1800

1900

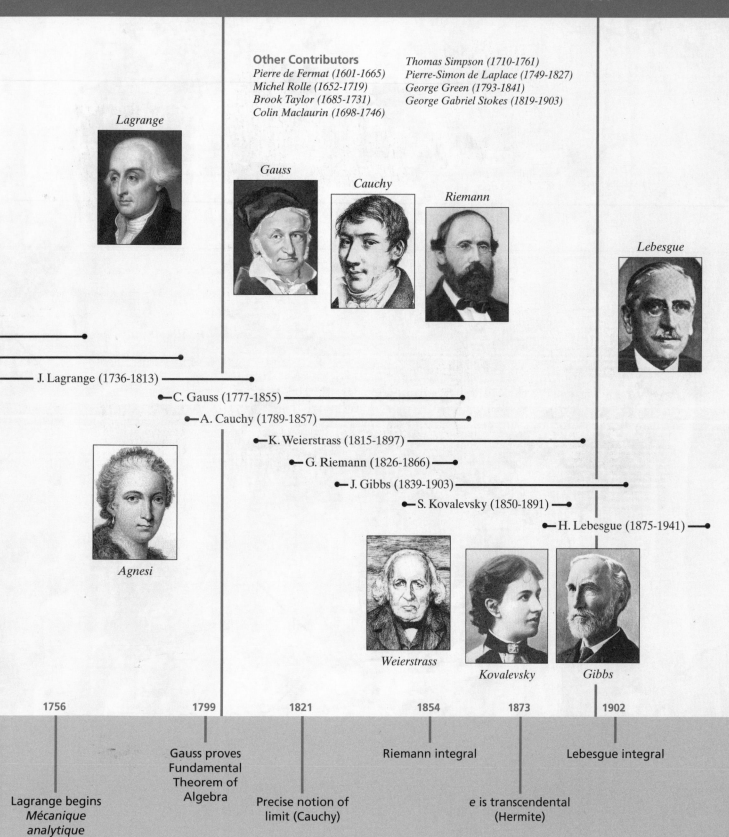

Other Contributors
Pierre de Fermat (1601-1665)
Michel Rolle (1652-1719)
Brook Taylor (1685-1731)
Colin Maclaurin (1698-1746)
Thomas Simpson (1710-1761)
Pierre-Simon de Laplace (1749-1827)
George Green (1793-1841)
George Gabriel Stokes (1819-1903)

Lagrange

Gauss

Cauchy

Riemann

Lebesgue

J. Lagrange (1736-1813)
C. Gauss (1777-1855)
A. Cauchy (1789-1857)
K. Weierstrass (1815-1897)
G. Riemann (1826-1866)
J. Gibbs (1839-1903)
S. Kovalevsky (1850-1891)
H. Lebesgue (1875-1941)

Agnesi

Weierstrass

Kovalevsky

Gibbs

1756 — Lagrange begins *Mécanique analytique*

1799 — Gauss proves Fundamental Theorem of Algebra

1821 — Precise notion of limit (Cauchy)

1854 — Riemann integral

1873 — e is transcendental (Hermite)

1902 — Lebesgue integral

FORMULAS FROM GEOMETRY

Triangle

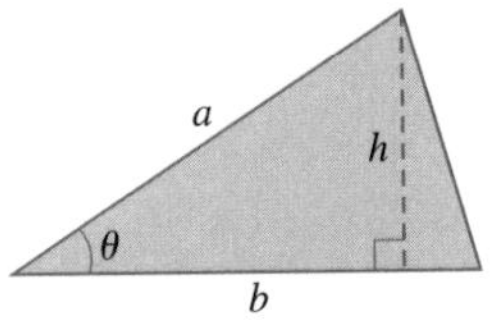

Area $= \frac{1}{2}bh$

Area $= \frac{1}{2}ab\sin\theta$

Parallelogram

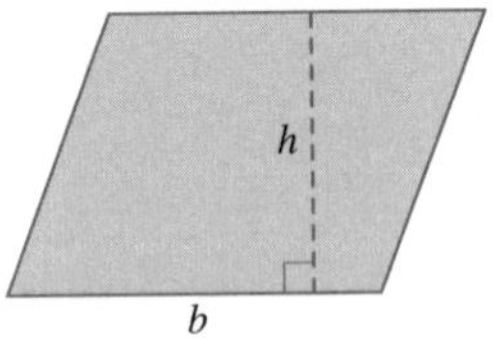

Area $= bh$

Trapezoid

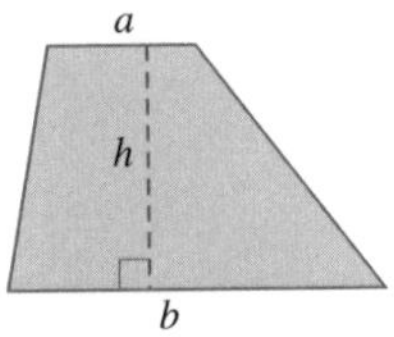

Area $= \frac{a+b}{2}h$

Circle

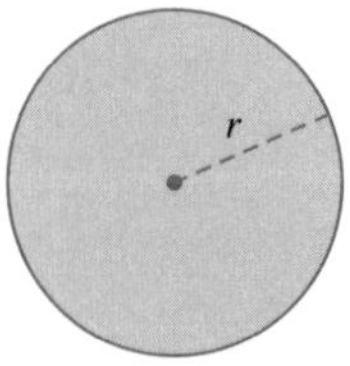

Circumference $= 2\pi r$

Area $= \pi r^2$

Sector of Circle

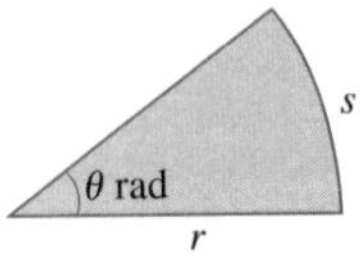

Arc length $s = r\theta$

Area $= \frac{1}{2}r^2\theta$

Polar Rectangle

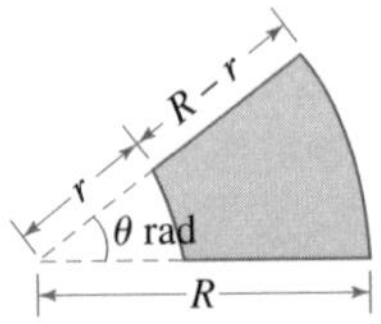

Area $= \frac{R+r}{2}(R-r)\theta$

Right Circular Cylinder

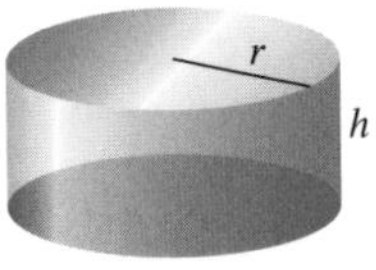

Lateral area $= 2\pi rh$

Volume $= \pi r^2 h$

Sphere

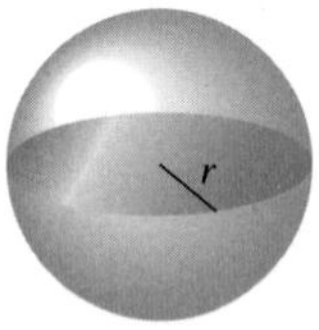

Area $= 4\pi r^2$

Volume $= \frac{4}{3}\pi r^3$

Right Circular Cone

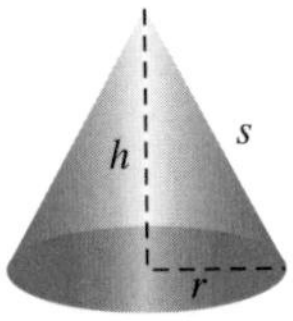

Lateral area $= \pi rs$

Volume $= \frac{1}{3}\pi r^2 h$

Frustum of Right Circular Cone

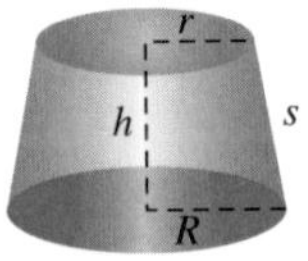

Lateral area $= \pi s(r + R)$

Volume $= \frac{1}{3}\pi(r^2 + rR + R^2)h$

General Cone

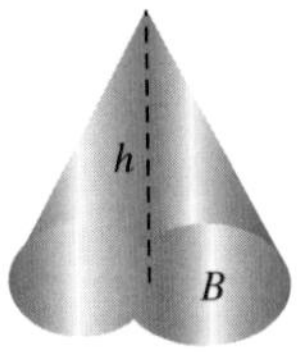

Volume $= \frac{1}{3}(\text{area } B)h$

Wedge

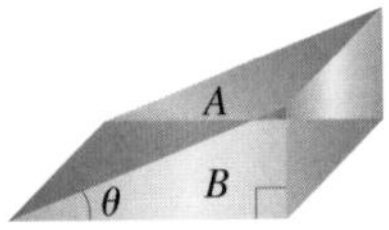

Area $A = (\text{area } B)\sec\theta$

Calculus

NINTH EDITION

Dale Varberg
Hamline University

Edwin J. Purcell
University of Arizona

Steven E. Rigdon
Southern Illinois University Edwardsville

PEARSON
Prentice Hall

Upper Saddle River, New Jersey 07458

Acquisitions Editor: *Adam Jaworski*
Editor-in-Chief: *Sally Yagan*
Project Manager: *Dawn Murrin*
Production Editor: *Debbie Ryan*
Assistant Managing Editor: *Bayani Mendoza de Leon*
Senior Managing Editor: *Linda Mihatov Behrens*
Executive Managing Editor: *Kathleen Schiaparelli*
Manufacturing Buyer: *Lisa McDowell*
Manufacturing Manager: *Alexis Heydt-Long*
Director of Marketing: *Patrice Jones*
Executive Marketing Manager: *Halee Dinsey*
Marketing Assistant: *Joon Won Moon*
Development Editor: *Frank Purcell*
Editor-in-Chief, Development: *Carol Trueheart*
Art Director: *Heather Scott*
Interior Designer: *Judith Matz-Coniglio*
Cover Designer: *Tamara Newnam*
Art Editor: *Thomas Benfatti*
Creative Director: *Juan R. López*
Director of Creative Services: *Paul Belfanti*
Manager, Cover Visual Research & Permissions: *Karen Sanatar*
Director, Image Resource Center: *Melinda Reo*
Manager, Rights and Permissions: *Zina Arabia*
Manager, Visual Research: *Beth Brenzel*
Image Permission: *Vickie Menanteaux*
Cover Photo: *Massimo Listri/Corbis; Interior view of Burj Al Arab Hotel, Dubai, United Arab Emirates*

Pearson Prentice Hall
Pearson Education, Inc.
Upper Saddle River, NJ 07458

Printed in the United States of America
10 9 8 7 6 5 4 3 2 1

ISBN 0-13-146968-1 (Instructor's Edition)
ISBN 0-13-142924-8 (Student Edition)

Pearson Education, Ltd., *London*
Pearson Education Australia PTY. Limited, *Sydney*
Pearson Education Singapore, Pte., Ltd
Pearson Education North Asia Ltd, *Hong Kong*
Pearson Education Canada, Ltd., *Toronto*
Pearson Education de Mexico, S.A. de C.V.
Pearson Education—Japan, *Tokyo*
Pearson Education Malaysia, Pte. Ltd

To
Pat, Chris, Mary, and Emily

Contents

Preface ix

0 Preliminaries 1

0.1 Real Numbers, Estimation, and Logic 1
0.2 Inequalities and Absolute Values 8
0.3 The Rectangular Coordinate System 16
0.4 Graphs of Equations 24
0.5 Functions and Their Graphs 29
0.6 Operations on Functions 35
0.7 Trigonometric Functions 41
0.8 Chapter Review 51
Review and Preview Problems 54

1 Limits 55

1.1 Introduction to Limits 55
1.2 Rigorous Study of Limits 61
1.3 Limit Theorems 68
1.4 Limits Involving Trigonometric Functions 73
1.5 Limits at Infinity; Infinite Limits 77
1.6 Continuity of Functions 82
1.7 Chapter Review 90
Review and Preview Problems 92

2 The Derivative 93

2.1 Two Problems with One Theme 93
2.2 The Derivative 100
2.3 Rules for Finding Derivatives 107
2.4 Derivatives of Trigonometric Functions 114
2.5 The Chain Rule 118
2.6 Higher-Order Derivatives 125
2.7 Implicit Differentiation 130
2.8 Related Rates 135
2.9 Differentials and Approximations 142
2.10 Chapter Review 147
Review and Preview Problems 150

3 Applications of the Derivative 151

3.1 Maxima and Minima 151
3.2 Monotonicity and Concavity 155
3.3 Local Extrema and Extrema on Open Intervals 162
3.4 Practical Problems 167
3.5 Graphing Functions Using Calculus 178
3.6 The Mean Value Theorem for Derivatives 185
3.7 Solving Equations Numerically 190
3.8 Antiderivatives 197
3.9 Introduction to Differential Equations 203
3.10 Chapter Review 209
Review and Preview Problems 214

4 The Definite Integral 215

4.1 Introduction to Area 215
4.2 The Definite Integral 224
4.3 The First Fundamental Theorem of Calculus 232
4.4 The Second Fundamental Theorem of Calculus and the Method of Substitution 243
4.5 The Mean Value Theorem for Integrals and the Use of Symmetry 253
4.6 Numerical Integration 260
4.7 Chapter Review 270
Review and Preview Problems 274

5 Applications of the Integral 275

5.1 The Area of a Plane Region 275
5.2 Volumes of Solids: Slabs, Disks, Washers 281
5.3 Volumes of Solids of Revolution: Shells 288
5.4 Length of a Plane Curve 294
5.5 Work and Fluid Force 301
5.6 Moments and Center of Mass 308
5.7 Probability and Random Variables 316
5.8 Chapter Review 322
Review and Preview Problems 324

6 Transcendental Functions 325

6.1 The Natural Logarithm Function 325
6.2 Inverse Functions and Their Derivatives 331
6.3 The Natural Exponential Function 337
6.4 General Exponential and Logarithmic Functions 342
6.5 Exponential Growth and Decay 347
6.6 First-Order Linear Differential Equations 355
6.7 Approximations for Differential Equations 359
6.8 The Inverse Trigonometric Functions and Their Derivatives 365
6.9 The Hyperbolic Functions and Their Inverses 374
6.10 Chapter Review 380
Review and Preview Problems 382

7 Techniques of Integration 383

7.1 Basic Integration Rules 383
7.2 Integration by Parts 387
7.3 Some Trigonometric Integrals 393
7.4 Rationalizing Substitutions 399
7.5 Integration of Rational Functions Using Partial Fractions 404
7.6 Strategies for Integration 411
7.7 Chapter Review 419
Review and Preview Problems 422

8 Indeterminate Forms and Improper Integrals 423

8.1 Indeterminate Forms of Type 0/0 423
8.2 Other Indeterminate Forms 428
8.3 Improper Integrals: Infinite Limits of Integration 433

8.4 Improper Integrals: Infinite Integrands 442
8.5 Chapter Review 446
Review and Preview Problems 448

9 Infinite Series 449

9.1 Infinite Sequences 449
9.2 Infinite Series 455
9.3 Positive Series: The Integral Test 463
9.4 Positive Series: Other Tests 468
9.5 Alternating Series, Absolute Convergence, and Conditional Convergence 474
9.6 Power Series 479
9.7 Operations on Power Series 484
9.8 Taylor and Maclaurin Series 489
9.9 The Taylor Approximation to a Function 497
9.10 Chapter Review 504
Review and Preview Problems 508

10 Conics and Polar Coordinates 509

10.1 The Parabola 509
10.2 Ellipses and Hyperbolas 513
10.3 Translation and Rotation of Axes 523
10.4 Parametric Representation of Curves in the Plane 530
10.5 The Polar Coordinate System 537
10.6 Graphs of Polar Equations 542
10.7 Calculus in Polar Coordinates 547
10.8 Chapter Review 552
Review and Preview Problems 554

11 Geometry in Space and Vectors 555

11.1 Cartesian Coordinates in Three-Space 555
11.2 Vectors 560
11.3 The Dot Product 566
11.4 The Cross Product 574
11.5 Vector-Valued Functions and Curvilinear Motion 579
11.6 Lines and Tangent Lines in Three-Space 589
11.7 Curvature and Components of Acceleration 593
11.8 Surfaces in Three-Space 603
11.9 Cylindrical and Spherical Coordinates 609
11.10 Chapter Review 613
Review and Preview Problems 616

12 Derivatives for Functions of Two or More Variables 617

12.1 Functions of Two or More Variables 617
12.2 Partial Derivatives 624
12.3 Limits and Continuity 629
12.4 Differentiability 635
12.5 Directional Derivatives and Gradients 641
12.6 The Chain Rule 647
12.7 Tangent Planes and Approximations 652
12.8 Maxima and Minima 657

12.9 The Method of Lagrange Multipliers 666
12.10 Chapter Review 672
Review and Preview Problems 674

13 Multiple Integrals 675

13.1 Double Integrals over Rectangles 675
13.2 Iterated Integrals 680
13.3 Double Integrals over Nonrectangular Regions 684
13.4 Double Integrals in Polar Coordinates 691
13.5 Applications of Double Integrals 696
13.6 Surface Area 700
13.7 Triple Integrals in Cartesian Coordinates 706
13.8 Triple Integrals in Cylindrical and Spherical Coordinates 713
13.9 Change of Variables in Multiple Integrals 718
13.10 Chapter Review 728
Review and Preview Problems 730

14 Vector Calculus 731

14.1 Vector Fields 731
14.2 Line Integrals 735
14.3 Independence of Path 742
14.4 Green's Theorem in the Plane 749
14.5 Surface Integrals 755
14.6 Gauss's Divergence Theorem 764
14.7 Stokes's Theorem 770
14.8 Chapter Review 773

15 Differential Equations 775*

15.1 Linear Homogeneous Equations
15.2 Nonhomogeneous Equations
15.3 Applications of Second-Order Equations
15.4 Chapter Review

*(This chapter can be found online at **www.prenhall.com**. Select "Browse our catalog" then click on "Mathematics," select your course and choose your text. Under "Resources" on the left side, select "Instructor" and choose this chapter for download.)

Appendix A-1

A.1 Mathematical Induction A-1
A.2 Proofs of Several Theorems A-3

Answers to Odd-Numbered Problems A-7

Index I-1

Photo Credits P-1

Teaching Outline T-1

Preface

The ninth edition of *Calculus* is again a modest revision. Some topics have been added, and some of the topics have been rearranged, but the spirit of the book has remained unchanged. Users of previous editions have reported success, and we have no intention of overhauling a workable text.

To many, this book would still be considered a traditional text. Most theorems are proved, left as an exercise, or left unproved when the proof is too difficult. When a proof is difficult, we try to give an intuitive explanation to make the result plausible before going on to the next topic. In some cases, we give a sketch of a proof, in which case we explain why it is a sketch and not a rigorous proof. The focus is still on understanding the concepts of calculus. While some see the emphasis on clear, rigorous presentation as being a distraction to understanding calculus, we see the two as complementary. Students are more likely to grasp the concepts of calculus if terms are clearly defined and theorems are clearly stated and proved.

A Brief Text The ninth edition continues to be the briefest of all the successful mainstream calculus texts. We have tried to prevent the text from ballooning upward with new topics and alternative approaches. In less than 800 pages, we cover the major topics of calculus, including a preliminary chapter, and the material from limits to vector calculus. In the last few decades, students have developed some bad habits. They prefer not to read the textbook. They want to find the appropriate worked-out example so it can be matched to their homework problem. Our goal with this text continues to be to keep calculus as a course focused on some few basic ideas centered around words, formulas, and graphs. Solving problem sets, while crucial to developing mathematical and problem-solving skills, should not overshadow the goal of understanding calculus.

Concepts Review Problems To encourage students to read the textbook with understanding, we begin every problem set with four fill-in-the-blank items. These test the mastery of the basic vocabulary, understanding of theorems, and ability to apply the concepts in the simplest settings. Students should respond to these items before proceeding to the later problems. We encourage this by giving immediate feedback; the correct answers are given at the end of the problem set. These items also make good quiz questions to see whether students have done the required reading and have prepared for class.

Review and Preview Problems We have also included a set of Review and Preview Problems between the end of one chapter and the beginning of the next. Many of these problems force students to review past topics before starting the new chapter. For example,

- Chapter 3, Applications of Derivatives: Students are asked to solve inequalities like the ones that arise when we ask where a function is increasing/decreasing or concave up/down.
- Chapter 7, Techniques of Integration: Students are asked to evaluate a number of integrals involving the method of substitution, the only substantive technique they have learned up to this point. Lacking skill using this technique would spell disaster in Chapter 7.
- Chapter 13, Multiple Integration: Students are asked to sketch the graphs of equations in Cartesian, cylindrical, and spherical coordinates. Visualizing regions in two- and three-space is key to understanding multiple integration.

Other Review and Preview Problems ask the student to use what they already know to get a head start on the upcoming chapter. For example,

- Chapter 5, Applications of Integration: Students are asked to find the length of a line segment between two functions, exactly the skill required to perform the *slice, approximate*, and *integrate* in the chapter. Also, students are asked to find the volume of a small disk, washer, and shell. Having worked these out before beginning the chapter would make the students better prepared to understand the idea of *slice, approximate*, and *integrate* as it applies to finding volumes of solids of revolution.

- Chapter 8, Indeterminate Forms and Improper Integrals: Students are asked to find the value of an integral like $\int_0^a e^{-x}\,dx$, for $a = 1, 2, 4, 8, 16$. We hope that students will work a problem like this and realize that as a grows, the value of the integral gets close to 1, thereby setting up the idea of improper integrals. There are similar problems involving sums before the chapter on infinite series.

Number Sense Number sense continues to play an important role in the book. All calculus students make numerical mistakes in solving problems, but the ones with the number sense recognize an absurd answer and rework the problem. To encourage and develop this important ability, we have emphasized the estimation process. We suggest how to make mental estimates and how to arrive at ballpark numerical answers. We have increased our own use of this in the text, using the symbol $\boxed{\approx}$ where we make a ballpark estimate. We hope students do the same, especially in problems with the $\boxed{\approx}$ mark.

Use of Technology Many problems in the ninth edition are flagged with one of these symbols:

$\boxed{\text{C}}$ indicates that an ordinary scientific calculator will be helpful

$\boxed{\text{GC}}$ indicates that a graphing calculator is required

$\boxed{\text{CAS}}$ indicates that a computer algebra system is required

The Technology Projects that were at the end of the chapters in the eighth edition are now available on the Web in pdf files.

Changes in the Ninth Edition The basic structure, and the overriding spirit, of the text has remained unchanged. Here are the most significant changes in the ninth edition:

- There is a set of Review and Preview Problems between the end of one chapter and the beginning of the next.
- The preliminary chapter, now called Chapter 0, has been condensed. The "precalculus" topics (that were in the beginning of Chapter 2 of the eighth edition) are now placed in Chapter 0. In the ninth edition, Chapter 1 begins with limits. How much of Chapter 0 needs to be covered depends on the background of the students and will vary from institution to institution.
- The sections on antiderivatives and an introduction to differential equations have been moved to Chapter 3. This allows a clear break between "rate of change" concepts and "accumulation" concepts, because Chapter 4 now begins with area, followed immediately by the definite integral and the fundamental theorems of calculus. "It has been the author's experience that many first-year students of calculus fail to make a clear distinction between the very different concepts of the indefinite integral (or antiderivative) and the definite integral as the limit of a sum." That was from the first edition, published in 1965, and it is just as true today. We hope that separating these topics will draw attention to the distinction.

- Probability and fluid pressure have been added to the Chapter 5, Applications of Integration. We emphasize that probability problems are treated much like mass problems along a line. The center of mass is the integral of x times the density, and the expectation in probability is the integral of x times the (probability) density.
- Material on conic sections has been condensed from five sections into three sections. Students have seen much (but not all) of this material in their precalculus courses.
- Vectors have been consolidated into a single chapter. In the eighth edition, we covered plane vectors in Chapter 13 and space vectors in Chapter 14. With this approach, we ended up repeating a number of topics, such as the dot product and curvature, in Chapter 14. The approach in the ninth edition is to cover vectors once. Most of the presentation is in terms of vectors in space, but we point out how plane vectors work. The context of a problem should dictate whether plane vectors or space vectors are needed.
- There are examples and an exercise on Kepler's Laws of Planetary Motion. The material on vectors culminates in the derivation of Kepler's laws from Newton's Law of Gravitation. We derive Kepler's second and third laws in examples, leaving the first law as an exercise. In this exercise, students are guided through the steps, (a) through (l), of the derivation.
- Chapter 13, Multiple Integration, now ends with a section on change of variables in multiple integrals using the Jacobian.
- The sections on numerical methods have been placed in appropriate places throughout the text. For example, the section on solving equations numerically has become Section 3.7; numerical integration has become Section 4.6; approximations for differential equations has become Section 6.7; and the Taylor approximation to a function has become Section 9.9.
- The chapter on differential equation has been removed, but it is available to users on the Web. The text already contains numerous sections on differential equations, including slope fields and Euler's method.
- The number of conceptual questions has increased significantly. Many more problems ask the student for graphs. We have also increased the use of numerical methods, such as Newton's method and numerical integration, in problems that cannot be treated analytically.

Acknowledgements I would like to thank the staff at Prentice Hall, including Adam Jaworski, Eric Frank, Dawn Murrin, Debbie Ryan, Bayani deLeon, Sally Yagan, Halee Dinsey, Patrice Jones, Heather Scott, and Thomas Benfatti for their encouragement and patience. I would also like to thank those who read the manuscript carefully, including Frank Purcell, Brad Davis, Pat Daly (Paley Company), and Edith Baker (Writewith, Inc.). I owe a great debt of gratitude to Kevin Bodden and Christopher Rigdon, who worked tirelessly preparing the solutions manuals, and to Barbara Kniepkamp and Brian Rife for preparing the back-of-the-book answers. I would also like to thank the faculty at Southern Illinois University Edwardsville (and elsewhere), especially George Pelekanos, Rahim Karimpour, Krzysztof Jarosz, Alan Wheeler, and Paul Phillips, for helpful comments.

I also thank the following faculty for their careful review and helpful comments during the preparation of the ninth edition.

Fritz Keinert, Iowa State University
Michael Martin, Johnson County Community College
Christopher Johnston, University of Missouri-Columbia
Nakhle Asmar, University of Missouri-Columbia
Zhonghai Ding, University of Nevada Las Vegas
Joel Foisy, SUNY Potsdam
Wolfe Snow, Brooklyn College
Ioana Mihaila, California State Polytechnic University, Pomona

Hasan Celik, California State Polytechnic University
Jeffrey Stopple, University of California, Santa Barbara
Jason Howell, Clemson University
John Goulet, Worcester Polytechnic Institute
Ryan Berndt, The Ohio State University
Douglas Meade, University of South Carolina
Elgin Johnston, Iowa State University
Brian Snyder, Lake Superior State University
Bruce Wenner, University of Missouri—Kansas City
Linda Kilgariff, University of North Carolina at Greensboro
Joel Robbin, University of Wisconsin—Madison
John Johnson, George Fox University
Julie Connolly, Wake Forest University
Chris Peterson, Colorado State University
Blake Thornton, Washington University in St. Louis
Sue Goodman, University of North Carolina—Chapel Hill
John Santomos, Villanova University

Finally, I would like to thank my wife Pat, and children Chris, Mary, and Emily for tolerating the many nights and weekends that I spent at the office.

S. E. R.
srigdon@siue.edu
Southern Illinois University Edwardsville

STUDENT RESOURCES

Student Study Pack

Everything a student needs to succeed in one place. It is packaged with the book, or can be available for purchase stand-alone. Study Pack contains:

- ***Student Solutions Manual***
  Fully worked solutions to odd-numbered exercises.
- ***Pearson Tutor Center***
  Tutors provide one-on-one tutoring for any problem with an answer at the back of the book. Students access the Tutor Center via toll-free phone, fax, or email. Available only to college students in the U.S. and Canada.
- ***CD Lecture Series***
  A comprehensive set of CD-ROMs, tied to the textbook, containing short video clips of an instructor working key book examples.

INSTRUCTOR RESOURCES

Instructor Resource Distribution

All instructor resources can be downloaded from the web site, **www.prenhall.com**. Select "Browse our catalog," then, click on "Mathematics;" select your course and choose your text. Under "Resources," on the left side, select "instructor" and choose the supplement you need to download. You will be required to run through a one time registration before you can complete this process.

- ***TestGen***
  Easily create tests from textbook section objectives. Questions are algorithmically generated allowing for unlimited versions. Edit problems or create your own.
- ***Test Item File***
  A printed test bank derived from TestGen.
- ***PowerPoint Lecture Slides***
  Fully editable slides that follow the textbook. Project in class or post to a website in an online course.
- ***Instructor Solutions Manual***
  Fully worked solutions to all textbook exercises and chapter projects.
- ***Technology Projects***
- ***Chapter 15, Differential Equations***
  The entire chapter is available in pdf for download.

Instructor's Edition

Provides answers to odd-numbered problems in the back of the text.

Includes a Teaching Outline organized by section, which includes topics to cover, suggested homework assignments, and various teaching tips.

MathXL®

MathXL® is a powerful online homework, tutorial, and assessment system that accompanies your textbook. Instructors can create, edit, and assign online homework and tests using algorithmically generated exercises correlated at the objective level to the textbook. Student work is tracked in an online gradebook. Students can take chapter tests and receive personalized study plans based on their results. The study plan diagnoses weaknesses and links students to tutorial exercises for objectives they need to study. Students can also access video clips from selected exercises. MathXL® is available to qualified adopters. For more information, visit our website at **www.mathxl.com**, or contact your Prentice Hall sales representative for a demonstration.

MyMathLab

MyMathLab is a text-specific, customizable online course for your textbooks. MyMathLab is powered by CourseCompass™—Pearson Education's online teaching and learning environment—and by MathXL®—our online homework, tutorial, and assessment system. MyMathLab gives you the tools you need to deliver all or a portion of your course online, whether your students are in a lab setting or working from home.

MyMathLab provides a rich and flexible set of course materials, featuring free-response exercises that are algorithmically generated for unlimited practice. Students can use online tools such as video lectures and a multimedia textbook to improve their performance. Instructors can use MyMathLab's homework and test managers to select and assign online exercises correlated to the textbook, and can import TestGen tests for added flexibility. The only gradebook—designed specifically for mathematics—automatically tracks students' homework and test results and gives the instructor control over how to calculate final grades. MyMathLab is available to qualified adopters. For more information, visit our website at **www.mymathlab.com** or contact your Prentice Hall sales representative for a product demonstration.

CHAPTER 0 Preliminaries

0.1 Real Numbers, Estimation, and Logic
0.2 Inequalities and Absolute Values
0.3 The Rectangular Coordinate System
0.4 Graphs of Equations
0.5 Functions and Their Graphs
0.6 Operations on Functions
0.7 Trigonometric Functions

0.1 Real Numbers, Estimation, and Logic

Calculus is based on the real number system and its properties. But what are the real numbers and what are their properties? To answer, we start with some simpler number systems.

The Integers and the Rational Numbers The simplest numbers of all are the **natural numbers,**

$$1, 2, 3, 4, 5, 6, \ldots$$

With them we can *count:* our books, our friends, and our money. If we include their negatives and zero, we obtain the **integers**

$$\ldots, -3, -2, -1, 0, 1, 2, 3, \ldots$$

When we *measure* length, weight, or voltage, the integers are inadequate. They are spaced too far apart to give sufficient precision. We are led to consider quotients (ratios) of integers (Figure 1), numbers such as

$$\frac{3}{4}, \frac{-7}{8}, \frac{21}{5}, \frac{19}{-2}, \frac{16}{2}, \text{ and } \frac{-17}{1}$$

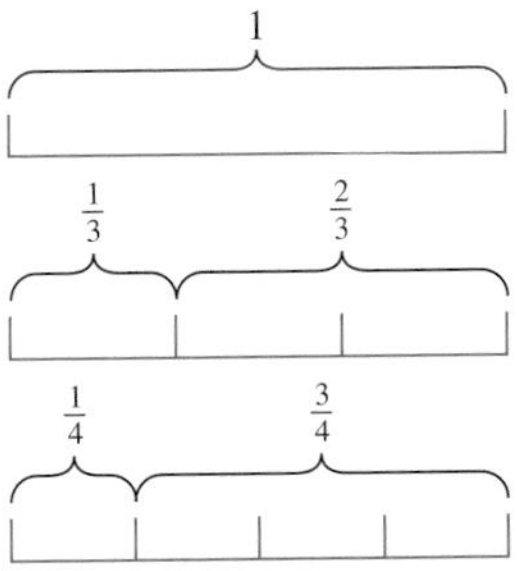

Figure 1

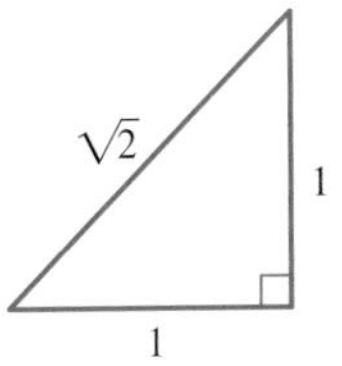

Figure 2

Note that we included $\frac{16}{2}$ and $\frac{-17}{1}$, though we would normally write them as 8 and -17 since they are equal to the latter by the ordinary meaning of division. We did not include $\frac{5}{0}$ or $\frac{-9}{0}$ since it is impossible to make sense out of these symbols (see Problem 30). Remember always that division by 0 is never allowed. Numbers that can be written in the form m/n, where m and n are integers with $n \neq 0$, are called **rational numbers.**

Do the rational numbers serve to measure all lengths? No. This surprising fact was discovered by the ancient Greeks in about the fifth century B.C. They showed that while $\sqrt{2}$ measures the hypotenuse of a right triangle with legs of length 1 (Figure 2), $\sqrt{2}$ cannot be written as a quotient of two integers (see Problem 77). Thus, $\sqrt{2}$ is an **irrational** (not rational) number. So are $\sqrt{3}$, $\sqrt{5}$, $\sqrt[3]{7}$, π, and a host of other numbers.

The Real Numbers Consider all numbers (rational and irrational) that can measure lengths, together with their negatives and zero. We call these numbers the **real numbers.**

The real numbers may be viewed as labels for points along a horizontal line. There they measure the distance to the right or left (the **directed distance**) from a

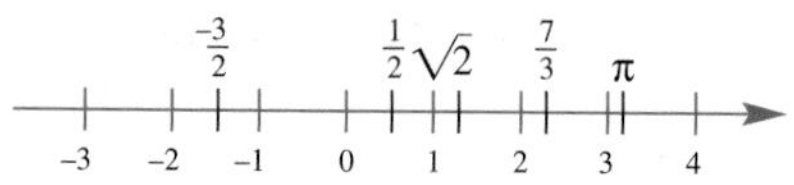

Figure 3

fixed point called the **origin** and labeled 0 (Figure 3). Though we cannot possibly show all the labels, each point does have a unique real number label. This number is called the **coordinate** of the point, and the resulting coordinate line is referred to as the **real line.** Figure 4 suggests the relationships among the sets of numbers discussed so far.

You may remember that the real number system can be enlarged still more—to the **complex numbers.** These are numbers of the form $a + bi$, where a and b are real numbers and $i = \sqrt{-1}$. Complex numbers will rarely be used in this book. In fact, if we say or suggest *number* without any qualifying adjective, you can assume that we mean real number. The real numbers are the principal characters in calculus.

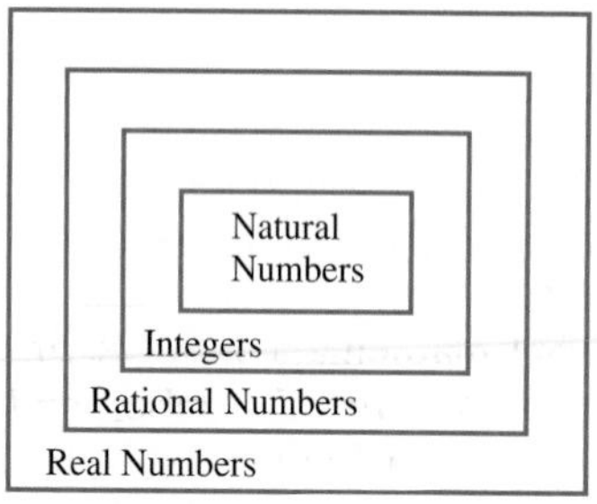

Figure 4

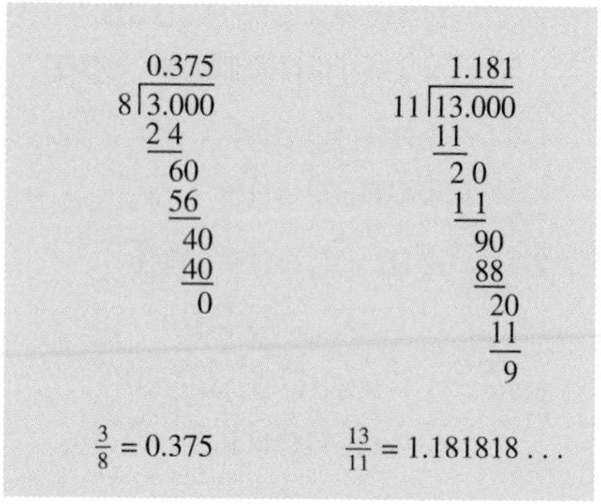

Figure 5

Repeating and Nonrepeating Decimals Every rational number can be written as a decimal, since by definition it can always be expressed as the quotient of two integers; if we divide the denominator into the numerator, we obtain a decimal (Figure 5). For example,

$$\frac{1}{2} = 0.5 \qquad \frac{3}{8} = 0.375 \qquad \frac{3}{7} = 0.428571428571428571\ldots$$

Irrational numbers, too, can be expressed as decimals. For instance,

$$\sqrt{2} = 1.4142135623\ldots, \qquad \pi = 3.1415926535\ldots$$

The decimal representation of a rational number either terminates (as in $\frac{3}{8} = 0.375$) or else repeats in regular cycles forever (as in $\frac{13}{11} = 1.181818\ldots$). A little experimenting with the long division algorithm will show you why. (Note that there can be only a finite number of different remainders.) A terminating decimal can be regarded as a repeating decimal with repeating zeros. For instance,

$$\frac{3}{8} = 0.375 = 0.3750000\ldots$$

Thus, every rational number can be written as a repeating decimal. In other words, if x is a rational number, then x can be written as a repeating decimal. It is a remarkable fact that the converse is also true; if x can be written as a repeating decimal, then x is a rational number. This is obvious in the case of a terminating decimal (for instance, $3.137 = 3137/1000$), and it is easy to show for the case of a nonterminating repeating decimal.

EXAMPLE 1 **(Repeating decimals are rational.)** Show that $x = 0.136136136\ldots$ represents a rational number.

SOLUTION We subtract x from $1000x$ and then solve for x.

$$\begin{aligned} 1000x &= 136.136136\ldots \\ x &= 0.136136\ldots \\ \hline 999x &= 136 \\ x &= \frac{136}{999} \end{aligned}$$

■

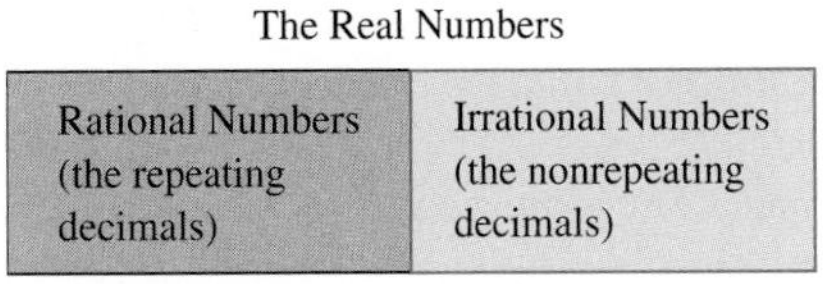

Figure 6

The decimal representations of irrational numbers do not repeat in cycles. Conversely, a nonrepeating decimal must represent an irrational number. Thus, for example,

$$0.101001000100001\ldots$$

must represent an irrational number (note the pattern of more and more 0s between the 1s). The diagram in Figure 6 summarizes what we have said.

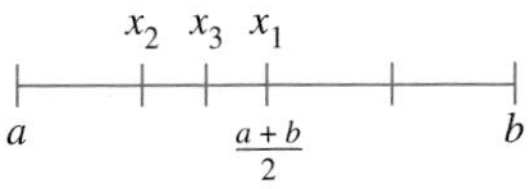

Figure 7

Denseness Between any two different real numbers a and b, no matter how close together, there is another real number. In particular, the number $x_1 = (a + b)/2$ is a real number that is midway between a and b (Figure 7). Since there is another real number, x_2, between a and x_1, and another real number, x_3, between x_1 and x_2, and since this argument can be repeated ad infinitum, we conclude that there are infinitely many real numbers between a and b. Thus, there is no such thing as "the real number just larger than 3."

Actually, we can say more. Between any two distinct real numbers, there are both a rational number and an irrational number. (In Problem 57 you are asked to show that there is a rational number between any two real numbers.) Hence, by the preceding argument, there are infinitely many of each.

One way that mathematicians describe the situation we have been discussing is to say that both the rational numbers and the irrational numbers are **dense** along the real line. Every number has both rational and irrational neighbors arbitrarily close to it.

Figure 8

One consequence of the density property is that any irrational number can be approximated as closely as we please by a rational number—in fact, by a rational number with a terminating decimal representation. Take $\sqrt{2}$ as an example. The sequence of rational numbers 1, 1.4, 1.41, 1.414, 1.4142, 1.41421, 1.414213, ... marches steadily and inexorably toward $\sqrt{2}$ (Figure 8). By going far enough along in this sequence, we can get as near to $\sqrt{2}$ as we wish.

Calculators and Computers Today many calculators are capable of performing numerical, graphical, and symbolic operations. For decades now, calculators have been able to perform numerical operations such as giving decimal approximations to $\sqrt{12.2}$ and $1.25 \sin 22°$. By the early 1990s calculators could display the graph of almost any algebraic, trigonometric, exponential, or logarithmic function. Recent advances allow calculators to perform many symbolic operations, such as expanding $(x - 3y)^{12}$ or solving $x^3 - 2x^2 + x = 0$. Computer software such as *Mathematica* or *Maple* can perform symbolic operations like these, as well as a great many others.

Our recommendations regarding the use of a calculator are these:

1. Know when your calculator or computer gives you an exact answer and when it gives you an approximation. For example, if you ask for $\sin 60°$, your calculator may give the exact answer, $\sqrt{3}/2$, or it may give you a decimal approximation, 0.8660254.
2. In most cases, an exact answer is preferred. This is especially true when you must use the result in further calculations. For example, if you subsequently need to square the result of $\sin 60°$, it is easier, as well as being more accurate, to compute $\left(\sqrt{3}/2\right)^2 = 3/4$ than it is to compute 0.8660254^2.
3. In an applied problem, give an exact answer, if possible, as well as an approximation. You can often check whether your answer is reasonable, as it relates to the description of the problem, by looking at your numerical approximation to the solution.

> Many problems in this book are marked with a special sysmbol.
>
> C means use a calculator.
>
> GC means use a graphing calculator.
>
> CAS means use a computer algebra system.
>
> EXPL means the problem asks you to explore and go beyond the explanations given in the book.

Estimation Given a complicated arithmetic problem, a careless student might quickly press a few keys on a calculator and report the answer, not realizing that a missed parenthesis or a slip of the finger has given an incorrect result. A careful student with a feeling for numbers will press the same keys, immediately recognize

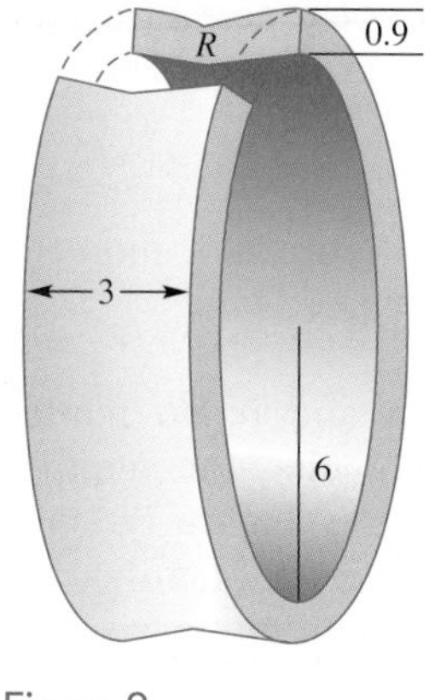

Figure 9

that the answer is wrong if it is far too big or far too small, and recalculate it correctly. It is important to know how to make a mental estimate.

EXAMPLE 2 Calculate $\left(\sqrt{430} + 72 + \sqrt[3]{7.5}\right)/2.75$.

SOLUTION A wise student approximated this as $(20 + 72 + 2)/3$ and said that the answer should be in the neighborhood of 30. Thus, when her calculator gave 93.448 for an answer, she was suspicious (she had actually calculated $\sqrt{430} + 72 + \sqrt[3]{7.5}/2.75$).

On recalculating, she got the correct answer: 34.434. ■

EXAMPLE 3 Suppose that the shaded region R shown in Figure 9 is revolved about the x-axis. Estimate the volume of the resulting solid ring S.

SOLUTION The region R is about 3 units long and 0.9 units high. We estimate its area as $3(0.9) \approx 3$ square units. Imagine the solid ring S to be slit open and laid out flat, forming a box about $2\pi r \approx 2(3)(6) = 36$ units long. The volume of a box is its cross-sectional area times its length. Thus, we estimate the volume of the box to be $3(36) = 108$ cubic units. If you calculate it to be 1000 cubic units, you need to check your work. ■

≈

In Example 3, we have used ≈ to mean "approximately equal." Use this symbol in your scratch work when making an approximation. In more formal work, never use this symbol without knowing how large the error could be.

Many problems are marked with this symbol.

≈ means make an estimate of the answer before working the problem; then check your answer against this estimate.

The process of *estimation* is just ordinary common sense combined with reasonable numerical approximations. We urge you to use it frequently, especially on word problems. Before you attempt to get a precise answer, make an estimate. If your answer is close to your estimate, there is no guarantee that your answer is correct. On the other hand, if your answer and your estimate are far apart, you should check your work. There is probably an error in your answer or in your approximation. Remember that $\pi \approx 3$, $\sqrt{2} \approx 1.4$, $2^{10} \approx 1000$, 1 foot $\approx$ 10 inches, 1 mile $\approx$ 5000 feet, and so on.

A central theme in this text is number sense. By this, we mean the ability to work through a problem and tell whether your solution is a reasonable one for the stated problem. A student with good number sense will immediately recognize and correct an answer that is obviously unreasonable. For many of the examples worked out in the text, we provide an initial estimate of the solution before proceeding to find the exact solution.

A Bit of Logic

Important results in mathematics are called **theorems;** you will find many theorems in this book. The most important ones occur with the label *Theorem* and are usually given names (e.g., the Pythagorean Theorem). Others occur in the problem sets and are introduced with the words *show that* or *prove that*. In contrast to axioms or definitions, which are taken for granted, theorems require proof.

Many theorems are stated in the form "If P then Q" or they can be restated in this form. We often abbreviate the statement "If P then Q" by $P \Rightarrow Q$, which is also read "P implies Q." We call P the *hypothesis* and Q the *conclusion* of the theorem. A proof consists of showing that Q must be true whenever P is true.

Beginning students (and some mature ones) may confuse $P \Rightarrow Q$ with its **converse,** $Q \Rightarrow P$. These two statements are not equivalent. "If John is a Missourian, then John is an American" is a true statement, but its converse "If John is an American, then John is a Missourian" may not be true.

The **negation** of the statement P is written $\sim P$. For example, if P is the statement "It is raining," then $\sim P$ is the statement "It is not raining." The statement $\sim Q \Rightarrow \sim P$ is called the **contrapositive** of the statement $P \Rightarrow Q$ and it is equivalent to $P \Rightarrow Q$. By "equivalent" we mean that $P \Rightarrow Q$ and $\sim Q \Rightarrow \sim P$ are either both true or both false. For our example about John, the contrapositive of "If John is a Missourian, then John is an American" is "If John is not an American, then John is not a Missourian."

Because a statement and its contrapositive are equivalent, we can prove a theorem of the form "If P then Q" by proving its contrapositive "If $\sim Q$ then $\sim P$."

Thus, to prove $P \Rightarrow Q$, we can assume $\sim Q$ and try to deduce $\sim P$. Here is a simple example.

> **Proof by Contradiction**
>
> Proof by contradiction also goes by the name *reductio ad absurdum.* Here is what the great mathematician G. H. Hardy had to say about it.
>
> "Reductio ad absurdum, which Euclid loved so much, is one of a mathematician's finest weapons. It is a far finer gambit than any chess gambit; a chess player may offer the sacrifice of a pawn or even a piece, but a mathematician offers the game."

EXAMPLE 4 Prove that if n^2 is even, then n is even.

Proof The contrapositive of this sentence is "If n is not even, then n^2 is not even," which is equivalent to "If n is odd, then n^2 is odd." We will prove the contrapositive. If n is odd, then there exists an integer k such that $n = 2k + 1$. Then

$$n^2 = (2k + 1)^2 = 4k^2 + 4k + 1 = 2(2k^2 + 2k) + 1$$

Therefore, n^2 is equal to one more than twice an integer. Hence n^2 is odd. ■

The *Law of the Excluded Middle* says: Either R or $\sim R$, but not both. Any proof that begins by assuming the conclusion of a theorem is false and proceeds to show this assumption leads to a contradiction is called a **proof by contradiction.**

Occasionally, we will need another type of proof called **mathematical induction.** It would take us too far afield to describe this now, but we have given a complete discussion in Appendix A.1.

Sometimes both the statements $P \Rightarrow Q$ (if P then Q) and $Q \Rightarrow P$ (if Q then P) are true. In this case we write $P \Leftrightarrow Q$, which is read "P if and only if Q." In Example 4 we showed that "If n^2 is even, then n is even," but the converse "If n is even, then n^2 is even" is also true. Thus, we would say "n is even if and only if n^2 is even."

> **Order on the Real Line**
>
> To say that $x < y$ means that x is to the left of y on the real line.
>
> 
>

Order

The nonzero real numbers separate nicely into two disjoint sets—the positive real numbers and the negative real numbers. This fact allows us to introduce the order relation $<$ (read "is less than") by

$$x < y \Leftrightarrow y - x \text{ is positive}$$

We agree that $x < y$ and $y > x$ shall mean the same thing. Thus, $3 < 4$, $4 > 3$, $-3 < -2$, and $-2 > -3$.

The order relation $\leq$ (read "is less than or equal to") is a first cousin of $<$. It is defined by

$$x \leq y \Leftrightarrow y - x \text{ is positive or zero}$$

Order properties 2, 3, and 4 in the margin box hold when the symbols $<$ and $>$ are replaced by $\leq$ and $\geq$.

> **The Order Properties**
>
> 1. **Trichotomy.** If x and y are numbers, then exactly one of the following holds:
>
> $x < y$ or $x = y$ or $x > y$
> 2. **Transitivity.** $x < y$ and $y < z \Rightarrow x < z$.
> 3. **Addition.** $x < y \Leftrightarrow x + z < y + z$.
> 4. **Multiplication.** When z is positive, $x < y \Leftrightarrow xz < yz$. When z is negative, $x < y \Leftrightarrow xz > yz$.

Quantifiers

Many mathematical statements involve a variable x, and the truth of the statement depends on the value of x. For example, the statement "$\sqrt{x}$ is a rational number" depends on the value of x; it is true for some values of x, such as $x = 1, 4, 9, \frac{4}{9}$, and $\frac{10{,}000}{49}$, and false for other values of x, such as $x = 2, 3, 77$, and π. Some statements, such as "$x^2 \geq 0$," are true for all real numbers x, and other statements, such as "x is an even integer greater than 2 and x is a prime number," are always false. We will let $P(x)$ denote a statement whose truth depends on the value of x.

We say "For all x, $P(x)$" or "For every x, $P(x)$" when the statement $P(x)$ is true for every value of x. When there is at least one value of x for which $P(x)$ is true, we say "There exists an x such that $P(x)$." The two important *quantifiers* are "for all" and "there exists."

EXAMPLE 5 Which of the following statements are true?

(a) For all x, $x^2 > 0$.
(b) For all x, $x < 0 \Rightarrow x^2 > 0$.
(c) For every x, there exists a y such that $y > x$.
(d) There exists a y such that, for all x, $y > x$.

SOLUTION

(a) False. If we choose $x = 0$, then it is not true that $x^2 > 0$.
(b) True. If x is negative, then x^2 will be positive.
(c) True. This statement contains two quantifiers, "for every" and "there exists." To read the statement correctly, we must apply them in the right order. The statement begins "for every," so if the statement is true, then what follows must be true for every value of x that we choose. If you are not sure whether the whole statement is true, try a few values of x and see whether the second part of the statement is true or false. For example, we might choose $x = 100$; given this choice, does there exist a y that is greater than x? In other words, is there a number greater than 100? Yes, of course. The number 101 would do. Next choose another value for x, say $x = 1{,}000{,}000$. Does there exist a y that is greater than this value of x? Again, yes; in this case the number 1,000,001 would do. Now, ask yourself: "If I let x be any real number, will I be able to find a y that is larger than x?" The answer is yes. Just choose y to be $x + 1$.
(d) False. This statement says that there is a real number that is larger than every other real number. In other words, there is a largest real number. This is false; here is a proof by contradiction. Suppose that there exists a largest real number y. Let $x = y + 1$. Then $x > y$, which is contrary to the assumption that y is the largest real number. ■

The **negation** of the statement P is the statement "not P." (The statement "not P" is true provided P is false.) Consider the negation of the statement "for all x, $P(x)$." If this negated statement is true, then there must be at least one value of x for which $P(x)$ is false; in other words, there exists an x such that "not $P(x)$." Now consider the negation of the statement "there exists an x such that $P(x)$." If this negated statement is true, then there is not a single x for which $P(x)$ is true. This means that $P(x)$ is false no matter what the value of x. In other words, "for all x, not $P(x)$." In summary,

The negation of "for all x, $P(x)$" is "there exists an x such that not $P(x)$."

The negation of "there exists an x such that $P(x)$" is "for every x, not $P(x)$."

Concepts Review

1. Numbers that can be written as the ratio of two integers are called _____.

2. Between any two real numbers, there is another real number. This is what it means to say that the real numbers are _____.

3. The contrapositive of "If P then Q" is _____.

4. Axioms and definitions are taken for granted, but _____ require proof.

Problem Set 0.1

In Problems 1–16, simplify as much as possible. Be sure to remove all parentheses and reduce all fractions.

1. $4 - 2(8 - 11) + 6$

2. $3[2 - 4(7 - 12)]$

3. $-4[5(-3 + 12 - 4) + 2(13 - 7)]$

4. $5[-1(7 + 12 - 16) + 4] + 2$

5. $\frac{5}{7} - \frac{1}{13}$

6. $\frac{3}{4 - 7} + \frac{3}{21} - \frac{1}{6}$

7. $\frac{1}{3}\left[\frac{1}{2}\left(\frac{1}{4} - \frac{1}{3}\right) + \frac{1}{6}\right]$

8. $-\frac{1}{3}\left[\frac{2}{5} - \frac{1}{2}\left(\frac{1}{3} - \frac{1}{5}\right)\right]$

9. $\frac{14}{21}\left(\frac{2}{5 - \frac{1}{3}}\right)^2$

10. $\left(\frac{2}{7} - 5\right)/\left(1 - \frac{1}{7}\right)$

11. $\dfrac{\frac{11}{7} - \frac{12}{21}}{\frac{11}{7} + \frac{12}{21}}$

12. $\dfrac{\frac{1}{2} - \frac{3}{4} + \frac{7}{8}}{\frac{1}{2} + \frac{3}{4} - \frac{7}{8}}$

13. $1 - \dfrac{1}{1 + \frac{1}{2}}$

14. $2 + \dfrac{3}{1 + \frac{5}{2}}$

15. $\left(\sqrt{5} + \sqrt{3}\right)\left(\sqrt{5} - \sqrt{3}\right)$

16. $\left(\sqrt{5} - \sqrt{3}\right)^2$

In Problems 17–28, perform the indicated operations and simplify.

17. $(3x - 4)(x + 1)$

18. $(2x - 3)^2$

19. $(3x - 9)(2x + 1)$

20. $(4x - 11)(3x - 7)$

21. $(3t^2 - t + 1)^2$

22. $(2t + 3)^3$

23. $\dfrac{x^2 - 4}{x - 2}$

24. $\dfrac{x^2 - x - 6}{x - 3}$

25. $\dfrac{t^2 - 4t - 21}{t + 3}$

26. $\dfrac{2x - 2x^2}{x^3 - 2x^2 + x}$

27. $\dfrac{12}{x^2 + 2x} + \dfrac{4}{x} + \dfrac{2}{x + 2}$

28. $\dfrac{2}{6y - 2} + \dfrac{y}{9y^2 - 1}$

29. Find the value of each of the following; if undefined, say so.

(a) $0 \cdot 0$ (b) $\frac{0}{0}$ (c) $\frac{0}{17}$

(d) $\frac{3}{0}$ (e) 0^5 (f) 17^0

30. Show that division by 0 is meaningless as follows: Suppose that $a \neq 0$. If $a/0 = b$, then $a = 0 \cdot b = 0$, which is a contradiction. Now find a reason why $0/0$ is also meaningless.

In Problems 31–36, change each rational number to a decimal by performing long division.

31. $\frac{1}{12}$

32. $\frac{2}{7}$

33. $\frac{3}{21}$

34. $\frac{5}{17}$

35. $\frac{11}{3}$

36. $\frac{11}{13}$

In Problems 37–42, change each repeating decimal to a ratio of two integers (see Example 1).

37. 0.123123123...

38. 0.217171717...

39. 2.56565656...

40. 3.929292...

41. 0.199999...

42. 0.399999...

43. Since 0.199999... = 0.200000... and 0.399999... = 0.400000... (see Problems 41 and 42), we see that certain rational numbers have two different decimal expansions. Which rational numbers have this property?

44. Show that any rational number p/q, for which the prime factorization of q consists entirely of 2s and 5s, has a terminating decimal expansion.

45. Find a positive rational number and a positive irrational number both smaller than 0.00001.

46. What is the smallest positive integer? The smallest positive rational number? The smallest positive irrational number?

47. Find a rational number between 3.14159 and π. Note that $\pi = 3.141592...$

48. Is there a number between 0.9999... (repeating 9s) and 1? How do you resolve this with the statement that between any two different real numbers there is another real number?

49. Is 0.12345678910111213 14... rational or irrational? (You should see a pattern in the given sequence of digits.)

50. Find two irrational numbers whose sum is rational.

≈ *In Problems 51–56, find the best decimal approximation that your calculator allows. Begin by making a mental estimate.*

51. $(\sqrt{3} + 1)^3$

52. $(\sqrt{2} - \sqrt{3})^4$

53. $\sqrt[4]{1.123} - \sqrt[3]{1.09}$

54. $(3.1415)^{-1/2}$

55. $\sqrt{8.9\pi^2 + 1} - 3\pi$

56. $\sqrt[4]{(6\pi^2 - 2)\pi}$

57. Show that between any two different real numbers there is a rational number. (*Hint:* If $a < b$, then $b - a > 0$, so there is a natural number n such that $1/n < b - a$. Consider the set $\{k : k/n > b\}$ and use the fact that a set of integers that is bounded from below contains a least element.) Show that between any two different real numbers there are infinitely many rational numbers.

≈ **58.** Estimate the number of cubic inches in your head.

≈ **59.** Estimate the length of the equator in feet. Assume the radius of the earth to be 4000 miles.

≈ **60.** About how many times has your heart beat by your twentieth birthday?

≈ **61.** The General Sherman tree in California is about 270 feet tall and averages about 16 feet in diameter. Estimate the number of board feet (1 board foot equals 1 inch by 12 inches by 12 inches) of lumber that could be made from this tree, assuming no waste and ignoring the branches.

≈ **62.** Assume that the General Sherman tree (Problem 61) produces an annual growth ring of thickness 0.004 foot. Estimate the resulting increase in the volume of its trunk each year.

63. Write the converse and the contrapositive to the following statements.

(a) If it rains today, then I will stay home from work.

(b) If the candidate meets all the qualifications, then she will be hired.

64. Write the converse and the contrapositive to the following statements.

(a) If I get an A on the final exam, I will pass the course.

(b) If I finish my research paper by Friday, then I will take off next week.

65. Write the converse and the contrapositive to the following statements.

(a) (Let a, b, and c be the lengths of sides of a triangle.) If $a^2 + b^2 = c^2$, then the triangle is a right triangle.

(b) If angle ABC is acute, then its measure is greater than 0° and less than 90°.

66. Write the converse and the contrapositive to the following statements.

(a) If the measure of angle ABC is 45°, then angle ABC is an acute angle.

(b) If $a < b$ then $a^2 < b^2$.

67. Consider the statements in Problem 65 along with their converses and contrapositives. Which are true?

68. Consider the statements in Problem 66 along with their converses and contrapositives. Which are true?

69. Use the rules regarding the negation of statements involving quantifiers to write the negation of the following statements. Which is true, the original statement or its negation?

(a) Every isosceles triangle is equilateral.

(b) There is a real number that is not an integer.

(c) Every natural number is less than or equal to its square.

70. Use the rules regarding the negation of statements involving quantifiers to write the negation of the following statements. Which is true, the original statement or its negation?

(a) Every natural number is rational.

(b) There is a circle whose area is larger than 9π.

(c) Every real number is larger than its square.

71. Which of the following are true? Assume that x and y are real numbers.

(a) For every x, $x > 0 \Rightarrow x^2 > 0$.

(b) For every x, $x > 0 \Leftrightarrow x^2 > 0$.

(c) For every x, $x^2 > x$.

(d) For every x, there exists a y such that $y > x^2$.

(e) For every positive number y, there exists another positive number x such that $0 < x < y$.

72. Which of the following are true? Unless it is stated otherwise, assume that x, y, and ε are real numbers.

(a) For every x, $x < x + 1$.

(b) There exists a natural number N such that all prime numbers are less than N. (A **prime number** is a natural number whose only factors are 1 and itself.)

(c) For every $x > 0$, there exists a y such that $y > \frac{1}{x}$.

(d) For every positive x, there exists a natural number n such that $\frac{1}{n} < x$.

(e) For every positive ε, there exists a natural number n such that $\frac{1}{2^n} < \varepsilon$.

73. Prove the following statements.

(a) If n is odd, then n^2 is odd. (*Hint:* If n is odd, then there exists an integer k such that $n = 2k + 1$.)

(b) If n^2 is odd, then n is odd. (*Hint:* Prove the contrapositive.)

74. Prove that n is odd if and only if n^2 is odd. (See Problem 73.)

75. According to the **Fundamental Theorem of Arithmetic,** every natural number greater than 1 can be written as the product of primes in a unique way, except for the order of the factors. For example, $45 = 3 \cdot 3 \cdot 5$. Write each of the following as a product of primes.

(a) 243 (b) 124 (c) 5100

76. Use the Fundamental Theorem of Arithmetic (Problem 75) to show that the square of any natural number greater than 1 can be written as the product of primes in a unique way, except for the order of the factors, with each prime occurring an *even* number of times. For example, $(45)^2 = 3 \cdot 3 \cdot 3 \cdot 3 \cdot 5 \cdot 5$.

77. Show that $\sqrt{2}$ is irrational. *Hint:* Try a proof by contradiction. Suppose that $\sqrt{2} = p/q$, where p and q are natural numbers (necessarily different from 1). Then $2 = p^2/q^2$, and so $2q^2 = p^2$. Now use Problem 76 to get a contradiction.

78. Show that $\sqrt{3}$ is irrational (see Problem 77).

79. Show that the sum of two rational numbers is rational.

80. Show that the product of a rational number (other than 0) and an irrational number is irrational. *Hint:* Try proof by contradiction.

81. Which of the following are rational and which are irrational?

(a) $-\sqrt{9}$ (b) 0.375

(c) $(3\sqrt{2})(5\sqrt{2})$ (d) $(1 + \sqrt{3})^2$

82. A number b is called an **upper bound** for a set S of numbers if $x \le b$ for all x in S. For example 5, 6.5, and 13 are upper bounds for the set $S = \{1, 2, 3, 4, 5\}$. The number 5 is the **least upper bound** for S (the smallest of all upper bounds). Similarly, 1.6, 2, and 2.5 are upper bounds for the infinite set $T = \{1.4, 1.49, 1.499, 1.4999, \ldots\}$, whereas 1.5 is its least upper bound. Find the least upper bound of each of the following sets.

(a) $S = \{-10, -8, -6, -4, -2\}$

(b) $S = \{-2, -2.1, -2.11, -2.111, -2.1111, \ldots\}$

(c) $S = \{2.4, 2.44, 2.444, 2.4444, \ldots\}$

(d) $S = \{1 - \frac{1}{2}, 1 - \frac{1}{3}, 1 - \frac{1}{4}, 1 - \frac{1}{5}, \ldots\}$

(e) $S = \{x : x = (-1)^n + 1/n, n \text{ a positive integer}\}$; that is, S is the set of all numbers x that have the form $x = (-1)^n + 1/n$, where n is a positive integer.

(f) $S = \{x : x^2 < 2, x \text{ a rational number}\}$

EXPL **83. The Axiom of Completeness** for the real numbers says: Every set of real numbers that has an upper bound has a *least* upper bound that is a real number.

(a) Show that the italicized statement is false if the word *real* is replaced by *rational*.

(b) Would the italicized statement be true or false if the word *real* were replaced by *natural*?

Answers to Concepts Review: 1. rational numbers **2.** dense **3.** "If not Q then not P." **4.** theorems

0.2 Inequalities and Absolute Values

Solving equations (for instance, $3x - 17 = 6$ or $x^2 - x - 6 = 0$) is one of the traditional tasks of mathematics; it will be important in this course and we assume that you remember how to do it. But of almost equal significance in calculus is the notion of solving an inequality (e.g., $3x - 17 < 6$ or $x^2 - x - 6 \ge 0$). To **solve** an inequality is to find the set of all real numbers that make the inequality true. In contrast to an equation, whose solution set normally consists of one number or perhaps a finite set of numbers, the solution set of an inequality is usually an entire interval of numbers or, in some cases, the union of such intervals.

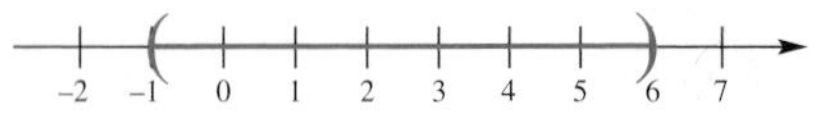

$(-1, 6) = \{x : -1 < x < 6\}$

Figure 1

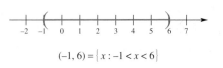

Intervals Several kinds of intervals will arise in our work and we introduce special terminology and notation for them. The inequality $a < x < b$, which is actually two inequalities, $a < x$ and $x < b$, describes the **open interval** consisting of all numbers between a and b, not including the end points a and b. We denote this interval by the symbol (a, b) (Figure 1). In contrast, the inequality $a \le x \le b$ describes the corresponding **closed interval,** which does include the end points a and

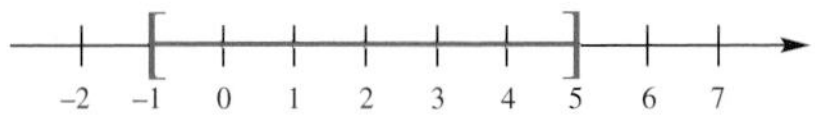

$[-1, 5] = \{x : -1 \le x \le 5\}$

Figure 2

b. This interval is denoted by $[a, b]$ (Figure 2). The table indicates the wide variety of possibilities and introduces our notation.

Set Notation	Interval Notation	Graph
$\{x: a < x < b\}$	(a, b)	
$\{x: a \le x \le b\}$	$[a, b]$	
$\{x: a \le x < b\}$	$[a, b)$	
$\{x: a < x \le b\}$	$(a, b]$	
$\{x: x \le b\}$	$(-\infty, b]$	
$\{x: x < b\}$	$(-\infty, b)$	
$\{x: x \ge a\}$	$[a, \infty)$	
$\{x: x > a\}$	(a, ∞)	
$\mathbb{R}$	$(-\infty, \infty)$	

Solving Inequalities As with equations, the procedure for solving an inequality consists of transforming the inequality one step at a time until the solution set is obvious. We may perform certain operations on both sides of an inequality without changing its solution set. In particular,

1. We may add the same number to both sides of an inequality.
2. We may multiply both sides of an inequality by the same positive number.
3. We may multiply both sides by the same negative number, but then we must reverse the direction of the inequality sign.

EXAMPLE 1 Solve the inequality $2x - 7 < 4x - 2$ and show the graph of its solution set.

SOLUTION

$$\begin{aligned} 2x - 7 &< 4x - 2 & \\ 2x &< 4x + 5 & \text{(adding 7)} \\ -2x &< 5 & \text{(adding } -4x\text{)} \\ x &> -\tfrac{5}{2} & \text{(multiplying by } -\tfrac{1}{2}\text{)} \end{aligned}$$

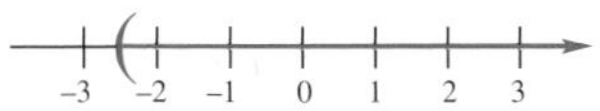

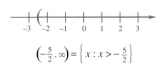

$\left(-\frac{5}{2}, \infty\right) = \left\{x : x > -\frac{5}{2}\right\}$

Figure 3

The graph appears in Figure 3. ■

EXAMPLE 2 Solve $-5 \le 2x + 6 < 4$.

SOLUTION

$$\begin{aligned} -5 \le 2x &+ 6 < 4 & \\ -11 \le 2x &\quad < -2 & \text{(adding } -6\text{)} \\ -\tfrac{11}{2} \le x &\quad < -1 & \text{(multiplying by } \tfrac{1}{2}\text{)} \end{aligned}$$

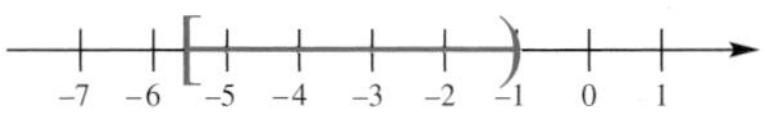

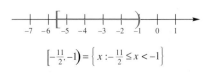

$\left[-\frac{11}{2}, -1\right) = \left\{x : -\frac{11}{2} \le x < -1\right\}$

Figure 4

Figure 4 shows the corresponding graph. ■

Before tackling a quadratic inequality, we point out that a linear factor of the form $x - a$ is positive for $x > a$ and negative for $x < a$. If follows that a product $(x - a)(x - b)$ can change from being positive to negative, or vice versa, only at a or b. These points, where a factor is zero, are called **split points.** They are the keys to determining the solution sets of quadratic and other more complicated inequalities.

EXAMPLE 3 Solve the quadratic inequality $x^2 - x < 6$.

SOLUTION As with quadratic equations, we move all nonzero terms to one side and factor.

$$x^2 - x < 6$$

$$x^2 - x - 6 < 0 \qquad \text{(adding } -6\text{)}$$

$$(x - 3)(x + 2) < 0 \qquad \text{(factoring)}$$

Test Point	Sign of $(x - 3)$	Sign of $(x + 2)$	Sign of $(x - 3)(x + 2)$
−3	−	−	+
0	−	+	−
5	+	+	+

We see that −2 and 3 are the split points; they divide the real line into the three intervals $(-\infty, -2)$, $(-2, 3)$, and $(3, \infty)$. On each of these intervals, $(x - 3)(x + 2)$ is of one sign; that is, it is either always positive or always negative. To find this sign in each interval, we use the **test points** −3, 0, and 5 (any points in the three intervals would do). Our results are shown in the margin.

The information we have obtained is summarized in the top half of Figure 5. We conclude that the solution set for $(x - 3)(x + 2) < 0$ is the interval $(-2, 3)$. Its graph is shown in the bottom half of Figure 5. ■

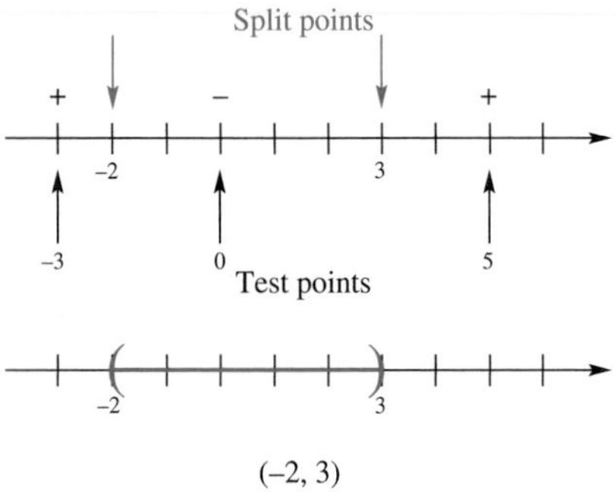

Figure 5

EXAMPLE 4 Solve $3x^2 - x - 2 > 0$.

SOLUTION Since

$$3x^2 - x - 2 = (3x + 2)(x - 1) = 3(x - 1)\left(x + \tfrac{2}{3}\right)$$

the split points are $-\frac{2}{3}$ and 1. These points, together with the test points −2, 0, and 2, establish the information shown in the top part of Figure 6. We conclude that the solution set of the inequality consists of the points in either $\left(-\infty, -\frac{2}{3}\right)$ or $(1, \infty)$. In set language, the solution set is the **union** (symbolized by ∪) of these two intervals; that is, it is $\left(-\infty, -\frac{2}{3}\right) \cup (1, \infty)$. ■

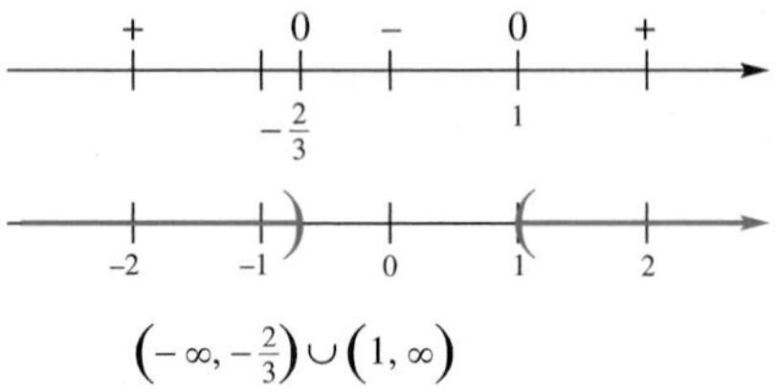

Figure 6

EXAMPLE 5 Solve $\dfrac{x - 1}{x + 2} \geq 0$.

SOLUTION Our inclination to multiply both sides by $x + 2$ leads to an immediate dilemma, since $x + 2$ may be either positive or negative. Should we reverse the inequality sign or leave it alone? Rather than try to untangle this problem (which would require breaking it into two cases), we observe that the quotient $(x - 1)/(x + 2)$ can change sign only at the split points of the numerator and denominator, that is, at 1 and −2. The test points −3, 0, and 2 yield the information displayed in the top part of Figure 7. The symbol u indicates that the quotient is undefined at −2. We conclude that the solution set is $(-\infty, -2) \cup [1, \infty)$. Note that −2 is not in the solution set because the quotient is undefined there. On the other hand, 1 is included because the inequality is true when $x = 1$. ■

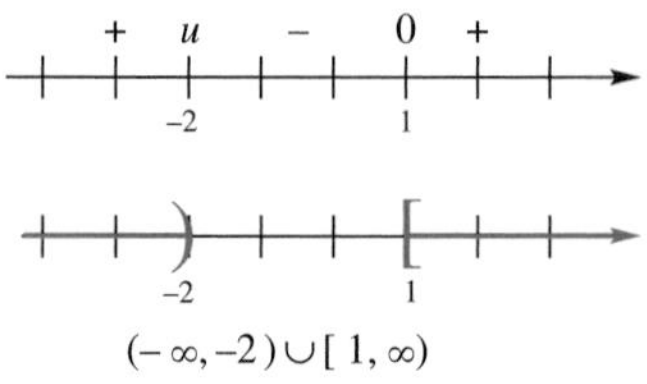

Figure 7

EXAMPLE 6 Solve $(x + 1)(x - 1)^2(x - 3) \leq 0$.

SOLUTION The split points are −1, 1 and 3, which divide the real line into four intervals, as shown in Figure 8. After testing these intervals, we conclude that the solution set is $[-1, 1] \cup [1, 3]$, which is the interval $[-1, 3]$. ■

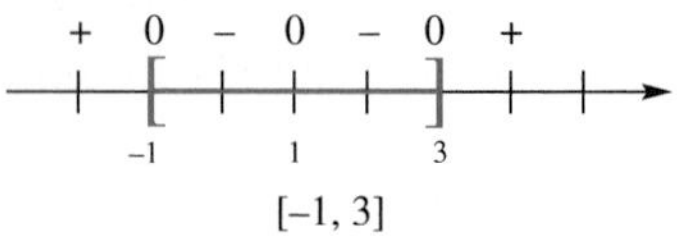

Figure 8

EXAMPLE 7 Solve $2.9 < \dfrac{1}{x} < 3.1$.

SOLUTION It is tempting to multiply through by x, but this again brings up the dilemma that x may be positive or negative. In this case, however, $\frac{1}{x}$ must be between 2.9 and 3.1, which guarantees that x is positive. It is therefore permissible to multiply by x and not reverse the inequalities. Thus,

$$2.9x < 1 < 3.1x$$

At this point, we must break this compound inequality into two inequalities, which we solve separately.

$$2.9x < 1 \quad \text{and} \quad 1 < 3.1x$$

$$x < \frac{1}{2.9} \quad \text{and} \quad \frac{1}{3.1} < x$$

Any value of x that satisfies the original inequality must satisfy both of these inequalities. The solution set thus consists of those values of x satisfying

$$\frac{1}{3.1} < x < \frac{1}{2.9}$$

This inequality can be written as

$$\frac{10}{31} < x < \frac{10}{29}$$

The interval $\left(\frac{10}{31}, \frac{10}{29}\right)$ is shown in Figure 9. ■

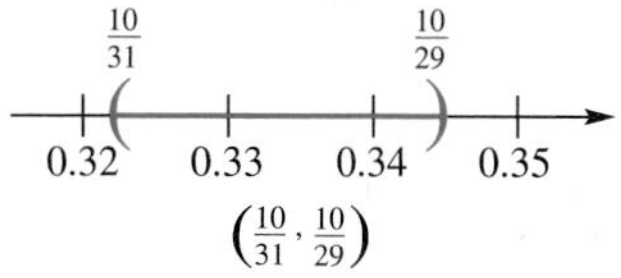

Figure 9

Absolute Values

The concept of absolute value is extremely useful in calculus, and the reader should acquire skill in working with it. The **absolute value** of a real number x, denoted by $|x|$, is defined by

$$|x| = x \quad \text{if } x \geq 0$$
$$|x| = -x \quad \text{if } x < 0$$

For example, $|6| = 6$, $|0| = 0$, and $|-5| = -(-5) = 5$. This two-pronged definition merits careful study. Note that it does not say that $|-x| = x$ (try $x = -5$ to see why). It is true that $|x|$ is always nonnegative; it is also true that $|-x| = |x|$.

One of the best ways to think of the absolute value of a number is as an undirected distance. In particular, $|x|$ is the distance between x and the origin. Similarly, $|x - a|$ is the distance between x and a (Figure 10).

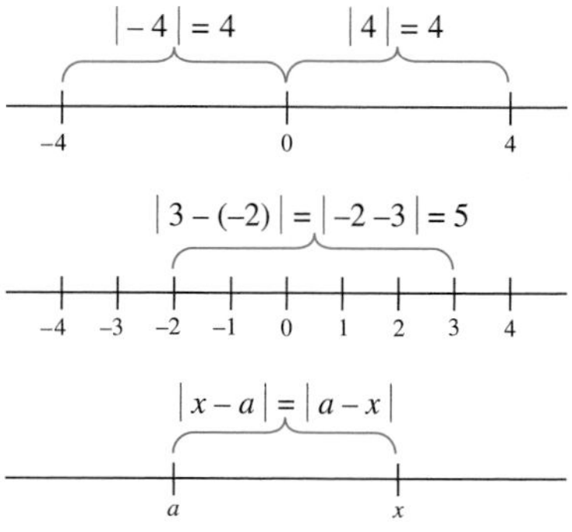

Figure 10

Properties

Absolute values behave nicely under multiplication and division, but not so well under addition and subtraction.

Properties of Absolute Values

1. $|ab| = |a||b|$
2. $\left|\frac{a}{b}\right| = \frac{|a|}{|b|}$
3. $|a + b| \leq |a| + |b|$ (Triangle Inequality)
4. $|a - b| \geq \big||a| - |b|\big|$

Inequalities Involving Absolute Values

If $|x| < 3$, then the distance between x and the origin must be less than 3. In other words, x must be simultaneously less than 3 *and* greater than -3; that is, $-3 < x < 3$. On the other hand, if $|x| > 3$, then the distance between x and the origin must be at least 3. This can happen when $x > 3$ *or* $x < -3$ (Figure 11). These are special cases of the following general statements that hold when $a > 0$.

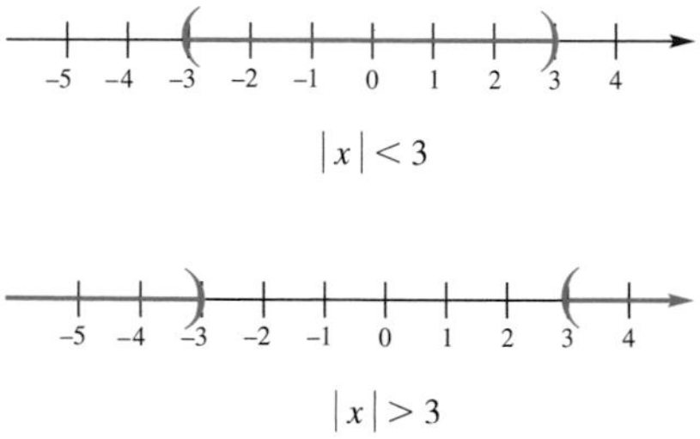

Figure 11

(1) $$|x| < a \Leftrightarrow -a < x < a$$
$$|x| > a \Leftrightarrow x < -a \quad \text{or} \quad x > a$$

We can use these facts to solve inequalities involving absolute values, since they provide a way of removing absolute value signs.

EXAMPLE 8 Solve the inequality $|x - 4| < 2$ and show the solution set on the real line. Interpret the absolute value as a distance.

SOLUTION From the equations in (1), with x replaced by $|x - 4|$, we see that

$$|x - 4| < 2 \Leftrightarrow -2 < x - 4 < 2$$

When we add 4 to all three members of this latter inequality, we obtain $2 < x < 6$. The graph is shown in Figure 12.

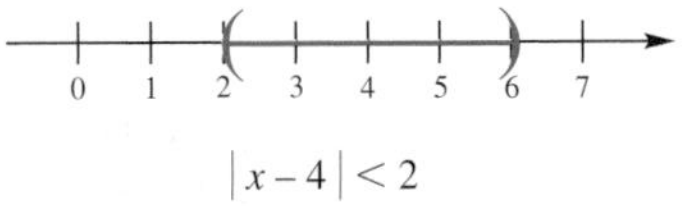

Figure 12

In terms of distance, the symbol $|x - 4|$ represents the distance between x and 4. The inequality says that the distance between x and 4 is less than 2. The numbers x with this property are the numbers between 2 and 6; that is, $2 < x < 6$. ■

The statements in the equations just before Example 8 are valid with $<$ and $>$ replaced by $\le$ and $\ge$, respectively. We need the second statement in this form in our next example.

EXAMPLE 9 Solve the inequality $|3x - 5| \ge 1$ and show its solution set on the real line.

SOLUTION The given inequality may be written successively as

$$\begin{aligned} 3x - 5 \le -1 \quad &\text{or} \quad 3x - 5 \ge 1 \\ 3x \le 4 \quad &\text{or} \quad 3x \ge 6 \\ x \le \tfrac{4}{3} \quad &\text{or} \quad x \ge 2 \end{aligned}$$

The solution set is the union of two intervals, $\left(-\infty, \frac{4}{3}\right] \cup [2, \infty)$, and is shown in Figure 13. ■

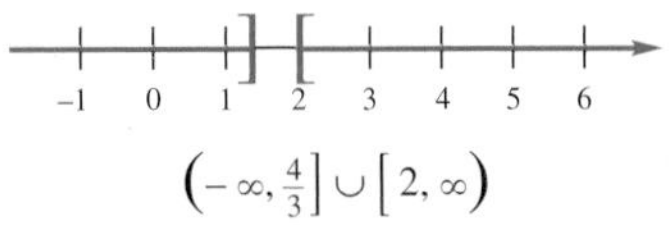

Figure 13

In Chapter 1, we will need to make the kind of manipulations illustrated by the next two examples. Delta (δ) and epsilon (ε) are the fourth and fifth letters, respectively, of the Greek alphabet and are traditionally used to stand for small positive numbers.

EXAMPLE 10 Let ε (epsilon) be a positive number. Show that

$$|x - 2| < \frac{\varepsilon}{5} \Leftrightarrow |5x - 10| < \varepsilon$$

In terms of distance, this says that the distance between x and 2 is less than $\varepsilon/5$ if and only if the distance between $5x$ and 10 is less than ε.

SOLUTION

$$\begin{aligned} |x - 2| < \frac{\varepsilon}{5} \quad &\Leftrightarrow \quad 5|x - 2| < \varepsilon && \text{(multiplying by 5)} \\ &\Leftrightarrow \quad |5|\,|(x - 2)| < \varepsilon && (|5| = 5) \\ &\Leftrightarrow \quad |5(x - 2)| < \varepsilon && (|a||b| = |ab|) \\ &\Leftrightarrow \quad |5x - 10| < \varepsilon \end{aligned}$$

■

> **Finding Delta**
>
> Note two facts about our solution to Example 11.
>
> 1. The value we find for δ must depend on ε. Our choice is $\delta = \varepsilon/6$.
> 2. Any positive δ smaller than $\varepsilon/6$ is acceptable. For example $\delta = \varepsilon/7$ or $\delta = \varepsilon/(2\pi)$ are other correct choices.

EXAMPLE 11 Let ε be a positive number. Find a positive number δ (delta) such that

$$|x - 3| < \delta \Rightarrow |6x - 18| < \varepsilon$$

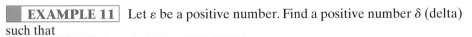

SOLUTION

$$\begin{aligned} |6x - 18| < \varepsilon &\Leftrightarrow |6(x - 3)| < \varepsilon \\ &\Leftrightarrow 6|x - 3| < \varepsilon \qquad (|ab| = |a||b|) \\ &\Leftrightarrow |x - 3| < \frac{\varepsilon}{6} \qquad \left(\text{multiplying by } \frac{1}{6}\right) \end{aligned}$$

Therefore, we choose $\delta = \varepsilon/6$. Following the implications backward, we see that

$$|x - 3| < \delta \Rightarrow |x - 3| < \frac{\varepsilon}{6} \Rightarrow |6x - 18| < \varepsilon$$ ■

Here is a practical problem that uses the same type of reasoning.

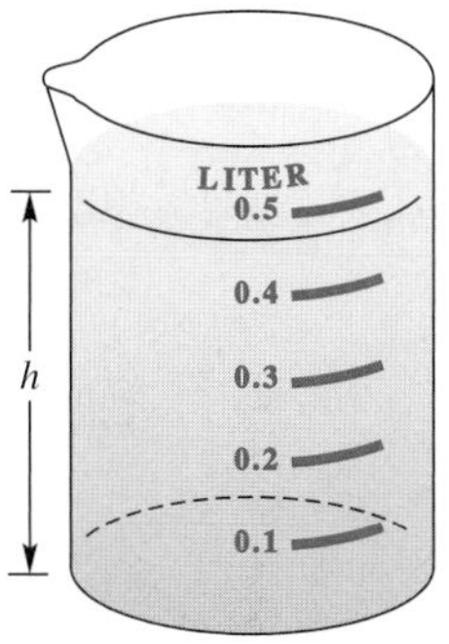

Figure 14

EXAMPLE 12 A $\frac{1}{2}$-liter (500 cubic centimeter) glass beaker has an inner radius of 4 centimeters. How closely must we measure the height h of water in the beaker to be sure that we have $\frac{1}{2}$ liter of water with an error of less than 1%, that is, an error of less than 5 cubic centimeters? See Figure 14.

SOLUTION The volume V of water in the glass is given by the formula $V = 16\pi h$. We want $|V - 500| < 5$ or, equivalently, $|16\pi h - 500| < 5$. Now

$$\begin{aligned} |16\pi h - 500| < 5 &\Leftrightarrow \left|16\pi\left(h - \frac{500}{16\pi}\right)\right| < 5 \\ &\Leftrightarrow 16\pi\left|h - \frac{500}{16\pi}\right| < 5 \\ &\Leftrightarrow \left|h - \frac{500}{16\pi}\right| < \frac{5}{16\pi} \\ &\Leftrightarrow |h - 9.947| < 0.09947 \approx 0.1 \end{aligned}$$

Thus, we must measure the height to an accuracy of about 0.1 centimeter, or 1 millimeter. ■

Notation for Square Roots

Every positive number has two square roots. For example, the two square roots of 9 are 3 and -3. We sometimes represent these two numbers as ± 3. For $a \geq 0$, the symbol $\sqrt{a}$, called the **principal square root** of a, denotes the nonnegative square root of a. Thus, $\sqrt{9} = 3$ and $\sqrt{121} = 11$. It is incorrect to write $\sqrt{16} = \pm 4$ because $\sqrt{16}$ means the nonnegative square root of 16, that is, 4. The number 7 has two square roots, which are written as $\pm\sqrt{7}$, but $\sqrt{7}$ represents a single real number. Just remember this:

$$a^2 = 16$$

has two solutions, $a = -4$ and $a = 4$, but

$$\sqrt{16} = 4$$

Quadratic Formula

Most students will recall the **Quadratic Formula.** The solutions to the quadratic equation $ax^2 + bx + c = 0$ are given by

$$x = \frac{-b \pm \sqrt{b^2 - 4ac}}{2a}$$

The number $d = b^2 - 4ac$ is called the **discriminant** of the quadratic equation. The equation $ax^2 + bx + c = 0$ has two real solutions if $d > 0$, one real solution if $d = 0$, and no real solutions if $d < 0$. With the Quadratic Formula, we can easily solve quadratic inequalities even if they do not factor by inspection.

EXAMPLE 13 Solve $x^2 - 2x - 4 \leq 0$.

SOLUTION The two solutions of $x^2 - 2x - 4 = 0$ are

$$x_1 = \frac{-(-2) - \sqrt{4 + 16}}{2} = 1 - \sqrt{5} \approx -1.24$$

and

$$x_2 = \frac{-(-2) + \sqrt{4 + 16}}{2} = 1 + \sqrt{5} \approx 3.24$$

Thus,

$$x^2 - 2x - 4 = (x - x_1)(x - x_2) = \left(x - 1 + \sqrt{5}\right)\left(x - 1 - \sqrt{5}\right)$$

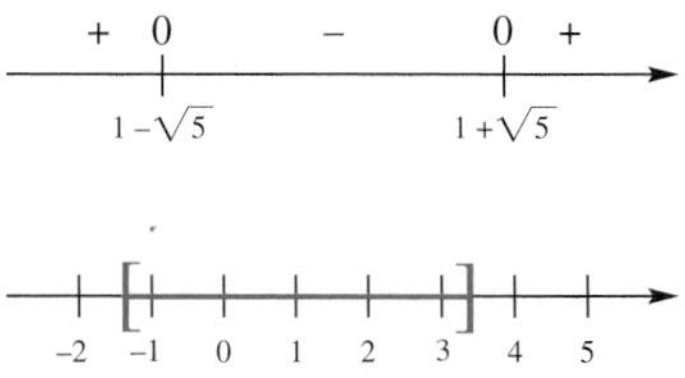

Figure 15

The split points $1 - \sqrt{5}$ and $1 + \sqrt{5}$ divide the real line into three intervals (Figure 15). When we test them with the test points -2, 0, and 4, we conclude that the solution set for $x^2 - 2x - 4 \leq 0$ is $\left[1 - \sqrt{5}, 1 + \sqrt{5}\right]$. ■

Squares

Turning to squares, we notice that

$$|x|^2 = x^2 \quad \text{and} \quad |x| = \sqrt{x^2}$$

> **Notation for Roots**
>
> If n is even and $a \geq 0$ the symbol $\sqrt[n]{a}$ denotes the nonnegative nth root of a. When n is odd, there is only one real nth root of a, denoted by the symbol $\sqrt[n]{a}$. Thus, $\sqrt[4]{16} = 2$, $\sqrt[3]{27} = 3$, and $\sqrt[3]{-8} = -2$.

These follow from the property $|a||b| = |ab|$.

Does the squaring operation preserve inequalities? In general, the answer is no. For instance, $-3 < 2$, but $(-3)^2 > 2^2$. On the other hand, $2 < 3$ and $2^2 < 3^2$. If we are dealing with nonnegative numbers, then $a < b \Leftrightarrow a^2 < b^2$. A useful variant of this (see Problem 63) is

$$|x| < |y| \Leftrightarrow x^2 < y^2$$

EXAMPLE 14 Solve the inequality $|3x + 1| < 2|x - 6|$.

SOLUTION This inequality is more difficult to solve than our earlier examples, because there are two sets of absolute value signs. We can remove both of them by using the last boxed result.

$$\begin{aligned} |3x + 1| < 2|x - 6| &\Leftrightarrow |3x + 1| < |2x - 12| \\ &\Leftrightarrow (3x + 1)^2 < (2x - 12)^2 \\ &\Leftrightarrow 9x^2 + 6x + 1 < 4x^2 - 48x + 144 \\ &\Leftrightarrow 5x^2 + 54x - 143 < 0 \\ &\Leftrightarrow (x + 13)(5x - 11) < 0 \end{aligned}$$

The split points for this quadratic inequality are -13 and $\frac{11}{5}$; they divide the real line into the three intervals: $(-\infty, -13)$, $\left(-13, \frac{11}{5}\right)$, and $\left(\frac{11}{5}, \infty\right)$. When we use the test points -14, 0, and 3, we discover that only the points in $\left(-13, \frac{11}{5}\right)$ satisfy the inequality. ■

Concepts Review

1. The set $\{x\colon -1 \leq x < 5\}$ is written in interval notation as _____ and the set $\{x\colon x \leq -2\}$ is written as _____.

2. If $a/b < 0$, then either $a < 0$ and _____ or $a > 0$ and _____.

3. Which of the following are always true?

(a) $|-x| = x$ (b) $|x|^2 = x^2$

(c) $|xy| = |x||y|$ (d) $\sqrt{x^2} = x$

4. The inequality $|x - 2| \leq 3$ is equivalent to _____ $\leq x \leq$ _____.

Problem Set 0.2

1. Show each of the following intervals on the real line.

(a) $[-1, 1]$ (b) $(-4, 1]$

(c) $(-4, 1)$ (d) $[1, 4]$

(e) $[-1, \infty)$ (f) $(-\infty, 0]$

2. Use the notation of Problem 1 to describe the following intervals.

(a)

(b)

(c)

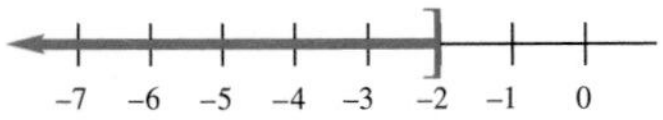

(d)

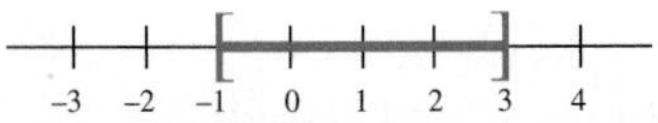

In each of Problems 3–26, express the solution set of the given inequality in interval notation and sketch its graph.

3. $x - 7 < 2x - 5$

4. $3x - 5 < 4x - 6$

5. $7x - 2 \leq 9x + 3$

6. $5x - 3 > 6x - 4$

7. $-4 < 3x + 2 < 5$

8. $-3 < 4x - 9 < 11$

9. $-3 < 1 - 6x \leq 4$

10. $4 < 5 - 3x < 7$

11. $x^2 + 2x - 12 < 0$

12. $x^2 - 5x - 6 > 0$

13. $2x^2 + 5x - 3 > 0$

14. $4x^2 - 5x - 6 < 0$

15. $\dfrac{x + 4}{x - 3} \leq 0$

16. $\dfrac{3x - 2}{x - 1} \geq 0$

17. $\dfrac{2}{x} < 5$

18. $\dfrac{7}{4x} \leq 7$

19. $\dfrac{1}{3x - 2} \leq 4$

20. $\dfrac{3}{x + 5} > 2$

21. $(x + 2)(x - 1)(x - 3) > 0$

22. $(2x + 3)(3x - 1)(x - 2) < 0$

23. $(2x - 3)(x - 1)^2(x - 3) \geq 0$

24. $(2x - 3)(x - 1)^2(x - 3) > 0$

25. $x^3 - 5x^2 - 6x < 0$ **26.** $x^3 - x^2 - x + 1 > 0$

27. Tell whether each of the following is true or false.

(a) $-3 < -7$ (b) $-1 > -17$ (c) $-3 < -\frac{22}{7}$

28. Tell whether each of the following is true or false.

(a) $-5 > -\sqrt{26}$ (b) $\frac{6}{7} < \frac{34}{39}$ (c) $-\frac{5}{7} < -\frac{44}{59}$

29. Assume that $a > 0, b > 0$. Prove each statement. *Hint:* Each part requires two proofs: one for $\Rightarrow$ and one for $\Leftarrow$.

(a) $a < b \Leftrightarrow a^2 < b^2$ (b) $a < b \Leftrightarrow \frac{1}{a} > \frac{1}{b}$

30. Which of the following are true if $a \leq b$?

(a) $a^2 \leq ab$ (b) $a - 3 \leq b - 3$

(c) $a^3 \leq a^2 b$ (d) $-a \leq -b$

31. Find all values of x that satisfy both inequalities simultaneously.

(a) $3x + 7 > 1$ and $2x + 1 < 3$

(b) $3x + 7 > 1$ and $2x + 1 > -4$

(c) $3x + 7 > 1$ and $2x + 1 < -4$

32. Find all the values of x that satisfy at least one of the two inequalities.

(a) $2x - 7 > 1$ or $2x + 1 < 3$

(b) $2x - 7 \leq 1$ or $2x + 1 < 3$

(c) $2x - 7 \leq 1$ or $2x + 1 > 3$

33. Solve for x, expressing your answer in interval notation.

(a) $(x + 1)(x^2 + 2x - 7) \geq x^2 - 1$

(b) $x^4 - 2x^2 \geq 8$

(c) $(x^2 + 1)^2 - 7(x^2 + 1) + 10 < 0$

34. Solve each inequality. Express your solution in interval notation.

(a) $1.99 < \frac{1}{x} < 2.01$ (b) $2.99 < \frac{1}{x + 2} < 3.01$

In Problems 35–44, find the solution sets of the given inequalities.

35. $|x - 2| \geq 5$ **36.** $|x + 2| < 1$

37. $|4x + 5| \leq 10$ **38.** $|2x - 1| > 2$

39. $\left|\frac{2x}{7} - 5\right| \geq 7$ **40.** $\left|\frac{x}{4} + 1\right| < 1$

41. $|5x - 6| > 1$ **42.** $|2x - 7| > 3$

43. $\left|\frac{1}{x} - 3\right| > 6$ **44.** $\left|2 + \frac{5}{x}\right| > 1$

In Problems 45–48, solve the given quadratic inequality using the Quadratic Formula.

45. $x^2 - 3x - 4 \geq 0$ **46.** $x^2 - 4x + 4 \leq 0$

47. $3x^2 + 17x - 6 > 0$ **48.** $14x^2 + 11x - 15 \leq 0$

In Problems 49–52, show that the indicated implication is true.

49. $|x - 3| < 0.5 \Rightarrow |5x - 15| < 2.5$

50. $|x + 2| < 0.3 \Rightarrow |4x + 8| < 1.2$

51. $|x - 2| < \frac{\varepsilon}{6} \Rightarrow |6x - 12| < \varepsilon$

52. $|x + 4| < \frac{\varepsilon}{2} \Rightarrow |2x + 8| < \varepsilon$

In Problems 53–56, find δ (depending on ε) so that the given implication is true.

53. $|x - 5| < \delta \Rightarrow |3x - 15| < \varepsilon$

54. $|x - 2| < \delta \Rightarrow |4x - 8| < \varepsilon$

55. $|x + 6| < \delta \Rightarrow |6x + 36| < \varepsilon$

56. $|x + 5| < \delta \Rightarrow |5x + 25| < \varepsilon$

57. On a lathe, you are to turn out a disk (thin right circular cylinder) of circumference 10 inches. This is done by continually measuring the diameter as you make the disk smaller. How closely must you measure the diameter if you can tolerate an error of at most 0.02 inch in the circumference?

58. Fahrenheit temperatures and Celsius temperatures are related by the formula $C = \frac{5}{9}(F - 32)$. An experiment requires that a solution be kept at 50°C with an error of at most 3% (or 1.5°). You have only a Fahrenheit thermometer. What error are you allowed on it?

In Problems 59–62, solve the inequalities.

59. $|x - 1| < 2|x - 3|$ **60.** $|2x - 1| \geq |x + 1|$

61. $2|2x - 3| < |x + 10|$ **62.** $|3x - 1| < 2|x + 6|$

63. Prove that $|x| < |y| \Leftrightarrow x^2 < y^2$ by giving a reason for each of these steps:

$$|x| < |y| \Rightarrow |x||x| \leq |x||y| \quad \text{and} \quad |x||y| < |y||y|$$
$$\Rightarrow |x|^2 < |y|^2$$
$$\Rightarrow x^2 < y^2$$

Conversely,

$$x^2 < y^2 \Rightarrow |x|^2 < |y|^2$$
$$\Rightarrow |x|^2 - |y|^2 < 0$$
$$\Rightarrow (|x| - |y|)(|x| + |y|) < 0$$
$$\Rightarrow |x| - |y| < 0$$
$$\Rightarrow |x| < |y|$$

64. Use the result of Problem 63 to show that

$$0 < a < b \Rightarrow \sqrt{a} < \sqrt{b}$$

65. Use the properties of the absolute value to show that each of the following is true.

(a) $|a - b| \leq |a| + |b|$ (b) $|a - b| \geq |a| - |b|$

(c) $|a + b + c| \leq |a| + |b| + |c|$

66. Use the Triangle Inequality and the fact that $0 < |a| < |b| \Rightarrow 1/|b| < 1/|a|$ to establish the following chain of inequalities.

$$\left|\frac{1}{x^2 + 3} - \frac{1}{|x| + 2}\right| \leq \frac{1}{x^2 + 3} + \frac{1}{|x| + 2} \leq \frac{1}{3} + \frac{1}{2}$$

67. Show that (see Problem 66)

$$\left|\frac{x - 2}{x^2 + 9}\right| \leq \frac{|x| + 2}{9}$$

68. Show that

$$|x| \leq 2 \Rightarrow \left|\frac{x^2 + 2x + 7}{x^2 + 1}\right| \leq 15$$

69. Show that

$$|x| \le 1 \Rightarrow |x^4 + \tfrac{1}{2}x^3 + \tfrac{1}{4}x^2 + \tfrac{1}{8}x + \tfrac{1}{16}| < 2$$

70. Show each of the following:

(a) $x < x^2$ for $x < 0$ or $x > 1$

(b) $x^2 < x$ for $0 < x < 1$

71. Show that $a \ne 0 \Rightarrow a^2 + 1/a^2 \ge 2$. *Hint:* Consider $(a - 1/a)^2$.

72. The number $\frac{1}{2}(a + b)$ is called the average, or **arithmetic mean,** of a and b. Show that the arithmetic mean of two numbers is between the two numbers; that is, prove that

$$a < b \Rightarrow a < \frac{a + b}{2} < b$$

73. The number $\sqrt{ab}$ is called the **geometric mean** of two positive numbers a and b. Prove that

$$0 < a < b \Rightarrow a < \sqrt{ab} < b$$

74. For two positive numbers a and b, prove that

$$\sqrt{ab} \le \tfrac{1}{2}(a + b)$$

This is the simplest version of a famous inequality called the **geometric mean–arithmetic mean inequality.**

75. Show that, among all rectangles with given perimeter p, the square has the largest area. *Hint:* If a and b denote the lengths of adjacent sides of a rectangle of perimeter p, then the area is ab, and for the square the area is $a^2 = [(a + b)/2]^2$. Now see Problem 74.

76. Solve $1 + x + x^2 + x^3 + \cdots + x^{99} \le 0$.

77. The formula $\frac{1}{R} = \frac{1}{R_1} + \frac{1}{R_2} + \frac{1}{R_3}$ gives the total resistance R in an electric circuit due to three resistances, R_1, R_2, and R_3, connected in parallel. If $10 \le R_1 \le 20$, $20 \le R_2 \le 30$, and $30 \le R_3 \le 40$, find the range of values for R.

78. The radius of a sphere is measured to be about 10 inches. Determine a tolerance δ in this measurement that will ensure an error of less than 0.01 square inch in the calculated value of the surface area of the sphere.

Answers to Concepts Review **1.** $[-1, 5)$; $(-\infty, -2]$ **2.** $b > 0$; $b < 0$ **3.** (b) and (c) **4.** $-1 \le x \le 5$

0.3 The Rectangular Coordinate System

In the plane, produce two copies of the real line, one horizontal and the other vertical, so that they intersect at the zero points of the two lines. The two lines are called **coordinate axes;** their intersection is labeled O and is called the **origin.** By convention, the horizontal line is called the ***x*-axis** and the vertical line is called the ***y*-axis.** The positive half of the x-axis is to the right; the positive half of the y-axis is upward. The coordinate axes divide the plane into four regions, called **quadrants,** labeled I, II, III, and IV, as shown in Figure 1.

Each point P in the plane can now be assigned a pair of numbers, called its **Cartesian coordinates.** If vertical and horizontal lines through P intersect the x- and y-axes at a and b, respectively, then P has coordinates (a, b) (see Figure 2). We call (a, b) an **ordered pair** of numbers because it makes a difference which number is first. The first number a is the ***x*-coordinate;** the second number b is the ***y*-coordinate.**

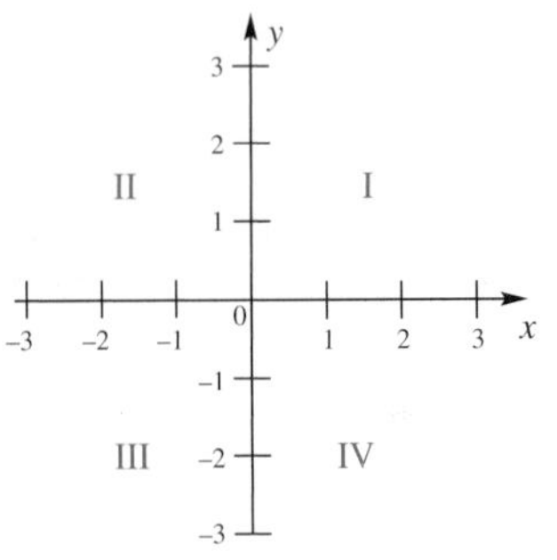

Figure 1

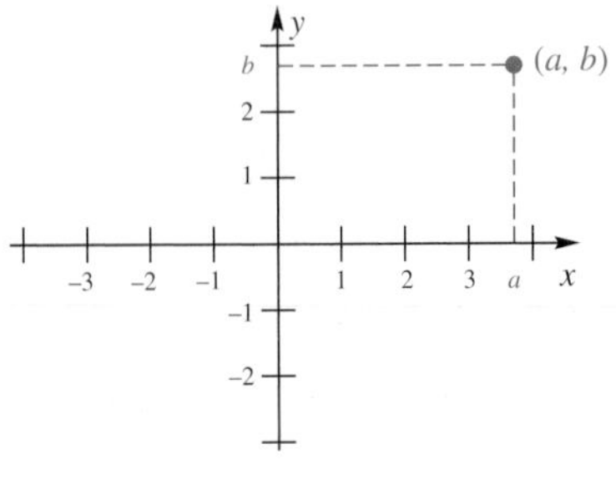

Figure 2

The Distance Formula With coordinates in hand, we can introduce a simple formula for the distance between any two points in the plane. It is based on the **Pythagorean Theorem,** which says that, if a and b measure the two legs of a right triangle and c measures its hypotenuse (Figure 3), then

$$a^2 + b^2 = c^2$$

Conversely, this relationship between the three sides of a triangle holds only for a right triangle.

Now consider any two points P and Q, with coordinates (x_1, y_1) and (x_2, y_2), respectively. Together with R, the point with coordinates (x_2, y_1), P and Q are vertices of a right triangle (Figure 4). The lengths of PR and RQ are $|x_2 - x_1|$ and $|y_2 - y_1|$, respectively. When we apply the Pythagorean Theorem and take the principal square root of both sides, we obtain the following expression for the **Distance Formula**

$$d(P, Q) = \sqrt{(x_2 - x_1)^2 + (y_2 - y_1)^2}$$

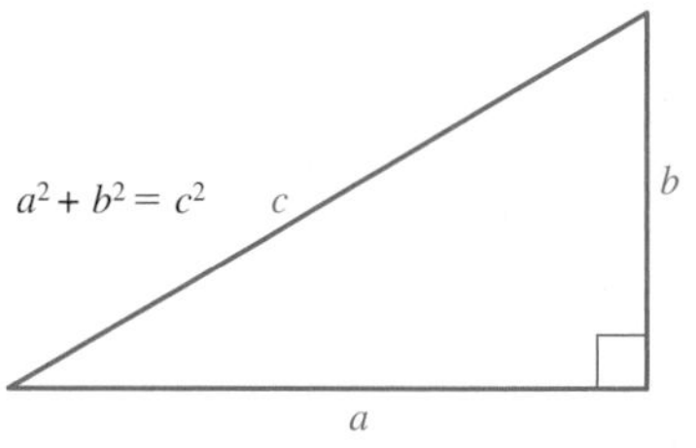

Figure 3

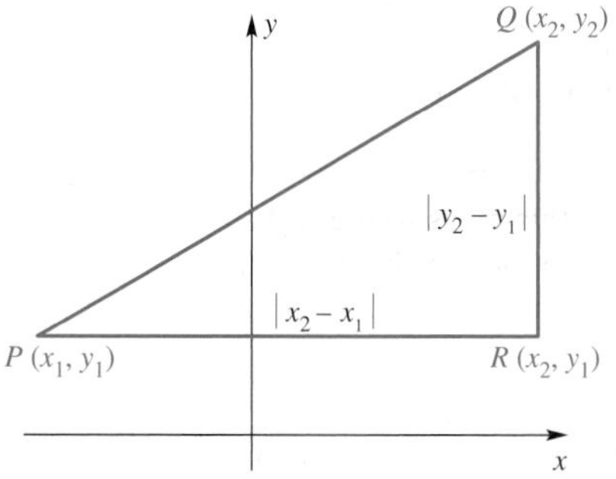

Figure 4

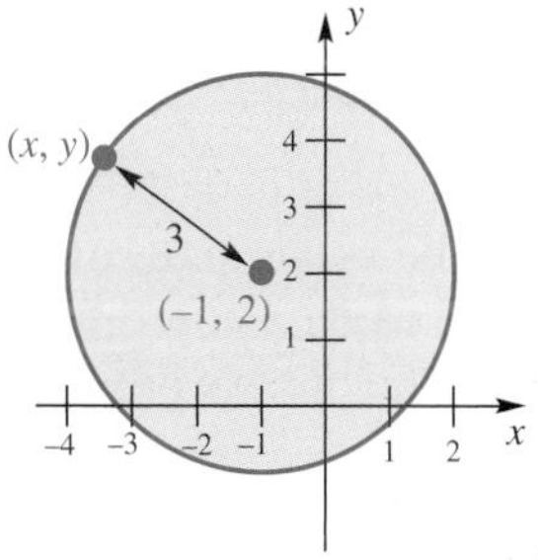

Figure 5

EXAMPLE 1 Find the distance between

(a) $P(-2, 3)$ and $Q(4, -1)$ (b) $P(\sqrt{2}, \sqrt{3})$ and $Q(\pi, \pi)$

SOLUTION

(a) $d(P, Q) = \sqrt{(4 - (-2))^2 + (-1 - 3)^2} = \sqrt{36 + 16} = \sqrt{52} \approx 7.21$

(b) $d(P, Q) = \sqrt{(\pi - \sqrt{2})^2 + (\pi - \sqrt{3})^2} \approx \sqrt{4.971} \approx 2.23$ ■

The formula holds even if the two points lie on the same horizontal line or the same vertical line. Thus, the distance between $P(-2, 2)$ and $Q(6, 2)$ is

$$\sqrt{(6-(-2))^2 + (2 - 2)^2} = \sqrt{64} = 8$$

The Equation of a Circle It is a small step from the distance formula to the equation of a circle. A **circle** is the set of points that lie at a fixed distance (the *radius*) from a fixed point (the *center*). Consider, for example, the circle of radius 3 with center at $(-1, 2)$ (Figure 5). Let (x, y) denote any point on this circle. By the Distance Formula,

$$\sqrt{(x + 1)^2 + (y - 2)^2} = 3$$

When we square both sides, we obtain

$$(x + 1)^2 + (y - 2)^2 = 9$$

which we call the equation of this circle.

More generally, the circle of radius r and center (h, k) has the equation

$$(x - h)^2 + (y - k)^2 = r^2 \qquad (1)$$

We call this the **standard equation of a circle.**

EXAMPLE 2 Find the standard equation of a circle of radius 5 and center $(1, -5)$. Also find the y-coordinates of the two points on this circle with x-coordinate 2.

SOLUTION The desired equation is

$$(x - 1)^2 + (y + 5)^2 = 25$$

To accomplish the second task, we substitute $x = 2$ in the equation and solve for y.

$$(2 - 1)^2 + (y + 5)^2 = 25$$
$$(y + 5)^2 = 24$$
$$y + 5 = \pm\sqrt{24}$$
$$y = -5 \pm \sqrt{24} = -5 \pm 2\sqrt{6}$$ ■

If we expand the two squares in the boxed equation (1) and combine the constants, then the equation takes the form

$$x^2 + ax + y^2 + by = c$$

This suggests asking whether every equation of the latter form is the equation of a circle. The answer is yes, with some obvious exceptions.

Circle ↔ Equation

To say that

$$(x + 1)^2 + (y - 2)^2 = 9$$

is the equation of the circle of radius 3 with center $(-1, 2)$ means two things:

1. If a point is on this circle, then its coordinates (x, y) satisfy the equation.
2. If x and y are numbers that satisfy the equation, then they are the coordinates of a point on the circle.

EXAMPLE 3 Show that the equation

$$x^2 - 2x + y^2 + 6y = -6$$

represents a circle, and find its center and radius.

SOLUTION We need to *complete the square*, a process important in many contexts. To complete the square of $x^2 \pm bx$, add $(b/2)^2$. Thus, we add $(-2/2)^2 = 1$ to $x^2 - 2x$ and $(6/2)^2 = 9$ to $y^2 + 6y$, and of course we must add the same numbers to the right side of the equation, to obtain

$$x^2 - 2x + 1 + y^2 + 6y + 9 = -6 + 1 + 9$$

$$(x - 1)^2 + (y + 3)^2 = 4$$

The last equation is in standard form. It is the equation of a circle with center $(1, -3)$ and radius 2. If, as a result of this process, we had come up with a negative number on the right side of the final equation, the equation would not have represented any curve. If we had come up with zero, the equation would have represented the single point $(1, -3)$. ■

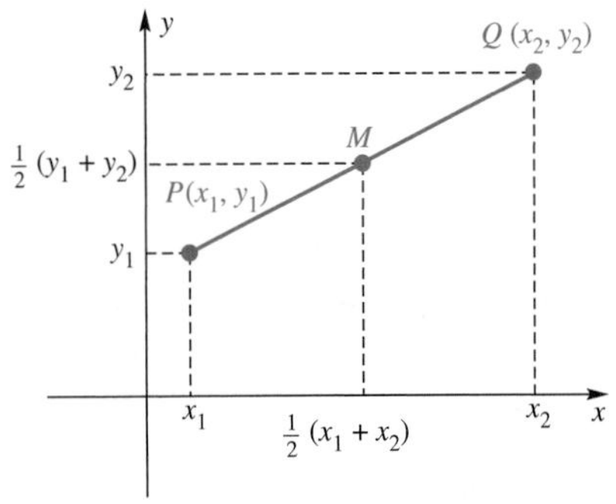

Figure 6

The Midpoint Formula Consider two points $P(x_1, y_1)$ and $Q(x_2, y_2)$ with $x_1 \le x_2$ and $y_1 \le y_2$, as in Figure 6. The distance between x_1 and x_2 is $x_2 - x_1$. When we add half this distance, $\frac{1}{2}(x_2 - x_1)$, to x_1, we should get the number midway between x_1 and x_2.

$$x_1 + \frac{1}{2}(x_2 - x_1) = x_1 + \frac{1}{2}x_2 - \frac{1}{2}x_1 = \frac{1}{2}x_1 + \frac{1}{2}x_2 = \frac{x_1 + x_2}{2}$$

Thus, the point $(x_1 + x_2)/2$ is midway between x_1 and x_2 on the x-axis and, consequently, the midpoint M of the segment PQ has $(x_1 + x_2)/2$ as its x-coordinate. Similarly, we can show that $(y_1 + y_2)/2$ is the y-coordinate of M. Thus, we have the **Midpoint Formula.**

> The midpoint of the line segment joining $P(x_1, y_1)$ and $Q(x_2, y_2)$ is
>
> $$\left(\frac{x_1 + x_2}{2}, \frac{y_1 + y_2}{2}\right)$$

EXAMPLE 4 Find the equation of the circle having the segment from $(1, 3)$ to $(7, 11)$ as a diameter.

SOLUTION The center of the circle is at the midpoint of the diameter; thus, the center has coordinates $(1 + 7)/2 = 4$ and $(3 + 11)/2 = 7$. The length of the diameter, obtained from the distance formula, is

$$\sqrt{(7 - 1)^2 + (11 - 3)^2} = \sqrt{36 + 64} = 10$$

and so the radius of the circle is 5. The equation of the circle is

$$(x - 4)^2 + (y - 7)^2 = 25$$ ■

Lines Consider the line in Figure 7. From point A to point B, there is a **rise** (vertical change) of 2 units and a **run** (horizontal change) of 5 units. We say that the line has a slope of $\frac{2}{5}$. In general (Figure 8), for a line through $A(x_1, y_1)$ and $B(x_2, y_2)$, where $x_1 \neq x_2$, we define the **slope** m of that line by

$$m = \frac{\text{rise}}{\text{run}} = \frac{y_2 - y_1}{x_2 - x_1}$$

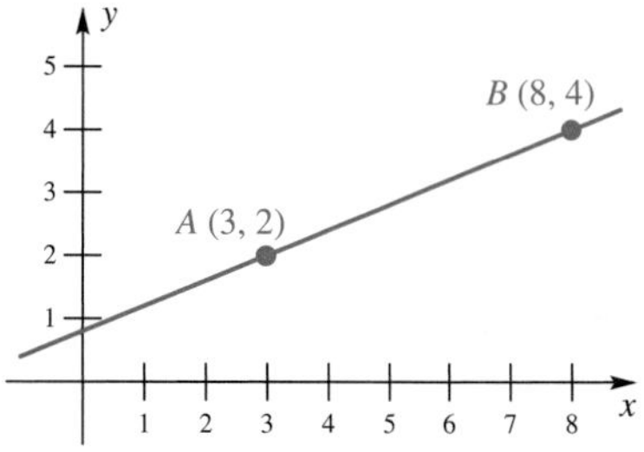

Figure 7

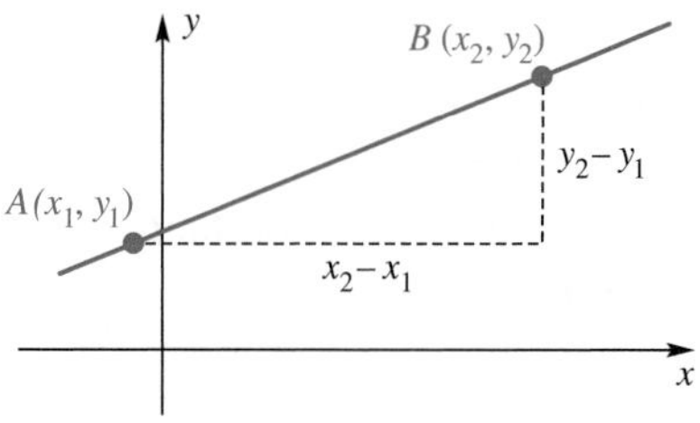

Figure 8

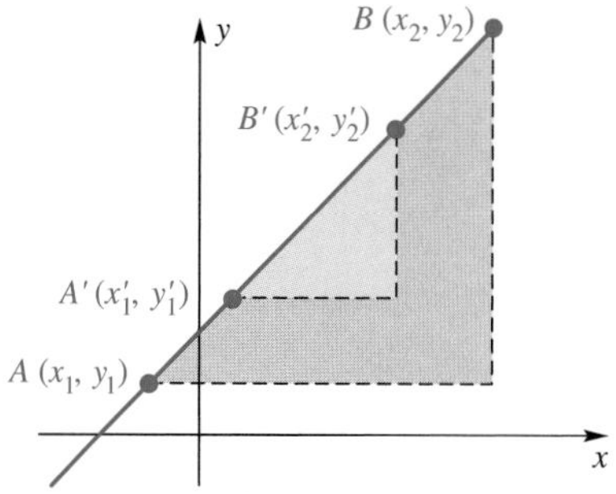

Figure 9

Does the value we get for the slope depend on which pair of points we use for A and B? The similar triangles in Figure 9 show us that

$$\frac{y_2' - y_1'}{x_2' - x_1'} = \frac{y_2 - y_1}{x_2 - x_1}$$

Thus, points A' and B' would do just as well as A and B. It does not even matter whether A is to the left or right of B, since

$$\frac{y_1 - y_2}{x_1 - x_2} = \frac{y_2 - y_1}{x_2 - x_1}$$

All that matters is that we subtract the coordinates in the same order in the numerator and the denominator.

The slope m is a measure of the steepness of a line, as Figure 10 illustrates. Notice that a horizontal line has zero slope, a line that rises to the right has positive slope, and a line that falls to the right has negative slope. The larger the absolute value of the slope is, the steeper the line. The concept of slope for a vertical line makes no sense, since it would involve division by zero. Therefore, slope for a vertical line is left undefined.

Grade and Pitch

The international symbol for the slope of a road (called the grade) is shown below. The grade is given as a percentage. A grade of 10% corresponds to a slope of ± 0.10.

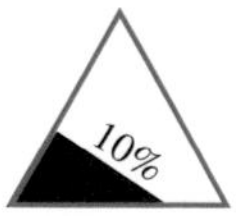

Carpenters use the term *pitch*. A 9:12 pitch corresponds to a slope of $\frac{9}{12}$.

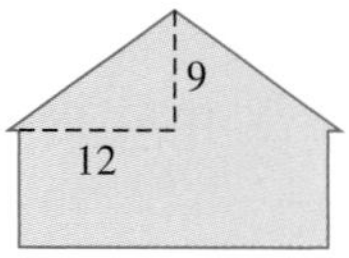

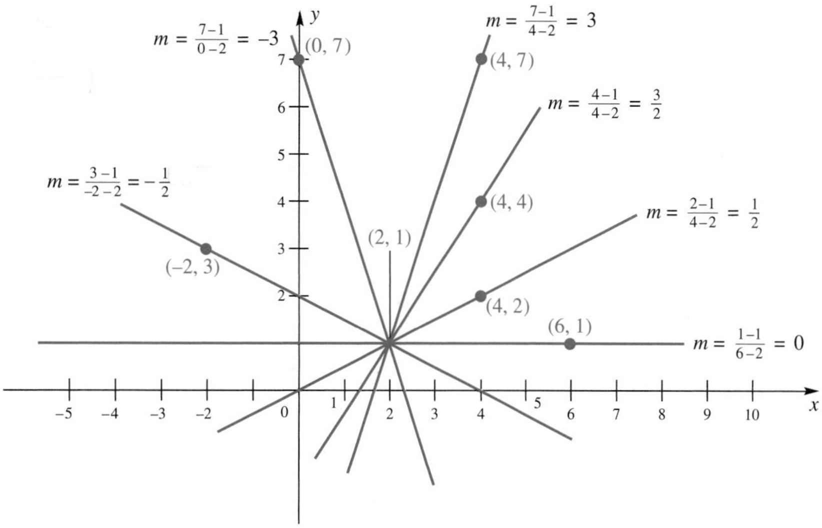

Lines of various slopes

Figure 10

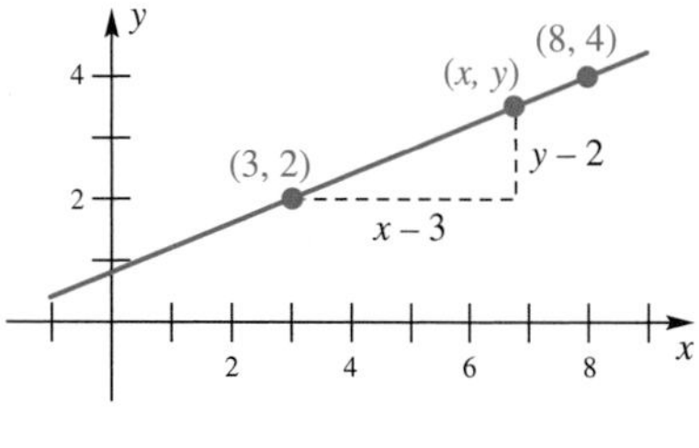

Figure 11

The Point–Slope Form Consider again the line of our opening discussion; it is reproduced in Figure 11. We know that this line

1. passes through $(3, 2)$ and
2. has slope $\frac{2}{5}$.

Take any other point on this line, such as one with coordinates (x, y). If we use this point and the point $(3, 2)$ to measure slope, we must get $\frac{2}{5}$, that is,

$$\frac{y-2}{x-3} = \frac{2}{5}$$

or, after multiplying by $x - 3$,

$$y - 2 = \tfrac{2}{5}(x - 3)$$

Notice that this last equation is satisfied by all points on the line, even by (3, 2). Moreover, none of the points not on the line can satisfy this equation.

What we have just done in an example can be done in general. The line passing through the (fixed) point (x_1, y_1) with slope m has equation

$$\boxed{y - y_1 = m(x - x_1)}$$

We call this the **point–slope** form of the equation of a line.

Consider once more the line of our example. That line passes through (8, 4) as well as (3, 2). If we use (8, 4) as (x_1, y_1), we get the equation

$$y - 4 = \tfrac{2}{5}(x - 8)$$

which looks quite different from $y - 2 = \frac{2}{5}(x - 3)$. However, both can be simplified to $5y - 2x = 4$; they are equivalent.

EXAMPLE 5 Find an equation of the line through $(-4, 2)$ and $(6, -1)$.

SOLUTION The slope is $m = (-1 - 2)/(6 + 4) = -\frac{3}{10}$. Thus, using $(-4, 2)$ as the fixed point, we obtain the equation

$$y - 2 = -\tfrac{3}{10}(x + 4)$$ ■

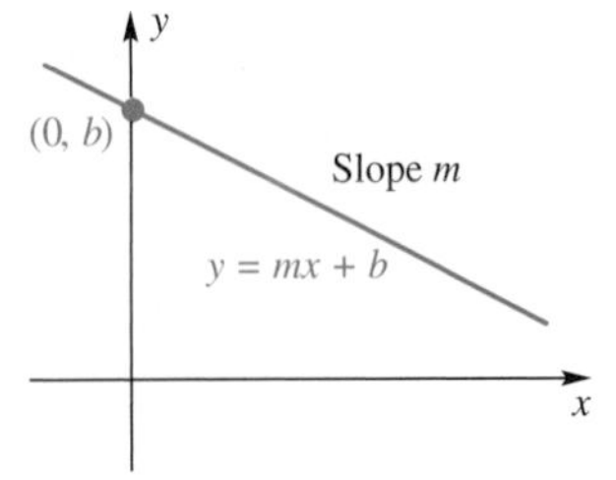

Figure 12

The Slope–Intercept Form The equation of a line can be expressed in various forms. Suppose that we are given the slope m for a line and the y-intercept b (i.e., the line intersects the y-axis at $(0, b)$), as shown in Figure 12. Choosing $(0, b)$ as (x_1, y_1) and applying the point-slope form, we get

$$y - b = m(x - 0)$$

which we can rewrite as

$$\boxed{y = mx + b}$$

The latter is called the **slope–intercept** form. Any time we see an equation written this way, we recognize it as a line and can immediately read its slope and y-intercept. For example, consider the equation

$$3x - 2y + 4 = 0$$

If we solve for y, we get

$$y = \tfrac{3}{2}x + 2$$

It is the equation of a line with slope $\frac{3}{2}$ and y-intercept 2.

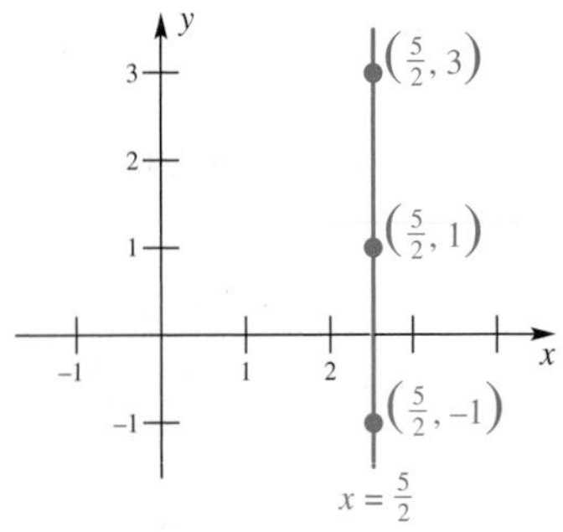

Figure 13

Equation of a Vertical Line Vertical lines do not fit within the preceding discussion since the concept of slope is not defined for them. But they do have equations, very simple ones. The line in Figure 13 has equation $x = \frac{5}{2}$, since a point is on the line if and only if it satisfies this equation. The equation of any vertical line can be put in the form $x = k$, where k is a constant. It should be noted that the equation of a horizontal line can be written in the form $y = k$.

The Form $Ax + By + C = 0$ It would be nice to have a form that covered all lines, including vertical lines. Consider, for example,

Summary: Equations of Lines
Vertical line: $x = k$
Horizontal line: $y = k$
Point–slope form: $y - y_1 = m(x - x_1)$
Slope–intercept form: $y = mx + b$
General linear equation: $Ax + By + C = 0$

$$y - 2 = -4(x + 2)$$

$$y = 5x - 3$$

$$x = 5$$

These can be rewritten (by taking everything to the left-hand side) as follows:

$$4x + y + 6 = 0$$

$$-5x + y + 3 = 0$$

$$x + 0y - 5 = 0$$

All are of the form

$$Ax + By + C = 0, \qquad A \text{ and } B \text{ not both } 0$$

which we call the **general linear equation.** It takes only a moment's thought to see that the equation of any line can be put in this form. Conversely, the graph of the general linear equation is always a line.

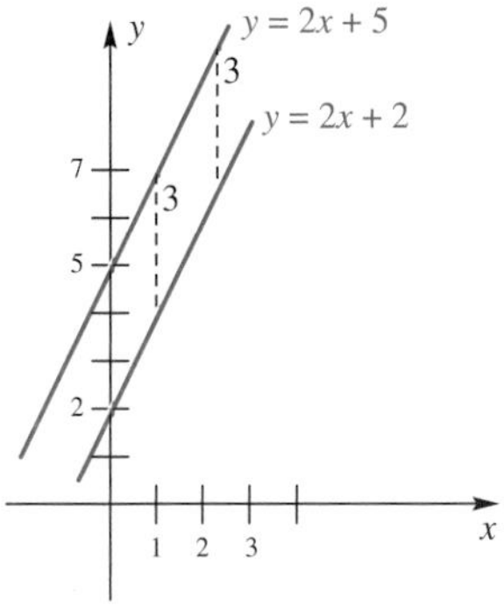

Figure 14

Parallel Lines Two lines that have no points in common are said to be parallel. For example, the lines whose equations are $y = 2x + 2$ and $y = 2x + 5$ are parallel because, for every value of x, the second line is three units above the first (see Figure 14). Similarly, the lines with equations $-2x + 3y + 12 = 0$ and $4x - 6y = 5$ are parallel. To see this, solve each equation for y (i.e., put each in the slope–intercept form). This gives $y = \frac{2}{3}x - 4$ and $y = \frac{2}{3}x - \frac{5}{6}$, respectively. Again, because the slopes are equal, one line will be a fixed number of units above or below the other, so the lines will never intersect. If two lines have the same slope *and* the same y-intercept, then the two lines are the same, and they are not parallel.

We summarize by stating that two nonvertical lines are parallel if and only if they have the same slope and different y-intercepts. Two vertical lines are parallel if and only if they are distinct lines.

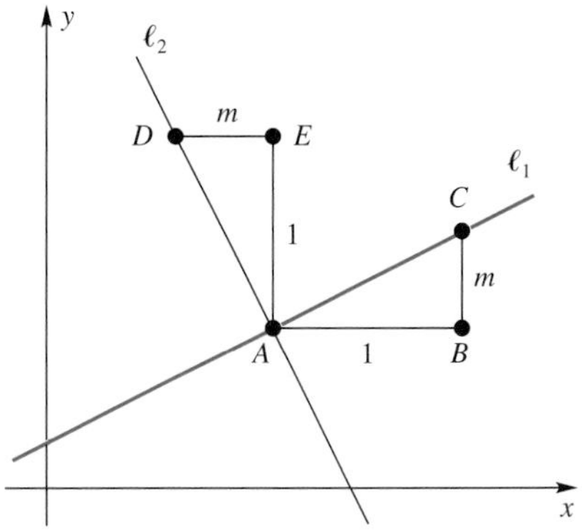

Figure 15

EXAMPLE 6 Find the equation of the line through (6, 8) that is parallel to the line with equation $3x - 5y = 11$.

SOLUTION When we solve $3x - 5y = 11$ for y, we obtain $y = \frac{3}{5}x - \frac{11}{5}$, from which we read the slope of the line to be $\frac{3}{5}$. The equation of the desired line is

$$y - 8 = \tfrac{3}{5}(x - 6)$$

or, equivalently, $y = \frac{3}{5}x + \frac{22}{5}$. We know that these lines are distinct because the y-intercepts are different. ■

Perpendicular Lines Is there a simple slope condition that characterizes perpendicular lines? Yes; *two nonvertical lines are perpendicular if and only if their slopes are negative reciprocals of each other.* To see why this is true, consider Figure 15. This picture tells almost the whole story; it is left as an exercise (Problem 57) to construct a geometric proof that the two (nonvertical) lines are perpenicular if and only if $m_2 = -1/m_1$.

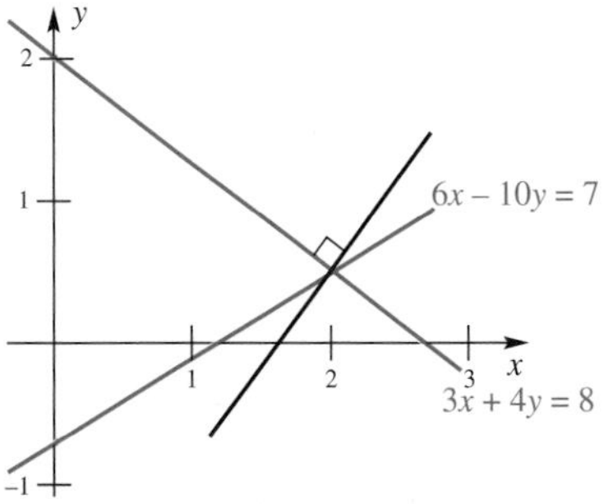

Figure 16

EXAMPLE 7 Find the equation of the line through the point of intersection of the lines with equations $3x + 4y = 8$ and $6x - 10y = 7$ that is perpendicular to the first of these two lines (Figure 16).

SOLUTION To find the point of intersection of the two lines, we multiply the first equation by -2 and add it to the second equation.

$$\begin{aligned} -6x - 8y &= -16 \\ 6x - 10y &= 7 \\ \hline -18y &= -9 \\ y &= \frac{1}{2} \end{aligned}$$

Substituting $y = \frac{1}{2}$ in either of the original equations yields $x = 2$. The point of intersection is $\left(2, \frac{1}{2}\right)$. When we solve the first equation for y (to put it in slope-intercept form), we get $y = -\frac{3}{4}x + 2$. A line perpendicular to it has slope $\frac{4}{3}$. The equation of the required line is

$$y - \frac{1}{2} = \frac{4}{3}(x - 2)$$ ■

Concepts Review

1. The distance between the points $(-2, 3)$ and (x, y) is ______.

2. The equation of the circle of radius 5 and center $(-4, 2)$ is ______.

3. The midpoint of the line segment joining $(-2, 3)$ and $(5, 7)$ is ______.

4. The line through (a, b) and (c, d) has slope $m =$ ______ provided $a \neq c$.

Problem Set 0.3

In Problems 1–4, plot the given points in the coordinate plane and then find the distance between them.

1. $(3, 1), (1, 1)$
2. $(-3, 5), (2, -2)$
3. $(4, 5), (5, -8)$
4. $(-1, 5), (6, 3)$

5. Show that the triangle whose vertices are $(5, 3)$, $(-2, 4)$, and $(10, 8)$ is isosceles.

6. Show that the triangle whose vertices are $(2, -4)$, $(4, 0)$, and $(8, -2)$ is a right triangle.

7. The points $(3, -1)$ and $(3, 3)$ are two vertices of a square. Give three other pairs of possible vertices.

8. Find the point on the x-axis that is equidistant from $(3, 1)$ and $(6, 4)$.

9. Find the distance between $(-2, 3)$ and the midpoint of the segment joining $(-2, -2)$ and $(4, 3)$.

10. Find the length of the line segment joining the midpoints of the segments AB and CD, where $A = (1, 3)$, $B = (2, 6)$, $C = (4, 7)$, and $D = (3, 4)$.

In Problems 11–16, find the equation of the circle satisfying the given conditions.

11. Center $(1, 1)$, radius 1
12. Center $(-2, 3)$, radius 4
13. Center $(2, -1)$, goes through $(5, 3)$
14. Center $(4, 3)$, goes through $(6, 2)$
15. Diameter AB, where $A = (1, 3)$ and $B = (3, 7)$
16. Center $(3, 4)$ and tangent to x-axis

In Problems 17–22, find the center and radius of the circle with the given equation.

17. $x^2 + 2x + 10 + y^2 - 6y - 10 = 0$
18. $x^2 + y^2 - 6y = 16$
19. $x^2 + y^2 - 12x + 35 = 0$
20. $x^2 + y^2 - 10x + 10y = 0$
21. $4x^2 + 16x + 15 + 4y^2 + 6y = 0$
22. $x^2 + 16x + \frac{105}{16} + 4y^2 + 3y = 0$

In Problems 23–28, find the slope of the line containing the given two points.

23. $(1, 1)$ and $(2, 2)$
24. $(3, 5)$ and $(4, 7)$
25. $(2, 3)$ and $(-5, -6)$
26. $(2, -4)$ and $(0, -6)$
27. $(3, 0)$ and $(0, 5)$
28. $(-6, 0)$ and $(0, 6)$

In Problems 29–34, find an equation for each line. Then write your answer in the form $Ax + By + C = 0$.

29. Through $(2, 2)$ with slope -1
30. Through $(3, 4)$ with slope -1
31. With y-intercept 3 and slope 2
32. With y-intercept 5 and slope 0
33. Through $(2, 3)$ and $(4, 8)$
34. Through $(4, 1)$ and $(8, 2)$

In Problems 35–38, find the slope and y-intercept of each line.

35. $3y = -2x + 1$
36. $-4y = 5x - 6$

37. $6 - 2y = 10x - 2$ **38.** $4x + 5y = -20$

39. Write an equation for the line through $(3, -3)$ that is
(a) parallel to the line $y = 2x + 5$;
(b) perpendicular to the line $y = 2x + 5$;
(c) parallel to the line $2x + 3y = 6$;
(d) perpendicular to the line $2x + 3y = 6$;
(e) parallel to the line through $(-1, 2)$ and $(3, -1)$;
(f) parallel to the line $x = 8$;
(g) perpendicular to the line $x = 8$.

40. Find the value of c for which the line $3x + cy = 5$
(a) passes through the point $(3, 1)$;
(b) is parallel to the y-axis;
(c) is parallel to the line $2x + y = -1$;
(d) has equal x- and y-intercepts;
(e) is perpendicular to the line $y - 2 = 3(x + 3)$.

41. Write the equation for the line through $(-2, -1)$ that is perpendicular to the line $y + 3 = -\frac{2}{3}(x - 5)$.

42. Find the value of k such that the line $kx - 3y = 10$
(a) is parallel to the line $y = 2x + 4$;
(b) is perpendicular to the line $y = 2x + 4$;
(c) is perpendicular to the line $2x + 3y = 6$.

43. Does $(3, 9)$ lie above or below the line $y = 3x - 1$?

44. Show that the equation of the line with x-intercept $a \neq 0$ and y-intercept $b \neq 0$ can be written as

$$\frac{x}{a} + \frac{y}{b} = 1$$

In Problems 45–48, find the coordinates of the point of intersection. Then write an equation for the line through that point perpendicular to the line given first.

45. $2x + 3y = 4$, $-3x + y = 5$ **46.** $4x - 5y = 8$, $2x + y = -10$

47. $3x - 4y = 5$, $2x + 3y = 9$ **48.** $5x - 2y = 5$, $2x + 3y = 6$

49. The points $(2, 3)$, $(6, 3)$, $(6, -1)$, and $(2, -1)$ are corners of a square. Find the equations of the inscribed and circumscribed circles.

≈ **50.** A belt fits tightly around the two circles, with equations $(x - 1)^2 + (y + 2)^2 = 16$ and $(x + 9)^2 + (y - 10)^2 = 16$. How long is this belt?

51. Show that the midpoint of the hypotenuse of any right triangle is equidistant from the three vertices.

52. Find the equation of the circle circumscribed about the right triangle whose vertices are $(0, 0)$, $(8, 0)$, and $(0, 6)$.

53. Show that the two circles $x^2 + y^2 - 4x - 2y - 11 = 0$ and $x^2 + y^2 + 20x - 12y + 72 = 0$ do not intersect. *Hint:* Find the distance between their centers.

54. What relationship between a, b, and c must hold if $x^2 + ax + y^2 + by + c = 0$ is the equation of a circle?

55. The ceiling of an attic makes an angle of 30° with the floor. A pipe of radius 2 inches is placed along the edge of the attic in such a way that one side of the pipe touches the ceiling and another side touches the floor (see Figure 17). What is the distance d from the edge of the attic to where the pipe touches the floor?

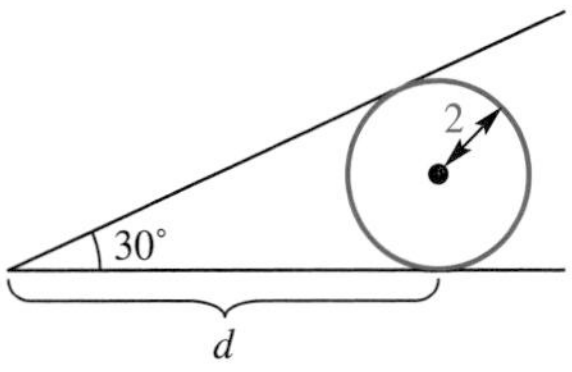

Figure 17

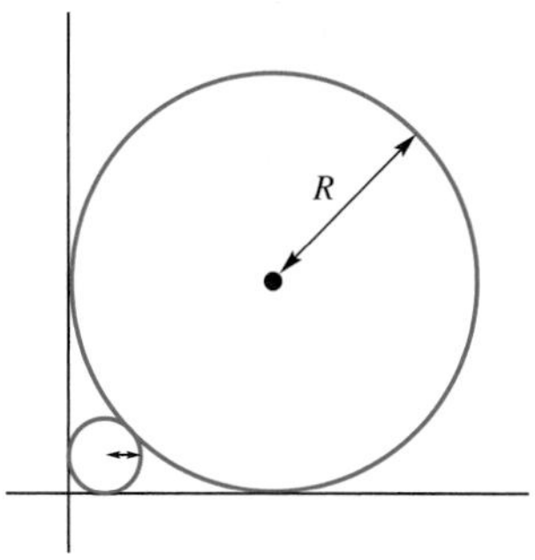

Figure 18

56. A circle of radius R is placed in the first quadrant as shown in Figure 18. What is the radius r of the largest circle that can be placed between the original circle and the origin?

57. Construct a geometric proof using Figure 15 that shows two lines are perpendicular if and only if their slopes are negative reciprocals of one another.

58. Show that the set of points that are twice as far from $(3, 4)$ as from $(1, 1)$ form a circle. Find its center and radius.

59. The Pythagorean Theorem says that the areas A, B, and C of the squares in Figure 19 satisfy $A + B = C$. Show that semicircles and equilateral triangles satisfy the same relation and then guess what a very general theorem says.

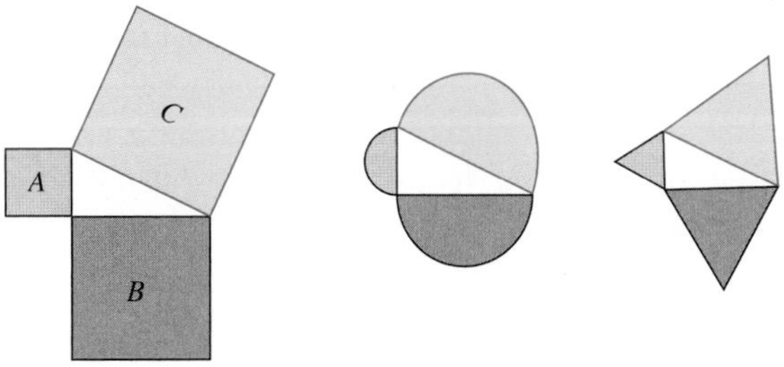

Figure 19

60. Consider a circle C and a point P exterior to the circle. Let line segment PT be tangent to C at T, and let the line through P and the center of C intersect C at M and N. Show that $(PM)(PN) = (PT)^2$.

≈ **61.** A belt fits around the three circles $x^2 + y^2 = 4$, $(x - 8)^2 + y^2 = 4$, and $(x - 6)^2 + (y - 8)^2 = 4$, as shown in Figure 20. Find the length of this belt.

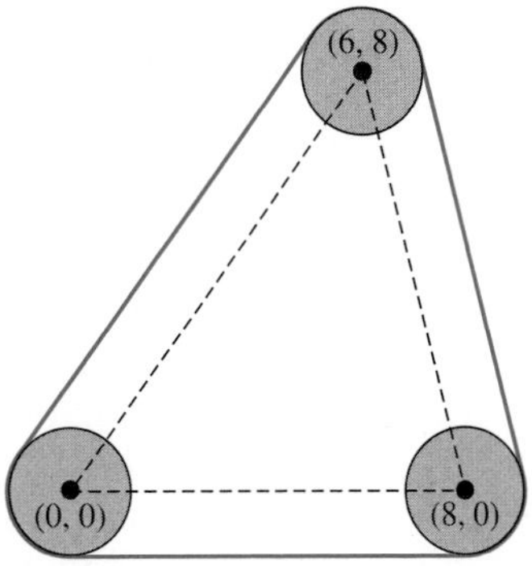

Figure 20

62. Study Problems 50 and 61. Consider a set of nonintersecting circles of radius r with centers at the vertices of a convex n-sided polygon having sides of lengths $d_1, d_2, \ldots, d_n$. How long is the belt that fits around these circles (in the manner of Figure 20)?

It can be shown that the distance d from the point (x_1, y_1) to the line $Ax + By + C = 0$ is

$$d = \frac{|Ax_1 + By_1 + C|}{\sqrt{A^2 + B^2}}$$

Use this result to find the distance from the given point to the given line.

63. $(-3, 2); 3x + 4y = 6$

64. $(4, -1); 2x - 2y + 4 = 0$

65. $(-2, -1); 5y = 12x + 1$

66. $(3, -1); y = 2x - 5$

In Problems 67 and 68, find the (perpendicular) distance between the given parallel lines. Hint: First find a point on one of the lines.

67. $2x + 4y = 7, 2x + 4y = 5$

68. $7x - 5y = 6, 7x - 5y = -1$

69. Find the equation for the line that bisects the line segment from $(-2, 3)$ to $(1, -2)$ and is at right angles to this line segment.

70. The center of the circumscribed circle of a triangle lies on the perpendicular bisectors of the sides. Use this fact to find the center of the circle that circumscribes the triangle with vertices $(0, 4), (2, 0)$, and $(4, 6)$.

71. Find the radius of the circle that is inscribed in a triangle with sides of lengths 3, 4, and 5 (see Figure 21).

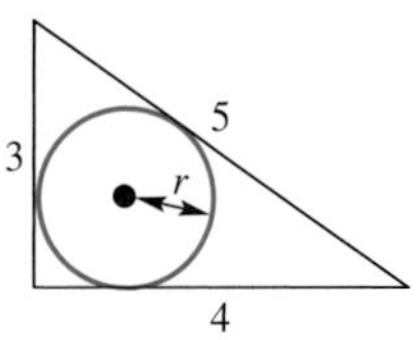

Figure 21

72. Suppose that (a, b) is on the circle $x^2 + y^2 = r^2$. Show that the line $ax + by = r^2$ is tangent to the circle at (a, b).

73. Find the equations of the two tangent lines to the circle $x^2 + y^2 = 36$ that go through $(12, 0)$. *Hint:* See Problem 72.

74. Express the perpendicular distance between the parallel lines $y = mx + b$ and $y = mx + B$ in terms of m, b, and B. *Hint:* The required distance is the same as that between $y = mx$ and $y = mx + B - b$.

75. Show that the line through the midpoints of two sides of a triangle is parallel to the third side. *Hint:* You may assume that the triangle has vertices at $(0, 0)$, $(a, 0)$, and (b, c).

76. Show that the line segments joining the midpoints of adjacent sides of any quadrilateral (four-sided polygon) form a parallelogram.

77. A wheel whose rim has equation $x^2 + (y - 6)^2 = 25$ is rotating rapidly in the counterclockwise direction. A speck of dirt on the rim came loose at the point $(3, 2)$ and flew toward the wall $x = 11$. About how high up on the wall did it hit? *Hint:* The speck of dirt flies off on a tangent so fast that the effects of gravity are negligible by the time it has hit the wall.

Answers to Concepts Review: **1.** $\sqrt{(x + 2)^2 + (y - 3)^2}$ **2.** $(x + 4)^2 + (y - 2)^2 = 25$ **3.** $(1.5, 5)$ **4.** $(d - b)/(c - a)$

0.4 Graphs of Equations

The use of coordinates for points in the plane allows us to describe a curve (a geometric object) by an equation (an algebraic object). We saw how this was done for circles and lines in the previous section. Now we want to consider the reverse process: graphing an equation. The **graph of an equation** in x and y consists of those points in the plane whose coordinates (x, y) satisfy the equation, that is, make it a true equality.

The Graphing Procedure To graph an equation, for example, $y = 2x^3 - x + 19$, by hand, we can follow a simple three-step procedure:

Step 1: Obtain the coordinates of a few points that satisfy the equation.

Step 2: Plot these points in the plane.

Step 3: Connect the points with a smooth curve.

This simplistic method will have to suffice until Chapter 3 when we use more advanced methods to graph equations. The best way to do Step 1 is to make a table of values. Assign values to one of the variables, such as x, and determine the corresponding values of the other variable, listing the results in tabular form.

A graphing calculator or a computer algebra system will follow much the same procedure, although its procedure is transparent to the user. A user simply defines the function and asks the graphing calculator or computer to plot it.

EXAMPLE 1 Graph the equation $y = x^2 - 3$.

SOLUTION The three-step procedure is shown in Figure 1.

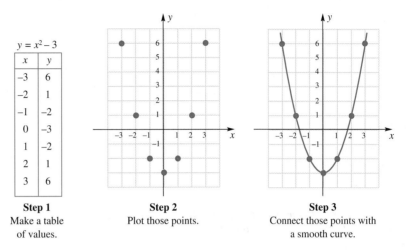

$y = x^2 - 3$

x	y
–3	6
–2	1
–1	–2
0	–3
1	–2
2	1
3	6

Step 1
Make a table of values.

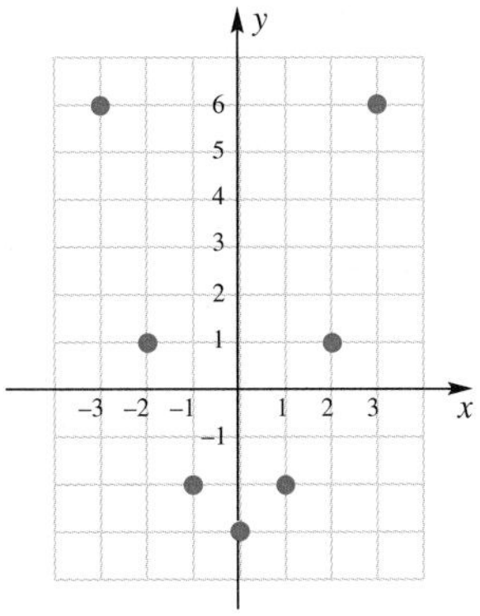

Step 2
Plot those points.

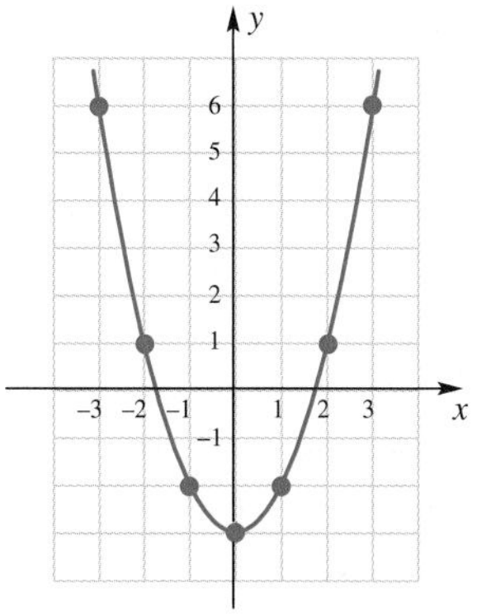

Step 3
Connect those points with a smooth curve.

Figure 1

■

Of course, you need to use common sense and even a little faith. When you have a point that seems out of place, check your calculations. When you connect the points you have plotted with a smooth curve, you are assuming that the curve behaves nicely between consecutive points, which is faith. This is why you should plot enough points so that the outline of the curve seems very clear; the more points you plot, the less faith you will need. Also, you should recognize that you can seldom display the whole curve. In our example, the curve has infinitely long arms, opening wider and wider. But our graph does show the essential features. This is our goal in graphing. Show enough of the graph so that the essential features are visible. Later (Section 3.5) we will use the tools of calculus to refine and improve our understanding of graphs.

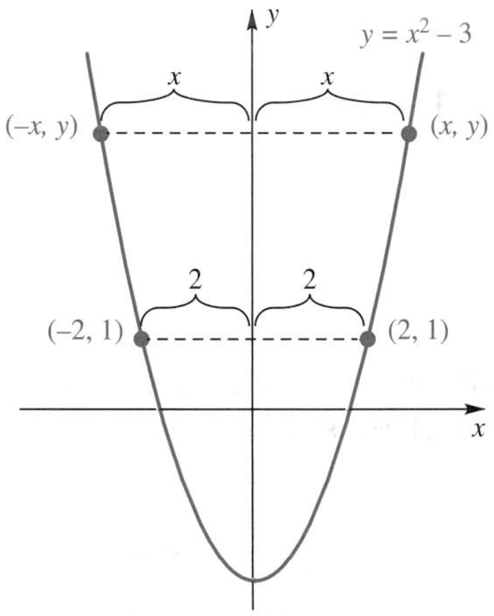

Symmetry with respect to the y-axis

Figure 2

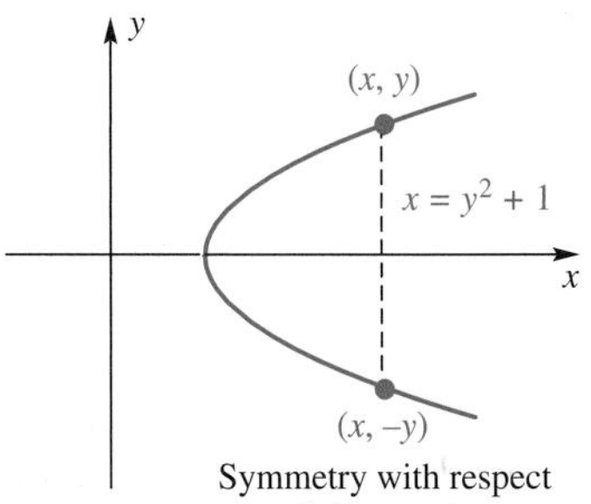

Symmetry with respect to the x-axis

Figure 3

Symmetry of a Graph

We can sometimes cut our graphing effort in half by recognizing certain symmetries of the graph as revealed by its equation. Look at the graph of $y = x^2 - 3$, drawn above and again in Figure 2. If the coordinate plane is folded along the y-axis, the two branches of the graph will coincide. For example, $(3, 6)$ will coincide with $(-3, 6)$, $(2, 1)$ will coincide with $(-2, 1)$, and, more generally, (x, y) will coincide with $(-x, y)$. Algebraically, this corresponds to the fact that replacing x by $-x$ in the equation $y = x^2 - 3$ results in an equivalent equation.

Consider an arbitrary graph. It is **symmetric with respect to the *y*-axis** if, whenever (x, y) is on the graph, $(-x, y)$ is also on the graph (Figure 2). Similarly, it is **symmetric with respect to the *x*-axis** if, whenever (x, y) is on the graph, $(x, -y)$ is also on the graph (Figure 3). Finally, a graph is **symmetric with respect to the origin** if, whenever (x, y) is on the graph, $(-x, -y)$ is also on the graph (see Example 2).

In terms of equations, we have three simple tests. The graph of an equation is

1. symmetric with respect to the y-axis if replacing x by $-x$ gives an equivalent equation (e.g., $y = x^2$);
2. symmetric with respect to the x-axis if replacing y by $-y$ gives an equivalent equation (e.g., $x = y^2 + 1$);
3. symmetric with respect to the origin if replacing x by $-x$ and y by $-y$ gives an equivalent equation ($y = x^3$ is a good example since $-y = (-x)^3$ is equivalent to $y = x^3$).

$y = x^3$

x	y
0	0
1	1
2	8
3	27
4	64

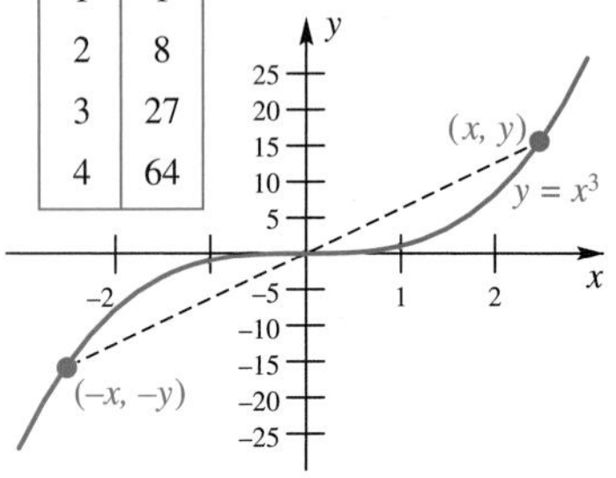

Symmetry with respect to the origin

Figure 4

EXAMPLE 2 Sketch the graph of $y = x^3$.

SOLUTION We note, as pointed out above, that the graph will be symmetric with respect to the origin, so we need only get a table of values for nonnegative x's; we can find matching points by symmetry. For example, $(2, 8)$ being on the graph tells us that $(-2, -8)$ is on the graph, $(3, 27)$ being on the graph tells us that $(-3, -27)$ is on the graph, and so on. See Figure 4. ■

In graphing $y = x^3$, we used a different scale on the y-axis than on the x-axis. This made it possible to show a larger portion of the graph (it also distorted the graph by flattening it). When you graph by hand we suggest that before putting scales on the two axes you should examine your table of values. Choose scales so that all or most of your points can be plotted and still keep your graph of reasonable size. A graphing calculator or a CAS will often choose the scale for the y's once you have chosen the x's to be used. The first choice you make, therefore, is the x values to plot. Most graphing calculators and CASs allow you to override the automatic y-axis scaling. In some cases you may want to use this option.

Intercepts The points where the graph of an equation crosses the two coordinate axes play a significant role in many problems. Consider, for example,

$$y = x^3 - 2x^2 - 5x + 6 = (x + 2)(x - 1)(x - 3)$$

Notice that $y = 0$ when $x = -2, 1, 3$. The numbers $-2, 1,$ and 3 are called ***x*-intercepts.** Similarly, $y = 6$ when $x = 0$, and so 6 is called the ***y*-intercept.**

Graphing Calculators
If you have a graphing calculator, use it whenever possible to reproduce the plots shown in the figures.

EXAMPLE 3 Find all intercepts of the graph of $y^2 - x + y - 6 = 0$.

SOLUTION Putting $y = 0$ in the given equation, we get $x = -6$, and so the x-intercept is -6. Putting $x = 0$ in the equation, we find that $y^2 + y - 6 = 0$, or $(y + 3)(y - 2) = 0$; the y-intercepts are -3 and 2. A check on symmetries indicates that the graph has none of the three types discussed earlier. The graph is displayed in Figure 5. ■

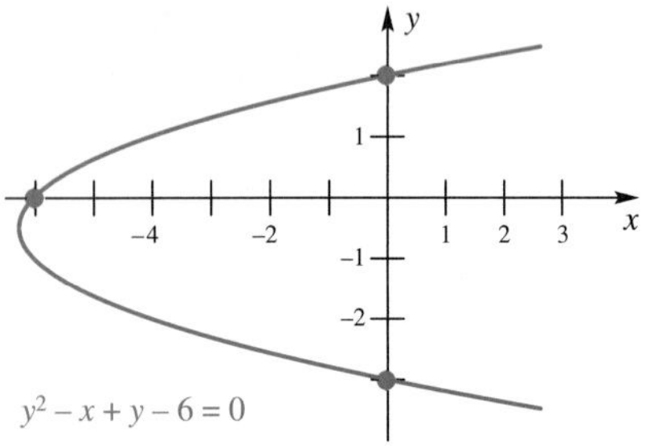

Figure 5

Since quadratic and cubic equations will often be used as examples in later work, we display their typical graphs in Figure 6.

The graphs of quadratic equations are cup-shaped curves called **parabolas.** If an equation has the form $y = ax^2 + bx + c$ or $x = ay^2 + by + c$ with $a \neq 0$, its graph is a parabola. In the first case, the graph opens up if $a > 0$ and opens down if $a < 0$. In the second case, the graph opens right if $a > 0$ and opens left if $a < 0$. Note that the equation of Example 3 can be put in the form $x = y^2 + y - 6$.

Intersections of Graphs Frequently, we need to know the points of intersection of two graphs. These points are found by solving the two equations for the graphs simultaneously, as illustrated in the next example.

EXAMPLE 4 Find the points of intersection of the line $y = -2x + 2$ and the parabola $y = 2x^2 - 4x - 2$, and sketch both graphs on the same coordinate plane.

SOLUTION We must solve the two equations simultaneously. This is easy to do by substituting the expression for y from the first equation into the second equation and then solving the resulting equation for x.

$$\begin{aligned} -2x + 2 &= 2x^2 - 4x - 2 \\ 0 &= 2x^2 - 2x - 4 \\ 0 &= 2(x + 1)(x - 2) \\ x &= -1, \qquad x = 2 \end{aligned}$$

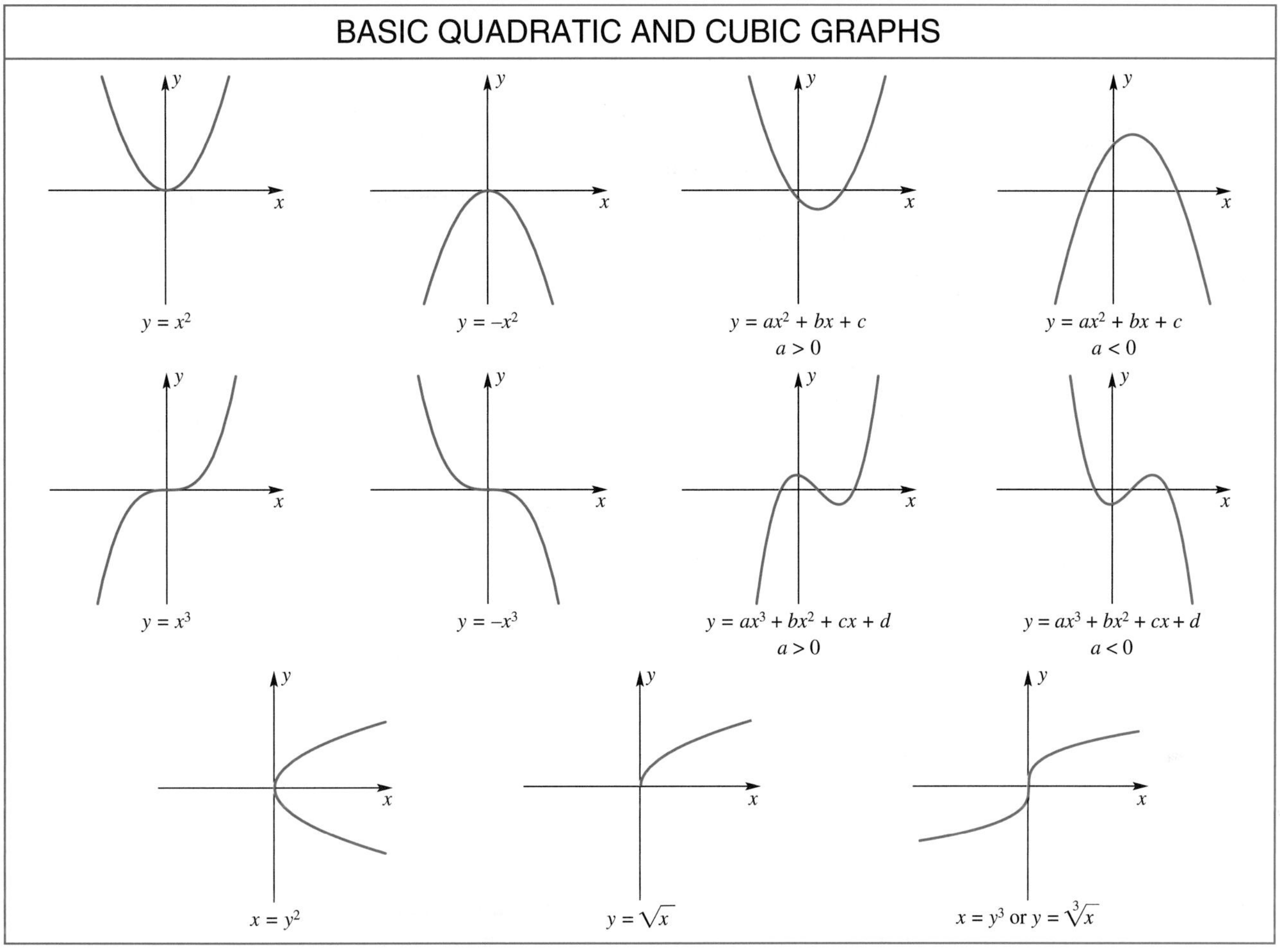

Figure 6

By substitution, we find the corresponding values of y to be 4 and -2; the intersection points are therefore $(-1, 4)$ and $(2, -2)$. The two graphs are shown in Figure 7.

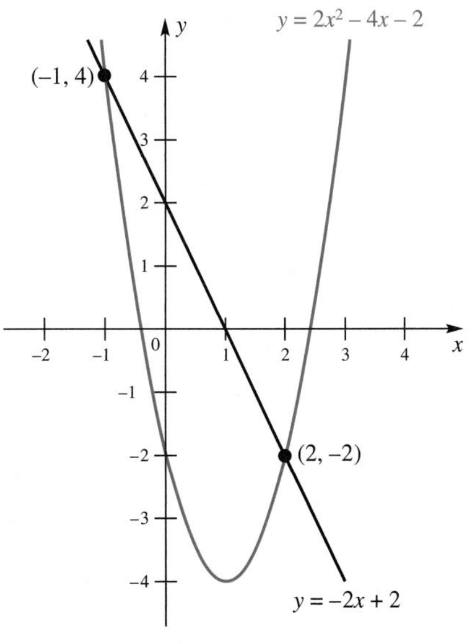

Figure 7 ■

Concepts Review

1. If whenever (x, y) is on a graph, $(-x, y)$ is also on the graph, then the graph is symmetric with respect to the _____.

2. If $(-4, 2)$ is on a graph that is symmetric with respect to the origin, then _____ is also on the graph.

3. The graph of $y = (x + 2)(x - 1)(x - 4)$ has y-intercept _____ and x-intercepts _____.

4. The graph of $y = ax^2 + bx + c$ is a _____ if $a = 0$ and a _____ if $a \neq 0$.

Problem Set 0.4

In Problems 1–30, plot the graph of each equation. Begin by checking for symmetries and be sure to find all x- and y-intercepts.

1. $y = -x^2 + 1$

2. $x = -y^2 + 1$

3. $x = -4y^2 - 1$

4. $y = 4x^2 - 1$

5. $x^2 + y = 0$

6. $y = x^2 - 2x$

7. $7x^2 + 3y = 0$

8. $y = 3x^2 - 2x + 2$

9. $x^2 + y^2 = 4$

10. $3x^2 + 4y^2 = 12$

11. $y = -x^2 - 2x + 2$

12. $4x^2 + 3y^2 = 12$

13. $x^2 - y^2 = 4$

14. $x^2 + (y - 1)^2 = 9$

15. $4(x - 1)^2 + y^2 = 36$

16. $x^2 - 4x + 3y^2 = -2$

17. $x^2 + 9(y + 2)^2 = 36$

GC **18.** $x^4 + y^4 = 1$

GC **19.** $x^4 + y^4 = 16$

GC **20.** $y = x^3 - x$

GC **21.** $y = \dfrac{1}{x^2 + 1}$

GC **22.** $y = \dfrac{x}{x^2 + 1}$

GC **23.** $2x^2 - 4x + 3y^2 + 12y = -2$

GC **24.** $4(x - 5)^2 + 9(y + 2)^2 = 36$

GC **25.** $y = (x - 1)(x - 2)(x - 3)$

GC **26.** $y = x^2(x - 1)(x - 2)$

GC **27.** $y = x^2(x - 1)^2$

GC **28.** $y = x^4(x - 1)^4(x + 1)^4$

GC **29.** $|x| + |y| = 1$

GC **30.** $|x| + |y| = 4$

GC *In Problems 31–38, plot the graphs of both equations on the same coordinate plane. Find and label the points of intersection of the two graphs (see Example 4).*

31. $y = -x + 1$, $y = (x + 1)^2$

32. $y = 2x + 3$, $y = -(x - 1)^2$

33. $y = -2x + 3$, $y = -2(x - 4)^2$

34. $y = -2x + 3$, $y = 3x^2 - 3x + 12$

35. $y = x$, $x^2 + y^2 = 4$

36. $y = x - 1$, $2x^2 + 3y^2 = 12$

37. $y - 3x = 1$, $x^2 + 2x + y^2 = 15$

38. $y = 4x + 3$, $x^2 + y^2 = 81$

39. Choose the equation that corresponds to each graph in Figure 8.

(a) $y = ax^2$, with $a > 0$

(b) $y = ax^3 + bx^2 + cx + d$, with $a > 0$

(c) $y = ax^3 + bx^2 + cx + d$, with $a < 0$

(d) $y = ax^3$, with $a > 0$

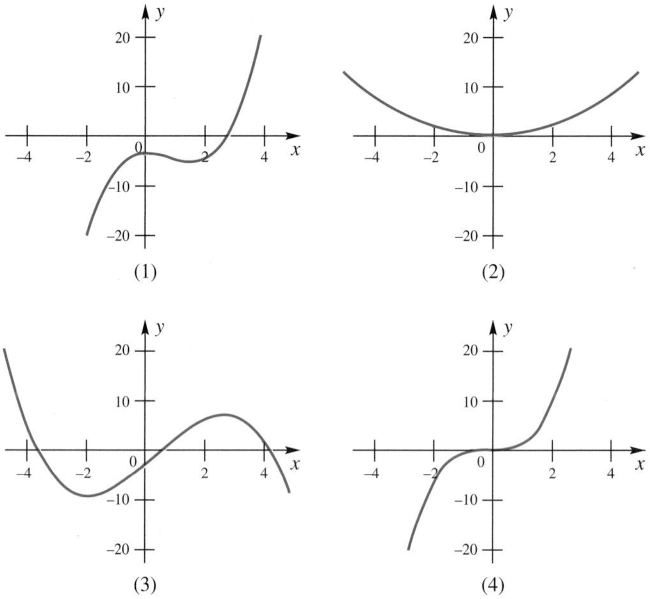

Figure 8

≈ **40.** Find the distance between the points on the circle $x^2 + y^2 = 13$ with the x-coordinates -2 and 2. How many such distances are there?

≈ **41.** Find the distance between the points on the circle $x^2 + 2x + y^2 - 2y = 20$ with the x-coordinates -2 and 2. How many such distances are there?

Answers to Concepts Review: **1.** y-axis **2.** $(4, -2)$ **3.** 8; $-2, 1, 4$ **4.** line; parabola

0.5 Functions and Their Graphs

The concept of function is one of the most basic in all mathematics, and it plays an indispensable role in calculus.

> **Definition**
>
> A **function** f is a rule of correspondence that associates with each object x in one set, called the **domain,** a single value $f(x)$ from a second set. The set of all values so obtained is called the **range** of the function. (See Figure 1.)

Think of a function as a machine that takes as its input a value x and produces an output $f(x)$. (See Figure 2.) Each input value is matched with a *single* output value. It can, however, happen that several different input values give the same output value.

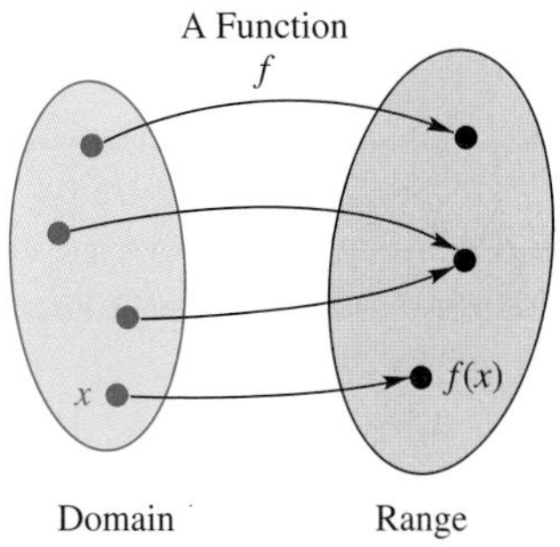

Figure 1

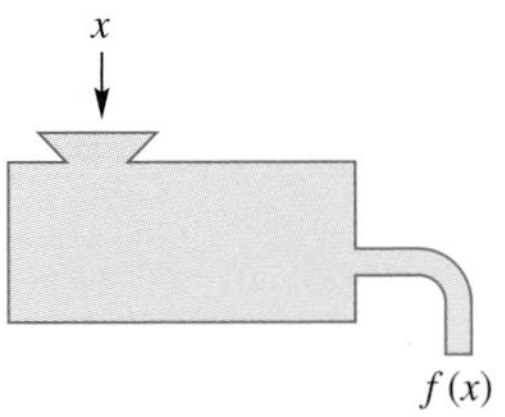

Figure 2

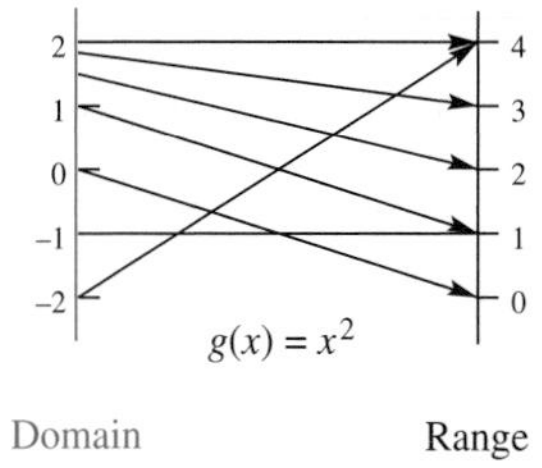

Figure 3

The definition puts no restriction on the domain and range sets. The domain might consist of the set of people in your calculus class, the range the set of grades $\{A, B, C, D, F\}$ that will be given, and the rule of correspondence the assignment of grades. Nearly all functions you encounter in this book will be functions of one or more real numbers. For example, the function g might take a real number x and square it, producing the real number x^2. In this case we have a formula that gives the rule of correspondence, that is, $g(x) = x^2$. A schematic diagram for this function is shown in Figure 3.

Function Notation A single letter like f (or g or F) is used to name a function. Then $f(x)$, read "f of x" or "f at x," denotes the value that f assigns to x. Thus, if $f(x) = x^3 - 4$, then

$$f(2) = 2^3 - 4 = 4$$
$$f(a) = a^3 - 4$$
$$f(a + h) = (a + h)^3 - 4 = a^3 + 3a^2h + 3ah^2 + h^3 - 4$$

Study the following examples carefully. Although some of these examples may look odd now, they will play an important role in Chapter 2.

EXAMPLE 1 For $f(x) = x^2 - 2x$, find and simplify

(a) $f(4)$ (b) $f(4 + h)$
(c) $f(4 + h) - f(4)$ (d) $[f(4 + h) - f(4)]/h$

SOLUTION

(a) $f(4) = 4^2 - 2 \cdot 4 = 8$

(b) $f(4 + h) = (4 + h)^2 - 2(4 + h) = 16 + 8h + h^2 - 8 - 2h$
$= 8 + 6h + h^2$

(c) $f(4 + h) - f(4) = 8 + 6h + h^2 - 8 = 6h + h^2$

(d) $\dfrac{f(4 + h) - f(4)}{h} = \dfrac{6h + h^2}{h} = \dfrac{h(6 + h)}{h} = 6 + h$ ■

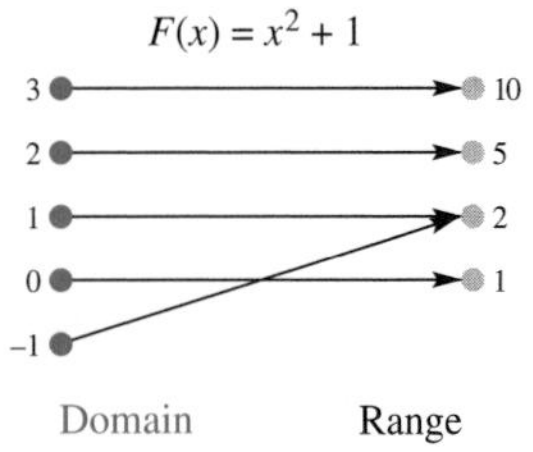

Figure 4

Domain and Range To specify a function completely, we must state, in addition to the rule of correspondence, the domain of the function. For example, if F is the function defined by $F(x) = x^2 + 1$ with domain $\{-1, 0, 1, 2, 3\}$ (Figure 4), then the range is $\{1, 2, 5, 10\}$. The rule of correspondence, together with the domain, determines the range.

When no domain is specified for a function, we assume that it is the largest set of real numbers for which the rule for the function makes sense. This is called the **natural domain.** Numbers that you should remember to exclude from the natural domain are those values that would cause division by zero or the square root of a negative number.

EXAMPLE 2 Find the natural domains for

(a) $f(x) = 1/(x - 3)$ (b) $g(t) = \sqrt{9 - t^2}$

(c) $h(w) = 1/\sqrt{9 - w^2}$

SOLUTION

(a) We must exclude 3 from the domain because it would require division by zero. Thus, the natural domain is $\{x: x \neq 3\}$. This may be read "the set of x's such that x is not equal to 3."

(b) To avoid the square root of a negative number, we must choose t so that $9 - t^2 \geq 0$. Thus, t must satisfy $|t| \leq 3$. The natural domain is therefore $\{t: |t| \leq 3\}$, which we can write using interval notation as $[-3, 3]$.

(c) Now we must avoid division by zero *and* square roots of negative numbers, so we must exclude -3 and 3 from the natural domain. The natural domain is therefore the interval $(-3, 3)$. ■

When the rule for a function is given by an equation of the form $y = f(x)$, we call x the **independent variable** and y the **dependent variable.** *Any* value in the domain may be substituted for the independent variable. Once selected, this value of x completely determines the corresponding value of the dependent variable y.

The input for a function need not be a single real number. In many important applications, a function depends on more than one independent variable. For example, the amount A of a monthly car payment depends on the loan's principal P, the rate of interest r, and the required number n of monthly payments. We could write such a function as $A(P, r, n)$. The value of $A(16000, 0.07, 48)$, that is, the required monthly payment to retire a \$16,000 loan in 48 months at an annual interest rate of 7%, is \$383.14. In this situation, there is no simple mathematical formula that gives the output A in terms of the input variables P, r, and n.

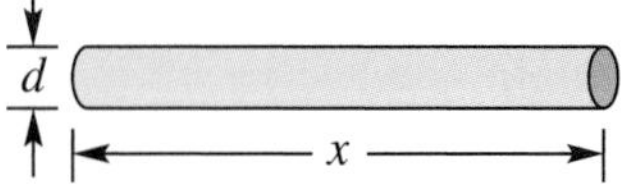

Figure 5

EXAMPLE 3 Let $V(x, d)$ denote the volume of a cylindrical rod of length x and diameter d. (See Figure 5.) Find

(a) a formula for $V(x, d)$
(b) the domain and range of V
(c) $V(4, 0.1)$

SOLUTION

(a) $V(x, d) = x \cdot \pi\left(\dfrac{d}{2}\right)^2 = \dfrac{\pi x d^2}{4}$

(b) Because the length and diameter of the rod must be positive, the domain is the set of all ordered pairs (x, d) where $x > 0$ and $d > 0$. Any positive volume is possible so the range is $(0, \infty)$.

(c) $V(4, 0.1) = \dfrac{\pi \cdot 4 \cdot 0.1^2}{4} = 0.01\pi$ ■

Chapters 1 through 11 will deal mostly with functions of a single independent variable. Beginning in Chapter 12, we will study properties of functions of two or more independent variables.

Graphing Calculator

Remember, use your graphing calculator to reproduce the figures in this book. Experiment with various graphing windows until you are convinced that you understand all important aspects of the graph.

Graphs of Functions When both the domain and range of a function are sets of real numbers, we can picture the function by drawing its graph on a coordinate plane. The **graph of a function** f is simply the graph of the equation $y = f(x)$.

EXAMPLE 4 Sketch the graphs of

(a) $f(x) = x^2 - 2$ (b) $g(x) = 2/(x - 1)$

SOLUTION The natural domains of f and g are, respectively, all real numbers and all real numbers except 1. Following the procedure described in Section 0.4 (make a table of values, plot the corresponding points, connect these points with a smooth curve), we obtain the two graphs shown in Figures 6 and 7a. ■

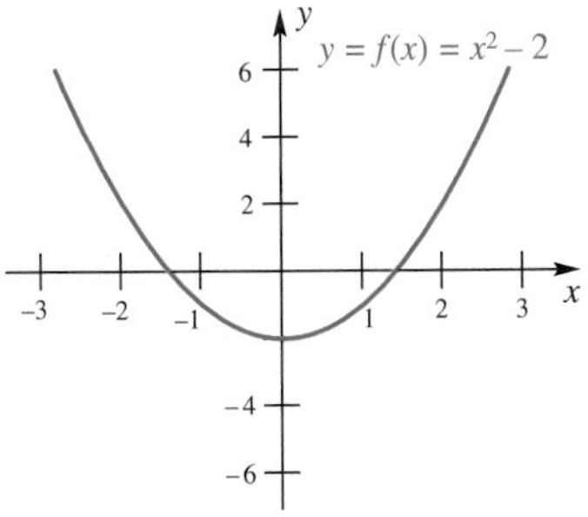

Figure 6

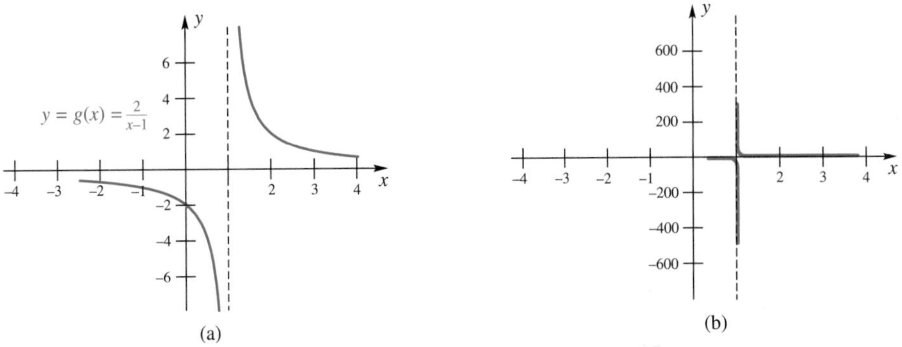

Figure 7

Pay special attention to the graph of g; it points to an oversimplification that we have made and now need to correct. When connecting the plotted points by a smooth curve, do not do so in a mechanical way that ignores special features that may be apparent from the formula for the function. In the case of $g(x) = 2/(x - 1)$, something dramatic happens as x nears 1. In fact, the values of $|g(x)|$ increase without bound; for example, $g(0.99) = 2/(0.99 - 1) = -200$ and $g(1.001) = 2000$. We have indicated this by drawing a dashed vertical line, called an **asymptote,** at $x = 1$. As x approaches 1, the graph gets closer and closer to this line, though this line itself is not part of the graph. Rather, it is a guideline. Notice that the graph of g also has a horizontal asymptote, the x-axis.

Functions like $g(x) = 2/(x - 1)$ can even cause problems when you graph them on a CAS. For example, *Maple*, when asked to plot $g(x) = 2/(x - 1)$ over the domain $[-4, 4]$ responded with the graph shown in Figure 7b. Computer Algebra Systems use an algorithm much like that described in Section 0.4; they choose a number of x-values over the stated domain, find the corresponding y-values, and plot these points with connecting lines. When *Maple* chose a number near 1, the resulting output was large, leading to the y-axis scaling in the figure. *Maple* also connected the points right across the break at $x = 1$. Always be cautious and careful when you use a graphing calculator or a CAS to plot functions.

The domains and ranges for the functions f and g are shown in the table below.

Function	Domain	Range
$f(x) = x^2 - 2$	all real numbers	$\{y: y \geq -2\}$
$g(x) = \dfrac{2}{x - 1}$	$\{x: x \neq 1\}$	$\{y: y \neq 0\}$

Even and Odd Functions We can often predict the symmetries of the graph of a function by inspecting the formula for the function. If $f(-x) = f(x)$ for all x, then the graph is symmetric with respect to the y-axis. Such a function is called an

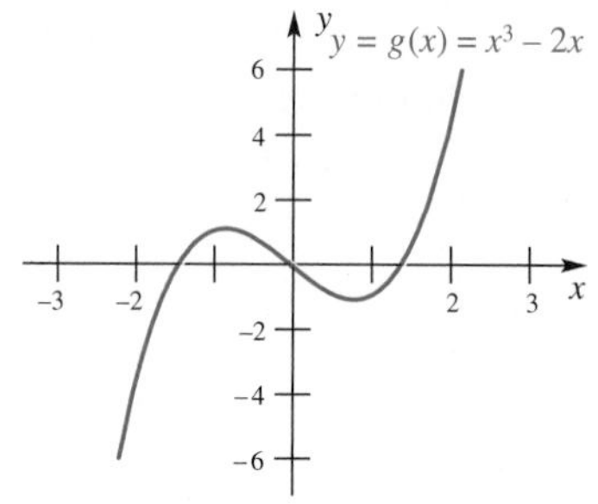

Figure 8

even function, probably because a function that specifies $f(x)$ as a sum of only even powers of x is even. The function $f(x) = x^2 - 2$ (graphed in Figure 6) is even; so are $f(x) = 3x^6 - 2x^4 + 11x^2 - 5$, $f(x) = x^2/(1 + x^4)$, and $f(x) = (x^3 - 2x)/3x$.

If $f(-x) = -f(x)$ for all x, the graph is symmetric with respect to the origin. We call such a function an **odd function.** A function that gives $f(x)$ as a sum of only odd powers of x is odd. Thus, $g(x) = x^3 - 2x$ (graphed in Figure 8) is odd. Note that

$$g(-x) = (-x)^3 - 2(-x) = -x^3 + 2x = -(x^3 - 2x) = -g(x)$$

Consider the function $g(x) = 2/(x - 1)$ from Example 4, which we graphed in Figure 7. It is neither even nor odd. To see this, observe that $g(-x) = 2/(-x - 1)$, which is not equal to either $g(x)$ or $-g(x)$. Note that the graph of $y = g(x)$ is neither symmetric with respect to the y-axis nor the origin.

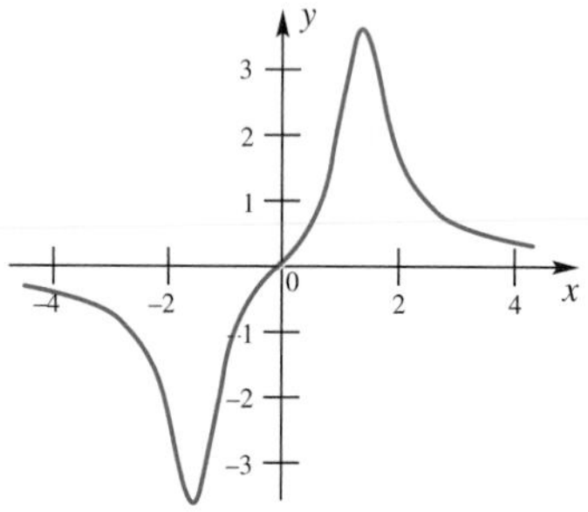

Figure 9

EXAMPLE 5 Is $f(x) = \dfrac{x^3 + 3x}{x^4 - 3x^2 + 4}$ even, odd, or neither?

SOLUTION Since

$$f(-x) = \frac{(-x)^3 + 3(-x)}{(-x)^4 - 3(-x)^2 + 4} = \frac{-(x^3 + 3x)}{x^4 - 3x^2 + 4} = -f(x)$$

f is an odd function. The graph of $y = f(x)$ (Figure 9) is symmetric with respect to the origin. ■

Two Special Functions Among the functions that will often be used as examples are two very special ones: the **absolute value function,** $|\ \ |$, and the **greatest integer function,** $[\![\ \ ]\!]$. They are defined by

$$|x| = \begin{cases} x & \text{if } x \geq 0 \\ -x & \text{if } x < 0 \end{cases}$$

and

$$[\![x]\!] = \text{the greatest integer less than or equal to } x$$

Thus, $|-3.1| = |3.1| = 3.1$, while $[\![-3.1]\!] = -4$ and $[\![3.1]\!] = 3$. We show the graphs of these two functions in Figures 10 and 11. The absolute value function is even, since $|-x| = |x|$. The greatest integer function is neither even nor odd, as you can see from its graph.

We will often appeal to the following special features of these graphs. The graph of $|x|$ has a sharp corner at the origin, while the graph of $[\![x]\!]$ takes a jump at each integer.

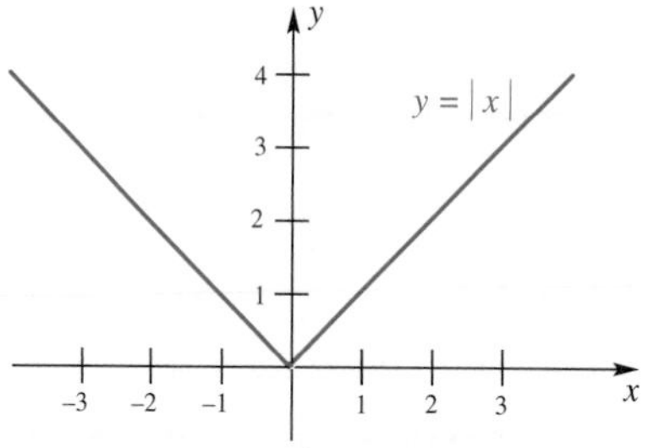

Figure 10

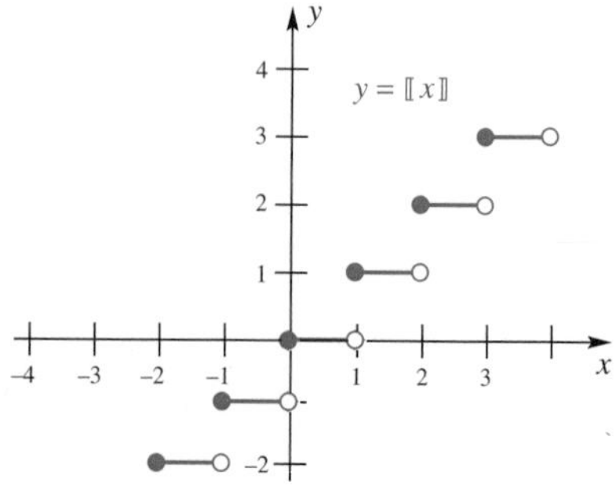

Figure 11

Concepts Review

1. The set of allowable inputs for a function is called the _____ of the function; the set of outputs that are obtained is called the _____ of the function.

2. If $f(x) = 3x^2$, then $f(2u) =$ _____ and $f(x + h) =$ _____.

3. If $f(x)$ gets closer and closer to L as $|x|$ increases indefinitely, then the line $y = L$ is a(an) _____ for the graph of f.

4. If $f(-x) = f(x)$ for all x in the domain of f, then f is called a(an) _____ function; if $f(-x) = -f(x)$ for all x in the domain of f, then f is called a(an) _____ function. In the first case, the graph of f is symmetric with respect to the _____; in the second case, it is symmetric with respect to the _____.

Problem Set 0.5

1. For $f(x) = 1 - x^2$, find each value.

(a) $f(1)$ (b) $f(-2)$ (c) $f(0)$
(d) $f(k)$ (e) $f(-5)$ (f) $f\left(\frac{1}{4}\right)$
(g) $f(1 + h)$ (h) $f(1 + h) - f(1)$
(i) $f(2 + h) - f(2)$

2. For $F(x) = x^3 + 3x$, find each value.

(a) $F(1)$ (b) $F(\sqrt{2})$ (c) $F\left(\frac{1}{4}\right)$
(d) $F(1 + h)$ (e) $F(1 + h) - F(1)$
(f) $F(2 + h) - F(2)$

3. For $G(y) = 1/(y - 1)$, find each value.

(a) $G(0)$ (b) $G(0.999)$ (c) $G(1.01)$
(d) $G(y^2)$ (e) $G(-x)$ (f) $G\left(\frac{1}{x^2}\right)$

4. For $\Phi(u) = \dfrac{u + u^2}{\sqrt{u}}$, find each value. ($\Phi$ is the uppercase Greek letter phi.)

(a) $\Phi(1)$ (b) $\Phi(-t)$ (c) $\Phi\left(\frac{1}{2}\right)$
(d) $\Phi(u + 1)$ (e) $\Phi(x^2)$ (f) $\Phi(x^2 + x)$

5. For

$$f(x) = \frac{1}{\sqrt{x - 3}}$$

find each value.

(a) $f(0.25)$ (b) $f(\pi)$ (c) $f(3 + \sqrt{2})$

C **6.** For $f(x) = \sqrt{x^2 + 9}/(x - \sqrt{3})$, find each value.

(a) $f(0.79)$ (b) $f(12.26)$ (c) $f(\sqrt{3})$

7. Which of the following determine a function f with formula $y = f(x)$? For those that do, find $f(x)$. *Hint:* Solve for y in terms of x and note that the definition of a function requires a single y for each x.

(a) $x^2 + y^2 = 1$ (b) $xy + y + x = 1, x \neq -1$
(c) $x = \sqrt{2y + 1}$ (d) $x = \dfrac{y}{y + 1}$

8. Which of the graphs in Figure 12 are graphs of functions?

This problem suggests a rule: *For a graph to be the graph of a function, each vertical line must meet the graph in at most one point.*

9. For $f(x) = 2x^2 - 1$, find and simplify $[f(a + h) - f(a)]/h$.

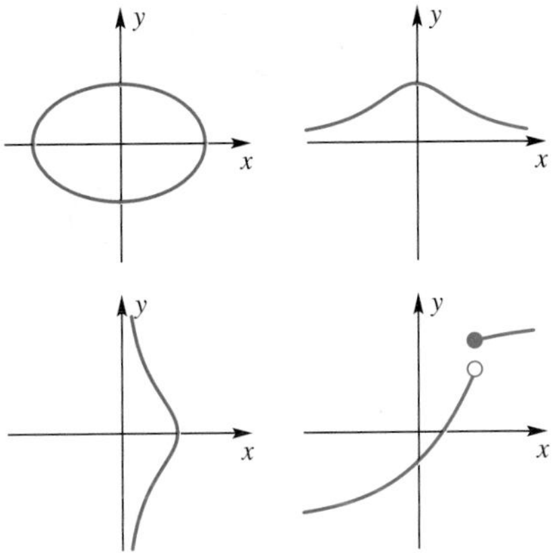

Figure 12

10. For $F(t) = 4t^3$, find and simplify $[F(a + h) - F(a)]/h$.

11. For $g(u) = 3/(u - 2)$, find and simplify $[g(x + h) - g(x)]/h$.

12. For $G(t) = t/(t + 4)$, find and simplify $[G(a + h) - G(a)]/h$.

13. Find the natural domain for each of the following.

(a) $F(z) = \sqrt{2z + 3}$ (b) $g(v) = 1/(4v - 1)$
(c) $\psi(x) = \sqrt{x^2 - 9}$ (d) $H(y) = -\sqrt{625 - y^4}$

14. Find the natural domain in each case.

(a) $f(x) = \dfrac{4 - x^2}{x^2 - x - 6}$ (b) $G(y) = \sqrt{(y + 1)^{-1}}$
(c) $\phi(u) = |2u + 3|$ (d) $F(t) = t^{2/3} - 4$

In Problems 15–30, specify whether the given function is even, odd, or neither, and then sketch its graph.

15. $f(x) = -4$ **16.** $f(x) = 3x$

17. $F(x) = 2x + 1$ **18.** $F(x) = 3x - \sqrt{2}$

19. $g(x) = 3x^2 + 2x - 1$ **20.** $g(u) = \dfrac{u^3}{8}$

21. $g(x) = \dfrac{x}{x^2 - 1}$ **22.** $\phi(z) = \dfrac{2z + 1}{z - 1}$

23. $f(w) = \sqrt{w - 1}$ **24.** $h(x) = \sqrt{x^2 + 4}$

25. $f(x) = |2x|$ **26.** $F(t) = -|t + 3|$

27. $g(x) = \left[\!\left[\dfrac{x}{2}\right]\!\right]$ **28.** $G(x) = [\![2x - 1]\!]$

29. $g(t) = \begin{cases} 1 & \text{if } t \le 0 \\ t+1 & \text{if } 0 < t < 2 \\ t^2 - 1 & \text{if } t \ge 2 \end{cases}$

30. $h(x) = \begin{cases} -x^2 + 4 & \text{if } x \le 1 \\ 3x & \text{if } x > 1 \end{cases}$

31. A plant has the capacity to produce from 0 to 100 computers per day. The daily overhead for the plant is \$5000, and the direct cost (labor and materials) of producing one computer is \$805. Write a formula for $T(x)$, the total cost of producing x computers in one day, and also for the unit cost $u(x)$ (average cost per computer). What are the domains of these functions?

C 32. It costs the ABC Company $400 + 5\sqrt{x(x-4)}$ dollars to make x toy stoves that sell for \$6 each.

(a) Find a formula for $P(x)$, the total profit in making x stoves.

(b) Evaluate $P(200)$ and $P(1000)$.

(c) How many stoves does ABC have to make to just break even?

C 33. Find the formula for the amount $E(x)$ by which a number x exceeds its square. Plot a graph of $E(x)$ for $0 \le x \le 1$. Use the graph to estimate the positive number less than or equal to 1 that exceeds its square by the maximum amount.

34. Let p denote the perimeter of an equilateral triangle. Find a formula for $A(p)$, the area of such a triangle.

35. A right triangle has a fixed hypotenuse of length h and one leg that has length x. Find a formula for the length $L(x)$ of the other leg.

36. A right triangle has a fixed hypotenuse of length h and one leg that has length x. Find a formula for the area $A(x)$ of the triangle.

37. The Acme Car Rental Agency charges \$24 a day for the rental of a car plus \$0.40 per mile.

(a) Write a formula for the total rental expense $E(x)$ for one day, where x is the number of miles driven.

(b) If you rent a car for one day, how many miles can you drive for \$120?

38. A right circular cylinder of radius r is inscribed in a sphere of radius $2r$. Find a formula for $V(r)$, the volume of the cylinder, in terms of r.

39. A 1-mile track has parallel sides and equal semicircular ends. Find a formula for the area enclosed by the track, $A(d)$, in terms of the diameter d of the semicircles. What is the natural domain for this function?

40. Let $A(c)$ denote the area of the region bounded from above by the line $y = x + 1$, from the left by the y-axis, from below by the x-axis, and from the right by the line $x = c$. Such a function is called an **accumulation function.** (See Figure 13.) Find

(a) $A(1)$ (b) $A(2)$

(c) $A(0)$ (d) $A(c)$

(e) Sketch the graph of $A(c)$.

(f) What are the domain and range of A?

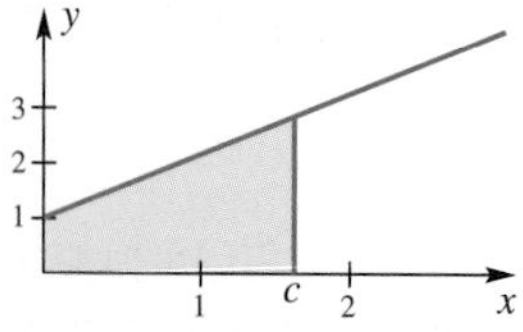

Figure 13

41. Let $B(c)$ denote the area of the region bounded from above by the graph of the curve $y = x(1-x)$, from below by the x-axis, and from the right by the line $x = c$. The domain of B is the interval $[0, 1]$. (See Figure 14.) Given that $B(1) = \frac{1}{6}$,

(a) Find $B(0)$ (b) Find $B\left(\frac{1}{2}\right)$

(c) As best you can, sketch a graph of $B(c)$.

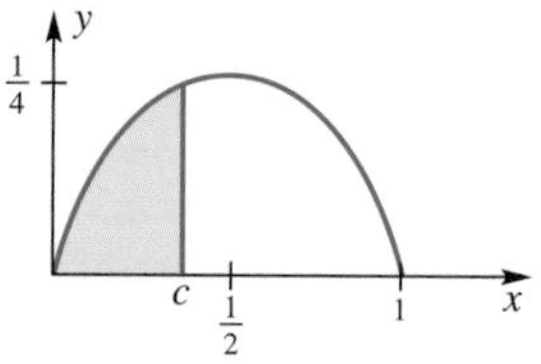

Figure 14

42. Which of the following functions satisfies $f(x+y) = f(x) + f(y)$ for all real numbers x and y?

(a) $f(t) = 2t$ (b) $f(t) = t^2$

(c) $f(t) = 2t + 1$ (d) $f(t) = -3t$

43. Let $f(x+y) = f(x) + f(y)$ for all x and y. Prove that there is a number m such that $f(t) = mt$ for all rational numbers t. *Hint:* First decide what m has to be. Then proceed in steps, starting with $f(0) = 0$, $f(p) = mp$ for a natural number p, $f(1/p) = m/p$, and so on.

44. A baseball diamond is a square with sides of 90 feet. A player, after hitting a home run, loped around the diamond at 10 feet per second. Let s represent the player's distance from home plate after t seconds.

(a) Express s as a function of t by means of a four-part formula.

(b) Express s as a function of t by means of a three-part formula.

GC *To use technology effectively, you need to discover its capabilities, its strengths, and its weaknesses. We urge you to practice graphing functions of various types using your own computer package or calculator. Problems 45–50 are designed for this purpose.*

45. Let $f(x) = (x^3 + 3x - 5)/(x^2 + 4)$.

(a) Evaluate $f(1.38)$ and $f(4.12)$.

(b) Construct a table of values for this function corresponding to $x = -4, -3, \ldots, 3, 4$.

46. Follow the instructions in Problem 45 for $f(x) = (\sin^2 x - 3\tan x)/\cos x$.

47. Draw the graph of $f(x) = x^3 - 5x^2 + x + 8$ on the domain $[-2, 5]$.

(a) Determine the range of f.

(b) Where on this domain is $f(x) \ge 0$?

48. Superimpose the graph of $g(x) = 2x^2 - 8x - 1$ with domain $[-2, 5]$ on the graph of $f(x)$ of Problem 47.

(a) Estimate the x-values where $f(x) = g(x)$.

(b) Where on $[-2, 5]$ is $f(x) \ge g(x)$?

(c) Estimate the largest value of $|f(x) - g(x)|$ on $[-2, 5]$.

49. Graph $f(x) = (3x - 4)/(x^2 + x - 6)$ on the domain $[-6, 6]$.

(a) Determine the x- and y-intercepts.

(b) Determine the range of f for the given domain.

(c) Determine the vertical asymptotes of the graph.

(d) Determine the horizontal asymptote for the graph when the domain is enlarged to the natural domain.

50. Follow the directions in Problem 49 for the function $g(x) = (3x^2 - 4)/(x^2 + x - 6)$

Answers to Concepts Review: **1.** domain; range **2.** $12u^2$; $3(x + h)^2 = 3x^2 + 6xh + 3h^2$ **3.** asymptote **4.** even; odd; y-axis; origin

0.6 Operations on Functions

Just as two numbers a and b can be added to produce a new number $a + b$, so two functions f and g can be added to produce a new function $f + g$. This is just one of several operations on functions that we will describe in this section.

Sums, Differences, Products, Quotients, and Powers Consider functions f and g with formulas

$$f(x) = \frac{x - 3}{2}, \qquad g(x) = \sqrt{x}$$

We can make a new function $f + g$ by having it assign to x the value $f(x) + g(x) = (x - 3)/2 + \sqrt{x}$; that is,

$$(f + g)(x) = f(x) + g(x) = \frac{x - 3}{2} + \sqrt{x}$$

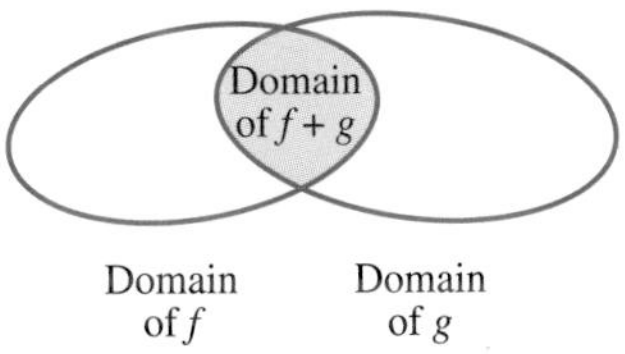

Figure 1

Of course, we must be a little careful about domains. Clearly, x must be a number on which both f and g can work. In other words, the domain of $f + g$ is the intersection (common part) of the domains of f and g (Figure 1).

The functions $f - g$, $f \cdot g$, and f/g are introduced in a completely analogous way. Assuming that f and g have their natural domains, we have the following:

Formula	Domain
$(f + g)(x) = f(x) + g(x) = \frac{x - 3}{2} + \sqrt{x}$	$[0, \infty)$
$(f - g)(x) = f(x) - g(x) = \frac{x - 3}{2} - \sqrt{x}$	$[0, \infty)$
$(f \cdot g)(x) = f(x) \cdot g(x) = \frac{x - 3}{2}\sqrt{x}$	$[0, \infty)$
$\left(\frac{f}{g}\right)(x) = \frac{f(x)}{g(x)} = \frac{x - 3}{2\sqrt{x}}$	$(0, \infty)$

We had to exclude 0 from the domain of f/g to avoid division by 0.

We may also raise a function to a power. By f^n, we mean the function that assigns to x the value $[f(x)]^n$. Thus,

$$g^3(x) = [g(x)]^3 = \left(\sqrt{x}\right)^3 = x^{3/2}$$

There is one exception to the above agreement on exponents, namely, when $n = -1$. We reserve the symbol f^{-1} for the inverse function, which will be discussed in Section 6.2. Thus, f^{-1} does not mean $1/f$.

EXAMPLE 1 Let $F(x) = \sqrt[4]{x + 1}$ and $G(x) = \sqrt{9 - x^2}$, with respective natural domains $[-1, \infty)$ and $[-3, 3]$. Find formulas for $F + G$, $F - G$, $F \cdot G$, F/G, and F^5 and give their natural domains.

SOLUTION

Formula	Domain
$(F+G)(x) = F(x) + G(x) = \sqrt[4]{x+1} + \sqrt{9-x^2}$	$[-1, 3]$
$(F-G)(x) = F(x) - G(x) = \sqrt[4]{x+1} - \sqrt{9-x^2}$	$[-1, 3]$
$(F \cdot G)(x) = F(x) \cdot G(x) = \sqrt[4]{x+1}\sqrt{9-x^2}$	$[-1, 3]$
$\left(\frac{F}{G}\right)(x) = \frac{F(x)}{G(x)} = \frac{\sqrt[4]{x+1}}{\sqrt{9-x^2}}$	$[-1, 3)$
$F^5(x) = [F(x)]^5 = \left(\sqrt[4]{x+1}\right)^5 = (x+1)^{5/4}$	$[-1, \infty)$ ■

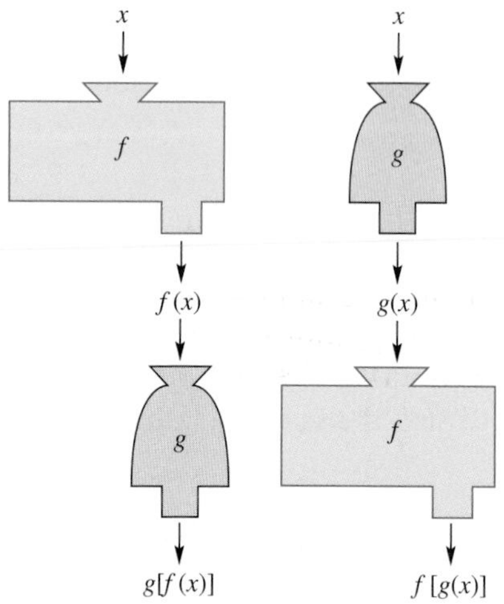

Figure 2

Composition of Functions Earlier, we asked you to think of a function as a machine. It accepts x as input, works on x, and produces $f(x)$ as output. Two machines may often be put together in tandem to make a more complicated machine; so may two functions f and g (Figure 2). If f works on x to produce $f(x)$ and g then works on $f(x)$ to produce $g(f(x))$, we say that we have *composed* g with f. The resulting function, called the **composition** of g with f, is denoted by $g \circ f$. Thus,

$$\boxed{(g \circ f)(x) = g(f(x))}$$

In our previous examples we had $f(x) = (x-3)/2$ and $g(x) = \sqrt{x}$. We may compose these functions in two ways:

$$(g \circ f)(x) = g(f(x)) = g\left(\frac{x-3}{2}\right) = \sqrt{\frac{x-3}{2}}$$

$$(f \circ g)(x) = f(g(x)) = f\left(\sqrt{x}\right) = \frac{\sqrt{x}-3}{2}$$

Right away we notice that $g \circ f$ does not equal $f \circ g$. Thus, we say that the composition of functions is not commutative.

We must be careful in describing the domain of a composite function. The domain of $g \circ f$ is equal to the set of those values x that satisfy the following properties:

1. x is in the domain of f.
2. $f(x)$ is in the domain of g.

In other words, x must be a valid input for f, and $f(x)$ must be a valid input for g. In our example, the value $x = 2$ is in the domain of f, but it is not in the domain of $g \circ f$ because this would lead to the square root of a negative number:

$$g(f(2)) = g((2-3)/2) = g\left(-\frac{1}{2}\right) = \sqrt{-\frac{1}{2}}$$

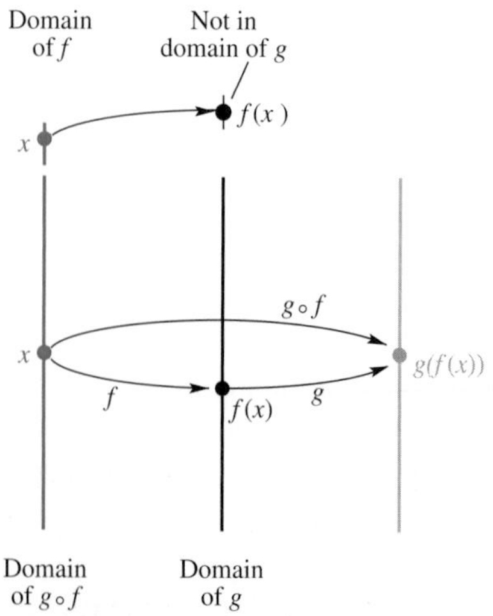

Figure 3

The domain for $g \circ f$ is the interval $[3, \infty)$ because $f(x)$ is nonnegative on this interval, and the input to g must be nonnegative. The domain for $f \circ g$ is the interval $[0, \infty)$ (why?), so we see that the domains of $g \circ f$ and $f \circ g$ can be different. Figure 3 shows how the domain of $g \circ f$ excludes those values of x for which $f(x)$ is not in the domain of g.

EXAMPLE 2 Let $f(x) = 6x/(x^2 - 9)$ and $g(x) = \sqrt{3x}$, with their natural domains. First, find $(f \circ g)(12)$; then find $(f \circ g)(x)$ and give its domain.

SOLUTION

$$(f \circ g)(12) = f(g(12)) = f\left(\sqrt{36}\right) = f(6) = \frac{6 \cdot 6}{6^2 - 9} = \frac{4}{3}$$

$$(f \circ g)(x) = f(g(x)) = f\left(\sqrt{3x}\right) = \frac{6\sqrt{3x}}{\left(\sqrt{3x}\right)^2 - 9}$$

The expression $\sqrt{3x}$ appears in both the numerator and denominator. Any negative number for x will lead to the square root of a negative number. Thus, all negative numbers must be excluded from the domain of $f \circ g$. For $x \geq 0$, we have $(\sqrt{3x})^2 = 3x$, allowing us to write

$$(f \circ g)(x) = \frac{6\sqrt{3x}}{3x - 9} = \frac{2\sqrt{3x}}{x - 3}$$

We must also exclude $x = 3$ from the domain of $f \circ g$ because $g(3)$ is not in the domain of f. (It would cause division by 0.) Thus, the domain of $f \circ g$ is $[0, 3) \cup (3, \infty)$. ■

In calculus, we will often need to take a given function and write it as the composition of two simpler functions. Usually, this can be done in a number of ways. For example, $p(x) = \sqrt{x^2 + 4}$ can be written as

$$p(x) = g(f(x)), \quad \text{where } g(x) = \sqrt{x} \text{ and } f(x) = x^2 + 4$$

or as

$$p(x) = g(f(x)), \quad \text{where } g(x) = \sqrt{x + 4} \text{ and } f(x) = x^2$$

(You should check that both of these compositions give $p(x) = \sqrt{x^2 + 4}$ with domain $(-\infty, \infty)$.) The decomposition $p(x) = g(f(x))$ with $f(x) = x^2 + 4$ and $g(x) = \sqrt{x}$ is regarded as simpler and is usually preferred. We can therefore view $p(x) = \sqrt{x^2 + 4}$ as the square root of a function of x. This way of looking at functions will be important in Chapter 2.

EXAMPLE 3 Write the function $p(x) = (x + 2)^5$ as a composite function $g \circ f$.

SOLUTION The most obvious way to decompose p is to write

$$p(x) = g(f(x)), \quad \text{where } g(x) = x^5 \text{ and } f(x) = x + 2$$

We thus view $p(x) = (x + 2)^5$ as the fifth power of a function of x. ■

Translations Observing how a function is built up from simpler ones can be a big aid in graphing. We may ask this question: How are the graphs of

$$y = f(x) \qquad y = f(x - 3) \qquad y = f(x) + 2 \qquad y = f(x - 3) + 2$$

related to each other? Consider $f(x) = |x|$ as an example. The corresponding four graphs are displayed in Figure 4.

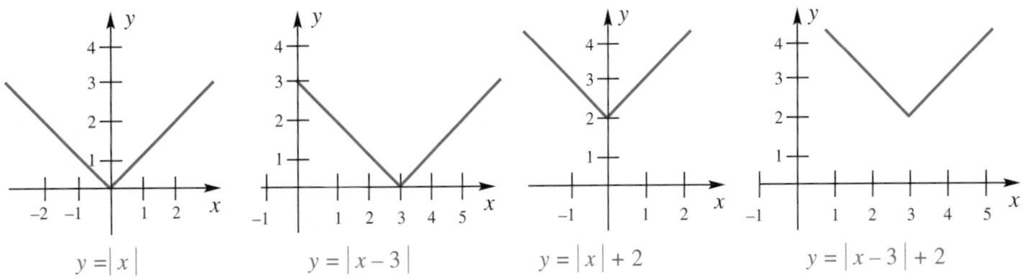

Figure 4

Notice that all four graphs have the same shape; the last three are just translations of the first. Replacing x by $x - 3$ translates the graph 3 units to the right; adding 2 translates it upward by 2 units.

What happened with $f(x) = |x|$ is typical. Figure 5 offers an illustration for the function $f(x) = x^3 + x^2$.

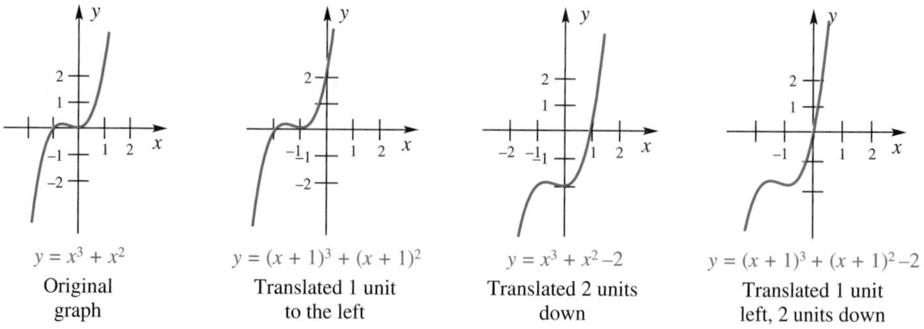

Figure 5

Exactly the same principles apply in the general situation. They are illustrated in Figure 6 with both h and k positive. If $h < 0$, the translation is to the left; if $k < 0$, the translation is downward.

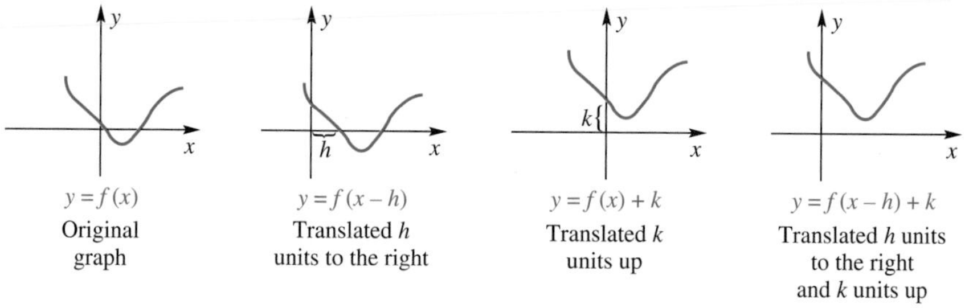

Figure 6

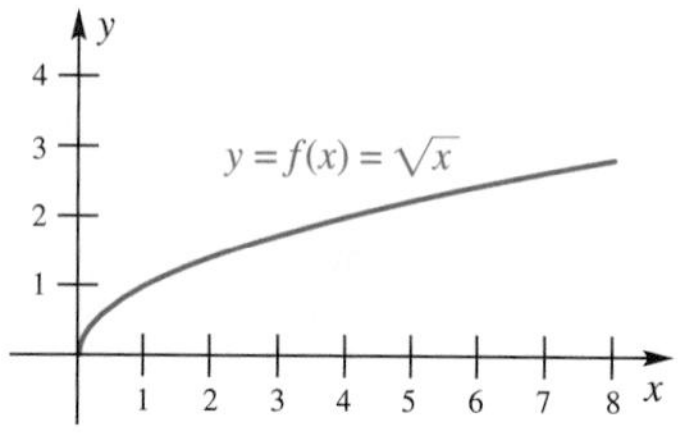

Figure 7

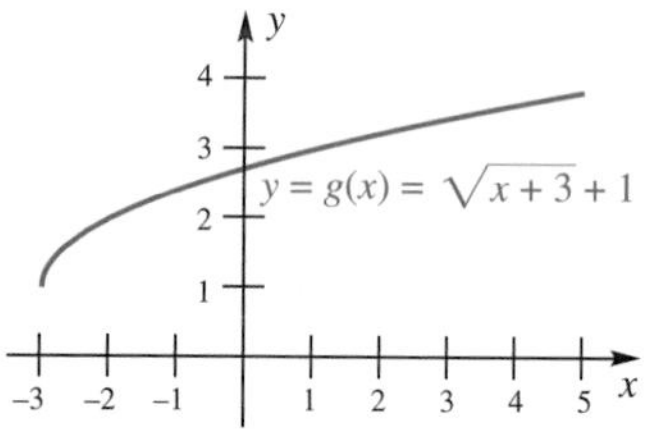

Figure 8

EXAMPLE 4 Sketch the graph of $g(x) = \sqrt{x + 3} + 1$ by first graphing $f(x) = \sqrt{x}$ and then making appropriate translations.

SOLUTION By translating the graph of f (Figure 7) 3 units left and 1 unit up, we obtain the graph of g (Figure 8). ■

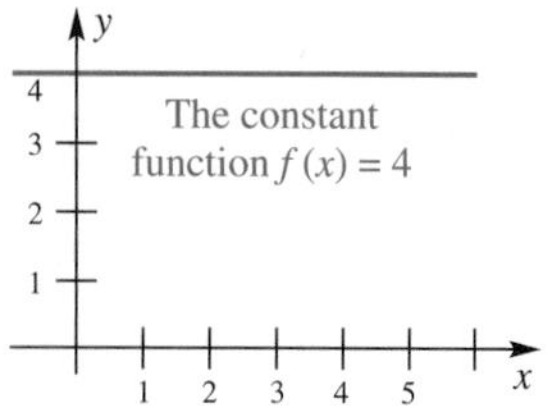

Figure 9

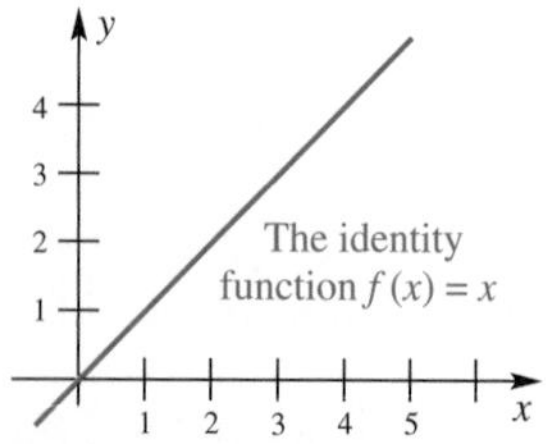

Figure 10

Partial Catalog of Functions A function of the form $f(x) = k$, where k is a constant (real number), is called a **constant function.** Its graph is a horizontal line (Figure 9). The function $f(x) = x$ is called the **identity function.** Its graph is a line through the origin having slope 1 (Figure 10). From these simple functions, we can build many important functions.

Any function that can be obtained from the constant functions and the identity function by use of the operations of addition, subtraction, and multiplication is called a **polynomial function.** This amounts to saying that f is a polynomial function if it is of the form

$$f(x) = a_n x^n + a_{n-1} x^{n-1} + \cdots + a_1 x + a_0$$

where the a's are real numbers and n is a nonnegative integer. If $a_n \neq 0$, n is the **degree** of the polynomial function. In particular, $f(x) = ax + b$ is a first-degree polynomial function, or **linear function,** and $f(x) = ax^2 + bx + c$ is a second-degree polynomial function, or **quadratic function.**

Quotients of polynomial functions are called **rational functions.** Thus, f is a rational function if it is of the form

$$f(x) = \frac{a_n x^n + a_{n-1}x^{n-1} + \cdots + a_1 x + a_0}{b_m x^m + b_{m-1}x^{m-1} + \cdots + b_1 x + b_0}$$

The domain of a rational function consists of those real numbers for which the denominator is nonzero.

An **explicit algebraic function** is one that can be obtained from the constant functions and the identity function via the five operations of addition, subtraction, multiplication, division, and root extraction. Examples are

$$f(x) = 3x^{2/5} = 3\sqrt[5]{x^2} \qquad g(x) = \frac{(x+2)\sqrt{x}}{x^3 + \sqrt[3]{x^2 - 1}}$$

The functions listed so far, together with the trigonometric, inverse trigonometric, exponential, and logarithmic functions (to be introduced later), are the basic raw materials for calculus.

Concepts Review

1. If $f(x) = x^2 + 1$, then $f^3(x) =$ ____.

2. The value of the composite function $f \circ g$ at x is given by $(f \circ g)(x) =$ ____.

3. Compared to the graph of $y = f(x)$, the graph of $y = f(x + 2)$ is translated ____ units to the ____.

4. A rational function is defined as ____.

Problem Set 0.6

1. For $f(x) = x + 3$ and $g(x) = x^2$, find each value (if possible).
(a) $(f + g)(2)$ (b) $(f \cdot g)(0)$ (c) $(g/f)(3)$
(d) $(f \circ g)(1)$ (e) $(g \circ f)(1)$ (f) $(g \circ f)(-8)$

2. For $f(x) = x^2 + x$ and $g(x) = 2/(x + 3)$, find each value.
(a) $(f - g)(2)$ (b) $(f/g)(1)$ (c) $g^2(3)$
(d) $(f \circ g)(1)$ (e) $(g \circ f)(1)$ (f) $(g \circ g)(3)$

3. For $\Phi(u) = u^3 + 1$ and $\Psi(v) = 1/v$, find each value.
(a) $(\Phi + \Psi)(t)$ (b) $(\Phi \circ \Psi)(r)$
(c) $(\Psi \circ \Phi)(r)$ (d) $\Phi^3(z)$
(e) $(\Phi - \Psi)(5t)$ (f) $((\Phi - \Psi) \circ \Psi)(t)$

4. If $f(x) = \sqrt{x^2 - 1}$ and $g(x) = 2/x$, find formulas for the following and state their domains.
(a) $(f \cdot g)(x)$ (b) $f^4(x) + g^4(x)$
(c) $(f \circ g)(x)$ (d) $(g \circ f)(x)$

5. If $f(s) = \sqrt{s^2 - 4}$ and $g(w) = |1 + w|$, find formulas for $(f \circ g)(x)$ and $(g \circ f)(x)$.

6. If $g(x) = x^2 + 1$, find formulas for $g^3(x)$ and $(g \circ g \circ g)(x)$.

C 7. Calculate $g(3.141)$ if $g(u) = \dfrac{\sqrt{u^3 + 2u}}{2 + u}$.

C 8. Calculate $g(2.03)$ if $g(x) = \dfrac{(\sqrt{x} - \sqrt[3]{x})^4}{1 - x + x^2}$.

C 9. Calculate $[g^2(\pi) - g(\pi)]^{1/3}$ if $g(v) = |11 - 7v|$.

C 10. Calculate $[g^3(\pi) - g(\pi)]^{1/3}$ if $g(x) = 6x - 11$.

11. Find f and g so that $F = g \circ f$. (See Example 3.)
(a) $F(x) = \sqrt{x + 7}$ (b) $F(x) = (x^2 + x)^{15}$

12. Find f and g so that $p = f \circ g$.
(a) $p(x) = \dfrac{2}{(x^2 + x + 1)^3}$ (b) $p(x) = \dfrac{1}{x^3 + 3x}$

13. Write $p(x) = 1/\sqrt{x^2 + 1}$ as a composite of three functions in two different ways.

14. Write $p(x) = 1/\sqrt{x^2 + 1}$ as a composite of four functions.

15. Sketch the graph of $f(x) = \sqrt{x - 2} - 3$ by first sketching $g(x) = \sqrt{x}$ and then translating. (See Example 4.)

16. Sketch the graph of $g(x) = |x + 3| - 4$ by first sketching $h(x) = |x|$ and then translating.

17. Sketch the graph of $f(x) = (x - 2)^2 - 4$ using translations.

18. Sketch the graph of $g(x) = (x + 1)^3 - 3$ using translations.

19. Sketch the graphs of $f(x) = (x - 3)/2$ and $g(x) = \sqrt{x}$ using the same coordinate axes. Then sketch $f + g$ by adding y-coordinates.

20. Follow the directions of Problem 19 for $f(x) = x$ and $g(x) = |x|$.

21. Sketch the graph of $F(t) = \dfrac{|t| - t}{t}$.

22. Sketch the graph of $G(t) = t - [\![t]\!]$.

23. State whether each of the following is an odd function, an even function, or neither. Prove your statements.

(a) The sum of two even functions
(b) The sum of two odd functions
(c) The product of two even functions
(d) The product of two odd functions
(e) The product of an even function and an odd function

24. Let F be any function whose domain contains $-x$ whenever it contains x. Prove each of the following.

(a) $F(x) - F(-x)$ is an odd function.
(b) $F(x) + F(-x)$ is an even function.
(c) F can always be expressed as the sum of an odd and an even function.

25. Is every polynomial of even degree an even function? Is every polynomial of odd degree an odd function? Explain.

26. Classify each of the following as a PF (polynomial function), RF (rational function but not a polynomial function), or neither.

(a) $f(x) = 3x^{1/2} + 1$ (b) $f(x) = 3$
(c) $f(x) = 3x^2 + 2x^{-1}$ (d) $f(x) = \pi x^3 - 3\pi$
(e) $f(x) = \dfrac{1}{x + 1}$ (f) $f(x) = \dfrac{x + 1}{\sqrt{x + 3}}$

27. The relationship between the unit price P (in cents) for a certain product and the demand D (in thousands of units) appears to satisfy

$$P = \sqrt{29 - 3D + D^2}$$

On the other hand, the demand has risen over the t years since 1970 according to $D = 2 + \sqrt{t}$.

(a) Express P as a function of t.
(b) Evaluate P when $t = 15$.

28. After being in business for t years, a manufacturer of cars is making $120 + 2t + 3t^2$ units per year. The sales price in dollars per unit has risen according to the formula $6000 + 700t$. Write a formula for the manufacturer's yearly revenue $R(t)$ after t years.

29. Starting at noon, airplane A flies due north at 400 miles per hour. Starting 1 hour later, airplane B flies due east at 300 miles per hour. Neglecting the curvature of the Earth and assuming that they fly at the same altitude, find a formula for $D(t)$, the distance between the two airplanes t hours after noon. *Hint:* There will be two formulas for $D(t)$, one if $0 < t < 1$ and the other if $t \geq 1$.

≈ C **30.** Find the distance between the airplanes of Problem 29 at 2:30 p.m.

31. Let $f(x) = \dfrac{ax + b}{cx - a}$. Show that $f(f(x)) = x$, provided $a^2 + bc \neq 0$ and $x \neq a/c$.

32. Let $f(x) = \dfrac{x - 3}{x + 1}$. Show that $f(f(f(x))) = x$, provided $x \neq \pm 1$.

33. Let $f(x) = \dfrac{x}{x - 1}$. Find and simplify each value.

(a) $f(1/x)$ (b) $f(f(x))$ (c) $f(1/f(x))$

34. Let $f(x) = \dfrac{x}{\sqrt{x} - 1}$. Find and simplify.

(a) $f\left(\dfrac{1}{x}\right)$ (b) $f(f(x))$

35. Prove that the operation of composition of functions is associative; that is, $f_1 \circ (f_2 \circ f_3) = (f_1 \circ f_2) \circ f_3$.

36. Let $f_1(x) = x$, $f_2(x) = 1/x$, $f_3(x) = 1 - x$, $f_4(x) = 1/(1 - x)$, $f_5(x) = (x - 1)/x$, and $f_6(x) = x/(x - 1)$. Note that $f_3(f_4(x)) = f_3(1/(1 - x)) = 1 - 1/(1 - x) = x/(x - 1) = f_6(x)$; that is, $f_3 \circ f_4 = f_6$. In fact, the composition of any two of these functions is another one in the list. Fill in the composition table in Figure 11.

$\circ$	f_1	f_2	f_3	f_4	f_5	f_6
f_1						
f_2						
f_3				f_6		
f_4						
f_5						
f_6						

Figure 11

Then use this table to find each of the following. From Problem 35, you know that the associative law holds.

(a) $f_3 \circ f_3 \circ f_3 \circ f_3 \circ f_3$ (b) $f_1 \circ f_2 \circ f_3 \circ f_4 \circ f_5 \circ f_6$
(c) F if $F \circ f_6 = f_1$ (d) G if $G \circ f_3 \circ f_6 = f_1$
(e) H if $f_2 \circ f_5 \circ H = f_5$

GC *Use a computer or a graphing calculator in Problems 37–40.*

37. Let $f(x) = x^2 - 3x$. Using the same axes, draw the graphs of $y = f(x)$, $y = f(x - 0.5) - 0.6$, and $y = f(1.5x)$, all on the domain $[-2, 5]$.

38. Let $f(x) = |x^3|$. Using the same axes, draw the graphs of $y = f(x)$, $y = f(3x)$, and $y = f(3(x - 0.8))$, all on the domain $[-3, 3]$.

39. Let $f(x) = 2\sqrt{x} - 2x + 0.25x^2$. Using the same axes, draw the graphs of $y = f(x)$, $y = f(1.5x)$, and $y = f(x - 1) + 0.5$, all on the domain $[0, 5]$.

40. Let $f(x) = 1/(x^2 + 1)$. Using the same axes, draw the graphs of $y = f(x)$, $y = f(2x)$, and $y = f(x - 2) + 0.6$, all on the domain $[-4, 4]$.

CAS **41.** Your computer algebra system (CAS) may allow the use of parameters in defining functions. In each case, draw the graph of $y = f(x)$ for the specified values of the parameter k, using the same axes and $-5 \leq x \leq 5$.

(a) $f(x) = |kx|^{0.7}$ for $k = 1, 2, 0.5$, and 0.2.
(b) $f(x) = |x - k|^{0.7}$ for $k = 0, 2, -0.5$, and -3.
(c) $f(x) = |x|^k$ for $k = 0.4, 0.7, 1$, and 1.7.

CAS **42.** Using the same axes, draw the graph of $f(x) = |k(x - c)|^n$ for the following choices of parameters.

(a) $c = -1, k = 1.4, n = 0.7$ (b) $c = 2, k = 1.4, n = 1$

(c) $c = 0, k = 0.9, n = 0.6$

Answers to Concepts Review: **1.** $(x^2 + 1)^3$ **2.** $f(g(x))$ **3.** 2; left **4.** a quotient of two polynomial functions

0.7 Trigonometric Functions

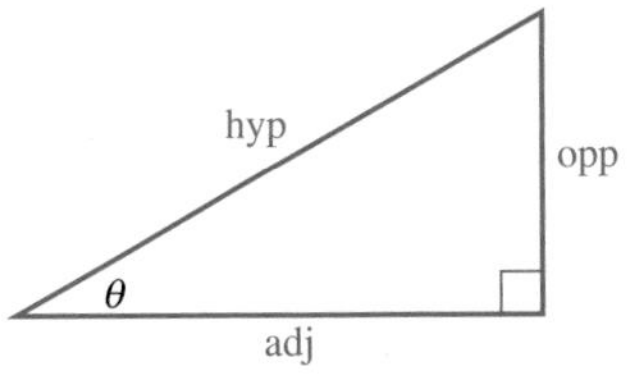

$\sin\theta = \dfrac{\text{opp}}{\text{hyp}}$ $\cos\theta = \dfrac{\text{adj}}{\text{hyp}}$ $\tan\theta = \dfrac{\text{opp}}{\text{adj}}$

Figure 1

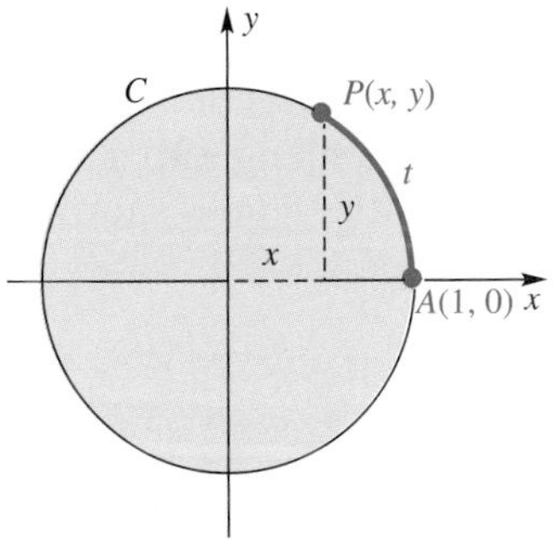

The unit circle

Figure 2

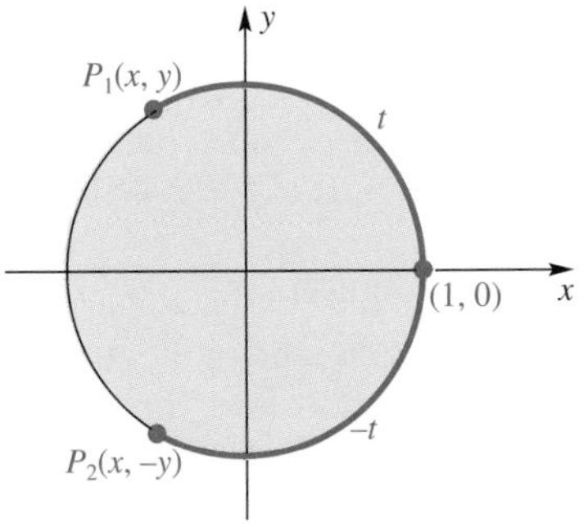

Figure 3

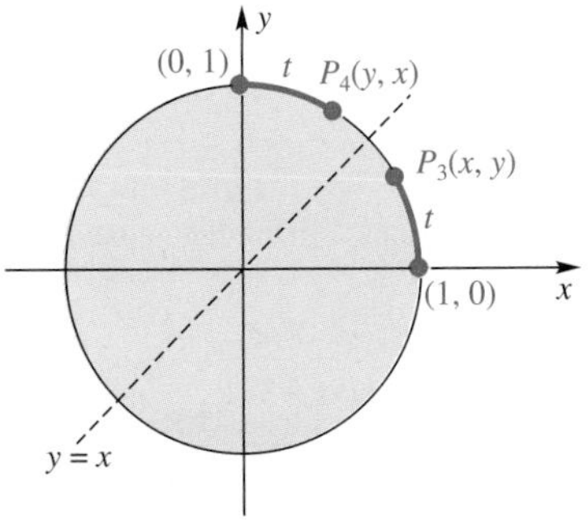

Figure 4

You have probably seen the definitions of the trigonometric functions based on right triangles. Figure 1 summarizes the definitions of the sine, cosine, and tangent functions. You should review Figure 1 carefully, because these concepts are needed for many applications later in this book.

More generally, we define the trigonometric functions based on the unit circle. The unit circle, which we denote by C, is the circle with radius 1 and center at the origin; it has equation $x^2 + y^2 = 1$. Let A be the point $(1, 0)$ and let t be a positive number. There is a single point P on the circle C such that the distance, measured in the *counterclockwise* direction around the arc AP, is equal to t. (See Figure 2.) Recall that the circumference of a circle with radius r is $2\pi r$, so the circumference of C is 2π. Thus, if $t = \pi$, then the point P is exactly halfway around the circle from the point A; in this case, P is the point $(-1, 0)$. If $t = 3\pi/2$, then P is the point $(0, -1)$, and if $t = 2\pi$, then P is the point A. If $t > 2\pi$, then it will take more than one complete circuit of the circle C to trace the arc AP.

When $t < 0$, we trace the circle in a *clockwise* direction. There will be a single point P on the circle C such that the arc length measured in the clockwise direction from A is t. Thus, for every real number t, we can associate a unique point $P(x, y)$ on the unit circle. This allows us to make the key definitions of the sine and cosine functions. The functions sine and cosine are written as sin and cos, rather than as a single letter such as f or g. Parentheses around the independent variable are usually omitted unless there is some ambiguity.

Definition Sine and Cosine Functions

Let t be a real number that determines the point $P(x, y)$ as indicated above. Then

$$\sin t = y \quad \text{and} \quad \cos t = x$$

Basic Properties of Sine and Cosine A number of facts follow almost immediately from the definitions given above. First, since t can be any real number, the domain for both the sine and cosine functions is $(-\infty, \infty)$. Second, x and y are always between -1 and 1. Thus, the range for both the sine and cosine functions is the interval $[-1, 1]$.

Because the unit circle has circumference 2π, the values t and $t + 2\pi$ determine the *same* point $P(x, y)$. Thus,

$$\sin(t + 2\pi) = \sin t \quad \text{and} \quad \cos(t + 2\pi) = \cos t$$

(Notice that parentheses are needed to make it clear that we mean $\sin(t + 2\pi)$, rather than $(\sin t) + 2\pi$. The expression $\sin t + 2\pi$ would be ambiguous.)

The points P_1 and P_2 that correspond to t and $-t$, respectively, are symmetric about the x-axis (Figure 3). Thus, the x-coordinates for P_1 and P_2 are the same, and the y-coordinates differ only in sign. Consequently,

$$\sin(-t) = -\sin t \quad \text{and} \quad \cos(-t) = \cos t$$

In other words, sine is an odd function and cosine is an even function.

The points P_3 and P_4 corresponding to t and $\pi/2 - t$, respectively, are symmetric with respect to the line $y = x$ and thus they have their coordinates interchanged (Figure 4). This means that

$$\sin\left(\frac{\pi}{2} - t\right) = \cos t \quad \text{and} \quad \cos\left(\frac{\pi}{2} - t\right) = \sin t$$

Finally, we mention an important identity connecting the sine and cosine functions:

$$\sin^2 t + \cos^2 t = 1$$

for every real number t. This identity follows from the fact that since the point (x, y) is on the unit circle, x and y satisfy $x^2 + y^2 = 1$.

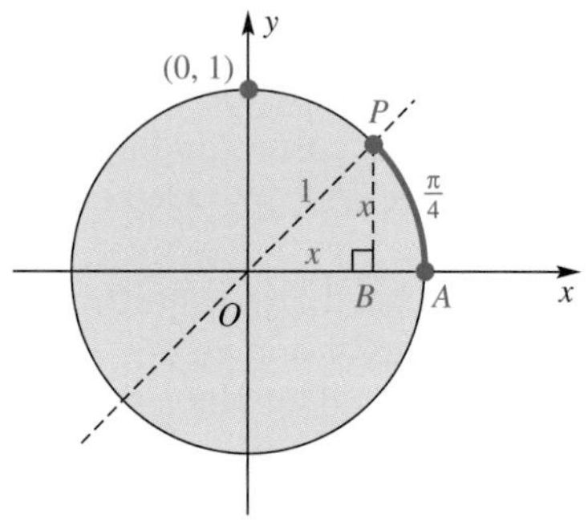

Figure 5

Graphs of Sine and Cosine To graph $y = \sin t$ and $y = \cos t$, we follow our usual procedure of making a table of values, plotting the corresponding points, and connecting these points with a smooth curve. So far, however, we know the values of sine and cosine for only a few values of t. A number of other values can be determined from geometric arguments. For example, if $t = \pi/4$, then t determines the point half of the way counterclockwise around the unit circle between the points $(1, 0)$ and $(0, 1)$. By symmetry, x and y will be on the line $y = x$, so $y = \sin t$ and $x = \cos t$ will be equal. Thus, the two legs of the right triangle OBP are equal, and the hypotenuse is 1 (Figure 5). The Pythagorean Theorem can be applied to give

$$1 = x^2 + x^2 = \cos^2\frac{\pi}{4} + \cos^2\frac{\pi}{4}$$

From this we conclude that $\cos(\pi/4) = 1/\sqrt{2} = \sqrt{2}/2$. Similarly, $\sin(\pi/4) = \sqrt{2}/2$. We can determine $\sin t$ and $\cos t$ for a number of other values of t. Some of these are shown in the table in the margin. Using these results, along with a number of results from a calculator (in radian mode), we obtain the graphs shown in Figure 6.

t	$\sin t$	$\cos t$
0	0	1
$\pi/6$	$1/2$	$\sqrt{3}/2$
$\pi/4$	$\sqrt{2}/2$	$\sqrt{2}/2$
$\pi/3$	$\sqrt{3}/2$	$1/2$
$\pi/2$	1	0
$2\pi/3$	$\sqrt{3}/2$	$-1/2$
$3\pi/4$	$\sqrt{2}/2$	$-\sqrt{2}/2$
$5\pi/6$	$1/2$	$-\sqrt{3}/2$
π	0	-1

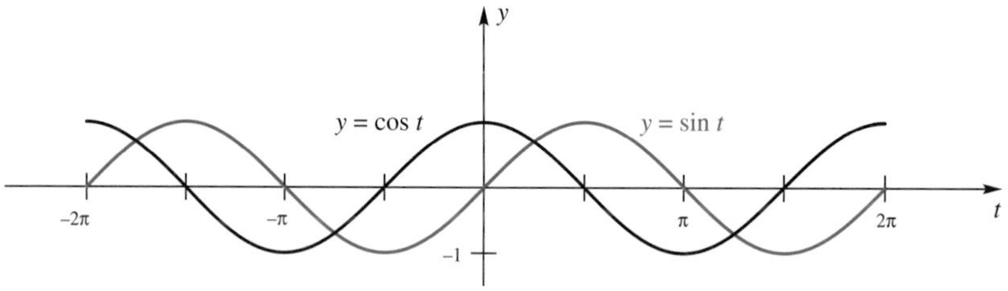

Figure 6

Four things are noticeable from these graphs:

1. Both $\sin t$ and $\cos t$ range from -1 to 1.
2. Both graphs repeat themselves on adjacent intervals of length 2π.
3. The graph of $y = \sin t$ is symmetric about the origin, and $y = \cos t$ is symmetric about the y-axis. (Thus, the sine function is odd and the cosine function is even.)
4. The graph of $y = \sin t$ is the same as that of $y = \cos t$, but translated $\pi/2$ units to the right.

The next example deals with functions of the form $\sin(at)$ or $\cos(at)$, which occur frequently in applications.

EXAMPLE 1 Sketch the graphs of

(a) $y = \sin(2\pi t)$ (b) $y = \cos(2t)$

SOLUTION

(a) As t goes from 0 to 1, the argument $2\pi t$ goes from 0 to 2π. Thus, the graph of this function will repeat itself on adjacent intervals of length 1. From the entries in the following table, we can sketch a graph of $y = \sin(2\pi t)$.

t	$\sin(2\pi t)$	t	$\sin(2\pi t)$
0	$\sin(2\pi\cdot 0) = 0$	$\frac{5}{8}$	$\sin\left(2\pi\cdot\frac{5}{8}\right) = -\frac{\sqrt{2}}{2}$
$\frac{1}{8}$	$\sin\left(2\pi\cdot\frac{1}{8}\right) = \frac{\sqrt{2}}{2}$	$\frac{3}{4}$	$\sin\left(2\pi\cdot\frac{3}{4}\right) = -1$
$\frac{1}{4}$	$\sin\left(2\pi\cdot\frac{1}{4}\right) = 1$	$\frac{7}{8}$	$\sin\left(2\pi\cdot\frac{7}{8}\right) = -\frac{\sqrt{2}}{2}$
$\frac{3}{8}$	$\sin\left(2\pi\cdot\frac{3}{8}\right) = \frac{\sqrt{2}}{2}$	1	$\sin(2\pi\cdot 1) = 0$
$\frac{1}{2}$	$\sin\left(2\pi\cdot\frac{1}{2}\right) = 0$	$\frac{9}{8}$	$\sin\left(2\pi\cdot\frac{9}{8}\right) = \frac{\sqrt{2}}{2}$

Figure 7 shows a sketch of the graph of $y = \sin(2\pi t)$.

(b) As t goes from 0 to π, the argument $2t$ goes from 0 to 2π. Thus, the graph of $y = \cos(2t)$ will repeat itself on adjacent intervals of length π. Once we construct a table we can sketch a plot of $y = \cos(2t)$. Figure 8 shows the graph.

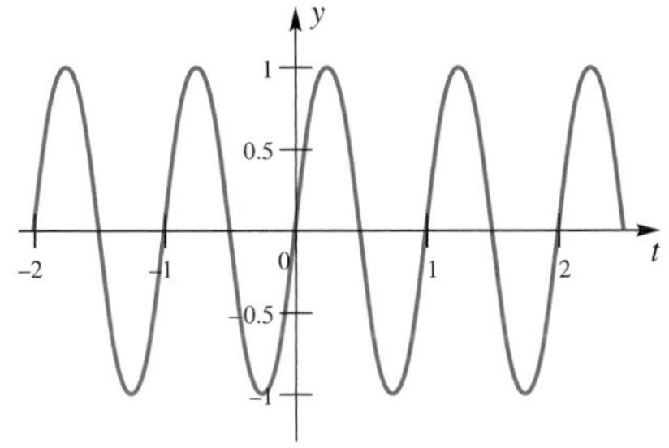

Figure 7

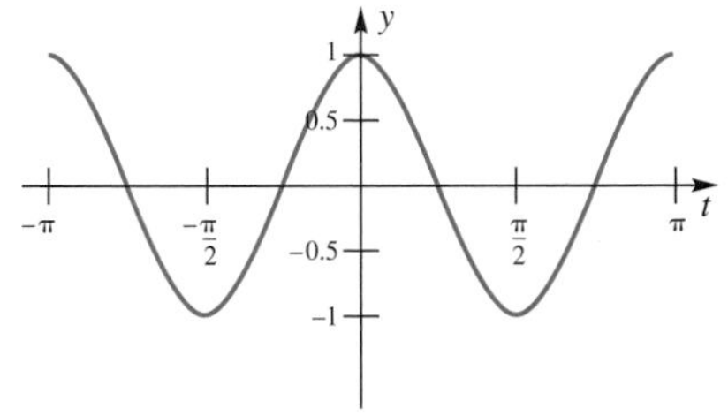

Figure 8

t	$\cos(2t)$	t	$\cos(2t)$
0	$\cos(2\cdot 0) = 1$	$\frac{5\pi}{8}$	$\cos\left(2\cdot\frac{5\pi}{8}\right) = -\frac{\sqrt{2}}{2}$
$\frac{\pi}{8}$	$\cos\left(2\cdot\frac{\pi}{8}\right) = \frac{\sqrt{2}}{2}$	$\frac{3\pi}{4}$	$\cos\left(2\cdot\frac{3\pi}{4}\right) = 0$
$\frac{\pi}{4}$	$\cos\left(2\cdot\frac{\pi}{4}\right) = 0$	$\frac{7\pi}{8}$	$\cos\left(2\cdot\frac{7\pi}{8}\right) = \frac{\sqrt{2}}{2}$
$\frac{3\pi}{8}$	$\cos\left(2\cdot\frac{3\pi}{8}\right) = -\frac{\sqrt{2}}{2}$	π	$\cos(2\cdot\pi) = 1$
$\frac{\pi}{2}$	$\cos\left(2\cdot\frac{\pi}{2}\right) = -1$	$\frac{9\pi}{8}$	$\cos\left(2\cdot\frac{9\pi}{8}\right) = \frac{\sqrt{2}}{2}$

■

Period and Amplitude of the Trigonometric Functions A function f is **periodic** if there is a positive number p such that

$$f(x + p) = f(x)$$

for all real numbers x in the domain of f. The smallest such positive number p is called the **period** of f. The sine function is periodic because $\sin(x + 2\pi) = \sin x$ for all x. It is also true that

$$\sin(x + 4\pi) = \sin(x - 2\pi) = \sin(x + 12\pi) = \sin x$$

for all x. Thus, $4\pi, -2\pi$, and 12π are all numbers p with the property $\sin(x + p) = \sin x$. The period is defined to be the *smallest* such positive number p. For the sine function, the smallest positive p with the property that $\sin(x + p) = \sin x$ is $p = 2\pi$. We therefore say that the sine function is periodic with period 2π. The cosine function is also periodic with period 2π.

The function sin (at) has period $2\pi/a$ since

$$\sin\left[a\left(t + \frac{2\pi}{a}\right)\right] = \sin[at + 2\pi] = \sin(at)$$

The period of the function $\cos(at)$ is also $2\pi/a$.

EXAMPLE 2 What are the periods of the following functions?

(a) $\sin(2\pi t)$ (b) $\cos(2t)$ (c) $\sin(2\pi t/12)$

SOLUTION

(a) Because the function $\sin(2\pi t)$ is of the form $\sin(at)$ with $a = 2\pi$, its period is $p = \frac{2\pi}{2\pi} = 1$.

(b) The function $\cos(2t)$ is of the form $\cos(at)$ with $a = 2$. Thus, the period of $\cos(2t)$ is $p = \frac{2\pi}{2} = \pi$.

(c) The function $\sin(2\pi t/12)$ has period $p = \frac{2\pi}{2\pi/12} = 12$. ■

If the periodic function f attains a minimum and a maximum, we define the **amplitude** A as half the vertical distance between the highest point and the lowest point on the graph.

EXAMPLE 3 Find the amplitude of the following periodic functions.

(a) $\sin(2\pi t/12)$ (b) $3\cos(2t)$

(c) $50 + 21\sin(2\pi t/12 + 3)$

SOLUTION

(a) Since the range of the function $\sin(2\pi t/12)$ is $[-1, 1]$, its amplitude is $A = 1$.

(b) The function $3\cos(2t)$ will take on values from -3 (which occurs when $t = \pm\frac{\pi}{2}, \pm\frac{3\pi}{2}, \ldots$) to 3 (which occurs when $t = 0, \pm\pi, \pm 2\pi, \ldots$). The amplitude is therefore $A = 3$.

(c) The function $21\sin(2\pi t/12 + 3)$ takes on values from -21 to 21. Thus, $50 + 21\sin(2\pi t/12 + 3)$ takes on values from $50 - 21 = 29$ to $50 + 21 = 71$. The amplitude is therefore 21. ■

In general, for $a > 0$ and $A > 0$,

$$C + A\sin(a(t + b)) \text{ and } C + A\cos(a(t + b)) \text{ have period } \frac{2\pi}{a} \text{ and amplitude } A.$$

Trigonometric functions can be used to model a number of physical phenomena, including daily tide levels and yearly temperatures.

EXAMPLE 4 The normal high temperature for St. Louis, Missouri, ranges from 37°F for January 15 to 89°F for July 15. The normal high temperature follows roughly a sinusoidal curve.

(a) Find values of C, A, a, and b such that

$$T(t) = C + A\sin(a(t + b))$$

where t, expressed in months since January 1, is a reasonable model for the normal high temperature.

(b) Use this model to approximate the normal high temperature for May 15.

SOLUTION

(a) The required function must have period $t = 12$ since the seasons repeat every 12 months. Thus, $\frac{2\pi}{a} = 12$, so we have $a = \frac{2\pi}{12}$. The amplitude is half the difference between the lowest and highest points; in this case,

$A = \frac{1}{2}(89 - 37) = 26$. The value of C is equal to the midpoint of the low and high temperatures, so $C = \frac{1}{2}(89 + 37) = 63$. The function $T(t)$ must therefore be of the form

$$T(t) = 63 + 26 \sin\left(\frac{2\pi}{12}(t + b)\right)$$

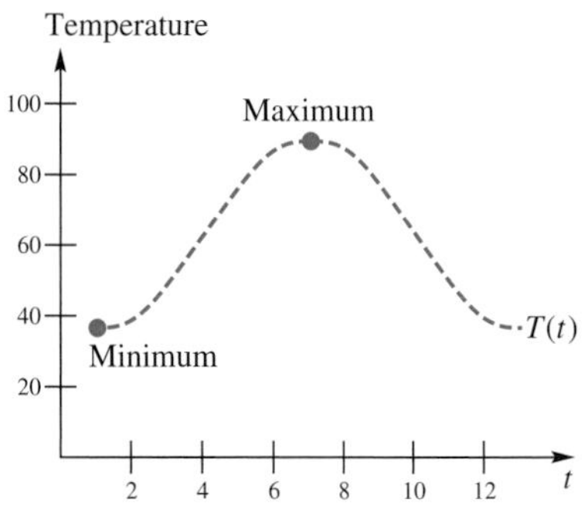

Figure 9

The only constant left to find is b. The lowest normal high temperature is 37, which occurs on January 15, roughly in the middle of January. Thus, our function must satisfy $T(1/2) = 37$, and the function must reach its minimum of 37 when $t = 1/2$. Figure 9 summarizes the information that we have so far. The function $63 + 26\sin(2\pi t/12)$ reaches its minimum when $2\pi t/12 = -\pi/2$, that is, when $t = -3$. We must therefore translate the curve defined by $y = 63 + 26\sin(2\pi t/12)$ to the right by the amount $1/2 - (-3) = 7/2$. In Section 0.6, we showed that replacing x with $x - c$ translates the graph of $y = f(x)$ to the right by c units. Thus, in order to translate the graph of $y = 63 + 26\sin(2\pi t/12)$ to the right by $7/2$ units, we must replace t with $t - 7/2$. Thus,

$$T(t) = 63 + 26 \sin\left(\frac{2\pi}{12}\left(t - \frac{7}{2}\right)\right)$$

Figure 10 shows a plot of the normal high temperature T as a function of time t, where t is given in months.

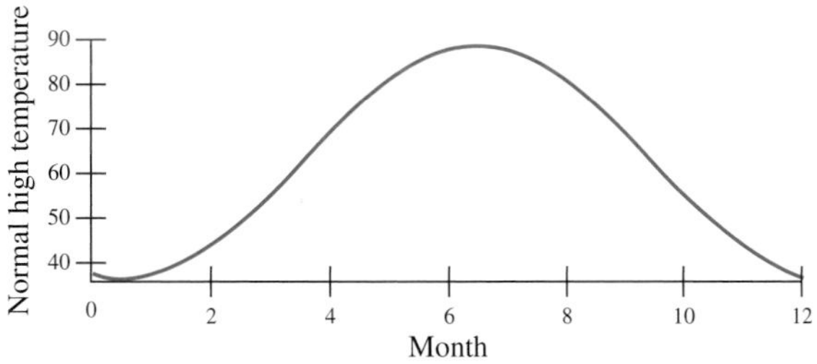

Figure 10

(b) To estimate the normal high temperature for May 15, we substitute $t = 4.5$ (because the middle of May is four and one-half months into the year) and obtain

$$T(4.5) = 63 + 26\sin(2\pi(4.5 - 3.5)/12) = 76$$

The normal high temperature for St. Louis on May 15 is actually 75°F. Thus, our model overpredicts by 1°, which is remarkably accurate considering how little information was given. ■

> **Models and Modeling**
>
> It is important to keep in mind that all models such as this are simplifications of reality. (That is why they are called *models*.) Although such models are inherently simplifications of reality, many of them are still useful for prediction.

Four Other Trigonometric Functions We could get by with just the sine and cosine functions, but it is convenient to introduce four additional trigonometric functions: tangent, cotangent, secant, and cosecant.

$$\tan t = \frac{\sin t}{\cos t} \qquad \cot t = \frac{\cos t}{\sin t}$$

$$\sec t = \frac{1}{\cos t} \qquad \csc t = \frac{1}{\sin t}$$

What we know about sine and cosine will automatically give us knowledge about these four new functions.

EXAMPLE 5 Show that tangent is an odd function.

SOLUTION

$$\tan(-t) = \frac{\sin(-t)}{\cos(-t)} = \frac{-\sin t}{\cos t} = -\tan t$$ ■

EXAMPLE 6 Verify that the following are identities.

$$1 + \tan^2 t = \sec^2 t \qquad 1 + \cot^2 t = \csc^2 t$$

SOLUTION

$$1 + \tan^2 t = 1 + \frac{\sin^2 t}{\cos^2 t} = \frac{\cos^2 t + \sin^2 t}{\cos^2 t} = \frac{1}{\cos^2 t} = \sec^2 t$$

$$1 + \cot^2 t = 1 + \frac{\cos^2 t}{\sin^2 t} = \frac{\sin^2 t + \cos^2 t}{\sin^2 t} = \frac{1}{\sin^2 t} = \csc^2 t$$ ■

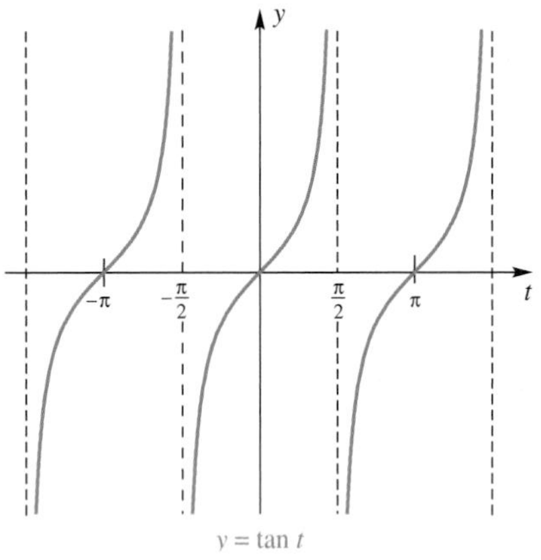

Figure 11

When we study the tangent function (Figure 11), we are in for two minor surprises. First, we notice that there are vertical asymptotes at $\pm\pi/2, \pm 3\pi/2, \ldots$. We should have anticipated this since $\cos t = 0$ at these values of t, which means that $\sin t/\cos t$ would involve a division by zero. Second, it appears that the tangent is periodic (which we expected), but with period π (which we might not have expected). You will see the analytic reason for this in Problem 33.

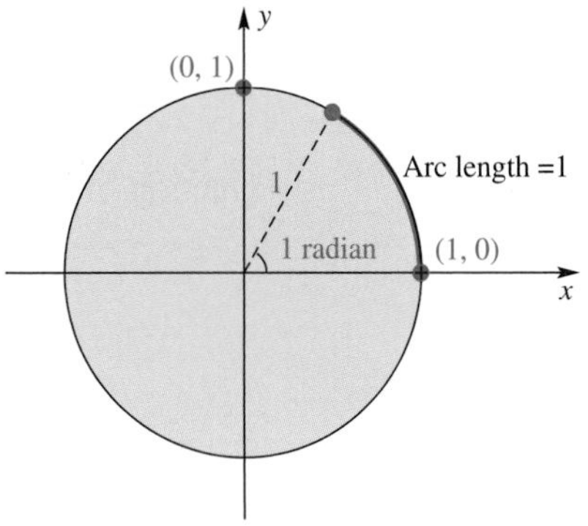

Figure 12

Relation to Angle Trigonometry

Angles are commonly measured either in degrees or in radians. One radian is by definition the angle corresponding to an arc of length 1 on the unit circle. See Figure 12. The angle corresponding to a complete revolution measures 360°, but only 2π radians. Equivalently, a straight angle measures 180° or π radians, a fact worth remembering.

$$180° = \pi \text{ radians} \approx 3.1415927 \text{ radians}$$

This leads to the results

$$1 \text{ radian} \approx 57.29578° \qquad 1° \approx 0.0174533 \text{ radian}$$

Degrees	Radians
0	0
30	π/6
45	π/4
60	π/3
90	π/2
120	2π/3
135	3π/4
150	5π/6
180	π
360	2π

Figure 13

Figure 13 shows some other common conversions between degrees and radians.

The division of a revolution into 360 parts is quite arbitrary (due to the ancient Babylonians, who liked multiples of 60). The division into 2π parts is more fundamental and lies behind the almost universal use of radian measure in calculus. Notice, in particular, that the length s of the arc cut off on a circle of radius r by a central angle of t radians satisfies (see Figure 14)

$$\frac{s}{2\pi r} = \frac{t}{2\pi}$$

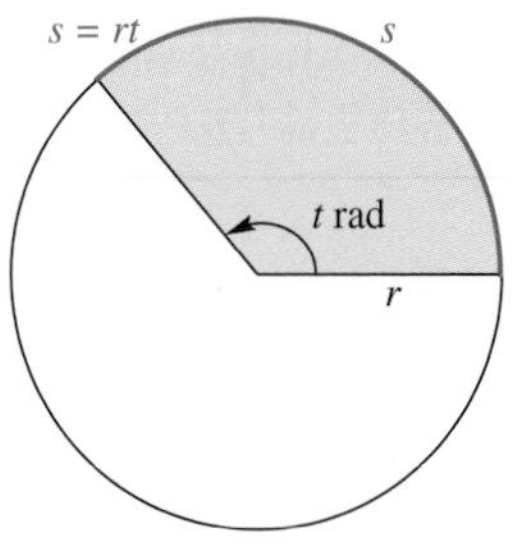

Figure 14

That is, the fraction of the total circumference $2\pi r$ corresponding to an angle t is the same as the fraction of the unit circle corresponding to the same angle t. This implies that $s = rt$.

When $r = 1$, this gives $s = t$. This means that *the length of the arc on the unit circle cut off by a central angle of t radians is t*. This is correct even if t is negative, provided that we interpret length to be negative when measured in the clockwise direction.

EXAMPLE 7 Find the distance traveled by a bicycle with wheels of radius 30 centimeters when the wheels turn through 100 revolutions.

Another View

We have based our discussion of trigonometry on the unit circle. We could as well have used a circle of radius r.

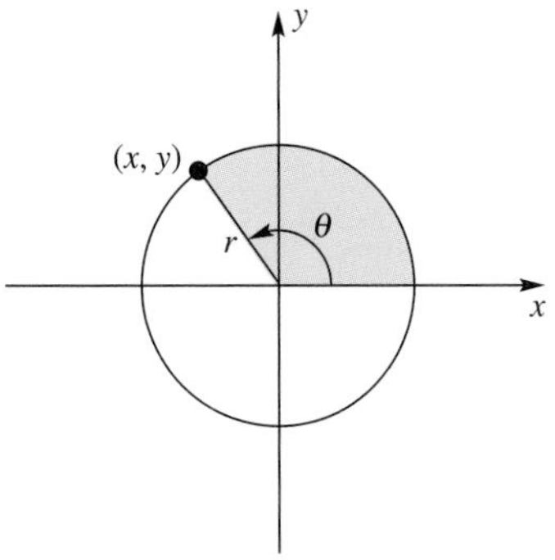

Then

$$\sin\theta = \frac{y}{r}$$

$$\cos\theta = \frac{x}{r}$$

SOLUTION We use the fact that $s = rt$, recognizing that 100 revolutions correspond to $100 \cdot (2\pi)$ radians.

$$s = (30)(100)(2\pi) = 6000\pi \approx 18{,}849.6 \text{ centimeters} \approx 188.5 \text{ meters} \quad \blacksquare$$

Now we can make the connection between angle trigonometry and unit circle trigonometry. If θ is an angle measuring t radians, that is, if θ is an angle that cuts off an arc of length t from the unit circle, then

$$\sin\theta = \sin t \qquad \cos\theta = \cos t$$

In calculus, when we meet an angle measured in degrees, we almost always change it to radians before doing any calculations. For example,

$$\sin 31.6^\circ = \sin\left(31.6 \cdot \frac{\pi}{180}\text{radian}\right) \approx \sin 0.552$$

List of Important Identities We will not take space to verify all the following identities. We simply assert their truth and suggest that most of them will be needed somewhere in this book.

Trigonometric Identities The following are true for all x and y, provided that both sides are defined at the chosen x and y.

Odd–even identities

$$\sin(-x) = -\sin x$$

$$\cos(-x) = \cos x$$

$$\tan(-x) = -\tan x$$

Cofunction identities

$$\sin\left(\frac{\pi}{2} - x\right) = \cos x$$

$$\cos\left(\frac{\pi}{2} - x\right) = \sin x$$

$$\tan\left(\frac{\pi}{2} - x\right) = \cot x$$

Pythagorean identities

$$\sin^2 x + \cos^2 x = 1$$

$$1 + \tan^2 x = \sec^2 x$$

$$1 + \cot^2 x = \csc^2 x$$

Addition identities

$$\sin(x + y) = \sin x \cos y + \cos x \sin y$$

$$\cos(x + y) = \cos x \cos y - \sin x \sin y$$

$$\tan(x + y) = \frac{\tan x + \tan y}{1 - \tan x \tan y}$$

Double-angle identities

$$\sin 2x = 2\sin x \cos x$$

$$\cos 2x = \cos^2 x - \sin^2 x$$

$$= 2\cos^2 x - 1$$

$$= 1 - 2\sin^2 x$$

Half-angle identities

$$\sin\left(\frac{x}{2}\right) = \pm\sqrt{\frac{1 - \cos x}{2}}$$

$$\cos\left(\frac{x}{2}\right) = \pm\sqrt{\frac{1 + \cos x}{2}}$$

Sum identities

$$\sin x + \sin y = 2\sin\left(\frac{x + y}{2}\right)\cos\left(\frac{x - y}{2}\right)$$

$$\cos x + \cos y = 2\cos\left(\frac{x + y}{2}\right)\cos\left(\frac{x - y}{2}\right)$$

Product identities

$$\sin x \sin y = -\tfrac{1}{2}[\cos(x + y) - \cos(x - y)]$$

$$\cos x \cos y = \tfrac{1}{2}[\cos(x + y) + \cos(x - y)]$$

$$\sin x \cos y = \tfrac{1}{2}[\sin(x + y) + \sin(x - y)]$$

Concepts Review

1. The natural domain of the sine function is _____; its range is _____.

2. The period of the cosine function is _____; the period of the sine function is _____; the period of the tangent function is _____.

3. Since $\sin(-x) = -\sin x$, the sine function is _____, and since $\cos(-x) = \cos x$, the cosine function is _____.

4. If $(-4, 3)$ lies on the terminal side of an angle θ whose vertex is at the origin and initial side is along the positive x-axis, then $\cos\theta =$ _____.

Problem Set 0.7

1. Convert the following degree measures to radians (leave π in your answer).

(a) $30°$ (b) $45°$ (c) $-60°$
(d) $240°$ (e) $-370°$ (f) $10°$

2. Convert the following radian measures to degrees.

(a) $\frac{7}{6}\pi$ (b) $\frac{3}{4}\pi$ (c) $-\frac{1}{3}\pi$
(d) $\frac{4}{3}\pi$ (e) $-\frac{35}{18}\pi$ (f) $\frac{3}{18}\pi$

C **3.** Convert the following degree measures to radians ($1° = \pi/180 \approx 1.7453 \times 10^{-2}$ radian).

(a) $33.3°$ (b) $46°$ (c) $-66.6°$
(d) $240.11°$ (e) $-369°$ (f) $11°$

C **4.** Convert the following radian measures to degrees ($1 \text{ radian} = 180/\pi \approx 57.296$ degrees).

(a) 3.141 (b) 6.28 (c) 5.00
(d) 0.001 (e) -0.1 (f) 36.0

C **5.** Calculate (be sure that your calculator is in radian or degree mode as needed).

(a) $\dfrac{56.4 \tan 34.2°}{\sin 34.1°}$ (b) $\dfrac{5.34 \tan 21.3°}{\sin 3.1° + \cot 23.5°}$

(c) $\tan 0.452$ (d) $\sin(-0.361)$

C **6.** Calculate.

(a) $\dfrac{234.1 \sin 1.56}{\cos 0.34}$ (b) $\sin^2 2.51 + \sqrt{\cos 0.51}$

C **7.** Calculate.

(a) $\dfrac{56.3 \tan 34.2°}{\sin 56.1°}$ (b) $\left(\dfrac{\sin 35°}{\sin 26° + \cos 26°}\right)^3$

8. Verify the values of $\sin t$ and $\cos t$ in the table used to construct Figure 6.

9. Evaluate without using a calculator.

(a) $\tan\frac{\pi}{6}$ (b) $\sec\pi$ (c) $\sec\frac{3\pi}{4}$

(d) $\csc\frac{\pi}{2}$ (e) $\cot\frac{\pi}{4}$ (f) $\tan\left(-\frac{\pi}{4}\right)$

10. Evaluate without using a calculator.

(a) $\tan\frac{\pi}{3}$ (b) $\sec\frac{\pi}{3}$ (c) $\cot\frac{\pi}{3}$

(d) $\csc\frac{\pi}{4}$ (e) $\tan\left(-\frac{\pi}{6}\right)$ (f) $\cos\left(-\frac{\pi}{3}\right)$

11. Verify that the following are identities (see Example 6).

(a) $(1 + \sin z)(1 - \sin z) = \dfrac{1}{\sec^2 z}$

(b) $(\sec t - 1)(\sec t + 1) = \tan^2 t$

(c) $\sec t - \sin t \tan t = \cos t$

(d) $\dfrac{\sec^2 t - 1}{\sec^2 t} = \sin^2 t$

12. Verify that the following are identities (see Example 6).

(a) $\sin^2 v + \dfrac{1}{\sec^2 v} = 1$

(b) $\cos 3t = 4\cos^3 t - 3\cos t$ *Hint:* Use a double-angle identity.

(c) $\sin 4x = 8 \sin x \cos^3 x - 4 \sin x \cos x$ *Hint:* Use a double-angle identity twice.

(d) $(1 + \cos\theta)(1 - \cos\theta) = \sin^2\theta$

13. Verify the following are identities.

(a) $\dfrac{\sin u}{\csc u} + \dfrac{\cos u}{\sec u} = 1$

(b) $(1 - \cos^2 x)(1 + \cot^2 x) = 1$

(c) $\sin t(\csc t - \sin t) = \cos^2 t$

(d) $\dfrac{1 - \csc^2 t}{\csc^2 t} = \dfrac{-1}{\sec^2 t}$

14. Sketch the graphs of the following on $[-\pi, 2\pi]$.

(a) $y = \sin 2x$ (b) $y = 2 \sin t$

(c) $y = \cos\left(x - \frac{\pi}{4}\right)$ (d) $y = \sec t$

15. Sketch the graphs of the following on $[-\pi, 2\pi]$.

(a) $y = \csc t$ (b) $y = 2 \cos t$

(c) $y = \cos 3t$ (d) $y = \cos\left(t + \frac{\pi}{3}\right)$

Determine the period, amplitude, and shifts (both horizontal and vertical) and draw a graph over the interval $-5 \le x \le 5$ for the functions listed in Problems 16–23.

16. $y = 3\cos\frac{x}{2}$ **17.** $y = 2\sin 2x$

18. $y = \tan x$ **19.** $y = 2 + \frac{1}{6}\cot 2x$

20. $y = 3 + \sec(x - \pi)$ **21.** $y = 21 + 7\sin(2x + 3)$

22. $y = 3\cos\left(x - \frac{\pi}{2}\right) - 1$ **23.** $y = \tan\left(2x - \frac{\pi}{3}\right)$

24. Which of the following represent the same graph? Check your result analytically using trigonometric identities.

(a) $y = \sin\left(x + \frac{\pi}{2}\right)$ (b) $y = \cos\left(x + \frac{\pi}{2}\right)$

(c) $y = -\sin(x + \pi)$ (d) $y = \cos(x - \pi)$

(e) $y = -\sin(\pi - x)$ (f) $y = \cos\left(x - \frac{\pi}{2}\right)$

(g) $y = -\cos(\pi - x)$ (h) $y = \sin\left(x - \frac{\pi}{2}\right)$

25. Which of the following are odd functions? Even functions? Neither?

(a) $t \sin t$ (b) $\sin^2 t$ (c) $\csc t$

(d) $|\sin t|$ (e) $\sin(\cos t)$ (f) $x + \sin x$

26. Which of the following are odd functions? Even functions? Neither?

(a) $\cot t + \sin t$ (b) $\sin^3 t$ (c) $\sec t$

(d) $\sqrt{\sin^4 t}$ (e) $\cos(\sin t)$ (f) $x^2 + \sin x$

Find the exact values in Problems 27–31. Hint: Half-angle identities may be helpful.

27. $\cos^2\frac{\pi}{3}$ **28.** $\sin^2\frac{\pi}{6}$

29. $\sin^3\frac{\pi}{6}$ **30.** $\cos^2\frac{\pi}{12}$

31. $\sin^2\frac{\pi}{8}$

32. Find identities analogous to the addition identities for each expression.

(a) $\sin(x - y)$ (b) $\cos(x - y)$ (c) $\tan(x - y)$

33. Use the addition identity for the tangent to show that $\tan(t + \pi) = \tan t$ for all t in the domain of $\tan t$.

34. Show that $\cos(x - \pi) = -\cos x$ for all x.

≈ C **35.** Suppose that a tire on a truck has an outer radius of 2.5 feet. How many revolutions per minute does the tire make when the truck is traveling 60 miles per hour?

≈ **36.** How far does a wheel of radius 2 feet roll along level ground in making 150 revolutions?

≈ C **37.** A belt passes around two wheels, as shown in Figure 15. How many revolutions per second does the small wheel make when the large wheel makes 21 revolutions per second?

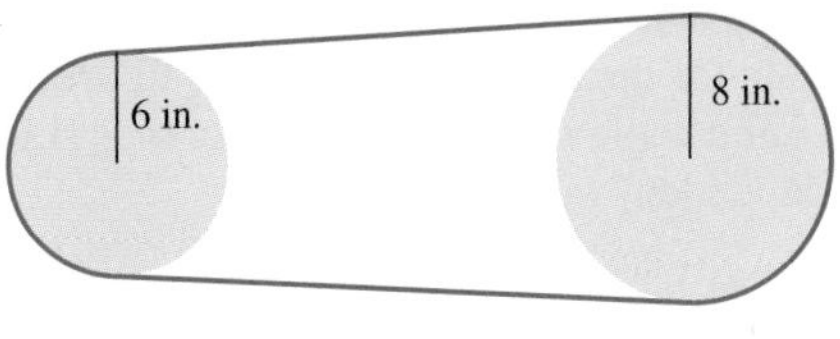

Figure 15

38. The **angle of inclination** α of a line is the smallest positive angle from the positive x-axis to the line ($\alpha = 0$ for a horizontal line). Show that the slope m of the line is equal to $\tan\alpha$.

39. Find the angle of inclination of the following lines (see Problem 38).

(a) $y = \sqrt{3}x - 7$ (b) $\sqrt{3}x + 3y = 6$

40. Let ℓ_1 and ℓ_2 be two nonvertical intersecting lines with slopes m_1 and m_2, respectively. If θ, the angle from ℓ_1 to ℓ_2, is not a right angle, then

$$\tan\theta = \frac{m_2 - m_1}{1 + m_1 m_2}$$

Show this using the fact that $\theta = \theta_2 - \theta_1$ in Figure 16.

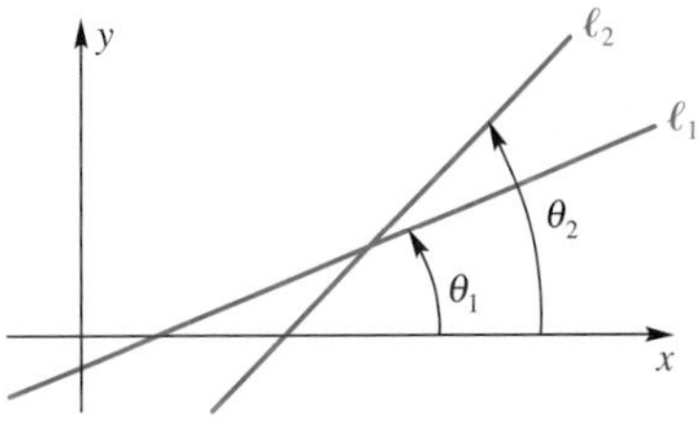

Figure 16

C **41.** Find the angle (in radians) from the first line to the second (see Problem 40).

(a) $y = 2x, y = 3x$ (b) $y = \frac{x}{2}, y = -x$

(c) $2x - 6y = 12, 2x + y = 0$

42. Derive the formula $A = \frac{1}{2}r^2 t$ for the area of a sector of a circle. Here r is the radius and t is the radian measure of the central angle (see Figure 17).

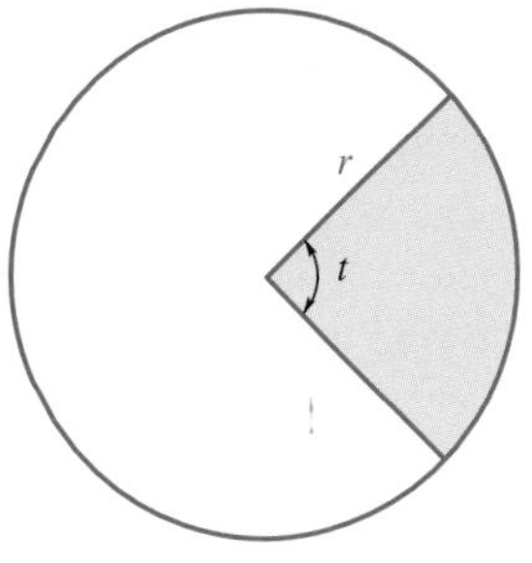

Figure 17

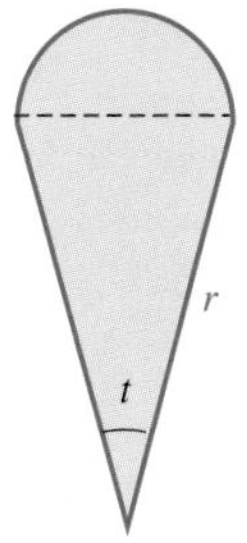

Figure 18

43. Find the area of the sector of a circle of radius 5 centimeters and central angle 2 radians (see Problem 42).

44. A regular polygon of n sides is inscribed in a circle of radius r. Find formulas for the perimeter, P, and area, A, of the polygon in terms of n and r.

45. An isosceles triangle is topped by a semicircle, as shown in Figure 18. Find a formula for the area A of the whole figure in terms of the side length r and angle t (radians). (We say that A is a function of the two independent variables r and t.)

46. From a product identity, we obtain

$$\cos\frac{x}{2}\cos\frac{x}{4} = \frac{1}{2}\left[\cos\left(\frac{3}{4}x\right) + \cos\left(\frac{1}{4}x\right)\right]$$

Find the corresponding sum of cosines for

$$\cos\frac{x}{2}\cos\frac{x}{4}\cos\frac{x}{8}\cos\frac{x}{16}$$

Do you see a generalization?

47. The normal high temperature for Las Vegas, Nevada, is 55°F for January 15 and 105° for July 15. Assuming that these are the extreme high and low temperatures for the year, use this information to approximate the average high temperature for November 15.

48. Tides are often measured by arbitrary height markings at some location. Suppose that a high tide occurs at noon when the water level is at 12 feet. Six hours later, a low tide with a water level of 5 feet occurs, and by midnight another high tide with a water level of 12 feet occurs. Assuming that the water level is periodic, use this information to find a formula that gives the water level as a function of time. Then use this function to approximate the water level at 5:30 P.M.

EXPL **49.** Circular motion can be modeled by using the parametric representations of the form $x(t) = \sin t$ and $y(t) = \cos t$. (A *parametric representation* means that a variable, t in this case, determines both $x(t)$ and $y(t)$.) This will give the full circle for $0 \le t \le 2\pi$. If we consider a 4-foot-diameter wheel making one complete rotation clockwise once every 10 seconds, show that the motion of a point on the rim of the wheel can be represented by $x(t) = 2\sin(\pi t/5)$ and $y(t) = 2\cos(\pi t/5)$.

(a) Find the positions of the point on the rim of the wheel when $t = 2$ seconds, 6 seconds, and 10 seconds. Where was this point when the wheel started to rotate at $t = 0$?

(b) How will the formulas giving the motion of the point change if the wheel is rotating *counterclockwise*.

(c) At what value of t is the point at $(2, 0)$ for the first time?

EXPL **50.** The circular frequency v of oscillation of a point is given by $v = \frac{2\pi}{\text{period}}$. What happens when you add two motions that have the same frequency or period? To investigate, we can graph the functions $y(t) = 2\sin(\pi t/5)$ and $y(t) = \sin(\pi t/5) + \cos(\pi t/5)$ and look for similarities. Armed with this information, we can investigate by graphing the following functions over the interval $[-5, 5]$:

(a) $y(t) = 3\sin(\pi t/5) - 5\cos(\pi t/5) + 2\sin((\pi t/5) - 3)$

(b) $y(t) = 3\cos(\pi t/5 - 2) + \cos(\pi t/5) + \cos((\pi t/5) - 3)$

EXPL **51.** We now explore the relationship between $A\sin(\omega t) + B\cos(\omega t)$ and $C\sin(\omega t + \phi)$.

(a) By expanding $\sin(\omega t + \phi)$ using the sum of the angles formula, show that the two expressions are equivalent if $A = C\cos\phi$ and $B = C\sin\phi$.

(b) Consequently, show that $A^2 + B^2 = C^2$ and that ϕ then satisfies the equation $\tan\phi = \frac{B}{A}$.

(c) Generalize your result to state a proposition about $A_1\sin(\omega t + \phi_1) + A_2\sin(\omega t + \phi_2) + A_3\sin(\omega t + \phi_3)$.

(d) Write an essay, in your own words, that expresses the importance of the identity between $A\sin(\omega t) + B\cos(\omega t)$ and $C\sin(\omega t + \phi)$. Be sure to note that $|C| \ge \max(|A|, |B|)$ and that the identity holds only when you are forming a linear combination (adding and/or subtracting multiples of single powers) of sine and cosine of the same frequency.

Trigonometric functions that have high frequencies pose special problems for graphing. We now explore how to plot such functions.

GC **52.** Graph the function $f(x) = \sin 50x$ using the window given by a y range of $-1.5 \le y \le 1.5$ and the x range given by

(a) $[-15, 15]$ (b) $[-10, 10]$ (c) $[-8, 8]$
(d) $[-1, 1]$ (e) $[-0.25, 0.25]$

Indicate briefly which x-window shows the true behavior of the function, and discuss reasons why the other x-windows give results that look different.

GC **53.** Graph the function $f(x) = \cos x + \frac{1}{50}\sin 50x$ using the windows given by the following ranges of x and y.

(a) $-5 \le x \le 5, -1 \le y \le 1$

(b) $-1 \le x \le 1, 0.5 \le y \le 1.5$

(c) $-0.1 \le x \le 0.1, 0.9 \le y \le 1.1$

Indicate briefly which (x, y)-window shows the true behavior of the function, and discuss reasons why the other (x, y)-windows give results that look different. In this case, is it true that only one window gives the important behavior, or do we need more than one window to graphically communicate the behavior of this function?

GC EXPL **54.** Let $f(x) = \frac{3x + 2}{x^2 + 1}$ and $g(x) = \frac{1}{100}\cos(100x)$.

(a) Use functional composition to form $h(x) = (f \circ g)(x)$, as well as $j(x) = (g \circ f)(x)$.

(b) Find the appropriate window or windows that give a clear picture of $h(x)$.

(c) Find the appropriate window or windows that give a clear picture of $j(x)$.

55. Suppose that a continuous function is periodic with period 1 and is linear between 0 and 0.25 and linear between -0.75 and 0. In addition, it has the value 1 at 0 and 2 at 0.25. Sketch the function over the domain $[-1, 1]$, and give a piecewise definition of the function.

56. Suppose that a continuous function is periodic with period 2 and is quadratic between -0.25 and 0.25 and linear between -1.75 and -0.25. In addition, it has the value 0 at 0 and 0.0625 at ± 0.25. Sketch the function over the domain $[-2, 2]$, and give a piecewise definition of the function.

Answers to Concepts Review: **1.** $(-\infty, \infty)$; $[-1, 1]$ **2.** 2π; 2π; π **3.** odd; even **4.** $-4/5$

0.8 Chapter Review

Concepts Test

Respond with true or false to each of the following assertions. Be prepared to justify your answer. Normally, this means that you should supply a reason if you answer true and provide a counter-example if you answer false.

1. Any number that can be written as a fraction p/q is rational.

2. The difference of any two rational numbers is rational.

3. The difference of any two irrational numbers is irrational.

4. Between two distinct irrational numbers, there is always another irrational number.

5. $0.999\ldots$ (repeating 9s) is less than 1.

6. The operation of exponentiation is commutative; that is, $(a^m)^n = (a^n)^m$.

7. The operation $*$ defined by $m*n = m^n$ is associative.

8. The inequalities $x \le y$, $y \le z$, and $z \le x$ together imply that $x = y = z$.

9. If $|x| < \varepsilon$ for every positive number ε, then $x = 0$.

10. If x and y are real numbers, then $(x - y)(y - x) \le 0$.

11. If $a < b < 0$, then $1/a > 1/b$.

12. It is possible for two closed intervals to have exactly one point in common.

13. If two open intervals have a point in common, then they have infinitely many points in common.

14. If $x < 0$, then $\sqrt{x^2} = -x$.

15. If x is a real number, then $|-x| = x$.

16. If $|x| < |y|$, then $x < y$.

17. If $|x| < |y|$, then $x^4 < y^4$.

18. If x and y are both negative, then $|x + y| = |x| + |y|$.

19. If $|r| < 1$, then $\dfrac{1}{1 + |r|} \le \dfrac{1}{1 - r} \le \dfrac{1}{1 - |r|}$.

20. If $|r| > 1$, then $\dfrac{1}{1 - |r|} \le \dfrac{1}{1 - r} \le \dfrac{1}{1 + |r|}$.

21. It is always true that $||x| - |y|| \le |x + y|$.

22. For every positive real number y, there exists a real number x such that $x^2 = y$.

23. For every real number y, there exists a real number x such that $x^3 = y$.

24. It is possible to have an inequality whose solution set consists of exactly one number.

25. The equation $x^2 + y^2 + ax + y = 0$ represents a circle for every real number a.

26. The equation $x^2 + y^2 + ax + by = c$ represents a circle for all real numbers a, b, c.

27. If (a, b) is on a line with slope $\frac{3}{4}$, then $(a + 4, b + 3)$ is also on that line.

28. If (a, b), (c, d) and (e, f) are on the same line, then $\dfrac{a - c}{b - d} = \dfrac{a - e}{b - f} = \dfrac{e - c}{f - d}$ provided all three points are different.

29. If $ab > 0$, then (a, b) lies in either the first or third quadrant.

30. For every $\varepsilon > 0$, there exists a positive number x such that $x < \varepsilon$.

31. If $ab = 0$, then (a, b) lies on either the x-axis or the y-axis.

32. If $\sqrt{(x_2 - x_1)^2 + (y_2 - y_1)^2} = |x_2 - x_1|$, then (x_1, y_1) and (x_2, y_2) lie on the same horizontal line.

33. The distance between $(a + b, a)$ and $(a - b, a)$ is $|2b|$.

34. The equation of every line can be written in point–slope form.

35. The equation of every line can be written in the general linear form $Ax + By + C = 0$.

36. If two nonvertical lines are parallel, they have the same slope.

37. It is possible for two lines to have positive slopes and be perpendicular.

38. If the x- and y-intercepts of a line are rational and nonzero, then the slope of the line is rational.

39. The lines $ax + y = c$ and $ax - y = c$ are perpendicular.

40. $(3x - 2y + 4) + m(2x + 6y - 2) = 0$ is the equation of a line for each real number m.

41. The natural domain of

$$f(x) = \sqrt{-(x^2 + 4x + 3)}$$

is the interval $-3 \le x \le -1$.

42. The natural domain of $T(\theta) = \sec(\theta) + \cos(\theta)$ is $(-\infty, \infty)$.

43. The range of $f(x) = x^2 - 6$ is the interval $[-6, \infty)$.

44. The range of the function $f(x) = \tan x - \sec x$ is the set $(-\infty, -1] \cup [1, \infty)$.

45. The range of the function $f(x) = \csc x - \sec x$ is the set $(-\infty, -1] \cup [1, \infty)$.

46. The sum of two even functions is an even function.

47. The sum of two odd functions is an odd function.

48. The product of two odd functions is an odd function.

49. The product of an even function with an odd function is an odd function.

50. The composition of an even function with an odd function is an odd function.

51. The composition of two odd functions is an even function.

52. The function $f(x) = (2x^3 + x)/(x^2 + 1)$ is odd.

53. The function

$$f(t) = \frac{(\sin t)^2 + \cos t}{\tan t \csc t}$$

is even.

54. If the range of a function consists of just one number, then its domain also consists of just one number.

55. If the domain of a function contains at least two numbers then the range also contains at least two numbers.

56. If $g(x) = [\![x/2]\!]$, then $g(-1.8) = -1$.

57. If $f(x) = x^2$ and $g(x) = x^3$, then $f \circ g = g \circ f$.

58. If $f(x) = x^2$ and $g(x) = x^3$, then $(f \circ g)(x) = f(x) \cdot g(x)$.

59. If f and g have the same domain, then f/g also has that domain.

60. If the graph of $y = f(x)$ has an x-intercept at $x = a$, then the graph of $y = f(x + h)$ has an x-intercept at $x = a - h$.

61. The cotangent is an odd function.

62. The natural domain of the tangent function is the set of all real numbers.

63. If $\cos s = \cos t$, then $s = t$.

Sample Test Problems

1. Calculate each value for $n = 1, 2$, and -2.

(a) $\left(n + \frac{1}{n}\right)^n$ (b) $(n^2 - n + 1)^2$

(c) $4^{3/n}$ (d) $\sqrt[n]{\left|\frac{1}{n}\right|}$

2. Simplify.

(a) $\left(1 + \frac{1}{m} + \frac{1}{n}\right)\left(1 - \frac{1}{m} + \frac{1}{n}\right)^{-1}$

(b) $\dfrac{\frac{2}{x+1} - \frac{x}{x^2 - x - 2}}{\frac{3}{x+1} - \frac{2}{x-2}}$

(c) $\dfrac{t^3 - 1}{t - 1}$

3. Show that the average of two rational numbers is a rational number.

4. Write the repeating decimal 4.1282828... as a ratio of two integers.

5. Find an irrational number between $\frac{1}{2}$ and $\frac{13}{25}$.

C **6.** Calculate $\left(\sqrt[3]{8.15 \times 10^4} - 1.32\right)^2/3.24$.

C **7.** Calculate $\left(\pi - \sqrt{2.0}\right)^{2.5} - \sqrt[3]{2.0}$.

C **8.** Calculate $\sin^2(2.45) + \cos^2(2.40) - 1.00$.

In Problems 9–18, find the solution set, graph this set on the real line, and express this set in interval notation.

9. $1 - 3x > 0$ **10.** $6x + 3 > 2x - 5$

11. $3 - 2x \le 4x + 1 \le 2x + 7$

12. $2x^2 + 5x - 3 < 0$ **13.** $21t^2 - 44t + 12 \le -3$

14. $\dfrac{2x - 1}{x - 2} > 0$

15. $(x + 4)(2x - 1)^2(x - 3) \le 0$

16. $|3x - 4| < 6$

17. $\dfrac{3}{1 - x} \le 2$

18. $|12 - 3x| \ge |x|$

19. Find a value of x for which $|-x| \ne x$.

20. For what values of x does the equation $|-x| = x$ hold?

21. For what values of t does the equation $|t - 5| = 5 - t$ hold?

22. For what values of a and t does the equation $|t - a| = a - t$ hold?

23. Suppose $|x| \le 2$. Use properties of absolute values to show that

$$\left|\frac{2x^2 + 3x + 2}{x^2 + 2}\right| \le 8$$

24. Write a sentence involving the word *distance* to express the following algebraic sentences:

(a) $|x - 5| = 3$ (b) $|x + 1| \le 2$

(c) $|x - a| > b$

25. Sketch the triangle with vertices $A(-2, 6)$, $B(1, 2)$, and $C(5, 5)$, and show that it is a right triangle.

26. Find the distance from $(3, -6)$ to the midpoint of the line segment from $(1, 2)$ to $(7, 8)$.

27. Find the equation of the circle with diameter AB if $A = (2, 0)$ and $B = (10, 4)$.

28. Find the center and radius of the circle with equation $x^2 + y^2 - 8x + 6y = 0$.

29. Find the distance between the centers of the circles with equations

$$x^2 - 2x + y^2 + 2y = 2 \quad \text{and} \quad x^2 + 6x + y^2 - 4y = -7$$

30. Find the equation of the line through the indicated point that is parallel to the indicated line, and sketch both lines.

(a) $(3, 2)$: $3x + 2y = 6$ (b) $(1, -1)$: $y = \frac{2}{3}x + 1$

(c) $(5, 9)$: $y = 10$ (d) $(-3, 4)$: $x = -2$

31. Write the equation of the line through $(-2, 1)$ that

(a) goes through $(7, 3)$;

(b) is parallel to $3x - 2y = 5$;

(c) is perpendicular to $3x + 4y = 9$;

(d) is perpendicular to $y = 4$;

(e) has y-intercept 3.

32. Show that $(2, -1)$, $(5, 3)$, and $(11, 11)$ are on the same line.

33. Figure 1 can be represented by which equation?

(a) $y = x^3$ (b) $x = y^3$

(c) $y = x^2$ (d) $x = y^2$

34. Figure 2 can be represented by which equation?

(a) $y = ax^2 + bx + c$, with $a > 0, b > 0$, and $c > 0$

(b) $y = ax^2 + bx + c$, with $a < 0, b > 0$, and $c > 0$

(c) $y = ax^2 + bx + c$, with $a < 0, b > 0$, and $c < 0$

(d) $y = ax^2 + bx + c$, with $a > 0, b > 0$, and $c < 0$

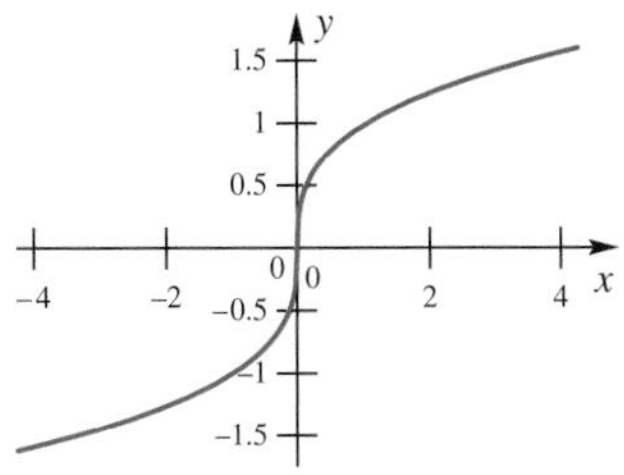

Figure 1

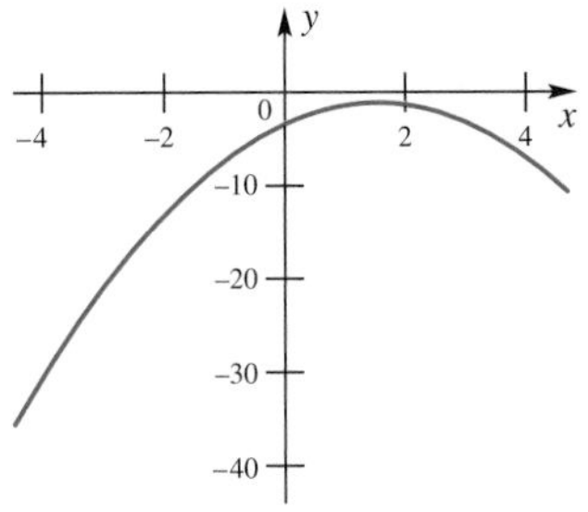

Figure 2

In Problems 35–38, sketch the graph of each equation.

35. $3y - 4x = 6$ **36.** $x^2 - 2x + y^2 = 3$

GC **37.** $y = \dfrac{2x}{x^2 + 2}$ GC **38.** $x = y^2 - 3$

GC **39.** Find the points of intersection of the graphs of $y = x^2 - 2x + 4$ and $y - x = 4$.

40. Among all lines perpendicular to $4x - y = 2$, find the equation of the one that, together with the positive x- and y-axes, forms a triangle of area 8.

41. For $f(x) = 1/(x + 1) - 1/x$, find each value (if possible).

(a) $f(1)$ (b) $f(-\frac{1}{2})$ (c) $f(-1)$

(d) $f(t - 1)$ (e) $f\left(\dfrac{1}{t}\right)$

42. For $g(x) = (x + 1)/x$, find and simplify each value.

(a) $g(2)$ (b) $g(\frac{1}{2})$

(c) $\dfrac{g(2 + h) - g(2)}{h}$

43. Describe the natural domain of each function.

(a) $f(x) = \dfrac{x}{x^2 - 1}$ (b) $g(x) = \sqrt{4 - x^2}$

44. Which of the following functions are odd? Even? Neither even nor odd?

(a) $f(x) = \dfrac{3x}{x^2 + 1}$ (b) $g(x) = |\sin x| + \cos x$

(c) $h(x) = x^3 + \sin x$ (d) $k(x) = \dfrac{x^2 + 1}{|x| + x^4}$

45. Sketch the graph of each function.

(a) $f(x) = x^2 - 1$ (b) $g(x) = \dfrac{x}{x^2 + 1}$

(c) $h(x) = \begin{cases} x^2 & \text{if } 0 \le x \le 2 \\ 6 - x & \text{if } x > 2 \end{cases}$

46. Suppose that f is an even function satisfying $f(x) = -1 + \sqrt{x}$ for $x \ge 0$. Sketch the graph of f for $-4 \le x \le 4$.

47. An open box is made by cutting squares of side x inches from the four corners of a sheet of cardboard 24 inches by 32 inches and then turning up the sides. Express the volume $V(x)$ in terms of x. What is the domain for this function?

48. Let $f(x) = x - 1/x$ and $g(x) = x^2 + 1$. Find each value.

(a) $(f + g)(2)$ (b) $(f \cdot g)(2)$ (c) $(f \circ g)(2)$

(d) $(g \circ f)(2)$ (e) $f^3(-1)$

(f) $f^2(2) + g^2(2)$

49. Sketch the graph of each of the following, making use of translations.

(a) $y = \frac{1}{4}x^2$ (b) $y = \frac{1}{4}(x + 2)^2$

(c) $y = -1 + \frac{1}{4}(x + 2)^2$

50. Let $f(x) = \sqrt{16 - x}$ and $g(x) = x^4$. What is the domain of each of the following?

(a) f (b) $f \circ g$ (c) $g \circ f$

51. Write $F(x) = \sqrt{1 + \sin^2 x}$ as the composite of four functions, $f \circ g \circ h \circ k$.

52. Calculate each of the following without using a calculator.

(a) $\sin 570°$ (b) $\cos \dfrac{9\pi}{2}$

(c) $\cos\left(\dfrac{-13\pi}{6}\right)$

53. If $\sin t = 0.8$ and $\cos t < 0$, find each value.

(a) $\sin(-t)$ (b) $\cos t$ (c) $\sin 2t$

(d) $\tan t$ (e) $\cos\left(\dfrac{\pi}{2} - t\right)$ (f) $\sin(\pi + t)$

54. Write $\sin 3t$ in terms of $\sin t$. *Hint:* $3t = 2t + t$.

55. A fly sits on the rim of a wheel spinning at the rate of 20 revolutions per minute. If the radius of the wheel is 9 inches, how far does the fly travel in 1 second?

REVIEW & PREVIEW PROBLEMS

1. Solve the following inequalities:

(a) $1 < 2x + 1 < 5$ (b) $-3 < \frac{x}{2} < 8$

2. Solve the following inequalities:

(a) $14 < 2x + 1 < 15$ (b) $-3 < 1 - \frac{x}{2} < 8$

3. Solve $|x - 7| = 3$ for x.

4. Solve $|x + 3| = 2$ for x.

5. The distance along the number line between x and 7 is equal to 3. What are the possible values for x?

6. The distance along the number line between x and 7 is equal to d. What are the possible values for x?

7. Solve the following inequalities:

(a) $|x - 7| < 3$ (b) $|x - 7| \le 3$

(c) $|x - 7| \le 1$ (d) $|x - 7| < 0.1$

8. Solve the following inequalities:

(a) $|x - 2| < 1$ (b) $|x - 2| \ge 1$

(c) $|x - 2| < 0.1$ (d) $|x - 2| < 0.01$

9. What are the natural domains of the following functions?

(a) $f(x) = \frac{x^2 - 1}{x - 1}$ (b) $g(x) = \frac{x^2 - 2x + 1}{2x^2 - x - 1}$

10. What are the natural domains of the following functions?

(a) $F(x) = \frac{|x|}{x}$ (b) $G(x) = \frac{\sin x}{x}$

11. Evaluate the functions $f(x)$ and $g(x)$ from Problem 9 at the following values of x: $0, 0.9, 0.99, 0.999, 1.001, 1.01, 1.1, 2$.

12. Evaluate the functions $F(x)$ and $G(x)$ from Problem 10 at the following values of x: $-1, -0.1, -0.01, -0.001, 0.001, 0.01, 0.1, 1$.

13. The distance between x and 5 is less than 0.1. What are the possible values for x?

14. The distance between x and 5 is less than ε, where ε is a positive number. What are the possible values for x?

15. True or false. Assume that a, x, and y are real numbers and n is a natural number.

(a) For every $x > 0$, there exists a y such that $y > x$.

(b) For every $a \ge 0$, there exists an n such that $\frac{1}{n} < a$.

(c) For every $a > 0$, there exists an n such that $\frac{1}{n} < a$.

(d) For every circle C in the plane, there exists an n such that the circle C and its interior are all within n units of the origin.

16. Use the Addition Identity for the sine function to find $\sin(c + h)$ in terms of $\sin c$, $\sin h$, $\cos c$, and $\cos h$.

CHAPTER 1 Limits

1.1 Introduction to Limits
1.2 Rigorous Study of Limits
1.3 Limit Theorems
1.4 Limits Involving Trigonometric Functions
1.5 Limits at Infinity; Infinite Limits
1.6 Continuity of Functions

1.1 Introduction to Limits

The topics discussed in the previous chapter are part of what is called *precalculus*. They provide the foundation for calculus, but they are not calculus. Now we are ready for an important new idea, the notion of *limit*. It is this idea that distinguishes calculus from other branches of mathematics. In fact, we define calculus this way:

Calculus is the study of limits.

Problems Leading to the Limit Concept The concept of **limit** is central to many problems in the physical, engineering, and social sciences. Basically the question is this: what happens to the function $f(x)$ as x gets close to some constant c? There are variations on this theme, but the basic idea is the same in many circumstances.

Suppose that as an object steadily moves forward we know its position at any given time. We denote the position at time t by $s(t)$. How fast is the object moving at time $t = 1$? We can use the formula "distance equals rate times time" to find the speed (rate of change of position) over any interval of time; in other words

$$\text{speed} = \frac{\text{distance}}{\text{time}}$$

We call this the "average" speed over the interval since, no matter how small the interval is, we never know whether the speed is constant over this interval. For example, over the interval $[1, 2]$, the average speed is $\frac{s(2) - s(1)}{2 - 1}$; over the interval $[1, 1.2]$, the average speed is $\frac{s(1.2) - s(1)}{1.2 - 1}$; over the interval $[1, 1.02]$, the average speed is $\frac{s(1.02) - s(1)}{1.02 - 1}$, etc. How fast is the object traveling at time $t = 1$? To give meaning to this "instantaneous" velocity we must talk about the *limit* of the average speed over smaller and smaller intervals.

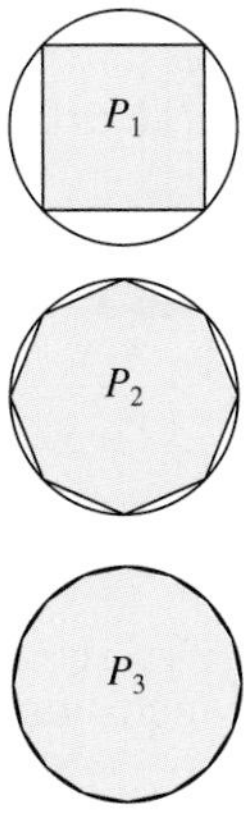

Figure 1

We can find areas of rectangles and triangles using formulas from geometry, but what about regions with curved boundaries, such as a circle? Archimedes had this idea over two thousand years ago. Imagine regular polygons inscribed in a circle as shown in Figure 1. Archimedes was able to find the area of a regular polygon with n sides, and by taking the regular polygon with more and more sides, he was able to approximate the area of a circle to any desired level of accuracy. In other words, the area of the circle is the *limit* of the areas of the inscribed polygons as n (the number of sides in the polygon) increases without bound.

Consider the graph of the function $y = f(x)$ for $a \le x \le b$. If the graph is a straight line, the length of the curve is easy to find using the distance formula. But what if the graph is curved? We can find numerous points along the curve and connect them with line segments as shown in Figure 2. If we add up the lengths of these line segments we should get a sum that is approximately the length of the curve. In fact, by "length of the curve" we mean the *limit* of the sum of the lengths of these line segments as the number of line segments increases without bound.

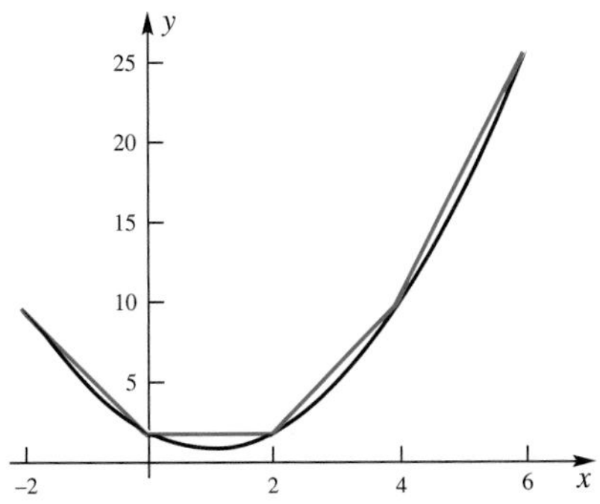

Figure 2

The last three paragraphs describe situations that lead to the concept of *limit*. There are many others, and we will study them throughout this book. We begin with an intuitive explanation of limits. The precise definition is given in the next section.

An Intuitive Understanding Consider the function defined by

$$f(x) = \frac{x^3 - 1}{x - 1}$$

Note that it is not defined at $x = 1$ since at this point $f(x)$ has the form $\frac{0}{0}$, which is meaningless. We can, however, still ask what is happening to $f(x)$ as x approaches 1. More precisely, is $f(x)$ approaching some specific number as x approaches 1? To get at the answer, we can do three things. We can calculate some values of $f(x)$ for x near 1, we can show these values in a schematic diagram, and we can sketch the graph of $y = f(x)$. All this has been done, and the results are shown in Figure 3.

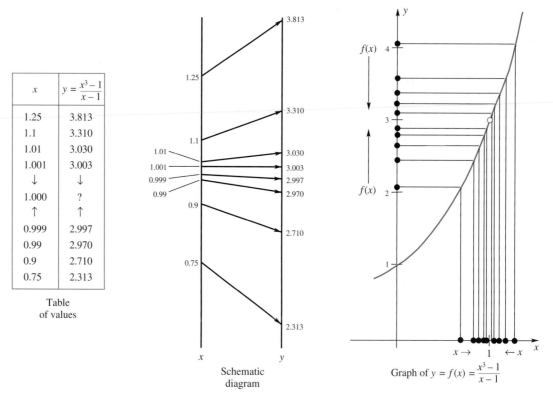

x	$y = \frac{x^3 - 1}{x - 1}$
1.25	3.813
1.1	3.310
1.01	3.030
1.001	3.003
↓	↓
1.000	?
↑	↑
0.999	2.997
0.99	2.970
0.9	2.710
0.75	2.313

Table of values

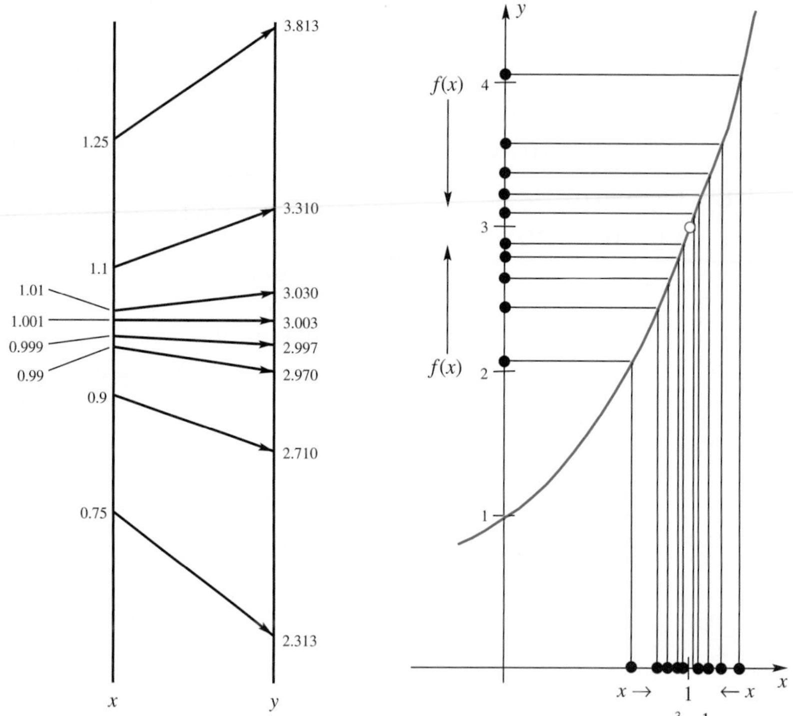

Schematic diagram

Graph of $y = f(x) = \frac{x^3 - 1}{x - 1}$

Figure 3

All the information we have assembled seems to point to the same conclusion: $f(x)$ approaches 3 as x approaches 1. In mathematical symbols, we write

$$\lim_{x \to 1} \frac{x^3 - 1}{x - 1} = 3$$

This is read "the limit as x approaches 1 of $(x^3 - 1)/(x - 1)$ is 3."

Being good algebraists (thus knowing how to factor the difference of cubes), we can provide more and better evidence.

$$\lim_{x \to 1} \frac{x^3 - 1}{x - 1} = \lim_{x \to 1} \frac{(x - 1)(x^2 + x + 1)}{x - 1}$$

$$= \lim_{x \to 1} (x^2 + x + 1) = 1^2 + 1 + 1 = 3$$

Note that $(x - 1)/(x - 1) = 1$ as long as $x \neq 1$. This justifies the second step. The third step should seem reasonable; a rigorous justification will come later.

To be sure that we are on the right track, we need to have a clearly understood meaning for the word *limit*. Here is our first attempt at a definition.

> **Definition** Intuitive Meaning of Limit
>
> To say that $\lim_{x \to c} f(x) = L$ means that when x is near but different from c then $f(x)$ is near L.

Notice that we do not require anything *at c*. The function f need not even be defined at c; it was not in the example $f(x) = (x^3 - 1)/(x - 1)$ just considered. The notion of limit is associated with the behavior of a function *near c*, not *at c*.

A cautious reader is sure to object to our use of the word *near*. What does *near* mean? How near is near? For precise answers, you will have to study the next section; however, some further examples will help to clarify the idea.

More Examples Our first example is almost trivial, but nonetheless important.

EXAMPLE 1 Find $\lim_{x \to 3} (4x - 5)$.

SOLUTION When x is near 3, $4x - 5$ is near $4 \cdot 3 - 5 = 7$. We write

$$\lim_{x \to 3} (4x - 5) = 7$$

■

EXAMPLE 2 Find $\lim_{x \to 3} \frac{x^2 - x - 6}{x - 3}$.

SOLUTION Note that $(x^2 - x - 6)/(x - 3)$ is not defined at $x = 3$, but this is all right. To get an idea of what is happening as x approaches 3, we could use a calculator to evaluate the given expression, for example, at 3.1, 3.01, 3.001, and so on. But it is much better to use a little algebra to simplify the problem.

$$\lim_{x \to 3} \frac{x^2 - x - 6}{x - 3} = \lim_{x \to 3} \frac{(x - 3)(x + 2)}{x - 3} = \lim_{x \to 3} (x + 2) = 3 + 2 = 5$$

The cancellation of $x - 3$ in the second step is legitimate because the definition of limit ignores the behavior *at* $x = 3$. Remember, $\frac{x - 3}{x - 3} = 1$ as long as x is not equal to 3. ■

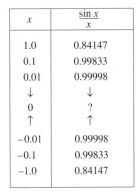

x	$\frac{\sin x}{x}$
1.0	0.84147
0.1	0.99833
0.01	0.99998
↓	↓
0	?
↑	↑
−0.01	0.99998
−0.1	0.99833
−1.0	0.84147

Figure 4

EXAMPLE 3 Find $\lim_{x \to 0} \frac{\sin x}{x}$.

SOLUTION No algebraic trick will simplify our task; certainly, we cannot cancel the x's. A calculator will help us to get an idea of the limit. Use your own calculator (radian mode) to check the values in the table of Figure 4. Figure 5 shows a plot of $y = (\sin x)/x$. Our conclusion, though we admit it is a bit shaky, is that

$$\lim_{x \to 0} \frac{\sin x}{x} = 1$$

We will give a rigorous demonstration in Section 1.4. ■

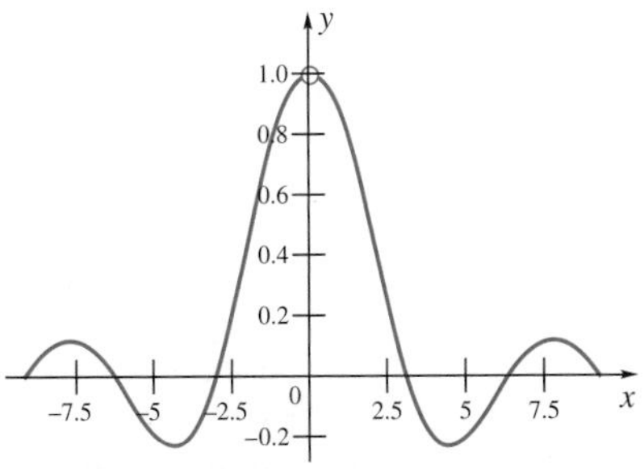

Figure 5

Some Warning Flags Things are not quite as simple as they may appear. Calculators may mislead us; so may our own intuition. The examples that follow suggest some possible pitfalls.

x	$x^2 - \frac{\cos x}{10{,}000}$
± 1	0.99995
± 0.5	0.24991
± 0.1	0.00990
± 0.01	0.000000005
↓	↓
0	?

Figure 6

EXAMPLE 4 **(Your calculator may fool you.)** Find $\lim_{x \to 0}\left[x^2 - \frac{\cos x}{10{,}000}\right]$.

SOLUTION Following the procedure used in Example 3, we construct the table of values shown in Figure 6. The conclusion it suggests is that the desired limit is 0. But this is wrong. If we recall the graph of $y = \cos x$, we realize that $\cos x$ approaches 1 as x approaches 0. Thus,

$$\lim_{x \to 0}\left[x^2 - \frac{\cos x}{10{,}000}\right] = 0^2 - \frac{1}{10{,}000} = -\frac{1}{10{,}000}$$ ■

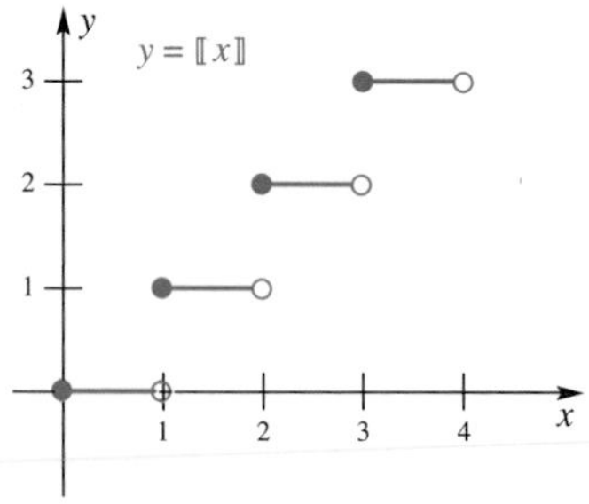

Figure 7

EXAMPLE 5 **(No limit at a jump)** Find $\lim_{x \to 2} [\![x]\!]$.

SOLUTION Recall that $[\![x]\!]$ denotes the greatest integer less than or equal to x (see Section 0.5). The graph of $y = [\![x]\!]$ is shown in Figure 7. For all numbers x less than 2 but near 2, $[\![x]\!] = 1$, but for all numbers x greater than 2 but near 2, $[\![x]\!] = 2$. Is $[\![x]\!]$ near a single number L when x is near 2? No. No matter what number we propose for L, there will be x's arbitrarily close to 2 on one side or the other, where $[\![x]\!]$ differs from L by at least $\frac{1}{2}$. Our conclusion is that $\lim_{x \to 2} [\![x]\!]$ does not exist. If you check back, you will see that we have not claimed that every limit we can write must exist. ■

EXAMPLE 6 **(Too many wiggles)** Find $\lim_{x \to 0} \sin(1/x)$.

SOLUTION This example poses the most subtle limit question asked yet. Since we do not want to make too big a story out of it, we ask you to do two things. First, pick a sequence of x-values approaching 0. Use your calculator to evaluate $\sin(1/x)$ at these x's. Unless you happen on some very lucky choices, your values will oscillate wildly.

Second, consider trying to graph $y = \sin(1/x)$. No one will ever do this very well, but the table of values in Figure 8 gives a good clue about what is happening. In any neighborhood of the origin, the graph wiggles up and down between -1 and 1 infinitely many times (Figure 9). Clearly, $\sin(1/x)$ is not near a single number L when x is near 0. We conclude that $\lim_{x \to 0} \sin(1/x)$ does not exist. ■

x	$\sin\frac{1}{x}$
$2/\pi$	1
$2/(2\pi)$	0
$2/(3\pi)$	−1
$2/(4\pi)$	0
$2/(5\pi)$	1
$2/(6\pi)$	0
$2/(7\pi)$	−1
$2/(8\pi)$	0
$2/(9\pi)$	1
$2/(10\pi)$	0
$2/(11\pi)$	−1
$2/(12\pi)$	0
↓	↓
0	?

Figure 8

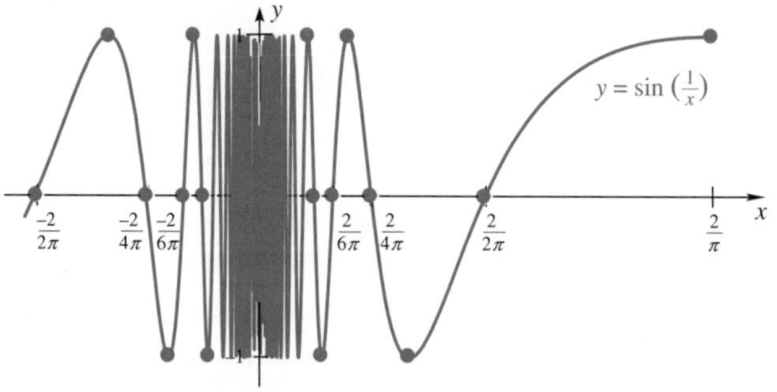

Figure 9

One-Sided Limits When a function takes a jump (as does $[\![x]\!]$ at each integer in Example 5), then the limit does not exist at the jump points. Such functions suggest the introduction of **one-sided limits.** Let the symbol $x \to c^+$ mean that x approaches c from the right, and let $x \to c^-$ mean that x approaches c from the left.

Definition **Right- and Left-Hand Limits**

To say that $\lim_{x \to c^+} f(x) = L$ means that when x is near but to the right of c then $f(x)$ is near L. Similarly, to say that $\lim_{x \to c^-} f(x) = L$ means that when x is near but to the left of c then $f(x)$ is near L.

Thus, while $\lim_{x\to 2} [\![x]\!]$ does not exist, it is correct to write (look at the graph in Figure 7)

$$\lim_{x\to 2^-} [\![x]\!] = 1 \quad \text{and} \quad \lim_{x\to 2^+} [\![x]\!] = 2$$

We believe that you will find the following theorem quite reasonable.

Theorem A

$\lim_{x\to c} f(x) = L$ if and only if $\lim_{x\to c^-} f(x) = L$ and $\lim_{x\to c^+} f(x) = L$.

Figure 10 should give additional insight. Two of the limits do not exist, although all but one of the one-sided limits exist.

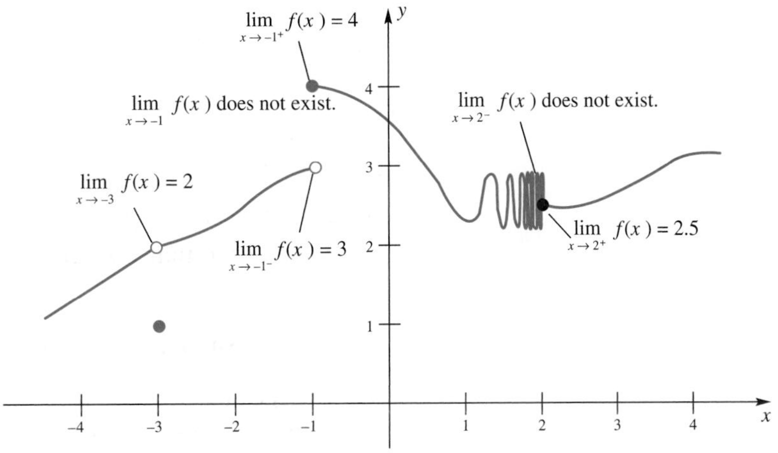

Figure 10

Concepts Review

1. $\lim_{x\to c} f(x) = L$ means that $f(x)$ gets close to ______ when x gets sufficiently close to (but is different from) ______.

2. Let $f(x) = (x^2 - 9)/(x - 3)$ and note that $f(3)$ is undefined. Nevertheless, $\lim_{x\to 3} f(x) =$ ______.

3. $\lim_{x\to c^+} f(x) = L$ means that $f(x)$ gets near to ______ when x approaches c from the ______.

4. If both $\lim_{x\to c^-} f(x) = M$ and $\lim_{x\to c^+} f(x) = M$, then ______.

Problem Set 1.1

In Problems 1–6, find the indicated limit.

1. $\lim_{x\to 3}(x - 5)$

2. $\lim_{t\to -1}(1 - 2t)$

3. $\lim_{x\to -2}(x^2 + 2x - 1)$

4. $\lim_{x\to -2}(x^2 + 2t - 1)$

5. $\lim_{t\to -1}(t^2 - 1)$

6. $\lim_{t\to -1}(t^2 - x^2)$

In Problems 7–18, find the indicated limit. In most cases, it will be wise to do some algebra first (see Example 2).

7. $\lim_{x\to 2} \dfrac{x^2 - 4}{x - 2}$

8. $\lim_{t\to -7} \dfrac{t^2 + 4t - 21}{t + 7}$

9. $\lim_{x\to -1} \dfrac{x^3 - 4x^2 + x + 6}{x + 1}$

10. $\lim_{x\to 0} \dfrac{x^4 + 2x^3 - x^2}{x^2}$

11. $\lim_{x\to -t} \dfrac{x^2 - t^2}{x + t}$

12. $\lim_{x\to 3} \dfrac{x^2 - 9}{x - 3}$

13. $\lim_{t\to 2} \dfrac{\sqrt{(t + 4)(t - 2)^4}}{(3t - 6)^2}$

14. $\lim_{t\to 7^+} \dfrac{\sqrt{(t - 7)^3}}{t - 7}$

15. $\lim_{x\to 3} \dfrac{x^4 - 18x^2 + 81}{(x - 3)^2}$

16. $\lim_{u\to 1} \dfrac{(3u + 4)(2u - 2)^3}{(u - 1)^2}$

17. $\lim_{h\to 0} \dfrac{(2 + h)^2 - 4}{h}$

18. $\lim_{h\to 0} \dfrac{(x + h)^2 - x^2}{h}$

GC *In Problems 19–28, use a calculator to find the indicated limit. Use a graphing calculator to plot the function near the limit point.*

19. $\lim_{x\to 0} \dfrac{\sin x}{2x}$

20. $\lim_{t\to 0} \dfrac{1 - \cos t}{2t}$

21. $\lim_{x\to 0} \dfrac{(x - \sin x)^2}{x^2}$

22. $\lim_{x\to 0} \dfrac{(1 - \cos x)^2}{x^2}$

23. $\lim_{t\to 1} \dfrac{t^2 - 1}{\sin(t - 1)}$

24. $\lim_{x\to 3} \dfrac{x - \sin(x - 3) - 3}{x - 3}$

25. $\lim_{x\to \pi} \dfrac{1 + \sin(x - 3\pi/2)}{x - \pi}$

26. $\lim_{t\to 0} \dfrac{1 - \cot t}{1/t}$

27. $\lim_{x\to \pi/4} \dfrac{(x - \pi/4)^2}{(\tan x - 1)^2}$

28. $\lim_{u\to \pi/2} \dfrac{2 - 2\sin u}{3u}$

29. For the function f graphed in Figure 11, find the indicated limit or function value, or state that it does not exist.

(a) $\lim_{x\to-3} f(x)$ (b) $f(-3)$ (c) $f(-1)$
(d) $\lim_{x\to-1} f(x)$ (e) $f(1)$ (f) $\lim_{x\to1} f(x)$
(g) $\lim_{x\to1^-} f(x)$ (h) $\lim_{x\to1^+} f(x)$ (i) $\lim_{x\to-1^+} f(x)$

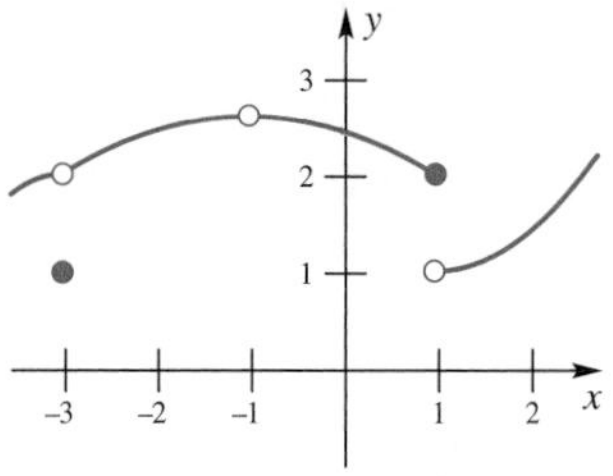

Figure 11

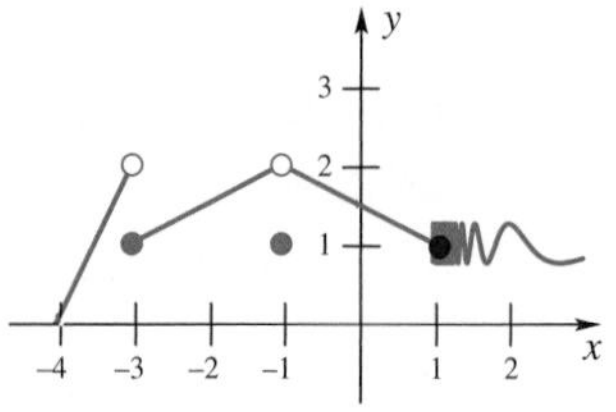

Figure 12

30. Follow the directions of Problem 29 for the function f graphed in Figure 12.

31. For the function f graphed in Figure 13, find the indicated limit or function value, or state that it does not exist.

(a) $f(-3)$ (b) $f(3)$ (c) $\lim_{x\to-3^-} f(x)$
(d) $\lim_{x\to-3^+} f(x)$ (e) $\lim_{x\to-3} f(x)$ (f) $\lim_{x\to3^+} f(x)$

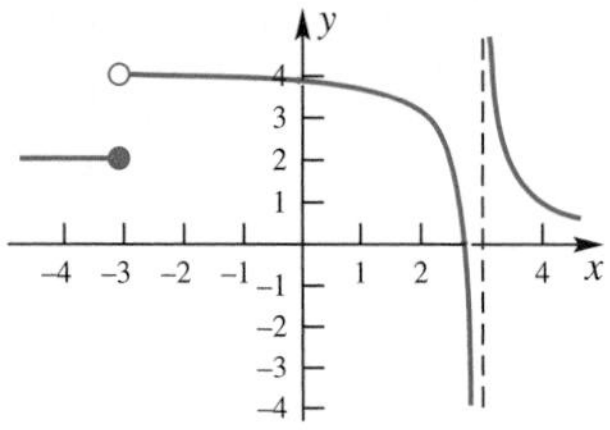

Figure 13

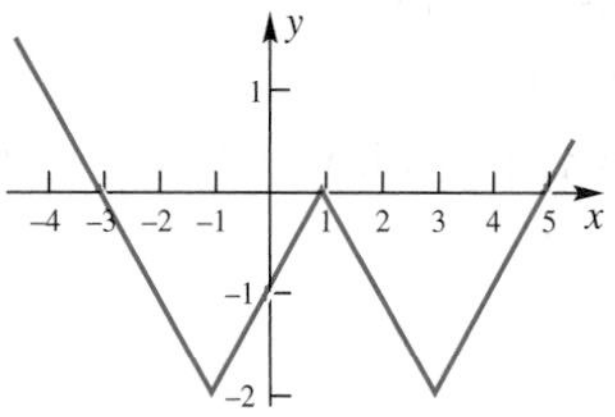

Figure 14

32. For the function f graphed in Figure 14, find the indicated limit or function value, or state that it does not exist.

(a) $\lim_{x\to-1^-} f(x)$ (b) $\lim_{x\to-1^+} f(x)$ (c) $\lim_{x\to-1} f(x)$
(d) $f(-1)$ (e) $\lim_{x\to1} f(x)$ (f) $f(1)$

33. Sketch the graph of

$$f(x) = \begin{cases} -x & \text{if } x < 0 \\ x & \text{if } 0 \le x < 1 \\ 1 + x & \text{if } x \ge 1 \end{cases}$$

Then find each of the following or state that it does not exist.

(a) $\lim_{x\to0} f(x)$ (b) $\lim_{x\to1} f(x)$
(c) $f(1)$ (d) $\lim_{x\to1^+} f(x)$

34. Sketch the graph of

$$g(x) = \begin{cases} -x + 1 & \text{if } x < 1 \\ x - 1 & \text{if } 1 < x < 2 \\ 5 - x^2 & \text{if } x \ge 2 \end{cases}$$

Then find each of the following or state that it does not exist.

(a) $\lim_{x\to1} g(x)$ (b) $g(1)$
(c) $\lim_{x\to2} g(x)$ (d) $\lim_{x\to2^+} g(x)$

35. Sketch the graph of $f(x) = x - [\![x]\!]$; then find each of the following or state that it does not exist.

(a) $f(0)$ (b) $\lim_{x\to0} f(x)$
(c) $\lim_{x\to0^-} f(x)$ (d) $\lim_{x\to1/2} f(x)$

36. Follow the directions of Problem 35 for $f(x) = x/|x|$.

37. Find $\lim_{x\to1}(x^2 - 1)/|x - 1|$ or state that it does not exist.

38. Evaluate $\lim_{x\to0}\left(\sqrt{x+2} - \sqrt{2}\right)/x$. *Hint:* Rationalize the numerator by multiplying the numerator and denominator by $\sqrt{x+2} + \sqrt{2}$.

39. Let

$$f(x) = \begin{cases} x & \text{if } x \text{ is rational} \\ -x & \text{if } x \text{ is irrational} \end{cases}$$

Find each value, if possible.

(a) $\lim_{x\to1} f(x)$ (b) $\lim_{x\to0} f(x)$

40. Sketch, as best you can, the graph of a function f that satisfies all the following conditions.

(a) Its domain is the interval $[0, 4]$.
(b) $f(0) = f(1) = f(2) = f(3) = f(4) = 1$
(c) $\lim_{x\to1} f(x) = 2$ (d) $\lim_{x\to2} f(x) = 1$
(e) $\lim_{x\to3^-} f(x) = 2$ (f) $\lim_{x\to3^+} f(x) = 1$

41. Let

$$f(x) = \begin{cases} x^2 & \text{if } x \text{ is rational} \\ x^4 & \text{if } x \text{ is irrational} \end{cases}$$

For what values of a does $\lim_{x\to a} f(x)$ exist?

42. The function $f(x) = x^2$ had been carefully graphed, but during the night a mysterious visitor changed the values of f at a million different places. Does this affect the value of $\lim_{x\to a} f(x)$ at any a? Explain.

43. Find each of the following limits or state that it does not exist.

(a) $\lim_{x\to1} \dfrac{|x-1|}{x-1}$ (b) $\lim_{x\to1^-} \dfrac{|x-1|}{x-1}$
(c) $\lim_{x\to1^-} \dfrac{x^2 - |x-1| - 1}{|x-1|}$ (d) $\lim_{x\to1^-}\left[\dfrac{1}{x-1} - \dfrac{1}{|x-1|}\right]$

44. Find each of the following limits or state that it does not exist.

(a) $\lim_{x\to1^+} \sqrt{x - [\![x]\!]}$ (b) $\lim_{x\to0^+} [\![1/x]\!]$
(c) $\lim_{x\to0^+} x(-1)^{[\![1/x]\!]}$ (d) $\lim_{x\to0^+} [\![x]\!](-1)^{[\![1/x]\!]}$

45. Find each of the following limits or state that it does not exist.

(a) $\lim_{x\to0^+} x[\![1/x]\!]$ (b) $\lim_{x\to0^+} x^2[\![1/x]\!]$
(c) $\lim_{x\to3^-} ([\![x]\!] + [\![-x]\!])$ (d) $\lim_{x\to3^+} ([\![x]\!] + [\![-x]\!])$

46. Find each of the following limits or state that it does not exist.

(a) $\lim_{x\to3} [\![x]\!]/x$ (b) $\lim_{x\to0^+} [\![x]\!]/x$
(c) $\lim_{x\to1.8} [\![x]\!]$ (d) $\lim_{x\to1.8} [\![x]\!]/x$

CAS *Many software packages have programs for calculating limits, although you should be warned that they are not infallible. To develop confidence in your program, use it to recalculate some of the limits in Problems 1–28. Then for each of the following, find the limit or state that it does not exist.*

47. $\lim_{x\to0} \sqrt{x}$ **48.** $\lim_{x\to0^+} x^x$

49. $\lim_{x\to0} \sqrt{|x|}$ **50.** $\lim_{x\to0} |x|^x$

51. $\lim_{x\to 0} (\sin 2x)/4x$

52. $\lim_{x\to 0} (\sin 5x)/3x$

53. $\lim_{x\to 0} \cos(1/x)$

54. $\lim_{x\to 0} x \cos(1/x)$

55. $\lim_{x\to 1} \dfrac{x^3 - 1}{\sqrt{2x + 2} - 2}$

56. $\lim_{x\to 0} \dfrac{x \sin 2x}{\sin(x^2)}$

57. $\lim_{x\to 2^-} \dfrac{x^2 - x - 2}{|x - 2|}$

58. $\lim_{x\to 1^+} \dfrac{2}{1 + 2^{1/(x-1)}}$

CAS **59.** Since calculus software packages find $\lim_{x\to a} f(x)$ by sampling a few values of $f(x)$ for x near a, they can be fooled. Find a function f for which $\lim_{x\to 0} f(x)$ fails to exist but for which your software gives a value for the limit.

Answers to Concepts Review: **1.** $L; c$ **2.** 6 **3.** L; right **4.** $\lim_{x\to c} f(x) = M$

1.2 Rigorous Study of Limits

We gave an informal definition of *limit* in the previous section. Here is a slightly better, but still informal, rewording of that definition. To say that $\lim_{x\to c} f(x) = L$ means that $f(x)$ can be made to be as close as we like to L provided x is close enough, but not equal to c. The first example illustrates this point.

EXAMPLE 1 Use a plot of $y = f(x) = 3x^2$ to determine how close x must be to 2 to guarantee that $f(x)$ is within 0.05 of 12.

SOLUTION In order for $f(x)$ to be within 0.05 of 12, we must have $11.95 < f(x) < 12.05$. The lines $y = 11.95$ and $y = 12.05$ have been drawn in Figure 1. If we solve $y = 3x^2$ for x we get $x = \sqrt{y/3}$. Thus $f\left(\sqrt{11.95/3}\right) = 11.95$ and $f\left(\sqrt{12.05/3}\right) = 12.05$. Figure 1 indicates that if $\sqrt{11.95/3} < x < \sqrt{12.05/3}$ then $f(x)$ satisfies $11.95 < f(x) < 12.05$. This interval for x is approximately $1.99583 < x < 2.00416$. Of the two endpoints of this interval, the upper one, 2.00416, is closer to 2 and it is within 0.00416 of 2. Thus, if x is within 0.00416 of 2 then $f(x)$ is within 0.05 of 12. ■

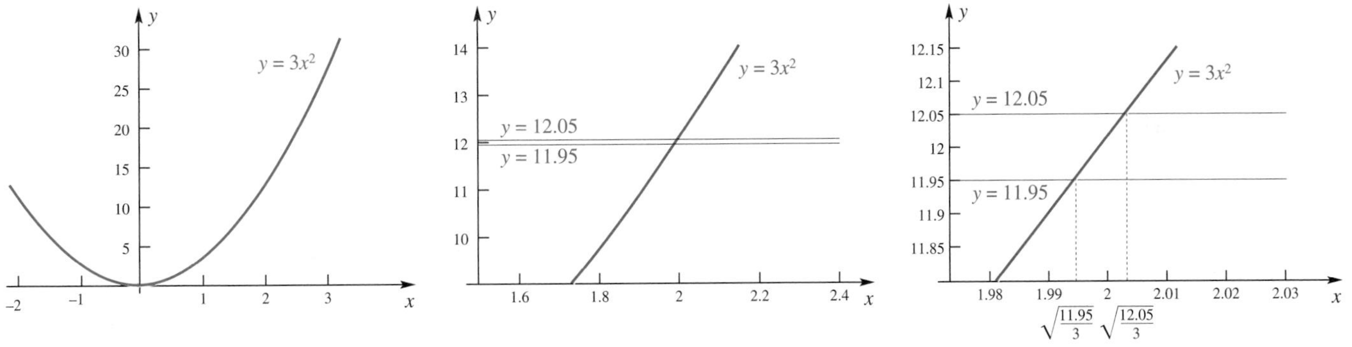

Figure 1

Absolute Value as Distance

Think of two points a and b on a number line. What is the distance between them? If $a < b$, then $b - a$ is the distance, but if $b < a$ then $a - b$ is the distance. We can combine these statements into one by saying that the distance is $|b - a|$. This geometric interpretation of the absolute value of a difference as the distance between two points on a number line is important in understanding our definition of the limit.

If we now asked how close x would have to be to 2 to guarantee that $f(x)$ is within 0.01 of 12, the solution would proceed along the same lines, and we would find that x would have to be in a smaller interval than we obtained above. If we wanted $f(x)$ to be within 0.001 of 12, we would require an interval that is narrower still. In this example, it seems plausible that no matter how close we want $f(x)$ to be to 12, we can accomplish this by taking x sufficiently close to 2.

We now make the definition of the limit precise.

Making the Definition Precise We follow the tradition in using the Greek letters ε (epsilon) and δ (delta) to stand for (usually small) arbitrary positive numbers.

To say that $f(x)$ is within ε of L means that $L - \varepsilon < f(x) < L + \varepsilon$, or equivalently, $|f(x) - L| < \varepsilon$. This means that $f(x)$ lies in the open interval $(L - \varepsilon, L + \varepsilon)$ shown on the graph in Figure 2.

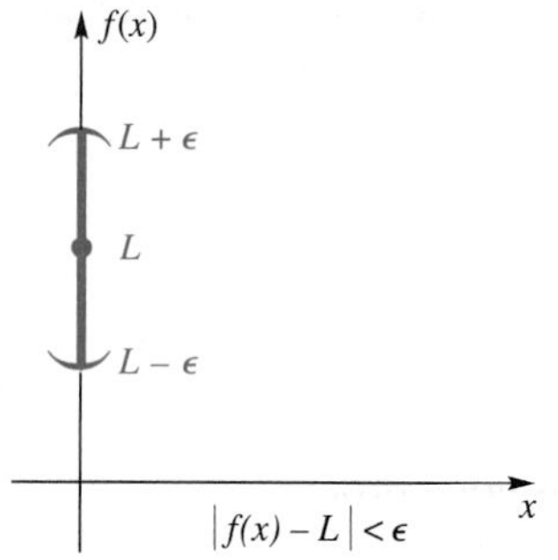

Figure 2

Next, to say that x is sufficiently close to but different from c is to say that, for some δ, x is in the open interval $(c - \delta, c + \delta)$ with c deleted. Perhaps the best way to say this is to write

$$0 < |x - c| < \delta$$

Note that $|x - c| < \delta$ would describe the interval $c - \delta < x < c + \delta$, while $0 < |x - c|$ requires that $x = c$ be excluded. The interval that we are describing is shown in Figure 3.

We are now ready for what some have called the most important definition in calculus.

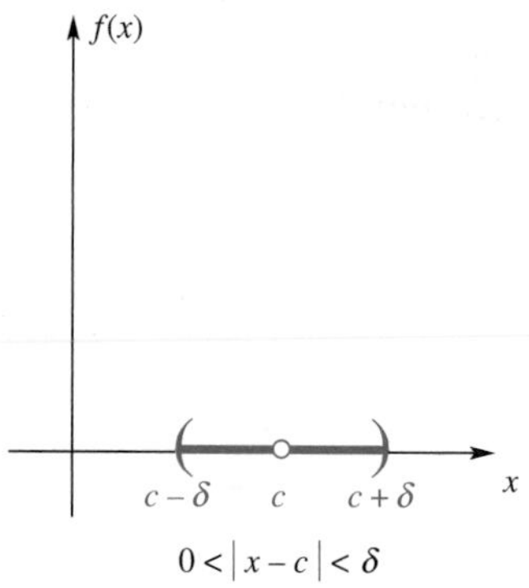

Figure 3

Definition **Precise Meaning of Limit**

To say that $\lim_{x \to c} f(x) = L$ means that for each given $\varepsilon > 0$ (no matter how small) there is a corresponding $\delta > 0$ such that $|f(x) - L| < \varepsilon$, provided that $0 < |x - c| < \delta$; that is,

$$0 < |x - c| < \delta \Rightarrow |f(x) - L| < \varepsilon$$

The pictures in Figure 4 may help you absorb this definition.

We must emphasize that the real number ε must be given *first*; the number δ is to be produced, and it will usually depend on ε. Suppose that David wishes to prove to Emily that $\lim_{x \to c} f(x) = L$. Emily can challenge David with any particular

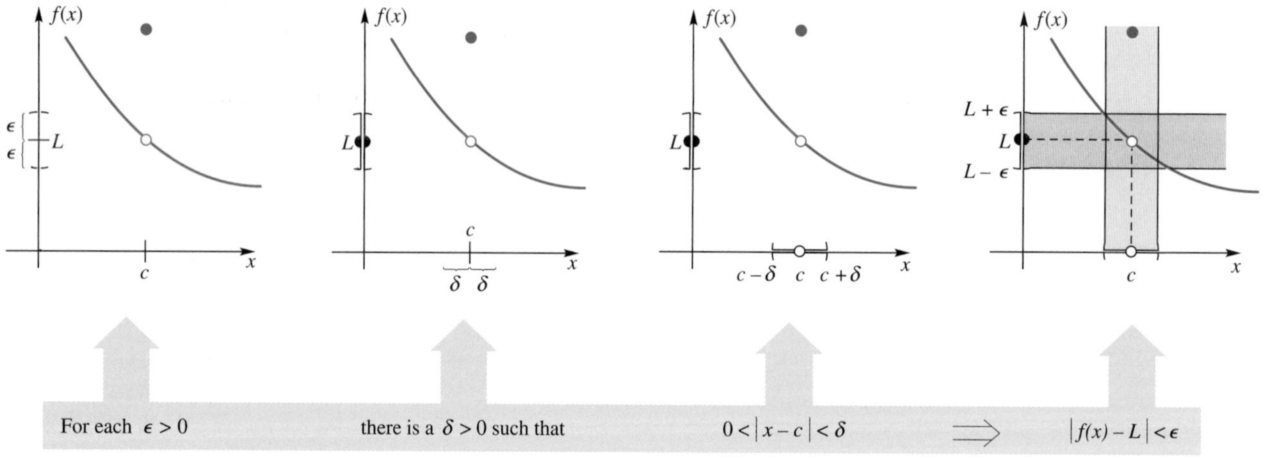

Figure 4

ε she chooses (e.g., $\varepsilon = 0.01$) and demand that David produce a corresponding δ. Let's apply David's reasoning to the limit $\lim_{x \to 3}(2x + 1)$. By inspection, David would conjecture that the limit is 7. Now, can David find a δ such that $|(2x + 1) - 7| < 0.01$ whenever $0 < |x - 3| < \delta$? A little algebra shows that

$$|(2x + 1) - 7| < 0.01 \Leftrightarrow 2|x - 3| < 0.01$$

$$\Leftrightarrow |x - 3| < \frac{0.01}{2}$$

Thus, the answer to the question is yes! David can choose $\delta = 0.01/2$ (or any smaller value) and this will guarantee that $|(2x + 1) - 7| < 0.01$ whenever $0 < |x - 3| < 0.01/2$. In other words, David can make $2x + 1$ within 0.01 of 7, provided that x is within 0.01/2 of 3.

Now suppose that Emily challenges David again, but this time she wants $|(2x + 1) - 7| < 0.000002$. Can David find a δ for this value of ε? Following the reasoning used above,

$$|(2x + 1) - 7| < 0.000002 \Leftrightarrow 2|x - 3| < 0.000002$$
$$\Leftrightarrow |x - 3| < \frac{0.000002}{2}$$

Thus, $|(2x + 1) - 7| < 0.000002$ whenever $|x - 3| < 0.000002/2$.

This kind of reasoning, while it may convince some, is not a proof that the limit is 7. The definition says that we must be able to find a δ for *every* $\varepsilon > 0$ (not for *some* $\varepsilon > 0$). Emily could challenge David repeatedly, but they would never *prove* that the limit is 7. David must be able to produce a δ for *every* positive ε (no matter how small).

David opts to take things into his own hands and proposes to let ε be any positive real number. He follows the same reasoning as above, but this time he uses ε instead of 0.000002.

$$|(2x + 1) - 7| < \varepsilon \Leftrightarrow 2|x - 3| < \varepsilon$$
$$\Leftrightarrow |x - 3| < \frac{\varepsilon}{2}$$

David can choose $\delta = \varepsilon/2$, and it follows that $|(2x + 1) - 7| < \varepsilon$ whenever $|x - 3| < \varepsilon/2$. In other words, he can make $2x + 1$ within ε of 7 provided x is within $\varepsilon/2$ of 3. Now David has met the requirements of the definition of the limit and has therefore verified that the limit is 7, as suspected.

> **Two Different Limits?**
>
> A natural question to ask is "Can a function have two different limits at c?" The obvious intuitive answer is no. If a function is getting closer and closer to L as $x \to c$, it cannot also be getting closer and closer to a different number M. You are asked to show this rigorously in Problem 23.

Some Limit Proofs In each of the following examples, we begin with what we call a preliminary analysis. We include it so that our choice of δ in each proof does not seem to suggest incredible insight on our part. It shows the kind of work you need to do on scratch paper in order to construct the proof. Once you feel that you grasp an example, take another look at it, but cover up the preliminary analysis and note how elegant, but mysterious, the proof seems to be.

EXAMPLE 2 Prove that $\lim_{x \to 4} (3x - 7) = 5$.

PRELIMINARY ANALYSIS Let ε be any positive number. We must produce a $\delta > 0$ such that

$$0 < |x - 4| < \delta \Rightarrow |(3x - 7) - 5| < \varepsilon$$

Consider the inequality on the right.

$$|(3x - 7) - 5| < \varepsilon \Leftrightarrow |3x - 12| < \varepsilon$$
$$\Leftrightarrow |3(x - 4)| < \varepsilon$$
$$\Leftrightarrow |3||(x - 4)| < \varepsilon$$
$$\Leftrightarrow |x - 4| < \frac{\varepsilon}{3}$$

Now we see how to choose δ; that is, $\delta = \varepsilon/3$. Of course, any smaller δ would work.

FORMAL PROOF Let $\varepsilon > 0$ be given. Choose $\delta = \varepsilon/3$. Then $0 < |x - 4| < \delta$ implies that

$$|(3x - 7) - 5| = |3x - 12| = |3(x - 4)| = 3|x - 4| < 3\delta = \varepsilon$$

If you read this chain of equalities and an inequality from left to right and use the transitive properties of $=$ and $<$, you see that

$$|(3x - 7) - 5| < \varepsilon$$

Now, David knows a rule for choosing the value of δ given Emily's challenge. If Emily were to challenge David with $\varepsilon = 0.01$, then David would respond with $\delta = 0.01/3$. If Emily said $\varepsilon = 0.000003$, then David would say $\delta = 0.000001$. If he gave a smaller value for δ, that would be fine, too.

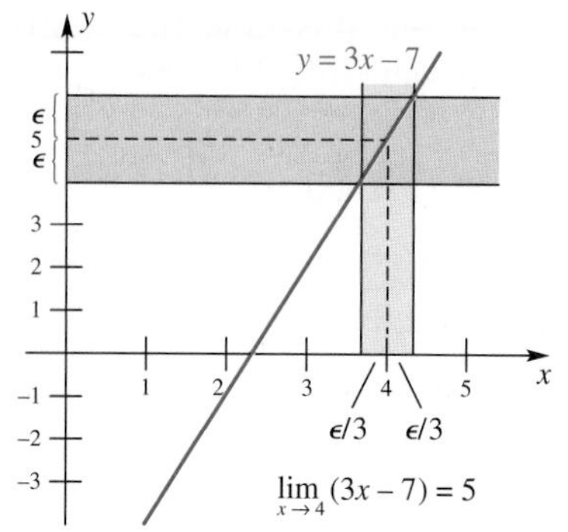

Figure 5

Of course, if you think about the graph of $y = 3x - 7$ (a line with slope 3, as in Figure 5), you know that to force $3x - 7$ to be close to 5 you had better make x even closer (closer by a factor of one-third) to 4. ■

Now look at Figure 6 and convince yourself that $\delta = 2\varepsilon$ would be an appropriate choice for δ in showing that $\lim_{x\to 4}\left(\frac{1}{2}x + 3\right) = 5$.

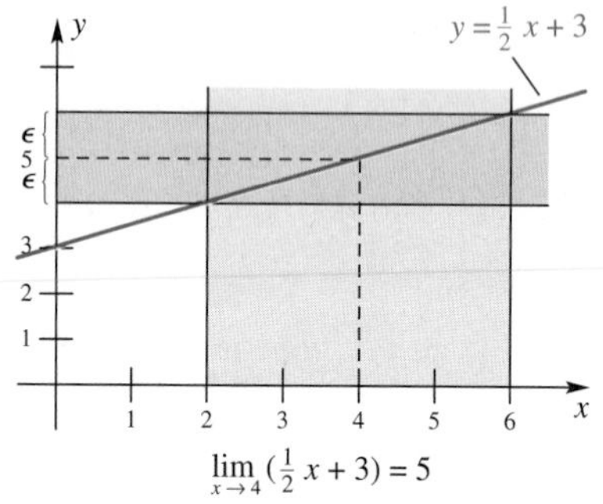

Figure 6

EXAMPLE 3 Prove that $\lim_{x\to 2}\dfrac{2x^2 - 3x - 2}{x - 2} = 5$.

PRELIMINARY ANALYSIS We are looking for a δ such that

$$0 < |x - 2| < \delta \Rightarrow \left|\frac{2x^2 - 3x - 2}{x - 2} - 5\right| < \varepsilon$$

Now, for $x \neq 2$,

$$\begin{aligned}\left|\frac{2x^2 - 3x - 2}{x - 2} - 5\right| < \varepsilon &\Leftrightarrow \left|\frac{(2x + 1)(x - 2)}{x - 2} - 5\right| < \varepsilon\\ &\Leftrightarrow |(2x + 1) - 5| < \varepsilon\\ &\Leftrightarrow |2(x - 2)| < \varepsilon\\ &\Leftrightarrow |2||x - 2| < \varepsilon\\ &\Leftrightarrow |x - 2| < \frac{\varepsilon}{2}\end{aligned}$$

This indicates that $\delta = \varepsilon/2$ will work (see Figure 7).

FORMAL PROOF Let $\varepsilon > 0$ be given. Choose $\delta = \varepsilon/2$. Then $0 < |x - 2| < \delta$ implies that

$$\begin{aligned}\left|\frac{2x^2 - 3x - 2}{x - 2} - 5\right| &= \left|\frac{(2x + 1)(x - 2)}{x - 2} - 5\right| = |2x + 1 - 5|\\ &= |2(x - 2)| = 2|x - 2| < 2\delta = \varepsilon\end{aligned}$$

The cancellation of the factor $x - 2$ is legitimate because $0 < |x - 2|$ implies that $x \neq 2$, and $\dfrac{x - 2}{x - 2} = 1$ as long as $x \neq 2$. ■

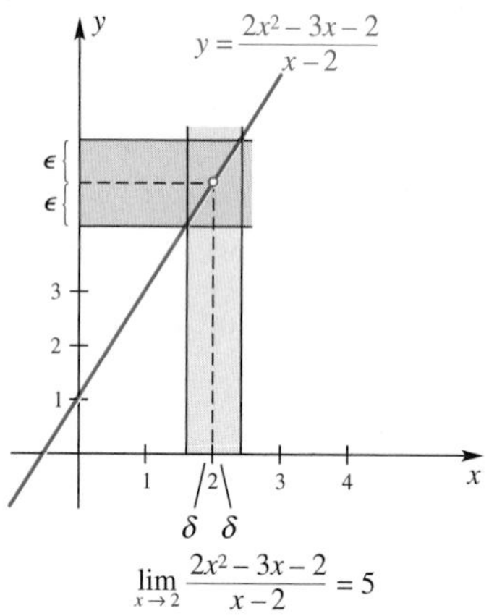

Figure 7

EXAMPLE 4 Prove that $\lim_{x\to c}(mx + b) = mc + b$.

PRELIMINARY ANALYSIS We want to find δ such that

$$0 < |x - c| < \delta \Rightarrow |(mx + b) - (mc + b)| < \varepsilon$$

Now

$$|(mx + b) - (mc + b)| = |mx - mc| = |m(x - c)| = |m||x - c|$$

It appears that $\delta = \varepsilon/|m|$ should do as long as $m \neq 0$. (Note that m could be positive or negative, so we need to keep the absolute value bars. Recall from Chapter 0 that $|ab| = |a||b|$.)

FORMAL PROOF Let $\varepsilon > 0$ be given. Choose $\delta = \varepsilon/|m|$. Then $0 < |x - c| < \delta$ implies that

$$|(mx + b) - (mc + b)| = |mx - mc| = |m||x - c| < |m|\delta = \varepsilon$$

And in case $m = 0$, any δ will do just fine since

$$|(0x + b) - (0c + b)| = |0| = 0$$

The latter is less than ε for all x. ■

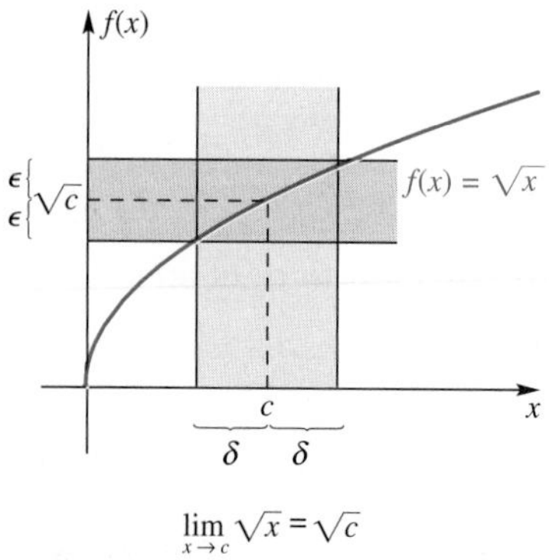

Figure 8

EXAMPLE 5 Prove that if $c > 0$ then $\lim_{x\to c}\sqrt{x} = \sqrt{c}$.

PRELIMINARY ANALYSIS Refer to Figure 8. We must find δ such that

$$0 < |x - c| < \delta \Rightarrow |\sqrt{x} - \sqrt{c}| < \varepsilon$$

Now

$$|\sqrt{x} - \sqrt{c}| = \left|\frac{(\sqrt{x} - \sqrt{c})(\sqrt{x} + \sqrt{c})}{\sqrt{x} + \sqrt{c}}\right| = \left|\frac{x - c}{\sqrt{x} + \sqrt{c}}\right|$$
$$= \frac{|x - c|}{\sqrt{x} + \sqrt{c}} \le \frac{|x - c|}{\sqrt{c}}$$

To make the latter less than ε requires that we have $|x - c| < \varepsilon\sqrt{c}$.

FORMAL PROOF Let $\varepsilon > 0$ be given. Choose $\delta = \varepsilon\sqrt{c}$. Then $0 < |x - c| < \delta$ implies that

$$|\sqrt{x} - \sqrt{c}| = \left|\frac{(\sqrt{x} - \sqrt{c})(\sqrt{x} + \sqrt{c})}{\sqrt{x} + \sqrt{c}}\right| = \left|\frac{x - c}{\sqrt{x} + \sqrt{c}}\right|$$
$$= \frac{|x - c|}{\sqrt{x} + \sqrt{c}} \le \frac{|x - c|}{\sqrt{c}} < \frac{\delta}{\sqrt{c}} = \varepsilon$$

There is one technical point here. We began with $c > 0$, but it could happen that c sits very close to 0 on the x-axis. We should insist that $\delta \le c$, for then $|x - c| < \delta$ implies that $x > 0$ so that $\sqrt{x}$ is defined. Thus, for absolute rigor, choose δ to be the smaller of c and $\varepsilon\sqrt{c}$. ■

Our demonstration in Example 5 depended on *rationalizing the numerator*, a trick frequently useful in calculus.

EXAMPLE 6 Prove that $\lim_{x\to 3}(x^2 + x - 5) = 7$.

PRELIMINARY ANALYSIS Our task is to find δ such that

$$0 < |x - 3| < \delta \Rightarrow |(x^2 + x - 5) - 7| < \varepsilon$$

Now

$$|(x^2 + x - 5) - 7| = |x^2 + x - 12| = |x + 4||x - 3|$$

The factor $|x - 3|$ can be made as small as we wish, and we know that $|x + 4|$ will be about 7. We therefore seek an upper bound for $|x + 4|$. To do this, we first agree to make $\delta \le 1$. Then $|x - 3| < \delta$ implies that

$$\begin{aligned} |x + 4| &= |x - 3 + 7| \\ &\le |x - 3| + |7| \qquad \text{(Triangle Inequality)} \\ &< 1 + 7 = 8 \end{aligned}$$

(Figure 9 offers an alternative demonstration of this fact.) If we also require that $\delta \le \varepsilon/8$, then the product $|x + 4||x - 3|$ will be less than ε.

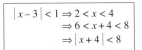

$$\begin{aligned} |x-3| < 1 &\Rightarrow 2 < x < 4 \\ &\Rightarrow 6 < x+4 < 8 \\ &\Rightarrow |x+4| < 8 \end{aligned}$$

Figure 9

FORMAL PROOF Let $\varepsilon > 0$ be given. Choose $\delta = \min\{1, \varepsilon/8\}$; that is, choose δ to be the smaller of 1 and $\varepsilon/8$. Then $0 < |x - 3| < \delta$ implies that

$$|(x^2 + x - 5) - 7| = |x^2 + x - 12| = |x + 4||x - 3| < 8\cdot\frac{\varepsilon}{8} = \varepsilon \qquad ■$$

EXAMPLE 7 Prove that $\lim_{x\to c} x^2 = c^2$.

PROOF We mimic the proof in Example 6. Let $\varepsilon > 0$ be given. Choose $\delta = \min\{1, \varepsilon/(1 + 2|c|)\}$. Then $0 < |x - c| < \delta$ implies that

$$\begin{aligned} |x^2 - c^2| &= |x + c||x - c| = |x - c + 2c||x - c| \\ &\le (|x - c| + 2|c|)|x - c| \qquad \text{(Triangle Inequality)} \\ &< (1 + 2|c|)|x - c| < \frac{(1 + 2|c|)\cdot\varepsilon}{1 + 2|c|} = \varepsilon \end{aligned} \qquad ■$$

Although appearing incredibly insightful, we did not pull δ "out of the air" in Example 7. We simply did not show you the preliminary analysis this time.

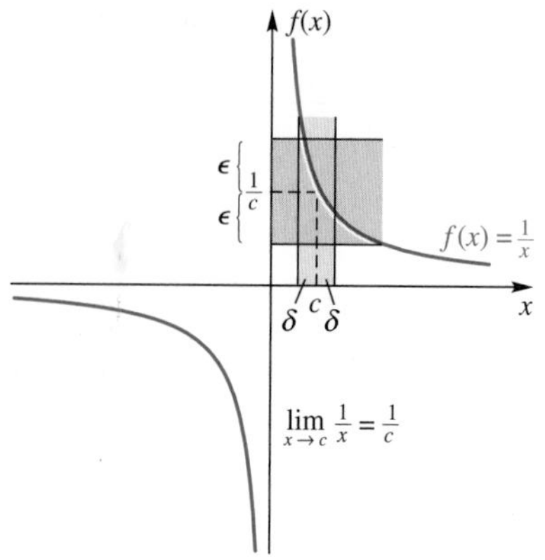

Figure 10

EXAMPLE 8 Prove that $\lim_{x\to c} \frac{1}{x} = \frac{1}{c}$, $c \neq 0$.

PRELIMINARY ANALYSIS Study Figure 10. We must find δ such that

$$0 < |x - c| < \delta \Rightarrow \left|\frac{1}{x} - \frac{1}{c}\right| < \varepsilon$$

Now

$$\left|\frac{1}{x} - \frac{1}{c}\right| = \left|\frac{c - x}{xc}\right| = \frac{1}{|x|}\cdot\frac{1}{|c|}\cdot|x - c|$$

The factor $1/|x|$ is troublesome, especially if x is near 0. We can bound this factor if we can keep x away from 0. To that end, note that

$$|c| = |c - x + x| \leq |c - x| + |x|$$

so

$$|x| \geq |c| - |x - c|$$

Thus, if we choose $\delta \leq |c|/2$, we succeed in making $|x| \geq |c|/2$. Finally, if we also require $\delta \leq \varepsilon c^2/2$, then

$$\frac{1}{|x|}\cdot\frac{1}{|c|}\cdot|x - c| < \frac{1}{|c|/2}\cdot\frac{1}{|c|}\cdot\frac{\varepsilon c^2}{2} = \varepsilon$$

FORMAL PROOF Let $\varepsilon > 0$ be given. Choose $\delta = \min\{|c|/2, \varepsilon c^2/2\}$. Then $0 < |x - c| < \delta$ implies

$$\left|\frac{1}{x} - \frac{1}{c}\right| = \left|\frac{c - x}{xc}\right| = \frac{1}{|x|}\cdot\frac{1}{|c|}\cdot|x - c| < \frac{1}{|c|/2}\cdot\frac{1}{|c|}\cdot\frac{\varepsilon c^2}{2} = \varepsilon \quad \blacksquare$$

One-Sided Limits It does not take much imagination to give the ε–δ definitions of right- and left-hand limits.

Definition Right-Hand Limit

To say $\lim_{x\to c^+} f(x) = L$ means that for each $\varepsilon > 0$ there is a corresponding $\delta > 0$ such that

$$0 < x - c < \delta \Rightarrow |f(x) - L| < \varepsilon$$

We leave the ε–δ definition for the left-hand limit to the reader. (See Problem 5.)

The ε–δ concept presented in this section is probably the most intricate and elusive topic in a calculus course. It may take you some time to grasp this concept, but it is worth the effort. Calculus is the study of limits, so a clear understanding of the concept of limit is a worthy goal.

The discovery of calculus is usually attributed to Isaac Newton (1642–1727) and Gottfried Wilhelm von Leibniz (1646–1716), who worked independently in the late 1600s. Although Newton and Leibniz, along with their successors, discovered a number of properties of calculus, and calculus was found to have many applications in the physical sciences, it was not until the nineteenth century that a precise definition of a limit was proposed. Augustin Louis Cauchy (1789–1857), a French engineer and mathematician, gave this definition: "If the successive values attributed to the same variable approach indefinitely a fixed value, such that they finally differ from it by as little as one wishes, this latter is called the limit of all the others." Even Cauchy, a master at rigor, was somewhat vague in his definition of a limit. What are "successive values," and what does it mean to "finally differ"? The phrase "finally differ from it by as little as one wishes" contains the seed of the ε–δ

definition, because for the first time it indicates that the difference between $f(x)$ and its limit L can be made smaller than any given number, the number we labeled ε. The German mathematician Karl Weierstrass (1815–1897) first put together the definition that is equivalent to our ε–δ definition of a limit.

Concepts Review

1. The inequality $|f(x) - L| < \varepsilon$ is equivalent to _____ $< f(x) <$ _____.

2. The precise meaning of $\lim_{x \to a} f(x) = L$ is this: Given any positive number ε, there is a corresponding positive number δ such that _____ implies _____.

3. To be sure that $|3x - 3| < \varepsilon$, we would require that $|x - 1| <$ _____.

4. $\lim_{x \to a}(mx + b) =$ _____.

Problem Set 1.2

In Problems 1–6, give the appropriate ε–δ definition of each statement.

1. $\lim_{t \to a} f(t) = M$
2. $\lim_{u \to b} g(u) = L$
3. $\lim_{z \to d} h(z) = P$
4. $\lim_{y \to e} \phi(y) = B$
5. $\lim_{x \to c^-} f(x) = L$
6. $\lim_{t \to a^+} g(t) = D$

In Problems 7–10, plot the function $f(x)$ over the interval $[1.5, 2.5]$. Zoom in on the graph of each function to determine how close x must be to 2 in order that $f(x)$ is within 0.002 of 4. Your answer should be of the form "If x is within _____ of 2, then $f(x)$ is within 0.002 of 4."

7. $f(x) = 2x$
8. $f(x) = x^2$
9. $f(x) = \sqrt{8x}$
10. $f(x) = \dfrac{8}{x}$

In Problems 11–22, give an ε–δ proof of each limit fact.

11. $\lim_{x \to 0}(2x - 1) = -1$
12. $\lim_{x \to -21}(3x - 1) = -64$
13. $\lim_{x \to 5} \dfrac{x^2 - 25}{x - 5} = 10$
14. $\lim_{x \to 0} \left(\dfrac{2x^2 - x}{x}\right) = -1$
15. $\lim_{x \to 5} \dfrac{2x^2 - 11x + 5}{x - 5} = 9$
16. $\lim_{x \to 1} \sqrt{2x} = \sqrt{2}$
17. $\lim_{x \to 4} \dfrac{\sqrt{2x - 1}}{\sqrt{x - 3}} = \sqrt{7}$
18. $\lim_{x \to 1} \dfrac{14x^2 - 20x + 6}{x - 1} = 8$
19. $\lim_{x \to 1} \dfrac{10x^3 - 26x^2 + 22x - 6}{(x - 1)^2} = 4$
20. $\lim_{x \to 1}(2x^2 + 1) = 3$
21. $\lim_{x \to -1}(x^2 - 2x - 1) = 2$
22. $\lim_{x \to 0} x^4 = 0$

23. Prove that if $\lim_{x \to c} f(x) = L$ and $\lim_{x \to c} f(x) = M$, then $L = M$.

24. Let F and G be functions such that $0 \le F(x) \le G(x)$ for all x near c, except possibly at c. Prove that if $\lim_{x \to c} G(x) = 0$, then $\lim_{x \to c} F(x) = 0$.

25. Prove that $\lim_{x \to 0} x^4 \sin^2(1/x) = 0$. *Hint*: Use Problems 22 and 24.

26. Prove that $\lim_{x \to 0^+} \sqrt{x} = 0$.

27. By considering left- and right-hand limits, prove that $\lim_{x \to 0} |x| = 0$.

28. Prove that if $|f(x)| < B$ for $|x - a| < 1$ and $\lim_{x \to a} g(x) = 0$, then $\lim_{x \to a} f(x)g(x) = 0$.

29. Suppose that $\lim_{x \to a} f(x) = L$ and that $f(a)$ exists (though it may be different from L). Prove that f is bounded on some interval containing a; that is, show that there is an interval (c, d) with $c < a < d$ and a constant M such that $|f(x)| \le M$ for all x in (c, d).

30. Prove that if $f(x) \le g(x)$ for all x in some deleted interval about a and if $\lim_{x \to a} f(x) = L$ and $\lim_{x \to a} g(x) = M$, then $L \le M$.

31. Which of the following are equivalent to the definition of limit?

(a) For some $\varepsilon > 0$ and every $\delta > 0$, $0 < |x - c| < \delta \Rightarrow |f(x) - L| < \varepsilon$.

(b) For every $\delta > 0$, there is a corresponding $\varepsilon > 0$ such that
$$0 < |x - c| < \varepsilon \Rightarrow |f(x) - L| < \delta$$

(c) For every positive integer N, there is a corresponding positive integer M such that $0 < |x - c| < 1/M \Rightarrow |f(x) - L| < 1/N$.

(d) For every $\varepsilon > 0$, there is a corresponding $\delta > 0$ such that $0 < |x - c| < \delta$ and $|f(x) - L| < \varepsilon$ for some x.

32. State in ε–δ language what it means to say $\lim_{x \to c} f(x) \ne L$.

GC 33. Suppose we wish to give an ε–δ proof that
$$\lim_{x \to 3} \frac{x + 6}{x^4 - 4x^3 + x^2 + x + 6} = -1$$
We begin by writing $\dfrac{x + 6}{x^4 - 4x^3 + x^2 + x + 6} + 1$ in the form $(x - 3)g(x)$.

(a) Determine $g(x)$.

(b) Could we choose $\delta = \min(1, \varepsilon/n)$ for some n? Explain.

(c) If we choose $\delta = \min\left(\frac{1}{4}, \varepsilon/m\right)$, what is the smallest integer m that we could use?

Answers to Concepts Review 1. $L - \varepsilon$; $L + \varepsilon$ 2. $0 < |x - a| < \delta$; $|f(x) - L| < \varepsilon$ 3. $\varepsilon/3$ 4. $ma + b$

1.3 Limit Theorems

Most readers will agree that proving the existence and values of limits using the ε–δ definition of the preceding section is both time consuming and difficult. That is why the theorems of this section are so welcome. Our first theorem is the big one. With it, we can handle most limit problems that we will face for quite some time.

One-Sided Limits

Although stated in terms of two-sided limits, Theorem A remains true for both left- and right-hand limits.

Theorem A Main Limit Theorem

Let n be a positive integer, k be a constant, and f and g be functions that have limits at c. Then

1. $\lim_{x\to c} k = k$;
2. $\lim_{x\to c} x = c$;
3. $\lim_{x\to c} kf(x) = k \lim_{x\to c} f(x)$;
4. $\lim_{x\to c} [f(x) + g(x)] = \lim_{x\to c} f(x) + \lim_{x\to c} g(x)$;
5. $\lim_{x\to c} [f(x) - g(x)] = \lim_{x\to c} f(x) - \lim_{x\to c} g(x)$;
6. $\lim_{x\to c} [f(x)\cdot g(x)] = \lim_{x\to c} f(x)\cdot \lim_{x\to c} g(x)$;
7. $\lim_{x\to c} \dfrac{f(x)}{g(x)} = \dfrac{\lim_{x\to c} f(x)}{\lim_{x\to c} g(x)}$, provided $\lim_{x\to c} g(x) \neq 0$;
8. $\lim_{x\to c} [f(x)]^n = \left[\lim_{x\to c} f(x)\right]^n$;
9. $\lim_{x\to c} \sqrt[n]{f(x)} = \sqrt[n]{\lim_{x\to c} f(x)}$, provided $\lim_{x\to c} f(x) > 0$ when n is even.

These important results are remembered best if learned in words. For example, Statement 4 translates as *The limit of a sum is the sum of the limits.*

Of course, Theorem A needs to be proved. We postpone that job till the end of the section, choosing first to show how this multipart theorem is used.

Applications of the Main Limit Theorem In the next examples, the circled numbers refer to the numbered statements from Theorem A. Each equality is justified by the indicated statement.

EXAMPLE 1 Find $\lim_{x\to 3} 2x^4$.

$$\lim_{x\to 3} 2x^4 \overset{③}{=} 2 \lim_{x\to 3} x^4 \overset{⑧}{=} 2\left[\lim_{x\to 3} x\right]^4 \overset{②}{=} 2[3]^4 = 162$$

■

EXAMPLE 2 Find $\lim_{x\to 4}(3x^2 - 2x)$.

SOLUTION

$$\lim_{x\to 4}(3x^2 - 2x) \overset{⑤}{=} \lim_{x\to 4} 3x^2 - \lim_{x\to 4} 2x \overset{③}{=} 3\lim_{x\to 4} x^2 - 2\lim_{x\to 4} x$$

$$\overset{⑧}{=} 3\left(\lim_{x\to 4} x\right)^2 - 2\lim_{x\to 4} x \overset{②}{=} 3(4)^2 - 2(4) = 40$$

■

EXAMPLE 3 Find $\lim_{x \to 4} \frac{\sqrt{x^2 + 9}}{x}$.

SOLUTION

$$\lim_{x \to 4} \frac{\sqrt{x^2 + 9}}{x} \overset{(7)}{=} \frac{\lim_{x \to 4} \sqrt{x^2 + 9}}{\lim_{x \to 4} x} \overset{(9,2)}{=} \frac{\sqrt{\lim_{x \to 4} (x^2 + 9)}}{4} \overset{(4)}{=} \frac{1}{4}\sqrt{\lim_{x \to 4} x^2 + \lim_{x \to 4} 9}$$

$$\overset{(8,1)}{=} \frac{1}{4}\sqrt{\left[\lim_{x \to 4} x\right]^2 + 9} \overset{(2)}{=} \frac{1}{4}\sqrt{4^2 + 9} = \frac{5}{4}$$

■

EXAMPLE 4 If $\lim_{x \to 3} f(x) = 4$ and $\lim_{x \to 3} g(x) = 8$, find

$$\lim_{x \to 3}\left[f^2(x) \cdot \sqrt[3]{g(x)}\right]$$

SOLUTION

$$\lim_{x \to 3} [f^2(x) \cdot \sqrt[3]{g(x)}] \overset{(6)}{=} \lim_{x \to 3} f^2(x) \cdot \lim_{x \to 3} \sqrt[3]{g(x)}$$

$$\overset{(8,9)}{=} \left[\lim_{x \to 3} f(x)\right]^2 \cdot \sqrt[3]{\lim_{x \to 3} g(x)}$$

$$= [4]^2 \cdot \sqrt[3]{8} = 32$$

■

Recall that a polynomial function f has the form

$$f(x) = a_n x^n + a_{n-1}x^{n-1} + \cdots + a_1 x + a_0$$

whereas a rational function f is the quotient of two polynomial functions, that is,

$$f(x) = \frac{a_n x^n + a_{n-1}x^{n-1} + \cdots + a_1 x + a_0}{b_m x^m + b_{m-1}x^{m-1} + \cdots + b_1 x + b_0}$$

Theorem B Substitution Theorem

If f is a polynomial function or a rational function, then

$$\lim_{x \to c} f(x) = f(c)$$

provided $f(c)$ is defined. In the case of a rational function, this means that the value of the denominator at c is not zero.

Evaluating a Limit "by Substitution"

When we apply Theorem B, the Substitution Theorem, we say we evaluate the limit *by substitution*. Not all limits can be evaluated by substitution; consider $\lim_{x \to 1} \frac{x^2 - 1}{x - 1}$. The Substitution Theorem does not apply here because the denominator is 0 when $x = 1$, but the limit does exist.

The proof of Theorem B follows from repeated applications of Theorem A. Note that Theorem B allows us to find limits for polynomial and rational functions by simply substituting c for x throughout, provided the denominator of the rational function is not zero at c.

EXAMPLE 5 Find $\lim_{x \to 2} \frac{7x^5 - 10x^4 - 13x + 6}{3x^2 - 6x - 8}$.

SOLUTION

$$\lim_{x\to 2}\frac{7x^5 - 10x^4 - 13x + 6}{3x^2 - 6x - 8} = \frac{7(2)^5 - 10(2)^4 - 13(2) + 6}{3(2)^2 - 6(2) - 8} = -\frac{11}{2}$$ ■

EXAMPLE 6 Find $\lim_{x\to 1}\frac{x^3 + 3x + 7}{x^2 - 2x + 1} = \lim_{x\to 1}\frac{x^3 + 3x + 7}{(x - 1)^2}$.

SOLUTION Neither Theorem B nor Statement 7 of Theorem A applies, since the limit of the denominator is 0. However, since the limit of the numerator is 11, we see that as x nears 1 we are dividing a number near 11 by a positive number near 0. The result is a large positive number. In fact, the resulting number can be made as large as you like by letting x get close enough to 1. We say that the limit does not exist. (Later in this chapter (see Section 1.5) we will allow ourselves to say that the limit is $+\infty$.) ■

In many cases, Theorem B cannot be applied because substitution of c causes the denominator to be 0. In cases like this, it sometimes happens that the function can be simplified, for example by factoring. For example, we can write

$$\frac{x^2 + 3x - 10}{x^2 + x - 6} = \frac{(x - 2)(x + 5)}{(x - 2)(x + 3)} = \frac{x + 5}{x + 3}$$

We have to be careful with this last step. The fraction $(x + 5)/(x + 3)$ is equal to the one on the left side of the equal sign only if x is not equal to 2. If $x = 2$, the left side is undefined (because the denominator is 0), whereas the right side is equal to $(2 + 5)/(2 + 3) = 7/5$. This brings up the question about whether the limits

$$\lim_{x\to 2}\frac{x^2 + 3x - 10}{x^2 + x - 6} \quad\text{and}\quad \lim_{x\to 2}\frac{x + 5}{x + 3}$$

are equal. The answer is contained in the following theorem.

Theorem C

If $f(x) = g(x)$ for all x in an open interval containing the number c, except possibly at the number c itself, and if $\lim_{x\to c} g(x)$ exists, then $\lim_{x\to c} f(x)$ exists and $\lim_{x\to c} f(x) = \lim_{x\to c} g(x)$.

EXAMPLE 7 Find $\lim_{x\to 1}\frac{x - 1}{\sqrt{x} - 1}$.

SOLUTION

$$\lim_{x\to 1}\frac{x - 1}{\sqrt{x} - 1} = \lim_{x\to 1}\frac{\left(\sqrt{x} - 1\right)\left(\sqrt{x} + 1\right)}{\sqrt{x} - 1} = \lim_{x\to 1}\left(\sqrt{x} + 1\right) = \sqrt{1} + 1 = 2$$ ■

EXAMPLE 8 Find $\lim_{x\to 2}\frac{x^2 + 3x - 10}{x^2 + x - 6}$.

SOLUTION Theorem B does not apply because the denominator is 0 when $x = 2$. When we substitute $x = 2$ in the numerator we also get 0, so the quotient takes on the meaningless form 0/0 at $x = 2$. When this happens we should look for some sort of simplification such as factoring.

$$\lim_{x\to 2}\frac{x^2 + 3x - 10}{x^2 + x - 6} = \lim_{x\to 2}\frac{(x - 2)(x + 5)}{(x - 2)(x + 3)} = \lim_{x\to 2}\frac{x + 5}{x + 3} = \frac{7}{5}$$

> **Optional?**
>
> How much theorem proving should be done in a first calculus course? Mathematics teachers argue long and hard about this and about the right balance between
>
> - logic and intuition
> - proof and explanation
> - theory and application
>
> A great scientist of long ago had some sage advice.
>
> "He who loves practice without theory is like the sailor who boards ship without a rudder and compass and never knows where he may cast."
>
> *Leonardo da Vinci*

The second to last equality is justified by Theorem C since

$$\frac{(x-2)(x+5)}{(x-2)(x+3)} = \frac{x+5}{x+3}$$

for all x except $x = 2$. Once we apply Theorem C, we can evaluate the limit by substitution (i.e., by applying Theorem B). ■

Proof of Theorem A (Optional) You should not be too surprised when we say that the proofs of some parts of Theorem A are quite sophisticated. Because of this, we prove only the first five parts here, deferring the others to the Appendix (Section A.2, Theorem A). To get your feet wet, you might try Problems 35 and 36.

Proofs of Statements 1 and 2 These statements result from $\lim_{x \to c}(mx + b) = mc + b$ (Example 4 of Section 1.2) using first $m = 0$ and then $m = 1, b = 0$. ■

Proof of Statement 3 If $k = 0$, the result is trivial, so we suppose that $k \neq 0$. Let $\varepsilon > 0$ be given. By hypothesis, $\lim_{x \to c} f(x)$ exists; call its value L. By definition of limit, there is a number δ such that

$$0 < |x - c| < \delta \Rightarrow |f(x) - L| < \frac{\varepsilon}{|k|}$$

Someone is sure to complain that we put $\varepsilon/|k|$ rather than ε at the end of the inequality above. Well, isn't $\varepsilon/|k|$ a positive number? Yes. Doesn't the definition of limit require that for *any* positive number there be a corresponding δ? Yes.

Now, for δ so determined (again by a preliminary analysis that we have not shown here), we assert that $0 < |x - c| < \delta$ implies that

$$|kf(x) - kL| = |k||f(x) - L| < |k|\frac{\varepsilon}{|k|} = \varepsilon$$

This shows that

$$\lim_{x \to c} kf(x) = kL = k \lim_{x \to c} f(x)$$ ■

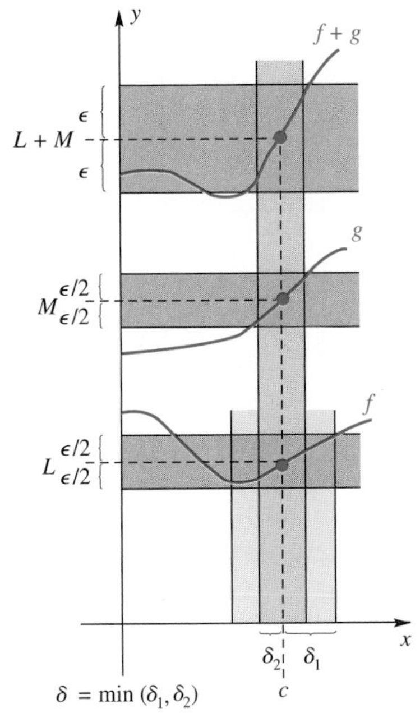

Figure 1

Proof of Statement 4 Refer to Figure 1. Let $\lim_{x \to c} f(x) = L$ and $\lim_{x \to c} g(x) = M$. If ε is any given positive number, then $\varepsilon/2$ is positive. Since $\lim_{x \to c} f(x) = L$, there is a positive number δ_1 such that

$$0 < |x - c| < \delta_1 \Rightarrow |f(x) - L| < \frac{\varepsilon}{2}$$

Since $\lim_{x \to c} g(x) = M$, there is a positive number δ_2 such that

$$0 < |x - c| < \delta_2 \Rightarrow |g(x) - M| < \frac{\varepsilon}{2}$$

Choose $\delta = \min\{\delta_1, \delta_2\}$; that is, choose δ to be the smaller of δ_1 and δ_2. Then $0 < |x - c| < \delta$ implies that

$$\begin{aligned} |f(x) + g(x) - (L + M)| &= |[f(x) - L] + [g(x) - M]| \\ &\leq |f(x) - L| + |g(x) - M| \\ &< \frac{\varepsilon}{2} + \frac{\varepsilon}{2} = \varepsilon \end{aligned}$$

In this chain, the first inequality is the Triangle Inequality (Section 0.2); the second results from the choice of δ. We have just shown that

$$0 < |x - c| < \delta \Rightarrow |f(x) + g(x) - (L + M)| < \varepsilon$$

Thus,

$$\lim_{x \to c} [f(x) + g(x)] = L + M = \lim_{x \to c} f(x) + \lim_{x \to c} g(x)$$ ■

Proof of Statement 5

$$\begin{aligned}\lim_{x\to c}[f(x) - g(x)] &= \lim_{x\to c}[f(x) + (-1)g(x)] \\ &= \lim_{x\to c} f(x) + \lim_{x\to c}(-1)g(x) \\ &= \lim_{x\to c} f(x) + (-1)\lim_{x\to c} g(x) \\ &= \lim_{x\to c} f(x) - \lim_{x\to c} g(x)\end{aligned}$$ ■

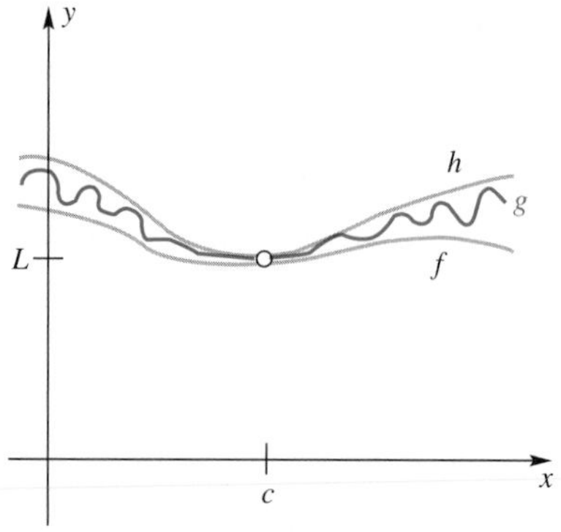

Figure 2

The Squeeze Theorem You have likely heard someone say, "I was caught between a rock and a hard place." This is what happens to g in the following theorem (see Figure 2).

> **Theorem D** **Squeeze Theorem**
>
> Let f, g, and h be functions satisfying $f(x) \le g(x) \le h(x)$ for all x near c, except possibly at c. If $\lim_{x\to c} f(x) = \lim_{x\to c} h(x) = L$, then $\lim_{x\to c} g(x) = L$.

Proof (Optional) Let $\varepsilon > 0$ be given. Choose δ_1 such that

$$0 < |x - c| < \delta_1 \Rightarrow L - \varepsilon < f(x) < L + \varepsilon$$

and δ_2 such that

$$0 < |x - c| < \delta_2 \Rightarrow L - \varepsilon < h(x) < L + \varepsilon$$

Choose δ_3 so that

$$0 < |x - c| < \delta_3 \Rightarrow f(x) \le g(x) \le h(x)$$

Let $\delta = \min\{\delta_1, \delta_2, \delta_3\}$. Then

$$0 < |x - c| < \delta \Rightarrow L - \varepsilon < f(x) \le g(x) \le h(x) < L + \varepsilon$$

We conclude that $\lim_{x\to c} g(x) = L$. ■

EXAMPLE 9 Assume that we have proved $1 - x^2/6 \le (\sin x)/x \le 1$ for all x near but different from 0. What can we conclude about $\lim_{x\to 0}\frac{\sin x}{x}$?

SOLUTION Let $f(x) = 1 - x^2/6$, $g(x) = (\sin x)/x$, and $h(x) = 1$. It follows that $\lim_{x\to 0} f(x) = 1 = \lim_{x\to 0} h(x)$ and so, by Theorem D,

$$\lim_{x\to 0}\frac{\sin x}{x} = 1$$ ■

Concepts Review

1. If $\lim_{x\to 3} f(x) = 4$, then $\lim_{x\to 3}(x^2 + 3)f(x) =$ ____.

2. If $\lim_{x\to 2} g(x) = -2$, then $\lim_{x\to 2}\sqrt{g^2(x) + 12} =$ ____.

3. If $\lim_{x\to c} f(x) = 4$ and $\lim_{x\to c} g(x) = -2$, then $\lim_{x\to c}\frac{f^2(x)}{g(x)} =$ ____ and $\lim_{x\to c}\left[g(x)\sqrt{f(x)} + 5x\right] =$ ____.

4. If $\lim_{x\to c} f(x) = L$ and $\lim_{x\to c} g(x) = L$, then $\lim_{x\to c}[f(x) - L]g(x) =$ ____.

Problem Set 1.3

In Problems 1–12, use Theorem A to find each of the limits. Justify each step by appealing to a numbered statement, as in Examples 1–4.

1. $\lim_{x\to 1}(2x + 1)$

2. $\lim_{x\to -1}(3x^2 - 1)$

3. $\lim_{x\to 0}[(2x + 1)(x - 3)]$

4. $\lim_{x\to\sqrt{2}}[(2x^2 + 1)(7x^2 + 13)]$

5. $\lim_{x\to 2}\frac{2x + 1}{5 - 3x}$

6. $\lim_{x\to -3}\frac{4x^3 + 1}{7 - 2x^2}$

7. $\lim_{x \to 3} \sqrt{3x - 5}$

8. $\lim_{x \to -3} \sqrt{5x^2 + 2x}$

9. $\lim_{t \to -2} (2t^3 + 15)^{13}$

10. $\lim_{w \to -2} \sqrt{-3w^3 + 7w^2}$

11. $\lim_{y \to 2} \left(\dfrac{4y^3 + 8y}{y + 4} \right)^{1/3}$

12. $\lim_{w \to 5} (2w^4 - 9w^3 + 19)^{-1/2}$

In Problems 13–24, find the indicated limit or state that it does not exist. In many cases, you will want to do some algebra before trying to evaluate the limit.

13. $\lim_{x \to 2} \dfrac{x^2 - 4}{x^2 + 4}$

14. $\lim_{x \to 2} \dfrac{x^2 - 5x + 6}{x - 2}$

15. $\lim_{x \to -1} \dfrac{x^2 - 2x - 3}{x + 1}$

16. $\lim_{x \to -1} \dfrac{x^2 + x}{x^2 + 1}$

17. $\lim_{x \to -1} \dfrac{x^3 - 6x^2 + 11x - 6}{x^3 + 4x^2 - 19x + 14}$

18. $\lim_{x \to 2} \dfrac{x^2 + 7x + 10}{x + 2}$

19. $\lim_{x \to 1} \dfrac{x^2 + x - 2}{x^2 - 1}$

20. $\lim_{x \to -3} \dfrac{x^2 - 14x - 51}{x^2 - 4x - 21}$

21. $\lim_{u \to -2} \dfrac{u^2 - ux + 2u - 2x}{u^2 - u - 6}$

22. $\lim_{x \to 1} \dfrac{x^2 + ux - x - u}{x^2 + 2x - 3}$

23. $\lim_{x \to \pi} \dfrac{2x^2 - 6x\pi + 4\pi^2}{x^2 - \pi^2}$

24. $\lim_{w \to -2} \dfrac{(w + 2)(w^2 - w - 6)}{w^2 + 4w + 4}$

In Problems 25–30, find the limits if $\lim_{x \to a} f(x) = 3$ and $\lim_{x \to a} g(x) = -1$ (see Example 4).

25. $\lim_{x \to a} \sqrt{f^2(x) + g^2(x)}$

26. $\lim_{x \to a} \dfrac{2f(x) - 3g(x)}{f(x) + g(x)}$

27. $\lim_{x \to a} \sqrt[3]{g(x)}\left[f(x) + 3\right]$

28. $\lim_{x \to a} \left[f(x) - 3\right]^4$

29. $\lim_{t \to a} \left[|f(t)| + |3g(t)|\right]$

30. $\lim_{u \to a} \left[f(u) + 3g(u)\right]^3$

In Problems 31–34, find $\lim_{x \to 2} [f(x) - f(2)]/(x - 2)$ for each given function f.

31. $f(x) = 3x^2$

32. $f(x) = 3x^2 + 2x + 1$

33. $f(x) = \dfrac{1}{x}$

34. $f(x) = \dfrac{3}{x^2}$

35. Prove Statement 6 of Theorem A. *Hint:*

$$\begin{aligned} |f(x)g(x) - LM| &= |f(x)g(x) - Lg(x) + Lg(x) - LM| \\ &= |g(x)[f(x) - L] + L[g(x) - M]| \\ &\le |g(x)||f(x) - L| + |L||g(x) - M| \end{aligned}$$

Now show that if $\lim_{x \to c} g(x) = M$, then there is a number δ_1 such that

$$0 < |x - c| < \delta_1 \Rightarrow |g(x)| < |M| + 1$$

36. Prove Statement 7 of Theorem A by first giving an ε–δ proof that $\lim_{x \to c} [1/g(x)] = 1/\left[\lim_{x \to c} g(x)\right]$ and then applying Statement 6.

37. Prove that $\lim_{x \to c} f(x) = L \Leftrightarrow \lim_{x \to c} [f(x) - L] = 0$.

38. Prove that $\lim_{x \to c} f(x) = 0 \Leftrightarrow \lim_{x \to c} |f(x)| = 0$.

39. Prove that $\lim_{x \to c} |x| = |c|$.

40. Find examples to show that if

(a) $\lim_{x \to c} \left[f(x) + g(x)\right]$ exists, this does not imply that either $\lim_{x \to c} f(x)$ or $\lim_{x \to c} g(x)$ exists;

(b) $\lim_{x \to c} \left[f(x) \cdot g(x)\right]$ exists, this does not imply that either $\lim_{x \to c} f(x)$ or $\lim_{x \to c} g(x)$ exists.

In Problems 41–48, find each of the right-hand and left-hand limits or state that they do not exist.

41. $\lim_{x \to -3^+} \dfrac{\sqrt{3 + x}}{x}$

42. $\lim_{x \to -\pi^+} \dfrac{\sqrt{\pi^3 + x^3}}{x}$

43. $\lim_{x \to 3^+} \dfrac{x - 3}{\sqrt{x^2 - 9}}$

44. $\lim_{x \to 1^-} \dfrac{\sqrt{1 + x}}{4 + 4x}$

45. $\lim_{x \to 2^+} \dfrac{(x^2 + 1)[\![x]\!]}{(3x - 1)^2}$

46. $\lim_{x \to 3^-} (x - [\![x]\!])$

47. $\lim_{x \to 0^-} \dfrac{x}{|x|}$

48. $\lim_{x \to 3^+} [\![x^2 + 2x]\!]$

49. Suppose that $f(x)g(x) = 1$ for all x and $\lim_{x \to a} g(x) = 0$. Prove that $\lim_{x \to a} f(x)$ does not exist.

50. Let R be the rectangle joining the midpoints of the sides of the quadrilateral Q having vertices $(\pm x, 0)$ and $(0, \pm 1)$. Calculate

$$\lim_{x \to 0^+} \frac{\text{perimeter of } R}{\text{perimeter of } Q}$$

51. Let $y = \sqrt{x}$ and consider the points M, N, O, and P with coordinates $(1, 0)$, $(0, 1)$, $(0, 0)$, and (x, y) on the graph of $y = \sqrt{x}$, respectively. Calculate

(a) $\lim_{x \to 0^+} \dfrac{\text{perimeter of } \Delta NOP}{\text{perimeter of } \Delta MOP}$

(b) $\lim_{x \to 0^+} \dfrac{\text{area of } \Delta NOP}{\text{area of } \Delta MOP}$

Answers to Concepts Review: **1.** 48 **2.** 4 **3.** −8; −4 + 5c **4.** 0

1.4 Limits Involving Trigonometric Functions

Theorem B of the previous section says that limits of polynomial functions can always be found by substitution, and limits of rational functions can be found by substitution as long as the denominator is not zero at the limit point. This substitution rule applies to the trigonometric functions as well. This result is stated next.

Theorem A Limits of Trigonometric Functions

For every real number c in the function's domain,

1. $\lim_{t\to c} \sin t = \sin c$
2. $\lim_{t\to c} \cos t = \cos c$
3. $\lim_{t\to c} \tan t = \tan c$
4. $\lim_{t\to c} \cot t = \cot c$
5. $\lim_{t\to c} \sec t = \sec c$
6. $\lim_{t\to c} \csc t = \csc c$

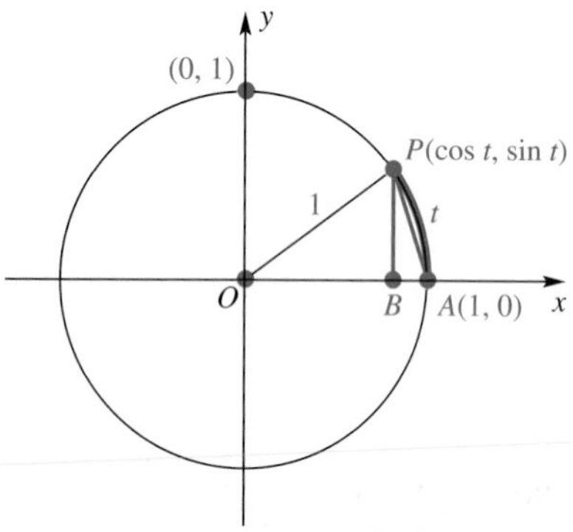

Figure 1

Proof of Statement 1 We first establish the special case in which $c = 0$. Suppose that $t > 0$ and let points A, B, and P be defined as in Figure 1. Then

$$0 < |BP| < |AP| < \text{arc}(AP)$$

But $|BP| = \sin t$ and arc $(AP) = t$, so

$$0 < \sin t < t$$

If $t < 0$, then $t < \sin t < 0$. We can thus apply the Squeeze Theorem (Theorem 1.3D) and conclude that $\lim_{t\to 0} \sin t = 0$. To complete the proof, we will also need the result that $\lim_{t\to 0} \cos t = 1$. This follows by applying a trigonometric identity and Theorem 1.3A:

$$\lim_{t\to 0} \cos t = \lim_{t\to 0} \sqrt{1 - \sin^2 t} = \sqrt{1 - \left(\lim_{t\to 0} \sin t\right)^2} = \sqrt{1 - 0^2} = 1$$

Now, to show that $\lim_{t\to c} \sin t = \sin c$, we first let $h = t - c$ so that $h \to 0$ as $t \to c$. Then

$$\begin{aligned}\lim_{t\to c} \sin t &= \lim_{h\to 0} \sin(c + h)\\ &= \lim_{h\to 0}(\sin c \cos h + \cos c \sin h) \qquad \text{(Addition Identity)}\\ &= (\sin c)\left(\lim_{h\to 0} \cos h\right) + (\cos c)\left(\lim_{h\to 0} \sin h\right)\\ &= (\sin c)(1) + (\cos c)(0) = \sin c\end{aligned}$$

■

Proof of Statement 2 We use another identity along with Theorem 1.3A. If $\cos c > 0$, then for t near c we have $\cos t = \sqrt{1 - \sin^2 t}$. Thus,

$$\lim_{t\to c} \cos t = \lim_{t\to c} \sqrt{1 - \sin^2 t} = \sqrt{1 - \left(\lim_{t\to c} \sin t\right)^2} = \sqrt{1 - \sin^2 c} = \cos c$$

On the other hand, if $\cos c < 0$, then for t near c we have $\cos t = -\sqrt{1 - \sin^2 t}$. In this case,

$$\begin{aligned}\lim_{t\to c} \cos t &= \lim_{t\to c}\left(-\sqrt{1 - \sin^2 t}\right) = -\sqrt{1 - \left(\lim_{t\to c} \sin t\right)^2} = -\sqrt{1 - \sin^2 c}\\ &= -\sqrt{\cos^2 c} = -|\cos c| = \cos c\end{aligned}$$

The case $c = 0$ was handled in the proof of Statement 1. ■

The proofs for the other statements are left as exercises. (See Problems 21 and 22.) Theorem A can be used along with Theorem 1.3A to evaluate other limits.

EXAMPLE 1 Find $\lim_{t\to 0} \dfrac{t^2 \cos t}{t + 1}$.

SOLUTION

$$\lim_{t\to 0} \frac{t^2 \cos t}{t + 1} = \left(\lim_{t\to 0} \frac{t^2}{t + 1}\right)\left(\lim_{t\to 0} \cos t\right) = 0 \cdot 1 = 0$$

■

Two important limits that we cannot evaluate by substitution are

$$\lim_{t\to 0} \frac{\sin t}{t} \quad \text{and} \quad \lim_{t\to 0} \frac{1 - \cos t}{t}$$

We met the first of these limits in Section 1.1, where we conjectured that the limit was 1. Now we prove that 1 is indeed the limit.

Theorem B **Special Trigonometric Limits**

1. $\lim_{t \to 0} \frac{\sin t}{t} = 1$
2. $\lim_{t \to 0} \frac{1 - \cos t}{t} = 0$

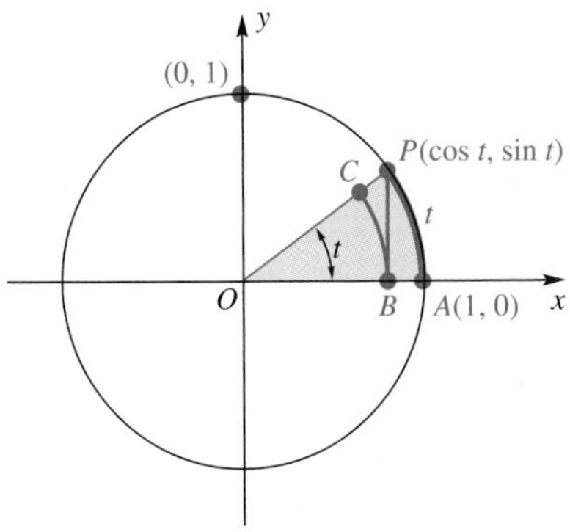

Figure 2

Proof of Statement 1 In the proof of Theorem A of this section, we showed that

$$\lim_{t \to 0} \cos t = 1 \quad \text{and} \quad \lim_{t \to 0} \sin t = 0$$

For $-\pi/2 \le t \le \pi/2, t \ne 0$ (remember, it does not matter what happens at $t = 0$), draw the vertical line segment BP and the circular arc BC, as shown in Figure 2. (If $t < 0$, then think of the shaded region as being reflected across the x-axis.) It is evident from Figure 2 that

$$\text{area (sector } OBC) \le \text{ area } (\Delta OBP) \le \text{ area (sector } OAP)$$

The area of a triangle is one-half its base times the height, and the area of a circular sector with central angle t and radius r is $\frac{1}{2}r^2|t|$ (see Problem 42 of Section 0.7). Applying these results to the three regions gives

$$\frac{1}{2}(\cos t)^2|t| \le \frac{1}{2}\cos t\,|\sin t| \le \frac{1}{2}1^2|t|$$

which, after multiplying by 2 and dividing by the positive number $|t|\cos t$, yields

$$\cos t \le \frac{|\sin t|}{|t|} \le \frac{1}{\cos t}$$

Since the expression $(\sin t)/t$ is positive for $-\pi/2 \le t \le \pi/2, t \ne 0$, we have $|\sin t|/|t| = (\sin t)/t$. Therefore,

$$\cos t \le \frac{\sin t}{t} \le \frac{1}{\cos t}$$

Since we are after the limit of the middle function and we know the limit of each "outside" function, this double inequality begs for the Squeeze Theorem. When we apply it, we get

$$\lim_{t \to 0} \frac{\sin t}{t} = 1$$ ■

Proof of Statement 2 The second limit follows easily from the first. Just multiply the numerator and denominator by $(1 + \cos t)$; this gives

$$\lim_{t \to 0} \frac{1 - \cos t}{t} = \lim_{t \to 0} \frac{1 - \cos t}{t} \cdot \frac{1 + \cos t}{1 + \cos t} = \lim_{t \to 0} \frac{1 - \cos^2 t}{t(1 + \cos t)}$$

$$= \lim_{t \to 0} \frac{\sin^2 t}{t(1 + \cos t)}$$

$$= \left(\lim_{t \to 0} \frac{\sin t}{t}\right) \frac{\lim_{t \to 0} \sin t}{\lim_{t \to 0}(1 + \cos t)} = 1 \cdot \frac{0}{2} = 0$$ ■

We will make explicit use of these two limit statements in Chapter 2. Right now, we can use them to evaluate other limits.

EXAMPLE 2 Find each limit.

(a) $\lim_{x \to 0} \frac{\sin 3x}{x}$ (b) $\lim_{t \to 0} \frac{1 - \cos t}{\sin t}$ (c) $\lim_{x \to 0} \frac{\sin 4x}{\tan x}$

SOLUTION

(a) $\lim_{x\to 0}\frac{\sin 3x}{x} = \lim_{x\to 0} 3\frac{\sin 3x}{3x} = 3\lim_{x\to 0}\frac{\sin 3x}{3x}$

Here the argument to the sine function is $3x$, not simply x as required by Theorem B. Let $y = 3x$. Then $y \to 0$ if and only if $x \to 0$, so

$$\lim_{x\to 0}\frac{\sin 3x}{3x} = \lim_{y\to 0}\frac{\sin y}{y} = 1$$

Thus,

$$\lim_{x\to 0}\frac{\sin 3x}{x} = 3\lim_{x\to 0}\frac{\sin 3x}{3x} = 3$$

(b) $$\lim_{t\to 0}\frac{1-\cos t}{\sin t} = \lim_{t\to 0}\frac{\dfrac{1-\cos t}{t}}{\dfrac{\sin t}{t}} = \frac{\lim_{t\to 0}\dfrac{1-\cos t}{t}}{\lim_{t\to 0}\dfrac{\sin t}{t}} = \frac{0}{1} = 0$$

(c) $$\lim_{x\to 0}\frac{\sin 4x}{\tan x} = \lim_{x\to 0}\frac{\dfrac{4\sin 4x}{4x}}{\dfrac{\sin x}{x\cos x}}$$

$$= \frac{4\lim_{x\to 0}\dfrac{\sin 4x}{4x}}{\left(\lim_{x\to 0}\dfrac{\sin x}{x}\right)\left(\lim_{x\to 0}\dfrac{1}{\cos x}\right)} = \frac{4}{1\cdot 1} = 4$$ ■

EXAMPLE 3 Sketch the graphs of $u(x) = |x|$, $l(x) = -|x|$, and $f(x) = x\cos(1/x)$. Use these graphs along with the Squeeze Theorem (Theorem D of Section 1.3) to determine $\lim_{x\to 0} f(x)$.

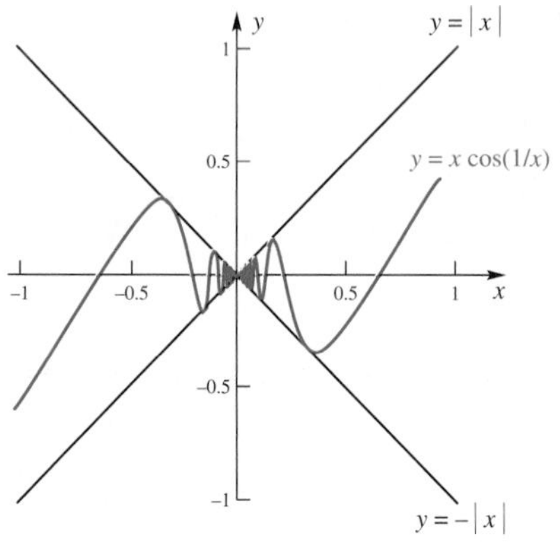

Figure 3

SOLUTION Note that $\cos(1/x)$ is always between -1 and 1 and $f(x) = x\cos(1/x)$. Thus, $x\cos(1/x)$ will always be between $-x$ and x if x is positive and between x and $-x$ if x is negative. In other words, the graph of $y = x\cos(1/x)$ is between the graphs of $y = |x|$ and $y = -|x|$, as shown in Figure 3. We know that $\lim_{x\to 0}|x| = \lim_{x\to 0}(-|x|) = 0$ (see Problem 27 of Section 1.2) and since the graph of $y = f(x) = x\cos(1/x)$ is "squeezed" between the graphs of $u(x) = |x|$ and $l(x) = -|x|$, both of which go to 0 as $x \to 0$, we can apply the Squeeze Theorem to conclude that $\lim_{x\to 0} f(x) = 0$. ■

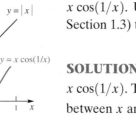

Concepts Review

1. $\lim_{t\to 0}\sin t =$ _____.

2. $\lim_{t\to \pi/4}\tan t =$ _____.

3. The limit $\lim_{t\to 0}\frac{\sin t}{t}$ cannot be evaluated by substitution because _____.

4. $\lim_{t\to 0}\frac{\sin t}{t} =$ _____.

Problem Set 1.4

In Problems 1–14, evaluate each limit.

1. $\lim_{x\to 0} \dfrac{\cos x}{x+1}$

2. $\lim_{\theta\to \pi/2} \theta \cos \theta$

3. $\lim_{t\to 0} \dfrac{\cos^2 t}{1+\sin t}$

4. $\lim_{x\to 0} \dfrac{3x \tan x}{\sin x}$

5. $\lim_{x\to 0} \dfrac{\sin x}{2x}$

6. $\lim_{\theta\to 0} \dfrac{\sin 3\theta}{2\theta}$

7. $\lim_{\theta\to 0} \dfrac{\sin 3\theta}{\tan \theta}$

8. $\lim_{\theta\to 0} \dfrac{\tan 5\theta}{\sin 2\theta}$

9. $\lim_{\theta\to 0} \dfrac{\cot(\pi\theta)\sin\theta}{2\sec\theta}$

10. $\lim_{t\to 0} \dfrac{\sin^2 3t}{2t}$

11. $\lim_{t\to 0} \dfrac{\tan^2 3t}{2t}$

12. $\lim_{t\to 0} \dfrac{\tan 2t}{\sin 2t - 1}$

13. $\lim_{t\to 0} \dfrac{\sin 3t + 4t}{t \sec t}$

14. $\lim_{\theta\to 0} \dfrac{\sin^2 \theta}{\theta^2}$

In Problems 15–19, plot the functions $u(x)$, $l(x)$, and $f(x)$. Then use these graphs along with the Squeeze Theorem to determine $\lim_{x\to 0} f(x)$.

15. $u(x) = |x|, l(x) = -|x|, f(x) = x\sin(1/x)$

16. $u(x) = |x|, l(x) = -|x|, f(x) = x\sin(1/x^2)$

17. $u(x) = |x|, l(x) = -|x|, f(x) = (1-\cos^2 x)/x$

18. $u(x) = 1, l(x) = 1 - x^2, f(x) = \cos^2 x$

19. $u(x) = 2, l(x) = 2 - x^2, f(x) = 1 + \dfrac{\sin x}{x}$

20. Prove that $\lim_{t\to c} \cos t = \cos c$ using an argument similar to the one used in the proof that $\lim_{t\to c} \sin t = \sin c$.

21. Prove statements 3 and 4 of Theorem A using Theorem 1.3A.

22. Prove statements 5 and 6 of Theorem A using Theorem 1.3A.

23. From area $(OBP) \le$ area (sector OAP) $\le$ area (OBP) + area $(ABPQ)$ in Figure 4, show that

$$\cos t \le \frac{t}{\sin t} \le 2 - \cos t$$

and thus obtain another proof that $\lim_{t\to 0^+} (\sin t)/t = 1$.

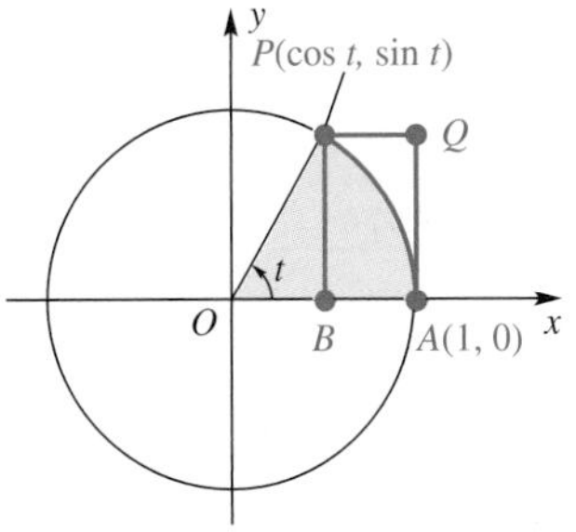

Figure 4

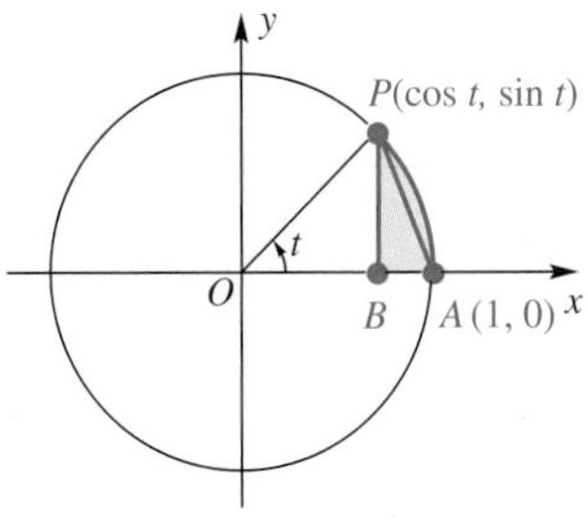

Figure 5

24. In Figure 5, let D be the area of triangle ABP and E the area of the shaded region.

(a) Guess the value of $\lim_{t\to 0^+} \dfrac{D}{E}$ by looking at the figure.

(b) Find a formula for D/E in terms of t.

C (c) Use a calculator to get an accurate estimate of $\lim_{t\to 0^+} \dfrac{D}{E}$.

Answers to Concepts Review: **1.** 0 **2.** 1 **3.** the denominator is zero when $t = 0$ **4.** 1

1.5 Limits at Infinity; Infinite Limits

The deepest problems and most profound paradoxes of mathematics are often intertwined with the use of the concept of the infinite. Yet mathematical progress can in part be measured in terms of our understanding the concept of infinity. We have already used the symbols ∞ and $-\infty$ in our notation for certain intervals. Thus, $(3, \infty)$ is our way of denoting the set of all real numbers greater than 3. Note that we have never referred to ∞ as a number. For example, we have never added it to a number or divided it by a number. We will use the symbols ∞ and $-\infty$ in a new way in this section, but they will still not represent numbers.

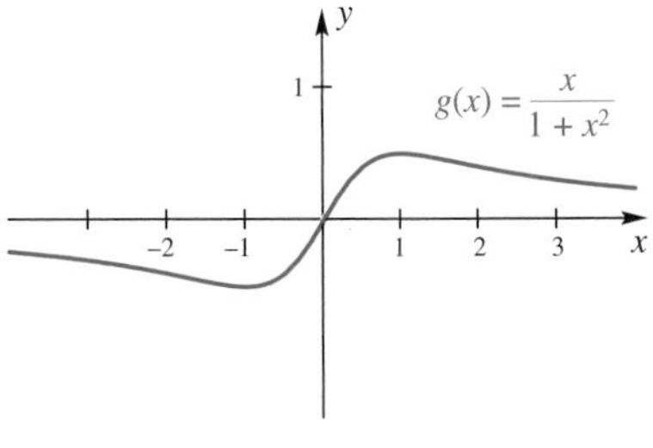

Figure 1

Limits at Infinity Consider the function $g(x) = x/(1+x^2)$ whose graph is shown in Figure 1. We ask this question: What happens to $g(x)$ as x gets larger and larger? In symbols, we ask for the value of $\lim_{x\to\infty} g(x)$.

When we write $x \to \infty$, we are *not* implying that somewhere far, far to the right on the x-axis there is a number—bigger than all other numbers—that x is approaching. Rather, we use $x \to \infty$ as a shorthand way of saying that x gets larger and larger without bound.

In the table in Figure 2, we have listed values of $g(x) = x/(1+x^2)$ for several values of x. It appears that $g(x)$ gets smaller and smaller as x gets larger and larger. We write

x	$\frac{x}{1+x^2}$
10	0.099
100	0.010
1000	0.001
10000	0.0001
$\downarrow$	$\downarrow$
∞	?

Figure 2

$$\lim_{x\to\infty} \frac{x}{1 + x^2} = 0$$

Experimenting with negative numbers far to the left of zero on the real number line would lead us to write

$$\lim_{x\to-\infty} \frac{x}{1 + x^2} = 0$$

Rigorous Definitions of Limits as $x \to \pm\infty$ In analogy with our ε–δ definition for ordinary limits, we make the following definition.

> **Definition** Limit as $x \to \infty$
>
> Let f be defined on $[c, \infty)$ for some number c. We say that $\lim_{x\to\infty} f(x) = L$ if for each $\varepsilon > 0$ there is a corresponding number M such that
>
> $$x > M \Rightarrow |f(x) - L| < \varepsilon$$

You will note that M can, and usually does, depend on ε. In general, the smaller ε is, the larger M will have to be. The graph in Figure 3 may help you to understand what we are saying.

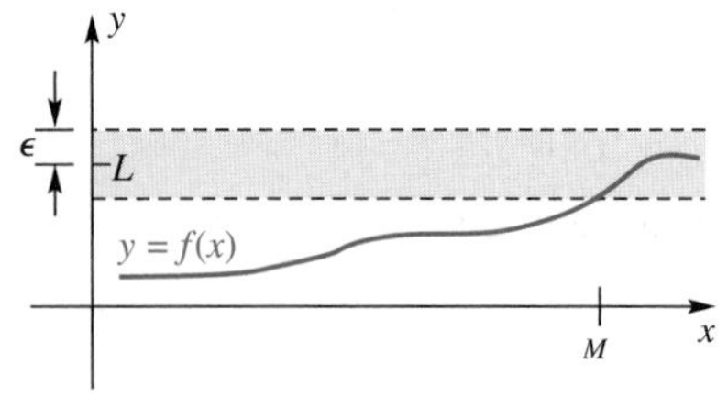

Figure 3

> **Definition** Limit as $x \to -\infty$
>
> Let f be defined on $(-\infty, c]$ for some number c. We say that $\lim_{x\to-\infty} f(x) = L$ if for each $\varepsilon > 0$ there is a corresponding number M such that
>
> $$x < M \Rightarrow |f(x) - L| < \varepsilon$$

EXAMPLE 1 Show that if k is a positive integer, then

$$\lim_{x\to\infty} \frac{1}{x^k} = 0 \quad \text{and} \quad \lim_{x\to-\infty} \frac{1}{x^k} = 0$$

SOLUTION Let $\varepsilon > 0$ be given. After a preliminary analysis (as in Section 1.2), we chose $M = \sqrt[k]{1/\varepsilon}$. Then $x > M$ implies that

$$\left|\frac{1}{x^k} - 0\right| = \frac{1}{x^k} < \frac{1}{M^k} = \varepsilon$$

The proof of the second statement is similar. ■

Having given the definitions of these new kinds of limits, we must face the question of whether the Main Limit Theorem (Theorem 1.3A) holds for them. The answer is yes, and the proof is similar to the original one. Note how we use this theorem in the following examples.

EXAMPLE 2 Prove that $\lim_{x\to\infty} \frac{x}{1 + x^2} = 0$.

SOLUTION Here we use a standard trick: divide the numerator and denominator by the highest power of x that appears in the denominator, that is, x^2.

$$\lim_{x\to\infty} \frac{x}{1 + x^2} = \lim_{x\to\infty} \frac{\frac{x}{x^2}}{\frac{1 + x^2}{x^2}} = \lim_{x\to\infty} \frac{\frac{1}{x}}{\frac{1}{x^2} + 1}$$

$$= \frac{\lim_{x\to\infty} \frac{1}{x}}{\lim_{x\to\infty} \frac{1}{x^2} + \lim_{x\to\infty} 1} = \frac{0}{0 + 1} = 0$$

■

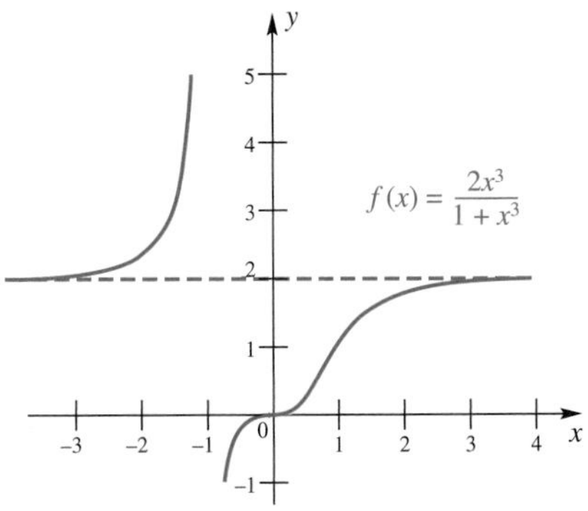

Figure 4

EXAMPLE 3 Find $\lim_{x\to-\infty} \frac{2x^3}{1+x^3}$.

SOLUTION The graph of $f(x) = 2x^3/(1+x^3)$ is shown in Figure 4. To find the limit, divide both the numerator and denominator by x^3.

$$\lim_{x\to-\infty} \frac{2x^3}{1+x^3} = \lim_{x\to-\infty} \frac{2}{1/x^3+1} = \frac{2}{0+1} = 2$$ ■

Limits of Sequences The domain for some functions is the set of natural numbers $\{1, 2, 3, \ldots\}$. In this situation, we usually write a_n rather than $a(n)$ to denote the nth term of the sequence, or $\{a_n\}$ to denote the whole sequence. For example, we might define the sequence by $a_n = n/(n+1)$. Let's consider what happens as n gets large. A little calculation shows that

$$a_1 = \frac{1}{2}, \quad a_2 = \frac{2}{3}, \quad a_3 = \frac{3}{4}, \quad a_4 = \frac{4}{5}, \quad \ldots, \quad a_{100} = \frac{100}{101}, \quad \ldots$$

It looks as if these values are approaching 1, so it seems reasonable to say that for this sequence $\lim_{n\to\infty} a_n = 1$. The next definition gives meaning to this idea of the limit of a sequence.

> **Definition** Limit of a Sequence
>
> Let a_n be defined for all natural numbers greater than or equal to some number c. We say that $\lim_{n\to\infty} a_n = L$ if for each $\varepsilon > 0$ there is a corresponding natural number M such that
>
> $$n > M \Rightarrow |a_n - L| < \varepsilon$$

Notice that this definition is nearly identical to the definition of $\lim_{x\to\infty} f(x)$. The only difference is that now we are requiring that the argument to the function be a natural number. As we might expect, the Main Limit Theorem (Theorem 1.3A) holds for sequences.

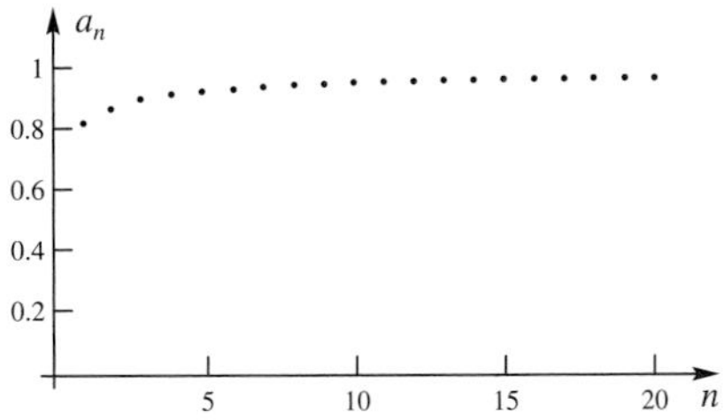

Figure 5

EXAMPLE 4 Find $\lim_{n\to\infty} \sqrt{\frac{n+1}{n+2}}$.

SOLUTION Figure 5 shows a graph of $a_n = \sqrt{\frac{n+1}{n+2}}$. Applying Theorem 1.3A gives

$$\lim_{n\to\infty} \sqrt{\frac{n+1}{n+2}} = \left(\lim_{n\to\infty} \frac{n+1}{n+2}\right)^{1/2} = \left(\lim_{n\to\infty} \frac{1+1/n}{1+2/n}\right)^{1/2} = \left(\frac{1+0}{1+0}\right)^{1/2} = 1$$ ■

We will need the concept of the limit of a sequence in Section 3.7 and in Chapter 4. Sequences are covered more thoroughly in Chapter 9.

Infinite Limits Consider the function $f(x) = 1/(x-2)$, which is graphed in Figure 6. As x gets close to 2 from the left, the function seems to decrease without bound. Similarly, as x approaches 2 from the right, the function seems to increase without bound. It therefore makes no sense to talk about $\lim_{x\to 2} 1/(x-2)$, but we think it is reasonable to write

$$\lim_{x\to 2^-} \frac{1}{x-2} = -\infty \quad \text{and} \quad \lim_{x\to 2^+} \frac{1}{x-2} = \infty$$

Here is the precise definition.

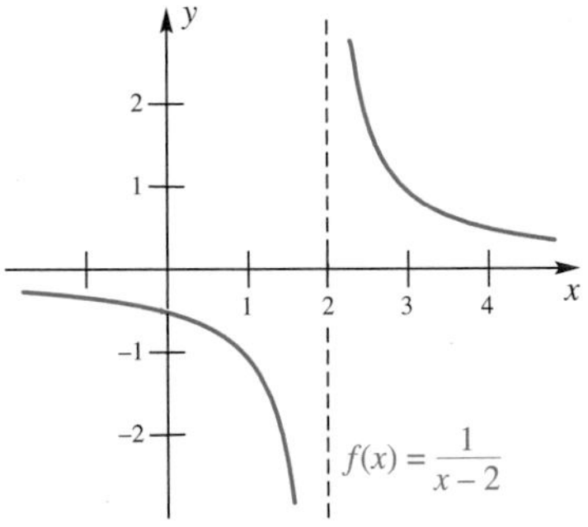

Figure 6

> **Definition** Infinite Limit
>
> We say that $\lim_{x\to c^+} f(x) = \infty$ if for every positive number M, there exists a corresponding $\delta > 0$ such that
>
> $$0 < x - c < \delta \Rightarrow f(x) > M$$

In other words, $f(x)$ can be made as large as we wish (greater than any M that we choose) by taking x to be sufficiently close to but to the right of c. There are corresponding definitions of

$$\lim_{x\to c^+} f(x) = -\infty \quad \lim_{x\to c^-} f(x) = \infty \quad \lim_{x\to c^-} f(x) = -\infty$$
$$\lim_{x\to\infty} f(x) = \infty \quad \lim_{x\to\infty} f(x) = -\infty \quad \lim_{x\to-\infty} f(x) = \infty \quad \lim_{x\to-\infty} f(x) = -\infty$$

(See Problems 51 and 52.)

EXAMPLE 5 Find $\lim_{x\to 1^-} \frac{1}{(x-1)^2}$ and $\lim_{x\to 1^+} \frac{1}{(x-1)^2}$.

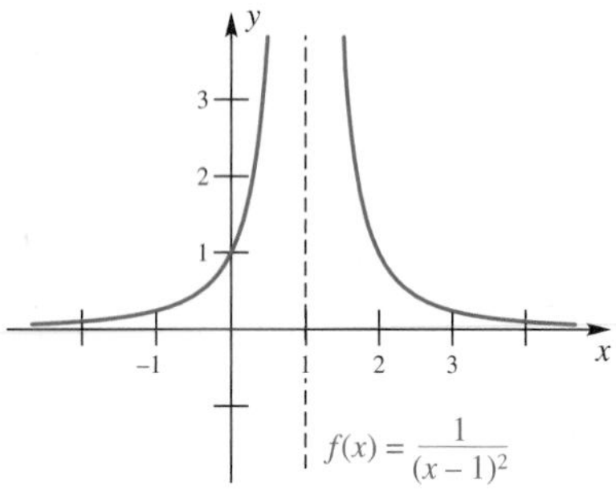

Figure 7

SOLUTION The graph of $f(x) = 1/(x-1)^2$ is shown in Figure 7. As $x \to 1^+$, the denominator remains positive but goes to zero, while the numerator is 1 for all x. Thus, the ratio $1/(x-1)^2$ can be made arbitrarily large by restricting x to be near, but to the right of, 1. Similarly, as $x \to 1^-$, the denominator is positive and can be made arbitrarily close to 0. Thus $1/(x-1)^2$ can be made arbitrarily large by restricting x to be near, but to the left of, 1. We therefore conclude that

$$\lim_{x\to 1^+} \frac{1}{(x-1)^2} = \infty \quad \text{and} \quad \lim_{x\to 1^-} \frac{1}{(x-1)^2} = \infty$$

Since both limits are ∞, we could also write

$$\lim_{x\to 1} \frac{1}{(x-1)^2} = \infty$$

■

EXAMPLE 6 Find $\lim_{x\to 2^+} \frac{x+1}{x^2-5x+6}$.

SOLUTION

$$\lim_{x\to 2^+} \frac{x+1}{x^2-5x+6} = \lim_{x\to 2^+} \frac{x+1}{(x-3)(x-2)}$$

As $x \to 2^+$ we see that $x + 1 \to 3$, $x - 3 \to -1$, and $x - 2 \to 0^+$; thus, the numerator is approaching 3, but the denominator is negative and approaching 0. We conclude that

$$\lim_{x\to 2^+} \frac{x+1}{(x-3)(x-2)} = -\infty$$

■

Relation to Asymptotes Asymptotes were discussed briefly in Section 0.5, but now we can say more about them. The line $x = c$ is a **vertical asymptote** of the graph of $y = f(x)$ if any of the following four statements is true.

1. $\lim_{x\to c^+} f(x) = \infty$
2. $\lim_{x\to c^+} f(x) = -\infty$
3. $\lim_{x\to c^-} f(x) = \infty$
4. $\lim_{x\to c^-} f(x) = -\infty$

Thus, in Figure 6, the line $x = 2$ is a vertical asymptote. Likewise, the lines $x = 2$ and $x = 3$, although not shown graphically, are vertical asymptotes in Example 6.

In a similar vein, the line $y = b$ is a **horizontal asymptote** of the graph of $y = f(x)$ if either

$$\lim_{x\to\infty} f(x) = b \quad \text{or} \quad \lim_{x\to-\infty} f(x) = b$$

The line $y = 0$ is a horizontal asymptote in both Figures 6 and 7.

EXAMPLE 7 Find the vertical and horizontal asymptotes of the graph of $y = f(x)$ if

$$f(x) = \frac{2x}{x-1}$$

> **Do Infinite Limits Exist?**
>
> In previous sections we required that a limit be equal to a real number. For example, we said that $\lim_{x\to 2^+} \frac{1}{x-2}$ does not exist because $1/(x-2)$ does not approach a real number as x approaches 2 from the right. Many mathematicians maintain that this limit does not exist even though we write $\lim_{x\to 2^+} \frac{1}{x-2} = \infty$; to say that the limit is ∞ is to describe the particular way in which the limit does not exist. Here we will use the phrase "exists in the infinite sense" to describe such limits.

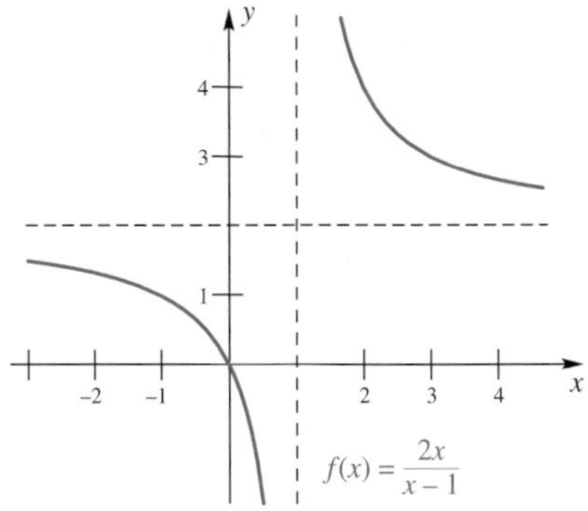

Figure 8

SOLUTION We often have a vertical asymptote at a point where the denominator is zero, and in this case we do because

$$\lim_{x\to 1^+} \frac{2x}{x-1} = \infty \quad \text{and} \quad \lim_{x\to 1^-} \frac{2x}{x-1} = -\infty$$

On the other hand,

$$\lim_{x\to\infty} \frac{2x}{x-1} = \lim_{x\to\infty} \frac{2}{1 - 1/x} = 2 \quad \text{and} \quad \lim_{x\to-\infty} \frac{2x}{x-1} = 2$$

and so $y = 2$ is a horizontal asymptote. The graph of $y = 2x/(x-1)$ is shown in Figure 8. ■

Concepts Review

1. To say that $x \to \infty$ means that _____; to say that $\lim_{x\to\infty} f(x) = L$ means that _____. Give your answers in informal language.

2. To say that $\lim_{x\to c^+} f(x) = \infty$ means that _____; to say that $\lim_{x\to c^-} f(x) = -\infty$ means that _____. Give your answers in informal language.

3. If $\lim_{x\to\infty} f(x) = 6$, then the line _____ is a _____ asymptote of the graph of $y = f(x)$.

4. If $\lim_{x\to 6^+} f(x) = \infty$, then the line _____ is a _____ asymptote of the graph of $y = f(x)$.

Problem Set 1.5

In Problems 1–42, find the limits.

1. $\lim_{x\to\infty} \frac{x}{x-5}$

2. $\lim_{x\to\infty} \frac{x^2}{5-x^3}$

3. $\lim_{t\to-\infty} \frac{t^2}{7-t^2}$

4. $\lim_{t\to-\infty} \frac{t}{t-5}$

5. $\lim_{x\to\infty} \frac{x^2}{(x-5)(3-x)}$

6. $\lim_{x\to\infty} \frac{x^2}{x^2-8x+15}$

7. $\lim_{x\to\infty} \frac{x^3}{2x^3-100x^2}$

8. $\lim_{\theta\to-\infty} \frac{\pi\theta^5}{\theta^5-5\theta^4}$

9. $\lim_{x\to\infty} \frac{3x^3-x^2}{\pi x^3-5x^2}$

10. $\lim_{\theta\to\infty} \frac{\sin^2\theta}{\theta^2-5}$

11. $\lim_{x\to\infty} \frac{3\sqrt{x^3}+3x}{\sqrt{2x^3}}$

12. $\lim_{x\to\infty} \sqrt[3]{\frac{\pi x^3+3x}{\sqrt{2x^3}+7x}}$

13. $\lim_{x\to\infty} \sqrt[3]{\frac{1+8x^2}{x^2+4}}$

14. $\lim_{x\to\infty} \sqrt{\frac{x^2+x+3}{(x-1)(x+1)}}$

15. $\lim_{n\to\infty} \frac{n}{2n+1}$

16. $\lim_{n\to\infty} \frac{n^2}{n^2+1}$

17. $\lim_{n\to\infty} \frac{n^2}{n+1}$

18. $\lim_{n\to\infty} \frac{n}{n^2+1}$

19. $\lim_{x\to\infty} \frac{2x+1}{\sqrt{x^2+3}}$. *Hint:* Divide numerator and denominator by x. Note that, for $x > 0$, $\sqrt{x^2+3}/x = \sqrt{(x^2+3)/x^2}$.

20. $\lim_{x\to\infty} \frac{\sqrt{2x+1}}{x+4}$

21. $\lim_{x\to\infty} \left(\sqrt{2x^2+3} - \sqrt{2x^2-5}\right)$. *Hint:* Multiply and divide by $\sqrt{2x^2+3} + \sqrt{2x^2-5}$.

22. $\lim_{x\to\infty} \left(\sqrt{x^2+2x} - x\right)$

23. $\lim_{y\to-\infty} \frac{9y^3+1}{y^2-2y+2}$. *Hint:* Divide numerator and denominator by y^2.

24. $\lim_{x\to\infty} \frac{a_0x^n + a_1x^{n-1} + \cdots + a_{n-1}x + a_n}{b_0x^n + b_1x^{n-1} + \cdots + b_{n-1}x + b_n}$, where $a_0 \neq 0$, $b_0 \neq 0$, and n is a natural number.

25. $\lim_{n\to\infty} \frac{n}{\sqrt{n^2+1}}$

26. $\lim_{n\to\infty} \frac{n^2}{\sqrt{n^3+2n+1}}$

27. $\lim_{x\to 4^+} \frac{x}{x-4}$

28. $\lim_{t\to -3^+} \frac{t^2-9}{t+3}$

29. $\lim_{t\to 3^-} \frac{t^2}{9-t^2}$

30. $\lim_{x\to\sqrt[3]{5}^+} \frac{x^2}{5-x^3}$

31. $\lim_{x\to 5^-} \frac{x^2}{(x-5)(3-x)}$

32. $\lim_{\theta\to\pi^+} \frac{\theta^2}{\sin\theta}$

33. $\lim_{x\to 3^-} \frac{x^3}{x-3}$

34. $\lim_{\theta\to(\pi/2)^+} \frac{\pi\theta}{\cos\theta}$

35. $\lim_{x\to 3^-} \frac{x^2-x-6}{x-3}$

36. $\lim_{x\to 2^+} \frac{x^2+2x-8}{x^2-4}$

37. $\lim_{x\to 0^+} \frac{[\![x]\!]}{x}$

38. $\lim_{x\to 0^-} \frac{[\![x]\!]}{x}$

39. $\lim_{x \to 0^-} \frac{|x|}{x}$

40. $\lim_{x \to 0^+} \frac{|x|}{x}$

41. $\lim_{x \to 0^-} \frac{1 + \cos x}{\sin x}$

42. $\lim_{x \to \infty} \frac{\sin x}{x}$

GC *In Problems 43–48, find the horizontal and vertical asymptotes for the graphs of the indicated functions. Then sketch their graphs.*

43. $f(x) = \frac{3}{x + 1}$

44. $f(x) = \frac{3}{(x + 1)^2}$

45. $F(x) = \frac{2x}{x - 3}$

46. $F(x) = \frac{3}{9 - x^2}$

47. $g(x) = \frac{14}{2x^2 + 7}$

48. $g(x) = \frac{2x}{\sqrt{x^2 + 5}}$

49. The line $y = ax + b$ is called an **oblique asymptote** to the graph of $y = f(x)$ if either $\lim_{x \to \infty} [f(x) - (ax + b)] = 0$ or $\lim_{x \to -\infty} [f(x) - (ax + b)] = 0$. Find the oblique asymptote for

$$f(x) = \frac{2x^4 + 3x^3 - 2x - 4}{x^3 - 1}$$

Hint: Begin by dividing the denominator into the numerator.

50. Find the oblique asymptote for

$$f(x) = \frac{3x^3 + 4x^2 - x + 1}{x^2 + 1}$$

51. Using the symbols M and δ, give precise definitions of each expression.

(a) $\lim_{x \to c^+} f(x) = -\infty$ (b) $\lim_{x \to c^-} f(x) = \infty$

52. Using the symbols M and N, give precise definitions of each expression.

(a) $\lim_{x \to \infty} f(x) = \infty$ (b) $\lim_{x \to -\infty} f(x) = \infty$

53. Give a rigorous proof that if $\lim_{x \to \infty} f(x) = A$ and $\lim_{x \to \infty} g(x) = B$, then

$$\lim_{x \to \infty} [f(x) + g(x)] = A + B$$

54. We have given meaning to $\lim_{x \to A} f(x)$ for $A = a, a^-, a^+, -\infty, \infty$. Moreover, in each case, this limit may be L (finite), $-\infty$, ∞, or may fail to exist in any sense. Make a table illustrating each of the 20 possible cases.

55. Find each of the following limits or indicate that it does not exist even in the infinite sense.

(a) $\lim_{x \to \infty} \sin x$ (b) $\lim_{x \to \infty} \sin \frac{1}{x}$

(c) $\lim_{x \to \infty} x \sin \frac{1}{x}$ (d) $\lim_{x \to \infty} x^{3/2} \sin \frac{1}{x}$

(e) $\lim_{x \to \infty} x^{-1/2} \sin x$ (f) $\lim_{x \to \infty} \sin\left(\frac{\pi}{6} + \frac{1}{x}\right)$

(g) $\lim_{x \to \infty} \sin\left(x + \frac{1}{x}\right)$ (h) $\lim_{x \to \infty} \left[\sin\left(x + \frac{1}{x}\right) - \sin x\right]$

56. Einstein's Special Theory of Relativity says that the mass $m(v)$ of an object is related to its velocity v by

$$m(v) = \frac{m_0}{\sqrt{1 - v^2/c^2}}$$

Here m_0 is the rest mass and c is the velocity of light. What is $\lim_{v \to c^-} m(v)$?

GC *Use a computer or a graphing calculator to find the limits in Problems 57–64. Begin by plotting the function in an appropriate window.*

57. $\lim_{x \to \infty} \frac{3x^2 + x + 1}{2x^2 - 1}$

58. $\lim_{x \to -\infty} \sqrt{\frac{2x^2 - 3x}{5x^2 + 1}}$

59. $\lim_{x \to -\infty} \left(\sqrt{2x^2 + 3x} - \sqrt{2x^2 - 5}\right)$

60. $\lim_{x \to \infty} \frac{2x + 1}{\sqrt{3x^2 + 1}}$

61. $\lim_{x \to \infty} \left(1 + \frac{1}{x}\right)^{10}$

62. $\lim_{x \to \infty} \left(1 + \frac{1}{x}\right)^{x}$

63. $\lim_{x \to \infty} \left(1 + \frac{1}{x}\right)^{x^2}$

64. $\lim_{x \to \infty} \left(1 + \frac{1}{x}\right)^{\sin x}$

CAS *Find the one-sided limits in Problems 65–71. Begin by plotting the function in an appropriate window. Your computer may indicate that some of these limits do not exist, but, if so, you should be able to interpret the answer as either ∞ or $-\infty$.*

65. $\lim_{x \to 3^-} \frac{\sin|x - 3|}{x - 3}$

66. $\lim_{x \to 3^-} \frac{\sin|x - 3|}{\tan(x - 3)}$

67. $\lim_{x \to 3^-} \frac{\cos(x - 3)}{x - 3}$

68. $\lim_{x \to \frac{\pi}{2}^+} \frac{\cos x}{x - \pi/2}$

69. $\lim_{x \to 0^+} \left(1 + \sqrt{x}\right)^{1/\sqrt{x}}$

70. $\lim_{x \to 0^+} \left(1 + \sqrt{x}\right)^{1/x}$

71. $\lim_{x \to 0^+} \left(1 + \sqrt{x}\right)^{x}$

Answers to Concepts Review: **1.** x increases without bound; $f(x)$ gets close to L as x increases without bound **2.** $f(x)$ increases without bound as x approaches c from the right; $f(x)$ decreases without bound as x approaches c from the left **3.** $y = 6$; horizontal **4.** $x = 6$; vertical

1.6 Continuity of Functions

In mathematics and science, we use the word *continuous* to describe a process that goes on without abrupt changes. In fact, our experience leads us to assume that this is an essential feature of many natural processes. It is this notion as it pertains to functions that we now want to make precise. In the three graphs shown in Figure 1, only the third graph exhibits continuity at c. In the first two graphs, either $\lim_{x \to c} f(x)$ does not exist, or it exists but does not equal $f(c)$. Only in the third graph does $\lim_{x \to c} f(x) = f(c)$.

A Discontinuous Machine

A good example of a discontinuous machine is the postage machine, which (in 2005) charged \$0.37 for a 1-ounce letter but \$0.60 for a letter the least little bit over 1 ounce.

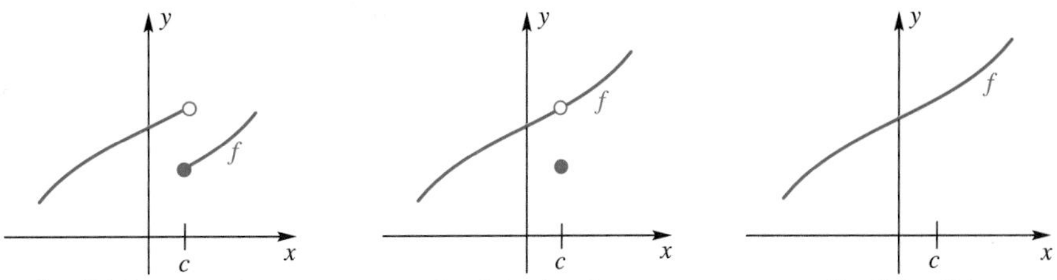

$\lim_{x \to c} f(x)$ does not exist.

$\lim_{x \to c} f(x)$ exists, but $\lim_{x \to c} f(x) \neq f(c)$.

$\lim_{x \to c} f(x) = f(c)$

Figure 1

Here is the formal definition.

Definition Continuity at a Point

Let f be defined on an open interval containing c. We say that f is **continuous** at c if

$$\lim_{x \to c} f(x) = f(c)$$

We mean by this definition to require three things:

1. $\lim_{x \to c} f(x)$ exists,
2. $f(c)$ exists (i.e., c is in the domain of f), and
3. $\lim_{x \to c} f(x) = f(c)$.

If any one of these three fails, then f is **discontinuous** at c. Thus, the functions represented by the first and second graphs of Figure 1 are discontinuous at c. They do appear, however, to be continuous at other points of their domains.

EXAMPLE 1 Let $f(x) = \dfrac{x^2 - 4}{x - 2}$, $x \neq 2$. How should f be defined at $x = 2$ in order to make it continuous there?

SOLUTION

$$\lim_{x \to 2} \frac{x^2 - 4}{x - 2} = \lim_{x \to 2} \frac{(x - 2)(x + 2)}{x - 2} = \lim_{x \to 2}(x + 2) = 4$$

Therefore, we define $f(2) = 4$. The graph of the resulting function is shown in Figure 2. In fact, we see that $f(x) = x + 2$ for all x. ■

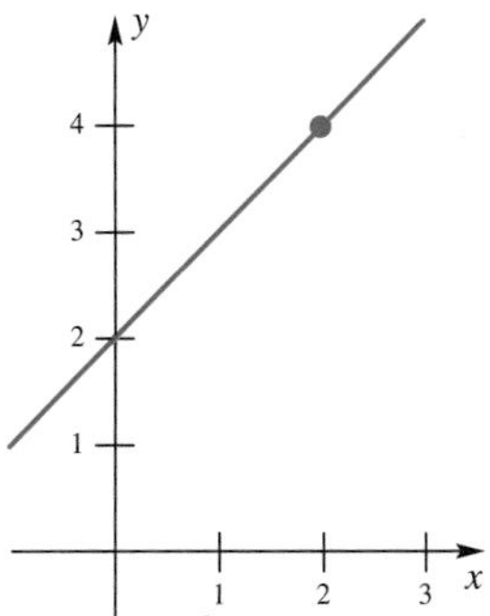

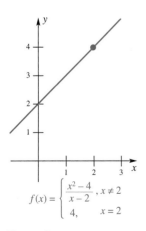

$$f(x) = \begin{cases} \dfrac{x^2 - 4}{x - 2}, & x \neq 2 \\ 4, & x = 2 \end{cases}$$

Figure 2

A point of discontinuity c is called **removable** if the function can be defined or redefined at c so as to make the function continuous. Otherwise, a point of discontinuity is called **nonremovable.** The function f in Example 1 has a removable discontinuity at 2 because we could define $f(2) = 4$ and the function would be continuous there.

Continuity of Familiar Functions

Most functions that we will meet in this book are either (1) continuous everywhere or (2) continuous everywhere except at a few points. In particular, Theorem 1.3B implies the following result.

Theorem A Continuity of Polynomial and Rational Functions

A polynomial function is continuous at every real number c. A rational function is continuous at every real number c in its domain, that is, everywhere except where its denominator is zero.

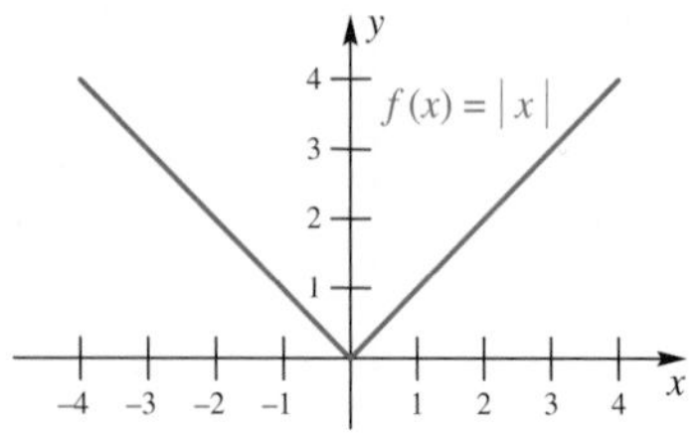

Figure 3

Recall the absolute value function $f(x) = |x|$; its graph is shown in Figure 3. For $x < 0, f(x) = -x$, a polynomial; for $x > 0, f(x) = x$, another polynomial. Thus, $|x|$ is continuous at all numbers different from 0 by Theorem A. But

$$\lim_{x \to 0} |x| = 0 = |0|$$

(see Problem 27 of Section 1.2). Therefore, $|x|$ is also continuous at 0; it is continuous everywhere.

By the Main Limit Theorem (Theorem 1.3A)

$$\lim_{x \to c} \sqrt[n]{x} = \sqrt[n]{\lim_{x \to c} x} = \sqrt[n]{c}$$

provided $c > 0$ when n is even. This means that $f(x) = \sqrt[n]{x}$ is continuous at each point where it makes sense to talk about continuity. In particular, $f(x) = \sqrt{x}$ is continuous at each real number $c > 0$ (Figure 4). We summarize.

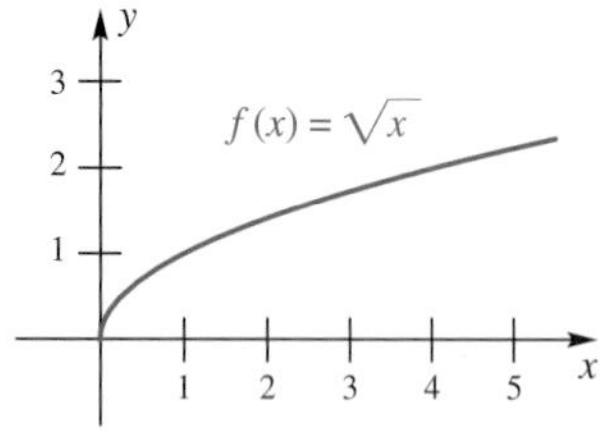

Figure 4

Theorem B Continuity of Absolute Value and *n*th Root Functions

The absolute value function is continuous at every real number c. If n is odd, the nth root function is continuous at every real number c; if n is even, the nth-root function is continuous at every positive real number c.

Continuity under Function Operations Do the standard function operations preserve continuity? Yes, according to the next theorem. In it, f and g are functions, k is a constant, and n is a positive integer.

Theorem C Continuity under Function Operations

If f and g are continuous at c, then so are $kf, f + g, f - g, f \cdot g, f/g$ (provided that $g(c) \neq 0$), f^n, and $\sqrt[n]{f}$ (provided that $f(c) > 0$ if n is even).

Proof All these results are easy consequences of the corresponding facts for limits from Theorem 1.3A. For example, that theorem, combined with the fact that f and g are continuous at c, gives

$$\lim_{x \to c} f(x)g(x) = \lim_{x \to c} f(x) \cdot \lim_{x \to c} g(x) = f(c)g(c)$$

This is precisely what it means to say that $f \cdot g$ is continuous at c. ■

EXAMPLE 2 At what numbers is $F(x) = (3|x| - x^2)/\left(\sqrt{x} + \sqrt[3]{x}\right)$ continuous?

SOLUTION We need not even consider nonpositive numbers, since F is not defined at such numbers. For any positive number, the functions $\sqrt{x}, \sqrt[3]{x}, |x|$, and x^2 are all continuous (Theorems A and B). It follows from Theorem C that $3|x|$, $3|x| - x^2$, $\sqrt{x} + \sqrt[3]{x}$, and finally,

$$\frac{(3|x| - x^2)}{\left(\sqrt{x} + \sqrt[3]{x}\right)}$$

are continuous at each positive number. ■

The continuity of the trigonometric functions follows from Theorem 1.4A.

Theorem D Continuity of Trigonometric Functions

The sine and cosine functions are continuous at every real number c. The functions $\tan x$, $\cot x$, $\sec x$, and $\csc x$ are continuous at every real number c in their domains.

Proof Theorem 1.4A says that for every real number c in the function's domain, $\lim_{x \to c} \sin x = \sin c$, $\lim_{x \to c} \cos x = \cos c$, and so forth, for all six of the trigonometric functions. These are exactly the conditions required for these functions to be continuous at every real number in their respective domains. ■

EXAMPLE 3 Determine all points of discontinuity of $f(x) = \dfrac{\sin x}{x(1 - x)}$, $x \neq 0, 1$. Classify each point of discontinuity as removable or nonremovable.

SOLUTION By Theorem D, the numerator is continuous at every real number. The denominator is also continuous at every real number, but when $x = 0$ or $x = 1$, the denominator is 0. Thus, by Theorem C, f is continuous at every real number except $x = 0$ and $x = 1$. Since

$$\lim_{x \to 0} \frac{\sin x}{x(1 - x)} = \lim_{x \to 0} \frac{\sin x}{x} \cdot \lim_{x \to 0} \frac{1}{(1 - x)} = (1)(1) = 1$$

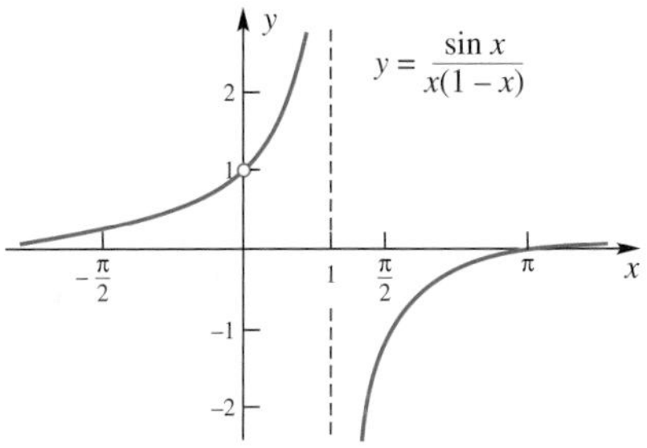

Figure 5

we could define $f(0) = 1$ and the function would continuous there. Thus, $x = 0$ is a removable discontinuity. Also, since

$$\lim_{x \to 1^+} \frac{\sin x}{x(1 - x)} = -\infty \quad \text{and} \quad \lim_{x \to 1^-} \frac{\sin x}{x(1 - x)} = \infty$$

there is no way to define $f(1)$ to make f continuous at $x = 1$. Thus $x = 1$ is a nonremovable discontinuity. A graph of $y = f(x)$ is shown in Figure 5. ■

There is another functional operation, composition, that will be very important in later work. It, too, preserves continuity.

Theorem E **Composite Limit Theorem**

If $\lim_{x \to c} g(x) = L$ and if f is continuous at L, then

$$\lim_{x \to c} f(g(x)) = f\left(\lim_{x \to c} g(x)\right) = f(L)$$

In particular, if g is continuous at c and f is continuous at $g(c)$, then the composite $f \circ g$ is continuous at c.

Proof of Theorem E (Optional)

Proof Let $\varepsilon > 0$ be given. Since f is continuous at L, there is a corresponding $\delta_1 > 0$ such that

$$|t - L| < \delta_1 \Rightarrow |f(t) - f(L)| < \varepsilon$$

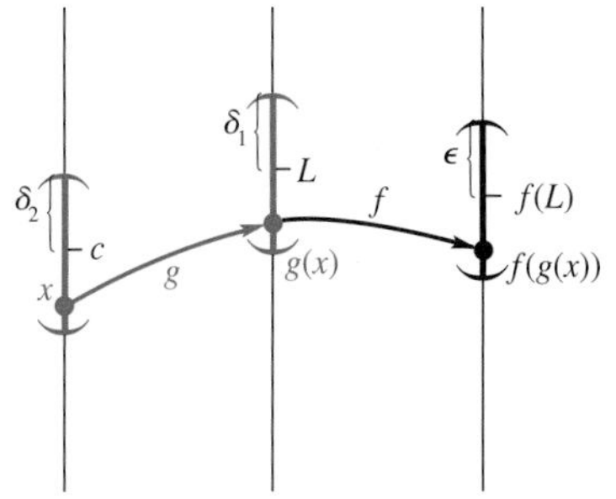

Figure 6

and so (see Figure 6)

$$|g(x) - L| < \delta_1 \Rightarrow |f(g(x)) - f(L)| < \varepsilon$$

But because $\lim_{x \to c} g(x) = L$, for a given $\delta_1 > 0$ there is a corresponding $\delta_2 > 0$ such that

$$0 < |x - c| < \delta_2 \Rightarrow |g(x) - L| < \delta_1$$

When we put these two facts together, we have

$$0 < |x - c| < \delta_2 \Rightarrow |f(g(x)) - f(L)| < \varepsilon$$

This shows that

$$\lim_{x \to c} f(g(x)) = f(L)$$

The second statement in Theorem E follows from the observation that if g is continuous at c then $L = g(c)$. ■

EXAMPLE 4 Show that $h(x) = |x^2 - 3x + 6|$ is continuous at each real number.

SOLUTION Let $f(x) = |x|$ and $g(x) = x^2 - 3x + 6$. Both are continuous at each real number, and so their composite

$$h(x) = f(g(x)) = |x^2 - 3x + 6|$$

is also. ■

EXAMPLE 5 Show that

$$h(x) = \sin \frac{x^4 - 3x + 1}{x^2 - x - 6}$$

is continuous except at 3 and -2.

SOLUTION $x^2 - x - 6 = (x - 3)(x + 2)$. Thus, the rational function

$$g(x) = \frac{x^4 - 3x + 1}{x^2 - x - 6}$$

is continuous except at 3 and -2 (Theorem A). We know from Theorem D that the sine function is continuous at every real number. Thus, from Theorem E, we conclude that, since $h(x) = \sin(g(x))$, h is also continuous except at 3 and -2. ■

Continuity on an Interval So far, we have been discussing continuity at a point. We now wish to discuss continuity on an interval. Continuity on an interval ought to mean continuity at each point of that interval. This is exactly what it does mean for an *open* interval.

When we consider a closed interval $[a, b]$, we face a problem. It might be that f is not even defined to the left of a (e.g., this occurs for $f(x) = \sqrt{x}$ at $a = 0$), so, strictly speaking, $\lim_{x \to a} f(x)$ does not exist. We choose to get around this problem by calling f continuous on $[a, b]$ if it is continuous at each point of (a, b) and if $\lim_{x \to a^+} f(x) = f(a)$ and $\lim_{x \to b^-} f(x) = f(b)$. We summarize in a formal definition.

Definition Continuity on an Interval

The function f is **right continuous** at a if $\lim_{x \to a^+} f(x) = f(a)$ and **left continuous** at b if $\lim_{x \to b^-} f(x) = f(b)$.

We say f is **continuous on an open interval** if it is continuous at each point of that interval. It is **continuous on the closed interval** $[a, b]$ if it is continuous on (a, b), right continuous at a, and left continuous at b.

For example, it is correct to say that $f(x) = 1/x$ is continuous on $(0, 1)$ and that $g(x) = \sqrt{x}$ is continuous on $[0, 1]$.

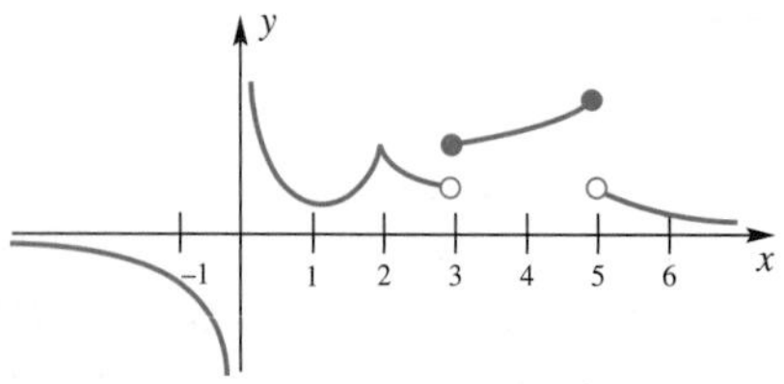

Figure 7

EXAMPLE 6 Using the definition above, describe the continuity properties of the function whose graph is sketched in Figure 7.

SOLUTION The function appears to be continuous on the open intervals $(-\infty, 0)$ $(0, 3)$, and $(5, \infty)$, and also on the closed interval $[3, 5]$. ■

EXAMPLE 7 What is the largest interval over which the function defined by $g(x) = \sqrt{4 - x^2}$ is continuous?

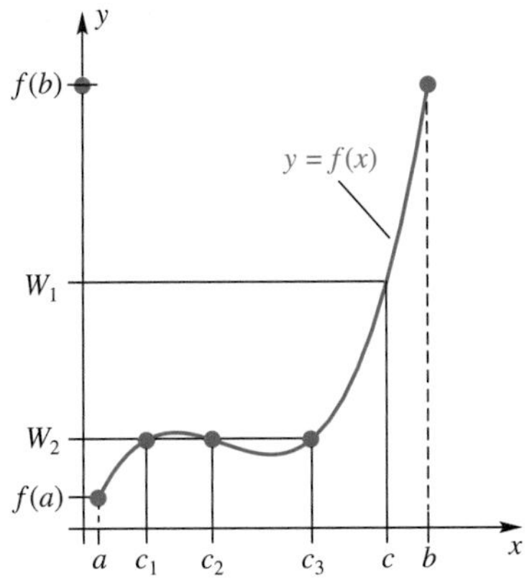

Figure 8

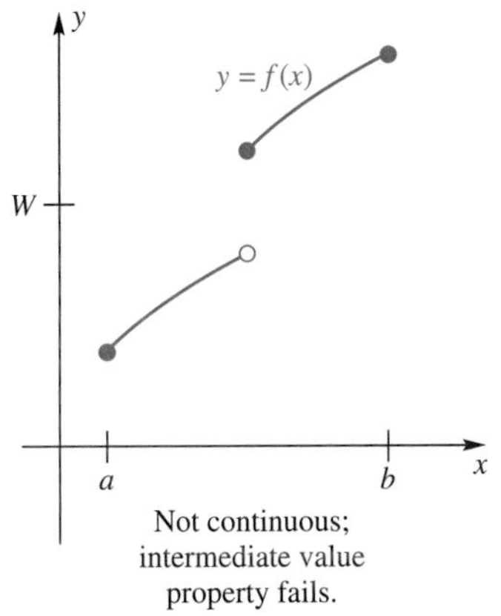

Figure 9

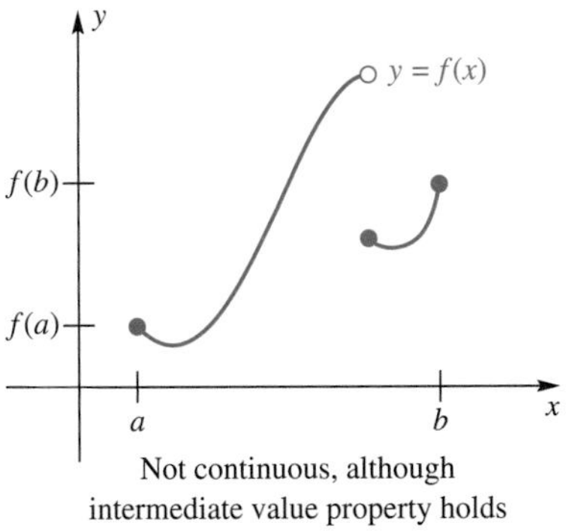

Figure 10

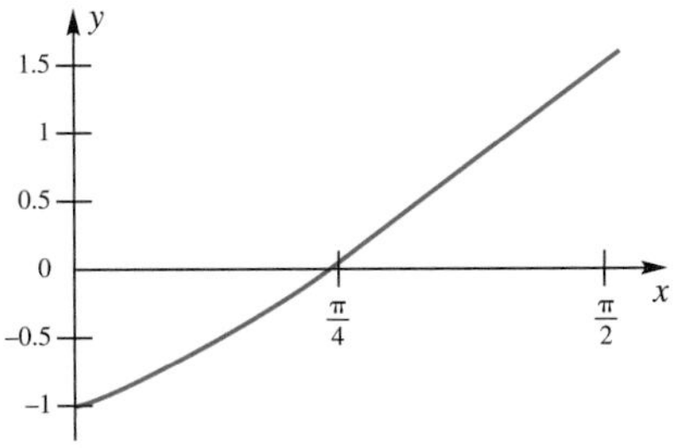

Figure 11

SOLUTION The domain of g is the interval $[-2, 2]$. If c is in the open interval $(-2, 2)$, then g is continuous at c by Theorem E; hence, g is continuous on $(-2, 2)$. The one-sided limits are

$$\lim_{x\to -2^+} \sqrt{4 - x^2} = \sqrt{4 - \left(\lim_{x\to -2^+} x\right)^2}\sqrt{4 - 4} = 0 = g(-2)$$

and

$$\lim_{x\to 2^-} \sqrt{4 - x^2} = \sqrt{4 - \left(\lim_{x\to 2^-} x\right)^2} = \sqrt{4 - 4} = 0 = g(2)$$

This implies that g is right continuous at -2 and left continuous at 2. Thus, g is continuous on its domain, the closed interval $[-2, 2]$. ■

Intuitively, for f to be continuous on $[a, b]$ means that the graph of f on $[a, b]$ should have no jumps, so we should be able to "draw" the graph of f from the point $(a, f(a))$ to the point $(b, f(b))$ without lifting our pencil from the paper. Thus, the function f should take on every value between $f(a)$ and $f(b)$. This property is stated more precisely in Theorem F.

Theorem F Intermediate Value Theorem

Let f be a function defined on $[a, b]$ and let W be a number between $f(a)$ and $f(b)$. If f is continuous on $[a, b]$, then there is at least one number c between a and b such that $f(c) = W$.

Figure 8 shows the graph of a function $f(x)$ that is continuous on $[a, b]$. The Intermediate Value Theorem says that for every W in $(f(a), f(b))$ there must be a c in $[a, b]$ such that $f(c) = W$. In other words, f takes on every value between $f(a)$ and $f(b)$. Continuity is needed for this theorem, for otherwise it is possible to find a function f and a number W between $f(a)$ and $f(b)$ such that there is no c in $[a, b]$ that satisfies $f(c) = W$. Figure 9 shows an example of such a function.

It seems clear that continuity is sufficient, although a formal proof of this result turns out to be difficult. We leave the proof to more advanced works.

The converse of this theorem, which is not true in general, says that if f takes on every value between $f(a)$ and $f(b)$ then f is continuous. Figures 8 and 10 show functions that take on all values between $f(a)$ and $f(b)$, but the function in Figure 10 is not continuous on $[a, b]$. Just because a function has the intermediate value property does not mean that it must be continuous.

The Intermediate Value Theorem can be used to tell us something about the solutions of equations, as the next example shows.

EXAMPLE 8 Use the Intermediate Value Theorem to show that the equation $x - \cos x = 0$ has a solution between $x = 0$ and $x = \pi/2$.

SOLUTION Let $f(x) = x - \cos x$, and let $W = 0$. Then $f(0) = 0 - \cos 0 = -1$ and $f(\pi/2) = \pi/2 - \cos \pi/2 = \pi/2$. Since f is continuous on $[0, \pi/2]$ and since $W = 0$ is between $f(0)$ and $f(\pi/2)$, the Intermediate Value Theorem implies the existence of a c in the interval $(0, \pi/2)$ with the property that $f(c) = 0$. Such a c is a solution to the equation $x - \cos x = 0$. Figure 11 suggests that there is exactly one such c.

We can go one step further. The midpoint of the interval $[0, \pi/2]$ is the point $x = \pi/4$. When we evaluate $f(\pi/4)$, we get

$$f(\pi/4) = \frac{\pi}{4} - \cos\frac{\pi}{4} = \frac{\pi}{4} - \frac{\sqrt{2}}{2} \approx 0.0782914$$

which is greater than 0. Thus, $f(0) < 0$ and $f(\pi/4) > 0$, so another application of the Intermediate Value Theorem tells us that there exists a c between 0 and $\pi/4$ such that $f(c) = 0$. We have thus narrowed down the interval containing the desired c from $[0, \pi/2]$ to $[0, \pi/4]$. There is nothing stopping us from selecting the midpoint of $[0, \pi/4]$ and evaluating f at that point, thereby narrowing even further the interval containing c. This process could be continued indefinitely until we find that c is in a sufficiently small interval. This method of zeroing in on a solution is called the *bisection method*, and we will study it further in Section 3.7. ■

The Intermediate Value Theorem can also lead to some surprising results.

EXAMPLE 9 Use the Intermediate Value Theorem to show that on a circular wire ring there are always two points opposite from each other with the same temperature.

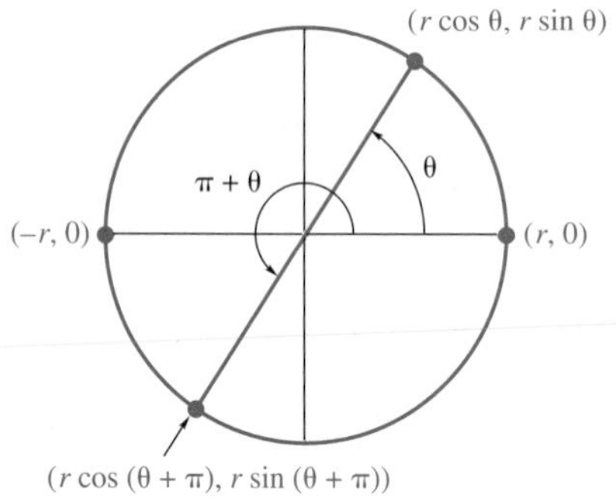

Figure 12

SOLUTION Choose coordinates for this problem so that the center of the ring is the origin, and let r be the radius of the ring. (See Figure 12.) Define $T(x, y)$ to be the temperature at the point (x, y). Consider a diameter of the circle that makes an angle θ with the x-axis, and define $f(\theta)$ to be the temperature difference between the points that make angles of θ and $\theta + \pi$; that is,

$$f(\theta) = T(r\cos\theta, r\sin\theta) - T(r\cos(\theta + \pi), r\sin(\theta + \pi))$$

With this definition

$$f(0) = T(r, 0) - T(-r, 0)$$
$$f(\pi) = T(-r, 0) - T(r, 0) = -\left[T(r, 0) - T(-r, 0)\right] = -f(0)$$

Thus, either $f(0)$ and $f(\pi)$ are both zero, or one is positive and the other is negative. If both are zero, then we have found the required two points. Otherwise, we can apply the Intermediate Value Theorem. Assuming that temperature varies continuously, we conclude that there exists a c between 0 and π such that $f(c) = 0$. Thus, for the two points at the angles c and $c + \pi$, the temperatures are the same. ■

Concepts Review

1. A function f is continuous at c if _____ $= f(c)$.

2. The function $f(x) = [\![x]\!]$ is discontinuous at _____.

3. A function f is said to be continuous on a closed interval $[a, b]$ if it is continuous at every point of (a, b) and if _____ and _____.

4. The Intermediate Value Theorem says that if a function f is continuous on $[a, b]$ and W is a number between $f(a)$ and $f(b)$, then there is a number c between _____ and _____ such that _____.

Problem Set 1.6

In Problems 1–15, state whether the indicated function is continuous at 3. If it is not continuous, tell why.

1. $f(x) = (x - 3)(x - 4)$

2. $g(x) = x^2 - 9$

3. $h(x) = \dfrac{3}{x - 3}$

4. $g(t) = \sqrt{t - 4}$

5. $h(t) = \dfrac{|t - 3|}{t - 3}$

6. $h(t) = \dfrac{|\sqrt{(t - 3)^4}|}{t - 3}$

7. $f(t) = |t|$

8. $g(t) = |t - 2|$

9. $h(x) = \dfrac{x^2 - 9}{x - 3}$

10. $f(x) = \dfrac{21 - 7x}{x - 3}$

11. $r(t) = \begin{cases} \dfrac{t^3 - 27}{t - 3} & \text{if } t \neq 3 \\ 27 & \text{if } t = 3 \end{cases}$

12. $r(t) = \begin{cases} \dfrac{t^3 - 27}{t - 3} & \text{if } t \neq 3 \\ 23 & \text{if } t = 3 \end{cases}$

13. $f(t) = \begin{cases} t - 3 & \text{if } t \leq 3 \\ 3 - t & \text{if } t > 3 \end{cases}$

14. $f(t) = \begin{cases} t^2 - 9 & \text{if } t \leq 3 \\ (3 - t)^2 & \text{if } t > 3 \end{cases}$

15. $f(x) = \begin{cases} -3x + 7 & \text{if } x \leq 3 \\ -2 & \text{if } x > 3 \end{cases}$

16. From the graph of g (see Figure 13), indicate the values where g is discontinuous. For each of these values state whether g is continuous from the right, left, or neither.

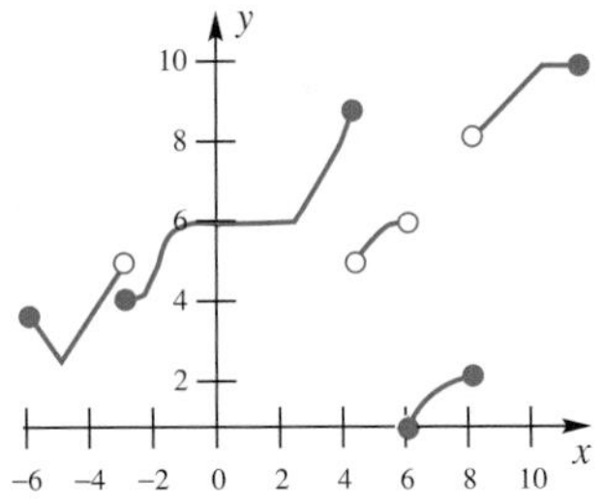

Figure 13

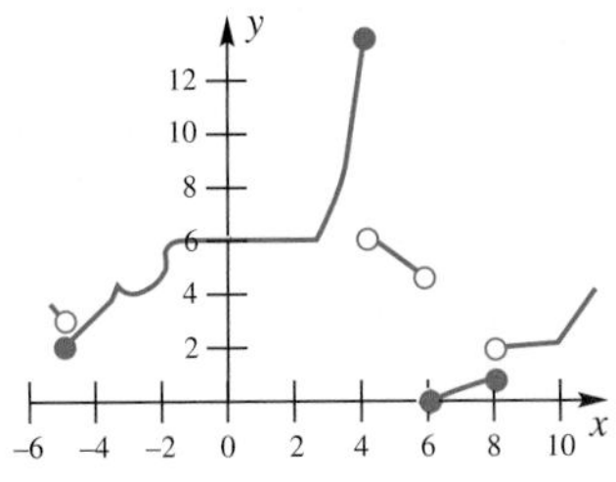

Figure 14

17. From the graph of h given in Figure 14, indicate the intervals on which h is continuous.

In Problems 18–23, the given function is not defined at a certain point. How should it be defined in order to make it continuous at that point? (See Example 1.)

18. $f(x) = \dfrac{x^2 - 49}{x - 7}$ **19.** $f(x) = \dfrac{2x^2 - 18}{3 - x}$

20. $g(\theta) = \dfrac{\sin\theta}{\theta}$ **21.** $H(t) = \dfrac{\sqrt{t} - 1}{t - 1}$

22. $\phi(x) = \dfrac{x^4 + 2x^2 - 3}{x + 1}$ **23.** $F(x) = \sin\dfrac{x^2 - 1}{x + 1}$

In Problems 24–35, at what points, if any, are the functions discontinuous?

24. $f(x) = \dfrac{3x + 7}{(x - 30)(x - \pi)}$

25. $f(x) = \dfrac{33 - x^2}{x\pi + 3x - 3\pi - x^2}$

26. $h(\theta) = |\sin\theta + \cos\theta|$ **27.** $r(\theta) = \tan\theta$

28. $f(u) = \dfrac{2u + 7}{\sqrt{u + 5}}$ **29.** $g(u) = \dfrac{u^2 + |u - 1|}{\sqrt[3]{u + 1}}$

30. $F(x) = \dfrac{1}{\sqrt{4 + x^2}}$ **31.** $G(x) = \dfrac{1}{\sqrt{4 - x^2}}$

32. $f(x) = \begin{cases} x & \text{if } x < 0 \\ x^2 & \text{if } 0 \le x \le 1 \\ 2 - x & \text{if } x > 1 \end{cases}$

33. $g(x) = \begin{cases} x^2 & \text{if } x < 0 \\ -x & \text{if } 0 \le x \le 1 \\ x & \text{if } x > 1 \end{cases}$

34. $f(t) = [\![t]\!]$ **35.** $g(t) = [\![t + \frac{1}{2}]\!]$

36. Sketch the graph of a function f that satisfies all the following conditions.

(a) Its domain is $[-2, 2]$.

(b) $f(-2) = f(-1) = f(1) = f(2) = 1$.

(c) It is discontinuous at -1 and 1.

(d) It is right continuous at -1 and left continuous at 1.

37. Sketch the graph of a function that has domain $[0, 2]$ and is continuous on $[0, 2)$ but not on $[0, 2]$.

38. Sketch the graph of a function that has domain $[0, 6]$ and is continuous on $[0, 2]$ and $(2, 6]$ but is not continuous on $[0, 6]$.

39. Sketch the graph of a function that has domain $[0, 6]$ and is continuous on $(0, 6)$ but not on $[0, 6]$.

40. Let

$$f(x) = \begin{cases} x & \text{if } x \text{ is rational} \\ -x & \text{if } x \text{ is irrational} \end{cases}$$

Sketch the graph of this function as best you can and decide where it is continuous.

In Problems 41–48, determine whether the function is continuous at the given point c. If the function is not continuous, determine whether the discontinuity is removable or nonremovable.

41. $f(x) = \sin x;\ c = 0$ **42.** $f(x) = \dfrac{x^2 - 100}{x - 10};\ c = 10$

43. $f(x) = \dfrac{\sin x}{x};\ c = 0$ **44.** $f(x) = \dfrac{\cos x}{x};\ c = 0$

45. $g(x) = \begin{cases} \dfrac{\sin x}{x}, & x \ne 0 \\ 0, & x = 0 \end{cases}$ **46.** $F(x) = x\sin\dfrac{1}{x};\ c = 0$

47. $f(x) = \sin\dfrac{1}{x};\ c = 0$ **48.** $f(x) = \dfrac{4 - x}{2 - \sqrt{x}};\ c = 4$

49. A cell phone company charges \$0.12 for connecting a call plus \$0.08 per minute or any part thereof (e.g., a phone call lasting 2 minutes and 5 seconds costs $\$0.12 + 3 \times \$0.08$). Sketch a graph of the cost of making a call as a function of the length of time t that the call lasts. Discuss the continuity of this function.

50. A rental car company charges \$20 for one day, allowing up to 200 miles. For each additional 100 miles, or any fraction thereof, the company charges \$18. Sketch a graph of the cost for renting a car for one day as a function of the miles driven. Discuss the continuity of this function.

51. A cab company charges \$2.50 for the first $\frac{1}{4}$ mile and \$0.20 for each additional $\frac{1}{8}$ mile. Sketch a graph of the cost of a cab ride as a function of the number of miles driven. Discuss the continuity of this function.

52. Use the Intermediate Value Theorem to prove that $x^3 + 3x - 2 = 0$ has a real solution between 0 and 1.

53. Use the Intermediate Value Theorem to prove that $(\cos t)t^3 + 6\sin^5 t - 3 = 0$ has a real solution between 0 and 2π.

GC **54.** Use the Intermediate Value Theorem to show that $x^3 - 7x^2 + 14x - 8 = 0$ has at least one solution in the interval $[0, 5]$. Sketch the graph of $y = x^3 - 7x^2 + 14x - 8$ over $[0, 5]$. How many solutions does this equation really have?

GC **55.** Use the Intermediate Value Theorem to show that $\sqrt{x} - \cos x = 0$ has a solution between 0 and $\pi/2$. Zoom in on the graph of $y = \sqrt{x} - \cos x$ to find an interval having length 0.1 that contains this solution.

56. Show that the equation $x^5 + 4x^3 - 7x + 14 = 0$ has at least one real solution.

57. Prove that f is continuous at c if and only if $\lim_{t \to 0} f(c + t) = f(c)$.

58. Prove that if f is continuous at c and $f(c) > 0$ there is an interval $(c - \delta, c + \delta)$ such that $f(x) > 0$ on this interval.

59. Prove that if f is continuous on $[0, 1]$ and satisfies $0 \le f(x) \le 1$ there, then f has a *fixed point*; that is, there is a number c in $[0, 1]$ such that $f(c) = c$. *Hint:* Apply the Intermediate Value Theorem to $g(x) = x - f(x)$.

60. Find the values of a and b so that the following function is continuous everywhere.

$$f(x) = \begin{cases} x + 1 & \text{if } x < 1 \\ ax + b & \text{if } 1 \le x < 2 \\ 3x & \text{if } x \ge 2 \end{cases}$$

61. A stretched elastic string covers the interval $[0, 1]$. The ends are released and the string contracts so that it covers the interval $[a, b]$, $a \ge 0, b \le 1$. Prove that this results in at least one point of the string being where it was originally. See Problem 59.

62. Let $f(x) = \frac{1}{x - 1}$. Then $f(-2) = -\frac{1}{3}$ and $f(2) = 1$. Does the Intermediate Value Theorem imply the existence of a number c between -2 and 2 such that $f(c) = 0$? Explain.

63. Starting at 4 A.M., a hiker slowly climbed to the top of a mountain, arriving at noon. The next day, he returned along the same path, starting at 5 A.M. and getting to the bottom at 11 A.M. Show that at some point along the path his watch showed the same time on both days.

64. Let D be a bounded, but otherwise arbitrary, region in the first quadrant. Given an angle θ, $0 \le \theta \le \pi/2$, D can be circumscribed by a rectangle whose base makes angle θ with the x-axis as shown in Figure 15. Prove that at some angle this rectangle is a square. (This means that *any* bounded region can be circumscribed by a *square*.)

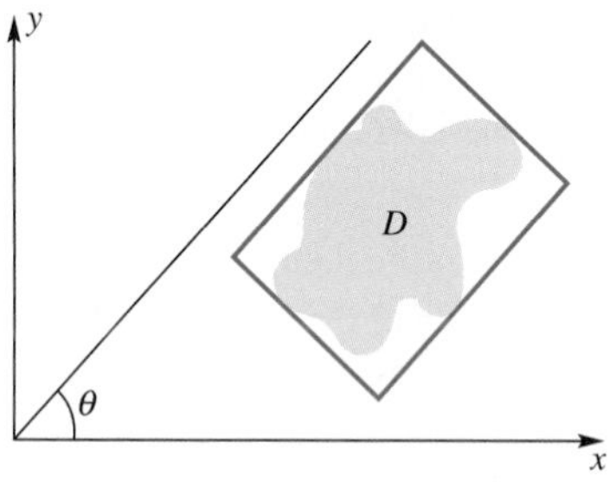

Figure 15

65. The gravitational force exerted by the earth on an object having mass m that is a distance r from the center of the earth is

$$g(r) = \begin{cases} \dfrac{GMmr}{R^3}, & \text{if } r < R \\ \dfrac{GMm}{r^2}, & \text{if } r \ge R \end{cases}$$

Here G is the gravitational constant, M is the mass of the earth, and R is the earth's radius. Is g a continuous function of r?

66. Suppose that f is continuous on $[a, b]$ and it is never zero there. Is it possible that f changes sign on $[a, b]$? Explain.

67. Let $f(x + y) = f(x) + f(y)$ for all x and y and suppose that f is continuous at $x = 0$.

(a) Prove that f is continuous everywhere.

(b) Prove that there is a constant m such that $f(t) = mt$ for all t (see Problem 43 of Section 0.5).

68. Prove that if $f(x)$ is a continuous function on an interval then so is the function $|f(x)| = \sqrt{(f(x))^2}$.

69. Show that if $g(x) = |f(x)|$ is continuous it is not necessarily true that $f(x)$ is continuous.

70. Let $f(x) = 0$ if x is irrational and let $f(x) = 1/q$ if x is the rational number p/q in reduced form $(q > 0)$.

(a) Sketch (as best you can) the graph of f on $(0, 1)$.

(b) Show that f is continuous at each irrational number in $(0, 1)$, but is discontinuous at each rational number in $(0, 1)$.

71. A thin equilateral triangular block of side length 1 unit has its face in the vertical xy-plane with a vertex V at the origin. Under the influence of gravity, it will rotate about V until a side hits the x-axis floor (Figure 16). Let x denote the initial x-coordinate of the midpoint M of the side opposite V, and let $f(x)$ denote the final x-coordinate of this point. Assume that the block balances when M is directly above V.

(a) Determine the domain and range of f.

(b) Where on this domain is f discontinuous?

(c) Identify any fixed points of f (see Problem 59).

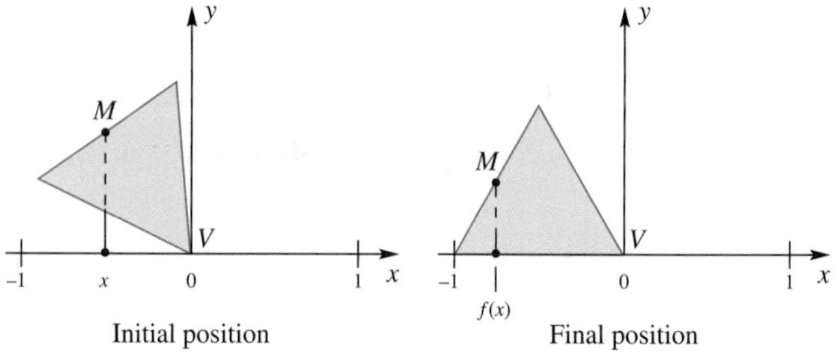

Figure 16

Answers to Concepts Review: **1.** $\lim_{x \to c} f(x)$ **2.** every integer **3.** $\lim_{x \to a^+} f(x) = f(a)$; $\lim_{x \to b^-} f(x) = f(b)$ **4.** a; b; $f(c) = W$

1.7 Chapter Review

Concepts Test

Respond with true or false to each of the following assertions. Be prepared to justify your answer.

1. If $f(c) = L$, then $\lim_{x \to c} f(x) = L$.

2. If $\lim_{x \to c} f(x) = L$, then $f(c) = L$.

3. If $\lim_{x \to c} f(x)$ exists, then $f(c)$ exists.

4. If $\lim_{x \to 0} f(x) = 0$, then for every $\varepsilon > 0$ there exists a $\delta > 0$ such that $0 < |x| < \delta$ implies $|f(x)| < \varepsilon$.

5. If $f(c)$ is undefined, then $\lim_{x \to c} f(x)$ does not exist.

6. The coordinates of the hole in the graph of $y = \frac{x^2 - 25}{x - 5}$ are $(5, 10)$.

7. If $p(x)$ is a polynomial, then $\lim_{x \to c} p(x) = p(c)$.

8. $\lim_{x \to 0} \frac{\sin x}{x}$ does not exist.

9. For every real number c, $\lim_{x \to c} \tan x = \tan c$.

10. $\tan x$ is continuous at every point of its domain.

11. The function $f(x) = 2\sin^2 x - \cos x$ is continuous at every real number.

12. If f is continuous at c, then $f(c)$ exists.

13. If f is continuous on the interval $(1, 3)$, then f is continuous at 2.

14. If f is continuous on $[0, 4]$, then $\lim_{x\to 0} f(x)$ exists.

15. If f is a continuous function such that $A \le f(x) \le B$ for all x, then $\lim_{x\to\infty} f(x)$ exists and it satisfies $A \le \lim_{x\to\infty} f(x) \le B$.

16. If f is continuous on (a, b) then $\lim_{x\to c} f(x) = f(c)$ for all c in (a, b).

17. $\lim_{x\to\infty} \frac{\sin x}{x} = 1$

18. If the line $y = 2$ is a horizontal asymptote of the graph of $y = f(x)$, then $\lim_{x\to\infty} f(x) = 2$.

19. The graph of $y = \tan x$ has many horizontal asymptotes.

20. The graph of $y = \frac{1}{x^2 - 4}$ has two vertical asymptotes.

21. $\lim_{t\to 1^+} \frac{2t}{t - 1} = \infty$.

22. If $\lim_{x\to c^-} f(x) = \lim_{x\to c^+} f(x)$, then f is continuous at $x = c$.

23. If $\lim_{x\to c} f(x) = f\left(\lim_{x\to c} x\right)$, then f is continuous at $x = c$.

24. The function $f(x) = [\![x/2]\!]$ is continuous at $x = 2.3$.

25. If $\lim_{x\to 2} f(x) = f(2) > 0$, then $f(x) < 1.001f(2)$ for all x in some interval containing 2.

26. If $\lim_{x\to c} [f(x) + g(x)]$ exists, then $\lim_{x\to c} f(x)$ and $\lim_{x\to c} g(x)$ both exist.

27. If $0 \le f(x) \le 3x^2 + 2x^4$ for all x, then $\lim_{x\to 0} f(x) = 0$.

28. If $\lim_{x\to a} f(x) = L$ and $\lim_{x\to a} f(x) = M$, then $L = M$.

29. If $f(x) \ne g(x)$ for all x, then $\lim_{x\to c} f(x) \ne \lim_{x\to c} g(x)$.

30. If $f(x) < 10$ for all x and $\lim_{x\to 2} f(x)$ exists, then $\lim_{x\to 2} f(x) < 10$.

31. If $\lim_{x\to a} f(x) = b$, then $\lim_{x\to a} |f(x)| = |b|$.

32. If f is continuous and positive on $[a, b]$, then $1/f$ must assume every value between $1/f(a)$ and $1/f(b)$.

Sample Test Problems

In Problems 1–22, find the indicated limit or state that it does not exist.

1. $\lim_{x\to 2} \frac{x - 2}{x + 2}$

2. $\lim_{u\to 1} \frac{u^2 - 1}{u + 1}$

3. $\lim_{u\to 1} \frac{u^2 - 1}{u - 1}$

4. $\lim_{u\to 1} \frac{u + 1}{u^2 - 1}$

5. $\lim_{x\to 2} \frac{1 - 2/x}{x^2 - 4}$

6. $\lim_{z\to 2} \frac{z^2 - 4}{z^2 + z - 6}$

7. $\lim_{x\to 0} \frac{\tan x}{\sin 2x}$

8. $\lim_{y\to 1} \frac{y^3 - 1}{y^2 - 1}$

9. $\lim_{x\to 4} \frac{x - 4}{\sqrt{x} - 2}$

10. $\lim_{x\to 0} \frac{\cos x}{x}$

11. $\lim_{x\to 0^-} \frac{|x|}{x}$

12. $\lim_{x\to 1/2^+} [\![4x]\!]$

13. $\lim_{t\to 2^-} ([\![t]\!] - t)$

14. $\lim_{x\to 1^-} \frac{|x - 1|}{x - 1}$

15. $\lim_{x\to 0} \frac{\sin 5x}{3x}$

16. $\lim_{x\to 0} \frac{1 - \cos 2x}{3x}$

17. $\lim_{x\to\infty} \frac{x - 1}{x + 2}$

18. $\lim_{t\to\infty} \frac{\sin t}{t}$

19. $\lim_{t\to 2} \frac{t + 2}{(t - 2)^2}$

20. $\lim_{x\to 0^+} \frac{\cos x}{x}$

21. $\lim_{x\to \pi/4^-} \tan 2x$

22. $\lim_{x\to 0^+} \frac{1 + \sin x}{x}$

23. Prove using an ε–δ argument that $\lim_{x\to 3}(2x + 1) = 7$.

24. Let $f(x) = \begin{cases} x^3 & \text{if } x < -1 \\ x & \text{if } -1 < x < 1 \\ 1 - x & \text{if } x \ge 1 \end{cases}$

Find each value.

(a) $f(1)$ (b) $\lim_{x\to 1^+} f(x)$

(c) $\lim_{x\to 1^-} f(x)$ (d) $\lim_{x\to -1} f(x)$

25. Refer to f of Problem 24. (a) What are the values of x at which f is discontinuous? (b) How should f be defined at $x = -1$ to make it continuous there?

26. Give the ε–δ definition in each case.

(a) $\lim_{u\to a} g(u) = M$ (b) $\lim_{x\to a^-} f(x) = L$

27. If $\lim_{x\to 3} f(x) = 3$ and $\lim_{x\to 3} g(x) = -2$ and if g is continuous at $x = 3$, find each value.

(a) $\lim_{x\to 3} [2f(x) - 4g(x)]$ (b) $\lim_{x\to 3} g(x)\frac{x^2 - 9}{x - 3}$

(c) $g(3)$ (d) $\lim_{x\to 3} g(f(x))$

(e) $\lim_{x\to 3} \sqrt{f^2(x) - 8g(x)}$ (f) $\lim_{x\to 3} \frac{|g(x) - g(3)|}{f(x)}$

28. Sketch the graph of a function f that satisfies all the following conditions.

(a) Its domain is $[0, 6]$.

(b) $f(0) = f(2) = f(4) = f(6) = 2$.

(c) f is continuous except at $x = 2$.

(d) $\lim_{x\to 2^-} f(x) = 1$ and $\lim_{x\to 5^+} f(x) = 3$.

29. Let $f(x) = \begin{cases} -1 & \text{if } x \le 0 \\ ax + b & \text{if } 0 < x < 1 \\ 1 & \text{if } x \ge 1 \end{cases}$

Determine a and b so that f is continuous everywhere.

30. Use the Intermediate Value Theorem to prove that the equation $x^5 - 4x^3 - 3x + 1 = 0$ has at least one solution between $x = 2$ and $x = 3$.

In Problems 31–36, find the equations of all vertical and horizontal asymptotes for the given function.

31. $f(x) = \frac{x}{x^2 + 1}$

32. $g(x) = \frac{x^2}{x^2 + 1}$

33. $F(x) = \frac{x^2}{x^2 - 1}$

34. $G(x) = \frac{x^3}{x^2 - 4}$

35. $h(x) = \tan 2x$

36. $H(x) = \frac{\sin x}{x^2}$

REVIEW & PREVIEW PROBLEMS

1. Let $f(x) = x^2$. Find and simplify each of the following.

(a) $f(2)$

(b) $f(2.1)$

(c) $f(2.1) - f(2)$

(d) $\dfrac{f(2.1) - f(2)}{2.1 - 2}$

(e) $f(a + h)$

(f) $f(a + h) - f(a)$

(g) $\dfrac{f(a + h) - f(a)}{(a + h) - a}$

(h) $\lim_{h \to 0} \dfrac{f(a + h) - f(a)}{(a + h) - a}$

2. Repeat (a) through (h) of Problem 1 for the function $f(x) = 1/x$.

3. Repeat (a) through (h) of Problem 1 for the function $f(x) = \sqrt{x}$.

4. Repeat (a) through (h) of Problem 1 for the function $f(x) = x^3 + 1$.

5. Write the first two terms in the expansions of the following:

(a) $(a + b)^3$

(b) $(a + b)^4$

(c) $(a + b)^5$

6. Based on your results from Problem 5, make a conjecture about the first two terms in the expansion of $(a + b)^n$ for an arbitrary n.

7. Use a trigonometric identity to write $\sin(x + h)$ in terms of $\sin x$, $\sin h$, $\cos x$, and $\cos h$.

8. Use a trigonometric identity to write $\cos(x + h)$ in terms of $\cos x$, $\cos h$, $\sin x$, and $\sin h$.

9. A wheel centered at the origin and of radius 10 centimeters is rotating counterclockwise at a rate of 4 revolutions per second. A point P on the rim of the wheel is at position $(10, 0)$ at time $t = 0$.

(a) What are the coordinates of P at times $t = 1, 2, 3$?

(b) At what time does the point P first return to the starting position $(10, 0)$?

10. Assume that a soap bubble retains its spherical shape as it expands. At time $t = 0$ the soap bubble has radius 2 centimeters. At time $t = 1$, the radius has increased to 2.5 centimeters. How much has the volume changed in this 1 second interval?

11. One airplane leaves an airport at noon flying north at 300 miles per hour. Another leaves the same airport one hour later and flies east at 400 miles per hour.

(a) What are the positions of the airplanes at 2:00 P.M.?

(b) What is the distance between the two planes at 2:00 P.M.?

(c) What is the distance between the two planes at 2:15 P.M.?

CHAPTER 2 The Derivative

2.1 Two Problems with One Theme

2.2 The Derivative

2.3 Rules for Finding Derivatives

2.4 Derivatives of Trigonometric Functions

2.5 The Chain Rule

2.6 Higher-Order Derivatives

2.7 Implicit Differentiation

2.8 Related Rates

2.9 Differentials and Approximations

2.1 Two Problems with One Theme

Our first problem is very old; it dates back to the great Greek scientist Archimedes (287–212 B.C.). We refer to the problem of the *slope of the tangent line.* Our second problem is newer. It grew out of attempts by Kepler (1571–1630), Galileo (1564–1642), Newton (1642–1727), and others to describe the speed of a moving body. It is the problem of *instantaneous velocity*.

The two problems, one geometric and the other mechanical, appear to be quite unrelated. In this case, appearances are deceptive. The two problems are identical twins.

The Tangent Line Euclid's notion of a tangent as a line touching a curve at just one point is all right for circles (Figure 1) but completely unsatisfactory for most other curves (Figure 2). The idea of a tangent to a curve at P as the line that best approximates the curve near P is better, but is still too vague for mathematical precision. The concept of limit provides a way of getting the best description.

Let P be a point on a curve and let Q be a nearby *movable point* on that curve. Consider the line through P and Q, called a **secant line.** The **tangent line** at P is the limiting position (if it exists) of the secant line as Q moves toward P along the curve (Figure 3).

Suppose that the curve is the graph of the equation $y = f(x)$. Then P has coordinates $(c, f(c))$, a nearby point Q has coordinates $(c + h, f(c + h))$, and the secant line through P and Q has slope m_{sec} given by (Figure 4):

$$m_{\text{sec}} = \frac{f(c + h) - f(c)}{h}$$

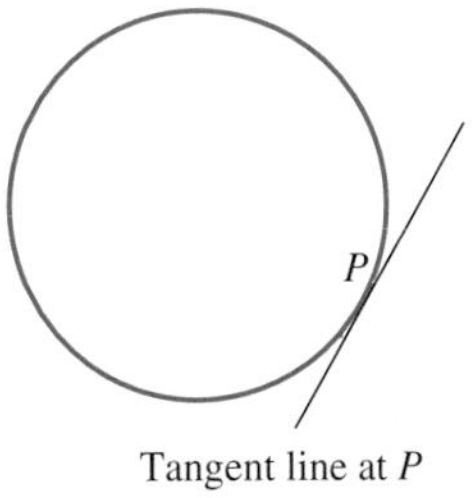

Figure 1

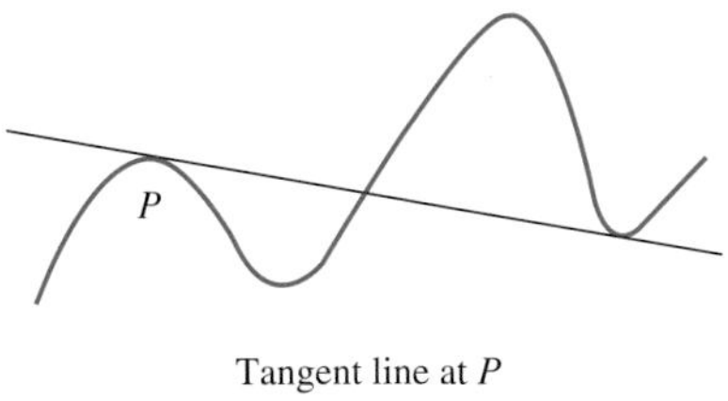

Figure 2

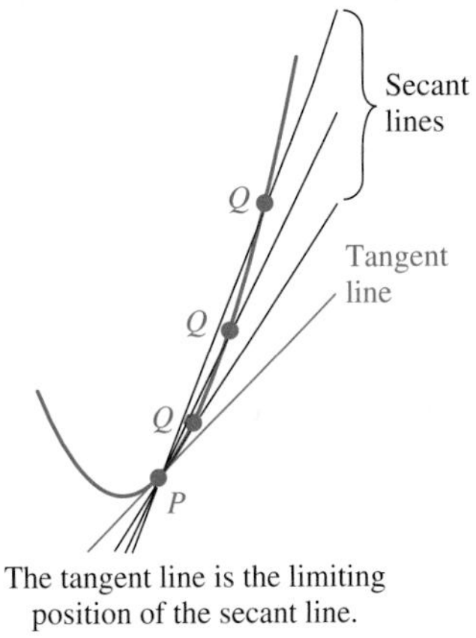

Figure 3

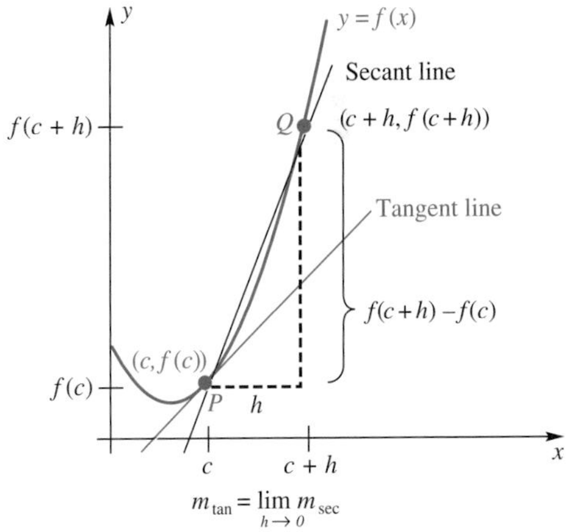

Figure 4

Using the concept of limit, which we studied in the last chapter, we can now give a formal definition of the tangent line.

> **Definition** Tangent Line
>
> The **tangent line** to the curve $y = f(x)$ at the point $P(c, f(c))$ is that line through P with slope
>
> $$m_{\tan} = \lim_{h \to 0} m_{\text{sec}} = \lim_{h \to 0} \frac{f(c + h) - f(c)}{h}$$
>
> provided that this limit exists and is not ∞ or $-\infty$.

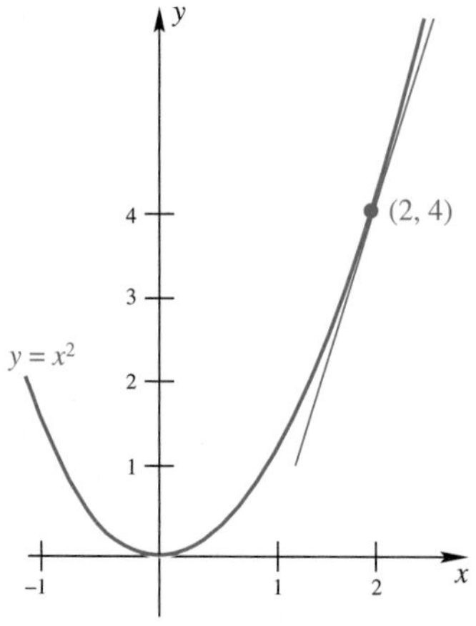

Figure 5

EXAMPLE 1 Find the slope of the tangent line to the curve $y = f(x) = x^2$ at the point (2, 4).

SOLUTION The line whose slope we are seeking is shown in Figure 5. Clearly it has a large positive slope.

$$\begin{aligned} m_{\tan} &= \lim_{h\to 0} \frac{f(2+h) - f(2)}{h} \\ &= \lim_{h\to 0} \frac{(2+h)^2 - 2^2}{h} \\ &= \lim_{h\to 0} \frac{4 + 4h + h^2 - 4}{h} \\ &= \lim_{h\to 0} \frac{\cancel{h}(4+h)}{\cancel{h}} \\ &= 4 \end{aligned}$$

■

EXAMPLE 2 Find the slopes of the tangent lines to the curve $y = f(x) = -x^2 + 2x + 2$ at the points with x-coordinates $-1, \frac{1}{2}, 2,$ and 3.

SOLUTION Rather than make four separate calculations, it seems wise to calculate the slope at the point with x-coordinate c and then obtain the four desired answers by substitution.

$$\begin{aligned} m_{\tan} &= \lim_{h\to 0} \frac{f(c+h) - f(c)}{h} \\ &= \lim_{h\to 0} \frac{-(c+h)^2 + 2(c+h) + 2 - (-c^2 + 2c + 2)}{h} \\ &= \lim_{h\to 0} \frac{-c^2 - 2ch - h^2 + 2c + 2h + 2 + c^2 - 2c - 2}{h} \\ &= \lim_{h\to 0} \frac{\cancel{h}(-2c - h + 2)}{\cancel{h}} \\ &= -2c + 2 \end{aligned}$$

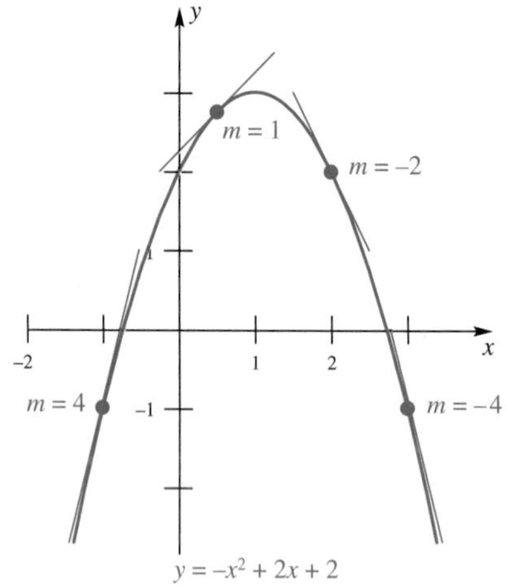

Figure 6

The four desired slopes (obtained by letting $c = -1, \frac{1}{2}, 2, 3$) are 4, 1, −2, and −4. These answers do appear to be consistent with the graph in Figure 6. ■

EXAMPLE 3 Find the equation of the tangent line to the curve $y = 1/x$ at $\left(2, \frac{1}{2}\right)$ (see Figure 7).

SOLUTION Let $f(x) = 1/x$.

$$\begin{aligned} m_{\tan} &= \lim_{h\to 0} \frac{f(2+h) - f(2)}{h} \\ &= \lim_{h\to 0} \frac{\dfrac{1}{2+h} - \dfrac{1}{2}}{h} \\ &= \lim_{h\to 0} \frac{\dfrac{2}{2(2+h)} - \dfrac{2+h}{2(2+h)}}{h} \\ &= \lim_{h\to 0} \frac{2 - (2+h)}{2(2+h)h} \\ &= \lim_{h\to 0} \frac{-h}{2(2+h)h} \\ &= \lim_{h\to 0} \frac{-1}{2(2+h)} = -\frac{1}{4} \end{aligned}$$

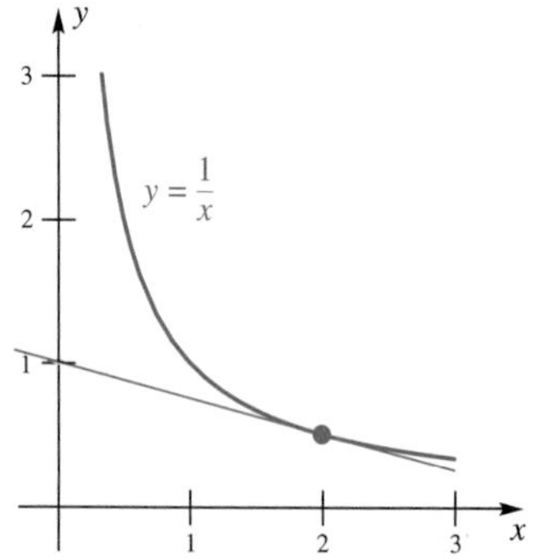

Figure 7

Knowing that the slope of the line is $-\frac{1}{4}$ and that the point $\left(2, \frac{1}{2}\right)$ is on it, we can easily write its equation by using the point-slope form $y - y_0 = m(x - x_0)$. The result is $y - \frac{1}{2} = -\frac{1}{4}(x - 2)$, or equivalently, $y = 1 - \frac{1}{4}x$. ■

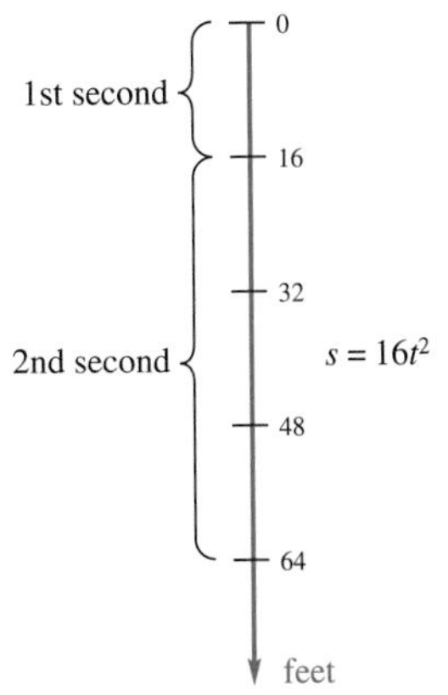

Figure 8

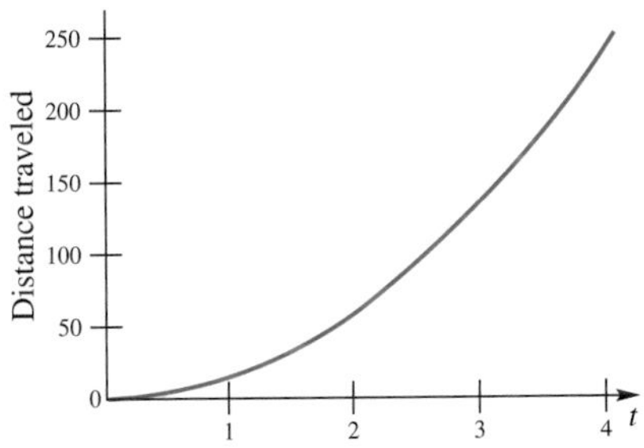

Figure 9

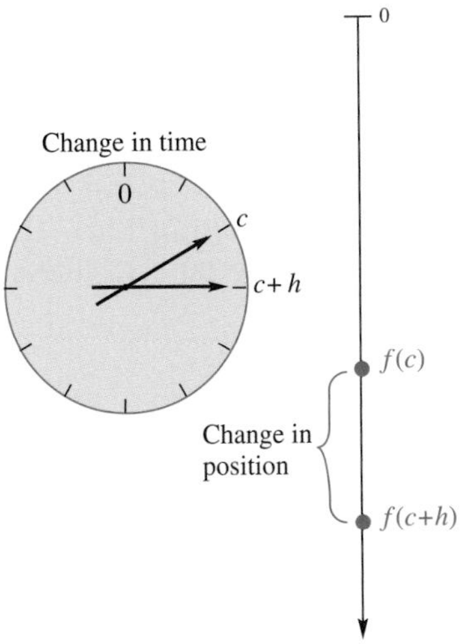

Figure 10

Average Velocity and Instantaneous Velocity

If we drive an automobile from one town to another 80 miles away in 2 hours, our average velocity is 40 miles per hour. *Average velocity* is the distance from the first position to the second position divided by the elapsed time.

But during our trip the speedometer reading was often different from 40. At the start, it registered 0; at times it rose as high as 57; at the end it fell back to 0 again. Just what does the speedometer measure? Certainly, it does not indicate average velocity.

Consider the more precise example of an object P falling in a vacuum. Experiment shows that if it starts from rest, P falls $16t^2$ feet in t seconds. Thus, it falls 16 feet in the first second and 64 feet during the first 2 seconds (Figure 8); clearly, it falls faster and faster as time goes on. Figure 9 shows the distance traveled (on the vertical axis) as a function of time (on the horizontal axis).

During the second second (i.e., in the time interval from $t = 1$ to $t = 2$), P fell $64 - 16 = 48$ feet. Its average velocity was

$$v_{\text{avg}} = \frac{64 - 16}{2 - 1} = 48 \text{ feet per second}$$

During the time interval from $t = 1$ to $t = 1.5$, it fell $16(1.5)^2 - 16 = 20$ feet. Its average velocity was

$$v_{\text{avg}} = \frac{16(1.5)^2 - 16}{1.5 - 1} = \frac{20}{0.5} = 40 \text{ feet per second}$$

Similarly, on the time intervals $t = 1$ to $t = 1.1$ and $t = 1$ to $t = 1.01$, we calculate the respective average velocities to be

$$v_{\text{avg}} = \frac{16(1.1)^2 - 16}{1.1 - 1} = \frac{3.36}{0.1} = 33.6 \text{ feet per second}$$

$$v_{\text{avg}} = \frac{16(1.01)^2 - 16}{1.01 - 1} = \frac{0.3216}{0.01} = 32.16 \text{ feet per second}$$

What we have done is to calculate the average velocity over shorter and shorter time intervals, each starting at $t = 1$. The shorter the time interval is, the better we should approximate the *instantaneous velocity* at the instant $t = 1$. Looking at the numbers 48, 40, 33.6, and 32.16, we might guess 32 feet per second to be the instantaneous velocity.

But let us be more precise. Suppose that an object P moves along a coordinate line so that its position at time t is given by $s = f(t)$. At time c the object is at $f(c)$; at the nearby time $c + h$, it is at $f(c + h)$ (see Figure 10). Thus the **average velocity** on this interval is

$$v_{\text{avg}} = \frac{f(c + h) - f(c)}{h}$$

We can now define instantaneous velocity.

Definition Instantaneous Velocity

If an object moves along a coordinate line with position function $f(t)$, then its **instantaneous velocity** at time c is

$$v = \lim_{h \to 0} v_{\text{avg}} = \lim_{h \to 0} \frac{f(c + h) - f(c)}{h}$$

provided that the limit exists and is not ∞ or $-\infty$.

In the case where $f(t) = 16t^2$, the instantaneous velocity at $t = 1$ is

$$\begin{aligned} v &= \lim_{h\to 0} \frac{f(1+h) - f(1)}{h} \\ &= \lim_{h\to 0} \frac{16(1+h)^2 - 16}{h} \\ &= \lim_{h\to 0} \frac{16 + 32h + 16h^2 - 16}{h} \\ &= \lim_{h\to 0} (32 + 16h) = 32 \end{aligned}$$

This confirms our earlier guess.

Two Problems with One Theme

Now you see why we called this section "Two Problems with One Theme." Look at the definitions of *slope of the tangent line* and *instantaneous velocity*. They give different names for the same mathematical concept.

EXAMPLE 4 An object, initially at rest, falls due to gravity. Find its instantaneous velocity at $t = 3.8$ seconds and at $t = 5.4$ seconds.

SOLUTION We calculate the instantaneous velocity at $t = c$ seconds. Since $f(t) = 16t^2$,

$$\begin{aligned} v &= \lim_{h\to 0} \frac{f(c+h) - f(c)}{h} \\ &= \lim_{h\to 0} \frac{16(c+h)^2 - 16c^2}{h} \\ &= \lim_{h\to 0} \frac{16c^2 + 32ch + 16h^2 - 16c^2}{h} \\ &= \lim_{h\to 0} (32c + 16h) = 32c \end{aligned}$$

Thus, the instantaneous velocity at 3.8 seconds is $32(3.8) = 121.6$ feet per second; at 5.4 seconds, it is $32(5.4) = 172.8$ feet per second. ■

EXAMPLE 5 How long will it take the falling object of Example 4 to reach an instantaneous velocity of 112 feet per second?

SOLUTION We learned in Example 4 that the instantaneous velocity after c seconds is $32c$. Thus, we must solve the equation $32c = 112$. The solution is $c = \frac{112}{32} = 3.5$ seconds. ■

EXAMPLE 6 A particle moves along a coordinate line and s, its directed distance in centimeters from the origin after t seconds, is given by $s = f(t) = \sqrt{5t + 1}$. Find the instantaneous velocity of the particle after 3 seconds.

SOLUTION Figure 11 shows the distance traveled as a function of time. The instantaneous velocity at time $t = 3$ is equal to the slope of the tangent line at $t = 3$.

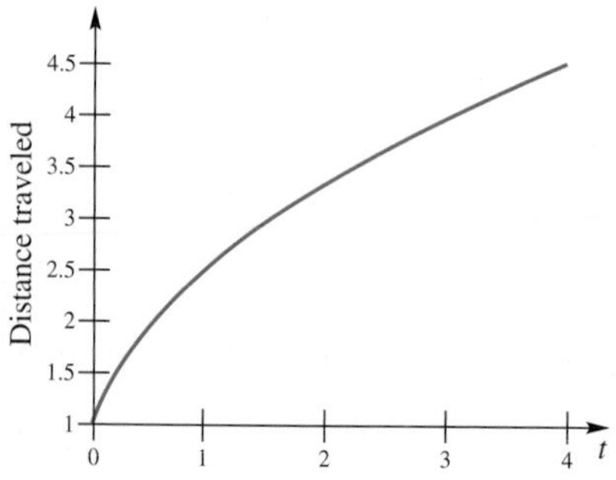

Figure 11

$$\begin{aligned} v &= \lim_{h\to 0} \frac{f(3+h) - f(3)}{h} \\ &= \lim_{h\to 0} \frac{\sqrt{5(3+h)+1} - \sqrt{5(3)+1}}{h} \\ &= \lim_{h\to 0} \frac{\sqrt{16+5h} - 4}{h} \end{aligned}$$

To evaluate this limit, we rationalize the numerator by multiplying the numerator and denominator by $\sqrt{16 + 5h} + 4$. We obtain

$$v = \lim_{h \to 0}\left(\frac{\sqrt{16 + 5h} - 4}{h} \cdot \frac{\sqrt{16 + 5h} + 4}{\sqrt{16 + 5h} + 4}\right)$$

$$= \lim_{h \to 0} \frac{16 + 5h - 16}{h\left(\sqrt{16 + 5h} + 4\right)}$$

$$= \lim_{h \to 0} \frac{5}{\sqrt{16 + 5h} + 4} = \frac{5}{8}$$

We conclude that the instantaneous velocity after 3 seconds is $\frac{5}{8}$ centimeter per second. ■

> **Velocity or Speed**
>
> For the time being, we will use the terms *velocity* and *speed* interchangeably. Later in this chapter, we will distinguish between these two words.

Rates of Change Velocity is only one of many rates of change that will be important in this course; it is the rate of change of distance with respect to time. Other rates of change that will interest us are density for a wire (the rate of change of mass with respect to distance), marginal revenue (the rate of change of revenue with respect to the number of items produced), and current (the rate of change of electrical charge with respect to time). These rates and many more are discussed in the problem set. In each case, we must distinguish between an *average* rate of change on an interval and an *instantaneous* rate of change at a point. The phrase *rate of change* without an adjective will mean instantaneous rate of change.

Concepts Review

1. The line that most closely approximates a curve near the point P is the _____ through that point.

2. More precisely, the tangent line to a curve at P is the limiting position of the _____ line through P and Q as Q approaches P along the curve.

3. The slope $m_{\tan}$ of the tangent line to the curve $y = f(x)$ at $(c, f(c))$ is given by $m_{\tan} = \lim_{h \to 0}$ _____.

4. The instantaneous velocity of a point P (moving along a line) at time c is the limit of the _____ on the time interval c to $c + h$ as h approaches zero.

Problem Set 2.1

In Problems 1 and 2, a tangent line to a curve is drawn. Estimate its slope (slope = rise/run). Be careful to note the difference in scales on the two axes.

1.

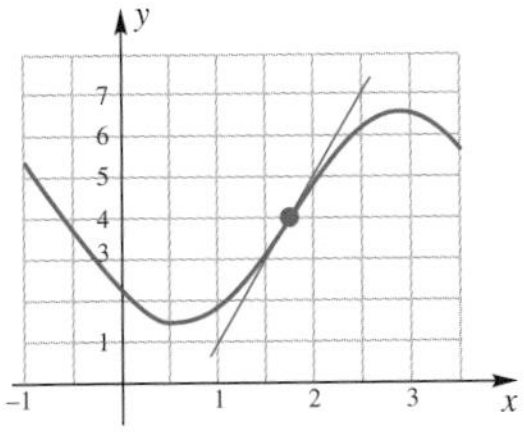

2.

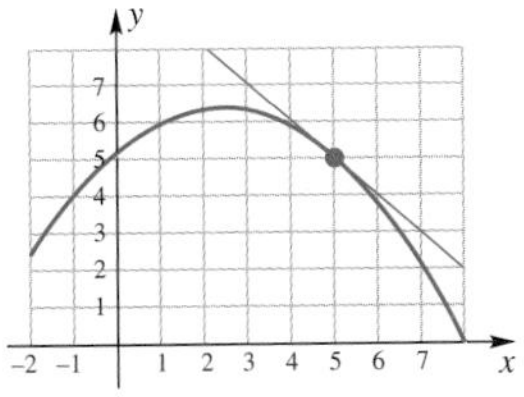

In Problems 3–6, draw the tangent line to the curve through the indicated point and estimate its slope.

3.

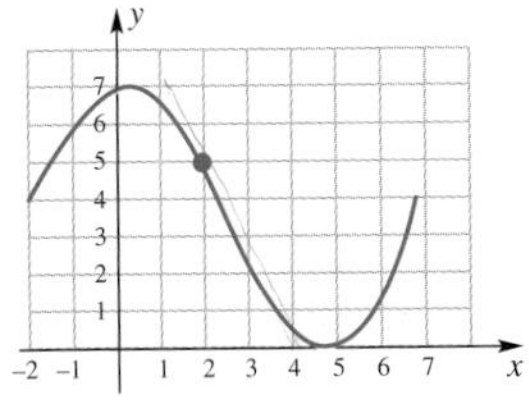

4.

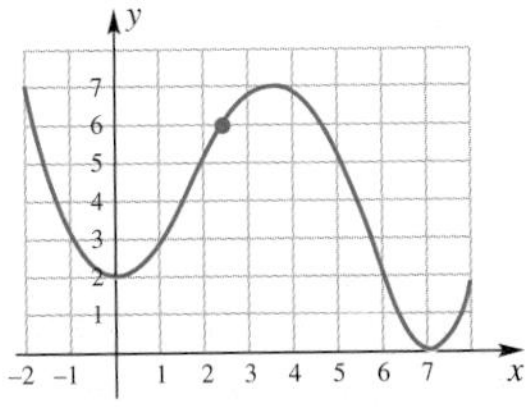

5.

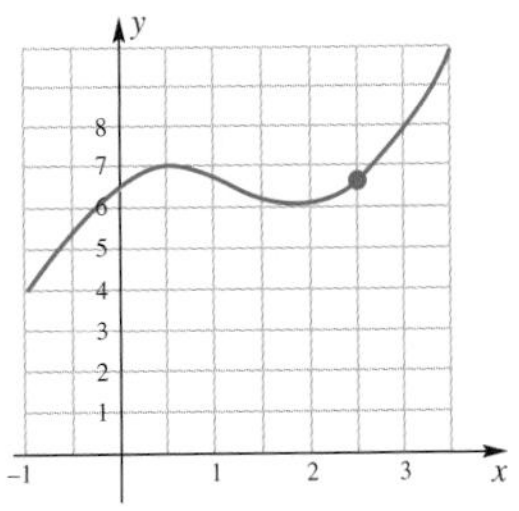

6.

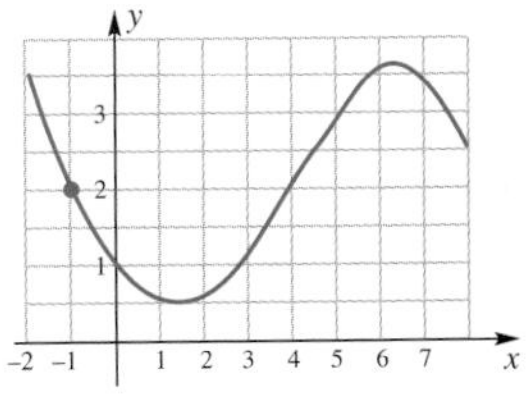

7. Consider $y = x^2 + 1$.

(a) Sketch its graph as carefully as you can.

(b) Draw the tangent line at $(1, 2)$.

≈ (c) Estimate the slope of this tangent line.

C (d) Calculate the slope of the secant line through $(1, 2)$ and $(1.01, (1.01)^2 + 1.0)$.

(e) Find by the limit process (see Example 1) the slope of the tangent line at $(1, 2)$.

8. Consider $y = x^3 - 1$.

(a) Sketch its graph as carefully as you can.

(b) Draw the tangent line at $(2, 7)$.

(c) Estimate the slope of this tangent line.

(d) Calculate the slope of the secant line through $(2, 7)$ and $(2.01, (2.01)^3 - 1.0)$.

(e) Find by the limit process (see Example 1) the slope of the tangent line at $(2, 7)$.

9. Find the slopes of the tangent lines to the curve $y = x^2 - 1$ at the points where $x = -2, -1, 0, 1, 2$ (see Example 2).

10. Find the slopes of the tangent lines to the curve $y = x^3 - 3x$ at the points where $x = -2, -1, 0, 1, 2$.

11. Sketch the graph of $y = 1/(x + 1)$ and then find the equation of the tangent line at $\left(1, \frac{1}{2}\right)$ (see Example 3).

12. Find the equation of the tangent line to $y = 1/(x - 1)$ at $(0, -1)$.

13. Experiment suggests that a falling body will fall approximately $16t^2$ feet in t seconds.

(a) How far will it fall between $t = 0$ and $t = 1$?

(b) How far will it fall between $t = 1$ and $t = 2$?

(c) What is its average velocity on the interval $2 \le t \le 3$?

(d) What is its average velocity on the interval $3 \le t \le 3.01$?

(e) Find its instantaneous velocity at $t = 3$ (see Example 4).

14. An object travels along a line so that its position s is $s = t^2 + 1$ meters after t seconds.

(a) What is its average velocity on the interval $2 \le t \le 3$?

(b) What is its average velocity on the interval $2 \le t \le 2.003$?

(c) What is its average velocity on the interval $2 \le t \le 2 + h$?

(d) Find its instantaneous velocity at $t = 2$.

15. Suppose that an object moves along a coordinate line so that its directed distance from the origin after t seconds is $\sqrt{2t + 1}$ feet.

(a) Find its instantaneous velocity at $t = \alpha, \alpha > 0$.

(b) When will it reach a velocity of $\frac{1}{2}$ foot per second? (see Example 5.)

16. If a particle moves along a coordinate line so that its directed distance from the origin after t seconds is $(-t^2 + 4t)$ feet, when did the particle come to a momentary stop (i.e., when did its instantaneous velocity become zero)?

17. A certain bacterial culture is growing so that it has a mass of $\frac{1}{2}t^2 + 1$ grams after t hours.

(a) How much did it grow during the interval $2 \le t \le 2.01$?

(b) What was its average growth rate during the interval $2 \le t \le 2.01$?

(c) What was its instantaneous growth rate at $t = 2$?

18. A business is prospering in such a way that its total (accumulated) profit after t years is $1000t^2$ dollars.

(a) How much did the business make during the third year (between $t = 2$ and $t = 3$)?

(b) What was its average rate of profit during the first half of the third year, between $t = 2$ and $t = 2.5$? (The rate will be in dollars per year.)

(c) What was its instantaneous rate of profit at $t = 2$?

19. A wire of length 8 centimeters is such that the mass between its left end and a point x centimeters to the right is x^3 grams (Figure 12).

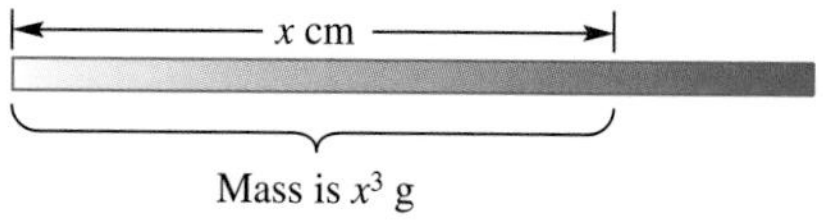

Figure 12

(a) What is the average density of the middle 2-centimeter segment of this wire? *Note*: Average density equals mass/length.

(b) What is the actual density at the point 3 centimeters from the left end?

20. Suppose that the revenue $R(n)$ in dollars from producing n computers is given by $R(n) = 0.4n - 0.001n^2$. Find the instantaneous rates of change of revenue when $n = 10$ and $n = 100$. (The instantaneous rate of change of revenue with respect to the amount of product produced is called the *marginal revenue*.)

21. The rate of change of velocity with respect to time is called **acceleration.** Suppose that the velocity at time t of a particle is given by $v(t) = 2t^2$. Find the instantaneous acceleration when $t = 1$ second.

22. A city is hit by an Asian flu epidemic. Officials estimate that t days after the beginning of the epidemic the number of persons sick with the flu is given by $p(t) = 120t^2 - 2t^3$, when $0 \le t \le 40$. At what rate is the flu spreading at time $t = 10$; $t = 20$; $t = 40$?

23. The graph in Figure 13 shows the amount of water in a city water tank during one day when no water was pumped into the tank. What was the average rate of water usage during the day? How fast was water being used at 8 A.M.?

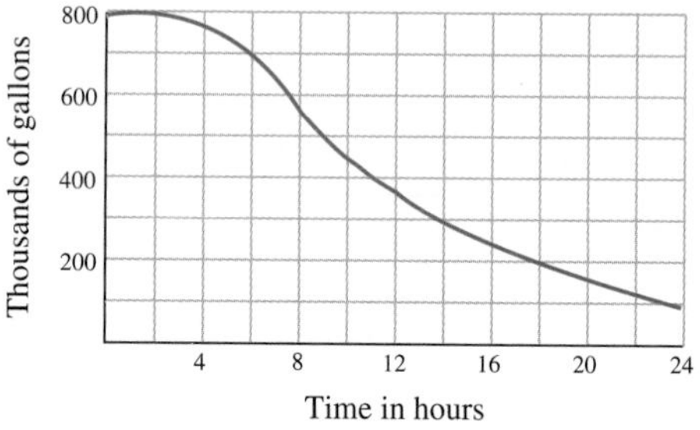

Figure 13

24. Passengers board an elevator at the ground floor (i.e., the zeroth floor) and take it to the seventh floor, which is 84 feet above the ground floor. The elevator's position s as a function of time t (measured in seconds) is shown in Figure 14.

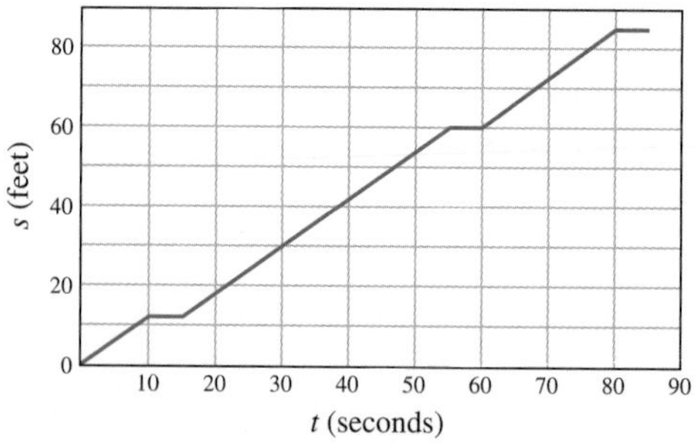

Figure 14

(a) What is the average velocity of the elevator from the time the elevator began moving until the time that it reached the seventh floor?

(b) What was the elevator's approximate velocity at time $t = 20$?

(c) How many stops did the elevator make between the ground floor and the seventh floor (excluding the ground and seventh floors)? On which floors do you think the elevator stopped?

25. Figure 15 shows the normal high temperature for St. Louis, Missouri, as a function of time (measured in days beginning January 1).

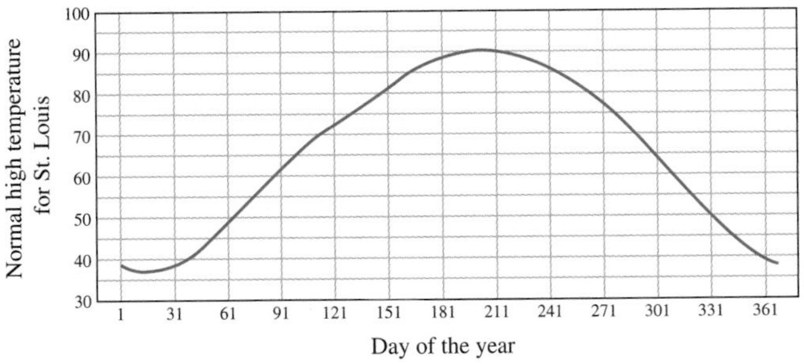

Figure 15

(a) What is the approximate rate of change in the normal high temperature on March 2 (i.e., on day number 61)? What are the units of this rate of change?

(b) What is the approximate rate of change in the normal high temperature on July 10 (i.e., on day number 191)?

(c) In what months is there a moment when the rate of change is equal to 0?

(d) In what months is the absolute value of the rate of change the greatest?

26. Figure 16 shows the population in millions of a developing country for the years 1900 to 1999. What is the approximate rate of change of the population in 1930? In 1990? The percentage growth is often a more appropriate measure of population growth. This is the rate of growth divided by the population size at that time. For this population, what was the approximate percentage growth in 1930? In 1990?

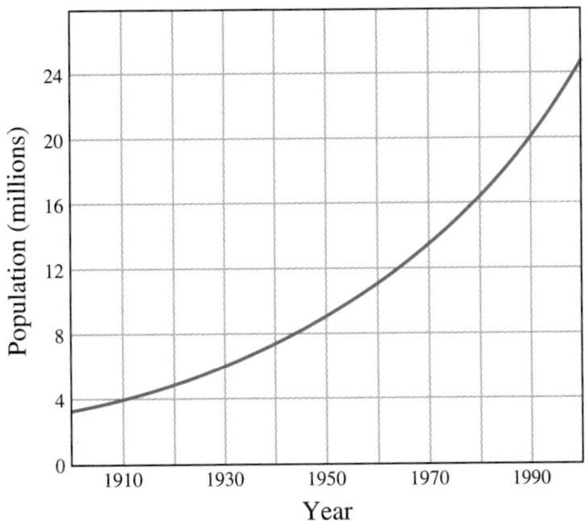

Figure 16

27. Figures 17a and 17b show the position s as a function of time t for two particles that are moving along a line. For each particle, is the velocity increasing or decreasing? Explain.

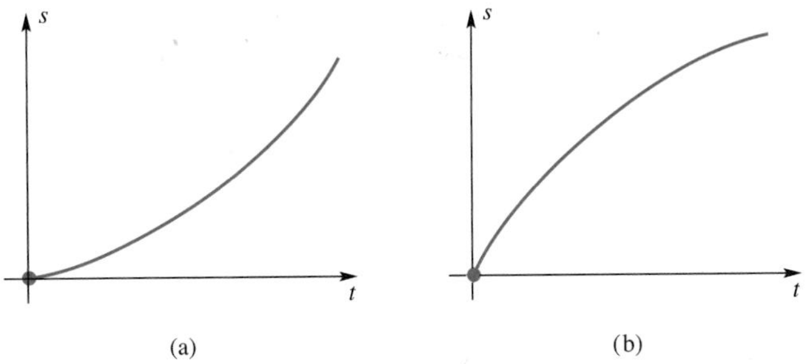

Figure 17

28. The rate of change of electric charge with respect to time is called **current.** Suppose that $\frac{1}{3}t^3 + t$ coulombs of charge flow through a wire in t seconds. Find the current in amperes (coulombs per second) after 3 seconds. When will a 20-ampere fuse in the line blow?

29. The radius of a circular oil spill is growing at a constant rate of 2 kilometers per day. At what rate is the area of the spill growing 3 days after it began?

30. The radius of a spherical balloon is increasing at the rate of 0.25 inch per second. If the radius is 0 at time $t = 0$, find the rate of change in the volume at time $t = 3$.

GC *Use a graphing calculator or a CAS to do Problems 31–34.*

31. Draw the graph of $y = f(x) = x^3 - 2x^2 + 1$. Then find the slope of the tangent line at

(a) -1 (b) 0 (c) 1 (d) 3.2

32. Draw the graph of $y = f(x) = \sin x \sin^2 2x$. Then find the slope of the tangent line at

(a) $\pi/3$ (b) 2.8 (c) π (d) 4.2

33. If a point moves along a line so that its distance s (in feet) from 0 is given by $s = t + t\cos^2 t$ at time t seconds, find its instantaneous velocity at $t = 3$.

34. If a point moves along a line so that its distance s (in meters) from 0 is given by $s = (t + 1)^3/(t + 2)$ at time t minutes, find its instantaneous velocity at $t = 1.6$.

Answers to Concepts Review: **1.** tangent line **2.** secant **3.** $[f(c + h) - f(c)]/h$ **4.** average velocity

2.2 The Derivative

We have seen that *slope of the tangent line* and *instantaneous velocity* are manifestations of the same basic idea. Rate of growth of an organism (biology), marginal profit (economics), density of a wire (physics), and dissolution rates (chemistry) are other versions of the same basic concept. Good mathematical sense suggests that we study this concept independently of these specialized vocabularies and diverse applications. We choose the neutral name *derivative*. Add it to *function* and *limit* as one of the key words in calculus.

Definition Derivative

The **derivative** of a function f is another function f' (read "f prime") whose value at any number x is

$$f'(x) = \lim_{h\to 0} \frac{f(x+h) - f(x)}{h}$$

If this limit does exist, we say that f is **differentiable** at x. Finding a derivative is called **differentiation**; the part of calculus associated with the derivative is called **differential calculus.**

Finding Derivatives

We illustrate with several examples.

EXAMPLE 1 Let $f(x) = 13x - 6$. Find $f'(4)$.

SOLUTION

$$f'(4) = \lim_{h\to 0} \frac{f(4+h) - f(4)}{h} = \lim_{h\to 0} \frac{[13(4+h) - 6] - [13(4) - 6]}{h}$$

$$= \lim_{h\to 0} \frac{13h}{h} = \lim_{h\to 0} 13 = 13$$

■

EXAMPLE 2 If $f(x) = x^3 + 7x$, find $f'(x)$.

SOLUTION

$$\begin{aligned} f'(x) &= \lim_{h\to 0} \frac{f(x+h) - f(x)}{h} \\ &= \lim_{h\to 0} \frac{\left[(x+h)^3 + 7(x+h)\right] - \left[x^3 + 7x\right]}{h} \\ &= \lim_{h\to 0} \frac{3x^2h + 3xh^2 + h^3 + 7h}{h} \\ &= \lim_{h\to 0} (3x^2 + 3xh + h^2 + 7) \\ &= 3x^2 + 7 \end{aligned}$$

■

EXAMPLE 3 If $f(x) = 1/x$, find $f'(x)$.

SOLUTION

$$f'(x) = \lim_{h\to 0} \frac{f(x+h) - f(x)}{h} = \lim_{h\to 0} \frac{\dfrac{1}{x+h} - \dfrac{1}{x}}{h}$$

$$= \lim_{h\to 0} \left[\frac{x - (x+h)}{(x+h)x} \cdot \frac{1}{h}\right] = \lim_{h\to 0} \left[\frac{-h}{(x+h)x} \cdot \frac{1}{h}\right]$$

$$= \lim_{h\to 0} \frac{-1}{(x+h)x} = -\frac{1}{x^2}$$

Thus, f' is the function given by $f'(x) = -1/x^2$. Its domain is all real numbers except $x = 0$. ■

EXAMPLE 4 Find $F'(x)$ if $F(x) = \sqrt{x}$, $x > 0$.

SOLUTION

$$F'(x) = \lim_{h \to 0} \frac{F(x+h) - F(x)}{h} = \lim_{h \to 0} \frac{\sqrt{x+h} - \sqrt{x}}{h}$$

By this time you will have noticed that finding a derivative always involves taking the limit of a quotient where both numerator and denominator are approaching zero. Our task is to simplify this quotient so that we can cancel a factor h from the numerator and denominator, thereby allowing us to evaluate the limit by substitution. In the present example, this can be accomplished by rationalizing the numerator.

$$\begin{aligned} F'(x) &= \lim_{h \to 0} \left[\frac{\sqrt{x+h} - \sqrt{x}}{h} \cdot \frac{\sqrt{x+h} + \sqrt{x}}{\sqrt{x+h} + \sqrt{x}} \right] \\ &= \lim_{h \to 0} \frac{x + h - x}{h\left(\sqrt{x+h} + \sqrt{x}\right)} \\ &= \lim_{h \to 0} \frac{h}{h\left(\sqrt{x+h} + \sqrt{x}\right)} \\ &= \lim_{h \to 0} \frac{1}{\sqrt{x+h} + \sqrt{x}} \\ &= \frac{1}{\sqrt{x} + \sqrt{x}} = \frac{1}{2\sqrt{x}} \end{aligned}$$

Thus, F', the derivative of F, is given by $F'(x) = 1/\left(2\sqrt{x}\right)$. Its domain is $(0, \infty)$. ■

Equivalent Forms for the Derivative There is nothing sacred about use of the letter h in defining $f'(c)$. Notice, for example, that

$$\begin{aligned} f'(c) &= \lim_{h \to 0} \frac{f(c+h) - f(c)}{h} \\ &= \lim_{p \to 0} \frac{f(c+p) - f(c)}{p} \\ &= \lim_{s \to 0} \frac{f(c+s) - f(c)}{s} \end{aligned}$$

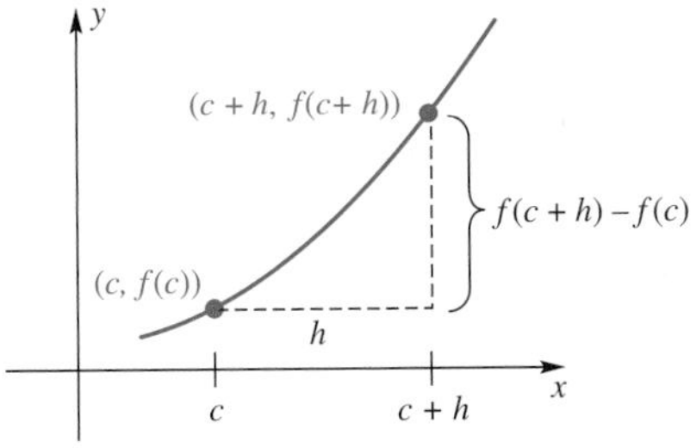

Figure 1

A more radical change, but still just a change of notation, may be understood by comparing Figures 1 and 2. Note how x takes the place of $c + h$, and so $x - c$ replaces h. Thus,

$$f'(c) = \lim_{x \to c} \frac{f(x) - f(c)}{x - c}$$

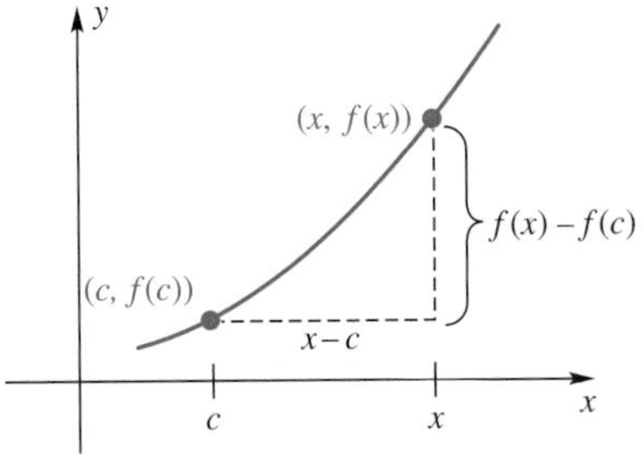

Figure 2

Note that in all cases the number at which f' is evaluated is held fixed during the limit operation.

EXAMPLE 5 Use the last boxed result to find $g'(c)$ if $g(x) = 2/(x+3)$.

SOLUTION

$$g'(c) = \lim_{x\to c} \frac{g(x) - g(c)}{x - c} = \lim_{x\to c} \frac{\dfrac{2}{x+3} - \dfrac{2}{c+3}}{x - c}$$

$$= \lim_{x\to c} \left[\frac{2(c+3) - 2(x+3)}{(x+3)(c+3)} \cdot \frac{1}{x-c}\right]$$

$$= \lim_{x\to c} \left[\frac{-2(x-c)}{(x+3)(c+3)} \cdot \frac{1}{x-c}\right]$$

$$= \lim_{x\to c} \frac{-2}{(x+3)(c+3)} = \frac{-2}{(c+3)^2}$$

Here we manipulated the quotient until we could cancel a factor of $x - c$ from the numerator and denominator. Then we could evaluate the limit. ■

EXAMPLE 6 Each of the following is a derivative, but of what function and at what point?

(a) $\displaystyle\lim_{h\to 0} \frac{(4+h)^2 - 16}{h}$ (b) $\displaystyle\lim_{x\to 3} \frac{\dfrac{2}{x} - \dfrac{2}{3}}{x - 3}$

SOLUTION

(a) This is the derivative of $f(x) = x^2$ at $x = 4$.

(b) This is the derivative of $f(x) = 2/x$ at $x = 3$. ■

Differentiability Implies Continuity If a curve has a tangent line at a point, then that curve cannot take a jump or wiggle too badly at the point. The precise formulation of this fact is an important theorem.

Theorem A Differentiability Implies Continuity

If $f'(c)$ exists, then f is continuous at c.

Proof We need to show that $\lim_{x\to c} f(x) = f(c)$. We begin by writing $f(x)$ in a fancy way.

$$f(x) = f(c) + \frac{f(x) - f(c)}{x - c} \cdot (x - c), \qquad x \neq c$$

Therefore,

$$\lim_{x\to c} f(x) = \lim_{x\to c} \left[f(c) + \frac{f(x) - f(c)}{x - c} \cdot (x - c)\right]$$

$$= \lim_{x\to c} f(c) + \lim_{x\to c} \frac{f(x) - f(c)}{x - c} \cdot \lim_{x\to c} (x - c)$$

$$= f(c) + f'(c) \cdot 0$$

$$= f(c)$$ ■

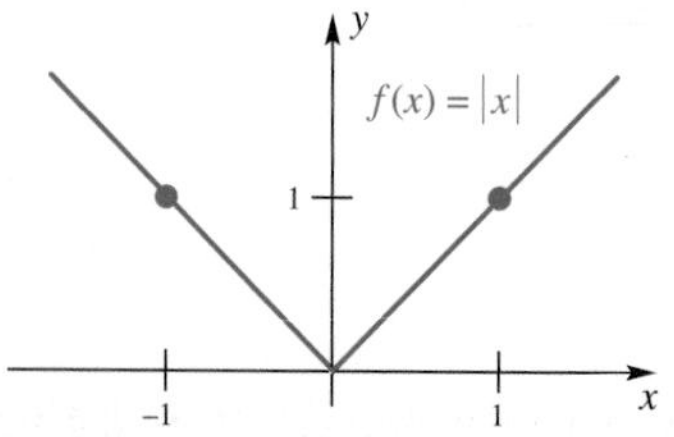

Figure 3

The converse of this theorem is false. If a function f is continuous at c, it does not follow that f has a derivative at c. This is easily seen by considering $f(x) = |x|$ at the origin (Figure 3). This function is certainly continuous at zero. However, it does not have a derivative there, as we now show. Note that

$$\frac{f(0+h)-f(0)}{h}=\frac{|0+h|-|0|}{h}=\frac{|h|}{h}$$

Thus,

$$\lim_{h\to 0^+}\frac{f(0+h)-f(0)}{h}=\lim_{h\to 0^+}\frac{|h|}{h}=\lim_{h\to 0^+}\frac{h}{h}=1$$

whereas

$$\lim_{h\to 0^-}\frac{f(0+h)-f(0)}{h}=\lim_{h\to 0^-}\frac{|h|}{h}=\lim_{h\to 0^-}\frac{-h}{h}=-1$$

Since the right- and left-hand limits are different,

$$\lim_{h\to 0}\frac{f(0+h)-f(0)}{h}$$

does not exist. Therefore, $f'(0)$ does not exist.

A similar argument shows that at any point where the graph of a continuous function has a sharp corner the function is not differentiable. The graph in Figure 4 indicates a number of ways for a function to be nondifferentiable at a point.

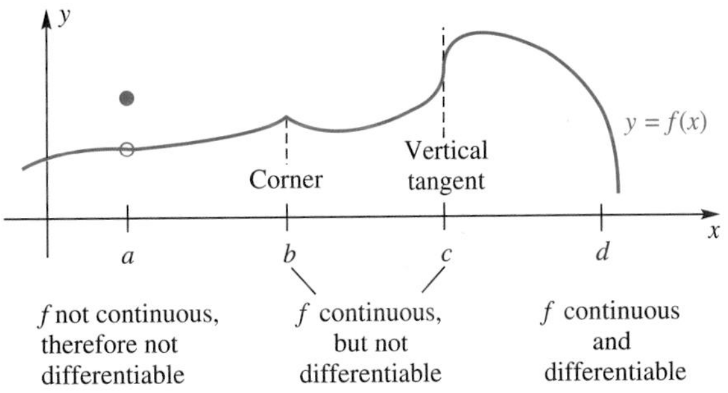

Figure 4

For the function shown in Figure 4 the derivative does not exist at the point c where the tangent line is vertical. This is because

$$\lim_{h\to 0}\frac{f(c+h)-f(c)}{h}=\infty$$

This corresponds to the fact that the slope of a vertical line is undefined.

Increments If the value of a variable x changes from x_1 to x_2, then $x_2 - x_1$, the change in x, is called an **increment** of x and is commonly denoted by Δx (read "delta x"). Note that Δx does *not* mean Δ times x. If $x_1 = 4.1$ and $x_2 = 5.7$, then

$$\Delta x = x_2 - x_1 = 5.7 - 4.1 = 1.6$$

If $x_1 = c$ and $x_2 = c + h$, then

$$\Delta x = x_2 - x_1 = c + h - c = h$$

Suppose next that $y = f(x)$ determines a function. If x changes from x_1 to x_2, then y changes from $y_1 = f(x_1)$ to $y_2 = f(x_2)$. Thus, corresponding to the increment $\Delta x = x_2 - x_1$ in x, there is an increment in y given by

$$\Delta y = y_2 - y_1 = f(x_2) - f(x_1)$$

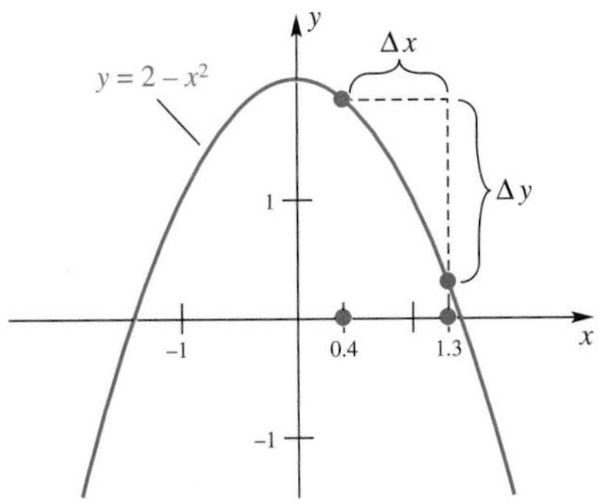

Figure 5

EXAMPLE 7 Let $y = f(x) = 2 - x^2$. Find Δy when x changes from 0.4 to 1.3 (see Figure 5).

SOLUTION

$$\Delta y = f(1.3) - f(0.4) = \left[2 - (1.3)^2\right] - \left[2 - (0.4)^2\right] = -1.53 \quad \blacksquare$$

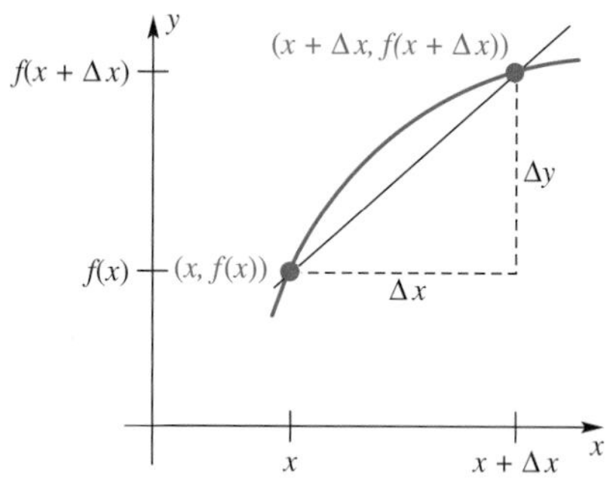

Figure 6

Leibniz Notation for the Derivative Suppose now that the independent variable changes from x to $x + \Delta x$. The corresponding change in the dependent variable, y, will be

$$\Delta y = f(x + \Delta x) - f(x)$$

and the ratio

$$\frac{\Delta y}{\Delta x} = \frac{f(x + \Delta x) - f(x)}{\Delta x}$$

represents the slope of a secant line through $(x, f(x))$, as shown in Figure 6. As $\Delta x \to 0$, the slope of this secant line approaches that of the tangent line, and for this latter slope we use the symbol dy/dx. Thus,

$$\boxed{\frac{dy}{dx} = \lim_{\Delta x \to 0} \frac{\Delta y}{\Delta x} = \lim_{\Delta x \to 0} \frac{f(x + \Delta x) - f(x)}{\Delta x} = f'(x)}$$

Gottfried Wilhelm Leibniz, a contemporary of Isaac Newton, called dy/dx a quotient of two infinitesimals. The meaning of the word *infinitesimal* is vague, and we will not use it. However, dy/dx is a standard symbol for the derivative and we will use it frequently from now on.

The Graph of the Derivative The derivative $f'(x)$ gives the slope of the tangent line to the graph of $y = f(x)$ at the value of x. Thus, when the tangent line is sloping up to the right, the derivative is positive, and when the tangent line is sloping down to the right, the derivative is negative. We can therefore get a rough picture of the derivative given just the graph of the function.

EXAMPLE 8 Given the graph of $y = f(x)$ shown in the first part of Figure 7, sketch a graph of the derivative $f'(x)$.

SOLUTION For $x < 0$, the tangent line to the graph of $y = f(x)$ has positive slope. A rough calculation from the plot suggests that when $x = -2$, the slope is about 3. As we move from left to right along the curve in Figure 7, we see that the slope is still positive (for a while) but that the tangent lines become flatter and flatter. When $x = 0$, the tangent line is horizontal, telling us that $f'(0) = 0$. For x between 0 and 2, the tangent lines have negative slope, indicating that the derivative will be negative over this interval. When $x = 2$, we are again at a point where the tangent line is horizontal, so the derivative is equal to zero when $x = 2$. For $x > 2$, the tangent line again has positive slope. The graph of the derivative $f'(x)$ is shown in the last part of Figure 7. $\blacksquare$

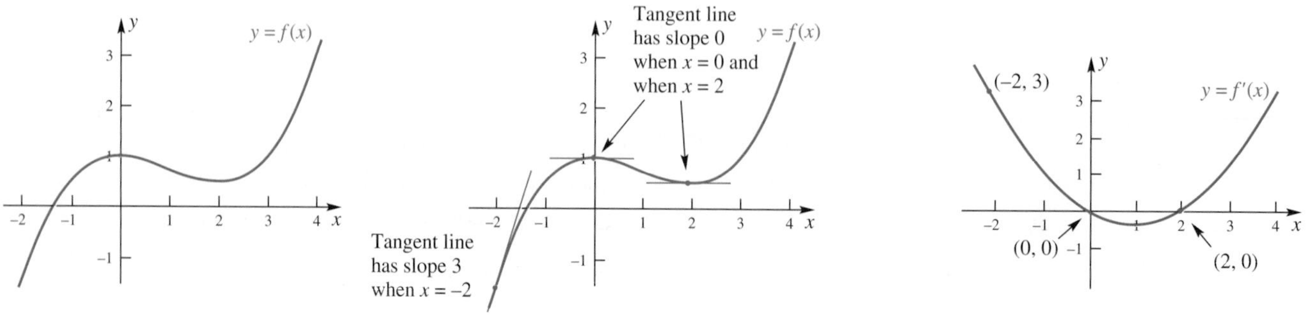

Figure 7

Concepts Review

1. The derivative of f at x is given by $f'(x) = \lim_{h\to 0}$ ____. Equivalently, $f'(x) = \lim_{t\to x}$ ____.

2. The slope of the tangent line to the graph of $y = f(x)$ at the point $(c, f(c))$ is ____.

3. If f is differentiable at c, then f is ____ at c. The converse is false, as is shown by the example $f(x) =$ ____.

4. If $y = f(x)$, we now have two different symbols for the derivative of y with respect to x. They are ____ and ____.

Problem Set 2.2

In Problems 1–4, use the definition

$$f'(c) = \lim_{h\to 0} \frac{f(c+h) - f(c)}{h}$$

to find the indicated derivative.

1. $f'(1)$ if $f(x) = x^2$

2. $f'(2)$ if $f(t) = (2t)^2$

3. $f'(3)$ if $f(t) = t^2 - t$

4. $f'(4)$ if $f(s) = \dfrac{1}{s-1}$

In Problems 5–22, use $f'(x) = \lim_{h\to 0} [f(x+h) - f(x)]/h$ *to find the derivative at* x.

5. $s(x) = 2x + 1$

6. $f(x) = \alpha x + \beta$

7. $r(x) = 3x^2 + 4$

8. $f(x) = x^2 + x + 1$

9. $f(x) = ax^2 + bx + c$

10. $f(x) = x^4$

11. $f(x) = x^3 + 2x^2 + 1$

12. $g(x) = x^4 + x^2$

13. $h(x) = \dfrac{2}{x}$

14. $S(x) = \dfrac{1}{x+1}$

15. $F(x) = \dfrac{6}{x^2+1}$

16. $F(x) = \dfrac{x-1}{x+1}$

17. $G(x) = \dfrac{2x-1}{x-4}$

18. $G(x) = \dfrac{2x}{x^2-x}$

19. $g(x) = \sqrt{3x}$

20. $g(x) = \dfrac{1}{\sqrt{3x}}$

21. $H(x) = \dfrac{3}{\sqrt{x-2}}$

22. $H(x) = \sqrt{x^2+4}$

In Problems 23–26, use $f'(x) = \lim_{t\to x} [f(t) - f(x)]/[t - x]$ *to find* $f'(x)$ *(see Example 5).*

23. $f(x) = x^2 - 3x$

24. $f(x) = x^3 + 5x$

25. $f(x) = \dfrac{x}{x-5}$

26. $f(x) = \dfrac{x+3}{x}$

In Problems 27–36, the given limit is a derivative, but of what function and at what point? (See Example 6.)

27. $\displaystyle\lim_{h\to 0} \frac{2(5+h)^3 - 2(5)^3}{h}$

28. $\displaystyle\lim_{h\to 0} \frac{(3+h)^2 + 2(3+h) - 15}{h}$

29. $\displaystyle\lim_{x\to 2} \frac{x^2 - 4}{x - 2}$

30. $\displaystyle\lim_{x\to 3} \frac{x^3 + x - 30}{x - 3}$

31. $\displaystyle\lim_{t\to x} \frac{t^2 - x^2}{t - x}$

32. $\displaystyle\lim_{p\to x} \frac{p^3 - x^3}{p - x}$

33. $\displaystyle\lim_{x\to t} \frac{\frac{2}{x} - \frac{2}{t}}{x - t}$

34. $\displaystyle\lim_{x\to y} \frac{\sin x - \sin y}{x - y}$

35. $\displaystyle\lim_{h\to 0} \frac{\cos(x+h) - \cos x}{h}$

36. $\displaystyle\lim_{h\to 0} \frac{\tan(t+h) - \tan t}{h}$

In Problems 37–44, the graph of a function $y = f(x)$ *is given. Use this graph to sketch the graph of* $y = f'(x)$.

37.

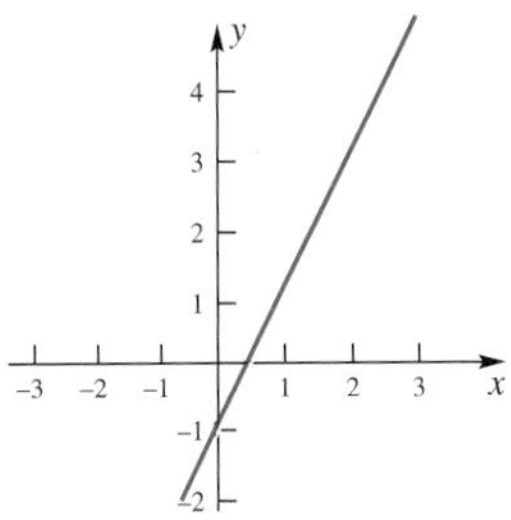

38.

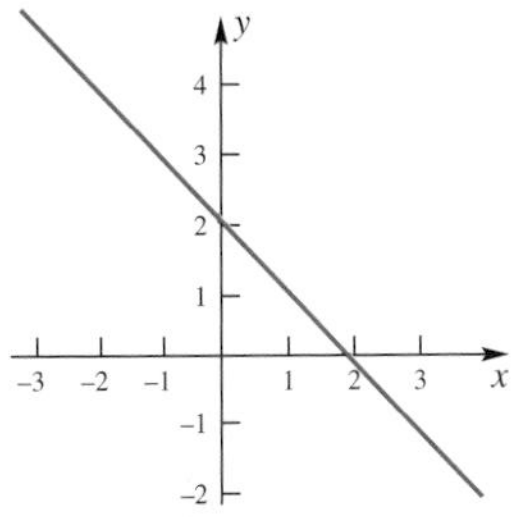

39.

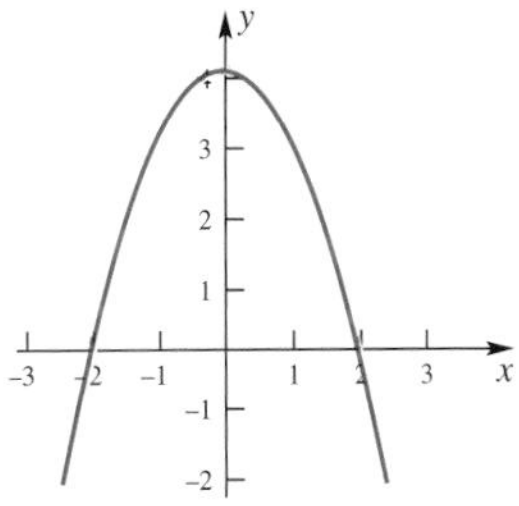

40.

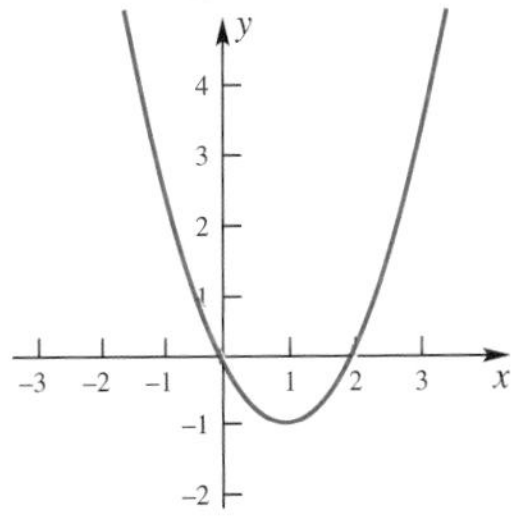

41.

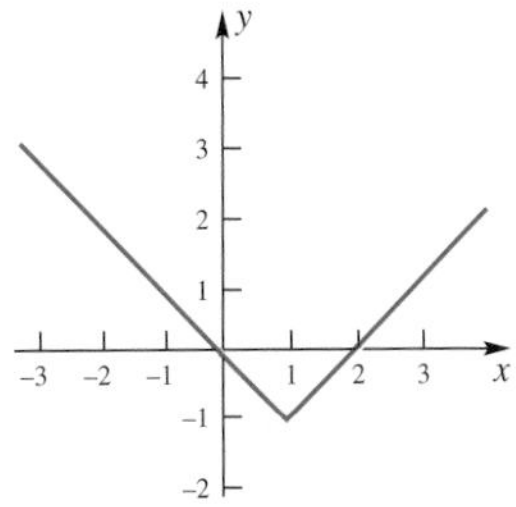

42.

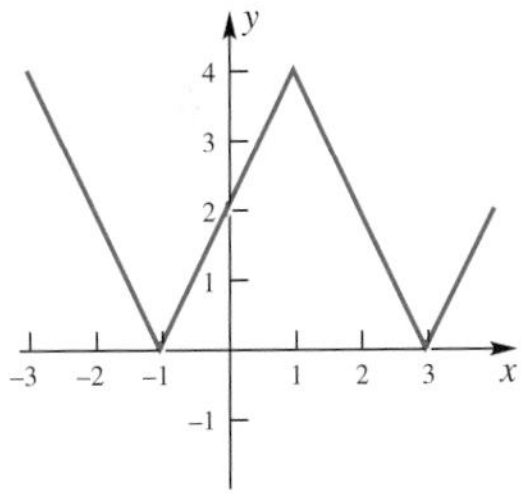

43.

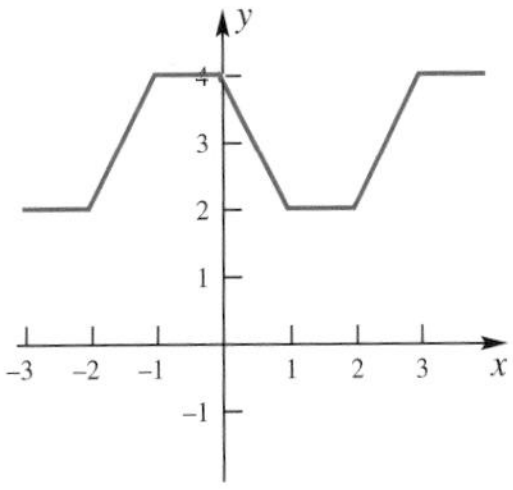

44.

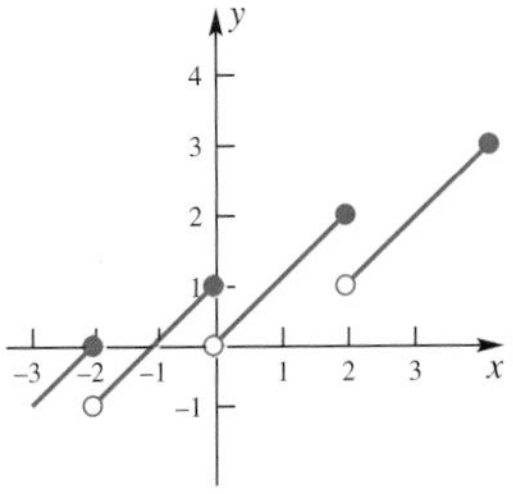

In Problems 45–50, find Δy *for the given values of* x_1 *and* x_2 *(see Example 7).*

45. $y = 3x + 2$, $x_1 = 1$, $x_2 = 1.5$

46. $y = 3x^2 + 2x + 1$, $x_1 = 0.0$, $x_2 = 0.1$

47. $y = \dfrac{1}{x}$, $x_1 = 1.0$, $x_2 = 1.2$

48. $y = \dfrac{2}{x+1}$, $x_1 = 0$, $x_2 = 0.1$

C **49.** $y = \dfrac{3}{x+1}$, $x_1 = 2.34$, $x_2 = 2.31$

C **50.** $y = \cos 2x$, $x_1 = 0.571$, $x_2 = 0.573$

In Problems 51–56, first find and simplify

$$\frac{\Delta y}{\Delta x} = \frac{f(x + \Delta x) - f(x)}{\Delta x}$$

Then find dy/dx by taking the limit of your answer as $\Delta x \to 0$.

51. $y = x^2$

52. $y = x^3 - 3x^2$

53. $y = \dfrac{1}{x+1}$

54. $y = 1 + \dfrac{1}{x}$

55. $y = \dfrac{x-1}{x+1}$

56. $y = \dfrac{x^2 - 1}{x}$

57. From Figure 8, estimate $f'(0)$, $f'(2)$, $f'(5)$, and $f'(7)$.

58. From Figure 9, estimate $g'(-1)$, $g'(1)$, $g'(4)$, and $g'(6)$.

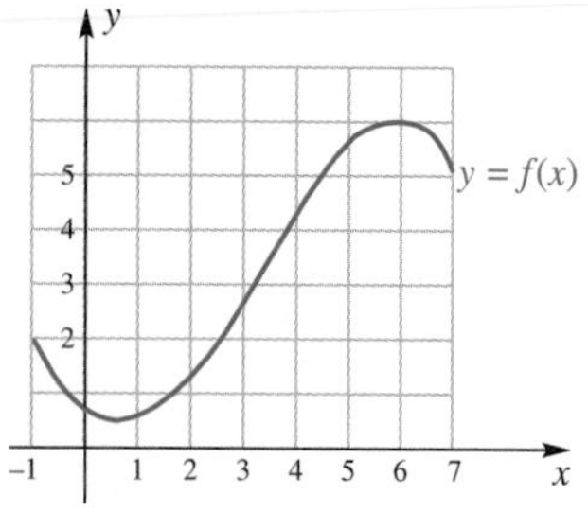

Figure 8

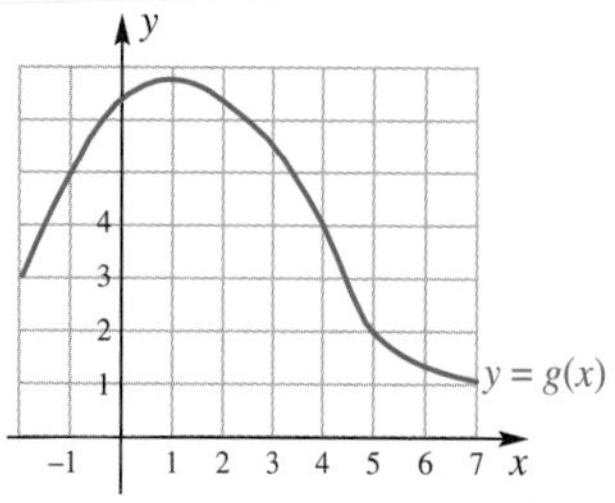

Figure 9

59. Sketch the graph of $y = f'(x)$ on $-1 < x < 7$ for the function f in Figure 8.

60. Sketch the graph of $y = g'(x)$ on $-1 < x < 7$ for the function g in Figure 9.

61. Consider the function $y = f(x)$, whose graph is sketched in Figure 10.

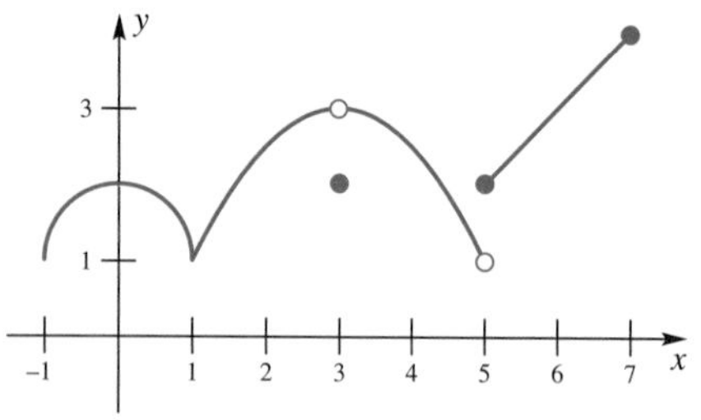

Figure 10

(a) Estimate $f(2)$, $f'(2)$, $f(0.5)$, and $f'(0.5)$.

(b) Estimate the average rate of change in f on the interval $0.5 \le x \le 2.5$.

(c) Where on the interval $-1 < x < 7$ does $\lim_{u \to x} f(u)$ fail to exist?

(d) Where on the interval $-1 < x < 7$ does f fail to be continuous?

(e) Where on the interval $-1 < x < 7$ does f fail to have a derivative?

(f) Where on the interval $-1 < x < 7$ is $f'(x) = 0$?

(g) Where on the interval $-1 < x < 7$ is $f'(x) = 1$?

62. Figure 14 in Section 2.1 shows the position s of an elevator as a function of time t. At what points does the derivative exist? Sketch the derivative of s.

63. Figure 15 in Section 2.1 shows the normal high temperature for St. Louis, Missouri. Sketch the derivative.

64. Figure 11 shows two functions. One is the function f, and the other is its derivative f'. Which one is which?

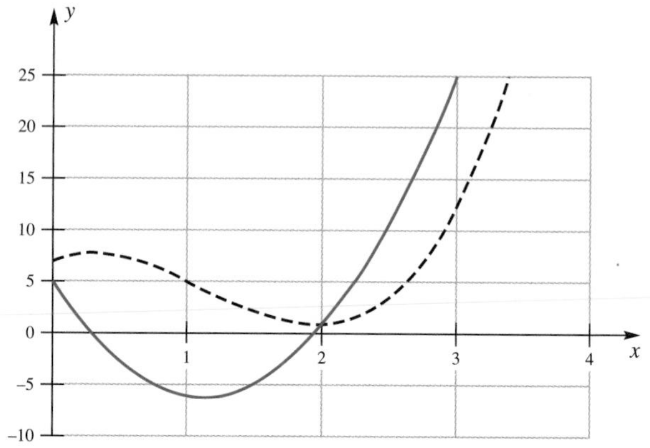

Figure 11

65. Figure 12 shows three functions. One is the function f; another is its derivative f', which we will call g; and the third is the derivative of g. Which one is which?

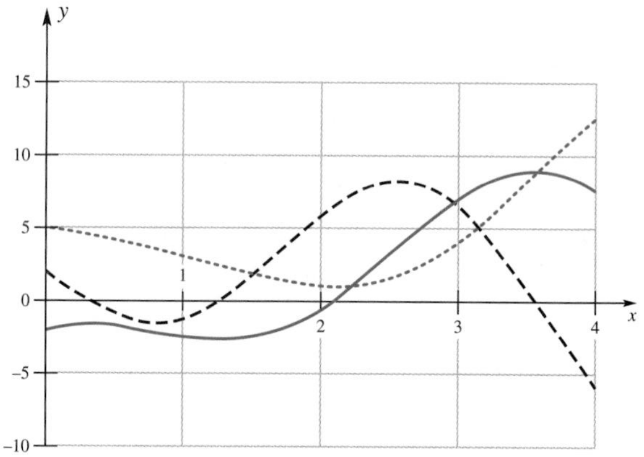

Figure 12

EXPL **66.** Suppose that $f(x + y) = f(x)f(y)$ for all x and y. Show that if $f'(0)$ exists then $f'(a)$ exists and $f'(a) = f(a)f'(0)$.

67. Let $f(x) = \begin{cases} mx + b & \text{if } x < 2 \\ x^2 & \text{if } x \ge 2 \end{cases}$

Determine m and b so that f is differentiable everywhere.

EXPL **68.** The **symmetric derivative** $f_s(x)$ is defined by

$$f_s(x) = \lim_{h \to 0} \frac{f(x + h) - f(x - h)}{2h}$$

Show that if $f'(x)$ exists then $f_s(x)$ exists, but that the converse is false.

69. Let f be differentiable and let $f'(x_0) = m$. Find $f'(-x_0)$ if

(a) f is an odd function.

(b) f is an even function.

70. Prove that the derivative of an odd function is an even function and that the derivative of an even function is an odd function.

CAS *Use a CAS to do Problems 71 and 72.*

EXPL **71.** Draw the graphs of $f(x) = x^3 - 4x^2 + 3$ and its derivative $f'(x)$ on the interval $[-2, 5]$ using the same axes.

(a) Where on this interval is $f'(x) < 0$?

(b) Where on this interval is $f(x)$ decreasing as x increases?

(c) Make a conjecture. Experiment with other intervals and other functions to support this conjecture.

EXPL **72.** Draw the graphs of $f(x) = \cos x - \sin(x/2)$ and its derivative $f'(x)$ on the interval $[0, 9]$ using the same axes.

(a) Where on this interval is $f'(x) > 0$?

(b) Where on this interval is $f(x)$ increasing as x increases?

(c) Make a conjecture. Experiment with other intervals and other functions to support this conjecture.

Answers to Concepts Review: **1.** $[f(x + h) - f(x)]/h$; $[f(t) - f(x)]/(t - x)$ **2.** $f'(c)$ **3.** continuous; $|x|$ **4.** $f'(x)$; $\dfrac{dy}{dx}$

2.3 Rules for Finding Derivatives

The process of finding the derivative of a function directly from the definition of the derivative, that is, by setting up the difference quotient

$$\frac{f(x + h) - f(x)}{h}$$

and evaluating its limit, can be time consuming and tedious. We are going to develop tools that will allow us to shortcut this lengthy process—that will, in fact, allow us to find derivatives of the most complicated looking functions.

Recall that the derivative of a function f is another function f'. We saw in the previous section that, if $f(x) = x^3 + 7x$ is the formula for f, then $f'(x) = 3x^2 + 7$ is the formula for f'. When we take the derivative of f, we say that we are differentiating f. The derivative *operates* on f to produce f'. We often use the symbol D_x to indicate the operation of differentiating (Figure 1). The D_x symbol says that we are to take the derivative (with respect to the variable x) of what follows. Thus, we write $D_x f(x) = f'(x)$ or (in the case just mentioned) $D_x(x^3 + 7x) = 3x^2 + 7$. This D_x is an example of an **operator.** As Figure 1 suggests, an operator is a function whose input is a function and whose output is another function.

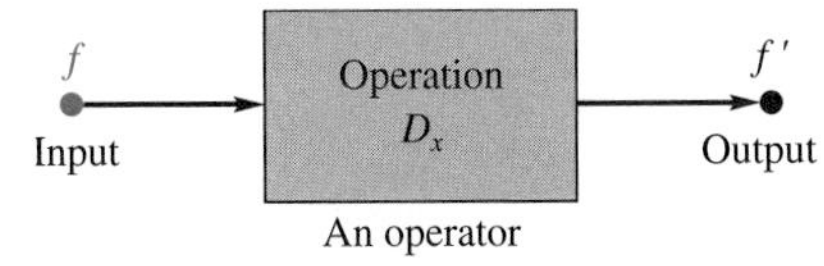

Figure 1

With Leibniz notation, introduced in the last section, we now have three notations for the derivative. If $y = f(x)$, we can denote the derivative of f by

$$f'(x) \quad \text{or} \quad D_x f(x) \quad \text{or} \quad \frac{dy}{dx}$$

We will use the notation $\dfrac{d}{dx}$ to mean the same as the operator D_x.

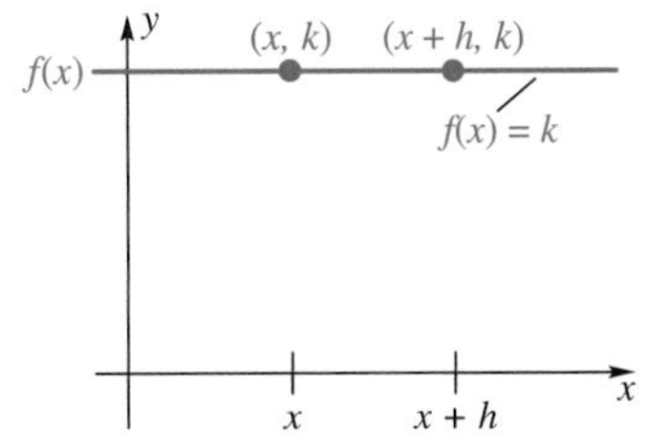

Figure 2

The Constant and Power Rules The graph of the constant function $f(x) = k$ is a horizontal line (Figure 2), which therefore has slope zero everywhere. This is one way to understand our first theorem.

Theorem A Constant Function Rule

If $f(x) = k$, where k is a constant, then for any x, $f'(x) = 0$; that is,

$$D_x(k) = 0$$

Proof

$$f'(x) = \lim_{h \to 0} \frac{f(x + h) - f(x)}{h} = \lim_{h \to 0} \frac{k - k}{h} = \lim_{h \to 0} 0 = 0$$

■

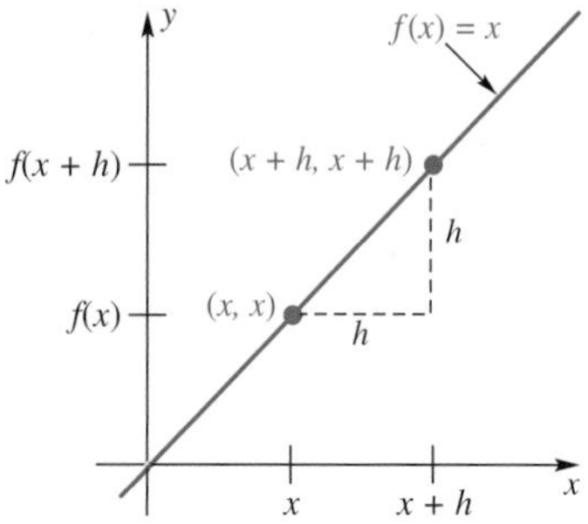

Figure 3

The graph of $f(x) = x$ is a line through the origin with slope 1 (Figure 3); so we should expect the derivative of this function to be 1 for all x.

Theorem B Identity Function Rule

If $f(x) = x$, then $f'(x) = 1$; that is,

$$D_x(x) = 1$$

Proof

$$f'(x) = \lim_{h \to 0} \frac{f(x+h) - f(x)}{h} = \lim_{h \to 0} \frac{x + h - x}{h} = \lim_{h \to 0} \frac{h}{h} = 1 \qquad \blacksquare$$

Before stating our next theorem, we recall something from algebra: how to raise a binomial to a power.

$$(a+b)^2 = a^2 + 2ab + b^2$$
$$(a+b)^3 = a^3 + 3a^2b + 3ab^2 + b^3$$
$$(a+b)^4 = a^4 + 4a^3b + 6a^2b^2 + 4ab^3 + b^4$$
$$\vdots$$
$$(a+b)^n = a^n + na^{n-1}b + \frac{n(n-1)}{2}a^{n-2}b^2 + \cdots + nab^{n-1} + b^n$$

Theorem C Power Rule

If $f(x) = x^n$, where n is a positive integer, then $f'(x) = nx^{n-1}$; that is,

$$D_x(x^n) = nx^{n-1}$$

Proof

$$f'(x) = \lim_{h \to 0} \frac{f(x+h) - f(x)}{h} = \lim_{h \to 0} \frac{(x+h)^n - x^n}{h}$$

$$= \lim_{h \to 0} \frac{x^n + nx^{n-1}h + \frac{n(n-1)}{2}x^{n-2}h^2 + \cdots + nxh^{n-1} + h^n - x^n}{h}$$

$$= \lim_{h \to 0} \frac{\not{h}\left[nx^{n-1} + \frac{n(n-1)}{2}x^{n-2}h + \cdots + nxh^{n-2} + h^{n-1}\right]}{\not{h}}$$

Within the brackets, all terms except the first have h as a factor, and so for every value of x, each of these terms has limit zero as h approaches zero. Thus,

$$f'(x) = nx^{n-1} \qquad \blacksquare$$

As illustrations of Theorem C, note that

$$D_x(x^3) = 3x^2 \qquad D_x(x^9) = 9x^8 \qquad D_x(x^{100}) = 100x^{99}$$

D_x Is a Linear Operator The operator D_x behaves very well when applied to constant multiples of functions or to sums of functions.

Theorem D Constant Multiple Rule

If k is a constant and f is a differentiable function, then $(kf)'(x) = k \cdot f'(x)$; that is,

$$D_x[k \cdot f(x)] = k \cdot D_x f(x)$$

In words, *a constant multiplier k can be passed across the operator D_x.*

Proof Let $F(x) = k \cdot f(x)$. Then

$$F'(x) = \lim_{h \to 0} \frac{F(x+h) - F(x)}{h} = \lim_{h \to 0} \frac{k \cdot f(x+h) - k \cdot f(x)}{h}$$

$$= \lim_{h \to 0} k \cdot \frac{f(x+h) - f(x)}{h} = k \cdot \lim_{h \to 0} \frac{f(x+h) - f(x)}{h}$$

$$= k \cdot f'(x)$$

The next-to-last step was the critical one. We could shift k past the limit sign because of the Main Limit Theorem Part 3. ■

Examples that illustrate this result are

$$D_x(-7x^3) = -7D_x(x^3) = -7 \cdot 3x^2 = -21x^2$$

and

$$D_x\left(\tfrac{4}{3}x^9\right) = \tfrac{4}{3}D_x(x^9) = \tfrac{4}{3} \cdot 9x^8 = 12x^8$$

Theorem E Sum Rule

If f and g are differentiable functions, then $(f + g)'(x) = f'(x) + g'(x)$; that is,

$$D_x\left[f(x) + g(x)\right] = D_x f(x) + D_x g(x)$$

In words, *the derivative of a sum is the sum of the derivatives.*

Proof Let $F(x) = f(x) + g(x)$. Then

$$F'(x) = \lim_{h \to 0} \frac{\left[f(x+h) + g(x+h)\right] - \left[f(x) + g(x)\right]}{h}$$

$$= \lim_{h \to 0} \left[\frac{f(x+h) - f(x)}{h} + \frac{g(x+h) - g(x)}{h}\right]$$

$$= \lim_{h \to 0} \frac{f(x+h) - f(x)}{h} + \lim_{h \to 0} \frac{g(x+h) - g(x)}{h}$$

$$= f'(x) + g'(x)$$

Again, the next-to-last step was the critical one. It is justified by the Main Limit Theorem Part 4. ■

Any operator L with the properties stated in Theorems D and E is called *linear*; that is, L is a **linear operator** if for all functions f and g:

1. $L(kf) = kL(f)$, for every constant k;
2. $L(f + g) = L(f) + L(g)$.

Linear operators will appear again and again in this book; D_x is a particularly important example. A linear operator always satisfies the difference rule $L(f - g) = L(f) - L(g)$ stated next for D_x.

Theorem F Difference Rule

If f and g are differentiable functions, then $(f - g)'(x) = f'(x) - g'(x)$; that is,

$$D_x\left[f(x) - g(x)\right] = D_x f(x) - D_x g(x)$$

The proof of Theorem F is left as an exercise (Problem 54).

Linear Operator

The fundamental meaning of the word *linear*, as used in mathematics, is that given in this section. An operator L is linear if it satisfies the two key conditions:

- $L(ku) = kL(u)$
- $L(u + v) = L(u) + L(v)$

Linear operators play a central role in the *linear algebra* course, which many readers of this book will take.

Functions of the form $f(x) = mx + b$ are called *linear functions* because of their connections with lines. This terminology can be confusing because linear functions are not linear in the operator sense. To see this, note that

$$f(kx) = m(kx) + b$$

whereas

$$kf(x) = k(mx + b)$$

Thus, $f(kx) \neq kf(x)$ unless b happens to be zero.

EXAMPLE 1 Find the derivatives of $5x^2 + 7x - 6$ and $4x^6 - 3x^5 - 10x^2 + 5x + 16$.

SOLUTION

$$\begin{aligned} D_x(5x^2 + 7x - 6) &= D_x(5x^2 + 7x) - D_x(6) && \text{(Theorem F)} \\ &= D_x(5x^2) + D_x(7x) - D_x(6) && \text{(Theorem E)} \\ &= 5D_x(x^2) + 7D_x(x) - D_x(6) && \text{(Theorem D)} \\ &= 5 \cdot 2x + 7 \cdot 1 - 0 && \text{(Theorems C, B, A)} \\ &= 10x + 7 \end{aligned}$$

To find the next derivative, we note that the theorems on sums and differences extend to any finite number of terms. Thus,

$$\begin{aligned} &D_x(4x^6 - 3x^5 - 10x^2 + 5x + 16) \\ &\quad = D_x(4x^6) - D_x(3x^5) - D_x(10x^2) + D_x(5x) + D_x(16) \\ &\quad = 4D_x(x^6) - 3D_x(x^5) - 10D_x(x^2) + 5D_x(x) + D_x(16) \\ &\quad = 4(6x^5) - 3(5x^4) - 10(2x) + 5(1) + 0 \\ &\quad = 24x^5 - 15x^4 - 20x + 5 \end{aligned}$$

■

The method of Example 1 allows us to find the derivative of any polynomial. If you know the Power Rule and do what comes naturally, you are almost sure to get the right result. Also, with practice, you will find that you can write the derivative immediately, without having to write any intermediate steps.

Product and Quotient Rules Now we are in for a surprise. So far, we have seen that the limit of a sum or difference is equal to the sum or difference of the limits (Theorem 1.3A, Parts 4 and 5), the limit of a product or quotient is the product or quotient of the limits (Theorem 1.3A, Parts 6 and 7), and the derivative of a sum or difference is the sum or difference of the derivatives (Theorems E and F). So what could be more natural than to have the derivative of a product be the product of the derivatives?

This may seem natural, but it is wrong. To see why, let's look at the following example.

EXAMPLE 2 Let $g(x) = x, h(x) = 1 + 2x,$ and $f(x) = g(x) \cdot h(x) = x(1 + 2x)$. Find $D_x f(x), D_x g(x),$ and $D_x h(x)$, and show that $D_x f(x) \neq [D_x g(x)][D_x h(x)]$.

SOLUTION

$$\begin{aligned} D_x f(x) &= D_x[x(1 + 2x)] \\ &= D_x(x + 2x^2) \\ &= 1 + 4x \\ D_x g(x) &= D_x x = 1 \\ D_x h(x) &= D_x(1 + 2x) = 2 \end{aligned}$$

Notice that

$$D_x(g(x))D_x(h(x)) = 1 \cdot 2 = 2$$

whereas

$$D_x f(x) = D_x[g(x)h(x)] = 1 + 4x$$

Thus, $D_x f(x) \neq [D_x g(x)][D_x h(x)]$. ■

That the derivative of a product should be the product of the derivatives seemed so natural that it even fooled Gottfried Wilhelm von Leibniz, one of the discoverers of calculus. In a manuscript of November 11, 1675, he computed the derivative of the product of two functions and said (without checking) that it was equal to the product of the derivatives. Ten days later, he caught the error and gave the correct product rule, which we present as Theorem G.

> **Memorization**
>
> Some people say that memorization is passé, that only logical reasoning is important in mathematics. They are wrong. Some things (including the rules of this section) must become so much a part of our mental apparatus that we can use them without stopping to reflect.
>
> "Civilization advances by extending the number of important operations which we can perform without thinking about them."
>
> *Alfred N. Whitehead*

Theorem G Product Rule

If f and g are differentiable functions, then

$$(f \cdot g)'(x) = f(x)g'(x) + g(x)f'(x)$$

That is,

$$D_x[f(x)g(x)] = f(x)D_xg(x) + g(x)D_xf(x)$$

This rule should be memorized in words as follows: *The derivative of a product of two functions is the first times the derivative of the second plus the second times the derivative of the first.*

Proof Let $F(x) = f(x)g(x)$. Then

$$\begin{aligned} F'(x) &= \lim_{h\to0} \frac{F(x+h) - F(x)}{h} \\ &= \lim_{h\to0} \frac{f(x+h)g(x+h) - f(x)g(x)}{h} \\ &= \lim_{h\to0} \frac{f(x+h)g(x+h) - f(x+h)g(x) + f(x+h)g(x) - f(x)g(x)}{h} \\ &= \lim_{h\to0}\left[f(x+h)\cdot\frac{g(x+h) - g(x)}{h} + g(x)\cdot\frac{f(x+h) - f(x)}{h}\right] \\ &= \lim_{h\to0} f(x+h)\cdot\lim_{h\to0}\frac{g(x+h) - g(x)}{h} + g(x)\cdot\lim_{h\to0}\frac{f(x+h) - f(x)}{h} \\ &= f(x)g'(x) + g(x)f'(x) \end{aligned}$$

The derivation just given relies first on the trick of adding and subtracting the same thing, that is, $f(x+h)g(x)$. Second, at the very end, we use the fact that

$$\lim_{h\to0} f(x+h) = f(x)$$

This is just an application of Theorem 2.2A (which says that differentiability at a point implies continuity there) and the definition of continuity at a point. ■

EXAMPLE 3 Find the derivative of $(3x^2 - 5)(2x^4 - x)$ by use of the Product Rule. Check the answer by doing the problem a different way.

SOLUTION

$$\begin{aligned} D_x[(3x^2 - 5)(2x^4 - x)] &= (3x^2 - 5)D_x(2x^4 - x) + (2x^4 - x)D_x(3x^2 - 5) \\ &= (3x^2 - 5)(8x^3 - 1) + (2x^4 - x)(6x) \\ &= 24x^5 - 3x^2 - 40x^3 + 5 + 12x^5 - 6x^2 \\ &= 36x^5 - 40x^3 - 9x^2 + 5 \end{aligned}$$

To check, we first multiply and then take the derivative.

$$(3x^2 - 5)(2x^4 - x) = 6x^6 - 10x^4 - 3x^3 + 5x$$

Thus,

$$D_x\left[(3x^2 - 5)(2x^4 - x)\right] = D_x(6x^6) - D_x(10x^4) - D_x(3x^3) + D_x(5x)$$
$$= 36x^5 - 40x^3 - 9x^2 + 5$$ ■

Theorem H **Quotient Rule**

Let f and g be differentiable functions with $g(x) \neq 0$. Then

$$\left(\frac{f}{g}\right)'(x) = \frac{g(x)f'(x) - f(x)g'(x)}{g^2(x)}$$

That is,

$$D_x\left(\frac{f(x)}{g(x)}\right) = \frac{g(x)D_xf(x) - f(x)D_xg(x)}{g^2(x)}$$

We strongly urge you to memorize this in words, as follows: *The derivative of a quotient is equal to the denominator times the derivative of the numerator minus the numerator times the derivative of the denominator, all divided by the square of the denominator.*

Proof Let $F(x) = f(x)/g(x)$. Then

$$F'(x) = \lim_{h\to 0} \frac{F(x+h) - F(x)}{h}$$

$$= \lim_{h\to 0} \frac{\dfrac{f(x+h)}{g(x+h)} - \dfrac{f(x)}{g(x)}}{h}$$

$$= \lim_{h\to 0} \frac{g(x)f(x+h) - f(x)g(x+h)}{h} \cdot \frac{1}{g(x)g(x+h)}$$

$$= \lim_{h\to 0} \left[\frac{g(x)f(x+h) - g(x)f(x) + f(x)g(x) - f(x)g(x+h)}{h} \cdot \frac{1}{g(x)g(x+h)}\right]$$

$$= \lim_{h\to 0} \left\{\left[g(x)\frac{f(x+h) - f(x)}{h} - f(x)\frac{g(x+h) - g(x)}{h}\right]\frac{1}{g(x)g(x+h)}\right\}$$

$$= \left[g(x)f'(x) - f(x)g'(x)\right]\frac{1}{g(x)g(x)}$$ ■

EXAMPLE 4 Find $\dfrac{d}{dx}\dfrac{(3x-5)}{(x^2+7)}$.

SOLUTION

$$\frac{d}{dx}\left[\frac{3x-5}{x^2+7}\right] = \frac{(x^2+7)\dfrac{d}{dx}(3x-5) - (3x-5)\dfrac{d}{dx}(x^2+7)}{(x^2+7)^2}$$

$$= \frac{(x^2+7)(3) - (3x-5)(2x)}{(x^2+7)^2}$$

$$= \frac{-3x^2 + 10x + 21}{(x^2+7)^2}$$ ■

EXAMPLE 5 Find $D_x y$ if $y = \frac{2}{x^4 + 1} + \frac{3}{x}$.

SOLUTION

$$\begin{aligned} D_x y &= D_x\left(\frac{2}{x^4 + 1}\right) + D_x\left(\frac{3}{x}\right) \\ &= \frac{(x^4 + 1)D_x(2) - 2D_x(x^4 + 1)}{(x^4 + 1)^2} + \frac{xD_x(3) - 3D_x(x)}{x^2} \\ &= \frac{(x^4 + 1)(0) - (2)(4x^3)}{(x^4 + 1)^2} + \frac{(x)(0) - (3)(1)}{x^2} \\ &= \frac{-8x^3}{(x^4 + 1)^2} - \frac{3}{x^2} \end{aligned}$$

■

EXAMPLE 6 Show that the Power Rule holds for negative integral exponents; that is,

$$D_x\left(x^{-n}\right) = -nx^{-n-1}$$

$$D_x(x^{-n}) = D_x\left(\frac{1}{x^n}\right) = \frac{x^n \cdot 0 - 1 \cdot nx^{n-1}}{x^{2n}} = \frac{-nx^{n-1}}{x^{2n}} = -nx^{-n-1}$$

■

We saw as part of Example 5 that $D_x(3/x) = -3/x^2$. Now we have another way to see the same thing.

Concepts Review

1. The derivative of a product of two functions is the first times _____ plus the _____ times the derivative of the first. In symbols, $D_x\left[f(x)g(x)\right] =$ _____.

2. The derivative of a quotient is the _____ times the derivative of the numerator minus the numerator times the derivative of the _____, all divided by the _____. In symbols, $D_x\left[f(x)/g(x)\right] =$ _____.

3. The second term (the term involving h) in the expansion of $(x + h)^n$ is _____. It is this fact that leads to the formula $D_x\left[x^n\right] =$ _____.

4. L is called a linear operator if $L(kf) =$ _____ and $L(f + g) =$ _____. The derivative operator denoted by _____ is such an operator.

Problem Set 2.3

In Problems 1–44, find $D_x y$ using the rules of this section.

1. $y = 2x^2$
2. $y = 3x^3$
3. $y = \pi x$
4. $y = \pi x^3$
5. $y = 2x^{-2}$
6. $y = -3x^{-4}$
7. $y = \frac{\pi}{x}$
8. $y = \frac{\alpha}{x^3}$
9. $y = \frac{100}{x^5}$
10. $y = \frac{3\alpha}{4x^5}$
11. $y = x^2 + 2x$
12. $y = 3x^4 + x^3$
13. $y = x^4 + x^3 + x^2 + x + 1$
14. $y = 3x^4 - 2x^3 - 5x^2 + \pi x + \pi^2$
15. $y = \pi x^7 - 2x^5 - 5x^{-2}$
16. $y = x^{12} + 5x^{-2} - \pi x^{-10}$
17. $y = \frac{3}{x^3} + x^{-4}$
18. $y = 2x^{-6} + x^{-1}$
19. $y = \frac{2}{x} - \frac{1}{x^2}$
20. $y = \frac{3}{x^3} - \frac{1}{x^4}$
21. $y = \frac{1}{2x} + 2x$
22. $y = \frac{2}{3x} - \frac{2}{3}$
23. $y = x(x^2 + 1)$
24. $y = 3x(x^3 - 1)$
25. $y = (2x + 1)^2$
26. $y = (-3x + 2)^2$
27. $y = (x^2 + 2)(x^3 + 1)$
28. $y = (x^4 - 1)(x^2 + 1)$
29. $y = (x^2 + 17)(x^3 - 3x + 1)$
30. $y = (x^4 + 2x)(x^3 + 2x^2 + 1)$
31. $y = (5x^2 - 7)(3x^2 - 2x + 1)$
32. $y = (3x^2 + 2x)(x^4 - 3x + 1)$

33. $y = \dfrac{1}{3x^2 + 1}$

34. $y = \dfrac{2}{5x^2 - 1}$

35. $y = \dfrac{1}{4x^2 - 3x + 9}$

36. $y = \dfrac{4}{2x^3 - 3x}$

37. $y = \dfrac{x - 1}{x + 1}$

38. $y = \dfrac{2x - 1}{x - 1}$

39. $y = \dfrac{2x^2 - 1}{3x + 5}$

40. $y = \dfrac{5x - 4}{3x^2 + 1}$

41. $y = \dfrac{2x^2 - 3x + 1}{2x + 1}$

42. $y = \dfrac{5x^2 + 2x - 6}{3x - 1}$

43. $y = \dfrac{x^2 - x + 1}{x^2 + 1}$

44. $y = \dfrac{x^2 - 2x + 5}{x^2 + 2x - 3}$

45. If $f(0) = 4, f'(0) = -1, g(0) = -3$, and $g'(0) = 5$, find
(a) $(f \cdot g)'(0)$ (b) $(f + g)'(0)$ (c) $(f/g)'(0)$

46. If $f(3) = 7, f'(3) = 2, g(3) = 6$, and $g'(3) = -10$, find
(a) $(f - g)'(3)$ (b) $(f \cdot g)'(3)$ (c) $(g/f)'(3)$

47. Use the Product Rule to show that $D_x[f(x)]^2 = 2 \cdot f(x) \cdot D_x f(x)$.

EXPL **48.** Develop a rule for $D_x[f(x)g(x)h(x)]$.

49. Find the equation of the tangent line to $y = x^2 - 2x + 2$ at the point $(1, 1)$.

50. Find the equation of the tangent line to $y = 1/(x^2 + 4)$ at the point $(1, 1/5)$.

51. Find all points on the graph of $y = x^3 - x^2$ where the tangent line is horizontal.

52. Find all points on the graph of $y = \frac{1}{3}x^3 + x^2 - x$ where the tangent line has slope 1.

53. Find all points on the graph of $y = 100/x^5$ where the tangent line is perpendicular to the line $y = x$.

54. Prove Theorem F in two ways.

55. The height s in feet of a ball above the ground at t seconds is given by $s = -16t^2 + 40t + 100$.
(a) What is its instantaneous velocity at $t = 2$?
(b) When is its instantaneous velocity 0?

56. A ball rolls down a long inclined plane so that its distance s from its starting point after t seconds is $s = 4.5t^2 + 2t$ feet. When will its instantaneous velocity be 30 feet per second?

≈ **57.** There are two tangent lines to the curve $y = 4x - x^2$ that go through $(2, 5)$. Find the equations of both of them. *Hint*: Let (x_0, y_0) be a point of tangency. Find two conditions that (x_0, y_0) must satisfy. See Figure 4.

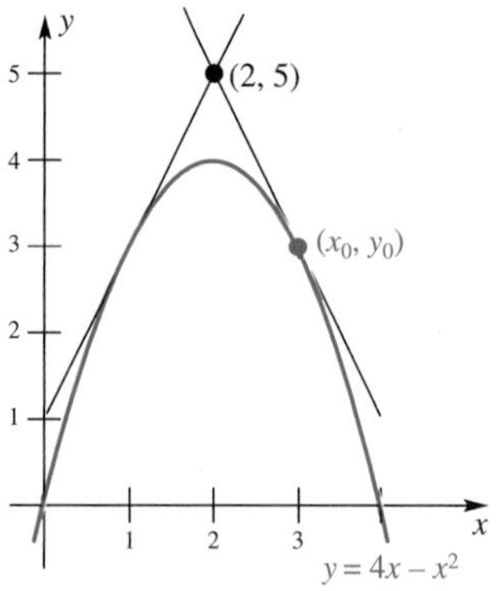

Figure 4

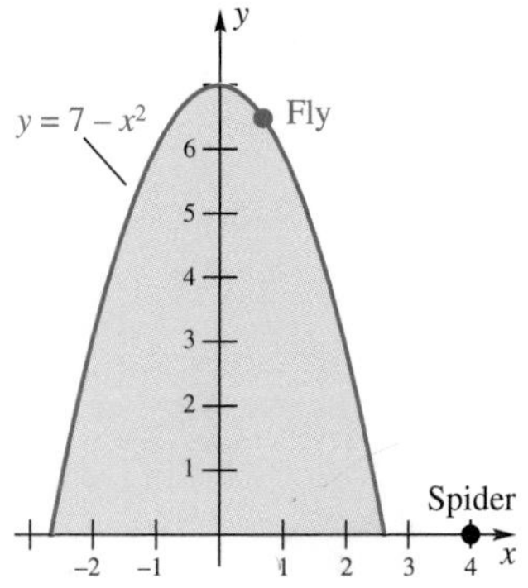

Figure 5

≈ **58.** A space traveler is moving from left to right along the curve $y = x^2$. When she shuts off the engines, she will continue traveling along the tangent line at the point where she is at that time. At what point should she shut off the engines in order to reach the point $(4, 15)$?

≈ **59.** A fly is crawling from left to right along the top of the curve $y = 7 - x^2$ (Figure 5). A spider waits at the point $(4, 0)$. Find the distance between the two insects when they first see each other.

60. Let $P(a, b)$ be a point on the first quadrant portion of the curve $y = 1/x$ and let the tangent line at P intersect the x-axis at A. Show that triangle AOP is isosceles and determine its area.

61. The radius of a spherical watermelon is growing at a constant rate of 2 centimeters per week. The thickness of the rind is always one-tenth of the radius. How fast is the volume of the rind growing at the end of the fifth week? Assume that the radius is initially 0.

CAS **62.** Redo Problems 29–44 on a computer and compare your answers with those you get by hand.

Answers to Concepts Review: **1.** the derivative of the second; second; $f(x)D_x g(x) + g(x)D_x f(x)$ **2.** denominator; denominator; square of the denominator; $[g(x)D_x f(x) - f(x)D_x g(x)]/g^2(x)$ **3.** $nx^{n-1}h$; nx^{n-1} **4.** $kL(f)$; $L(f) + L(g)$; D_x

2.4 Derivatives of Trigonometric Functions

Figure 1 reminds us of the definition of the sine and cosine functions. In what follows, t should be thought of as a number measuring the length of an arc on the unit circle or, equivalently, as the number of radians in the corresponding angle. Thus, $f(t) = \sin t$ and $g(t) = \cos t$ are functions for which both domain and range are sets of real numbers. We may consider the problem of finding their derivatives.

The Derivative Formulas We choose to use x rather than t as our basic variable. To find $D_x(\sin x)$, we appeal to the definition of derivative and use the Addition Identity for $\sin(x + h)$.

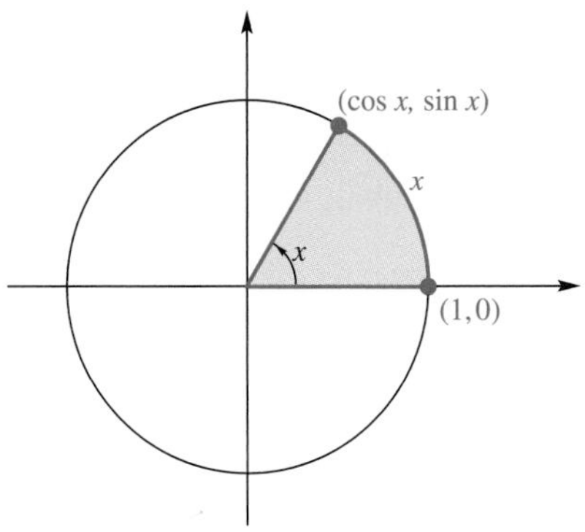

Figure 1

$$D_x(\sin x) = \lim_{h\to 0} \frac{\sin(x+h) - \sin x}{h}$$
$$= \lim_{h\to 0} \frac{\sin x \cos h + \cos x \sin h - \sin x}{h}$$
$$= \lim_{h\to 0} \left(-\sin x \frac{1-\cos h}{h} + \cos x \frac{\sin h}{h}\right)$$
$$= (-\sin x)\left[\lim_{h\to 0} \frac{1-\cos h}{h}\right] + (\cos x)\left[\lim_{h\to 0} \frac{\sin h}{h}\right]$$

Notice that the two limits in this last expression are exactly the limits we studied in Section 1.4. In Theorem 1.4B we proved that

$$\lim_{h\to 0} \frac{\sin h}{h} = 1 \quad \text{and} \quad \lim_{h\to 0} \frac{1-\cos h}{h} = 0$$

Thus,

$$D_x(\sin x) = (-\sin x)\cdot 0 + (\cos x)\cdot 1 = \cos x$$

Similarly,

$$D_x(\cos x) = \lim_{h\to 0} \frac{\cos(x+h) - \cos x}{h}$$
$$= \lim_{h\to 0} \frac{\cos x \cos h - \sin x \sin h - \cos x}{h}$$
$$= \lim_{h\to 0} \left(-\cos x \frac{1-\cos h}{h} - \sin x \frac{\sin h}{h}\right)$$
$$= (-\cos x)\cdot 0 - (\sin x)\cdot 1$$
$$= -\sin x$$

We summarize these results in an important theorem.

Theorem A

The functions $f(x) = \sin x$ and $g(x) = \cos x$ are both differentiable and,

$$D_x(\sin x) = \cos x \qquad D_x(\cos x) = -\sin x$$

Could You Have Guessed?

The solid curve below is the graph of $y = \sin x$. Note that the slope is 1 at 0, 0 at $\pi/2$, -1 at π, and so on. When we graph the slope function (the derivative), we obtain the dashed curve. Could you have guessed that $D_x \sin x = \cos x$?

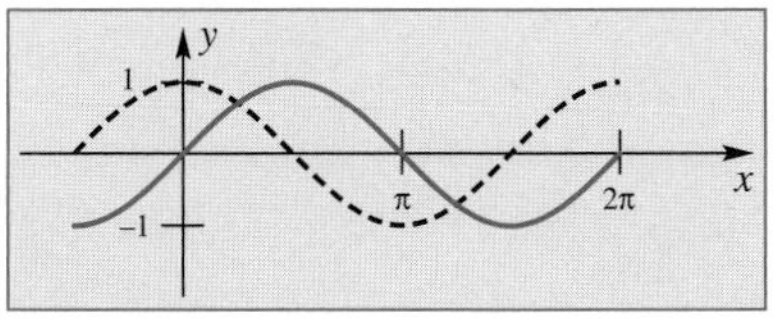

Try plotting these two functions in the same window on your CAS or graphing calculator.

EXAMPLE 1 Find $D_x(3 \sin x - 2 \cos x)$.

SOLUTION

$$D_x(3\sin x - 2\cos x) = 3D_x(\sin x) - 2D_x(\cos x)$$
$$= 3\cos x + 2\sin x$$

■

EXAMPLE 2 Find the equation of the tangent line to the graph of $y = 3 \sin x$ at the point $(\pi, 0)$. (See Figure 2.)

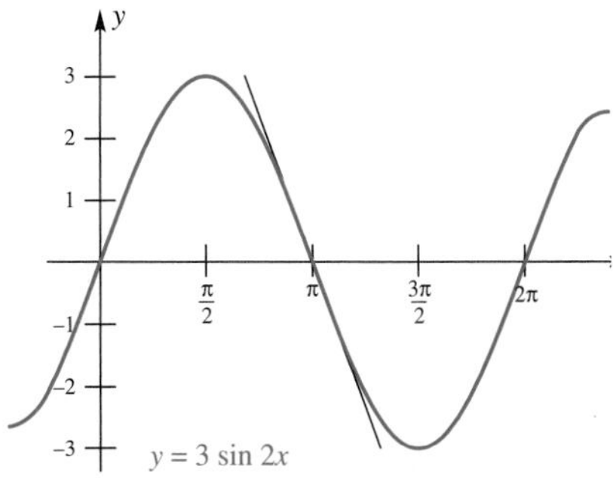

Figure 2

SOLUTION The derivative is $\frac{dy}{dx} = 3 \cos x$, so when $x = \pi$, the slope is $3 \cos \pi = -3$. Using the point-slope form for a line we find that the equation of the tangent line is

$$y - 0 = -3(x - \pi)$$
$$y = -3x + 3\pi$$ ■

The Product and Quotient Rules are useful when evaluating derivatives of functions involving the trigonometric functions.

EXAMPLE 3 Find $D_x(x^2 \sin x)$.

SOLUTION The Product Rule is needed here.

$$D_x(x^2 \sin x) = x^2 D_x(\sin x) + \sin x(D_x x^2) = x^2 \cos x + 2x \sin x$$ ■

EXAMPLE 4 Find $\frac{d}{dx}\left(\frac{1 + \sin x}{\cos x}\right)$.

SOLUTION For this problem, the Quotient Rule is needed.

$$\frac{d}{dx}\left(\frac{1 + \sin x}{\cos x}\right) = \frac{\cos x\left(\frac{d}{dx}(1 + \sin x)\right) - (1 + \sin x)\left(\frac{d}{dx}\cos x\right)}{\cos^2 x}$$
$$= \frac{\cos^2 x + \sin x + \sin^2 x}{\cos^2 x}$$
$$= \frac{1 + \sin x}{\cos^2 x}$$ ■

EXAMPLE 5 At time t seconds, the center of a bobbing cork is $y = 2 \sin t$ centimeters above (or below) water level. What is the velocity of the cork at $t = 0, \pi/2, \pi$?

SOLUTION The velocity is the derivative of position, and $\frac{dy}{dt} = 2 \cos t$. Thus, when $t = 0, \frac{dy}{dt} = 2 \cos 0 = 2$, when $t = \pi/2, \frac{dy}{dt} = 2 \cos \frac{\pi}{2} = 0$, and when $t = \pi, \frac{dy}{dt} = 2 \cos \pi = -2$. ■

Since the tangent, cotangent, secant, and cosecant functions are defined in terms of the sine and cosine functions, the derivatives of these functions can be obtained from Theorem A by applying the Quotient Rule. The results are summarized in Theorem B; for proofs, see Problems 5–8.

Theorem B

For all points x in the function's domain,

$$D_x \tan x = \sec^2 x \qquad D_x \cot x = -\csc^2 x$$
$$D_x \sec x = \sec x \tan x \qquad D_x \csc x = -\csc x \cot x$$

EXAMPLE 6 Find $D_x(x^n \tan x)$ for $n \geq 1$.

SOLUTION We apply the Product Rule along with Theorem B.

$$D_x(x^n \tan x) = x^n D_x(\tan x) + \tan x(D_x x^n)$$
$$= x^n \sec^2 x + nx^{n-1} \tan x$$ ■

EXAMPLE 7 Find the equation of the tangent line to the graph of $y = \tan x$ at the point $(\pi/4, 1)$.

SOLUTION The derivative of $y = \tan x$ is $\dfrac{dy}{dx} = \sec^2 x$. When $x = \pi/4$, the derivative is equal to $\sec^2 \dfrac{\pi}{4} = \left(\dfrac{2}{\sqrt{2}}\right)^2 = 2$. Thus the required line has slope 2 and passes through $(\pi/4, 1)$. Thus

$$y - 1 = 2\left(x - \frac{\pi}{4}\right)$$

$$y = 2x - \frac{\pi}{2} + 1$$ ■

EXAMPLE 8 Find all points on the graph of $y = \sin^2 x$ where the tangent line is horizontal.

SOLUTION The tangent line is horizontal when the derivative is equal to zero. To get the derivative of $\sin^2 x$, we use the Product Rule.

$$\frac{d}{dx}\sin^2 x = \frac{d}{dx}(\sin x \sin x) = \sin x \cos x + \sin x \cos x = 2 \sin x \cos x$$

The product of $\sin x$ and $\cos x$ is equal to zero when either $\sin x$ or $\cos x$ is equal to zero; that is, at $x = 0, \pm\dfrac{\pi}{2}, \pm\pi, \pm\dfrac{3\pi}{2}, \ldots$. ■

Concepts Review

1. By definition, $D_x(\sin x) = \lim_{h\to 0}$ ______.

2. To evaluate the limit in the preceding statement, we first use the Addition Identity for the sine function and then do a little algebra to obtain

$$D_x(\sin x) = (-\sin x)\left(\lim_{h\to 0}\frac{1 - \cos h}{h}\right) + (\cos x)\left(\lim_{h\to 0}\frac{\sin h}{h}\right)$$

The two displayed limits have the values ______ and ______, respectively.

3. The result of the calculation in the preceding statement is the important derivative formula $D_x(\sin x) =$ ______. The corresponding derivative formula $D_x(\cos x) =$ ______ is obtained in a similar manner.

4. At $x = \pi/3$, $D_x(\sin x)$ has the value ______. Thus, the equation of the tangent line to $y = \sin x$ at $x = \pi/3$ is ______.

Problem Set 2.4

In Problems 1–18, find $D_x y$.

1. $y = 2 \sin x + 3 \cos x$
2. $y = \sin^2 x$
3. $y = \sin^2 x + \cos^2 x$
4. $y = 1 - \cos^2 x$
5. $y = \sec x = 1/\cos x$
6. $y = \csc x = 1/\sin x$
7. $y = \tan x = \dfrac{\sin x}{\cos x}$
8. $y = \cot x = \dfrac{\cos x}{\sin x}$
9. $y = \dfrac{\sin x + \cos x}{\cos x}$
10. $y = \dfrac{\sin x + \cos x}{\tan x}$
11. $y = \sin x \cos x$
12. $y = \sin x \tan x$
13. $y = \dfrac{\sin x}{x}$
14. $y = \dfrac{1 - \cos x}{x}$
15. $y = x^2 \cos x$
16. $y = \dfrac{x \cos x + \sin x}{x^2 + 1}$
17. $y = \tan^2 x$
18. $y = \sec^3 x$

C 19. Find the equation of the tangent line to $y = \cos x$ at $x = 1$.

20. Find the equation of the tangent line to $y = \cot x$ at $x = \dfrac{\pi}{4}$.

21. Use the trigonometric identity $\sin 2x = 2 \sin x \cos x$ along with the Product Rule to find $D_x \sin 2x$.

22. Use the trigonometric identity $\cos 2x = 2\cos^2 x - 1$ along with the Product Rule to find $D_x \cos 2x$.

23. A Ferris wheel of radius 30 feet is rotating counterclockwise with an angular velocity of 2 radians per second. How fast is a seat on the rim rising (in the vertical direction) when it is 15 feet above the horizontal line through the center of the wheel? *Hint:* Use the result of Problem 21.

24. A Ferris wheel of radius 20 feet is rotating counterclockwise with an angular velocity of 1 radian per second. One seat on the rim is at $(20, 0)$ at time $t = 0$.

(a) What are its coordinates at $t = \pi/6$?

(b) How fast is it rising (vertically) at $t = \pi/6$?

(c) How fast is it rising when it is rising at the fastest rate?

25. Find the equation of the tangent line to $y = \tan x$ at $x = 0$.

26. Find all points on the graph of $y = \tan^2 x$ where the tangent line is horizontal.

27. Find all points on the graph of $y = 9 \sin x \cos x$ where the tangent line is horizontal.

28. Let $f(x) = x - \sin x$. Find all points on the graph of $y = f(x)$ where the tangent line is horizontal. Find all points on the graph of $y = f(x)$ where the tangent line has slope 2.

29. Show that the curves $y = \sqrt{2} \sin x$ and $y = \sqrt{2} \cos x$ intersect at right angles at a certain point with $0 < x < \pi/2$.

30. At time t seconds, the center of a bobbing cork is $3 \sin 2t$ centimeters above (or below) water level. What is the velocity of the cork at $t = 0, \pi/2, \pi$?

31. Use the definition of the derivative to show that $D_x(\sin x^2) = 2x \cos x^2$.

32. Use the definition of the derivative to show that $D_x(\sin 5x) = 5 \cos 5x$.

GC *Problems 33 and 34 are computer or graphing calculator exercises.*

33. Let $f(x) = x \sin x$.

(a) Draw the graphs of $f(x)$ and $f'(x)$ on $[\pi, 6\pi]$.

(b) How many solutions does $f(x) = 0$ have on $[\pi, 6\pi]$? How many solutions does $f'(x) = 0$ have on this interval?

(c) What is wrong with the following conjecture? If f and f' are both continuous and differentiable on $[a, b]$, if $f(a) = f(b) = 0$, and if $f(x) = 0$ has exactly n solutions on $[a, b]$, then $f'(x) = 0$ has exactly $n - 1$ solutions on $[a, b]$.

(d) Determine the maximum value of $|f(x) - f'(x)|$ on $[\pi, 6\pi]$.

34. Let $f(x) = \cos^3 x - 1.25 \cos^2 x + 0.225$. Find $f'(x_0)$ at that point x_0 in $[\pi/2, \pi]$ where $f(x_0) = 0$.

Answers to Concepts Review: **1.** $[\sin(x + h) - \sin x]/h$ **2.** 0; 1 **3.** $\cos x$; $-\sin x$ **4.** $\frac{1}{2}$; $y - \sqrt{3}/2 = \frac{1}{2}(x - \pi/3)$

2.5 The Chain Rule

Imagine trying to find the derivative of

$$F(x) = (2x^2 - 4x + 1)^{60}$$

We could find the derivative, but we would first have to multiply together the 60 quadratic factors of $2x^2 - 4x + 1$ and then differentiate the resulting polynomial. Or, how about trying to find the derivative of

$$G(x) = \sin 3x$$

We might be able to use some trigonometric identities to reduce it to something that depends on $\sin x$ and $\cos x$ and then use the rules from the previous section.

Fortunately, there is a better way. After learning the *Chain Rule*, we will be able to write the answers

$$F'(x) = 60(2x^2 - 4x + 1)^{59}(4x - 4)$$

and

$$G'(x) = 3 \cos 3x$$

The Chain Rule is so important that we will seldom again differentiate any function without using it.

Differentiating a Composite Function If David can type twice as fast as Mary and Mary can type three times as fast as Joe, then David can type $2 \times 3 = 6$ times as fast as Joe.

Consider the composite function $y = f(g(x))$. If we let $u = g(x)$, we can then think of f as a function of u. Suppose that $f(u)$ changes twice as fast as u, and $u = g(x)$ changes three times as fast as x. How fast is y changing? The statements

"$y = f(u)$ changes twice as fast as u" and "$u = g(x)$ changes three times as fast as x" can be restated as

$$\frac{dy}{du} = 2 \quad \text{and} \quad \frac{du}{dx} = 3$$

Just as in the previous paragraph, it seems as if the rates should multiply; that is, the rate of change of y with respect to x should equal the rate of change of y with respect to u times the rate of change of u with respect to x. In other words,

$$\frac{dy}{dx} = \frac{dy}{du} \times \frac{du}{dx}$$

This is in fact true, and we will sketch the proof at the end of this section. The result is called the **Chain Rule.**

Theorem A Chain Rule

Let $y = f(u)$ and $u = g(x)$. If g is differentiable at x and f is differentiable at $u = g(x)$, then the composite function $f \circ g$, defined by $(f \circ g)(x) = f(g(x))$, is differentiable at x and

$$(f \circ g)'(x) = f'(g(x))g'(x)$$

That is,

$$D_x(f(g(x)) = f'(g(x))g'(x)$$

or

$$\frac{dy}{dx} = \frac{dy}{du}\frac{du}{dx}$$

You can remember the Chain Rule this way: *The derivative of a composite function is the derivative of the outer function evaluated at the inner function, times the derivative of the inner function.*

Applications of the Chain Rule We begin with the example $(2x^2 - 4x + 1)^{60}$ introduced at the beginning of this section.

EXAMPLE 1 If $y = (2x^2 - 4x + 1)^{60}$, find $D_x y$.

SOLUTION We think of y as the 60th power of a function of x; that is

$$y = u^{60} \quad \text{and} \quad u = 2x^2 - 4x + 1$$

The outer function is $f(u) = u^{60}$ and the inner function is $u = g(x) = 2x^2 - 4x + 1$. Thus,

$$\begin{aligned} D_x y &= D_x f(g(x)) \\ &= f'(u)g'(x) \\ &= (60u^{59})(4x - 4) \\ &= 60(2x^2 - 4x + 1)^{59}(4x - 4) \end{aligned}$$

■

EXAMPLE 2 If $y = 1/(2x^5 - 7)^3$, find $\frac{dy}{dx}$.

SOLUTION Think of it this way.

$$y = \frac{1}{u^3} = u^{-3} \quad \text{and} \quad u = 2x^5 - 7$$

Thus,

$$\begin{aligned}\frac{dy}{dx} &= \frac{dy}{du}\frac{du}{dx} \\ &= (-3u^{-4})(10x^4) \\ &= \frac{-3}{u^4} \cdot 10x^4 \\ &= \frac{-30x^4}{(2x^5 - 7)^4}\end{aligned}$$ ■

> **The Last First**
>
> Here is an informal rule that may help you in using the derivative rules.
>
> *The last step in calculation corresponds to the first step in differentiation.*
>
> For example, the last step in calculating $(2x + 1)^3$ is to cube $2x + 1$, so you would first apply the Chain Rule to the cube function. The last step in calculating
>
> $$\frac{x^2 - 1}{x^2 + 1}$$
>
> is to take the quotient, so the first rule to use in differentiating is the Quotient Rule.

EXAMPLE 3 Find $D_t\left(\frac{t^3 - 2t + 1}{t^4 + 3}\right)^{13}$.

SOLUTION The last step in calculating this expression would be to raise the expression on the inside to the power 13. Thus, we begin by applying the Chain Rule to the function $y = u^{13}$, where $u = (t^3 - 2t + 1)/(t^4 + 3)$. The Chain Rule followed by the Quotient Rule gives

$$\begin{aligned}D_t\left(\frac{t^3 - 2t + 1}{t^4 + 3}\right)^{13} &= 13\left(\frac{t^3 - 2t + 1}{t^4 + 3}\right)^{13-1} D_t\left(\frac{t^3 - 2t + 1}{t^4 + 3}\right) \\ &= 13\left(\frac{t^3 - 2t + 1}{t^4 + 3}\right)^{12} \frac{(t^4 + 3)(3t^2 - 2) - (t^3 - 2t + 1)(4t^3)}{(t^4 + 3)^2} \\ &= 13\left(\frac{t^3 - 2t + 1}{t^4 + 3}\right)^{12} \frac{-t^6 + 6t^4 - 4t^3 + 9t^2 - 6}{(t^4 + 3)^2}\end{aligned}$$ ■

The Chain Rule simplifies computation of many derivatives involving the trigonometric functions. Although it is possible to differentiate $y = \sin 2x$ using trigonometric identities (see Problem 21 of the previous section), it is much easier to use the Chain Rule.

EXAMPLE 4 If $y = \sin 2x$, find $\frac{dy}{dx}$.

SOLUTION The last step in calculating this expression would be to take the sine of the quantity $2x$. Thus we use the Chain Rule on the function $y = \sin u$ where $u = 2x$.

$$\frac{dy}{dx} = (\cos 2x)\left(\frac{d}{dx}2x\right) = 2\cos 2x$$ ■

EXAMPLE 5 Find $F'(y)$ where $F(y) = y \sin y^2$.

SOLUTION The last step in calculating this expression would be to multiply y and $\sin y^2$, so we begin by applying the Product Rule. The Chain Rule is needed when we differentiate $\sin y^2$.

$$\begin{aligned}F'(y) &= yD_y[\sin y^2] + (\sin y^2)D_y(y) \\ &= y(\cos y^2)D_y(y^2) + (\sin y^2)(1) \\ &= 2y^2 \cos y^2 + \sin y^2\end{aligned}$$ ■

EXAMPLE 6 Find $D_x\left(\dfrac{x^2(1-x)^3}{1+x}\right)$.

SOLUTION The last step in calculating this expression would be to take the quotient. Thus, the Quotient Rule is the first to be applied. But notice that when we take the derivative of the numerator, we must apply the Product Rule and then the Chain Rule.

$$\begin{aligned}
D_x\left(\frac{x^2(1-x)^3}{1+x}\right) &= \frac{(1+x)D_x(x^2(1-x)^3) - x^2(1-x)^3D_x(1+x)}{(1+x)^2} \\
&= \frac{(1+x)[x^2D_x(1-x)^3 + (1-x)^3D_x(x^2)] - x^2(1-x)^3(1)}{(1+x)^2} \\
&= \frac{(1+x)[x^2(3(1-x)^2(-1)) + (1-x)^3(2x)] - x^2(1-x)^3}{(1+x)^2} \\
&= \frac{(1+x)[-3x^2(1-x)^2 + 2x(1-x)^3] - x^2(1-x)^3}{(1+x)^2} \\
&= \frac{(1+x)(1-x)^2x(2-5x) - x^2(1-x)^3}{(1+x)^2}
\end{aligned}$$

■

EXAMPLE 7 Find $\dfrac{d}{dx}\dfrac{1}{(2x-1)^3}$.

SOLUTION

$$\frac{d}{dx}\frac{1}{(2x-1)^3} = \frac{d}{dx}(2x-1)^{-3} = -3(2x-1)^{-3-1}\frac{d}{dx}(2x-1) = -\frac{6}{(2x-1)^4}$$

■

In this last example we were able to avoid use of the Quotient Rule. If you use the Quotient Rule, you would notice that the derivative of the numerator is 0, which simplifies the calculation. (You should check that the Quotient Rule gives the same answer as above.) As a general rule, if the numerator of a fraction is a constant, then do not use the Quotient Rule; instead write the quotient as the product of the constant and the expression in the denominator raised to a negative power, and then use the Chain Rule.

> **Notations for the Derivative**
>
> In this section, we have used all the various notations for the derivative, namely,
>
> $$f'(x)$$
>
> $$\frac{dy}{dx}$$
>
> and
>
> $$D_x f(x)$$
>
> You should by now be familiar with all of these notations. They will all be used in the remainder of the book.

EXAMPLE 8 Express the following derivatives in terms of the function $F(x)$. Assume that F is differentiable.

$$\text{(a) } D_x(F(x^3)) \quad \text{and} \quad \text{(b) } D_x[(F(x))^3]$$

SOLUTION

(a) The last step in calculating this expression would be to apply the function F. (Here the inner function is $u = x^3$ and the outer function is $F(u)$.) Thus

$$D_x(F(x^3)) = F'(x^3)D_x(x^3) = 3x^2\,F'(x^3)$$

(b) For this expression we would first evaluate $F(x)$ and then cube the result. (Here the inner function is $u = F(x)$ and the outer function is u^3.) Thus we apply the Power Rule first, then the Chain Rule.

$$D_x[(F(x))^3] = 3[F(x)]^2D_x(F(x)) = 3[F(x)]^2F'(x)$$

■

Applying the Chain Rule More than Once Sometimes when we apply the Chain Rule to a composite function we find that differentiation of the inner function also requires the Chain Rule. In cases like this, we simply have to use the Chain Rule a second time.

EXAMPLE 9 Find $D_x \sin^3(4x)$.

SOLUTION Remember, $\sin^3(4x) = [\sin(4x)]^3$, so we view this as the cube of a function of x. Thus, using our rule "derivative of the outer function evaluated at the inner function times the derivative of the inner function," we have

$$D_x \sin^3(4x) = D_x[\sin(4x)]^3 = 3[\sin(4x)]^{3-1} D_x[\sin(4x)]$$

Now we apply the Chain Rule once again for the derivative of the inner function.

$$\begin{aligned} D_x \sin^3(4x) &= 3[\sin(4x)]^{3-1} D_x \sin(4x) \\ &= 3[\sin(4x)]^2 \cos(4x) D_x(4x) \\ &= 3[\sin(4x)]^2 \cos(4x)(4) \\ &= 12 \cos(4x) \sin^2(4x) \end{aligned}$$

■

EXAMPLE 10 Find $D_x \sin[\cos(x^2)]$.

SOLUTION

$$\begin{aligned} D_x \sin[\cos(x^2)] &= \cos[\cos(x^2)] \cdot [-\sin(x^2)] \cdot 2x \\ &= -2x \sin(x^2) \cos[\cos(x^2)] \end{aligned}$$

■

EXAMPLE 11 Suppose that the graphs of $y = f(x)$ and $y = g(x)$ are as shown in Figure 1. Use these graphs to approximate (a) $(f - g)'(2)$ and (b) $(f \circ g)'(2)$.

SOLUTION

(a) By Theorem 2.3F, $(f - g)'(2) = f'(2) - g'(2)$. From Figure 1, we can determine that $f'(2) \approx 1$ and $g'(2) \approx -\frac{1}{2}$. Thus,

$$(f - g)'(2) \approx 1 - \left(-\frac{1}{2}\right) = \frac{3}{2}.$$

(b) From Figure 1 we can determine that $f'(1) \approx \frac{1}{2}$. Thus, by the Chain Rule,

$$(f \circ g)'(2) = f'(g(2))g'(2) = f'(1)g'(2) \approx \frac{1}{2}\left(-\frac{1}{2}\right) = -\frac{1}{4}$$

■

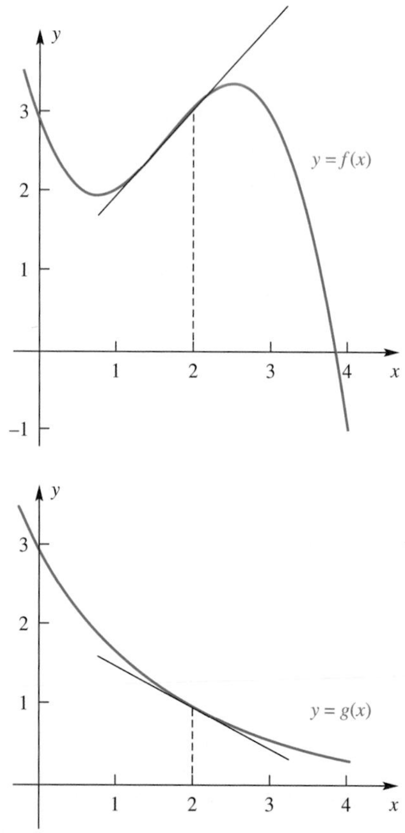

Figure 1

A Partial Proof of the Chain Rule

We can now give a sketch of the proof of the Chain Rule.

Proof We suppose that $y = f(u)$ and $u = g(x)$, that g is differentiable at x, and that f is differentiable at $u = g(x)$. When x is given an increment Δx, there are corresponding increments in u and y given by

$$\begin{aligned} \Delta u &= g(x + \Delta x) - g(x) \\ \Delta y &= f(g(x + \Delta x)) - f(g(x)) \\ &= f(u + \Delta u) - f(u) \end{aligned}$$

Thus,

$$\begin{aligned} \frac{dy}{dx} &= \lim_{\Delta x \to 0} \frac{\Delta y}{\Delta x} = \lim_{\Delta x \to 0} \frac{\Delta y}{\Delta u} \frac{\Delta u}{\Delta x} \\ &= \lim_{\Delta x \to 0} \frac{\Delta y}{\Delta u} \cdot \lim_{\Delta x \to 0} \frac{\Delta u}{\Delta x} \end{aligned}$$

Since g is differentiable at x, it is continuous there (Theorem 2.2A), and so $\Delta x \to 0$ forces $\Delta u \to 0$. Hence,

$$\frac{dy}{dx} = \lim_{\Delta u \to 0} \frac{\Delta y}{\Delta u} \cdot \lim_{\Delta x \to 0} \frac{\Delta u}{\Delta x} = \frac{dy}{du} \cdot \frac{du}{dx}$$

This proof was very slick, but unfortunately it contains a subtle flaw. There are functions $u = g(x)$ that have the property that $\Delta u = 0$ for some points in every neighborhood of x (the constant function $g(x) = k$ is a good example). This means the division by Δu at our first step might not be legal. There is no simple way to get around this difficulty, though the Chain Rule is valid even in this case. We give a complete proof of the Chain Rule in the appendix (Section A.2, Theorem B). ■

Concepts Review

1. If $y = f(u)$, where $u = g(t)$, then $D_t y = D_u y \cdot$ _____. In function notation, $(f \circ g)'(t) =$ _____ _____.

2. If $w = G(v)$, where $v = H(s)$, then $D_s w =$ _____ $D_s v$. In function notation $(G \circ H)'(s) =$ _____ _____.

3. $D_x \cos[(f(x))^2] = -\sin($_____$) \cdot D_x($_____$)$.

4. If $y = (2x + 1)^3 \sin(x^2)$, then $D_x y = (2x + 1)^3 \cdot$ _____ $+ \sin(x^2) \cdot$ _____.

Problem Set 2.5

In Problems 1–20, find $D_x y$.

1. $y = (1 + x)^{15}$

2. $y = (7 + x)^5$

3. $y = (3 - 2x)^5$

4. $y = (4 + 2x^2)^7$

5. $y = (x^3 - 2x^2 + 3x + 1)^{11}$

6. $y = (x^2 - x + 1)^{-7}$

7. $y = \dfrac{1}{(x + 3)^5}$

8. $y = \dfrac{1}{(3x^2 + x - 3)^9}$

9. $y = \sin(x^2 + x)$

10. $y = \cos(3x^2 - 2x)$

11. $y = \cos^3 x$

12. $y = \sin^4(3x^2)$

13. $y = \left(\dfrac{x + 1}{x - 1}\right)^3$

14. $y = \left(\dfrac{x - 2}{x - \pi}\right)^{-3}$

15. $y = \cos\left(\dfrac{3x^2}{x + 2}\right)$

16. $y = \cos^3\left(\dfrac{x^2}{1 - x}\right)$

17. $y = (3x - 2)^2(3 - x^2)^2$

18. $y = (2 - 3x^2)^4(x^7 + 3)^3$

19. $y = \dfrac{(x + 1)^2}{3x - 4}$

20. $y = \dfrac{2x - 3}{(x^2 + 4)^2}$

In Problems 21–28, find the indicated derivative.

21. y' where $y = (x^2 + 4)^2$

22. y' where $y = (x + \sin x)^2$

23. $D_t\left(\dfrac{3t - 2}{t + 5}\right)^3$

24. $D_s\left(\dfrac{s^2 - 9}{s + 4}\right)$

25. $\dfrac{d}{dt}\left(\dfrac{(3t - 2)^3}{t + 5}\right)$

26. $\dfrac{d}{d\theta}(\sin^3 \theta)$

27. $\dfrac{dy}{dx}$, where $y = \left(\dfrac{\sin x}{\cos 2x}\right)^3$

28. $\dfrac{dy}{dt}$, where $y = [\sin t \tan(t^2 + 1)]$

In Problems 29–32, evaluate the indicated derivative.

29. $f'(3)$ if $f(x) = \left(\dfrac{x^2 + 1}{x + 2}\right)^3$

30. $G'(1)$ if $G(t) = (t^2 + 9)^3(t^2 - 2)^4$

C **31.** $F'(1)$ if $F(t) = \sin(t^2 + 3t + 1)$

32. $g'\left(\frac{1}{2}\right)$ if $g(s) = \cos \pi s \sin^2 \pi s$

In Problems 33–40, apply the Chain Rule more than once to find the indicated derivative.

33. $D_x[\sin^4(x^2 + 3x)]$

34. $D_t[\cos^5(4t - 19)]$

35. $D_t[\sin^3(\cos t)]$

36. $D_u\left[\cos^4\left(\dfrac{u + 1}{u - 1}\right)\right]$

37. $D_\theta[\cos^4(\sin \theta^2)]$

38. $D_x[x \sin^2(2x)]$

39. $\dfrac{d}{dx}\{\sin[\cos(\sin 2x)]\}$

40. $\dfrac{d}{dt}\{\cos^2[\cos(\cos t)]\}$

In Problems 41–46, use Figures 2 and 3 to approximate the indicated expressions.

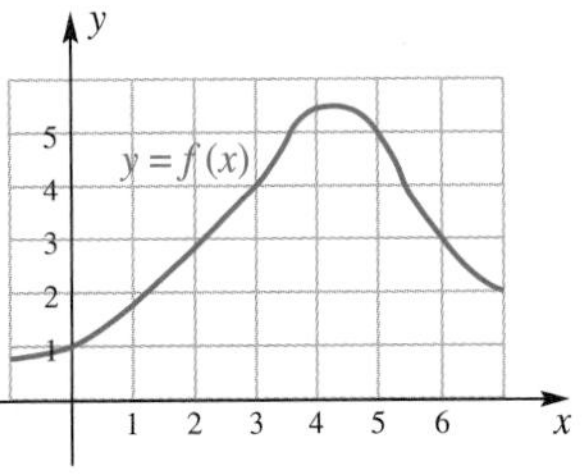

Figure 2

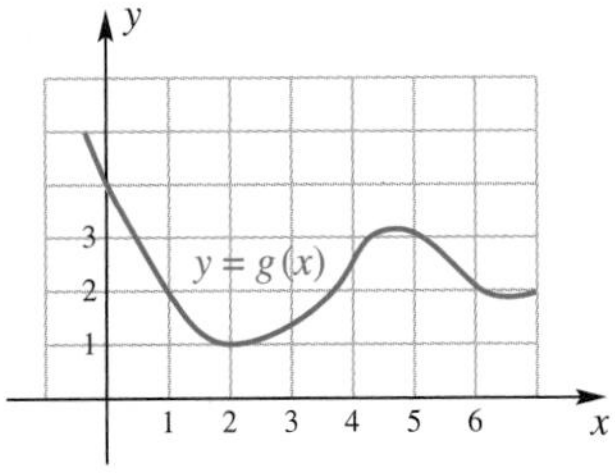

Figure 3

41. $(f + g)'(4)$

42. $(f - 2g)'(2)$

43. $(fg)'(2)$

44. $(f/g)'(2)$

45. $(f \circ g)'(6)$

46. $(g \circ f)'(3)$

In Problems 47–58, express the indicated derivative in terms of the function $F(x)$. Assume that F is differentiable.

47. $D_x(F(2x))$

48. $D_x(F(x^2 + 1))$

49. $D_t((F(t))^{-2})$

50. $\frac{d}{dz}\left(\frac{1}{(F(z))^2}\right)$

51. $\frac{d}{dz}(1 + (F(2z)))^2$

52. $\frac{d}{dy}\left(y^2 + \frac{1}{F(y^2)}\right)$

53. $\frac{d}{dx}F(\cos x)$

54. $\frac{d}{dx}\cos F(x)$

55. $D_x \tan F(2x)$

56. $\frac{d}{dx}g(\tan 2x)$

57. $D_x(F(x)\sin^2 F(x))$

58. $D_x \sec^3 F(x)$

59. Given that $f(0) = 1$ and $f'(0) = 2$, find $g'(0)$ where $g(x) = \cos f(x)$.

60. Given that $F(0) = 2$ and $F'(0) = -1$, find $G'(0)$ where $G(x) = \frac{x}{1 + \sec F(2x)}$.

61. Given that $f(1) = 2, f'(1) = -1, g(1) = 0$ and $g'(1) = 1$, find $F'(1)$ where $F(x) = f(x)\cos g(x)$.

62. Find the equation of the tangent line to the graph of $y = 1 + x\sin 3x$ at $\left(\frac{\pi}{3}, 1\right)$. Where does this line cross the x-axis?

63. Find all points on the graph of $y = \sin^2 x$ where the tangent line has slope 1.

64. Find the equation of the tangent line to $y = (x^2 + 1)^3(x^4 + 1)^2$ at $(1, 32)$.

65. Find the equation of the tangent line to $y = (x^2 + 1)^{-2}$ at $\left(1, \frac{1}{4}\right)$.

66. Where does the tangent line to $y = (2x + 1)^3$ at $(0, 1)$ cross the x-axis?

67. Where does the tangent line to $y = (x^2 + 1)^{-2}$ at $\left(1, \frac{1}{4}\right)$ cross the x-axis?

68. A point P is moving in the plane so that its coordinates after t seconds are $(4\cos 2t, 7\sin 2t)$, measured in feet.

(a) Show that P is following an elliptical path. *Hint:* Show that $(x/4)^2 + (y/7)^2 = 1$, which is an equation of an ellipse.

(b) Obtain an expression for L, the distance of P from the origin at time t.

(c) How fast is the distance between P and the origin changing when $t = \pi/8$? You will need the fact that $D_u(\sqrt{u}) = 1/(2\sqrt{u})$ (see Example 4 of Section 2.2).

69. A wheel centered at the origin and of radius 10 centimeters is rotating counterclockwise at a rate of 4 revolutions per second. A point P on the rim is at $(10, 0)$ at $t = 0$.

(a) What are the coordinates of P at time t seconds?

(b) At what rate is P rising (or falling) at time $t = 1$?

70. Consider the wheel-piston device in Figure 4. The wheel has radius 1 foot and rotates counterclockwise at 2 radians per second. The connecting rod is 5 feet long. The point P is at $(1, 0)$ at time $t = 0$.

(a) Find the coordinates of P at time t.

(b) Find the y-coordinate of Q at time t (the x-coordinate is always zero).

(c) Find the velocity of Q at time t. You will need the fact that $D_u(\sqrt{u}) = 1/(2\sqrt{u})$.

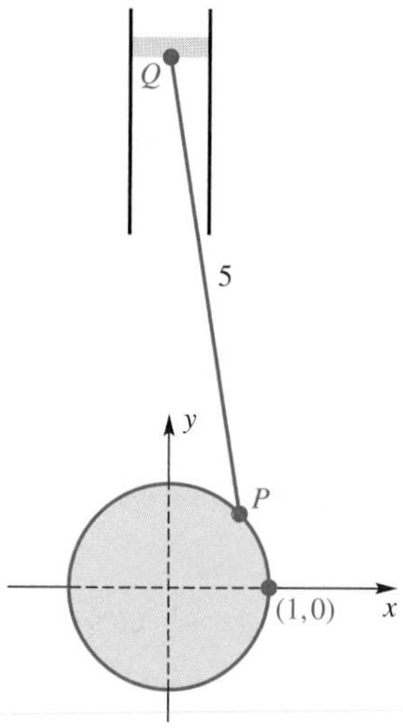

Figure 4

71. Do Problem 70, assuming that the wheel is rotating at 60 revolutions per minute and t is measured in seconds.

72. The dial of a standard clock has a 10-centimeter radius. One end of an elastic string is attached to the rim at 12 and the other to the tip of the 10-centimeter minute hand. At what rate is the string stretching at 12:15 (assuming that the clock is not slowed down by this stretching)?

C **73.** The hour and minute hands of a clock are 6 and 8 inches long, respectively. How fast are the tips of the hands separating at 12:20 (see Figure 5). *Hint:* Law of Cosines.

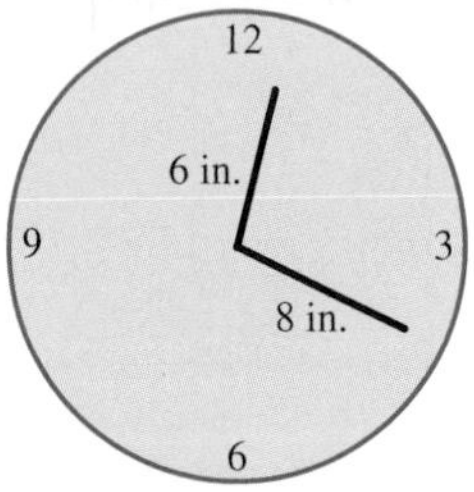

Figure 5

≈ GC **74.** Find the approximate time between 12:00 and 1:00 when the distance s between the tips of the hands of the clock of Figure 5 is increasing most rapidly, that is, when the derivative ds/dt is largest.

75. Let x_0 be the smallest positive value of x at which the curves $y = \sin x$ and $y = \sin 2x$ intersect. Find x_0 and also the acute angle at which the two curves intersect at x_0 (see Problem 40 of Section 0.7).

76. An isosceles triangle is topped by a semicircle, as shown in Figure 6. Let D be the area of triangle AOB and E be the area of the shaded region. Find a formula for D/E in terms of t and then calculate

$$\lim_{t\to 0^+}\frac{D}{E} \quad \text{and} \quad \lim_{t\to \pi^-}\frac{D}{E}$$

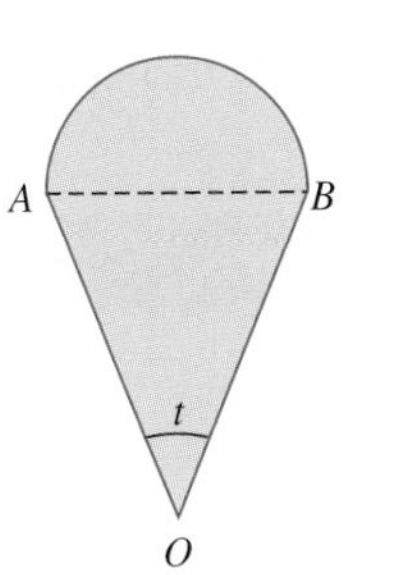

Figure 6

77. Show that $D_x|x| = |x|/x, x \neq 0$. *Hint:* Write $|x| = \sqrt{x^2}$ and use the Chain Rule with $u = x^2$.

78. Apply the result of Problem 77 to find $D_x|x^2 - 1|$.

79. Apply the result of Problem 77 to find $D_x|\sin x|$.

80. In Chapter 6 we will study a function L satisfying $L'(x) = 1/x$. Find each of the following derivatives.

(a) $D_x(L(x^2))$ (b) $D_x(L(\cos^4 x))$

81. Let $f(0) = 0$ and $f'(0) = 2$. Find the derivative of $f(f(f(f(x))))$ at $x = 0$.

82. Suppose that f is a differentiable function.

(a) Find $\frac{d}{dx}f(f(x))$. (b) Find $\frac{d}{dx}f(f(f(x)))$.

(c) Let $f^{[n]}$ denote the function defined as follows: $f^{[1]} = f$ and $f^{[n]} = f \circ f^{[n-1]}$ for $n \geq 2$. Thus $f^{[2]} = f \circ f$, $f^{[3]} = f \circ f \circ f$, etc. Based on your results from parts (a) and (b), make a conjecture regarding $\frac{d}{dx}f^{[n]}$. Prove your conjecture.

83. Give a second proof of the Quotient Rule. Write

$$D_x\left(\frac{f(x)}{g(x)}\right) = D_x\left(f(x)\frac{1}{g(x)}\right)$$

and use the Product Rule and the Chain Rule.

84. Suppose that f is differentiable and that there are real numbers x_1 and x_2 such that $f(x_1) = x_2$ and $f(x_2) = x_1$. Let $g(x) = f(f(f(f(x))))$. Show that $g'(x_1) = g'(x_2)$.

Answers to Concepts Review: 1. $D_t u$; $f'(g(t))g'(t)$ **2.** $D_v w$; $G'(H(s))H'(s)$ **3.** $(f(x))^2$; $(f(x))^2$ **4.** $2x\cos(x^2)$; $6(2x + 1)^2$

2.6 Higher-Order Derivatives

The operation of differentiation takes a function f and produces a new function f'. If we now differentiate f', we produce still another function, denoted by f'' (read "f double prime") and called the **second derivative** of f. It in turn, may be differentiated, thereby producing f''', which is called the **third derivative** of f, and so on. The **fourth derivative** is denoted $f^{(4)}$, the **fifth derivative** is denoted $f^{(5)}$, and so on.

If, for example

$$f(x) = 2x^3 - 4x^2 + 7x - 8$$

then

$$\begin{aligned} f'(x) &= 6x^2 - 8x + 7 \\ f''(x) &= 12x - 8 \\ f'''(x) &= 12 \\ f^{(4)}(x) &= 0 \end{aligned}$$

Since the derivative of the zero function is zero, the fourth derivative and all *higher-order derivatives* of f will be zero.

We have introduced three notations for the derivative (now also called the *first derivative*) of $y = f(x)$. They are

$$f'(x) \qquad D_x y \qquad \frac{dy}{dx}$$

called, respectively, the *prime notation*, the *D notation*, and the *Leibniz notation*. There is a variation of the prime notation, y', that we will also use occasionally. All these notations have extensions for higher-order derivates, as shown in the accompanying table. Note especially the Leibniz notation, which, though complicated, seemed most appropriate to Leibniz. What, thought he, is more natural than to write

$$\frac{d}{dx}\left(\frac{dy}{dx}\right) \quad \text{as} \quad \frac{d^2y}{dx^2}$$

Leibniz's notation for the second derivative is read *the second derivative of y with respect to x.*

Notations for Derivatives of $y = f(x)$

Derivative	f' Notation	y' Notation	D Notation	Leibniz Notation
First	$f'(x)$	y'	$D_x y$	$\dfrac{dy}{dx}$
Second	$f''(x)$	y''	$D_x^2 y$	$\dfrac{d^2y}{dx^2}$
Third	$f'''(x)$	y'''	$D_x^3 y$	$\dfrac{d^3y}{dx^3}$
Fourth	$f^{(4)}(x)$	$y^{(4)}$	$D_x^4 y$	$\dfrac{d^4y}{dx^4}$
$\vdots$	$\vdots$	$\vdots$	$\vdots$	$\vdots$
nth	$f^{(n)}(x)$	$y^{(n)}$	$D_x^n y$	$\dfrac{d^ny}{dx^n}$

EXAMPLE 1 If $y = \sin 2x$, find d^3y/dx^3, d^4y/dx^4, and $d^{12}y/dx^{12}$.

SOLUTION

$$\frac{dy}{dx} = 2\cos 2x$$
$$\frac{d^2y}{dx^2} = -2^2 \sin 2x$$
$$\frac{d^3y}{dx^3} = -2^3 \cos 2x$$
$$\frac{d^4y}{dx^4} = 2^4 \sin 2x$$
$$\frac{d^5y}{dx^5} = 2^5 \cos 2x$$
$$\vdots$$
$$\frac{d^{12}y}{dx^{12}} = 2^{12} \sin 2x$$

■

Velocity and Acceleration In Section 2.1, we used the notion of instantaneous velocity to motivate the definition of the derivative. Let's review this notion by means of an example. Also, from now on we will use the single word *velocity* in place of the more cumbersome phrase *instantaneous velocity*.

EXAMPLE 2 An object moves along a coordinate line so that its position s satisfies $s = 2t^2 - 12t + 8$, where s is measured in centimeters and t in seconds with $t \geq 0$. Determine the velocity of the object when $t = 1$ and when $t = 6$. When is the velocity 0? When is it positive?

SOLUTION If we use the symbol $v(t)$ for the velocity at time t, then

$$v(t) = \frac{ds}{dt} = 4t - 12$$

Thus,

$$v(1) = 4(1) - 12 = -8 \text{ centimeters per second}$$
$$v(6) = 4(6) - 12 = 12 \text{ centimeters per second}$$

The velocity is 0 when $4t - 12 = 0$, that is, when $t = 3$. The velocity is positive when $4t - 12 > 0$, or when $t > 3$. All this is shown schematically in Figure 1.

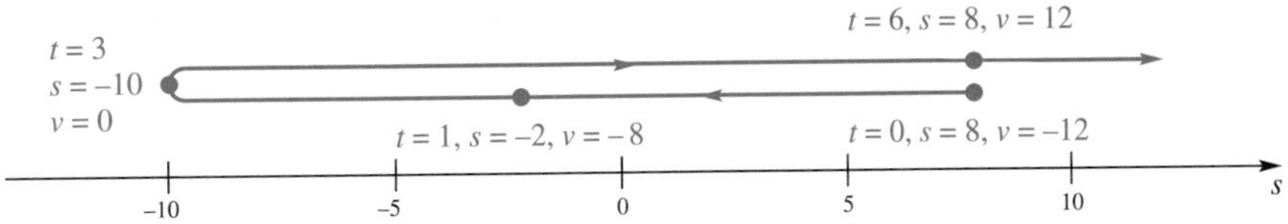

Figure 1

The object is, of course, moving along the s-axis, not on the colored path above it. But the colored path shows what happens to the object. Between $t = 0$ and $t = 3$, the velocity is negative; the object is moving to the left (backing up). By the time $t = 3$, it has "slowed" to a zero velocity. It then starts moving to the right as its velocity becomes positive. Thus, negative velocity corresponds to moving in the direction of decreasing s; positive velocity corresponds to moving in the direction of increasing s. A rigorous discussion of these points will be given in Chapter 3. ■

There is a technical distinction between the words *velocity* and *speed*. Velocity has a sign associated with it; it may be positive or negative. **Speed** is defined to be the absolute value of the velocity. Thus, in the example above, the speed at $t = 1$ is $|-8| = 8$ centimeters per second. The meter in most cars is a *speed*ometer; it always gives nonnegative values.

Now we want to give a physical interpretation of the second derivative d^2s/dt^2. It is, of course, just the first derivative of the velocity. Thus, it measures the rate of change of velocity with respect to time, which has the name **acceleration.** If it is denoted by a, then

$$a = \frac{dv}{dt} = \frac{d^2s}{dt^2}$$

Measuring Time

If $t = 0$ corresponds to the present moment, then $t < 0$ corresponds to the past, and $t > 0$ to the future. In many problems, it will be obvious that we are concerned only with the future. However, since the statement of Example 3 does not specify this, it seems reasonable to allow t to have negative as well as positive values.

In Example 2, $s = 2t^2 - 12t + 8$. Thus,

$$v = \frac{ds}{dt} = 4t - 12$$

$$a = \frac{d^2s}{dt^2} = 4$$

This means that the velocity is increasing at a constant rate of 4 centimeters per second every second, which we write as 4 centimeters per second per second, or as 4 cm/sec^2.

EXAMPLE 3 An object moves along a horizontal coordinate line in such a way that its position at time t is specified by

$$s = t^3 - 12t^2 + 36t - 30$$

Here s is measured in feet and t in seconds.

(a) When is the velocity 0?
(b) When is the velocity positive?
(c) When is the object moving to the left (that is, in the negative direction)?
(d) When is the acceleration positive?

SOLUTION

(a) $v = ds/dt = 3t^2 - 24t + 36 = 3(t - 2)(t - 6)$. Thus, $v = 0$ at $t = 2$ and at $t = 6$.

(b) $v > 0$ when $(t - 2)(t - 6) > 0$. We learned how to solve quadratic inequalities in Section 0.2. The solution is $\{t: t < 2 \text{ or } t > 6\}$ or, in interval notation, $(-\infty, 2) \cup (6, \infty)$; see Figure 2.

Figure 2

(c) The object is moving to the left when $v < 0$; that is, when $(t - 2)(t - 6) < 0$. This inequality has as its solution the interval $(2, 6)$.
(d) $a = dv/dt = 6t - 24 = 6(t - 4)$. Thus, $a > 0$ when $t > 4$. The motion of the object is shown schematically in Figure 3.

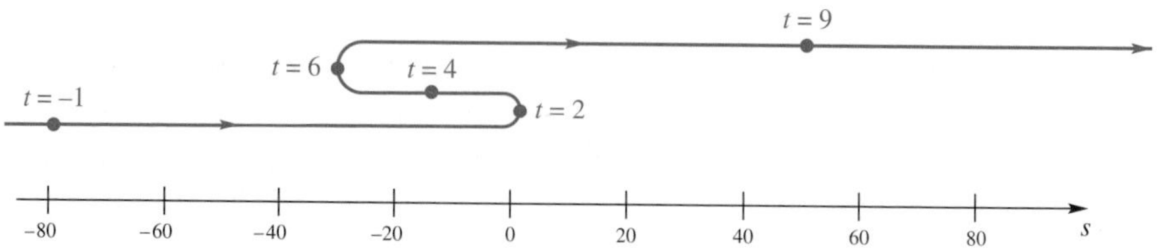

Figure 3

■

Falling-Body Problems If an object is thrown straight upward (or downward) from an initial height of s_0 feet with an initial velocity of v_0 feet per second and if s is its height above the ground in feet after t seconds, then

$$s = -16t^2 + v_0t + s_0$$

This assumes that the experiment takes place near sea level and that air resistance can be neglected. The diagram in Figure 4 portrays the situation we have in mind. Notice that positive velocity means that the object is moving upward.

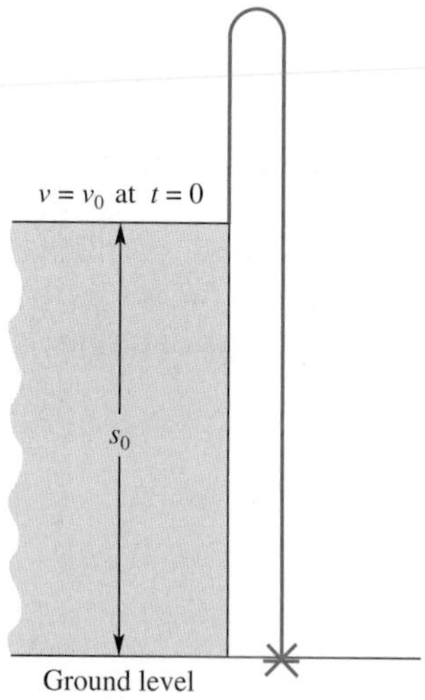

Figure 4

EXAMPLE 4 From the top of a building 160 feet high, a ball is thrown upward with an initial velocity of 64 feet per second.

(a) When does it reach its maximum height?
(b) What is its maximum height?
(c) When does it hit the ground?
(d) With what speed does it hit the ground?
(e) What is its acceleration at $t = 2$?

SOLUTION Let $t = 0$ correspond to the instant when the ball was thrown. Then $s_0 = 160$ and $v_0 = 64$ (v_0 is positive because the ball was thrown *upward*). Thus,

$$s = -16t^2 + 64t + 160$$

$$v = \frac{ds}{dt} = -32t + 64$$

$$a = \frac{dv}{dt} = -32$$

(a) The ball reached its maximum height at the time its velocity was 0, that is, when $-32t + 64 = 0$ or when $t = 2$ seconds.
(b) At $t = 2$, $s = -16(2)^2 + 64(2) + 160 = 224$ feet.
(c) The ball hit the ground when $s = 0$, that is, when

$$-16t^2 + 64t + 160 = 0$$

Dividing by -16 yields

$$t^2 - 4t - 10 = 0$$

The quadratic formula then gives

$$t = \frac{4 \pm \sqrt{16 + 40}}{2} = \frac{4 \pm 2\sqrt{14}}{2} = 2 \pm \sqrt{14}$$

Only the positive answer makes sense. Thus, the ball hit the ground at $t = 2 + \sqrt{14} \approx 5.74$ seconds.
(d) At $t = 2 + \sqrt{14}$, $v = -32\left(2 + \sqrt{14}\right) + 64 \approx -119.73$. Thus, the ball hit the ground with a speed of 119.73 feet per second.

(e) The acceleration is always -32 feet per second per second. This is the acceleration of gravity near sea level. ■

Concepts Review

1. If $y = f(x)$, then the third derivative of y with respect to x can be denoted by any one of the following four symbols: ______.

2. If $s = f(t)$ denotes the position of a particle on a coordinate line at time t, then its velocity is given by ______, its speed is given by ______, and its acceleration is given by ______.

3. If $s = f(t)$ denotes the position of an object at time t, then the object is moving to the right if ______.

4. Assume that an object is thrown straight upward so that its height s at time t is given by $s = f(t)$. The object reaches its maximum height when $ds/dt =$ ______, after which, ds/dt ______.

Problem Set 2.6

In Problems 1–8, find d^3y/dx^3.

1. $y = x^3 + 3x^2 + 6x$

2. $y = x^5 + x^4$

3. $y = (3x + 5)^3$

4. $y = (3 - 5x)^5$

5. $y = \sin(7x)$

6. $y = \sin(x^3)$

7. $y = \dfrac{1}{x - 1}$

8. $y = \dfrac{3x}{1 - x}$

In Problems 9–16, find $f''(2)$.

9. $f(x) = x^2 + 1$

10. $f(x) = 5x^3 + 2x^2 + x$

11. $f(t) = \dfrac{2}{t}$

12. $f(u) = \dfrac{2u^2}{5 - u}$

13. $f(\theta) = (\cos \theta\pi)^{-2}$

14. $f(t) = t \sin(\pi/t)$

15. $f(s) = s(1 - s^2)^3$

16. $f(x) = \dfrac{(x + 1)^2}{x - 1}$

17. Let $n! = n(n - 1)(n - 2)\cdots 3\cdot 2\cdot 1$. Thus, $4! = 4\cdot 3\cdot 2\cdot 1 = 24$ and $5! = 5\cdot 4\cdot 3\cdot 2\cdot 1$. We give $n!$ the name ***n* factorial.** Show that $D_x^n(x^n) = n!$.

18. Find a formula for

$$D_x^n(a_{n-1}x^{n-1} + \cdots + a_1x + a_0)$$

19. Without doing any calculating, find each derivative.

(a) $D_x^4(3x^3 + 2x - 19)$ (b) $D_x^{12}(100x^{11} - 79x^{10})$

(c) $D_x^{11}(x^2 - 3)^5$

20. Find a formula for $D_x^n(1/x)$.

21. If $f(x) = x^3 + 3x^2 - 45x - 6$, find the value of f'' at each zero of f', that is, at each point c where $f'(c) = 0$.

22. Suppose that $g(t) = at^2 + bt + c$ and $g(1) = 5$, $g'(1) = 3$, and $g''(1) = -4$. Find a, b, and c.

In Problems 23–28, an object is moving along a horizontal coordinate line according to the formula $s = f(t)$, where s, the directed distance from the origin, is in feet and t is in seconds. In each case, answer the following questions (see Examples 2 and 3).

(a) What are $v(t)$ and $a(t)$, the velocity and acceleration, at time t?

(b) When is the object moving to the right?

(c) When is it moving to the left?

(d) When is its acceleration negative?

(e) Draw a schematic diagram that shows the motion of the object.

23. $s = 12t - 2t^2$

24. $s = t^3 - 6t^2$

25. $s = t^3 - 9t^2 + 24t$

26. $s = 2t^3 - 6t + 5$

27. $s = t^2 + \dfrac{16}{t}, t > 0$

28. $s = t + \dfrac{4}{t}, t > 0$

29. If $s = \frac{1}{2}t^4 - 5t^3 + 12t^2$, find the velocity of the moving object when its acceleration is zero.

30. If $s = \frac{1}{10}(t^4 - 14t^3 + 60t^2)$, find the velocity of the moving object when its acceleration is zero.

31. Two objects move along a coordinate line. At the end of t seconds their directed distances from the origin, in feet, are given by $s_1 = 4t - 3t^2$ and $s_2 = t^2 - 2t$, respectively.

(a) When do they have the same velocity?

(b) When do they have the same speed?

(c) When do they have the same position?

32. The positions of two objects, P_1 and P_2, on a coordinate line at the end of t seconds are given by $s_1 = 3t^3 - 12t^2 + 18t + 5$ and $s_2 = -t^3 + 9t^2 - 12t$, respectively. When do the two objects have the same velocity?

33. An object thrown directly upward is at a height of $s = -16t^2 + 48t + 256$ feet after t seconds (see Example 4).

(a) What is its initial velocity?

(b) When does it reach its maximum height?

(c)What is its maximum height?

C (d) When does it hit the ground?

C (e) With what speed does it hit the ground?

34. An object thrown directly upward from ground level with an initial velocity of 48 feet per second is $s = 48t - 16t^2$ feet high at the end of t seconds.

(a) What is the maximum height attained?

(b) How fast is the object moving, and in which direction, at the end of 1 second?

(c) How long does it take to return to its original position?

C **35.** A projectile is fired directly upward from the ground with an initial velocity of v_0 feet per second. Its height in t seconds is given by $s = v_0t - 16t^2$ feet. What must its initial velocity be for the projectile to reach a maximum height of 1 mile?

36. An object thrown directly downward from the top of a cliff with an initial velocity of v_0 feet per second falls $s = v_0t + 16t^2$ feet in t seconds. If it strikes the ocean below in 3 seconds with a speed of 140 feet per second, how high is the cliff?

37. An object moves along a horizontal coordinate line in such a way that its position at time t is specified by $s = t^3 - 3t^2 - 24t - 6$. Here s is measured in centimeters and t in seconds. When is the object slowing down; that is, when is its *speed* decreasing?

38. Explain why an object moving along a line is slowing down when its velocity and acceleration have opposite signs (see Problem 37).

EXPL **39.** Leibniz obtained a general formula for $D_x^n(uv)$, where u and v are both functions of x. See if you can find it. *Hint:* Begin by considering the cases $n = 1$, $n = 2$, and $n = 3$.

40. Use the formula of Problem 39 to find $D_x^4(x^4 \sin x)$.

GC **41.** Let $f(x) = x[\sin x - \cos(x/2)]$.

(a) Draw the graphs of $f(x), f'(x), f''(x)$, and $f'''(x)$ on $[0, 6]$ using the same axes.

(b) Evaluate $f'''(2.13)$.

GC **42.** Repeat Problem 41 for $f(x) = (x + 1)/(x^2 + 2)$.

Answers to Concepts Review: **1.** $f'''(x)$; $D_x^3 y$; d^3y/dx^3; y''' **2.** ds/dt; $|ds/dt|$; d^2s/dt^2 **3.** $f'(t) > 0$ **4.** 0; <0

2.7 Implicit Differentiation

In the equation

$$y^3 + 7y = x^3$$

we cannot solve for y in terms of x. It still may be the case, however, that there is exactly one y corresponding to each x. For example, we may ask what y-values (if any) correspond to $x = 2$. To answer this question, we must solve

$$y^3 + 7y = 8$$

Certainly, $y = 1$ is one solution, and it turns out that $y = 1$ is the *only* real solution. Given $x = 2$, the equation $y^3 + 7y = x^3$ determines a corresponding y-value. We say that the equation defines y as an **implicit** function of x. The graph of this equation, shown in Figure 1, certainly looks like the graph of a differentiable function. The new element is that we do not have an equation of the form $y = f(x)$. Based on the graph, we assume that y is some unknown function of x. If we denote this function by $y(x)$, we can write the equation as

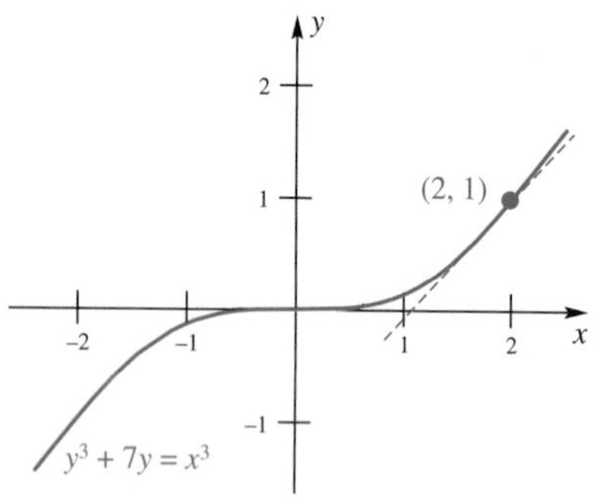

Figure 1

$$[y(x)]^3 + 7y(x) = x^3$$

Even though we do not have a formula for $y(x)$, we can nevertheless get a relation between x, $y(x)$, and $y'(x)$, by differentiating both sides of the equation with respect to x. Remembering to apply the Chain Rule, we get

$$\frac{d}{dx}(y^3) + \frac{d}{dx}(7y) = \frac{d}{dx}x^3$$

$$3y^2\frac{dy}{dx} + 7\frac{dy}{dx} = 3x^2$$

$$\frac{dy}{dx}(3y^2 + 7) = 3x^2$$

$$\frac{dy}{dx} = \frac{3x^2}{3y^2 + 7}$$

Note that our expression for dy/dx involves both x and y, a fact that is often a nuisance. But if we wish only to find a slope at a point where we know both coordinates, no difficulty exists. At (2, 1),

$$\frac{dy}{dx} = \frac{3(2)^2}{3(1)^2 + 7} = \frac{12}{10} = \frac{6}{5}$$

The slope is $\frac{6}{5}$.

The method just illustrated for finding dy/dx without first solving the given equation for y explicitly in terms of x is called **implicit differentiation.** But is the method legitimate—does it give the right answer?

An Example That Can Be Checked To give some evidence for the correctness of the method, consider the following example, which can be worked two ways.

EXAMPLE 1 Find dy/dx if $4x^2y - 3y = x^3 - 1$.

SOLUTION

Method 1 We can solve the given equation explicitly for y as follows:

$$y(4x^2 - 3) = x^3 - 1$$
$$y = \frac{x^3 - 1}{4x^2 - 3}$$

Thus,

$$\frac{dy}{dx} = \frac{(4x^2 - 3)(3x^2) - (x^3 - 1)(8x)}{(4x^2 - 3)^2} = \frac{4x^4 - 9x^2 + 8x}{(4x^2 - 3)^2}$$

Method 2 Implicit Differentiation We equate the derivatives of the two sides.

$$\frac{d}{dx}(4x^2y - 3y) = \frac{d}{dx}(x^3 - 1)$$

We obtain, after using the Product Rule on the first term,

$$4x^2 \cdot \frac{dy}{dx} + y \cdot 8x - 3\frac{dy}{dx} = 3x^2$$
$$\frac{dy}{dx}(4x^2 - 3) = 3x^2 - 8xy$$
$$\frac{dy}{dx} = \frac{3x^2 - 8xy}{4x^2 - 3}$$

These two answers look different. For one thing, the answer obtained from Method 1 involves x only, whereas the answer from Method 2 involves both x and y. Remember, however, that the original equation could be solved for y in terms of x to give $y = (x^3 - 1)/(4x^2 - 3)$. When we substitute $y = (x^3 - 1)/(4x^2 - 3)$ into the expression just obtained for dy/dx, we get the following:

$$\frac{dy}{dx} = \frac{3x^2 - 8xy}{4x^2 - 3} = \frac{3x^2 - 8x\dfrac{x^3 - 1}{4x^2 - 3}}{4x^2 - 3}$$
$$= \frac{12x^4 - 9x^2 - 8x^4 + 8x}{(4x^2 - 3)^2} = \frac{4x^4 - 9x^2 + 8x}{(4x^2 - 3)^2}$$ ■

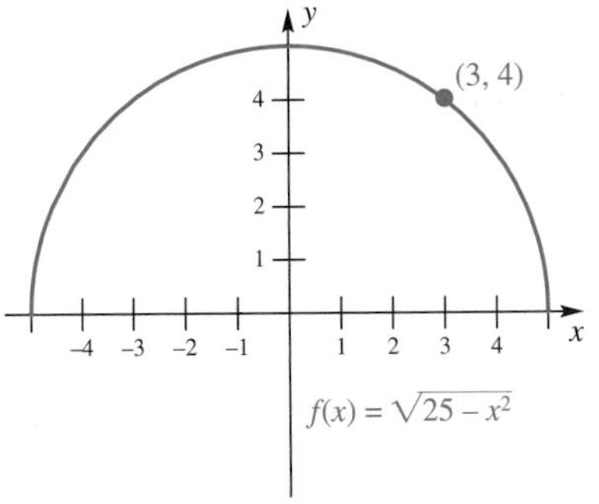

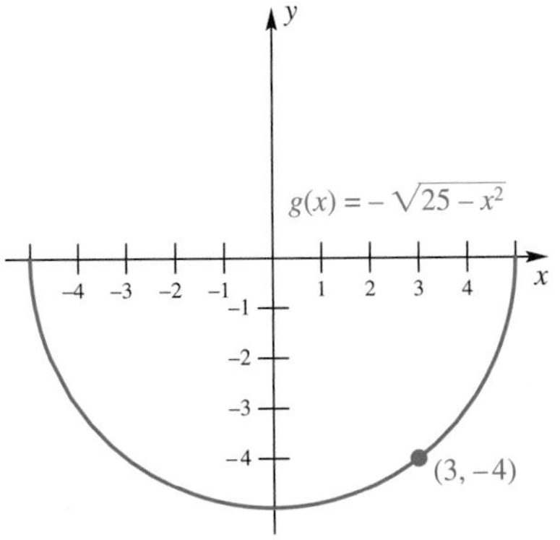

Figure 2

Some Subtle Difficulties If an equation in x and y determines a function $y = f(x)$ and if this function is differentiable, then the method of implicit differentiation will yield a correct expression for dy/dx. But notice there are two big *ifs* in this statement.

Consider the equation

$$x^2 + y^2 = 25$$

which determines both the function $y = f(x) = \sqrt{25 - x^2}$ and the function $y = g(x) = -\sqrt{25 - x^2}$. Their graphs are shown in Figure 2.

Happily, both of these functions are differentiable on $(-5, 5)$. Consider f first. It satisfies

$$x^2 + [f(x)]^2 = 25$$

When we differentiate implicitly and solve for $f'(x)$, we obtain

$$2x + 2f(x)f'(x) = 0$$

$$f'(x) = -\frac{x}{f(x)} = -\frac{x}{\sqrt{25 - x^2}}$$

A similar treatment of $g(x)$ yields

$$g'(x) = -\frac{x}{g(x)} = \frac{x}{\sqrt{25 - x^2}}$$

For practical purposes, we can obtain both of these results simultaneously by implicit differentiation of $x^2 + y^2 = 25$. This gives

$$2x + 2y\frac{dy}{dx} = 0$$

$$\frac{dy}{dx} = -\frac{x}{y} = \begin{cases} \dfrac{-x}{\sqrt{25 - x^2}} & \text{if } y = f(x) \\ \dfrac{-x}{-\sqrt{25 - x^2}} & \text{if } y = g(x) \end{cases}$$

Naturally, the results are identical with those obtained above.

Note that it is often enough to know that $dy/dx = -x/y$ in order to apply our results. Suppose we want to know the slopes of the tangent lines to the circle $x^2 + y^2 = 25$ when $x = 3$. For $x = 3$, the corresponding y-values are 4 and -4. The slopes at $(3, 4)$ and $(3, -4)$, obtained by substituting in $-x/y$, are $-\frac{3}{4}$ and $\frac{3}{4}$, respectively (see Figure 2).

To complicate matters, we point out that

$$x^2 + y^2 = 25$$

determines many other functions. For example, consider the function h defined by

$$h(x) = \begin{cases} \sqrt{25 - x^2} & \text{if } -5 \le x \le 3 \\ -\sqrt{25 - x^2} & \text{if } 3 < x \le 5 \end{cases}$$

It too satisfies $x^2 + y^2 = 25$, since $x^2 + [h(x)]^2 = 25$. But it is not even continuous at $x = 3$, so it certainly does not have a derivative there (see Figure 3).

While the subject of implicit functions leads to difficult technical questions (treated in advanced calculus), the problems we study have straightforward solutions.

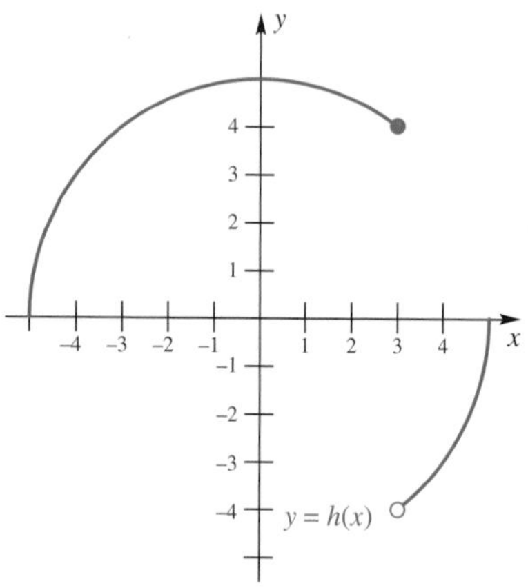

Figure 3

More Examples In the examples that follow, we assume that the given equation determines one or more differentiable functions whose derivatives can be found by implicit differentiation. Note that in each case we begin by taking the derivative of each side of the given equation with respect to the appropriate variable. Then we use the Chain Rule as needed.

EXAMPLE 2 Find dy/dx if $x^2 + 5y^3 = x + 9$.

SOLUTION

$$\frac{d}{dx}(x^2 + 5y^3) = \frac{d}{dx}(x + 9)$$

$$2x + 15y^2\frac{dy}{dx} = 1$$

$$\frac{dy}{dx} = \frac{1 - 2x}{15y^2}$$

■

EXAMPLE 3 Find the equation of the tangent line to the curve

$$y^3 - xy^2 + \cos xy = 2$$

at the point (0, 1).

SOLUTION For simplicity, let us use the notation y' for dy/dx. When we differentiate both sides and equate the results, we obtain

$$3y^2y' - x(2yy') - y^2 - (\sin xy)(xy' + y) = 0$$

$$y'(3y^2 - 2xy - x\sin xy) = y^2 + y\sin xy$$

$$y' = \frac{y^2 + y\sin xy}{3y^2 - 2xy - x\sin xy}$$

At (0, 1), $y' = \frac{1}{3}$. Thus, the equation of the tangent line at (0, 1) is

$$y - 1 = \tfrac{1}{3}(x - 0)$$

or

$$y = \tfrac{1}{3}x + 1$$

■

The Power Rule Again We have learned that $D_x(x^n) = nx^{n-1}$, where n is any nonzero integer. We now extend this to the case where n is any nonzero rational number.

Theorem A Power Rule

Let r be any nonzero rational number. Then, for $x > 0$,

$$D_x(x^r) = rx^{r-1}$$

If r can be written in lowest terms as $r = p/q$, where q is odd, then $D_x(x^r) = rx^{r-1}$ for all x.

Proof Since r is rational, r may be written as p/q, where p and q are integers with $q > 0$. Let

$$y = x^r = x^{p/q}$$

Then

$$y^q = x^p$$

and, by implicit differentiation,

$$qy^{q-1}D_xy = px^{p-1}$$

Thus,

$$D_xy = \frac{px^{p-1}}{qy^{q-1}} = \frac{p}{q}\frac{x^{p-1}}{(x^{p/q})^{q-1}} = \frac{p}{q}\frac{x^{p-1}}{x^{p-p/q}}$$

$$= \frac{p}{q}x^{p-1-p+p/q} = \frac{p}{q}x^{p/q-1} = rx^{r-1}$$

We have obtained the desired result, but, to be honest, we must point out a flaw in our argument. In the implicit differentiation step, we assumed that D_xy exists, that is, that $y = x^{p/q}$ is differentiable. We can fill this gap, but since it is hard work we relegate the complete proof to the appendix (Section A.2, Theorem C). ■

EXAMPLE 4 If $y = 2x^{5/3} + \sqrt{x^2 + 1}$, find D_xy.

SOLUTION Using Theorem A and the Chain Rule, we have

$$D_x y = 2D_x x^{5/3} + D_x(x^2 + 1)^{1/2}$$
$$= 2 \cdot \frac{5}{3} x^{5/3-1} + \frac{1}{2}(x^2 + 1)^{1/2-1} \cdot (2x)$$
$$= \frac{10}{3} x^{2/3} + \frac{x}{\sqrt{x^2 + 1}}$$ ■

Concepts Review

1. The implicit relation $yx^3 - 3y = 9$ can be solved explicitly for y giving $y =$ ______.

2. Implicit differentiation of $y^3 + x^3 = 2x$ with respect to x gives ______ $+ 3x^2 = 2$.

3. Implicit differentiation of $xy^2 + y^3 - y = x^3$ with respect to x gives ______ $=$ ______.

4. The Power Rule with rational exponents says that $D_x(x^{p/q}) =$ ______. This rule, together with the Chain Rule, implies that $D_x[(x^2 - 5x)^{5/3}] =$ ______.

Problem Set 2.7

Assuming that each equation in Problems 1–12 defines a differentiable function of x, find $D_x y$ by implicit differentiation.

1. $y^2 - x^2 = 1$

2. $9x^2 + 4y^2 = 36$

3. $xy = 1$

4. $x^2 + \alpha^2 y^2 = 4\alpha^2$, where α is a constant.

5. $xy^2 = x - 8$

6. $x^2 + 2x^2 y + 3xy = 0$

7. $4x^3 + 7xy^2 = 2y^3$

8. $x^2 y = 1 + y^2 x$

9. $\sqrt{5xy} + 2y = y^2 + xy^3$

10. $x\sqrt{y + 1} = xy + 1$

11. $xy + \sin(xy) = 1$

12. $\cos(xy^2) = y^2 + x$

In Problems 13–18, find the equation of the tangent line at the indicated point (see Example 3).

13. $x^3 y + y^3 x = 30$; $(1, 3)$

14. $x^2 y^2 + 4xy = 12y$; $(2, 1)$

15. $\sin(xy) = y$; $(\pi/2, 1)$

16. $y + \cos(xy^2) + 3x^2 = 4$; $(1, 0)$

17. $x^{2/3} - y^{2/3} - 2y = 2$; $(1, -1)$

18. $\sqrt{y} + xy^2 = 5$; $(4, 1)$

In Problems 19–32, find dy/dx.

19. $y = 3x^{5/3} + \sqrt{x}$

20. $y = \sqrt[3]{x} - 2x^{7/2}$

21. $y = \sqrt[3]{x} + \dfrac{1}{\sqrt[3]{x}}$

22. $y = \sqrt[4]{2x + 1}$

23. $y = \sqrt[4]{3x^2 - 4x}$

24. $y = (x^3 - 2x)^{1/3}$

25. $y = \dfrac{1}{(x^3 + 2x)^{2/3}}$

26. $y = (3x - 9)^{-5/3}$

27. $y = \sqrt{x^2 + \sin x}$

28. $y = \sqrt{x^2 \cos x}$

29. $y = \dfrac{1}{\sqrt[3]{x^2 \sin x}}$

30. $y = \sqrt[4]{1 + \sin 5x}$

31. $y = \sqrt[4]{1 + \cos(x^2 + 2x)}$

32. $y = \sqrt{\tan^2 x + \sin^2 x}$

33. If $s^2 t + t^3 = 1$, find ds/dt and dt/ds.

34. If $y = \sin(x^2) + 2x^3$, find dx/dy.

35. Sketch the graph of the circle $x^2 + 4x + y^2 + 3 = 0$ and then find equations of the two tangent lines that pass through the origin.

36. Find the equation of the **normal line** (line perpendicular to the tangent line) to the curve $8(x^2 + y^2)^2 = 100(x^2 - y^2)$ at $(3, 1)$.

37. Suppose that $xy + y^3 = 2$. Then implicit differentiation twice with respect to x yields in turn:

(a) $xy' + y + 3y^2 y' = 0$;

(b) $xy'' + y' + y' + 3y^2 y'' + 6y(y')^2 = 0$.

Solve (a) for y' and substitute in (b), and then solve for y''.

38. Find y'' if $x^3 - 4y^2 + 3 = 0$ (see Problem 37).

39. Find y'' at $(2, 1)$ if $2x^2 y - 4y^3 = 4$ (see Problem 37).

40. Use implicit differentiation twice to find y'' at $(3, 4)$ if $x^2 + y^2 = 25$.

41. Show that the normal line to $x^3 + y^3 = 3xy$ at $\left(\frac{3}{2}, \frac{3}{2}\right)$ passes through the origin.

42. Show that the hyperbolas $xy = 1$ and $x^2 - y^2 = 1$ intersect at right angles.

43. Show that the graphs of $2x^2 + y^2 = 6$ and $y^2 = 4x$ intersect at right angles.

44. Suppose that curves C_1 and C_2 intersect at (x_0, y_0) with slopes m_1 and m_2, respectively, as in Figure 4. Then (see Problem 40 of Section 0.7) the positive angle θ from C_1 (i.e., from the tangent line to C_1 at (x_0, y_0)) to C_2 satisfies

$$\tan \theta = \frac{m_2 - m_1}{1 + m_1 m_2}$$

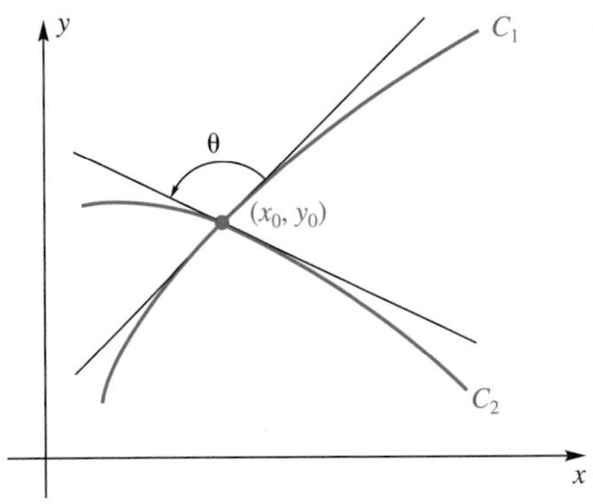

Figure 4

Find the angles from the circle $x^2 + y^2 = 1$ to the circle $(x - 1)^2 + y^2 = 1$ at the two points of intersection.

45. Find the angle from the line $y = 2x$ to the curve $x^2 - xy + 2y^2 = 28$ at their point of intersection in the first quadrant (see Problem 44).

46. A particle of mass m moves along the x-axis so that its position x and velocity $v = dx/dt$ satisfy

$$m(v^2 - v_0^2) = k(x_0^2 - x^2)$$

where v_0, x_0, and k are constants. Show by implicit differentiation that

$$m\frac{dv}{dt} = -kx$$

whenever $v \neq 0$.

47. The curve $x^2 - xy + y^2 = 16$ is an ellipse centered at the origin and with the line $y = x$ as its major axis. Find the equations of the tangent lines at the two points where the ellipse intersects the x-axis.

48. Find all points on the curve $x^2y - xy^2 = 2$ where the tangent line is vertical, that is, where $dx/dy = 0$.

≈ **49.** How high h must the light bulb in Figure 5 be if the point $(1.25, 0)$ is on the edge of the illuminated region?

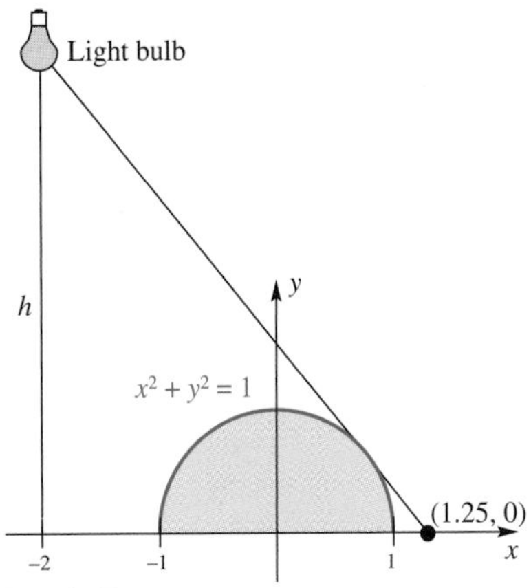

Figure 5

Answers to Concepts Review: **1.** $9/(x^3 - 3)$ **2.** $3y^2\frac{dy}{dx}$ **3.** $x \cdot 2y\frac{dy}{dx} + y^2 + 3y^2\frac{dy}{dx} - \frac{dy}{dx} = 3x^2$ **4.** $\frac{p}{q}x^{p/q-1}$; $\frac{5}{3}(x^2 - 5x)^{2/3}(2x - 5)$

2.8 Related Rates

If a variable y depends on time t, then its derivative dy/dt is called a **time rate of change.** Of course, if y measures distance, then this time rate of change is also called velocity. We are interested in a wide variety of time rates: the rate at which water is flowing into a bucket, the rate at which the area of an oil spill is growing, the rate at which the value of a piece of real estate is increasing, and so on. If y is given explicitly in terms of t, the problem is simple; we just differentiate and then evaluate the derivative at the required time.

It may be that, in place of knowing y explicitly in terms of t, we know a relationship that connects y and another variable x, and that we also know something about dx/dt. We may still be able to find dy/dt, since dy/dt and dx/dt are **related rates.** This will usually require implicit differentiation.

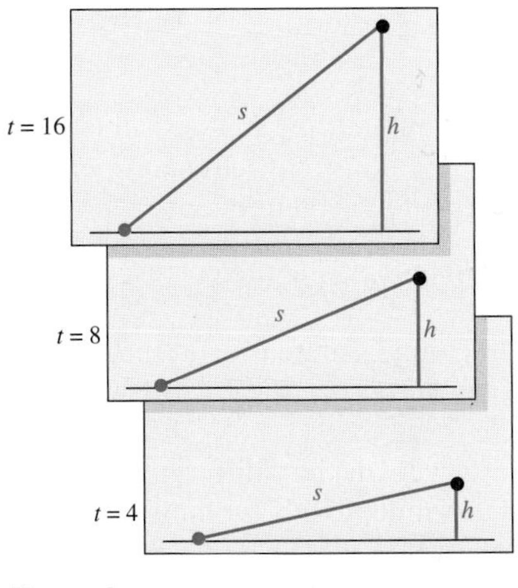

Figure 1

Two Simple Examples

In preparation for outlining a systematic procedure for solving related rate problems, we discuss two examples.

EXAMPLE 1 A small balloon is released at a point 150 feet away from an observer, who is on level ground. If the balloon goes straight up at a rate of 8 feet per second, how fast is the distance from the observer to the balloon increasing when the balloon is 50 feet high?

SOLUTION Let t denote the number of seconds after the balloon is released. Let h denote the height of the balloon and s its distance from the observer (see Figure 1). Both h and s are variables that depend on t; however, the base of the

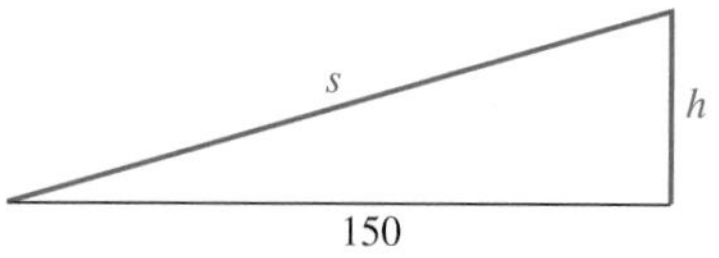

Figure 2

triangle (the distance from the observer to the point of release) remains unchanged as t increases. Figure 2 shows the key quantities in one simple diagram.

≈ Before going farther, we pick up a theme discussed earlier in the book, *estimating the answer*. Note that, initially, s changes hardly at all ($ds/dt \approx 0$), but eventually s changes about as fast as h changes ($ds/dt \approx dh/dt = 8$). An estimate for ds/dt when $h = 50$ might be about one-third to one-half of dh/dt, or 3. If we get an answer far from this value, we will know we have made a mistake. For example, answers such as 17 and even 7 are clearly wrong.

We continue with the exact solution. For emphasis, we ask and answer three fundamental questions.

(a) What is given? *Answer*: $dh/dt = 8$.

(b) What do we want to know? *Answer*: We want to know ds/dt at the instant when $h = 50$.

(c) How are s and h related? *Answer*: The variables s and h change with time (they are implicit functions of t), but they are always related by the Pythagorean equation

$$s^2 = h^2 + (150)^2$$

If we differentiate implicitly with respect to t and use the Chain Rule, we obtain

$$2s\frac{ds}{dt} = 2h\frac{dh}{dt}$$

or

$$s\frac{ds}{dt} = h\frac{dh}{dt}$$

This relationship holds for all $t > 0$.

Now, and *not before now*, we turn to the specific instant when $h = 50$. From the Pythagorean Theorem, we see that, when $h = 50$,

$$s = \sqrt{(50)^2 + (150)^2} = 50\sqrt{10}$$

Substituting in $s(ds/dt) = h(dh/dt)$ yields

$$50\sqrt{10}\frac{ds}{dt} = 50(8)$$

or

$$\frac{ds}{dt} = \frac{8}{\sqrt{10}} \approx 2.53$$

At the instant when $h = 50$, the distance between the balloon and the observer is increasing at the rate of 2.53 feet per second. ■

Similar Triangles

Two triangles are similar if their corresponding angles are congruent.

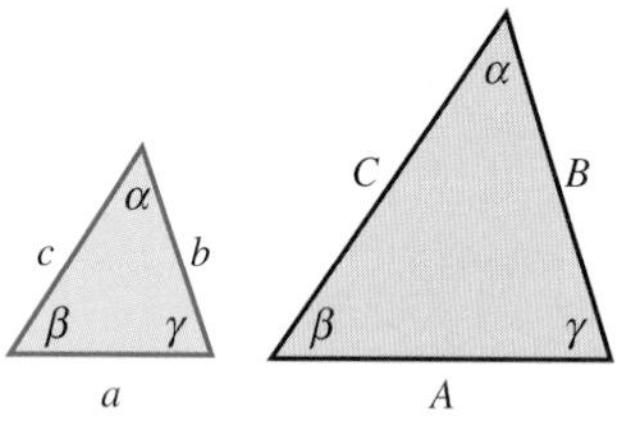

From geometry, we learn that ratios of corresponding sides of similar triangles are equal. For example,

$$\frac{b}{a} = \frac{B}{A}$$

This fact, used in Example 2, will be needed often in the problem set.

EXAMPLE 2 Water is pouring into a conical tank at the rate of 8 cubic feet per minute. If the height of the tank is 12 feet and the radius of its circular opening is 6 feet, how fast is the water level rising when the water is 4 feet deep?

SOLUTION Denote the depth of the water by h and let r be the corresponding radius of the surface of the water (see Figure 3).

We are *given* that the volume, V, of water in the tank is increasing at the rate of 8 cubic feet per minute; that is, $dV/dt = 8$. We *want to know* how fast the water is rising (that is, dh/dt) at the instant when $h = 4$.

We need to find an equation relating V and h; we will then differentiate it to get a relationship between dV/dt and dh/dt. The formula for the volume of water in the tank, $V = \frac{1}{3}\pi r^2 h$, contains the unwanted variable r; it is unwanted because we do not know its rate dr/dt. However, by similar triangles (see the marginal box), we have $r/h = 6/12$, so $r = h/2$. Substituting this in $V = \frac{1}{3}\pi r^2 h$ gives

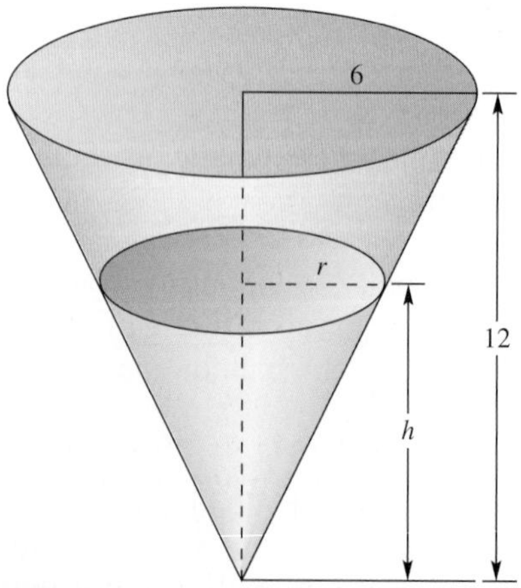

Figure 3

$$V = \frac{1}{3}\pi\left(\frac{h}{2}\right)^2 h = \frac{\pi h^3}{12}$$

Now we differentiate implicitly, keeping in mind that both V and h depend on t. We obtain

$$\frac{dV}{dt} = \frac{3\pi h^2}{12}\frac{dh}{dt} = \frac{\pi h^2}{4}\frac{dh}{dt}$$

Now that we have a relationship between dV/dt and dh/dt, and not earlier, we consider the situation when $h = 4$. Substituting $h = 4$ and $dV/dt = 8$, we obtain

$$8 = \frac{\pi(4)^2}{4}\frac{dh}{dt}$$

from which

$$\frac{dh}{dt} = \frac{2}{\pi} \approx 0.637$$

When the depth of the water is 4 feet, the water level is rising at 0.637 foot per minute. ■

If you think about Example 2 for a moment, you realize that the water level will rise more and more slowly as time goes on. For example, when $h = 10$

$$8 = \frac{\pi(10)^2}{4}\frac{dh}{dt}$$

so $dh/dt = 32/100\pi \approx 0.102$ foot per minute.

What we are really saying is that the acceleration d^2h/dt^2 is negative. We can calculate an expression for it. At any time t,

$$8 = \frac{\pi h^2}{4}\frac{dh}{dt}$$

so

$$\frac{32}{\pi} = h^2\frac{dh}{dt}$$

If we differentiate implicitly again, we get

$$0 = h^2\frac{d^2h}{dt^2} + \frac{dh}{dt}\left(2h\frac{dh}{dt}\right)$$

from which

$$\frac{d^2h}{dt^2} = \frac{-2\left(\frac{dh}{dt}\right)^2}{h}$$

This is clearly negative.

A Systematic Procedure Examples 1 and 2 suggest the following method for solving a related rates problem.

Step 1: Let t denote the elapsed time. Draw a diagram that is valid for all $t > 0$. Label those quantities whose values do not change as t increases with their given *constant* values. Assign letters to the quantities that vary with t, and label the appropriate parts of the figure with these variables.

Step 2: State what is given about the variables and what information is wanted about them. This information will be in the form of derivatives with respect to t.

Step 3: Relate the variables by writing an equation that is valid at all times $t > 0$, not just at some particular instant.

Step 4: Differentiate the equation found in Step 3 implicitly with respect to t. The resulting equation, containing derivatives with respect to t, is true for all $t > 0$.

Step 5: At this point, and not earlier, substitute in the equation found in Step 4 all data that are valid *at the particular instant* for which the answer to the problem is required. Solve for the desired derivative.

EXAMPLE 3 An airplane flying north at 640 miles per hour passes over a certain town at noon. A second airplane going east at 600 miles per hour is directly over the same town 15 minutes later. If the airplanes are flying at the same altitude, how fast will they be separating at 1:15 P.M.?

SOLUTION

Step 1: Let t denote the number of hours after 12:15 P.M., y the distance in miles flown by the northbound airplane after 12:15 P.M., x the distance flown by the eastbound airplane after 12:15 P.M., and s the distance between the airplanes. In the 15 minutes from noon to 12:15 P.M. the northbound airplane will have flown $\frac{640}{4} = 160$ miles, so the distance from the town to the northbound airplane at time t will be $y + 160$. (See Figure 4.)

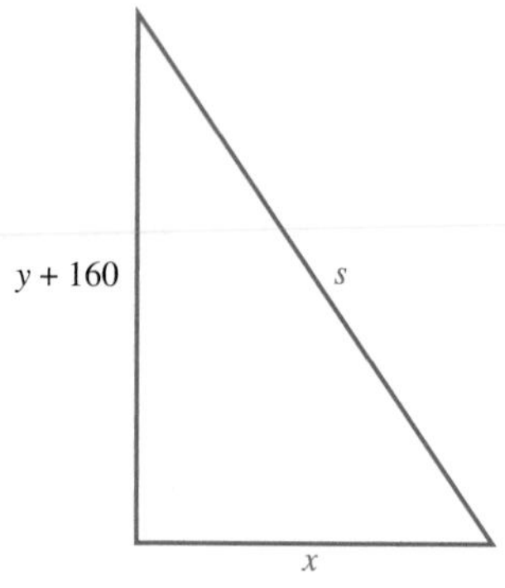

Figure 4

Step 2: We are given that, for all $t > 0$, $dy/dt = 640$ and $dx/dt = 600$. We want to know ds/dt at $t = 1$, that is, at 1:15 P.M.

Step 3: By the Pythagorean Theorem,

$$s^2 = x^2 + (y + 160)^2$$

Step 4: Differentiating implicitly with respect to t and using the Chain Rule, we have

$$2s\frac{ds}{dt} = 2x\frac{dx}{dt} + 2(y + 160)\frac{dy}{dt}$$

or

$$s\frac{ds}{dt} = x\frac{dx}{dt} + (y + 160)\frac{dy}{dt}$$

Step 5: For all $t > 0$, $dx/dt = 600$ and $dy/dt = 640$, while at the particular instant $t = 1$, $x = 600$, $y = 640$, and $s = \sqrt{(600)^2 + (640 + 160)^2} = 1000$. When we substitute these data in the equation of Step 4, we obtain

$$1000\frac{ds}{dt} = (600)(600) + (640 + 160)(640)$$

from which

$$\frac{ds}{dt} = 872$$

At 1:15 P.M., the airplanes are separating at 872 miles per hour.

≈ Now let's see if our answer makes sense. Look at Figure 4 again. Clearly, s is increasing faster than either x or y is increasing, so ds/dt exceeds 640. On the other hand, s is surely increasing more slowly than the sum of x and y; that is, $ds/dt < 600 + 640 = 1240$. Our answer, $ds/dt = 872$, is reasonable. ■

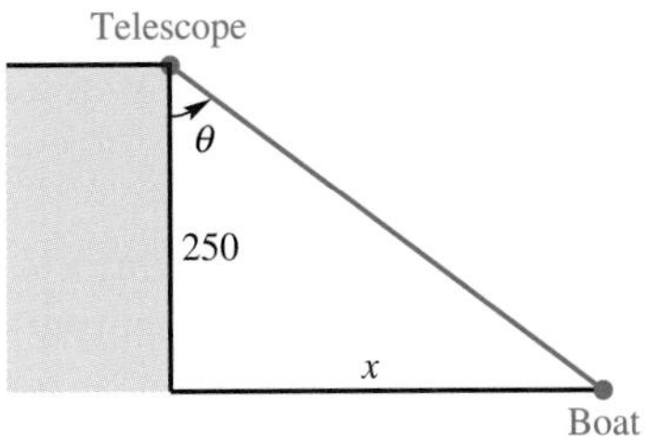

Figure 5

EXAMPLE 4 A woman standing on a cliff is watching a motorboat through a telescope as the boat approaches the shoreline directly below her. If the telescope is 250 feet above the water level and if the boat is approaching at 20 feet per second, at what rate is the angle of the telescope changing when the boat is 250 feet from the shore?

SOLUTION

Step 1: We draw a figure (Figure 5) and introduce variables x and θ, as shown.

Step 2: We are given that $dx/dt = -20$; the sign is negative because x is decreasing with time. We want to know $d\theta/dt$ at the instant when $x = 250$.

Step 3: From trigonometry,

$$\tan\theta = \frac{x}{250}$$

Step 4: We differentiate implicitly using the fact that $D_\theta \tan\theta = \sec^2\theta$ (Theorem 2.4B). This gives

$$\sec^2\theta \frac{d\theta}{dt} = \frac{1}{250}\frac{dx}{dt}$$

Step 5: At the instant when $x = 250$, θ is $\pi/4$ radians and $\sec^2\theta = \sec^2(\pi/4) = 2$. Thus,

$$2\frac{d\theta}{dt} = \frac{1}{250}(-20)$$

or

$$\frac{d\theta}{dt} = \frac{-1}{25} = -0.04$$

The angle is changing at -0.04 radian per second. The negative sign shows that θ is decreasing with time. ■

EXAMPLE 5 As the sun sets behind a 120-foot building, the building's shadow grows. How fast is the shadow growing (in feet per second) when the sun's rays make an angle of 45° (or $\pi/4$ radians).

SOLUTION

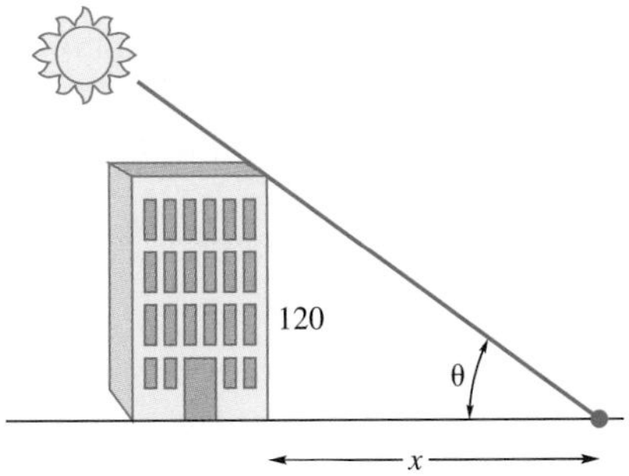

Figure 6

Step 1: Let t denote time in seconds since midnight. Let x denote the length of the shadow in feet, and let θ denote the angle of the sun's ray. See Figure 6.

Step 2: Since the earth rotates once every 24 hours, or 86,400 seconds, we know that $d\theta/dt = -2\pi/86{,}400$. (The negative sign is needed because θ *decreases* as the sun sets.) We want to know dx/dt when $\theta = \pi/4$.

Step 3: Figure 6 indicates that the quantities x and θ satisfy $\cot\theta = x/120$, so $x = 120\cot\theta$.

Step 4: Differentiating both sides of $x = 120\cot\theta$ with respect to t gives

$$\frac{dx}{dt} = 120(-\csc^2\theta)\frac{d\theta}{dt} = -120(\csc^2\theta)\left(-\frac{2\pi}{86{,}400}\right) = \frac{\pi}{360}\csc^2\theta$$

Step 5: When $\theta = \pi/4$, we have

$$\frac{dx}{dt} = \frac{\pi}{360}\csc^2\frac{\pi}{4} = \frac{\pi}{360}(\sqrt{2})^2 = \frac{\pi}{180} \approx 0.0175\frac{\text{ft}}{\text{sec}}$$

Notice that as the sun sets, θ is decreasing (hence $d\theta/dt$ is negative), while the shadow x is increasing (hence dx/dt is positive). ■

A Graphical Related Rates Problem Often in a real-life situation, we do not know a formula for a certain function, but rather have an empirically determined graph for it. We may still be able to answer questions about rates.

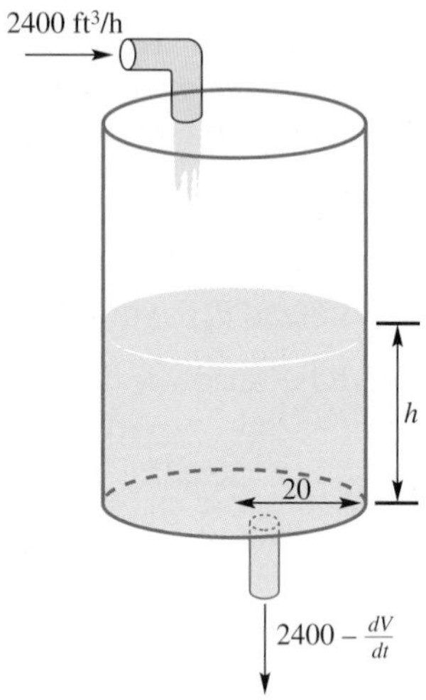

Figure 7

EXAMPLE 6 Webster City monitors the height of the water in its cylindrical water tank with an automatic recording device. Water is constantly pumped into the tank at a rate of 2400 cubic feet per hour, as shown in Figure 7. During a certain 12-hour period (beginning at midnight), the water level rose and fell according to the graph in Figure 8. If the radius of the tank is 20 feet, at what rate was water being used at 7:00 A.M.?

SOLUTION Let t denote the number of hours past midnight, h the height of the water in the tank at time t, and V the volume of water in the tank at that time (see Figure 7). Then dV/dt is the rate in minus the rate out, so $2400 - dV/dt$ is the rate at which water is being used at any time t. Since the slope of the tangent line at $t = 7$ is approximately -3 (Figure 8), we conclude that $dh/dt \approx -3$ at that time.

For a cylinder, $V = \pi r^2 h$, and so

$$V = \pi(20)^2 h$$

from which

$$\frac{dV}{dt} = 400\pi \frac{dh}{dt}$$

At $t = 7$,

$$\frac{dV}{dt} \approx 400\pi(-3) \approx -3770$$

Thus Webster City residents were using water at the rate of $2400 + 3770 = 6170$ cubic feet per hour at 7:00 A.M. ■

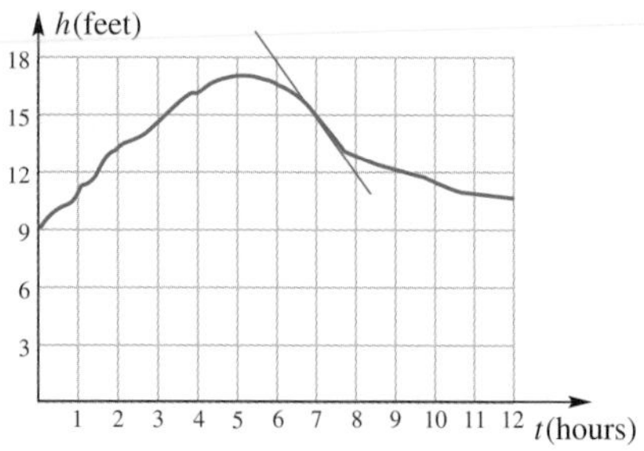

Figure 8

Concepts Review

1. To ask how fast u is changing with respect to time t after 2 hours is to ask the value of _____ at _____.

2. An airplane with a constant speed of 400 miles per hour flew directly over an observer. The distance between the observer and plane grew at an increasing rate, eventually approaching a rate of _____.

3. If dh/dt is decreasing as time t increases, then d^2h/dt^2 is _____.

4. If water is pouring into a spherical tank at a constant rate, then the height of the water grows at a variable and positive rate dh/dt, but d^2h/dt^2 is _____ until h reaches half the height of the tank, after which d^2h/dt^2 becomes _____.

Problem Set 2.8

1. Each edge of a variable cube is increasing at a rate of 3 inches per second. How fast is the volume of the cube increasing when an edge is 12 inches long?

2. Assuming that a soap bubble retains its spherical shape as it expands, how fast is its radius increasing when its radius is 3 inches if air is blown into it at a rate of 3 cubic inches per second?

≈ 3. An airplane, flying horizontally at an altitude of 1 mile, passes directly over an observer. If the constant speed of the airplane is 400 miles per hour, how fast is its distance from the observer increasing 45 seconds later? *Hint:* Note that in 45 seconds $\left(\frac{3}{4} \cdot \frac{1}{60} = \frac{1}{80} \text{ hour}\right)$, the airplane goes 5 miles.

4. A student is using a straw to drink from a conical paper cup, whose axis is vertical, at a rate of 3 cubic centimeters per second. If the height of the cup is 10 centimeters and the diameter of its opening is 6 centimeters, how fast is the level of the liquid falling when the depth of the liquid is 5 centimeters?

≈ 5. An airplane flying west at 300 miles per hour goes over the control tower at noon, and a second airplane at the same altitude, flying north at 400 miles per hour, goes over the tower an hour later. How fast is the distance between the airplanes changing at 2:00 P.M.? *Hint:* See Example 3.

≈ 6. A woman on a dock is pulling in a rope fastened to the bow of a small boat. If the woman's hands are 10 feet higher than the point where the rope is attached to the boat and if she is retrieving the rope at a rate of 2 feet per second, how fast is the boat approaching the dock when 25 feet of rope is still out?

≈ 7. A 20-foot ladder is leaning against a building. If the bottom of the ladder is sliding along the level pavement directly away from the building at 1 foot per second, how fast is the top of the ladder moving down when the foot of the ladder is 5 feet from the wall?

8. We assume that an oil spill is being cleaned up by deploying bacteria that consume the oil at 4 cubic feet per hour. The oil spill itself is modeled in the form of a very thin cyclinder whose height is the thickness of the oil slick. When the thickness of the slick is 0.001 foot, the cylinder is 500 feet in diameter. If the height is decreasing at 0.0005 foot per hour, at what rate is the area of the slick changing?

9. Sand is pouring from a pipe at the rate of 16 cubic feet per second. If the falling sand forms a conical pile on the ground whose altitude is always $\frac{1}{4}$ the diameter of the base, how fast is the altitude increasing when the pile is 4 feet high? *Hint:* Refer to Figure 9 and use the fact that $V = \frac{1}{3}\pi r^2 h$.

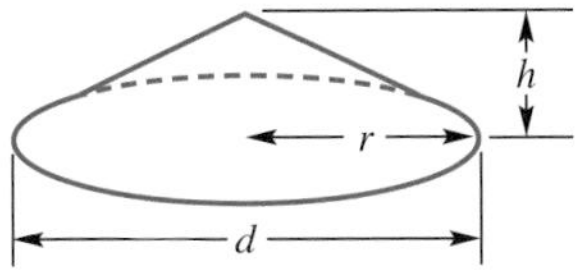

Figure 9

≈ **10.** A child is flying a kite. If the kite is 90 feet above the child's hand level and the wind is blowing it on a horizontal course at 5 feet per second, how fast is the child paying out cord when 150 feet of cord is out? (Assume that the cord remains straight from hand to kite, actually an unrealistic assumption.)

11. A rectangular swimming pool is 40 feet long, 20 feet wide, 8 feet deep at the deep end, and 3 feet deep at the shallow end (see Figure 10). If the pool is filled by pumping water into it at the rate of 40 cubic feet per minute, how fast is the water level rising when it is 3 feet deep at the deep end?

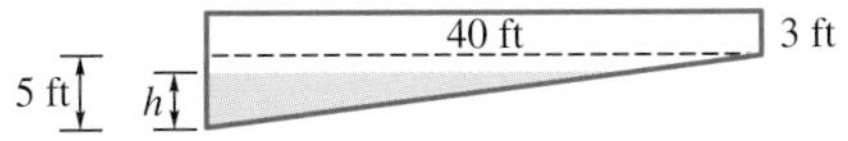

Figure 10

≈ **12.** A particle P is moving along the graph of $y = \sqrt{x^2 - 4}$, $x \geq 2$, so that the x-coordinate of P is increasing at the rate of 5 units per second. How fast is the y-coordinate of P increasing when $x = 3$?

13. A metal disk expands during heating. If its radius increases at the rate of 0.02 inch per second, how fast is the area of one of its faces increasing when its radius is 8.1 inches?

≈ **14.** Two ships sail from the same island port, one going north at 24 knots (24 nautical miles per hour) and the other east at 30 knots. The northbound ship departed at 9:00 A.M. and the eastbound ship left at 11:00 A.M. How fast is the distance between them increasing at 2:00 P.M.? *Hint:* Let $t = 0$ at 11:00 A.M.

15. A light in a lighthouse 1 kilometer offshore from a straight shoreline is rotating at 2 revolutions per minute. How fast is the beam moving along the shoreline when it passes the point $\frac{1}{2}$ kilometer from the point opposite the lighthouse?

[C] **16.** An aircraft spotter observes a plane flying at a constant altitude of 4000 feet toward a point directly above her head. She notes that when the angle of elevation is $\frac{1}{2}$ radian it is increasing at a rate of $\frac{1}{10}$ radian per second. What is the speed of the airplane?

17. Chris, who is 6 feet tall, is walking away from a street light pole 30 feet high at a rate of 2 feet per second.

(a) How fast is his shadow increasing in length when Chris is 24 feet from the pole? 30 feet?

(b) How fast is the tip of his shadow moving?

(c) To follow the tip of his shadow, at what angular rate must Chris be lifting his eyes when his shadow is 6 feet long?

18. The vertex angle θ opposite the base of an isosceles triangle with equal sides of length 100 centimeters is increasing at $\frac{1}{10}$ radian per minute. How fast is the area of the triangle increasing when the vertex angle measures $\pi/6$ radians? *Hint:* $A = \frac{1}{2}ab\sin\theta$.

≈ **19.** A long, level highway bridge passes over a railroad track that is 100 feet below it and at right angles to it. If an automobile traveling 45 miles per hour (66 feet per second) is directly above a train engine going 60 miles per hour (88 feet per second), how fast will they be separating 10 seconds later?

20. Water is pumped at a uniform rate of 2 liters (1 liter = 1000 cubic centimeters) per minute into a tank shaped like a frustum of a right circular cone. The tank has altitude 80 centimeters and lower and upper radii of 20 and 40 centimeters, respectively (Figure 11). How fast is the water level rising when the depth of the water is 30 centimeters? *Note:* The volume, V, of a frustum of a right circular cone of altitude h and lower and upper radii a and b is $V = \frac{1}{3}\pi h \cdot (a^2 + ab + b^2)$.

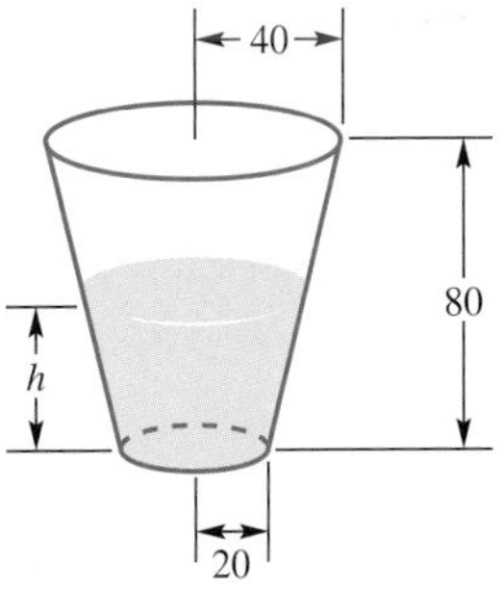

Figure 11

21. Water is leaking out the bottom of a hemispherical tank of radius 8 feet at a rate of 2 cubic feet per hour. The tank was full at a certain time. How fast is the water level changing when its height h is 3 feet? *Note:* The volume of a segment of height h in a hemisphere of radius r is $\pi h^2[r - (h/3)]$. (See Figure 12.)

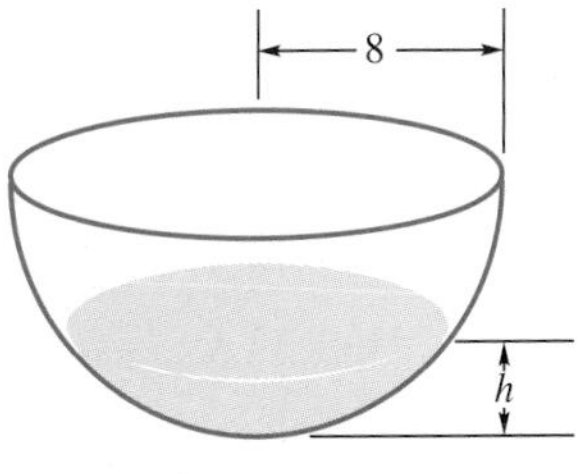

Figure 12

22. The hands on a clock are of length 5 inches (minute hand) and 4 inches (hour hand). How fast is the distance between the tips of the hands changing at 3:00?

23. A steel ball will drop $16t^2$ feet in t seconds. Such a ball is dropped from a height of 64 feet at a horizontal distance 10 feet from a 48-foot street light. How fast is the ball's shadow moving when the ball hits the ground?

24. Rework Example 6 assuming that the water tank is a sphere of radius 20 feet. (See Problem 21 for the volume of a spherical segment.)

25. Rework Example 6 assuming that the water tank is in the shape of an upper hemisphere of radius 20 feet. (See Problem 21 for the volume of a spherical segment.)

26. Refer to Example 6. How much water did Webster City use during this 12-hour period from midnight to noon? *Hint:* This is not a differentiation problem.

≈ **27.** An 18-foot ladder leans against a 12-foot vertical wall, its top extending over the wall. The bottom end of the ladder is pulled along the ground away from the wall at 2 feet per second.

(a) Find the vertical velocity of the top end when the ladder makes an angle of 60° with the ground.

(b) Find the vertical acceleration at the same instant.

28. A spherical steel ball rests at the bottom of the tank of Problem 21. Answer the question posed there if the ball has radius

(a) 6 inches, and (b) 2 feet.

(Assume that the ball does not affect the flow from the tank.)

29. A snowball melts at a rate proportional to its surface area.

(a) Show that its radius shrinks at a constant rate.

(b) If it melts to $\frac{8}{27}$ its original volume in one hour, how long will it take to melt completely?

30. A right circular cylinder with a piston at one end is filled with gas. Its volume is continually changing because of the movement of the piston. If the temperature of the gas is kept constant, then, by **Boyle's Law,** $PV = k$, where P is the pressure (pounds per square inch), V is the volume (cubic inches), and k is a constant. The pressure was monitored by a recording device over one 10-minute period. The results are shown in Figure 13. Approximately how fast was the volume changing at $t = 6.5$ if its volume was 300 cubic inches at that instant? (See Example 6.)

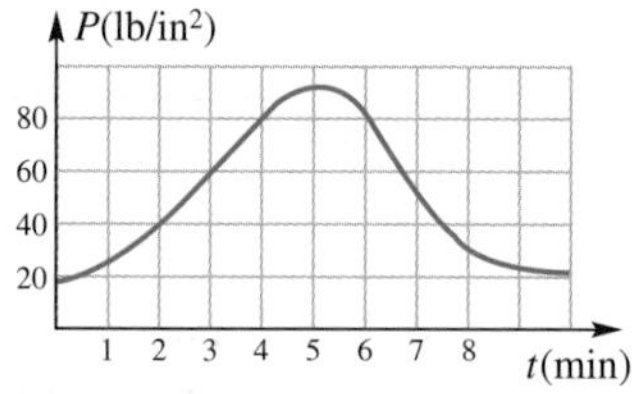

Figure 13

31. A girl 5 feet tall walks toward a street light 20 feet high at a rate of 4 feet per second. Her little brother, 3 feet tall, follows at a constant distance of 4 feet directly behind her (Figure 14).

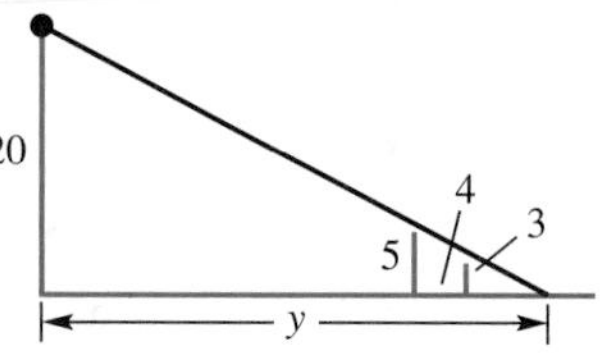

Figure 14

Determine how fast the tip of the shadow is moving, that is, determine dy/dt. *Note:* When the girl is far from the light, she controls the tip of the shadow, whereas her brother controls it near the light.

Answers to Concepts Review: **1.** du/dt; $t = 2$ **2.** 400 mi/h **3.** negative **4.** negative; positive

2.9 Differentials and Approximations

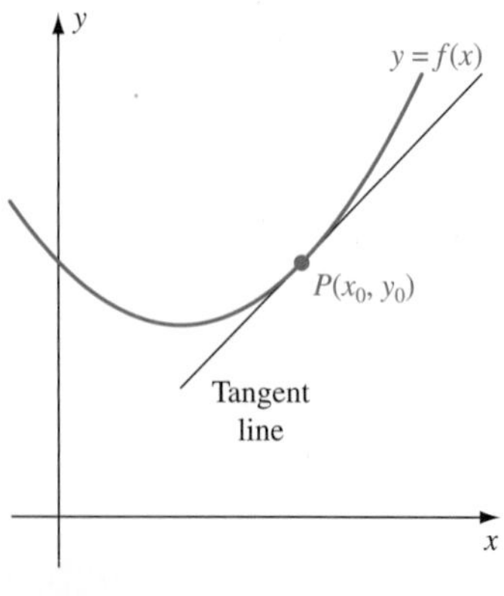

Figure 1

The Leibniz notation dy/dx has been used to mean the derivative of y with respect to x. The notation d/dx has been used as an operator to mean the derivative (of whatever follows d/dx) with respect to x. Thus, d/dx and D_x are synonymous. Up to now, we have treated dy/dx (or d/dx) as a *single* symbol and have not tried to give separate meanings to the symbols dy and dx. In this section we will give meanings to dy and to dx.

Let f be a differentiable function. To motivate our definitions, let $P(x_0, y_0)$ be a point on the graph of $y = f(x)$ as shown in Figure 1. Since f is differentiable,

$$\lim_{\Delta x \to 0} \frac{f(x_0 + \Delta x) - f(x_0)}{\Delta x} = f'(x_0)$$

Thus, if Δx is small, the quotient $[f(x_0 + \Delta x) - f(x_0)]/\Delta x$ will be approximately $f'(x_0)$, so

$$f(x_0 + \Delta x) - f(x_0) \approx \Delta x\, f'(x_0)$$

The left side of this expression is called Δy; this is the *actual* change in y as x changes from x_0 to $x_0 + \Delta x$. The right side is called dy, and it serves as an

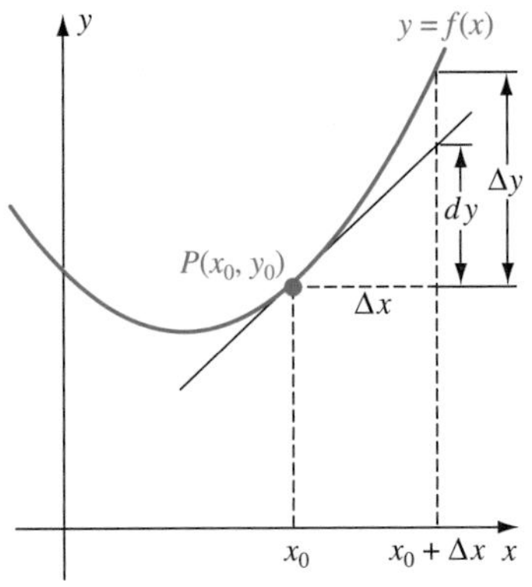

Figure 2

approximation to Δy. As Figure 2 indicates, the quantity dy is equal to the change in the tangent line to the curve at P as x changes from x_0 to $x_0 + \Delta x$. When Δx is small, we expect dy to be a good approximation to Δy, and being just a constant times Δx, it is usually easier to calculate.

Differentials Defined Here are the formal definitions of the differentials dx and dy.

> **Definition Differentials**
>
> Let $y = f(x)$ be a differentiable function of the independent variable x.
>
> Δx is an arbitrary increment in the independent variable x.
>
> dx, called the **differential of the independent variable** x, is equal to Δx.
>
> Δy is the actual change in the variable y as x changes from x to $x + \Delta x$; that is, $\Delta y = f(x + \Delta x) - f(x)$.
>
> dy, called the **differential of the dependent variable** y, is defined by $dy = f'(x)\,dx$.

EXAMPLE 1 Find dy if

(a) $y = x^3 - 3x + 1$ (b) $y = \sqrt{x^2 + 3x}$

(c) $y = \sin(x^4 - 3x^2 + 11)$

SOLUTION If we know how to calculate derivatives, we know how to calculate differentials. We simply calculate the derivative and multiply it by dx.

(a) $dy = (3x^2 - 3)\,dx$

(b) $dy = \frac{1}{2}(x^2 + 3x)^{-1/2}(2x + 3)\,dx = \dfrac{2x + 3}{2\sqrt{x^2 + 3x}}dx$

(c) $dy = \cos(x^4 - 3x^2 + 11)\cdot(4x^3 - 6x)\,dx$ ■

We ask you to note two things. First, since $dy = f'(x)\,dx$, division of both sides by dx yields

$$f'(x) = \frac{dy}{dx}$$

and we can, if we wish, interpret the derivative as a quotient of two differentials.

Second, corresponding to every derivative rule, there is a differential rule obtained from the former by "multiplying" through by dx. We illustrate the major rules in the following table.

> **Distinguish between Derivatives and Differentials**
>
> Derivatives and differentials are not the same. When you write $D_x y$ or dy/dx, you are using a symbol for the derivative; when you write dy, you are denoting a differential. Do not be sloppy and write dy when you mean to label a derivative. This will lead to boundless confusion.

Derivative Rule	Differential Rule
1. $\dfrac{dk}{dx} = 0$	1. $dk = 0$
2. $\dfrac{d(ku)}{dx} = k\dfrac{du}{dx}$	2. $d(ku) = k\,du$
3. $\dfrac{d(u + v)}{dx} = \dfrac{du}{dx} + \dfrac{dv}{dx}$	3. $d(u + v) = du + dv$
4. $\dfrac{d(uv)}{dx} = u\dfrac{dv}{dx} + v\dfrac{du}{dx}$	4. $d(uv) = u\,dv + v\,du$
5. $\dfrac{d(u/v)}{dx} = \dfrac{v(du/dx) - u(dv/dx)}{v^2}$	5. $d\left(\dfrac{u}{v}\right) = \dfrac{v\,du - u\,dv}{v^2}$
6. $\dfrac{d(u^n)}{dx} = nu^{n-1}\dfrac{du}{dx}$	6. $d(u^n) = nu^{n-1}\,du$

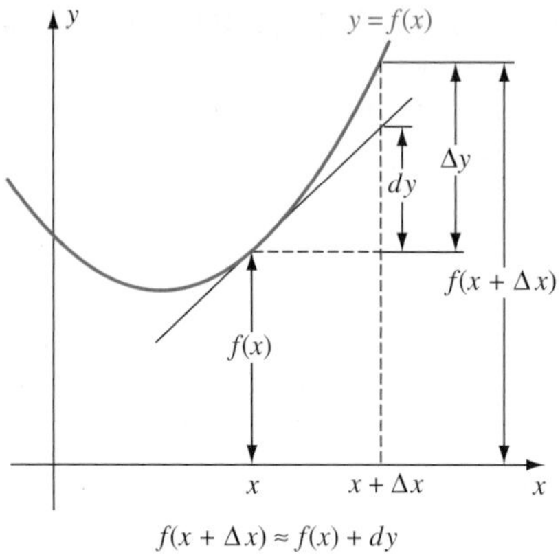

Figure 3

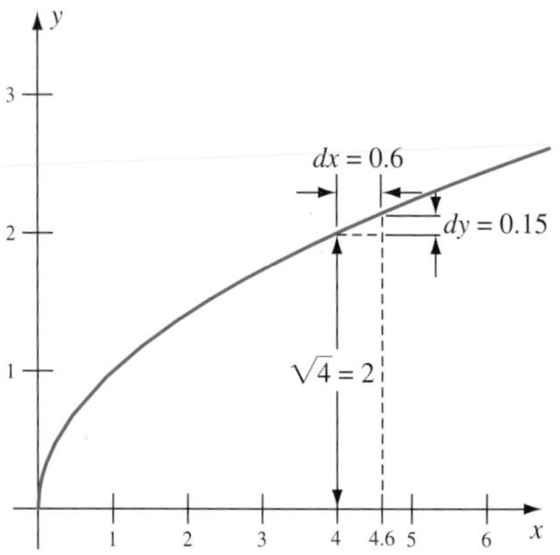

Figure 4

Approximations Differentials will play several roles in this book, but for now their chief use is in providing approximations. We hinted at this earlier.

Suppose that $y = f(x)$, as shown in Figure 3. An increment Δx produces a corresponding increment Δy in y, which can be approximated by dy. Thus, $f(x + \Delta x)$ is approximated by

$$\boxed{f(x + \Delta x) \approx f(x) + dy = f(x) + f'(x)\,\Delta x}$$

This is the basis for the solutions to all the examples that follow.

EXAMPLE 2 Suppose you need good approximations to $\sqrt{4.6}$ and $\sqrt{8.2}$, but your calculator has died. What might you do?

SOLUTION Consider the graph of $y = \sqrt{x}$ sketched in Figure 4. When x changes from 4 to 4.6, $\sqrt{x}$ changes from $\sqrt{4} = 2$ to (approximately) $\sqrt{4} + dy$. Now

$$dy = \frac{1}{2}x^{-1/2}\,dx = \frac{1}{2\sqrt{x}}dx$$

which, at $x = 4$ and $dx = 0.6$, has the value

$$dy = \frac{1}{2\sqrt{4}}(0.6) = \frac{0.6}{4} = 0.15$$

Thus,

$$\sqrt{4.6} \approx \sqrt{4} + dy = 2 + 0.15 = 2.15$$

Similarly, at $x = 9$ and $dx = -0.8$,

$$dy = \frac{1}{2\sqrt{9}}(-0.8) = \frac{-0.8}{6} \approx -0.133$$

Hence,

$$\sqrt{8.2} \approx \sqrt{9} + dy \approx 3 - 0.133 = 2.867$$

Note that both dx and dy were negative in this case.

The approximate values 2.15 and 2.867 may be compared to the true values (to four decimal places) of 2.1448 and 2.8636. ■

EXAMPLE 3 Use differentials to approximate the increase in the area of a soap bubble when its radius increases from 3 inches to 3.025 inches.

SOLUTION The area of a spherical soap bubble is given by $A = 4\pi r^2$. We may approximate the exact change, ΔA, by the differential dA, where

$$dA = 8\pi r\,dr$$

At $r = 3$ and $dr = \Delta r = 0.025$,

$$dA = 8\pi(3)(0.025) \approx 1.885 \text{ square inches}$$ ■

Estimating Errors Here is a typical problem in science. A researcher measures a certain variable x to have a value x_0 with a possible error of size $\pm\Delta x$. The value x_0 is then used to calculate a value y_0 for y that depends on x. The value y_0 is contaminated by the error in x, but how badly? The standard procedure is to estimate this error by means of differentials.

EXAMPLE 4 The side of a cube is measured as 11.4 centimeters with a possible error of ± 0.05 centimeter. Evaluate the volume of the cube and give an estimate for the possible error in this value.

SOLUTION The volume V of a cube of side x is $V = x^3$. Thus, $dV = 3x^2\,dx$. If $x = 11.4$ and $dx = 0.05$, then $V = (11.4)^3 \approx 1482$ and

$$\Delta V \approx dV = 3(11.4)^2(0.05) \approx 19$$

Thus, we might report the volume of the cube as 1482 ± 19 cubic centimeters. ■

The quantity ΔV in Example 4 is called the **absolute error.** Another measure of error is the **relative error,** which is found by dividing the absolute error by the total volume. We can approximate the relative error $\Delta V/V$ by dV/V. In Example 4, the relative error is

$$\frac{\Delta V}{V} \approx \frac{dV}{V} \approx \frac{19}{1482} \approx 0.0128$$

The relative error is often expressed in terms of a percentage. Thus, we say that for the cube in Example 4 the relative error is approximately 1.28%.

EXAMPLE 5 Poiseuille's Law for blood flow says that the volume flowing through an artery is proportional to the fourth power of the radius, that is, $V = kR^4$. By how much must the radius be increased in order to increase the blood flow by 50%?

SOLUTION The differentials satisfy $dV = 4kR^3\,dR$. The relative change in the volume is

$$\frac{\Delta V}{V} \approx \frac{dV}{V} = \frac{4kR^3\,dR}{kR^4} = 4\frac{dR}{R}$$

so for a 50% change in volume,

$$0.5 \approx \frac{dV}{V} = 4\frac{dR}{R}$$

The relative change in R must be

$$\frac{\Delta R}{R} \approx \frac{dR}{R} \approx \frac{0.5}{4} = 0.125$$

Thus, just a 12.5% increase in the radius of an artery will increase the blood flow by about 50%. ■

Linear Approximation If f is differentiable at a, then from the point-slope form of a line, the tangent line to f at $(a, f(a))$ is given by $y = f(a) + f'(a)(x - a)$. The function

$$L(x) = f(a) + f'(a)(x - a)$$

is called the **linear approximation** to the function f at a, and it is often a very good approximation to f when x is close to a.

EXAMPLE 6 Find and plot the linear approximation to $f(x) = 1 + \sin 2x$ at $x = \pi/2$.

SOLUTION: The derivative of f is $f'(x) = 2\cos 2x$, so the linear approximation is

$$\begin{aligned} L(x) &= f(\pi/2) + f'(\pi/2)(x - \pi/2) \\ &= (1 + \sin\pi) + (2\cos\pi)(x - \pi/2) \\ &= 1 - 2(x - \pi/2) = (1 + \pi) - 2x \end{aligned}$$

Figure 5a shows both the graph of the function f and the linear approximation L over the interval $[0, \pi]$. We can see that the approximation is good near $\pi/2$, but not so good as you move away from $\pi/2$. Figures 5b and c also show plots of the functions L and f over smaller and smaller intervals. For values of x close to $\pi/2$, we see that the linear approximation is very close to the function f. ■

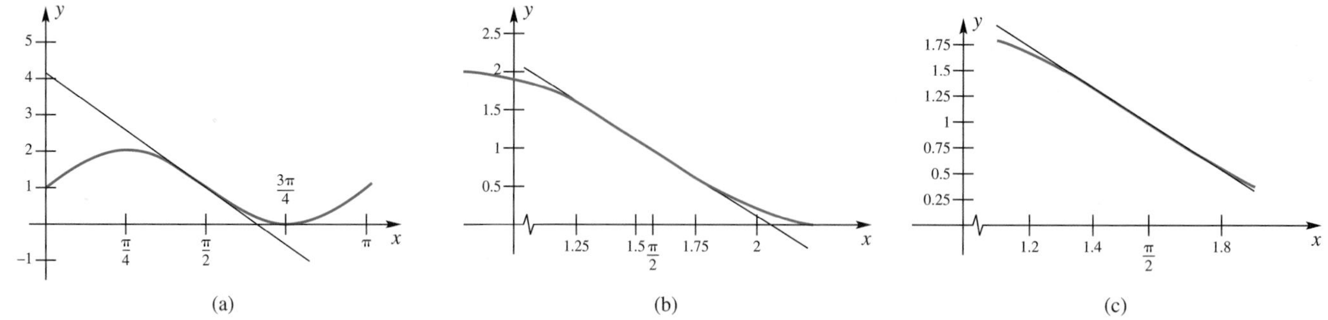

Figure 5

Concepts Review

1. Let $y = f(x)$. The differential of y in terms of dx is defined by $dy =$ _____.

2. Consider the curve $y = f(x)$ and suppose that x is given an increment Δx. The corresponding change in y on the curve is denoted by _____, whereas the corresponding change in y on the tangent line is denoted by _____.

3. We can expect dy to be a good approximation to Δy, provided that _____.

4. On the curve $y = \sqrt{x}$, we should expect dy to be close to Δy, but always _____ than Δy. On the curve $y = x^2, x \ge 0$, we should expect dy to be _____ than Δy.

Problem Set 2.9

In Problems 1–8, find dy.

1. $y = x^2 + x - 3$ **2.** $y = 7x^3 + 3x^2 + 1$

3. $y = (2x + 3)^{-4}$ **4.** $y = (3x^2 + x + 1)^{-2}$

5. $y = (\sin x + \cos x)^3$ **6.** $y = (\tan x + 1)^3$

7. $y = (7x^2 + 3x - 1)^{-3/2}$ **8.** $y = \left(x^{10} + \sqrt{\sin 2x}\right)^2$

9. If $s = \sqrt{(t^2 - \cot t + 2)^3}$, find ds.

10. Let $y = f(x) = x^3$. Find the value of dy in each case.
(a) $x = 0.5, dx = 1$ (b) $x = -1, dx = 0.75$

11. For the function defined in Problem 10, make a careful drawing of the graph of f for $-1.5 \le x \le 1.5$ and the tangents to the curve at $x = 0.5$ and $x = -1$; on this drawing label dy and dx for each of the given sets of data in parts (a) and (b).

12. Let $y = 1/x$. Find the value of dy in each case.
(a) $x = 1, dx = 0.5$ (b) $x = -2, dx = 0.75$

13. For the function defined in Problem 12, make a careful drawing (as in Problem 11) for $-3 \le x < 0$ and $0 < x \le 3$.

C **14.** For the data of Problem 10, find the actual changes in y, that is, Δy.

C **15.** For the data of Problem 12, find the actual changes in y, that is, Δy.

16. If $y = x^2 - 3$, find the values of Δy and dy in each case.
(a) $x = 2$ and $dx = \Delta x = 0.5$
C (b) $x = 3$ and $dx = \Delta x = -0.12$

17. If $y = x^4 + 2x$, find the values of Δy and dy in each case.
(a) $x = 2$ and $dx = \Delta x = 1$
C (b) $x = 2$ and $dx = \Delta x = 0.005$

In Problems 18–20, use differentials to approximate the given number (see Example 2). Compare with calculator values.

18. $\sqrt{402}$ **19.** $\sqrt{35.9}$

20. $\sqrt[3]{26.91}$

C **21.** Approximate the volume of material in a spherical shell of inner radius 5 centimeters and outer radius 5.125 centimeters (see Example 3).

C **22.** All six sides of a cubical metal box are 0.25 inch thick, and the volume of the interior of the box is 40 cubic inches. Use differentials to find the approximate volume of metal used to make the box.

23. The outside diameter of a thin spherical shell is 12 feet. If the shell is 0.3 inch thick, use differentials to approximate the volume of the region interior to the shell.

24. The interior of an open cylindrical tank is 12 feet in diameter and 8 feet deep. The bottom is copper and the sides are steel. Use differentials to find approximately how many gallons of waterproofing paint are needed to apply a 0.05-inch coat to the steel part of the inside of the tank (1 gallon $\approx$ 231 cubic inches).

25. Assuming that the equator is a circle whose radius is approximately 4000 miles, how much longer than the equator would a concentric, coplanar circle be if each point on it were 2 feet above the equator? Use differentials.

26. The period of a simple pendulum of length L feet is given by $T = 2\pi\sqrt{L/g}$ seconds. We assume that g, the acceleration due to gravity on (or very near) the surface of the earth, is 32 feet per second per second. If the pendulum is that of a clock that keeps good time when $L = 4$ feet, how much time will the clock gain in 24 hours if the length of the pendulum is decreased to 3.97 feet?

27. The diameter of a sphere is measured as 20 ± 0.1 centimeters. Calculate the volume and estimate the absolute error and the relative error (see Example 4).

28. A cylindrical roller is exactly 12 inches long and its diameter is measured as 6 ± 0.005 inches. Calculate its volume with an estimate for the absolute error and the relative error.

C **29.** The angle θ between the two equal sides of an isosceles triangle measures 0.53 ± 0.005 radian. The two equal sides are exactly 151 centimeters long. Calculate the length of the third side with an estimate for the absolute error and the relative error.

C **30.** Calculate the area of the triangle of Problem 29 with an estimate for the absolute error and the relative error. *Hint:* $A = \frac{1}{2}ab \sin \theta$.

31. It can be shown that if $|d^2y/dx^2| \le M$ on a closed interval with c and $c + \Delta x$ as end points, then

$$|\Delta y - dy| \le \tfrac{1}{2}M(\Delta x)^2$$

Find, using differentials, the change in $y = 3x^2 - 2x + 11$ when x increases from 2 to 2.001 and then give a bound for the error that you have made by using differentials.

32. Suppose that f is a function satisfying $f(1) = 10$, and $f'(1.02) = 12$. Use this information to approximate $f(1.02)$.

33. Suppose f is a function satisfying $f(3) = 8$ and $f'(3.05) = \frac{1}{4}$. Use this information to approximate $f(3.05)$.

34. A conical cup, 10 centimeters high and 8 centimeters wide at the top, is filled with water to a depth of 9 centimeters. An ice cube 3 centimeters on a side is about to be dropped in. Use differentials to decide whether the cup will overflow.

35. A tank has the shape of a cylinder with hemispherical ends. If the cylindrical part is 100 centimeters long and has an outside diameter of 20 centimeters, about how much paint is required to coat the outside of the tank to a thickness of 1 millimeter?

C **36.** Einstein's Special Theory of Relativity says that an object's mass m is related to its velocity v by the formula

$$m = \frac{m_0}{\sqrt{1 - v^2/c^2}} = m_0\left(1 - \frac{v^2}{c^2}\right)^{-1/2}$$

Here m_0 is the rest mass and c is the speed of light. Use differentials to determine the percent increase in mass of an object when its velocity increases from $0.9c$ to $0.92c$.

In Problems 37–44, find the linear approximation to the given functions at the specified points. Plot the function and its linear approximation over the indicated interval.

37. $f(x) = x^2$ at $a = 2$, $[0, 3]$

38. $g(x) = x^2 \cos x$ at $a = \pi/2$, $[0, \pi]$

39. $h(x) = \sin x$ at $a = 0$, $[-\pi, \pi]$

40. $F(x) = 3x + 4$ at $a = 3$, $[0, 6]$

41. $f(x) = \sqrt{1 - x^2}$ at $a = 0$, $[-1, 1]$

42. $g(x) = x/(1 - x^2)$ at $a = \frac{1}{2}$, $[0, 1)$

43. $h(x) = x \sec x$ at $a = 0$, $(-\pi/2, \pi/2)$

44. $G(x) = x + \sin 2x$, at $a = \pi/2$, $[0, \pi]$

45. Find the linear approximation to $f(x) = mx + b$ at an arbitrary a. What is the relationship between $f(x)$ and $L(x)$?

46. Show that for every $a > 0$ the linear approximation $L(x)$ to the function $f(x) = \sqrt{x}$ at a satisfies $f(x) \le L(x)$ for all $x > 0$.

47. Show that for every a the linear approximation $L(x)$ to the function $f(x) = x^2$ at a satisfies $L(x) \le f(x)$ for all x.

EXPL **48.** Find a linear approximation to $f(x) = (1 + x)^\alpha$ at $x = 0$, where α is any number. For various values of α, plot $f(x)$ and its linear approximation $L(x)$. For what values of α does the linear approximation always overestimate $f(x)$? For what values of α does the linear approximation always underestimate $f(x)$?

EXPL **49.** Suppose f is differentiable. If we use the approximation $f(x + h) \approx f(x) + f'(x)\,h$ the error is $\varepsilon(h) = f(x + h) - f(x) - f'(x)\,h$. Show that

(a) $\lim_{h\to 0} \varepsilon(h) = 0$ and (b) $\lim_{h\to 0} \dfrac{\varepsilon(h)}{h} = 0$.

Answers to Concepts Review: **1.** $f'(x)\,dx$ **2.** Δy; dy **3.** Δx is small **4.** larger; smaller

2.10 Chapter Review

Concepts Test

Respond with true or false to each of the following assertions. Be prepared to justify your answer.

1. The tangent line to a curve at a point cannot cross the curve at that point.

2. The tangent line to a curve can touch the curve at only one point.

3. The slope of the tangent line to the curve $y = x^4$ is different at every point of the curve.

4. The slope of the tangent line to the curve $y = \cos x$ is different at every point on the curve.

5. It is possible for the velocity of an object to be increasing while its speed is decreasing.

6. It is possible for the speed of an object to be increasing while its velocity is decreasing.

7. If the tangent line to the graph of $y = f(x)$ is horizontal at $x = c$, then $f'(c) = 0$.

8. If $f'(x) = g'(x)$ for all x, then $f(x) = g(x)$ for all x.

9. If $g(x) = x$, then $f'(g(x)) = D_x f(g(x))$.

10. If $y = \pi^5$, then $D_x y = 5\pi^4$.

11. If $f'(c)$ exists, then f is continuous at c.

12. The graph of $y = \sqrt[3]{x}$ has a tangent line at $x = 0$ and yet $D_x y$ does not exist there.

13. The derivative of a product is always the product of the derivatives.

14. If the acceleration of an object is negative, then its velocity is decreasing.

15. If x^3 is a factor of the differentiable function $f(x)$, then x^2 is a factor of its derivative.

16. The equation of the line tangent to the graph of $y = x^3$ at $(1, 1)$ is $y - 1 = 3x^2(x - 1)$.

17. If $y = f(x)g(x)$, then $D_x^2 y = f(x)g''(x) + g(x)f''(x)$.

18. If $y = (x^3 + x)^8$, then $D_x^{25} y = 0$.

19. The derivative of a polynomial is a polynomial.

20. The derivative of a rational function is a rational function.

21. If $f'(c) = g'(c) = 0$ and $h(x) = f(x)g(x)$, then $h'(c) = 0$.

22. The expression

$$\lim_{x \to \pi/2} \frac{\sin x - 1}{x - \pi/2}$$

is the derivative of $f(x) = \sin x$ at $x = \pi/2$.

23. The operator D^2 is linear.

24. If $h(x) = f(g(x))$ where both f and g are differentiable, then $g'(c) = 0$ implies that $h'(c) = 0$.

25. If $f'(2) = g'(2) = g(2) = 2$, then $(f \circ g)'(2) = 4$.

26. If f is differentiable and increasing and if $dx = \Delta x > 0$, then $\Delta y > dy$.

27. If the radius of a sphere is increasing at 3 feet per second, then its volume is increasing at 27 cubic feet per second.

28. If the radius of a circle is increasing at 4 feet per second, then its circumference is increasing at 8π feet per second.

29. $D_x^{n+4}(\sin x) = D_x^n(\sin x)$ for every positive integer n.

30. $D_x^{n+3}(\cos x) = -D_x^n(\sin x)$ for every positive integer n.

31. $\lim_{x \to 0} \frac{\tan x}{3x} = \frac{1}{3}$.

32. If $s = 5t^3 + 6t - 300$ gives the position of an object on a horizontal coordinate line at time t, then that object is always moving to the right (the direction of increasing s).

33. If air is being pumped into a spherical rubber balloon at a constant rate of 3 cubic inches per second, then the radius will increase, but at a slower and slower rate.

34. If water is being pumped into a spherical tank of fixed radius at a rate of 3 gallons per second, the height of the water in the tank will increase more and more rapidly as the tank nears being full.

35. If an error Δr is made in measuring the radius of a sphere, the corresponding error in the calculated volume will be approximately $S \cdot \Delta r$, where S is the surface area of the sphere.

36. If $y = x^5$, then $dy \geq 0$.

37. The linear approximation to the function defined by $f(x) = \cos x$ at $x = 0$ has positive slope.

Sample Test Problems

1. Use $f'(x) = \lim_{h \to 0} [f(x + h) - f(x)]/h$ to find the derivative of each of the following.

(a) $f(x) = 3x^3$ (b) $f(x) = 2x^5 + 3x$

(c) $f(x) = \frac{1}{3x}$ (d) $f(x) = \frac{1}{3x^2 + 2}$

(e) $f(x) = \sqrt{3x}$ (f) $f(x) = \sin 3x$

(g) $f(x) = \sqrt{x^2 + 5}$ (h) $f(x) = \cos \pi x$

2. Use $g'(x) = \lim_{t \to x} \frac{g(t) - g(x)}{t - x}$ to find $g'(x)$ in each case.

(a) $g(x) = 2x^2$ (b) $g(x) = x^3 + x$

(c) $g(x) = \frac{1}{x}$ (d) $g(x) = \frac{1}{x^2 + 1}$

(e) $g(x) = \sqrt{x}$ (f) $g(x) = \sin \pi x$

(g) $g(x) = \sqrt{x^3 + C}$ (h) $g(x) = \cos 2x$

3. The given limit is a derivative, but of what function f and at what point?

(a) $\lim_{h \to 0} \frac{3(1 + h) - 3}{h}$ (b) $\lim_{h \to 0} \frac{4(2 + h)^3 - 4(2)^3}{h}$

(c) $\lim_{\Delta x \to 0} \frac{\sqrt{(1 + \Delta x)^3} - 1}{\Delta x}$ (d) $\lim_{\Delta x \to 0} \frac{\sin(\pi + \Delta x)}{\Delta x}$

(e) $\lim_{t \to x} \frac{4/t - 4/x}{t - x}$ (f) $\lim_{t \to x} \frac{\sin 3x - \sin 3t}{t - x}$

(g) $\lim_{h \to 0} \frac{\tan(\pi/4 + h) - 1}{h}$ (h) $\lim_{h \to 0} \left(\frac{1}{\sqrt{5 + h}} - \frac{1}{\sqrt{5}} \right) \frac{1}{h}$

4. Use the sketch of $s = f(t)$ in Figure 1 to approximate each of the following.

(a) $f'(2)$ (b) $f'(6)$

(c) v_{avg} on $[3, 7]$ (d) $\frac{d}{dt} f(t^2)$ at $t = 2$

(e) $\frac{d}{dt} [f^2(t)]$ at $t = 2$ (f) $\frac{d}{dt}(f(f(t)))$ at $t = 2$

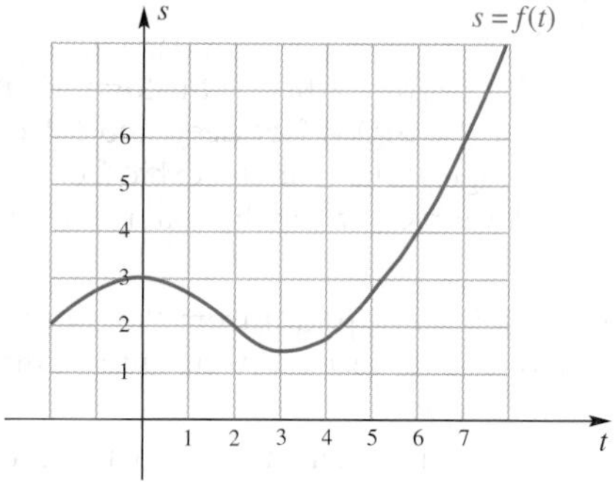

Figure 1

In Problems 5–29, find the indicated derivative by using the rules that we have developed.

5. $D_x(3x^5)$

6. $D_x(x^3 - 3x^2 + x^{-2})$

7. $D_z(z^3 + 4z^2 + 2z)$

8. $D_x\left(\frac{3x - 5}{x^2 + 1}\right)$

9. $D_t\left(\frac{4t - 5}{6t^2 + 2t}\right)$

10. $D_x^2(3x + 2)^{2/3}$

11. $\frac{d}{dx}\left(\frac{4x^2 - 2}{x^3 + x}\right)$

12. $D_t\left(t\sqrt{2t + 6}\right)$

13. $\frac{d}{dx}\left(\frac{1}{\sqrt{x^2 + 4}}\right)$

14. $\frac{d}{dx}\sqrt{\frac{x^2 - 1}{x^3 - x}}$

15. $D_\theta^2(\sin \theta + \cos^3 \theta)$

16. $\frac{d}{dt}[\sin(t^2) - \sin^2(t)]$

17. $D_\theta(\sin(\theta^2))$

18. $\frac{d}{dx}(\cos^3 5x)$

19. $\frac{d}{d\theta}[\sin^2(\sin(\pi\theta))]$

20. $\frac{d}{dt}[\sin^2(\cos 4t)]$

21. $D_\theta \tan 3\theta$

22. $\frac{d}{dx}\left(\frac{\sin 3x}{\cos 5x^2}\right)$

23. $f'(2)$ if $f(x) = (x^2 - 1)^2(3x^3 - 4x)$

24. $g''(0)$ if $g(x) = \sin 3x + \sin^2 3x$

25. $\frac{d}{dx}\left(\frac{\cot x}{\sec x^2}\right)$

26. $D_t\left(\frac{4t \sin t}{\cos t - \sin t}\right)$

27. $f'(2)$ if $f(x) = (x - 1)^3(\sin \pi x - x)^2$

28. $h''(0)$ if $h(t) = (\sin 2t + \cos 3t)^5$

29. $g'''(1)$ if $g(r) = \cos^3 5r$

In Problems 30–33, assume that all the functions given are differentiable, and find the indicated derivative.

30. $f'(t)$ if $f(t) = h(g(t)) + g^2(t)$

31. $G''(x)$ if $G(x) = F(r(x) + s(x)) + s(x)$

32. If $F(x) = Q(R(x))$, $R(x) = \cos x$, and $Q(R) = R^3$, find $F'(x)$.

33. If $F(z) = r(s(z))$, $r(x) = \sin 3x$, and $s(t) = 3t^3$, find $F'(z)$.

34. Find the coordinates of the point on the curve $y = (x - 2)^2$ where there is a tangent line that is perpendicular to the line $2x - y + 2 = 0$.

35. A spherical balloon is expanding from the sun's heat. Find the rate of change of the volume of the balloon with respect to its radius when the radius is 5 meters.

36 If the volume of the balloon of Problem 35 is increasing at a constant rate of 10 cubic meters per hour, how fast is its radius increasing when the radius is 5 meters?

37. A trough 12 feet long has a cross section in the form of an isosceles triangle (with base at the top) 4 feet deep and 6 feet across. If water is filling the trough at the rate of 9 cubic feet per minute, how fast is the water level rising when the water is 3 feet deep?

38. An object is projected directly upward from the ground with an initial velocity of 128 feet per second. Its height s at the end of t seconds is $s = 128t - 16t^2$ feet.

(a) When does it reach its maximum height and what is this height?

(b) When does it hit the ground and with what velocity?

39. An object moves on a horizontal coordinate line. Its directed distance s from the origin at the end of t seconds is $s = t^3 - 6t^2 + 9t$ feet.

(a) When is the object moving to the left?

(b) What is its acceleration when its velocity is zero?

(c) When is its acceleration positive?

40. Find $D_x^{20}y$ in each case.

(a) $y = x^{19} + x^{12} + x^5 + 10$

(b) $y = x^{20} + x^{19} + x^{18}$

(c) $y = 7x^{21} + 3x^{20}$

(d) $y = \sin x + \cos x$

(e) $y = \sin 2x$

(f) $y = \frac{1}{x}$

41. Find dy/dx in each case.

(a) $(x - 1)^2 + y^2 = 5$

(b) $xy^2 + yx^2 = 1$

(c) $x^3 + y^3 = x^3y^3$

(d) $x \sin(xy) = x^2 + 1$

(e) $x \tan(xy) = 2$

42. Show that the tangent lines to the curves $y^2 = 4x^3$ and $2x^2 + 3y^2 = 14$ at (1, 2) are perpendicular to each other. *Hint:* Use implicit differentiation.

43. Let $y = \sin(\pi x) + x^2$. If x changes from 2 to 2.01, approximately how much does y change?

44. Let $xy^2 + 2y(x + 2)^2 + 2 = 0$.

(a) If x changes from -2.00 to -2.01 and $y > 0$, approximately how much does y change?

(b) If x changes from -2.00 to -2.01 and $y < 0$, approximately how much does y change?

45. Suppose that $f(2) = 3$, $f'(2) = 4$, $f''(2) = -1$, $g(2) = 2$, and $g'(2) = 5$. Find each value.

(a) $\frac{d}{dx}[f^2(x) + g^3(x)]$ at $x = 2$

(b) $\frac{d}{dx}[f(x)g(x)]$ at $x = 2$

(c) $\frac{d}{dx}[f(g(x))]$ at $x = 2$

(d) $D_x^2[f^2(x)]$ at $x = 2$

46. A 13-foot ladder is leaning against a vertical wall. If the bottom of the ladder is being pulled away from the wall at a constant rate of 2 feet per second how fast is the top end of the ladder moving down the wall when it is 5 feet above the ground?

47. An airplane is climbing at a 15° angle to the horizontal. How fast is it gaining altitude if its speed is 400 miles per hour?

48. Given that $D_x|x| = \frac{|x|}{x}$, $x \neq 0$, find a formula for

(a) $D_x(|x|^2)$

(b) $D_x^2|x|$

(c) $D_x^3|x|$

(d) $D_x^2(|x|^2)$

49. Given that $D_t|t| = \frac{|t|}{t}$, $t \neq 0$, find a formula for

(a) $D_\theta|\sin \theta|$

(b) $D_\theta|\cos \theta|$

50. Find the linear approximation to the following functions at the given points.

(a) $\sqrt{x + 1}$ at $a = 3$

(b) $x \cos x$ at $a = 1$

REVIEW & PREVIEW PROBLEMS

In Problems 1–6, solve the given inequalities. (See Section 0.2.)

1. $(x - 2)(x - 3) < 0$

2. $x^2 - x - 6 > 0$

3. $x(x - 1)(x - 2) \leq 0$

4. $x^3 + 3x^2 + 2x \geq 0$

5. $\dfrac{x(x-2)}{x^2 - 4} \geq 0$

6. $\dfrac{x^2 - 9}{x^2 + 2} > 0$

In Problems 7–14, find the derivative $f'(x)$ of the given function.

7. $f(x) = (2x + 1)^4$

8. $f(x) = \sin \pi x$

9. $f(x) = (x^2 - 1)\cos 2x$

10. $f(x) = \dfrac{\sec x}{x}$

11. $f(x) = \tan^2 3x$

12. $f(x) = \sqrt{1 + \sin^2 x}$

13. $f(x) = \sin\sqrt{x}$

14. $f(x) = \sqrt{\sin 2x}$

15. Find all points on the graph of $y = \tan^2 x$ where the tangent line is horizontal.

16. Find all points on the graph of $y = x + \sin x$ where the tangent line is horizontal.

17. Find all points on the graph of $y = x + \sin x$ where the tangent line is parallel to the line $y = 2 + x$.

18. A rectangular box is to be made from a piece of cardboard 24 inches long and 9 inches wide by cutting out identical squares from the four corners and turning up the sides as in Figure 1. If x is the length of the side of one of the squares that is cut out, what is the volume of the resulting box?

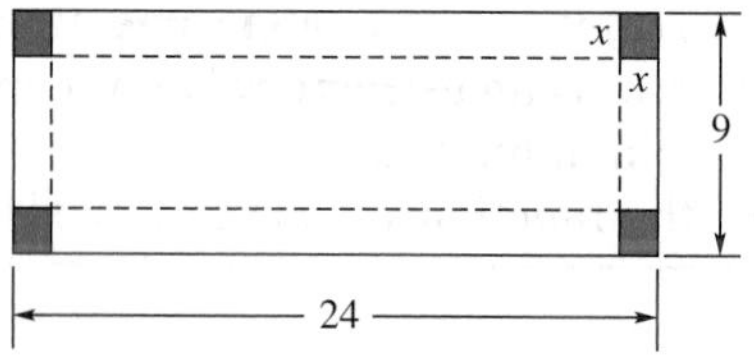

Figure 1

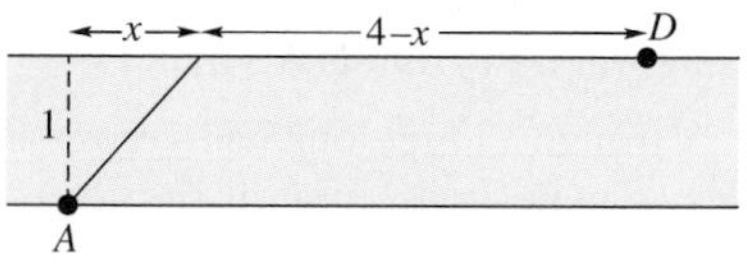

Figure 2

19. Andy wants to cross a river that is 1 kilometer wide and get to a point 4 kilometers downstream. (See Figure 2.) He can swim at 4 kilometers per hour and run 10 kilometers per hour. Assuming that he begins by swimming and that he swims toward a point x kilometers downstream from his initial starting point A, how long will it take him to reach his destination D?

20. Let $f(x) = x - \cos x$.

(a) Does the equation $x - \cos x = 0$ have a solution between $x = 0$ and $x = \pi$? How do you know?

(b) Find the equation of the tangent line at $x = \pi/2$.

(c) Where does the tangent line from part (b) intersect the x-axis?

21. Find a function whose derivative is

(a) $2x$ (b) $\sin x$ (c) $x^2 + x + 1$

22. Add 1 to each answer from Problem 21. Are these functions also solutions to Problem 21? Explain.

CHAPTER 3

Applications of the Derivative

3.1 Maxima and Minima

3.2 Monotonicity and Concavity

3.3 Local Extrema and Extrema on Open Intervals

3.4 Practical Problems

3.5 Graphing Functions Using Calculus

3.6 The Mean Value Theorem for Derivatives

3.7 Solving Equations Numerically

3.8 Antiderivatives

3.9 Introduction to Differential Equations

3.1 Maxima and Minima

Often in life, we are faced with the problem of finding the *best* way to do something. For example, a farmer wants to choose the mix of crops that is likely to produce the largest profit. A doctor wishes to select the smallest dosage of a drug that will cure a certain disease. A manufacturer would like to minimize the cost of distributing its products. Often such a problem can be formulated so that it involves maximizing or minimizing a function over a specified set. If so, the methods of calculus provide a powerful tool for solving the problem.

Suppose then that we are given a function $f(x)$ and a domain S as in Figure 1. We now pose three questions:

1. Does $f(x)$ have a maximum or minimum value on S?
2. If it does have a maximum or a minimum, where are they attained?
3. If they exist, what are the maximum and minimum values?

Answering these questions is the principal goal of this section. We begin by introducing a precise vocabulary.

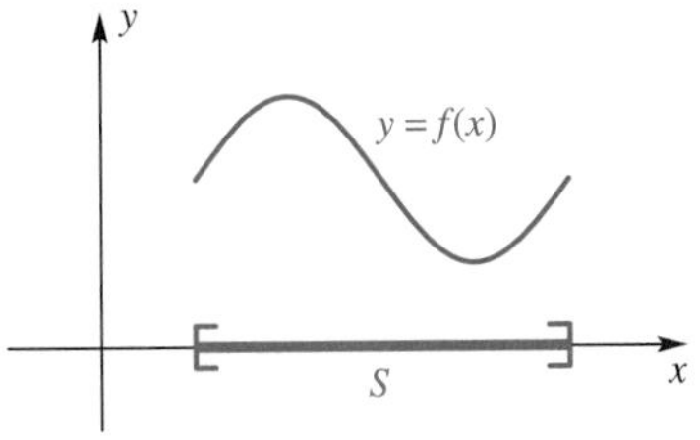

Figure 1

Definition

Let S, the domain of f, contain the point c. We say that

(i) $f(c)$ is the **maximum value** of f on S if $f(c) \geq f(x)$ for all x in S;
(ii) $f(c)$ is the **minimum value** of f on S if $f(c) \leq f(x)$ for all x in S;
(iii) $f(c)$ is an **extreme value** of f on S if it is either the maximum value or the minimum value;
(iv) the function we want to maximize or minimize is the **objective function.**

The Existence Question *Does f have a maximum (or minimum) value on S?* The answer depends first of all on the set S. Consider $f(x) = 1/x$ on $S = (0, \infty)$; it has neither a maximum value nor a minimum value (Figure 2). On the other hand, the same function on $S = [1, 3]$ has the maximum value of $f(1) = 1$ and the minimum value of $f(3) = \frac{1}{3}$. On $S = (1, 3]$, f has no maximum value and the minimum value of $f(3) = \frac{1}{3}$.

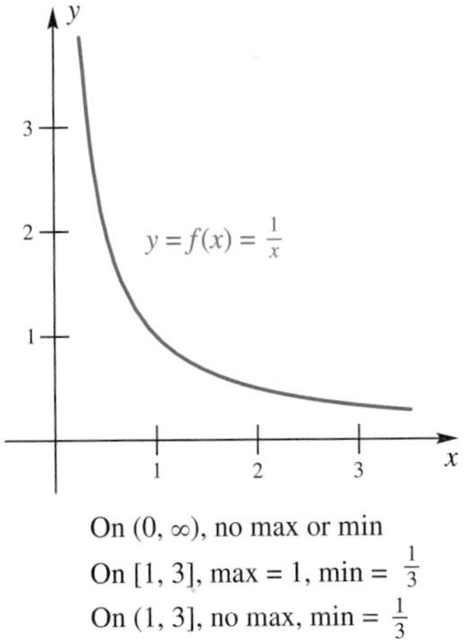

Figure 2

The answer also depends on the type of function. Consider the discontinuous function g (Figure 3) defined by

$$g(x) = \begin{cases} x & \text{if } 1 \leq x < 2 \\ x - 2 & \text{if } 2 \leq x \leq 3 \end{cases}$$

On $S = [1, 3]$, g has no maximum value (it gets arbitrarily close to 2 but never attains it). However, g has the minimum value $g(2) = 0$.

There is a nice theorem that answers the existence question for many of the problems that come up in practice. Though it is intuitively obvious, a rigorous proof is quite difficult; we leave that for more advanced textbooks.

Theorem A **Max–Min Existence Theorem**

If f is continuous on a closed interval $[a, b]$, then f attains both a maximum value and a minimum value there.

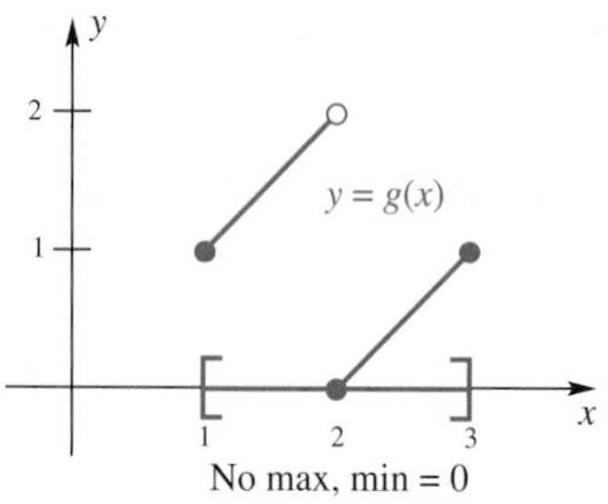

Figure 3

Note the key words in Theorem A; f is required to be *continuous* and the set S is required to be a *closed interval.*

Where Do Extreme Values Occur? Usually, the objective function will have an interval I as its domain. But this interval may be any of the nine types discussed in Section 0.2. Some of them contain their end points; some do not. For instance, $I = [a, b]$ contains both its end points; $[a, b)$ contains only its left end point; (a, b) contains neither end point. Extreme values of functions defined on closed intervals often occur at end points (see Figure 4).

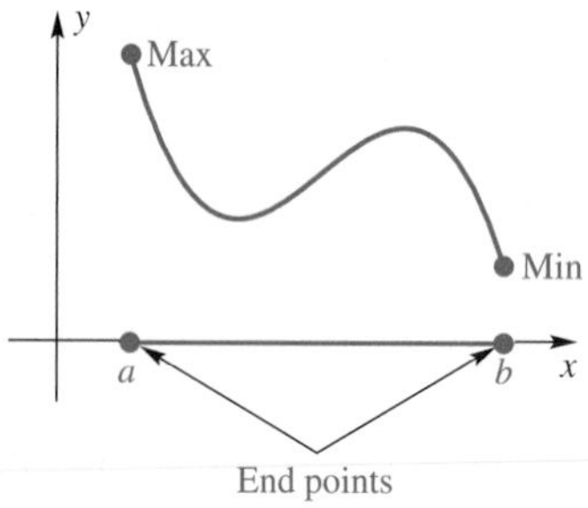

Figure 4

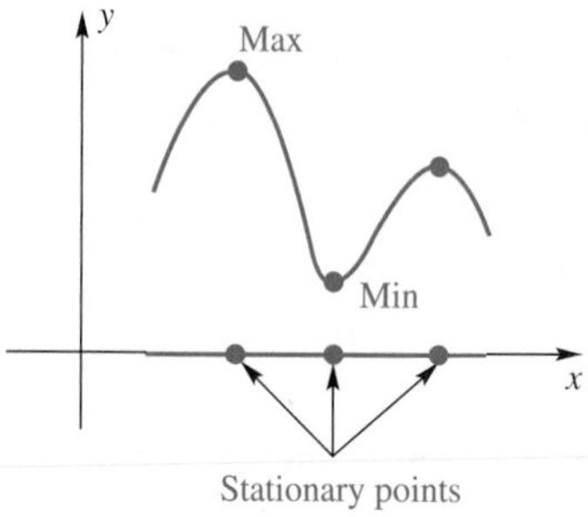

Figure 5

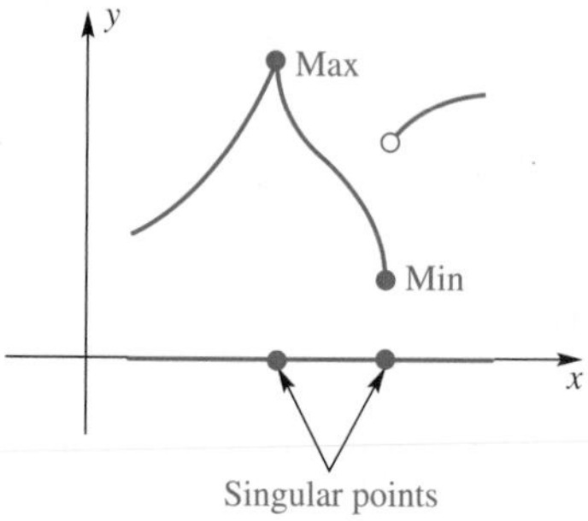

Figure 6

If c is a point at which $f'(c) = 0$, we call c a **stationary point.** The name derives from the fact that at a stationary point the graph of f levels off, since the tangent line is horizontal. Extreme values often occur at stationary points (see Figure 5).

Finally, if c is an interior point of I where f' fails to exist, we call c a **singular point.** It is a point where the graph of f has a sharp corner, a vertical tangent, or perhaps takes a jump, or near where the graph wiggles very badly. Extreme values can occur at singular points (Figure 6), though in practical problems this is quite rare.

These three kinds of points (end points, stationary points, and singular points) are the key points of max–min theory. Any point of one of these three types in the domain of a function f is called a **critical point** of f.

EXAMPLE 1 Find the critical points of $f(x) = -2x^3 + 3x^2$ on $\left[-\frac{1}{2}, 2\right]$.

SOLUTION The end points are $-\frac{1}{2}$ and 2. To find the stationary points, we solve $f'(x) = -6x^2 + 6x = 0$ for x, obtaining 0 and 1. There are no singular points. Thus, the critical points are $-\frac{1}{2}$, 0, 1, and 2. ■

Theorem B Critical Point Theorem

Let f be defined on an interval I containing the point c. If $f(c)$ is an extreme value, then c must be a critical point; that is, either c is

(i) an end point of I;
(ii) a stationary point of f; that is, a point where $f'(c) = 0$; or
(iii) a singular point of f; that is, a point where $f'(c)$ does not exist.

Proof Consider first the case where $f(c)$ is the maximum value of f on I and suppose that c is neither an end point nor a singular point. We must show that c is a stationary point.

Now, since $f(c)$ is the maximum value, $f(x) \le f(c)$ for all x in I; that is,

$$f(x) - f(c) \le 0$$

Thus, if $x < c$, so that $x - c < 0$, then

$$\frac{f(x) - f(c)}{x - c} \ge 0 \qquad (1)$$

whereas if $x > c$, then

$$(2) \qquad \frac{f(x) - f(c)}{x - c} \leq 0$$

But $f'(c)$ exists because c is not a singular point. Consequently, when we let $x \to c^-$ in (1) and $x \to c^+$ in (2), we obtain, respectively, $f'(c) \geq 0$ and $f'(c) \leq 0$. We conclude that $f'(c) = 0$, as desired.

The case where $f(c)$ is the minimum value is handled similarly. ■

In the proof just given, we used the fact that the inequality $\leq$ is preserved under the operation of taking limits.

What Are the Extreme Values? In view of Theorems A and B, we can now state a very simple procedure for finding the maximum value and minimum value of a continuous function f on a *closed interval I.*

Step 1: Find the critical points of f on I.

Step 2: Evaluate f at each of these critical points. The largest of these values is the maximum value; the smallest is the minimum value.

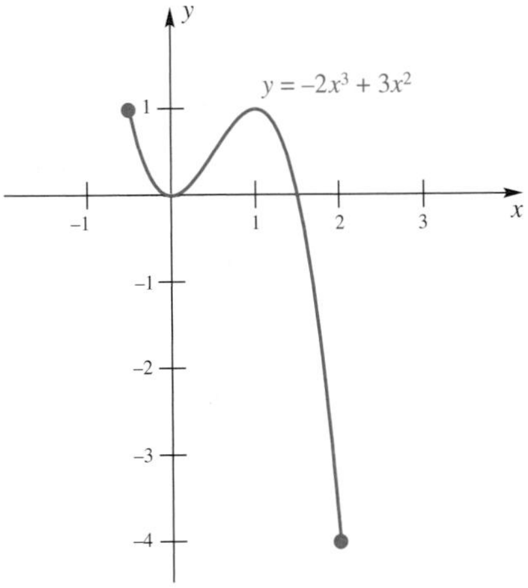

Figure 7

EXAMPLE 2 Find the maximum and minimum values of $f(x) = x^3$ on $[-2, 2]$.

SOLUTION The derivative is $f'(x) = 3x^2$, which is defined on $(-2, 2)$ and is zero only when $x = 0$. The critical points are therefore $x = 0$ and the end points $x = -2$ and $x = 2$. Evaluating f at the critical points yields $f(-2) = -8$, $f(0) = 0$, and $f(2) = 8$. Thus, the maximum value of f is 8 (attained at $x = 2$) and the minimum is -8 (attained at $x = -2$). ■

Terminology
Notice the way that terms are used in Example 3. The maximum is 1, which is equal to $f\left(-\frac{1}{2}\right)$ and $f(1)$. We say that the maximum is attained at $-\frac{1}{2}$ and at 1. Similarly, the minimum is -4, which is attained at 2.

Notice that in Example 2, $f'(0) = 0$, but f did not attain a minimum or a maximum at $x = 0$. This does not contradict Theorem B. Theorem B does not say that if c is a critical point then $f(c)$ is a minimum or maximum; it says that if $f(c)$ is a minimum or a maximum, then c is a critical point.

EXAMPLE 3 Find the maximum and minimum values of

$$f(x) = -2x^3 + 3x^2$$

on $\left[-\frac{1}{2}, 2\right]$.

SOLUTION In Example 1, we identified $-\frac{1}{2}$, 0, 1, and 2 as the critical points. Now $f\left(-\frac{1}{2}\right) = 1$, $f(0) = 0$, $f(1) = 1$, and $f(2) = -4$. Thus, the maximum value is 1 (attained at both $x = -\frac{1}{2}$ and $x = 1$), and the minimum value is -4 (attained at $x = 2$). The graph of f is shown in Figure 7. ■

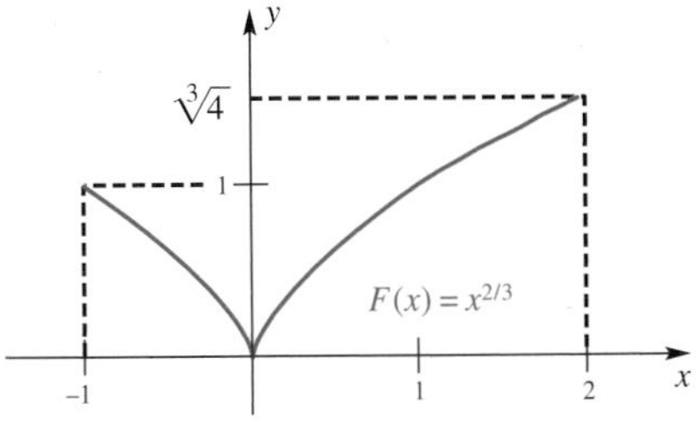

Figure 8

EXAMPLE 4 The function $F(x) = x^{2/3}$ is continuous everywhere. Find its maximum and minimum values on $[-1, 2]$.

SOLUTION $F'(x) = \frac{2}{3}x^{-1/3}$, which is never 0. However, $F'(0)$ does not exist, so 0 is a critical point, as are the end points -1 and 2. Now $F(-1) = 1$, $F(0) = 0$, and $F(2) = \sqrt[3]{4} \approx 1.59$. Thus, the maximum value is $\sqrt[3]{4}$; the minimum value is 0. The graph is shown in Figure 8. ■

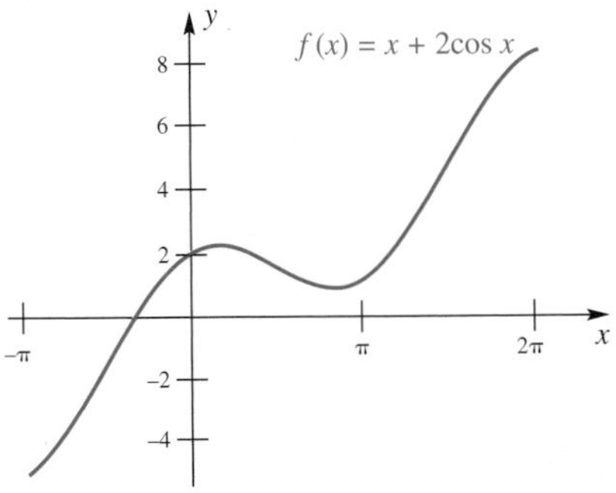

Figure 9

EXAMPLE 5 Find the maximum and minimum values of $f(x) = x + 2\cos x$ on $[-\pi, 2\pi]$.

SOLUTION Figure 9 shows a plot of $y = f(x)$. The derivative is $f'(x) = 1 - 2\sin x$, which is defined on $(-\pi, 2\pi)$ and is zero when $\sin x = 1/2$. The only values of x in the interval $[-\pi, 2\pi]$ that satisfy $\sin x = 1/2$ are $x = \pi/6$ and $x = 5\pi/6$. These two numbers, together with the end points $-\pi$ and 2π, are the critical points. Now, evaluate f at each critical point:

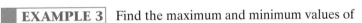

$$f(-\pi) = -2 - \pi \approx -5.14 \qquad f(\pi/6) = \sqrt{3} + \frac{\pi}{6} \approx 2.26$$

$$f(5\pi/6) = -\sqrt{3} + \frac{5\pi}{6} \approx 0.89 \qquad f(2\pi) = 2 + 2\pi \approx 8.28$$

Thus, $-2 - \pi$ is the minimum (attained at $x = -\pi$), and the maximum is $2 + 2\pi$ (attained at $x = 2\pi$). ■

Concepts Review

1. A _____ function on a _____ interval will always have both a maximum value and a minimum value on that interval.

2. The term _____ value denotes either a maximum or a minimum value.

3. A function can attain an extreme value only at a critical point. Critical points are of three types: _____, _____, and _____.

4. A stationary point for f is a number c such that _____; a singular point for f is a number c such that _____.

Problem Set 3.1

In Problems 1–4, find all critical points and find the minimum and maximum of the function. Each function has domain $[-2, 4]$.

1.

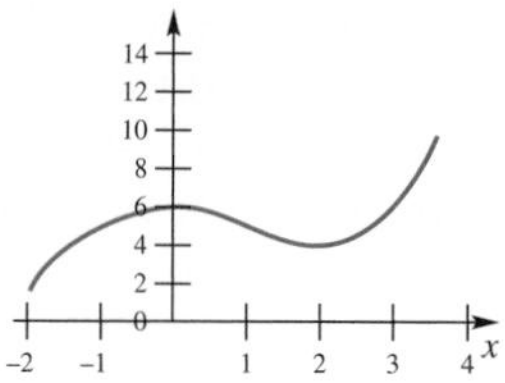

2.

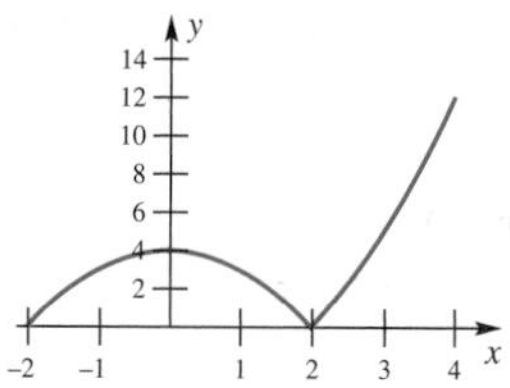

3.

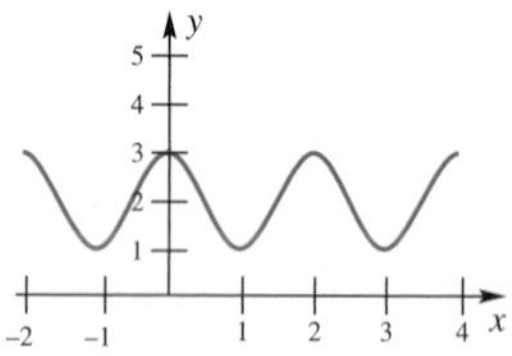

4.

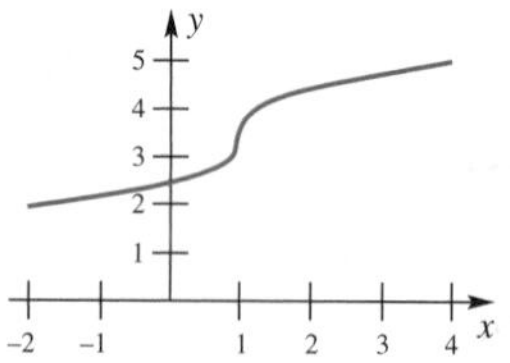

In Problems 5–26, identify the critical points and find the maximum value and minimum value on the given interval.

5. $f(x) = x^2 + 4x + 4$; $I = [-4, 0]$

6. $h(x) = x^2 + x$; $I = [-2, 2]$

7. $\Psi(x) = x^2 + 3x$; $I = [-2, 1]$

8. $G(x) = \frac{1}{5}(2x^3 + 3x^2 - 12x)$; $I = [-3, 3]$

9. $f(x) = x^3 - 3x + 1$; $I = \left(-\frac{3}{2}, 3\right)$ *Hint:* Sketch the graph.

10. $f(x) = x^3 - 3x + 1$; $I = \left[-\frac{3}{2}, 3\right]$

11. $h(r) = \frac{1}{r}$; $I = [-1, 3]$

12. $g(x) = \frac{1}{1 + x^2}$; $I = [-3, 1]$

13. $f(x) = x^4 - 2x^2 + 2$; $I = [-2, 2]$

14. $f(x) = x^5 - \frac{25}{3}x^3 + 20x - 1$; $I = [-3, 2]$

15. $g(x) = \frac{1}{1 + x^2}$; $I = (-\infty, \infty)$ *Hint:* Sketch the graph.

16. $f(x) = \frac{x}{1 + x^2}$; $I = [-1, 4]$

17. $r(\theta) = \sin\theta$; $I = \left[-\frac{\pi}{4}, \frac{\pi}{6}\right]$

18. $s(t) = \sin t - \cos t$; $I = [0, \pi]$

19. $a(x) = |x - 1|$; $I = [0, 3]$

20. $f(s) = |3s - 2|$; $I = [-1, 4]$

21. $g(x) = \sqrt[3]{x}$; $I = [-1, 27]$

22. $s(t) = t^{2/5}$; $I = [-1, 32]$

23. $H(t) = \cos t$; $I = [0, 8\pi]$

24. $g(x) = x - 2\sin x$; $I = [-2\pi, 2\pi]$

25. $g(\theta) = \theta^2 \sec\theta$; $I = \left[-\frac{\pi}{4}, \frac{\pi}{4}\right]$

26. $h(t) = \frac{t^{5/3}}{2 + t}$; $I = [-1, 8]$

GC **27.** Identify the critical points and find the extreme values on the interval $[-1, 5]$ for each function:

(a) $f(x) = x^3 - 6x^2 + x + 2$ (b) $g(x) = |f(x)|$

GC **28.** Identify the critical points and find the extreme values on the interval $[-1, 5]$ for each function:

(a) $f(x) = \cos x + x\sin x + 2$ (b) $g(x) = |f(x)|$

In Problems 29–36, sketch the graph of a function with the given properties.

29. f is differentiable, has domain $[0, 6]$, reaches a maximum of 6 (attained when $x = 3$) and a minimum of 0 (attained when $x = 0$). Additionally, $x = 5$ is a stationary point.

30. f is differentiable, has domain $[0, 6]$, reaches a maximum of 4 (attained when $x = 6$) and a minimum of -2 (attained when $x = 1$). Additionally, $x = 2, 3, 4, 5$ are stationary points.

31. f is continuous, but not necessarily differentiable, has domain $[0, 6]$, reaches a maximum of 6 (attained when $x = 5$), and a minimum of 2 (attained when $x = 3$). Additionally, $x = 1$ and $x = 5$ are the only stationary points.

32. f is continuous, but not necessarily differentiable, has domain $[0, 6]$, reaches a maximum of 4 (attained when $x = 4$), and a minimum of 2 (attained when $x = 2$). Additionally, f has no stationary points.

33. f is differentiable, has domain $[0, 6]$, reaches a maximum of 4 (attained at two different values of x, neither of which is an end point), and a minimum of 1 (attained at three different values of x, exactly one of which is an end point.)

34. f is continuous but not necessarily differentiable, has domain $[0, 6]$, reaches a maximum of 6 (attained when $x = 0$) and a minimum of 0 (attained when $x = 6$). Additionally, f has two stationary points and two singular points in $(0, 6)$.

35. f has domain $[0, 6]$, but is not necessarily continuous, and f does not attain a maximum.

36. f has domain $[0, 6]$, but is not necessarily continuous, and f attains neither a maximum nor a minimum.

Answers to Concepts Review: **1.** continuous; closed **2.** extreme **3.** end points; stationary points; singular points **4.** $f'(c) = 0$; $f'(c)$ does not exist

3.2 Monotonicity and Concavity

Consider the graph in Figure 1. No one will be surprised when we say that f is decreasing to the left of c and increasing to the right of c. But to make sure that we agree on terminology, we give precise definitions.

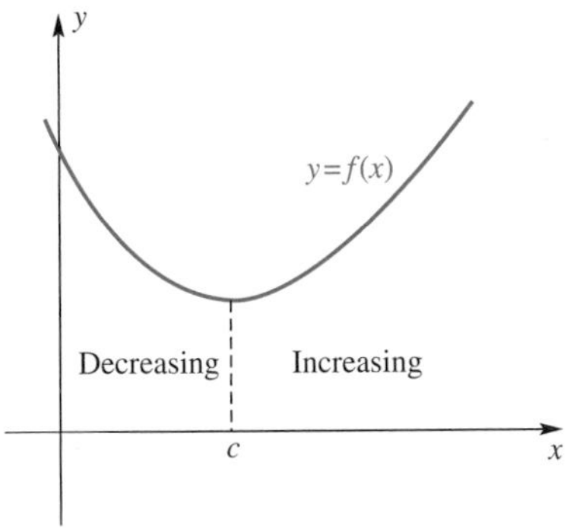

Figure 1

Definition

Let f be defined on an interval I (open, closed, or neither). We say that

(i) f is **increasing** on I if, for every pair of numbers x_1 and x_2 in I,

$$x_1 < x_2 \Rightarrow f(x_1) < f(x_2)$$

(ii) f is **decreasing** on I if, for every pair of numbers x_1 and x_2 in I,

$$x_1 < x_2 \Rightarrow f(x_1) > f(x_2)$$

(iii) f is **strictly monotonic** on I if it is either increasing on I or decreasing on I.

How shall we decide where a function is increasing? We could draw its graph and look at it, but a graph is usually drawn by plotting a few points and connecting those points with a smooth curve. Who can be sure that the graph does not wiggle between the plotted points? Even computer algebra systems and graphing calculators plot by simply connecting points. We need a better procedure.

The First Derivative and Monotonicity Recall that the first derivative $f'(x)$ gives us the slope of the tangent line to the graph of f at the point x. Thus, if $f'(x) > 0$, then the tangent line is rising to the right, suggesting that f is increasing. (See Figure 2.) Similarly, if $f'(x) < 0$, then the tangent line is falling to the right, suggesting that f is decreasing. We can also look at this in terms of motion along a line. Suppose an object is at position $s(t)$ at time t and that its velocity is always positive, that is, $s'(t) = ds/dt > 0$. Then it seems reasonable that the object will continue to move to the right as long as the derivative stays positive. In other words, $s(t)$ will be an *increasing* function of t. These observations do not prove Theorem A, but they make the result plausible. We postpone a rigorous proof until Section 3.6.

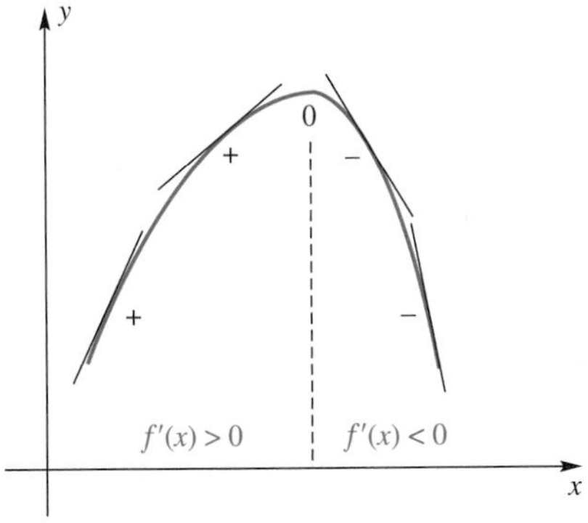

Figure 2

Theorem A Monotonicity Theorem

Let f be continuous on an interval I and differentiable at every interior point of I.

(i) If $f'(x) > 0$ for all x interior to I, then f is increasing on I.
(ii) If $f'(x) < 0$ for all x interior to I, then f is decreasing on I.

This theorem usually allows us to determine precisely where a differentiable function increases and where it decreases. It is a matter of solving two inequalities.

EXAMPLE 1 If $f(x) = 2x^3 - 3x^2 - 12x + 7$, find where f is increasing and where it is decreasing.

SOLUTION We begin by finding the derivative of f.

$$f'(x) = 6x^2 - 6x - 12 = 6(x + 1)(x - 2)$$

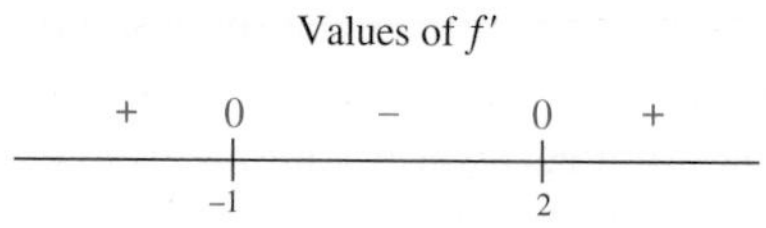

Figure 3

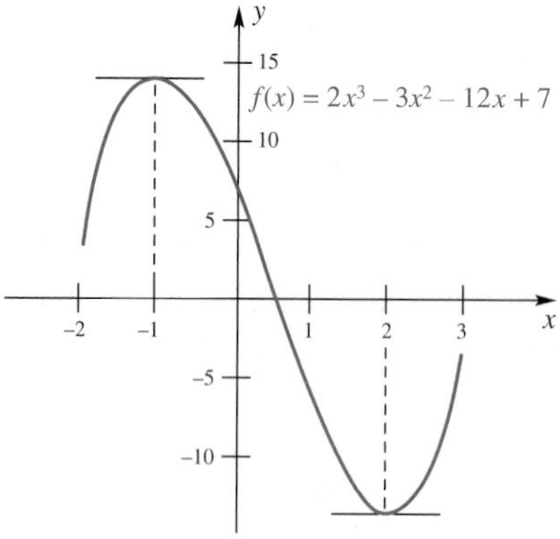

Figure 4

We need to determine where

$$(x + 1)(x - 2) > 0$$

and also where

$$(x + 1)(x - 2) < 0$$

This problem was discussed in great detail in Section 0.2, a section worth reviewing now. The split points are -1 and 2; they split the x-axis into three intervals: $(-\infty, -1)$, $(-1, 2)$, and $(2, \infty)$. Using the test points -2, 0, and 3, we conclude that $f'(x) > 0$ on the first and last of these intervals and that $f'(x) < 0$ on the middle interval (Figure 3). Thus, by Theorem A, f is increasing on $(-\infty, -1]$ and $[2, \infty)$; it is decreasing on $[-1, 2]$. Note that the theorem allows us to include the end points of these intervals, even though $f'(x) = 0$ at those points. The graph of f is shown in Figure 4. ■

EXAMPLE 2 Determine where $g(x) = x/(1 + x^2)$ is increasing and where it is decreasing.

SOLUTION

$$g'(x) = \frac{(1 + x^2) - x(2x)}{(1 + x^2)^2} = \frac{1 - x^2}{(1 + x^2)^2} = \frac{(1 - x)(1 + x)}{(1 + x^2)^2}$$

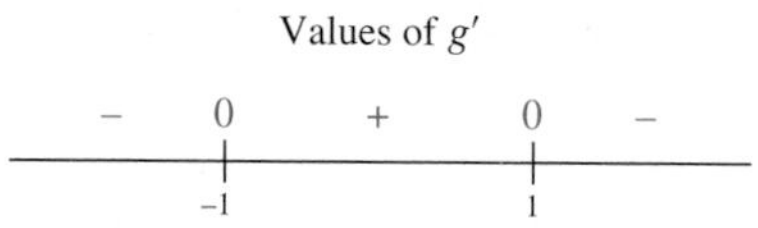

Figure 5

Since the denominator is always positive, $g'(x)$ has the same sign as the numerator $(1 - x)(1 + x)$. The split points, -1 and 1, determine the three intervals $(-\infty, -1)$, $(-1, 1)$, and $(1, \infty)$. When we test them, we find that $g'(x) < 0$ on the first and last of these intervals and that $g'(x) > 0$ on the middle one (Figure 5). We conclude from Theorem A that g is decreasing on $(-\infty, -1]$ and $[1, \infty)$ and that it is increasing on $[-1, 1]$. We postpone graphing g until later, but if you want to see the graph, turn to Figure 11 and Example 4. ■

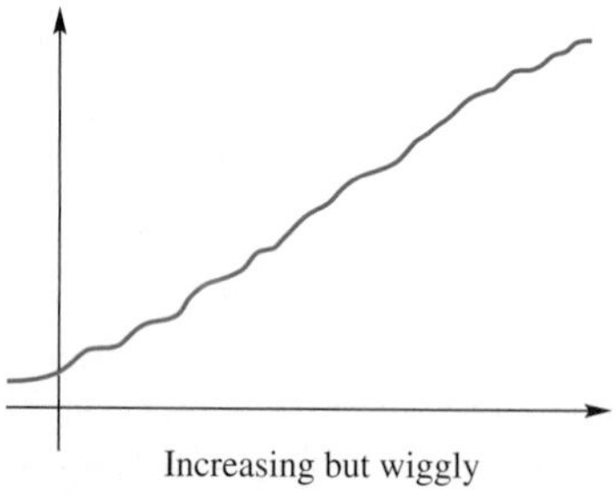

Figure 6

The Second Derivative and Concavity A function may be increasing and still have a very wiggly graph (Figure 6). To analyze wiggles, we need to study how the tangent line turns as we move from left to right along the graph. If the tangent line turns steadily in the counterclockwise direction, we say that the graph is *concave up*; if the tangent turns in the clockwise direction, the graph is *concave down*. Both definitions are better stated in terms of functions and their derivatives.

> **Definition**
>
> Let f be differentiable on an open interval I. We say that f (as well as its graph) is **concave up** on I if f' is increasing on I, and we say that f is **concave down** on I if f' is decreasing on I.

The diagrams in Figure 7 will help to clarify these notions. Note that a curve that is concave *up* is shaped like a *cup*.

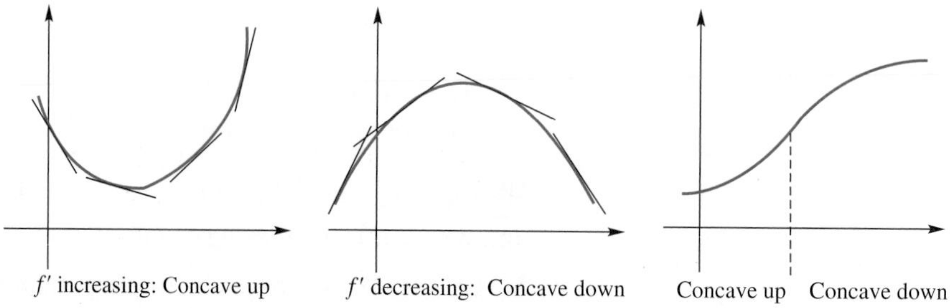

Figure 7

> **Conditions in Theorems A and B**
>
> The conditions regarding the derivatives in Theorems A and B are sufficient to guarantee the conclusions stated. These conditions are not, however, necessary. It is possible that a function is increasing on some interval even though the derivative isn't always positive on that interval. If we consider the function $f(x) = x^3$ over the interval $[-4, 4]$ we note that it is increasing but its derivative is not always positive on that interval ($f'(0) = 0$). The function $g(x) = x^4$ is concave up on the interval $[-4, 4]$, but the second derivative, $g''(x) = 12x^2$, is not always positive on that interval.

In view of Theorem A, we have a simple criterion for deciding where a curve is concave up and where it is concave down. We simply keep in mind that the second derivative of f is the first derivative of f'. Thus, f' is increasing if f'' is positive; it is decreasing if f'' is negative.

> **Theorem B** **Concavity Theorem**
>
> Let f be twice differentiable on the open interval I.
>
> (i) If $f''(x) > 0$ for all x in I, then f is concave up on I.
>
> (ii) If $f''(x) < 0$ for all x in I, then f is concave down on I.

For most functions, this theorem reduces the problem of determining concavity to the problem of solving inequalities. By now we are experts at this.

EXAMPLE 3 Where is $f(x) = \frac{1}{3}x^3 - x^2 - 3x + 4$ increasing, decreasing, concave up, and concave down?

SOLUTION

$$f'(x) = x^2 - 2x - 3 = (x + 1)(x - 3)$$

$$f''(x) = 2x - 2 = 2(x - 1)$$

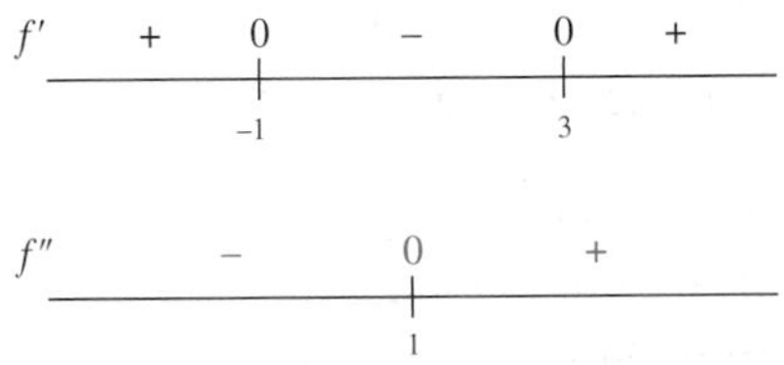

Figure 8

By solving the inequalities $(x + 1)(x - 3) > 0$ and its opposite, $(x + 1)(x - 3) < 0$, we conclude that f is increasing on $(-\infty, -1]$ and $[3, \infty)$ and decreasing on $[-1, 3]$ (Figure 8). Similarly, solving $2(x - 1) > 0$ and $2(x - 1) < 0$ shows that f is concave up on $(1, \infty)$ and concave down on $(-\infty, 1)$. The graph of f is shown in Figure 9. ■

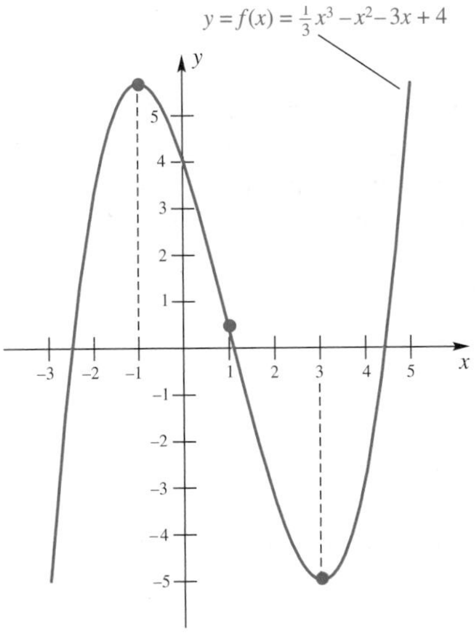

Figure 9

EXAMPLE 4 Where is $g(x) = x/(1 + x^2)$ concave up and where is it concave down? Sketch the graph of g.

SOLUTION We began our study of this function in Example 2. There we learned that g is decreasing on $(-\infty, -1]$ and $[1, \infty)$ and increasing on $[-1, 1]$. To analyze concavity, we calculate g''.

$$g'(x) = \frac{1 - x^2}{(1 + x^2)^2}$$

$$\begin{aligned} g''(x) &= \frac{(1 + x^2)^2(-2x) - (1 - x^2)(2)(1 + x^2)(2x)}{(1 + x^2)^4} \\ &= \frac{(1 + x^2)[(1 + x^2)(-2x) - (1 - x^2)(4x)]}{(1 + x^2)^4} \\ &= \frac{2x^3 - 6x}{(1 + x^2)^3} \\ &= \frac{2x(x^2 - 3)}{(1 + x^2)^3} \end{aligned}$$

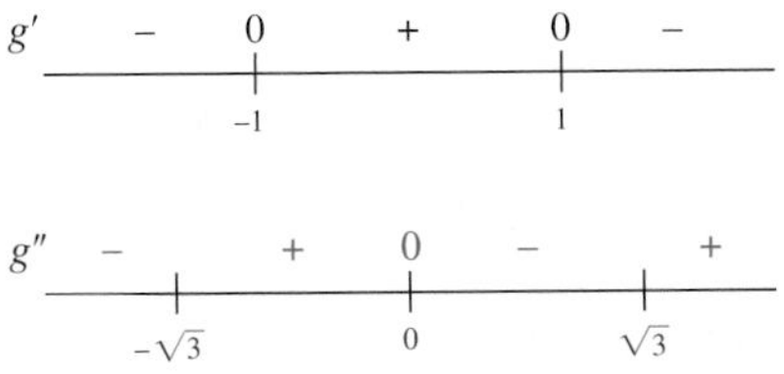

Figure 10

Since the denominator is always positive, we need only solve $x(x^2 - 3) > 0$ and its opposite. The split points are $-\sqrt{3}$, 0, and $\sqrt{3}$. These three split points determine four intervals. After testing them (Figure 10), we conclude that g is concave up on $\left(-\sqrt{3}, 0\right)$ and $\left(\sqrt{3}, \infty\right)$ and that it is concave down on $\left(-\infty, -\sqrt{3}\right)$ and $\left(0, \sqrt{3}\right)$.

To sketch the graph of g, we make use of all the information obtained so far, plus the fact that g is an odd function whose graph is symmetric with respect to the origin (Figure 11).

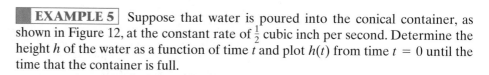

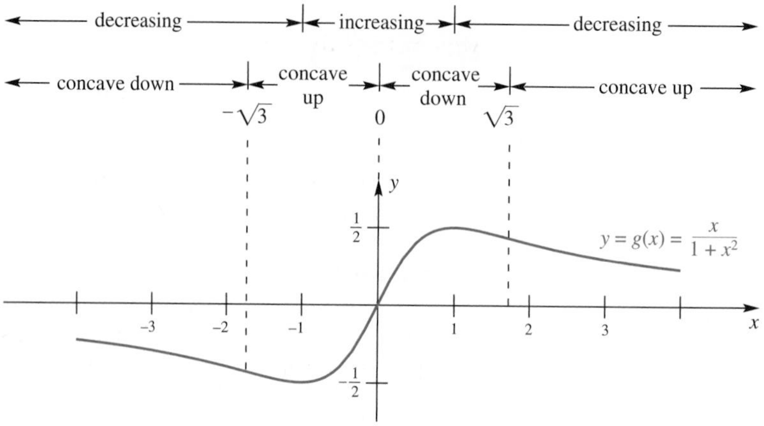

Figure 11

■

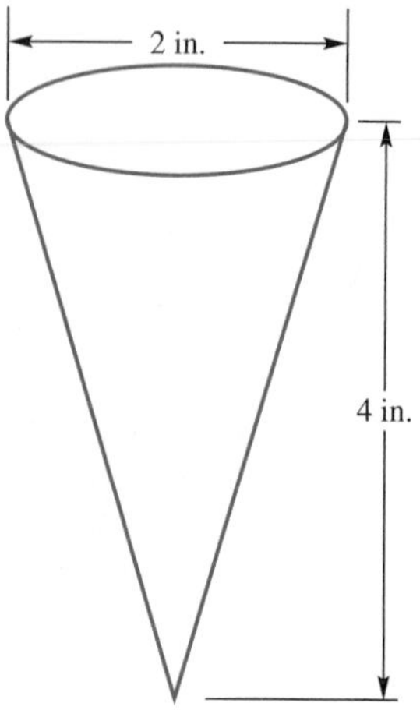

Figure 12

EXAMPLE 5 Suppose that water is poured into the conical container, as shown in Figure 12, at the constant rate of $\frac{1}{2}$ cubic inch per second. Determine the height h of the water as a function of time t and plot $h(t)$ from time $t = 0$ until the time that the container is full.

SOLUTION ≈ Before we solve this problem, let's think about what the graph will look like. At first, the height will increase rapidly, since it takes very little water to fill the bottom of the cone. As the container fills up, the height will increase less rapidly. What do these statements say about the function $h(t)$, its derivative $h'(t)$, and its second derivative $h''(t)$? Since the water is steadily pouring in, the height will always increase, so $h'(t)$ will be positive. The height will increase more slowly as the water level rises. Thus, the function $h'(t)$ is decreasing so $h''(t)$ is negative. The graph of $h(t)$ is therefore increasing (because $h'(t)$ is positive) and concave down (because $h''(t)$ is negative).

Now, once we have an intuitive idea about what the graph should look like (increasing and concave down), let's solve the problem analytically. The volume of a right circular cone is $V = \frac{1}{3}\pi r^2 h$, where V, r, and h are all functions of time. The functions h and r are related; notice the similar triangles in Figure 13. Using properties of similar triangles, we have

$$\frac{r}{h} = \frac{1}{4}$$

Thus, $r = h/4$. The volume of the water inside the cone is thus

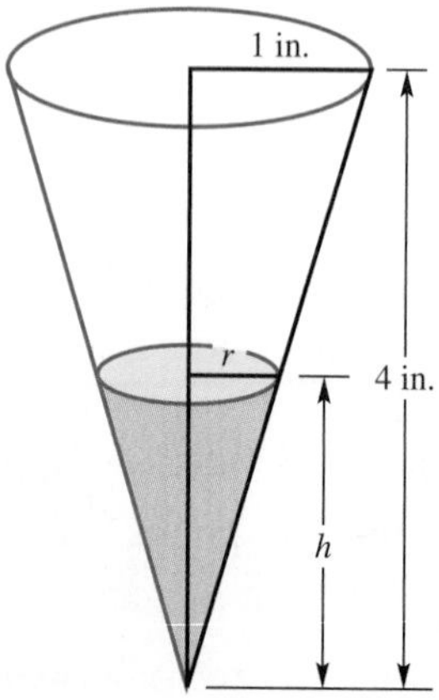

Figure 13

$$V = \frac{1}{3}\pi r^2 h = \frac{\pi}{3}\left(\frac{h}{4}\right)^2 h = \frac{\pi}{48}h^3$$

On the other hand, since water is flowing into the container at the rate of $\frac{1}{2}$ cubic inch per second, the volume at time t is $V = \frac{1}{2}t$, where t is measured in seconds. Equating these two expressions for V gives

$$\frac{1}{2}t = \frac{\pi}{48}h^3$$

When $h = 4$, we have $t = \frac{2\pi}{48}4^3 = \frac{8}{3}\pi \approx 8.4$; thus, it takes about 8.4 seconds to fill the container. Now solve for h in the above equation relating h and t to obtain

$$h(t) = \sqrt[3]{\frac{24}{\pi}t}$$

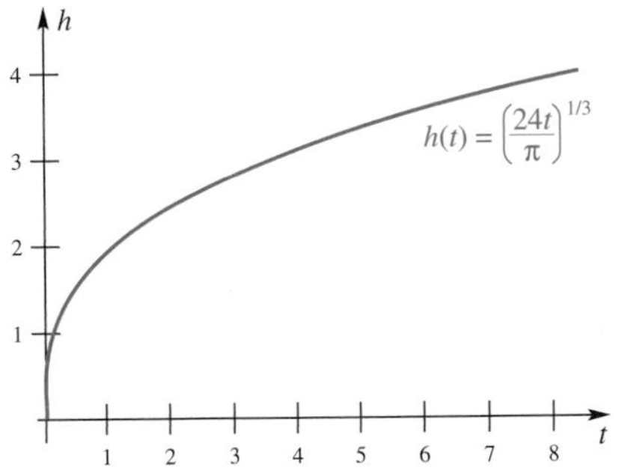

Figure 14

The first and second derivatives of h are

$$h'(t) = D_t \sqrt[3]{\frac{24}{\pi}t} = \frac{8}{\pi}\left(\frac{24}{\pi}t\right)^{-2/3} = \frac{2}{\sqrt[3]{9\pi t^2}}$$

which is positive, and

$$h''(t) = D_t \frac{2}{\sqrt[3]{9\pi t^2}} = -\frac{4}{3\sqrt[3]{9\pi t^5}}$$

which is negative. The graph of $h(t)$ is shown in Figure 14. As expected, the graph of h is increasing and concave down. ■

EXAMPLE 6 A news agency reported in May 2004 that unemployment in eastern Asia was continuing to increase at an increasing rate. On the other hand, the price of food was increasing, but at a slower rate than before. Interpret these statements in terms of increasing/decreasing functions and concavity.

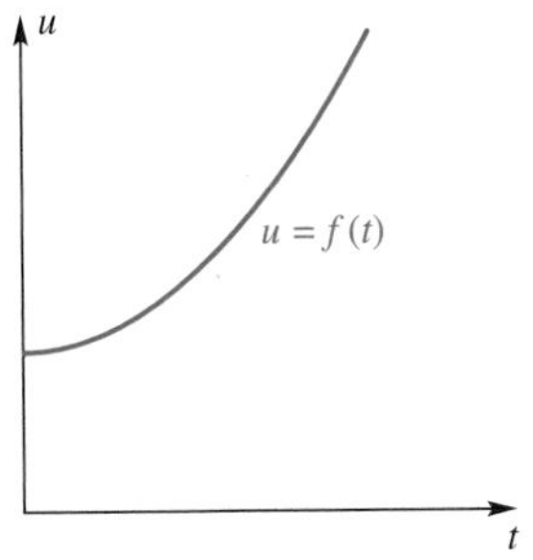

Figure 15

SOLUTION Let $u = f(t)$ denote the number of people unemployed at time t. Although u actually jumps by unit amounts, we will follow standard practice in representing u by a smooth curve as in Figure 15. To say unemployment is increasing is to say that $du/dt > 0$. To say that it is increasing at an increasing rate is to say that the function du/dt is *increasing*; but this means that the derivative of du/dt must be positive. Thus, $d^2u/dt^2 > 0$. In Figure 15, notice that the slope of the tangent line increases as t increases. Unemployment is increasing and concave up.

Similarly, if $p = g(t)$ represents the price of food (e.g., the typical cost of one day's groceries for one person) at time t, then dp/dt is positive but *decreasing*. Thus, the derivative of dp/dt is negative, so $d^2p/dt^2 < 0$. In Figure 16, notice that the slope of the tangent line decreases as t increases. The price of food is increasing but concave down. ■

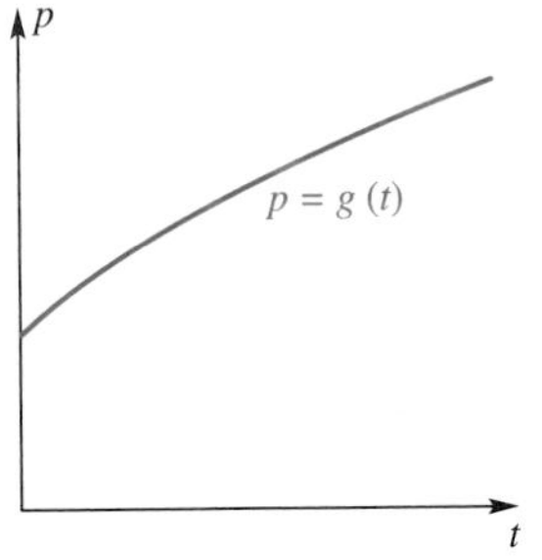

Figure 16

Inflection Points Let f be continuous at c. We call $(c, f(c))$ an **inflection point** of the graph of f if f is concave up on one side of c and concave down on the other side. The graph in Figure 17 indicates a number of possibilities.

> **Terminology**
>
> While a function's minimum or maximum is a *number*, an inflection point is always an *ordered pair* $(c, f(c))$.

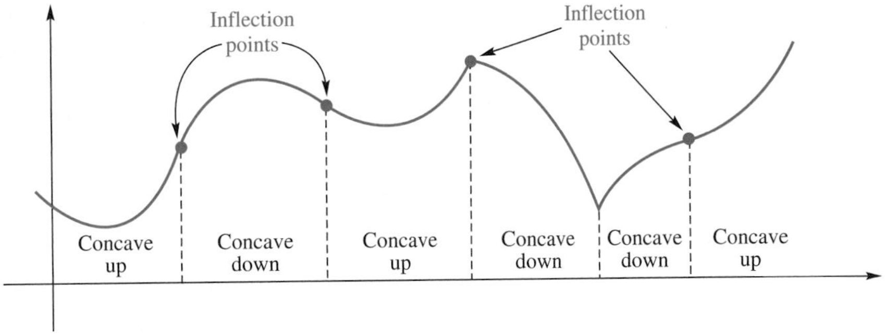

Figure 17

As you might guess, *points where $f''(x) = 0$ or where $f''(x)$ does not exist are the candidates for points of inflection.* We use the word *candidate* deliberately. Just as a candidate for political office may fail to be elected, so, for example, may a point where $f''(x) = 0$ fail to be a point of inflection. Consider $f(x) = x^4$, which has the graph shown in Figure 18. It is true that $f''(0) = 0$; yet the origin is not a point of inflection. Therefore, in searching for inflection points, we begin by identifying those points where $f''(x) = 0$ (and where $f''(x)$ does not exist). Then we check to see if they really are inflection points.

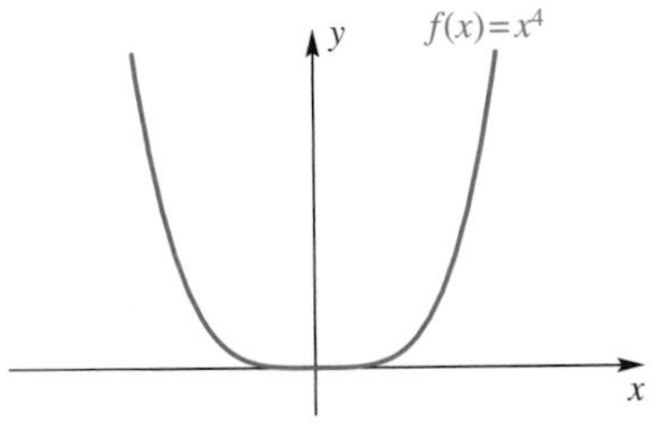

Figure 18

Look back at the graph in Example 4. You will see that it has three inflection points. They are $\left(-\sqrt{3}, -\sqrt{3}/4\right)$, $(0, 0)$, and $\left(\sqrt{3}, \sqrt{3}/4\right)$.

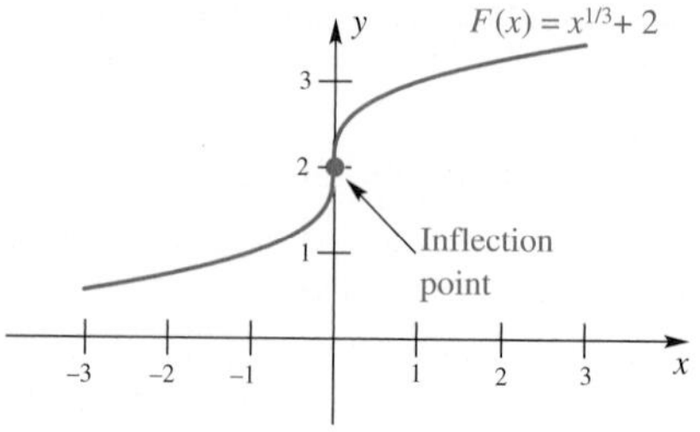

Figure 19

EXAMPLE 7 Find all points of inflection of $F(x) = x^{1/3} + 2$.

SOLUTION

$$F'(x) = \frac{1}{3x^{2/3}}, \qquad F''(x) = \frac{-2}{9x^{5/3}}$$

The second derivative, $F''(x)$, is never 0; however, it fails to exist at $x = 0$. The point (0, 2) is an inflection point since $F''(x) > 0$ for $x < 0$ and $F''(x) < 0$ for $x > 0$. The graph is sketched in Figure 19. ■

Concepts Review

1. If $f'(x) > 0$ everywhere, then f is _____ everywhere; if $f''(x) > 0$ everywhere, then f is _____ everywhere.

2. If _____ and _____ on an open interval I, then f is both increasing and concave down on I.

3. A point on the graph of a continuous function where the concavity changes is called _____.

4. In trying to locate the inflection points for the graph of a function f, we should look at numbers c, where either _____ or _____.

Problem Set 3.2

In Problems 1–10, use the Monotonicity Theorem to find where the given function is increasing and where it is decreasing.

1. $f(x) = 3x + 3$
2. $g(x) = (x + 1)(x - 2)$
3. $h(t) = t^2 + 2t - 3$
4. $f(x) = x^3 - 1$
5. $G(x) = 2x^3 - 9x^2 + 12x$
6. $f(t) = t^3 + 3t^2 - 12$
7. $h(z) = \frac{z^4}{4} - \frac{4z^3}{6}$
8. $f(x) = \frac{x - 1}{x^2}$
9. $H(t) = \sin t, 0 \le t \le 2\pi$
10. $R(\theta) = \cos^2 \theta, 0 \le \theta \le 2\pi$

In Problems 11–18, use the Concavity Theorem to determine where the given function is concave up and where it is concave down. Also find all inflection points.

11. $f(x) = (x - 1)^2$
12. $G(w) = w^2 - 1$
13. $T(t) = 3t^3 - 18t$
14. $f(z) = z^2 - \frac{1}{z^2}$
15. $q(x) = x^4 - 6x^3 - 24x^2 + 3x + 1$
16. $f(x) = x^4 + 8x^3 - 2$
17. $F(x) = 2x^2 + \cos^2 x$
18. $G(x) = 24x^2 + 12 \sin^2 x$

In Problems 19–28, determine where the graph of the given function is increasing, decreasing, concave up, and concave down. Then sketch the graph (see Example 4).

19. $f(x) = x^3 - 12x + 1$
20. $g(x) = 4x^3 - 3x^2 - 6x + 12$
21. $g(x) = 3x^4 - 4x^3 + 2$
22. $F(x) = x^6 - 3x^4$
23. $G(x) = 3x^5 - 5x^3 + 1$
24. $H(x) = \frac{x^2}{x^2 + 1}$
25. $f(x) = \sqrt{\sin x}$ on $[0, \pi]$
26. $g(x) = x\sqrt{x - 2}$
27. $f(x) = x^{2/3}(1 - x)$
28. $g(x) = 8x^{1/3} + x^{4/3}$

In Problems 29–34, sketch the graph of a continuous function f on [0, 6] that satisfies all the stated conditions.

29. $f(0) = 1; f(6) = 3$; increasing and concave down on $(0, 6)$

30. $f(0) = 8; f(6) = -2$; decreasing on $(0, 6)$; inflection point at the ordered pair $(2, 3)$, concave up on $(2, 6)$

31. $f(0) = 3; f(3) = 0; f(6) = 4$;
$f'(x) < 0$ on $(0, 3)$; $f'(x) > 0$ on $(3, 6)$;
$f''(x) > 0$ on $(0, 5)$; $f''(x) < 0$ on $(5, 6)$

32. $f(0) = 3; f(2) = 2; f(6) = 0$;
$f'(x) < 0$ on $(0, 2) \cup (2, 6)$; $f'(2) = 0$;
$f''(x) < 0$ on $(0, 1) \cup (2, 6)$; $f''(x) > 0$ on $(1, 2)$

33. $f(0) = f(4) = 1; f(2) = 2; f(6) = 0$;
$f'(x) > 0$ on $(0, 2)$; $f'(x) < 0$ on $(2, 4) \cup (4, 6)$;
$f'(2) = f'(4) = 0$; $f''(x) > 0$ on $(0, 1) \cup (3, 4)$;
$f''(x) < 0$ on $(1, 3) \cup (4, 6)$

34. $f(0) = f(3) = 3; f(2) = 4; f(4) = 2; f(6) = 0$;
$f'(x) > 0$ on $(0, 2)$; $f'(x) < 0$ on $(2, 4) \cup (4, 5)$;
$f'(2) = f'(4) = 0$; $f'(x) = -1$ on $(5, 6)$;
$f''(x) < 0$ on $(0, 3) \cup (4, 5)$; $f''(x) > 0$ on $(3, 4)$

35. Prove that a quadratic function has no point of inflection.

36. Prove that a cubic function has exactly one point of inflection.

37. Prove that, if $f'(x)$ exists and is continuous on an interval I and if $f'(x) \neq 0$ at all interior points of I, then either f is

increasing throughout I or decreasing throughout I. *Hint:* Use the Intermediate Value Theorem to show that there cannot be two points x_1 and x_2 of I where f' has opposite signs.

38. Suppose that f is a function whose derivative is $f'(x) = (x^2 - x + 1)/(x^2 + 1)$. Use Problem 37 to prove that f is increasing everywhere.

39. Use the Monotonicity Theorem to prove each statement if $0 < x < y$.

(a) $x^2 < y^2$ (b) $\sqrt{x} < \sqrt{y}$ (c) $\dfrac{1}{x} > \dfrac{1}{y}$

40. What conditions on a, b, and c will make $f(x) = ax^3 + bx^2 + cx + d$ always increasing?

41. Determine a and b so that $f(x) = a\sqrt{x} + b/\sqrt{x}$ has the point (4, 13) as an inflection point.

42. Suppose that the cubic function $f(x)$ has three real zeros, r_1, r_2, and r_3. Show that its inflection point has x-coordinate $(r_1 + r_2 + r_3)/3$. *Hint:* $f(x) = a(x - r_1)(x - r_2)(x - r_3)$.

43. Suppose that $f'(x) > 0$ and $g'(x) > 0$ for all x. What simple additional conditions (if any) are needed to guarantee that:

(a) $f(x) + g(x)$ is increasing for all x;

(b) $f(x) \cdot g(x)$ is increasing for all x;

(c) $f(g(x))$ is increasing for all x?

44. Suppose that $f''(x) > 0$ and $g''(x) > 0$ for all x. What simple additional conditions (if any) are needed to guarantee that

(a) $f(x) + g(x)$ is concave up for all x;

(b) $f(x) \cdot g(x)$ is concave up for all x;

(c) $f(g(x))$ is concave up for all x?

GC *Use a graphing calculator or a computer to do Problems 45–48.*

45. Let $f(x) = \sin x + \cos(x/2)$ on the interval $I = (-2, 7)$.

(a) Draw the graph of f on I.

(b) Use this graph to estimate where $f'(x) < 0$ on I.

(c) Use this graph to estimate where $f''(x) < 0$ on I.

(d) Plot the graph of f' to confirm your answer to part (b).

(e) Plot the graph of f'' to confirm your answer to part (c).

46. Repeat Problem 45 for $f(x) = x\cos^2(x/3)$ on (0, 10).

47. Let $f'(x) = x^3 - 5x^2 + 2$ on $I = [-2, 4]$. Where on I is f increasing?

48. Let $f''(x) = x^4 - 5x^3 + 4x^2 + 4$ on $I = [-2, 3]$. Where on I is f concave down?

49. Translate each of the following into the language of derivatives of distance with respect to time. For each part, sketch a plot of the car's position s against time t, and indicate the concavity.

(a) The speed of the car is proportional to the distance it has traveled.

(b) The car is speeding up.

(c) I didn't say the car was slowing down; I said its rate of increase in speed was slowing down.

(d) The car's speed is increasing 10 miles per hour every minute.

(e) The car is slowing very gently to a stop.

(f) The car always travels the same distance in equal time intervals.

50. Translate each of the following into the language of derivatives, sketch a plot of the appropriate function and indicate the concavity.

(a) Water is evaporating from the tank at a constant rate.

(b) Water is being poured into the tank at 3 gallons per minute but is also leaking out at $\frac{1}{2}$ gallon per minute.

(c) Since water is being poured into the conical tank at a constant rate, the water level is rising at a slower and slower rate.

(d) Inflation held steady this year but is expected to rise more and more rapidly in the years ahead.

(e) At present the price of oil is dropping, but this trend is expected to slow and then reverse direction in 2 years.

(f) David's temperature is still rising, but the penicillin seems to be taking effect.

51. Translate each of the following statements into mathematical language, sketch a plot of the appropriate function, and indicate the concavity.

(a) The cost of a car continues to increase and at a faster and faster rate.

(b) During the last 2 years, the United States has continued to cut its consumption of oil, but at a slower and slower rate.

(c) World population continues to grow, but at a slower and slower rate.

(d) The angle that the Leaning Tower of Pisa makes with the vertical is increasing more and more rapidly.

(e) Upper Midwest firm's profit growth slows.

(f) The XYZ Company has been losing money, but will soon turn this situation around.

52. Translate each statement from the following newspaper column into a statement about derivatives.

(a) In the United States, the ratio R of government debt to national income remained unchanged at around 28% up to 1981, but

(b) then it began to increase more and more sharply, reaching 36% during 1983.

53. Coffee is poured into the cup shown in Figure 20 at the rate of 2 cubic inches per second. The top diameter is 3.5 inches, the bottom diameter is 3 inches, and the height of the cup is 5 inches. This cup holds about 23 fluid ounces. Determine the height h of the coffee as a function of time t, and sketch the graph of $h(t)$ from time $t = 0$ until the time that the cup is full.

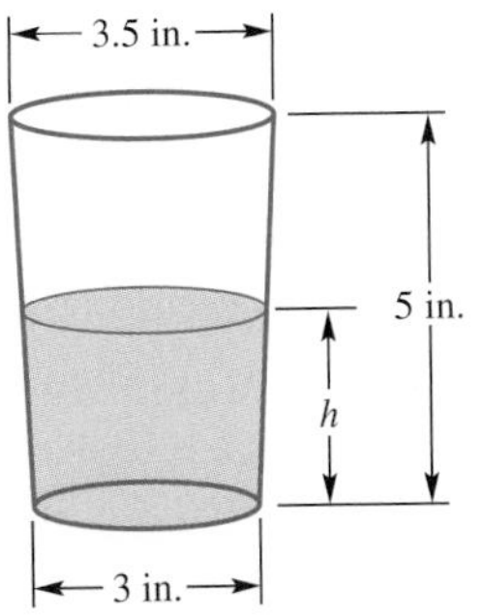

Figure 20

54. Water is being pumped into a cylindrical tank at a constant rate of 5 gallons per minute, as shown in Figure 21. The tank has diameter 3 feet and length 9.5 feet. The volume of the tank is $\pi r^2 l = \pi \times 1.5^2 \times 9.5 \approx 67.152$ cubic feet ≈ 500 gallons. Without doing any calculations, sketch a graph of the height h of the water as a function of time t (see Example 6). Where is h concave up? Concave down?

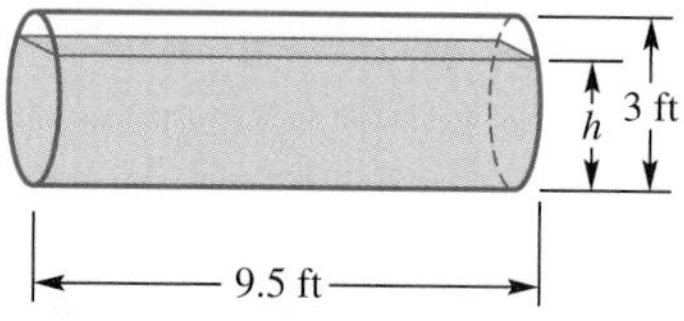

Figure 21

55. A liquid is poured into the container shown in Figure 22 at the rate of 3 cubic inches per second. The container holds about 24 cubic inches. Sketch a graph of the height h of the liquid as a function of time t. In your graph, pay special attention to the concavity of h.

56. A 20-gallon barrel, as shown in Figure 23, leaks at the constant rate of 0.1 gallon per day. Sketch a plot of the height h of the water as a function of time t, assuming that the barrel is full at time $t = 0$. In your graph, pay special attention to the concavity of h.

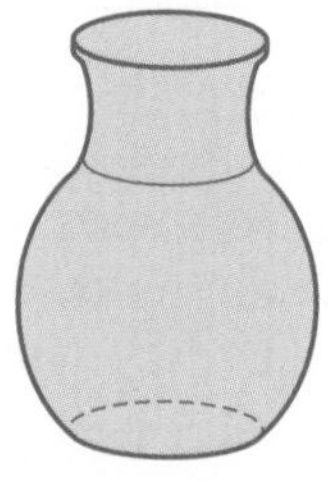

Figure 22

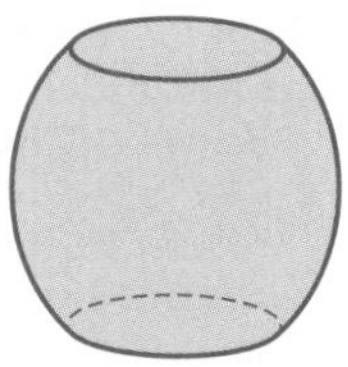

Figure 23

57. What are you able to deduce about the shape of a vase based on each of the following tables, which give measurements of the volume of the water as a function of the depth.

(a)

Depth	1	2	3	4	5	6
Volume	4	8	11	14	20	28

(b)

Depth	1	2	3	4	5	6
Volume	4	9	12	14	20	28

Answers to Concepts Review: 1. increasing; concave up **2.** $f'(x) > 0; f''(x) < 0$ **3.** an inflection point **4.** $f''(c) = 0; f''(c)$ does not exist

3.3 Local Extrema and Extrema on Open Intervals

We recall from Section 3.1 that the maximum value (if it exists) of a function f on a set S is the largest value that f attains on the whole set S. It is sometimes referred to as the **global maximum value,** or the *absolute maximum value* of f. Thus, for the function f with domain $S = [a, b]$ whose graph is sketched in Figure 1, $f(a)$ is the global maximum value. But what about $f(c)$? It may not be king of the country, but at least it is chief of its own locality. We call it a **local maximum value,** or a *relative maximum value*. Of course, a global maximum value is automatically a local maximum value. Figure 2 illustrates a number of possibilities. Note that the global maximum value (if it exists) is simply the largest of the local maximum values. Similarly, the global minimum value is the smallest of the local minimum values.

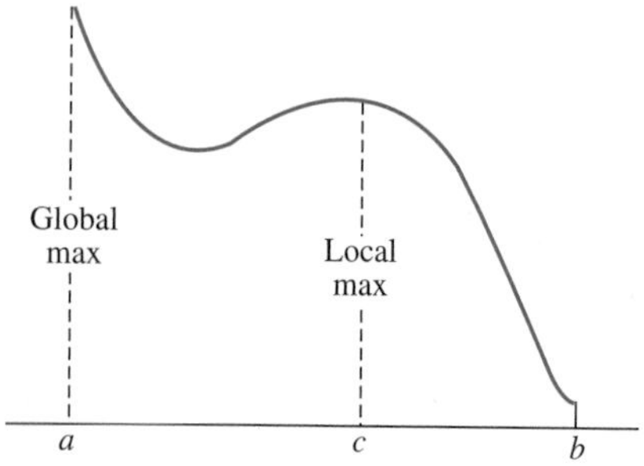

Figure 1

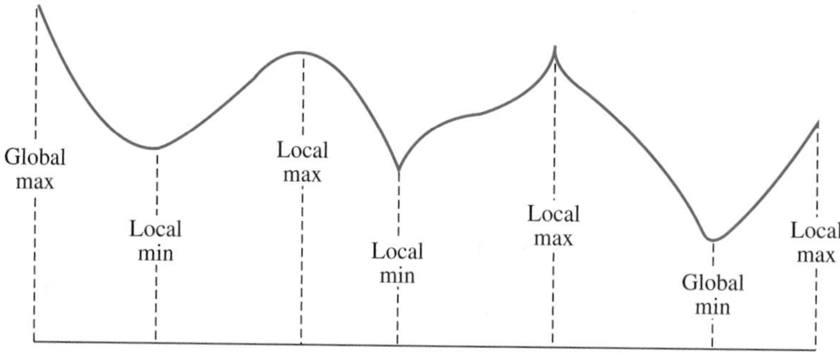

Figure 2

Here is the formal definition of local maxima and local minima. Recall that the symbol $\cap$ denotes the intersection (common part) of two sets.

Definition

Let S, the domain of f, contain the point c. We say that

(i) $f(c)$ is a **local maximum value** of f if there is an interval (a, b) containing c such that $f(c)$ is the maximum value of f on $(a, b) \cap S$;

(ii) $f(c)$ is a **local minimum value** of f if there is an interval (a, b) containing c such that $f(c)$ is the minimum value of f on $(a, b) \cap S$;

(iii) $f(c)$ is a **local extreme value** of f if it is either a local maximum or a local minimum value.

Where Do Local Extreme Values Occur? The Critical Point Theorem (Theorem 3.1B) holds with the phrase *extreme value* replaced by *local extreme value*; the proof is essentially the same. Thus, the critical points (end points, stationary points, and singular points) are the candidates for points where local extrema may occur. We say *candidates* because we are not claiming that there must be a local extremum at every critical point. The left graph in Figure 3 makes this clear. However, if the derivative is positive on one side of the critical point and negative on the other (and if the function is continuous), then we have a local extremum, as shown in the middle and right graphs of Figure 3.

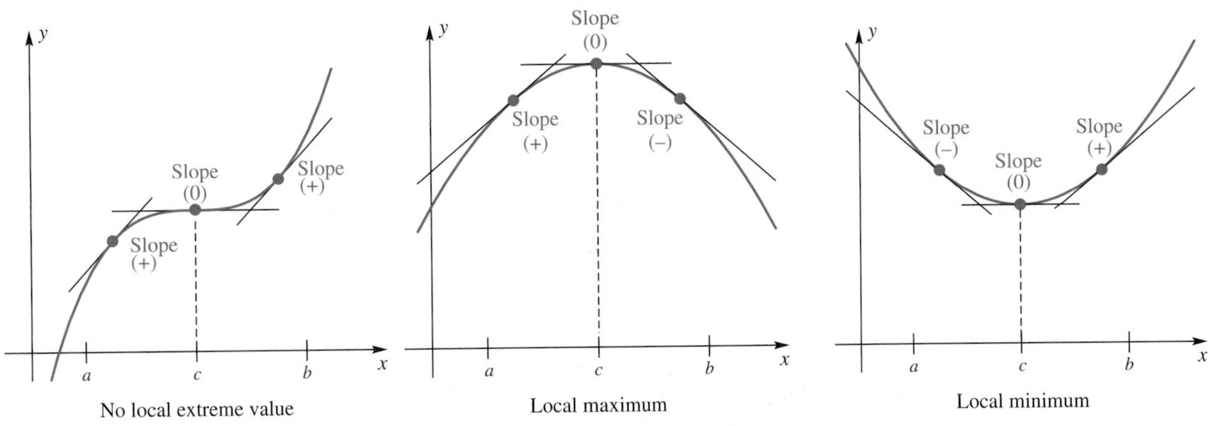

Figure 3

Theorem A	First Derivative Test

Let f be continuous on an open interval (a, b) that contains a critical point c.

(i) If $f'(x) > 0$ for all x in (a, c) and $f'(x) < 0$ for all x in (c, b), then $f(c)$ is a local maximum value of f.

(ii) If $f'(x) < 0$ for all x in (a, c) and $f'(x) > 0$ for all x in (c, b), then $f(c)$ is a local minimum value of f.

(iii) If $f'(x)$ has the same sign on both sides of c, then $f(c)$ is not a local extreme value of f.

Proof of (i) Since $f'(x) > 0$ for all x in (a, c), f is increasing on $(a, c]$ by the Monotonicity Theorem. Again, since $f'(x) < 0$ for all x in (c, b), f is decreasing on $[c, b)$. Thus, $f(x) < f(c)$ for all x in (a, b), except of course at $x = c$. We conclude that $f(c)$ is a local maximum.

The proofs of (ii) and (iii) are similar. ■

EXAMPLE 1 Find the local extreme values of the function $f(x) = x^2 - 6x + 5$ on $(-\infty, \infty)$.

SOLUTION The polynomial function f is continuous everywhere, and its derivative, $f'(x) = 2x - 6$, exists for all x. Thus, the only critical point for f is the single solution of $f'(x) = 0$; that is, $x = 3$.

Since $f'(x) = 2(x - 3) < 0$ for $x < 3$, f is decreasing on $(-\infty, 3]$; and because $2(x - 3) > 0$ for $x > 3$, f is increasing on $[3, \infty)$. Therefore, by the First Derivative Test, $f(3) = -4$ is a local minimum value of f. Since 3 is the only critical point, there are no other extreme values. The graph of f is shown in Figure 4. Note that $f(3)$ is actually the (global) minimum value in this case. ■

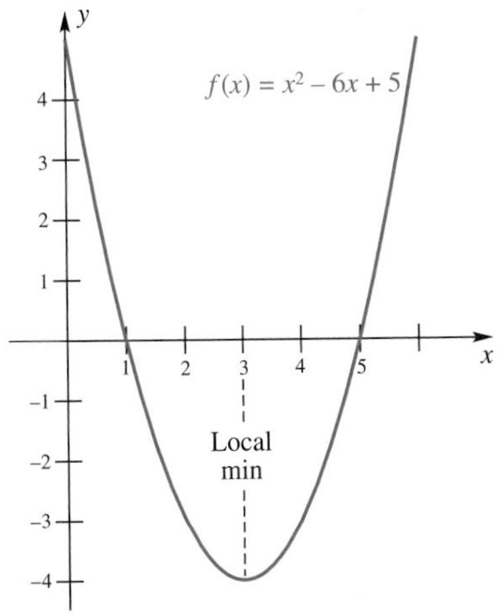

Figure 4

EXAMPLE 2 Find the local extreme values of $f(x) = \frac{1}{3}x^3 - x^2 - 3x + 4$ on $(-\infty, \infty)$.

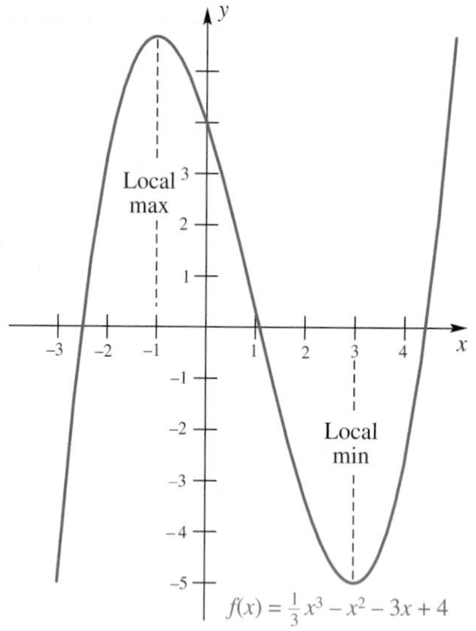

Figure 5

SOLUTION Since $f'(x) = x^2 - 2x - 3 = (x + 1)(x - 3)$, the only critical points of f are -1 and 3. When we use the test points -2, 0, and 4, we learn that $(x + 1)(x - 3) > 0$ on $(-\infty, -1)$ and $(3, \infty)$ and $(x + 1)(x - 3) < 0$ on $(-1, 3)$. By the First Derivative Test, we conclude that $f(-1) = \frac{17}{3}$ is a local maximum value and that $f(3) = -5$ is a local minimum value (Figure 5). ■

EXAMPLE 3 Find the local extreme values of $f(x) = (\sin x)^{2/3}$ on $(-\pi/6, 2\pi/3)$.

SOLUTION

$$f'(x) = \frac{2\cos x}{3(\sin x)^{1/3}}, \qquad x \neq 0$$

The points 0 and $\pi/2$ are critical points, since $f'(0)$ does not exist and $f'(\pi/2) = 0$. Now $f'(x) < 0$ on $(-\pi/6, 0)$ and on $(\pi/2, 2\pi/3)$, while $f'(x) > 0$ on $(0, \pi/2)$. By the First Derivative Test, we conclude that $f(0) = 0$ is a local minimum value and that $f(\pi/2) = 1$ is a local maximum value. The graph of f is shown in Figure 6. ■

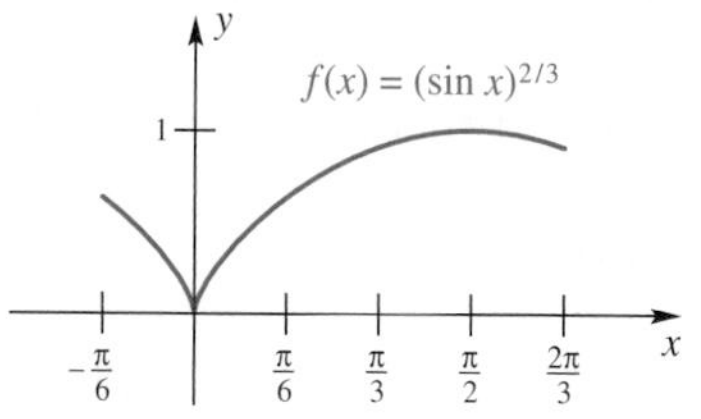

Figure 6

The Second Derivative Test There is another test for local maxima and minima that is sometimes easier to apply than the First Derivative Test. It involves evaluating the second derivative at the stationary points. It does not apply to singular points.

Theorem B Second Derivative Test

Let f' and f'' exist at every point in an open interval (a, b) containing c, and suppose that $f'(c) = 0$.

(i) If $f''(c) < 0$, then $f(c)$ is a local maximum value of f.
(ii) If $f''(c) > 0$, then $f(c)$ is a local minimum value of f.

Proof of (i) It is tempting to say that, since $f''(c) < 0$, f is concave downward near c and to claim that this proves (i). However, to be sure that f is concave downward in a neighborhood of c, we need $f''(x) < 0$ in that neighborhood (not just at c), and nothing in our hypothesis guarantees that. We must be a bit more careful. By definition and hypothesis,

$$f''(c) = \lim_{x \to c} \frac{f'(x) - f'(c)}{x - c} = \lim_{x \to c} \frac{f'(x) - 0}{x - c} < 0$$

so we can conclude that there is a (possibly small) interval (α, β) around c where

$$\frac{f'(x)}{x - c} < 0, \qquad x \neq c$$

But this inequality implies that $f'(x) > 0$ for $\alpha < x < c$ and $f'(x) < 0$ for $c < x < \beta$. Thus, by the First Derivative Test, $f(c)$ is a local maximum value.

The proof of (ii) is similar. ■

EXAMPLE 4 For $f(x) = x^2 - 6x + 5$, use the Second Derivative Test to identify local extrema.

SOLUTION This is the function of Example 1. Note that

$$f'(x) = 2x - 6 = 2(x - 3)$$
$$f''(x) = 2$$

Thus, $f'(3) = 0$ and $f''(3) > 0$. Therefore, by the Second Derivative Test, $f(3)$ is a local minimum value. ■

EXAMPLE 5 For $f(x) = \frac{1}{3}x^3 - x^2 - 3x + 4$, use the Second Derivative Test to identify local extrema.

SOLUTION This is the function of Example 2.

$$f'(x) = x^2 - 2x - 3 = (x + 1)(x - 3)$$

$$f''(x) = 2x - 2$$

The critical points are -1 and 3 ($f'(-1) = f'(3) = 0$). Since $f''(-1) = -4$ and $f''(3) = 4$, we conclude by the Second Derivative Test that $f(-1)$ is a local maximum value and that $f(3)$ is a local minimum value. ■

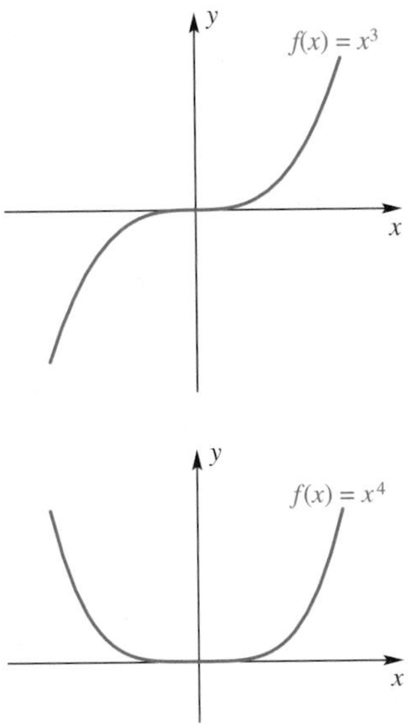

Figure 7

Unfortunately, the Second Derivative Test sometimes fails, since $f''(x)$ may be 0 at a stationary point. For both $f(x) = x^3$ and $f(x) = x^4$, $f'(0) = 0$ and $f''(0) = 0$ (see Figure 7). The first does not have a local maximum or minimum value at 0; the second has a local minimum there. This shows that if $f''(x) = 0$ at a stationary point we are unable to draw a conclusion about maxima or minima without more information.

Extrema on Open Intervals The problems that we studied in this section and in Section 3.1 often assumed that the set on which we wanted to maximize or minimize a function was a *closed* interval. However, the intervals that arise in practice are not always closed; they are sometimes open, or even open on one end and closed on the other. We can still handle these problems if we correctly apply the theory developed in this section. Keep in mind that maximum (minimum) with no qualifying adjective means global maximum (minimum).

EXAMPLE 6 Find (if any exist) the minimum and maximum values of $f(x) = x^4 - 4x$ on $(-\infty, \infty)$.

SOLUTION

$$f'(x) = 4x^3 - 4 = 4(x^3 - 1) = 4(x - 1)(x^2 + x + 1)$$

Since $x^2 + x + 1 = 0$ has no real solutions (quadratic formula), there is only one critical point, $x = 1$. For $x < 1, f'(x) < 0$, whereas for $x > 1, f'(x) > 0$. We conclude that $f(1) = -3$ is a local minimum value for f; and since f is decreasing on the left of 1 and increasing on the right of 1, it must actually be the minimum value of f.

The facts stated above imply that f cannot have a maximum value. The graph of f is shown in Figure 8. ■

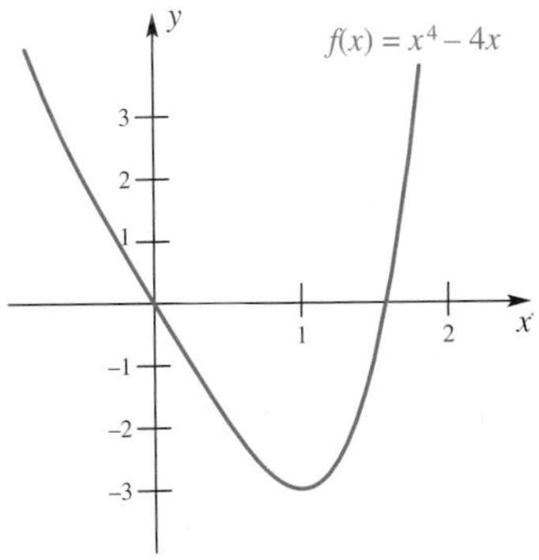

Figure 8

EXAMPLE 7 Find (if any exist) the maximum and minimum values of

$$G(p) = \frac{1}{p(1 - p)}$$

on $(0, 1)$.

SOLUTION

$$G'(p) = \frac{d}{dp}[p(1 - p)]^{-1} = \frac{2p - 1}{p^2(1 - p)^2}$$

The only critical point is $p = 1/2$. For every value of p in the interval $(0, 1)$ the denominator is positive; thus, the numerator determines the sign. If p is in the interval $(0, 1/2)$, then the numerator is negative; hence, $G'(p) < 0$. Similarly, if p is in the interval $(1/2, 1)$, $G'(p) > 0$. Thus, by the First Derivative Test, $G(1/2) = 4$ is a local minimum. Since there are no end points or singular points to check, $G(1/2)$ is a global minimum. There is no maximum. The graph of $y = G(p)$ is shown in Figure 9. ■

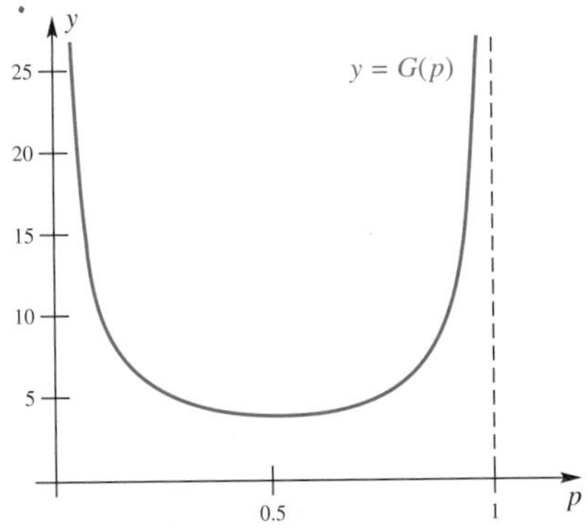

Figure 9

Concepts Review

1. If f is continuous at c, $f'(x) > 0$ near to c on the left, and $f'(x) < 0$ near to c on the right, then $f(c)$ is a local _____ value for f.

2. If $f'(x) = (x + 2)(x - 1)$, then $f(-2)$ is a local _____ value for f and $f(1)$ is a local _____ value for f.

3. If $f'(c) = 0$ and $f''(c) < 0$, we expect to find a local _____ value for f at c.

4. If $f(x) = x^3$, then $f(0)$ is neither a _____ nor a _____, even though $f''(0) =$ _____.

Problem Set 3.3

In Problems 1–10, identify the critical points. Then use (a) the First Derivative Test and (if possible) (b) the Second Derivative Test to decide which of the critical points give a local maximum and which give a local minimum.

1. $f(x) = x^3 - 6x^2 + 4$
2. $f(x) = x^3 - 12x + \pi$
3. $f(\theta) = \sin 2\theta, 0 < \theta < \dfrac{\pi}{4}$
4. $f(x) = \frac{1}{2}x + \sin x, 0 < x < 2\pi$
5. $\Psi(\theta) = \sin^2 \theta, -\pi/2 < \theta < \pi/2$
6. $r(z) = z^4 + 4$
7. $f(x) = \dfrac{x}{x^2 + 4}$
8. $g(z) = \dfrac{z^2}{1 + z^2}$
9. $h(y) = y^2 - \dfrac{1}{y}$
10. $f(x) = \dfrac{3x + 1}{x^2 + 1}$

In Problems 11–20, find the critical points and use the test of your choice to decide which critical points give a local maximum value and which give a local minimum value. What are these local maximum and minimum values?

11. $f(x) = x^3 - 3x$
12. $g(x) = x^4 + x^2 + 3$
13. $H(x) = x^4 - 2x^3$
14. $f(x) = (x - 2)^5$
15. $g(t) = \pi - (t - 2)^{2/3}$
16. $r(s) = 3s + s^{2/5}$
17. $f(t) = t - \dfrac{1}{t}, t \neq 0$
18. $f(x) = \dfrac{x^2}{\sqrt{x^2 + 4}}$
19. $\Lambda(\theta) = \dfrac{\cos \theta}{1 + \sin \theta}, 0 < \theta < 2\pi$
20. $g(\theta) = |\sin \theta|, 0 < \theta < 2\pi$

In Problems 21–30, find, if possible, the (global) maximum and minimum values of the given function on the indicated interval.

21. $f(x) = \sin^2 2x$ on $[0, 2]$
22. $f(x) = \dfrac{2x}{x^2 + 4}$ on $[0, \infty)$
23. $g(x) = \dfrac{x^2}{x^3 + 32}$ on $[0, \infty)$
24. $h(x) = \dfrac{1}{x^2 + 4}$ on $[0, \infty)$
25. $F(x) = 6\sqrt{x} - 4x$ on $[0, 4]$
26. $F(x) = 6\sqrt{x} - 4x$ on $[0, \infty)$
27. $f(x) = \dfrac{64}{\sin x} + \dfrac{27}{\cos x}$ on $(0, \pi/2)$
28. $g(x) = x^2 + \dfrac{16x^2}{(8 - x)^2}$ on $(8, \infty)$
29. $H(x) = |x^2 - 1|$ on $[-2, 2]$
30. $h(t) = \sin t^2$ on $[0, \pi]$

In Problems 31–36, the first derivative f' is given. Find all values of x that make the function f (a) a local minimum and (b) a local maximum.

31. $f'(x) = x^3(1 - x)^2$
32. $f'(x) = -(x - 1)(x - 2)(x - 3)(x - 4)$
33. $f'(x) = (x - 1)^2(x - 2)^2(x - 3)(x - 4)$
34. $f'(x) = (x - 1)^2(x - 2)^2(x - 3)^2(x - 4)^2$
35. $f'(x) = (x - A)^2(x - B)^2, A \neq B$
36. $f'(x) = x(x - A)(x - B), 0 < A < B$

In Problems 37–42, sketch a graph of a function with the given properties. If it is impossible to graph such a function, then indicate this and justify your answer.

37. f is differentiable, has domain $[0, 6]$, and has two local maxima and two local minima on $(0, 6)$.

38. f is differentiable, has domain $[0, 6]$, and has three local maxima and two local minima on $(0, 6)$.

39. f is continuous, but not necessarily differentiable, has domain $[0, 6]$, and has one local minimum and one local maximum on $(0, 6)$.

40. f is continuous, but not necessarily differentiable, has domain $[0, 6]$, and has one local minimum and no local maximum on $(0, 6)$.

41. f has domain $[0, 6]$, but is not necessarily continuous, and has three local maxima and no local minimum on $(0, 6)$.

42. f has domain $[0, 6]$, but is not necessarily continuous, and has two local maxima and no local minimum on $(0, 6)$.

43. Consider $f(x) = Ax^2 + Bx + C$ with $A > 0$. Show that $f(x) \geq 0$ for all x if and only if $B^2 - 4AC \leq 0$.

44. Consider $f(x) = Ax^3 + Bx^2 + Cx + D$ with $A > 0$. Show that f has one local maximum and one local minimum if and only if $B^2 - 3AC > 0$.

45. What conclusions can you draw about f from the information that $f'(c) = f''(c) = 0$ and $f'''(c) > 0$?

Answers to Concepts Review: **1.** maximum **2.** maximum; minimum **3.** maximum **4.** local maximum; local minimum; 0

3.4 Practical Problems

Based on the examples and the theory developed in the first three sections of this chapter, we suggest the following step-by-step method that can be applied to many practical optimization problems. Do not follow it slavishly; common sense may sometimes suggest an alternative approach or omission of some steps.

Step 1: Draw a picture for the problem and assign appropriate variables to the important quantities.

Step 2: Write a formula for the objective function Q to be maximized or minimized in terms of the variables from step 1.

Step 3: Use the conditions of the problem to eliminate all but one of these variables, and thereby express Q as a function of a single variable.

Step 4: Find the critical points (end points, stationary points, singular points).

Step 5: Either substitute the critical values into the objective function or use the theory from the last section (i.e., the First and Second Derivative Tests) to determine the maximum or minimum.

Throughout, use your intuition to get some idea of what the solution of the problem should be. For many physical problems you can get a "ballpark" estimate of the optimal value before you begin to carry out the details.

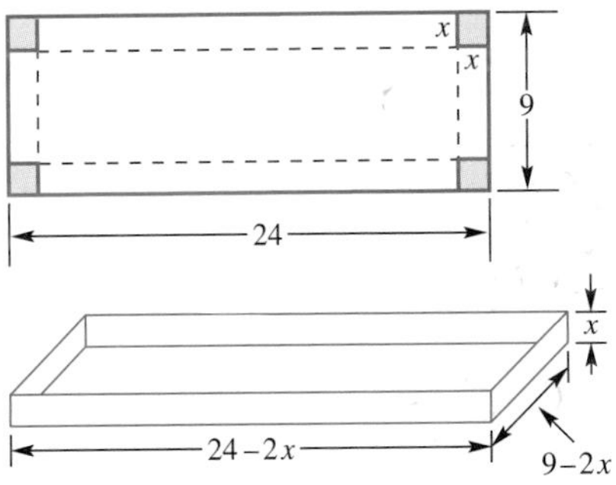

Figure 1

EXAMPLE 1 A rectangular box is to be made from a piece of cardboard 24 inches long and 9 inches wide by cutting out identical squares from the four corners and turning up the sides, as in Figure 1. Find the dimensions of the box of maximum volume. What is this volume?

SOLUTION Let x be the width of the square to be cut out and V the volume of the resulting box. Then

$$V = x(9 - 2x)(24 - 2x) = 216x - 66x^2 + 4x^3$$

Now x cannot be less than 0 or more than 4.5. Thus, our problem is to maximize V on $[0, 4.5]$. The stationary points are found by setting dV/dx equal to 0 and solving the resulting equation:

$$\frac{dV}{dx} = 216 - 132x + 12x^2 = 12(18 - 11x + x^2) = 12(9 - x)(2 - x) = 0$$

This gives $x = 2$ or $x = 9$, but 9 is not in the interval $[0, 4.5]$. We see that there are only three critical points, 0, 2, and 4.5. At the end points 0 and 4.5, $V = 0$; at 2, $V = 200$. We conclude that the box has a maximum volume of 200 cubic inches if $x = 2$, that is, if the box is 20 inches long, 5 inches wide, and 2 inches deep. ■

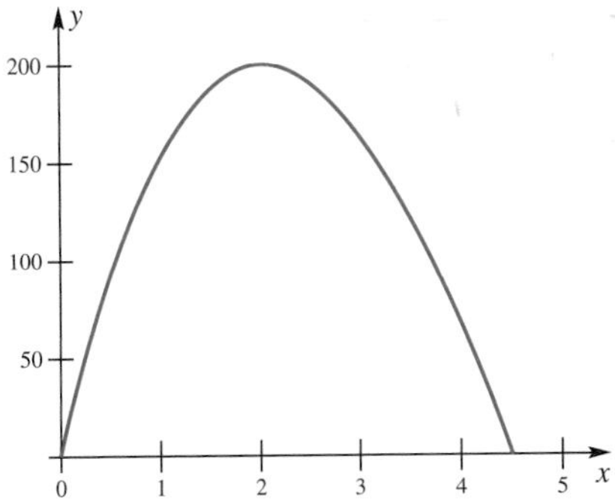

Figure 2

It is often helpful to plot the objective function. Plotting functions can be done easily with a graphing calculator or a CAS. Figure 2 shows a plot of the function $V(x) = 216x - 66x^2 + 4x^3$. When $x = 0$, $V(x)$ is equal to zero. In the context of folding the box, this means that when the width of the cut-out corner is zero there is nothing to fold up, so the volume is zero. Also, when $x = 4.5$, the cardboard gets folded in half, so there is no base to the box; this box will also have zero volume. Thus, $V(0) = 0$ and $V(4.5) = 0$. The greatest volume must be attained for some value of x between 0 and 4.5. The graph suggests that the maximum volume occurs when x is about 2; by using calculus, we can determine that the *exact* value of x that maximizes the volume of the box is $x = 2$.

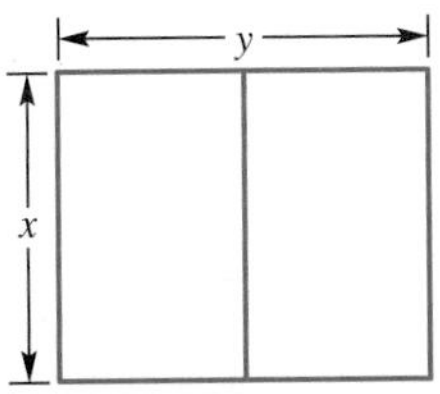

Figure 3

EXAMPLE 2 A farmer has 100 meters of wire fence with which he plans to build two identical adjacent pens, as shown in Figure 3. What are the dimensions of the enclosure that has maximum area?

SOLUTION Let x be the width and y the length of the total enclosure, both in meters. Because there are 100 meters of fence, $3x + 2y = 100$; that is,

$$y = 50 - \tfrac{3}{2}x$$

The total area A is given by

$$A = xy = 50x - \tfrac{3}{2}x^2$$

Since there must be three sides of length x, we see that $0 \le x \le \frac{100}{3}$. Thus, our problem is to maximize A on $\left[0, \frac{100}{3}\right]$. Now

$$\frac{dA}{dx} = 50 - 3x$$

When we set $50 - 3x$ equal to 0 and solve, we get $x = \frac{50}{3}$ as a stationary point. Thus, there are three critical points: $0, \frac{50}{3}$, and $\frac{100}{3}$. The two end points 0 and $\frac{100}{3}$ give $A = 0$, while $x = \frac{50}{3}$ yields $A \approx 416.67$. The desired dimensions are $x = \frac{50}{3} \approx 16.67$ meters and $y = 50 - \frac{3}{2}\left(\frac{50}{3}\right) = 25$ meters.

≈ Is this answer sensible? Yes. We should expect to use more of the given fence in the y-direction than the x-direction because the former is fenced only twice, whereas the latter is fenced three times. ■

EXAMPLE 3 Find the dimensions of the right circular cylinder of greatest volume that can be inscribed in a given right circular cone.

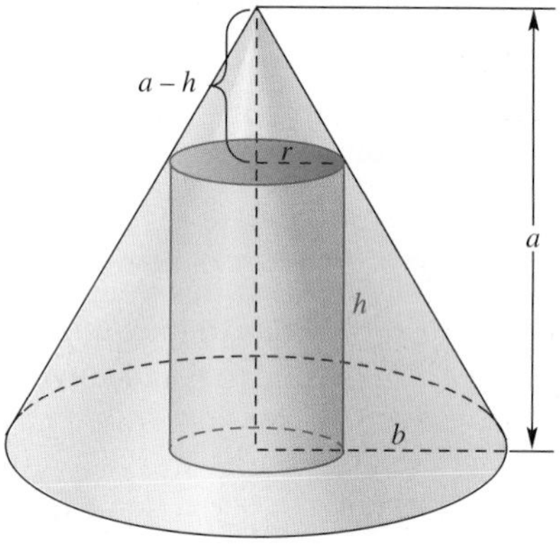

Figure 4

SOLUTION Let a be the altitude and b the radius of the base of the given cone (both constants). Denote by h, r, and V the altitude, radius, and volume, respectively, of an inscribed cylinder (see Figure 4).

≈ Before proceeding, let's apply some intuition. If the cylinder's radius is close to the radius of the cone's base, then the cylinder's volume would be close to zero. Now, imagine inscribed cylinders with increasing height, but decreasing radius. Initially, the volumes would increase from zero, but then they would decrease to zero as the cylinders' heights get close to the cone's height. Intuitively, the volume should peak for some cylinder. Since the radius is squared in the volume formula, it counts more than the height and we would expect $r > h$ at the maximum.

The volume of the inscribed cylinder is

$$V = \pi r^2 h$$

From similar triangles,

$$\frac{a - h}{r} = \frac{a}{b}$$

which gives

$$h = a - \frac{a}{b}r$$

When we substitute this expression for h in the formula for V, we obtain

$$V = \pi r^2\left(a - \frac{a}{b}r\right) = \pi a r^2 - \pi\frac{a}{b}r^3$$

We wish to maximize V for r in the interval $[0, b]$. Now

$$\frac{dV}{dr} = 2\pi a r - 3\pi\frac{a}{b}r^2 = \pi a r\left(2 - \frac{3}{b}r\right)$$

This yields the stationary points $r = 0$ and $r = 2b/3$, giving us three critical points on $[0, b]$ to consider: 0, $2b/3$, and b. As expected, $r = 0$ and $r = b$ both give a volume of 0. Thus, $r = 2b/3$ has to give the maximum volume. When we substitute

Algebra and Geometry

Whenever possible, try to view a problem from both a geometric and an algebraic point of view. Example 3 is a good example for which this kind of thinking lends insight into the problem.

this value for r in the equation connecting r and h, we find that $h = a/3$. In other words, the inscribed cylinder has greatest volume when its radius is two-thirds the radius of the cone's base and its height is one-third the altitude of the cone. ■

EXAMPLE 4 Suppose that a fish swims upstream with velocity v relative to the water and that the current of the river has velocity $-v_c$ (the negative sign indicates that the current's velocity is in the direction opposite that of the fish). The energy expended in traveling a distance d up the river is directly proportional to the time required to travel the distance d and the cube of the velocity. What velocity v minimizes the energy expended in swimming this distance?

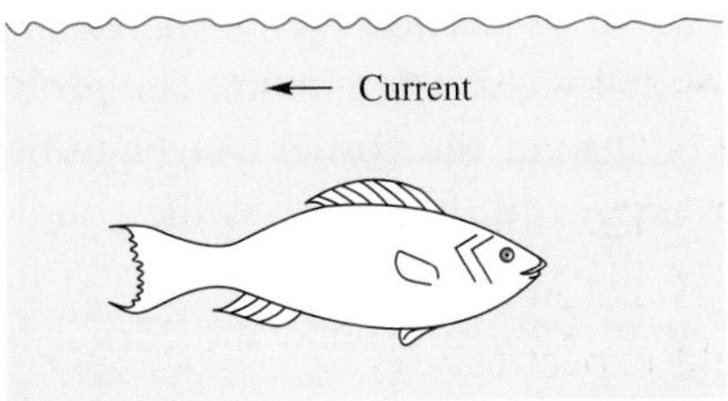

Figure 5

SOLUTION Figure 5 illustrates the situation. Since the fish's velocity up the stream (i.e., relative to the banks of the stream) is $v - v_c$, we have $d = (v - v_c)t$, where t is the required time. Thus $t = d/(v - v_c)$. For a fixed value of v, the energy required for the fish to travel the distance d is therefore

$$E(v) = k\frac{d}{v - v_c}v^3 = kd\frac{v^3}{v - v_c}$$

The domain for the function E is the open interval (v_c, ∞). To find the value of v that minimizes the required energy we set $E'(v) = 0$ and solve for v:

$$E'(v) = kd\frac{(v - v_c)3v^2 - v^3(1)}{(v - v_c)^2} = \frac{kd}{(v - v_c)^2}v^2(2v - 3v_c) = 0$$

The only critical point in the interval (v_0, ∞) is found by solving $2v - 3v_c = 0$, which leads to $v = \frac{3}{2}v_c$. The interval is open so there are no end points to check. The sign of $E'(v)$ depends entirely on the expression $2v - 3v_c$, since all the other expressions are positive. If $v < \frac{3}{2}v_c$, then $2v - 3v_c < 0$ so E is decreasing to the left of $\frac{3}{2}v_c$. If $v > \frac{3}{2}v_c$, then $2v - 3v_c > 0$ so E is increasing to the right of $\frac{3}{2}v_c$. Thus, by the First Derivative Test, $v = \frac{3}{2}v_c$ yields a local minimum. Since this is the only critical point on the interval (v_0, ∞), this must give a global minimum. The velocity that minimizes the expended energy is therefore one and a half times the speed of the current. ■

EXAMPLE 5 A 6-foot-wide hallway makes a right-angle turn. What is the length of the longest thin rod that can be carried around the corner assuming you cannot tilt the rod?

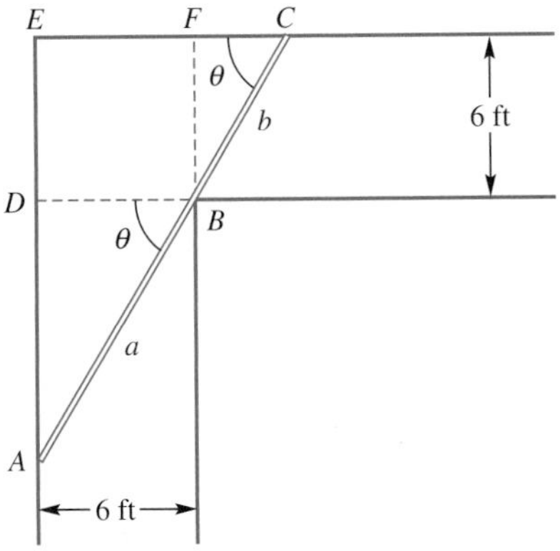

Figure 6

SOLUTION The rod that barely fits around the corner will touch the outside walls as well as the inside corner. As suggested in Figure 6, let a and b represent the lengths of the segments AB and BC, and let θ denote the angles $\angle DBA$ and $\angle FCB$. Consider the two similar right triangles, $\triangle ADB$ and $\triangle BFC$; these have hypotenuses a and b, respectively. A little trigonometry applied to these angles gives

$$a = \frac{6}{\cos\theta} = 6\sec\theta \quad \text{and} \quad b = \frac{6}{\sin\theta} = 6\csc\theta$$

Note that the angle θ determines the position of the rod. The total length of the rod in Figure 6 is thus

$$L(\theta) = a + b = 6\sec\theta + 6\csc\theta$$

The domain for θ is the open interval $(0, \pi/2)$. The derivative of L is

$$L'(\theta) = 6 \sec\theta \tan\theta - 6 \csc\theta \cot\theta$$
$$= 6\left(\frac{\sin\theta}{\cos^2\theta} - \frac{\cos\theta}{\sin^2\theta}\right)$$
$$= 6\frac{\sin^3\theta - \cos^3\theta}{\sin^2\theta\cos^2\theta}$$

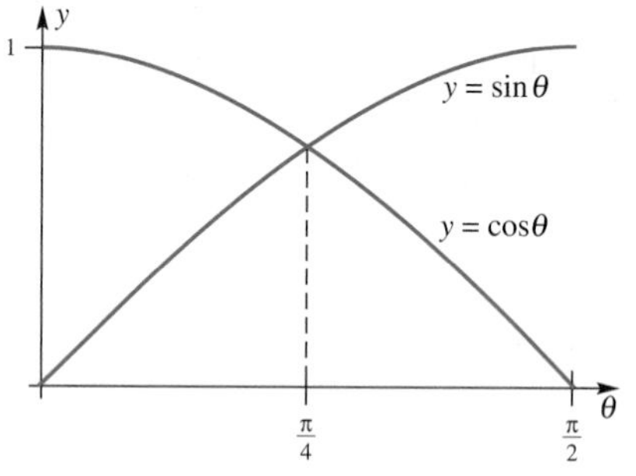

Figure 7

Thus $L'(\theta) = 0$ provided $\sin^3\theta - \cos^3\theta = 0$. This leads to $\sin\theta = \cos\theta$. The only angle in $(0, \pi/2)$ for which $\sin\theta = \cos\theta$ is the angle $\pi/4$ (see Figure 7). We again apply the First Derivative Test. If $0 < \theta < \pi/4$, then $\sin\theta < \cos\theta$ (see Figure 7 again) so $\sin^3\theta - \cos^3\theta < 0$. Thus, $L(\theta)$ is decreasing on $(0, \pi/4)$. If $\pi/4 < \theta < \pi/2$ then $\sin\theta > \cos\theta$ so $\sin^3\theta - \cos^3\theta > 0$. Thus, $L(\theta)$ is increasing on $(\pi/4, \pi/2)$. By the First Derivative Test, $\theta = \pi/4$ yields a minimum. The problem, however, asks for the *longest* rod that fits around the corner. As Figure 8 below indicates, we are actually finding the *smallest* rod that satisfies the conditions in Figure 6; in other words, we are finding the smallest rod that doesn't fit around the corner. Therefore, the longest rod that does fit around the corner is $L(\pi/4) = 6 \sec\pi/4 + 6 \csc\pi/4 = 12\sqrt{2} \approx 16.97$ feet. ■

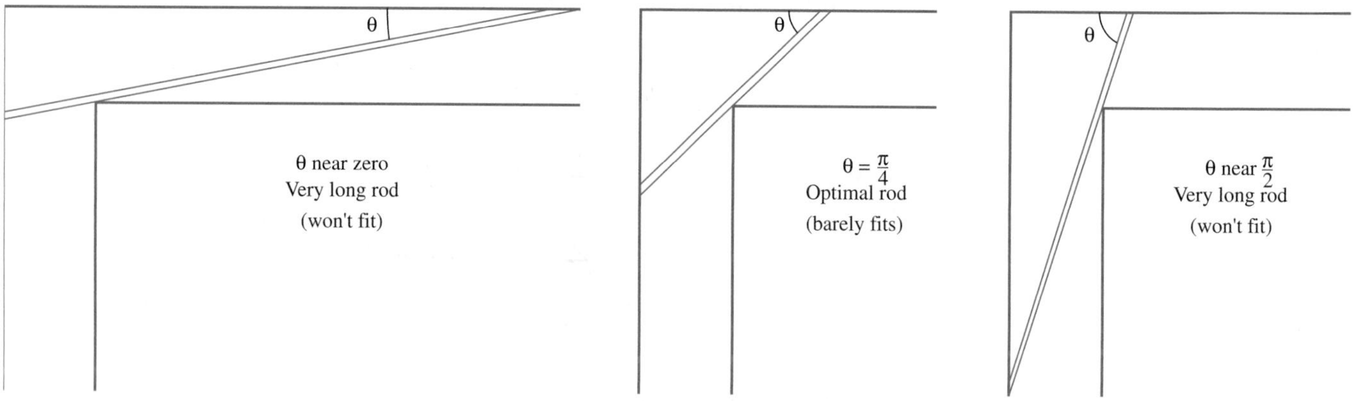

Figure 8

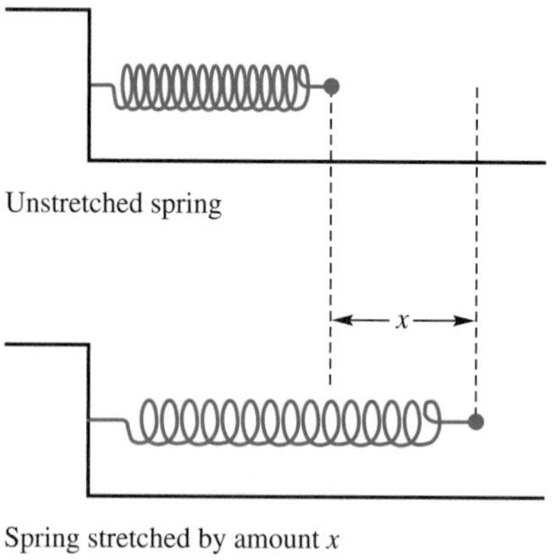

Figure 9

Distance Stretched, x (meters)	Force y Exerted by Spring (newtons)
0.005	8
0.010	17
0.015	22
0.020	32
0.025	36

Figure 10

Least Squares (Optional) There are a number of physical, economic, and social phenomena in which one variable is proportional to another. For example, Newton's Second Law says that the force F on an object of mass m is proportional to its acceleration a ($F = ma$). Hooke's Law says that the force exerted by a spring is proportional to the distance it is stretched ($F = kx$). (Hooke's Law is often given as $F = -kx$, with the negative sign indicating that the force is in the direction opposite the stretch. For now, we will ignore the sign of the force.) Manufacturing costs are proportional to the number of units produced. The number of traffic accidents is proportional to the volume of traffic. These are *models*, and in an experiment we will rarely find that the observed data fit the model exactly.

Suppose that we observe the force exerted by a spring when it is stretched by x centimeters (Figure 9). For example, when we stretch the spring by 0.5 centimeter (0.005 meter), we observe a force of 8 newtons, when we stretch the spring by 1.0 centimeter, we observe a force of 17 newtons, and so on. Figure 10 shows additional observations, and Figure 11 shows a plot of the ordered pairs (x_i, y_i), where x_i is the distance stretched and y_i is the force exerted on the spring. A plot of ordered pairs like this is called a **scatter plot.**

Let's generalize the problem to one in which we are given n points $(x_1, y_1), (x_2, y_2), \ldots, (x_n, y_n)$. Our goal is to find a line through the origin that *best fits* these points. Before proceeding, we must introduce sigma (Σ) notation.

The symbol $\sum_{i=1}^{n} a_i$ means the sum of the numbers $a_1, a_2, \ldots, a_n$. For example,

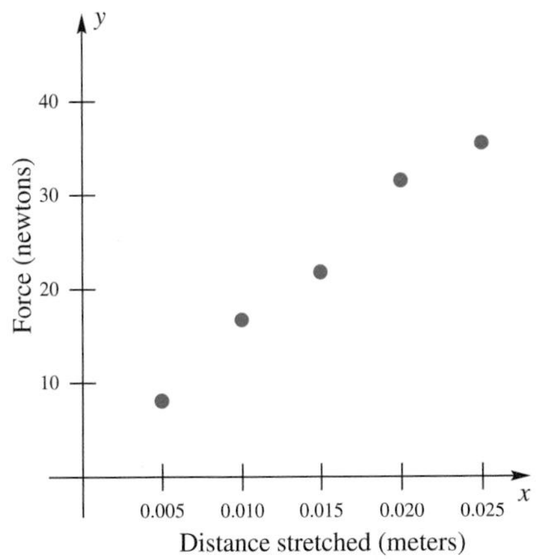

Figure 11

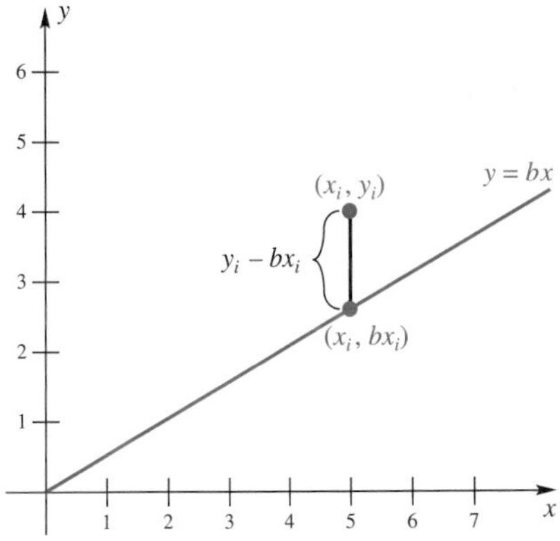

Figure 12

$$\sum_{i=1}^{3} i^2 = 1^2 + 2^2 + 3^2 = 14 \qquad \text{and} \qquad \sum_{i=1}^{n} x_i y_i = x_1 y_1 + x_2 y_2 + \cdots + x_n y_n$$

In the second case, we multiply x_i and y_i first and then sum.

To find the line that best fits these n points, we must be specific about how we measure the fit. Our *best-fit* line through the origin is defined to be the one that minimizes the sum of the squared vertical deviations between (x_i, y_i) and the line $y = bx$. If (x_i, y_i) is a point in the data set, then (x_i, bx_i) is the point on the line $y = bx$ that is directly above or below (x_i, y_i). The vertical deviation between (x_i, y_i) and (x_i, bx_i) is therefore $y_i - bx_i$. (See Figure 12.) The squared deviation is thus $(y_i - bx_i)^2$. The problem is to find the value of b that minimizes the sum of these squared deviations. If we define

$$S = \sum_{i=1}^{n} (y_i - bx_i)^2$$

then we must find the value of b that *minimizes* S. This is a minimization problem like the ones encountered before. Keep in mind, however, that the ordered pairs (x_i, y_i), $i = 1, 2, \ldots, n$ are *fixed*; the variable in this problem is b.

We proceed as before by finding dS/db, setting the result equal to 0, and solving for b. Since the derivative is a linear operator, we have

$$\begin{aligned}
\frac{dS}{db} &= \frac{d}{db}\sum_{i=1}^{n}(y_i - bx_i)^2 \\
&= \sum_{i=1}^{n}\frac{d}{db}(y_i - bx_i)^2 \\
&= \sum_{i=1}^{n}2(y_i - bx_i)\left(\frac{d}{db}(y_i - bx_i)\right) \\
&= -2\sum_{i=1}^{n}x_i(y_i - bx_i)
\end{aligned}$$

Setting this result equal to zero and solving yields

$$\begin{aligned}
0 &= -2\sum_{i=1}^{n}x_i(y_i - bx_i) \\
0 &= \sum_{i=1}^{n}x_i y_i - b\sum_{i=1}^{n}x_i^2 \\
b &= \frac{\sum_{i=1}^{n}x_i y_i}{\sum_{i=1}^{n}x_i^2}
\end{aligned}$$

To see that this yields a minimum value for S, we note that

$$\frac{d^2S}{db^2} = 2\sum_{i=1}^{n}x_i^2$$

which is always positive. There are no end points to check. Thus, by the Second Derivative Test, we conclude that the line $y = bx$, with $b = \sum_{i=1}^{n} x_i y_i \Big/ \sum_{i=1}^{n} x_1^2$, is the best-fit line, in the sense of minimizing S. The line $y = bx$ is called the **least-squares line through the origin.**

EXAMPLE 6 Find the least-squares line through the origin for the spring data in Figure 10.

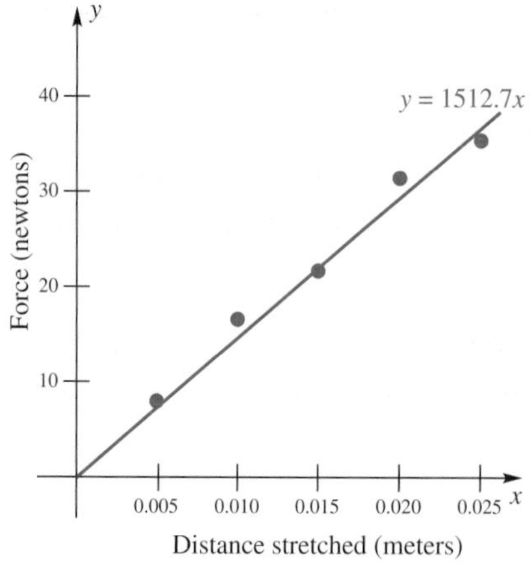

Figure 13

SOLUTION

$$b = \frac{0.005 \cdot 8 + 0.010 \cdot 17 + 0.015 \cdot 22 + 0.020 \cdot 32 + 0.025 \cdot 36}{0.005^2 + 0.010^2 + 0.015^2 + 0.020^2 + 0.025^2}$$

$$\approx 1512.7$$

The least-squares line through the origin is therefore $y = 1512.7x$ and is shown in Figure 13. The estimate of the spring constant is therefore $k = 1512.7$. ■

For most line-fitting problems, it is unreasonable to assume that the line passes through the origin. A more reasonable assumption is that y is related to x by $y = a + bx$. In this case, however, the sum of squares is a function of both a and b so we are faced with the problem of minimizing a function of two variables, a problem we address in Chapter 12.

Economic Applications (Optional) Consider a typical company, the ABC Company. For simplicity, assume that ABC produces and markets a single product; it might be television sets, car batteries, or bars of soap. If it sells x units of the product in a fixed period of time (e.g., a year), it will be able to charge a **price,** $p(x)$, for each unit. In other words, $p(x)$ is the price required to attract a demand for x units. The **total revenue** that ABC can expect is given by $R(x) = xp(x)$, the number of units times the price per unit.

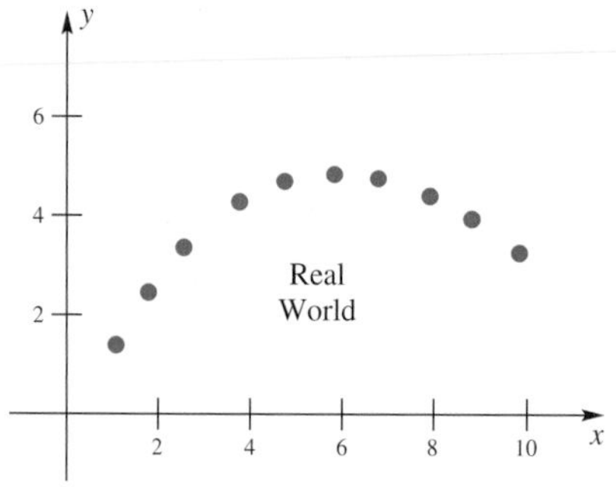

Figure 14

To produce and market x units, ABC will have a total cost, $C(x)$. This is normally the sum of a **fixed cost** (office utilities, real estate taxes, and so on) plus a **variable cost,** which depends directly on the number of units produced.

The key concept for a company is the **total profit,** $P(x)$. It is just the difference between revenue and cost; that is,

$$P(x) = R(x) - C(x) = xp(x) - C(x)$$

Generally, a company seeks to maximize its total profit.

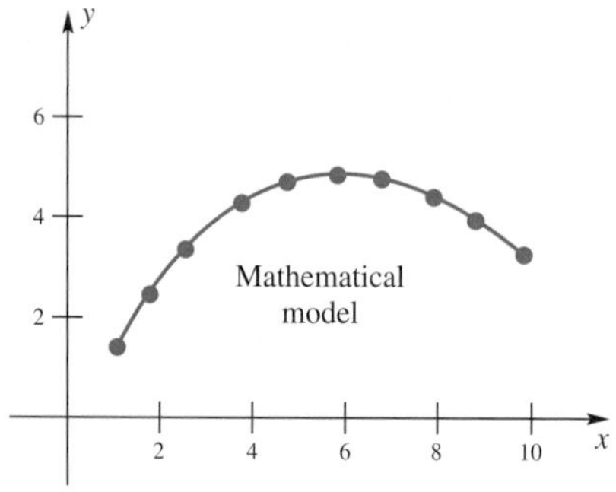

Figure 15

There is a feature that tends to distinguish problems in economics from those in the physical sciences. In most cases, ABC's product will be in discrete units (you can't make or sell 8.23 television sets or π car batteries). Thus, the functions $R(x)$, $C(x)$, and $P(x)$ are usually defined only for $x = 0, 1, 2, \ldots$ and, consequently, their graphs consist of discrete points (Figure 14). In order to make the tools of calculus available, we connect these points with a smooth curve (Figure 15), thereby pretending that R, C, and P are nice differentiable functions. This illustrates an aspect of *mathematical modeling* that is almost always necessary, especially in economics. To model a real-world problem, we must make some simplifying assumptions. This means that the answers we get are only approximations of the answers that we seek—one of the reasons economics is a less than perfect science. A well-known statistician once said: No model is accurate, but many models are useful.

A related problem for an economist is how to obtain formulas for the functions $C(x)$ and $p(x)$. In a simple case, $C(x)$ might have the form

$$C(x) = 10{,}000 + 50x$$

If so, \$10,000 is the **fixed cost** and $\$50x$ is the **variable cost,** based on a \$50 direct cost for each unit produced. Perhaps a more typical situation is

$$C_1(x) = 10{,}000 + 45x + 100\sqrt{x}$$

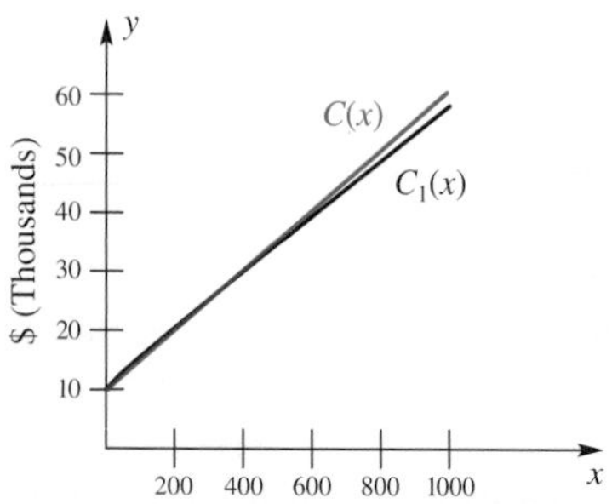

Figure 16

Both cost functions are shown in Figure 16.

The cost function $C(x)$ indicates that the cost of making an additional unit is the same regardless of how many units have been made. On the other hand, the cost function $C_1(x)$ indicates that the cost of making additional units increases but at a decreasing rate. Thus, $C_1(x)$ allows for what economists call economies of scale.

Selecting appropriate functions to model cost and price is a nontrivial task. Occasionally, they can be inferred from basic assumptions. In other cases, a careful

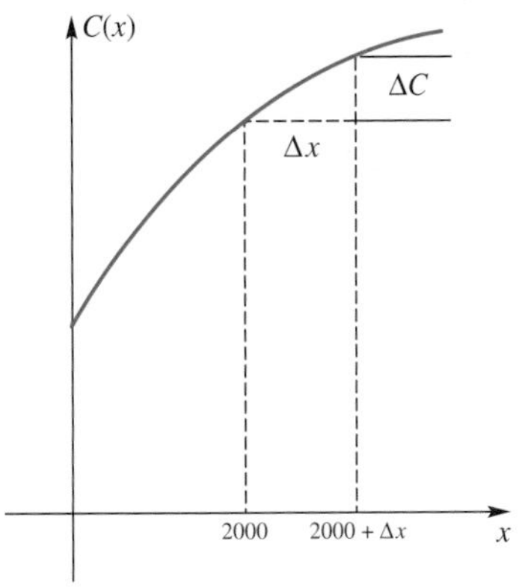

Figure 17

study of the history of the firm will suggest reasonable choices. Sometimes, we must simply make intelligent guesses.

Use of the Word *Marginal* Suppose that ABC knows its cost function $C(x)$ and that it has tentatively planned to produce 2000 units this year. We would like to determine the additional cost per unit if ABC increased production slightly. Would it, for example, be less than the additional revenue per unit? If so, it would make good economic sense to increase production.

If the cost function is the one shown in Figure 17, we are asking for the value of $\Delta C/\Delta x$ when $\Delta x = 1$. But we expect that this will be very close to the value of

$$\lim_{\Delta x \to 0} \frac{\Delta C}{\Delta x}$$

when $x = 2000$. This limit is called the **marginal cost.** We mathematicians recognize it as dC/dx, the derivative of C with respect to x.

In a similar vein, we define **marginal price** as dp/dx, **marginal revenue** as dR/dx, and **marginal profit** as dP/dx.

We now illustrate how to solve a wide variety of economic problems.

> **Economic Vocabulary**
>
> Because economics tends to be a study of discrete phenomena, your economics professor may define marginal cost at x as the cost of producing one additional unit, that is, as
>
> $$C(x + 1) - C(x)$$
>
> In the mathematical model, this number will be very close in value to dC/dx, and since the latter is a principal concept in calculus, we choose to take it as the definition of marginal cost. Similar statements hold for marginal revenue and marginal profit.

EXAMPLE 7 Suppose that $C(x) = 8300 + 3.25x + 40\sqrt[3]{x}$ dollars. Find the average cost per unit and the marginal cost, and then evaluate them when $x = 1000$.

SOLUTION

$$\text{Average cost:} \quad \frac{C(x)}{x} = \frac{8300 + 3.25x + 40x^{1/3}}{x}$$

$$\text{Marginal cost:} \quad \frac{dC}{dx} = 3.25 + \frac{40}{3}x^{-2/3}$$

At $x = 1000$, these have the values 11.95 and 3.38, respectively. This means that it costs, on the average, \$11.95 per unit to produce the first 1000 units; to produce one additional unit beyond 1000 costs only about \$3.38. ■

EXAMPLE 8 In manufacturing and selling x units of a certain commodity, the price function p and the cost function C (in dollars) are given by

$$p(x) = 5.00 - 0.002x$$

$$C(x) = 3.00 + 1.10x$$

Find expressions for the marginal revenue, marginal cost, and marginal profit. Determine the production level that will produce the maximum total profit.

SOLUTION

$$R(x) = xp(x) = 5.00x - 0.002x^2$$

$$P(x) = R(x) - C(x) = -3.00 + 3.90x - 0.002x^2$$

Thus, we have the following derivatives:

$$\text{Marginal revenue:} \quad \frac{dR}{dx} = 5 - 0.004x$$

$$\text{Marginal cost:} \quad \frac{dC}{dx} = 1.1$$

$$\text{Marginal profit:} \quad \frac{dP}{dx} = \frac{dR}{dx} - \frac{dC}{dx} = 3.9 - 0.004x$$

To maximize profit, we set $dP/dx = 0$ and solve. This gives $x = 975$ as the only critical point to consider. It does provide a maximum, as may be checked by the First Derivative Test. The maximum profit is $P(975) = \$1898.25$. ■

Note that at $x = 975$ both the marginal revenue and the marginal cost are \$1.10. In general, a company should expect to be at a maximum profit level when the cost of producing an additional unit equals the revenue from that unit.

Concepts Review

1. If a rectangle of area 100 has length x and width y, then the allowable values for x are _____.

2. The perimeter of the rectangle in Question 1, expressed in terms of x (only), is _____.

3. The least squares line through the origin minimizes $S = \sum_{i=1}^{n} (\_\_\_\_\_)^2$

4. In economics, $\frac{dR}{dx}$ is called _____ and $\frac{dC}{dx}$ is called _____.

Problem Set 3.4

1. Find two numbers whose product is -16 and the sum of whose squares is a minimum.

2. For what number does the principal square root exceed eight times the number by the largest amount?

3. For what number does the principal fourth root exceed twice the number by the largest amount?

4. Find two numbers whose product is -12 and the sum of whose squares is a minimum.

5. Find the points on the parabola $y = x^2$ that are closest to the point $(0, 5)$. *Hint:* Minimize the square of the distance between (x, y) and $(0, 5)$.

6. Find the points on the parabola $x = 2y^2$ that are closest to the point $(10, 0)$. *Hint:* Minimize the square of the distance between (x, y) and $(10, 0)$.

7. What number exceeds its square by the maximum amount? Begin by convincing yourself that this number is on the interval $[0, 1]$.

8. Show that for a rectangle of given perimeter K the one with maximum area is a square.

9. Find the volume of the largest open box that can be made from a piece of cardboard 24 inches square by cutting equal squares from the corners and turning up the sides (see Example 1).

≈ **10.** A farmer has 80 feet of fence with which he plans to enclose a rectangular pen along one side of his 100-foot barn, as shown in Figure 18 (the side along the barn needs no fence). What are the dimensions of the pen that has maximum area?

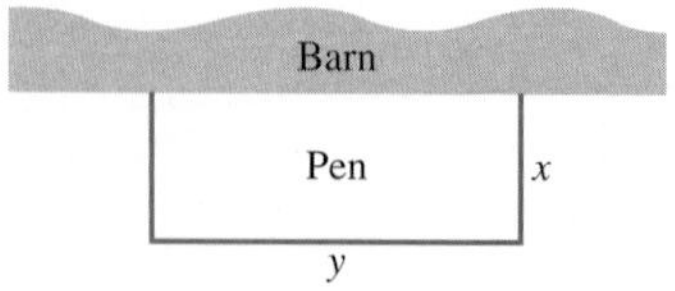

Figure 18

≈ **11.** The farmer of Problem 10 decides to make three identical pens with his 80 feet of fence, as shown in Figure 19. What dimensions for the total enclosure make the area of the pens as large as possible?

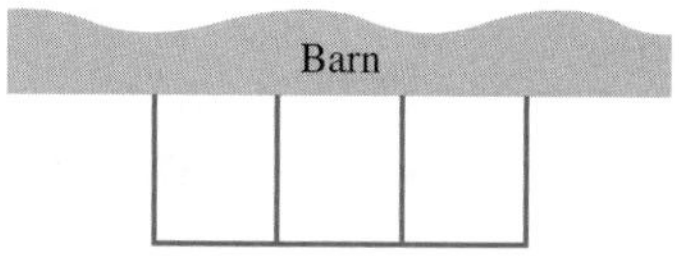

Figure 19

12. Suppose that the farmer of Problem 10 has 180 feet of fence and wants the pen to adjoin to the whole side of the 100-foot barn, as shown in Figure 20. What should the dimensions be for maximum area? Note that $0 \le x \le 40$ in this case.

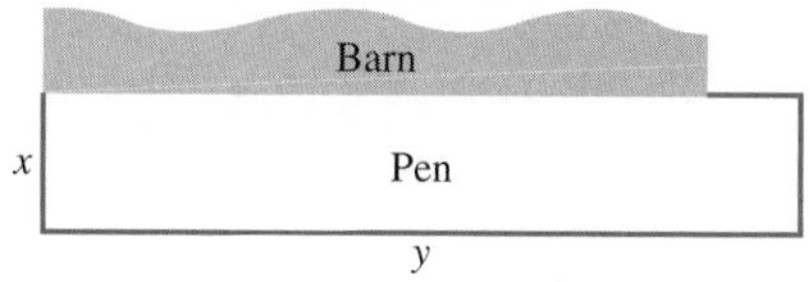

Figure 20

13. A farmer wishes to fence off two identical adjoining rectangular pens, each with 900 square feet of area, as shown in Figure 21. What are x and y so that the least amount of fence is required?

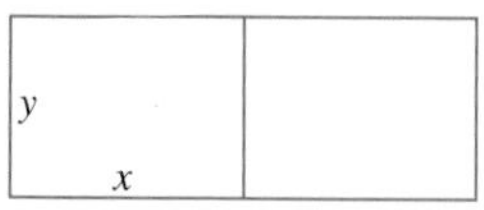

Figure 21

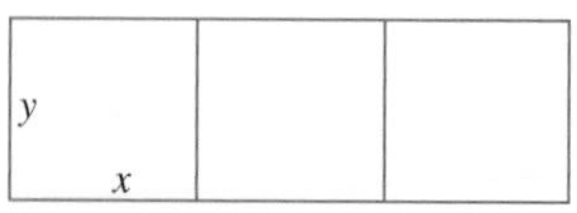

Figure 22

14. A farmer wishes to fence off three identical adjoining rectangular pens (see Figure 22), each with 300 square feet of area. What should the width and length of each pen be so that the least amount of fence is required?

15. Suppose that the outer boundary of the pens in Problem 14 requires heavy fence that costs \$3 per foot, but that the two internal partitions require fence costing only \$2 per foot. What dimensions x and y will produce the least expensive cost for the pens?

16. Solve Problem 14 assuming that the area of each pen is 900 square feet. Study the solution to this problem and to Problem 14 and make a conjecture about the ratio of x/y in all problems of this type. Try to prove your conjecture.

17. Find the points P and Q on the curve $y = x^2/4$, $0 \le x \le 2\sqrt{3}$, that are closest to and farthest from the point $(0, 4)$. *Hint:* The algebra is simpler if you consider the square of the required distance rather than the distance itself.

18. A right circular cone is to be inscribed in another right circular cone of given volume, with the same axis and with the vertex of the inner cone touching the base of the outer cone. What must be the ratio of their altitudes for the inscribed cone to have maximum volume?

19. A small island is 2 miles from the nearest point P on the straight shoreline of a large lake. If a woman on the island can row a boat 3 miles per hour and can walk 4 miles per hour, where should the boat be landed in order to arrive at a town 10 miles down the shore from P in the least time?

20. In Problem 19, suppose that the woman will be picked up by a car that will average 50 miles per hour when she gets to the shore. Then where should she land?

21. In Problem 19, suppose that the woman uses a motorboat that goes 20 miles per hour. Then where should she land?

22. A powerhouse is located on one bank of a straight river that is w feet wide. A factory is situated on the opposite bank of the river, L feet downstream from the point A directly opposite the powerhouse. What is the most economical path for a cable connecting the powerhouse to the factory if it costs a dollars per foot to lay the cable under water and b dollars per foot on land ($a > b$)?

23. At 7:00 A.M. one ship was 60 miles due east from a second ship. If the first ship sailed west at 20 miles per hour and the second ship sailed southeast at 30 miles per hour, when were they closest together?

24. Find the equation of the line that is tangent to the ellipse $b^2x^2 + a^2y^2 = a^2b^2$ in the first quadrant and forms with the coordinate axes the triangle with smallest possible area (a and b are positive constants).

25. Find the greatest volume that a right circular cylinder can have if it is inscribed in a sphere of radius r.

26. Show that the rectangle with maximum perimeter that can be inscribed in a circle is a square.

27. What are the dimensions of the right circular cylinder with greatest curved surface area that can be inscribed in a sphere of radius r?

28. The illumination at a point is inversely proportional to the square of the distance of the point from the light source and directly proportional to the intensity of the light source. If two light sources are s feet apart and their intensities are I_1 and I_2, respectively, at what point between them will the sum of their illuminations be a minimum?

29. A wire of length 100 centimeters is cut into two pieces; one is bent to form a square, and the other is bent to form an equilateral triangle. Where should the cut be made if (a) the sum of the two areas is to be a minimum; (b) a maximum? (Allow the possibility of no cut.)

30. A closed box in the form of a rectangular parallelepiped with a square base is to have a given volume. If the material used in the bottom costs 20% more per square inch than the material in the sides, and the material in the top costs 50% more per square inch than that of the sides, find the most economical proportions for the box.

31. An observatory is to be in the form of a right circular cylinder surmounted by a hemispherical dome. If the hemispherical dome costs twice as much per square foot as the cylindrical wall, what are the most economical proportions for a given volume?

32. A weight connected to a spring moves along the x-axis so that its x-coordinate at time t is

$$x = \sin 2t + \sqrt{3}\cos 2t$$

What is the farthest that the weight gets from the origin?

33. A flower bed will be in the shape of a sector of a circle (a pie-shaped region) of radius r and vertex angle θ. Find r and θ if its area is a constant A and the perimeter is a minimum.

34. A fence h feet high runs parallel to a tall building and w feet from it (Figure 23). Find the length of the shortest ladder that will reach from the ground across the top of the fence to the wall of the building.

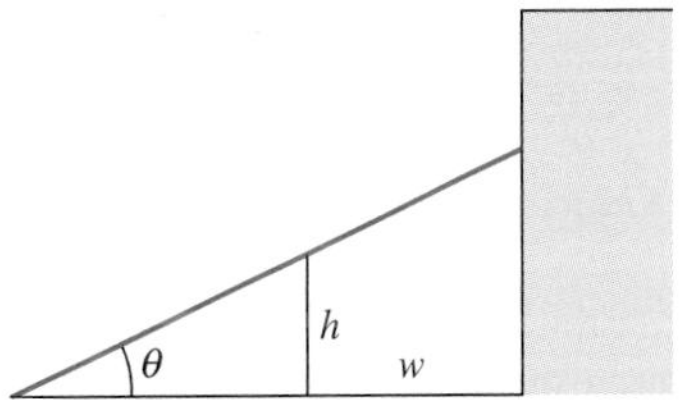

Figure 23

35. A rectangle has two corners on the x-axis and the other two on the parabola $y = 12 - x^2$, with $y \ge 0$ (Figure 24). What are the dimensions of the rectangle of this type with maximum area?

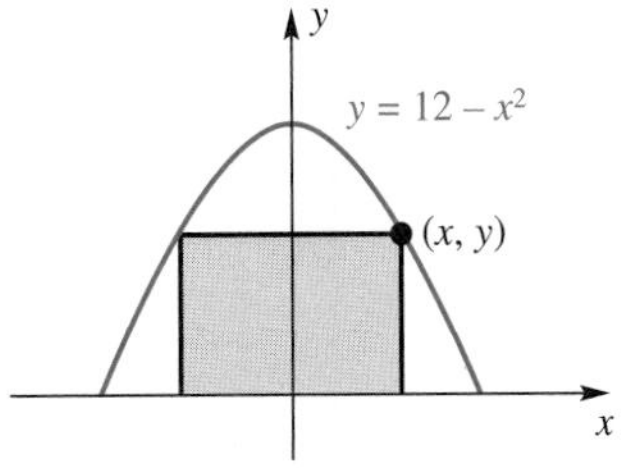

Figure 24

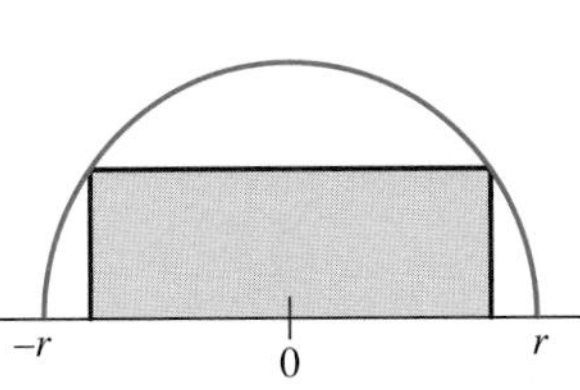

Figure 25

36. A rectangle is to be inscribed in a semicircle of radius r, as shown in Figure 25. What are the dimensions of the rectangle if its area is to be maximized?

37. Of all right circular cylinders with a given surface area, find the one with the maximum volume. *Note:* The ends of the cylinders are closed.

38. Find the dimensions of the rectangle of greatest area that can be inscribed in the ellipse $x^2/a^2 + y^2/b^2 = 1$.

39. Of all rectangles with a given diagonal, find the one with the maximum area.

40. A humidifier uses a rotating disk of radius r, which is partially submerged in water. The most evaporation occurs when the exposed wetted region (shown as the upper shaded region in Figure 26) is maximized. Show that this happens when h (the distance from the center to the water) is equal to $r/\sqrt{1+\pi^2}$.

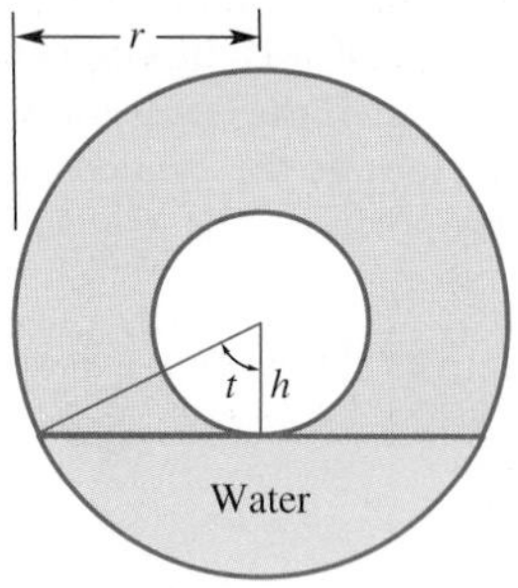

Figure 26

41. A metal rain gutter is to have 3-inch sides and a 3-inch horizontal bottom, the sides making an equal angle θ with the bottom (Figure 27). What should θ be in order to maximize the carrying capacity of the gutter? *Note:* $0 \le \theta \le \pi/2$.

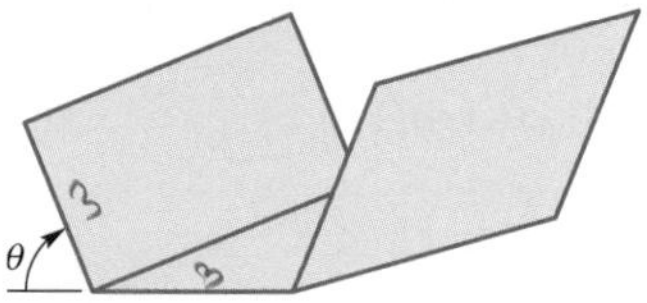

Figure 27

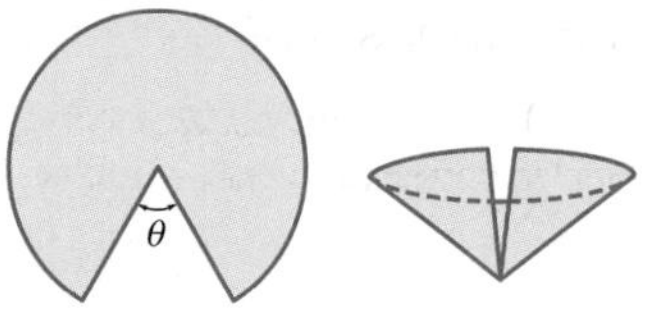

Figure 28

42. A huge conical tank is to be made from a circular piece of sheet metal of radius 10 meters by cutting out a sector with vertex angle θ and then welding together the straight edges of the remaining piece (Figure 28). Find θ so that the resulting cone has the largest possible volume.

43. A covered box is to be made from a rectangular sheet of cardboard measuring 5 feet by 8 feet. This is done by cutting out the shaded regions of Figure 29 and then folding on the dotted lines. What are the dimensions x, y, and z that maximize the volume?

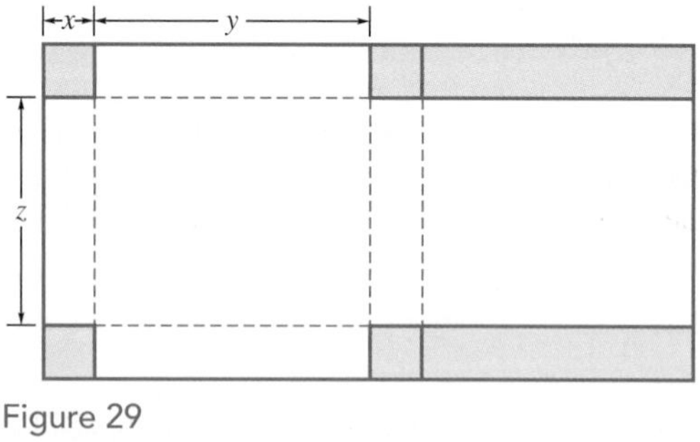

Figure 29

44. I have enough pure silver to coat 1 square meter of surface area. I plan to coat a sphere and a cube. What dimensions should they be if the total volume of the silvered solids is to be a maximum? A minimum? (Allow the possibility of all the silver going onto one solid.)

45. One corner of a long narrow strip of paper is folded over so that it just touches the opposite side, as shown in Figure 30. With parts labeled as indicated, determine x in order to

(a) maximize the area of triangle A;

(b) minimize the area of triangle B;

(c) minimize the length z.

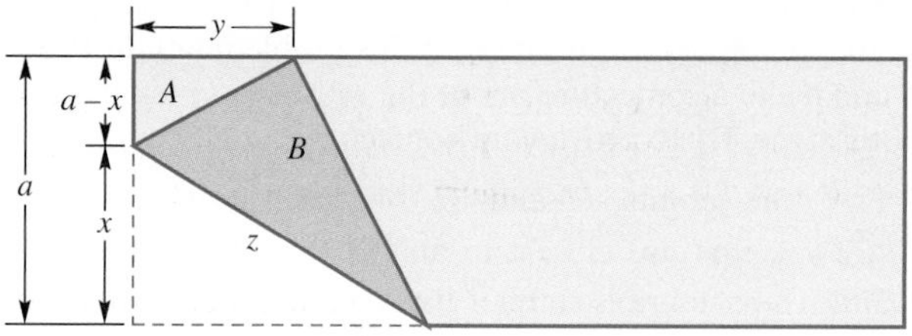

Figure 30

46. Determine θ so that the area of the symmetric cross shown in Figure 31 is maximized. Then find this maximum area.

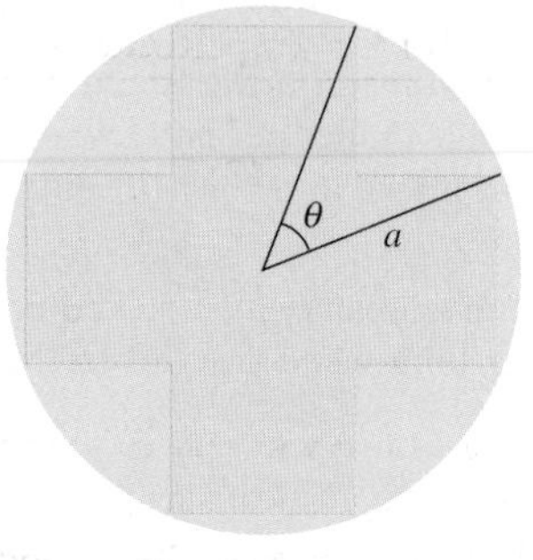

Figure 31

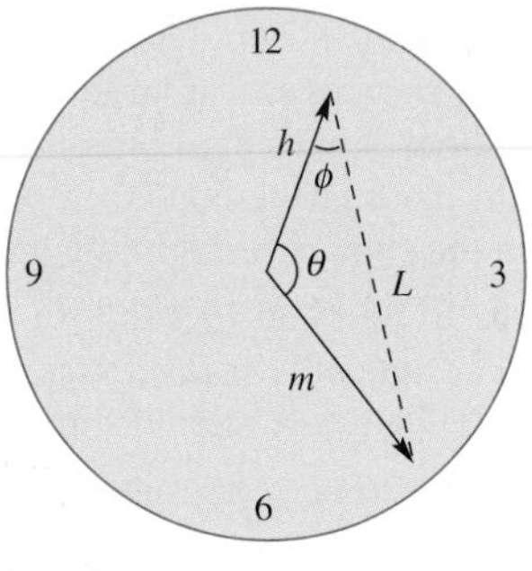

Figure 32

CAS **47.** A clock has hour and minute hands of lengths h and m, respectively, with $h \le m$. We wish to study this clock at times between 12:00 and 12:30. Let θ, ϕ, and L be as in Figure 32 and note that θ increases at a constant rate. By the Law of Cosines, $L = L(\theta) = (h^2 + m^2 - 2hm\cos\theta)^{1/2}$, and so

$$L'(\theta) = hm(h^2 + m^2 - 2hm\cos\theta)^{-1/2}\sin\theta$$

(a) For $h = 3$ and $m = 5$, determine L', L, and ϕ at the instant when L' is largest.

(b) Rework part (a) when $h = 5$ and $m = 13$.

(c) Based on parts (a) and (b), make conjectures about the values of L', L, and ϕ at the instant when the tips of the hands are separating most rapidly.

(d) Try to prove your conjectures.

≈ C **48.** An object thrown from the edge of a 100-foot cliff follows the path given by $y = -\frac{x^2}{10} + x + 100$. An observer stands 2 feet from the bottom of the cliff.

(a) Find the position of the object when it is closest to the observer.

(b) Find the position of the object when it is farthest from the observer.

≈ CAS **49.** The earth's position in the solar system at time t can be described approximately by $P(93\cos(2\pi t), 93\sin(2\pi t))$, where the sun is at the origin and distances are measured in millions of miles. Suppose that an asteroid has position $Q(60\cos[2\pi(1.51t - 1)], 120\sin[2\pi(1.51t - 1)])$. When, over the time period $[0, 20]$ (i.e., over the next 20 years), does the asteroid come closest to the earth? How close does it come?

50. An advertising flyer is to contain 50 square inches of printed matter, with 2-inch margins at the top and bottom and 1-inch margins on each side. What dimensions for the flyer would use the least paper?

≈ **51.** One end of a 27-foot ladder rests on the ground and the other end rests on the top of an 8-foot wall. As the bottom end is pushed along the ground toward the wall, the top end extends beyond the wall. Find the maximum horizontal overhang of the top end.

C **52.** Brass is produced in long rolls of a thin sheet. To monitor the quality, inspectors select at random a piece of the sheet, measure its area, and count the number of surface imperfections on that piece. The area varies from piece to piece. The following table gives data on the area (in square feet) of the selected piece and the number of surface imperfections found on that piece.

Piece	Area in Square Feet	Number of Surface Imperfections
1	1.0	3
2	4.0	12
3	3.6	9
4	1.5	5
5	3.0	8

(a) Make a scatter plot with area on the horizontal axis and number of surface imperfections on the vertical axis.

(b) Does it look like a line through the origin would be a good model for these data? Explain.

(c) Find the equation of the least-squares line through the origin.

(d) Use the result of part (c) to predict how many surface imperfections there would be on a sheet with area 2.0 square feet.

C **53.** Suppose that every customer order taken by the XYZ Company requires exacty 5 hours of labor for handling the paperwork; this length of time is *fixed* and does not vary from lot to lot. The total number of hours y required to manufacture and sell a lot of size x would then be

$$y = (\text{number of hours to produce a lot of size } x) + 5$$

Some data on XYZ's bookcases are given in the following table.

Order	Lot Size x	Total Labor Hours y
1	11	38
2	16	52
3	8	29
4	7	25
5	10	38

(a) From the description of the problem, the least-squares line should have 5 as its y-intercept. Find a formula for the value of the slope b that minimizes the sum of squares

$$S = \sum_{i=1}^{n}[y_i - (5 + bx_i)]^2$$

(b) Use this formula to estimate the slope b.

(c) Use your least-squares line to predict the total number of labor hours to produce a lot consisting of 15 bookcases.

54. The fixed monthly cost of operating a plant that makes Zbars is \$7000, while the cost of manufacturing each unit is \$100. Write an expression for $C(x)$, the total cost of making x Zbars in a month.

55. The manufacturer of Zbars estimates that 100 units per month can be sold if the unit price is \$250 and that sales will increase by 10 units for each \$5 decrease in price. Write an expression for the price $p(n)$ and the revenue $R(n)$ if n units are sold in one month, $n \geq 100$

56. Use the information in Problems 54 and 55 to write an expression for the total monthly profit $P(n)$, $n \geq 100$.

57. Sketch the graph of $P(n)$ of Problem 56, and from it estimate the value of n that maximizes P. Find this n exactly by the methods of calculus.

C **58.** The total cost of producing and selling x units of Xbars per month is $C(x) = 100 + 3.002x - 0.0001x^2$. If the production level is 1600 units per month, find the average cost, $C(x)/x$, of each unit and the marginal cost.

59. The total cost of producing and selling n units of a certain commodity per week is $C(n) = 1000 + n^2/1200$. Find the average cost, $C(n)/n$, of each unit and the marginal cost at a production level of 800 units per week.

60. The total cost of producing and selling $100x$ units of a particular commodity per week is

$$C(x) = 1000 + 33x - 9x^2 + x^3$$

Find (a) the level of production at which the marginal cost is a minimum, and (b) the minimum marginal cost.

61. A price function, p, is defined by

$$p(x) = 20 + 4x - \frac{x^2}{3}$$

where $x \geq 0$ is the number of units.

(a) Find the total revenue function and the marginal revenue function.

(b) On what interval is the total revenue increasing?

(c) For what number x is the marginal revenue a maximum?

C **62.** For the price function defined by

$$p(x) = (182 - x/36)^{1/2}$$

find the number of units x_1 that makes the total revenue a maximum and state the maximum possible revenue. What is the marginal revenue when the optimum number of units, x_1, is sold?

63. For the price function given by

$$p(x) = 800/(x + 3) - 3$$

find the number of units x_1 that makes the total revenue a maximum and state the maximum possible revenue. What is the marginal revenue when the optimum number of units, x_1, is sold?

64. A riverboat company offers a fraternal organization a Fourth of July excursion with the understanding that there will be at least 400 passengers. The price of each ticket will be \$12.00, and the company agrees to discount the price by \$0.20 for each 10 passengers in excess of 400. Write an expression for the price function $p(x)$ and find the number x_1 of passengers that makes the total revenue a maximum.

65. The XYZ Company manufactures wicker chairs. With its present machines, it has a maximum yearly output of 500 units. If it makes x chairs, it can set a price of $p(x) = 200 - 0.15x$ dollars each and will have a total yearly cost of $C(x) = 5000 + 6x - 0.002x^2$ dollars. The company has the opportunity to buy a new machine for \$4000 with which the company can make up to an additional 250 chairs per year. The cost function for values of x between 500 and 750 is thus $C(x) = 9000 + 6x - 0.002x^2$. Basing your analysis on the profit for the next year, answer the following questions.

(a) Should the company purchase the additional machine?

(b) What should be the level of production?

66. Repeat Problem 65, assuming that the additional machine costs \$3000.

C **67.** The ZEE Company makes zingos, which it markets at a price of $p(x) = 10 - 0.001x$ dollars, where x is the number produced each month. Its total monthly cost is $C(x) = 200 + 4x - 0.01x^2$. At peak production, it can make 300 units. What is its maximum monthly profit and what level of production gives this profit?

C **68.** If the company of Problem 67 expands its facilities so that it can produce up to 450 units each month, its monthly cost function takes the form $C(x) = 800 + 3x - 0.01x^2$ for $300 < x \le 450$. Find the production level that maximizes monthly profit and evaluate this profit. Sketch the graph of the monthly profit function $P(x)$ on $0 \le x \le 450$.

EXPL **69.** The arithmetic mean of the numbers a and b is $(a + b)/2$, and the geometric mean of two positive numbers a and b is $\sqrt{ab}$. Suppose that $a > 0$ and $b > 0$.

(a) Show that $\sqrt{ab} \le (a + b)/2$ holds by squaring both sides and simplifying.

(b) Use calculus to show that $\sqrt{ab} \le (a + b)/2$. *Hint:* Consider a to be fixed. Square both sides of the inequality and divide through by b. Define the function $F(b) = (a + b)^2/4b$. Show that F has its minimum at a.

(c) The geometric mean of three positive numbers a, b, and c is $(abc)^{1/3}$. Show that the analogous inequality holds:

$$(abc)^{1/3} \le \frac{a + b + c}{3}$$

Hint: Consider a and c to be fixed and define $F(b) = (a + b + c)^3/27b$. Show that F has a minimum at $b = (a + c)/2$ and that this minimum is $[(a + c)/2]^2$. Then use the result from (b).

EXPL **70.** Show that of all three-dimensional boxes with a given surface area, the cube has the greatest volume. *Hint:* The surface area is $S = 2(lw + lh + hw)$ and the volume is $V = lwh$. Let $a = lw$, $b = lh$, and $c = hw$. Use the previous problem to show that $(V^2)^{1/3} \le S/6$. When does equality hold?

Answers to Concepts Review: 1. $0 < x < \infty$ **2.** $2x + 200/x$ **3.** $y_i - bx_i$ **4.** marginal revenue; marginal cost

3.5 Graphing Functions Using Calculus

Our treatment of graphing in Section 0.4 was elementary. We proposed plotting enough points so that the essential features of the graph were clear. We mentioned that symmetries of the graph could reduce the effort involved. We suggested that one should be alert to possible asymptotes. But if the equation to be graphed is complicated or if we want a very accurate graph, the techniques of that section are inadequate.

Calculus provides a powerful tool for analyzing the fine structure of a graph, especially in identifying those points where the character of the graph changes. We can locate local maximum points, local minimum points, and inflection points; we can determine precisely where the graph is increasing or where it is concave up. Inclusion of all these ideas in our graphing procedure is the program for this section.

Polynomial Functions A polynomial function of degree 1 or 2 is easy to graph by hand; one of degree 50 could be next to impossible. If the degree is of modest size, such as 3 to 6, we can use the tools of calculus to great advantage.

EXAMPLE 1 Sketch the graph of $f(x) = \dfrac{3x^5 - 20x^3}{32}$.

SOLUTION Since $f(-x) = -f(x)$, f is an odd function, and therefore its graph is symmetric with respect to the origin. Setting $f(x) = 0$, we find the x-intercepts to be 0 and $\pm\sqrt{20/3} \approx \pm 2.6$. We can go this far without calculus.

When we differentiate f, we obtain

$$f'(x) = \frac{15x^4 - 60x^2}{32} = \frac{15x^2(x - 2)(x + 2)}{32}$$

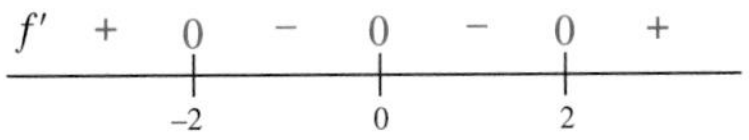

Figure 1

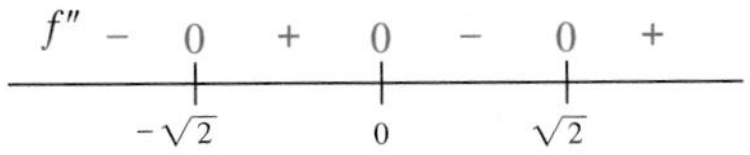

Figure 2

Thus, the critical points are $-2, 0$, and 2; we quickly discover (Figure 1) that $f'(x) > 0$ on $(-\infty, -2)$ and $(2, \infty)$ and that $f'(x) < 0$ on $(-2, 0)$ and $(0, 2)$. These facts tell us where f is increasing and where it is decreasing; they also confirm that $f(-2) = 2$ is a local maximum value and that $f(2) = -2$ is a local minimum value.

Differentiating again, we get

$$f''(x) = \frac{60x^3 - 120x}{32} = \frac{15x\left(x - \sqrt{2}\right)\left(x + \sqrt{2}\right)}{8}$$

By studying the sign of $f''(x)$ (Figure 2), we deduce that f is concave upward on $\left(-\sqrt{2}, 0\right)$ and $\left(\sqrt{2}, \infty\right)$ and concave downward on $\left(-\infty, -\sqrt{2}\right)$ and $\left(0, \sqrt{2}\right)$. Thus, there are three points of inflection: $\left(-\sqrt{2}, 7\sqrt{2}/8\right) \approx (-1.4, 1.2)$, $(0, 0)$, and $\left(\sqrt{2}, -7\sqrt{2}/8\right) \approx (1.4, -1.2)$.

Much of this information is collected at the top of Figure 3, which we use to sketch the graph directly below it.

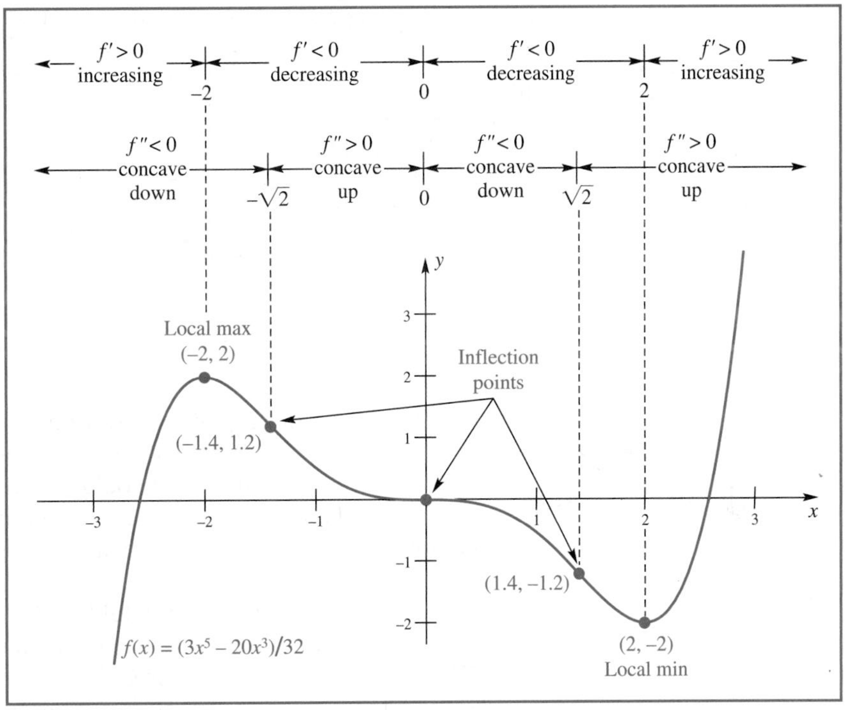

Figure 3

■

Rational Functions A rational function, being the quotient of two polynomial functions, is considerably more complicated to graph than a polynomial. In particular, we can expect dramatic behavior near where the denominator would be zero.

EXAMPLE 2 Sketch the graph of $f(x) = \dfrac{x^2 - 2x + 4}{x - 2}$.

SOLUTION This function is neither even nor odd, so we do not have any of the usual symmetries. There are no x-intercepts, since the solutions to $x^2 - 2x + 4 = 0$ are not real numbers. The y-intercept is -2. We anticipate a vertical asymptote at $x = 2$. In fact,

$$\lim_{x \to 2^-} \frac{x^2 - 2x + 4}{x - 2} = -\infty \quad \text{and} \quad \lim_{x \to 2^+} \frac{x^2 - 2x + 4}{x - 2} = \infty$$

Differentiation twice gives

$$f'(x) = \frac{x(x-4)}{(x-2)^2} \quad \text{and} \quad f''(x) = \frac{8}{(x-2)^3}$$

The stationary points are therefore $x = 0$ and $x = 4$.

Thus, $f'(x) > 0$ on $(-\infty, 0) \cup (4, \infty)$ and $f'(x) < 0$ on $(0, 2) \cup (2, 4)$. (Remember, $f'(x)$ does not exist when $x = 2$.) Also, $f''(x) > 0$ on $(2, \infty)$ and $f''(x) < 0$ on $(-\infty, 2)$. Since $f''(x)$ is never 0, there are no inflection points. On the other hand, $f(0) = -2$ and $f(4) = 6$ give local maximum and minimum values, respectively.

It is a good idea to check on the behavior of $f(x)$ for large $|x|$. Since

$$f(x) = \frac{x^2 - 2x + 4}{x - 2} = x + \frac{4}{x - 2}$$

the graph of $y = f(x)$ gets closer and closer to the line $y = x$ as $|x|$ gets larger and larger. We call the line $y = x$ an **oblique asymptote** for the graph of f (see Problem 49 of Section 1.5).

With all this information, we are able to sketch a rather accurate graph (Figure 4).

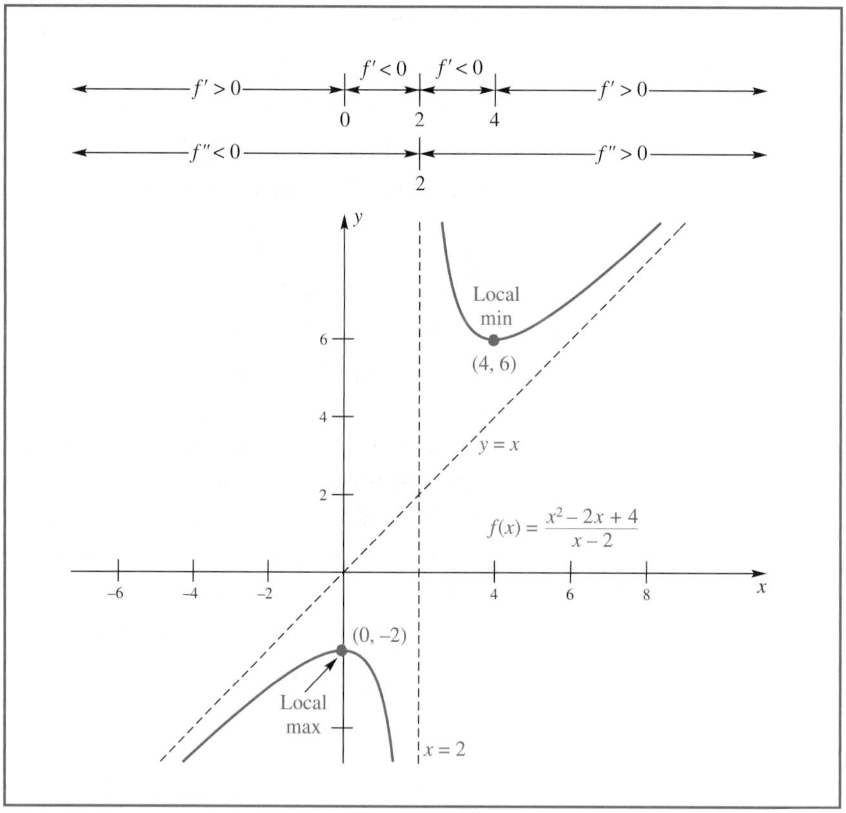

Figure 4

■

Functions Involving Roots There is an endless variety of functions involving roots. Here is one example.

EXAMPLE 3 Analyze the function

$$F(x) = \frac{\sqrt{x}(x-5)^2}{4}$$

and sketch its graph.

SOLUTION The domain of F is $[0, \infty)$ and the range is $[0, \infty)$, so the graph of F is confined to the first quadrant and the positive coordinate axes. The x-intercepts are 0 and 5; the y-intercept is 0. From

$$F'(x) = \frac{5(x-1)(x-5)}{8\sqrt{x}}, \qquad x > 0$$

we find the stationary points 1 and 5. Since $F'(x) > 0$ on $(0, 1)$ and $(5, \infty)$, while $F'(x) < 0$ on $(1, 5)$, we conclude that $F(1) = 4$ is a local maximum value and $F(5) = 0$ is a local minimum value.

So far, it has been clear sailing. But on calculating the second derivative, we obtain

$$F''(x) = \frac{5(3x^2 - 6x - 5)}{16x^{3/2}}, \qquad x > 0$$

which is quite complicated. However, $3x^2 - 6x - 5 = 0$ has one solution in $(0, \infty)$, namely $1 + 2\sqrt{6}/3 \approx 2.6$.

Using the test points 1 and 3, we conclude that $f''(x) < 0$ on $(0, 1 + 2\sqrt{6}/3$ and $f''(x) > 0$ on $(1 + 2\sqrt{6}/3, \infty)$. It then follows that the point $\left(1 + 2\sqrt{6}/3, F(1 + 2\sqrt{6}/3)\right)$, which is approximately (2.6, 2.3), is an inflection point.

As x grows large, $F(x)$ grows without bound and much faster than any linear function; there are no asymptotes. The graph is sketched in Figure 5. ■

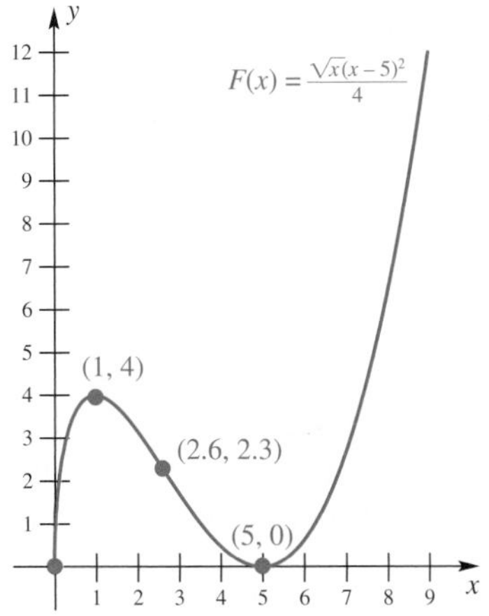

Figure 5

Summary of the Method In graphing functions, there is no substitute for common sense. However, the following procedure will be helpful in most cases.

Step 1: Precalculus analysis.

(a) Check the *domain* and *range* of the function to see if any regions of the plane are excluded.

(b) Test for *symmetry* with respect to the y-axis and the origin. (Is the function even or odd?)

(c) Find the *intercepts*.

Step 2: Calculus analysis.

(a) Use the first derivative to find the critical points and to find out where the graph is *increasing* and *decreasing*.

(b) Test the critical points for *local maxima* and *minima*.

(c) Use the second derivative to find out where the graph is *concave upward* and *concave downward* and to locate *inflection points*.

(d) Find the *asymptotes*.

Step 3: Plot a few points (including all critical points and inflection points).

Step 4: Sketch the graph.

EXAMPLE 4 Sketch the graphs of $f(x) = x^{1/3}$ and $g(x) = x^{2/3}$ and their derivatives.

SOLUTION The domain for both functions is $(-\infty, \infty)$. (Remember, the cube root exists for every real number.) The range for $f(x)$ is $(-\infty, \infty)$ since every real number is the cube root of some other number. Writing $g(x)$ as $g(x) = x^{2/3} = (x^{1/3})^2$, we see that $g(x)$ must be nonnegative; it's range is $[0, \infty)$. Since $f(-x) = (-x)^{1/3} = -x^{1/3} = -f(x)$, we see that f is an odd function. Similarly, since $g(-x) = (-x)^{2/3} = ((-x)^2)^{1/3} = (x^2)^{1/3} = g(x)$, we see that g is an even function. The first derivatives are

$$f'(x) = \frac{1}{3}x^{-2/3} = \frac{1}{3x^{2/3}}$$

and

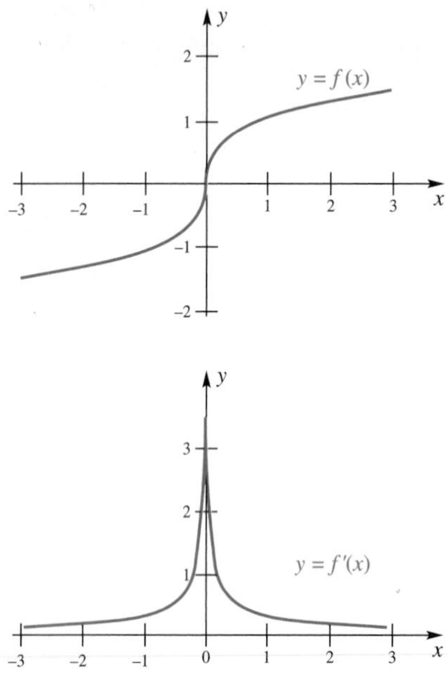

Figure 6

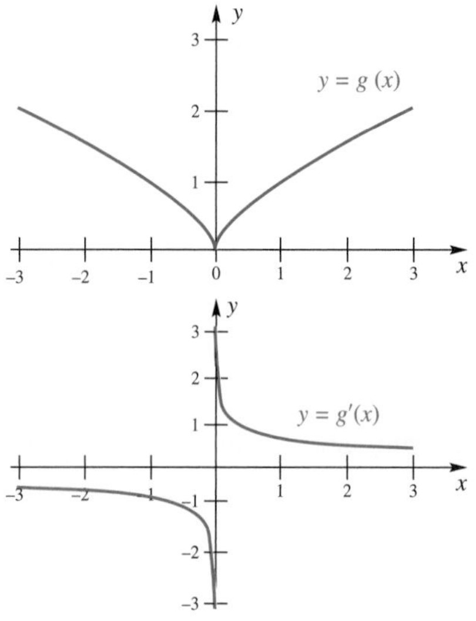

Figure 7

$$g'(x) = \frac{2}{3}x^{-1/3} = \frac{2}{3x^{1/3}}$$

and the second derivatives are

$$f''(x) = -\frac{2}{9}x^{-5/3} = -\frac{2}{9x^{5/3}}$$

and

$$g''(x) = -\frac{2}{9}x^{-4/3} = -\frac{2}{9x^{4/3}}$$

For both functions the only critical point, in this case a point where the derivative doesn't exist, is $x = 0$.

Note that $f'(x) > 0$ for all x, except $x = 0$. Thus, f is increasing on $(-\infty, 0]$ and also on $[0, \infty)$, but because f is continuous on $(-\infty, \infty)$, we can conclude that f is always increasing. Consequently, f has no local maxima or minima. Since $f''(x)$ is positive when x is negative and negative when x is positive (and undefined when $x = 0$), we conclude that f is concave up on $(-\infty, 0)$ and concave down on $(0, \infty)$. The point $(0, 0)$ is an inflection point because that is where the concavity changes.

Now consider $g(x)$. Note that $g'(x)$ is negative when x is negative and positive when x is positive. Since g is decreasing on $(-\infty, 0]$ and increasing on $[0, \infty)$, $g(0) = 0$ is a local mimimum. Note also that $g''(x)$ is negative as long as $x \neq 0$. Thus g is concave down on $(-\infty, 0)$ and concave down on $(0, \infty)$, so $(0, 0)$ is not an inflection point. The graphs of $f(x)$, $f'(x)$, $g(x)$ and $g'(x)$ are shown in Figures 6 and 7. ■

Note that in the above example both functions had one critical point, $x = 0$, where the derivative was undefined. Yet the graphs of the functions are fundamentally different. The graph of $y = f(x)$ has a tangent line at all points, but it is vertical when $x = 0$. (If the tangent line is vertical, then the derivative doesn't exist at that point.) The graph of $y = g(x)$ has a sharp point, called a **cusp,** at $x = 0$.

Using the Derivative's Graph to Graph a Function Knowing just a function's derivative can tell us a lot about the function itself and what its graph looks like.

EXAMPLE 5 Figure 8 shows a plot of $y = f'(x)$. Find all local extrema and points of inflection of f on the interval $[-1, 3]$. Given that $f(1) = 0$, sketch the graph of $y = f(x)$.

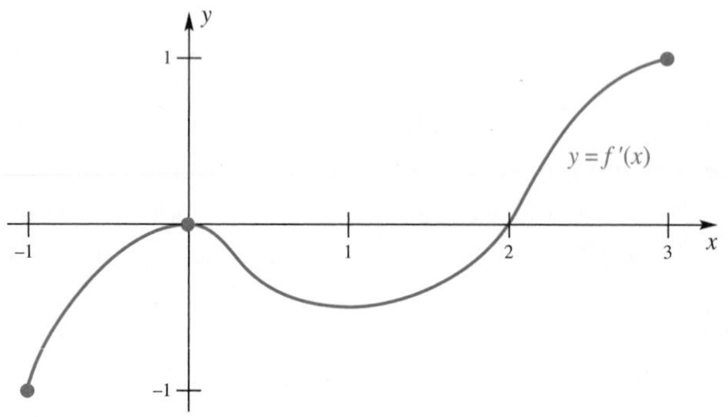

Figure 8

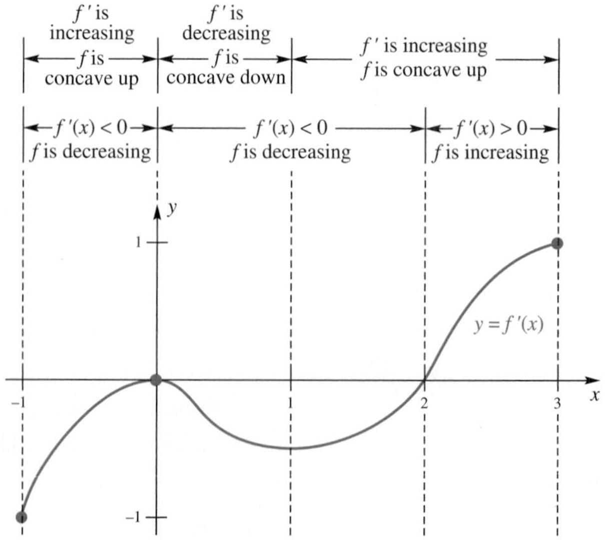

Figure 9

$f(-1)$	Local maximum
$f(2)$	Local minimum
$f(3)$	Local maximum
$(0, f(0))$	Inflection point
$(1, f(1))$	Inflection point

SOLUTION The derivative is negative on the intervals $(-1, 0)$ and $(0, 2)$, and positive on the interval $(2, 3)$. Thus, f is decreasing on $[-1, 0]$ and on $[0, 2]$ so there is a local maximum at the left end point $x = -1$. Since $f'(x)$ is positive on $(2, 3)$, f is increasing on $[2, 3]$ so there is a local maximum at the right end point $x = 3$. Since f is decreasing on $[-1, 2]$ and increasing on $[2, 3]$, there is a local minimum at $x = 2$. Figure 9 summarizes this information.

Inflection points for f occur when the concavity of f changes. Since f' is increasing on $(-1, 0)$ and on $(1, 3)$, f is concave up on $(-1, 0)$ and on $(1, 3)$. Since f' is decreasing on $(0, 1)$, f is concave down on $(0, 1)$. Thus, f changes concavity at $x = 0$ and $x = 1$. The inflection points are therefore $(0, f(0))$ and $(1, f(1))$.

The information given above, together with the fact that $f(1) = 0$, can be used to sketch the graph of $y = f(x)$. (The sketch cannot be too precise because we still have limited information about f.) A sketch is shown in Figure 10.

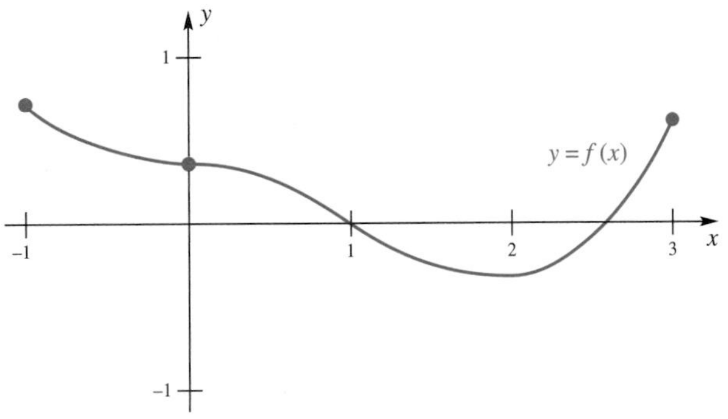

Figure 10

Concepts Review

1. The graph of f is symmetric with respect to the y-axis if $f(-x) =$ _____ for every x; the graph is symmetric with respect to the origin if $f(-x) =$ _____ for every x.

2. If $f'(x) < 0$ and $f''(x) > 0$ for all x in an interval I, then the graph of f is both _____ and _____ on I.

3. The graph of $f(x) = x^3/[(x + 1)(x - 2)(x - 3)]$ has as vertical asymptotes the lines _____ and as a horizontal asymptote the line _____.

4. We call $f(x) = 3x^5 - 2x^2 + 6$ a(n) _____ function, and we call $g(x) = (3x^5 - 2x^2 + 6)/(x^2 - 4)$ a(n) _____ function.

Problem Set 3.5

In Problems 1–27, make an analysis as suggested in the summary above and then sketch the graph.

1. $f(x) = x^3 - 3x + 5$
2. $f(x) = 2x^3 - 3x - 10$
3. $f(x) = 2x^3 - 3x^2 - 12x + 3$
4. $f(x) = (x - 1)^3$
5. $G(x) = (x - 1)^4$
6. $H(t) = t^2(t^2 - 1)$
7. $f(x) = x^3 - 3x^2 + 3x + 10$
8. $F(s) = \dfrac{4s^4 - 8s^2 - 12}{3}$
9. $g(x) = \dfrac{x}{x + 1}$
10. $g(s) = \dfrac{(s - \pi)^2}{s}$
11. $f(x) = \dfrac{x}{x^2 + 4}$
12. $\Lambda(\theta) = \dfrac{\theta^2}{\theta^2 + 1}$
13. $h(x) = \dfrac{x}{x - 1}$
14. $P(x) = \dfrac{1}{x^2 + 1}$
15. $f(x) = \dfrac{(x - 1)(x - 3)}{(x + 1)(x - 2)}$
16. $w(z) = \dfrac{z^2 + 1}{z}$
17. $g(x) = \dfrac{x^2 + x - 6}{x - 1}$
18. $f(x) = |x|^3$ *Hint:* $\dfrac{d}{dx}|x| = \dfrac{x}{|x|}$
19. $R(z) = z|z|$
20. $H(q) = q^2|q|$
21. $g(x) = \dfrac{|x| + x}{2}(3x + 2)$
22. $h(x) = \dfrac{|x| - x}{2}(x^2 - x + 6)$
23. $f(x) = |\sin x|$
24. $f(x) = \sqrt{\sin x}$
25. $h(t) = \cos^2 t$
26. $g(t) = \tan^2 t$
27. [C] $f(x) = \dfrac{5.235x^3 - 1.245x^2}{7.126x - 3.141}$

28. Sketch the graph of a function f that has the following properties:

(a) f is everywhere continuous; (b) $f(0) = 0, f(1) = 2$;
(c) f is an even function; (d) $f'(x) > 0$ for $x > 0$;
(e) $f''(x) > 0$ for $x > 0$.

29. Sketch the graph of a function f that has the following properties:

(a) f is everywhere continuous; (b) $f(2) = -3, f(6) = 1$;
(c) $f'(2) = 0, f'(x) > 0$ for $x \neq 2, f'(6) = 3$;
(d) $f''(6) = 0, f''(x) > 0$ for $2 < x < 6, f''(x) < 0$ for $x > 6$.

30. Sketch the graph of a function g that has the following properties:

(a) g is everywhere *smooth* (continuous with a continuous first derivative);
(b) $g(0) = 0$; (c) $g'(x) < 0$ for all x;
(d) $g''(x) < 0$ for $x < 0$ and $g''(x) > 0$ for $x > 0$.

31. Sketch the graph of a function f that has the following properties:

(a) f is everywhere continuous;
(b) $f(-3) = 1$;
(c) $f'(x) < 0$ for $x < -3, f'(x) > 0$ for $x > -3, f''(x) < 0$ for $x \neq -3$.

32. Sketch the graph of a function f that has the following properties:

(a) f is everywhere continuous;
(b) $f(-4) = -3, f(0) = 0, f(3) = 2$;
(c) $f'(-4) = 0, f'(3) = 0, f'(x) > 0$ for $x < -4, f'(x) > 0$ for $-4 < x < 3, f'(x) < 0$ for $x > 3$;
(d) $f''(-4) = 0, f''(0) = 0, f''(x) < 0$ for $x < -4, f''(x) > 0$ for $-4 < x < 0, f''(x) < 0$ for $x > 0$.

33. Sketch the graph of a function f that

(a) has a continuous first derivative;
(b) is decreasing and concave up for $x < 3$;
(c) has an extremum at $(3, 1)$;
(d) is increasing and concave up for $3 < x < 5$;
(e) has an inflection point at $(5, 4)$;
(f) is increasing and concave down for $5 < x < 6$;
(g) has an extremum at $(6, 7)$;
(h) is decreasing and concave down for $x > 6$.

GC *Linear approximations provide particularly good approximations near points of inflection. Using a graphing calculator, investigate this behavior in Problems 34–36.*

34. Graph $y = \sin x$ and its linear approximation $L(x) = x$ at $x = 0$.

35. Graph $y = \cos x$ and its linear approximation $L(x) = -x + \pi/2$ at $x = \pi/2$.

36. Find the linear approximation to the curve $y = (x - 1)^5 + 3$ at its point of inflection. Graph both the function and its linear approximation in the neighborhood of the inflection point.

37. Suppose $f'(x) = (x - 2)(x - 3)(x - 4)$ and $f(2) = 2$. Sketch a graph of $y = f(x)$.

38. Suppose $f'(x) = (x - 3)(x - 2)^2(x - 1)$ and $f(2) = 0$. Sketch a graph of $y = f(x)$.

39. Suppose $h'(x) = x^2(x - 1)^2(x - 2)$ and $h(0) = 0$. Sketch a graph of $y = h(x)$.

40. Consider a general quadratic curve $y = ax^2 + bx + c$. Show that such a curve has no inflection points.

41. Show that the curve $y = ax^3 + bx^2 + cx + d$ where $a \neq 0$, has exactly one inflection point.

42. Consider a general quartic curve $y = ax^4 + bx^3 + cx^2 + dx + e$, where $a \neq 0$. What is the maximum number of inflection points that such a curve can have?

EXPL CAS *In Problems 43–47, the graph of $y = f(x)$ depends on a parameter c. Using a CAS, investigate how the extremum and inflection points depend on the value of c. Identify the values of c at which the basic shape of the curve changes.*

43. $f(x) = x^2\sqrt{x^2 - c^2}$ **44.** $f(x) = \dfrac{cx}{4 + (cx)^2}$

45. $f(x) = \dfrac{1}{(cx^2 - 4)^2 + cx^2}$ **46.** $f(x) = \dfrac{1}{x^2 + 4x + c}$

47. $f(x) = c + \sin cx$

48. What conclusions can you draw about f from the information that $f'(c) = f''(c) = 0$ and $f'''(c) > 0$?

49. Let $g(x)$ be a function that has two derivatives and satisfies the following properties:

(a) $g(1) = 1$;
(b) $g'(x) > 0$ for all $x \neq 1$;
(c) g is concave down for all $x < 1$ and concave up for all $x > 1$;
(d) $f(x) = g(x^4)$;

Sketch a possible graph of $f(x)$ and justify your answer.

50. Let $H(x)$ have three continuous derivatives, and be such that $H(1) = H'(1) = H''(1) = 0$, but $H'''(1) \neq 0$. Does $H(x)$ have a local maximum, local minimum, or a point of inflection at $x = 1$? Justify your answer.

51. In each case, is it possible for a function F with two continuous derivatives to satisfy the following properties? If so sketch such a function. If not, justify your answer.

(a) $F'(x) > 0, F''(x) > 0$, while $F(x) < 0$ for all x.
(b) $F''(x) < 0$, while $F(x) > 0$.
(c) $F''(x) < 0$, while $F'(x) > 0$.

GC **52.** Use a graphing calculator or a CAS to plot the graphs of each of the following functions on the indicated interval. Determine the coordinates of any of the global extrema and any inflection points. You should be able to give answers that are accurate to at least one decimal place. Restrict the y-axis window to $-5 \leq y \leq 5$.

(a) $f(x) = x^2 \tan x; \left(-\dfrac{\pi}{2}, \dfrac{\pi}{2}\right)$
(b) $f(x) = x^3 \tan x; \left(-\dfrac{\pi}{2}, \dfrac{\pi}{2}\right)$
(c) $f(x) = 2x + \sin x; [-\pi, \pi]$
(d) $f(x) = x - \dfrac{\sin x}{2}; [-\pi, \pi]$

GC **53.** Each of the following functions is periodic. Use a graphing calculator or a CAS to plot the graph of each of the following functions over one full period with the center of the interval located at the origin. Determine the coordinates of any of the global extrema and any inflection points. You should be able to give answers that are accurate to at least one decimal place.

(a) $f(x) = 2 \sin x + \cos^2 x$ (b) $f(x) = 2 \sin x + \sin^2 x$
(c) $f(x) = \cos 2x - 2 \cos x$ (d) $f(x) = \sin 3x - \sin x$
(e) $f(x) = \sin 2x - \cos 3x$

54. Let f be a continuous function with $f(-3) = f(0) = 2$. If the graph of $y = f'(x)$ is as shown in Figure 11, sketch a possible graph for $y = f(x)$.

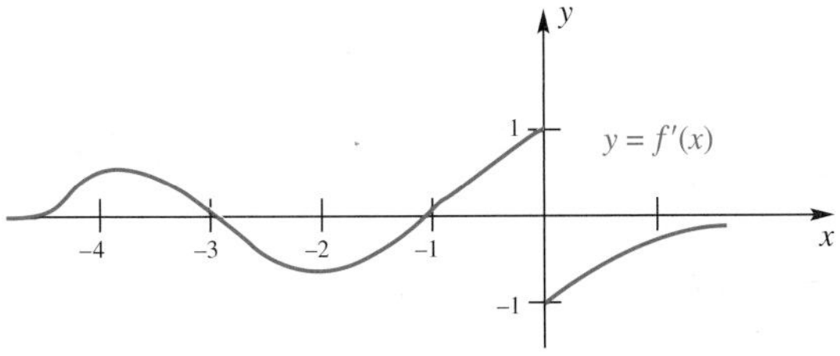

Figure 11

55. Let f be a continuous function and let f' have the graph shown in Figure 12. Sketch a possible graph for f and answer the following questions.

(a) Where is f increasing? Decreasing?
(b) Where is f concave up? Concave down?
(c) Where does f attain a local maximum? A local minimum?
(d) Where are there inflection points for f?

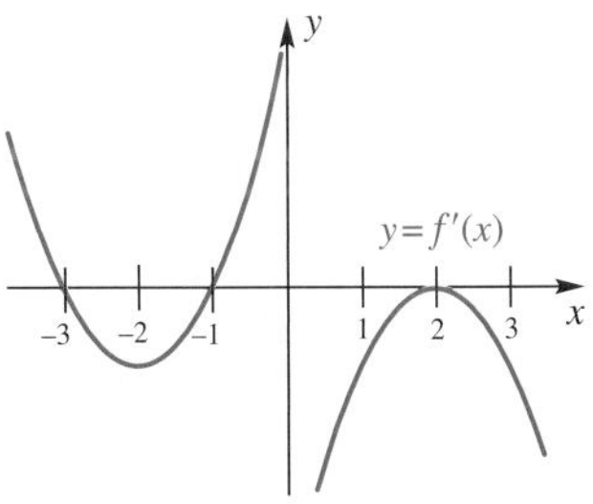

Figure 12

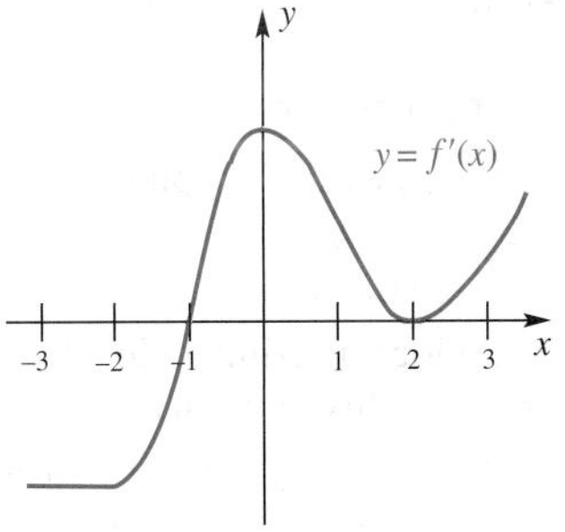

Figure 13

56. Repeat Problem 55 for Figure 13.

57. Let f be a continuous function with $f(0) = f(2) = 0$. If the graph of $y = f'(x)$ is as shown in Figure 14, sketch a possible graph for $y = f(x)$.

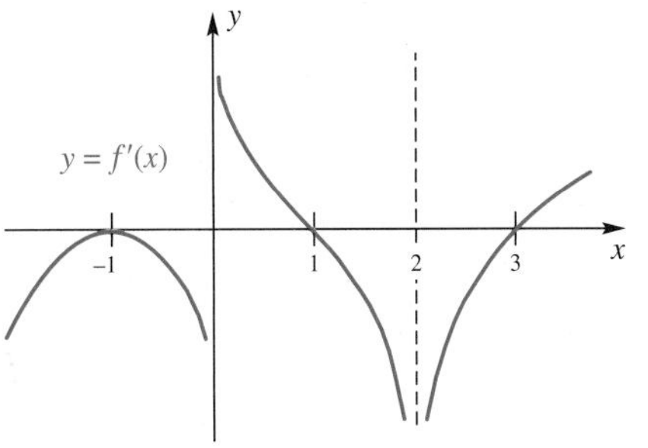

Figure 14

58. Suppose that $f'(x) = (x - 3)(x - 1)^2(x + 2)$ and $f(1) = 2$. Sketch a possible graph of f.

GC **59.** Use a graphing calculator or a CAS to plot the graph of each of the following functions on $[-1, 7]$. Determine the coordinates of any global extrema and any inflection points. You should be able to give answers that are accurate to at least one decimal place.

(a) $f(x) = x\sqrt{x^2 - 6x + 40}$
(b) $f(x) = \sqrt{|x|}(x^2 - 6x + 40)$
(c) $f(x) = \sqrt{x^2 - 6x + 40}/(x - 2)$
(d) $f(x) = \sin[(x^2 - 6x + 40)/6]$

GC **60.** Repeat Problem 59 for the following functions.

(a) $f(x) = x^3 - 8x^2 + 5x + 4$
(b) $f(x) = |x^3 - 8x^2 + 5x + 4|$
(c) $f(x) = (x^3 - 8x^2 + 5x + 4)/(x - 1)$
(d) $f(x) = (x^3 - 8x^2 + 5x + 4)/(x^3 + 1)$

Answers to Concepts Review: **1.** $f(x); -f(x)$ **2.** decreasing; concave up **3.** $x = -1, x = 2, x = 3; y = 1$ **4.** polynomial; rational

3.6 The Mean Value Theorem for Derivatives

In geometric language, the Mean Value Theorem is easy to state and understand. It says that, if the graph of a continuous function has a nonvertical tangent line at every point between A and B, then there is at least one point C on the graph between A and B at which the tangent line is parallel to the secant line AB. In Figure 1, there is just one such point C; in Figure 2, there are several. First we state the theorem in the language of functions; then we prove it.

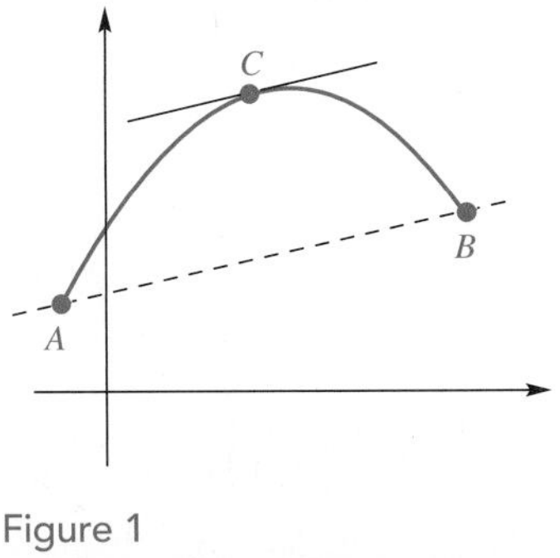

Figure 1

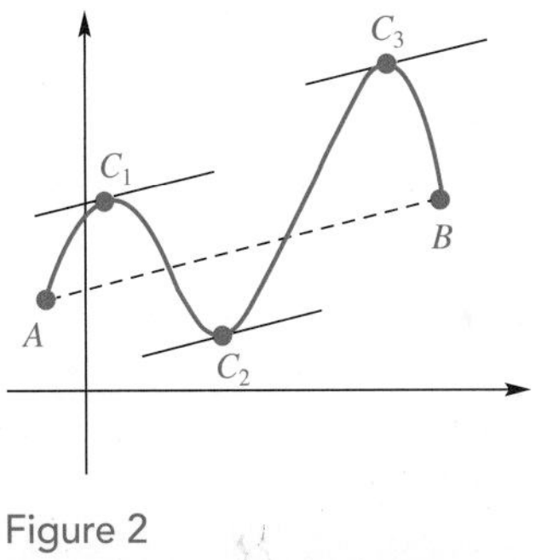

Figure 2

Theorem A Mean Value Theorem for Derivatives

If f is continuous on a closed interval $[a, b]$ and differentiable on its interior (a, b), then there is at least one number c in (a, b) where

$$\frac{f(b) - f(a)}{b - a} = f'(c)$$

or, equivalently, where

$$f(b) - f(a) = f'(c)(b - a)$$

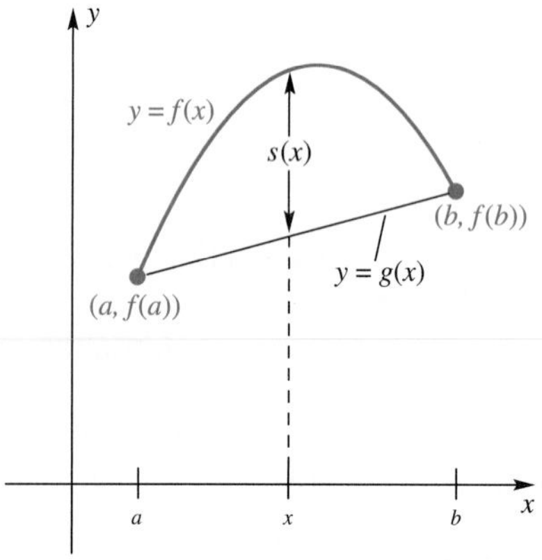

Figure 3

Proof Our proof rests on a careful analysis of the function $s(x) = f(x) - g(x)$, introduced in Figure 3. Here $y = g(x)$ is the equation of the line through $(a, f(a))$ and $(b, f(b))$. Since this line has slope $[f(b) - f(a)]/(b - a)$ and goes through $(a, f(a))$, the point-slope form for its equation is

$$g(x) - f(a) = \frac{f(b) - f(a)}{b - a}(x - a)$$

This, in turn, yields a formula for $s(x)$:

$$s(x) = f(x) - g(x) = f(x) - f(a) - \frac{f(b) - f(a)}{b - a}(x - a)$$

Note immediately that $s(b) = s(a) = 0$ and that, for x in (a, b),

$$s'(x) = f'(x) - \frac{f(b) - f(a)}{b - a}$$

The Key to a Proof

The key to this proof is that c is the value at which $f'(c) = \dfrac{f(b) - f(a)}{b - a}$ *and* $s'(c) = 0$. Many proofs have one or two key ideas; if you understand the key, you will understand the proof.

Now we make a crucial observation. If we knew that there was a number c in (a, b) satisfying $s'(c) = 0$, we would be all done. For then the last equation would say that

$$0 = f'(c) - \frac{f(b) - f(a)}{b - a}$$

which is equivalent to the conclusion of the theorem.

To see that $s'(c) = 0$ for some c in (a, b), reason as follows. Clearly, s is continuous on $[a, b]$, being the difference of two continuous functions. Thus, by the Max–Min Existence Theorem (Theorem 3.1A), s must attain both a maximum and a minimum value on $[a, b]$. If both of these values happen to be 0, then $s(x)$ is identically 0 on $[a, b]$, and consequently $s'(x) = 0$ for all x in (a, b), much more than we need.

If either the maximum value or the minimum value is different from 0, then that value is attained at an interior point c, since $s(a) = s(b) = 0$. Now s has a derivative at each point of (a, b), and so, by the Critical Point Theorem (Theorem 3.1B), $s'(c) = 0$. That is all we needed to know. ■

The Theorem Illustrated

EXAMPLE 1 Find the number c guaranteed by the Mean Value Theorem for $f(x) = 2\sqrt{x}$ on $[1, 4]$.

SOLUTION

$$f'(x) = 2 \cdot \frac{1}{2}x^{-1/2} = \frac{1}{\sqrt{x}}$$

and

$$\frac{f(4) - f(1)}{4 - 1} = \frac{4 - 2}{3} = \frac{2}{3}$$

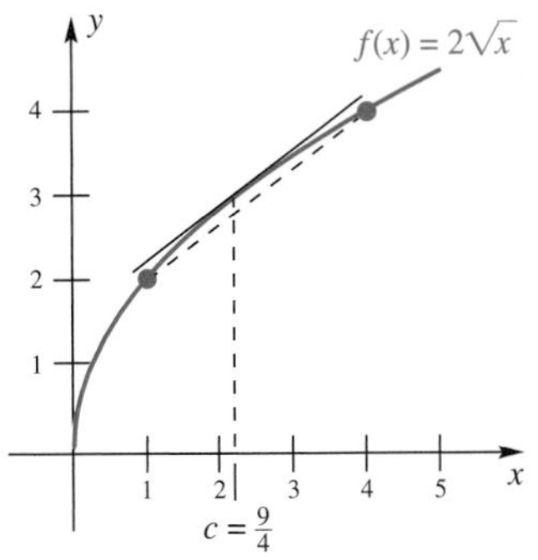

Figure 4

Thus, we must solve

$$\frac{1}{\sqrt{c}} = \frac{2}{3}$$

The single solution is $c = \frac{9}{4}$ (Figure 4). ■

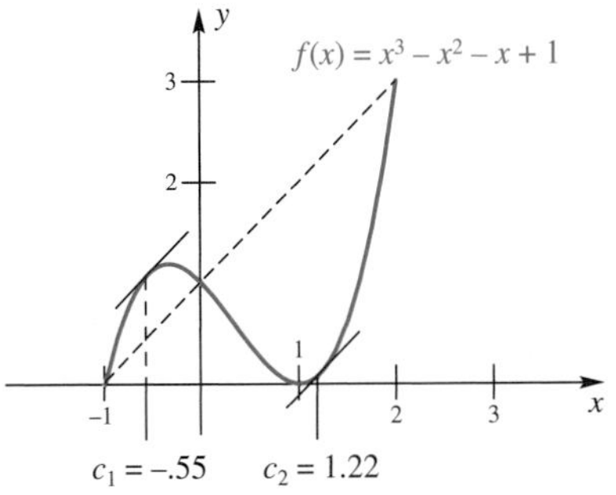

Figure 5

EXAMPLE 2 Let $f(x) = x^3 - x^2 - x + 1$ on $[-1, 2]$. Find all numbers c satisfying the conclusion to the Mean Value Theorem.

SOLUTION Figure 5 shows a graph of the function f. From this graph, it appears that there are two numbers c_1 and c_2 with the required property. We now find

$$f'(x) = 3x^2 - 2x - 1$$

and

$$\frac{f(2) - f(-1)}{2 - (-1)} = \frac{3 - 0}{3} = 1$$

Therefore, we must solve

$$3c^2 - 2c - 1 = 1$$

or, equivalently,

$$3c^2 - 2c - 2 = 0$$

By the Quadratic Formula, there are two solutions, $\left(2 \pm \sqrt{4 + 24}\right)/6$, which correspond to $c_1 \approx -0.55$ and $c_2 \approx 1.22$. Both numbers are in the interval $(-1, 2)$. ■

EXAMPLE 3 Let $f(x) = x^{2/3}$ on $[-8, 27]$. Show that the conclusion to the Mean Value Theorem fails and figure out why.

SOLUTION

$$f'(x) = \frac{2}{3}x^{-1/3}, \qquad x \neq 0$$

and

$$\frac{f(27) - f(-8)}{27 - (-8)} = \frac{9 - 4}{35} = \frac{1}{7}$$

We must solve

$$\frac{2}{3}c^{-1/3} = \frac{1}{7}$$

which gives

$$c = \left(\frac{14}{3}\right)^3 \approx 102$$

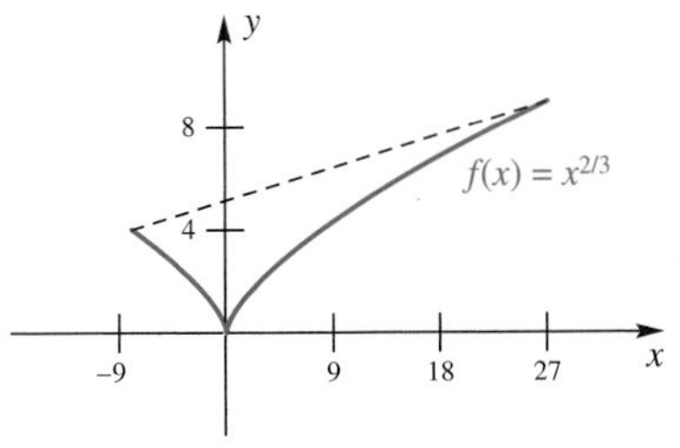

Figure 6

But $c = 102$ is not in the interval $(-8, 27)$ as required. As the graph of $y = f(x)$ suggests (Figure 6), $f'(0)$ fails to exist, so the problem is that $f(x)$ is not everywhere differentiable on $(-8, 27)$. ■

If the function $s(t)$ represents the position of an object at time t, then the Mean Value Theorem states that over any interval of time, there is some time for which the instantaneous velocity equals the average velocity.

EXAMPLE 4 Suppose that an object has position function $s(t) = t^2 - t - 2$. Find the average velocity over the interval [3, 6] and find the time at which the instantaneous velocity equals the average velocity.

SOLUTION The average velocity over the interval $[3, 6]$ is equal to $(s(6) - s(3))/(6 - 3) = 8$. The instantaneous velocity is $s'(t) = 2t - 1$. To find the point where average velocity equals instantaneous velocity, we equate $8 = 2t - 1$ and solve to get $t = 9/2$. ■

The Theorem Used In Section 3.2, we promised a rigorous proof of the Monotonicity Theorem (Theorem 3.2A). This is the theorem that relates the sign of the derivative of a function to whether that function is increasing or decreasing.

Proof of the Monotonicity Theorem We suppose that f is continuous on I and that $f'(x) > 0$ at each point x in the interior of I. Consider any two points x_1 and x_2 of I with $x_1 < x_2$. By the Mean Value Theorem applied to the interval $[x_1, x_2]$, there is a number c in (x_1, x_2) satisfying

$$f(x_2) - f(x_1) = f'(c)(x_2 - x_1)$$

Since $f'(c) > 0$, we see that $f(x_2) - f(x_1) > 0$; that is, $f(x_2) > f(x_1)$. This is what we mean when we say that f is increasing on I.

The case where $f'(x) < 0$ on I is handled similarly. ■

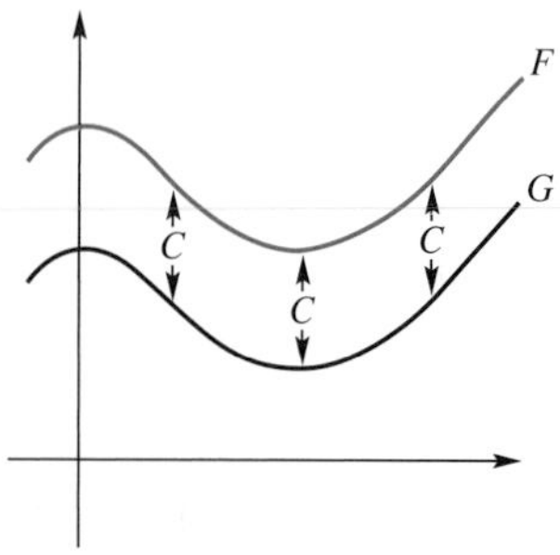

Figure 7

Our next theorem will be used repeatedly in this and the next chapter. In words, it says that *two functions with the same derivative differ by a constant*, possibly the zero constant (see Figure 7).

Theorem B

If $F'(x) = G'(x)$ for all x in (a, b), then there is a constant C such that

$$F(x) = G(x) + C$$

for all x in (a, b).

Geometry and Algebra

As with most topics in this book, you should try to see things from an algebraic and a geometrical point of view. Geometrically, Theorem B says that if F and G have the same derivative then the graph of G is a vertical translation of the graph of F.

Proof Let $H(x) = F(x) - G(x)$. Then

$$H'(x) = F'(x) - G'(x) = 0$$

for all x in (a, b). Choose x_1, as some (fixed) point in (a, b), and let x be any other point there. The function H satisfies the hypotheses of the Mean Value Theorem on the closed interval with end points x_1, and x. Thus, there is a number c between x_1 and x such that

$$H(x) - H(x_1) = H'(c)(x - x_1)$$

But $H'(c) = 0$ by hypothesis. Therefore, $H(x) - H(x_1) = 0$ or, equivalently, $H(x) = H(x_1)$ for all x in (a, b). Since $H(x) = F(x) - G(x)$, we conclude that $F(x) - G(x) = H(x_1)$. Now let $C = H(x_1)$, and we have the conclusion $F(x) = G(x) + C$. ■

Concepts Review

1. The Mean Value Theorem for Derivatives says that if f is ____ on $[a, b]$ and differentiable on ____ then there is a point c in (a, b) such that ____.

2. The function, $f(x) = |\sin x|$ would satisfy the hypotheses of the Mean Value Theorem on the interval $[0, 1]$ but would not satisfy them on the interval $[-1, 1]$ because ____.

3. If two functions F and G have the same derivative on the interval (a, b), then there is a constant C such that ____.

4. Since $D_x(x^4) = 4x^3$, it follows that every function F that satisfies $F'(x) = 4x^3$ has the form $F(x) =$ ____.

Problem Set 3.6

In each of the Problems 1–21, a function is defined and a closed interval is given. Decide whether the Mean Value Theorem applies to the given function on the given interval. If it does, find all possible values of c; if not, state the reason. In each problem, sketch the graph of the given function on the given interval.

1. $f(x) = |x|; [1, 2]$ **2.** $g(x) = |x|; [-2, 2]$

3. $f(x) = x^2 + x; [-2, 2]$ **4.** $g(x) = (x + 1)^3; [-1, 1]$

5. $H(s) = s^2 + 3s - 1; [-3, 1]$

6. $F(x) = \frac{x^3}{3}; [-2, 2]$

7. $f(z) = \frac{1}{3}(z^3 + z - 4); [-1, 2]$

8. $F(t) = \frac{1}{t - 1}; [0, 2]$ **9.** $h(x) = \frac{x}{x - 3}; [0, 2]$

10. $f(x) = \frac{x - 4}{x - 3}; [0, 4]$ **11.** $h(t) = t^{2/3}; [0, 2]$

12. $h(t) = t^{2/3}; [-2, 2]$ **13.** $g(x) = x^{5/3}; [0, 1]$

14. $g(x) = x^{5/3}; [-1, 1]$ **15.** $S(\theta) = \sin\theta; [-\pi, \pi]$

16. $C(\theta) = \csc\theta; [-\pi, \pi]$ **17.** $T(\theta) = \tan\theta; [0, \pi]$

18. $f(x) = x + \frac{1}{x}; \left[-1, \frac{1}{2}\right]$ **19.** $f(x) = x + \frac{1}{x}; [1, 2]$

20. $f(x) = [\![x]\!]; [1, 2]$ **21.** $f(x) = x + |x|; [-2, 1]$

22. (Rolle's Theorem) *If f is continuous on [a, b] and differentiable on (a, b), and if f(a) = f(b), then there is at least one number c in (a, b) such that f′(c) = 0.* Show that Rolle's Theorem is just a special case of the Mean Value Theorem. (Michel Rolle (1652–1719) was a French mathematician.)

23. For the function graphed in Figure 8, find (approximately) all points c that satisfy the conclusion to the Mean Value Theorem for the interval $[0, 8]$.

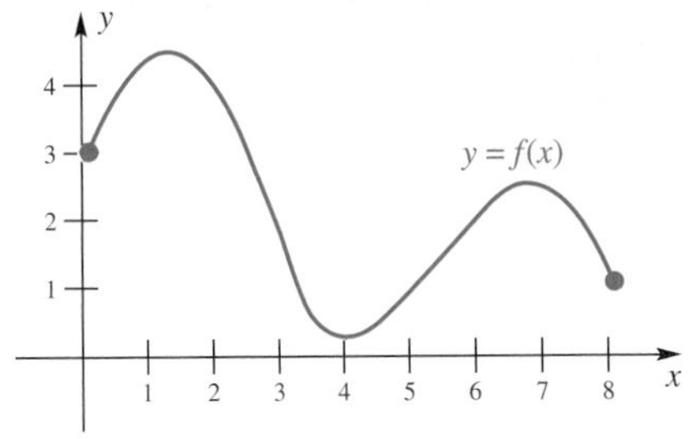

Figure 8

24. Show that if f is the quadratic function defined by $f(x) = \alpha x^2 + \beta x + \gamma, \alpha \neq 0$, then the number c of the Mean Value Theorem is always the midpoint of the given interval $[a, b]$.

25. Prove: If f is continuous on (a, b) and if $f'(x)$ exists and satisfies $f'(x) > 0$ except at one point x_0 in (a, b), then f is increasing on (a, b). *Hint:* Consider f on each of the intervals $(a, x_0]$ and $[x_0, b)$ separately.

26. Use Problem 25 to show that each of the following is increasing on $(-\infty, \infty)$.

(a) $f(x) = x^3$ (b) $f(x) = x^5$

(c) $f(x) = \begin{cases} x^3, & x \leq 0 \\ x, & x > 0 \end{cases}$

27. Use the Mean Value Theorem to show that $s = 1/t$ decreases on any interval over which it is defined.

28. Use the Mean Value Theorem to show that $s = 1/t^2$ decreases on any interval to the right of the origin.

29. Prove that if $F'(x) = 0$ for all x in (a, b) then there is a constant C such that $F(x) = C$ for all x in (a, b). *Hint:* Let $G(x) = 0$ and apply Theorem B.

30. Suppose that you know that $\cos(0) = 1$, $\sin(0) = 0$, $D_x \cos x = -\sin x$, and $D_x \sin x = \cos x$, but nothing else about the sine and cosine functions. Show that $\cos^2 x + \sin^2 x = 1$. *Hint:* Let $F(x) = \cos^2 x + \sin^2 x$ and use Problem 29.

31. Prove that if $F'(x) = D$ for all x in (a, b) then there is a constant C such that $F(x) = Dx + C$ for all x in (a, b). *Hint:* Let $G(x) = Dx$ and apply Theorem B.

32. Suppose that $F'(x) = 5$ and $F(0) = 4$. Find a formula for $F(x)$. *Hint:* See Problem 31.

33. Prove: Let f be continuous on $[a, b]$ and differentiable on (a, b). If $f(a)$ and $f(b)$ have opposite signs and if $f'(x) \neq 0$ for all x in (a, b), then the equation $f(x) = 0$ has one and only one solution between a and b. *Hint:* Use the Intermediate Value Theorem and Rolle's Theorem (Problem 22).

34. Show that $f(x) = 2x^3 - 9x^2 + 1 = 0$ has exactly one solution on each of the intervals $(-1, 0)$, $(0, 1)$, and $(4, 5)$. *Hint:* Apply Problem 33.

35. Let f have a derivative on an interval I. Prove that between successive distinct zeros of f' there can be at most one zero of f. *Hint:* Try a proof by contradiction and use Rolle's Theorem (Problem 22).

36. Let g be continuous on $[a, b]$ and suppose that $g''(x)$ exists for all x in (a, b). Prove that if there are three values of x in $[a, b]$ for which $g(x) = 0$ then there is at least one value of x in (a, b) such that $g''(x) = 0$.

37. Let $f(x) = (x - 1)(x - 2)(x - 3)$. Prove by using Problem 36 that there is at least one value in the interval $[0, 4]$ where $f''(x) = 0$ and two values in the same interval where $f'(x) = 0$.

38. Prove that if $|f'(x)| \leq M$ for all x in (a, b) and if x_1 and x_2 are any two points in (a, b) then

$$|f(x_2) - f(x_1)| \leq M|x_2 - x_1|$$

Note: A function satisfying the above inequality is said to satisfy a *Lipschitz condition* with constant M. (Rudolph Lipschitz (1832–1903) was a German mathematician.)

39. Show that $f(x) = \sin 2x$ satisfies a Lipschitz condition with constant 2 on the interval $(-\infty, \infty)$. See Problem 38.

40. A function f is said to be **nondecreasing** on an interval I if $x_1 < x_2 \Rightarrow f(x_1) \leq f(x_2)$ for x_1 and x_2 in I. Similarly, f is **nonincreasing** on I if $x_1 < x_2 \Rightarrow f(x_1) \geq f(x_2)$ for x_1 and x_2 in I.

(a) Sketch the graph of a function that is nondecreasing but not increasing.

(b) Sketch the graph of a function that is nonincreasing but not decreasing.

41. Prove that, if f is continuous on I and if $f'(x)$ exists and satisfies $f'(x) \geq 0$ on the interior of I, then f is nondecreasing on I. Similarly, if $f'(x) \leq 0$, then f is nonincreasing on I.

42. Prove that if $f(x) \geq 0$ and $f'(x) \geq 0$ on I, then f^2 is nondecreasing on I.

43. Prove that if $g'(x) \leq h'(x)$ for all x in (a, b) then

$$x_1 < x_2 \Rightarrow g(x_2) - g(x_1) \leq h(x_2) - h(x_1)$$

for all x_1 and x_2 in (a, b). *Hint:* Apply Problem 41 with $f(x) = h(x) - g(x)$.

44. Use the Mean Value Theorem to prove that

$$\lim_{x\to\infty} \left(\sqrt{x + 2} - \sqrt{x}\right) = 0$$

45. Use the Mean Value Theorem to show that

$$|\sin x - \sin y| \leq |x - y|$$

46. Suppose that in a race, horse A and horse B begin at the same point and finish in a dead heat. Prove that their speeds were identical at some instant of the race.

47. In Problem 46, suppose that the two horses crossed the finish line together at the same speed. Show that they had the same acceleration at some instant.

48. Use the Mean Value Theorem to show that the graph of a concave up function f is always above its tangent line; that is, show that

$$f(x) > f(c) + f'(c)(x - c), \qquad x \neq c$$

49. Prove that if $|f(y) - f(x)| \leq M(y - x)^2$ for all x and y then f is a constant function.

50. Give an example of a function f that is continuous on $[0, 1]$, differentiable on $(0, 1)$, and *not* differentiable on $[0, 1]$, and has a tangent line at every point of $[0, 1]$.

51. John traveled 112 miles in 2 hours and claimed that he never exceeded 55 miles per hour. Use the Mean Value Theorem to disprove John's claim. *Hint:* Let $f(t)$ be the distance traveled in time t.

52. A car is stationary at a toll booth. Eighteen minutes later at a point 20 miles down the road the car is clocked at 60 miles per hour. Sketch a possible graph of v versus t. Sketch a possible graph of the distance traveled s against t. Use the Mean Value Theorem to show that the car must have exceeded the 60 mile per hour speed limit at some time after leaving the toll booth, but before the car was clocked at 60 miles per hour.

53. A car is stationary at a toll booth. Twenty minutes later at a point 20 miles down the road the car is clocked at 60 miles per hour. Explain why the car must have exceeded 60 miles per hour at some time after leaving the toll booth, but before the car was clocked at 60 miles per hour.

54. Show that if an object's position function is given by $s(t) = at^2 + bt + c$, then the average velocity over the interval $[A, B]$ is equal to the instantaneous velocity at the midpoint of $[A, B]$.

Answers to Concepts Review: **1.** continuous; (a, b); $f(b) - f(a) = f'(c)(b - a)$ **2.** $f'(0)$ does not exist **3.** $F(x) = G(x) + C$ **4.** $x^4 + C$

3.7 Solving Equations Numerically

In mathematics and science, we often need to find the roots (solutions) of an equation $f(x) = 0$. To be sure, if $f(x)$ is a linear or quadratic polynomial, formulas for writing exact solutions exist and are well known. But for other algebraic equations, and certainly for equations involving transcendental functions, formulas for exact solutions are rarely available. What can be done in such cases?

There is a general method of solving problems known to all resourceful people. Given a cup of tea, we add sugar a bit at a time until it tastes just right. Given a stopper too large for a hole, we whittle it down until it fits. We change the solution a bit at a time, improving the accuracy, until we are satisfied. Mathematicians call this the *method of successive approximations*, or the *method of iterations*.

In this section, we present three such methods for solving equations: the Bisection Method, Newton's Method, and the Fixed-Point Method. All are designed to approximate the real roots of $f(x) = 0$, and they all require many computations. You will want to keep your calculator handy.

The Bisection Method In Example 7 of Section 1.6 we saw how to use the Intermediate Value Theorem to approximate a solution of $f(x) = 0$ by successively bisecting an interval known to contain a solution. This Bisection Method has two great virtues—simplicity and reliability. It also has a major vice—the large number of steps needed to achieve the desired accuracy (otherwise known as slowness of convergence).

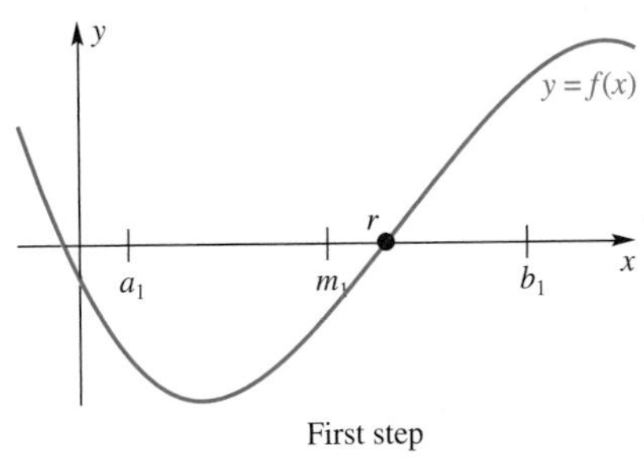

Figure 1

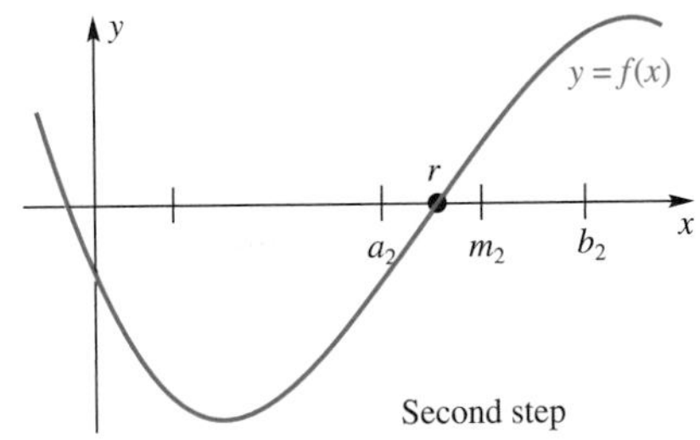

Figure 2

Begin the process by sketching the graph of f, which is assumed to be a continuous function (see Figure 1). A real root r of $f(x) = 0$ is a point (technically, the x-coordinate of a point) where the graph crosses the x-axis. As a first step in pinning down this point, locate two points, $a_1 < b_1$, at which you are sure that f has opposite signs; if f has opposite signs at a_1 and b_1, then the product $f(a_1) \cdot f(b_1)$ will be negative. (Try choosing a_1 and b_1 on opposite sides of your best guess at r.) The Intermediate Value Theorem guarantees the existence of a root between a_1 and b_1. Now evaluate f at the midpoint $m_1 = (a_1 + b_1)/2$ of $[a_1, b_1]$. The number m_1 is our first approximation to r.

Either $f(m_1) = 0$, in which case we are done, or $f(m_1)$ differs in sign from $f(a_1)$ or $f(b_1)$. Denote the one of the subintervals $[a_1, m_1]$ or $[m_1, b_1]$ on which the sign change occurs by the symbol $[a_2, b_2]$, and evaluate f at its midpoint $m_2 = (a_2 + b_2)/2$ (Figure 2). The number m_2 is our second approximation to r.

Repeat the process, thus determining a sequence of approximations $m_1, m_2, m_3, \ldots$, and subintervals $[a_1, b_1], [a_2, b_2], [a_3, b_3], \ldots$, each subinterval containing the root r and each half the length of its predecessor. Stop when r is determined to the desired accuracy, that is, when $(b_n - a_n)/2$ is less than the allowable error, which we will denote by E.

Algorithm **Bisection Method**

Let $f(x)$ be a continuous function, and let a_1 and b_1 be numbers satisfying $a_1 < b_1$ and $f(a_1) \cdot f(b_1) < 0$. Let E denote the desired bound for the error $|r - m_n|$. Repeat steps 1 to 5 for $n = 1, 2, \ldots$ until $h_n < E$:

1. Calculate $m_n = (a_n + b_n)/2$.
2. Calculate $f(m_n)$, and if $f(m_n) = 0$, then STOP.
3. Calculate $h_n = (b_n - a_n)/2$.
4. If $f(a_n) \cdot f(m_n) < 0$, then set $a_{n+1} = a_n$ and $b_{n+1} = m_n$.
5. If $f(a_n) \cdot f(m_n) > 0$, then set $a_{n+1} = m_n$ and $b_{n+1} = b_n$.

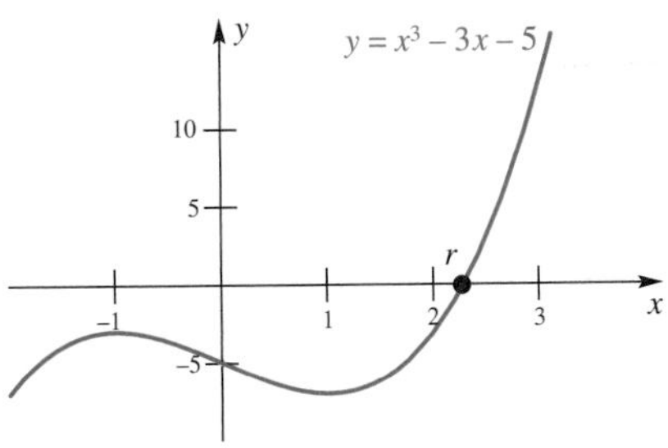

Figure 3

EXAMPLE 1 Determine the real root of $f(x) = x^3 - 3x - 5 = 0$ to accuracy within 0.0000001.

SOLUTION We first sketch the graph of $y = x^3 - 3x - 5$ (Figure 3) and, noting that it crosses the x-axis between 2 and 3, we begin with $a_1 = 2$ and $b_1 = 3$.

Step 1: $m_1 = (a_1 + b_1)/2 = (2 + 3)/2 = 2.5$

Step 2: $f(m_1) = f(2.5) = 2.5^3 - 3 \cdot 2.5 - 5 = 3.125$

Step 3: $h_1 = (b_1 - a_1)/2 = (3 - 2)/2 = 0.5$

Step 4: Since

$$f(a_1) \cdot f(m_1) = f(2)f(2.5) = (-3)(3.125) = -9.375 < 0$$

we set $a_2 = a_1 = 2$ and $b_2 = m_1 = 2.5$.

Step 5: The condition $f(a_n) \cdot f(m_n) > 0$ is false.

Next we increment n so that it has the value 2 and repeat these steps. We can continue this process to obtain the entries in the following table:

n	h_n	m_n	$f(m_n)$
1	0.5	2.5	3.125
2	0.25	2.25	−0.359
3	0.125	2.375	1.271
4	0.0625	2.3125	0.429
5	0.03125	2.28125	0.02811
6	0.015625	2.265625	−0.16729
7	0.0078125	2.2734375	−0.07001
8	0.0039063	2.2773438	−0.02106
9	0.0019531	2.2792969	0.00350
10	0.0009766	2.2783203	−0.00878
11	0.0004883	2.2788086	−0.00264
12	0.0002441	2.2790528	0.00043
13	0.0001221	2.2789307	−0.00111
14	0.0000610	2.2789918	−0.00034
15	0.0000305	2.2790224	0.00005
16	0.0000153	2.2790071	−0.00015
17	0.0000076	2.2790148	−0.00005
18	0.0000038	2.2790187	−0.000001
19	0.0000019	2.2790207	0.000024
20	0.0000010	2.2790197	0.000011
21	0.0000005	2.2790192	0.000005
22	0.0000002	2.2790189	0.0000014
23	0.0000001	2.2790187	−0.0000011
24	0.0000001	2.2790188	0.0000001

We conclude that $r = 2.2790188$ with an error of at most 0.0000001. ■

Example 1 illustrates the shortcoming of the Bisection Method. The approximations $m_1, m_2, m_3, \ldots$ converge very slowly to the root r. But they do converge; that is, $\lim_{n\to\infty} m_n = r$. The method works, and we have at step n a good bound for the error $E_n = r - m_n$, namely, $|E_n| \le h_n$.

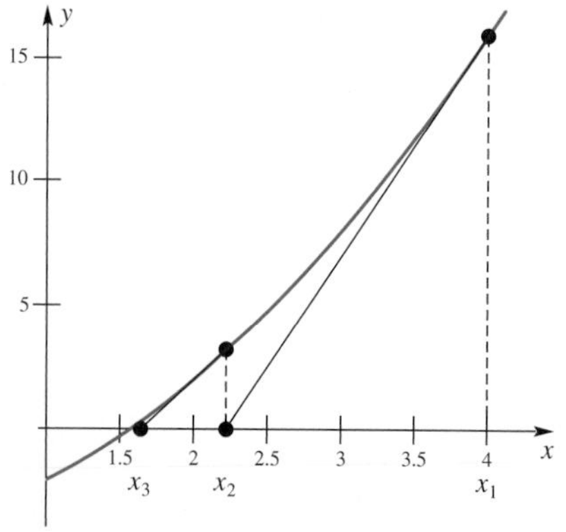

Figure 4

Newton's Method We are still considering the problem of solving the equation $f(x) = 0$ for a root r. Suppose that f is differentiable, so the graph of $y = f(x)$ has a tangent line at each point. If we can find a first approximation x_1 to r by graphing or any other means (see Figure 4), then a better approximation x_2 ought to lie at the intersection of the tangent at $(x_1, f(x_1))$ with the x-axis. Using x_2 as an approximation, we can then find a still better approximation x_3, and so on.

The process can be mechanized so that it is easy to do on a calculator. The equation of the tangent line at $(x_1, f(x_1))$ is

$$y - f(x_1) = f'(x_1)(x - x_1)$$

and its x-intercept x_2 is found by setting $y = 0$ and solving for x. The result is

$$x_2 = x_1 - \frac{f(x_1)}{f'(x_1)}$$

More generally, we have the following algorithm, also called a *recursion formula* or an *iteration scheme*.

Algorithm **Newton's Method**

Let $f(x)$ be a differentiable function and let x_1 be an initial approximation to the root r of $f(x) = 0$. Let E denote a bound for the error $|r - x_n|$.

Repeat the following step for $n = 1, 2, \ldots$ until $|x_{n+1} - x_n| < E$:

1. $$x_{n+1} = x_n - \frac{f(x_n)}{f'(x_n)}$$

EXAMPLE 2 Use Newton's Method to find the real root r of $f(x) = x^3 - 3x - 5 = 0$ to seven decimal places.

SOLUTION This is the same equation considered in Example 1. Let's use $x_1 = 2.5$ as our first approximation to r, as we did there. Since $f(x) = x^3 - 3x - 5$ and $f'(x) = 3x^2 - 3$, the algorithm is

$$x_{n+1} = x_n - \frac{x_n^3 - 3x_n - 5}{3x_n^2 - 3} = \frac{2x_n^3 + 5}{3x_n^2 - 3}$$

We obtain the following table.

n	x_n
1	2.5
2	2.30
3	2.2793
4	2.2790188
5	2.2790188

After just four steps, we get a repetition of the first eight digits. We feel confident in reporting that $r \approx 2.2790188$, with perhaps some question about the last digit. ■

Algorithms

Algorithms have been part of mathematics since people first learned to do long division, but it is computer science that has given algorithmic thinking its present popularity. What is an algorithm? Donald Knuth, dean of computer scientists, responds,

"An algorithm is a precisely defined sequence of rules telling how to produce specified output information from given input information in a finite number of steps."

And what is computer science? According to Knuth,

"It is the study of algorithms."

EXAMPLE 3 Use Newton's Method to find the positive real root r of $f(x) = 2 - x + \sin x = 0$.

SOLUTION The graph of $y = 2 - x + \sin x$ is shown in Figure 5. We will use the starting value $x_1 = 2$. Since $f'(x) = -1 + \cos x$, the iteration becomes

$$x_{n+1} = x_n - \frac{2 - x_n + \sin x_n}{-1 + \cos x_n}$$

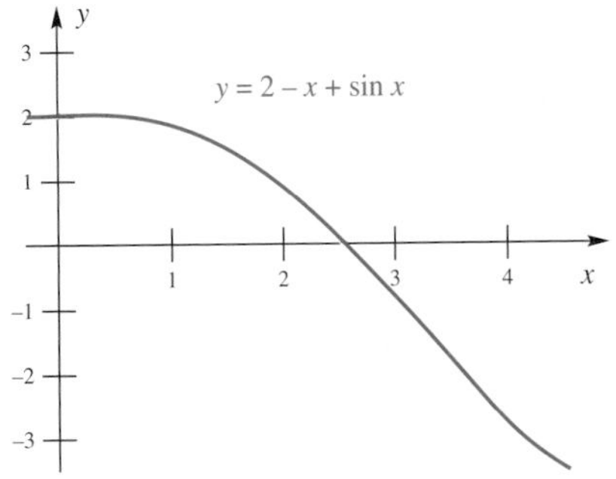

Figure 5

which leads to the following table:

n	x_n
1	2.0
2	2.6420926
3	2.5552335
4	2.5541961
5	2.5541960
6	2.5541960

After just five steps, we get a repetition of the seven digits after the decimal point. We conclude that $r \approx 2.5541960$. ■

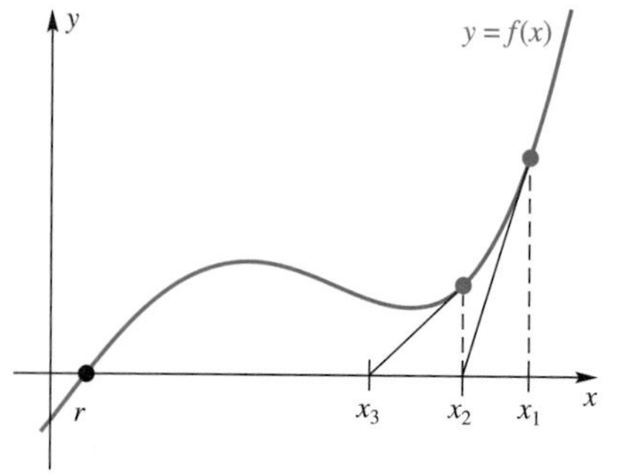

Figure 6

Newton's Method creates a *sequence* of successive approximations to the root. (We mentioned sequences briefly in Section 1.5.) It is often the case that Newton's Method produces a sequence $\{x_n\}$ that converges to the root of $f(x) = 0$, that is, $\lim_{n\to\infty} x_n = r$. This is not always the case, however. Figure 6 illustrates what can go wrong (see also Problem 22). For the function in Figure 6, the difficulty is that x_1 is not close enough to r to get a convergent process started. Other difficulties arise if $f'(x)$ is zero or undefined at or near r. When Newton's Method fails to produce approximations that converge to the solution, then you can retry Newton's Method with a different starting point, or use a different method such as the Bisection Method.

The Fixed-Point Algorithm The Fixed-Point Algorithm is simple and straightforward, but it often works.

Suppose that an equation can be written in the form $x = g(x)$. To solve this equation is to find a number r that is unchanged by the function g. We call such a number a **fixed point** of g. To find this number, we propose the following algorithm. Make a first guess x_1. Then let $x_2 = g(x_1)$, $x_3 = g(x_2)$, and so on. If we are lucky, x_n will converge to the root r as $n \to \infty$.

Algorithm **Fixed-Point Algorithm**

Let $g(x)$ be a continuous function, and let x_1 be an initial approximation to the root r of $x = g(x)$. Let E denote a bound for the error $|r - x_n|$.

Repeat the following step for $n = 1, 2, \ldots$ until $|x_{n+1} - x_n| < E$:

1. $x_{n+1} = g(x_n)$

EXAMPLE 4 Use the Fixed-Point Algorithm to approximate the solution of $f(x) = x^2 - 2\sqrt{x+1} = 0$.

SOLUTION We write $x^2 = 2\sqrt{x+1}$, which leads to $x = \pm\left(2\sqrt{x+1}\right)^{1/2}$. Since we know the solution is positive, we take the positive square root and write the iteration as

$$x_{n+1} = \left(2\sqrt{x_n+1}\right)^{1/2} = \sqrt{2}\,(x_n+1)^{1/4}$$

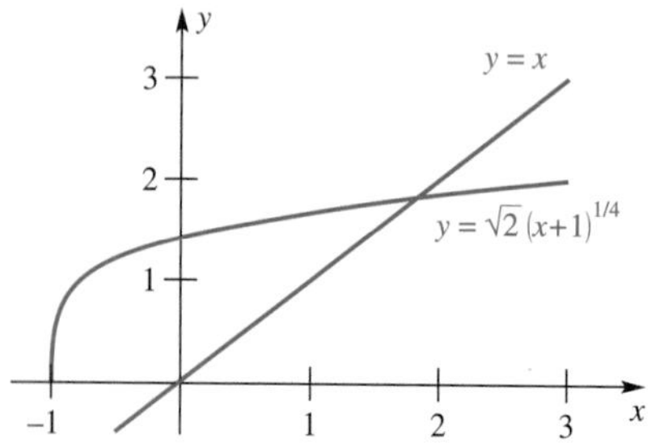

Figure 7

Figure 7 suggests that the point of intersection of the curves $y = x$ and $y = \sqrt{2}\left(x+1\right)^{1/4}$ occurs between 1 and 2, probably closer to 2, so we take $x_1 = 2$ as our starting point. This leads to the following table. The solution is approximately 1.8350867.

n	x_n	n	x_n
1	2.0	7	1.8350896
2	1.8612097	8	1.8350871
3	1.8392994	9	1.8350868
4	1.8357680	10	1.8350867
5	1.8351969	11	1.8350867
6	1.8351045	12	1.8350867

■

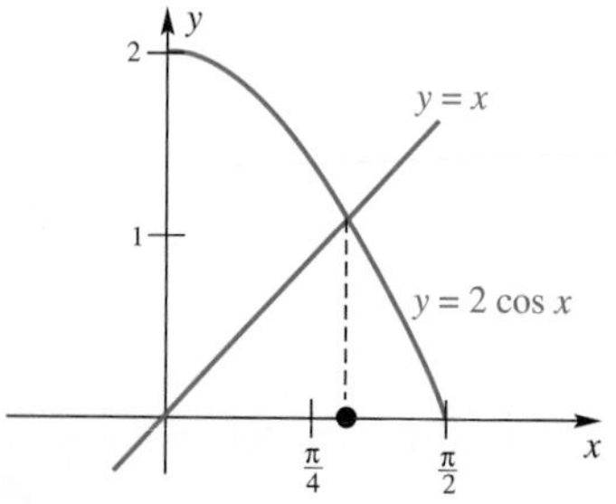

Figure 8

EXAMPLE 5 Solve $x = 2\cos x$ using the Fixed-Point Algorithm.

SOLUTION Note first that solving this equation is equivalent to solving the pair of equations $y = x$ and $y = 2\cos x$. Thus, to get our initial value, we graph these two equations (Figure 8) and observe that the two curves cross at approximately $x = 1$. Taking $x_1 = 1$ and applying the algorithm $x_{n+1} = 2\cos x_n$, we obtain the results in the following table.

n	x_n	n	x_n
1	1	6	1.4394614
2	1.0806046	7	0.2619155
3	0.9415902	8	1.9317916
4	1.1770062	9	−0.7064109
5	0.7673820	10	1.5213931

Quite clearly the process is unstable, even though our initial guess is very close to the actual root.

Let's take a different tack. Rewrite the equation $x = 2\cos x$ as $x = (x + 2\cos x)/2$ and use the algorithm

$$x_{n+1} = \frac{x_n + 2\cos x_n}{2}$$

This process produces a convergent sequence, shown in the following table. (The oscillation in the last digit is probably due to round-off errors.)

n	x_n	n	x_n	n	x_n
1	1	7	1.0298054	13	1.0298665
2	1.0403023	8	1.0298883	14	1.0298666
3	1.0261107	9	1.0298588	15	1.0298665
4	1.0312046	10	1.0298693	16	1.0298666
5	1.0293881	11	1.0298655		
6	1.0300374	12	1.0298668		

■

Now we raise an obvious question. Why did the second algorithm yield a convergent sequence, whereas the first one failed to do so? Whether or not the Fixed-Point Algorithm works depends on two factors. One is the formulation of the equation $x = g(x)$. Example 5 demonstrates that an equation such as $x = 2\cos x$ can be rewritten in a form that yields a different sequence of approximations. In Example 5 the reformulation was $x = (x + 2\cos x)/2$. In general, there may be many ways to write the equation and the trick is to find one that works. Another factor that affects whether the Fixed-Point Algorithm converges is the closeness of the starting point x_1 to the root r. As we suggested for Newton's Method, if the Fixed-Point Algorithm fails with one starting point, you can try a different one.

Concepts Review

1. The virtues of the Bisection Method are its simplicity and reliability; its vice is its _____.

2. If f is continuous on $[a, b]$, and $f(a)$ and $f(b)$ have opposite signs, then there is a _____ of $f(x) = 0$ between a and b. This follows from the _____ Theorem.

3. The Bisection Method, Newton's Method, and the Fixed-Point Algorithm are examples of _____; that is, they provide a finite sequence of steps that, if followed, will produce a root of an equation to desired accuracy.

4. A point x satisfying $g(x) = x$ is called a _____ of g.

Problem Set 3.7

C *In Problems 1–4, use the Bisection Method to approximate the real root of the given equation on the given interval. Each answer should be accurate to two decimal places.*

1. $x^3 + 2x - 6 = 0; [1, 2]$ **2.** $x^4 + 5x^3 + 1 = 0; [-1, 0]$

3. $2\cos x - \sin x = 0; [1, 2]$

4. $x - 2 + 2\cos x = 0; [1, 2]$

C *In Problems 5–14, use Newton's Method to approximate the indicated root of the given equation accurate to five decimal places. Begin by sketching a graph.*

5. The largest root of $x^3 + 6x^2 + 9x + 1 = 0$

6. The real root of $7x^3 + x - 5 = 0$

7. The largest root of $x - 2 + 2\cos x = 0$ (see Problem 4)

8. The smallest positive root of $2 \cos x - \sin x = 0$ (see Problem 3)

9. The root of $\cos x = 2x$

10. The root of $2x - \sin x = 1$

11. All real roots of $x^4 - 8x^3 + 22x^2 - 24x + 8 = 0$

12. All real roots of $x^4 + 6x^3 + 2x^2 + 24x - 8 = 0$

13. The positive root of $2x^2 - \sin x = 0$

14. The smallest positive root of $2 \cot x = x$

C **15.** Use Newton's Method to calculate $\sqrt[3]{6}$ to five decimal places. *Hint:* Solve $x^3 - 6 = 0$.

C **16.** Use Newton's Method to calculate $\sqrt[4]{47}$ to five decimal places.

GC *In Problems 17–20, approximate the values of x that give maximum and minimum values of the function on the indicated intervals.*

17. $f(x) = x^4 + x^3 + x^2 + x; [-1, 1]$

18. $f(x) = \dfrac{x^3 + 1}{x^4 + 1}; [-4, 4]$

19. $f(x) = \dfrac{\sin x}{x}; [\pi, 3\pi]$

20. $f(x) = x^2 \sin\dfrac{x}{2}; [0, 4\pi]$

C **21.** Kepler's equation $x = m + E \sin x$ is important in astronomy. Use the Fixed-Point Algorithm to solve this equation for x when $m = 0.8$ and $E = 0.2$.

22. Sketch the graph of $y = x^{1/3}$. Obviously, its only x-intercept is zero. Convince yourself that Newton's Method fails to converge. Explain this failure.

23. In installment buying, one would like to figure out the real interest rate (effective rate), but unfortunately this involves solving a complicated equation. If one buys an item worth \$$P$ today and agrees to pay for it with payments of \$$R$ at the end of each month for k months, then

$$P = \frac{R}{i}\left[1 - \frac{1}{(1 + i)^k}\right]$$

where i is the interest rate per month. Tom bought a used car for \$2000 and agreed to pay for it with \$100 payments at the end of each of the next 24 months.

(a) Show that i satisfies the equation

$$20i(1 + i)^{24} - (1 + i)^{24} + 1 = 0$$

(b) Show that Newton's Method for this equation reduces to

$$i_{n+1} = i_n - \left[\frac{20i_n^2 + 19i_n - 1 + (1 + i_n)^{-23}}{500i_n - 4}\right]$$

C (c) Find i accurate to five decimal places starting with $i = 0.012$, and then give the annual rate r as a percent ($r = 1200i$).

24. In applying Newton's Method to solve $f(x) = 0$, one can usually tell by simply looking at the numbers $x_1, x_2, x_3, \ldots$ whether the sequence is converging. But even if it converges, say to $\bar{x}$, can we be sure that $\bar{x}$ is a solution? Show that the answer is yes provided f and f' are continuous at $\bar{x}$ and $f'(\bar{x}) \neq 0$.

C *In Problems 25–28, use the Fixed-Point Algorithm with x_1 as indicated to solve the equations to five decimal places.*

25. $x = \frac{3}{2}\cos x; x_1 = 1$

26. $x = 2 - \sin x; x_1 = 2$

27. $x = \sqrt{2.7 + x}; x_1 = 1$

28. $x = \sqrt{3.2 + x}; x_1 = 47$

GC **29.** Consider the equation $x = 2(x - x^2) = g(x)$.

(a) Sketch the graph of $y = x$ and $y = g(x)$ using the same coordinate system, and thereby approximately locate the positive root of $x = g(x)$.

(b) Try solving the equation by the Fixed-Point Algorithm starting with $x_1 = 0.7$.

(c) Solve the equation algebraically.

GC **30.** Follow the directions of Problem 29 for $x = 5(x - x^2) = g(x)$.

C **31.** Consider $x = \sqrt{1 + x}$.

(a) Apply the Fixed-Point Algorithm starting with $x_1 = 0$ to find x_2, x_3, x_4, and x_5.

(b) Algebraically solve for x in $x = \sqrt{1 + x}$.

(c) Evaluate $\sqrt{1 + \sqrt{1 + \sqrt{1 + \cdots}}}$.

C **32.** Consider $x = \sqrt{5 + x}$.

(a) Apply the Fixed-Point Algorithm starting with $x_1 = 0$ to find x_2, x_3, x_4, and x_5.

(b) Algebraically solve for x in $x = \sqrt{5 + x}$.

(c) Evaluate $\sqrt{5 + \sqrt{5 + \sqrt{5 + \cdots}}}$.

C **33.** Consider $x = 1 + \dfrac{1}{x}$.

(a) Apply the Fixed-Point Algorithm starting with $x_1 = 1$ to find x_2, x_3, x_4, and x_5.

(b) Algebraically solve for x in $x = 1 + \dfrac{1}{x}$.

(c) Evaluate the following expression. (An expression like this is called a **continued fraction.**)

$$1 + \cfrac{1}{1 + \cfrac{1}{1 + \cfrac{1}{1 + \cdots}}}$$

EXPL **34.** Consider the equation $x = x - f(x)/f'(x)$ and suppose that $f'(x) \neq 0$ in an interval $[a, b]$.

(a) Show that if r is in $[a, b]$ then r is a root of the equation $x = x - f(x)/f'(x)$ if and only if $f(r) = 0$.

(b) Show that Newton's Method is a special case of the Fixed-Point Algorithm, in which $g'(r) = 0$.

35. Experiment with the algorithm

$$x_{n+1} = 2x_n - ax_n^2$$

using several different values of a.

(a) Make a conjecture about what this algorithm computes.

(b) Prove your conjecture.

C *After differentiating and setting the result equal to zero, many practical max–min problems lead to an equation that cannot be solved exactly. For the following problems, use a numerical method to approximate the solution to the problem.*

36. A rectangle has two corners on the x-axis and the other two on the curve $y = \cos x$, with $-\pi/2 < x < \pi/2$. What are the dimensions of the rectangle of this type with maximum area? (See Figure 24 of Section 3.4.)

37. Two hallways meet in a right angle as shown in Figure 6 of Section 3.4, except the widths of the hallways are 8.6 feet and 6.2 feet. What is the length of the longest thin rod that can be carried around the corner?

38. An 8-foot-wide hallway makes a turn as shown in Figure 9. What is the length of the longest thin rod that can be carried around the corner?

39. An object thrown from the edge of a 42-foot cliff follows the path given by $y = -\frac{2x^2}{25} + x + 42$ (Figure 10). An observer stands 3 feet from the bottom of the cliff.

(a) Find the position of the object when it is closest to the observer.

(b) Find the position of the object when it is farthest from the observer.

Answers to Concepts Review: **1.** slowness of convergence **2.** root: Intermediate Value **3.** algorithms **4.** fixed point

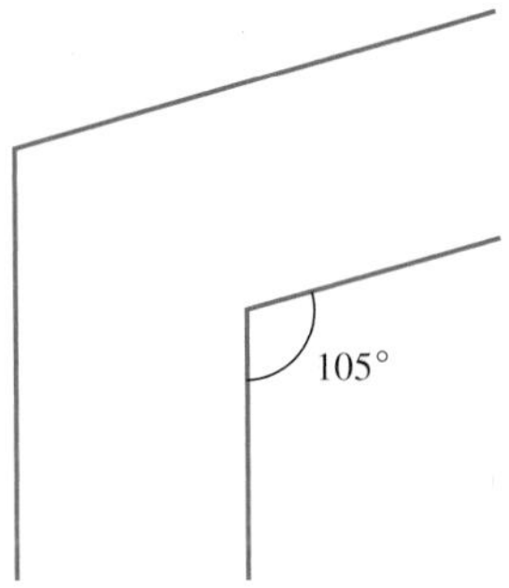

Figure 9

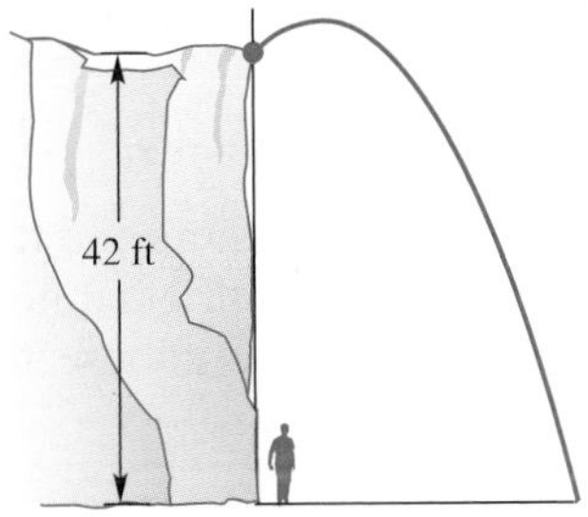

Figure 10

3.8 Antiderivatives

Most of the mathematical operations that we work with come in inverse pairs: addition and subtraction, multiplication and division, and exponentiation and root taking. In each case, the second operation undoes the first, and vice versa. One reason for our interest in inverse operations is their usefulness in solving equations. For example, solving $x^3 = 8$ involves taking roots. We have been studying differentiation in this chapter and the previous one. If we want to solve equations involving derivatives we will need its inverse, called *antidifferentiation* or *integration*.

> **Definition**
>
> We call F an **antiderivative** of f on the interval I if $D_xF(x) = f(x)$ on I, that is, if $F'(x) = f(x)$ for all x in I.

We said *an* antiderivative rather than *the* antiderivative in our definition. You will soon see why.

EXAMPLE 1 Find an antiderivative of the function $f(x) = 4x^3$ on $(-\infty, \infty)$.

SOLUTION We seek a function F satisfying $F'(x) = 4x^3$ for all real x. From our experience with differentiation, we know that $F(x) = x^4$ is one such function. ■

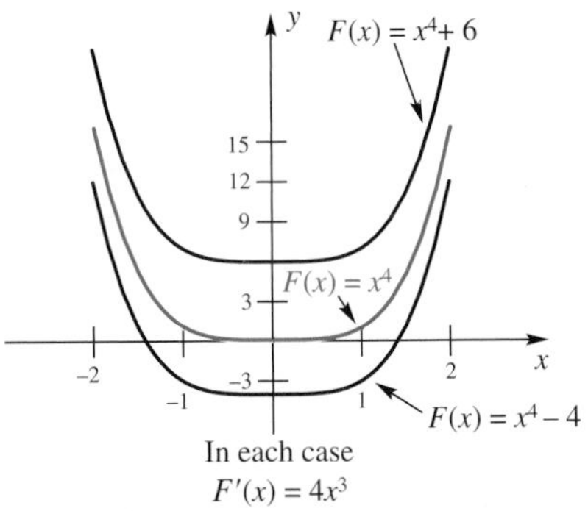

Figure 1

A moment's thought will suggest other solutions to Example 1. The function $F(x) = x^4 + 6$ also satisfies $F'(x) = 4x^3$; it too is an antiderivative of $f(x) = 4x^3$. In fact, $F(x) = x^4 + C$, where C is any constant, is an antiderivative of $4x^3$ on $(-\infty, \infty)$ (see Figure 1).

Now we pose an important question. Is *every* antiderivative of $f(x) = 4x^3$ of the form $F(x) = x^4 + C$? The answer is yes. This follows from Theorem 3.6B which says that if two functions have the same derivative, they must differ by a constant.

Our conclusion is this. If a function f has an antiderivative, it will have a whole family of them, and each member of this family can be obtained from one of them by the addition of an appropriate constant. We call this family of functions the **general antiderivative** of f. After we get used to this notion, we will often omit the adjective *general.*

EXAMPLE 2 Find the general antiderivative of $f(x) = x^2$ on $(-\infty, \infty)$.

SOLUTION The function $F(x) = x^3$ will not do since its derivative is $3x^2$. But this suggests $F(x) = \frac{1}{3}x^3$, which satisfies $F'(x) = \frac{1}{3}\cdot 3x^2 = x^2$. However, the general antiderivative is $\frac{1}{3}x^3 + C$. ■

Notation for Antiderivatives Since we used the symbol D_x for the operation of taking a derivative, it would be natural to use A_x for the operation of finding the antiderivative. Thus,

$$A_x(x^2) = \frac{1}{3}x^3 + C$$

This is the notation used by several authors and it was, in fact, used in earlier editions of this book. However, Leibniz's original notation continues to enjoy overwhelming popularity, and we therefore choose to follow him. Rather than A_x, Leibniz used the symbol $\int \ldots dx$. He wrote

$$\int x^2\,dx = \frac{1}{3}x^3 + C$$

and

$$\int 4x^3\,dx = x^4 + C$$

Leibniz chose to use the elongated s, $\int$, and the dx for reasons that will not become apparent until the next chapter. For the moment, simply think of $\int \ldots dx$ as indicating the antiderivative with respect to x, just as D_x indicates the derivative with respect to x. Note that

$$D_x \int f(x)\,dx = f(x) \quad \text{and} \quad \int D_x f(x)\,dx = f(x) + C$$

> **Proving Rules for Antiderivatives**
>
> To establish any result of the form
>
> $$\int f(x)\,dx = F(x) + C$$
>
> all we have to do is show that
>
> $$D_x[F(x) + C] = f(x)$$

> **Theorem A** **Power Rule**
>
> If r is any rational number except -1, then
>
> $$\int x^r\,dx = \frac{x^{r+1}}{r+1} + C$$

Proof The derivative of the right side is

$$D_x\left[\frac{x^{r+1}}{r+1} + C\right] = \frac{1}{r+1}(r+1)x^r = x^r$$ ■

We make two comments about Theorem A. First, it is meant to include the case $r = 0$; that is,

$$\boxed{\int 1\,dx = x + C}$$

Second, since no interval I is specified, the conclusion is understood to be valid only on intervals on which x^r is defined. In particular, we must exclude any interval containing the origin if $r < 0$.

Following Leibniz, we shall often use the term **indefinite integral** in place of antiderivative. To antidifferentiate is also to **integrate.** In the symbol $\int f(x)\,dx$, $\int$ is called the **integral sign** and $f(x)$ is called the **integrand.** Thus, we integrate the integrand and thereby evaluate the indefinite integral. Perhaps Leibniz used the adjective *indefinite* to suggest that the indefinite integral always involves an arbitrary constant.

EXAMPLE 3 Find the general antiderivative of $f(x) = x^{4/3}$.

SOLUTION

$$\int x^{4/3}\,dx = \frac{x^{7/3}}{\frac{7}{3}} + C = \tfrac{3}{7}x^{7/3} + C \qquad \blacksquare$$

Note that *to integrate a power of x, we increase the exponent by* 1 *and divide by the new exponent.*

Antiderivative formulas for the sine and cosine functions follow directly from the derivative.

Theorem B

$$\int \sin x\,dx = -\cos x + C \quad \text{and} \quad \int \cos x\,dx = \sin x + C$$

Proof Simply note that $D_x(-\cos x + C) = \sin x$ and $D_x(\sin x + C) = \cos x$. ■

The Indefinite Integral Is Linear Recall from Chapter 2 that D_x is a linear operator. This means two things.

1. $D_x[kf(x)] = kD_xf(x)$
2. $D_x[f(x) + g(x)] = D_xf(x) + D_xg(x)$

From these two properties, a third follows automatically.

3. $D_x[f(x) - g(x)] = D_xf(x) - D_xg(x)$

It turns out that $\int \ldots dx$ also has these properties of a linear operator.

Theorem C Indefinite Integral Is a Linear Operator

Let f and g have antiderivatives (indefinite integrals) and let k be a constant. Then

(i) $\displaystyle\int kf(x)\,dx = k\int f(x)\,dx$;

(ii) $\displaystyle\int [f(x) + g(x)]\,dx = \int f(x)\,dx + \int g(x)\,dx$;

(iii) $\displaystyle\int [f(x) - g(x)]\,dx = \int f(x)\,dx - \int g(x)\,dx$.

Proof To show (i) and (ii), we simply differentiate the right side and observe that we get the integrand of the left side.

$$D_x\left[k\int f(x)\,dx\right] = kD_x\int f(x)\,dx = kf(x)$$

$$D_x\left[\int f(x)\,dx + \int g(x)\,dx\right] = D_x\int f(x)\,dx + D_x\int g(x)\,dx$$

$$= f(x) + g(x)$$

Property (iii) follows from (i) and (ii). ■

EXAMPLE 4 Using the linearity of $\int$, evaluate

(a) $\int (3x^2 + 4x)\,dx$ (b) $\int (u^{3/2} - 3u + 14)\,du$ (c) $\int \left(1/t^2 + \sqrt{t}\right) dt$

SOLUTION

(a)

$$\begin{aligned}\int (3x^2 + 4x)\,dx &= \int 3x^2\,dx + \int 4x\,dx \\ &= 3\int x^2\,dx + 4\int x\,dx \\ &= 3\left(\frac{x^3}{3} + C_1\right) + 4\left(\frac{x^2}{2} + C_2\right) \\ &= x^3 + 2x^2 + (3C_1 + 4C_2) \\ &= x^3 + 2x^2 + C\end{aligned}$$

Two arbitrary constants C_1 and C_2 appeared, but they were combined into one constant, C, a practice we consistently follow.

(b) Note the use of the variable u rather than x. This is fine as long as the corresponding differential symbol is du, since we then have a complete change of notation.

$$\begin{aligned}\int (u^{3/2} - 3u + 14)\,du &= \int u^{3/2}\,du - 3\int u\,du + 14\int 1\,du \\ &= \tfrac{2}{5}u^{5/2} - \tfrac{3}{2}u^2 + 14u + C\end{aligned}$$

(c)

$$\begin{aligned}\int \left(\frac{1}{t^2} + \sqrt{t}\right) dt &= \int (t^{-2} + t^{1/2})\,dt = \int t^{-2}\,dt + \int t^{1/2}\,dt \\ &= \frac{t^{-1}}{-1} + \frac{t^{3/2}}{\frac{3}{2}} + C = -\frac{1}{t} + \frac{2}{3}t^{3/2} + C\end{aligned}$$

■

Generalized Power Rule Recall the Chain Rule as applied to a power of a function. If $u = g(x)$ is a differentiable function and r is a rational number $(r \neq -1)$, then

$$D_x\left[\frac{u^{r+1}}{r+1}\right] = u^r \cdot D_x u$$

or, in functional notation,

$$D_x\left(\frac{[g(x)]^{r+1}}{r+1}\right) = [g(x)]^r \cdot g'(x)$$

From this we obtain an important rule for indefinite integrals.

Theorem D Generalized Power Rule

Let g be a differentiable function and r a rational number different from -1. Then

$$\int [g(x)]^r g'(x)\,dx = \frac{[g(x)]^{r+1}}{r+1} + C$$

To apply Theorem D, we must be able to recognize the functions g and g' in the integrand.

EXAMPLE 5 Evaluate

(a) $\int (x^4 + 3x)^{30}(4x^3 + 3)\,dx$ (b) $\int \sin^{10} x \cos x\,dx$

SOLUTION

(a) Let $g(x) = x^4 + 3x$; then $g'(x) = 4x^3 + 3$. Thus, by Theorem D,

$$\int (x^4 + 3x)^{30}(4x^3 + 3)\,dx = \int [g(x)]^{30}g'(x)\,dx = \frac{[g(x)]^{31}}{31} + C$$

$$= \frac{(x^4 + 3x)^{31}}{31} + C$$

(b) Let $g(x) = \sin x$; then $g'(x) = \cos x$. Thus,

$$\int \sin^{10} x \cos x\,dx = \int [g(x)]^{10}g'(x)\,dx = \frac{[g(x)]^{11}}{11} + C$$

$$= \frac{\sin^{11} x}{11} + C$$

■

Example 5 shows why Leibniz used the differential dx in his notation $\int \ldots dx$. If we let $u = g(x)$, then $du = g'(x)\,dx$. The conclusion of Theorem D is therefore

$$\int u^r\,du = \frac{u^{r+1}}{r + 1} + C, \qquad r \neq -1$$

which is the ordinary power rule with u as the variable. Thus, the generalized power rule is just the ordinary power rule applied to functions. But, in applying it, we must always make sure that we have du to go with u^r. The following examples illustrate what we mean.

EXAMPLE 6 Evaluate

(a) $\int (x^3 + 6x)^5(6x^2 + 12)\,dx$ (b) $\int (x^2 + 4)^{10}x\,dx$

SOLUTION

(a) Let $u = x^3 + 6x$; then $du = (3x^2 + 6)\,dx$. Thus, $(6x^2 + 12)\,dx = 2(3x^2 + 6)\,dx = 2\,du$, and so

$$\int (x^3 + 6x)^5(6x^2 + 12)\,dx = \int u^5\,2\,du$$

$$= 2\int u^5\,du$$

$$= 2\left[\frac{u^6}{6} + C\right]$$

$$= \frac{u^6}{3} + 2C$$

$$= \frac{(x^3 + 6x)^6}{3} + K$$

Two things should be noted about our solution. First, the fact that $(6x^2 + 12)\,dx$ is $2du$ instead of du caused no trouble; the factor 2 could be moved in front of the integral sign by linearity. Second, we wound up with an arbitrary constant of $2C$. This is still an arbitrary constant; we called it K.

(b) Let $u = x^2 + 4$; then $du = 2x\,dx$. Thus,

$$\begin{aligned}\int (x^2+4)^{10} x\,dx &= \int (x^2+4)^{10} \cdot \frac{1}{2} \cdot 2x\,dx \\ &= \frac{1}{2}\int u^{10}\,du \\ &= \frac{1}{2}\left(\frac{u^{11}}{11} + C\right) \\ &= \frac{(x^2+4)^{11}}{22} + K\end{aligned}$$

■

Concepts Review

1. The Power Rule for derivatives says that $d(x^r)/dx =$ ______. The Power Rule for integrals says that $\int x^r\,dx =$ ______.

2. The Generalized Power Rule for derivatives says that $d[f(x)]^r/dx =$ ______. The Generalized Power Rule for integrals says that $\int$ ______ $dx = [f(x)]^{r+1}/(r+1) + C, r \neq -1$.

3. $\int (x^4 + 3x^2 + 1)^8 (4x^3 + 6x)\,dx =$ ______.

4. By linearity, $\int [c_1 f(x) + c_2 g(x)]\,dx =$ ______.

Problem Set 3.8

Find the general antiderivative $F(x) + C$ for each of the following.

1. $f(x) = 5$
2. $f(x) = x - 4$
3. $f(x) = x^2 + \pi$
4. $f(x) = 3x^2 + \sqrt{3}$
5. $f(x) = x^{5/4}$
6. $f(x) = 3x^{2/3}$
7. $f(x) = 1/\sqrt[3]{x^2}$
8. $f(x) = 7x^{-3/4}$
9. $f(x) = x^2 - x$
10. $f(x) = 3x^2 - \pi x$
11. $f(x) = 4x^5 - x^3$
12. $f(x) = x^{100} + x^{99}$
13. $f(x) = 27x^7 + 3x^5 - 45x^3 + \sqrt{2}x$
14. $f(x) = x^2(x^3 + 5x^2 - 3x + \sqrt{3})$
15. $f(x) = \dfrac{3}{x^2} - \dfrac{2}{x^3}$
16. $f(x) = \dfrac{\sqrt{2x}}{x} + \dfrac{3}{x^5}$
17. $f(x) = \dfrac{4x^6 + 3x^4}{x^3}$
18. $f(x) = \dfrac{x^6 - x}{x^3}$

In Problems 19–26, evaluate the indicated indefinite integrals.

19. $\int (x^2 + x)\,dx$
20. $\int (x^3 + \sqrt{x})\,dx$
21. $\int (x + 1)^2\,dx$
22. $\int (z + \sqrt{2}z)^2\,dz$
23. $\int \dfrac{(z^2 + 1)^2}{\sqrt{z}}\,dz$
24. $\int \dfrac{s(s + 1)^2}{\sqrt{s}}\,ds$
25. $\int (\sin\theta - \cos\theta)\,d\theta$
26. $\int (t^2 - 2\cos t)\,dt$

In Problems 27–36, use the methods of Examples 5 and 6 to evaluate the indefinite integrals.

27. $\int (\sqrt{2}x + 1)^3 \sqrt{2}\,dx$
28. $\int (\pi x^3 + 1)^4 3\pi x^2\,dx$
29. $\int (5x^2 + 1)(5x^3 + 3x - 8)^6\,dx$
30. $\int (5x^2 + 1)\sqrt{5x^3 + 3x - 2}\,dx$
31. $\int 3t\sqrt[3]{2t^2 - 11}\,dt$
32. $\int \dfrac{3y}{\sqrt{2y^2 + 5}}\,dy$
33. $\int x^2\sqrt{x^3 + 4}\,dx$
34. $\int (x^3 + x)\sqrt{x^4 + 2x^2}\,dx$
35. $\int \sin x\,(1 + \cos x)^4\,dx$
36. $\int \sin x \cos x\,\sqrt{1 + \sin^2 x}\,dx$

In Problems 37–42, $f''(x)$ is given. Find $f(x)$ by antidifferentiating twice. Note that in this case your answer should involve two arbitrary constants, one from each antidifferentiation. For example, if $f''(x) = x$, then $f'(x) = x^2/2 + C_1$ and $f(x) = x^3/6 + C_1x + C_2$. The constants C_1 and C_2 cannot be combined because C_1x is not a constant.

37. $f''(x) = 3x + 1$
38. $f''(x) = -2x + 3$

39. $f''(x) = \sqrt{x}$

40. $f''(x) = x^{4/3}$

41. $f''(x) = \dfrac{x^4 + 1}{x^3}$

42. $f''(x) = 2\sqrt[3]{x + 1}$

43. Prove the formula

$$\int [f(x)g'(x) + g(x)f'(x)]\,dx = f(x)g(x) + C$$

Hint: See the box in the margin next to Theorem A.

44. Prove the formula

$$\int \frac{g(x)f'(x) - f(x)g'(x)}{g^2(x)}\,dx = \frac{f(x)}{g(x)} + C$$

45. Use the formula from Problem 43 to find

$$\int \left[\frac{x^2}{2\sqrt{x - 1}} + 2x\sqrt{x - 1}\right] dx$$

46. Use the formula from Problem 43 to find

$$\int \left[\frac{-x^3}{(2x + 5)^{3/2}} + \frac{3x^2}{\sqrt{2x + 5}}\right] dx$$

47. Find $\int f''(x)\,dx$ if $f(x) = x\sqrt{x^3 + 1}$.

48. Prove the formula

$$\int \frac{2g(x)f'(x) - f(x)g'(x)}{2[g(x)]^{3/2}} = \frac{f(x)}{\sqrt{g(x)}} + C$$

49. Prove the formula

$$\int f^{m-1}(x)g^{n-1}(x)[nf(x)g'(x) + mg(x)f'(x)]\,dx = f^m(x)g^n(x) + C$$

50. Evaluate the indefinite integral

$$\int \sin^3[(x^2 + 1)^4]\cos[(x^2 + 1)^4](x^2 + 1)^3 x\,dx$$

Hint: Let $u = \sin(x^2 + 1)^4$.

51. Evaluate $\int |x|\,dx$.

52. Evaluate $\int \sin^2 x\,dx$.

CAS **53.** Some software packages can evaluate indefinite integrals. Use your software on each of the following.

(a) $\int 6\sin(3(x - 2))\,dx$

(b) $\int \sin^3(x/6)\,dx$

(c) $\int (x^2\cos 2x + x\sin 2x)\,dx$

EXPL CAS **54.** Let $F_0(x) = x\sin x$ and $F_{n+1}(x) = \int F_n(x)\,dx$.

(a) Determine $F_1(x)$, $F_2(x)$, $F_3(x)$, and $F_4(x)$.

(b) On the basis of part (a), conjecture the form of $F_{16}(x)$.

Answers to Concepts Review: **1.** rx^{r-1}; $x^{r+1}/(r + 1) + C, r \neq -1$ **2.** $r[f(x)]^{r-1}f'(x)$; $[f(x)]^r f'(x)$ **3.** $(x^4 + 3x^2 + 1)^9/9 + C$ **4.** $c_1\int f(x)\,dx + c_2\int g(x)\,dx$

3.9 Introduction to Differential Equations

In the previous section, our task was to antidifferentiate (integrate) a function f to obtain a new function F. We wrote

$$\boxed{\int f(x)\,dx = F(x) + C}$$

and this was correct by definition provided $F'(x) = f(x)$. Now $F'(x) = f(x)$ in the language of derivatives is equivalent to $dF(x) = f(x)\,dx$ in differential language (Section 2.9). Thus, we may look on the boxed formula as saying that

$$\boxed{\int dF(x) = F(x) + C}$$

From this perspective, we integrate the differential of a function to obtain the function (plus a constant). This was Leibniz's viewpoint; adopting it will help us to solve *differential equations*.

What Is a Differential Equation? To motivate our answer, we begin with a simple example.

EXAMPLE 1 Find the xy-equation of the curve that passes through $(-1, 2)$ and whose slope at any point on the curve is equal to twice the x-coordinate of that point.

SOLUTION The condition that must hold at each point (x, y) on the curve is

$$\frac{dy}{dx} = 2x$$

We are looking for a function $y = f(x)$ that satisfies this equation and the additional condition that $y = 2$ when $x = -1$. We suggest two ways of looking at this problem.

Method 1 When an equation has the form $dy/dx = g(x)$, we observe that y must be an antiderivative of $g(x)$; that is,

$$y = \int g(x)\, dx$$

In our case,

$$y = \int 2x\, dx = x^2 + C$$

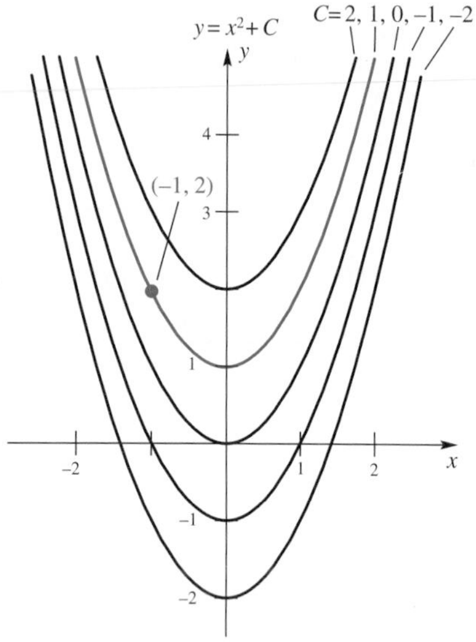

Figure 1

Method 2 Think of dy/dx as a quotient of two differentials. When we multiply both sides of $dy/dx = 2x$ by dx, we get

$$dy = 2x\, dx$$

Next we integrate the differentials on both sides, equate the results, and simplify.

$$\int dy = \int 2x\, dx$$

$$y + C_1 = x^2 + C_2$$

$$y = x^2 + C_2 - C_1$$

$$y = x^2 + C$$

The second method works in a wide variety of problems that are not of the simple form $dy/dx = g(x)$, as we shall see.

The solution $y = x^2 + C$ represents the family of curves illustrated in Figure 1. From this family, we must choose the one for which $y = 2$ when $x = -1$; thus, we want

$$2 = (-1)^2 + C$$

We conclude that $C = 1$ and therefore that $y = x^2 + 1$. ■

The equations $dy/dx = 2x$ and $dy = 2x\, dx$ are called *differential equations*. Other examples are

$$\frac{dy}{dx} = 2xy + \sin x$$

$$y\, dy = (x^3 + 1)\, dx$$

$$\frac{d^2y}{dx^2} + 3\frac{dy}{dx} - 2xy = 0$$

Any equation in which the unknown is a function and that involves derivatives (or differentials) of this unknown function is called a **differential equation.** A function that, when substituted in the differential equation yields an equality, is called a **solution** of the differential equation. Thus, to solve a differential equation is to find an unknown *function*. In general, this is a difficult job and one about which many thick books have been written. Here we consider only the simplest type, **first-order separable** differential equations. These are equations involving just the first derivative of the unknown function and are such that the variables can be separated, one on each side of the equation.

Separation of Variables Consider the differential equation

$$\frac{dy}{dx} = \frac{x + 3x^2}{y^2}$$

If we multiply both sides by $y^2\,dx$, we obtain

$$y^2\,dy = (x + 3x^2)\,dx$$

In this form, the differential equation has its variables separated; that is, the y terms are on one side of the equation and the x terms are on the other. In separated form, we can solve the differential equation using Method 2 (integrate both sides, equate the results, and simplify), as we now illustrate.

EXAMPLE 2 Solve the differential equation

$$\frac{dy}{dx} = \frac{x + 3x^2}{y^2}$$

Then find that solution for which $y = 6$ when $x = 0$.

SOLUTION As noted earlier, the given equation is equivalent to

$$y^2\,dy = (x + 3x^2)\,dx$$

Thus,

$$\int y^2\,dy = \int (x + 3x^2)\,dx$$

$$\frac{y^3}{3} + C_1 = \frac{x^2}{2} + x^3 + C_2$$

$$y^3 = \frac{3x^2}{2} + 3x^3 + (3C_2 - 3C_1)$$

$$= \frac{3x^2}{2} + 3x^3 + C$$

$$y = \sqrt[3]{\frac{3x^2}{2} + 3x^3 + C}$$

To find the constant C, we use the condition $y = 6$ when $x = 0$. This gives

$$6 = \sqrt[3]{C}$$

$$216 = C$$

Thus,

$$y = \sqrt[3]{\frac{3x^2}{2} + 3x^3 + 216}$$

To check our work we can substitute this result in both sides of the original differential equation to see that it gives an equality. We should also check that $y = 6$ when $x = 0$.

Substituting in the left side, we get

$$\frac{dy}{dx} = \frac{1}{3}\left(\frac{3x^2}{2} + 3x^3 + 216\right)^{-2/3}(3x + 9x^2)$$

$$= \frac{x + 3x^2}{\left(\frac{3}{2}x^2 + 3x^3 + 216\right)^{2/3}}$$

On the right side, we get

$$\frac{x + 3x^2}{y^2} = \frac{x + 3x^2}{\left(\frac{3}{2}x^2 + 3x^3 + 216\right)^{2/3}}$$

As expected, the two expressions are equal. When $x = 0$, we have

$$y = \sqrt[3]{\frac{3 \cdot 0^2}{2} + 3 \cdot 0^3 + 216} = \sqrt[3]{216} = 6$$

Thus, $y = 6$ when $x = 0$, as we expected. ■

Motion Problems Recall that if $s(t)$, $v(t)$, and $a(t)$ represent the position, velocity, and acceleration, respectively, at time t of an object moving along a coordinate line then

$$v(t) = s'(t) = \frac{ds}{dt}$$

$$a(t) = v'(t) = \frac{dv}{dt} = \frac{d^2s}{dt^2}$$

In some earlier work (Section 2.6), we assumed that $s(t)$ was known, and from this we calculated $v(t)$ and $a(t)$. Now we want to consider the reverse process: given the acceleration $a(t)$, find the velocity $v(t)$ and the position $s(t)$.

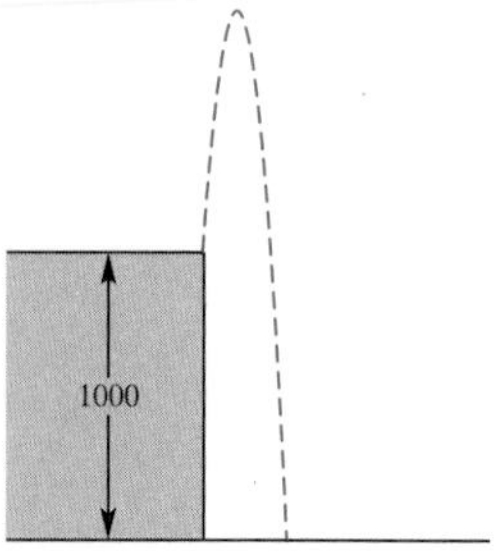

Figure 2

EXAMPLE 3 **Falling-Body Problem**

Near the surface of the earth, the acceleration of a falling body due to gravity is 32 feet per second per second, provided that air resistance is neglected. If an object is thrown upward from an initial height of 1000 feet (Figure 2) with a velocity of 50 feet per second, find its velocity and height 4 seconds later.

SOLUTION Let us assume that the height s is measured positively in the upward direction. Then $v = ds/dt$ is initially positive (s is increasing), but $a = dv/dt$ is negative. (The pull of gravity is downward, thus decreasing v). Hence, we start our analysis with the differential equation $dv/dt = -32$, with the additional conditions that $v = 50$ and $s = 1000$ when $t = 0$. Either Method 1 (direct antidifferentiation) or Method 2 (separation of variables) works well.

$$\frac{dv}{dt} = -32$$

$$v = \int -32\, dt = -32t + C$$

Since $v = 50$ at $t = 0$, we find that $C = 50$, and so

$$\boxed{v = -32t + 50}$$

Now $v = ds/dt$, and so we have another differential equation,

$$\frac{ds}{dt} = -32t + 50$$

When we integrate, we obtain

$$s = \int (-32t + 50)\, dt$$

$$= -16t^2 + 50t + K$$

Since $s = 1000$ at $t = 0$, $K = 1000$ and

$$\boxed{s = -16t^2 + 50t + 1000}$$

Finally, at $t = 4$,

$$v = -32(4) + 50 = -78 \text{ feet per second}$$

$$s = -16(4)^2 + 50(4) + 1000 = 944 \text{ feet}$$ ■

We remark that if $v = v_0$ and $s = s_0$ at $t = 0$, the procedure of Example 3 leads to the well-known falling-body formulas:

$$a = -32$$
$$v = -32t + v_0$$
$$s = -16t^2 + v_0 t + s_0$$

EXAMPLE 4 The acceleration of an object moving along a coordinate line is given by $a(t) = (2t + 3)^{-3}$ in meters per second per second. If the velocity at $t = 0$ is 4 meters per second, find the velocity 2 seconds later.

SOLUTION We begin with the differential equation shown in the first line below. To perform the integration in the second line, we multiply and divide by 2, thus preparing the integral for the Generalized Power Rule.

$$\frac{dv}{dt} = (2t + 3)^{-3}$$
$$v = \int (2t + 3)^{-3}\, dt = \frac{1}{2}\int (2t + 3)^{-3}\, 2\, dt$$
$$= \frac{1}{2}\frac{(2t + 3)^{-2}}{-2} + C = -\frac{1}{4(2t + 3)^2} + C$$

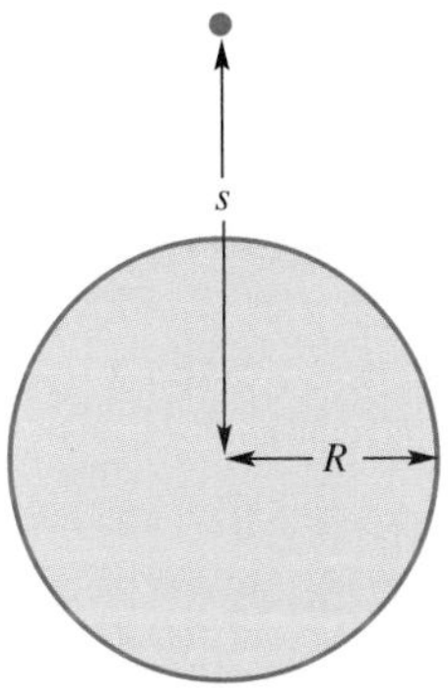

Figure 3

Since $v = 4$ at $t = 0$,

$$4 = -\frac{1}{4(3)^2} + C$$

which gives $C = \frac{145}{36}$. Thus,

$$v = -\frac{1}{4(2t + 3)^2} + \frac{145}{36}$$

At $t = 2$,

$$v = -\frac{1}{4(49)} + \frac{145}{36} \approx 4.023 \text{ meters per second}$$ ■

EXAMPLE 5 **Escape Velocity (Optional)**

The gravitational attraction F exerted by the earth on an object of mass m at a distance s from the center of the earth is given by $F = -mgR^2/s^2$, where $-g$ ($g \approx 32$ feet per second per second) is the acceleration of gravity at the surface of the earth and R ($R \approx 3960$ miles) is the radius of the earth (Figure 3). Show that an object launched outward from the earth with an initial velocity $v_0 \geq \sqrt{2gR} \approx 6.93$ miles per second will not fall back to the earth. Neglect air resistance in making this calculation.

SOLUTION According to Newton's Second Law, $F = ma$; that is,

$$F = m\frac{dv}{dt} = m\frac{dv}{ds}\frac{ds}{dt} = m\frac{dv}{ds}v$$

Thus,

$$mv\frac{dv}{ds} = -mg\frac{R^2}{s^2}$$

Separating variables gives

$$v\, dv = -gR^2 s^{-2}\, ds$$
$$\int v\, dv = -gR^2 \int s^{-2}\, ds$$
$$\frac{v^2}{2} = \frac{gR^2}{s} + C$$

Now $v = v_0$ when $s = R$, and so $C = \frac{1}{2}v_0^2 - gR$. Consequently,

$$v^2 = \frac{2gR^2}{s} + v_0^2 - 2gR$$

Finally, since $2gR^2/s$ gets small with increasing s, we see that v remains positive if and only if $v_0 \geq \sqrt{2gR}$. ■

Concepts Review

1. $dy/dx = 3x^2 + 1$ and $dy/dx = x/y^2$ are examples of what is called a _____.

2. To solve the differential equation $dy/dx = g(x, y)$ is to find the _____ that, when substituted for y, yields an equality.

3. To solve the differential equation $dy/dx = x^2y^3$, the first step would be to _____.

4. To solve a falling-body problem near the surface of the earth, we start with the experimental fact that the acceleration a of gravity is -32 feet per second per second; that is, $a = dv/dt = -32$. Solving this differential equation gives $v = ds/dt =$ _____, and solving the resulting differential equation gives $s =$ _____.

Problem Set 3.9

In Problems 1–4, show that the indicated function is a solution of the given differential equation; that is, substitute the indicated function for y to see that it produces an equality.

1. $\dfrac{dy}{dx} + \dfrac{x}{y} = 0;\ y = \sqrt{1 - x^2}$

2. $-x\dfrac{dy}{dx} + y = 0;\ y = Cx$

3. $\dfrac{d^2y}{dx^2} + y = 0;\ y = C_1 \sin x + C_2 \cos x$

4. $\left(\dfrac{dy}{dx}\right)^2 + y^2 = 1;\ y = \sin(x + C)$ and $y = \pm 1$

In Problems 5–14, first find the general solution (involving a constant C) for the given differential equation. Then find the particular solution that satisfies the indicated condition. (See Example 2.)

5. $\dfrac{dy}{dx} = x^2 + 1;\ y = 1$ at $x = 1$

6. $\dfrac{dy}{dx} = x^{-3} + 2;\ y = 3$ at $x = 1$

7. $\dfrac{dy}{dx} = \dfrac{x}{y};\ y = 1$ at $x = 1$

8. $\dfrac{dy}{dx} = \sqrt{\dfrac{x}{y}};\ y = 4$ at $x = 1$

9. $\dfrac{dz}{dt} = t^2z^2;\ z = 1/3$ at $t = 1$

10. $\dfrac{dy}{dt} = y^4;\ y = 1$ at $t = 0$

11. $\dfrac{ds}{dt} = 16t^2 + 4t - 1;\ s = 100$ at $t = 0$

12. $\dfrac{du}{dt} = u^3(t^3 - t);\ u = 4$ at $t = 0$

13. $\dfrac{dy}{dx} = (2x + 1)^4;\ y = 6$ at $x = 0$

14. $\dfrac{dy}{dx} = -y^2x(x^2 + 2)^4;\ y = 1$ at $x = 0$

15. Find the xy-equation of the curve through $(1, 2)$ whose slope at any point is three times its x-coordinate (see Example 1).

16. Find the xy-equation of the curve through $(1, 2)$ whose slope at any point is three times the square of its y-coordinate.

In Problems 17–20, an object is moving along a coordinate line subject to the indicated acceleration a (in centimeters per second per second) with the initial velocity v_0 (in centimeters per second) and directed distance s_0 (in centimeters). Find both the velocity v and directed distance s after 2 seconds (see Example 4).

17. $a = t;\ v_0 = 3,\ s_0 = 0$

18. $a = (1 + t)^{-4};\ v_0 = 0,\ s_0 = 10$

C **19.** $a = \sqrt[3]{2t + 1};\ v_0 = 0,\ s_0 = 10$

C **20.** $a = (3t + 1)^{-3};\ v_0 = 4,\ s_0 = 0$

21. A ball is thrown upward from the surface of the earth with an initial velocity of 96 feet per second. What is the maximum height that it reaches? (See Example 3.)

22. A ball is thrown upward from the surface of a planet where the acceleration of gravity is k (a negative constant) feet per second per second. If the initial velocity is v_0, show that the maximum height is $-v_0^2/2k$.

C **23.** On the surface of the moon, the acceleration of gravity is -5.28 feet per second per second. If an object is thrown upward from an initial height of 1000 feet with a velocity of 56 feet per second, find its velocity and height 4.5 seconds later.

C **24.** What is the maximum height that the object of Problem 23 reaches?

25. The rate of change of volume V of a melting snowball is proportional to the surface area S of the ball; that is,

$dV/dt = -kS$, where k is a positive constant. If the radius of the ball at $t = 0$ is $r = 2$ and at $t = 10$ is $r = 0.5$, show that $r = -\frac{3}{20}t + 2$.

26. From what height above the earth must a ball be dropped in order to strike the ground with a velocity of -136 feet per second?

C **27.** Determine the *escape velocity* for an object launched from each of the following celestial bodies (see Example 5). Here $g \approx 32$ feet per second per second.

	Acceleration of Gravity	Radius (miles)
Moon	$-0.165g$	1,080
Venus	$-0.85g$	3,800
Jupiter	$-2.6g$	43,000
Sun	$-28g$	432,000

28. If the brakes of a car, when fully applied, produce a constant deceleration of 11 feet per second per second, what is the shortest distance in which the car can be braked to a halt from a speed of 60 miles per hour?

29. What constant acceleration will cause a car to increase its velocity from 45 to 60 miles per hour in 10 seconds?

30. A block slides down an inclined plane with a constant acceleration of 8 feet per second per second. If the inclined plane is 75 feet long and the block reaches the bottom in 3.75 seconds, what was the initial velocity of the block?

31. A certain rocket, initially at rest, is shot straight up with an acceleration of $6t$ meters per second per second during the first 10 seconds after blast-off, after which the engine cuts out and the rocket is subject only to gravitational acceleration of -10 meters per second per second. How high will the rocket go?

32. Starting at station A, a commuter train accelerates at 3 meters per second per second for 8 seconds, then travels at constant speed v_m for 100 seconds, and finally brakes (decelerates) to a stop at station B at 4 meters per second per second. Find (a) v_m and (b) the distance between A and B.

33. Starting from rest, a bus increases speed at constant acceleration a_1, then travels at constant speed v_m, and finally brakes to a stop at constant acceleration a_2 ($a_2 < 0$). It took 4 minutes to travel the 2 miles between stop C and stop D and then 3 minutes to go the 1.4 miles between stop D and stop E.

(a) Sketch the graph of the velocity v as a function of time t, $0 \le t \le 7$.

(b) Find the maximum speed v_m.

(c) If $a_1 = -a_2 = a$, evaluate a.

34. A hot-air balloon left the ground rising at 4 feet per second. Sixteen seconds later, Victoria threw a ball straight up to her friend Colleen in the balloon. At what speed did she throw the ball if it just made it to Colleen?

35. According to Torricelli's Law, the time rate of change of the volume V of water in a draining tank is proportional to the square root of the water's depth. A cylindrical tank of radius $10/\sqrt{\pi}$ centimeters and height 16 centimeters, which was full initially, took 40 seconds to drain.

(a) Write the differential equation for V at time t and the two corresponding conditions.

(b) Solve the differential equation.

(c) Find the volume of water after 10 seconds.

C **36.** The wolf population P in a certain state has been growing at a rate proportional to the cube root of the population size. The population was estimated at 1000 in 1980 and at 1700 in 1990.

(a) Write the differential equation for P at time t with the two corresponding conditions.

(b) Solve the differential equation.

(c) When will the wolf population reach 4000?

37. At $t = 0$, a ball was dropped from a height of 16 feet. It hit the floor and rebounded to a height of 9 feet (Figure 4).

(a) Find a two-part formula for the velocity $v(t)$ that is valid until the ball hits the floor a second time.

(b) At what two times was the ball at height 9 feet?

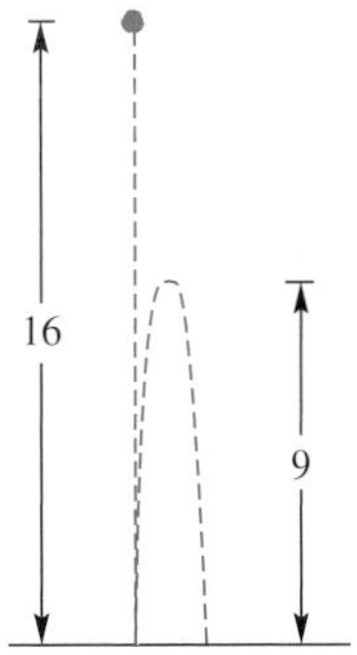

Figure 4

Answers to Concepts Review: **1.** differential equation **2.** function **3.** separate variables **4.** $-32t + v_0$; $-16t^2 + v_0t + s_0$

3.10 Chapter Review

Concepts Test

Respond with true or false to each of the following assertions. Be prepared to justify your answer.

1. A continuous function defined on a closed interval must attain a maximum value on that interval.

2. If a differentiable function f attains a maximum value at an interior point c of its domain, then $f'(c) = 0$.

3. It is possible for a function to have an infinite number of critical points.

4. A continuous function that is increasing on $(-\infty, \infty)$ must be differentiable everywhere.

5. If $f(x) = 3x^6 + 4x^4 + 2x^2$, then the graph of f is concave up on the whole real line.

6. If f is an increasing differentiable function on an interval I, then $f'(x) > 0$ for all x in I.

7. If $f'(x) > 0$, for all x in I, then f is increasing on I.

8. If $f''(c) = 0$, then f has an inflection point at $(c, f(c))$.

9. A quadratic function has no inflection points.

10. If $f'(x) > 0$ for all x in $[a, b]$, then f attains its maximum value on $[a, b]$ at b.

11. The function $y = \tan^2 x$ has no minimum value.

12. The function $y = 2x^3 + x$ has no maximum or minimum values.

13. The function $y = 2x^3 + x + \tan x$ has no maximum or minimum values.

14. The graph of $y = \dfrac{x^2 - x - 6}{x - 3} = \dfrac{(x + 2)(x - 3)}{x - 3}$ has a vertical asymptote at $x = 3$.

15. The graph of $y = \dfrac{x^2 + 1}{1 - x^2}$ has a horizontal asymptote of $y = -1$.

16. The graph of $y = \dfrac{3x^2 + 2x + \sin x}{x}$ has an oblique asymptote of $y = 3x + 2$.

17. The function $f(x) = \sqrt{x}$ satisfies the hypotheses of the Mean Value Theorem on $[0, 2]$.

18. The function $f(x) = |x|$ satisfies the hypotheses of the Mean Value Theorem on $[-1, 1]$.

19. On the interval $[-1, 1]$, there will be just one point where the tangent line to $y = x^3$ is parallel to the secant line.

20. If $f'(x) = 0$ for all x in (a, b), then f is constant on this interval.

21. If $f'(c) = f''(c) = 0$, then $f(c)$ is neither a maximum nor minimum value.

22. The graph of $y = \sin x$ has infinitely many points of inflection.

23. Among rectangles of fixed area K, the one with maximum perimeter is a square.

24. If the graph of a differentiable function has three x-intercepts, then it must have at least two points where the tangent line is horizontal.

25. The sum of two increasing functions is an increasing function.

26. The product of two increasing functions is an increasing function.

27. If $f'(0) = 0$ and $f''(x) > 0$ for $x \geq 0$, then f is increasing on $[0, \infty)$.

28. If $f'(x) \leq 2$ for all x on the interval $[0, 3]$ and $f(0) = 1$, then $f(3) < 4$.

29. If f is a differentiable function, then f is nondecreasing on (a, b) if and only if $f'(x) \geq 0$ on (a, b).

30. Two differentiable functions have the same derivative on (a, b) if and only if they differ by a constant on (a, b).

31. If $f''(x) > 0$ for all x, then the graph of $y = f(x)$ cannot have a horizontal asymptote.

32. A global maximum value is always a local maximum value.

33. A cubic function $f(x) = ax^3 + bx^2 + cx + d, a \neq 0$, can have at most one local maximum value on any open interval.

34. The linear function $f(x) = ax + b, a \neq 0$, has no minimum value on any open interval.

35. If f is continuous on $[a, b]$ and $f(a)f(b) < 0$, then $f(x) = 0$ has a root between a and b.

36. One of the virtues of the Bisection Method is its rapid convergence.

37. Newton's Method will produce a convergent sequence for the function $f(x) = x^{1/3}$.

38. If Newton's Method fails to converge for one starting value, then it will fail to converge for every starting value.

39. If g is continuous on $[a, b]$ and if $a < g(a) < g(b) < b$, then g has a fixed point between a and b.

40. One of the virtues of the Bisection Method is that it always converges.

41. The indefinite integral is a linear operator.

42. $\int [f(x)g'(x) + g(x)f'(x)]\,dx = f(x)g(x) + C$.

43. $y = \cos x$ is a solution to the differential equation $(dy/dx)^2 = 1 - y^2$.

44. All functions that are antiderivatives must have derivatives.

45. If the second derivatives of two functions are equal, then the functions differ at most by a constant.

46. $\int f'(x)\,dx = f(x)$ for every differentiable function f.

47. If $s = -16t^2 + v_0 t$ gives the height at time t of a ball thrown straight up from the surface of the earth, then the ball will hit the ground with velocity $-v_0$.

Sample Test Problems

In Problems 1–12, a function f and its domain are given. Determine the critical points, evaluate f at these points, and find the (global) maximum and minimum values.

1. $f(x) = x^2 - 2x$; $[0, 4]$

2. $f(t) = \dfrac{1}{t}$; $[1, 4]$

3. $f(z) = \dfrac{1}{z^2}$; $\left[-2, -\frac{1}{2}\right]$

4. $f(x) = \dfrac{1}{x^2}$; $[-2, 0)$

5. $f(x) = |x|$; $\left[-\frac{1}{2}, 1\right]$

6. $f(s) = s + |s|; [-1, 1]$

7. $f(x) = 3x^4 - 4x^3; [-2, 3]$

8. $f(u) = u^2(u - 2)^{1/3}; [-1, 3]$

9. $f(x) = 2x^5 - 5x^4 + 7; [-1, 3]$

10. $f(x) = (x - 1)^3(x + 2)^2; [-2, 2]$

11. $f(\theta) = \sin \theta; [\pi/4, 4\pi/3]$

12. $f(\theta) = \sin^2 \theta - \sin \theta; [0, \pi]$

In Problems 13–19, a function f is given with domain $(-\infty, \infty)$. Indicate where f is increasing and where it is concave down.

13. $f(x) = 3x - x^2$

14. $f(x) = x^9$

15. $f(x) = x^3 - 3x + 3$

16. $f(x) = -2x^3 - 3x^2 + 12x + 1$

17. $f(x) = x^4 - 4x^5$

18. $f(x) = x^3 - \frac{6}{5}x^5$

19. $f(x) = x^3 - x^4$

20. Find where the function g, defined by $g(t) = t^3 + 1/t$, is increasing and where it is decreasing. Find the local extreme values of g. Find the point of inflection. Sketch the graph.

21. Find where the function f, defined by $f(x) = x^2(x - 4)$, is increasing and where it is decreasing. Find the local extreme values of f. Find the point of inflection. Sketch the graph.

22. Find the maximum and minimum values, if they exist, of the function defined by

$$f(x) = \frac{4}{x^2 + 1} + 2$$

In Problems 23–30, sketch the graph of the given function f, labeling all extrema (local and global) and the inflection points and showing any asymptotes. Be sure to make use of f' and f''.

23. $f(x) = x^4 - 2x$

24. $f(x) = (x^2 - 1)^2$

25. $f(x) = x\sqrt{x - 3}$

26. $f(x) = \dfrac{x - 2}{x - 3}$

27. $f(x) = 3x^4 - 4x^3$

28. $f(x) = \dfrac{x^2 - 1}{x}$

29. $f(x) = \dfrac{3x^2 - 1}{x}$

30. $f(x) = \dfrac{2}{(x + 1)^2}$

In Problems 31–36, sketch the graph of the given function f in the region $(-\pi, \pi)$, unless otherwise indicated, labeling all extrema (local and global) and the inflection points and showing any asymptotes. Be sure to make use of f' and f''.

31. $f(x) = \cos x - \sin x$

32. $f(x) = \sin x - \tan x$

33. $f(x) = x \tan x; (-\pi/2, \pi/2)$

34. $f(x) = 2x - \cot x; (0, \pi)$

35. $f(x) = \sin x - \sin^2 x$

36. $f(x) = 2 \cos x - 2 \sin x$

37. Sketch the graph of a function F that has all the following properties:

(a) F is everywhere continuous;

(b) $F(-2) = 3, F(2) = -1$;

(c) $F'(x) = 0$ for $x > 2$;

(d) $F''(x) < 0$ for $x < 2$.

38. Sketch the graph of a function F that has all the following properties:

(a) F is everywhere continuous;

(b) $F(-1) = 6, F(3) = -2$;

(c) $F'(x) < 0$ for $x < -1, F'(-1) = F'(3) = -2, F'(7) = 0$;

(d) $F''(x) < 0$ for $x < -1, F''(x) = 0$ for $-1 < x < 3$, $F''(x) > 0$ for $x > 3$.

39. Sketch the graph of a function F that has the following properties:

(a) F is everywhere continuous;

(b) F has period π;

(c) $0 \le F(x) \le 2, F(0) = 0, F\left(\dfrac{\pi}{2}\right) = 2$;

(d) $F'(x) > 0$ for $0 < x < \dfrac{\pi}{2}, F'(x) < 0$ for $\dfrac{\pi}{2} < x < \pi$;

(e) $F''(x) < 0$ for $0 < x < \pi$.

40. A long sheet of metal, 16 inches wide, is to be turned up at both sides to make a horizontal gutter with vertical sides. How many inches should be turned up at each side for maximum carrying capacity?

41. A fence, 8 feet high, is parallel to the wall of a building and 1 foot from the building. What is the shortest plank that can go over the fence, from the level ground, to prop the wall?

42. A page of a book is to contain 27 square inches of print. If the margins at the top, bottom, and one side are 2 inches and the margin at the other side is 1 inch, what size page would use the least paper?

43. A metal water trough with equal semicircular ends and open top is to have a capacity of 128π cubic feet (Figure 1). Determine its radius r and length h if the trough is to require the least material for its construction.

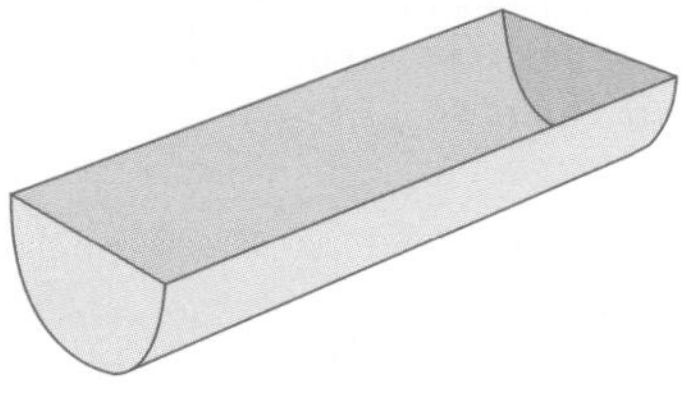

Figure 1

44. Find the maximum and the minimum of the function defined on the closed interval $[-2, 2]$ by

$$f(x) = \begin{cases} \frac{1}{4}(x^2 + 6x + 8), & \text{if } -2 \le x \le 0 \\ -\frac{1}{6}(x^2 + 4x - 12), & \text{if } 0 \le x \le 2 \end{cases}$$

Find where the graph is concave up and where it is concave down. Sketch the graph.

45. For each of the following functions, decide whether the Mean Value Theorem applies on the indicated interval I. If so, find all possible values of c, if not, tell why. Make a sketch.

(a) $f(x) = \dfrac{x^3}{3}$; $I = [-3, 3]$

(b) $F(x) = x^{3/5} + 1$; $I = [-1, 1]$

(c) $g(x) = \dfrac{x+1}{x-1}$; $I = [2, 3]$

46. Find the equations of the tangent lines at the inflection points of the graph of

$$y = x^4 - 6x^3 + 12x^2 - 3x + 1$$

47. Let f be a continuous function with $f(1) = -1/4$, $f(2) = 0$, and $f(3) = -1/4$. If the graph of $y = f'(x)$ is as shown in Figure 2, sketch a possible graph for $y = f(x)$.

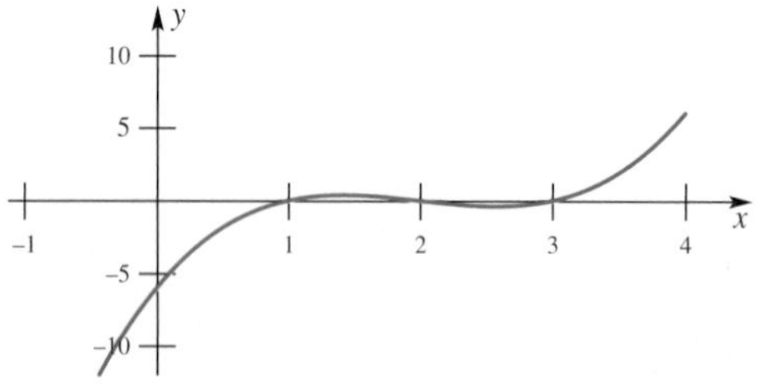

Figure 2

48. Sketch the graph of a function G with all the following properties:

(a) $G(x)$ is continuous and $G''(x) > 0$ for all x in $(-\infty, 0) \cup (0, \infty)$;

(b) $G(-2) = G(2) = 3$;

(c) $\lim_{x\to-\infty} G(x) = 2$, $\lim_{x\to\infty} [G(x) - x] = 0$;

(d) $\lim_{x\to 0^+} G(x) = \lim_{x\to 0^-} G(x) = \infty$.

C **49.** Use the Bisection Method to solve $3x - \cos 2x = 0$ accurate to six decimal places. Use $a_1 = 0$ and $b_1 = 1$.

C **50.** Use Newton's Method to solve $3x - \cos 2x = 0$ accurate to six decimal places. Use $x_1 = 0.5$.

C **51.** Use the Fixed-Point Algorithm to solve $3x - \cos 2x = 0$, starting with $x_1 = 0.5$.

C **52.** Use Newton's Method to find the solution of $x - \tan x = 0$ in the interval $(\pi, 2\pi)$ accurate to four decimal places. *Hint:* Sketch graphs of $y = x$ and $y = \tan x$ using the same axes to get a good initial guess for x_1.

In Problems 53–67, evaluate the indicated integrals.

53. $\displaystyle\int \left(x^3 - 3x^2 + 3\sqrt{x}\right) dx$

54. $\displaystyle\int \frac{2x^4 - 3x^2 + 1}{x^2}\, dx$

55. $\displaystyle\int \frac{y^3 - 9y\sin y + 26y^{-1}}{y}\, dy$

56. $\displaystyle\int y\sqrt{y^2 - 4}\, dy$

57. $\displaystyle\int z(2z^2 - 3)^{1/3}\, dz$

58. $\displaystyle\int \cos^4 x \sin x\, dx$

59. $\displaystyle\int (x+1)\tan^2(3x^2 + 6x)\sec^2(3x^2 + 6x)\, dx$

60. $\displaystyle\int \frac{t^3}{\sqrt{t^4 + 9}}\, dt$

61. $\displaystyle\int t^4(t^5 + 5)^{2/3}\, dt$

62. $\displaystyle\int \frac{x}{\sqrt{x^2 + 4}}\, dx$

63. $\displaystyle\int \frac{x^2}{\sqrt{x^3 + 9}}\, dx$

64. $\displaystyle\int \frac{1}{(y+1)^2}\, dy$

65. $\displaystyle\int \frac{2}{(2y-1)^3}\, dy$

66. $\displaystyle\int \frac{y^2 - 1}{(y^3 - 3y)^2}\, dy$

67. $\displaystyle\int \frac{(y^2 + y + 1)}{\sqrt[5]{2y^3 + 3y^2 + 6y}}\, dy$

In Problems 68–74, solve the differential equation subject to the indicated condition.

68. $\frac{dy}{dx} = \sin x$; $y = 2$ at $x = 0$

69. $\frac{dy}{dx} = \frac{1}{\sqrt{x+1}}$; $y = 18$ at $x = 3$

70. $\frac{dy}{dx} = \csc y$; $y = \pi$ at $x = 0$

71. $\frac{dy}{dt} = \sqrt{2t-1}$; $y = -1$ at $t = \frac{1}{2}$

72. $\frac{dy}{dt} = t^2y^4$; $y = 1$ at $t = 1$

73. $\frac{dy}{dx} = \frac{6x - x^3}{2y}$; $y = 3$ at $x = 0$

74. $\frac{dy}{dx} = x \sec y$; $y = \pi$ at $x = 0$

75. A ball is thrown directly upward from a tower 448 feet high with an initial velocity of 48 feet per second. In how many seconds will it strike the ground and with what velocity? Assume that $g = 32$ feet per second per second and neglect air resistance.

REVIEW & PREVIEW PROBLEMS

In Problems 1–12, find the area of the shaded region.

1.

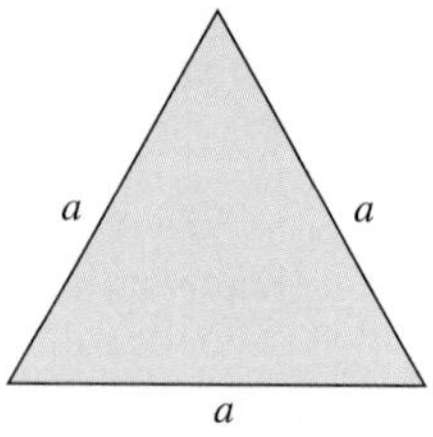

2.

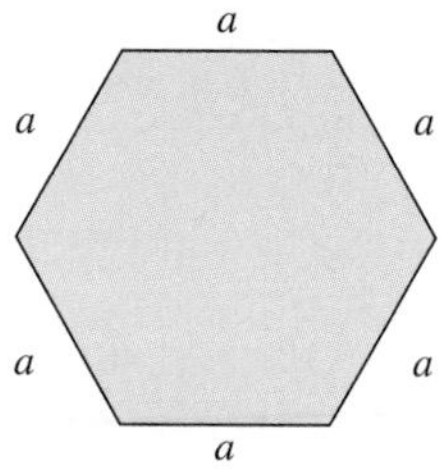

3.

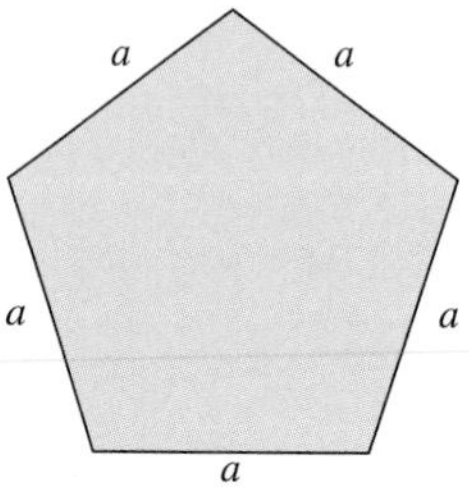

4.

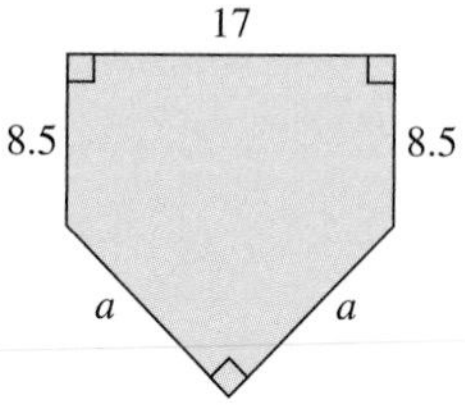

5.

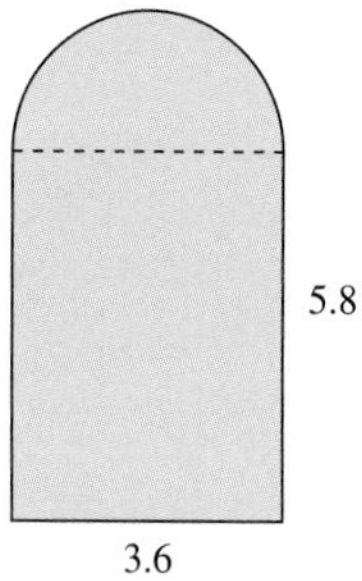

6.

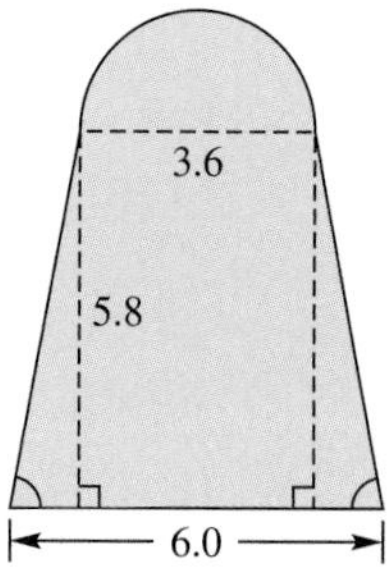

7.

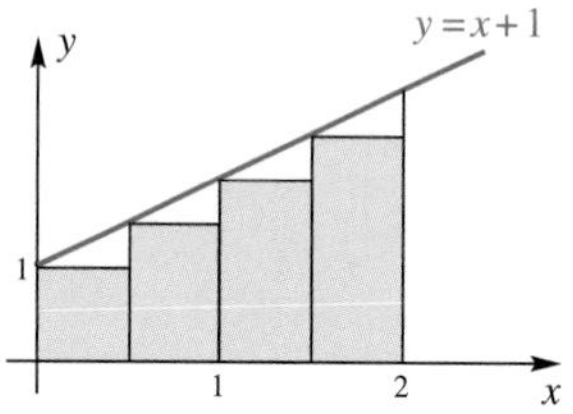

8.

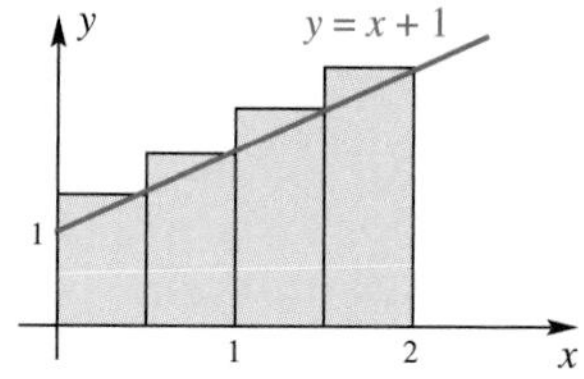

9.

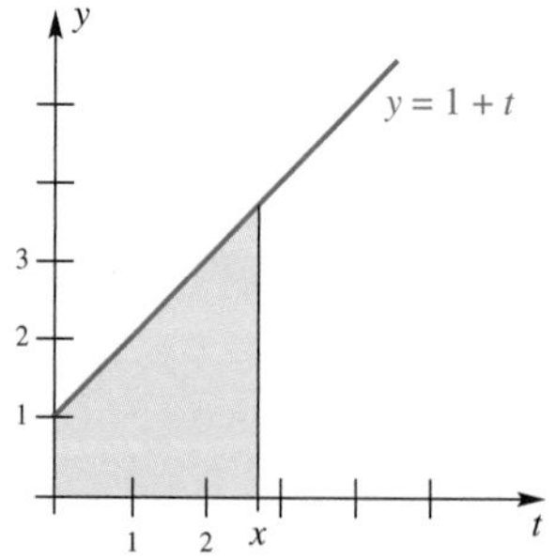

10.

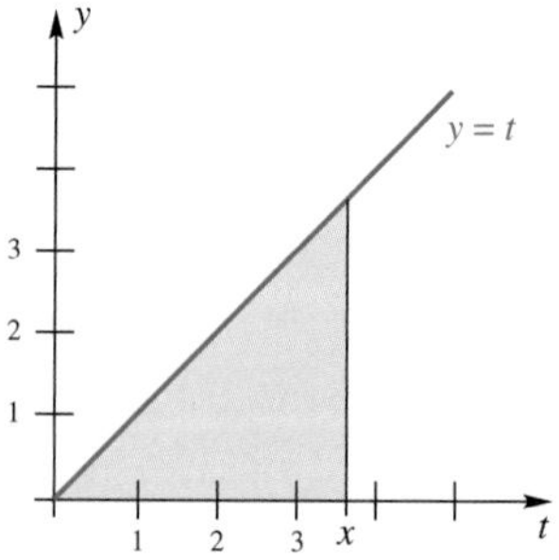

11.

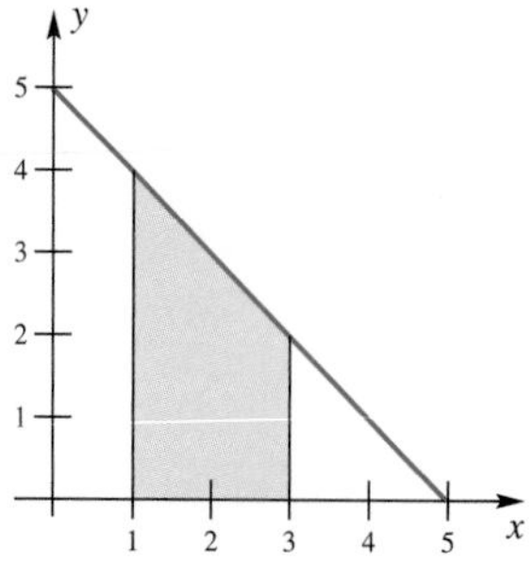

12.

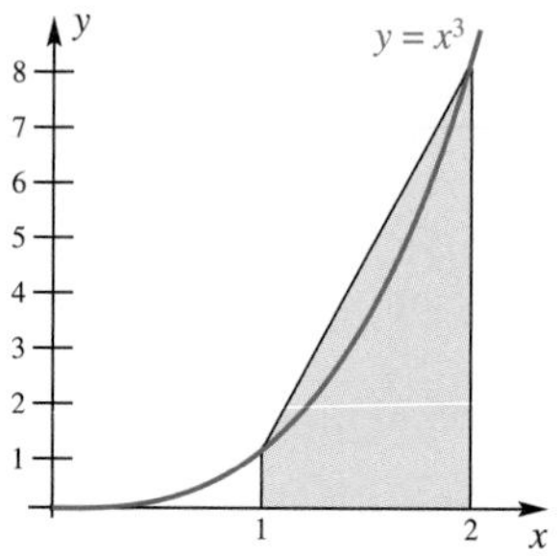

CHAPTER 4 The Definite Integral

4.1 Introduction to Area
4.2 The Definite Integral
4.3 The First Fundamental Theorem of Calculus
4.4 The Second Fundamental Theorem of Calculus and the Method of Substitution
4.5 The Mean Value Theorem for Integrals and the Use of Symmetry
4.6 Numerical Integration

4.1 Introduction to Area

Two problems, both from geometry, motivate the two most important ideas in calculus. The problem of finding the tangent line led us to the *derivative*. The problem of finding area will lead us to the *definite integral*.

For polygons (closed plane regions bounded by line segments), the problem of finding area is hardly a problem at all. We start by defining the area of a rectangle to be the familiar length times width, and from this we successively derive the formulas for the area of a parallelogram, a triangle, and any polygon. The sequence of figures in Figure 1 suggests how this is done.

Even in this simple setting, it is clear that area should satisfy five properties.

1. The area of a plane region is a nonnegative (real) number.
2. The area of a rectangle is the product of its length and width (both measured in the same units). The result is in square units, for example, square feet or square centimeters.
3. Congruent regions have equal areas.
4. The area of the union of two regions that overlap only in a line segment is the sum of the areas of the two regions.
5. If one region is contained in a second region, then the area of the first region is less than or equal to that of the second.

When we consider a region with a curved boundary, the problem of assigning area is more difficult. However, over 2000 years ago, Archimedes provided the key to a solution. Consider a sequence of inscribed polygons that approximate the curved region with greater and greater accuracy. For example, for the circle of

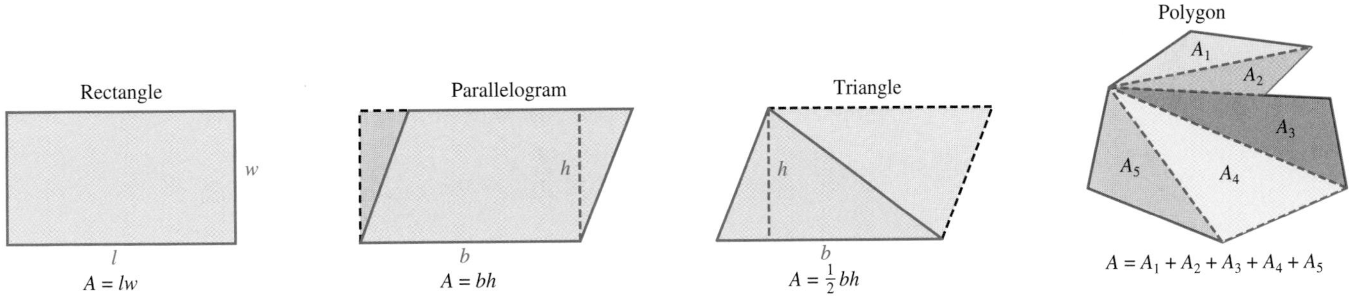

Figure 1

radius 1, consider regular *inscribed* polygons $P_1, P_2, P_3, \ldots$ with 4 sides, 8 sides, 16 sides, ..., as shown in Figure 2. The area of the circle is the limit as $n \to \infty$ of the areas of P_n. Thus, if $A(F)$ denotes the area of a region F, then

$$A(\text{circle}) = \lim_{n\to\infty} A(P_n)$$

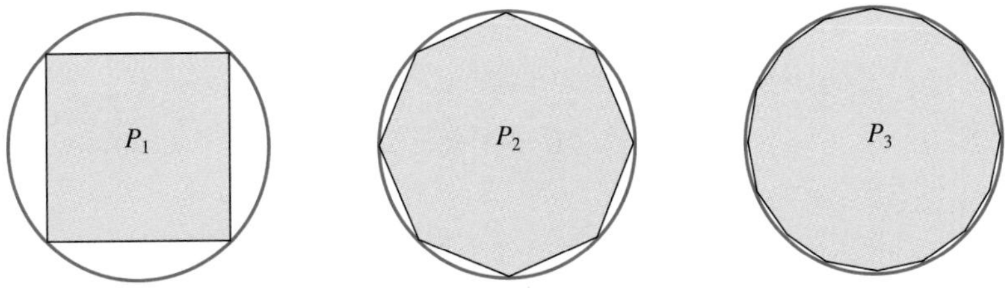

Figure 2

Use and Abuse of Language

Following common usage, we allow ourselves a certain abuse of language. The words *triangle, rectangle, polygon,* and *circle* will be used to denote both two-dimensional regions of the indicated shape and also their one-dimensional boundaries. Note that regions have areas, whereas curves have lengths. When we say that a circle has area πr^2 and circumference $2\pi r$, the context should make clear whether *circle* means the region or the boundary.

Archimedes went further, considering also *circumscribed* polygons $T_1, T_2, T_3, \ldots$ (Figure 3). He showed that you get the same value for the area of the circle of radius 1 (what we call π) whether you use inscribed or circumscribed polygons. It is just a small step from what he did to our modern treatment of area.

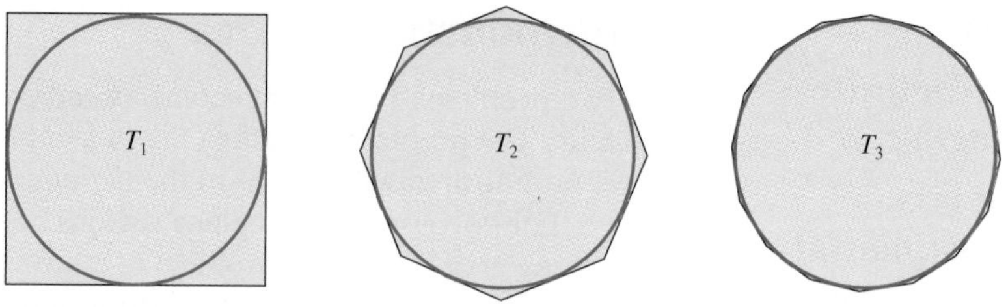

Figure 3

Sigma Notation Our approach to finding the area of a curved region R will involve the following steps:

1. Approximate the region R by n rectangles where the n rectangles taken together either contain R, producing a **circumscribed polygon,** or are contained in R, producing an **inscribed polygon.**
2. Find the area of each rectangle.
3. Sum the areas of the n rectangles.
4. Take the limit as $n \to \infty$.

If the limit of areas of inscribed and circumscribed polygons is the same, we call this limit the area of the region R.

Step 3, involving summing the areas of rectangles, requires us to have a notation for summation, as well as some of its properties. Consider, for example, the following sums:

$$1^2 + 2^2 + 3^2 + 4^2 + \cdots + 100^2$$

and

$$a_1 + a_2 + a_3 + a_4 + \cdots + a_n$$

To indicate these sums in a compact way, we write these sums as

$$\sum_{i=1}^{100} i^2 \quad \text{and} \quad \sum_{i=1}^{n} a_i$$

respectively. Here Σ (capital Greek sigma), which corresponds to the English S, says that we are to sum (add) all numbers of the form indicated as the *index i* runs through the positive integers, starting with the integer shown below Σ and ending with the integer above Σ. Thus,

$$\sum_{i=2}^{4} a_i b_i = a_2 b_2 + a_3 b_3 + a_4 b_4$$

$$\sum_{j=1}^{n} \frac{1}{j} = \frac{1}{1} + \frac{1}{2} + \frac{1}{3} + \cdots + \frac{1}{n}$$

$$\sum_{k=1}^{4} \frac{k}{k^2+1} = \frac{1}{1^2+1} + \frac{2}{2^2+1} + \frac{3}{3^2+1} + \frac{4}{4^2+1}$$

If all the c_i in $\sum_{i=1}^{n} c_i$ have the same value, say c, then

$$\sum_{i=1}^{n} c_i = \underbrace{c + c + c + \cdots + c}_{n \text{ terms}}$$

As a result,

$$\sum_{i=1}^{n} c = nc$$

In particular,

$$\sum_{i=1}^{5} 2 = 5(2) = 10 \quad \text{and} \quad \sum_{i=1}^{100} (-4) = 100(-4) = -400$$

Properties of $\sum$ Thought of as an operator, Σ operates on sequences, and it does so in a linear way.

Theorem A Linearity of $\sum$

If c is a constant, then

(i) $\sum_{i=1}^{n} ca_i = c\sum_{i=1}^{n} a_i$;

(ii) $\sum_{i=1}^{n} (a_i + b_i) = \sum_{i=1}^{n} a_i + \sum_{i=1}^{n} b_i$;

(iii) $\sum_{i=1}^{n} (a_i - b_i) = \sum_{i=1}^{n} a_i - \sum_{i=1}^{n} b_i$.

Proof The proofs are easy; we consider only (i).

$$\sum_{i=1}^{n} ca_i = ca_1 + ca_2 + \cdots + ca_n = c(a_1 + a_2 + \cdots + a_n) = c\sum_{i=1}^{n} a_i$$ ■

EXAMPLE 1 Suppose that $\sum_{i=1}^{100} a_i = 60$ and $\sum_{i=1}^{100} b_i = 11$. Calculate

$$\sum_{i=1}^{100} (2a_i - 3b_i + 4)$$

SOLUTION

$$\begin{aligned}\sum_{i=1}^{100} (2a_i - 3b_i + 4) &= \sum_{i=1}^{100} 2a_i - \sum_{i=1}^{100} 3b_i + \sum_{i=1}^{100} 4 \\ &= 2\sum_{i=1}^{100} a_i - 3\sum_{i=1}^{100} b_i + \sum_{i=1}^{100} 4 \\ &= 2(60) - 3(11) + 100(4) = 487\end{aligned}$$ ■

EXAMPLE 2 **Collapsing Sums**

Show that:

(a) $\sum_{i=1}^{n} (a_{i+1} - a_i) = a_{n+1} - a_1$

(b) $\sum_{i=1}^{n} [(i + 1)^2 - i^2] = (n + 1)^2 - 1$

SOLUTION

(a) Here we should resist our inclination to apply linearity and instead write out the sum, hoping for some nice cancellations.

$$\begin{aligned}\sum_{i=1}^{n}(a_{i+1} - a_i) &= (a_2 - a_1) + (a_3 - a_2) + (a_4 - a_3) + \cdots + (a_{n+1} - a_n)\\ &= -a_1 + a_2 - a_2 + a_3 - a_3 + a_4 - \cdots - a_n + a_{n+1}\\ &= -a_1 + a_{n+1} = a_{n+1} - a_1\end{aligned}$$

(b) This follows immediately from part (a). ■

The symbol used for the index does not matter. Thus,

$$\sum_{i=1}^{n} a_i = \sum_{j=1}^{n} a_j = \sum_{k=1}^{n} a_k$$

and all of these are equal to $a_1 + a_2 + \cdots + a_n$. For this reason, the index is sometimes called a **dummy index.**

Some Special Sum Formulas When finding areas of regions we will often need to consider the sum of the first n positive integers, as well as the sums of their squares, cubes, and so on. There are nice formulas for these; proofs are discussed after Example 4.

1. $\displaystyle\sum_{i=1}^{n} i = 1 + 2 + 3 + \cdots + n = \frac{n(n+1)}{2}$
2. $\displaystyle\sum_{i=1}^{n} i^2 = 1^2 + 2^2 + 3^2 + \cdots + n^2 = \frac{n(n+1)(2n+1)}{6}$
3. $\displaystyle\sum_{i=1}^{n} i^3 = 1^3 + 2^3 + 3^3 + \cdots + n^3 = \left[\frac{n(n+1)}{2}\right]^2$
4. $\displaystyle\sum_{i=1}^{n} i^4 = 1^4 + 2^4 + 3^4 + \cdots + n^4 = \frac{n(n+1)(2n+1)(3n^2+3n-1)}{30}$

EXAMPLE 3 Find a formula for $\displaystyle\sum_{j=1}^{n}(j+2)(j-5)$.

SOLUTION We make use of linearity and Formulas 1 and 2 from above.

$$\begin{aligned}\sum_{j=1}^{n}(j+2)(j-5) &= \sum_{j=1}^{n}(j^2 - 3j - 10) = \sum_{j=1}^{n} j^2 - 3\sum_{j=1}^{n} j - \sum_{j=1}^{n} 10\\ &= \frac{n(n+1)(2n+1)}{6} - 3\frac{n(n+1)}{2} - 10n\\ &= \frac{n}{6}[2n^2 + 3n + 1 - 9n - 9 - 60]\\ &= \frac{n(n^2 - 3n - 34)}{3}\end{aligned}$$

■

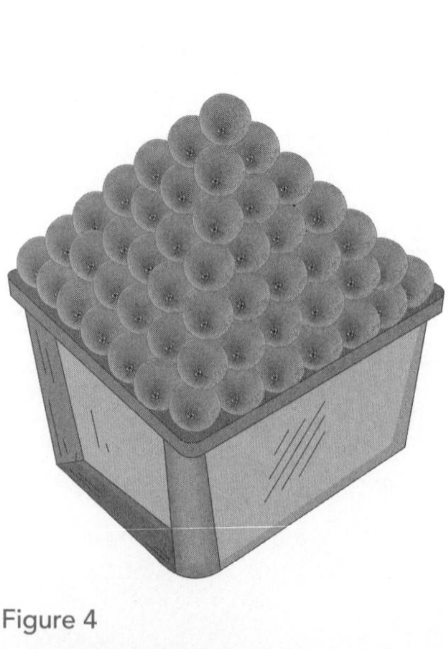

Figure 4

EXAMPLE 4 How many oranges are in the pyramid shown in Figure 4?

SOLUTION $1^2 + 2^2 + 3^2 + \cdots + 7^2 = \displaystyle\sum_{i=1}^{7} i^2 = \frac{7(8)(15)}{6} = 140$ ■

Proofs of Special Sum Formulas To prove Special Sum Formula 1, we start with the identity $(i+1)^2 - i^2 = 2i + 1$, sum both sides, apply Example 2 on the left, and use linearity on the right.

$$(i + 1)^2 - i^2 = 2i + 1$$

$$\sum_{i=1}^{n}[(i + 1)^2 - i^2] = \sum_{i=1}^{n}(2i + 1)$$

$$(n + 1)^2 - 1^2 = 2\sum_{i=1}^{n} i + \sum_{i=1}^{n} 1$$

$$n^2 + 2n = 2\sum_{i=1}^{n} i + n$$

$$\frac{n^2 + n}{2} = \sum_{i=1}^{n} i$$

Almost the same technique works to establish Formulas 2, 3, and 4 (Problems 29–31).

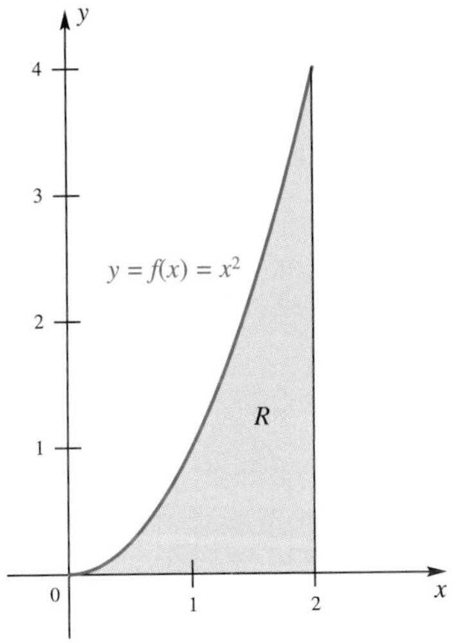

Figure 5

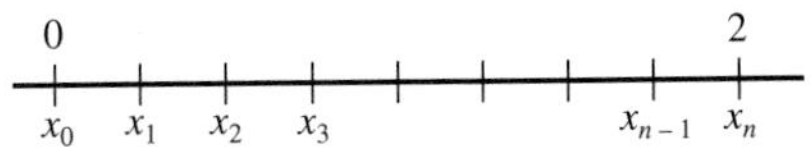

Figure 6

Area by Inscribed Polygons Consider the region R bounded by the parabola $y = f(x) = x^2$, the x-axis, and the vertical line $x = 2$ (Figure 5). We refer to R as the region under the curve $y = x^2$ between $x = 0$ and $x = 2$. Our aim is to calculate its area $A(R)$.

Partition (as in Figure 6) the interval [0, 2] into n subintervals, each of length $\Delta x = 2/n$, by means of the $n + 1$ points

$$0 = x_0 < x_1 < x_2 < \cdots < x_{n-1} < x_n = 2$$

Thus,

$$x_0 = 0$$

$$x_1 = \Delta x = \frac{2}{n}$$

$$x_2 = 2 \cdot \Delta x = \frac{4}{n}$$

$$\vdots$$

$$x_i = i \cdot \Delta x = \frac{2i}{n}$$

$$\vdots$$

$$x_{n-1} = (n - 1) \cdot \Delta x = \frac{(n - 1)2}{n}$$

$$x_n = n \cdot \Delta x = n\left(\frac{2}{n}\right) = 2$$

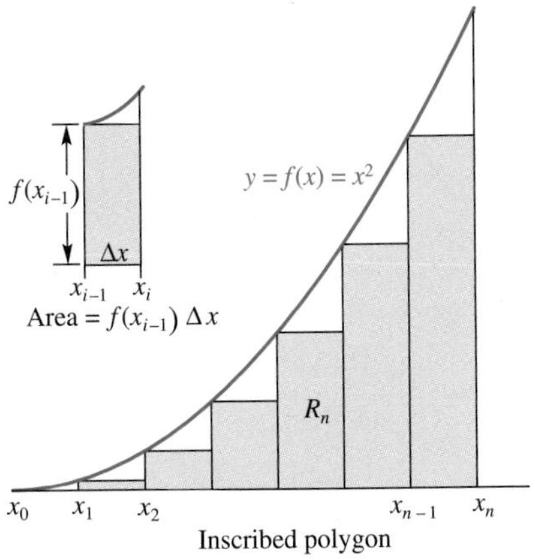

Figure 7

Consider the typical rectangle with base $[x_{i-1}, x_i]$ and height $f(x_{i-1}) = x_{i-1}^2$. Its area is $f(x_{i-1})\,\Delta x$ (see the upper-left part of Figure 7). The union R_n of all such rectangles forms the inscribed polygon shown in the lower-right part of Figure 7.

The area $A(R_n)$ can be calculated by summing the areas of these rectangles.

$$A(R_n) = f(x_0)\,\Delta x + f(x_1)\,\Delta x + f(x_2)\,\Delta x + \cdots + f(x_{n-1})\,\Delta x$$

Now

$$f(x_i)\,\Delta x = x_i^2\,\Delta x = \left(\frac{2i}{n}\right)^2 \cdot \frac{2}{n} = \left(\frac{8}{n^3}\right)i^2$$

Thus,

$$A(R_n) = \left[\frac{8}{n^3}(0^2) + \frac{8}{n^3}(1^2) + \frac{8}{n^3}(2^2) + \cdots + \frac{8}{n^3}(n - 1)^2\right]$$

$$= \frac{8}{n^3}[1^2 + 2^2 + \cdots + (n - 1)^2]$$

$$= \frac{8}{n^3}\left[\frac{(n-1)n(2n-1)}{6}\right] \qquad \text{(Special Sum Formula 2, with } n-1 \text{ replacing } n\text{)}$$

$$= \frac{8}{6}\left(\frac{2n^3 - 3n^2 + n}{n^3}\right)$$

$$= \frac{4}{3}\left(2 - \frac{3}{n} + \frac{1}{n^2}\right)$$

$$= \frac{8}{3} - \frac{4}{n} + \frac{4}{3n^2}$$

We conclude that

$$A(R) = \lim_{n\to\infty} A(R_n) = \lim_{n\to\infty}\left(\frac{8}{3} - \frac{4}{n} + \frac{4}{3n^2}\right) = \frac{8}{3}$$

The diagrams in Figure 8 should help you to visualize what is happening as n gets larger and larger.

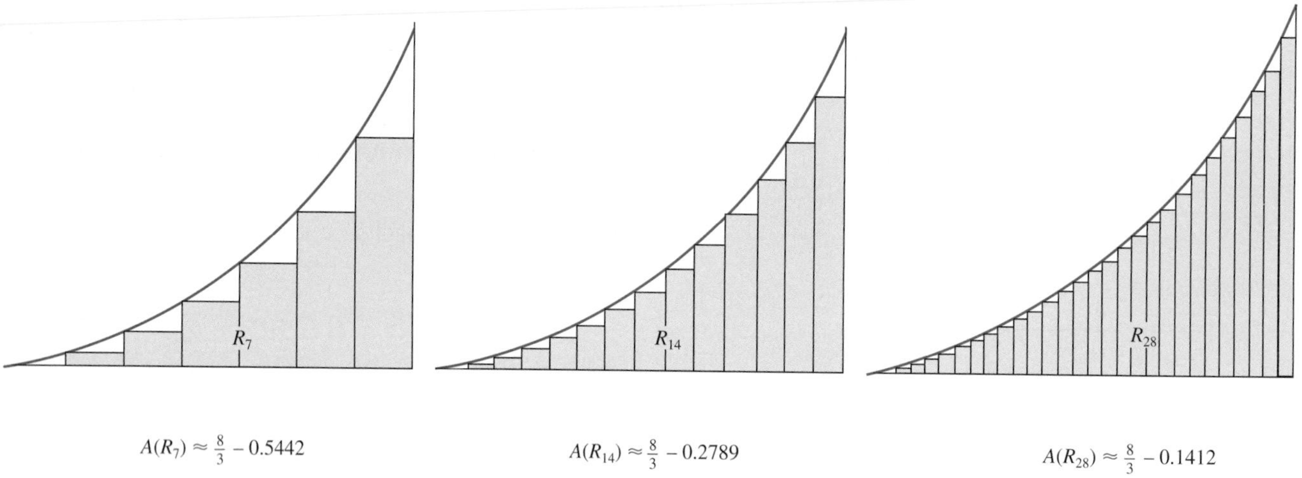

Figure 8

Area by Circumscribed Polygons Perhaps you are still not convinced that $A(R) = \frac{8}{3}$. We can give more evidence. Consider the rectangle with base $[x_{i-1}, x_i]$ and height $f(x_i) = x_i^2$ (shown at the upper left in Figure 9). Its area is $f(x_i)\,\Delta x$. The union S_n of all such rectangles forms a circumscribed polygon for the region R, as shown at the lower right in Figure 9.

The area $A(S_n)$ is calculated in analogy with the calculation of $A(R_n)$.

$$A(S_n) = f(x_1)\,\Delta x + f(x_2)\,\Delta x + \cdots + f(x_n)\,\Delta x$$

As before, $f(x_i)\,\Delta x = x_i^2\,\Delta x = (8/n^3)i^2$, and so

$$A(S_n) = \left[\frac{8}{n^3}(1^2) + \frac{8}{n^3}(2^2) + \cdots + \frac{8}{n^3}(n^2)\right]$$

$$= \frac{8}{n^3}[1^2 + 2^2 + \cdots + n^2]$$

$$= \frac{8}{n^3}\left[\frac{n(n+1)(2n+1)}{6}\right] \qquad \text{(Special Sum Formula 2)}$$

$$= \frac{8}{6}\left[\frac{2n^3 + 3n^2 + n}{n^3}\right]$$

$$= \frac{8}{3} + \frac{4}{n} + \frac{4}{3n^2}$$

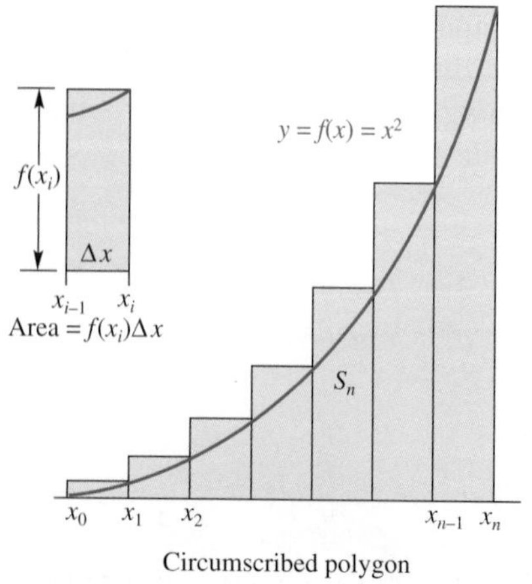

Figure 9

Again, we conclude that

$$A(R) = \lim_{n\to\infty} A(S_n) = \lim_{n\to\infty}\left(\frac{8}{3} + \frac{4}{n} + \frac{4}{3n^2}\right) = \frac{8}{3}$$

Another Problem—Same Theme Suppose that an object is traveling along the x-axis in such a way that its velocity at time t is given by $v = f(t) = \frac{1}{4}t^3 + 1$ feet per second. How far did it travel between $t = 0$ and $t = 3$? This problem can be solved by the method of differential equations (Section 3.9), but we have something else in mind.

Our starting point is the familiar fact that, if an object travels at constant velocity k over a time interval of length Δt, then the distance traveled is $k\,\Delta t$. But this is just the area of a rectangle, the one shown in Figure 10.

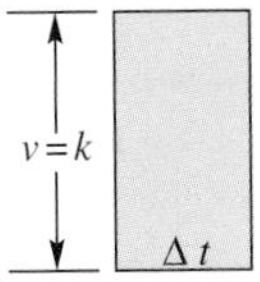

Distance = $k\,\Delta t$

Figure 10

Next consider the given problem, where $v = f(t) = \frac{1}{4}t^3 + 1$. The graph is shown in the top half of Figure 11. Partition the interval $[0, 3]$ into n subintervals of length $\Delta t = 3/n$ by means of points $0 = t_0 < t_1 < t_2 < \cdots < t_n = 3$. Then consider the corresponding circumscribed polygon S_n, displayed in the bottom half of Figure 11 (we could as well have considered the inscribed polygon). Its area, $A(S_n)$, should be a good approximation of the distance traveled, especially if Δt is small, since on each subinterval the actual velocity is almost equal to a constant (the value of v at the end of the subinterval). Moreover, this approximation should get better and better as n gets larger. We are led to the conclusion that the exact distance traveled is $\lim_{n\to\infty} A(S_n)$; that is, it is the area of the region under the velocity curve between $t = 0$ and $t = 3$.

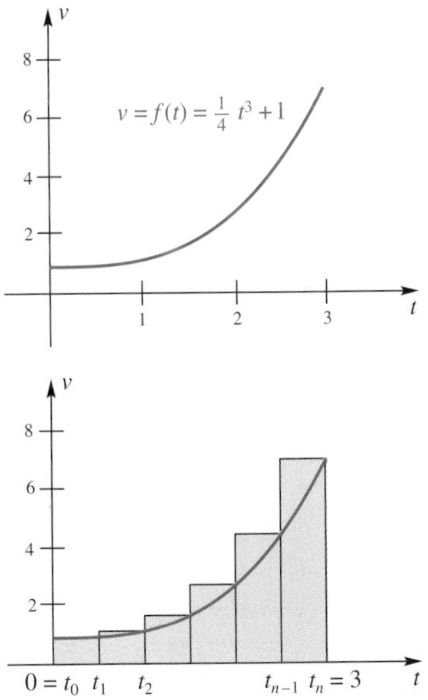

Figure 11

To calculate $A(S_n)$, note that $t_i = 3i/n$, and so the area of the ith rectangle is

$$f(t_i)\,\Delta t = \left[\frac{1}{4}\left(\frac{3i}{n}\right)^3 + 1\right]\frac{3}{n} = \frac{81}{4n^4}i^3 + \frac{3}{n}$$

Thus,

$$\begin{aligned}
A(S_n) &= f(t_1)\,\Delta t + f(t_2)\,\Delta t + \cdots + f(t_n)\,\Delta t \\
&= \sum_{i=1}^{n} f(t_i)\,\Delta t \\
&= \sum_{i=1}^{n}\left(\frac{81}{4n^4}i^3 + \frac{3}{n}\right) \\
&= \frac{81}{4n^4}\sum_{i=1}^{n} i^3 + \sum_{i=1}^{n}\frac{3}{n} \\
&= \frac{81}{4n^4}\left[\frac{n(n+1)}{2}\right]^2 + \frac{3}{n}\cdot n \qquad \text{(Special Sum Formula 3)} \\
&= \frac{81}{16}\left[n^2\frac{(n^2+2n+1)}{n^4}\right] + 3 \\
&= \frac{81}{16}\left(1 + \frac{2}{n} + \frac{1}{n^2}\right) + 3
\end{aligned}$$

We conclude that

$$\lim_{n\to\infty} A(S_n) = \frac{81}{16} + 3 = \frac{129}{16} \approx 8.06$$

The object traveled about 8.06 feet between $t = 0$ and $t = 3$.

What was true in this example is true for any object moving with positive velocity. *The distance traveled is the area of the region under the velocity curve.*

Concepts Review

1. The value of $\sum_{i=1}^{5} 2i$ is _____, and the value of $\sum_{i=1}^{5} 2$ is _____.

2. If $\sum_{i=1}^{10} a_i = 9$ and $\sum_{i=1}^{10} b_i = 7$, then the value of $\sum_{i=1}^{10} (3a_i - 2b_i) =$ _____ and the value of $\sum_{i=1}^{10} (a_i + 4) =$ _____.

3. The area of a(n) _____ polygon underestimates the area of a region, whereas the area of a(n) _____ polygon overestimates this area.

4. The exact area of the region under the curve $y = [\![x]\!]$ between 0 and 4 is _____.

Problem Set 4.1

In Problems 1–8, find the value of the indicated sum.

1. $\sum_{k=1}^{6} (k - 1)$

2. $\sum_{i=1}^{6} i^2$

3. $\sum_{k=1}^{7} \frac{1}{k + 1}$

4. $\sum_{l=3}^{8} (l + 1)^2$

5. $\sum_{m=1}^{8} (-1)^m 2^{m-2}$

6. $\sum_{k=3}^{7} \frac{(-1)^k 2^k}{(k + 1)}$

7. $\sum_{n=1}^{6} n \cos(n\pi)$

8. $\sum_{k=-1}^{6} k \sin(k\pi/2)$

In Problems 9–14, write the indicated sum in sigma notation.

9. $1 + 2 + 3 + \cdots + 41$

10. $2 + 4 + 6 + 8 + \cdots + 50$

11. $1 + \frac{1}{2} + \frac{1}{3} + \cdots + \frac{1}{100}$

12. $1 - \frac{1}{2} + \frac{1}{3} - \frac{1}{4} + \cdots - \frac{1}{100}$

13. $a_1 + a_3 + a_5 + a_7 + \cdots + a_{99}$

14. $f(w_1)\,\Delta x + f(w_2)\,\Delta x + \cdots + f(w_n)\,\Delta x$

In Problems 15–18, suppose that $\sum_{i=1}^{10} a_i = 40$ and $\sum_{i=1}^{10} b_i = 50$. Calculate each of the following (see Example 1).

15. $\sum_{i=1}^{10} (a_i + b_i)$

16. $\sum_{n=1}^{10} (3a_n + 2b_n)$

17. $\sum_{p=0}^{9} (a_{p+1} - b_{p+1})$

18. $\sum_{q=1}^{10} (a_q - b_q - q)$

In Problems 19–24, use Special Sum Formulas 1–4 to find each sum.

19. $\sum_{i=1}^{100} (3i - 2)$

20. $\sum_{i=1}^{10} [(i - 1)(4i + 3)]$

21. $\sum_{k=1}^{10} (k^3 - k^2)$

22. $\sum_{k=1}^{10} 5k^2(k + 4)$

23. $\sum_{i=1}^{n} (2i^2 - 3i + 1)$

24. $\sum_{i=1}^{n} (2i - 3)^2$

25. Add both sides of the two equalities below, solve for S, and thereby give another proof of Formula 1.

$$S = 1 + 2 + 3 + \cdots + (n - 2) + (n - 1) + n$$

$$S = n + (n - 1) + (n - 2) + \cdots + 3 + 2 + 1$$

26. Prove the following formula for a **geometric sum:**

$$\sum_{k=0}^{n} ar^k = a + ar + ar^2 + \cdots + ar^n = \frac{a - ar^{n+1}}{1 - r} \quad (r \neq 1)$$

Hint: Let $S = a + ar + \cdots + ar^n$. Simplify $S - rS$ and solve for S.

27. Use Problem 26 to calculate each sum.

(a) $\sum_{k=1}^{10} \left(\frac{1}{2}\right)^k$ (b) $\sum_{k=1}^{10} 2^k$

28. Use a derivation like that in Problem 25 to obtain a formula for the **arithmetic sum:**

$$\sum_{k=0}^{n} (a + kd) = a + (a + d) + (a + 2d) + \cdots + (a + nd)$$

29. Use the identity $(i + 1)^3 - i^3 = 3i^2 + 3i + 1$ to prove Special Sum Formula 2.

30. Use the identity $(i + 1)^4 - i^4 = 4i^3 + 6i^2 + 4i + 1$ to prove Special Sum Formula 3.

31. Use the identity $(i + 1)^5 - i^5 = 5i^4 + 10i^3 + 10i^2 + 5i + 1$ to prove Special Sum Formula 4.

32. Use the diagrams in Figure 12 to establish Formulas 1 and 3.

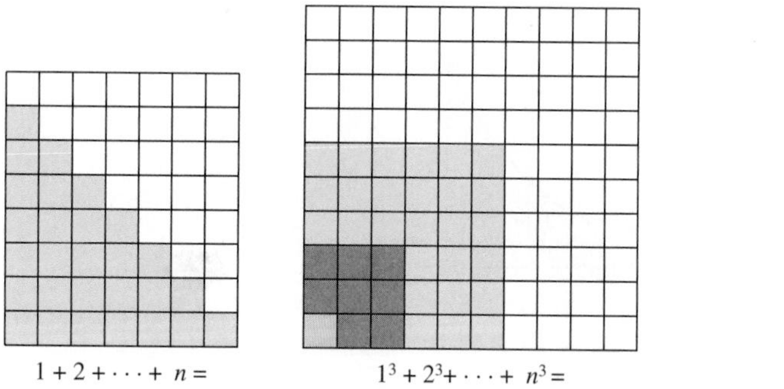

Figure 12

C 33. In statistics we define the *mean* $\bar{x}$ and the *variance* s^2 of a sequence of numbers $x_1, x_2, \ldots, x_n$ by

$$\bar{x} = \frac{1}{n} \sum_{i=1}^{n} x_i, \qquad s^2 = \frac{1}{n} \sum_{i=1}^{n} (x_i - \bar{x})^2$$

Find $\bar{x}$ and s^2 for the sequence of numbers 2, 5, 7, 8, 9, 10, 14.

34. Using the definitions in Problem 33, find $\bar{x}$ and s^2 for each sequence of numbers.

(a) 1, 1, 1, 1, 1 (b) 1001, 1001, 1001, 1001, 1001

(c) 1, 2, 3 (d) 1,000,001; 1,000,002; 1,000,003

35. Use the definitions in Problem 33 to show that each is true.

(a) $\sum_{i=1}^{n} (x_i - \bar{x}) = 0$ (b) $s^2 = \left(\frac{1}{n} \sum_{i=1}^{n} x_i^2\right) - \bar{x}^2$

36. Based on your response to parts (a) and (b) of Problem 34, make a conjecture about the variance of n identical numbers. Prove your conjecture.

37. Let $x_1, x_2, \ldots, x_n$ be any real numbers. Find the value of c that minimizes $\sum_{i=1}^{n}(x_i - c)^2$.

38. In the song *The Twelve Days of Christmas*, my true love gave me 1 gift on the first day, $1 + 2$ gifts on the second day, $1 + 2 + 3$ gifts on the third day, and so on for 12 days.

(a) Find the total number of gifts given in 12 days.

(b) Find a simple formula for T_n, the total number of gifts given during a Christmas of n days.

39. A grocer stacks oranges in a pyramidlike pile. If the bottom layer is rectangular with 10 rows of 16 oranges and the top layer has a single row of oranges, how many oranges are in the stack?

40. Answer the same question in Problem 39 if the bottom layer has 50 rows of 60 oranges.

41. Generalize the result of Problems 39 and 40 to the case of m rows of n oranges.

42. Find a nice formula for the sum

$$\frac{1}{1\cdot 2} + \frac{1}{2\cdot 3} + \frac{1}{3\cdot 4} + \cdots + \frac{1}{n(n+1)}$$

Hint: $\dfrac{1}{i(i+1)} = \dfrac{1}{i} - \dfrac{1}{i+1}$.

In Problems 43–48, find the area of the indicated inscribed or circumscribed polygon.

43.

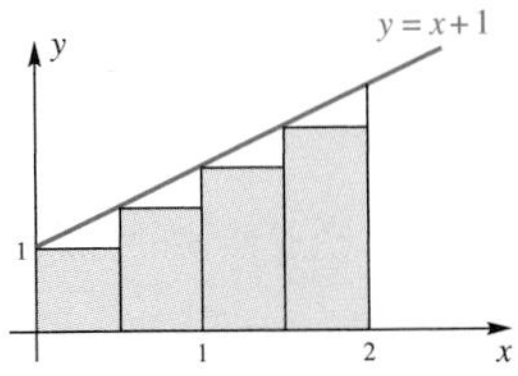

44.

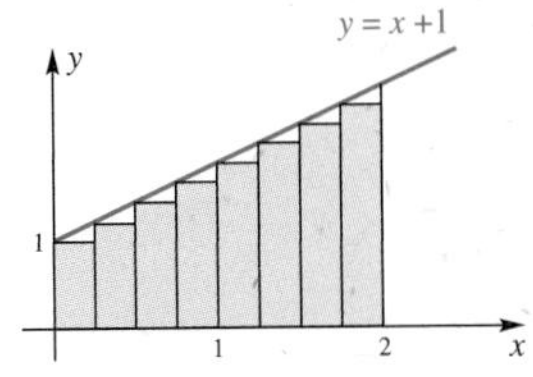

45.

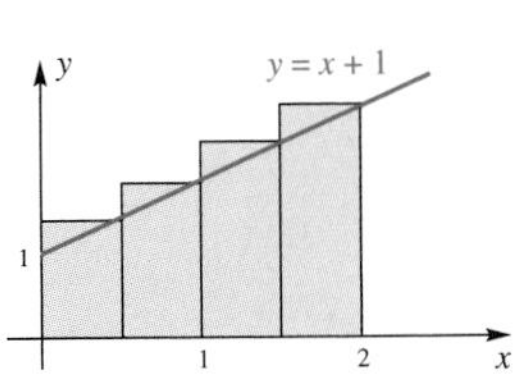

46.

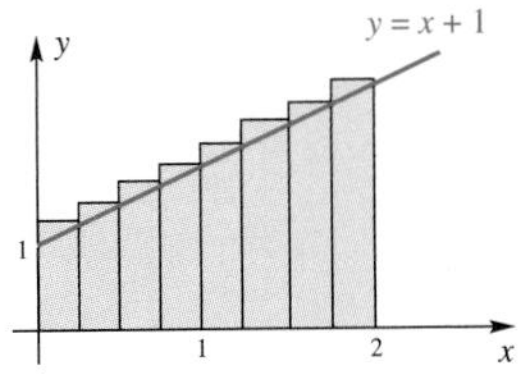

47.

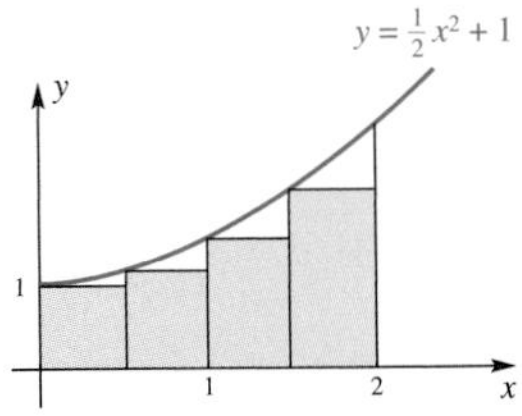

48.

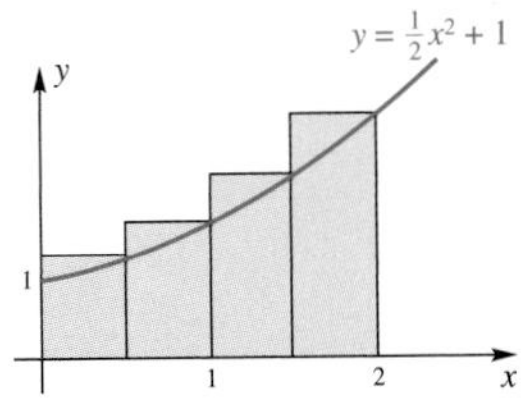

In Problems 49–52, sketch the graph of the given function over the interval $[a, b]$; then divide $[a, b]$ into n equal subintervals. Finally, calculate the area of the corresponding circumscribed polygon.

49. $f(x) = x + 1; a = -1, b = 2, n = 3$

50. $f(x) = 3x - 1; a = 1, b = 3, n = 4$

C **51.** $f(x) = x^2 - 1; a = 2, b = 3, n = 6$

C **52.** $f(x) = 3x^2 + x + 1; a = -1, b = 1, n = 10$

In Problems 53–58, find the area of the region under the curve $y = f(x)$ over the interval $[a, b]$. To do this, divide the interval $[a, b]$ into n equal subintervals, calculate the area of the corresponding circumscribed polygon, and then let $n \to \infty$. (See the example for $y = x^2$ in the text.)

53. $y = x + 2; a = 0, b = 1$

54. $y = \frac{1}{2}x^2 + 1; a = 0, b = 1$

55. $y = 2x + 2; a = -1, b = 1$. *Hint:* $x_i = -1 + \dfrac{2i}{n}$

56. $y = x^2; a = -2, b = 2$

≈ **57.** $y = x^3; a = 0, b = 1$

≈ **58.** $y = x^3 + x; a = 0, b = 1$

59. Suppose that an object is traveling along the x-axis in such a way that its velocity at time t seconds is $v = t + 2$ feet per second. How far did it travel between $t = 0$ and $t = 1$? *Hint:* See the discussion of the velocity problem at the end of this section and use the result of Problem 53.

60. Follow the directions of Problem 59 given that $v = \frac{1}{2}t^2 + 2$. You may use the result of Problem 54.

61. Let A_a^b denote the area under the curve $y = x^2$ over the interval $[a, b]$.

(a) Prove that $A_0^b = b^3/3$. *Hint:* $\Delta x = b/n$, so $x_i = ib/n$; use circumscribed polygons.

(b) Show that $A_a^b = b^3/3 - a^3/3$. Assume that $a \geq 0$.

62. Suppose that an object, moving along the x-axis, has velocity $v = t^2$ meters per second at time t seconds. How far did it travel between $t = 3$ and $t = 5$? See Problem 61.

63. Use the results of Problem 61 to calculate the area under the curve $y = x^2$ over each of the following intervals.

(a) $[0, 5]$ (b) $[1, 4]$ (c) $[2, 5]$

64. From Special Sum Formulas 1–4 you might guess that

$$1^m + 2^m + 3^m + \cdots + n^m = \frac{n^{m+1}}{m+1} + C_n$$

where C_n is a polynomial in n of degree m. Assume that this is true (which it is) and, for $a \geq 0$, let $A_a^b(x^m)$ be the area under the curve $y = x^m$ over the interval $[a, b]$.

(a) Prove that $A_0^b(x^m) = \dfrac{b^{m+1}}{(m+1)}$.

(b) Show that $A_a^b(x^m) = \dfrac{b^{m+1}}{m+1} - \dfrac{a^{m+1}}{m+1}$.

65. Use the results of Problem 64 to calculate each of the following areas.

(a) $A_0^2(x^3)$ (b) $A_1^2(x^3)$ (c) $A_1^2(x^5)$ (d) $A_0^2(x^9)$

66. Derive the formulas $A_n = \frac{1}{2}nr^2\sin(2\pi/n)$ and $B_n = nr^2\tan(\pi/n)$ for the areas of the inscribed and circumscribed regular n-sided polygons for a circle of radius r. Then show that $\lim_{n\to\infty} A_n$ and $\lim_{n\to\infty} B_n$ are both πr^2.

Answers to Concepts Review: **1.** 30; 10 **2.** 13; 49 **3.** inscribed; circumscribed **4.** 6

4.2 The Definite Integral

All the preparations have been made; we are ready to define the definite integral. Both Newton and Leibniz introduced early versions of this concept. However, it was Georg Friedrich Bernhard Riemann (1826–1866) who gave us the modern definition. In formulating this definition, we are guided by the ideas we discussed in the previous section. The first notion is that of a Riemann sum.

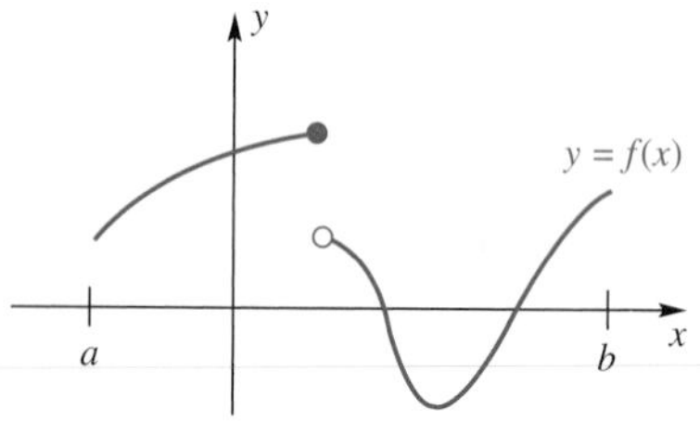

Figure 1

Riemann Sums Consider a function f defined on a closed interval $[a, b]$. It may have both positive and negative values on the interval, and it does not even need to be continuous. Its graph might look something like the one in Figure 1.

Consider a partition P of the interval $[a, b]$ into n subintervals (not necessarily of equal length) by means of points $a = x_0 < x_1 < x_2 < \cdots < x_{n-1} < x_n = b$, and let $\Delta x_i = x_i - x_{i-1}$. On each subinterval $[x_{i-1}, x_i]$, pick an arbitrary point $\bar{x}_i$ (which may be an end point); we call it a *sample point* for the ith subinterval. An example of these constructions is shown in Figure 2 for $n = 6$.

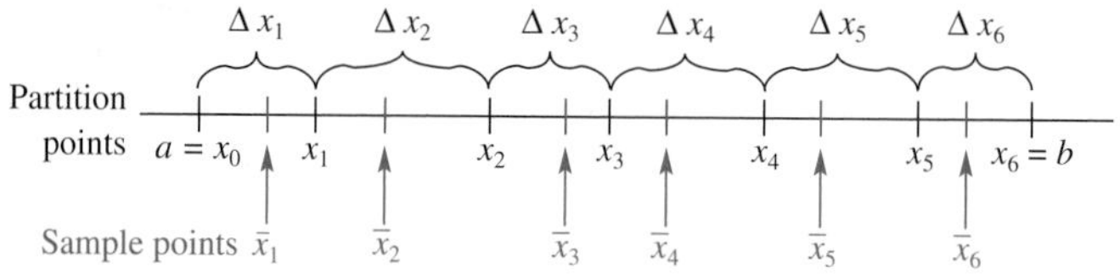

A Partition of $[a, b]$ with Sample Points $\bar{x}_i$

Figure 2

We call the sum

$$R_P = \sum_{i=1}^{n} f(\bar{x}_i)\,\Delta x_i$$

a **Riemann sum** for f corresponding to the partition P. Its geometric interpretation is shown in Figure 3.

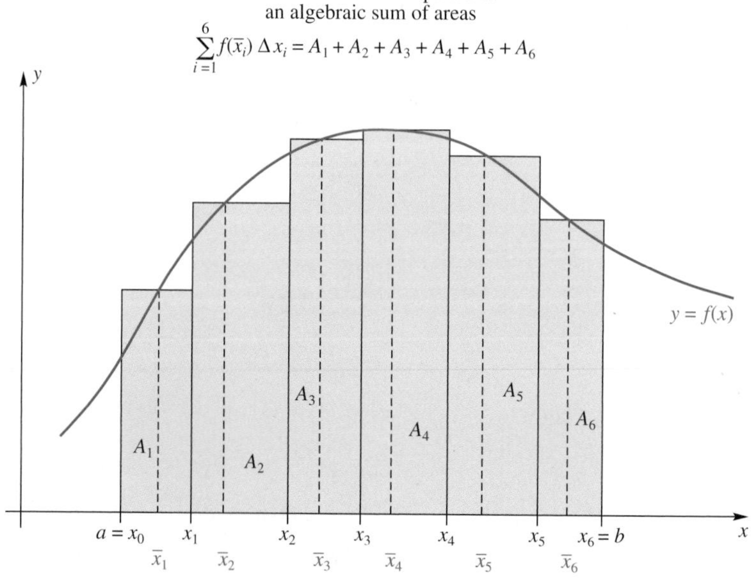

Figure 3

EXAMPLE 1 Evaluate the Riemann sum for $f(x) = x^2 + 1$ on the interval $[-1, 2]$ using the equally spaced partition points $-1 < -0.5 < 0 < 0.5 < 1 < 1.5 < 2$, with the sample point $\bar{x}_i$ being the midpoint of the ith subinterval.

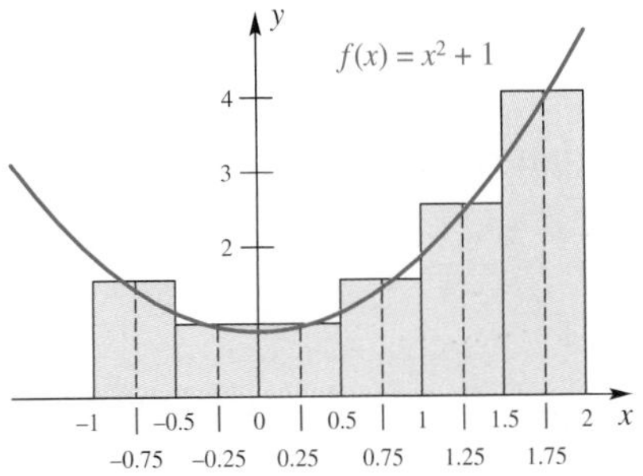

Figure 4

SOLUTION Note the picture in Figure 4.

$$\begin{aligned} R_P &= \sum_{i=1}^{6} f(\bar{x}_i)\,\Delta x_i \\ &= \left[f(-0.75) + f(-0.25) + f(0.25) + f(0.75) + f(1.25) + f(1.75)\right](0.5) \\ &= [1.5625 + 1.0625 + 1.0625 + 1.5625 + 2.5625 + 4.0625](0.5) \\ &= 5.9375 \end{aligned}$$

■

The functions in Figures 3 and 4 were positive. As a consequence of this, the Riemann sum is simply the sum of the areas of the rectangles. But what if f is negative? In this case, a sample point $\bar{x}_i$ with the property that $f(\bar{x}_i) < 0$ will lead to a rectangle that is entirely below the x-axis, and the product $f(\bar{x}_i)\,\Delta x_i$ will be negative. This means that the contribution of such a rectangle to the Riemann sum is negative. Figure 5 illustrates this.

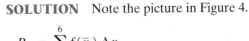

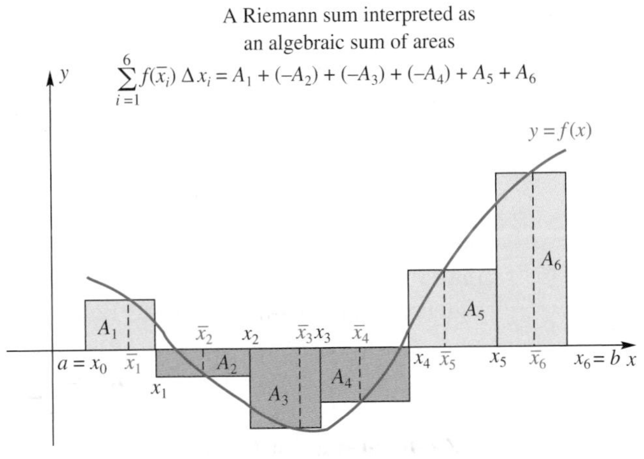

Figure 5

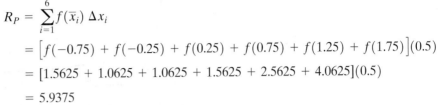

EXAMPLE 2 Evaluate the Riemann sum R_P for

$$f(x) = (x + 1)(x - 2)(x - 4) = x^3 - 5x^2 + 2x + 8$$

on the interval $[0, 5]$ using the partition P with partition points $0 < 1.1 < 2 < 3.2 < 4 < 5$ and the corresponding sample points $\bar{x}_1 = 0.5, \bar{x}_2 = 1.5, \bar{x}_3 = 2.5, \bar{x}_4 = 3.6$, and $\bar{x}_5 = 5$.

SOLUTION

$$\begin{aligned} R_P &= \sum_{i=1}^{5} f(\bar{x}_i)\,\Delta x_i \\ &= f(\bar{x}_1)\,\Delta x_1 + f(\bar{x}_2)\,\Delta x_2 + f(\bar{x}_3)\,\Delta x_3 + f(\bar{x}_4)\,\Delta x_4 + f(\bar{x}_5)\,\Delta x_5 \\ &= f(0.5)(1.1 - 0) + f(1.5)(2 - 1.1) + f(2.5)(3.2 - 2) \\ &\quad + f(3.6)(4 - 3.2) + f(5)(5 - 4) \\ &= (7.875)(1.1) + (3.125)(0.9) + (-2.625)(1.2) + (-2.944)(0.8) + 18(1) \\ &= 23.9698 \end{aligned}$$

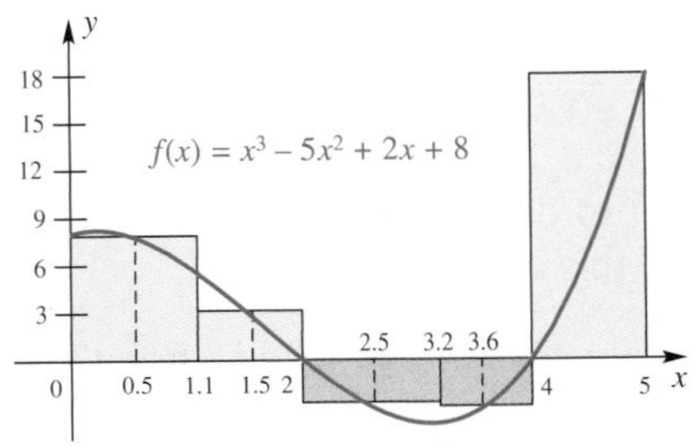

Figure 6

The corresponding geometric picture appears in Figure 6. ■

Definition of the Definite Integral Suppose now that P, Δx_i, and $\overline{x}_i$ have the meanings discussed above. Also let $\|P\|$, called the **norm** of P, denote the length of the longest of the subintervals of the partition P. For instance, in Example 1, $\|P\| = 0.5$; in Example 2, $\|P\| = 3.2 - 2 = 1.2$.

> **Notation for Integrals**
>
> We have chosen as our symbol for the definite integral the same elongated "S" as we did for the antiderative in the last chapter. The "S", of course, stands for "sum" since the definite integral is the limit of a particular type of sum, the Riemann sum.
>
> The connection between the antiderivative from Chapter 3 and the definite integral in this section will become clear in Section 4.4 when we present the Second Fundamental Theorem of Calculus.

> **Definition** Definite Integral
>
> Let f be a function that is defined on the closed interval $[a, b]$. If
>
> $$\lim_{\|P\|\to 0} \sum_{i=1}^{n} f(\overline{x}_i)\,\Delta x_i$$
>
> exists, we say f that is **integrable** on $[a, b]$. Moreover, $\int_a^b f(x)\,dx$, called the **definite integral** (or Riemann integral) of f from a to b, is then given by
>
> $$\int_a^b f(x)\,dx = \lim_{\|P\|\to 0} \sum_{i=1}^{n} f(\overline{x}_i)\,\Delta x_i$$

The heart of the definition is the final line. The concept captured in that equation grows out of our discussion of area in the previous section. However, we have considerably modified the notion presented there. For example, we now allow f to be negative on part or all of $[a, b]$, we use partitions with subintervals that may be of unequal length, and we allow $\overline{x}_i$ to be *any* point on the ith subinterval. Since we have made these changes, it is important to state precisely how the definite integral relates to area. In general, $\int_a^b f(x)\,dx$ gives the *signed area* of the region trapped between the curve $y = f(x)$ and the x-axis on the interval $[a, b]$, meaning that a positive sign is attached to areas of parts above the x-axis, and a negative sign is attached to areas of parts below the x-axis. In symbols,

$$\int_a^b f(x)\,dx = A_{\text{up}} - A_{\text{down}}$$

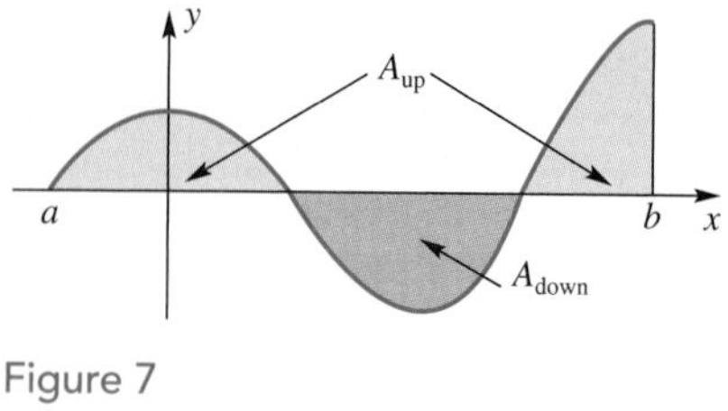

Figure 7

where A_{up} and A_{down} are as shown in Figure 7.

The meaning of the word *limit* in the definition of the definite integral is more general than in earlier usage and should be explained. The equality

$$\lim_{\|P\|\to 0} \sum_{i=1}^{n} f(\overline{x}_i)\,\Delta x_i = L$$

means that, corresponding to each $\varepsilon > 0$, there is a $\delta > 0$ such that

$$\left| \sum_{i=1}^{n} f(\overline{x}_i)\,\Delta x_i - L \right| < \varepsilon$$

for all Riemann sums $\sum_{i=1}^{n} f(\overline{x}_i)\,\Delta x_i$ for f on $[a, b]$ for which the norm $\|P\|$ of the associated partition is less than δ. In this case, we say that the indicated limit exists and has the value L.

That was a mouthful, and we are not going to digest it just now. We simply assert that the usual limit theorems also hold for this kind of limit.

Returning to the symbol $\int_a^b f(x)\,dx$, we might call a the lower end point and b the upper end point for the integral. However, most authors use the terminology **lower limit** of integration and **upper limit** of integration, which is fine provided we realize that this usage of the word *limit* has nothing to do with its more technical meaning.

In our definition of $\int_a^b f(x)\,dx$, we implicitly assumed that $a < b$. We remove that restriction with the following definitions.

$$\int_a^a f(x)\,dx = 0$$

$$\int_a^b f(x)\,dx = -\int_b^a f(x)\,dx, \quad a > b$$

Thus,

$$\int_2^2 x^3\,dx = 0, \qquad \int_6^2 x^3\,dx = -\int_2^6 x^3\,dx$$

Finally, we point out that x is a **dummy variable** in the symbol $\int_a^b f(x)\,dx$. By this we mean that x can be replaced by any other letter (provided, of course, that it is replaced in each place where it occurs). Thus,

$$\int_a^b f(x)\,dx = \int_a^b f(t)\,dt = \int_a^b f(u)\,du$$

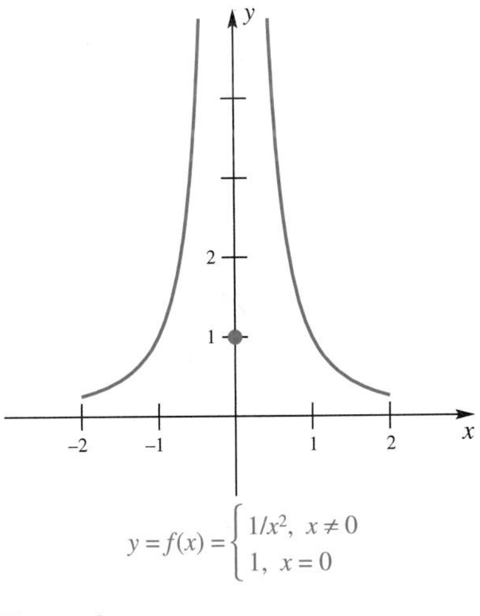

Figure 8

What Functions Are Integrable?

Not every function is integrable on a closed interval $[a, b]$. For example, the unbounded function

$$f(x) = \begin{cases} \dfrac{1}{x^2} & \text{if } x \neq 0 \\ 1 & \text{if } x = 0 \end{cases}$$

which is graphed in Figure 8, is not integrable on $[-2, 2]$. It can be shown that for this unbounded function, the Riemann sum can be made arbitrarily large. Therefore, the limit of the Riemann sum over $[-2, 2]$ does not exist.

Even some bounded functions can fail to be integrable, but they have to be pretty complicated (see Problem 39 for one example). Theorem A (below) is the most important theorem about integrability. Unfortunately, it is too difficult to prove here; we leave that for advanced calculus books.

Theorem A Integrability Theorem

If f is bounded on $[a, b]$ and if it is continuous there except at a finite number of points, then f is integrable on $[a, b]$. In particular, if f is continuous on the whole interval $[a, b]$, it is integrable on $[a, b]$.

As a consequence of this theorem, the following functions are integrable on every closed interval $[a, b]$.

1. Polynomial functions
2. Sine and cosine functions
3. Rational functions, provided that the interval $[a, b]$ contains no points where the denominator is 0

Calculating Definite Integrals

Knowing that a function is integrable allows us to calculate its integral by using a **regular partition** (i.e., a partition with equal-length subintervals) and by picking the sample points $\overline{x}_i$ in any way that is convenient for us. Examples 3 and 4 involve polynomials, which we just learned are integrable.

EXAMPLE 3 Evaluate $\int_{-2}^{3} (x + 3)\, dx$.

SOLUTION Partition the interval $[-2, 3]$ into n equal subintervals, each of length $\Delta x = 5/n$. In each subinterval $[x_{i-1}, x_i]$, use $\overline{x}_i = x_i$ as the sample point. Then

$$\begin{aligned}
x_0 &= -2\\
x_1 &= -2 + \Delta x = -2 + \frac{5}{n}\\
x_2 &= -2 + 2\,\Delta x = -2 + 2\left(\frac{5}{n}\right)\\
&\vdots\\
x_i &= -2 + i\,\Delta x = -2 + i\left(\frac{5}{n}\right)\\
&\vdots\\
x_n &= -2 + n\,\Delta x = -2 + n\left(\frac{5}{n}\right) = 3
\end{aligned}$$

Thus, $f(x_i) = x_i + 3 = 1 + i(5/n)$, and so

$$\begin{aligned}
\sum_{i=1}^{n} f(\overline{x}_i)\,\Delta x_i &= \sum_{i=1}^{n} f(x_i)\,\Delta x\\
&= \sum_{i=1}^{n}\left[1 + i\left(\frac{5}{n}\right)\right]\frac{5}{n}\\
&= \frac{5}{n}\sum_{i=1}^{n} 1 + \frac{25}{n^2}\sum_{i=1}^{n} i\\
&= \frac{5}{n}(n) + \frac{25}{n^2}\left[\frac{n(n+1)}{2}\right] \qquad \text{(Special Sum Formula 1)}\\
&= 5 + \frac{25}{2}\left(1 + \frac{1}{n}\right)
\end{aligned}$$

Since P is a regular partition, $\|P\| \to 0$ is equivalent to $n \to \infty$. We conclude that

$$\begin{aligned}
\int_{-2}^{3} (x + 3)\, dx &= \lim_{\|P\|\to 0} \sum_{i=1}^{n} f(\overline{x}_i)\,\Delta x_i\\
&= \lim_{n\to\infty}\left[5 + \frac{25}{2}\left(1 + \frac{1}{n}\right)\right]\\
&= \frac{35}{2}
\end{aligned}$$

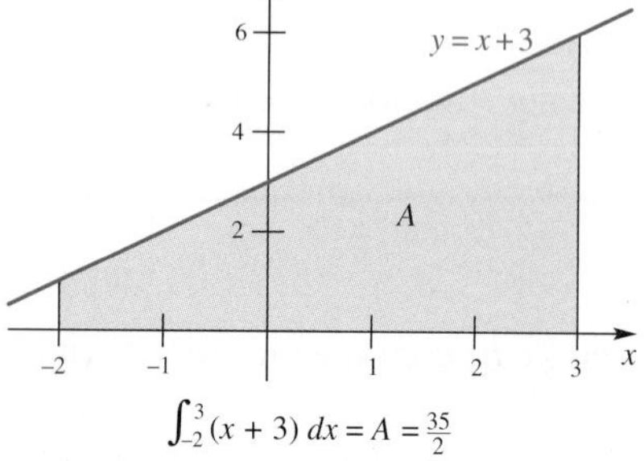

Figure 9

We can easily check our answer, since the required integral gives the area of the trapezoid in Figure 9. The familiar trapezoidal area formula $A = \frac{1}{2}(a + b)h$ gives $\frac{1}{2}(1 + 6)5 = 35/2$. ■

EXAMPLE 4 Evaluate $\int_{-1}^{3} (2x^2 - 8)\, dx$.

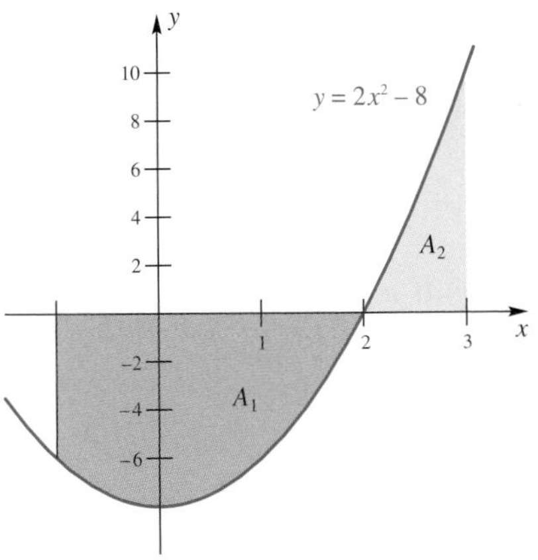

$\int_{-1}^{3}(2x^2 - 8)\,dx = -A_1 + A_2 = -\frac{40}{3}$

Figure 10

SOLUTION No formulas from elementary geometry will help here. Figure 10 suggests that the integral is equal to $-A_1 + A_2$, where A_1 and A_2 are the areas of the regions below and above the x-axis.

Let P be a regular partition of $[-1, 3]$ into n equal subintervals, each of length $\Delta x = 4/n$. In each subinterval $[x_{i-1}, x_i]$, choose $\overline{x}_i$ to be the right end point, so $\overline{x}_i = x_i$. Then

$$x_i = -1 + i\,\Delta x = -1 + i\left(\frac{4}{n}\right)$$

and

$$\begin{aligned} f(x_i) &= 2x_i^2 - 8 = 2\left[-1 + i\left(\frac{4}{n}\right)\right]^2 - 8 \\ &= -6 - \frac{16i}{n} + \frac{32i^2}{n^2} \end{aligned}$$

Consequently,

$$\begin{aligned} \sum_{i=1}^{n} f(\overline{x}_i)\,\Delta x_i &= \sum_{i=1}^{n} f(x_i)\,\Delta x \\ &= \sum_{i=1}^{n}\left[-6 - \frac{16}{n}\,i + \frac{32}{n^2}\,i^2\right]\frac{4}{n} \\ &= -\frac{24}{n}\sum_{i=1}^{n}1 - \frac{64}{n^2}\sum_{i=1}^{n}i + \frac{128}{n^3}\sum_{i=1}^{n}i^2 \\ &= -\frac{24}{n}(n) - \frac{64}{n^2}\frac{n(n+1)}{2} + \frac{128}{n^3}\frac{n(n+1)(2n+1)}{6} \\ &= -24 - 32\left(1 + \frac{1}{n}\right) + \frac{128}{6}\left(2 + \frac{3}{n} + \frac{1}{n^2}\right) \end{aligned}$$

We conclude that

$$\begin{aligned} \int_{-1}^{3}(2x^2 - 8)\,dx &= \lim_{\|P\|\to 0}\sum_{i=1}^{n} f(\overline{x}_i)\,\Delta x_i \\ &= \lim_{n\to\infty}\left[-24 - 32\left(1 + \frac{1}{n}\right) + \frac{128}{6}\left(2 + \frac{3}{n} + \frac{1}{n^2}\right)\right] \\ &= -24 - 32 + \frac{128}{3} = -\frac{40}{3} \end{aligned}$$

That the answer is negative is not surprising, since the region below the x-axis appears to be larger than the one above the x-axis (Figure 10). Our answer is close to the estimate given in the margin note COMMON SENSE; this reassures us that our answer is likely to be correct. ■

> **Common Sense**
>
> Given the graph of a function, we can always make a rough estimate for the value of a definite integral by using the fact that it is the signed area
>
> $$A_{\text{up}} - A_{\text{down}}$$
>
> Thus, in Example 4, we might estimate the value of the integral by pretending that the part above the x-axis is a triangle and the part below is a rectangle. Our estimate is
>
> $$\tfrac{1}{2}(1)(10) - (3)(6) = -13$$

The Interval Additive Property Our definition of the definite integral was motivated by the problem of area for curved regions. Consider the two curved regions R_1 and R_2 in Figure 11 and let $R = R_1 \cup R_2$. It is clear that

$$A(R) = A(R_1 \cup R_2) = A(R_1) + A(R_2)$$

which suggests that

$$\int_a^c f(x)\,dx = \int_a^b f(x)\,dx + \int_b^c f(x)\,dx$$

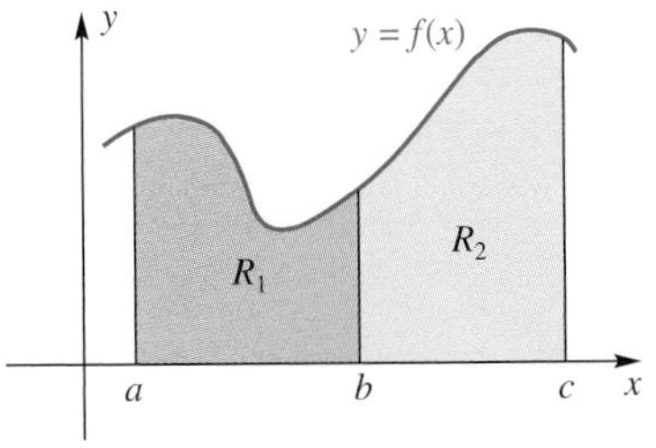

Figure 11

We quickly point out that this does not constitute a proof of this fact about integrals, since, first of all, our discussion of area in Section 4.1 was rather informal and, second, our diagram supposes that f is positive, which it need not be. Nevertheless, definite integrals do satisfy this interval additive property, and they do it no matter how the three points a, b, and c are arranged. We leave the rigorous proof to more advanced works.

Theorem B Interval Additive Property

If f is integrable on an interval containing the points a, b, and c, then

$$\int_a^c f(x)\,dx = \int_a^b f(x)\,dx + \int_b^c f(x)\,dx$$

no matter what the order of a, b, and c.

For example,

$$\int_0^2 x^2\,dx = \int_0^1 x^2\,dx + \int_1^2 x^2\,dx$$

which most people readily believe. But it is also true that

$$\int_0^2 x^2\,dx = \int_0^3 x^2\,dx + \int_3^2 x^2\,dx$$

which may seem surprising. If you mistrust the theorem, you might actually evaluate each of the above integrals to see that the equality holds.

Velocity and Position Near the end of Section 4.1 we explained how the area under the velocity curve is equal to the distance traveled, provided the velocity function $v(t)$ is positive. In general, the position (which could be positive or negative) is equal to the definite integral of the velocity function (which could be positive or negative). More specifically, if $v(t)$ is the velocity of an object at time t, where $t \geq 0$, and if the object is at position 0 at time 0, then the position of the object at time a is $\int_0^a v(t)\,dt$.

EXAMPLE 5 An object at the origin at time $t = 0$ has velocity, measured in meters per second,

$$v(t) = \begin{cases} t/20, & \text{if } 0 \leq t \leq 40 \\ 2, & \text{if } 40 < t \leq 60 \\ 5 - t/20 & \text{if } t > 60 \end{cases}$$

Sketch the velocity curve. Express the object's position at $t = 140$ as a definite integral and evaluate it using formulas from plane geometry.

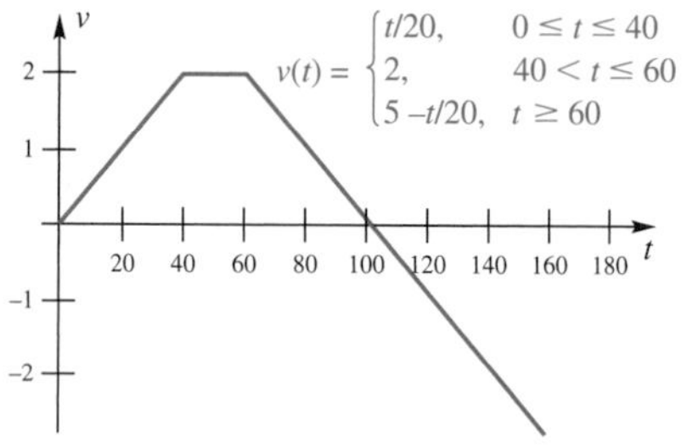

Figure 12

SOLUTION Figure 12 shows the velocity curve. The position at time 140 is equal to the definite integral $\int_0^{140} v(t)\,dt$, which we can evaluate using formulas for the area of a triangle and a rectangle and using the Interval Additive Property (Theorem B):

$$\int_0^{140} v(t)\,dt = \int_0^{40} \frac{t}{20}\,dt + \int_{40}^{60} 2\,dt + \int_{60}^{140}\left(5 - \frac{t}{20}\right)dt$$
$$= 40 + 40 + 40 - 40 = 80$$

■

Concepts Review

1. A sum of the form $\sum_{i=1}^{n} f(\overline{x}_i)\,\Delta x_i$ is called a _____.

2. The limit of the sum above for f defined on $[a, b]$ is called a _____ and is symbolized by _____.

3. Geometrically, the definite integral corresponds to a signed area. In terms of A_{up} and A_{down}, $\int_a^b f(x)\,dx =$ _____.

4. Thus, the value of $\int_{-1}^{4} x\,dx$ is _____.

Problem Set 4.2

In Problems 1 and 2, calculate the Riemann sum suggested by each figure.

1.

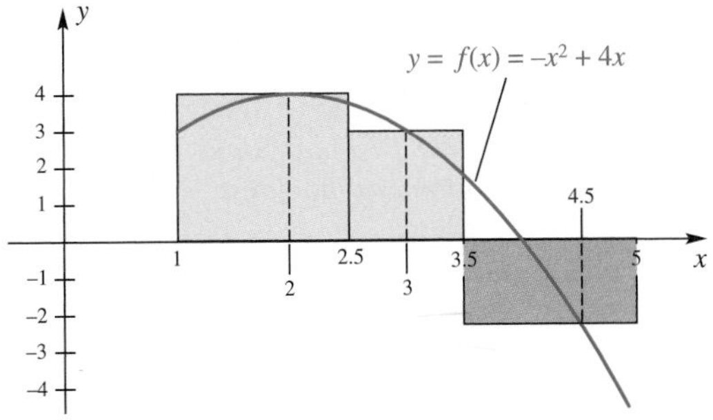

2.

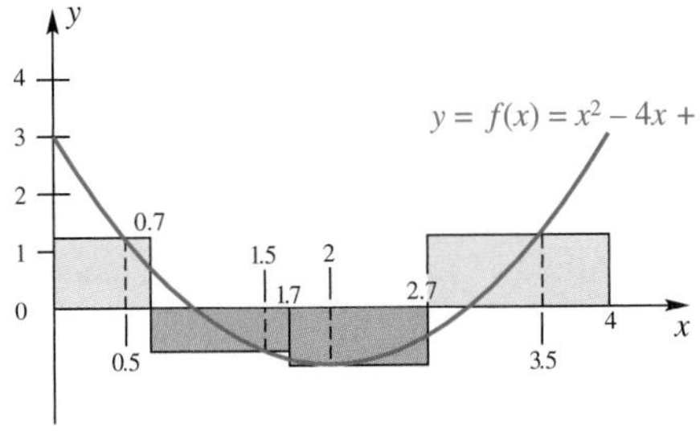

In Problems 3–6, calculate the Riemann sum $\sum_{i=1}^{n} f(\bar{x}_i)\,\Delta x_i$ *for the given data.*

3. $f(x) = x - 1$; $P: 3 < 3.75 < 4.25 < 5.5 < 6 < 7$; $\bar{x}_1 = 3, \bar{x}_2 = 4, \bar{x}_3 = 4.75, \bar{x}_4 = 6, \bar{x}_5 = 6.5$

4. $f(x) = -x/2 + 3$; $P: -3 < -1.3 < 0 < 0.9 < 2$; $\bar{x}_1 = -2, \bar{x}_2 = -0.5, \bar{x}_3 = 0, \bar{x}_4 = 2$

C **5.** $f(x) = x^2/2 + x$; $[-2, 2]$ is divided into eight equal subintervals, $\bar{x}_i$ is the midpoint.

C **6.** $f(x) = 4x^3 + 1$; $[0, 3]$ is divided into six equal subintervals, $\bar{x}_i$ is the right end point.

In Problems 7–10, use the given values of a and b and express the given limit as a definite integral.

7. $\lim_{\|P\|\to 0} \sum_{i=1}^{n} (\bar{x}_i)^3\,\Delta x_i$; $a = 1, b = 3$

8. $\lim_{\|P\|\to 0} \sum_{i=1}^{n} (\bar{x}_i + 1)^3\,\Delta x_i$; $a = 0, b = 2$

9. $\lim_{\|P\|\to 0} \sum_{i=1}^{n} \frac{\bar{x}_i^2}{1 + \bar{x}_i}\,\Delta x_i$; $a = -1, b = 1$

10. $\lim_{\|P\|\to 0} \sum_{i=1}^{n} (\sin \bar{x}_i)^2\,\Delta x_i$; $a = 0, b = \pi$

In Problems 11–16, evaluate the definite integrals using the definition, as in Examples 3 and 4.

11. $\int_0^2 (x + 1)\,dx$

12. $\int_0^2 (x^2 + 1)\,dx$

Hint: Use $\bar{x}_i = 2i/n$.

13. $\int_{-2}^{1} (2x + \pi)\,dx$

14. $\int_{-2}^{1} (3x^2 + 2)\,dx$

Hint: Use $\bar{x}_i = -2 + 3i/n$.

15. $\int_0^5 (x + 1)\,dx$

16. $\int_{-10}^{10} (x^2 + x)\,dx$

In Problems 17–22, calculate $\int_a^b f(x)\,dx$*, where a and b are the left and right end points for which f is defined, by using the Interval Additive Property and the appropriate area formulas from plane geometry. Begin by graphing the given function.*

17. $f(x) = \begin{cases} 2x & \text{if } 0 \le x \le 1 \\ 2 & \text{if } 1 < x \le 2 \\ x & \text{if } 2 < x \le 5 \end{cases}$

18. $f(x) = \begin{cases} 3x & \text{if } 0 \le x \le 1 \\ 2(x-1) + 2 & \text{if } 1 < x \le 2 \end{cases}$

19. $f(x) = \begin{cases} \sqrt{1 - x^2} & \text{if } 0 \le x \le 1 \\ x - 1 & \text{if } 1 < x \le 2 \end{cases}$

20. $f(x) = \begin{cases} -\sqrt{4 - x^2} & \text{if } -2 \le x \le 0 \\ -2x - 2 & \text{if } 0 < x \le 2 \end{cases}$

21. $f(x) = \sqrt{A^2 - x^2}$; $-A \le x \le A$

22. $f(x) = 4 - |x|$, $-4 \le x \le 4$

In Problems 23–26, the velocity function for an object is given. Assuming that the object is at the origin at time t = 0, find the position at time t = 4.

23. $v(t) = t/60$

24. $v(t) = 1 + 2t$

25. $v(t) = \begin{cases} t/2 & \text{if } 0 \le t \le 2 \\ 1 & \text{if } 2 < t \le 4 \end{cases}$

26. $v(t) = \begin{cases} \sqrt{4 - t^2} & \text{if } 0 \le t \le 2 \\ 0 & \text{if } 2 < t \le 4 \end{cases}$

In Problems 27–30, an object's velocity function is graphed. Use this graph to determine the object's position at times t = 20, 40, 60, 80, 100, and 120 assuming the object is at the origin at time t = 0.

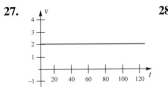

27.

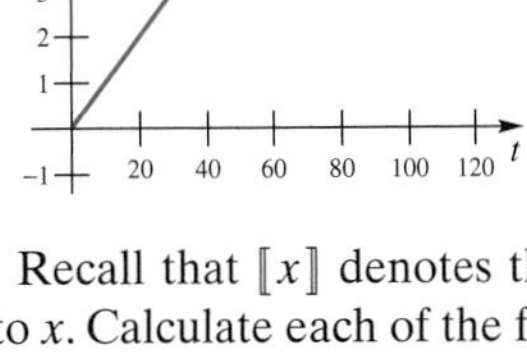

28.

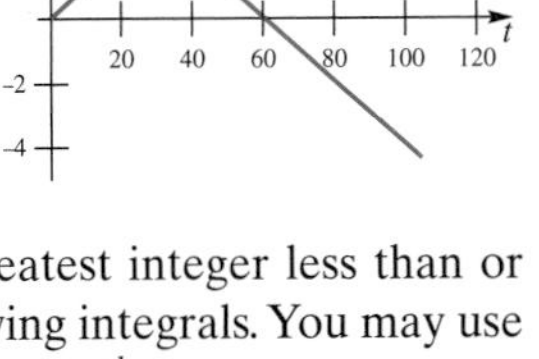

29.

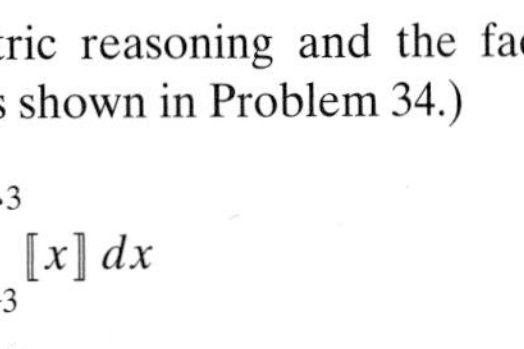

30.

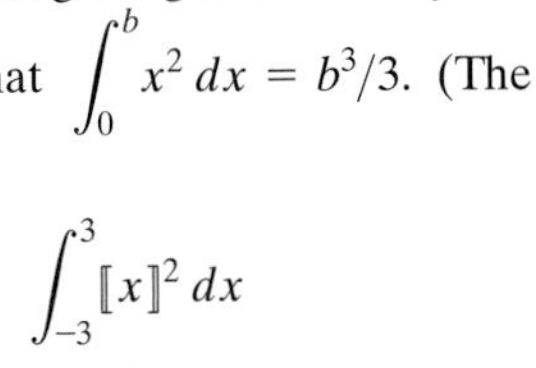

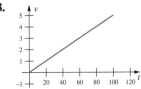

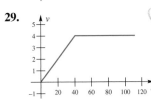

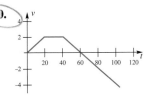

31. Recall that $[\![x]\!]$ denotes the greatest integer less than or equal to x. Calculate each of the following integrals. You may use geometric reasoning and the fact that $\int_0^b x^2\,dx = b^3/3$. (The latter is shown in Problem 34.)

(a) $\int_{-3}^{3} [\![x]\!]\,dx$

(b) $\int_{-3}^{3} [\![x]\!]^2\,dx$

(c) $\int_{-3}^{3} (x - [\![x]\!])\,dx$

(d) $\int_{-3}^{3} (x - [\![x]\!])^2\,dx$

(e) $\int_{-3}^{3} |x|\,dx$ (f) $\int_{-3}^{3} x|x|\,dx$

(g) $\int_{-1}^{2} |x|[\![x]\!]\,dx$ (h) $\int_{-1}^{2} x^2[\![x]\!]\,dx$

32. Let f be an odd function and g be an even function, and suppose that $\int_0^1 |f(x)|\,dx = \int_0^1 g(x)\,dx = 3$. Use geometric reasoning to calculate each of the following:

(a) $\int_{-1}^{1} f(x)\,dx$ (b) $\int_{-1}^{1} g(x)\,dx$

(c) $\int_{-1}^{1} |f(x)|\,dx$ (d) $\int_{-1}^{1} [-g(x)]\,dx$

(e) $\int_{-1}^{1} xg(x)\,dx$ (f) $\int_{-1}^{1} f^3(x)g(x)\,dx$

33. Show that $\int_a^b x\,dx = \frac{1}{2}(b^2 - a^2)$ by completing the following argument. For the partition $a = x_0 < x_1 < \cdots < x_n = b$, choose $\overline{x}_i = \frac{1}{2}(x_{i-1} + x_i)$. Then $R_P = \sum_{i=1}^{n} \overline{x}_i\,\Delta x_i = \frac{1}{2}\sum_{i=1}^{n}(x_i + x_{i-1})(x_i - x_{i-1})$. Now simplify R_P (collapsing sum) and take a limit.

34. Show that $\int_a^b x^2\,dx = \frac{1}{3}(b^3 - a^3)$ by an argument like that in Problem 33, but using $\overline{x}_i = \left[\frac{1}{3}(x_{i-1}^2 + x_{i-1}x_i + x_i^2)\right]^{1/2}$. Assume that $0 \le a < b$.

CAS *Many computer algebra systems permit the evaluation of Riemann sums for left end point, right end point, or midpoint evaluations of the function. Using such a system in Problems 35–38, evaluate the 10-subinterval Riemann sums using left end point, right end point, and midpoint evaluations.*

35. $\int_0^2 (x^3 + 1)\,dx$ **36.** $\int_0^1 \tan x\,dx$

37. $\int_0^1 \cos x\,dx$ **38.** $\int_1^3 (1/x)\,dx$

39. Prove that the function f defined by

$$f(x) = \begin{cases} 1 & \text{if } x \text{ is rational} \\ 0 & \text{if } x \text{ is irrational} \end{cases}$$

is not integrable on $[0, 1]$. *Hint:* Show that no matter how small the norm of the partition, $\|P\|$, the Riemann sum can be made to have value either 0 or 1.

Answers to Concepts Review: **1.** Riemann sum **2.** definite integral; $\int_a^b f(x)\,dx$ **3.** $A_{\text{up}} - A_{\text{down}}$ **4.** $\frac{15}{2}$

4.3 The First Fundamental Theorem of Calculus

Calculus is the study of limits, and the two most important limits that you have studied so far are the derivative and the definite integral. The derivative of a function f is

$$f'(x) = \lim_{h\to 0} \frac{f(x+h) - f(x)}{h}$$

and the definite integral is

$$\int_a^b f(x)\,dx = \lim_{\|P\|\to 0} \sum_{i=1}^{n} f(\overline{x}_i)\,\Delta x_i$$

These two kinds of limits appear to have no connection whatsoever. There is, however, a very close connection, as we shall see in this section.

Newton and Leibniz are usually credited with the simultaneous but independent discovery of calculus. Yet, the concepts of the slope of a tangent line (which led to the derivative) were known earlier, having been studied by Blaise Pascal and Isaac Barrow years before Newton and Leibniz. And Archimedes had studied areas of curved regions 1800 years earlier, in the third century B.C. Why then do Newton and Leibniz get the credit? They understood and exploited the intimate relationship between antiderivatives and definite integrals. This important relationship is called the *First Fundamental Theorem of Calculus*.

The First Fundamental Theorem You have likely met several "fundamental theorems" in your mathematical career. The Fundamental Theorem of Arithmetic says that a whole number factors uniquely into a product of primes. The Fundamental Theorem of Algebra says that an nth-degree polynomial has n roots, counting complex roots and multiplicities. Any "fundamental theorem" should be studied carefully, and then permanently committed to memory.

Near the end of Section 4.1, we studied a problem in which the velocity of an object at time t is given by $v = f(t) = \frac{1}{4}t^3 + 1$. We found that the distance traveled from time $t = 0$ to time $t = 3$ is equal to

$$\lim_{n\to\infty} \sum_{i=1}^{n} f(t_i)\,\Delta t = \frac{129}{16}$$

Using the terminology from Section 4.2, we now see that the distance traveled from time $t = 0$ to time $t = 3$ is equal to the definite integral

$$\lim_{n\to\infty} \sum_{i=1}^{n} f(t_i)\,\Delta t = \int_0^3 f(t)\,dt$$

(Since the velocity is positive for all $t \geq 0$, the distance traveled through time t is equal to the position of the object at time t. If the velocity were negative for some value of t, then the object would be traveling backward at that time t; in such a case, distance traveled would not equal position.) We could use the same reasoning to find that the distance s traveled from time $t = 0$ to time $t = x$ is

$$s(x) = \int_0^x f(t)\,dt$$

The question we now pose is this: What is the derivative of s?

Since the derivative of distance traveled (as long as the velocity is always positive) is the velocity, we have

$$s'(x) = v = f(x)$$

In other words,

$$\boxed{\frac{d}{dx}s(x) = \frac{d}{dx}\int_0^x f(t)\,dt = f(x)}$$

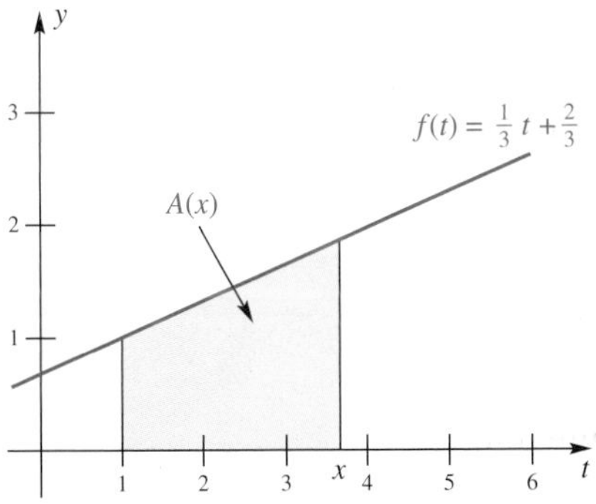

Figure 1

Now, define $A(x)$ to be the area under the graph of $y = \frac{1}{3}t + \frac{2}{3}$, above the t-axis, and between the vertical lines $t = 1$ and $t = x$, where $x \geq 1$ (see Figure 1). A function such as this is called an **accumulation function** because it accumulates area under a curve from a fixed value ($t = 1$ in this case) to a variable value ($t = x$ in this case). What is the derivative of A?

The area $A(x)$ is equal to the definite integral

$$A(x) = \int_1^x \left(\frac{2}{3} + \frac{1}{3}t\right) dt$$

In this case we can evaluate this definite integral using a geometrical argument; $A(x)$ is just the area of a trapezoid, so

$$A(x) = (x - 1)\frac{1 + \left(\frac{2}{3} + \frac{1}{3}x\right)}{2} = \frac{1}{6}x^2 + \frac{2}{3}x - \frac{5}{6}$$

With this done, we see that the derivative of A is

$$A'(x) = \frac{d}{dx}\left(\frac{1}{6}x^2 + \frac{2}{3}x - \frac{5}{6}\right) = \frac{1}{3}x + \frac{2}{3}$$

In other words,

$$\boxed{\frac{d}{dx}\int_1^x \left(\frac{2}{3} + \frac{1}{3}t\right) dt = \frac{2}{3} + \frac{1}{3}x}$$

Let's define another accumulation function B as the area under the curve $y = t^2$, above the t-axis, to the right of the origin, and to the left of the line $t = x$,

Terminology

- The indefinite integral $\int f(x)\,dx$ is a *family of functions* of x.
- The definite integral $\int_a^b f(x)\,dx$ is a *number*, provided that a and b are fixed.
- If the upper limit in a definite integral is a variable x, then the definite integral [e.g., $\int_a^x f(t)\,dt$] is a *function of* x.
- A function of the form $F(x) = \int_a^x f(t)\,dt$ is called an *accumulation function*.

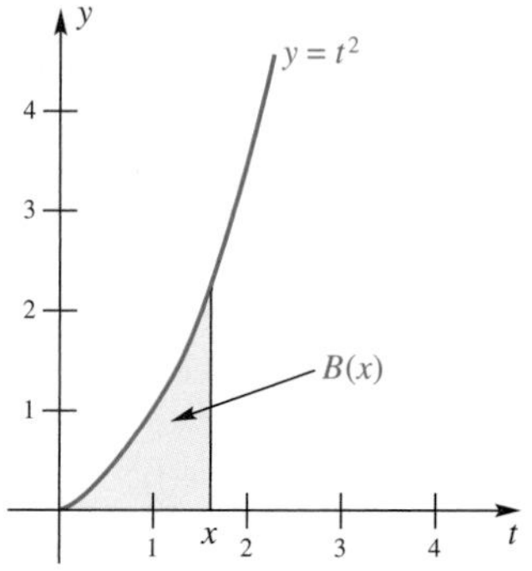

Figure 2

where $x \geq 0$ (see Figure 2). This area is given by the definite integral $\int_0^x t^2\,dt$. To find this area, we first construct a Riemann sum. We use a regular partition of $[0, x]$ and evaluate the function at the right end point of each subinterval. Then $\Delta t = x/n$ and the right end point of the ith interval is $t_i = 0 + i\Delta t = ix/n$. The Riemann sum is therefore

$$\begin{aligned}\sum_{i=1}^{n} f(t_i)\,\Delta t &= \sum_{i=1}^{n} f\left(\frac{ix}{n}\right)\frac{x}{n}\\ &= \frac{x}{n}\sum_{i=1}^{n}\left(\frac{ix}{n}\right)^2\\ &= \frac{x^3}{n^3}\sum_{i=1}^{n} i^2\\ &= \frac{x^3}{n^3}\,\frac{n(n+1)(2n+1)}{6}\end{aligned}$$

The definite integral is the limit of these Riemann sums.

$$\begin{aligned}\int_0^x t^2\,dt &= \lim_{n\to\infty}\sum_{i=1}^{n} f(t_i)\,\Delta t\\ &= \lim_{n\to\infty}\frac{x^3}{n^3}\,\frac{n(n+1)(2n+1)}{6}\\ &= \frac{x^3}{6}\lim_{n\to\infty}\frac{2n^3+3n^2+n}{n^3}\\ &= \frac{x^3}{6}\cdot 2 = \frac{x^3}{3}\end{aligned}$$

Thus, $B(x) = x^3/3$, so the derivative of B is

$$B'(x) = \frac{d}{dx}\frac{x^3}{3} = x^2$$

In other words,

$$\boxed{\frac{d}{dx}\int_0^x t^2\,dt = x^2}$$

The results of the last three boxed equations suggest that the derivative of an accumulation function is equal to the function being accumulated. But is this *always* the case? And *why* should this be the case?

Suppose that we are using a "retractable" paintbrush to paint the region under a curve. (By retractable, we mean that the brush becomes wider or narrower as it moves left to right so that it just covers the height to be painted. The brush is wide when the integrand values are large and narrow when the integrand values are small. See Figure 3.) With this analogy, the accumulated area is the painted area, and the rate of accumulation is the rate at which the paint is being applied. But the rate at which paint is being applied is equal to the width of the brush, in effect, the height of the function. We can restate this result as follows.

> *The rate of accumulation at $t = x$ is equal to the value of the function being accumulated at $t = x$.*

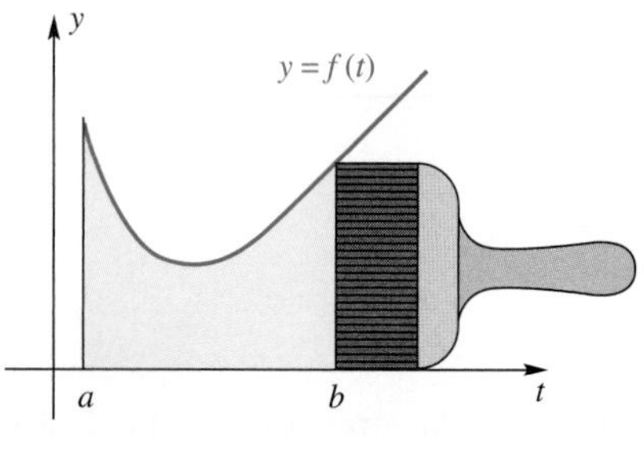

Figure 3

This, in a nutshell, is the First Fundamental Theorem of Calculus. It is *fundamental* because it links the derivative and the definite integral, the most important kinds of limits you have studied so far.

Theorem A First Fundamental Theorem of Calculus

Let f be continuous on the closed interval $[a, b]$ and let x be a (variable) point in (a, b). Then

$$\frac{d}{dx}\int_a^x f(t)\,dt = f(x)$$

Sketch of Proof For now, we present a sketch of the proof. This sketch shows the important features of the proof, but a complete proof must wait until after we have established a few other results. For x in $[a, b]$, define $F(x) = \int_a^x f(t)\,dt$. Then for x in (a, b)

$$\begin{aligned}\frac{d}{dx}\int_a^x f(t)\,dt &= F'(x)\\ &= \lim_{h\to 0}\frac{F(x+h)-F(x)}{h}\\ &= \lim_{h\to 0}\frac{1}{h}\left[\int_a^{x+h} f(t)\,dt - \int_a^x f(t)\,dt\right]\\ &= \lim_{h\to 0}\frac{1}{h}\int_x^{x+h} f(t)\,dt\end{aligned}$$

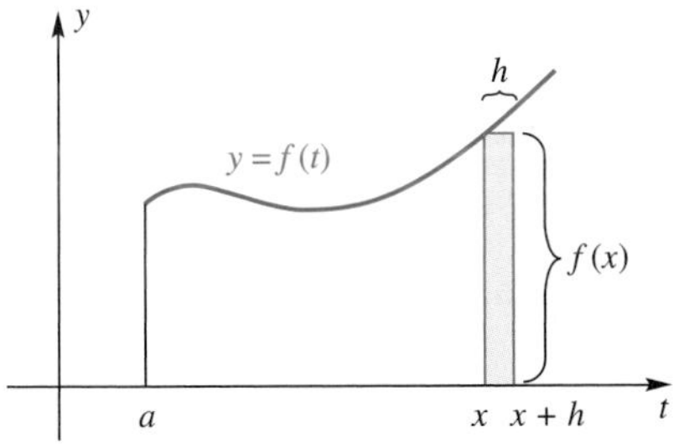

Figure 4

The last line follows from the Interval Additive Property (Theorem 4.2B). Now, when h is small, f does not change much over the interval $[x, x+h]$. On this interval, f is roughly equal to $f(x)$, the value of f evaluated at the left end point of the interval (see Figure 4). The area under the curve $y = f(t)$ from x to $x+h$ is approximately equal to the area of the rectangle with width h and height $f(x)$; that is, $\int_x^{x+h} f(t)\,dt \approx hf(x)$. Therefore,

$$\frac{d}{dx}\int_a^x f(t)\,dt \approx \lim_{h\to 0}\frac{1}{h}[hf(x)] = f(x) \qquad \blacksquare$$

Of course, the flaw in this argument is that h is never 0, so we cannot claim that f does not change over the interval $[x, x+h]$. We will give a complete proof later in this section.

Comparison Properties Consideration of the areas of the regions R_1 and R_2 in Figure 5 suggests another property of definite integrals.

Theorem B Comparison Property

If f and g are integrable on $[a, b]$ and if $f(x) \le g(x)$ for all x in $[a, b]$, then

$$\int_a^b f(x)\,dx \le \int_a^b g(x)\,dx$$

In informal but descriptive language, we say that the definite integral preserves inequalities.

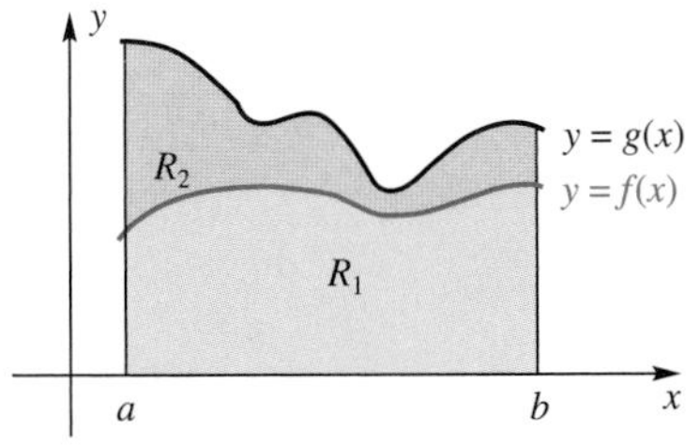

Figure 5

Proof Let $P\colon a = x_0 < x_1 < x_2 < \cdots < x_n = b$ be an arbitrary partition of $[a, b]$, and for each i let $\bar{x}_i$ be any sample point on the ith subinterval $[x_{i-1}, x_i]$. We may conclude successively that

$$f(\overline{x}_i) \le g(\overline{x}_i)$$
$$f(\overline{x}_i)\,\Delta x_i \le g(\overline{x}_i)\,\Delta x_i$$
$$\sum_{i=1}^{n} f(\overline{x}_i)\,\Delta x_i \le \sum_{i=1}^{n} g(\overline{x}_i)\,\Delta x_i$$
$$\lim_{\|P\|\to 0} \sum_{i=1}^{n} f(\overline{x}_i)\,\Delta x_i \le \lim_{\|P\|\to 0} \sum_{i=1}^{n} g(\overline{x}_i)\,\Delta x_i$$
$$\int_a^b f(x)\,dx \le \int_a^b g(x)\,dx$$
■

Theorem C Boundedness Property

If f is integrable on $[a, b]$ and $m \le f(x) \le M$ for all x in $[a, b]$, then

$$m(b - a) \le \int_a^b f(x)\,dx \le M(b - a)$$

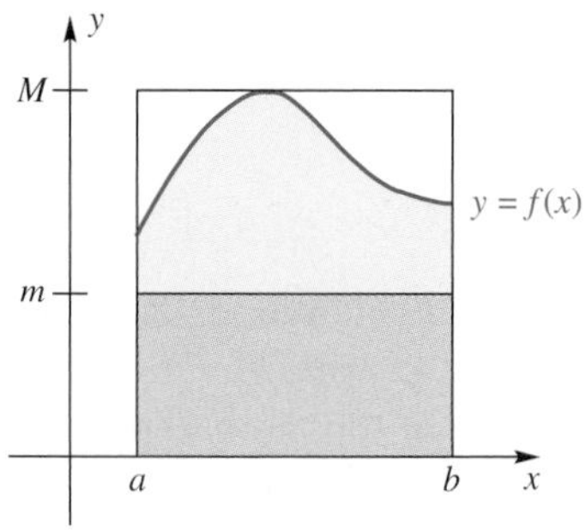

Figure 6

Proof The picture in Figure 6 helps us to understand the theorem. Note that $m(b - a)$ is the area of the lower, small rectangle, $M(b - a)$ is the area of the large rectangle, and $\int_a^b f(x)\,dx$ is the area under the curve.

To prove the right-hand inequality, let $g(x) = M$ for all x in $[a, b]$. Then, by Theorem B,

$$\int_a^b f(x)\,dx \le \int_a^b g(x)\,dx$$

However, $\int_a^b g(x)\,dx$ is equal to the area of a rectangle with width $b - a$ and height M. Thus,

$$\int_a^b g(x)\,dx = M(b - a)$$

The left-hand inequality is handled similarly. ■

The Definite Integral Is a Linear Operator

Earlier we learned that D_x, $\int \cdots dx$, and Σ are linear operators. You can add $\int_a^b \cdots dx$ to the list.

Theorem D Linearity of the Definite Integral

Suppose that f and g are integrable on $[a, b]$ and that k is a constant. Then kf and $f + g$ are integrable and

(i) $\int_a^b kf(x)\,dx = k\int_a^b f(x)\,dx$;

(ii) $\int_a^b [f(x) + g(x)]\,dx = \int_a^b f(x)\,dx + \int_a^b g(x)\,dx$; and

(iii) $\int_a^b [f(x) - g(x)]\,dx = \int_a^b f(x)\,dx - \int_a^b g(x)\,dx$.

Proof The proofs of (i) and (ii) depend on the linearity of Σ and the properties of limits. We show (ii).

$$\begin{aligned}\int_a^b [f(x) + g(x)]\, dx &= \lim_{\|P\|\to 0} \sum_{i=1}^{n} [f(\overline{x}_i) + g(\overline{x}_i)]\Delta x_i \\ &= \lim_{\|P\|\to 0} \left[\sum_{i=1}^{n} f(\overline{x}_i)\, \Delta x_i + \sum_{i=1}^{n} g(\overline{x}_i)\, \Delta x_i \right] \\ &= \lim_{\|P\|\to 0} \sum_{i=1}^{n} f(\overline{x}_i)\, \Delta x_i + \lim_{\|P\|\to 0} \sum_{i=1}^{n} g(\overline{x}_i)\, \Delta x_i \\ &= \int_a^b f(x)\, dx + \int_a^b g(x)\, dx\end{aligned}$$

Part (iii) follows from (i) and (ii) on writing $f(x) - g(x)$ as $f(x) + (-1)g(x)$. ■

Proof of the First Fundamental Theorem of Calculus With these results in hand, we are now ready to prove the First Fundamental Theorem of Calculus.

Proof In the sketch of the proof presented earlier, we defined $F(x) = \int_a^x f(t)\, dt$, and we established the fact that

$$F(x + h) - F(x) = \int_x^{x+h} f(t)\, dt$$

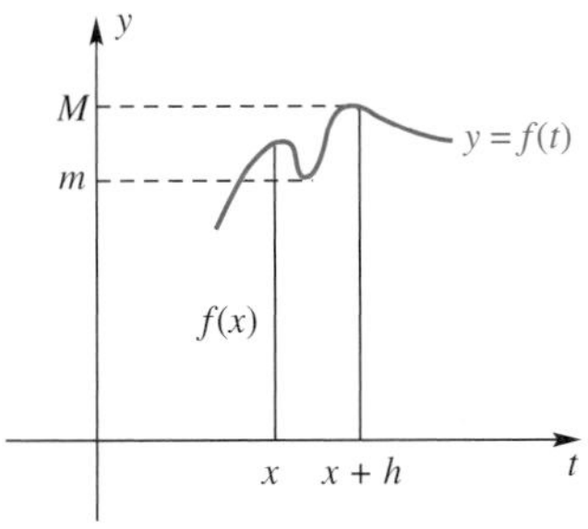

Figure 7

Assume for the moment that $h > 0$ and let m and M be the minimum value and maximum value, respectively, of f on the interval $[x, x + h]$ (Figure 7). By Theorem C,

$$mh \le \int_x^{x+h} f(t)\, dt \le Mh$$

or

$$mh \le F(x + h) - F(x) \le Mh$$

Dividing by h, we obtain

$$m \le \frac{F(x + h) - F(x)}{h} \le M$$

Now m and M really depend on h. Moreover, since f is continuous, both m and M must approach $f(x)$ as $h \to 0$. Thus, by the Squeeze Theorem,

$$\lim_{h\to 0} \frac{F(x + h) - F(x)}{h} = f(x)$$

The case where $h < 0$ is handled similarly. ■

One theoretical consequence of this theorem is that every continuous function f has an antiderivative F given by the accumulation function

$$F(x) = \int_a^x f(t)\, dt$$

However, this fact is not helpful in getting a simple formula for any particular antiderivative. Section 7.6 gives several examples of important functions that are defined as accumulation functions. In Chapter 6 we will define the natural logarithm function as an accumulation function.

EXAMPLE 1 Find $\frac{d}{dx}\left[\int_1^x t^3\,dt\right]$.

SOLUTION By the First Fundamental Theorem of Calculus,

$$\frac{d}{dx}\left[\int_1^x t^3\,dt\right] = x^3$$

■

EXAMPLE 2 Find $\frac{d}{dx}\left[\int_2^x \frac{t^{3/2}}{\sqrt{t^2+17}}\,dt\right]$.

SOLUTION We challenge anyone to do this example by first evaluating the integral. However, by the First Fundamental Theorem of Calculus, it is a trivial problem.

$$\frac{d}{dx}\left[\int_2^x \frac{t^{3/2}}{\sqrt{t^2+17}}\,dt\right] = \frac{x^{3/2}}{\sqrt{x^2+17}}$$

■

EXAMPLE 3 Find $\frac{d}{dx}\left[\int_x^4 \tan^2 u \cos u\,du\right], \frac{\pi}{2} < x < \frac{3\pi}{2}$.

SOLUTION Use of the dummy variable u rather than t should not bother anyone. However, the fact that x is the lower limit, rather than the upper limit, is troublesome. Here is how we handle this difficulty.

$$\begin{aligned}\frac{d}{dx}\left[\int_x^4 \tan^2 u \cos u\,du\right] &= \frac{d}{dx}\left[-\int_4^x \tan^2 u \cos u\,du\right]\\ &= -\frac{d}{dx}\left[\int_4^x \tan^2 u \cos u\,du\right] = -\tan^2 x \cos x\end{aligned}$$

The interchange of the upper and lower limits is allowed if we prefix a negative sign. (Recall that by definition $\int_b^a f(x)\,dx = -\int_a^b f(x)\,dx$.) ■

EXAMPLE 4 Find $D_x\left[\int_1^{x^2} (3t-1)\,dt\right]$ in two ways.

SOLUTION One way to find this derivative is to apply the First Fundamental Theorem of Calculus, although now we have a new complication; the upper limit is x^2 rather than x. This problem is handled by the Chain Rule. We may think of the expression in brackets as

$$\int_1^u (3t-1)\,dt \qquad \text{where } u = x^2$$

By the Chain Rule, the derivative with respect to x of this composite function is

$$D_u\left[\int_1^u (3t-1)\,dt\right]\cdot D_x u = (3u-1)(2x) = (3x^2-1)(2x) = 6x^3 - 2x$$

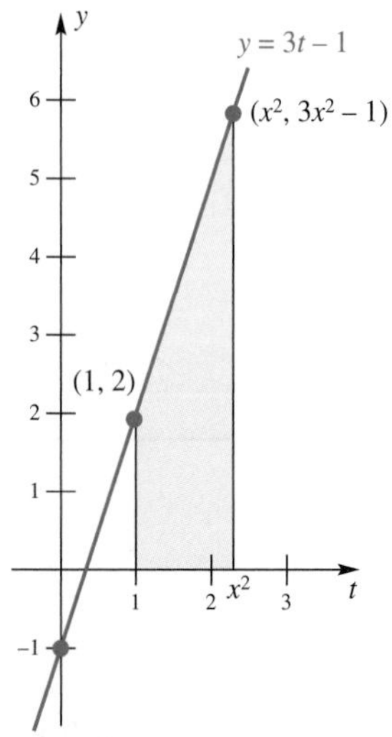

Figure 8

Another way to find this derivative is to evaluate the definite integral first, and then use our rules for derivatives. The definite integral $\int_1^{x^2} (3t-1)\,dt$ is the area below the line $y = 3t - 1$ between $t = 1$ and $t = x^2$ (see Figure 8). Since the area of this trapezoid is $\frac{x^2-1}{2}[2 + (3x^2 - 1)] = \frac{3}{2}x^4 - x^2 - \frac{1}{2}$,

$$\int_1^{x^2} (3t - 1)\, dt = \frac{3}{2}x^4 - x^2 - \frac{1}{2}$$

Thus,

$$D_x \int_1^{x^2} (3t - 1)\, dt = D_x\left(\frac{3}{2}x^4 - x^2 - \frac{1}{2}\right) = 6x^3 - 2x$$

■

Position as Accumulated Velocity In the last section we saw how the position of an object, initially at the origin, is equal to the definite integral of the velocity function. This often leads to accumulation functions, as the next example illustrates.

EXAMPLE 5 An object at the origin at time $t = 0$ has velocity, measured in meters per second,

$$v(t) = \begin{cases} t/20 & \text{if } 0 \le t \le 40 \\ 2 & \text{if } 40 < t \le 60 \\ 5 - t/20 & \text{if } t > 60 \end{cases}$$

When, if ever, does the object return to the origin?

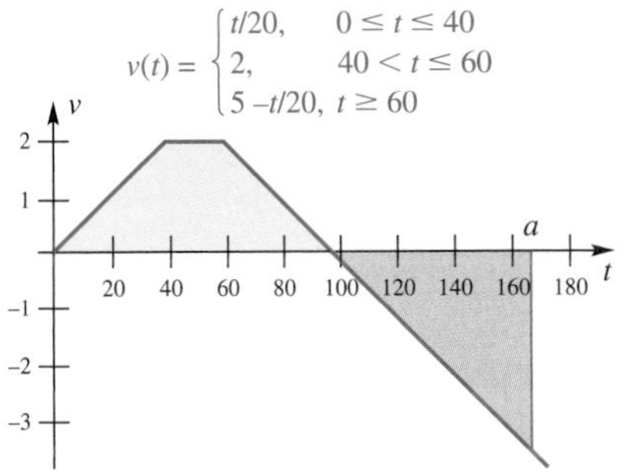

Figure 9

SOLUTION Let $F(a) = \int_0^a v(t)\, dt$ denote the position of the object at time a. The accumulation is illustrated in Figure 9. If the object returns to the origin at some time a, then a must satisfy $F(a) = 0$. The required value of a is certainly greater than 100 because the area below the curve between 0 and 100 must exactly equal the area above the curve and below the x-axis between 100 and a. Therefore,

$$\begin{aligned} F(a) &= \int_0^a v(t)\, dt = \int_0^{100} v(t)\, dt + \int_{100}^a v(t)\, dt \\ &= \frac{1}{2}40 \cdot 2 + 20 \cdot 2 + \frac{1}{2}40 \cdot 2 + \int_{100}^a (5 - t/20)\, dt \\ &= 120 + \frac{1}{2}(a - 100)(5 - a/20) \\ &= -130 + 5a - \frac{1}{40}a^2 \end{aligned}$$

We must then set $F(a) = 0$. The two solutions to this quadratic equation are $a = 100 \pm 40\sqrt{3}$. Taking the minus sign gives a value less than 100, which cannot be the solution so, we discard it. The other solution is $100 + 40\sqrt{3} \approx 169.3$. Let's check this solution:

$$\begin{aligned} F(a) &= \int_0^{100+40\sqrt{3}} v(t)\, dt \\ &= \int_0^{100} v(t)\, dt + \int_{100}^{100+40\sqrt{3}} v(t)\, dt \\ &= 120 + \frac{1}{2}\left(100 + 40\sqrt{3} - 100\right)\left(5 - \left(100 + 40\sqrt{3}\right)/20\right) \\ &= 0 \end{aligned}$$

Thus, the object returns to the origin at time $t = 100 + 40\sqrt{3} \approx 169.3$ seconds.

■

A Way to Evaluate Definite Integrals The next example shows a way (admittedly a rather awkward way) to evaluate a definite integral. If this method seems long and cumbersome, be patient. The next section deals with efficient ways to evaluate definite integrals.

EXAMPLE 6 Let $A(x) = \int_1^x t^3\,dt$.

(a) Let $y = A(x)$ and show that $dy/dx = x^3$.

(b) Find the solution of the differential equation $dy/dx = x^3$ that satisfies $y = 0$ when $x = 1$.

(c) Find $\int_1^4 t^3\,dt$.

SOLUTION

(a) By the First Fundamental Theorem of Calculus,

$$\frac{dy}{dx} = A'(x) = x^3$$

(b) Since the differential equation $dy/dx = x^3$ is separable, we can write

$$dy = x^3\,dx$$

Integrating both sides gives

$$y = \int x^3\,dx = \frac{x^4}{4} + C$$

When $x = 1$, we must have $y = A(1) = \int_1^1 t^3\,dt = 0$. Thus, we choose C so that

$$0 = A(1) = \frac{1^4}{4} + C$$

Therefore, $C = -1/4$. The solution to the differential equation is thus $y = x^4/4 - 1/4$.

(c) Since $y = A(x) = x^4/4 - 1/4$, we have

$$\int_1^4 t^3\,dt = A(4) = \frac{4^4}{4} - \frac{1}{4} = 64 - \frac{1}{4} = \frac{255}{4}$$ ■

Concepts Review

1. Since $4 \le x^2 \le 16$ for all x in $[2, 4]$, the Boundedness Property of the definite integral allows us to say that ____ $\le \int_2^4 x^2\,dx \le$ ____.

2. $\frac{d}{dx}\left[\int_1^x \sin^3 t\,dt\right] =$ ____.

3. By linearity, $\int_1^4 cf(x)\,dx = c \cdot$ ____ and $\int_2^5 (x + \sqrt{x})\,dx = \int_2^5 x\,dx +$ ____.

4. If $\int_1^4 f(x)\,dx = 5$ and if $g(x) \le f(x)$ for all x in $[1, 4]$, then the Comparison Property allows us to say that $\int_1^4 g(x)\,dx \le$ ____.

Problem Set 4.3

In Problems 1–8, find a formula for and graph the accumulation function $A(x)$ that is equal to the indicated area.

1.

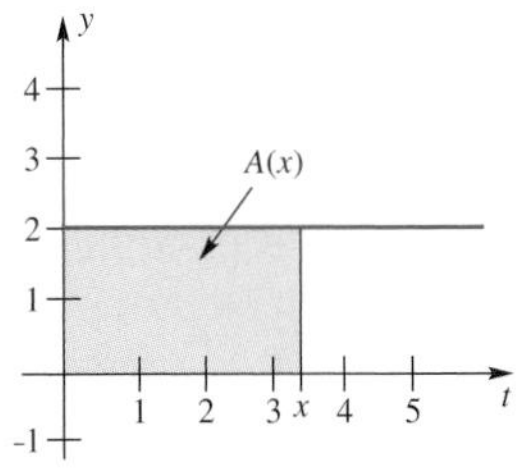

2.

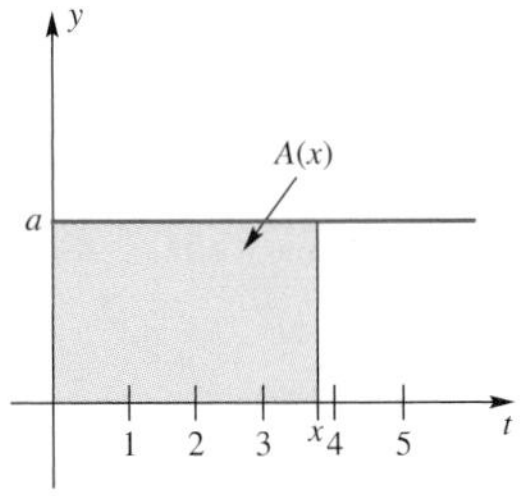

3.

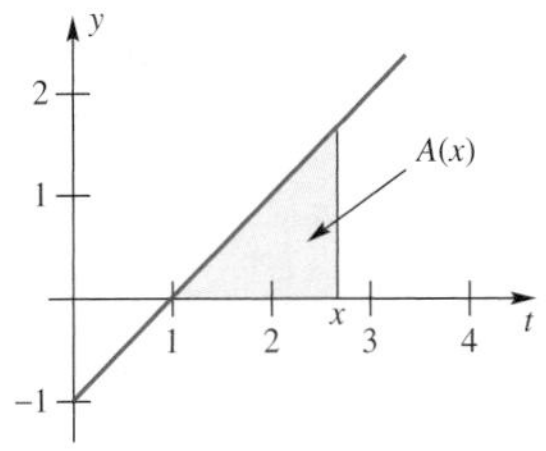

4.

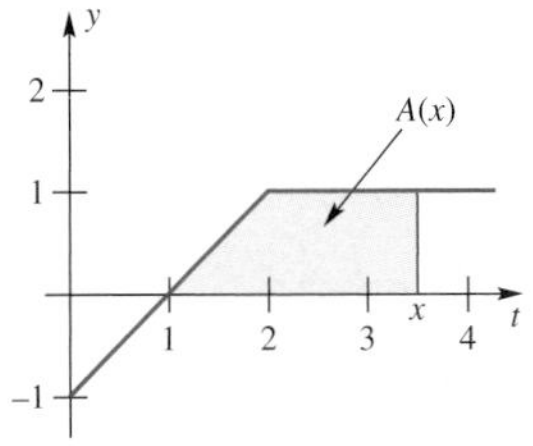

5.

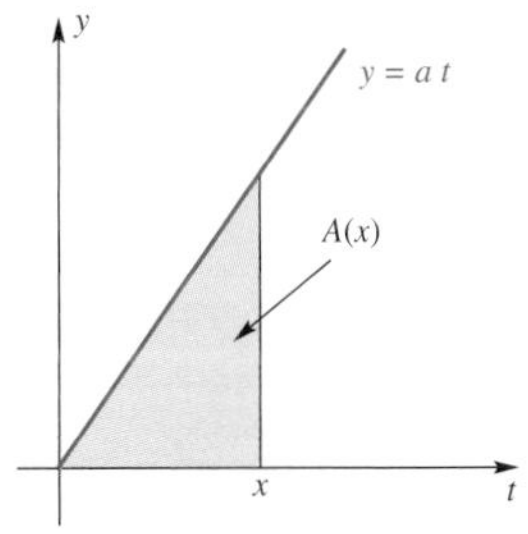

6.

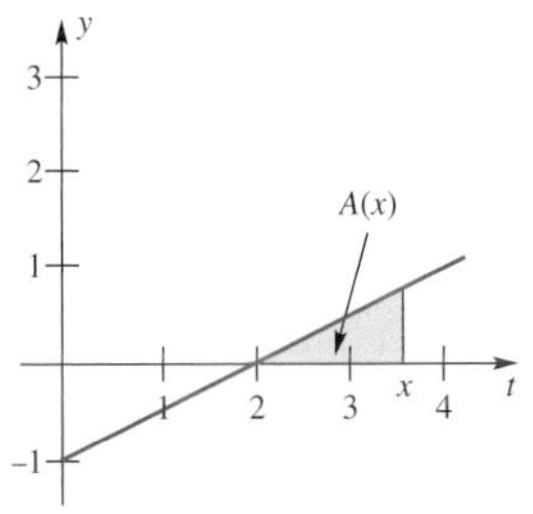

7.

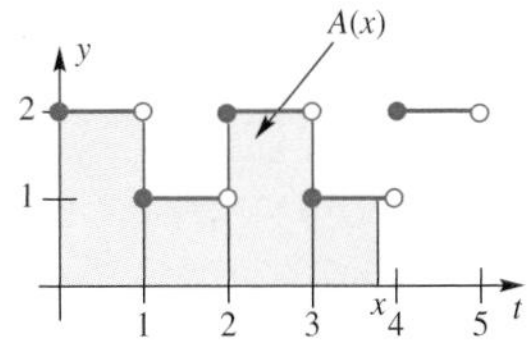

8.

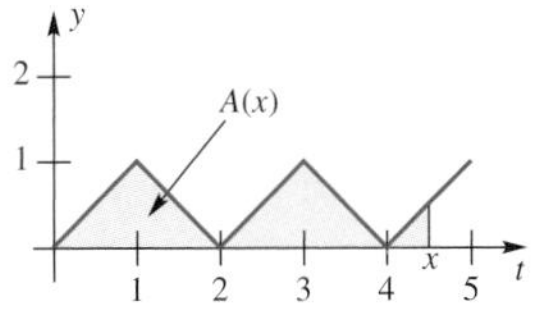

Suppose that $\int_0^1 f(x)\,dx = 2$, $\int_1^2 f(x)\,dx = 3$, $\int_0^1 g(x)\,dx = -1$, *and* $\int_0^2 g(x)\,dx = 4$. *Use properties of definite integrals (linearity, interval additivity, and so on) to calculate each of the integrals in Problems 9–16.*

9. $\int_1^2 2f(x)\,dx$

10. $\int_0^2 2f(x)\,dx$

11. $\int_0^2 [2f(x) + g(x)]\,dx$

12. $\int_0^1 [2f(s) + g(s)]\,ds$

13. $\int_2^1 [2f(s) + 5g(s)]\,ds$

14. $\int_1^1 [3f(x) + 2g(x)]\,dx$

15. $\int_0^2 [3f(t) + 2g(t)]\,dt$

16. $\int_0^2 \left[\sqrt{3}f(t) + \sqrt{2}g(t) + \pi\right] dt$

In Problems 17–26, find $G'(x)$.

17. $G(x) = \int_1^x 2t\,dt$

18. $G(x) = \int_x^1 2t\,dt$

19. $G(x) = \int_0^x \left(2t^2 + \sqrt{t}\right) dt$

20. $G(x) = \int_1^x \cos^3 2t \tan t\,dt;\ -\pi/2 < x < \pi/2$

21. $G(x) = \int_x^{\pi/4} (s - 2) \cot 2s\,ds;\ 0 < x < \pi/2$

22. $G(x) = \int_1^x xt\,dt$ (Be careful.)

23. $G(x) = \int_1^{x^2} \sin t\,dt$

24. $G(x) = \int_1^{x^2+x} \sqrt{2z + \sin z}\,dz$

25. $G(x) = \int_{-x^2}^x \frac{t^2}{1 + t^2}\,dt$ *Hint:* $\int_{-x^2}^x = \int_{-x^2}^0 + \int_0^x$

26. $G(x) = \int_{\cos x}^{\sin x} t^5\,dt$

In Problems 27–32, find the interval(s) on which the graph of $y = f(x)$, $x \geq 0$, is (a) increasing, and (b) concave up.

27. $f(x) = \int_0^x \frac{s}{\sqrt{1 + s^2}}\,ds$

28. $f(x) = \int_0^x \frac{1 + t}{1 + t^2}\,dt$

29. $f(x) = \int_0^x \cos u\,du$

30. $f(x) = \int_0^x (t + \sin t)\,dt$

31. $f(x) = \int_1^x \frac{1}{\theta}\,d\theta$

32. $f(x)$ is the accumulation function $A(x)$ in Problem 8.

In Problems 33–36, use the Interval Additive Property and linearity to evaluate $\int_0^4 f(x)\,dx$. Begin by drawing a graph of f.

33. $f(x) = \begin{cases} 2 & \text{if } 0 \leq x < 2 \\ x & \text{if } 2 \leq x \leq 4 \end{cases}$

34. $f(x) = \begin{cases} 1 & \text{if } 0 \leq x < 1 \\ x & \text{if } 1 \leq x < 2 \\ 4 - x & \text{if } 2 \leq x \leq 4 \end{cases}$

35. $f(x) = |x - 2|$

36. $f(x) = 3 + |x - 3|$

37. Consider the function $G(x) = \int_0^x f(t)\,dt$, where $f(t)$ oscillates about the line $y = 2$ over the x-region $[0, 10]$ and is given by Figure 10.

(a) At what values of x over this region do the local maxima and minima of $G(x)$ occur?

(b) Where does $G(x)$ attain its absolute maximum and absolute minimum?

(c) On what intervals is $G(x)$ concave down?

(d) Sketch a graph of $G(x)$.

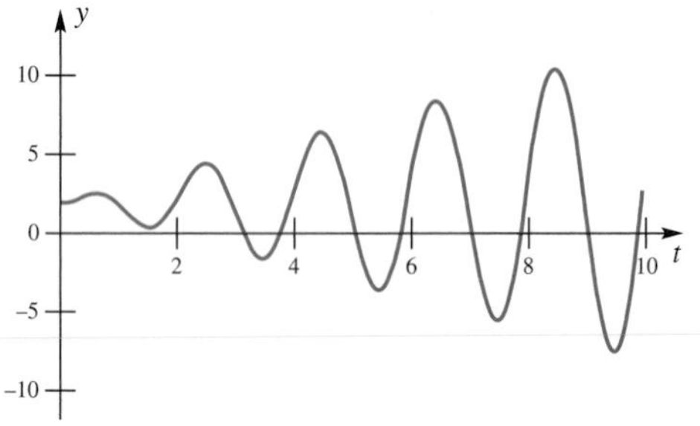

Figure 10

38. Perform the same analysis as you did in Problem 37 for the function $G(x) = \int_0^x f(t)\,dt$ given by Figure 11, where $f(t)$ oscillates about the line $y = 2$ for the interval $[0, 10]$.

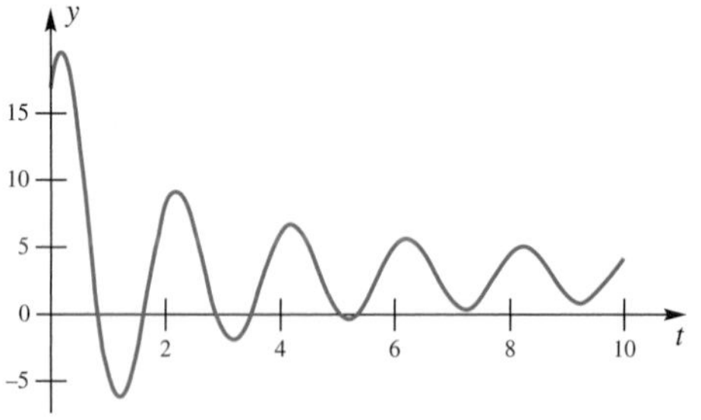

Figure 11

39. Let $F(x) = \int_0^x (t^4 + 1)\,dt$.

(a) Find $F(0)$.

(b) Let $y = F(x)$. Apply the First Fundamental Theorem of Calculus to obtain $dy/dx = F'(x) = x^4 + 1$. Solve the differential equation $dy/dx = x^4 + 1$.

(c) Find the solution to this differential equation that satisfies $y = F(0)$ when $x = 0$.

(d) Show that $\int_0^1 (x^4 + 1)\,dx = \frac{6}{5}$.

40. Let $G(x) = \int_0^x \sin t\,dt$.

(a) Find $G(0)$ and $G(2\pi)$.

(b) Let $y = G(x)$. Apply the First Fundamental Theorem of Calculus to obtain $dy/dx = G'(x) = \sin x$. Solve the differential equation $dy/dx = \sin x$.

(c) Find the solution to this differential equation that satisfies $y = G(0)$ when $x = 0$.

(d) Show that $\int_0^\pi \sin x\,dx = 2$.

(e) Find all relative extrema and inflection points of G on the interval $[0, 4\pi]$.

(f) Plot a graph of $y = G(x)$ over the interval $[0, 4\pi]$.

41. Show that $1 \le \int_0^1 \sqrt{1 + x^4}\,dx \le \frac{6}{5}$. *Hint:* Explain why $1 \le \sqrt{1 + x^4} \le 1 + x^4$ for x in the closed interval $[0, 1]$; then use the Comparison Property (Theorem B) and the result of Problem 39d.

42. Show that $2 \le \int_0^1 \sqrt{4 + x^4} \le \frac{21}{5}$. (See the hint for Problem 41.)

GC *In Problems 43–48, use a graphing calculator to graph each integrand. Then use the Boundedness Property (Theorem C) to find a lower bound and an upper bound for each definite integral.*

43. $\int_0^4 (5 + x^3)\,dx$

44. $\int_2^4 (x + 6)^5\,dx$

45. $\int_1^5 \left(3 + \frac{2}{x}\right) dx$

46. $\int_{10}^{20} \left(1 + \frac{1}{x}\right)^5 dx$

47. $\int_{4\pi}^{8\pi} \left(5 + \frac{1}{20}\sin^2 x\right) dx$

48. $\int_{0.2}^{0.4} (0.002 + 0.0001\cos^2 x)\,dx$

49. Find $\lim_{x\to 0} \frac{1}{x}\int_0^x \frac{1 + t}{2 + t}\,dt$.

50. Find $\lim_{x\to 1} \frac{1}{x - 1}\int_1^x \frac{1 + t}{2 + t}\,dt$.

51. Find $f(x)$ if $\int_1^x f(t)\,dt = 2x - 2$.

52. Find $f(x)$ if $\int_0^x f(t)\,dt = x^2$.

53. Find $f(x)$ if $\int_0^{x^2} f(t)\,dt = \frac{1}{3}x^3$.

54. Does there exist a function f such that $\int_0^x f(t)\,dt = x + 1$? Explain.

In Problems 55–60, decide whether the given statement is true or false. Then justify your answer.

55. If f is continuous and $f(x) \ge 0$ for all x in $[a, b]$, then $\int_a^b f(x)\,dx \ge 0$.

56. If $\int_a^b f(x)\,dx \ge 0$, then $f(x) \ge 0$ for all x in $[a, b]$.

57. If $\int_a^b f(x)\,dx = 0$, then $f(x) = 0$ for all x in $[a, b]$.

58. If $f(x) \ge 0$ and $\int_a^b f(x)\,dx = 0$, then $f(x) = 0$ for all x in $[a, b]$.

59. If $\int_a^b f(x)\,dx > \int_a^b g(x)\,dx$, then

$$\int_a^b [f(x) - g(x)]\,dx > 0$$

60. If f and g are continuous and $f(x) > g(x)$ for all x in $[a, b]$, then $\left|\int_a^b f(x)\,dx\right| > \left|\int_a^b g(x)\,dx\right|$.

61. The velocity of an object is $v(t) = 2 - |t - 2|$. Assuming that the object is at the origin at time 0, find a formula for its position at time t. (*Hint:* You will have to consider separately the intervals $0 \le t \le 2$, and $t > 2$.) When, if ever, does the object return to the origin?

62. The velocity of an object is

$$v(t) = \begin{cases} 5 & \text{if } 0 \le t \le 100 \\ 6 - t/100 & \text{if } 100 < t \le 700 \\ -1 & \text{if } t > 700 \end{cases}$$

(a) Assuming that the object is at the origin at time 0, find a formula for its position at time $t (t \ge 0)$.

(b) What is the farthest to the right of the origin that this object ever gets?

(c) When, if ever, does the object return to the origin?

63. Let f be continuous on $[a, b]$ and thus integrable there. Show that

$$\left|\int_a^b f(x)\,dx\right| \le \int_a^b |f(x)|\,dx$$

Hint: $-|f(x)| \le f(x) \le |f(x)|$; use Theorem B.

64. Suppose that f' is integrable and $|f'(x)| \le M$ for all x. Prove that $|f(x)| \le |f(a)| + M|x - a|$ for every a.

Answers to Concepts Review: **1.** 8; 32 **2.** $\sin^3 x$ **3.** $\int_1^4 f(x)\,dx;\ \int_2^5 \sqrt{x}\,dx$ **4.** 5

4.4 The Second Fundamental Theorem of Calculus and the Method of Substitution

The First Fundamental Theorem of Calculus, given in the previous section, gives the inverse relationship between definite integrals and derivatives. Although it is not yet apparent, this relationship gives us a powerful tool for evaluating definite integrals. This tool is called the Second Fundamental Theorem of Calculus, and we will apply it much more often than the First Fundamental Theorem of Calculus.

Is It Fundamental?

The Second Fundamental Theorem of Calculus is important in providing a powerful tool for evaluating definite integrals. But its deepest significance lies in the link it makes between differentiation and integration, between derivatives and integrals. This link appears in sparkling clarity when we rewrite the conclusion to the theorem with $f(x)$ replaced by $g'(x)$.

$$\int_a^b g'(x)\,dx = g(b) - g(a)$$

Theorem A Second Fundamental Theorem of Calculus

Let f be continuous (hence integrable) on $[a, b]$, and let F be any antiderivative of f on $[a, b]$. Then

$$\int_a^b f(x)\,dx = F(b) - F(a)$$

Proof For x in the interval $[a, b]$, define $G(x) = \int_a^x f(t)\,dt$. Then, by the First Fundamental Theorem of Calculus, $G'(x) = f(x)$ for all x in (a, b). Thus, G is an antiderivative of f; but F is also an antiderivative of f. From Theorem 3.6B, we conclude that since $F'(x) = G'(x)$, the functions F and G differ by a constant. Thus, for all x in (a, b)

$$F(x) = G(x) + C$$

Since the functions F and G are continuous on the closed interval $[a, b]$ (Problem 77), we have $F(a) = G(a) + C$ and $F(b) = G(b) + C$. Thus, $F(x) = G(x) + C$ on the *closed* interval $[a, b]$.

Since $G(a) = \int_a^a f(t)\,dt = 0$, we have

$$F(a) = G(a) + C = 0 + C = C$$

Therefore,

$$F(b) - F(a) = [G(b) + C] - C = G(b) = \int_a^b f(t)\,dt$$

■

In Section 3.8, we defined the *indefinite* integral as an antiderivative. In Section 4.2, we defined the *definite* integral as the limit of a Riemann sum. We used the same word (integral) in both cases, although at the time there seemed to be little in common between the two. Theorem A is fundamental because it shows how indefinite integration (antidifferentiation) and definite integration (signed area) are related. Before going on to examples, ask yourself why we can use the word *any*, in the statement of the theorem.

EXAMPLE 1 Show that $\int_a^b k\,dx = k(b - a)$, where k is a constant.

SOLUTION $F(x) = kx$ is an antiderivative of $f(x) = k$. Thus, by the Second Fundamental Theorem of Calculus,

$$\int_a^b k\,dx = F(b) - F(a) = kb - ka = k(b - a)$$ ■

EXAMPLE 2 Show that $\int_a^b x\,dx = \frac{b^2}{2} - \frac{a^2}{2}$.

SOLUTION $F(x) = x^2/2$ is an antiderivative of $f(x) = x$. Therefore,

$$\int_a^b x\,dx = F(b) - F(a) = \frac{b^2}{2} - \frac{a^2}{2}$$ ■

EXAMPLE 3 Show that if r is a rational number different from -1, then

$$\int_a^b x^r\,dx = \frac{b^{r+1}}{r+1} - \frac{a^{r+1}}{r+1}$$

SOLUTION $F(x) = x^{r+1}/(r + 1)$ is an antiderivative of $f(x) = x^r$. Thus, by the Second Fundamental Theorem of Calculus,

$$\int_a^b x^r\,dx = F(b) - F(a) = \frac{b^{r+1}}{r+1} - \frac{a^{r+1}}{r+1}$$

If $r < 0$, we require that 0 not be in $[a, b]$. Why? ■

It is convenient to introduce a special symbol for $F(b) - F(a)$. We write

$$F(b) - F(a) = \left[F(x)\right]_a^b$$

With this notation,

$$\int_2^5 x^2\,dx = \left[\frac{x^3}{3}\right]_2^5 = \frac{125}{3} - \frac{8}{3} = \frac{117}{3} = 39$$

EXAMPLE 4 Evaluate $\int_{-1}^2 (4x - 6x^2)\,dx$

(a) using the Second Fundamental Theorem of Calculus directly, and
(b) using linearity (Theorem 4.3D) first.

SOLUTION

(a) $$\int_{-1}^2 (4x - 6x^2)\,dx = \left[2x^2 - 2x^3\right]_{-1}^2$$

$$= (8 - 16) - (2 + 2) = -12$$

(b) Using linearity first, we have

$$\begin{aligned}\int_{-1}^{2}(4x - 6x^2)\,dx &= 4\int_{-1}^{2} x\,dx - 6\int_{-1}^{2} x^2\,dx \\ &= 4\left[\frac{x^2}{2}\right]_{-1}^{2} - 6\left[\frac{x^3}{3}\right]_{-1}^{2} \\ &= 4\left(\frac{4}{2} - \frac{1}{2}\right) - 6\left(\frac{8}{3} + \frac{1}{3}\right) \\ &= -12\end{aligned}$$

■

EXAMPLE 5 Evaluate $\int_{1}^{8}(x^{1/3} + x^{4/3})\,dx$.

SOLUTION

$$\begin{aligned}\int_{1}^{8}(x^{1/3} + x^{4/3})\,dx &= \left[\tfrac{3}{4}x^{4/3} + \tfrac{3}{7}x^{7/3}\right]_{1}^{8} \\ &= \left(\tfrac{3}{4}\cdot 16 + \tfrac{3}{7}\cdot 128\right) - \left(\tfrac{3}{4}\cdot 1 + \tfrac{3}{7}\cdot 1\right) \\ &= \tfrac{45}{4} + \tfrac{381}{7} \approx 65.68\end{aligned}$$

■

EXAMPLE 6 Find $D_x \int_{0}^{x} 3\sin t\,dt$ in two ways.

SOLUTION The easy way is to apply the First Fundamental Theorem of Calculus.

$$D_x \int_{0}^{x} 3\sin t\,dt = 3\sin x$$

A second way to do this problem is to apply the Second Fundamental Theorem of Calculus to evaluate the integral from 0 to x; then apply the rules of derivatives.

$$\int_{0}^{x} 3\sin t\,dt = [-3\cos t]_{0}^{x} = -3\cos x - (-3\cos 0) = -3\cos x + 3$$

Then

$$D_x \int_{0}^{x} 3\sin t\,dt = D_x(-3\cos x + 3) = 3\sin x$$

■

In terms of the symbol for the indefinite integral, we may write the conclusion of the Second Fundamental Theorem of Calculus as

$$\int_{a}^{b} f(x)\,dx = \left[\int f(x)\,dx\right]_{a}^{b}$$

The nontrivial part of applying the theorem is always to find the indefinite integral $\int f(x)\,dx$. One of the most powerful techniques for doing this is the method of substitution

The Method of Substitution In Section 3.8, we introduced the method of substitution for the power rule. This rule can be extended to a more general case as the following theorem shows. An astute reader will see that the substitution rule is nothing more than the Chain Rule in reverse.

Using the Second Fundamental Theorem of Calculus

The way to use the Second Fundamental Theorem of Calculus to evaluate a definite integral such as $\int_a^b f(x)\,dx$, is to

(1) find an antiderivative $F(x)$ of the integrand $f(x)$, and
(2) substitute the limits and compute $F(b) - F(a)$

This all hinges on being able to find an antiderivative. It is for this reason that we return briefly to the evaluation of *indefinite* integrals.

Theorem B Substitution Rule for Indefinite Integrals

Let g be a differentiable function and suppose that F is an antiderivative of f. Then

$$\int f(g(x))g'(x)\,dx = F(g(x)) + C$$

Proof All we need to do to prove this result is to show that the derivative of the right side is equal to the integrand of the integral on the left. This is a simple application of the Chain Rule.

$$D_x[F(g(x)) + C] = F'(g(x))g'(x) = f(g(x))g'(x) \qquad \blacksquare$$

We normally apply Theorem B as follows. In an integral such as $\int f(g(x))g'(x)\,dx$ we let $u = g(x)$, so that $du/dx = g'(x)$. Thus, $du = g'(x)\,dx$. The integral then becomes

$$\int f(\underbrace{g(x)}_{u})\underbrace{g'(x)\,dx}_{du} = \int f(u)\,du = F(u) + C = F(g(x)) + C$$

Thus, if we can find an antiderivative for $f(x)$, we can evaluate $\int f(g(x))g'(x)\,dx$. The trick to applying the method of substitution is to choose the right substitution to make. In some cases this substitution is obvious; in other cases it is not so obvious. Proficiency in applying the method of substitution comes from practice.

EXAMPLE 7 Evaluate $\int \sin 3x\,dx$.

SOLUTION The obvious substitution here is $u = 3x$, so that $du = 3\,dx$. Thus

$$\int \sin 3x\,dx = \int \frac{1}{3}\sin(\underbrace{3x}_{u})\underbrace{3\,dx}_{du}$$

$$= \frac{1}{3}\int \sin u\,du = -\frac{1}{3}\cos u + C = -\frac{1}{3}\cos 3x + C$$

Notice how we had to multiply by $\frac{1}{3}\cdot 3$ in order to have the expression $3\,dx = du$ in the integral. ■

EXAMPLE 8 Evaluate $\int x \sin x^2\,dx$.

SOLUTION Here the appropriate substitution is $u = x^2$. This gives us $\sin x^2 = \sin u$ in the integrand, but more importantly, the extra x in the integrand can be put with the differential, because $du = 2x\,dx$. Thus

$$\int x\sin x^2\,dx = \int \frac{1}{2}\sin(\underbrace{x^2}_{u})\underbrace{2x\,dx}_{du}$$

$$= \frac{1}{2}\int \sin u\,du = -\frac{1}{2}\cos u + C = -\frac{1}{2}\cos x^2 + C \qquad ■$$

No law says that you have to write out the u-substitution. If you can do the substitution mentally, that is fine. Here is an illustration.

EXAMPLE 9 Evaluate $\int x^3\sqrt{x^4 + 11}\, dx$.

SOLUTION Mentally, substitute $u = x^4 + 11$.

$$\int x^3\sqrt{x^4+11}\,dx = \frac{1}{4}\int (x^4+11)^{1/2}(4x^3\,dx)$$

$$= \frac{1}{6}(x^4+11)^{3/2} + C \qquad \blacksquare$$

What Makes This Substitution Work?

Note that in Example 10 the derivative of u is precisely $2x + 1$. This is what makes the substition work. If the expression in parentheses were $3x + 1$ rather than $2x + 1$, the Substitution Rule would not apply and we would have a much more difficult problem.

EXAMPLE 10 Evaluate $\int_0^4 \sqrt{x^2 + x}\,(2x + 1)\, dx$.

SOLUTION Let $u = x^2 + x$; then $du = (2x + 1)\,dx$. Thus,

$$\int \underbrace{\sqrt{x^2+x}}_{u}\,\underbrace{(2x+1)\,dx}_{du} = \int u^{1/2}\,du = \tfrac{2}{3}u^{3/2} + C$$

$$= \tfrac{2}{3}(x^2+x)^{3/2} + C$$

Therefore, by the Second Fundamental Theorem of Calculus,

$$\int_0^4 \sqrt{x^2+x}\,(2x+1)\,dx = \left[\tfrac{2}{3}(x^2+x)^{3/2} + C\right]_0^4$$

$$= \left[\tfrac{2}{3}(20)^{3/2} + C\right] - [0 + C]$$

$$= \tfrac{2}{3}(20)^{3/2} \approx 59.63 \qquad \blacksquare$$

Note that the C of the indefinite integration cancels out, as it always will, in the definite integration. That is why in the statement of the Second Fundamental Theorem we could use the phrase *any antiderivative*. In particular, we may always choose $C = 0$ in applying the Second Fundamental Theorem.

EXAMPLE 11 Evaluate $\int_0^{\pi/4} \sin^3 2x \cos 2x\, dx$.

SOLUTION Let $u = \sin 2x$; then $du = 2 \cos 2x\, dx$. Thus,

$$\int \sin^3 2x \cos 2x\,dx = \frac{1}{2}\int \underbrace{(\sin 2x)}_{u}{}^3\underbrace{(2\cos 2x)\,dx}_{du} = \frac{1}{2}\int u^3\,du$$

$$= \frac{1}{2}\frac{u^4}{4} + C = \frac{\sin^4 2x}{8} + C$$

Therefore, by the Second Fundamental Theorem of Calculus,

$$\int_0^{\pi/4} \sin^3 2x\cos 2x\,dx = \left[\frac{\sin^4 2x}{8}\right]_0^{\pi/4} = \frac{1}{8} - 0 = \frac{1}{8} \qquad \blacksquare$$

Note that in the two-step procedure illustrated in Examples 10 and 11, we must be sure to express the indefinite integral in terms of x before we apply the Second Fundamental Theorem. This is because the limits, 0 and 4 in Example 10, and 0 and $\pi/4$ in Example 11, apply to x, not to u. But what if, in making the substitution $u = \sin 2x$ in Example 11, we also made the corresponding changes in the limits of integration to u?

If $x = 0$, then $u = \sin(2\cdot 0) = 0$.

If $x = \pi/4$, then $u = \sin(2(\pi/4)) = \sin(\pi/2) = 1$.

Could we then finish the integration with the definite integral in terms of u? The answer is *yes*.

$$\int_0^{\pi/4} \sin^3 2x \cos 2x\, dx = \left[\frac{1}{2}\frac{u^4}{4}\right]_0^1 = \frac{1}{8} - 0 = \frac{1}{8}$$

Here is the general result, which lets us substitute the limits of integration, thereby producing a procedure with fewer steps.

Substitution in Definite Integrals

To make a substitution in a definite integral, three changes are required:

1. Make the substitution in the integrand.
2. Make the appropriate change in the differential.
3. Change the limits from a and b to $g(a)$ and $g(b)$.

Theorem C Substitution Rule for Definite Integrals

Let g have a continuous derivative on $[a, b]$, and let f be continuous on the range of g. Then

$$\int_a^b f(g(x))g'(x)\, dx = \int_{g(a)}^{g(b)} f(u)\, du$$

where $u = g(x)$.

Proof Let F be an antiderivative of f (the existence of F is guaranteed by Theorem 4.3A). Then, by the Second Fundamental Theorem of Calculus,

$$\int_{g(a)}^{g(b)} f(u)\, du = \left[F(u)\right]_{g(a)}^{g(b)} = F(g(b)) - F(g(a))$$

On the other hand, by the Substitution Rule for Indefinite Integrals (Theorem B),

$$\int f(g(x))g'(x)\, dx = F(g(x)) + C$$

and so, again by the Second Fundamental Theorem of Calculus,

$$\int_a^b f(g(x))g'(x)\, dx = \left[F(g(x))\right]_a^b = F(g(b)) - F(g(a))$$ ■

EXAMPLE 12 Evaluate $\displaystyle\int_0^1 \frac{x+1}{(x^2+2x+6)^2}\, dx$.

SOLUTION Let $u = x^2 + 2x + 6$, so $du = (2x+2)\, dx = 2(x+1)\, dx$, and note that $u = 6$ when $x = 0$ and $u = 9$ when $x = 1$. Thus,

$$\begin{aligned}\int_0^1 \frac{x+1}{(x^2+2x+6)^2}\, dx &= \frac{1}{2}\int_0^1 \frac{2(x+1)}{(x^2+2x+6)^2}\, dx \\ &= \frac{1}{2}\int_6^9 u^{-2}\, du = \left[-\frac{1}{2}\frac{1}{u}\right]_6^9 \\ &= -\frac{1}{18} - \left(-\frac{1}{12}\right) = \frac{1}{36}\end{aligned}$$ ■

EXAMPLE 13 Evaluate $\displaystyle\int_{\pi^2/9}^{\pi^2/4} \frac{\cos\sqrt{x}}{\sqrt{x}}\, dx$.

SOLUTION Let $u = \sqrt{x}$, so $du = dx/(2\sqrt{x})$. Thus,

$$\begin{aligned}\int_{\pi^2/9}^{\pi^2/4} \frac{\cos\sqrt{x}}{\sqrt{x}}\, dx &= 2\int_{\pi^2/9}^{\pi^2/4} \cos\sqrt{x}\cdot\frac{1}{2\sqrt{x}}\, dx \\ &= 2\int_{\pi/3}^{\pi/2} \cos u\, du \\ &= \left[2\sin u\right]_{\pi/3}^{\pi/2} = 2 - \sqrt{3}\end{aligned}$$

The change in the limits of integration occurred at the second equality. When $x = \pi^2/9$, $u = \sqrt{\pi^2/9} = \pi/3$; when $x = \pi^2/4$, $u = \pi/2$. ■

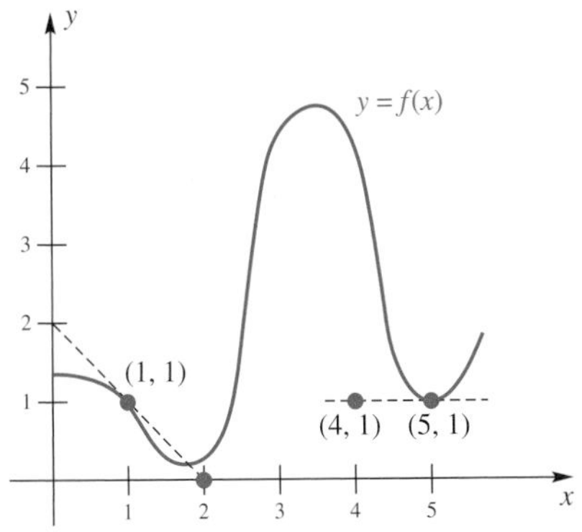

Figure 1

EXAMPLE 14 Figure 1 shows the graph of a function f that has a continuous third derivative. The dashed lines are tangent to the graph of $y = f(x)$ at $(1, 1)$ and $(5, 1)$. Based on what is shown, tell, if possible, whether the following integrals are positive, negative, or zero.

(a) $\int_1^5 f(x)\,dx$ (b) $\int_1^5 f'(x)\,dx$

(c) $\int_1^5 f''(x)\,dx$ (d) $\int_1^5 f'''(x)\,dx$

SOLUTION

(a) The function f is positive for all x in the interval $[1, 5]$, and the graph indicates that there is some area above the x-axis. Thus, $\int_1^5 f(x)\,dx > 0$.

(b) By the Second Fundamental Theorem of Calculus,

$$\int_1^5 f'(x)\,dx = f(5) - f(1) = 1 - 1 = 0$$

(c) Again using the Second Fundamental Theorem of Calculus (this time with f' being an antiderivative of f''), we see that

$$\int_1^5 f''(x)\,dx = f'(5) - f'(1) = 0 - (-1) = 1$$

(d) The function f is concave up at $x = 5$, so $f''(5) > 0$, and it is concave down at $x = 1$, so $f''(1) < 0$. Thus,

$$\int_1^5 f'''(x)\,dx = f''(5) - f''(1) > 0$$

■

This example illustrates the remarkable property that to evaluate a definite integral all we need to know are the values of an antiderivative at the end points a and b. For example, to evaluate $\int_1^5 f''(x)\,dx$, all we needed to know was $f'(5)$ and $f'(1)$; we did not need to know f' or f'' at any of the points in the open interval (a, b).

Accumulated Rate of Change The Second Fundamental Theorem of Calculus can be restated in this way:

$$\int_a^b F'(t)\,dt = F(b) - F(a)$$

If $F(t)$ measures the amount of some quantity at time t, then the Second Fundamental Theorem of Calculus says that the accumulated rate of change from time $t = a$ to time $t = b$ is equal to the net change in that quantity over the interval $[a, b]$, that is, the amount present at time $t = b$ minus the amount present at time $t = a$.

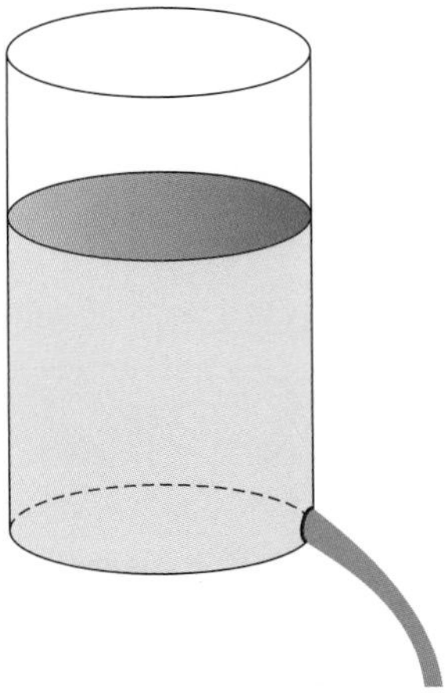
Figure 2

EXAMPLE 15 Water leaks out of a 55-gallon tank at the rate $V'(t) = 11 - 1.1t$ where t is measured in hours and V in gallons. (See Figure 2.) Initially, the tank is full. (a) How much water leaks out of the tank between $t = 3$ and $t = 5$ hours? (b) How long does it take until there are just 5 gallons remaining in the tank?

SOLUTION $V(t)$ represents the amount of water that has leaked out through time t.

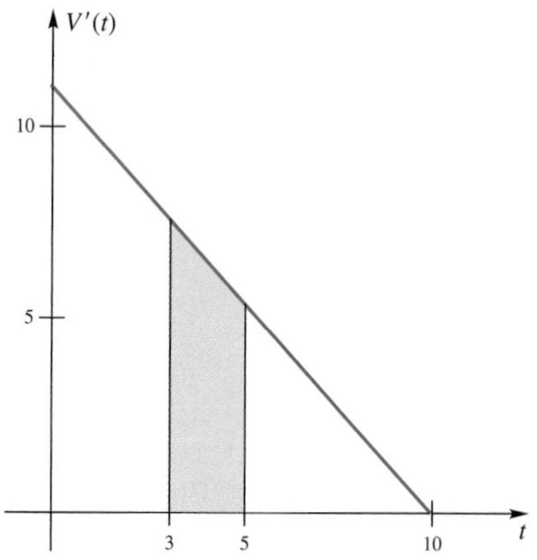

Figure 3

(a) The amount that has leaked out between $t = 3$ and $t = 5$ hours is equal to the area under the $V'(t)$ curve from 3 to 5 (Figure 3). Thus

$$V(5) - V(3) = \int_3^5 V'(t)\,dt = \int_3^5 (11 - 1.1t)\,dt = \left[11t - \frac{1.1}{2}t^2\right]_3^5 = 13.2$$

Thus, 13.2 gallons leaked in the two hours between time $t = 3$ and $t = 5$.

(b) Let t_1 denote the time when 5 gallons remain in the tank. Then the amount that has leaked out is equal to 50, so $V(t_1) = 50$. Since the tank was initially full (i.e., nothing has leaked out), we have $V(0) = 0$. Thus,

$$V(t_1) - V(0) = \int_0^{t_1} (11 - 1.1t)\,dt$$

$$50 - 0 = \left[11t - \frac{1.1}{2}t^2\right]_0^{t_1}$$

$$0 = -50 + 11t_1 - 0.55t_1^2$$

The solutions of this last equation are $10\left(11 \pm \sqrt{11}\right)/11$, approximately 6.985 and 13.015. Note that since $\int_0^{10}(11 - 1.1t)\,dt = 55$, the entire tank is drained by time $t = 10$, leading us to discard the latter solution. Thus, 5 gallons remain after 6.985 hours. ■

Concepts Review

1. If f is continuous on $[a, b]$ and if F is any ____ of f there, then $\int_a^b f(x)\,dx =$ ____.

2. The symbol $\left[F(x)\right]_a^b$ stands for the expression ____.

3. By the Second Fundamental Theorem of Calculus, $\int_c^d F'(x)\,dx =$ ____.

4. Under the substitution $u = x^3 + 1$, the definite integral $\int_0^1 x^2(x^3 + 1)^4\,dx$ transforms to the new definite integral ____.

Problem Set 4.4

In Problems 1–14, use the Second Fundamental Theorem of Calculus to evaluate each definite integral.

1. $\int_0^2 x^3\,dx$
2. $\int_{-1}^2 x^4\,dx$
3. $\int_{-1}^2 (3x^2 - 2x + 3)\,dx$
4. $\int_1^2 (4x^3 + 7)\,dx$
5. $\int_1^4 \frac{1}{w^2}\,dw$
6. $\int_1^3 \frac{2}{t^3}\,dt$
7. $\int_0^4 \sqrt{t}\,dt$
8. $\int_1^8 \sqrt[3]{w}\,dw$
9. $\int_{-4}^{-2} \left(y^2 + \frac{1}{y^3}\right) dy$
10. $\int_1^4 \frac{s^4 - 8}{s^2}\,ds$
11. $\int_0^{\pi/2} \cos x\,dx$
12. $\int_{\pi/6}^{\pi/2} 2\sin t\,dt$
13. $\int_0^1 (2x^4 - 3x^2 + 5)\,dx$
14. $\int_0^1 (x^{4/3} - 2x^{1/3})\,dx$

In Problems 15–34, use the method of substitution to find each of the following indefinite integrals.

15. $\int \sqrt{3x + 2}\,dx$
16. $\int \sqrt[3]{2x - 4}\,dx$
17. $\int \cos(3x + 2)\,dx$
18. $\int \sin(2x - 4)\,dx$
19. $\int \sin(6x - 7)\,dx$
20. $\int \cos\left(\pi v - \sqrt{7}\right) dv$
21. $\int x\sqrt{x^2 + 4}\,dx$
22. $\int x^2(x^3 + 5)^9\,dx$
23. $\int x(x^2 + 3)^{-12/7}\,dx$
24. $\int v\left(\sqrt{3}v^2 + \pi\right)^{7/8} dv$
25. $\int x\sin(x^2 + 4)\,dx$
26. $\int x^2\cos(x^3 + 5)\,dx$
27. $\int \frac{x\sin\sqrt{x^2 + 4}}{\sqrt{x^2 + 4}}\,dx$
28. $\int \frac{z\cos\left(\sqrt[3]{z^2 + 3}\right)}{\left(\sqrt[3]{z^2 + 3}\right)^2}\,dz$

29. $\int x^2(x^3 + 5)^8 \cos[(x^3 + 5)^9]\, dx$

30. $\int x^6(7x^7 + \pi)^8 \sin[(7x^7 + \pi)^9]\, dx$

31. $\int x \cos(x^2 + 4)\sqrt{\sin(x^2 + 4)}\, dx$

32. $\int x^6 \sin(3x^7 + 9)\sqrt[3]{\cos(3x^7 + 9)}\, dx$

33. $\int x^2 \sin(x^3 + 5) \cos^9(x^3 + 5)\, dx$

34. $\int x^{-4} \sec^2(x^{-3} + 1)\sqrt[5]{\tan(x^{-3} + 1)}\, dx$

Hint: $D_x \tan x = \sec^2 x$

In Problems 35–58, use the Substitution Rule for Definite Integrals to evaluate each definite integral.

35. $\int_0^1 (x^2 + 1)^{10}(2x)\, dx$

36. $\int_{-1}^0 \sqrt{x^3 + 1}(3x^2)\, dx$

37. $\int_{-1}^3 \frac{1}{(t + 2)^2}\, dt$

38. $\int_2^{10} \sqrt{y - 1}\, dy$

39. $\int_5^8 \sqrt{3x + 1}\, dx$

40. $\int_1^7 \frac{1}{\sqrt{2x + 2}}\, dx$

41. $\int_{-3}^3 \sqrt{7 + 2t^2}\,(8t)\, dt$

42. $\int_1^3 \frac{x^2 + 1}{\sqrt{x^3 + 3x}}\, dx$

43. $\int_0^{\pi/2} \cos^2 x \sin x\, dx$

44. $\int_0^{\pi/2} \sin^2 3x \cos 3x\, dx$

45. $\int_0^1 (x + 1)(x^2 + 2x)^2\, dx$

46. $\int_1^4 \frac{(\sqrt{x} - 1)^3}{\sqrt{x}}\, dx$

47. $\int_0^{\pi/6} \sin^3 \theta \cos \theta\, d\theta$

48. $\int_0^{\pi/6} \frac{\sin \theta}{\cos^3 \theta}\, d\theta$

49. $\int_0^1 \cos(3x - 3)\, dx$

50. $\int_0^{1/2} \sin(2\pi x)\, dx$

51. $\int_0^1 x \sin(\pi x^2)\, dx$

52. $\int_0^{\pi} x^4 \cos(2x^5)\, dx$

53. $\int_0^{\pi/4} (\cos 2x + \sin 2x)\, dx$

54. $\int_{-\pi/2}^{\pi/2} (\cos 3x + \sin 5x)\, dx$

55. $\int_0^{\pi/2} \sin x \sin(\cos x)\, dx$

56. $\int_{-\pi/2}^{\pi/2} \cos \theta \cos(\pi \sin \theta)\, d\theta$

57. $\int_0^1 x \cos^3(x^2) \sin(x^2)\, dx$

58. $\int_{-\pi/2}^{\pi/2} x^2 \sin^2(x^3) \cos(x^3)\, dx$

59. Figure 4 shows the graph of a function f that has a continuous third derivative. The dashed lines are tangent to the graph of $y = f(x)$ at the points $(0, 2)$ and $(3, 0)$. Based on what is shown, tell, if possible, whether the following integrals are positive, negative, or zero.

(a) $\int_0^3 f(x)\, dx$ (b) $\int_0^3 f'(x)\, dx$

(c) $\int_0^3 f''(x)\, dx$ (d) $\int_0^3 f'''(x)\, dx$

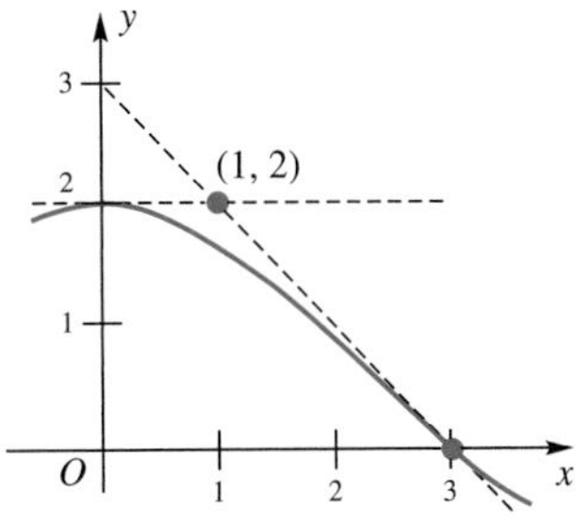

Figure 4

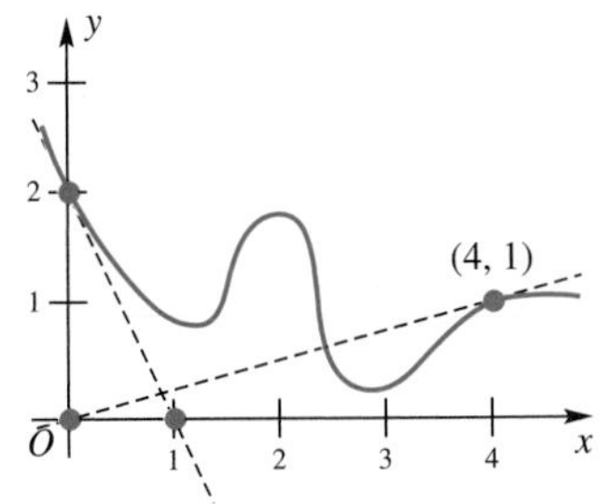

Figure 5

60. Figure 5 shows the graph of a function f that has a continuous third derivative. The dashed lines are tangent to the graph of $y = f(x)$ at the points $(0, 2)$ and $(4, 1)$. Based on what is shown, tell, if possible, whether the following integrals are positive, negative, or zero.

(a) $\int_0^4 f(x)\, dx$ (b) $\int_0^4 f'(x)\, dx$

(c) $\int_0^4 f''(x)\, dx$ (d) $\int_0^4 f'''(x)\, dx$

61. Water leaks out of a 200-gallon storage tank (initially full) at the rate $V'(t) = 20 - t$, where t is measured in hours and V in gallons. How much water leaked out between 10 and 20 hours? How long will it take the tank to drain completely?

62. Oil is leaking at the rate of $V'(t) = 1 - t/110$ from a storage tank that is initially full of 55 gallons. How much leaks out during the first hour? During the tenth hour? How long until the entire tank is drained?

63. The water usage in a small town is measured in gallons per hour. A plot of this rate of usage is shown in Figure 6 for the hours midnight through noon for a particular day. Estimate the total amount of water used during this 12-hour period.

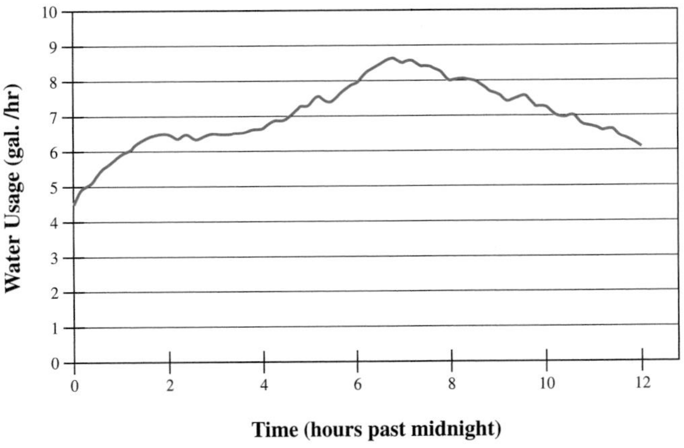

Figure 6

64. Figure 7 shows the rate of oil consumption in million barrels per year for the United States from 1973 to 2003. Approximately how many barrels of oil were consumed between 1990 and 2000?

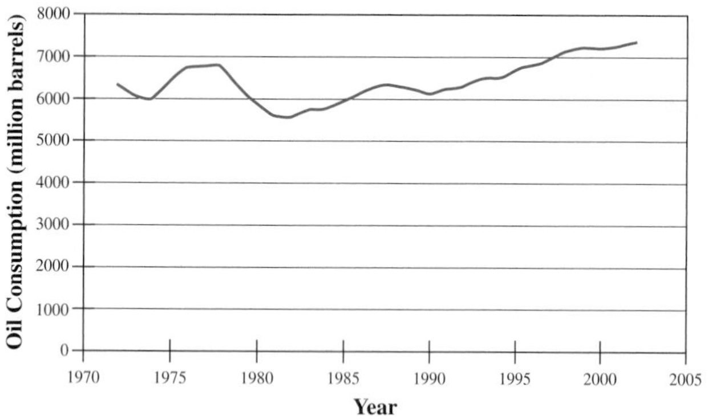

Figure 7

65. Figure 8 shows the power usage, measured in megawatts, for a small town for one day (measured from midnight to midnight). Estimate the energy usage for the day measured in megawatt-hours. *Hint:* Power is the derivative of energy.

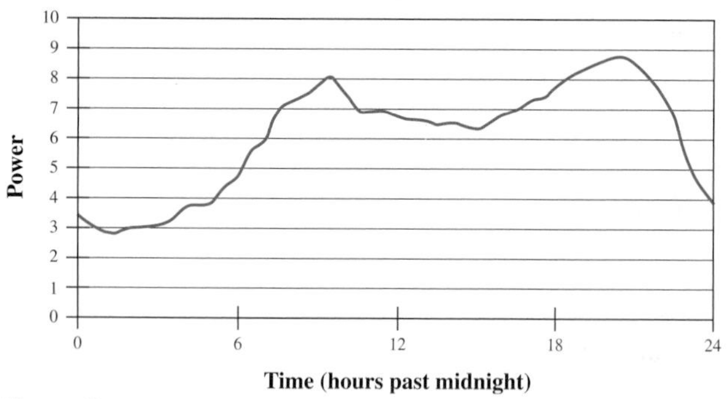

Figure 8

66. The mass, in kilograms, of a rod measured from the left endpoint to the point x meters away is $m(x) = x + x^2/8$. What is the density $\delta(x)$ of the rod, measured in kilograms per meter? Assuming that the rod is 2 meters long, express the total mass of the rod in terms of its density.

67. We claim that

$$\int_a^b x^n\,dx + \int_{a^n}^{b^n} \sqrt[n]{y}\,dy = b^{n+1} - a^{n+1}$$

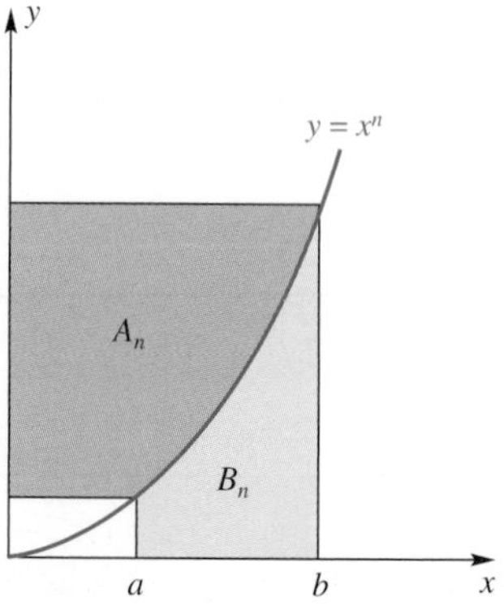

Figure 9

(a) Use Figure 9 to justify this by a geometric argument.

(b) Prove the result using the Second Fundamental Theorem of Calculus.

(c) Show that $A_n = nB_n$.

68. Prove the Second Fundamental Theorem of Calculus following the method suggested in Example 6 of Section 4.3.

In Problems 69–72, first recognize the given limit as a definite integral and then evaluate that integral by the Second Fundamental Theorem of Calculus.

69. $\lim_{n\to\infty} \sum_{i=1}^{n} \left(\frac{3i}{n}\right)^2 \frac{3}{n}$

70. $\lim_{n\to\infty} \sum_{i=1}^{n} \left(\frac{2i}{n}\right)^3 \frac{2}{n}$

71. $\lim_{n\to\infty} \sum_{i=1}^{n} \left[\sin\left(\frac{\pi i}{n}\right)\right] \frac{\pi}{n}$

72. $\lim_{n\to\infty} \sum_{i=1}^{n} \left[1 + \frac{2i}{n} + \left(\frac{2i}{n}\right)^2\right] \frac{2}{n}$

C **73.** Explain why $(1/n^3)\sum_{i=1}^{n} i^2$ should be a good approximation to $\int_0^1 x^2\,dx$ for large n. Now calculate the summation expression for $n = 10$, and evaluate the integral by the Second Fundamental Theorem of Calculus. Compare their values.

74. Evaluate $\int_{-2}^{4} (2[\![x]\!] - 3|x|)\,dx$.

75. Show that $\frac{1}{2}x|x|$ is an antiderivative of $|x|$, and use this fact to get a simple formula for $\int_a^b |x|\,dx$.

76. Find a nice formula for $\int_0^b [\![x]\!]\,dx, b > 0$.

77. Suppose that f is continuous on $[a, b]$.

(a) Let $G(x) = \int_a^x f(t)\,dt$. Show that G is continuous on $[a, b]$.

(b) Let $F(x)$ be any antiderivative of f on $[a, b]$. Show that F is continuous on $[a, b]$.

78. Give an example to show that the accumulation function $G(x) = \int_a^x f(x)\,dx$ can be continuous even if f is not continuous.

Answers to Concepts Review: **1.** antiderivative; $F(b) - F(a)$ **2.** $F(b) - F(a)$ **3.** $F(d) - F(c)$ **4.** $\int_1^2 \frac{1}{3}u^4\,du$

4.5 The Mean Value Theorem for Integrals and the Use of Symmetry

We know what is meant by the average of a set of n numbers, $y_1, y_2, \ldots, y_n$; we simply add them up and divide by n

$$\bar{y} = \frac{y_1 + y_2 + \cdots + y_n}{n}$$

Can we give meaning to the concept of the average of a function f over an interval $[a, b]$? Well, suppose we take a regular partition of $[a, b]$, say $P: a = x_0 < x_1 < x_2 < \cdots < x_{n-1} < x_n = b$, with $\Delta x = (b - a)/n$. The average of the n values $f(x_1), f(x_2), \ldots, f(x_n)$ is

$$\begin{aligned}\frac{f(x_1) + f(x_2) + \cdots + f(x_n)}{n} &= \frac{1}{n}\sum_{i=1}^{n} f(x_i) \\ &= \sum_{i=1}^{n} f(x_i)\frac{b-a}{n}\frac{1}{b-a} \\ &= \frac{1}{b-a}\sum_{i=1}^{n} f(x_i)\,\Delta x\end{aligned}$$

This last sum is a Riemann sum for f on $[a, b]$ and therefore

$$\begin{aligned}\lim_{n\to\infty}\frac{f(x_1) + f(x_2) + \cdots + f(x_n)}{n} &= \frac{1}{b-a}\lim_{n\to\infty}\sum_{i=1}^{n} f(x_i)\,\Delta x \\ &= \frac{1}{b-a}\int_a^b f(x)\,dx\end{aligned}$$

This suggests the following definition.

Definition **Average Value of a Function**

If f is integrable on the interval $[a, b]$, then the **average value** of f on $[a, b]$ is

$$\frac{1}{b-a}\int_a^b f(x)\,dx$$

EXAMPLE 1 Find the average value of the function defined by $f(x) = x \sin x^2$ on the interval $\left[0, \sqrt{\pi}\right]$. (See Figure 1.)

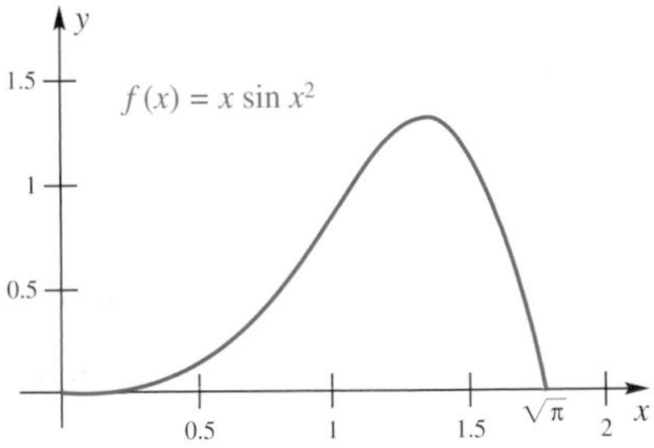

Figure 1

SOLUTION The average value is

$$\frac{1}{\sqrt{\pi} - 0}\int_0^{\sqrt{\pi}} x \sin x^2\,dx$$

To evaluate this integral, we make the substitution $u = x^2$, so that $du = 2x\,dx$. When $x = 0$, $u = 0$ and when $x = \sqrt{\pi}$, $u = \pi$. Thus,

$$\frac{1}{\sqrt{\pi}}\int_0^{\sqrt{\pi}} x \sin x^2\,dx = \frac{1}{\sqrt{\pi}}\int_0^{\pi}\frac{1}{2}\sin u\,du = \frac{1}{2\sqrt{\pi}}\left[-\cos u\right]_0^{\pi} = \frac{1}{2\sqrt{\pi}}(2) = \frac{1}{\sqrt{\pi}} \blacksquare$$

EXAMPLE 2 Suppose the temperature in degrees Fahrenheit of a metal bar of length 2 feet depends on the position x according to the function $T(x) = 40 + 20x(2 - x)$. Find the average temperature in the bar. Is there a point where the actual temperature equals the average temperature?

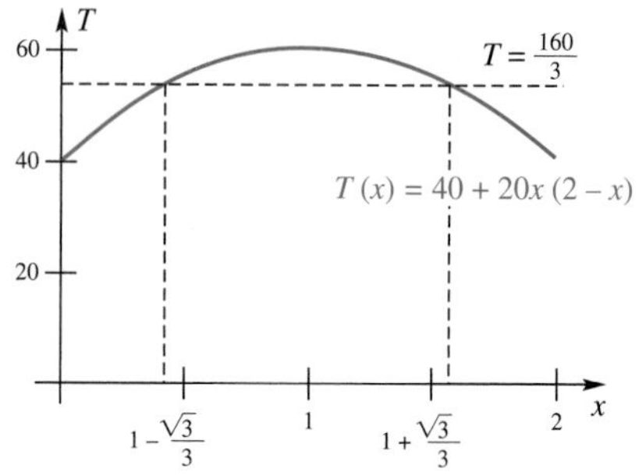

Figure 2

SOLUTION The average temperature is

$$\frac{1}{2}\int_0^2 [40 + 20x(2 - x)]\,dx = \int_0^2 (20 + 20x - 10x^2)\,dx$$
$$= \left[20x + 10x^2 - \frac{10}{3}x^3\right]_0^2$$
$$= \left(40 + 40 - \frac{80}{3}\right) = \frac{160}{3}\ \text{°F}$$

Figure 2, which shows the temperature T as a function of x, indicates that we should expect two points at which the actual temperature equals the average temperature. To find these points, we set $T(x)$ equal to 160/3 and try to solve for x.

$$40 + 20x(2 - x) = \frac{160}{3}$$
$$3x^2 - 6x + 2 = 0$$

The Quadratic Formula gives

$$x = \frac{1}{3}\left(3 - \sqrt{3}\right) \approx 0.42265 \quad \text{and} \quad x = \frac{1}{3}\left(3 + \sqrt{3}\right) \approx 1.5774$$

Both solutions are between 0 and 2, so there are two points at which the actual temperature equals the average temperature. ■

It seems as if there should always be a value of x with the property that $f(x)$ equals the average value of the function. This is true provided only that the function f is continuous.

The Two Mean Value Theorems

The Mean Value Theorem for Derivatives says that there is some point c in the interval $[a, b]$ at which the average rate of change of f, $(f(b) - f(a))/(b - a)$, equals the instantaneous rate of change, $f'(c)$.

The Mean Value Theorem for Integrals says that there is some point c in the interval $[a, b]$ at which the average value of a function $\frac{1}{b-a}\int_a^b f(t)\,dt$ is equal to the actual value of the function, $f(c)$.

Theorem A Mean Value Theorem for Integrals

If f is continuous on $[a, b]$, then there is a number c between a and b such that

$$f(c) = \frac{1}{b-a}\int_a^b f(t)\,dt$$

Proof For $a \le x \le b$ define $G(x) = \int_a^x f(t)\,dt$. By the Mean Value Theorem for Derivatives (applied to G) there is a c in the interval (a, b) such that

$$G'(c) = \frac{G(b) - G(a)}{b - a}$$

Since $G(a) = \int_a^a f(t)\,dt = 0$, $G(b) = \int_a^b f(t)\,dt$, and $G'(c) = f(c)$, this leads to

$$G'(c) = f(c) = \frac{1}{b-a}\int_a^b f(t)\,dt \qquad ■$$

The Mean Value Theorem for Integrals is often expressed as follows: If f is integrable on $[a, b]$, then there exists a c in (a, b) such that

$$\int_a^b f(t)\,dt = (b - a)\,f(c)$$

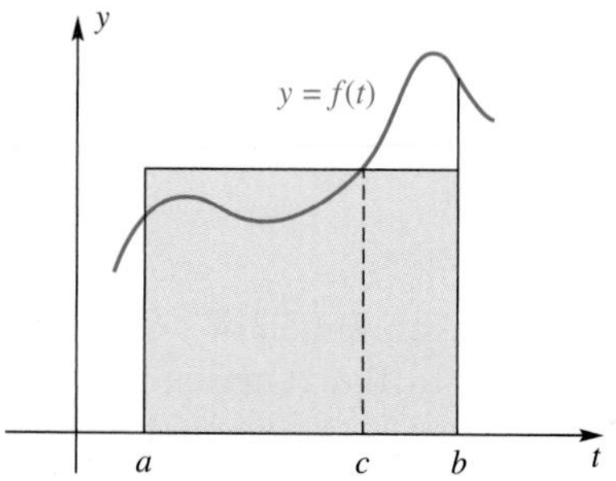

Figure 3

When viewed this way, the Mean Value Theorem for Integrals says that there is some c in the interval $[a, b]$ such that the area of the rectangle with height $f(c)$ and width $b - a$ is equal to the area under the curve. In Figure 3, the area under the curve is equal to the area of the rectangle.

> **Estimating Integrals**
>
> This version of the Mean Value Theorem for Integrals with the accompanying Figure 3 suggests a good way to estimate the value of a definite integral. The area of the region under a curve is equal to the area of a rectangle. One can make a good guess at this rectangle by simply "eyeballing" the region. In Figure 3, the area of the shaded part *above* the curve should match the area of the white part *below* the curve.

EXAMPLE 3 Find all values of c that satisfy the Mean Value Theorem for Integrals for $f(x) = x^2$ on the interval $[-3, 3]$.

SOLUTION The graph of $f(x)$ shown in Figure 4 indicates that there could be two values of c that satisfy the Mean Value Theorem for Integrals. The average value of the function is

$$\frac{1}{3-(-3)}\int_{-3}^{3} x^2\,dx = \frac{1}{6}\left[\frac{x^3}{3}\right]_{-3}^{3} = \frac{1}{18}[27-(-27)] = 3$$

To find the values of c, we solve

$$3 = f(c) = c^2$$
$$c = \pm\sqrt{3}$$

Both $-\sqrt{3}$ and $\sqrt{3}$ are in the interval $[-3, 3]$, so both values satisfy the Mean Value Theorem for Integrals. ■

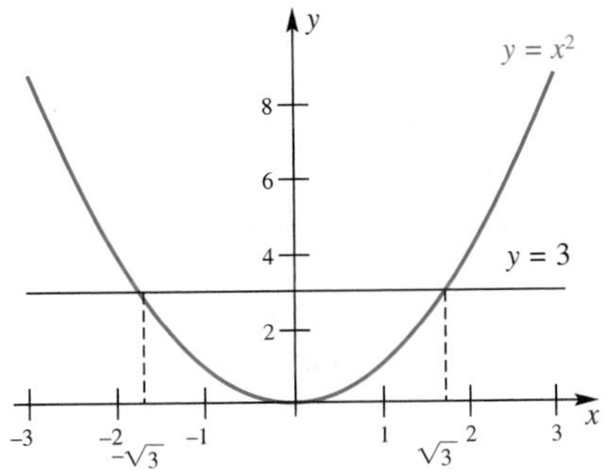

Figure 4

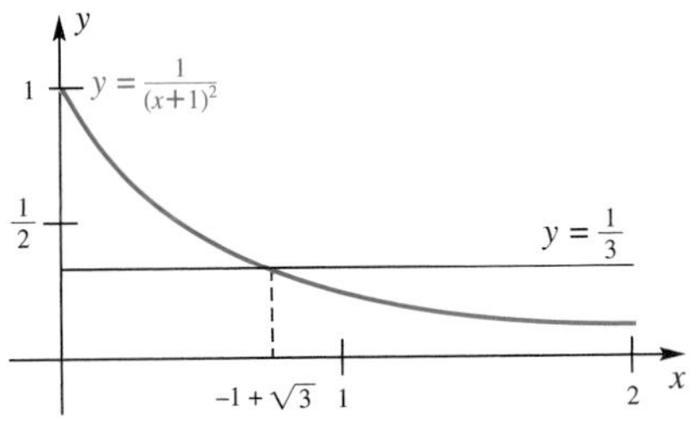

Figure 5

EXAMPLE 4 Find all values of c that satisfy the Mean Value Theorem for Integrals for $f(x) = \dfrac{1}{(x+1)^2}$ on the interval $[0, 2]$.

SOLUTION The graph of $f(x)$ shown in Figure 5 indicates that there should be one value of c that satisfies the Mean Value Theorem for Integrals. The average value of the function is found by making the substitution $u = x + 1$, $du = dx$, where when $x = 0$, $u = 1$ and when $x = 2$, $u = 3$:

$$\frac{1}{2-0}\int_0^2 \frac{1}{(x+1)^2}\,dx = \frac{1}{2}\int_1^3 \frac{1}{u^2}\,du = \frac{1}{2}\left[-u^{-1}\right]_1^3 = \frac{1}{2}\left(-\frac{1}{3}+1\right) = \frac{1}{3}$$

To find the value of c we solve

$$\frac{1}{3} = f(c) = \frac{1}{(c+1)^2}$$

$$c^2 + 2c + 1 = 3$$

$$c = \frac{-2 \pm \sqrt{2^2 - 4(1)(-2)}}{2} = -1 \pm \sqrt{3}$$

Note that $-1 - \sqrt{3} \approx -2.7321$ and $-1 + \sqrt{3} \approx 0.73205$. The only one of these two solutions that is in the interval $[0, 2]$ is $c = -1 + \sqrt{3}$; thus, this is the only value of c that satisfies the Mean Value Theorem for Integrals. ■

The Use of Symmetry in Evaluating Definite Integrals

Recall that an even function is one satisfying $f(-x) = f(x)$, whereas an odd function satisfies

$f(-x) = -f(x)$. The graph of the former is symmetric with respect to the y-axis; the graph of the latter is symmetric with respect to the origin. Here is a useful integration theorem for such functions.

Theorem B **Symmetry Theorem**

If f is an even function, then

$$\int_{-a}^{a} f(x)\,dx = 2\int_{0}^{a} f(x)\,dx$$

If f is an odd function, then

$$\int_{-a}^{a} f(x)\,dx = 0$$

Proof for Even Functions The geometric interpretation of this theorem is shown in Figures 6 and 7. To justify the results analytically, we first write

$$\int_{-a}^{a} f(x)\,dx = \int_{-a}^{0} f(x)\,dx + \int_{0}^{a} f(x)\,dx$$

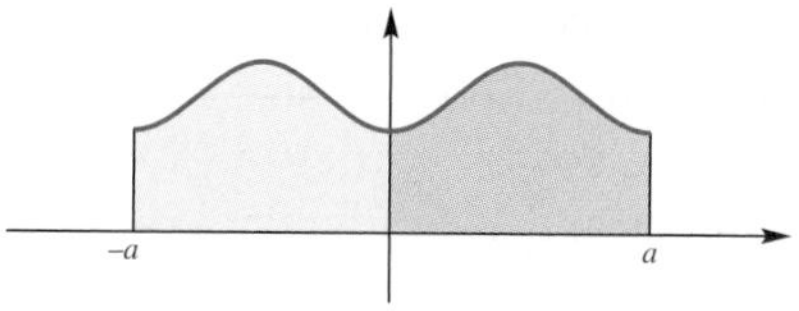

Even function
Left area = right area

Figure 6

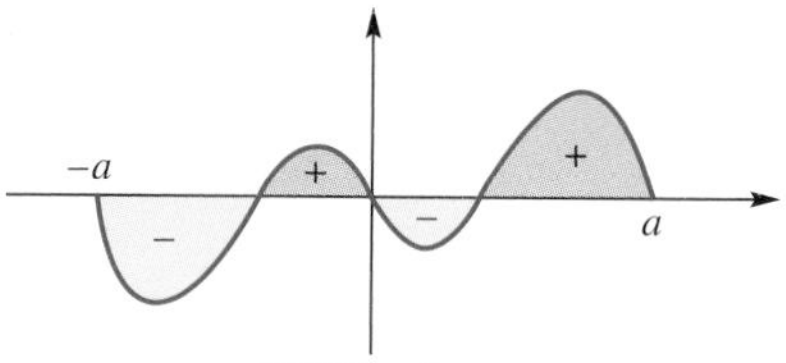

Odd function
Left area neutralizes right area

Figure 7

In the first of the integrals on the right, we make the substitution $u = -x$, $du = -dx$. If f is even, $f(u) = f(-x) = f(x)$ and

$$\int_{-a}^{0} f(x)\,dx = -\int_{-a}^{0} f(-x)(-dx) = -\int_{a}^{0} f(u)\,du = \int_{0}^{a} f(u)\,du = \int_{0}^{a} f(x)\,dx$$

Therefore,

$$\int_{-a}^{a} f(x)\,dx = \int_{0}^{a} f(x)\,dx + \int_{0}^{a} f(x)\,dx = 2\int_{0}^{a} f(x)\,dx$$

The proof for odd functions is left as an exercise (Problem 60). ■

EXAMPLE 5 Evaluate $\int_{-\pi}^{\pi} \cos\left(\frac{x}{4}\right) dx$.

Be sure to note the hypotheses of the Symmetry Theorem. The integrand must be even or odd and the interval of integration must be symmetric about the origin. These are restrictive conditions, but it is surprising how often they hold in applications. When they do hold, they can greatly simplify integrations.

SOLUTION Since $\cos(-x/4) = \cos(x/4)$, $f(x) = \cos(x/4)$ is an even function. Thus,

$$\int_{-\pi}^{\pi} \cos\left(\frac{x}{4}\right) dx = 2\int_{0}^{\pi} \cos\left(\frac{x}{4}\right) dx = 8\int_{0}^{\pi} \cos\left(\frac{x}{4}\right)\cdot\frac{1}{4}dx$$

$$= 8\int_{0}^{\pi/4} \cos u\,du = \left[8\sin u\right]_{0}^{\pi/4} = 4\sqrt{2}$$ ■

EXAMPLE 6 Evaluate $\int_{-5}^{5} \frac{x^5}{x^2+4}\,dx$.

SOLUTION $f(x) = x^5/(x^2+4)$ is an odd function. Thus, the above integral has the value 0. ■

EXAMPLE 7 Evaluate $\int_{-2}^{2} (x \sin^4 x + x^3 - x^4)\,dx$.

SOLUTION The first two terms in the integrand are odd, and the last is even. Thus, we may write the integral as

$$\int_{-2}^{2} (x \sin^4 x + x^3)\,dx - \int_{-2}^{2} x^4\,dx = 0 - 2\int_{0}^{2} x^4\,dx$$
$$= \left[-2\frac{x^5}{5}\right]_0^2 = -\frac{64}{5}$$ ■

EXAMPLE 8 Evaluate $\int_{-\pi}^{\pi} \sin^3 x \cos^5 x\,dx$.

SOLUTION The function $\sin x$ is odd and $\cos x$ is even. An odd function raised to an odd power is odd, so $\sin^3 x$ is odd. An even function raised to any integer power is even, so $\cos^5 x$ is even. An odd function times an even function is odd. Thus the integrand in this integral is an odd function and the interval is symmetric about 0, so the value of this integral is 0. ■

Use of Periodicity Recall that a function f is *periodic* if there is a number p such that $f(x+p) = f(x)$ for all x in the domain of f. The smallest such positive number p is called the **period** of f. The trigonometric functions are examples of periodic functions.

Theorem C

If f is periodic with period p, then

$$\int_{a+p}^{b+p} f(x)\,dx = \int_{a}^{b} f(x)\,dx$$

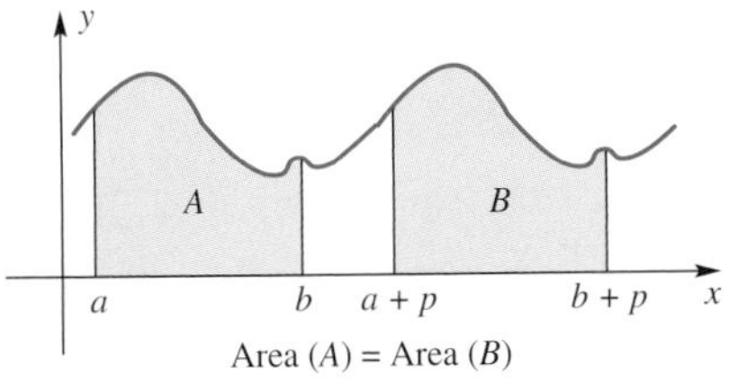

Figure 8

Proof The geometric interpretation can be seen in Figure 8. To prove the result, let $u = x - p$ so that $x = u + p$ and $du = dx$. Then

$$\int_{a+p}^{b+p} f(x)\,dx = \int_{a}^{b} f(u+p)\,du = \int_{a}^{b} f(u)\,du = \int_{a}^{b} f(x)\,dx$$

We could replace $f(u+p)$ by $f(u)$ because f is periodic. ■

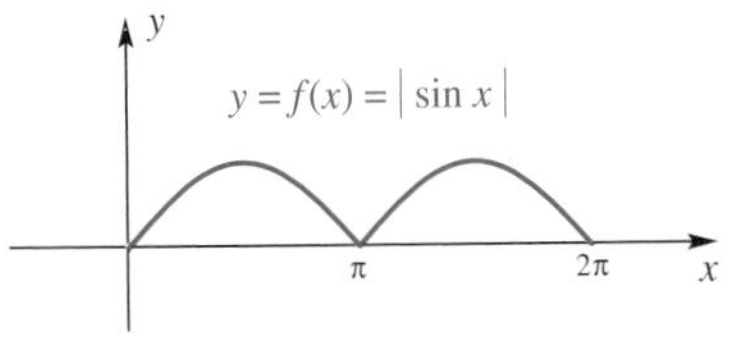

Figure 9

EXAMPLE 9 Evaluate (a) $\int_{0}^{2\pi} |\sin x|\,dx$ and (b) $\int_{0}^{100\pi} |\sin x|\,dx$.

SOLUTION

(a) Note that $f(x) = |\sin x|$ is periodic with period π (Figure 9). The integral in (a) is thus

$$\int_0^{2\pi} |\sin x|\, dx = \int_0^{\pi} |\sin x|\, dx + \int_{\pi}^{2\pi} |\sin x|\, dx$$
$$= \int_0^{\pi} |\sin x|\, dx + \int_0^{\pi} |\sin x|\, dx$$
$$= 2\int_0^{\pi} \sin x\, dx = 2\left[-\cos x\right]_0^{\pi} = 2[1 - (-1)] = 4$$

(b) The integral in (b) is

$$\int_0^{100\pi} |\sin x|\, dx = \underbrace{\int_0^{\pi} |\sin x|\, dx + \int_{\pi}^{2\pi} |\sin x|\, dx + \cdots + \int_{99\pi}^{100\pi} |\sin x|\, dx}_{100 \text{ integrals each equal to } \int_0^{\pi} \sin x\, dx}$$
$$= 100\int_0^{\pi} \sin x\, dx = 100\left[-\cos x\right]_0^{\pi} = 100(2) = 200 \quad ■$$

Note that in Example 9, we had to use symmetry because we can't find an antiderivative for $|\sin x|$ over the interval $[0, 100\pi]$.

Concepts Review

1. The average value of a function f on the interval $[a, b]$ is ______.

2. The Mean Value Theorem for Integrals says there exists a c in the interval (a, b) such that the average value of the function on $[a, b]$ is equal to ______.

3. If f is an odd function, $\int_{-2}^{2} f(x)\, dx =$ ______; if f is an even function, $\int_{-2}^{2} f(x)\, dx =$ ______.

4. The function f is periodic if there is a number p such that ______ for all x in the domain of f. The smallest such positive number p is called the ______ of the function.

Problem Set 4.5

In Problems 1–14, find the average value of the function on the given interval.

1. $f(x) = 4x^3;\ [1, 3]$
2. $f(x) = 5x^2;\ [1, 4]$
3. $f(x) = \dfrac{x}{\sqrt{x^2 + 16}};\ [0, 3]$
4. $f(x) = \dfrac{x^2}{\sqrt{x^3 + 16}};\ [0, 2]$
5. $f(x) = 2 + |x|;\ [-2, 1]$
6. $f(x) = x + |x|;\ [-3, 2]$
7. $f(x) = \cos x;\ [0, \pi]$
8. $f(x) = \sin x;\ [0, \pi]$
9. $f(x) = x \cos x^2;\ \left[0, \sqrt{\pi}\right]$
10. $f(x) = \sin^2 x \cos x;\ [0, \pi/2]$
11. $F(y) = y(1 + y^2)^3;\ [1, 2]$
12. $g(x) = \tan x \sec^2 x;\ [0, \pi/4]$
13. $h(z) = \dfrac{\sin\sqrt{z}}{\sqrt{z}};\ [\pi/4, \pi/2]$
14. $G(v) = \dfrac{\sin v \cos v}{\sqrt{1 + \cos^2 v}};\ [0, \pi/2]$

In Problems 15–28, find all values of c that satisfy the Mean Value Theorem for Integrals on the given interval.

15. $f(x) = \sqrt{x + 1};\ [0, 3]$
16. $f(x) = x^2;\ [-1, 1]$
17. $f(x) = 1 - x^2;\ [-4, 3]$
18. $f(x) = x(1 - x);\ [0, 1]$
19. $f(x) = |x|;\ [0, 2]$
20. $f(x) = |x|;\ [-2, 2]$
21. $H(z) = \sin z;\ [-\pi, \pi]$
22. $g(y) = \cos 2y;\ [0, \pi]$
23. $R(v) = v^2 - v;\ [0, 2]$
24. $T(x) = x^3;\ [0, 2]$
25. $f(x) = ax + b;\ [1, 4]$
26. $S(y) = y^2;\ [0, b]$
27. $f(x) = ax + b;\ [A, B]$
28. $q(y) = ay^2;\ [0, b]$

GC ≈ *Use a graphing calculator to plot the graph of the integrand in Problems 29–32. Then estimate the integral as suggested in the margin note accompanying Theorem B.*

29. $\int_0^{2\pi} (5 + \sin x)^4\, dx$
30. $\int_0^{2} \left[3 + \sin(x^2)\right] dx$
31. $\int_{-1}^{1} \dfrac{2}{1 + x^2}\, dx$
32. $\int_{10}^{20} \left(1 + \dfrac{1}{x}\right)^5 dx$

33. Figure 10 shows the relative humidity H as a function of time t (measured in days since Sunday) for an office building. Approximate the average relative humidity for the week.

34. Figure 11 shows temperature T as a function of time t (measured in hours past midnight) for one day in St. Louis, Missouri.

(a) Approximate the average temperature for the day.

(b) Must there be a time when the temperature is equal to the average temperature for the day? Explain.

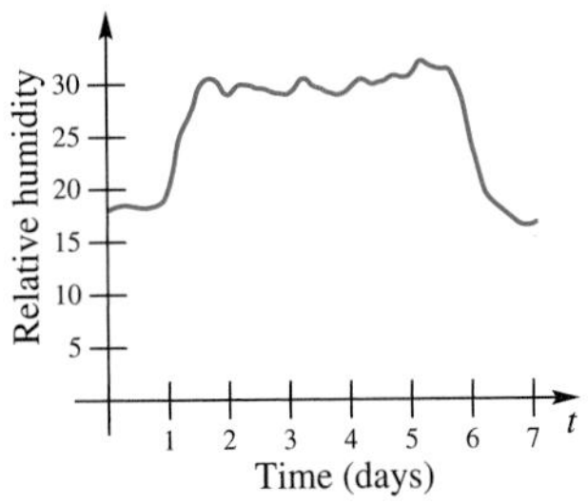

Figure 10

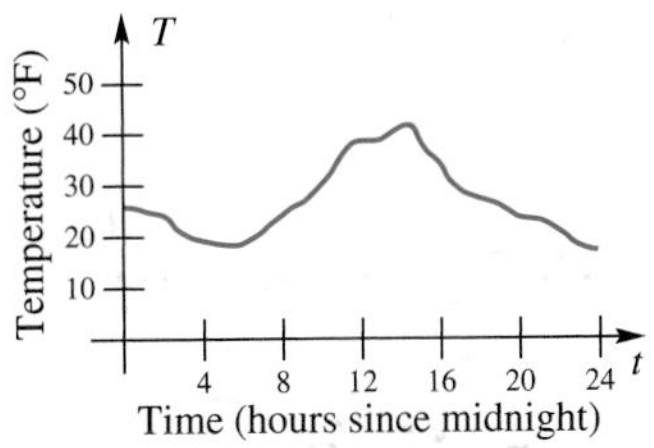

Figure 11

In Problems 35–44, use symmetry to help you evaluate the given integral.

35. $\int_{-\pi}^{\pi} (\sin x + \cos x)\, dx$ **36.** $\int_{-1}^{1} \frac{x^3}{(1 + x^2)^4}\, dx$

37. $\int_{-\pi/2}^{\pi/2} \frac{\sin x}{1 + \cos x}\, dx$ **38.** $\int_{-\sqrt[3]{\pi}}^{\sqrt[3]{\pi}} x^2 \cos(x^3)\, dx$

39. $\int_{-\pi}^{\pi} (\sin x + \cos x)^2\, dx$ **40.** $\int_{-\pi/2}^{\pi/2} z \sin^2(z^3) \cos(z^3)\, dz$

41. $\int_{-1}^{1} (1 + x + x^2 + x^3)\, dx$

42. $\int_{-100}^{100} (v + \sin v + v \cos v + \sin^3 v)^5\, dv$

43. $\int_{-1}^{1} (|x^3| + x^3)\, dx$

44. $\int_{-\pi/4}^{\pi/4} (|x| \sin^5 x + |x|^2 \tan x)\, dx$

45. How does $\int_{-b}^{-a} f(x)\, dx$ compare with $\int_{a}^{b} f(x)\, dx$ when f is an even function? An odd function?

46. Prove (by a substitution) that

$$\int_{a}^{b} f(-x)\, dx = \int_{-b}^{-a} f(x)\, dx$$

47. Use periodicity to calculate $\int_{0}^{4\pi} |\cos x|\, dx$.

48. Calculate $\int_{0}^{4\pi} |\sin 2x|\, dx$.

49. If f is periodic with period p, then

$$\int_{a}^{a+p} f(x)\, dx = \int_{0}^{p} f(x)\, dx$$

Convince yourself that this is true by drawing a picture and then use the result to calculate $\int_{1}^{1+\pi} |\sin x|\, dx$.

50. Use the result in Problem 49 to calculate

$$\int_{2}^{2+\pi/2} |\sin 2x|\, dx.$$

51. Calculate $\int_{1}^{1+\pi} |\cos x|\, dx$.

52. Prove or disprove that the integral of the average value equals the integral of the function on the interval: $\int_{a}^{b} \overline{f}\, dx = \int_{a}^{b} f(x)\, dx$, where $\overline{f}$ is the average value of the function f over the interval $[a, b]$.

EXPL **53.** Assuming that u and v can be integrated over the interval $[a, b]$ and that the average values over the interval are denoted by $\overline{u}$ and $\overline{v}$, prove or disprove that

(a) $\overline{u} + \overline{v} = \overline{u + v}$;

(b) $k\overline{u} = \overline{ku}$, where k is any constant;

(c) if $u \le v$ then $\overline{u} \le \overline{v}$.

54. Household electric current can be modeled by the voltage $V = \hat{V} \sin(120\pi t + \phi)$, where t is measured in seconds, $\hat{V}$ is the maximum value that V can attain, and ϕ is the phase angle. Such a voltage is usually said to be 60-cycle, since in 1 second the voltage goes through 60 oscillations. The root-mean-square voltage, usually denoted by V_{rms} is defined to be the square root of the average of V^2. Hence

$$V_{rms} = \sqrt{\int_{\phi}^{1+\phi} (\hat{V} \sin(120\pi t + \phi))^2\, dt}$$

A good measure of how much heat a given voltage can produce is given by V_{rms}.

(a) Compute the average voltage over 1 second.

(b) Compute the average voltage over 1/60 of a second.

(c) Show that $V_{rms} = \frac{\hat{V}\sqrt{2}}{2}$ by computing the integral for V_{rms}.

Hint: $\int \sin^2 t\, dt = -\frac{1}{2} \cos t \sin t + \frac{1}{2} t + C.$

(d) If the V_{rms} for household current is usually 120 volts, what is the value $\hat{V}$ in this case?

55. Give a proof of the Mean Value Theorem for Integrals (Theorem A) that does not use the First Fundamental Theorem of Calculus. *Hint:* Apply the Max–Min Existence Theorem and the Intermediate Value Theorem.

56. Integrals that occur frequently in applications are $\int_{0}^{2\pi} \cos^2 x\, dx$ and $\int_{0}^{2\pi} \sin^2 x\, dx$.

(a) Using a trigonometric identity, show that

$$\int_{0}^{2\pi} (\sin^2 x + \cos^2 x)\, dx = 2\pi$$

(b) Show from graphical considerations that

$$\int_{0}^{2\pi} \cos^2 x\, dx = \int_{0}^{2\pi} \sin^2 x\, dx$$

(c) Conclude that $\int_{0}^{2\pi} \cos^2 x\, dx = \int_{0}^{2\pi} \sin^2 x\, dx = \pi$.

GC **57.** Let $f(x) = |\sin x| \sin(\cos x)$.

(a) Is f even, odd, or neither?

(b) Note that f is periodic. What is its period?

(c) Evaluate the definite integral of f for each of the following intervals: $[0, \pi/2]$, $[-\pi/2, \pi/2]$, $[0, 3\pi/2]$, $[-3\pi/2, 3\pi/2]$, $[0, 2\pi]$, $[\pi/6, 13\pi/6]$, $[\pi/6, 4\pi/3]$, $[13\pi/6, 10\pi/3]$.

58. Repeat Problem 57 for $f(x) = \sin x\, |\sin(\sin x)|$.

59. Complete the generalization of the Pythagorean Theorem begun in Problem 59 of Section 0.3 by showing that $A + B = C$ in Figure 12, these being the areas of similar regions built on the two legs and the hypotenuse of a right triangle.

(a) Convince yourself that similarity means

$$g(x) = \frac{a}{c} f\left(\frac{c}{a} x\right) \quad \text{and} \quad h(x) = \frac{b}{c} f\left(\frac{c}{b} x\right)$$

(b) Show that $\displaystyle\int_0^a g(x)\,dx + \int_0^b h(x)\,dx = \int_0^c f(x)\,dx$.

60. Prove the Symmetry Theorem for the case of odd functions.

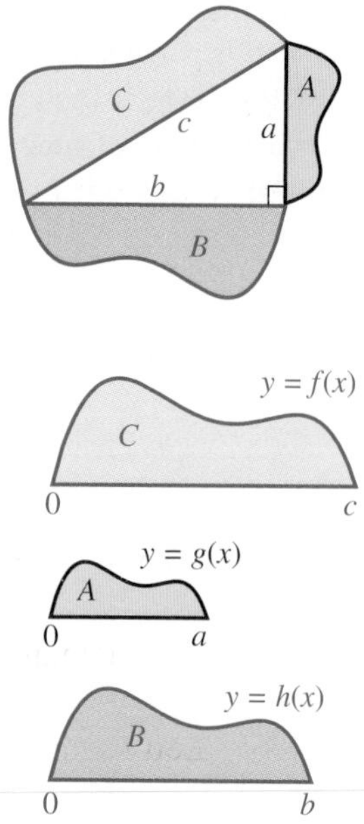

Figure 12

Answers to Concepts Review: **1.** $\displaystyle\frac{1}{b-a}\int_a^b f(x)\,dx$ **2.** $f(c)$ **3.** $0; 2\displaystyle\int_0^2 f(x)\,dx$ **4.** $f(x + p) = f(x)$; period

4.6 Numerical Integration

We know that if f is continuous on a closed interval $[a, b]$, then the definite integral $\displaystyle\int_a^b f(x)\,dx$ must exist. Existence is one thing; evaluation is a very different matter. There are many definite integrals that cannot be evaluated by the methods that we have learned, that is, by use of the Second Fundamental Theorem of Calculus. For example, the indefinite integrals

$$\int \sin(x^2)\,dx, \quad \int \sqrt{1 - x^4}\,dx, \quad \int \frac{\sin x}{x}\,dx$$

cannot be expressed algebraically in terms of elementary functions, that is, in terms of functions studied in a first calculus course. Even when elementary indefinite integrals can be found, it is often advantageous to use the approximation methods of this section, since they lead to efficient algorithms that can be directly programmed on a calculator or computer. In Section 4.2 we saw how Riemann sums can be used to approximate a definite integral. In this section we review these Riemann sums and we present two additional methods: the Trapezoidal Rule and the Parabolic Rule.

Riemann Sums In Section 4.2 we introduced the concept of a Riemann sum. Suppose f is defined on $[a, b]$ and we partition the interval $[a, b]$ into n smaller intervals with end points $a = x_0 < x_1 < \cdots < x_{n-1} < x_n = b$. The Riemann sum is then defined to be

$$\sum_{i=1}^{n} f(\bar{x}_i)\,\Delta x_i$$

where $\bar{x}_i$ is some point (possibly even an end point) in the interval $[x_{i-1}, x_i]$, and $\Delta x_i = x_i - x_{i-1}$. For now, we will assume that the partition is *regular*, that is,

$\Delta x_i = (b - a)/n$ for all i. Riemann sums were introduced in Section 4.2 with the goal of defining the definite integral as the limit of the Riemann sum. Here we look at the Riemann sum as a way to approximate a definite integral.

We consider the three cases: where the sample point $\overline{x}_i$ is the left end point, the right end point, or the midpoint of $[x_{i-1}, x_i]$. The left end point, right end point, and midpoint of the interval $[x_{i-1}, x_i]$ are

$$\text{left end point} = x_{i-1} = a + (i - 1)\frac{b - a}{n}$$

$$\text{right end point} = x_i = a + i\,\frac{b - a}{n}$$

$$\text{midpoint} = \frac{x_{i-1} + x_i}{2} = \frac{a + (i - 1)\frac{b - a}{n} + a + i\frac{b - a}{n}}{2} = a + \left(i - \tfrac{1}{2}\right)\frac{b - a}{n}$$

For a left Riemann sum, we take $\overline{x}_i$ to be x_{i-1}, the left end point:

$$\text{Left Riemann Sum} = \sum_{i=1}^{n} f(\overline{x}_i)\,\Delta x_i = \frac{b - a}{n}\sum_{i=1}^{n} f\left(a + (i - 1)\frac{b - a}{n}\right)$$

For a right Riemann sum, we take $\overline{x}_i$ to be x_i, the right end point:

$$\text{Right Riemann Sum} = \sum_{i=1}^{n} f(\overline{x}_i)\,\Delta x_i = \frac{b - a}{n}\sum_{i=1}^{n} f\left(a + i\,\frac{b - a}{n}\right)$$

For a midpoint Riemann sum, we take $\overline{x}_i$ to be $(x_{i-1} + x_i)/2$, the midpoint of the interval $[x_{i-1}, x_i]$:

$$\text{Midpoint Riemann Sum} = \sum_{i=1}^{n} f(\overline{x}_i)\,\Delta x_i = \frac{b - a}{n}\sum_{i=1}^{n} f\left(a + \left(i - \tfrac{1}{2}\right)\frac{b - a}{n}\right)$$

The figures in the large table on the next page illustrate how these approximations (and two others we will introduce later in this section) work.

EXAMPLE 1 Approximate the definite integral $\int_1^3 \sqrt{4 - x}\,dx$ using left, right, and midpoint Riemann sums with $n = 4$.

SOLUTION Let $f(x) = \sqrt{4 - x}$. We have $a = 1, b = 3$, and $n = 4$, so $(b - a)/n = 0.5$. The values of x_i and $f(x_i)$ are

$$\begin{aligned}
x_0 &= 1.0 & f(x_0) &= f(1.0) = \sqrt{4 - 1} \approx 1.7321\\
x_1 &= 1.5 & f(x_1) &= f(1.5) = \sqrt{4 - 1.5} \approx 1.5811\\
x_2 &= 2.0 & f(x_2) &= f(2.0) = \sqrt{4 - 2} \approx 1.4142\\
x_3 &= 2.5 & f(x_3) &= f(2.5) = \sqrt{4 - 2.5} \approx 1.2247\\
x_4 &= 3.0 & f(x_4) &= f(3.0) = \sqrt{4 - 3} = 1.0000
\end{aligned}$$

Using the left Riemann sum, we have the following approximation:

$$\begin{aligned}
\int_1^3 \sqrt{4 - x}\,dx &\approx \text{Left Riemann Sum}\\
&= \frac{b - a}{n}[f(x_0) + f(x_1) + f(x_2) + f(x_3)]\\
&= 0.5[f(1.0) + f(1.5) + f(2.0) + f(2.5)]\\
&\approx 0.5(1.7321 + 1.5811 + 1.4142 + 1.2247)\\
&\approx 2.9761
\end{aligned}$$

Methods for Approximating $\int_a^b f(x)\,dx$

1. Left Riemann Sum

$$\text{Area of } i\text{th rectangle} = f(x_{i-1})\,\Delta x_i = \frac{b-a}{n} f\left(a + (i-1)\frac{b-a}{n}\right)$$

$$\int_a^b f(x)\,dx \approx \frac{b-a}{n}\sum_{i=1}^{n} f\left(a + (i-1)\frac{b-a}{n}\right)$$

$$E_n = \frac{(b-a)^2}{2n} f'(c) \text{ for some } c \text{ in } [a, b]$$

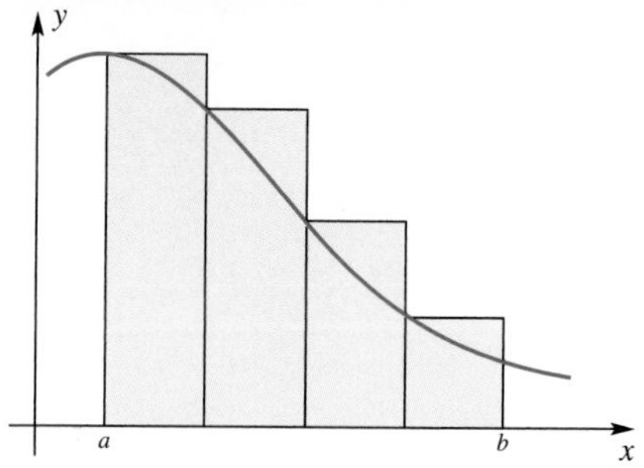

2. Right Riemann Sum

$$\text{Area of } i\text{th rectangle} = f(x_i)\,\Delta x_i = \frac{b-a}{n} f\left(a + i\frac{b-a}{n}\right)$$

$$\int_a^b f(x)\,dx \approx \frac{b-a}{n}\sum_{i=1}^{n} f\left(a + i\frac{b-a}{n}\right)$$

$$E_n = -\frac{(b-a)^2}{2n} f'(c) \text{ for some } c \text{ in } [a, b]$$

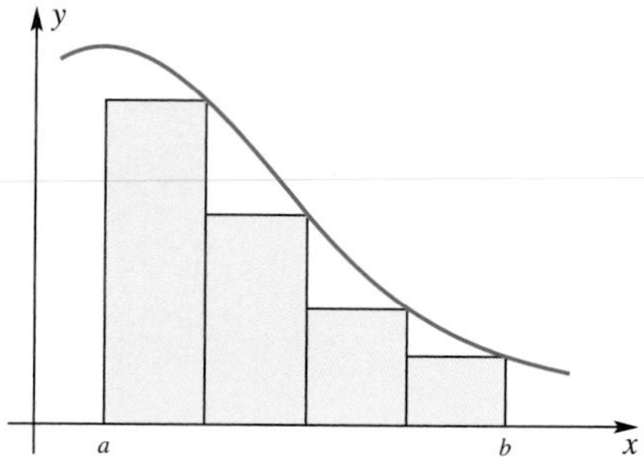

3. Midpoint Riemann Sum

$$\text{Area of } i\text{th rectangle} = f\left(\frac{x_{i-1}+x_i}{2}\right)\Delta x_i = \frac{b-a}{n} f\left(a + \left(i - \frac{1}{2}\right)\frac{b-a}{n}\right)$$

$$\int_a^b f(x)\,dx \approx \frac{b-a}{n}\sum_{i=1}^{n} f\left(a + \left(i - \frac{1}{2}\right)\frac{b-a}{n}\right)$$

$$E_n = \frac{(b-a)^3}{24n^2} f''(c) \text{ for some } c \text{ in } [a, b]$$

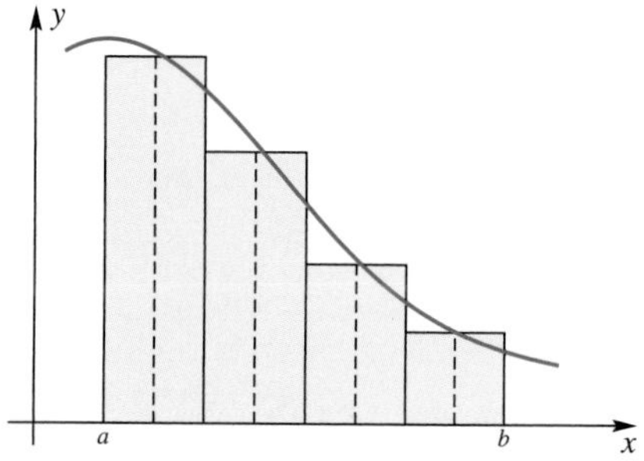

4. Trapezoidal Rule

$$\text{Area of } i\text{th trapezoid} = \frac{b-a}{n}\frac{f(x_{i-1}) + f(x_i)}{2}$$

$$\int_a^b f(x)\,dx \approx \frac{b-a}{n}\sum_{i=1}^{n}\frac{f(x_{i-1}) + f(x_i)}{2}$$

$$= \frac{b-a}{2n}\left[f(a) + 2\sum_{i=1}^{n-1} f\left(a + i\frac{b-a}{n}\right) + f(b)\right]$$

$$E_n = -\frac{(b-a)^3}{12n^2} f''(c) \text{ for some } c \text{ in } [a, b]$$

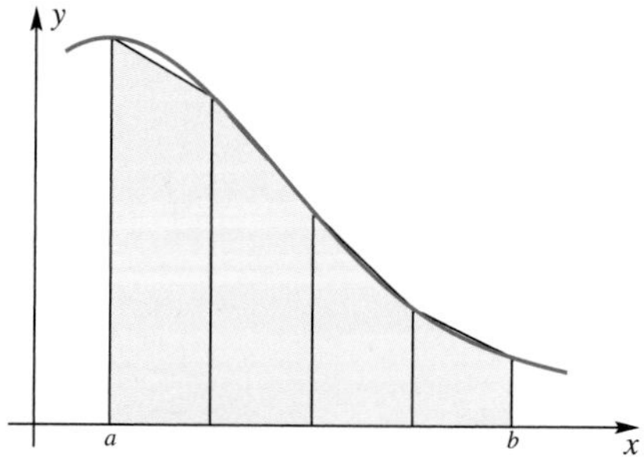

5. Parabolic Rule (n must be even)

$$\int_a^b f(x)\,dx \approx \frac{b-a}{3n}[f(x_0) + 4f(x_1) + 2f(x_2) + 4f(x_3) + 2f(x_4) + \cdots + 4f(x_{n-3}) + 2f(x_{n-2}) + 4f(x_{n-1}) + f(x_n)]$$

$$= \frac{b-a}{3n}\left[f(a) + 4\sum_{i=1}^{n/2} f\left(a + (2i-1)\frac{b-a}{n}\right) + 2\sum_{i=1}^{n/2-1} f\left(a + 2i\frac{b-a}{n}\right) + f(b)\right]$$

$$E_n = -\frac{(b-a)^5}{180n^4} f^{(4)}(c) \text{ for some } c \text{ in } [a, b]$$

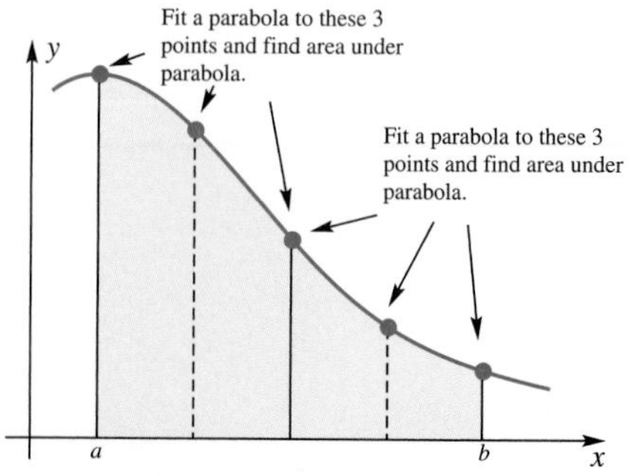

The right Riemann sum leads to the following approximation:

$$\begin{aligned}\int_1^3 \sqrt{4-x}\,dx &\approx \text{Right Riemann Sum}\\ &= \frac{b-a}{n}[f(x_1)+f(x_2)+f(x_3)+f(x_4)]\\ &= 0.5[f(1.5)+f(2.0)+f(2.5)+f(3.0)]\\ &\approx 0.5(1.5811+1.4142+1.2247+1.0000)\\ &\approx 2.6100\end{aligned}$$

Finally, the midpoint Riemann sum approximation to the definite integral is

$$\begin{aligned}&\int_1^3 \sqrt{4-x}\,dx \approx \text{Midpoint Riemann Sum}\\ &\quad= \frac{b-a}{n}\left[f\left(\frac{x_0+x_1}{2}\right)+f\left(\frac{x_1+x_2}{2}\right)+f\left(\frac{x_2+x_3}{2}\right)+f\left(\frac{x_3+x_4}{2}\right)\right]\\ &\quad= 0.5\,[f(1.25)+f(1.75)+f(2.25)+f(2.75)]\\ &\quad\approx 0.5(1.6583+1.5000+1.3229+1.1180)\\ &\quad\approx 2.7996\end{aligned}$$

■

In this last example, approximations were not needed because we could have evaluated this integral using the Second Fundamental Theorem of Calculus:

$$\begin{aligned}\int_1^3 \sqrt{4-x}\,dx &= \left[-\frac{2}{3}(4-x)^{3/2}\right]_1^3 = -\frac{2}{3}(4-3)^{3/2}+\frac{2}{3}(4-1)^{3/2}\\ &= 2\sqrt{3}-\frac{2}{3} \approx 2.7974\end{aligned}$$

The midpoint Riemann sum approximation turned out to be the closest. The figures in the large table on the previous page suggest that this will often be the case.

The next example is more realistic, in the sense that it is not possible to apply the Second Fundamental Theorem of Calculus.

EXAMPLE 2 Approximate the definite integral $\int_0^2 \sin x^2\,dx$ using a right Riemann sum with $n = 8$.

SOLUTION Let $f(x) = \sin x^2$. We have $a = 0, b = 2$, and $n = 8$, so $(b-a)/n = 0.25$. Using the right Riemann sum, we have the following approximation:

$$\begin{aligned}\int_0^2 \sin x^2\,dx &\approx \text{Right Riemann Sum}\\ &= \frac{b-a}{n}\left[\sum_{i=1}^{8} f\left(a+i\frac{b-a}{n}\right)\right]\\ &= 0.25(\sin 0.25^2+\sin 0.5^2+\sin 0.75^2+\sin 1^2\\ &\quad+\sin 1.25^2+\sin 1.5^2+\sin 1.75^2+\sin 2^2)\\ &\approx 0.69622\end{aligned}$$

■

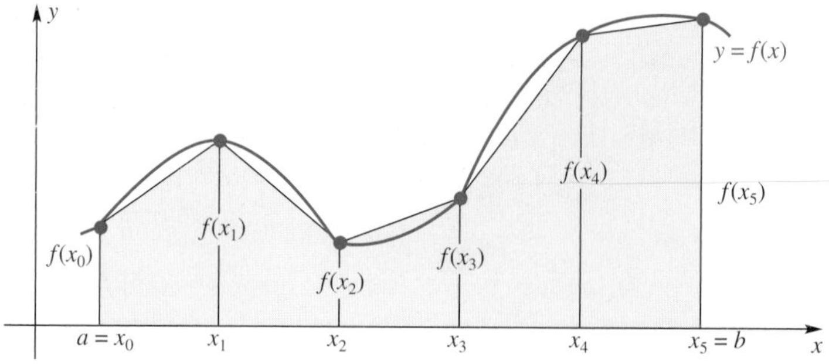

Figure 1

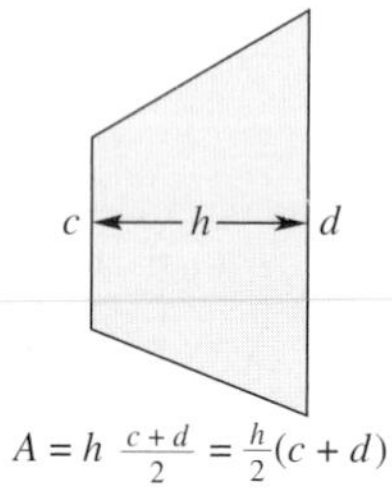

Figure 2

The Trapezoidal Rule Suppose we join the pairs of points $(x_{i-1}, f(x_{i-1}))$ and $(x_i, f(x_i))$ by line segments as shown in Figure 1, thus forming n trapezoids. Then instead of approximating the area under the curve by summing the areas of *rectangles*, we approximate it by summing the areas of the *trapezoids*. This method is called the **Trapezoidal Rule.**

Recalling the area formula shown in Figure 2, we can write the area of the ith trapezoid as

$$A_i = \frac{h}{2}[f(x_{i-1}) + f(x_i)]$$

More accurately, we should say *signed* area, since A_i will be negative for a subinterval where f is negative. The definite integral $\int_a^b f(x)\,dx$ is approximately equal to $A_1 + A_2 + \cdots + A_n$, that is, to

$$\frac{h}{2}[f(x_0) + f(x_1)] + \frac{h}{2}[f(x_1) + f(x_2)] + \cdots + \frac{h}{2}[f(x_{n-1}) + f(x_n)]$$

This simplifies to the **Trapezoidal Rule:**

Trapezoidal Rule

$$\int_a^b f(x)\,dx \approx \frac{h}{2}[f(x_0) + 2f(x_1) + 2f(x_2) + \cdots + 2f(x_{n-1}) + f(x_n)]$$

$$= \frac{b-a}{2n}\left[f(a) + 2\sum_{i=1}^{n-1} f\left(a + i\frac{b-a}{n}\right) + f(b)\right]$$

EXAMPLE 3 Approximate the definite integral $\int_0^2 \sin x^2\,dx$ using the Trapezoidal Rule with $n = 8$.

SOLUTION This is the same integrand and interval as in Example 2.

$$\int_0^2 \sin x^2\,dx \approx \frac{b-a}{2n}\left[f(a) + 2\sum_{i=1}^{7} f\left(a + i\frac{b-a}{n}\right) + f(b)\right]$$

$$= 0.125\left[\sin 0^2 + 2(\sin 0.25^2 + \sin 0.5^2 + \sin 0.75^2 + \sin 1^2 + \sin 1.25^2 + \sin 1.5^2 + \sin 1.75^2) + \sin 2^2\right]$$

$$\approx 0.79082$$

■

Presumably we could get a better approximation by taking n larger; this would be easy to do using a computer. However, while taking n larger reduces the error of the method, it at least potentially increases the error of calculation. It would be unwise, for example, to take $n = 1{,}000{,}000$, since the potential round-off errors would more than compensate for the fact that the error of the method would be minuscule. We will have more to say about errors shortly.

The Parabolic Rule (Simpson's Rule) In the Trapezoidal Rule, we approximated the curve $y = f(x)$ by line segments. It seems likely that we could do better using parabolic segments. Just as before, partition the interval $[a, b]$ into n subintervals of length $h = (b - a)/n$, but this time with n an *even* number. Then fit parabolic segments to neighboring triples of points, as shown in Figure 3.

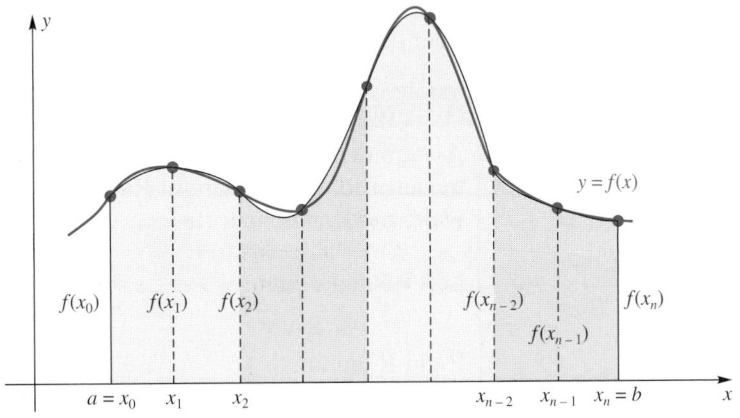

Figure 3

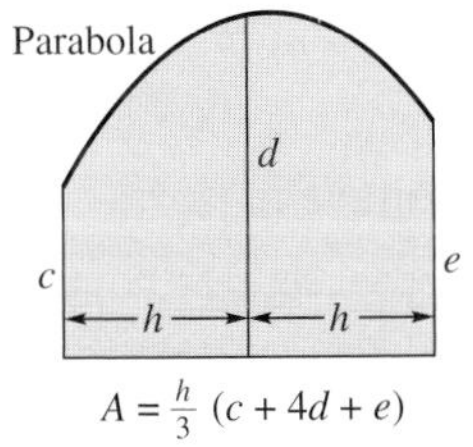

Figure 4

Using the area formula in Figure 4 (see Problem 17 for the derivation) leads to an approximation called the **Parabolic Rule.** It is also called **Simpson's Rule,** after the English mathematician Thomas Simpson (1710–1761).

Parabolic Rule (n even)

$$\int_a^b f(x)\,dx \approx \frac{h}{3}[f(x_0) + 4f(x_1) + 2f(x_2) + \cdots + 4f(x_{n-1}) + f(x_n)]$$

$$= \frac{b-a}{3n}\left[f(a) + 4\sum_{i=1}^{n/2} f\left(a + (2i-1)\frac{b-a}{n}\right) + 2\sum_{i=1}^{n/2-1} f\left(a + 2i\frac{b-a}{n}\right) + f(b)\right]$$

The pattern of coefficients is 1, 4, 2, 4, 2, 4, 2, . . . , 2, 4, 1.

EXAMPLE 4 Approximate the definite integral $\int_0^3 \frac{1}{1+x^2}\,dx$ using the Parabolic Rule with $n = 6$.

SOLUTION Let $f(x) = 1/(1 + x^2)$, $a = 0$, $b = 3$, and $n = 6$. The x_i's are $x_0 = 0, x_1 = 0.5, x_2 = 1.0, \ldots, x_6 = 3.0$

$$\int_0^3 \frac{1}{1+x^2}\,dx \approx \frac{3-0}{3\cdot 6}[f(0) + 4f(0.5) + 2f(1.0) + 4f(1.5) + 2f(2.0) + 4f(2.5) + f(3.0)]$$

$$= \frac{1}{6}(1 + 4\cdot 0.8 + 2\cdot 0.5 + 4\cdot 0.30769 + 2\cdot 0.2 + 4\cdot 0.13793 + 0.1)$$

$$= 1.2471$$

■

Error Analysis In any practical use of the approximation methods described in this section, we need to have some idea of the size of the error involved. Fortunately, the methods described in this section have fairly simple error formulas, provided the integrand possesses sufficiently many derivatives. We call E_n the error if it satisfies

$$\int_a^b f(x)\,dx = \text{approximation based on } n \text{ subintervals} + E_n$$

The error formulas are given in the next theorem. The proofs of these results are rather difficult and we omit them here.

Theorem A

Assuming that the required derivatives exist on the interval $[a, b]$, the errors for the left Riemann sum, right Riemann sum, midpoint Riemann sum, Trapezoidal Rule, and Parabolic Rule are

Left Riemann Sum: $$E_n = \frac{(b-a)^2}{2n} f'(c) \text{ for some } c \text{ in } [a, b]$$

Right Riemann Sum: $$E_n = -\frac{(b-a)^2}{2n} f'(c) \text{ for some } c \text{ in } [a, b]$$

Midpoint Riemann Sum: $$E_n = \frac{(b-a)^3}{24n^2} f''(c) \text{ for some } c \text{ in } [a, b]$$

Trapezoidal Rule: $$E_n = -\frac{(b-a)^3}{12n^2} f''(c) \text{ for some } c \text{ in } [a, b]$$

Parabolic Rule: $$E_n = -\frac{(b-a)^5}{180n^4} f^{(4)}(c) \text{ for some } c \text{ in } [a, b]$$

The most important thing to notice about these error formulas is the position of n, the number of subintervals. In all cases, the n occurs raised to some power in the *denominator*. Thus, as n increases, the error decreases. Also, the larger the exponent on n, the faster the error term will go to zero. For example, the error term for the Parabolic Rule involves an n^4 in the denominator. Since n^4 grows much faster than n^2, the error term for the Parabolic Rule will go to zero faster than the error term for the Trapezoidal Rule or the midpoint Riemann sum rule. Similarly, the error term for the Trapezoidal Rule will go to zero faster than the error term for the left or right Riemann sum rules. One other thing to notice about these error formulas is that they hold "for some c in $[a, b]$." In most practical situations we can never tell what the value of c is. All we can hope to do is obtain an upper bound on how large the error could be. The next example illustrates this.

EXAMPLE 5 Approximate the definite integral $\int_1^4 \frac{1}{1+x}\,dx$ using the Parabolic Rule with $n = 6$ and give a bound for the absolute value of the error.

SOLUTION Let $f(x) = \frac{1}{1+x}$, $a = 1$, $b = 4$, and $n = 6$. Then

$$\begin{aligned}\int_1^4 \frac{1}{1+x}\,dx &\approx \frac{b-a}{3n}[f(x_0) + 4f(x_1) + 2f(x_2) + 4f(x_3) + 2f(x_4) + \\ &\qquad 4f(x_5) + f(x_6)] \\ &= \frac{3}{3(6)}[f(1.0) + 4f(1.5) + 2f(2.0) + 4f(2.5) + 2f(3.0) + \\ &\qquad 4f(3.5) + f(4.0)] \\ &\approx \frac{1}{6}(5.4984) \approx 0.9164\end{aligned}$$

The error term for the Parabolic Rule involves the fourth derivative of the integrand:

$$f'(x) = -\frac{1}{(1+x)^2}$$

$$f''(x) = \frac{2}{(1+x)^3}$$

$$f'''(x) = -\frac{6}{(1+x)^4}$$

$$f^{(4)}(x) = \frac{24}{(1+x)^5}$$

The question we now face is, how large can $|f^{(4)}(x)|$ be on the interval $[1, 4]$? It is clear that $f^{(4)}(x)$ is a nonnegative decreasing function on this interval, so its absolute value achieves its largest value at the left endpoint, that is, when $x = 1$. The value of the fourth derivative at $x = 1$ is $f^{(4)}(1) = 24/(1+1)^5 = 3/4$. Thus

$$|E_6| = \left|-\frac{(b-a)^5}{180n^4} f^{(4)}(c)\right| = \frac{(4-1)^5}{180 \cdot 6^4}|f^{(4)}(c)| \le \frac{(4-1)^5}{180 \cdot 6^4}\,\frac{3}{4} \approx 0.00078$$

The error is therefore no larger than 0.00078. ■

In the next example, we turn things around. Rather than specifying n and asking for the error, we give the desired error and ask how large n must be.

EXAMPLE 6 How large must n be in order to guarantee that the absolute value of the error is less than 0.00001 when we use (a) the right Riemann sum, (b) the Trapezoidal Rule, and (c) the Parabolic Rule to estimate $\int_1^4 \frac{1}{1+x}\,dx$?

SOLUTION The derivatives of the integrand $f(x) = 1/(1+x)$ are given in the previous example.

(a) The absolute value of the error term for the right Riemann sum is

$$|E_n| = \left|-\frac{(4-1)^2}{2n} f'(c)\right| = \frac{3^2}{2n}\left|\frac{1}{(1+c)^2}\right| \le \frac{9}{2n}\,\frac{1}{(1+1)^2} = \frac{9}{8n}$$

We want $|E_n| \le 0.00001$, so we require

$$\frac{9}{8n} \le 0.00001$$

$$n \ge \frac{9}{8 \cdot 0.00001} = 112{,}500$$

(b) For the Trapezoidal Rule we have

$$|E_n| = \left|-\frac{(4-1)^3}{12n^2} f''(c)\right| = \frac{3^3}{12n^2}\left|\frac{2}{(1+c)^3}\right| \le \frac{54}{12n^2(1+1)^3} = \frac{9}{16n^2}$$

We want $|E_n| \le 0.00001$, so n must satisfy

$$\frac{9}{16n^2} \le 0.00001$$

$$n^2 \ge \frac{9}{16 \cdot 0.00001} = 56{,}250$$

$$n \ge \sqrt{56{,}250} \approx 237.17$$

Thus, $n = 238$ should do it.

(c) For the Parabolic Rule,

$$|E_n| = \left| -\frac{(b-a)^5}{180n^4} f^{(4)}(c) \right| = \frac{3^5}{180n^4} \left| \frac{24}{(1+c)^5} \right| \le \frac{3^5 \cdot 24}{180n^4(1+1)^5} = \frac{81}{80n^4}$$

We want $|E_n| \le 0.00001$, so

$$\frac{81}{80n^4} \le 0.00001$$

$$n^4 \ge \frac{81}{80 \cdot 0.00001} \approx 101{,}250$$

$$n \ge 101{,}250^{1/4} \approx 17.8$$

We must round up to the next even integer (since n must be even for the Parabolic Rule). Thus we require $n = 18$. ■

Notice how much different the answers were for the three parts in the previous example. Eighteen subintervals for the Parabolic Rule will give about the same accuracy as over 100,000 subintervals for the right Riemann sum! The Parabolic Rule is indeed a powerful method for approximating definite integrals.

Functions Defined by a Table In all the previous examples, the function we integrated was defined over the whole interval of integration. There are many situations where this is not the case. For example, speed is measured every minute, water flow from a tank is measured every 10 seconds, and cross-sectional area is measured every 0.1 millimeter. In all of these cases, the integral has a clearly defined meaning. Although we cannot obtain the integral exactly, we can use the methods of this section to approximate the integral.

Minutes	Speed
0	0
10	55
20	57
30	60
40	70
50	70
60	70
70	70
80	19
90	0
100	59
110	63
120	65
130	62
140	0
150	0
160	0
170	22
180	38
190	35
200	25
210	0

EXAMPLE 7 While his father drove from St. Louis to Jefferson City, Chris noted the speed of the car every 10 minutes, that is, every one-sixth of an hour. The table to the left shows these speedometer readings. Use the Trapezoidal Rule to approximate how far they drove.

SOLUTION Let $v(t)$ denote the velocity of the car at time t, where t is measured in hours since the beginning of the trip. If we knew $v(t)$ for all t in the interval $[0, 3.5]$, we could find the distance traveled by taking $\int_0^{3.5} v(t)\,dt$. The problem is, we know $v(t)$ only for 22 values of t: $t_k = k/6$, where $k = 0, 1, 2, \ldots, 21$. Figure 5 shows a graph of the information we are given. We partition the interval $[0, 3.5]$ into 21 intervals of width $\frac{1}{6}$ (since 10 minutes is one-sixth of an hour). The Trapezoidal Rule then gives

$$\int_0^{3.5} v(t)\,dt \approx \frac{3.5-0}{2\cdot 21}\left[v(0) + 2\sum_{i=1}^{20} v\left(0 + i\frac{3.5-0}{21}\right) + v(21)\right]$$

$$= \frac{3.5}{42}[0 + 2(55 + 57 + 60 + 70 + 70 + 70 + 70 + 19 + 0 + 59$$

$$+ 63 + 65 + 62 + 0 + 0 + 0 + 22 + 38 + 35 + 25) + 0]$$

$$= 140$$

They drove approximately 140 miles. ■

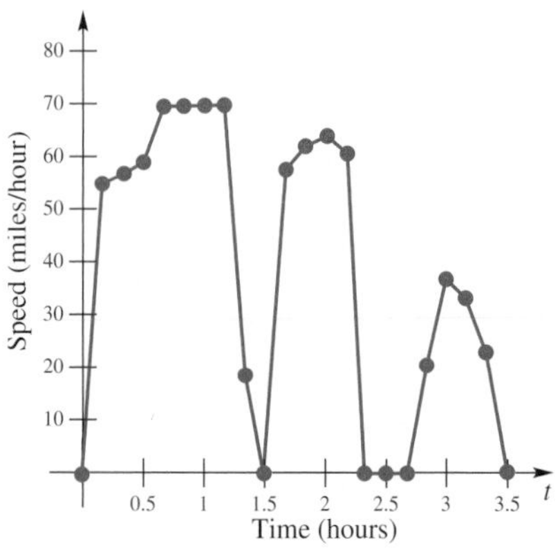

Figure 5

Concepts Review

1. The pattern of coefficients in the Trapezoidal Rule is ______.

2. The pattern of coefficients in the Parabolic Rule is ______.

3. The error in the Trapezoidal Rule has n^2 in the denominator, whereas the error in the Parabolic Rule has ______ in the denominator, so we expect the latter to give a better approximation to a definite integral.

4. If f is positive and concave up, then the Trapezoidal Rule will always give a value for $\int_a^b f(x)\,dx$ that is too ______.

Problem Set 4.6

C *In Problems 1–6, use the methods of (1) left Riemann sum, (2) right Riemann sum, (3) Trapezoidal Rule, (4) Parabolic Rule with $n = 8$ to approximate the definite integral. Then use the Second Fundamental Theorem of Calculus to find the exact value of each integral.*

1. $\int_1^3 \frac{1}{x^2}\,dx$ **2.** $\int_1^3 \frac{1}{x^3}\,dx$

3. $\int_0^2 \sqrt{x}\,dx$ **4.** $\int_1^3 x\sqrt{x^2+1}\,dx$

5. $\int_0^1 x(x^2+1)^5\,dx$ **6.** $\int_1^4 (x+1)^{3/2}\,dx$

C *In Problems 7–10, use the methods of (1) left Riemann sum, (2) right Riemann sum, (3) midpoint Riemann sum, (4) Trapezoidal Rule, (5) Parabolic Rule with $n = 4, 8, 16$. (Note that none of these can be evaluated using the Second Fundamental Theorem of Calculus with the techniques you have learned so far.) Present your approximations in a table like this:*

	LRS	RRS	MRS	Trap	Parabolic
$n = 4$					
$n = 8$					
$n = 16$					

7. $\int_1^3 \frac{1}{1+x^2}\,dx$ **8.** $\int_1^3 \frac{1}{x}\,dx$

9. $\int_0^2 \sqrt{x^2+1}\,dx$ **10.** $\int_1^3 x\sqrt{x^3+1}\,dx$

C *In Problems 11–14, determine an n so that the Trapezoidal Rule will approximate the integral with an error E_n satisfying $|E_n| \le 0.01$. Then, using that n, approximate the integral.*

11. $\int_1^3 \frac{1}{x}\,dx$ **12.** $\int_1^3 \frac{1}{1+x}\,dx$

13. $\int_1^4 \sqrt{x}\,dx$ **14.** $\int_1^3 \sqrt{x+1}\,dx$

C *In Problems 15–16, determine an n so that the Parabolic Rule will approximate the integral with an error E_n satisfying $|E_n| \le 0.01$. Then, using that n, approximate the integral.*

15. $\int_1^3 \frac{1}{x}\,dx$ **16.** $\int_4^8 \sqrt{x+1}\,dx$

17. Let $f(x) = ax^2 + bx + c$. Show that

$$\int_{m-h}^{m+h} f(x)\,dx \quad\text{and}\quad \frac{h}{3}[f(m-h) + 4f(m) + f(m+h)]$$

both have the value $(h/3)[a(6m^2 + 2h^2) + b(6m) + 6c]$. This establishes the area formula on which the Parabolic Rule is based.

18. Show that the Parabolic Rule is exact for any cubic polynomial in two different ways.

(a) By direct calculation.

(b) By showing that $E_n = 0$.

Justify your answers to Problems 19–22 two ways: (1) using the properties of the graph of the function, and (2) using the error formulas from Theorem A.

19. If a function f is increasing on $[a, b]$, will the left Riemann sum be larger or smaller than $\int_a^b f(x)\,dx$?

20. If a function f is increasing on $[a, b]$, will the right Riemann sum be larger or smaller than $\int_a^b f(x)\,dx$?

21. If a function f is concave down on $[a, b]$, will the midpoint Riemann sum be larger or smaller than $\int_a^b f(x)\,dx$?

22. If a function f is concave down on $[a, b]$, will the Trapezoidal Rule approximation be larger or smaller than $\int_a^b f(x)\,dx$?

23. Show that the Parabolic Rule gives the exact value of $\int_{-a}^{a} x^k\,dx$ provided that k is odd.

24. It is interesting that a modified version of the Trapezoidal Rule turns out to be in general more accurate than the Parabolic Rule. This version says that

$$\int_a^b f(x)\,dx \approx T - \frac{[f'(b) - f'(a)]h^2}{12}$$

where T is the standard trapezoidal estimate.

(a) Use this formula with $n = 8$ to estimate $\int_1^3 x^4\,dx$ and note its remarkable accuracy.

(b) Use this formula with $n = 12$ to estimate $\int_0^{\pi} \sin x\,dx$.

25. Without doing any calculations, rank from smallest to largest the approximations of $\int_0^1 \sqrt{x^2+1}\,dx$ for the following methods: left Riemann sum, right Riemann sum, midpoint Riemann sum, Trapezoidal Rule.

26. Without doing any calculations, rank from smallest to largest the approximations of $\int_1^3 (x^3+x^2+x+1)\,dx$ for the following methods: left Riemann sum, right Riemann sum, Trapezoidal Rule, Parabolic Rule.

27. Use the Trapezoidal Rule to approximate the area of the lakeside lot shown in Figure 6. Dimensions are in feet.

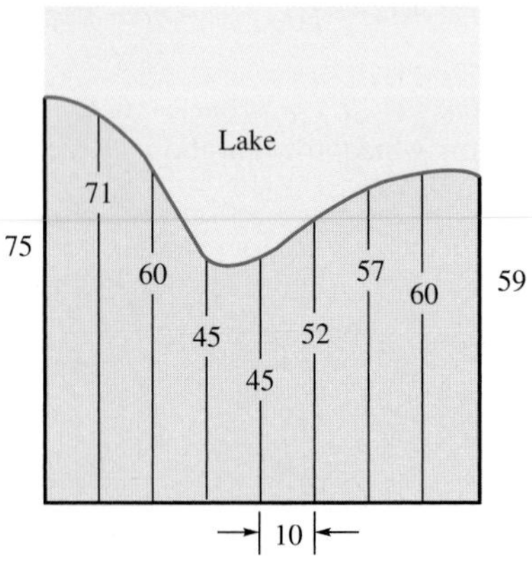

Figure 6

28. Use the Parabolic Rule to approximate the amount of water required to fill a pool shaped like Figure 7 to a depth of 6 feet. All dimensions are in feet.

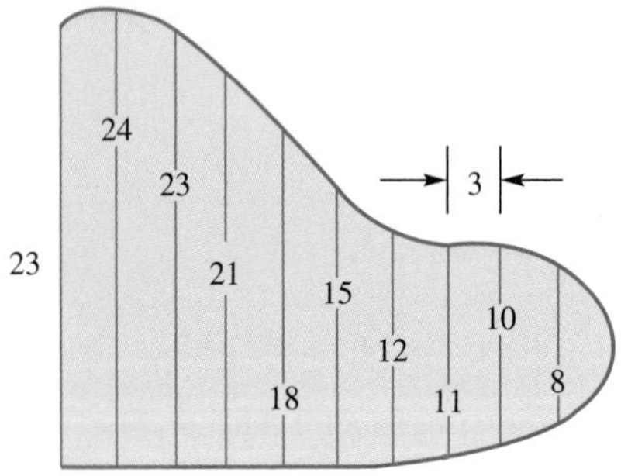

Figure 7

C **29.** Figure 8 shows the depth in feet of the water in a river measured at 20-foot intervals across the width of the river. If the river flows at 4 miles per hour, how much water (in cubic feet) flows past the place where these measurements were taken in one day? Use the Parabolic Rule.

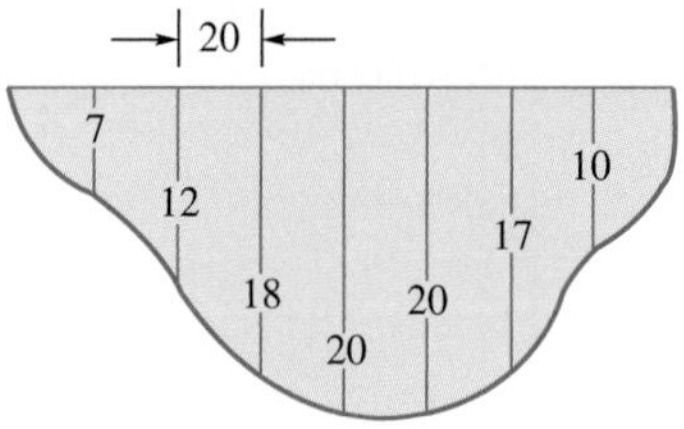

Figure 8

30. On her way to work, Teri noted her speed every 3 minutes. The results are shown in the table below. How far did she drive?

Time (minutes)	0	3	6	9	12	15	18	21	24
Speed (mi/h)	0	31	54	53	52	35	31	28	0

31. Every 12 minutes between 4:00 P.M. and 6:00 P.M., the rate (in gallons per minute) at which water flowed out of a town's water tank was measured. The results are shown in the table below. How much water was used in this 2-hour span?

Time	4:00	4:12	4:24	4:36	4:48	5:00
Flow (gal/min)	65	71	68	78	105	111

Time	5:12	5:24	5:36	5:48	6:00
Flow (gal/min)	108	144	160	152	148

Answers to Concepts Review: **1.** 1, 2, 2, ..., 2, 1 **2.** 1, 4, 2, 4, 2, ..., 4, 1 **3.** n^4 **4.** large

4.7 Chapter Review

Concepts Test

Respond with true or false to each of the following assertions. Be prepared to justify your answer.

1. The indefinite integral is a linear operator.

2. $\int [f(x)g'(x) + g(x)f'(x)]\,dx = f(x)g(x) + C.$

3. All functions that are antiderivatives must have derivatives.

4. If the second derivatives of two functions are equal, then the functions differ at most by a constant.

5. $\int f'(x)\,dx = f(x)$ for every differentiable function f.

6. If $s = -16t^2 + v_0 t$ gives the height at time t of a ball thrown straight up from the surface of the earth with velocity v_0 at time $t = 0$, then the ball will hit the ground with velocity $-v_0$.

7. $\sum_{i=1}^{n}(a_i + a_{i-1}) = a_0 + a_n + 2\sum_{i=1}^{n-1} a_i.$

8. $\sum_{i=1}^{100}(2i-1) = 10{,}000$.

9. If $\sum_{i=1}^{10} a_i^2 = 100$ and $\sum_{i=1}^{10} a_i = 20$, then $\sum_{i=1}^{10}(a_i+1)^2 = 150$.

10. If f is bounded on $[a, b]$, then f is integrable there.

11. $\int_a^a f(x)\,dx = 0$.

12. If $\int_a^b f(x)\,dx = 0$, then $f(x) = 0$ for all x in $[a, b]$.

13. If $\int_a^b [f(x)]^2\,dx = 0$, then $f(x) = 0$ for all x in $[a, b]$.

14. If $a > x$ and $G(x) = \int_a^x f(z)\,dz$, then $G'(x) = -f(x)$.

15. The value of $\int_x^{x+2\pi} (\sin t + \cos t)\,dt$ is independent of x.

16. The operator lim is linear.

17. $\int_{-\pi}^{\pi} \sin^{13} x\,dx = 0$.

18. $\int_1^5 \sin^2 x\,dx = \int_1^7 \sin^2 x\,dx + \int_7^5 \sin^2 x\,dx$.

19. If f is continuous and positive everywhere, then $\int_c^d f(x)\,dx$ is positive.

20. $D_x\left[\int_0^{x^2} \frac{1}{1+t^2}\,dt\right] = \frac{1}{1+x^4}$.

21. $\int_0^{2\pi} |\sin x|\,dx = \int_0^{2\pi} |\cos x|\,dx$.

22. $\int_0^{2\pi} |\sin x|\,dx = 4\int_0^{\pi/2} \sin x\,dx$.

23. The antiderivatives of odd functions are even functions.

24. If $F(x)$ is an antiderivative of $f(x)$, then $F(5x)$ is an antiderivative of $f(5x)$.

25. If $F(x)$ is an antiderivative of $f(x)$, then $F(2x+1)$ is an antiderivative of $f(2x+1)$.

26. If $F(x)$ is an antiderivative of $f(x)$, then $F(x)+1$ is an antiderivative of $f(x)+1$.

27. If $F(x)$ is an antiderivative of $f(x)$, then

$$\int f(v(x))\,dx = F(v(x)) + C$$

28. If $F(x)$ is an antiderivative of $f(x)$, then

$$\int f^2(x)\,dx = \tfrac{1}{3}F^3(x) + C$$

29. If $F(x)$ is an antiderivative of $f(x)$, then

$$\int f(x)\frac{df}{dx}\,dx = \tfrac{1}{2}F^2(x) + C$$

30. If $f(x) = 4$ on $[0, 3]$, then every Riemann sum for f on the given interval has the value 12.

31. If $F'(x) = G'(x)$ for all x in $[a, b]$, then $F(b) - F(a) = G(b) - G(a)$.

32. If $f(x) = f(-x)$ for all x in $[-a, a]$, then $\int_{-a}^{a} f(x)\,dx = 0$.

33. If $\bar{z} = \frac{1}{2}\int_{-1}^{1} z(t)\,dt$, then $z(t) - \bar{z}$ is an odd function for $-1 \le t \le 1$.

34. If $F'(x) = f(x)$ for all x in $[0, b]$, then $\int_0^b f(x)\,dx = F(b)$.

35. $\int_{-99}^{99} (ax^3 + bx^2 + cx)\,dx = 2\int_0^{99} bx^2\,dx$.

36. If $f(x) \le g(x)$ on $[a. b]$, then $\int_a^b |f(x)|\,dx \le \int_a^b |g(x)|\,dx$.

37. If $f(x) \le g(x)$ on $[a. b]$, then $\left|\int_a^b f(x)\,dx\right| \le \left|\int_a^b g(x)\,dx\right|$.

38. $\left|\sum_{i=1}^{n} a_i\right| \le \sum_{i=1}^{n} |a_i|$.

39. If f is continuous on $[a. b]$, then $\left|\int_a^b f(x)\,dx\right| \le \int_a^b |f(x)|\,dx$.

40. $\lim_{n\to\infty} \sum_{i=1}^{n} \sin\left(\frac{2i}{n}\right)\cdot\frac{2}{n} = \int_0^2 \sin x\,dx$.

41. If $\|P\| \to 0$, then the number of subintervals in the partition P tends to ∞.

42. We can always express the indefinite integral of an elementary function in terms of elementary functions.

43. For an increasing function, the left Riemann sum will always be less than the right Riemann sum.

44. For a linear function $f(x)$, the midpoint Riemann sum will give the exact value of $\int_a^b f(x)\,dx$ no matter what n is.

45. The Trapezoidal Rule with $n = 10$ will give an estimate for $\int_0^5 x^3\,dx$ that is smaller than the true value.

46. The Parabolic Rule with $n = 10$ will give the exact value of $\int_0^5 x^3\,dx$.

Sample Test Problems

In Problems 1–12, evaluate the indicated integrals.

1. $\int_0^1 \left(x^3 - 3x^2 + 3\sqrt{x}\right)dx$

2. $\int_1^2 \frac{2x^4 - 3x^2 + 1}{x^2}\,dx$

3. $\int_1^{\pi} \frac{y^3 - 9y\sin y + 26y^{-1}}{y}\,dy$

4. $\int_4^9 y\sqrt{y^2 - 4}\,dy$

5. $\int_2^8 z(2z^2 - 3)^{1/3}\,dz$

6. $\int_0^{\pi/2} \cos^4 x \sin x\,dx$

7. $\int_0^{\pi} (x+1)\tan^2(3x^2 + 6x)\sec^2(3x^2 + 6x)\,dx$

8. $\int_0^2 \frac{t^3}{\sqrt{t^4 + 9}}\,dt$

9. $\int_1^2 t^4(t^5 + 5)^{2/3}\,dt$

10. $\displaystyle\int_2^3 \frac{y^2 - 1}{(y^3 - 3y)^2}\, dy$

11. $\displaystyle\int (x + 1) \sin (x^2 + 2x + 3)\, dx$

12. $\displaystyle\int_1^5 \frac{(y^2 + y + 1)}{\sqrt[5]{2y^3 + 3y^2 + 6y}}\, dy$

13. Let P be a regular partition of the interval $[0, 2]$ into four equal subintervals, and let $f(x) = x^2 - 1$. Write out the Riemann sum for f on P, in which $\bar{x}_i$ is the right end point of each subinterval of $P, i = 1, 2, 3, 4$. Find the value of this Riemann sum and make a sketch.

14. If $f(x) = \displaystyle\int_{-2}^{x} \frac{1}{t + 3}\, dt, -2 \le x$, find $f'(7)$.

15. Evaluate $\displaystyle\int_0^3 (2 - \sqrt{x + 1})^2\, dx$.

16. If $f(x) = 3x^2\sqrt{x^3 - 4}$, find the average value of f on $[2, 5]$.

17. Evaluate $\displaystyle\int_2^4 \frac{5x^2 - 1}{x^2}\, dx$.

18. Evaluate $\displaystyle\sum_{i=1}^{n} (3^i - 3^{i-1})$.

19. Evaluate $\displaystyle\sum_{i=1}^{10} (6i^2 - 8i)$.

20. Evaluate each sum.

(a) $\displaystyle\sum_{m=2}^{4} \left(\frac{1}{m}\right)$ (b) $\displaystyle\sum_{i=1}^{6} (2 - i)$ (c) $\displaystyle\sum_{k=0}^{4} \cos\left(\frac{k\pi}{4}\right)$

21. Write in sigma notation.

(a) $\displaystyle\frac{1}{2} + \frac{1}{3} + \frac{1}{4} + \cdots + \frac{1}{78}$

(b) $x^2 + 2x^4 + 3x^6 + 4x^8 + \cdots + 50x^{100}$

22. Sketch the region under the curve $y = 16 - x^2$ between $x = 0$ and $x = 3$, showing the inscribed polygon corresponding to a regular partition of $[0, 3]$ into n subintervals. Find a formula for the area of this polygon and then find the area under the curve by taking a limit.

23. If $\displaystyle\int_0^1 f(x)\, dx = 4$, $\displaystyle\int_0^2 f(x)\, dx = 2$, and $\displaystyle\int_0^2 g(x)\, dx = -3$, evaluate each integral.

(a) $\displaystyle\int_1^2 f(x)\, dx$ (b) $\displaystyle\int_1^0 f(x)\, dx$

(c) $\displaystyle\int_0^2 3f(u)\, du$ (d) $\displaystyle\int_0^2 [2g(x) - 3f(x)]\, dx$

(e) $\displaystyle\int_0^{-2} f(-x)\, dx$

24. Evaluate each integral.

(a) $\displaystyle\int_0^4 |x - 1|\, dx$ (b) $\displaystyle\int_0^4 [\![x]\!]\, dx$

(c) $\displaystyle\int_0^4 (x - [\![x]\!])\, dx$

Hint: In parts (a) and (b), first sketch a graph.

25. Suppose that $f(x) = f(-x)$, $f(x) \le 0$, $g(-x) = -g(x)$, $\displaystyle\int_0^2 f(x)\, dx = -4$, and $\displaystyle\int_0^2 g(x)\, dx = 5$. Evaluate each integral.

(a) $\displaystyle\int_{-2}^{2} f(x)\, dx$ (b) $\displaystyle\int_{-2}^{2} |f(x)|\, dx$

(c) $\displaystyle\int_{-2}^{2} g(x)\, dx$ (d) $\displaystyle\int_{-2}^{2} [f(x) + f(-x)]\, dx$

(e) $\displaystyle\int_0^2 [2g(x) + 3f(x)]\, dx$ (f) $\displaystyle\int_{-2}^{0} g(x)\, dx$

26. Evaluate $\displaystyle\int_{-100}^{100} (x^3 + \sin^5 x)\, dx$.

27. Find c of the Mean Value Theorem for Integrals for $f(x) = 3x^2$ on $[-4, -1]$.

28. Find $G'(x)$ for each function G.

(a) $G(x) = \displaystyle\int_1^x \frac{1}{t^2 + 1}\, dt$

(b) $G(x) = \displaystyle\int_1^{x^2} \frac{1}{t^2 + 1}\, dt$

(c) $G(x) = \displaystyle\int_x^{x^3} \frac{1}{t^2 + 1}\, dt$

29. Find $G'(x)$ for each function G.

(a) $G(x) = \displaystyle\int_1^x \sin^2 z\, dz$

(b) $G(x) = \displaystyle\int_x^{x+1} f(z)\, dz$

(c) $G(x) = \displaystyle\frac{1}{x}\int_0^x f(z)\, dz$

(d) $G(x) = \displaystyle\int_0^x \left(\int_0^u f(t)\, dt\right) du$

(e) $G(x) = \displaystyle\int_0^{g(x)} \left(\frac{d}{du} g(u)\right) du$

(f) $G(x) = \displaystyle\int_0^{-x} f(-t)\, dt$

30. Evaluate each of the following limits by recognizing it as a definite integral.

(a) $\displaystyle\lim_{n\to\infty} \sum_{i=1}^{n} \sqrt{\frac{4i}{n}} \cdot \frac{4}{n}$ (b) $\displaystyle\lim_{n\to\infty} \sum_{i=1}^{n} \left(1 + \frac{2i}{n}\right)^2 \frac{2}{n}$

31. Show that if $f(x) = \displaystyle\int_{2x}^{5x} \frac{1}{t}\, dt$, then f is a constant function on $(0, \infty)$.

32. Approximate $\displaystyle\int_1^2 \frac{1}{1 + x^4}\, dx$ using left, right, and midpoint Riemann sums with $n = 8$.

33. Approximate $\displaystyle\int_1^2 \frac{1}{1 + x^4}\, dx$ using the Trapezoidal Rule with $n = 8$, and give an upper bound for the absolute value of the error.

34. Approximate $\int_0^4 \frac{1}{1+2x}\,dx$ using the Parabolic Rule with $n = 8$, and give an upper bound for the absolute value of the error.

35. How large must n be for the Trapezoidal Rule in order to approximate $\int_1^2 \frac{1}{1+x^4}\,dx$ with an error no larger then 0.0001?

36. How large must n be for the Parabolic Rule in order to approximate $\int_0^4 \frac{1}{1+2x}\,dx$ with an error no larger then 0.0001?

37. Without doing any calculations, rank from smallest to largest the approximations of $\int_1^6 \frac{1}{x}\,dx$ for the following methods: left Riemann sum, midpoint Riemann sum, Trapezoidal rule.

REVIEW & PREVIEW PROBLEMS

In Problems 1–6, find the length of the solid green line.

1.
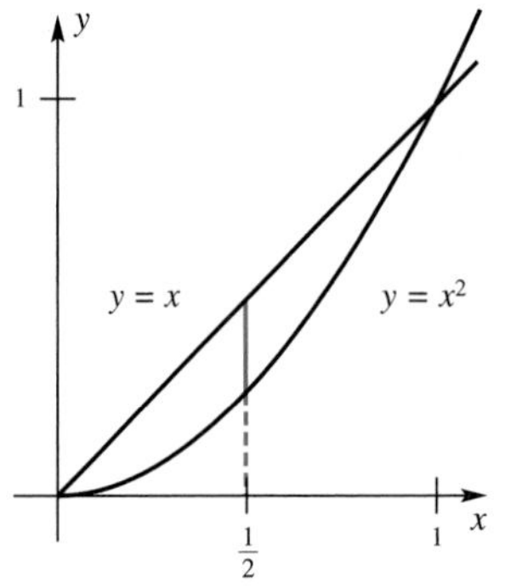

2.
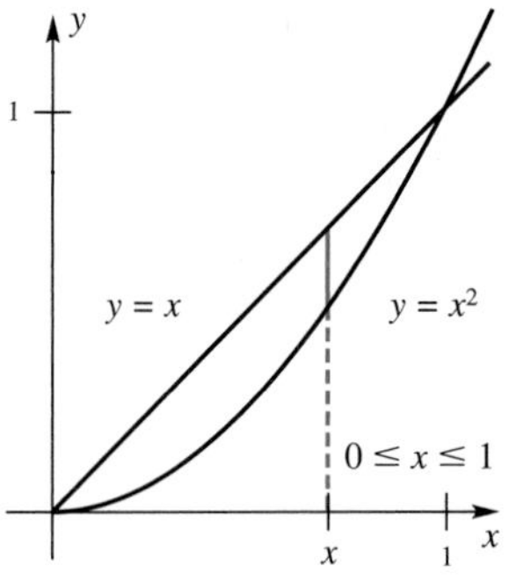

3.
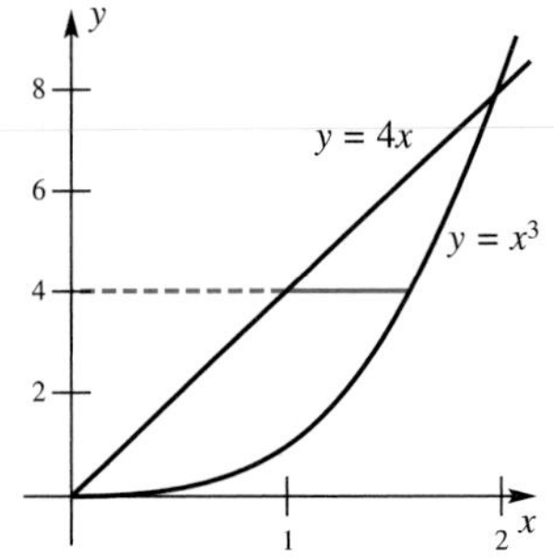

4.
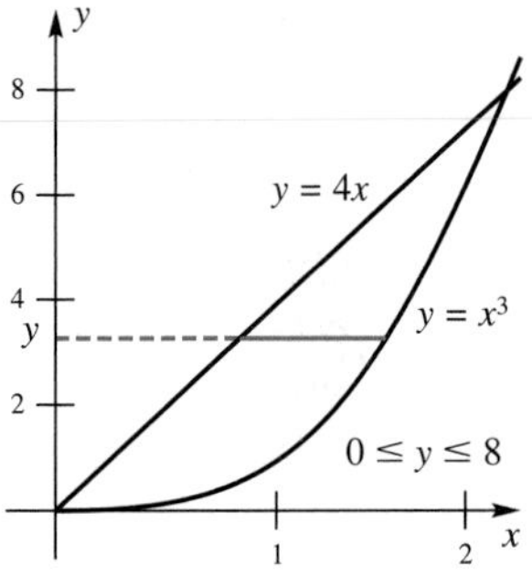

5.
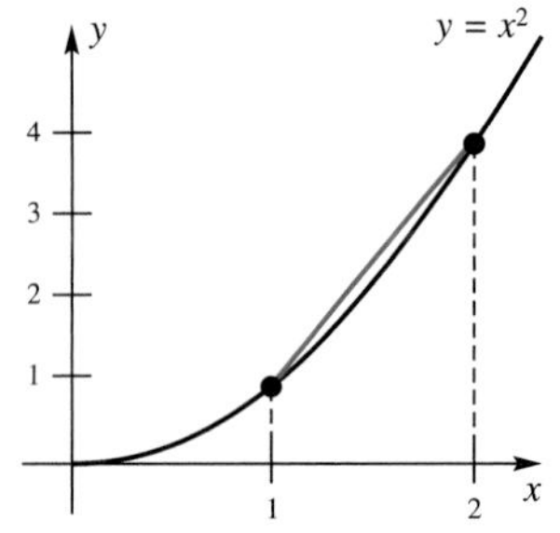

6.
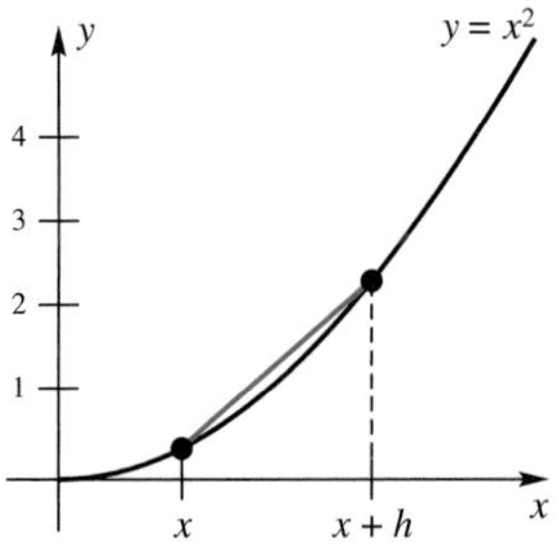

For each of the following figures, the volume of the solid is equal to the base area times the height. Give the volume of each of these solids.

7.
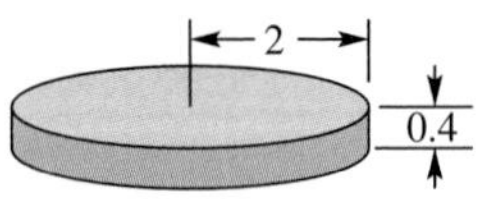

8.
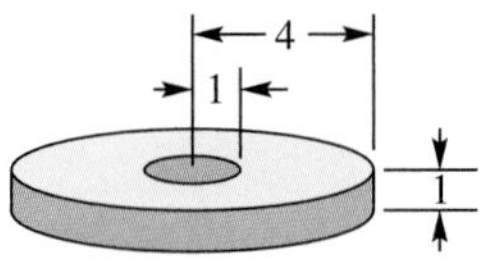

9.
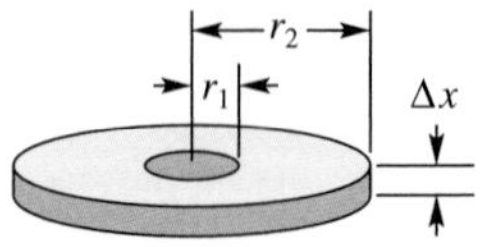

10.
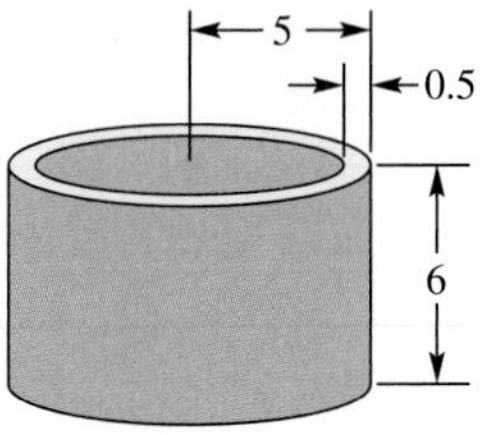

Evaluate each of the following definite integrals.

11. $\int_{-1}^{2} (x^4 - 2x^3 + 2)\, dx$

12. $\int_{0}^{3} y^{2/3}\, dy$

13. $\int_{0}^{2} \left(1 - \frac{x^2}{2} + \frac{x^4}{16}\right) dx$

14. $\int_{1}^{4} \sqrt{1 + \frac{9}{4}x}\, dx$

CHAPTER 5 Applications of the Integral

5.1 The Area of a Plane Region
5.2 Volumes of Solids: Slabs, Disks, Washers
5.3 Volumes of Solids of Revolution: Shells
5.4 Length of a Plane Curve
5.5 Work and Fluid Force
5.6 Moments and Center of Mass
5.7 Probability and Random Variables

5.1 The Area of a Plane Region

The brief discussion of area in Section 4.1 served to motivate the definition of the definite integral. With the latter notion now firmly established, we use the definite integral to calculate areas of regions of more and more complicated shapes. As is our practice, we begin with simple cases.

A Region above the x-Axis Let $y = f(x)$ determine a curve in the xy-plane and suppose that f is continuous and nonnegative on the interval $a \le x \le b$ (as in Figure 1). Consider the region R bounded by the graphs of $y = f(x)$, $x = a$, $x = b$, and $y = 0$. We refer to R as the region under $y = f(x)$ between $x = a$ and $x = b$. Its area $A(R)$ is given by

$$A(R) = \int_a^b f(x)\,dx$$

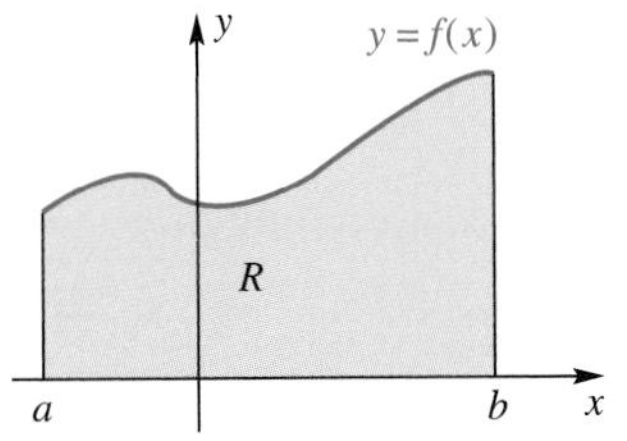

Figure 1

EXAMPLE 1 Find the area of the region R under $y = x^4 - 2x^3 + 2$ between $x = -1$ and $x = 2$.

SOLUTION The graph of R is shown in Figure 2. A reasonable estimate for the area of R is its base times an average height, say $(3)(2) = 6$. The exact value is

$$\begin{aligned} A(R) &= \int_{-1}^{2} (x^4 - 2x^3 + 2)\,dx = \left[\frac{x^5}{5} - \frac{x^4}{2} + 2x\right]_{-1}^{2} \\ &= \left(\frac{32}{5} - \frac{16}{2} + 4\right) - \left(-\frac{1}{5} - \frac{1}{2} - 2\right) = \frac{51}{10} = 5.1 \end{aligned}$$

The calculated value 5.1 is close enough to our estimate, 6, to give us confidence in its correctness. ■

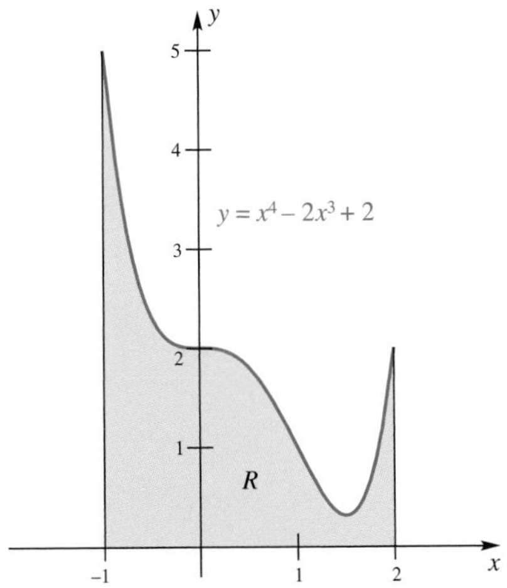

Figure 2

A Region Below the x-Axis Area is a nonnegative number. If the graph of $y = f(x)$ is below the x-axis, then $\int_a^b f(x)\,dx$ is a negative number and therefore cannot be an area. However, it is just the negative of the area of the region bounded by $y = f(x)$, $x = a$, $x = b$, and $y = 0$.

EXAMPLE 2 Find the area of the region R bounded by $y = x^2/3 - 4$, the x-axis, $x = -2$, and $x = 3$.

SOLUTION The region R is shown in Figure 3. Our preliminary estimate for its area is $(5)(3) = 15$. The exact value is

$$\begin{aligned} A(R) &= -\int_{-2}^{3}\left(\frac{x^2}{3} - 4\right)dx = \int_{-2}^{3}\left(-\frac{x^2}{3} + 4\right)dx \\ &= \left[-\frac{x^3}{9} + 4x\right]_{-2}^{3} = \left(-\frac{27}{9} + 12\right) - \left(\frac{8}{9} - 8\right) = \frac{145}{9} \approx 16.11 \end{aligned}$$

We are reassured by the nearness of 16.11 to our estimate. ■

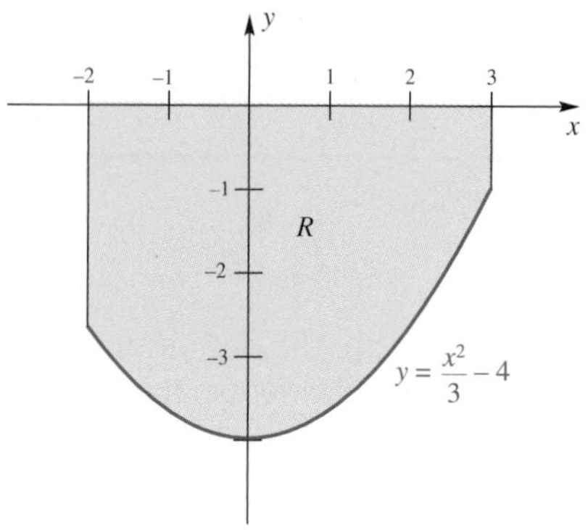

Figure 3

EXAMPLE 3 Find the area of the region R bounded by $y = x^3 - 3x^2 - x + 3$, the segment of the x-axis between $x = -1$ and $x = 2$, and the line $x = 2$.

SOLUTION The region R is shaded in Figure 4. Note that part of it is above the x-axis and part is below. The areas of these two parts, R_1 and R_2, must be calculated separately. You can check that the curve crosses the x-axis at -1, 1, and 3. Thus,

$$\begin{aligned} A(R) &= A(R_1) + A(R_2) \\ &= \int_{-1}^{1} (x^3 - 3x^2 - x + 3)\,dx - \int_{1}^{2} (x^3 - 3x^2 - x + 3)\,dx \\ &= \left[\frac{x^4}{4} - x^3 - \frac{x^2}{2} + 3x\right]_{-1}^{1} - \left[\frac{x^4}{4} - x^3 - \frac{x^2}{2} + 3x\right]_{1}^{2} \\ &= 4 - \left(-\frac{7}{4}\right) = \frac{23}{4} \end{aligned}$$

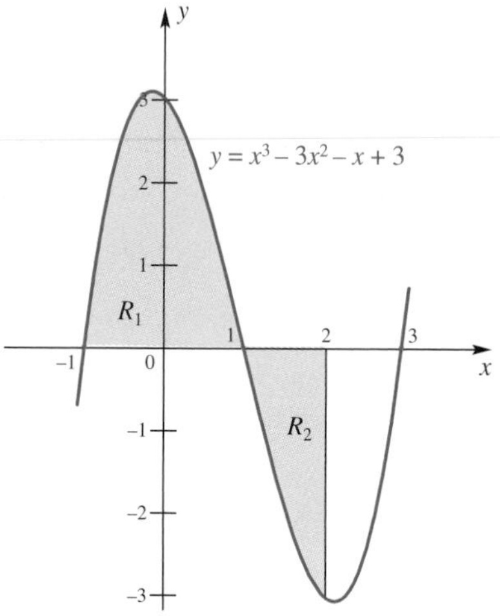

Figure 4

Notice that we could have written this area as one integral using the absolute value symbol,

$$A(R) = \int_{-1}^{2} |x^3 - 3x^2 - x + 3|\,dx$$

but this is no real simplification since, in order to evaluate this integral, we would have to split it into two parts, just as we did above. ■

A Helpful Way of Thinking For simple regions of the type considered above, it is quite easy to write down the correct integral. When we consider more complicated regions (e.g., regions between two curves), the task of selecting the right integral is more difficult. However, there is a way of thinking that can be very helpful. It goes back to the definition of area and of the definite integral. Here it is in five steps.

Step 1: Sketch the region.

Step 2: Slice it into thin pieces (strips); label a typical piece.

Step 3: Approximate the area of this typical piece as if it were a rectangle.

Step 4: Add up the approximations to the areas of the pieces.

Step 5: Take the limit as the width of the pieces approaches zero, thus getting a definite integral.

To illustrate, we consider yet another simple example.

EXAMPLE 4 Set up the integral for the area of the region under $y = 1 + \sqrt{x}$ between $x = 0$ and $x = 4$ (Figure 5).

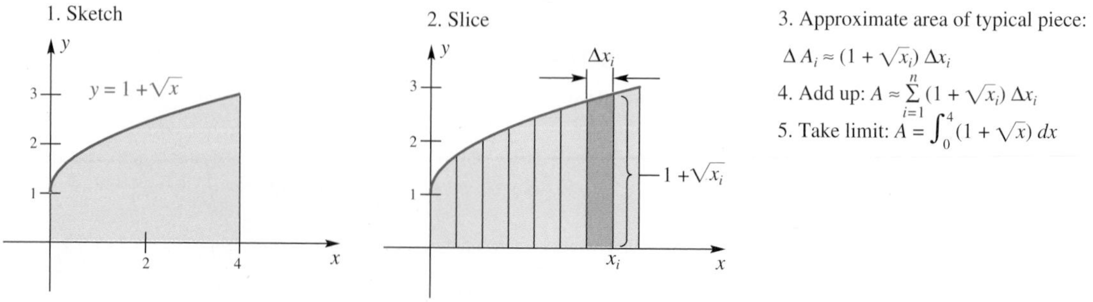

Figure 5

SOLUTION Once we understand this five-step procedure, we can abbreviate it to three: *slice, approximate, integrate*. Think of the word *integrate* as incorporating two steps: (1) add the areas of the pieces and (2) take the limit as the piece width tends to zero. In this process, $\Sigma \ldots \Delta x$ transforms into $\int \ldots dx$ as we take the limit. Figure 6 gives the abbreviated form for the same problem.

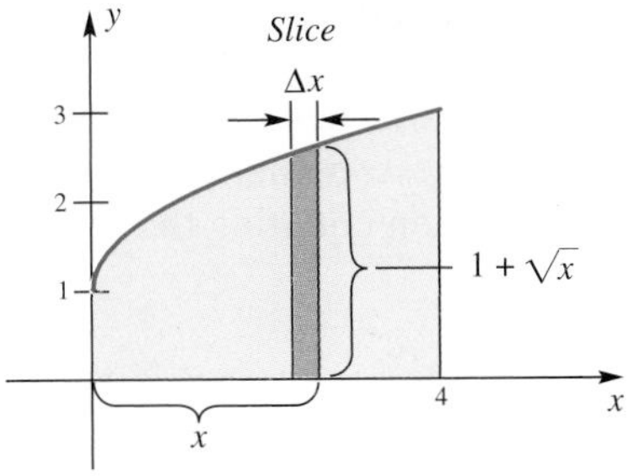

Approximate

$$\Delta A \approx (1 + \sqrt{x})\,\Delta x$$

Integrate

$$A = \int_0^4 (1 + \sqrt{x})\,dx$$

Figure 6

■

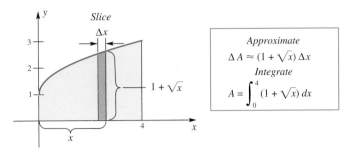

A Region Between Two Curves Consider curves $y = f(x)$ and $y = g(x)$ with $g(x) \le f(x)$ on $a \le x \le b$. They determine the region shown in Figure 7. We use the *slice, approximate, integrate* method to find its area. Be sure to note that $f(x) - g(x)$ gives the correct height for the thin slice, even when the graph of g goes below the x-axis. In this case $g(x)$ is negative; so subtracting $g(x)$ is the same as adding a positive number. You can check that $f(x) - g(x)$ also gives the correct height, even when both $f(x)$ and $g(x)$ are negative.

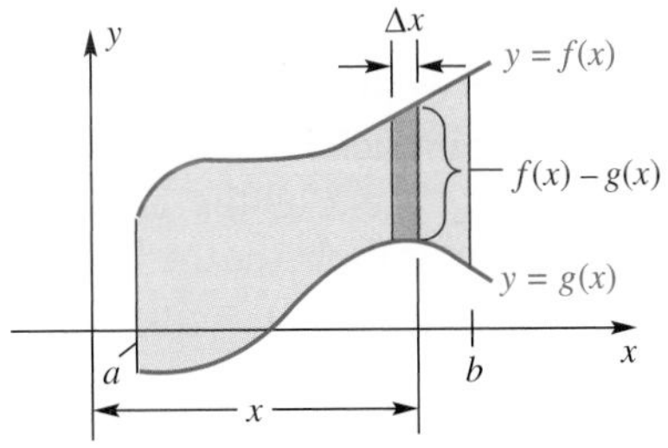

$$\Delta A \approx [f(x) - g(x)]\,\Delta x$$

$$A = \int_a^b [f(x) - g(x)]\,dx$$

Figure 7

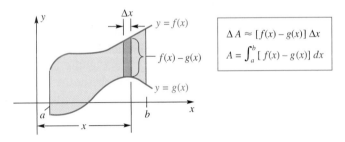

EXAMPLE 5 Find the area of the region between the curves $y = x^4$ and $y = 2x - x^2$.

SOLUTION We start by finding where the two curves intersect. To do this, we need to solve $2x - x^2 = x^4$, a fourth-degree equation, which would usually be difficult to solve. However, in this case $x = 0$ and $x = 1$ are rather obvious solutions. Our sketch of the region, together with the appropriate approximation and the corresponding integral, is shown in Figure 8.

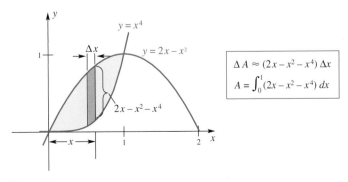

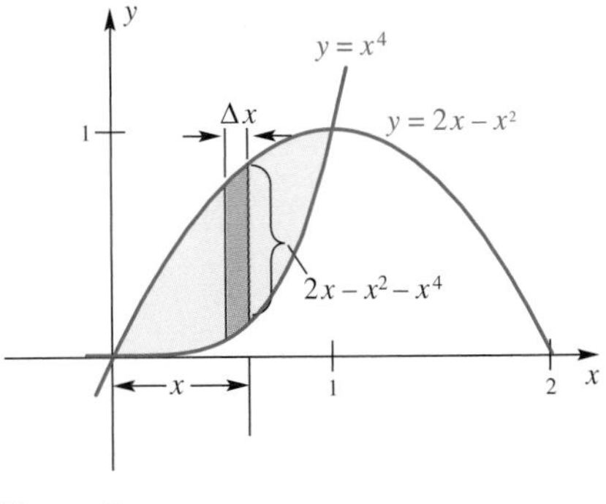

$$\Delta A \approx (2x - x^2 - x^4)\,\Delta x$$

$$A = \int_0^1 (2x - x^2 - x^4)\,dx$$

Figure 8

One job remains: to evaluate the integral.

$$\int_0^1 (2x - x^2 - x^4)\,dx = \left[x^2 - \frac{x^3}{3} - \frac{x^5}{5}\right]_0^1 = 1 - \frac{1}{3} - \frac{1}{5} = \frac{7}{15} \qquad ■$$

EXAMPLE 6 **Horizontal Slicing** Find the area of the region between the parabola $y^2 = 4x$ and the line $4x - 3y = 4$.

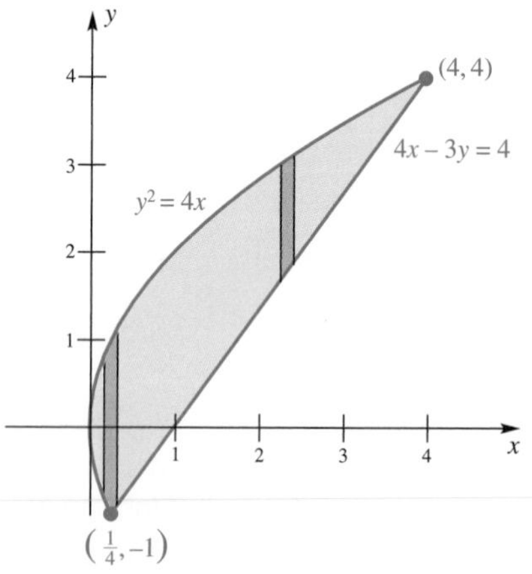

Figure 9

SOLUTION We will need the points of intersection of these two curves. The y-coordinates of these points can be found by writing the second equation as $4x = 3y + 4$ and then equating the two expressions for $4x$.

$$\begin{aligned} y^2 &= 3y + 4 \\ y^2 - 3y - 4 &= 0 \\ (y - 4)(y + 1) &= 0 \\ y &= 4, -1 \end{aligned}$$

When $y = 4$, $x = 4$ and when $y = -1$, $x = \frac{1}{4}$, so we conclude that the points of intersection are $(4, 4)$ and $\left(\frac{1}{4}, -1\right)$. The region between the curves is shown in Figure 9.

Now imagine slicing this region vertically. We face a problem, because the lower boundary consists of two different curves. Slices at the extreme left extend from the lower branch of the parabola to its upper branch. For the rest of the region, slices extend from the line to the parabola. To solve the problem with vertical slices requires that we first split our region into two parts, set up an integral for each part, and then evaluate both integrals.

A far better approach is to slice the region horizontally as shown in Figure 10, thus using y rather than x as the integration variable. Note that horizontal slices always go from the parabola (at the left) to the line (at the right). The length of such a slice is the larger x-value $\left(x = \frac{1}{4}(3y + 4)\right)$ minus the smaller x-value $\left(x = \frac{1}{4}y^2\right)$.

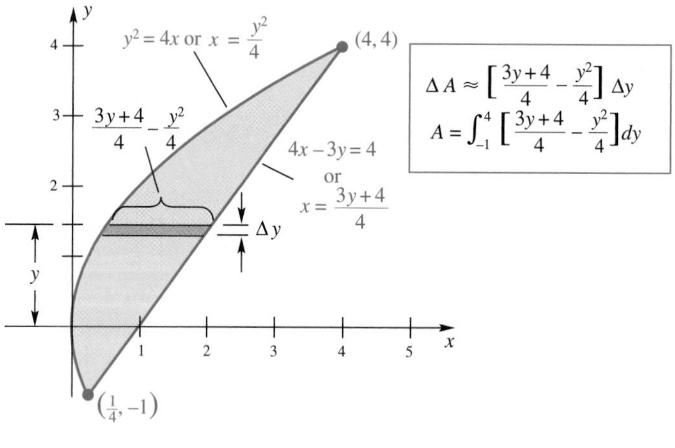

Figure 10

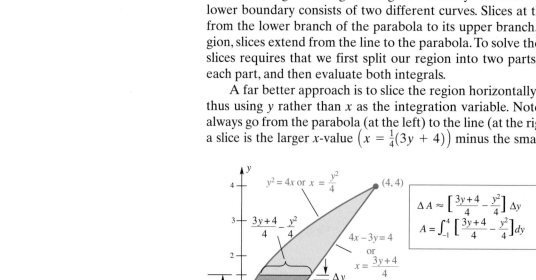

$$\begin{aligned} A &= \int_{-1}^{4}\left[\frac{3y + 4 - y^2}{4}\right]dy = \frac{1}{4}\int_{-1}^{4}(3y + 4 - y^2)\,dy \\ &= \frac{1}{4}\left[\frac{3y^2}{2} + 4y - \frac{y^3}{3}\right]_{-1}^{4} \\ &= \frac{1}{4}\left[\left(24 + 16 - \frac{64}{3}\right) - \left(\frac{3}{2} - 4 + \frac{1}{3}\right)\right] \\ &= \frac{125}{24} \approx 5.21 \end{aligned}$$

There are two items to note: (1) The integrand resulting from a horizontal slicing involves y, not x; and (2) to get the integrand, solve both equations for x and subtract the smaller x-value from the larger. ■

Distance and Displacement Consider an object moving along a straight line with velocity $v(t)$ at time t. If $v(t) \geq 0$, then $\int_a^b v(t)\,dt$ gives the distance traveled during the time interval $a \leq t \leq b$. However, if $v(t)$ is sometimes negative (which corresponds to the object moving in reverse), then

$$\int_a^b v(t)\,dt = s(b) - s(a)$$

measures the **displacement** of the object, that is, the directed distance from its starting position $s(a)$ to its ending position $s(b)$. To get the **total distance** that the object traveled during $a \leq t \leq b$, we must calculate $\int_a^b |v(t)|\,dt$, the area between the velocity curve and the t-axis.

EXAMPLE 7 An object is at position $s = 3$ at time $t = 0$. Its velocity at time t is $v(t) = 5 \sin 6\pi t$. What is the position of the object at time $t = 2$, and how far did it travel during this time?

SOLUTION The object's displacement, that is, change in position, is

$$s(2) - s(0) = \int_0^2 v(t)\,dt = \int_0^2 5 \sin 6\pi t\,dt = \left[-\frac{5}{6\pi}\cos 6\pi t\right]_0^2 = 0$$

Thus, $s(2) = s(0) + 0 = 3 + 0 = 3$. The object is at position 3 at time $t = 2$. The total distance traveled is

$$\int_0^2 |v(t)|\,dt = \int_0^2 |5 \sin 6\pi t|\,dt$$

To perform this integration we make use of symmetry (see Figure 11). Thus

$$\int_0^2 |v(t)|\,dt = 12\int_0^{2/12} 5 \sin 6\pi t\,dt = 60\left[-\frac{1}{6\pi}\cos 6\pi t\right]_0^{1/6} = \frac{20}{\pi} \approx 6.3662$$ ■

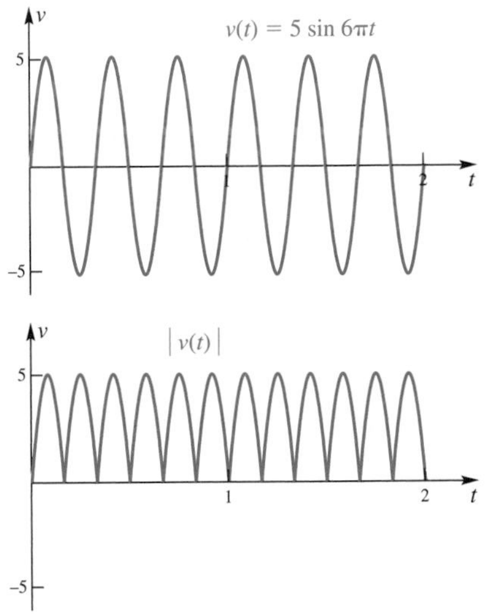

Figure 11

Concepts Review

1. Let R be the region between the curve $y = f(x)$ and the x-axis on the interval $[a, b]$. If $f(x) \geq 0$ for all x in $[a, b]$, then $A(R) =$ _____, but if $f(x) \leq 0$ for all x in $[a, b]$, then $A(R) =$ _____.

2. To find the area of the region between two curves, it is wise to think of the following three-word motto: _____.

3. Suppose that the curves $y = f(x)$ and $y = g(x)$ bound a region R on which $f(x) \leq g(x)$. Then the area of R is given by $A(R) = \int_a^b$ _____ dx, where a and b are determined by solving the equation _____.

4. If $p(y) \leq q(y)$ for all y in $[c, d]$, then the area $A(R)$ of the region R bounded by the curves $x = p(y)$ and $x = q(y)$ between $y = c$ and $y = d$ is given by $A(R) =$ _____.

Problem Set 5.1

In Problems 1–10, use the three-step procedure (slice, approximate, integrate) to set up and evaluate an integral (or integrals) for the area of the indicated region.

1.

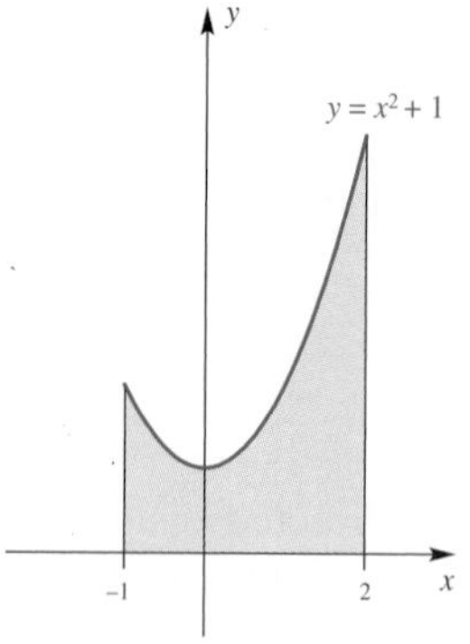

2.

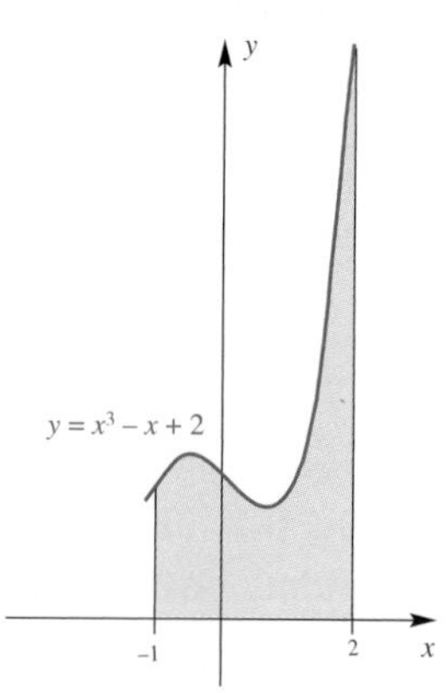

3.

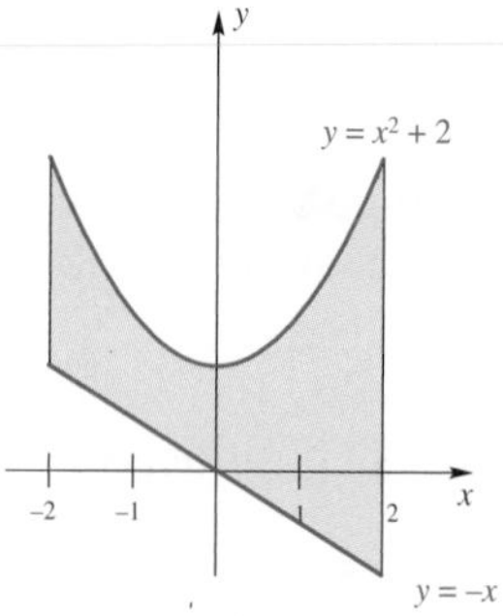

4.

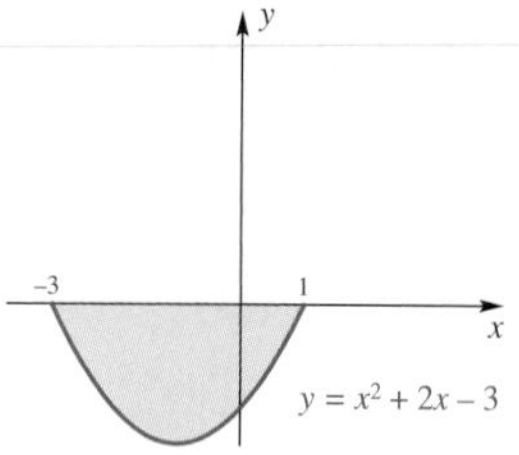

5.

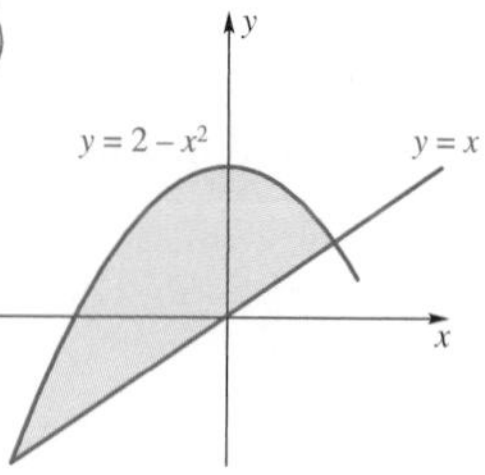

6.

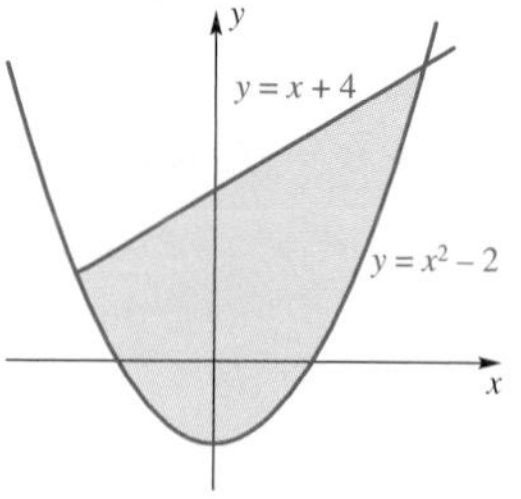

7.

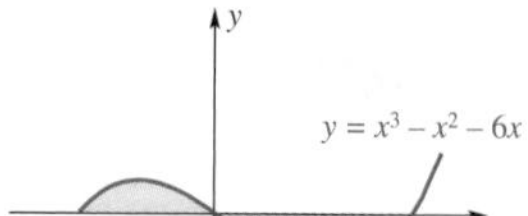

8.

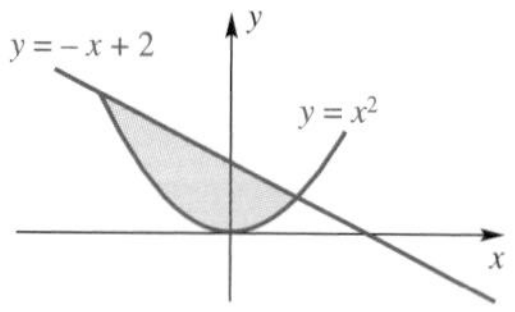

9.

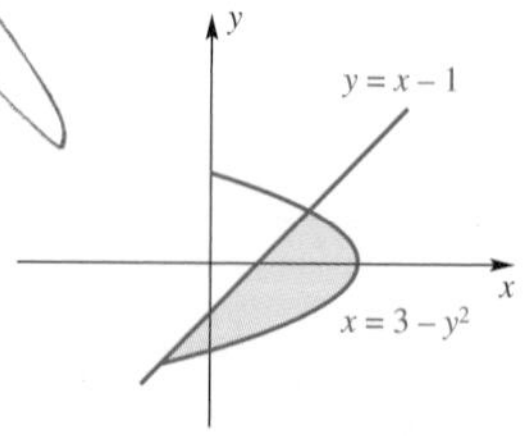

10.

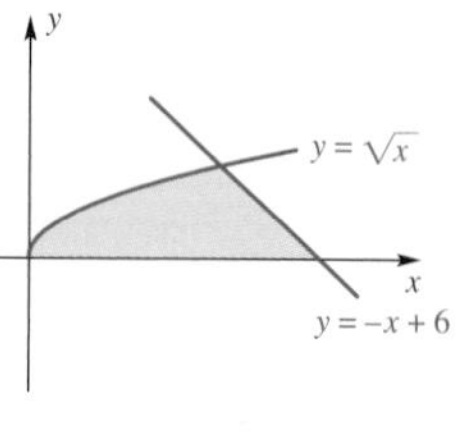

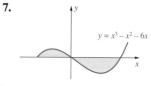

In Problems 11–28, sketch the region bounded by the graphs of the given equations, show a typical slice, approximate its area, set up an integral, and calculate the area of the region. Make an estimate of the area to confirm your answer.

11. $y = 3 - \frac{1}{3}x^2$, $y = 0$, between $x = 0$ and $x = 3$

12. $y = 5x - x^2$, $y = 0$, between $x = 1$ and $x = 3$

13. $y = (x - 4)(x + 2)$, $y = 0$, between $x = 0$ and $x = 3$

14. $y = x^2 - 4x - 5$, $y = 0$, between $x = -1$ and $x = 4$

15. $y = \frac{1}{4}(x^2 - 7)$, $y = 0$, between $x = 0$ and $x = 2$

16. $y = x^3$, $y = 0$, between $x = -3$ and $x = 3$

17. $y = \sqrt[3]{x}$, $y = 0$, between $x = -2$ and $x = 2$

18. $y = \sqrt{x} - 10$, $y = 0$, between $x = 0$ and $x = 9$

19. $y = (x - 3)(x - 1)$, $y = x$

20. $y = \sqrt{x}$, $y = x - 4$, $x = 0$

21. $y = x^2 - 2x$, $y = -x^2$

22. $y = x^2 - 9$, $y = (2x - 1)(x + 3)$

23. $x = 8y - y^2$, $x = 0$

24. $x = (3 - y)(y + 1)$, $x = 0$

25. $x = -6y^2 + 4y$, $x + 3y - 2 = 0$

26. $x = y^2 - 2y$, $x - y - 4 = 0$

27. $4y^2 - 2x = 0$, $4y^2 + 4x - 12 = 0$

28. $x = 4y^4$, $x = 8 - 4y^4$

29. Sketch the region R bounded by $y = x + 6$, $y = x^3$, and $2y + x = 0$. Then find its area. *Hint:* Divide R into two pieces.

30. Find the area of the triangle with vertices at $(-1, 4)$, $(2, -2)$, and $(5, 1)$ by integration.

31. An object moves along a line so that its velocity at time t is $v(t) = 3t^2 - 24t + 36$ feet per second. Find the displacement and total distance traveled by the object for $-1 \le t \le 9$.

32. Follow the directions of Problem 31 if $v(t) = \frac{1}{2} + \sin 2t$ and the interval is $0 \le t \le 3\pi/2$.

33. Starting at $s = 0$ when $t = 0$, an object moves along a line so that its velocity at time t is $v(t) = 2t - 4$ centimeters per second. How long will it take to get to $s = 12$? To travel a total distance of 12 centimeters?

34. Consider the curve $y = 1/x^2$ for $1 \le x \le 6$.

(a) Calculate the area under this curve.

(b) Determine c so that the line $x = c$ bisects the area of part (a).

(c) Determine d so that the line $y = d$ bisects the area of part (a).

35. Calculate areas A, B, C, and D in Figure 12. Check by calculating $A + B + C + D$ in one integration.

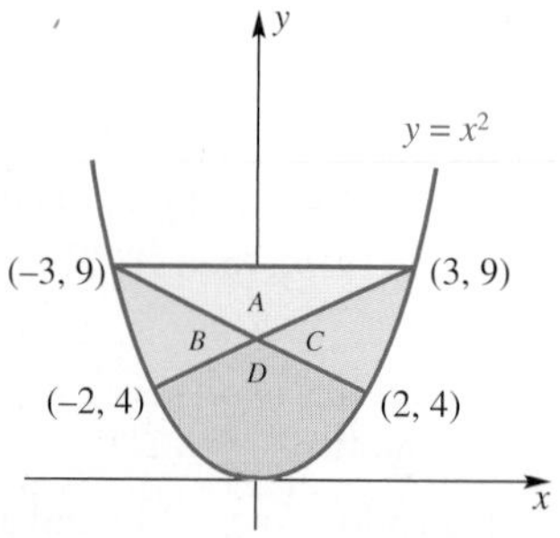

Figure 12

36. Prove Cavalieri's Principle. (Bonaventura Cavalieri (1598–1647) developed this principle in 1635.) If two regions have the same height at every x in $[a, b]$, then they have the same area (see Figure 13).

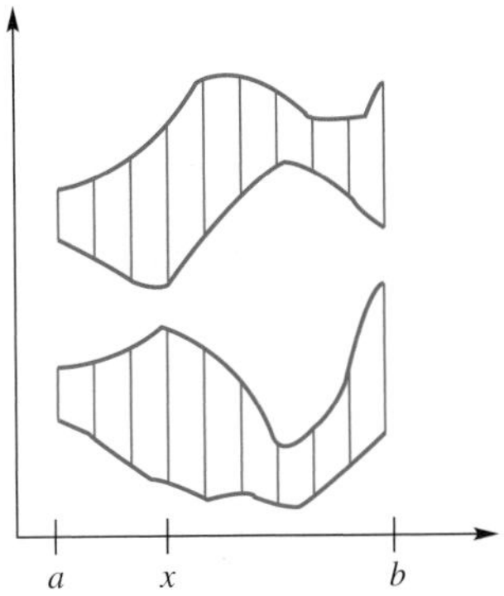

Figure 13

37. Use Cavalieri's Principle (not integration; see Problem 36) to show that the shaded regions in Figure 14 have the same area.

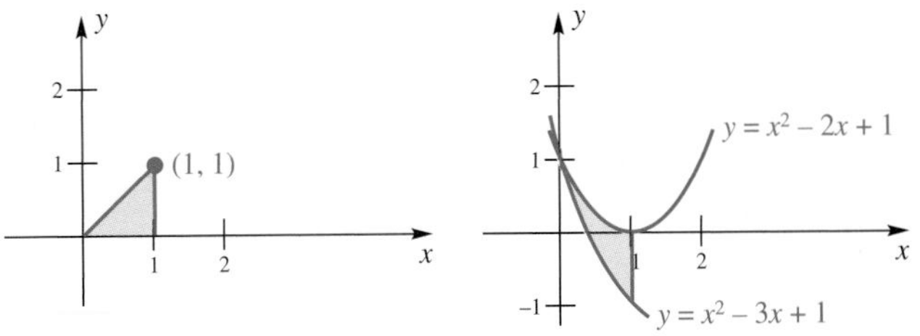

Figure 14

38. Find the area of the region trapped between $y = \sin x$ and $y = \frac{1}{2}, 0 \leq x \leq 17\pi/6$.

Answers to Concepts Review: **1.** $\int_a^b f(x)\,dx; -\int_a^b f(x)\,dx$ **2.** slice, approximate, integrate **3.** $[g(x) - f(x)]; f(x) = g(x)$ **4.** $\int_c^d [q(y) - p(y)]\,dy$

5.2 Volumes of Solids: Slabs, Disks, Washers

That the definite integral can be used to calculate *areas* is not surprising; it was invented for that purpose. But uses of the integral go far beyond that application. Many quantities can be thought of as a result of slicing something into small pieces, approximating each piece, adding up, and taking the limit as the pieces shrink in size. This method of slice, approximate, and integrate can be used to find the *volumes* of solids provided that the volume of each slice is easy to approximate.

What is volume? We start with simple solids called *right cylinders*, four of which are shown in Figure 1. In each case, the solid is generated by moving a plane region (the base) through a distance h in a direction perpendicular to that region. And in each case, the volume of the solid is defined to be the area A of the base times the height h; that is,

$$V = A \cdot h$$

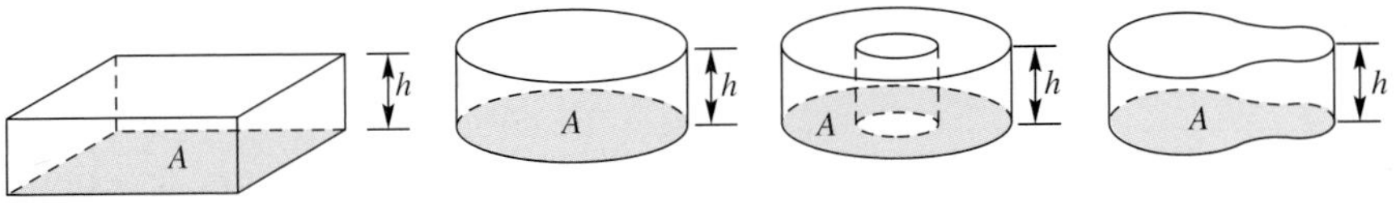

Figure 1

The Volume of a Coin

Consider an ordinary coin, say a quarter.

A quarter has a radius of about 1 centimeter and a thickness of about 0.2 centimeter. Its volume is the area of the base, $A = \pi(1^2)$, times the thickness $h = 0.2$; that is,

$$V = (1\pi)(0.2) \approx 0.63$$

cubic centimeters.

Next consider a solid with the property that cross sections perpendicular to a given line have known area. In particular, suppose that the line is the x-axis and that the area of the cross section at x is $A(x)$, $a \leq x \leq b$ (Figure 2). We partition the interval $[a, b]$ by inserting points $a = x_0 < x_1 < x_2 < \cdots < x_n = b$. We then pass planes through these points perpendicular to the x-axis, thus slicing the solid into thin **slabs** (Figure 3). The *volume* ΔV_i of a slab should be approximately the volume of a cylinder; that is,

$$\Delta V_i \approx A(\overline{x}_i)\,\Delta x_i$$

(Recall that $\overline{x}_i$, called a *sample point*, is any number in the interval $[x_{i-1}, x_i]$.)

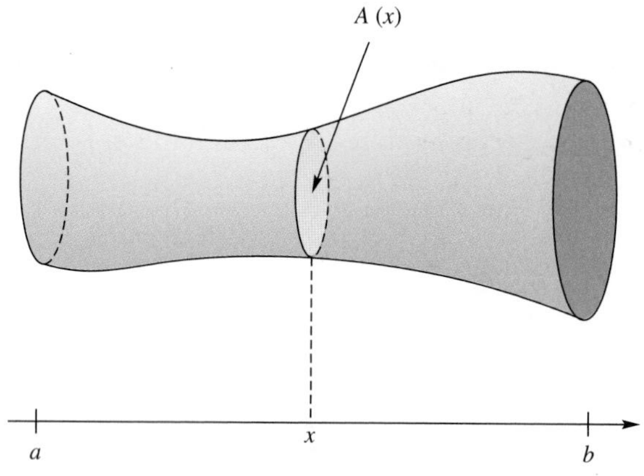

Figure 2

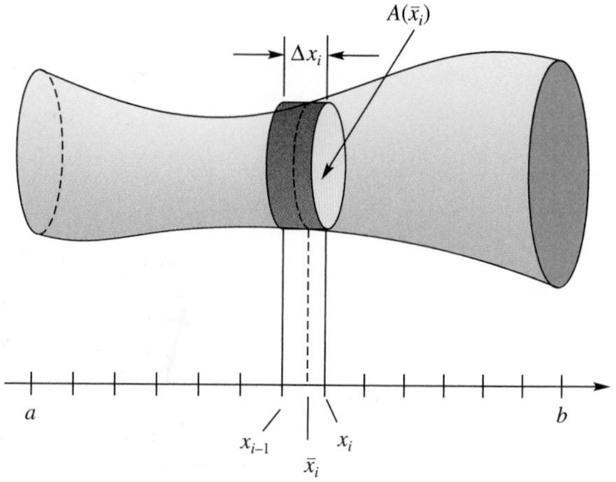

Figure 3

The "volume" V of the solid should be given approximately by the Riemann sum

$$V \approx \sum_{i=1}^{n} A(\overline{x}_i)\, \Delta x_i$$

When we let the norm of the partition approach zero, we obtain a definite integral; this integral is defined to be the **volume** of the solid.

$$V = \int_a^b A(x)\, dx$$

Rather than routinely applying the boxed formula to obtain volumes, we suggest that in each problem you go through the process that led to it. Just as for areas, we call this process *slice, approximate, integrate*. It is illustrated in the examples that follow.

Solids of Revolution: Method of Disks When a plane region, lying entirely on one side of a fixed line in its plane, is revolved about that line, it generates a **solid of revolution.** The fixed line is called the **axis** of the solid of revolution.

As an illustration, if the region bounded by a semicircle and its diameter is revolved about that diameter, it sweeps out a spherical solid (Figure 4). If the region inside a right triangle is revolved about one of its legs, it generates a conical solid (Figure 5). When a circular region is revolved about a line in its plane that does not intersect the circle (Figure 6), it sweeps out a torus (doughnut). In each case, it is possible to represent the volume as a definite integral.

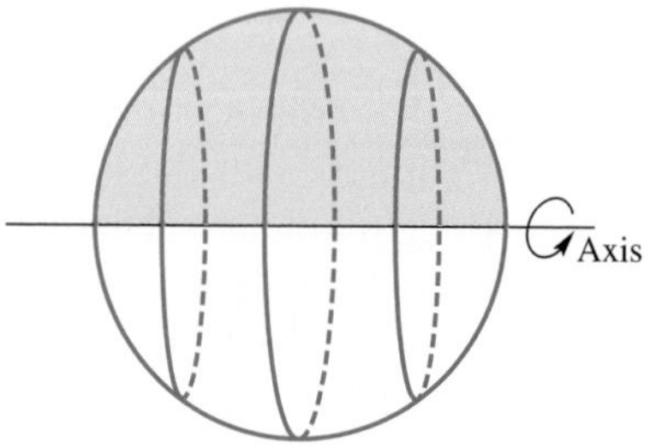

Figure 4

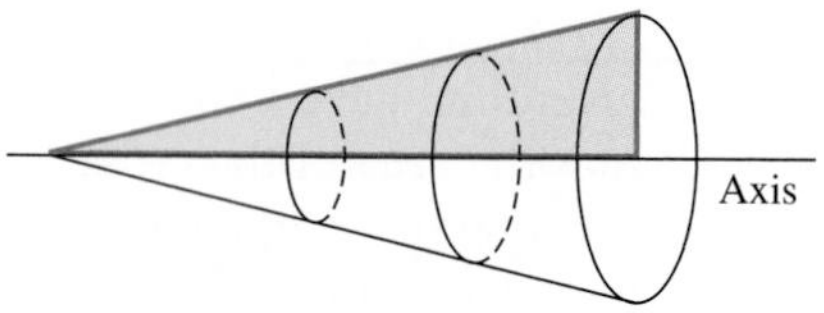

Figure 5

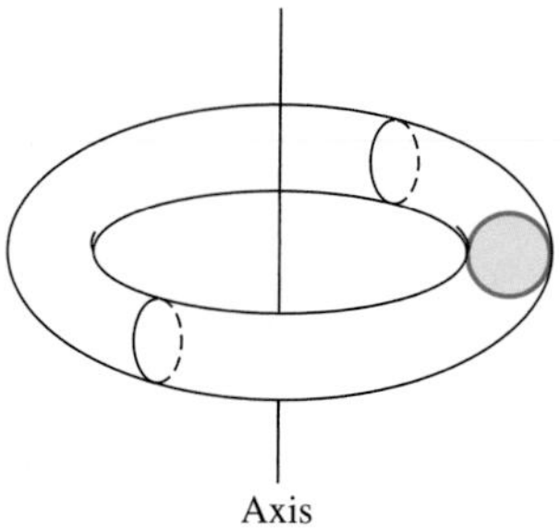

Figure 6

EXAMPLE 1 Find the volume of the solid of revolution obtained by revolving the plane region R bounded by $y = \sqrt{x}$, the x-axis, and the line $x = 4$ about the x-axis.

SOLUTION The region R, with a typical slice, is displayed as the left part of Figure 7. When revolved about the x-axis, this region generates a solid of revolution and the slice generates a disk, a thin coin-shaped object.

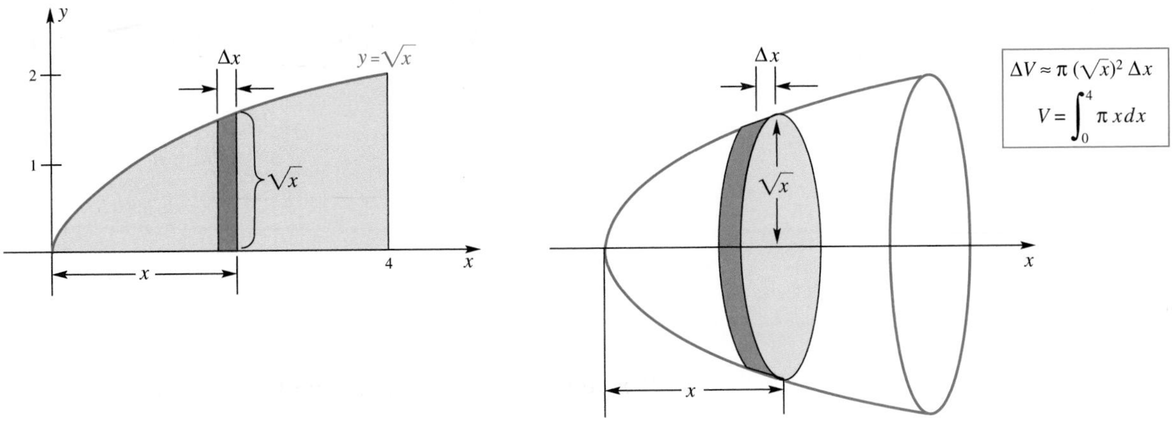

Figure 7

Recalling that the volume of a circular cylinder is $\pi r^2 h$, we approximate the volume ΔV of this disk with $\Delta V \approx \pi\left(\sqrt{x}\right)^2 \Delta x$ and then integrate.

$$V = \pi \int_0^4 x\,dx = \pi\left[\frac{x^2}{2}\right]_0^4 = \pi\frac{16}{2} = 8\pi \approx 25.13$$

≈ Is this answer reasonable? The right circular cylinder that contains the solid has volume $V = \pi 2^2 \cdot 4 = 16\pi$. Half this number seems reasonable. ■

EXAMPLE 2 Find the volume of the solid generated by revolving the region bounded by the curve $y = x^3$, the y-axis, and the line $y = 3$ about the y-axis (Figure 8).

SOLUTION Here we slice horizontally, which makes y the choice for the integration variable. Note that $y = x^3$ is equivalent to $x = \sqrt[3]{y}$ and $\Delta V \approx \pi\left(\sqrt[3]{y}\right)^2 \Delta y$.

The volume is therefore

$$V = \pi \int_0^3 y^{2/3}\,dy = \pi\left[\frac{3}{5}y^{5/3}\right]_0^3 = \pi\frac{9\sqrt[3]{9}}{5} \approx 11.76$$

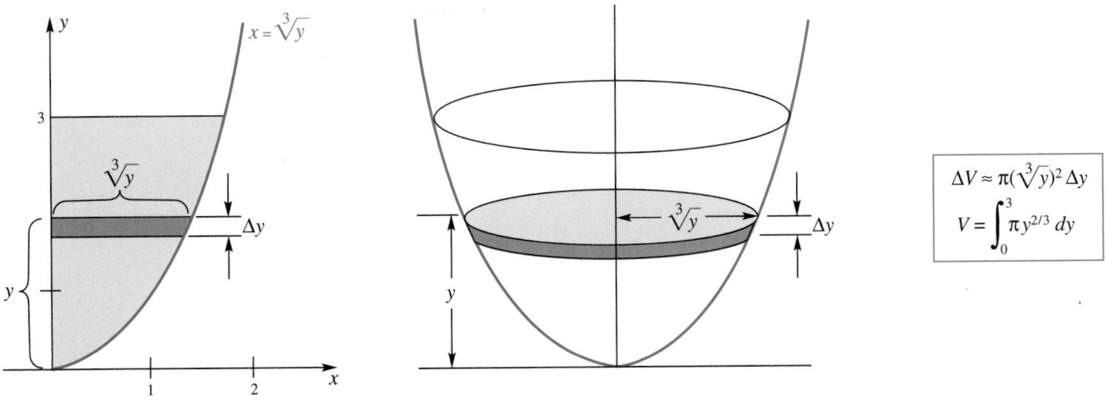

Figure 8

■

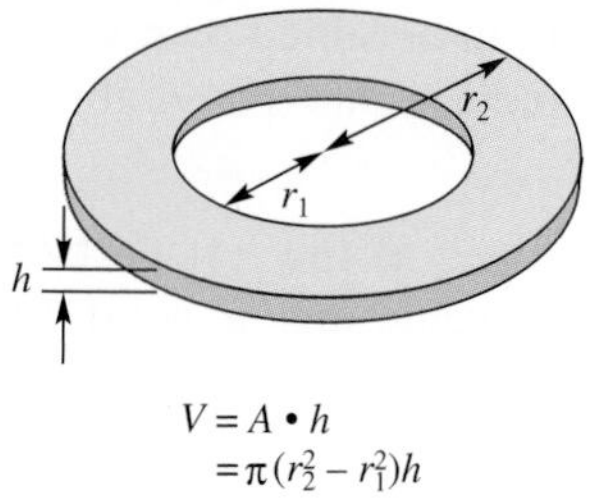

Figure 9

Method of Washers Sometimes, slicing a solid of revolution results in disks with holes in the middle. We call them **washers.** See the diagram and accompanying volume formula shown in Figure 9.

EXAMPLE 3 Find the volume of the solid generated by revolving the region bounded by the parabolas $y = x^2$ and $y^2 = 8x$ about the x-axis.

SOLUTION The key words are still *slice, approximate, integrate* (see Figure 10).

$$V = \pi \int_0^2 (8x - x^4)\, dx = \pi \left[\frac{8x^2}{2} - \frac{x^5}{5} \right]_0^2 = \frac{48\pi}{5} \approx 30.16$$

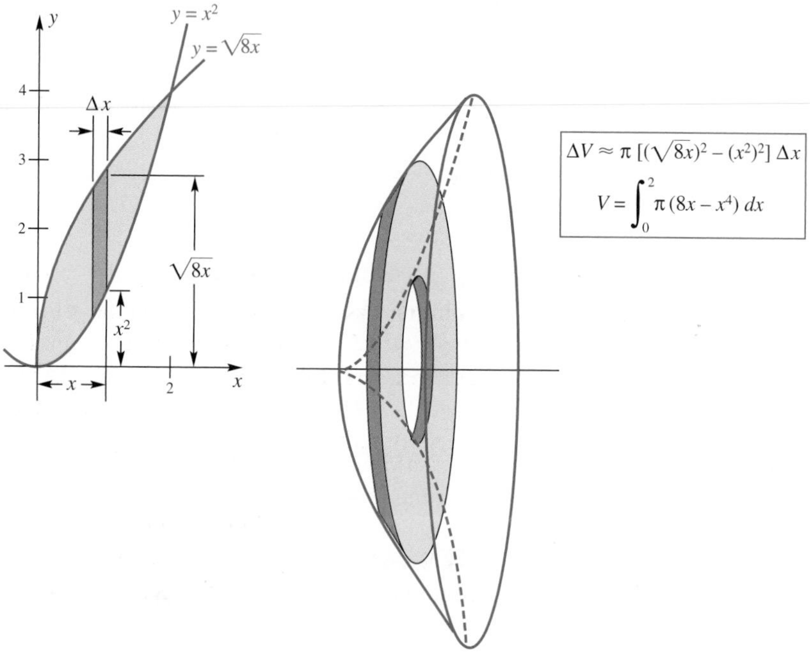

Figure 10

■

EXAMPLE 4 The semicircular region bounded by the curve $x = \sqrt{4 - y^2}$ and the y-axis is revolved about the line $x = -1$. Set up the integral that represents its volume.

SOLUTION Here the outer radius of the washer is $1 + \sqrt{4 - y^2}$ and the inner radius is 1. Figure 11 exhibits the solution. The integral can be simplified. The part above the x-axis has the same volume as the part below it (which manifests itself in an even integrand). Thus, we may integrate from 0 to 2 and double the result.

$$V = \pi \int_{-2}^{2} \left[\left(1 + \sqrt{4 - y^2}\right)^2 - 1^2 \right] dy$$

$$= 2\pi \int_0^2 \left[2\sqrt{4 - y^2} + 4 - y^2 \right] dy$$

Now see Problem 35 for a way to evaluate this integral.

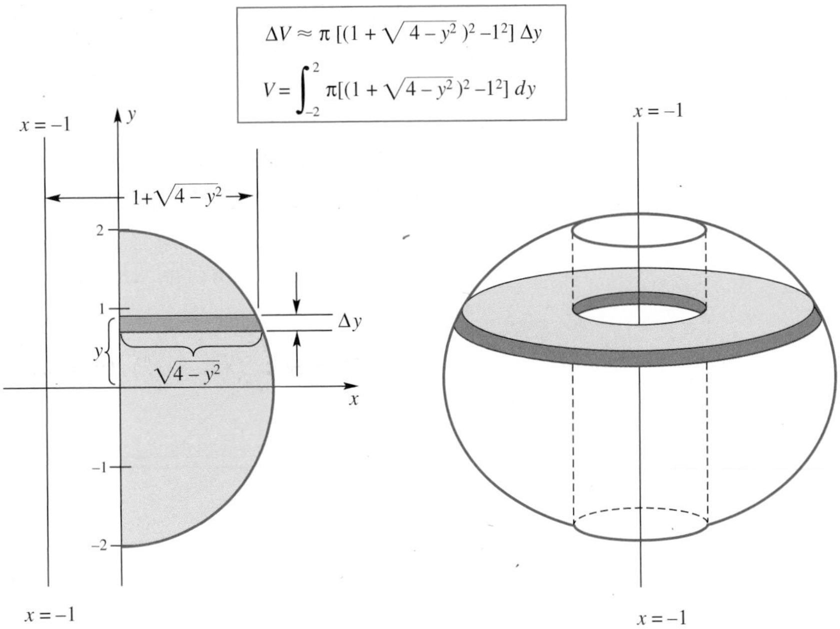

Figure 11

Other Solids with Known Cross Sections So far, our solids have had circular cross sections. However, the method for finding volume works just as well for solids whose cross sections are squares or triangles. In fact, all that is really needed is that the areas of the cross sections can be determined, since, in this case, we can also approximate the volume of the slice—a slab—with this cross section. The volume is then found by integrating.

EXAMPLE 5 Let the base of a solid be the first quadrant plane region bounded by $y = 1 - x^2/4$, the x-axis, and the y-axis. Suppose that cross sections perpendicular to the x-axis are squares. Find the volume of the solid.

SOLUTION When we slice this solid perpendicularly to the x-axis, we get thin square boxes (Figure 12), like slices of cheese.

$$V = \int_0^2 \left(1 - \frac{x^2}{4}\right)^2 dx = \int_0^2 \left(1 - \frac{x^2}{2} + \frac{x^4}{16}\right) dx = \left[x - \frac{x^3}{6} + \frac{x^5}{80}\right]_0^2$$

$$= 2 - \frac{8}{6} + \frac{32}{80} = \frac{16}{15} \approx 1.07$$

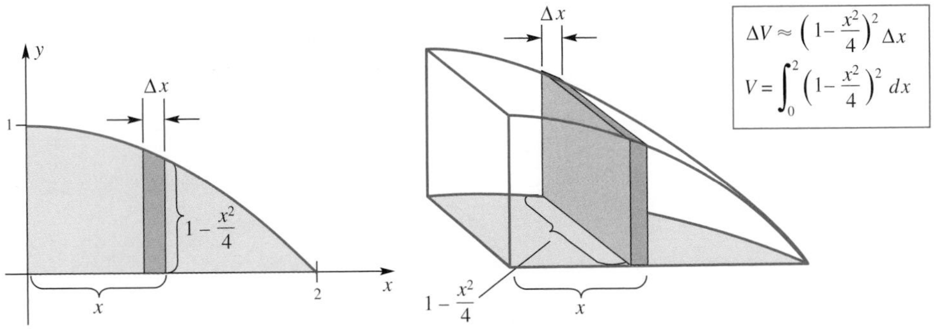

Figure 12

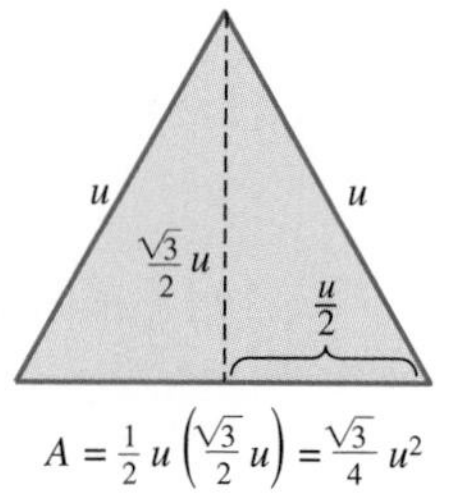

Figure 13

EXAMPLE 6 The base of a solid is the region between one arch of $y = \sin x$ and the x-axis. Each cross section perpendicular to the x-axis is an equilateral triangle sitting on this base. Find the volume of the solid.

SOLUTION We need the fact that the area of an equilateral triangle of side u is $\sqrt{3}\,u^2/4$ (see Figure 13). We proceed as shown in Figure 14.

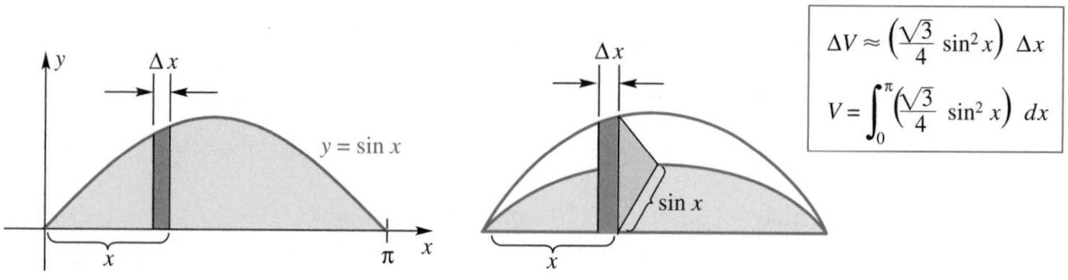

Figure 14

To perform the indicated integration, we use the half-angle formula $\sin^2 x = (1 - \cos 2x)/2$.

$$V = \frac{\sqrt{3}}{4}\int_0^{\pi}\frac{1 - \cos 2x}{2}\,dx = \frac{\sqrt{3}}{8}\int_0^{\pi}(1 - \cos 2x)\,dx$$

$$= \frac{\sqrt{3}}{8}\left[\int_0^{\pi} 1\,dx - \frac{1}{2}\int_0^{\pi}\cos 2x \cdot 2\,dx\right]$$

$$= \frac{\sqrt{3}}{8}\left[x - \frac{1}{2}\sin 2x\right]_0^{\pi} = \frac{\sqrt{3}}{8}\pi \approx 0.68$$

■

Concepts Review

1. The volume of a disk of radius r and thickness h is ______.

2. The volume of a washer of inner radius r, outer radius R, and thickness h is ______.

3. If the region R bounded by $y = x^2$, $y = 0$, and $x = 3$ is revolved about the x-axis, the disk at x will have volume $\Delta V \approx$ ______.

4. If the region R of Question 3 is revolved about the line $y = -2$, the washer at x will have volume $\Delta V \approx$ ______.

Problem Set 5.2

In Problems 1–4, find the volume of the solid generated when the indicated region is revolved about the specified axis; slice, approximate, integrate.

1. x-axis

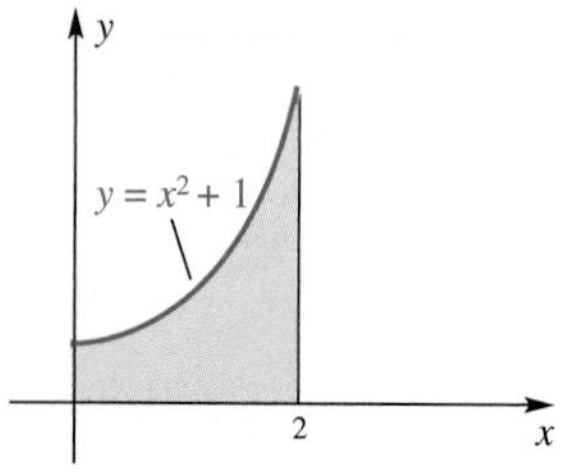

2. x-axis

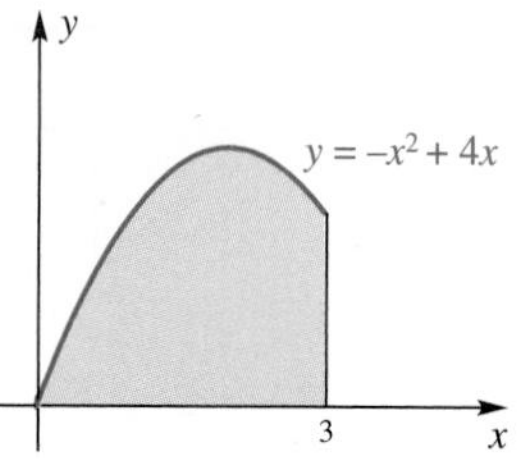

3. (a) x-axis
(b) y-axis

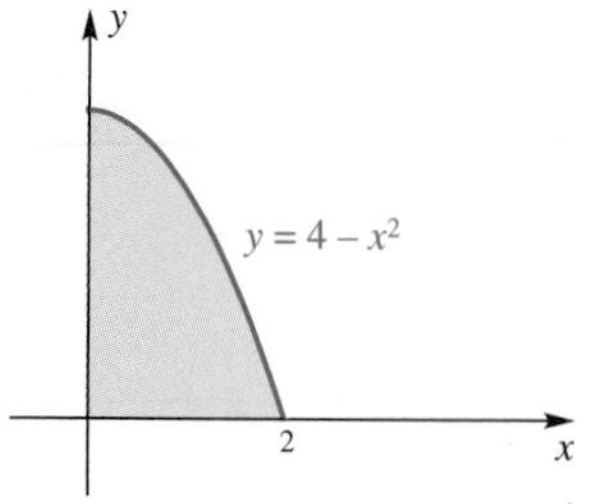

4. (a) x-axis
(b) y-axis

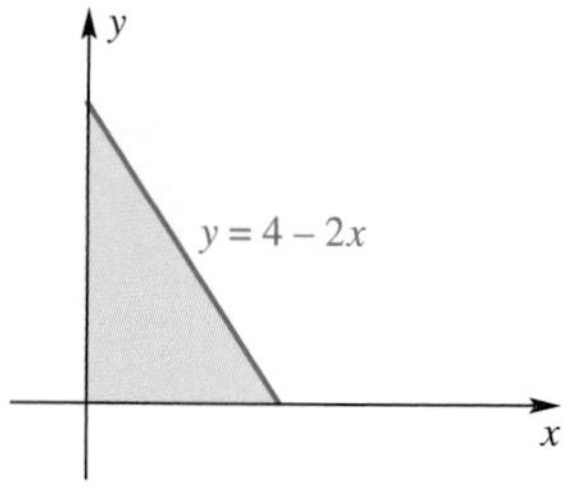

In Problems 5–10, sketch the region R bounded by the graphs of the given equations, and show a typical vertical slice. Then find the volume of the solid generated by revolving R about the x-axis.

5. $y = \dfrac{x^2}{\pi}, x = 4, y = 0$

6. $y = x^3, x = 3, y = 0$

7. $y = \dfrac{1}{x}, x = 2, x = 4, y = 0$

8. $y = x^{3/2}, y = 0$, between $x = 2$ and $x = 3$

9. $y = \sqrt{9 - x^2}, y = 0$, between $x = -2$ and $x = 3$

10. $y = x^{2/3}, y = 0$, between $x = 1$ and $x = 27$

In Problems 11–16, sketch the region R bounded by the graphs of the given equations and show a typical horizontal slice. Find the volume of the solid generated by revolving R about the y-axis.

11. $x = y^2, x = 0, y = 3$

12. $x = \dfrac{2}{y}, y = 2, y = 6, x = 0$

13. $x = 2\sqrt{y}, y = 4, x = 0$

14. $x = y^{2/3}, y = 27, x = 0$

15. $x = y^{3/2}, y = 9, x = 0$

16. $x = \sqrt{4 - y^2}, x = 0$

17. Find the volume of the solid generated by revolving about the x-axis the region bounded by the upper half of the ellipse

$$\frac{x^2}{a^2} + \frac{y^2}{b^2} = 1$$

and the x-axis, and thus find the volume of a *prolate spheroid*. Here a and b are positive constants, with $a > b$.

18. Find the volume of the solid generated by revolving about the x-axis the region bounded by the line $y = 6x$ and the parabola $y = 6x^2$.

19. Find the volume of the solid generated by revolving about the x-axis the region bounded by the line $x - 2y = 0$ and the parabola $y^2 = 4x$.

20. Find the volume of the solid generated by revolving about the x-axis the region in the first quadrant bounded by the circle $x^2 + y^2 = r^2$, the x-axis, and the line $x = r - h, 0 < h < r$, and thus find the volume of a *spherical segment* of height h, of a sphere of radius r.

21. Find the volume of the solid generated by revolving about the y-axis the region bounded by the line $y = 4x$ and the parabola $y = 4x^2$.

22. Find the volume of the solid generated by revolving about the line $y = 2$ the region in the first quadrant bounded by the parabolas $3x^2 - 16y + 48 = 0$ and $x^2 - 16y + 80 = 0$ and the y-axis.

23. The base of a solid is the region inside the circle $x^2 + y^2 = 4$. Find the volume of the solid if every cross section by a plane perpendicular to the x-axis is a square. *Hint:* See Examples 5 and 6.

24. Do Problem 23 assuming that every cross section by a plane perpendicular to the x-axis is an isosceles triangle with base on the xy-plane and altitude 4. *Hint:* To complete the evaluation, interpret $\int_{-2}^{2} \sqrt{4 - x^2}\, dx$ as the area of a semicircle.

25. The base of a solid is bounded by one arch of $y = \sqrt{\cos x}, -\pi/2 \le x \le \pi/2$, and the x-axis. Each cross section perpendicular to the x-axis is a square sitting on this base. Find the volume of the solid.

26. The base of a solid is the region bounded by $y = 1 - x^2$ and $y = 1 - x^4$. Cross sections of the solid that are perpendicular to the x-axis are squares. Find the volume of the solid.

27. Find the volume of one octant (one-eighth) of the solid region common to two right circular cylinders of radius 1 whose axes intersect at right angles. *Hint:* Horizontal cross sections are squares. See Figure 15.

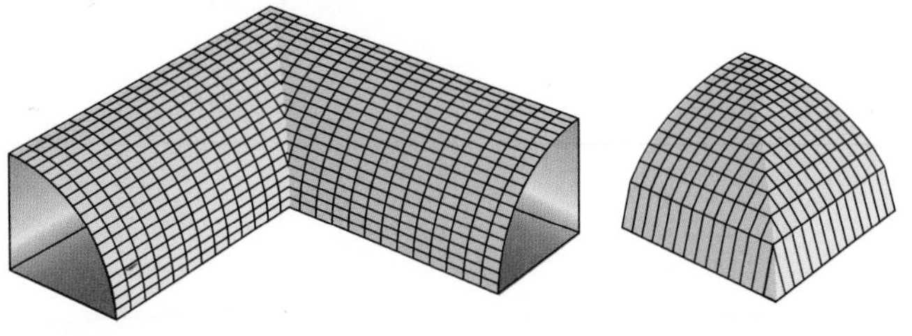

Figure 15

28. Find the volume inside the "+" shown in Figure 16. Assume that both cylinders have radius 2 inches and length 12 inches. *Hint:* The volume is equal to the volume of the first cylinder plus the volume of the second cylinder minus the volume of the region common to both. Use the result of Problem 27.

29. Find the volume inside the "+" in Figure 16, assuming that both cylinders have radius r and length L.

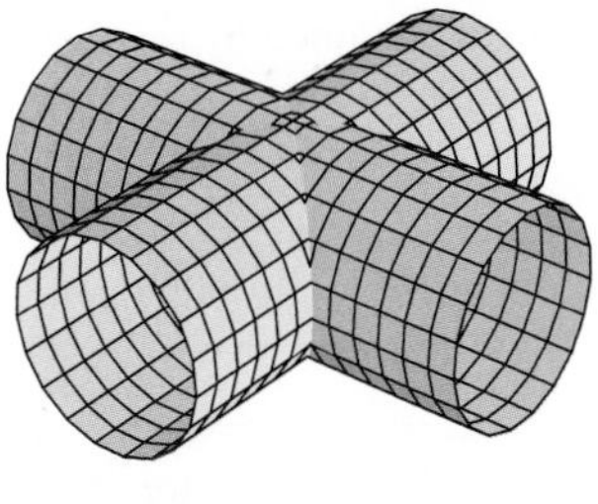

Figure 16

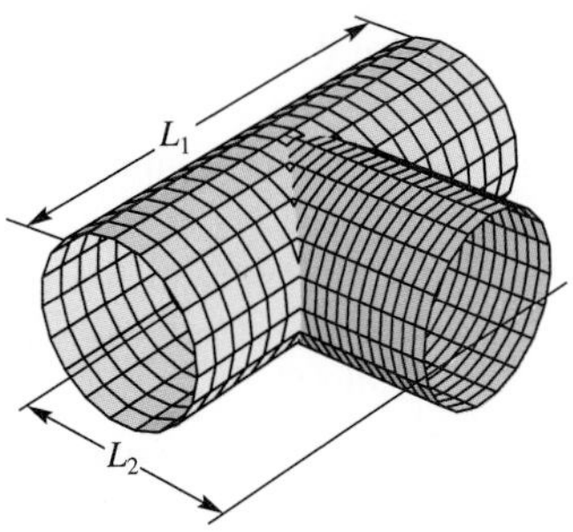

Figure 17

30. Find the volume inside the "T" in Figure 17, assuming that each cylinder has radius $r = 2$ inches and that the lengths are $L_1 = 12$ inches and $L_2 = 8$ inches.

31. Repeat Problem 30 for arbitrary r, L_1, and L_2.

32. The base of a solid is the region R bounded by $y = \sqrt{x}$ and $y = x^2$. Each cross section perpendicular to the x-axis is a semicircle with diameter extending across R. Find the volume of the solid.

33. Find the volume of the solid generated by revolving the region in the first quadrant bounded by the curve $y^2 = x^3$, the line $x = 4$, and the x-axis:

(a) about the line $x = 4$; (b) about the line $y = 8$.

34. Find the volume of the solid generated by revolving the region bounded by the curve $y^2 = x^3$, the line $y = 8$, and the y-axis:

(a) about the line $x = 4$; (b) about the line $y = 8$.

35. Complete the evaluation of the integral in Example 4 by noting that

$$\int_0^2 \left[2\sqrt{4-y^2} + 4 - y^2\right] dy$$
$$= 2\int_0^2 \sqrt{4-y^2}\, dy + \int_0^2 (4-y^2)\, dy$$

Now interpret the first integral as the area of a quarter circle.

36. An open barrel of radius r and height h is initially full of water. It is tilted and water pours out until the water level coincides with a diameter of the base and just touches the rim of the top. Find the volume of water left in the barrel. See Figure 18.

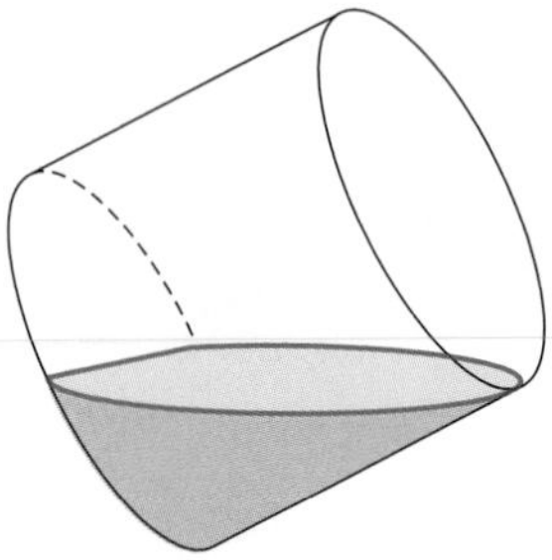

Figure 18

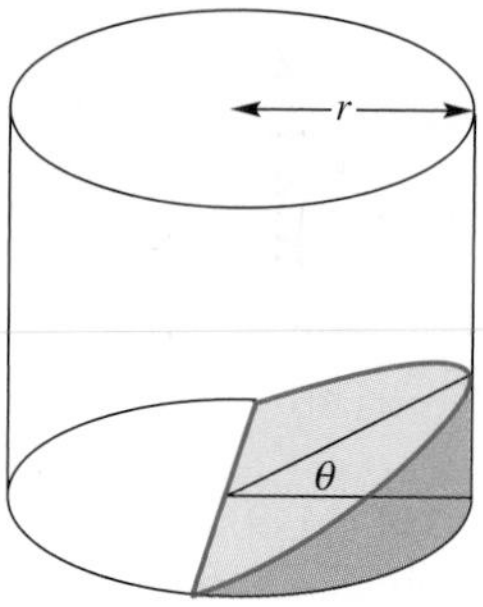

Figure 19

37. A wedge is cut from a right circular cylinder of radius r (Figure 19). The upper surface of the wedge is in a plane through a diameter of the circular base and makes an angle θ with the base. Find the volume of the wedge.

38. (The Water Clock) A water tank is obtained by revolving the curve $y = kx^4$, $k > 0$, about the y-axis.

(a) Find $V(y)$, the volume of water in the tank as a function of its depth y.

(b) Water drains through a small hole according to Torricelli's Law $\left(dV/dt = -m\sqrt{y}\right)$. Show that the water level falls at a constant rate.

39. Show that the volume of a general cone (Figure 20) is $\frac{1}{3}Ah$, where A is the area of the base and h is the height. Use this result to give the formula for the volume of

(a) a right circular cone of radius r and height h;

(b) a regular tetrahedron with edge length r.

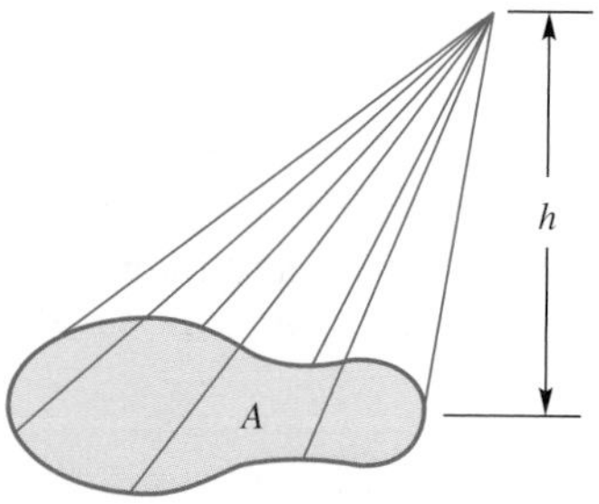

Figure 20

40. State the version of Cavalieri's Principle for volume (see Problem 36 of Section 5.1).

41. Apply Cavalieri's Principle for volumes to the two solids shown in Figure 21. (One is a hemisphere of radius r; the other is a cylinder of radius r and height r with a right circular cone of radius r and height r removed.) Assuming that the volume of a right circular cone is $\frac{1}{3}\pi r^2 h$, find the volume of a hemisphere of radius r.

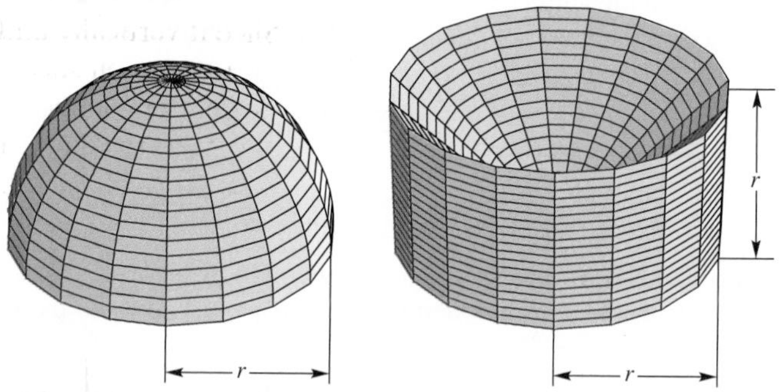

Figure 21

Answers to Concepts Review: **1.** $\pi r^2 h$ **2.** $\pi(R^2 - r^2)h$ **3.** $\pi x^4 \,\Delta x$ **4.** $\pi[(x^2+2)^2 - 4]\,\Delta x$

5.3 Volumes of Solids of Revolution: Shells

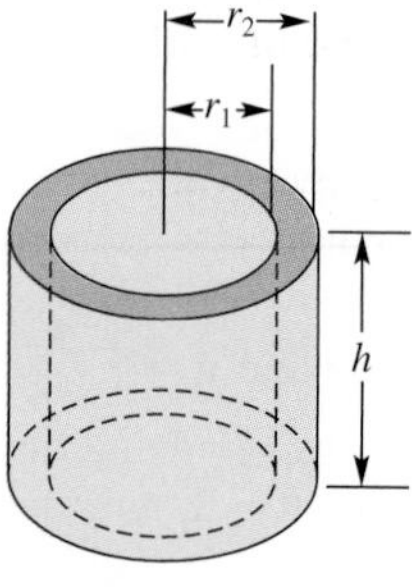

Figure 1

There is another method for finding the volume of a solid of revolution: the method of cylindrical shells. For many problems, it is easier to apply than the methods of disks or washers.

A cylindrical shell is a solid bounded by two concentric right circular cylinders (Figure 1). If the inner radius is r_1, the outer radius is r_2, and the height is h, then its volume is given by

$$\begin{aligned} V &= (\text{area of base}) \cdot (\text{height}) \\ &= (\pi r_2^2 - \pi r_1^2)h \\ &= \pi(r_2 + r_1)(r_2 - r_1)h \\ &= 2\pi\left(\frac{r_2 + r_1}{2}\right)h(r_2 - r_1) \end{aligned}$$

The expression $(r_1 + r_2)/2$, which we will denote by r, is the average of r_1 and r_2. Thus,

$$V = 2\pi \cdot (\text{average radius}) \cdot (\text{height}) \cdot (\text{thickness})$$
$$= 2\pi rh\, \Delta r$$

Here is a good way to remember this formula: If the shell were very thin and flexible (like paper), we could slit it down the side, open it up to form a rectangular sheet, and then calculate its volume by pretending that this sheet forms a thin box of length $2\pi r$, height h, and thickness Δr (Figure 2).

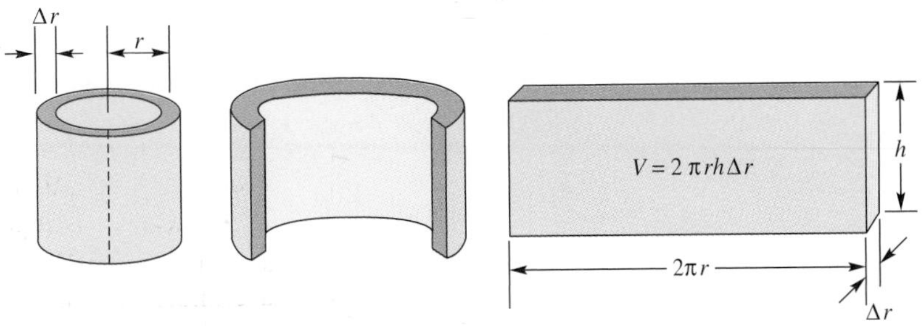

Figure 2

The Method of Shells Consider now a region of the type shown in Figure 3. Slice it vertically and revolve it about the y-axis. It will generate a solid of revolution, and each slice will generate a piece that is approximately a cylindrical shell. To get the volume of this solid, we calculate the volume ΔV of a typical shell, add, and take the limit as the thickness of the shells tends to zero. The latter is, of course, an integral. Again, the strategy is *slice, approximate, integrate*.

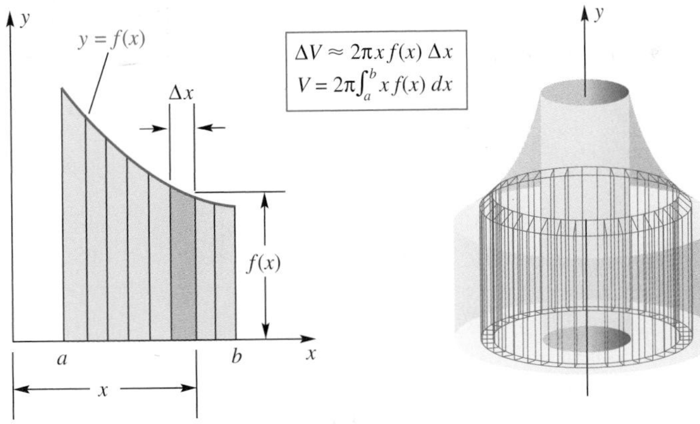

Figure 3

EXAMPLE 1 The region bounded by $y = 1/\sqrt{x}$, the x-axis, $x = 1$, and $x = 4$ is revolved about the y-axis. Find the volume of the resulting solid.

SOLUTION From Figure 3 we see that the volume of the shell generated by the slice is

$$\Delta V \approx 2\pi x f(x)\, \Delta x$$

which, for $f(x) = 1/\sqrt{x}$, becomes

$$\Delta V \approx 2\pi x \frac{1}{\sqrt{x}} \Delta x$$

The volume is then found by integrating.

$$V = 2\pi \int_1^4 x \frac{1}{\sqrt{x}}\, dx = 2\pi \int_1^4 x^{1/2}\, dx$$

$$= 2\pi \left[\frac{2}{3} x^{3/2}\right]_1^4 = 2\pi\left(\frac{2}{3}\cdot 8 - \frac{2}{3}\cdot 1\right) = \frac{28\pi}{3} \approx 29.32$$

■

EXAMPLE 2 The region bounded by the line $y = (r/h)x$, the x-axis, and $x = h$ is revolved about the x-axis, thereby generating a cone (assume that $r > 0, h > 0$). Find its volume by the disk method and by the shell method.

SOLUTION

Disk Method Follow the steps suggested by Figure 4; that is, *slice, approximate, integrate.*

$$V = \pi \frac{r^2}{h^2} \int_0^h x^2\, dx = \pi \frac{r^2}{h^2}\left[\frac{x^3}{3}\right]_0^h = \frac{\pi r^2 h^3}{3h^2} = \frac{1}{3}\pi r^2 h$$

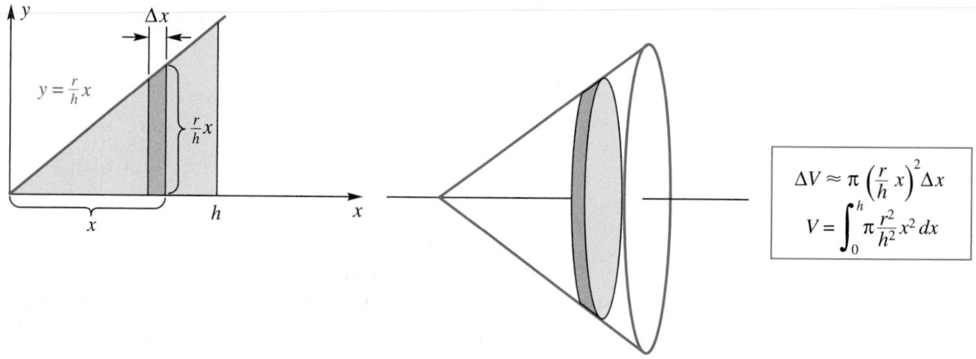

Figure 4

Shell Method Follow the steps suggested by Figure 5. The volume is then

$$V = \int_0^r 2\pi y\left(h - \frac{h}{r}y\right) dy = 2\pi h \int_0^r \left(y - \frac{1}{r}y^2\right) dy$$

$$= 2\pi h \left[\frac{y^2}{2} - \frac{y^3}{3r}\right]_0^r = 2\pi h\left[\frac{r^2}{2} - \frac{r^2}{3}\right] = \frac{1}{3}\pi r^2 h$$

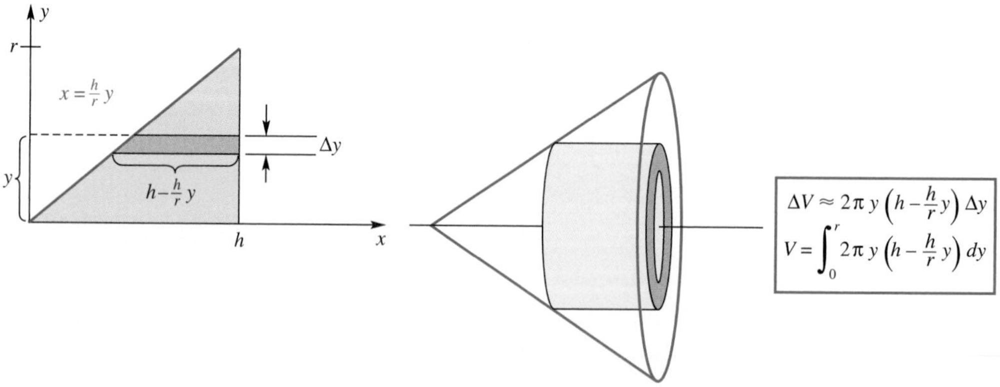

Figure 5

As should be expected, both methods yield the well-known formula for the volume of a right circular cone. ■

EXAMPLE 3 Find the volume of the solid generated by revolving the region in the first quadrant that is above the parabola $y = x^2$ and below the parabola $y = 2 - x^2$ about the y-axis.

SOLUTION One look at the region (left part of Figure 6) should convince you that horizontal slices leading to the disk method are not the best choice (because the right boundary consists of parts of two curves, making it necessary to use two integrals). However, vertical slices, resulting in cylindrical shells, will work fine.

$$V = \int_0^1 2\pi x(2 - 2x^2)\,dx = 4\pi \int_0^1 (x - x^3)\,dx$$

$$= 4\pi\left[\frac{x^2}{2} - \frac{x^4}{4}\right]_0^1 = 4\pi\left[\frac{1}{2} - \frac{1}{4}\right] = \pi \approx 3.14$$

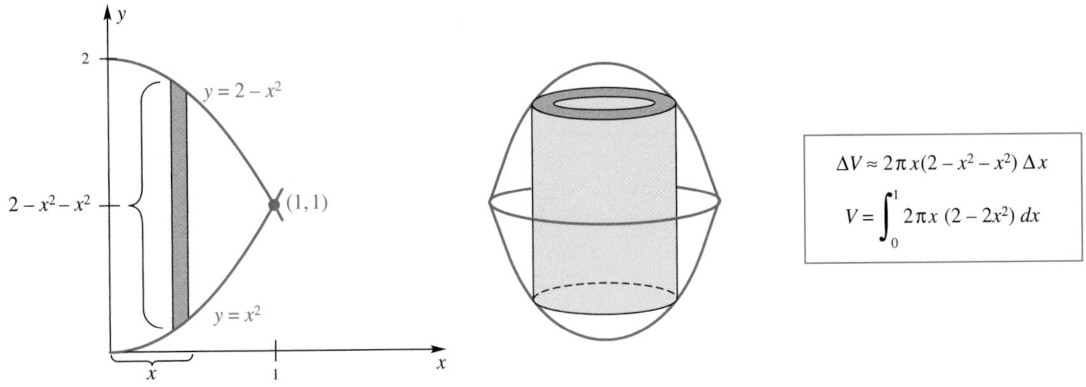

Figure 6

■

Putting It All Together Although most of us can draw a reasonably good plane figure, some of us do less well at drawing three-dimensional solids. But no law says that we have to draw a solid in order to calculate its volume. Usually, a plane figure will do, provided we can visualize the corresponding solid in our minds. In the next example, we are going to imagine revolving the region R of Figure 7 about various axes. Our job is to set up and evaluate an integral for the volume of the resulting solid, and we are going to do it by looking at a plane figure.

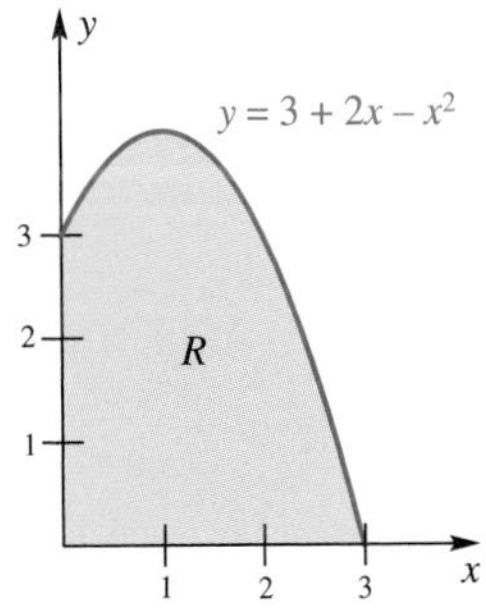

Figure 7

EXAMPLE 4 Set up and evaluate an integral for the volume of the solid that results when the region R shown in Figure 7 is revolved about

(a) the x-axis,
(b) the y-axis,
(c) the line $y = -1$,
(d) the line $x = 4$.

SOLUTION

(a)

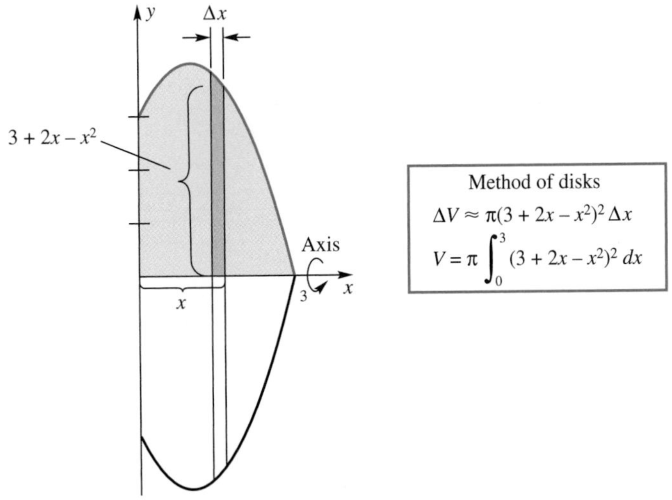

$$V = \pi \int_0^3 (3 + 2x - x^2)^2\,dx = \frac{153}{5}\pi \approx 96.13$$

(b)

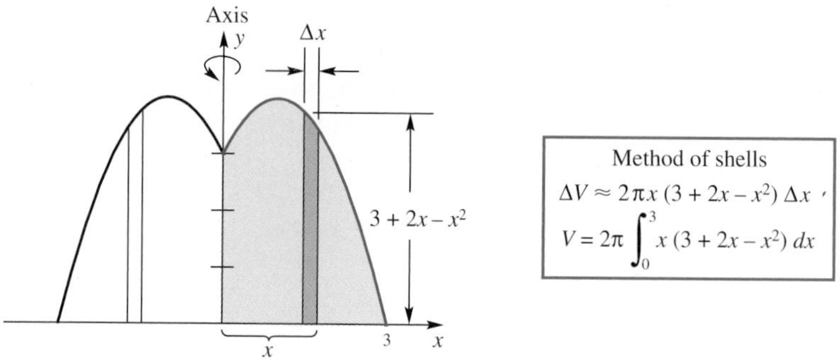

$$V = 2\pi \int_0^3 x(3 + 2x - x^2)\,dx = \frac{45}{2}\pi \approx 70.69$$

> **Technology**
>
> In all four parts of this example, the integrand turned out to be a polynomial, but finding the polynomial involved some messy expansions. Once the integrals are set up, evaluating them is an ideal task for a CAS.

(c)

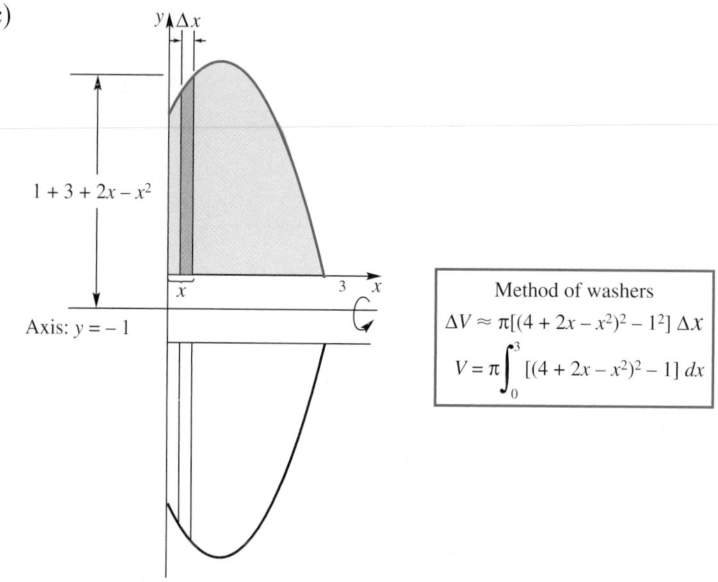

$$V = \pi \int_0^3 [(4 + 2x - x^2)^2 - 1]\,dx = \frac{243}{5}\pi \approx 152.68$$

(d)

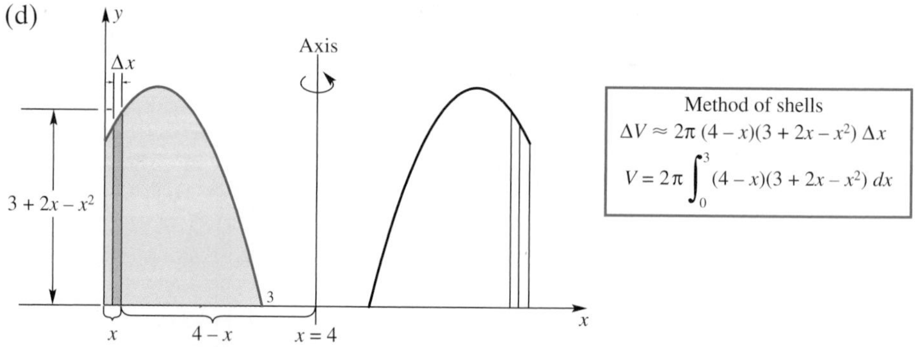

$$V = 2\pi \int_0^3 (4 - x)(3 + 2x - x^2)\,dx = \frac{99}{2}\pi \approx 155.51$$

Note that in all four cases the limits of integration are the same; it is the original plane region that determines these limits. ■

Concepts Review

1. The volume ΔV of a thin cylindrical shell of radius x, height $f(x)$, and thickness Δx is given by $\Delta V \approx$ _____.

2. The triangular region R bounded by $y = x$, $y = 0$, and $x = 2$ is revolved about the y-axis, generating a solid. The method of shells gives the integral _____ as its volume; the method of washers gives the integral _____ as its volume.

3. The region R of Question 2 is revolved about the line $x = -1$, generating a solid. The method of shells gives the integral _____ as its volume.

4. The region R of Question 2 is revolved about the line $y = -1$, generating a solid. The method of shells gives the integral _____ as its volume.

Problem Set 5.3

In Problems 1–12, find the volume of the solid generated when the region R bounded by the given curves is revolved about the indicated axis. Do this by performing the following steps.

(a) Sketch the region R.

(b) Show a typical rectangular slice properly labeled.

(c) Write a formula for the approximate volume of the shell generated by this slice.

(d) Set up the corresponding integral.

(e) Evaluate this integral.

1. $y = \frac{1}{x}$, $x = 1$, $x = 4$, $y = 0$; about the y-axis

2. $y = x^2$, $x = 1$, $y = 0$; about the y-axis

3. $y = \sqrt{x}$, $x = 3$, $y = 0$; about the y-axis

4. $y = 9 - x^2 (x \geq 0)$, $x = 0$, $y = 0$; about the y-axis

5. $y = \sqrt{x}$, $x = 5$, $y = 0$; about the line $x = 5$

6. $y = 9 - x^2 (x \geq 0)$, $x = 0$, $y = 0$; about the line $x = 3$

7. $y = \frac{1}{4}x^3 + 1$, $y = 1 - x$, $x = 1$; about the y-axis

8. $y = x^2$, $y = 3x$; about the y-axis

9. $x = y^2$, $y = 1$, $x = 0$; about the x-axis

10. $x = \sqrt{y} + 1$, $y = 4$, $x = 0$, $y = 0$; about the x-axis

11. $x = y^2$, $y = 2$, $x = 0$; about the line $y = 2$

12. $x = \sqrt{2y} + 1$, $y = 2$, $x = 0$, $y = 0$; about the line $y = 3$

13. Consider the region R (Figure 8). Set up an integral for the volume of the solid obtained when R is revolved about the given line using the indicated method.

(a) The x-axis (washers) (b) The y-axis (shells)

(c) The line $x = a$ (shells) (d) The line $x = b$ (shells)

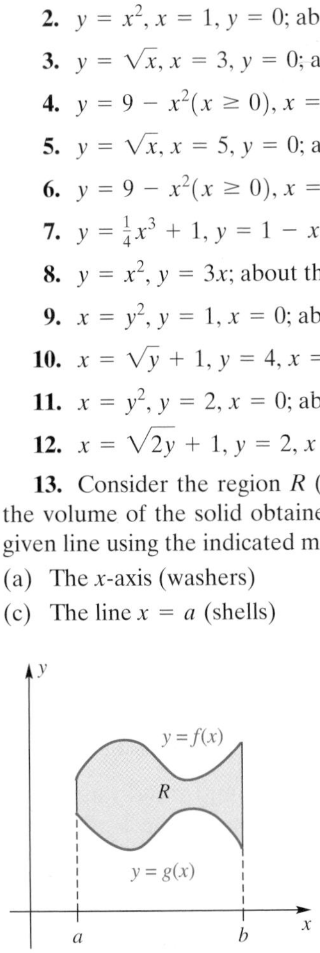

Figure 8

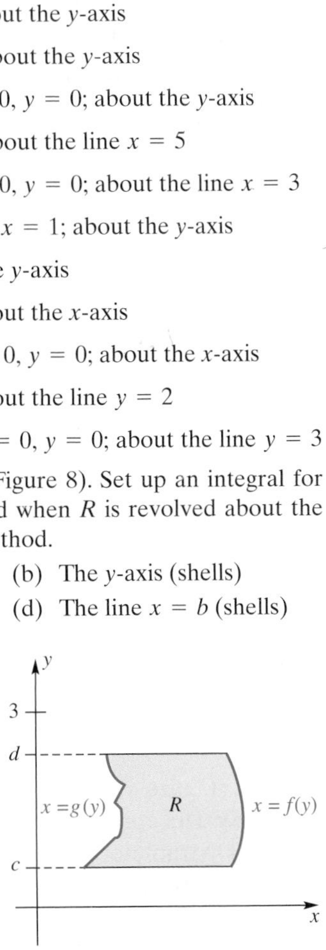

Figure 9

14. A region R is shown in Figure 9. Set up an integral for the volume of the solid obtained when R is revolved about each of the following lines. Use the indicated method.

(a) The y-axis (washers) (b) The x-axis (shells)

(c) The line $y = 3$ (shells)

15. Sketch the region R bounded by $y = 1/x^3$, $x = 1$, $x = 3$, and $y = 0$. Set up (but do not evaluate) integrals for each of the following.

(a) Area of R

(b) Volume of the solid obtained when R is revolved about the y-axis

(c) Volume of the solid obtained when R is revolved about $y = -1$

(d) Volume of the solid obtained when R is revolved about $x = 4$

16. Follow the directions of Problem 15 for the region R bounded by $y = x^3 + 1$ and $y = 0$ and between $x = 0$ and $x = 2$.

17. Find the volume of the solid generated by revolving the region R bounded by the curves $x = \sqrt{y}$ and $x = y^3/32$ about the x-axis.

18. Follow the directions of Problem 17, but revolve R about the line $y = 4$.

19. A round hole of radius a is drilled through the center of a solid sphere of radius b (assume that $b > a$). Find the volume of the solid that remains.

20. Set up the integral (using shells) for the volume of the torus obtained by revolving the region inside the circle $x^2 + y^2 = a^2$ about the line $x = b$, where $b > a$. Then evaluate this integral. *Hint:* As you simplify, it may help to think of part of this integral as an area.

21. The region in the first quadrant bounded by $x = 0$, $y = \sin(x^2)$, and $y = \cos(x^2)$ is revolved about the y-axis. Find the volume of the resulting solid.

22. The region bounded by $y = 2 + \sin x$, $y = 0$, $x = 0$, and $x = 2\pi$ is revolved about the y-axis. Find the volume that results.

Hint: $\int x \sin x \, dx = \sin x - x \cos x + C$.

23. Let R be the region bounded by $y = x^2$ and $y = x$. Find the volume of the solid that results when R is revolved around:

(a) the x-axis; (b) the y-axis; (c) the line $y = x$.

24. Suppose that we know the formula $S = 4\pi r^2$ for the surface area of a sphere, but do not know the corresponding formula for its volume V. Obtain this formula by slicing the solid sphere into thin concentric *spherical shells* (Figure 10). *Hint:* The volume ΔV of a thin spherical shell of outer radius x is $\Delta V \approx 4\pi x^2 \Delta x$.

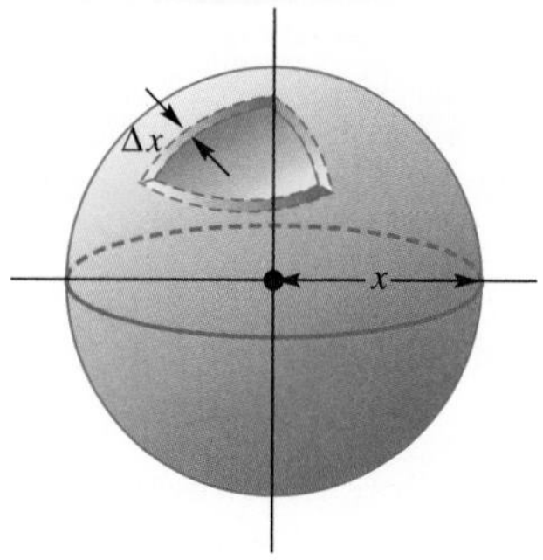

Figure 10

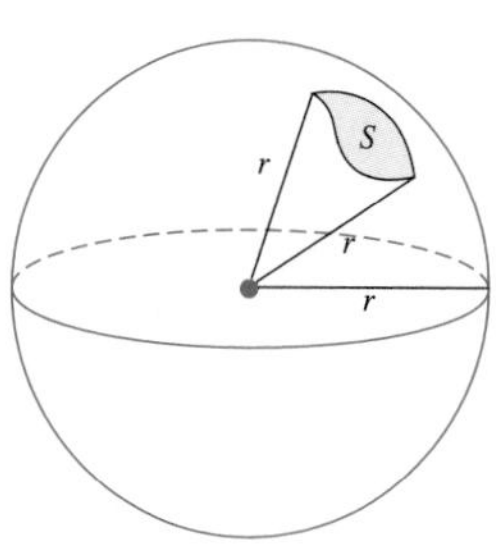

Figure 11

25. Consider a region of area S on the surface of a sphere of radius r. Find the volume of the solid that results when each point of this region is connected to the center of the sphere by a line segment (Figure 11). *Hint:* Use the method of spherical shells mentioned in Problem 24.

Answers to Concepts Review: **1.** $2\pi x f(x)\,\Delta x$ **2.** $2\pi\int_0^2 x^2\,dx;\ \pi\int_0^2 (4-y^2)\,dy$ **3.** $2\pi\int_0^2 (1+x)x\,dx$ **4.** $2\pi\int_0^2 (1+y)(2-y)\,dy$

5.4 Length of a Plane Curve

How long is the spiral curve shown in Figure 1? If it were a piece of string, most of us would stretch it taut and measure it with a ruler. But if it is the graph of an equation, this is a little hard to do.

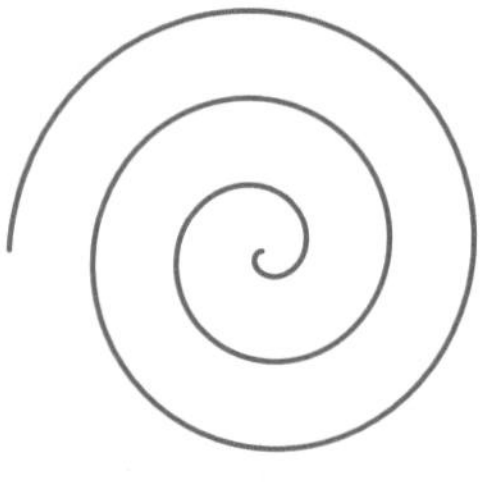

Figure 1

A little reflection suggests a prior question. What is a plane curve? We have used the term *curve* informally until now, often in reference to the graph of a function. Now it is time to be more precise, even for curves that are not graphs of functions. We begin with several examples.

The graph of $y = \sin x, 0 \le x \le \pi$, is a plane curve (Figure 2). So is the graph of $x = y^2, -2 \le y \le 2$ (Figure 3). In both cases, the curve is the graph of a function, the first of the form $y = f(x)$, the second of the form $x = g(y)$. However, the spiral curve does not fit either pattern. Neither does the circle $x^2 + y^2 = a^2$, though in this case we could think of it as the combined graph of the two functions $y = f(x) = \sqrt{a^2 - x^2}$ and $y = g(x) = -\sqrt{a^2 - x^2}$.

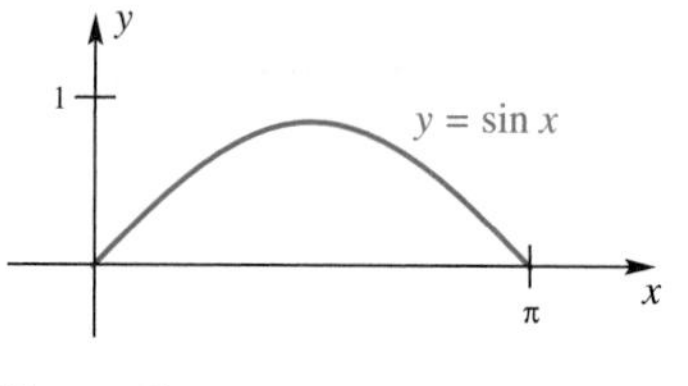

Figure 2

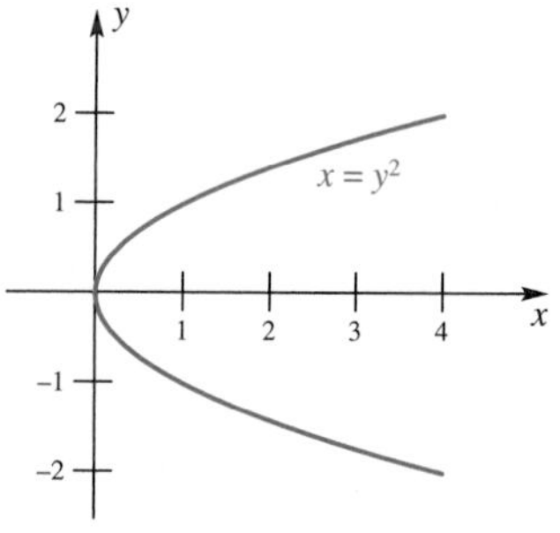

Figure 3

The circle suggests another way of thinking about curves. Recall from trigonometry that

$$x = a\cos t, \quad y = a\sin t, \qquad 0 \le t \le 2\pi$$

describe the circle $x^2 + y^2 = a^2$ (Figure 4). Think of t as time and x and y as giving the position of a particle at time t. The variable t is called a **parameter.** Both x and y are expressed in terms of this parameter. We say that $x = a\cos t,\ y = a\sin t,\ 0 \le t \le 2\pi$, are **parametric equations** describing the circle.

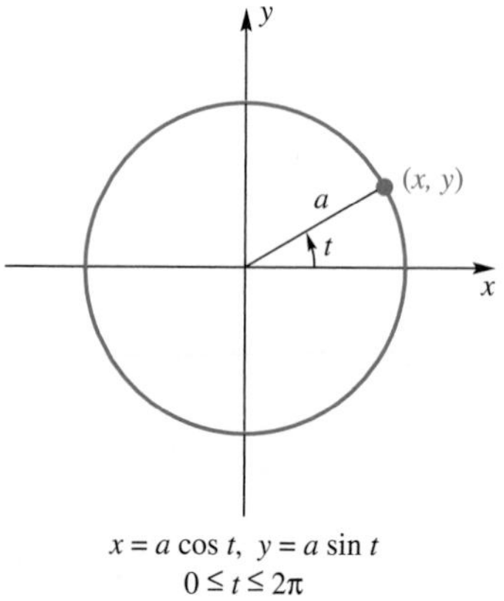

Figure 4

If we were to graph the parametric equations $x = t\cos t,\ y = t\sin t,\ 0 \le t \le 5\pi$, we would get a curve something like the spiral with which we started. And we can even think of the sine curve (Figure 2) and the parabola (Figure 3) in parametric form. We write

$$x = t, \qquad y = \sin t, \qquad 0 \le t \le \pi$$

$x = 2t + 1, y = t^2 - 1$
$0 \le t \le 3$

t	x	y
0	1	−1
1	3	0
2	5	3
3	7	8

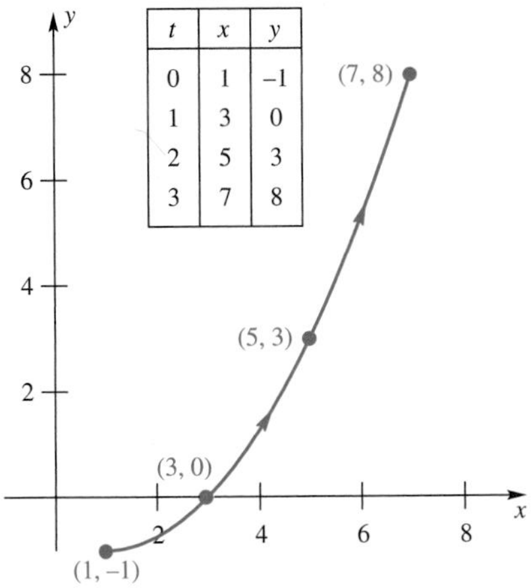

Figure 5

and

$$x = t^2, \qquad y = t, \qquad -2 \le t \le 2$$

Thus, for us, a **plane curve** is determined by a pair of parametric equations $x = f(t), y = g(t), a \le t \le b$, where we assume that f and g are continuous on the given interval. As t increases from a to b, the point (x, y) traces out a curve in the plane. Here is another example.

EXAMPLE 1 Sketch the curve determined by the parametric equations $x = 2t + 1, y = t^2 - 1, 0 \le t \le 3$.

SOLUTION We make a three-column table of values, then plot the ordered pairs (x, y), and finally connect these points in the order of increasing t, as shown in Figure 5. A graphing calculator or a CAS can be used to produce such a graph. Such software usually produces a graph by creating a table, just as we did, and connecting the points. ■

Actually, the definition we have given is too broad for the purposes we have in mind, so we immediately restrict it to what is called a *smooth curve*. The adjective *smooth* is chosen to indicate that as an object moves along the curve so that its position at time t is (x, y) it suffers no sudden changes of direction (continuity of f' and g' ensures this) and does not stop or double back ($f'(t)$ and $g'(t)$ not simultaneously zero ensures this).

Definition

A plane curve is **smooth** if it is determined by a pair of parametric equations $x = f(t), y = g(t), a \le t \le b$, where f' and g' exist and are continuous on $[a, b]$, and $f'(t)$ and $g'(t)$ are not simultaneously zero on (a, b).

The way a curve is parametrized, that is, the way the functions $f(t)$ and $g(t)$, and the domain for t are chosen, determines a *positive direction*. For example, when $t = 0$ in Example 1 (Figure 5), the curve is at the point $(1, -1)$, and when $t = 1$, the curve is at $(3, 0)$. As t increases from $t = 0$ to $t = 3$, the curve traces a path from $(1, -1)$ to $(7, 8)$. This direction, which is often indicated by an arrow on the curve as shown in Figure 5, is called the **orientation** of the curve. The orientation of a curve is irrelevant as far as determining its length goes, but in problems that we will encounter later in this book the orientation does matter.

$x = t - \sin t, y = 1 - \cos t$
$0 \le t \le 4\pi$

t	$x(t)$	$y(t)$
0	0.00	0
$\pi/2$	0.57	1
π	3.14	2
$3\pi/2$	5.71	1
2π	6.28	0
$5\pi/2$	6.85	1
3π	9.42	2
$7\pi/2$	10.00	1
4π	12.57	0

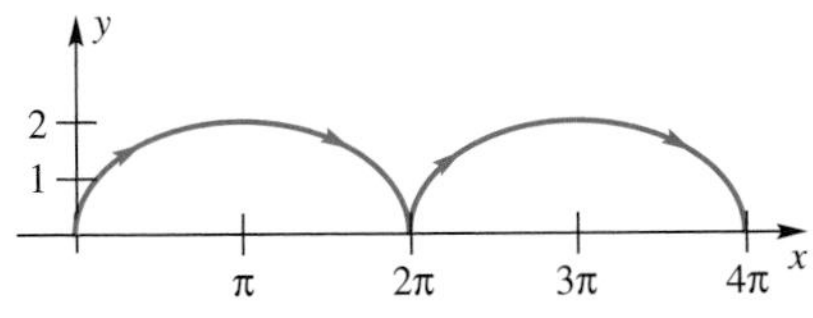

Figure 6

EXAMPLE 2 Sketch the curve determined by the parametric equations $x = t - \sin t, y = 1 - \cos t, 0 \le t \le 4\pi$. Indicate the orientation. Is this curve smooth?

SOLUTION The table, which shows the values of x and y for several values of t from 0 to 4π, leads to the graph in Figure 6. This curve is not smooth even though x and y are both differentiable functions of t. The problem is that $dx/dt = 1 - \cos t$ and $dy/dt = \sin t$ are simultaneously 0 when $t = 2\pi$. The object slows down to a stop at time $t = 2\pi$, then starts up in a new direction. ■

The curve described in Example 2 is called the **cycloid**. It describes the path of a fixed point on the rim of a wheel of radius 1 as the wheel rolls along the x-axis. (See Problem 18 for a derivation of this result.)

Arc Length Finally, we are ready for the main question. What is meant by the length of the smooth curve given parametrically by $x = f(t), \; y = g(t), \; a \le t \le b$?

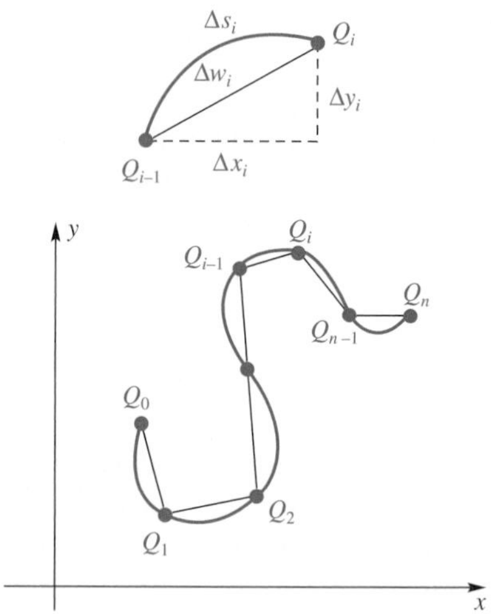

Figure 7

Partition the interval $[a, b]$ into n subintervals by means of points t_i:

$$a = t_0 < t_1 < t_2 < \cdots < t_n = b$$

This cuts the curve into n pieces with corresponding end points $Q_0, Q_1, Q_2, \ldots, Q_{n-1}, Q_n$, as shown in Figure 7.

Our idea is to approximate the curve by the indicated polygonal line segments, calculate their total length, and then take the limit as the norm of the partition approaches zero. In particular, we approximate the *length* Δs_i of the ith segment (see Figure 7) by

$$\begin{aligned}\Delta s_i \approx \Delta w_i &= \sqrt{(\Delta x_i)^2 + (\Delta y_i)^2} \\ &= \sqrt{[f(t_i) - f(t_{i-1})]^2 + [g(t_i) - g(t_{i-1})]^2}\end{aligned}$$

From the Mean Value Theorem for Derivatives (Theorem 3.6A), we know that there are points $\bar{t}_i$ and $\hat{t}_i$ in (t_{i-1}, t_i) such that

$$\begin{aligned}f(t_i) - f(t_{i-1}) &= f'(\bar{t}_i)\,\Delta t_i \\ g(t_i) - g(t_{i-1}) &= g'(\hat{t}_i)\,\Delta t_i\end{aligned}$$

where $\Delta t_i = t_i - t_{i-1}$. Thus,

$$\begin{aligned}\Delta w_i &= \sqrt{[f'(\bar{t}_i)\,\Delta t_i]^2 + [g'(\hat{t}_i)\,\Delta t_i]^2} \\ &= \sqrt{[f'(\bar{t}_i)]^2 + [g'(\hat{t}_i)]^2}\,\Delta t_i\end{aligned}$$

and the total length of the polygonal line segments is

$$\sum_{i=1}^{n} \Delta w_i = \sum_{i=1}^{n} \sqrt{[f'(\bar{t}_i)]^2 + [g'(\hat{t}_i)]^2}\,\Delta t_i$$

The latter expression is almost a Riemann sum, the only difficulty being that $\bar{t}_i$ and $\hat{t}_i$ are not likely to be the same point. However, it is shown in advanced calculus books that in the limit (as the norm of the partition goes to 0), this makes no difference. Thus, we may define the **arc length** L of the curve to be the limit of the expression above as the norm of the partition approaches zero; that is,

$$L = \int_a^b \sqrt{[f'(t)]^2 + [g'(t)]^2}\,dt = \int_a^b \sqrt{\left(\frac{dx}{dt}\right)^2 + \left(\frac{dy}{dt}\right)^2}\,dt$$

Two special cases are of great interest. If the curve is given by $y = f(x)$, $a \le x \le b$, we treat x as the parameter and the boxed result takes the form

$$L = \int_a^b \sqrt{1 + \left(\frac{dy}{dx}\right)^2}\,dx$$

Similarly, if the curve is given by $x = g(y)$, $c \le y \le d$, we treat y as the parameter, obtaining

$$L = \int_c^d \sqrt{1 + \left(\frac{dx}{dy}\right)^2}\,dy$$

These formulas yield the familiar results for circles and line segments, as the following two examples illustrate.

EXAMPLE 3 Find the circumference of the circle $x^2 + y^2 = a^2$.

SOLUTION We write the equation of the circle in parametric form: $x = a \cos t$, $y = a \sin t$, $0 \le t \le 2\pi$. Then $dx/dt = -a \sin t$, $dy/dt = a \cos t$, and, by the first of our formulas,

$$L = \int_0^{2\pi} \sqrt{a^2 \sin^2 t + a^2 \cos^2 t}\, dt = \int_0^{2\pi} a\, dt = \Big[at\Big]_0^{2\pi} = 2\pi a$$

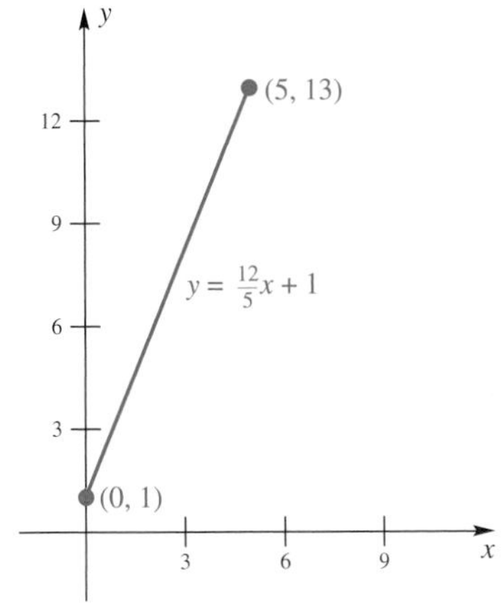

Figure 8

EXAMPLE 4 Find the length of the line segment from $A(0, 1)$ to $B(5, 13)$.

SOLUTION The given line segment is shown in Figure 8. Note that the equation of the corresponding line is $y = \frac{12}{5}x + 1$, so $dy/dx = \frac{12}{5}$; and so, by the second of the three length formulas,

$$L = \int_0^5 \sqrt{1 + \left(\frac{12}{5}\right)^2}\, dx = \int_0^5 \sqrt{\frac{5^2 + 12^2}{5^2}}\, dx = \frac{13}{5}\int_0^5 1\, dx$$

$$= \left[\frac{13}{5}x\right]_0^5 = 13$$

This agrees with the result obtained by use of the distance formula.

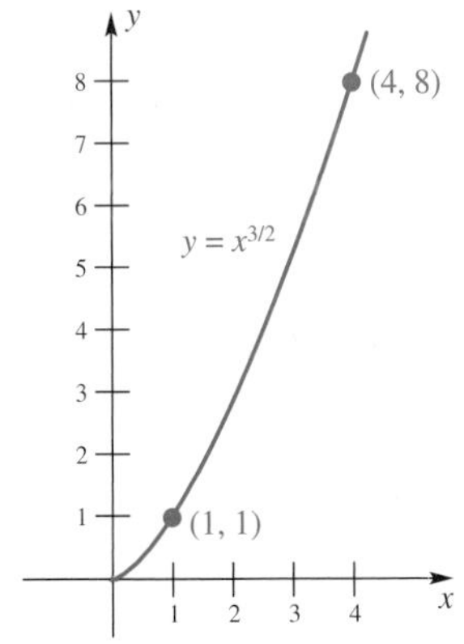

Figure 9

EXAMPLE 5 Find the length of the arc of the curve $y = x^{3/2}$ from the point $(1, 1)$ to the point $(4, 8)$ (see Figure 9).

≈ **SOLUTION** We begin by estimating this length by finding the length of the segment from (1, 1) to (4, 8): $\sqrt{(4-1)^2 + (8-1)^2} = \sqrt{58} \approx 7.6$. The actual length should be slightly larger.

For the exact calculation, we note that $dy/dx = \frac{3}{2}x^{1/2}$, so

$$L = \int_1^4 \sqrt{1 + \left(\frac{3}{2}x^{1/2}\right)^2}\, dx = \int_1^4 \sqrt{1 + \frac{9}{4}x}\, dx$$

Let $u = 1 + \frac{9}{4}x$; then $du = \frac{9}{4}dx$. Hence,

$$\int \sqrt{1 + \frac{9}{4}x}\, dx = \frac{4}{9}\int \sqrt{u}\, du = \frac{4}{9}\frac{2}{3}u^{3/2} + C$$

$$= \frac{8}{27}\left(1 + \frac{9}{4}x\right)^{3/2} + C$$

Therefore,

$$\int_1^4 \sqrt{1 + \frac{9}{4}x}\, dx = \left[\frac{8}{27}\left(1 + \frac{9}{4}x\right)^{3/2}\right]_1^4 = \frac{8}{27}\left(10^{3/2} - \frac{13^{3/2}}{8}\right) \approx 7.63$$

For most arc length problems it is easy to set up the definite integral that gives the length. This is just a matter of substituting the required derivatives in the formula. However, it is often difficult or impossible to evaluate these integrals using the Second Fundamental Theorem of Calculus because of the difficulty of finding antiderivatives. For many problems we must resort to using a numerical technique such as the Parabolic Rule described in Section 4.6 in order to obtain an approximation to the definite integral.

EXAMPLE 6 Sketch the graph of the curve given parametrically by $x = 2 \cos t$, $y = 4 \sin t$, $0 \le t \le \pi$. Set up a definite integral that gives the arc length of the curve and approximate this definite integral using the Parabolic Rule with $n = 8$.

$x = 2\cos t,\ y = 4\sin t$
$0 \le t \le \pi$

t	x	y
0	2	0
$\pi/6$	$\sqrt{3}$	2
$\pi/3$	1	$2\sqrt{3}$
$\pi/2$	0	4
$2\pi/3$	-1	$2\sqrt{3}$
$5\pi/6$	$-\sqrt{3}$	2
π	-2	0

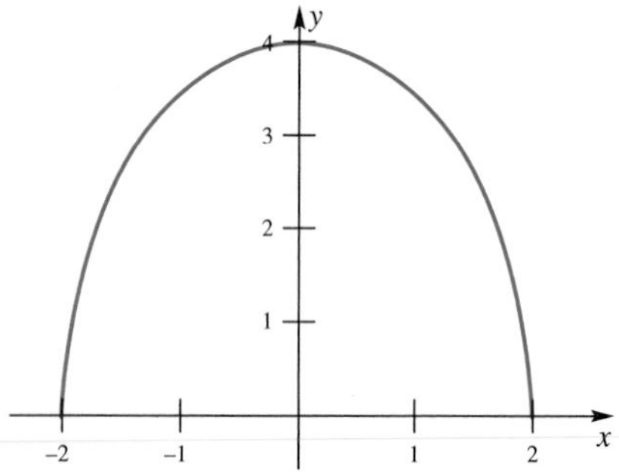

Figure 10

SOLUTION The graph (Figure 10) is drawn, as in previous examples, by first making a three-column table of values. The definite integral that gives the arc length is

$$\begin{aligned} L &= \int_0^{\pi} \sqrt{\left(\frac{dx}{dt}\right)^2 + \left(\frac{dy}{dt}\right)^2}\, dt \\ &= \int_0^{\pi} \sqrt{(-2\sin t)^2 + (4\cos t)^2}\, dt \\ &= \int_0^{\pi} 2\sqrt{\sin^2 t + 4\cos^2 t}\, dt \\ &= 2\int_0^{\pi} \sqrt{1 + 3\cos^2 t}\, dt \end{aligned}$$

This definite integral cannot be evaluated using the Second Fundamental Theorem of Calculus. Let $f(t) = \sqrt{1 + 3\cos^2 t}$. The approximation using the Parabolic Rule with $n = 8$ is

$$\begin{aligned} L &\approx 2\frac{\pi - 0}{3 \cdot 8}\left[f(0) + 4f\left(\frac{\pi}{8}\right) + 2f\left(\frac{2\pi}{8}\right) + 4f\left(\frac{3\pi}{8}\right) + 2f\left(\frac{4\pi}{8}\right)\right. \\ &\qquad \left. + 4f\left(\frac{5\pi}{8}\right) + 2f\left(\frac{6\pi}{8}\right) + 4f\left(\frac{7\pi}{8}\right) + f(\pi)\right] \\ &\approx 2\frac{\pi}{24}[2.0 + 4 \cdot 1.8870 + 2 \cdot 1.5811 + 4 \cdot 1.1997 + 2 \cdot 1.0 \\ &\qquad + 4 \cdot 1.1997 + 2 \cdot 1.5811 + 4 \cdot 1.8870 + 2.0)] \\ &\approx 9.6913 \end{aligned}$$

■

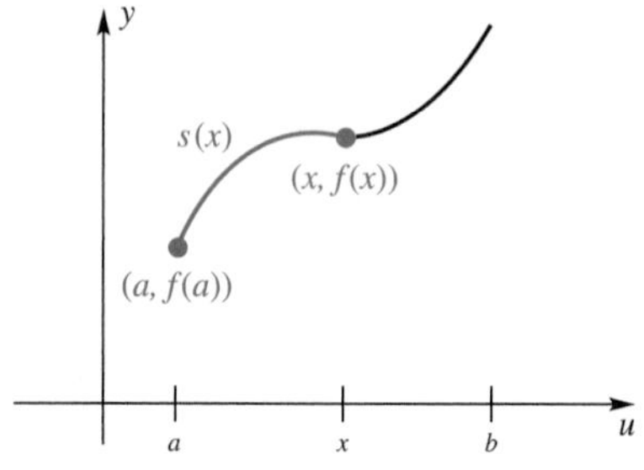

Figure 11

Differential of Arc Length Let f be continuously differentiable on $[a, b]$. For each x in (a, b), define $s(x)$ by

$$s(x) = \int_a^x \sqrt{1 + [f'(u)]^2}\, du$$

Then $s(x)$ gives the arc length of the curve $y = f(u)$ from the point $(a, f(a))$ to $(x, f(x))$ (see Figure 11). By the First Fundamental Theorem of Calculus (Theorem 4.3A),

$$s'(x) = \frac{ds}{dx} = \sqrt{1 + [f'(x)]^2} = \sqrt{1 + \left(\frac{dy}{dx}\right)^2}$$

Thus, ds, the **differential of arc length,** can be written as

$$ds = \sqrt{1 + \left(\frac{dy}{dx}\right)^2}\, dx$$

In fact, depending on how a graph is parametrized, we are led to three formulas for ds:

$$ds = \sqrt{1 + \left(\frac{dy}{dx}\right)^2}\, dx = \sqrt{1 + \left(\frac{dx}{dy}\right)^2}\, dy = \sqrt{\left(\frac{dx}{dt}\right)^2 + \left(\frac{dy}{dt}\right)^2}\, dt$$

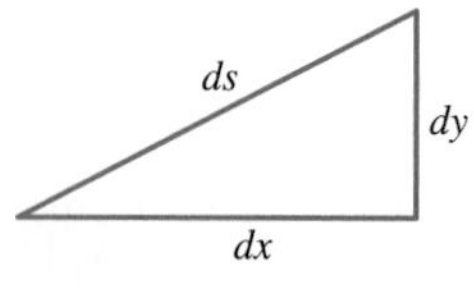

Figure 12

Some people prefer to remember these formulas by writing (see Figure 12)

$$(ds)^2 = (dx)^2 + (dy)^2$$

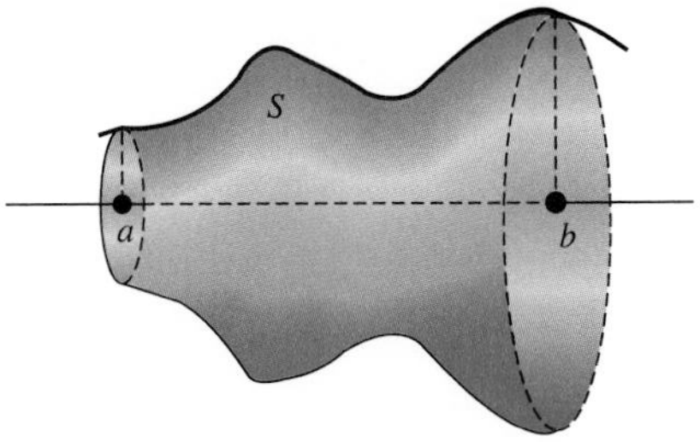

Figure 13

The three forms arise by dividing and multiplying the right-hand side by $(dx)^2$, $(dy)^2$, and $(dt)^2$, respectively. For example,

$$(ds)^2 = \left[\frac{(dx)^2}{(dx)^2} + \frac{(dy)^2}{(dx)^2}\right](dx)^2 = \left[1 + \left(\frac{dy}{dx}\right)^2\right](dx)^2$$

which gives the first of the three formulas.

Area of a Surface of Revolution If a smooth plane curve is revolved about an axis in its plane, it generates a surface of revolution as illustrated in Figure 13. Our aim is to determine the area of such a surface.

To get started, we introduce the formula for the area of the frustum of a cone. A **frustum** of a cone is that part of the surface of a cone between two planes perpendicular to the axis of the cone (shaded in Figure 14). If a frustum has base radii r_1 and r_2 and slant height ℓ, then its area A is given by

$$A = 2\pi\left(\frac{r_1 + r_2}{2}\right)\ell = 2\pi(\text{average radius})\cdot(\text{slant height})$$

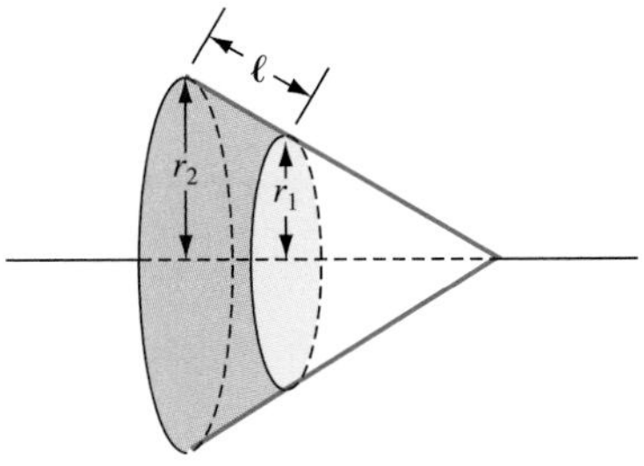

Figure 14

The derivation of this result depends only on the formula for the area of a circle (see Problem 31).

Let $y = f(x)$, $a \le x \le b$, determine a smooth curve in the upper half of the xy-plane, as shown in Figure 15. Partition the interval $[a, b]$ into n pieces by means of points $a = x_0 < x_1 < \cdots < x_n = b$, thereby also dividing the curve into n pieces. Let Δs_i denote the length of the ith piece, and let y_i be the y-coordinate of a point on this piece. When the curve is revolved about the x-axis, it generates a surface, and the typical piece generates a narrow band. The "area" of this band ought to be approximately that of a frustum, that is, approximately $2\pi y_i\,\Delta s_i$. When we add the contributions of all the pieces and take the limit as the norm of the partition approaches zero, we get what we define to be the area of the surface of revolution. All this is indicated in Figure 16. The surface area is thus

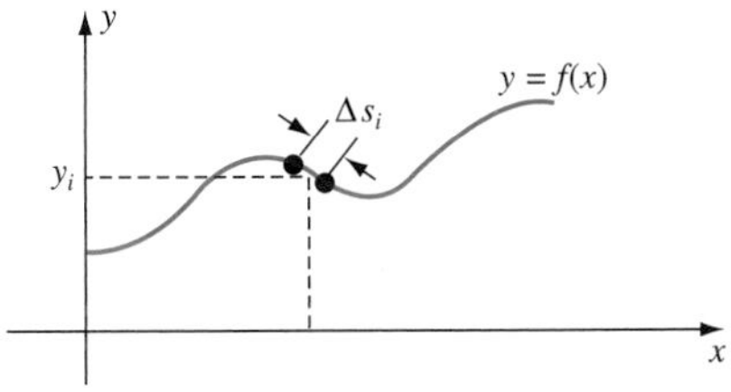

Figure 15

$$\begin{aligned} A &= \lim_{\|P\|\to 0}\sum_{i=1}^{n} 2\pi y_i\,\Delta s_i \\ &= 2\pi\int_a^b y\,ds \\ &= 2\pi\int_a^b f(x)\sqrt{1 + [f'(x)]^2}\,dx \end{aligned}$$

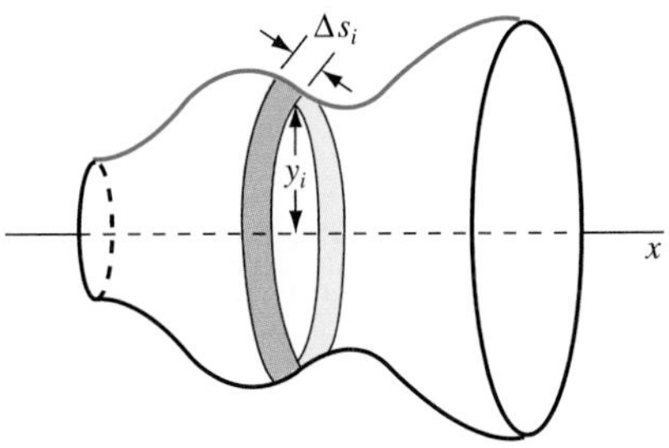

Figure 16

EXAMPLE 7 Find the area of the surface of revolution generated by revolving the curve $y = \sqrt{x}$, $0 \le x \le 4$, about the x-axis (Figure 17).

SOLUTION Here, $f(x) = \sqrt{x}$ and $f'(x) = 1/(2\sqrt{x})$. Thus,

$$\begin{aligned} A &= 2\pi\int_0^4 \sqrt{x}\sqrt{1 + \frac{1}{4x}}\,dx = 2\pi\int_0^4 \sqrt{x}\sqrt{\frac{4x+1}{4x}}\,dx \\ &= \pi\int_0^4 \sqrt{4x+1}\,dx = \left[\pi\cdot\frac{1}{4}\cdot\frac{2}{3}(4x+1)^{3/2}\right]_0^4 \\ &= \frac{\pi}{6}(17^{3/2} - 1^{3/2}) \approx 36.18 \end{aligned}$$

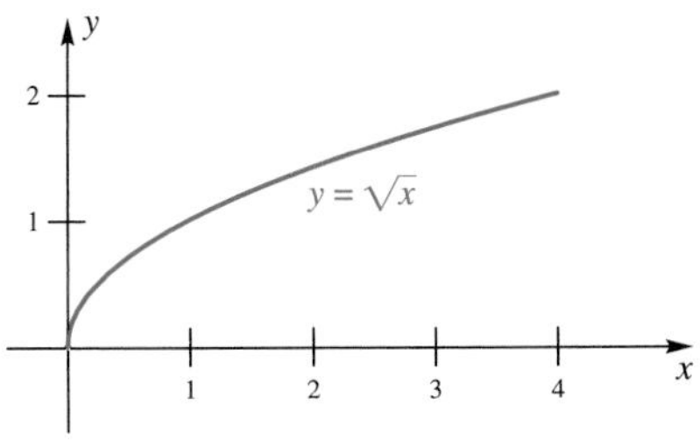

Figure 17

If the curve is given parametrically by $x = f(t)$, $y = g(t)$, $a \le t \le b$, then the surface area formula becomes

$$A = 2\pi\int_a^b y\,ds = 2\pi\int_a^b g(t)\sqrt{[f'(t)]^2 + [g'(t)]^2}\,dt$$

■

Concepts Review

1. The graph of the parametric equations $x = 4\cos t$, $y = 4\sin t$, $0 \le t \le 2\pi$, is a curve called a _____.

2. The curve determined by $y = x^2 + 1$, $0 \le x \le 4$, can be put in parametric form using x as the parameter by writing $x =$ _____, $y =$ _____.

3. The formula for the length L of the curve $x = f(t)$, $y = g(t)$, $a \le t \le b$, is $L =$ _____.

4. The proof of the formula for the length of a curve depends strongly on an earlier theorem with the name _____.

Problem Set 5.4

In Problems 1–6, find the length of the indicated curve.

1. $y = 4x^{3/2}$ between $x = 1/3$ and $x = 5$

2. $y = \frac{2}{3}(x^2 + 1)^{3/2}$ between $x = 1$ and $x = 2$

3. $y = (4 - x^{2/3})^{3/2}$ between $x = 1$ and $x = 8$

4. $y = (x^4 + 3)/(6x)$ between $x = 1$ and $x = 3$

5. $x = y^4/16 + 1/(2y^2)$ between $y = -3$ and $y = -2$ *Hint:* Watch signs; $\sqrt{u^2} = -u$ when $u < 0$.

6. $30xy^3 - y^8 = 15$ between $y = 1$ and $y = 3$

In Problems 7–10, sketch the graph of the given parametric equation and find its length.

7. $x = t^3/3$, $y = t^2/2$; $0 \le t \le 1$

8. $x = 3t^2 + 2$, $y = 2t^3 - 1/2$; $1 \le t \le 4$

9. $x = 4\sin t$, $y = 4\cos t - 5$; $0 \le t \le \pi$

10. $x = \sqrt{5}\sin 2t - 2$, $y = \sqrt{5}\cos 2t - \sqrt{3}$; $0 \le t \le \pi/4$

11. Use an x-integration to find the length of the segment of the line $y = 2x + 3$ between $x = 1$ and $x = 3$. Check by using the distance formula.

12. Use a y-integration to find the length of the segment of the line $2y - 2x + 3 = 0$ between $y = 1$ and $y = 3$. Check by using the distance formula.

In Problems 13–16, set up a definite integral that gives the arc length of the given curve. Approximate the integral using the Parabolic Rule with $n = 8$.

13. $x = t$, $y = t^2$; $0 \le t \le 2$

14. $x = t^2$, $y = \sqrt{t}$; $1 \le t \le 4$

15. $x = \sin t$, $y = \cos 2t$; $0 \le t \le \pi/2$

16. $x = t$, $y = \tan t$; $0 \le t \le \pi/4$

17. Sketch the graph of the four-cusped *hypocycloid* $x = a\sin^3 t$, $y = a\cos^3 t$, $0 \le t \le 2\pi$, and find its length. *Hint:* By symmetry, you can quadruple the length of the first quadrant portion.

18. A point P on the rim of a wheel of radius a is initially at the origin. As the wheel rolls to the right along the x-axis, P traces out a curve called a *cycloid* (see Figure 18). Derive parametric equations for the cycloid as follows. The parameter is θ.

(a) Show that $\overline{OT} = a\theta$.

(b) Convince yourself that $\overline{PQ} = a\sin\theta$, $\overline{QC} = a\cos\theta$, $0 \le \theta \le \pi/2$.

(c) Show that $x = a(\theta - \sin\theta)$, $y = a(1 - \cos\theta)$.

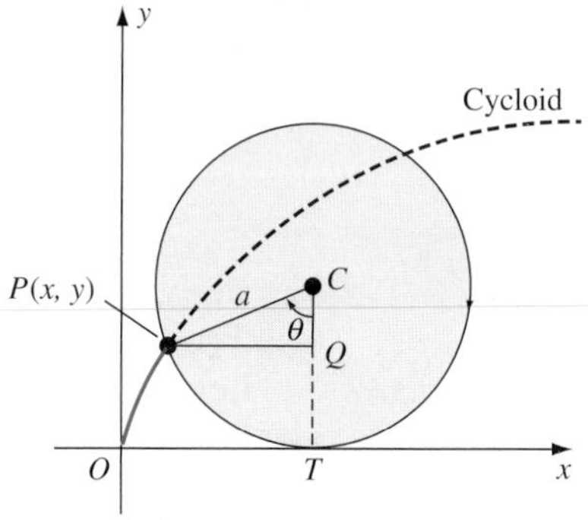

Figure 18

19. Find the length of one arch of the cycloid of Problem 18. *Hint:* First show that

$$\left(\frac{dx}{d\theta}\right)^2 + \left(\frac{dy}{d\theta}\right)^2 = 4a^2\sin^2\left(\frac{\theta}{2}\right)$$

20. Suppose that the wheel of Problem 18 turns at a constant rate $\omega = d\theta/dt$, where t is time. Then $\theta = \omega t$.

(a) Show that the speed ds/dt of P along the cycloid is

$$\frac{ds}{dt} = 2a\omega\left|\sin\frac{\omega t}{2}\right|$$

(b) When is the speed a maximum and when is it a minimum?

(c) Explain why a bug on a wheel of a car going 60 miles per hour is itself sometimes traveling at 120 miles per hour.

21. Find the length of each curve.

(a) $y = \int_1^x \sqrt{u^3 - 1}\,du$, $1 \le x \le 2$

(b) $x = t - \sin t$, $y = 1 - \cos t$, $0 \le t \le 4\pi$

22. Find the length of each curve.

(a) $y = \int_{\pi/6}^x \sqrt{64\sin^2 u\cos^4 u - 1}\,du$, $\frac{\pi}{6} \le x \le \frac{\pi}{3}$

(b) $x = a\cos t + at\sin t$, $y = a\sin t - at\cos t$, $-1 \le t \le 1$

In Problems 23–30, find the area of the surface generated by revolving the given curve about the x-axis.

23. $y = 6x$, $0 \le x \le 1$

24. $y = \sqrt{25 - x^2}$, $-2 \le x \le 3$

25. $y = x^3/3$, $1 \le x \le \sqrt{7}$

26. $y = (x^6 + 2)/(8x^2)$, $1 \le x \le 3$

27. $x = t$, $y = t^3$, $0 \le t \le 1$

28. $x = 1 - t^2, y = 2t, 0 \le t \le 1$

29. $y = \sqrt{r^2 - x^2}, -r \le x \le r$

30. $x = r \cos t, y = r \sin t, 0 \le t \le \pi$

31. If the surface of a cone of slant height ℓ and base radius r is cut along a lateral edge and laid flat, it becomes the sector of a circle of radius ℓ and central angle θ (see Figure 19).

(a) Show that $\theta = 2\pi r/\ell$ radians.

(b) Use the formula $\frac{1}{2}\ell^2\theta$ for the area of a sector of radius ℓ and central angle θ to show that the lateral surface area of a cone is $\pi r \ell$.

(c) Use the result of part (b) to obtain the formula $A = 2\pi[(r_1 + r_2)/2]\ell$ for the lateral area of a frustum of a cone with base radii r_1 and r_2 and slant height ℓ.

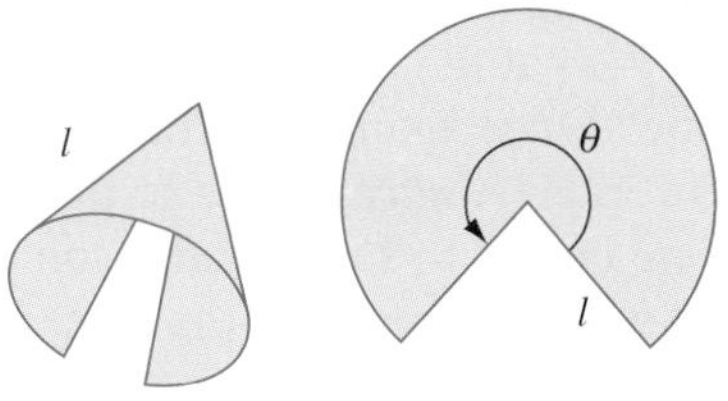

Figure 19

32. Show that the area of the part of the surface of a sphere of radius a between two parallel planes h units apart ($h < 2a$) is $2\pi ah$. Thus, show that if a right circular cylinder is circumscribed about a sphere then two planes parallel to the base of the cylinder bound regions of the same area on the sphere and the cylinder.

33. Figure 20 shows one arch of a cycloid. Its parametric equations (see Problem 18) are given by

$$x = a(t - \sin t), \quad y = a(1 - \cos t), \quad 0 \le t \le 2\pi$$

(a) Show that the area of the surface generated when this curve is revolved about the x-axis is

$$A = 2\sqrt{2}\pi a^2 \int_0^{2\pi} (1 - \cos t)^{3/2}\, dt$$

(b) With the help of the half-angle formula $1 - \cos t = 2\sin^2(t/2)$, evaluate A.

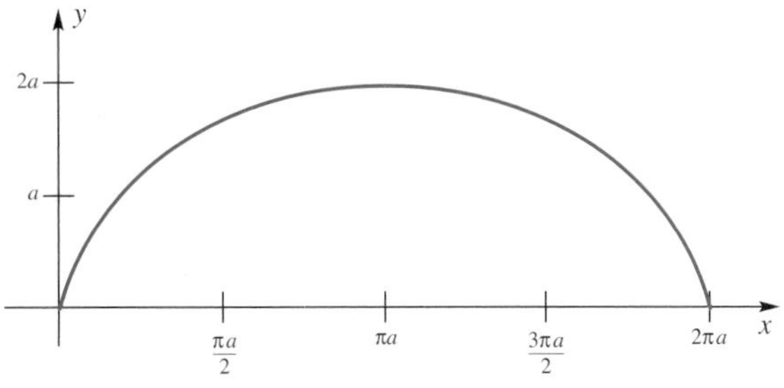

Figure 20

34. The circle $x = a \cos t, y = a \sin t, 0 \le t \le 2\pi$, is revolved about the line $x = b, 0 < a < b$, thus generating a torus (doughnut). Find its surface area.

GC **35.** Sketch the graphs of each of the following parametric equations.

(a) $x = 3 \cos t, y = 3 \sin t, 0 \le t \le 2\pi$

(b) $x = 3 \cos t, y = \sin t, 0 \le t \le 2\pi$

(c) $x = t \cos t, y = t \sin t, 0 \le t \le 6\pi$

(d) $x = \cos t, y = \sin 2t, 0 \le t \le 2\pi$

(e) $x = \cos 3t, y = \sin 2t, 0 \le t \le 2\pi$

(f) $x = \cos t, y = \sin \pi t, 0 \le t \le 40$

CAS **36.** Find the lengths of each of the curves in Problem 35. You will first have to set up the appropriate integral and then use a computer to evaluate it.

CAS **37.** Using the same axes, draw the graphs of $y = x^n$ on $[0, 1]$ for $n = 1, 2, 4, 10$, and 100. Find the length of each of these curves. Guess at the length when $n = 10{,}000$.

Answers to Concepts Review: **1.** circle **2.** $x; x^2 + 1$ **3.** $\int_a^b \sqrt{[f'(t)]^2 + [g'(t)]^2}\, dt$ **4.** Mean Value Theorem for Derivatives

5.5 Work and Fluid Force

In physics, we learn that if an object moves a distance d along a line while subjected to a *constant* force F in the direction of the motion then the **work done by the force** is

$$\text{Work} = (\text{Force}) \cdot (\text{Distance})$$

That is,

$$W = F \cdot D$$

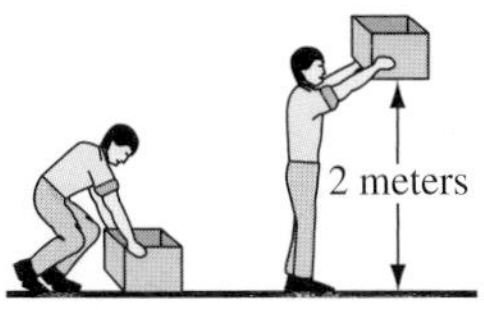

Figure 1

If force is measured in newtons (the force required to give a mass of 1 kilogram an acceleration of 1 meter per second per second), then work is in *newton-meters*, also called *joules*. If force is measured in pounds and distance in feet, then work is in *foot-pounds*. For example, a person lifting a weight (force) of 3 newtons a distance of 2 meters does $3 \cdot 2 = 6$ joules of work (Figure 1). (Strictly speaking, a force slightly greater than 3 newtons is required for a short distance to get the

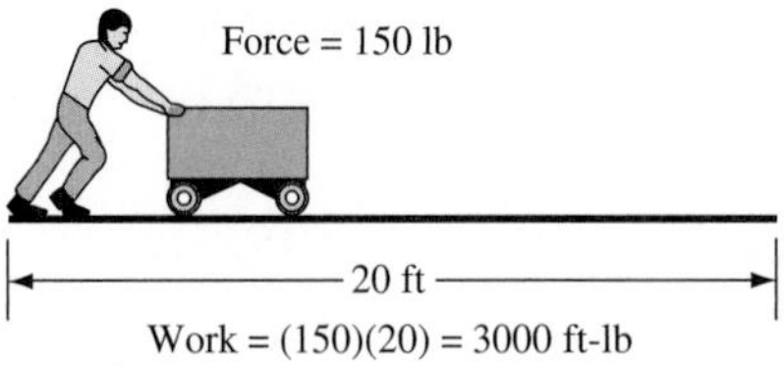

Figure 2

package moving upward, and as the package nears the height of 2 meters, a force slightly less than 3 newtons is required for a short distance to get it stopped. Even in this case, the work is 6 newtons, but this is harder to show.) Similarly, a worker pushing a cart with a constant force of 150 pounds (in order to overcome friction) a distance of 20 feet does $150 \cdot 20 = 3000$ foot-pounds of work (Figure 2).

In many practical situations, force is not constant, but rather varies as the object moves along the line. Suppose that the object is being moved along the x-axis from a to b subject to a variable force of magnitude $F(x)$ at the point x, where F is a continuous function. Then how much work is done? Once again, the strategy of *slice*, *approximate*, and *integrate* leads us to an answer. Here *slice* means to partition the interval $[a, b]$ into small pieces. *Approximate* means to suppose that, on a typical piece from x to $x + \Delta x$, the force is constant with value $F(x)$. If the force is constant (with value $F(x_i)$) over the interval $[x_{i-1}, x_i]$, then the work required to move the object from x_{i-1} to x_i is $F(x_i)(x_i - x_{i-1})$ (Figure 3). *Integrate* means to add up all the bits of work and then take the limit as the length of the pieces approaches zero. Thus, the work done in moving the object from a to b is

$$W = \lim_{\Delta x \to 0} \sum_{i=1}^{n} F(x_i)\,\Delta x = \int_a^b F(x)\,dx$$

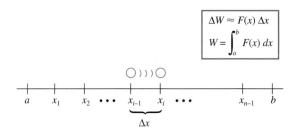

$$\Delta W \approx F(x)\,\Delta x$$
$$W = \int_a^b F(x)\,dx$$

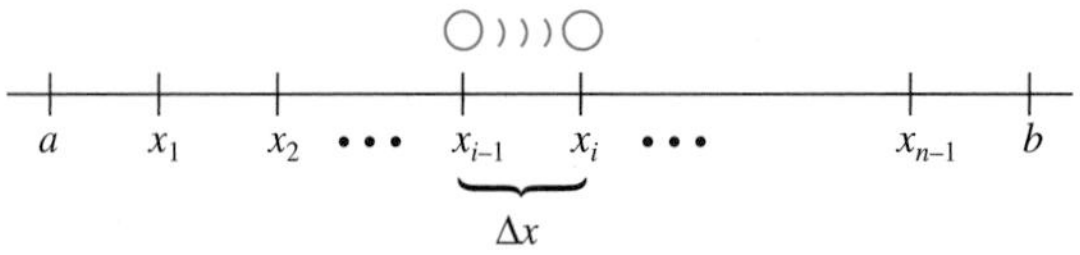

Figure 3

Application to Springs According to Hooke's Law in physics, the force $F(x)$ necessary to keep a spring stretched (or compressed) x units beyond (or short of) its natural length (Figure 4) is given by

$$F(x) = kx$$

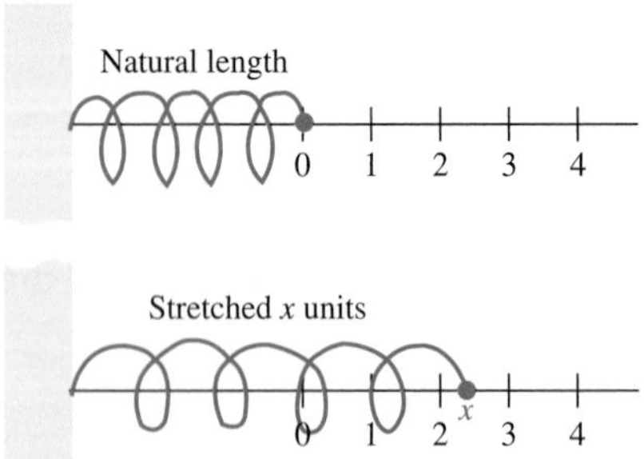

Figure 4

Here, the constant k, the *spring constant*, is positive and depends on the particular spring under consideration. The stiffer the spring is, the greater the value of k.

EXAMPLE 1 If the natural length of a spring is 0.2 meter, and if it takes a force of 12 newtons to keep it extended 0.04 meter, find the work done in stretching the spring from its natural length to a length of 0.3 meter.

SOLUTION By Hooke's Law, the force $F(x)$ required to keep the spring stretched x meters is given by $F(x) = kx$. To determine the spring constant k for this particular spring, we note that $F(0.04) = 12$. Thus, $k \cdot 0.04 = 12$, or $k = 300$, and so

$$F(x) = 300x$$

Figure 5

When the spring is at its natural length of 0.2 meter, $x = 0$; when it is 0.3 meter long, $x = 0.1$. Therefore, the work done in stretching the spring is

$$W = \int_0^{0.1} 300x\,dx = \left[150x^2\right]_0^{0.1} = 1.5 \text{ joules} \qquad \blacksquare$$

Application to Pumping a Liquid To pump water out of a tank requires work, as anyone who has ever tried a hand pump will know (Figure 5). But how much work? The answer to this question rests on the same basic principles presented in the previous discussion.

EXAMPLE 2 A tank in the shape of a right circular cone (Figure 6) is full of water. If the height of the tank is 10 feet and the radius of its top is 4 feet, find the work done in (a) pumping the water over the top edge of the tank, and (b) pumping the water to a height 10 feet above the top of the tank.

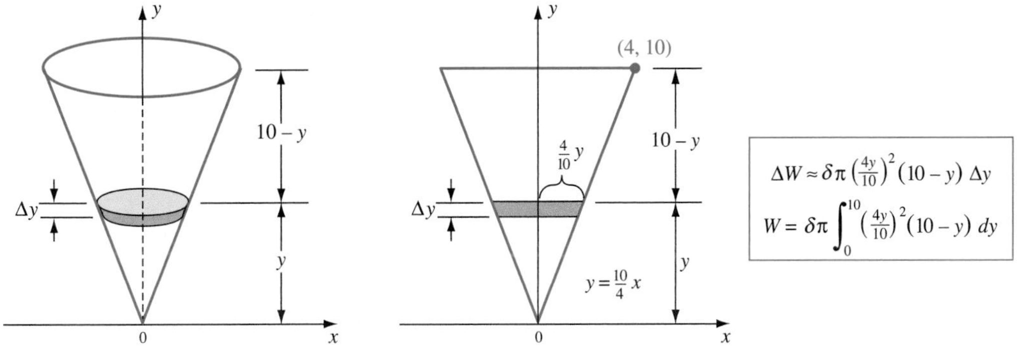

Figure 6

SOLUTION

(a) Position the tank in a coordinate system, as shown in Figure 6. Both a three-dimensional view and a two-dimensional cross section are shown. Imagine slicing the water into thin horizontal disks, each of which must be lifted over the edge of the tank. A disk of thickness Δy at height y has radius $4y/10$ (by similar triangles). Thus, its volume is approximately $\pi(4y/10)^2\,\Delta y$ cubic feet, and its weight is about $\delta\pi(4y/10)^2\,\Delta y$, where $\delta = 62.4$ is the (weight) density of water in pounds per cubic foot. The force necessary to lift this disk of water is its weight, and the disk must be lifted a distance $10 - y$ feet. Thus, the work ΔW done on this disk is approximately

$$\Delta W = (\text{force})\cdot(\text{distance}) \approx \delta\pi\left(\frac{4y}{10}\right)^2 \Delta y\cdot(10 - y)$$

Thus,

$$W = \int_0^{10} \delta\pi\left(\frac{4y}{10}\right)^2 (10 - y)\,dy = \delta\pi\frac{4}{25}\int_0^{10} (10y^2 - y^3)\,dy$$

$$= \frac{(4\pi)(62.4)}{25}\left[\frac{10y^3}{3} - \frac{y^4}{4}\right]_0^{10} \approx 26{,}138 \text{ foot-pounds}$$

(b) Part (b) is just like part (a), except that each disk of water must now be lifted a distance $20 - y$, rather than $10 - y$. Thus,

$$W = \delta\pi\int_0^{10} \left(\frac{4y}{10}\right)^2 (20 - y)\,dy = \delta\pi\frac{4}{25}\int_0^{10} (20y^2 - y^3)\,dy$$

$$= \frac{(4\pi)(62.4)}{25}\left[\frac{20y^3}{3} - \frac{y^4}{4}\right]_0^{10} \approx 130{,}690 \text{ foot-pounds}$$

Note that the limits are still 0 and 10 (not 0 and 20). Why? ■

EXAMPLE 3 Find the work done in pumping the water over the rim of a tank that is 50 feet long and has a semicircular end of radius 10 feet if the tank is filled to a depth of 7 feet (Figure 7).

SOLUTION We position the end of the tank in a coordinate system, as shown in Figure 8. A typical horizontal slice is shown both on this two-dimensional picture and on the three-dimensional one in Figure 7. This slice is approximately a thin box, so we calculate its volume by multiplying length, width, and thickness. Its weight is its density $\delta = 62.4$ times its volume. Finally, we note that this slice must be lifted through a distance $-y$ (the minus sign results from the fact that y is negative in our diagram).

$$\begin{aligned} W &= \delta \int_{-10}^{-3} 100\sqrt{100 - y^2}(-y)\, dy \\ &= 50\delta \int_{-10}^{-3} (100 - y^2)^{1/2}(-2y)\, dy \\ &= \left[(50\delta)\left(\tfrac{2}{3}\right)(100 - y^2)^{3/2}\right]_{-10}^{-3} \\ &= \tfrac{100}{3}(91)^{3/2}\delta \approx 1{,}805{,}616 \text{ foot-pounds} \end{aligned}$$

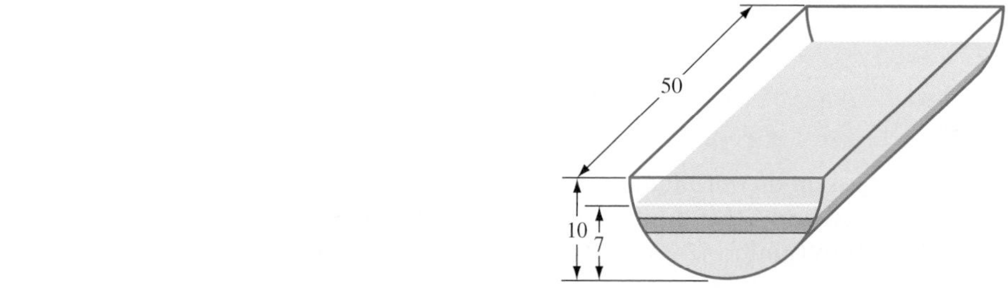

Figure 7

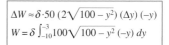

$\Delta W \approx \delta \cdot 50\,(2\sqrt{100 - y^2})\,(\Delta y)\,(-y)$
$W = \delta \int_{-10}^{-3} 100\sqrt{100 - y^2}\,(-y)\,dy$

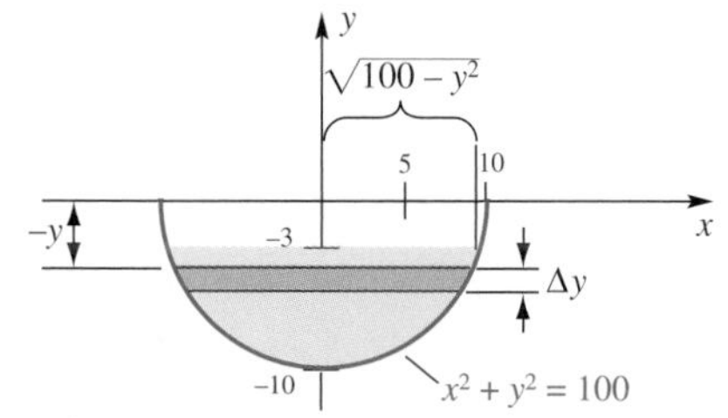

Figure 8 ■

Fluid Force Imagine the tank shown in Figure 9 to be filled to a depth h with a fluid of density δ. Then the force exerted by the fluid on a horizontal rectangle of area A on the bottom is equal to the weight of the column of fluid that stands directly over that rectangle (Figure 10), that is, $F = \delta h A$.

It is a fact, first stated by Blaise Pascal (1623–1662), that the pressure (force per unit area) exerted by a fluid is the same in all directions. Thus, the pressure at all points on a surface, whether that surface is horizontal, vertical, or at some other angle, is the same—provided the points are at the same depth. In particular, the force against each of the three small rectangles in Figure 9 is approximately the

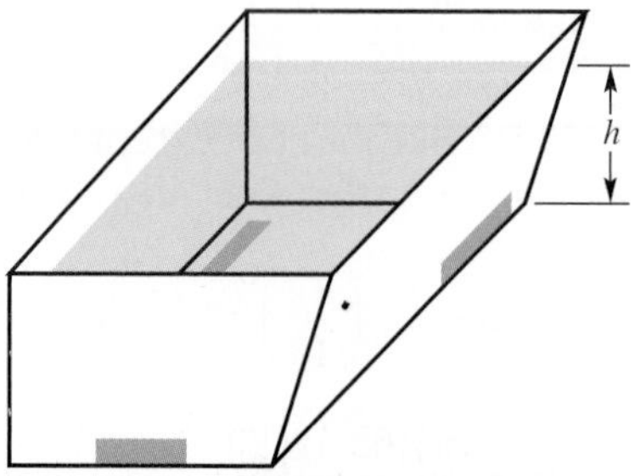

Figure 9

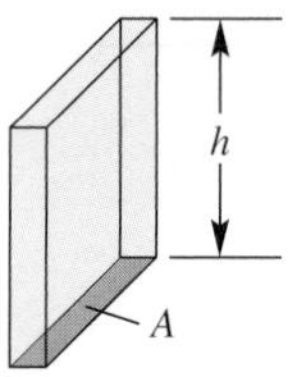

$F = \delta h A$

Figure 10

The condition for balance at the origin is that $M = 0$. Of course, we should not expect balance at the origin except in special circumstances. But surely any system of masses will balance somewhere. The question is where. What is the x-coordinate of the point where the fulcrum should be placed to make the system in Figure 4 balance?

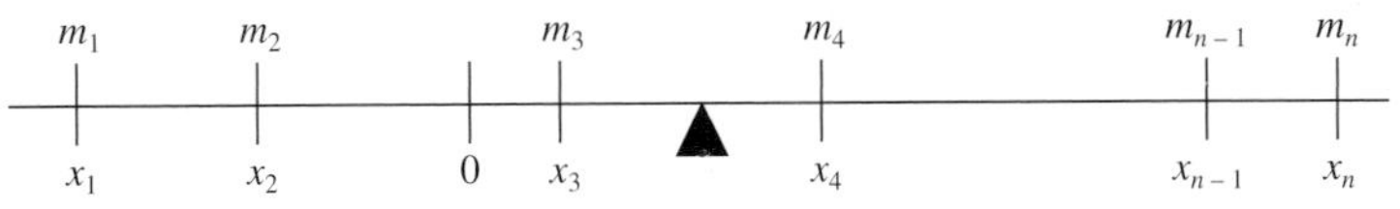

Figure 4

Call the desired coordinate $\bar{x}$. The total moment with respect to it should be zero; that is,

$$(x_1 - \bar{x})m_1 + (x_2 - \bar{x})m_2 + \cdots + (x_n - \bar{x})m_n = 0$$

or

$$x_1 m_1 + x_2 m_2 + \cdots + x_n m_n = \bar{x}m_1 + \bar{x}m_2 + \cdots + \bar{x}m_n$$

When we solve for $\bar{x}$, we obtain

$$\bar{x} = \frac{M}{m} = \frac{\sum_{i=1}^{n} x_i m_i}{\sum_{i=1}^{n} m_i}$$

The point $\bar{x}$, called the **center of mass,** is the balance point. Notice that it is just the total moment with respect to the origin divided by the total mass.

EXAMPLE 1 Masses of 4, 2, 6, and 7 kilograms are located at points 0, 1, 2, and 4, respectively, along the x-axis (Figure 5). Find the center of mass.

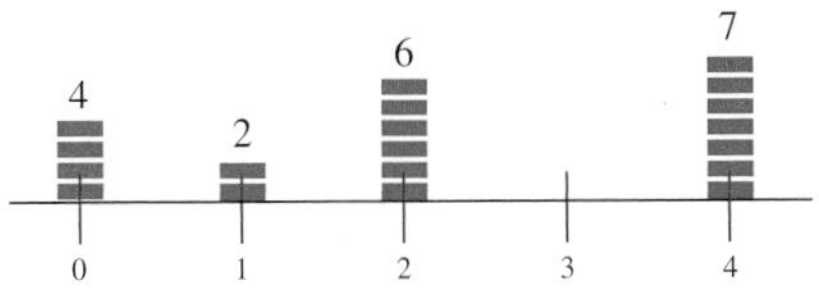

Figure 5

SOLUTION

$$\bar{x} = \frac{(0)(4) + (1)(2) + (2)(6) + (4)(7)}{4 + 2 + 6 + 7} = \frac{42}{19} \approx 2.21$$

≈ Your intuition should confirm that $x = 2.21$ is about right for the balance point. ■

Continuous Mass Distribution along a Line Consider now a straight segment of thin wire of varying density (mass per unit length) for which we desire to find the balance point. We impose a coordinate line along the wire and follow our usual procedure of *slice, approximate*, and *integrate*. Supposing that the density at x is $\delta(x)$, we first obtain the total mass m and then the total moment M with respect to the origin (Figure 6). This leads to the formula

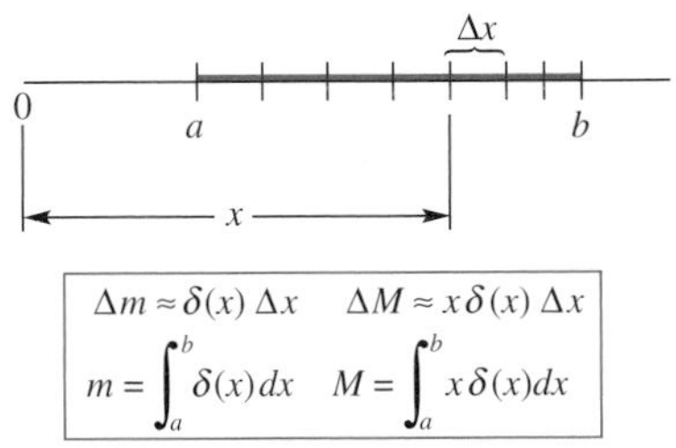

Figure 6

$$\bar{x} = \frac{M}{m} = \frac{\int_a^b x\delta(x)\,dx}{\int_a^b \delta(x)\,dx}$$

Two comments are in order. First, remember this formula by analogy with the formula for point masses:

$$\frac{\sum x_i m_i}{\sum m_i} \sim \frac{\sum x\,\Delta m}{\sum \Delta m} \sim \frac{\int x\,\delta(x)\,dx}{\int \delta(x)\,dx}$$

Second, note that we have assumed that moments of small pieces of wire add together to give the total moment, just as was the case for point masses. This should seem reasonable to you if you imagine the mass of the typical piece of length Δx to be concentrated at the point x.

EXAMPLE 2 The density $\delta(x)$ of a wire at the point x centimeters from one end is given by $\delta(x) = 3x^2$ grams per centimeter. Find the center of mass of the piece between $x = 0$ and $x = 10$.

Figure 7

SOLUTION We expect $\bar{x}$ to be nearer 10 than 0, since the wire is much heavier (denser) toward the right end (Figure 7).

$$\bar{x} = \frac{\int_0^{10} x \cdot 3x^2\,dx}{\int_0^{10} 3x^2\,dx} = \frac{\left[3x^4/4\right]_0^{10}}{\left[x^3\right]_0^{10}} = \frac{7500}{1000} = 7.5\text{ cm}$$ ■

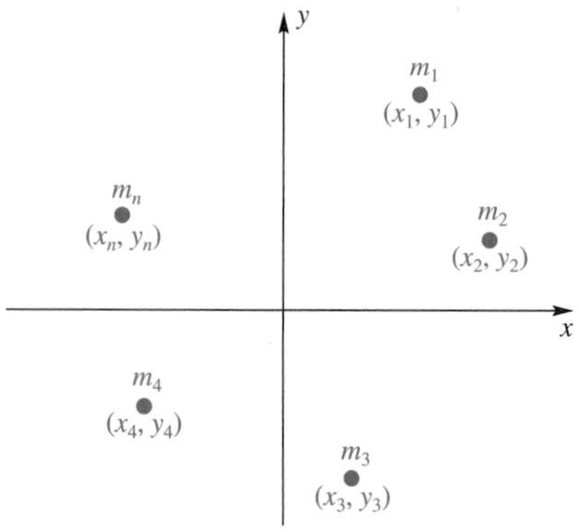

Figure 8

Mass Distributions in the Plane Consider n point masses of sizes $m_1, m_2, \ldots, m_n$ situated at points $(x_1, y_1), (x_2, y_2), \ldots, (x_n, y_n)$ in the coordinate plane (Figure 8). Then the total moments M_y and M_x with respect to the y-axis and x-axis, respectively, are given by

$$M_y = \sum_{i=1}^{n} x_i m_i \qquad M_x = \sum_{i=1}^{n} y_i m_i$$

The coordinates $(\bar{x}, \bar{y})$ of the center of mass (balance point) are

$$\bar{x} = \frac{M_y}{m} = \frac{\sum_{i=1}^{n} x_i m_i}{\sum_{i=1}^{n} m_i} \qquad \bar{y} = \frac{M_x}{m} = \frac{\sum_{i=1}^{n} y_i m_i}{\sum_{i=1}^{n} m_i}$$

EXAMPLE 3 Five particles, having masses 1, 4, 2, 3, and 2 units, are located at points $(6, -1)$, $(2, 3)$, $(-4, 2)$, $(-7, 4)$, and $(2, -2)$, respectively. Find the center of mass.

SOLUTION

$$\bar{x} = \frac{(6)(1) + (2)(4) + (-4)(2) + (-7)(3) + (2)(2)}{1 + 4 + 2 + 3 + 2} = -\frac{11}{12}$$

$$\bar{y} = \frac{(-1)(1) + (3)(4) + (2)(2) + (4)(3) + (-2)(2)}{1 + 4 + 2 + 3 + 2} = \frac{23}{12}$$ ■

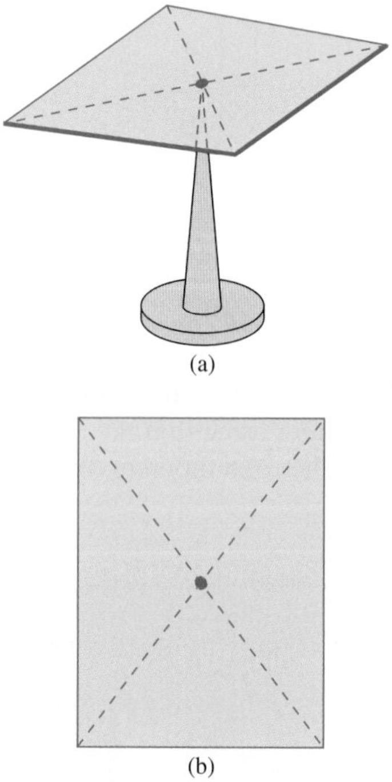

Figure 9

We next consider the problem of finding the center of mass of a *lamina* (thin planar sheet). For simplicity, we suppose that it is homogeneous; that is, it has constant mass density δ. For a homogeneous rectangular sheet, the center of mass (sometimes called the center of gravity) is at the geometric center, as diagrams (a) and (b) in Figure 9 suggest.

Consider the homogeneous lamina bounded by $x = a$, $x = b$, $y = f(x)$, and $y = g(x)$, with $g(x) \le f(x)$. *Slice* this lamina into narrow strips parallel to the y-axis, which are therefore nearly rectangular in shape, and imagine the mass of each strip to be concentrated at its geometric center. Then *approximate* and *integrate* (Figure 10). From this we can calculate the coordinates $(\overline{x}, \overline{y})$ of the center of mass using the formulas

$$\overline{x} = \frac{M_y}{m} \qquad \overline{y} = \frac{M_x}{m}$$

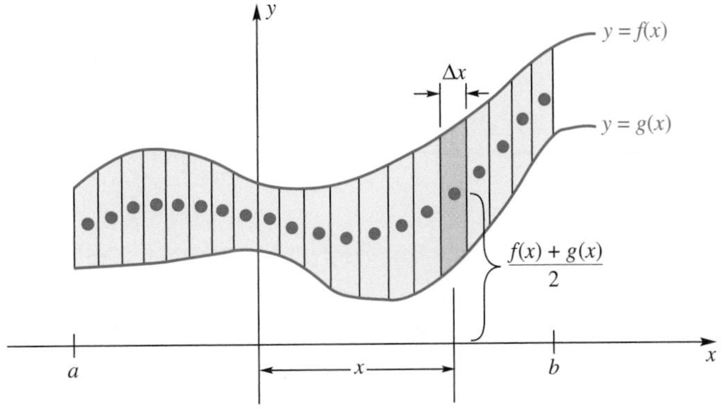

$\Delta m \approx \delta\,[f(x) - g(x)]\,\Delta x$	$\Delta M_y \approx x\,\delta\,[f(x) - g(x)]\,\Delta x$	$\Delta M_x \approx \frac{\delta}{2}\,[(f(x))^2 - (g(x))^2]\,\Delta x$
$m = \delta \int_a^b [f(x) - g(x)]\,dx$	$M_y = \delta \int_a^b x\,[f(x) - g(x)]\,dx$	$M_x = \frac{\delta}{2} \int_a^b [f^2(x) - g^2(x)]\,dx$

Figure 10

When we do, the factor δ cancels between numerator and denominator, and we obtain

$$\overline{x} = \frac{\displaystyle\int_a^b x[f(x) - g(x)]\,dx}{\displaystyle\int_a^b [f(x) - g(x)]\,dx}$$

$$\overline{y} = \frac{\displaystyle\int_a^b \frac{f(x) + g(x)}{2}[f(x) - g(x)]\,dx}{\displaystyle\int_a^b [f(x) - g(x)]\,dx} = \frac{\displaystyle\frac{1}{2}\int_a^b [(f(x))^2 - (g(x))^2]\,dx}{\displaystyle\int_a^b [f(x) - g(x)]\,dx}$$

Sometimes, slicing parallel to the x-axis works better than slicing parallel to the y-axis. This leads to formulas for $\overline{x}$ and $\overline{y}$ in which y is the variable of integration. Do not try to memorize all these formulas. It is much better to remember how they were derived.

The center of mass of a homogeneous lamina does not depend on its density or its mass, but only on the shape of the corresponding region in the plane. Thus, our problem becomes a geometric problem rather than a physical one. Accordingly, we often speak of the **centroid** of a planar region, rather than the center of mass of a homogeneous lamina.

EXAMPLE 4 Find the centroid of the region bounded by the curves $y = x^3$ and $y = \sqrt{x}$.

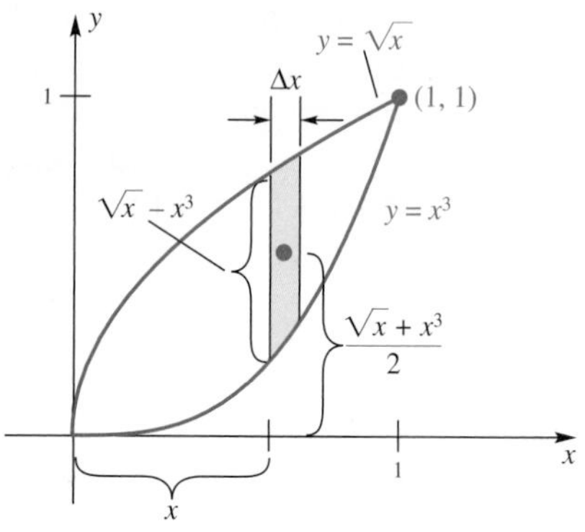

Figure 11

SOLUTION Note the diagram in Figure 11.

$$\bar{x} = \frac{\int_0^1 x(\sqrt{x} - x^3)\,dx}{\int_0^1 (\sqrt{x} - x^3)\,dx} = \frac{\left[\frac{2}{5}x^{5/2} - \frac{x^5}{5}\right]_0^1}{\left[\frac{2}{3}x^{3/2} - \frac{x^4}{4}\right]_0^1} = \frac{\frac{1}{5}}{\frac{5}{12}} = \frac{12}{25}$$

$$\bar{y} = \frac{\int_0^1 \frac{1}{2}\left(\sqrt{x} + x^3\right)\left(\sqrt{x} - x^3\right)dx}{\int_0^1 \left(\sqrt{x} - x^3\right)dx} = \frac{\frac{1}{2}\int_0^1 \left[\left(\sqrt{x}\right)^2 - (x^3)^2\right]dx}{\int_0^1 \left(\sqrt{x} - x^3\right)dx}$$

$$= \frac{\frac{1}{2}\left[\frac{x^2}{2} - \frac{x^7}{7}\right]_0^1}{\frac{5}{12}} = \frac{\frac{5}{28}}{\frac{5}{12}} = \frac{3}{7}$$

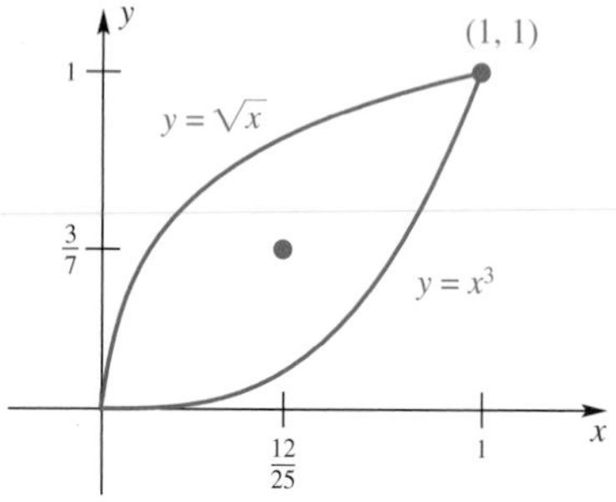

Figure 12

The centroid is shown in Figure 12. ■

EXAMPLE 5 Find the centroid of the region under the curve $y = \sin x$, $0 \le x \le \pi$ (Figure 13).

SOLUTION This region is symmetric about the line $x = \pi/2$, from which we conclude (without an integration) that $\bar{x} = \pi/2$. In fact, it is both intuitively obvious and true that if a region has a vertical or horizontal line of symmetry then the centroid will lie on that line.

Your intuition should also tell you that $\bar{y}$ will be less than $\frac{1}{2}$, since more of the region is below $\frac{1}{2}$ than above it. But to find this number exactly, we must calculate

$$\bar{y} = \frac{\int_0^\pi \frac{1}{2}\sin x \cdot \sin x\,dx}{\int_0^\pi \sin x\,dx} = \frac{\frac{1}{2}\int_0^\pi \sin^2 x\,dx}{\int_0^\pi \sin x\,dx}$$

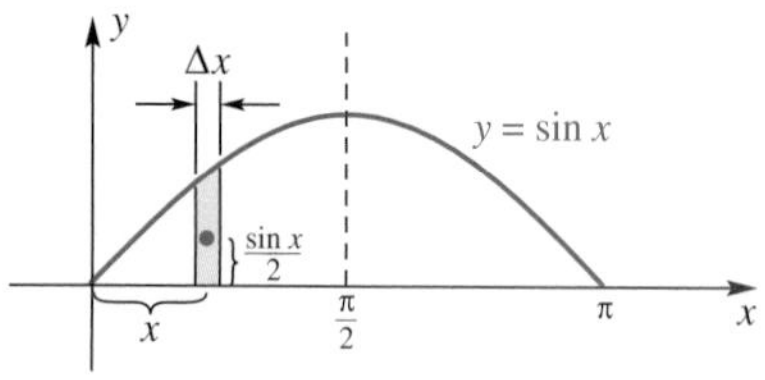

Figure 13

The denominator is easy to calculate; it has value 2. To calculate the numerator, we use the half-angle formula $\sin^2 x = (1 - \cos 2x)/2$.

$$\int_0^\pi \sin^2 x\,dx = \frac{1}{2}\left(\int_0^\pi 1\,dx - \int_0^\pi \cos 2x\,dx\right)$$

$$= \frac{1}{2}\left[x - \frac{1}{2}\sin 2x\right]_0^\pi = \frac{\pi}{2}$$

Thus,

$$\bar{y} = \frac{\frac{1}{2}\cdot\frac{\pi}{2}}{2} = \frac{\pi}{8} \approx 0.39$$ ■

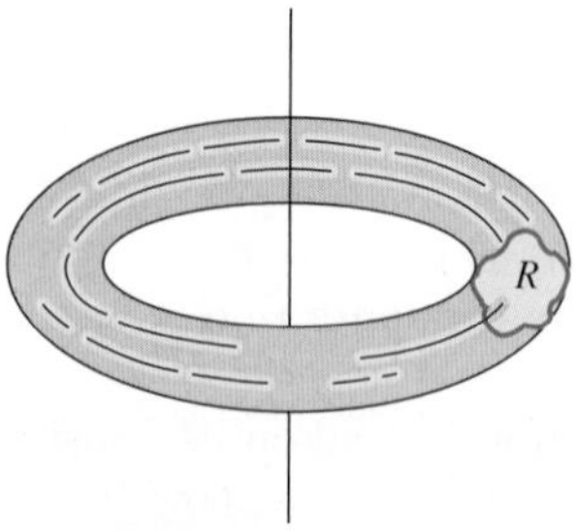

Figure 14

The Theorem of Pappus About A.D. 300, the Greek geometer Pappus stated a novel result, which connects centroids with volumes of solids of revolution (Figure 14).

Theorem A Pappus's Theorem

If a region R, lying on one side of a line in its plane, is revolved about that line, then the volume of the resulting solid is equal to the area of R multiplied by the distance traveled by its centroid.

All in all, we would expect $0 + 3750 + 7500 + 3750 = 15{,}000$ heads. Thus, we expect $15{,}000/10{,}000 = 1.5$ heads per trial (tossing three coins). A little reflection on the calculations suggests that the 10,000 is arbitrary and that it washes out anyway. We multiplied each probability by 10,000 to get the expected frequency, but then we divided by 10,000. That is,

$$\begin{aligned} 1.5 &= \frac{15{,}000}{10{,}000} \\ &= \frac{1}{10{,}000}[0P(X = 0)\ 10{,}000 + 1P(X = 1)\ 10{,}000 \\ &\qquad + 2P(X = 2)\ 10{,}000 + 3P(X = 3)\ 10{,}000] \\ &= 0P(X = 0) + 1P(X = 1) + 2P(X = 2) + 3P(X = 3) \end{aligned}$$

This last line is what we mean by the expectation.

Definition Expectation of a Random Variable

If X is a random variable with probability distribution

x	x_1	x_2	$\cdots$	x_n
$P(X = x)$	p_1	p_2	$\cdots$	p_n

then the **expectation** of X, denoted $E(X)$, also called the **mean** of X and denoted μ, is

$$\mu = E(X) = x_1p_1 + x_2p_2 + \cdots + x_np_n = \sum_{i=1}^{n} x_ip_i$$

Since $\sum_{i=1}^{n} p_i$ (all probabilities must sum to one), the formula for $E(X)$ is the same as the formula for the center of mass of a finite set of particles having masses $p_1, p_2, \ldots, p_n$ located at positions $x_1, x_2, \ldots, x_n$:

$$\text{Center of Mass} = \frac{M}{m} = \frac{\sum_{i=1}^{n} x_ip_i}{\sum_{i=1}^{n} p_i} = \frac{\sum_{i=1}^{n} x_ip_i}{1} = \sum_{i=1}^{n} x_ip_i = E(X)$$

EXAMPLE 1 Plastic parts are made 20 at a time by injecting plastic into a mold. The twenty parts are inspected for defects such as voids (bubbles inside the part) and cracks. Suppose that the probability distribution for the number of defective parts out of the 20 is given in the table below.

x_i	0	1	2	3
p_i	0.90	0.06	0.03	0.01

Find (a) the probability that a batch of 20 parts contains at least one defective part, and (b) the expected number of defective parts per batch of 20.

SOLUTION

(a) $P(X \geq 1) = P(X = 1) + P(X = 2) + P(X = 3)$
$= 0.06 + 0.03 + 0.01 = 0.10$

(b) The expected value for the number of defective parts is

$$E(X) = 0 \cdot 0.90 + 1 \cdot 0.06 + 2 \cdot 0.03 + 3 \cdot 0.01 = 0.15$$

Thus, on average, we would expect 0.15 defective part per batch. ■

> **Models**
>
> A famous statistician once said, "No models are correct, but many are useful." Probability models like those in this section should be regarded as approximations to the real world, not as perfectly accurate representations of the real world.

So far in this section we have dealt with random variables where the number of possible values is finite; this situation is analogous to having point masses in the previous section. There are other situations where there are infinitely many possible outcomes. If the set of possible values of a random variable X is finite, such as $\{x_1, x_2, \ldots, x_n\}$, or is infinite, but can be put in a list such as $\{x_1, x_2, \ldots\}$, then the random variable X is said to be **discrete.** If a random variable X can take on any value in some interval of real numbers then we say that X is a **continuous** random variable. There are many situations where, theoretically at least, the outcome can be any real number in an interval: for example, waiting time for a stop light, mass of a plastic molded part, or lifetime of a battery. Of course, in practice, every measurement is rounded, for example to the nearest second, milligram, day, etc. In situations like this the random variable is actually discrete (with very many possible outcomes), but a continuous random variable is often a good approximation.

Continuous random variables are studied in a manner analogous to the continuous distribution of mass in the previous section. For a continuous random variable X, we must specify the probability density function (PDF). A PDF for a random variable X that takes on values in the interval $[A, B]$ is a function satisfying

1. $f(x) \geq 0$
2. $\int_A^B f(x)\,dx = 1$
3. $P(a \leq X \leq b) = \int_a^b f(x)\,dx$ for all a, b $(a \leq b)$ in the interval $[A, B]$

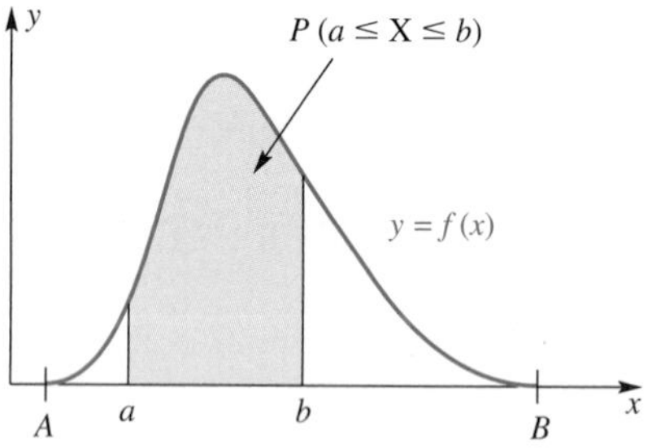

Figure 1

The third property says that we can find probabilities for a continuous random variable by finding areas under the PDF (see Figure 1). It is customary to define the PDF to be zero outside of the interval $[A, B]$.

The **expected value,** or **mean,** of a continuous random variable X is

$$\mu = E(X) = \int_A^B x\,f(x)\,dx$$

Just as for the case of discrete random variables, this is analogous to the center of mass of an object with variable density:

$$\text{Center of Mass} = \frac{M}{m} = \frac{\int_A^B x\,f(x)\,dx}{\int_A^B f(x)\,dx} = \frac{\int_A^B x\,f(x)\,dx}{1} = \int_A^B xf(x)\,dx = E(X)$$

EXAMPLE 2 A continuous random variable X has PDF

$$f(x) = \begin{cases} \frac{1}{10}, & \text{if } 0 \leq x \leq 10 \\ 0, & \text{otherwise} \end{cases}$$

Find (a) $P(1 \leq X \leq 9)$ (b) $P(X \geq 4)$ (c) $E(X)$.

SOLUTION The random variable X takes on values in $[0, 10]$.

(a) $P(1 \leq X \leq 9) = \int_1^9 \frac{1}{10}\,dx = \frac{1}{10} \cdot 8 = \frac{4}{5}$

(b) $P(X \geq 4) = \int_4^{10} \frac{1}{10}\,dx = \frac{1}{10} \cdot 6 = \frac{3}{5}$

(c) $E(X) = \int_0^{10} x\frac{1}{10}\,dx = \left[\frac{x^2}{20}\right]_0^{10} = 5$

≈ Are these answers reasonable? The random variable X is uniformly distributed on the interval $[0, 10]$, so 80% of the probability should be between 1 and 9, just as 80% of the mass of a uniform rod would be between 1 and 9. By symmetry, we would expect the mean, or expectation, of X to be 5, just as we would expect the center of mass of a uniform bar of length 10 to be 5 units from either side. ■

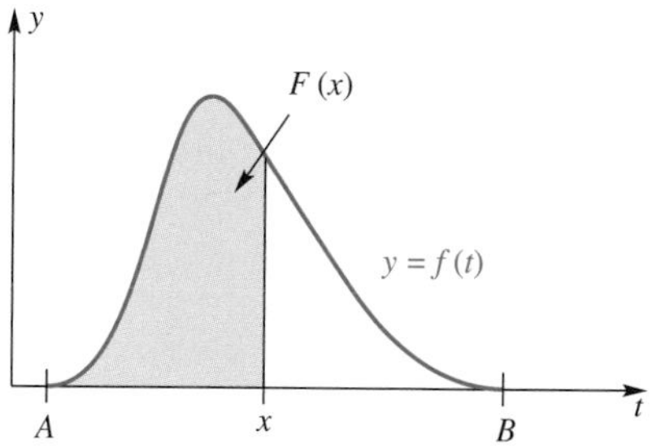

Figure 2

A function closely related to the PDF is the **cumulative distribution function** (CDF), which, for a random variable X, is the function F defined by

$$F(x) = P(X \leq x)$$

This function is defined for both discrete and continuous random variables. For a discrete random variable like the one given in Example 1, the CDF is a step function that takes a jump of $p_i = P(X = x_i)$ at the value x_i (see Problem 33). For a continuous random variable X that takes on values on the interval $[A, B]$ and having PDF $f(x)$, the CDF is equal to the definite integral (see Figure 2).

$$F(x) = \int_A^x f(t)\,dt, \quad A \leq x \leq B$$

For $x < A$, the CDF $F(x)$ is zero, since the probability of being less than or equal to a value less than A is zero. Similarly, for $x > B$, the CDF is one, since the probability of being less than or equal to a value that is greater than B is one.

In Chapter 4 we used the term *accumulation function* to refer to a function defined this way. The CDF is defined as the accumulated area under the PDF, so it is an accumulation function. The next theorem gives several properties of the CDF. The proofs are easy and are left as exercises. (See Problem 19.)

Theorem A

Let X be a continuous random variable taking on values in the interval $[A, B]$ and having PDF $f(x)$ and CDF $F(x)$. Then

1. $F'(x) = f(x)$
2. $F(A) = 0$ and $F(B) = 1$
3. $P(a \leq X \leq b) = F(b) - F(a)$

EXAMPLE 3 In reliability theory, the random variable is often the lifetime of some item, such as a laptop computer battery. The PDF can be used to find probabilities and expectations about the lifetime. Suppose then that the lifetime in hours of a battery is a continuous random variable X having PDF

$$f(x) = \begin{cases} \frac{12}{625}x^2(5 - x), & \text{if } 0 \leq x \leq 5 \\ 0, & \text{otherwise} \end{cases}$$

(a) Verify that this is a valid PDF and sketch its graph.
(b) Find the probability that the battery lasts at least three hours.
(c) Find the expected value of the lifetime.
(d) Find and sketch a graph of the CDF.

SOLUTION A graphing calculator or a CAS may be helpful in evaluating the integrals for this problem.

(a) For all x, $f(x)$ is nonnegative and

$$\int_0^5 \frac{12}{625}x^2(5 - x)\,dx = \frac{12}{625}\int_0^5 (5x^2 - x^3)\,dx$$
$$= \frac{12}{625}\left[\frac{5}{3}x^3 - \frac{1}{4}x^4\right]_0^5 = 1$$

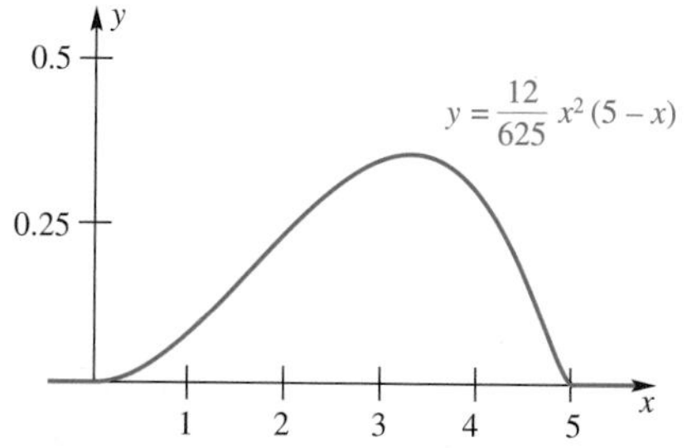

Figure 3

A graph of the PDF is given in Figure 3.

(b) The probability is found by integrating:

$$P(X \geq 3) = \int_3^5 \frac{12}{625}x^2(5 - x)\,dx$$
$$= \frac{12}{625}\left[\frac{5}{3}x^3 - \frac{1}{4}x^4\right]_3^5$$
$$= \frac{328}{625} = 0.5248$$

(c) The expected lifetime is

$$E(X) = \int_0^5 x\left[\frac{12}{625}x^2(5 - x)\right] dx$$

$$= \frac{12}{625}\int_0^5 (5x^3 - x^4)\, dx$$

$$= \frac{12}{625}\left[\frac{5}{4}x^4 - \frac{1}{5}x^5\right]_0^5 = 3 \text{ hours}$$

(d) For x between 0 and 5,

$$F(x) = \int_0^x \frac{12}{625}t^2(5 - t)\, dt$$

$$= \frac{4}{125}x^3 - \frac{3}{625}x^4$$

For $x < 0$, $F(x) = 0$, and for $x > 5$, $F(x) = 1$. A graph is given in Figure 4. ■

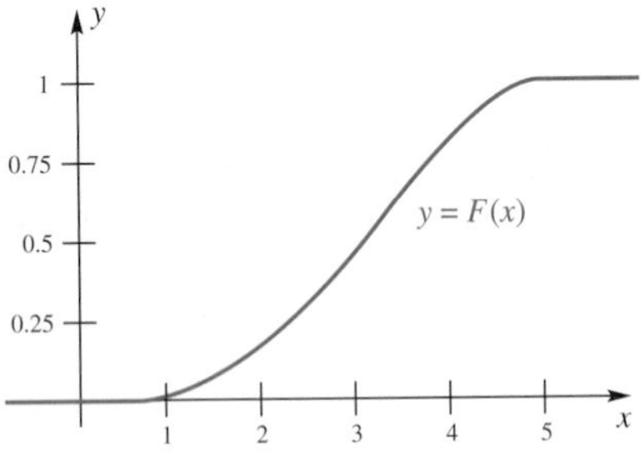

Figure 4

Concepts Review

1. A random variable whose set of possible outcomes is finite or can be put into an infinite list is called a _____ random variable. A random variable whose set of possible outcomes makes up an interval of real numbers is called a _____ random variable.

2. For discrete random variables, probabilities and expectations are found by evaluating a (an) _____, whereas for continuous random variables, probabilities and expectations are found by evaluating a (an) _____.

3. If a continuous random variable X takes on values in [0, 20] and has PDF $f(x)$, then $P(X \leq 5)$ is found by evaluating _____.

4. The accumulation function $\int_A^x f(t)\, dt$, which accumulates the probability (area under the PDF), is called the _____.

Problem Set 5.7

In Problems 1–8, a discrete probability distribution for a random variable X is given. Use the given distribution to find (a) $P(X \geq 2)$ and (b) $E(X)$.

1.

x_i	0	1	2	3
p_i	0.80	0.10	0.05	0.05

2.

x_i	0	1	2	3	4
p_i	0.70	0.15	0.05	0.05	0.05

3.

x_i	−2	−1	0	1	2
p_i	0.2	0.2	0.2	0.2	0.2

4.

x_i	−2	−1	0	1	2
p_i	0.1	0.2	0.4	0.2	0.1

5.

x_i	1	2	3	4
p_i	0.4	0.2	0.2	0.2

6.

x_i	−0.1	100	1000
p_i	0.980	0.018	0.002

7. $p_i = (5 - i)/10$, $x_i = i$, $i = 1, 2, 3, 4$

8. $p_i = (2 - i)^2/10$, $x_i = i$, $i = 0, 1, 2, 3, 4$

In Problems 9–18, a PDF for a continuous random variable X is given. Use the PDF to find (a) $P(X \geq 2)$, (b) $E(X)$, and (c) the CDF.

9. $f(x) = \begin{cases} \frac{1}{20}, & \text{if } 0 \leq x \leq 20 \\ 0, & \text{otherwise} \end{cases}$

10. $f(x) = \begin{cases} \frac{1}{40}, & \text{if } -20 \leq x \leq 20 \\ 0, & \text{otherwise} \end{cases}$

11. $f(x) = \begin{cases} \frac{3}{256}x(8 - x), & \text{if } 0 \leq x \leq 8 \\ 0, & \text{otherwise} \end{cases}$

12. $f(x) = \begin{cases} \frac{3}{4000}x(20 - x), & \text{if } 0 \leq x \leq 20 \\ 0, & \text{otherwise} \end{cases}$

13. $f(x) = \begin{cases} \frac{3}{64}x^2(4 - x), & \text{if } 0 \leq x \leq 4 \\ 0, & \text{otherwise} \end{cases}$

14. $f(x) = \begin{cases} (8 - x)/32, & \text{if } 0 \leq x \leq 8 \\ 0, & \text{otherwise} \end{cases}$

15. $f(x) = \begin{cases} \frac{\pi}{8}\sin(\pi x/4), & \text{if } 0 \leq x \leq 4 \\ 0, & \text{otherwise} \end{cases}$

16. $f(x) = \begin{cases} \frac{\pi}{8}\cos(\pi x/8), & \text{if } 0 \leq x \leq 4 \\ 0, & \text{otherwise} \end{cases}$

17. $f(x) = \begin{cases} \frac{4}{3}x^{-2}, & \text{if } 1 \le x \le 4 \\ 0, & \text{otherwise} \end{cases}$

18. $f(x) = \begin{cases} \frac{81}{40}x^{-3}, & \text{if } 1 \le x \le 9 \\ 0, & \text{otherwise} \end{cases}$

19. Prove the three properties of the CDF in Theorem A.

20. A continuous random variable X is said to have a **uniform distribution** on the interval $[a, b]$ if the PDF has the form

$$f(x) = \begin{cases} \dfrac{1}{b-a}, & \text{if } a \le x \le b \\ 0, & \text{otherwise} \end{cases}$$

(a) Find the probability that the value of X is closer to a than it is to b.

(b) Find the expected value of X.

(c) Find the CDF of X.

21. The **median** of a continuous random variable X is a value x_0 such that $P(X \le x_0) = 0.5$. Find the median of a uniform random variable on the interval $[a, b]$.

22. Without doing any integration, find the median of the random variable that has PDF $f(x) = \frac{15}{512}x^2(4-x)^2$, $0 \le x \le 4$. *Hint:* Use symmetry.

23. Find the value of k that makes $f(x) = kx(5-x)$, $0 \le x \le 5$, a valid PDF. *Hint:* The PDF must integrate to 1.

24. Find the value of k that makes $f(x) = kx^2(5-x)^2$, $0 \le x \le 5$, a valid PDF.

25. The time in minutes that it takes a worker to complete a task is a random variable with PDF $f(x) = k(2 - |x-2|)$, $0 \le x \le 4$.

(a) Find the value of k that makes this a valid PDF.

(b) What is the probability that it takes more than 3 minutes to complete the task?

(c) Find the expected value of the time to complete the task.

(d) Find the CDF $F(x)$.

(e) Let Y denote the time in seconds required to complete the task. What is the CDF of Y? *Hint:* $P(Y \le y) = P(60X \le y)$.

26. The daily summer air quality index (AQI) in St. Louis is a random variable whose PDF is $f(x) = kx^2(180-x)$, $0 \le x \le 180$.

(a) Find the value of k that makes this a valid PDF.

(b) A day is an "orange alert" day if the AQI is between 100 and 150. What is the probability that a summer day is an orange alert day?

(c) Find the expected value of the summer AQI.

CAS **27.** Holes drilled by a machine have a diameter, measured in millimeters, that is a random variable with PDF $f(x) = kx^6(0.6-x)^8$, $0 \le x \le 0.6$.

(a) Find the value of k that makes this a valid PDF.

(b) Specifications require that the hole's diameter be between 0.35 and 0.45 mm. Those units not meeting this requirement are scrapped. What is the probability that a unit is scrapped?

(c) Find the expected value of the hole's diameter.

(d) Find the CDF $F(x)$.

(e) Let Y denote the hole's diameter in inches. (1 inch = 25.4 mm.) What is the CDF of Y?

CAS **28.** A company monitors the total impurities in incoming batches of chemicals. The PDF for total impurity X in a batch, measured in parts per million (PPM), has PDF $f(x) = kx^2(200-x)^8$, $0 \le x \le 200$.

(a) Find the value of k that makes this a valid PDF.

(b) The company does not accept batches whose total impurity is 100 or above. What is the probability that a batch is not accepted?

(c) Find the expected value of the total impurity in PPM.

(d) Find the CDF $F(x)$.

(e) Let Y denote the total impurity in percent, rather than in PPM. What is the CDF of Y?

29. Suppose that X is a random variable that has a uniform distribution on the interval $[0, 1]$. (See Problem 20.) The point $(1, X)$ is plotted in the plane. Let Y be the distance from $(1, X)$ to the origin. Find the CDF and the PDF of the random variable Y. *Hint:* Find the CDF first.

30. Suppose that X is a continuous random variable. Explain why $P(X = x) = 0$. Which of the following probabilities are the same? Explain.

$$P(a < X < b), \quad P(a \le X \le b),$$
$$P(a < X \le b), \quad P(a \le X < b)$$

31. Show that if A^c is the complement of A, that is, the set of all outcomes in the sample space S that are not in A, then $P(A^c) = 1 - P(A)$.

32. Use the result in Problem 31 to find $P(X \ge 1)$ in Problems 1, 2, and 5.

33. If X is a discrete random variable then the CDF is a step function. By considering values of x less than zero, between 0 and 1, etc., find and graph the CDF for the random variable X in Problem 1.

34. Find and graph the CDF of the random variable X in Problem 2.

35. Suppose a random variable Y has CDF

$$F(y) = \begin{cases} 0, & \text{if } y < 0 \\ 2y/(y+1), & \text{if } 0 \le y \le 1 \\ 1, & \text{if } y > 1 \end{cases}$$

Find each of the following:

(a) $P(Y < 2)$

(b) $P(0.5 < Y < 0.6)$

(c) the PDF of Y

(d) Use the Parabolic Rule with $n = 8$ to approximate $E(Y)$.

36. Suppose a random variable Z has CDF

$$F(z) = \begin{cases} 0, & \text{if } z < 0 \\ z^2/9, & \text{if } 0 \le z \le 3 \\ 1, & \text{if } z > 3 \end{cases}$$

Find each of the following:

(a) $P(Z > 1)$

(b) $P(1 < Z < 2)$

(c) the PDF of Z

(d) $E(Z)$

CAS **37.** The expected value of a function $g(X)$ of a continuous random variable X having PDF $f(x)$ is defined to be $E[g(X)] = \int_A^B g(x)f(x)\,dx$. If the PDF of X is $f(x) = \frac{15}{512}x^2(4-x)^2$, $0 \le x \le 4$, find $E(X)$ and $E(X^2)$.

CAS **38.** A continuous random variable X has PDF $f(x) = \frac{3}{256}x(8-x), 0 \le x \le 8$. Find $E(X^2)$ and $E(X^3)$.

CAS **39.** The **variance** of a continuous random variable, denoted $V(X)$ or σ^2, is defined to be $V(X) = E[(X-\mu)^2]$, where μ is the expected value, or mean, of the random variable X. Find the variance of the random variable in Problem 37.

CAS **40.** Find the variance of the random variable in Problem 38.

41. Show that the variance of a random variable is equal to $E(X^2) - \mu^2$ and use this result to find the variance of the random variable in Problem 37.

Answers to Concepts Review: **1.** discrete; continuous **2.** sum; integral **3.** $\int_0^5 f(x)\,dx$ **4.** cumulative distribution function

5.8 Chapter Review

Concepts Test

Respond with true or false to each of the following assertions. Be prepared to justify your answer.

1. The area of the region bounded by $y = \cos x$, $y = 0$, $x = 0$, and $x = \pi$ is $\int_0^{\pi} \cos x\,dx$.

2. The area of a circle of radius a is $4\int_0^a \sqrt{a^2 - x^2}\,dx$.

3. The area of the region bounded by $y = f(x)$, $y = g(x)$, $x = a$, and $x = b$ is either $\int_a^b [f(x) - g(x)]\,dx$ or its negative.

4. All right cylinders whose bases have the same area and whose heights are the same have identical volumes.

5. If two solids with bases in the same plane have cross sections of the same area in all planes parallel to their bases, then they have the same volume.

6. If the radius of the base of a cone is doubled while the height is halved, the volume will remain the same.

7. To calculate the volume of the solid obtained by revolving the region bounded by $y = -x^2 + x$ and $y = 0$ about the y-axis, one should use the method of washers in preference to the method of shells.

8. The solids obtained by revolving the region of Problem 7 about $x = 0$ and $x = 1$ have the same volume.

9. Any smooth curve in the plane that lies entirely within the unit circle will have finite length.

10. The work required to stretch a spring 2 inches beyond its natural length is twice that required to stretch it 1 inch (assume Hooke's Law).

11. It will require the same amount of work to empty a cone-shaped tank and a cylindrical tank of water by pumping it over the rim if both tanks have the same height and volume.

12. A boat contains circular windows of radius 6 inches that are below the surface of the water. The force exerted by the water on a window is the same regardless of the depth.

13. If $\bar{x}$ is the center of mass of a system of masses $m_1, m_2, \ldots, m_n$ distributed along a line at points with coordinates $x_1, x_2, \ldots, x_n$, respectively, then $\sum_{i=1}^{n}(x_i - \bar{x})m_i = 0$.

14. The centroid of the region bounded by $y = \cos x$, $y = 0$, $x = 0$, and $x = 2\pi$ is at $(\pi, 0)$.

15. According to Pappus's Theorem, the volume of the solid obtained by revolving the region (of area 2) bounded by $y = \sin x$, $y = 0$, $x = 0$, and $x = \pi$ about the y-axis is $2(2\pi)\left(\frac{\pi}{2}\right) = 2\pi^2$.

16. The area of the region bounded by $y = \sqrt{x}$, $y = 0$, and $x = 9$ is $\int_0^3 (9 - y^2)\,dy$.

17. If the density of a wire is proportional to the square of the distance from its midpoint, then its center of mass is at the midpoint.

18. The centroid of a triangle with base on the x-axis has y-coordinate equal to one-third the altitude of the triangle.

19. A random variable that can take on only a finite number of values is a discrete random variable.

20. Consider a wire with density $\delta(x), 0 \le x \le a$ and a random variable with PDF $f(x), 0 \le x \le a$. If $\delta(x) = f(x)$ for all x in $[0, a]$, then the center of mass of the wire will equal the expectation of the random variable.

21. A random variable that takes on the value 5 with probability one will have expectation equal to 5.

22. If $F(x)$ is the CDF of a continuous random variable X, then $F'(x)$ is equal to the PDF $f(x)$.

23. If X is a continuous random variable, then $P(X = 1) = 0$.

Sample Test Problems

Problems 1–7 refer to the plane region R bounded by the curve $y = x - x^2$ and the x-axis (Figure 1).

1. Find the area of R.

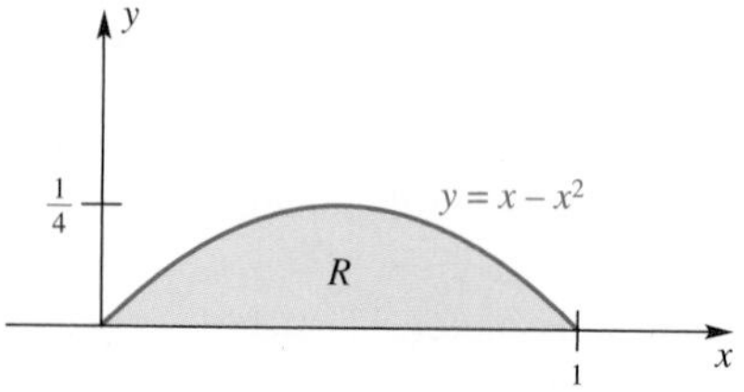

Figure 1

2. Find the volume of the solid S_1 generated by revolving the region R about the x-axis.

3. Use the shell method to find the volume of the solid S_2 generated by revolving R about the y-axis.

4. Find the volume of the solid S_3 generated by revolving R about the line $y = -2$.

5. Find the volume of the solid S_4 generated by revolving R about the line $x = 3$.

6. Find the coordinates of the centroid of R.

7. Use Pappus's Theorem and Problems 1 and 6 to find the volumes of the solids S_1, S_2, S_3, and S_4, above.

8. The natural length of a certain spring is 16 inches, and a force of 8 pounds is required to keep it stretched 8 inches. Find the work done in each case.

(a) Stretching it from a length of 18 inches to a length of 24 inches.

(b) Compressing it from its natural length to a length of 12 inches.

9. An upright cylindrical tank is 10 feet in diameter and 10 feet high. If water in the tank is 6 feet deep, how much work is done in pumping all the water over the top edge of the tank?

10. An object weighing 200 pounds is suspended from the top of a building by a uniform cable. If the cable is 100 feet long and weighs 120 pounds, how much work is done in pulling the object and the cable to the top?

11. A region R is bounded by the line $y = 4x$ and the parabola $y = x^2$. Find the area of R by

(a) taking x as the integration variable, and

(b) taking y as the integration variable.

12. Find the centroid of R in Problem 11.

13. Find the volume of the solid of revolution generated by revolving the region R of Problem 11 about the x-axis. Check by using Pappus's Theorem.

14. Find the total force exerted by the water in a right circular cylinder with height 3 feet and radius 8 feet

(a) on the lateral surface, and

(b) on the bottom surface.

15. Find the length of the arc of the curve $y = x^3/3 + 1/(4x)$ from $x = 1$ to $x = 3$.

16. Sketch the graph of the parametric equations

$$x = t^2, \qquad y = \tfrac{1}{3}(t^3 - 3t)$$

Then find the length of the loop of the resulting curve.

17. A solid with the semicircular base bounded by $y = \sqrt{9 - x^2}$ and $y = 0$ has cross sections perpendicular to the x-axis that are squares. Find the volume of this solid.

In Problems 18–23, write an expression involving integrals that represents the required concept. Refer to Figure 2.

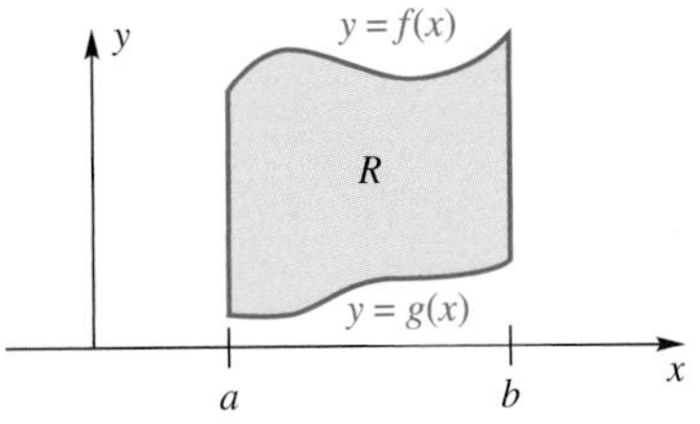

Figure 2

18. The area of R.

19. The volume of the solid obtained when R is revolved about the x-axis.

20. The volume of the solid obtained when R is revolved about $x = a$.

21. The moments M_x and M_y of a homogeneous lamina with shape R, assuming that its density is δ.

22. The *total* length of the boundary of R.

23. The *total* surface area of the solid of Problem 19.

24. Let X be a continuous random variable with PDF

$$f(x) = \begin{cases} \dfrac{8 - x^3}{12}, & \text{if } 0 \le x \le 2 \\ 0, & \text{otherwise} \end{cases}$$

(a) Find $P(X \ge 1)$.

(b) Find the probability that X is closer to 0 than it is to 1.

(c) Find $E(X)$.

(d) Find the CDF of X.

25. A random variable X has CDF

$$F(x) = \begin{cases} 0, & \text{if } x < 0 \\ 1 - \dfrac{(6 - x)^2}{36}, & \text{if } 0 \le x \le 6 \\ 1, & \text{if } x > 6 \end{cases}$$

(a) Find $P(X \le 3)$.

(b) Find the PDF $f(x)$.

(c) Find $E(X)$.

REVIEW & PREVIEW PROBLEMS

Find the following antiderivatives.

1. $\int \frac{1}{x^2}\,dx$

2. $\int \frac{1}{x^{1.5}}\,dx$

3. $\int \frac{1}{x^{1.01}}\,dx$

4. $\int \frac{1}{x^{0.99}}\,dx$

For Problems 5–8, let $F(x) = \int_1^x \frac{1}{t}\,dt$ *and find the following.*

5. $F(1)$

6. $F'(x)$

7. $D_x F(x^2)$

8. $D_x F(x^3)$

In Problems 9–12, evaluate the expressions at the given values.

9. $(1 + h)^{1/h}; h = 1, \frac{1}{5}, \frac{1}{10}, \frac{1}{50}, \frac{1}{100}$

10. $\left(1 + \frac{1}{n}\right)^n; n = 1, 10, 100, 1000$

11. $\left(1 + \frac{h}{2}\right)^{2/h}; h = 1, \frac{1}{5}, \frac{1}{10}, \frac{1}{50}, \frac{1}{100}$

12. $\left(1 + \frac{2}{n}\right)^{n/2}; n = 1, 10, 100, 1000$

In Problems 13–16, find all values of x that satisfy the given relationship.

13. $\sin x = \frac{1}{2}$

14. $\cos x = -1$

15. $\tan x = 1$

16. $\sec x = 0$

For the triangles shown in Problems 17–20, find all of the following in terms of x: $\sin\theta, \cos\theta, \tan\theta, \cot\theta, \sec\theta,$ *and* $\csc\theta$.

17.

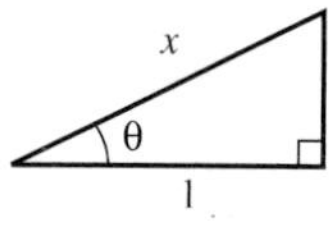

18.

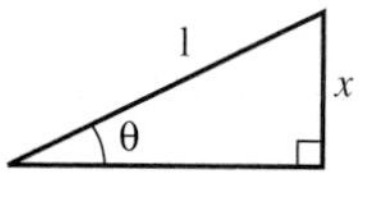

19.

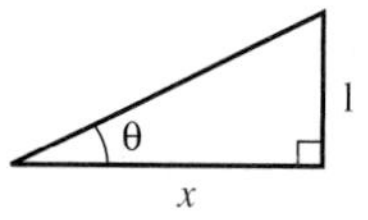

20.

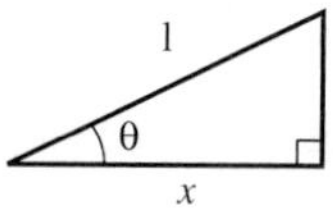

In Problems 21–22, solve the differential equation subject to the given condition

21. $y' = xy^2$, $y = 1$ when $x = 0$

22. $y' = \frac{\cos x}{y}$, $y = 4$ when $x = 0$

CHAPTER 6 Transcendental Functions

6.1 The Natural Logarithm Function

6.2 Inverse Functions and Their Derivatives

6.3 The Natural Exponential Function

6.4 General Exponential and Logarithmic Functions

6.5 Exponential Growth and Decay

6.6 First-Order Linear Differential Equations

6.7 Approximations for Differential Equations

6.8 The Inverse Trigonometric Functions and Their Derivatives

6.9 The Hyperbolic Functions and Their Inverses

6.1 The Natural Logarithm Function

The power of calculus, both that of derivatives and integrals, has already been amply demonstrated. Yet we have only scratched the surface of potential applications. To dig deeper, we need to expand the class of functions with which we can work. That is the object of this chapter.

We begin by asking you to notice a peculiar gap in our knowledge of derivatives.

$$D_x\left(\frac{x^2}{2}\right) = x^1,\ D_x(x) = x^0,\ D_x(??) = x^{-1},\ D_x\left(-\frac{1}{x}\right) = x^{-2},\ D_x\left(-\frac{x^{-2}}{2}\right) = x^{-3}$$

Is there a function whose derivative is $1/x$? In other words, is there an antiderivative $\int 1/x\,dx$? The First Fundamental Theorem of Calculus states that the accumulation function

$$F(x) = \int_a^x f(t)\,dt$$

is a function whose derivative is $f(x)$, provided that f is continuous on an interval I that contains a and x. In this sense, we can find an antiderivative of *any* continuous function. The existence of an antiderivative does not mean that the antiderivative can be expressed in terms of functions that we have studied so far. In this chapter we will introduce and study a number of new functions.

Our first new function is chosen to fill the gap noticed above. We call it the **natural logarithm function,** and it does have something to do with the logarithm studied in algebra, but this relationship will only appear later. For now, just accept the fact that we are going to define a *new* function and study its properties.

> **Definition** Natural Logarithm Function
>
> The **natural logarithm function,** denoted by ln, is defined by
>
> $$\ln x = \int_1^x \frac{1}{t}\,dt, \qquad x > 0$$
>
> The domain of the natural logarithm function is the set of positive real numbers.

The diagrams in Figure 1 indicate the geometric meaning of ln x. The natural logarithm (or natural log) function measures the area under the curve $y = 1/t$ between 1 and x if $x > 1$ and the negative of the area if $0 < x < 1$. The natural logarithm is an accumulation function because it accumulates area under the curve

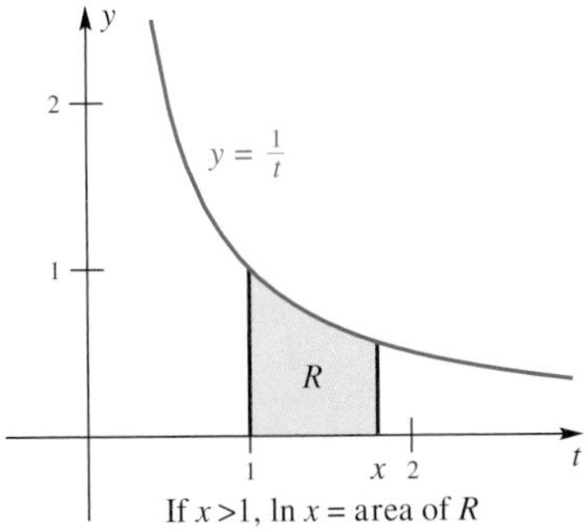

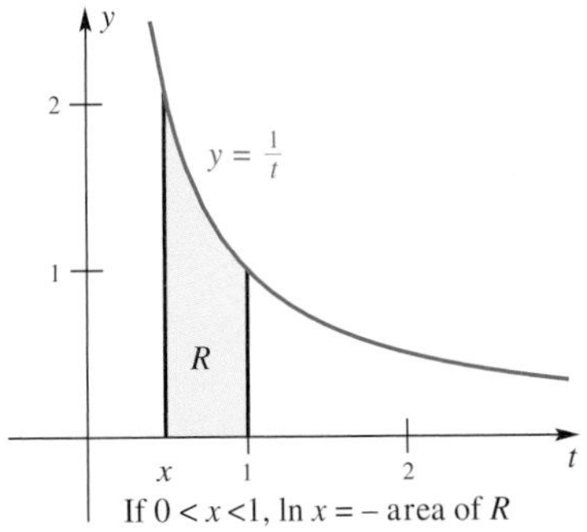

Figure 1

$y = 1/t$. Clearly, $\ln x$ is well defined for $x > 0$; $\ln x$ is not defined for $x \leq 0$ because this definite integral does not exist over an interval that includes 0.

And what is the derivative of this new function? Just exactly what we want.

Derivative of the Natural Logarithm Function From the First Fundamental Theorem of Calculus, we have

$$D_x \int_1^x \frac{1}{t}\, dt = D_x \ln x = \frac{1}{x}, \qquad x > 0$$

This can be combined with the Chain Rule. If $u = f(x) > 0$ and if f is differentiable, then

$$D_x \ln u = \frac{1}{u} D_x u$$

EXAMPLE 1 Find $D_x \ln \sqrt{x}$.

SOLUTION Let $u = \sqrt{x} = x^{1/2}$. Then

$$D_x \ln \sqrt{x} = \frac{1}{x^{1/2}} \cdot D_x(x^{1/2}) = \frac{1}{x^{1/2}} \cdot \frac{1}{2} x^{-1/2} = \frac{1}{2x}$$

■

EXAMPLE 2 Find $D_x \ln(x^2 - x - 2)$.

SOLUTION This problem makes sense, provided that $x^2 - x - 2 > 0$. Now $x^2 - x - 2 = (x - 2)(x + 1)$, which is positive provided that $x < -1$ or $x > 2$. Thus, the domain of $\ln(x^2 - x - 2)$ is $(-\infty, -1) \cup (2, \infty)$. On this domain,

$$D_x \ln(x^2 - x - 2) = \frac{1}{x^2 - x - 2} D_x(x^2 - x - 2) = \frac{2x - 1}{x^2 - x - 2}$$

■

EXAMPLE 3 Show that

$$D_x \ln|x| = \frac{1}{x}, \qquad x \neq 0$$

SOLUTION Two cases are to be considered. If $x > 0$, $|x| = x$, and

$$D_x \ln|x| = D_x \ln x = \frac{1}{x}$$

If $x < 0$, $|x| = -x$, and so

$$D_x \ln|x| = D_x \ln(-x) = \frac{1}{-x} D_x(-x) = \left(\frac{1}{-x}\right)(-1) = \frac{1}{x}$$

■

We know that for every differentiation formula there is a corresponding integration formula. Thus, Example 3 implies that

$$\int \frac{1}{x}\, dx = \ln|x| + C, \qquad x \neq 0$$

or, with u replacing x,

$$\int \frac{1}{u}\, du = \ln|u| + C, \qquad u \neq 0$$

This fills the long-standing gap in the Power Rule: $\int u^r\, du = u^{r+1}/(r + 1)$, from which we had to exclude the exponent $r = -1$.

EXAMPLE 4 Find $\int \frac{5}{2x+7}\,dx$.

SOLUTION Let $u = 2x + 7$, so $du = 2\,dx$. Then

$$\int \frac{5}{2x+7}\,dx = \frac{5}{2}\int \frac{1}{2x+7}\,2\,dx = \frac{5}{2}\int \frac{1}{u}\,du$$

$$= \frac{5}{2}\ln|u| + C = \frac{5}{2}\ln|2x+7| + C \quad \blacksquare$$

EXAMPLE 5 Evaluate $\int_{-1}^{3} \frac{x}{10-x^2}\,dx$.

SOLUTION Let $u = 10 - x^2$, so $du = -2x\,dx$. Then

$$\int \frac{x}{10-x^2}\,dx = -\frac{1}{2}\int \frac{-2x}{10-x^2}\,dx = -\frac{1}{2}\int \frac{1}{u}\,du$$

$$= -\frac{1}{2}\ln|u| + C = -\frac{1}{2}\ln|10-x^2| + C$$

Thus, by the Second Fundamental Theorem of Calculus,

$$\int_{-1}^{3} \frac{x}{10-x^2}\,dx = \left[-\frac{1}{2}\ln|10-x^2|\right]_{-1}^{3} = -\frac{1}{2}\ln 1 + \frac{1}{2}\ln 9 = \frac{1}{2}\ln 9$$

For the above calculation to be valid, $10 - x^2$ must never be 0 on the interval $[-1, 3]$. It is easy to see that this is true. ■

When the integrand is the quotient of two polynomials (that is, a rational function) and the numerator is of equal or greater degree than the denominator, always *divide the denominator into the numerator first.*

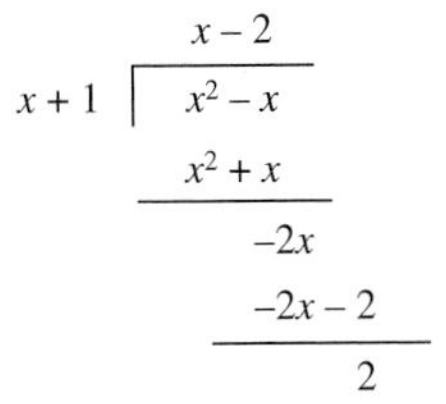

Figure 2

EXAMPLE 6 Find $\int \frac{x^2 - x}{x+1}\,dx$.

SOLUTION By long division (Figure 2),

$$\frac{x^2 - x}{x+1} = x - 2 + \frac{2}{x+1}$$

Hence,

$$\int \frac{x^2-x}{x+1}\,dx = \int (x-2)\,dx + 2\int \frac{1}{x+1}\,dx$$

$$= \frac{x^2}{2} - 2x + 2\int \frac{1}{x+1}\,dx$$

$$= \frac{x^2}{2} - 2x + 2\ln|x+1| + C \quad \blacksquare$$

Common Logarithms

Properties (ii) and (iii) for **common logarithms** (base 10 logarithms) were what motivated the invention of logarithms. John Napier (1550–1617) wanted to simplify the complicated calculations that arose in astronomy and navigation. To replace multiplication by addition and division by subtraction was his goal—exactly what (ii) and (iii) accomplish. For over 350 years, common logarithms were an essential aid in computation, but today we use calculators and computers for this purpose. However, natural logarithms retain their importance for other reasons, as you will see.

Properties of the Natural Logarithm The next theorem lists several important properties of the natural log function.

Theorem A

If a and b are positive numbers and r is any rational number, then

(i) $\ln 1 = 0$; (ii) $\ln ab = \ln a + \ln b$;

(iii) $\ln \frac{a}{b} = \ln a - \ln b$; (iv) $\ln a^r = r \ln a$.

Proof

(i) $\ln 1 = \int_1^1 \frac{1}{t}\,dt = 0.$

(ii) Since, for $x > 0$,

$$D_x \ln ax = \frac{1}{ax} \cdot a = \frac{1}{x}$$

and

$$D_x \ln x = \frac{1}{x}$$

it follows from the theorem about two functions with the same derivative (Theorem 3.6B) that

$$\ln ax = \ln x + C$$

To determine C, let $x = 1$, obtaining $\ln a = C$. Thus,

$$\ln ax = \ln x + \ln a$$

Finally, let $x = b$.

(iii) Replace a by $1/b$ in (ii) to obtain

$$\ln\frac{1}{b} + \ln b = \ln\left(\frac{1}{b} \cdot b\right) = \ln 1 = 0$$

Thus,

$$\ln\frac{1}{b} = -\ln b$$

Applying (ii) again, we get

$$\ln\frac{a}{b} = \ln\left(a \cdot \frac{1}{b}\right) = \ln a + \ln\frac{1}{b} = \ln a - \ln b$$

(iv) Since, for $x > 0$,

$$D_x(\ln x^r) = \frac{1}{x^r} \cdot rx^{r-1} = \frac{r}{x}$$

and

$$D_x(r \ln x) = r \cdot \frac{1}{x} = \frac{r}{x}$$

it follows by Theorem 3.6B, which we used in (ii), that

$$\ln x^r = r \ln x + C$$

Let $x = 1$, which gives $C = 0$. Thus,

$$\ln x^r = r \ln x$$

Finally, let $x = a$. ■

EXAMPLE 7 Find dy/dx if $y = \ln \sqrt[3]{(x-1)/x^2}$, $x > 1$.

SOLUTION Our task is easier if we first use the properties of natural logarithms to simplify y.

$$y = \ln\left(\frac{x-1}{x^2}\right)^{1/3} = \frac{1}{3}\ln\left(\frac{x-1}{x^2}\right)$$

$$= \frac{1}{3}\left[\ln(x-1) - \ln x^2\right] = \frac{1}{3}\left[\ln(x-1) - 2\ln x\right]$$

Thus,

$$\frac{dy}{dx} = \frac{1}{3}\left[\frac{1}{x-1} - \frac{2}{x}\right] = \frac{2-x}{3x(x-1)}$$ ■

Logarithmic Differentiation The labor of differentiating expressions involving quotients, products, or powers can often be substantially reduced by first applying the natural logarithm function and using its properties. This method, called **logarithmic differentiation,** is illustrated in Example 8.

EXAMPLE 8 Differentiate $y = \dfrac{\sqrt{1-x^2}}{(x+1)^{2/3}}$.

SOLUTION First we take natural logarithms; then we differentiate implicitly with respect to x (recall Section 2.7).

$$\ln y = \frac{1}{2}\ln(1-x^2) - \frac{2}{3}\ln(x+1)$$

$$\frac{1}{y}\frac{dy}{dx} = \frac{-2x}{2(1-x^2)} - \frac{2}{3(x+1)} = \frac{-(x+2)}{3(1-x^2)}$$

Thus,

$$\frac{dy}{dx} = \frac{-y(x+2)}{3(1-x^2)} = \frac{-\sqrt{1-x^2}\,(x+2)}{3(x+1)^{2/3}(1-x^2)}$$

$$= \frac{-(x+2)}{3(x+1)^{2/3}(1-x^2)^{1/2}}$$ ■

Example 8 could have been done directly, without first taking logarithms, and we suggest you try it. You should be able to make the two answers agree.

The Graph of the Natural Logarithm The domain of $\ln x$ consists of the set of all positive real numbers, so the graph of $y = \ln x$ is in the right half-plane. Also, for $x > 0$,

$$D_x \ln x = \frac{1}{x} > 0$$

and

$$D_x^2 \ln x = -\frac{1}{x^2} < 0$$

The first formula tells us that the natural log function is continuous (why?) and rises as x increases; the second tells us that the graph is everywhere concave down. In Problems 43 and 44, you are asked to show that

$$\lim_{x\to\infty} \ln x = \infty$$

and

$$\lim_{x\to 0^+} \ln x = -\infty$$

Finally, $\ln 1 = 0$. These facts imply that the graph of $y = \ln x$ is similar in shape to that shown in Figure 3.

If your calculator has an ln button, values for the natural logarithm are at your fingertips. For example,

$$\ln 2 \approx 0.6931$$

$$\ln 3 \approx 1.0986$$

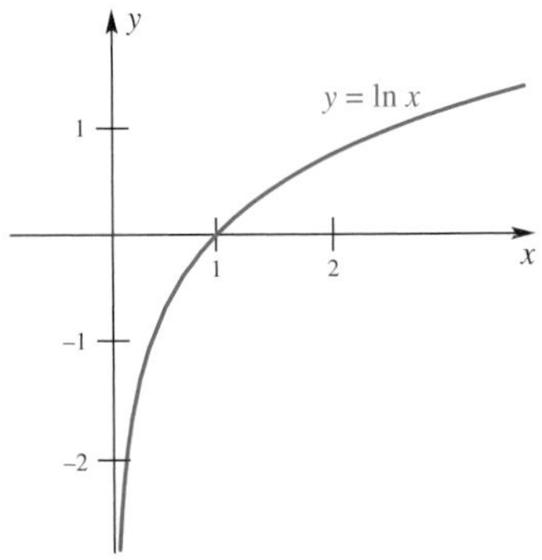

Figure 3

Trigonometric Integrals Some trigonometric integrals can be evaluated using the natural log function.

EXAMPLE 9 Evaluate $\int \tan x\, dx$.

SOLUTION Since $\tan x = \dfrac{\sin x}{\cos x}$ we can make the substitution $u = \cos x$, $du = -\sin x\, dx$, to obtain

$$\int \tan x\, dx = \int \frac{\sin x}{\cos x}\, dx = \int \frac{-1}{\cos x}(-\sin x\, dx) = -\ln|\cos x| + C \quad \blacksquare$$

Similarly, $\int \cot x\, dx = \ln|\sin x|$.

EXAMPLE 10 Evaluate $\int \sec x \csc x\, dx$.

SOLUTION For this one we use the trig identity $\sec x \csc x = \tan x + \cot x$. Then

$$\int \sec x \csc x\, dx = \int (\tan x + \cot x)\, dx = -\ln|\cos x| + \ln|\sin x| + C \quad \blacksquare$$

Concepts Review

1. The function ln is defined by $\ln x =$ ______. The domain of this function is ______ and its range is ______.

2. From the preceding definition, it follows that $D_x \ln x =$ ______ for $x > 0$.

3. More generally, for $x \neq 0$, $D_x \ln|x| =$ ______ and so $\int (1/x)\, dx =$ ______.

4. Some common properties of ln are $\ln(xy) =$ ______, $\ln(x/y) =$ ______, and $\ln(x^r) =$ ______.

Problem Set 6.1

1. Use the approximations $\ln 2 \approx 0.693$ and $\ln 3 \approx 1.099$ together with the properties stated in Theorem A to calculate approximations to each of the following. For example, $\ln 6 = \ln(2 \cdot 3) = \ln 2 + \ln 3 \approx 0.693 + 1.099 = 1.792$.

(a) $\ln 6$ (b) $\ln 1.5$ (c) $\ln 81$

(d) $\ln \sqrt{2}$ (e) $\ln\left(\frac{1}{36}\right)$ (f) $\ln 48$

2. Use your calculator to make the computations in Problem 1 directly.

In Problems 3–14, find the indicated derivative (see Examples 1 and 2). Assume in each case that x is restricted so that ln is defined.

3. $D_x \ln(x^2 + 3x + \pi)$
4. $D_x \ln(3x^3 + 2x)$
5. $D_x \ln(x - 4)^3$
6. $D_x \ln \sqrt{3x - 2}$
7. $\dfrac{dy}{dx}$ if $y = 3 \ln x$
8. $\dfrac{dy}{dx}$ if $y = x^2 \ln x$
9. $\dfrac{dz}{dx}$ if $z = x^2 \ln x^2 + (\ln x)^3$
10. $\dfrac{dr}{dx}$ if $r = \dfrac{\ln x}{x^2 \ln x^2} + \left(\ln \dfrac{1}{x}\right)^3$
11. $g'(x)$ if $g(x) = \ln\left(x + \sqrt{x^2 + 1}\right)$
12. $h'(x)$ if $h(x) = \ln\left(x + \sqrt{x^2 - 1}\right)$
13. $f'(81)$ if $f(x) = \ln \sqrt[3]{x}$
14. $f'\left(\dfrac{\pi}{4}\right)$ if $f(x) = \ln(\cos x)$

In Problems 15–26, find the integrals (see Examples 4, 5, and 6).

15. $\displaystyle\int \frac{1}{2x + 1}\, dx$
16. $\displaystyle\int \frac{1}{1 - 2x}\, dx$
17. $\displaystyle\int \frac{6v + 9}{3v^2 + 9v}\, dv$
18. $\displaystyle\int \frac{z}{2z^2 + 8}\, dz$
19. $\displaystyle\int \frac{2 \ln x}{x}\, dx$
20. $\displaystyle\int \frac{-1}{x(\ln x)^2}\, dx$
21. $\displaystyle\int_0^3 \frac{x^4}{2x^5 + \pi}\, dx$
22. $\displaystyle\int_0^1 \frac{t + 1}{2t^2 + 4t + 3}\, dt$
23. $\displaystyle\int \frac{x^2}{x - 1}\, dx$
24. $\displaystyle\int \frac{x^2 + x}{2x - 1}\, dx$
25. $\displaystyle\int \frac{x^4}{x + 4}\, dx$
26. $\displaystyle\int \frac{x^3 + x^2}{x + 2}\, dx$

In Problems 27–30, use Theorem A to write the expressions as the logarithm of a single quantity.

27. $2\ln(x + 1) - \ln x$
28. $\frac{1}{2}\ln(x - 9) + \frac{1}{2}\ln x$
29. $\ln(x - 2) - \ln(x + 2) + 2 \ln x$
30. $\ln(x^2 - 9) - 2\ln(x - 3) - \ln(x + 3)$

In Problems 31–34, find dy/dx by logarithmic differentiation (see Example 8).

31. $y = \dfrac{x + 11}{\sqrt{x^3 - 4}}$
32. $y = (x^2 + 3x)(x - 2)(x^2 + 1)$
33. $y = \dfrac{\sqrt{x + 13}}{(x - 4)\sqrt[3]{2x + 1}}$
34. $y = \dfrac{(x^2 + 3)^{2/3}(3x + 2)^2}{\sqrt{x + 1}}$

In Problems 35–38, make use of the known graph of $y = \ln x$ to sketch the graphs of the equations.

35. $y = \ln|x|$

36. $y = \ln\sqrt{x}$

37. $y = \ln\left(\frac{1}{x}\right)$

38. $y = \ln(x - 2)$

39. Sketch the graph of $y = \ln\cos x + \ln\sec x$ on $(-\pi/2, \pi/2)$, but think before you begin.

40. Explain why $\lim_{x\to 0} \ln\frac{\sin x}{x} = 0$.

41. Find all local extreme values of $f(x) = 2x^2 \ln x - x^2$ on its domain.

42. The rate of transmission in a telegraph cable is observed to be proportional to $x^2 \ln(1/x)$, where x is the ratio of the radius of the core to the thickness of the insulation $(0 < x < 1)$. What value of x gives the maximum rate of transmission?

43. Use the fact that $\ln 4 > 1$ to show that $\ln 4^m > m$ for $m > 1$. Conclude that $\ln x$ can be made as large as desired by choosing x sufficiently large. What does this imply about $\lim_{x\to\infty} \ln x$?

44. Use the fact that $\ln x = -\ln(1/x)$ and Problem 43 to show that $\lim_{x\to 0^+} \ln x = -\infty$.

45. Solve for x: $\int_{1/3}^{x} \frac{1}{t}\,dt = 2\int_{1}^{x} \frac{1}{t}\,dt$.

46. Prove the following statements.

(a) Since $1/t < 1/\sqrt{t}$ for $t > 1$, $\ln x < 2(\sqrt{x} - 1)$ for $x > 1$.

(b) $\lim_{x\to\infty} (\ln x)/x = 0$.

47. Calculate

$$\lim_{n\to\infty}\left[\frac{1}{n+1} + \frac{1}{n+2} + \cdots + \frac{1}{2n}\right]$$

by writing the expression in brackets as

$$\left[\frac{1}{1+1/n} + \frac{1}{1+2/n} + \cdots + \frac{1}{1+n/n}\right]\frac{1}{n}$$

and recognizing the latter as a Riemann sum.

C **48.** A famous theorem (the Prime Number Theorem) says that the number of primes less than n for large n is approximately $n/(\ln n)$. About how many primes are there less than 1,000,000?

49. Find and simplify $f'(1)$.

(a) $f(x) = \ln\left(\frac{ax - b}{ax + b}\right)^c$, where $c = \frac{a^2 - b^2}{2ab}$.

(b) $f(x) = \int_1^u \cos^2 t\,dt$, where $u = \ln(x^2 + x - 1)$.

50. Evaluate $\int_0^{\pi/3} \tan x\,dx$.

51. Evaluate $\int_{\pi/4}^{\pi/3} \sec x \csc x\,dx$.

52. Evaluate $\int \frac{\cos x}{1 + \sin x}\,dx$.

53. The region bounded by $y = (x^2 + 4)^{-1}$, $y = 0$, $x = 1$, and $x = 4$, is revolved about the y-axis, generating a solid. Find its volume.

54. Find the length of the curve $y = x^2/4 - \ln\sqrt{x}$, $1 \le x \le 2$.

55. By appealing to the graph of $y = 1/x$, show that

$$\frac{1}{2} + \frac{1}{3} + \cdots + \frac{1}{n} < \ln n < 1 + \frac{1}{2} + \frac{1}{3} + \cdots + \frac{1}{n-1}$$

56. Prove **Napier's Inequality,** which says that, for $0 < x < y$,

$$\frac{1}{y} < \frac{\ln y - \ln x}{y - x} < \frac{1}{x}$$

CAS **57.** Let $f(x) = \ln(1.5 + \sin x)$.

(a) Find the absolute extreme points on $[0, 3\pi]$.

(b) Find any inflection points on $[0, 3\pi]$.

(c) Evaluate $\int_0^{3\pi} \ln(1.5 + \sin x)\,dx$.

CAS **58.** Let $f(x) = \cos(\ln x)$.

(a) Find the absolute extreme points on $[0.1, 20]$.

(b) Find the absolute extreme points on $[0.01, 20]$.

(c) Evaluate $\int_{0.1}^{20} \cos(\ln x)\,dx$.

CAS **59.** Draw the graphs of $f(x) = x\ln(1/x)$ and $g(x) = x^2\ln(1/x)$ on $(0, 1]$.

(a) Find the area of the region between these curves on $(0, 1]$.

(b) Find the absolute maximum value of $|f(x) - g(x)|$ on $(0, 1]$.

CAS **60.** Follow the directions of Problem 59 for $f(x) = x\ln x$ and $g(x) = \sqrt{x}\ln x$.

Answers to Concepts Review: **1.** $\int_1^x (1/t)\,dt$; $(0, \infty)$; $(-\infty, \infty)$ **2.** $1/x$ **3.** $1/x$; $\ln|x| + C$ **4.** $\ln x + \ln y$; $\ln x - \ln y$; $r\ln x$

6.2 Inverse Functions and Their Derivatives

Our stated aim for this chapter is to expand the number of functions in our repertoire. One way to manufacture new functions is to take old ones and "reverse" them. When we do this for the natural logarithm function, we will be led to the natural exponential function, the subject of Section 6.3. In this section, we study the general problem of reversing (or inverting) a function. Here is the idea.

A function f takes a number x from its domain D and assigns to it a single value y from its range R. If we are lucky, as in the case of the two functions graphed in Figures 1 and 2, we can reverse f; that is, for any given y in R, we can unambiguously go back and find the x from which it came. This new function that takes y and assigns x to it is denoted by f^{-1}. Note that its domain is R and its range is D. It is called the **inverse** of f, or simply f-inverse. Here we are using the superscript -1 in a new way. The symbol f^{-1} does not denote $1/f$, as you might have expected. We, and all mathematicians, use it to name the inverse function.

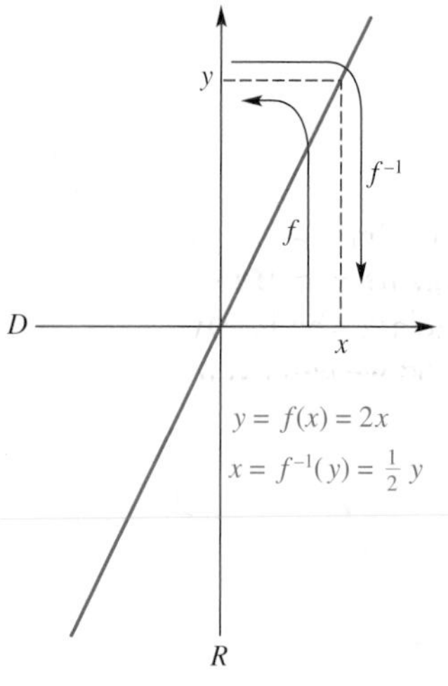

Figure 1

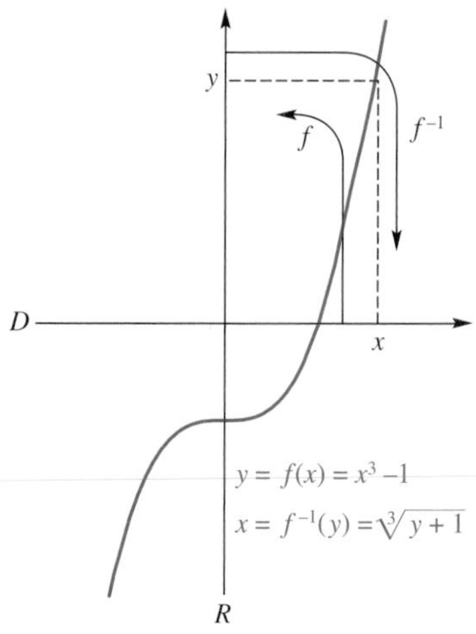

Figure 2

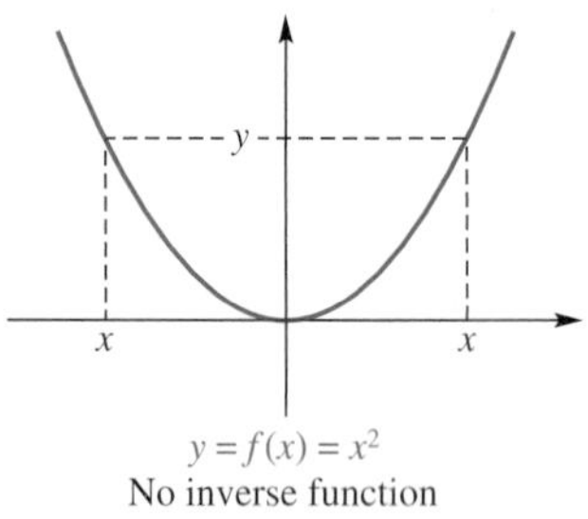

Figure 3

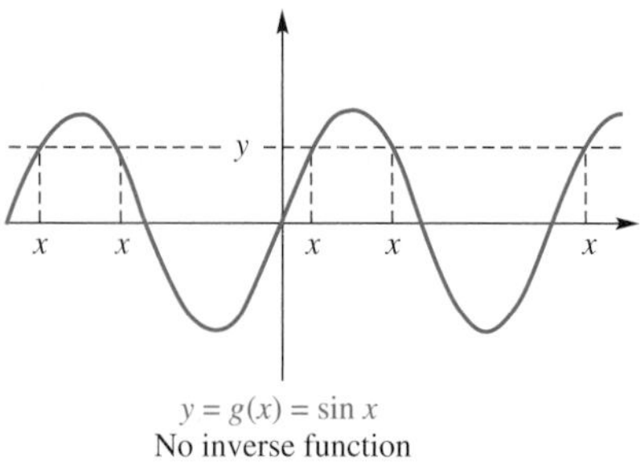

Figure 4

Sometimes, we can give a formula for f^{-1}. If $y = f(x) = 2x$, then $x = f^{-1}(y) = \frac{1}{2}y$ (see Figure 1). Similarly, if $y = f(x) = x^3 - 1$, then $x = f^{-1}(y) = \sqrt[3]{y + 1}$ (Figure 2). In each case, we simply solve the equation that determines f for x in terms of y. The result is $x = f^{-1}(y)$.

But life is more complicated than these two examples indicate. Not every function can be reversed in an unambiguous way. Consider $y = f(x) = x^2$, for example. For a given y there are *two* x's that correspond to it (Figure 3). The function $y = g(x) = \sin x$ is even worse. For each y, there are infinitely many x's that correspond to it (Figure 4). Such functions do not have inverses; at least, they do not unless we somehow restrict the set of x-values, a subject we will take up later.

Existence of Inverse Functions It would be nice to have a simple criterion for deciding whether a function f has an inverse. One such criterion is that the function be **one-to-one;** that is, $x_1 \neq x_2$ implies $f(x_1) \neq f(x_2)$. This is equivalent to the geometric condition that every horizontal line meet the graph of $y = f(x)$ in at most one point. But, in a given situation, this criterion may be very hard to apply, since it demands that we have complete knowledge of the graph. A more practical criterion that covers most examples that arise in this book is that a function be **strictly monotonic.** By this we mean that it is either increasing or decreasing on its domain. (See the definitions in Section 3.2.)

Theorem A

If f is strictly monotonic on its domain, then f has an inverse.

Proof Let x_1 and x_2 be distinct numbers in the domain of f, with $x_1 < x_2$. Since f is monotonic, $f(x_1) < f(x_2)$ or $f(x_1) > f(x_2)$. Either way, $f(x_1) \neq f(x_2)$. Thus, $x_1 \neq x_2$ implies $f(x_1) \neq f(x_2)$, which means that f is one-to-one and therefore has an inverse. ■

This is a practical result, because we have an easy way of deciding whether a differentiable function f is strictly monotonic. We simply examine the sign of f'.

EXAMPLE 1 Show that $f(x) = x^5 + 2x + 1$ has an inverse.

SOLUTION $f'(x) = 5x^4 + 2 > 0$ for all x. Thus, f is increasing on the whole real line and so it has an inverse there. ■

We do not claim that we can always give a formula for f^{-1}. In the example just considered, this would require that we be able to solve $y = x^5 + 2x + 1$ for x. Although we could use a CAS or a graphing calculator to solve this equation for x for a particular value of y, there is no simple formula that would give us x in terms of y for an arbitrary y.

There is a way of salvaging the notion of inverse for functions that do not have inverses on their natural domain. We simply *restrict the domain* to a set on which the graph is either increasing or decreasing. Thus, for $y = f(x) = x^2$, we may restrict the domain to $x \geq 0$ ($x \leq 0$ would also work). For $y = g(x) = \sin x$, we restrict the domain to the interval $[-\pi/2, \pi/2]$. Then both functions have inverses (see Figure 5), and we can even give a formula for the first one: $f^{-1}(y) = \sqrt{y}$.

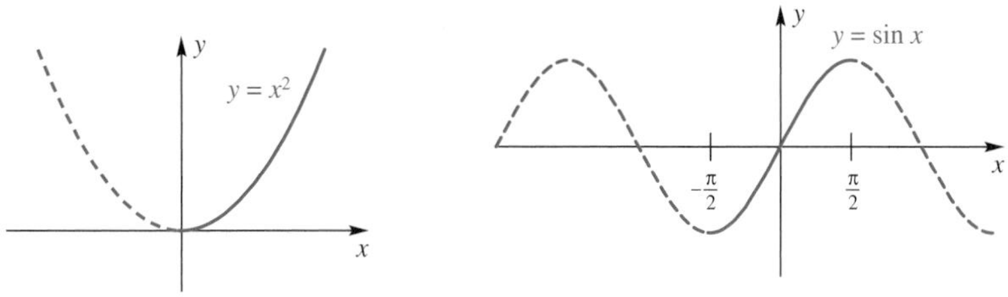

Domain restricted to $x \geq 0$ Domain restricted to $[-\frac{\pi}{2}, \frac{\pi}{2}]$

Figure 5

If f has an inverse f^{-1}, then f^{-1} also has an inverse, namely, f. Thus, we may call f and f^{-1} a pair of inverse functions. One function undoes (or reverses) what the other did; that is,

$$f^{-1}(f(x)) = x \quad \text{and} \quad f(f^{-1}(y)) = y$$

Undoing Machines

We may view a function as a machine that accepts an input and produces an output. If the f machine and the f^{-1} machine are hooked together in tandem, they undo each other.

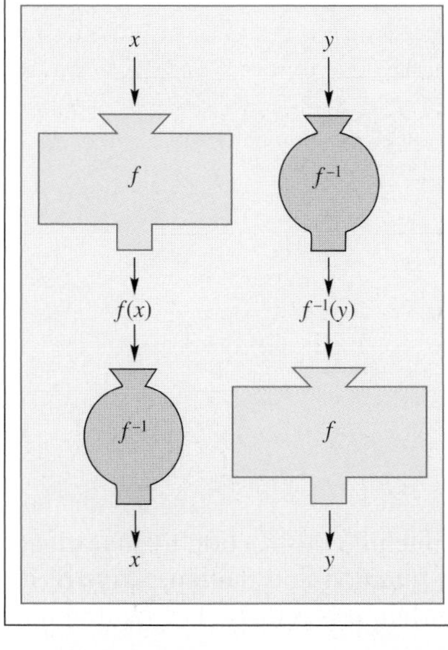

EXAMPLE 2 Show that $f(x) = 2x + 6$ has an inverse, find a formula for $f^{-1}(y)$, and verify the results in the box above.

SOLUTION Since f is an increasing function, it has an inverse. To find $f^{-1}(y)$, we solve $y = 2x + 6$ for x, which gives $x = (y - 6)/2 = f^{-1}(y)$. Finally, note that

$$f^{-1}(f(x)) = f^{-1}(2x + 6) = \frac{(2x + 6) - 6}{2} = x$$

and

$$f(f^{-1}(y)) = f\left(\frac{y - 6}{2}\right) = 2\left(\frac{y - 6}{2}\right) + 6 = y$$ ■

The Graph of $y = f^{-1}(x)$

Suppose that f has an inverse. Then

$$x = f^{-1}(y) \Leftrightarrow y = f(x)$$

Consequently, $y = f(x)$ and $x = f^{-1}(y)$ determine the same (x, y) pairs and so have identical graphs. However, it is conventional to use x as the domain variable for functions, so we now inquire about the graph of $y = f^{-1}(x)$ (note that we have interchanged the roles of x and y). A little thought convinces us that to interchange the roles of x and y on a graph is to reflect the graph across the line $y = x$.

Thus the graph of $y = f^{-1}(x)$ *is just the reflection of the graph of* $y = f(x)$ *across the line* $y = x$ (Figure 6).

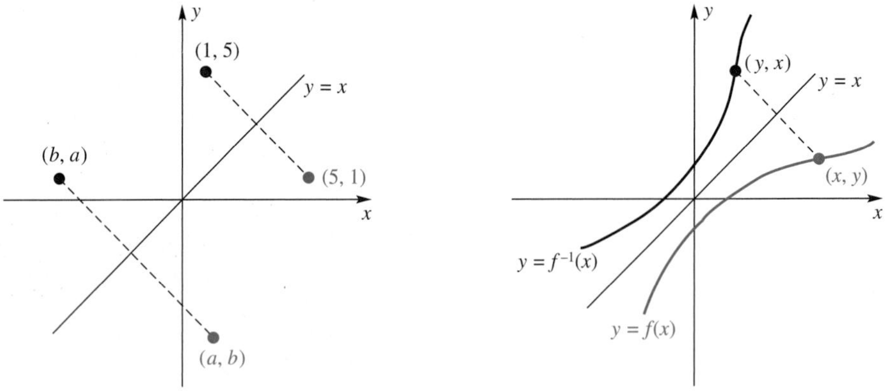

Figure 6

A related matter is that of finding a formula for $f^{-1}(x)$. To do it, we first find $f^{-1}(y)$ and then replace y by x in the resulting formula. Thus, we propose the following three-step process for finding $f^{-1}(x)$.

Step 1: Solve the equation $y = f(x)$ for x in terms of y.

Step 2: Use $f^{-1}(y)$ to name the resulting expression in y.

Step 3: Replace y by x to get the formula for $f^{-1}(x)$.

Before trying the three-step process on a particular function f, you might think we should first verify that f has an inverse. However, if we can actually carry out the first step and get a single x for each y, then f^{-1} does exist. (Note that when we try this for $y = f(x) = x^2$ we get $x = \pm\sqrt{y}$, which immediately shows that f^{-1} does not exist, unless, of course, we have restricted the domain to eliminate one of the two signs, + or −.)

EXAMPLE 3 Find a formula for $f^{-1}(x)$ if $y = f(x) = x/(1 - x)$.

SOLUTION Here are the three steps for this example.

Step 1:

$$\begin{aligned} y &= \frac{x}{1 - x} \\ (1 - x)y &= x \\ y - xy &= x \\ x + xy &= y \\ x(1 + y) &= y \\ x &= \frac{y}{1 + y} \end{aligned}$$

Step 2: $f^{-1}(y) = \dfrac{y}{1 + y}$

Step 3: $f^{-1}(x) = \dfrac{x}{1 + x}$ ■

Derivatives of Inverse Functions We conclude this section by investigating the relationship between the derivative of a function and the derivative of its inverse. Consider first what happens to a line l_1 when it is reflected across the line

$y = x$. As the left half of Figure 7 makes clear, l_1 is reflected into a line l_2; moreover, their respective slopes m_1 and m_2 are related by $m_2 = 1/m_1$, provided $m_1 \neq 0$. If l_1 happens to be the tangent line to the graph of f at the point (c, d), then l_2 is the tangent line to the graph of f^{-1} at the point (d, c) (see the right half of Figure 7). We are led to the conclusion that

$$(f^{-1})'(d) = m_2 = \frac{1}{m_1} = \frac{1}{f'(c)}$$

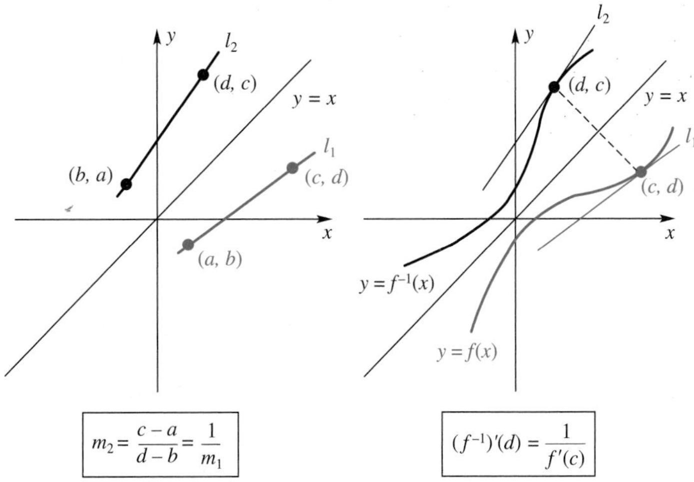

Figure 7

Pictures are sometimes deceptive, so we claim only to have made the following result plausible. For a formal proof, see any advanced calculus book.

Theorem B Inverse Function Theorem

Let f be differentiable and strictly monotonic on an interval I. If $f'(x) \neq 0$ at a certain x in I, then f^{-1} is differentiable at the corresponding point $y = f(x)$ in the range of f and

$$(f^{-1})'(y) = \frac{1}{f'(x)}$$

The conclusion to Theorem B is often written symbolically as

$$\frac{dx}{dy} = \frac{1}{dy/dx}$$

EXAMPLE 4 Let $y = f(x) = x^5 + 2x + 1$, as in Example 1. Find $(f^{-1})'(4)$.

SOLUTION Even though we cannot find a formula for f^{-1} in this case, we note that $y = 4$ corresponds to $x = 1$, and, since $f'(x) = 5x^4 + 2$,

$$(f^{-1})'(4) = \frac{1}{f'(1)} = \frac{1}{5+2} = \frac{1}{7}$$ ■

Concepts Review

1. A f function is one-to-one if $x_1 \neq x_2$ implies _____.

2. A one-to-one function f has an inverse f^{-1} satisfying $f^{-1}(f(x)) =$ _____ and $f($_____$) = y$.

3. A useful criterion for f to be one-to-one (and so have an inverse) on a domain is that f be strictly _____ there. This means that f is either _____ or _____.

4. Let $y = f(x)$, where f has the inverse f^{-1}. The relation connecting the derivatives of f and f^{-1} is _____.

Problem Set 6.2

In Problems 1–6, the graph of $y = f(x)$ is shown. In each case, decide whether f has an inverse and, if so, estimate $f^{-1}(2)$.

1.

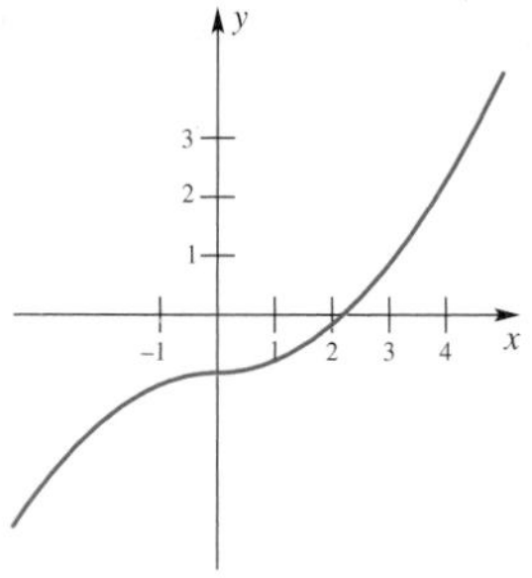

2.

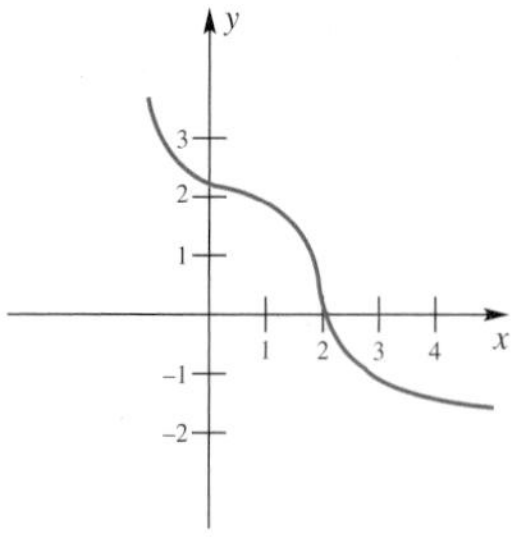

3.

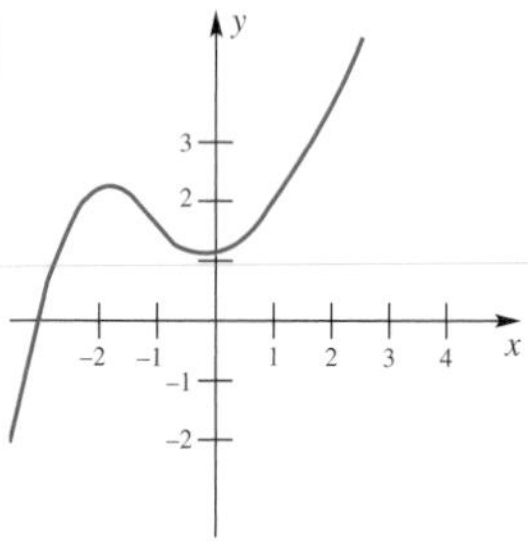

4.

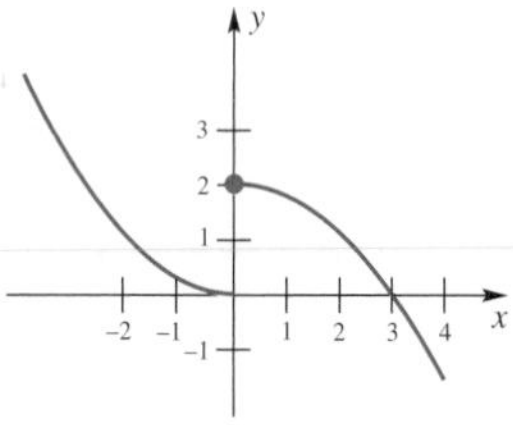

5.

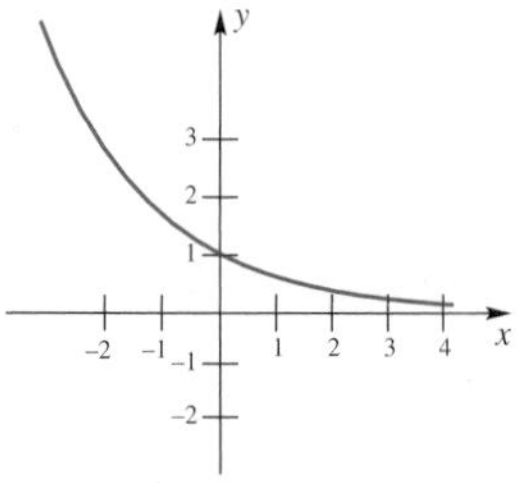

6.

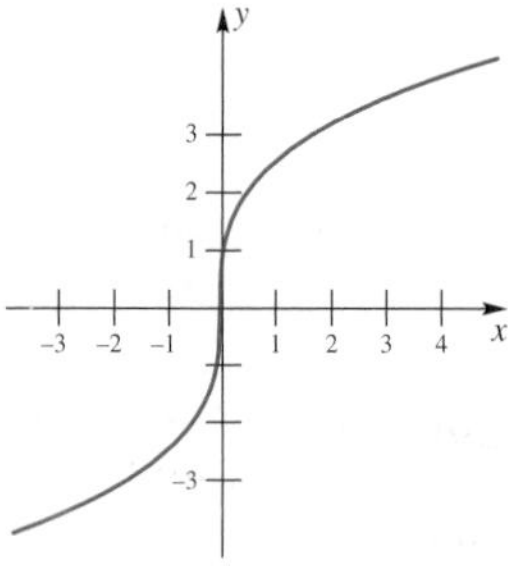

In Problems 7–14, show that f has an inverse by showing that it is strictly monotonic (see Example 1).

7. $f(x) = -x^5 - x^3$

8. $f(x) = x^7 + x^5$

9. $f(\theta) = \cos\theta, 0 \le \theta \le \pi$

10. $f(x) = \cot x = \dfrac{\cos x}{\sin x}, 0 < x < \dfrac{\pi}{2}$

11. $f(z) = (z - 1)^2, z \ge 1$

12. $f(x) = x^2 + x - 6, x \ge 2$

13. $f(x) = \displaystyle\int_0^x \sqrt{t^4 + t^2 + 10}\, dt$

14. $f(r) = \displaystyle\int_r^1 \cos^4 t\, dt$

In Problems 15–28, find a formula for $f^{-1}(x)$ and then verify that $f^{-1}(f(x)) = x$ and $f(f^{-1}(x)) = x$ (see Examples 2 and 3).

15. $f(x) = x + 1$

16. $f(x) = -\dfrac{x}{3} + 1$

17. $f(x) = \sqrt{x + 1}$

18. $f(x) = -\sqrt{1 - x}$

19. $f(x) = -\dfrac{1}{x - 3}$

20. $f(x) = \sqrt{\dfrac{1}{x - 2}}$

21. $f(x) = 4x^2, x \le 0$

22. $f(x) = (x - 3)^2, x \ge 3$

23. $f(x) = (x - 1)^3$

24. $f(x) = x^{5/2}, x \ge 0$

25. $f(x) = \dfrac{x - 1}{x + 1}$

26. $f(x) = \left(\dfrac{x - 1}{x + 1}\right)^3$

27. $f(x) = \dfrac{x^3 + 2}{x^3 + 1}$

28. $f(x) = \left(\dfrac{x^3 + 2}{x^3 + 1}\right)^5$

29. Find the volume V of water in the conical tank of Figure 8 as a function of the height h. Then find the height h as a function of volume V.

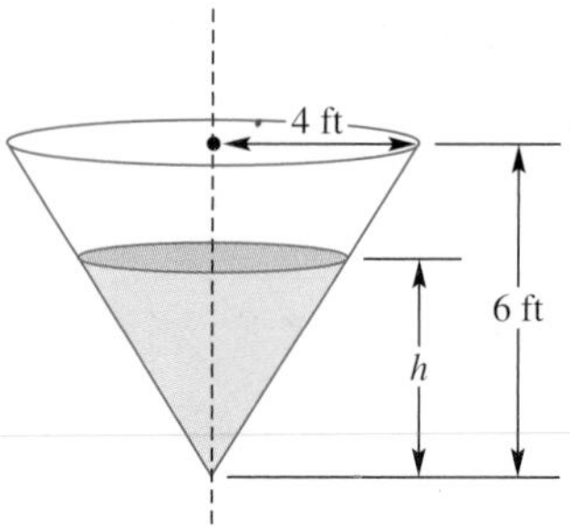

Figure 8

30. A ball is thrown vertically upward with velocity v_0. Find the maximum height H of the ball as a function of v_0. Then find the velocity v_0 required to achieve a height of H.

In Problems 31 and 32, restrict the domain of f so that f has an inverse, yet keeping its range as large as possible. Then find $f^{-1}(x)$. Suggestion: First graph f.

31. $f(x) = 2x^2 + x - 4$

32. $f(x) = x^2 - 3x + 1$

In each of Problems 33–36, the graph of $y = f(x)$ is shown. Sketch the graph of $y = f^{-1}(x)$ and estimate $(f^{-1})'(3)$.

33.

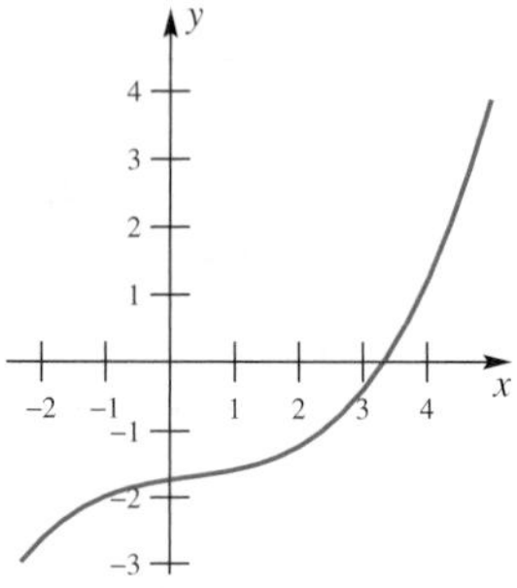

34.

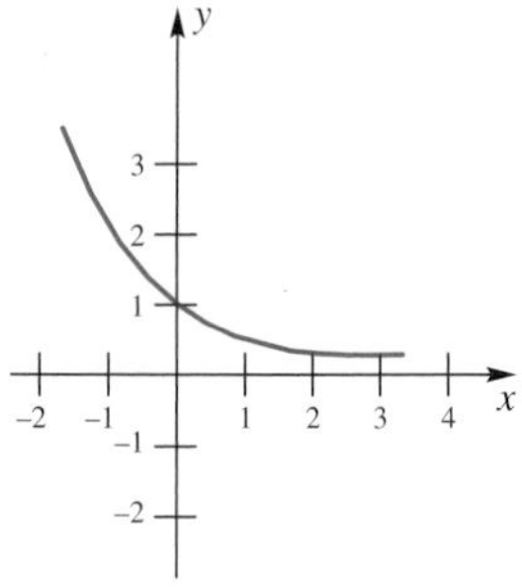

35.

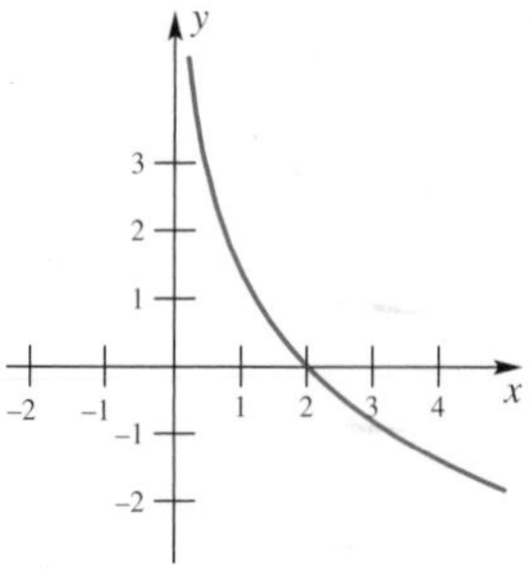

36.

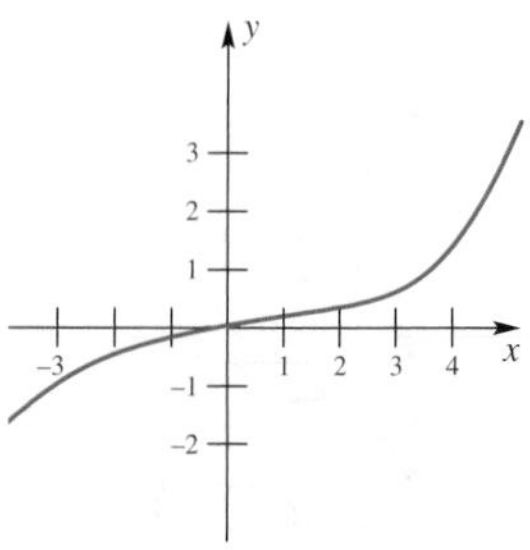

In Problems 37–40, find $(f^{-1})'(2)$ by using Theorem B (see Example 4). Note that you can find the x corresponding to $y = 2$ by inspection.

37. $f(x) = 3x^5 + x - 2$

38. $f(x) = x^5 + 5x - 4$

39. $f(x) = 2 \tan x, -\frac{\pi}{2} < x < \frac{\pi}{2}$

40. $f(x) = \sqrt{x + 1}$

41. Suppose that both f and g have inverses and that $h(x) = (f \circ g)(x) = f(g(x))$. Show that h has an inverse given by $h^{-1} = g^{-1} \circ f^{-1}$.

42. Verify the result of Problem 41 for $f(x) = 1/x, g(x) = 3x + 2$.

43. If $f(x) = \int_0^x \sqrt{1 + \cos^2 t}\, dt$, then f has an inverse. (Why?) Let $A = f(\pi/2)$ and $B = f(5\pi/6)$. Find

(a) $(f^{-1})'(A)$, (b) $(f^{-1})'(B)$,

(c) $(f^{-1})'(0)$.

44. Let $f(x) = \frac{ax + b}{cx + d}$ and assume $bc - ad \neq 0$.

(a) Find the formula for $f^{-1}(x)$.

(b) Why is the condition $bc - ad \neq 0$ needed?

(c) What condition on a, b, c, and d will make $f = f^{-1}$?

45. Suppose that f is continuous and strictly increasing on $[0, 1]$ with $f(0) = 0$ and $f(1) = 1$. If $\int_0^1 f(x)\, dx = \frac{2}{5}$, calculate $\int_0^1 f^{-1}(y)\, dy$. *Hint:* Draw a picture.

EXPL **46.** Let f be continuous and strictly increasing on $[0, \infty)$ with $f(0) = 0$ and $f(x) \to \infty$ as $x \to \infty$. Use geometric reasoning to establish **Young's Inequality.** For $a > 0, b > 0$,

$$ab \leq \int_0^a f(x)\, dx + \int_0^b f^{-1}(y)\, dy$$

What is the condition for equality?

EXPL **47.** Let $p > 1, q > 1$, and $1/p + 1/q = 1$. Show that the inverse of $f(x) = x^{p-1}$ is $f^{-1}(y) = y^{q-1}$ and use this together with Problem 46 to prove **Minkowski's Inequality:**

$$ab \leq \frac{a^p}{p} + \frac{b^q}{q}, \qquad a > 0, b > 0$$

Answers to Concepts Review: **1.** $f(x_1) \neq f(x_2)$ **2.** x; $f^{-1}(y)$ **3.** monotonic; increasing; decreasing **4.** $(f^{-1})'(y) = 1/f'(x)$

6.3 The Natural Exponential Function

The graph of $y = f(x) = \ln x$ was obtained at the end of Section 6.1 and is reproduced in Figure 1. The natural logarithm function is differentiable (hence continuous) and increasing on its domain $D = (0, \infty)$; its range is $R = (-\infty, \infty)$. It is, in fact, precisely the kind of function studied in Section 6.2, and therefore has an inverse $\ln^{-1}$ with domain $(-\infty, \infty)$ and range $(0, \infty)$. This function is so important that it is given a special name and a special symbol.

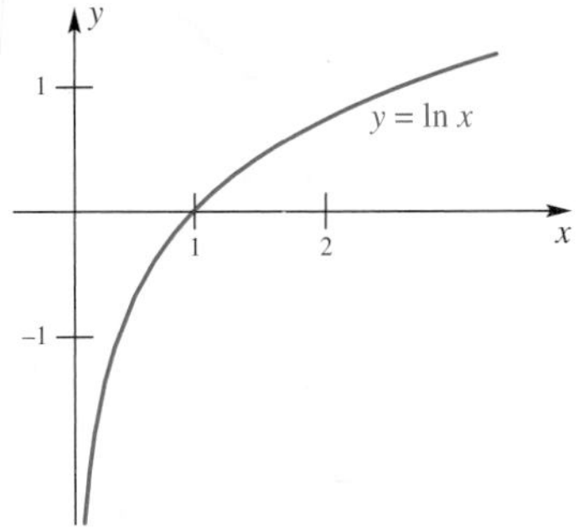

Figure 1

Definition

The inverse of ln is called the **natural exponential function** and is denoted by exp. Thus,

$$x = \exp y \Leftrightarrow y = \ln x$$

It follows immediately from this definition that

1. $\exp(\ln x) = x, \quad x > 0$
2. $\ln(\exp y) = y, \quad$ for all y

Since exp and ln are inverse functions, the graph of $y = \exp x$ is just the graph of $y = \ln x$ reflected across the line $y = x$ (Figure 2).

But why the name *exponential function*? You will see.

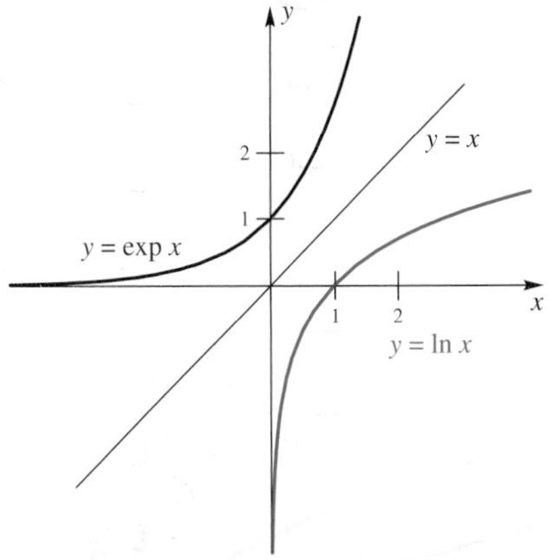

Figure 2

Properties of the Exponential Function We begin by introducing a new number, which, like π, is so important in mathematics that it gets a special symbol, e. The letter e is appropriate since Leonhard Euler first recognized the significance of this number.

Definition

The letter e denotes the unique positive real number such that $\ln e = 1$.

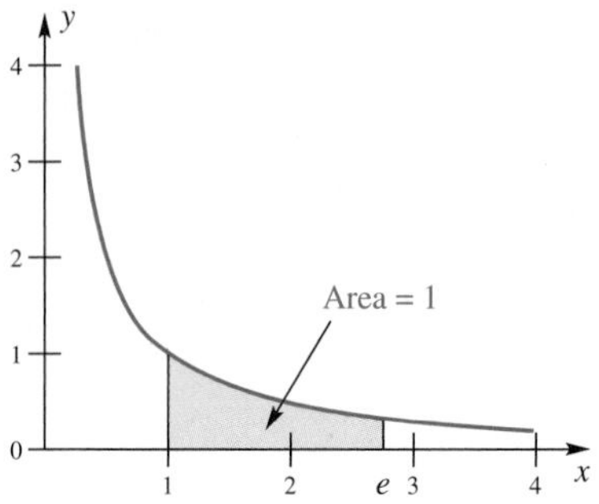

Figure 3

Figure 3 illustrates this definition; the area under the graph of $y = 1/x$ between $x = 1$ and $x = e$ is 1. Since $\ln e = 1$, it is also true that $\exp 1 = e$. The number e, like π, is irrational. Its decimal expansion is known to thousands of places; the first few digits are

$$e \approx 2.718281828459045$$

Now we make a crucial observation, one that depends only on facts already demonstrated: (1) above and Theorem 6.1A. If r is any rational number,

$$e^r = \exp(\ln e^r) = \exp(r \ln e) = \exp r$$

Let us emphasize the result. For rational r, $\exp r$ *is identical with* e^r. What was introduced in the most abstract way as the inverse of the natural logarithm, which itself was defined by an integral, has turned out to be a simple power.

But what if r is irrational? Here we remind you of a gap in all elementary algebra books. Never are irrational powers defined in anything approaching a rigorous manner. What is meant by $e^{\sqrt{2}}$? You will have a hard time pinning that number down, based on elementary algebra. But we must pin it down if we are to talk of such things as $D_x e^x$. Guided by what we learned above, we simply define e^x for all x (rational or irrational) by

$$e^x = \exp x$$

Note that (1) and (2) at the beginning of this section now take the following form:

$$(1)' \quad e^{\ln x} = x, \qquad x > 0$$

$$(2)' \quad \ln(e^y) = y, \quad \text{for all } y$$

Note also that $(1)'$ says that $\ln x$ is the *exponent* you need to put on e to get x. This is just the usual definition of the logarithm to the base e, as given in most precalculus books.

We can now easily prove two of the familiar laws of exponents.

Theorem A

Let a and b be any real numbers. Then $e^a e^b = e^{a+b}$ and $e^a/e^b = e^{a-b}$.

Proof To prove the first, we write

$$\begin{aligned} e^a e^b &= \exp(\ln e^a e^b) && \text{(by (1))} \\ &= \exp(\ln e^a + \ln e^b) && \text{(Theorem 6.1A)} \\ &= \exp(a + b) && \text{(by (2)')} \\ &= e^{a+b} && \text{(since } \exp x = e^x\text{)} \end{aligned}$$

The second fact is proved similarly. ■

The Derivative of e^x Since exp and ln are inverses, we know from Theorem 6.2B that $\exp x = e^x$ is differentiable. To find a formula for $D_x e^x$, we could use that theorem. Alternatively, let $y = e^x$ so that

$$x = \ln y$$

Now differentiate both sides with respect to x. Using the Chain Rule, we obtain

$$1 = \frac{1}{y} D_x y$$

Thus,

$$D_x y = y = e^x$$

Definitions of e

Authors choose different ways to define e.

1. $e = \ln^{-1} 1$ (our definition)
2. $e = \lim_{h \to 0} (1 + h)^{1/h}$
3. $e = \lim_{n \to \infty} \left(1 + \frac{1}{1!} + \frac{1}{2!} + \cdots + \frac{1}{n!}\right)$

In our book, definitions 2 and 3 become theorems (see Section 6.5, Theorem A, and Section 9.7, Example 3).

Phoenix

The number e appears throughout mathematics, but its importance rests most securely on its use as the base for the natural exponential function. And what makes this function so significant?

"Who has not been amazed to learn that the function $y = e^x$, like a phoenix rising again from its own ashes, is its own derivative?"

François Le Lionnais

We have proved the remarkable fact that e^x is its own derivative; that is,

$$D_x e^x = e^x$$

Thus, $y = e^x$ is a solution of the differential equation $y' = y$.

If $u = f(x)$ is differentiable, then the Chain Rule yields

$$D_x e^u = e^u D_x u$$

EXAMPLE 1 Find $D_x e^{\sqrt{x}}$.

SOLUTION Using $u = \sqrt{x}$, we obtain

$$D_x e^{\sqrt{x}} = e^{\sqrt{x}} D_x \sqrt{x} = e^{\sqrt{x}} \cdot \frac{1}{2} x^{-1/2} = \frac{e^{\sqrt{x}}}{2\sqrt{x}}$$ ■

EXAMPLE 2 Find $D_x e^{x^2 \ln x}$.

SOLUTION

$$\begin{aligned} D_x e^{x^2 \ln x} &= e^{x^2 \ln x} D_x(x^2 \ln x) \\ &= e^{x^2 \ln x}\left(x^2 \cdot \frac{1}{x} + 2x \ln x\right) \\ &= xe^{x^2 \ln x}(1 + \ln x^2) \end{aligned}$$ ■

EXAMPLE 3 Let $f(x) = xe^{x/2}$. Find where f is increasing and decreasing, and where it is concave upward and downward. Also, identify all extreme values and points of inflection. Then, sketch the graph of f.

SOLUTION

$$f'(x) = \frac{xe^{x/2}}{2} + e^{x/2} = e^{x/2}\left(\frac{x + 2}{2}\right)$$

and

$$f''(x) = \frac{e^{x/2}}{2} + \left(\frac{x + 2}{2}\right)\frac{e^{x/2}}{2} = e^{x/2}\left(\frac{x + 4}{4}\right)$$

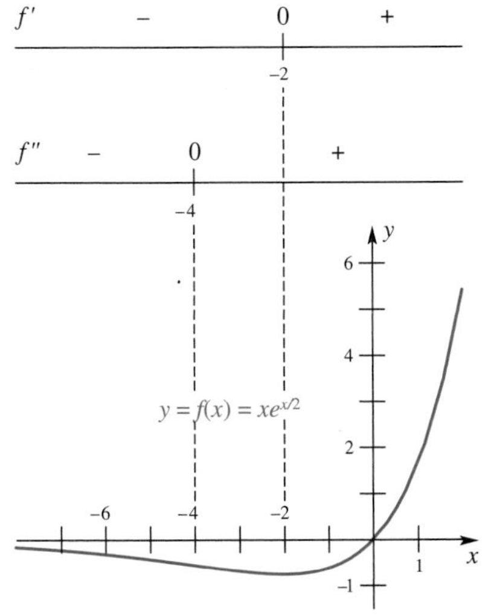

Figure 4

Keeping in mind that $e^{x/2} > 0$ for all x, we see that $f'(x) < 0$ for $x < -2$, $f'(-2) = 0$, and $f'(x) > 0$ for all $x > -2$. Thus, f is decreasing on $(-\infty, -2]$ and increasing on $[-2, \infty)$, and has its minimum value at $x = -2$ of $f(-2) = -2/e \approx -0.7$.

Also, $f''(x) < 0$ for $x < -4$, $f''(-4) = 0$, and $f''(x) > 0$ for $x > -4$; so the graph of f is concave downward on $(-\infty, -4)$ and concave upward on $(-4, \infty)$, and has a point of inflection at $(-4, -4e^{-2}) \approx (-4, -0.54)$. Since $\lim_{x \to -\infty} xe^{x/2} = 0$, the line $y = 0$ is a horizontal asymptote. This information supports the graph in Figure 4. ■

The derivative formula $D_x e^x = e^x$ automatically yields the integral formula $\int e^x\, dx = e^x + C$, or, with u replacing x,

$$\int e^u\, du = e^u + C$$

EXAMPLE 4 Evaluate $\int e^{-4x}\,dx$.

SOLUTION Let $u = -4x$, so $du = -4\,dx$. Then

$$\int e^{-4x}\,dx = -\frac{1}{4}\int e^{-4x}(-4\,dx) = -\frac{1}{4}\int e^{u}\,du = -\frac{1}{4}e^{u} + C = -\frac{1}{4}e^{-4x} + C \quad \blacksquare$$

EXAMPLE 5 Evaluate $\int x^2 e^{-x^3}\,dx$.

SOLUTION Let $u = -x^3$, so $du = -3x^2\,dx$. Then

$$\begin{aligned}\int x^2 e^{-x^3}\,dx &= -\frac{1}{3}\int e^{-x^3}(-3x^2\,dx)\\ &= -\frac{1}{3}\int e^{u}\,du = -\frac{1}{3}e^{u} + C\\ &= -\frac{1}{3}e^{-x^3} + C\end{aligned} \quad \blacksquare$$

EXAMPLE 6 Evaluate $\int_1^3 xe^{-3x^2}\,dx$.

SOLUTION Let $u = -3x^2$, so $du = -6x\,dx$. Then

$$\begin{aligned}\int xe^{-3x^2}\,dx &= -\frac{1}{6}\int e^{-3x^2}(-6x\,dx) = -\frac{1}{6}\int e^{u}\,du\\ &= -\frac{1}{6}e^{u} + C = -\frac{1}{6}e^{-3x^2} + C\end{aligned}$$

Thus, by the Second Fundamental Theorem of Calculus,

$$\int_1^3 xe^{-3x^2}\,dx = \left[-\frac{1}{6}e^{-3x^2}\right]_1^3 = -\frac{1}{6}(e^{-27} - e^{-3}) = \frac{e^{-3} - e^{-27}}{6} \approx 0.0082978$$

The last result can be obtained directly with a calculator. ■

EXAMPLE 7 Evaluate $\int \frac{6e^{1/x}}{x^2}\,dx$.

SOLUTION Think of $\int e^{u}\,du$. Let $u = 1/x$, so $du = (-1/x^2)\,dx$. Then

$$\begin{aligned}\int \frac{6e^{1/x}}{x^2}\,dx &= -6\int e^{1/x}\left(\frac{-1}{x^2}\,dx\right) = -6\int e^{u}\,du\\ &= -6e^{u} + C = -6e^{1/x} + C\end{aligned} \quad \blacksquare$$

Although the symbol e^y will largely supplant exp y throughout the rest of this book, exp occurs frequently in scientific writing, especially when the exponent y is complicated. For example, in statistics, one often encounters the normal probability density function, which is

$$f(x) = \frac{1}{\sigma\sqrt{2\pi}}\exp\left[-\frac{(x-\mu)^2}{2\sigma^2}\right]$$

Concepts Review

1. The function ln is _____ on $(0, \infty)$ and so has an inverse denoted by $\ln^{-1}$ or by _____.

2. The number e is defined in terms of ln by _____; its value to two decimal places is _____.

3. Since $e^x = \exp x = \ln^{-1} x$, it follows that $e^{\ln x} =$ _____ for $x > 0$, and $\ln(e^x) =$ _____.

4. Two remarkable facts about e^x are that $D_x(e^x) =$ _____ and $\int e^x\,dx =$ _____.

Problem Set 6.3

C **1.** Use your calculator to calculate each of the following. *Note:* On some calculators there is an e^x button. On others you must press the INV (or 2nd) and ln x buttons.

(a) e^3 (b) $e^{2.1}$

(c) $e^{\sqrt{2}}$ (d) $e^{\cos(\ln 4)}$

C **2.** Calculate the following and explain why your answers are not surprising.

(a) $e^{3 \ln 2}$ (b) $e^{(\ln 64)/2}$

In Problems 3–10, simplify the given expression.

3. $e^{3 \ln x}$ **4.** $e^{-2 \ln x}$

5. $\ln e^{\cos x}$ **6.** $\ln e^{-2x-3}$

7. $\ln(x^3 e^{-3x})$ **8.** $e^{x - \ln x}$

9. $e^{\ln 3 + 2 \ln x}$ **10.** $e^{\ln x^2 - y \ln x}$

In Problems 11–22, find $D_x y$ (see Examples 1 and 2).

11. $y = e^{x+2}$ **12.** $y = e^{2x^2 - x}$

13. $y = e^{\sqrt{x+2}}$ **14.** $y = e^{-1/x^2}$

15. $y = e^{2 \ln x}$ **16.** $y = e^{x/\ln x}$

17. $y = x^3 e^x$ **18.** $y = e^{x^3 \ln x}$

19. $y = \sqrt{e^{x^2}} + e^{\sqrt{x^2}}$ **20.** $y = e^{1/x^2} + 1/e^{x^2}$

21. $e^{xy} + xy = 2$ *Hint:* Use implicit differentiation.

22. $e^{x+y} = 4 + x + y$

23. Use your knowledge of the graph of $y = e^x$ to sketch the graphs of (a) $y = -e^x$ and (b) $y = e^{-x}$.

24. Explain why $a < b \Rightarrow e^{-a} > e^{-b}$.

In Problems 25–36, first find the domain of the given function f and then find where it is increasing and decreasing, and also where it is concave upward and downward. Identify all extreme values and points of inflection. Then sketch the graph of $y = f(x)$.

25. $f(x) = e^{2x}$ **26.** $f(x) = e^{-x/2}$

27. $f(x) = xe^{-x}$ **28.** $f(x) = e^x + x$

29. $f(x) = \ln(x^2 + 1)$ **30.** $f(x) = \ln(2x - 1)$

31. $f(x) = \ln(1 + e^x)$ **32.** $f(x) = e^{1-x^2}$

33. $f(x) = e^{-(x-2)^2}$ **34.** $f(x) = e^x - e^{-x}$

35. $f(x) = \int_0^x e^{-t^2}\, dt$ **36.** $f(x) = \int_0^x te^{-t}\, dt$

In Problems 37–44, find each integral.

37. $\int e^{3x+1}\, dx$ **38.** $\int xe^{x^2-3}\, dx$

39. $\int (x + 3)e^{x^2+6x}\, dx$ **40.** $\int \frac{e^x}{e^x - 1}\, dx$

41. $\int \frac{e^{-1/x}}{x^2}\, dx$ **42.** $\int e^{x+e^x}\, dx$

43. $\int_0^1 e^{2x+3}\, dx$ **44.** $\int_1^2 \frac{e^{3/x}}{x^2}\, dx$

45. Find the volume of the solid generated by revolving the region bounded by $y = e^x$, $y = 0$, $x = 0$, and $x = \ln 3$ about the x-axis.

46. The region bounded by $y = e^{-x^2}$, $y = 0$, $x = 0$, and $x = 1$ is revolved about the y-axis. Find the volume of the resulting solid.

47. Find the area of the region bounded by the curve $y = e^{-x}$ and the line through the points $(0, 1)$ and $(1, 1/e)$.

48. Show that $f(x) = \frac{x}{e^x - 1} - \ln(1 - e^{-x})$ is decreasing for $x > 0$.

C **49. Stirling's Formula** says that for large n we can approximate $n! = 1 \cdot 2 \cdot 3 \cdots n$ by

$$n! \approx \sqrt{2\pi n}\left(\frac{n}{e}\right)^n$$

(a) Calculate 10! exactly and then approximately using the above formula.

(b) Approximate 60!.

C **50.** It will be shown later (Section 9.9) that for small x

$$e^x \approx 1 + x + \frac{x^2}{2!} + \frac{x^3}{3!} + \frac{x^4}{4!}$$

Use this result to approximate $e^{0.3}$ and compare your result with what you get by calculating it directly. (Computers and calculators use sums like this to approximate e^x.)

51. Find the length of the curve given parametrically by $x = e^t \sin t$, $y = e^t \cos t$, $0 \le t \le \pi$.

C **52.** If customers arrive at a check-out counter at the average rate of k per minute, then (see books on probability theory) the probability that exactly n customers will arrive in a period of x minutes is given by the formula

$$P_n(x) = \frac{(kx)^n e^{-kx}}{n!}, \quad n = 0, 1, 2, \ldots$$

Find the probability that exactly 8 customers will arrive during a 30-minute period if the average arrival rate for this check-out counter is 1 customer every 4 minutes.

53. Let $f(x) = \frac{\ln x}{1 + (\ln x)^2}$ for x in $(0, \infty)$. Find

(a) $\lim_{x \to 0^+} f(x)$ and $\lim_{x \to \infty} f(x)$;

(b) the maximum and minimum values of $f(x)$;

(c) $F'(\sqrt{e})$ if $F(x) = \int_1^{x^2} f(t)\, dt$.

54. Let R be the region bounded by $x = 0$, $y = e^x$, and the tangent line to $y = e^x$ that goes through the origin. Find

(a) the area of R;

(b) the volume of the solid obtained when R is revolved about the x-axis.

GC *Use a graphing calculator or a CAS to do Problems 55–60.*

55. Evaluate.

(a) $\int_{-3}^{3} \exp(-1/x^2)\, dx$ (b) $\int_0^{8\pi} e^{-0.1x} \sin x\, dx$

56. Evaluate.

(a) $\lim_{x \to 0} (1 + x)^{1/x}$ (b) $\lim_{x \to 0} (1 + x)^{-1/x}$

57. Find the area of the region between the graphs of $y = f(x) = \exp(-x^2)$ and $y = f''(x)$ on $[-3, 3]$.

EXPL **58.** Draw the graphs of $y = x^p e^{-x}$ for various positive values of p using the same axes. Make conjectures about

(a) $\lim_{x\to\infty} x^p e^{-x}$,

(b) the x-coordinate of the maximum point for $f(x) = x^p e^{-x}$.

59. Describe the behavior of $\ln(x^2 + e^{-x})$ for large negative x. For large positive x.

60. Draw the graphs of f and f', where $f(x) = 1/(1 + e^{1/x})$. Then determine each of the following:

(a) $\lim_{x\to 0^+} f(x)$ (b) $\lim_{x\to 0^-} f(x)$

(c) $\lim_{x\to \pm\infty} f(x)$ (d) $\lim_{x\to 0} f'(x)$

(e) The maximum and minimum values of f (if they exist).

Answers to Concepts Review: **1.** increasing; exp **2.** $\ln e = 1$; 2.72 **3.** x; x **4.** e^x; $e^x + C$

6.4 General Exponential and Logarithmic Functions

We defined $e^{\sqrt{2}}$, e^{π}, and all other irrational powers of e in the previous section. But what about $2^{\sqrt{2}}$, π^{π}, π^{e}, and similar irrational powers of other numbers? In fact, we want to give meaning to a^x for $a > 0$ and x any real number. Now, if $r = p/q$ is a rational number, then $a^r = \left(\sqrt[q]{a}\right)^p$. But we also know that

$$a^r = \exp(\ln a^r) = \exp(r \ln a) = e^{r \ln a}$$

This suggests the definition of the **exponential function to the base *a*.**

Definition

For $a > 0$ and any real number x,

$$a^x = e^{x \ln a}$$

Of course, this definition will be appropriate only if the usual properties of exponents are valid for it, a matter we take up shortly. To shore up our confidence in the definition, we use it to calculate 3^2 (with a little help from our calculator):

$$3^2 = e^{2 \ln 3} \approx e^{2(1.0986123)} \approx 9.000000$$

Your calculator may give a result that differs slightly from 9. Calculators use approximations for e^x and $\ln x$, and they round to a fixed number of decimal places (usually about 8).

Now we can fill a small gap in the properties of the natural logarithm left over from Section 6.1.

$$\ln(a^x) = \ln(e^{x \ln a}) = x \ln a$$

Thus, Property (iv) of Theorem 6.1A holds for all real x, not just rational x as claimed there. We will need this fact in the proof of Theorem A below.

What is 2^{π}?

In algebra, 2^n is first defined for positive integers n. Thus, $2^1 = 2$ and $2^4 = 2\cdot 2\cdot 2\cdot 2$. Next, we define 2^n for zero,

$$2^0 = 1$$

and for negative integers:

$$2^{-n} = 1/2^n \quad \text{if} \quad n > 0$$

This means that $2^{-3} = 1/2^3 = 1/8$. Finally, we used root functions to define 2^r for rational numbers r. Thus,

$$2^{7/3} = \sqrt[3]{2^7}$$

Calculus is required to extend the definition of 2^x to the set of real numbers. One way to define 2^{π} would be to say that it is the limit of the sequence

$$2^3, 2^{3.1}, 2^{3.14}, 2^{3.141}, \ldots$$

The definition we use is

$$2^{\pi} = e^{\pi \ln 2}$$

This definition involves calculus, because our definition of the natural log function involved the definite integral.

Properties of a^x Theorem A summarizes the familiar properties of exponents, which can all be proved now in a completely rigorous manner. Theorem B shows us how to differentiate and integrate a^x.

Theorem A Properties of Exponents

If $a > 0$, $b > 0$, and x and y are real numbers, then

(i) $a^x a^y = a^{x+y}$; (ii) $\dfrac{a^x}{a^y} = a^{x-y}$;

(iii) $(a^x)^y = a^{xy}$; (iv) $(ab)^x = a^x b^x$;

(v) $\left(\dfrac{a}{b}\right)^x = \dfrac{a^x}{b^x}$.

Proof We will prove (ii) and (iii), leaving the others for you.

(ii) $\frac{a^x}{a^y} = e^{\ln(a^x/a^y)} = e^{\ln a^x - \ln a^y}$

$= e^{x \ln a - y \ln a} = e^{(x-y)\ln a} = a^{x-y}$

(iii) $(a^x)^y = e^{y \ln a^x} = e^{yx \ln a} = a^{yx} = a^{xy}$ ■

Theorem B **Exponential Function Rules**

$$D_x a^x = a^x \ln a$$

$$\int a^x\,dx = \left(\frac{1}{\ln a}\right)a^x + C, \qquad a \neq 1$$

Proof

$$D_x a^x = D_x(e^{x \ln a}) = e^{x \ln a} D_x(x \ln a)$$

$$= a^x \ln a$$

The integral formula follows immediately from the derivative formula. ■

EXAMPLE 1 Find $D_x\left(3^{\sqrt{x}}\right)$.

SOLUTION We use the Chain Rule with $u = \sqrt{x}$.

$$D_x\left(3^{\sqrt{x}}\right) = 3^{\sqrt{x}} \ln 3 \cdot D_x\sqrt{x} = \frac{3^{\sqrt{x}} \ln 3}{2\sqrt{x}}$$ ■

EXAMPLE 2 Find dy/dx if $y = (x^4 + 2)^5 + 5^{x^4+2}$.

SOLUTION

$$\frac{dy}{dx} = 5(x^4 + 2)^4 \cdot 4x^3 + 5^{x^4+2} \ln 5 \cdot 4x^3$$

$$= 4x^3[5(x^4 + 2)^4 + 5^{x^4+2} \ln 5]$$

$$= 20x^3[(x^4 + 2)^4 + 5^{x^4+1} \ln 5]$$ ■

EXAMPLE 3 Find $\int 2^{x^3} x^2\,dx$.

SOLUTION Let $u = x^3$, so $du = 3x^2\,dx$. Then

$$\int 2^{x^3} x^2\,dx = \frac{1}{3}\int 2^{x^3}(3x^2\,dx) = \frac{1}{3}\int 2^u\,du$$

$$= \frac{1}{3}\frac{2^u}{\ln 2} + C = \frac{2^{x^3}}{3 \ln 2} + C$$ ■

Why Other Bases?

Are bases other than e really needed? No. The formulas

$$a^x = e^{x \ln a}$$

and

$$\log_a x = \frac{\ln x}{\ln a}$$

allow us to turn any problem involving exponential functions or logarithmic functions with base a to corresponding functions with base e. This supports our terminology: *natural* exponential and *natural* logarithmic functions. It also explains the universal use of the latter functions in advanced work.

The Function $\log_a$

Finally, we are ready to make a connection with the logarithms that you studied in algebra. We note that if $0 < a < 1$ then $f(x) = a^x$ is a decreasing function; if $a > 1$, it is an increasing function, as you may check by considering the derivative. In either case, f has an inverse. We call this inverse the **logarithmic function to the base *a*.** This is equivalent to the following definition.

Definition

Let a be a positive number different from 1. Then

$$y = \log_a x \Leftrightarrow x = a^y$$

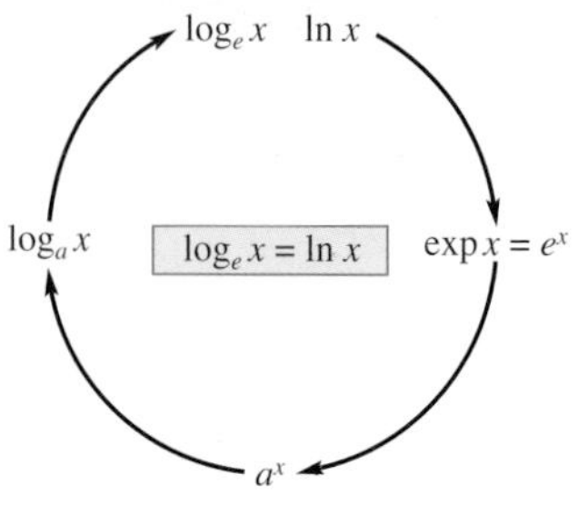

Figure 1

Historically, the most commonly used base was base 10, and the resulting logarithms were called **common logarithms.** But in calculus and all of advanced mathematics, the significant base is e. Notice that $\log_e$, being the inverse of $f(x) = e^x$, is just another symbol for ln; that is,

$$\log_e x = \ln x$$

We have come full circle (Figure 1). The function ln, which we introduced in Section 6.1, has turned out to be an ordinary logarithm, but to a rather special base, e.

Now observe that if $y = \log_a x$, so that $x = a^y$, then

$$\ln x = y \ln a$$

from which we conclude that

$$\boxed{\log_a x = \frac{\ln x}{\ln a}}$$

From this, it follows that $\log_a$ satisfies the usual properties associated with logarithms (see Theorem 6.1A). Also,

$$\boxed{D_x \log_a x = \frac{1}{x \ln a}}$$

EXAMPLE 4 If $y = \log_{10}(x^4 + 13)$, find $\frac{dy}{dx}$.

SOLUTION Let $u = x^4 + 13$ and apply the Chain Rule.

$$\frac{dy}{dx} = \frac{1}{(x^4 + 13) \ln 10} \cdot 4x^3 = \frac{4x^3}{(x^4 + 13) \ln 10}$$ ■

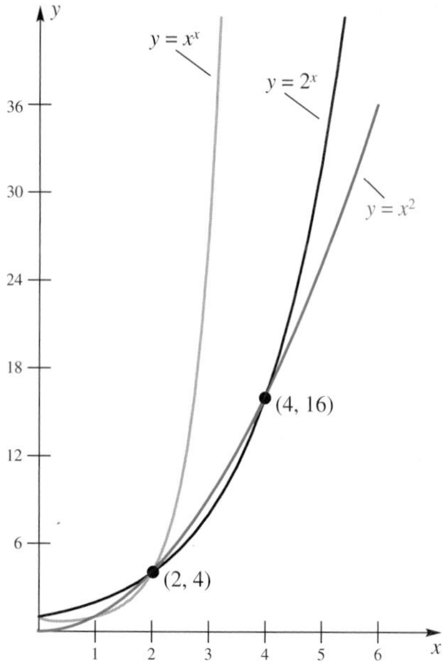

Figure 2

The Functions a^x, x^a, and x^x Begin by comparing the three graphs in Figure 2. More generally, let a be a constant. Do not confuse $f(x) = a^x$, an *exponential function*, with $g(x) = x^a$, a *power function*. And do not confuse their derivatives. We have just learned that

$$\boxed{D_x(a^x) = a^x \ln a}$$

What about $D_x(x^a)$? For a rational, we proved the Power Rule in Chapter 2, which says that

$$D_x(x^a) = ax^{a-1}$$

Now we assert that this is true even if a is irrational. To see this, write

$$D_x(x^a) = D_x(e^{a \ln x}) = e^{a \ln x} \cdot \frac{a}{x}$$

$$= x^a \cdot \frac{a}{x} = ax^{a-1}$$

The corresponding rule for integrals also holds even if a is irrational.

$$\boxed{\int x^a \, dx = \frac{x^{a+1}}{a + 1} + C, \qquad a \neq -1}$$

Finally, we consider $f(x) = x^x$, a variable to a variable power. There is a formula for $D_x(x^x)$, but we do not recommend that you memorize it. Rather, we suggest that you learn two methods for finding it, as illustrated below.

EXAMPLE 5 If $y = x^x$, $x > 0$, find $D_x y$ by two different methods.

SOLUTION

Method 1 We may write

$$y = x^x = e^{x \ln x}$$

Thus, using the Chain Rule and the Product Rule,

$$D_x y = e^{x \ln x} D_x(x \ln x) = x^x\left(x \cdot \frac{1}{x} + \ln x\right) = x^x(1 + \ln x)$$

Method 2 Recall the *logarithmic differentiation* technique from Section 6.1.

$$y = x^x$$

$$\ln y = x \ln x$$

$$\frac{1}{y} D_x y = x \cdot \frac{1}{x} + \ln x$$

$$D_x y = y(1 + \ln x) = x^x(1 + \ln x)$$

■

EXAMPLE 6 If $y = (x^2 + 1)^\pi + \pi^{\sin x}$, find dy/dx.

SOLUTION

$$\frac{dy}{dx} = \pi(x^2 + 1)^{\pi - 1}(2x) + \pi^{\sin x} \ln \pi \cdot \cos x$$

■

From a^x to $[f(x)]^{g(x)}$

Note the increasing complexity of the functions that we have considered. The progression a^x to x^a to x^x is one chain. A more complex chain is $a^{f(x)}$ to $[f(x)]^a$ to $[f(x)]^{g(x)}$. We now know how to find the derivatives of all these functions. Finding the derivative of the last of these is best accomplished by logarithmic differentiation, a technique introduced in Section 6.1 and illustrated in Examples 5 and 7.

EXAMPLE 7 If $y = (x^2 + 1)^{\sin x}$, find $\frac{dy}{dx}$.

SOLUTION We use logarithmic differentiation.

$$\ln y = (\sin x) \ln(x^2 + 1)$$

$$\frac{1}{y}\frac{dy}{dx} = (\sin x)\frac{2x}{x^2 + 1} + (\cos x) \ln(x^2 + 1)$$

$$\frac{dy}{dx} = (x^2 + 1)^{\sin x}\left[\frac{2x \sin x}{x^2 + 1} + (\cos x) \ln(x^2 + 1)\right]$$

■

EXAMPLE 8 Evaluate $\int_{1/2}^{1} \frac{5^{1/x}}{x^2}\,dx$.

SOLUTION Let $u = 1/x$, so $du = (-1/x^2)\,dx$. Then

$$\int \frac{5^{1/x}}{x^2}\,dx = -\int 5^{1/x}\left(-\frac{1}{x^2}\,dx\right) = -\int 5^u\,du$$

$$= -\frac{5^u}{\ln 5} + C = -\frac{5^{1/x}}{\ln 5} + C$$

Thus, by the Second Fundamental Theorem of Calculus,

$$\int_{1/2}^{1} \frac{5^{1/x}}{x^2}\,dx = \left[-\frac{5^{1/x}}{\ln 5}\right]_{1/2}^{1} = \frac{1}{\ln 5}(5^2 - 5)$$

$$= \frac{20}{\ln 5} \approx 12.43$$

■

Concepts Review

1. In terms of e and ln, $\pi^{\sqrt{3}} =$ ____. More generally $a^x =$ ____.

2. $\ln x = \log_a x$, where $a =$ ____.

3. $\log_a x$ can be expressed in terms of ln by $\log_a x =$ ____.

4. The derivative of the power function $f(x) = x^a$ is $f'(x) =$ ____; the derivative of the exponential function $g(x) = a^x$ is $g'(x) =$ ____.

Problem Set 6.4

In Problems 1–8, solve for x. Hint: $\log_a b = c \Leftrightarrow a^c = b$.

1. $\log_2 8 = x$

2. $\log_5 x = 2$

3. $\log_4 x = \frac{3}{2}$

4. $\log_x 64 = 4$

5. $2 \log_9\left(\frac{x}{3}\right) = 1$

6. $\log_4\left(\frac{1}{2x}\right) = 3$

7. $\log_2(x + 3) - \log_2 x = 2$

8. $\log_5(x + 3) - \log_5 x = 1$

C *Use* $\log_a x = (\ln x)/(\ln a)$ *to calculate each of the logarithms in Problems 9–12.*

9. $\log_5 12$

10. $\log_7(0.11)$

11. $\log_{11}(8.12)^{1/5}$

12. $\log_{10}(8.57)^7$

C *In Problems 13–16, use natural logarithms to solve each of the exponential equations. Hint: To solve* $3^x = 11$, *take ln of both sides, obtaining* $x \ln 3 = \ln 11$; *then* $x = (\ln 11)/(\ln 3) \approx 2.1827$.

13. $2^x = 17$

14. $5^x = 13$

15. $5^{2s-3} = 4$

16. $12^{1/(\theta-1)} = 4$

In Problems 17–26, find the indicated derivative or integral.

17. $D_x(6^{2x})$

18. $D_x(3^{2x^2-3x})$

19. $D_x \log_3 e^x$

20. $D_x \log_{10}(x^3 + 9)$

21. $D_z[3^z \ln(z + 5)]$

22. $D_\theta \sqrt{\log_{10}(3^{\theta^2-\theta})}$

23. $\int x2^{x^2}\,dx$

24. $\int 10^{5x-1}\,dx$

25. $\int_1^4 \frac{5^{\sqrt{x}}}{\sqrt{x}}\,dx$

26. $\int_0^1 (10^{3x} + 10^{-3x})\,dx$

In Problems 27–32, find dy/dx. Note: You must distinguish among problems of the type a^x, x^a, *and* x^x *as in Examples 5–7.*

27. $y = 10^{(x^2)} + (x^2)^{10}$

28. $y = \sin^2 x + 2^{\sin x}$

29. $y = x^{\pi+1} + (\pi + 1)^x$

30. $y = 2^{(e^x)} + (2^e)^x$

31. $y = (x^2 + 1)^{\ln x}$

32. $y = (\ln x^2)^{2x+3}$

33. If $f(x) = x^{\sin x}$, find $f'(1)$.

C **34.** Let $f(x) = \pi^x$ and $g(x) = x^\pi$. Which is larger, $f(e)$ or $g(e)$? $f'(e)$ or $g'(e)$?

In Problems 35–40, first find the domain of the given function f and then find where it is increasing and decreasing, and also where it is concave upward and downward. Identify all extreme values and points of inflection. Then sketch the graph of $y = f(x)$.

35. $f(x) = 2^{-x}$

36. $f(x) = x2^{-x}$

37. $f(x) = \log_2(x^2 + 1)$

38. $f(x) = x \log_3(x^2 + 1)$

39. $f(x) = \int_1^x 2^{-t^2}\,dt$

40. $f(x) = \int_0^x \log_{10}(t^2 + 1)\,dt$

41. How are $\log_{1/2} x$ and $\log_2 x$ related?

42. Sketch the graphs of $\log_{1/3} x$ and $\log_3 x$ using the same coordinate axes.

C **43.** The magnitude M of an earthquake on the **Richter scale** is

$$M = 0.67 \log_{10}(0.37E) + 1.46$$

where E is the energy of the earthquake in kilowatt-hours. Find the energy of an earthquake of magnitude 7. Of magnitude 8.

C **44.** The loudness of sound is measured in decibels in honor of Alexander Graham Bell (1847–1922), inventor of the telephone. If the variation in pressure is P pounds per square inch, then the loudness L in decibels is

$$L = 20 \log_{10}(121.3P)$$

Find the variation in pressure caused by music at 115 decibels.

C **45.** In the equally tempered scale to which keyed instruments have been tuned since the days of J.S. Bach (1685–1750), the frequencies of successive notes C, C#, D, D#, E, F, F#, G, G#, A, A#, B, $\overline{\text{C}}$ form a geometric sequence (progression), with $\overline{\text{C}}$ having twice the frequency of C (C# is read C sharp). What is the ratio r between the frequencies of successive notes? If the frequency of A is 440, find the frequency of $\overline{\text{C}}$.

46. Prove that $\log_2 3$ is irrational. *Hint:* Use proof by contradiction.

GC **47.** You suspect that the xy-data that you collect lie on either an exponential curve $y = Ab^x$ or a power curve $y = Cx^d$. To check, you plot $\ln y$ against x on one graph and $\ln y$ against $\ln x$ on another graph. (Graphing calculators and CASs have options to make the vertical axis, or both the vertical and horizontal axes, a logarithmic scale.) Explain how these graphs will help you to come to a conclusion.

48. (An Amusement) Given the problem of finding y' if $y = x^x$, student A did the following:

Wrong 1

$$y = x^x$$
$$y' = x \cdot x^{x-1} \cdot 1 \qquad \text{(misapplying the Power Rule)}$$
$$= x^x$$

Student B did this:

Wrong 2

$$y = x^x$$
$$y' = x^x \cdot \ln x \cdot 1 \qquad \text{(misapplying the Exponential Function Rule)}$$
$$= x^x \ln x$$

The sum $x^x + x^x \ln x$ is correct (Example 5), so

WRONG 1 + WRONG 2 = RIGHT

Show that the same procedure yields a correct answer for finding the derivative of $y = f(x)^{g(x)}$.

49. Convince yourself that $f(x) = (x^x)^x$ and $g(x) = x^{(x^x)}$ are not the same function. Then find $f'(x)$ and $g'(x)$. *Note:* When mathematicians write x^{x^x}, they mean $x^{(x^x)}$.

50. Consider $f(x) = \dfrac{a^x - 1}{a^x + 1}$ for fixed a, $a > 0$, $a \neq 1$. Show that f has an inverse and find a formula for $f^{-1}(x)$.

51. For a fixed $a > 1$, let $f(x) = x^a/a^x$ on $[0, \infty)$. Show:

(a) $\lim_{x\to\infty} f(x) = 0$ *Hint:* Study $\ln f(x)$;

(b) $f(x)$ is maximized at $x_0 = a/\ln a$;

(c) $x^a = a^x$ has two positive solutions if $a \neq e$ and only one such solution if $a = e$;

(d) $\pi^e < e^\pi$.

52. Let $f_u(x) = x^u e^{-x}$ for $x \geq 0$. Show that for any fixed $u > 0$:

(a) $f_u(x)$ attains its maximum at $x_0 = u$;

(b) $f_u(u) > f_u(u+1)$ and $f_{u+1}(u+1) > f_{u+1}(u)$ imply

$$\left(\frac{u+1}{u}\right)^u < e < \left(\frac{u+1}{u}\right)^{u+1}$$

(c) $\dfrac{u}{u+1}e < \left(\dfrac{u+1}{u}\right)^u < e$.

Conclude from part (c) that $\lim_{u\to\infty}\left(1 + \dfrac{1}{u}\right)^u = e$.

GC **53.** Find $\lim_{x\to 0^+} x^x$. Also find the coordinates of the minimum point for $f(x) = x^x$ on $[0, 4]$.

GC **54.** Draw the graphs of $y = x^3$ and $y = 3^x$ using the same axes and find all their intersection points.

CAS **55.** Evaluate $\displaystyle\int_0^{4\pi} x^{\sin x}\,dx$.

CAS **56.** Refer to Problem 49. Draw the graphs of f and g using the same axes. Then draw the graphs of f' and g' using the same axes.

Our graphing experience so far has been restricted to using standard (linear) coordinate spacings. When working with exponential and logarithmic functions it may be more instructive to use logarithmic and log–log scales. We explore these techniques in Problems 57 and 58.

GC **57.** On a single set of axes, use your calculator to draw the graphs of $y = 2^x$, $y = 3^x$, and $y = 4^x$ over the interval $0 < x < 4$. Do the same for the inverse functions $y = \log_2 x$, $y = \log_3 x$, and $y = \log_4 x$. If we use a computer graphing program that permits the use of semilog axes (a logarithmic scale on the y-axis and a normal scale on the x-axis) to graph the functions $y = 2^x$, $y = 3^x$, and $y = 4^x$ over the region $-5 < x < 5$ (Figure 3), we get three lines.

(a) Identify each of the lines in Figure 3.

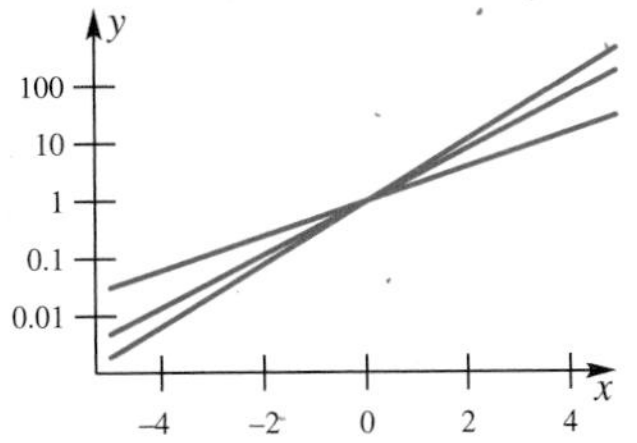

Figure 3

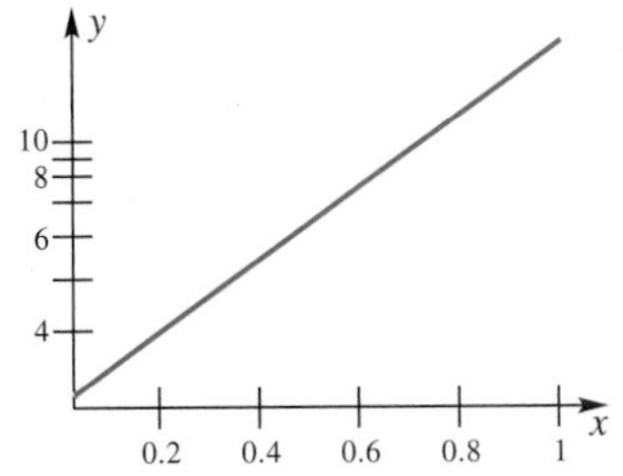

Figure 4

(b) Noting that, if $y = Cb^x$ then $\ln y = \ln C + x \ln b$, explain why all the curves in Figure 3 are lines through the point $(0, 1)$.

(c) Based on the semilog plot given by Figure 4, determine the C and b in the equation $y = Cb^x$.

58. If we use log scaling for the x-axis as well as the y-axis (called a log–log plot) and graph several power functions, we will also get lines. Using the result that, upon taking logs, $y = Cx^r$ becomes $\log y = \log C + r \log x$, identify the equations that are graphed in Figure 5.

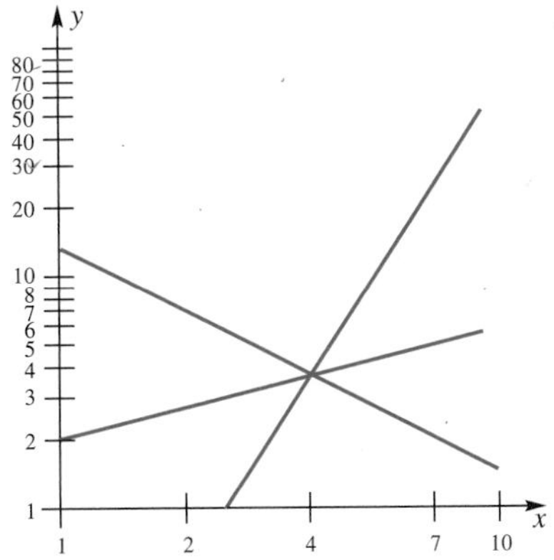

Figure 5

Answers to Concepts Review: **1.** $e^{\sqrt{3}\ln\pi}$; $e^{x\ln a}$ **2.** e **3.** $(\ln x)/(\ln a)$ **4.** ax^{a-1}; $a^x \ln a$

6.5 Exponential Growth and Decay

At the beginning of 2004, the world's population was about 6.4 billion. It is said that by the year 2020 it will reach 7.9 billion. How are such predictions made?

To treat the problem mathematically, let $y = f(t)$ denote the size of the population at time t, where t is the number of years after 2004. Actually, $f(t)$ is an integer, and its graph "jumps" when someone is born or someone dies. However, for a large population, these jumps are so small relative to the total population that we will not go far wrong if we pretend that f is a nice differentiable function.

It seems reasonable to suppose that the increase Δy in population (births minus deaths) during a short time period Δt is proportional to the size of the population at the beginning of the period, and to the length of that period. Thus, $\Delta y = ky\,\Delta t$, or

$$\frac{\Delta y}{\Delta t} = ky$$

In its limiting form, this gives the differential equation

$$\boxed{\frac{dy}{dt} = ky}$$

If $k > 0$, the population is growing; if $k < 0$, it is shrinking. For world population, history indicates that k is about 0.0132 (assuming that t is measured in years), though some agencies report a different figure.

Solving the Differential Equation We began our study of differential equations in Section 3.9, and you might refer to that section now. We want to solve $dy/dt = ky$ subject to the condition that $y = y_0$ when $t = 0$. Separating variables and integrating, we obtain

$$\frac{dy}{y} = k\,dt$$

$$\int \frac{dy}{y} = \int k\,dt$$

$$\ln y = kt + C$$

The condition $y = y_0$ at $t = 0$ gives $C = \ln y_0$. Thus,

$$\ln y - \ln y_0 = kt$$

or

$$\ln \frac{y}{y_0} = kt$$

Changing to exponential form yields

$$\frac{y}{y_0} = e^{kt}$$

or, finally,

$$\boxed{y = y_0\,e^{kt}}$$

When $k > 0$, this type of growth is called **exponential growth,** and when $k < 0$, it is called **exponential decay.**

Returning to the problem of world population, we choose to measure time t in years after January 1, 2004, and y in billions of people. Thus, $y_0 = 6.4$ and, since $k = 0.0132$,

$$y = 6.4e^{0.0132t}$$

By the year 2020, when $t = 16$, we can predict that y will be about

$$y = 6.4e^{0.0132(16)} \approx 7.9 \text{ billion}$$

EXAMPLE 1 How long will it take world population to double under the assumptions above?

SOLUTION The question is equivalent to asking "In how many years after 2004 will the population reach 12.8 billion?" We need to solve

$$12.8 = 6.4e^{0.0132t}$$
$$2 = e^{0.0132t}$$

for t. Taking logarithms of both sides gives

$$\ln 2 = 0.0132t$$
$$t = \frac{\ln 2}{0.0132} \approx 53 \text{ years}$$ ■

If world population will double in the first 53 years after 2004, it will double in any 53-year period; so, for example, it will quadruple in 106 years. More generally, if an exponentially growing quantity doubles from y_0 to $2y_0$ in an initial interval of length T, it will double in *any* interval of length T, since

$$\frac{y(t+T)}{y(t)} = \frac{y_0 e^{k(t+T)}}{y_0 e^{kt}} = \frac{y_0 e^{kT}}{y_0} = \frac{2y_0}{y_0} = 2$$

We call the number T the **doubling time.**

EXAMPLE 2 The number of bacteria in a rapidly growing culture was estimated to be 10,000 at noon and 40,000 after 2 hours. Predict how many bacteria there will be at 5 P.M.

SOLUTION We assume that the differential equation $dy/dt = ky$ is applicable, so $y = y_0e^{kt}$. Now we have two conditions ($y_0 = 10{,}000$ and $y = 40{,}000$ at $t = 2$), from which we conclude that

$$40{,}000 = 10{,}000e^{k(2)}$$

or

$$4 = e^{2k}$$

Taking logarithms yields

$$\ln 4 = 2k$$

or

$$k = \tfrac{1}{2}\ln 4 = \ln \sqrt{4} = \ln 2$$

Thus,

$$y = 10{,}000e^{(\ln 2)t}$$

and, at $t = 5$, this gives

$$y = 10{,}000e^{0.693(5)} \approx 320{,}000$$ ■

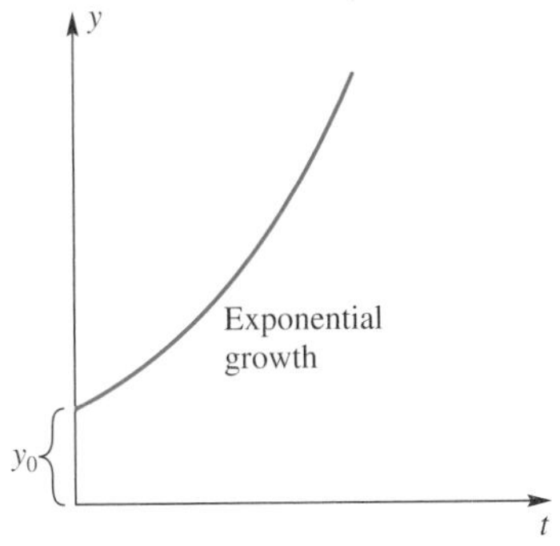

Figure 1

The exponential model $y = y_0e^{kt}$, $k > 0$, for population growth is flawed since it projects faster and faster growth indefinitely far into the future (Figure 1). In most cases (including that of world population), the limited amount of space and resources will eventually force a slowing of the growth rate. This suggests another model for population growth, called the **logistic model,** in which we assume that the rate of growth is proportional both to the population size y and to the difference $L - y$, where L is the maximum population that can be supported. This leads to the differential equation

$$\frac{dy}{dt} = ky(L - y)$$

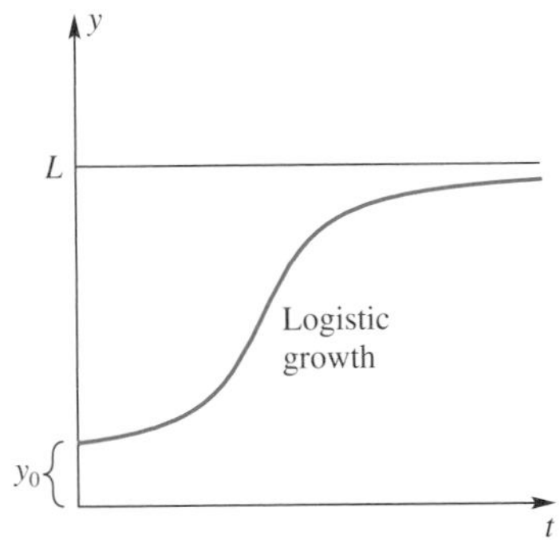

Figure 2

Note that for small y, $dy/dt \approx kLy$, which suggests exponential-type growth. But as y nears L, growth is curtailed and dy/dt gets smaller and smaller, producing a growth curve like Figure 2. This model is explored in Problems 34, 35, and 49 of this section and again in Section 7.5.

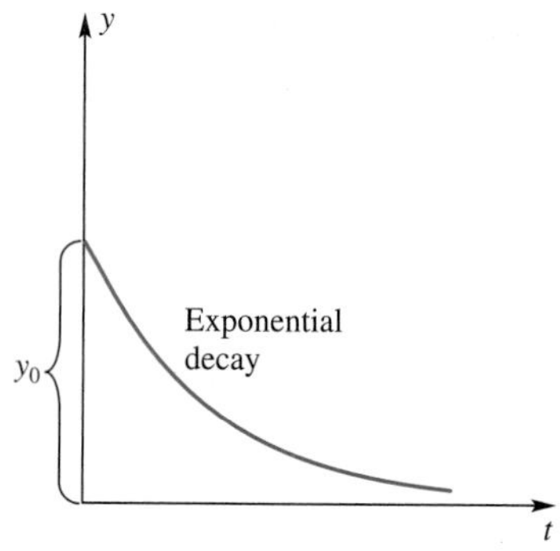

Figure 3

Radioactive Decay Not everything grows; some things decrease over time. For example, radioactive elements *decay*, and they do it at a rate proportional to the amount present. Thus, their change rates also satisfy the differential equation

$$\frac{dy}{dt} = ky$$

but now with k negative. It is still true that $y = y_0e^{kt}$ is the solution to this equation. A typical graph appears in Figure 3.

EXAMPLE 3 Carbon 14, an isotope of carbon, is radioactive and decays at a rate proportional to the amount present. Its **half-life** is 5730 years; that is, it takes 5730 years for a given amount of carbon 14 to decay to one-half its original size. If there were 10 grams present originally, how much would be left after 2000 years?

SOLUTION The half-life of 5730 allows us to determine k, since it implies that

$$\frac{1}{2} = 1e^{k(5730)}$$

or, after taking logarithms,

$$-\ln 2 = 5730k$$

$$k = \frac{-\ln 2}{5730} \approx -0.000121$$

Thus,

$$y = 10e^{-0.000121t}$$

At $t = 2000$, this gives

$$y = 10e^{-0.000121(2000)} \approx 7.85 \text{ grams}$$ ■

In Problem 17, we show how Example 3 may be used to determine the age of fossils and other once-living things.

Newton's Law of Cooling Newton's Law of Cooling states that the rate at which an object cools (or warms) is proportional to the difference in temperature between the object and the surrounding medium. To be specific, suppose that an object initially at temperature T_0 is placed in a room where the temperature is T_1. If $T(t)$ represents the temperature of the object at time t, then Newton's Law of Cooling says that

$$\frac{dT}{dt} = k(T - T_1)$$

This differential equation is separable and can be solved like the growth and decay problems in this section.

EXAMPLE 4 An object is taken from an oven at 350°F and left to cool in a room at 70°F. If the temperature fell to 250°F in one hour, what would its temperature be three hours after it was removed from the oven?

SOLUTION The differential equation can be written as

$$\frac{dT}{dt} = k(T - 70)$$

$$\frac{dT}{T - 70} = k\,dt$$

$$\int \frac{dT}{T - 70} = \int k\,dt$$

$$\ln|T - 70| = kt + C$$

Since the initial temperature is greater than 70, it seems reasonable that the object's temperature will decrease toward 70; thus $T - 70$ will be positive and the absolute value is unnecessary. This leads to

$$T - 70 = e^{kt+C}$$
$$T = 70 + C_1 e^{kt}$$

where $C_1 = e^C$. Now we apply the initial condition, $T(0) = 350$ to find C_1:

$$350 = T(0) = 70 + C_1 e^{k \cdot 0}$$
$$280 = C_1$$

Thus, the solution of the differential equation is

$$T(t) = 70 + 280e^{kt}$$

To find k we apply the condition that at time $t = 1$ the temperature was $T(1) = 250$.

$$250 = T(1) = 70 + 280e^{k \cdot 1}$$
$$280e^k = 180$$
$$e^k = \frac{180}{280}$$
$$k = \ln\frac{180}{280} \approx -0.44183$$

This gives

$$T(t) = 70 + 280e^{-0.44183t}$$

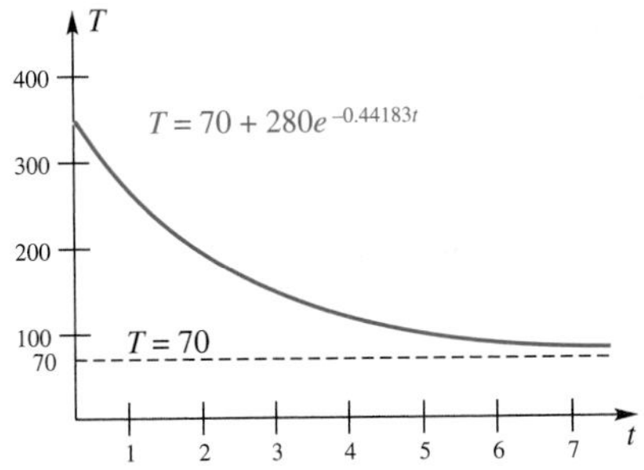

Figure 4

See Figure 4. After 3 hours, the temperature is

$$T(3) = 70 + 280e^{-0.44183 \cdot 3} \approx 144.4°\text{F}$$ ■

Compound Interest If we put \$100 in the bank at 12% interest compounded monthly, it will be worth \$100(1.01) at the end of 1 month, $\$100(1.01)^2$ at the end of 2 months, and $\$100(1.01)^{12}$ at the end of 12 months, or 1 year. More generally, if we put A_0 dollars in the bank at $100r$ percent compounded n times per year, it will be worth $A(t)$ dollars at the end of t years, where

$$A(t) = A_0\left(1 + \frac{r}{n}\right)^{nt}$$

EXAMPLE 5 Suppose that Catherine put \$500 in the bank at 4% interest, compounded daily. How much will it be worth at the end of 3 years?

SOLUTION Here $r = 0.04$ and $n = 365$, so

$$A = 500\left(1 + \frac{0.04}{365}\right)^{365(3)} \approx \$563.74$$ ■

Now let us consider what happens when interest is **compounded continuously,** that is, when n, the number of compounding periods in a year, tends to infinity. Then we claim that

$$A(t) = \lim_{n\to\infty} A_0\left(1 + \frac{r}{n}\right)^{nt} = A_0 \lim_{n\to\infty}\left[\left(1 + \frac{r}{n}\right)^{n/r}\right]^{rt}$$
$$= A_0\left[\lim_{h\to 0}(1 + h)^{1/h}\right]^{rt} = A_0 e^{rt}$$

Here we replaced r/n by h and noted that $n \to \infty$ corresponds to $h \to 0$. But the big step is knowing that the expression in brackets is the number e. This result is important enough to be called a theorem.

Theorem A

$$\lim_{h \to 0}(1 + h)^{1/h} = e$$

Another View of Continuity

Recall that to say a function is continuous at x_0 means that

$$\lim_{x \to x_0} f(x) = f(x_0)$$

That is,

$$\lim_{x \to x_0} f(x) = f\left(\lim_{x \to x_0} x\right)$$

Thus, continuity for a function means that we can pass a limit inside the function. This is what we did for the function $f(x) = \exp(x)$ near the end of the proof of Theorem A.

Proof First recall that if $f(x) = \ln x$, then $f'(x) = 1/x$ and, in particular, $f'(1) = 1$. Then, from the definition of the derivative and properties of ln, we get

$$1 = f'(1) = \lim_{h \to 0} \frac{f(1 + h) - f(1)}{h} = \lim_{h \to 0} \frac{\ln(1 + h) - \ln 1}{h}$$

$$= \lim_{h \to 0} \frac{1}{h} \ln(1 + h) = \lim_{h \to 0} \ln(1 + h)^{1/h}$$

Thus, $\lim_{h \to 0} \ln(1 + h)^{1/h} = 1$, a result we will use in a moment. Now, $g(x) = e^x = \exp x$ is a continuous function, and it therefore follows that we can pass the limit inside the exponential function in the following argument:

$$\lim_{h \to 0}(1 + h)^{1/h} = \lim_{h \to 0} \exp[\ln(1 + h)^{1/h}] = \exp\left[\lim_{h \to 0} \ln(1 + h)^{1/h}\right]$$

$$= \exp 1 = e$$ ■

For another proof of Theorem A, see Problem 52 of Section 6.4.

EXAMPLE 6 Suppose that the bank of Example 5 compounded interest continuously. How much would Catherine then have at the end of 3 years?

SOLUTION

$$A(t) = A_0 e^{rt} = 500e^{(0.04)(3)} \approx \$563.75$$

Note that, though some banks try to get advertising mileage out of offering continuous compounding of interest, the difference in yields between continuous and daily compounding (which many banks offer) is miniscule. ■

Here is another approach to the problem of continuous compounding of interest. Let A be the value at time t of A_0 dollars invested at the interest rate r. To say that interest is compounded continuously is to say that the instantaneous rate of change of A with respect to time is rA; that is,

$$\frac{dA}{dt} = rA$$

This differential equation was solved at the beginning of the section; its solution is $A = A_0 e^{rt}$.

Concepts Review

1. The rate of change dy/dt of a quantity y growing exponentially satisfies the differential equation $dy/dt =$ _____. In contrast, if y is growing logistically toward an upper bound L, $dy/dt =$ _____.

2. If a quantity growing exponentially doubles after T years, it will be _____ times as large after $3T$ years.

3. The time for an exponentially decaying quantity y to go from size y_0 to size $y_0/2$ is called the _____.

4. The number e can be expressed as a limit by $e = \lim_{h \to 0}$ _____.

Problem Set 6.5

In Problems 1–4, solve the given differential equation subject to the given condition. Note that $y(a)$ denotes the value of y at $t = a$.

1. $\frac{dy}{dt} = -6y,\ y(0) = 4$
2. $\frac{dy}{dt} = 6y,\ y(0) = 1$
3. $\frac{dy}{dt} = 0.005y,\ y(10) = 2$
4. $\frac{dy}{dt} = -0.003y,\ y(-2) = 3$

5. A bacterial population grows at a rate proportional to its size. Initially, it is 10,000, and after 10 days it is 20,000. What is the population after 25 days? See Example 2.

6. How long will it take the population of Problem 5 to double? See Example 1.

7. How long will it take the population of Problem 5 to triple? See Example 1.

8. The population of the United States was 3.9 million in 1790 and 178 million in 1960. If the rate of growth is assumed proportional to the number present, what estimate would you give for the population in 2000? (Compare your answer with the actual 2000 population, which was 275 million.)

9. The population of a certain country is growing at 3.2% per year; that is, if it is A at the beginning of a year, it is $1.032A$ at the end of that year. Assuming that it is 4.5 million now, what will it be at the end of 1 year? 2 years? 10 years? 100 years?

10. Determine the proportionality constant k in $dy/dt = ky$ for Problem 9. Then use $y = 4.5e^{kt}$ to find the population after 100 years.

11. A population is growing at a rate proportional to its size. After 5 years, the population size was 164,000. After 12 years, the population size was 235,000. What was the original population size?

12. The mass of a tumor grows at a rate proportional to its size. The first measurement of its size was 4.0 grams. Four months later its mass was 6.76 grams. How large was the tumor six months before the first measurement? If the instrument can detect tumors of mass 1 gram or greater, would the tumor have been detected at that time?

13. A radioactive substance has a half-life of 700 years. If there were 10 grams initially, how much would be left after 300 years?

14. If a radioactive substance loses 15% of its radioactivity in 2 days, what is its half-life?

15. Cesium 137 and strontium 90 are two radioactive chemicals that were released at the Chernobyl nuclear reactor in April 1986. The half-life of cesium 137 is 30.22 years, and that of strontium 90 is 28.8 years. In what year will the amount of cesium 137 be equal to 1% of what was released? Answer this question for strontium 90.

16. An unknown amount of a radioactive substance is being studied. After two days, the mass is 15.231 grams. After eight days, the mass is 9.086 grams. How much was there initially? What is the half-life of this substance?

17. **(Carbon Dating)** All living things contain carbon 12, which is stable, and carbon 14, which is radioactive. While a plant or animal is alive, the ratio of these two isotopes of carbon remains unchanged, since the carbon 14 is constantly renewed; after death, no more carbon 14 is absorbed. The half-life of carbon 14 is 5730 years. If charred logs of an old fort show only 70% of the carbon 14 expected in living matter, when did the fort burn down? Assume that the fort burned soon after it was built of freshly cut logs.

18. Human hair from a grave in Africa proved to have only 51% of the carbon 14 of living tissue. When was the body buried?

19. An object is taken from an oven at 300°F and left to cool in a room at 75°F. If the temperature fell to 200°F in $\frac{1}{2}$ hour, what will it be after 3 hours?

20. A thermometer registered −20°C outside and then was brought into a house where the temperature was 24°C. After 5 minutes it registered 0°C. When will it register 20°C?

21. An object initially at 26°C is placed in water having temperature 90°C. If the temperature of the object rises to 70°C in 5 minutes, what will be the temperature after 10 minutes?

22. A batch of brownies is taken from a 350°F oven and placed in a refrigerator at 40°F and left to cool. After 15 minutes, the brownies have cooled to 250°F. When will the temperature of the brownies be 110°F?

23. A dead body is found at 10 P.M. to have temperature 82°F. One hour later the temperature was 76°F. The temperature of the room was a constant 70°F. Assuming that the temperature of the body was 98.6°F when it was alive, estimate the time of death.

24. Solve the differential equation for Newton's Law of Cooling for an arbitrary T_0, T_1, and k, assuming that $T_0 > T_1$. Show that $\lim_{t\to\infty} T(t) = T_1$.

25. If \$375 is put in the bank today, what will it be worth at the end of 2 years if interest is 3.5% and is compounded as specified?

(a) Annually (b) Monthly
(c) Daily (d) Continuously

26. Do Problem 25 assuming that the interest rate is 4.6%.

27. How long does it take money to double in value for the specified interest rate?

(a) 6% compounded monthly
(b) 6% compounded continuously

28. Inflation between 1999 and 2004 ran at about 2.5% per year. On this basis, what would you expect a car that would have cost \$20,000 in 1999 to cost in 2004?

29. Manhattan Island is said to have been bought by Peter Minuit in 1626 for \$24. Suppose that Minuit had instead put the \$24 in the bank at 6% interest compounded continuously. What would that \$24 have been worth in 2000?

30. If Methuselah's parents had put \$100 in the bank for him at birth and he left it there, what would Methuselah have had at his death (969 years later) if interest was 4% compounded annually?

31. Find the value of \$1000 at the end of 1 year when the interest is compounded continuously at 5%. This is called the **future value.**

32. Suppose that after 1 year you have \$1000 in the bank. If the interest was compounded continuously at 5%, how much money did you put in the bank one year ago? This is called the **present value.**

33. It will be shown later for small x that $\ln(1 + x) \approx x$. Use this fact to show that the doubling time for money invested at p percent compounded annually is about $70/p$ years.

34. The equation for logistic growth is

$$\frac{dy}{dt} = ky(L - y)$$

Show that this differential equation has the solution

$$y = \frac{Ly_0}{y_0 + (L - y_0)e^{-Lkt}}$$

Hint: $\frac{1}{y(L-y)} = \frac{1}{Ly} + \frac{1}{L(L-y)}$.

35. Sketch the graph of the solution in Problem 34 when $y_0 = 6.4$, $L = 16$, and $k = 0.00186$ (a *logistic model* for world population; see the discussion at the beginning of this section). Note that $\lim_{t\to\infty} y = 16$.

36. Find each of the following limits.

(a) $\lim_{x\to 0}(1 + x)^{1000}$ (b) $\lim_{x\to 0}(1)^{1/x}$

(c) $\lim_{x\to 0^+}(1 + \varepsilon)^{1/x}, \varepsilon > 0$ (d) $\lim_{x\to 0^-}(1 + \varepsilon)^{1/x}, \varepsilon > 0$

(e) $\lim_{x\to 0}(1 + x)^{1/x}$

37. Use the fact that $e = \lim_{h\to 0}(1 + h)^{1/h}$ to find each limit.

(a) $\lim_{x\to 0}(1 - x)^{1/x}$ *Hint:* $(1 - x)^{1/x} = [(1 - x)^{1/(-x)}]^{-1}$

(b) $\lim_{x\to 0}(1 + 3x)^{1/x}$ (c) $\lim_{n\to\infty}\left(\frac{n+2}{n}\right)^n$

(d) $\lim_{n\to\infty}\left(\frac{n-1}{n}\right)^{2n}$

38. Show that the differential equation

$$\frac{dy}{dt} = ay + b$$

has solution

$$y = \left(y_0 + \frac{b}{a}\right)e^{at} - \frac{b}{a}$$

Assume that $a \neq 0$.

39. Consider a country with a population of 10 million in 1985, a growth rate of 1.2% per year, and immigration from other countries of 60,000 per year. Use the differential equation of Problem 38 to model this situation and predict the population in 2010. Take $a = 0.012$.

40. Important news is said to diffuse through an adult population of fixed size L at a time rate proportional to the number of people who have not heard the news. Five days after a scandal in City Hall was reported, a poll showed that half the people had heard it. How long will it take for 99% of the people to hear it?

EXPL *Besides providing an easy way to differentiate products, logarithmic differentiation also provides a measure of the **relative** or **fractional rate of change**, defined as y'/y. We explore this concept in Problems 41–44.*

41. Show that the relative rate of change of e^{kt} as a function of t is k.

42. Show that the relative rate of change of any polynomial approaches zero as the independent variable approaches infinity.

43. Prove that if the relative rate of change is a positive constant then the function must represent exponential growth.

44. Prove that if the relative rate of change is a negative constant then the function must represent exponential decay.

45. Assume that (1) world population continues to grow exponentially with growth constant $k = 0.0132$, (2) it takes $\frac{1}{2}$ acre of land to supply food for one person, and (3) there are 13,500,000 square miles of arable land in the world. How long will it be before the world reaches the maximum population? *Note:* There were 6.4 billion people in 2004 and 1 square mile is 640 acres.

GC **46.** The Census Bureau estimates that the growth rate k of the world population will decrease by roughly 0.0002 per year for the next few decades. In 2004, k was 0.0132.

(a) Express k as a function of time t, where t is measured in years since 2004.

(b) Find a differential equation that models the population y for this problem.

(c) Solve the differential equation with the additional condition that the population in 2004 ($t = 0$) was 6.4 billion.

(d) Graph the population y for the next 300 years.

(e) With this model, when will the population reach a maximum? When will the population drop below the 2004 level?

GC **47.** Repeat Exercise 46 under the assumption that k will decrease by 0.0001 per year.

EXPL **48.** Let E be a differentiable function satisfying $E(u + v) = E(u)E(v)$ for all u and v. Find a formula for $E(x)$. *Hint:* First find $E'(x)$.

GC **49.** Using the same axes, draw the graphs for $0 \le t \le 100$ of the following two models for the growth of world population (both described in this section).

(a) Exponential growth: $y = 6.4e^{0.0132t}$

(b) Logistic growth: $y = 102.4/(6 + 10e^{-0.030t})$

Compare what the two models predict for world population in 2010, 2040, and 2090. *Note:* Both models assume that world population was 6.4 billion in 2004 ($t = 0$).

GC **50.** Evaluate:

(a) $\lim_{x\to 0}(1 + x)^{1/x}$ (b) $\lim_{x\to 0}(1 - x)^{1/x}$

The limit in part (a) determines e. What special number does the limit in part (b) determine?

Answers to Concepts Review: **1.** ky; $ky(L - y)$ **2.** 8 **3.** half-life **4.** $(1 + h)^{1/h}$

6.6 First-Order Linear Differential Equations

We first solved differential equations in Section 3.9. There we developed the method of separation of variables for finding a solution. In the previous section we used the method of separation of variables to solve differential equations involving growth and decay.

Not all differential equations are *separable*. For example, in the differential equation

$$\frac{dy}{dx} = 2x - 3y$$

there is no way to separate the variables in such a way as to have dy and all expressions involving y on one side, and dx and all expressions involving x on the other side. This equation can, however, be put in the form

$$\frac{dy}{dx} + P(x)y = Q(x)$$

where $P(x)$ and $Q(x)$ are functions of x only. A differential equation of this form is said to be a **first-order linear differential equation.** *First-order* refers to the fact that the only derivative is a first derivative. *Linear* refers to fact that the equation can be written in the form $D_x y + P(x)Iy = Q(x)$, where D_x is the derivative operator, and I is the identity operator (that is $Iy = y$). Both D_x and I are *linear operators.*

The family of all solutions of a differential equation is called the **general solution.** Many problems require that the solution satisfy the condition $y = b$ when $x = a$, where a and b are given. Such a condition is called an **initial condition,** and a function that satisfies the differential equation and the initial condition is called a **particular solution.**

Solving First-Order Linear Equations To solve the first-order linear differential equation, we first multiply both sides by the **integrating factor**

$$e^{\int P(x)\,dx}$$

(The reason for this step will become clear shortly.) The differential equation is then

$$e^{\int P(x)\,dx}\frac{dy}{dx} + e^{\int P(x)\,dx}P(x)y = e^{\int P(x)\,dx}Q(x)$$

The left side is the derivative of the product $y \cdot e^{\int P(x)\,dx}$, so the equation takes the form

$$\frac{d}{dx}\left(y \cdot e^{\int P(x)\,dx}\right) = e^{\int P(x)\,dx}Q(x)$$

Integration of both sides yields

$$ye^{\int P(x)\,dx} = \int \left(Q(x)e^{\int P(x)\,dx}\right) dx$$

The general solution is thus

$$y = e^{-\int P(x)\,dx}\int \left(Q(x)e^{\int P(x)\,dx}\right) dx$$

It is not worth memorizing this final result; the process of getting there is easily recalled and that is what we illustrate.

EXAMPLE 1 Solve

$$\frac{dy}{dx} + \frac{2}{x}y = \frac{\sin 3x}{x^2}$$

SOLUTION Our integrating factor is

$$e^{\int P(x)\,dx} = e^{\int (2/x)\,dx} = e^{2\ln|x|} = e^{\ln x^2} = x^2$$

(We have taken the arbitrary constant from the integration $\int P(x)\,dx$ to be 0. The choice for the constant does not affect the answer. See Problems 27 and 28.) Multiplying both sides of the original equation by x^2, we obtain

$$x^2\frac{dy}{dx} + 2xy = \sin 3x$$

The left side of this equation is the derivative of the product x^2y. Thus,

$$\frac{d}{dx}(x^2y) = \sin 3x$$

Integration of both sides yields

$$x^2y = \int \sin 3x\,dx = -\tfrac{1}{3}\cos 3x + C$$

or

$$y = \left(-\tfrac{1}{3}\cos 3x + C\right)x^{-2}$$ ■

EXAMPLE 2 Find the particular solution of

$$\frac{dy}{dx} - 3y = xe^{3x}$$

that satisfies $y = 4$ when $x = 0$.

SOLUTION The appropriate integrating factor is

$$e^{\int (-3)\,dx} = e^{-3x}$$

Upon multiplication by this factor, our equation takes the form

$$\frac{d}{dx}(e^{-3x}y) = x$$

or

$$e^{-3x}y = \int x\,dx = \frac{1}{2}x^2 + C$$

Thus, the general solution is

$$y = \frac{1}{2}x^2e^{3x} + Ce^{3x}$$

Substitution of $y = 4$ when $x = 0$ makes $C = 4$. The desired particular solution is

$$y = \frac{1}{2}x^2e^{3x} + 4e^{3x}$$ ■

Applications We begin with a mixture problem, typical of many problems that arise in chemistry.

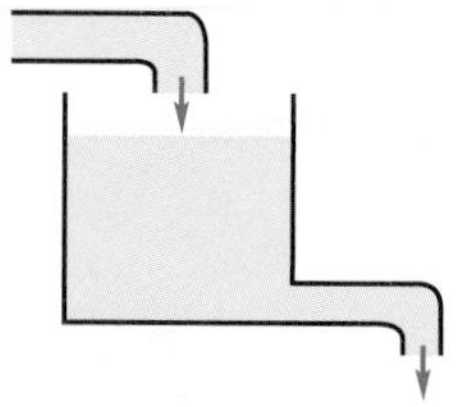
Figure 1

EXAMPLE 3 A tank initially contains 120 gallons of brine, holding 75 pounds of dissolved salt in solution. Salt water containing 1.2 pounds of salt per gallon is entering the tank at the rate of 2 gallons per minute and brine flows out at the same rate (Figure 1). If the mixture is kept uniform by constant stirring, find the amount of salt in the tank at the end of 1 hour.

SOLUTION Let y be the number of pounds of salt in the tank at the end of t minutes. From the brine flowing in, the tank gains 2.4 pounds of salt per minute; from that flowing out, it loses $\frac{2}{120}y$ pounds per minute. Thus,

$$\frac{dy}{dt} = 2.4 - \frac{1}{60}y$$

subject to the condition $y = 75$ when $t = 0$. The equivalent equation

$$\frac{dy}{dt} + \frac{1}{60}y = 2.4$$

has the integrating factor $e^{t/60}$, and so

$$\frac{d}{dt}[ye^{t/60}] = 2.4e^{t/60}$$

We conclude that

$$ye^{t/60} = \int 2.4e^{t/60}\,dt = (60)(2.4)e^{t/60} + C$$

Substituting $y = 75$ when $t = 0$ yields $C = -69$, and so

$$y = e^{-t/60}[144e^{t/60} - 69] = 144 - 69e^{-t/60}$$

At the end of 1 hour ($t = 60$),

$$y = 144 - 69e^{-1} \approx 118.62 \text{ pounds}$$

Note that the limiting value for y as $t \to \infty$ is 144. This corresponds to the fact that the tank will ultimately take on the complexion of the brine entering the tank. One hundred twenty gallons of brine with a concentration of 1.2 pounds of salt per gallon will contain 144 pounds of salt. ■

> **A General Principle**
>
> In flow problems such as Example 3, we apply a general principle. Let y measure the quantity of interest that is in the tank at time t. Then the rate of change of y with respect to time is the input rate minus the output rate; that is
>
> $$\frac{dy}{dt} = \text{rate in} - \text{rate out}$$

We turn next to an example from electricity. According to **Kirchhoff's Law,** a simple electrical circuit (Figure 2) containing a resistor with a resistance of R ohms and an inductor with an inductance of L henrys in series with a source of electromotive force (a battery or generator) that supplies a voltage of $E(t)$ volts at time t satisfies

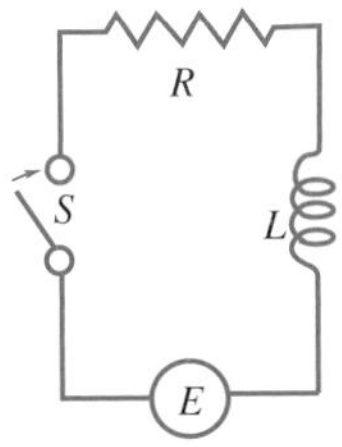

Figure 2

$$L\frac{dI}{dt} + RI = E(t)$$

where I is the current measured in amperes. This is a linear equation, easily solved by the method of this section.

EXAMPLE 4 Consider a circuit (Figure 2) with $L = 2$ henrys, $R = 6$ ohms, and a battery supplying a constant voltage of 12 volts. If $I = 0$ at $t = 0$ (when the switch S is closed), find I at time t.

SOLUTION The differential equation is

$$2\frac{dI}{dt} + 6I = 12 \quad \text{or} \quad \frac{dI}{dt} + 3I = 6$$

Following our standard procedure (multiply by the integrating factor e^{3t}, integrate, and multiply by e^{-3t}), we obtain

$$I = e^{-3t}(2e^{3t} + C) = 2 + Ce^{-3t}$$

The initial condition, $I = 0$ at $t = 0$, gives $C = -2$; hence

$$I = 2 - 2e^{-3t}$$

As t increases, the current tends toward a current of 2 amps. ■

Concepts Review

1. The general first-order linear differential equation has the form $dy/dx + P(x)y = Q(x)$. An integrating factor for this equation is ______.

2. Multiplying both sides of the first-order linear differential equation in Question 1 by its integrating factor makes the left side $\frac{d}{dx}$(______).

3. The integrating factor for $dy/dx - (1/x)y = x$, where $x > 0$, is ______. When we multiply both sides by this factor, the equation takes the form ______. The general solution to this equation is $y =$ ______.

4. The solution to the differential equation in Question 1 satisfying $y(a) = b$ is called a ______ solution.

Problem Set 6.6

In Problems 1–14, solve each differential equation.

1. $\frac{dy}{dx} + y = e^{-x}$

2. $(x + 1)\frac{dy}{dx} + y = x^2 - 1$

3. $(1 - x^2)\frac{dy}{dx} + xy = ax, |x| < 1$

4. $y' + y \tan x = \sec x$

5. $\frac{dy}{dx} - \frac{y}{x} = xe^x$

6. $y' - ay = f(x)$

7. $\frac{dy}{dx} + \frac{y}{x} = \frac{1}{x}$

8. $y' + \frac{2y}{x+1} = (x + 1)^3$

9. $y' + yf(x) = f(x)$

10. $\frac{dy}{dx} + 2y = x$ *Hint:* $\int xe^{2x}\, dx = \frac{x}{2}e^{2x} - \frac{1}{4}e^{2x} + C$

11. $\frac{dy}{dx} - \frac{y}{x} = 3x^3$; $y = 3$ when $x = 1$.

12. $y' = e^{2x} - 3y$; $y = 1$ when $x = 0$.

13. $xy' + (1 + x)y = e^{-x}$; $y = 0$ when $x = 1$.

14. $\sin x \frac{dy}{dx} + 2y \cos x = \sin 2x$; $y = 2$ when $x = \frac{\pi}{6}$.

15. A tank contains 20 gallons of a solution, with 10 pounds of chemical A in the solution. At a certain instant, we begin pouring in a solution containing the same chemical in a concentration of 2 pounds per gallon. We pour at a rate of 3 gallons per minute while simultaneously draining off the resulting (well-stirred) solution at the same rate. Find the amount of chemical A in the tank after 20 minutes.

16. A tank initially contains 200 gallons of brine, with 50 pounds of salt in solution. Brine containing 2 pounds of salt per gallon is entering the tank at the rate of 4 gallons per minute and is flowing out at the same rate. If the mixture in the tank is kept uniform by constant stirring, find the amount of salt in the tank at the end of 40 minutes.

17. A tank initially contains 120 gallons of pure water. Brine with 1 pound of salt per gallon flows into the tank at 4 gallons per minute, and the well-stirred solution runs out at 6 gallons per minute. How much salt is in the tank after t minutes, $0 \le t \le 60$?

18. A tank initially contains 50 gallons of brine, with 30 pounds of salt in solution. Water runs into the tank at 3 gallons per minute and the well-stirred solution runs out at 2 gallons per minute. How long will it be until there are 25 pounds of salt in the tank?

19. Find the current I as a function of time for the circuit of Figure 3 if the switch S is closed and $I = 0$ at $t = 0$.

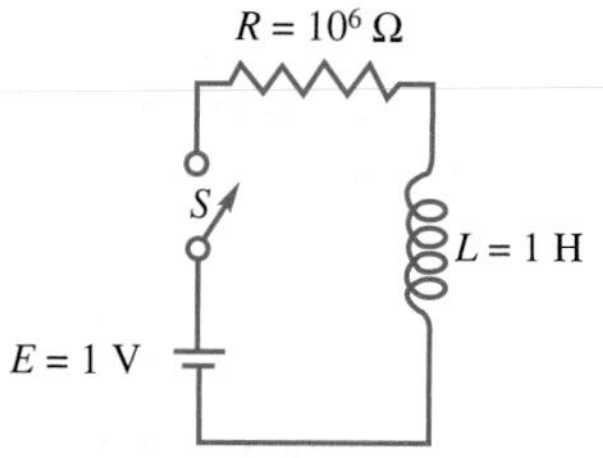

Figure 3

20. Find I as a function of time for the circuit of Figure 4, assuming that the switch is closed and $I = 0$ at $t = 0$.

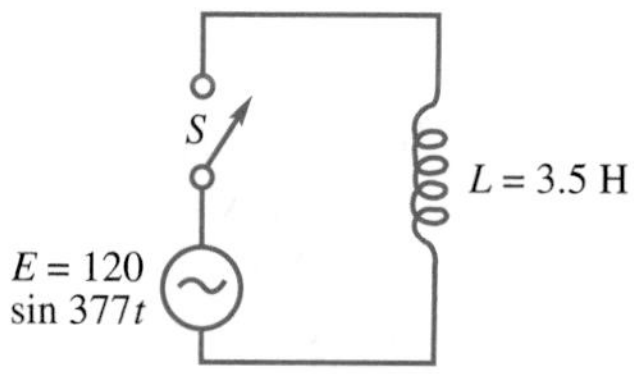

Figure 4

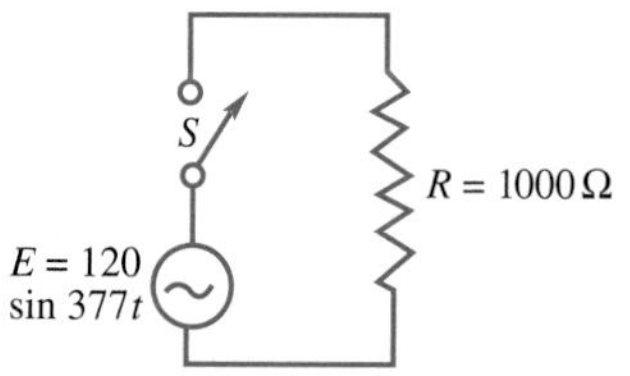

Figure 5

21. Find I as a function of time for the circuit of Figure 5, assuming that the switch is closed and $I = 0$ at $t = 0$.

22. Suppose that tank 1 initially contains 100 gallons of solution, with 50 pounds of dissolved salt, and tank 2 contains 200 gallons, with 150 pounds of dissolved salt. Pure water flows into tank 1 at 2 gallons per minute, the well-mixed solution flows out and into tank 2 at the same rate, and finally, the solution in tank 2 drains away also at the same rate. Let $x(t)$ and $y(t)$ denote the amounts of salt in tanks 1 and 2, respectively, at time t. Find $y(t)$. *Hint:* First find $x(t)$ and use it in setting up the differential equation for tank 2.

23. A tank of capacity 100 gallons is initially full of pure alcohol. The flow rate of the drain pipe is 5 gallons per minute; the flow rate of the filler pipe can be adjusted to c gallons per minute. An unlimited amount of 25% alcohol solution can be brought in through the filler pipe. Our goal is to reduce the amount of alcohol in the tank so that it will contain 100 gallons of 50% solution. Let T be the number of minutes required to accomplish the desired change.

(a) Evaluate T if $c = 5$ and both pipes are opened.

(b) Evaluate T if $c = 5$ and we first drain away a sufficient amount of the pure alcohol and then close the drain and open the filler pipe.

(c) For what values of c (if any) would strategy (b) give a faster time than (a)?

(d) Suppose that $c = 4$. Determine the equation for T if we initially open both pipes and then close the drain.

EXPL **24.** The differential equation for a falling body near the earth's surface with air resistance proportional to the velocity v is $dv/dt = -g - av$, where $g = 32$ feet per second per second is the acceleration of gravity and $a > 0$ is the *drag coefficient*. Show each of the following:

(a) $v(t) = (v_0 - v_\infty)e^{-at} + v_\infty$, where $v_0 = v(0)$, and

$$v_\infty = -g/a = \lim_{t\to\infty} v(t)$$

the so-called terminal velocity.

(b) If $y(t)$ denotes the altitude, then

$$y(t) = y_0 + tv_\infty + (1/a)(v_0 - v_\infty)(1 - e^{-at})$$

25. A ball is thrown straight up from ground level with an initial velocity $v_0 = 120$ feet per second. Assuming a drag coefficient of $a = 0.05$, determine each of the following:

(a) the maximum altitude

(b) an equation for T, the time when the ball hits the ground

26. Mary bailed out of her plane at an altitude of 8000 feet, fell freely for 15 seconds, and then opened her parachute. Assume that the drag coefficients are $a = 0.10$ for free fall and $a = 1.6$ with the parachute. When did she land?

27. For the differential equation $\frac{dy}{dx} - \frac{y}{x} = x^2$, $x > 0$, the integrating factor is $e^{\int(-1/x)\,dx}$. The general antiderivative $\int\left(-\frac{1}{x}\right)dx$ is equal to $-\ln x + C$.

(a) Multiply both sides of the differential equation by $\exp\left(\int\left(-\frac{1}{x}\right)dx\right) = \exp(-\ln x + C)$, and show that $\exp(-\ln x + C)$ is an integrating factor for every value of C.

(b) Solve the resulting equation for y, and show that the solution agrees with the solution obtained when we assumed that $C = 0$ in the integrating factor.

28. Multiply both sides of the equation $\frac{dy}{dx} + P(x)y = Q(x)$ by the factor $e^{\int P(x)\,dx + C}$.

(a) Show that $e^{\int P(x)\,dx + C}$ is an integrating factor for every value of C.

(b) Solve the resulting equation for y, and show that it agrees with the general solution given before Example 1.

Answers to Concepts Review: **1.** $\exp\left(\int P(x)\,dx\right)$ **2.** $y\exp\left(\int P(x)\,dx\right)$ **3.** $1/x$; $\frac{d}{dx}\left(\frac{y}{x}\right) = 1$; $x^2 + Cx$ **4.** particular

---

6.7 Approximations for Differential Equations

In the previous section we studied a number of differential equations that arise from physical applications. For each equation, we were always able to find an **analytic solution**; that is, we found an explicit function that satisfies the equation. Many differential equations do not have such analytic solutions, so for these equations we must settle for approximations. In this section, we will study two ways to approximate a solution to a differential equation; one method is graphical and the other is numerical.

> **A Function of Two Variables**
>
> The function f depends on two variables. Since $y'(x) = f(x, y)$, the slope of a solution depends on *both* the x- and y-coordinates. Functions of two or more variables were introduced in Section 0.5. We will study them further in Chapter 12.

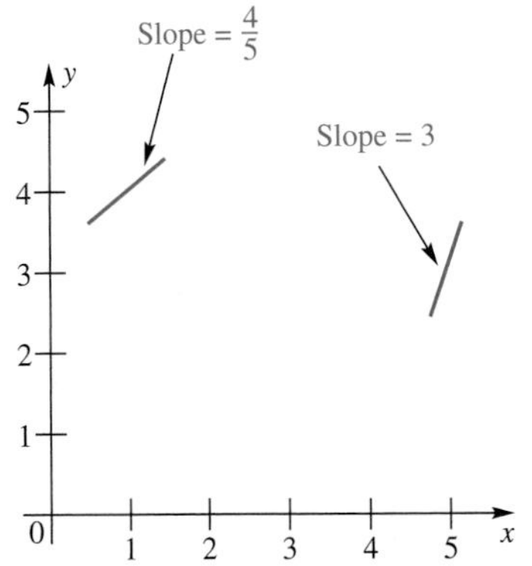

Figure 1

Slope Fields Consider a first-order differential equation of the form

$$y' = f(x, y)$$

This equation says that at the point (x, y) the slope of a solution is given by $f(x, y)$. For example, the differential equation $y' = y$ says that the slope of the curve passing through the point (x, y) is equal to y.

For the differential equation $y' = \frac{1}{5}xy$, at the point $(5, 3)$ the slope of the solution is $y' = \frac{1}{5}\cdot 5\cdot 3 = 3$; at the point $(1, 4)$ the slope is $y' = \frac{1}{5}\cdot 1\cdot 4 = \frac{4}{5}$. We can indicate graphically this latter result by drawing a small line segment through the point $(1, 4)$ having slope $\frac{4}{5}$ (see Figure 1).

If we repeat this process for a number of ordered pairs (x, y), we obtain a **slope field**. Since plotting a slope field is a tedious job if done by hand, the task is best suited for computers; *Mathematica* and *Maple* are capable of plotting slope fields. Figure 2 shows a slope field for the differential equation $y' = \frac{1}{5}xy$. Given an initial condition, we can follow the slopes to get at least a rough approximation to the particular solution. We can often see from the slope field the behavior of all solutions to the differential equation.

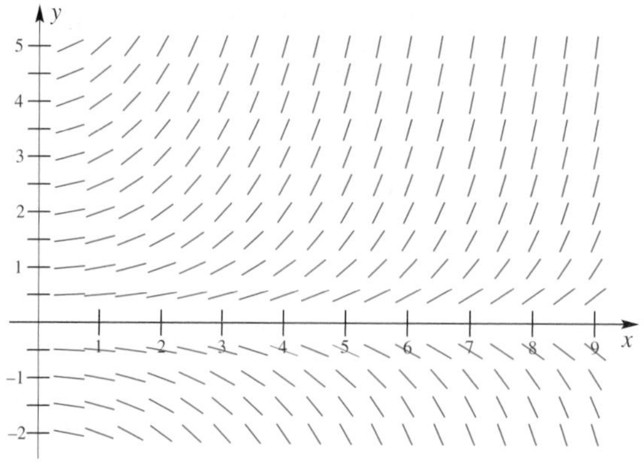

Figure 2

EXAMPLE 1 Suppose that the size y of a population satisfies the differential equation $y' = 0.2y(16 - y)$. The slope field for this differential equation is shown in Figure 3.

(a) Sketch the solution that satisfies the initial condition $y(0) = 3$.

Describe the behavior of solutions when

(b) $y(0) > 16$, and (c) $0 < y(0) < 16$.

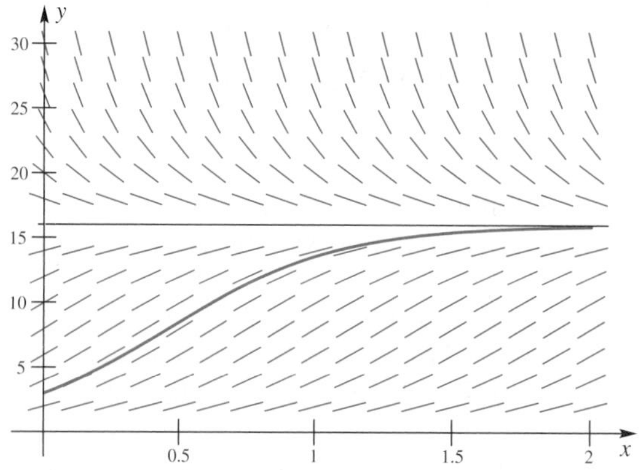

Figure 3

SOLUTION

(a) The solution that satisfies the initial condition $y(0) = 3$ contains the point $(0, 3)$. From that point to the right, the solution follows the slope lines. The curve in Figure 3 shows a graph of the solution.

(b) If $y(0) > 16$, then the solution decreases toward the horizontal asymptote $y = 16$.

(c) If $0 < y(0) < 16$, then the solution increases toward the horizontal asymptote $y = 16$.

Parts (b) and (c) indicate that the size of the population will converge toward the value 16 for any initial population size. ■

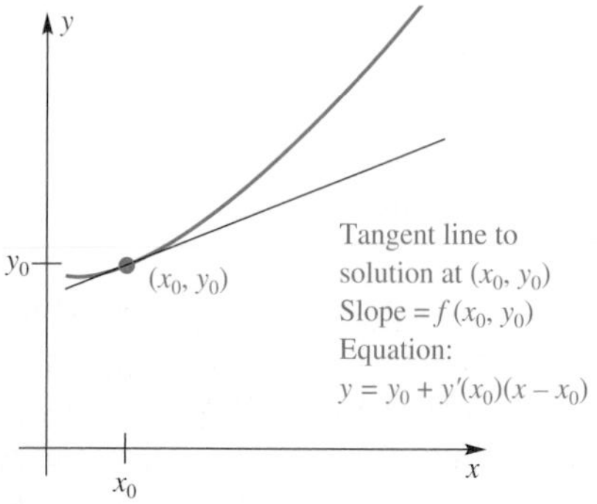

Figure 4

Euler's Method We again consider differential equations of the form $y' = f(x, y)$ with initial condition $y(x_0) = y_0$. Keep in mind that y is a function of x, whether we write it explicitly or not. The initial condition $y(x_0) = y_0$ tells us that the ordered pair (x_0, y_0) is a point on the graph of the solution. We also know just a bit more about the unknown solution: the slope of the tangent line to the solution at x_0 is $f(x_0, y_0)$. This information is summarized in Figure 4.

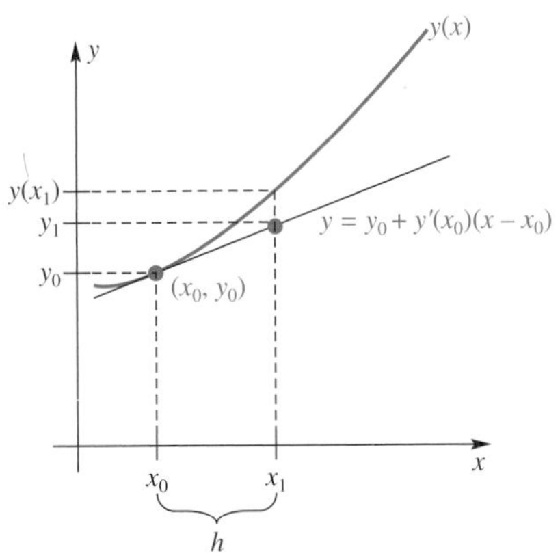

Figure 5

If h is positive, but small, we would expect the tangent line, which has equation

$$P_1(x) = y_0 + y'(x_0)(x - x_0) = y_0 + f(x_0, y_0)(x - x_0)$$

to be close to the solution $y(x)$ over the interval $[x_0, x_0 + h]$. Let $x_1 = x_0 + h$. Then, at x_1, we have

$$P_1(x_1) = y_0 + hy'(x_0) = y_0 + hf(x_0, y_0)$$

Setting $y_1 = y_0 + hf(x_0, y_0)$, we now have an approximation for the solution at x_1. Figure 5 illustrates the method we have just described.

Since $y' = f(x, y)$, we know that the slope of the solution when $x = x_1$ is $f(x_1, y(x_1))$. At this point, we do not know $y(x_1)$, but we do have the approximation y_1 for it. Thus, we repeat the process to obtain the estimate $y_2 = y_1 + hf(x_1, y_1)$ for the solution at the point $x_2 = x_1 + h$. This process, when continued in this fashion, is called **Euler's Method**, named after the Swiss mathematician Leonhard Euler (1707–1783). (Euler is pronounced "oiler.") The parameter h is often called the **step size**

Algorithm Euler's Method

To approximate the solution of the differential equation $y' = f(x, y)$ with initial condition $y(x_0) = y_0$, choose a step size h and repeat the following steps for $n = 1, 2, \ldots$.

1. Set $x_n = x_{n-1} + h$.
2. Set $y_n = y_{n-1} + hf(x_{n-1}, y_{n-1})$.

Remember, the solution to a differential equation is a *function*. Euler's Method, however, does not yield a function; rather, it gives a set of ordered pairs (x_i, y_i) that approximates the solution y. Often, this set of ordered pairs is enough to describe the solution to the differential equation.

Notice the difference between $y(x_n)$ and y_n; $y(x_n)$ (usually unknown) is the value of the exact solution at x_n, and y_n is our approximation to the exact solution at x_n. In other words, y_n is our approximation to $y(x_n)$.

EXAMPLE 2 Use Euler's Method with $h = 0.2$ to approximate the solution to

$$y' = y, \qquad y(0) = 1$$

over the interval $[0, 1]$.

SOLUTION For this problem, $f(x, y) = y$. Beginning with $x_0 = 0$ and $y_0 = 1$, we have

$$y_1 = y_0 + hf(x_0, y_0) = 1 + 0.2 \cdot 1 = 1.2$$

$$y_2 = 1.2 + 0.2 \cdot 1.2 = 1.44$$

$$y_3 = 1.44 + 0.2 \cdot 1.44 = 1.728$$

$$y_4 = 1.728 + 0.2 \cdot 1.728 = 2.0736$$

$$y_5 = 2.0736 + 0.2 \cdot 2.0736 = 2.48832$$ ■

n	x_n	y_n	e^{x_n}
0	0.0	1.0	1.00000
1	0.2	1.2	1.22140
2	0.4	1.44	1.49182
3	0.6	1.728	1.82212
4	0.8	2.0736	2.22554
5	1.0	2.48832	2.71828

The differential equation $y' = y$ says that y is its own derivative. Thus, we know that a solution is $y(x) = e^x$, and in fact $y(x) = e^x$ is *the* solution, since we are told that $y(0)$ must be 1. In this case, we can compare the five estimated y-values from Euler's Method with the exact y-values as shown in the table in the margin. Figure 6a shows the five approximations (x_i, y_i), $i = 1, 2, 3, 4, 5$, to the solution y; Figure 6 also shows the exact solution $y(x) = e^x$. Choosing a smaller h

will usually result in a more accurate approximation. Of course, a smaller h means that it will take more steps to get to $x = 1$.

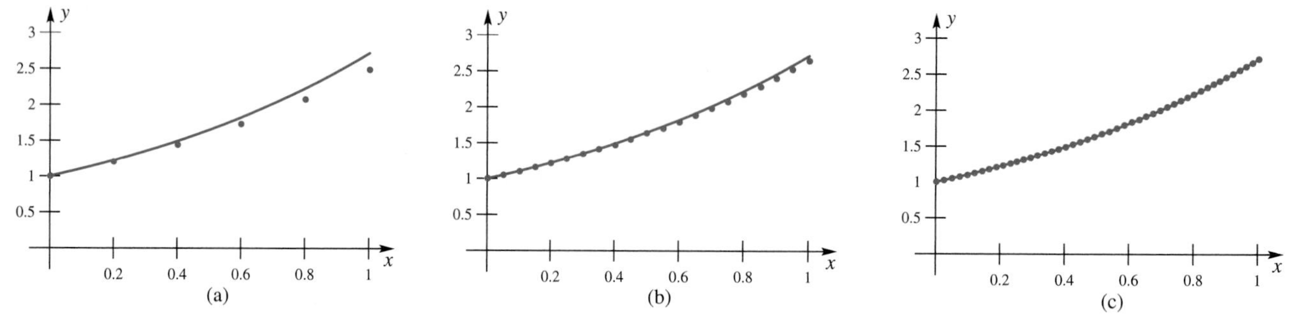

Figure 6

EXAMPLE 3 Use Euler's Method with $h = 0.05$ and $h = 0.01$ to approximate the solution to

$$y' = y, \quad y(0) = 1$$

over the interval [0, 1].

SOLUTION We proceed as in Example 1, but shrink the step size h to 0.05 and get the following table:

n	x_n	y_n	n	x_n	y_n
0	0.00	1.000000	11	0.55	1.710339
1	0.05	1.050000	12	0.60	1.795856
2	0.10	1.102500	13	0.65	1.885649
3	0.15	1.157625	14	0.70	1.979932
4	0.20	1.215506	15	0.75	2.078928
5	0.25	1.276282	16	0.80	2.182875
6	0.30	1.340096	17	0.85	2.292018
7	0.35	1.407100	18	0.90	2.406619
8	0.40	1.477455	19	0.95	2.526950
9	0.45	1.551328	20	1.00	2.653298
10	0.50	1.628895			

Figure 6b shows the approximation to the solution when Euler's Method with $h = 0.05$ is used.

Computations proceed similarly for the case when $h = 0.01$. The results are summarized in the table in the margin and in Figure 6c. ■

n	x_n	y_n
0	0.00	1.000000
1	0.01	1.010000
2	0.02	1.020100
3	0.03	1.030301
⋮	⋮	⋮
99	0.99	2.678033
100	1.00	2.704814

Notice in Example 3 that as the step size h decreases, the approximation to $y(1)$ (which in this case is $e^1 \approx 2.718282$) improves. When $h = 0.2$, the error is approximately $e - y_5 = 2.718282 - 2.488320 = 0.229962$. Approximations to the error for other step sizes are shown in the following table:

h	Euler Approximation of $y(1)$	Error = Exact − Estimate
0.2	2.488320	0.229962
0.1	2.593742	0.124540
0.05	2.653298	0.064984
0.01	2.704814	0.013468
0.005	2.711517	0.006765

Note in the table that as the step size h is *halved* the error is approximately *halved*. The error at a given point is therefore roughly proportional to the step size h. We found a similar result with numerical integration in Section 4.6. There we saw that the error for the left or right Riemann Sum Rule is proportional to $h = 1/n$ and that the error for the Trapezoidal Rule is proportional to $h^2 = 1/n^2$. The Parabolic Rule is even better, having an error proportional to $h^4 = 1/n^4$. This raises the question of whether there are better methods for approximating the solution of $y' = f(x, y)$, $y(x_0) = y_0$. In fact, there are a number of methods that are better then Euler's Method, in the sense that the error is proportional to a higher power of h. These methods are conceptually similar to Euler's Method: they are "step methods," that is, they begin with the initial condition and successively approximate the solution at each of a number of steps to the right. One method, the **Fourth-Order Runge-Kutta Method**, has an error that is proportional to $h^4 = 1/n^4$.

Concepts Review

1. For the differential equation $y' = f(x, y)$, a plot of line segments whose slopes equal $f(x, y)$ is called a _____.

2. The basis for Euler's Method is that the _____ to the solution at x_0 will be a good approximation to the solution over the interval $[x_0, x_0 + h]$.

3. The recursive formula for the approximation to the solution of a differential equation using Euler's Method is $y_n =$ _____.

4. If the solution of a differential equation is concave up, then Euler's Method will _____ (underestimate or overestimate) the solution.

Problem Set 6.7

In Problems 1–4, a slope field is given for a differential equation of the form $y' = f(x, y)$. Use the slope field to sketch the solution that satisfies the given initial condition. In each case, find $\lim_{x \to \infty} y(x)$ and approximate $y(2)$.

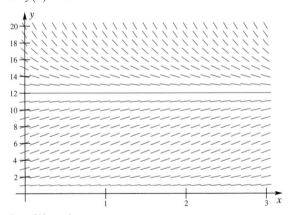

1. $y(0) = 5$

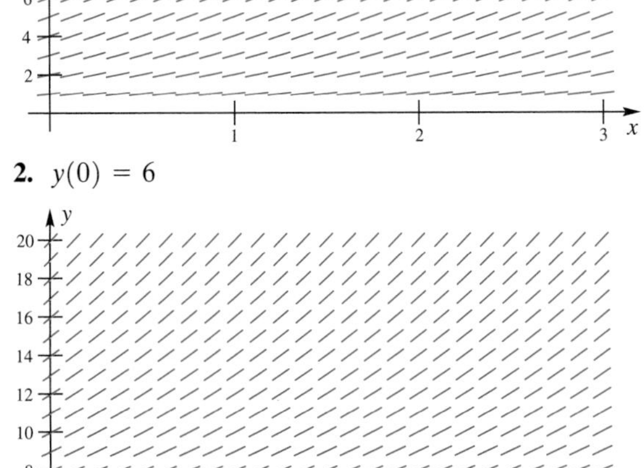

2. $y(0) = 6$

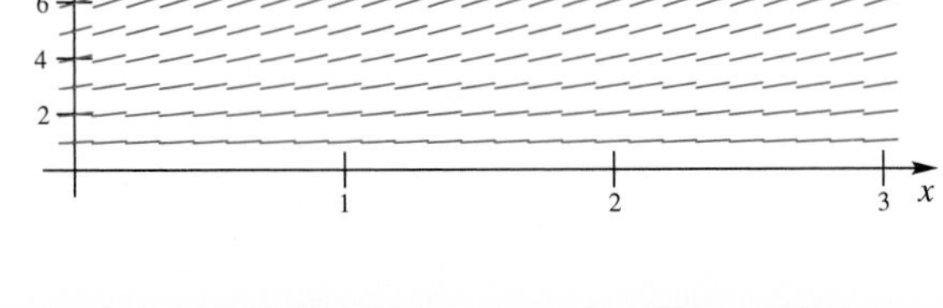

3. $y(0) = 16$

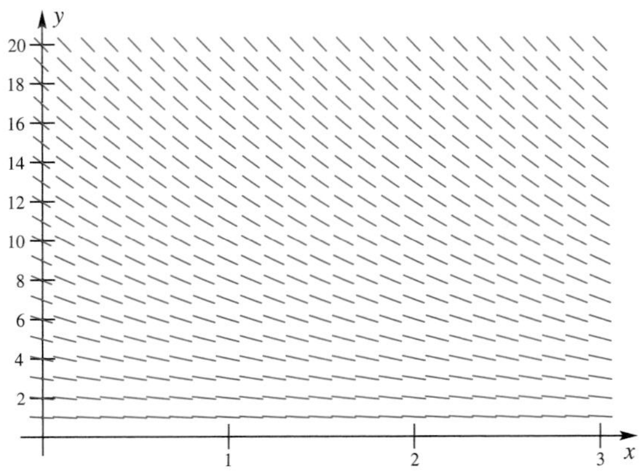

4. $y(1) = 3$

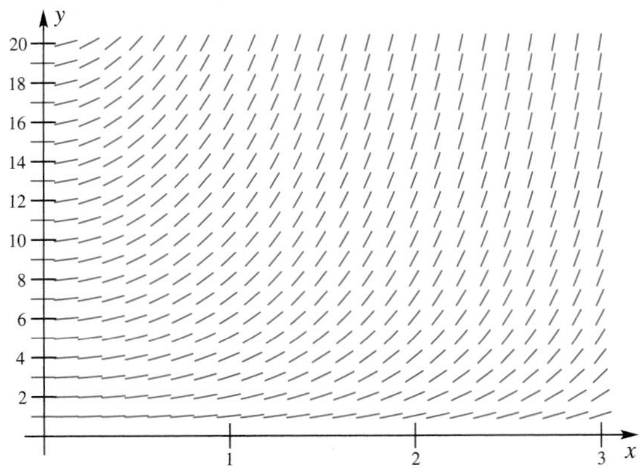

In Problems 5 and 6, a slope field is given for a differential equation of the form $y' = f(x, y)$. In both cases, every solution has the

same oblique asymptote (see Section 3.5). Sketch the solution that satisfies the given initial condition, and find the equation of the oblique asymptote.

5. $y(0) = 6$

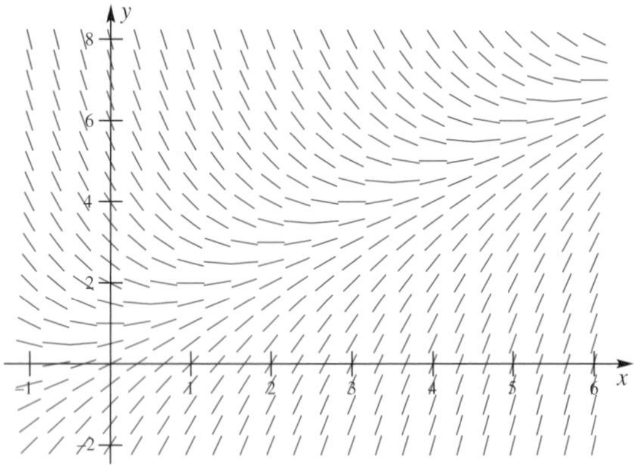

6. $y(0) = 8$

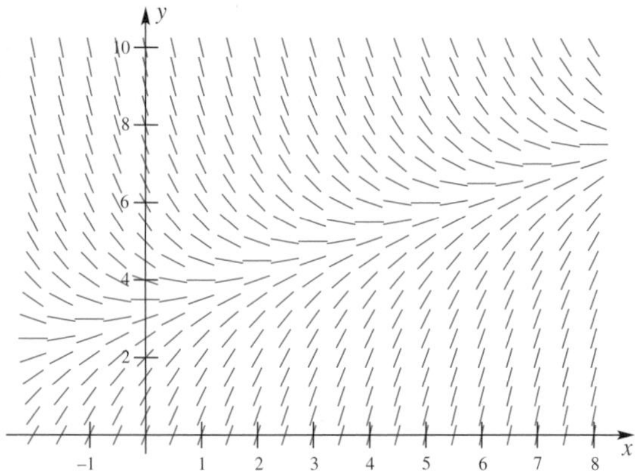

CAS *In Problems 7–10, plot a slope field for each differential equation. Use the method of separation of variables (Section 3.9) or an integrating factor (Section 6.6) to find a particular solution of the differential equation that satisfies the given initial condition, and plot the particular solution.*

7. $y' = \frac{1}{2}y;\ y(0) = \frac{1}{2}$

8. $y' = -y;\ y(0) = 4$

9. $y' = x - y + 2;\ y(0) = 4$

10. $y' = 2x - y + \frac{3}{2};\ y(0) = 3$

C *In Problems 11–16, use Euler's Method with $h = 0.2$ to approximate the solution over the indicated interval.*

11. $y' = 2y,\ y(0) = 3,\ [0, 1]$

12. $y' = -y,\ y(0) = 2,\ [0, 1]$

13. $y' = x,\ y(0) = 0,\ [0, 1]$

14. $y' = x^2,\ y(0) = 0,\ [0, 1]$

15. $y' = xy,\ y(1) = 1,\ [1, 2]$

16. $y' = -2xy,\ y(1) = 2,\ [1, 2]$

17. Apply Euler's Method to the equation $y' = y$, $y(0) = 1$ with an arbitrary step size $h = 1/N$ where N is a positive integer.

(a) Derive the relationship $y_n = y_0(1 + h)^n$.

(b) Explain why y_N is an approximation to e.

18. Suppose that the function $f(x, y)$ depends only on x. The differential equation $y' = f(x, y)$ can then be written as

$$y' = f(x), \qquad y(x_0) = 0$$

Explain how to apply Euler's Method to this differential equation if $y_0 = 0$.

19. Consider the differential equation $y' = f(x)$, $y(x_0) = 0$ of Problem 18. For this problem, let $f(x) = \sin x^2$, $x_0 = 0$, and $h = 0.1$.

(a) Integrate both sides of the equation from x_0 to $x_1 = x_0 + h$. To approximate the integral, use a Riemann sum with a single interval, evaluating the integrand at the left end point.

(b) Integrate both sides from x_0 to $x_2 = x_0 + 2h$. Again, to approximate the integral use a left end point Riemann sum, but with two intervals.

(c) Continue the process described in parts (a) and (b) until $x_n = 1$. Use a left end point Riemann sum with ten intervals to approximate the integral.

(d) Describe how this method is related to Euler's Method.

20. Repeat parts (a) through (c) of Problem 19 for the differential equation $y' = \sqrt{x + 1}$, $y(0) = 0$.

21. (Improved Euler Method) Consider the change Δy in the solution between x_0 and x_1. One approximation is obtained from Euler's Method: $\frac{\Delta y}{\Delta x} = \frac{y(x_1) - y_0}{h} \approx \frac{\hat{y}_1 - y_0}{h} = f(x_0, y_0)$. (Here we have used $\hat{y}_1$ to indicate Euler's approximation to the solu-tion at x_1.) Another approximation is obtained by finding an approximation to the slope of the solution at x_1:

$$\frac{\Delta y}{\Delta x} = \frac{y(x_1) - y_0}{h} \approx f(x_1, y_1) \approx f(x_1, \hat{y}_1)$$

(a) Average these two solutions to get a single approximation for $\Delta y/\Delta x$.

(b) Solve for $y_1 = y(x_1)$ to obtain

$$y_1 = y_0 + \frac{h}{2}[f(x_0, y_0) + f(x_1, \hat{y}_1)]$$

(c) This is the first step in the Improved Euler Method. Additional steps follow the same pattern. Fill in the blanks for the following three-step algorithm that yields the Improved Euler Method:

1. Set $x_n =$ ______________
2. Set $\hat{y}_n =$ ______________
3. Set $y_n =$ ______________

C *For Problems 22–27, use the Improved Euler Method with $h = 0.2$ on the equations in Problems 11–16. Compare your answer with those obtained using Euler's Method.*

CAS **28.** Apply the Improved Euler Method to the equation $y' = y$, $y(0) = 1$, with $h = 0.2, 0.1, 0.05, 0.01, 0.005$ to approximate the solution on the interval $[0, 1]$. (Note that the exact solution is $y = e^x$, so $y(1) = e$.) Compute the error in approximating $y(1)$ (see Example 3 and the subsequent discussion) and fill in the following table. For the Improved Euler Method, is the error proportional to h, h^2, or some other power of h?

h	Error from Euler's Method	Error from Improved Euler Method
0.2	0.229962	0.015574
0.1	0.124540	
0.05	0.064984	0.001091
0.01	0.013468	0.000045
0.005	0.006765	

Answers to Concepts Review: **1.** slope field **2.** tangent line **3.** $y_{n-1} + hf(x_{n-1}, y_{n-1})$ **4.** underestimate

6.8 The Inverse Trigonometric Functions and Their Derivatives

The six basic trigonometric functions (sine, cosine, tangent, cotangent, secant, and cosecant) were defined in Section 0.7, and we have used them occasionally in examples and problems. With respect to the notion of inverse, they are miserable functions, since for each y in their range there are infinitely many x's that correspond to it (Figure 1). Nonetheless, we are going to introduce a notion of inverse for them. That this is possible rests on a procedure called **restricting the domain,** which was discussed briefly in Section 6.2.

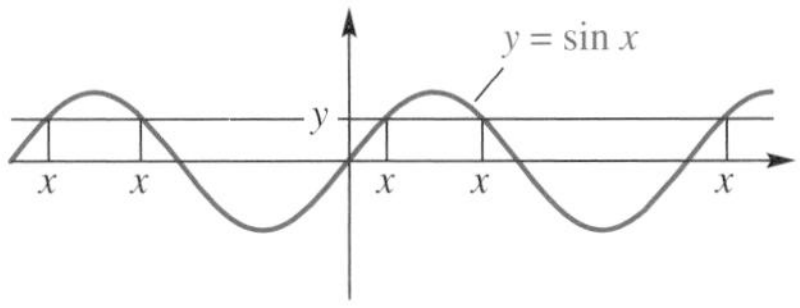

Figure 1

Inverse Sine and Inverse Cosine In the case of sine and cosine, we restrict the domain, keeping the range as large as possible while insisting that the resulting function have an inverse. This can be done in many ways, but the agreed procedure is suggested by Figures 2 and 3. We also show the graph of the corresponding inverse function, obtained, as usual, by reflecting across the line $y = x$.

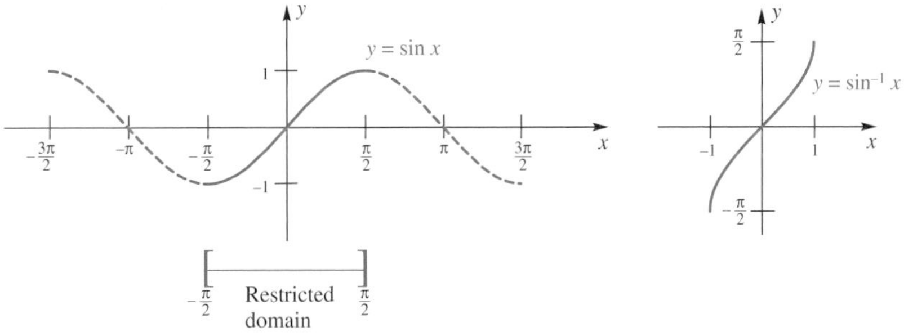

Figure 2

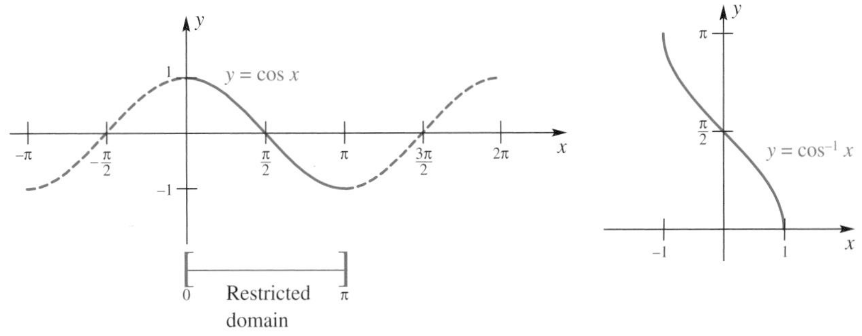

Figure 3

We formalize what we have shown in a definition.

Definition

To obtain inverses for sine and cosine, we restrict their domains to $[-\pi/2, \pi/2]$ and $[0, \pi]$, respectively. Thus,

$$x = \sin^{-1} y \iff y = \sin x, -\frac{\pi}{2} \le x \le \frac{\pi}{2}$$

$$x = \cos^{-1} y \iff y = \cos x, 0 \le x \le \pi$$

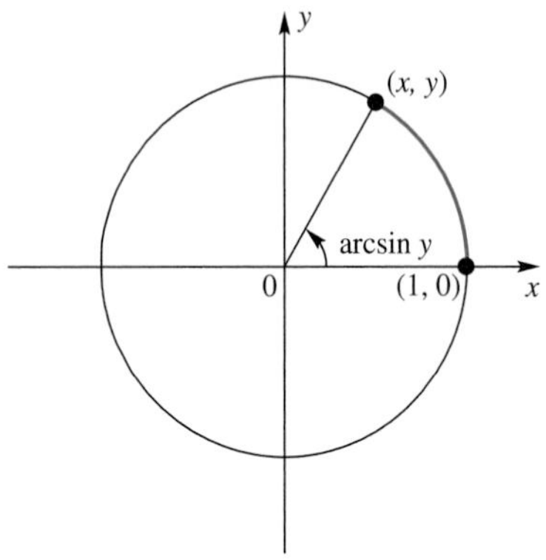

Figure 4

The symbol arcsin is often used for $\sin^{-1}$, and arccos is similarly used for $\cos^{-1}$. Think of arcsin as meaning "the arc whose sine is" or "the angle whose sine is" (Figure 4). We will use both forms throughout the rest of this book.

EXAMPLE 1 Calculate

(a) $\sin^{-1}(\sqrt{2}/2)$, (b) $\cos^{-1}\left(-\frac{1}{2}\right)$,

(c) $\cos(\cos^{-1} 0.6)$, and (d) $\sin^{-1}(\sin 3\pi/2)$

SOLUTION

(a) $\sin^{-1}\left(\frac{\sqrt{2}}{2}\right) = \frac{\pi}{4}$ (b) $\cos^{-1}\left(-\frac{1}{2}\right) = \frac{2\pi}{3}$

(c) $\cos(\cos^{-1} 0.6) = 0.6$ (d) $\sin^{-1}\left(\sin\frac{3\pi}{2}\right) = -\frac{\pi}{2}$

The only one of these that is tricky is (d). Note that it would be wrong to give $3\pi/2$ as the answer, since $\sin^{-1} y$ is always in the interval $[-\pi/2, \pi/2]$. Work the problem in steps, as follows.

$$\sin^{-1}\left(\sin\frac{3\pi}{2}\right) = \sin^{-1}(-1) = -\pi/2$$

■

EXAMPLE 2 Use a calculator to find

(a) $\cos^{-1}(-0.61)$, (b) $\sin^{-1}(1.21)$, (c) $\sin^{-1}(\sin 4.13)$

SOLUTION Use a calculator in radian mode. It has been programmed to give answers that are consistent with the definitions that we have given.

(a) $\cos^{-1}(-0.61) = 2.2268569$

(b) Your calculator should indicate an error, since $\sin^{-1}(1.21)$ does not exist.

(c) $\sin^{-1}(\sin 4.13) = -0.9884073$ ■

Inverse Tangent and Inverse Secant In Figure 5, we show the graph of the tangent function, its restricted domain, and the graph of $y = \tan^{-1} x$.

There is a standard way to restrict the domain of the cotangent function, that is, to $(0, \pi)$, so that it has an inverse. However, this function does not play a significant role in calculus.

Another Way To Say It

$\sin^{-1} y$

is the number in the interval $[-\pi/2, \pi/2]$ whose sine is y.

$\cos^{-1} y$

is the number in the interval $[0, \pi]$ whose cosine is y.

$\tan^{-1} y$

is the number in the interval $(-\pi/2, \pi/2)$ whose tangent is y.

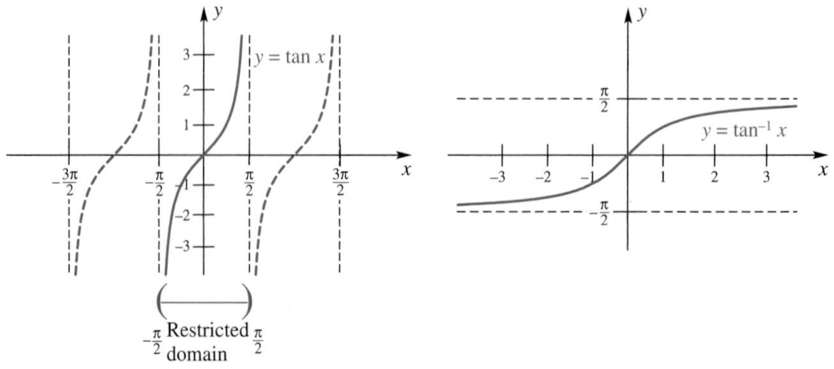

Figure 5

To obtain an inverse for secant, we graph $y = \sec x$, restrict its domain appropriately, and then graph $y = \sec^{-1} x$ (Figure 6).

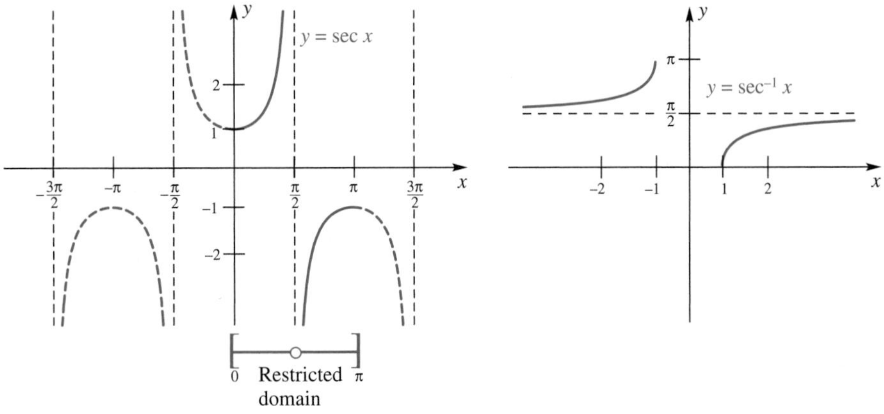

Figure 6

> **Definition**
>
> To obtain inverses for tangent and secant, we restrict their domains to $(-\pi/2, \pi/2)$ and $[0, \pi/2) \cup (\pi/2, \pi]$, respectively. Thus,
>
> $$x = \tan^{-1} y \quad \Leftrightarrow \quad y = \tan x, -\frac{\pi}{2} < x < \frac{\pi}{2}$$
>
> $$x = \sec^{-1} y \quad \Leftrightarrow \quad y = \sec x, 0 \le x \le \pi, x \neq \frac{\pi}{2}$$

Some authors restrict the domain of the secant in a different way. Thus, if you refer to another book, you must check that author's definition. We will have no need to define $\csc^{-1}$, though this can also be done.

EXAMPLE 3 Calculate

(a) $\tan^{-1}(1)$, (b) $\tan^{-1}(-\sqrt{3})$,

(c) $\tan^{-1}(\tan 5.236)$, (d) $\sec^{-1}(-1)$,

(e) $\sec^{-1}(2)$, and (f) $\sec^{-1}(-1.32)$

SOLUTION

(a) $\tan^{-1}(1) = \frac{\pi}{4}$ (b) $\tan^{-1}(-\sqrt{3}) = -\frac{\pi}{3}$

(c) $\tan^{-1}(\tan 5.236) = -1.0471853$

Most of us have trouble remembering our secants; moreover, most calculators fail to have a secant button. Therefore, we suggest that you remember that $\sec x = 1/\cos x$. From this, it follows that

$$\boxed{\sec^{-1} y = \cos^{-1}\left(\frac{1}{y}\right)}$$

and this allows us to use known facts about the cosine.

(d) $\sec^{-1}(-1) = \cos^{-1}(-1) = \pi$

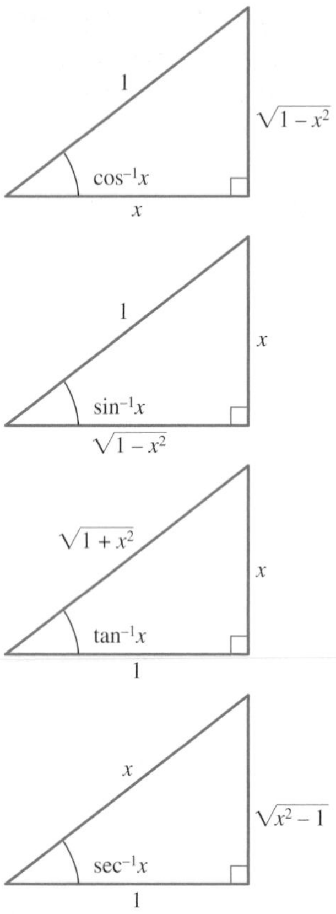

Figure 7

(e) $\sec^{-1}(2) = \cos^{-1}\left(\frac{1}{2}\right) = \frac{\pi}{3}$

(f) $\sec^{-1}(-1.32) = \cos^{-1}\left(-\frac{1}{1.32}\right) = \cos^{-1}(0.7575758)$

$= 2.4303875$ ■

Four Useful Identities Theorem A gives some useful identities. You can recall them by reference to the triangles in Figure 7.

Theorem A

(i) $\sin(\cos^{-1} x) = \sqrt{1 - x^2}$

(ii) $\cos(\sin^{-1} x) = \sqrt{1 - x^2}$

(iii) $\sec(\tan^{-1} x) = \sqrt{1 + x^2}$

(iv) $\tan(\sec^{-1} x) = \begin{cases} \sqrt{x^2 - 1}, & \text{if } x \geq 1 \\ -\sqrt{x^2 - 1}, & \text{if } x \leq -1 \end{cases}$

Proof To prove (i), recall that $\sin^2 \theta + \cos^2 \theta = 1$. If $0 \leq \theta \leq \pi$, then

$$\sin \theta = \sqrt{1 - \cos^2 \theta}$$

Now apply this with $\theta = \cos^{-1} x$ and use the fact that $\cos(\cos^{-1} x) = x$ to get

$$\sin(\cos^{-1} x) = \sqrt{1 - \cos^2(\cos^{-1} x)} = \sqrt{1 - x^2}$$

Identity (ii) is proved in a completely similar manner. To prove (iii) and (iv), use the identity $\sec^2 \theta = 1 + \tan^2 \theta$ in place of $\sin^2 \theta + \cos^2 \theta = 1$. ■

EXAMPLE 4 Calculate $\sin\left[2 \cos^{-1}\left(\frac{2}{3}\right)\right]$.

SOLUTION Recall the double-angle identity $\sin 2\theta = 2 \sin \theta \cos \theta$. Thus,

$$\sin\left[2 \cos^{-1}\left(\frac{2}{3}\right)\right] = 2 \sin\left[\cos^{-1}\left(\frac{2}{3}\right)\right] \cos\left[\cos^{-1}\left(\frac{2}{3}\right)\right]$$

$$= 2 \cdot \sqrt{1 - \left(\frac{2}{3}\right)^2} \cdot \frac{2}{3} = \frac{4\sqrt{5}}{9}$$ ■

Derivatives of Trigonometric Functions We learned in Section 2.4 the derivative formulas for the six trigonometric functions. They should be memorized.

$D_x \sin x = \cos x$	$D_x \cos x = -\sin x$
$D_x \tan x = \sec^2 x$	$D_x \cot x = -\csc^2 x$
$D_x \sec x = \sec x \tan x$	$D_x \csc x = -\csc x \cot x$

We can combine the rules above with the Chain Rule. For example, if $u = f(x)$ is differentiable, then

$$D_x \sin u = \cos u \cdot D_x u$$

Inverse Trigonometric Functions From the Inverse Function Theorem (Theorem 6.2B), we conclude that $\sin^{-1}$, $\cos^{-1}$, $\tan^{-1}$, and $\sec^{-1}$ are differentiable. Our aim is to find formulas for their derivatives. We state the results and then show how they can be derived.

Theorem B Derivatives of Four Inverse Trigonometric Functions

(i) $D_x \sin^{-1} x = \dfrac{1}{\sqrt{1-x^2}}, \quad -1 < x < 1$

(ii) $D_x \cos^{-1} x = -\dfrac{1}{\sqrt{1-x^2}}, \quad -1 < x < 1$

(iii) $D_x \tan^{-1} x = \dfrac{1}{1+x^2}$

(iv) $D_x \sec^{-1} x = \dfrac{1}{|x|\sqrt{x^2-1}}, \quad |x| > 1$

Proof Our proofs follow the same pattern in each case. To prove (i), let $y = \sin^{-1} x$, so that

$$x = \sin y$$

Now differentiate both sides with respect to x, using the Chain Rule on the right-hand side. Then

$$\begin{aligned} 1 &= \cos y \, D_x y = \cos(\sin^{-1} x)\, D_x(\sin^{-1} x) \\ &= \sqrt{1-x^2}\, D_x(\sin^{-1} x) \end{aligned}$$

At the last step, we used Theorem A(ii). We conclude that $D_x(\sin^{-1} x) = 1/\sqrt{1-x^2}$.

Results (ii), (iii), and (iv) are proved similarly, but (iv) has a little twist. Let $y = \sec^{-1} x$, so

$$x = \sec y$$

Differentiating both sides with respect to x and using Theorem A(iv), we obtain

$$\begin{aligned} 1 &= \sec y \tan y \, D_x y \\ &= \sec(\sec^{-1} x)\tan(\sec^{-1} x)\, D_x(\sec^{-1} x) \\ &= \begin{cases} x\sqrt{x^2-1}\, D_x(\sec^{-1} x), & \text{if } x \ge 1 \\ x\left(-\sqrt{x^2-1}\right) D_x(\sec^{-1} x), & \text{if } x \le -1 \end{cases} \\ &= |x|\sqrt{x^2-1}\, D_x(\sec^{-1} x) \end{aligned}$$

The desired result follows immediately. ■

> **$D_x \sec^{-1} x$**
>
> Here is another way to derive the formula for the derivative of $\sec^{-1} x$.
>
> $$\begin{aligned} D_x \sec^{-1} x &= D_x \cos^{-1}\left(\frac{1}{x}\right) \\ &= \frac{-1}{\sqrt{1-1/x^2}} \cdot \frac{-1}{x^2} \\ &= \frac{1}{\sqrt{x^2-1}} \cdot \frac{\sqrt{x^2}}{x^2} \\ &= \frac{1}{\sqrt{x^2-1}} \cdot \frac{|x|}{x^2} \\ &= \frac{1}{|x|\sqrt{x^2-1}} \end{aligned}$$

EXAMPLE 5 Find $D_x \sin^{-1}(3x-1)$.

SOLUTION We use Theorem B(i) and the Chain Rule.

$$\begin{aligned} D_x \sin^{-1}(3x-1) &= \frac{1}{\sqrt{1-(3x-1)^2}} D_x(3x-1) \\ &= \frac{3}{\sqrt{-9x^2+6x}} \end{aligned}$$

Of course, every differentiation formula leads to an integration formula, a matter we will say much more about in the next chapter. In particular,

1. $\displaystyle\int \frac{1}{\sqrt{1-x^2}}\,dx = \sin^{-1} x + C$

2. $\displaystyle\int \frac{1}{1+x^2}\,dx = \tan^{-1} x + C$

3. $\displaystyle\int \frac{1}{x\sqrt{x^2-1}}\,dx = \sec^{-1}|x| + C$

These integration formulas can be generalized slightly (see Problems 81–84) to the following:

1′. $\displaystyle\int \frac{1}{\sqrt{a^2-x^2}}\,dx = \sin^{-1}\left(\frac{x}{a}\right) + C$

2′. $\displaystyle\int \frac{1}{a^2+x^2}\,dx = \frac{1}{a}\tan^{-1}\left(\frac{x}{a}\right) + C$

3′. $\displaystyle\int \frac{1}{x\sqrt{x^2-a^2}}\,dx = \frac{1}{a}\sec^{-1}\left(\frac{|x|}{a}\right) + C$

EXAMPLE 6 Evaluate $\displaystyle\int_0^1 \frac{1}{\sqrt{4-x^2}}\,dx$.

SOLUTION

$$\int_0^1 \frac{1}{\sqrt{4-x^2}}\,dx = \left[\sin^{-1}\left(\frac{x}{2}\right)\right]_0^1 = \sin^{-1}\frac{1}{2} - \sin^{-1} 0 = \frac{\pi}{6} - 0 = \frac{\pi}{6}$$ ■

EXAMPLE 7 Evaluate $\displaystyle\int \frac{3}{\sqrt{5-9x^2}}\,dx$.

SOLUTION Think of $\displaystyle\int \frac{du}{\sqrt{a^2-u^2}}$. Let $u = 3x$, so $du = 3\,dx$. Then

$$\int \frac{3}{\sqrt{5-9x^2}}\,dx = \int \frac{1}{\sqrt{5-u^2}}\,du = \sin^{-1}\left(\frac{u}{\sqrt{5}}\right) + C$$
$$= \sin^{-1}\left(\frac{3x}{\sqrt{5}}\right) + C$$ ■

EXAMPLE 8 Evaluate $\displaystyle\int \frac{e^x}{4+9e^{2x}}\,dx$.

SOLUTION Think of $\displaystyle\int \frac{1}{a^2+u^2}\,du$. Let $u = 3e^x$, so $du = 3e^x\,dx$. Then

$$\int \frac{e^x}{4+9e^{2x}}\,dx = \frac{1}{3}\int \frac{1}{4+9e^{2x}}(3e^x\,dx) = \frac{1}{3}\int \frac{1}{4+u^2}\,du$$
$$= \frac{1}{3}\cdot\frac{1}{2}\tan^{-1}\left(\frac{u}{2}\right) + C = \frac{1}{6}\tan^{-1}\left(\frac{3e^x}{2}\right) + C$$ ■

EXAMPLE 9 Evaluate $\displaystyle\int_6^{18} \frac{1}{x\sqrt{x^2-9}}\,dx$

SOLUTION

$$\int_6^{18} \frac{1}{x\sqrt{x^2-9}}\,dx = \frac{1}{3}\left[\sec^{-1}\frac{|x|}{3}\right]_6^{18}$$
$$= \frac{1}{3}\left(\sec^{-1}\frac{|18|}{3} - \sec^{-1}\frac{|6|}{3}\right)$$
$$= \frac{1}{3}\left(\sec^{-1} 6 - \frac{\pi}{3}\right) \approx 0.1187$$ ■

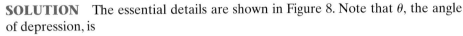

EXAMPLE 10 A man standing on top of a vertical cliff is 200 feet above a lake. As he watches, a motorboat moves directly away from the foot of the cliff at a rate of 25 feet per second. How fast is the angle of depression of his line of sight changing when the boat is 150 feet from the foot of the cliff?

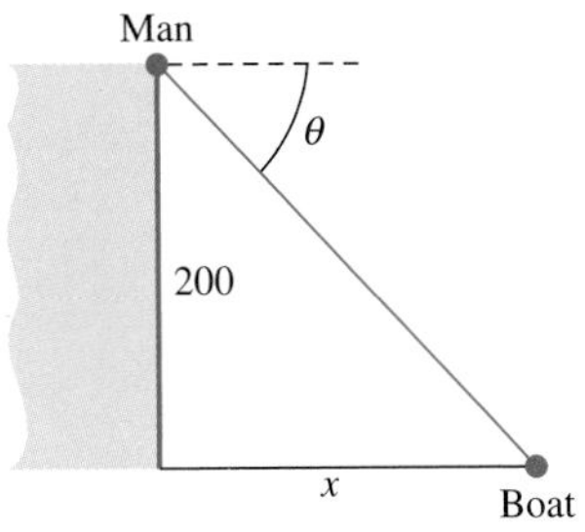

Figure 8

SOLUTION The essential details are shown in Figure 8. Note that θ, the angle of depression, is

$$\theta = \tan^{-1}\left(\frac{200}{x}\right)$$

Thus,

$$\frac{d\theta}{dt} = \frac{1}{1 + (200/x)^2} \cdot \frac{-200}{x^2} \cdot \frac{dx}{dt} = \frac{-200}{x^2 + 40{,}000} \cdot \frac{dx}{dt}$$

When we substitute $x = 150$ and $dx/dt = 25$, we obtain $d\theta/dt = -0.08$ radian per second. ■

Manipulating the Integrand Before you make a substitution, you may find it helpful to rewrite the integrand in a more convenient form. Integrals with quadratic expressions in the denominator can often be reduced to standard forms by *completing the square*. Recall that $x^2 + bx$ becomes a perfect square by the addition of $(b/2)^2$.

EXAMPLE 11 Evaluate $\displaystyle\int \frac{7}{x^2 - 6x + 25}\,dx$.

SOLUTION

$$\begin{aligned}\int \frac{7}{x^2 - 6x + 25}\,dx &= \int \frac{7}{x^2 - 6x + 9 + 16}\,dx \\ &= 7\int \frac{1}{(x-3)^2 + 4^2}\,dx \\ &= \frac{7}{4}\tan^{-1}\left(\frac{x-3}{4}\right) + C\end{aligned}$$

We made the mental substitution $u = x - 3$ at the final stage. ■

Concepts Review

1. To obtain an inverse for the sine function, we restrict its domain to _____. The resulting inverse function is denoted by $\sin^{-1}$ or by _____.

2. To obtain an inverse for the tangent function, we restrict the domain to _____. The resulting inverse function is denoted by $\tan^{-1}$ or by _____.

3. $D_x \sin(\arcsin x) =$ _____.

4. Since $D_x \arctan x = 1/(1 + x^2)$, it follows that $4\int_0^1 1/(1 + x^2)\,dx =$ _____.

Problem Set 6.8

In Problems 1–10, find the exact value without using a calculator.

1. $\arccos\left(\frac{\sqrt{2}}{2}\right)$

2. $\arcsin\left(-\frac{\sqrt{3}}{2}\right)$

3. $\sin^{-1}\left(-\frac{\sqrt{3}}{2}\right)$

4. $\sin^{-1}\left(-\frac{\sqrt{2}}{2}\right)$

5. $\arctan\left(\sqrt{3}\right)$

6. $\text{arcsec}\,(2)$

7. $\arcsin\left(-\frac{1}{2}\right)$

8. $\tan^{-1}\left(-\frac{\sqrt{3}}{3}\right)$

9. $\sin(\sin^{-1} 0.4567)$

10. $\cos(\sin^{-1} 0.56)$

In Problems 11–18, approximate each value.

11. $\sin^{-1}(0.1113)$ **12.** $\arccos(0.6341)$

13. $\cos(\text{arccot } 3.212)$ **14.** $\sec(\arccos 0.5111)$

15. $\sec^{-1}(-2.222)$ **16.** $\tan^{-1}(-60.11)$

17. $\cos(\sin(\tan^{-1} 2.001))$ **18.** $\sin^2(\ln(\cos 0.5555))$

In Problems 19–24, express θ in terms of x using the inverse trigonometric functions $\sin^{-1}$, $\cos^{-1}$, $\tan^{-1}$, and $\sec^{-1}$.

19.

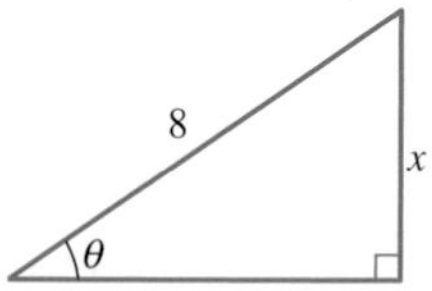

20.

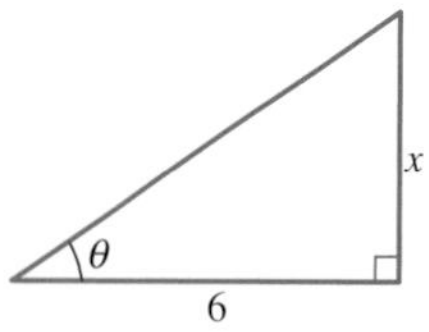

21.

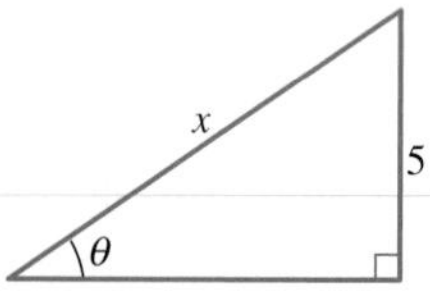

22.

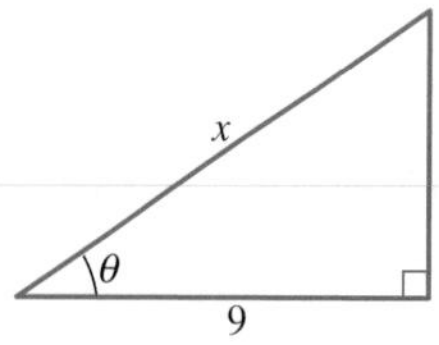

23.

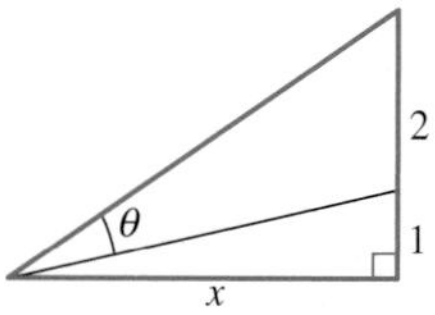

24.

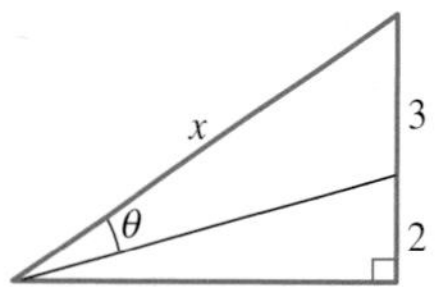

In Problems 25–28, find each value without using a calculator (see Example 4).

25. $\cos\left[2\sin^{-1}\left(-\frac{2}{3}\right)\right]$ **26.** $\tan\left[2\tan^{-1}\left(\frac{1}{3}\right)\right]$

27. $\sin\left[\cos^{-1}\left(\frac{3}{5}\right) + \cos^{-1}\left(\frac{5}{13}\right)\right]$

28. $\cos\left[\cos^{-1}\left(\frac{4}{5}\right) + \sin^{-1}\left(\frac{12}{13}\right)\right]$

In Problems 29–32, show that each equation is an identity.

29. $\tan(\sin^{-1} x) = \dfrac{x}{\sqrt{1 - x^2}}$

30. $\sin(\tan^{-1} x) = \dfrac{x}{\sqrt{1 + x^2}}$

31. $\cos(2\sin^{-1} x) = 1 - 2x^2$

32. $\tan(2\tan^{-1} x) = \dfrac{2x}{1 - x^2}$

33. Find each limit.

(a) $\lim_{x\to\infty} \tan^{-1} x$ (b) $\lim_{x\to-\infty} \tan^{-1} x$

34. Find each limit.

(a) $\lim_{x\to\infty} \sec^{-1} x$ (b) $\lim_{x\to-\infty} \sec^{-1} x$

35. Find each limit.

(a) $\lim_{x\to 1^-} \sin^{-1} x$ (b) $\lim_{x\to -1^+} \sin^{-1} x$

36. Does $\lim_{x\to 1} \sin^{-1} x$ exist? Explain.

37. Describe what happens to the slope of the tangent line to the graph of $y = \sin^{-1} x$ at the point c if c approaches 1 from the left.

38. Sketch the graph of $y = \cot^{-1} x$, assuming that it has been obtained by restricting the domain of the cotangent to $(0, \pi)$.

In Problems 39–54, find dy/dx.

39. $y = \ln(2 + \sin x)$ **40.** $y = e^{\tan x}$

41. $y = \ln(\sec x + \tan x)$ **42.** $y = -\ln(\csc x + \cot x)$

43. $y = \sin^{-1}(2x^2)$ **44.** $y = \arccos(e^x)$

45. $y = x^3 \tan^{-1}(e^x)$ **46.** $y = e^x \arcsin x^2$

47. $y = (\tan^{-1} x)^3$ **48.** $y = \tan(\cos^{-1} x)$

49. $y = \sec^{-1}(x^3)$ **50.** $y = (\sec^{-1} x)^3$

51. $y = (1 + \sin^{-1} x)^3$ **52.** $y = \sin^{-1}\left(\dfrac{1}{x^2 + 4}\right)$

53. $y = \tan^{-1}(\ln x^2)$ **54.** $y = x\,\text{arcsec}(x^2 + 1)$

In Problems 55–72, evaluate each integral.

55. $\int \cos 3x\, dx$ **56.** $\int x \sin(x^2)\, dx$

57. $\int \sin 2x \cos 2x\, dx$ **58.** $\int \tan x\, dx = \int \dfrac{\sin x}{\cos x}\, dx$

59. $\int_0^1 e^{2x}\cos(e^{2x})\, dx$ **60.** $\int_0^{\pi/2} \sin^2 x \cos x\, dx$

61. $\int_0^{\sqrt{2}/2} \dfrac{1}{\sqrt{1 - x^2}}\, dx$ **62.** $\int_{\sqrt{2}}^{2} \dfrac{dx}{x\sqrt{x^2 - 1}}$

63. $\int_{-1}^{1} \dfrac{1}{1 + x^2}\, dx$ **64.** $\int_0^{\pi/2} \dfrac{\sin\theta}{1 + \cos^2\theta}\, d\theta$

65. $\int \dfrac{1}{1 + 4x^2}\, dx$ **66.** $\int \dfrac{e^x}{1 + e^{2x}}\, dx$

67. $\int \dfrac{1}{\sqrt{12 - 9x^2}}\, dx$ **68.** $\int \dfrac{x}{\sqrt{12 - 9x^2}}\, dx$

69. $\int \dfrac{1}{x^2 - 6x + 13}\, dx$ **70.** $\int \dfrac{1}{2x^2 + 8x + 25}\, dx$

71. $\int \dfrac{1}{x\sqrt{4x^2 - 9}}\, dx$ **72.** $\int \dfrac{x + 1}{\sqrt{4 - 9x^2}}\, dx$

C **73.** A picture 5 feet in height is hung on a wall so that its bottom is 8 feet from the floor, as shown in Figure 9. A viewer with eye level at 5.4 feet stands b feet from the wall. Express θ, the vertical angle subtended by the picture at her eye, in terms of b, and then find θ if $b = 12.9$ feet.

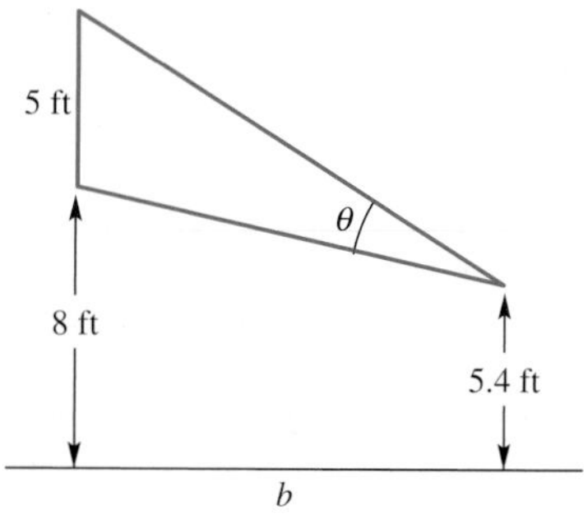

Figure 9

74. Find formulas for $f^{-1}(x)$ for each of the following functions f, first indicating how you would restrict the domain so that f has an inverse. For example, if $f(x) = 3 \sin 2x$ and we restrict the domain to $-\pi/4 \le x \le \pi/4$, then $f^{-1}(x) = \frac{1}{2}\sin^{-1}(x/3)$.

(a) $f(x) = 3 \cos 2x$ (b) $f(x) = 2 \sin 3x$

(c) $f(x) = \frac{1}{2} \tan x$ (d) $f(x) = \sin \dfrac{1}{x}$

75. By repeated use of the addition formula

$$\tan(x + y) = (\tan x + \tan y)/(1 - \tan x \tan y)$$

show that

$$\frac{\pi}{4} = 3 \tan^{-1}\left(\frac{1}{4}\right) + \tan^{-1}\left(\frac{5}{99}\right)$$

76. Verify that

$$\frac{\pi}{4} = 4 \tan^{-1}\left(\frac{1}{5}\right) - \tan^{-1}\left(\frac{1}{239}\right)$$

a result discovered by John Machin in 1706 and used by him to calculate the first 100 decimal places of π.

77. Without using calculus, find a formula for the area of the shaded region in Figure 10 in terms of a and b. Note that the center of the larger circle is on the rim of the smaller.

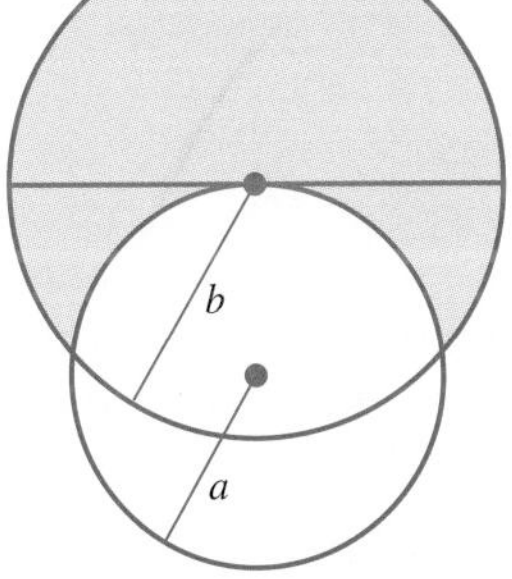

Figure 10

GC **78.** Draw the graphs of

$$y = \arcsin x \quad \text{and} \quad y = \arctan\left(x/\sqrt{1 - x^2}\right)$$

using the same axes. Make a conjecture. Prove it.

GC **79.** Draw the graph of $y = \pi/2 - \arcsin x$. Make a conjecture. Prove it.

GC **80.** Draw the graph of $y = \sin(\arcsin x)$ on $[-1, 1]$. Then draw the graph of $y = \arcsin(\sin x)$ on $[-2\pi, 2\pi]$. Explain the differences that you observe.

81. Show that

$$\int \frac{dx}{\sqrt{a^2 - x^2}} = \sin^{-1}\frac{x}{a} + C, \qquad a > 0$$

by writing $a^2 - x^2 = a^2[1 - (x/a)^2]$ and making the substitution $u = x/a$.

82. Show the result in Problem 81 by differentiating the right side to get the integrand.

83. Show that

$$\int \frac{dx}{a^2 + x^2} = \frac{1}{a}\tan^{-1}\frac{x}{a} + C, \qquad a \ne 0$$

84. Show that

$$\int \frac{dx}{x\sqrt{x^2 - a^2}} = \frac{1}{a}\sec^{-1}\frac{|x|}{a} + C, \qquad a > 0$$

85. Show, by differentiating the right side, that

$$\int \sqrt{a^2 - x^2}\, dx = \frac{x}{2}\sqrt{a^2 - x^2} + \frac{a^2}{2}\sin^{-1}\frac{x}{a} + C, \qquad a > 0$$

86. Use the result of Problem 85 to show that

$$\int_{-a}^{a} \sqrt{a^2 - x^2}\, dx = \frac{\pi a^2}{2}$$

Why is this result expected?

87. The lower edge of a wall hanging, 10 feet in height, is 2 feet above the observer's eye level. Find the ideal distance b to stand from the wall for viewing the hanging; that is, find b that maximizes the angle subtended at the viewer's eye. (See Problem 73.)

88. Express $d\theta/dt$ in terms of x, dx/dt, and the constants a and b.

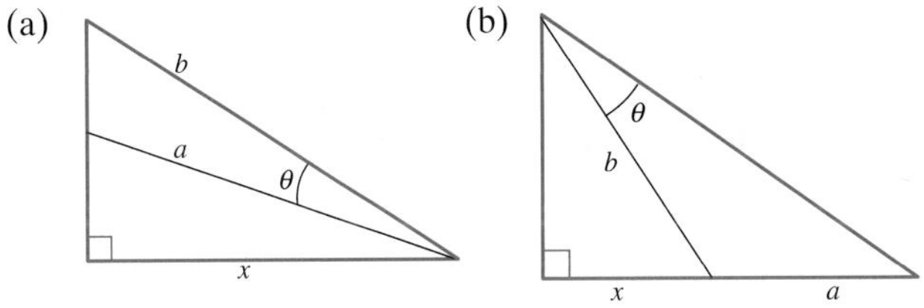

89. The structural steel work of a new office building is finished. Across the street, 60 feet from the ground floor of the freight elevator shaft in the building, a spectator is standing and watching the freight elevator ascend at a constant rate of 15 feet per second. How fast is the angle of elevation of the spectator's line of sight to the elevator increasing 6 seconds after his line of sight passes the horizontal?

90. An airplane is flying at a constant altitude of 2 miles and a constant speed of 600 miles per hour on a straight course that will take it directly over an observer on the ground. How fast is the angle of elevation of the observer's line of sight increasing when the distance from her to the plane is 3 miles? Give your result in radians per minute.

91. A revolving beacon light is located on an island and is 2 miles away from the nearest point P of the straight shoreline of the mainland. The beacon throws a spot of light that moves along the shoreline as the beacon revolves. If the speed of the spot of light on the shoreline is 5π miles per minute when the spot is 1 mile from P, how fast is the beacon revolving?

92. A man on a dock is pulling in a rope attached to a rowboat at a rate of 5 feet per second. If the man's hands are 8 feet higher than the point where the rope is attached to the boat, how fast is the angle of depression of the rope changing when there are still 17 feet of rope out?

C **93.** A visitor from outer space is approaching the earth (radius = 6376 kilometers) at 2 kilometers per second. How fast is the angle θ subtended by the earth at her eye increasing when she is 3000 kilometers from the surface?

Answers to Concepts Review: **1.** $[-\pi/2, \pi/2]$; arcsin **2.** $(-\pi/2, \pi/2)$; arctan **3.** 1 **4.** π

6.9 The Hyperbolic Functions and Their Inverses

In both mathematics and science, certain combinations of e^x and e^{-x} occur so often that they are given special names.

Definition Hyperbolic Functions

The hyperbolic sine, hyperbolic cosine, and four related functions are defined by

$$\sinh x = \frac{e^x - e^{-x}}{2} \qquad \cosh x = \frac{e^x + e^{-x}}{2}$$

$$\tanh x = \frac{\sinh x}{\cosh x} \qquad \coth x = \frac{\cosh x}{\sinh x}$$

$$\operatorname{sech} x = \frac{1}{\cosh x} \qquad \operatorname{csch} x = \frac{1}{\sinh x}$$

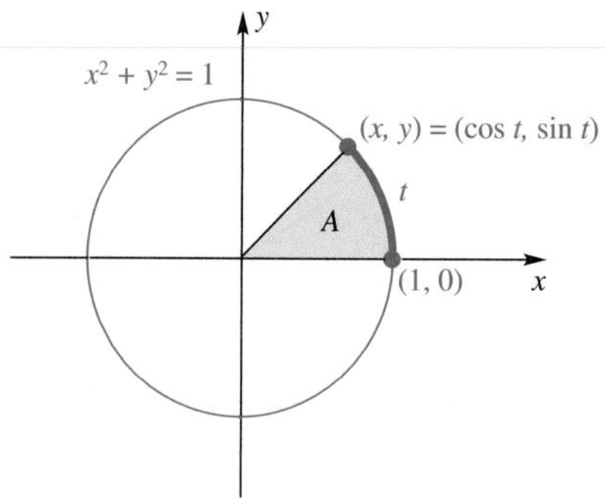

Figure 1

The terminology suggests that there must be some connection with the trigonometric functions; there is. First, the fundamental identity for the hyperbolic functions (reminiscent of $\cos^2 x + \sin^2 x = 1$ in trigonometry) is

$$\cosh^2 x - \sinh^2 x = 1$$

To verify it, we write

$$\cosh^2 x - \sinh^2 x = \frac{e^{2x} + 2 + e^{-2x}}{4} - \frac{e^{2x} - 2 + e^{-2x}}{4} = 1$$

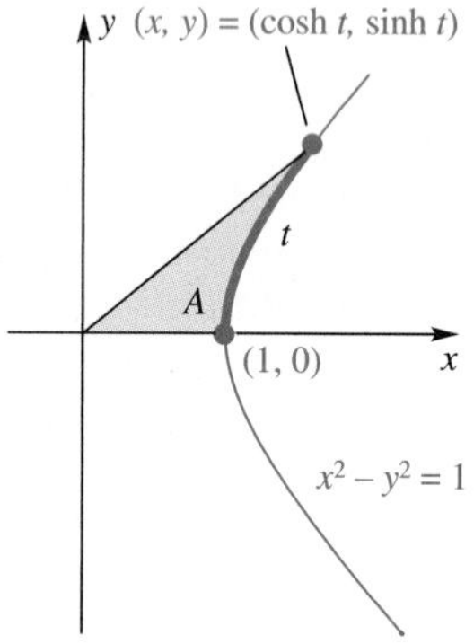

Figure 2

Second, recall that the trigonometric functions are intimately related to the unit circle (Figure 1), so much so that they are sometimes called the *circular functions*. In fact, the parametric equations $x = \cos t$, $y = \sin t$ describe the unit circle. In parallel fashion, the parametric equations $x = \cosh t$, $y = \sinh t$ describe the right branch of the unit hyperbola $x^2 - y^2 = 1$ (Figure 2). Moreover, in both cases the parameter t is related to the shaded area A by $t = 2A$, though this is not obvious in the second case (Problem 56).

Since $\sinh(-x) = -\sinh x$, sinh is an odd function; $\cosh(-x) = \cosh x$, so cosh is an even function. Correspondingly, the graph of $y = \sinh x$ is symmetric with respect to the origin and the graph of $y = \cosh x$ is symmetric with respect to the y-axis. Similarly, tanh is an odd function and sech is an even function. The graphs are shown in Figure 3.

Derivatives of Hyperbolic Functions We can find $D_x \sinh x$ and $D_x \cosh x$ directly from the definitions.

$$D_x \sinh x = D_x\left(\frac{e^x - e^{-x}}{2}\right) = \frac{e^x + e^{-x}}{2} = \cosh x$$

and

$$D_x \cosh x = D_x\left(\frac{e^x + e^{-x}}{2}\right) = \frac{e^x - e^{-x}}{2} = \sinh x$$

Note that these facts confirm the character of the graphs in Figure 3. For example, since $D_x(\sinh x) = \cosh x > 0$, the graph of hyperbolic sine is always increasing. Similarly, $D_x^2(\cosh x) = \cosh x > 0$, which means that the graph of hyperbolic cosine is concave upward.

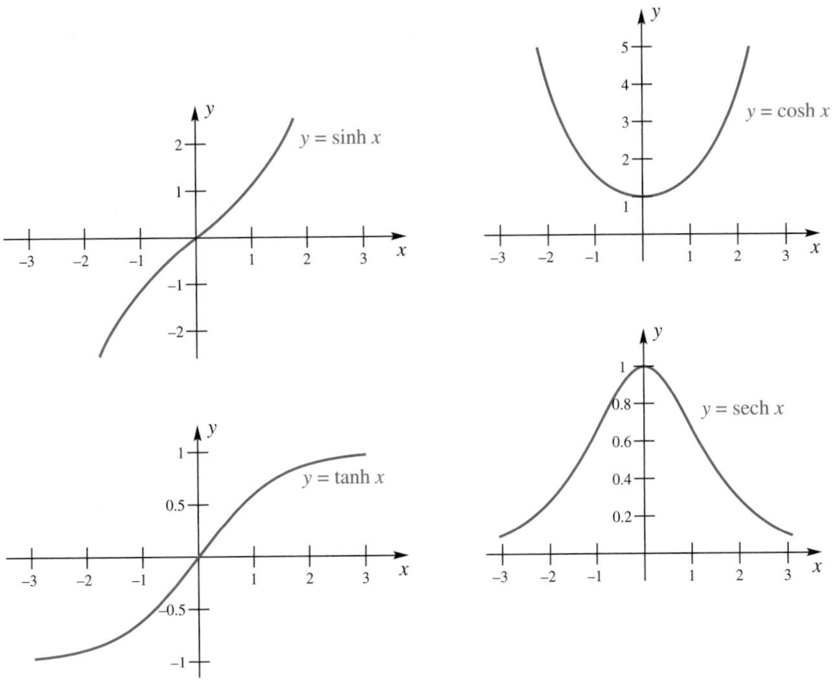

Figure 3

The derivatives of the other four hyperbolic functions follow from those for the first two, combined with the Quotient Rule. The results are summarized in Theorem A.

Theorem A **Derivatives of Hyperbolic Functions**

$D_x \sinh x = \cosh x$	$D_x \cosh x = \sinh x$
$D_x \tanh x = \operatorname{sech}^2 x$	$D_x \coth x = -\operatorname{csch}^2 x$
$D_x \operatorname{sech} x = -\operatorname{sech} x \tanh x$	$D_x \operatorname{csch} x = -\operatorname{csch} x \coth x$

Another way that the trigonometric and hyperbolic functions are connected concerns differential equations. The functions $\sin x$ and $\cos x$ are solutions to the second-order differential equation $y'' = -y$, and $\sinh x$ and $\cosh x$ are solutions to the differential equation $y'' = y$.

EXAMPLE 1 Find $D_x \tanh(\sin x)$.

SOLUTION

$$\begin{aligned} D_x \tanh(\sin x) &= \operatorname{sech}^2(\sin x)\, D_x(\sin x) \\ &= \cos x \cdot \operatorname{sech}^2(\sin x) \end{aligned}$$

■

EXAMPLE 2 Find $D_x \cosh^2(3x - 1)$.

SOLUTION We apply the Chain Rule twice.

$$\begin{aligned} D_x \cosh^2(3x - 1) &= 2\cosh(3x - 1)\, D_x \cosh(3x - 1) \\ &= 2\cosh(3x - 1)\sinh(3x - 1)\, D_x(3x - 1) \\ &= 6\cosh(3x - 1)\sinh(3x - 1) \end{aligned}$$

■

EXAMPLE 3 Find $\int \tanh x\, dx$.

SOLUTION Let $u = \cosh x$, so $du = \sinh x\, dx$.

$$\int \tanh x\, dx = \int \frac{\sinh x}{\cosh x}\, dx = \int \frac{1}{u}\, du$$

$$= \ln|u| + C = \ln|\cosh x| + C = \ln(\cosh x) + C$$

We could drop the absolute value signs because $\cosh x > 0$. ■

Inverse Hyperbolic Functions Since hyperbolic sine and hyperbolic tangent have positive derivatives, they are increasing functions and automatically have inverses. To obtain inverses for hyperbolic cosine and hyperbolic secant, we restrict their domains to $x \geq 0$. Thus,

$$x = \sinh^{-1} y \quad \Leftrightarrow \quad y = \sinh x$$

$$x = \cosh^{-1} y \quad \Leftrightarrow \quad y = \cosh x \quad \text{and} \quad x \geq 0$$

$$x = \tanh^{-1} y \quad \Leftrightarrow \quad y = \tanh x$$

$$x = \operatorname{sech}^{-1} y \quad \Leftrightarrow \quad y = \operatorname{sech} x \quad \text{and} \quad x \geq 0$$

Since the hyperbolic functions are defined in terms of e^x and e^{-x}, it is not too surprising that the inverse hyperbolic functions can be expressed in terms of the natural logarithm. For example, consider $y = \cosh x$ for $x \geq 0$; that is, consider

$$y = \frac{e^x + e^{-x}}{2}, \qquad x \geq 0$$

Our goal is to solve this equation for x, which will give $\cosh^{-1} y$. Multiplying both sides by $2e^x$, we get $2ye^x = e^{2x} + 1$, or

$$(e^x)^2 - 2ye^x + 1 = 0, \qquad x \geq 0$$

If we solve this quadratic equation in e^x, we obtain

$$e^x = \frac{2y + \sqrt{(2y)^2 - 4}}{2} = y + \sqrt{y^2 - 1}$$

The Quadratic Formula gives two solutions, the one given above and $\left(2y - \sqrt{(2y)^2 - 4}\right)/2$. This latter solution is extraneous because it is less than 1, whereas e^x is greater than 1 for all $x > 0$. Thus, $x = \ln\left(y + \sqrt{y^2 - 1}\right)$, so

$$x = \cosh^{-1} y = \ln\left(y + \sqrt{y^2 - 1}\right)$$

Similar arguments apply to each of the inverse hyperbolic functions. We obtain the following results (note that the roles of x and y have been interchanged). Figure 3 suggests the necessary domain restrictions. Graphs of the inverse hyperbolic functions are shown in Figure 4.

$$\sinh^{-1} x = \ln\left(x + \sqrt{x^2 + 1}\right)$$

$$\cosh^{-1} x = \ln\left(x + \sqrt{x^2 - 1}\right), \qquad x \geq 1$$

$$\tanh^{-1} x = \frac{1}{2}\ln\frac{1 + x}{1 - x}, \qquad -1 < x < 1$$

$$\operatorname{sech}^{-1} x = \ln\left(\frac{1 + \sqrt{1 - x^2}}{x}\right), \qquad 0 < x \leq 1$$

Each of these functions is differentiable. In fact,

$$D_x \sinh^{-1} x = \frac{1}{\sqrt{x^2+1}}$$

$$D_x \cosh^{-1} x = \frac{1}{\sqrt{x^2-1}}, \qquad x > 1$$

$$D_x \tanh^{-1} x = \frac{1}{1-x^2}, \qquad -1 < x < 1$$

$$D_x \operatorname{sech}^{-1} x = \frac{-1}{x\sqrt{1-x^2}}, \qquad 0 < x < 1$$

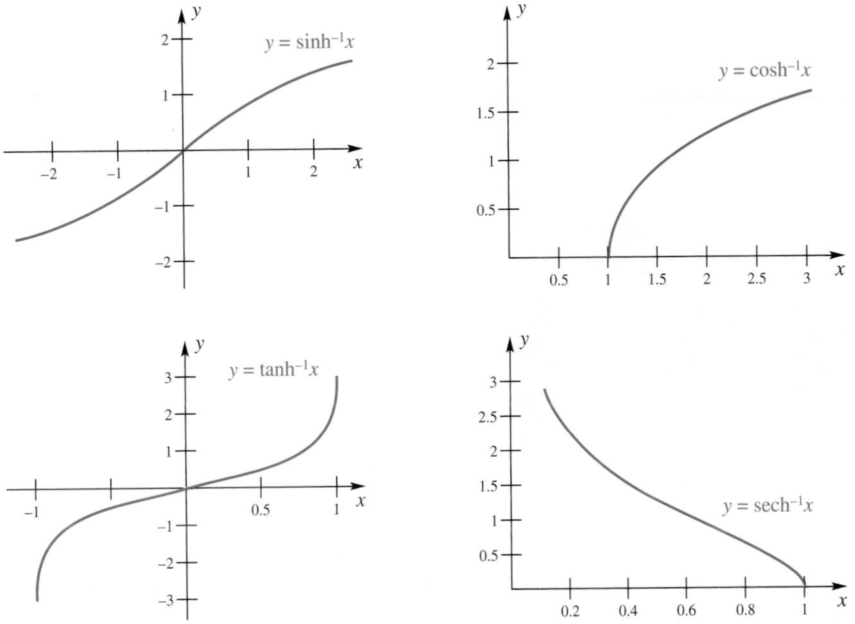

Figure 4

EXAMPLE 4 Show that $D_x \sinh^{-1} x = 1/\sqrt{x^2+1}$ by two different methods.

SOLUTION

Method 1 Let $y = \sinh^{-1} x$, so

$$x = \sinh y$$

Now differentiate both sides with respect to x.

$$1 = (\cosh y)\, D_x y$$

Thus,

$$D_x y = D_x(\sinh^{-1} x) = \frac{1}{\cosh y} = \frac{1}{\sqrt{1+\sinh^2 y}} = \frac{1}{\sqrt{1+x^2}}$$

Method 2 Use the logarithmic expression for $\sinh^{-1} x$.

$$\begin{aligned} D_x(\sinh^{-1} x) &= D_x \ln\left(x + \sqrt{x^2+1}\right) \\ &= \frac{1}{x+\sqrt{x^2+1}} D_x\left(x + \sqrt{x^2+1}\right) \\ &= \frac{1}{x+\sqrt{x^2+1}}\left(1 + \frac{x}{\sqrt{x^2+1}}\right) \\ &= \frac{1}{\sqrt{x^2+1}} \end{aligned}$$

■

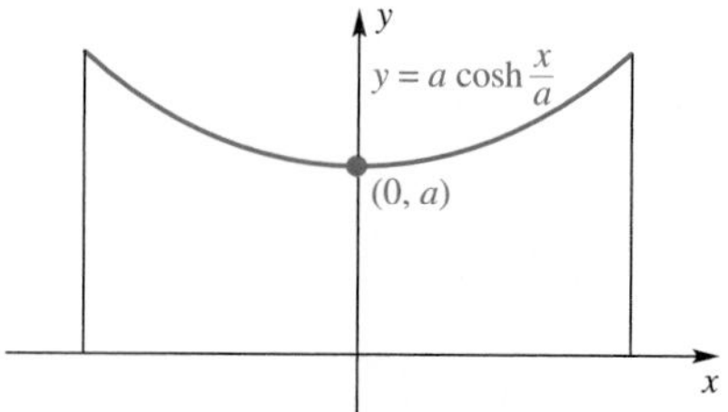

The catenary

Figure 5

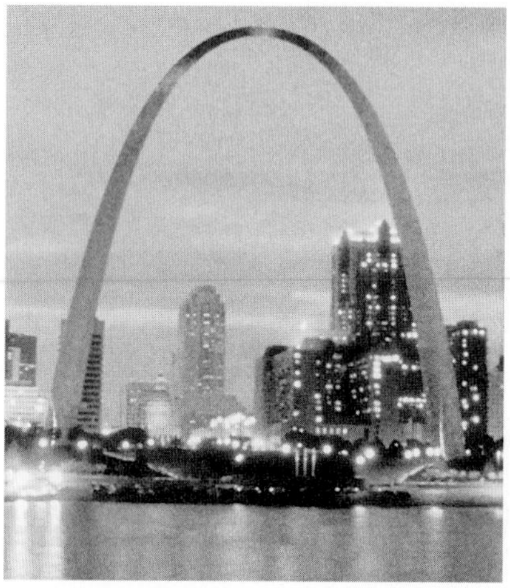
An Inverted Catenary

Applications: The Catenary If a homogeneous flexible cable or chain is suspended between two fixed points at the same height, it forms a curve called a **catenary** (Figure 5). Furthermore (see Problem 53), a catenary can be placed in a coordinate system so that its equation takes the form

$$y = a \cosh \frac{x}{a}$$

EXAMPLE 5 Find the length of the catenary $y = a\cosh(x/a)$ between $x = -a$ and $x = a$.

SOLUTION The desired length (see Section 5.4) is given by

$$\begin{aligned}
\int_{-a}^{a} \sqrt{1 + \left(\frac{dy}{dx}\right)^2}\, dx &= \int_{-a}^{a} \sqrt{1 + \sinh^2\left(\frac{x}{a}\right)}\, dx \\
&= \int_{-a}^{a} \sqrt{\cosh^2\left(\frac{x}{a}\right)}\, dx \\
&= 2\int_{0}^{a} \cosh\left(\frac{x}{a}\right) dx \\
&= 2a\int_{0}^{a} \cosh\left(\frac{x}{a}\right)\left(\frac{1}{a}\, dx\right) \\
&= \left[2a \sinh \frac{x}{a}\right]_0^a \\
&= 2a \sinh 1 \approx 2.35a
\end{aligned}$$

■

Concepts Review

1. sinh and cosh are defined by $\sinh x =$ _____ and $\cosh x =$ _____.

2. In *hyperbolic* trigonometry, the identity corresponding to $\sin^2 x + \cos^2 x = 1$ is _____.

3. Because of the identity in Question 2, the graph of the parametric equations $x = \cosh t$, $y = \sinh t$ is _____.

4. The graph of $y = a\cosh(x/a)$ is a curve called a _____; this curve is important as a model for _____.

Problem Set 6.9

In Problems 1–12, verify that the given equations are identities.

1. $e^x = \cosh x + \sinh x$

2. $e^{2x} = \cosh 2x + \sinh 2x$

3. $e^{-x} = \cosh x - \sinh x$

4. $e^{-2x} = \cosh 2x - \sinh 2x$

5. $\sinh(x + y) = \sinh x \cosh y + \cosh x \sinh y$

6. $\sinh(x - y) = \sinh x \cosh y - \cosh x \sinh y$

7. $\cosh(x + y) = \cosh x \cosh y + \sinh x \sinh y$

8. $\cosh(x - y) = \cosh x \cosh y - \sinh x \sinh y$

9. $\tanh(x + y) = \dfrac{\tanh x + \tanh y}{1 + \tanh x \tanh y}$

10. $\tanh(x - y) = \dfrac{\tanh x - \tanh y}{1 - \tanh x \tanh y}$

11. $\sinh 2x = 2 \sinh x \cosh x$

12. $\cosh 2x = \cosh^2 x + \sinh^2 x$

In Problems 13–36, find $D_x y$.

13. $y = \sinh^2 x$

14. $y = \cosh^2 x$

15. $y = 5\sinh^2 x$

16. $y = \cosh^3 x$

17. $y = \cosh(3x + 1)$

18. $y = \sinh(x^2 + x)$

19. $y = \ln(\sinh x)$

20. $y = \ln(\coth x)$

21. $y = x^2 \cosh x$

22. $y = x^{-2} \sinh x$

23. $y = \cosh 3x \sinh x$

24. $y = \sinh x \cosh 4x$

25. $y = \tanh x \sinh 2x$

26. $y = \coth 4x \sinh x$

27. $y = \sinh^{-1}(x^2)$

28. $y = \cosh^{-1}(x^3)$

29. $y = \tanh^{-1}(2x - 3)$

30. $y = \coth^{-1}(x^5)$

31. $y = x \cosh^{-1}(3x)$

32. $y = x^2 \sinh^{-1}(x^5)$

33. $y = \ln(\cosh^{-1} x)$

34. $y = \cosh^{-1}(\cos x)$

35. $y = \tanh(\cot x)$

36. $y = \coth^{-1}(\tanh x)$

37. Find the area of the region bounded by $y = \cosh 2x$, $y = 0$, $x = 0$, and $x = \ln 3$.

In Problems 38–45, evaluate each integral.

38. $\int \sinh(3x + 2)\, dx$

39. $\int x \cosh(\pi x^2 + 5)\, dx$

40. $\int \frac{\cosh \sqrt{z}}{\sqrt{z}}\, dz$

41. $\int \frac{\sinh(2z^{1/4})}{\sqrt[4]{z^3}}\, dz$

42. $\int e^x \sinh e^x\, dx$

43. $\int \cos x \sinh(\sin x)\, dx$

44. $\int \tanh x \ln(\cosh x)\, dx$

45. $\int x \coth x^2 \ln(\sinh x^2)\, dx$

46. Find the area of the region bounded by $y = \cosh 2x$, $y = 0$, $x = -\ln 5$, and $x = \ln 5$.

47. Find the area of the region bounded by $y = \sinh x$, $y = 0$, and $x = \ln 2$.

48. Find the area of the region bounded by $y = \tanh x$, $y = 0$, $x = -8$, and $x = 8$.

49. The region bounded by $y = \cosh x$, $y = 0$, $x = 0$, and $x = 1$ is revolved about the x-axis. Find the volume of the resulting solid. *Hint:* $\cosh^2 x = (1 + \cosh 2x)/2$.

50. The region bounded by $y = \sinh x$, $y = 0$, $x = 0$, and $x = \ln 10$ is revolved about the x-axis. Find the volume of the resulting solid.

51. The curve $y = \cosh x$, $0 \le x \le 1$, is revolved about the x-axis. Find the area of the resulting surface.

52. The curve $y = \sinh x$, $0 \le x \le 1$, is revolved about the x-axis. Find the area of the resulting surface.

53. To derive the equation of a hanging cable (catenary), we consider the section AP from the lowest point A to a general point $P(x, y)$ (see Figure 6) and imagine the rest of the cable to have been removed.

The forces acting on the cable are

1. H = horizontal tension pulling at A;
2. T = tangential tension pulling at P;
3. $W = \delta s$ = weight of s feet of cable of density δ pounds per foot.

To be in equilibrium, the horizontal and vertical components of T must just balance H and W, respectively. Thus, $T \cos \phi = H$ and $T \sin \phi = W = \delta s$, and so

$$\frac{T \sin \phi}{T \cos \phi} = \tan \phi = \frac{\delta s}{H}$$

But since $\tan \phi = dy/dx$, we get

$$\frac{dy}{dx} = \frac{\delta s}{H}$$

and therefore

$$\frac{d^2y}{dx^2} = \frac{\delta}{H}\frac{ds}{dx} = \frac{\delta}{H}\sqrt{1 + \left(\frac{dy}{dx}\right)^2}$$

Now show that $y = a \cosh(x/a) + C$ satisfies this differential equation with $a = H/\delta$.

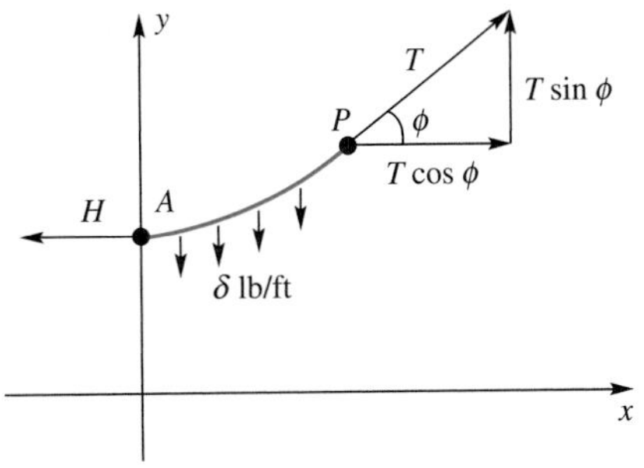

Figure 6

C **54.** Call the graph of $y = b - a\cosh(x/a)$ an inverted catenary and imagine it to be an arch sitting on the x-axis. Show that if the width of this arch along the x-axis is $2a$ then each of the following is true.

(a) $b = a \cosh 1 \approx 1.54308a$.

(b) The height of the arch is approximately $0.54308a$.

(c) The height of an arch of width 48 is approximately 13.

C **55.** A farmer built a large hayshed of length 100 feet and width 48 feet. A cross section has the shape of an inverted catenary (see Problem 54) with equation $y = 37 - 24\cosh(x/24)$.

(a) Draw a picture of this shed.

(b) Find the volume of the shed.

(c) Find the surface area of the roof of the shed.

56. Show that $A = t/2$, where A denotes the area in Figure 2 of this section. *Hint:* At some point you will need to use Formula 44 from the back of the book.

57. Demonstrate that for every real number r:

(a) $(\sinh x + \cosh x)^r = \sinh rx + \cosh rx$

(b) $(\cosh x - \sinh x)^r = \cosh rx - \sinh rx$

(c) $(\cos x + i \sin x)^r = \cos rx + i \sin rx$

(d) $(\cos x - i \sin x)^r = \cos rx - i \sin rx$

58. The **gudermannian** of t is defined by

$$\text{gd}(t) = \tan^{-1}(\sinh t)$$

Show that

(a) gd is odd and increasing with an inflection point at the origin;

(b) $\text{gd}(t) = \sin^{-1}(\tanh t) = \int_0^t \text{sech}\, u\, du$.

59. Show that the area under the curve $y = \cosh t$, $0 \le t \le x$, is numerically equal to its arc length.

60. Find the equation of the Gateway Arch in St. Louis, Missouri, given that it is an inverted catenary (see Problem 54). Assume that it stands on the x-axis, that it is symmetric with respect to the y-axis, and that it is 630 feet wide at the base and 630 feet high at the center.

GC **61.** Draw the graphs of $y = \sinh x$, $y = \ln\left(x + \sqrt{x^2 + 1}\right)$, and $y = x$ using the same axes and scaled so that $-3 \le x \le 3$ and $-3 \le y \le 3$. What does this demonstrate?

CAS **62.** Refer to Problem 58. Derive a formula for $\mathrm{gd}^{-1}(x)$. Draw its graph and also that of $\mathrm{gd}(x)$ using the same axes and thereby confirm your formula.

Answers to Concepts Review: **1.** $(e^x - e^{-x})/2$; $(e^x + e^{-x})/2$ **2.** $\cosh^2 x - \sinh^2 x = 1$ **3.** the graph of $x^2 - y^2 = 1$ (a hyperbola) **4.** catenary; a hanging cable

6.10 Chapter Review

Concepts Test

Respond with true or false to each of the following assertions. Be prepared to justify your answer.

1. $\ln|x|$ is defined for all real x.

2. The graph of $y = \ln x$ has no inflection points.

3. $\int_1^{e^3} \frac{1}{t}\,dt = 3$

4. The graph of an invertible function $y = f(x)$ is intersected exactly once by every horizontal line.

5. The domain of $\ln^{-1}$ is the set of all real numbers.

6. $\ln x/\ln y = \ln x - \ln y$

7. $(\ln x)^4 = 4 \ln x$

8. $\ln(2e^{x+1}) - \ln(2e^x) = 1$ for all real numbers x.

9. The functions $f(x) = 4 + e^x$ and $g(x) = \ln(x - 4)$ are inverses of each other.

10. $\exp x + \exp y = \exp(x + y)$.

11. If $x > y > 0$, then $\ln x > \ln y$.

12. If $a \ln x < b \ln x$, then $a < b$.

13. If $a < b$, then $ae^x < be^x$.

14. If $a < b$, then $e^a < e^b$.

15. $\lim_{x\to 0^+} (\ln \sin x - \ln x) = 0$.

16. $\pi^{\sqrt{2}} = e^{\sqrt{2}\ln \pi}$

17. $\frac{d}{dx}(\ln \pi) = \frac{1}{\pi}$

18. $\int \frac{1}{x}\,dx = \ln 3|x| + C$

19. $D_x(x^e) = ex^{e-1}$

20. If $f(x) \cdot \exp[g(x)] = 0$ for $x = x_0$, then $f(x_0) = 0$.

21. $D_x(x^x) = x^x \ln x$.

22. $y = \tan x + \sec x$ is a solution of $2y' - y^2 = 1$.

23. An integrating factor for $y' + \frac{4}{x}y = e^x$ is x^4.

24. The solution to the differential equation $y' = 2y$ that passes through the point $(2, 1)$ has slope 2 at that point.

25. Euler's Method will always overestimate the solution of the differential equation $y' = 2y$ with initial condition $y(0) = 1$.

26. $\sin(\arcsin x) = x$ for all real numbers x.

27. $\arcsin(\sin x) = x$ for all real numbers x.

28. If $a < b$, then $\sinh a < \sinh b$.

29. If $a < b$, then $\cosh a < \cosh b$.

30. $\cosh x \le e^{|x|}$

31. $|\sinh x| \le e^{|x|}/2$

32. $\tan^{-1} x = \sin^{-1} x/\cos^{-1} x$

33. $\cosh(\ln 3) = \frac{5}{6}$

34. $\lim_{x\to 0} \ln\left(\frac{\sin x}{x}\right) = 1$

35. $\lim_{x\to -\infty} \tan^{-1} x = -\frac{\pi}{2}$

36. $\sin^{-1}(\cosh x)$ is defined for all real numbers x.

37. $f(x) = \tanh x$ is an odd function.

38. Both $y = \sinh x$ and $y = \cosh x$ satisfy the differential equation $y'' + y = 0$.

39. $\ln(3^{100}) > 100$.

40. $\ln(2x^2 - 18) - \ln(x - 3) - \ln(x + 3) = \ln 2$ for all real numbers x.

41. If y is growing exponentially and if y triples between $t = 0$ and $t = t_1$, then y will also triple between $t = 2t_1$ and $t = 3t_1$.

42. The time necessary for $x(t) = Ce^{-kt}$ to drop to half its value is $\frac{\ln 2}{\ln k}$.

43. If $y'(t) = ky(t)$ and $z'(t) = kz(t)$, then $(y(t) + z(t))' = k(y(t) + z(t))$.

44. If $y_1(t)$ and $y_2(t)$ both satisfy $y'(t) = ky(t) + C$, then so does $(y_1(t) + y_2(t))$.

45. $\lim_{h\to 0}(1 - h)^{-1/h} = e^{-1}$.

46. It is to a saver's advantage to have money invested at 5% compounded continuously rather than 6% compounded monthly.

47. If $D_x(a^x) = a^x$ with $a > 0$, then $a = e$.

Sample Test Problems

In Problems 1–24, differentiate each function.

1. $\ln\frac{x^4}{2}$

2. $\sin^2(x^3)$

3. e^{x^2-4x}

4. $\log_{10}(x^5 - 1)$

5. $\tan(\ln e^x)$

6. $e^{\ln \cot x}$

7. $2 \tanh \sqrt{x}$

8. $\tanh^{-1}(\sin x)$

9. $\sinh^{-1}(\tan x)$

10. $2 \sin^{-1}\sqrt{3x}$

11. $\sec^{-1} e^x$

12. $\ln \sin^2\left(\frac{x}{2}\right)$

13. $3 \ln(e^{5x} + 1)$

14. $\ln(2x^3 - 4x + 5)$

15. $\cos e^{\sqrt{x}}$

16. $\ln(\tanh x)$

17. $2 \cos^{-1}\sqrt{x}$

18. $4^{3x} + (3x)^4$

19. $2 \csc e^{\ln \sqrt{x}}$

20. $(\log_{10} 2x)^{2/3}$

21. $4 \tan 5x \sec 5x$

22. $x \tan^{-1}\frac{x^2}{2}$

23. x^{1+x}

24. $(1 + x^2)^e$

In Problems 25–34, find the antiderivative of each function and verify your result by differentiation.

25. e^{3x-1}

26. $6 \cot 3x$

27. $e^x \sin e^x$

28. $\dfrac{6x + 3}{x^2 + x - 5}$

29. $\dfrac{e^{x+2}}{e^{x+3} + 1}$

30. $4x \cos x^2$

31. $\dfrac{4}{\sqrt{1 - 4x^2}}$

32. $\dfrac{\cos x}{1 + \sin^2 x}$

33. $\dfrac{-1}{x + x(\ln x)^2}$

34. $\operatorname{sech}^2(x - 3)$

In Problems 35 and 36, find the intervals on which f is increasing and the intervals on which f is decreasing. Find where the graph of f is concave upward and where it is concave downward. Find any extreme values and points of inflection. Then sketch the graph of f.

35. $f(x) = \sin x + \cos x, -\dfrac{\pi}{2} \le x \le \dfrac{\pi}{2}$

36. $f(x) = \dfrac{x^2}{e^x}, -\infty < x < \infty$

37. Let $f(x) = x^5 + 2x^3 + 4x, -\infty < x < \infty$.

(a) Prove that f has an inverse $g = f^{-1}$.

(b) Evaluate $g(7) = f^{-1}(7)$.

(c) Evaluate $g'(7)$.

38. A certain radioactive substance has a half-life of 10 years. How long will it take for 100 grams to decay to 1 gram?

C **39.** Use Euler's Method with $h = 0.2$ to approximate the solution to the differential equation $y' = xy$ with initial condition $y(1) = 2$ over the interval $[1, 2]$.

40. An airplane is flying horizontally at an altitude of 500 feet with a speed of 300 feet per second directly away from a searchlight on the ground. The searchlight is kept directed at the plane. At what rate is the angle between the light beam and the ground changing when this angle is 30°?

41. Find the equation of the tangent line to $y = (\cos x)^{\sin x}$ at $(0, 1)$.

42. The population of a town grew exponentially from 10,000 in 1990 to 14,000 in 2000. Assuming that the same type of growth continues, what will the population be in 2010?

In Problems 43–47, solve each differential equation.

43. $\dfrac{dy}{dx} + \dfrac{y}{x} = 0$

44. $\dfrac{dy}{dx} - \dfrac{x^2 - 2y}{x} = 0$

45. $\dfrac{dy}{dx} + 2x(y - 1) = 0$; $y = 3$ when $x = 0$

46. $\dfrac{dy}{dx} - ay = e^{ax}$

47. $\dfrac{dy}{dx} - 2y = e^x$

48. Suppose that glucose is infused into the bloodstream of a patient at the rate of 3 grams per minute, but that the patient's body converts and removes glucose from its blood at a rate proportional to the amount present (with constant of proportionality 0.02). Let $Q(t)$ be the amount present at time t, with $Q(0) = 120$.

(a) Write the differential equation for Q.

(b) Solve this differential equation.

(c) Determine what happens to Q in the long run.

REVIEW & PREVIEW PROBLEMS

Evaluate the integrals in Problems 1–8.

1. $\int \sin 2x \, dx$

2. $\int e^{3t} \, dt$

3. $\int x \sin x^2 \, dx$

4. $\int x e^{3x^2} \, dx$

5. $\int \frac{\sin t}{\cos t} \, dt$

6. $\int \sin^2 x \cos x \, dx$

7. $\int x\sqrt{x^2 + 2} \, dx$

8. $\int \frac{x}{x^2 + 1} \, dx$

Find and simplify the derivatives of the functions in Problems 9–12.

9. $f(x) = x \ln x - x$

10. $f(x) = x \arcsin x + \sqrt{1 - x^2}$

11. $f(x) = -x^2 \cos x + 2x \sin x + 2 \cos x$

12. $f(x) = e^x(\sin x - \cos x)$

13. Use one of the double-angle identities (from Section 0.7) to find an expression for $\sin^2 x$ that involves $\cos 2x$.

14. Use one of the double-angle identities to find an expression for $\cos^2 x$ that involves $\cos 2x$.

15. Use one of the double-angle identities to find an expression for $\cos^4 x$ that involves $\cos 2x$.

16. Use one of the product identities (from Section 0.7) to express $\sin 3x \cos 4x$ in terms of the sine function only, in such a way that no two trigonometric functions are multiplied together.

17. Use one of the product identities to express $\cos 3x \cos 5x$ in terms of the cosine function only, in such a way that no two trigonometric functions are multiplied together.

18. Use one of the product identities to express $\sin 2x \sin 3x$ in terms of the cosine function only, in such a way that no two trigonometric functions are multiplied together.

19. Evaluate $\sqrt{a^2 - x^2}$ when $x = a \sin t$, if $-\pi/2 \le t \le \pi/2$.

20. Evaluate $\sqrt{a^2 + x^2}$ when $x = a \tan t$, if $-\pi/2 < t < \pi/2$.

21. Evaluate $\sqrt{x^2 - a^2}$ when $x = a \sec t$, if $0 \le t \le \pi$ and $t \ne \pi/2$.

22. Solve for a in the equation $\int_0^a e^{-x} \, dx = \frac{1}{2}$.

In Problems 23–26, find a common denominator, add the two fractions, and simplify.

23. $\frac{1}{1 - x} - \frac{1}{x}$

24. $\frac{7/5}{x + 2} + \frac{8/5}{x - 3}$

25. $-\frac{1}{x} - \frac{1/2}{x + 1} + \frac{3/2}{x - 3}$

26. $\frac{1}{y} + \frac{1}{2000 - y}$

CHAPTER 7 Techniques of Integration

7.1 Basic Integration Rules
7.2 Integration by Parts
7.3 Some Trigonometric Integrals
7.4 Rationalizing Substitutions
7.5 Integration of Rational Functions Using Partial Fractions
7.6 Strategies for Integration

7.1 Basic Integration Rules

Our repertoire of functions now includes all the elementary functions. These are the constant functions, the power functions, the algebraic functions, the logarithmic and exponential functions, the trigonometric and inverse trigonometric functions, and all functions obtained from them by addition, subtraction, multiplication, division, and composition. Thus,

$$f(x) = \frac{e^x + e^{-x}}{2} = \cosh x$$

$$g(x) = (1 + \cos^4 x)^{1/2}$$

$$h(x) = \frac{3^{x^2-2x}}{\ln(x^2 + 1)} - \sin[\cos(\cosh x)]$$

are elementary functions.

Differentiation of an elementary function is straightforward, requiring only a systematic use of the rules that we have learned. And the result is always an elementary function. Integration (antidifferentiation) is a far different matter. It involves a few techniques and a large bag of tricks; what is worse, it does not always yield an elementary function. For example, it is known that the antiderivatives of e^{-x^2} and $(\sin x)/x$ are not elementary functions.

The two principal techniques for integration are *substitution* and *integration by parts*. The method of substitution was introduced in Section 4.4; we have used it occasionally in the intervening chapters.

Standard Forms Effective use of the method of substitution and integration by parts depends on the ready availability of a list of known integrals. One such list (but too long to memorize) appears inside the back cover of this book. The short list shown below is so useful that we think that every calculus student should memorize it.

Standard Integral Forms

Constants, Powers

1. $\int k\, du = ku + C$

2. $\int u^r\, du = \begin{cases} \dfrac{u^{r+1}}{r+1} + C & r \neq -1 \\ \ln|u| + C & r = -1 \end{cases}$

Exponentials

3. $\int e^u\, du = e^u + C$

4. $\int a^u\, du = \dfrac{a^u}{\ln a} + C, a \neq 1, a > 0$

Trigonometric Functions

5. $\int \sin u\, du = -\cos u + C$

6. $\int \cos u\, du = \sin u + C$

7. $\int \sec^2 u\, du = \tan u + C$

8. $\int \csc^2 u\, du = -\cot u + C$

9. $\int \sec u \tan u\, du = \sec u + C$

10. $\int \csc u \cot u\, du = -\csc u + C$

11. $\int \tan u\, du = -\ln|\cos u| + C$

12. $\int \cot u\, du = \ln|\sin u| + C$

Algebraic Functions

13. $\int \dfrac{du}{\sqrt{a^2 - u^2}} = \sin^{-1}\left(\dfrac{u}{a}\right) + C$

14. $\int \dfrac{du}{a^2 + u^2} = \dfrac{1}{a}\tan^{-1}\left(\dfrac{u}{a}\right) + C$

15. $\displaystyle\int \frac{du}{u\sqrt{u^2 - a^2}} = \frac{1}{a}\sec^{-1}\left(\frac{|u|}{a}\right) + C = \frac{1}{a}\cos^{-1}\left(\frac{a}{|u|}\right) + C$

Hyperbolic Functions 16. $\displaystyle\int \sinh u\, du = \cosh u + C$ 17. $\displaystyle\int \cosh u\, du = \sinh u + C$

Substitution in Indefinite Integrals Suppose that you face an indefinite integral. If it is a standard form, simply write the answer. If not, look for a substitution that will change it to a standard form. If the first substitution that you try does not work, try another. Skill at this, like most worthwhile activities, depends on practice.

The method of substitution was given in Theorem 4.4B and is restated here for easy reference.

Theorem A Substitution in Indefinite Integrals

Let g be a differentiable function and suppose that F is an antiderivative of f. Then, if $u = g(x)$,

$$\int f(g(x))g'(x)\, dx = \int f(u)\, du = F(u) + C = F(g(x)) + C$$

EXAMPLE 1 Find $\displaystyle\int \frac{x}{\cos^2(x^2)}\, dx$.

SOLUTION Look at this integral for a few moments. Since $1/\cos^2 x = \sec^2 x$, you may be reminded of the standard form $\int \sec^2 u\, du$. Let $u = x^2$, $du = 2x\, dx$. Then

$$\int \frac{x}{\cos^2(x^2)}\, dx = \frac{1}{2}\int \frac{1}{\cos^2(x^2)} \cdot 2x\, dx = \frac{1}{2}\int \sec^2 u\, du$$

$$= \frac{1}{2}\tan u + C = \frac{1}{2}\tan(x^2) + C$$ ■

EXAMPLE 2 Find $\displaystyle\int \frac{3}{\sqrt{5 - 9x^2}}\, dx$.

SOLUTION Think of $\displaystyle\int \frac{du}{\sqrt{a^2 - u^2}}$. Let $u = 3x$, so $du = 3\, dx$. Then

$$\int \frac{3}{\sqrt{5 - 9x^2}}\, dx = \int \frac{1}{\sqrt{5 - u^2}}\, du = \sin^{-1}\left(\frac{u}{\sqrt{5}}\right) + C$$

$$= \sin^{-1}\left(\frac{3x}{\sqrt{5}}\right) + C$$ ■

EXAMPLE 3 Find $\displaystyle\int \frac{6e^{1/x}}{x^2}\, dx$.

SOLUTION Think of $\int e^u\, du$. Let $u = 1/x$, so $du = (-1/x^2)\, dx$. Then

$$\int \frac{6e^{1/x}}{x^2}\, dx = -6\int e^{1/x}\left(\frac{-1}{x^2}\, dx\right) = -6\int e^u\, du$$

$$= -6e^u + C = -6e^{1/x} + C$$ ■

EXAMPLE 4 Find $\int \frac{e^x}{4 + 9e^{2x}}\, dx$.

SOLUTION Think of $\int \frac{1}{a^2 + u^2}\, du$. Let $u = 3e^x$, so $du = 3e^x\, dx$. Then

$$\int \frac{e^x}{4 + 9e^{2x}}\, dx = \frac{1}{3}\int \frac{1}{4 + 9e^{2x}}(3e^x\, dx) = \frac{1}{3}\int \frac{1}{4 + u^2}\, du$$

$$= \frac{1}{3}\cdot\frac{1}{2}\tan^{-1}\left(\frac{u}{2}\right) + C = \frac{1}{6}\tan^{-1}\left(\frac{3e^x}{2}\right) + C$$ ■

No law says that you have to write out the u-substitution. If you can do the substitution mentally, that is fine. Here are two illustrations.

EXAMPLE 5 Find $\int x \cos x^2\, dx$

SOLUTION Mentally substitute $u = x^2$.

$$\int x \cos x^2\, dx = \frac{1}{2}\int (\cos x^2)(2x\, dx) = \frac{1}{2}\sin x^2 + C$$ ■

EXAMPLE 6 Find $\int \frac{a^{\tan t}}{\cos^2 t}\, dt$.

SOLUTION Mentally, substitute $u = \tan t$.

$$\int \frac{a^{\tan t}}{\cos^2 t}\, dt = \int a^{\tan t}(\sec^2 t\, dt) = \frac{a^{\tan t}}{\ln a} + C$$ ■

Substitution in Definite Integrals This topic was covered in Section 4.4. It is just like substitution in indefinite integrals, but we must remember to make the appropriate change in the limits of integration.

EXAMPLE 7 Evaluate $\int_2^5 t\sqrt{t^2 - 4}\, dt$.

SOLUTION Let $u = t^2 - 4$, so $du = 2t\, dt$; note that when $t = 2, u = 0$, and when $t = 5, u = 21$. Thus,

$$\int_2^5 t\sqrt{t^2 - 4}\, dt = \frac{1}{2}\int_2^5 (t^2 - 4)^{1/2}(2t\, dt)$$

$$= \frac{1}{2}\int_0^{21} u^{1/2}\, du$$

$$= \left[\frac{1}{3}u^{3/2}\right]_0^{21} = \frac{1}{3}(21)^{3/2} \approx 32.08$$ ■

EXAMPLE 8 Find $\int_1^3 x^3\sqrt{x^4 + 11}\, dx$.

SOLUTION Mentally substitute $u = x^4 + 11$.

$$\int_1^3 x^3\sqrt{x^4 + 11}\, dx = \frac{1}{4}\int_1^3 (x^4 + 11)^{1/2}(4x^3\, dx)$$

$$= \left[\frac{1}{6}(x^4 + 11)^{3/2}\right]_1^3$$

$$= \frac{1}{6}[92^{3/2} - 12^{3/2}] \approx 140.144$$ ■

Concepts Review

1. Differentiation of an elementary function is straightforward, but there are cases where the antiderivative of an elementary function cannot be expressed as a(an) ____.

2. The substitution $u = 1 + x^3$ transforms $\int 3x^2(1 + x^3)^5\,dx$ to ____.

3. The substitution $u =$ ____ transforms $\int e^x/(4 + e^{2x})\,dx$ to $\int 1/(4 + u^2)\,du$.

4. The substitution $u = 1 + \sin x$ transforms $\int_0^{\pi/2} (1 + \sin x)^3 \cos x\,dx$ to ____.

Problem Set 7.1

In Problems 1–54, perform the indicated integrations.

1. $\int (x - 2)^5\,dx$

2. $\int \sqrt{3x}\,dx$

3. $\int_0^2 x(x^2 + 1)^5\,dx$

4. $\int_0^1 x\sqrt{1 - x^2}\,dx$

5. $\int \frac{dx}{x^2 + 4}$

6. $\int \frac{e^x}{2 + e^x}\,dx$

7. $\int \frac{x}{x^2 + 4}\,dx$

8. $\int \frac{2t^2}{2t^2 + 1}\,dt$

9. $\int 6z\sqrt{4 + z^2}\,dz$

10. $\int \frac{5}{\sqrt{2t + 1}}\,dt$

11. $\int \frac{\tan z}{\cos^2 z}\,dz$

12. $\int e^{\cos z} \sin z\,dz$

13. $\int \frac{\sin\sqrt{t}}{\sqrt{t}}\,dt$

14. $\int \frac{2x\,dx}{\sqrt{1 - x^4}}$

15. $\int_0^{\pi/4} \frac{\cos x}{1 + \sin^2 x}\,dx$

16. $\int_0^{3/4} \frac{\sin\sqrt{1 - x}}{\sqrt{1 - x}}\,dx$

17. $\int \frac{3x^2 + 2x}{x + 1}\,dx$

18. $\int \frac{x^3 + 7x}{x - 1}\,dx$

19. $\int \frac{\sin(\ln 4x^2)}{x}\,dx$

20. $\int \frac{\sec^2(\ln x)}{2x}\,dx$

21. $\int \frac{6e^x}{\sqrt{1 - e^{2x}}}\,dx$

22. $\int \frac{x}{x^4 + 4}\,dx$

23. $\int \frac{3e^{2x}}{\sqrt{1 - e^{2x}}}\,dx$

24. $\int \frac{x^3}{x^4 + 4}\,dx$

25. $\int_0^1 t\,3^{t^2}\,dt$

26. $\int_0^{\pi/6} 2^{\cos x} \sin x\,dx$

27. $\int \frac{\sin x - \cos x}{\sin x}\,dx$

28. $\int \frac{\sin(4t - 1)}{1 - \sin^2(4t - 1)}\,dt$

29. $\int e^x \sec e^x\,dx$ *Hint:* See Problem 56.

30. $\int e^x \sec^2(e^x)\,dx$

31. $\int \frac{\sec^3 x + e^{\sin x}}{\sec x}\,dx$

32. $\int \frac{(6t - 1)\sin\sqrt{3t^2 - t - 1}}{\sqrt{3t^2 - t - 1}}\,dt$

33. $\int \frac{t^2\cos(t^3 - 2)}{\sin^2(t^3 - 2)}\,dt$

34. $\int \frac{1 + \cos 2x}{\sin^2 2x}\,dx$

35. $\int \frac{t^2\cos^2(t^3 - 2)}{\sin^2(t^3 - 2)}\,dt$

36. $\int \frac{\csc^2 2t}{\sqrt{1 + \cot 2t}}\,dt$

37. $\int \frac{e^{\tan^{-1} 2t}}{1 + 4t^2}\,dt$

38. $\int (t + 1)e^{-t^2 - 2t - 5}\,dt$

39. $\int \frac{y}{\sqrt{16 - 9y^4}}\,dy$

40. $\int \cosh 3x\,dx$

41. $\int x^2 \sinh x^3\,dx$

42. $\int \frac{5}{\sqrt{9 - 4x^2}}\,dx$

43. $\int \frac{e^{3t}}{\sqrt{4 - e^{6t}}}\,dt$

44. $\int \frac{dt}{2t\sqrt{4t^2 - 1}}$

45. $\int_0^{\pi/2} \frac{\sin x}{16 + \cos^2 x}\,dx$

46. $\int_0^1 \frac{e^{2x} - e^{-2x}}{e^{2x} + e^{-2x}}\,dx$

47. $\int \frac{1}{x^2 + 2x + 5}\,dx$

48. $\int \frac{1}{x^2 - 4x + 9}\,dx$

49. $\int \frac{dx}{9x^2 + 18x + 10}$

50. $\int \frac{dx}{\sqrt{16 + 6x - x^2}}$

51. $\int \frac{x + 1}{9x^2 + 18x + 10}\,dx$

52. $\int \frac{3 - x}{\sqrt{16 + 6x - x^2}}\,dx$

53. $\int \frac{dt}{t\sqrt{2t^2 - 9}}$

54. $\int \frac{\tan x}{\sqrt{\sec^2 x - 4}}\,dx$

55. Find the length of the curve $y = \ln(\cos x)$ between $x = 0$ and $x = \pi/4$.

56. Establish the identity

$$\sec x = \frac{\sin x}{\cos x} + \frac{\cos x}{1 + \sin x}$$

and then use it to derive the formula

$$\int \sec x\,dx = \ln|\sec x + \tan x| + C$$

57. Evaluate $\int_0^{2\pi} \frac{x|\sin x|}{1 + \cos^2 x}\,dx$. *Hint:* Make the substitution $u = x - \pi$ in the definite integral and then use symmetry properties.

58. Let R be the region bounded by $y = \sin x$ and $y = \cos x$ between $x = -\pi/4$ and $x = 3\pi/4$. Find the volume of the solid obtained when R is revolved about $x = -\pi/4$. *Hint:* Use cylindrical shells to write a single integral, make the substitution $u = x - \pi/4$, and apply symmetry properties.

Answers to Concepts Review: 1. elementary function 2. $\int u^5\,du$ 3. e^x 4. $\int_1^2 u^3\,du$

7.2 Integration by Parts

If integration by substitution fails, it may be possible to use a double substitution, better known as *integration by parts*. This method is based on the integration of the formula for the derivative of a product of two functions.

Let $u = u(x)$ and $v = v(x)$. Then

$$D_x[u(x)v(x)] = u(x)v'(x) + v(x)u'(x)$$

or

$$u(x)v'(x) = D_x[u(x)v(x)] - v(x)u'(x)$$

By integrating both sides of this equation, we obtain

$$\int u(x)v'(x)\,dx = u(x)v(x) - \int v(x)u'(x)\,dx$$

Since $dv = v'(x)\,dx$ and $du = u'(x)\,dx$, the preceding equation is usually written symbolically as follows:

Integration by Parts: Indefinite Integrals

$$\int u\,dv = uv - \int v\,du$$

The corresponding formula for definite integrals is

$$\int_a^b u(x)v'(x)\,dx = \Big[u(x)v(x)\Big]_a^b - \int_a^b v(x)u'(x)\,dx$$

Integration by Parts

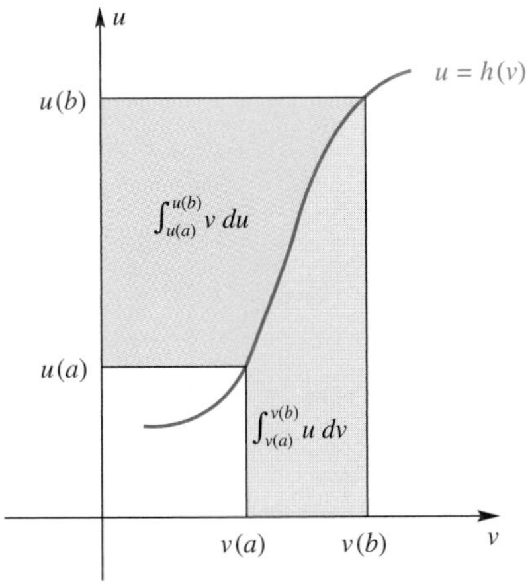

$\int_{v(a)}^{v(b)} u\,dv = u(b)v(b) - u(a)v(a) - \int_{u(a)}^{u(b)} v\,du$

Figure 1

Figure 1 illustrates a geometric interpretation of integration by parts for definite integrals. We abbreviate this as follows:

Integration by Parts: Definite Integrals

$$\int_a^b u\,dv = \Big[uv\Big]_a^b - \int_a^b v\,du$$

These formulas allow us to shift the problem of integrating $u\,dv$ to that of integrating $v\,du$. Success depends on the proper choice of u and dv, which comes with practice.

EXAMPLE 1 Find $\displaystyle\int x \cos x\,dx$.

SOLUTION We wish to write $x \cos x\,dx$ as $u\,dv$. One possibility is to let $u = x$ and $dv = \cos x\,dx$. Then $du = dx$ and $v = \int \cos x\,dx = \sin x$ (we can omit the arbitrary constant at this stage). Here is a summary of this double substitution in a convenient format.

$$u = x \qquad dv = \cos x\,dx$$
$$du = dx \qquad v = \sin x$$

The formula for integration by parts gives

$$\int \underbrace{x}_{u}\,\underbrace{\cos x\,dx}_{dv} = \underbrace{x}_{u}\,\underbrace{\sin x}_{v} - \int \underbrace{\sin x}_{v}\,\underbrace{dx}_{du}$$
$$= x \sin x + \cos x + C$$

We were successful on our first try. Another substitution would be

$$u = \cos x \qquad dv = x\,dx$$

$$du = -\sin x\,dx \qquad v = \frac{x^2}{2}$$

This time the formula for integration by parts gives

$$\int \underbrace{(\cos x)}_{u}\,\underbrace{x\,dx}_{dv} = \underbrace{(\cos x)}_{u}\,\underbrace{\frac{x^2}{2}}_{v} - \int \underbrace{\frac{x^2}{2}}_{v}\,\underbrace{(-\sin x\,dx)}_{du}$$

which is correct but not helpful. The new integral on the right-hand side is more complicated than the original one. Thus, we see the importance of a wise choice for u and dv. ■

EXAMPLE 2 Find $\int_1^2 \ln x\,dx$.

SOLUTION We make the following substitutions:

$$u = \ln x \qquad dv = dx$$

$$du = \left(\frac{1}{x}\right)dx \qquad v = x$$

Then

$$\begin{aligned}\int_1^2 \ln x\,dx &= [x \ln x]_1^2 - \int_1^2 x\frac{1}{x}dx \\ &= 2\ln 2 - \int_1^2 dx \\ &= 2\ln 2 - 1 \approx 0.386\end{aligned}$$ ■

EXAMPLE 3 Find $\int \arcsin x\,dx$.

SOLUTION We make the substitutions

$$u = \arcsin x \qquad dv = dx$$

$$du = \frac{1}{\sqrt{1 - x^2}}dx \qquad v = x$$

Then

$$\begin{aligned}\int \arcsin x\,dx &= x \arcsin x - \int \frac{x}{\sqrt{1 - x^2}}dx \\ &= x \arcsin x + \frac{1}{2}\int (1 - x^2)^{-1/2}(-2x\,dx) \\ &= x \arcsin x + \frac{1}{2}\cdot 2(1 - x^2)^{1/2} + C \\ &= x \arcsin x + \sqrt{1 - x^2} + C\end{aligned}$$ ■

EXAMPLE 4 Find $\int_1^2 t^6 \ln t\,dt$.

SOLUTION We make the following substitutions

$$u = \ln t \qquad dv = t^6\, dt$$
$$du = \frac{1}{t}\,dt \qquad v = \frac{1}{7}t^7$$

Then

$$\int_1^2 t^6 \ln t\, dt = \left[\frac{1}{7}t^7 \ln t\right]_1^2 - \int_1^2 \frac{1}{7}t^7\left(\frac{1}{t}dt\right)$$
$$= \frac{1}{7}(128 \ln 2 - \ln 1) - \frac{1}{7}\int_1^2 t^6\, dt$$
$$= \frac{128}{7}\ln 2 - \frac{1}{49}[t^7]_1^2$$
$$= \frac{128}{7}\ln 2 - \frac{127}{49} \approx 10.083$$

■

Repeated Integration by Parts Sometimes it is necessary to apply integration by parts several times.

EXAMPLE 5 Find $\int x^2 \sin x\, dx$.

SOLUTION Let

$$u = x^2 \qquad dv = \sin x\, dx$$
$$du = 2x\, dx \qquad v = -\cos x$$

Then

$$\int x^2 \sin x\, dx = -x^2 \cos x + 2\int x \cos x\, dx$$

We have improved our situation (the exponent on x has gone from 2 to 1), which suggests reapplying integration by parts to the integral on the right. Actually, we did this integration in Example 1, so we will make use of the result obtained there.

$$\int x^2 \sin x\, dx = -x^2 \cos x + 2(x \sin x + \cos x + C)$$
$$= -x^2 \cos x + 2x \sin x + 2\cos x + K$$

■

EXAMPLE 6 Find $\int e^x \sin x\, dx$.

SOLUTION Take $u = e^x$ and $dv = \sin x\, dx$. Then $du = e^x\, dx$ and $v = -\cos x$. Thus,

$$\int e^x \sin x\, dx = -e^x \cos x + \int e^x \cos x\, dx$$

which does not seem to have improved things—but does not leave us any worse off either. So, let's not give up and try integration by parts again. In the integral on the right, let $u = e^x$ and $dv = \cos x\, dx$, so $du = e^x\, dx$ and $v = \sin x$. Then

$$\int e^x \cos x\, dx = e^x \sin x - \int e^x \sin x\, dx$$

When we substitute this in our first result, we get

$$\int e^x \sin x\,dx = -e^x \cos x + e^x \sin x - \int e^x \sin x\,dx$$

By moving the last term to the left side and combining terms, we obtain

$$2\int e^x \sin x\,dx = e^x(\sin x - \cos x) + C$$

from which

$$\int e^x \sin x\,dx = \frac{1}{2}e^x(\sin x - \cos x) + K \qquad \blacksquare$$

The fact that the integral we wanted to find reappeared on the right side is what made Example 6 work.

Reduction Formulas A formula of the form

$$\int f^n(x)g(x)\,dx = h(x) + \int f^k(x)\,g(x)\,dx$$

where $k < n$, is called a **reduction formula** (the exponent on f is reduced). Such formulas can often be obtained via integration by parts.

EXAMPLE 7 Derive a reduction formula for $\int \sin^n x\,dx$.

SOLUTION Let $u = \sin^{n-1} x$ and $dv = \sin x\,dx$. Then

$$du = (n-1)\sin^{n-2} x \cos x\,dx \quad \text{and} \quad v = -\cos x$$

from which

$$\int \sin^n x\,dx = -\sin^{n-1} x \cos x + (n-1)\int \sin^{n-2} x \cos^2 x\,dx$$

If we replace $\cos^2 x$ by $1 - \sin^2 x$ in the last integral, we obtain

$$\int \sin^n x\,dx = -\sin^{n-1} x \cos x + (n-1)\int \sin^{n-2} x\,dx - (n-1)\int \sin^n x\,dx$$

After combining the first and last integrals above and solving for $\int \sin^n x\,dx$, we get the reduction formula (valid for $n \geq 2$)

$$\int \sin^n x\,dx = \frac{-\sin^{n-1} x \cos x}{n} + \frac{n-1}{n}\int \sin^{n-2} x\,dx \qquad \blacksquare$$

EXAMPLE 8 Use the reduction formula above to evaluate $\int_0^{\pi/2} \sin^8 x\,dx$.

SOLUTION Note first that

$$\int_0^{\pi/2} \sin^n x\,dx = \left[\frac{-\sin^{n-1} x \cos x}{n}\right]_0^{\pi/2} + \frac{n-1}{n}\int_0^{\pi/2} \sin^{n-2} x\,dx$$

$$= 0 + \frac{n-1}{n}\int_0^{\pi/2} \sin^{n-2} x\,dx$$

Thus,

$$\begin{aligned}\int_0^{\pi/2} \sin^8 x\,dx &= \frac{7}{8}\int_0^{\pi/2} \sin^6 x\,dx \\ &= \frac{7}{8}\cdot\frac{5}{6}\int_0^{\pi/2} \sin^4 x\,dx \\ &= \frac{7}{8}\cdot\frac{5}{6}\cdot\frac{3}{4}\int_0^{\pi/2} \sin^2 x\,dx \\ &= \frac{7}{8}\cdot\frac{5}{6}\cdot\frac{3}{4}\cdot\frac{1}{2}\int_0^{\pi/2} 1\,dx \\ &= \frac{7}{8}\cdot\frac{5}{6}\cdot\frac{3}{4}\cdot\frac{1}{2}\cdot\frac{\pi}{2} = \frac{35}{256}\pi\end{aligned}$$

■

The general formula for $\int_0^{\pi/2} \sin^n x\,dx$ can be found in a similar way (Formula 113 at the back of the book).

Concepts Review

1. The integration-by-parts formula says that $\int u\,dv =$ _____.

2. To apply this formula to $\int x \sin x\,dx$, let $u =$ _____ and $dv =$ _____.

3. Applying the integration-by-parts formula yields the value _____ for $\int_0^{\pi/2} x \sin x\,dx$.

4. A formula that expresses $\int f^n(x)\,g(x)\,dx$ in terms of $\int f^k(x)\,g(x)\,dx$, where $k < n$, is called a _____ formula.

Problem Set 7.2

In Problems 1–36, use integration by parts to evaluate each integral.

1. $\int xe^x\,dx$

2. $\int xe^{3x}\,dx$

3. $\int te^{5t+\pi}\,dt$

4. $\int (t+7)e^{2t+3}\,dt$

5. $\int x \cos x\,dx$

6. $\int x \sin 2x\,dx$

7. $\int (t-3)\cos(t-3)\,dt$

8. $\int (x-\pi)\sin x\,dx$

9. $\int t\sqrt{t+1}\,dt$

10. $\int t\sqrt[3]{2t+7}\,dt$

11. $\int \ln 3x\,dx$

12. $\int \ln(7x^5)\,dx$

13. $\int \arctan x\,dx$

14. $\int \arctan 5x\,dx$

15. $\int \frac{\ln x}{x^2}\,dx$

16. $\int_2^3 \frac{\ln 2x^5}{x^2}\,dx$

17. $\int_1^e \sqrt{t}\ln t\,dt$

18. $\int_1^5 \sqrt{2x}\ln x^3\,dx$

19. $\int z^3 \ln z\,dz$

20. $\int t \arctan t\,dt$

21. $\int \arctan(1/t)\,dt$

22. $\int t^5 \ln(t^7)\,dt$

23. $\int_{\pi/6}^{\pi/2} x\csc^2 x\,dx$

24. $\int_{\pi/6}^{\pi/4} x\sec^2 x\,dx$

25. $\int x^5\sqrt{x^3+4}\,dx$

26. $\int x^{13}\sqrt{x^7+1}\,dx$

27. $\int \frac{t^7}{(7-3t^4)^{3/2}}\,dt$

28. $\int x^3\sqrt{4-x^2}\,dx$

29. $\int \frac{z^7}{(4-z^4)^2}\,dz$

30. $\int x\cosh x\,dx$

31. $\int x\sinh x\,dx$

32. $\int \frac{\ln x}{\sqrt{x}}\,dx$

33. $\int x(3x+10)^{49}\,dx$

34. $\int_0^1 t(t-1)^{12}\,dt$

35. $\int x\,2^x\,dx$

36. $\int z\,a^z\,dz$

In Problems 37–48, apply integration by parts twice to evaluate each integral (see Examples 5 and 6).

37. $\int x^2 e^x\,dx$

38. $\int x^5 e^{x^2}\,dx$

39. $\int \ln^2 z\,dz$

40. $\int \ln^2 x^{20}\,dx$

41. $\int e^t \cos t\,dt$

42. $\int e^{at}\sin t\,dt$

43. $\int x^2 \cos x\,dx$

44. $\int r^2 \sin r\,dr$

45. $\int \sin(\ln x)\,dx$

46. $\int \cos(\ln x)\,dx$

47. $\int (\ln x)^3\,dx$ *Hint:* Use Problem 39.

48. $\int (\ln x)^4\,dx$ *Hint:* Use Problems 39 and 47.

In Problems 49–54, use integration by parts to derive the given formula.

49. $\int \sin x \sin 3x\,dx =$

$$-\tfrac{3}{8}\sin x\cos 3x + \tfrac{1}{8}\cos x \sin 3x + C$$

50. $\int \cos 5x \sin 7x\,dx =$

$$-\tfrac{7}{24}\cos 5x\cos 7x - \tfrac{5}{24}\sin 5x\sin 7x + C$$

51. $\int e^{\alpha z}\sin \beta z\,dz = \dfrac{e^{\alpha z}(\alpha\sin\beta z - \beta\cos\beta z)}{\alpha^2+\beta^2} + C$

52. $\int e^{\alpha z}\cos \beta z\,dz = \dfrac{e^{\alpha z}(\alpha\cos\beta z + \beta\sin\beta z)}{\alpha^2+\beta^2} + C$

53. $\int x^\alpha \ln x\,dx = \dfrac{x^{\alpha+1}}{\alpha+1}\ln x - \dfrac{x^{\alpha+1}}{(\alpha+1)^2} + C,\ \alpha \neq -1$

54. $\int x^\alpha(\ln x)^2\,dx = \dfrac{x^{\alpha+1}}{\alpha+1}(\ln x)^2$

$$-\,2\frac{x^{\alpha+1}}{(\alpha+1)^2}\ln x + 2\frac{x^{\alpha+1}}{(\alpha+1)^3} + C,\ \alpha\neq -1$$

In Problems 55–61, derive the given reduction formula using integration by parts.

55. $\int x^\alpha e^{\beta x}\,dx = \dfrac{x^\alpha e^{\beta x}}{\beta} - \dfrac{\alpha}{\beta}\int x^{\alpha-1}e^{\beta x}\,dx$

56. $\int x^\alpha \sin\beta x\,dx = -\dfrac{x^\alpha\cos\beta x}{\beta} + \dfrac{\alpha}{\beta}\int x^{\alpha-1}\cos\beta x\,dx$

57. $\int x^\alpha \cos\beta x\,dx = \dfrac{x^\alpha\sin\beta x}{\beta} - \dfrac{\alpha}{\beta}\int x^{\alpha-1}\sin\beta x\,dx$

58. $\int (\ln x)^\alpha\,dx = x(\ln x)^\alpha - \alpha\int(\ln x)^{\alpha-1}\,dx$

59. $\int (a^2-x^2)^\alpha\,dx =$

$$x(a^2-x^2)^\alpha + 2\alpha\int x^2(a^2-x^2)^{\alpha-1}\,dx$$

60. $\int \cos^\alpha x\,dx = \dfrac{\cos^{\alpha-1}x\sin x}{\alpha} + \dfrac{\alpha-1}{\alpha}\int\cos^{\alpha-2}x\,dx$

61. $\int \cos^\alpha \beta x\,dx =$

$$\frac{\cos^{\alpha-1}\beta x\sin\beta x}{\alpha\beta} + \frac{\alpha-1}{\alpha}\int\cos^{\alpha-2}\beta x\,dx$$

62. Use Problem 55 to derive

$$\int x^4e^{3x}\,dx = \tfrac{1}{3}x^4e^{3x} - \tfrac{4}{9}x^3e^{3x} + \tfrac{4}{9}x^2e^{3x} - \tfrac{8}{27}xe^{3x} + \tfrac{8}{81}e^{3x} + C$$

63. Use Problems 56 and 57 to derive

$$\int x^4\cos 3x\,dx = \tfrac{1}{3}x^4\sin 3x + \tfrac{4}{9}x^3\cos 3x - \tfrac{4}{9}x^2\sin 3x - \tfrac{8}{27}x\cos 3x + \tfrac{8}{81}\sin 3x + C.$$

64. Use Problem 61 to derive

$$\int\cos^6 3x\,dx = \tfrac{1}{18}\sin 3x\cos^5 3x + \tfrac{5}{72}\sin 3x\cos^3 3x + \tfrac{5}{48}\sin 3x\cos 3x + \tfrac{5}{16}x + C.$$

65. Find the area of the region bounded by the curve $y = \ln x$, the x-axis, and the line $x = e$.

66. Find the volume of the solid generated by revolving the region of Problem 65 about the x-axis.

67. Find the area of the region bounded by the curves $y = 3e^{-x/3}$, $y = 0$, $x = 0$, and $x = 9$. Make a sketch.

68. Find the volume of the solid generated by revolving the region described in Problem 67 about the x-axis.

69. Find the area of the region bounded by the graphs of $y = x\sin x$ and $y = x\cos x$ from $x = 0$ to $x = \pi/4$.

70. Find the volume of the solid obtained by revolving the region under the graph of $y = \sin(x/2)$ from $x = 0$ to $x = 2\pi$ about the y-axis.

71. Find the centroid (see Section 5.6) of the region bounded by $y = \ln x^2$ and the x-axis from $x = 1$ to $x = e$.

72. Evaluate the integral $\int\cot x\csc^2 x\,dx$ by parts in two different ways:

(a) By differentiating $\cot x$ (b) By differentiating $\csc x$

(c) Show that the two results are equivalent up to a constant.

73. If $p(x)$ is a polynomial of degree n and $G_1, G_2, \ldots, G_{n+1}$, are successive antiderivatives of a function g, then, by repeated integration by parts,

$$\int p(x)g(x)\,dx = p(x)G_1(x) - p'(x)G_2(x) + p''(x)G_3(x) - \cdots + (-1)^n p^{(n)}(x)G_{n+1}(x) + C$$

Use this result to find each of the following:

(a) $\int(x^3-2x)e^x\,dx$ (b) $\int(x^2-3x+1)\sin x\,dx$

74. The graph of $y = x\sin x$ for $x \geq 0$ is sketched in Figure 2.

(a) Find a formula for the area of the nth arch.

(b) The second arch is revolved about the y-axis. Find the volume of the resulting solid.

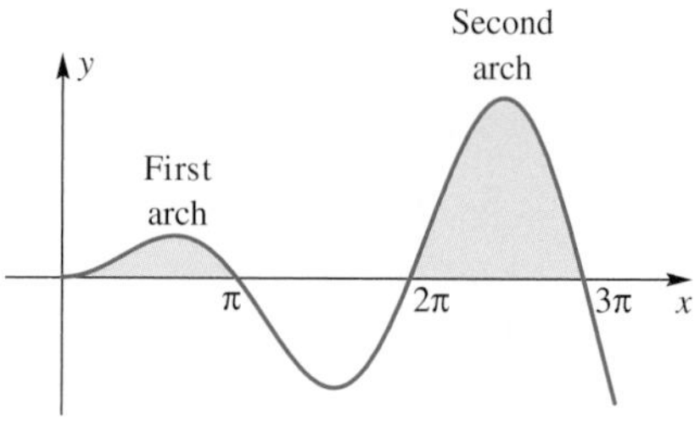

Figure 2

75. The quantity $a_n = \dfrac{1}{\pi}\int_{-\pi}^{\pi} f(x)\sin nx\,dx$ plays an important role in applied mathematics. Show that if $f'(x)$ is continuous on $[-\pi, \pi]$, then $\lim_{n\to\infty} a_n = 0$. *Hint:* Integration by parts.

76. Let $G_n = \sqrt[n]{(n+1)(n+2)\cdots(n+n)}$. Show that $\lim_{n\to\infty}(G_n/n) = 4/e$. *Hint:* Consider $\ln(G_n/n)$, recognize it as a Riemann sum, and use Example 2.

77. Find the error in the following "proof" that $0 = 1$. In $\int(1/t)\,dt$, set $u = 1/t$ and $dv = dt$. Then $du = -t^{-2}\,dt$ and $uv = 1$. Integration by parts gives

$$\int(1/t)\,dt = 1 - \int(-1/t)\,dt$$

or $0 = 1$.

78. Suppose that you want to evaluate the integral

$$\int e^{5x}(4\cos 7x + 6\sin 7x)\,dx$$

and you know from experience that the result will be of the form $e^{5x}(C_1\cos 7x + C_2\sin 7x) + C_3$. Compute C_1 and C_2 by differentiating the result and setting it equal to the integrand.

Many surprising theoretical results can be derived through the use of integration by parts. In all cases, one starts with an integral. We explore two of these results here.

79. Show that

$$\int_a^b f(x)\,dx = [xf(x)]_a^b - \int_a^b xf'(x)\,dx$$
$$= [(x-a)f(x)]_a^b - \int_a^b (x-a)f'(x)\,dx$$

80. Using Problem 79 and replacing f by f', show that

$$f(b) - f(a) = \int_a^b f'(x)\,dx$$
$$= f'(b)(b-a) - \int_a^b (x-a)f''(x)\,dx$$
$$= f'(a)(b-a) - \int_a^b (x-b)f''(x)\,dx$$

81. Show that

$$f(t) = f(a) + \sum_{i=1}^{n} \frac{f^{(i)}(a)}{i!}(t-a)^i + \int_a^t \frac{(t-x)^n}{n!} f^{(n+1)}(x)\,dx,$$

provided that f can be differentiated $n+1$ times.

82. The *Beta function*, which is important in many branches of mathematics, is defined as

$$B(\alpha, \beta) = \int_0^1 x^{\alpha-1}(1-x)^{\beta-1}\,dx,$$

with the condition that $\alpha \geq 1$ and $\beta \geq 1$.

(a) Show by a change of variables that

$$B(\alpha, \beta) = \int_0^1 x^{\beta-1}(1-x)^{\alpha-1}\,dx = B(\beta, \alpha)$$

(b) Integrate by parts to show that

$$B(\alpha, \beta) = \frac{\alpha-1}{\beta}B(\alpha-1, \beta+1) = \frac{\beta-1}{\alpha}B(\alpha+1, \beta-1)$$

(c) Assume now that $\alpha = n$ and $\beta = m$, and that n and m are positive integers. By using the result in part (b) repeatedly, show that

$$B(n, m) = \frac{(n-1)!\,(m-1)!}{(n+m-1)!}$$

This result is valid even in the case where n and m are not integers, provided that we can give meaning to $(n-1)!$, $(m-1)!$, and $(n+m-1)!$.

83. Suppose that $f(t)$ has the property that $f'(a) = f'(b) = 0$ and that $f(t)$ has two continuous derivatives. Use integration by parts to prove that $\int_a^b f''(t)f(t)\,dt \leq 0$. *Hint:* Use integration by parts by differentiating $f(t)$ and integrating $f''(t)$. This result has many applications in the field of applied mathematics.

84. Derive the formula

$$\int_0^x \left(\int_0^t f(z)\,dz \right) dt = \int_0^x f(t)(x-t)\,dt$$

using integration by parts.

85. Generalize the formula given in Problem 84 to one for an n-fold iterated integral

$$\int_0^x \int_0^{t_1} \cdots \int_0^{t_{n-1}} f(t_n)\,dt_n \ldots dt_1 = \frac{1}{(n-1)!}\int_0^x f(t_1)(x-t_1)^{n-1}\,dt_1$$

86. If $P_n(x)$ is a polynomial of degree n, show that

$$\int e^x P_n(x)\,dx = e^x \sum_{j=0}^{n} (-1)^j \frac{d^j P_n(x)}{dx^j}$$

87. Use the result from Problem 86 to evaluate

$$\int (3x^4 + 2x^2)e^x\,dx$$

Answers to Concepts Review: **1.** $uv - \int v\,du$ **2.** x; $\sin x\,dx$ **3.** 1 **4.** reduction

7.3 Some Trigonometric Integrals

When we combine the method of substitution with a clever use of trigonometric identities, we can integrate a wide variety of trigonometric forms. We consider five commonly encountered types.

1. $\int \sin^n x\,dx$ and $\int \cos^n x\,dx$
2. $\int \sin^m x \cos^n x\,dx$
3. $\int \sin mx \cos nx\,dx$, $\int \sin mx \sin nx\,dx$, $\int \cos mx \cos nx\,dx$
4. $\int \tan^n x\,dx$, $\int \cot^n x\,dx$
5. $\int \tan^m x \sec^n x\,dx$, $\int \cot^m x \csc^n x\,dx$

Useful Identities

Some trigonometric identities needed in this section are the following.

Pythagorean Identities

$\sin^2 x + \cos^2 x = 1$

$1 + \tan^2 x = \sec^2 x$

$1 + \cot^2 x = \csc^2 x$

Half-Angle Identities

$\sin^2 x = \dfrac{1 - \cos 2x}{2}$

$\cos^2 x = \dfrac{1 + \cos 2x}{2}$

Type 1 $\left(\int \sin^n x\, dx, \int \cos^n x\, dx\right)$ Consider first the case where n is an odd positive integer. After taking out either the factor $\sin x$ or $\cos x$, use the identity $\sin^2 x + \cos^2 x = 1$.

EXAMPLE 1 **(n Odd)** Find $\int \sin^5 x\, dx$.

SOLUTION

$$\begin{aligned}
\int \sin^5 x\, dx &= \int \sin^4 x \sin x\, dx \\
&= \int (1 - \cos^2 x)^2 \sin x\, dx \\
&= \int (1 - 2\cos^2 x + \cos^4 x) \sin x\, dx \\
&= -\int (1 - 2\cos^2 x + \cos^4 x)(-\sin x\, dx) \\
&= -\cos x + \tfrac{2}{3}\cos^3 x - \tfrac{1}{5}\cos^5 x + C
\end{aligned}$$

■

EXAMPLE 2 **(n Even)** Find $\int \sin^2 x\, dx$ and $\int \cos^4 x\, dx$.

SOLUTION Here we make use of half-angle identities.

$$\begin{aligned}
\int \sin^2 x\, dx &= \int \frac{1 - \cos 2x}{2}\, dx \\
&= \frac{1}{2}\int dx - \frac{1}{4}\int (\cos 2x)(2\, dx) \\
&= \frac{1}{2}x - \frac{1}{4}\sin 2x + C
\end{aligned}$$

$$\begin{aligned}
\int \cos^4 x\, dx &= \int \left(\frac{1 + \cos 2x}{2}\right)^2 dx \\
&= \frac{1}{4}\int (1 + 2\cos 2x + \cos^2 2x)\, dx \\
&= \frac{1}{4}\int dx + \frac{1}{4}\int (\cos 2x)(2)\, dx + \frac{1}{8}\int (1 + \cos 4x)\, dx \\
&= \frac{3}{8}\int dx + \frac{1}{4}\int \cos 2x(2\, dx) + \frac{1}{32}\int \cos 4x(4\, dx) \\
&= \frac{3}{8}x + \frac{1}{4}\sin 2x + \frac{1}{32}\sin 4x + C
\end{aligned}$$

■

Type 2 $\left(\int \sin^m x \cos^n x\, dx\right)$ If either m or n is an odd positive integer and the other exponent is any number, we factor out $\sin x$ or $\cos x$ and use the identity $\sin^2 x + \cos^2 x = 1$.

EXAMPLE 3 **(m or n Odd)** Find $\int \sin^3 x \cos^{-4} x\, dx$.

SOLUTION

$$\begin{aligned}\int \sin^3 x \cos^{-4} x\,dx &= \int (1 - \cos^2 x)(\cos^{-4} x)(\sin x)\,dx \\ &= -\int (\cos^{-4} x - \cos^{-2} x)(-\sin x\,dx) \\ &= -\left[\frac{(\cos x)^{-3}}{-3} - \frac{(\cos x)^{-1}}{-1}\right] + C \\ &= \frac{1}{3}\sec^3 x - \sec x + C\end{aligned}$$

■

If both m and n are even positive integers, we use half-angle identities to reduce the degree of the integrand. Example 4 gives an illustration.

EXAMPLE 4 **(Both m and n Even)** Find $\int \sin^2 x \cos^4 x\,dx$.

SOLUTION

$$\begin{aligned}&\int \sin^2 x \cos^4 x\,dx \\ &= \int \left(\frac{1 - \cos 2x}{2}\right)\left(\frac{1 + \cos 2x}{2}\right)^2 dx \\ &= \frac{1}{8}\int (1 + \cos 2x - \cos^2 2x - \cos^3 2x)\,dx \\ &= \frac{1}{8}\int \left[1 + \cos 2x - \frac{1}{2}(1 + \cos 4x) - (1 - \sin^2 2x)\cos 2x\right] dx \\ &= \frac{1}{8}\int \left[\frac{1}{2} - \frac{1}{2}\cos 4x + \sin^2 2x \cos 2x\right] dx \\ &= \frac{1}{8}\left[\int \frac{1}{2}dx - \frac{1}{8}\int \cos 4x(4\,dx) + \frac{1}{2}\int \sin^2 2x(2\cos 2x\,dx)\right] \\ &= \frac{1}{8}\left[\frac{1}{2}x - \frac{1}{8}\sin 4x + \frac{1}{6}\sin^3 2x\right] + C\end{aligned}$$

■

Are They Different?

Indefinite integrations may lead to different looking answers. By one method,

$$\begin{aligned}\int \sin x \cos x\,dx &= -\int \cos x(-\sin x)\,dx \\ &= -\tfrac{1}{2}\cos^2 x + C\end{aligned}$$

By a second method,

$$\begin{aligned}\int \sin x \cos x\,dx &= \int \sin x(\cos x)\,dx \\ &= \tfrac{1}{2}\sin^2 x + C\end{aligned}$$

But two such answers should differ by at most a constant. Note, however, that

$$\begin{aligned}\tfrac{1}{2}\sin^2 x + C &= \tfrac{1}{2}(1 - \cos^2 x) + C \\ &= -\tfrac{1}{2}\cos^2 x + \left(\tfrac{1}{2} + C\right)\end{aligned}$$

Now reconcile these answers with a third answer.

$$\begin{aligned}\int \sin x \cos x\,dx &= \tfrac{1}{2}\int \sin 2x\,dx \\ &= -\tfrac{1}{4}\cos 2x + C\end{aligned}$$

Type 3 $\left(\int \sin mx \cos nx\,dx, \int \sin mx \sin nx\,dx, \int \cos mx \cos nx\,dx\right)$ Integrals of this type occur in many physics and engineering applications. To handle these integrals, we use the product identities.

1. $\sin mx \cos nx = \frac{1}{2}[\sin(m + n)x + \sin(m - n)x]$
2. $\sin mx \sin nx = -\frac{1}{2}[\cos(m + n)x - \cos(m - n)x]$
3. $\cos mx \cos nx = \frac{1}{2}[\cos(m + n)x + \cos(m - n)x]$

EXAMPLE 5 Find $\int \sin 2x \cos 3x\,dx$.

SOLUTION Apply product identity 1.

$$\int \sin 2x \cos 3x\,dx = \frac{1}{2}\int [\sin 5x + \sin(-x)]\,dx$$
$$= \frac{1}{10}\int \sin 5x(5\,dx) - \frac{1}{2}\int \sin x\,dx$$
$$= -\frac{1}{10}\cos 5x + \frac{1}{2}\cos x + C$$

■

EXAMPLE 6 If m and n are positive integers, show that

$$\int_{-\pi}^{\pi} \sin mx \sin nx\,dx = \begin{cases} 0 & \text{if } m \neq n \\ \pi & \text{if } m = n \end{cases}$$

SOLUTION Apply product identity 2. If $m \neq n$, then

$$\int_{-\pi}^{\pi} \sin mx \sin nx\,dx = -\frac{1}{2}\int_{-\pi}^{\pi} [\cos(m+n)x - \cos(m-n)x]\,dx$$
$$= -\frac{1}{2}\left[\frac{1}{m+n}\sin(m+n)x - \frac{1}{m-n}\sin(m-n)x\right]_{-\pi}^{\pi}$$
$$= 0$$

If $m = n$,

$$\int_{-\pi}^{\pi} \sin mx \sin nx\,dx = -\frac{1}{2}\int_{-\pi}^{\pi} [\cos 2mx - 1]\,dx$$
$$= -\frac{1}{2}\left[\frac{1}{2m}\sin 2mx - x\right]_{-\pi}^{\pi}$$
$$= -\frac{1}{2}[-2\pi] = \pi$$

■

EXAMPLE 7 If m and n are positive integers, find

$$\int_{-L}^{L} \sin\frac{m\pi x}{L} \sin\frac{n\pi x}{L}\,dx$$

SOLUTION Let $u = \pi x/L$, $du = \pi dx/L$. If $x = -L$, then $u = -\pi$, and if $x = L$, then $u = \pi$. Thus,

$$\int_{-L}^{L} \sin\frac{m\pi x}{L} \sin\frac{n\pi x}{L}\,dx = \frac{L}{\pi}\int_{-\pi}^{\pi} \sin mu \sin nu\,du$$
$$= \begin{cases} \dfrac{L}{\pi}\cdot 0 & \text{if } m \neq n \\ \dfrac{L}{\pi}\cdot \pi & \text{if } m = n \end{cases}$$
$$= \begin{cases} 0 & \text{if } m \neq n \\ L & \text{if } m = n \end{cases}$$

Here we have used the result of Example 6. ■

A number of times in this book we have suggested that you should view things from both an algebraic and a geometric point of view. So far, this section has been entirely algebraic, but with definite integrals such as those in Examples 6 and 7, we have an opportunity to view things geometrically.

Figure 1 shows graphs of $y = \sin 3x \sin 2x$ and $y = \sin(3\pi x/10)\sin(2\pi x/10)$. The graphs suggest that the areas above and below the x-axis are the same, leaving $A_{\text{up}} - A_{\text{down}} = 0$. Examples 6 and 7 confirm this.

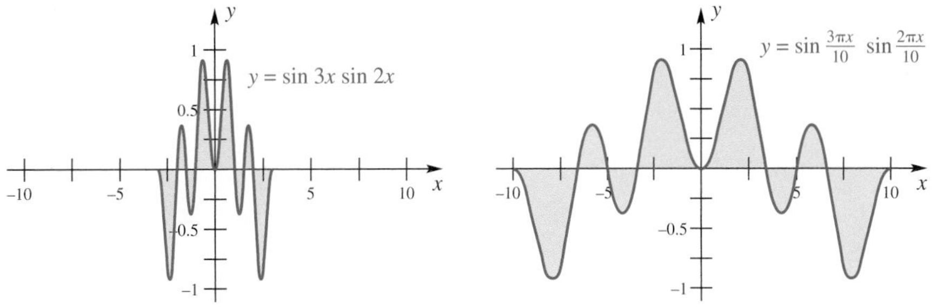

Figure 1

Figure 2 shows graphs of $y = \sin 2x \sin 2x = \sin^2 2x$, $-\pi \le x \le \pi$, and $y = \sin(2\pi x/10)\sin(2\pi x/10) = \sin^2(2\pi x/10)$, $-10 \le x \le 10$. These two graphs look the same, except the one on the right has been stretched horizontally by the factor $10/\pi$. Does it then make sense that the area will increase by this same factor? That would make the shaded area in the figure on the right equal to $10/\pi$ times the shaded area in the figure on the left; that is, the area on the right should be $(10/\pi)\cdot\pi = 10$, which corresponds to the result of Example 7 with $L = 10$.

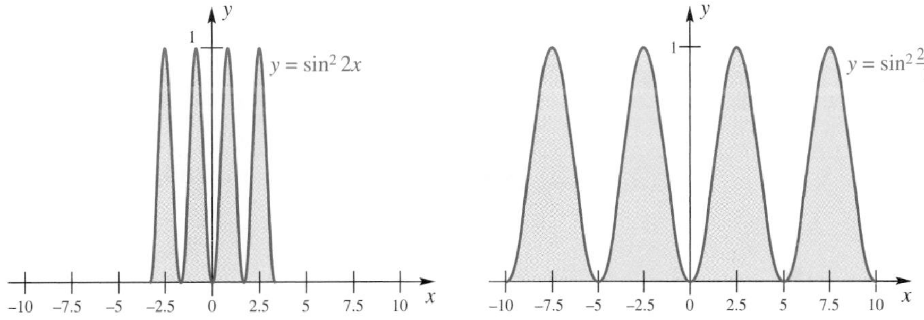

Figure 2

Type 4 $\left(\int \tan^n x\,dx, \int \cot^n x\,dx\right)$ In the tangent case, factor out $\tan^2 x = \sec^2 x - 1$; in the cotangent case, factor out $\cot^2 x = \csc^2 x - 1$.

EXAMPLE 8 Find $\int \cot^4 x\,dx$.

SOLUTION

$$\begin{aligned}\int \cot^4 x\,dx &= \int \cot^2 x\,(\csc^2 x - 1)\,dx \\ &= \int \cot^2 x \csc^2 x\,dx - \int \cot^2 x\,dx \\ &= -\int \cot^2 x\,(-\csc^2 x\,dx) - \int (\csc^2 x - 1)\,dx \\ &= -\tfrac{1}{3}\cot^3 x + \cot x + x + C\end{aligned}$$

■

EXAMPLE 9 Find $\int \tan^5 x\,dx$.

SOLUTION

$$\int \tan^5 x\,dx = \int \tan^3 x\,(\sec^2 x - 1)\,dx$$
$$= \int \tan^3 x \sec^2 x\,dx - \int \tan^3 x\,dx$$
$$= \int \tan^3 x\,(\sec^2 x\,dx) - \int \tan x\,(\sec^2 x - 1)\,dx$$
$$= \int \tan^3 x\,(\sec^2 x\,dx) - \int \tan x\,(\sec^2 x\,dx) + \int \tan x\,dx$$
$$= \tfrac{1}{4}\tan^4 x - \tfrac{1}{2}\tan^2 x - \ln|\cos x| + C$$ ■

Type 5 $\left(\int \tan^m x \sec^n x\,dx,\ \int \cot^m x \csc^n x\,dx\right)$

EXAMPLE 10 **(n Even, m Any Number)** Find $\int \tan^{-3/2} x \sec^4 x\,dx$.

SOLUTION

$$\int \tan^{-3/2} x \sec^4 x\,dx = \int (\tan^{-3/2} x)(1 + \tan^2 x)\sec^2 x\,dx$$
$$= \int (\tan^{-3/2} x)\sec^2 x\,dx + \int (\tan^{1/2} x)\sec^2 x\,dx$$
$$= -2\tan^{-1/2} x + \tfrac{2}{3}\tan^{3/2} x + C$$ ■

EXAMPLE 11 **(m Odd, n Any Number)** Find $\int \tan^3 x \sec^{-1/2} x\,dx$.

SOLUTION

$$\int \tan^3 x \sec^{-1/2} x\,dx = \int (\tan^2 x)(\sec^{-3/2} x)(\sec x \tan x)\,dx$$
$$= \int (\sec^2 x - 1)\sec^{-3/2} x\,(\sec x \tan x\,dx)$$
$$= \int \sec^{1/2} x\,(\sec x \tan x\,dx) - \int \sec^{-3/2} x\,(\sec x \tan x\,dx)$$
$$= \tfrac{2}{3}\sec^{3/2} x + 2\sec^{-1/2} x + C$$ ■

Concepts Review

1. To handle $\int \cos^2 x\,dx$, we first rewrite it as _____.

2. To handle $\int \cos^3 x\,dx$, we first rewrite it as _____.

3. To handle $\int \sin^2 x \cos^3 x\,dx$, we first rewrite it as _____.

4. To handle $\int_{-\pi}^{\pi} \cos mx \cos nx\,dx$, where $m \neq n$, we use the trigonometric identity _____.

Problem Set 7.3

In Problems 1–28, perform the indicated integrations.

1. $\int \sin^2 x\, dx$

2. $\int \sin^4 6x\, dx$

3. $\int \sin^3 x\, dx$

4. $\int \cos^3 x\, dx$

5. $\int_0^{\pi/2} \cos^5 \theta\, d\theta$

6. $\int_0^{\pi/2} \sin^6 \theta\, d\theta$

7. $\int \sin^5 4x \cos^2 4x\, dx$

8. $\int (\sin^3 2t)\sqrt{\cos 2t}\, dt$

9. $\int \cos^3 3\theta \sin^{-2} 3\theta\, d\theta$

10. $\int \sin^{1/2} 2z \cos^3 2z\, dz$

11. $\int \sin^4 3t \cos^4 3t\, dt$

12. $\int \cos^6 \theta \sin^2 \theta\, d\theta$

13. $\int \sin 4y \cos 5y\, dy$

14. $\int \cos y \cos 4y\, dy$

15. $\int \sin^4\left(\frac{w}{2}\right)\cos^2\left(\frac{w}{2}\right) dw$

16. $\int \sin 3t \sin t\, dt$

17. $\int x \cos^2 x \sin x\, dx$ *Hint:* Use integration by parts.

18. $\int x \sin^3 x \cos x\, dx$

19. $\int \tan^4 x\, dx$

20. $\int \cot^4 x\, dx$

21. $\int \tan^3 x\, dx$

22. $\int \cot^3 2t\, dt$

23. $\int \tan^5\left(\frac{\theta}{2}\right) d\theta$

24. $\int \cot^5 2t\, dt$

25. $\int \tan^{-3} x \sec^4 x\, dx$

26. $\int \tan^{-3/2} x \sec^4 x\, dx$

27. $\int \tan^3 x \sec^2 x\, dx$

28. $\int \tan^3 x \sec^{-1/2} x\, dx$

29. Find $\int_{-\pi}^{\pi} \cos mx \cos nx\, dx, m \neq n; m, n$ integers.

30. Find $\int_{-L}^{L} \cos\frac{m\pi x}{L}\cos\frac{n\pi x}{L} dx, m \neq n, m, n$ integers.

≈ **31.** The region bounded by $y = x + \sin x, y = 0, x = \pi$, is revolved about the x-axis. Find the volume of the resulting solid.

≈ **32.** The region bounded by $y = \sin^2(x^2), y = 0$, and $x = \sqrt{\pi/2}$ is revolved about the y-axis. Find the volume of the resulting solid.

33. Let $f(x) = \sum_{n=1}^{N} a_n \sin(nx)$. Use Example 6 to show each of the following for a positive integer m.

(a) $\frac{1}{\pi}\int_{-\pi}^{\pi} f(x)\sin(mx)\, dx = \begin{cases} a_m & \text{if } m \leq N \\ 0 & \text{if } m > N \end{cases}$

(b) $\frac{1}{\pi}\int_{-\pi}^{\pi} f^2(x)\, dx = \sum_{n=1}^{N} a_n^2$

Note: Integrals of this type occur in a subject called *Fourier series*, which has applications to heat, vibrating strings, and other physical phenomena.

34. Show that

$$\lim_{n\to\infty} \cos\frac{x}{2}\cos\frac{x}{4}\cos\frac{x}{8}\cdots\cos\frac{x}{2^n} = \frac{\sin x}{x}$$

by completing the following steps.

(a) $\cos\frac{x}{2}\cos\frac{x}{4}\cdots\cos\frac{x}{2^n} =$

$$\left[\cos\frac{1}{2^n}x + \cos\frac{3}{2^n}x + \cdots + \cos\frac{2^n-1}{2^n}x\right]\frac{1}{2^{n-1}}$$

(See Problem 46 of Section 0.7.)

(b) Recognize a Riemann sum leading to a definite integral.

(c) Evaluate the definite integral.

35. Use the result of Problem 34 to obtain the famous formula of François Viète (1540–1603):

$$\frac{2}{\pi} = \frac{\sqrt{2}}{2}\cdot\frac{\sqrt{2+\sqrt{2}}}{2}\cdot\frac{\sqrt{2+\sqrt{2+\sqrt{2}}}}{2}\cdots$$

36. The shaded region (Figure 3) between one arch of $y = \sin x, 0 \leq x \leq \pi$, and the line $y = k, 0 \leq k \leq 1$, is revolved about the line $y = k$, generating a solid S. Determine k so that S has

(a) minimum volume and (b) maximum volume.

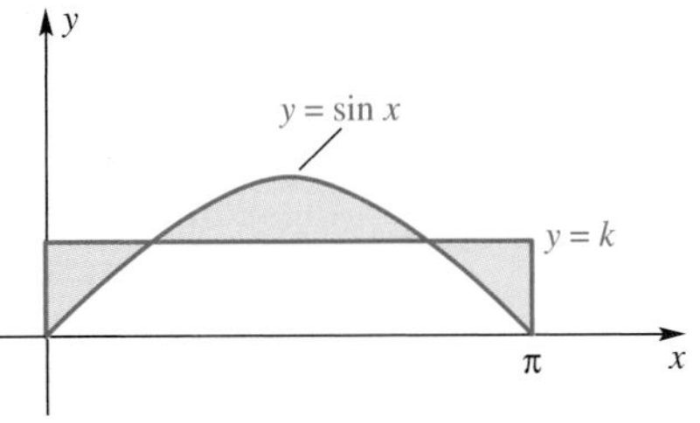

Figure 3

Answers to Concepts Review: **1.** $\int[(1 + \cos 2x)/2]\, dx$ **2.** $\int(1 - \sin^2 x)\cos x\, dx$ **3.** $\int \sin^2 x(1 - \sin^2 x)\cos x\, dx$ **4.** $\cos mx \cos nx = \frac{1}{2}[\cos(m+n)x + \cos(m-n)x]$

7.4 Rationalizing Substitutions

Radicals in an integrand are often troublesome and we usually try to get rid of them. Often an appropriate substitution will rationalize the integrand.

Integrands Involving $\sqrt[n]{ax+b}$ If $\sqrt[n]{ax+b}$ appears in an integral, the substitution $u = \sqrt[n]{ax+b}$ will eliminate the radical.

EXAMPLE 1 Find $\displaystyle\int \frac{dx}{x-\sqrt{x}}$.

SOLUTION Let $u = \sqrt{x}$, so $u^2 = x$ and $2u\,du = dx$. Then

$$\int \frac{dx}{x-\sqrt{x}} = \int \frac{2u}{u^2-u}\,du = 2\int \frac{1}{u-1}\,du$$
$$= 2\ln|u-1| + C = 2\ln|\sqrt{x}-1| + C \quad ■$$

EXAMPLE 2 Find $\displaystyle\int x\sqrt[3]{x-4}\,dx$.

SOLUTION Let $u = \sqrt[3]{x-4}$, so $u^3 = x-4$ and $3u^2\,du = dx$. Then

$$\int x\sqrt[3]{x-4}\,dx = \int (u^3+4)u\cdot(3u^2\,du) = 3\int (u^6+4u^3)\,du$$
$$= 3\left[\frac{u^7}{7} + u^4\right] + C = \frac{3}{7}(x-4)^{7/3} + 3(x-4)^{4/3} + C \quad ■$$

EXAMPLE 3 Find $\displaystyle\int x\sqrt[5]{(x+1)^2}\,dx$.

SOLUTION Let $u = (x+1)^{1/5}$, so $u^5 = x+1$ and $5u^4\,du = dx$. Then

$$\int x(x+1)^{2/5}\,dx = \int (u^5-1)u^2\cdot 5u^4\,du$$
$$= 5\int (u^{11}-u^6)\,du = \tfrac{5}{12}u^{12} - \tfrac{5}{7}u^7 + C$$
$$= \tfrac{5}{12}(x+1)^{12/5} - \tfrac{5}{7}(x+1)^{7/5} + C \quad ■$$

Integrands Involving $\sqrt{a^2-x^2}$, $\sqrt{a^2+x^2}$ and $\sqrt{x^2-a^2}$ To rationalize these three expressions, we may assume that a is positive and make the following trigonometric substitutions.

Radical	Substitution	Restriction on t
1. $\sqrt{a^2-x^2}$	$x = a\sin t$	$-\pi/2 \le t \le \pi/2$
2. $\sqrt{a^2+x^2}$	$x = a\tan t$	$-\pi/2 < t < \pi/2$
3. $\sqrt{x^2-a^2}$	$x = a\sec t$	$0 \le t \le \pi, t \ne \pi/2$

Now note the simplifications that these substitutions achieve.

1. $\sqrt{a^2-x^2} = \sqrt{a^2-a^2\sin^2 t} = \sqrt{a^2\cos^2 t} = |a\cos t| = a\cos t$
2. $\sqrt{a^2+x^2} = \sqrt{a^2+a^2\tan^2 t} = \sqrt{a^2\sec^2 t} = |a\sec t| = a\sec t$
3. $\sqrt{x^2-a^2} = \sqrt{a^2\sec^2 t - a^2} = \sqrt{a^2\tan^2 t} = |a\tan t| = \pm a\tan t$

The restrictions on t allowed us to remove the absolute value signs in the first two cases, but they also achieved something else. These restrictions are exactly the ones we introduced in Section 6.7 in order to make sine, tangent, and secant invertible functions. This means that we can solve the substitution equations for t in each case, and this will allow us to write our final answers in the following examples in terms of x.

EXAMPLE 4 Find $\int \sqrt{a^2 - x^2}\, dx$.

SOLUTION We make the substitution

$$x = a \sin t, \qquad -\frac{\pi}{2} \le t \le \frac{\pi}{2}$$

Then $dx = a \cos t\, dt$ and $\sqrt{a^2 - x^2} = a \cos t$. Thus,

$$\int \sqrt{a^2 - x^2}\, dx = \int a \cos t \cdot a \cos t\, dt = a^2 \int \cos^2 t\, dt$$
$$= \frac{a^2}{2} \int (1 + \cos 2t)\, dt$$
$$= \frac{a^2}{2}\left(t + \frac{1}{2}\sin 2t\right) + C$$
$$= \frac{a^2}{2}(t + \sin t \cos t) + C$$

Now, $x = a \sin t$ is equivalent to $x/a = \sin t$ and, since t was restricted to make the sine function invertible,

$$t = \sin^{-1}\left(\frac{x}{a}\right)$$

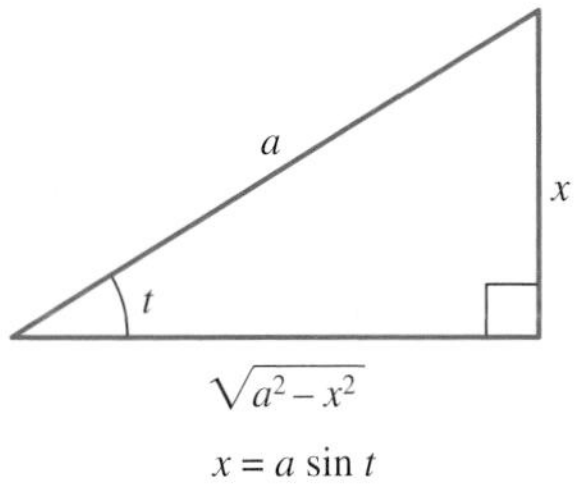

Figure 1

Using the right triangle in Figure 1 (as we did in Section 6.8), we see that

$$\cos t = \cos\left[\sin^{-1}\left(\frac{x}{a}\right)\right] = \sqrt{1 - \frac{x^2}{a^2}} = \frac{1}{a}\sqrt{a^2 - x^2}$$

Thus,

$$\int \sqrt{a^2 - x^2}\, dx = \frac{a^2}{2}\sin^{-1}\left(\frac{x}{a}\right) + \frac{x}{2}\sqrt{a^2 - x^2} + C$$

■

The result in Example 4 allows us to calculate the following definite integral, which represents the area of a semicircle (Figure 2). Thus, calculus confirms a result that we already know.

$$\int_{-a}^{a} \sqrt{a^2 - x^2}\, dx = \left[\frac{a^2}{2}\sin^{-1}\left(\frac{x}{a}\right) + \frac{x}{2}\sqrt{a^2 - x^2}\right]_{-a}^{a} = \frac{a^2}{2}\left[\frac{\pi}{2} + \frac{\pi}{2}\right] = \frac{\pi a^2}{2}$$

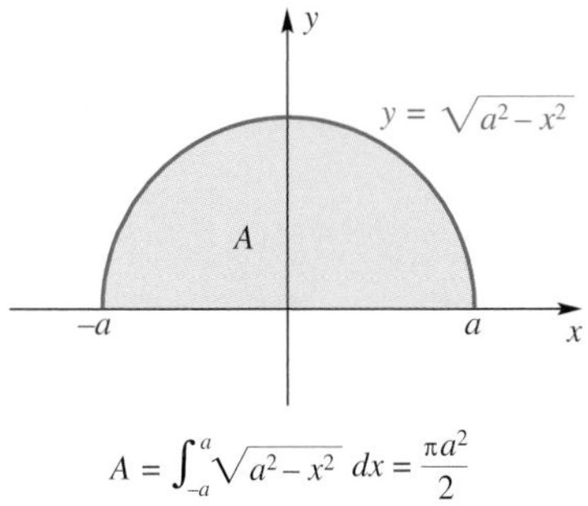

Figure 2

EXAMPLE 5 Find $\int \frac{dx}{\sqrt{9 + x^2}}$.

SOLUTION Let $x = 3 \tan t, -\pi/2 < t < \pi/2$. Then $dx = 3 \sec^2 t\, dt$ and $\sqrt{9 + x^2} = 3 \sec t$.

$$\int \frac{dx}{\sqrt{9 + x^2}} = \int \frac{3 \sec^2 t}{3 \sec t}\, dt = \int \sec t\, dt$$
$$= \ln|\sec t + \tan t| + C$$

The last step, the integration of $\sec t$, was handled in Problem 56 of Section 7.1. Now $\tan t = x/3$, which suggests the triangle in Figure 3, from which we conclude that $\sec t = \sqrt{9 + x^2}/3$. Thus,

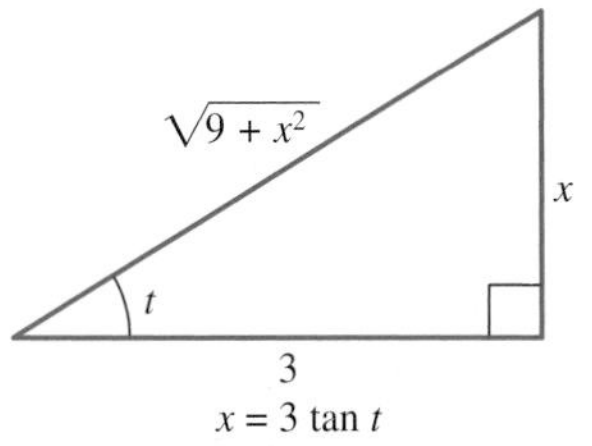

Figure 3

$$\int \frac{dx}{\sqrt{9 + x^2}} = \ln\left|\frac{\sqrt{9 + x^2} + x}{3}\right| + C$$
$$= \ln|\sqrt{9 + x^2} + x| - \ln 3 + C$$
$$= \ln|\sqrt{9 + x^2} + x| + K$$

■

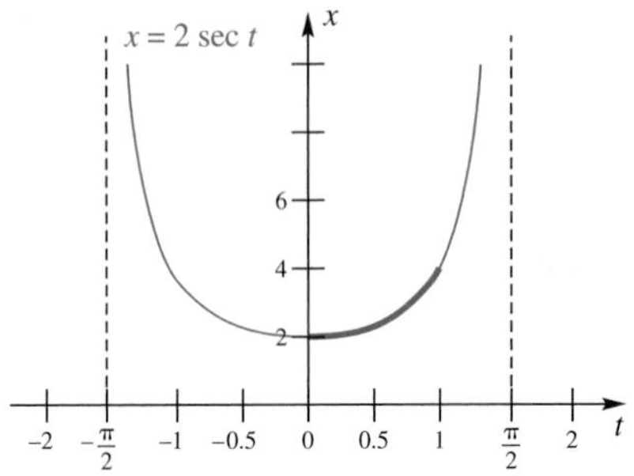

Figure 4

EXAMPLE 6 Calculate $\displaystyle\int_2^4 \frac{\sqrt{x^2-4}}{x}\,dx$.

SOLUTION Let $x = 2\sec t$, where $0 \le t < \pi/2$. Note that the restriction of t to this interval is acceptable, since x is in the interval $2 \le x \le 4$ (see Figure 4). This is important because it allows us to remove the absolute value sign that normally appears when we simplify $\sqrt{x^2-a^2}$. In our case,

$$\sqrt{x^2-4} = \sqrt{4\sec^2 t - 4} = \sqrt{4\tan^2 t} = 2|\tan t| = 2\tan t$$

We now use the theorem on substitution in a definite integral (which requires changing the limits of integration) to write

$$\begin{aligned}\int_2^4 \frac{\sqrt{x^2-4}}{x}\,dx &= \int_0^{\pi/3} \frac{2\tan t}{2\sec t}\, 2\sec t\tan t\,dt\\ &= \int_0^{\pi/3} 2\tan^2 t\,dt = 2\int_0^{\pi/3}(\sec^2 t - 1)\,dt\\ &= 2\Big[\tan t - t\Big]_0^{\pi/3} = 2\sqrt{3} - \frac{2\pi}{3} \approx 1.37\end{aligned}$$

■

Completing the Square When a quadratic expression of the type $x^2 + Bx + C$ appears under a radical, completing the square will prepare it for a trigonometric substitution.

EXAMPLE 7 Find (a) $\displaystyle\int \frac{dx}{\sqrt{x^2+2x+26}}$ and (b) $\displaystyle\int \frac{2x}{\sqrt{x^2+2x+26}}\,dx$.

SOLUTION

(a) $x^2 + 2x + 26 = x^2 + 2x + 1 + 25 = (x+1)^2 + 25$. Let $u = x + 1$ and $du = dx$. Then

$$\int \frac{dx}{\sqrt{x^2+2x+26}} = \int \frac{du}{\sqrt{u^2+25}}$$

Next let $u = 5\tan t$, $-\pi/2 < t < \pi/2$. Then $du = 5\sec^2 t\,dt$ and $\sqrt{u^2+25} = \sqrt{25(\tan^2 t + 1)} = 5\sec t$, so

$$\begin{aligned}\int \frac{du}{\sqrt{u^2+25}} &= \int \frac{5\sec^2 t\,dt}{5\sec t} = \int \sec t\,dt\\ &= \ln|\sec t + \tan t| + C\\ &= \ln\left|\frac{\sqrt{u^2+25}}{5} + \frac{u}{5}\right| + C \qquad \text{(by Figure 5)}\\ &= \ln|\sqrt{u^2+25} + u| - \ln 5 + C\\ &= \ln|\sqrt{x^2+2x+26} + x + 1| + K\end{aligned}$$

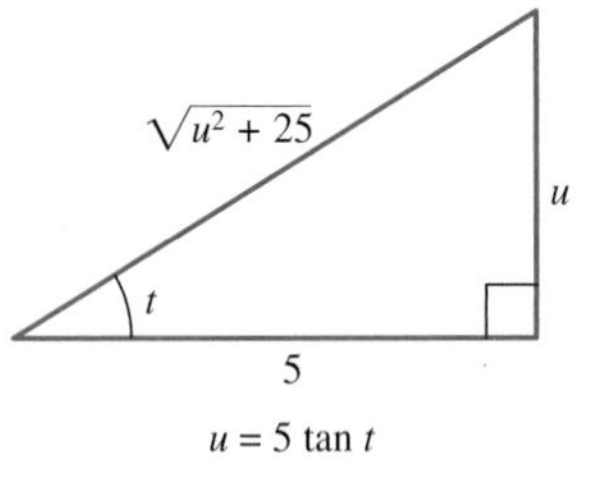

Figure 5

(b) To handle the second integral, we write

$$\int \frac{2x}{\sqrt{x^2+2x+26}}\,dx = \int \frac{2x+2}{\sqrt{x^2+2x+26}}\,dx - 2\int \frac{1}{\sqrt{x^2+2x+26}}\,dx$$

The first of the integrals on the right is handled by the substitution $u = x^2 + 2x + 26$; the second was just done. We obtain

$$\int \frac{2x}{\sqrt{x^2+2x+26}}\,dx =$$

$$2\sqrt{x^2+2x+26} - 2\ln|\sqrt{x^2+2x+26} + x + 1| + K$$

■

Concepts Review

1. To handle $\int x\sqrt{x-3}\,dx$, make the substitution $u =$ ______.

2. To handle an integral involving $\sqrt{4-x^2}$, make the substitution $x =$ ______.

3. To handle an integral involving $\sqrt{4+x^2}$, make the substitution $x =$ ______.

4. To handle an integral involving $\sqrt{x^2-4}$, make the substitution $x =$ ______.

Problem Set 7.4

In Problems 1–16, perform the indicated integrations.

1. $\int x\sqrt{x+1}\,dx$

2. $\int x\sqrt[3]{x+\pi}\,dx$

3. $\int \frac{t\,dt}{\sqrt{3t+4}}$

4. $\int \frac{x^2+3x}{\sqrt{x+4}}\,dx$

5. $\int_1^2 \frac{dt}{\sqrt{t}+e}$

6. $\int_0^1 \frac{\sqrt{t}}{t+1}\,dt$

7. $\int t(3t+2)^{3/2}\,dt$

8. $\int x(1-x)^{2/3}\,dx$

9. $\int \frac{\sqrt{4-x^2}}{x}\,dx$

10. $\int \frac{x^2\,dx}{\sqrt{16-x^2}}$

11. $\int \frac{dx}{(x^2+4)^{3/2}}$

12. $\int_2^3 \frac{dt}{t^2\sqrt{t^2-1}}$

13. $\int_{-2}^{-3} \frac{\sqrt{t^2-1}}{t^3}\,dt$

14. $\int \frac{t}{\sqrt{1-t^2}}\,dt$

15. $\int \frac{2z-3}{\sqrt{1-z^2}}\,dz$

16. $\int_0^{\pi} \frac{\pi x-1}{\sqrt{x^2+\pi^2}}\,dx$

In Problems 17–26, use the method of completing the square, along with a trigonometric substitution if needed, to evaluate each integral.

17. $\int \frac{dx}{\sqrt{x^2+2x+5}}$

18. $\int \frac{dx}{\sqrt{x^2+4x+5}}$

19. $\int \frac{3x}{\sqrt{x^2+2x+5}}\,dx$

20. $\int \frac{2x-1}{\sqrt{x^2+4x+5}}\,dx$

21. $\int \sqrt{5-4x-x^2}\,dx$

22. $\int \frac{dx}{\sqrt{16+6x-x^2}}$

23. $\int \frac{dx}{\sqrt{4x-x^2}}$

24. $\int \frac{x}{\sqrt{4x-x^2}}\,dx$

25. $\int \frac{2x+1}{x^2+2x+2}\,dx$

26. $\int \frac{2x-1}{x^2-6x+18}\,dx$

27. The region bounded by $y = 1/(x^2+2x+5)$, $y = 0$, $x = 0$, and $x = 1$, is revolved about the x-axis. Find the volume of the resulting solid.

28. The region of Problem 27 is revolved about the y-axis. Find the volume of the resulting solid.

29. Find $\int \frac{x\,dx}{x^2+9}$ by

(a) an algebraic substitution and

(b) a trigonometric substitution. Then reconcile your answers.

30. Find $\int_0^3 \frac{x^3\,dx}{\sqrt{9+x^2}}$ by making the substitutions

$$u = \sqrt{9+x^2},\quad u^2 = 9+x^2,\quad 2u\,du = 2x\,dx$$

31. Find $\int \frac{\sqrt{4-x^2}}{x}\,dx$ by

(a) the substitution $u = \sqrt{4-x^2}$ and

(b) a trigonometric substitution. Then reconcile your answers.

Hint: $\int \csc x\,dx = \ln|\csc x - \cot x| + C$.

32. Two circles of radius b intersect as shown in Figure 6 with their centers $2a$ apart $(0 \le a \le b)$. Find the area of the region of their overlap.

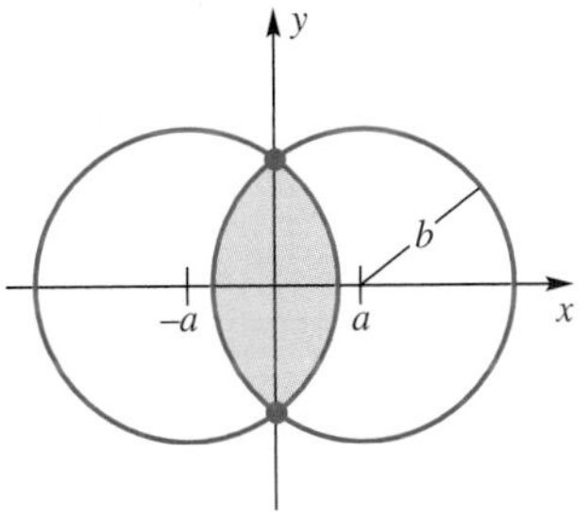

Figure 6

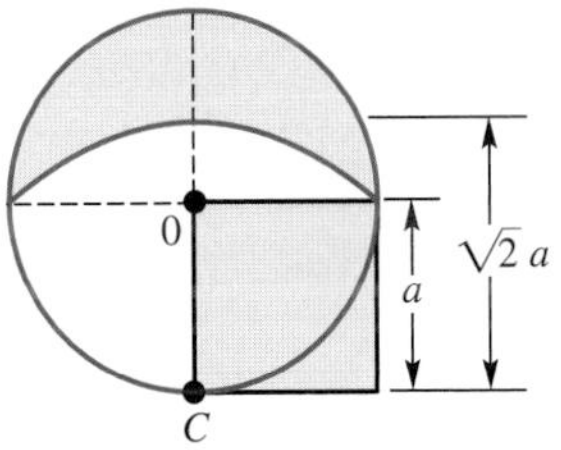

Figure 7

33. Hippocrates of Chios (ca. 430 B.C.) showed that the two shaded regions in Figure 7 have the same area (he squared the lune). Note that C is the center of the lower arc of the lune. Show Hippocrates' result

(a) using calculus and (b) without calculus.

34. Generalize the idea in Problem 33 by finding a formula for the area of the shaded lune shown in Figure 8.

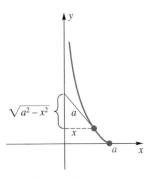

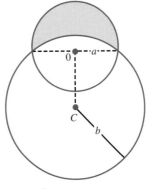

Figure 8

Figure 9

35. Starting at $(a, 0)$, an object is pulled along by a string of length a with the pulling end moving along the positive y-axis (Figure 9). The path of the object is a curve called a **tractrix**

and has the property that the string is always tangent to the curve. Set up a differential equation for the curve and solve it.

Answers to Concepts Review: **1.** $\sqrt{x-3}$ **2.** $2 \sin t$ **3.** $2 \tan t$ **4.** $2 \sec t$

7.5 Integration of Rational Functions Using Partial Fractions

A **rational function** is by definition the quotient of two polynomial functions. Examples are

$$f(x) = \frac{2}{(x+1)^3}, \qquad g(x) = \frac{2x+2}{x^2-4x+8}, \qquad h(x) = \frac{x^5+2x^3-x+1}{x^3+5x}$$

Of these, f and g are **proper rational functions,** meaning that the degree of the numerator is less than that of the denominator. An improper (not proper) rational function can always be written as a sum of a polynomial function and a proper rational function. Thus, for example,

$$h(x) = \frac{x^5+2x^3-x+1}{x^3+5x} = x^2 - 3 + \frac{14x+1}{x^3+5x}$$

$$\begin{array}{r|l} & x^2 - 3 \\ \hline x^3+5x & x^5 + 2x^3 - x + 1 \\ & \underline{x^5 + 5x^3} \\ & -3x^3 - x \\ & \underline{-3x^3 - 15x} \\ & 14x + 1 \end{array}$$

Figure 1

a result obtained by long division (Figure 1). Since polynomials are easy to integrate, the problem of integrating rational functions is really that of integrating proper rational functions. But can we always integrate proper rational functions? In theory, the answer is yes, though the practical details may be messy. Consider first the integrals of f and g above.

EXAMPLE 1 Find $\int \frac{2}{(x+1)^3}\,dx$.

SOLUTION Think of the substitution $u = x + 1$.

$$\int \frac{2}{(x+1)^3}\,dx = 2\int (x+1)^{-3}\,dx = \frac{2(x+1)^{-2}}{-2} + C$$

$$= -\frac{1}{(x+1)^2} + C \qquad \blacksquare$$

EXAMPLE 2 Find $\int \frac{2x+2}{x^2-4x+8}\,dx$.

SOLUTION Think first of the substitution $u = x^2 - 4x + 8$ for which $du = (2x-4)\,dx$. Then write the given integral as a sum of two integrals.

$$\int \frac{2x+2}{x^2-4x+8}\,dx = \int \frac{2x-4}{x^2-4x+8}\,dx + \int \frac{6}{x^2-4x+8}\,dx$$

$$= \ln|x^2-4x+8| + 6\int \frac{1}{x^2-4x+8}\,dx$$

In the second integral, complete the square.

$$\int \frac{1}{x^2-4x+8}\,dx = \int \frac{1}{x^2-4x+4+4}\,dx = \int \frac{1}{(x-2)^2+4}\,dx$$

$$= \int \frac{1}{(x-2)^2+4}\,dx = \frac{1}{2}\tan^{-1}\left(\frac{x-2}{2}\right) + C$$

We conclude that

$$\int \frac{2x+2}{x^2-4x+8}\,dx = \ln|x^2-4x+8| + 3\tan^{-1}\left(\frac{x-2}{2}\right) + K \qquad \blacksquare$$

It is a remarkable fact that any proper rational function can be written as a sum of *simple* proper rational functions like those illustrated in Examples 1 and 2.

Partial Fraction Decomposition (Linear Factors) To add fractions is a standard algebraic exercise: find a common denominator and add. For example,

$$\frac{2}{x-1} + \frac{3}{x+1} = \frac{2(x+1) + 3(x-1)}{(x-1)(x+1)} = \frac{5x-1}{(x-1)(x+1)} = \frac{5x-1}{x^2-1}$$

It is the reverse process of decomposing a fraction into a sum of simpler fractions that interests us now. We focus on the denominator and consider cases.

EXAMPLE 3 **Distinct Linear Factors** Decompose $(3x - 1)/(x^2 - x - 6)$ and then find its indefinite integral.

SOLUTION Since the denominator factors as $(x + 2)(x - 3)$, it seems reasonable to hope for a decomposition of the following form:

(1) $$\frac{3x-1}{(x+2)(x-3)} = \frac{A}{x+2} + \frac{B}{x-3}$$

Our job is, of course, to determine A and B so that (1) is an identity, a task that we find easier after we have multiplied both sides by $(x + 2)(x - 3)$. We obtain

(2) $$3x - 1 = A(x - 3) + B(x + 2)$$

or, equivalently,

(3) $$3x - 1 = (A + B)x + (-3A + 2B)$$

However, (3) is an identity if and only if coefficients of like powers of x on both sides are equal; that is,

$$A + B = 3$$
$$-3A + 2B = -1$$

By solving this pair of equations for A and B, we obtain $A = \frac{7}{5}$, $B = \frac{8}{5}$. Consequently,

$$\frac{3x-1}{x^2-x-6} = \frac{3x-1}{(x+2)(x-3)} = \frac{\frac{7}{5}}{x+2} + \frac{\frac{8}{5}}{x-3}$$

and

$$\int \frac{3x-1}{x^2-x-6}\,dx = \frac{7}{5}\int \frac{1}{x+2}\,dx + \frac{8}{5}\int \frac{1}{x-3}\,dx$$
$$= \frac{7}{5}\ln|x+2| + \frac{8}{5}\ln|x-3| + C$$ ■

If there was anything difficult about this process, it was the determination of A and B. We found their values by "brute force," but there is an easier way. In (2), which we wish to be an identity (that is, true for *every* value of x), substitute the convenient values $x = 3$ and $x = -2$, obtaining

$$8 = A \cdot 0 + B \cdot 5$$
$$-7 = A \cdot (-5) + B \cdot 0$$

This immediately gives $B = \frac{8}{5}$ and $A = \frac{7}{5}$.

You have just witnessed an odd, but correct, mathematical maneuver. Equation (1) turns out to be an identity (true for all x except -2 and 3) if and only if the essentially equivalent equation (2) is true precisely at -2 and 3. Ask yourself why

Solve This D.E.

"Often, there is little resemblance between a differential equation and its solution. Who would suppose that an expression as simple as

$$\frac{dy}{dx} = \frac{1}{a^2 - x^2}$$

could be transformed into

$$y = \frac{1}{2a}\log_e\left(\frac{a+x}{a-x}\right) + C$$

This resembles the transformation of a chrysalis into a butterfly."

Silvanus P. Thompson

The method of partial fractions makes this an easy transformation. Do you see how it is done?

this is so. Ultimately it depends on the fact that the two sides of equation (2), both linear polynomials, are identical if they have the same values at any two points.

EXAMPLE 4 **Distinct Linear Factors** Find $\int \frac{5x+3}{x^3-2x^2-3x}\,dx$.

SOLUTION Since the denominator factors as $x(x+1)(x-3)$, we write

$$\frac{5x+3}{x(x+1)(x-3)} = \frac{A}{x} + \frac{B}{x+1} + \frac{C}{x-3}$$

and seek to determine A, B, and C. Clearing the fractions gives

$$5x + 3 = A(x+1)(x-3) + Bx(x-3) + Cx(x+1)$$

Substitution of the values $x = 0$, $x = -1$, and $x = 3$ results in

$$3 = A(-3)$$
$$-2 = B(4)$$
$$18 = C(12)$$

or $A = -1$, $B = -\frac{1}{2}$, $C = \frac{3}{2}$. Thus,

$$\int \frac{5x+3}{x^3-2x^2-3x}\,dx = -\int \frac{1}{x}\,dx - \frac{1}{2}\int \frac{1}{x+1}\,dx + \frac{3}{2}\int \frac{1}{x-3}\,dx$$
$$= -\ln|x| - \frac{1}{2}\ln|x+1| + \frac{3}{2}\ln|x-3| + C$$

■

EXAMPLE 5 **Repeated Linear Factors** Find $\int \frac{x}{(x-3)^2}\,dx$.

SOLUTION Now the decomposition takes the form

$$\frac{x}{(x-3)^2} = \frac{A}{x-3} + \frac{B}{(x-3)^2}$$

with A and B to be determined. After clearing the fractions, we get

$$x = A(x-3) + B$$

If we now substitute the convenient value $x = 3$ and any other value, such as $x = 0$, we obtain $B = 3$ and $A = 1$. Thus,

$$\int \frac{x}{(x-3)^2}\,dx = \int \frac{1}{x-3}\,dx + 3\int \frac{1}{(x-3)^2}\,dx$$
$$= \ln|x-3| - \frac{3}{x-3} + C$$

■

EXAMPLE 6 **Some Distinct, Some Repeated Linear Factors** Find

$$\int \frac{3x^2-8x+13}{(x+3)(x-1)^2}\,dx$$

SOLUTION We decompose the integrand in the following way:

$$\frac{3x^2-8x+13}{(x+3)(x-1)^2} = \frac{A}{x+3} + \frac{B}{x-1} + \frac{C}{(x-1)^2}$$

Clearing the fractions changes this to

$$3x^2 - 8x + 13 = A(x-1)^2 + B(x+3)(x-1) + C(x+3)$$

Substitution of $x = 1$, $x = -3$, and $x = 0$ yields $C = 2$, $A = 4$, and $B = -1$. Thus,

$$\int \frac{3x^2 - 8x + 13}{(x+3)(x-1)^2}\,dx = 4\int \frac{dx}{x+3} - \int \frac{dx}{x-1} + 2\int \frac{dx}{(x-1)^2}$$

$$= 4\ln|x+3| - \ln|x-1| - \frac{2}{x-1} + C \quad \blacksquare$$

Be sure to note the inclusion of the two fractions $B/(x-1)$ and $C/(x-1)^2$ in the decomposition above. The general rule for decomposing fractions with repeated linear factors in the denominator is this: for each factor $(ax+b)^k$ of the denominator, there are k terms in the partial fraction decomposition:

$$\frac{A_1}{ax+b} + \frac{A_2}{(ax+b)^2} + \frac{A_3}{(ax+b)^3} + \cdots + \frac{A_k}{(ax+b)^k}$$

Partial Fraction Decomposition (Quadratic Factors) In factoring the denominator of a fraction, we may well get some quadratic factors, such as $x^2 + 1$, that cannot be factored into linear factors without introducing complex numbers.

EXAMPLE 7 **A Single Quadratic Factor** Decompose $\dfrac{6x^2 - 3x + 1}{(4x+1)(x^2+1)}$ and then find its indefinite integral.

SOLUTION The best we can hope for is a decomposition of the form

$$\frac{6x^2 - 3x + 1}{(4x+1)(x^2+1)} = \frac{A}{4x+1} + \frac{Bx + C}{x^2+1}$$

To determine the constants A, B, and C, we multiply both sides by $(4x+1)(x^2+1)$ and obtain

$$6x^2 - 3x + 1 = A(x^2+1) + (Bx + C)(4x+1)$$

Substitution of $x = -\frac{1}{4}$, $x = 0$, and $x = 1$ yields

$$\frac{6}{16} + \frac{3}{4} + 1 = A\left(\tfrac{17}{16}\right) \quad \Rightarrow \quad A = 2$$

$$1 = 2 + C \quad \Rightarrow \quad C = -1$$

$$4 = 4 + (B-1)5 \quad \Rightarrow \quad B = 1$$

Thus,

$$\int \frac{6x^2 - 3x + 1}{(4x+1)(x^2+1)}\,dx = \int \frac{2}{4x+1}\,dx + \int \frac{x-1}{x^2+1}\,dx$$

$$= \frac{1}{2}\int \frac{4\,dx}{4x+1} + \frac{1}{2}\int \frac{2x\,dx}{x^2+1} - \int \frac{dx}{x^2+1}$$

$$= \frac{1}{2}\ln|4x+1| + \frac{1}{2}\ln(x^2+1) - \tan^{-1} x + C \quad \blacksquare$$

EXAMPLE 8 **A Repeated Quadratic Factor** Find $\displaystyle\int \frac{6x^2 - 15x + 22}{(x+3)(x^2+2)^2}\,dx.$

SOLUTION Here the appropriate decomposition is

$$\frac{6x^2 - 15x + 22}{(x+3)(x^2+2)^2} = \frac{A}{x+3} + \frac{Bx + C}{x^2+2} + \frac{Dx + E}{(x^2+2)^2}$$

After considerable work, we discover that $A = 1$, $B = -1$, $C = 3$, $D = -5$, and $E = 0$. Thus,

$$\int \frac{6x^2 - 15x + 22}{(x+3)(x^2+2)^2}\,dx$$
$$= \int \frac{dx}{x+3} - \int \frac{x-3}{x^2+2}\,dx - 5\int \frac{x}{(x^2+2)^2}\,dx$$
$$= \int \frac{dx}{x+3} - \frac{1}{2}\int \frac{2x}{x^2+2}\,dx + 3\int \frac{dx}{x^2+2} - \frac{5}{2}\int \frac{2x\,dx}{(x^2+2)^2}$$
$$= \ln|x+3| - \frac{1}{2}\ln(x^2+2) + \frac{3}{\sqrt{2}}\tan^{-1}\left(\frac{x}{\sqrt{2}}\right) + \frac{5}{2(x^2+2)} + C$$ ■

Summary To decompose a rational function $f(x) = p(x)/q(x)$ into partial fractions, proceed as follows:

Step 1: If $f(x)$ is improper, that is, if $p(x)$ is of degree at least that of $q(x)$, divide $p(x)$ by $q(x)$, obtaining

$$f(x) = \text{a polynomial} + \frac{N(x)}{D(x)}$$

Step 2: Factor $D(x)$ into a product of linear and irreducible quadratic factors with real coefficients. By a theorem of algebra, this is always (theoretically) possible.

Step 3: For each factor of the form $(ax+b)^k$, expect the decomposition to have the terms

$$\frac{A_1}{(ax+b)} + \frac{A_2}{(ax+b)^2} + \cdots + \frac{A_k}{(ax+b)^k}$$

Step 4: For each factor of the form $(ax^2+bx+c)^m$, expect the decomposition to have the terms

$$\frac{B_1x + C_1}{ax^2+bx+c} + \frac{B_2x+C_2}{(ax^2+bx+c)^2} + \cdots + \frac{B_mx+C_m}{(ax^2+bx+c)^m}$$

Step 5: Set $N(x)/D(x)$ equal to the sum of all the terms found in Steps 3 and 4. The number of constants to be determined should equal the degree of the denominator, $D(x)$.

Step 6: Multiply both sides of the equation found in Step 5 by $D(x)$ and solve for the unknown constants. This can be done by either of two methods: (1) Equate coefficients of like-degree terms or (2) assign convenient values to the variable x.

The Logistic Differential Equation In the last chapter, we saw that the assumption that the rate of growth of a population is proportional to its size, that is, $y' = ky$, leads to exponential growth. This assumption may be realistic until the available resources in the system are unable to sustain the population. In such a case, more reasonable assumptions are that there is a maximum capacity L that the system can sustain, and that the rate of growth is proportional to the product of the population size y and the "available room" $L - y$. These assumptions lead to the differential equation

$$y' = ky(L - y)$$

This is called the **logistic differential equation.** It is separable and now that we have covered the method of partial fractions, we can perform the necessary integration to solve it.

A Bound for the Answer

The initial population size is 800, and the rate of change in population size, y', is positive, so the population grows. As it nears 2000, the rate of change gets close to zero, so as $t \to \infty$, we have $y \to 2000$. The population at time $t = 2$ should be somewhere between 800 and 2000.

EXAMPLE 9 A population grows according to the logistic differential equation $y' = 0.0003y(2000 - y)$. The initial population size is 800. Solve this differential equation and use the solution to predict the population size at time $t = 2$.

SOLUTION Writing y' as dy/dt, we see that the differential equation can be written as

$$\frac{dy}{dt} = 0.0003y(2000 - y)$$

$$\frac{dy}{y(2000 - y)} = 0.0003\,dt$$

$$\int \frac{dy}{y(2000 - y)} = \int 0.0003\,dt$$

The integral on the left can be evaluated using the method of partial fractions. We write

$$\frac{1}{y(2000 - y)} = \frac{A}{y} + \frac{B}{2000 - y}$$

which leads to

$$1 = A(2000 - y) + By$$

Substituting $y = 0$ and $y = 2000$ yields

$$1 = 2000A$$
$$1 = 2000B$$

Thus, $A = \frac{1}{2000}$ and $B = \frac{1}{2000}$, leading to

$$\int \left(\frac{1}{2000y} + \frac{1}{2000(2000 - y)} \right) dy = 0.0003t + C$$

$$\frac{1}{2000}\ln y - \frac{1}{2000}\ln(2000 - y) = 0.0003t + C$$

$$\ln \frac{y}{2000 - y} = 0.6t + 2000C$$

$$\frac{y}{2000 - y} = e^{0.6t + 2000C}$$

$$\frac{y}{2000 - y} = C_1 e^{0.6t}$$

Here, $C_1 = e^{2000C}$. At this point we can use the initial condition $y(0) = 800$ to determine C_1.

$$\frac{800}{2000 - 800} = C_1 e^{0.6 \cdot 0}$$

$$C_1 = \frac{800}{1200} = \frac{2}{3}$$

Thus

$$\frac{y}{2000 - y} = \frac{2}{3}e^{0.6t}$$

$$y = \frac{2}{3}(2000 - y)e^{0.6t}$$

$$y + \frac{2}{3}ye^{0.6t} = \frac{4000}{3}e^{0.6t}$$

$$y = \frac{(4000/3)e^{0.6t}}{1 + (2/3)e^{0.6t}} = \frac{4000/3}{2/3 + e^{-0.6t}}$$

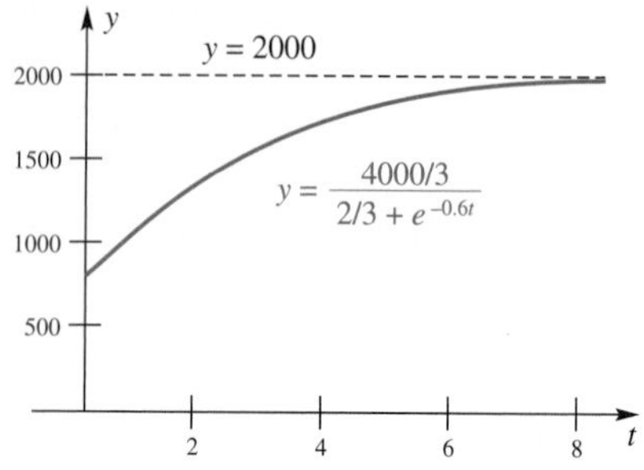

Figure 2

The population at time $t = 2$ is thus

$$y = \frac{4000/3}{2/3 + e^{-0.6 \cdot 2}} \approx 1378$$

A sketch of the population size as a function of t is given in Figure 2. ■

Concepts Review

1. If the degree of the polynomial $p(x)$ is less than the degree of $q(x)$, then $f(x) = p(x)/q(x)$ is called a _____ rational function.

2. To integrate the improper rational function $f(x) = (x^2 + 4)/(x + 1)$, we first rewrite it as $f(x) =$ _____.

3. If $(x - 1)(x + 1) + 3x + x^2 = ax^2 + bx + c$, then $a =$ _____, $b =$ _____, and $c =$ _____.

4. $(3x + 1)/[(x - 1)^2(x^2 + 1)]$ can be decomposed in the form _____.

Problem Set 7.5

In Problems 1–40, use the method of partial fraction decomposition to perform the required integration.

1. $\displaystyle\int \frac{1}{x(x + 1)}\,dx$

2. $\displaystyle\int \frac{2}{x^2 + 3x}\,dx$

3. $\displaystyle\int \frac{3}{x^2 - 1}\,dx$

4. $\displaystyle\int \frac{5x}{2x^3 + 6x^2}\,dx$

5. $\displaystyle\int \frac{x - 11}{x^2 + 3x - 4}\,dx$

6. $\displaystyle\int \frac{x - 7}{x^2 - x - 12}\,dx$

7. $\displaystyle\int \frac{3x - 13}{x^2 + 3x - 10}\,dx$

8. $\displaystyle\int \frac{x + \pi}{x^2 - 3\pi x + 2\pi^2}\,dx$

9. $\displaystyle\int \frac{2x + 21}{2x^2 + 9x - 5}\,dx$

10. $\displaystyle\int \frac{2x^2 - x - 20}{x^2 + x - 6}\,dx$

11. $\displaystyle\int \frac{17x - 3}{3x^2 + x - 2}\,dx$

12. $\displaystyle\int \frac{5 - x}{x^2 - x(\pi + 4) + 4\pi}\,dx$

13. $\displaystyle\int \frac{2x^2 + x - 4}{x^3 - x^2 - 2x}\,dx$

14. $\displaystyle\int \frac{7x^2 + 2x - 3}{(2x - 1)(3x + 2)(x - 3)}\,dx$

15. $\displaystyle\int \frac{6x^2 + 22x - 23}{(2x - 1)(x^2 + x - 6)}\,dx$

16. $\displaystyle\int \frac{x^3 - 6x^2 + 11x - 6}{4x^3 - 28x^2 + 56x - 32}\,dx$

17. $\displaystyle\int \frac{x^3}{x^2 + x - 2}\,dx$

18. $\displaystyle\int \frac{x^3 + x^2}{x^2 + 5x + 6}\,dx$

19. $\displaystyle\int \frac{x^4 + 8x^2 + 8}{x^3 - 4x}\,dx$

20. $\displaystyle\int \frac{x^6 + 4x^3 + 4}{x^3 - 4x^2}\,dx$

21. $\displaystyle\int \frac{x + 1}{(x - 3)^2}\,dx$

22. $\displaystyle\int \frac{5x + 7}{x^2 + 4x + 4}\,dx$

23. $\displaystyle\int \frac{3x + 2}{x^3 + 3x^2 + 3x + 1}\,dx$

24. $\displaystyle\int \frac{x^6}{(x - 2)^2(1 - x)^5}\,dx$

25. $\displaystyle\int \frac{3x^2 - 21x + 32}{x^3 - 8x^2 + 16x}\,dx$

26. $\displaystyle\int \frac{x^2 + 19x + 10}{2x^4 + 5x^3}\,dx$

27. $\displaystyle\int \frac{2x^2 + x - 8}{x^3 + 4x}\,dx$

28. $\displaystyle\int \frac{3x + 2}{x(x + 2)^2 + 16x}\,dx$

29. $\displaystyle\int \frac{2x^2 - 3x - 36}{(2x - 1)(x^2 + 9)}\,dx$

30. $\displaystyle\int \frac{1}{x^4 - 16}\,dx$

31. $\displaystyle\int \frac{1}{(x - 1)^2(x + 4)^2}\,dx$

32. $\displaystyle\int \frac{x^3 - 8x^2 - 1}{(x + 3)(x^2 - 4x + 5)}\,dx$

33. $\displaystyle\int \frac{(\sin^3 t - 8\sin^2 t - 1)\cos t}{(\sin t + 3)(\sin^2 t - 4\sin t + 5)}\,dt$

34. $\displaystyle\int \frac{\cos t}{\sin^4 t - 16}\,dt$

35. $\displaystyle\int \frac{x^3 - 4x}{(x^2 + 1)^2}\,dx$

36. $\displaystyle\int \frac{(\sin t)(4\cos^2 t - 1)}{(\cos t)(1 + 2\cos^2 t + \cos^4 t)}\,dt$

37. $\displaystyle\int \frac{2x^3 + 5x^2 + 16x}{x^5 + 8x^3 + 16x}\,dx$

38. $\displaystyle\int_4^6 \frac{x - 17}{x^2 + x - 12}\,dx$

39. $\displaystyle\int_0^{\pi/4} \frac{\cos\theta}{(1 - \sin^2\theta)(\sin^2\theta + 1)^2}\,d\theta$

40. $\displaystyle\int_1^5 \frac{3x + 13}{x^2 + 4x + 3}\,dx$

In Problems 41–44, solve the logistic differential equation representing population growth with the given initial condition. Then use the solution to predict the population size at time $t = 3$.

41. $y' = y(1 - y),\ y(0) = 0.5$

42. $y' = \frac{1}{10}y(12 - y),\ y(0) = 2$

43. $y' = 0.0003y\,(8000 - y),\ y(0) = 1000$

44. $y' = 0.001y\,(4000 - y),\ y(0) = 100$

45. Solve the logistic differential equation for an arbitrary constant of proportionality k, capacity L, and initial condition $y(0) = y_0$.

46. Explain what happens to the solution of the logistic differential equation if the initial population size is *larger* than the maximum capacity.

47. Without solving the logistic equation or referring to its solution, explain how you know that if $y_0 < L$, then the population size is increasing.

48. Assuming $y_0 < L$, for what values of t is the graph of the population size $y(t)$ concave up?

49. Suppose that the earth will not support a population of more than 16 billion and that there were 2 billion people in 1925 and 4 billion people in 1975. Then, if y is the population t years after 1925, an appropriate model is the logistic differential equation

$$\frac{dy}{dt} = ky(16 - y)$$

(a) Solve this differential equation.

(b) Predict the population in 2015.

(c) When will the population be 9 billion?

50. Do Problem 49 assuming that the upper limit for the population is 10 billion.

51. The Law of Mass Action in chemistry results in the differential equation

$$\frac{dx}{dt} = k(a - x)(b - x), \qquad k > 0,\ \ a > 0,\ \ b > 0$$

where x is the amount of a substance at time t resulting from the reaction of two others. Assume that $x = 0$ when $t = 0$.

(a) Solve this differential equation in the case $b > a$.

(b) Show that $x \to a$ as $t \to \infty$ (if $b > a$).

(c) Suppose that $a = 2$ and $b = 4$, and that 1 gram of the substance is formed in 20 minutes. How much will be present in 1 hour?

(d) Solve the differential equation if $a = b$.

52. The differential equation

$$\frac{dy}{dt} = k(y - m)(M - y)$$

with $k > 0$ and $0 \le m < y_0 < M$ is used to model some growth problems. Solve the equation and find $\lim_{t\to\infty} y$.

53. As a model for the production of trypsin from trypsinogen in digestion, biochemists have proposed the model

$$\frac{dy}{dt} = k(A - y)(B + y)$$

where $k > 0$, A is the initial amount of trypsinogen, and B is the original amount of trypsin. Solve this differential equation.

54. Evaluate

$$\int_{\pi/6}^{\pi/2} \frac{\cos x}{\sin x(\sin^2 x + 1)^2}\,dx$$

Answers to Concepts Review: **1.** proper **2.** $x - 1 + \frac{5}{x + 1}$ **3.** 2; 3; −1 **4.** $\frac{A}{x - 1} + \frac{B}{(x - 1)^2} + \frac{Cx + D}{x^2 + 1}$

7.6 Strategies for Integration

Throughout this chapter we have presented a number of techniques for finding an antiderivative (or indefinite integral) of a given function. By now it should be clear that while differentiation is a straightforward process, antidifferentiation is not. The Sum Rule, Product Rule, Quotient Rule, and Chain Rule can be used to find the derivative of almost any function, but there is no sure-fire method for finding antiderivatives. There is only a set of techniques that one might apply. Thus, to a large extent, antidifferentiation is a trial and error process; when one method fails, look for another. This being said, however, we can give the following strategies for finding antiderivatives.

1. Look for a *substitution* that makes the integral look like one of the basic integration formulas from the first section of this chapter. For example,

$$\int \sin 2x\,dx,\ \int xe^{-x^2}\,dx,\ \int x\sqrt{x^2 - 1}\,dx$$

can be evaluated using simple substitutions.

2. Look for situations where you have the product of two functions, where the derivative of one of them times the antiderivative of the other is one of the basic integration formulas from Section 7.1. *Integration by parts* can be used

for these integrals. For example $\int xe^x\,dx$ and $\int x\sinh x\,dx$ can both be evaluated using integration by parts.

3. *Trigonometric Substitutions*
   If the integrand contains $\sqrt{a^2 - x^2}$, consider the substitution $x = a\sin t$.
   If the integrand contains $\sqrt{x^2 + a^2}$, consider the substitution $x = a\tan t$.
   If the integrand contains $\sqrt{x^2 - a^2}$, consider the substitution $x = a\sec t$.
4. If the integrand is a proper rational function, that is, the degree of the numerator is less than that of the denominator, then decompose the integrand using the method of *partial fractions*. Often the terms in the resulting sum can be integrated one at a time. If the integrand is an improper rational function, apply long division to write it as the sum of a polynomial and a proper rational function. Then apply the method of partial fractions to the proper rational function.

These suggestions, along with a bit of ingenuity, will go a long way in evaluating antiderivatives.

Tables of Integrals The inside back cover of the book contains 110 (indefinite) integration formulas. There are larger tables, such as those found in the *CRC Standard Mathematical Tables and Formulae* (published by CRC Press) and *Handbook of Mathematical Functions* (edited by Abramowitz and Stegun, published by Dover), but the list of 110 will suffice for our purposes. The important thing to keep in mind is that you must often use these formulas along with the method of substitution to evaluate an indefinite integral. This is why many tables of integrals, including those at the end of this book, use u for the variable of integration, rather than x. You should think of u as being some function of x (maybe just x itself). The next example shows how one formula can be used to evaluate several integrals using the method of substitution.

EXAMPLE 1 Use Formula (54)

$$(54)\quad \int \sqrt{a^2 - u^2}\,du = \frac{u}{2}\sqrt{a^2 - u^2} + \frac{a^2}{2}\sin^{-1}\frac{u}{a} + C$$

to evaluate the following integrals:

(a) $\int \sqrt{9 - x^2}\,dx$ (b) $\int \sqrt{16 - 4y^2}\,dy$

(c) $\int y\sqrt{1 - 4y^4}\,dy$ (d) $\int e^t\sqrt{100 - e^{2t}}\,dt$

SOLUTION

(a) In this integral we have $a = 3$ and $u = x$, so

$$\int \sqrt{9 - x^2}\,dx = \frac{x}{2}\sqrt{9 - x^2} + \frac{9}{2}\sin^{-1}\frac{x}{3} + C$$

For part (b), we have to recognize $4y^2$ as $(2y)^2$, so the appropriate substitution is $u = 2y$ and $du = 2\,dy$. Thus

$$\begin{aligned}\int \sqrt{16 - 4y^2}\,dy &= \frac{1}{2}\int \sqrt{4^2 - (2y)^2}\,(2\,dy)\\ &= \frac{1}{2}\left(\frac{2y}{2}\sqrt{4^2 - (2y)^2} + \frac{4^2}{2}\sin^{-1}\frac{2y}{4}\right) + C\\ &= \frac{y}{2}\sqrt{16 - 4y^2} + 4\sin^{-1}\frac{y}{2} + C\end{aligned}$$

Part (c) requires a little foresight. We might be tempted to make the substitution $u = 1 - 4y^4$, but then $du = -16y^3\,dy$. The presence of y^3 in the expression du is troubling because we have just y in the remainder of the integrand. For this part we must see the radical as $\sqrt{1 - (2y^2)^2}$ making Formula (54) applicable with the substitution $u = 2y^2$ and $du = 4y\,dy$. Thus

$$\begin{aligned}\int y\sqrt{1 - 4y^4}\,dy &= \frac{1}{4}\int \sqrt{1 - (2y^2)^2}\,(4y\,dy)\\ &= \frac{1}{4}\left(\frac{2y^2}{2}\sqrt{1 - (2y^2)^2} + \frac{1}{2}\sin^{-1}\frac{2y^2}{1}\right) + C\\ &= \frac{y^2}{4}\sqrt{1 - 4y^4} + \frac{1}{8}\sin^{-1} 2y^2 + C\end{aligned}$$

For part (d) we must recognize that the radical can be written as $\sqrt{100 - (e^t)^2}$ and that we should make the substitution $u = e^t$ and $du = e^t\,dt$. Thus

$$\begin{aligned}\int e^t\sqrt{100 - e^{2t}}\,dt &= \int \sqrt{10^2 - (e^t)^2}\,(e^t\,dt)\\ &= \frac{e^t}{2}\sqrt{10^2 - (e^t)^2} + \frac{10^2}{2}\sin^{-1}\frac{e^t}{10} + C\\ &= \frac{e^t}{2}\sqrt{100 - e^{2t}} + 50\sin^{-1}\frac{e^t}{10} + C\end{aligned}$$

■

Tables and Substitution

In Example 1 we were able to evaluate four seemingly unrelated integrals using the same formula from the table of integrals. Each required a different substitution. When you use a table of integrals to help you evaluate an integral, keep in the back of your mind the method of substitution.

Computer Algebra Systems and Calculators Today a computer algebra system such as *Maple*, *Mathematica*, or *Derive*, can be used to evaluate indefinite or definite integrals. Many calculators are also capable of evaluating integrals. If such systems are used to evaluate *definite integrals*, it is important to distinguish whether the system is giving you an *exact* answer, usually obtained by applying the Second Fundamental Theorem of Calculus, or whether it is giving an *approximation* (using something similar to, but probably a little more sophisticated than, the Parabolic Method of Section 4.6). It might seem like the two are equally good in practical situations, and if we were presented with just one integral to evaluate, this may well be correct. However, in many cases, the result of the definite integral will be used in subsequent calculations. In a case like this it is more accurate, and often easier, to find the exact answer and then use the exact answer in further calculations. For example, if $\int_0^1 \frac{1}{1 + x^2}\,dx$ is needed in subsequent calculations, it would be better to find an antiderivative and use the Second Fundamental Theorem of Calculus to obtain

$$\int_0^1 \frac{1}{1 + x^2}\,dx = [\tan^{-1} x]_0^1 = \tan^{-1} 1 - \tan^{-1} 0 = \frac{\pi}{4}$$

Using $\pi/4$ in subsequent calculations would be preferable to 0.785398, which is what *Mathematica* gives for a numerical approximation to the integral.

In some cases, however, it is not possible to evaluate a definite integral by applying the Second Fundamental Theorem of Calculus, because some functions do not have antiderivatives that can be expressed in terms of elementary functions. Our inability to find a simple formula for an antiderivative does not absolve us of the task of finding the value of the definite integral. It just means that we must use a numerical method to *approximate* the definite integral. Many practical problems lead to just this situation, where the necessary integral is intractable and we must resort to a numerical method. We discussed numerical integration in Section 4.6. Usually a CAS will use a method similar to the Parabolic Rule, but a little more sophisticated.

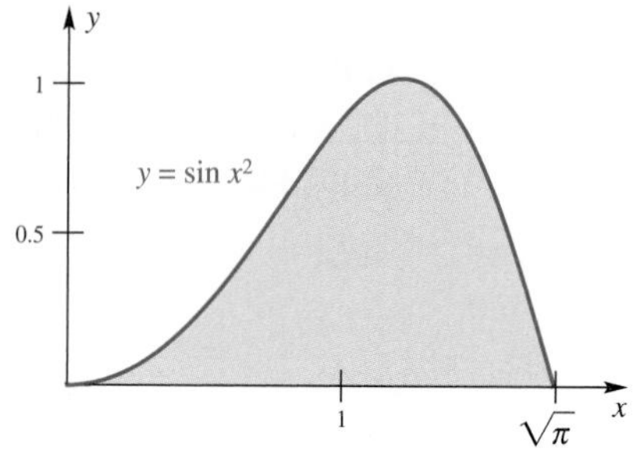

Figure 1

EXAMPLE 2 Find the center of mass of the homogeneous lamina shown in Figure 1.

SOLUTION Using the formulas from Section 5.6, we have

$$m = \delta \int_0^{\sqrt{\pi}} \sin x^2\, dx$$

$$M_y = \delta \int_0^{\sqrt{\pi}} x \sin x^2\, dx$$

$$M_x = \frac{\delta}{2} \int_0^{\sqrt{\pi}} \sin^2 x^2\, dx$$

Among these integrals, only the second can be evaluated using the Second Fundamental Theorem of Calculus. For the first and the third, there is no antiderivative that can be expressed in terms of elementary functions. We must therefore resort to an approximation for the integrals. A CAS gives the following values for these integrals

$$m = \delta \int_0^{\sqrt{\pi}} \sin x^2\, dx \approx 0.89483\,\delta$$

$$M_y = \delta \int_0^{\sqrt{\pi}} x \sin x^2\, dx = \delta\left[-\frac{1}{2}\cos x^2\right]_0^{\sqrt{\pi}} = \delta$$

$$M_x = \frac{\delta}{2} \int_0^{\sqrt{\pi}} \sin^2 x^2\, dx \approx 0.33494\,\delta$$

Notice that the CAS was able to give an exact value for the second integral and approximations for the first and third. From these results we can calculate

$$\overline{x} = \frac{M_y}{m} \approx \frac{\delta}{0.89483\,\delta} \approx 1.1175$$

$$\overline{y} = \frac{M_x}{m} \approx \frac{0.33494\,\delta}{0.89483\,\delta} \approx 0.3743$$

■

There are also situations where the upper limit of an integral is an unknown. If this is the case, then use of the Second Fundamental Theorem of Calculus is preferred over the use of a numerical approximation. The next two examples illustrate this. The two problems are in principle the same, but the methods of solution are, of necessity, different.

EXAMPLE 3 A rod has density equal to $\delta(x) = \exp(-x/4)$ for $x > 0$. Where should the rod be cut off so that the mass from 0 to the cut is equal to one?

SOLUTION Let a denote the cut-off point. We then require

$$1 = \int_0^a \delta(x)\, dx = \int_0^a \exp(-x/4)\, dx = 4 - 4e^{-a/4}$$

Solving for a gives

$$1 = 4 - 4e^{-a/4}$$

$$4e^{-a/4} = 3$$

$$a = -4 \ln\frac{3}{4} \approx 1.1507$$

Here we obtained the exact answer, $a = -4\ln(3/4)$, which we could approximate as 1.1507 if we needed an approximation. ■

An Approximate Answer

The mass is the integral of the density, so the mass can be thought of as the area under the density curve. At position $x = 0$, the density is 1, and it decreases slowly as x increases. In order to make the area under the density curve equal to 1, we would expect to have to choose the cut-off point to be slightly larger than 1.

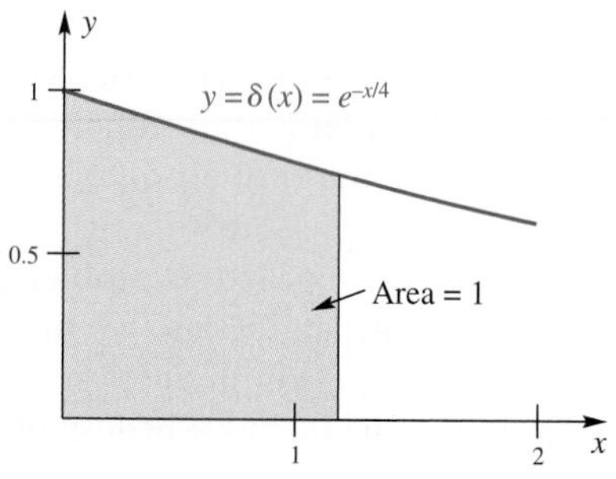

Another Approximation

Using the fact that mass is area under the density curve, we see from the figure below that the cut-off must be somewhere between 0.5 and 1. This give us a starting point to approximate the answer.

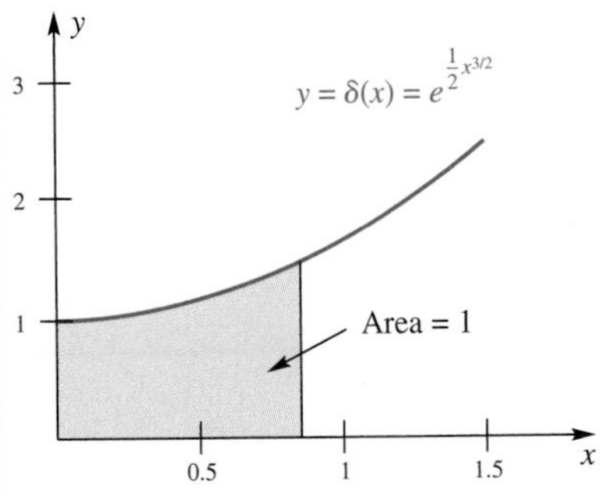

EXAMPLE 4 A rod has density equal to $\delta(x) = \exp\left(\frac{1}{2}x^{3/2}\right)$ for $x > 0$. Where should the rod be cut off so that the mass from 0 to the cut is equal to one? Use the Bisection Method to approximate the cut-off point accurate to two significant places.

SOLUTION Again, let a denote the position of the cut. We then require that

$$1 = \int_0^a \delta(x)\,dx = \int_0^a \exp\left(\frac{1}{2}x^{3/2}\right)dx$$

The antiderivative of $\exp\left(\frac{1}{2}x^{3/2}\right)$ cannot be expressed in terms of elementary functions, so we cannot use the Second Fundamental Theorem of Calculus to evaluate the definite integral. We are forced to approximate the integral using numerical methods. The problem is that we must have fixed upper and lower limits for the integral in order to approximate it, but in this case the upper limit is the variable a. A little trial and error, along with a program to approximate definite integrals, leads to the following:

$a = 1;$ $\quad \int_0^1 \exp\left(\frac{1}{2}x^{3/2}\right)dx \approx 1.2354$ $\quad a = 1$ is too large

$a = 0.5;$ $\quad \int_0^{0.5} \exp\left(\frac{1}{2}x^{3/2}\right)dx \approx 0.5374$ $\quad a = 0.5$ is too small

At this point we know that the desired value of a is between 0.5 and 1.0. The midpoint of [0.5, 1.0] is 0.75, so we try 0.75:

$a = 0.75;$ $\quad \int_0^{0.75} \exp\left(\frac{1}{2}x^{3/2}\right)dx \approx 0.85815$ $\quad a = 0.75$ is too small

Continuing in this manner,

$a = 0.875;$ $\quad \int_0^{0.875} \exp\left(\frac{1}{2}x^{3/2}\right)dx \approx 1.0385$ $\quad a = 0.875$ is too large

$a = 0.8125;$ $\quad \int_0^{0.8125} \exp\left(\frac{1}{2}x^{3/2}\right)dx \approx 0.94643$ $\quad a = 0.8125$ is too small

$a = 0.84375;$ $\quad \int_0^{0.84375} \exp\left(\frac{1}{2}x^{3/2}\right)dx \approx 0.99198$ $\quad a = 0.84375$ is too small

$a = 0.859375;$ $\quad \int_0^{0.859375} \exp\left(\frac{1}{2}x^{3/2}\right)dx \approx 1.0151$ $\quad a = 0.859375$ is too large

$a = 0.8515625;$ $\quad \int_0^{0.8515625} \exp\left(\frac{1}{2}x^{3/2}\right)dx \approx 1.0035$ $\quad a = 0.8515625$ is too large

$a = 0.84765625;$ $\quad \int_0^{0.84765625} \exp\left(\frac{1}{2}x^{3/2}\right)dx \approx 0.99775$ $\quad a = 0.84765625$ is too small

At this point, we have trapped a between 0.84765625 and 0.8515625, so correct to two places, the cut-off point should be $a = 0.85$. ■

EXAMPLE 5 Use Newton's Method to approximate the solution of the equation in Example 4.

SOLUTION The equation to be solved can be written as

$$\int_0^a \exp\left(\frac{1}{2}x^{3/2}\right) dx - 1 = 0$$

Let $F(a)$ denote the left side of this equation. We are then asking for an approximation to the solution of $F(a) = 0$. Recall that Newton's Method is an iterative method defined by

$$a_{n+1} = a_n - \frac{F(a_n)}{F'(a_n)}$$

In this case we can use the First Fundamental Theorem of Calculus to obtain

$$F'(a) = \exp\left(\frac{1}{2}a^{3/2}\right)$$

We begin with $a_1 = 1$ as our initial guess (which we know from the solution of Example 3 is on the high side). Then

$$a_2 = 1 - \frac{\int_0^1 \exp\left(\frac{1}{2}x^{3/2}\right) dx - 1}{\exp\left(\frac{1}{2}1^{3/2}\right)} \approx 0.857197$$

$$a_3 = 0.857197 - \frac{\int_0^{0.857197} \exp\left(\frac{1}{2}x^{3/2}\right) dx - 1}{\exp\left(\frac{1}{2}0.857197^{3/2}\right)} \approx 0.849203$$

$$a_4 = 0.849203 - \frac{\int_0^{0.849203} \exp\left(\frac{1}{2}x^{3/2}\right) dx - 1}{\exp\left(\frac{1}{2}0.849203^{3/2}\right)} \approx 0.849181$$

$$a_5 = 0.849181 - \frac{\int_0^{0.849181} \exp\left(\frac{1}{2}x^{3/2}\right) dx - 1}{\exp\left(\frac{1}{2}0.849181^{3/2}\right)} \approx 0.849181$$

Our approximation for the cut-off point is 0.849181. Notice that Newton's Method required less work and gave a more accurate answer. ■

Functions Defined by Tables It is now common to have computers collect data from a system at periodic points in time, often very frequently, such as once per second. When the data collected represent a function which must be integrated, we really cannot use the Second Fundamental Theorem of Calculus. Instead, we must apply a numerical method that uses just the sampled points.

EXAMPLE 6 Cars are often equipped with devices that monitor instantaneous fuel consumption (measured in miles per gallon). Suppose a computer is hooked up to the car so that it collects the instantaneous fuel consumption as well as the instantaneous speed. A graph showing both speed (in miles per hour) and fuel consumption (in miles per gallon) are shown in Figure 2 for a two-hour trip. The top (black) curve shows speed, and the bottom curve shows fuel consumption. The fuel consumption varies quite a lot, depending mainly on whether the car is going up or down a hill. Part of the data are shown in the table below. How far did the car travel in this two-hour trip, and how much fuel was consumed?

Time (Minutes)	Speed (miles/hr)	Fuel Consumption (miles/gal)	Speed/ Fuel Cons.
0	36	20.00	1.80
1	37	22.35	1.66
2	36	23.67	1.52
3	36	28.75	1.25
⋮	⋮	⋮	⋮
118	42	24.30	1.73
119	40	24.83	1.61
120	41	26.19	1.57

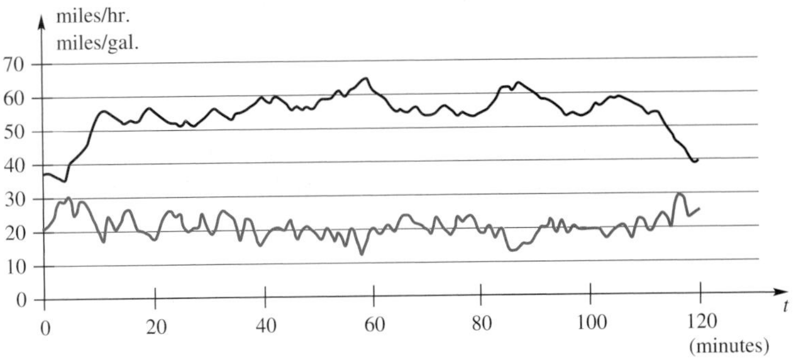

Figure 2

A Rough Approximation

The figure suggests that the average fuel consumption is about 20 miles per gallon and that the average velocity is about 50 miles per hour. After 2 hours (120 minutes) the car would have traveled about 100 miles, and, at roughly 20 miles per gallon, the fuel consumed would be $\frac{100 \text{ miles}}{20 \text{ miles per gallon}} = 5$ gallons. We expect our answer to be near 5 gallons.

SOLUTION We will use the Trapezoidal Rule to approximate the integrals. The distance traveled is the definite integral of instantaneous speed, so

$$D = \int_0^2 \frac{ds}{dt}\,dt \approx \frac{2-0}{2\cdot 120}[36 + 2(37 + 36 + \cdots + 40) + 41] = 109.4 \text{ miles}$$

The total amount of fuel consumed is the integral of $\frac{df}{dt}$, where $f(t)$ is the amount of fuel in the car's tank at time t. Notice that fuel consumption is given in miles per gallon, which is ds/df. The last column in the above table is the speed ds/dt divided by ds/df. The fuel consumed is therefore

$$\int_0^2 \frac{df}{dt}\,dt = \int_0^2 \frac{ds/dt}{ds/df}\,dt$$
$$\approx \frac{2-0}{2\cdot 120}[1.80 + 2(1.66 + 1.52 + \cdots + 1.61) + 1.57]$$
$$\approx 5.30 \text{ gallons}$$

■

Special Functions Many definite integrals that cannot be evaluated using the Second Fundamental Theorem of Calculus arise so often in applied mathematics that they are given special names. Here are some of these accumulation functions, along with their common names and abbreviations:

the error function	$\operatorname{erf}(x) = \frac{2}{\sqrt{\pi}}\int_0^x e^{-t^2}\,dt$
the sine integral	$\operatorname{Si}(x) = \int_0^x \frac{\sin t}{t}\,dt$
the Fresnel sine integral	$S(x) = \int_0^x \sin\left(\frac{\pi t^2}{2}\right)dt$
the Fresnel cosine integral	$C(x) = \int_0^x \cos\left(\frac{\pi t^2}{2}\right)dt$

There are numerous others; see *Handbook of Mathematical Functions* for many more. Algorithms, often involving infinite series (a topic we will take up in Chapter 9), have been developed to approximate these functions. These algorithms are usually accurate and efficient. In fact, it is no more difficult (for a computer, anyway) to approximate the Fresnel integral $S(1)$ than it is to approximate the sine of 1. Since many practical problems work out to involve such functions, it is important to know that they exist and how to find approximations for them.

EXAMPLE 7 Express the mass of the lamina from Example 2 in terms of the Fresnel sine integral.

SOLUTION The mass was found to be

$$m = \delta \int_0^{\sqrt{\pi}} \sin x^2 \, dx$$

If we make the substitution $x = t\sqrt{\pi/2}$, then $x^2 = t^2\pi/2$ and $dx = \sqrt{\pi/2}\, dt$. The limits on the definite integral must also be transformed

$$x = 0 \Rightarrow t = 0$$
$$x = \sqrt{\pi} \Rightarrow t = \sqrt{2}$$

Thus,

$$m = \delta \int_0^{\sqrt{2}} \sin\left(\frac{t^2\pi}{2}\right)\sqrt{\frac{\pi}{2}}\, dt$$
$$= \delta\sqrt{\frac{\pi}{2}} \int_0^{\sqrt{2}} \sin\left(\frac{\pi t^2}{2}\right) dt$$
$$= \delta\sqrt{\frac{\pi}{2}}\, S\left(\sqrt{2}\right) \approx 0.895\,\delta$$

■

Concepts Review

1. Tables of Integrals are most helpful when used in conjunction with the method of _____.

2. Both $\int (x^2+9)^{3/2}\,dx$ and $\int (\sin^2 x + 1)^{3/2} \cos x\, dx$ can be evaluated using Formula Number _____.

3. When using a CAS to evaluate a definite integral it is important to know whether the system is giving us an exact answer or a(n) _____.

4. The sine integral evaluated at $t = 0$ is $S(0) =$ _____.

Problem Set 7.6

In Problems 1–12, evaluate the given integral.

1. $\int xe^{-5x}\, dx$

2. $\int \frac{x}{x^2+9}\, dx$

3. $\int_1^2 \frac{\ln x}{x}\, dx$

4. $\int \frac{x}{x^2 - 5x + 6}\, dx$

5. $\int \cos^4 2x\, dx$

6. $\int \sin^3 x \cos x\, dx$

7. $\int_1^2 \frac{1}{x^2 + 6x + 8}\, dx$

8. $\int_0^{1/2} \frac{1}{1 - t^2}\, dt$

9. $\int_0^5 x\sqrt{x+2}\, dx$

10. $\int_3^4 \frac{1}{t - \sqrt{2t}}\, dt$

11. $\int_{-\pi/2}^{\pi/2} \cos^2 x \sin x\, dx$

12. $\int_0^{2\pi} |\sin 2x|\, dx$

In Problems 13–30, use the table of integrals on the inside back cover, perhaps combined with a substitution, to evaluate the given integrals.

13. (a) $\int x\sqrt{3x+1}\, dx$ (b) $\int e^x\sqrt{3e^x+1}\, e^x\, dx$

14. (a) $\int 2t\sqrt{3-4t}\, dt$ (b) $\int \cos t\sqrt{3 - 4\cos t}\, \sin t\, dt$

15. (a) $\int \frac{dx}{9 - 16x^2}$ (b) $\int \frac{e^x}{9 - 16e^{2x}}\, dx$

16. (a) $\int \frac{dx}{5x^2 - 11}$ (b) $\int \frac{x}{5x^4 - 11}\, dx$

17. (a) $\int x^2\sqrt{9 - 2x^2}\, dx$

(b) $\int \sin^2 x \cos x\sqrt{9 - 2\sin^2 x}\, dx$

18. (a) $\int \frac{\sqrt{16 - 3t^2}}{t}\, dt$ (b) $\int \frac{\sqrt{16 - 3t^6}}{t}\, dt$

19. (a) $\int \frac{dx}{\sqrt{5 + 3x^2}}$ (b) $\int \frac{x}{\sqrt{5 + 3x^4}}\, dx$

20. (a) $\int t^2\sqrt{3 + 5t^2}\, dt$ (b) $\int t^8\sqrt{3 + 5t^6}\, dt$

21. (a) $\int \frac{dt}{\sqrt{t^2 + 2t - 3}}$ (b) $\int \frac{dt}{\sqrt{t^2 + 3t - 5}}$

22. (a) $\int \frac{\sqrt{x^2 + 2x - 3}}{x + 1}\, dx$ (b) $\int \frac{\sqrt{x^2 - 4x}}{x - 2}\, dx$

23. (a) $\int \frac{y}{\sqrt{3y + 5}}\, dy$ (b) $\int \frac{\sin t \cos t}{\sqrt{3\sin t + 5}}\, dt$

24. (a) $\int \frac{dz}{z\sqrt{5 - 4z}}$ (b) $\int \frac{\sin x}{\cos x\sqrt{5 - 4\cos x}}\, dx$

25. $\int \sinh^2 3t\, dt$

26. $\int \frac{\operatorname{sech}\sqrt{x}}{\sqrt{x}}\, dx$

27. $\displaystyle\int \frac{\cos t \sin t}{\sqrt{2}\cos t + 1}\,dt$

28. $\displaystyle\int \cos t \sin t \sqrt{4\cos t - 1}\,dt$

29. $\displaystyle\int \frac{\cos^2 t \sin t}{\sqrt{\cos t + 1}}\,dt$

30. $\displaystyle\int \frac{1}{(9 + x^2)^3}\,dx$

Use a CAS to evaluate the definite integrals in Problems 31–40. If the CAS does not give an exact answer in terms of elementary functions, give a numerical approximation.

31. $\displaystyle\int_0^{\pi} \frac{\cos^2 x}{1 + \sin x}\,dx$

32. $\displaystyle\int_0^{1} \operatorname{sech}\sqrt[3]{x}\,dx$

33. $\displaystyle\int_0^{\pi/2} \sin^{12} x\,dx$

34. $\displaystyle\int_0^{\pi} \cos^4 \frac{x}{2}\,dx$

35. $\displaystyle\int_1^{4} \frac{\sqrt{t}}{1 + t^8}\,dt$

36. $\displaystyle\int_0^{3} x^4 e^{-x/2}\,dx$

37. $\displaystyle\int_0^{\pi/2} \frac{1}{1 + 2\cos^5 x}\,dx$

38. $\displaystyle\int_{-\pi/4}^{\pi/4} \frac{x^3}{4 + \tan x}\,dx$

39. $\displaystyle\int_2^{3} \frac{x^2 + 2x - 1}{x^2 - 2x + 1}\,dx$

40. $\displaystyle\int_1^{3} \frac{du}{u\sqrt{2u - 1}}$

In Problems 41–48, the density of a rod is given. Find c so that the mass from 0 to c is equal to 1. Whenever possible find an exact solution. If this is not possible, find an approximation for c. (See Examples 4 and 5).

41. $\delta(x) = \dfrac{1}{x + 1}$

42. $\delta(x) = \dfrac{2}{x^2 + 1}$

43. $\delta(x) = \ln(x + 1)$

44. $\delta(x) = \dfrac{x}{x^2 + 1}$

45. $\delta(x) = 2e^{-x^{3/2}}$

46. $\delta(x) = \ln(x^3 + 1)$

47. $\delta(x) = 6\cos\left(\dfrac{x^2}{2}\right)$

48. $\delta(x) = 4\dfrac{\sin x}{x}$

49. Find c so that $\displaystyle\int_0^{c} \frac{1}{3\sqrt{2\pi}} x^{3/2} e^{-x/2}\,dx = 0.90$.

50. Find c so that $\displaystyle\int_{-c}^{c} \frac{1}{\sqrt{2\pi}} e^{-x^2/2}\,dx = 0.95$. *Hint:* Use symmetry.

In Problems 51–54, the graph of $y = f(x)$ is given along with the graph of a line. Find c so that the x component of the center of mass of the shaded homogeneous lamina is equal to 2.

51.

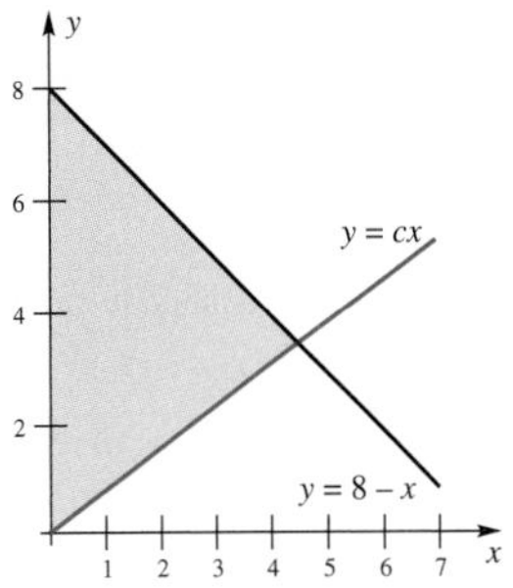

52.

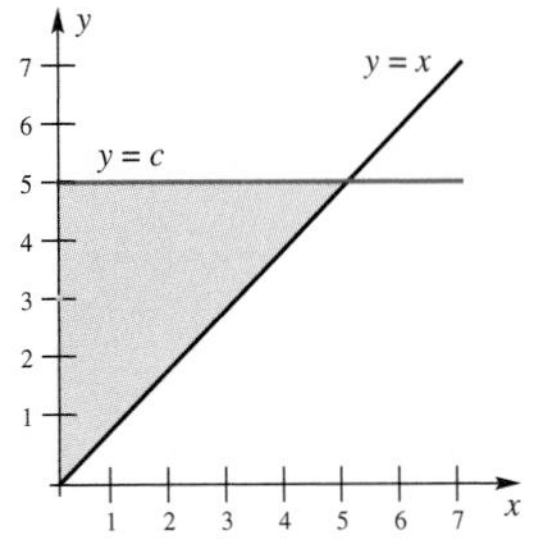

53.

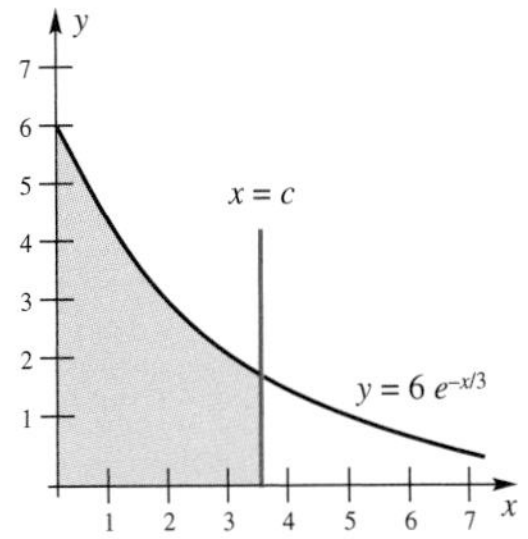

54.

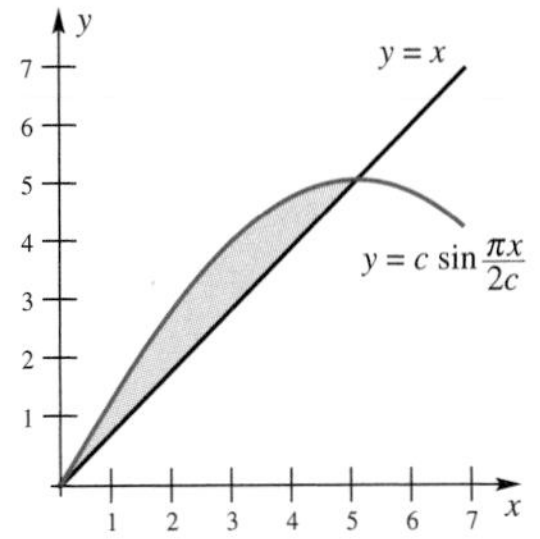

55. Find the following derivatives.

(a) $\dfrac{d}{dx}\operatorname{erf}(x)$ (b) $\dfrac{d}{dx}\operatorname{Si}(x)$

56. Find the derivatives of the Fresnel functions

(a) $\dfrac{d}{dx}S(x)$ (b) $\dfrac{d}{dx}C(x)$

57. Over what intervals (on the nonnegative side of the number line) is the error function increasing? Concave up?

58. Over what subintervals of $[0, 2]$ is the Fresnel function $S(x)$ increasing? Concave up?

59. Over what subintervals of $[0, 2]$ is the Fresnel function $C(x)$ increasing? Concave up?

60. Find the coordinates of the first inflection point of the Fresnel function $S(x)$ that is to the right of the origin.

Answers to Concepts Review: **1.** substitution **2.** 53 **3.** approximation **4.** 0

7.7 Chapter Review

Concepts Test

Respond with true or false to each of the following assertions. Be prepared to justify your answer.

1. To evaluate $\displaystyle\int x\sin(x^2)\,dx$, make the substitution $u = x^2$.

2. To evaluate $\displaystyle\int \frac{x}{9 + x^4}\,dx$, make the substitution $u = x^2$.

3. To evaluate $\displaystyle\int \frac{x^3}{9 + x^4}\,dx$, make the substitution $u = x^2$.

4. To evaluate $\displaystyle\int \frac{2x - 3}{x^2 - 3x + 5}\,dx$, begin by completing the square of the denominator.

5. To evaluate $\displaystyle\int \frac{3}{x^2 - 3x + 5}\,dx$, begin by completing the square of the denominator.

6. To evaluate $\displaystyle\int \frac{1}{\sqrt{4 - 5x^2}}\,dx$, make the substitution $u = \sqrt{5}x$.

7. To evaluate $\int \frac{t+2}{t^3-9t}\,dt$, use partial fractions.

8. To evaluate $\int \frac{t^4}{t^2-1}\,dt$, use integration by parts.

9. To evaluate $\int \sin^6 x \cos^2 x\,dx$, use half-angle formulas.

10. To evaluate $\int \frac{e^x}{1+e^x}\,dx$, use integration by parts.

11. To evaluate $\int \frac{x+2}{\sqrt{-x^2-4x}}\,dx$, use a trigonometric substitution.

12. To evaluate $\int x^2\sqrt[3]{3-2x}\,dx$, let $u = \sqrt[3]{3-2x}$.

13. To evaluate $\int \sin^2 x \cos^5 x\,dx$, rewrite the integrand as $\sin^2 x(1-\sin^2 x)^2 \cos x$.

14. To evaluate $\int \frac{1}{x^2\sqrt{9-x^2}}\,dx$, make a trigonometric substitution.

15. To evaluate $\int x^2 \ln x\,dx$, use integration by parts.

16. To evaluate $\int \sin 2x \cos 4x\,dx$, use half-angle formulas.

17. $\frac{x^2}{x^2-1}$ can be expressed in the form $\frac{A}{x-1} + \frac{B}{x+1}$.

18. $\frac{x^2+2}{x(x^2-1)}$ can be expressed in the form

$$\frac{A}{x} + \frac{B}{x-1} + \frac{C}{x+1}$$

19. $\frac{x^2+2}{x(x^2+1)}$ can be expressed in the form $\frac{A}{x} + \frac{Bx+C}{x^2+1}$.

20. $\frac{x+2}{x^2(x^2-1)}$ can be expressed in the form

$$\frac{A}{x^2} + \frac{B}{x-1} + \frac{C}{x+1}$$

21. To complete the square of $ax^2 + bx$, add $(b/2)^2$.

22. Any polynomial with real coefficients can be factored into a product of linear polynomials with real coefficients.

23. Two polynomials in x have the same values for all x if and only if the coefficients of like-degree terms are identical.

24. The integral $\int x^2\sqrt{25-4x^2}\,dx$ can be evaluated using Formula 57 from the table of integrals along with an appropriate substitution.

25. The integral $\int x\sqrt{25-4x^2}\,dx$ can be evaluated using Formula 57 from the table of integrals along with an appropriate substitution.

26. $\mathrm{erf}(0) < \mathrm{erf}(1)$

27. If $C(x) = \int_0^x \cos\left(\frac{\pi t^2}{2}\right) dt$, then $C'(x) = \cos\left(\frac{\pi x^2}{2}\right)$.

28. The sine integral $\mathrm{Si}(x) = \int_0^x \frac{\sin t}{t}\,dt$ is an increasing function on the interval $[0, \infty)$.

Sample Test Problems

In Problems 1–42, evaluate each integral.

1. $\int_0^4 \frac{t}{\sqrt{9+t^2}}\,dt$

2. $\int \cot^2(2\theta)\,d\theta$

3. $\int_0^{\pi/2} e^{\cos x}\sin x\,dx$

4. $\int_0^{\pi/4} x\sin 2x\,dx$

5. $\int \frac{y^3+y}{y+1}\,dy$

6. $\int \sin^3(2t)\,dt$

7. $\int \frac{y-2}{y^2-4y+2}\,dy$

8. $\int_0^{3/2} \frac{dy}{\sqrt{2y+1}}$

9. $\int \frac{e^{2t}}{e^t-2}\,dt$

10. $\int \frac{\sin x + \cos x}{\tan x}\,dx$

11. $\int \frac{dx}{\sqrt{16+4x-2x^2}}$

12. $\int x^2 e^x\,dx$

13. $\int \frac{dy}{\sqrt{2+3y^2}}$

14. $\int \frac{w^3}{1-w^2}\,dw$

15. $\int \frac{\tan x}{\ln|\cos x|}\,dx$

16. $\int \frac{3\,dt}{t^3-1}$

17. $\int \sinh x\,dx$

18. $\int \frac{(\ln y)^5}{y}\,dy$

19. $\int x\cot^2 x\,dx$

20. $\int \frac{\sin\sqrt{x}}{\sqrt{x}}\,dx$

21. $\int \frac{\ln t^2}{t}\,dt$

22. $\int \ln(y^2+9)\,dy$

23. $\int e^{t/3}\sin 3t\,dt$

24. $\int \frac{t+9}{t^3+9t}\,dt$

25. $\int \sin\frac{3x}{2}\cos\frac{x}{2}\,dx$

26. $\int \cos^4\left(\frac{x}{2}\right)dx$

27. $\int \tan^3 2x \sec 2x\,dx$

28. $\int \frac{\sqrt{x}}{1+\sqrt{x}}\,dx$

29. $\int \tan^{3/2} x \sec^4 x\,dx$

30. $\int \frac{dt}{t(t^{1/6}+1)}$

31. $\int \frac{e^{2y}\,dy}{\sqrt{9-e^{2y}}}$

32. $\int \cos^5 x\sqrt{\sin x}\,dx$

33. $\int e^{\ln(3\cos x)}\,dx$

34. $\int \frac{\sqrt{9-y^2}}{y}\,dy$

35. $\int \frac{e^{4x}}{1+e^{8x}}\,dx$

36. $\int \frac{\sqrt{x^2+a^2}}{x^4}\,dx$

37. $\int \frac{w}{\sqrt{w+5}}\,dw$

38. $\int \frac{\sin t\,dt}{\sqrt{1+\cos t}}$

39. $\int \frac{\sin y\cos y}{9+\cos^4 y}\,dy$

40. $\int \frac{dx}{\sqrt{1-6x-x^2}}$

41. $\int \frac{4x^2+3x+6}{x^2(x^2+3)}\,dx$

42. $\int \frac{dx}{(16+x^2)^{3/2}}$

43. Express the partial fraction decomposition of each rational function without computing the exact coefficients. For example,

$$\frac{3x+1}{(x-1)^2} = \frac{A}{(x-1)} + \frac{B}{(x-1)^2}$$

(a) $\dfrac{3-4x^2}{(2x+1)^3}$ (b) $\dfrac{7x-41}{(x-1)^2(2-x)^3}$

(c) $\dfrac{3x+1}{(x^2+x+10)^2}$ (d) $\dfrac{(x+1)^2}{(x^2-x+10)^2(1-x^2)^2}$

(e) $\dfrac{x^5}{(x+3)^4(x^2+2x+10)^2}$ (f) $\dfrac{(3x^2+2x-1)^2}{(2x^2+x+10)^3}$

44. Find the volume of the solid generated by revolving the region under the graph of

$$y = \frac{1}{\sqrt{3x-x^2}}$$

from $x = 1$ to $x = 2$ about

(a) the x-axis; (b) the y-axis.

45. Find the length of the curve $y = x^2/16$ from $x = 0$ to $x = 4$.

46. The region under the curve

$$y = \frac{1}{x^2+5x+6}$$

from $x = 0$ to $x = 3$ is rotated about the x-axis. Compute the volume of the solid that is generated.

47. If the region given in Problem 46 is rotated about the y-axis, find the volume of the solid.

48. Find the volume of the solid created by revolving the region bounded by the x-axis and the curve $y = 4x\sqrt{2-x}$ about the y-axis.

49. Find the volume when the region created by the x-axis, y-axis, the curve $y = 2(e^x - 1)$ and the curve $x = \ln 3$ is revolved about the line $x = \ln 3$.

50. Find the area of the region bounded by the x-axis, the curve $y = 18/\left(x^2\sqrt{x^2+9}\right)$, and the lines $x = \sqrt{3}$ and $x = 3\sqrt{3}$.

51. Find the area of the region bounded by the curve $s = t/(t-1)^2$, $s = 0$, $t = -6$, and $t = 0$.

≈ **52.** Find the volume of the solid generated by revolving the region

$$\left\{(x, y)\colon -3 \le x \le -1, \frac{6}{x\sqrt{x+4}} \le y \le 0\right\}$$

about the x-axis. Make a sketch.

≈ **53.** Find the length of the segment of the curve $y = \ln(\sin x)$ from $x = \pi/6$ to $x = \pi/3$.

54. Use the table of integrals to evaluate the following integrals:

(a) $\displaystyle\int \frac{\sqrt{81-4x^2}}{x}\,dx$ (b) $\displaystyle\int e^x(9-e^{2x})^{3/2}\,dx$

55. Use the table of integrals to evaluate the following integrals:

(a) $\displaystyle\int \cos x\sqrt{\sin^2 x + 4}\,dx$ (b) $\displaystyle\int \frac{1}{1-4x^2}\,dx$

56. Evaluate the first two derivatives of the sine integral $\mathrm{Si}(x) = \displaystyle\int_0^x \frac{\sin t}{t}\,dt$.

57. A rod has density $\delta(x) = \dfrac{1}{1+x^3}$. Use Newton's method to find the value of c so that the mass of the rod from 0 to c is 0.5.

REVIEW & PREVIEW PROBLEMS

Evaluate the limits in Problems 1–14.

1. $\lim_{x \to 2} \frac{x^2 + 1}{x^2 - 1}$

2. $\lim_{x \to 3} \frac{2x + 1}{x + 5}$

3. $\lim_{x \to 3} \frac{x^2 - 9}{x - 3}$

4. $\lim_{x \to 2} \frac{x^2 - 5x + 6}{x - 2}$

5. $\lim_{x \to 0} \frac{\sin 2x}{x}$

6. $\lim_{x \to 0} \frac{\tan 3x}{x}$

7. $\lim_{x \to \infty} \frac{x^2 + 1}{x^2 - 1}$

8. $\lim_{x \to \infty} \frac{2x + 1}{x + 5}$

9. $\lim_{x \to \infty} e^{-x}$

10. $\lim_{x \to \infty} e^{-x^2}$

11. $\lim_{x \to \infty} e^{2x}$

12. $\lim_{x \to -\infty} e^{-2x}$

13. $\lim_{x \to \infty} \tan^{-1} x$

14. $\lim_{x \to \infty} \sec^{-1} x$

Plot the functions given in Problems 15–18 on the domain $0 \le x \le 10$ and make a conjecture about $\lim_{x \to \infty} f(x)$.

15. $f(x) = xe^{-x}$

16. $f(x) = x^2e^{-x}$

17. $f(x) = x^3e^{-x}$

18. $f(x) = x^4e^{-x}$

19. Plot a graph of $y = x^{10}e^{-x}$ over some domain that allows you to make a conjecture about $\lim_{x \to \infty} x^{10}e^{-x}$.

20. Experiment with several positive integers n and make a conjecture about $\lim_{x \to \infty} x^n e^{-x}$.

Evaluate the integrals in Problems 21–28 for the indicated values of a.

21. $\int_0^a e^{-x}\,dx$; $a = 1, 2, 4, 8, 16$

22. $\int_0^a xe^{-x^2}\,dx$; $a = 1, 2, 4, 8, 16$

23. $\int_0^a \frac{x}{1 + x^2}\,dx$; $a = 1, 2, 4, 8, 16$

24. $\int_0^a \frac{1}{1 + x}\,dx$; $a = 1, 2, 4, 8, 16$

25. $\int_1^a \frac{1}{x^2}\,dx$; $a = 2, 4, 8, 16$

26. $\int_1^a \frac{1}{x^3}\,dx$; $a = 2, 4, 8, 16$

27. $\int_a^4 \frac{1}{\sqrt{x}}\,dx$; $a = 1, \frac{1}{2}, \frac{1}{4}, \frac{1}{8}, \frac{1}{16}$

28. $\int_a^4 \frac{1}{x}\,dx$; $a = 1, \frac{1}{2}, \frac{1}{4}, \frac{1}{8}, \frac{1}{16}$

CHAPTER 8

Indeterminate Forms and Improper Integrals

8.1 Indeterminate Forms of Type 0/0
8.2 Other Indeterminate Forms
8.3 Improper Integrals: Infinite Limits of Integration
8.4 Improper Integrals: Infinite Integrands

8.1 Indeterminate Forms of Type 0/0

Here are three familiar limit problems:

$$\lim_{x \to 0} \frac{\sin x}{x}, \qquad \lim_{x \to 3} \frac{x^2 - 9}{x^2 - x - 6}, \qquad \lim_{x \to a} \frac{f(x) - f(a)}{x - a}$$

The first was treated at length in Section 1.4, and the third actually defines the derivative $f'(a)$. The three limits have a common feature. In each case, a quotient is involved and, in each case, both numerator and denominator have 0 as their limits. An attempt to apply part 7 of the Main Limit Theorem (Theorem 1.3A), which says that the limit of a quotient is equal to the quotient of the limits, leads to the nonsensical result 0/0. In fact, the theorem does not apply, since it requires that the limit of the denominator be different from 0. We are not saying that these limits do not exist, only that the Main Limit Theorem will not determine them.

You may recall that an intricate geometric argument led us to the conclusion $\lim_{x \to 0}(\sin x)/x = 1$ (Theorem 1.4B). On the other hand, the algebraic technique of factoring yields

$$\lim_{x \to 3} \frac{x^2 - 9}{x^2 - x - 6} = \lim_{x \to 3} \frac{(x - 3)(x + 3)}{(x - 3)(x + 2)} = \lim_{x \to 3} \frac{x + 3}{x + 2} = \frac{6}{5}$$

Would it not be nice to have a standard procedure for handling all problems for which the limits of the numerator and denominator are both 0? That is too much to hope for. However, there is a simple rule that works beautifully on a wide variety of such problems.

L'Hôpital's Rule In 1696, Guillaume François Antoine de l'Hôpital published the first textbook on differential calculus; it included the following rule, which he had learned from his teacher Johann Bernoulli.

Theorem A L'Hôpital's Rule for forms of type 0/0

Suppose that $\lim_{x \to u} f(x) = \lim_{x \to u} g(x) = 0$. If $\lim_{x \to u}[f'(x)/g'(x)]$ exists in either the finite or infinite sense (i.e., if this limit is a finite number or $-\infty$ or $+\infty$), then

$$\lim_{x \to u} \frac{f(x)}{g(x)} = \lim_{x \to u} \frac{f'(x)}{g'(x)}$$

Before attempting to prove this theorem, we illustrate it. Note that l'Hôpital's Rule allows us to replace one limit by another, which may be simpler and, in particular, may not have the 0/0 form.

EXAMPLE 1 Use l'Hôpital's Rule to show that

$$\lim_{x \to 0} \frac{\sin x}{x} = 1 \quad \text{and} \quad \lim_{x \to 0} \frac{1 - \cos x}{x} = 0$$

Geometric Interpretation of l'Hôpital's Rule

Study the diagrams below. They should make l'Hôpital's Rule seem quite reasonable. (See Problems 38–42.)

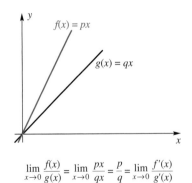

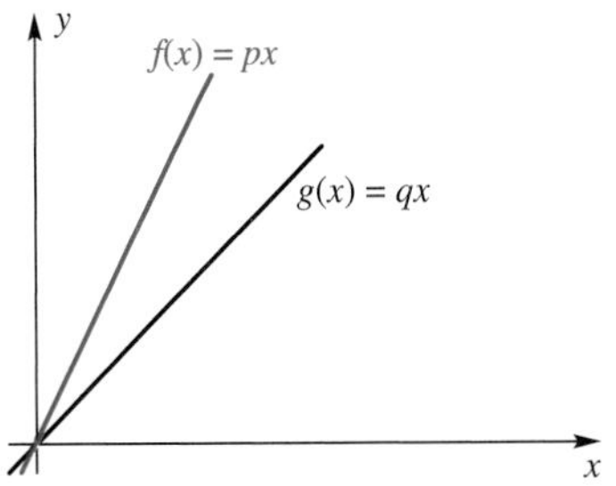

$$\lim_{x \to 0} \frac{f(x)}{g(x)} = \lim_{x \to 0} \frac{px}{qx} = \frac{p}{q} = \lim_{x \to 0} \frac{f'(x)}{g'(x)}$$

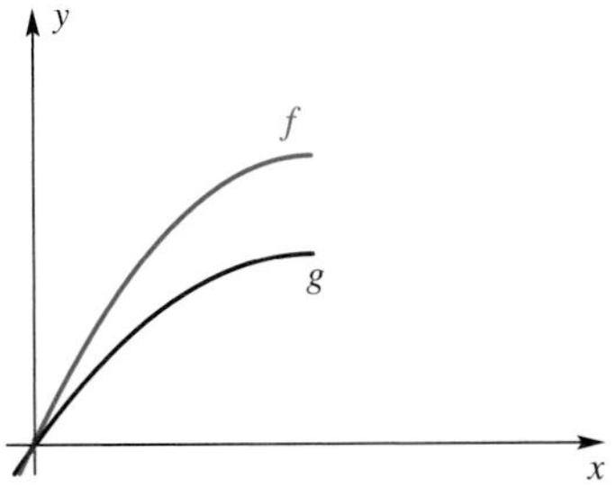

$$\lim_{x \to 0} \frac{f(x)}{g(x)} = \lim_{x \to 0} \frac{f'(x)}{g'(x)}$$

SOLUTION We worked pretty hard to demonstrate these two facts in Section 1.4. After noting that trying to evaluate both limits by substitution leads to the form 0/0, we can now establish the desired results in two lines (but see Problem 25). By l'Hôpital's Rule,

$$\lim_{x\to 0}\frac{\sin x}{x} = \lim_{x\to 0}\frac{D_x \sin x}{D_x x} = \lim_{x\to 0}\frac{\cos x}{1} = 1$$

$$\lim_{x\to 0}\frac{1-\cos x}{x} = \lim_{x\to 0}\frac{D_x(1-\cos x)}{D_x x} = \lim_{x\to 0}\frac{\sin x}{1} = 0$$ ■

EXAMPLE 2 Find $\lim_{x\to 3}\frac{x^2-9}{x^2-x-6}$ and $\lim_{x\to 2^+}\frac{x^2+3x-10}{x^2-4x+4}$.

SOLUTION Both limits have the 0/0 form, so, by l'Hôpital's Rule,

$$\lim_{x\to 3}\frac{x^2-9}{x^2-x-6} = \lim_{x\to 3}\frac{2x}{2x-1} = \frac{6}{5}$$

$$\lim_{x\to 2^+}\frac{x^2+3x-10}{x^2-4x+4} = \lim_{x\to 2^+}\frac{2x+3}{2x-4} = \infty$$

The first of these limits was handled at the beginning of this section by factoring and simplifying. Of course, we get the same answer either way. ■

EXAMPLE 3 Find $\lim_{x\to 0}\frac{\tan 2x}{\ln(1+x)}$.

SOLUTION Both numerator and denominator have limit 0. Hence,

$$\lim_{x\to 0}\frac{\tan 2x}{\ln(1+x)} = \lim_{x\to 0}\frac{2\sec^2 2x}{1/(1+x)} = \frac{2}{1} = 2$$ ■

Sometimes $\lim f'(x)/g'(x)$ also has the indeterminate form 0/0. Then we may apply l'Hôpital's Rule again, as we now illustrate. Each application of l'Hôpital's Rule is flagged with the symbol Ⓛ

EXAMPLE 4 Find $\lim_{x\to 0}\frac{\sin x - x}{x^3}$.

SOLUTION By l'Hôpital's Rule applied three times in succession,

$$\lim_{x\to 0}\frac{\sin x - x}{x^3} \overset{\text{Ⓛ}}{=} \lim_{x\to 0}\frac{\cos x - 1}{3x^2}$$

$$\overset{\text{Ⓛ}}{=} \lim_{x\to 0}\frac{-\sin x}{6x}$$

$$\overset{\text{Ⓛ}}{=} \lim_{x\to 0}\frac{-\cos x}{6} = -\frac{1}{6}$$ ■

Just because we have an elegant rule does not mean that we should use it indiscriminately. In particular, we must always make sure that it applies; that is, we must make sure that the limit has the indeterminate form 0/0. Otherwise, we will be led into all kinds of errors, as we now illustrate.

EXAMPLE 5 Find $\lim_{x\to 0} \frac{1 - \cos x}{x^2 + 3x}$.

SOLUTION We might be tempted to write

$$\lim_{x\to 0} \frac{1-\cos x}{x^2+3x} \overset{\text{L}}{=} \lim_{x\to 0} \frac{\sin x}{2x+3} \overset{\text{L}}{=} \lim_{x\to 0} \frac{\cos x}{2} = \frac{1}{2} \qquad \text{WRONG}$$

The first application of l'Hôpital's Rule was correct; the second was not, since at that stage, the limit did not have the 0/0 form. Here is what we should have done.

$$\lim_{x\to 0} \frac{1-\cos x}{x^2+3x} \overset{\text{L}}{=} \lim_{x\to 0} \frac{\sin x}{2x+3} = 0 \qquad \text{RIGHT}$$

We stop differentiating as soon as either the numerator or denominator has a nonzero limit. ■

Even if the conditions of l'Hôpital's Rule hold, an application of l'Hôpital's Rule may not help us; witness the following example.

EXAMPLE 6 Find $\lim_{x\to\infty} \frac{e^{-x}}{x^{-1}}$.

SOLUTION Since the numerator and denominator both tend to 0, the limit is indeterminate of the form 0/0. Thus, the conditions of Theorem A hold. We may apply l'Hôpital's Rule indefinitely.

$$\lim_{x\to\infty} \frac{e^{-x}}{x^{-1}} \overset{\text{L}}{=} \lim_{x\to\infty} \frac{e^{-x}}{x^{-2}} \overset{\text{L}}{=} \lim_{x\to\infty} \frac{e^{-x}}{2x^{-3}} = \cdots$$

Clearly, we are only complicating the problem. A better approach is to do a bit of algebra first.

$$\lim_{x\to\infty} \frac{e^{-x}}{x^{-1}} = \lim_{x\to\infty} \frac{x}{e^x}$$

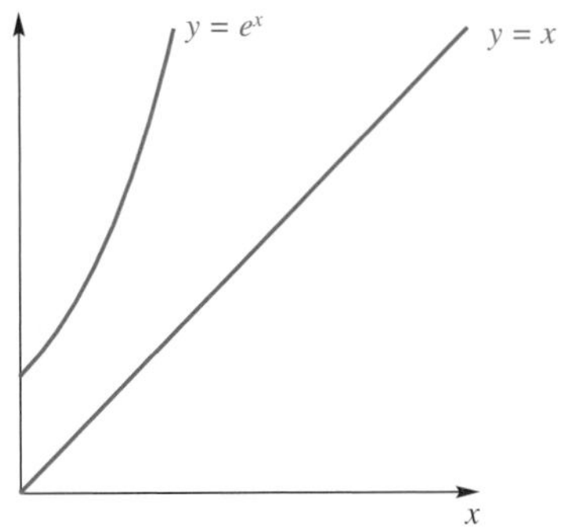

Figure 1

Written this way, the limit is indeterminate of the form ∞/∞, the subject of the next section. However, you should be able to guess that the limit is 0 by considering how much faster e^x grows than x (see Figure 1). A rigorous demonstration will come later (Example 1 of Section 8.2). ■

Cauchy's Mean Value Theorem The proof of l'Hôpital's Rule depends on an extension of the Mean Value Theorem for Derivatives due to Augustin Louis Cauchy (1789–1857).

Theorem B Cauchy's Mean Value Theorem

Let the functions f and g be differentiable on (a, b) and continuous on $[a, b]$. If $g'(x) \neq 0$ for all x in (a, b), then there exists a number c in (a, b) such that

$$\frac{f(b) - f(a)}{g(b) - g(a)} = \frac{f'(c)}{g'(c)}$$

Note that this theorem reduces to the ordinary Mean Value Theorem for Derivatives (Theorem 3.6A) when $g(x) = x$.

Proof It is tempting to apply the ordinary Mean Value Theorem to both numerator and denominator of the left side of the conclusion. If we do this, we obtain

$$f(b) - f(a) = f'(c_1)(b - a) \tag{1}$$

and

$$g(b) - g(a) = g'(c_2)(b - a) \tag{2}$$

for appropriate choices of c_1 and c_2. If only c_1 and c_2 were equal, we could divide the first equality by the second and be done; but there is no reason to hope for such a coincidence. However, this attempt is not a complete failure since (2) yields the valuable information that $g(b) - g(a) \neq 0$, a fact we will need later (this follows from the hypothesis that $g'(x) \neq 0$ for all x in (a, b)).

Recall that the proof of the Mean Value Theorem for Derivatives (Theorem 3.6A) rested on the introduction of an auxiliary function s. If we try to mimic that proof, we are led to the following choice for $s(x)$. Let

$$s(x) = f(x) - f(a) - \frac{f(b) - f(a)}{g(b) - g(a)}[g(x) - g(a)]$$

No division by zero is involved since we earlier established that $g(b) - g(a) \neq 0$. Note further that $s(a) = 0 = s(b)$. Also, s is continuous on $[a, b]$ and differentiable on (a, b), this following from the corresponding facts for f and g. Thus, by the Mean Value Theorem for Derivatives, there is a number c in (a, b) such that

$$s'(c) = \frac{s(b) - s(a)}{b - a} = \frac{0 - 0}{b - a} = 0$$

But

$$s'(c) = f'(c) - \frac{f(b) - f(a)}{g(b) - g(a)}g'(c) = 0$$

so

$$\frac{f'(c)}{g'(c)} = \frac{f(b) - f(a)}{g(b) - g(a)}$$

which is what we wished to prove. ■

Proof of L'Hôpital's Rule

Proof Refer back to Theorem A, which actually states several theorems at once. We will prove only the case where L is finite and the limit is the one-sided limit $\lim_{x \to a^+}$.

The hypotheses for Theorem A imply more than they say explicitly. In particular, the existence of $\lim_{x \to a^+}[f'(x)/g'(x)]$ implies that both $f'(x)$ and $g'(x)$ exist in at least a small interval $(a, b]$ and that $g'(x) \neq 0$ there. At a, we do not even know that f and g are defined, but we do know that $\lim_{x \to a^+} f(x) = 0$ and $\lim_{x \to a^+} g(x) = 0$. Thus, we may define (or redefine if necessary) both $f(a)$ and $g(a)$ to be zero, thereby making both f and g (right) continuous at a. All this is to say that f and g satisfy the hypotheses of Cauchy's Mean Value Theorem on $[a, b]$. Consequently, there is a number c in (a, b) such that

$$\frac{f(b) - f(a)}{g(b) - g(a)} = \frac{f'(c)}{g'(c)}$$

or, since $f(a) = 0 = g(a)$,

$$\frac{f(b)}{g(b)} = \frac{f'(c)}{g'(c)}$$

When we let $b \to a^+$, thereby forcing $c \to a^+$, we obtain

$$\lim_{b \to a^+} \frac{f(b)}{g(b)} = \lim_{c \to a^+} \frac{f'(c)}{g'(c)}$$

which is equivalent to what we wanted to prove.

A very similar proof works for the case of left-hand limits and thus for two-sided limits. The proofs for limits at infinity and infinite limits are harder, and we omit them. ■

Concepts Review

1. L'Hôpital's Rule is useful in finding $\lim_{x \to a}[f(x)/g(x)]$, where both _____ and _____ are zero.

2. L'Hôpital's Rule says that under appropriate conditions $\lim_{x \to a} f(x)/g(x) = \lim_{x \to a}$ _____.

3. From l'Hôpital's Rule, we can conclude that $\lim_{x \to 0}(\tan x)/x = \lim_{x \to 0}$ _____ $=$ _____, but l'Hôpital's Rule gives us no information about $\lim_{x \to 0}(\cos x)/x$ because _____.

4. The proof of l'Hôpital's Rule depends on _____ Theorem.

Problem Set 8.1

In Problems 1–24, find the indicated limit. Make sure that you have an indeterminate form before you apply l'Hôpital's Rule.

1. $\lim_{x \to 0} \frac{2x - \sin x}{x}$

2. $\lim_{x \to \pi/2} \frac{\cos x}{\frac{1}{2}\pi - x}$

3. $\lim_{x \to 0} \frac{x - \sin 2x}{\tan x}$

4. $\lim_{x \to 0} \frac{\tan^{-1} 3x}{\sin^{-1} x}$

5. $\lim_{x \to -2} \frac{x^2 + 6x + 8}{x^2 - 3x - 10}$

6. $\lim_{x \to 0} \frac{x^3 - 3x^2 + x}{x^3 - 2x}$

7. $\lim_{x \to 1^-} \frac{x^2 - 2x + 2}{x^2 - 1}$

8. $\lim_{x \to 1} \frac{\ln x^2}{x^2 - 1}$

9. $\lim_{x \to \pi/2} \frac{\ln(\sin x)^3}{\frac{1}{2}\pi - x}$

10. $\lim_{x \to 0} \frac{e^x - e^{-x}}{2 \sin x}$

11. $\lim_{t \to 1} \frac{\sqrt{t} - t^2}{\ln t}$

12. $\lim_{x \to 0^+} \frac{7^{\sqrt{x}} - 1}{2^{\sqrt{x}} - 1}$

13. $\lim_{x \to 0} \frac{\ln \cos 2x}{7x^2}$

14. $\lim_{x \to 0^-} \frac{3 \sin x}{\sqrt{-x}}$

15. $\lim_{x \to 0} \frac{\tan x - x}{\sin 2x - 2x}$

16. $\lim_{x \to 0} \frac{\sin x - \tan x}{x^2 \sin x}$

17. $\lim_{x \to 0^+} \frac{x^2}{\sin x - x}$

18. $\lim_{x \to 0} \frac{e^x - \ln(1 + x) - 1}{x^2}$

19. $\lim_{x \to 0} \frac{\tan^{-1} x - x}{8x^3}$

20. $\lim_{x \to 0} \frac{\cosh x - 1}{x^2}$

21. $\lim_{x \to 0^+} \frac{1 - \cos x - x \sin x}{2 - 2\cos x - \sin^2 x}$

22. $\lim_{x \to 0^-} \frac{\sin x + \tan x}{e^x + e^{-x} - 2}$

23. $\lim_{x \to 0} \frac{\int_0^x \sqrt{1 + \sin t}\, dt}{x}$

24. $\lim_{x \to 0^+} \frac{\int_0^x \sqrt{t} \cos t\, dt}{x^2}$

25. In Section 1.4, we worked very hard to prove that $\lim_{x \to 0}(\sin x)/x = 1$; l'Hôpital's Rule allows us to show this in one line. However, even if we had l'Hôpital's Rule, say at the end of Section 1.3, it would not have helped us. Explain why. (We really did need to establish $\lim_{x \to 0} \frac{\sin x}{x} = 1$ the way we did in Section 1.4.)

26. Find $\lim_{x \to 0} \frac{x^2 \sin(1/x)}{\tan x}$.

Hint: Begin by deciding why l'Hôpital's Rule is not applicable. Then find the limit by other means.

27. For Figure 2, compute the following limits.

(a) $\lim_{t \to 0^+} \frac{\text{area of triangle } ABC}{\text{area of curved region } ABC}$

(b) $\lim_{t \to 0^+} \frac{\text{area of curved region } BCD}{\text{area of curved region } ABC}$

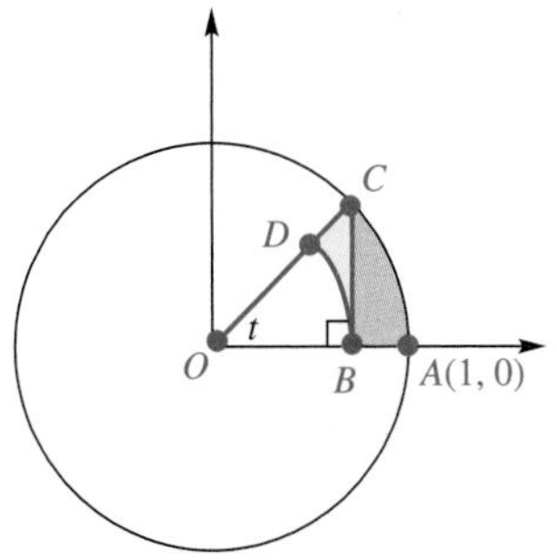

Figure 2

28. In Figure 3, $CD = DE = DF = t$. Find each limit.

(a) $\lim_{t \to 0^+} y$ (b) $\lim_{t \to 0^+} x$

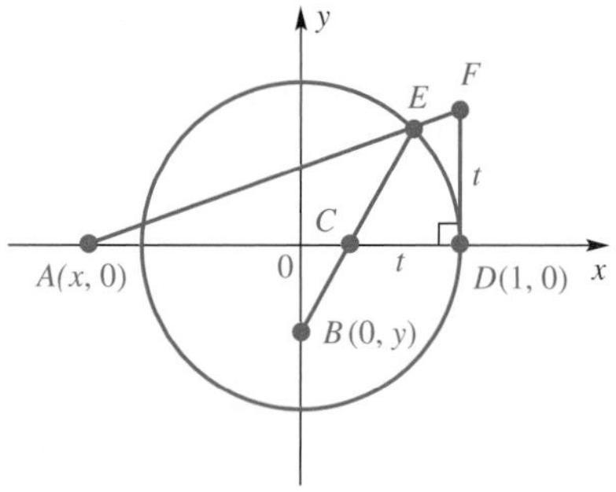

Figure 3

29. Let

$$f(x) = \begin{cases} \dfrac{e^x - 1}{x}, & \text{if } x \neq 0 \\ c, & \text{if } x = 0 \end{cases}$$

What value of c makes $f(x)$ continuous at $x = 0$?

30. Let

$$f(x) = \begin{cases} \dfrac{\ln x}{x - 1}, & \text{if } x \neq 1 \\ c, & \text{if } x = 1 \end{cases}$$

What value of c makes $f(x)$ continuous at $x = 1$?

31. Using the concepts of Section 5.4, you can show that the surface area of the prolate spheroid gotten by rotating the ellipse $x^2/a^2 + y^2/b^2 = 1$ $(a > b)$ about the x-axis is

$$A = 2\pi b^2 + 2\pi ab\left[\frac{a}{\sqrt{a^2 - b^2}} \arcsin \frac{\sqrt{a^2 - b^2}}{a}\right]$$

What should A approach as $a \to b^+$? Use l'Hôpital's Rule to show that this does happen.

32. Determine constants a, b, and c so that

$$\lim_{x \to 1} \frac{ax^4 + bx^3 + 1}{(x - 1) \sin \pi x} = c$$

33. L'Hôpital's Rule in its 1696 form said this: If $\lim_{x \to a} f(x) = \lim_{x \to a} g(x) = 0$, then $\lim_{x \to a} f(x)/g(x) = f'(a)/g'(a)$, provided that $f'(a)$ and $g'(a)$ both exist and $g'(a) \neq 0$. Prove this result without recourse to Cauchy's Mean Value Theorem.

CAS *Use a CAS to evaluate the limits in Problems 34–37.*

34. $\displaystyle\lim_{x \to 0} \frac{\cos x - 1 + x^2/2}{x^4}$

35. $\displaystyle\lim_{x \to 0} \frac{e^x - 1 - x - x^2/2 - x^3/6}{x^4}$

36. $\displaystyle\lim_{x \to 0} \frac{1 - \cos(x^2)}{x^3 \sin x}$ **37.** $\displaystyle\lim_{x \to 0} \frac{\tan x - x}{\arcsin x - x}$

GC *For Problems 38–41, plot the numerator $f(x)$ and the denominator $g(x)$ in the same graph window for each of these domains: $-1 \leq x \leq 1$, $-0.1 \leq x \leq 0.1$, and $-0.01 \leq x \leq 0.01$. From the plot, estimate the values of $f'(x)$ and $g'(x)$ and use these to approximate the given limit.*

38. $\displaystyle\lim_{x \to 0} \frac{3x - \sin x}{x}$ **39.** $\displaystyle\lim_{x \to 0} \frac{\sin x/2}{x}$

40. $\displaystyle\lim_{x \to 0} \frac{x}{e^{2x} - 1}$ **41.** $\displaystyle\lim_{x \to 0} \frac{e^x - 1}{e^{-x} - 1}$

EXPL **42.** Use the concept of the **linear approximation** to a function (Section 2.9) to explain the geometric interpretation of l'Hôpital's Rule in the marginal box next to Theorem A.

Answers to Concepts Review: **1.** $\lim_{x \to a} f(x)$; $\lim_{x \to a} g(x)$ **2.** $f'(x)/g'(x)$ **3.** $\sec^2 x$; 1; $\lim_{x \to 0} \cos x \neq 0$ **4.** Cauchy's Mean Value

---

8.2 Other Indeterminate Forms

In the solution to Example 6 of the previous section, we faced the following limit problem.

$$\lim_{x \to \infty} \frac{x}{e^x}$$

This is typical of a class of problems of the form $\lim_{x \to \infty} f(x)/g(x)$, where both numerator and denominator are growing indefinitely large; we call it an indeterminate form of type ∞/∞. It turns out that l'Hôpital's Rule also applies in this situation; that is,

$$\lim_{x \to \infty} \frac{f(x)}{g(x)} = \lim_{x \to \infty} \frac{f'(x)}{g'(x)}$$

A rigorous proof is quite difficult, but there is an intuitive way of seeing that the result has to be true. Imagine that $f(t)$ and $g(t)$ represent the positions of two cars on the t-axis at time t (Figure 1). These two cars, the f-car and the g-car, are on endless journeys with respective velocities $f'(t)$ and $g'(t)$. Now, if

$$\lim_{t \to \infty} \frac{f'(t)}{g'(t)} = L$$

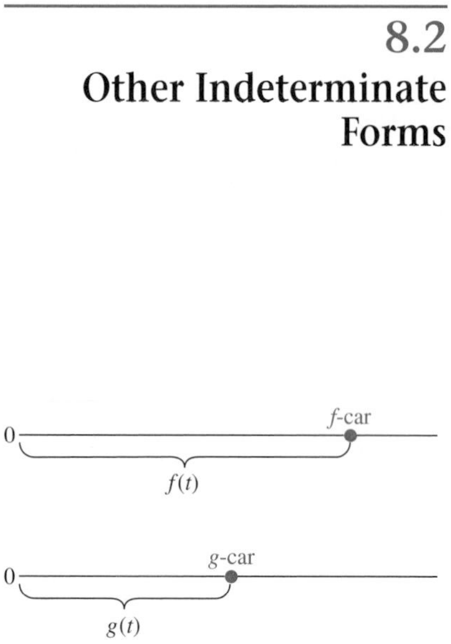

Figure 1

then ultimately the f-car travels about L times as fast as the g-car. It is therefore reasonable to say that, in the long run, it will travel about L times as far; that is,

$$\lim_{t \to \infty} \frac{f(t)}{g(t)} = L$$

We do not call this a proof, but it does lend plausibility to a result that we now state formally.

Theorem A **L'Hôpital's Rule for Forms of Type ∞/∞**

Suppose that $\lim_{x \to u} |f(x)| = \lim_{x \to u} |g(x)| = \infty$. If $\lim_{x \to u} [f'(x)/g'(x)]$ exists in either the finite or infinite sense, then

$$\lim_{x \to u} \frac{f(x)}{g(x)} = \lim_{x \to u} \frac{f'(x)}{g'(x)}$$

Here u may stand for any of the symbols a, a^-, a^+, $-\infty$, or ∞.

The Indeterminate Form ∞/∞ We use Theorem A to finish Example 6 of the previous section.

EXAMPLE 1 Find $\lim_{x \to \infty} \frac{x}{e^x}$.

SOLUTION Both x and e^x tend to ∞ as $x \to \infty$. Hence, by l'Hôpital's Rule,

$$\lim_{x \to \infty} \frac{x}{e^x} = \lim_{x \to \infty} \frac{D_x x}{D_x e^x} = \lim_{x \to \infty} \frac{1}{e^x} = 0$$ ■

Here is a general result of the same type.

EXAMPLE 2 Show that, if a is any positive real number, $\lim_{x \to \infty} \frac{x^a}{e^x} = 0$.

SOLUTION Suppose as a special case that $a = 2.5$. Then three applications of l'Hôpital's Rule give

$$\lim_{x \to \infty} \frac{x^{2.5}}{e^x} \overset{\text{L}}{=} \lim_{x \to \infty} \frac{2.5x^{1.5}}{e^x} \overset{\text{L}}{=} \lim_{x \to \infty} \frac{(2.5)(1.5)x^{0.5}}{e^x} \overset{\text{L}}{=} \lim_{x \to \infty} \frac{(2.5)(1.5)(0.5)}{x^{0.5}e^x} = 0$$

A similar argument works for any $a > 0$. Let m denote the greatest integer less than a. Then $m + 1$ applications of l'Hôpital's Rule give

$$\lim_{x \to \infty} \frac{x^a}{e^x} \overset{\text{L}}{=} \lim_{x \to \infty} \frac{ax^{a-1}}{e^x} \overset{\text{L}}{=} \lim_{x \to \infty} \frac{a(a-1)x^{a-2}}{e^x} \overset{\text{L}}{=} \cdots \overset{\text{L}}{=} \lim_{x \to \infty} \frac{a(a-1)\cdots(a-m)}{x^{m+1-a}e^x} = 0$$

■

EXAMPLE 3 Show that, if a is any positive real number, $\lim_{x \to \infty} \frac{\ln x}{x^a} = 0$.

See How They Grow

In computer science, one pays careful attention to the amount of time needed to perform a task. For example, to sort x items using the "bubble sort" algorithm takes time proportional to x^2, whereas the "quick sort" algorithm does the same task in time proportional to $x \ln x$, a major improvement. Here is a chart illustrating how some common functions grow as x increases from 10 to 100 to 1000.

$\ln x$	2.3	4.6	6.9
$\sqrt{x}$	3.2	10	31.6
x	10	100	1000
$x \ln x$	23	461	6908
x^2	100	10000	10^6
e^x	2.2×10^4	2.7×10^{43}	10^{434}

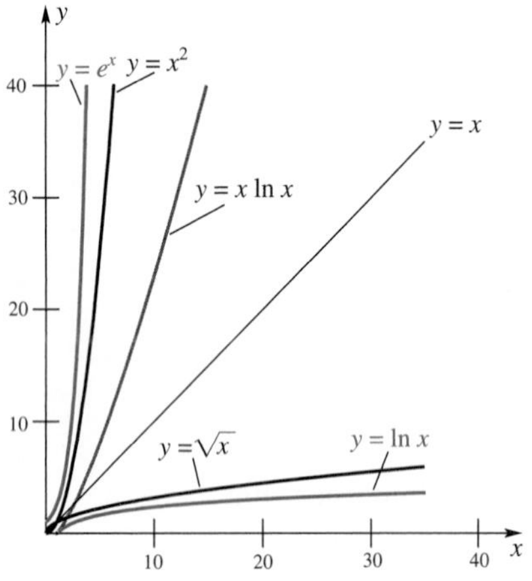

Figure 2

SOLUTION Both $\ln x$ and x^a tend to ∞ as $x \to \infty$. Hence, by one application of l'Hôpital's Rule,

$$\lim_{x \to \infty} \frac{\ln x}{x^a} \stackrel{L}{=} \lim_{x \to \infty} \frac{1/x}{ax^{a-1}} = \lim_{x \to \infty} \frac{1}{ax^a} = 0$$

■

≈ Examples 2 and 3 say something that is worth remembering: *for sufficiently large x, e^x grows faster as x increases than any constant power of x, whereas $\ln x$ grows more slowly than any constant power of x.* For example, when x is sufficiently large, e^x grows faster than x^{100} and $\ln x$ grows more slowly than $\sqrt[100]{x}$. The chart in the margin and Figure 2 offer additional illustration.

EXAMPLE 4 Find $\lim_{x \to 0^+} \frac{\ln x}{\cot x}$.

SOLUTION As $x \to 0^+$, $\ln x \to -\infty$ and $\cot x \to \infty$, so l'Hôpital's Rule applies.

$$\lim_{x \to 0^+} \frac{\ln x}{\cot x} \stackrel{L}{=} \lim_{x \to 0^+} \left[\frac{1/x}{-\csc^2 x}\right]$$

This is still indeterminate as it stands, but rather than apply l'Hôpital's Rule again (which only makes things worse), we rewrite the expression in brackets as

$$\frac{1/x}{-\csc^2 x} = -\frac{\sin^2 x}{x} = -\sin x \frac{\sin x}{x}$$

Thus,

$$\lim_{x \to 0^+} \frac{\ln x}{\cot x} = \lim_{x \to 0^+} \left[-\sin x \frac{\sin x}{x}\right] = 0 \cdot 1 = 0$$

■

The Indeterminate Forms $0 \cdot \infty$ and $\infty - \infty$ Suppose that $A(x) \to 0$, but $B(x) \to \infty$. What is going to happen to the product $A(x)B(x)$? Two competing forces are at work, tending to pull the product in opposite directions. Which will win this battle, A or B or neither? It depends on whether one is stronger (i.e., doing its job at a faster rate) or whether they are evenly matched. L'Hôpital's Rule will help us to decide, but only after we have transformed the problem to a $0/0$ or ∞/∞ form.

EXAMPLE 5 Find $\lim_{x \to \pi/2} (\tan x \cdot \ln \sin x)$.

SOLUTION Since $\lim_{x \to \pi/2} \ln \sin x = 0$ and $\lim_{x \to \pi/2} |\tan x| = \infty$, this is a $0 \cdot \infty$ indeterminate form. We can rewrite it as a $0/0$ form by the simple device of changing $\tan x$ to $1/\cot x$. Thus,

$$\lim_{x \to \pi/2} (\tan x \cdot \ln \sin x) = \lim_{x \to \pi/2} \frac{\ln \sin x}{\cot x}$$

$$\stackrel{L}{=} \lim_{x \to \pi/2} \frac{\frac{1}{\sin x} \cdot \cos x}{-\csc^2 x}$$

$$= \lim_{x \to \pi/2} (-\cos x \cdot \sin x) = 0$$

■

EXAMPLE 6 Find $\lim_{x \to 1^+} \left(\frac{x}{x-1} - \frac{1}{\ln x} \right)$.

SOLUTION The first term is growing without bound; so is the second. We say that the limit is an $\infty - \infty$ indeterminate form. L'Hôpital's Rule will determine the result, but only after we rewrite the problem in a form for which the rule applies. In this case, the two fractions must be combined, a procedure that changes the problem to a 0/0 form. Two applications of l'Hôpital's Rule yield

$$\lim_{x \to 1^+} \left(\frac{x}{x-1} - \frac{1}{\ln x} \right) = \lim_{x \to 1^+} \frac{x \ln x - x + 1}{(x-1)\ln x} \overset{\text{(L)}}{=} \lim_{x \to 1^+} \frac{x \cdot 1/x + \ln x - 1}{(x-1)(1/x) + \ln x}$$

$$= \lim_{x \to 1^+} \frac{x \ln x}{x - 1 + x \ln x} \overset{\text{(L)}}{=} \lim_{x \to 1^+} \frac{1 + \ln x}{2 + \ln x} = \frac{1}{2}$$

■

The Indeterminate Forms 0^0, ∞^0, 1^∞ We turn now to three indeterminate forms of exponential type. Here the trick is to consider not the original expression, but rather its logarithm. Usually, l'Hôpital's Rule will apply to the logarithm.

EXAMPLE 7 Find $\lim_{x \to 0^+} (x + 1)^{\cot x}$.

SOLUTION This takes the indeterminate form 1^∞. Let $y = (x + 1)^{\cot x}$, so

$$\ln y = \cot x \ln(x + 1) = \frac{\ln(x + 1)}{\tan x}$$

Using l'Hôpital's Rule for 0/0 forms, we obtain

$$\lim_{x \to 0^+} \ln y = \lim_{x \to 0^+} \frac{\ln(x + 1)}{\tan x} \overset{\text{(L)}}{=} \lim_{x \to 0^+} \frac{\frac{1}{x + 1}}{\sec^2 x} = 1$$

Now $y = e^{\ln y}$, and since the exponential function $f(x) = e^x$ is continuous,

$$\lim_{x \to 0^+} y = \lim_{x \to 0^+} \exp(\ln y) = \exp\left(\lim_{x \to 0^+} \ln y \right) = \exp 1 = e$$

■

EXAMPLE 8 Find $\lim_{x \to \pi/2^-} (\tan x)^{\cos x}$.

SOLUTION This has the indeterminate form ∞^0. Let $y = (\tan x)^{\cos x}$, so

$$\ln y = \cos x \cdot \ln \tan x = \frac{\ln \tan x}{\sec x}$$

Then

$$\lim_{x \to \pi/2^-} \ln y = \lim_{x \to \pi/2^-} \frac{\ln \tan x}{\sec x} \overset{\text{(L)}}{=} \lim_{x \to \pi/2^-} \frac{\frac{1}{\tan x} \cdot \sec^2 x}{\sec x \tan x}$$

$$= \lim_{x \to \pi/2^-} \frac{\sec x}{\tan^2 x} = \lim_{x \to \pi/2^-} \frac{\cos x}{\sin^2 x} = 0$$

Therefore,

$$\lim_{x\to\pi/2^-} y = e^0 = 1$$ ■

Summary We have classified certain limit problems as indeterminate forms, using the seven symbols $0/0$, ∞/∞, $0\cdot\infty$, $\infty - \infty$, 0^0, ∞^0, and 1^∞. Each involves a competition of opposing forces, which means that the result is not obvious. However, with the help of l'Hôpital's Rule, which applies directly only to the $0/0$ and ∞/∞ forms, we can usually determine the limit.

There are many other possibilities symbolized by, for example, $0/\infty$, $\infty/0$, $\infty + \infty$, $\infty\cdot\infty$, 0^∞, and ∞^∞. Why don't we call these indeterminate forms? Because, in each of these cases, the forces work together, not in competition.

EXAMPLE 9 Find $\lim_{x\to0^+}(\sin x)^{\cot x}$.

SOLUTION We might call this a 0^∞ form, but it is not indeterminate. Note that $\sin x$ is approaching zero, and raising it to the exponent $\cot x$, an increasingly large number, serves only to make it approach zero faster. Thus,

$$\lim_{x\to0^+}(\sin x)^{\cot x} = 0$$ ■

Concepts Review

1. If $\lim_{x\to a} f(x) = \lim_{x\to a} g(x) = \infty$, then l'Hôpital's Rule says that $\lim_{x\to a} f(x)/g(x) = \lim_{x\to a}$ _____.

2. If $\lim_{x\to a} f(x) = 0$ and $\lim_{x\to a} g(x) = \infty$, then $\lim_{x\to a} f(x)g(x)$ is an indeterminate form. To apply l'Hôpital's Rule, we may rewrite this latter limit as _____.

3. Seven indeterminate forms are discussed in this book. They are symbolized by $0/0$, ∞/∞, $0\cdot\infty$, and _____.

4. e^x grows faster than any power of x, but _____ grows more slowly than any power of x.

Problem Set 8.2

Find each limit in Problems 1–40. Be sure you have an indeterminate form before applying l'Hôpital's Rule.

1. $\lim_{x\to\infty}\frac{\ln x^{10000}}{x}$
2. $\lim_{x\to\infty}\frac{(\ln x)^2}{2^x}$
3. $\lim_{x\to\infty}\frac{x^{10000}}{e^x}$
4. $\lim_{x\to\infty}\frac{3x}{\ln(100x + e^x)}$
5. $\lim_{x\to\pi/2}\frac{3\sec x + 5}{\tan x}$
6. $\lim_{x\to0^+}\frac{\ln\sin^2 x}{3\ln\tan x}$
7. $\lim_{x\to\infty}\frac{\ln(\ln x^{1000})}{\ln x}$
8. $\lim_{x\to(1/2)^-}\frac{\ln(4-8x)^2}{\tan\pi x}$
9. $\lim_{x\to0^+}\frac{\cot x}{\sqrt{-\ln x}}$
10. $\lim_{x\to0}\frac{2\csc^2 x}{\cot^2 x}$
11. $\lim_{x\to0}(x\ln x^{1000})$
12. $\lim_{x\to0} 3x^2\csc^2 x$
13. $\lim_{x\to0}(\csc^2 x - \cot^2 x)$
14. $\lim_{x\to\pi/2}(\tan x - \sec x)$
15. $\lim_{x\to0^+}(3x)^{x^2}$
16. $\lim_{x\to0}(\cos x)^{\csc x}$
17. $\lim_{x\to(\pi/2)^-}(5\cos x)^{\tan x}$
18. $\lim_{x\to0}\left(\csc^2 x - \frac{1}{x^2}\right)^2$
19. $\lim_{x\to0}(x + e^{x/3})^{3/x}$
20. $\lim_{x\to(\pi/2)^-}(\cos 2x)^{x-\pi/2}$
21. $\lim_{x\to\pi/2}(\sin x)^{\cos x}$
22. $\lim_{x\to\infty} x^x$
23. $\lim_{x\to\infty} x^{1/x}$
24. $\lim_{x\to0}(\cos x)^{1/x^2}$
25. $\lim_{x\to0^+}(\tan x)^{2/x}$
26. $\lim_{x\to-\infty}(e^{-x} - x)$
27. $\lim_{x\to0^+}(\sin x)^x$
28. $\lim_{x\to0}(\cos x - \sin x)^{1/x}$
29. $\lim_{x\to0}\left(\csc x - \frac{1}{x}\right)$
30. $\lim_{x\to\infty}\left(1 + \frac{1}{x}\right)^x$
31. $\lim_{x\to0^+}(1 + 2e^x)^{1/x}$
32. $\lim_{x\to1}\left(\frac{1}{x-1} - \frac{x}{\ln x}\right)$
33. $\lim_{x\to0}(\cos x)^{1/x}$
34. $\lim_{x\to0^+}(x^{1/2}\ln x)$
35. $\lim_{x\to\infty} e^{\cos x}$
36. $\lim_{x\to\infty}[\ln(x+1) - \ln(x-1)]$
37. $\lim_{x\to0^+}\frac{x}{\ln x}$
38. $\lim_{x\to0^+}(\ln x\cot x)$
39. $\lim_{x\to\infty}\frac{\int_1^x\sqrt{1+e^{-t}}\,dt}{x}$
40. $\lim_{x\to1^+}\frac{\int_1^x\sin t\,dt}{x-1}$

41. Find each limit. *Hint:* Transform to problems involving a continuous variable x. Assume that $a > 0$.

(a) $\lim_{n\to\infty}\sqrt[n]{a}$ (b) $\lim_{n\to\infty}\sqrt[n]{n}$

(c) $\lim_{n\to\infty} n(\sqrt[n]{a} - 1)$ (d) $\lim_{n\to\infty} n(\sqrt[n]{n} - 1)$

42. Find each limit.

(a) $\lim_{x\to 0^+} x^x$ (b) $\lim_{x\to 0^+} (x^x)^x$

(c) $\lim_{x\to 0^+} x^{(x^x)}$ (d) $\lim_{x\to 0^+} ((x^x)^x)^x$

(e) $\lim_{x\to 0^+} x^{(x^{(x^x)})}$

43. Graph $y = x^{1/x}$ for $x > 0$. Show what happens for very small x and very large x. Indicate the maximum value.

44. Find each limit.

(a) $\lim_{x\to 0^+} (1^x + 2^x)^{1/x}$ (b) $\lim_{x\to 0^-} (1^x + 2^x)^{1/x}$

(c) $\lim_{x\to \infty} (1^x + 2^x)^{1/x}$ (d) $\lim_{x\to -\infty} (1^x + 2^x)^{1/x}$

45. For $k \geq 0$, find

$$\lim_{n\to\infty} \frac{1^k + 2^k + \cdots + n^k}{n^{k+1}}$$

Hint: Though this has the ∞/∞ form, l'Hôpital's Rule is not helpful. Think of a Riemann sum.

46. Let $c_1, c_2, \ldots, c_n$ be positive constants with $\sum_{i=1}^{n} c_i = 1$, and let $x_1, x_2, \ldots, x_n$ be positive numbers. Take natural logarithms and then use l'Hôpital's Rule to show that

$$\lim_{t\to 0^+} \left(\sum_{i=1}^{n} c_i x_i^t \right)^{1/t} = x_1^{c_1} x_2^{c_2} \cdots x_n^{c_n} = \prod_{i=1}^{n} x_i^{c_i}$$

Here $\prod$ means product; that is, $\prod_{i=1}^{n} a_i$ means $a_1 \cdot a_2 \cdot \cdots \cdot a_n$. In particular, if a, b, x, and y are positive and $a + b = 1$, then

$$\lim_{t\to 0^+} (ax^t + by^t)^{1/t} = x^a y^b$$

47. Verify the last statement in Problem 46 by calculating each of the following.

(a) $\lim_{t\to 0^+} \left(\frac{1}{2}2^t + \frac{1}{2}5^t\right)^{1/t}$ (b) $\lim_{t\to 0^+} \left(\frac{1}{5}2^t + \frac{4}{5}5^t\right)^{1/t}$

(c) $\lim_{t\to 0^+} \left(\frac{1}{10}2^t + \frac{9}{10}5^t\right)^{1/t}$

48. Consider $f(x) = n^2 x e^{-nx}$.

(a) Graph $f(x)$ for $n = 1, 2, 3, 4, 5, 6$ on $[0, 1]$ in the same graph window.

(b) For $x > 0$, find $\lim_{n\to\infty} f(x)$.

(c) Evaluate $\int_0^1 f(x)\,dx$ for $n = 1, 2, 3, 4, 5, 6$.

(d) Guess at $\lim_{n\to\infty} \int_0^1 f(x)\,dx$. Then justify your answer rigorously.

CAS **49.** Find the absolute maximum and minimum points (if they exist) for $f(x) = (x^{25} + x^3 + 2^x)e^{-x}$ on $[0, \infty)$.

Answers to Concepts Review: **1.** $f'(x)/g'(x)$ **2.** $\lim_{x\to a} f(x)/[1/g(x)]$ or $\lim_{x\to a} g(x)/[1/f(x)]$ **3.** $\infty - \infty, 0^0, \infty^0, 1^\infty$ **4.** $\ln x$

8.3 Improper Integrals: Infinite Limits of Integration

In the definition of $\int_a^b f(x)\,dx$, it was assumed that the interval $[a, b]$ was finite. However, in many applications in physics, economics, and probability we wish to allow a or b (or both) to be ∞ or $-\infty$. We must therefore find a way to give meaning to symbols like

$$\int_0^\infty \frac{1}{1 + x^2}\,dx, \qquad \int_{-\infty}^{-1} xe^{-x^2}\,dx, \qquad \int_{-\infty}^{\infty} x^2 e^{-x^2}\,dx$$

These integrals are called **improper integrals** with infinite limits.

One Infinite Limit Consider the function $f(x) = xe^{-x}$. It makes perfectly good sense to ask for $\int_0^1 xe^{-x}\,dx$ or $\int_0^2 xe^{-x}\,dx$, or indeed for $\int_0^b xe^{-x}\,dx$, where b is any positive number. As the table on the next page indicates, as we increase the upper limit in the definite integral the value of the integral (the area under the curve) increases, but apparently not without bound (in this example, at least). To give meaning to $\int_0^\infty xe^{-x}\,dx$ we begin by integrating from 0 to an arbitrary upper limit, say b, which using integration by parts gives

$$\int_0^b xe^{-x}\,dx = [-xe^{-x}]_0^b - \int_0^b (-e^{-x})\,dx = 1 - e^{-b} - be^{-b}$$

Now, imagine that the value of b marches off to infinity. (See the accompanying table.) As the above calculation shows, if we let $b \to \infty$, the value of the definite integral converges to 1. Thus, it seems natural to define

$$\int_0^\infty xe^{-x}\,dx = \lim_{b\to\infty} \int_0^b xe^{-x}\,dx = \lim_{b\to\infty} (1 - e^{-b} - be^{-b}) = 1$$

Integral	Picture	Exact Value	Numerical Approximation
$\int_0^1 xe^{-x}\,dx$	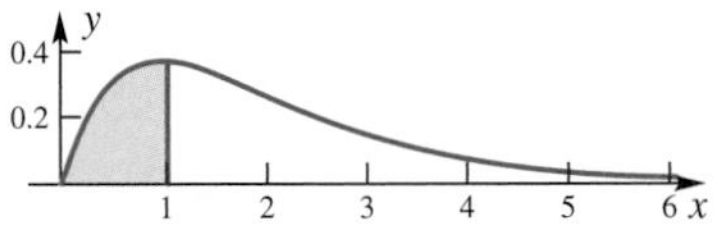	$1 - e^{-1} - 1e^{-1}$	0.2642
$\int_0^2 xe^{-x}\,dx$	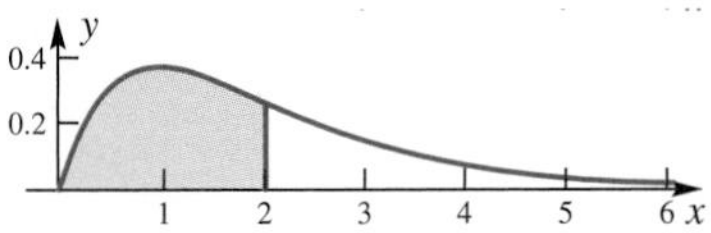	$1 - e^{-2} - 2e^{-2}$	0.5940
$\int_0^3 xe^{-x}\,dx$	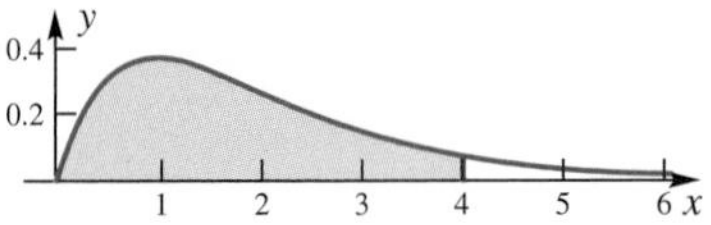	$1 - e^{-3} - 3e^{-3}$	0.8009
$\int_0^b xe^{-x}\,dx$	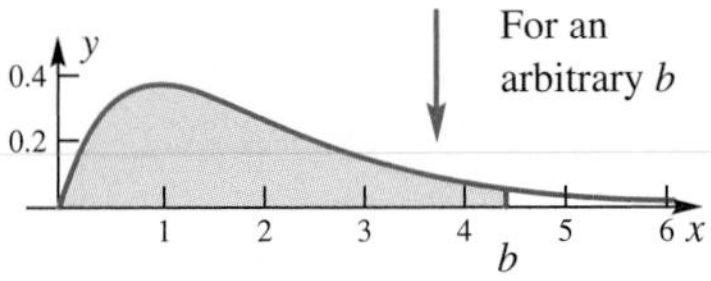	$1 - e^{-b} - be^{-b}$	
$\int_0^\infty xe^{-x}\,dx$	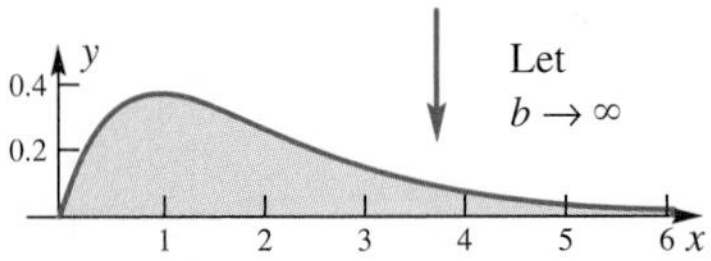	$\lim_{b\to\infty}\left[1 - e^{-b} - be^{-b}\right] = 1$	

Here is the general definition.

Definition

$$\int_{-\infty}^{b} f(x)\,dx = \lim_{a\to-\infty} \int_a^b f(x)\,dx$$

$$\int_{a}^{\infty} f(x)\,dx = \lim_{b\to\infty} \int_a^b f(x)\,dx$$

If the limits on the right exist and have finite values, then we say that the corresponding improper integrals **converge** and have those values. Otherwise, the integrals are said to **diverge.**

EXAMPLE 1 Find, if possible, $\int_{-\infty}^{-1} xe^{-x^2}\,dx$.

SOLUTION

$$\int_a^{-1} xe^{-x^2}\,dx = -\frac{1}{2}\int_a^{-1} e^{-x^2}(-2x\,dx) = \left[-\frac{1}{2}e^{-x^2}\right]_a^{-1}$$

$$= -\frac{1}{2}e^{-1} + \frac{1}{2}e^{-a^2}$$

Thus,

$$\int_{-\infty}^{-1} xe^{-x^2}\,dx = \lim_{a\to-\infty}\left[-\frac{1}{2}e^{-1} + \frac{1}{2}e^{-a^2}\right] = -\frac{1}{2e}$$

We say the integral converges and has value $-1/2e$. ■

EXAMPLE 2 Find, if possible, $\int_0^{\infty} \sin x\, dx$.

SOLUTION

$$\int_0^{\infty} \sin x\, dx = \lim_{b\to\infty} \int_0^{b} \sin x\, dx = \lim_{b\to\infty} [-\cos x]_0^b$$
$$= \lim_{b\to\infty} [1 - \cos b]$$

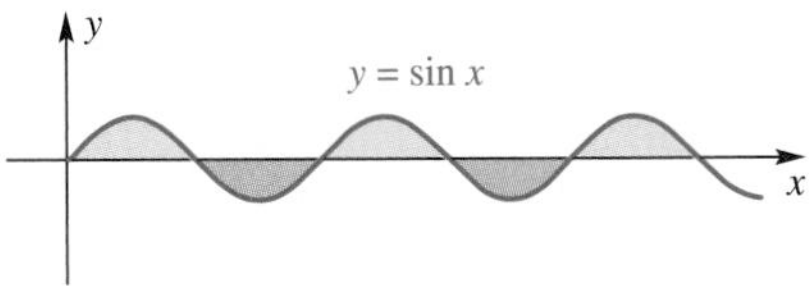

Figure 1

The latter limit does not exist; we conclude that the given integral diverges. Think about the geometric meaning of $\int_0^{\infty} \sin x\, dx$ to support this result (Figure 1). ■

EXAMPLE 3 According to Newton's Inverse Square Law, the force exerted by the earth on a space capsule is $-k/x^2$, where x is the distance (in miles, for instance) from the capsule to the center of the earth (Figure 2). The force $F(x)$ required to lift the capsule is therefore $F(x) = k/x^2$. How much work is done in propelling a 1000-pound capsule out of the earth's gravitational field?

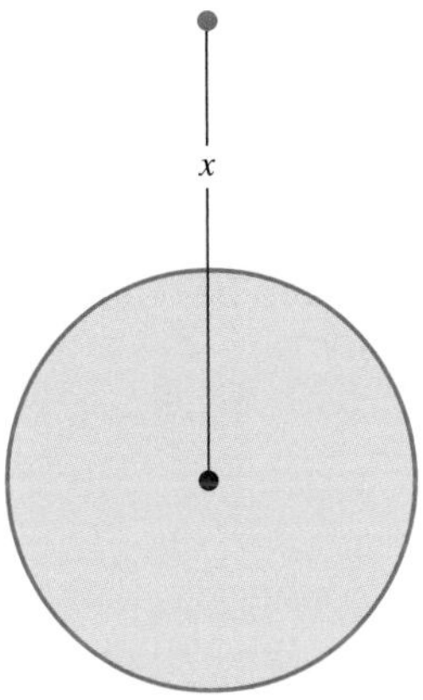

Figure 2

SOLUTION We can evaluate k by noting that at $x = 3960$ miles (the radius of the earth) $F = 1000$ pounds. This yields $k = 1000(3960)^2 \approx 1.568 \times 10^{10}$. The work done in mile-pounds is therefore

$$1.568 \times 10^{10} \int_{3960}^{\infty} \frac{1}{x^2}\, dx = \lim_{b\to\infty} 1.568 \times 10^{10}\left[-\frac{1}{x}\right]_{3960}^{b}$$
$$= \lim_{b\to\infty} 1.568 \times 10^{10}\left[-\frac{1}{b} + \frac{1}{3960}\right]$$
$$= \frac{1.568 \times 10^{10}}{3960} \approx 3.96 \times 10^6$$ ■

Both Limits Infinite

We can now give a definition for $\int_{-\infty}^{\infty} f(x)\, dx$.

Definition

If both $\int_{-\infty}^{0} f(x)\, dx$ and $\int_0^{\infty} f(x)\, dx$ converge, then $\int_{-\infty}^{\infty} f(x)\, dx$ is said to converge and have value

$$\int_{-\infty}^{\infty} f(x)\, dx = \int_{-\infty}^{0} f(x)\, dx + \int_0^{\infty} f(x)\, dx$$

Otherwise, $\int_{-\infty}^{\infty} f(x)\, dx$ diverges.

EXAMPLE 4 Evaluate $\int_{-\infty}^{\infty} \frac{1}{1 + x^2}\, dx$ or state that it diverges.

SOLUTION

$$\int_0^{\infty} \frac{1}{1 + x^2}\, dx = \lim_{b\to\infty} \int_0^{b} \frac{1}{1 + x^2}\, dx$$
$$= \lim_{b\to\infty} [\tan^{-1} x]_0^b$$
$$= \lim_{b\to\infty} [\tan^{-1} b - \tan^{-1} 0] = \frac{\pi}{2}$$

Since the integrand is an even function,

$$\int_{-\infty}^{0} \frac{1}{1+x^2}\,dx = \int_{0}^{\infty} \frac{1}{1+x^2}\,dx = \frac{\pi}{2}$$

Therefore,

$$\int_{-\infty}^{\infty} \frac{1}{1+x^2}\,dx = \int_{-\infty}^{0} \frac{1}{1+x^2}\,dx + \int_{0}^{\infty} \frac{1}{1+x^2}\,dx = \frac{\pi}{2} + \frac{\pi}{2} = \pi$$ ■

We will use the notation $[F(x)]_a^{\infty}$ to mean $\lim_{b\to\infty} F(b) - F(a)$. Similar definitions apply to $[F(x)]_{-\infty}^{a}$ and to $[F(x)]_{-\infty}^{\infty}$. Note that in none of these cases are we "substituting" infinity. Each is defined as a limit, which agrees with our approach to determining improper integrals.

Probability Density Functions When we first introduced random variables and probability density functions back in Section 5.7 we had to restrict attention to cases where the set of possible outcomes was bounded. In many situations, there is no upper (or lower) limit for the set of possible outcomes. For example, there is no upper bound on how long a battery will last, or how strong a mix of concrete is. Now that we have covered improper integrals, we can dispense with this restriction.

If the PDF $f(x)$ of a continuous random variable X is defined to be 0 outside the set of possible outcomes, then the requirements for a PDF are

1. $f(x) \geq 0$
2. $\int_{-\infty}^{\infty} f(x)\,dx = 1$

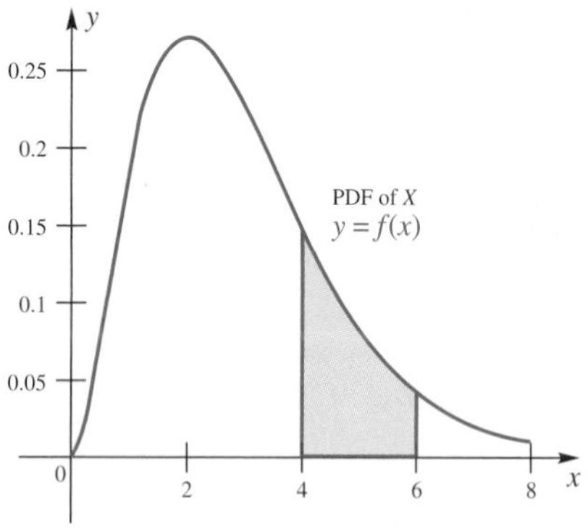

Figure 3

The PDF of a random variable allows us to find probabilities by integration; for example, Figure 3 illustrates the probability that X is between 4 and 6.

The mean and variance of a random variable are then defined by

$$\mu = E(X) = \int_{-\infty}^{\infty} x\,f(x)\,dx$$

$$\sigma^2 = V(X) = \int_{-\infty}^{\infty} (x-\mu)^2 f(x)\,dx$$

The variance σ^2 of a random variable is a measure of the dispersion, or "spread-out-ness" of the probability, and it can be computed from (see Problem 41 of Section 5.7)

$$\sigma^2 = E(X^2) - \mu^2$$

When σ^2 is small, the distribution of probability is, roughly speaking, clustered very closely around the mean; when σ^2 is large, the probability is more spread out.

The next two examples, and some of the exercises, introduce several useful families of probability distributions.

EXAMPLE 5 The **exponential distribution,** which is sometimes used to model the lifetimes of electrical or mechanical components, has PDF

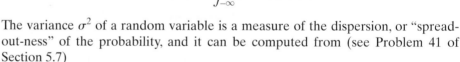

$$f(x) = \begin{cases} \lambda e^{-\lambda x}, & \text{if } 0 \leq x \\ 0, & \text{otherwise} \end{cases}$$

where λ is some positive constant.

(a) Show that this is a valid PDF.
(b) Find the mean μ and the variance σ^2.
(c) Find the cumulative distribution function (CDF) $F(x)$.
(d) If a component's lifetime X measured in hours is a random variable having an exponential distribution with $\lambda = 0.01$, what is the probability that the component works for at least 20 hours?

SOLUTION

(a) The function f is always nonnegative and

$$\begin{aligned}\int_{-\infty}^{\infty} f(x)\,dx &= \int_{-\infty}^{0} 0\,dx + \int_{0}^{\infty} \lambda e^{-\lambda x}\,dx \\ &= 0 + [-e^{-\lambda x}]_0^{\infty} \\ &= 1\end{aligned}$$

so $f(x)$ is a valid PDF.

(b)

$$\begin{aligned}E(X) &= \int_{-\infty}^{\infty} x f(x)\,dx \\ &= \int_{-\infty}^{0} x \cdot 0\,dx + \int_{0}^{\infty} x\lambda e^{-\lambda x}\,dx\end{aligned}$$

We apply integration by parts in the second integral: $u = x$, $dv = \lambda e^{-\lambda x}\,dx$, so that $du = dx$, $v = -e^{-\lambda x}$. Thus

$$\begin{aligned}E(X) &= [-x\lambda e^{-\lambda x}]_0^{\infty} - \int_0^{\infty} (-e^{-\lambda x})\,dx \\ &= (-0 + 0) + \left[-\frac{1}{\lambda}e^{-\lambda x}\right]_0^{\infty} \\ &= \frac{1}{\lambda}\end{aligned}$$

The variance is

$$\begin{aligned}\sigma^2 &= E(X^2) - \mu^2 \\ &= \int_{-\infty}^{\infty} x^2 f(x)\,dx - \left(\frac{1}{\lambda}\right)^2 \\ &= \int_{-\infty}^{0} x^2 \cdot 0\,dx + \int_0^{\infty} x^2\lambda e^{-\lambda x}\,dx - \frac{1}{\lambda^2} \\ &= [-x^2 e^{-\lambda x}]_0^{\infty} - \int_0^{\infty} (-e^{-\lambda x})\,2x\,dx - \frac{1}{\lambda^2} \\ &= (-0 + 0) + 2\int_0^{\infty} x e^{-\lambda x}\,dx - \frac{1}{\lambda^2} \\ &= 2\frac{1}{\lambda^2} - \frac{1}{\lambda^2} = \frac{1}{\lambda^2}\end{aligned}$$

(c) For $x < 0$, the CDF is $F(x) = P(X \le x) = 0$. For $x \ge 0$,

$$\begin{aligned}F(x) &= \int_{-\infty}^{x} f(t)\,dt \\ &= \int_{-\infty}^{0} 0\,dx + \int_0^{x} \lambda e^{-\lambda t}\,dt \\ &= 0 + [-e^{-\lambda t}]_0^{x} \\ &= 1 - e^{-\lambda x}\end{aligned}$$

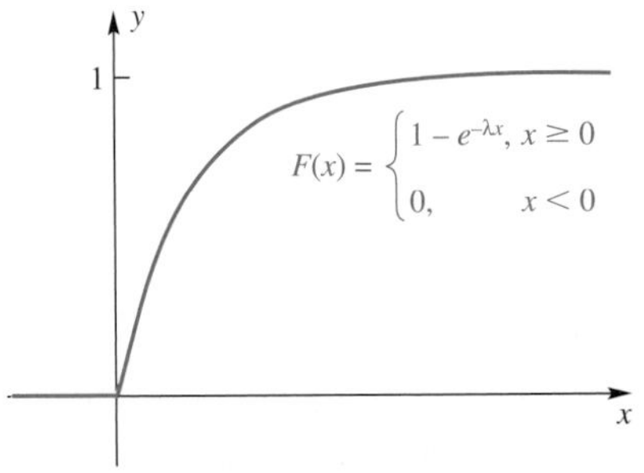

Figure 4

A graph of the CDF is shown in Figure 4.

(d) Set $\lambda = 0.01$. The probability that the component works for at least 20 hours is the probability that the lifetime is 20 hours or greater:

$$\begin{aligned} P(X > 20) &= \int_{20}^{\infty} 0.01e^{-0.01x}\,dx \\ &= [-e^{-0.01x}]_{20}^{\infty} \\ &= 0 - (-e^{-0.01 \cdot 20}) \\ &= e^{-0.2} \\ &\approx 0.819 \end{aligned}$$

■

The **normal distribution** is the familiar bell-shaped curve. It is really a family of distributions since the mean μ can be any number and the variance can be any positive number σ^2. The normal distribution with parameters μ and σ^2 has PDF

$$f(x) = \frac{1}{\sqrt{2\pi}\,\sigma} \exp[-(x - \mu)^2/2\sigma^2]$$

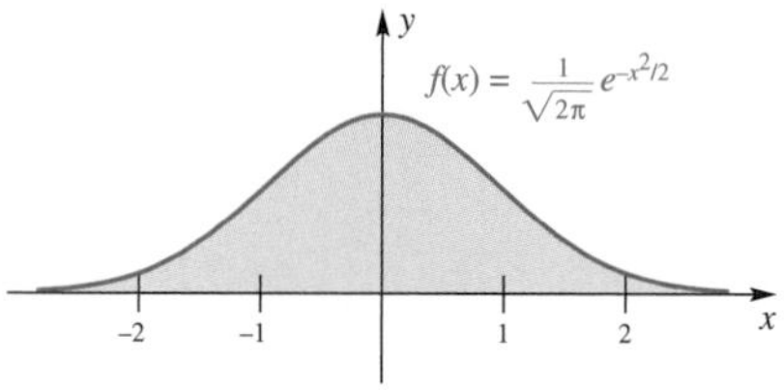

Figure 5

(The parameters μ and σ^2 turn out to be equal to the mean and variance, respectively, thus justifying the use of the Greek letters μ and σ.) Figure 5 shows a plot of the PDF for the normal distribution with mean $\mu = 0$ and variance $\sigma^2 = 1$. It is surprisingly difficult to show that

$$\int_{-\infty}^{\infty} \frac{1}{\sqrt{2\pi}\,\sigma} \exp[-(x - \mu)^2/2\sigma^2]\,dx = 1$$

although we will do so later (Section 13.4). Other properties of the normal distribution include the following:

(a) its graph is symmetric about the line $x = \mu$;
(b) it has a maximum at $x = \mu$;
(c) it has inflection points when $x = \mu \pm \sigma$;
(d) the mean is μ;
(e) the variance is σ^2.

Problem 33 involves some other properties of the normal PDF. The normal distribution with $\mu = 0$ and $\sigma^2 = 1$ is called the **standard normal distribution**. This is the normal distribution that is graphed in Figure 5.

EXAMPLE 6 Show that

(a) $$\frac{1}{\sqrt{2\pi}} \int_{-\infty}^{\infty} xe^{-x^2/2}\,dx = 0$$

(b) $$\frac{1}{\sqrt{2\pi}} \int_{-\infty}^{\infty} x^2e^{-x^2/2}\,dx = 1$$

SOLUTION

(a) $$\frac{1}{\sqrt{2\pi}} \int_{0}^{\infty} xe^{-x^2/2}\,dx = \lim_{b\to\infty}\left[-\frac{1}{\sqrt{2\pi}} \int_{0}^{b} e^{-x^2/2}(-x\,dx)\right]$$

$$= \lim_{b\to\infty}\left[-\frac{1}{\sqrt{2\pi}}e^{-x^2/2}\right]_0^b$$

$$= \frac{1}{\sqrt{2\pi}}$$

Since $xe^{-x^2/2}$ is an odd function,

$$\frac{1}{\sqrt{2\pi}}\int_{-\infty}^{0} xe^{-x^2/2}\,dx = -\frac{1}{\sqrt{2\pi}}\int_0^{\infty} xe^{-x^2/2}\,dx = -\frac{1}{\sqrt{2\pi}}$$

Thus,

$$\frac{1}{\sqrt{2\pi}}\int_{-\infty}^{\infty} xe^{-x^2/2}\,dx = \frac{1}{\sqrt{2\pi}}\int_{-\infty}^{0} xe^{-x^2/2}\,dx + \frac{1}{\sqrt{2\pi}}\int_0^{\infty} xe^{-x^2/2}\,dx$$

$$= -\frac{1}{\sqrt{2\pi}} + \frac{1}{\sqrt{2\pi}} = 0$$

(b) Since $e^{-x^2/2}$ is an even function and since $\int_{-\infty}^{\infty}\frac{1}{\sqrt{2\pi}}e^{-x^2/2}\,dx = 1$,

$$\frac{1}{\sqrt{2\pi}}\int_0^{\infty} e^{-x^2/2}\,dx = \frac{1}{2}$$

We then apply integration by parts and l'Hôpital's Rule.

$$\frac{1}{\sqrt{2\pi}}\int_0^{\infty} x^2e^{-x^2/2}\,dx = \lim_{b\to\infty}\frac{1}{\sqrt{2\pi}}\int_0^{b}(x)(e^{-x^2/2}x)\,dx$$

$$= \lim_{b\to\infty}\frac{1}{\sqrt{2\pi}}\left(\left[-xe^{-x^2/2}\right]_0^b + \int_0^b e^{-x^2/2}\,dx\right)$$

$$= \frac{1}{\sqrt{2\pi}}\left(0 + \int_0^{\infty} e^{-x^2/2}\,dx\right) = \frac{1}{2}$$

Since $x^2e^{-x^2/2}$ is an even function, we get the same contribution to the left of zero, and so

$$\frac{1}{\sqrt{2\pi}}\int_{-\infty}^{\infty} x^2e^{-x^2/2}\,dx = \frac{1}{2} + \frac{1}{2} = 1$$

■

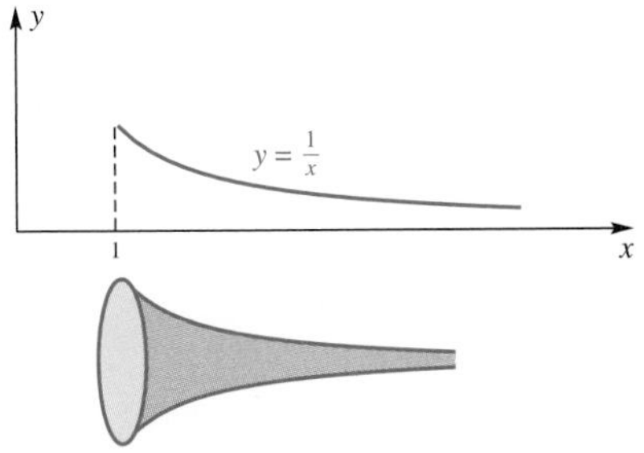

Figure 6

The Paradox of Gabriel's Horn Let the curve $y = 1/x$ on $[1, \infty)$ be revolved about the x-axis, thereby generating a surface called Gabriel's horn (Figure 6). We claim that

1. the volume V of this horn is finite;
2. the surface area A of the horn is infinite.

To put the results in practical terms, they seem to say that the horn can be filled with a finite amount of paint, and yet there is not enough to paint its inside surface. Before we try to unravel this paradox, let us establish (1) and (2). We use results for volume from Section 5.2 and for surface area from Section 5.4.

Hans Memling (1425/40–1494). Last Judgement, detail of right panel: angel blowing a trumpet and the damned falling into Hell. Pomorskie Museum, Gdansk, Poland. Scala/Art Resource, N.Y.

$$V = \int_1^\infty \pi\left(\frac{1}{x}\right)^2 dx = \lim_{b\to\infty} \pi \int_1^b x^{-2}\, dx$$

$$= \lim_{b\to\infty}\left[-\frac{\pi}{x}\right]_1^b = \pi$$

$$A = \int_1^\infty 2\pi y\, ds = \int_1^\infty 2\pi y \sqrt{1 + \left(\frac{dy}{dx}\right)^2}\, dx$$

$$= 2\pi \int_1^\infty \frac{1}{x}\sqrt{1 + \left(\frac{-1}{x^2}\right)^2}\, dx$$

$$= \lim_{b\to\infty} 2\pi \int_1^b \frac{\sqrt{x^4+1}}{x^3}\, dx$$

Now,

$$\frac{\sqrt{x^4+1}}{x^3} > \frac{\sqrt{x^4}}{x^3} = \frac{1}{x}$$

Thus,

$$\int_1^b \frac{\sqrt{x^4+1}}{x^3}\, dx > \int_1^b \frac{1}{x}\, dx = \ln b$$

and since $\ln b \to \infty$ as $b \to \infty$, we conclude that A is infinite.

Is something wrong with our mathematics? No. Imagine the horn to be slit along the side, opened up, and laid flat. Given a finite amount of paint, we could not possibly paint this surface with a paint coat of *uniform* thickness. However, we could do it if we allow the paint coat to get thinner and thinner as we move farther and farther from the horn's fat end. And, of course, that is exactly what happens when we fill the unslit horn with π cubic units of paint. (Imaginary paint can be spread to arbitrary thinness.)

This problem involved the study of two integrals of the form $\int_1^\infty 1/x^p\, dx$. For later reference, we now analyze this integral for all values of p.

> **Gabriel Paves a Street**
>
> When told to pave the infinite street $0 \le x < \infty, 0 \le y \le 1$ with pure gold, Gabriel obeyed but made the thickness h of the gold at x satisfy
>
> $$h = e^{-x}$$
>
> How much gold did it take?
>
> $$V = \int_0^\infty e^{-x}\, dx = \lim_{b\to\infty}\int_0^b e^{-x}\, dx$$
>
> $$= \lim_{b\to\infty}[-e^{-x}]_0^b = 1$$
>
> Just 1 cubic unit.

EXAMPLE 7 Show that $\int_1^\infty 1/x^p\, dx$ diverges for $p \le 1$ and converges for $p > 1$.

SOLUTION We showed in our solution to Gabriel's horn that the integral diverges for $p = 1$. If $p \ne 1$,

$$\int_1^\infty \frac{1}{x^p}\, dx = \lim_{b\to\infty}\int_1^b x^{-p}\, dx = \lim_{b\to\infty}\left[\frac{x^{-p+1}}{-p+1}\right]_1^b$$

$$= \lim_{b\to\infty}\left[\frac{1}{1-p}\right]\left[\frac{1}{b^{p-1}} - 1\right] = \begin{cases} \infty & \text{if } p < 1 \\ \dfrac{1}{p-1} & \text{if } p > 1 \end{cases}$$

The conclusion follows. ■

Concepts Review

1. $\int_a^\infty f(x)\, dx$ is said to ____ if $\lim_{b\to\infty}\int_a^b f(x)\, dx$ exists and is finite.

2. $\int_0^\infty \cos x\, dx$ does not converge because ____ does not exist.

3. $\int_{-\infty}^\infty f(x)\, dx$ is said to diverge if either ____ or ____ diverges.

4. $\int_1^\infty (1/x^p)\, dx$ converges if and only if ____.

Problem Set 8.3

In Problems 1–24, evaluate each improper integral or show that it diverges.

1. $\int_{100}^{\infty} e^x \, dx$

2. $\int_{-\infty}^{-5} \frac{dx}{x^4}$

3. $\int_{1}^{\infty} 2xe^{-x^2} \, dx$

4. $\int_{-\infty}^{1} e^{4x} \, dx$

5. $\int_{9}^{\infty} \frac{x \, dx}{\sqrt{1 + x^2}}$

6. $\int_{1}^{\infty} \frac{dx}{\sqrt{\pi x}}$

7. $\int_{1}^{\infty} \frac{dx}{x^{1.00001}}$

8. $\int_{10}^{\infty} \frac{x}{1 + x^2} \, dx$

9. $\int_{1}^{\infty} \frac{dx}{x^{0.99999}}$

10. $\int_{1}^{\infty} \frac{x}{(1 + x^2)^2} \, dx$

11. $\int_{e}^{\infty} \frac{1}{x \ln x} \, dx$

12. $\int_{e}^{\infty} \frac{\ln x}{x} \, dx$

13. $\int_{2}^{\infty} \frac{\ln x}{x^2} \, dx$

14. $\int_{1}^{\infty} xe^{-x} \, dx$

15. $\int_{-\infty}^{1} \frac{dx}{(2x - 3)^3}$

16. $\int_{4}^{\infty} \frac{dx}{(\pi - x)^{2/3}}$

17. $\int_{-\infty}^{\infty} \frac{x}{\sqrt{x^2 + 9}} \, dx$

18. $\int_{-\infty}^{\infty} \frac{dx}{(x^2 + 16)^2}$

19. $\int_{-\infty}^{\infty} \frac{1}{x^2 + 2x + 10} \, dx$

20. $\int_{-\infty}^{\infty} \frac{x}{e^{2|x|}} \, dx$

21. $\int_{-\infty}^{\infty} \operatorname{sech} x \, dx$ *Hint:* Use a table of integrals or a CAS.

22. $\int_{1}^{\infty} \operatorname{csch} x \, dx$

23. $\int_{0}^{\infty} e^{-x} \cos x \, dx$ *Hint:* Use table of integrals or a CAS.

24. $\int_{0}^{\infty} e^{-x} \sin x \, dx$

25. Find the area of the region under the curve $y = 2/(4x^2 - 1)$ to the right of $x = 1$. *Hint:* Use partial fractions.

26. Find the area of the region under the curve $y = 1/(x^2 + x)$ to the right of $x = 1$.

27. Suppose that Newton's law for the force of gravity had the form $-k/x$ rather than $-k/x^2$ (see Example 3). Show that it would then be impossible to send anything out of the earth's gravitational field.

28. If a 1000-pound capsule weighs only 165 pounds on the moon (radius 1080 miles), how much work is done in propelling this capsule out of the moon's gravitational field (see Example 3)?

29. Suppose that a company expects its annual profits t years from now to be $f(t)$ dollars and that interest is considered to be compounded continuously at an annual rate r. Then the present value of all future profits can be shown to be

$$FP = \int_{0}^{\infty} e^{-rt} f(t) \, dt$$

Find FP if $r = 0.08$ and $f(t) = 100{,}000$.

30. Do Problem 29 assuming that $f(t) = 100{,}000 + 1000t$.

31. A continuous random variable X has a **uniform distribution** if it has a probability density function of the form

$$f(x) = \begin{cases} \dfrac{1}{b - a} & \text{if } a < x < b \\ 0 & \text{if } x \le a \text{ or } x \ge b \end{cases}$$

(a) Show that $\int_{-\infty}^{\infty} f(x) \, dx = 1$.

(b) Find the mean μ and variance σ^2 of the uniform distribution.

(c) If $a = 0$ and $b = 10$, find the probability that X is less than 2.

32. A random variable X has a **Weibull distribution** if it has probability density function

$$f(x) = \begin{cases} \dfrac{\beta}{\theta}\left(\dfrac{x}{\theta}\right)^{\beta-1} e^{-(x/\theta)^{\beta}} & \text{if } x > 0 \\ 0 & \text{if } x \le 0 \end{cases}$$

(a) Show that $\int_{-\infty}^{\infty} f(x) \, dx = 1$. (Assume $\beta > 1$.)

(b) If $\theta = 3$ and $\beta = 2$, find the mean μ and the variance σ^2.

(c) If the lifetime of a computer monitor is a random variable X that has a Weibull distribution with $\theta = 3$ and $\beta = 2$ (where age is measured in years) find the probability that a monitor fails before two years.

33. Sketch the graph of the normal probability density function

$$f(x) = \frac{1}{\sigma\sqrt{2\pi}} e^{-(x-\mu)^2/2\sigma^2}$$

and show, using calculus, that σ is the distance from the mean μ to the x-coordinate of one of the inflection points.

CAS **34.** The Pareto probability density function has the form

$$f(x) = \begin{cases} \dfrac{CM^k}{x^{k+1}} & \text{if } x \ge M, \\ 0 & \text{if } x < M \end{cases}$$

where k and M are positive constants.

(a) Find the value of C that makes $f(x)$ a probability density function.

(b) For the value of C found in part (a), find the value of the mean μ. Is the mean finite for all positive k? If not, how does the mean depend on k?

(c) For the value of C found in part (a), find the variance σ^2. How does the variance depend on k?

35. The Pareto distribution is often used to model income distribution. Suppose that in some economy the income distribution does follow a Pareto distribution with $k = 3$. Suppose that the mean income is \$20,000.

(a) Find M and C.

(b) Find the variance σ^2.

(c) Find the fraction of income earners who earn more than \$100,000. (*Note:* This is the same as asking what is the probability that a randomly chosen person has an income of more than \$100,000.)

36. In electromagnetic theory, the magnetic potential u at a point on the axis of a circular coil is given by

$$u = Ar\int_a^{\infty} \frac{dx}{(r^2 + x^2)^{3/2}}$$

where A, r, and a are constants. Evaluate u.

37. There is a subtlety in the definition of $\int_{-\infty}^{\infty} f(x)\,dx$ that is illustrated by the following: Show that

(a) $\int_{-\infty}^{\infty} \sin x\,dx$ diverges and

(b) $\lim_{a\to\infty} \int_{-a}^{a} \sin x\,dx = 0$.

38. Consider an infinitely long wire coinciding with the positive x-axis and having mass density $\delta(x) = (1 + x^2)^{-1}$, $0 \le x < \infty$.

(a) Calculate the total mass of the wire.

(b) Show that this wire does not have a center of mass.

39. Give an example of a region in the first quadrant that gives a solid of finite volume when revolved about the x-axis, but gives a solid of infinite volume when revolved about the y-axis.

40. Let f be a nonnegative continuous function defined on $0 \le x < \infty$ with $\int_0^{\infty} f(x)\,dx < \infty$. Show that

(a) if $\lim_{x\to\infty} f(x)$ exists it must be 0;

(b) it is possible that $\lim_{x\to\infty} f(x)$ does not exist.

CAS **41.** We can use a computer to approximate $\int_1^{\infty} f(x)\,dx$ by taking b very large in $\int_1^{b} f(x)\,dx$ *provided* we know that the first integral converges. Calculate $\int_1^{100} (1/x^p)\,dx$ for $p = 2, 1.1, 1.01,$ 1, and 0.99. Note that this gives no hint that the integral $\int_1^{\infty} (1/x^p)\,dx$ converges for $p > 1$ and diverges for $p \le 1$.

CAS **42.** Approximate $\int_0^{a} \frac{1}{\pi}(1 + x^2)^{-1}\,dx$ for $a = 10$, 50, and 100.

CAS **43.** Approximate $\int_{-a}^{a} \frac{1}{\sqrt{2\pi}} \exp(-x^2/2)\,dx$ for $a = 1, 2, 3,$ and 4.

Answers to Concepts Review: **1.** converge **2.** $\lim_{b\to\infty} \int_0^{b} \cos x\,dx$ **3.** $\int_{-\infty}^{0} f(x)\,dx$; $\int_0^{\infty} f(x)\,dx$ **4.** $p > 1$

8.4 Improper Integrals: Infinite Integrands

Considering the many complicated integrations that we have done, here is one that looks simple enough but is incorrect.

$$\int_{-2}^{1} \frac{1}{x^2}\,dx = \left[-\frac{1}{x}\right]_{-2}^{1} = -1 - \frac{1}{2} = -\frac{3}{2} \quad \text{Wrong}$$

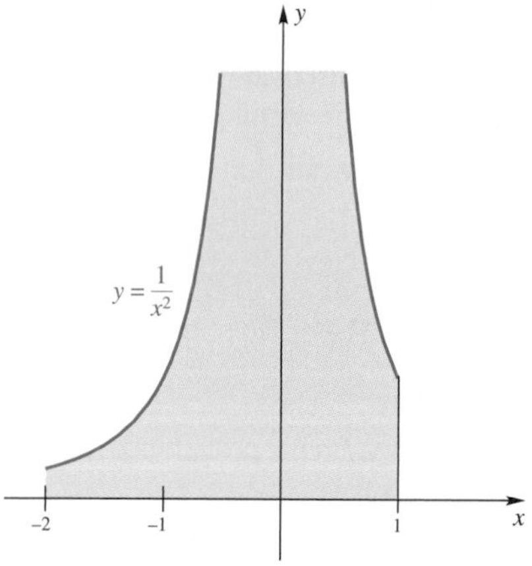

Figure 1

One glance at Figure 1 tells us that something is terribly wrong. The value of the integral (if there is one) has to be a positive number. (Why?)

Where is our mistake? To answer, we refer back to Section 4.2. Recall that for a function to be integrable in the standard (or proper) sense it must be bounded. Our function, $f(x) = 1/x^2$, is not bounded, so it is not integrable in the proper sense. We say that $\int_{-2}^{1} x^{-2}\,dx$ is an improper integral with an infinite integrand (*unbounded integrand* is a more accurate but less colorful term).

Until now, we have carefully avoided infinite integrands in all our examples and problems. We could continue to do this, but this would be to avoid a kind of integral that has important applications. Our task for this section is to define and analyze this new kind of integral.

Integrands That Are Infinite at an End Point We give the definition for the case where f tends to infinity at the right end point of the interval of integration. There is a completely analogous definition for the case where f tends to infinity at the left end point.

Definition

Let f be continuous on the half-open interval $[a, b)$ and suppose that $\lim_{x\to b^-} |f(x)| = \infty$. Then

$$\int_a^{b} f(x)\,dx = \lim_{t\to b^-} \int_a^{t} f(x)\,dx$$

provided that this limit exists and is finite, in which case we say that the integral converges. Otherwise, we say that the integral diverges.

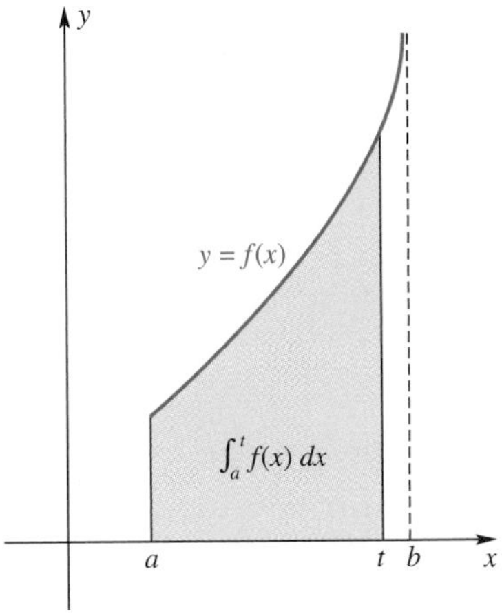

Figure 2

Note the geometric interpretation shown in Figure 2.

EXAMPLE 1 Evaluate, if possible, the improper integral $\int_0^2 \frac{dx}{\sqrt{4-x^2}}$.

SOLUTION Note that the integrand tends to infinity at 2.

$$\int_0^2 \frac{dx}{\sqrt{4-x^2}} = \lim_{t\to 2^-}\int_0^t \frac{dx}{\sqrt{4-x^2}} = \lim_{t\to 2^-}\left[\sin^{-1}\left(\frac{x}{2}\right)\right]_0^t$$

$$= \lim_{t\to 2^-}\left[\sin^{-1}\left(\frac{t}{2}\right) - \sin^{-1}\left(\frac{0}{2}\right)\right] = \frac{\pi}{2}$$ ■

EXAMPLE 2 Evaluate, if possible, $\int_0^{16} \frac{1}{\sqrt[4]{x}}\,dx$.

SOLUTION

$$\int_0^{16} x^{-1/4}\,dx = \lim_{t\to 0^+}\int_t^{16} x^{-1/4}\,dx = \lim_{t\to 0^+}\left[\frac{4}{3}x^{3/4}\right]_t^{16}$$

$$= \lim_{t\to 0^+}\left[\frac{32}{3} - \frac{4}{3}t^{3/4}\right] = \frac{32}{3}$$ ■

Two Key Examples

From Example 7 of Section 8.3, we learned that

$$\int_1^\infty \frac{1}{x^p}\,dx$$

converges if and only if $p > 1$. From Example 4 of the present section, we learn that

$$\int_0^1 \frac{1}{x^p}\,dx$$

converges if and only if $p < 1$. The first has an infinite limit of integration; the second has an infinite integrand. If you feel at home with these two integrals, you should also be at ease with any other improper integrals that you may meet.

EXAMPLE 3 Evaluate, if possible, $\int_0^1 \frac{1}{x}\,dx$.

SOLUTION

$$\int_0^1 \frac{1}{x}\,dx = \lim_{t\to 0^+}\int_t^1 \frac{1}{x}\,dx = \lim_{t\to 0^+}[\ln x]_t^1$$

$$= \lim_{t\to 0^+}[-\ln t] = \infty$$

We conclude that the integral diverges. ■

EXAMPLE 4 Show that $\int_0^1 \frac{1}{x^p}\,dx$ converges if $p < 1$, but diverges if $p \ge 1$.

SOLUTION Example 3 took care of the case $p = 1$. If $p \ne 1$,

$$\int_0^1 \frac{1}{x^p}\,dx = \lim_{t\to 0^+}\int_t^1 x^{-p}\,dx = \lim_{t\to 0^+}\left[\frac{x^{-p+1}}{-p+1}\right]_t^1$$

$$= \lim_{t\to 0^+}\left[\frac{1}{1-p} - \frac{1}{1-p}\cdot\frac{1}{t^{p-1}}\right] = \begin{cases} \dfrac{1}{1-p} & \text{if } p < 1 \\ \infty & \text{if } p > 1 \end{cases}$$ ■

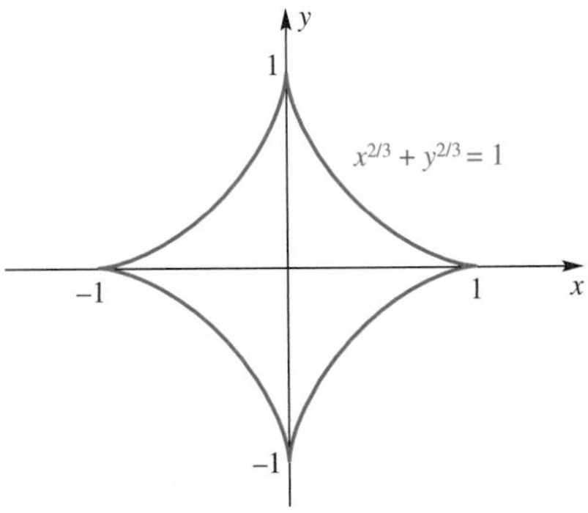

Figure 3

EXAMPLE 5 Sketch the graph of the hypocycloid of four cusps, $x^{2/3} + y^{2/3} = 1$, and find its perimeter.

≈ **SOLUTION** The graph is shown in Figure 3. To find the perimeter, it is enough to find the length L of the first quadrant portion and quadruple it. We estimate L to be a bit more than $\sqrt{2} \approx 1.4$. Its exact value (see Section 5.4) is

$$L = \int_0^1 \sqrt{1 + (y')^2}\,dx$$

By implicit differentiation of $x^{2/3} + y^{2/3} = 1$, we obtain

$$\frac{2}{3}x^{-1/3} + \frac{2}{3}y^{-1/3}y' = 0$$

or

$$y' = -\frac{y^{1/3}}{x^{1/3}}$$

Thus,

$$1 + (y')^2 = 1 + \frac{y^{2/3}}{x^{2/3}} = 1 + \frac{1 - x^{2/3}}{x^{2/3}} = \frac{1}{x^{2/3}}$$

and so

$$L = \int_0^1 \sqrt{1 + (y')^2}\, dx = \int_0^1 \frac{1}{x^{1/3}}\, dx$$

The value of this improper integral can be read from the solution to Example 4; it is $L = 1/\left(1 - \frac{1}{3}\right) = \frac{3}{2}$. We conclude that the hypocycloid has perimeter $4L = 6$. ■

Integrands That Are Infinite at an Interior Point The integral $\int_{-2}^{1} 1/x^2\, dx$ of our introduction has an integrand that tends to infinity at $x = 0$, an interior point of the interval $[-2, 1]$. Here is the appropriate definition to give meaning to such an integral.

Definition

Let f be continuous on $[a, b]$ except at a number c, where $a < c < b$, and suppose that $\lim_{x \to c} |f(x)| = \infty$. Then we define

$$\int_a^b f(x)\, dx = \int_a^c f(x)\, dx + \int_c^b f(x)\, dx$$

provided both integrals on the right converge. Otherwise, we say that $\int_a^b f(x)\, dx$ diverges.

EXAMPLE 6 Show that $\int_{-2}^{1} 1/x^2\, dx$ diverges.

SOLUTION

$$\int_{-2}^{1} \frac{1}{x^2}\, dx = \int_{-2}^{0} \frac{1}{x^2}\, dx + \int_{0}^{1} \frac{1}{x^2}\, dx$$

The second of the integrals on the right diverges by Example 4. This is enough to give the conclusion. ■

EXAMPLE 7 Evaluate, if possible, the improper integral $\int_0^3 \frac{dx}{(x-1)^{2/3}}$.

SOLUTION The integrand tends to infinity at $x = 1$ (see Figure 4). Thus,

$$\begin{aligned}
\int_0^3 \frac{dx}{(x-1)^{2/3}} &= \int_0^1 \frac{dx}{(x-1)^{2/3}} + \int_1^3 \frac{dx}{(x-1)^{2/3}} \\
&= \lim_{t \to 1^-} \int_0^t \frac{dx}{(x-1)^{2/3}} + \lim_{s \to 1^+} \int_s^3 \frac{dx}{(x-1)^{2/3}} \\
&= \lim_{t \to 1^-} \left[3(x-1)^{1/3}\right]_0^t + \lim_{s \to 1^+} \left[3(x-1)^{1/3}\right]_s^3 \\
&= 3 \lim_{t \to 1^-} \left[(t-1)^{1/3} + 1\right] + 3 \lim_{s \to 1^+} \left[2^{1/3} - (s-1)^{1/3}\right] \\
&= 3 + 3(2^{1/3}) \approx 6.78
\end{aligned}$$

■

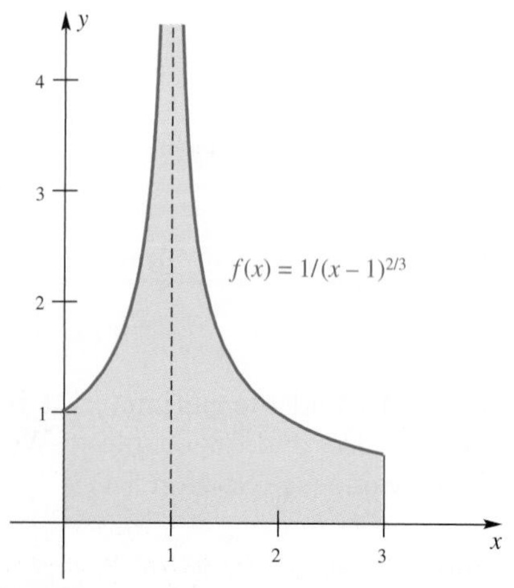

Figure 4

Concepts Review

1. The integral $\int_0^1 (1/\sqrt{x})\,dx$ does not exist in the proper sense because the function $f(x) = 1/\sqrt{x}$ is _____ on the interval $(0, 1]$.

2. Considered as an improper integral, $\int_0^1 (1/\sqrt{x})\,dx = \lim_{a \to 0^+} \int_a^1 x^{-1/2}\,dx =$ _____.

3. The improper integral $\int_0^4 \left(1/\sqrt{4 - x}\right) dx$ is defined by _____.

4. The improper integral $\int_0^1 (1/x^p)\,dx$ converges if and only if _____.

Problem Set 8.4

In Problems 1–32, evaluate each improper integral or show that it diverges.

1. $\int_1^3 \frac{dx}{(x-1)^{1/3}}$

2. $\int_1^3 \frac{dx}{(x-1)^{4/3}}$

3. $\int_3^{10} \frac{dx}{\sqrt{x-3}}$

4. $\int_0^9 \frac{dx}{\sqrt{9-x}}$

5. $\int_0^1 \frac{dx}{\sqrt{1-x^2}}$

6. $\int_{100}^{\infty} \frac{x}{\sqrt{1+x^2}}\,dx$

7. $\int_{-1}^3 \frac{1}{x^3}\,dx$

8. $\int_5^{-5} \frac{1}{x^{2/3}}\,dx$

9. $\int_{-1}^{128} x^{-5/7}\,dx$

10. $\int_0^1 \frac{x}{\sqrt[3]{1-x^2}}\,dx$

11. $\int_0^4 \frac{dx}{(2-3x)^{1/3}}$

12. $\int_{\sqrt{5}}^{\sqrt{8}} \frac{x}{(16-2x^2)^{2/3}}\,dx$

13. $\int_0^{-4} \frac{x}{16-2x^2}\,dx$

14. $\int_0^3 \frac{x}{\sqrt{9-x^2}}\,dx$

15. $\int_{-2}^{-1} \frac{dx}{(x+1)^{4/3}}$

16. $\int_0^3 \frac{dx}{x^2+x-2}$

17. $\int_0^3 \frac{dx}{x^3-x^2-x+1}$

18. $\int_0^{27} \frac{x^{1/3}}{x^{2/3}-9}\,dx$

19. $\int_0^{\pi/4} \tan 2x\,dx$

20. $\int_0^{\pi/2} \csc x\,dx$

21. $\int_0^{\pi/2} \frac{\sin x}{1-\cos x}\,dx$

22. $\int_0^{\pi/2} \frac{\cos x}{\sqrt[3]{\sin x}}\,dx$

23. $\int_0^{\pi/2} \tan^2 x \sec^2 x\,dx$

24. $\int_0^{\pi/4} \frac{\sec^2 x}{(\tan x - 1)^2}\,dx$

25. $\int_0^{\pi} \frac{dx}{\cos x - 1}$

26. $\int_{-3}^{-1} \frac{dx}{x\sqrt{\ln(-x)}}$

27. $\int_0^{\ln 3} \frac{e^x\,dx}{\sqrt{e^x-1}}$

28. $\int_2^4 \frac{dx}{\sqrt{4x-x^2}}$

29. $\int_1^e \frac{dx}{x \ln x}$

30. $\int_1^{10} \frac{dx}{x \ln^{100} x}$

31. $\int_{2c}^{4c} \frac{dx}{\sqrt{x^2-4c^2}}$

32. $\int_c^{2c} \frac{x\,dx}{\sqrt{x^2+xc-2c^2}}, c > 0$

33. It is often possible to change an improper integral into a proper one by using integration by parts. Consider $\lim_{c \to 0^+} \int_c^1 \frac{dx}{\sqrt{x}(1+x)}$. Use integration by parts on the interval $[c, 1]$ where $0 < c < 1$ to show that

$$\int_c^1 \frac{dx}{\sqrt{x}(1+x)} = 1 - \frac{2\sqrt{c}}{c+1} + 2\int_c^1 \frac{\sqrt{x}\,dx}{(1+x)^2}$$

and thus conclude that upon taking the limit as $c \to 0$ an improper integral can be turned into a proper integral.

34. Use integration by parts and the technique of Problem 33 to transform the improper integral $\int_0^1 \frac{dx}{\sqrt{x}(1+x)}$ into a proper integral.

35. If $f(x)$ tends to infinity at both a and b, then we define

$$\int_a^b f(x)\,dx = \int_a^c f(x)\,dx + \int_c^b f(x)\,dx,$$

where c is any point between a and b, provided of course that both the latter integrals converge. Otherwise, we say that the given integral diverges. Use this to evaluate $\int_{-3}^3 \frac{x}{\sqrt{9-x^2}}\,dx$ or show that it diverges.

36. Evaluate $\int_{-3}^3 \frac{x}{9-x^2}\,dx$ or show that it diverges. See Problem 35.

37. Evaluate $\int_{-4}^4 \frac{1}{16-x^2}\,dx$ or show that it diverges. See Problem 35.

38. Evaluate $\int_{-1}^1 \frac{1}{x\sqrt{-\ln|x|}}\,dx$ or show that it diverges.

39. If $\lim_{x \to 0^+} f(x) = \infty$, we define

$$\int_0^{\infty} f(x)\,dx = \lim_{c \to 0^+} \int_c^1 f(x)\,dx + \lim_{b \to \infty} \int_1^b f(x)\,dx$$

provided both limits exist. Otherwise, we say that $\int_0^{\infty} f(x)\,dx$ diverges. Show that $\int_0^{\infty} \frac{1}{x^p}\,dx$ diverges for all p.

40. Suppose that f is continuous on $[0, \infty)$ except at $x = 1$, where $\lim_{x \to 1} |f(x)| = \infty$. How would you define $\int_0^{\infty} f(x)\,dx$?

41. Find the area of the region between the curves $y = (x - 8)^{-2/3}$ and $y = 0$ for $0 \le x < 8$.

42. Find the area of the region between the curves $y = 1/x$ and $y = 1/(x^3 + x)$ for $0 < x \le 1$.

43. Let R be the region in the first quadrant below the curve $y = x^{-2/3}$ and to the left of $x = 1$.

(a) Show that the area of R is finite by finding its value.

(b) Show that the volume of the solid generated by revolving R about the x-axis is infinite.

44. Find b so that $\int_0^b \ln x \, dx = 0$.

45. Is $\int_0^1 \frac{\sin x}{x} dx$ an improper integral? Explain.

EXPL **46. (Comparison Test)** If $0 \le f(x) \le g(x)$ on $[a, \infty)$, it can be shown that the convergence of $\int_a^\infty g(x)\, dx$ implies the convergence of $\int_a^\infty f(x)\, dx$, and the divergence of $\int_a^\infty f(x)\, dx$ implies the divergence of $\int_a^\infty g(x)\, dx$. Use this to show that $\int_1^\infty \frac{1}{x^4(1 + x^4)} dx$ converges.
Hint: On $[1, \infty)$, $1/[x^4(1 + x^4)] \le 1/x^4$.

47. Use the Comparison Test of Problem 46 to show that $\int_1^\infty e^{-x^2} dx$ converges. *Hint:* $e^{-x^2} \le e^{-x}$ on $[1, \infty)$.

48. Use the Comparison Test of Problem 46 to show that $\int_2^\infty \frac{1}{\sqrt{x + 2} - 1} dx$ diverges.

49. Use the Comparison Test of Problem 46 to determine whether $\int_1^\infty \frac{1}{x^2 \ln(x + 1)} dx$ converges or diverges.

50. Formulate a comparison test for improper integrals with infinite integrands.

51. (a) Use Example 2 of Section 8.2 to show that for any positive number n there is a number M such that

$$0 < \frac{x^{n-1}}{e^x} \le \frac{1}{x^2} \text{ for } x \ge M$$

(b) Use part (a) and Problem 46 to show that $\int_1^\infty x^{n-1}e^{-x}\, dx$ converges.

52. Using Problem 50, prove that $\int_0^1 x^{n-1}e^{-x}\, dx$ converges for $n > 0$.

EXPL **53. (Gamma Function)** Let $\Gamma(n) = \int_0^\infty x^{n-1}e^{-x}\, dx$, $n > 0$. This integral converges by Problems 51 and 52. Show each of the following (note that the gamma function is defined for every positive real number n):

(a) $\Gamma(1) = 1$ (b) $\Gamma(n + 1) = n\Gamma(n)$

(c) $\Gamma(n + 1) = n!$, if n is a positive integer.

CAS **54.** Evaluate $\int_0^\infty x^{n-1}e^{-x}\, dx$ for $n = 1, 2, 3, 4,$ and 5, thereby confirming Problem 53(c).

55. The **gamma** probability density function is

$$f(x) = \begin{cases} Cx^{\alpha-1}e^{-\beta x}, & \text{if } x > 0 \\ 0, & \text{if } x \le 0 \end{cases}$$

where α and β are positive constants. (Both the gamma and the Weibull distributions are used to model lifetimes of people, animals, and equipment.)

(a) Find the value of C, depending on both α and β, that makes $f(x)$ a probability density function.

(b) For the value of C found in part (a), find the value of the mean μ.

(c) For the value of C found in part (a), find the variance σ^2.

EXPL **56.** The **Laplace transform,** named after the French mathematician Pierre-Simon de Laplace (1749–1827), of a function $f(x)$ is given by $L\{f(t)\}(s) = \int_0^\infty f(t)e^{-st}\, dt$. Laplace transforms are useful for solving differential equations.

(a) Show that the Laplace transform of t^α is given by $\Gamma(\alpha + 1)/s^{\alpha+1}$ and is defined for $s > 0$.

(b) Show that the Laplace transform of $e^{\alpha t}$ is given by $1/(s - \alpha)$ and is defined for $s > \alpha$.

(c) Show that the Laplace transform of $\sin(\alpha t)$ is given by $\alpha/(s^2 + \alpha^2)$ and is defined for $s > 0$.

57. By interpreting each of the following integrals as an area and then calculating this area by a y-integration, evaluate:

(a) $\int_0^1 \sqrt{\frac{1 - x}{x}}\, dx$ (b) $\int_{-1}^1 \sqrt{\frac{1 + x}{1 - x}}\, dx$

EXPL **58.** Suppose that $0 < p < q$ and $\int_0^\infty \frac{1}{x^p + x^q} dx$ converges. What can you say about p and q?

Answers to Concepts Review: **1.** unbounded **2.** 2 **3.** $\lim_{b \to 4^-} \int_0^b \left(1/\sqrt{4 - x}\right) dx$ **4.** $p < 1$

8.5 Chapter Review

Concepts Test

Respond with true or false to each of the following assertions. Be prepared to justify your answer.

1. $\lim_{x \to \infty} \frac{x^{100}}{e^x} = 0$

2. $\lim_{x \to \infty} \frac{x^{1/10}}{\ln x} = \infty$

3. $\lim_{x \to \infty} \frac{1000x^4 + 1000}{0.001x^4 + 1} = \infty$

4. $\lim_{x \to \infty} xe^{-1/x} = 0$

5. If $\lim_{x \to a} f(x) = \lim_{x \to a} g(x) = \infty$, then $\lim_{x \to a} \frac{f(x)}{g(x)} = 1$.

6. If $\lim_{x \to a} f(x) = 1$ and $\lim_{x \to a} g(x) = \infty$, then $\lim_{x \to a} [f(x)]^{g(x)} = 1$.

7. If $\lim_{x \to a} f(x) = 1$, then $\lim_{n \to \infty} \left\{\lim_{x \to a} [f(x)]^n\right\} = 1$.

8. If $\lim_{x \to a} f(x) = 0$ and $\lim_{x \to a} g(x) = \infty$, then $\lim_{x \to a} [f(x)]^{g(x)} = 0$. (Assume $f(x) \geq 0$ for $x \neq a$.)

9. If $\lim_{x \to a} f(x) = -1$ and $\lim_{x \to a} g(x) = \infty$, then $\lim_{x \to a} [f(x)g(x)] = -\infty$.

10. If $\lim_{x \to a} f(x) = 0$ and $\lim_{x \to a} g(x) = \infty$, then $\lim_{x \to a} [f(x)g(x)] = 0$.

11. If $\lim_{x \to \infty} \dfrac{f(x)}{g(x)} = 3$, then $\lim_{x \to \infty} [f(x) - 3g(x)] = 0$.

12. If $\lim_{x \to a} f(x) = 2$ and $\lim_{x \to a} g(x) = 0$, then $\lim_{x \to a} \dfrac{f(x)}{|g(x)|} = \infty$. (Assume $g(x) \neq 0$ for $x \neq a$.)

13. If $\lim_{x \to \infty} \ln f(x) = 2$, then $\lim_{x \to \infty} f(x) = e^2$.

14. If $f(x) \neq 0$ for $x \neq a$ and $\lim_{x \to a} f(x) = 0$, then $\lim_{x \to a} [1 + f(x)]^{1/f(x)} = e$.

15. If $p(x)$ is a polynomial, then $\lim_{x \to \infty} \dfrac{p(x)}{e^x} = 0$.

16. If $p(x)$ is a polynomial, then $\lim_{x \to 0} \dfrac{p(x)}{e^x} = p(0)$.

17. If $f(x)$ and $g(x)$ are both differentiable and $\lim_{x \to 0} \dfrac{f'(x)}{g'(x)} = L$, then $\lim_{x \to 0} \dfrac{f(x)}{g(x)} = L$.

18. $\displaystyle\int_0^1 \frac{1}{x^{1.001}}\, dx$ converges.

19. $\displaystyle\int_0^\infty \frac{1}{x^p}\, dx$ diverges for all $p > 0$.

20. If f is continuous on $[0, \infty)$ and $\lim_{x \to \infty} f(x) = 0$, then $\displaystyle\int_0^\infty f(x)\, dx$ converges.

21. If f is an even function and $\displaystyle\int_0^\infty f(x)\, dx$ converges, then $\displaystyle\int_{-\infty}^\infty f(x)\, dx$ converges.

22. If $\displaystyle\lim_{b \to \infty} \int_{-b}^{b} f(x)\, dx$ exists and is finite, then $\displaystyle\int_{-\infty}^\infty f(x)\, dx$ converges.

23. If f' is continuous on $[0, \infty)$ and $\lim_{x \to \infty} f(x) = 0$, then $\displaystyle\int_0^\infty f'(x)\, dx$ converges.

24. If $0 \leq f(x) \leq e^{-x}$ on $[0, \infty)$, then $\displaystyle\int_0^\infty f(x)\, dx$ converges.

25. $\displaystyle\int_0^{\pi/4} \frac{\tan x}{x}\, dx$ is an improper integral.

Sample Test Problems

Find each limit in Problems 1–18.

1. $\lim_{x \to 0} \dfrac{4x}{\tan x}$

2. $\lim_{x \to 0} \dfrac{\tan 2x}{\sin 3x}$

3. $\lim_{x \to 0} \dfrac{\sin x - \tan x}{\frac{1}{3}x^2}$

4. $\lim_{x \to 0} \dfrac{\cos x}{x^2}$

5. $\lim_{x \to 0} 2x \cot x$

6. $\lim_{x \to 1^-} \dfrac{\ln(1 - x)}{\cot \pi x}$

7. $\lim_{t \to \infty} \dfrac{\ln t}{t^2}$

8. $\lim_{x \to \infty} \dfrac{2x^3}{\ln x}$

9. $\lim_{x \to 0^+} (\sin x)^{1/x}$

10. $\lim_{x \to 0^+} x \ln x$

11. $\lim_{x \to 0^+} x^x$

12. $\lim_{x \to 0} (1 + \sin x)^{2/x}$

13. $\lim_{x \to 0^+} \sqrt{x} \ln x$

14. $\lim_{t \to \infty} t^{1/t}$

15. $\lim_{x \to 0^+} \left(\dfrac{1}{\sin x} - \dfrac{1}{x} \right)$

16. $\lim_{x \to \pi/2} \dfrac{\tan 3x}{\tan x}$

17. $\lim_{x \to \pi/2} (\sin x)^{\tan x}$

18. $\lim_{x \to \pi/2} \left(x \tan x - \dfrac{\pi}{2} \sec x \right)$

In Problems 19–38, evaluate the given improper integral or show that it diverges.

19. $\displaystyle\int_0^\infty \frac{dx}{(x + 1)^2}$

20. $\displaystyle\int_0^\infty \frac{dx}{1 + x^2}$

21. $\displaystyle\int_{-\infty}^1 e^{2x}\, dx$

22. $\displaystyle\int_{-1}^1 \frac{dx}{1 - x}$

23. $\displaystyle\int_0^\infty \frac{dx}{x + 1}$

24. $\displaystyle\int_{1/2}^2 \frac{dx}{x(\ln x)^{1/5}}$

25. $\displaystyle\int_1^\infty \frac{dx}{x^2 + x^4}$

26. $\displaystyle\int_{-\infty}^1 \frac{dx}{(2 - x)^2}$

27. $\displaystyle\int_{-2}^0 \frac{dx}{2x + 3}$

28. $\displaystyle\int_1^4 \frac{dx}{\sqrt{x - 1}}$

29. $\displaystyle\int_2^\infty \frac{dx}{x(\ln x)^2}$

30. $\displaystyle\int_0^\infty \frac{dx}{e^{x/2}}$

31. $\displaystyle\int_3^5 \frac{dx}{(4 - x)^{2/3}}$

32. $\displaystyle\int_2^\infty xe^{-x^2}\, dx$

33. $\displaystyle\int_{-\infty}^\infty \frac{x}{x^2 + 1}\, dx$

34. $\displaystyle\int_{-\infty}^\infty \frac{x}{1 + x^4}\, dx$

35. $\displaystyle\int_0^\infty \frac{e^x}{e^{2x} + 1}\, dx$

36. $\displaystyle\int_{-\infty}^\infty x^2 e^{-x^3}\, dx$

37. $\displaystyle\int_{-3}^3 \frac{x}{\sqrt{9 - x^2}}\, dx$

38. $\displaystyle\int_{\pi/3}^{\pi/2} \frac{\tan x}{(\ln \cos x)^2}\, dx$

39. For what values of p does the integral $\displaystyle\int_1^\infty \frac{1}{x^p}\, dx$ converge and for what values does it diverge?

40. For what values of p does the integral $\displaystyle\int_0^1 \frac{1}{x^p}\, dx$ converge and for what values does it diverge?

In Problems 41–44, use a comparison test (see Problem 46 of Section 8.4) to decide whether each integral converges or diverges.

41. $\displaystyle\int_1^\infty \frac{dx}{\sqrt{x^6 + x}}$

42. $\displaystyle\int_1^\infty \frac{\ln x}{e^{2x}}\, dx$

43. $\displaystyle\int_3^\infty \frac{\ln x}{x}\, dx$

44. $\displaystyle\int_1^\infty \frac{\ln x}{x^3}\, dx$

REVIEW & PREVIEW PROBLEMS

Recall from Section 0.1 that the converse of an implication $P \Rightarrow Q$ is $Q \Rightarrow P$, and the contrapositive is not $Q \Rightarrow$ not P. In Problems 1–8, give the converse and the contrapositive of the given statement. Which, among the original statement, its converse, and its contrapositive, are alays true?

1. If $x > 0$, then $x^2 > 0$.

2. If $x^2 > 0$, then $x > 0$.

3. If f is differentiable at c, then f is continuous at c.

4. If f is continuous at c, then f is differentiable at c.

5. If f is right continuous at c, then f is continuous at c.

6. If the derivative of f is always zero, then f is a constant function. [Assume that f is differentiable for all x.]

7. If $f(x) = x^2$, then $f'(x) = 2x$.

8. If $a < b$, then $a^2 < b^2$.

In Problems 9–12, evaluate the given sum.

9. $1 + \frac{1}{2} + \frac{1}{4}$

10. $1 + \frac{1}{2} + \frac{1}{4} + \frac{1}{8} + \frac{1}{16} + \frac{1}{32}$

11. $\sum_{i=1}^{4} \frac{1}{i}$

12. $\sum_{k=1}^{4} \frac{(-1)^k}{2^k}$

Evaluate the following limits.

13. $\lim_{x\to\infty} \frac{x}{2x + 1}$

14. $\lim_{n\to\infty} \frac{n^2}{2n^2 + 1}$

15. $\lim_{x\to\infty} \frac{x^2}{e^x}$

16. $\lim_{n\to\infty} \frac{n^2}{e^n}$

Which of the improper integrals converge?

17. $\int_1^\infty \frac{1}{x}\, dx$

18. $\int_1^\infty \frac{1}{x^2}\, dx$

19. $\int_1^\infty \frac{1}{x^{1.001}}\, dx$

20. $\int_1^\infty \frac{x}{x^2 + 1}\, dx$

21. $\int_2^\infty \frac{1}{x \ln x}\, dx$

22. $\int_2^\infty \frac{1}{x(\ln x)^2}\, dx$

CHAPTER 9

Infinite Series

9.1 Infinite Sequences
9.2 Infinite Series
9.3 Positive Series: The Integral Test
9.4 Positive Series: Other Tests
9.5 Alternating Series, Absolute Convergence, and Conditional Convergence
9.6 Power Series
9.7 Operations on Power Series
9.8 Taylor and Maclaurin Series
9.9 The Taylor Approximation to a Function

Someone is sure to argue that there are many different sequences that begin

$$1, 4, 7, 10, 13$$

and we agree. For example, the formula

$$3n - 2 + (n - 1)\cdot(n - 2)\cdots(n - 5)$$

generates those five numbers. Who but an expert would think of this formula? When we ask you to look for a pattern, we mean a simple and obvious pattern.

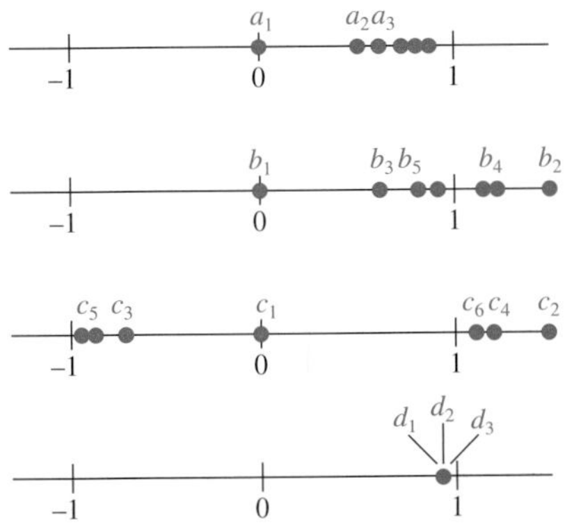

Figure 1

9.1 Infinite Sequences

In simple language, a sequence

$$a_1, a_2, a_3, a_4, \ldots$$

is an ordered arrangement of real numbers, one for each positive integer. More formally, an **infinite sequence** is a function whose domain is the set of positive integers and whose range is a set of real numbers. We may denote a sequence by $a_1, a_2, a_3, \ldots$, by $\{a_n\}_{n=1}^{\infty}$, or simply by $\{a_n\}$. Occasionally, we will extend the notion slightly by allowing the domain to consist of all integers greater than or equal to a specified integer, as in $b_0, b_1, b_2, \ldots$ and $c_8, c_9, c_{10}, \ldots$, which are also denoted by $\{b_n\}_{n=0}^{\infty}$ and $\{c_n\}_{n=8}^{\infty}$, respectively.

A sequence may be specified by giving enough initial terms to establish a pattern, as in

$$1, 4, 7, 10, 13, \ldots$$

by an **explicit formula** for the nth term, as in

$$a_n = 3n - 2, \qquad n \geq 1$$

or by a **recursion formula**

$$a_1 = 1, \qquad a_n = a_{n-1} + 3, \qquad n \geq 2$$

Note that each of our three illustrations describes the same sequence. Here are four more explicit formulas and the first few terms of the sequences that they generate.

(1) $a_n = 1 - \dfrac{1}{n}$, $\quad n \geq 1$: $\quad 0, \dfrac{1}{2}, \dfrac{2}{3}, \dfrac{3}{4}, \dfrac{4}{5}, \ldots$

(2) $b_n = 1 + (-1)^n \dfrac{1}{n}$, $\quad n \geq 1$: $\quad 0, \dfrac{3}{2}, \dfrac{2}{3}, \dfrac{5}{4}, \dfrac{4}{5}, \dfrac{7}{6}, \dfrac{6}{7}, \ldots$

(3) $c_n = (-1)^n + \dfrac{1}{n}$, $\quad n \geq 1$: $\quad 0, \dfrac{3}{2}, -\dfrac{2}{3}, \dfrac{5}{4}, -\dfrac{4}{5}, \dfrac{7}{6}, -\dfrac{6}{7}, \ldots$

(4) $d_n = 0.999$, $\quad n \geq 1$: $\quad 0.999, 0.999, 0.999, 0.999, \ldots$

Convergence Consider the four sequences just defined. Each has values that pile up near 1 (see the diagrams in Figure 1). But do they all *converge* to 1? The correct response is that sequences $\{a_n\}$ and $\{b_n\}$ converge to 1, but $\{c_n\}$ and $\{d_n\}$ do not.

For a sequence to converge to 1 means first that values of the sequence should get close to 1. But they must do more than get close; they must *remain* close for all n beyond a certain value. This rules out sequence $\{c_n\}$. And close means arbitrarily close, that is, within *any* specified nonzero distance from 1, which rules out sequence $\{d_n\}$. While sequence $\{d_n\}$ does not converge to 1, it is correct to say that it converges to 0.999. Sequence $\{c_n\}$ does not converge at all; we say it diverges.

Here is the formal definition, which we first introduced in Section 1.5.

Definition

The sequence $\{a_n\}$ is said to **converge** to L, and we write

$$\lim_{n\to\infty} a_n = L$$

if for each positive number ε there is a corresponding positive number N such that

$$n \geq N \Rightarrow |a_n - L| < \varepsilon$$

A sequence that fails to converge to any finite number L is said to **diverge,** or to be divergent.

To see a relationship with limits at infinity (Section 1.5), consider graphing $a_n = 1 - 1/n$ and $a(x) = 1 - 1/x$. The only difference is that in the sequence case the domain is restricted to the positive integers. In the first case, we write $\lim_{n\to\infty} a_n = 1$; in the second, $\lim_{x\to\infty} a(x) = 1$. Note the interpretations of ε and N in the diagrams in Figure 2.

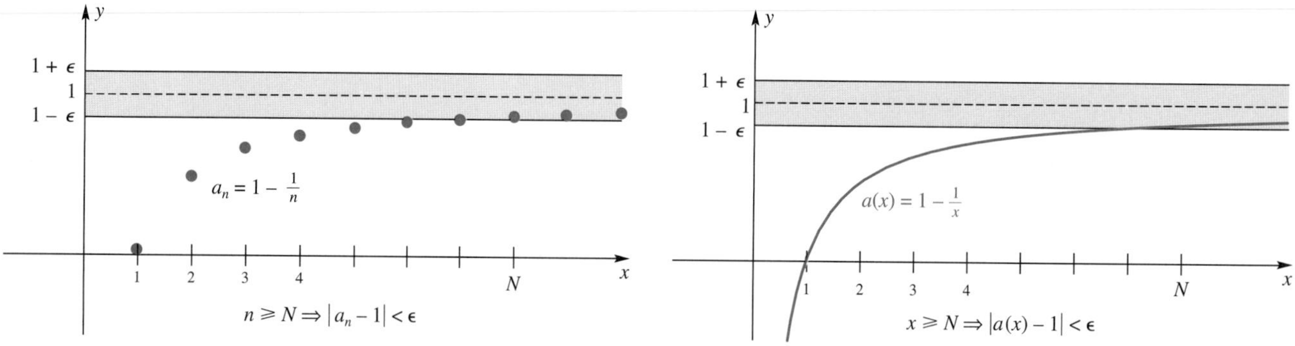

Figure 2

EXAMPLE 1 Show that if p is a positive integer, then

$$\lim_{n\to\infty} \frac{1}{n^p} = 0$$

SOLUTION This is almost obvious from earlier work, but we can give a formal demonstration. Let an arbitrary $\varepsilon > 0$ be given. Choose N to be any number greater than $\sqrt[p]{1/\varepsilon}$. Then $n \geq N$ implies that

$$|a_n - L| = \left|\frac{1}{n^p} - 0\right| = \frac{1}{n^p} \leq \frac{1}{N^p} < \frac{1}{\left(\sqrt[p]{1/\varepsilon}\right)^p} = \varepsilon \quad \blacksquare$$

All the familiar limit theorems hold for convergent sequences. We state them without proof.

Theorem A Properties of Limits of Sequences

Let $\{a_n\}$ and $\{b_n\}$ be convergent sequences and k a constant. Then

(i) $\lim_{n\to\infty} k = k$;

(ii) $\lim_{n\to\infty} ka_n = k \lim_{n\to\infty} a_n$;

(iii) $\lim_{n\to\infty} (a_n \pm b_n) = \lim_{n\to\infty} a_n \pm \lim_{n\to\infty} b_n$;

(iv) $\lim_{n\to\infty} (a_n \cdot b_n) = \lim_{n\to\infty} a_n \cdot \lim_{n\to\infty} b_n$;

(v) $\lim_{n\to\infty} \dfrac{a_n}{b_n} = \dfrac{\lim_{n\to\infty} a_n}{\lim_{n\to\infty} b_n}$, provided that $\lim_{n\to\infty} b_n \neq 0$.

EXAMPLE 2 Find $\lim_{n\to\infty} \frac{3n^2}{7n^2 + 1}$.

SOLUTION To decide what is happening to a quotient of two polynomials in n as n gets large, it is wise to divide the numerator and denominator by the largest power of n that occurs in the denominator. This justifies our first step below; the others are justified by appealing to statements from Theorem A as indicated by the circled numbers.

$$\begin{aligned}
\lim_{n\to\infty} \frac{3n^2}{n^2+1} &= \lim_{n\to\infty} \frac{3}{7+(1/n^2)} \\
&\overset{(5)}{=} \frac{\lim_{n\to\infty} 3}{\lim_{n\to\infty} [7+(1/n^2)]} \\
&\overset{(3)}{=} \frac{\lim_{n\to\infty} 3}{\lim_{n\to\infty} 7 + \lim_{n\to\infty} 1/n^2} \\
&\overset{(1)}{=} \frac{3}{7 + \lim_{n\to\infty} 1/n^2} = \frac{3}{7+0} = \frac{3}{7}
\end{aligned}$$

By this time, the limit theorems are so familiar that we will normally jump directly from the first step to the final result. ■

EXAMPLE 3 Does the sequence $\{(\ln n)/e^n\}$ converge and, if so, to what number?

SOLUTION Here, and in many sequence problems, it is convenient to use the following almost obvious fact (see Figure 2).

> If $\lim_{x\to\infty} f(x) = L$, then $\lim_{n\to\infty} f(n) = L$.

This is convenient because we can apply l'Hôpital's Rule to the continuous variable problem. In particular, by l'Hôpital's Rule,

$$\lim_{x\to\infty} \frac{\ln x}{e^x} = \lim_{x\to\infty} \frac{1/x}{e^x} = 0$$

Thus,

$$\lim_{n\to\infty} \frac{\ln n}{e^n} = 0$$

That is, $\{(\ln n)/e^n\}$ converges to 0. ■

> **As Expected**
>
> In Chapter 8, we showed using L'Hôpital's Rule that e^x grows *faster* than any power of x and $\ln x$ grows *slower* than any power of x. Thus we would expect that $\ln x/e^x$ and $\ln n/e^n$ both go to 0.

Here is another theorem that we have seen before in a slightly different guise (Theorem 1.3D).

Theorem B Squeeze Theorem

Suppose that $\{a_n\}$ and $\{c_n\}$ both converge to L and that $a_n \le b_n \le c_n$ for $n \ge K$ (K a fixed integer). Then $\{b_n\}$ also converges to L.

EXAMPLE 4 Show that $\lim_{n\to\infty} \frac{\sin^3 n}{n} = 0$.

SOLUTION For $n \ge 1$, $-1/n \le (\sin^3 n)/n \le 1/n$. Since $\lim_{n\to\infty}(-1/n) = 0$ and $\lim_{n\to\infty}(1/n) = 0$, the result follows by the Squeeze Theorem. ■

For sequences of variable sign, it is helpful to have the following result.

Theorem C

If $\lim_{n\to\infty} |a_n| = 0$, then $\lim_{n\to\infty} a_n = 0$.

Proof Since $-|a_n| \le a_n \le |a_n|$, the result follows from the Squeeze Theorem. ■

What happens to the numbers in the sequence $\{0.999^n\}$ as $n \to \infty$? We suggest that you calculate 0.999^n for $n = 10, 100, 1000$, and $10{,}000$ on your calculator to make a good guess. Then note the following example.

EXAMPLE 5 Show that if $-1 < r < 1$ then $\lim_{n\to\infty} r^n = 0$.

SOLUTION If $r = 0$, the result is trivial, so suppose otherwise. Then $1/|r| > 1$, and so $1/|r| = 1 + p$ for some number $p > 0$. By the Binomial Formula,

$$\frac{1}{|r|^n} = (1 + p)^n = 1 + pn + (\text{positive terms}) \ge pn$$

Thus,

$$0 \le |r|^n \le \frac{1}{pn}$$

Since $\lim_{n\to\infty} (1/pn) = (1/p) \lim_{n\to\infty} (1/n) = 0$, it follows from the Squeeze Theorem that $\lim_{n\to\infty} |r|^n = 0$ or, equivalently, $\lim_{n\to\infty} |r^n| = 0$. By Theorem C, $\lim_{n\to\infty} r^n = 0$. ■

What if $r > 1$; for example, $r = 1.5$? Then r^n will march off toward ∞. In this case, we write

$$\lim_{n\to\infty} r^n = \infty, \qquad r > 1$$

However, we say that the sequence $\{r^n\}$ diverges. To converge, a sequence must approach a *finite* limit. The sequence $\{r^n\}$ also diverges when $r \le -1$.

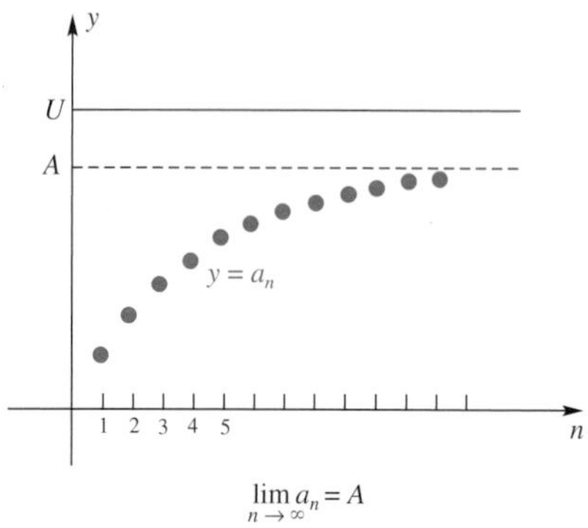

Figure 3

Monotonic Sequences

Consider now an arbitrary **nondecreasing sequence** $\{a_n\}$, by which we mean $a_n \le a_{n+1}$, $n \ge 1$. One example is the sequence $a_n = n^2$; another is $a_n = 1 - 1/n$. If you think about it a little, you may convince yourself that such a sequence can do one of only two things. Either it marches off to infinity or, if it cannot do that because it is bounded above, then it must approach a lid (see Figure 3). Here is the formal statement of this very important result.

Theorem D Monotonic Sequence Theorem

If U is an upper bound for a nondecreasing sequence $\{a_n\}$, then the sequence converges to a limit A that is less than or equal to U. Similarly, if L is a lower bound for a nonincreasing sequence $\{b_n\}$, then the sequence $\{b_n\}$ converges to a limit B that is greater than or equal to L.

The expression **monotonic sequence** is used to describe either a nondecreasing or nonincreasing sequence; hence the name for this theorem.

Theorem D describes a very deep property of the real number system. It is equivalent to the *completeness property* of the real numbers, which in simple language says that the real line has no "holes" in it (see Problems 47 and 48). It is this property that distinguishes the real number line from the rational number line (which is full of holes). A great deal more could be said about this topic; we hope Theorem D appeals to your intuition and that you will accept it on faith until you take a more advanced course.

58. $\lim_{n\to\infty}\left(\frac{2+n^2}{3+n^2}\right)^n$

59. $\lim_{n\to\infty}\left(\frac{2+n^2}{3+n^2}\right)^{n^2}$

Answers to Concepts Review: **1.** a sequence **2.** $\lim_{n\to\infty} a_n$ exists (finite sense) **3.** bounded above **4.** $-1; 1$

9.2 Infinite Series

When you ask a computer or a calculator for the sine of an angle, or e to some power, it is using an algorithm to make this approximation. Many such algorithms are based on **infinite series,** the topic of this and subsequent sections in this chapter. We will see in Section 9.8 that

$$\sin x = x - \frac{x^3}{3!} + \frac{x^5}{5!} - \frac{x^7}{7!} + \cdots$$

$$e^x = 1 + x + \frac{x^2}{2!} + \frac{x^3}{3!} + \frac{x^4}{4!} + \cdots$$

The topic of the last section of this chapter is the use of such series (actually, using just a finite number of terms in the series) to approximate a given function to a desired accuracy. But there is some work to do before reaching that point. We must give meaning to an "infinite" series like one of those above. We begin with the issue of an infinite series of numbers, as opposed to an infinite series of powers of x, called a power series (which we will address in Section 9.6). Consider for illustration the series

$$\frac{1}{2} + \frac{1}{4} + \frac{1}{8} + \cdots$$

If we include just the first term, we have a "sum" of $\frac{1}{2}$. If we include the first two terms, we have a sum of $\frac{1}{2} + \frac{1}{4} = \frac{3}{4}$; if we include the first three terms, we have a sum of $\frac{1}{2} + \frac{1}{4} + \frac{1}{8} = \frac{7}{8}$; and so on. This is the idea of a **partial sum,** that is the sum of a finite number of terms at the beginning of the series. We denote the nth partial sum, that is, the sum of the first n terms, by S_n. For this series the partial sums are

$$S_1 = \frac{1}{2}$$

$$S_2 = \frac{1}{2} + \frac{1}{4} = \frac{3}{4}$$

$$S_3 = \frac{1}{2} + \frac{1}{4} + \frac{1}{8} = \frac{7}{8}$$

$$\vdots$$

$$S_n = \frac{1}{2} + \frac{1}{4} + \frac{1}{8} + \cdots + \frac{1}{2^n} = 1 - \frac{1}{2^n}$$

Clearly, these partial sums get increasingly close to 1. In fact,

$$\lim_{n\to\infty} S_n = \lim_{n\to\infty}\left(1 - \frac{1}{2^n}\right) = 1$$

The infinite sum is then defined to be the limit of the partial sum S_n.

More generally, consider the **infinite series**

$$a_1 + a_2 + a_3 + a_4 + \cdots$$

which is also denoted by $\sum_{k=1}^{\infty} a_k$ or $\sum a_k$. Then S_n, the ***n*th partial sum,** is given by

$$S_n = a_1 + a_2 + a_3 + \cdots + a_n = \sum_{k=1}^{n} a_k$$

We make the following formal definition.

Definition

The infinite series $\sum_{k=1}^{\infty} a_k$ **converges** and has **sum** S if the sequence of partial sums $\{S_n\}$ converges to S. If $\{S_n\}$ diverges, then the series **diverges**. A divergent series has no sum.

Geometric Series A series of the form

$$\sum_{k=1}^{\infty} ar^{k-1} = a + ar + ar^2 + ar^3 + \cdots$$

where $a \neq 0$, is called a **geometric series.**

EXAMPLE 1 Show that a geometric series converges, and has sum $S = a/(1 - r)$ if $|r| < 1$, but diverges if $|r| \geq 1$.

SOLUTION Let $S_n = a + ar + ar^2 + \cdots + ar^{n-1}$. If $r = 1$, $S_n = na$, which grows without bound, and so $\{S_n\}$ diverges. If $r \neq 1$, we may write

$$S_n - rS_n = (a + ar + \cdots + ar^{n-1}) - (ar + ar^2 + \cdots + ar^n) = a - ar^n$$

and so

$$S_n = \frac{a - ar^n}{1 - r} = \frac{a}{1 - r} - \frac{a}{1 - r}r^n$$

If $|r| < 1$, then $\lim_{n\to\infty} r^n = 0$ (Section 9.1, Example 5), and thus

$$S = \lim_{n\to\infty} S_n = \frac{a}{1 - r}$$

If $|r| > 1$ or $r = -1$, the sequence $\{r^n\}$ diverges, and consequently so does $\{S_n\}$. ■

EXAMPLE 2 Use the result of Example 1 to find the sum of the following two geometric series.

(a) $\frac{4}{3} + \frac{4}{9} + \frac{4}{27} + \frac{4}{81} + \cdots$

(b) $0.515151\ldots = \frac{51}{100} + \frac{51}{10{,}000} + \frac{51}{1{,}000{,}000} + \cdots$

SOLUTION

(a) $S = \frac{a}{1 - r} = \frac{\frac{4}{3}}{1 - \frac{1}{3}} = \frac{\frac{4}{3}}{\frac{2}{3}} = 2$ (b) $S = \frac{\frac{51}{100}}{1 - \frac{1}{100}} = \frac{\frac{51}{100}}{\frac{99}{100}} = \frac{51}{99} = \frac{17}{33}$

Incidently, the procedure in part (b) suggests how to show that any repeating decimal represents a rational number. ■

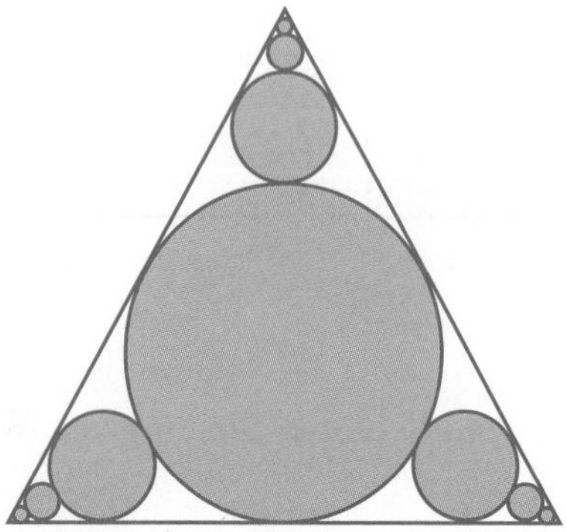

Figure 1

EXAMPLE 3 The diagram in Figure 1 represents an equilateral triangle containing infinitely many circles, tangent to the triangle and to neighboring circles, and reaching into the corners. What fraction of the area of the triangle is occupied by the circles?

SOLUTION Suppose for convenience that the large circle has radius 1, which makes the triangle have sides of length $2\sqrt{3}$. Concentrate attention on the vertical stack of circles. With a bit of geometric reasoning (the center of the large circle is two-thirds of the way from the upper vertex to the base), we see that the radii of these circles are $1, \frac{1}{3}, \frac{1}{9}, \ldots$ and conclude that the vertical stack has area

$$\pi\left[1^2 + \left(\frac{1}{3}\right)^2 + \left(\frac{1}{9}\right)^2 + \left(\frac{1}{27}\right)^2 + \cdots\right]$$
$$= \pi\left[1 + \frac{1}{9} + \frac{1}{81} + \frac{1}{729} + \cdots\right] = \pi\left[\frac{1}{1 - \frac{1}{9}}\right] = \frac{9\pi}{8}$$

The total area of all the circles is three times this number minus twice the area of the big circle, that is, $27\pi/8 - 2\pi$, or $11\pi/8$. Since the triangle has area $3\sqrt{3}$, the fraction of this area occupied by the circles is

$$\frac{11\pi}{24\sqrt{3}} \approx 0.83$$

■

EXAMPLE 4 Suppose that Peter and Paul take turns tossing a fair coin until one of them tosses a head. If Peter goes first, what is the probability that he wins?

SOLUTION Peter can win by tossing a head on the first toss, which happens with probability $\frac{1}{2}$. Or he can win if these three events occur in succession: Peter tosses a tail, Paul tosses a tail, and Peter tosses a head. Each of these events has probability $\frac{1}{2}$, so they all occur with probability $\frac{1}{2} \times \frac{1}{2} \times \frac{1}{2} = \frac{1}{8}$. Another way Peter can win is for the first four tosses to be all tails, while Peter's third toss (the fifth overall) is a head. This occurs with probability $\frac{1}{2} \times \frac{1}{2} \times \frac{1}{2} \times \frac{1}{2} \times \frac{1}{2} = \frac{1}{32}$. This process continues, so that the probability that Peter wins is the sum of the geometric series

$$\frac{1}{2} + \frac{1}{8} + \frac{1}{32} + \frac{1}{128} + \cdots = \frac{1/2}{1 - 1/4} = \frac{2}{3}$$

Paul therefore wins with probability $1 - \frac{2}{3} = \frac{1}{3}$. Peter has the greater chance of winning because he goes first. ■

A General Test for Divergence Consider the geometric series $a + ar + ar^2 + \cdots + ar^{n-1} + \cdots$ once more. Its nth term a_n is given by $a_n = ar^{n-1}$. Example 1 shows that a geometric series converges *if and only if* $\lim_{n\to\infty} a_n = 0$.

Could this possibly be true of all series? The answer is no, although half of the statement (the "only-if" half) is correct. This leads to an important divergence test for series.

> **Logic**
>
> Consider these two statements:
>
> 1. If $\sum_{k=1}^{\infty} a_n$ converges, then $\lim_{n\to\infty} a_n = 0$.
> 2. If $\lim_{n\to\infty} a_n = 0$, then $\sum_{n=1}^{\infty} a_n$ converges.
>
> The first statement is true for any sequence $\{a_n\}$; the second is not. This provides another example of a true statement (the first) whose *converse* is false.
>
> Recall that the *contrapositive* of a statement is true whenever the statement is true. The contrapositive of the first statement is
>
> 3. If $\lim_{n\to\infty} a_n \neq 0$, then $\sum_{n=1}^{\infty} a_n$ diverges.

Theorem A ***n*th-Term Test for Divergence**

If the series $\sum_{n=1}^{\infty} a_n$ converges, then $\lim_{n\to\infty} a_n = 0$. Equivalently, if $\lim_{n\to\infty} a_n \neq 0$ or if $\lim_{n\to\infty} a_n$ does not exist, then the series diverges.

Proof Let S_n be the nth partial sum and $S = \lim_{n\to\infty} S_n$. Note that $a_n = S_n - S_{n-1}$. Since $\lim_{n\to\infty} S_{n-1} = \lim_{n\to\infty} S_n = S$, it follows that

$$\lim_{n\to\infty} a_n = \lim_{n\to\infty} S_n - \lim_{n\to\infty} S_{n-1} = S - S = 0$$

■

EXAMPLE 5 Show that $\sum_{n=1}^{\infty} \frac{n^3}{3n^3 + 2n^2}$ diverges.

SOLUTION

$$\lim_{n \to \infty} a_n = \lim_{n \to \infty} \frac{n^3}{3n^3 + 2n^2} = \lim_{n \to \infty} \frac{1}{3 + 2/n} = \frac{1}{3}$$

Thus, by the nth-Term Test, the series diverges. ■

The Harmonic Series Students invariably want to turn Theorem A around and make it say that $a_n \to 0$ implies convergence of Σa_n. The **harmonic series**

$$\sum_{n=1}^{\infty} \frac{1}{n} = 1 + \frac{1}{2} + \frac{1}{3} + \cdots + \frac{1}{n} + \cdots$$

shows that this is false. Clearly, $\lim_{n \to \infty} a_n = \lim_{n \to \infty} (1/n) = 0$. However, the series diverges, as we now show.

EXAMPLE 6 Show that the harmonic series diverges.

SOLUTION We show that S_n grows without bound. Imagine n to be large and write

$$\begin{aligned} S_n &= 1 + \frac{1}{2} + \frac{1}{3} + \frac{1}{4} + \frac{1}{5} + \cdots + \frac{1}{n} \\ &= 1 + \frac{1}{2} + \left(\frac{1}{3} + \frac{1}{4}\right) + \left(\frac{1}{5} + \frac{1}{6} + \frac{1}{7} + \frac{1}{8}\right) + \left(\frac{1}{9} + \cdots + \frac{1}{16}\right) + \cdots + \frac{1}{n} \\ &> 1 + \frac{1}{2} + \frac{2}{4} + \frac{4}{8} + \frac{8}{16} + \cdots + \frac{1}{n} \\ &= 1 + \frac{1}{2} + \frac{1}{2} + \frac{1}{2} + \frac{1}{2} + \cdots + \frac{1}{n} \end{aligned}$$

It is clear that by taking n sufficiently large we can introduce as many $\frac{1}{2}$'s into the last expression as we wish. Thus, S_n grows without bound, and so $\{S_n\}$ diverges. Hence, the harmonic series diverges. ■

Collapsing Series A geometric series is one of the few series where we can actually give an explicit formula for S_n; a **collapsing series** is another (see Example 2 of Section 4.1).

EXAMPLE 7 Show that the following series converges and find its sum.

$$\sum_{k=1}^{\infty} \frac{1}{(k+2)(k+3)}$$

SOLUTION Use a partial fraction decomposition to write

$$\frac{1}{(k+2)(k+3)} = \frac{1}{k+2} - \frac{1}{k+3}$$

Then

$$\begin{aligned} S_n &= \sum_{k=1}^{n} \left(\frac{1}{k+2} - \frac{1}{k+3}\right) = \left(\frac{1}{3} - \frac{1}{4}\right) + \left(\frac{1}{4} - \frac{1}{5}\right) + \cdots + \left(\frac{1}{n+2} - \frac{1}{n+3}\right) \\ &= \frac{1}{3} - \frac{1}{n+3} \end{aligned}$$

Therefore,

$$\lim_{n \to \infty} S_n = \frac{1}{3}$$

The series converges and has sum $\frac{1}{3}$. ■

A Note on Terminology

This theorem introduces a subtle shift in terminology. The symbol $\sum_{k=1}^{\infty} a_k$ is now being used both for the infinite series $a_1 + a_2 + \cdots$ and for the sum of this series, which is a number.

Properties of Convergent Series Convergent series behave much like finite sums; what you expect to be true often *is* true.

Theorem B Linearity of Convergent Series

If $\sum_{k=1}^{\infty} a_k$ and $\sum_{k=1}^{\infty} b_k$ both converge, and if c is a constant, then $\sum_{k=1}^{\infty} ca_k$ and $\sum_{k=1}^{\infty} (a_k + b_k)$ also converge, and

(i) $\sum_{k=1}^{\infty} ca_k = c\sum_{k=1}^{\infty} a_k$;

(ii) $\sum_{k=1}^{\infty} (a_k + b_k) = \sum_{k=1}^{\infty} a_k + \sum_{k=1}^{\infty} b_k$.

Proof By hypothesis, $\lim_{n\to\infty} \sum_{k=1}^{n} a_k$ and $\lim_{n\to\infty} \sum_{k=1}^{n} b_k$ both exist. Thus, use the properties of sums with finitely many terms and the properties of limits of sequences.

$$\text{(i)}\quad \sum_{k=1}^{\infty} ca_k = \lim_{n\to\infty} \sum_{k=1}^{n} ca_k = \lim_{n\to\infty} c\sum_{k=1}^{n} a_k$$

$$= c \lim_{n\to\infty} \sum_{k=1}^{n} a_k = c\sum_{k=1}^{\infty} a_k$$

$$\text{(ii)}\quad \sum_{k=1}^{\infty} (a_k + b_k) = \lim_{n\to\infty} \sum_{k=1}^{n} (a_k + b_k) = \lim_{n\to\infty}\left[\sum_{k=1}^{n} a_k + \sum_{k=1}^{n} b_k\right]$$

$$= \lim_{n\to\infty} \sum_{k=1}^{n} a_k + \lim_{n\to\infty} \sum_{k=1}^{n} b_k = \sum_{k=1}^{\infty} a_k + \sum_{k=1}^{\infty} b_k$$ ■

EXAMPLE 8 Calculate $\sum_{k=1}^{\infty}\left[3\left(\frac{1}{8}\right)^k - 5\left(\frac{1}{3}\right)^k\right]$.

SOLUTION By Theorem B and Example 1,

$$\sum_{k=1}^{\infty}\left[3\left(\frac{1}{8}\right)^k - 5\left(\frac{1}{3}\right)^k\right] = 3\sum_{k=1}^{\infty}\left(\frac{1}{8}\right)^k - 5\sum_{k=1}^{\infty}\left(\frac{1}{3}\right)^k$$

$$= 3\frac{\frac{1}{8}}{1-\frac{1}{8}} - 5\frac{\frac{1}{3}}{1-\frac{1}{3}} = \frac{3}{7} - \frac{5}{2} = -\frac{29}{14}$$ ■

Theorem C

If $\sum_{k=1}^{\infty} a_k$ diverges and $c \neq 0$, then $\sum_{k=1}^{\infty} ca_k$ diverges.

We leave the proof of this theorem to you (Problem 41). It implies, for example, that

$$\sum_{k=1}^{\infty} \frac{1}{3k} = \sum_{k=1}^{\infty} \frac{1}{3}\cdot\frac{1}{k}$$

diverges, since we know that the harmonic series diverges.

The associative law of addition allows us to group terms in a *finite* sum in any way that we please. For example,

$$2 + 7 + 3 + 4 + 5 = (2 + 7) + (3 + 4) + 5 = 2 + (7 + 3) + (4 + 5)$$

But sometimes we lose sight of the definition of an *infinite* series as the limit of a sequence of partial sums, and we let our intuition guide us into a paradox. For example, the series

$$1 - 1 + 1 - 1 + \cdots + (-1)^{n+1} + \cdots$$

has partial sums

$$\begin{aligned} S_1 &= 1 \\ S_2 &= 1 - 1 = 0 \\ S_3 &= 1 - 1 + 1 = 1 \\ S_4 &= 1 - 1 + 1 - 1 = 0 \\ &\vdots \end{aligned}$$

The sequence of partial sums, $1, 0, 1, 0, 1, \ldots$, diverges; thus the series $1 - 1 + 1 - 1 + \cdots$ diverges. We might, however, view the series as

$$(1 - 1) + (1 - 1) + \cdots$$

and claim that the sum is 0. Alternatively, we might view the series as

$$1 - (1 - 1) - (1 - 1) - \cdots$$

and claim that the sum is 1. The sum of the series cannot be equal to both 0 and 1. It turns out that grouping of terms in a series is acceptable provided that the series is convergent; in such a case we can group terms in any way that we wish.

Theorem D Grouping Terms in an Infinite Series

The terms of a convergent series can be grouped in any way (provided that the order of the terms is maintained), and the new series will converge with the same sum as the original series.

Proof Let Σa_n be the original convergent series and let $\{S_n\}$ be its sequence of partial sums. If Σb_m is a series formed by grouping the terms of Σa_n and if $\{T_m\}$ is its sequence of partial sums, then each T_m is one of the S_n's. For example, T_4 might be

$$T_4 = a_1 + (a_2 + a_3) + (a_4 + a_5 + a_6) + (a_7 + a_8)$$

in which case $T_4 = S_8$. Thus, $\{T_m\}$ is a "subsequence" of $\{S_n\}$. A moment's thought should convince you that if $S_n \to S$ then $T_m \to S$. ■

Concepts Review

1. An expression of the form $a_1 + a_2 + a_3 + \cdots$ is called ______.

2. A series $a_1 + a_2 + \cdots$ is said to converge if the sequence $\{S_n\}$ converges, where $S_n =$ ______.

3. The geometric series $a + ar + ar^2 + \cdots$ converges if ______; in this case the sum of the series is ______.

4. If $\lim_{n\to\infty} a_n \neq 0$, we can be sure that the series $\sum_{n=1}^{\infty} a_n$ ______.

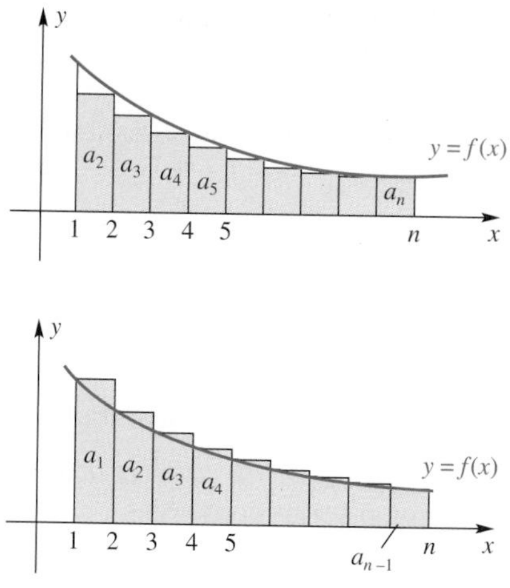

Figure 1

Proof The diagrams in Figure 1 indicate how we may interpret the partial sums of the series Σa_k as areas and thereby relate the series to a corresponding integral. Note that the area of each rectangle is equal to its height, since the width is 1 in each case. From these diagrams, we easily see that

$$\sum_{k=2}^{n} a_k \leq \int_1^n f(x)\,dx \leq \sum_{k=1}^{n-1} a_k$$

Now suppose that $\int_1^\infty f(x)\,dx$ converges. Then, by the left inequality above,

$$S_n = a_1 + \sum_{k=2}^{n} a_k \leq a_1 + \int_1^n f(x)\,dx \leq a_1 + \int_1^\infty f(x)\,dx$$

Therefore, by the Bounded Sum Test, $\sum_{k=1}^{\infty} a_k$ converges.

On the other hand, suppose that $\sum_{k=1}^{\infty} a_k$ converges. Then, by the right inequality above, if $t \leq n$,

$$\int_1^t f(x)\,dx \leq \int_1^n f(x)\,dx \leq \sum_{k=1}^{n-1} a_k \leq \sum_{k=1}^{\infty} a_k$$

Since $\int_1^t f(x)\,dx$ increases with t and is bounded above, $\lim_{t\to\infty} \int_1^t f(x)\,dx$ must exist; that is, $\int_1^\infty f(x)\,dx$ converges. ■

The conclusion to Theorem B is often stated this way. *The series* $\sum_{k=1}^{\infty} f(k)$ *and the improper integral* $\int_1^\infty f(x)\,dx$ *converge or diverge together.* You should see that this is equivalent to our statement.

EXAMPLE 2 **(*p*-Series Test)** The series

$$\sum_{k=1}^{\infty} \frac{1}{k^p} = 1 + \frac{1}{2^p} + \frac{1}{3^p} + \frac{1}{4^p} + \cdots$$

where p is a constant, is called a ***p*-series.** Show each of the following:

(a) The p-series converges if $p > 1$.
(b) The p-series diverges if $p \leq 1$.

SOLUTION If $p \geq 0$, the function $f(x) = 1/x^p$ is continuous, positive, and nonincreasing on $[1, \infty)$ and $f(k) = 1/k^p$. Thus, by the Integral Test, $\Sigma(1/k^p)$ converges if and only if $\lim_{t\to\infty} \int_1^t x^{-p}\,dx$ exists (as a finite number).

If $p \neq 1$,

$$\int_1^t x^{-p}\,dx = \left[\frac{x^{1-p}}{1-p}\right]_1^t = \frac{t^{1-p} - 1}{1 - p}$$

If $p = 1$,

$$\int_1^t x^{-1}\,dx = [\ln x]_1^t = \ln t$$

Since $\lim_{t\to\infty} t^{1-p} = 0$ if $p > 1$ and $\lim_{t\to\infty} t^{1-p} = \infty$ if $p < 1$, and since $\lim_{t\to\infty} \ln t = \infty$, we conclude that the p-series converges if $p > 1$ and diverges if $0 \leq p \leq 1$.

The Tail of a Series

The beginning of a series plays no role in its convergence or divergence. Only the tail is important (the tail really does wag the dog). By the *tail* of a series, we mean

$$a_N + a_{N+1} + a_{N+2} + \cdots$$

where N denotes an arbitrarily large number. Hence, in testing for convergence or divergence of a series, we can ignore the beginning terms or even change them. Clearly, however, the sum of a series does depend on all its terms, including the initial ones.

We still have the case $p < 0$ to consider. In this case, the nth term of $\Sigma(1/k^p)$, that is, $1/n^p$, does not even tend toward 0. Thus, by the nth-Term Test, the series diverges.

Note that the case $p = 1$ gives the harmonic series, which was treated in Section 9.2. Our results here and there are consistent. The harmonic series diverges. ■

EXAMPLE 3 Does $\sum_{k=4}^{\infty} \frac{1}{k^{1.001}}$ converge or diverge?

SOLUTION By the p-Series Test, $\sum_{k=1}^{\infty}(1/k^{1.001})$ converges. *The insertion or removal of a finite number of terms in a series cannot affect its convergence or divergence (though it may affect the sum).* Thus, the given series converges. ■

EXAMPLE 4 Determine whether $\sum_{k=2}^{\infty} \frac{1}{k \ln k}$ converges or diverges.

SOLUTION The hypotheses of the Integral Test are satisfied for $f(x) = 1/(x \ln x)$ on $[2, \infty)$. That the interval is $[2, \infty)$ rather than $[1, \infty)$ is inconsequential, as we noted right after Theorem B. Now,

$$\int_2^{\infty} \frac{1}{x \ln x}\, dx = \lim_{t\to\infty} \int_2^t \frac{1}{\ln x}\left(\frac{1}{x}\, dx\right) = \lim_{t\to\infty} [\ln \ln x]_2^t = \infty$$

Thus, $\Sigma 1/(k \ln k)$ diverges. ■

Approximating the Sum of a Series

So far we have been concerned with whether a series converges or diverges. Except for a few special cases, such as the geometric series, or a collapsing series, we have not addressed the question of what a series converges to, if it converges. This is, in general, a difficult question, but at this point we can use the method suggested by the integral test to *approximate* the sum of a series.

If we use the nth partial sum S_n to approximate the sum of the series

$$S = a_1 + a_2 + a_3 + \cdots$$

then the error we make is

$$E_n = S - S_n = a_{n+1} + a_{n+2} + \cdots$$

Let $f(x)$ be a function with the properties that $a_n = f(n)$, and f is positive, continuous, and nonincreasing on $[1, \infty)$; these are the conditions of Theorem B. Under these conditions

$$E_n = a_{n+1} + a_{n+2} + \cdots < \int_n^{\infty} f(x)\, dx$$

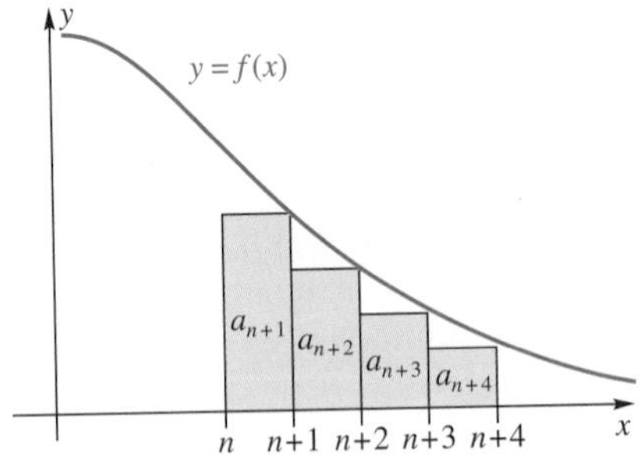

Figure 2

(see Figure 2). We can use this result to find an upper bound on the error involved in using the first n terms to approximate the sum S of the series, and we can use it to determine how large n must be to approximate S to a desired accuracy.

EXAMPLE 5 Find an upper bound for the error in using the sum of the first twenty terms to approximate the sum of the convergent series $S = \sum_{k=1}^{\infty} \frac{1}{k^{3/2}}$.

SOLUTION The obvious choice for $f(x)$ is $f(x) = 1/x^{3/2}$; this function is positive, continuous, and nonincreasing on $[1, \infty)$. The error satisfies

$$E_{20} = \sum_{k=20+1}^{\infty} \frac{1}{k^{3/2}} < \int_{20}^{\infty} \frac{1}{x^{3/2}}\, dx = \lim_{A\to\infty}\left[-2x^{-1/2}\right]_{20}^{A} = \frac{2}{\sqrt{20}} \approx 0.44721$$

Even with twenty terms, the error is somewhat large. ■

$$\sum_{n=1}^{\infty} \frac{1}{n^p} \quad \text{converges if } p > 1, \text{ diverges otherwise}$$

In the first we have found what the series converges to, provided that it converges; in the second, we have not. These series provide standards, or models, against which we can measure other series. Keep in mind that we are still considering series whose terms are positive (or at least nonnegative).

Comparing One Series with Another A series with terms less than the corresponding terms of a convergent series ought to converge; a series with terms greater than the corresponding terms of a divergent series ought to diverge. What ought to be true is true.

Theorem A Ordinary Comparison Test

Suppose that $0 \le a_n \le b_n$ for $n \ge N$.

(i) If Σb_n converges, so does Σa_n.

(ii) If Σa_n diverges, so does Σb_n.

Proof We suppose that $N = 1$; the case $N > 1$ is only slightly harder. To prove (i), let $S_n = a_1 + a_2 + \cdots + a_n$ and note that $\{S_n\}$ is a nondecreasing sequence. If Σb_n converges, for instance, with sum B, then

$$S_n \le b_1 + b_2 + \cdots + b_n \le \sum_{n=1}^{\infty} b_n = B$$

By the Bounded Sum Test (Theorem 9.3A), Σa_n converges.

Property (ii) follows from (i); for if Σb_n converged, then Σa_n would have to converge. ■

EXAMPLE 1 Does $\displaystyle\sum_{n=1}^{\infty} \frac{n}{5n^2 - 4}$ converge or diverge?

SOLUTION A good guess would be that it diverges, since the nth term behaves like $1/5n$ for large n. In fact,

$$\frac{n}{5n^2 - 4} > \frac{n}{5n^2} = \frac{1}{5} \cdot \frac{1}{n}$$

We know that $\displaystyle\sum_{n=1}^{\infty} \frac{1}{5} \cdot \frac{1}{n}$ diverges since it is one-fifth of the harmonic series (Theorem 9.2C). Thus, by the Ordinary Comparison Test, the given series also diverges. ■

EXAMPLE 2 Does $\displaystyle\sum_{n=1}^{\infty} \frac{n}{2^n(n + 1)}$ converge or diverge?

SOLUTION A good guess would be that it converges, since the nth term behaves like $(1/2)^n$ for large n. To substantiate our guess, we note that

$$\frac{n}{2^n(n + 1)} = \left(\frac{1}{2}\right)^n \frac{n}{n + 1} < \left(\frac{1}{2}\right)^n$$

Since $\Sigma\left(\frac{1}{2}\right)^n$ converges (it is a geometric series with $r = \frac{1}{2}$), we conclude that the given series converges. ■

If there is a problem in applying the Ordinary Comparison Test, it is in finding exactly the right known series with which to compare the series to be tested. Suppose that we wish to determine the convergence or divergence of

$$\sum_{n=3}^{\infty} \frac{1}{(n - 2)^2} = \sum_{n=3}^{\infty} \frac{1}{n^2 - 4n + 4}$$

We suspect convergence, so our inclination is to compare $1/(n-2)^2$ with $1/n^2$, but, unfortunately,

$$\frac{1}{(n-2)^2} > \frac{1}{n^2}$$

which gives no test at all (the inequality goes the wrong way for what we want). After some experimenting, we discover that

$$\frac{1}{(n-2)^2} \le \frac{9}{n^2}$$

for $n \ge 3$; since $\Sigma 9/n^2$ converges, so does $\Sigma 1/(n-2)^2$.

Can we avoid these contortions with inequalities? Our intuition tells us that Σa_n and Σb_n converge or diverge together, provided that a_n and b_n are approximately the same size for large n (give or take a multiplicative constant). This is the essential content of our next theorem.

Theorem B Limit Comparison Test

Suppose that $a_n \ge 0$, $b_n > 0$, and

$$\lim_{n\to\infty} \frac{a_n}{b_n} = L$$

If $0 < L < \infty$, then Σa_n and Σb_n converge or diverge together. If $L = 0$ and Σb_n converges, then Σa_n converges.

Proof Begin by taking $\varepsilon = L/2$ in the definition of limit of a sequence (Section 9.1). There is a number N such that $n \ge N \Rightarrow |(a_n/b_n) - L| < L/2$; that is,

$$-\frac{L}{2} < \frac{a_n}{b_n} - L < \frac{L}{2}$$

This inequality is equivalent (by adding L throughout) to

$$\frac{L}{2} < \frac{a_n}{b_n} < \frac{3L}{2}$$

Hence, for $n \ge N$,

$$b_n < \frac{2}{L} a_n \quad \text{and} \quad a_n < \frac{3L}{2} b_n$$

These two inequalities, together with the Ordinary Comparison Test, show that Σa_n and Σb_n converge or diverge together. We leave the proof of the final statement of the theorem to the reader (Problem 37). ■

EXAMPLE 3 Determine the convergence or divergence of each series.

(a) $\displaystyle\sum_{n=1}^{\infty} \frac{3n-2}{n^3 - 2n^2 + 11}$ (b) $\displaystyle\sum_{n=1}^{\infty} \frac{1}{\sqrt{n^2 + 19n}}$

SOLUTION We apply the Limit Comparison Test, but we still must decide to what we should compare the nth term. We see what the nth term is like for large n by looking at the largest-degree terms in the numerator and denominator. In the first case, the nth term is like $3/n^2$; in the second, it is like $1/n$.

(a) $$\lim_{n\to\infty} \frac{a_n}{b_n} = \lim_{n\to\infty} \frac{(3n-2)/(n^3 - 2n^2 + 11)}{3/n^2} = \lim_{n\to\infty} \frac{3n^3 - 2n^2}{3n^3 - 6n^2 + 33} = 1$$

(b) $$\lim_{n\to\infty} \frac{a_n}{b_n} = \lim_{n\to\infty} \frac{1/\sqrt{n^2 + 19n}}{1/n} = \lim_{n\to\infty} \sqrt{\frac{n^2}{n^2 + 19n}} = 1$$

Since $\Sigma 3/n^2$ converges and $\Sigma 1/n$ diverges, we conclude that the series in (a) converges and the series in (b) diverges. ■

EXAMPLE 4 Does $\sum_{n=1}^{\infty} \frac{\ln n}{n^2}$ converge or diverge?

SOLUTION To what shall we compare $(\ln n)/n^2$? If we try $1/n^2$, we get

$$\lim_{n\to\infty} \frac{a_n}{b_n} = \lim_{n\to\infty} \frac{\ln n}{n^2} \div \frac{1}{n^2} = \lim_{n\to\infty} \ln n = \infty$$

The test fails because its conditions are not satisfied. On the other hand, if we use $1/n$, we get

$$\lim_{n\to\infty} \frac{a_n}{b_n} = \lim_{n\to\infty} \frac{\ln n}{n^2} \div \frac{1}{n} = \lim_{n\to\infty} \frac{\ln n}{n} = 0$$

Again, the test fails. Possibly something between $1/n^2$ and $1/n$ will work, such as $1/n^{3/2}$.

$$\lim_{n\to\infty} \frac{a_n}{b_n} = \lim_{n\to\infty} \frac{\ln n}{n^2} \div \frac{1}{n^{3/2}} = \lim_{n\to\infty} \frac{\ln n}{\sqrt{n}} = 0$$

(The last equality follows from l'Hôpital's Rule.) We conclude from the second part of the Limit Comparison Test that $\Sigma(\ln n)/n^2$ converges (since $\Sigma 1/n^{3/2}$ converges by the p-Series Test). ■

Comparing a Series with Itself Getting useful results from the comparison tests requires insight or perseverance. We must choose wisely among known series to find one that is just right for comparison with the series that we wish to test. Wouldn't it be nice if we could somehow compare a series with itself and thereby determine convergence or divergence? Roughly speaking, this is what we do in the Ratio Test.

Theorem C Ratio Test

Let Σa_n be a series of positive terms and suppose that

$$\lim_{n\to\infty} \frac{a_{n+1}}{a_n} = \rho$$

(i) If $\rho < 1$, the series converges.

(ii) If $\rho > 1$ or if $\lim_{n\to\infty} a_{n+1}/a_n = \infty$, the series diverges.

(iii) If $\rho = 1$, the test is inconclusive.

Proof Here is what is behind the Ratio Test. Since $\lim_{n\to\infty} a_{n+1}/a_n = \rho$, $a_{n+1} \approx \rho a_n$; that is, the series behaves like a geometric series with ratio ρ. A geometric series converges when its ratio is less than 1, and diverges when its ratio is greater than 1. Tying down this argument is the task before us.

(i) Since $\rho < 1$, we may choose a number r such that $\rho < r < 1$, for example, $r = (\rho + 1)/2$. Next choose N so large that $n \geq N$ implies that $a_{n+1}/a_n < r$. (This can be done since $\lim_{n\to\infty} a_{n+1}/a_n = \rho < r$.)

Then

$$a_{N+1} < ra_N$$
$$a_{N+2} < ra_{N+1} < r^2 a_N$$
$$a_{N+3} < ra_{N+2} < r^3 a_N$$
$$\vdots$$

Since $ra_N + r^2a_N + r^3a_N + \cdots$ is a geometric series with $0 < r < 1$, it converges. By the Ordinary Comparison Test, $\sum_{n=N+1}^{\infty} a_n$ converges, and hence so does $\sum_{n=1}^{\infty} a_n$.

(ii) Since $\rho > 1$, there is a number N such that $a_{n+1}/a_n > 1$ for all $n \geq N$. Thus,

$$a_{N+1} > a_N$$
$$a_{N+2} > a_{N+1} > a_N$$
$$\vdots$$

Hence, $a_n > a_N > 0$ for all $n > N$, which means that $\lim_{n\to\infty} a_n$ cannot be zero. By the nth-Term Test for Divergence, Σa_n diverges.

(iii) We know that $\Sigma 1/n$ diverges, whereas $\Sigma 1/n^2$ converges. For the first series,

$$\lim_{n\to\infty} \frac{a_{n+1}}{a_n} = \lim_{n\to\infty} \frac{1}{n+1} \div \frac{1}{n} = \lim_{n\to\infty} \frac{n}{n+1} = 1$$

For the second series,

$$\lim_{n\to\infty} \frac{a_{n+1}}{a_n} = \lim_{n\to\infty} \frac{1}{(n+1)^2} \div \frac{1}{n^2} = \lim_{n\to\infty} \frac{n^2}{(n+1)^2} = 1$$

Thus, the Ratio Test does not distinguish between convergence and divergence when $\rho = 1$. ■

The Ratio Test will always be inconclusive for any series whose nth term is a rational expression in n, since in this case $\rho = 1$ (the cases $a_n = 1/n$ and $a_n = 1/n^2$ were treated above). However, for a series whose nth term involves $n!$ or r^n, the Ratio Test usually works beautifully.

EXAMPLE 5 Test for convergence or divergence: $\sum_{n=1}^{\infty} \frac{2^n}{n!}$.

SOLUTION

$$\rho = \lim_{n\to\infty} \frac{a_{n+1}}{a_n} = \lim_{n\to\infty} \frac{2^{n+1}}{(n+1)!} \frac{n!}{2^n} = \lim_{n\to\infty} \frac{2}{n+1} = 0$$

We conclude by the Ratio Test that the series converges. ■

EXAMPLE 6 Test for convergence or divergence: $\sum_{n=1}^{\infty} \frac{2^n}{n^{20}}$.

SOLUTION

$$\rho = \lim_{n\to\infty} \frac{a_{n+1}}{a_n} = \lim_{n\to\infty} \frac{2^{n+1}}{(n+1)^{20}} \frac{n^{20}}{2^n}$$
$$= \lim_{n\to\infty} \left(\frac{n}{n+1}\right)^{20} \cdot 2 = 2$$

We conclude that the given series diverges. ■

EXAMPLE 7 Test for convergence or divergence: $\sum_{n=1}^{\infty} \frac{n!}{n^n}$.

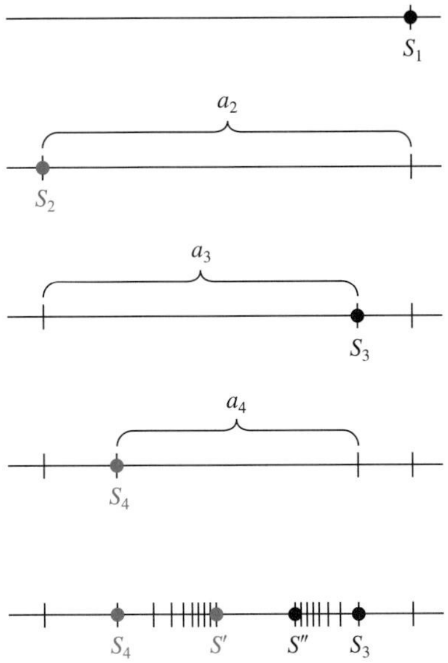

Figure 1

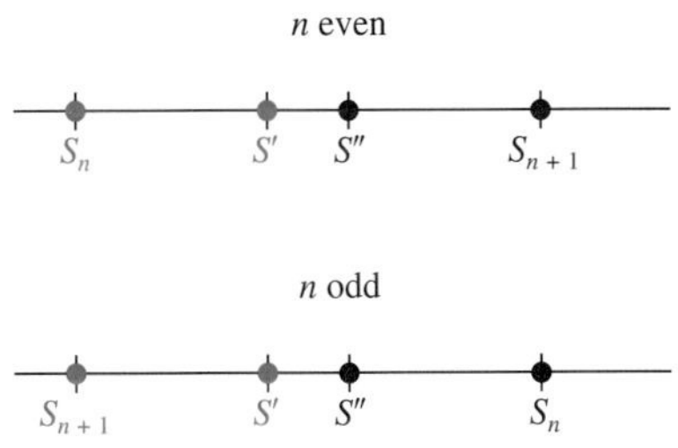

Figure 2

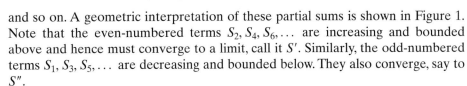

A Convergence Test Let us suppose that the sequence $\{a_n\}$ is decreasing; that is, $a_{n+1} < a_n$ for all n. Also, let S_n have its usual meaning. Thus, for the alternating series $a_1 - a_2 + a_3 - a_4 + \cdots$, we have

$$S_1 = a_1$$
$$S_2 = a_1 - a_2 = S_1 - a_2$$
$$S_3 = a_1 - a_2 + a_3 = S_2 + a_3$$
$$S_4 = a_1 - a_2 + a_3 - a_4 = S_3 - a_4$$

and so on. A geometric interpretation of these partial sums is shown in Figure 1. Note that the even-numbered terms $S_2, S_4, S_6, \ldots$ are increasing and bounded above and hence must converge to a limit, call it S'. Similarly, the odd-numbered terms $S_1, S_3, S_5, \ldots$ are decreasing and bounded below. They also converge, say to S''.

Both S' and S'' are between S_n and S_{n+1} for all n (see Figure 2), and so

$$|S'' - S'| \le |S_{n+1} - S_n| = a_{n+1}$$

Thus, the condition $a_{n+1} \to 0$ as $n \to \infty$ will guarantee that $S' = S''$ and, consequently, the convergence of the series to their common value, which we call S. Finally, we note that, since S is between S_n and S_{n+1},

$$|S - S_n| \le |S_{n+1} - S_n| = a_{n+1}$$

That is, the error made by using S_n as an approximation to the sum S of the whole series is not more than the magnitude of the first neglected term. We have proved the following theorem.

Theorem A Alternating Series Test

Let

$$a_1 - a_2 + a_3 - a_4 + \cdots$$

be an alternating series with $a_n > a_{n+1} > 0$. If $\lim_{n\to\infty} a_n = 0$, then the series converges. Moreover, the error made by using the sum S_n of the first n terms to approximate the sum S of the series is not more than a_{n+1}.

EXAMPLE 1 Show that the alternating harmonic series

$$1 - \tfrac{1}{2} + \tfrac{1}{3} - \tfrac{1}{4} + \cdots$$

converges. How many terms of this series would we need to take in order to get a partial sum S_n within 0.01 of the sum S of the whole series?

SOLUTION The alternating harmonic series satisfies the hypotheses of Theorem A and so converges. We want $|S - S_n| \le 0.01$, and this will hold if $a_{n+1} \le 0.01$. Since $a_{n+1} = 1/(n + 1)$, we require $1/(n + 1) \le 0.01$, which is satisfied if $n \ge 99$. Thus, we need to take 99 terms to make sure that we have the desired accuracy. This gives you an idea of how slowly the alternating harmonic series converges. (See Problem 45 for a clever way to find the exact sum of this series.) ■

EXAMPLE 2 Show that

$$\frac{1}{1!} - \frac{1}{2!} + \frac{1}{3!} - \frac{1}{4!} + \cdots$$

converges. Calculate S_5 and estimate the error made by using this as a value for the sum of the whole series.

SOLUTION The Alternating Series Test (Theorem A) applies and guarantees convergence.

$$S_5 = 1 - \frac{1}{2} + \frac{1}{6} - \frac{1}{24} + \frac{1}{120} \approx 0.6333$$

$$|S - S_5| \le a_6 = \frac{1}{6!} \approx 0.0014$$

■

EXAMPLE 3 Show that $\sum_{n=1}^{\infty} (-1)^{n-1} \frac{n^2}{2^n}$ converges.

SOLUTION To get a feeling for this series, we write the first few terms:

$$\tfrac{1}{2} - 1 + \tfrac{9}{8} - 1 + \tfrac{25}{32} - \tfrac{36}{64} + \cdots$$

The series is alternating and $\lim_{n\to\infty} n^2/2^n = 0$ (l'Hôpital's Rule), but unfortunately the terms are not decreasing initially. However, they do appear to be decreasing after the first two terms; this is good enough, since what happens at the beginning of a series never affects convergence or divergence. To show that the sequence $\{n^2/2^n\}$ is decreasing from the third term on, consider the function

$$f(x) = \frac{x^2}{2^x}$$

Note that if $x \ge 3$ the derivative

$$f'(x) = \frac{2x \cdot 2^x - x^2 2^x \ln 2}{2^{2x}} = \frac{x2^x(2 - x \ln 2)}{2^{2x}}$$

$$\approx \frac{x(2 - 0.69x)}{2^x} < 0$$

Thus, f is decreasing on $[3, \infty)$, and so $\{n^2/2^n\}$ is decreasing for $n \ge 3$. For a different demonstration of this last fact, see Example 6 of Section 9.1. ■

Absolute Convergence

Does a series such as

$$1 + \tfrac{1}{4} - \tfrac{1}{9} + \tfrac{1}{16} + \tfrac{1}{25} - \tfrac{1}{36} + \cdots$$

in which there is a pattern of two positive terms followed by one negative term, converge or diverge? The Alternating Series Test does not apply. However, since the corresponding series of all positive terms

$$1 + \tfrac{1}{4} + \tfrac{1}{9} + \tfrac{1}{16} + \tfrac{1}{25} + \tfrac{1}{36} + \cdots$$

converges (p-series with $p = 2$), it seems plausible to think that the same series with some terms negative should converge (even better). This is the content of our next theorem.

Theorem B **Absolute Convergence Test**

If $\Sigma |u_n|$ converges, then Σu_n converges.

Proof We use a trick. Let $v_n = u_n + |u_n|$, so

$$u_n = v_n - |u_n|$$

Now $0 \le v_n \le 2|u_n|$, and so Σv_n converges by the Ordinary Comparison Test. It follows from the Linearity Theorem (Theorem 9.2B) that $\Sigma u_n = \Sigma(v_n - |u_n|)$ converges. ■

A series Σu_n is said to **converge absolutely** if $\Sigma|u_n|$ converges. Theorem B asserts that absolute convergence implies convergence. All our tests for convergence of series of positive terms are automatically tests for the absolute convergence of a series in which some terms are negative. In particular, this is true of the Ratio Test, which we now restate.

Theorem C **Absolute Ratio Test**

Let Σu_n be a series of nonzero terms and suppose that

$$\lim_{n\to\infty} \frac{|u_{n+1}|}{|u_n|} = \rho$$

(i) If $\rho < 1$, the series converges absolutely (hence converges).
(ii) If $\rho > 1$, the series diverges.
(iii) If $\rho = 1$, the test is inconclusive.

Proof Proofs of (i) and (iii) are direct results of the Ratio Test. For (ii), we could conclude from the original Ratio Test that $\Sigma|u_n|$ diverges, but here we are claiming more, that Σu_n diverges. Since

$$\lim_{n\to\infty} \frac{|u_{n+1}|}{|u_n|} > 1$$

it follows that for n sufficiently large, say $n \ge N$, $|u_{n+1}| > |u_n|$. This, in turn, implies that $|u_n| > |u_N| > 0$ for all $n \ge N$, and so $\lim_{n\to\infty} u_n$ cannot be 0. We conclude by the nth-Term Test that Σu_n diverges. ■

EXAMPLE 4 Show that $\sum_{n=1}^{\infty}(-1)^{n+1}\frac{3^n}{n!}$ converges absolutely.

SOLUTION

$$\rho = \lim_{n\to\infty} \frac{|u_{n+1}|}{|u_n|} = \lim_{n\to\infty} \frac{3^{n+1}}{(n+1)!} \div \frac{3^n}{n!}$$

$$= \lim_{n\to\infty} \frac{3}{n+1} = 0$$

We conclude from the Absolute Ratio Test that the series converges absolutely (and therefore converges). ■

EXAMPLE 5 Test for the convergence or divergence of $\sum_{n=1}^{\infty} \frac{\cos(n!)}{n^2}$.

SOLUTION If you write out the first 100 terms of this series, you will discover that the signs of the terms vary in a rather random way. The series is in fact a difficult one to analyze directly. However,

$$\left|\frac{\cos(n!)}{n^2}\right| \le \frac{1}{n^2}$$

and so the series converges absolutely by the Ordinary Comparison Test. We conclude from the Absolute Convergence Test (Theorem B) that the series converges. ■

Conditional Convergence A common error is to try to turn Theorem B around. It does *not* say that convergence implies absolute convergence. That is clearly false; witness the alternating harmonic series. We know that

$$1 - \tfrac{1}{2} + \tfrac{1}{3} - \tfrac{1}{4} + \cdots$$

converges, but that

$$1 + \tfrac{1}{2} + \tfrac{1}{3} + \tfrac{1}{4} + \cdots$$

diverges. A series Σu_n is called **conditionally convergent** if Σu_n converges but $\Sigma |u_n|$ diverges. The alternating harmonic series is the premier example of a conditionally convergent series, but there are many others.

EXAMPLE 6 Show that $\sum_{n=1}^{\infty} (-1)^{n+1} \frac{1}{\sqrt{n}}$ is conditionally convergent.

SOLUTION $\sum_{n=1}^{\infty} (-1)^{n+1}[1/\sqrt{n}\,]$ converges by the Alternating Series Test. However, $\sum_{n=1}^{\infty} 1/\sqrt{n}$ diverges, since it is a p-series with $p = \frac{1}{2}$. ■

Absolutely convergent series behave much better than do conditionally convergent ones. Here is a nice theorem about absolutely convergent series. It is spectacularly false for conditionally convergent series (see Problems 35–38). The proof is difficult, so we do not include it here.

Theorem D Rearrangement Theorem

The terms of an absolutely convergent series can be rearranged without affecting either the convergence or the sum of the series.

For example, the series

$$1 + \tfrac{1}{4} - \tfrac{1}{9} + \tfrac{1}{16} + \tfrac{1}{25} - \tfrac{1}{36} + \tfrac{1}{49} + \tfrac{1}{64} - \tfrac{1}{81} + \cdots$$

converges absolutely. The rearrangement

$$1 + \tfrac{1}{4} + \tfrac{1}{16} - \tfrac{1}{9} + \tfrac{1}{25} + \tfrac{1}{49} + \tfrac{1}{64} - \tfrac{1}{36} + \cdots$$

converges and has the same sum as the original series.

Concepts Review

1. If $a_n \geq 0$ for all n, the alternating series $a_1 - a_2 + a_3 - \cdots$ will converge provided that the terms are decreasing in size and ______.

2. If $\Sigma |u_k|$ converges, we say that the series Σu_k converges ______; if Σu_k converges, but $\Sigma |u_k|$ diverges, we say that Σu_k converges ______.

3. The premier example of a conditionally convergent series is ______.

4. The terms of an absolutely convergent series may be ______ at will without affecting its convergence or its sum.

Problem Set 9.5

In Problems 1–6, show that each alternating series converges, and then estimate the error made by using the partial sum S_9 as an approximation to the sum S of the series (see Examples 1–3).

1. $\sum_{n=1}^{\infty} (-1)^{n+1} \frac{2}{3n + 1}$

2. $\sum_{n=1}^{\infty} (-1)^{n+1} \frac{1}{\sqrt{n}}$

3. $\sum_{n=1}^{\infty} (-1)^{n+1} \frac{1}{\ln(n + 1)}$

4. $\sum_{n=1}^{\infty} (-1)^{n+1} \frac{n}{n^2 + 1}$

5. $\sum_{n=1}^{\infty} (-1)^{n+1} \frac{\ln n}{n}$

6. $\sum_{n=1}^{\infty} (-1)^{n+1} \frac{\ln n}{\sqrt{n}}$

In Problems 7–12, show that each series converges absolutely.

7. $\sum_{n=1}^{\infty} \left(-\tfrac{3}{4}\right)^n$

8. $\sum_{n=1}^{\infty} (-1)^n \frac{1}{n\sqrt{n}}$

9. $\sum_{n=1}^{\infty} (-1)^{n+1} \frac{n}{2^n}$

10. $\sum_{n=1}^{\infty} (-1)^{n+1} \frac{n^2}{e^n}$

11. $\sum_{n=1}^{\infty} (-1)^{n+1} \frac{1}{n(n + 1)}$

12. $\sum_{n=1}^{\infty} (-1)^{n+1} \frac{2^n}{n!}$

In Problems 13–30, classify each series as absolutely convergent, conditionally convergent, or divergent.

13. $\sum_{n=1}^{\infty}(-1)^{n+1}\frac{1}{5n}$

14. $\sum_{n=1}^{\infty}(-1)^{n+1}\frac{1}{5n^{1.1}}$

15. $\sum_{n=1}^{\infty}(-1)^{n+1}\frac{n}{10n+1}$

16. $\sum_{n=1}^{\infty}(-1)^{n+1}\frac{n}{10n^{1.1}+1}$

17. $\sum_{n=2}^{\infty}(-1)^{n}\frac{1}{n\ln n}$

18. $\sum_{n=1}^{\infty}(-1)^{n+1}\frac{1}{n(1+\sqrt{n})}$

19. $\sum_{n=1}^{\infty}(-1)^{n+1}\frac{n^4}{2^n}$

20. $\sum_{n=2}^{\infty}(-1)^{n}\frac{1}{\sqrt{n^2-1}}$

21. $\sum_{n=1}^{\infty}(-1)^{n+1}\frac{n}{n^2+1}$

22. $\sum_{n=1}^{\infty}(-1)^{n+1}\frac{n-1}{n}$

23. $\sum_{n=1}^{\infty}\frac{\cos n\pi}{n}$

24. $\sum_{n=1}^{\infty}\frac{\sin(n\pi/2)}{n^2}$

25. $\sum_{n=1}^{\infty}(-1)^{n}\frac{\sin n}{n\sqrt{n}}$

26. $\sum_{n=1}^{\infty}n\sin\left(\frac{1}{n}\right)$

27. $\sum_{n=1}^{\infty}(-1)^{n+1}\frac{1}{\sqrt{n(n+1)}}$

28. $\sum_{n=1}^{\infty}\frac{(-1)^{n+1}}{\sqrt{n+1}+\sqrt{n}}$

29. $\sum_{n=1}^{\infty}\frac{(-3)^{n+1}}{n^2}$

30. $\sum_{n=1}^{\infty}(-1)^{n+1}\sin\frac{\pi}{n}$

31. Prove that if Σa_n diverges, so does $\Sigma|a_n|$.

32. Give an example of two series Σa_n and Σb_n, both convergent, such that $\Sigma a_n b_n$ diverges.

33. Show that the positive terms of the alternating harmonic series form a divergent series. Show the same for the negative terms.

34. Show that the results in Problem 33 hold for any conditionally convergent series.

35. Show that the alternating harmonic series

$$1-\tfrac{1}{2}+\tfrac{1}{3}-\tfrac{1}{4}+\tfrac{1}{5}-\tfrac{1}{6}+\cdots$$

(whose sum is actually $\ln 2 \approx 0.69$) can be rearranged to converge to 1.3 by using the following steps.

(a) Take enough of the positive terms $1+\frac{1}{3}+\frac{1}{5}+\cdots$ to just exceed 1.3.

(b) Now add enough of the negative terms $-\frac{1}{2}-\frac{1}{4}-\frac{1}{6}-\cdots$ so that the partial sum S_n falls just below 1.3.

(c) Add just enough more positive terms to again exceed 1.3, and so on.

C **36.** Use your calculator to help you find the first 20 terms of the series described in Problem 35. Calculate S_{20}.

37. Explain why a conditionally convergent series can be rearranged to converge to any given number.

38. Show that a conditionally convergent series can be rearranged so as to diverge.

39. Show that $\lim_{n\to\infty} a_n = 0$ is not sufficient to guarantee the convergence of the alternating series $\Sigma(-1)^{n+1}a_n$. *Hint:* Alternate the terms of $\Sigma 1/n$ and $\Sigma(-1/n^2)$.

40. Discuss the convergence or divergence of

$$\frac{1}{\sqrt{2}-1}-\frac{1}{\sqrt{2}+1}+\frac{1}{\sqrt{3}-1}-\frac{1}{\sqrt{3}+1}+\frac{1}{\sqrt{4}-1}-\frac{1}{\sqrt{4}+1}+\cdots$$

41. Prove that if $\sum_{k=1}^{\infty}a_k^2$ and $\sum_{k=1}^{\infty}b_k^2$ both converge then $\sum_{k=1}^{\infty}a_k b_k$ converges absolutely. *Hint:* First show that $2|a_k b_k| \le a_k^2 + b_k^2$.

42. Sketch the graph of $y = (\sin x)/x$ and then show that $\int_0^{\infty}(\sin x)/x\,dx$ converges.

43. Show that $\int_0^{\infty}|\sin x|/x\,dx$ diverges.

44. Show that the graph of $y = x\sin\frac{\pi}{x}$ on $(0, 1]$ has infinite length.

45. Note that

$$1-\frac{1}{2}+\frac{1}{3}-\frac{1}{4}+\cdots-\frac{1}{2n}$$
$$=1+\frac{1}{2}+\frac{1}{3}+\cdots+\frac{1}{2n}-\left(1+\frac{1}{2}+\frac{1}{3}+\cdots+\frac{1}{n}\right)$$
$$=\frac{1}{n+1}+\frac{1}{n+2}+\cdots+\frac{1}{2n}$$

Recognize the latter expression as a Riemann sum and use it to find the sum of the alternating harmonic series.

Answers to Concepts Review: **1.** $\lim_{n\to\infty} a_n = 0$ **2.** absolutely; conditionally **3.** the alternating harmonic series **4.** rearranged

9.6 Power Series

So far we have been studying what might be called *series of constants*, that is, series of the form Σu_n, where each u_n is a number. Now we consider *series of functions*, series of the form $\Sigma u_n(x)$. A typical example of such a series is

$$\sum_{n=1}^{\infty}\frac{\sin nx}{n^2}=\frac{\sin x}{1}+\frac{\sin 2x}{4}+\frac{\sin 3x}{9}+\cdots$$

Of course, as soon as we substitute a value for x (such as $x = 2.1$), we are back to familiar territory; we have a series of constants.

There are two important questions to ask about a series of functions.

1. For what x's does the series converge?
2. To what function does it converge; that is, what is the sum $S(x)$ of the series?

Fourier Series

The series of sine functions mentioned in the introduction is an example of a *Fourier series*, named after Jean Baptiste Joseph Fourier (1768–1830). Fourier series are of immense importance in the study of wave phenomena, since they allow us to represent a complicated wave as a sum of its fundamental components (called the pure tones in the case of sound waves). It is a large field, which we leave to other authors and other books.

The general situation is a proper subject for an advanced calculus course. However, even in elementary calculus, we can learn a good deal about the special case of a power series. A **power series in *x*** has the form

$$\sum_{n=0}^{\infty} a_n x^n = a_0 + a_1 x + a_2 x^2 + \cdots$$

(Here we interpret $a_0 x^0$ to be a_0 even if $x = 0$.) We can immediately answer our two questions for one such power series.

EXAMPLE 1 For what x's does the power series

$$\sum_{n=0}^{\infty} a x^n = a + ax + ax^2 + ax^3 + \cdots$$

converge and what is its sum? Assume that $a \neq 0$.

SOLUTION We actually studied this series in Section 9.2 (with r in place of x) and called it a geometric series. It converges for $-1 < x < 1$ and has sum $S(x)$ given by

$$S(x) = \frac{a}{1 - x}, \qquad -1 < x < 1$$

■

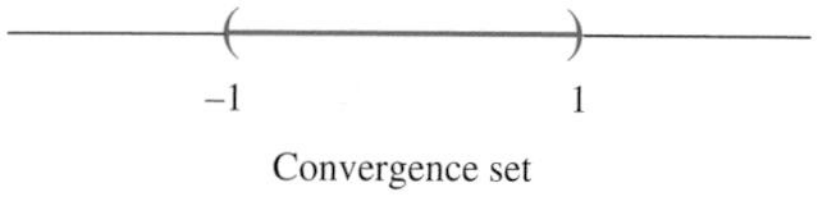

Figure 1

The Convergence Set We call the set on which a power series converges its **convergence set**. What kind of set can be a convergence set? Example 1 suggests that it can be an open interval (see Figure 1). Are there other possibilities?

EXAMPLE 2 What is the convergence set for

$$\sum_{n=0}^{\infty} \frac{x^n}{(n+1)2^n} = 1 + \frac{1}{2}\frac{x}{2} + \frac{1}{3}\frac{x^2}{2^2} + \frac{1}{4}\frac{x^3}{2^3} + \cdots$$

SOLUTION Note that some of the terms may be negative (if x is negative). Let's test for absolute convergence using the Absolute Ratio Test (Theorem 9.5C).

$$\rho = \lim_{n\to\infty} \left| \frac{x^{n+1}}{(n+2)2^{n+1}} \div \frac{x^n}{(n+1)2^n} \right| = \lim_{n\to\infty} \frac{|x|}{2} \cdot \frac{n+1}{n+2} = \frac{|x|}{2}$$

The series converges absolutely (hence converges) when $\rho = |x|/2 < 1$ and diverges when $|x|/2 > 1$. Consequently, it converges when $|x| < 2$ and diverges when $|x| > 2$.

If $x = 2$ or $x = -2$, the Ratio Test fails. However, when $x = 2$, the series is the harmonic series, which diverges; and when $x = -2$, it is the alternating harmonic series, which converges. We conclude that the convergence set for the given series is the interval $-2 \leq x < 2$ (Figure 2). ■

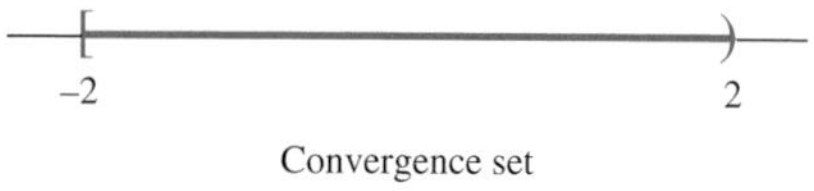

Figure 2

EXAMPLE 3 Find the convergence set for $\sum_{n=0}^{\infty} \frac{x^n}{n!}$.

SOLUTION

$$\rho = \lim_{n\to\infty} \left| \frac{x^{n+1}}{(n+1)!} \div \frac{x^n}{n!} \right| = \lim_{n\to\infty} \frac{|x|}{n+1} = 0$$

We conclude from the Absolute Ratio Test that the series converges for all x (Figure 3). ■

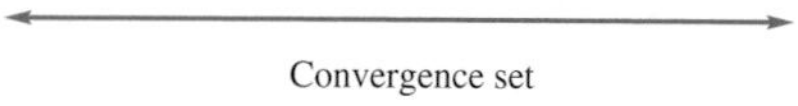

Figure 3

EXAMPLE 4 Find the convergence set for $\sum_{n=0}^{\infty} n!x^n$.

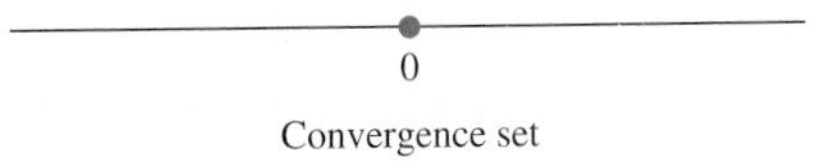

Figure 4

SOLUTION

$$\rho = \lim_{n\to\infty}\left|\frac{(n+1)!x^{n+1}}{n!x^n}\right| = \lim_{n\to\infty}(n+1)|x| = \begin{cases} 0 & \text{if } x = 0 \\ \infty & \text{if } x \neq 0 \end{cases}$$

We conclude that the series converges only at $x = 0$ (Figure 4). ■

In each of our examples, the convergence set was an interval (a degenerate interval in the last example). This will always be the case. For example, it is impossible for a power series to have a convergence set consisting of two disconnected parts (like $[0, 1] \cup [2, 3]$). Our next theorem tells the whole story.

Theorem A

The convergence set for a power series $\Sigma a_n x^n$ is always an interval of one of the following three types:

(i) The single point $x = 0$.
(ii) An interval $(-R, R)$, plus possibly one or both end points.
(iii) The whole real line.

In (i), (ii), and (iii), the series is said to have **radius of convergence** 0, R, and ∞, respectively.

Proof Suppose that the series converges at $x = x_1 \neq 0$. Then $\lim_{n\to\infty} a_n x_1^n = 0$, and so there is certainly a number N such that $|a_n x_1^n| < 1$ for $n \geq N$. Then, for any x for which $|x| < |x_1|$,

$$|a_n x^n| = |a_n x_1^n|\left|\frac{x}{x_1}\right|^n < \left|\frac{x}{x_1}\right|^n$$

this holding for $n \geq N$. Now $\Sigma|x/x_1|^n$ converges, since it is a geometric series with ratio less than 1. Thus, by the Ordinary Comparison Test (Theorem 9.4A), $\Sigma|a_n x^n|$ converges. We have shown that if a power series converges at x_1 it converges (absolutely) for all x such that $|x| < |x_1|$.

On the other hand, suppose that a power series diverges at x_2. Then it must diverge for all x for which $|x| > |x_2|$. For if it converged at x_1 such that $|x_1| > |x_2|$, then, by what we have already shown, it would converge at x_2, contrary to hypothesis.

These two paragraphs together eliminate all possible types of convergence sets except the three types mentioned in the theorem. ■

Actually we have proved slightly more than we have claimed in Theorem A, and it is worth stating this as another theorem.

Theorem B

A power series $\Sigma a_n x^n$ converges absolutely on the interior of its interval of convergence.

Of course, it might even converge absolutely at the end points of the interval of convergence, but of that we cannot be sure; witness Example 2.

Power Series in $x - a$ A series of the form

$$\sum a_n(x-a)^n = a_0 + a_1(x-a) + a_2(x-a)^2 + \cdots$$

is called a **power series in $x - a$.** All that we have said about power series in x applies equally well for series in $x - a$. In particular, its convergence set is always one of the following kinds of intervals:

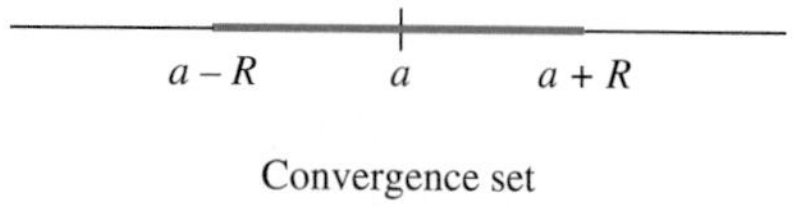

Figure 5

1. The single point $x = a$.
2. An interval $(a - R, a + R)$, plus possibly one or both end points (Figure 5).
3. The whole real line.

EXAMPLE 5 Find the convergence set for $\sum_{n=0}^{\infty} \frac{(x-1)^n}{(n+1)^2}$.

SOLUTION We apply the Absolute Ratio Test.

$$\rho = \lim_{n\to\infty}\left|\frac{(x-1)^{n+1}}{(n+2)^2} \div \frac{(x-1)^n}{(n+1)^2}\right| = \lim_{n\to\infty}|x-1|\frac{(n+1)^2}{(n+2)^2}$$
$$= |x-1|$$

Thus, the series converges if $|x-1| < 1$, that is, if $0 < x < 2$; it diverges if $|x-1| > 1$. It also converges (even absolutely) at both of the end points 0 and 2, as we see by substitution of these values. The convergence set is the closed interval $[0, 2]$ (Figure 6). ■

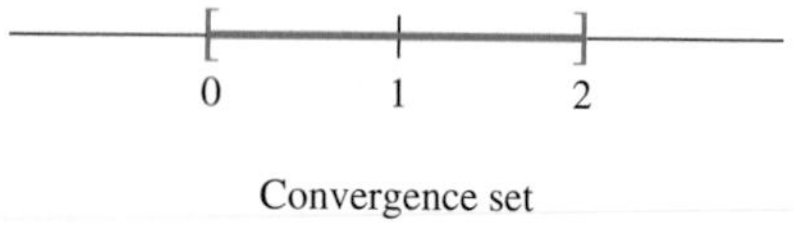

Figure 6

EXAMPLE 6 Determine the convergence set for

$$\frac{(x+2)^2 \ln 2}{2\cdot 9} + \frac{(x+2)^3 \ln 3}{3\cdot 27} + \frac{(x+2)^4 \ln 4}{4\cdot 81} + \cdots$$

SOLUTION The nth term is $u_n = \frac{(x+2)^n \ln n}{n\cdot 3^n}$, $n \ge 2$. Thus,

$$\rho = \lim_{n\to\infty}\left|\frac{(x+2)^{n+1}\ln(n+1)}{(n+1)3^{n+1}} \cdot \frac{n3^n}{(x+2)^n \ln n}\right|$$
$$= \frac{|x+2|}{3}\lim_{n\to\infty}\frac{n}{n+1}\,\frac{\ln(n+1)}{\ln n} = \frac{|x+2|}{3}$$

We know that the series converges when $\rho < 1$, that is, when $|x+2| < 3$ or, equivalently, $-5 < x < 1$, but we must check the end points -5 and 1.

At $x = -5$,

$$u_n = \frac{(-3)^n \ln n}{n3^n} = (-1)^n\frac{\ln n}{n}$$

and $\Sigma(-1)^n(\ln n)/n$ converges by the Alternating Series Test.

At $x = 1$, $u_n = (\ln n)/n$ and $\Sigma(\ln n)/n$ diverges by comparison with the harmonic series.

We conclude that the given series converges on the interval $-5 \le x < 1$. ■

Concepts Review

1. A series of the form $a_0 + a_1x + a_2x^2 + \cdots$ is called a ______.

2. Rather than asking whether a power series converges, we should ask ______.

3. A power series always converges on a(n) ______, which may or may not include its ______.

4. The series $5 + x + x^2 + x^3 + \cdots$ converges on the interval ______.

Problem Set 9.6

In Problems 1–8, find the convergence set for the given power series.

1. $\sum_{n=1}^{\infty} \frac{x^n}{(n-1)!}$

2. $\sum_{n=1}^{\infty} \frac{x^n}{3^n}$

3. $\sum_{n=1}^{\infty} \frac{x^n}{n^2}$

4. $\sum_{n=1}^{\infty} nx^n$

5. $\sum_{n=1}^{\infty} (-1)^{n+1} \frac{x^n}{n^2}$

6. $\sum_{n=1}^{\infty} (-1)^n \frac{x^n}{n}$

7. $\sum_{n=1}^{\infty} (-1)^n \frac{(x-2)^n}{n}$

8. $\sum_{n=1}^{\infty} \frac{(x+1)^n}{n!}$

In Problems 9–28, find the convergence set for the given power series. Hint: First find a formula for the nth term; then use the Absolute Ratio Test.

9. $\frac{x}{1 \cdot 2} - \frac{x^2}{2 \cdot 3} + \frac{x^3}{3 \cdot 4} - \frac{x^4}{4 \cdot 5} + \frac{x^5}{5 \cdot 6} - \cdots$

10. $1 + x + \frac{x^2}{2!} + \frac{x^3}{3!} + \frac{x^4}{4!} + \cdots$

11. $x - \frac{x^3}{3!} + \frac{x^5}{5!} - \frac{x^7}{7!} + \frac{x^9}{9!} - \cdots$

12. $1 - \frac{x^2}{2!} + \frac{x^4}{4!} - \frac{x^6}{6!} + \frac{x^8}{8!} - \frac{x^{10}}{10!} + \cdots$

13. $x + 2x^2 + 3x^3 + 4x^4 + \cdots$

14. $x + 2^2x^2 + 3^2x^3 + 4^2x^4 + \cdots$

15. $1 - x + \frac{x^2}{2} - \frac{x^3}{3} + \frac{x^4}{4} - \cdots$

16. $1 + x + \frac{x^2}{\sqrt{2}} + \frac{x^3}{\sqrt{3}} + \frac{x^4}{\sqrt{4}} + \frac{x^5}{\sqrt{5}} + \cdots$

17. $1 - \frac{x}{1 \cdot 3} + \frac{x^2}{2 \cdot 4} - \frac{x^3}{3 \cdot 5} + \frac{x^4}{4 \cdot 6} - \cdots$

18. $\frac{x}{2^2 - 1} + \frac{x^2}{3^2 - 1} + \frac{x^3}{4^2 - 1} + \frac{x^4}{5^2 - 1} + \cdots$

19. $1 - \frac{x}{2} + \frac{x^2}{2^2} - \frac{x^3}{2^3} + \frac{x^4}{2^4} - \cdots$

20. $1 + 2x + 2^2x^2 + 2^3x^3 + 2^4x^4 + \cdots$

21. $1 + 2x + \frac{2^2x^2}{2!} + \frac{2^3x^3}{3!} + \frac{2^4x^4}{4!} + \cdots$

22. $\frac{x}{2} + \frac{2x^2}{3} + \frac{3x^3}{4} + \frac{4x^4}{5} + \frac{5x^5}{6} + \cdots$

23. $\frac{x-1}{1} + \frac{(x-1)^2}{2} + \frac{(x-1)^3}{3} + \frac{(x-1)^4}{4} + \cdots$

24. $1 + (x+2) + \frac{(x+2)^2}{2!} + \frac{(x+2)^3}{3!} + \cdots$

25. $1 + \frac{x+1}{2} + \frac{(x+1)^2}{2^2} + \frac{(x+1)^3}{2^3} + \cdots$

26. $\frac{x-2}{1^2} + \frac{(x-2)^2}{2^2} + \frac{(x-2)^3}{3^2} + \frac{(x-2)^4}{4^2} + \cdots$

27. $\frac{x+5}{1 \cdot 2} + \frac{(x+5)^2}{2 \cdot 3} + \frac{(x+5)^3}{3 \cdot 4} + \frac{(x+5)^4}{4 \cdot 5} + \cdots$

28. $(x+3) - 2(x+3)^2 + 3(x+3)^3 - 4(x+3)^4 + \cdots$

29. From Example 3, we know that $\Sigma x^n/n!$ converges for all x. Why can we conclude that $\lim_{n\to\infty} x^n/n! = 0$ for all x?

30. Let k be an arbitrary number and $-1 < x < 1$. Prove that

$$\lim_{n\to\infty} \frac{k(k-1)(k-2)\cdots(k-n)}{n!} x^n = 0$$

Hint: See Problem 29.

31. Find the radius of convergence of

$$\sum_{n=1}^{\infty} \frac{1 \cdot 2 \cdot 3 \cdots n}{1 \cdot 3 \cdot 5 \cdots (2n-1)} x^{2n+1}$$

32. Find the radius of convergence of

$$\sum_{n=0}^{\infty} \frac{(pn)!}{(n!)^p} x^n$$

where p is a positive integer.

33. Find the sum $S(x)$ of $\sum_{n=0}^{\infty} (x-3)^n$. What is the convergence set?

34. Suppose that $\sum_{n=0}^{\infty} a_n(x-3)^n$ converges at $x = -1$. Why can you conclude that it converges at $x = 6$? Can you be sure that it converges at $x = 7$? Explain.

35. Find the convergence set for each series.

(a) $\sum_{n=1}^{\infty} \frac{(3x+1)^n}{n \cdot 2^n}$ (b) $\sum_{n=1}^{\infty} (-1)^n \frac{(2x-3)^n}{4^n\sqrt{n}}$

36. Refer to Problem 52 of Section 9.1, where the Fibonacci sequence $f_1, f_2, f_3, \ldots$ was defined. Find the radius of convergence of $\sum_{n=1}^{\infty} f_n x^n$.

37. Suppose that $a_{n+3} = a_n$ and let $S(x) = \sum_{n=0}^{\infty} a_n x^n$. Show that the series converges for $|x| < 1$ and give a formula for $S(x)$.

38. Follow the directions of Problem 37 for the case where $a_{n+p} = a_n$ for some fixed positive integer p.

Answers to Concepts Review: **1.** power series **2.** where it converges **3.** interval; end points **4.** $(-1, 1)$

9.7 Operations on Power Series

We know from the previous section that the convergence set of a power series $\Sigma a_n x^n$ is an interval I. This interval is the domain for a new function $S(x)$, the sum of the series. The most obvious question to ask about $S(x)$ is whether we can give a simple formula for it. We have done this for one series, a geometric series.

$$\sum_{n=0}^{\infty} ax^n = \frac{a}{1-x}, \qquad -1 < x < 1$$

Actually, there is little reason to hope that the sum of an arbitrarily given power series will be one of the elementary functions studied earlier in this book, though we will make a little progress in that direction in this section and more in Section 9.8.

A better question to ask now is whether we can say anything about the properties of $S(x)$. For example, is it differentiable? Is it integrable? The answer to both questions is yes.

Term-by-Term Differentiation and Integration Think of a power series as a polynomial with infinitely many terms. It behaves like a polynomial under both differentiation and integration; these operations can be performed term by term, as follows.

Theorem A

Suppose that $S(x)$ is the sum of a power series on an interval I; that is,

$$S(x) = \sum_{n=0}^{\infty} a_n x^n = a_0 + a_1 x + a_2 x^2 + a_3 x^3 + \cdots$$

Then, if x is interior to I,

(i) $$S'(x) = \sum_{n=0}^{\infty} D_x(a_n x^n) = \sum_{n=1}^{\infty} n a_n x^{n-1}$$
$$= a_1 + 2a_2 x + 3a_3 x^2 + \cdots$$

(ii) $$\int_0^x S(t)\,dt = \sum_{n=0}^{\infty} \int_0^x a_n t^n\,dt = \sum_{n=0}^{\infty} \frac{a_n}{n+1} x^{n+1}$$
$$= a_0 x + \frac{1}{2} a_1 x^2 + \frac{1}{3} a_2 x^3 + \frac{1}{4} a_3 x^4 + \cdots$$

The theorem entails several things. It asserts that S is both differentiable and integrable, it shows how the derivative and integral may be calculated, and it implies that the radius of convergence of both the differentiated and integrated series is the same as for the original series (though it says nothing about the end points of the interval of convergence). The theorem is hard to prove. We leave the proof to more advanced books.

A nice consequence of Theorem A is that we can apply it to a power series with a known sum formula to obtain sum formulas for other series.

EXAMPLE 1 Apply Theorem A to the geometric series

$$\frac{1}{1-x} = 1 + x + x^2 + x^3 + \cdots, \qquad -1 < x < 1$$

to obtain formulas for two new series.

SOLUTION Differentiating term by term yields

$$\frac{1}{(1-x)^2} = 1 + 2x + 3x^2 + 4x^3 + \cdots, \qquad -1 < x < 1$$

Integrating term by term gives

$$\int_0^x \frac{1}{1-t}\,dt = \int_0^x 1\,dt + \int_0^x t\,dt + \int_0^x t^2\,dt + \cdots$$

That is,

$$-\ln(1-x) = x + \frac{x^2}{2} + \frac{x^3}{3} + \cdots, \qquad -1 < x < 1$$

If we replace x by $-x$ in the latter and multiply both sides by -1, we obtain

$$\boxed{\ln(1+x) = x - \frac{x^2}{2} + \frac{x^3}{3} - \frac{x^4}{4} + \cdots, \qquad -1 < x < 1}$$

From Problem 45 of Section 9.5, we learn that this result is valid at the end point $x = 1$ (also see the note in the margin). ■

An End Point Result

The question of what is true at an end point of the interval of convergence of a power series is tricky. One result is due to Norway's greatest mathematician, Niels Henrik Abel (1802–1829). Suppose that

$$f(x) = \sum_{n=0}^{\infty} a_n x^n$$

for $|x| < R$. If f is continuous at an end point (R or $-R$) and if the series converges there, then the formula also holds at that end point.

EXAMPLE 2 Find the power series representation for $\tan^{-1} x$.

SOLUTION Recall that

$$\tan^{-1} x = \int_0^x \frac{1}{1+t^2}\,dt$$

From the geometric series for $1/(1-x)$, with x replaced by $-t^2$, we get

$$\frac{1}{1+t^2} = 1 - t^2 + t^4 - t^6 + \cdots, \qquad -1 < t < 1$$

Thus,

$$\tan^{-1} x = \int_0^x (1 - t^2 + t^4 - t^6 + \cdots)\,dt$$

That is,

$$\boxed{\tan^{-1} x = x - \frac{x^3}{3} + \frac{x^5}{5} - \frac{x^7}{7} + \cdots, \qquad -1 < x < 1}$$

(By the note in the margin, this also holds at $x = \pm 1$.) ■

EXAMPLE 3 Find a formula for the sum of the series

$$S(x) = 1 + x + \frac{x^2}{2!} + \frac{x^3}{3!} + \cdots$$

SOLUTION We saw earlier (Section 9.6, Example 3) that this series converges for all x. Differentiating term by term, we obtain

$$S'(x) = 1 + x + \frac{x^2}{2!} + \frac{x^3}{3!} + \cdots$$

That is, $S'(x) = S(x)$ for all x. Furthermore, $S(0) = 1$. This differential equation has the unique solution $S(x) = e^x$ (see Section 6.5). Thus,

$$\boxed{e^x = 1 + x + \frac{x^2}{2!} + \frac{x^3}{3!} + \cdots}$$

■

EXAMPLE 4 Obtain the power series representation for e^{-x^2}.

SOLUTION Simply substitute $-x^2$ for x in the series for e^x.

$$e^{-x^2} = 1 - x^2 + \frac{x^4}{2!} - \frac{x^6}{3!} + \cdots$$ ■

Algebraic Operations Convergent power series can be added and subtracted term by term (Theorem 9.2B). In that sense they behave like polynomials. Convergent power series can also be multiplied and divided in a manner suggested by the multiplication and "long" division of polynomials.

EXAMPLE 5 Multiply and divide the power series for $\ln(1 + x)$ by that for e^x.

SOLUTION We refer to Examples 1 and 3 for the required series. The key to multiplication is to first find the constant term, then the x-term, then the x^2-term, and so on. We arrange our work as follows.

$$0 + x - \frac{x^2}{2} + \frac{x^3}{3} - \frac{x^4}{4} + \cdots$$

$$1 + x + \frac{x^2}{2!} + \frac{x^3}{3!} + \frac{x^4}{4!} + \cdots$$

$$0 + (0 + 1)x + \left(0 + 1 - \frac{1}{2}\right)x^2 + \left(0 + \frac{1}{2!} - \frac{1}{2} + \frac{1}{3}\right)x^3$$

$$+ \left(0 + \frac{1}{3!} - \frac{1}{2!2} + \frac{1}{3} - \frac{1}{4}\right)x^4 + \cdots$$

$$= 0 + x + \frac{1}{2}x^2 + \frac{1}{3}x^3 + 0 \cdot x^4 + \cdots$$

Here is how division is done.

$$\begin{array}{r|l}
 & x - \frac{3}{2}x^2 + \frac{4}{3}x^3 - x^4 + \cdots \\
\hline
1 + x + \frac{1}{2}x^2 + \frac{1}{6}x^3 + \cdots & x - \frac{1}{2}x^2 + \frac{1}{3}x^3 - \frac{1}{4}x^4 + \cdots \\
 & x + x^2 + \frac{1}{2}x^3 + \frac{1}{6}x^4 + \cdots \\
\cline{2-2}
 & -\frac{3}{2}x^2 - \frac{1}{6}x^3 - \frac{5}{12}x^4 + \cdots \\
 & -\frac{3}{2}x^2 - \frac{3}{2}x^3 - \frac{3}{4}x^4 + \cdots \\
\cline{2-2}
 & \frac{4}{3}x^3 + \frac{1}{3}x^4 + \cdots \\
 & \frac{4}{3}x^3 + \frac{4}{3}x^4 + \cdots \\
\cline{2-2}
 & -x^4 + \cdots
\end{array}$$ ■

The real question relative to Example 5 is whether the two series that we have obtained converge to $[\ln(1 + x)]e^x$ and $[\ln(1 + x)]/e^x$, respectively. Our next theorem, stated without proof, answers this question.

Theorem B

Let $f(x) = \Sigma a_n x^n$ and $g(x) = \Sigma b_n x^n$, with both of these series converging at least for $|x| < r$. If the operations of addition, subtraction, and multiplication are performed on these series as if they were polynomials, the resulting series will converge for $|x| < r$ and represent $f(x) + g(x)$, $f(x) - g(x)$, and $f(x) \cdot g(x)$, respectively. If $b_0 \neq 0$, the corresponding result holds for division, but we can guarantee its validity only for $|x|$ sufficiently small.

We mention that the operation of substituting one power series into another is also legitimate for $|x|$ sufficiently small, provided that the constant term of the substituted series is zero. Here is an illustration.

S. Ramanujan (1887–1920)

One of the most remarkable people of the early 20th century was the Indian mathematician Srinivasa Ramanujan. Largely self-educated, Ramanujan left at his death a number of notebooks in which he had recorded his discoveries. These notebooks are only now being thoroughly studied. In them are many strange and wonderful formulas, some for the sums of infinite series. Here is one.

$$\frac{1}{\pi} = \frac{\sqrt{8}}{9801}\sum_{n=0}^{\infty}\frac{(4n)![1103 + 26{,}390n]}{(n!)^4(396)^{4n}}$$

Formulas like this were used in 1989 to calculate the decimal expansion of π to over 1 billion places. (See Problem 35.)

EXAMPLE 6 Find the power series for $e^{\tan^{-1} x}$ through terms of degree 4.

SOLUTION Since

$$e^u = 1 + u + \frac{u^2}{2!} + \frac{u^3}{3!} + \frac{u^4}{4!} + \cdots$$

$$e^{\tan^{-1} x} = 1 + \tan^{-1} x + \frac{(\tan^{-1} x)^2}{2!} + \frac{(\tan^{-1} x)^3}{3!} + \frac{(\tan^{-1} x)^4}{4!} + \cdots$$

Now substitute the series for $\tan^{-1} x$ from Example 2 and combine like terms.

$$e^{\tan^{-1} x} = 1 + \left(x - \frac{x^3}{3} + \cdots\right) + \frac{\left(x - \frac{x^3}{3} + \cdots\right)^2}{2!} + \frac{\left(x - \frac{x^3}{3} + \cdots\right)^3}{3!}$$
$$+ \frac{\left(x - \frac{x^3}{3} + \cdots\right)^4}{4!} + \cdots$$
$$= 1 + \left(x - \frac{x^3}{3} + \cdots\right) + \frac{\left(x^2 - \frac{2}{3}x^4 + \cdots\right)}{2} + \frac{(x^3 + \cdots)}{6}$$
$$+ \frac{(x^4 + \cdots)}{24} + \cdots$$
$$= 1 + x + \frac{x^2}{2} - \frac{x^3}{6} - \frac{7x^4}{24} + \cdots$$

■

Power Series in $x - a$ We have stated the theorems of this section for power series in x, but with obvious modifications they are equally valid for power series in $x - a$.

Concepts Review

1. A power series may be differentiated or _____ term by term on the _____ of its interval of convergence.

2. The first five terms in the power series expansion for $\ln(1 - x)$ are _____.

3. The first four terms in the power series expansion for $\exp(x^2)$ are _____.

4. The first five terms in the power series expansion for $\exp(x^2) - \ln(1 - x)$ are _____.

Problem Set 9.7

In Problems 1–10, find the power series representation for $f(x)$ and specify the radius of convergence. Each is somehow related to a geometric series (see Examples 1 and 2).

1. $f(x) = \dfrac{1}{1 + x}$

2. $f(x) = \dfrac{1}{(1 + x)^2}$ *Hint:* Differentiate Problem 1.

3. $f(x) = \dfrac{1}{(1 - x)^3}$

4. $f(x) = \dfrac{x}{(1 + x)^2}$

5. $f(x) = \dfrac{1}{2 - 3x} = \dfrac{\frac{1}{2}}{1 - \frac{3}{2}x}$

6. $f(x) = \dfrac{1}{3 + 2x}$

7. $f(x) = \dfrac{x^2}{1 - x^4}$

8. $f(x) = \dfrac{x^3}{2 - x^3}$

9. $f(x) = \displaystyle\int_0^x \ln(1 + t)\, dt$

10. $f(x) = \displaystyle\int_0^x \tan^{-1} t\, dt$

11. Obtain the power series in x for $\ln[(1 + x)/(1 - x)]$ and specify its radius of convergence. *Hint:*

$$\ln[(1 + x)/(1 - x)] = \ln(1 + x) - \ln(1 - x)$$

12. Show that any positive number M can be represented by $(1 + x)/(1 - x)$, where x lies within the interval of convergence of the series of Problem 11. Hence conclude that the natural logarithm of any positive number can be found by means of this series. Find ln 8 this way to three decimal places.

In Problems 13–16, use the result of Example 3 to find the power series in x for the given functions.

13. $f(x) = e^{-x}$

14. $f(x) = xe^{x^2}$

15. $f(x) = e^x + e^{-x}$

16. $f(x) = e^{2x} - 1 - 2x$

In Problems 17–24, use the methods of Example 5 to find power series in x for each function f.

17. $f(x) = e^{-x} \cdot \frac{1}{1-x}$

18. $f(x) = e^x \tan^{-1} x$

19. $f(x) = \frac{\tan^{-1} x}{e^x}$

20. $f(x) = \frac{e^x}{1 + \ln(1 + x)}$

21. $f(x) = (\tan^{-1} x)(1 + x^2 + x^4)$

22. $f(x) = \frac{\tan^{-1} x}{1 + x^2 + x^4}$

23. $f(x) = \int_0^x \frac{e^t}{1+t}\,dt$

24. $f(x) = \int_0^x \frac{\tan^{-1} t}{t}\,dt$

25. Find the sum of each of the following series by recognizing how it is related to something familiar.

(a) $x - x^2 + x^3 - x^4 + x^5 - \cdots$

(b) $\frac{1}{2!} + \frac{x}{3!} + \frac{x^2}{4!} + \frac{x^3}{5!} + \cdots$

(c) $2x + \frac{4x^2}{2} + \frac{8x^3}{3} + \frac{16x^4}{4} + \cdots$

26. Follow the directions of Problem 25.

(a) $1 + x^2 + x^4 + x^6 + x^8 + \cdots$

(b) $\cos x + \cos^2 x + \cos^3 x + \cos^4 x + \cdots$

(c) $\frac{x^2}{2} + \frac{x^4}{4} + \frac{x^6}{6} + \frac{x^8}{8} + \cdots$

27. Find the sum of $\sum_{n=1}^{\infty} nx^n$.

28. Find the sum of $\sum_{n=1}^{\infty} n(n+1)x^n$.

29. Use the method of substitution (Example 6) to find power series through terms of degree 3.

(a) $\tan^{-1}(e^x - 1)$

(b) $e^{e^x - 1}$

30. Suppose that $f(x) = \sum_{n=0}^{\infty} a_n x^n = \sum_{n=0}^{\infty} b_n x^n$ for $|x| < R$. Show that $a_n = b_n$ for all n. *Hint:* Let $x = 0$; then differentiate and let $x = 0$ again. Continue.

31. Find the power series representation of $x/(x^2 - 3x + 2)$. *Hint:* Use partial fractions.

32. Let $y = y(x) = x - \frac{x^3}{3!} + \frac{x^5}{5!} - \frac{x^7}{7!} + \cdots$. Show that y satisfies the differential equation $y'' + y = 0$ with the conditions $y(0) = 0$ and $y'(0) = 1$. From this, guess a simple formula for y.

33. Let $\{f_n\}$ be the Fibonacci sequence defined by

$$f_0 = 0, \quad f_1 = 1, \quad f_{n+2} = f_{n+1} + f_n$$

(See Problem 52 of Section 9.1 and Problem 36 of Section 9.6.) If $F(x) = \sum_{n=0}^{\infty} f_n x^n$, show that

$$F(x) - xF(x) - x^2F(x) = x$$

and then use this fact to obtain a simple formula for $F(x)$.

34. Let $y = y(x) = \sum_{n=0}^{\infty} \frac{f_n}{n!} x^n$, where f_n is as in Problem 33. Show that y satisfies the differential equation $y'' - y' - y = 0$.

C **35.** Did you ever wonder how people find the decimal expansion of π to a large number of places? One method depends on the following identity (see Problem 76 of Section 6.8).

$$\pi = 16\tan^{-1}\left(\frac{1}{5}\right) - 4\tan^{-1}\left(\frac{1}{239}\right)$$

Find the first 6 digits of π using this identity and the series for $\tan^{-1} x$. (You will need terms through $x^9/9$ for $\tan^{-1}(\frac{1}{5})$, but only the first term for $\tan^{-1}(1/239)$.) In 1706, John Machin used this method to calculate the first 100 digits of π, while in 1973, Jean Guilloud and Martine Bouyer found the first 1 million digits using the related identity

$$\pi = 48\tan^{-1}\left(\frac{1}{18}\right) + 32\tan^{-1}\left(\frac{1}{57}\right) - 20\tan^{-1}\left(\frac{1}{239}\right)$$

In 1983, π was calculated to over 16 million digits by a somewhat different method. Of course, computers were used in these recent calculations.

36. The number e is readily calculated to as many digits as desired using the rapidly converging series

$$e = 1 + 1 + \frac{1}{2!} + \frac{1}{3!} + \frac{1}{4!} + \cdots$$

This series can also be used to show that e is irrational. Do so by completing the following argument. Suppose that $e = p/q$, where p and q are positive integers. Choose $n > q$ and let

$$M = n!\left(e - 1 - 1 - \frac{1}{2!} - \frac{1}{3!} - \cdots - \frac{1}{n!}\right)$$

Now M is a positive integer. (Why?) Also,

$$M = n!\left[\frac{1}{(n+1)!} + \frac{1}{(n+2)!} + \frac{1}{(n+3)!} + \cdots\right]$$

$$= \frac{1}{n+1} + \frac{1}{(n+1)(n+2)} + \frac{1}{(n+1)(n+2)(n+3)} + \cdots$$

$$< \frac{1}{n+1} + \frac{1}{(n+1)^2} + \frac{1}{(n+1)^3} + \cdots$$

$$= \frac{1}{n}$$

which gives a contradiction (to what?).

Answers to Concepts Review: **1.** integrated; interior **2.** $-x - \frac{1}{2}x^2 - \frac{1}{3}x^3 - \frac{1}{4}x^4 - \frac{1}{5}x^5$ **3.** $1 + x^2 + \frac{1}{2}x^4 + \frac{1}{6}x^6$ **4.** $1 + x + \frac{3}{2}x^2 + \frac{1}{3}x^3 + \frac{3}{4}x^4$

9.8 Taylor and Maclaurin Series

The major question still dangling is this: Given a function f (e.g., $\sin x$ or $\ln(\cos^2 x)$), can we represent it as a power series in x, or more generally, in $x - a$? More precisely, can we find numbers $c_0, c_1, c_2, c_3, \ldots$ such that

$$f(x) = c_0 + c_1(x - a) + c_2(x - a)^2 + c_3(x - a)^3 + \cdots$$

for x belonging to some interval around a?

Suppose that such a representation exists. Then, by the theorem on differentiating series (Theorem 9.7A),

$$f'(x) = c_1 + 2c_2(x - a) + 3c_3(x - a)^2 + 4c_4(x - a)^3 + \cdots$$
$$f''(x) = 2!c_2 + 3!c_3(x - a) + 4 \cdot 3c_4(x - a)^2 + \cdots$$
$$f'''(x) = 3!c_3 + 4!c_4(x - a) + 5 \cdot 4 \cdot 3c_5(x - a)^2 + \cdots$$
$$\vdots$$

When we substitute $x = a$ and solve for c_n, we get

$$c_0 = f(a)$$
$$c_1 = f'(a)$$
$$c_2 = \frac{f''(a)}{2!}$$
$$c_3 = \frac{f'''(a)}{3!}$$

and, more generally,

$$c_n = \frac{f^{(n)}(a)}{n!}$$

(To make this valid for $n = 0$, we define $f^{(0)}(a)$ to mean $f(a)$ and $0!$ to be 1.) Thus, the coefficients c_n are determined by the function f. This also shows that a function f cannot be represented by two different power series in $x - a$, an important point that we have glossed over until now. We summarize in the following theorem.

Theorem A Uniqueness Theorem

Suppose that f satisfies

$$f(x) = c_0 + c_1(x - a) + c_2(x - a)^2 + c_3(x - a)^3 + \cdots$$

for all x in some interval around a. Then

$$c_n = \frac{f^{(n)}(a)}{n!}$$

Thus, a function cannot be represented by more than one power series in $x - a$. The power series representation of a function in $x - a$ is called its **Taylor series** after the English mathematician Brook Taylor (1685–1731). If $a = 0$, the corresponding series is called the **Maclaurin series** after the Scottish mathematician Colin Maclaurin (1698–1746).

Convergence of Taylor Series But the existence question remains. Given a function f, can we represent it in a power series in $x - a$ (which must necessarily be the Taylor series)? The next two theorems give the answer.

Theorem B **Taylor's Formula with Remainder**

Let f be a function whose $(n+1)$st derivative $f^{(n+1)}(x)$ exists for each x in an open interval I containing a. Then, for each x in I,

$$f(x) = f(a) + f'(a)(x-a) + \frac{f''(a)}{2!}(x-a)^2 + \cdots + \frac{f^{(n)}(a)}{n!}(x-a)^n + R_n(x)$$

where the remainder (or error) $R_n(x)$ is given by the formula

$$R_n(x) = \frac{f^{(n+1)}(c)}{(n+1)!}(x-a)^{n+1}$$

and c is some point between x and a.

Proof We will prove the theorem for the special case of $n = 4$; the proof for an arbitrary n follows the same structure and is left as an exercise. (See Problem 37.) First define the function $R_4(x)$ on I by

$$R_4(x) = f(x) - f(a) - f'(a)(x-a) - \frac{f''(a)}{2!}(x-a)^2 - \frac{f'''(a)}{3!}(x-a)^3 - \frac{f^{(4)}(a)}{4!}(x-a)^4$$

Now think of x and a as constants, and define a new function g on I by

$$g(t) = f(x) - f(t) - f'(t)(x-t) - \frac{f''(t)(x-t)^2}{2!} - \frac{f'''(t)(x-t)^3}{3!} - \frac{f^{(4)}(t)(x-t)^4}{4!} - R_4(x)\frac{(x-t)^5}{(x-a)^5}$$

Clearly, $g(x) = 0$ (remember, x is considered fixed) and

$$\begin{aligned} g(a) &= f(x) - f(a) - f'(a)(x-a) - \frac{f''(a)(x-a)^2}{2!} - \frac{f'''(a)(x-a)^3}{3!} - \frac{f^{(4)}(a)(x-a)^4}{4!} - R_4(x)\frac{(x-a)^5}{(x-a)^5} \\ &= R_4(x) - R_4(x) \\ &= 0 \end{aligned}$$

Since a and x are points in I with the property that $g(a) = g(x) = 0$, we can apply the Mean Value Theorem for Derivatives. There exists, therefore, a real number c between a and x such that $g'(c) = 0$. To obtain the derivative of g, we must repeatedly apply the product rule.

$$\begin{aligned} g'(t) &= 0 - f'(t) - [f'(t)(-1) + (x-t)f''(t)] - \frac{1}{2!}[f''(t)2(x-t)(-1) + (x-t)^2 f'''(t)] \\ &\quad - \frac{1}{3!}[f'''(t)3(x-t)^2(-1) + (x-t)^3 f^{(4)}(t)] \\ &\quad - \frac{1}{4!}[f^{(4)}(t)4(x-t)^3(-1) + (x-t)^4 f^{(5)}(t)] - R_4(x)\frac{5(x-t)^4(-1)}{(x-a)^5} \\ &= -\frac{1}{4!}(x-t)^4 f^{(5)}(t) + 5R_4(x)\frac{(x-t)^4}{(x-a)^5} \end{aligned}$$

Thus, by the Mean Value Theorem for Derivatives, there is some c between x and a such that,

$$0 = g'(c) = -\frac{1}{4!}(x-c)^4 f^{(5)}(c) + 5R_4(x)\frac{(x-c)^4}{(x-a)^5}$$

This leads to

$$\frac{1}{4!}(x-c)^4 f^{(5)}(c) = 5R_4(x)\frac{(x-c)^4}{(x-a)^5}$$

$$R_4(x) = \frac{f^{(5)}(c)}{5!}(x-a)^5$$ ■

This theorem tells us what the error can be when we approximate a function with a finite number of terms of its Taylor series. In the next section, we will further exploit the relationship given in Theorem B.

We now—finally—answer the question about whether a function f can be represented by a power series in $x - a$.

Theorem C Taylor's Theorem

Let f be a function with derivatives of all orders in some interval $(a - r, a + r)$. The Taylor series

$$f(a) + f'(a)(x-a) + \frac{f''(a)}{2!}(x-a)^2 + \frac{f'''(a)}{3!}(x-a)^3 + \cdots$$

represents the function f on the interval $(a - r, a + r)$ if and only if

$$\lim_{n\to\infty} R_n(x) = 0$$

where $R_n(x)$ is the remainder in Taylor's Formula,

$$R_n(x) = \frac{f^{(n+1)}(c)}{(n+1)!}(x-a)^{n+1}$$

and c is some point in $(a - r, a + r)$.

Proof We need only recall Taylor's Formula with Remainder (Theorem B),

$$f(x) = f(a) + f'(a)(x-a) + \cdots + \frac{f^{(n)}(a)}{n!}(x-a)^n + R_n(x)$$

and the result follows. ■

Note that if $a = 0$, we get the Maclaurin series

$$f(0) + f'(0)x + \frac{f''(0)}{2!}x^2 + \frac{f'''(0)}{3!}x^3 + \cdots$$

EXAMPLE 1 Find the Maclaurin series for $\sin x$ and prove that it represents $\sin x$ for all x.

Warning

Here is a fact that surprises many students. It is possible that the Taylor series for $f(x)$ converges on an interval but does not represent $f(x)$ there. This is shown by example in Problem 40. Of course,

$$\lim_{n\to\infty} R_n(x) \neq 0$$

in this example.

SOLUTION

$$\begin{aligned} f(x) &= \sin x & f(0) &= 0 \\ f'(x) &= \cos x & f'(0) &= 1 \\ f''(x) &= -\sin x & f''(0) &= 0 \\ f'''(x) &= -\cos x & f'''(0) &= -1 \\ f^{(4)}(x) &= \sin x & f^{(4)}(0) &= 0 \\ &\vdots & &\vdots \end{aligned}$$

Thus,

$$\sin x = x - \frac{x^3}{3!} + \frac{x^5}{5!} - \frac{x^7}{7!} + \cdots$$

and this is valid for all x, provided we can show that

$$\lim_{n\to\infty} R_n(x) = \lim_{n\to\infty} \frac{f^{(n+1)}(c)}{(n+1)!} x^{n+1} = 0$$

Now, $|f^{(n+1)}(x)| = |\cos x|$ or $|f^{(n+1)}(x)| = |\sin x|$, and so

$$|R_n(x)| \le \frac{|x|^{n+1}}{(n+1)!}$$

But $\lim_{n\to\infty} x^n/n! = 0$ for all x, since $x^n/n!$ is the nth term of a convergent series (see Example 3 and Problem 29 of Section 9.6). As a consequence, we see that $\lim_{n\to\infty} R_n(x) = 0$. ■

EXAMPLE 2 Find the Maclaurin series for $\cos x$ and show that it represents $\cos x$ for all x.

SOLUTION We could proceed as in Example 1. However, it is easier to get the result by differentiating the series of that example (a valid procedure according to Theorem 9.7A). We obtain

$$\cos x = 1 - \frac{x^2}{2!} + \frac{x^4}{4!} - \frac{x^6}{6!} + \cdots$$

■

EXAMPLE 3 Find the Maclaurin series for $f(x) = \cosh x$ in two different ways, and show that it represents $\cosh x$ for all x.

SOLUTION

Method 1. This is the direct method.

$$\begin{aligned} f(x) &= \cosh x & f(0) &= 1 \\ f'(x) &= \sinh x & f'(0) &= 0 \\ f''(x) &= \cosh x & f''(0) &= 1 \\ f'''(x) &= \sinh x & f'''(0) &= 0 \\ &\vdots & &\vdots \end{aligned}$$

Thus,

$$\cosh x = 1 + \frac{x^2}{2!} + \frac{x^4}{4!} + \frac{x^6}{6!} + \cdots$$

provided we can show that $\lim_{n\to\infty} R_n(x) = 0$ for all x.

Now let B be an arbitrary number and suppose that $|x| \le B$. Then

$$|\cosh x| = \left|\frac{e^x + e^{-x}}{2}\right| \le \frac{e^x}{2} + \frac{e^{-x}}{2} \le \frac{e^B}{2} + \frac{e^B}{2} = e^B$$

By similar reasoning, $|\sinh x| \le e^B$. Since $f^{(n+1)}(x)$ is either $\cosh x$ or $\sinh x$, we conclude that

$$|R_n(x)| = \left|\frac{f^{(n+1)}(c)x^{n+1}}{(n+1)!}\right| \le \frac{e^B|x|^{n+1}}{(n+1)!}$$

The latter expression tends to zero as $n \to \infty$, just as in Example 1.

Method 2. We use the fact that $\cosh x = (e^x + e^{-x})/2$. From Example 3 of Section 9.7,

$$e^x = 1 + x + \frac{x^2}{2!} + \frac{x^3}{3!} + \frac{x^4}{4!} + \cdots$$

$$e^{-x} = 1 - x + \frac{x^2}{2!} - \frac{x^3}{3!} + \frac{x^4}{4!} - \cdots$$

The previously obtained result follows by adding these two series and dividing by 2. ■

EXAMPLE 4 Find the Maclaurin series for $\sinh x$ and show that it represents $\sinh x$ for all x.

SOLUTION We do both jobs at once when we differentiate the series for $\cosh x$ (Example 3) term by term and use Theorem 9.7A.

$$\sinh x = x + \frac{x^3}{3!} + \frac{x^5}{5!} + \frac{x^7}{7!} + \cdots$$

■

The Binomial Series We are all familiar with the Binomial Formula. For a positive integer p,

$$(1 + x)^p = 1 + \binom{p}{1}x + \binom{p}{2}x^2 + \cdots + \binom{p}{p}x^p$$

where

$$\binom{p}{k} = \frac{p!}{k!(p-k)!} = \frac{p(p-1)(p-2)\cdots(p-k+1)}{k!}$$

Note that if we redefine $\binom{p}{k}$ to be

$$\binom{p}{k} = \frac{p(p-1)(p-2)\cdots(p-k+1)}{k!}$$

then $\binom{p}{k}$ makes sense for *any* real number p, provided that k is a positive integer. Of course, if p is a positive integer, then our new definition reduces to $p!/[k!(p-k)!]$.

Theorem D Binomial Series

For any real number p and for $|x| < 1$,

$$(1 + x)^p = 1 + \binom{p}{1}x + \binom{p}{2}x^2 + \binom{p}{3}x^3 + \cdots$$

Partial Proof Let $f(x) = (1 + x)^p$. Then

$$\begin{aligned} f(x) &= (1+x)^p & f(0) &= 1 \\ f'(x) &= p(1+x)^{p-1} & f'(0) &= p \\ f''(x) &= p(p-1)(1+x)^{p-2} & f''(0) &= p(p-1) \\ f'''(x) &= p(p-1)(p-2)(1+x)^{p-3} & f'''(0) &= p(p-1)(p-2) \\ &\vdots & &\vdots \end{aligned}$$

Thus, the Maclaurin series for $(1 + x)^p$ is as indicated in the theorem. To show that it represents $(1 + x)^p$, we need to show that $\lim_{n\to\infty} R_n(x) = 0$. This unfortunately, is difficult, and we leave it for more advanced courses. (See Problem 38 for a completely different way to prove Theorem D.) ■

If p is a positive integer, $\binom{p}{k} = 0$ for $k > p$, and so the Binomial Series collapses to a series with finitely many terms, the usual Binomial Formula.

EXAMPLE 5 Represent $(1 - x)^{-2}$ in a Maclaurin series for $-1 < x < 1$.

SOLUTION By Theorem D,

$$\begin{aligned} (1+x)^{-2} &= 1 + (-2)x + \frac{(-2)(-3)}{2!}x^2 + \frac{(-2)(-3)(-4)}{3!}x^3 + \cdots \\ &= 1 - 2x + 3x^2 - 4x^3 + \cdots \end{aligned}$$

Thus,

$$(1 - x)^{-2} = 1 + 2x + 3x^2 + 4x^3 + \cdots$$

Naturally, this agrees with a result we obtained by a different method in Example 1 of Section 9.7. ■

EXAMPLE 6 Represent $\sqrt{1 + x}$ in a Maclaurin series and use it to approximate $\sqrt{1.1}$ to five decimal places.

SOLUTION For $|x| < 1$ we have from Theorem D,

$$\begin{aligned} (1+x)^{1/2} &= 1 + \frac{1}{2}x + \frac{\left(\frac{1}{2}\right)\left(-\frac{1}{2}\right)}{2!}x^2 + \frac{\left(\frac{1}{2}\right)\left(-\frac{1}{2}\right)\left(-\frac{3}{2}\right)}{3!}x^3 \\ &\quad + \frac{\left(\frac{1}{2}\right)\left(-\frac{1}{2}\right)\left(-\frac{3}{2}\right)\left(-\frac{5}{2}\right)}{4!}x^4 + \cdots \\ &= 1 + \frac{1}{2}x - \frac{1}{8}x^2 + \frac{1}{16}x^3 - \frac{5}{128}x^4 + \cdots \end{aligned}$$

Since $|0.1| < 1$, we conclude that

$$\begin{aligned} \sqrt{1.1} = (1 + 0.1)^{1/2} &= 1 + \frac{0.1}{2} - \frac{0.01}{8} + \frac{0.001}{16} - \frac{5(0.0001)}{128} + \cdots \\ &\approx 1.04881 \end{aligned}$$

■

EXAMPLE 7 Compute $\int_0^{0.4} \sqrt{1 + x^4}\,dx$ to five decimal places.

SOLUTION From Example 6,

$$\sqrt{1 + x^4} = 1 + \frac{1}{2}x^4 - \frac{1}{8}x^8 + \frac{1}{16}x^{12} - \frac{5}{128}x^{16} + \cdots$$

Thus,

$$\int_0^{0.4} \sqrt{1 + x^4}\, dx = \left[x + \frac{x^5}{10} - \frac{x^9}{72} + \frac{x^{13}}{208} + \cdots\right]_0^{0.4} \approx 0.40102$$ ■

Summary We conclude our discussion of series with a list of the important Maclaurin series we have found. These series will be useful in doing the problem set, but, what is more significant, they find application throughout mathematics and science.

Important Maclaurin Series

1. $\dfrac{1}{1 - x} = 1 + x + x^2 + x^3 + x^4 + \cdots$ $\qquad -1 < x < 1$

2. $\ln(1 + x) = x - \dfrac{x^2}{2} + \dfrac{x^3}{3} - \dfrac{x^4}{4} + \dfrac{x^5}{5} - \cdots$ $\qquad -1 < x \le 1$

3. $\tan^{-1} x = x - \dfrac{x^3}{3} + \dfrac{x^5}{5} - \dfrac{x^7}{7} + \dfrac{x^9}{9} + \cdots$ $\qquad -1 \le x \le 1$

4. $e^x = 1 + x + \dfrac{x^2}{2!} + \dfrac{x^3}{3!} + \dfrac{x^4}{4!} + \cdots$

5. $\sin x = x - \dfrac{x^3}{3!} + \dfrac{x^5}{5!} - \dfrac{x^7}{7!} + \dfrac{x^9}{9!} - \cdots$

6. $\cos x = 1 - \dfrac{x^2}{2!} + \dfrac{x^4}{4!} - \dfrac{x^6}{6!} + \dfrac{x^8}{8!} - \cdots$

7. $\sinh x = x + \dfrac{x^3}{3!} + \dfrac{x^5}{5!} + \dfrac{x^7}{7!} + \dfrac{x^9}{9!} + \cdots$

8. $\cosh x = 1 + \dfrac{x^2}{2!} + \dfrac{x^4}{4!} + \dfrac{x^6}{6!} + \dfrac{x^8}{8!} + \cdots$

9. $(1 + x)^p = 1 + \binom{p}{1}x + \binom{p}{2}x^2 + \binom{p}{3}x^3 + \binom{p}{4}x^4 + \cdots$ $\qquad -1 < x < 1$

Concepts Review

1. If a function $f(x)$ is represented by the power series $\Sigma c_k x^k$, then $c_k =$ ______.

2. The Taylor series for a function will represent the function for those x for which the remainder $R_n(x)$ in Taylor's Formula satisfies ______.

3. The Maclaurin series for $\sin x$ represents $\sin x$ for ______ $< x <$ ______.

4. The first four terms in the Maclaurin series for $(1 + x)^{1/3}$ are ______.

Problem Set 9.8

In Problems 1–18, find the terms through x^5 in the Maclaurin series for $f(x)$. Hint: It may be easiest to use known Maclaurin series and then perform multiplications, divisions, and so on. For example, $\tan x = (\sin x)/(\cos x)$.

1. $f(x) = \tan x$

2. $f(x) = \tanh x$

3. $f(x) = e^x \sin x$

4. $f(x) = e^{-x} \cos x$

5. $f(x) = (\cos x)\ln(1 + x)$

6. $f(x) = (\sin x)\sqrt{1 + x}$

7. $f(x) = e^x + x + \sin x$

8. $f(x) = \dfrac{\cos x - 1 + x^2/2}{x^4}$

9. $f(x) = \dfrac{1}{1 - x}\cosh x$

10. $f(x) = \dfrac{1}{1 + x}\ln\left(\dfrac{1}{1 + x}\right) = \dfrac{-\ln(1 + x)}{1 + x}$

11. $f(x) = \dfrac{1}{1 + x + x^2}$

12. $f(x) = \dfrac{1}{1 - \sin x}$

13. $f(x) = \sin^3 x$

14. $f(x) = x(\sin 2x + \sin 3x)$

15. $f(x) = x\sec(x^2) + \sin x$ 16. $f(x) = \dfrac{\cos x}{\sqrt{1 + x}}$

17. $f(x) = (1 + x)^{3/2}$ 18. $f(x) = (1 - x^2)^{2/3}$

In Problems 19–24, find the Taylor series in $x - a$ through the term $(x - a)^3$.

19. $e^x, a = 1$ 20. $\sin x, a = \dfrac{\pi}{6}$

21. $\cos x, a = \dfrac{\pi}{3}$ 22. $\tan x, a = \dfrac{\pi}{4}$

23. $1 + x^2 + x^3, a = 1$

24. $2 - x + 3x^2 - x^3, a = -1$

25. Let $f(x) = \Sigma a_n x^n$ be an even function $(f(-x) = f(x))$ for x in $(-R, R)$. Prove that $a_n = 0$ if n is odd. *Hint:* Use the Uniqueness Theorem.

26. State and prove a theorem analogous to that in Problem 25 for odd functions.

27. Recall that

$$\sin^{-1} x = \int_0^x \frac{1}{\sqrt{1 - t^2}}\, dt$$

Find the first four nonzero terms in the Maclaurin series for $\sin^{-1} x$.

28. Given that

$$\sinh^{-1} x = \int_0^x \frac{1}{\sqrt{1 + t^2}}\, dt$$

find the first four nonzero terms in the Maclaurin series for $\sinh^{-1} x$.

[C] 29. Calculate, accurate to four decimal places,

$$\int_0^1 \cos(x^2)\, dx$$

[C] 30. Calculate, accurate to five decimal places,

$$\int_0^{0.5} \sin\sqrt{x}\, dx$$

31. By writing $1/x = 1/[1 - (1 - x)]$ and using the known expansion of $1/(1 - x)$, find the Taylor series for $1/x$ in powers of $x - 1$.

32. Let $f(x) = (1 + x)^{1/2} + (1 - x)^{1/2}$. Find the Maclaurin series for f and use it to find $f^{(4)}(0)$ and $f^{(51)}(0)$.

33. In each case, find the Maclaurin series for $f(x)$ by use of known series and then use it to calculate $f^{(4)}(0)$.

(a) $f(x) = e^{x+x^2}$

(b) $f(x) = e^{\sin x}$

(c) $f(x) = \displaystyle\int_0^x \frac{e^{t^2} - 1}{t^2}\, dt$

(d) $f(x) = e^{\cos x} = e \cdot e^{\cos x - 1}$

(e) $f(x) = \ln(\cos^2 x)$

34. One can sometimes find a Maclaurin series by the *method of equating coefficients*. For example, let

$$\tan x = \frac{\sin x}{\cos x} = a_0 + a_1 x + a_2 x^2 + \cdots$$

Then multiply by $\cos x$ and replace $\sin x$ and $\cos x$ by their series to obtain

$$x - \frac{x^3}{6} + \cdots = (a_0 + a_1 x + a_2 x^2 + \cdots)\left(1 - \frac{x^2}{2} + \cdots\right)$$
$$= a_0 + a_1 x + \left(a_2 - \frac{a_0}{2}\right)x^2 + \left(a_3 - \frac{a_1}{2}\right)x^3 + \cdots$$

Thus,

$$a_0 = 0, \quad a_1 = 1, \quad a_2 - \frac{a_0}{2} = 0, \quad a_3 - \frac{a_1}{2} = -\frac{1}{6}, \quad \cdots$$

so

$$a_0 = 0, \quad a_1 = 1, \quad a_2 = 0, \quad a_3 = \tfrac{1}{3}, \quad \cdots$$

and therefore

$$\tan x = 0 + x + 0 + \tfrac{1}{3}x^3 + \cdots$$

which agrees with Problem 1. Use this method to find the terms through x^4 in the series for $\sec x$.

35. Use the method of Problem 34 to find the terms through x^5 in the Maclaurin series for $\tanh x$.

36. Use the method of Problem 34 to find the terms through x^4 in the series for $\operatorname{sech} x$.

37. Prove Theorem B for

(a) the special case of $n = 3$, and

(b) an arbitrary n.

38. Prove Theorem D as follows: Let

$$f(x) = 1 + \sum_{n=1}^{\infty} \binom{p}{n} x^n$$

(a) Show that the series converges for $|x| < 1$.

(b) Show that $(1 + x)f'(x) = pf(x)$ and $f(0) = 1$.

(c) Solve this differential equation to get $f(x) = (1 + x)^p$.

39. Let

$$f(t) = \begin{cases} 0 & \text{if } t < 0 \\ t^4 & \text{if } t \ge 0 \end{cases}$$

Explain why $f(t)$ cannot be represented by a Maclaurin series. Also show that, if $g(t)$ gives the distance traveled by a car that is stationary for $t < 0$ and moving ahead for $t \ge 0$, $g(t)$ cannot be represented by a Maclaurin series.

40. Let

$$f(x) = \begin{cases} e^{-1/x^2} & \text{if } x \ne 0 \\ 0 & \text{if } x = 0 \end{cases}$$

(a) Show that $f'(0) = 0$ by using the definition of the derivative.

(b) Show that $f''(0) = 0$.

(c) Assuming the known fact that $f^{(n)}(0) = 0$ for all n, find the Maclaurin series for $f(x)$.

(d) Does the Maclaurin series represent $f(x)$?

(e) When $a = 0$, the formula in Theorem B is called **Maclaurin's Formula.** What is the remainder in Maclaurin's Formula for $f(x)$?

This shows that a Maclaurin series may exist and yet not represent the given function (the remainder does not tend to 0 as $n \to \infty$).

CAS *Use a CAS to find the first four nonzero terms in the Maclaurin series for each of the following. Check Problems 43–48 to see that you get the same answers using the methods of Section 9.7.*

41. $\sin x$ **42.** $\exp x$

43. $3 \sin x - 2 \exp x$ **44.** $\exp(x^2)$

45. $\sin(\exp x - 1)$ **46.** $\exp(\sin x)$

47. $(\sin x)(\exp x)$ **48.** $(\sin x)/(\exp x)$

Answers to Concepts Review: **1.** $f^{(k)}(0)/k!$ **2.** $\lim_{n\to\infty} R_n(x) = 0$ **3.** $-\infty; \infty$ **4.** $1 + \frac{1}{3}x - \frac{1}{9}x^2 + \frac{5}{81}x^3$

9.9 The Taylor Approximation to a Function

The Taylor and Maclaurin series introduced in the previous section cannot be used directly to approximate a function such as e^x or $\tan x$. However, truncating a Taylor or Maclaurin series, that is, cutting off the series after a finite number of terms, leads to a polynomial that we *can* use to approximate a function. Such polynomials are called Taylor or Maclaurin polynomials.

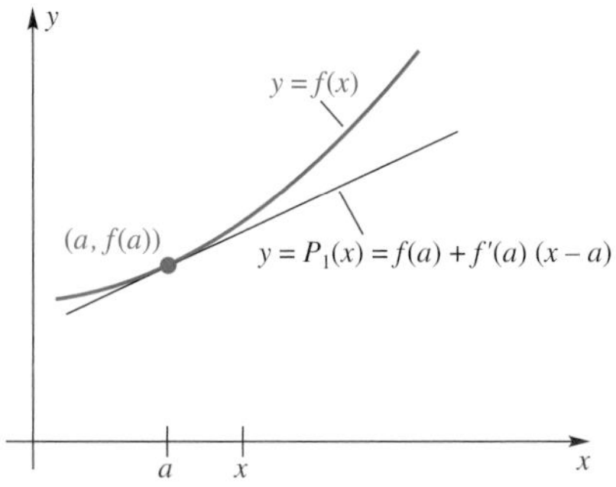

Figure 1

The Taylor Polynomial of Order 1 In Section 2.9 we emphasized that a function f can be approximated near a point a by its tangent line through the point $(a, f(a))$ (see Figure 1). We called such a line the linear approximation to f near a and we found it to be

$$P_1(x) = f(a) + f'(a)(x - a)$$

After studying Taylor series in Section 9.8, you should recognize that $P_1(x)$ is composed of the first two terms, that is, the terms of *order* 0 and 1, of the Taylor series of f expanded about a. We therefore call P_1 the **Taylor polynomial of order 1 based at *a*.** As Figure 1 suggests, we can expect $P_1(x)$ to be a good approximation to $f(x)$ only near $x = a$.

EXAMPLE 1 Find $P_1(x)$ based at $a = 1$ for $f(x) = \ln x$ and use it to approximate $\ln 0.9$ and $\ln 1.5$.

SOLUTION Since $f(x) = \ln x, f'(x) = 1/x$; thus, $f(1) = 0$ and $f'(1) = 1$. Therefore,

$$P_1(x) = 0 + 1(x - 1) = x - 1$$

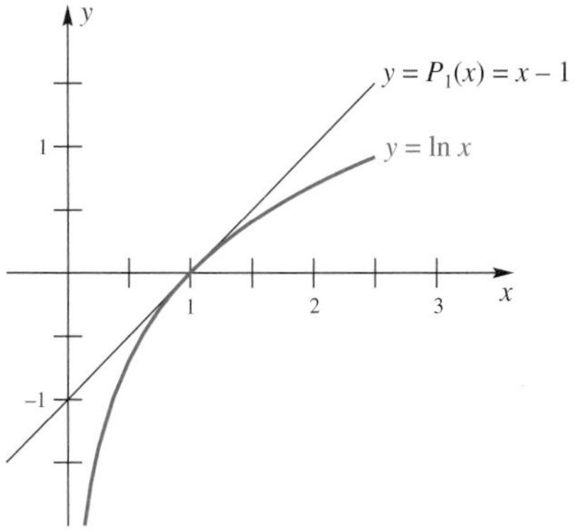

Figure 2

Consequently (see Figure 2), for x near 1,

$$\ln x \approx x - 1$$

and

$$\ln 0.9 \approx 0.9 - 1 = -0.1$$

$$\ln 1.5 \approx 1.5 - 1 = 0.5$$

The correct four-place values of $\ln 0.9$ and $\ln 1.5$ are -0.1054 and 0.4055. As expected, the approximation is much better for $\ln 0.9$ than for $\ln 1.5$, since 0.9 is closer to 1 than is 1.5. ■

The Taylor Polynomial of Order *n* The linear approximation $P_1(x)$ works well when x is near a, but less so when x is not close to a. As you might expect, summing to higher-order terms in the Taylor series will usually give a better approximation. Thus, the quadratic polynomial

$$P_2(x) = f(a) + f'(a)(x - a) + \frac{f''(a)}{2}(x - a)^2$$

which is composed of the first three terms of the Taylor series for f, will give a better approximation to f than the linear approximation $P_1(x)$. The **Taylor polynomial of order n based at a** is

$$P_n(x) = f(a) + f'(a)(x - a) + \frac{f''(a)}{2!}(x - a)^2 + \cdots + \frac{f^{(n)}(a)}{n!}(x - a)^n$$

EXAMPLE 2 Find $P_2(x)$ based at $a = 1$ for $f(x) = \ln x$ and use it to approximate $\ln 0.9$ and $\ln 1.5$.

SOLUTION Here $f(x) = \ln x$, $f'(x) = 1/x$, $f''(x) = -1/x^2$, and so $f(1) = 0$, $f'(1) = 1$, and $f''(1) = -1$. Therefore,

$$P_2(x) = 0 + 1(x - 1) - \frac{1}{2}(x - 1)^2$$

Consequently, for x near 1,

$$\ln x \approx (x - 1) - \frac{1}{2}(x - 1)^2$$

and

$$\ln 0.9 \approx (0.9 - 1) - \frac{1}{2}(0.9 - 1)^2 = -0.1050$$

$$\ln 1.5 \approx (1.5 - 1) - \frac{1}{2}(1.5 - 1)^2 = 0.3750$$

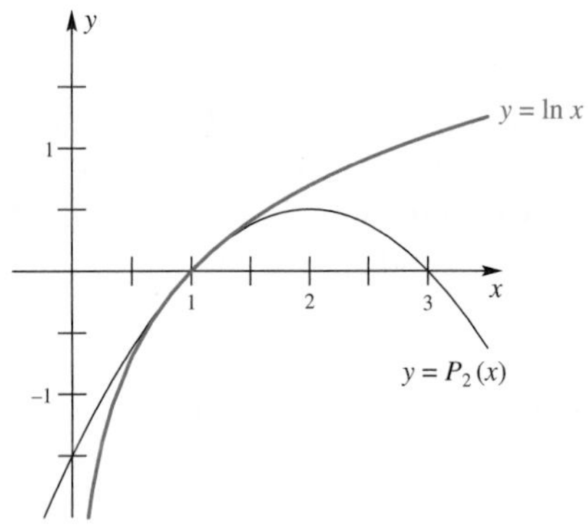

Figure 3

As expected, these are better approximations than we got using the linear approximation $P_1(x)$ (Example 1). Figure 3 shows the graph of $y = \ln x$ and the approximation $P_2(x)$. ■

Maclaurin Polynomials When $a = 0$, the Taylor polynomial of order n simplifies to the **Maclaurin polynomial of order n,** which gives a particularly useful approximation near $x = 0$:

$$f(x) \approx P_n(x) = f(0) + f'(0)x + \frac{f''(0)}{2!}x^2 + \cdots + \frac{f^{(n)}(0)}{n!}x^n$$

Order versus Degree

We have chosen the terminology Taylor (and Maclaurin) polynomial of *order n* because the highest-order derivative involved in its construction is of order n. Note that this polynomial can have *degree* less than n if $f^{(n)}(a) = 0$. If n is odd in Example 3, then the Maclaurin polynomial of order n for $\cos x$ will be of degree $n - 1$. For example, the Maclaurin polynomial of order 5 for $\cos x$ is

$$1 - \tfrac{1}{2}x^2 + \tfrac{1}{24}x^4$$

a polynomial of degree 4.

EXAMPLE 3 Find the Maclaurin polynomials of order n for e^x and $\cos x$. Then approximate $e^{0.2}$ and $\cos(0.2)$ using $n = 4$.

SOLUTION The calculation of the required derivatives is shown in the table.

n			At $x = 0$		At $x = 0$
0	$f(x)$	e^x	1	$\cos x$	1
1	$f'(x)$	e^x	1	$-\sin x$	0
2	$f''(x)$	e^x	1	$-\cos x$	-1
3	$f^{(3)}(x)$	e^x	1	$\sin x$	0
4	$f^{(4)}(x)$	e^x	1	$\cos x$	1
5	$f^{(5)}(x)$	e^x	1	$-\sin x$	0
⋮	⋮	⋮	⋮	⋮	⋮

It follows that

$$e^x \approx 1 + x + \frac{1}{2!}x^2 + \frac{1}{3!}x^3 + \frac{1}{4!}x^4 + \cdots + \frac{1}{n!}x^n$$

$$\cos x \approx 1 - \frac{1}{2!}x^2 + \frac{1}{4!}x^4 - \cdots + (-1)^{n/2}\frac{1}{n!}x^n \quad (n \text{ even})$$

Thus, using $n = 4$ and $x = 0.2$, we obtain

$$e^{0.2} \approx 1 + 0.2 + \frac{(0.2)^2}{2} + \frac{(0.2)^3}{6} + \frac{(0.2)^4}{24} = 1.2214000$$

$$\cos(0.2) \approx 1 - \frac{(0.2)^2}{2} + \frac{(0.2)^4}{24} = 0.9800667$$

Compare these results with the correct seven-place values of 1.2214028 and 0.9800666. ■

For a visual idea of how the Maclaurin polynomials provide approximations to $\cos x$, we have sketched the graphs of $P_1(x)$ through $P_5(x)$ and $P_8(x)$, along with the graph of $\cos x$, in Figure 4.

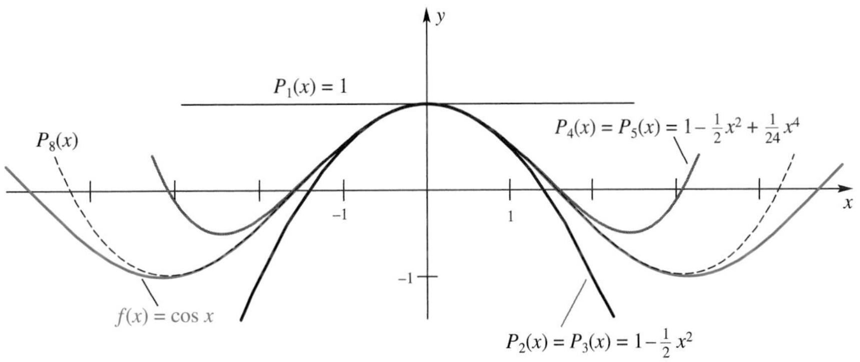

Maclaurin approximations to $f(x) = \cos x$

Figure 4

In Example 3 we used the Maclaurin polynomial of order 4 to approximate $\cos(0.2)$ as follows:

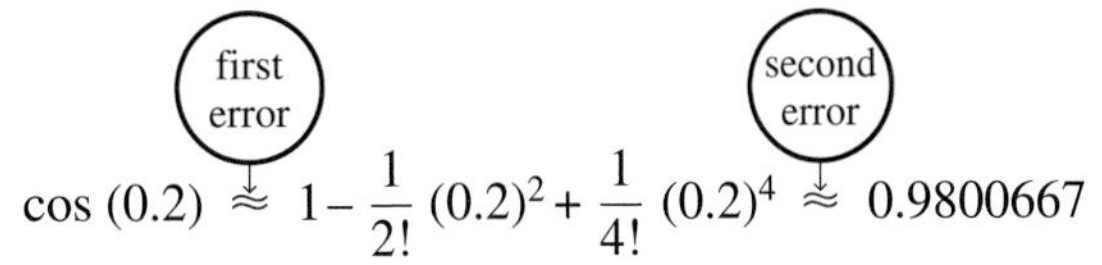

$$\cos(0.2) \approx 1 - \frac{1}{2!}(0.2)^2 + \frac{1}{4!}(0.2)^4 \approx 0.9800667$$

This example illustrates the two kinds of errors that occur in approximation processes. First, there is the **error of the method.** In this case, we approximated $\cos x$ by a fourth-degree polynomial instead of evaluating the *exact* sum of the series. Second, there is the **error of calculation.** This includes errors due to rounding, as when we replaced the unending decimal 0.9800666 ... by 0.9800667 in the last term above.

Now notice a sad fact of the numerical analyst's life. We can reduce the error of the method by using Maclaurin polynomials of higher order. But using polynomials of higher order means more calculations, which potentially increases the error of calculation. To be a good numerical analyst is to know how to compromise between these two types of error. Unfortunately, this is more of an art than a science. However, we can say something definite about the first type of error, the subject to which we now turn.

The Error in the Method In Section 9.8 we gave a formula for the error of approximating a function by its Taylor polynomial. Taylor's Formula with Remainder is

$$\begin{aligned} f(x) &= f(a) + f'(a)(x-a) + \frac{f''(a)}{2!}(x-a)^2 + \cdots \\ &\quad + \frac{f^{(n)}(a)}{n!}(x-a)^n + R_n(x) \\ &= P_n(x) + R_n(x) \end{aligned}$$

The error, or remainder, $R_n(x)$ is given by

$$R_n(x) = \frac{f^{(n+1)}(c)}{(n+1)!}(x-a)^{n+1}$$

where c is some real number between a and x. This formula for the error is due to the French–Italian mathematician Joseph Louis Lagrange (1736–1813) and is often called the Lagrange error bound for Taylor polynomials. When $a = 0$, Taylor's Formula is called **Maclaurin's Formula.**

One problem that you might foresee at this point is that we do not know what c is; all we know is that it is some real number between a and x. For most problems we must settle for a bound on the remainder using the known bounds on c. The next example illustrates this point.

EXAMPLE 4 Approximate $e^{0.8}$ with an error of less than 0.001.

SOLUTION For $f(x) = e^x$, Maclaurin's Formula gives the remainder

$$R_n(x) = \frac{f^{(n+1)}(c)}{(n+1)!}x^{n+1} = \frac{e^c}{(n+1)!}x^{n+1}$$

and so

$$R_n(0.8) = \frac{e^c}{(n+1)!}(0.8)^{n+1}$$

where $0 < c < 0.8$. Our goal is to choose n large enough so that $|R_n(0.8)| < 0.001$. Now, $e^c < e^{0.8} < 3$ and $(0.8)^{n+1} < (1)^{n+1}$, and so

$$|R_n(0.8)| < \frac{3(1)^{n+1}}{(n+1)!} = \frac{3}{(n+1)!}$$

It is easy to check that $3/(n+1)! < 0.001$ when $n \ge 6$, and so we can obtain the desired accuracy by using the Maclaurin polynomial of order 6:

$$e^{0.8} \approx 1 + (0.8) + \frac{(0.8)^2}{2!} + \frac{(0.8)^3}{3!} + \frac{(0.8)^4}{4!} + \frac{(0.8)^5}{5!} + \frac{(0.8)^6}{6!}$$

Our calculator gives 2.2254948 for this sum.

Can we be sure that this value is within 0.001 of the true result? Certainly the error of the method is less than 0.001. But could the error of calculation have distorted our answer? Possibly so; however, so few calculations are involved that we feel confident in reporting an answer of 2.2255 accurate within 0.001. ■

Useful Tools for Bounding $|R_n|$ The precise value of R_n is almost never obtainable, since we do not know c, only that c lies on a certain interval. Our task is therefore to find the maximum possible value of $|R_n|$ for c in the given interval. To do this exactly is often difficult, so we usually content ourselves with getting a "good" upper bound for $|R_n|$. This involves a sensible use of inequalities. Our chief tools are the triangle inequality, $|a \pm b| \le |a| + |b|$, and the fact that a fraction gets larger when we make its numerator larger or its denominator smaller.

EXAMPLE 5 If c is known to be in $[2, 4]$, give a good bound for the maximum value of

$$\left|\frac{c^2 - \sin c}{c}\right|$$

SOLUTION

$$\left|\frac{c^2 - \sin c}{c}\right| = \frac{|c^2 - \sin c|}{|c|} \le \frac{|c^2| + |\sin c|}{|c|} \le \frac{4^2 + 1}{2} = 8.5$$

A different and better bound is obtained as follows:

$$\left|\frac{c^2 - \sin c}{c}\right| = \left|c - \frac{\sin c}{c}\right| \le |c| + \left|\frac{\sin c}{c}\right| \le 4 + \frac{1}{2} = 4.5$$

■

EXAMPLE 6 Use a Taylor polynomial of order 2 to approximate cos 62° and then give a bound for the error of the approximation.

SOLUTION Since 62° is near 60° (whose cosine and sine are known), we use radian measure and the Taylor polynomial based at $a = \pi/3$.

$$f(x) = \cos x \qquad f\left(\frac{\pi}{3}\right) = \frac{1}{2}$$

$$f'(x) = -\sin x \qquad f'\left(\frac{\pi}{3}\right) = -\frac{\sqrt{3}}{2}$$

$$f''(x) = -\cos x \qquad f''\left(\frac{\pi}{3}\right) = -\frac{1}{2}$$

$$f'''(x) = \sin x \qquad f'''(c) = \sin c$$

Now

$$62^\circ = \frac{\pi}{3} + \frac{\pi}{90} \text{ radians}$$

Thus,

$$\cos x = \frac{1}{2} - \frac{\sqrt{3}}{2}\left(x - \frac{\pi}{3}\right) - \frac{1}{4}\left(x - \frac{\pi}{3}\right)^2 + R_2(x)$$

and

$$\cos\left(\frac{\pi}{3} + \frac{\pi}{90}\right) = \frac{1}{2} - \frac{\sqrt{3}}{2}\left(\frac{\pi}{90}\right) - \frac{1}{4}\left(\frac{\pi}{90}\right)^2 + R_2\left(\frac{\pi}{3} + \frac{\pi}{90}\right)$$

$$\approx 0.4694654 + R_2$$

and

$$|R_2| = \left|\frac{\sin c}{3!}\left(\frac{\pi}{90}\right)^3\right| < \frac{1}{6}\left(\frac{\pi}{90}\right)^3 \approx 0.0000071$$

Again the number of calculations is small, so we feel safe in reporting cos 62° = 0.4694654 with an error of less than 0.0000071. ■

The Error of Calculation In all our examples so far, we have assumed that the error of calculation is small enough so that it can be ignored. We will ordinarily make that assumption in this book, since our problems will always involve a small number of calculations. We feel obligated, however, to warn you that when computers are used to do thousands or millions of operations, these errors of calculation may well accumulate and distort an answer.

There are two sources of calculation errors that may be significant even in using a calculator. Consider calculating

$$a + b_1 + b_2 + b_3 + \cdots + b_m$$

where a is very much larger than any of the b's; for example, $a = 10{,}000{,}000$ and $b_i = 0.4$, $i = 1, 2, \ldots, m$. If we use eight-digit floating-point arithmetic and proceed from left to right, first adding b_1 to a, then adding b_2 to the result, and so on, we will simply get 10,000,000 at each stage. Yet a sum of just 25 of the b's ought to affect the seventh digit of the overall sum. The moral here is that in adding a large number of small terms to one or more large ones, it is wise to find the sum of the small terms first. Whenever possible, add the numbers from smallest to largest.

A more likely source of calculation error is due to the loss of significant digits in a subtraction of nearly equal numbers. For example, subtracting 0.823421 from 0.823445, each with six significant digits, results in 0.000024, which has only two significant digits. That this can cause trouble is easily illustrated by calculating a numerical approximation to a derivative.

Consider calculating $f'(2)$ for $f(x) = x^4$ by using the difference quotient

$$f'(2) \approx \frac{f(2 + h) - f(2)}{h} = \frac{(2 + 10^{-n})^4 - 2^4}{10^{-n}}$$

Theoretically, as n increases (and $h = 10^{-n}$ correspondingly decreases) the result should get closer and closer to the correct value, 32. But note what happens on one eight-digit calculator when n gets too large. Problems like this arise even if we use 16-digit or 32-digit floating-point arithmetic. Regardless of the number of significant digits used in the calculations, the difference quotient in the following table will be 0 for sufficiently large n.

n	$(2 + 10^{-n})^4 - 2^4$	$[(2 + 10^{-n})^4 - 2^4]/10^{-n}$
2	0.32240801	32.240801
3	0.03202401	32.024010
4	0.00320024	32.002400
5	0.00032000	32.000000
6	0.00003200	32.000000
7	0.00000320	32.000000
8	0.00000032	32.000000
9	0.00000003	30.000000
10	0.00000000	0.000000
⋮	⋮	⋮

Concepts Review

1. If $P_2(x)$ is the Taylor polynomial of order 2 based at 1 for $f(x)$, then $P_2(1) =$ _____, $P_2'(1) =$ _____, and $P_2''(1) =$ _____.

2. The coefficient of x^6 in the Maclaurin polynomial of order 9 for $f(x)$ is _____.

3. The two types of errors that arise in approximation theory are called _____ and _____.

4. Calculation errors in using Taylor's Formula tend to _____ as n increases, whereas errors of the method tend to _____ as n increases.

Problem Set 9.9

C *In Problems 1–8, find the Maclaurin polynomial of order 4 for $f(x)$ and use it to approximate $f(0.12)$.*

1. $f(x) = e^{2x}$

2. $f(x) = e^{-3x}$

3. $f(x) = \sin 2x$

4. $f(x) = \tan x$

5. $f(x) = \ln(1 + x)$

6. $f(x) = \sqrt{1 + x}$

7. $f(x) = \tan^{-1} x$

8. $f(x) = \sinh x$

C *In Problems 9–14, find the Taylor polynomial of order 3 based at a for the given function.*

9. e^x; $a = 1$

10. $\sin x$; $a = \dfrac{\pi}{4}$

11. $\tan x$; $a = \dfrac{\pi}{6}$

12. $\sec x$; $a = \dfrac{\pi}{4}$

13. $\cot^{-1} x$; $a = 1$

14. $\sqrt{x}$; $a = 2$

15. Find the Taylor polynomial of order 3 based at 1 for $f(x) = x^3 - 2x^2 + 3x + 5$ and show that it is an exact representation of $f(x)$.

16. Find the Taylor polynomial of order 4 based at 2 for $f(x) = x^4$ and show that it represents $f(x)$ exactly.

17. Find the Maclaurin polynomial of order n for $f(x) = 1/(1 - x)$. Then use it with $n = 4$ to approximate each of the following.

(a) $f(0.1)$ (b) $f(0.5)$ (c) $f(0.9)$ (d) $f(2)$

C **18.** Find the Maclaurin polynomial of order n (n odd) for $\sin x$. Then use it with $n = 5$ to approximate each of the following. (This example should convince you that the Maclaurin approximation can be exceedingly poor if x is far from zero.) Compare your answers with those given by your calculator. What conclusion do you draw?

(a) $\sin(0.1)$ (b) $\sin(0.5)$ (c) $\sin(1)$ (d) $\sin(10)$

CAS *In Problems 19–28, plot on the same axes the given function along with the Maclaurin polynomials of orders 1, 2, 3, and 4.*

19. $\cos 2x$

20. $\sin x$

21. $\sin x^2$

22. $\cos(x - \pi)$

23. e^{-x^2}

24. $e^{\sin x}$

25. $\sin e^x$

26. $\sin(\ln(1 + x))$

27. $\dfrac{\sin x}{2 + \sin x}$

28. $\dfrac{1}{1 + x^2}$

In Problems 29–36, find a good bound for the maximum value of the given expression, given that c is in the stated interval. Answers may vary depending on the technique used. (See Example 5.)

29. $|e^{2c} + e^{-2c}|$; $[0, 3]$

30. $|\tan c + \sec c|$; $\left[0, \dfrac{\pi}{4}\right]$

31. $\left|\dfrac{4c}{\sin c}\right|$; $\left[\dfrac{\pi}{4}, \dfrac{\pi}{2}\right]$

32. $\left|\dfrac{4c}{c + 4}\right|$; $[0, 1]$

33. $\left|\dfrac{e^c}{c + 5}\right|$; $[-2, 4]$

34. $\left|\dfrac{\cos c}{c + 2}\right|$; $\left[0, \dfrac{\pi}{4}\right]$

35. $\left|\dfrac{c^2 + \sin c}{10 \ln c}\right|$; $[2, 4]$

36. $\left|\dfrac{c^2 - c}{\cos c}\right|$; $\left[0, \dfrac{\pi}{4}\right]$

In Problems 37–42, find a formula for $R_6(x)$, the remainder for the Taylor polynomial of order 6 based at a. Then obtain a good bound for $|R_6(0.5)|$. See Examples 4 and 6.

37. $\ln(2 + x)$; $a = 0$

38. e^{-x}; $a = 1$

39. $\sin x$; $a = \pi/4$

40. $\dfrac{1}{x - 3}$; $a = 1$

41. $\dfrac{1}{x}$; $a = 1$

42. $\dfrac{1}{x^2}$; $a = 1$

43. Determine the order n of the Maclaurin polynomial for e^x that is required to approximate e to five decimal places, that is, so that $|R_n(1)| \le 0.000005$ (see Example 4).

44. Determine the order n of the Maclaurin polynomial for $4 \tan^{-1} x$ that is required to approximate $\pi = 4 \tan^{-1} 1$ to five decimal places, that is, so that $|R_n(1)| \le 0.000005$.

45. Find the third-order Maclaurin polynomial for $(1 + x)^{1/2}$ and bound the error $R_3(x)$ for $-0.5 \le x \le 0.5$.

46. Find the third-order Maclaurin polynomial for $(1 + x)^{3/2}$ and bound the error $R_3(x)$ if $-0.1 \le x \le 0$.

47. Find the third-order Maclaurin polynomial for

$$(1 + x)^{-1/2}$$

and bound the error $R_3(x)$ if $-0.05 \le x \le 0.05$.

48. Find the fourth-order Maclaurin polynomial for

$$\ln[(1 + x)/(1 - x)]$$

and bound the error $R_4(x)$ for $-0.5 \le x \le 0.5$.

49. Note that the fourth-order Maclaurin polynomial for $\sin x$ is really of third degree since the coefficient of x^4 is 0. Thus,

$$\sin x = x - \frac{x^3}{6} + R_4(x)$$

Show that if $0 \le x \le 0.5$, $|R_4(x)| \le 0.0002605$. Use this result to approximate $\int_0^{0.5} \sin x\, dx$ and give a bound for the error.

50. In analogy with Problem 49,

$$\cos x = 1 - \frac{x^2}{2} + \frac{x^4}{24} + R_5(x)$$

If $0 \le x \le 1$, give a good bound for $|R_5(x)|$. Then use your result to approximate $\int_0^1 \cos x\, dx$ and give a bound for the error.

51. Problem 49 suggests that if n is odd, then the nth order Maclaurin polynomial for $\sin x$ is also the $(n + 1)$st order polynomial, so the error can be calculated using R_{n+1}. Use this result to find how large n must be so that $|R_{n+1}(x)|$ is less than 0.00005 for all x in the interval $0 \le x \le \pi/2$. (Note, n must be odd.)

52. Problem 50 suggests that if n is even, then the nth order Maclaurin polynomial for $\cos x$ is also the $(n + 1)$st order polynomial, so the error can be calculated using R_{n+1}. Use this result to find how large n must be so that $|R_{n+1}(x)|$ is less than 0.00005 for all x in the interval $0 \le x \le \pi/2$. (Note, n must be even.)

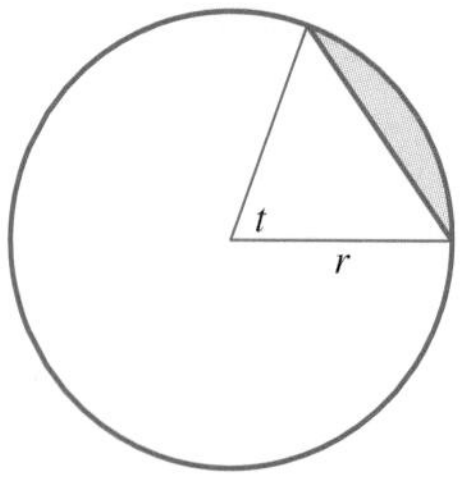

Figure 5

53. Use a Maclaurin polynomial to obtain the approximation $A \approx r^2t^3/12$ for the area of the shaded region in Figure 5. First express A exactly, then approximate.

54. If an object of rest mass m_0 has velocity v, then (according to the theory of relativity) its mass m is given by $m = m_0/\sqrt{1 - v^2/c^2}$, where c is the velocity of light. Explain how physicists get the approximation

$$m \approx m_0 + \frac{m_0}{2}\left(\frac{v}{c}\right)^2$$

55. If money is invested at an interest rate of r compounded monthly, it will double in n years, where n satisfies

$$\left(1 + \frac{r}{12}\right)^{12n} = 2$$

(a) Show that

$$n = \ln 2\left[\frac{1}{12\ln(1 + r/12)}\right]$$

(b) Use the Maclaurin polynomial of order 2 for $\ln(1 + x)$ and a partial fraction decomposition to obtain the approximation

$$n \approx \frac{0.693}{r} + 0.029$$

C (c) Some people use the *Rule of 72*, $n \approx 72/(100r)$, to approximate n. Fill in the table to compare the values obtained from these three formulas.

r	n (*Exact*)	n (*Approximation*)	n (*Rule of 72*)
0.05			
0.10			
0.15			
0.20			

56. The author of a biology text claimed that the smallest positive solution to $x = 1 - e^{-(1+k)x}$ is approximately $x = 2k$, provided k is very small. Show how she reached this conclusion and check it for $k = 0.01$.

57. Expand $x^4 - 3x^3 + 2x^2 + x - 2$ in a Taylor polynomial of order 4 based at 1 and show that $R_4(x) = 0$ for all x.

58. Let $f(x)$ be a function that possesses at least n derivatives at $x = a$ and let $P_n(x)$ be the Taylor polynomial of order n based at a. Show that

$$P_n(a) = f(a),\quad P_n'(a) = f'(a),\quad P_n''(a) = f''(a),\quad \ldots,\quad P_n^{(n)}(a) = f^{(n)}(a)$$

C **59.** Calculate $\sin 43° = \sin(43\pi/180)$ by using the Taylor polynomial of order 3 based at $\pi/4$ for $\sin x$. Then obtain a good bound for the error made. See Example 6.

C **60.** Calculate $\cos 63°$ by the method illustrated in Example 6. Choose n large enough so that $|R_n| \le 0.0005$.

C **61.** Show that if x is in $[0, \pi/2]$ the error in using

$$\sin x \approx x - \frac{x^3}{3!} + \frac{x^5}{5!} - \frac{x^7}{7!} + \frac{x^9}{9!}$$

is less than 5×10^{-5} and therefore, that this formula is good enough to build a four-place sine table.

62. Use Maclaurin's Formula, rather than l'Hôpital's Rule, to find

(a) $\displaystyle\lim_{x\to 0}\frac{\sin x - x + x^3/6}{x^5}$

(b) $\displaystyle\lim_{x\to 0}\frac{\cos x - 1 + x^2/2 - x^4/24}{x^6}$

EXPL **63.** Let $g(x) = p(x) + x^{n+1}f(x)$, where $p(x)$ is a polynomial of degree at most n and f has derivatives through order n. Show that $p(x)$ is the Maclaurin polynomial of order n for g.

EXPL **64.** Recall that the Second Derivative Test for Local Extrema (Section 3.3) does not apply when $f''(c) = 0$. Prove the following generalization, which may help determine a maximum or a minimum when $f''(c) = 0$. Suppose that

$$f'(c) = f''(c) = f'''(c) = \cdots = f^{(n)}(c) = 0$$

where n is odd and $f^{(n+1)}(x)$ is continuous near c.

1. If $f^{(n+1)}(c) < 0$, then $f(c)$ is a local maximum value.
2. If $f^{(n+1)}(c) > 0$, then $f(c)$ is a local minimum value.

Test this result on $f(x) = x^4$.

Answers to Concepts Review: **1.** $f(1); f'(1); f''(1)$ **2.** $f^{(6)}(0)/6!$ **3.** error of the method; error of calculation **4.** increase; decrease

9.10 Chapter Review

Concepts Test

Respond with true or false to each of the following assertions. Be prepared to justify your answer.

1. If $0 \le a_n \le b_n$ for all natural numbers n and $\lim_{n\to\infty} b_n$ exists, then $\lim_{n\to\infty} a_n$ exists.

2. For every positive integer n, $n! \le n^n \le (2n - 1)!$.

3. If $\lim_{n\to\infty} a_n = L$, then $\lim_{n\to\infty} a_{3n+4} = L$.

4. If $\lim_{n\to\infty} a_{2n} = L$ and $\lim_{n\to\infty} a_{3n} = L$, then $\lim_{n\to\infty} a_n = L$.

5. If $\lim_{n\to\infty} a_{mn} = L$ for every positive integer $m \ge 2$, then $\lim_{n\to\infty} a_n = L$.

6. If $\lim_{n\to\infty} a_{2n} = L$ and $\lim_{n\to\infty} a_{2n+1} = L$, then $\lim_{n\to\infty} a_n = L$.

7. If $\lim_{n\to\infty}(a_n - a_{n+1}) = 0$, then $\lim_{n\to\infty} a_n$ exists and is finite.

8. If $\{a_n\}$ and $\{b_n\}$ both diverge, then $\{a_n + b_n\}$ diverges.

9. If $\{a_n\}$ converges, then $\{a_n/n\}$ converges to 0.

10. If $\sum_{n=1}^{\infty} a_n$ converges, so does $\sum_{n=1}^{\infty} a_n^2$.

11. If $0 < a_{n+1} < a_n$ for all natural numbers n, and if $\lim_{n\to\infty} a_n = 0$, then $\sum_{n=1}^{\infty}(-1)^{n+1}a_n$ converges and has sum S satisfying $0 < S < a_1$.

12. $\sum_{n=1}^{\infty}\left(\frac{1}{n}\right)^n$ converges and has sum S satisfying $1 < S < 2$.

13. If a series $\sum a_n$ diverges, then its sequence of partial sums is unbounded.

14. If $0 \le a_n \le b_n$ for all natural numbers n, and if $\sum_{n=1}^{\infty} b_n$ diverges, then $\sum_{n=1}^{\infty} a_n$ diverges.

15. The Ratio Test will not help in determining the convergence or divergence of $\sum_{n=1}^{\infty}\frac{2n+3}{3n^4+2n^3+3n+1}$.

16. If $a_n > 0$ for all natural numbers n and $\sum_{n=1}^{\infty} a_n$ converges, then $\lim_{n\to\infty}(a_{n+1}/a_n) < 1$.

17. $\sum_{n=1}^{\infty}\left(1 - \frac{1}{n}\right)^n$ converges.

18. $\sum_{n=1}^{\infty}\frac{1}{\ln(n^4+1)}$ converges.

19. $\sum_{n=2}^{\infty}\frac{n+1}{(n\ln n)^2}$ converges.

20. $\sum_{n=1}^{\infty}\frac{\sin^2(n\pi/2)}{n}$ converges.

21. If $0 \le a_{n+100} \le b_n$ for all natural numbers n and if $\sum_{n=1}^{\infty} b_n$ converges, then $\sum_{n=1}^{\infty} a_n$ converges.

22. If, for some $c > 0$, $ca_n \ge 1/n$ for all natural numbers n, then $\sum_{n=1}^{\infty} a_n$ diverges.

23. $\frac{1}{3} + \left(\frac{1}{3}\right)^2 + \left(\frac{1}{3}\right)^3 + \cdots + \left(\frac{1}{3}\right)^{1000} < \frac{1}{2}$.

24. If $\sum_{n=1}^{\infty} a_n$ converges, then $\sum_{n=1}^{\infty}(-1)^n a_n$ converges.

25. If $b_n \le a_n \le 0$ for all natural numbers n, and if $\sum_{n=1}^{\infty} b_n$ converges, then $\sum_{n=1}^{\infty} a_n$ converges.

26. If $0 \le a_n$ for all natural numbers n, and if $\sum_{n=1}^{\infty} a_n$ converges, then $\sum_{n=1}^{\infty}(-1)^n a_n$ converges.

27. $\left|\sum_{n=1}^{\infty}(-1)^{n+1}\frac{1}{n} - \sum_{n=1}^{99}(-1)^{n+1}\frac{1}{n}\right| < 0.01$.

28. If $\sum_{n=1}^{\infty} a_n$ diverges, then $\sum_{n=1}^{\infty}|a_n|$ diverges.

29. If the power series $\sum_{n=0}^{\infty} a_n(x-3)^n$ converges at $x = -1.1$, then it also converges at $x = 7$.

30. If $\sum_{n=0}^{\infty} a_n x^n$ converges at $x = -2$, then it also converges at $x = 2$.

31. If $f(x) = \sum_{n=0}^{\infty} a_n x^n$ and if the series converges at $x = 1.5$, then $\int_0^1 f(x)\,dx = \sum_{n=0}^{\infty} a_n/(n+1)$.

32. Every power series converges for at least two values of the variable.

33. If $f(0), f'(0), f''(0), \ldots$ all exist, then the Maclaurin series for $f(x)$ converges to $f(x)$ in a neighborhood of $x = 0$.

34. The function $f(x) = 1 + x + x^2 + x^3 + \cdots$ satisfies the differential equation $y' = y^2$ on the interval $(-1, 1)$.

35. The function $f(x) = \sum_{n=0}^{\infty}(-1)^n x^n/n!$ satisfies the differential equation $y' + y = 0$ on the whole real line.

36. If $P(x)$ is the Maclaurin polynomial of order 2 for $f(x)$, then $P(0) = f(0)$, $P'(0) = f'(0)$, and $P''(0) = f''(0)$.

37. The Taylor polynomial of order n based at a for $f(x)$ is unique; that is, $f(x)$ has only one such polynomial.

38. $f(x) = x^{5/2}$ has a second-order Maclaurin polynomial.

39. The Maclaurin polynomial of order 3 for $f(x) = 2x^3 - x^2 + 7x - 11$ is an exact representation of $f(x)$.

40. The Maclaurin polynomial of order 16 for $\cos x$ involves only even powers of x.

41. If $f'(0)$ exists for an even function, then $f'(0) = 0$.

42. Taylor's Formula with Remainder contains the Mean Value Theorem for Derivatives as a special case.

Sample Test Problems

In Problems 1–8, determine whether the given sequence converges or diverges and, if it converges, find $\lim_{n\to\infty} a_n$.

1. $a_n = \frac{9n}{\sqrt{9n^2+1}}$

2. $a_n = \frac{\ln n}{\sqrt{n}}$

3. $a_n = \left(1 + \frac{4}{n}\right)^n$

4. $a_n = \frac{n!}{3^n}$

5. $a_n = \sqrt[n]{n}$

6. $a_n = \frac{1}{\sqrt[3]{n}} + \frac{1}{\sqrt[n]{3}}$

7. $a_n = \frac{\sin^2 n}{\sqrt{n}}$

8. $a_n = \cos\left(\frac{n\pi}{6}\right)$

In Problems 9–18, determine whether the given series converges or diverges and, if it converges, find its sum.

9. $\sum_{k=1}^{\infty}\left(\frac{1}{\sqrt{k}} - \frac{1}{\sqrt{k+1}}\right)$

10. $\sum_{k=1}^{\infty}\left(\frac{1}{k} - \frac{1}{k+2}\right)$

11. $\ln\frac{1}{2} + \ln\frac{2}{3} + \ln\frac{3}{4} + \cdots$

12. $\sum_{k=0}^{\infty} \cos k\pi$

13. $\sum_{k=0}^{\infty} e^{-2k}$

14. $\sum_{k=0}^{\infty}\left(\frac{3}{2^k} + \frac{4}{3^k}\right)$

15. $0.91919191\ldots = \sum_{k=1}^{\infty} 91\left(\frac{1}{100}\right)^k$

16. $\sum_{k=1}^{\infty}\left(\frac{1}{\ln 2}\right)^k$

17. $1 - \frac{2^2}{2!} + \frac{2^4}{4!} - \frac{2^6}{6!} + \cdots$

18. $1 - \frac{1}{1!} + \frac{1}{2!} - \frac{1}{3!} + \frac{1}{4!} - \cdots$

In Problems 19–32, indicate whether the given series converges or diverges and give a reason for your conclusion.

19. $\sum_{n=1}^{\infty} \frac{n}{1 + n^2}$

20. $\sum_{n=1}^{\infty} \frac{n + 5}{1 + n^3}$

21. $\sum_{n=1}^{\infty} (-1)^{n+1}\frac{1}{\sqrt[3]{n}}$

22. $\sum_{n=1}^{\infty} (-1)^{n+1}\frac{1}{\sqrt[n]{3}}$

23. $\sum_{n=1}^{\infty} \frac{2^n + 3^n}{4^n}$

24. $\sum_{n=1}^{\infty} \frac{n}{e^{n^2}}$

25. $\sum_{n=1}^{\infty} (-1)^{n+1}\frac{n + 1}{10n + 12}$

26. $\sum_{n=1}^{\infty} \frac{\sqrt{n}}{n^2 + 7}$

27. $\sum_{n=1}^{\infty} \frac{n^2}{n!}$

28. $\sum_{n=1}^{\infty} \frac{n^3 3^n}{(n + 1)!}$

29. $\sum_{n=1}^{\infty} \frac{2^n n!}{(n + 2)!}$

30. $\sum_{n=2}^{\infty} \left(1 - \frac{1}{n}\right)^n$

31. $\sum_{n=1}^{\infty} n^2\left(\frac{2}{3}\right)^n$

32. $\sum_{n=1}^{\infty} \frac{(-1)^n}{1 + \ln n}$

In Problems 33–36, state whether the given series is absolutely convergent, conditionally convergent, or divergent.

33. $\sum_{n=1}^{\infty} (-1)^n \frac{1}{3n - 1}$

34. $\sum_{n=1}^{\infty} \frac{(-1)^n n^3}{2^n}$

35. $\sum_{n=1}^{\infty} (-1)^n \frac{3^n}{2^{n+8}}$

36. $\sum_{n=2}^{\infty} \frac{(-1)^n \sqrt[n]{n}}{\ln n}$

In Problems 37–42, find the convergence set for the power series.

37. $\sum_{n=0}^{\infty} \frac{x^n}{n^3 + 1}$

38. $\sum_{n=0}^{\infty} \frac{(-2)^{n+1}x^n}{2n + 3}$

39. $\sum_{n=0}^{\infty} \frac{(-1)^n(x - 4)^n}{n + 1}$

40. $\sum_{n=0}^{\infty} \frac{3^n x^{3n}}{(3n)!}$

41. $\sum_{n=0}^{\infty} \frac{(x - 3)^n}{2^n + 1}$

42. $\sum_{n=0}^{\infty} \frac{n!(x + 1)^n}{3^n}$

43. By differentiating the geometric series

$$\frac{1}{1 + x} = 1 - x + x^2 - x^3 + x^4 - \cdots, \quad |x| < 1,$$

find a power series that represents $1/(1 + x)^2$. What is its interval of convergence?

44. Find a power series that represents $1/(1 + x)^3$ on the interval $(-1, 1)$.

45. Find the Maclaurin series for $\sin^2 x$. For what values of x does the series represent the function?

46. Find the first five terms of the Taylor series for e^x based at the point $x = 2$.

47. Write the Maclaurin series for $f(x) = \sin x + \cos x$. For what values of x does it represent f?

48. Determine how large n must be so that using the nth partial sum to approximate the series $\sum_{k=1}^{\infty} \frac{1}{9 + k^2}$ gives an error of no more than 0.00005.

49. Determine how large n must be so that using the nth partial sum to approximate the series $\sum_{k=1}^{\infty} \frac{k}{e^{k^2}}$ gives an error of no more than 0.000005.

50. How many terms do we have to take in the convergent series

$$1 - \frac{1}{\sqrt{2}} + \frac{1}{\sqrt{3}} - \frac{1}{\sqrt{4}} + \frac{1}{\sqrt{5}} - \frac{1}{\sqrt{6}} + \cdots$$

to be sure that we have approximated its sum to within 0.001?

51. Use the simplest method you can think of to find the first three nonzero terms of the Maclaurin series for each of the following:

(a) $\frac{1}{1 - x^3}$ (b) $\sqrt{1 + x^2}$

(c) $e^{-x} - 1 + x$ (d) $x \sec x$

(e) $e^{-x}\sin x$ (f) $\frac{1}{1 + \sin x}$

52. Find the Maclaurin polynomial of order 2 for $f(x) = \cos x$ and use it to approximate $\cos 0.1$.

C **53.** Find the Maclaurin polynomial of order 1 for $f(x) = x\cos x^2$ and use it to approximate $f(0.2)$.

C **54.** Find the Maclaurin polynomial of order 4 for $f(x)$, and use it to approximate $f(0.1)$.

(a) $f(x) = xe^x$

(b) $f(x) = \cosh x$

55. Find the Taylor polynomial of order 3 based at 2 for $g(x) = x^3 - 2x^2 + 5x - 7$, and show that it is an exact representation of $g(x)$.

56. Use the result of Problem 55 to calculate $g(2.1)$.

57. Find the Taylor polynomial of order 4 based at 1 for $f(x) = 1/(x + 1)$.

58. Obtain an expression for the error term $R_4(x)$ in Problem 57, and find a bound for it if $x = 1.2$.

59. Find the Maclaurin polynomial of order 4 for $f(x) = \sin^2 x = \frac{1}{2}(1 - \cos 2x)$, and find a bound for the error $R_4(x)$ if $|x| \le 0.2$. *Note:* A better bound is obtained if you observe that $R_4(x) = R_5(x)$ and then bound $R_5(x)$.

C **60.** If $f(x) = \ln x$, then $f^{(n)}(x) = (-1)^{n-1}(n-1)!/x^n$. Thus, the Taylor polynomial of order n based at 1 for $\ln x$ is

$$\ln x = (x-1) - \frac{1}{2}(x-1)^2 + \frac{1}{3}(x-1)^3 + \cdots + \frac{(-1)^{n-1}}{n}(x-1)^n + R_n(x)$$

How large would n have to be for us to know that $|R_n(x)| \le 0.00005$ if $0.8 \le x \le 1.2$?

C **61.** Refer to Problem 60. Use the Taylor polynomial of order 5 based at 1 to approximate

$$\int_{0.8}^{1.2} \ln x \, dx$$

and give a good bound for the error that is made.

REVIEW & PREVIEW PROBLEMS

1. For the graph of $y = x^2/4$, find the equation of the tangent line and the normal line (i.e., the line perpendicular to the tangent line) that pass through the point $(2, 1)$.

2. Find all points on the graph of $y = x^2/4$ (a) where the tangent line is parallel to the line $y = x$ and (b) where the normal line is parallel to the line $y = x$.

3. Find all points of intersection of $\frac{x^2}{16} + \frac{y^2}{9} = 1$ and $\frac{x^2}{9} + \frac{y^2}{16} = 1$.

4. Find all points of intersection of $\frac{x^2}{16} + \frac{y^2}{9} = 1$ and $x^2 + y^2 = 9$.

5. Use implicit differentiation to find the equation of the tangent line to the curve $x^2 + y^2/4 = 1$ at the point $\left(-\frac{\sqrt{3}}{2}, 1\right)$.

6. Use implicit differentiation to find the equation of the tangent line to the curve $\frac{x^2}{9} - \frac{y^2}{16} = 1$ at the point $(9, 8\sqrt{2})$.

7. Find all points of intersection of $\frac{x^2}{100} + \frac{y^2}{64} = 1$ and $\frac{x^2}{9} - \frac{y^2}{27} = 1$. For the point of intersection that is in the first quadrant, use implicit differentiation to find the equations of the tangent lines to both curves. Find all angles between these two tangent lines.

8. Suppose that $x = 2 \cos t$ and $y = 2 \sin t$. Fill in the table below and plot the ordered pairs (x, y).

t	$x = 2 \cos t$	$y = 2 \sin t$
0		
$\pi/6$		
$\pi/4$		
$\pi/3$		
$\pi/2$		
π		
$3\pi/2$		
2π		

In Problems 9–10, determine the values of r and θ.

9.

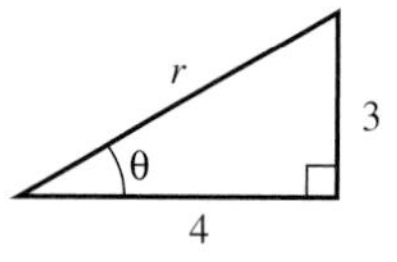

10.

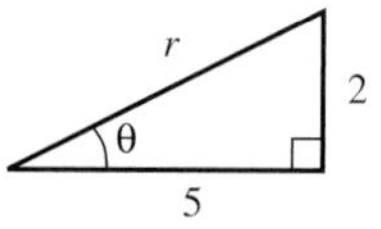

In Problems 11–12, determine the values of x and y.

11.

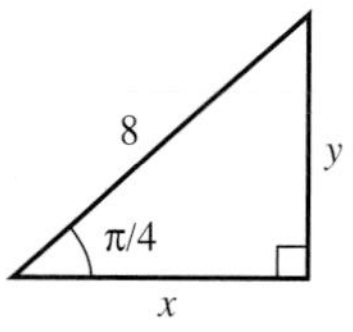

12.

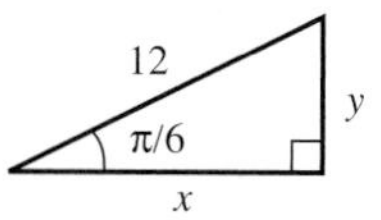

CHAPTER 10 Conics and Polar Coordinates

10.1 The Parabola
10.2 Ellipses and Hyperbolas
10.3 Translation and Rotation of Axes
10.4 Parametric Representation of Curves in the Plane
10.5 The Polar Coordinate System
10.6 Graphs of Polar Equations
10.7 Calculus in Polar Coordinates

10.1 The Parabola

Take a right circular cone with two nappes and pass planes through it at various angles, as shown in Figure 1. As sections, you will obtain curves called, respectively, an ellipse, a parabola, and a hyperbola. (You may also obtain various limiting forms: a circle, a point, a pair of intersecting lines, and one line.) These curves are called *conic sections,* or simply *conics*. This definition, which is due to the Greeks, is cumbersome and we shall immediately adopt a different one. It can be shown that the two notions are consistent.

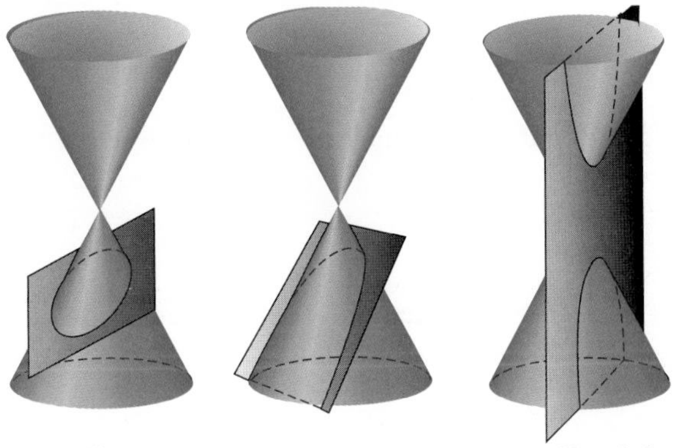

Figure 1

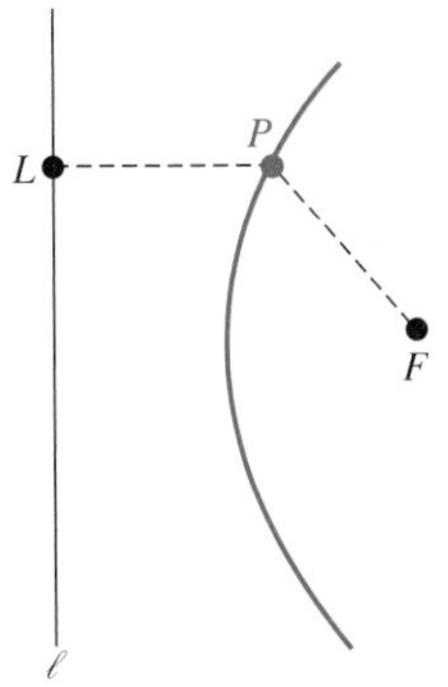

Figure 2

In the plane let ℓ be a fixed line (the **directrix**) and F be a fixed point (the **focus**) not on the line, as in Figure 2. The set of points P for which the ratio of the distance $|PF|$ from the focus to the distance $|PL|$ from the line is a positive constant e (the **eccentricity**), that is, the set of points P that satisfy

$$|PF| = e|PL|$$

is called a **conic.** If $0 < e < 1$, the conic is an **ellipse**; if $e = 1$, it is a **parabola**; if $e > 1$, it is a **hyperbola.**

When we draw the curves corresponding to $e = \frac{1}{2}$, $e = 1$, and $e = 2$, we get the three curves shown in Figure 3.

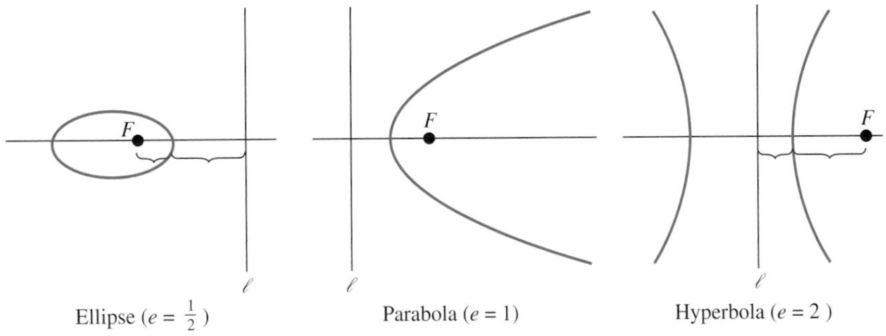

Figure 3

In each case, the curves are symmetric with respect to the line through the focus perpendicular to the directrix. We call this line the **major axis** (or simply the

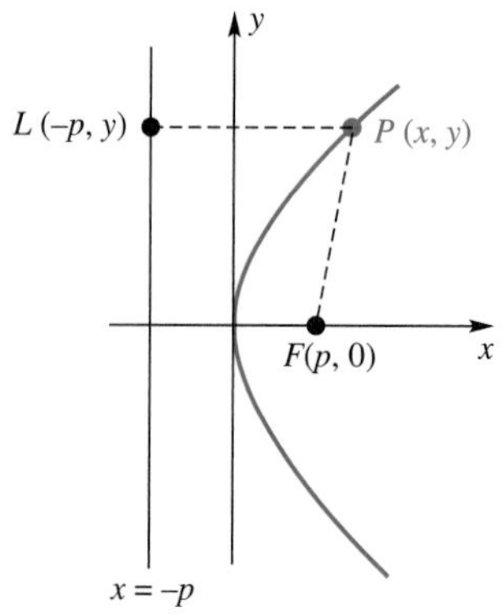

Figure 4

axis) of the conic. A point where the conic crosses the axis is called a **vertex.** The parabola has one vertex, while the ellipse and hyperbola have two vertices.

The Parabola ($e = 1$) A **parabola** is the set of points P that are equidistant from the directrix ℓ and the focus F, that is, points that satisfy

$$|PF| = |PL|$$

From this definition, we wish to derive the xy-equation, and we want it to be as simple as possible. The position of the coordinate axes has no effect on the curve, but it does affect the simplicity of the curve's equation. Since a parabola is symmetric with respect to its axis, it is natural to place one of the coordinate axes, for instance, the x-axis, along the axis. Let the focus F be to the right of the origin, say at $(p, 0)$, and the directrix to the left with equation $x = -p$. Then the vertex is at the origin. All this is shown in Figure 4.

From the condition $|PF| = |PL|$ and the distance formula, we get

$$\sqrt{(x-p)^2 + (y-0)^2} = \sqrt{(x+p)^2 + (y-y)^2}$$

After squaring both sides and simplifying, we obtain

$$\boxed{y^2 = 4px}$$

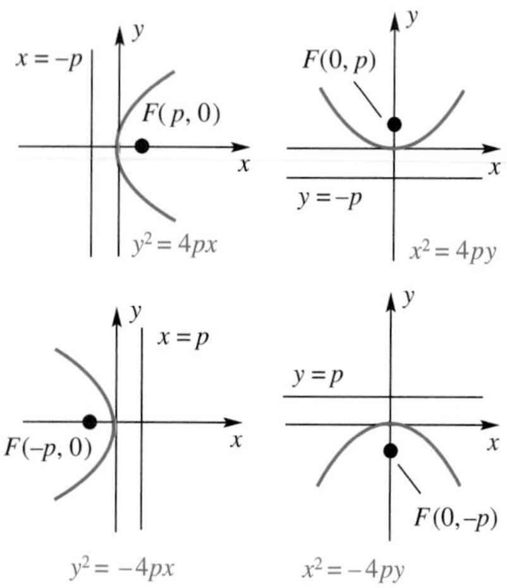

Figure 5

This is called the **standard equation** of a horizontal parabola (horizontal axis) opening to the right. Note that $p > 0$ and that p is the distance from the focus to the vertex.

EXAMPLE 1 Find the focus and directrix of the parabola with equation $y^2 = 12x$.

SOLUTION Since $y^2 = 4(3)x$, we see that $p = 3$. The focus is at (3, 0); the directrix is the line $x = -3$. ■

There are three variants of the standard equation. If we interchange the roles of x and y, we obtain the equation $x^2 = 4py$. It is the equation of a vertical parabola with focus at $(0, p)$ and directrix $y = -p$. Finally, introducing a minus sign on one side of the equation causes the parabola to open in the opposite direction. All four cases are shown in Figure 5.

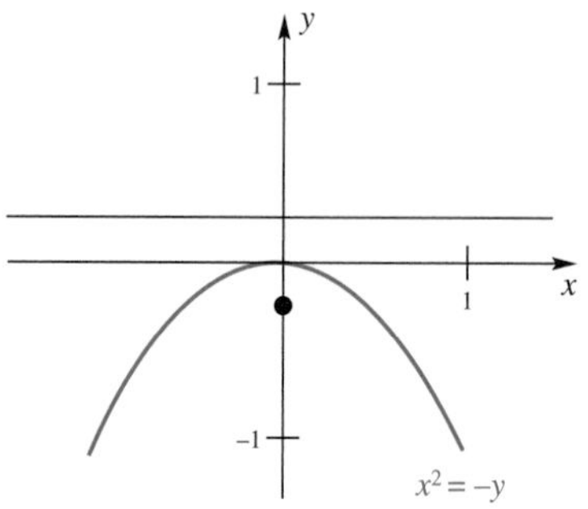

Figure 6

EXAMPLE 2 Determine the focus and directrix of the parabola $x^2 = -y$ and sketch the graph.

SOLUTION We write $x^2 = -4\left(\frac{1}{4}\right)y$, from which we conclude that $p = \frac{1}{4}$. The form of the equation tells us that the parabola is vertical and opens down. The focus is at $\left(0, -\frac{1}{4}\right)$; the directrix is the line $y = \frac{1}{4}$. The graph is shown in Figure 6. ■

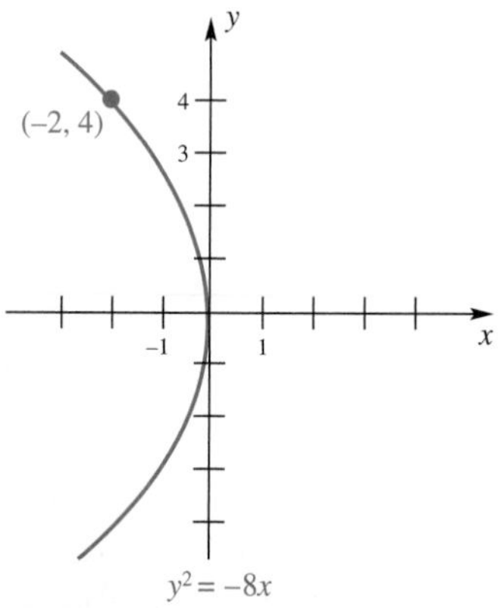

Figure 7

EXAMPLE 3 Find the equation of the parabola with vertex at the origin and focus at $(0, 5)$.

SOLUTION The parabola opens up and $p = 5$. The equation is $x^2 = 4(5)y$, that is, $x^2 = 20y$. ■

EXAMPLE 4 Find the equation of the parabola with vertex at the origin that goes through $(-2, 4)$ and opens left. Sketch the graph.

SOLUTION The equation has the form $y^2 = -4px$. Because $(-2, 4)$ is on the graph, $(4)^2 = -4p(-2)$, from which $p = 2$. The desired equation is $y^2 = -8x$ and its graph is sketched in Figure 7. ■

The Optical Property A simple geometric property of a parabola is the basis of many important applications. If F is the focus and P is any point on the parabola, the tangent line at P makes equal angles with FP and the line GP, which is parallel to the axis of the parabola (see Figure 8). A principle from physics says that when a light ray strikes a reflecting surface the angle of incidence is equal to the angle of reflection. It follows that, if a parabola is revolved about its axis to form a hollow reflecting shell, all light rays from the focus after hitting the shell are reflected outward parallel to the axis. This property of the parabola is used in designing searchlights, with the light source placed at the focus. Conversely, it is used in certain telescopes and satellite dishes in which incoming parallel rays from a distant star or satellite are focused at a single point.

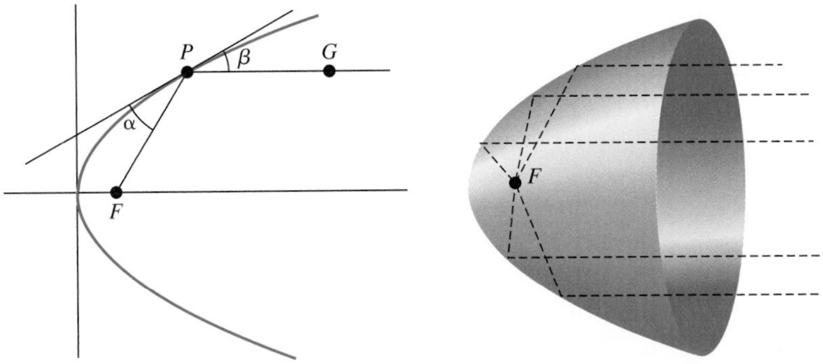

Figure 8

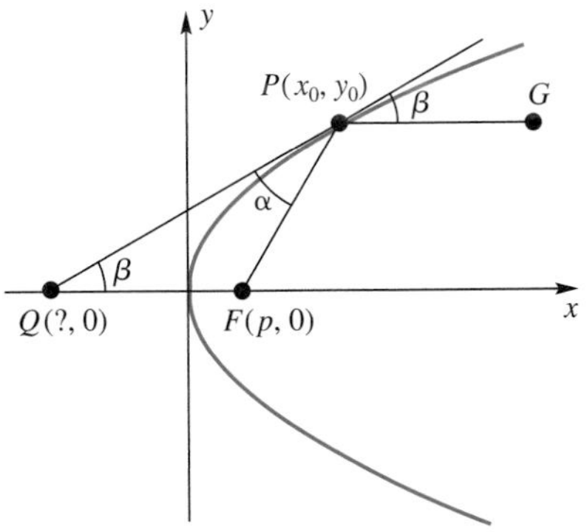

Figure 9

EXAMPLE 5 Prove the optical property of the parabola.

SOLUTION In Figure 9, let QP be the tangent line at P and let GP be the line through P parallel to the x-axis. We must show that $\alpha = \beta$. After noting that $\angle FQP = \beta$, we reduce the problem to showing that triangle FQP is isosceles.

First, we obtain the x-coordinate of Q. Differentiating $y^2 = 4px$ implicitly gives $2y'y = 4p$, from which we conclude that the slope of the tangent line at $P(x_0, y_0)$ is $2p/y_0$. The equation of this line is

$$y - y_0 = \frac{2p}{y_0}(x - x_0)$$

Setting $y = 0$ and solving for x gives $-y_0 = (2p/y_0)(x - x_0)$, or $x - x_0 = -y_0^2/2p$. Now $y_0^2 = 4px_0$, which gives $x - x_0 = -2x_0$, that is, $x = -x_0$; Q has coordinates $(-x_0, 0)$.

To show that the segments FP and FQ have equal length, we use the distance formula

$$|FP| = \sqrt{(x_0 - p)^2 + y_0^2} = \sqrt{x_0^2 - 2x_0p + p^2 + 4px_0}$$

$$= \sqrt{x_0^2 + 2x_0p + p^2} = x_0 + p = |FQ|$$

■

Sound obeys the same laws of reflection as light, and parabolic microphones are used to pick up and concentrate sounds from, for example, a distant part of a football stadium. Radar and radio telescopes are also based on the same principle.

There are many other applications of parabolas. For example, the path of a projectile is a parabola if air resistance and other minor factors are neglected. The cable of an evenly loaded suspension bridge takes the form of a parabola. Arches are often parabolic. The paths of some comets are parabolic.

Concepts Review

1. The set of points P satisfying $|PF| = e|PL|$ (i.e., distance to the focus equals e times distance to the directrix) is an ellipse if _____, a parabola if _____, and a hyperbola if _____.

2. The standard equation of a parabola, vertex at the origin and opening right, is _____.

3. The parabola $y = \frac{1}{4}x^2$ has focus _____ and directrix _____.

4. The rays from a light source at the focus of a parabolic mirror will be reflected in a direction _____.

Problem Set 10.1

In Problems 1–8, find the coordinates of the focus and the equation of the directrix for each parabola. Make a sketch showing the parabola, its focus, and its directrix.

1. $y^2 = 4x$

2. $y^2 = -12x$

3. $x^2 = -12y$

4. $x^2 = -16y$

5. $y^2 = x$

6. $y^2 + 3x = 0$

7. $6y - 2x^2 = 0$

8. $3x^2 - 9y = 0$

In Problems 9–14, find the standard equation of each parabola from the given information. Assume that the vertex is at the origin.

9. Focus is at $(2, 0)$

10. Directrix is $x = 3$

11. Directrix is $y - 2 = 0$

12. Focus is $\left(0, -\frac{1}{9}\right)$

13. Focus is $(-4, 0)$

14. Directrix is $y = \frac{7}{2}$

15. Find the equation of the parabola with vertex at the origin and axis along the x-axis if the parabola passes through the point $(3, -1)$. Make a sketch.

16. Find the equation of the parabola through the point $(-2, 4)$ if its vertex is at the origin and its axis is along the x-axis. Make a sketch.

17. Find the equation of the parabola through the point $(6, -5)$ if its vertex is at the origin and its axis is along the y-axis. Make a sketch.

18. Find the equation of the parabola whose vertex is the origin and whose axis is the y-axis if the parabola passes through the point $(-3, 5)$. Make a sketch.

In Problems 19–26, find the equations of the tangent and the normal lines to the given parabola at the given point. Sketch the parabola, the tangent line, and the normal line.

19. $y^2 = 16x, (1, -4)$

20. $x^2 = -10y, \left(2\sqrt{5}, -2\right)$

21. $x^2 = 2y, (4, 8)$

22. $y^2 = -9x, (-1, -3)$

23. $y^2 = -15x, \left(-3, -3\sqrt{5}\right)$

24. $x^2 = 4y, (4, 4)$

25. $x^2 = -6y, \left(3\sqrt{2}, -3\right)$

26. $y^2 = 20x, \left(2, -2\sqrt{10}\right)$

27. The slope of the tangent line to the parabola $y^2 = 5x$ at a certain point on the parabola is $\sqrt{5}/4$. Find the coordinates of that point. Make a sketch.

28. The slope of the tangent line to the parabola $x^2 = -14y$ at a certain point on the parabola is $-2\sqrt{7}/7$. Find the coordinates of that point.

29. Find the equation of the tangent line to the parabola $y^2 = -18x$ that is parallel to the line $3x - 2y + 4 = 0$.

30. Any line segment through the focus of a parabola, with end points on the parabola, is a **focal chord.** Prove that the tangent lines to a parabola at the end points of any focal chord intersect on the directrix.

31. Prove that the tangents to a parabola at the extremities of any focal chord are perpendicular to each other (see Problem 30).

32. A chord of a parabola that is perpendicular to the axis and 1 unit from the vertex has length 1 unit. How far is it from the vertex to the focus?

33. Prove that the vertex is the point on a parabola closest to the focus.

34. An asteroid from deep space is sighted from the earth moving on a parabolic path with the earth at the focus. When the line from the earth to the asteroid first makes an angle of 90° with the axis of the parabola, the asteroid is measured to be 40 million miles away. How close will the asteroid come to the earth (see Problem 33)? Treat the earth as a point.

C **35.** Work Problem 34, assuming that the angle is 75° rather than 90°.

36. The cables for the central span of a suspension bridge take the shape of a parabola (see Problem 41). If the towers are 800 meters apart and the cables are attached to them at points 400 meters above the floor of the bridge, how long must the vertical strut be that is 100 meters from the tower? Assume that the cable touches the bridge deck at the midpoint of the bridge (Figure 10).

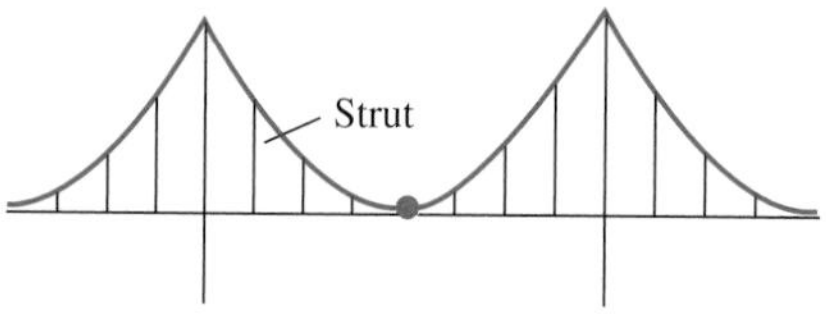

Figure 10

37. The focal chord that is perpendicular to the axis of a parabola is called the **latus rectum.** For the parabola $y^2 = 4px$ in Figure 11, let F be the focus, R be any point on the parabola to the left of the latus rectum, and G be the intersection of the latus rectum with the line through R parallel to the axis. Find $|FR| + |RG|$ and note that it is a constant.

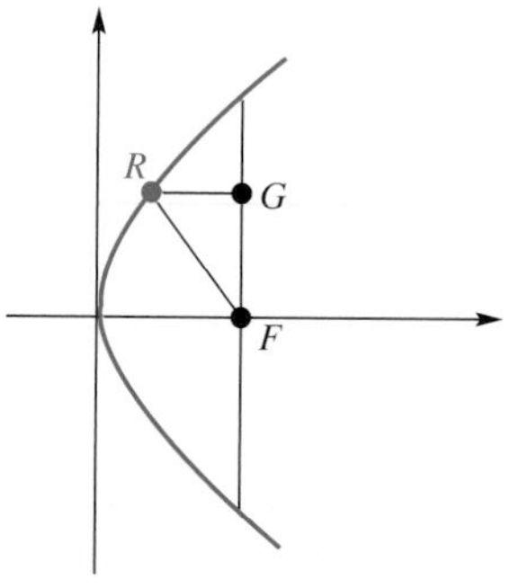

Figure 11

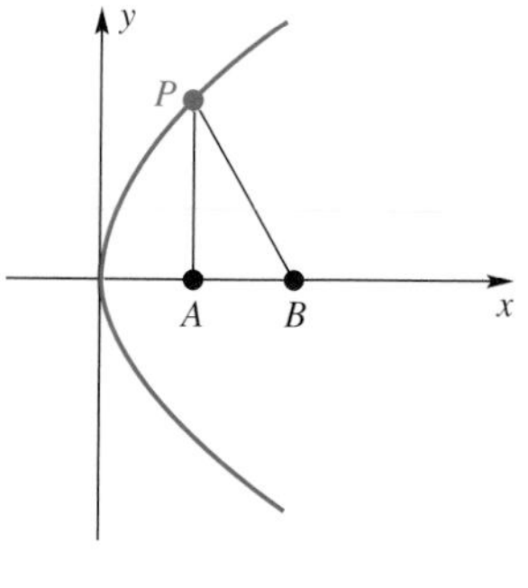

Figure 12

38. For the parabola $y^2 = 4px$ in Figure 12, P is any of its points except the vertex, PB is the normal line at P, PA is perpendicular to the axis of the parabola, and A and B are on the axis. Find $|AB|$ and note that it is a constant.

39. Show that the focal chord of the parabola $y^2 = 4px$ with end points (x_1, y_1) and (x_2, y_2) has length $x_1 + x_2 + 2p$. Specialize to find the length L of the latus rectum.

40. Show that the set of points equidistant from a circle and a line outside the circle is a parabola.

EXPL **41.** Consider a bridge deck weighing δ pounds per linear foot and supported by a cable, which is assumed to be of negligible weight compared to the bridge deck. The cable section OP from the lowest point (the origin) to a general point $P(x, y)$ is shown in Figure 13.

The forces acting on this section of cable are

H = horizontal tension pulling at O

T = tangential tension pulling at P

$W = \delta x$ = weight of x feet of bridge deck

For equilibrium, the horizontal and vertical components of T must balance H and W, respectively. Thus,

$$\frac{T \sin \phi}{T \cos \phi} = \tan \phi = \frac{\delta x}{H}$$

That is,

$$\frac{dy}{dx} = \frac{\delta x}{H}, \qquad y(0) = 0$$

Solve this differential equation to show that the cable hangs in the shape of a parabola. (Compare this result with that for the unloaded hanging cable of Problem 53 of Section 6.9.)

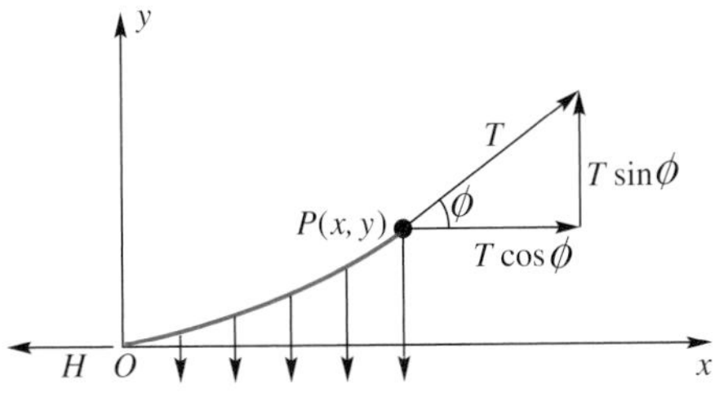

Figure 13

EXPL **42.** Consider the parabola $y = x^2$ over the interval $[a, b]$, and let $c = (a + b)/2$ be the midpoint of $[a, b]$, d be the midpoint of $[a, c]$, and e be the midpoint of $[c, b]$. Let T_1 be the triangle with vertices on the parabola at a, c, and b, and let T_2 be the union of the two triangles with vertices on the parabola at a, d, c and c, e, b, respectively (Figure 14). Continue to build triangles on triangles in this manner, thus obtaining sets $T_3, T_4, \ldots$.

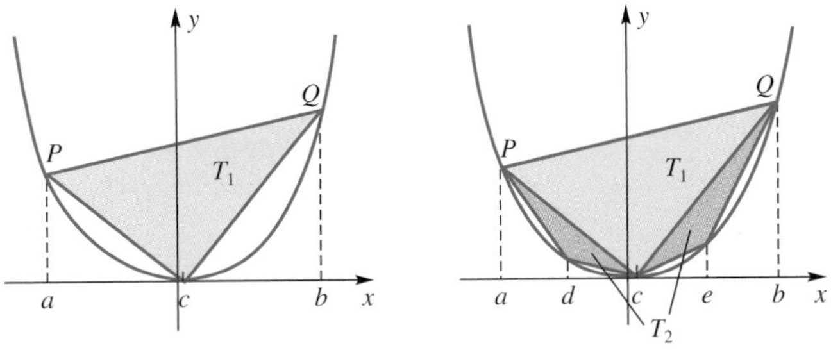

Figure 14

(a) Show that $A(T_1) = (b - a)^3/8$.

(b) Show that $A(T_2) = A(T_1)/4$.

(c) Let S be the parabolic segment cut off by the chord PQ. Show that the area of S satisfies

$$A(S) = A(T_1) + A(T_2) + A(T_3) + \cdots = \frac{4}{3}A(T_1)$$

This is a famous result of Archimedes, which he obtained without coordinates.

(d) Use these results to show that the area under $y = x^2$ between a and b is $b^3/3 - a^3/3$.

CAS **43.** Illustrate Problems 30 and 31 for the parabola $y = \frac{1}{4}x^2 + 2$ by plotting (in the same graph window) the parabola, its directrix, its focal chord parallel to the x-axis, and the tangent lines at the ends of the focal chord.

CAS **44.** In Problem 60 of Section 6.9 you were asked to find the equation of the Gateway Arch in St. Louis, Missouri.

(a) Find the equation of a parabola with the properties that its vertex is at $(0, 630)$ and it intersects the x-axis at ± 315.

(b) In the same graphing window, plot the catenary which is the Gateway Arch and the parabola you found in (a).

(c) Approximate (to the nearest foot) the largest vertical distance between the catenary and the parabola.

Answers to Concepts Review: **1.** $e < 1$; $e = 1$; $e > 1$ **2.** $y^2 = 4px$ **3.** $(0, 1)$; $y = -1$ **4.** parallel to the axis

10.2 Ellipses and Hyperbolas

Recall that the conic determined by the condition $|PF| = e|PL|$ is an **ellipse** if $0 < e < 1$ and a **hyperbola** if $e > 1$ (see the introduction to Section 10.1). In either case, the conic has two vertices, which we label A' and A. Call the point on the major axis midway between A' and A the **center** of the conic. Ellipses and hyperbolas are symmetric with respect to their centers (as we shall demonstrate soon) and are, therefore, called **central conics.**

To derive the equation of a central conic, place the x-axis along the major axis with the origin at the center. We may suppose the focus to be $F(c, 0)$, the directrix $x = k$, and the vertices $A'(-a, 0)$ and $A(a, 0)$, with c, k, and a all positive. It is

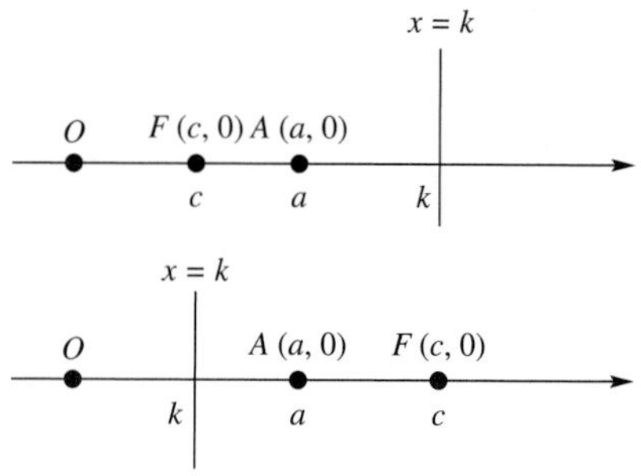

Figure 1

clear that A must lie between F and the line $x = k$. The two possible arrangements are shown in Figure 1. In the first case, applying $|PF| = e|PL|$ to the point $P = A$ gives

$$(1) \qquad a - c = e(k - a) = ek - ea$$

In the second case, applying $|PF| = e|PL|$ to the point $P = A$ gives

$$c - a = e(a - k) = ea - ek$$

which, when both sides are multiplied by -1, is seen to be equivalent to (1). Next, apply the relationship $|PF| = e|PL|$ to the points $A'(-a, 0)$ and $F'(-c, 0)$, and the line $x = -k$. This leads to

$$(2) \qquad a + c = e(k + a) = ek + ea$$

When equations (1) and (2) are solved for c and k, we get

$$c = ea \quad \text{and} \quad k = \frac{a}{e}$$

If $0 < e < 1$ then $c = ea < a$ and $k = a/e > a$. Thus for the case of an ellipse, the focus F lies to the left of the vertex A and the directrix $x = k$ lies to the right of A. On the other hand, if $e > 1$, then $c = ea > a$ and $k = a/e < a$. For the case of a hyperbola, the directrix $x = k$ lies to the left of A, and the focus F lies to the right of A. The two situations are shown in Figures 2 and 3.

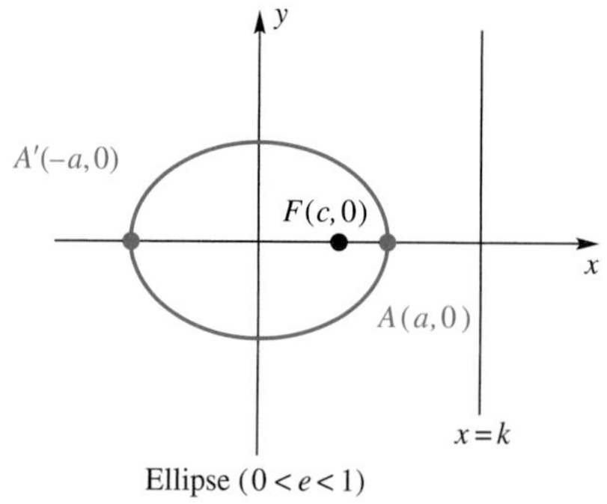

Figure 2

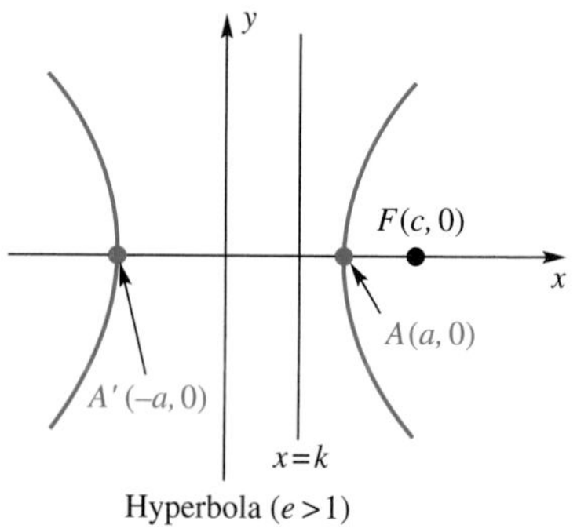

Figure 3

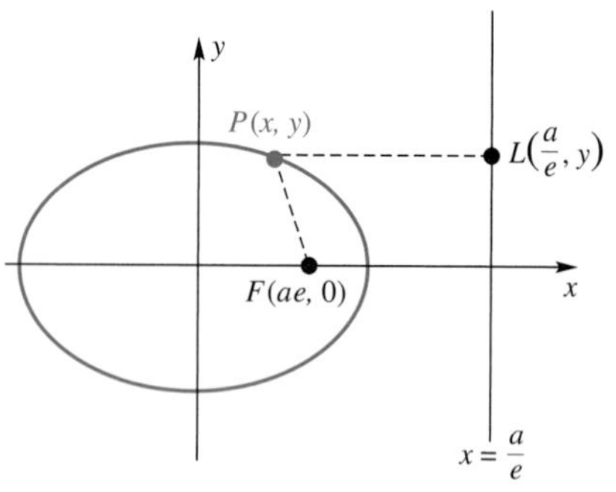

Figure 4

Now let $P(x, y)$ be any point on the ellipse (or hyperbola). Then $L(a/e, y)$ is its projection on the directrix (see Figure 4 for the case of the ellipse). The condition $|PF| = e|PL|$ becomes

$$\sqrt{(x - ae)^2 + y^2} = e\sqrt{\left(x - \frac{a}{e}\right)^2}$$

Squaring both members and collecting terms, we obtain the equivalent equation

$$x^2 - 2aex + a^2e^2 + y^2 = e^2\left(x^2 - \frac{2a}{e}x + \frac{a^2}{e^2}\right)$$

or

$$(1 - e^2)x^2 + y^2 = a^2(1 - e^2)$$

or

$$\frac{x^2}{a^2} + \frac{y^2}{a^2(1 - e^2)} = 1$$

Because this last equation contains x and y only to even powers, it corresponds to a curve that is symmetric with respect to both the x- and y-axes and the origin.

Also, because of this symmetry, there must be a second focus at $(-ae, 0)$ and a second directrix at $x = -a/e$. The axis containing the two vertices (and the two foci) is the **major axis,** and the axis perpendicular to it (through the center) is the **minor axis.**

Standard Equation of the Ellipse For the ellipse, $0 < e < 1$, and so $(1 - e^2)$ is positive. To simplify notation, let $b = a\sqrt{1 - e^2}$. Then the equation derived above takes the form

$$\boxed{\frac{x^2}{a^2} + \frac{y^2}{b^2} = 1}$$

which is called the **standard equation of an ellipse.** Since $c = ae$, the numbers a, b, and c satisfy the Pythagorean relationship $a^2 = b^2 + c^2$. In Figure 5, the shaded right triangle captures the condition $a^2 = b^2 + c^2$. Thus, the number $2a$ is the **major diameter,** whereas $2b$ is the **minor diameter.**

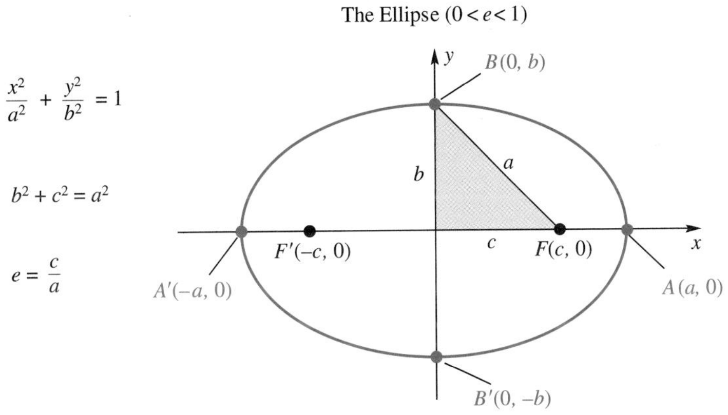

Figure 5

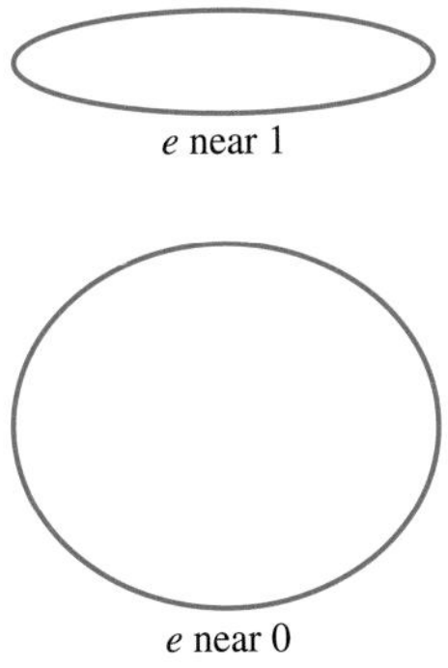

Figure 6

Consider now the effect of changing the value of e. If e is near 1, then $b = a\sqrt{1 - e^2}$ is small relative to a; the ellipse is thin and very *eccentric*. On the other hand, if e is near 0 (near zero eccentricity), b is almost as large as a; the ellipse is fat and well rounded (Figure 6). In the limiting case where $b = a$, the equation takes the form

$$\frac{x^2}{a^2} + \frac{y^2}{a^2} = 1$$

which is equivalent to $x^2 + y^2 = a^2$. This is the equation of a circle of radius a centered at the origin.

EXAMPLE 1 Sketch the graph of

$$\frac{x^2}{36} + \frac{y^2}{4} = 1$$

and determine its foci and eccentricity.

SOLUTION Since $a = 6$ and $b = 2$, we calculate

$$c = \sqrt{a^2 - b^2} = \sqrt{36 - 4} = 4\sqrt{2} \approx 5.66$$

The foci are at $(\pm c, 0) = \left(\pm 4\sqrt{2}, 0\right)$, and $e = c/a \approx 0.94$. The graph is sketched in Figure 7. ■

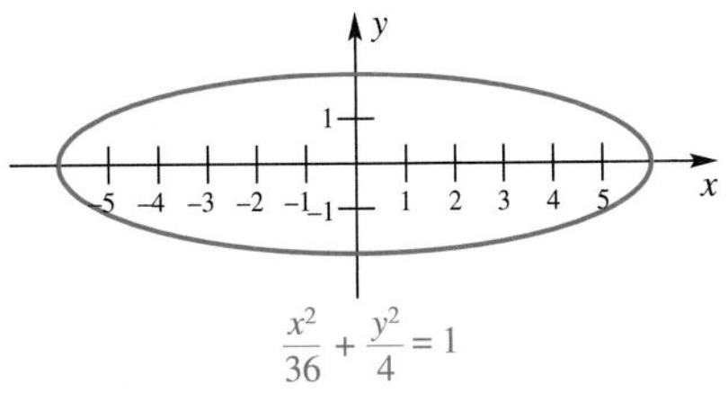

Figure 7

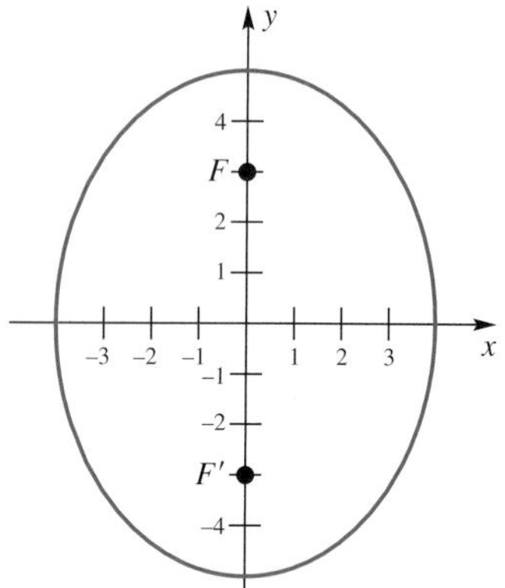

Figure 8

We call the ellipses sketched so far *horizontal ellipses* because the major axis is the x-axis. If we interchange the roles of x and y, we have the equation of a vertical ellipse:

$$\frac{y^2}{a^2} + \frac{x^2}{b^2} = 1 \quad \text{or} \quad \frac{x^2}{b^2} + \frac{y^2}{a^2} = 1$$

EXAMPLE 2 Sketch the graph of

$$\frac{x^2}{16} + \frac{y^2}{25} = 1$$

and determine its foci and eccentricity.

SOLUTION The larger square is now under y^2, which tells us that the major axis is vertical. Noting that $a = 5$ and $b = 4$, we conclude that $c = \sqrt{25 - 16} = 3$. Thus, the foci are $(0, \pm 3)$, and $e = c/a = \frac{3}{5} = 0.6$ (Figure 8). ■

Standard Equation of the Hyperbola

For the hyperbola, $e > 1$ and so $e^2 - 1$ is positive. If we let $b = a\sqrt{e^2 - 1}$, then the equation $x^2/a^2 + y^2/(1 - e^2)a^2 = 1$, which was derived earlier, takes the form

$$\boxed{\frac{x^2}{a^2} - \frac{y^2}{b^2} = 1}$$

This is called the **standard equation of a hyperbola.** Since $c = ae$, we now obtain $c^2 = a^2 + b^2$. (Note how this differs from the corresponding relationship for an ellipse.)

To interpret b, observe that if we solve for y in terms of x we get

$$y = \pm\frac{b}{a}\sqrt{x^2 - a^2}$$

For large x, $\sqrt{x^2 - a^2}$ behaves like x (i.e., $\left(\sqrt{x^2 - a^2} - x\right) \to 0$ as $x \to \infty$; see Problem 70) and hence y behaves like

$$y = \frac{b}{a}x \quad \text{or} \quad y = -\frac{b}{a}x$$

More precisely, the graph of the given hyperbola has these two lines as asymptotes.

The important facts for the hyperbola are summarized in Figure 9. As with the ellipse, there is an important right triangle (shaded in the diagram) that has legs a and b. This **fundamental triangle** determines the rectangle centered at the origin having sides of length $2a$ and $2b$. The extended diagonals of this rectangle are the asymptotes mentioned above.

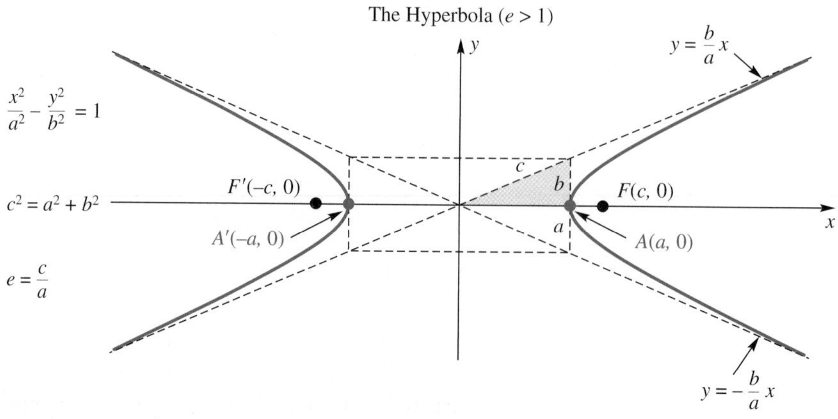

Figure 9

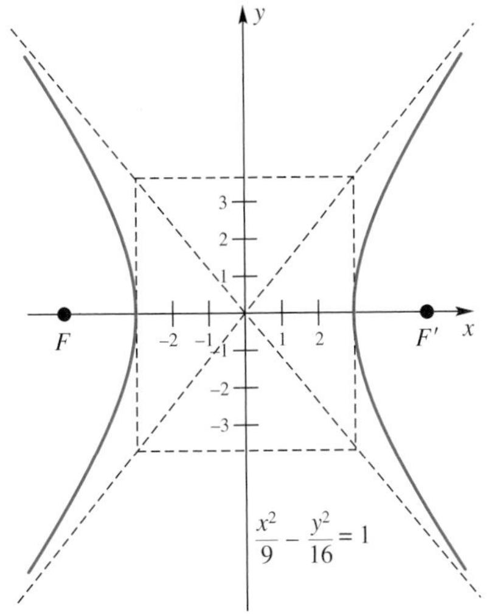

Figure 10

EXAMPLE 3 Sketch the graph of

$$\frac{x^2}{9} - \frac{y^2}{16} = 1$$

showing the asymptotes. What are the equations of the asymptotes? What are the foci?

SOLUTION We begin by determining the fundamental triangle; it has horizontal leg 3 and vertical leg 4. After drawing it, we can indicate the asymptotes and sketch the graph (Figure 10). The asymptotes are $y = \frac{4}{3}x$ and $y = -\frac{4}{3}x$. Since $c = \sqrt{a^2 + b^2} = \sqrt{9 + 16} = 5$, the foci are at $(\pm 5, 0)$. ■

Again, we should consider the effect of interchanging the roles of x and y. The equation takes the form

$$\frac{y^2}{a^2} - \frac{x^2}{b^2} = 1$$

This is the equation of a vertical hyperbola (vertical major axis). Its vertices are at $(0, \pm a)$; its foci are at $(0, \pm c)$.

For both the ellipse and the hyperbola, a is always the distance from the center to a vertex. For the ellipse, $a > b$; for the hyperbola, there is no such requirement.

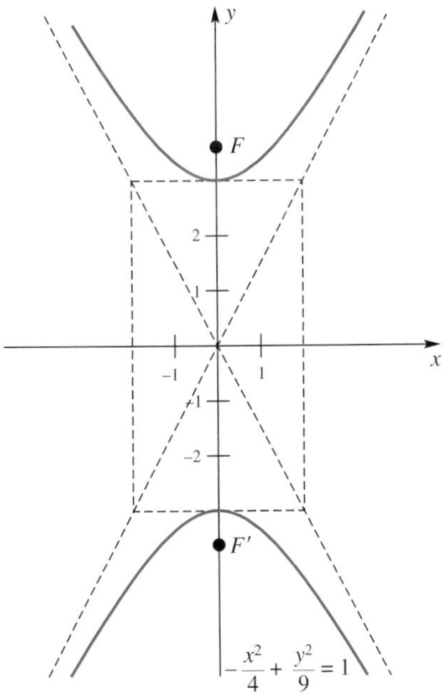

Figure 11

EXAMPLE 4 Determine the foci of

$$-\frac{x^2}{4} + \frac{y^2}{9} = 1$$

and sketch its graph.

SOLUTION We note immediately that this is a vertical hyperbola, which is determined by the fact that the plus sign is associated with the y^2 term. Thus, $a = 3$, $b = 2$, and $c = \sqrt{9 + 4} = \sqrt{13} \approx 3.61$. The foci are at $\left(0, \pm\sqrt{13}\right)$ (Figure 11). ■

EXAMPLE 5 According to Johannes Kepler (1571–1630), the planets revolve around the sun in elliptical orbits, with the sun at one focus. The earth's maximum distance from the sun is 94.56 million miles, and its minimum distance is 91.45 million miles. What is the eccentricity of the orbit, and what are the major and minor diameters?

SOLUTION Using the notation in Figure 12, we see that

$$a + c = 94.56 \qquad a - c = 91.45$$

When we solve these equations for a and c, we obtain $a = 93.01$ and $c = 1.56$. Thus,

$$e = \frac{c}{a} = \frac{1.56}{93.01} \approx 0.017$$

and the major diameter and minor diameter (in millions of miles) are, respectively,

$$2a \approx 186.02 \qquad 2b = 2\sqrt{a^2 - c^2} \approx 185.99$$ ■

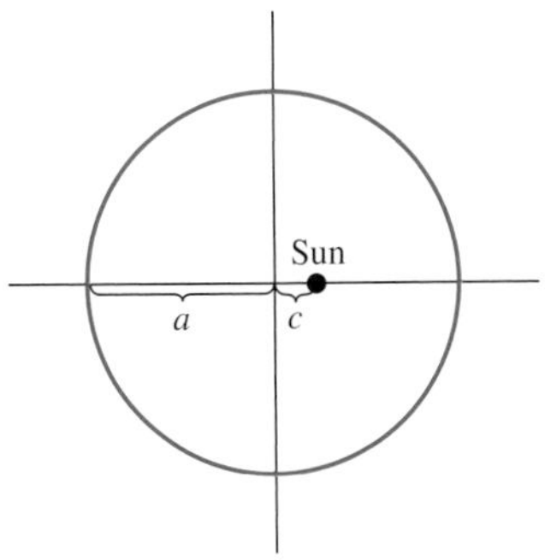

Figure 12

String Properties of the Ellipse and Hyperbola We have chosen to define conic sections in terms of the condition $|PF| = e|PL|$, where the figure is an ellipse if $0 < e < 1$ and a hyperbola if $e > 1$. This approach allows us to treat all conics in a unified way. Many authors prefer to define the ellipse and hyperbola via the following definitions.

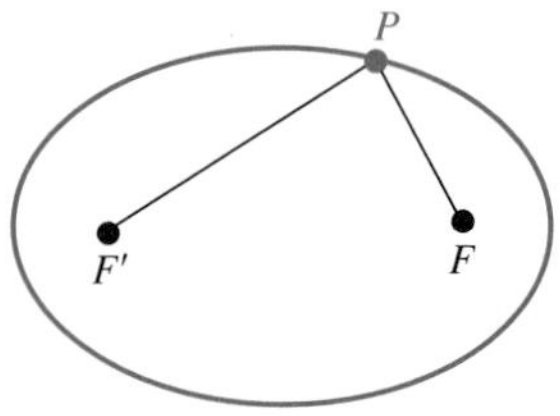

Figure 13

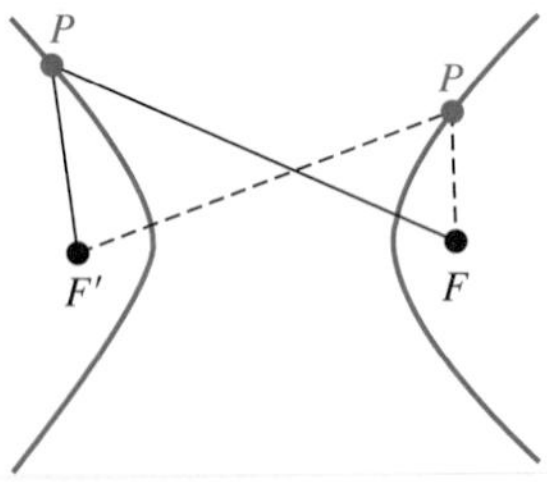

Figure 14

An ellipse is the set of points in the plane, the sum of whose distances from two fixed points (the foci) is a given constant $2a$. A hyperbola is the set of points in the plane, the difference of whose distances from two fixed points (the foci) is a given positive constant $2a$. (Here *difference* is taken to mean the larger minus the smaller distance.)

These definitions are illustrated in Figure 13 and 14. For the ellipse, imagine a string of length $2a$ tacked down at its two endpoints. If a pencil is used to stretch the string at point P, then the ellipse can be traced by moving the pencil around. These properties, called the *string properties*, should be consequences of our eccentricity definitions. We derive them now.

Suppose a and e are given. We know that the foci are $(\pm ae, 0)$ and the directrices are $x = \pm a/e$. The situations for the ellipse and hyperbola are illustrated in Figure 15.

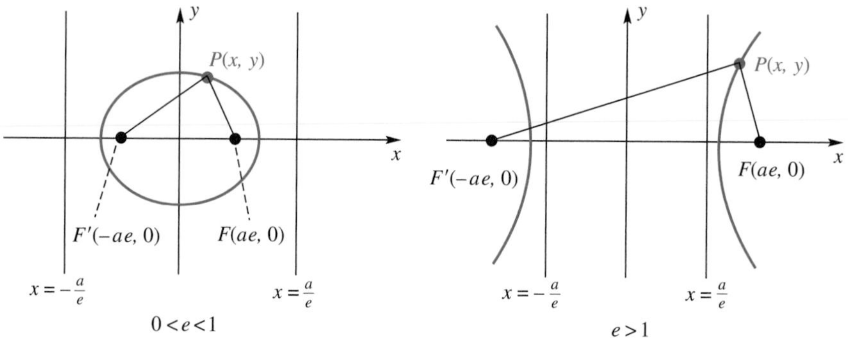

Figure 15

If we take an arbitrary point $P(x, y)$ on the ellipse, then, from the condition $|PF| = e|PL|$ applied first to the left focus and directrix and then to the right ones, we get

$$|PF'| = e\left(x + \frac{a}{e}\right) = ex + a \qquad |PF| = e\left(\frac{a}{e} - x\right) = a - ex$$

and so

$$\boxed{|PF'| + |PF| = 2a}$$

Next consider the hyperbola with $P(x, y)$ on its right branch, as shown in the right part of Figure 15. Then

$$|PF'| = e\left(x + \frac{a}{e}\right) = ex + a \qquad |PF| = e\left(x - \frac{a}{e}\right) = ex - a$$

and so $|PF'| - |PF| = 2a$. If $P(x, y)$ had been on the left branch, we would have gotten $-2a$ in place of $2a$. In either case,

$$\boxed{||PF'| - |PF|| = 2a}$$

EXAMPLE 6 Find the equation of the set of points the sum of whose distances from $(\pm 3, 0)$ is equal to 10.

SOLUTION This is a horizontal ellipse with $a = 5$ and $c = 3$. Thus, $b = \sqrt{a^2 - c^2} = 4$, and the equation is

$$\frac{x^2}{25} + \frac{y^2}{16} = 1$$

■

EXAMPLE 7 Find the equation of the set of points the difference of whose distances from $(0, \pm 6)$ is equal to 4.

SOLUTION This is a vertical hyperbola with $a = 2$ and $c = 6$. Thus, $b = \sqrt{c^2 - a^2} = \sqrt{32} = 4\sqrt{2}$, and the equation is

$$-\frac{x^2}{32} + \frac{y^2}{4} = 1$$

■

> **Lenses**
>
> The optical properties of the conics have been used in the grinding of lenses for hundreds of years. A recent innovation is the introduction of variable lenses to replace bifocal lenses in eyeglasses. Starting from the top, these lenses are ground so that the eccentricity varies continuously from $e < 1$ to $e = 1$ to $e > 1$, thus producing horizontal cross sections from ellipses to parabolas to hyperbolas and presumably allowing perfect viewing of objects at any distance by an appropriate tilt of the head.

Optical Properties Consider two mirrors, one with the shape of an ellipse and the other with the shape of a hyperbola. If a light ray emanating from one focus strikes the mirror, it will be reflected back to the other focus in the case of the ellipse and directly away from the other focus in the case of the hyperbola. These facts are shown in Figure 16.

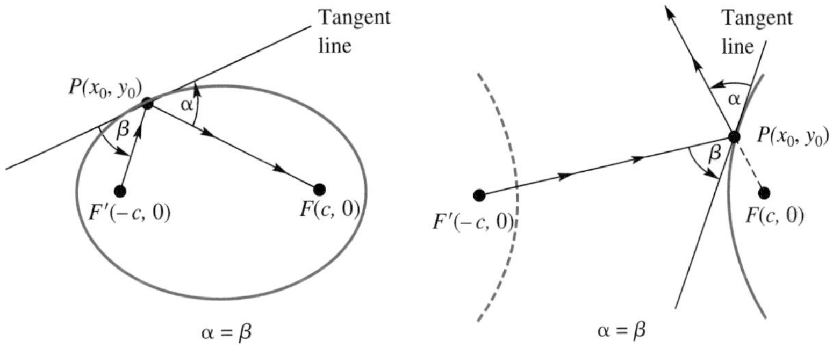

Figure 16

To demonstrate these optical properties (i.e., to show that $\alpha = \beta$ in both parts of Figure 16), we suppose the curves to be in standard position so that their equations are $x^2/a^2 + y^2/b^2 = 1$ and $x^2/a^2 - y^2/b^2 = 1$, respectively. For the ellipse, we differentiate implicitly to find the slope of the tangent line.

$$\frac{2x}{a^2} + \frac{2yy'}{b^2} = 0$$

$$y' = -\frac{b^2}{a^2}\frac{x}{y}$$

The slope of the tangent line at (x_0, y_0) is $m = -b^2x_0/(a^2y_0)$. Thus, the equation of the tangent line may be written successively as

$$y - y_0 = -\frac{b^2x_0}{a^2y_0}(x - x_0)$$

$$\frac{x_0}{a^2}(x - x_0) + \frac{y_0}{b^2}(y - y_0) = 0$$

$$\frac{x_0x}{a^2} + \frac{y_0y}{b^2} = \frac{x_0^2}{a^2} + \frac{y_0^2}{b^2} = 1$$

To calculate $\tan\alpha$ for the ellipse, we recall (Problem 40 of Section 0.7) a formula for the tangent of the counterclockwise angle from one line ℓ_1 to another ℓ in terms of their respective slopes m_1 and m:

$$\tan\alpha = \frac{m - m_1}{1 + mm_1}$$

Now refer to Figure 16 and let ℓ_1 be the line FP and ℓ be the tangent line at P. Then

$$\tan\alpha = \frac{\dfrac{-b^2x_0}{a^2y_0} - \dfrac{y_0 - 0}{x_0 - c}}{1 + \left(\dfrac{-b^2x_0}{a^2y_0}\right)\left(\dfrac{y_0 - 0}{x_0 - c}\right)} = \frac{-b^2x_0(x_0 - c) - a^2y_0^2}{a^2y_0(x_0 - c) - b^2x_0y_0}$$

$$= \frac{b^2cx_0 - (b^2x_0^2 + a^2y_0^2)}{(a^2 - b^2)x_0y_0 - a^2cy_0} = \frac{b^2cx_0 - a^2b^2}{c^2x_0y_0 - a^2cy_0}$$

$$= \frac{b^2(cx_0 - a^2)}{cy_0(cx_0 - a^2)} = \frac{b^2}{cy_0}$$

The same calculation with c replaced by $-c$ gives

$$\tan(-\beta) = \frac{b^2}{-cy_0}$$

and so $\tan\beta = b^2/cy_0$. We conclude that $\tan\alpha = \tan\beta$, and consequently $\alpha = \beta$.

A similar derivation establishes the corresponding result for the hyperbola.

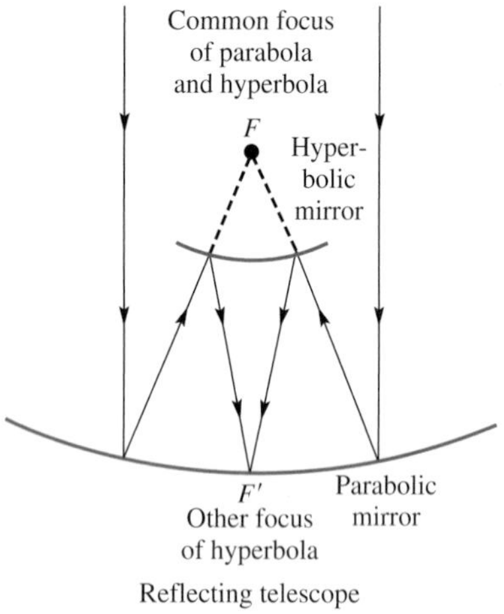

Figure 17

Applications The reflecting property of the ellipse is the basis of the *whispering gallery* effect that can be observed, for example, in the U.S. Capitol, the Mormon Tabernacle, and many science museums. A speaker standing at one focus can be heard whispering by a listener at the other focus, even though his or her voice is inaudible in other parts of the room.

The optical properties of the parabola and hyperbola are combined in one design for a reflecting telescope (Figure 17). The parallel rays from a star are finally focused at the eyepiece at F'.

The string property of the hyperbola is used in navigation. A ship at sea can determine the difference $2a$ in its distance from two fixed transmitters by measuring the difference in reception times of synchronized radio signals. This puts its path on a hyperbola, with the two transmitters F and F' as foci (see Figure 18). If another pair of transmitters G and G' are used, the ship must lie at the intersection of the two corresponding hyperbolas (see Figure 19). LORAN, a system of long-range navigation, is based on this principle.

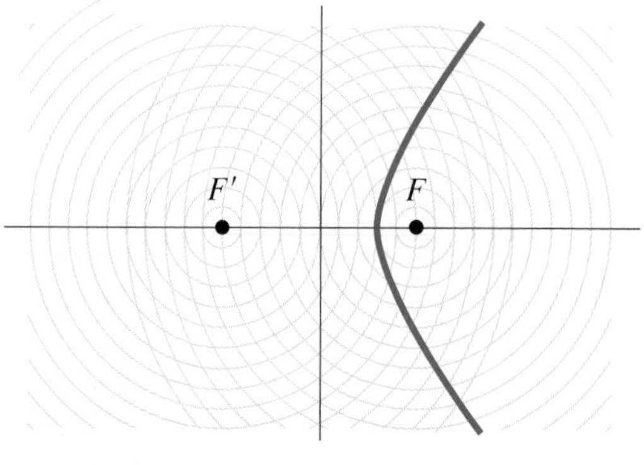

Figure 18

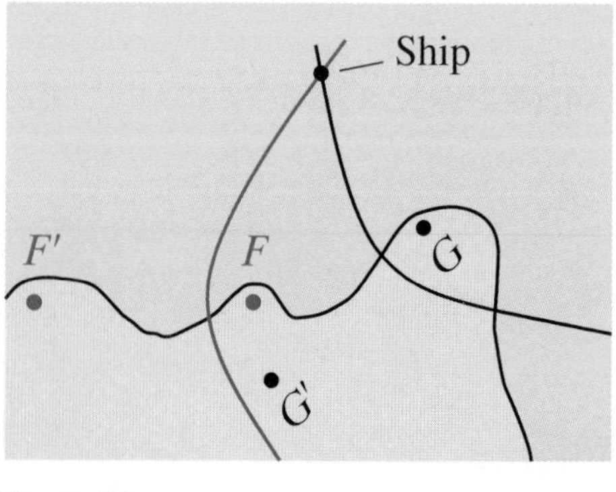

Figure 19

Concepts Review

1. The standard equation of the horizontal ellipse centered at $(0, 0)$ is ______.

2. The xy-equation of the vertical ellipse centered at $(0, 0)$ that has major diameter 8 and minor diameter 6 is ______.

3. An ellipse is the set of points P satisfying $|PF| + |PF'| = 2a$, where F and F' are fixed points called the ______ of the ellipse.

4. A ray from a light source at one focus of an elliptical mirror will be reflected ______ whereas a ray from a light source at one focus of a hyperbolic mirror will be reflected ______.

Problem Set 10.2

In Problems 1–8, name the conic (horizontal ellipse, vertical hyperbola, and so on) corresponding to the given equation.

1. $\frac{x^2}{9} + \frac{y^2}{4} = 1$ **2.** $\frac{x^2}{9} - \frac{y^2}{4} = 1$

3. $\frac{-x^2}{9} + \frac{y^2}{4} = 1$ **4.** $\frac{-x^2}{9} + \frac{y^2}{4} = -1$

5. $\frac{-x^2}{9} + \frac{y}{4} = 0$ **6.** $\frac{-x^2}{9} = \frac{y}{4}$

7. $9x^2 + 4y^2 = 9$ **8.** $x^2 - 4y^2 = 4$

In Problems 9–16, sketch the graph of the given equation, indicating vertices, foci, and asymptotes (if it is a hyperbola).

9. $\frac{x^2}{16} + \frac{y^2}{4} = 1$ **10.** $\frac{x^2}{16} - \frac{y^2}{4} = 1$

11. $\frac{-x^2}{9} + \frac{y^2}{4} = 1$ **12.** $\frac{x^2}{7} + \frac{y^2}{4} = 1$

13. $16x^2 + 4y^2 = 32$ **14.** $4x^2 + 25y^2 = 100$

15. $10x^2 - 25y^2 = 100$ **16.** $x^2 - 4y^2 = 8$

In Problems 17–30, find the equation of the given central conic.

17. Ellipse with a focus at $(-3, 0)$ and a vertex at $(6, 0)$

18. Ellipse with a focus at $(6, 0)$ and eccentricity $\frac{2}{3}$

19. Ellipse with a focus at $(0, -5)$ and eccentricity $\frac{1}{3}$

20. Ellipse with a focus at $(0, 3)$ and minor diameter 8

21. Ellipse with a vertex at $(5, 0)$ and passing through $(2, 3)$

22. Hyperbola with a focus at $(5, 0)$ and a vertex at $(4, 0)$

23. Hyperbola with a vertex at $(0, -4)$ and a focus at $(0, -5)$

24. Hyperbola with a vertex at $(0, -3)$ and eccentricity $\frac{3}{2}$

25. Hyperbola with asymptotes $2x \pm 4y = 0$ and a vertex at $(8, 0)$

26. Vertical hyperbola with eccentricity $\sqrt{6}/2$ that passes through $(2, 4)$

27. Ellipse with foci $(\pm 2, 0)$ and directrices $x = \pm 8$

28. Hyperbola with foci $(\pm 4, 0)$ and directrices $x = \pm 1$

29. Hyperbola whose asymptotes are $x \pm 2y = 0$ and that goes through the point $(4, 3)$

30. Horizontal ellipse that goes through $(-5, 1)$ and $(-4, -2)$

In Problems 31–34, find the equation of the set of points P satisfying the given conditions.

31. The sum of the distances of P from $(0, \pm 9)$ is 26.

32. The sum of the distances of P from $(\pm 4, 0)$ is 14.

33. The difference of the distances of P from $(\pm 7, 0)$ is 12.

34. The difference of the distances of P from $(0, \pm 6)$ is 10.

In Problems 35–42, find the equation of the tangent line to the given curve at the given point.

35. $\frac{x^2}{27} + \frac{y^2}{9} = 1$ at $(3, \sqrt{6})$

36. $\frac{x^2}{24} + \frac{y^2}{16} = 1$ at $(3\sqrt{2}, -2)$

37. $\frac{x^2}{27} + \frac{y^2}{9} = 1$ at $(3, -\sqrt{6})$

38. $\frac{x^2}{2} - \frac{y^2}{4} = 1$ at $(\sqrt{3}, \sqrt{2})$

39. $x^2 + y^2 = 169$ at $(5, 12)$

40. $x^2 - y^2 = -1$ at $(\sqrt{2}, \sqrt{3})$

41. The curve of Problem 31 at $(0, 13)$

42. The curve of Problem 32 at $(7, 0)$

43. A doorway in the shape of an elliptical arch (a half-ellipse) is 10 feet wide and 4 feet high at the center. A box 2 feet high is to be pushed through the doorway. How wide can the box be?

44. How high is the arch of Problem 43 at a distance 2 feet to the right of the center?

45. How long is the *latus rectum* (chord through the focus perpendicular to the major axis) for the ellipse $x^2/a^2 + y^2/b^2 = 1$?

46. Determine the length of the latus rectum (see Problem 45) of the hyperbola $x^2/a^2 - y^2/b^2 = 1$.

C **47.** Halley's comet has an elliptical orbit with major and minor diameters of 36.18 AU and 9.12 AU, respectively (1 AU is 1 astronomical unit, the earth's mean distance from the sun). What is its minimum distance from the sun (assuming the sun is at a focus)?

C **48.** The orbit of the comet Kahoutek is an ellipse with eccentricity $e = 0.999925$ with the sun at a focus. If its minimum distance to the sun is 0.13 AU, what is its maximum distance from the sun? See Problem 47.

C **49.** In 1957, Russia launched Sputnik I. Its elliptical orbit around the earth reached maximum and minimum distances from the earth of 583 miles and 132 miles, respectively. Assuming that the center of the earth is one focus and that the earth is a sphere of radius 4000 miles, find the eccentricity of the orbit.

50. The orbit of the planet Pluto has an eccentricity 0.249. The closest that Pluto comes to the sun is 29.65 AU, and the farthest is 49.31 AU. Find the major and minor diameters.

51. If two tangent lines to the ellipse $9x^2 + 4y^2 = 36$ intersect the y-axis at $(0, 6)$, find the points of tangency.

52. If the tangent lines to the hyperbola $9x^2 - y^2 = 36$ intersect the y-axis at $(0, 6)$, find the points of tangency.

53. The slope of the tangent line to the hyperbola

$$2x^2 - 7y^2 - 35 = 0$$

at two points on the hyperbola is $-\frac{2}{3}$. What are the coordinates of the points of tangency?

54. Find the equations of the tangent lines to the ellipse $x^2 + 2y^2 - 2 = 0$ that are parallel to the line

$$3x - 3\sqrt{2}y - 7 = 0$$

55. Find the area of the ellipse $b^2x^2 + a^2y^2 = a^2b^2$.

56. Find the volume of the solid obtained by revolving the ellipse $b^2x^2 + a^2y^2 = a^2b^2$ about the y-axis.

57. The region bounded by the hyperbola

$$b^2x^2 - a^2y^2 = a^2b^2$$

and a vertical line through a focus is revolved about the x-axis. Find the volume of the resulting solid.

58. If the ellipse of Problem 56 is revolved about the x-axis, find the volume of the resulting solid.

59. Find the dimensions of the rectangle having the greatest possible area that can be inscribed in the ellipse $b^2x^2 + a^2y^2 = a^2b^2$. Assume that the sides of the rectangle are parallel to the axes of the ellipse.

60. Show that the point of contact of any tangent line to a hyperbola is midway between the points in which the tangent intersects the asymptotes.

61. Find the point in the first quadrant where the two hyperbolas $25x^2 - 9y^2 = 225$ and $-25x^2 + 18y^2 = 450$ intersect.

62. Find the points of intersection of $x^2 + 4y^2 = 20$ and $x + 2y = 6$.

63. Sketch a design for a reflecting telescope that uses a parabola and an ellipse rather than a parabola and a hyperbola as described in the text and shown in Figure 17.

64. A ball placed at a focus of an elliptical billiard table is shot with tremendous force so that it continues to bounce off the cushions indefinitely. Describe its ultimate path? *Hint:* Draw a picture.

65. If the ball of Problem 64 is initially on the major axis between a focus and the neighboring vertex, what can you say about its path?

66. Show that an ellipse and a hyperbola with the same foci intersect at right angles. *Hint:* Draw a picture and use the optical properties.

67. Describe a string apparatus for constructing a hyperbola. (There are several possibilities.)

68. Sound travels at u feet per second and a rifle bullet at $v > u$ feet per second. The sound of the firing of a rifle and the impact of the bullet hitting the target were heard simultaneously. If the rifle was at $A(-c, 0)$, the target was at $B(c, 0)$, and the listener was at $P(x, y)$, find the equation of the curve on which P lies (in terms of u, v, and c).

69. Listeners $A(-8, 0)$, $B(8, 0)$, and $C(8, 10)$ recorded the exact times at which they heard an explosion. If B and C heard the explosion at the same time and A heard it 12 seconds later, where was the explosion? Assume that distances are in kilometers and that sound travels $\frac{1}{3}$ kilometer per second.

70. Show that $\left(\sqrt{x^2 - a^2} - x\right) \to 0$ as $x \to \infty$. *Hint:* Rationalize the numerator.

71. For an ellipse, let p and q be the distances from a focus to the two vertices. Show that $b = \sqrt{pq}$, with $2b$ being the minor diameter.

72. The wheel in Figure 20 is rotating at 1 radian per second so that Q has coordinates $(a \cos t, a \sin t)$. Find the coordinates (x, y) of R at time t and show that it is traveling in an elliptical path. *Note: PQR* is a right triangle when $P \neq R$ and $R \neq Q$.

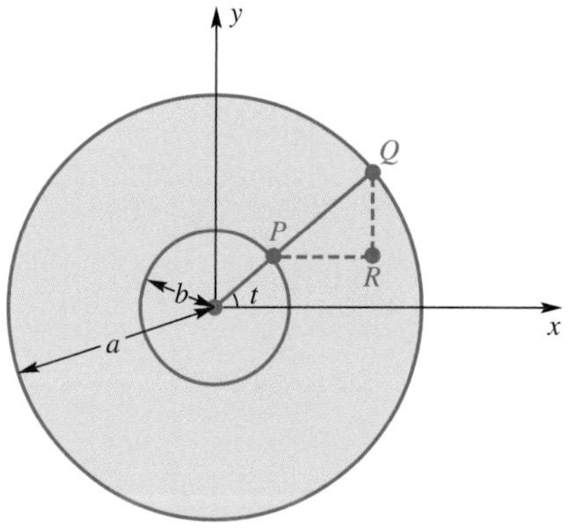

Figure 20

73. Let P be a point on a ladder of length $a + b$, P being a units from the top end. As the ladder slides with its top end on the y-axis and its bottom end on the x-axis, P traces out a curve. Find the equation of this curve.

74. Show that a line through a focus of a hyperbola and perpendicular to an asymptote intersects that asymptote on the directrix nearest the focus.

75. If a horizontal hyperbola and a vertical hyperbola have the same asymptotes, show that their eccentricities e and E satisfy $e^{-2} + E^{-2} = 1$.

76. Let C be the curve of intersection of a right circular cylinder and a plane making an angle ϕ $(0 < \phi < \pi/2)$ with the axis of the cylinder. Show that C is an ellipse.

GC EXPL **77.** Using the same axes, draw the conics $y = \pm(ax^2 + 1)^{1/2}$ for $-2 \le x \le 2$ and $-2 \le y \le 2$ using $a = -2, -1, -0.5, -0.1, 0, 0.1, 0.6, 1$. Make a conjecture about how the shape of the figure depends on a.

Answers to Concepts Review: **1.** $x^2/a^2 + y^2/b^2 = 1$ **2.** $x^2/9 + y^2/16 = 1$ **3.** foci **4.** to the other focus; directly away from the other focus

10.3 Translation and Rotation of Axes

So far we have placed the conics in the coordinate system in very special ways—always with the major axis along one of the coordinate axes and either the vertex (in the case of a parabola) or the center (in the case of an ellipse or hyperbola) at the origin. Now we place our conics in a more general position, though we still require that the major axis be parallel to one of the coordinate axes. Even this restriction will be removed later in this section.

The case of a circle is instructive. The circle of radius 5 centered at (2, 3) has equation

$$(x - 2)^2 + (y - 3)^2 = 25$$

or, in equivalent expanded form,

$$x^2 + y^2 - 4x - 6y = 12$$

The same circle with its center at the origin of the uv-coordinate system (Figure 1) has the simple equation

$$u^2 + v^2 = 25$$

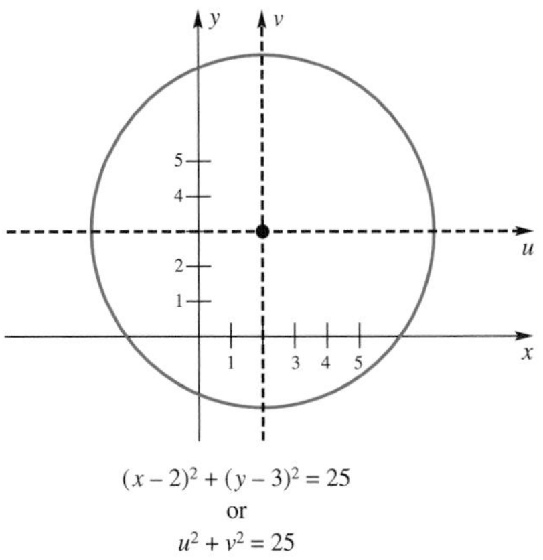

Figure 1

The introduction of new axes does not change the shape or size of a curve, but it may greatly simplify its equation. It is this *translation* of axes and the corresponding change of variables in an equation that we wish to investigate.

Translations If new axes are chosen in the plane, every point will have two sets of coordinates, the old ones, (x, y), relative to the old axes and the new ones, (u, v), relative to the new axes. The original coordinates are said to undergo a **transformation.** If the new axes are parallel, respectively, to the original axes and have the same directions and scales, then the transformation is called a **translation of axes.**

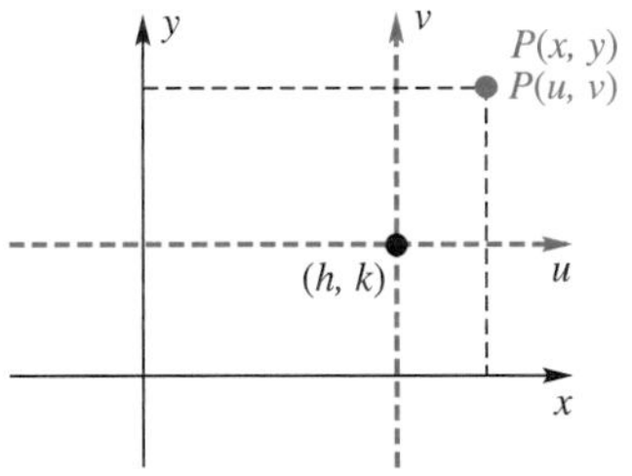

Figure 2

From Figure 2, it is easy to see how the new coordinates (u, v) relate to the old ones (x, y). Let (h, k) be the old coordinates of the new origin. Then

$$u = x - h, \qquad v = y - k$$

or, equivalently,

$$x = u + h, \qquad y = v + k$$

EXAMPLE 1 Find the new coordinates of $P(-6, 5)$ after a translation of axes to a new origin at $(2, -4)$.

SOLUTION Since $h = 2$ and $k = -4$, it follows that

$$u = x - h = -6 - 2 = -8 \qquad v = y - k = 5 - (-4) = 9$$

The new coordinates are $(-8, 9)$. ■

EXAMPLE 2 Given the equation $4x^2 + y^2 + 40x - 2y + 97 = 0$, find the equation of its graph after a translation with new origin $(-5, 1)$.

SOLUTION In the equation, we replace x by $u + h = u - 5$ and y by $v + k = v + 1$. We obtain

$$4(u - 5)^2 + (v + 1)^2 + 40(u - 5) - 2(v + 1) + 97 = 0$$

or

$$4u^2 - 40u + 100 + v^2 + 2v + 1 + 40u - 200 - 2v - 2 + 97 = 0$$

This simplifies to

$$4u^2 + v^2 = 4$$

or

$$u^2 + \frac{v^2}{4} = 1$$

which we recognize as the equation of an ellipse. ■

Completing the Square Given a complicated second-degree equation, how do we know what translation will simplify the equation and bring it to a recognizable form? We can complete the square to eliminate the first-degree terms of any expression of the form

$$Ax^2 + Cy^2 + Dx + Ey + F = 0, \qquad A \neq 0, \quad C \neq 0$$

EXAMPLE 3 Make a translation that will eliminate the first-degree terms of

$$4x^2 + 9y^2 + 8x - 90y + 193 = 0$$

and use this information to sketch the graph of the given equation.

SOLUTION Recall that to complete the square of $x^2 + ax$ we must add $a^2/4$ (the square of half the coefficient of x). Using this, we rewrite the given equation by adding the same numbers to both sides.

$$4(x^2 + 2x \qquad) + 9(y^2 - 10y \qquad) = -193$$
$$4(x^2 + 2x + 1) + 9(y^2 - 10y + 25) = -193 + 4 + 225$$
$$4(x + 1)^2 + 9(y - 5)^2 = 36$$
$$\frac{(x + 1)^2}{9} + \frac{(y - 5)^2}{4} = 1$$

The translation $u = x + 1$ and $v = y - 5$ transforms this to

$$\frac{u^2}{9} + \frac{v^2}{4} = 1$$

which is the standard form of a horizontal ellipse. The graph is shown in Figure 3. ■

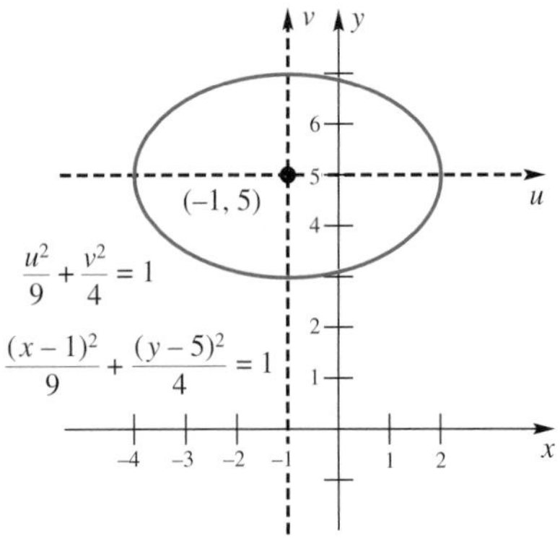

Figure 3

EXAMPLE 4 Use a translation to simplify

$$y^2 - 4x - 12y + 28 = 0$$

Then determine which conic it represents, list the important characteristics of this conic, and sketch its graph.

SOLUTION We complete the square.

$$y^2 - 12y = 4x - 28$$
$$y^2 - 12y + 36 = 4x - 28 + 36$$
$$(y - 6)^2 = 4(x + 2)$$

The translation $u = x + 2$, $v = y - 6$ transforms this to $v^2 = 4u$, which we recognize as a horizontal parabola opening right with $p = 1$ (Figure 4). ■

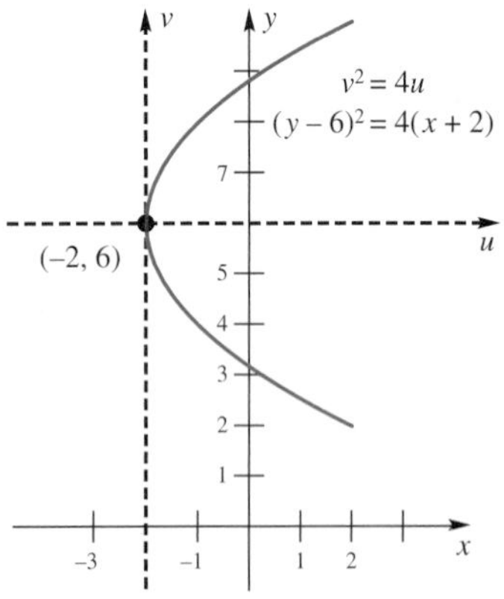

Figure 4

General Second-Degree Equations Now we ask an important question. Is the graph of an equation of the form

$$Ax^2 + Cy^2 + Dx + Ey + F = 0$$

always a conic? The answer is no, unless we admit certain limiting forms. The following table indicates the possibilities with a sample equation for each.

Thus, the graphs of the general quadratic equation above fall into three general categories, but yield nine different possibilities, including limiting forms.

Conics		Limiting Forms	
1. $(AC = 0)$ Parabola: $y^2 = 4x$	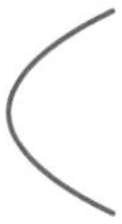	Parallel lines: $y^2 = 4$ Single line: $y^2 = 0$ Empty set: $y^2 = -1$	
2. $(AC > 0)$ Ellipse: $\frac{x^2}{9} + \frac{y^2}{4} = 1$		Circle: $x^2 + y^2 = 4$ Point: $2x^2 + y^2 = 0$ Empty set: $2x^2 + y^2 = -1$	 •
3. $(AC < 0)$ Hyperbola: $\frac{x^2}{9} - \frac{y^2}{4} = 1$		Intersecting lines: $x^2 - y^2 = 0$	

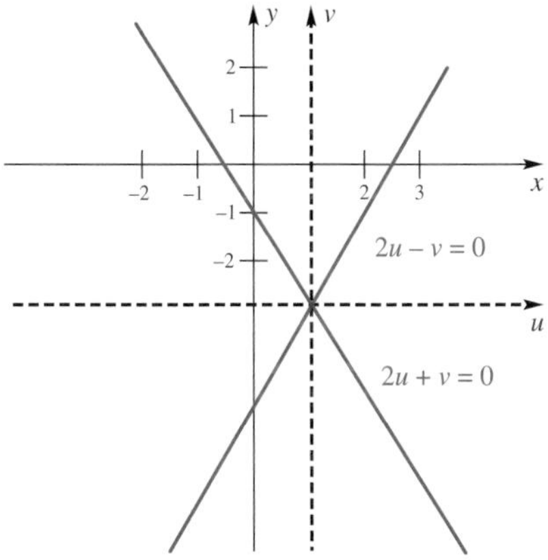

Figure 5

EXAMPLE 5 Use a translation to simplify

$$4x^2 - y^2 - 8x - 6y - 5 = 0$$

and sketch its graph.

SOLUTION We rewrite the equation as follows:

$$4(x^2 - 2x \qquad) - (y^2 + 6y \qquad) = 5$$
$$4(x^2 - 2x + 1) - (y^2 + 6y + 9) = 5 + 4 - 9$$
$$4(x - 1)^2 - (y + 3)^2 = 0$$

Let $u = x - 1$ and $v = y + 3$, which results in

$$4u^2 - v^2 = 0$$

or

$$(2u - v)(2u + v) = 0$$

This is the equation of two intersecting lines (Figure 5). ■

EXAMPLE 6 Write the equation of a hyperbola with foci at (1, 1) and (1, 11) and vertices at (1, 3) and (1, 9).

SOLUTION The center is (1, 6), midway between the vertices on a vertical major axis. Thus, $a = 3$ and $c = 5$, and so $b = \sqrt{c^2 - a^2} = 4$. The equation is

$$\frac{(y - 6)^2}{9} - \frac{(x - 1)^2}{16} = 1$$ ■

The General Equation of a Conic Section

Consider the equation

$$Ax^2 + Cy^2 + Dx + Ey + F = 0$$

If both A and C are zero, we have the equation of a line (provided, of course, that D and E are not both zero). If at least one of A and C is different from zero, we may apply the process of completing the square. We obtain one of several forms, the most typical being

(1) $$(y - k)^2 = \pm 4p(x - h)$$

(2) $$\frac{(x - h)^2}{a^2} + \frac{(y - k)^2}{b^2} = 1$$

(3) $$\frac{(x - h)^2}{a^2} - \frac{(y - k)^2}{b^2} = 1$$

These can be recognized even in this form as the equations of a horizontal parabola with vertex at (h, k), a horizontal ellipse (if $a^2 > b^2$) with center at (h, k), and a horizontal hyperbola with center at (h, k).

In all of these cases we get a figure whose major and minor axes are parallel to the x- and y-axes. If we include the cross-product term Bxy, as in

$$Ax^2 + Bxy + Cx^2 + Dx + Ey + F = 0$$

we still get a conic section (or a limiting form of one), but one where the major and minor axes are parallel to a **rotation** of the x- and y-axes.

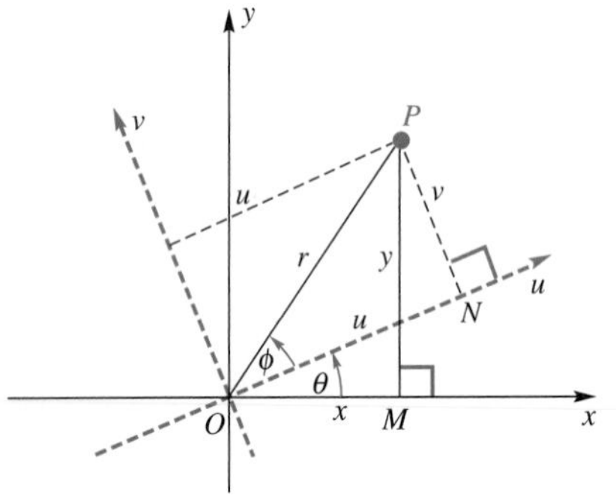

Figure 6

Rotations Introduce a new pair of coordinate axes, the u- and v-axes, with the same origin as the x- and y-axes but rotated through an angle θ, as shown in Figure 6. A point P then has two sets of coordinates: (x, y) and (u, v). How are they related?

Let r denote the length of OP, and let ϕ denote the angle from the positive u-axis to OP. Then x, y, u, and v have the geometric interpretations shown in the diagram.

Looking at the right triangle OPM, we see that

$$\cos(\phi + \theta) = \frac{x}{r}$$

so

$$x = r\cos(\phi + \theta) = r(\cos\phi\cos\theta - \sin\phi\sin\theta)$$
$$= (r\cos\phi)\cos\theta - (r\sin\phi)\sin\theta$$

Consideration of triangle OPN shows that $u = r\cos\phi$ and $v = r\sin\phi$. Thus,

$$\boxed{x = u\cos\theta - v\sin\theta}$$

Similar reasoning leads to

$$\boxed{y = u\sin\theta + v\cos\theta}$$

These formulas determine a transformation called a **rotation of axes.**

EXAMPLE 7 Find the new equation that results from $xy = 1$ after a rotation of axes through $\theta = \pi/4$. Sketch the graph.

SOLUTION The required substitutions are

$$x = u\cos\frac{\pi}{4} - v\sin\frac{\pi}{4} = \frac{\sqrt{2}}{2}(u - v)$$

$$y = u\sin\frac{\pi}{4} + v\cos\frac{\pi}{4} = \frac{\sqrt{2}}{2}(u + v)$$

The equation $xy = 1$ takes the form

$$\frac{\sqrt{2}}{2}(u - v)\frac{\sqrt{2}}{2}(u + v) = 1$$

which simplifies to

$$\frac{u^2}{2} - \frac{v^2}{2} = 1$$

This we recognize as the equation of a hyperbola with $a = b = \sqrt{2}$. Note that the cross-product term has disappeared as a result of the rotation. The choice of the angle $\theta = \pi/4$ was just right to make this happen. The graph is shown in Figure 7. ■

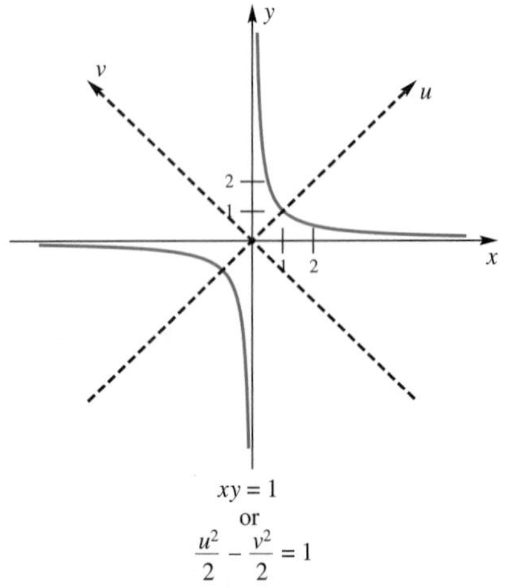

Figure 7

Determining the Angle θ How do we know what rotation to make in order to eliminate the cross-product term? Consider the equation

$$Ax^2 + Bxy + Cy^2 + Dx + Ey + F = 0$$

If we make the substitutions

$$x = u\cos\theta - v\sin\theta$$
$$y = u\sin\theta + v\cos\theta$$

this equation takes the form

$$au^2 + buv + cv^2 + du + ev + f = 0$$

where a, b, c, d, e, and f are numbers that depend on θ. We could find expressions for all of them, but we really care only about b. When we do the necessary algebra, we find that

$$\begin{aligned} b &= B(\cos^2\theta - \sin^2\theta) - 2(A - C)\sin\theta\cos\theta \\ &= B\cos 2\theta - (A - C)\sin 2\theta \end{aligned}$$

To make $b = 0$, we require that

$$B\cos 2\theta = (A - C)\sin 2\theta$$

or

$$\boxed{\cot 2\theta = \frac{A - C}{B}}$$

This formula answers our question. To eliminate the cross-product term, choose θ so that it satisfies this formula. In the equation $xy = 1$ of Example 7, $A = 0$, $B = 1$, and $C = 0$, so we choose θ satisfying $\cot 2\theta = 0$. One angle that works is $\theta = \pi/4$. We could also use $\theta = 3\pi/4$ or $\theta = -5\pi/4$, but it is customary to choose a first-quadrant angle; that is, we choose 2θ satisfying $0 \le 2\theta < \pi$ so that $0 \le \theta < \pi/2$.

EXAMPLE 8 Make a rotation of axes to eliminate the cross-product term in

$$4x^2 + 2\sqrt{3}xy + 2y^2 + 10\sqrt{3}x + 10y = 5$$

Then sketch the graph.

SOLUTION

$$\cot 2\theta = \frac{A - C}{B} = \frac{4 - 2}{2\sqrt{3}} = \frac{1}{\sqrt{3}}$$

which means that $2\theta = \pi/3$ and $\theta = \pi/6$. The appropriate substitutions are

$$x = u\frac{\sqrt{3}}{2} - v\frac{1}{2} = \frac{\sqrt{3}u - v}{2}$$

$$y = u\frac{1}{2} + v\frac{\sqrt{3}}{2} = \frac{u + \sqrt{3}v}{2}$$

Our equation transforms first to

$$4\frac{(\sqrt{3}u - v)^2}{4} + 2\sqrt{3}\frac{(\sqrt{3}u - v)(u + \sqrt{3}v)}{4}$$
$$+ 2\frac{(u + \sqrt{3}v)^2}{4} + 10\sqrt{3}\frac{\sqrt{3}u - v}{2} + 10\frac{u + \sqrt{3}v}{2} = 5$$

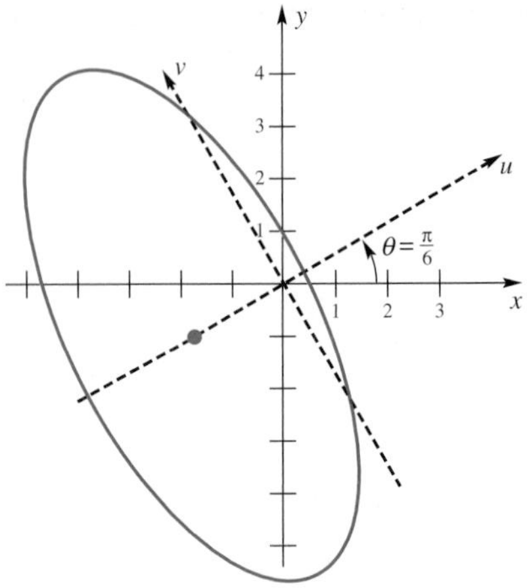

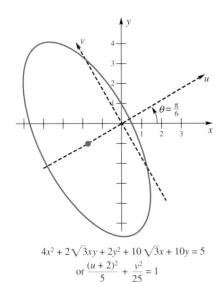

$4x^2 + 2\sqrt{3}xy + 2y^2 + 10\sqrt{3}x + 10y = 5$

or $\frac{(u+2)^2}{5} + \frac{v^2}{25} = 1$

Figure 8

and, after simplifying, to

$$5u^2 + v^2 + 20u = 5$$

To put this equation in recognizable form, we complete the square.

$$5(u^2 + 4u + 4) + v^2 = 5 + 20$$

$$\frac{(u+2)^2}{5} + \frac{v^2}{25} = 1$$

We identify the last equation as that of a vertical ellipse with center at $u = -2$ and $v = 0$ and with $a = 5$ and $b = \sqrt{5}$. This allows us to draw the graph shown in Figure 8. If we wanted to carry the simplifying process further, we would make the translation $r = u + 2$, $s = v$, which results in the standard equation $r^2/5 + s^2/25 = 1$. ■

Concepts Review

1. The quadratic form $x^2 + ax$ is made a square by adding ____.

2. $x^2 + 6x + 2(y^2 - 2y) = 3$ is (after completing the squares) equivalent to $(x + 3)^2 + 2(y - 1)^2 =$ ____, which is the equation of a(n) ____.

3. The cross-product term (the xy-term) can be eliminated by a rotation of axes through an angle θ satisfying $\cot 2\theta =$ ____.

4. To put a general second-degree equation in standard form, we first make a ____ of axes and then a ____ of axes.

Problem Set 10.3

In Problems 1–14, name the conic or limiting form represented by the given equation. Usually you will need to use the process of completing the square (see Examples 3–5).

1. $x^2 + y^2 - 2x + 2y + 1 = 0$

2. $x^2 + y^2 + 6x - 2y + 6 = 0$

3. $9x^2 + 4y^2 + 72x - 16y + 124 = 0$

4. $16x^2 - 9y^2 + 192x + 90y - 495 = 0$

5. $9x^2 + 4y^2 + 72x - 16y + 160 = 0$

6. $16x^2 + 9y^2 + 192x + 90y + 1000 = 0$

7. $y^2 - 5x - 4y - 6 = 0$

8. $4x^2 + 4y^2 + 8x - 28y - 11 = 0$

9. $3x^2 + 3y^2 - 6x + 12y + 60 = 0$

10. $4x^2 - 4y^2 - 2x + 2y + 1 = 0$

11. $4x^2 - 4y^2 + 8x + 12y - 5 = 0$

12. $4x^2 - 4y^2 + 8x + 12y - 6 = 0$

13. $4x^2 - 24x + 36 = 0$

14. $4x^2 - 24x + 35 = 0$

In Problems 15–28, sketch the graph of the given equation.

15. $\frac{(x+3)^2}{4} + \frac{(y+2)^2}{16} = 1$

16. $(x + 3)^2 + (y - 4)^2 = 25$

17. $\frac{(x+3)^2}{4} - \frac{(y+2)^2}{16} = 1$

18. $4(x + 3) = (y + 2)^2$

19. $(x + 2)^2 = 8(y - 1)$

20. $(x + 2)^2 = 4$

21. $(y - 1)^2 = 16$

22. $\frac{(x+3)^2}{4} + \frac{(y-2)^2}{8} = 0$

23. $x^2 + 4y^2 - 2x + 16y + 1 = 0$

24. $25x^2 + 9y^2 + 150x - 18y + 9 = 0$

25. $9x^2 - 16y^2 + 54x + 64y - 127 = 0$

26. $x^2 - 4y^2 - 14x - 32y - 11 = 0$

27. $4x^2 + 16x - 16y + 32 = 0$

28. $x^2 - 4x + 8y = 0$

29. Find the focus and directrix of the parabola

$$2y^2 - 4y - 10x = 0$$

30. Determine the distance between the vertices of

$$-9x^2 + 18x + 4y^2 + 24y = 9$$

31. Find the foci of the ellipse

$$16(x - 1)^2 + 25(y + 2)^2 = 400$$

32. Find the focus and directrix of the parabola

$$x^2 - 6x + 4y + 3 = 0$$

In Problems 33–42, find the equation of the given conic.

33. Horizontal ellipse with center $(5, 1)$, major diameter 10, minor diameter 8

34. Hyperbola with center $(2, -1)$, vertex at $(4, -1)$, and focus at $(5, -1)$

35. Parabola with vertex $(2, 3)$ and focus $(2, 5)$

36. Ellipse with center $(2, 3)$ passing through $(6, 3)$ and $(2, 5)$

37. Hyperbola with vertices at $(0, 0)$ and $(0, 6)$ and a focus at $(0, 8)$

38. Ellipse with foci at $(2, 0)$ and $(2, 12)$ and a vertex at $(2, 14)$

39. Parabola with focus $(2, 5)$ and directrix $x = 10$

40. Parabola with focus $(2, 5)$ and vertex $(2, 6)$

41. Ellipse with foci $(\pm 2, 2)$ that passes through the origin

42. Hyperbola with foci $(0, 0)$ and $(0, 4)$ that passes through $(12, 9)$

In Problems 43–48, eliminate the cross-product term by a suitable rotation of axes and then, if necessary, translate axes (complete the squares) to put the equation in standard form. Finally, graph the equation showing the rotated axes.

43. $x^2 + xy + y^2 = 6$

44. $3x^2 + 10xy + 3y^2 + 10 = 0$

45. $4x^2 + xy + 4y^2 = 56$

46. $4xy - 3y^2 = 64$

47. $-\frac{1}{2}x^2 + 7xy - \frac{1}{2}y^2 - 6\sqrt{2}x - 6\sqrt{2}y = 0$

48. $\frac{3}{2}x^2 + xy + \frac{3}{2}y^2 + \sqrt{2}x + \sqrt{2}y = 13$

In Problems 49–52, continue the directions for Problems 43–48. After finding $\cot 2\theta$, *you will need to use one of the identities* $\sin\theta = \pm\sqrt{(1 - \cos 2\theta)/2}$ *or* $\cos\theta = \pm\sqrt{(1 + \cos 2\theta)/2}$ *in order to determine* θ.

49. $4x^2 - 3xy = 18$

50. $11x^2 + 96xy + 39y^2 + 240x + 570y + 875 = 0$

51. $34x^2 + 24xy + 41y^2 + 250y = -325$

52. $16x^2 + 24xy + 9y^2 - 20x - 15y - 150 = 0$

53. A curve C goes through the three points, $(-1, 2)$, $(0, 0)$, and $(3, 6)$. Find an equation for C if C is

(a) a vertical parabola;

(b) a horizontal parabola;

(c) a circle.

54. The ends of an elastic string with a knot at $K(x, y)$ are attached to a fixed point $A(a, b)$ and a point P on the rim of a wheel of radius r centered at $(0, 0)$. As the wheel turns, K traces a curve C. Find the equation for C. Assume that the string stays taut and stretches uniformly (i.e., $\alpha = |KP|/|AP|$ is constant).

55. Name the conic $y^2 = Lx + Kx^2$ according to the value of K and then show that in every case $|L|$ is the length of the latus rectum of the conic. Assume that $L \neq 0$.

56. Show that the equations of the parabola and hyperbola with vertex $(a, 0)$ and focus $(c, 0)$, $c > a > 0$, can be written as $y^2 = 4(c - a)(x - a)$ and $y^2 = (b^2/a^2)(x^2 - a^2)$, respectively. Then use these expressions for y^2 to show that the parabola is always "inside" the right branch of the hyperbola.

57. The graph of $x\cos\alpha + y\sin\alpha = d$ is a line. Show that the perpendicular distance from the origin to this line is $|d|$ by making a rotation of axes through the angle α.

58. Transform the equation $x^{1/2} + y^{1/2} = a^{1/2}$ by a rotation of axes through 45° and then square twice to eliminate radicals on variables. Identify the corresponding curve.

59. Solve the rotation formulas for u and v in terms of x and y.

60. Use the results of Problem 59 to find the uv-coordinates corresponding to $(x, y) = (5, -3)$ after a rotation of axes through 60°.

61. Find the points of $x^2 + 14xy + 49y^2 = 100$ that are closest to the origin.

62. Recall that $Ax^2 + Bxy + Cy^2 + Dx + Ey + F = 0$ transforms to $au^2 + buv + cv^2 + du + ev + f = 0$ under a rotation of axes. Find formulas for a and c, and show that $a + c = A + C$.

63. Show that $b^2 - 4ac = B^2 - 4AC$ (see Problem 62).

64. Use the result of Problem 63 to convince yourself that the graph of the general second-degree equation will be

(a) a parabola if $B^2 - 4AC = 0$.

(b) an ellipse if $B^2 - 4AC < 0$,

(c) a hyperbola if $B^2 - 4AC > 0$,

or limiting forms of the above conics.

65. Let $Ax^2 + Bxy + Cy^2 = 1$ be transformed into $au^2 + cv^2 = 1$ by a rotation of axes, and suppose that $\Delta = 4AC - B^2 \neq 0$. Use Problems 62 and 63 to show that

(a) $1/ac = 4/\Delta$,

(b) $1/a + 1/c = 4(A + C)/\Delta$,

(c) $1/a$ and $1/c$ are the two values of

$$(2/\Delta)\left(A + C \pm \sqrt{(A - C)^2 + B^2}\right)$$

66. Show that, if $A + C$ and $\Delta = 4AC - B^2$ are both positive, then the graph of $Ax^2 + Bxy + Cy^2 = 1$ is an ellipse (or circle) with area $2\pi/\sqrt{\Delta}$. (Recall from Problem 55 of Section 10.2 that the area of the ellipse $x^2/p^2 + y^2/q^2 = 1$ is πpq.)

67. For what values of B is the graph of $x^2 + Bxy + y^2 = 1$

(a) an ellipse

(b) a circle

(c) a hyperbola

(d) two parallel lines

68. Use the results of Problems 65 and 66 to find the distance between the foci and the area of the ellipse

$$25x^2 + 8xy + y^2 = 1$$

69. Refer to Figure 6 and show that $y = u\sin\theta + v\cos\theta$.

Answers to Concepts Review: **1.** $a^2/4$ **2.** 14; ellipse **3.** $(A - C)/B$ **4.** rotation; translation

10.4 Parametric Representation of Curves in the Plane

We gave the general definition of a plane curve in Section 5.4 in connection with our derivation of the arc length formula. A **plane curve** is determined by a pair of parametric equations

$$x = f(t), \qquad y = g(t), \qquad t \text{ in } I$$

with f and g continuous on the interval I. Usually I is a closed interval $[a, b]$. Think of t, called the **parameter,** as measuring time. As t advances from a to b, the point (x, y) traces out the curve in the xy-plane. When I is the closed interval $[a, b]$, the points $P = (x(a), y(a))$ and $Q = (x(b), y(b))$ are called the **initial** and **final end points.** If the curve has end points that coincide, then we say that the curve is **closed.** If distinct values of t yield distinct points in the plane (except possibly for $t = a$ and $t = b$), we say the curve is a **simple** curve (Figure 1). The pair of relationships $x = f(t)$, $y = g(t)$, together with the interval I is called the **parametrization** of the curve.

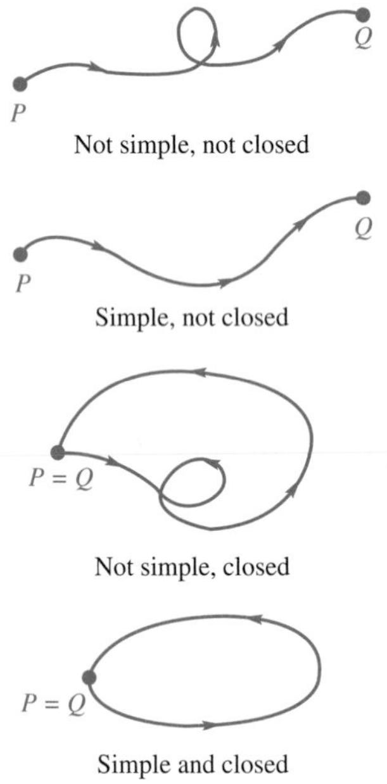

Figure 1

Eliminating the Parameter To recognize a curve given by parametric equations, it may be desirable to eliminate the parameter. Sometimes this can be accomplished by solving one equation for t and substituting in the other (Example 1). Often we can make use of a familiar identity, as in Example 2.

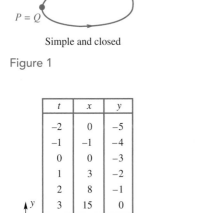

EXAMPLE 1 Eliminate the parameter in

$$x = t^2 + 2t, \qquad y = t - 3, \qquad -2 \le t \le 3$$

Then identify the corresponding curve and sketch its graph.

SOLUTION From the second equation, $t = y + 3$. Substituting this expression for t in the first equation gives

$$x = (y + 3)^2 + 2(y + 3) = y^2 + 8y + 15$$

or

$$x + 1 = (y + 4)^2$$

This we recognize as a parabola with vertex at $(-1, -4)$ and opening to the right.

In graphing the given equation, we must be careful to display only that part of the parabola corresponding to $-2 \le t \le 3$. A table of values and the graph are shown in Figure 2. The arrowhead indicates the curve's *orientation*, that is, the direction of increasing t. ■

t	x	y
−2	0	−5
−1	−1	−4
0	0	−3
1	3	−2
2	8	−1
3	15	0

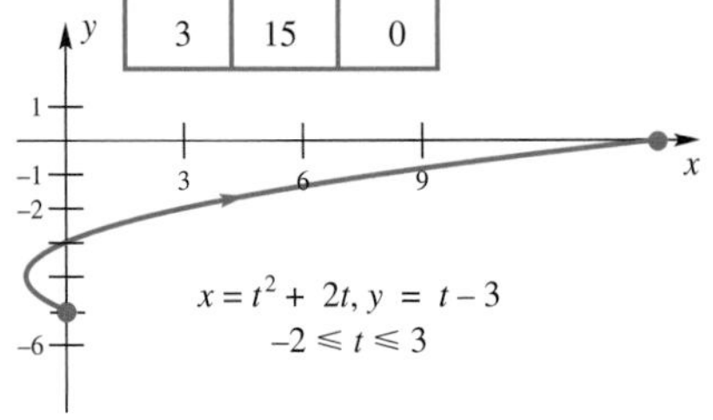

Figure 2

EXAMPLE 2 Show that

$$x = a \cos t, \qquad y = b \sin t, \qquad 0 \le t \le 2\pi$$

represents the ellipse shown in Figure 3.

SOLUTION We solve the equations for $\cos t$ and $\sin t$, then square, and add.

$$\left(\frac{x}{a}\right)^2 + \left(\frac{y}{b}\right)^2 = \cos^2 t + \sin^2 t = 1$$

$$\frac{x^2}{a^2} + \frac{y^2}{b^2} = 1$$

A quick check of a few values for t convinces us that we do get the complete ellipse. In particular, $t = 0$ and $t = 2\pi$ give the same point, namely, $(a, 0)$.

If $a = b$, we get the circle $x^2 + y^2 = a^2$. ■

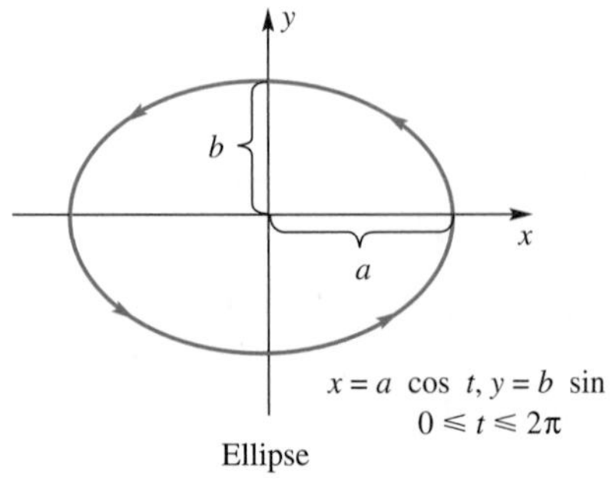

Figure 3

Different pairs of parametric equations may have the same graph. In other words, a given curve can have more than one parametrization.

EXAMPLE 3 Show that each of the following pairs of parametric equations has the same graph, namely, the semicircle shown in Figure 4.

(a) $x = \sqrt{1 - t^2},\ y = t,\ -1 \le t \le 1$

(b) $x = \cos t,\ y = \sin t,\ -\dfrac{\pi}{2} \le t \le \dfrac{\pi}{2}$

(c) $x = \dfrac{1 - t^2}{1 + t^2},\ y = \dfrac{2t}{1 + t^2},\ -1 \le t \le 1$

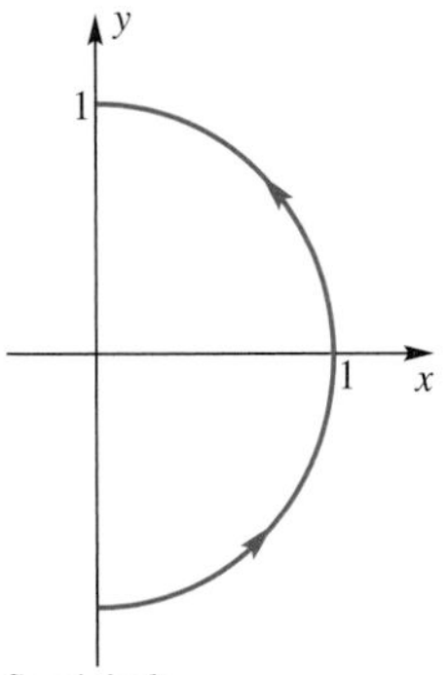

Figure 4

SOLUTION In each case, we discover that

$$x^2 + y^2 = 1$$

It is then just a matter of checking a few values of t to make sure that the given intervals for t yield the same section of the circle. ■

EXAMPLE 4 Show that each of the following pairs of parametric equations yields one branch of a hyperbola. Assume in both cases that $a > 0$ and $b > 0$.

(a) $x = a \sec t,\ y = b \tan t,\ -\dfrac{\pi}{2} < t < \dfrac{\pi}{2}$

(b) $x = a \cosh t,\ y = b \sinh t,\ -\infty < t < \infty$

SOLUTION

(a) In the first case,

$$\left(\frac{x}{a}\right)^2 - \left(\frac{y}{b}\right)^2 = \sec^2 t - \tan^2 t = 1$$

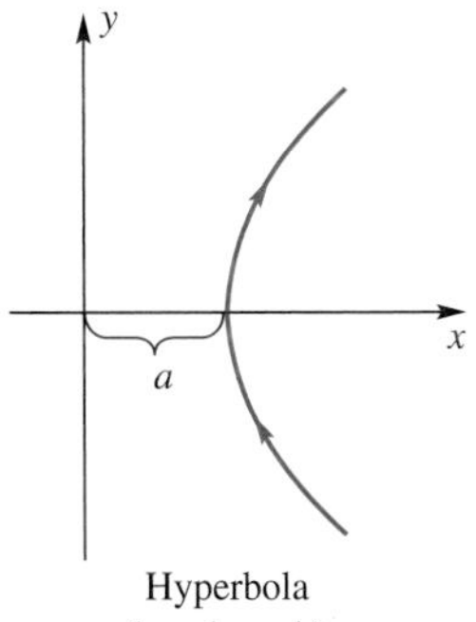

Figure 5

(b) In the second case,

$$\left(\frac{x}{a}\right)^2 - \left(\frac{y}{b}\right)^2 = \cosh^2 t - \sinh^2 t$$

$$= \left(\frac{e^t + e^{-t}}{2}\right)^2 - \left(\frac{e^t - e^{-t}}{2}\right)^2 = 1$$

Checking a few t-values shows that, in both cases, we obtain the branch of the hyperbola $x^2/a^2 - y^2/b^2 = 1$ shown in Figure 5. ■

Notice that in Example 4 we have in part (a) a parametric curve defined on the *open* interval $(-\pi/2, \pi/2)$, and in part (b) we have a curve defined on the *infinite* interval $(-\infty, \infty)$. Since the curve does not contain end points, it is not closed.

The Cycloid A **cycloid** is the curve traced by a point P on the rim of a wheel as the wheel rolls along a straight line without slipping (Figure 6). The Cartesian equation of a cycloid is quite complicated, but simple parametric equations are readily found, as shown in the next example.

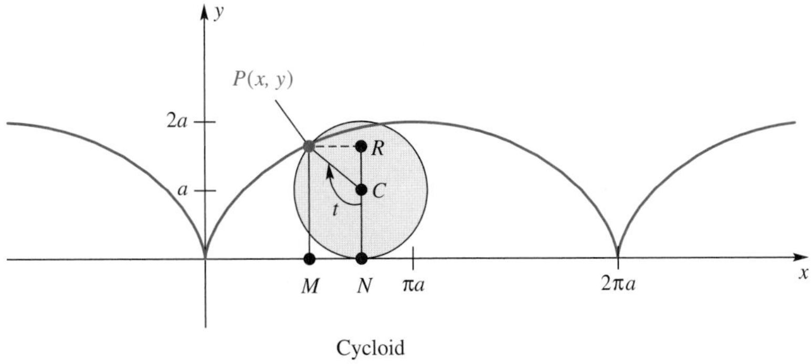

Figure 6

EXAMPLE 5 Find parametric equations for the cycloid.

SOLUTION Let the wheel roll along the x-axis with P initially at the origin. Denote the center of the wheel by C, and let a be its radius. Choose for a parameter the radian measure t of the clockwise angle through which the line segment CP has turned from its vertical position when P was at the origin. All of this is shown in Figure 6.

Since $|ON| = \text{arc } PN = at$,

$$x = |OM| = |ON| - |MN| = at - a\sin t = a(t - \sin t)$$

and

$$y = |MP| = |NR| = |NC| + |CR| = a - a\cos t = a(1 - \cos t)$$

Thus, the parametric equations for the cycloid are

$$x = a(t - \sin t), \qquad y = a(1 - \cos t), \qquad t > 0$$

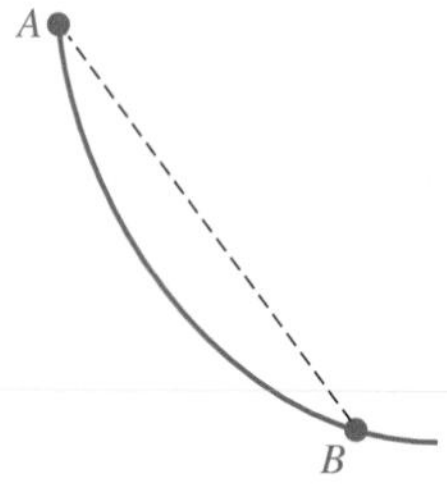

Figure 7

The cycloid has a number of interesting applications, especially in mechanics. It is the "curve of fastest descent." If a particle, acted on only by gravity, is allowed to slide down some curve from a point A to a lower point B not on the same vertical line, it completes its journey in the shortest time when the curve is an inverted cycloid (Figure 7). Of course, the shortest *distance* is along the straight line segment AB, but the *least time* is used when the path is along a cycloid; this is because the acceleration when it is released depends on the steepness of descent, and along a cycloid it builds up velocity much more quickly than it does along a straight line.

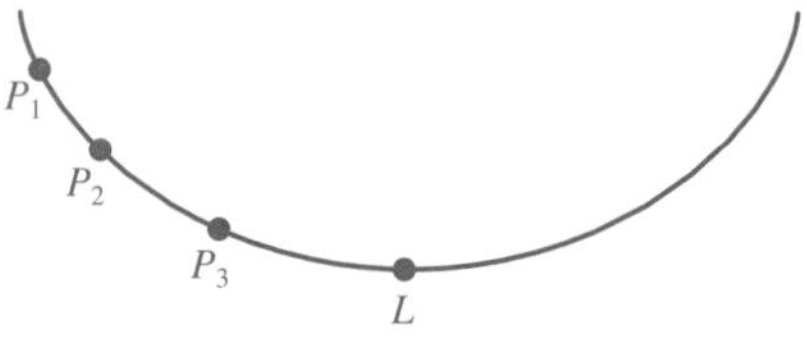

Figure 8

Another interesting property is this: If L is the lowest point on an arch of an inverted cycloid, the time that it takes a particle P to slide down the cycloid to L is the same no matter where P starts from on the inverted arch; thus, if several particles, P_1, P_2, and P_3, in different positions on the cycloid (Figure 8), start to slide at the same instant, all will reach the low point L at the same time.

In 1673, the Dutch astronomer Christian Huygens (1629–1695) published a description of an ideal pendulum clock. Because the bob swings between cycloidal "cheeks," the path of the bob is a cycloid (Figure 9). This means that the period of the swing is independent of the amplitude, and so the period does not change as the clock's spring unwinds.

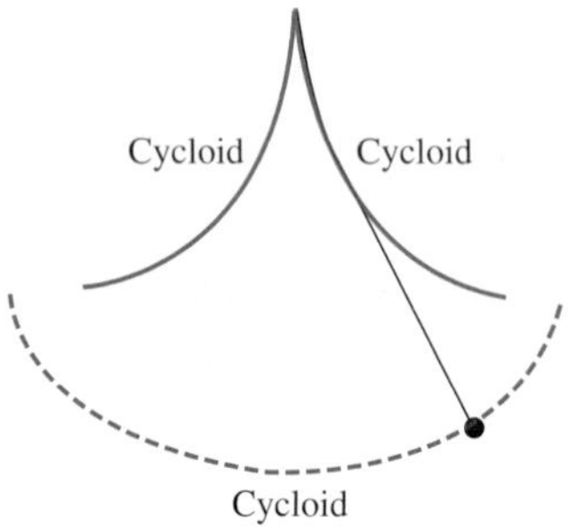

Figure 9

Calculus for Curves Defined Parametrically

Can we find the slope of the tangent line to a curve given parametrically without first eliminating the parameter? The answer is yes, according to the following theorem.

Theorem A

Let f and g be continuously differentiable with $f'(t) \neq 0$ on $\alpha < t < \beta$. Then the parametric equations

$$x = f(t), \qquad y = g(t)$$

define y as a differentiable function of x and

$$\frac{dy}{dx} = \frac{dy/dt}{dx/dt}$$

Proof Since $f'(t) \neq 0$ for $\alpha < t < \beta$, f is strictly monotonic and so has a differentiable inverse f^{-1} (see the Inverse Function Theorem (Theorem 6.2B)). Define F by $F = g \circ f^{-1}$ so that

$$y = g(t) = g(f^{-1}(x)) = F(x) = F(f(t))$$

Then, by the Chain Rule,

$$\frac{dy}{dt} = F'(f(t)) \cdot f'(t) = \frac{dy}{dx} \cdot \frac{dx}{dt}$$

Since $dx/dt \neq 0$, we have

$$\frac{dy}{dx} = \frac{dy/dt}{dx/dt}$$ ■

EXAMPLE 6 Find the first two derivatives dy/dx and d^2y/dx^2 for the function determined by

$$x = 5\cos t, \qquad y = 4\sin t, \qquad 0 < t < 3$$

and evaluate them at $t = \pi/6$ (see Example 2).

SOLUTION Let y' denote $\dfrac{dy}{dx}$. Then

$$\frac{dy}{dx} = \frac{dy/dt}{dx/dt} = \frac{4\cos t}{-5\sin t} = -\frac{4}{5}\cot t$$

$$\frac{d^2y}{dx^2} = \frac{dy'}{dx} = \frac{dy'/dt}{dx/dt} = \frac{\frac{4}{5}\csc^2 t}{-5\sin t} = -\frac{4}{25}\csc^3 t$$

At $t = \pi/6$,

$$\frac{dy}{dx} = \frac{-4\sqrt{3}}{5}, \qquad \frac{d^2y}{dx^2} = \frac{-4}{25}(8) = -\frac{32}{25}$$

The first value is the slope of the tangent line to the ellipse $x^2/25 + y^2/16 = 1$ at the point $\left(5\sqrt{3}/2, 2\right)$. You can check that this is so by implicit differentiation. ■

Sometimes a definite integral involves two variables, such as x and y, in the integrand and differential, and y may be defined as a function of x by equations that give x and y in terms of a parameter such as t. In such cases, it is often convenient to evaluate the definite integral by expressing the integrand and the differential in terms of t and dt, and adjusting the limits of integration before integrating with respect to t.

EXAMPLE 7 Evaluate (a) $\int_1^3 y\,dx$ and (b) $\int_1^3 xy^2\,dx$, where $x = 2t - 1$ and $y = t^2 + 2$.

SOLUTION From $x = 2t - 1$, we have $dx = 2\,dt$. When $x = 1, t = 1$ and when $x = 3, t = 2$.

(a) $$\int_1^3 y\,dx = \int_1^2 (t^2 + 2)2\,dt = 2\left[\frac{t^3}{3} + 2t\right]_1^2 = \frac{26}{3}$$

(b) $$\int_1^3 xy^2\,dx = \int_1^2 (2t - 1)(t^2 + 2)^2 2\,dt$$

$$= 2\int_1^2 (2t^5 - t^4 + 8t^3 - 4t^2 + 8t - 4)\,dt = 86\tfrac{14}{15}$$ ■

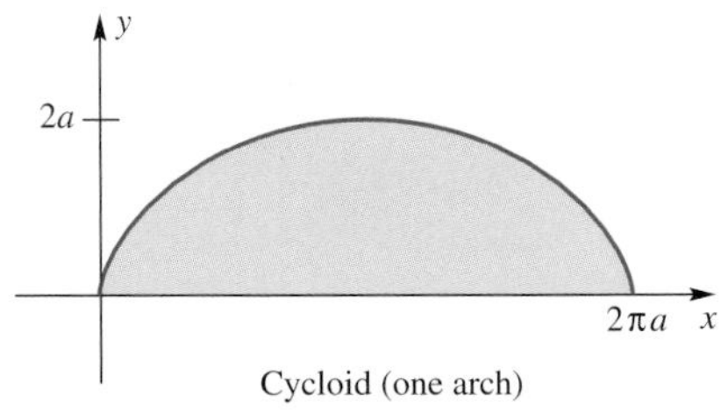

Figure 10

EXAMPLE 8 Find the area A under one arch of a cycloid (Figure 10) and the length L of this arch.

SOLUTION From Example 5, we know that we may represent one arch of the cycloid by

$$x = a(t - \sin t), \qquad y = a(1 - \cos t), \qquad 0 \le t \le 2\pi$$

Thus, $dx = a(1 - \cos t)\,dt$. The area A is therefore

$$\begin{aligned} A &= \int_0^{2\pi a} y\,dx \\ &= a^2 \int_0^{2\pi} (1 - \cos t)(1 - \cos t)\,dt \\ &= a^2 \int_0^{2\pi} (1 - 2\cos t + \cos^2 t)\,dt \\ &= a^2 \int_0^{2\pi} \left(1 - 2\cos t + \tfrac{1}{2} + \tfrac{1}{2}\cos 2t\right) dt \\ &= a^2\left[\tfrac{3}{2}t - 2\sin t + \tfrac{1}{4}\sin 2t\right]_0^{2\pi} = 3\pi a^2 \end{aligned}$$

To calculate L, we recall the arc-length formula from Section 5.4:

$$L = \int_\alpha^\beta \sqrt{\left(\frac{dx}{dt}\right)^2 + \left(\frac{dy}{dt}\right)^2}\,dt$$

In our case, this reduces to

$$\begin{aligned} L &= \int_0^{2\pi} \sqrt{a^2(1 - \cos t)^2 + a^2(\sin^2 t)}\,dt \\ &= a \int_0^{2\pi} \sqrt{2(1 - \cos t)}\,dt \\ &= a \int_0^{2\pi} \sqrt{4 \sin^2 \frac{t}{2}}\,dt \\ &= 2a \int_0^{2\pi} \sin \frac{t}{2}\,dt \\ &= \left[-4a \cos \frac{t}{2}\right]_0^{2\pi} = 8a \end{aligned}$$

■

Two Fleas on a Trike

Two fleas are arguing about who will get the longest ride when Jenny pedals her tricycle home from the park. Flea A will ride between the treads of the front tire; flea B will ride between the treads of one of the rear tires. Settle the argument by showing that their paths will have equal lengths. Example 8 should help.

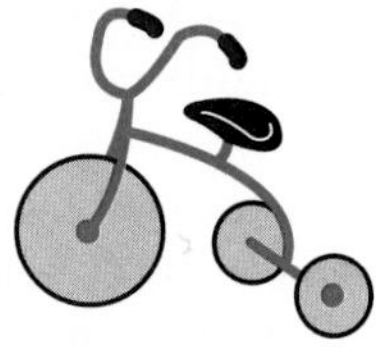

Concepts Review

1. A circle is a premier example of a curve that is both _____ and _____; a figure eight is an example of a closed curve that is not _____.

2. We call two equations $x = f(t)$ and $y = g(t)$ a _____ representation of a curve, and t is called a _____.

3. The path of a point on the rim of a rolling wheel is called a _____.

4. The formula for dy/dx, given the representation $x = f(t)$ and $y = g(t)$, is $dy/dx =$ _____.

Problem Set 10.4

In each of Problems 1–20, a parametric representation of a curve is given.
(a) *Graph the curve.*
(b) *Is the curve closed? Is it simple?*
(c) *Obtain the Cartesian equation of the curve by eliminating the parameter (see Examples 1–4).*

1. $x = 3t,\ y = 2t;\ -\infty < t < \infty$

2. $x = 2t,\ y = 3t;\ -\infty < t < \infty$

3. $x = 3t - 1,\ y = t;\ 0 \le t \le 4$

4. $x = 4t - 2,\ y = 2t;\ 0 \le t \le 3$

5. $x = 4 - t,\ y = \sqrt{t};\ 0 \le t \le 4$

6. $x = t - 3,\ y = \sqrt{2t};\ 0 \le t \le 8$

7. $x = \dfrac{1}{s},\ y = s;\ 1 \le s < 10$

8. $x = s,\ y = \dfrac{1}{s};\ 1 \le s \le 10$

9. $x = t^3 - 4t,\ y = t^2 - 4;\ -3 \le t \le 3$

10. $x = t^3 - 2t,\ y = t^2 - 2t;\ -3 \le t \le 3$

11. $x = 2\sqrt{t - 2},\ y = 3\sqrt{4 - t};\ 2 \le t \le 4$

12. $x = 3\sqrt{t-3},\ y = 2\sqrt{4-t};\ 3 \le t \le 4$

13. $x = 2\sin t,\ y = 3\cos t;\ 0 \le t \le 2\pi$

14. $x = 3\sin r,\ y = -2\cos r;\ 0 \le r \le 2\pi$

15. $x = -2\sin r,\ y = -3\cos r;\ 0 \le r \le 4\pi$

16. $x = 2\cos^2 r,\ y = 3\sin^2 r;\ 0 \le r \le 2\pi$

17. $x = 9\sin^2\theta,\ y = 9\cos^2\theta;\ 0 \le \theta \le \pi$

18. $x = 9\cos^2\theta,\ y = 9\sin^2\theta;\ 0 \le \theta \le \pi$

19. $x = \cos\theta,\ y = -2\sin^2 2\theta;\ -\infty < \theta < \infty$

20. $x = \sin\theta,\ y = 2\cos^2 2\theta;\ -\infty < \theta < \infty$

In Problems 21–30, find dy/dx and d^2y/dx^2 without eliminating the parameter.

21. $x = 3\tau^2,\ y = 4\tau^3;\ \tau \ne 0$

22. $x = 6s^2,\ y = -2s^3;\ s \ne 0$

23. $x = 2\theta^2,\ y = \sqrt{5}\theta^3;\ \theta \ne 0$

24. $x = \sqrt{3}\theta^2,\ y = -\sqrt{3}\theta^3;\ \theta \ne 0$

25. $x = 1 - \cos t,\ y = 1 + \sin t;\ t \ne n\pi$

26. $x = 3 - 2\cos t,\ y = -1 + 5\sin t;\ t \ne n\pi$

27. $x = 3\tan t - 1,\ y = 5\sec t + 2;\ t \ne \dfrac{(2n+1)\pi}{2}$

28. $x = \cot t - 2,\ y = -2\csc t + 5;\ 0 < t < \pi$

29. $x = \dfrac{1}{1+t^2},\ y = \dfrac{1}{t(1-t)};\ 0 < t < 1$

30. $x = \dfrac{2}{1+t^2},\ y = \dfrac{2}{t(1+t^2)};\ t \ne 0$

In Problems 31–34, find the equation of the tangent line to the given curve at the given value of t without eliminating the parameter. Make a sketch.

31. $x = t^2,\ y = t^3;\ t = 2$

32. $x = 3t,\ y = 8t^3;\ t = -\frac{1}{2}$

33. $x = 2\sec t,\ y = 2\tan t;\ t = -\dfrac{\pi}{6}$

34. $x = 2e^t,\ y = \frac{1}{3}e^{-t};\ t = 0$

In Problems 35–46, find the length of the parametric curve defined over the given interval.

35. $x = 2t - 1,\ y = 3t - 4;\ 0 \le t \le 3$

36. $x = 2 - t,\ y = 2t - 3;\ -3 \le t \le 3$

37. $x = t,\ y = t^{3/2};\ 0 \le t \le 3$

38. $x = 2\sin t,\ y = 2\cos t;\ 0 \le t \le \pi$

39. $x = 3t^2,\ y = t^3;\ 0 \le t \le 2$

40. $x = t + \dfrac{1}{t},\ y = \ln t^2;\ 1 \le t \le 4$

41. $x = 2e^t,\ y = 3e^{3t/2};\ \ln 3 \le t \le 2\ln 3$

42. $x = \sqrt{1-t^2},\ y = 1 - t;\ 0 \le t \le \dfrac{1}{4}$

43. $x = 4\sqrt{t},\ y = t^2 + \dfrac{1}{2t};\ \dfrac{1}{4} \le t \le 1$

44. $x = \tanh t,\ y = \ln(\cosh^2 t);\ -3 \le t \le 3$

45. $x = \cos t,\ y = \ln(\sec t + \tan t) - \sin t;\ 0 \le t \le \dfrac{\pi}{4}$

46. $x = \sin t - t\cos t,\ y = \cos t + t\sin t;\ \dfrac{\pi}{4} \le t \le \dfrac{\pi}{2}$

47. Find the length of the curve with the given parametric equations

(a) $x = \sin\theta,\ y = \cos\theta$ for $0 \le \theta \le 2\pi$

(b) $x = \sin 3\theta,\ y = \cos 3\theta$ for $0 \le \theta \le 2\pi$

(c) Explain why the lengths in parts (a) and (b) are not equal.

You can generate surfaces by revolving smooth curves, given parametrically, about a coordinate axis. As t increases from a to b, a smooth curve $x = F(t)$ and $y = G(t)$ is traced out exactly once. Revolving this curve about the x-axis for $y \ge 0$ gives the surface of revolution with surface area

$$S = \int_a^b 2\pi y\sqrt{\left(\frac{dx}{dt}\right)^2 + \left(\frac{dy}{dt}\right)^2}\,dt$$

See Section 5.4. Problems 48–54 relate to such surfaces.

48. Derive a formula for the surface area generated by the rotation of the curve $x = F(t),\ y = G(t)$ for $a \le t \le b$ about the y-axis for $x \ge 0$, and show that the result is given by

$$S = \int_a^b 2\pi x\sqrt{\left(\frac{dx}{dt}\right)^2 + \left(\frac{dy}{dt}\right)^2}\,dt$$

49. A parametrization of a circle of radius 1 centered at $(1, 0)$ in the xy-plane is given by $x = 1 + \cos t,\ y = \sin t$, for $0 \le t \le 2\pi$. Find the surface area when this curve is revolved about the y-axis.

50. Find the area of the surface generated by revolving the curve $x = \cos t,\ y = 3 + \sin t$, for $0 \le t \le 2\pi$ about the x-axis.

51. Find the area of the surface generated by revolving the curve $x = 2 + \cos t,\ y = 1 + \sin t$, for $0 \le t \le 2\pi$ about the x-axis.

52. Find the area of the surface generated by revolving the curve $x = (2/3)t^{3/2},\ y = 2\sqrt{t}$, for $0 \le t \le 2\sqrt{3}$ about the y-axis.

53. Find the area of the surface generated by revolving the curve $x = t + \sqrt{7},\ y = t^2/2 + \sqrt{7}t$, for $-\sqrt{7} \le t \le \sqrt{7}$ about the y-axis.

54. Find the area of the surface generated by revolving the curve $x = t^2/2 + at,\ y = t + a$, for $-\sqrt{a} \le t \le \sqrt{a}$ about the x-axis.

Evaluate the integrals in Problems 55 and 56.

55. $\displaystyle\int_0^1 (x^2 - 4y)\,dx$, where $x = t + 1,\ y = t^3 + 4$.

56. $\displaystyle\int_1^{\sqrt{3}} xy\,dy$, where $x = \sec t,\ y = \tan t$.

57. Find the area of the region between the curve $x = e^{2t},\ y = e^{-t}$, and the x-axis from $t = 0$ to $t = \ln 5$. Make a sketch.

58. The path of a projectile fired from level ground with a speed of v_0 feet per second at an angle α with the ground is given by the parametric equations

$$x = (v_0\cos\alpha)t, \qquad y = -16t^2 + (v_0\sin\alpha)t$$

(a) Show that the path is a parabola.

(b) Find the time of flight.

(c) Show that the range (horizontal distance traveled) is $(v_0^2/32)\sin 2\alpha$.

(d) For a given v_0, what value of α gives the largest possible range?

59. Modify the text discussion of the cycloid (and its accompanying diagram) to handle the case where the point P is $b < a$ units from the center of the wheel. Show that the corresponding parametric equations are

$$x = at - b\sin t, \qquad y = a - b\cos t$$

Sketch the graph of these equations (called a **curtate cycloid**) when $a = 8$ and $b = 4$.

60. Follow the instructions of Problem 59 for the case $b > a$ (a flanged wheel, as on a train), showing that you get the same parametric equations. Sketch the graph of these equations (called a **prolate cycloid**) when $a = 6$ and $b = 8$.

61. Let a circle of radius b roll, without slipping, inside a fixed circle of radius a, $a > b$. A point P on the rolling circle traces out a curve called a **hypocycloid.** Find parametric equations of the hypocycloid. *Hint:* Place the origin O of Cartesian coordinates at the center of the fixed, larger circle, and let the point $A(a, 0)$ be one position of the tracing point P. Denote by B the moving point of tangency of the two circles, and let t, the radian measure of the angle AOB, be the parameter (see Figure 11).

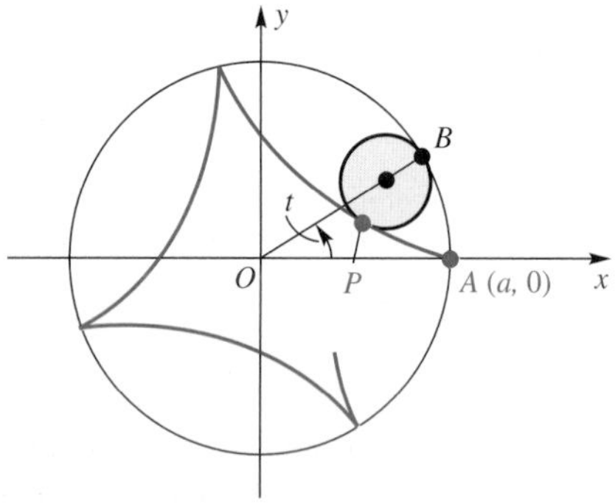

Figure 11

62. Show that if $b = a/4$ in Problem 61, the parametric equations of the hypocycloid may be simplified to

$$x = a\cos^3 t, \qquad y = a\sin^3 t$$

This is called a **hypocycloid of four cusps.** Sketch it carefully and show that its Cartesian equation is $x^{2/3} + y^{2/3} = a^{2/3}$.

63. The curve traced by a point on a circle of radius b as it rolls without slipping on the outside of a fixed circle of radius a is called an **epicycloid.** Show that it has parametric equations

$$x = (a + b)\cos t - b\cos\frac{a+b}{b}t$$

$$y = (a + b)\sin t - b\sin\frac{a+b}{b}t$$

(See the hint in Problem 61.)

64. If $b = a$, the equations in Problem 63 are

$$x = 2a\cos t - a\cos 2t$$

$$y = 2a\sin t - a\sin 2t$$

Find a Cartesian equation of the epicycloid by eliminating the parameter t between the equations.

65. If $b = a/3$ in Problem 61, we obtain a hypocycloid of three cusps, called a **deltoid,** with parametric equations

$$x = \left(\frac{a}{3}\right)(2\cos t + \cos 2t), \qquad y = \left(\frac{a}{3}\right)(2\sin t - \sin 2t)$$

Find the length of the deltoid.

66. Consider the ellipse $x^2/a^2 + y^2/b^2 = 1$.

(a) Show that its perimeter is

$$P = 4a\int_0^{\pi/2}\sqrt{1 - e^2\cos^2 t}\,dt,$$

where e is the eccentricity.

C (b) The integral in part (a) is called an *elliptic integral.* It has been studied at great length, and it is known that the integrand does not have an elementary antiderivative, so we must turn to approximate methods to evaluate P. Do so when $a = 1$ and $e = \frac{1}{4}$ using the Parabolic Rule with $n = 4$. (Your answer should be near 2π. Why?)

CAS (c) Repeat part (b) using $n = 20$.

CAS **67.** The parametric curve given by $x = \cos at$ and $y = \sin bt$ is known as a *Lissajous* figure. The x-coordinate oscillates a times between 1 and -1 as t goes from 0 to 2π, while the y-coordinate oscillates b times over the same t interval. This behavior is repeated over every interval of length 2π. The entire motion takes place in a unit square. Plot the following *Lissajous* figures for a range of t that ensures that the resulting figure is a closed curve. In each case, count the number of times that the curve touches the horizontal and vertical borders of the unit square.

(a) $x = \sin t$, $y = \cos t$
(b) $x = \sin 3t$, $y = \cos 5t$
(c) $x = \cos 5t$, $y = \sin 15t$
(d) $x = \sin 2t$, $y = \cos 9t$

CAS **68.** Plot the Lissajous figure defined by $x = \cos 2t$, $y = \sin 7t$, $0 \le t \le 2\pi$. Explain why this is a closed curve even though its graph does not look closed.

CAS **69.** Plot Lissajous figures for the following combinations of a and b for $0 \le t \le 2\pi$:

(a) $a = 1, b = 2$
(b) $a = 4, b = 8$
(c) $a = 5, b = 10$
(d) $a = 2, b = 3$
(e) $a = 6, b = 9$
(f) $a = 12, b = 18$

CAS **70.** Use the results from Problems 67–69 (and additional ones if necessary) to explain how the number of times the curve touches the sides or corners of the square for $0 \le t < 2\pi$ is related to the ratio a/b. *Hint:* If a curve touches a corner of a square, it counts as one-half a contact.

CAS **71.** Plot the following parametric curves. Describe in words how the point moves around the curve in each case.

(a) $x = \cos(t^2 - t)$, $y = \sin(t^2 - t)$
(b) $x = \cos(2t^2 + 3t + 1)$, $y = \sin(2t^2 + 3t + 1)$
(c) $x = \cos(-2\ln t)$, $y = \sin(-2\ln t)$
(d) $x = \cos(\sin t)$, $y = \sin(\sin t)$

CAS **72.** Using a computer algebra system, plot the following parametric curves for $0 \le t \le 2$. Describe the shape of the curve in each case and the similarities and differences among all the curves.

(a) $x = t, y = t^2$ (b) $x = t^3, y = t^6$
(c) $x = -t^4, y = -t^8$ (d) $x = t^5, y = t^{10}$

CAS EXPL **73.** Plot the graph of the hypocycloid (see Problem 61)

$$x = (a - b)\cos t + b\cos\frac{a-b}{b}t,$$

$$y = (a - b)\sin t - b\sin\frac{a-b}{b}t$$

for appropriate values of t in each of the following cases:
(a) $a = 4, b = 1$ (b) $a = 3, b = 1$
(c) $a = 5, b = 2$ (d) $a = 7, b = 4$

Experiment with other positive integer values of a and b and then make conjectures about the length of the t-interval required for the curve to return to its starting point and about the number of cusps. What can you say if a/b is irrational?

CAS EXPL **74.** Draw the graph of the epicycloid (see Problem 63)

$$x = (a + b)\cos t - b\cos\frac{a+b}{b}t,$$

$$y = (a + b)\sin t - b\sin\frac{a+b}{b}t$$

for various values of a and b. What conjectures can you make (see Problem 73)?

75. Draw the **Folium of Descartes** $x = 3t/(t^3 + 1)$, $y = 3t^2/(t^3 + 1)$. Then determine the values of t for which this graph is in each of the four quadrants.

Answers to Concepts Review: **1.** simple; closed; simple **2.** parametric; parameter **3.** cycloid **4.** $(dy/dt)/(dx/dt)$

10.5 The Polar Coordinate System

Two Frenchmen, Pierre de Fermat (1601–1665) and René Descartes (1596–1650), introduced what we now call the *Cartesian*, or *rectangular*, coordinate system. Their idea was to specify each point P in the plane by giving two numbers (x, y) the directed distances from a pair of perpendicular axes (Figure 1). This notion is by now so familiar that we use it almost without thinking. Yet it is the fundamental idea in analytic geometry and makes possible the development of calculus as we have given it so far.

Giving the directed distances from a pair of perpendicular axes is not the only way to specify a point. Another way to do this is by giving the *polar coordinates*.

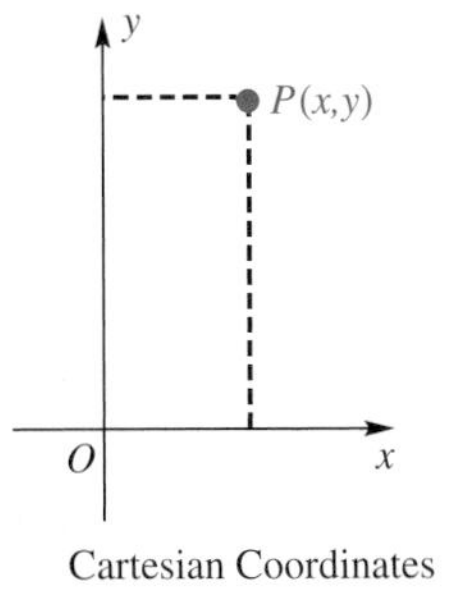

Figure 1

Polar Coordinates We start with a fixed half-line, called the **polar axis,** emanating from a fixed point O, called the **pole** or **origin** (see Figure 2). By custom, the polar axis is chosen to be horizontal and pointing to the right and may therefore be identified with the positive x-axis in the rectangular coordinate system. Any point P (other than the pole) is the intersection of a unique circle with center at O and a unique ray emanating from O. If r is the radius of the circle and θ is one of the angles that the ray makes with the polar axis, then (r, θ) is a pair of **polar coordinates** for P (Figure 2). Figure 3 shows several points plotted on a polar grid.

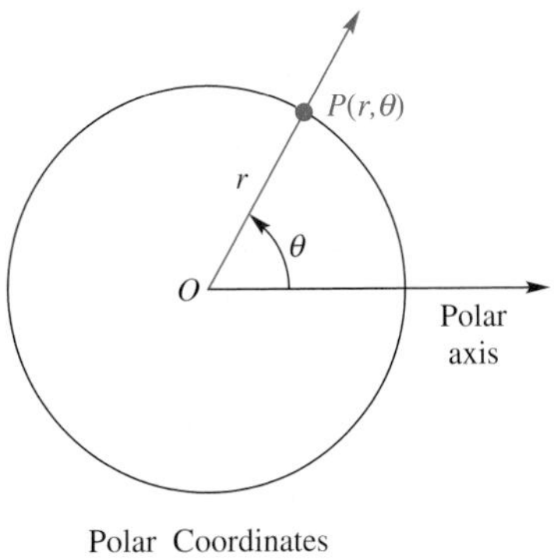

Figure 2

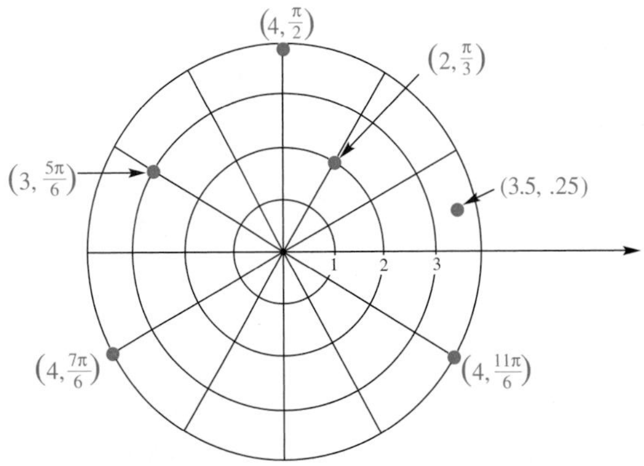

Figure 3

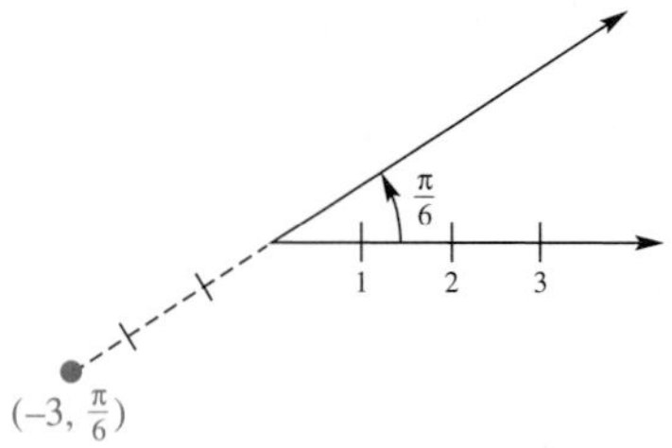

Figure 4

Notice a phenomenon that did not occur with Cartesian coordinates. Each point has infinitely many sets of polar coordinates, due to the fact that the angles $\theta + 2\pi n$, $n = 0, \pm 1, \pm 2, \ldots$, have the same terminal side. For example, the point with polar coordinates $(4, \pi/2)$ also has coordinates $(4, 5\pi/2)$, $(4, 9\pi/2)$, $(4, -3\pi/2)$, and so on. Additional representations occur because we allow r to be negative. In this case, (r, θ) is on the ray oppositely directed from the terminal side of θ and $|r|$ units from the origin. Thus, the point with polar coordinates $(-3, \pi/6)$ is as shown in Figure 4, and $(-4, 3\pi/2)$ is another set of coordinates for $(4, \pi/2)$. The origin has coordinates $(0, \theta)$, where θ is any angle.

Polar Equations Examples of polar equations are

$$r = 8 \sin \theta \quad \text{and} \quad r = \frac{2}{1 - \cos \theta}$$

Polar equations, like rectangular ones, are best visualized from their graphs. The **graph of a polar equation** is the set of points each of which has at least one pair of polar coordinates that satisfies the equation. The most basic way to sketch a graph is to construct a table of values, plot the corresponding points, and then connect these points. This is just what a graphing calculator or a CAS does to plot a polar equation.

EXAMPLE 1 Graph the polar equation $r = 8 \sin \theta$.

SOLUTION We substitute multiples of $\pi/6$ for θ and calculate the corresponding r-values. See the table in Figure 5. Note that as θ increases from 0 to 2π the graph in Figure 5 is traced twice. ■

EXAMPLE 2 Graph $r = \dfrac{2}{1 - \cos \theta}$.

SOLUTION See Figure 6. ■

Note a phenomenon that does not occur with rectangular coordinates. The coordinates $(-2, 3\pi/2)$ do not satisfy the equation in Example 2. Yet the point $P(-2, 3\pi/2)$ is on the graph, due to the fact that $(2, \pi/2)$ specifies the same point and does satisfy the equation. We conclude that *a set of coordinates having its corresponding point on the graph of an equation is no guarantee that these coordinates satisfy the equation*. This fact causes many difficulties; we must learn to live with them.

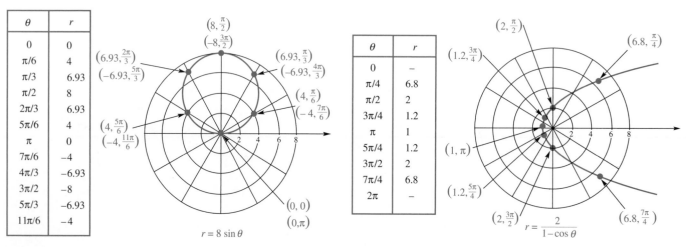

θ	r
0	0
$\pi/6$	4
$\pi/3$	6.93
$\pi/2$	8
$2\pi/3$	6.93
$5\pi/6$	4
π	0
$7\pi/6$	−4
$4\pi/3$	−6.93
$3\pi/2$	−8
$5\pi/3$	−6.93
$11\pi/6$	−4

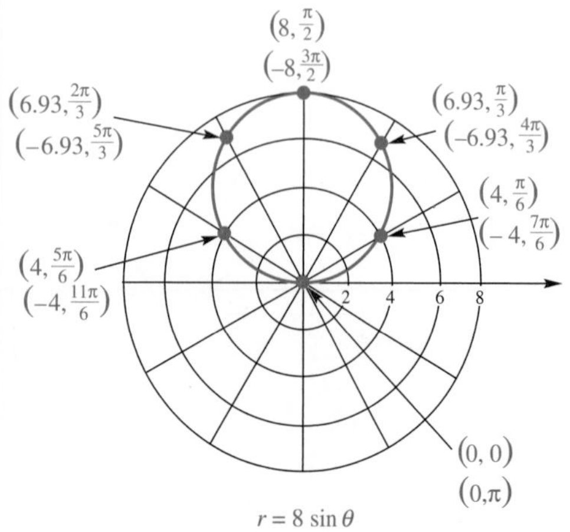

Figure 5

θ	r
0	–
$\pi/4$	6.8
$\pi/2$	2
$3\pi/4$	1.2
π	1
$5\pi/4$	1.2
$3\pi/2$	2
$7\pi/4$	6.8
2π	–

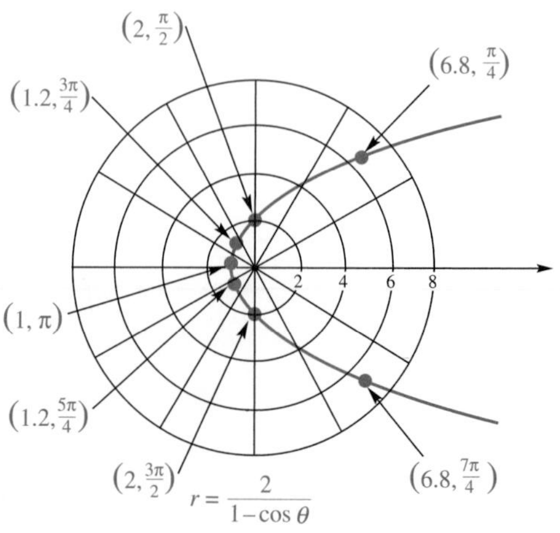

Figure 6

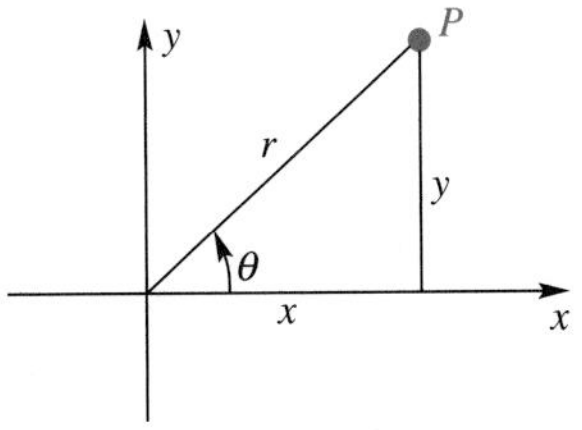

Figure 7

Relation to Cartesian Coordinates We suppose that the polar axis coincides with the positive x-axis of the Cartesian system. Then the polar coordinates (r, θ) of a point P and the Cartesian coordinates (x, y) of the same point are related by the equations

Polar to Cartesian	**Cartesian to Polar**
$x = r\cos\theta$	$r^2 = x^2 + y^2$
$y = r\sin\theta$	$\tan\theta = y/x$

That this is true for a point P in the first quadrant is clear from Figure 7 and is easy to show for points in the other quadrants.

EXAMPLE 3 Find the Cartesian coordinates corresponding to $(4, \pi/6)$ and polar coordinates corresponding to $\left(-3, \sqrt{3}\right)$.

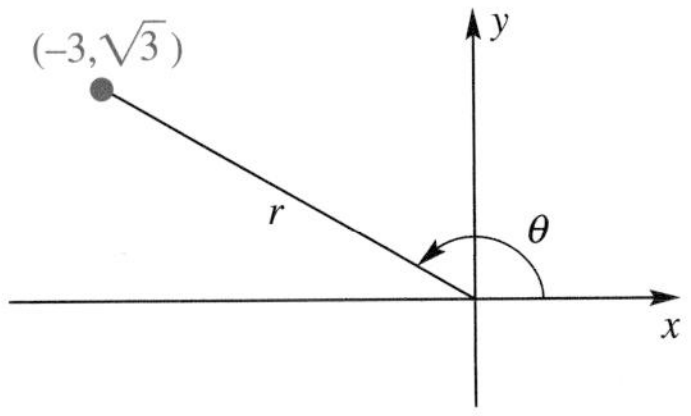

Figure 8

SOLUTION If $(r, \theta) = (4, \pi/6)$, then

$$x = 4\cos\frac{\pi}{6} = 4\cdot\frac{\sqrt{3}}{2} = 2\sqrt{3}$$

$$y = 4\sin\frac{\pi}{6} = 4\cdot\frac{1}{2} = 2$$

If $(x, y) = \left(-3, \sqrt{3}\right)$, then (see Figure 8)

$$r^2 = (-3)^2 + \left(\sqrt{3}\right)^2 = 12$$

$$\tan\theta = \frac{\sqrt{3}}{-3}$$

One value of (r, θ) is $\left(2\sqrt{3}, 5\pi/6\right)$. Another is $\left(-2\sqrt{3}, -\pi/6\right)$. ■

Sometimes we can identify the graph of a polar equation by finding its equivalent Cartesian form. Here is an illustration.

EXAMPLE 4 Show that the graph of $r = 8\sin\theta$ (Example 1) is a circle and that the graph of $r = 2/(1 - \cos\theta)$ (Example 2) is a parabola by changing to Cartesian coordinates.

Caution

Since r can be 0, there is a potential danger in multiplying both sides of a polar equation by r or in dividing both sides by r. In the first case, we might add the pole to the graph; in the second, we might delete the pole from the graph. In Example 4, we multiplied both sides of $r = 8\sin\theta$ by r, but no harm was done since the pole was already on the graph as the point with θ-coordinate 0.

SOLUTION If we multiply $r = 8\sin\theta$ by r, we get

$$r^2 = 8r\sin\theta$$

which, in Cartesian coordinates, is

$$x^2 + y^2 = 8y$$

and may be written successively as

$$x^2 + y^2 - 8y = 0$$
$$x^2 + y^2 - 8y + 16 = 16$$
$$x^2 + (y - 4)^2 = 16$$

The latter is the equation of a circle of radius 4 centered at $(0, 4)$.

The second equation is handled by the following steps.

$$r = \frac{2}{1 - \cos\theta}$$
$$r - r\cos\theta = 2$$
$$r - x = 2$$
$$r = x + 2$$
$$r^2 = x^2 + 4x + 4$$
$$x^2 + y^2 = x^2 + 4x + 4$$
$$y^2 = 4(x + 1)$$

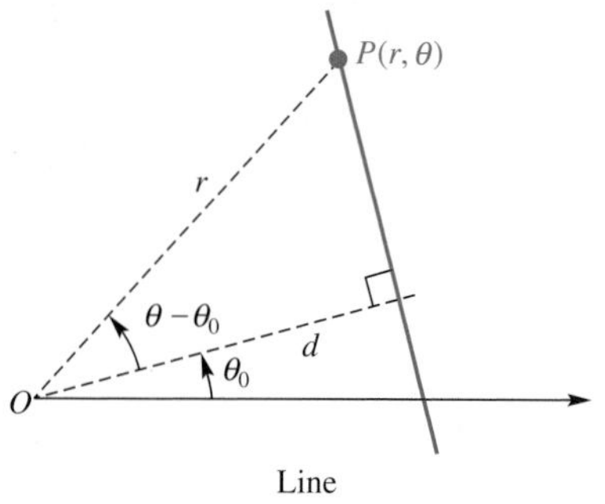

Figure 9

We recognize the last equation as that of a parabola with vertex at $(-1, 0)$ and focus at the origin. ■

Polar Equations for Lines, Circles, and Conics If a line passes through the pole, it has the simple equation $\theta = \theta_0$. If the line does not go through the pole, it is some distance $d > 0$ from it. Let θ_0 be the angle from the polar axis to the perpendicular from the pole to the given line (Figure 9). Then, if $P(r, \theta)$ is any point on the line, $\cos(\theta - \theta_0) = d/r$, or

$$\text{Line:} \quad r = \frac{d}{\cos(\theta - \theta_0)}$$

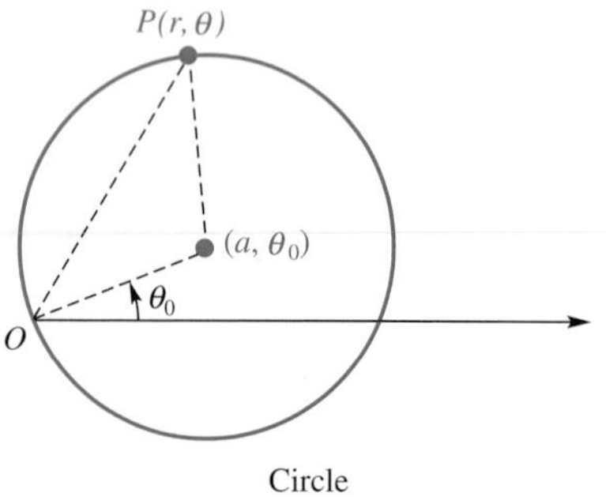

Figure 10

If a circle of radius a is centered at the pole, its equation is simply $r = a$. If it is centered at (r_0, θ_0), its equation is quite complicated unless we choose $r_0 = a$, as in Figure 10. Then, by the Law of Cosines, $a^2 = r^2 + a^2 - 2ra\cos(\theta - \theta_0)$, which simplifies to

$$\text{Circle:} \quad r = 2a\cos(\theta - \theta_0)$$

The cases $\theta_0 = 0$ and $\theta_0 = \pi/2$ are particularly nice. The first gives $r = 2a\cos\theta$; the second gives $r = 2a\cos(\theta - \pi/2)$; that is, $r = 2a\sin\theta$. The latter should be compared with Example 1.

Finally, if a conic (ellipse, parabola, or hyperbola) is placed so that its focus is at the pole and its directrix is d units away, as in Figure 11, then the familiar defining equation $|PF| = e|PL|$ takes the form

$$r = e[d - r\cos(\theta - \theta_0)]$$

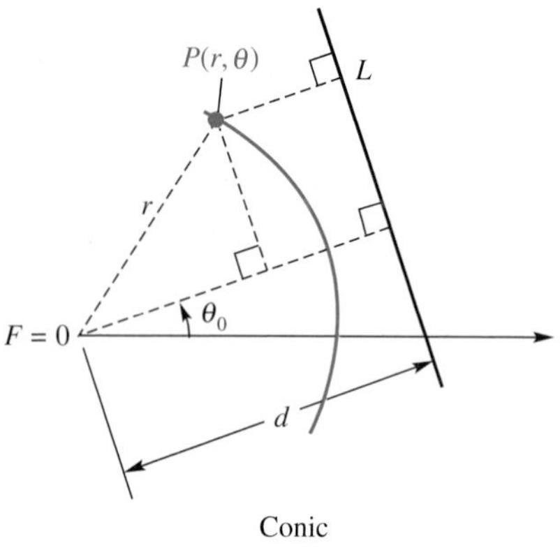

Figure 11

or, equivalently,

$$\text{Conic:} \quad r = \frac{ed}{1 + e\cos(\theta - \theta_0)}$$

Again, there is special interest in the cases $\theta_0 = 0$ and $\theta_0 = \pi/2$. Note in particular that if $e = 1$, $d = 2$, and $\theta_0 = 0$ we have the equation of Example 2.

Our results are summarized in the chart on the following page.

EXAMPLE 5 Find the equation of the horizontal ellipse with eccentricity $\frac{1}{2}$, focus at the pole, and vertical directrix 10 units to the right of the pole.

SOLUTION

$$r = \frac{\frac{1}{2} \cdot 10}{1 + \frac{1}{2}\cos\theta} = \frac{10}{2 + \cos\theta}$$ ■

EXAMPLE 6 Identify and sketch the graph of $r = \dfrac{7}{2 + 4\sin\theta}$.

SOLUTION The equation suggests a conic with vertical major axis. Putting it into the form shown in the polar equations chart gives

$$r = \frac{7}{2 + 4\sin\theta} = \frac{\frac{7}{2}}{1 + 2\sin\theta} = \frac{2\left(\frac{7}{4}\right)}{1 + 2\sin\theta}$$

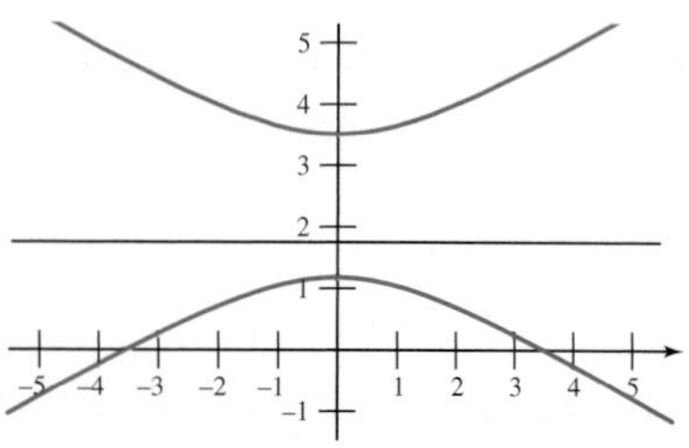

Figure 12

which we recognize as the polar equation of a hyperbola with $e = 2$, focus at the pole, and horizontal directrix $\frac{7}{4}$ units above the polar axis (Figure 12). ■

Summary of Polar Equations

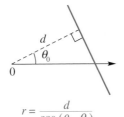

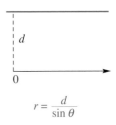

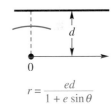

Type of Figure	General Case	$\theta_0 = 0$	$\theta_0 = \pi/2$
Line	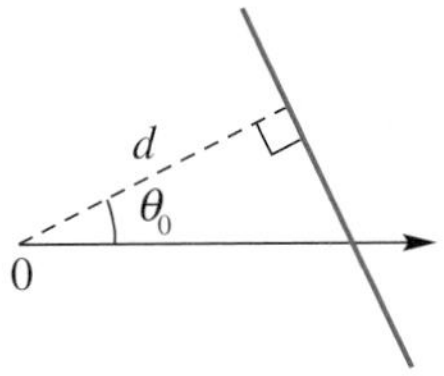 $r = \dfrac{d}{\cos(\theta - \theta_0)}$	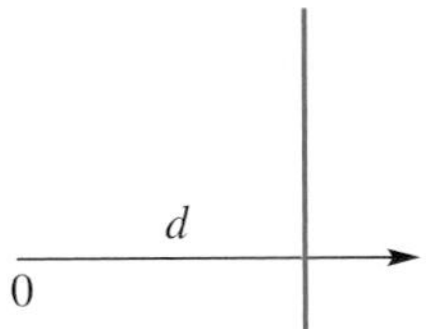 $r = \dfrac{d}{\cos\theta}$	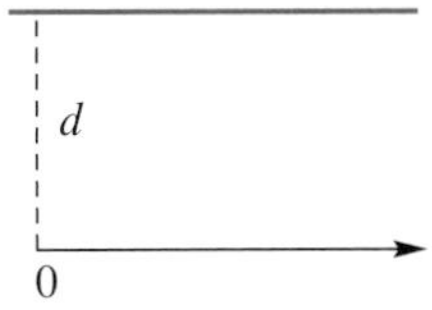 $r = \dfrac{d}{\sin\theta}$
Circle	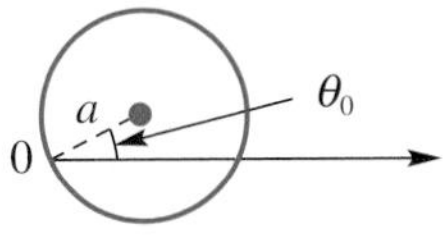 $r = 2a\cos(\theta - \theta_0)$	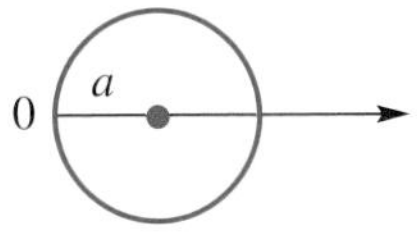 $r = 2a\cos\theta$	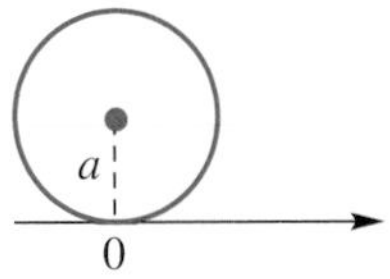 $r = 2a\sin\theta$
Ellipse $(0 < e < 1)$ **Parabola** $(e = 1)$ **Hyperbola** $(e > 1)$	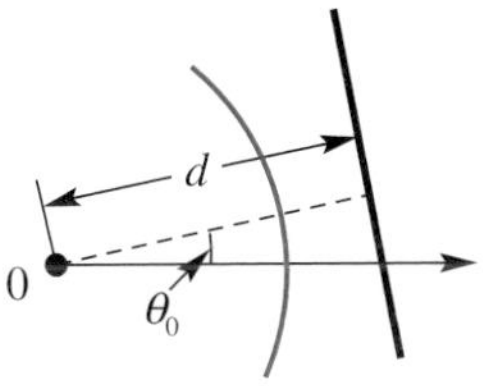 $r = \dfrac{ed}{1 + e\cos(\theta - \theta_0)}$	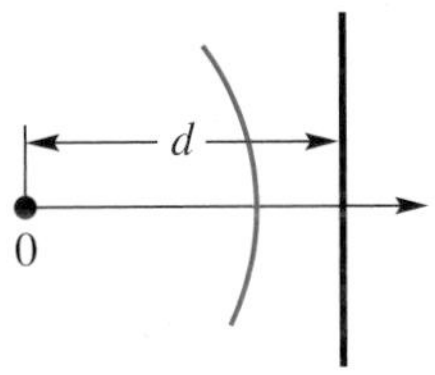 $r = \dfrac{ed}{1 + e\cos\theta}$	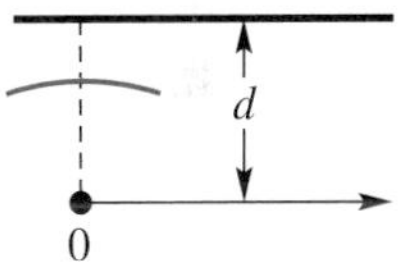 $r = \dfrac{ed}{1 + e\sin\theta}$

Concepts Review

1. Every point in the plane has a unique pair (x, y) of Cartesian coordinates, but _____ pairs (r, θ) of polar coordinates.

2. The relations $x =$ _____ and $y =$ _____ connect Cartesian and polar coordinates; also _____ $= x^2 + y^2$.

3. The graph of the polar equation $r = 5$ is a(n) _____; the graph of $\theta = 5$ is a(n) _____.

4. The graph of the polar equation $r = ed/(1 + e\cos\theta)$ is a(n) _____.

Problem Set 10.5

1. Plot the points whose polar coordinates are $\left(3, \frac{1}{3}\pi\right)$, $\left(1, \frac{1}{2}\pi\right)$, $\left(4, \frac{1}{3}\pi\right)$, $(0, \pi)$, $(1, 4\pi)$, $\left(3, \frac{11}{7}\pi\right)$, $\left(\frac{5}{3}, \frac{1}{2}\pi\right)$, and $(4, 0)$.

2. Plot the points whose polar coordinates are $(3, 2\pi)$, $\left(2, \frac{1}{2}\pi\right)$, $\left(4, -\frac{1}{3}\pi\right)$, $(0, 0)$, $(1, 54\pi)$, $\left(3, -\frac{1}{6}\pi\right)$, $\left(1, \frac{1}{2}\pi\right)$, and $\left(3, -\frac{3}{2}\pi\right)$.

3. Plot the points whose polar coordinates are $(3, 2\pi)$, $\left(-2, \frac{1}{3}\pi\right)$, $\left(-2, -\frac{1}{4}\pi\right)$, $(-1, 1)$, $(1, -4\pi)$, $\left(\sqrt{3}, -\frac{7}{6}\pi\right)$, $\left(-2, \frac{1}{4}\pi\right)$, and $\left(-1, -\frac{1}{2}\pi\right)$.

4. Plot the points whose polar coordinates are $\left(3, \frac{9}{4}\pi\right)$, $\left(-2, \frac{1}{2}\pi\right)$, $\left(-2, -\frac{1}{3}\pi\right)$, $(-1, -1)$, $(1, -7\pi)$, $\left(-3, -\frac{1}{6}\pi\right)$, $\left(-2, -\frac{1}{2}\pi\right)$, and $\left(3, -\frac{33}{2}\pi\right)$.

5. Plot the points whose polar coordinates follow. For each point, give four other pairs of polar coordinates, two with positive r and two with negative r.

(a) $\left(1, \frac{1}{2}\pi\right)$
(b) $\left(-1, \frac{1}{4}\pi\right)$
(c) $\left(\sqrt{2}, -\frac{1}{3}\pi\right)$
(d) $\left(-\sqrt{2}, \frac{5}{2}\pi\right)$

6. Plot the points whose polar coordinates follow. For each point, give four other pairs of polar coordinates, two with positive r and two with negative r.

(a) $\left(3\sqrt{2}, \frac{7}{2}\pi\right)$
(b) $\left(-1, \frac{15}{4}\pi\right)$
(c) $\left(-\sqrt{2}, -\frac{2}{3}\pi\right)$
(d) $\left(-2\sqrt{2}, \frac{29}{2}\pi\right)$

7. Find the Cartesian coordinates of the points in Problem 5.

8. Find the Cartesian coordinates of the points in Problem 6.

9. Find polar coordinates of the points whose Cartesian coordinates are given.

(a) $(3\sqrt{3}, 3)$ (b) $(-2\sqrt{3}, 2)$
(c) $(-\sqrt{2}, -\sqrt{2})$ (d) $(0, 0)$

10. Find polar coordinates of the points whose Cartesian coordinates are given.

(a) $(-3/\sqrt{3}, 1/\sqrt{3})$ (b) $(-\sqrt{3}/2, \sqrt{3}/2)$
(c) $(0, -2)$ (d) $(3, -4)$

In each of Problems 11–16, sketch the graph of the given Cartesian equation, and then find the polar equation for it.

11. $x - 3y + 2 = 0$ **12.** $x = 0$

13. $y = -2$ **14.** $x - y = 0$

15. $x^2 + y^2 = 4$ **16.** $x^2 = 4py$

In Problems 17–22, find the Cartesian equations of the graphs of the given polar equations.

17. $\theta = \frac{1}{2}\pi$ **18.** $r = 3$

19. $r \cos \theta + 3 = 0$ **20.** $r - 5 \cos \theta = 0$

21. $r \sin \theta - 1 = 0$

22. $r^2 - 6r \cos \theta - 4r \sin \theta + 9 = 0$

In Problems 23–36, name the curve with the given polar equation. If it is a conic, give its eccentricity. Sketch the graph.

23. $r = 6$ **24.** $\theta = \dfrac{2\pi}{3}$

25. $r = \dfrac{3}{\sin \theta}$ **26.** $r = \dfrac{-4}{\cos \theta}$

27. $r = 4 \sin \theta$ **28.** $r = -4 \cos \theta$

29. $r = \dfrac{4}{1 + \cos \theta}$ **30.** $r = \dfrac{4}{1 + 2 \sin \theta}$

31. $r = \dfrac{6}{2 + \sin \theta}$ **32.** $r = \dfrac{6}{4 - \cos \theta}$

33. $r = \dfrac{4}{2 + 2 \cos \theta}$ **34.** $r = \dfrac{4}{2 + 2 \cos(\theta - \pi/3)}$

35. $r = \dfrac{4}{\frac{1}{2} + \cos(\theta - \pi)}$ **36.** $r = \dfrac{4}{3 \cos(\theta - \pi/3)}$

37. Show that the polar equation of the circle with center (c, α) and radius a is $r^2 + c^2 - 2rc \cos(\theta - \alpha) = a^2$.

38. Prove that $r = a \sin \theta + b \cos \theta$ represents a circle and find its center and radius.

39. Find the length of the latus rectum for the general conic $r = ed/[1 + e \cos(\theta - \theta_0)]$ in terms of e and d.

40. Let r_1 and r_2 be the minimum and maximum distances (**perihelion** and **aphelion,** respectively) of the ellipse $r = ed/[1 + e \cos(\theta - \theta_0)]$ from a focus. Show that

(a) $r_1 = ed/(1 + e), r_2 = ed/(1 - e)$,

(b) major diameter $= 2ed/(1 - e^2)$ and minor diameter $= 2ed/\sqrt{1 - e^2}$.

41. The perihelion and aphelion for the orbit of the asteroid Icarus are 17 and 183 million miles, respectively. What is the eccentricity of its elliptical orbit?

42. Earth's orbit around the sun is an ellipse of eccentricity 0.0167 and major diameter 185.8 million miles. Find its perihelion.

43. The path of a certain comet is a parabola with the sun at the focus. The angle between the axis of the parabola and a ray from the sun to the comet is 120° (measured from the point of the perihelion to the sun to the comet) when the comet is 100 million miles from the sun. How close does the comet get to the sun?

44. The position of a comet with a highly eccentric elliptical orbit (e very near 1) is measured with respect to a fixed polar axis (sun is at a focus but the polar axis is not an axis of the ellipse) at two times, giving the two points $(4, \pi/2)$ and $(3, \pi/4)$ of the orbit. Here distances are measured in astronomical units (1 AU $\approx$ 93 million miles). For the part of the orbit near the sun, assume that $e = 1$, so the orbit is given by

$$r = \frac{d}{1 + \cos(\theta - \theta_0)}$$

(a) The two points give two conditions for d and θ_0. Use them to show that $4.24 \cos \theta_0 - 3.76 \sin \theta_0 - 2 = 0$.

(b) Solve for θ_0 using Newton's Method.

(c) How close does the comet get to the sun?

CAS **45.** In order to graph a polar equation such as $r = f(t)$ using a parametric equation grapher, you must replace this equation by $x = f(t) \cos t$ and $y = f(t) \sin t$. These equations can be obtained by multiplying $r = f(t)$ by $\cos t$ and $\sin t$, respectively. Confirm the discussions of conics in the text by graphing $r = 4e/(1 + e \cos t)$ for $e = 0.1, 0.5, 0.9, 1, 1.1$ and 1.3 on $[-\pi, \pi]$.

Answers to Concepts Review: **1.** infinitely many **2.** $r \cos \theta$; $r \sin \theta$; r^2 **3.** circle; line **4.** conic

10.6 Graphs of Polar Equations

The polar equations considered in the previous section led to familiar graphs, mainly lines, circles, and conics. Now we turn our attention to more exotic graphs—cardioids, limaçons, lemniscates, roses, and spirals. The polar equations for these curves are still rather simple; the corresponding Cartesian equations are quite complicated. Thus, we see one of the advantages of having available more than one coordinate system. Some curves have simple equations in one system; other curves have simple equations in the other system. We will exploit this later in the book when we often begin the solution of a problem by choosing a convenient coordinate system.

Symmetry can help us to understand a graph. Here are some *sufficient* tests for symmetry in polar coordinates. The diagrams in the margin will help you to establish their validity.

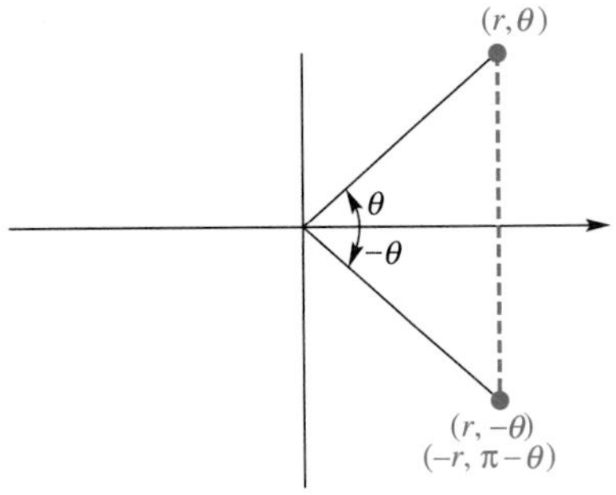

Figure 1

1. The graph of a polar equation is symmetric about the x-axis (the polar axis) if replacing (r, θ) by $(r, -\theta)$ (or by $(-r, \pi - \theta)$) produces an equivalent equation (Figure 1).
2. The graph of a polar equation is symmetric about the y-axis (the line $\theta = \pi/2$) if replacing (r, θ) by $(-r, -\theta)$ (or by $(r, \pi - \theta)$) produces an equivalent equation (Figure 2).
3. The graph of a polar equation is symmetric about the origin (pole) if replacing (r, θ) by $(-r, \theta)$ (or by $(r, \pi + \theta)$) produces an equivalent equation (Figure 3).

Because of the multiple representation of points in polar coordinates, symmetries may exist that are not identified by these three tests (see Problem 39).

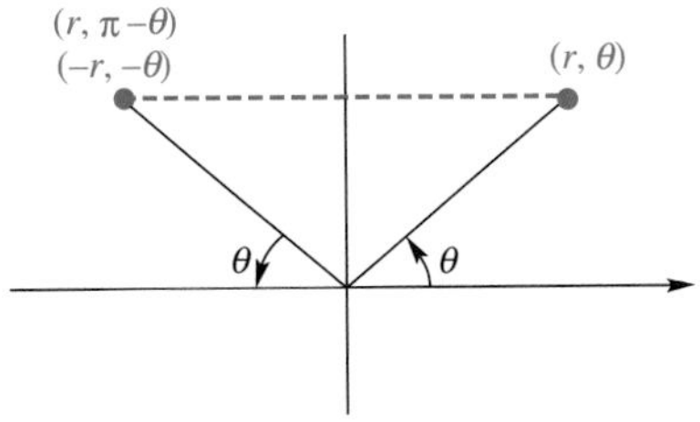

Figure 2

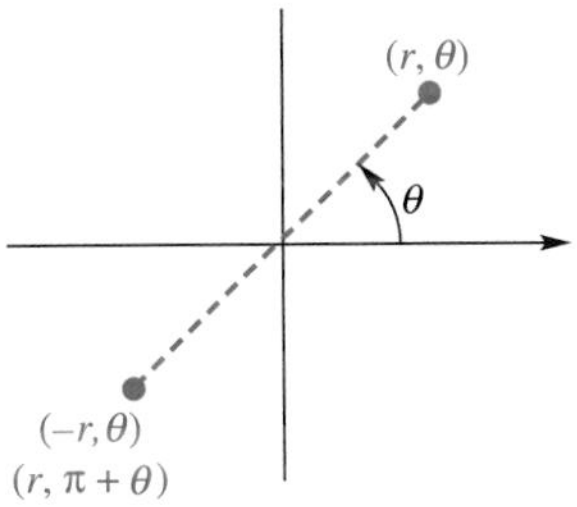

Figure 3

Cardioids and Limaçons We consider equations of the form

$$r = a \pm b \cos \theta \qquad r = a \pm b \sin \theta$$

with a and b positive. Their graphs are called **limaçons,** with the special cases in which $a = b$ referred to as **cardioids.** Typical graphs are shown in Figure 4.

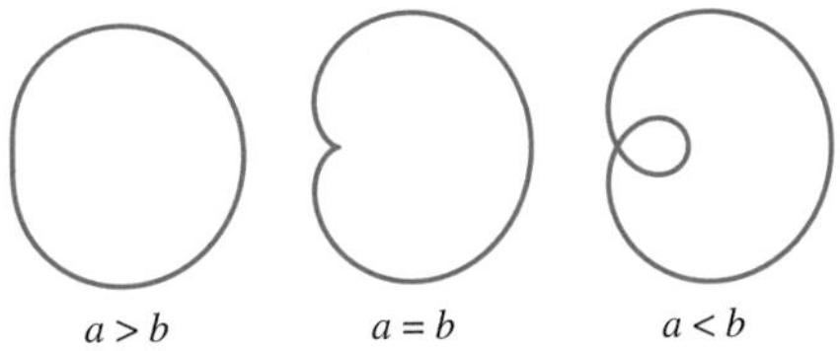

Figure 4

EXAMPLE 1 Analyze the equation $r = 2 + 4 \cos \theta$ for symmetry and sketch its graph.

SOLUTION Since cosine is an even function ($\cos(-\theta) = \cos \theta$), the graph is symmetric with respect to the x-axis. The other symmetry tests fail. A table of values and the graph appear in Figure 5. ■

θ	r
0	6
$\pi/6$	5.5
$\pi/3$	4
$\pi/2$	2
$7\pi/12$	1.0
$2\pi/3$	0
$3\pi/4$	−0.8
$5\pi/6$	−1.5
π	−2

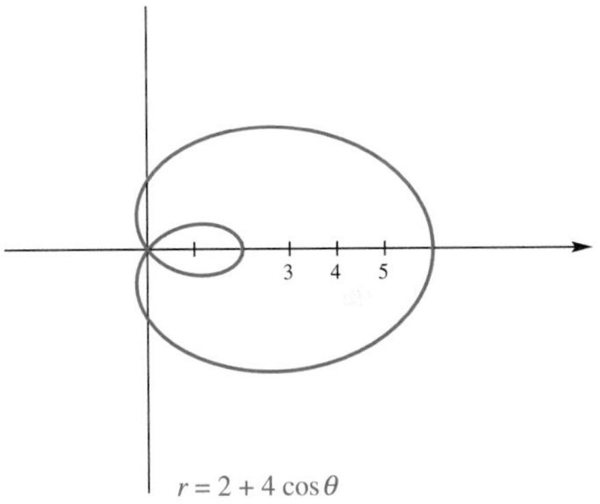

Figure 5

Lemniscates The graphs of

$$r^2 = \pm a \cos 2\theta \qquad r^2 = \pm a \sin 2\theta$$

are figure-eight-shaped curves called **lemniscates**.

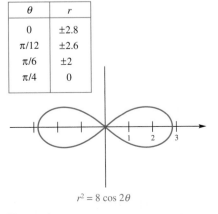

EXAMPLE 2 Analyze the equation $r^2 = 8 \cos 2\theta$ for symmetry and sketch its graph.

SOLUTION Since $\cos(-2\theta) = \cos 2\theta$ and

$$\cos[2(\pi - \theta)] = \cos(2\pi - 2\theta) = \cos(-2\theta) = \cos 2\theta$$

the graph is symmetric with respect to both axes. Clearly, it is also symmetric with respect to the origin. A table of values and the graph are shown in Figure 6. ■

θ	r
0	±2.8
$\pi/12$	±2.6
$\pi/6$	±2
$\pi/4$	0

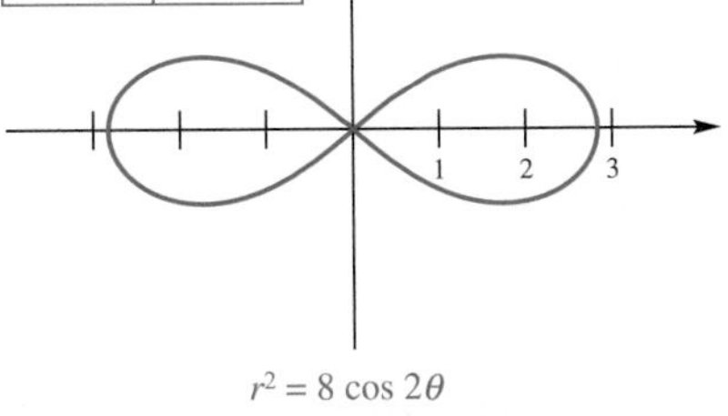

Figure 6

Roses Polar equations of the form

$$r = a \cos n\theta \qquad r = a \sin n\theta$$

represent flower-shaped curves called **roses**. The rose has n leaves if n is odd and $2n$ leaves if n is even.

EXAMPLE 3 Analyze $r = 4 \sin 2\theta$ for symmetry and sketch its graph.

SOLUTION You can check that $r = 4 \sin 2\theta$ satisfies all three symmetry tests. For example, it meets Test 1 since

$$\sin 2(\pi - \theta) = \sin(2\pi - 2\theta) = -\sin 2\theta$$

and so replacing (r, θ) by $(-r, \pi - \theta)$ produces an equivalent equation.

A rather extensive table of values for $0 \le \theta \le \pi/2$, a somewhat briefer one for $\pi/2 \le \theta \le 2\pi$, and the corresponding graph are shown in Figure 7. The arrows on the curve indicate the direction $P(r, \theta)$ moves as θ increases from 0 to 2π. ■

θ	r	θ	r
0	0	$2\pi/3$	–3.5
$\pi/12$	2	$5\pi/6$	–3.5
$\pi/8$	2.8	π	0
$\pi/6$	3.5	$7\pi/6$	3.5
$\pi/4$	4	$4\pi/3$	3.5
$\pi/3$	3.5	$3\pi/2$	0
$3\pi/8$	2.8	$5\pi/3$	–3.5
$5\pi/12$	2	$11\pi/6$	–3.5
$\pi/2$	0	2π	0

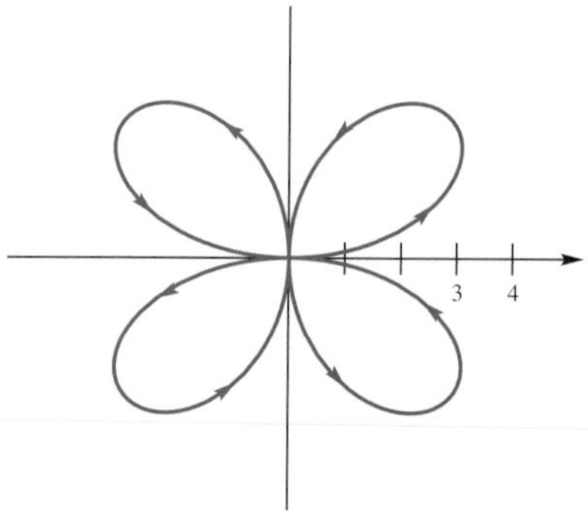

Figure 7

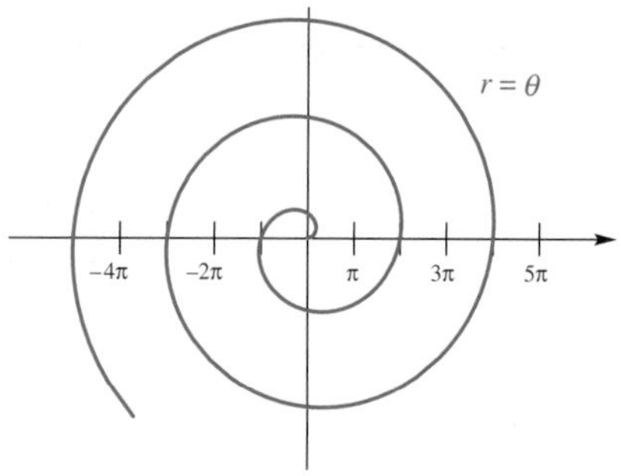

Figure 8

Spirals The graph of $r = a\theta$ is called a **spiral of Archimedes;** the graph of $r = ae^{b\theta}$ is called a **logarithmic spiral.**

EXAMPLE 4 Sketch the graph of $r = \theta$ for $\theta \ge 0$.

SOLUTION We omit a table of values, but note that the graph crosses the polar axis at $(0, 0)$, $(2\pi, 2\pi)$, $(4\pi, 4\pi)$, ... and crosses its extension to the left at (π, π), $(3\pi, 3\pi)$, $(5\pi, 5\pi)$, ..., as in Figure 8. ■

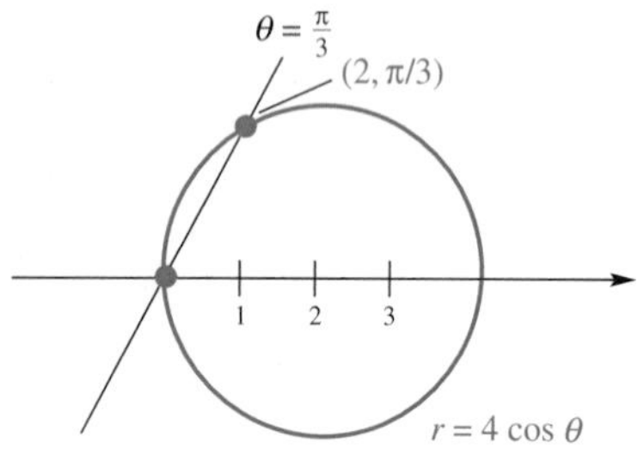

Figure 9

Intersection of Curves in Polar Coordinates In Cartesian coordinates, all points of intersection of two curves can be found by solving the equations of the curves simultaneously. But in polar coordinates, this is not always the case. This is because a point P has many pairs of polar coordinates, and one pair may satisfy the polar equation of one curve and a different pair may satisfy the polar equation of the other curve. For instance (see Figure 9), the circle $r = 4 \cos \theta$ intersects the line $\theta = \pi/3$ in two points, the pole and $(2, \pi/3)$ and yet only the latter is a common solution of the two equations. This happens because the coordinates of the pole that satisfy the equation of the line are $(0, \pi/3)$ and those that satisfy the equation of the circle are $(0, \pi/2 + n\pi)$.

Our conclusion is this. In order to find all intersections of two curves whose polar equations are given, solve the equations simultaneously; then graph the two equations carefully to discover other possible points of intersection.

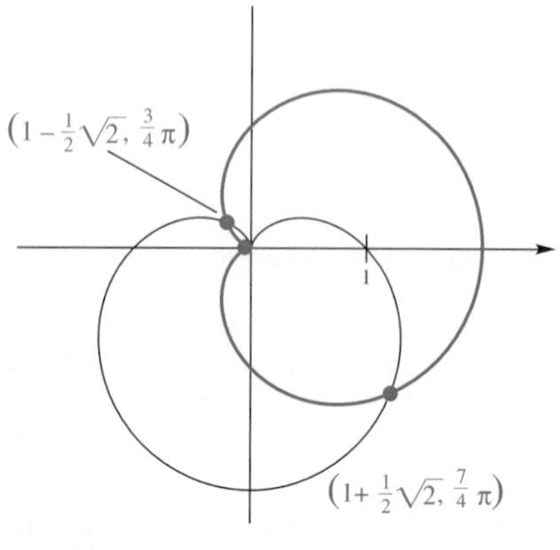

Figure 10

EXAMPLE 5 Find the points of intersection of the two cardioids $r = 1 + \cos \theta$ and $r = 1 - \sin \theta$.

SOLUTION If we eliminate r between the two equations, we get $1 + \cos \theta = 1 - \sin \theta$. Thus, $\cos \theta = -\sin \theta$, or $\tan \theta = -1$. We conclude that $\theta = \frac{3}{4}\pi$ or $\theta = \frac{7}{4}\pi$, which yields the two intersection points $\left(1 - \frac{1}{2}\sqrt{2}, \frac{3}{4}\pi\right)$ and $\left(1 + \frac{1}{2}\sqrt{2}, \frac{7}{4}\pi\right)$. The graphs in Figure 10 show, however, that we have missed a third intersection point, the pole. The reason we missed it is that $r = 0$ in $r = 1 + \cos \theta$ when $\theta = \pi$, but $r = 0$ in $r = 1 - \sin \theta$ when $\theta = \pi/2$. ■

Concepts Review

1. The graph of $r = 3 + 2\cos\theta$ is a(n) ______.

2. The graph of $r = 2 + 2\cos\theta$ is a(n) ______.

3. The graph of $r = 4\sin n\theta$ is a(n) ______ with n leaves if n is ______ and $2n$ leaves if n is ______.

4. The graph of $r = \theta/3$ is a(n) ______.

Problem Set 10.6

In Problems 1–32, sketch the graph of the given polar equation and verify its symmetry (see Examples 1–3).

1. $\theta^2 - \pi^2/16 = 0$

2. $(r - 3)\left(\theta - \frac{\pi}{4}\right) = 0$

3. $r\sin\theta + 4 = 0$

4. $r = -4\sec\theta$

5. $r = 2\cos\theta$

6. $r = 4\sin\theta$

7. $r = \dfrac{2}{1 - \cos\theta}$

8. $r = \dfrac{4}{1 + \sin\theta}$

9. $r = 3 - 3\cos\theta$ (cardioid)

10. $r = 5 - 5\sin\theta$ (cardioid)

11. $r = 1 - \sin\theta$ (cardioid)

12. $r = \sqrt{2} - \sqrt{2}\sin\theta$ (cardioid)

13. $r = 1 - 2\sin\theta$ (limaçon)

14. $r = 4 - 3\cos\theta$ (limaçon)

15. $r = 2 - 3\sin\theta$ (limaçon)

16. $r = 5 - 3\cos\theta$ (limaçon)

17. $r^2 = 4\cos 2\theta$ (lemniscate)

18. $r^2 = 9\sin 2\theta$ (lemniscate)

19. $r^2 = -9\cos 2\theta$ (lemniscate)

20. $r^2 = -16\cos 2\theta$ (lemniscate)

21. $r = 5\cos 3\theta$ (three-leaved rose)

22. $r = 3\sin 3\theta$ (three-leaved rose)

23. $r = 6\sin 2\theta$ (four-leaved rose)

24. $r = 4\cos 2\theta$ (four-leaved rose)

25. $r = 7\cos 5\theta$ (five-leaved rose)

26. $r = 3\sin 5\theta$ (five-leaved rose)

27. $r = \frac{1}{2}\theta, \theta \geq 0$ (spiral of Archimedes)

28. $r = 2\theta, \theta \geq 0$ (spiral of Archimedes)

29. $r = e^{\theta}, \theta \geq 0$ (logarithmic spiral)

30. $r = e^{\theta/2}, \theta \geq 0$ (logarithmic spiral)

31. $r = \dfrac{2}{\theta}, \theta > 0$ (reciprocal spiral)

32. $r = -\dfrac{1}{\theta}, \theta > 0$ (reciprocal spiral)

In Problems 33–38, sketch the given curves and find their points of intersection.

33. $r = 6, r = 4 + 4\cos\theta$

34. $r = 1 - \cos\theta, r = 1 + \cos\theta$

35. $r = 3\sqrt{3}\cos\theta, r = 3\sin\theta$

36. $r = 5, r = \dfrac{5}{1 - 2\cos\theta}$

37. $r = 6\sin\theta, r = \dfrac{6}{1 + 2\sin\theta}$

38. $r^2 = 4\cos 2\theta, r = 2\sqrt{2}\sin\theta$

39. The conditions for symmetry given in the text are sufficient conditions, not necessary conditions. Give an example of a polar equation $r = f(\theta)$ whose graph is symmetric with respect to the y-axis, even though replacing (r, θ) by either $(-r, -\theta)$ or $(r, \pi - \theta)$ fails to yield an equivalent equation.

40. Let a and b be fixed positive numbers and suppose that AP is part of the line that passes through $(0, 0)$, with A on the line $x = a$ and $|AP| = b$. Find both the polar equation and the rectangular equation for the set of points P (called a *conchoid*) and sketch its graph.

41. Let F and F' be fixed points with polar coordinates $(a, 0)$ and $(-a, 0)$, respectively. Show that the set of points P satisfying $|PF||PF'| = a^2$ is a lemniscate by finding its polar equation.

42. A line segment L of length $2a$ has its two end points on the x- and y-axes, respectively. The point P is on L and is such that OP is perpendicular to L. Show that the set of points P satisfying this condition is a four-leaved rose by finding its polar equation.

43. Find the polar equation for the curve described by the following Cartesian equations.

(a) $y = 45$ (b) $x^2 + y^2 = 36$

(c) $x^2 - y^2 = 1$ (d) $4xy = 1$

(e) $y = 3x + 2$ (f) $3x^2 + 4y = 2$

(g) $x^2 + 2x + y^2 - 4y - 25 = 0$

Computers and graphing calculators offer a wonderful opportunity to experiment with the graphing of polar equations of the form $r = f(\theta)$. In some cases these aids require that the equations be recast in a parametric form. Since $x = r\cos\theta = f(\theta)\cos\theta$ and $y = r\sin\theta = f(\theta)\sin\theta$, you can use the parametric graphing capabilities to graph $x = f(t)\cos t$ and $y = f(t)\sin t$ as a set of parametric equations.

GC 44. Graph the curve $r = \cos(8\theta/5)$ using the parametric graphing facility of a graphing calculator or computer. Notice that it is necessary to determine the proper domain for θ. Assuming that you start at $\theta = 0$, you have to determine the value of θ that makes the curve start to repeat itself. Explain why the correct domain is $0 \leq \theta \leq 10\pi$.

45. Match the polar equations to the graphs labeled I–VIII in Figure 11, giving reasons for your choices.

(a) $r = \cos(\theta/2)$ (b) $r = \sec(3\theta)$

(c) $r = 2 - 3\sin(5\theta)$ (d) $r = 1 - 2\sin(5\theta)$

(e) $r = 3 + 2\cos\theta$ (f) $r = \theta\cos\theta$

(g) $r = 1/\theta^{3/2}$ (h) $r = 2\cos 3\theta$

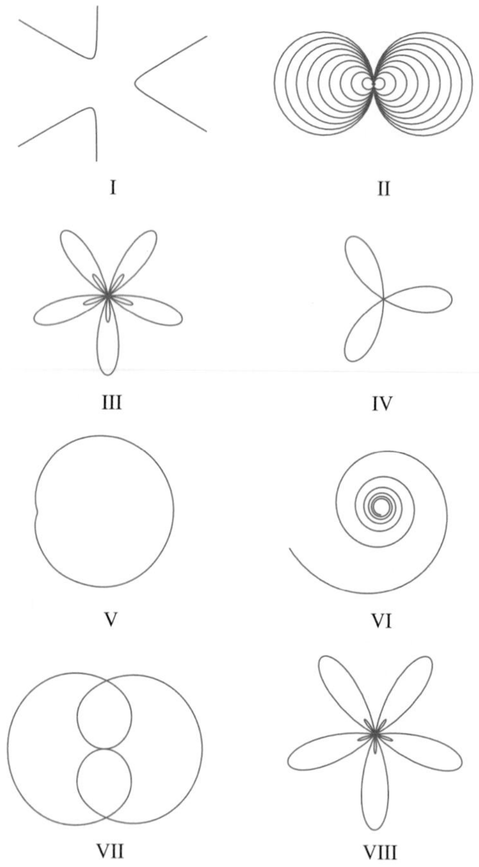

Figure 11

GC *In Problems 46–49, use a computer or graphing calculator to graph the given equation. Make sure that you choose a sufficiently large interval for the parameter so that the entire curve is drawn.*

46. $r = \sqrt{1 - 0.5\sin^2\theta}$ **47.** $r = \cos(13\theta/5)$

48. $r = \sin(5\theta/7)$ **49.** $r = 1 + 3\cos(\theta/3)$

GC EXPL **50.** In many cases, polar graphs are related to each other by rotation. We explore that concept here.

(a) How are the graphs of $r = 1 + \sin(\theta - \pi/3)$ and $r = 1 + \sin(\theta + \pi/3)$ related to the graph of $r = 1 + \sin\theta$?

(b) How is the graph of $r = 1 + \sin\theta$ related to the graph of $r = 1 - \sin\theta$?

(c) How is the graph of $r = 1 + \sin\theta$ related to the graph of $r = 1 + \cos\theta$?

(d) How is the graph of $r = f(\theta)$ related to the graph of $r = f(\theta - \alpha)$?

GC EXPL **51.** Investigate the family of curves given by $r = a + b\cos(n(\theta + \phi))$ where a, b, and ϕ are real numbers and n is a positive integer. As you answer the following questions, be sure that you graph a sufficient number of examples to justify your conclusions.

(a) How are the graphs for $\phi = 0$ related to those for which $\phi \neq 0$?

(b) How does the graph change as n increases?

(c) How do the relative magnitude and sign of a and b change the nature of the graph?

52. Investigate the family of curves defined by the polar equations $r = |\cos n\theta|$, where n is some positive integer. How do the number of leaves depend on n?

53. Polar graphs can be used to represent different spirals. The spirals can unwind clockwise or counterclockwise. Find the condition on c to make the *spiral of Archimedes*, $r = c\theta$, unwind clockwise and counterclockwise.

54. Sketch the *reciprocal spiral* given by $r = c/\theta$. For $c > 0$, does it unwind in the clockwise direction?

55. The following polar equations are represented by six graphs in Figure 12. Match each graph with its equation.

(a) $r = \sin 3\theta + \sin^2 2\theta$ (b) $r = \cos 2\theta + \cos^2 4\theta$

(c) $r = \sin 4\theta + \sin^2 5\theta$ (d) $r = \cos 2\theta + \cos^2 3\theta$

(e) $r = \cos 4\theta + \cos^2 4\theta$ (f) $r = \sin 4\theta + \sin^2 4\theta$

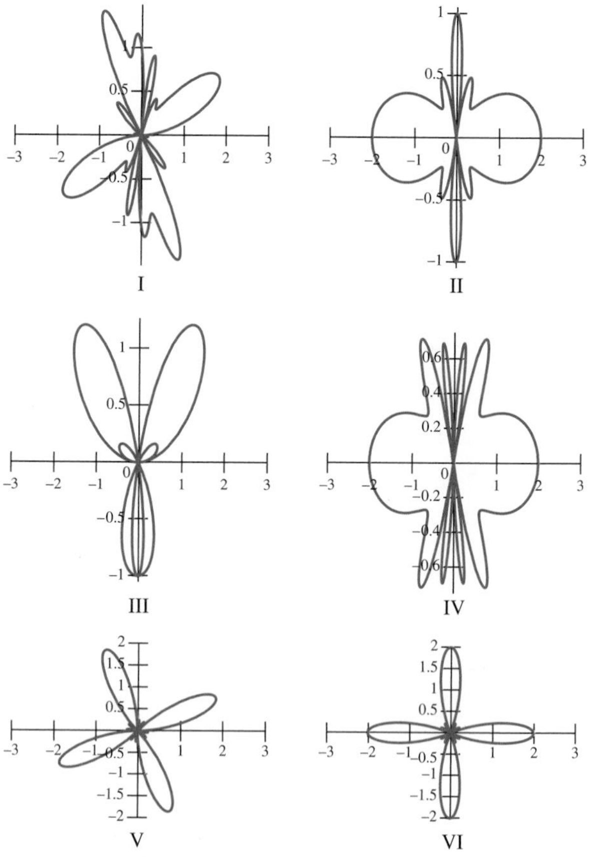

Figure 12

Answers to Concepts Review: **1.** limaçon **2.** cardioid **3.** rose; odd; even **4.** spiral

10.7 Calculus in Polar Coordinates

The two most basic problems in calculus are the determinations of the slope of a tangent line and the area of a curved region. Here we consider both problems, but in the context of polar coordinates. The area problem plays a larger role in the rest of the book, so we consider it first.

In Cartesian coordinates, the fundamental building block in area problems was the rectangle. In polar coordinates, it is the sector of a circle (a pie-shaped region like that in Figure 1). From the fact that the area of a circle is πr^2, we infer that the area of a sector with central angle θ radians is $(\theta/2\pi)\,\pi r^2$; that is,

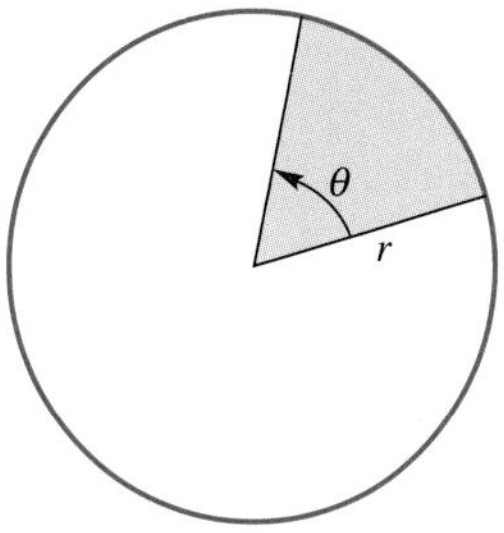

$A = \frac{1}{2}\theta r^2$

Figure 1

$$\text{Area of a sector:} \quad A = \frac{1}{2}\theta r^2$$

Area in Polar Coordinates To begin, let $r = f(\theta)$ determine a curve in the plane, where f is a continuous, nonnegative function for $\alpha \le \theta \le \beta$ and $\beta - \alpha \le 2\pi$. The curves $r = f(\theta)$, $\theta = \alpha$, and $\theta = \beta$ bound a region R (the one shown at the left in Figure 2), whose area $A(R)$ we wish to determine.

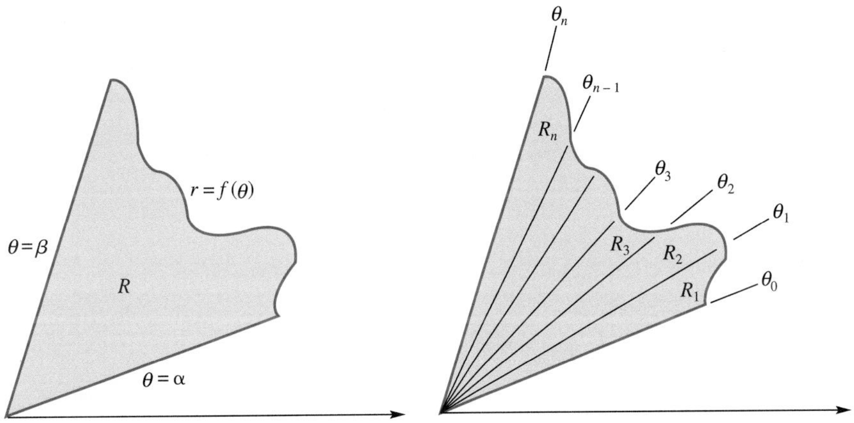

Figure 2

Partition the interval $[\alpha, \beta]$ into n subintervals by means of numbers $\alpha = \theta_0 < \theta_1 < \theta_2 < \cdots < \theta_n = \beta$, thereby slicing R into n smaller pie-shaped regions $R_1, R_2, \ldots, R_n$, as shown in the right half of Figure 2. Clearly, $A(R) = A(R_1) + A(R_2) + \cdots + A(R_n)$.

We approximate the area $A(R_i)$ of the ith slice; in fact, we do it in two ways. On the ith interval $[\theta_{i-1}, \theta_i]$, f achieves its minimum value and maximum value, for instance, at u_i and v_i, respectively (Figure 3). Thus, if $\Delta\theta_i = \theta_i - \theta_{i-1}$,

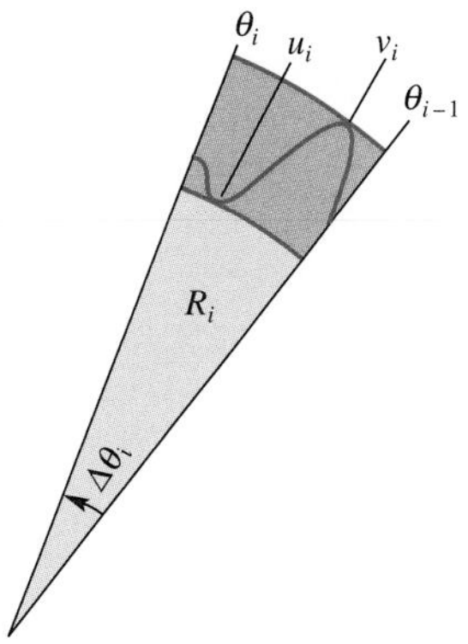

Figure 3

$$\tfrac{1}{2}[f(u_i)]^2\,\Delta\theta_i \le A(R_i) \le \tfrac{1}{2}[f(v_i)]^2\,\Delta\theta_i$$

and so

$$\sum_{i=1}^{n} \tfrac{1}{2}[f(u_i)]^2\,\Delta\theta_i \le \sum_{i=1}^{n} A(R_i) \le \sum_{i=1}^{n} \tfrac{1}{2}[f(v_i)]^2\,\Delta\theta_i$$

The first and third members of this inequality are Riemann sums for the same integral: $\int_\alpha^\beta \frac{1}{2}[f(\theta)]^2\,d\theta$. When we let the norm of the partition tend toward zero, we obtain (using the Squeeze Theorem) the area formula

$$A = \frac{1}{2}\int_\alpha^\beta [f(\theta)]^2\,d\theta$$

> **Equichordal Points**
>
> Limaçons share with circles the property of having an equichordal point (a point through which all chords have the same length). For the limaçon $r = 2 + \cos\theta$ of Example 1, all chords through the pole have length 4. Note that this limaçon has area $9\pi/2$, whereas the corresponding circle of diameter 4 has area 4π. Thus, having equal chords in all directions through a point is not enough to determine area.
>
> Here is a famous unsolved problem first posed in 1916. *Can a plane region have two equichordal points?* A correct answer to this question (either an example of such a region or a proof that no such region exists) would make you instantly famous. Still, we suggest you work on the problems at the end of the section before you tackle this challenge.

This formula can, of course, be memorized. We prefer that you remember how it was derived. In fact, you will note that the three familiar words *slice, approximate,* and *integrate* are also the key to area problems in polar coordinates. We illustrate what we mean.

EXAMPLE 1 Find the area of the region inside the limaçon $r = 2 + \cos\theta$.

SOLUTION The graph is sketched in Figure 4; note that θ varies from 0 to 2π.

≈ For a rough approximation, we might observe that the region looks much like a circle of radius 2. We therefore expect the answer to be approximately $\pi 2^2 = 4\pi$. To find the exact area, we slice, approximate, and integrate.

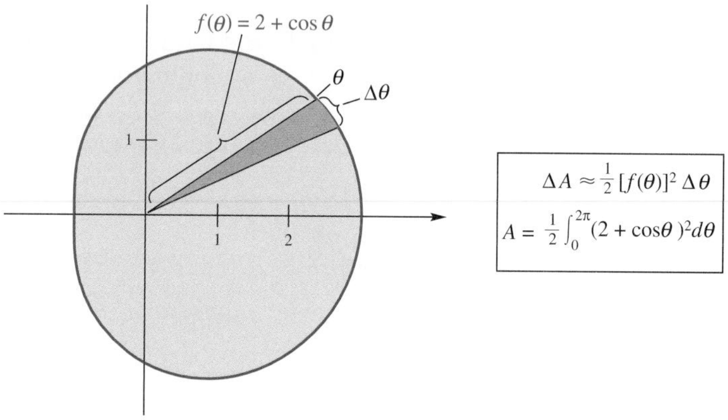

Figure 4

By symmetry, we can double the integral from 0 to π. Thus,

$$\begin{aligned}
A &= \int_0^{\pi}(2 + \cos\theta)^2\,d\theta = \int_0^{\pi}(4 + 4\cos\theta + \cos^2\theta)\,d\theta \\
&= \int_0^{\pi}4\,d\theta + 4\int_0^{\pi}\cos\theta\,d\theta + \frac{1}{2}\int_0^{\pi}(1 + \cos 2\theta)\,d\theta \\
&= \int_0^{\pi}\frac{9}{2}\,d\theta + 4\int_0^{\pi}\cos\theta\,d\theta + \frac{1}{4}\int_0^{\pi}\cos 2\theta\cdot 2\,d\theta \\
&= \left[\frac{9}{2}\theta\right]_0^{\pi} + [4\sin\theta]_0^{\pi} + \left[\frac{1}{4}\sin 2\theta\right]_0^{\pi} \\
&= \frac{9\pi}{2}
\end{aligned}$$

■

EXAMPLE 2 Find the area of one leaf of the four-leaved rose $r = 4\sin 2\theta$.

≈ **SOLUTION** The complete rose was sketched in Example 3 of the previous section. Here we show only the first-quadrant leaf (Figure 5). This leaf is 4 units long and averages about 1.5 units in width, giving 6 as an estimate for its area. The exact area A is given by

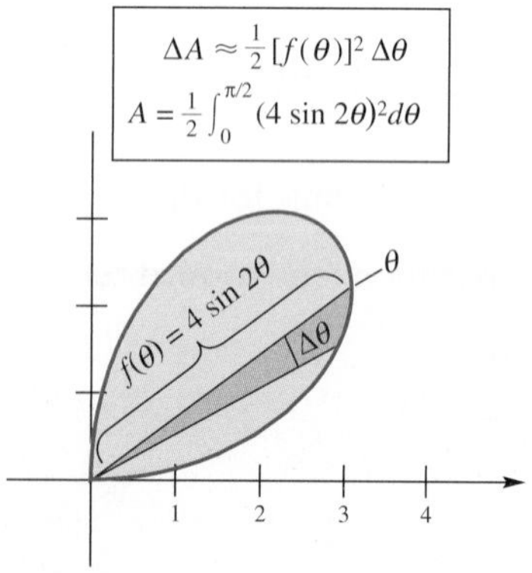

Figure 5

$$\begin{aligned}
A &= \frac{1}{2}\int_0^{\pi/2}16\sin^2 2\theta\,d\theta = 8\int_0^{\pi/2}\frac{1 - \cos 4\theta}{2}\,d\theta \\
&= 4\int_0^{\pi/2}d\theta - \int_0^{\pi/2}\cos 4\theta\cdot 4\,d\theta \\
&= [4\theta]_0^{\pi/2} - [\sin 4\theta]_0^{\pi/2} = 2\pi
\end{aligned}$$

■

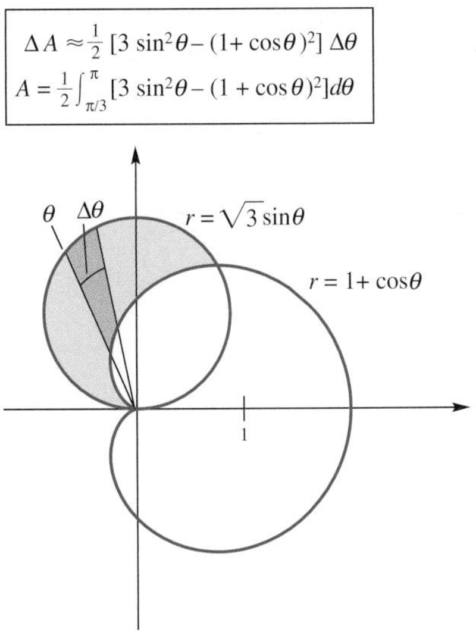

Figure 6

EXAMPLE 3 Find the area of the region outside the cardioid $r = 1 + \cos\theta$ and inside the circle $r = \sqrt{3}\sin\theta$.

SOLUTION The graphs of the two curves are sketched in Figure 6. We will need the θ-coordinates of the points of intersection. Let's try solving the two equations simultaneously.

$$1 + \cos\theta = \sqrt{3}\sin\theta$$
$$1 + 2\cos\theta + \cos^2\theta = 3\sin^2\theta$$
$$1 + 2\cos\theta + \cos^2\theta = 3(1 - \cos^2\theta)$$
$$4\cos^2\theta + 2\cos\theta - 2 = 0$$
$$2\cos^2\theta + \cos\theta - 1 = 0$$
$$(2\cos\theta - 1)(\cos\theta + 1) = 0$$
$$\cos\theta = \frac{1}{2} \quad \text{or} \quad \cos\theta = -1$$
$$\theta = \frac{\pi}{3} \quad \text{or} \quad \theta = \pi$$

Now slice, approximate, and integrate.

$$\begin{aligned}
A &= \frac{1}{2}\int_{\pi/3}^{\pi}[3\sin^2\theta - (1+\cos\theta)^2]\,d\theta \\
&= \frac{1}{2}\int_{\pi/3}^{\pi}[3\sin^2\theta - 1 - 2\cos\theta - \cos^2\theta]\,d\theta \\
&= \frac{1}{2}\int_{\pi/3}^{\pi}\left[\frac{3}{2}(1 - \cos 2\theta) - 1 - 2\cos\theta - \frac{1}{2}(1 + \cos 2\theta)\right]d\theta \\
&= \frac{1}{2}\int_{\pi/3}^{\pi}[-2\cos\theta - 2\cos 2\theta]\,d\theta \\
&= \frac{1}{2}[-2\sin\theta - \sin 2\theta]_{\pi/3}^{\pi} \\
&= \frac{1}{2}\left[2\frac{\sqrt{3}}{2} + \frac{\sqrt{3}}{2}\right] = \frac{3\sqrt{3}}{4} \approx 1.299
\end{aligned}$$

■

Tangents in Polar Coordinates In Cartesian coordinates, the slope m of the tangent line to a curve is given by $m = dy/dx$. We quickly reject $dr/d\theta$ as the corresponding slope formula in polar coordinates. Rather, if $r = f(\theta)$ determines the curve, we write

$$y = r\sin\theta = f(\theta)\sin\theta$$
$$x = r\cos\theta = f(\theta)\cos\theta$$

Thus,

$$\frac{dy}{dx} = \lim_{\Delta x\to 0}\frac{\Delta y}{\Delta x} = \lim_{\Delta\theta\to 0}\frac{\Delta y/\Delta\theta}{\Delta x/\Delta\theta} = \frac{dy/d\theta}{dx/d\theta}$$

That is,

$$m = \frac{f(\theta)\cos\theta + f'(\theta)\sin\theta}{-f(\theta)\sin\theta + f'(\theta)\cos\theta}$$

The formula just derived simplifies when the graph of $r = f(\theta)$ passes through the pole. For example, suppose for some angle α that $r = f(\alpha) = 0$ and $f'(\alpha) \neq 0$. Then (at the pole) our formula for m is

$$m = \frac{f'(\alpha)\sin\alpha}{f'(\alpha)\cos\alpha} = \tan\alpha$$

Since the line $\theta = \alpha$ also has slope $\tan \alpha$, we conclude that this line is tangent to the curve at the pole. We infer the useful fact that *tangent lines at the pole can be found by solving the equation* $f(\theta) = 0$. We illustrate this next.

EXAMPLE 4 Consider the polar equation $r = 4 \sin 3\theta$.
(a) Find the slope of the tangent line at $\theta = \pi/6$ and $\theta = \pi/4$.
(b) Find the tangent lines at the pole.
(c) Sketch the graph.
(d) Find the area of one leaf.

SOLUTION

(a) $$m = \frac{f(\theta)\cos\theta + f'(\theta)\sin\theta}{-f(\theta)\sin\theta + f'(\theta)\cos\theta} = \frac{4\sin 3\theta\cos\theta + 12\cos 3\theta\sin\theta}{-4\sin 3\theta\sin\theta + 12\cos 3\theta\cos\theta}$$

At $\theta = \pi/6$,

$$m = \frac{4\cdot 1\cdot\frac{\sqrt{3}}{2} + 12\cdot 0\cdot\frac{1}{2}}{-4\cdot 1\cdot\frac{1}{2} + 12\cdot 0\cdot\frac{\sqrt{3}}{2}} = -\sqrt{3}$$

At $\theta = \pi/4$,

$$m = \frac{4\cdot\frac{\sqrt{2}}{2}\cdot\frac{\sqrt{2}}{2} - 12\cdot\frac{\sqrt{2}}{2}\cdot\frac{\sqrt{2}}{2}}{-4\cdot\frac{\sqrt{2}}{2}\cdot\frac{\sqrt{2}}{2} - 12\cdot\frac{\sqrt{2}}{2}\cdot\frac{\sqrt{2}}{2}} = \frac{2-6}{-2-6} = \frac{1}{2}$$

θ	r
0	0
$\pi/12$	2.8
$\pi/6$	4
$\pi/4$	2.8
$\pi/3$	0
$5\pi/12$	−2.8
$\pi/2$	−4

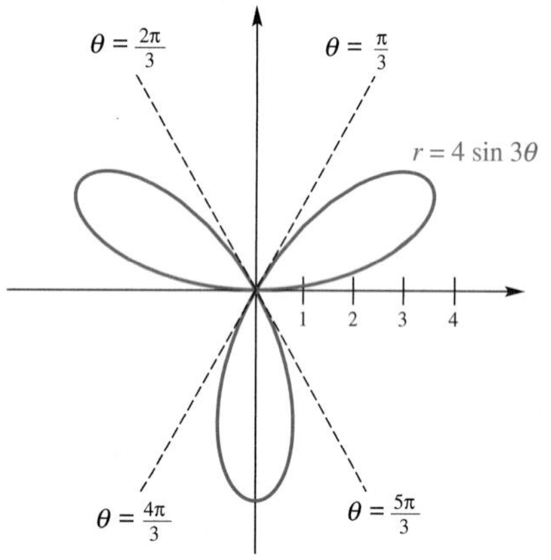

Figure 7

(b) We set $f(\theta) = 4\sin 3\theta = 0$ and solve. This yields $\theta = 0$, $\theta = \pi/3$, $\theta = 2\pi/3$, $\theta = \pi$, $\theta = 4\pi/3$, and $\theta = 5\pi/3$.

(c) After noting that

$$\sin 3(\pi - \theta) = \sin(3\pi - 3\theta) = \sin 3\pi\cos 3\theta - \cos 3\pi\sin 3\theta = \sin 3\theta$$

which implies symmetry with respect to the y-axis, we obtain a table of values and sketch the graph shown in Figure 7.

(d) $$A = \frac{1}{2}\int_0^{\pi/3}(4\sin 3\theta)^2\,d\theta = 8\int_0^{\pi/3}\sin^2 3\theta\,d\theta$$
$$= 4\int_0^{\pi/3}(1-\cos 6\theta)\,d\theta = 4\int_0^{\pi/3}d\theta - \frac{4}{6}\int_0^{\pi/3}\cos 6\theta\cdot 6\,d\theta$$
$$= \left[4\theta - \frac{2}{3}\sin 6\theta\right]_0^{\pi/3} = \frac{4\pi}{3}$$ ■

Concepts Review

1. The formula for the area A of a sector of a circle of radius r and angle θ (in radians) is $A =$ _____.

2. The formula in Question 1 leads to the formula for the area A of the region bounded by the curve $r = f(\theta)$ between $\theta = \alpha$ and $\theta = \beta$, that is, $A =$ _____.

3. From the formula of Question 2, we conclude that the area A of the region inside the cardioid $r = 2 + 2\cos\theta$ can be expressed as $A =$ _____.

4. The tangent lines to the polar curve $r = f(\theta)$ at the pole can be found by solving the equation _____.

Problem Set 10.7

In Problems 1–10, sketch the graph of the given equation and find the area of the region bounded by it.

1. $r = a, a > 0$

2. $r = 2a\cos\theta, a > 0$

3. $r = 2 + \cos\theta$

4. $r = 5 + 4\cos\theta$

5. $r = 3 - 3\sin\theta$

6. $r = 3 + 3\sin\theta$

7. $r = a(1 + \cos\theta), a > 0$

8. $r^2 = 6\cos 2\theta$

9. $r^2 = 9 \sin 2\theta$ **10.** $r^2 = a \cos 2\theta, a > 0$

11. Sketch the limaçon $r = 3 - 4 \sin \theta$, and find the area of the region inside its small loop.

12. Sketch the limaçon $r = 2 - 4 \cos \theta$, and find the area of the region inside its small loop.

13. Sketch the limaçon $r = 2 - 3 \cos \theta$, and find the area of the region inside its large loop.

14. Sketch one leaf of the four-leaved rose $r = 3 \cos 2\theta$, and find the area of the region enclosed by it.

15. Sketch the three-leaved rose $r = 4 \cos 3\theta$, and find the area of the total region enclosed by it.

16. Sketch the three-leaved rose $r = 2 \sin 3\theta$, and find the area of the region bounded by it.

17. Find the area of the region between the two concentric circles $r = 7$ and $r = 10$.

18. Sketch the region that is inside the circle $r = 3 \sin \theta$ and outside the cardioid $r = 1 + \sin \theta$, and find its area.

19. Sketch the region that is outside the circle $r = 2$ and inside the lemniscate $r^2 = 8 \cos 2\theta$, and find its area.

20. Sketch the limaçon $r = 3 - 6 \sin \theta$, and find the area of the region that is inside its large loop, but outside its small loop.

21. Sketch the region in the first quadrant that is inside the cardioid $r = 3 + 3 \cos \theta$ and outside the cardioid $r = 3 + 3 \sin \theta$, and find its area.

22. Sketch the region in the second quadrant that is inside the cardioid $r = 2 + 2 \sin \theta$ and outside the cardioid $r = 2 + 2 \cos \theta$, and find its area.

23. Find the slope of the tangent line to each of the following curves at $\theta = \pi/3$.

(a) $r = 2 \cos \theta$ (b) $r = 1 + \sin \theta$

(c) $r = \sin 2\theta$ (d) $r = 4 - 3 \cos \theta$

24. Find all points on the cardioid $r = a(1 + \cos \theta)$ where the tangent line is

(a) horizontal, and (b) vertical.

25. Find all points on the limaçon $r = 1 - 2 \sin \theta$ where the tangent line is horizontal.

26. Let $r = f(\theta)$, where f is continuous on the closed interval $[\alpha, \beta]$. Derive the following formula for the length L of the corresponding polar curve from $\theta = \alpha$ to $\theta = \beta$.

$$L = \int_{\alpha}^{\beta} \sqrt{[f(\theta)]^2 + [f'(\theta)]^2}\, d\theta$$

27. Use the formula of Problem 26 to find the perimeter of the cardioid $r = a(1 + \cos \theta)$.

28. Find the length of the logarithmic spiral $r = e^{\theta/2}$ from $\theta = 0$ to $\theta = 2\pi$.

29. Find the total area of the rose $r = a \cos n\theta$, where n is a positive integer.

30. Sketch the graph of the *strophoid* $r = \sec \theta - 2 \cos \theta$, and find the area of its loop.

31. Consider the two circles $r = 2a \sin \theta$ and $r = 2b \cos \theta$, with a and b positive.

(a) Find the area of the region inside both circles.

(b) Show that the two circles intersect at right angles.

32. Assume that a planet of mass m is revolving around the sun (located at the pole) with constant angular momentum $mr^2\, d\theta/dt$. Deduce Kepler's Second Law: The line from the sun to the planet sweeps out equal areas in equal times.

33. First Old Goat Problem A goat is tethered to the edge of a circular pond of radius a by a rope of length ka $(0 < k \le 2)$. Use the method of this section to find its grazing area (the shaded area in Figure 8). *Note:* We solved this problem once before (Problem 77 of Section 6.8); you should be able to make your answers agree.

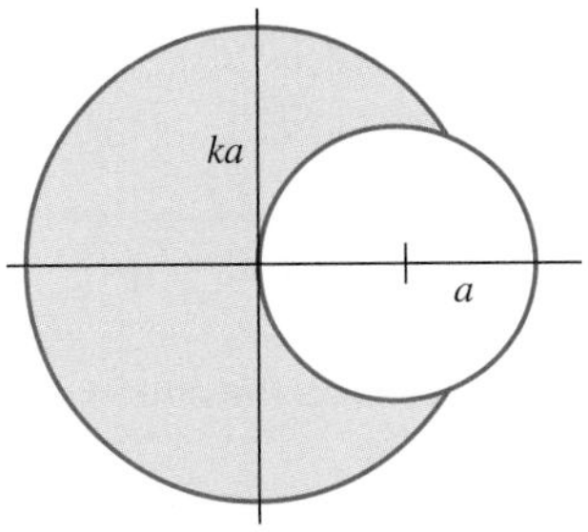

Figure 8

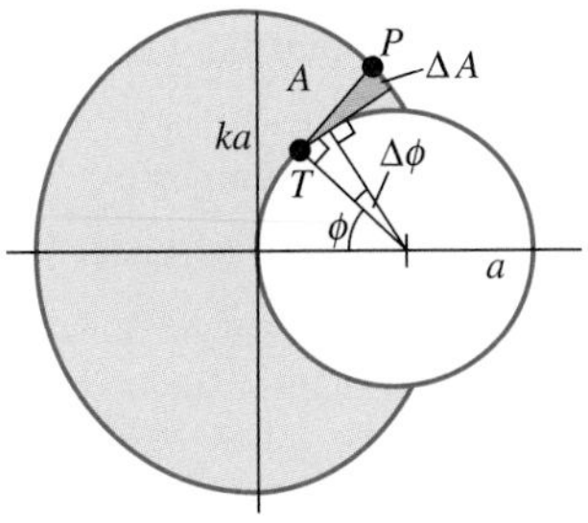

Figure 9

34. Second Old Goat Problem Do Problem 33 again, but assume that the pond has a fence around it so that, in forming the wedge A, the rope wraps around the fence (Figure 9). *Hint:* If you are exceedingly ambitious, try the method of this section. Better, note that in the wedge A,

$$\Delta A \approx \frac{1}{2}|PT|^2\, \Delta\phi$$

which leads to a Riemann sum for an integral. The final answer is $a^2(\pi k^2/2 + k^3/3)$, a result needed in Problem 35.

C **35. Third Old Goat Problem** An untethered goat grazes inside a yard enclosed by a circular fence of radius a; another grazes outside the fence tethered as in Problem 34. Find the length of the rope if the two goats have the same grazing area.

CAS *Use a computer to do Problems 36–39. In each case be sure to make a mental estimate first. Note the length formula in Problem 26.*

36. Find the lengths of the limaçons $r = 2 + \cos \theta$ and $r = 2 + 4 \cos \theta$ (see Example 1 of this section and Example 1 of Section 10.6).

37. Find the area and length of the three-leaved rose $r = 4 \sin 3\theta$ (see Example 4).

38. Find the area and length of the lemniscate $r^2 = 8 \cos 2\theta$ (see Example 2 of Section 10.6).

39. Plot the curve $r = 4 \sin(3\theta/2), 0 \le \theta \le 4\pi$, and then find its length.

Answers to Concepts Review: **1.** $\frac{1}{2}r^2\theta$

2. $\frac{1}{2}\int_{\alpha}^{\beta} [f(\theta)]^2\, d\theta$ **3.** $\frac{1}{2}\int_{0}^{2\pi} (2 + 2\cos\theta)^2\, d\theta$

4. $f(\theta) = 0$

10.8 Chapter Review

Concepts Test

Respond with true or false to each of the following assertions. Be prepared to justify your answer.

1. The graph of $y = ax^2 + bx + c$ is a parabola for all choices of a, b, and c.

2. The vertex of a parabola is midway between the focus and the directrix.

3. A vertex of an ellipse is closer to a directrix than to a focus.

4. The point on a parabola closest to its focus is the vertex.

5. The hyperbolas $x^2/a^2 - y^2/b^2 = 1$ and $y^2/b^2 - x^2/a^2 = 1$ have the same asymptotes.

6. The circumference C of the ellipse $x^2/a^2 + y^2/b^2 = 1$, with $b < a$, satisfies $2\pi b < C < 2\pi a$.

7. The smaller the eccentricity e of an ellipse, the more nearly circular the ellipse is.

8. The ellipse $6x^2 + 4y^2 = 24$ has its foci on the x-axis.

9. The equation $x^2 - y^2 = 0$ represents a hyperbola.

10. The equation $(y^2 - 4x + 1)^2 = 0$ represents a parabola.

11. If $k \neq 0$, $x^2/a^2 - y^2/b^2 = k$ is an equation of a hyperbola.

12. If $k \neq 0$, $x^2/a^2 + y^2/b^2 = k$ is an equation of an ellipse.

13. The distance between the foci of the graph of $x^2/a^2 + y^2/b^2 = 1$ is $2\sqrt{a^2 - b^2}$.

14. The graph of $x^2/9 - y^2/8 = -2$ does not intersect the x-axis.

15. Light emanating from a point between a focus and the nearest vertex of an elliptical mirror will be reflected beyond the other focus.

16. An ellipse that is drawn using a string of length 8 units attached to foci 2 units apart will have minor diameter of length $\sqrt{60}$ units.

17. The graph of $x^2 + y^2 + Cx + Dy + F = 0$ is either a circle, a point, or the empty set.

18. The graph of $2x^2 + y^2 + Cx + Dy + F = 0$ cannot be a single point.

19. The graph of $Ax^2 + Bxy + Cy^2 + Dyx + Ey + F = 0$ is the intersection of a plane with a cone of two nappes for all choices of A, B, C, D, E, and F.

20. In an appropriate coordinate system, the intersection of a plane with a cone of two nappes will have an equation of the form $Ax^2 + Cy^2 + Dx + Ey + F = 0$.

21. The graph of a hyperbola must enter all four quadrants.

22. If one of the conic sections passes through the four points $(1, 0)$, $(-1, 0)$, $(0, 1)$ and $(0, -1)$, it must be a circle.

23. The parametric representation of a curve is unique.

24. The graph of $x = 2t^3$, $y = t^3$ is a line.

25. If $x = f(t)$ and $y = g(t)$, then we can find a function h such that $y = h(x)$.

26. The curve with parametric representation $x = \ln t$ and $y = t^2 - 1$ passes through the origin.

27. If $x = f(t)$ and $y = g(t)$ and if both f'' and g'' exist, then $d^2y/dx^2 = g''(t)/f''(t)$ wherever $f''(t) \neq 0$.

28. A curve may have more than one tangent line at a point on the curve.

29. The graph of the polar equation $r = 4\cos(\theta - \pi/3)$ is a circle.

30. Every point in the plane has infinitely many sets of polar coordinates.

31. All points of intersection of the graphs of the polar equations $r = f(\theta)$ and $r = g(\theta)$ can be found by solving these two equations simultaneously.

32. If f is an odd function, then the graph of $r = f(\theta)$ is symmetric with respect to the y-axis (the line $\theta = \pi/2$).

33. If f is an even function, then the graph of $r = f(\theta)$ is symmetric with respect to the x-axis (the line $\theta = 0$).

34. The graph of $r = 4\cos 3\theta$ is a rose of three leaves whose area is less than half that of the circle $r = 4$.

Sample Test Problems

1. From the numbered list, pick the correct response to put in each blank that follows.

(1) no graph
(2) a single point
(3) a single line
(4) two parallel lines
(5) two intersecting lines
(6) a circle
(7) a parabola
(8) an ellipse
(9) a hyperbola
(10) none of the above

(a) ____ $x^2 - 4y^2 = 0$
(b) ____ $x^2 - 4y^2 = 0.01$
(c) ____ $x^2 - 4 = 0$
(d) ____ $x^2 - 4x + 4 = 0$
(e) ____ $x^2 + 4y^2 = 0$
(f) ____ $x^2 + 4y^2 = x$
(g) ____ $x^2 + 4y^2 = -x$
(h) ____ $x^2 + 4y^2 = -1$
(i) ____ $(x^2 + 4y - 1)^2 = 0$
(j) ____ $3x^2 + 4y^2 = -x^2 + 1$

In each of Problems 2–10, name the conic that has the given equation. Find its vertices and foci, and sketch its graph.

2. $y^2 - 6x = 0$

3. $9x^2 + 4y^2 - 36 = 0$

4. $25x^2 - 36y^2 + 900 = 0$

5. $x^2 + 9y = 0$

6. $x^2 - 4y^2 - 16 = 0$

7. $9x^2 + 25y^2 - 225 = 0$

8. $9x^2 + 9y^2 - 225 = 0$

9. $r = \dfrac{5}{2 + 2\sin\theta}$

10. $r(2 + \cos\theta) = 3$

In each of Problems 11–18, find the Cartesian equation of the conic with the given properties.

11. Vertices $(\pm 4, 0)$ and eccentricity $\frac{1}{2}$

12. Eccentricity 1, focus $(0, -3)$, and vertex $(0, 0)$

13. Eccentricity 1, vertex $(0, 0)$, symmetric with respect to the x-axis, and passing through the point $(-1, 3)$

14. Eccentricity $\frac{5}{3}$ and vertices $(0, \pm 3)$

15. Vertices $(\pm 2, 0)$ and asymptotes $x \pm 2y = 0$

16. Parabola with focus $(3, 2)$ and vertex $(3, 3)$

17. Ellipse with center $(1, 2)$, and focus $(4, 2)$, and major diameter 10

18. Hyperbola with vertices $(2, 0)$ and $(2, 6)$ and eccentricity $\frac{10}{3}$

In Problems 19–22, use the process of completing the square to transform the given equation to a standard form. Then name the corresponding curve and sketch its graph.

19. $4x^2 + 4y^2 - 24x + 36y + 81 = 0$

20. $4x^2 + 9y^2 - 24x - 36y + 36 = 0$

21. $x^2 + 8x + 6y + 28 = 0$

22. $3x^2 - 10y^2 + 36x - 20y + 68 = 0$

23. A rotation of axes through $\theta = 45°$ transforms $x^2 + 3xy + y^2 = 10$ into $ru^2 + sv^2 = 10$. Determine r and s, name the corresponding conic, and find the distance between its foci.

24. Determine the rotation angle θ needed to eliminate the cross-product term in $7x^2 + 8xy + y^2 = 9$. Then obtain the corresponding uv-equation and identify the conic that it represents.

In Problems 25–28, a parametric representation of a curve is given. Eliminate the parameter to obtain the corresponding Cartesian equation. Sketch the given curve.

25. $x = 6t + 2,\ y = 2t;\ -\infty < t < \infty$

26. $x = 4t^2,\ y = 4t;\ -1 \le t \le 2$

27. $x = 4 \sin t - 2,\ y = 3 \cos t + 1;\ 0 \le t \le 2\pi$

28. $x = 2 \sec t,\ y = \tan t;\ -\dfrac{\pi}{2} < t < \dfrac{\pi}{2}$

In Problems 29 and 30, find the equations of the tangent line at $t = 0$.

29. $x = 2t^3 - 4t + 7,\ y = t + \ln(t + 1)$

30. $x = 3e^{-t},\ y = \frac{1}{2}e^t$

31. Find the length of the curve $x = 1 + t^{3/2},\ y = 2 + t^{3/2}$, from $t = 0$ to $t = 9$.

32. Find the length of the curve $x = \cos t + t \sin t$, $y = \sin t - t \cos t$ from 0 to 2π. Make a sketch.

In Problems 33–44, analyze the given polar equation and sketch its graph.

33. $r = 6 \cos \theta$

34. $r = \dfrac{5}{\sin \theta}$

35. $r = \cos 2\theta$

36. $r = \dfrac{3}{\cos \theta}$

37. $r = 4$

38. $r = 5 - 5 \cos \theta$

39. $r = 4 - 3 \cos \theta$

40. $r = 2 - 3 \cos \theta$

41. $\theta = \frac{2}{3}\pi$

42. $r = 4 \sin 3\theta$

43. $r^2 = 16 \sin 2\theta$

44. $r = -\theta,\ \theta \ge 0$

45. Find a Cartesian equation of the graph of

$$r^2 - 6r(\cos \theta + \sin \theta) + 9 = 0$$

and then sketch the graph.

46. Find a Cartesian equation of the graph of $r^2 \cos 2\theta = 9$ and then sketch the graph.

47. Find the slope of the tangent line to the graph of $r = 3 + 3 \cos \theta$ at the point on the graph where $\theta = \frac{1}{6}\pi$.

48. Sketch the graphs of $r = 5 \sin \theta$ and $r = 2 + \sin \theta$ and find their points of intersection.

49. Find the area of the region bounded by the graph of $r = 5 - 5 \cos \theta$.

50. Find the area of the region that is outside the limaçon $r = 2 + \sin \theta$ and inside the circle $r = 5 \sin \theta$.

51. A racing car driving on the elliptical race track $x^2/400 + y^2/100 = 1$ went out of control at the point (16, 6) and thereafter continued on the tangent line until it hit a tree at $(14, k)$. Determine k.

52. Match each polar equation with its graph.

(a) $r = 1 - 2 \sin \theta$ (b) $r = 1 + \dfrac{\sin \theta}{2}$

(c) $r = 1 + 2 \cos \theta$ (d) $r = 1 + \dfrac{\cos \theta}{2}$

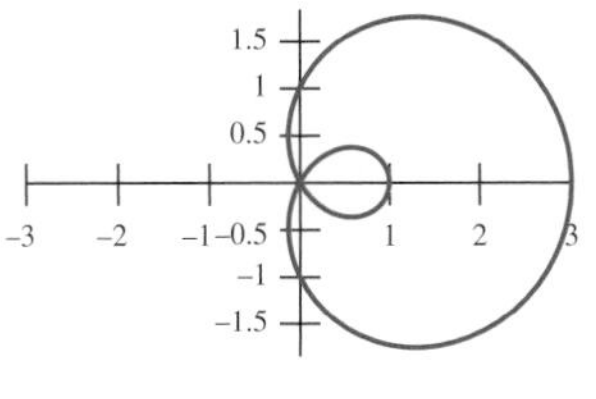

I

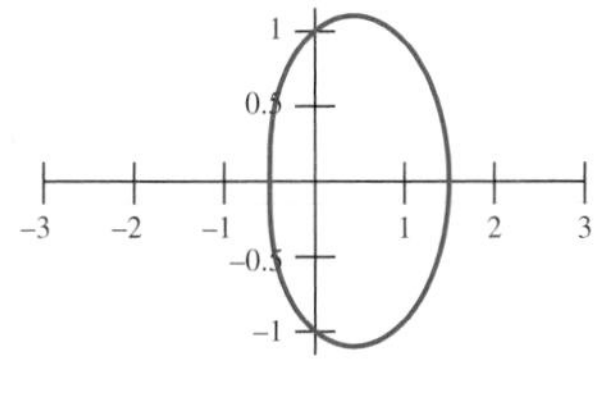

II

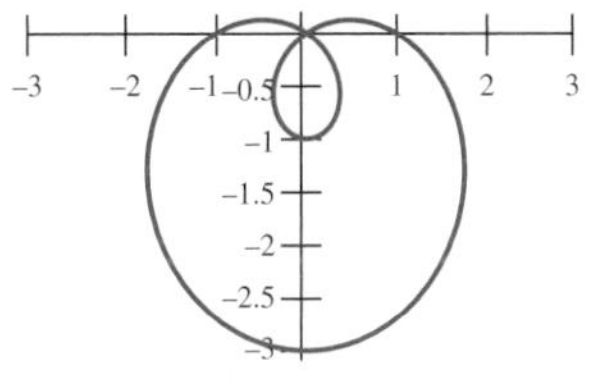

III

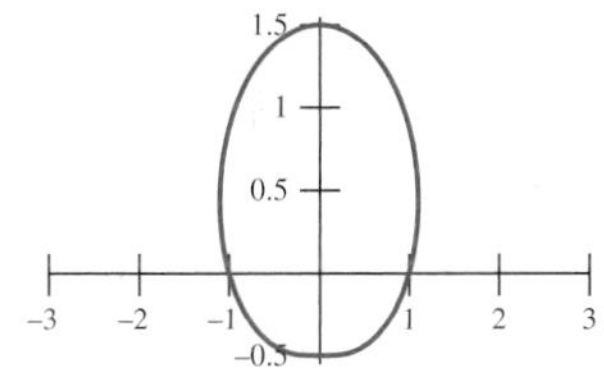

IV

53. Match each polar equation with its graph.

(a) $r = 4 \cos 2\theta$ (b) $r = 3 \cos 3\theta$

(c) $r = 5 \cos 5\theta$ (d) $r = 3 \sin 2\theta$

I

II

III

IV

REVIEW & PREVIEW PROBLEMS

In Problems 1–6, plot the curve whose parametric equation is given.

1. $x = 2t,\ y = t - 3;\ 1 \le t \le 4$

2. $x = t/2,\ y = t^2;\ -1 \le t \le 2$

3. $x = 2\cos t,\ y = 2\sin t;\ 0 \le t \le 2\pi$

4. $x = 2\sin t,\ y = -2\cos t;\ 0 \le t \le 2\pi$

5. $x = t,\ y = \tan 2t;\ -\pi/4 < t < \pi/4$

6. $x = \cosh t,\ y = \sinh t;\ -4 \le t \le 4$

In Problems 7–8, find expressions for x and y in terms of h and θ.

7.

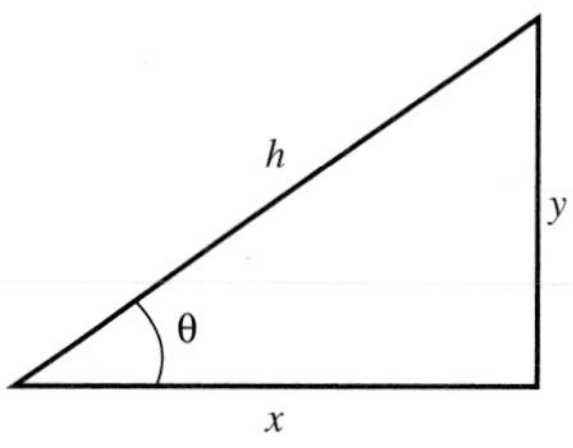

8.

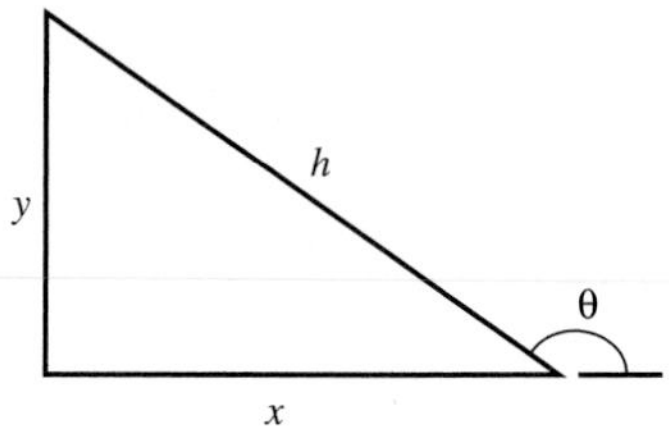

In Problems 9–12, find the length of the given curve.

9. $x = t,\ y = 3t^{3/2};\ 0 \le t \le 4$

10. $x = t + 2,\ y = 2t - 3;\ 1 \le t \le 5$

11. $x = a\cos 2t,\ y = a\sin 2t;\ 0 \le t \le \pi/2$

12. $x = \tanh t,\ y = \operatorname{sech} t;\ 0 \le t \le 4$

13. Find the point on the line $y = 2x + 1$ that is closest to the point $(0, 3)$. What is the minimum distance between the point and the line?

14. Find parametric equations of the form $x = a_1 t + b_1$ and $y = a_2 t + b_2$ for the line through points $(1, -1)$ and $(3, 3)$.

15. An object moving along the x-axis has position $s(t) = t^2 - 6t + 8$.

(a) Find the velocity and acceleration.

(b) When is the object moving forward?

16. An object initially at rest at position $x = 20$ has acceleration $a = 2$.

(a) Find the velocity and position.

(b) When will the object reach position 100?

In Problems 17–20, sketch a plot of the given conic section.

17. $8x = y^2$

18. $\dfrac{x^2}{4} + \dfrac{y^2}{9} = 1$

19. $x^2 - 4y^2 = 0$

20. $x^2 - y^2 = 4$

In Problems 21–24, sketch a graph of the given polar equation.

21. $r = 2$

22. $\theta = \pi/6$

23. $r = 4\sin\theta$

24. $r = \dfrac{1}{1 + \frac{1}{2}\cos\theta}$

CHAPTER 11

Geometry in Space and Vectors

11.1 Cartesian Coordinates in Three-Space
11.2 Vectors
11.3 The Dot Product
11.4 The Cross Product
11.5 Vector-Valued Functions and Curvilinear Motion
11.6 Lines and Tangent Lines in Three-Space
11.7 Curvature and Components of Acceleration
11.8 Surfaces in Three-Space
11.9 Cylindrical and Spherical Coordinates

11.1 Cartesian Coordinates in Three-Space

We have reached an important transition point in our study of calculus. Until now, we have been traveling across that broad flat expanse known as the Euclidean plane, or two-space. The concepts of calculus have been applied to functions of a single variable, functions whose graphs can be drawn in the plane. We are now going to study calculus in three dimensions. All the familiar ideas (such as limit, derivative, integral) are to be explored again from a loftier perspective.

To begin, consider three mutually perpendicular coordinate lines (the x-, y-, and z-axes) with their zero points at a common point O, called the *origin*. Although these lines can be oriented in any way one pleases, we follow a custom in thinking of the y- and z-axes as lying in the plane of the paper with their positive directions to the right and upward, respectively. The x-axis is then perpendicular to the paper, and we suppose its positive end to point toward us, thus forming a **right-handed system.** We call it right-handed because, if the fingers of the right hand are curled so that they curve from the positive x-axis toward the positive y-axis, the thumb points in the direction of the positive z-axis (Figure 1).

The three axes determine three planes, the yz-, xz-, and xy-planes, which divide space into eight octants (Figure 2). To each point P in space corresponds an ordered triple of numbers (x, y, z), its **Cartesian coordinates,** which measure its directed distances from the three planes (Figure 3).

Plotting points in the first octant (the octant where all three coordinates are positive) is relatively easy. In Figures 4 and 5, we illustrate something more difficult by plotting two points from other octants, the points $P(2, -3, 4)$ and $Q(-3, 2, -5)$.

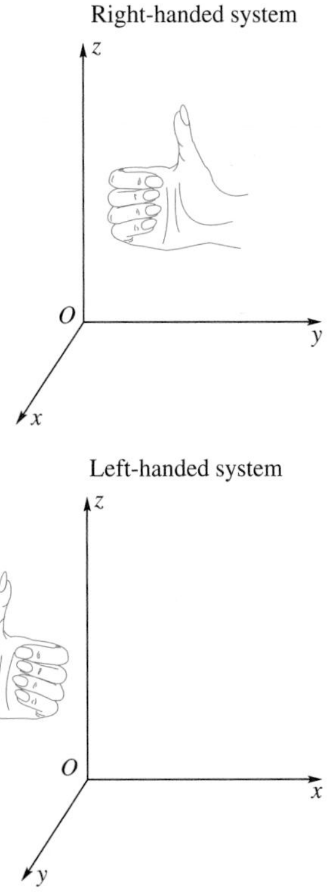

Figure 1

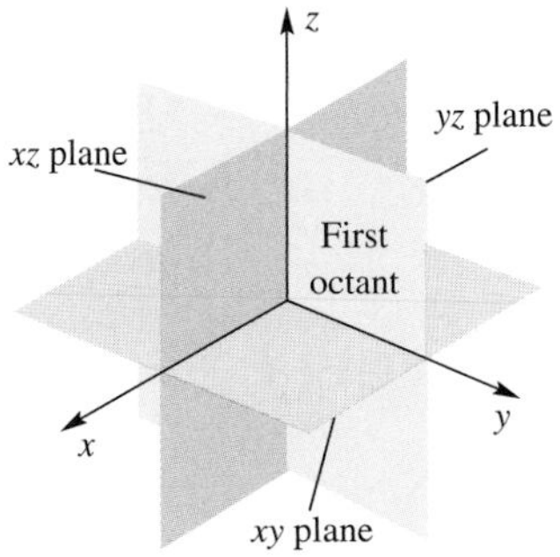

Figure 2

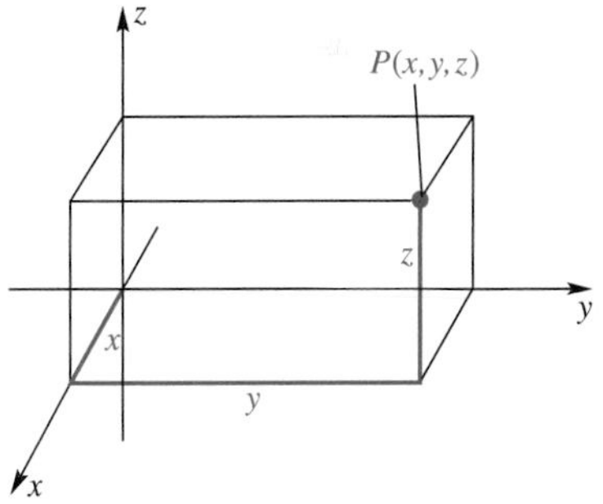

Figure 3

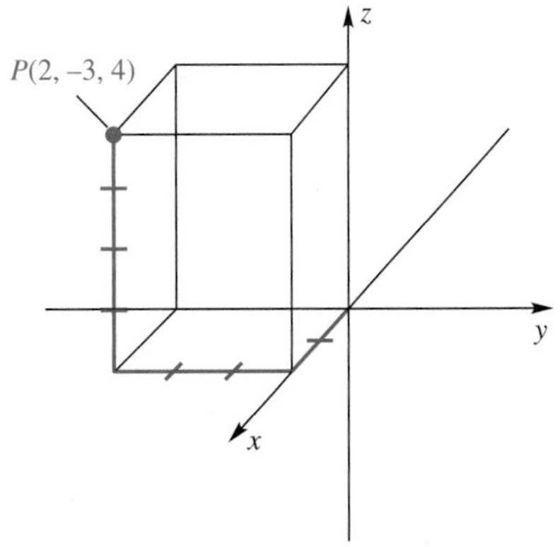

Figure 4

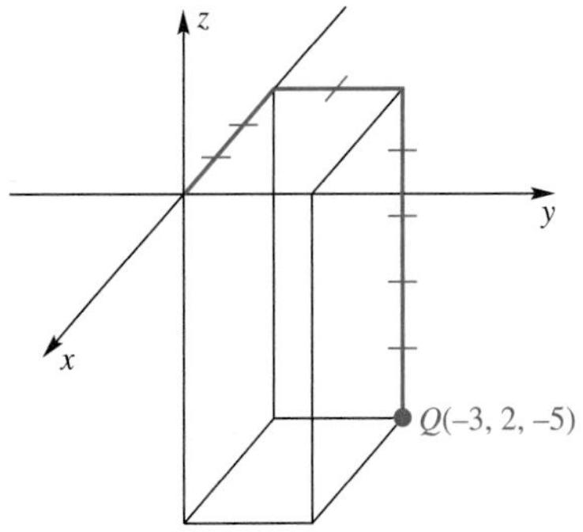

Figure 5

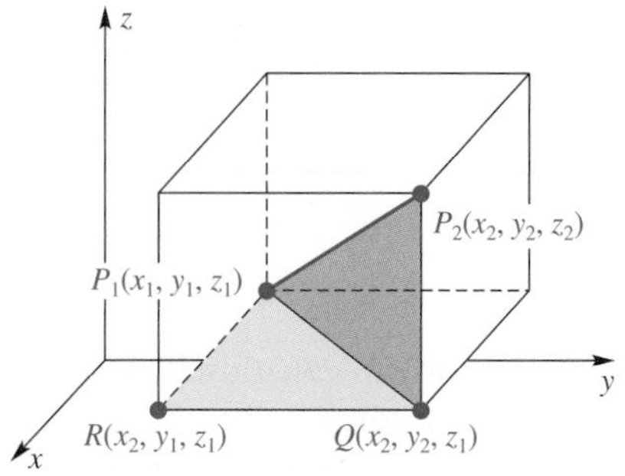

Figure 6

The Distance Formula Consider two points $P_1(x_1, y_1, z_1)$ and $P_2(x_2, y_2, z_2)$ in three-space ($x_1 \neq x_2, y_1 \neq y_2, z_1 \neq z_2$). They determine a **parallelepiped** (i.e., a rectangular box) with P_1 and P_2 as opposite vertices and with edges parallel to the coordinate axes (Figure 6). The triangles P_1QP_2 and P_1RQ are right triangles and, by the Pythagorean Theorem,

$$|P_1P_2|^2 = |P_1Q|^2 + |QP_2|^2$$

and

$$|P_1Q|^2 = |P_1R|^2 + |RQ|^2$$

Thus,

$$\begin{aligned}|P_1P_2|^2 &= |P_1R|^2 + |RQ|^2 + |QP_2|^2 \\ &= (x_2 - x_1)^2 + (y_2 - y_1)^2 + (z_2 - z_1)^2\end{aligned}$$

This gives us the **Distance Formula** in three-space, which applies even if some coordinates are identical.

$$|P_1P_2| = \sqrt{(x_2 - x_1)^2 + (y_2 - y_1)^2 + (z_2 - z_1)^2}$$

EXAMPLE 1 Find the distance between the points $P(2, -3, 4)$ and $Q(-3, 2, -5)$, which were plotted in Figures 4 and 5.

SOLUTION

$$|PQ| = \sqrt{(-3 - 2)^2 + (2 + 3)^2 + (-5 - 4)^2} = \sqrt{131} \approx 11.45 \quad \blacksquare$$

Spheres and Their Equations It is a small step from the Distance Formula to the equation of a sphere. By a **sphere,** we mean the set of all points in three-dimensional space that are a constant distance (the radius) from a fixed point (the center). (Recall that a circle is defined as the set of points *in a plane* that are a constant distance from a fixed point.) In fact, if (x, y, z) is a point on the sphere of radius r centered at (h, k, l), then (see Figure 7)

$$(x - h)^2 + (y - k)^2 + (z - l)^2 = r^2$$

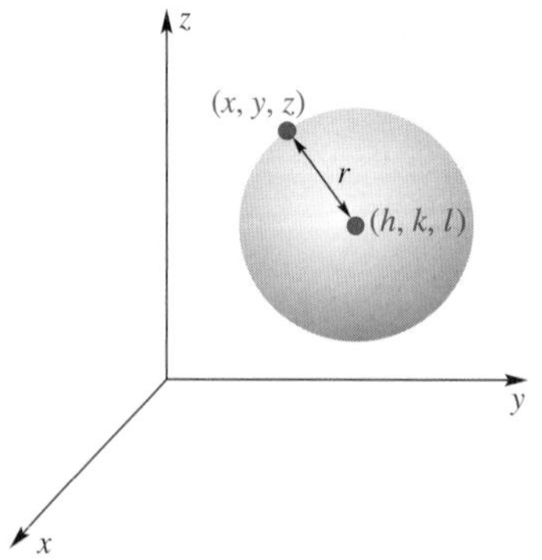

Figure 7

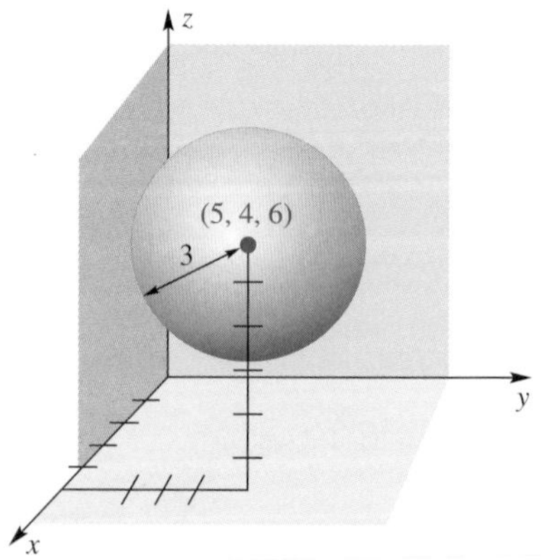

Figure 8

We call this the **standard equation of a sphere.**

In expanded form, the boxed equation may be written as

$$x^2 + y^2 + z^2 + Gx + Hy + Iz + J = 0$$

Conversely, the graph of any equation of this form is either a sphere, a point (a degenerate sphere), or the empty set. To see why, consider the following example.

What Is a Sphere?

We have defined a sphere to be the set of points a given distance away from some point, that is, those points (x, y, z) satisfying $(x - h)^2 + (y - k)^2 + (z - l)^2 = r^2$. Just as by "circle" we sometimes mean the points on and inside the circle's boundary (e.g., when we talk about the "area" of a circle being πr^2), there are times when by "sphere" we mean the boundary together with the interior. (This is sometimes called a ball or a solid sphere.) In other words, we sometimes mean the set of points satisfying $(x - h)^2 + (y - k)^2 + (z - l)^2 \leq r^2$. When we say that the volume of a sphere is $\frac{4}{3}\pi r^3$, we of course mean this latter interpretation. The context of a problem will usually dictate which "sphere" we are talking about.

EXAMPLE 2 Find the center and radius of the sphere with equation

$$x^2 + y^2 + z^2 - 10x - 8y - 12z + 68 = 0$$

and sketch its graph.

SOLUTION We use the process of completing the square.

$$(x^2 - 10x + \quad) + (y^2 - 8y + \quad) + (z^2 - 12z + \quad) = -68$$

$$(x^2 - 10x + 25) + (y^2 - 8y + 16) + (z^2 - 12z + 36) = -68 + 25 + 16 + 36$$

$$(x - 5)^2 + (y - 4)^2 + (z - 6)^2 = 9$$

Thus, the equation represents a sphere with center at (5, 4, 6) and radius 3. Its graph is shown in Figure 8. ■

If, after completing the square in Example 2, the equation had been

$$(x - 5)^2 + (y - 4)^2 + (z - 6)^2 = 0$$

then the graph would be the single point (5, 4, 6); if the right side were negative, the graph would be the empty set.

Another simple result that follows from the Distance Formula is the **Midpoint Formula.** If $P_1(x_1, y_1, z_1)$ and $P_2(x_2, y_2, z_2)$ are end points of a line segment, then the midpoint $M(m_1, m_2, m_3)$ has coordinates

$$m_1 = \frac{x_1 + x_2}{2}, \qquad m_2 = \frac{y_1 + y_2}{2}, \qquad m_3 = \frac{z_1 + z_2}{2}$$

In other words, to find the coordinates of the midpoint of a segment, simply take the average of corresponding coordinates of the end points.

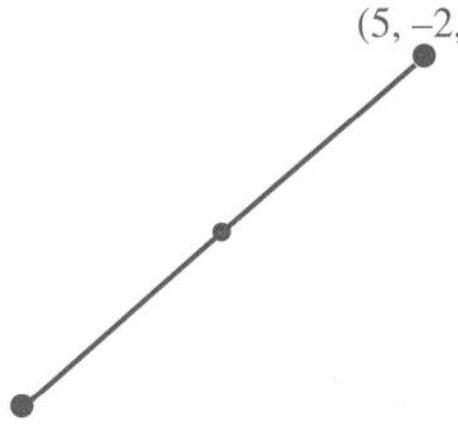

Figure 9

EXAMPLE 3 Find the equation of the sphere that has the line segment joining (−1, 2, 3) and (5, −2, 7) as a diameter (Figure 9).

SOLUTION The center of this sphere is at the midpoint of the segment, that is, at (2, 0, 5); the radius r satisfies

$$r^2 = (5 - 2)^2 + (-2 - 0)^2 + (7 - 5)^2 = 17$$

We conclude that the equation of the sphere is

$$(x - 2)^2 + y^2 + (z - 5)^2 = 17$$

■

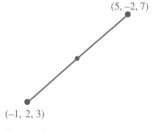

Graphs in Three-Space It was natural to consider a quadratic equation first because of its relation to the Distance Formula. But, presumably, a **linear equation** in x, y, and z, that is, an equation of the form

$$Ax + By + Cz = D, \quad A^2 + B^2 + C^2 \neq 0$$

should be even easier to analyze. (Note that $A^2 + B^2 + C^2 \neq 0$ is a compact way of saying that A, B, and C are not all zero.) As a matter of fact, we will show in Section 11.3 that the graph of a linear equation is a plane. Taking this for granted for now, let's consider how we might graph such an equation.

If, as will often be the case, the plane intersects the three axes, we begin by finding these intersection points; that is, we find the x-, y-, and z-intercepts. These three points determine the plane and allow us to draw the (coordinate-plane) **traces,** which are the lines of intersection of that plane with the coordinate planes. Then, with just a bit of artistry, we can shade in the plane.

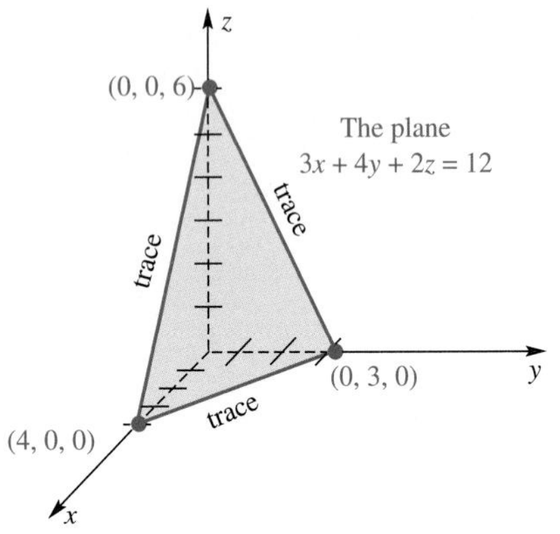

Figure 10

EXAMPLE 4 Sketch the graph of $3x + 4y + 2z = 12$.

SOLUTION To find the x-intercept, set y and z equal to zero and solve for x, obtaining $x = 4$. The corresponding point is $(4, 0, 0)$. Similarly, the y- and z-intercepts are $(0, 3, 0)$ and $(0, 0, 6)$. Next, connect these points by line segments to get the traces. Then shade in (the first octant part of) the plane, thereby obtaining the result shown in Figure 10. ■

What if the plane does not intersect all three axes? This will happen, for example, if one of the variables in the equation of the plane is missing (i.e., has a zero coefficient).

EXAMPLE 5 Sketch the graph of the linear equation

$$2x + 3y = 6$$

in three-space.

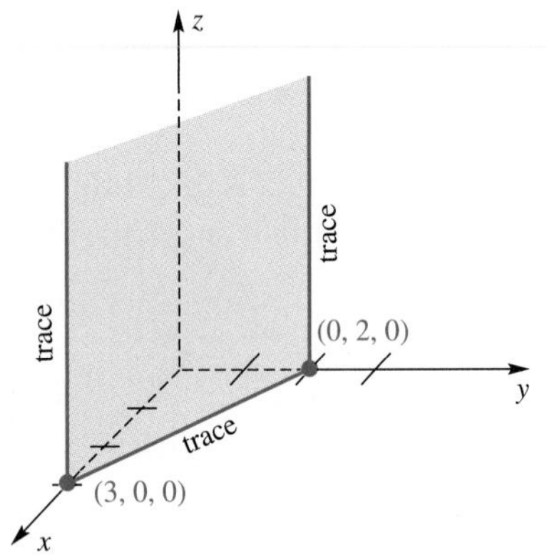

Figure 11

SOLUTION The x- and y-intercepts are $(3, 0, 0)$ and $(0, 2, 0)$, respectively, and these points determine the trace in the xy-plane. The plane never crosses the z-axis (x and y cannot both be 0), and so the plane is parallel to the z-axis. We have sketched the graph in Figure 11. ■

Notice that in each of our examples the graph of an equation in three-space was a *surface*. This contrasts with the two-space case, where the graph of an equation was usually a *curve*. We will have a good deal more to say about graphing equations and the corresponding surfaces in Section 11.8.

Curves in Three-Space

We saw parametrized curves in the plane in Section 5.4. This concept generalizes easily to three dimensions. A curve in three-space is determined by the parametric equations

$$x = f(t), \quad y = g(t), \quad z = h(t); \quad a \le t \le b$$

We say that a curve is **smooth** if $f'(t)$, $g'(t)$, and $h'(t)$ exist and are not simultaneously zero.

The concept of arc length also generalizes easily to curves in three-space. For the parametric curve defined above, the arc length is

$$L = \int_a^b \sqrt{[f'(t)]^2 + [g'(t)]^2 + [h'(t)]^2}\, dt$$

EXAMPLE 6 An object's position at time t is given by the parametrically defined curve $x = \cos t$, $y = \sin t$, $z = t/\pi$ for $0 \le t \le 2\pi$. Sketch this curve and find its arc length.

SOLUTION We begin by making a table of values of t, x, y, and z; then we connect the dots in three-space; the curve is shown in Figure 12. The arc length is

$$\begin{aligned} L &= \int_0^{2\pi} \sqrt{(-\sin t)^2 + (\cos t)^2 + (1/\pi)^2}\, dt \\ &= \int_0^{2\pi} \sqrt{\sin^2 t + \cos^2 t + 1/\pi^2}\, dt \\ &= \int_0^{2\pi} \sqrt{1 + 1/\pi^2}\, dt \\ &= 2\pi\sqrt{1 + 1/\pi^2} \end{aligned}$$

■

t	x	y	z
0	1	0	0
$\frac{\pi}{4}$	$\frac{\sqrt{2}}{2}$	$\frac{\sqrt{2}}{2}$	$\frac{1}{4}$
$\frac{\pi}{2}$	0	1	$\frac{1}{2}$
$\frac{3\pi}{4}$	$-\frac{\sqrt{2}}{2}$	$\frac{\sqrt{2}}{2}$	$\frac{3}{4}$
π	-1	0	1
$\frac{5\pi}{4}$	$-\frac{\sqrt{2}}{2}$	$-\frac{\sqrt{2}}{2}$	$\frac{5}{4}$
$\frac{3\pi}{2}$	0	-1	$\frac{3}{2}$
$\frac{7\pi}{4}$	$\frac{\sqrt{2}}{2}$	$-\frac{\sqrt{2}}{2}$	$\frac{7}{4}$
2π	1	0	2

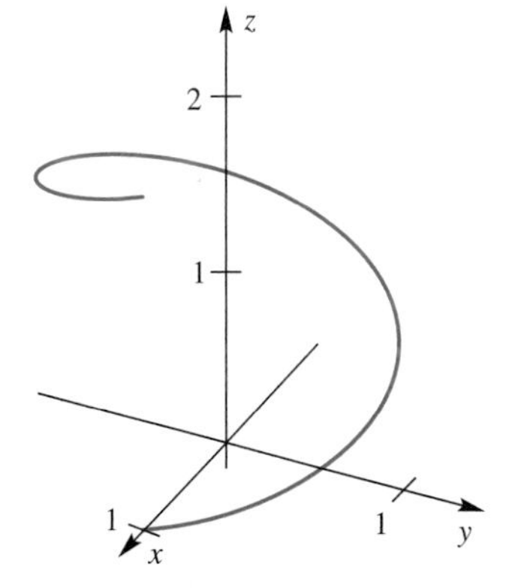

Figure 12

The curve in Example 6 is called a **helix.** Notice that if we ignore (for a moment) the motion in the z-dimension, the object is in uniform circular motion. Introducing back the motion in the z-dimension, which is up with constant speed, we see that the object is going around and around as it moves upward, much like a spiral staircase.

≈ Here is another way to obtain the length of this curve. The helix lies entirely on the surface of a right circular cylinder as shown in Figure 13. Now imagine that the cylinder is cut as indicated and that the cylinder is "peeled" back to make a rectangle. The helix will become a diagonal of the rectangle, so it will have length $\sqrt{4 + 4\pi^2} = \sqrt{4\pi^2(1 + 1/\pi^2)} = 2\pi\sqrt{1 + 1/\pi^2}$.

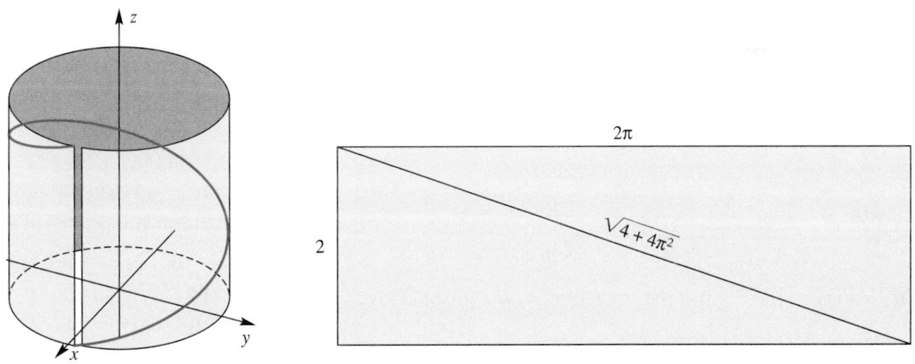

Figure 13

Concepts Review

1. The numbers $x, y,$ and z in (x, y, z) are called the _____ of a point in three-space.

2. The distance between the points $(-1, 3, 5)$ and (x, y, z) is _____.

3. The equation $(x + 1)^2 + (y - 3)^2 + (z - 5)^2 = 16$ determines a sphere with center _____ and radius _____.

4. The graph of $3x - 2y + 4z = 12$ is a _____ with x-intercept _____, y-intercept _____, and z-intercept _____.

Problem Set 11.1

1. Plot the points whose coordinates are $(1, 2, 3)$, $(2, 0, 1)$, $(-2, 4, 5)$, $(0, 3, 0)$, and $(-1, -2, -3)$. If appropriate, show the "box" as in Figures 4 and 5.

2. Follow the directions of Problem 1 for $(\sqrt{3}, -3, 3)$, $(0, \pi, -3)$, $(-2, \frac{1}{3}, 2)$, and $(0, 0, e)$.

3. What is peculiar to the coordinates of all points in the yz-plane? On the z-axis?

4. What is peculiar to the coordinates of all points in the xz-plane? On the y-axis?

5. Find the distance between the following pairs of points.

(a) $(6, -1, 0)$ and $(1, 2, 3)$

(b) $(-2, -2, 0)$ and $(2, -2, -3)$

(c) $(e, \pi, 0)$ and $(-\pi, -4, \sqrt{3})$

6. Show that $(4, 5, 3)$, $(1, 7, 4)$, and $(2, 4, 6)$ are vertices of an equilateral triangle.

7. Show that $(2, 1, 6)$, $(4, 7, 9)$, and $(8, 5, -6)$ are vertices of a right triangle. *Hint:* Only right triangles satisfy the Pythagorean Theorem.

8. Find the distance from $(2, 3, -1)$ to

(a) the xy-plane, (b) the y-axis, and

(c) the origin.

9. A rectangular box has its faces parallel to the coordinate planes and has $(2, 3, 4)$ and $(6, -1, 0)$ as the end points of a main diagonal. Sketch the box and find the coordinates of all eight vertices.

10. $P(x, 5, z)$ is on a line through $Q(2, -4, 3)$ that is parallel to one of the coordinate axes. Which axis must it be and what are x and z?

11. Write the equation of the sphere with the given center and radius.

(a) $(1, 2, 3); 5$ (b) $(-2, -3, -6); \sqrt{5}$

(c) $(\pi, e, \sqrt{2}); \sqrt{\pi}$

12. Find the equation of the sphere whose center is $(2, 4, 5)$ and that is tangent to the xy-plane.

In Problems 13–16, complete the squares to find the center and radius of the sphere whose equation is given (see Example 2).

13. $x^2 + y^2 + z^2 - 12x + 14y - 8z + 1 = 0$

14. $x^2 + y^2 + z^2 + 2x - 6y - 10z + 34 = 0$

15. $4x^2 + 4y^2 + 4z^2 - 4x + 8y + 16z - 13 = 0$

16. $x^2 + y^2 + z^2 + 8x - 4y - 22z + 77 = 0$

In Problems 17–24, sketch the graphs of the given equations. Begin by sketching the traces in the coordinate planes (see Examples 4 and 5).

17. $2x + 6y + 3z = 12$ **18.** $3x - 4y + 2z = 24$

19. $x + 3y - z = 6$ **20.** $-3x + 2y + z = 6$

21. $x + 3y = 8$ **22.** $3x + 4z = 12$

23. $x^2 + y^2 + z^2 = 9$ **24.** $(x - 2)^2 + y^2 + z^2 = 4$

In Problems 25–32, find the arc length of the given curve.

25. $x = t, y = t, z = 2t; 0 \leq t \leq 2$

26. $x = t/4, y = t/3, z = t/2; 1 \leq t \leq 3$

27. $x = t^{3/2}, y = 3t, z = 4t; 1 \leq t \leq 4$

28. $x = t^{3/2}, y = t^{3/2}, z = t; 2 \leq t \leq 4$

29. $x = t^2, y = (4/3)t^{3/2}, z = t; 0 \leq t \leq 8$

30. $x = t^2, y = \frac{4\sqrt{3}}{3}t^{3/2}, z = 3t; 1 \leq t \leq 4$

31. $x = 2\cos t, y = 2\sin t, z = 3t; -\pi \leq t \leq \pi$

32. $x = 2\cos t, y = 2\sin t, z = t/20; 0 \leq t \leq 8\pi$

CAS *In Problems 33–36, set up a definite integral for the arc length of the given curve. Use the Parabolic Rule with $n = 10$ or a CAS to approximate the integral.*

33. $x = \sqrt{t}, y = t, z = t; 1 \leq t \leq 6$

34. $x = t, y = t^2, z = t^3; 1 \leq t \leq 2$

35. $x = 2\cos t, y = \sin t, z = t; 0 \leq t \leq 6\pi$

36. $x = \sin t, y = \cos t, z = \sin t; 0 \leq t \leq 2\pi$

37. Find the equation of the sphere that has the line segment joining $(-2, 3, 6)$ and $(4, -1, 5)$ as a diameter (see Example 3).

38. Find the equations of the tangent spheres of equal radii whose centers are $(-3, 1, 2)$ and $(5, -3, 6)$.

39. Find the equation of the sphere that is tangent to the three coordinate planes if its radius is 6 and its center is in the first octant.

40. Find the equation of the sphere with center $(1, 1, 4)$ that is tangent to the plane $x + y = 12$.

41. Describe the graph in three-space of each equation.

(a) $z = 2$ (b) $x = y$

(c) $xy = 0$ (d) $xyz = 0$

(e) $x^2 + y^2 = 4$ (f) $z = \sqrt{9 - x^2 - y^2}$

42. The sphere $(x - 1)^2 + (y + 2)^2 + (z + 1)^2 = 10$ intersects the plane $z = 2$ in a circle. Find the circle's center and radius.

43. An object's position P changes so that its distance from $(1, 2, -3)$ is always twice its distance from $(1, 2, 3)$. Show that P is on a sphere and find its center and radius.

44. An object's position P changes so that its distance from $(1, 2, -3)$ always equals its distance from $(2, 3, 2)$. Find the equation of the plane on which P lies.

45. The solid spheres $(x - 1)^2 + (y - 2)^2 + (z - 1)^2 \leq 4$ and $(x - 2)^2 + (y - 4)^2 + (z - 3)^2 \leq 4$ intersect in a solid. Find its volume.

46. Do Problem 45 assuming that the second solid sphere is $(x - 2)^2 + (y - 4)^2 + (z - 3)^2 \leq 9$.

CAS **47.** The curve defined by $x = a\cos t, y = a\sin t, z = ct$ is a helix. Hold a fixed and use a CAS to obtain a parmetric plot of the helix for various values of c. What effect does c have on the curve?

CAS **48.** For the helix described in Problem 47, hold c fixed and use a CAS to obtain a parametric plot for various values of a. What effect does a have on the curve?

Answers to Concepts Review: **1.** coordinates **2.** $\sqrt{(x + 1)^2 + (y - 3)^2 + (z - 5)^2}$ **3.** $(-1, 3, 5); 4$ **4.** plane; 4; −6; 3

11.2 Vectors

Many quantities that occur in science (e.g., length, mass, volume, and electric charge) can be specified by giving a single number. These quantities (and the numbers that measure them) are called **scalars.** Other quantities, such as velocity, force, torque, and displacement, require both a magnitude and a direction for complete specification. We call such quantities **vectors** and represent them by arrows (directed line segments). The length of the arrow represents the **magnitude,** or length, of the vector; its direction is the **direction** of the vector. The vector in Figure 1 has length 2.3 units and direction 30° north of east (or 30° from the positive x-axis).

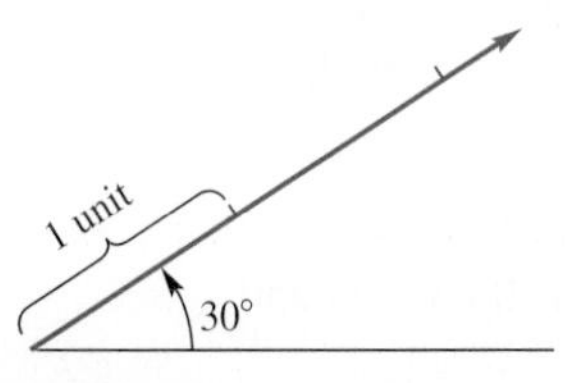

Figure 1

Arrows that we draw, like those shot from a bow, have two ends. There is the feather end (the initial point), called the **tail,** and the pointed end (the terminal

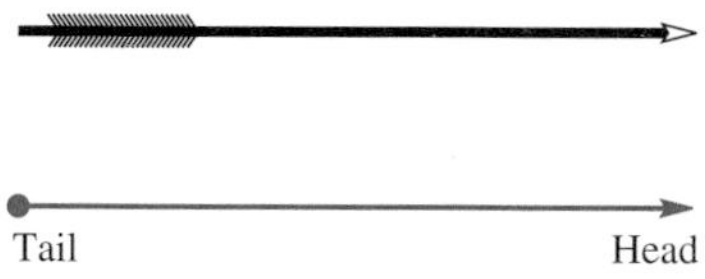

Figure 2

point), called the **head,** or tip (Figure 2). Two vectors are considered to be **equivalent** if they have the same magnitude and direction (Figure 3). We shall symbolize vectors by boldface letters, such as **u** and **v**. Since this is hard to accomplish in normal writing, you might use $\vec{u}$ and $\vec{v}$. The magnitude, or length, of a vector **u** is symbolized by $\|\mathbf{u}\|$.

In general, we think of vectors as being three-dimensional; that is, their initial and terminal points are points in three-space. There are many applications, however, where the vectors lie entirely in the *xy*-plane. The context of a problem should indicate whether the vectors are two- or three-dimensional.

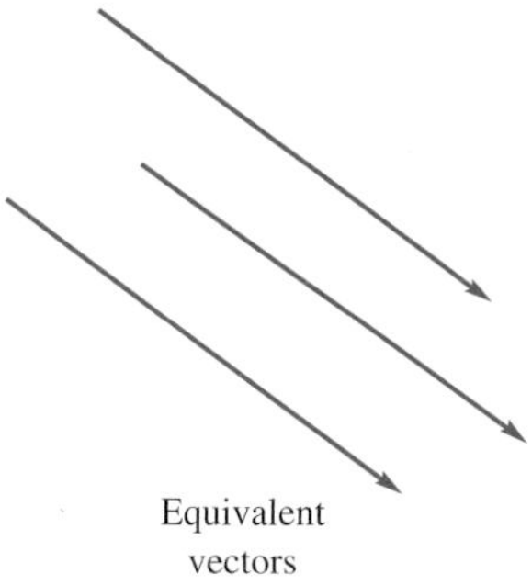

Figure 3

Operations on Vectors To find the **sum,** or **resultant,** of **u** and **v**, move **v** without changing its magnitude or direction until its tail coincides with the head of **u**. Then $\mathbf{u} + \mathbf{v}$ is the vector connecting the tail of **u** to the head of **v**. This method (called the *Triangle Law*) is illustrated in the left half of Figure 4.

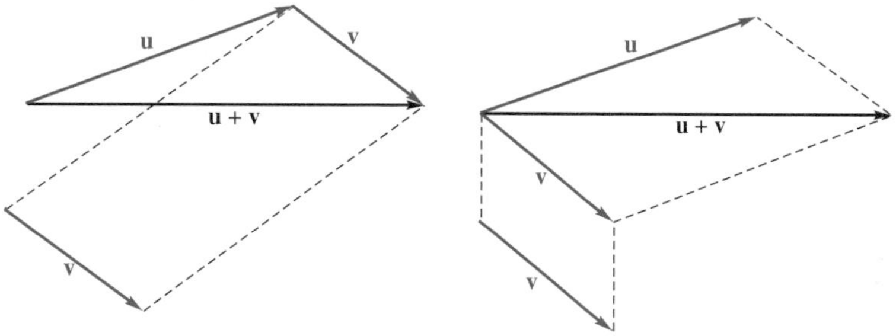

Two equivalent ways of adding vectors

Figure 4

As an alternative way to find $\mathbf{u} + \mathbf{v}$, move **v** so that its tail coincides with that of **u**. Then $\mathbf{u} + \mathbf{v}$ is the vector with this common tail and coinciding with the diagonal of the parallelogram that has **u** and **v** as sides. This method (called the *Parallelogram Law*) is illustrated on the right in Figure 4.

These two methods are equivalent ways to define what we mean by the sum of two vectors. You should convince yourself that vector addition is commutative and associative; that is,

$$\mathbf{u} + \mathbf{v} = \mathbf{v} + \mathbf{u}$$

$$(\mathbf{u} + \mathbf{v}) + \mathbf{w} = \mathbf{u} + (\mathbf{v} + \mathbf{w})$$

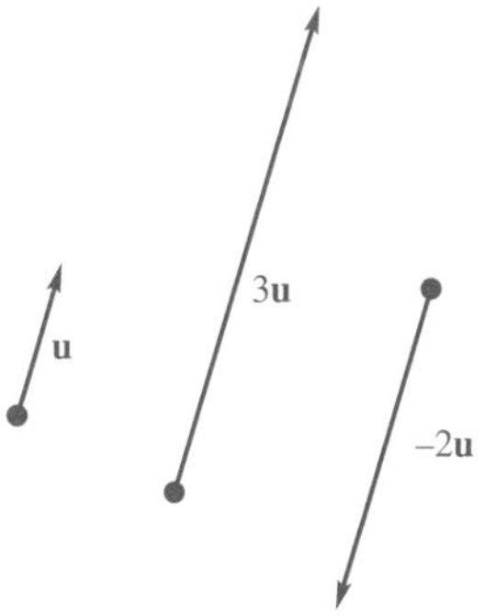

Figure 5

If **u** is a vector, then $3\mathbf{u}$ is the vector with the same direction as **u** but three times as long; $-2\mathbf{u}$ is twice as long but oppositely directed (Figure 5). In general, $c\mathbf{u}$, called a **scalar multiple** of **u**, has magnitude $|c|$ times that of **u** and is similarly or oppositely directed, depending on whether c is positive or negative. In particular, $(-1)\mathbf{u}$ (usually written $-\mathbf{u}$) has the same length as **u**, but opposite direction. It is called the **negative** of **u** because, when we add it to **u**, the result is a vector that is nothing more than a point. This latter vector (the only vector without a well-defined direction) is called the **zero vector** and is denoted by $\mathbf{0}$. It is the identity element for addition; that is, $\mathbf{u} + \mathbf{0} = \mathbf{0} + \mathbf{u} = \mathbf{u}$. Finally, subtraction is defined by

$$\mathbf{u} - \mathbf{v} = \mathbf{u} + (-\mathbf{v})$$

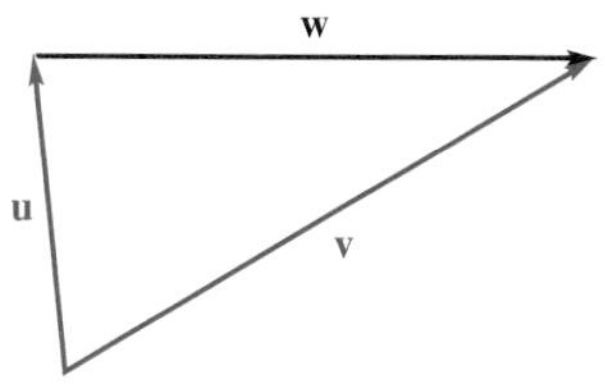

Figure 6

EXAMPLE 1 In Figure 6, express **w** in terms of **u** and **v**.

SOLUTION Since $\mathbf{u} + \mathbf{w} = \mathbf{v}$, it follows that

$$\mathbf{w} = \mathbf{v} - \mathbf{u}$$ ■

If P and Q are points in the plane, then $\overrightarrow{PQ}$ denotes the vector with tail at P and head at Q.

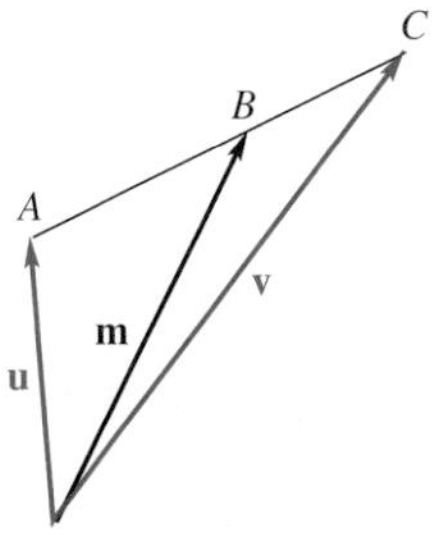

Figure 7

EXAMPLE 2 In Figure 7, $\overrightarrow{AB} = \frac{2}{3}\overrightarrow{AC}$. Express **m** in terms of **u** and **v**.

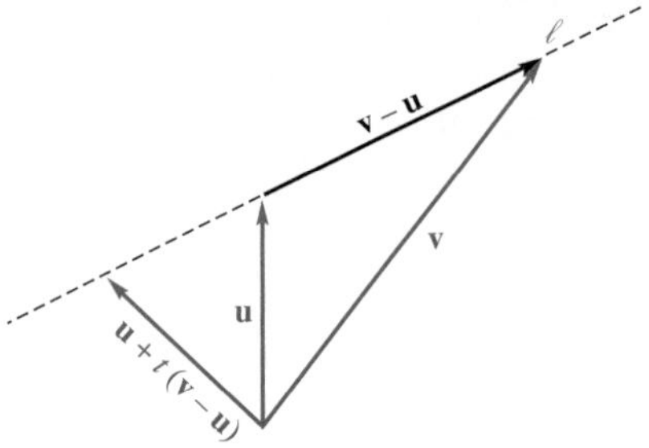

Figure 8

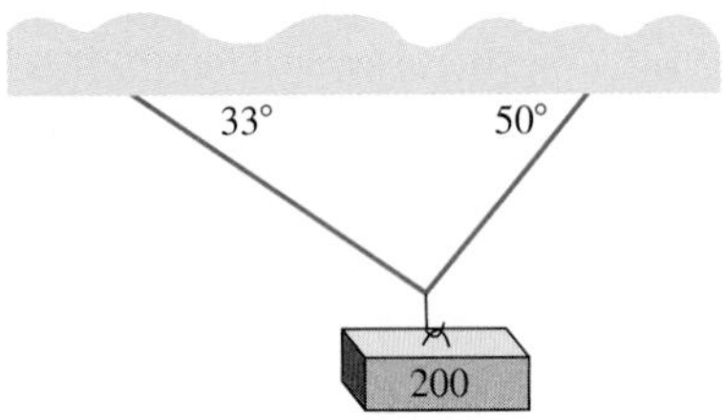

Figure 9

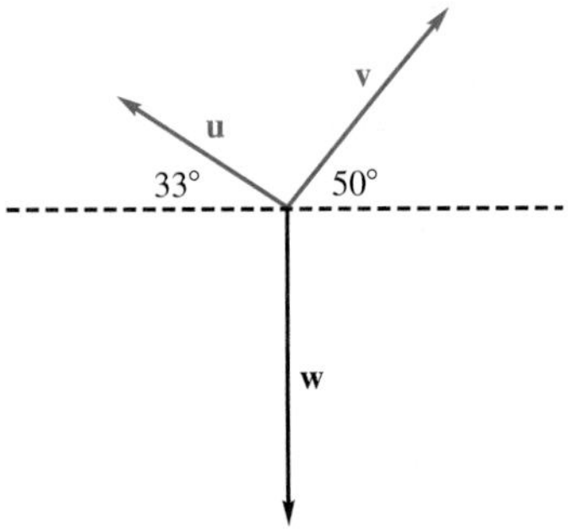

Figure 10

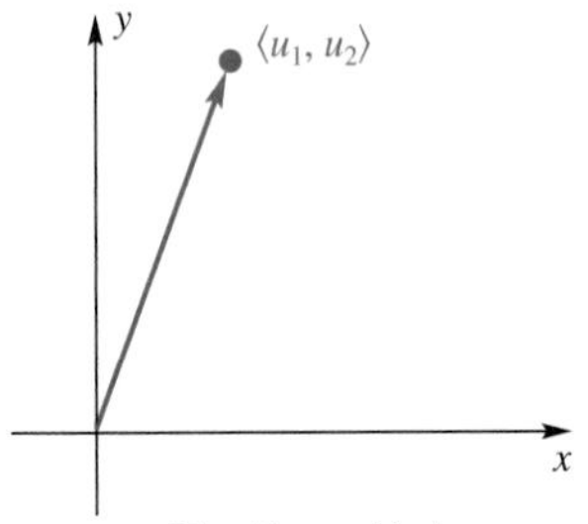

Identify **u** with the ordered pair $\langle u_1, u_2 \rangle$.

Figure 11

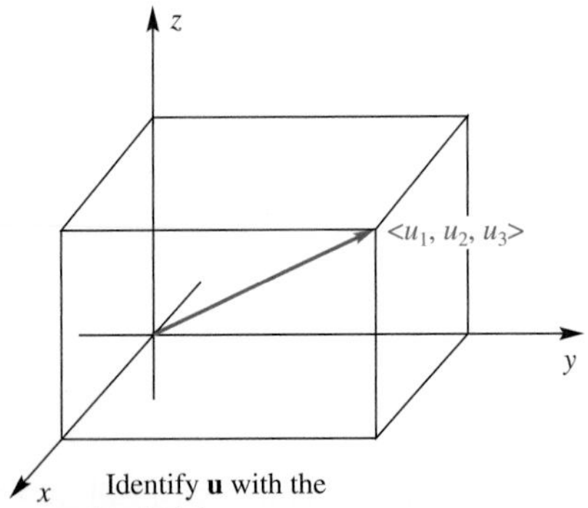

Identify **u** with the ordered triple $\langle u_1, u_2, u_3 \rangle$.

Figure 12

SOLUTION

$$\mathbf{m} = \mathbf{u} + \overrightarrow{AB} = \mathbf{u} + \tfrac{2}{3}\overrightarrow{AC} = \mathbf{u} + \tfrac{2}{3}(\mathbf{v} - \mathbf{u}) = \tfrac{1}{3}\mathbf{u} + \tfrac{2}{3}\mathbf{v}$$

More generally, if $\overrightarrow{AB} = t\overrightarrow{AC}$, where $0 < t < 1$, then

$$\mathbf{m} = (1 - t)\mathbf{u} + t\mathbf{v}$$

■

The expression just obtained for **m** can also be written as

$$\mathbf{u} + t(\mathbf{v} - \mathbf{u})$$

If we allow t to range over all scalars, we obtain the set of all vectors with tails at the same point as the tail of **u** and heads on the line ℓ (see Figure 8). This fact will be important to us later in describing lines using vector language.

An Application A force has both a magnitude and a direction. If two forces **u** and **v** act at a point, the resultant force at the point is the vector sum of the two forces.

EXAMPLE 3 A weight of 200 newtons is supported by two wires, as shown in Figure 9. Find the magnitude of the tension in each wire.

SOLUTION All forces are in one plane, so the vectors in this problem are two-dimensional. The weight **w** and the two tensions **u** and **v** are forces that behave as vectors (Figure 10). Each of these vectors can be expressed as a sum of a horizontal and a vertical component. The weight is in equilibrium, so (1) the magnitude of the leftward force must equal the magnitude of the rightward force, and (2) the magnitude of the upward force must equal the magnitude of the downward force. In other words, the net force is zero. Thus,

(1) $$\|\mathbf{u}\|\cos 33° = \|\mathbf{v}\|\cos 50°$$

(2) $$\|\mathbf{u}\|\sin 33° + \|\mathbf{v}\|\sin 50° = \|\mathbf{w}\| = 200$$

When we solve (1) for $\|\mathbf{v}\|$ and substitute in (2), we get

$$\|\mathbf{u}\|\sin 33° + \frac{\|\mathbf{u}\|\cos 33°}{\cos 50°}\sin 50° = 200$$

or

$$\|\mathbf{u}\| = \frac{200}{\sin 33° + \cos 33° \tan 50°} \approx 129.52 \text{ newtons}$$

Then

$$\|\mathbf{v}\| = \frac{\|\mathbf{u}\|\cos 33°}{\cos 50°} \approx \frac{129.52 \cos 33°}{\cos 50°} \approx 168.99 \text{ newtons}$$

■

Algebraic Approach to Vectors For a given vector **u** in the plane we choose as its representative the arrow that has its tail at the origin (Figure 11). This arrow is uniquely determined by the coordinates u_1 and u_2 of its head; that is, the vector **u** is completely described by the ordered pair $\langle u_1, u_2 \rangle$. The numbers u_1 and u_2 are called the **components** of the vector **u**. We write $\langle u_1, u_2 \rangle$ rather than (u_1, u_2) to distinguish the vector originating at the origin and terminating at the point with coordinates u_1 and u_2, from the point having coordinates u_1 and u_2.

For vectors in three-space, the generalization is straightforward. We represent the vector by an arrow starting at the origin and terminating at the point with coordinates u_1, u_2, and u_3, and we denote this vector by $\langle u_1, u_2, u_3 \rangle$ (Figure 12). In the remainder of this section, we develop the properties of vectors in three dimensions; the results for vectors in two dimensions should be obvious.

The vectors $\mathbf{u} = \langle u_1, u_2, u_3 \rangle$ and $\mathbf{v} = \langle v_1, v_2, v_3 \rangle$ are equal if and only if the corresponding components are equal; that is, $u_1 = v_1$, $u_2 = v_2$, and $u_3 = v_3$. To multiply a vector **u** by a scalar c, we multiply each component by c; that is,

$$c\mathbf{u} = \mathbf{u}c = \langle cu_1, cu_2, cu_3 \rangle$$

The notation $-\mathbf{u}$ indicates the vector $(-1)\mathbf{u} = \langle -u_1, -u_2, -u_3 \rangle$. The vector with all components equal to zero is called the zero vector; that is $\mathbf{0} = \langle 0, 0, 0 \rangle$. The sum of the two vectors **u** and **v** is

$$\mathbf{u} + \mathbf{v} = \langle u_1 + v_1, u_2 + v_2, u_3 + v_3 \rangle$$

The vector $\mathbf{u} - \mathbf{v}$ is defined to be

$$\mathbf{u} - \mathbf{v} = \mathbf{u} + (-1)\mathbf{v} = \langle u_1 - v_1, u_2 - v_2, u_3 - v_3 \rangle$$

Figure 13 indicates that these definitions are equivalent to the geometric ones given earlier in this section.

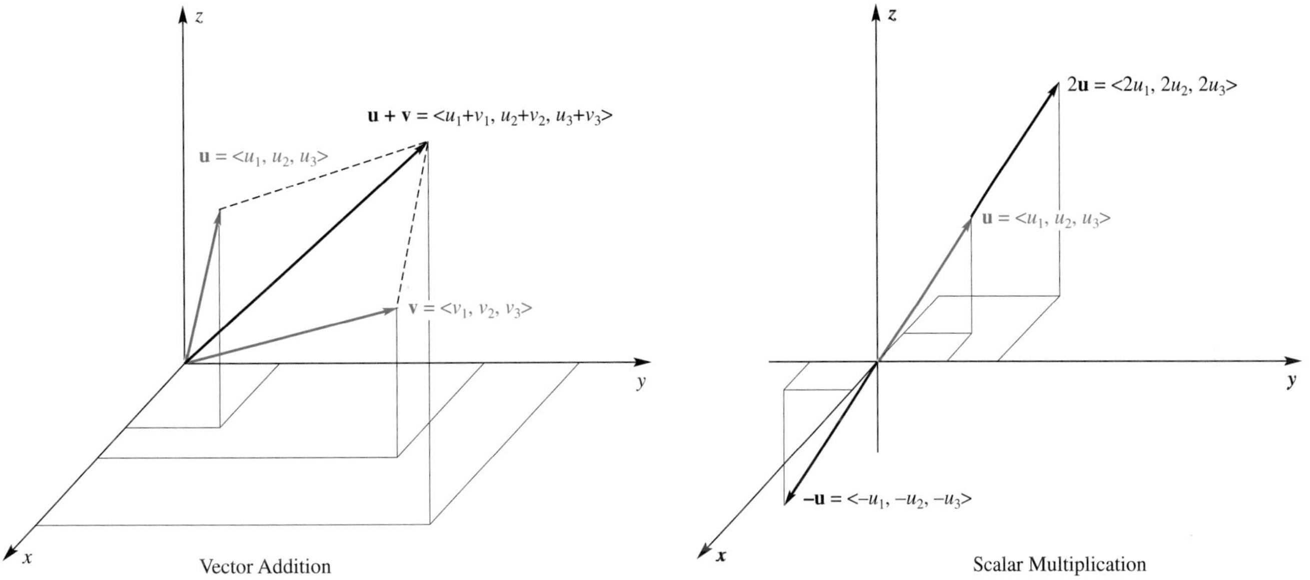

Vector Addition

Scalar Multiplication

Figure 13

Vectors in Two-Space

Here are the definitions for vectors in two-space. If $\mathbf{u} = \langle u_1, u_2 \rangle$, $\mathbf{v} = \langle v_1, v_2 \rangle$, and c is a scalar, then

$$c\mathbf{u} = \langle cu_1, cu_2 \rangle$$

$$\mathbf{u} + \mathbf{v} = \langle u_1 + v_1, u_2 + v_2 \rangle$$

$$\mathbf{u} - \mathbf{v} = \langle u_1 - v_1, u_2 - v_2 \rangle$$

$$\mathbf{i} = \langle 1, 0 \rangle; \quad \mathbf{j} = \langle 0, 1 \rangle$$

$$\|\mathbf{u}\| = \sqrt{u_1^2 + u_2^2}$$

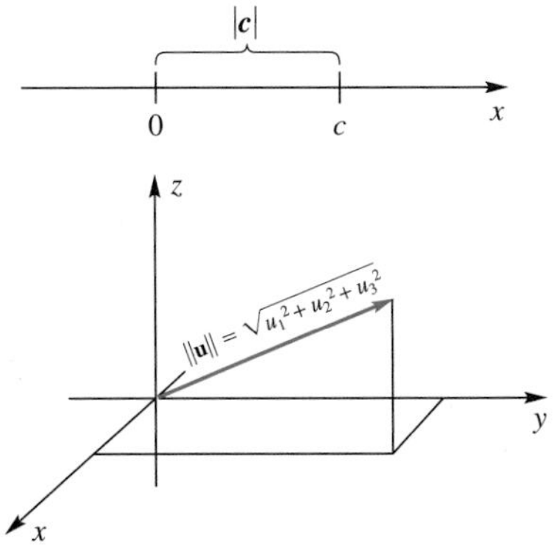

Figure 14

Three special vectors in three-space are $\mathbf{i} = \langle 1, 0, 0 \rangle$, $\mathbf{j} = \langle 0, 1, 0 \rangle$, and $\mathbf{k} = \langle 0, 0, 1 \rangle$. These are called the **standard unit vectors,** or **basis vectors.** Every vector $\mathbf{u} = \langle u_1, u_2, u_3 \rangle$ can be written in terms of **i**, **j**, and **k** as follows:

$$\mathbf{u} = \langle u_1, u_2, u_3 \rangle = u_1\mathbf{i} + u_2\mathbf{j} + u_3\mathbf{k}$$

The **magnitude** of a vector is just the length of the arrow that represents it. If the arrow begins at the origin and ends at (u_1, u_2, u_3), then its length can be readily determined from the distance formula:

$$\|\mathbf{u}\| = \sqrt{(u_1 - 0)^2 + (u_2 - 0)^2 + (u_3 - 0)^2} = \sqrt{u_1^2 + u_2^2 + u_3^2}$$

Just as $|c|$ gives the distance from the origin to a point c on the number line, $\|\mathbf{u}\|$ gives the distance from the origin to the point in space whose ordered triple is $\mathbf{u} = \langle u_1, u_2, u_3 \rangle$ (Figure 14). Using this algebraic interpretation of vectors, the following rules for operating with vectors can be be easily established.

Theorem A

For any vectors **u**, **v**, and **w**, and any scalars a and b, the following relationships hold.

1. $\mathbf{u} + \mathbf{v} = \mathbf{v} + \mathbf{u}$
2. $(\mathbf{u} + \mathbf{v}) + \mathbf{w} = \mathbf{u} + (\mathbf{v} + \mathbf{w})$
3. $\mathbf{u} + \mathbf{0} = \mathbf{0} + \mathbf{u} = \mathbf{u}$
4. $\mathbf{u} + (-\mathbf{u}) = \mathbf{0}$
5. $a(b\mathbf{u}) = (ab)\mathbf{u}$
6. $a(\mathbf{u} + \mathbf{v}) = a\mathbf{u} + a\mathbf{v}$
7. $(a + b)\mathbf{u} = a\mathbf{u} + b\mathbf{u}$
8. $1\mathbf{u} = \mathbf{u}$
9. $\|a\mathbf{u}\| = |a|\|\mathbf{u}\|$

Proof We illustrate the proof by demonstrating Rules 6 and 9 for the case of three-dimensional vectors.

$$\begin{aligned} a(\mathbf{u} + \mathbf{v}) &= a(\langle u_1, u_2, u_3\rangle + \langle v_1, v_2, v_3\rangle) \\ &= a\langle u_1 + v_1, u_2 + v_2, u_3 + v_3\rangle \\ &= \langle a(u_1 + v_1), a(u_2 + v_2), a(u_3 + v_3)\rangle \\ &= \langle au_1 + av_1, au_2 + av_2, au_3 + av_3\rangle \\ &= \langle au_1, au_2, au_3\rangle + \langle av_1, av_2, av_3\rangle \\ &= a\langle u_1, u_2, u_3\rangle + a\langle v_1, v_2, v_3\rangle \\ &= a\mathbf{u} + a\mathbf{v} \end{aligned}$$

This proves Rule 6. Now, for Rule 9,

$$\begin{aligned} \|a\mathbf{u}\| &= \|\langle au_1, au_2, au_3\rangle\| \\ &= \sqrt{(au_1)^2 + (au_2)^2 + (au_3)^2} \\ &= \sqrt{a^2(u_1^2 + u_2^2 + u_3^2)} \\ &= \sqrt{a^2}\sqrt{u_1^2 + u_2^2 + u_3^2} = |a|\ \|\mathbf{u}\| \end{aligned}$$

■

EXAMPLE 4 Let $\mathbf{u} = \langle 1, 1, 2\rangle$ and $\mathbf{v} = \langle 0, -1, 2\rangle$. Find (a) $\mathbf{u} + \mathbf{v}$, and (b) $\mathbf{u} - 2\mathbf{v}$, and express them in terms of **i**, **j**, and **k**. Find (c) $\|\mathbf{u}\|$, and (d) $\|-3\mathbf{u}\|$.

SOLUTION

(a) $\mathbf{u} + \mathbf{v} = \langle 1, 1, 2\rangle + \langle 0, -1, 2\rangle = \langle 1 + 0, 1 + (-1), 2 + 2\rangle$
$= \langle 1, 0, 4\rangle = 1\mathbf{i} + 0\mathbf{j} + 4\mathbf{k} = \mathbf{i} + 4\mathbf{k}$

(b) $\mathbf{u} - 2\mathbf{v} = \langle 1, 1, 2\rangle - 2\langle 0, -1, 2\rangle = \langle 1 - 0, 1 - (-2), 2 - 4\rangle$
$= \langle 1, 3, -2\rangle = \mathbf{i} + 3\mathbf{j} - 2\mathbf{k}$

(c) $\|\mathbf{u}\| = \sqrt{1^2 + 1^2 + 2^2} = \sqrt{6}$

(d) $\|-3\mathbf{u}\| = |-3|\|\mathbf{u}\| = 3\sqrt{6}$

■

Definition Unit Vector

A vector having length one is called a **unit vector.**

Dividing a Vector by a Scalar

We will often talk of "dividing" a vector **v** by a scalar c. By this we mean we multiply the vector by the reciprocal of c. That is,

$$\frac{\mathbf{v}}{c} = \frac{1}{c}\mathbf{v}$$

provided, of course, that $c \neq 0$. The expression on the right is simply a scalar, times a vector, which we defined earlier in this section. Dividing one vector by another is, of course, nonsense.

EXAMPLE 5 Let $\mathbf{v} = \langle 4, -3\rangle$. Find $\|\mathbf{v}\|$, and find a unit vector **u** with the same direction as **v**.

SOLUTION In this problem, all vectors are two-dimensional. The length, or magnitude, of **v** is $\|\mathbf{v}\| = \sqrt{4^2 + (-3)^2} = 5$. To find **u**, we divide **v** by its length $\|\mathbf{v}\|$; that is,

$$\mathbf{u} = \frac{\mathbf{v}}{\|\mathbf{v}\|} = \frac{\langle 4, -3\rangle}{\sqrt{4^2 + (-3)^2}} = \frac{\langle 4, -3\rangle}{5} = \frac{1}{5}\langle 4, -3\rangle = \left\langle \frac{4}{5}, -\frac{3}{5}\right\rangle$$

The length of **u** is then

$$\|\mathbf{u}\| = \left\|\frac{\mathbf{v}}{\|\mathbf{v}\|}\right\| = \left\|\frac{1}{\|\mathbf{v}\|}\mathbf{v}\right\| = \left|\frac{1}{\|\mathbf{v}\|}\right|\|\mathbf{v}\| = \frac{1}{\|\mathbf{v}\|}\|\mathbf{v}\| = 1$$

■

Concepts Review

1. Vectors are distinguished from scalars in that vectors have both _____ and _____.

2. Two vectors are considered to be equivalent if _____.

3. If the tail of **v** coincides with the head of **u**, then $\mathbf{u} + \mathbf{v}$ is the vector with tail at _____ and head at _____.

4. The vector $\mathbf{u} = \langle 6, 3, 3\rangle$ has length _____ times that of the vector $\mathbf{u} = \langle 2, 1, 1\rangle$.

Problem Set 11.2

In Problems 1–4, draw the vector **w**.

1. $\mathbf{w} = \mathbf{u} + \frac{3}{2}\mathbf{v}$ **2.** $\mathbf{w} = 2\mathbf{u} - 3\mathbf{v}$

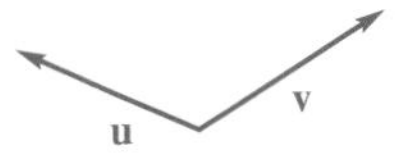

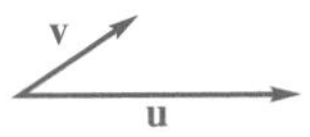

3. $\mathbf{w} = \mathbf{u}_1 + \mathbf{u}_2 + \mathbf{u}_3$ **4.** $\mathbf{w} = \mathbf{u}_1 + \mathbf{u}_2 + \mathbf{u}_3$

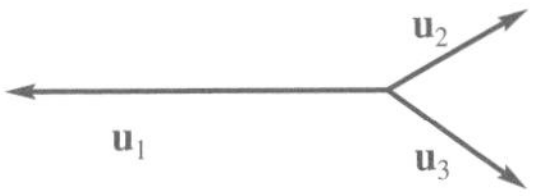

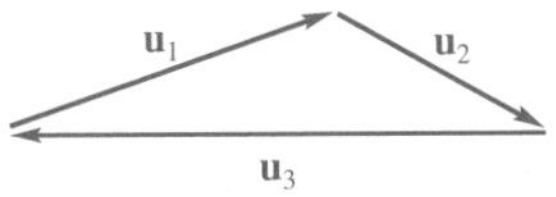

5. Figure 15 is a parallelogram. Express **w** in terms of **u** and **v**.

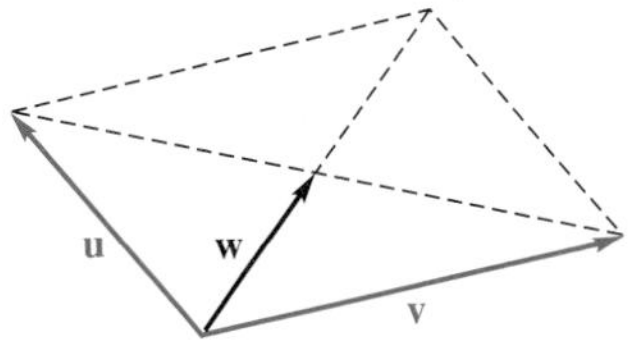

Figure 15

6. In the large triangle of Figure 16, **m** is a median (it bisects the side to which it is drawn). Express **m** and **n** in terms of **u** and **v**.

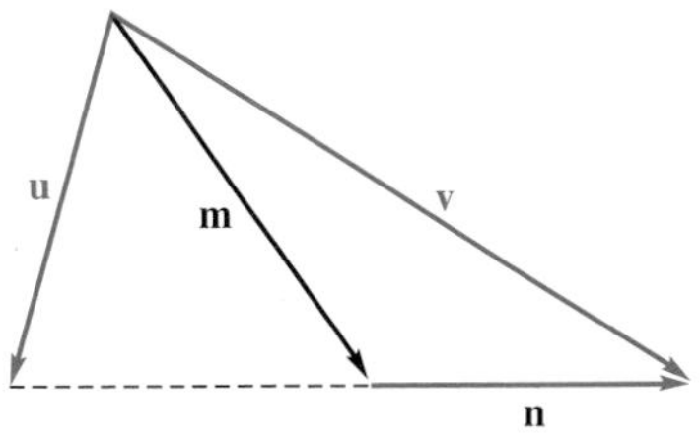

Figure 16

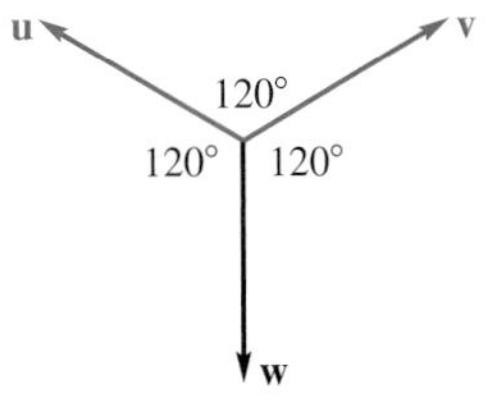

Figure 17

7. In Figure 17, $\mathbf{w} = -(\mathbf{u} + \mathbf{v})$ and $\|\mathbf{u}\| = \|\mathbf{v}\| = 1$. Find $\|\mathbf{w}\|$.

8. Do Problem 7 if the top angle is 90° and the two side angles are each 135°.

For the two-dimensional vectors **u** *and* **v** *in Problems 9–12, find the sum* $\mathbf{u} + \mathbf{v}$, *the difference* $\mathbf{u} - \mathbf{v}$, *and the magnitudes* $\|\mathbf{u}\|$ *and* $\|\mathbf{v}\|$.

9. $\mathbf{u} = \langle -1, 0 \rangle, \mathbf{v} = \langle 3, 4 \rangle$

10. $\mathbf{u} = \langle 0, 0 \rangle, \mathbf{v} = \langle -3, 4 \rangle$

11. $\mathbf{u} = \langle 12, 12 \rangle, \mathbf{v} = \langle -2, 2 \rangle$

12. $\mathbf{u} = \langle -0.2, 0.8 \rangle, \mathbf{v} = \langle -2.1, 1.3 \rangle$

For the three-dimensional vectors **u** *and* **v** *in Problems 13–16, find the sum* $\mathbf{u} + \mathbf{v}$, *the difference* $\mathbf{u} - \mathbf{v}$, *and the magnitudes* $\|\mathbf{u}\|$ *and* $\|\mathbf{v}\|$.

13. $\mathbf{u} = \langle -1, 0, 0 \rangle, \mathbf{v} = \langle 3, 4, 0 \rangle$

14. $\mathbf{u} = \langle 0, 0, 0 \rangle, \mathbf{v} = \langle -3, 3, 1 \rangle$

15. $\mathbf{u} = \langle 1, 0, 1 \rangle, \mathbf{v} = \langle -5, 0, 0 \rangle$

16. $\mathbf{u} = \langle 0.3, 0.3, 0.5 \rangle, \mathbf{v} = \langle 2.2, 1.3, -0.9 \rangle$

C **17.** In Figure 18, forces **u** and **v** each have magnitude 50 pounds. Find the magnitude and direction of the force **w** needed to counterbalance **u** and **v**.

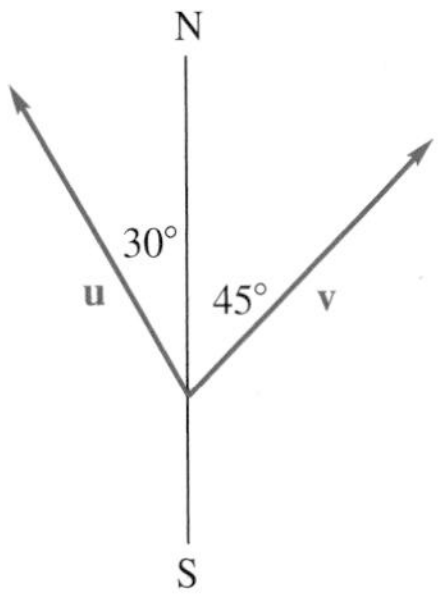

Figure 18

≈ **18.** Mark pushes on a post in the direction S 30° E (30° east of south) with a force of 60 pounds. Dan pushes on the same post in the direction S 60° W with a force of 80 pounds. What are the magnitude and direction of the resultant force?

≈ **19.** A 300-newton weight rests on a smooth (friction negligible) inclined plane that makes an angle of 30° with the horizontal. What force parallel to the plane will just keep the weight from sliding down the plane? *Hint:* Consider the downward force of 300 newtons to be the sum of two forces, one parallel to the plane and one perpendicular to it.

C **20.** An object weighing 258.5 pounds is held in equilibrium by two ropes that make angles of 27.34° and 39.22°, respectively, with the vertical. Find the magnitude of the force exerted on the object by each rope.

C **21.** A wind with velocity 45 miles per hour is blowing in the direction N 20° W. An airplane that flies at 425 miles per hour in still air is supposed to fly straight north. How should the airplane be headed and how fast will it then be flying with respect to the ground?

22. A ship is sailing due south at 20 miles per hour. A man walks west (i.e., at right angles to the side of the ship) across the deck at 3 miles per hour. What are the magnitude and direction of his velocity relative to the surface of the water?

23. Julie, flying in a wind blowing 40 miles per hour due south, discovers that she is heading due east when she points her airplane in the direction N 60° E. Find the airspeed (speed in still air) of the plane.

C **24.** What heading and airspeed are required for an airplane to fly 837 miles per hour due north if a wind of 63 miles per hour is blowing in the direction S 11.5° E?

25. Prove all parts of Theorem A for the case of two-dimensional vectors.

26. Prove parts 1–5 and 7–8 of Theorem A for the case of three-dimensional vectors.

27. Prove, using vector methods, that the line segment joining the midpoints of two sides of a triangle is parallel to the third side.

28. Prove that the midpoints of the four sides of an arbitrary quadrilateral are the vertices of a parallelogram.

29. Let $\mathbf{v}_1, \mathbf{v}_2, \ldots, \mathbf{v}_n$ be the edges of a polygon arranged in cyclic order as shown for the case $n = 7$ in Figure 19. Show that

$$\mathbf{v}_1 + \mathbf{v}_2 + \cdots + \mathbf{v}_n = \mathbf{0}$$

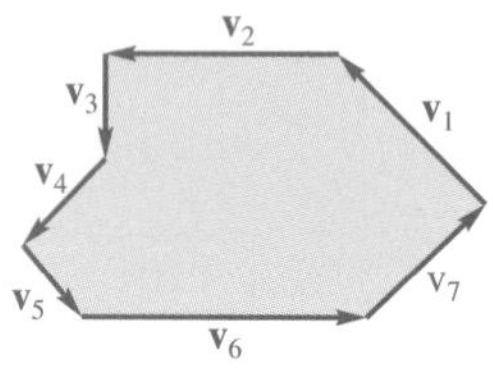

Figure 19

30. Let n points be equally spaced on a circle, and let $\mathbf{v}_1, \mathbf{v}_2, \ldots, \mathbf{v}_n$ be the vectors from the center of the circle to these n points. Show that $\mathbf{v}_1 + \mathbf{v}_2 + \cdots + \mathbf{v}_n = \mathbf{0}$.

31. Consider a horizontal triangular table with each vertex angle less than 120°. At the vertices are frictionless pulleys over which pass strings knotted at P, each with a weight W attached as shown in Figure 20. Show that at equilibrium the three angles at P are equal; that is, show that $\alpha + \beta = \alpha + \gamma = \beta + \gamma = 120°$.

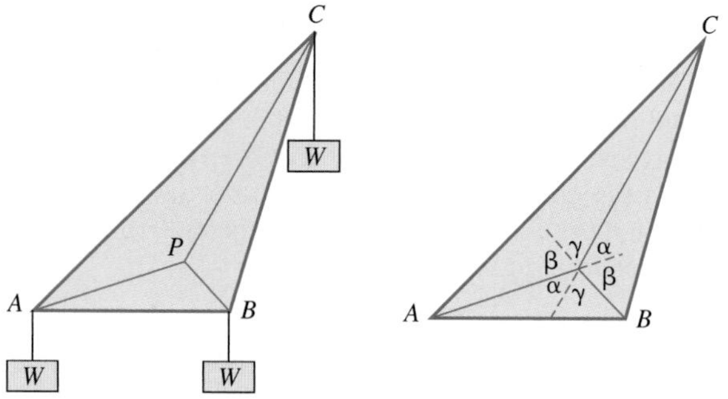

Figure 20

32. Show that the point P of the triangle of Problem 31 that minimizes $|AP| + |BP| + |CP|$ is the point where the three angles at P are equal. *Hint:* Let A', B', and C' be the points where the weights are attached. The center of gravity is then located $\frac{1}{3}(|AA'| + |BB'| + |CC'|)$ units below the plane of the triangle. The system is in equilibrium when the center of gravity of the three weights is lowest.

33. Let the weights at A, B, and C of Problem 31 be $3w$, $4w$, and $5w$, respectively. Determine the three angles at P at equilibrium. What geometric quantity (as in Problem 32) is now minimized?

34. A company will build a plant to manufacture refrigerators to be sold in cities A, B, and C in quantities a, b, and c, respectively, each year. Where is the best location for the plant, that is, the location that will minimize delivery costs (see Problem 33)?

35. A 100-pound chandelier is held in place by four wires attached to the ceiling at the four corners of a square. Each wire makes an angle of 45° with the horizontal. Find the magnitude of the tension in each wire.

36. Repeat Problem 35 for the case where there are *three* wires attached to the ceiling at the three corners of an equilateral triangle.

Answers to Concepts Review: **1.** magnitude; direction **2.** they have the same magnitude and direction **3.** the tail of **u**; the head of **v** **4.** 3

11.3 The Dot Product

We have discussed scalar multiplication, that is, the multiplication of a vector **u** by a scalar c. The result $c\mathbf{u}$ is always a vector. Now we introduce a multiplication for two vectors **u** and **v**. It is called the **dot product,** or **scalar product,** and is symbolized by $\mathbf{u} \cdot \mathbf{v}$. We define it for two-dimensional vectors as

$$\mathbf{u} \cdot \mathbf{v} = \langle u_1, u_2 \rangle \cdot \langle v_1, v_2 \rangle = u_1 v_1 + u_2 v_2$$

and for three-dimensional vectors as

$$\mathbf{u} \cdot \mathbf{v} = \langle u_1, u_2, u_3 \rangle \cdot \langle v_1, v_2, v_3 \rangle = u_1 v_1 + u_2 v_2 + u_3 v_3$$

EXAMPLE 1 Let $\mathbf{u} = \langle 0, 1, 1 \rangle$, $\mathbf{v} = \langle 2, -1, 1 \rangle$, and $\mathbf{w} = \langle 6, -3, 3 \rangle$. Compute each of the following if they are defined: (a) $\mathbf{u} \cdot \mathbf{v}$, (b) $\mathbf{v} \cdot \mathbf{u}$, (c) $\mathbf{v} \cdot \mathbf{w}$, (d) $\mathbf{u} \cdot \mathbf{u}$, and (e) $(\mathbf{u} \cdot \mathbf{v}) \cdot \mathbf{w}$.

SOLUTION

(a) $\mathbf{u} \cdot \mathbf{v} = \langle 0, 1, 1 \rangle \cdot \langle 2, -1, 1 \rangle = (0)(2) + (1)(-1) + (1)(1) = 0$

(b) $\mathbf{v} \cdot \mathbf{u} = \langle 2, -1, 1 \rangle \cdot \langle 0, 1, 1 \rangle = (2)(0) + (-1)(1) + (1)(1) = 0$

(c) $\mathbf{v} \cdot \mathbf{w} = \langle 2, -1, 1 \rangle \cdot \langle 6, -3, 3 \rangle = (2)(6) + (-1)(-3) + (1)(3) = 18$

(d) $\mathbf{u} \cdot \mathbf{u} = \langle 0, 1, 1 \rangle \cdot \langle 0, 1, 1 \rangle = 0^2 + 1^2 + 1^2 = 2$

(e) $(\mathbf{u} \cdot \mathbf{v}) \cdot \mathbf{w}$ is not defined. The quantity $\mathbf{u} \cdot \mathbf{v}$ is a *scalar*. A scalar dotted with a vector doesn't make sense. ■

The properties of the dot product are easy to establish. (See Problems 46–50.) Note that this theorem, as well as all others in this section, applies to both two- and three-dimensional vectors.

Theorem A Properties of the Dot Product

If **u**, **v**, and **w** are vectors, and c is a scalar, then

1. $\mathbf{u}\cdot\mathbf{v} = \mathbf{v}\cdot\mathbf{u}$
2. $\mathbf{u}\cdot(\mathbf{v}+\mathbf{w}) = \mathbf{u}\cdot\mathbf{v} + \mathbf{u}\cdot\mathbf{w}$
3. $c(\mathbf{u}\cdot\mathbf{v}) = (c\mathbf{u})\cdot\mathbf{v}$
4. $\mathbf{0}\cdot\mathbf{u} = 0$
5. $\mathbf{u}\cdot\mathbf{u} = \|\mathbf{u}\|^2$

To emphasize the significance of the dot product we offer the following alternative formula for it that involves the geometric properties of the vectors **u** and **v**.

Theorem B

If θ is the smallest nonnegative angle between the nonzero vectors **u** and **v**, then

$$\mathbf{u}\cdot\mathbf{v} = \|\mathbf{u}\|\|\mathbf{v}\|\cos\theta$$

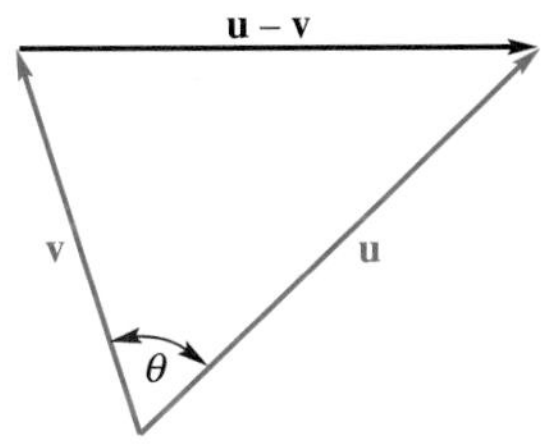

Figure 1

Proof To prove this result, apply the Law of Cosines to the triangle in Figure 1.

$$\|\mathbf{u}-\mathbf{v}\|^2 = \|\mathbf{u}\|^2 + \|\mathbf{v}\|^2 - 2\|\mathbf{u}\|\|\mathbf{v}\|\cos\theta$$

On the other hand, from the properties of the dot product stated in Theorem A,

$$\begin{aligned}\|\mathbf{u}-\mathbf{v}\|^2 &= (\mathbf{u}-\mathbf{v})\cdot(\mathbf{u}-\mathbf{v})\\ &= \mathbf{u}\cdot(\mathbf{u}-\mathbf{v}) - \mathbf{v}\cdot(\mathbf{u}-\mathbf{v})\\ &= \mathbf{u}\cdot\mathbf{u} - \mathbf{u}\cdot\mathbf{v} - \mathbf{v}\cdot\mathbf{u} + \mathbf{v}\cdot\mathbf{v}\\ &= \|\mathbf{u}\|^2 + \|\mathbf{v}\|^2 - 2\mathbf{u}\cdot\mathbf{v}\end{aligned}$$

Equating the two expressions for $\|\mathbf{u}-\mathbf{v}\|^2$ gives

$$\begin{aligned}\|\mathbf{u}\|^2 + \|\mathbf{v}\|^2 - 2\|\mathbf{u}\|\|\mathbf{v}\|\cos\theta &= \|\mathbf{u}\|^2 + \|\mathbf{v}\|^2 - 2\mathbf{u}\cdot\mathbf{v}\\ -2\|\mathbf{u}\|\|\mathbf{v}\|\cos\theta &= -2\mathbf{u}\cdot\mathbf{v}\\ \mathbf{u}\cdot\mathbf{v} &= \|\mathbf{u}\|\|\mathbf{v}\|\cos\theta\end{aligned}$$

■

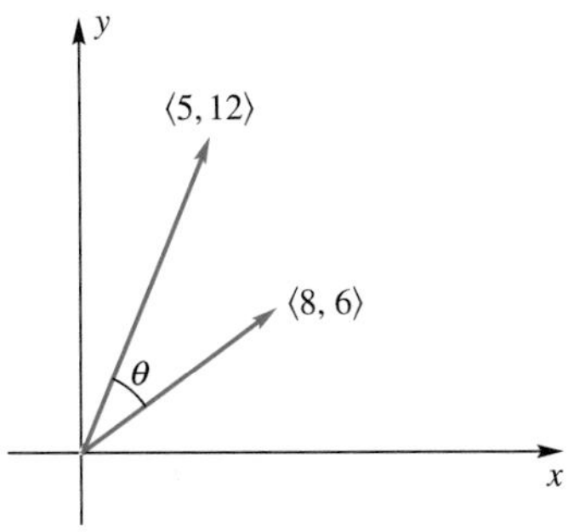

Figure 2

EXAMPLE 2 Find the angle between $\mathbf{u} = \langle 8, 6\rangle$ and $\mathbf{v} = \langle 5, 12\rangle$ (see Figure 2).

SOLUTION Solving for $\cos\theta$ in Theorem B gives

$$\cos\theta = \frac{\mathbf{u}\cdot\mathbf{v}}{\|\mathbf{u}\|\|\mathbf{v}\|} = \frac{(8)(5) + (6)(12)}{(10)(13)} = \frac{112}{130} \approx 0.862$$

Then

$$\theta \approx \cos^{-1}(0.862) \approx 0.532 \text{ (or } 30.5^\circ)$$

■

An important consequence of Theorem B is the following.

Theorem C Perpendicularity Criterion

Two vectors **u** and **v** are perpendicular if and only if their dot product, $\mathbf{u}\cdot\mathbf{v}$, is 0.

Proof Two nonzero vectors are perpendicular if and only if the smallest nonnegative angle θ between them is $\pi/2$; that is, if and only if $\cos\theta = 0$. But $\cos\theta = 0$ if and only if $\mathbf{u}\cdot\mathbf{v} = 0$. (This result is valid for zero vectors, provided that we agree that a zero vector is perpendicular to every other vector.) ■

Definition Orthogonal

Vectors that are perpendicular are said to be **orthogonal.**

EXAMPLE 3 Find the angles between each of the three pairs of vectors from Example 1. Which pairs are orthogonal?

SOLUTION For the vectors **u** and **v**, we have

$$\cos\theta_1 = \frac{\mathbf{u}\cdot\mathbf{v}}{\|\mathbf{u}\|\|\mathbf{v}\|} = \frac{(0)(2) + (1)(-1) + (1)(1)}{\|\langle 0, 1, 1\rangle\|\|\langle 2, -1, 1\rangle\|} = \frac{0}{\sqrt{2}\sqrt{6}} = 0$$

For the vectors **u** and **w**, we have

$$\cos\theta_2 = \frac{\mathbf{u}\cdot\mathbf{w}}{\|\mathbf{u}\|\|\mathbf{w}\|} = \frac{(0)(6) + (1)(-3) + (1)(3)}{\|\langle 0, 1, 1\rangle\|\|\langle 6, -3, 3\rangle\|} = \frac{0}{\sqrt{2}\left(3\sqrt{6}\right)} = 0$$

Finally, for the vectors **v** and **w**, we have

$$\cos\theta_3 = \frac{\mathbf{v}\cdot\mathbf{w}}{\|\mathbf{v}\|\|\mathbf{w}\|} = \frac{(2)(6) + (-1)(-3) + (1)(3)}{\|\langle 2, -1, 1\rangle\|\|\langle 6, -3, 3\rangle\|} = \frac{18}{\sqrt{6}\left(3\sqrt{6}\right)} = 1$$

Thus the pair **u** and **v** and the pair **u** and **w** are orthogonal, so $\theta_1 = \theta_2 = \pi/2$. Note that for the pair **v** and **w**, the cosine of the angle between them is 1, indicating that $\theta_3 = 0$; that is, the vectors point the same direction. ■

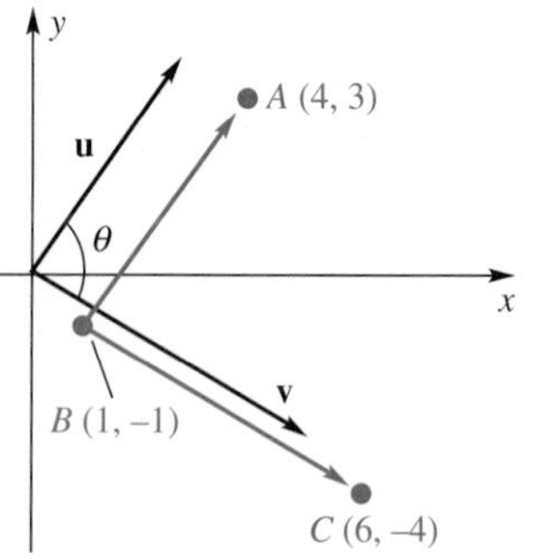

Figure 3

Recall that every vector **u** in the plane can be written as $\mathbf{u} = u_1\mathbf{i} + u_2\mathbf{j}$, where $\mathbf{i} = \langle 1, 0\rangle$ and $\mathbf{j} = \langle 0, 1\rangle$, and that every vector **v** in three space can be written as $\mathbf{v} = v_1\mathbf{i} + v_2\mathbf{j} + v_3\mathbf{k}$ where, in this case, $\mathbf{i} = \langle 1, 0, 0\rangle$, $\mathbf{j} = \langle 0, 1, 0\rangle$, and $\mathbf{k} = \langle 0, 0, 1\rangle$.

EXAMPLE 4 Find the measure of the angle ABC, where the three points are $A(4, 3)$, $B(1, -1)$, and $C(6, -4)$ as in Figure 3.

SOLUTION

$$\mathbf{u} = \overrightarrow{BA} = (4 - 1)\mathbf{i} + (3 + 1)\mathbf{j} = 3\mathbf{i} + 4\mathbf{j} = \langle 3, 4\rangle$$
$$\mathbf{v} = \overrightarrow{BC} = (6 - 1)\mathbf{i} + (-4 + 1)\mathbf{j} = 5\mathbf{i} - 3\mathbf{j} = \langle 5, -3\rangle$$
$$\|\mathbf{u}\| = \sqrt{3^2 + 4^2} = 5$$
$$\|\mathbf{v}\| = \sqrt{5^2 + (-3)^2} = \sqrt{34}$$
$$\mathbf{u}\cdot\mathbf{v} = (3)(5) + (4)(-3) = 3$$
$$\cos\theta = \frac{\mathbf{u}\cdot\mathbf{v}}{\|\mathbf{u}\|\|\mathbf{v}\|} = \frac{3}{5\sqrt{34}} \approx 0.1029$$
$$\theta \approx 1.468 \text{ (about } 84.09^\circ)$$
■

EXAMPLE 5 Find the measure of angle ABC if the points are $A(1, -2, 3)$, $B(2, 4, -6)$, and $C(5, -3, 2)$ (Figure 4).

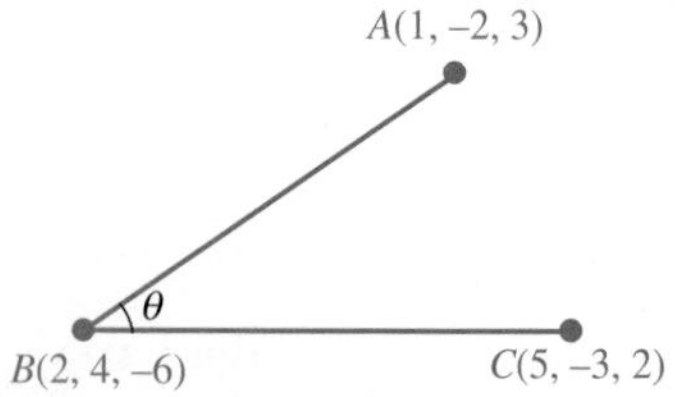

Figure 4

SOLUTION First we determine vectors **u** and **v** (emanating from the origin) equivalent to $\overrightarrow{BA}$ and $\overrightarrow{BC}$. This is done by subtracting the coordinates of the initial points from those of the terminal points, that is,

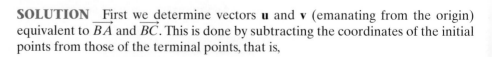

$$\mathbf{u} = \overrightarrow{BA} = \langle 1 - 2, -2 - 4, 3 + 6\rangle = \langle -1, -6, 9\rangle$$
$$\mathbf{v} = \overrightarrow{BC} = \langle 5 - 2, -3 - 4, 2 + 6\rangle = \langle 3, -7, 8\rangle$$

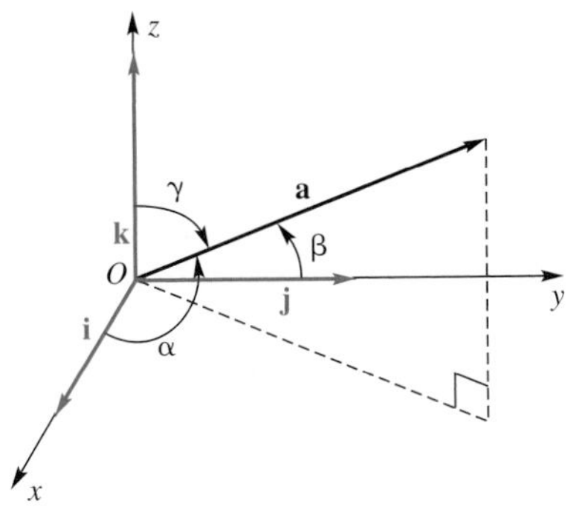

Figure 5

Thus,

$$\cos\theta = \frac{\mathbf{u}\cdot\mathbf{v}}{\|\mathbf{u}\|\|\mathbf{v}\|} = \frac{(-1)(3) + (-6)(-7) + (9)(8)}{\sqrt{1 + 36 + 81}\sqrt{9 + 49 + 64}} \approx 0.9251$$

$$\theta = 0.3894 \quad \text{(about } 22.31^\circ\text{)}$$

■

Direction Angles and Cosines The smallest nonnegative angles between a nonzero three-dimensional vector **a** and the basis vectors **i**, **j**, and **k** are called the **direction angles** of **a**; they are denoted α, β and γ, respectively, as shown in Figure 5. It is often more convenient to work with the **direction cosines:** $\cos\alpha$, $\cos\beta$, and $\cos\gamma$. If $\mathbf{a} = a_1\mathbf{i} + a_2\mathbf{j} + a_3\mathbf{k}$, then

$$\cos\alpha = \frac{\mathbf{a}\cdot\mathbf{i}}{\|\mathbf{a}\|\|\mathbf{i}\|} = \frac{a_1}{\|\mathbf{a}\|},$$

$$\cos\beta = \frac{\mathbf{a}\cdot\mathbf{j}}{\|\mathbf{a}\|\|\mathbf{j}\|} = \frac{a_2}{\|\mathbf{a}\|},$$

$$\cos\gamma = \frac{\mathbf{a}\cdot\mathbf{k}}{\|\mathbf{a}\|\|\mathbf{k}\|} = \frac{a_3}{\|\mathbf{a}\|}$$

Notice that

$$\cos^2\alpha + \cos^2\beta + \cos^2\gamma = \frac{a_1^2}{\|\mathbf{a}\|^2} + \frac{a_2^2}{\|\mathbf{a}\|^2} + \frac{a_3^2}{\|\mathbf{a}\|^2} = 1$$

The vector $\langle\cos\alpha, \cos\beta, \cos\gamma\rangle$ is a unit vector having the same direction as **a**.

EXAMPLE 6 Find the direction angles for the vector $\mathbf{a} = 4\mathbf{i} - 5\mathbf{j} + 3\mathbf{k}$.

SOLUTION Since $\|\mathbf{a}\| = \sqrt{4^2 + (-5)^2 + 3^2} = 5\sqrt{2}$,

$$\cos\alpha = \frac{4}{5\sqrt{2}} = \frac{2\sqrt{2}}{5}, \quad \cos\beta = -\frac{5}{5\sqrt{2}} = -\frac{\sqrt{2}}{2}, \quad \cos\gamma = \frac{3}{5\sqrt{2}} = \frac{3\sqrt{2}}{10}$$

and

$$\alpha \approx 55.55^\circ, \qquad \beta = 135^\circ, \qquad \gamma \approx 64.90^\circ$$

■

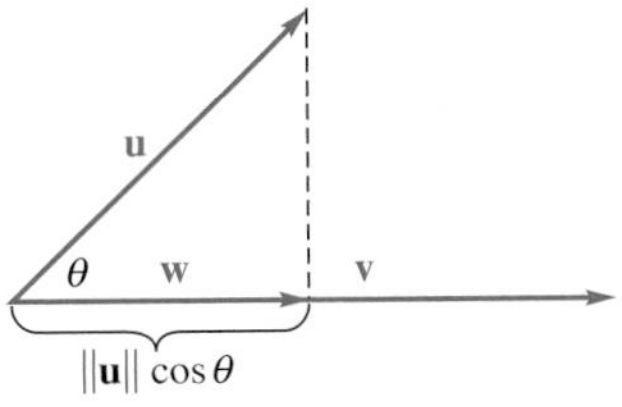

Figure 6

Projections Let **u** and **v** be vectors, and let θ be the angle between them. For now, we assume that $0 \le \theta \le \pi/2$. Let **w** be the vector in the direction of **v** that has magnitude $\|\mathbf{u}\|\cos\theta$ (see Figure 6). Since **w** has the same direction as **v**, we know that $\mathbf{w} = c\mathbf{v}$ for some nonnegative scalar c. On the other hand, the *magnitude* of **w** must be $\|\mathbf{u}\|\cos\theta$. Thus,

$$\|\mathbf{u}\|\cos\theta = \|\mathbf{w}\| = \|c\mathbf{v}\| = c\|\mathbf{v}\|$$

The constant c is therefore

$$c = \frac{\|\mathbf{u}\|}{\|\mathbf{v}\|}\cos\theta = \frac{\|\mathbf{u}\|}{\|\mathbf{v}\|}\frac{\mathbf{u}\cdot\mathbf{v}}{\|\mathbf{u}\|\|\mathbf{v}\|} = \frac{\mathbf{u}\cdot\mathbf{v}}{\|\mathbf{v}\|^2}$$

Thus,

$$\mathbf{w} = \left(\frac{\mathbf{u}\cdot\mathbf{v}}{\|\mathbf{v}\|^2}\right)\mathbf{v}$$

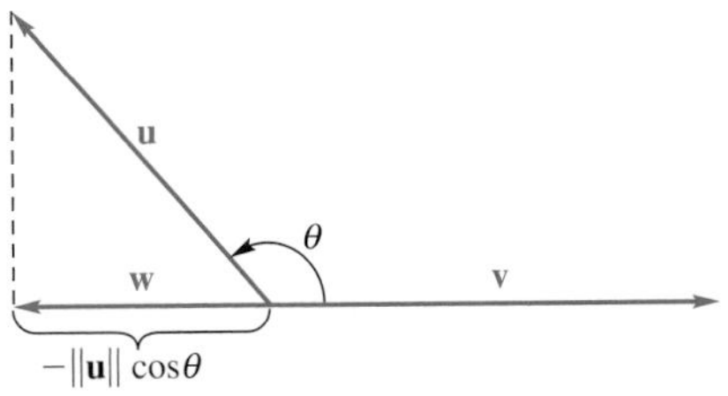

Figure 7

For $\pi/2 < \theta \le \pi$, we define **w** to be the vector in the line determined by **v**, but pointing in the direction opposite **v** (see Figure 7). The magnitude of this vector is $\|\mathbf{w}\| = -\|\mathbf{u}\|\cos\theta = c\|\mathbf{v}\|$ for some positive scalar c. Thus, $c = (-\|\mathbf{u}\|\cos\theta)/(\|\mathbf{v}\|) = -\mathbf{u}\cdot\mathbf{v}/\|\mathbf{v}\|^2$. Since **w** points in the direction *opposite* **v**, we have $\mathbf{w} = -c\mathbf{v} = (\mathbf{u}\cdot\mathbf{v}/\|\mathbf{v}\|^2)\mathbf{v}$. Thus, in both cases we have $\mathbf{w} = (\mathbf{u}\cdot\mathbf{v}/\|\mathbf{v}\|^2)\mathbf{v}$. The vector **w** is called

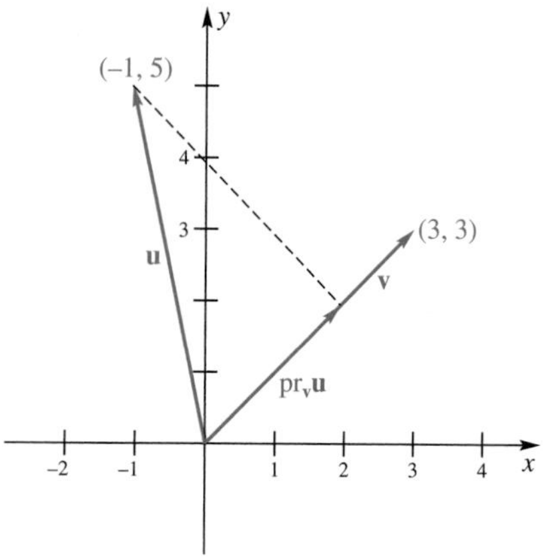

Figure 8

the **vector projection of u on v**, or sometimes just the **projection of u on v**, and is denoted $\text{pr}_\mathbf{v}\,\mathbf{u}$:

$$\text{pr}_\mathbf{v}\,\mathbf{u} = \left(\frac{\mathbf{u}\cdot\mathbf{v}}{\|\mathbf{v}\|^2}\right)\mathbf{v}$$

The **scalar projection of u on v** is defined to be $\|\mathbf{u}\|\cos\theta$. It is positive, zero, or negative, depending on whether θ is acute, right, or obtuse. When $0 \le \theta \le \pi/2$, the scalar projection is equal to the magnitude of $\text{pr}_\mathbf{v}\,\mathbf{u}$, and when $\pi/2 < \theta \le \pi$, the scalar projection is equal to the opposite of the magnitude of $\text{pr}_\mathbf{v}\,\mathbf{u}$.

EXAMPLE 7 Let $\mathbf{u} = \langle -1, 5\rangle$ and $\mathbf{v} = \langle 3, 3\rangle$. Find the vector projection of **u** on **v** and the scalar projection of **u** on **v**.

SOLUTION Figure 8 shows the two vectors. The vector projection is

$$\text{pr}_{\langle 3,3\rangle}\langle -1, 5\rangle = \left(\frac{\langle -1,5\rangle\cdot\langle 3,3\rangle}{\|\langle 3,3\rangle\|^2}\right)\langle 3,3\rangle = \frac{-3+15}{3^2+3^2}\langle 3,3\rangle = \langle 2,2\rangle = 2\mathbf{i} + 2\mathbf{j}$$

and the scalar projection is

$$\|\mathbf{u}\|\cos\theta = \|\langle -1,5\rangle\|\frac{\langle -1,5\rangle\cdot\langle 3,3\rangle}{\|\langle -1,5\rangle\|\|\langle 3,3\rangle\|} = \frac{-3+15}{\sqrt{3^2+3^2}} = 2\sqrt{2}$$ ■

The work done by a constant force **F** in moving an object along the line from P to Q is the magnitude of the force in the direction of the motion, times the distance moved. Thus, if **D** is the vector from P to Q, the work done is

$$(\text{Scalar projection of } \mathbf{F} \text{ on } \mathbf{D})\|\mathbf{D}\| = \big(\|\mathbf{F}\|\cos\theta\big)\|\mathbf{D}\|$$

That is,

$$\boxed{\text{Work} = \mathbf{F}\cdot\mathbf{D}}$$

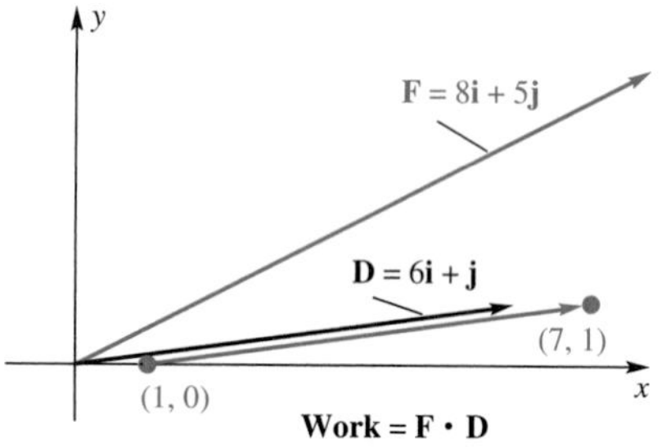

Figure 9

EXAMPLE 8 A force $\mathbf{F} = 8\mathbf{i} + 5\mathbf{j}$ in newtons moves an object from $(1, 0)$ to $(7, 1)$, where distance is measured in meters (Figure 9). How much work is done?

SOLUTION Let **D** be the vector from $(1, 0)$ to $(7, 1)$; that is, let $\mathbf{D} = 6\mathbf{i} + \mathbf{j}$. Then

$$\text{Work} = \mathbf{F}\cdot\mathbf{D} = (8)(6) + (5)(1) = 53 \text{ newton-meters} = 53 \text{ joules}$$ ■

Planes One fruitful way to describe a plane is by using vector language. Let $\mathbf{n} = \langle A, B, C\rangle$ be a fixed nonzero vector and $P_1(x_1, y_1, z_1)$ be a fixed point. The set of points $P(x, y, z)$ satisfying $\overrightarrow{P_1P}\cdot\mathbf{n} = 0$ is the **plane** through P_1 perpendicular to **n**. Since every plane contains a point and is perpendicular to some vector, a plane can be characterized in this way.

To get the Cartesian equation of the plane, write the vector $\overrightarrow{P_1P}$ in component form; that is,

$$\overrightarrow{P_1P} = \langle x - x_1, y - y_1, z - z_1\rangle$$

Then $\overrightarrow{P_1P}\cdot\mathbf{n} = 0$ is equivalent to

$$\boxed{A(x - x_1) + B(y - y_1) + C(z - z_1) = 0}$$

This equation (in which at least one of A, B, and C is different from zero) is called the **standard form for the equation of a plane.**

If we remove the parentheses and simplify, the boxed equation takes the form of the general linear equation

$$Ax + By + Cz = D, \qquad A^2 + B^2 + C^2 \neq 0$$

Thus, every plane has a linear equation. Conversely, the graph of a linear equation in three-space is always a plane. To see the latter, let (x_1, y_1, z_1) satisfy the equation; that is,

$$Ax_1 + By_1 + Cz_1 = D$$

When we subtract this equation from the one above, we have the boxed equation, which we know represents a plane.

EXAMPLE 9 Find the equation of the plane through $(5, 1, -2)$ perpendicular to $\mathbf{n} = \langle 2, 4, 3 \rangle$. Then find the angle between this plane and the one with equation $3x - 4y + 7z = 5$.

SOLUTION To perform the first task, simply apply the standard form for the equation of a plane to the problem at hand, which gives

$$2(x - 5) + 4(y - 1) + 3(z + 2) = 0$$

or, equivalently,

$$2x + 4y + 3z = 8$$

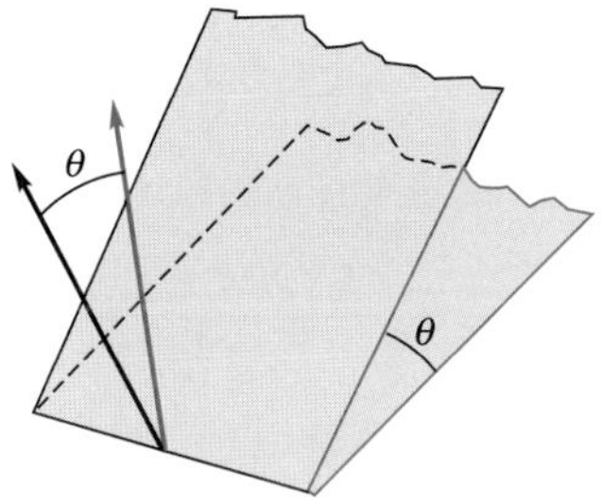

Figure 10

A vector $\mathbf{m}$ perpendicular, or normal, to the second plane is $\mathbf{m} = \langle 3, -4, 7 \rangle$. The angle θ between two planes is the angle between their normals (Figure 10). Thus,

$$\cos\theta = \frac{\mathbf{m} \cdot \mathbf{n}}{\|\mathbf{m}\|\|\mathbf{n}\|} = \frac{(3)(2) + (-4)(4) + (7)(3)}{\sqrt{9 + 16 + 49}\sqrt{4 + 16 + 9}} \approx 0.2375$$

$$\theta \approx 76.26°$$

Actually, there are two angles between two planes, but they are *supplementary*. The process just described will lead to one of them. The other, if desired, is obtained by subtracting the first value from 180°. In our case, it would be 103.74°. ■

EXAMPLE 10 Show that the distance L from the point (x_0, y_0, z_0) to the plane $Ax + By + Cz = D$ is given by the formula

$$L = \frac{|Ax_0 + By_0 + Cz_0 - D|}{\sqrt{A^2 + B^2 + C^2}}$$

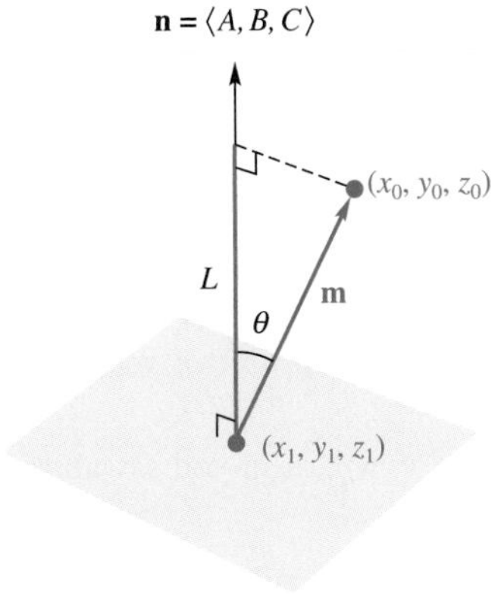

Figure 11

SOLUTION Let (x_1, y_1, z_1) be a point on the plane, and let $\mathbf{m} = \langle x_0 - x_1, y_0 - y_1, z_0 - z_1 \rangle$ be the vector from (x_1, y_1, z_1) to (x_0, y_0, z_0), as in Figure 11. Now $\mathbf{n} = \langle A, B, C \rangle$ is a vector perpendicular to the given plane, though it might point in the opposite direction of that in our figure. The number L that we seek is the length of the projection of $\mathbf{m}$ on $\mathbf{n}$. Thus,

$$L = |\|\mathbf{m}\|\cos\theta| = \frac{|\mathbf{m} \cdot \mathbf{n}|}{\|\mathbf{n}\|}$$

$$= \frac{|A(x_0 - x_1) + B(y_0 - y_1) + C(z_0 - z_1)|}{\sqrt{A^2 + B^2 + C^2}}$$

$$= \frac{|Ax_0 + By_0 + Cz_0 - (Ax_1 + By_1 + Cz_1)|}{\sqrt{A^2 + B^2 + C^2}}$$

But (x_1, y_1, z_1) is on the plane, and so

$$Ax_1 + By_1 + Cz_1 = D$$

Substitution of this result in the expression for L yields the desired formula. ■

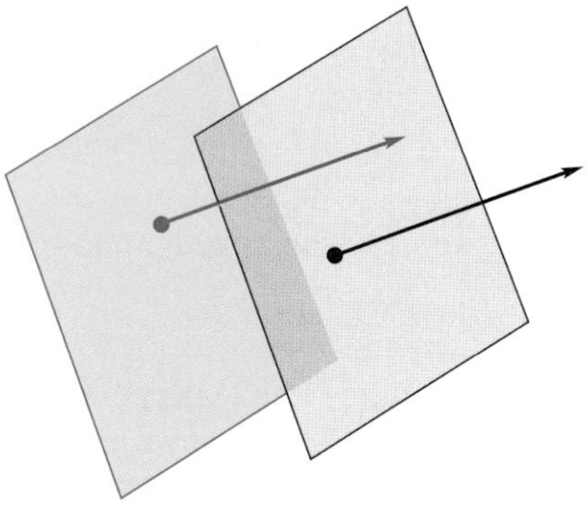
Figure 12

EXAMPLE 11 Find the distance between the parallel planes $3x - 4y + 5z = 9$ and $3x - 4y + 5z = 4$.

SOLUTION The planes are parallel, since the vector $\langle 3, -4, 5\rangle$ is perpendicular to both of them (Figure 12). The point $(1, 1, 2)$ is easily seen to be on the first plane. We find the distance L from $(1, 1, 2)$ to the second plane using the formula of Example 10.

$$L = \frac{|3(1) - 4(1) + 5(2) - 4|}{\sqrt{9 + 16 + 25}} = \frac{5}{5\sqrt{2}} \approx 0.7071 \quad \blacksquare$$

Concepts Review

1. The dot product of $\mathbf{u} = \langle u_1, u_2, u_3\rangle$ and $\mathbf{v} = \langle v_1, v_2, v_3\rangle$ is defined by _____. The corresponding geometric formulation for $\mathbf{u}\cdot\mathbf{v}$ is _____ where θ is the angle between **u** and **v**.

2. Two vectors **u** and **v** are orthogonal if and only if their dot product is _____.

3. The work done by a constant force **F** in moving an object along the vector **D** is given by _____.

4. A normal vector to the plane $Ax + By + Cz = D$ is _____.

Problem Set 11.3

1. Let $\mathbf{a} = -2\mathbf{i} + 3\mathbf{j}$, $\mathbf{b} = 2\mathbf{i} - 3\mathbf{j}$, and $\mathbf{c} = -5\mathbf{j}$. Find each of the following:

(a) $2\mathbf{a} - 4\mathbf{b}$ (b) $\mathbf{a}\cdot\mathbf{b}$
(c) $\mathbf{a}\cdot(\mathbf{b} + \mathbf{c})$ (d) $(-2\mathbf{a} + 3\mathbf{b})\cdot 5\mathbf{c}$
(e) $\|\mathbf{a}\|\mathbf{c}\cdot\mathbf{a}$ (f) $\mathbf{b}\cdot\mathbf{b} - \|\mathbf{b}\|$

2. Let $\mathbf{a} = \langle 3, -1\rangle$, $\mathbf{b} = \langle 1, -1\rangle$, and $\mathbf{c} = \langle 0, 5\rangle$. Find each of the following:

(a) $-4\mathbf{a} + 3\mathbf{b}$ (b) $\mathbf{b}\cdot\mathbf{c}$
(c) $(\mathbf{a} + \mathbf{b})\cdot\mathbf{c}$ (d) $2\mathbf{c}\cdot(3\mathbf{a} + 4\mathbf{b})$
(e) $\|\mathbf{b}\|\mathbf{b}\cdot\mathbf{a}$ (f) $\|\mathbf{c}\|^2 - \mathbf{c}\cdot\mathbf{c}$

3. Find the cosine of the angle between **a** and **b** and make a sketch.

(a) $\mathbf{a} = \langle 1, -3\rangle, \mathbf{b} = \langle -1, 2\rangle$ (b) $\mathbf{a} = \langle -1, -2\rangle, \mathbf{b} = \langle 6, 0\rangle$
(c) $\mathbf{a} = \langle 2, -1\rangle, \mathbf{b} = \langle -2, -4\rangle$
(d) $\mathbf{a} = \langle 4, -7\rangle, \mathbf{b} = \langle -8, 10\rangle$

4. Find the angle between **a** and **b** and make a sketch.

(a) $\mathbf{a} = 12\mathbf{i}, \mathbf{b} = -5\mathbf{i}$
(b) $\mathbf{a} = 4\mathbf{i} + 3\mathbf{j}, \mathbf{b} = -8\mathbf{i} - 6\mathbf{j}$
(c) $\mathbf{a} = -\mathbf{i} + 3\mathbf{j}, \mathbf{b} = 2\mathbf{i} - 6\mathbf{j}$
(d) $\mathbf{a} = \sqrt{3}\mathbf{i} + \mathbf{j}, \mathbf{b} = 3\mathbf{i} + \sqrt{3}\mathbf{j}$

5. Let $\mathbf{a} = \mathbf{i} + 2\mathbf{j} - \mathbf{k}$, $\mathbf{b} = \mathbf{j} + \mathbf{k}$, and $\mathbf{c} = -\mathbf{i} + \mathbf{j} + 2\mathbf{k}$. Find each of the following:

(a) $\mathbf{a}\cdot\mathbf{b}$ (b) $(\mathbf{a} + \mathbf{c})\cdot\mathbf{b}$
(c) $\mathbf{a}/\|\mathbf{a}\|$ (d) $(\mathbf{b} - \mathbf{c})\cdot\mathbf{a}$
(e) $\dfrac{\mathbf{a}\cdot\mathbf{b}}{\|\mathbf{a}\|\|\mathbf{b}\|}$ (f) $\mathbf{b}\cdot\mathbf{b} - \|\mathbf{b}\|^2$

6. Let $\mathbf{a} = \langle\sqrt{2}, \sqrt{2}, 0\rangle$, $\mathbf{b} = \langle 1, -1, 1\rangle$, and $\mathbf{c} = \langle -2, 2, 1\rangle$. Find each of the following:

(a) $\mathbf{a}\cdot\mathbf{c}$ (b) $(\mathbf{a} - \mathbf{c})\cdot\mathbf{b}$
(c) $\mathbf{a}/\|\mathbf{a}\|$ (d) $(\mathbf{b} - \mathbf{c})\cdot\mathbf{a}$
(e) $\dfrac{\mathbf{b}\cdot\mathbf{c}}{\|\mathbf{b}\|\|\mathbf{c}\|}$ (f) $\mathbf{a}\cdot\mathbf{a} - \|\mathbf{a}\|^2$

7. For the vectors **a**, **b**, and **c** from Problem 6, find the angle between each pair of vectors.

8. Let $\mathbf{a} = \langle\sqrt{3}/3, \sqrt{3}/3, \sqrt{3}/3\rangle$, $\mathbf{b} = \langle 1, -1, 0\rangle$, and $\mathbf{c} = \langle -2, -2, 1\rangle$. Find the angle between each pair of vectors.

9. For the vectors **a**, **b**, and **c** from Problem 6, find the direction cosines and the direction angles.

10. For the vectors **a**, **b**, and **c** from Problem 8, find the direction cosines and the direction angles.

11. Show that the vectors $\langle 6, 3\rangle$ and $\langle -1, 2\rangle$ are orthogonal.

12. Show that the vectors $\mathbf{a} = \langle 1, 1, 1\rangle$, $\mathbf{b} = \langle 1, -1, 0\rangle$, and $\mathbf{c} = \langle -1, -1, 2\rangle$ are mutually orthogonal, that is, each pair of vectors is orthogonal.

13. Show that the vectors $\mathbf{a} = \mathbf{i} - \mathbf{j}$, $\mathbf{b} = \mathbf{i} + \mathbf{j}$, and $\mathbf{c} = 2\mathbf{k}$ are mutually orthogonal, that is, each pair of vectors is orthogonal.

14. If $\mathbf{u} + \mathbf{v}$ is orthogonal to $\mathbf{u} - \mathbf{v}$, what can you say about the relative magnitudes of **u** and **v**?

15. Find two vectors of length 10, each of which is perpendicular to both $-4\mathbf{i} + 5\mathbf{j} + \mathbf{k}$ and $4\mathbf{i} + \mathbf{j}$.

16. Find all vectors perpendicular to both $\langle 1, -2, -3\rangle$ and $\langle -3, 2, 0\rangle$.

17. Find the angle ABC if the points are $A(1, 2, 3)$, $B(-4, 5, 6)$, and $C(1, 0, 1)$.

18. Show that the triangle ABC is a right triangle if the vertices are $A(6, 3, 3)$, $B(3, 1, -1)$, and $C(-1, 10, -2.5)$. *Hint:* Check the angle at B.

19. For what numbers c are $\langle c, 6\rangle$ and $\langle c, -4\rangle$ orthogonal?

20. For what numbers c are $2c\mathbf{i} - 8\mathbf{j}$ and $3\mathbf{i} + c\mathbf{j}$ orthogonal?

21. For what numbers c and d are $\mathbf{u} = c\mathbf{i} + \mathbf{j} + \mathbf{k}$ and $\mathbf{v} = 2\mathbf{j} + d\mathbf{k}$ orthogonal?

22. For what values of a, b, and c are the three vectors $\langle a, 0, 1\rangle$, $\langle 0, 2, b\rangle$, and $\langle 1, c, 1\rangle$ mutually orthogonal.

In Problems 23–28, find each of the given projections if $\mathbf{u} = \mathbf{i} + 2\mathbf{j}$, $\mathbf{v} = 2\mathbf{i} - \mathbf{j}$, *and* $\mathbf{w} = \mathbf{i} + 5\mathbf{j}$.

23. $\text{proj}_{\mathbf{v}}\,\mathbf{u}$

24. $\text{proj}_{\mathbf{u}}\,\mathbf{v}$

25. $\text{proj}_{\mathbf{u}}\,\mathbf{w}$

26. $\text{proj}_{\mathbf{u}}(\mathbf{w} - \mathbf{v})$

27. $\text{proj}_{\mathbf{j}}\,\mathbf{u}$

28. $\text{proj}_{\mathbf{i}}\,\mathbf{u}$

In Problems 29–34, find each of the given projections if $\mathbf{u} = 3\mathbf{i} + 2\mathbf{j} + \mathbf{k}$, $\mathbf{v} = 2\mathbf{i} - \mathbf{k}$, *and* $\mathbf{w} = \mathbf{i} + 5\mathbf{j} - 3\mathbf{k}$.

29. $\text{proj}_{\mathbf{v}}\,\mathbf{u}$

30. $\text{proj}_{\mathbf{u}}\,\mathbf{v}$

31. $\text{proj}_{\mathbf{u}}\,\mathbf{w}$

32. $\text{proj}_{\mathbf{u}}(\mathbf{w} + \mathbf{v})$

33. $\text{proj}_{\mathbf{k}}\,\mathbf{u}$

34. $\text{proj}_{\mathbf{i}}\,\mathbf{u}$

35. Find a simple expression for each of the following for an arbitrary vector $\mathbf{u}$.

(a) $\text{proj}_{\mathbf{u}}\,\mathbf{u}$ (b) $\text{proj}_{-\mathbf{u}}\,\mathbf{u}$

36. Find a simple expression for each of the following for an arbitrary vector $\mathbf{u}$.

(a) $\text{proj}_{\mathbf{u}}(-\mathbf{u})$ (b) $\text{proj}_{-\mathbf{u}}(-\mathbf{u})$

37. Find the scalar projection of $\mathbf{u} = -\mathbf{i} + 5\mathbf{j} + 3\mathbf{k}$ on $\mathbf{v} = -\mathbf{i} + \mathbf{j} - \mathbf{k}$.

38. Find the scalar projection of $\mathbf{u} = 5\mathbf{i} + 5\mathbf{j} + 2\mathbf{k}$ on $\mathbf{v} = -\sqrt{5}\mathbf{i} + \sqrt{5}\mathbf{j} + \mathbf{k}$.

39. A vector $\mathbf{u} = 2\mathbf{i} + 3\mathbf{j} + z\mathbf{k}$ emanating from the origin points into the first octant (i.e., that part of three-space where all components are positive). If $\|\mathbf{u}\| = 5$, find z.

40. If $\alpha = 46°$ and $\beta = 108°$ are direction angles for a vector $\mathbf{u}$, find two possible values for the third angle.

41. Find two perpendicular vectors $\mathbf{u}$ and $\mathbf{v}$ such that each is also perpendicular to $\mathbf{w} = \langle -4, 2, 5\rangle$.

42. Find the vector emanating from the origin whose terminal point is the midpoint of the segment joining $(3, 2, -1)$ and $(5, -7, 2)$.

43. Which of the following do not make sense?

(a) $\mathbf{u}\cdot(\mathbf{v}\cdot\mathbf{w})$ (b) $(\mathbf{u}\cdot\mathbf{w}) + \mathbf{w}$

(c) $\|\mathbf{u}\|(\mathbf{v}\cdot\mathbf{w})$ (d) $(\mathbf{u}\cdot\mathbf{v})\mathbf{w}$

44. Which of the following do not make sense?

(a) $\mathbf{u}\cdot(\mathbf{v} + \mathbf{w})$ (b) $(\mathbf{u}\cdot\mathbf{w})\|\mathbf{w}\|$

(c) $\|\mathbf{u}\|\cdot(\mathbf{v} + \mathbf{w})$ (d) $(\mathbf{u} + \mathbf{v})\mathbf{w}$

In Problems 45–50, give a proof of the indicated property for two-dimensional vectors. Use $\mathbf{u} = \langle u_1, u_2\rangle$, $\mathbf{v} = \langle v_1, v_2\rangle$, *and* $\mathbf{w} = \langle w_1, w_2\rangle$.

45. $(a + b)\mathbf{u} = a\mathbf{u} + b\mathbf{u}$

46. $\mathbf{u}\cdot\mathbf{v} = \mathbf{v}\cdot\mathbf{u}$

47. $c(\mathbf{u}\cdot\mathbf{v}) = (c\mathbf{u})\cdot\mathbf{v}$

48. $\mathbf{u}\cdot(\mathbf{v} + \mathbf{w}) = \mathbf{u}\cdot\mathbf{v} + \mathbf{u}\cdot\mathbf{w}$

49. $\mathbf{0}\cdot\mathbf{u} = 0$

50. $\mathbf{u}\cdot\mathbf{u} = \|\mathbf{u}\|^2$

51. Given the two nonparallel vectors $\mathbf{a} = 3\mathbf{i} - 2\mathbf{j}$ and $\mathbf{b} = -3\mathbf{i} + 4\mathbf{j}$ and another vector $\mathbf{r} = 7\mathbf{i} - 8\mathbf{j}$, find scalars k and m such that $\mathbf{r} = k\mathbf{a} + m\mathbf{b}$.

52. Given the two nonparallel vectors $\mathbf{a} = -4\mathbf{i} + 3\mathbf{j}$ and $\mathbf{b} = 2\mathbf{i} - \mathbf{j}$ and another vector $\mathbf{r} = 6\mathbf{i} - 7\mathbf{j}$, find scalars k and m such that $\mathbf{r} = k\mathbf{a} + m\mathbf{b}$.

53. Show that the vector $\mathbf{n} = a\mathbf{i} + b\mathbf{j}$ is perpendicular to the line with equation $ax + by = c$. *Hint:* Let $P_1(x_1, y_1)$ and $P_2(x_2, y_2)$ be two points on the line and show that $\mathbf{n}\cdot\overrightarrow{P_1P_2} = 0$.

54. Prove that $\|\mathbf{u} + \mathbf{v}\|^2 + \|\mathbf{u} - \mathbf{v}\|^2 = 2\|\mathbf{u}\|^2 + 2\|\mathbf{v}\|^2$.

55. Prove that $\mathbf{u}\cdot\mathbf{v} = \frac{1}{4}\|\mathbf{u} + \mathbf{v}\|^2 - \frac{1}{4}\|\mathbf{u} - \mathbf{v}\|^2$.

56. Find the angle between a main diagonal of a cube and one of its faces.

57. Find the smallest angle between the main diagonals of a rectangular box 4 feet by 6 feet by 10 feet.

58. Find the angles formed by the diagonals of a cube.

59. Find the work done by the force $\mathbf{F} = 3\mathbf{i} + 10\mathbf{j}$ newtons in moving an object 10 meters north (i.e., in the $\mathbf{j}$ direction).

60. Find the work done by a force of 100 newtons acting in the direction S 70° E in moving an object 30 meters east.

61. Find the work done by the force $\mathbf{F} = 6\mathbf{i} + 8\mathbf{j}$ pounds in moving an object from $(1, 0)$ to $(6, 8)$, where distance is in feet.

62. Find the work done by a force $\mathbf{F} = -5\mathbf{i} + 8\mathbf{j}$ newtons in moving an object 12 meters north.

63. Find the work done by a force $\mathbf{F} = -4\mathbf{k}$ newtons in moving an object from $(0, 0, 8)$ to $(4, 4, 0)$, where distance is in meters.

64. Find the work done by a force $\mathbf{F} = 3\mathbf{i} - 6\mathbf{j} + 7\mathbf{k}$ pounds in moving an object from $(2, 1, 3)$ to $(9, 4, 6)$, where distance is in feet.

In Problems 65–68, find the equation of the plane having the given normal vector $\mathbf{n}$ *and passing through the given point* P.

65. $\mathbf{n} = 2\mathbf{i} - 4\mathbf{j} + 3\mathbf{k}$; $P(1, 2, -3)$

66. $\mathbf{n} = 3\mathbf{i} - 2\mathbf{j} - 1\mathbf{k}$; $P(-2, -3, 4)$

67. $\mathbf{n} = \langle 1, 4, 4\rangle$; $P(1, 2, 1)$

68. $\mathbf{n} = \langle 0, 0, 1\rangle$; $P(1, 2, -3)$

69. Find the smaller of the angles between the two planes from Problems 65 and 66.

70. Find the equation of the plane through $(-1, 2, -3)$ and parallel to the plane $2x + 4y - z = 6$.

71. Find the equation of the plane passing through $(-4, -1, 2)$ and parallel to

(a) the xy-plane

(b) the plane $2x - 3y - 4z = 0$

72. Find the equation of the plane passing through the origin and parallel to

(a) the xy-plane

(b) the plane $x + y + z = 1$

73. Find the distance from $(1, -1, 2)$ to the plane $x + 3y + z = 7$.

74. Find the distance from $(2, 6, 3)$ to the plane $-3x + 2y + z = 9$.

75. Find the distance between the parallel planes $-3x + 2y + z = 9$ and $6x - 4y - 2z = 19$.

76. Find the distance between the parallel planes $5x - 3y - 2z = 5$ and $-5x + 3y + 2z = 7$.

77. Find the distance from the sphere $x^2 + y^2 + z^2 + 2x + 6y - 8z = 0$ to the plane $3x + 4y + z = 15$.

78. Find the equation of the plane each of whose points is equidistant from $(-2, 1, 4)$ and $(6, 1, -2)$.

79. Prove the **Cauchy-Schwarz Inequality** for two-dimensional vectors:

$$|\mathbf{u} \cdot \mathbf{v}| \le \|\mathbf{u}\|\|\mathbf{v}\|$$

80. Prove the **Triangle Inequality** (see Figure 13) for two-dimensional vectors:

$$\|\mathbf{u} + \mathbf{v}\| \le \|\mathbf{u}\| + \|\mathbf{v}\|$$

Hint: Use the dot product to compute $\|\mathbf{u} + \mathbf{v}\|$; then use the Cauchy-Schwarz Inequality from Problem 79.

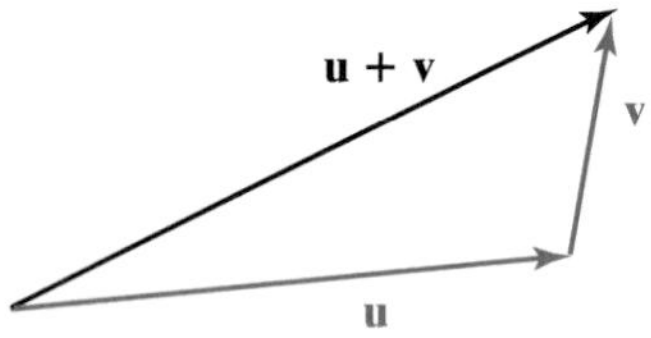

Figure 13

81. A weight of 30 pounds is suspended by three wires with resulting tensions $3\mathbf{i} + 4\mathbf{j} + 15\mathbf{k}$, $-8\mathbf{i} - 2\mathbf{j} + 10\mathbf{k}$, and $a\mathbf{i} + b\mathbf{j} + c\mathbf{k}$. Determine a, b, and c so that the net force is straight up.

82. Show that the work done by a constant force $\mathbf{F}$ on an object that moves completely around a closed polygonal path is 0.

83. Let $\mathbf{a} = \langle a_1, a_2, a_3 \rangle$ and $\mathbf{b} = \langle b_1, b_2, b_3 \rangle$ be fixed vectors. Show that $(\mathbf{x} - \mathbf{a}) \cdot (\mathbf{x} - \mathbf{b}) = 0$ is the equation of a sphere, and find its center and radius.

84. Refine the method of Example 10 by showing that the distance L between the parallel planes $Ax + By + Cz = D$ and $Ax + By + Cz = E$ is

$$L = \frac{|D - E|}{\sqrt{A^2 + B^2 + C^2}}$$

85. The medians of a triangle meet at a point P (the centroid by Problem 30 of Section 5.6) that is two-thirds of the way from a vertex to the midpoint of the opposite edge. Show that P is the head of the position vector $(\mathbf{a} + \mathbf{b} + \mathbf{c})/3$, where $\mathbf{a}$, $\mathbf{b}$, and $\mathbf{c}$ are the position vectors of the vertices, and use this to find P if the vertices are $(2, 6, 5)$, $(4, -1, 2)$, and $(6, 1, 2)$.

86. Let $\mathbf{a}$, $\mathbf{b}$, $\mathbf{c}$, and $\mathbf{d}$ be the position vectors of the vertices of a tetrahedron. Show that the lines joining the vertices to the centroids of the opposite faces meet in a point P, and give a nice vector formula for it, thus generalizing Problem 85.

87. Suppose that the three coordinate planes bounding the first octant are mirrors. A light ray with direction $a\mathbf{i} + b\mathbf{j} + c\mathbf{k}$ is reflected successively from the xy-plane, the xz-plane, and the yz-plane. Determine the direction of the ray after each reflection, and state a nice conclusion concerning the final reflected ray.

Answers to Concepts Review: **1.** $u_1v_1 + u_2v_2 + u_3v_3$; $\|\mathbf{u}\|\|\mathbf{v}\| \cos\theta$ **2.** 0 **3.** $\mathbf{F} \cdot \mathbf{D}$ **4.** $\langle A, B, C \rangle$

11.4 The Cross Product

The dot product of two vectors is a scalar. We have explored some of its uses in the previous section. Now we introduce the **cross product** (or vector product); it will also have many uses. The cross product $\mathbf{u} \times \mathbf{v}$ of $\mathbf{u} = \langle u_1, u_2, u_3 \rangle$ and $\mathbf{v} = \langle v_1, v_2, v_3 \rangle$ is defined by

$$\mathbf{u} \times \mathbf{v} = \langle u_2v_3 - u_3v_2, u_3v_1 - u_1v_3, u_1v_2 - u_2v_1 \rangle$$

In this form, the formula is hard to remember and its significance is not obvious. Note the one thing that is obvious. The cross product of two vectors is a vector.

To help us remember the formula for the cross product, we recall a subject from an earlier mathematics course, namely, *determinants*. First, the value of a 2×2 determinant is

$$\begin{vmatrix} a & b \\ c & d \end{vmatrix} = ad - bc$$

Then the value of a 3×3 determinant is (expanding along to the top row)

$$\begin{vmatrix} a_1 & a_2 & a_3 \\ b_1 & b_2 & b_3 \\ c_1 & c_2 & c_3 \end{vmatrix} = a_1\begin{vmatrix} a_1 & a_2 & a_3 \\ b_1 & b_2 & b_3 \\ c_1 & c_2 & c_3 \end{vmatrix} - a_2\begin{vmatrix} a_1 & a_2 & a_3 \\ b_1 & b_2 & b_3 \\ c_1 & c_2 & c_3 \end{vmatrix} + a_3\begin{vmatrix} a_1 & a_2 & a_3 \\ b_1 & b_2 & b_3 \\ c_1 & c_2 & c_3 \end{vmatrix}$$

$$= a_1\begin{vmatrix} b_2 & b_3 \\ c_2 & c_3 \end{vmatrix} - a_2\begin{vmatrix} b_1 & b_3 \\ c_1 & c_3 \end{vmatrix} + a_3\begin{vmatrix} b_1 & b_2 \\ c_1 & c_2 \end{vmatrix}$$

Using determinants, we can write the definition of $\mathbf{u} \times \mathbf{v}$ as

$$\mathbf{u} \times \mathbf{v} = \begin{vmatrix} \mathbf{i} & \mathbf{j} & \mathbf{k} \\ u_1 & u_2 & u_3 \\ v_1 & v_2 & v_3 \end{vmatrix} = \begin{vmatrix} u_2 & u_3 \\ v_2 & v_3 \end{vmatrix}\mathbf{i} - \begin{vmatrix} u_1 & u_3 \\ v_1 & v_3 \end{vmatrix}\mathbf{j} + \begin{vmatrix} u_1 & u_2 \\ v_1 & v_2 \end{vmatrix}\mathbf{k}$$

Note that the components of the left vector $\mathbf{u}$ go in the second row, and those of the right vector $\mathbf{v}$ go in the third row. This is important, because if we interchange the positions of $\mathbf{u}$ and $\mathbf{v}$, we interchange the second and third rows of the determinant, and this changes the sign of the determinant's value, as you may check. Thus,

$$\mathbf{u} \times \mathbf{v} = -(\mathbf{v} \times \mathbf{u})$$

which is sometimes called the *anticommutative law*.

Torque

The cross product plays an important role in mechanics. Let O be a *fixed point* in a body, and suppose that a force $\mathbf{F}$ is applied at another point P of the body. Then $\mathbf{F}$ tends to rotate the body about an axis through O and perpendicular to the plane of OP and $\mathbf{F}$. The vector

$$\tau = \overrightarrow{OP} \times \mathbf{F}$$

is called the **torque.** It points in the direction of the axis and has magnitude $\|\overrightarrow{OP}\|\|\mathbf{F}\| \sin \theta$, which is just the moment of force about the axis due to $\mathbf{F}$.

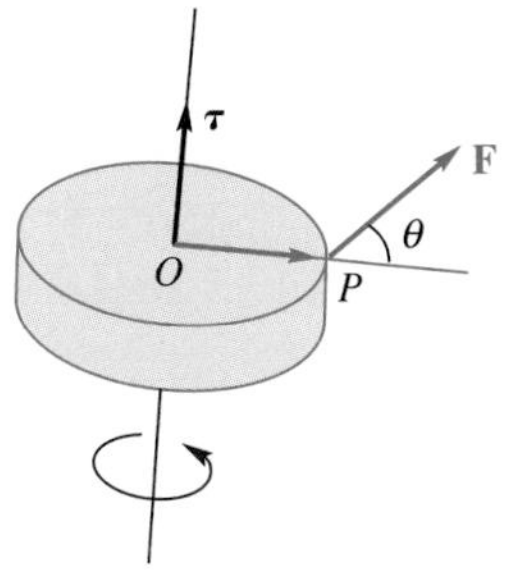

EXAMPLE 1 Let $\mathbf{u} = \langle 1, -2, -1 \rangle$ and $\mathbf{v} = \langle -2, 4, 1 \rangle$. Calculate $\mathbf{u} \times \mathbf{v}$ and $\mathbf{v} \times \mathbf{u}$ using the determinant definition.

SOLUTION

$$\mathbf{u} \times \mathbf{v} = \begin{vmatrix} \mathbf{i} & \mathbf{j} & \mathbf{k} \\ 1 & -2 & -1 \\ -2 & 4 & 1 \end{vmatrix} = \begin{vmatrix} -2 & -1 \\ 4 & 1 \end{vmatrix}\mathbf{i} - \begin{vmatrix} 1 & -1 \\ -2 & 1 \end{vmatrix}\mathbf{j} + \begin{vmatrix} 1 & -2 \\ -2 & 4 \end{vmatrix}\mathbf{k}$$

$$= 2\mathbf{i} + \mathbf{j} + 0\mathbf{k}$$

$$\mathbf{v} \times \mathbf{u} = \begin{vmatrix} \mathbf{i} & \mathbf{j} & \mathbf{k} \\ -2 & 4 & 1 \\ 1 & -2 & -1 \end{vmatrix} = \begin{vmatrix} 4 & 1 \\ -2 & -1 \end{vmatrix}\mathbf{i} - \begin{vmatrix} -2 & 1 \\ 1 & -1 \end{vmatrix}\mathbf{j} + \begin{vmatrix} -2 & 4 \\ 1 & -2 \end{vmatrix}\mathbf{k}$$

$$= -2\mathbf{i} - \mathbf{j} + 0\mathbf{k}$$ ■

Geometric Interpretation of $\mathbf{u} \times \mathbf{v}$ Like the dot product, the cross product gains significance from its geometric interpretation.

Theorem A

Let $\mathbf{u}$ and $\mathbf{v}$ be vectors in three-space and θ be the angle between them. Then

1. $\mathbf{u} \cdot (\mathbf{u} \times \mathbf{v}) = 0 = \mathbf{v} \cdot (\mathbf{u} \times \mathbf{v})$, that is, $\mathbf{u} \times \mathbf{v}$ is perpendicular to both $\mathbf{u}$ and $\mathbf{v}$;
2. $\mathbf{u}$, $\mathbf{v}$, and $\mathbf{u} \times \mathbf{v}$ form a right-handed triple;
3. $\|\mathbf{u} \times \mathbf{v}\| = \|\mathbf{u}\|\|\mathbf{v}\| \sin \theta$.

Proof Let $\mathbf{u} = \langle u_1, u_2, u_3 \rangle$ and $\mathbf{v} = \langle v_1, v_2, v_3 \rangle$.

(i) $\mathbf{u} \cdot (\mathbf{u} \times \mathbf{v}) = u_1(u_2v_3 - u_3v_2) + u_2(u_3v_1 - u_1v_3) + u_3(u_1v_2 - u_2v_1)$. When we remove parentheses, the six terms cancel in pairs, leaving a sum of 0. A similar event occur's when we expand $\mathbf{v} \cdot (\mathbf{u} \times \mathbf{v})$.

(ii) The meaning of right-handedness for the triple $\mathbf{u}$, $\mathbf{v}$, $\mathbf{u} \times \mathbf{v}$ is illustrated in Figure 1. There θ is the angle between $\mathbf{u}$ and $\mathbf{v}$, and the fingers of the right hand are curled in the direction of the rotation through θ that makes $\mathbf{u}$ coincide with $\mathbf{v}$. It is difficult to establish analytically that the indicated triple is right-handed, but you might check it with a few examples. Note in particular that $\mathbf{i} \times \mathbf{j} = \mathbf{k}$, and by definition we know that the triple $\mathbf{i}, \mathbf{j}, \mathbf{k}$ is right-handed.

(iii) We need Lagrange's Identity,

$$\|\mathbf{u} \times \mathbf{v}\|^2 = \|\mathbf{u}\|^2\|\mathbf{v}\|^2 - (\mathbf{u} \cdot \mathbf{v})^2$$

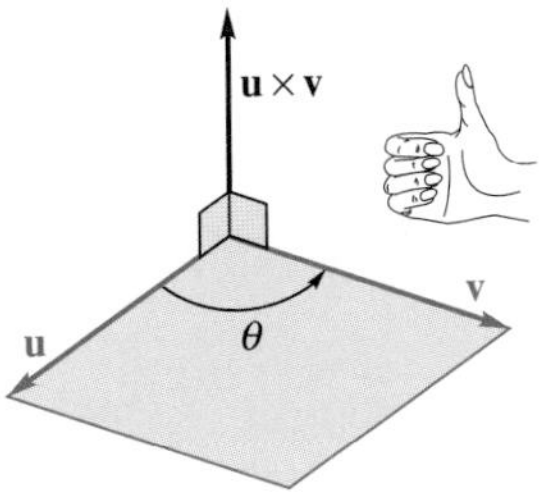

Figure 1

whose proof is a simple algebraic exercise (Problem 31). Using this identity, we may write

$$\begin{aligned}\|\mathbf{u}\times\mathbf{v}\|^2 &= \|\mathbf{u}\|^2\|\mathbf{v}\|^2 - (\|\mathbf{u}\|\|\mathbf{v}\|\cos\theta)^2\\ &= \|\mathbf{u}\|^2\|\mathbf{v}\|^2(1-\cos^2\theta)\\ &= \|\mathbf{u}\|^2\|\mathbf{v}\|^2\sin^2\theta\end{aligned}$$

Since $0 \le \theta \le \pi$, $\sin\theta \ge 0$. Taking the principal square root yields

$$\|\mathbf{u}\times\mathbf{v}\| = \|\mathbf{u}\|\|\mathbf{v}\|\sin\theta$$

■

It is important that we have geometric interpretations of both $\mathbf{u}\cdot\mathbf{v}$ and $\mathbf{u}\times\mathbf{v}$. While both products were originally defined in terms of components that depend on a choice of coordinate system, they are actually independent of coordinate systems. They are intrinsic geometric quantities, and you will get the same results for $\mathbf{u}\cdot\mathbf{v}$ and $\mathbf{u}\times\mathbf{v}$ no matter how you introduce the coordinates used to compute them.

Here is a simple consequence of Theorem A (part 3) and the fact that vectors are parallel if and only if the angle θ between them is either 0° or 180°.

Theorem B

Two vectors $\mathbf{u}$ and $\mathbf{v}$ in three-space are parallel if and only if $\mathbf{u}\times\mathbf{v} = \mathbf{0}$.

Applications Our first application is to find the equation of the plane through three noncollinear points.

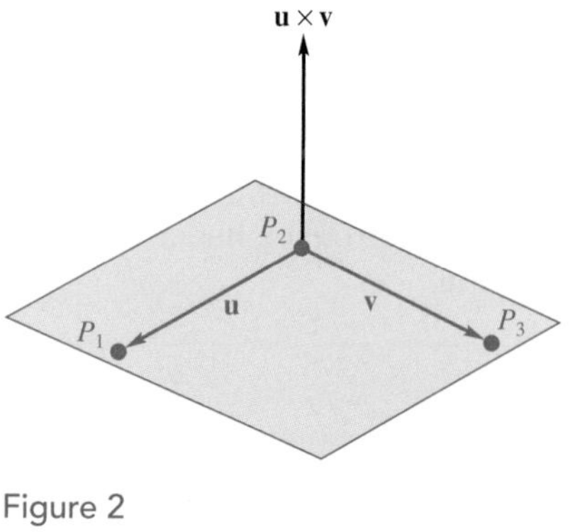

Figure 2

EXAMPLE 2 Find the equation of the plane (Figure 2) through the three points $P_1(1, -2, 3)$, $P_2(4, 1, -2)$, and $P_3(-2, -3, 0)$.

SOLUTION Let $\mathbf{u} = \overrightarrow{P_2P_1} = \langle -3, -3, 5\rangle$ and $\mathbf{v} = \overrightarrow{P_2P_3} = \langle -6, -4, 2\rangle$. From the first part of Theorem A we know that

$$\mathbf{u}\times\mathbf{v} = \begin{vmatrix}\mathbf{i} & \mathbf{j} & \mathbf{k}\\ -3 & -3 & 5\\ -6 & -4 & 2\end{vmatrix} = 14\mathbf{i} - 24\mathbf{j} - 6\mathbf{k}$$

is perpendicular to both $\mathbf{u}$ and $\mathbf{v}$ and thus to the plane containing them. The plane through $(4, 1, -2)$ with normal $14\mathbf{i} - 24\mathbf{j} - 6\mathbf{k}$ has equation (see Section 11.3)

$$14(x-4) - 24(y-1) - 6(z+2) = 0$$

or

$$14x - 24y - 6z = 44$$

■

EXAMPLE 3 Show that the area of a parallelogram with $\mathbf{a}$ and $\mathbf{b}$ as adjacent sides is $\|\mathbf{a}\times\mathbf{b}\|$.

SOLUTION Recall that the area of a parallelogram is the product of the base times the height. Now look at Figure 3 and use the fact that $\|\mathbf{a}\times\mathbf{b}\| = \|\mathbf{a}\|\|\mathbf{b}\|\sin\theta$. ■

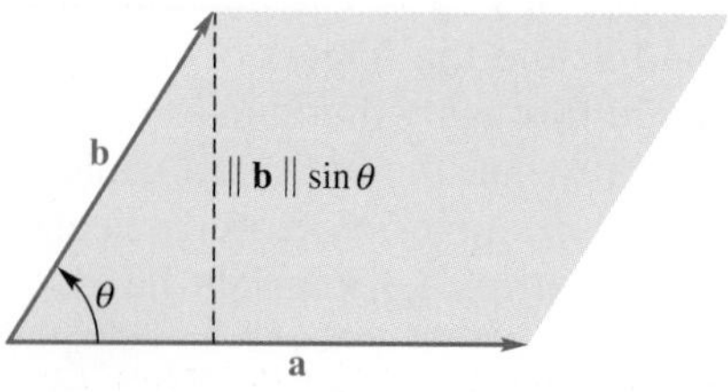

Figure 3

EXAMPLE 4 Show that the volume of the parallelepiped determined by the vectors $\mathbf{a}$, $\mathbf{b}$, and $\mathbf{c}$ is

$$V = |\mathbf{a}\cdot(\mathbf{b}\times\mathbf{c})| = \left|\begin{vmatrix}a_1 & a_2 & a_3\\ b_1 & b_2 & b_3\\ c_1 & c_2 & c_3\end{vmatrix}\right|$$

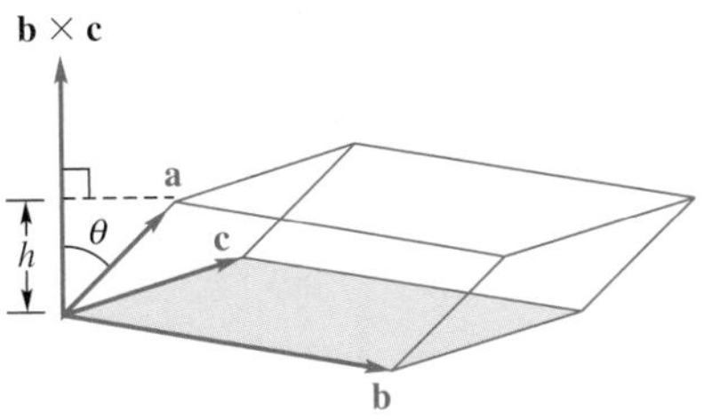

Figure 4

SOLUTION Refer to Figure 4 and regard the parallelogram determined by **b** and **c** as the base of the parallelepiped. The area of this base is $\|\mathbf{b}\times\mathbf{c}\|$ by Example 3; the height h of the parallelepiped is the absolute value of the scalar projection of **a** on $\mathbf{b}\times\mathbf{c}$. Thus,

$$h = \|\mathbf{a}\||\cos\theta| = \frac{\|\mathbf{a}\||\mathbf{a}\cdot(\mathbf{b}\times\mathbf{c})|}{\|\mathbf{a}\|\|\mathbf{b}\times\mathbf{c}\|} = \frac{|\mathbf{a}\cdot(\mathbf{b}\times\mathbf{c})|}{\|\mathbf{b}\times\mathbf{c}\|}$$

and

$$V = h\|\mathbf{b}\times\mathbf{c}\| = |\mathbf{a}\cdot(\mathbf{b}\times\mathbf{c})|$$

■

Suppose that the vectors **a**, **b**, and **c** from the previous example are in the *same* plane. In this case, the parallelepiped has height zero, so the volume should be zero. Does the formula for the volume yield $V = 0$? If **a** is in the plane determined by **b** and **c**, then any vector perpendicular to **b** and **c** will be perpendicular to **a** as well. The vector $\mathbf{b}\times\mathbf{c}$ is perpendicular to both **b** and **c**; hence $\mathbf{b}\times\mathbf{c}$ is perpendicular to **a**. Thus, $\mathbf{a}\cdot(\mathbf{b}\times\mathbf{c}) = 0$.

> **Check Extreme Cases**
>
> ≈ Never read a mathematics book passively; rather, ask questions as you go. In particular, you should look at extreme cases whenever possible. Here we look at the case where the vectors **a**, **b**, and **c** are in the *same* plane. The volume of the parallelepiped should be zero, and indeed the formula does give zero. What happens in Example 3 if the vectors **b** and **c** are parallel?

Algebraic Properties The rules for calculating with cross products are summarized in the following theorem. Proving this theorem is a matter of writing everything out in terms of components and will be left as an exercise.

Theorem C

If **u**, **v**, and **w** are vectors in three-space and k is a scalar, then

1. $\mathbf{u}\times\mathbf{v} = -(\mathbf{v}\times\mathbf{u})$ (anticommutative law);
2. $\mathbf{u}\times(\mathbf{v}+\mathbf{w}) = (\mathbf{u}\times\mathbf{v}) + (\mathbf{u}\times\mathbf{w})$ (left distributive law);
3. $k(\mathbf{u}\times\mathbf{v}) = (k\mathbf{u})\times\mathbf{v} = \mathbf{u}\times(k\mathbf{v})$;
4. $\mathbf{u}\times\mathbf{0} = \mathbf{0}\times\mathbf{u} = \mathbf{0}, \mathbf{u}\times\mathbf{u} = \mathbf{0}$;
5. $(\mathbf{u}\times\mathbf{v})\cdot\mathbf{w} = \mathbf{u}\cdot(\mathbf{v}\times\mathbf{w})$;
6. $\mathbf{u}\times(\mathbf{v}\times\mathbf{w}) = (\mathbf{u}\cdot\mathbf{w})\mathbf{v} - (\mathbf{u}\cdot\mathbf{v})\mathbf{w}$.

Once the rules in Theorem C are mastered, complicated calculations with vectors can be done with ease. We illustrate by calculating a cross product in a new way. We will need the following simple but important products.

$$\mathbf{i}\times\mathbf{j} = \mathbf{k}, \quad \mathbf{j}\times\mathbf{k} = \mathbf{i}, \quad \mathbf{k}\times\mathbf{i} = \mathbf{j}$$

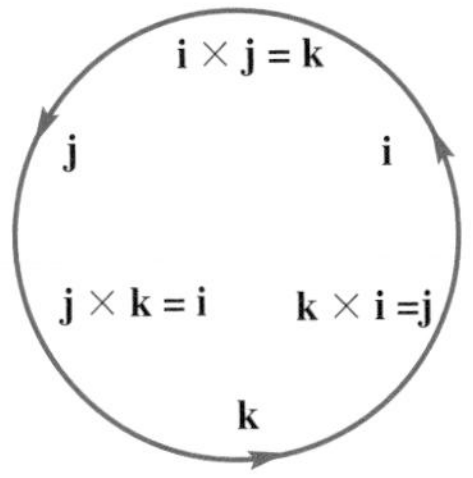

Figure 5

These results have a cyclic order, which can be remembered by appealing to Figure 5.

EXAMPLE 5 Calculate $\mathbf{u}\times\mathbf{v}$ if $\mathbf{u} = 3\mathbf{i} - 2\mathbf{j} + \mathbf{k}$ and $\mathbf{v} = 4\mathbf{i} + 2\mathbf{j} - 3\mathbf{k}$.

SOLUTION We appeal to Theorem C, especially the distributive law and the anticommutative law.

$$\begin{aligned}\mathbf{u}\times\mathbf{v} &= (3\mathbf{i} - 2\mathbf{j} + \mathbf{k})\times(4\mathbf{i} + 2\mathbf{j} - 3\mathbf{k}) \\ &= 12(\mathbf{i}\times\mathbf{i}) + 6(\mathbf{i}\times\mathbf{j}) - 9(\mathbf{i}\times\mathbf{k}) - 8(\mathbf{j}\times\mathbf{i}) - 4(\mathbf{j}\times\mathbf{j}) \\ &\quad + 6(\mathbf{j}\times\mathbf{k}) + 4(\mathbf{k}\times\mathbf{i}) + 2(\mathbf{k}\times\mathbf{j}) - 3(\mathbf{k}\times\mathbf{k}) \\ &= 12(\mathbf{0}) + 6(\mathbf{k}) - 9(-\mathbf{j}) - 8(-\mathbf{k}) - 4(\mathbf{0}) \\ &\quad + 6(\mathbf{i}) + 4(\mathbf{j}) + 2(-\mathbf{i}) - 3(\mathbf{0}) \\ &= 4\mathbf{i} + 13\mathbf{j} + 14\mathbf{k}\end{aligned}$$

Experts would do most of this in their heads; novices might find the determinant method easier.

■

Concepts Review

1. The cross product of $\mathbf{u} = \langle -1, 2, 1 \rangle$ and $\mathbf{v} = \langle 3, 1, -1 \rangle$ is given by a specific determinant; evaluation of this determinant gives $\mathbf{u} \times \mathbf{v} =$ _____.

2. Geometrically, $\mathbf{u} \times \mathbf{v}$ is a vector perpendicular to the plane of $\mathbf{u}$ and $\mathbf{v}$ and has length $\|\mathbf{u} \times \mathbf{v}\| =$ _____.

3. The cross product is anticommutative; that is, $\mathbf{u} \times \mathbf{v} =$ _____.

4. Two vectors are _____ if and only if their cross product is $\mathbf{0}$.

Problem Set 11.4

1. Let $\mathbf{a} = -3\mathbf{i} + 2\mathbf{j} - 2\mathbf{k}$, $\mathbf{b} = -\mathbf{i} + 2\mathbf{j} - 4\mathbf{k}$, and $\mathbf{c} = 7\mathbf{i} + 3\mathbf{j} - 4\mathbf{k}$. Find each of the following:

(a) $\mathbf{a} \times \mathbf{b}$ (b) $\mathbf{a} \times (\mathbf{b} + \mathbf{c})$

(c) $\mathbf{a} \cdot (\mathbf{b} + \mathbf{c})$ (d) $\mathbf{a} \times (\mathbf{b} \times \mathbf{c})$

2. If $\mathbf{a} = \langle 3, 3, 1 \rangle$, $\mathbf{b} = \langle -2, -1, 0 \rangle$, and $\mathbf{c} = \langle -2, -3, -1 \rangle$, find each of the following:

(a) $\mathbf{a} \times \mathbf{b}$ (b) $\mathbf{a} \times (\mathbf{b} + \mathbf{c})$

(c) $\mathbf{a} \cdot (\mathbf{b} \times \mathbf{c})$ (d) $\mathbf{a} \times (\mathbf{b} \times \mathbf{c})$

3. Find all vectors perpendicular to both of the vectors $\mathbf{a} = \mathbf{i} + 2\mathbf{j} + 3\mathbf{k}$ and $\mathbf{b} = -2\mathbf{i} + 2\mathbf{j} - 4\mathbf{k}$.

4. Find all vectors perpendicular to both of the vectors $\mathbf{a} = -2\mathbf{i} + 5\mathbf{j} - 2\mathbf{k}$ and $\mathbf{b} = 3\mathbf{i} - 2\mathbf{j} + 4\mathbf{k}$.

5. Find the unit vectors perpendicular to the plane determined by the three points $(1, 3, 5)$, $(3, -1, 2)$, and $(4, 0, 1)$.

6. Find the unit vectors perpendicular to the plane determined by the three points $(-1, 3, 0)$, $(5, 1, 2)$, and $(4, -3, -1)$.

7. Find the area of the parallelogram with $\mathbf{a} = -\mathbf{i} + \mathbf{j} - 3\mathbf{k}$ and $\mathbf{b} = 4\mathbf{i} + 2\mathbf{j} - 4\mathbf{k}$ as the adjacent sides.

8. Find the area of the parallelogram with $\mathbf{a} = 2\mathbf{i} + 2\mathbf{j} - \mathbf{k}$ and $\mathbf{b} = -\mathbf{i} + \mathbf{j} - 4\mathbf{k}$ as the adjacent sides.

9. Find the area of the triangle with $(3, 2, 1)$, $(2, 4, 6)$, and $(-1, 2, 5)$ as vertices.

10. Find the area of the triangle with $(1, 2, 3)$, $(3, 1, 5)$, and $(4, 5, 6)$ as vertices.

In Problems 11–14, find the equation of the plane through the given points.

11. $(1, 3, 2)$, $(0, 3, 0)$, and $(2, 4, 3)$

12. $(1, 1, 2)$, $(0, 0, 1)$, and $(-2, -3, 0)$

13. $(7, 0, 0)$, $(0, 3, 0)$, and $(0, 0, 5)$

14. $(a, 0, 0)$, $(0, b, 0)$, and $(0, 0, c)$, (None of a, b, and c is zero.)

15. Find the equation of the plane through $(2, 5, 1)$ that is parallel to the plane $x - y + 2z = 4$.

16. Find the equation of the plane through $(0, 0, 2)$ that is parallel to the plane $x + y + z = 1$.

17. Find the equation of the plane through $(-1, -2, 3)$ and perpendicular to both the planes $x - 3y + 2z = 7$ and $2x - 2y - z = -3$.

18. Find the equation of the plane through $(2, -1, 4)$ that is perpendicular to both the planes $x + y + z = 2$ and $x - y - z = 4$.

19. Find the equation of the plane through $(2, -3, 2)$ and parallel to the plane of the vectors $4\mathbf{i} + 3\mathbf{j} - \mathbf{k}$ and $2\mathbf{i} - 5\mathbf{j} + 6\mathbf{k}$.

20. Find the equation of the plane through the origin that is perpendicular to the xy-plane and the plane $3x - 2y + z = 4$.

21. Find the equation of the plane through $(6, 2, -1)$ and perpendicular to the line of intersection of the planes $4x - 3y + 2z + 5 = 0$ and $3x + 2y - z + 11 = 0$.

22. Let $\mathbf{a}$ and $\mathbf{b}$ be nonparallel vectors, and let $\mathbf{c}$ be any nonzero vector. Show that $(\mathbf{a} \times \mathbf{b}) \times \mathbf{c}$ is a vector in the plane of $\mathbf{a}$ and $\mathbf{b}$.

23. Find the volume of the parallelepiped with edges $\langle 2, 3, 4 \rangle$, $\langle 0, 4, -1 \rangle$, and $\langle 5, 1, 3 \rangle$ (see Example 4).

24. Find the volume of the parallelepiped with edges $3\mathbf{i} - 4\mathbf{j} + 2\mathbf{k}$, $-\mathbf{i} + 2\mathbf{j} + \mathbf{k}$, and $3\mathbf{i} - 2\mathbf{j} + 5\mathbf{k}$.

25. Let K be the parallelepiped determined by $\mathbf{u} = \langle 3, 2, 1 \rangle$, $\mathbf{v} = \langle 1, 1, 2 \rangle$, and $\mathbf{w} = \langle 1, 3, 3 \rangle$.

(a) Find the volume of K.

(b) Find the area of the face determined by $\mathbf{u}$ and $\mathbf{v}$.

(c) Find the angle between $\mathbf{u}$ and the plane containing the face determined by $\mathbf{v}$ and $\mathbf{w}$.

26. The formula for the volume of a parallelepiped derived in Example 4 should not depend on the choice of which one of the three vectors we call $\mathbf{a}$, which one we call $\mathbf{b}$, and which one we call $\mathbf{c}$. Use this result to explain why $|\mathbf{a} \cdot (\mathbf{b} \times \mathbf{c})| = |\mathbf{b} \cdot (\mathbf{a} \times \mathbf{c})| = |\mathbf{c} \cdot (\mathbf{a} \times \mathbf{b})|$.

27. Which of the following do *not* make sense?

(a) $\mathbf{u} \cdot (\mathbf{v} \times \mathbf{w})$ (b) $\mathbf{u} + (\mathbf{v} \times \mathbf{w})$

(c) $(\mathbf{a} \cdot \mathbf{b}) \times \mathbf{c}$ (d) $(\mathbf{a} \times \mathbf{b}) + k$

(e) $(\mathbf{a} \cdot \mathbf{b}) + k$ (f) $(\mathbf{a} + \mathbf{b}) \times (\mathbf{c} + \mathbf{d})$

(g) $(\mathbf{u} \times \mathbf{v}) \times \mathbf{w}$ (h) $(k\mathbf{u}) \times \mathbf{v}$

28. Show that if $\mathbf{a}$, $\mathbf{b}$, $\mathbf{c}$, and $\mathbf{d}$ all lie in the same plane then

$$(\mathbf{a} \times \mathbf{b}) \times (\mathbf{c} \times \mathbf{d}) = \mathbf{0}$$

29. The volume of a tetrahedron is known to be $\frac{1}{3}$(area of base)(height). From this, show that the volume of the tetrahedron with edges $\mathbf{a}$, $\mathbf{b}$, and $\mathbf{c}$ is $\frac{1}{6}|\mathbf{a} \cdot (\mathbf{b} \times \mathbf{c})|$.

30. Find the volume of the tetrahedron with vertices $(-1, 2, 3)$, $(4, -1, 2)$, $(5, 6, 3)$, and $(1, 1, -2)$ (see Problem 29).

31. Prove **Lagrange's Identity,**

$$\|\mathbf{u} \times \mathbf{v}\|^2 = \|\mathbf{u}\|^2\|\mathbf{v}\|^2 - (\mathbf{u} \cdot \mathbf{v})^2$$

without using Theorem A.

32. Prove the left distributive law,

$$\mathbf{u} \times (\mathbf{v} + \mathbf{w}) = (\mathbf{u} \times \mathbf{v}) + (\mathbf{u} \times \mathbf{w})$$

33. Use Problem 32 and the anticommutative law to prove the right distributive law.

34. If both $\mathbf{u} \times \mathbf{v} = \mathbf{0}$ and $\mathbf{u} \cdot \mathbf{v} = 0$, what can you conclude about $\mathbf{u}$ or $\mathbf{v}$?

35. Use Example 3 to develop a formula for the area of the triangle with vertices $P(a, 0, 0)$, $Q(0, b, 0)$, and $R(0, 0, c)$ shown in the left half of Figure 6.

36. Show that the triangle in the plane with vertices (x_1, y_1), (x_2, y_2), and (x_3, y_3) has area equal to one-half the absolute value of the determinant

$$\begin{vmatrix} x_1 & y_1 & 1 \\ x_2 & y_2 & 1 \\ x_3 & y_3 & 1 \end{vmatrix}$$

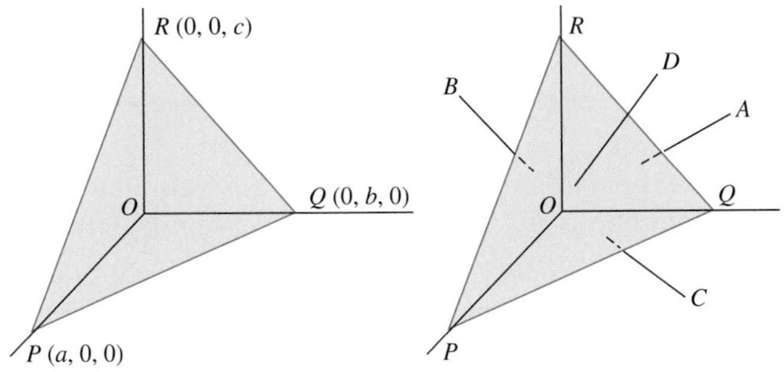

Figure 6

37. A Pythagorean Theorem in Three-Space As in Figure 6, let $P, Q, R,$ and O be the vertices of a (right-angled) tetrahedron, and let $A, B, C,$ and D be the areas of the opposite faces, respectively. Show that $A^2 + B^2 + C^2 = D^2$.

38. Let vectors **a**, **b**, and **c** with common initial point determine a tetrahedron, and let **m**, **n**, **p**, and **q** be vectors perpendicular to the four faces, pointing outward, and having length equal to the area of the corresponding face. Show that $\mathbf{m} + \mathbf{n} + \mathbf{p} + \mathbf{q} = \mathbf{0}$.

39. Let **a**, **b**, and $\mathbf{a} - \mathbf{b}$ denote the three edges of a triangle with lengths a, b, and c, respectively. Use Lagrange's Identity together with $2\mathbf{a} \cdot \mathbf{b} = \|\mathbf{a}\|^2 + \|\mathbf{b}\|^2 - \|\mathbf{a} - \mathbf{b}\|^2$ to prove **Heron's Formula** for the area A of a triangle,

$$A = \sqrt{s(s-a)(s-b)(s-c)}$$

where s is the semiperimeter $(a + b + c)/2$.

40. Use the method of Example 5 to show directly that, if $\mathbf{u} = u_1\mathbf{i} + u_2\mathbf{j} + u_3\mathbf{k}$ and $\mathbf{v} = v_1\mathbf{i} + v_2\mathbf{j} + v_3\mathbf{k}$, then

$$\mathbf{u} \times \mathbf{v} = (u_2v_3 - u_3v_2)\mathbf{i} + (u_3v_1 - u_1v_3)\mathbf{j} + (u_1v_2 - u_2v_1)\mathbf{k}$$

Answers to Concepts Review: **1.** $\langle -3, 2, -7 \rangle$ or $-3\mathbf{i} + 2\mathbf{j} - 7\mathbf{k}$ **2.** $\|\mathbf{u}\|\|\mathbf{v}\| \sin\theta$ **3.** $-(\mathbf{v} \times \mathbf{u})$ **4.** parallel

11.5 Vector-Valued Functions and Curvilinear Motion

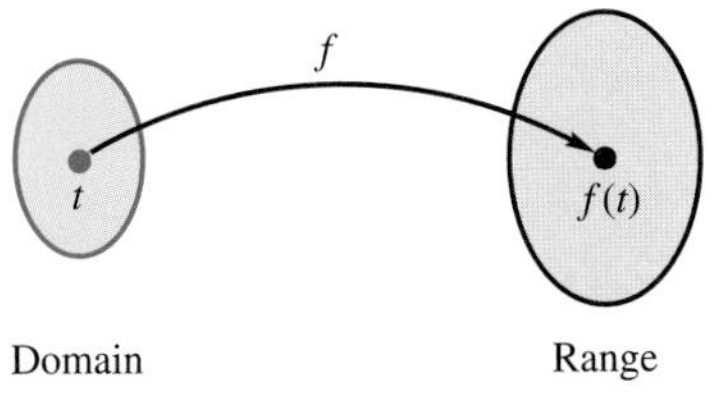

Figure 1

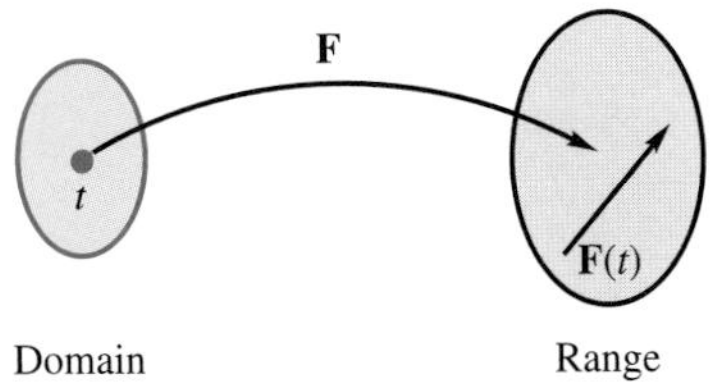

Figure 2

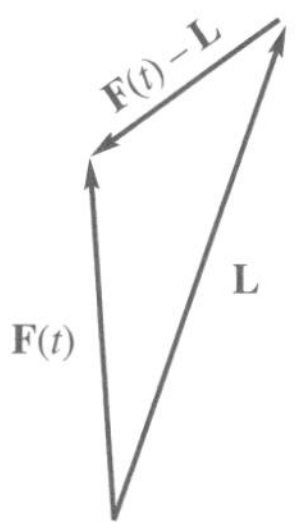

Figure 3

Recall that a function f is a rule that associates with each member t of one set (the domain) a unique value $f(t)$ from a second set (Figure 1). The set of values so obtained is the range of the function. So far in this book, our functions have been real-valued functions (scalar-valued functions) of a real variable; that is, both the domain and range have been sets of real numbers. A typical example is $f(t) = t^2$, which associates with each real number t the real number t^2.

Now we offer the first of many generalizations (Figure 2). A **vector-valued function F** of a real variable t associates with each real number t a vector $\mathbf{F}(t)$. Thus,

$$\mathbf{F}(t) = f(t)\mathbf{i} + g(t)\mathbf{j} + h(t)\mathbf{k} = \langle f(t), g(t), h(t) \rangle$$

where f, g, and h are ordinary real-valued functions. A typical example is

$$\mathbf{F}(t) = t^2\mathbf{i} + e^t\mathbf{j} + 2\mathbf{k} = \langle t^2, e^t, 2 \rangle$$

Note our use of a boldface letter; this helps us to distinguish between vector functions and scalar functions.

Calculus for Vector Functions

The most fundamental notion in calculus is that of limit. Intuitively, $\lim_{t \to c} \mathbf{F}(t) = \mathbf{L}$ means that the vector $\mathbf{F}(t)$ tends toward the vector $\mathbf{L}$ as t tends toward c. Alternatively, it means that the vector $\mathbf{F}(t) - \mathbf{L}$ approaches $\mathbf{0}$ as $t \to c$ (Figure 3). The precise ε-δ definition is nearly identical with that given for real-valued functions in Section 1.2.

Definition Limit of a Vector-Valued Function

To say that $\lim_{t \to c} \mathbf{F}(t) = \mathbf{L}$ means that, for each given $\varepsilon > 0$ (no matter how small), there is a corresponding $\delta > 0$ such that $\|\mathbf{F}(t) - \mathbf{L}\| < \varepsilon$, provided that $0 < |t - c| < \delta$; that is,

$$0 < |t - c| < \delta \Rightarrow \|\mathbf{F}(t) - \mathbf{L}\| < \varepsilon$$

The definition of $\lim_{t \to c} \mathbf{F}(t)$ is nearly the same as our definition of the limit from Chapter 1, once we interpret $\|\mathbf{F}(t) - \mathbf{L}\|$ as the length of the vector $\mathbf{F}(t) - \mathbf{L}$. Our definition says that we can make $\mathbf{F}(t)$ as close as we like (within ε) to $\mathbf{L}$ (here distance is measured in three-dimensional space), as long as we take t to be close enough (within δ) of c. The next theorem, which is proved for two-dimensional vectors in Appendix A.2, Theorem D, gives the relationship between the limit of $\mathbf{F}(t)$ and the limits of the components of $\mathbf{F}(t)$.

> **Vectors in Two Dimensions**
>
> The definitions and theorems in this section are given for three-dimensional vectors. The results for two-dimensional vectors should be obvious. For example, if $\mathbf{F}(t) = \langle f(t), g(t)\rangle = f(t)\mathbf{i} + g(t)\mathbf{j}$, then Theorem A says
>
> $$\lim_{t \to c} \mathbf{F}(t) = \left[\lim_{t \to c} f(t)\right]\mathbf{i} + \left[\lim_{t \to c} g(t)\right]\mathbf{j}$$

Theorem A

Let $\mathbf{F}(t) = f(t)\mathbf{i} + g(t)\mathbf{j} + h(t)\mathbf{k}$. Then $\mathbf{F}$ has a limit at c if and only if f, g, and h have limits at c. In this case,

$$\lim_{t \to c} \mathbf{F}(t) = \left[\lim_{t \to c} f(t)\right]\mathbf{i} + \left[\lim_{t \to c} g(t)\right]\mathbf{j} + \left[\lim_{t \to c} h(t)\right]\mathbf{k}$$

As you would expect, all the standard limit theorems hold. Also, continuity has its usual meaning; that is, $\mathbf{F}$ is **continuous** at c if $\lim_{t \to c} \mathbf{F}(t) = \mathbf{F}(c)$. From Theorem A, it is clear that $\mathbf{F}$ is continuous at c if and only if f, g, and h are all continuous there. Finally, the **derivative** $\mathbf{F}'(t)$ is defined just as for real-valued functions by

$$\mathbf{F}'(t) = \lim_{\Delta t \to 0} \frac{\mathbf{F}(t + \Delta t) - \mathbf{F}(t)}{\Delta t}$$

This can also be written in terms of components.

$$\begin{aligned}
\mathbf{F}'(t) &= \\
&\lim_{\Delta t \to 0} \frac{[f(t + \Delta t)\mathbf{i} + g(t + \Delta t)\mathbf{j} + h(t + \Delta t)\mathbf{k}] - [f(t)\mathbf{i} + g(t)\mathbf{j} + h(t)\mathbf{k}]}{\Delta t} \\
&= \lim_{\Delta t \to 0} \frac{f(t + \Delta t) - f(t)}{\Delta t}\mathbf{i} + \lim_{\Delta t \to 0} \frac{g(t + \Delta t) - g(t)}{\Delta t}\mathbf{j} \\
&\qquad + \lim_{\Delta t \to 0} \frac{h(t + \Delta t) - h(t)}{\Delta t}\mathbf{k} \\
&= f'(t)\mathbf{i} + g'(t)\mathbf{j} + h'(t)\mathbf{k}
\end{aligned}$$

In summary, if $\mathbf{F}(t) = f(t)\mathbf{i} + g(t)\mathbf{j} + h(t)\mathbf{k}$, then

$$\boxed{\mathbf{F}'(t) = f'(t)\mathbf{i} + g'(t)\mathbf{j} + h'(t)\mathbf{k} = \langle f'(t), g'(t), h'(t)\rangle}$$

EXAMPLE 1 If $\mathbf{F}(t) = (t^2 + t)\mathbf{i} + e^t\mathbf{j} + 2\mathbf{k}$, find $\mathbf{F}'(t)$, $\mathbf{F}''(t)$, and the angle θ between $\mathbf{F}'(0)$ and $\mathbf{F}''(0)$.

SOLUTION $\mathbf{F}'(t) = (2t + 1)\mathbf{i} + e^t\mathbf{j}$ and $\mathbf{F}''(t) = 2\mathbf{i} + e^t\mathbf{j}$. Thus, $\mathbf{F}'(0) = \mathbf{i} + \mathbf{j}$, $\mathbf{F}''(0) = 2\mathbf{i} + \mathbf{j}$, and

$$\cos\theta = \frac{\mathbf{F}'(0) \cdot \mathbf{F}''(0)}{\|\mathbf{F}'(0)\|\|\mathbf{F}''(0)\|} = \frac{(1)(2) + (1)(1) + (0)(0)}{\sqrt{1^2 + 1^2 + 0^2}\sqrt{2^2 + 1^2 + 0^2}} = \frac{3}{\sqrt{2}\sqrt{5}}$$

$$\theta \approx 0.3218 \quad \text{(about } 18.43^\circ\text{)}$$

■

Here are the rules for differentiation.

Theorem B **Differentiation Formulas**

Let $\mathbf{F}$ and $\mathbf{G}$ be differentiable, vector-valued functions, p a differentiable, real-valued function, and c a scalar. Then

1. $D_t[\mathbf{F}(t) + \mathbf{G}(t)] = \mathbf{F}'(t) + \mathbf{G}'(t)$
2. $D_t[c\mathbf{F}(t)] = c\mathbf{F}'(t)$
3. $D_t[p(t)\mathbf{F}(t)] = p(t)\mathbf{F}'(t) + p'(t)\mathbf{F}(t)$
4. $D_t[\mathbf{F}(t)\cdot\mathbf{G}(t)] = \mathbf{F}(t)\cdot\mathbf{G}'(t) + \mathbf{G}(t)\cdot\mathbf{F}'(t)$
5. $D_t[\mathbf{F}(t)\times\mathbf{G}(t)] = \mathbf{F}(t)\times\mathbf{G}'(t) + \mathbf{F}'(t)\times\mathbf{G}(t)$
6. $D_t[\mathbf{F}(p(t))] = \mathbf{F}'(p(t))p'(t)$ (Chain Rule)

Proof We prove formula 4 and leave the other parts to the reader. Let

$$\mathbf{F}(t) = f_1(t)\mathbf{i} + f_2(t)\mathbf{j} + f_3(t)\mathbf{k},$$
$$\mathbf{G}(t) = g_1(t)\mathbf{i} + g_2(t)\mathbf{j} + g_3(t)\mathbf{k}$$

Then

$$\begin{aligned} D_t[\mathbf{F}(t)\cdot\mathbf{G}(t)] &= D_t\left[f_1(t)g_1(t) + f_2(t)g_2(t) + f_3(t)g_3(t)\right] \\ &= f_1(t)g_1'(t) + g_1(t)f_1'(t) + f_2(t)g_2'(t) + g_2(t)f_2'(t) \\ &\qquad + f_3(t)g_3'(t) + g_3(t)f_3'(t) \\ &= \left[f_1(t)g_1'(t) + f_2(t)g_2'(t) + f_3(t)g_3'(t)\right] \\ &\qquad + \left[g_1(t)f_1'(t) + g_2(t)f_2'(t) + g_3(t)f_3'(t)\right] \\ &= \mathbf{F}(t)\cdot\mathbf{G}'(t) + \mathbf{G}(t)\cdot\mathbf{F}'(t) \end{aligned}$$ ■

Since derivatives of vector-valued functions are found by differentiating components, it is natural to define integration in terms of components; that is, if $\mathbf{F}(t) = f(t)\mathbf{i} + g(t)\mathbf{j} + h(t)\mathbf{k}$,

$$\int \mathbf{F}(t)\,dt = \left[\int f(t)\,dt\right]\mathbf{i} + \left[\int g(t)\,dt\right]\mathbf{j} + \left[\int h(t)\,dt\right]\mathbf{k}$$

$$\int_a^b \mathbf{F}(t)\,dt = \left[\int_a^b f(t)\,dt\right]\mathbf{i} + \left[\int_a^b g(t)\,dt\right]\mathbf{j} + \left[\int_a^b h(t)\,dt\right]\mathbf{k}$$

EXAMPLE 2 If $\mathbf{F}(t) = t^2\mathbf{i} + e^{-t}\mathbf{j} - 2\mathbf{k}$, find

(a) $D_t\left[t^3\mathbf{F}(t)\right]$ (b) $\int_0^1 \mathbf{F}(t)\,dt.$

SOLUTION

(a) $$\begin{aligned} D_t\left[t^3\mathbf{F}(t)\right] &= t^3(2t\mathbf{i} - e^{-t}\mathbf{j}) + 3t^2(t^2\mathbf{i} + e^{-t}\mathbf{j} - 2\mathbf{k}) \\ &= 5t^4\mathbf{i} + (3t^2 - t^3)e^{-t}\mathbf{j} - 6t^2\mathbf{k} \end{aligned}$$

(b) $$\begin{aligned} \int_0^1 \mathbf{F}(t)\,dt &= \left(\int_0^1 t^2\,dt\right)\mathbf{i} + \left(\int_0^1 e^{-t}\,dt\right)\mathbf{j} + \left(\int_0^1 (-2)\,dt\right)\mathbf{k} \\ &= \tfrac{1}{3}\mathbf{i} + (1 - e^{-1})\mathbf{j} - 2\mathbf{k} \end{aligned}$$ ■

Curvilinear Motion We are going to use the theory developed above for vector-valued functions to study the motion of a point in space. Let t measure time, and suppose that the coordinates of a moving point P are given by the parametric equations $x = f(t)$, $y = g(t)$, $z = h(t)$. Then the vector

$$\mathbf{r}(t) = f(t)\mathbf{i} + g(t)\mathbf{j} + h(t)\mathbf{k}$$

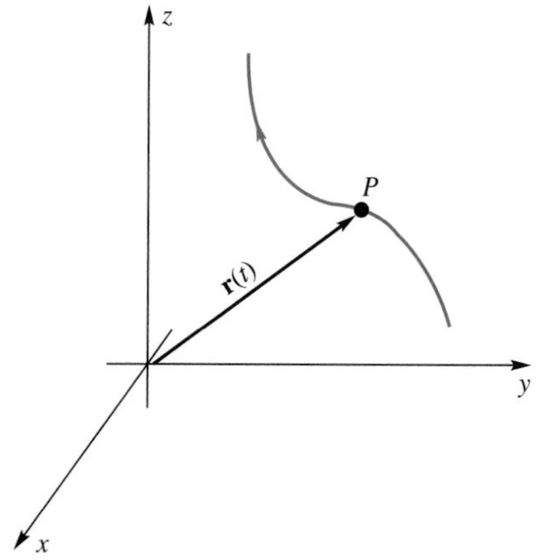

Figure 4

assumed to emanate from the origin, is called the **position vector** of the point. As t varies, the head of $\mathbf{r}(t)$ traces the path of the moving point P (Figure 4). This is a curve, and we call the corresponding motion **curvilinear motion.**

In analogy with linear (straight line) motion, we define the **velocity** $\mathbf{v}(t)$ and the **acceleration** $\mathbf{a}(t)$ of the moving point P by

$$\mathbf{v}(t) = \mathbf{r}'(t) = f'(t)\mathbf{i} + g'(t)\mathbf{j} + h'(t)\mathbf{k}$$

$$\mathbf{a}(t) = \mathbf{r}''(t) = f''(t)\mathbf{i} + g''(t)\mathbf{j} + h''(t)\mathbf{k}$$

Since

$$\mathbf{v}(t) = \lim_{\Delta t \to 0} \frac{\mathbf{r}(t + \Delta t) - \mathbf{r}(t)}{\Delta t}$$

it is clear (from Figure 5) that $\mathbf{v}(t)$ has the direction of the tangent line. The acceleration vector $\mathbf{a}(t)$ points to the **concave** side of the curve (i.e., the side toward which the curve is bending).

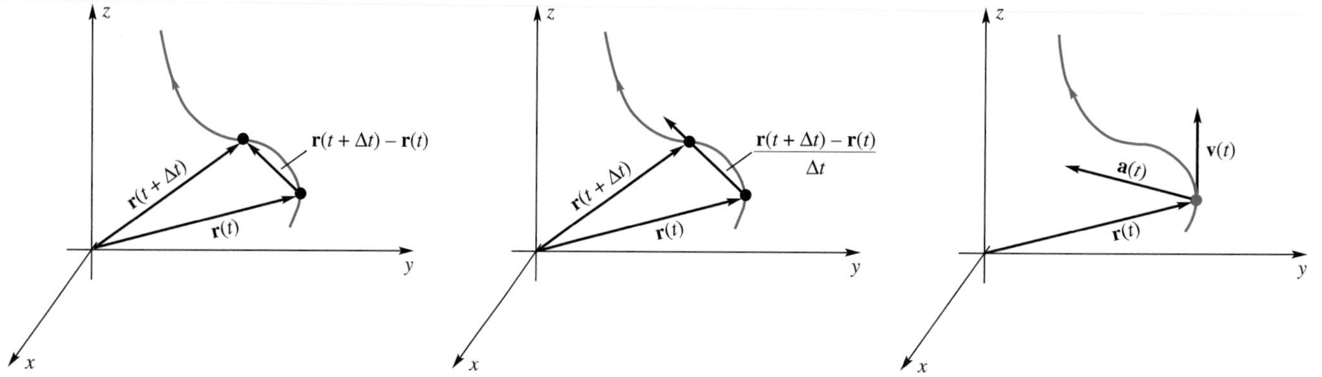

Figure 5

If $\mathbf{r}(t)$ is the position vector of an object, then the arc length of the path that it traces from time $t = a$ to time $t = b$ is

$$L = \int_a^b \sqrt{[f'(t)]^2 + [g'(t)]^2 + [h'(t)]^2}\, dt = \int_a^b \|\mathbf{r}'(t)\|\, dt$$

The accumulated arc length from time $t = a$ to an arbitrary time t is thus

$$s = \int_a^t \sqrt{[f'(u)]^2 + [g'(u)]^2 + [h'(u)]^2}\, du = \int_a^t \|\mathbf{r}'(u)\|\, du$$

By the First Fundamental Theorem of Calculus, the derivative of the accumulated arc length, ds/dt, is

$$\frac{ds}{dt} = \sqrt{[f'(t)]^2 + [g'(t)]^2 + [h'(t)]^2} = \|\mathbf{r}'(t)\|$$

But the derivative (i.e., rate of change) of accumulated arc length is what we think of as speed. Thus, the **speed** of an object is

$$\text{speed} = \frac{ds}{dt} = \|\mathbf{r}'(t)\| = \|\mathbf{v}(t)\|$$

Note that the speed of an object is a scalar quantity, whereas its velocity is a vector.

One of the most important applications of curvilinear motion, **uniform circular motion,** occurs in two dimensions. Suppose that an object moves in the xy-plane counterclockwise around a circle with center (0,0) and radius a at a constant angular speed of ω radians per second. If its initial position is $(a, 0)$, then its position vector is

$$\mathbf{r}(t) = a \cos \omega t\, \mathbf{i} + a \sin \omega t\, \mathbf{j}$$

EXAMPLE 3 Find the velocity, acceleration, and speed for uniform circular motion.

SOLUTION We differentiate the position vector $\mathbf{r}(t) = a \cos \omega t\, \mathbf{i} + a \sin \omega t\, \mathbf{j}$ to get $\mathbf{v}(t)$ and $\mathbf{a}(t)$.

$$\mathbf{v}(t) = \mathbf{r}'(t) = -a\omega \sin \omega t\, \mathbf{i} + a\omega \cos \omega t\, \mathbf{j}$$

$$\mathbf{a}(t) = \mathbf{v}'(t) = -a\omega^2 \cos \omega t\, \mathbf{i} - a\omega^2 \sin \omega t\, \mathbf{j}$$

The speed is

$$\frac{ds}{dt} = \|\mathbf{v}(t)\| = \sqrt{(-a\omega \sin \omega t)^2 + (a\omega \cos \omega t)^2}$$

$$= \sqrt{a^2\omega^2(\sin^2 \omega t + \cos^2 \omega t)} = a\omega$$

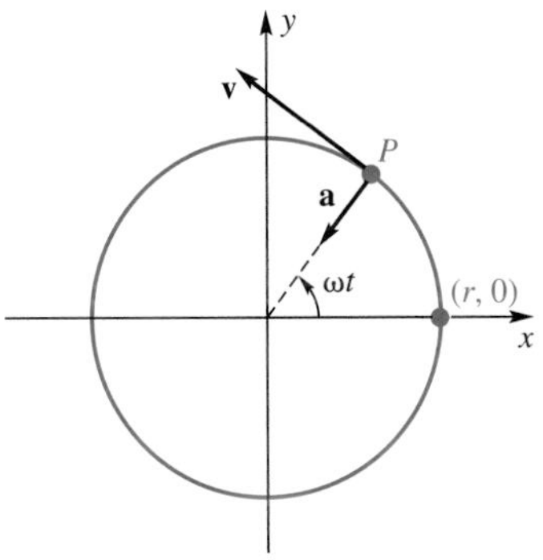

Figure 6

Note that if we think of **a** as being based at the object's location at point P, then **a** points directly toward the origin and is perpendicular to the velocity vector **v** (Figure 6). ■

We saw a particular case of a **helix** in Example 6 of Section 11.1. Here we generalize that concept a bit and say that the path traced out by an object whose position vector is given by

$$\mathbf{r}(t) = a \cos \omega t\, \mathbf{i} + a \sin \omega t\, \mathbf{j} + ct\, \mathbf{k}$$

is a **helix.** If we look at just the x- and y-components of motion, we see uniform circular motion, and if we look at just the z-component of motion, we see uniform straight line motion. When we put these two together, we see that the object spirals around and around as it moves higher and higher (assuming $c > 0$).

EXAMPLE 4 Find the velocity, acceleration, and speed for motion along a helix.

SOLUTION The velocity and acceleration vectors are

$$\mathbf{v}(t) = \mathbf{r}'(t) = -a\omega \sin \omega t\, \mathbf{i} + a\omega \cos \omega t\, \mathbf{j} + c\, \mathbf{k}$$

$$\mathbf{a}(t) = \mathbf{v}'(t) = -a\omega^2 \cos \omega t\, \mathbf{i} - a\omega^2 \sin \omega t\, \mathbf{j}$$

The speed is

$$\frac{ds}{dt} = \|\mathbf{v}(t)\| = \sqrt{(-a\omega \sin \omega t)^2 + (a\omega \cos \omega t)^2 + c^2} = \sqrt{a^2\omega^2 + c^2}$$ ■

EXAMPLE 5 Parametric equations for an object moving in the plane are $x = 3 \cos t$ and $y = 2 \sin t$, where t represents time and $0 \le t \le 2\pi$. Let P denote the object's position.

(a) Graph the path of P.
(b) Find expressions for the velocity $\mathbf{v}(t)$, speed $\|\mathbf{v}(t)\|$, and acceleration $\mathbf{a}(t)$.
(c) Find the maximum and minimum values of the speed and where they occur.
(d) Show that the acceleration vector based at P always points to the origin.

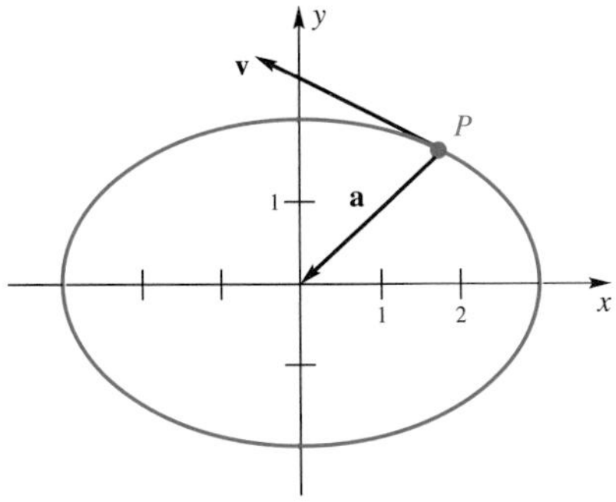

$\mathbf{r}(t) = 3 \cos t\, \mathbf{i} + 2 \sin t\, \mathbf{j}$

Figure 7

SOLUTION

(a) Since $x^2/9 + y^2/4 = 1$, the path is the ellipse shown in Figure 7.
(b) The position vector is

$$\mathbf{r}(t) = 3 \cos t\, \mathbf{i} + 2 \sin t\, \mathbf{j}$$

and so

$$\mathbf{v}(t) = -3\sin t\,\mathbf{i} + 2\cos t\,\mathbf{j}$$
$$\|\mathbf{v}(t)\| = \sqrt{9\sin^2 t + 4\cos^2 t} = \sqrt{5\sin^2 t + 4}$$
$$\mathbf{a}(t) = -3\cos t\,\mathbf{i} - 2\sin t\,\mathbf{j}$$

(c) Since the speed is given by $\sqrt{5\sin^2 t + 4}$, the maximum speed of 3 occurs when $\sin t = \pm 1$, that is, when $t = \pi/2$ or $3\pi/2$. This corresponds to the points $(0, \pm 2)$ on the ellipse. Similarly, the minimum speed of 2 occurs when $\sin t = 0$, which corresponds to the points $(\pm 3, 0)$.

(d) Note that $\mathbf{a}(t) = -\mathbf{r}(t)$. Thus, if we base $\mathbf{a}(t)$ at P, this vector will point to and exactly reach the origin. We conclude that $\|\mathbf{a}(t)\|$ is largest at $(\pm 3, 0)$ and smallest at $(0, \pm 2)$. ■

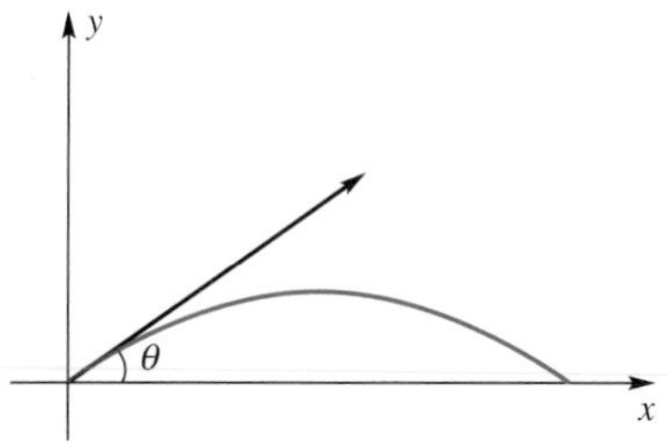

Figure 8

EXAMPLE 6 A projectile is shot from the origin at an angle θ from the positive x-axis with an initial speed of v_0 feet per second (Figure 8). Neglecting friction, find expressions for the velocity $\mathbf{v}(t)$ and position $\mathbf{r}(t)$, and show that the path is a parabola.

SOLUTION The acceleration due to gravity is $\mathbf{a}(t) = -32\mathbf{j}$ feet per second per second. The initial conditions are $\mathbf{r}(0) = \mathbf{0}$ and $\mathbf{v}(0) = v_0\cos\theta\,\mathbf{i} + v_0\sin\theta\,\mathbf{j}$. Starting with $\mathbf{a}(t) = -32\mathbf{j}$, we integrate twice.

$$\mathbf{v}(t) = \int \mathbf{a}(t)\,dt = \int (-32)\,dt\,\mathbf{j} = -32t\,\mathbf{j} + \mathbf{C}_1$$

The condition $\mathbf{v}(0) = v_0\cos\theta\,\mathbf{i} + v_0\sin\theta\,\mathbf{j}$ allows us to evaluate $\mathbf{C}_1$ and gives $\mathbf{C}_1 = v_0\cos\theta\,\mathbf{i} + v_0\sin\theta\,\mathbf{j}$. Thus,

$$\mathbf{v}(t) = (v_0\cos\theta)\mathbf{i} + (v_0\sin\theta - 32t)\mathbf{j}$$

and

$$\mathbf{r}(t) = \int \mathbf{v}(t)\,dt = (tv_0\cos\theta)\mathbf{i} + (tv_0\sin\theta - 16t^2)\mathbf{j} + \mathbf{C}_2$$

The condition $\mathbf{r}(0) = \mathbf{0}$ implies that $\mathbf{C}_2 = \mathbf{0}$, so

$$\mathbf{r}(t) = (tv_0\cos\theta)\mathbf{i} + (tv_0\sin\theta - 16t^2)\mathbf{j}$$

To find the equation of the path, we eliminate the parameter t in the equations

$$x = (v_0\cos\theta)t, \qquad y = (v_0\sin\theta)t - 16t^2$$

Specifically, we solve the first equation for t and substitute in the second, giving

$$y = (\tan\theta)x - \left(\frac{4}{v_0\cos\theta}\right)^2 x^2$$

This is the equation of a parabola. ■

EXAMPLE 7 A baseball is thrown with an initial velocity of 75 miles per hour (110 feet per second) 1 degree above horizontal in the direction of the positive x-axis from an initial height of 8 feet. The initial position is $\mathbf{r}(0) = 8\mathbf{k}$. In addition to acceleration due to gravity, the spin on the ball causes an acceleration of 2 feet per second per second in the positive y direction. What is the position of the ball when its x-component is 60.5 feet?

SOLUTION The initial position vector is $\mathbf{r}(0) = 8\mathbf{k}$, and the initial velocity vector is $\mathbf{v}(0) = 110\cos 1°\mathbf{i} + 110\sin 1°\mathbf{k}$. The acceleration vector is $\mathbf{a}(t) = 2\mathbf{j} - 32\mathbf{k}$. Proceeding as in the previous example, we have

$$\mathbf{v}(t) = \int \mathbf{a}(t)\,dt = \int (2\mathbf{j} - 32\mathbf{k})\,dt = 2t\,\mathbf{j} - 32t\,\mathbf{k} + \mathbf{C}_1$$

Since $110\cos 1°\mathbf{i} + 110\sin 1°\mathbf{k} = \mathbf{v}(0) = \mathbf{C}_1$, we have

$$\mathbf{v}(t) = 110\cos 1°\mathbf{i} + 2t\,\mathbf{j} + (110\sin 1° - 32t)\mathbf{k}$$

Integrating the velocity vector gives the position:

$$\mathbf{r}(t) = \int \mathbf{v}(t)\,dt = \int \left[110 \cos 1°\mathbf{i} + 2t\,\mathbf{j} + (110 \sin 1° - 32t)\mathbf{k}\right] dt$$

$$= 110(\cos 1°)t\mathbf{i} + t^2\mathbf{j} + \left[110(\sin 1°)t - 16t^2\right]\mathbf{k} + \mathbf{C}_2$$

The initial position $\mathbf{r}(0) = 8\mathbf{k}$ implies that $\mathbf{C}_2 = 8\mathbf{k}$. Thus

$$\mathbf{r}(t) = 110(\cos 1°)t\mathbf{i} + t^2\mathbf{j} + \left[8 + 110(\sin 1°)t - 16t^2\right]\mathbf{k}$$

Next, we must find the value of t for which the x-component is 60.5 feet. Setting $110(\cos 1°)t = 60.5$ yields $t = 60.5/(110 \cos 1°) \approx 0.55008$ second. The position of the ball at this time is

$$\begin{aligned}\mathbf{r}(0.55008) &= 110(\cos 1°)0.55008\mathbf{i} + (0.55008)^2\mathbf{j} \\ &\quad + \left[8 + 110(\sin 1°)0.55008 - 16(0.55008)^2\right]\mathbf{k} \\ &\approx 60.5\mathbf{i} + 0.303\mathbf{j} + 4.21\mathbf{k}\end{aligned}$$

If this pitch were thrown by a major league pitcher to a major league batter, the ball would be just above the waist (4.21 feet off the ground) and about 4 inches (0.303 feet) from the center of home plate. ■

Kepler's Laws of Planetary Motion (Optional) In the early part of the 17th century, Johannes Kepler inherited a collection of planetary data from the Danish nobleman Tycho Brahe. Kepler spent years studying the data and through trial and error, and a little luck, he formulated his three laws of planetary motion:

1. Planets move in elliptical orbits with the sun at one focus.
2. A line from the sun to the planet sweeps out equal areas in equal times.
3. The square of a planet's orbital period is proportional to the cube of its mean distance from the sun.

Only later was it discovered that Kepler's Laws of Planetary Motion are a consequence of Newton's Laws of Motion. Kepler's First Law can be stated as

$$r(\theta) = \frac{r_0(1 + e)}{1 + e \cos \theta}$$

which is the polar equation of an ellipse. Here $r(\theta)$ is the planet's distance from the sun for the angle θ, and e is the eccentricity of the ellipse. Problem 48, which guides the reader through the derivation of Kepler's First Law, shows that

$$e = \frac{r_0 v_0^2}{GM} - 1 = \frac{1}{r_0 GM}\left(2\frac{dA}{dt}\right)^2 - 1$$

where M is the sun's mass, G is the gravitational constant, r_0 is the shortest distance from the sun to the planet, v_0 is the planet's speed when it is closest to the sun, and dA/dt is the rate of change in the area swept out by a line segment joining the sun and planet (a constant by Kepler's Second Law). We will assume Kepler's First Law.

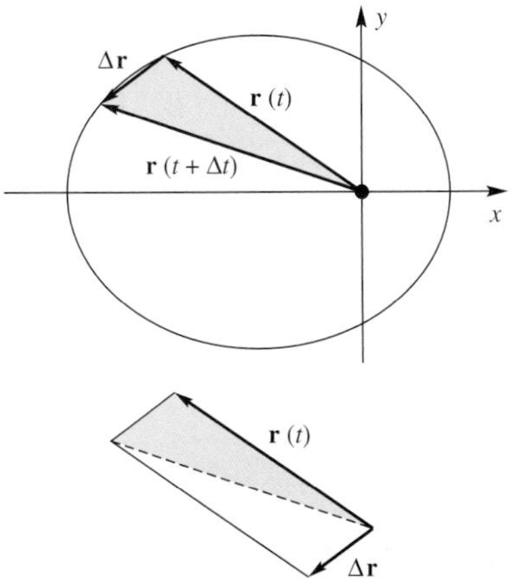

Figure 9

EXAMPLE 8 Derive Kepler's Second Law.

SOLUTION Let $\mathbf{r}(t)$ denote the position vector of a planet at time t, and let $\mathbf{r}(t + \Delta t)$ be its position Δt time units later (Figure 9). The area ΔA swept out in time Δt is approximately half the area of the parallelogram formed by $\mathbf{r}(t)$ and $\Delta\mathbf{r} = \mathbf{r}(t + \Delta t) - \mathbf{r}(t)$. Using the fact from the previous section that the area of a triangle formed by two vectors is half the magnitude of the cross product of the vectors, we have

$$\Delta A \approx \frac{1}{2}\|\mathbf{r}(t) \times \Delta\mathbf{r}\|$$

Thus

$$\frac{\Delta A}{\Delta t} \approx \frac{1}{2}\left\|\mathbf{r}(t) \times \frac{\Delta\mathbf{r}}{\Delta t}\right\|$$

so, letting $\Delta t \to 0$, we get

$$\frac{dA}{dt} = \frac{1}{2}\|\mathbf{r}(t) \times \mathbf{r}'(t)\|$$

The only force acting on the planet is the gravitational attraction of the sun which acts along the line from the sun to the planet and has magnitude $GMm/\|\mathbf{r}(t)\|^2$, where m is the planet's mass. Newton's Second Law ($F = ma$) implies

$$-\frac{GMm}{\|\mathbf{r}(t)\|^3}\mathbf{r}(t) = m\mathbf{a}(t) = m\mathbf{r}''(t)$$

Dividing both sides by m gives $\mathbf{r}''(t) = -(GM/\|\mathbf{r}(t)\|^3)\mathbf{r}(t)$.

In light of this, consider the vector $\mathbf{r}(t) \times \mathbf{r}'(t)$ in the above expression for dA/dt. Differentiating this vector using Property 5 of Theorem B gives

$$\begin{aligned}\frac{d}{dt}(\mathbf{r}(t) \times \mathbf{r}'(t)) &= \mathbf{r}(t) \times \mathbf{r}''(t) + \mathbf{r}'(t) \times \mathbf{r}'(t)\\ &= \mathbf{r}(t) \times \left(-\frac{GM}{\|\mathbf{r}(t)\|^3}\mathbf{r}(t)\right) + \mathbf{0}\\ &= \left(-\frac{GM}{\|\mathbf{r}(t)\|^3}\right)\mathbf{r}(t) \times \mathbf{r}(t) = \mathbf{0}\end{aligned}$$

This tells us that the vector $\mathbf{r}(t) \times \mathbf{r}'(t)$ is a constant and as a result, its magnitude $\|\mathbf{r}(t) \times \mathbf{r}'(t)\|$ is constant. Thus, dA/dt is a constant. ■

EXAMPLE 9 Derive Kepler's Third Law.

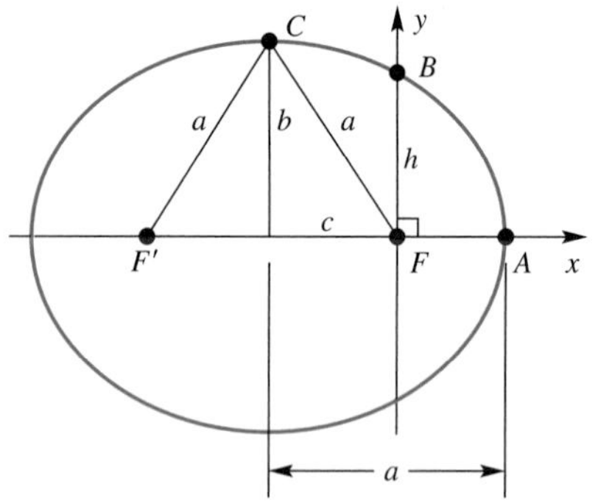

Figure 10

SOLUTION Place the sun at the origin, and the x-axis so that the planet's perihelion (point on the orbit closest to the sun) lies along the x-axis. The perihelion occurs at point A in Figure 10. Let C denote the point on the orbit that lies on the minor axis, and let B denote the point on the orbit that lies on a line perpendicular to the major axis at the origin as shown in Figure 10. Let a and b denote half the lengths of the major and minor axes of the ellipse, respectively, and let c denote the distance from the center of the two foci to a focus. The string property for ellipses says that the sum of the distances from the foci to any point on the ellipse is $2a$. Thus $\overline{F'C} + \overline{CF} = 2a$, and since $\overline{F'C} = \overline{CF}$, we conclude that $\overline{F'C} = \overline{CF} = a$. Another application of the string property to point B gives $\overline{F'B} + \overline{BF} = 2a$.

Using the Pythagorean Theorem we conclude $a^2 = b^2 + c^2$ and $(\overline{F'B})^2 = h^2 + (2c)^2$ (see Figure 11). From above, $\overline{F'B} = 2a - \overline{BF} = 2a - h$. Putting these results together gives

$$h^2 + (2c)^2 = (\overline{F'B})^2 = (2a - h)^2 = 4a^2 - 4ah + h^2$$

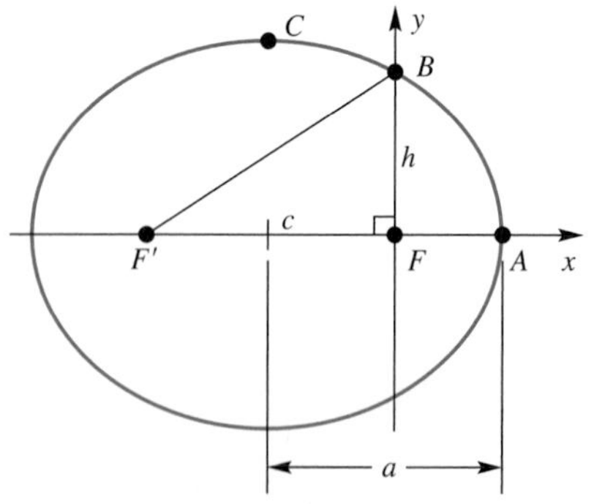

Figure 11

so we conclude that $4c^2 = 4a^2 - 4ah$, hence $c^2 = a^2 - ah$. Since $a^2 = b^2 + c^2$, we conclude that

$$a^2 - b^2 = c^2 = a^2 - ah$$

Thus, $b^2 = ah$.

The point B also occurs when the angle θ is $\pi/2$. Using Kepler's First Law,

$$h = r(\pi/2) = \frac{r_0(1 + e)}{1 + e\cos(\pi/2)} = \frac{1}{GM}\left(2\frac{dA}{dt}\right)^2$$

Let T denote the planet's period. Over one orbit about the sun, the area πab is swept out. The average rate at which area is swept out is thus $\pi ab/T$ but since dA/dt is constant (Kepler's Second Law), $dA/dt = \pi ab/T$. Thus

$$T = \frac{\pi ab}{dA/dt}$$

Now it all comes together. Using the relationships $b^2 = ah$ and $h = \left(2\frac{dA}{dt}\right)^2/GM$ from above, we have

$$T^2 = \left(\frac{\pi ab}{dA/dt}\right)^2 = \frac{\pi^2 a^2}{(dA/dt)^2} ah = \frac{\pi^2 a^3}{(dA/dt)^2}\frac{(2\,dA/dt)^2}{GM} = \frac{4\pi^2}{GM}a^3$$

The closest the planet gets to the sun is $a - c$ and the farthest is $a + c$. Kepler called the average of these two values, $(a - c + a + c)/2 = a$, the mean distance from the sun. The last formula thus implies that the square of the period is proportional to the cube of the mean distance from the sun. ■

Concepts Review

1. A function that associates with each real number a single vector is called a(n) _____.

2. The function $\mathbf{F}(t) = f(t)\mathbf{i} + g(t)\mathbf{j}$ is continuous at $t = c$ if and only if _____. The derivative of $\mathbf{F}$ is given in terms of f and g by $\mathbf{F}'(t) =$ _____.

3. If a point moves along a curve so that it is at point P at time t, then the vector $\mathbf{r}(t)$ from the origin to P is called the _____ vector of P.

4. In terms of $\mathbf{r}(t)$, the velocity is _____ and the acceleration is _____. The velocity vector at t is _____ to the curve, whereas the acceleration vector points to the _____ side of the curve.

Problem Set 11.5

In Problems 1–8, find the required limit or indicate that it does not exist.

1. $\lim_{t\to 1}[2t\mathbf{i} - t^2\mathbf{j}]$

2. $\lim_{t\to 3}[2(t-3)^2\mathbf{i} - 7t^3\mathbf{j}]$

3. $\lim_{t\to 1}\left[\frac{t-1}{t^2-1}\mathbf{i} - \frac{t^2+2t-3}{t-1}\mathbf{j}\right]$

4. $\lim_{t\to -2}\left[\frac{2t^2-10t-28}{t+2}\mathbf{i} - \frac{7t^3}{t-3}\mathbf{j}\right]$

5. $\lim_{t\to 0}\left[\frac{\sin t\cos t}{t}\mathbf{i} - \frac{7t^3}{e^t}\mathbf{j} + \frac{t}{t+1}\mathbf{k}\right]$

6. $\lim_{t\to\infty}\left[\frac{t\sin t}{t^2}\mathbf{i} - \frac{7t^3}{t^3-3t}\mathbf{j} - \frac{\sin t}{t}\mathbf{k}\right]$

7. $\lim_{t\to 0^+}\langle \ln(t^3), t^2\ln t, t\rangle$

8. $\lim_{t\to 0^-}\left\langle e^{-1/t^2}, \frac{t}{|t|}, |t|\right\rangle$

9. When no domain is given in the definition of a vector-valued function, it is to be understood that the domain is the set of all (real) scalars for which the rule for the function makes sense and gives real vectors (i.e., vectors with real components). Find the domain of each of the following vector-valued functions:

(a) $\mathbf{r}(t) = \frac{2}{t-4}\mathbf{i} + \sqrt{3-t}\,\mathbf{j} + \ln|4-t|\mathbf{k}$

(b) $\mathbf{r}(t) = [\![t^2]\!]\mathbf{i} - \sqrt{20-t}\,\mathbf{j} + 3\mathbf{k}$ ($[\![\ ]\!]$ denotes the greatest integer function.)

(c) $\mathbf{r}(t) = \cos t\,\mathbf{i} + \sin t\,\mathbf{j} + \sqrt{9-t^2}\,\mathbf{k}$

10. State the domain of each of the following vector-valued functions:

(a) $\mathbf{r}(t) = \ln(t-1)\mathbf{i} + \sqrt{20-t}\,\mathbf{j}$

(b) $\mathbf{r}(t) = \ln(t^{-1})\mathbf{i} + \tan^{-1}t\,\mathbf{j} + t\,\mathbf{k}$

(c) $\mathbf{r}(t) = \frac{1}{\sqrt{1-t^2}}\mathbf{j} + \frac{1}{\sqrt{9-t^2}}\mathbf{k}$

11. For what values of t is each function in Problem 9 continuous?

12. For what values of t is each function in Problem 10 continuous?

13. Find $D_t\mathbf{r}(t)$ and $D_t^2\mathbf{r}(t)$ for each of the following:

(a) $\mathbf{r}(t) = (3t+4)^3\mathbf{i} + e^{t^2}\mathbf{j} + \mathbf{k}$

(b) $\mathbf{r}(t) = \sin^2 t\,\mathbf{i} + \cos 3t\,\mathbf{j} + t^2\mathbf{k}$

14. Find $\mathbf{r}'(t)$ and $\mathbf{r}''(t)$ for each of the following:

(a) $\mathbf{r}(t) = (e^t + e^{-t^2})\mathbf{i} + 2^t\mathbf{j} + t\,\mathbf{k}$

(b) $\mathbf{r}(t) = \tan 2t\,\mathbf{i} + \arctan t\,\mathbf{j}$

15. If $\mathbf{r}(t) = e^{-t}\mathbf{i} - \ln(t^2)\mathbf{j}$, find $D_t[\mathbf{r}(t)\cdot\mathbf{r}''(t)]$.

16. If $\mathbf{r}(t) = \sin 3t\,\mathbf{i} - \cos 3t\,\mathbf{j}$, find $D_t[\mathbf{r}(t)\cdot\mathbf{r}'(t)]$.

17. If $\mathbf{r}(t) = \sqrt{t-1}\,\mathbf{i} + \ln(2t^2)\mathbf{j}$ and $h(t) = e^{-3t}$, find $D_t[h(t)\mathbf{r}(t)]$.

18. If $\mathbf{r}(t) = \sin 2t\,\mathbf{i} + \cosh t\,\mathbf{j}$ and $h(t) = \ln(3t-2)$, find $D_t[h(t)\mathbf{r}(t)]$.

In Problems 19–30, find the velocity $\mathbf{v}$, *acceleration* $\mathbf{a}$, *and speed* s *at the indicated time* $t = t_1$.

19. $\mathbf{r}(t) = 4t\mathbf{i} + 5(t^2-1)\mathbf{j} + 2t\mathbf{k};\ t_1 = 1$

20. $\mathbf{r}(t) = t\mathbf{i} + (t-1)^2\mathbf{j} + (t-3)^3\mathbf{k};\ t_1 = 0$

21. $\mathbf{r}(t) = (1/t)\mathbf{i} + (t^2-1)^{-1}\mathbf{j} + t^5\mathbf{k};\ t_1 = 2$

22. $\mathbf{r}(t) = t^6\mathbf{i} + (6t^2-5)^6\mathbf{j} + t\mathbf{k};\ t_1 = 1$

23. $\mathbf{r}(t) = \mathbf{i} + \left(\int_0^t x^2\,dx\right)\mathbf{j} + t^{2/3}\mathbf{k};\ t_1 = 2$

24. $\mathbf{r}(t) = \int_1^t [x^2\mathbf{i} + 5(x-1)^3\mathbf{j} + (\sin\pi x)\mathbf{k}]\,dx;\ t_1 = 2$

25. $\mathbf{r}(t) = \cos t\,\mathbf{i} + \sin t\,\mathbf{j} + t\,\mathbf{k};\ t_1 = \pi$

26. $\mathbf{r}(t) = \sin 2t\,\mathbf{i} + \cos 3t\,\mathbf{j} + \cos 4t\,\mathbf{k};\ t_1 = \frac{\pi}{2}$

27. $\mathbf{r}(t) = \tan t\,\mathbf{i} + 3e^t\,\mathbf{j} + \cos 4t\,\mathbf{k};\ t_1 = \frac{\pi}{4}$

28. $\mathbf{r}(t) = \left(\int_t^1 e^x\,dx\right)\mathbf{i} + \left(\int_t^\pi \sin\pi\theta\,d\theta\right)\mathbf{j} + t^{2/3}\mathbf{k};\ t_1 = 2$

29. $\mathbf{r}(t) = t\sin \pi t\,\mathbf{i} + t\cos \pi t\,\mathbf{j} + e^{-t}\mathbf{k};\ t_1 = 2$

30. $\mathbf{r}(t) = \ln t\,\mathbf{i} + \ln t^2\mathbf{j} + \ln t^3\mathbf{k};\ t_1 = 2$

31. Show that if the speed of a moving particle is constant its acceleration vector is always perpendicular to its velocity vector.

32. Prove that $\|\mathbf{r}(t)\|$ is constant if and only if $\mathbf{r}(t)\cdot\mathbf{r}'(t) = 0$.

In Problems 33–38, find the length of the curve with the given vector equation.

33. $\mathbf{r}(t) = t\,\mathbf{i} + \sin t\,\mathbf{j} + \cos t\,\mathbf{k};\ 0 \le t \le 2$

34. $\mathbf{r}(t) = t\cos t\,\mathbf{i} + t\sin t\,\mathbf{j} + \sqrt{2}t\,\mathbf{k};\ 0 \le t \le 2$

35. $\mathbf{r}(t) = \sqrt{6}t^2\mathbf{i} + \frac{2}{3}t^3\mathbf{j} + 6t\mathbf{k};\ 3 \le t \le 6$

36. $\mathbf{r}(t) = t^2\mathbf{i} - 2t^3\mathbf{j} + 6t^3\mathbf{k};\ 0 \le t \le 1$

37. $\mathbf{r}(t) = t^3\mathbf{i} - 2t^3\mathbf{j} + 6t^3\mathbf{k};\ 0 \le t \le 1$

38. $\mathbf{r}(t) = \sqrt{7}t^7\mathbf{i} - \sqrt{2}t^7\mathbf{j} + 6t^7\mathbf{k};\ 0 \le t \le 1$

In Problems 39 and 40, $\mathbf{F}(t) = \mathbf{f}(u(t))$. Find $\mathbf{F}'(t)$ in terms of t.

39. $\mathbf{f}(u) = \cos u\,\mathbf{i} + e^{3u}\mathbf{j}$ and $u(t) = 3t^2 - 4$

40. $\mathbf{f}(u) = u^2\mathbf{i} + \sin^2 u\,\mathbf{j}$ and $u(t) = \tan t$

Evaluate the integrals in Problems 41 and 42.

41. $\displaystyle\int_0^1 (e^t\mathbf{i} + e^{-t}\mathbf{j})\,dt$

42. $\displaystyle\int_{-1}^1 [(1+t)^{3/2}\mathbf{i} + (1-t)^{3/2}\mathbf{j}]\,dt$

43. A point moves around the circle $x^2 + y^2 = 25$ at constant angular speed of 6 radians per second starting at (5, 0). Find expressions for $\mathbf{r}(t)$, $\mathbf{v}(t)$, $\|\mathbf{v}(t)\|$, and $\mathbf{a}(t)$ (see Example 3).

44. Consider the motion of a particle along a helix given by $\mathbf{r}(t) = \sin t\,\mathbf{i} + \cos t\,\mathbf{j} + (t^2 - 3t + 2)\mathbf{k}$, where the $\mathbf{k}$ component measures the height in meters above the ground and $t \ge 0$.

(a) Does the particle ever move downward?

(b) Does the particle ever stop moving?

(c) At what times does it reach a position 12 meters above the ground?

(d) What is the velocity of the particle when it is 12 meters above the ground?

EXPL **45.** In many places in the solar system, a moon orbits a planet, which in turn orbits the sun. In some cases the orbits are very close to circular. We will assume that these orbits are circular with the sun at the center of the planet's orbit and the planet at the center of the moon's orbit. We will further assume that all motion is in a single xy-plane. Suppose that in the time the planet orbits the sun once the moon orbits the planet ten times.

(a) If the radius of the moon's orbit is R_m and the radius of the planet's orbit about the sun is R_p, show that the motion of the moon with respect to the sun at the origin could be given by

$$x = R_p\cos t + R_m\cos 10t, \qquad y = R_p\sin t + R_m\sin 10t$$

CAS (b) For $R_p = 1$ and $R_m = 0.1$, plot the path traced by the moon as the planet makes one revolution around the sun.

(c) Find one set of values for R_p, R_m and t so that at time t the moon is motionless with respect to the sun.

46. Assuming that the orbits of the earth about the sun and the moon about the earth lie in the same plane and are circular, we can represent the motion of the moon by

$$\mathbf{r}(t) = [93\cos(2\pi t) + 0.24\cos(26\pi t)]\mathbf{i} + [93\sin(2\pi t) + 0.24\sin(26\pi t)]\mathbf{j}$$

where $\mathbf{r}(t)$ is measured in millions of miles.

(a) What are the proper units for t?

CAS (b) Plot the path traced by the moon as the earth makes one revolution around the sun.

(c) What is the period of each of the two motions?

(d) What is the maximum distance that the moon is from the sun?

(e) What is the minimum distance that the moon is from the sun?

(f) Is there ever a time that the moon is stationary with respect to the sun?

(g) What are the velocity, speed, and acceleration of the moon when $t = 1/2$?

47. Describe in general terms the following "helical" type motions:

(a) $\mathbf{r}(t) = \sin t\,\mathbf{i} + \cos t\,\mathbf{j} + t\,\mathbf{k}$

(b) $\mathbf{r}(t) = \sin t^3\mathbf{i} + \cos t^3\mathbf{j} + t^3\mathbf{k}$

(c) $\mathbf{r}(t) = \sin(t^3 + \pi)\,\mathbf{i} + t^3\mathbf{j} + \cos(t^3 + \pi)\,\mathbf{k}$

(d) $\mathbf{r}(t) = t\sin t\,\mathbf{i} + t\cos t\,\mathbf{j} + t\,\mathbf{k}$

(e) $\mathbf{r}(t) = t^{-2}\sin t\,\mathbf{i} + t^{-2}\cos t\,\mathbf{j} + t\,\mathbf{k},\ t > 0$

(f) $\mathbf{r}(t) = t^2\sin(\ln t)\,\mathbf{i} + \ln t\,\mathbf{j} + t^2\cos(\ln t)\,\mathbf{k},\ t > 1$

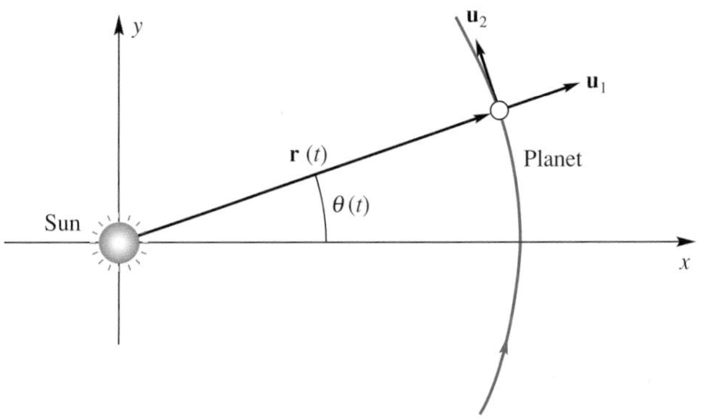

Figure 12

EXPL **48.** In this exercise you will derive Kepler's First Law, that planets travel in elliptical orbits. We begin with the notation. Place the coordinate system so that the sun is at the origin and the planet's closest approach to the sun (the perihelion) is on the positive x-axis and occurs at time $t = 0$. Let $\mathbf{r}(t)$ denote the position vector and let $r(t) = \|\mathbf{r}(t)\|$ denote the distance from the sun at time t. Also, let $\theta(t)$ denote the angle that the vector $\mathbf{r}(t)$ makes with the positive x-axis at time t. Thus, $(r(t), \theta(t))$ is the polar coordinate representation of the planet's position. Let $\mathbf{u}_1 = \mathbf{r}/r = (\cos\theta)\mathbf{i} + (\sin\theta)\mathbf{j}$ and $\mathbf{u}_2 = (-\sin\theta)\mathbf{i} + (\cos\theta)\mathbf{j}$. Vectors $\mathbf{u}_1$ and $\mathbf{u}_2$ are orthogonal unit vectors pointing in the directions of increasing r and increasing θ, respectively. Figure 12 summarizes this notation. We will often omit the argument t, but keep in mind that $\mathbf{r}$, θ, $\mathbf{u}_1$, and $\mathbf{u}_2$ are all functions of t. A prime indicates differentiation with respect to time t.

(a) Show that $\mathbf{u}_1' = \theta'\mathbf{u}_2$ and $\mathbf{u}_2' = -\theta'\mathbf{u}_1$.

(b) Show that the velocity and acceleration vectors satisfy

$$\mathbf{v} = r'\mathbf{u}_1 + r\theta'\mathbf{u}_2$$
$$\mathbf{a} = (r'' - r(\theta')^2)\mathbf{u}_1 + (2r'\theta' + r\theta'')\mathbf{u}_2$$

(c) Use the fact that the only force acting on the planet is the gravity of the sun to express $\mathbf{a}$ as a multiple of $\mathbf{u}_1$, then explain how we can conclude that

$$r'' - r(\theta')^2 = \frac{-GM}{r^2}$$
$$2r'\theta' + r\theta'' = 0$$

(d) Consider $\mathbf{r} \times \mathbf{r}'$, which we showed in Example 8 was a constant vector, say $\mathbf{D}$. Use the result from (b) to show that $\mathbf{D} = r^2\theta'\mathbf{k}$.

(e) Substitute $t = 0$ to get $\mathbf{D} = r_0 v_0 \mathbf{k}$, where $r_0 = r(0)$ and $v_0 = \|\mathbf{v}(0)\|$. Then argue that $r^2\theta' = r_0 v_0$ for all t.

(f) Make the substitution $q = r'$ and use the result from (e) to obtain the first-order (nonlinear) differential equation in q:

$$q\frac{dq}{dr} = \frac{r_0^2 v_0^2}{r^3} - \frac{GM}{r^2}$$

(g) Integrate with respect to r on both sides of the above equation and use an initial condition to obtain

$$q^2 = 2GM\left(\frac{1}{r} - \frac{1}{r_0}\right) + v_0^2\left(1 - \frac{r_0^2}{r^2}\right)$$

(h) Substitute $p = 1/r$ into the above equation to obtain

$$\frac{r_0^2 v_0^2}{(\theta')^2}\left(\frac{dp}{dt}\right)^2 = 2GM(p - p_0) + v_0^2\left(1 - \frac{p^2}{p_0^2}\right)$$

(i) Show that

$$\left(\frac{dp}{d\theta}\right)^2 = \left(p_0 - \frac{p_0^2 GM}{v_0^2}\right)^2 - \left(p - \frac{p_0^2 GM}{v_0^2}\right)^2$$

(j) From part (i) we can immediately conclude that

$$\frac{dp}{d\theta} = \pm\sqrt{\left(p_0 - \frac{p_0^2 GM}{v_0^2}\right)^2 - \left(p - \frac{p_0^2 GM}{v_0^2}\right)^2}$$

Explain why the minus sign is the correct sign in this case.

(k) Separate variables and integrate to obtain

$$\cos^{-1}\left(\frac{p - p_0^2 GM/v_0^2}{p_0 - p_0^2 GM/v_0^2}\right) = \theta$$

(l) Finally, obtain r as a function of θ:

$$r = \frac{r_0(1 + e)}{1 + e\cos\theta}$$

where $e = \dfrac{r_0 v_0^2}{GM} - 1$ is the eccentricity.

Answers to Concepts Review: **1.** vector-valued function of a real variable **2.** f and g are continuous at c; $f'(t)\mathbf{i} + g'(t)\mathbf{j}$ **3.** position **4.** $\mathbf{r}'(t)$; $\mathbf{r}''(t)$; tangent; concave

11.6 Lines and Tangent Lines in Three-Space

The simplest of all curves is a line. A line is determined by a fixed point P_0 and a fixed vector $\mathbf{v} = a\mathbf{i} + b\mathbf{j} + c\mathbf{k}$ called the **direction vector** for the line. It is the set of all points P such that $\overrightarrow{P_0P}$ is parallel to $\mathbf{v}$, that is, that satisfy

$$\overrightarrow{P_0P} = t\mathbf{v}$$

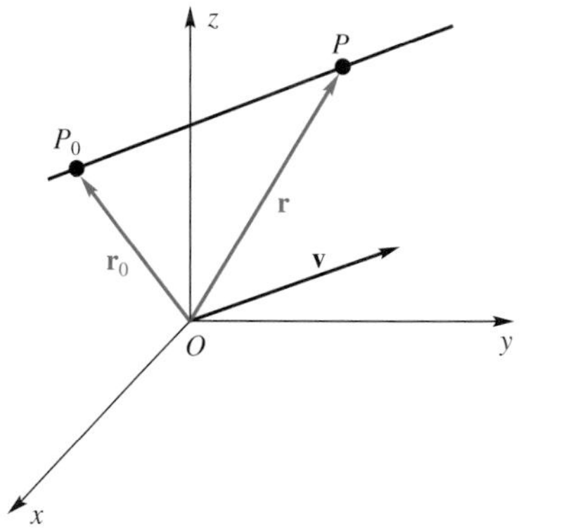

Figure 1

for some real number t (Figure 1). If $\mathbf{r} = \overrightarrow{OP}$ and $\mathbf{r}_0 = \overrightarrow{OP_0}$ are the position vectors of P and P_0, respectively, then $\overrightarrow{P_0P} = \mathbf{r} - \mathbf{r}_0$, and the equation of the line can thus be written

$$\mathbf{r} = \mathbf{r}_0 + t\mathbf{v}$$

If we write $\mathbf{r} = \langle x, y, z\rangle$ and $\mathbf{r}_0 = \langle x_0, y_0, z_0\rangle$ and equate components in the last equation above, we obtain

$$\boxed{x = x_0 + at, \quad y = y_0 + bt, \quad z = z_0 + ct}$$

These are **parametric equations** of the line through (x_0, y_0, z_0) and parallel to $\mathbf{v} = \langle a, b, c\rangle$. The numbers a, b, and c are called **direction numbers** for the line. They are not unique; any nonzero constant multiples ka, kb, and kc are also direction numbers.

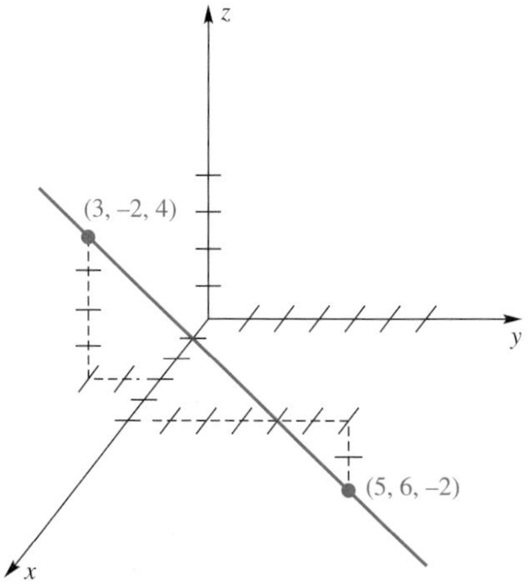

Figure 2

EXAMPLE 1 Find parametric equations for the line through $(3, -2, 4)$ and $(5, 6, -2)$ (see Figure 2).

SOLUTION A vector parallel to the given line is

$$\mathbf{v} = \langle 5 - 3, 6 + 2, -2 - 4\rangle = \langle 2, 8, -6\rangle$$

If we choose (x_0, y_0, z_0) as $(3, -2, 4)$, we obtain the parametric equations

$$x = 3 + 2t, \quad y = -2 + 8t, \quad z = 4 - 6t$$

Note that $t = 0$ determines the point $(3, -2, 4)$, whereas $t = 1$ gives $(5, 6, -2)$. In fact, $0 \le t \le 1$ corresponds to the segment joining these two points. ■

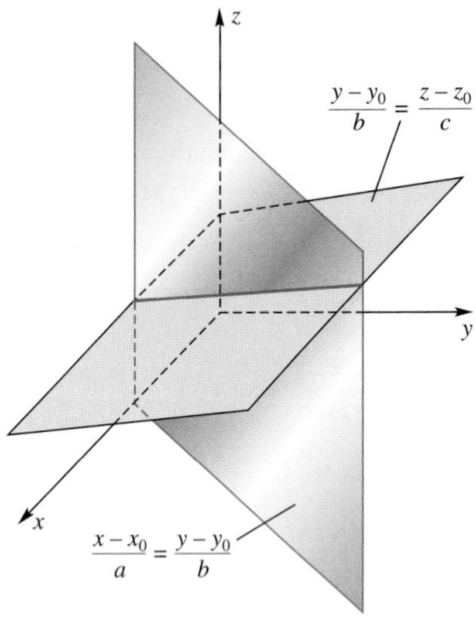

Figure 3

If we solve each of the parametric equations for t (assuming that a, b, and c are all different from zero) and equate the results, we obtain the **symmetric equations** for the line through (x_0, y_0, z_0) with direction numbers a, b, c; that is,

$$\frac{x - x_0}{a} = \frac{y - y_0}{b} = \frac{z - z_0}{c}$$

This is the conjunction of the two equations

$$\frac{x - x_0}{a} = \frac{y - y_0}{b} \quad \text{and} \quad \frac{y - y_0}{b} = \frac{z - z_0}{c}$$

both of which are the equations of planes (Figure 3); and, of course, the intersection of two planes is a line.

EXAMPLE 2 Find the symmetric equations of the line that is parallel to the vector $\langle 4, -3, 2\rangle$ and goes through $(2, 5, -1)$.

SOLUTION

$$\frac{x - 2}{4} = \frac{y - 5}{-3} = \frac{z + 1}{2}$$ ■

EXAMPLE 3 Find the symmetric equations of the line of intersection of the planes

$$2x - y - 5z = -14 \quad \text{and} \quad 4x + 5y + 4z = 28$$

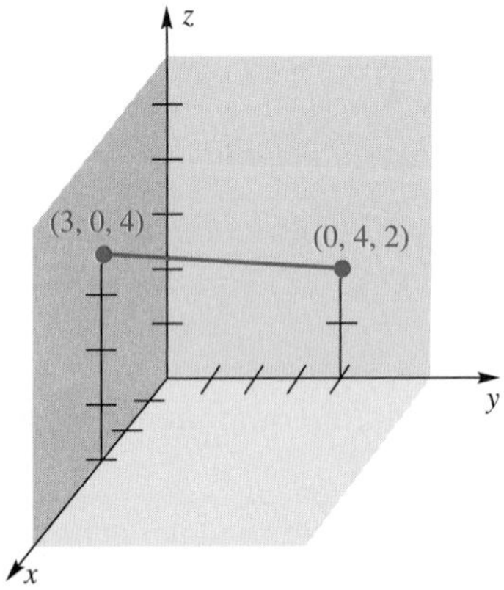

Figure 4

SOLUTION We begin by finding two points on the line. Any two points would do, but we choose to find the points where the line pierces the yz-plane and the xz-plane (Figure 4). The former is obtained by setting $x = 0$ and solving the resulting equations $-y - 5z = -14$ and $5y + 4z = 28$ simultaneously. This yields the point $(0, 4, 2)$. A similar procedure with $y = 0$ gives the point $(3, 0, 4)$. Consequently, a vector parallel to the required line is

$$\langle 3 - 0, 0 - 4, 4 - 2\rangle = \langle 3, -4, 2\rangle$$

Using $(3, 0, 4)$ for (x_0, y_0, z_0), we get

$$\frac{x - 3}{3} = \frac{y - 0}{-4} = \frac{z - 4}{2}$$

An alternative solution is based on the fact that the line of intersection of two planes is perpendicular to both of their normals. The vector $\mathbf{u} = \langle 2, -1, -5\rangle$ is normal to the first plane; $\mathbf{v} = \langle 4, 5, 4\rangle$ is normal to the second. Since

$$\mathbf{u} \times \mathbf{v} = \begin{vmatrix} \mathbf{i} & \mathbf{j} & \mathbf{k} \\ 2 & -1 & -5 \\ 4 & 5 & 4 \end{vmatrix} = 21\mathbf{i} - 28\mathbf{j} + 14\mathbf{k}$$

the vector $\mathbf{w} = \langle 21, -28, 14\rangle$ is parallel to the required line. This implies that $\frac{1}{7}\mathbf{w} = \langle 3, -4, 2\rangle$ also has this property. Next, find any point on the line of intersection, for example, $(3, 0, 4)$, and proceed as in the earlier solution. ■

EXAMPLE 4 Find parametric equations of the line through $(1, -2, 3)$ that is perpendicular to both the x-axis and the line

$$\frac{x - 4}{2} = \frac{y - 3}{-1} = \frac{z}{5}$$

SOLUTION The x-axis and the given line have directions $\mathbf{u} = \langle 1, 0, 0\rangle$ and $\mathbf{v} = \langle 2, -1, 5\rangle$, respectively. A vector perpendicular to both $\mathbf{u}$ and $\mathbf{v}$ is

$$\mathbf{u} \times \mathbf{v} = \begin{vmatrix} \mathbf{i} & \mathbf{j} & \mathbf{k} \\ 1 & 0 & 0 \\ 2 & -1 & 5 \end{vmatrix} = 0\mathbf{i} - 5\mathbf{j} - \mathbf{k}$$

The required line is parallel to $\langle 0, -5, -1\rangle$ and so also to $\langle 0, 5, 1\rangle$. Since the first direction number is zero, the line does not have symmetric equations. Its parametric equations are

$$x = 1, \qquad y = -2 + 5t, \qquad z = 3 + t$$ ■

Tangent Line to a Curve

Let

$$\mathbf{r} = \mathbf{r}(t) = f(t)\mathbf{i} + g(t)\mathbf{j} + h(t)\mathbf{k} = \langle f(t), g(t), h(t)\rangle$$

be the position vector determining a curve in three-space (Figure 5). The tangent line to the curve has direction vector

$$\mathbf{r}'(t) = f'(t)\mathbf{i} + g'(t)\mathbf{j} + h'(t)\mathbf{k} = \langle f'(t), g'(t), h'(t)\rangle$$

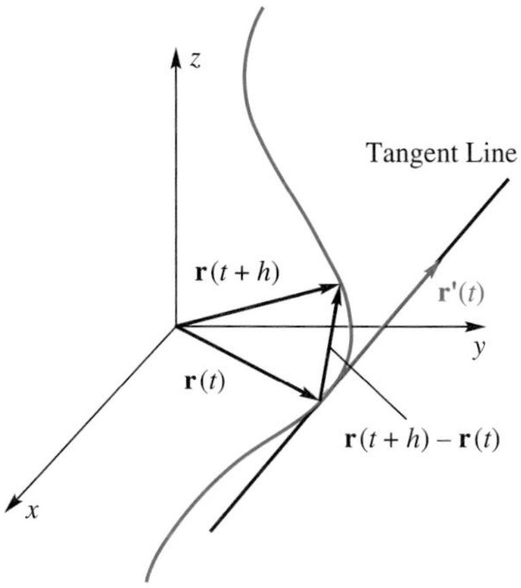

Figure 5

EXAMPLE 5 Find the parametric equations and symmetric equations for the tangent line to the curve determined by

$$\mathbf{r}(t) = t\,\mathbf{i} + \tfrac{1}{2}t^2\mathbf{j} + \tfrac{1}{3}t^3\mathbf{k}$$

at $P(2) = \left(2, 2, \frac{8}{3}\right)$.

SOLUTION

$$\mathbf{r}'(t) = \mathbf{i} + t\,\mathbf{j} + t^2\mathbf{k}$$

and

$$\mathbf{r}'(2) = \mathbf{i} + 2\mathbf{j} + 4\mathbf{k}$$

so the tangent line has direction vector $\langle 1, 2, 4\rangle$. Its symmetric equations are

$$\frac{x-2}{1} = \frac{y-2}{2} = \frac{z-\frac{8}{3}}{4}$$

The parametric equations are

$$x = 2 + t, \qquad y = 2 + 2t, \qquad z = \frac{8}{3} + 4t$$ ■

There is exactly one plane perpendicular to a smooth curve at a given point P. If we have a direction vector for the tangent line to the curve at P, then it is a normal vector for the plane (Figure 6). This, together with the given point, is enough to obtain the equation of the desired plane.

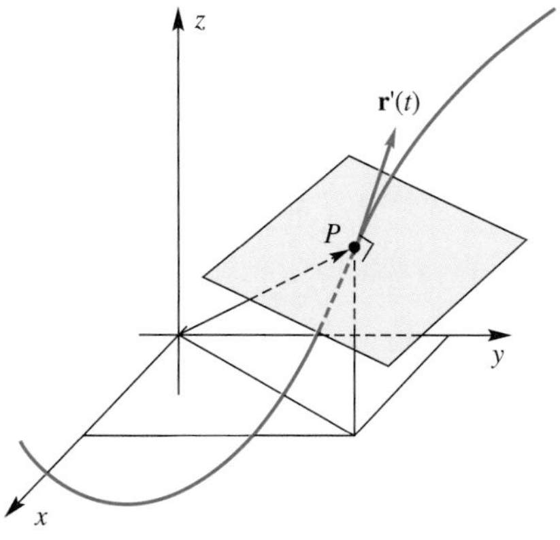

Figure 6

EXAMPLE 6 Find the equation of the plane perpendicular to the curve $\mathbf{r}(t) = 2\cos\pi t\,\mathbf{i} + \sin\pi t\,\mathbf{j} + t^3\mathbf{k}$ at $P\,(2, 0, 8)$.

SOLUTION The first issue to address is the value of t that yields the given point. Equating the z components gives $t^3 = 8$, leading to $t = 2$. A quick check verifies that $t = 2$ also yields the x- and y-components of P. Since $r'(t) = -2\pi\sin\pi t\,\mathbf{i} + \pi\cos\pi t\,\mathbf{j} + 3t^2\,\mathbf{k}$, we see that the direction vector for the tangent line at P, which is also a normal vector for the desired plane, is $r'(2) = \pi\mathbf{j} + 12\mathbf{k} = \langle 0, \pi, 12\rangle$. The equation of the plane is therefore

$$0x + \pi y + 12z = D$$

To determine D, we substitute $x = 2$, $y = 0$, and $z = 8$:

$$D = 0(2) + \pi(0) + 12(8) = 96$$

The equation of the desired plane is $\pi y + 12z = 96$. ■

Concepts Review

1. The parametric equations for a line through $(1, -3, 2)$ parallel to the vector $\langle 4, -2, -1\rangle$ are $x =$ _____, $y =$ _____, $z =$ _____.

2. The symmetric equations for the line of Question 1 are _____.

3. If $\mathbf{r}(t) = t^2\mathbf{i} - 3t\mathbf{j} + t^3\mathbf{k}$, then $\mathbf{r}'(t) =$ _____.

4. A vector parallel to the tangent line at $t = 1$ of the curve determined by the position vector $\mathbf{r}(t)$ of Question 3 is _____. This tangent line has symmetric equations _____.

Problem Set 11.6

In Problems 1–4, find the parametric equations of the line through the given pair of points.

1. $(1, -2, 3), (4, 5, 6)$

2. $(2, -1, -5), (7, -2, 3)$

3. $(4, 2, 3), (6, 2, -1)$

4. $(5, -3, -3), (5, 4, 2)$

In Problems 5–8, write both the parametric equations and the symmetric equations for the line through the given point parallel to the given vector.

5. $(4, 5, 6), \langle 3, 2, 1\rangle$

6. $(-1, 3, -6), \langle -2, 0, 5\rangle$

7. $(1, 1, 1), \langle -10, -100, -1000\rangle$

8. $(-2, 2, -2), \langle 7, -6, 3\rangle$

In Problems 9–12, find the symmetric equations of the line of intersection of the given pair of planes.

9. $4x + 3y - 7z = 1,\ 10x + 6y - 5z = 10$

10. $x + y - z = 2,\ 3x - 2y + z = 3$

11. $x + 4y - 2z = 13,\ 2x - y - 2z = 5$

12. $x - 3y + z = -1,\ 6x - 5y + 4z = 9$

13. Find the symmetric equations of the line through $(4, 0, 6)$ and perpendicular to the plane $x - 5y + 2z = 10$.

14. Find the symmetric equations of the line through $(-5, 7, -2)$ and perpendicular to both $\langle 2, 1, -3\rangle$ and $\langle 5, 4, -1\rangle$.

15. Find the parametric equations of the line through $(5, -3, 4)$ that intersects the z-axis at a right angle.

16. Find the symmetric equations of the line through $(2, -4, 5)$ that is parallel to the plane $3x + y - 2z = 5$ and perpendicular to the line

$$\frac{x+8}{2} = \frac{y-5}{3} = \frac{z-1}{-1}$$

17. Find the equation of the plane that contains the parallel lines

$$\begin{cases} x = -2 + 2t \\ y = 1 + 4t \\ z = 2 - t \end{cases} \quad \text{and} \quad \begin{cases} x = 2 - 2t \\ y = 3 - 4t \\ z = 1 + t \end{cases}$$

18. Show that the lines

$$\frac{x-1}{-4} = \frac{y-2}{3} = \frac{z-4}{-2}$$

and

$$\frac{x-2}{-1} = \frac{y-1}{1} = \frac{z+2}{6}$$

intersect, and find the equation of the plane that they determine.

19. Find the equation of the plane containing the line $x = 1 + 2t$, $y = -1 + 3t$, $z = 4 + t$ and the point $(1, -1, 5)$.

20. Find the equation of the plane containing the line $x = 3t$, $y = 1 + t$, $z = 2t$ and parallel to the intersection of the planes $2x - y + z = 0$ and $y + z + 1 = 0$.

21. Find the distance between the skew (nonintersecting and nonparallel) lines $x = 2 - t$, $y = 3 + 4t$, $z = 2t$ and $x = -1 + t$, $y = 2$, $z = -1 + 2t$ by using the following steps.

(a) Note by setting $t = 0$ that $(2, 3, 0)$ is on the first line.

(b) Find the equation of the plane π through $(2, 3, 0)$ parallel to both given lines (i.e., with normal perpendicular to both).

(c) Find a point Q on the second line.

(d) Find the distance from Q to the plane π. (See Example 10 of Section 11.3.)

See Problem 32 for another way to do this problem.

22. Find the distance between the skew lines $x = 1 + 2t$, $y = -3 + 4t$, $z = -1 - t$ and $x = 4 - 2t$, $y = 1 + 3t$, $z = 2t$ (see Problem 21).

23. Find the symmetric equations of the tangent line to the curve with equation

$$\mathbf{r}(t) = 2\cos t\,\mathbf{i} + 6\sin t\,\mathbf{j} + t\,\mathbf{k}$$

at $t = \pi/3$.

24. Find the parametric equations of the tangent line to the curve $x = 2t^2$, $y = 4t$, $z = t^3$ at $t = 1$.

25. Find the equation of the plane perpendicular to the curve $x = 3t$, $y = 2t^2$, $z = t^5$ at $t = -1$.

26. Find the equation of the plane perpendicular to the curve

$$\mathbf{r}(t) = t\sin t\,\mathbf{i} + 3t\,\mathbf{j} + 2t\cos t\,\mathbf{k}$$

at $t = \pi/2$.

27. Consider the curve

$$\mathbf{r}(t) = 2t\,\mathbf{i} + \sqrt{7t}\,\mathbf{j} + \sqrt{9 - 7t - 4t^2}\,\mathbf{k},\ 0 \le t \le \tfrac{1}{2}.$$

(a) Show that the curve lies on a sphere centered at the origin.

(b) Where does the tangent line at $t = \dfrac{1}{4}$ intersect the xz-plane?

28. Consider the curve $\mathbf{r}(t) = \sin t\cos t\,\mathbf{i} + \sin^2 t\,\mathbf{j} + \cos t\,\mathbf{k}$, $0 \le t \le 2\pi$.

(a) Show that the curve lies on a sphere centered at the origin.

(b) Where does the tangent line at $t = \pi/6$ intersect the xy-plane?

29. Consider the curve $\mathbf{r}(t) = 2t\,\mathbf{i} + t^2\mathbf{j} + (1 - t^2)\mathbf{k}$

(a) Show that this curve lies on a plane and find the equation of this plane.

(b) Where does the tangent line at $t = 2$ intersect the xy-plane?

30. Let P be a point on a plane with normal vector $\mathbf{n}$ and Q be a point off the plane. Show that the result of Example 10 of

Section 11.3, the distance d between the point Q and the plane, can be expressed as

$$d = \frac{|\overrightarrow{PQ} \cdot \mathbf{n}|}{\|\mathbf{n}\|}$$

and use this result to find the distance from $(4, -2, 3)$ to the plane $4x - 4y + 2z = 2$.

31. Point to Line Let P be a point on a line with direction $\mathbf{n}$ and Q a point off the line (Figure 7). Show that the distance d from Q to the line is given by

$$d = \frac{\|\overrightarrow{PQ} \times \mathbf{n}\|}{\|\mathbf{n}\|}$$

and use this result to find each distance in parts (a) and (b).

(a) From $Q(1, 0, -4)$ to the line $\frac{x-3}{2} = \frac{y+2}{-2} = \frac{z-1}{1}$

(b) From $Q(2, -1, 3)$ to the line $x = 1 + 2t$, $y = -1 + 3t$, $z = -6t$

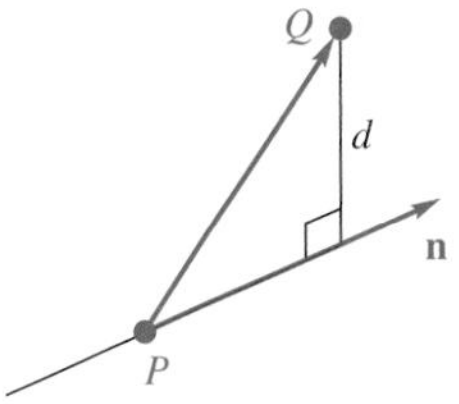

Figure 7

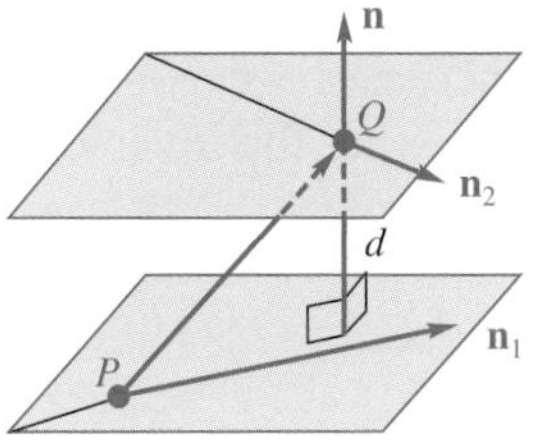

Figure 8

32. Line to Line Let P and Q be points on nonintersecting skew lines with directions $\mathbf{n}_1$ and $\mathbf{n}_2$, and let $\mathbf{n} = \mathbf{n}_1 \times \mathbf{n}_2$ (Figure 8). Show that the distance d between these lines is given by

$$d = \frac{|\overrightarrow{PQ} \cdot \mathbf{n}|}{\|\mathbf{n}\|}$$

and use this result to find the distance between each pair of lines in parts (a) and (b).

(a) $\frac{x-3}{1} = \frac{y+2}{1} = \frac{z-1}{2}$ and $\frac{x+4}{3} = \frac{y+5}{4} = \frac{z}{5}$

(b) $x = 1 + 2t$, $y = -2 + 3t$, $z = -4t$ and $x = 3t$, $y = 1 + t$, $z = -5t$

Answers to Concepts Review: **1.** $1 + 4t; -3 - 2t; 2 - t$
2. $\frac{x-1}{4} = \frac{y+3}{-2} = \frac{z-2}{-1}$ **3.** $2t\mathbf{i} - 3\mathbf{j} + 3t^2\mathbf{k}$
4. $\langle 2, -3, 3\rangle$; $\frac{x-1}{2} = \frac{y+3}{-3} = \frac{z-1}{3}$

11.7 Curvature and Components of Acceleration

We want to introduce a number, called the **curvature,** that measures how sharply a curve bends at a given point. A line should have curvature zero, and a curve that is turning sharply should have a large curvature (Figure 1).

Let $\mathbf{r}(t) = f(t)\mathbf{i} + g(t)\mathbf{j} + h(t)\mathbf{k}$ denote the position of an object at time t. We will assume that $\mathbf{r}'(t)$ is continuous and that $\mathbf{r}'(t)$ is never equal to the zero vector. This last condition assures that the accumulated arc length $s(t)$ increases as t increases. Our measure of curvature is going to involve how fast the tangent vector is changing. Rather than working with the tangent vector $\mathbf{r}'(t)$ we choose to work with the unit tangent vector (Figure 2)

$$\mathbf{T}(t) = \frac{\mathbf{r}'(t)}{\|\mathbf{r}'(t)\|} = \frac{\mathbf{v}(t)}{\|\mathbf{v}(t)\|}$$

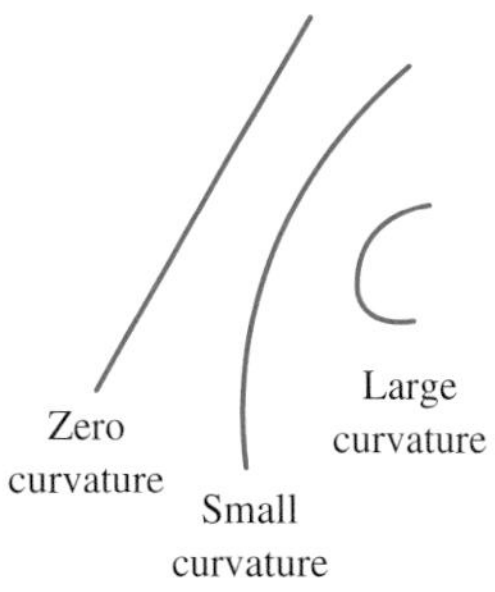

Figure 1

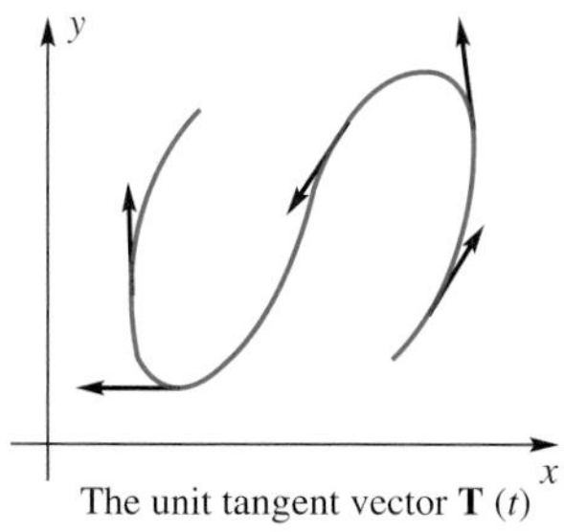

Figure 2

To accomplish the task of defining curvature, we consider the rate of change in the unit tangent vector. Figures 3 and 4 illustrate this concept for a given curve. As the object moves from points A to B (Figure 3) in time Δt, the unit tangent vector

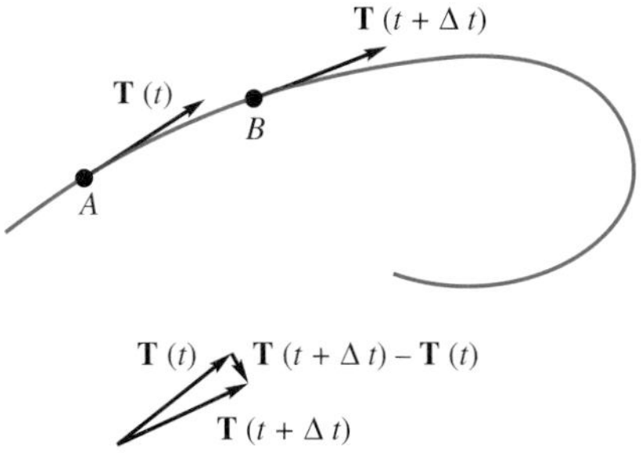

Figure 3

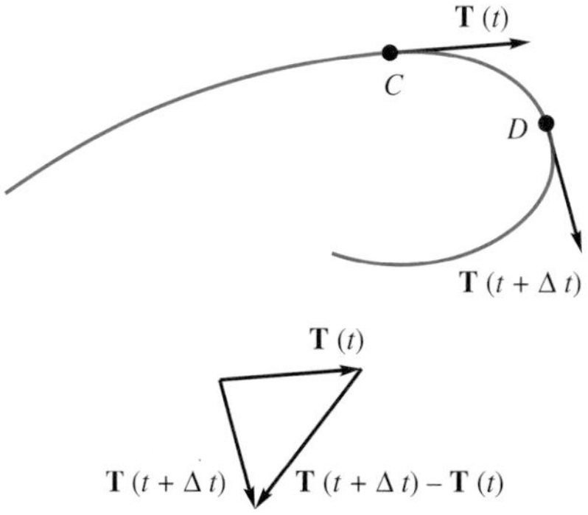

Figure 4

changed very little; in other words, the magnitude of $\mathbf{T}(t + \Delta t) - \mathbf{T}(t)$ is small. On the other hand, as the object moves from points C to D (Figure 4), also in time Δt, the unit tangent vector changed quite a bit; in other words, the magnitude of $\mathbf{T}(t + \Delta t) - \mathbf{T}(t)$ is large. Our definition of curvature κ is therefore the magnitude of the rate of change of the unit tangent vector with respect to arc length s; that is,

$$\kappa = \left\| \frac{d\mathbf{T}}{ds} \right\|$$

We differentiate with respect to arc length s rather than with respect to t because we want the curvature to be an intrinsic property of the curve, not how fast the object moves along the curve. (Imagine circular motion; the curvature of the circle should not depend on how fast the object travels around the curve.)

The definition of curvature given above does not help us to actually evaluate the curvature of a particular curve. To find a workable formula, we proceed as follows. In Section 11.5 we saw that the speed of an object could be expressed as

$$\text{speed} = \|\mathbf{v}(t)\| = \frac{ds}{dt}$$

Since s increases as t increases we can apply the Inverse Function Theorem (Theorem 6.2B) to conclude that the inverse of $s(t)$ exists and

$$\frac{dt}{ds} = \frac{1}{ds/dt} = \frac{1}{\|\mathbf{v}(t)\|}$$

This allows us to write

$$\kappa = \left\| \frac{d\mathbf{T}}{ds} \right\| = \left\| \frac{d\mathbf{T}}{dt}\frac{dt}{ds} \right\| = \left| \frac{dt}{ds} \right| \left\| \frac{d\mathbf{T}}{dt} \right\| = \frac{1}{\|\mathbf{v}(t)\|}\|\mathbf{T}'(t)\| = \frac{\|\mathbf{T}'(t)\|}{\|\mathbf{r}'(t)\|}$$

Some Important Examples To convince you that our definition of curvature is sensible, we illustrate with some familiar curves.

EXAMPLE 1 Show that the curvature of a line is identically zero.

SOLUTION For a line, the unit tangent vector is a constant, so its derivative is $\mathbf{0}$. But to illustrate vector methods, we give an algebraic demonstration. If motion is along the line whose parametric equation is given by

$$x = x_0 + at$$
$$y = y_0 + bt$$
$$z = z_0 + ct$$

then the position vector can be written as

$$\mathbf{r}(t) = \langle x_0, y_0, z_0 \rangle + t\langle a, b, c \rangle$$

Thus

$$\mathbf{v}(t) = \mathbf{r}'(t) = \langle a, b, c \rangle$$

$$\mathbf{T}(t) = \frac{\langle a, b, c \rangle}{\sqrt{a^2 + b^2 + c^2}}$$

$$\kappa = \frac{\|\mathbf{T}'(t)\|}{\|\mathbf{v}(t)\|} = \frac{\|\mathbf{0}\|}{\sqrt{a^2 + b^2 + c^2}} = 0$$

■

> **Curvature**
>
> Curvature is a relatively simple concept. However, the computations required for computing the curvature are often long and messy.

EXAMPLE 2 Find the curvature of a circle of radius a.

SOLUTION We assume that the circle lies in the xy-plane and is centered at the origin so that the position vector is

$$\mathbf{r}(t) = a \cos t\, \mathbf{i} + a \sin t\, \mathbf{j}$$

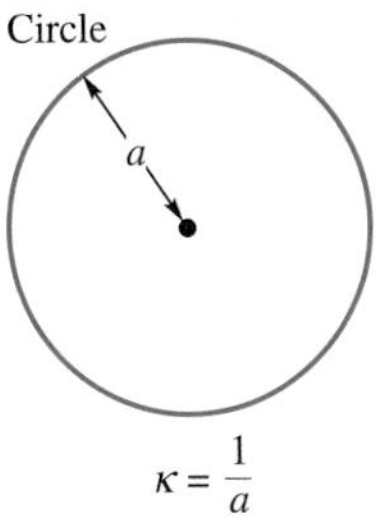

Figure 5

Thus,

$$\mathbf{v}(t) = \mathbf{r}'(t) = -a \sin t\, \mathbf{i} + a \cos t\, \mathbf{j}$$

$$\|\mathbf{v}(t)\| = \sqrt{a^2 \sin^2 t + a^2 \cos^2 t} = a$$

$$\mathbf{T}(t) = \frac{\mathbf{v}(t)}{\|\mathbf{v}(t)\|} = \frac{-a \sin t\, \mathbf{i} + a \cos t\, \mathbf{j}}{a} = -\sin t\, \mathbf{i} + \cos t\, \mathbf{j}$$

$$\kappa = \frac{\|\mathbf{T}'(t)\|}{\|\mathbf{v}(t)\|} = \frac{\|-\cos t\, \mathbf{i} - \sin t\, \mathbf{j}\|}{a} = \frac{1}{a}$$

Since κ is the reciprocal of the radius, small circles have large curvature, and large circles have small curvature. See Figure 5. ■

Checking Extreme Cases

Once again, it is instructive to check extreme cases. If $c = 0$, then motion is in a circle with radius a and the curvature is $a/(a^2 + 0^2) = 1/a$, which is the curvature of a circle with radius a. If $c \to \infty$, then we are stretching our helix vertically into a line and

$$\lim_{c \to \infty} \frac{a}{a^2 + c^2} = 0$$

which is the curvature of a line. Both results are as expected.

EXAMPLE 3 Find the curvature for the helix $\mathbf{r}(t) = a \cos t\, \mathbf{i} + a \sin t\, \mathbf{j} + ct\, \mathbf{k}$.

SOLUTION

$$\mathbf{v}(t) = \mathbf{r}'(t) = -a \sin t\, \mathbf{i} + a \cos t\, \mathbf{j} + c\, \mathbf{k}$$

$$\|\mathbf{v}(t)\| = \sqrt{a^2 \sin^2 t + a^2 \cos^2 t + c^2} = \sqrt{a^2 + c^2}$$

$$\mathbf{T}(t) = \frac{\mathbf{v}(t)}{\|\mathbf{v}(t)\|} = \frac{-a \sin t\, \mathbf{i} + a \cos t\, \mathbf{j} + c\, \mathbf{k}}{\sqrt{a^2 + c^2}}$$

$$\mathbf{T}'(t) = \frac{-a \cos t\, \mathbf{i} - a \sin t\, \mathbf{j}}{\sqrt{a^2 + c^2}}$$

$$\kappa = \frac{\|\mathbf{T}'(t)\|}{\|\mathbf{v}(t)\|} = \frac{\|(-a \cos t\, \mathbf{i} - a \sin t\, \mathbf{j})/\sqrt{a^2 + c^2}\|}{\sqrt{a^2 + c^2}} = \frac{a}{a^2 + c^2}$$ ■

For the three curves discussed so far, the line, circle, and helix, the curvature is a constant. This phenomenon occurs only for special curves. Normally the curvature is a function of t.

Circle of Curvature for Curves in Three-Space

The concepts of radius of curvature and circle of curvature apply more generally to curves in three-space. The radius of curvature is still $R = 1/\kappa$, but the circle of curvature is a more involved concept. The circle of curvature lies entirely in the **osculating plane** (defined near the end of this section). For a plane curve, the osculating plane *is* the plane containing the curve.

Radius and Center of Curvature for a Plane Curve Let P be a point on a plane curve (i.e., a curve lying entirely in the xy-plane) where the curvature is nonzero. Consider the circle that is tangent to the curve at P which has the same curvature there. Its center will lie on the concave side of the curve. This circle is called the **circle of curvature** or **osculating circle.** Its radius $R = 1/\kappa$ is called the **radius of curvature** and its center is the **center of curvature.** (See Figure 6.) These notions are illustrated in the next example.

EXAMPLE 4 Find the curvature and the radius of curvature of the curve traced by the position vector

$$\mathbf{r}(t) = 2t\, \mathbf{i} + t^2 \mathbf{j}$$

at the points $(0, 0)$ and $(2, 1)$.

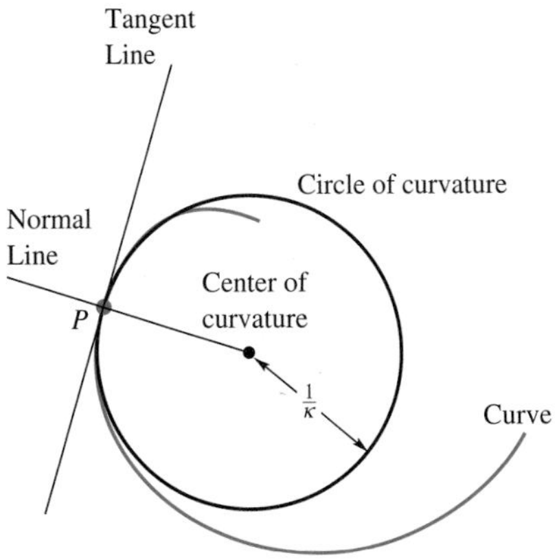

Figure 6

SOLUTION

$$\mathbf{v}(t) = \mathbf{r}'(t) = 2\, \mathbf{i} + 2t\, \mathbf{j}$$

$$\|\mathbf{v}(t)\| = \sqrt{2^2 + (2t)^2} = 2\sqrt{1 + t^2}$$

$$\mathbf{T}(t) = \frac{\mathbf{v}(t)}{\|\mathbf{v}(t)\|} = \frac{2\mathbf{i} + 2t\, \mathbf{j}}{2\sqrt{1 + t^2}} = \frac{1}{\sqrt{1 + t^2}}\, \mathbf{i} + \frac{t}{\sqrt{1 + t^2}}\, \mathbf{j}$$

$$\mathbf{T}'(t) = -\frac{t}{(1 + t^2)^{3/2}}\, \mathbf{i} + \frac{1}{(1 + t^2)^{3/2}}\, \mathbf{j}$$

$$\kappa(t) = \frac{\|\mathbf{T}'(t)\|}{\|\mathbf{v}(t)\|} = \frac{\sqrt{\frac{t^2}{(1 + t^2)^3} + \frac{1}{(1 + t^2)^3}}}{2\sqrt{1 + t^2}} = \frac{1}{2(1 + t^2)^{3/2}}$$

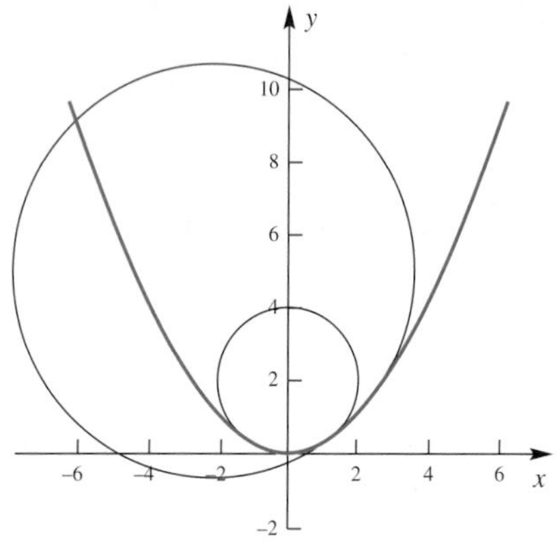

Figure 7

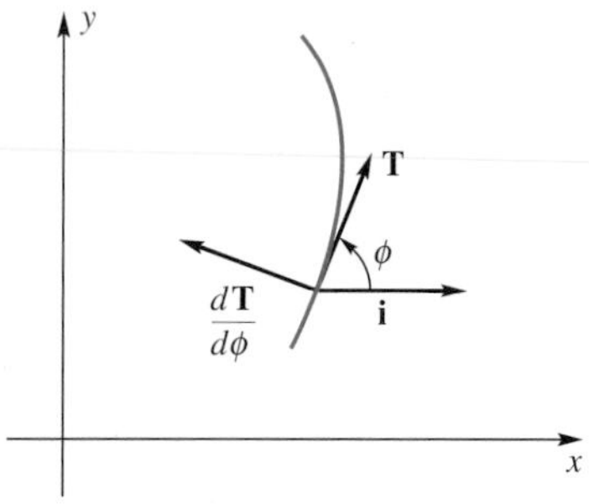

Figure 8

The points (0, 0) and (2, 1) occur when $t = 0$ and $t = 1$, respectively. Thus, the values of the curvature at these points are

$$\kappa(0) = \frac{1}{2(1 + 0^2)^{3/2}} = \frac{1}{2}$$

$$\kappa(1) = \frac{1}{2(1 + 1^2)^{3/2}} = \frac{\sqrt{2}}{8}$$

The two values for the radius of curvature are thus $1/\kappa(0) = 2$ and $1/\kappa(1) = 8/\sqrt{2} = 4\sqrt{2}$. The circles of curvature are shown in Figure 7. ■

Other Formulas for Curvature of a Plane Curve Let ϕ denote the angle measured counterclockwise from **i** to **T** (Figure 8). Then,

$$\mathbf{T} = \cos\phi\,\mathbf{i} + \sin\phi\,\mathbf{j}$$

and so

$$\frac{d\mathbf{T}}{d\phi} = -\sin\phi\,\mathbf{i} + \cos\phi\,\mathbf{j}$$

Now $d\mathbf{T}/d\phi$ is a unit vector (length 1) and $\mathbf{T}\cdot d\mathbf{T}/d\phi = 0$. Moreover,

$$\kappa = \left\|\frac{d\mathbf{T}}{ds}\right\| = \left\|\frac{d\mathbf{T}}{d\phi}\frac{d\phi}{ds}\right\| = \left\|\frac{d\mathbf{T}}{d\phi}\right\|\left|\frac{d\phi}{ds}\right| = \left|\frac{d\phi}{ds}\right|$$

This formula for κ helps our intuitive understanding of curvature (it measures the rate of change of ϕ with respect to s), and it also allows us to give a fairly simple proof of the following important theorem.

Theorem A

Consider a curve with vector equation $\mathbf{r}(t) = f(t)\mathbf{i} + g(t)\mathbf{j}$, that is, with parametric equations $x = f(t)$ and $y = g(t)$. Then

$$\kappa = \frac{|x'y'' - y'x''|}{\left[(x')^2 + (y')^2\right]^{3/2}}$$

In particular, if the curve is the graph of $y = g(x)$, then

$$\kappa = \frac{|y''|}{\left[1 + (y')^2\right]^{3/2}}$$

Primes indicate differentiation with respect to t in the first formula and with respect to x in the second formula.

Proof We might calculate κ directly from the formula $\kappa = \|\mathbf{T}'(t)\|/\|\mathbf{r}'(t)\|$, a task we propose in Problem 78. It is a good (but painful) exercise in differentiation and algebraic manipulation. Rather, we choose to use the formula $\kappa = |d\phi/ds|$ derived above. Refer to Figure 8, from which we see that

$$\tan\phi = \frac{dy}{dx} = \frac{dy/dt}{dx/dt} = \frac{y'}{x'}$$

Differentiate both sides of this equation with respect to t to obtain

$$\sec^2\phi\,\frac{d\phi}{dt} = \frac{x'y'' - y'x''}{(x')^2}$$

Then

$$\frac{d\phi}{dt} = \frac{x'y'' - y'x''}{(x')^2 \sec^2 \phi} = \frac{x'y'' - y'x''}{(x')^2(1 + \tan^2 \phi)}$$

$$= \frac{x'y'' - y'x''}{(x')^2(1 + (y')^2/(x')^2)} = \frac{x'y'' - y'x''}{(x')^2 + (y')^2}$$

But

$$\kappa = \left|\frac{d\phi}{ds}\right| = \left|\frac{d\phi}{dt}\frac{dt}{ds}\right| = \left|\frac{d\phi/dt}{ds/dt}\right| = \frac{|d\phi/dt|}{\left[(x')^2 + (y')^2\right]^{1/2}}$$

When we put these two results together, we obtain

$$\kappa = \frac{|x'y'' - y'x''|}{\left[(x')^2 + (y')^2\right]^{3/2}}$$

which is the first assertion of the theorem.

To obtain the second assertion, simply regard $y = g(x)$ as corresponding to the parametric equations $x = t$, $y = g(t)$ so that $x' = 1$ and $x'' = 0$. The conclusion follows. ■

EXAMPLE 5 Find the curvature of the ellipse

$$x = 3\cos t, \qquad y = 2\sin t$$

at the points corresponding to $t = 0$ and $t = \pi/2$, that is, at $(3, 0)$ and $(0, 2)$. Sketch the ellipse showing the corresponding circles of curvature.

SOLUTION From the given equations,

$$x' = -3\sin t, \qquad y' = 2\cos t$$

$$x'' = -3\cos t \qquad y'' = -2\sin t$$

Thus,

$$\kappa = \kappa(t) = \frac{|x'y'' - y'x''|}{\left[(x')^2 + (y')^2\right]^{3/2}} = \frac{6\sin^2 t + 6\cos^2 t}{\left[9\sin^2 t + 4\cos^2 t\right]^{3/2}}$$

$$= \frac{6}{\left[5\sin^2 t + 4\right]^{3/2}}$$

Consequently,

$$\kappa(0) = \frac{6}{4^{3/2}} = \frac{3}{4}$$

$$\kappa\left(\frac{\pi}{2}\right) = \frac{6}{9^{3/2}} = \frac{2}{9}$$

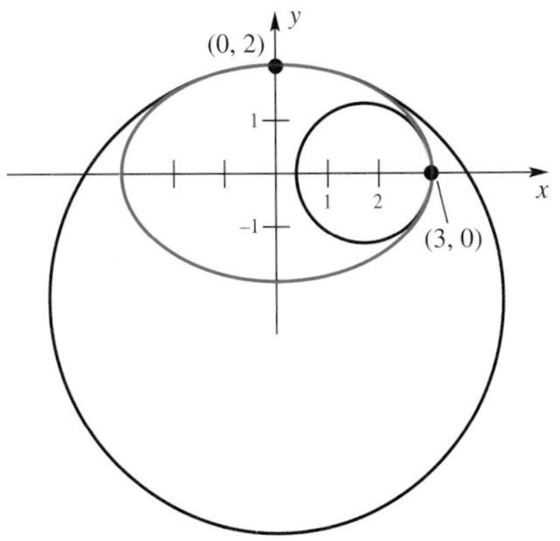

Figure 9

Note that $\kappa(0)$ is larger than $\kappa(\pi/2)$, as it should be. Figure 9 shows the circle of curvature at $(3, 0)$, which has radius $\frac{4}{3}$, and the one at $(0, 2)$, which has radius $\frac{9}{2}$. ■

EXAMPLE 6 Find the curvature of $y = \ln|\cos x|$ at $x = \pi/3$.

SOLUTION We employ the second formula of Theorem A, noting that the primes now indicate differentiation with respect to x. Since $y' = -\tan x$ and $y'' = -\sec^2 x$,

$$\kappa = \frac{|-\sec^2 x|}{(1 + \tan^2 x)^{3/2}} = \frac{\sec^2 x}{(\sec^2 x)^{3/2}} = |\cos x|$$

At $x = \pi/3$, $\kappa = \frac{1}{2}$. ■

Components of Acceleration For motion along the curve with position vector $\mathbf{r}(t)$, the unit tangent vector is $\mathbf{T}(t) = \mathbf{r}'(t)/\|\mathbf{r}'(t)\|$. This vector satisfies

$$\mathbf{T}(t) \cdot \mathbf{T}(t) = 1$$

for all t. Differentiating both sides with respect to t, and using the Product Rule on the left side, gives

$$\mathbf{T}(t) \cdot \mathbf{T}'(t) + \mathbf{T}(t) \cdot \mathbf{T}'(t) = 0$$

This reduces to $\mathbf{T}(t) \cdot \mathbf{T}'(t) = 0$ telling us that $\mathbf{T}(t)$ and $\mathbf{T}'(t)$ are perpendicular for all t. In general, $\mathbf{T}'$ is not a unit vector, so we define the **principal unit normal vector** to be

$$\mathbf{N}(t) = \frac{\mathbf{T}'(t)}{\|\mathbf{T}'(t)\|}$$

Now, imagine that you are riding in a car on a winding road. As the car accelerates you feel pushed in the opposite direction. If the car speeds up, you feel a push backwards, and when you are turning left, you feel a push to the right. These two kinds of acceleration are called the **tangential** and **normal components of acceleration,** respectively. What we would like to do is to express the acceleration vector $\mathbf{a}(t) = \mathbf{r}''(t)$ in terms of these two components, that is, in terms of the unit tangent vector $\mathbf{T}(t)$ and the unit normal vector $\mathbf{N}(t)$. Specifically, we would like to find scalars a_T and a_N so that

$$\mathbf{a} = a_T\mathbf{T} + a_N\mathbf{N}$$

To accomplish this we note that

$$\mathbf{T} = \frac{\mathbf{v}}{\|\mathbf{v}\|} = \frac{\mathbf{v}}{ds/dt}$$

so

$$\mathbf{v} = \frac{ds}{dt}\mathbf{T}$$

Differentiating both sides with respect to t, and using the Product Rule, gives

$$\mathbf{v}' = \frac{ds}{dt}\mathbf{T}' + \mathbf{T}\frac{d^2s}{dt^2}$$

Using the facts that $\mathbf{a} = \mathbf{v}'$, $\mathbf{T}' = \|\mathbf{T}'\|\mathbf{N}$, and $\|\mathbf{T}'\| = \kappa\dfrac{ds}{dt}$ we have

$$\mathbf{a} = \frac{d^2s}{dt^2}\mathbf{T} + \frac{ds}{dt}\|\mathbf{T}'\|\mathbf{N} = \frac{d^2s}{dt^2}\mathbf{T} + \left(\frac{ds}{dt}\right)^2\kappa\mathbf{N}$$

The tangential and normal components of acceleration are

$$a_T = \frac{d^2s}{dt^2}$$

and

$$a_N = \left(\frac{ds}{dt}\right)^2\kappa$$

These results make sense from a physical point of view. If you are speeding up on a straight road, then $a_T = \dfrac{d^2s}{dt^2} > 0$, and $\kappa = 0$ so $a_N = 0$. Thus, in this case you would feel a push backward and no push to either side. On the other hand, if you are going around a curve at a constant speed (i.e., ds/dt is constant) then

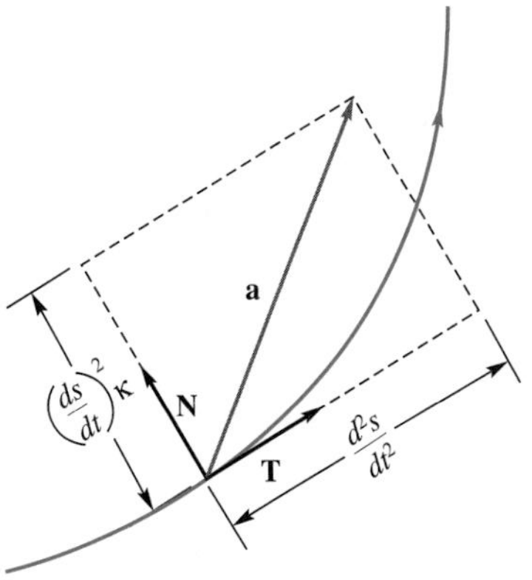

Figure 10

$a_T = \dfrac{d^2s}{dt^2} = 0$ and $\kappa > 0$, making a_N positive. Finally, imagine going around a curve while speeding up. In this case both a_T and a_N will be positive, and **a** will point inward and forward as shown in Figure 10. You would feel thrown back and to the right.

To calculate a_N, it appears that we must calculate the curvature κ. However, this can be avoided by noting that since **T** and **N** are orthogonal,

$$\|\mathbf{a}\|^2 = a_T^2 + a_N^2$$

so we can compute

$$a_N = \sqrt{\|\mathbf{a}\|^2 - a_T^2}$$

The vector **N** can be computed indirectly from

$$\mathbf{N} = \frac{\mathbf{a} - a_T\mathbf{T}}{a_N}$$

Vector Forms for the Components of Acceleration We can write the formulas for the components of acceleration in terms of the position vector **r**. We begin with

$$\mathbf{a} = a_T\mathbf{T} + a_N\mathbf{N}$$

and dot both sides by **T** to get

$$\mathbf{T}\cdot\mathbf{a} = \mathbf{T}\cdot(a_T\mathbf{T} + a_N\mathbf{N}) = a_T\mathbf{T}\cdot\mathbf{T} + a_N\mathbf{T}\cdot\mathbf{N} = a_T(1) + a_N(0) = a_T$$

Here we have used the facts that **T** is a unit vector and that **T** and **N** are orthogonal. Thus,

$$a_T = \mathbf{T}\cdot\mathbf{a} = \frac{\mathbf{r}'}{\|\mathbf{r}'\|}\cdot\mathbf{r}'' = \frac{\mathbf{r}'\cdot\mathbf{r}''}{\|\mathbf{r}'\|}$$

We can find a similar formula for a_N by crossing both sides by **T**:

$$\mathbf{T}\times\mathbf{a} = a_T\mathbf{T}\times\mathbf{T} + a_N\mathbf{T}\times\mathbf{N} = a_T\mathbf{0} + a_N(\mathbf{T}\times\mathbf{N}) = a_N(\mathbf{T}\times\mathbf{N})$$

Taking the magnitude of both sides gives

$$\|\mathbf{T}\times\mathbf{a}\| = |a_N|\|\mathbf{T}\times\mathbf{N}\| = a_N\|\mathbf{T}\|\|\mathbf{N}\|\sin\frac{\pi}{2} = a_N(1)(1)(1) = a_N$$

Notice that $a_N = (ds/dt)^2\kappa > 0$, so the absolute value bars are not needed for a_N. Thus,

$$a_N = \|\mathbf{T}\times\mathbf{a}\| = \left\|\frac{\mathbf{r}'}{\|\mathbf{r}'\|}\times\mathbf{r}''\right\| = \frac{\|\mathbf{r}'\times\mathbf{r}''\|}{\|\mathbf{r}'\|}$$

Finally, we can find a formula for the curvature κ:

$$\kappa = \frac{a_N}{(ds/dt)^2} = \frac{\|\mathbf{r}'\times\mathbf{r}''\|/\|\mathbf{r}'\|}{\|\mathbf{r}'\|^2} = \frac{\|\mathbf{r}'\times\mathbf{r}''\|}{\|\mathbf{r}'\|^3}$$

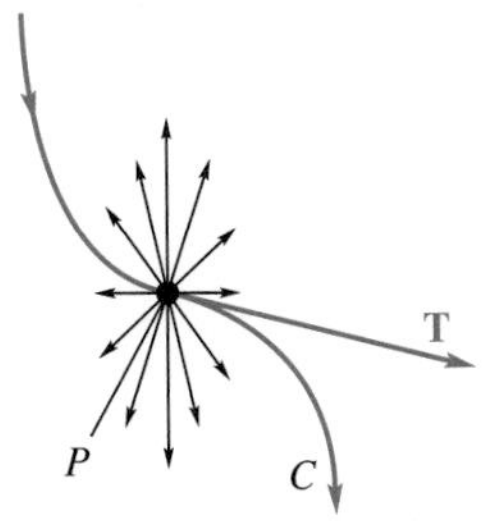

Figure 11

Binormal at P (Optional) Given a curve C and the unit tangent vector **T** at P, there are, of course, infinitely many unit vectors perpendicular to **T** at P (Figure 11). We picked one of them, $\mathbf{N} = \mathbf{T}'/\|\mathbf{T}'\|$, and called it the principal normal. The vector

$$\mathbf{B} = \mathbf{T}\times\mathbf{N}$$

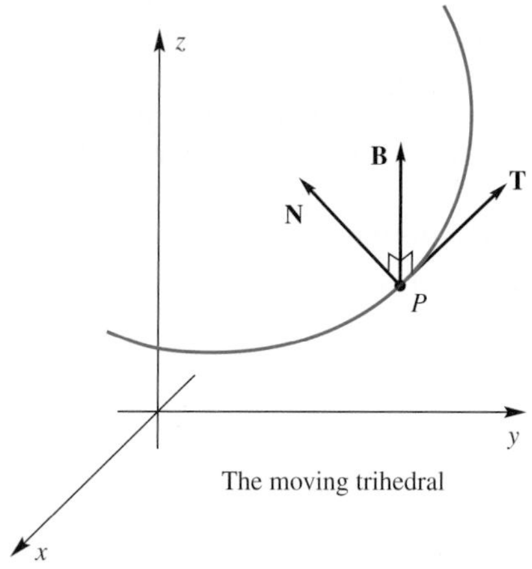

Figure 12

is called the **binormal.** It, too, is a unit vector and it is perpendicular to both **T** and **N**. (Why?)

If the unit tangent vector **T**, the principal normal **N**, and the binormal **B** have their initial points at P, they form a right-handed, mutually perpendicular triple of unit vectors known as the **trihedral** at P (Figure 12). This moving trihedral plays a crucial role in a subject called differential geometry. The plane of **T** and **N** is called the **osculating plane** at P.

EXAMPLE 7 Find **T**, **N**, and **B**, and the normal and tangential components of acceleration for uniform circular motion $\mathbf{r}(t) = a\cos\omega t\,\mathbf{i} + a\sin\omega t\,\mathbf{j}$.

SOLUTION

$$\mathbf{T} = \frac{\mathbf{r}'}{\|\mathbf{r}'\|} = \frac{-a\omega\sin\omega t\,\mathbf{i} + a\omega\cos\omega t\,\mathbf{j}}{\|-a\omega\sin\omega t\,\mathbf{i} + a\omega\cos\omega t\,\mathbf{j}\|} = -\sin\omega t\,\mathbf{i} + \cos\omega t\,\mathbf{j}$$

$$\mathbf{N} = \frac{\mathbf{T}'}{\|\mathbf{T}'\|} = \frac{-\omega\cos\omega t\,\mathbf{i} - \omega\sin\omega t\,\mathbf{j}}{\|-\omega\cos\omega t\,\mathbf{i} - \omega\sin\omega t\,\mathbf{j}\|} = -\cos\omega t\,\mathbf{i} - \sin\omega t\,\mathbf{j}$$

$$\mathbf{B} = \mathbf{T}\times\mathbf{N} = \begin{vmatrix} \mathbf{i} & \mathbf{j} & \mathbf{k} \\ -\sin\omega t & \cos\omega t & 0 \\ -\cos\omega t & -\sin\omega t & 0 \end{vmatrix} = \mathbf{k}$$

$$a_T = \frac{\mathbf{r}'\cdot\mathbf{r}''}{\|\mathbf{r}'\|} = \frac{(-a\omega\sin\omega t\mathbf{i} + a\omega\cos\omega t\mathbf{j})\cdot(-a\omega^2\cos\omega t\mathbf{i} - a\omega^2\sin\omega t\mathbf{j})}{a\omega} = 0$$

$$\mathbf{r}'\times\mathbf{r}'' = \begin{vmatrix} \mathbf{i} & \mathbf{j} & \mathbf{k} \\ -a\omega\sin\omega t & a\omega\cos\omega t & 0 \\ -a\omega^2\cos\omega t & -a\omega^2\sin\omega t & 0 \end{vmatrix} = a^2\omega^3\mathbf{k}$$

$$a_N = \frac{\|\mathbf{r}'\times\mathbf{r}''\|}{\|\mathbf{r}'\|} = \frac{a^2\omega^3}{a\omega} = a\omega^2$$

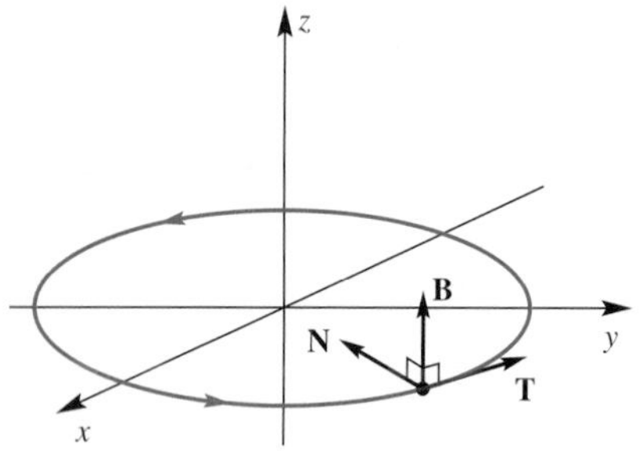

Figure 13

The tangential component of acceleration is 0 since the object is moving at uniform speed. The normal component of acceleration is equal to the magnitude of the acceleration vector. Figure 13 shows the vectors **T**, **N**, and **B**. ■

EXAMPLE 8 At the point $\left(1, 1, \frac{1}{3}\right)$, find **T**, **N**, **B**, a_T, a_N, and κ for the curvilinear motion

$$\mathbf{r}(t) = t\,\mathbf{i} + t^2\mathbf{j} + \tfrac{1}{3}t^3\mathbf{k}$$

SOLUTION

$$\mathbf{r}'(t) = \mathbf{i} + 2t\,\mathbf{j} + t^2\mathbf{k}$$
$$\mathbf{r}''(t) = 2\,\mathbf{j} + 2t\,\mathbf{k}$$

At $t = 1$, which gives the point $\left(1, 1, \frac{1}{3}\right)$, we have

$$\mathbf{r}' = \mathbf{i} + 2\mathbf{j} + \mathbf{k}$$
$$\mathbf{r}'' = 2\mathbf{j} + 2\mathbf{k}$$
$$\mathbf{T} = \frac{\mathbf{r}'}{\|\mathbf{r}'\|} = \frac{\mathbf{i} + 2\mathbf{j} + \mathbf{k}}{\sqrt{6}}$$
$$a_T = \frac{\mathbf{r}'\cdot\mathbf{r}''}{\|\mathbf{r}'\|} = \frac{6}{\sqrt{6}}$$

$$a_N = \frac{\|\mathbf{r}' \times \mathbf{r}''\|}{\|\mathbf{r}'\|} = \frac{1}{\sqrt{6}} \left\| \begin{vmatrix} \mathbf{i} & \mathbf{j} & \mathbf{k} \\ 1 & 2 & 1 \\ 0 & 2 & 2 \end{vmatrix} \right\| = \frac{1}{\sqrt{6}} \|2\mathbf{i} - 2\mathbf{j} + 2\mathbf{k}\| = \sqrt{2}$$

$$\mathbf{N} = \frac{\mathbf{a} - a_T \mathbf{T}}{a_N} = \frac{(2\mathbf{j} + 2\mathbf{k}) - (\mathbf{i} + 2\mathbf{j} + \mathbf{k})}{\sqrt{2}} = \frac{-\mathbf{i} + \mathbf{k}}{\sqrt{2}}$$

$$\mathbf{B} = \mathbf{T} \times \mathbf{N}$$

$$= \begin{vmatrix} \mathbf{i} & \mathbf{j} & \mathbf{k} \\ 1/\sqrt{6} & 2/\sqrt{6} & 1/\sqrt{6} \\ -1/\sqrt{2} & 0 & 1/\sqrt{2} \end{vmatrix}$$

$$= \frac{1}{\sqrt{3}}\mathbf{i} - \frac{1}{\sqrt{3}}\mathbf{j} + \frac{1}{\sqrt{3}}\mathbf{k}$$

$$\kappa = \frac{\|\mathbf{r}' \times \mathbf{r}''\|}{\|\mathbf{r}'\|^3} = \frac{a_N}{\|\mathbf{r}'\|^2} = \frac{\sqrt{2}}{6}$$ ■

Concepts Review

1. Curvature is defined to be the magnitude of the vector ______.

2. The curvature of a circle of radius a is constant and has value $\kappa =$ ______; the curvature of a line is ______.

3. The acceleration vector $\mathbf{a}$ can be expressed as $\mathbf{a} =$ ______ $\mathbf{T}$ + ______ $\mathbf{N}$.

4. For uniform circular motion in the plane, the tangential component of acceleration is ______.

Problem Set 11.7

In Problems 1–6, sketch the curve over the indicated domain for t. Find $\mathbf{v}$, $\mathbf{a}$, $\mathbf{T}$, *and* κ *at the point where* $t = t_1$.

1. $\mathbf{r}(t) = t\,\mathbf{i} + t^2\mathbf{j};\; 0 \le t \le 2;\, t_1 = 1$

2. $\mathbf{r}(t) = t^2\mathbf{i} + (2t + 1)\mathbf{j};\; 0 \le t \le 2;\, t_1 = 1$

3. $\mathbf{r}(t) = t\,\mathbf{i} + 2\cos t\,\mathbf{j} + 2\sin t\,\mathbf{k};\; 0 \le t \le 4\pi;\, t_1 = \pi$

4. $\mathbf{r}(t) = 5\cos t\,\mathbf{i} + 2t\,\mathbf{j} + 5\sin t\,\mathbf{k};\; 0 \le t \le 4\pi;\, t_1 = \pi$

5. $\mathbf{r}(t) = \frac{t^2}{8}\mathbf{i} + 5\cos t\,\mathbf{j} + 5\sin t\,\mathbf{k};\; 0 \le t \le 4\pi;\, t_1 = \pi$

6. $\mathbf{r}(t) = \frac{t^2}{4}\mathbf{i} + 2\cos t\,\mathbf{j} + 2\sin t\,\mathbf{k};\; 0 \le t \le 4\pi;\, t_1 = \pi$

In Problems 7–14, find the unit tangent vector $\mathbf{T}(t)$ *and the curvature* $\kappa(t)$ *at the point where* $t = t_1$. *For calculating* κ, *we suggest using Theorem A, as in Example 5.*

7. $\mathbf{u}(t) = 4t^2\mathbf{i} + 4t\,\mathbf{j};\, t_1 = \frac{1}{2}$

8. $\mathbf{r}(t) = \frac{1}{3}t^3\mathbf{i} + \frac{1}{2}t^2\mathbf{j};\, t_1 = 1$

9. $\mathbf{z}(t) = 3\cos t\,\mathbf{i} + 4\sin t\,\mathbf{j};\, t_1 = \pi/4$

10. $\mathbf{r}(t) = e^t\mathbf{i} + e^t\mathbf{j};\, t_1 = \ln 2$

11. $x(t) = 1 - t^2,\, y(t) = 1 - t^3;\, t_1 = 1$

12. $x(t) = \sinh t,\, y(t) = \cosh t;\, t_1 = \ln 3$

13. $x(t) = e^{-t}\cos t,\, y(t) = e^{-t}\sin t;\, t_1 = 0$

14. $\mathbf{r}(t) = t\cos t\,\mathbf{i} + t\sin t\,\mathbf{j};\, t_1 = 1$

In Problems 15–26, sketch the curve in the xy-plane. Then, for the given point, find the curvature and the radius of curvature. Finally, draw the circle of curvature at the point. Hint: For the curvature, you will use the second formula in Theorem A, as in Example 6.

15. $y = 2x^2, (1, 2)$

16. $y = x(x - 4)^2, (4, 0)$

17. $y = \sin x, \left(\frac{\pi}{4}, \frac{\sqrt{2}}{2}\right)$

18. $y^2 = x - 1, (1, 0)$

19. $y^2 - 4x^2 = 20, (2, 6)$

20. $y^2 - 4x^2 = 20, (2, -6)$

21. $y = \cos 2x, \left(\frac{1}{6}\pi, \frac{1}{2}\right)$

22. $y = e^{-x^2}, (1, 1/e)$

23. $y = \tan x, (\pi/4, 1)$

24. $y = \sqrt{x}, (1, 1)$

25. $y = \sqrt[3]{x}, (1, 1)$

26. $y = \tanh x, \left(\ln 2, \frac{3}{5}\right)$

In Problems 27–34, find the curvature κ, *the unit tangent vector* $\mathbf{T}$, *the unit normal vector* $\mathbf{N}$, *and the binormal vector* $\mathbf{B}$ *at* $t = t_1$.

27. $\mathbf{r}(t) = \frac{1}{2}t^2\mathbf{i} + t\,\mathbf{j} + \frac{1}{3}t^3\mathbf{k};\, t_1 = 2$

28. $x = \sin 3t,\, y = \cos 3t,\, z = t,\, t_1 = \pi/9$

29. $x = 7\sin 3t,\, y = 7\cos 3t,\, z = 14t,\, t_1 = \pi/3$

30. $\mathbf{r}(t) = \cos^3 t\,\mathbf{i} + \sin^3 t\,\mathbf{k};\, t_1 = \pi/2$

31. $\mathbf{r}(t) = 3\cosh(t/3)\,\mathbf{i} + t\,\mathbf{j};\, t_1 = 1$

32. $\mathbf{r}(t) = e^{7t}\cos 2t\,\mathbf{i} + e^{7t}\sin 2t\,\mathbf{j} + e^{7t}\mathbf{k};\, t_1 = \pi/3$

33. $\mathbf{r}(t) = e^{-2t}\mathbf{i} + e^{2t}\mathbf{j} + 2\sqrt{2}t\,\mathbf{k};\, t_1 = 0$

34. $x = \ln t,\, y = 3t,\, z = t^2;\, t_1 = 2$

In Problems 35–40, find the point of the curve at which the curvature is a maximum.

35. $y = \ln x$

36. $y = \sin x;\, -\pi \le x \le \pi$

37. $y = \cosh x$

38. $y = \sinh x$

39. $y = e^x$

40. $y = \ln \cos x$ for $-\pi/2 < x < \pi/2$

In Problems 41–52, find the tangential and normal components (a_T and a_N) of the acceleration vector at t. Then evaluate at $t = t_1$. See Examples 7 and 8.

41. $\mathbf{r}(t) = 3t\,\mathbf{i} + 3t^2\,\mathbf{j};\ t_1 = \frac{1}{3}$

42. $\mathbf{r}(t) = t^2\mathbf{i} + t\,\mathbf{j};\ t_1 = 1$

43. $\mathbf{r}(t) = (2t + 1)\mathbf{i} + (t^2 - 2)\mathbf{j};\ t_1 = -1$

44. $\mathbf{r}(t) = a \cos t\,\mathbf{i} + a \sin t\,\mathbf{j};\ t_1 = \pi/6$

45. $\mathbf{r}(t) = a \cosh t\,\mathbf{i} + a \sinh t\,\mathbf{j};\ t_1 = \ln 3$

46. $x(t) = 1 + 3t,\ y(t) = 2 - 6t;\ t_1 = 2$

47. $\mathbf{r}(t) = (t + 1)\mathbf{i} + 3t\,\mathbf{j} + t^2\mathbf{k};\ t_1 = 1$

48. $x = t,\ y = t^2,\ z = t^3;\ t_1 = 2$

49. $x = e^{-t},\ y = 2t,\ z = e^t;\ t_1 = 0$

50. $\mathbf{r}(t) = (t - 2)^2\mathbf{i} - t^2\mathbf{j} + t\,\mathbf{k};\ t_1 = 2$

51. $\mathbf{r}(t) = \left(t - \frac{1}{3}t^3\right)\mathbf{i} - \left(t + \frac{1}{3}t^3\right)\mathbf{j} + t\,\mathbf{k};\ t_1 = 3$

52. $\mathbf{r}(t) = t\,\mathbf{i} + \frac{1}{3}t^3\mathbf{j} + t^{-1}\mathbf{k},\ t > 0;\ t_1 = 1$

53. Sketch the path for a particle if its position vector is $\mathbf{r} = \sin t\,\mathbf{i} + \sin 2t\,\mathbf{j},\ 0 \le t \le 2\pi$ (you should get a figure eight). Where is the acceleration zero? Where does the acceleration vector point to the origin?

54. The position vector of a particle at time $t \ge 0$ is

$$\mathbf{r}(t) = (\cos t + t \sin t)\mathbf{i} + (\sin t - t \cos t)\mathbf{j}$$

(a) Show that the speed $ds/dt = t$.

(b) Show that $a_T = 1$ and $a_N = t$.

55. If, for a particle, $a_T = 0$ for all t, what can you conclude about its speed? If $a_N = 0$ for all t, what can you conclude about its curvature?

56. Find **N** for the ellipse

$$\mathbf{r}(t) = a \cos \omega t\,\mathbf{i} + b \sin \omega t\,\mathbf{j}$$

57. Consider the motion of a particle along a helix given by $\mathbf{r}(t) = \sin t\,\mathbf{i} + \cos t\,\mathbf{j} + (t^2 - 3t + 2)\mathbf{k}$, where the **k** component measures the height in meters above the ground and $t \ge 0$. If the particle leaves the helix and moves along the line tangent to the helix when it is 12 meters above the ground, give the direction vector for the line.

58. An object moves along the curve $y = \sin 2x$. Without doing any calculating, decide where $a_N = 0$.

59. A dog is running counterclockwise around the circle $x^2 + y^2 = 400$ (distances in feet). At the point $(-12, 16)$, it is running at 10 feet per second and is speeding up at 5 feet per second per second. Express its acceleration **a** at the point first in terms of **T** and **N**, and then in terms of **i** and **j**.

60. An object moves along the parabola $y = x^2$ with constant speed of 4. Express **a** at the point (x, x^2) in terms of **T** and **N**.

61. A car traveling at constant speed v rounds a level curve, which we take to be a circle of radius R. If the car is to avoid sliding outward, the horizontal frictional force F exerted by the road on the tires must at least balance the centrifugal force pulling outward. The force F satisfies $F = \mu m g$, where μ is the *coefficient of friction*, m is the mass of the car, and g is the acceleration of gravity. Thus, $\mu m g \ge m v^2/R$. Show that v_R, the speed beyond which skidding will occur, satisfies

$$v_R = \sqrt{\mu g R}$$

and use this to determine v_R for a curve with $R = 400$ feet and $\mu = 0.4$. Use $g = 32$ feet per second per second.

62. Consider again the car of Problem 61. Suppose that the curve is icy at its worst spot ($\mu = 0$), but is banked at angle θ from the horizontal (Figure 14). Let **F** be the force exerted by the road on the car. Then, at the critical speed v_R, $mg = \|\mathbf{F}\| \cos \theta$ and $m v_R^2/R = \|\mathbf{F}\| \sin \theta$.

(a) Show that $v_R = \sqrt{Rg \tan \theta}$.

(b) Find v_R for a curve with $R = 400$ feet and $\theta = 10°$.

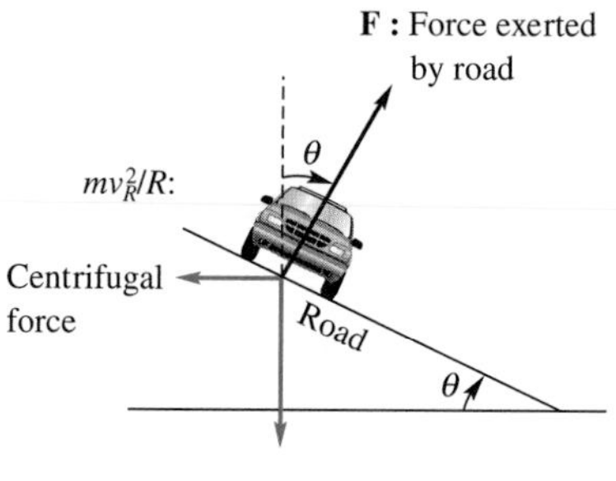

Figure 14

63. Demonstrate that the second formula in Theorem A can also be written as $\kappa = |y'' \cos^3 \phi|$, where ϕ is the angle of inclination of the tangent line to the graph of $y = f(x)$.

64. Show that for a plane curve **N** points to the concave side of the curve. *Hint:* One method is to show that

$$\mathbf{N} = (-\sin \phi\,\mathbf{i} + \cos \phi\,\mathbf{j})\frac{d\phi/ds}{|d\phi/ds|}$$

Then consider the cases $d\phi/ds > 0$ (curve bends to the left) and $d\phi/ds < 0$ (curve bends to the right).

65. Prove that $\mathbf{N} = \mathbf{B} \times \mathbf{T}$. Derive a similar result for **T** in terms of **N** and **B**.

66. Show that the curve

$$y = \begin{cases} 0 & \text{if } x \le 0 \\ x^3 & \text{if } x > 0 \end{cases}$$

has continuous first derivatives and curvature at all points.

EXPL 67. Find a curve given by a polynominal $P_5(x)$ that provides a smooth transition between two horizontal lines. That is, assume a function of the form $P_5(x) = a_0 + a_1x + a_2x^2 + a_3x^3 + a_4x^4 + a_5x^5$, which provides a smooth transition between $y = 0$ for $x \le 0$ and $y = 1$ for $x \ge 1$ in such a way that the function, its derivative, and curvature are all continuous for all values of x.

$$y = \begin{cases} 0 & \text{if } x \le 0 \\ P_5(x) & \text{if } 0 < x < 1 \\ 1 & \text{if } x \ge 1 \end{cases}$$

Hint: $P_5(x)$ must satisfy the six conditions $P_5(0) = 0$, $P_5'(0) = 0$, $P_5''(0) = 0$, $P_5(1) = 1$, $P_5'(1) = 0$, and $P_5''(1) = 0$. Use these six conditions to determine $a_0, \ldots, a_5$ uniquely and thus find $P_5(x)$.

68. Find a curve given by a polynomial $P_5(x)$ that provides a smooth transition between $y = 0$ for $x \le 0$ and $y = x$ for $x \ge 1$.

EXPL **69.** Derive the polar coordinate curvature formula

$$\kappa = \frac{|r^2 + 2(r')^2 - rr''|}{(r^2 + (r')^2)^{3/2}}$$

where the derivatives are with respect to θ.

In Problems 70–75, use the formula in Problem 69 to find the curvature κ of the following:

70. Circle: $r = 4\cos\theta$

71. Cardioid: $r = 1 + \cos\theta$ at $\theta = 0$

72. $r = \theta$ at $\theta = 1$

73. $r = 4(1 + \cos\theta)$ at $\theta = \pi/2$

74. $r = e^{3\theta}$ at $\theta = 1$

75. $r = 4(1 + \sin\theta)$ at $\theta = \pi/2$

76. Show that the curvature of the polar curve $r = e^{6\theta}$ is proportional to $1/r$.

77. Show that the curvature of the polar curve $r^2 = \cos 2\theta$ is directly proportional to r for $r > 0$.

78. Derive the first curvature formula in Theorem A by working directly with $\kappa = \|\mathbf{T}'(t)\|/\|\mathbf{r}'(t)\|$.

GC **79.** Draw the graph of $x = 4\cos t$, $y = 3\sin(t + 0.5)$, $0 \le t \le 2\pi$. Estimate its maximum and minimum curvature by looking at the graph (curvature is the reciprocal of the radius of curvature). Then use a graphing calculator or a CAS to approximate these two numbers to four decimal places.

80. Show that the unit binormal vector $\mathbf{B} = \mathbf{T} \times \mathbf{N}$ has the property that $\dfrac{d\mathbf{B}}{ds}$ is perpendicular to $\mathbf{B}$.

81. Show that the unit binormal vector $\mathbf{B} = \mathbf{T} \times \mathbf{N}$ has the property that $\dfrac{d\mathbf{B}}{ds}$ is perpendicular to $\mathbf{T}$.

82. Using the results obtained in Problems 80 and 81, show that $\dfrac{d\mathbf{B}}{ds}$ must be parallel to $\mathbf{N}$ and, consequently, there must be a number τ depending on s such that $\dfrac{d\mathbf{B}}{ds} = -\tau(s)\mathbf{N}$. The function $\tau(s)$ is called the torsion of the curve and measures the twist of the curve from the plane determined by $\mathbf{T}$ and $\mathbf{N}$.

83. Show that for a plane curve the torsion is $\tau(s) = 0$.

84. Show that for a straight line $\mathbf{r}(t) = \mathbf{r}_0 + a_0 t\,\mathbf{i} + b_0 t\,\mathbf{j} + c_0 t\,\mathbf{k}$ both κ and τ are zero.

C **85.** A fly is crawling along a wire helix so that its position vector is $\mathbf{r}(t) = 6\cos\pi t\,\mathbf{i} + 6\sin\pi t\,\mathbf{j} + 2t\,\mathbf{k}$, $t \ge 0$. At what point will the fly hit the sphere $x^2 + y^2 + z^2 = 100$, and how far did it travel in getting there (assuming that it started when $t = 0$)?

C **86.** The DNA molecule in humans is a double helix, each with about 2.9×10^8 complete turns. Each helix has radius about 10 angstroms and rises about 34 angstroms on each complete turn (an angstrom is 10^{-8} centimeter). What is the total length of such a helix?

Answers to Concepts Review: **1.** $\dfrac{d\mathbf{T}}{ds}$ **2.** $1/a$; 0 **3.** $\dfrac{d^2s}{dt^2}$; $\left(\dfrac{ds}{dt}\right)^2\kappa$ **4.** 0

11.8 Surfaces in Three-Space

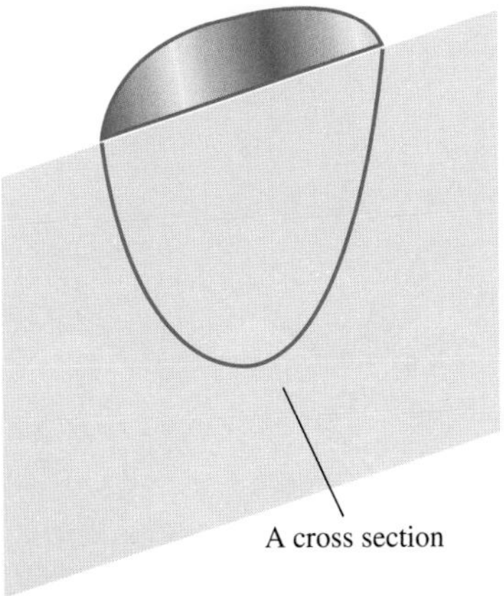

Figure 1

The graph of an equation in three variables is normally a surface. We have met two examples already. The graph of $Ax + By + Cz = D$ is a plane; the graph of $(x - h)^2 + (y - k)^2 + (z - l)^2 = r^2$ is a sphere. Graphing surfaces is best accomplished by finding the intersections of the surface with well-chosen planes. These intersections are called **cross sections** (Figure 1); those with the coordinate planes are also called **traces.**

EXAMPLE 1 Sketch the graph of

$$\frac{x^2}{16} + \frac{y^2}{25} + \frac{z^2}{9} = 1$$

SOLUTION To find the trace in the xy-plane, we set $z = 0$ in the given equation. The graph of the resulting equation

$$\frac{x^2}{16} + \frac{y^2}{25} = 1$$

is an ellipse. The traces in the xz-plane and the yz-plane (obtained by setting $y = 0$ and $x = 0$, respectively) are also ellipses. These three traces are shown in Figure 2 and help to provide a good visual image of the required surface (called an ellipsoid). ■

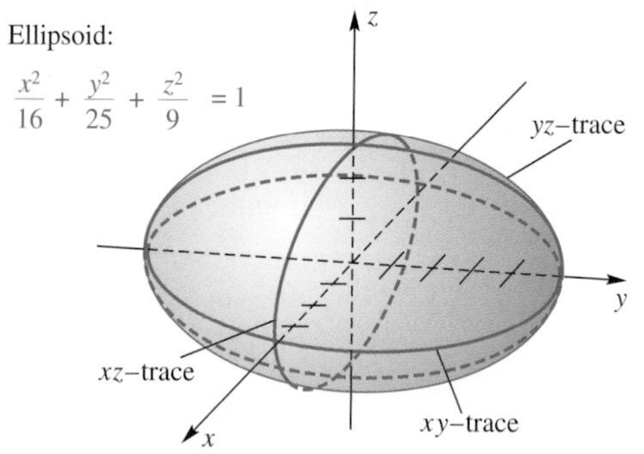

Figure 2

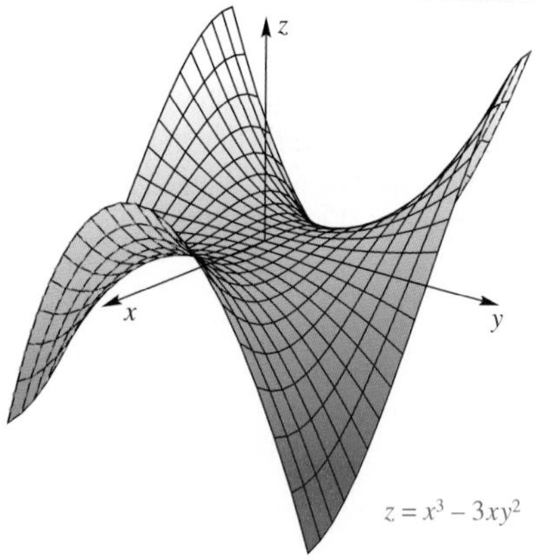

Figure 3

If the surface is very complicated, it may be useful to show the cross sections with many planes parallel to the coordinate planes. Here a computer with graphics capability can be very helpful. In Figure 3 we show a typical computer-generated graph, the graph of the "monkey saddle" $z = x^3 - 3xy^2$. We will have more to say about computer-generated graphs in the next chapter.

Cylinders

You should be familiar with right circular cylinders from high school geometry. Here the word *cylinder* will denote a much more extensive class of surfaces.

Let C be a plane curve, and let l be a line intersecting C that is not in the plane of C. The set of all points on lines that are parallel to l and that intersect C is called a **cylinder** (Figure 4).

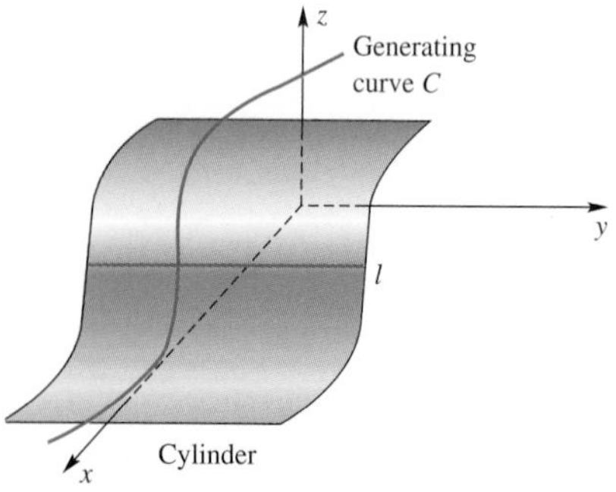

Figure 4

Cylinders occur naturally when we graph an equation in three-space that involves just two variables. Consider as a first example

$$\frac{y^2}{a^2} - \frac{x^2}{b^2} = 1$$

in which the variable z is missing. This equation determines a curve C in the xy-plane, a hyperbola. Moreover, if $(x_1, y_1, 0)$ satisfies the equation, so does (x_1, y_1, z). As z runs through all real values, the point (x_1, y_1, z) traces out a line parallel to the z-axis. We conclude that the graph of the given equation is a cylinder, a hyperbolic cylinder (Figure 5).

A second example is the graph of $z = \sin y$ (Figure 6).

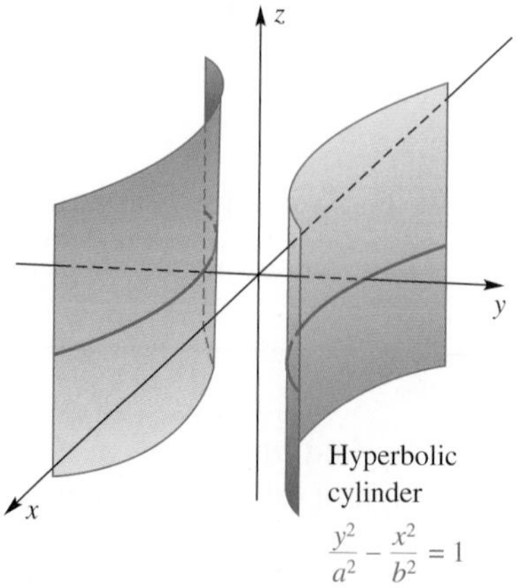

Figure 5

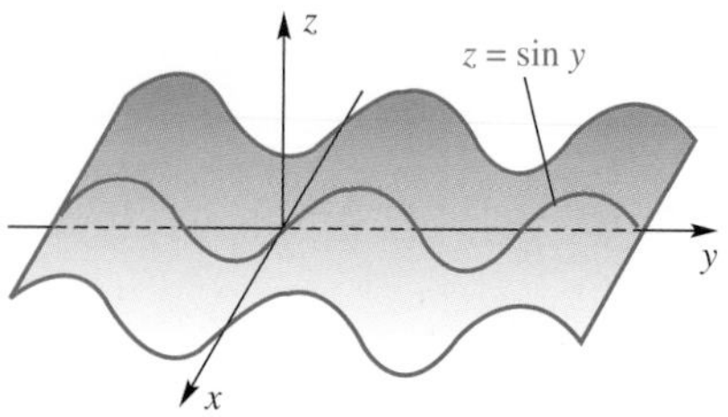

Figure 6

Quadric Surfaces If a surface is the graph in three-space of an equation of second degree, it is called a **quadric surface.** Plane sections of a quadric surface are conics.

The general second-degree equation has the form

$$Ax^2 + By^2 + Cz^2 + Dxy + Exz + Fyz + Gx + Hy + Iz + J = 0$$

It can be shown that any such equation can be reduced, by rotation and translation of coordinate axes, to one of the two forms

$$Ax^2 + By^2 + Cz^2 + J = 0$$

or

$$Ax^2 + By^2 + Iz = 0$$

The quadric surfaces represented by the first of these equations are symmetric with respect to the coordinate planes and the origin. They are called **central quadrics.**

In Figures 7 through 12, we show six general types of quadric surfaces. Study them carefully. The graphs were drawn by a technical artist; we do not expect that most of our readers will be able to duplicate them in doing the problems. A more reasonable drawing for most people to make is like the one that is shown in Figure 13 with our next example.

QUADRIC SURFACES

ELLIPSOID: $\dfrac{x^2}{a^2} + \dfrac{y^2}{b^2} + \dfrac{z^2}{c^2} = 1$

Plane	Cross Section
xy-plane	Ellipse
xz-plane	Ellipse
yz-plane	Ellipse
Parallel to xy-plane	Ellipse, point, or empty set
Parallel to xz-plane	Ellipse, point, or empty set
Parallel to yz-plane	Ellipse, point, or empty set

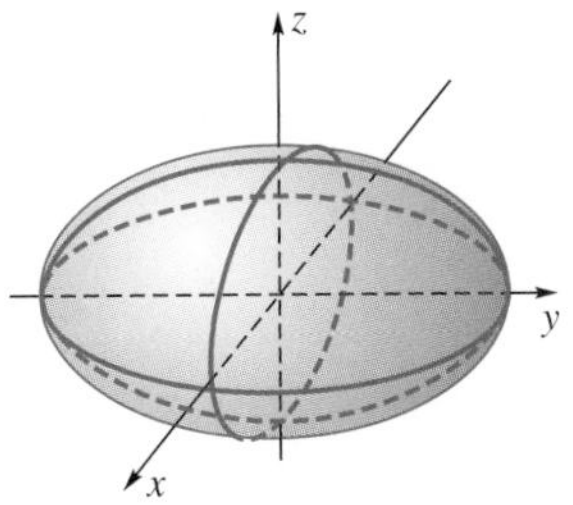

Figure 7

HYPERBOLOID OF ONE SHEET: $\dfrac{x^2}{a^2} + \dfrac{y^2}{b^2} - \dfrac{z^2}{c^2} = 1$

Plane	Cross Section
xy-plane	Ellipse
xz-plane	Hyperbola
yz-plane	Hyperbola
Parallel to xy-plane	Ellipse
Parallel to xz-plane	Hyperbola
Parallel to yz-plane	Hyperbola

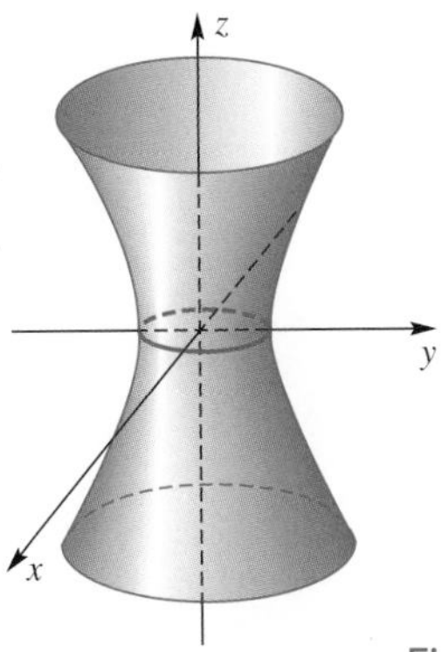

Figure 8

QUADRIC SURFACES (continued)

HYPERBOLOID OF TWO SHEETS: $\frac{x^2}{a^2} - \frac{y^2}{b^2} - \frac{z^2}{c^2} = 1$

Plane	Cross Section
xy-plane	Hyperbola
xz-plane	Hyperbola
yz-plane	Empty set
Parallel to *xy*-plane	Hyperbola
Parallel to *xz*-plane	Hyperbola
Parallel to *yz*-plane	Ellipse, point, or empty set

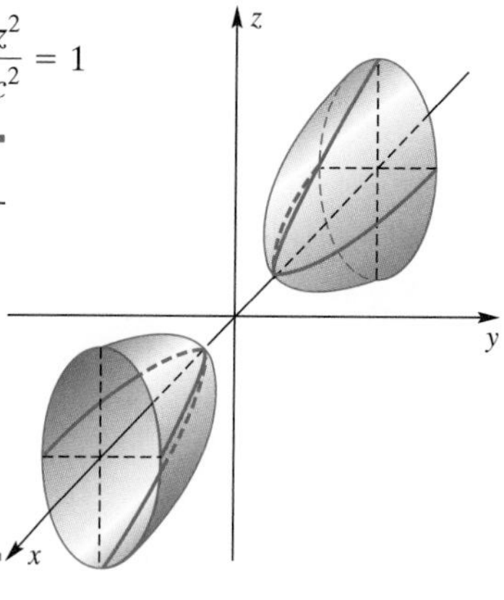

Figure 9

ELLIPTIC PARABOLOID: $z = \frac{x^2}{a^2} + \frac{y^2}{b^2}$

Plane	Cross Section
xy-plane	Point
xz-plane	Parabola
yz-plane	Parabola
Parallel to *xy*-plane	Ellipse, point, or empty set
Parallel to *xz*-plane	Parabola
Parallel to *yz*-plane	Parabola

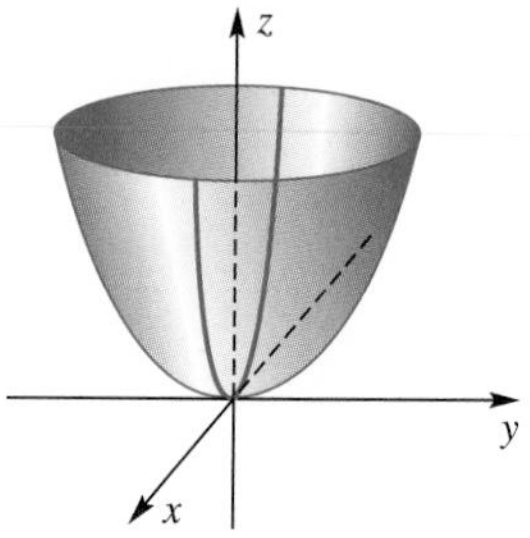

Figure 10

HYPERBOLIC PARABOLOID: $z = \frac{y^2}{b^2} - \frac{x^2}{a^2}$

Plane	Cross Section
xy-plane	Intersecting straight lines
xz-plane	Parabola
yz-plane	Parabola
Parallel to *xy*-plane	Hyperbola or intersecting straight lines
Parallel to *xz*-plane	Parabola
Parallel to *yz*-plane	Parabola

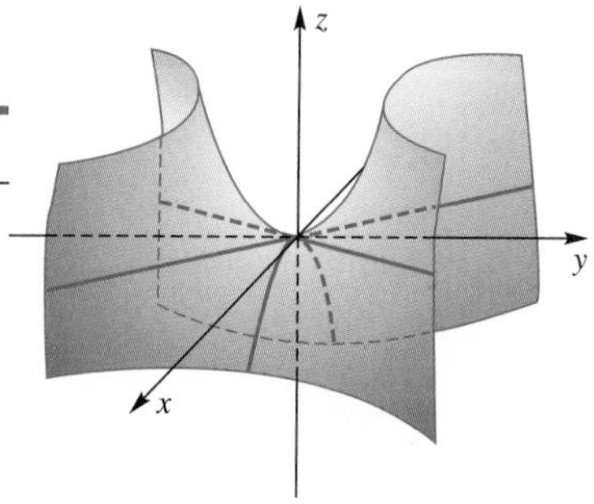

Figure 11

ELLIPTIC CONE: $\frac{x^2}{a^2} + \frac{y^2}{b^2} - \frac{z^2}{c^2} = 0$

Plane	Cross Section
xy-plane	Point
xz-plane	Intersecting straight lines
yz-plane	Intersecting straight lines
Parallel to *xy*-plane	Ellipse or point
Parallel to *xz*-plane	Hyperbola or intersecting straight lines
Parallel to *yz*-plane	Hyperbola or intersecting straight lines

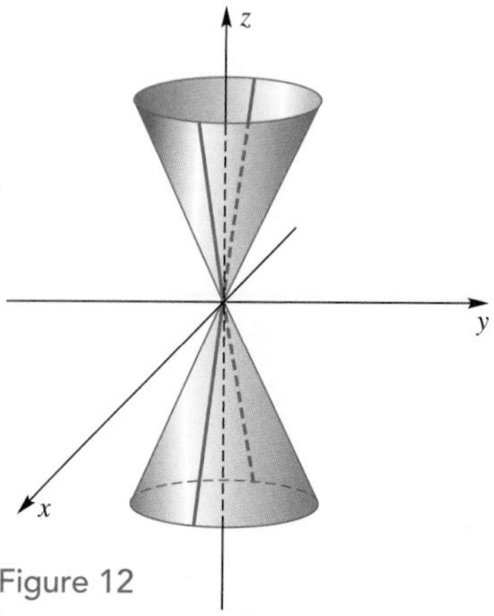

Figure 12

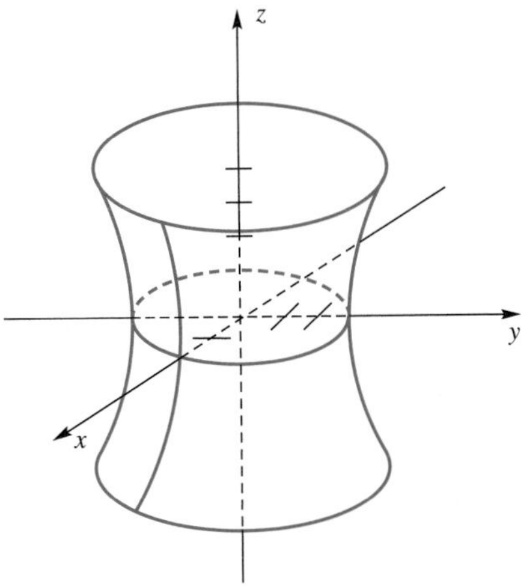

Figure 13

EXAMPLE 2 Analyze the equation

$$\frac{x^2}{4} + \frac{y^2}{9} - \frac{z^2}{16} = 1$$

and sketch its graph.

SOLUTION The traces in the three coordinate planes are obtained by setting $z = 0$, $y = 0$, and $x = 0$, respectively.

$$xy\text{-plane:} \quad \frac{x^2}{4} + \frac{y^2}{9} = 1, \quad \text{an ellipse}$$

$$xz\text{-plane:} \quad \frac{x^2}{4} - \frac{z^2}{16} = 1, \quad \text{a hyperbola}$$

$$yz\text{-plane:} \quad \frac{y^2}{9} - \frac{z^2}{16} = 1, \quad \text{a hyperbola}$$

These traces are graphed in Figure 13. We have also shown the cross sections in the planes $z = 4$ and $z = -4$. Note that when we substitute $z = \pm 4$ in the original equation we get

$$\frac{x^2}{4} + \frac{y^2}{9} - \frac{16}{16} = 1$$

which is equivalent to

$$\frac{x^2}{8} + \frac{y^2}{18} = 1$$

an ellipse. ■

EXAMPLE 3 Name the graph of each of the following equations:

(a) $4x^2 + 4y^2 - 25z^2 + 100 = 0$ (b) $y^2 + z^2 - 12y = 0$

(c) $x^2 - z^2 = 0$ (d) $9x^2 + 4z^2 - 36y = 0$

SOLUTION

(a) Dividing both sides of this equation by -100 gives the form

$$-\frac{x^2}{25} - \frac{y^2}{25} + \frac{z^2}{4} = 1$$

Its graph is a hyperboloid of two sheets. It does not intersect the xy-plane, but cross sections in planes parallel to this plane (and at least 2 units away) are circles.

(b) The variable x does not appear, so the graph is a cylinder parallel to the x-axis. Since the equation can be written in the form $(y - 6)^2 + z^2 = 36$, its graph is a circular cylinder.

(c) Since the variable y is missing, the graph is a cylinder. The given equation can be written $(x - z)(x + z) = 0$; so its graph consists of the two planes $x = z$ and $x = -z$.

(d) The equation can be rewritten as

$$\frac{x^2}{4} + \frac{z^2}{9} = y$$

which has an elliptic paraboloid as its graph. It is symmetric with respect to the y-axis. ■

Concepts Review

1. The intersections of a surface with the coordinate planes are called ______. More generally, intersections with any plane are called ______.

2. Equations involving just two variables when graphed in three-space generate surfaces called ______. In particular, the graph of $x^2 + y^2 = 1$ is an ordinary right circular cylinder whose axis is the ______.

3. The graph of $3x^2 + 2y^2 + 4z^2 = 12$ is a surface called a(n) ______.

4. The graph of $4z = x^2 + 2y^2$ is a surface called a(n) ______.

Problem Set 11.8

In Problems 1–20, name and sketch the graph of each of the following equations in three-space.

1. $4x^2 + 36y^2 = 144$ **2.** $y^2 + z^2 = 15$

3. $3x + 2z = 10$ **4.** $z^2 = 3y$

5. $x^2 + y^2 - 8x + 4y + 13 = 0$

6. $2x^2 - 16z^2 = 0$

7. $4x^2 + 9y^2 + 49z^2 = 1764$

8. $9x^2 - y^2 + 9z^2 - 9 = 0$

9. $4x^2 + 16y^2 - 32z = 0$ **10.** $-x^2 + y^2 + z^2 = 0$

11. $y = e^{2z}$ **12.** $6x - 3y = \pi$

13. $x^2 - z^2 + y = 0$ **14.** $x^2 + y^2 - 4z^2 + 4 = 0$

15. $9x^2 + 4z^2 - 36y = 0$

16. $9x^2 + 25y^2 + 9z^2 = 225$

17. $5x + 8y - 2z = 10$ **18.** $y = \cos x$

19. $z = \sqrt{16 - x^2 - y^2}$ **20.** $z = \sqrt{x^2 + y^2 + 1}$

21. The graph of an equation in x, y, and z is symmetric with respect to the xy-plane if replacing z by $-z$ results in an equivalent equation. What condition leads to a graph that is symmetric with respect to each of the following?

(a) yz-plane (b) z-axis (c) origin

22. What condition leads to a graph that is symmetric with respect to the following?

(a) xz-plane (b) y-axis (c) x-axis

23. Find the general equation of a central ellipsoid that is symmetric with respect to the following:

(a) origin (b) x-axis (c) xy-plane

24. Find the general equation of a central hyperboloid of one sheet that is symmetric with respect to the following:

(a) origin (b) y-axis (c) xy-plane

25. Find the general equation of a central hyperboloid of two sheets that is symmetric with respect to the following:

(a) origin (b) z-axis (c) yz-plane

26. Which of the equations in Problems 1–20 has a graph that is symmetric with respect to each of the following?

(a) xy-plane (b) z-axis

27. If the curve $z = x^2$ in the xz-plane is revolved about the z-axis, the resulting surface has equation $z = x^2 + y^2$, obtained as a result of replacing x by $\sqrt{x^2 + y^2}$. If $y = 2x^2$ in the xy-plane is revolved about the y-axis, what is the equation of the resulting surface?

28. Find the equation of the surface that results when the curve $z = 2y$ in the yz-plane is revolved about the z-axis.

29. Find the equation of the surface that results when the curve $4x^2 + 3y^2 = 12$ in the xy-plane is revolved about the y-axis.

30. Find the equation of the surface that results when the curve $4x^2 - 3y^2 = 12$ in the xy-plane is revolved about the x-axis.

31. Find the coordinates of the foci of the ellipse that is the intersection of $z = x^2/4 + y^2/9$ with the plane $z = 4$.

32. Find the coordinates of the focus of the parabola that is the intersection of $z = x^2/4 + y^2/9$ with $x = 4$.

33. Find the area of the elliptical cross section cut from the surface $x^2/a^2 + y^2/b^2 + z^2/c^2 = 1$ by the plane $z = h$, $-c < h < c$. *Recall:* The area of the ellipse $x^2/A^2 + y^2/B^2 = 1$ is πAB.

34. Show that the volume of the solid bounded by the elliptic paraboloid $x^2/a^2 + y^2/b^2 = h - z$, $h > 0$, and the xy-plane is $\pi abh^2/2$, that is, the volume is one-half the area of the base times the height. *Hint*: Use the method of slabs of Section 5.2.

35. Show that the projection in the xz-plane of the curve that is the intersection of the surfaces $y = 4 - x^2$ and $y = x^2 + z^2$ is an ellipse, and find its major and minor diameters.

36. Sketch the triangle in the plane $y = x$ that is above the plane $z = y/2$, below the plane $z = 2y$, and inside the cylinder $x^2 + y^2 = 8$. Then find the area of this triangle.

37. Show that the spiral $\mathbf{r} = t\cos t\,\mathbf{i} + t\sin t\,\mathbf{j} + t\,\mathbf{k}$ lies on the circular cone $x^2 + y^2 - z^2 = 0$. On what surface does the spiral $\mathbf{r} = 3t\cos t\,\mathbf{i} + t\sin t\,\mathbf{j} + t\,\mathbf{k}$ lie?

38. Show that the curve determined by $\mathbf{r} = t\,\mathbf{i} + t\,\mathbf{j} + t^2\mathbf{k}$ is a parabola, and find the coordinates of its focus.

Answers to Concepts Review: **1.** traces; cross sections **2.** cylinders; z-axis **3.** ellipsoid **4.** elliptic paraboloid

11.9 Cylindrical and Spherical Coordinates

Giving the Cartesian (rectangular) coordinates (x, y, z) is just one of many ways of specifying the position of a point in three-space. Two other kinds of coordinates that play a significant role in calculus are cylindrical coordinates (r, θ, z) and spherical coordinates (ρ, θ, ϕ). The meaning of the three kinds of coordinates is illustrated for the same point P in Figure 1.

The **cylindrical coordinate system** uses the polar coordinates r and θ (Section 10.5) in place of Cartesian coordinates x and y in the plane. The z-coordinate is the same as in Cartesian coordinates. We will usually require that $r \geq 0$, and we will restrict θ so that $0 \leq \theta < 2\pi$.

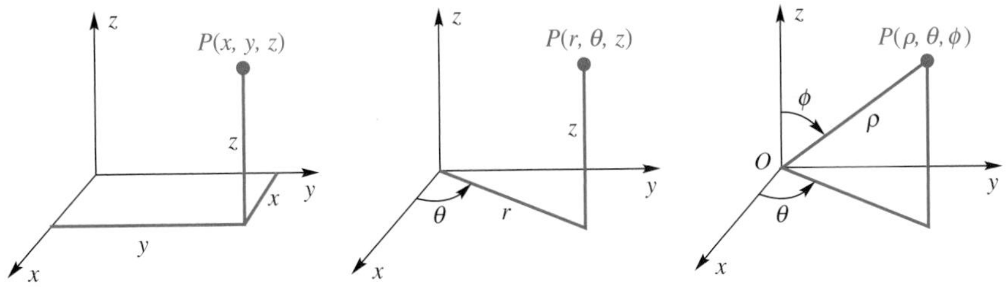

Figure 1

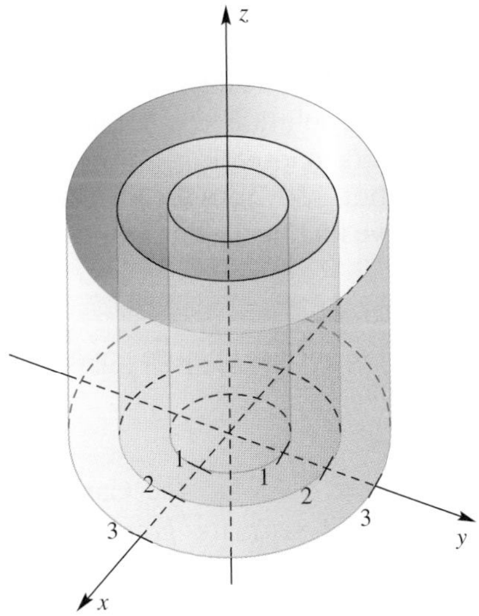

The cylinders $r = 1, r = 2, r = 3$

Figure 2

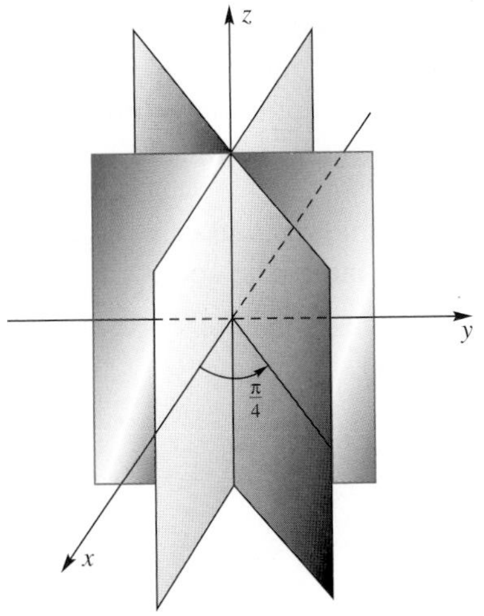

The planes $\theta = 0,\ \theta = \frac{\pi}{4},\ \theta = \frac{\pi}{2}$

Figure 3

A point P has **spherical coordinates** (ρ, θ, ϕ) if ρ (rho) is the distance $|OP|$ from the origin to P, θ is the polar angle associated with the projection P' of P onto the xy-plane, and ϕ is the angle between the positive z-axis and the line segment OP. We require that

$$\rho \geq 0, \qquad 0 \leq \theta < 2\pi, \qquad 0 \leq \phi \leq \pi$$

Cylindrical Coordinates

If a solid or a surface has an axis of symmetry, it is often wise to orient it so that this axis is the z-axis and then use cylindrical coordinates. Note in particular the simplicity of the equation of a circular cylinder with z-axis symmetry (Figure 2) and also of a plane containing the z-axis (Figure 3). In Figure 3, we have allowed $r < 0$.

Cylindrical and Cartesian coordinates are related by the following equations:

Cylindrical to Cartesian	Cartesian to Cylindrical
$x = r\cos\theta$	$r = \sqrt{x^2 + y^2}$
$y = r\sin\theta$	$\tan\theta = y/x$
$z = z$	$z = z$

With these relationships, we can go back and forth between the two coordinate systems.

EXAMPLE 1 Find

(a) the Cartesian coordinates of the point with cylindrical coordinates $(4, 2\pi/3, 5)$

(b) the cylindrical coordinates of the point with Cartesian coordinates $(-5, -5, 2)$.

SOLUTION

(a) $$x = 4\cos\frac{2\pi}{3} = 4\cdot\left(-\frac{1}{2}\right) = -2$$

$$y = 4\sin\frac{2\pi}{3} = 4\cdot\left(\frac{\sqrt{3}}{2}\right) = 2\sqrt{3}$$

$$z = 5$$

Thus, the Cartesian coordinates of $(4, 2\pi/3, 5)$ are $\left(-2, 2\sqrt{3}, 5\right)$.

(b) $$r = \sqrt{(-5)^2 + (-5)^2} = 5\sqrt{2}$$

$$\tan\theta = \frac{-5}{-5} = 1$$

$$z = 2$$

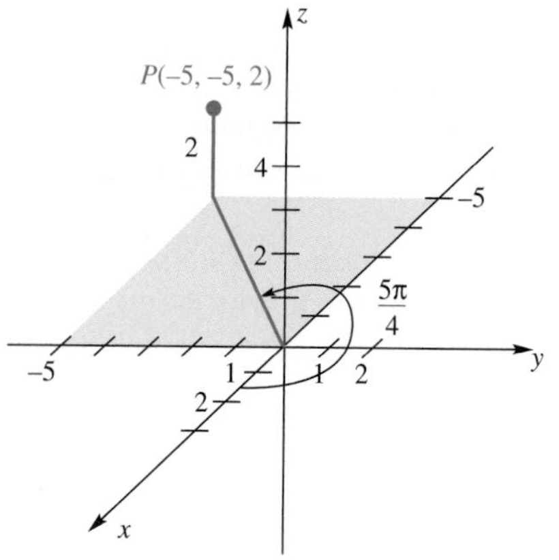

Figure 4

Figure 4 indicates that θ is between $\pi/2$ and π. Since $\tan\theta = 1$, we must have $\theta = 5\pi/4$. The cylindrical coordinates of $(-5, -5, 2)$ are $\left(5\sqrt{2}, 5\pi/4, 2\right)$. ■

EXAMPLE 2 Find the equations in cylindrical coordinates of the paraboloid and cylinder whose Cartesian equations are $x^2 + y^2 = 4 - z$ and $x^2 + y^2 = 2x$.

SOLUTION

Paraboloid: $r^2 = 4 - z$

Cylinder: $r^2 = 2r\cos\theta$ or (equivalently) $r = 2\cos\theta$

Division of an equation by a variable creates the potential for losing a solution. For example, dividing $x^2 = x$ by x gives $x = 1$ and loses the solution $x = 0$. Similarly, dividing $r^2 = 2r\cos\theta$ by r gives $r = 2\cos\theta$ and appears to lose the solution $r = 0$ (the origin). However, the origin satisfies the equation $r = 2\cos\theta$ with coordinates $(0, \pi/2)$. Thus, $r^2 = 2r\cos\theta$ and $r = 2\cos\theta$ have identical polar graphs (see CAUTION in the margin of Section 10.5). ■

EXAMPLE 3 Find the Cartesian equations of the surfaces whose equations in cylindrical coordinates are $r^2 + 4z^2 = 16$ and $r^2\cos 2\theta = z$.

SOLUTION Since $r^2 = x^2 + y^2$, the surface $r^2 + 4z^2 = 16$ has the Cartesian equation $x^2 + y^2 + 4z^2 = 16$ or $x^2/16 + y^2/16 + z^2/4 = 1$. Its graph is an ellipsoid.

Since $\cos 2\theta = \cos^2\theta - \sin^2\theta$, the second equation can be written $r^2\cos^2\theta - r^2\sin^2\theta = z$. In Cartesian coordinates it becomes $x^2 - y^2 = z$, the graph of which is a hyperbolic paraboloid. ■

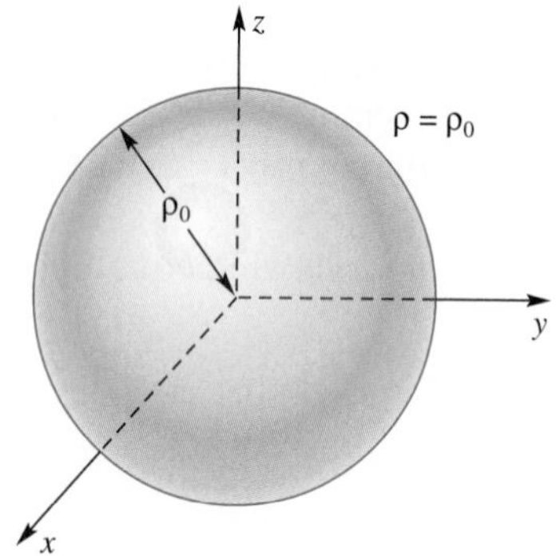

Figure 5

Spherical Coordinates When a solid or a surface is symmetric with respect to a point, spherical coordinates are likely to play a simplifying role. In particular, a sphere centered at the origin (Figure 5) has the simple equation $\rho = \rho_0$. Also note that the equation of a cone with axis along the z-axis and vertex at the origin (Figure 6) is $\phi = \phi_0$.

It is easy to determine the relationships between spherical and cylindrical coordinates and between spherical and Cartesian coordinates. The following table shows some of these relationships.

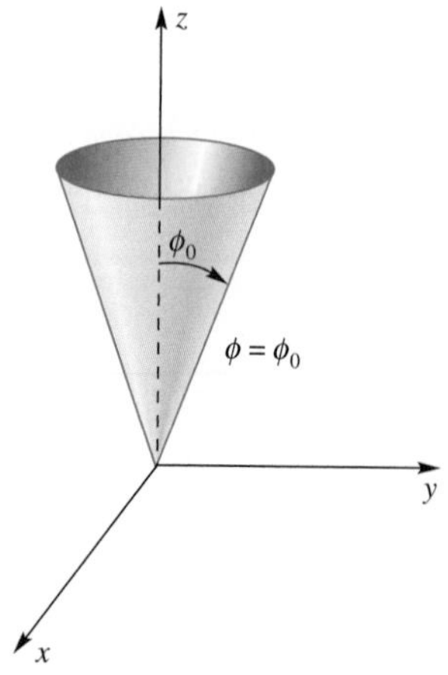

Figure 6

Spherical to Cartesian	**Cartesian to Spherical**
$x = \rho\sin\phi\cos\theta$	$\rho = \sqrt{x^2 + y^2 + z^2}$
$y = \rho\sin\phi\sin\theta$	$\tan\theta = y/x$
$z = \rho\cos\phi$	$\cos\phi = \dfrac{z}{\sqrt{x^2 + y^2 + z^2}}$

EXAMPLE 4 Find the Cartesian coordinates of the point P with spherical coordinates $(8, \pi/3, 2\pi/3)$.

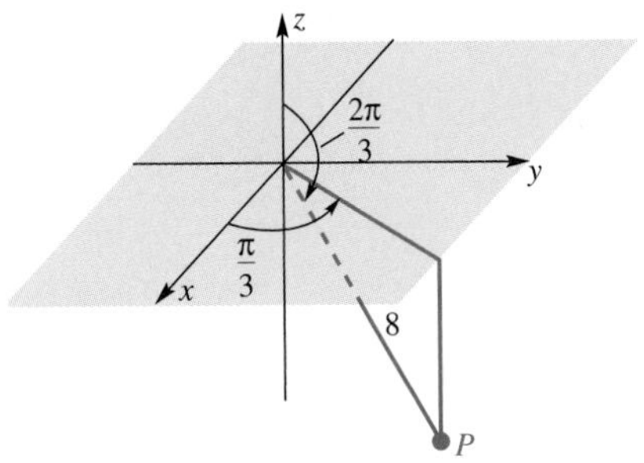

Figure 7

SOLUTION We have plotted the point P in Figure 7.

$$x = 8\sin\frac{2\pi}{3}\cos\frac{\pi}{3} = 8\frac{\sqrt{3}}{2}\frac{1}{2} = 2\sqrt{3}$$

$$y = 8\sin\frac{2\pi}{3}\sin\frac{\pi}{3} = 8\frac{\sqrt{3}}{2}\frac{\sqrt{3}}{2} = 6$$

$$z = 8\cos\frac{2\pi}{3} = 8\left(-\frac{1}{2}\right) = -4$$

Thus, P has Cartesian coordinates $\left(2\sqrt{3}, 6, -4\right)$. ■

EXAMPLE 5 Describe the graph of $\rho = 2\cos\phi$.

SOLUTION We change to Cartesian coordinates. Multiply both sides by ρ to obtain

$$\rho^2 = 2\rho\cos\phi$$

$$x^2 + y^2 + z^2 = 2z$$

$$x^2 + y^2 + (z-1)^2 = 1$$

The graph is a sphere of radius 1 centered at the point with Cartesian coordinates $(0, 0, 1)$. ■

EXAMPLE 6 Find the equation of the paraboloid $z = x^2 + y^2$ in spherical coordinates.

SOLUTION Substituting for x, y, and z yields

$$\rho\cos\phi = \rho^2\sin^2\phi\cos^2\theta + \rho^2\sin^2\phi\sin^2\theta$$

$$\rho\cos\phi = \rho^2\sin^2\phi(\cos^2\theta + \sin^2\theta)$$

$$\rho\cos\phi = \rho^2\sin^2\phi$$

$$\cos\phi = \rho\sin^2\phi$$

$$\rho = \cos\phi\csc^2\phi$$

Note that $\phi = \pi/2$ yields $\rho = 0$, which shows that we did not lose the origin when we canceled ρ at the fourth step. ■

Spherical Coordinates in Geography Geographers and navigators use a coordinate system very closely related to spherical coordinates, the longitude–latitude system. Suppose that the earth is a sphere with center at the origin, that the positive z-axis passes through the North Pole, and that the positive x-axis passes through the prime meridian (Figure 8). By convention, longitudes are specified in degrees east or west of the prime meridian and latitudes in degrees north or south of the equator. It is a simple matter to determine spherical coordinates from such data.

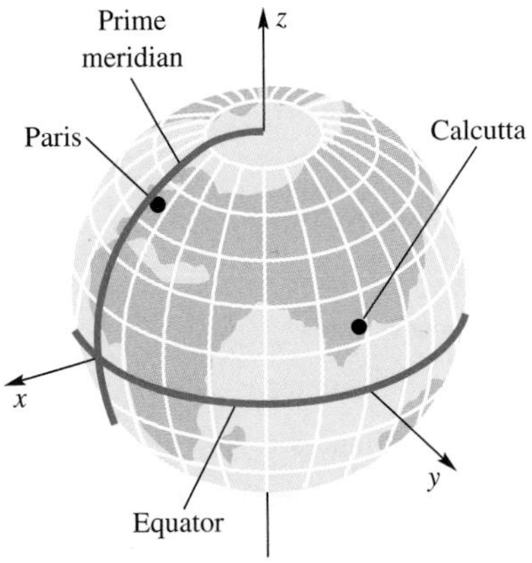

Figure 8

EXAMPLE 7 Assuming the earth to be a sphere of radius 3960 miles, find the great-circle distance from Paris (longitude 2.2° E, latitude 48.4° N) to Calcutta (longitude 88.2° E, latitude 22.3° N).

SOLUTION We first calculate the spherical angles θ and ϕ for the two cities.

Paris: $\theta = 2.2° \approx 0.0384$ radian

$\phi = 90° - 48.4° = 41.6° \approx 0.7261$ radian

Calcutta: $\theta = 88.2° \approx 1.5394$ radians

$\phi = 90° - 22.3° = 67.7° \approx 1.1816$ radians

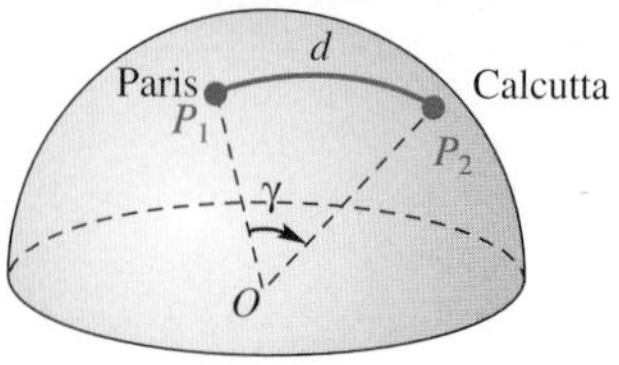

Figure 9

From these data and $\rho = 3960$ miles, we determine the Cartesian coordinates, as illustrated in Example 4.

Paris:	$P_1(2627.2, 100.9, 2961.3)$
Calcutta:	$P_2(115.1, 3662.0, 1502.6)$

Next, referring to Figure 9, we determine γ, the angle between $\overrightarrow{OP_1}$ and $\overrightarrow{OP_2}$.

$$\cos\gamma = \frac{\overrightarrow{OP_1}\cdot\overrightarrow{OP_2}}{\|\overrightarrow{OP_1}\|\|\overrightarrow{OP_2}\|} \approx \frac{(2627.2)(115.1) + (100.9)(3662) + (2961.3)(1502.6)}{(3960)(3960)}$$

$$\approx 0.3266$$

Thus, $\gamma \approx 1.2381$ radians and the great-circle distance d is

$$d = \rho\gamma \approx (3960)(1.2381) \approx 4903 \text{ miles}$$

■

Concepts Review

1. In cylindrical coordinates, the graph of $r = 6$ is a(n) _____; in spherical coordinates, the graph of $\rho = 6$ is a(n) _____.

2. In cylindrical coordinates, the graph of $\theta = \pi/6$ is a(n) _____; in spherical coordinates, the graph of $\phi = \pi/6$ is a(n) _____.

3. The equation _____ connects ρ with r and z.

4. The equation $\rho^2 = 4\rho\cos\phi$ in spherical coordinates becomes the equation _____ when written in rectangular coordinates.

Problem Set 11.9

1. Make a table, like the one just before Example 4, that gives the relationships between cylindrical and spherical coordinates.

2. Change the following from cylindrical to spherical coordinates.

(a) $(1, \pi/2, 1)$ (b) $(-2, \pi/4, 2)$

3. Change the following from cylindrical to Cartesian (rectangular) coordinates.

(a) $(6, \pi/6, -2)$ (b) $(4, 4\pi/3, -8)$

4. Change the following from spherical to Cartesian coordinates.

(a) $(8, \pi/4, \pi/6)$ (b) $(4, \pi/3, 3\pi/4)$

5. Change the following from Cartesian to spherical coordinates.

(a) $(2, -2\sqrt{3}, 4)$ (b) $(-\sqrt{2}, \sqrt{2}, 2\sqrt{3})$

6. Change the following from Cartesian to cylindrical coordinates.

(a) $(2, 2, 3)$ (b) $(4\sqrt{3}, -4, 6)$

In Problems 7–16, sketch the graph of the given cylindrical or spherical equation.

7. $r = 5$

8. $\rho = 5$

9. $\phi = \pi/6$

10. $\theta = \pi/6$

11. $r = 3\cos\theta$

12. $r = 2\sin 2\theta$

13. $\rho = 3\cos\phi$

14. $\rho = \sec\phi$

15. $r^2 + z^2 = 9$

16. $r^2\cos^2\theta + z^2 = 4$

In Problems 17–30, make the required change in the given equation.

17. $x^2 + y^2 = 9$ to cylindrical coordinates

18. $x^2 - y^2 = 25$ to cylindrical coordinates

19. $x^2 + y^2 + 4z^2 = 10$ to cylindrical coordinates

20. $x^2 + y^2 + 4z^2 = 10$ to spherical coordinates

21. $2x^2 + 2y^2 - 4z^2 = 0$ to spherical coordinates

22. $x^2 - y^2 - z^2 = 1$ to spherical coordinates

23. $r^2 + 2z^2 = 4$ to spherical coordinates

24. $\rho = 2\cos\phi$ to cylindrical coordinates

25. $x + y = 4$ to cylindrical coordinates

26. $x + y + z = 1$ to spherical coordinates

27. $x^2 + y^2 = 9$ to spherical coordinates

28. $r = 2\sin\theta$ to Cartesian coordinates

29. $r^2\cos 2\theta = z$ to Cartesian coordinates

30. $\rho\sin\phi = 1$ to Cartesian coordinates

31. The parabola $z = 2x^2$ in the xz-plane is revolved about the z-axis. Write the equation of the resulting surface in cylindrical coordinates.

32. The hyperbola $2x^2 - z^2 = 2$ in the xz-plane is revolved about the z-axis. Write the equation of the resulting surface in cylindrical coordinates.

[C] **33.** Find the great-circle distance from St. Paul (longitude 93.1° W, latitude 45° N) to Oslo (longitude 10.5° E, latitude 59.6° N). See Example 7.

[C] **34.** Find the great-circle distance from New York (longitude 74° W, latitude 40.4° N) to Greenwich (longitude 0°, latitude 51.3° N).

[C] **35.** Find the great-circle distance from St. Paul (longitude 93.1° W, latitude 45° N) to Turin, Italy (longitude 7.4° E, latitude 45° N).

[C] **36.** What is the distance along the 45° parallel between St. Paul and Turin? See Problem 35.

[C] **37.** How close does the great-circle route from St. Paul to Turin get to the North Pole? See Problem 35.

38. Let $(\rho_1, \theta_1, \phi_1)$ and $(\rho_2, \theta_2, \phi_2)$ be the spherical coordinates of two points, and let d be the straight-line distance between them. Show that

$$d = \{(\rho_1 - \rho_2)^2 + 2\rho_1\rho_2[1 - \cos(\theta_1 - \theta_2)\sin\phi_1\sin\phi_2 - \cos\phi_1\cos\phi_2]\}^{1/2}$$

39. Let (a, θ_1, ϕ_1) and (a, θ_2, ϕ_2) be two points on the sphere $\rho = a$. Show (using Problem 38) that the great-circle distance between these points is $a\gamma$, where $0 \le \gamma \le \pi$ and

$$\cos\gamma = \cos(\theta_1 - \theta_2)\sin\phi_1\sin\phi_2 + \cos\phi_1\cos\phi_2$$

40. As you may have guessed, there is a simple formula for expressing great-circle distance directly in terms of longitude and latitude. Let (α_1, β_1) and (α_2, β_2) be the longitude–latitude coordinates of two points on the surface of the earth, where we interpret N and E as positive and S and W as negative. Show that the great-circle distance between these points is 3960γ miles, where $0 \le \gamma \le \pi$ and

$$\cos\gamma = \cos(\alpha_1 - \alpha_2)\cos\beta_1\cos\beta_2 + \sin\beta_1\sin\beta_2$$

[C] **41.** Use Problem 40 to find the great-circle distance between each pair of places.

(a) New York and Greenwich (see Problem 34)

(b) St. Paul and Turin (see Problem 35)

(c) Turin and the South Pole (use $\alpha_1 = \alpha_2$)

(d) New York and Cape Town (18.4° E, 33.9° S)

(e) Two points on the equator with longitudes 100° E and 80° W, respectively

42. It is easy to see that the graph of $\rho = 2a\cos\phi$ is a sphere of radius a sitting on the xy-plane at the origin. But what is the graph of $\rho = 2a\sin\phi$?

Answers to Concepts Review: **1.** circular cylinder; sphere **2.** plane; cone **3.** $\rho^2 = r^2 + z^2$ **4.** $x^2 + y^2 + (z - 2)^2 = 4$

11.10 Chapter Review

Concepts Test

Respond with true or false to each of the following assertions. Be prepared to justify your answer.

1. Each point in three-space has a unique set of Cartesian coordinates.

2. The equation $x^2 + y^2 + z^2 - 4x + 9 = 0$ represents a sphere.

3. The linear equation $Ax + By + Cz = D$ represents a plane in three-space provided that A, B, and C are not all zero.

4. In three-space, the equation $Ax + By = C$ represents a line.

5. The planes $3x - 2y + 4z = 12$ and $3x - 2y + 4z = -12$ are parallel and 24 units apart.

6. The vector $\langle 1, -2, 3\rangle$ is parallel to the plane $2x - 4y + 6z = 5$.

7. The line $x = 2t - 1$, $y = 4t + 2$, $z = 6t - 5$ goes through the point $(0, 4, -2)$.

8. If $\mathbf{u} = a\mathbf{i} + b\mathbf{j} + c\mathbf{k}$ is a unit vector, then a, b, and c are direction cosines for $\mathbf{u}$.

9. The vectors $2\mathbf{i} - 3\mathbf{j}$ and $6\mathbf{i} + 4\mathbf{j}$ are perpendicular.

10. If $\mathbf{u}$ and $\mathbf{v}$ are unit vectors, then the angle θ between them satisfies $\cos\theta = \mathbf{u}\cdot\mathbf{v}$.

11. The dot product for vectors satisfies the associative law.

12. If $\mathbf{u}$ and $\mathbf{v}$ are any two vectors, then $\|\mathbf{u}\cdot\mathbf{v}\| \le \|\mathbf{u}\|\|\mathbf{v}\|$.

13. $\|\mathbf{u}\cdot\mathbf{v}\| = \|\mathbf{u}\|\|\mathbf{v}\|$ for nonzero vectors $\mathbf{u}$ and $\mathbf{v}$ if and only if $\mathbf{u}$ is a scalar multiple of $\mathbf{v}$.

14. If $\|\mathbf{u}\| = \|\mathbf{v}\| = \|\mathbf{u} + \mathbf{v}\|$, then $\mathbf{u} = \mathbf{v} = \mathbf{0}$.

15. If $\mathbf{u} + \mathbf{v}$ and $\mathbf{u} - \mathbf{v}$ are perpendicular, then $\|\mathbf{u}\| = \|\mathbf{v}\|$.

16. For any two vectors $\mathbf{u}$ and $\mathbf{v}$,

$$\|\mathbf{u} + \mathbf{v}\|^2 = \|\mathbf{u}\|^2 + \|\mathbf{v}\|^2 + 2\mathbf{u}\cdot\mathbf{v}.$$

17. The vector-valued function $\langle f(t), g(t), h(t)\rangle$ is continuous at $t = a$ if and only if f, g, and h are continuous at $t = a$.

18. $D_t[\mathbf{F}(t)\cdot\mathbf{F}(t)] = 2\mathbf{F}(t)\cdot\mathbf{F}'(t)$.

19. For every vector $\mathbf{u}$, $\|\,\|\mathbf{u}\|\mathbf{u}\,\| = \|\mathbf{u}\|^2$.

20. For every vector $\mathbf{u}$, $\|\mathbf{u}\|\cdot\mathbf{u} = \mathbf{u}\cdot\|\mathbf{u}\|$.

21. For all vectors $\mathbf{u}$ and $\mathbf{v}$, $\|\mathbf{u}\times\mathbf{v}\| = \|\mathbf{v}\times\mathbf{u}\|$.

22. If $\mathbf{u}$ is a scalar multiple of $\mathbf{v}$, then $\mathbf{u}\times\mathbf{v} = \mathbf{0}$.

23. The cross product of two unit vectors is a unit vector.

24. Multiplying each component of a vector $\mathbf{v}$ by the scalar a multiplies the length of $\mathbf{v}$ by a.

25. For any nonzero and nonperpendicular vectors $\mathbf{u}$ and $\mathbf{v}$ with angle θ between them, $\|\mathbf{u}\times\mathbf{v}\|/(\mathbf{u}\cdot\mathbf{v}) = \tan\theta$.

26. If $\mathbf{u}\cdot\mathbf{v} = 0$ and $\mathbf{u}\times\mathbf{v} = \mathbf{0}$, then $\mathbf{u}$ or $\mathbf{v}$ is $\mathbf{0}$.

27. The volume of the parallelepiped determined by $2\mathbf{i}$, $2\mathbf{j}$, and $\mathbf{j}\times\mathbf{i}$ is 4.

28. For all vectors **u**, **v**, and **w**,

$$\mathbf{u} \times (\mathbf{v} \times \mathbf{w}) = (\mathbf{u} \times \mathbf{v}) \times \mathbf{w}$$

29. If $a_1\mathbf{i} + a_2\mathbf{j} + a_3\mathbf{k}$ is a vector in the plane $b_1x + b_2y + b_3z = 0$, then $a_1b_1 + a_2b_2 + a_3b_3 = 0$.

30. Any line can be represented by both parametric equations and symmetric equations.

31. When $\kappa(t) = 0$ for all t, the path is a straight line.

32. An ellipse has its maximum curvature at points on the major axis.

33. The curvature depends on the shape of the curve and the speed with which you move along the curve.

34. The curvature of the curve determined by $x = 3t + 4$ and $y = 2t - 1$ is zero for all t.

35. The curvature of the curve determined by $x = 2 \cos t$ and $y = 2 \sin t$ is 2 for all t.

36. If $\mathbf{T} = \mathbf{T}(t)$ is a unit vector tangent to a smooth curve, then $\mathbf{T}(t)$ and $\mathbf{T}'(t)$ are perpendicular.

37. If $v = \|\mathbf{v}\|$ is the speed of a particle moving along a smooth curve, then $|dv/dt|$ is the magnitude of the acceleration.

38. If $y = f(x)$ and $y'' = 0$ everywhere, then the curvature of this curve is zero.

39. If $y = f(x)$ and y'' is a constant, then the curvature of this curve is a constant.

40. If $\mathbf{u} \cdot \mathbf{v} = 0$, then either $\mathbf{u} = \mathbf{0}$ or $\mathbf{v} = \mathbf{0}$, or both **u** and **v** are **0**.

41. If $\|\mathbf{r}(t)\| = 1$ for all t, then $\|\mathbf{r}'(t)\| =$ constant.

42. If $\mathbf{v}(t) \cdot \mathbf{v}(t) =$ constant, then $\mathbf{v}(t) \cdot \mathbf{v}'(t) = 0$.

43. For motion along a helix, **N** always points toward the z-axis.

44. If the velocity of the motion along the curve is of constant magnitude, then there can be no acceleration.

45. **T**, **N**, and **B** depend only on the shape of the curve and not on the speed of motion along the curve.

46. If **v** is perpendicular to **a**, then the speed of motion along the curve must be a constant.

47. If **v** is perpendicular to **a**, then the path of motion must be a circle.

48. The only curves with constant curvature are straight lines and circles.

49. The curves given by $\mathbf{r}_1(t) = \sin t\,\mathbf{i} + \cos t\,\mathbf{j} + t^3\mathbf{k}$ and $\mathbf{r}_2(t) = \sin t^3\mathbf{i} + \cos t^3\mathbf{j} + t^9\mathbf{k}$ for $0 \le t \le 1$ are identical.

50. The motions along the curves given by $\mathbf{r}_1(t) = \sin t\,\mathbf{i} + \cos t\,\mathbf{j} + t^3\mathbf{k}$ and $\mathbf{r}_2(t) = \sin t^3\mathbf{i} + \cos t^3\mathbf{j} + t^9\mathbf{k}$ for $0 \le t \le 1$ are identical.

51. The length of a given curve is independent of the parametrization used to describe the curve.

52. If a curve lies in a plane, then the binormal vector **B** must be a constant.

53. If $\|\mathbf{r}(t)\| =$ constant, then $\mathbf{r}'(t) = \mathbf{0}$.

54. The curve that is the intersection of the sphere $x^2 + y^2 + z^2 = 1$ and the plane $ax + by + cz = 0$ has constant curvature 1.

55. The graph of the equation $\phi = 0$ is the z-axis (here ϕ is a spherical coordinate).

56. The graph of $y = x^2$ in three-space is a paraboloid.

57. If we restrict ρ, θ, and ϕ by $\rho \ge 0, 0 \le \theta < 2\pi$, and $0 \le \phi \le \pi$, then each point in three-space has a unique set of spherical coordinates.

Sample Test Problems

1. Find the equation of the sphere that has $(-2, 3, 3)$ and $(4, 1, 5)$ as end points of a diameter.

2. Find the center and radius of the sphere with equation $x^2 + y^2 + z^2 - 6x + 2y - 8z = 0$.

3. Let $\mathbf{a} = \langle 2, -5 \rangle$, $\mathbf{b} = \langle 1, 1 \rangle$, and $\mathbf{c} = \langle -6, 0 \rangle$. Find each of the following:

(a) $3\mathbf{a} - 2\mathbf{b}$ (b) $\mathbf{a} \cdot \mathbf{b}$

(c) $\mathbf{a} \cdot (\mathbf{b} + \mathbf{c})$ (d) $(4\mathbf{a} + 5\mathbf{b}) \cdot 3\mathbf{c}$

(e) $\|\mathbf{c}\|\mathbf{c} \cdot \mathbf{b}$ (f) $\mathbf{c} \cdot \mathbf{c} - \|\mathbf{c}\|$

4. Find the cosine of the angle between **a** and **b** and make a sketch.

(a) $\mathbf{a} = 3\mathbf{i} + 2\mathbf{j}, \mathbf{b} = -\mathbf{i} + 4\mathbf{j}$

(b) $\mathbf{a} = -5\mathbf{i} - 3\mathbf{j}, \mathbf{b} = 2\mathbf{i} - \mathbf{j}$

(c) $\mathbf{a} = \langle 7, 0 \rangle, \mathbf{b} = \langle 5, 1 \rangle$

5. Let $\mathbf{a} = -\mathbf{i} + \mathbf{j} + 2\mathbf{k}$, $\mathbf{b} = \mathbf{j} - 2\mathbf{k}$, and $\mathbf{c} = 3\mathbf{i} - \mathbf{j} + 4\mathbf{k}$. Find each of the following if they are defined.

(a) $\mathbf{a} + \mathbf{b} + \mathbf{c}$ (b) $\mathbf{b} \cdot \mathbf{c}$

(c) $\mathbf{a} \cdot (\mathbf{b} \times \mathbf{c})$ (d) $\mathbf{a} \times (\mathbf{b} \cdot \mathbf{c})$

(e) $\|\mathbf{a} - \mathbf{b}\|$ (f) $\|\mathbf{b} \times \mathbf{c}\|$

6. Find the angle between each pair of vectors.

(a) $\mathbf{a} = \langle 1, 5, -1 \rangle, \mathbf{b} = \langle 0, 1, 3 \rangle$

(b) $\mathbf{a} = -\mathbf{i} + 2\mathbf{k}, \mathbf{b} = \mathbf{i} - \mathbf{j} + 3\mathbf{k}$

7. Sketch the two position vectors $\mathbf{a} = 2\mathbf{i} - \mathbf{j} + 2\mathbf{k}$ and $\mathbf{b} = 5\mathbf{i} + \mathbf{j} - 3\mathbf{k}$. Then find each of the following:

(a) their lengths

(b) their direction cosines

(c) the unit vector with the same direction as **a**

(d) the angle θ between **a** and **b**

8. Let $\mathbf{a} = 2\mathbf{i} - \mathbf{j} + \mathbf{k}$, $\mathbf{b} = -\mathbf{i} + 3\mathbf{j} + 2\mathbf{k}$, and $\mathbf{c} = \mathbf{i} + 2\mathbf{j} - \mathbf{k}$. Find each of the following:

(a) $\mathbf{a} \times \mathbf{b}$ (b) $\mathbf{a} \times (\mathbf{b} + \mathbf{c})$

(c) $\mathbf{a} \cdot (\mathbf{b} \times \mathbf{c})$ (d) $\mathbf{a} \times (\mathbf{b} \times \mathbf{c})$

9. Find all vectors that are perpendicular to both of the vectors $3\mathbf{i} + 3\mathbf{j} - \mathbf{k}$ and $-\mathbf{i} - 2\mathbf{j} + 4\mathbf{k}$.

10. Find the unit vectors that are perpendicular to the plane determined by the three points $(3, -6, 4)$, $(2, 1, 1)$, and $(5, 0, -2)$.

11. Write the equation of the plane through the point $(-5, 7, -2)$ that satisfies each condition.

(a) Parallel to the xz-plane

(b) Perpendicular to the x-axis

(c) Parallel to both the x- and y-axes

(d) Parallel to the plane $3x - 4y + z = 7$

12. A plane through the point $(2, -4, -5)$ is perpendicular to the line joining the points $(-1, 5, -7)$ and $(4, 1, 1)$.

(a) Write a vector equation of the plane.

(b) Find a Cartesian equation of the plane.
(c) Sketch the plane by drawing its traces.

13. Find the value of C if the plane $x + 5y + Cz + 6 = 0$ is perpendicular to the plane $4x - y + z - 17 = 0$.

14. Find a Cartesian equation of the plane through the three points $(2, 3, -1)$, $(-1, 5, 2)$, and $(-4, -2, 2)$.

15. Find parametric equations for the line through $(-2, 1, 5)$ and $(6, 2, -3)$.

16. Find the points where the line of intersection of the planes $x - 2y + 4z - 14 = 0$ and $-x + 2y - 5z + 30 = 0$ pierces the yz- and xz-planes.

17. Write the equation of the line in Problem 16 in parametric form.

18. Find symmetric equations of the line through $(4, 5, 8)$ and perpendicular to the plane $3x + 5y + 2z = 30$. Sketch the plane and the line.

19. Write a vector equation of the line through $(2, -2, 1)$ and $(-3, 2, 4)$.

20. Sketch the curve whose vector equation is $\mathbf{r}(t) = t\,\mathbf{i} + \frac{1}{2}t^2\,\mathbf{j} + \frac{1}{3}t^3\,\mathbf{k}$, $-2 \le t \le 3$.

21. Find the symmetric equations for the tangent line to the curve of Problem 20 at the point where $t = 2$. Also find the equation of the normal plane at this point.

22. Find $\mathbf{r}'(\pi/2)$, $\mathbf{T}(\pi/2)$, and $\mathbf{r}''(\pi/2)$ if

$$\mathbf{r}(t) = \langle t\cos t, t\sin t, 2t\rangle$$

23. Find the length of the curve

$$\mathbf{r}(t) = e^t \sin t\,\mathbf{i} + e^t \cos t\,\mathbf{j} + e^t\,\mathbf{k}, \qquad 1 \le t \le 5$$

24. Two forces $\mathbf{F}_1 = 2\mathbf{i} - 3\mathbf{j}$ and $\mathbf{F}_2 = 3\mathbf{i} + 12\mathbf{j}$ are applied at a point. What force $\mathbf{F}$ must be applied at the point to counteract the resultant of these two forces?

25. What heading and airspeed are required for an airplane to fly 450 miles per hour due north if a wind of 100 miles per hour is blowing in the direction N 60° E?

26. If $\mathbf{r}(t) = \langle e^{2t}, e^{-t}\rangle$ find each of the following:

(a) $\lim_{t\to 0} \mathbf{r}(t)$ (b) $\lim_{h\to 0} \dfrac{\mathbf{r}(0 + h) - \mathbf{r}(0)}{h}$
(c) $\int_0^{\ln 2} \mathbf{r}(t)\,dt$ (d) $D_t[t\mathbf{r}(t)]$
(e) $D_t[\mathbf{r}(3t + 10)]$ (f) $D_t[\mathbf{r}(t)\cdot\mathbf{r}'(t)]$

27. Find $\mathbf{r}'(t)$ and $\mathbf{r}''(t)$ for each of the following:
(a) $\mathbf{r}(t) = (\ln t)\mathbf{i} - 3t^2\,\mathbf{j}$ (b) $\mathbf{r}(t) = \sin t\,\mathbf{i} + \cos 2t\,\mathbf{j}$
(c) $\mathbf{r}(t) = \tan t\,\mathbf{i} - t^4\mathbf{j}$

28. Suppose that an object is moving so that its position vector at time t is

$$\mathbf{r}(t) = e^t\,\mathbf{i} + e^{-t}\,\mathbf{j} + 2t\,\mathbf{k}$$

Find $\mathbf{v}(t)$, $\mathbf{a}(t)$, and $\kappa(t)$ at $t = \ln 2$.

29. If $\mathbf{r}(t) = t\,\mathbf{i} + t^2\mathbf{j} + t^3\mathbf{k}$ is the position vector for a moving particle at time t, find the tangential and normal components, a_T and a_N, of the acceleration vector at $t = 1$.

For each equation in Problems 30–38, name and sketch the graph in three-space.

30. $x^2 + y^2 = 81$ **31.** $x^2 + y^2 + z^2 = 81$
32. $z^2 = 4y$ **33.** $x^2 + z^2 = 4y$
34. $3y - 6z - 12 = 0$ **35.** $3x + 3y - 6z - 12 = 0$
36. $x^2 + y^2 - z^2 - 1 = 0$
37. $3x^2 + 4y^2 + 9z^2 - 36 = 0$
38. $3x^2 + 4y^2 + 9z^2 + 36 = 0$

39. Write the following Cartesian equations in cylindrical coordinate form.
(a) $x^2 + y^2 = 9$ (b) $x^2 + 4y^2 = 16$
(c) $x^2 + y^2 = 9z$ (d) $x^2 + y^2 + 4z^2 = 10$

40. Find the Cartesian equation corresponding to each of the following cylindrical coordinate equations.
(a) $r^2 + z^2 = 9$ (b) $r^2\cos^2\theta + z^2 = 4$
(c) $r^2\cos 2\theta + z^2 = 1$

41. Write the following equations in spherical coordinate form.
(a) $x^2 + y^2 + z^2 = 4$ (b) $2x^2 + 2y^2 - 2z^2 = 0$
(c) $x^2 - y^2 - z^2 = 1$ (d) $x^2 + y^2 = z$

42. Find the (straight-line) distance between the points whose spherical coordinates are $(8, \pi/4, \pi/6)$ and $(4, \pi/3, 3\pi/4)$.

43. Find the distance between the parallel planes $2x - 3y + \sqrt{3}z = 4$ and $2x - 3y + \sqrt{3}z = 9$.

44. Find the acute angle between the planes $2x - 4y + z = 7$ and $3x + 2y - 5z = 9$.

45. Show that if the speed of a moving particle is constant then its velocity and acceleration vectors are orthogonal.

REVIEW & PREVIEW PROBLEMS

In Problems 1–4, sketch a graph of the cylinder or quadric surface.

1. $x^2 + y^2 + z^2 = 64$

2. $x^2 + z^2 = 4$

3. $z = x^2 + 4y^2$

4. $z = x^2 - y^2$

In Problems 5–8, find the indicated derivative.

5. (a) $\frac{d}{dx}2x^3$ (b) $\frac{d}{dx}5x^3$

(c) $\frac{d}{dx}kx^3$ (d) $\frac{d}{dx}ax^3$

6. (a) $\frac{d}{dx}\sin 2x$ (b) $\frac{d}{dt}\sin 17t$

(c) $\frac{d}{dt}\sin at$ (d) $\frac{d}{dt}\sin bt$

7. (a) $\frac{d}{da}\sin 2a$ (b) $\frac{d}{da}\sin 17a$

(c) $\frac{d}{da}\sin ta$ (d) $\frac{d}{da}\sin sa$

8. (a) $\frac{d}{dt}e^{4t+1}$ (b) $\frac{d}{dx}e^{-7x+4}$

(c) $\frac{d}{dx}e^{ax+b}$ (d) $\frac{d}{dx}e^{tx+s}$

In Problems 9–12, say whether the function is continuous and whether it is differentiable at the given point.

9. $f(x) = \frac{1}{x^2 - 1}$ at $x = 2$

10. $f(x) = \tan x$ at $x = \pi/2$

11. $f(x) = |x - 4|$ at $x = 4$

12. $f(x) = \begin{cases} \sin\frac{1}{x}, & \text{if } x \neq 0 \\ 0, & \text{if } x = 0 \end{cases}$ at $x = 0$

In Problems 13–14, find the maximum and minimum value of the function on the given interval. Use the Second Derivative Test to determine whether each stationary point is a maximum or a minimum.

13. $f(x) = 3x - (x - 1)^3$ on $[0, 4]$

14. $f(x) = x^4 - 18x^3 + 113x^2 - 288x + 252$ on $[2, 6]$

15. A storage can is to be made in the shape of a right circular cylinder of height h and radius r. Find the surface area of the container (including the circular top and bottom) as a function of r only, if the volume is to be 8 cubic feet.

16. A three-dimensional box without a lid is to be made of a material that costs \$1 per square foot for the sides and \$3 per square foot for the bottom. The box is to contain 27 cubic feet. Let l, w, and h denote, respectively, the length, width, and height of the box. Since the box must contain 27 cubic feet, we must have $lwh = 27$, or, equivalently, $h = 27/(lw)$. Use this expression for h to find a formula for the cost of a box having length l and width w.

CHAPTER 12 Derivatives for Functions of Two or More Variables

12.1 Functions of Two or More Variables
12.2 Partial Derivatives
12.3 Limits and Continuity
12.4 Differentiability
12.5 Directional Derivatives and Gradients
12.6 The Chain Rule
12.7 Tangent Planes and Approximations
12.8 Maxima and Minima
12.9 The Method of Lagrange Multipliers

12.1 Functions of Two or More Variables

Two kinds of functions have been emphasized so far. The first, typified by $f(x) = x^2$, associates with the real number x another real number $f(x)$. We call it a real-valued function of a real variable. The second type of function, illustrated by $\mathbf{f}(x) = \langle x^3, e^x \rangle$, associates with the real number x a vector $\mathbf{f}(x)$. We call it a vector-valued function of a real variable.

Our interest shifts now to a **real-valued function of two real variables,** that is, a function f (Figure 1) that assigns to each ordered pair (x, y) in some set D of the plane a (unique) real number $f(x, y)$. Examples are

(1) $$f(x, y) = x^2 + 3y^2$$

(2) $$g(x, y) = 2x\sqrt{y}$$

Note that $f(-1, 4) = (-1)^2 + 3(4)^2 = 49$ and $g(-1, 4) = 2(-1)\sqrt{4} = -4$.

The set D is called the **domain** of the function. If it is not specified, we take D to be the natural domain, that is, the set of all points (x, y) in the plane for which the function rule makes sense and gives a real number value. For $f(x, y) = x^2 + 3y^2$, the natural domain is the whole plane; for $g(x, y) = 2x\sqrt{y}$, it is $\{(x, y): -\infty < x < \infty, y \geq 0\}$. The **range** of a function is its set of values. If $z = f(x, y)$, we call x and y the **independent variables** and z the **dependent variable.**

All that we have said extends in a natural way to real-valued functions of three real variables (or even n real variables). We will feel free to use such functions without further comment.

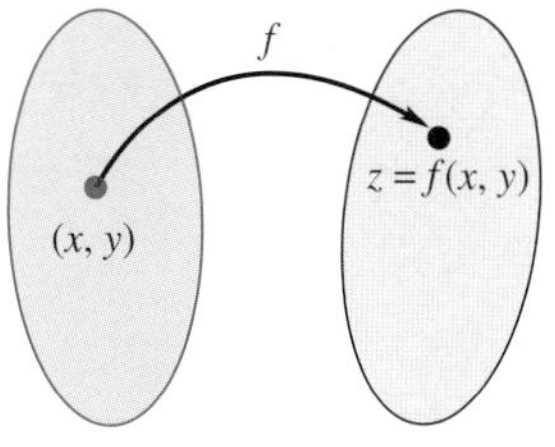

Figure 1

EXAMPLE 1 In the xy-plane, sketch the natural domain for

$$f(x, y) = \frac{\sqrt{y - x^2}}{x^2 + (y - 1)^2}$$

SOLUTION For this rule to make sense, we must exclude $\{(x, y): y < x^2\}$ and the point $(0, 1)$. The resulting domain is shown in Figure 2. ■

Graphs By the **graph** of a function f of two variables, we mean the graph of the equation $z = f(x, y)$. This graph will normally be a surface (Figure 3) and, since to each (x, y) in the domain there corresponds just one value z, each line perpendicular to the xy-plane intersects the surface in at most one point.

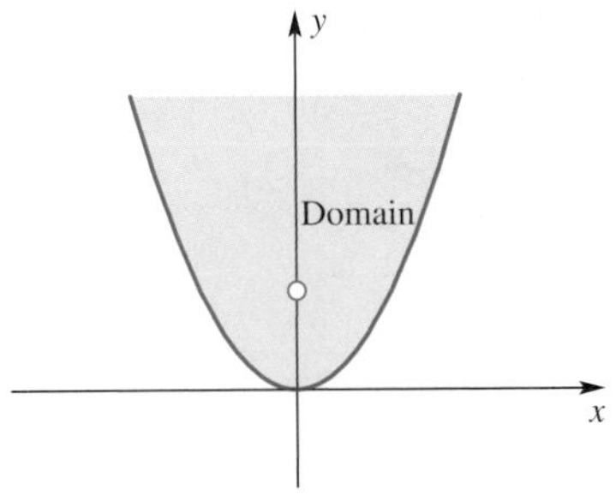

Figure 2

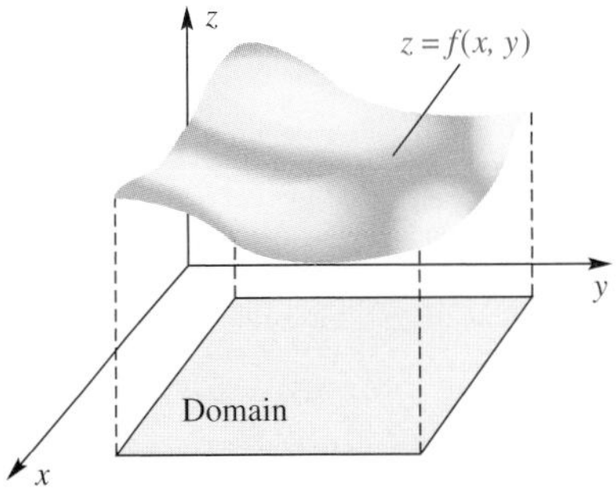

Figure 3

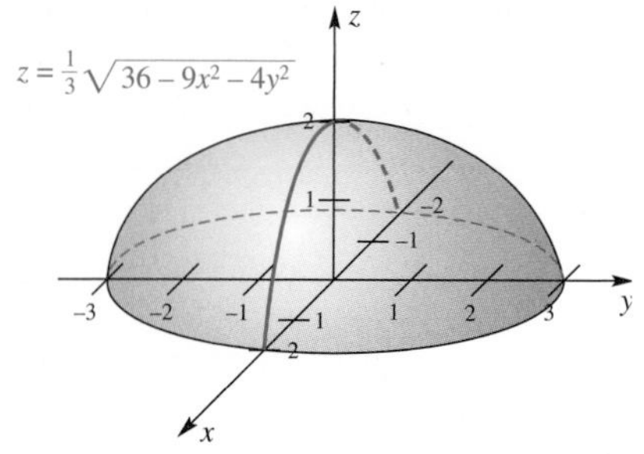

Figure 4

EXAMPLE 2 Sketch the graph of $f(x, y) = \frac{1}{3}\sqrt{36 - 9x^2 - 4y^2}$.

SOLUTION Let $z = \frac{1}{3}\sqrt{36 - 9x^2 - 4y^2}$ and note that $z \geq 0$. If we square both sides and simplify, we obtain the equation

$$9x^2 + 4y^2 + 9z^2 = 36$$

which we recognize as the equation of an ellipsoid (see Section 11.8). The graph of the given function is the upper half of this ellipsoid; it is shown in Figure 4. ■

EXAMPLE 3 Sketch the graph of $z = f(x, y) = y^2 - x^2$.

SOLUTION The graph is a hyperbolic paraboloid (see Section 11.8); it is graphed in Figure 5. ■

Computer Graphs A number of software packages, including *Maple* and *Mathematica*, can produce complicated three-dimensional graphs with ease. In Figures 6 through 9, we show four such graphs. Often, as in these four examples, we choose to show the graph with the y-axis pointing partially toward the viewer, rather than keeping it in the plane of the paper. Also, we often show the axes in a frame around the outside of the graph, rather than in the usual position, which could interfere with our view of the graph. The variable (x or y) indicating the axis is placed near the center of the axis that it represents.

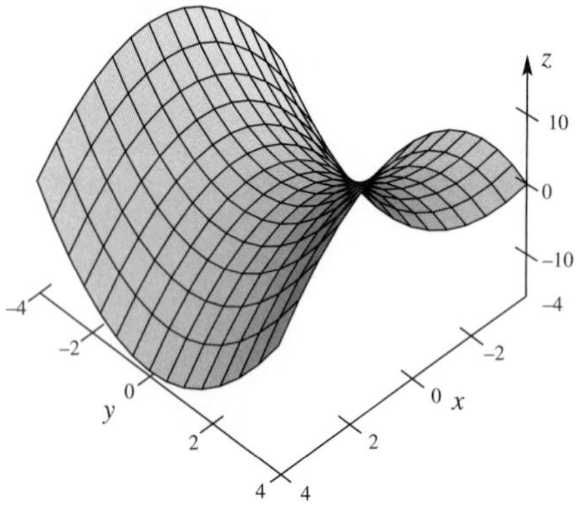

$z = y^2 - x^2$

Figure 5

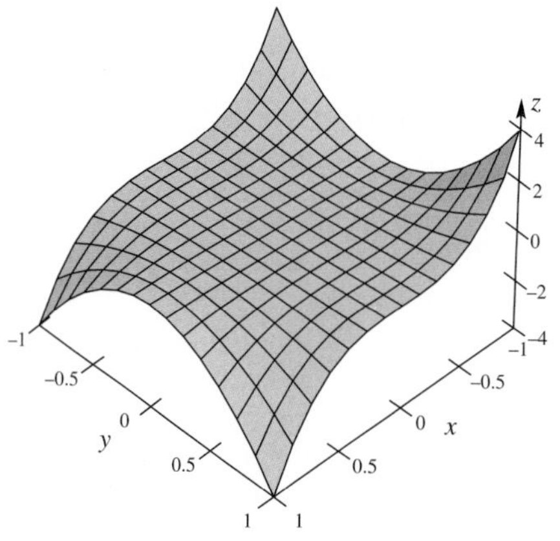

$z = -4x^3y^2$

Figure 6

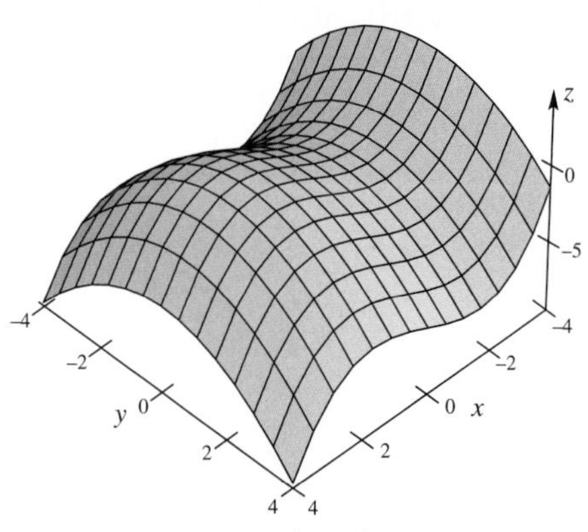

$z = x - \frac{1}{8}x^3 - \frac{1}{3}y^2$

Figure 7

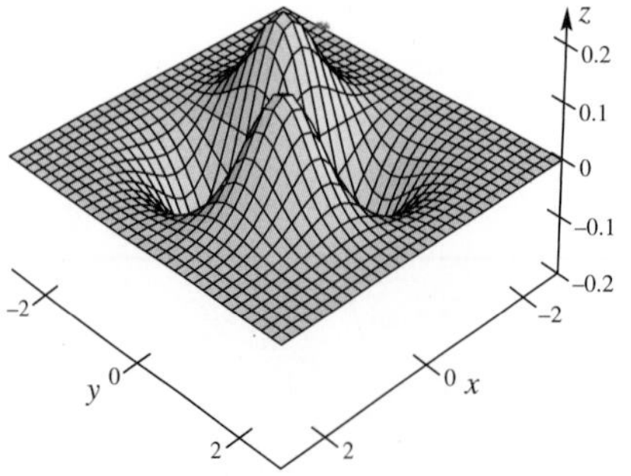

$z = xy\ \exp(-x^2 - y^2)$

Figure 8

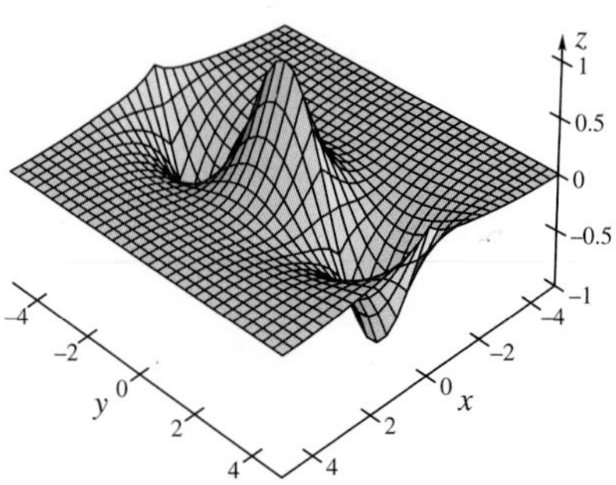

$z = e^{-|x|} \cos\sqrt{x^2 + y^2}$

Figure 9

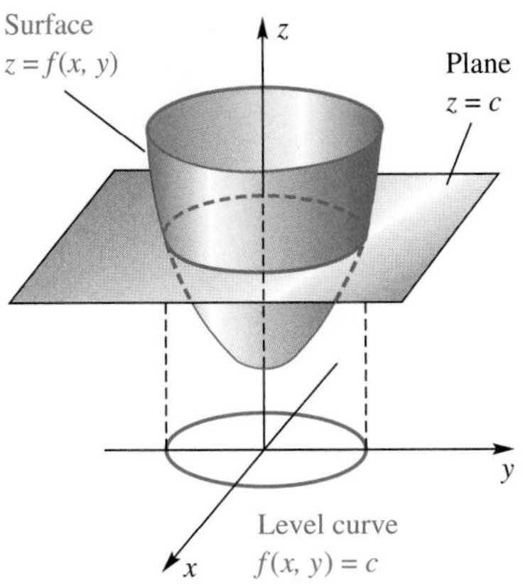

Figure 10

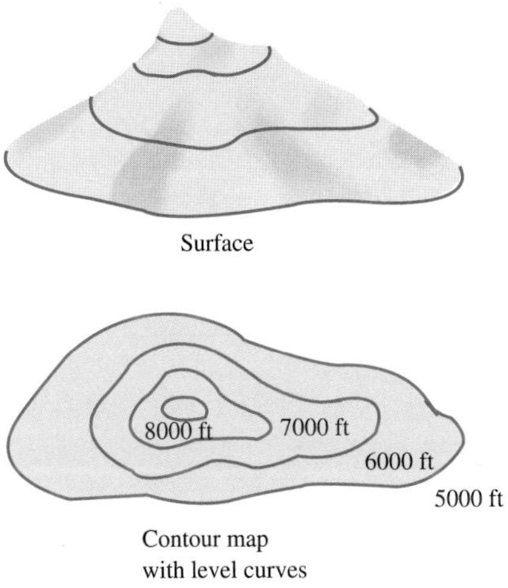

Figure 11

Level Curves

To sketch the surface corresponding to the graph of a function $z = f(x, y)$ of two variables is often very difficult. Map makers have given us another and usually simpler way to picture a surface: the contour map. Each horizontal plane $z = c$ intersects the surface in a curve. The projection of this curve on the xy-plane is called a **level curve** (Figure 10), and a collection of such curves is a **contour plot** or a **contour map.** We show a contour map for a hill-shaped surface in Figure 11.

We will often show contours on the three-dimensional graph itself, as is done in the top diagram in Figure 11. When this is done, we will usually make the y-axis go away from the viewer and the x-axis go to the right. This will help us to see the connection between the three-dimensional plot and the contour plot.

EXAMPLE 4 Draw contour maps for the surfaces corresponding to $z = \frac{1}{3}\sqrt{36 - 9x^2 - 4y^2}$ and $z = y^2 - x^2$ (see Examples 2 and 3, and Figures 4 and 5).

SOLUTION The level curves of $z = \frac{1}{3}\sqrt{36 - 9x^2 - 4y^2}$ corresponding to $z = 0, 1, 1.5, 1.75, 2$ are shown in Figure 12. They are ellipses. Similarly, in Figure 13, we show the level curves of $z = y^2 - x^2$ for $z = -5, -4, -3, \ldots, 2, 3, 4$. These curves are hyperbolas unless $z = 0$. The level curve for $z = 0$ is a pair of intersecting lines. ■

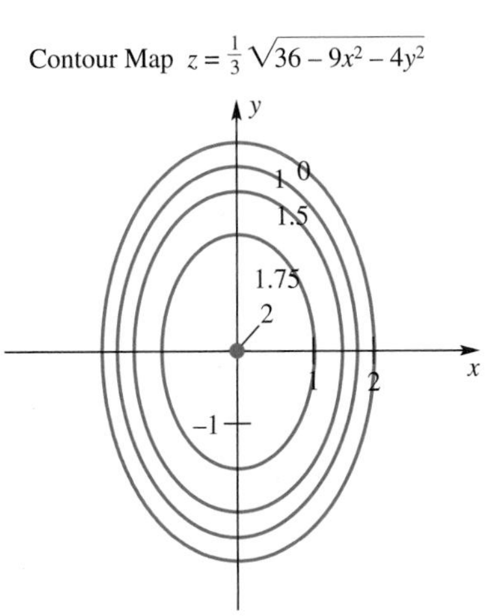

Figure 12

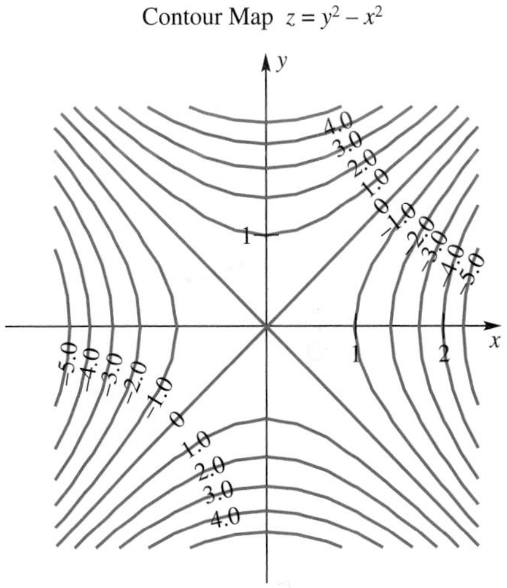

Figure 13

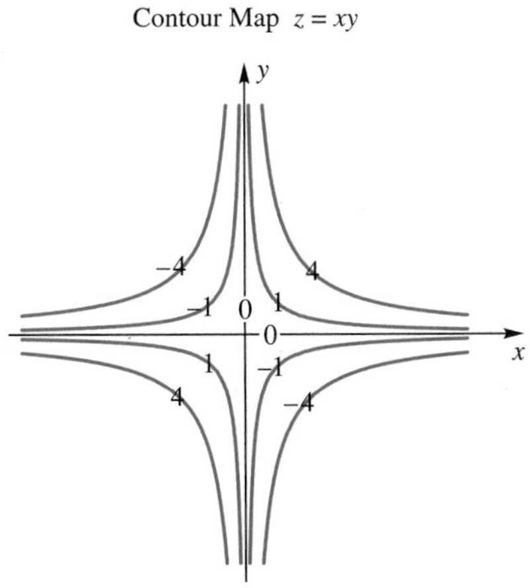

Figure 14

EXAMPLE 5 Sketch a contour map for $z = f(x, y) = xy$.

SOLUTION The level curves corresponding to $z = -4, -1, 0, 1, 4$ are shown in Figure 14. It can be shown that they are hyperbolas. Comparing the contour map of Figure 14 with that of Figure 13 suggests that the graph of $z = xy$ might be a hyperbolic paraboloid but with axes rotated through 45°. The suggestion is correct. ■

Computer Graphs and Level Curves

In Figures 15 through 19, we have drawn five more surfaces, but we now also show the corresponding level curves. A third plot is a three-dimensional plot with level curves on the surface. Note that we have rotated the xy-plane so that the x-axis points to the right, making it easier to relate the surface and the level curves.

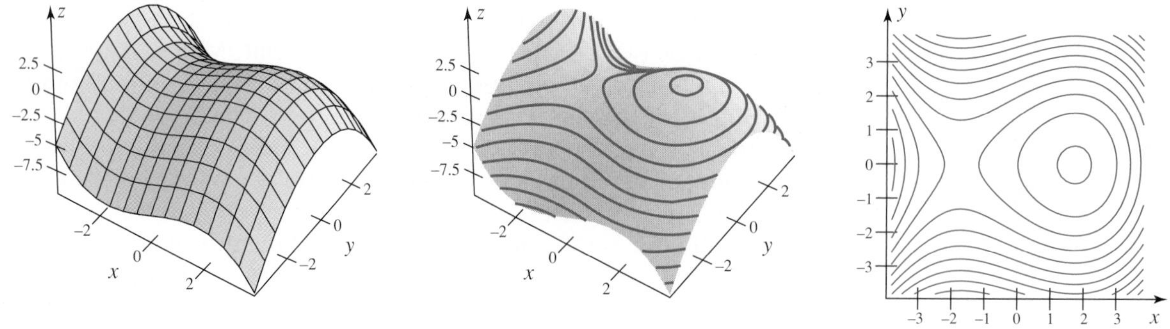

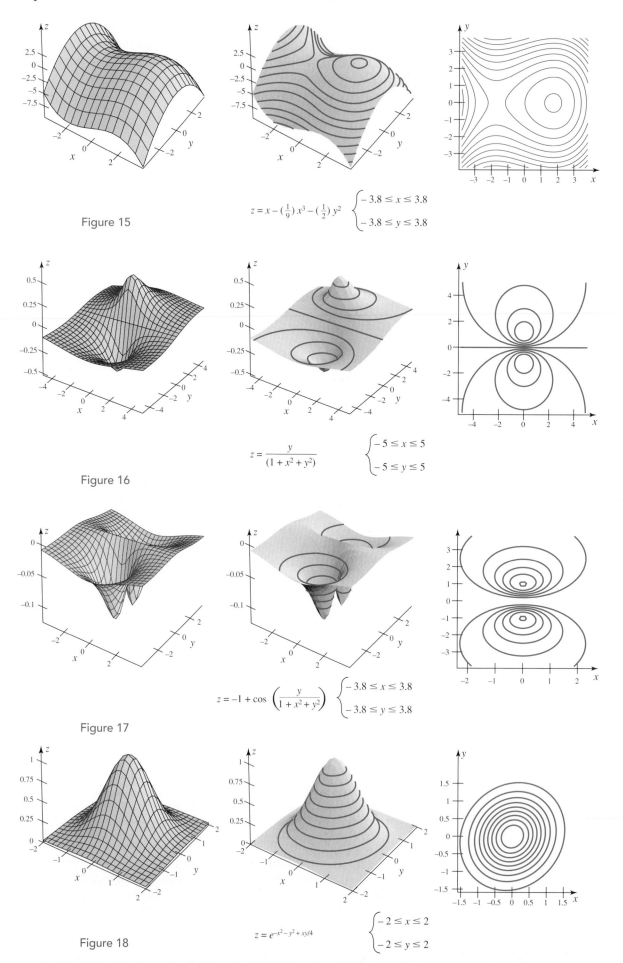

$$z = x - \left(\tfrac{1}{9}\right)x^3 - \left(\tfrac{1}{2}\right)y^2 \quad \begin{cases} -3.8 \le x \le 3.8 \\ -3.8 \le y \le 3.8 \end{cases}$$

Figure 15

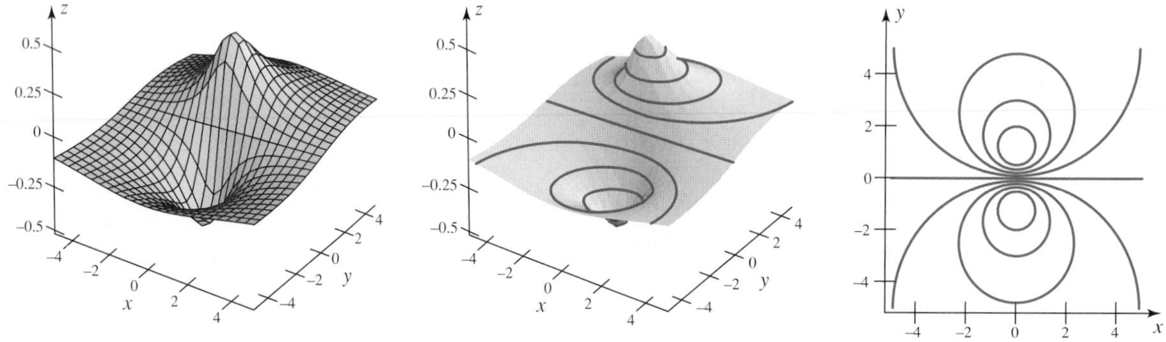

$$z = \frac{y}{(1 + x^2 + y^2)} \quad \begin{cases} -5 \le x \le 5 \\ -5 \le y \le 5 \end{cases}$$

Figure 16

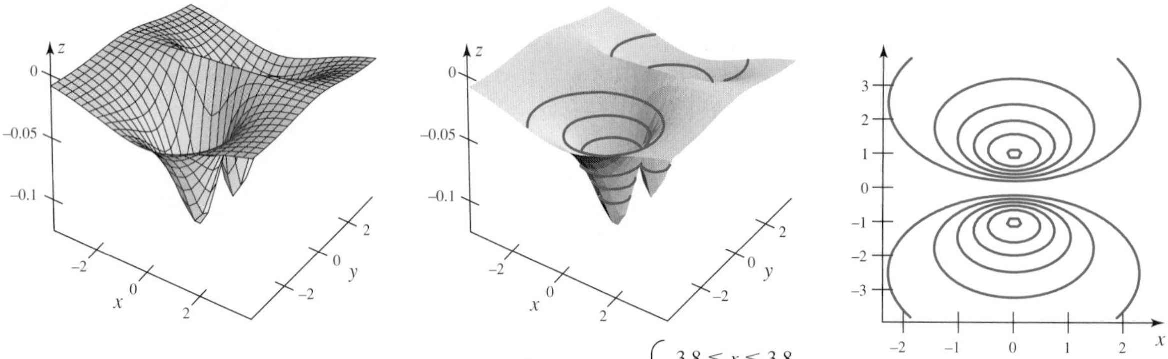

$$z = -1 + \cos\left(\frac{y}{1 + x^2 + y^2}\right) \quad \begin{cases} -3.8 \le x \le 3.8 \\ -3.8 \le y \le 3.8 \end{cases}$$

Figure 17

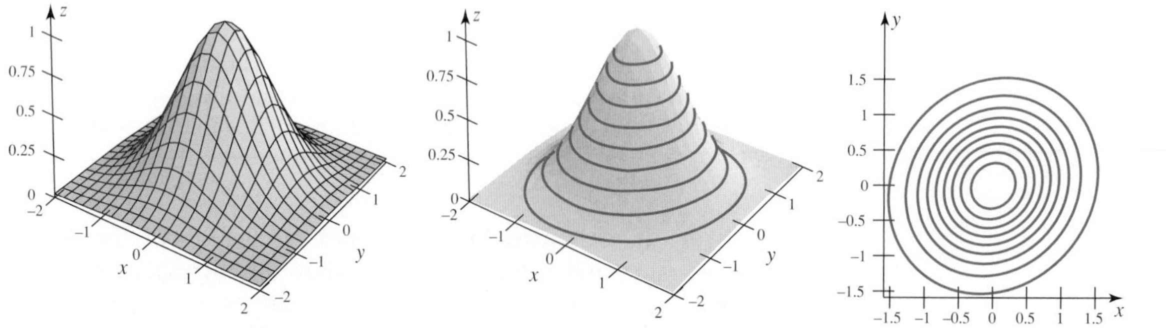

$$z = e^{-x^2 - y^2 + xy/4} \quad \begin{cases} -2 \le x \le 2 \\ -2 \le y \le 2 \end{cases}$$

Figure 18

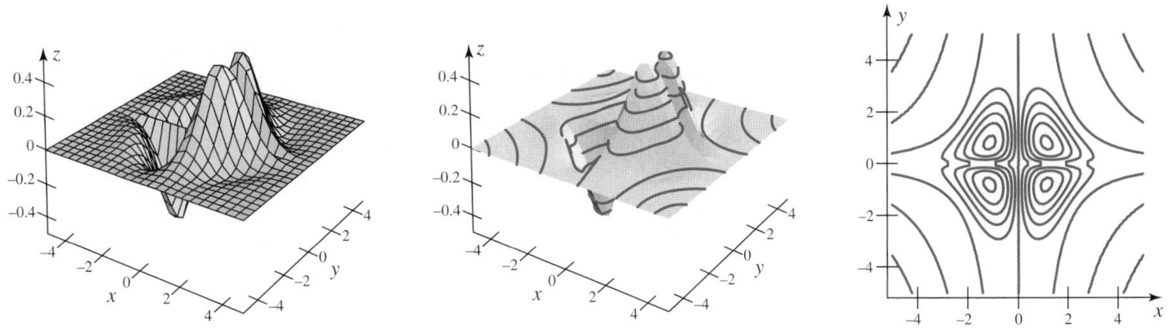

$z = e^{-(x^2+y^2)/4}\sin(x\sqrt{|y|})$ $\quad \begin{cases} -5 \le x \le 5 \\ -5 \le y \le 5 \end{cases}$

Figure 19

Applications of Contour Plots Contour maps are frequently used to show weather or other conditions at various points on a map. Temperature, for example, varies from place to place. We can envision $T(x, y)$ as being equal to the temperature at the location (x, y). Level curves for equal temperatures are called **isotherms** or **isothermal curves.** Figure 20 shows an isothermal map for the United States.

On April 9, 1917, a strong earthquake centered near the Mississippi river just south of St. Louis was felt as far north as Iowa and as far south as Mississippi. The intensity of an earthquake is measured from I to XII, with higher numbers corresponding to a more severe earthquake. A magnitude VI earthquake will cause physical damage to structures. Figure 21 shows an example of another type of contour map. If we envision the intensity I as a function of the location (x, y), then we can illustrate the earthquake's intensity using a map with level curves corresponding to equal intensity. Curves with constant intensity are called **isoseismic curves.** Figure 21 shows that the regions that experienced an intensity of VI include the St. Louis area and a strip in southeastern Missouri. Much of eastern Missouri and southwestern Illinois experienced an intensity between V and VI. Since Kansas City and Memphis are near the same isoseismic curve, the intensity was about the same in Kansas City and Memphis.

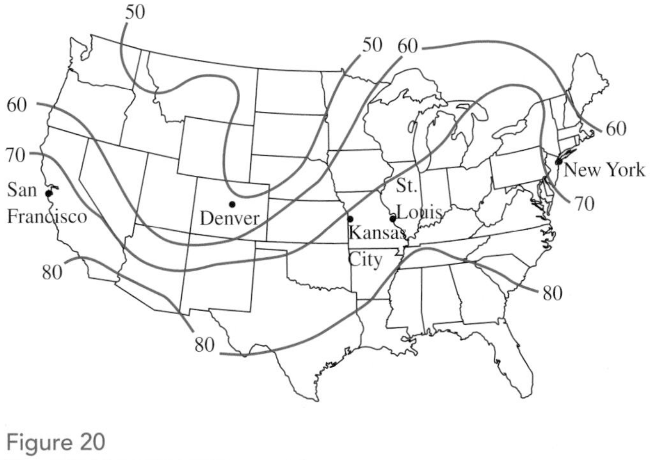

Figure 20

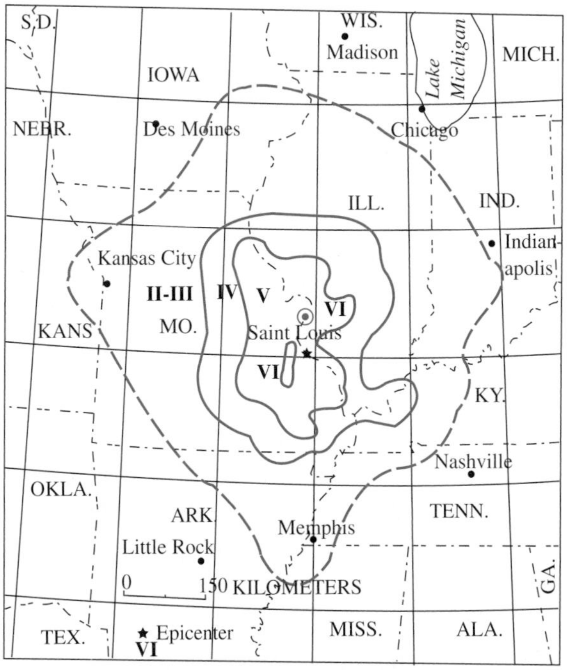

Figure 21

Functions of Three or More Variables A number of quantities depend on three or more variables. For example, the temperature in a large auditorium may depend on the location (x, y, z); this leads to the function $T(x, y, z)$. The velocity of a fluid may depend on the location (x, y, z), as well as on time t; this leads to the function $V(x, y, z, t)$. Finally, the average exam score in a class of 50 students depends on the 50 exam scores $x_1, x_2, \ldots, x_{50}$; this leads to the function $A(x_1, x_2, \ldots, x_{50})$.

We can visualize functions of three variables by plotting **level surfaces**, that is, surfaces in three-dimensional space that lead to a constant value for the function. Functions of four or more variables are much more difficult to visualize. The natural domain of a function of three or more variables is the set of all ordered triples (or quadruples, etc.) for which the function makes sense and gives a real number.

EXAMPLE 6 Find the domain of each function and describe the level surfaces for f.

(a) $f(x, y, z) = \sqrt{x^2 + y^2 + z^2 - 1}$

(b) $g(w, x, y, z) = \dfrac{1}{\sqrt{w^2 + x^2 + y^2 + z^2 - 1}}$

SOLUTION

(a) To avoid roots of negative numbers, the ordered triple (x, y, z) must satisfy $x^2 + y^2 + z^2 - 1 \geq 0$. Thus, the domain for f consists of all points (x, y, z) that are on or outside the unit sphere. Level surfaces for f are surfaces in three-space where $f(x, y, z) = \sqrt{x^2 + y^2 + z^2 - 1} = c$. As long as $c \geq 0$, this relationship leads to $x^2 + y^2 + z^2 = c + 1$, a sphere centered at the origin. Level surfaces are therefore concentric spheres centered at $(0, 0, 0)$.

(b) The ordered quadruple (w, x, y, z) must satisfy $w^2 + x^2 + y^2 + z^2 - 1 > 0$, since we must avoid roots of negative numbers and division by 0. ■

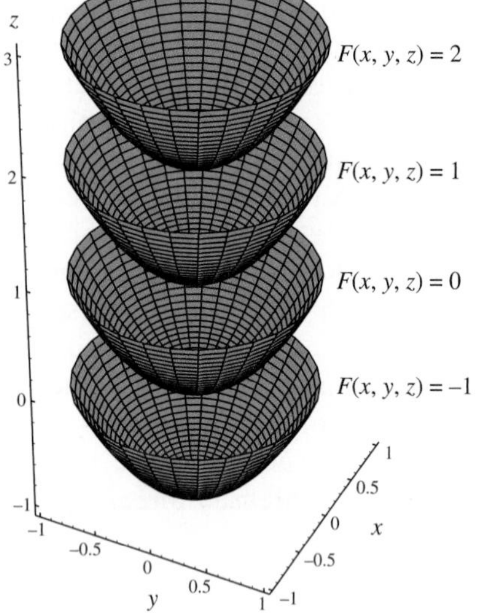

Figure 22

EXAMPLE 7 Let $F(x, y, z) = z - x^2 - y^2$. Describe the level surfaces for F and plot level surfaces for $-1, 0, 1$, and 2.

SOLUTION The relationship $F(x, y, z) = z - x^2 - y^2 = c$ leads to $z = c + x^2 + y^2$. This is a paraboloid opening upward having vertex at $(0, 0, c)$. The level surfaces are shown in Figure 22. ■

Concepts Review

1. A function f determined by $z = f(x, y)$ is called a(n) _____.

2. The projection of the curve $z = f(x, y) = c$ to the xy-plane is called a(n) _____, and a collection of such curves is called a(n) _____.

3. The contour map for $z = x^2 + y^2$ consists of _____.

4. The contour map for $z = x^2$ consists of _____.

Problem Set 12.1

1. Let $f(x, y) = x^2 y + \sqrt{y}$. Find each value.

(a) $f(2, 1)$ (b) $f(3, 0)$

(c) $f(1, 4)$ (d) $f(a, a^4)$

(e) $f(1/x, x^4)$ (f) $f(2, -4)$

What is the natural domain for this function?

2. Let $f(x, y) = y/x + xy$. Find each value.

(a) $f(1, 2)$ (b) $f\left(\frac{1}{4}, 4\right)$

(c) $f\left(4, \frac{1}{4}\right)$ (d) $f(a, a)$

(e) $f(1/x, x^2)$ (f) $f(0, 0)$

What is the natural domain for this function?

3. Let $g(x, y, z) = x^2 \sin yz$. Find each value.

(a) $g(1, \pi, 2)$ (b) $g(2, 1, \pi/6)$

(c) $g(4, 2, \pi/4)$ C (d) $g(\pi, \pi, \pi)$

4. Let $g(x, y, z) = \sqrt{x \cos y} + z^2$. Find each value.

(a) $g(4, 0, 2)$ (b) $g(-9, \pi, 3)$

(c) $g(2, \pi/3, -1)$ C (d) $g(3, 6, 1.2)$

5. Find $F(f(t), g(t))$ if $F(x, y) = x^2 y$ and $f(t) = t \cos t$, $g(t) = \sec^2 t$.

6. Find $F(f(t), g(t))$ if $F(x, y) = e^x + y^2$ and $f(t) = \ln t^2$, $g(t) = e^{t/2}$.

In Problems 7–16, sketch the graph of f.

7. $f(x, y) = 6$ **8.** $f(x, y) = 6 - x$

9. $f(x, y) = 6 - x - 2y$ **10.** $f(x, y) = 6 - x^2$

11. $f(x, y) = \sqrt{16 - x^2 - y^2}$

12. $f(x, y) = \sqrt{16 - 4x^2 - y^2}$

13. $f(x, y) = 3 - x^2 - y^2$ **14.** $f(x, y) = 2 - x - y^2$

15. $f(x, y) = e^{-(x^2+y^2)}$ **16.** $f(x, y) = x^2/y, y > 0$

In Problems 17–22, sketch the level curve $z = k$ for the indicated values of k.

17. $z = \frac{1}{2}(x^2 + y^2), k = 0, 2, 4, 6, 8$

18. $z = \dfrac{x}{y}, k = -2, -1, 0, 1, 2$

19. $z = \dfrac{x^2}{y}, k = -4, -1, 0, 1, 4$

20. $z = x^2 + y, k = -4, -1, 0, 1, 4$

21. $z = \dfrac{x^2 + 1}{x^2 + y^2}, k = 1, 2, 4$

22. $z = y - \sin x, k = -2, -1, 0, 1, 2$

23. Let $T(x, y)$ be the temperature at a point (x, y) in the plane. Draw the isothermal curves corresponding to $T = \frac{1}{10}, \frac{1}{5}, \frac{1}{2}, 0$ if

$$T(x, y) = \frac{x^2}{x^2 + y^2}$$

24. If $V(x, y)$ is the voltage at a point (x, y) in the plane, the level curves of V are called **equipotential curves.** Draw the equipotential curves corresponding to $V = \frac{1}{2}, 1, 2, 4$ for

$$V(x, y) = \frac{4}{\sqrt{(x - 2)^2 + (y + 3)^2}}$$

25. Figure 20 shows isotherms for the United States.

(a) Which of San Francisco, Denver, and New York had approximately the same temperature as St. Louis?

(b) If you were in Kansas City and wanted to drive toward cooler weather as quickly as possible, in which direction would you travel? What if you wanted to drive toward warmer weather?

(c) If you were leaving Kansas City, in which directions could you go and stay at approximately the same temperature?

26. Figure 23 shows a contour map for barometric pressure in millibars. Level curves for barometric pressure are called **isobars.**

(a) What part of the country had the lowest barometric pressure? The highest?

(b) If you were in St. Louis, in which direction would you have to travel to move as fast as possible toward lower barometric pressure? Higher barometric pressure?

(c) If you were leaving St. Louis, in which directions could you go in order to remain at approximately the same barometric pressure?

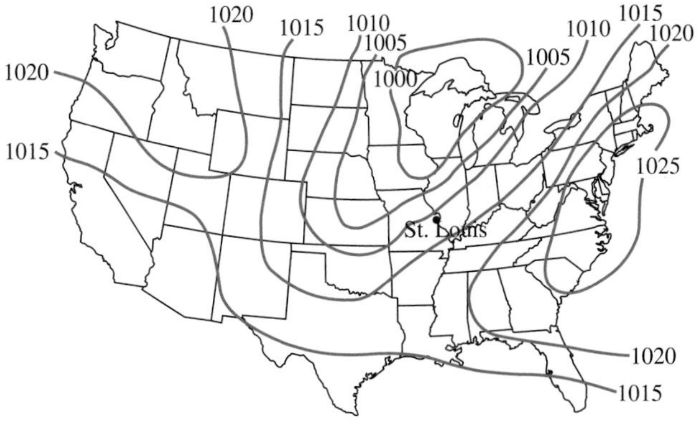

Figure 23

In Problems 27–32, describe geometrically the domain of each of the indicated functions of three variables.

27. $f(x, y, z) = \sqrt{x^2 + y^2 + z^2 - 16}$

28. $f(x, y, z) = \sqrt{x^2 + y^2 - z^2 - 9}$

29. $f(x, y, z) = \sqrt{144 - 16x^2 - 9y^2 - 144z^2}$

30. $f(x, y, z) = \dfrac{(144 - 16x^2 - 16y^2 + 9z^2)^{3/2}}{xyz}$

31. $f(x, y, z) = \ln(x^2 + y^2 + z^2)$

32. $f(x, y, z) = z \ln(xy)$

Describe geometrically the level surfaces for the functions defined in Problems 33–38.

33. $f(x, y, z) = x^2 + y^2 + z^2; k > 0$

34. $f(x, y, z) = 100x^2 + 16y^2 + 25z^2; k > 0$

35. $f(x, y, z) = 16x^2 + 16y^2 - 9z^2$

36. $f(x, y, z) = 9x^2 - 4y^2 - z^2$

37. $f(x, y, z) = 4x^2 - 9y^2$

38. $f(x, y, z) = e^{x^2+y^2+z^2}, k > 0$

39. Find the domain of each function.

(a) $f(w, x, y, z) = \dfrac{1}{\sqrt{w^2 + x^2 + y^2 + z^2}}$

(b) $g(x_1, x_2, \ldots, x_n) = \exp(-x_1^2 - x_2^2 - \cdots - x_n^2)$

(c) $h(x_1, x_2, \ldots, x_n) = \sqrt{1 - (x_1^2 + x_2^2 + \cdots + x_n^2)}$

40. Sketch (as best you can) the graph of the monkey saddle $z = x(x^2 - 3y^2)$. Begin by noting where $z = 0$.

41. The contour map in Figure 24 shows level curves for a mountain 3000 feet high.

(a) What is special about the path to the top labeled AC? What is special about BC?

(b) Make good estimates of the total lengths of path AC and path BC.

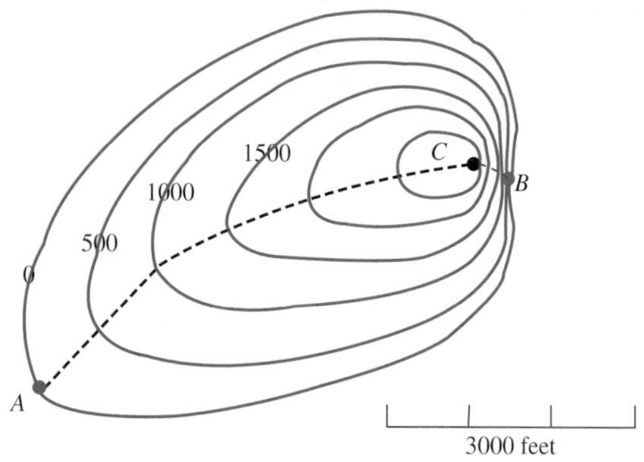

Figure 24

42. Identify the graph of $f(x, y) = x^2 - x + 3y^2 + 12y - 13$, state where it attains its minimum value, and find this minimum value.

CAS *For each of the functions in Problems 43–46, draw the graph and the corresponding contour plot.*

43. $f(x, y) = \sin\sqrt{2x^2 + y^2}; -2 \le x \le 2, -2 \le y \le 2$

44. $f(x, y) = \sin(x^2 + y^2)/(x^2 + y^2), f(0, 0) = 1;$ $-2 \le x \le 2, -2 \le y \le 2$

45. $f(x, y) = (2x - y^2)\exp(-x^2 - y^2); -2 \le x \le 2,$ $-2 \le y \le 2$

46. $f(x, y) = (\sin x \sin y)/(1 + x^2 + y^2); -2 \le x \le 2,$ $-2 \le y \le 2$

Answers to Concepts Review: 1. real-valued function of two real variables **2.** level curve; contour map **3.** concentric circles **4.** parallel lines

12.2 Partial Derivatives

Suppose that f is a function of two variables x and y. If y is held constant, say $y = y_0$, then $f(x, y_0)$ is a function of the single variable x. Its derivative at $x = x_0$ is called the **partial derivative of f with respect to x** at (x_0, y_0) and is denoted by $f_x(x_0, y_0)$. Thus,

$$f_x(x_0, y_0) = \lim_{\Delta x \to 0} \frac{f(x_0 + \Delta x, y_0) - f(x_0, y_0)}{\Delta x}$$

Similarly, the partial derivative of f with respect to y at (x_0, y_0) is denoted by $f_y(x_0, y_0)$ and is given by

$$f_y(x_0, y_0) = \lim_{\Delta y \to 0} \frac{f(x_0, y_0 + \Delta y) - f(x_0, y_0)}{\Delta y}$$

Rather than calculate $f_x(x_0, y_0)$ and $f_y(x_0, y_0)$ directly from the boxed definitions, we typically find $f_x(x, y)$ and $f_y(x, y)$ using the standard rules for derivatives; then we substitute $x = x_0$ and $y = y_0$. The key point here is that the rules for differentiating a function of one variable (Chapter 2) work for finding partial derivatives, as long as we hold one variable fixed.

EXAMPLE 1 Find $f_x(1, 2)$ and $f_y(1, 2)$ if $f(x, y) = x^2y + 3y^3$.

SOLUTION To find $f_x(x, y)$, we treat y as a constant and differentiate with respect to x, obtaining

$$f_x(x, y) = 2xy + 0$$

Thus,

$$f_x(1, 2) = 2 \cdot 1 \cdot 2 = 4$$

Similarly, we treat x as a constant and differentiate with respect to y, obtaining

$$f_y(x, y) = x^2 + 9y^2$$

and so

$$f_y(1, 2) = 1^2 + 9 \cdot 2^2 = 37$$

■

If $z = f(x, y)$, we use the following alternative notations:

$$f_x(x, y) = \frac{\partial z}{\partial x} = \frac{\partial f(x, y)}{\partial x} \qquad f_y(x, y) = \frac{\partial z}{\partial y} = \frac{\partial f(x, y)}{\partial y}$$

$$f_x(x_0, y_0) = \left.\frac{\partial z}{\partial x}\right|_{(x_0, y_0)} \qquad f_y(x_0, y_0) = \left.\frac{\partial z}{\partial y}\right|_{(x_0, y_0)}$$

The symbol ∂ is special to mathematics and is called the partial derivative sign. The symbols $\frac{\partial}{\partial x}$ and $\frac{\partial}{\partial y}$ represent linear operators, much like the linear operators D_x and $\frac{d}{dx}$ that we encountered in Chapter 2.

EXAMPLE 2 If $z = x^2 \sin(xy^2)$, find $\partial z/\partial x$ and $\partial z/\partial y$.

SOLUTION

$$\begin{aligned}\frac{\partial z}{\partial x} &= x^2 \frac{\partial}{\partial x}[\sin(xy^2)] + \sin(xy^2)\frac{\partial}{\partial x}(x^2) \\ &= x^2 \cos(xy^2)\frac{\partial}{\partial x}(xy^2) + \sin(xy^2)\cdot 2x \\ &= x^2\cos(xy^2)\cdot y^2 + 2x\sin(xy^2) \\ &= x^2y^2\cos(xy^2) + 2x\sin(xy^2)\end{aligned}$$

$$\frac{\partial z}{\partial y} = x^2\cos(xy^2)\cdot 2xy = 2x^3y\cos(xy^2)$$

■

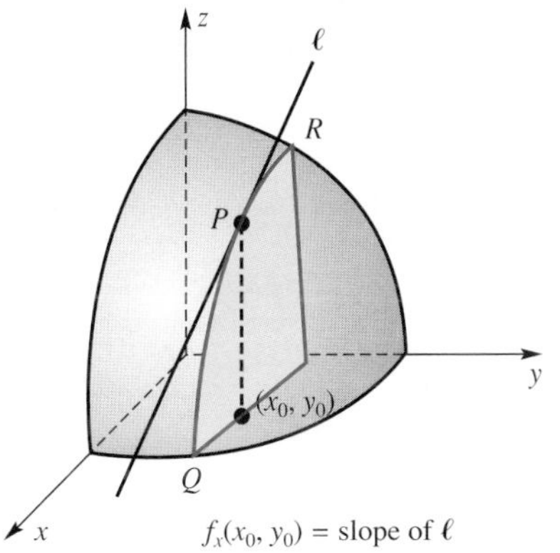

Figure 1

Geometric and Physical Interpretations Consider the surface whose equation is $z = f(x, y)$. The plane $y = y_0$ intersects this surface in the plane curve QPR (Figure 1), and the value of $f_x(x_0, y_0)$ is the slope of the tangent line to this curve at $P(x_0, y_0, f(x_0, y_0))$. Similarly, the plane $x = x_0$ intersects the surface in the plane curve LPM (Figure 2), and $f_y(x_0, y_0)$ is the slope of the tangent line to this curve at P.

Partial derivatives may also be interpreted as (instantaneous) rates of change. Suppose that a violin string is fixed at points A and B and vibrates in the xz-plane. Figure 3 shows the position of the string at a typical time t. If $z = f(x, t)$ denotes the height of the string at the point P with x-coordinate x at time t, then $\partial z/\partial x$ is the slope of the string at P, and $\partial z/\partial t$ is the time rate of change of height of P along the indicated vertical line. In other words, $\partial z/\partial t$ is the vertical velocity of P.

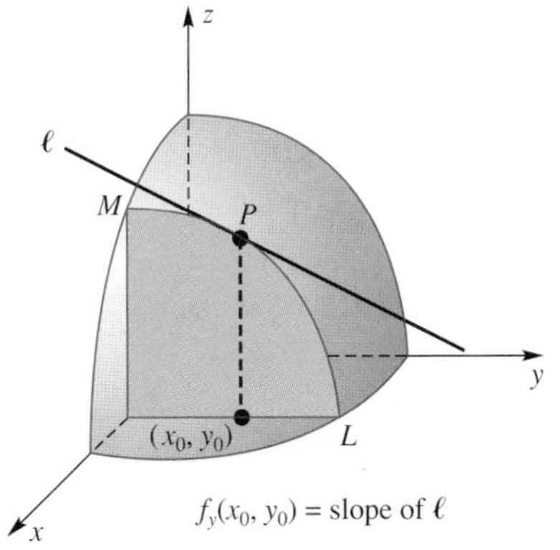

Figure 2

EXAMPLE 3 The surface $z = f(x, y) = \sqrt{9 - 2x^2 - y^2}$ and the plane $y = 1$ intersect in a curve as in Figure 1. Find parametric equations for the tangent line at $(\sqrt{2}, 1, 2)$.

SOLUTION

$$f_x(x, y) = \tfrac{1}{2}(9 - 2x^2 - y^2)^{-1/2}(-4x)$$

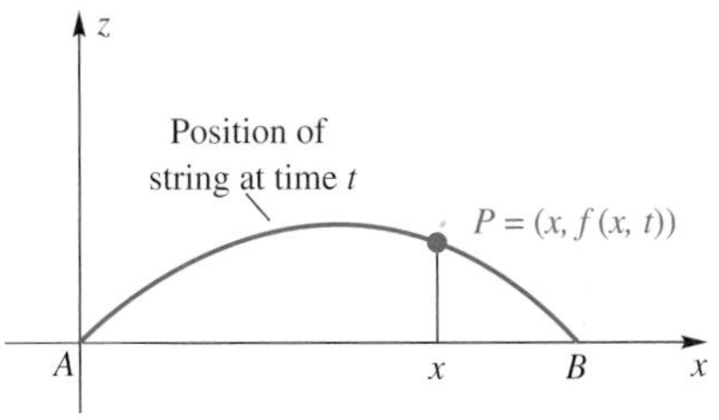

Figure 3

and so $f_x(\sqrt{2}, 1) = -\sqrt{2}$. This number is the slope of the tangent line to the curve at $(\sqrt{2}, 1, 2)$; that is, $-\sqrt{2}/1$ is the ratio of rise to run along the tangent line. It follows that this line has direction vector $\langle 1, 0, -\sqrt{2}\rangle$ and, since it goes through $(\sqrt{2}, 1, 2)$,

$$x = \sqrt{2} + t, \qquad y = 1, \qquad z = 2 - \sqrt{2}t$$

provide the required parametric equations. ■

EXAMPLE 4 The volume of a certain gas is related to its temperature T and its pressure P by the gas law $PV = 10T$, where V is measured in cubic inches, P in pounds per square inch, and T in degrees Kelvin. If V is kept constant at 50, what is the rate of change of pressure with respect to temperature when $T = 200$?

SOLUTION Since $P = 10T/V$,

$$\frac{\partial P}{\partial T} = \frac{10}{V}$$

Thus,

$$\left.\frac{\partial P}{\partial T}\right|_{T=200,\, V=50} = \frac{10}{50} = \frac{1}{5}$$

Thus, the pressure is increasing at the rate of $\frac{1}{5}$ pound per square inch per degree Kelvin. ■

Higher Partial Derivatives Since a partial derivative of a function of x and y is, in general, another function of these same two variables, it may be differentiated partially with respect to either x or y, resulting in four **second partial derivatives** of f.

$$f_{xx} = \frac{\partial}{\partial x}\left(\frac{\partial f}{\partial x}\right) = \frac{\partial^2 f}{\partial x^2} \qquad f_{yy} = \frac{\partial}{\partial y}\left(\frac{\partial f}{\partial y}\right) = \frac{\partial^2 f}{\partial y^2}$$

$$f_{xy} = (f_x)_y = \frac{\partial}{\partial y}\left(\frac{\partial f}{\partial x}\right) = \frac{\partial^2 f}{\partial y\,\partial x} \qquad f_{yx} = (f_y)_x = \frac{\partial}{\partial x}\left(\frac{\partial f}{\partial y}\right) = \frac{\partial^2 f}{\partial x\,\partial y}$$

EXAMPLE 5 Find the four second partial derivatives of

$$f(x, y) = xe^y - \sin(x/y) + x^3y^2$$

SOLUTION

$$f_x(x, y) = e^y - \frac{1}{y}\cos\left(\frac{x}{y}\right) + 3x^2y^2$$

$$f_y(x, y) = xe^y + \frac{x}{y^2}\cos\left(\frac{x}{y}\right) + 2x^3y$$

$$f_{xx}(x, y) = \frac{1}{y^2}\sin\left(\frac{x}{y}\right) + 6xy^2$$

$$f_{yy}(x, y) = xe^y + \frac{x^2}{y^4}\sin\left(\frac{x}{y}\right) - \frac{2x}{y^3}\cos\left(\frac{x}{y}\right) + 2x^3$$

$$f_{xy}(x, y) = e^y - \frac{x}{y^3}\sin\left(\frac{x}{y}\right) + \frac{1}{y^2}\cos\left(\frac{x}{y}\right) + 6x^2y$$

$$f_{yx}(x, y) = e^y - \frac{x}{y^3}\sin\left(\frac{x}{y}\right) + \frac{1}{y^2}\cos\left(\frac{x}{y}\right) + 6x^2y$$ ■

Notice that in Example 5, $f_{xy} = f_{yx}$, which is usually the case for the functions of two variables encountered in a first course. A criterion for this equality will be given in Section 12.3 (Theorem C).

Partial derivatives of the third and higher orders are defined analogously, and the notation for them is similar. Thus, if f is a function of the two variables x and y, the third partial derivative of f obtained by differentiating f partially, first with respect to x and then twice with respect to y, will be indicated by

$$\frac{\partial}{\partial y}\left[\frac{\partial}{\partial y}\left(\frac{\partial f}{\partial x}\right)\right] = \frac{\partial}{\partial y}\left(\frac{\partial^2 f}{\partial y\,\partial x}\right) = \frac{\partial^3 f}{\partial y^2\,\partial x} = f_{xyy}$$

Altogether, there are eight third partial derivatives.

More Than Two Variables Let f be a function of three variables, x, y, and z. The **partial derivative of f with respect to x** at (x, y, z) is denoted by $f_x(x, y, z)$ or $\partial f(x, y, z)/\partial x$ and is defined by

$$f_x(x, y, z) = \lim_{\Delta x \to 0} \frac{f(x + \Delta x, y, z) - f(x, y, z)}{\Delta x}$$

Thus, $f_x(x, y, z)$ may be obtained by treating y and z as constants and differentiating with respect to x.

The partial derivatives with respect to y and z are defined in an analogous way. Partial derivatives of functions of four or more variables are defined similarly (see Problem 49). Partial derivatives, such as f_{xy} and f_{xyz}, that involve differentiation with respect to more than one variable are called **mixed partial derivatives.**

EXAMPLE 6 If $f(x, y, z) = xy + 2yz + 3zx$, find f_x, f_y, and f_z.

SOLUTION To get f_x, we think of y and z as constants and differentiate with respect to the variable x. Thus,

$$f_x(x, y, z) = y + 3z$$

To find f_y, we treat x and z as constants and differentiate with respect to y:

$$f_y(x, y, z) = x + 2z$$

Similarly,

$$f_z(x, y, z) = 2y + 3x$$

■

EXAMPLE 7 If $T(w, x, y, z) = ze^{w^2+x^2+y^2}$, find all first partial derivatives and $\dfrac{\partial^2 T}{\partial w\,\partial x}$, $\dfrac{\partial^2 T}{\partial x\,\partial w}$, and $\dfrac{\partial^2 T}{\partial z^2}$.

SOLUTION The four first partials are

$$\frac{\partial T}{\partial w} = \frac{\partial}{\partial w}\left(ze^{w^2+x^2+y^2}\right) = 2wze^{w^2+x^2+y^2}$$

$$\frac{\partial T}{\partial x} = \frac{\partial}{\partial x}\left(ze^{w^2+x^2+y^2}\right) = 2xze^{w^2+x^2+y^2}$$

$$\frac{\partial T}{\partial y} = \frac{\partial}{\partial y}\left(ze^{w^2+x^2+y^2}\right) = 2yze^{w^2+x^2+y^2}$$

$$\frac{\partial T}{\partial z} = \frac{\partial}{\partial z}\left(ze^{w^2+x^2+y^2}\right) = e^{w^2+x^2+y^2}$$

The other partial derivatives are

$$\frac{\partial^2 T}{\partial w\,\partial x} = \frac{\partial^2}{\partial w\,\partial x}\left(ze^{w^2+x^2+y^2}\right) = \frac{\partial}{\partial w}\left(2xze^{w^2+x^2+y^2}\right) = 4wxze^{w^2+x^2+y^2}$$

$$\frac{\partial^2 T}{\partial x\,\partial w} = \frac{\partial^2}{\partial x\,\partial w}\left(ze^{w^2+x^2+y^2}\right) = \frac{\partial}{\partial x}\left(2wze^{w^2+x^2+y^2}\right) = 4wxze^{w^2+x^2+y^2}$$

$$\frac{\partial^2 T}{\partial z^2} = \frac{\partial^2}{\partial z^2}\left(ze^{w^2+x^2+y^2}\right) = \frac{\partial}{\partial z}\left(e^{w^2+x^2+y^2}\right) = 0$$

■

Concepts Review

1. As a limit, $f_x(x_0, y_0)$ is defined by ______ and is called the ______ at (x_0, y_0).

2. If $f(x, y) = x^3 + xy$, then $f_x(1, 2) =$ ______ and $f_y(1, 2) =$ ______.

3. Another notation for $f_{xy}(x, y)$ is ______.

4. If $f(x, y) = g(x) + h(y)$, then $f_{xy}(x, y) =$ ______.

Problem Set 12.2

In Problems 1–16, find all first partial derivatives of each function.

1. $f(x, y) = (2x - y)^4$

2. $f(x, y) = (4x - y^2)^{3/2}$

3. $f(x, y) = \dfrac{x^2 - y^2}{xy}$

4. $f(x, y) = e^x \cos y$

5. $f(x, y) = e^y \sin x$

6. $f(x, y) = (3x^2 + y^2)^{-1/3}$

7. $f(x, y) = \sqrt{x^2 - y^2}$

8. $f(u, v) = e^{uv}$

9. $g(x, y) = e^{-xy}$

10. $f(s, t) = \ln(s^2 - t^2)$

11. $f(x, y) = \tan^{-1}(4x - 7y)$

12. $F(w, z) = w \sin^{-1}\left(\dfrac{w}{z}\right)$

13. $f(x, y) = y \cos(x^2 + y^2)$

14. $f(s, t) = e^{t^2 - s^2}$

15. $F(x, y) = 2 \sin x \cos y$

16. $f(r, \theta) = 3r^3 \cos 2\theta$

In Problems 17–20, verify that

$$\frac{\partial^2 f}{\partial y\, \partial x} = \frac{\partial^2 f}{\partial x\, \partial y}$$

17. $f(x, y) = 2x^2y^3 - x^3y^5$

18. $f(x, y) = (x^3 + y^2)^5$

19. $f(x, y) = 3e^{2x} \cos y$

20. $f(x, y) = \tan^{-1} xy$

21. If $F(x, y) = \dfrac{2x - y}{xy}$, find $F_x(3, -2)$ and $F_y(3, -2)$.

22. If $F(x, y) = \ln(x^2 + xy + y^2)$, find $F_x(-1, 4)$ and $F_y(-1, 4)$.

23. If $f(x, y) = \tan^{-1}(y^2/x)$, find $f_x(\sqrt{5}, -2)$ and $f_y(\sqrt{5}, -2)$.

24. If $f(x, y) = e^y \cosh x$, find $f_x(-1, 1)$ and $f_y(-1, 1)$.

25. Find the slope of the tangent to the curve of intersection of the surface $36z = 4x^2 + 9y^2$ and the plane $x = 3$ at the point $(3, 2, 2)$.

26. Find the slope of the tangent to the curve of intersection of the surface $3z = \sqrt{36 - 9x^2 - 4y^2}$ and the plane $x = 1$ at the point $(1, -2, \sqrt{11}/3)$.

27. Find the slope of the tangent to the curve of intersection of the surface $2z = \sqrt{9x^2 + 9y^2 - 36}$ and the plane $y = 1$ at the point $(2, 1, \frac{3}{2})$.

28. Find the slope of the tangent to the curve of intersection of the cylinder $4z = 5\sqrt{16 - x^2}$ and the plane $y = 3$ at the point $(2, 3, 5\sqrt{3}/2)$.

29. The volume V of a right circular cylinder is given by $V = \pi r^2 h$, where r is the radius and h is the height. If h is held fixed at $h = 10$ inches, find the rate of change of V with respect to r when $r = 6$ inches.

30. The temperature in degrees Celsius on a metal plate in the xy-plane is given by $T(x, y) = 4 + 2x^2 + y^3$. What is the rate of change of temperature with respect to distance (measured in feet) if we start moving from $(3, 2)$ in the direction of the positive y-axis?

31. According to the ideal gas law, the pressure, temperature, and volume of a gas are related by $PV = kT$, where k is a constant. Find the rate of change of pressure (pounds per square inch) with respect to temperature when the temperature is 300°K if the volume is kept fixed at 100 cubic inches.

32. Show that, for the gas law of Problem 31,

$$V\frac{\partial P}{\partial V} + T\frac{\partial P}{\partial T} = 0 \quad \text{and} \quad \frac{\partial P}{\partial V}\frac{\partial V}{\partial T}\frac{\partial T}{\partial P} = -1$$

A function of two variables that satisfies **Laplace's Equation,**

$$\frac{\partial^2 f}{\partial x^2} + \frac{\partial^2 f}{\partial y^2} = 0$$

is said to be **harmonic.** *Show that the functions defined in Problems 33 and 34 are harmonic functions.*

33. $f(x, y) = x^3y - xy^3$

34. $f(x, y) = \ln(4x^2 + 4y^2)$

35. If $F(x, y) = 3x^4y^5 - 2x^2y^3$, find $\partial^3 F(x, y)/\partial y^3$.

36. If $f(x, y) = \cos(2x^2 - y^2)$, find $\partial^3 f(x, y)/\partial y\, \partial x^2$.

37. Express the following in ∂ notation.

(a) f_{yyy} (b) f_{xxy} (c) f_{xyyy}

38. Express the following in subscript notation.

(a) $\dfrac{\partial^3 f}{\partial x^2\, \partial y}$ (b) $\dfrac{\partial^4 f}{\partial x^2\, \partial y^2}$ (c) $\dfrac{\partial^5 f}{\partial x^3\, \partial y^2}$

39. If $f(x, y, z) = 3x^2y - xyz + y^2z^2$, find each of the following:

(a) $f_x(x, y, z)$ (b) $f_y(0, 1, 2)$ (c) $f_{xy}(x, y, z)$

40. If $f(x, y, z) = (x^3 + y^2 + z)^4$, find each of the following:

(a) $f_x(x, y, z)$ (b) $f_y(0, 1, 1)$ (c) $f_{zz}(x, y, z)$

41. If $f(x, y, z) = e^{-xyz} - \ln(xy - z^2)$, find $f_x(x, y, z)$.

42. If $f(x, y, z) = (xy/z)^{1/2}$, find $f_x(-2, -1, 8)$.

43. A bee was flying upward along the curve that is the intersection of $z = x^4 + xy^3 + 12$ with the plane $x = 1$. At the point $(1, -2, 5)$, it went off on the tangent line. Where did the bee hit the xz-plane? (See Example 3.)

44. Let $A(x, y)$ be the area of a nondegenerate rectangle of dimensions x and y, the rectangle being inside a circle of radius 10. Determine the domain and range for this function.

45. The interval $[0, 1]$ is to be separated into three pieces by making cuts at x and y. Let $A(x, y)$ be the area of any nondegenerate triangle that can be formed from these three pieces. Determine the domain and range for this function.

46. The **wave equation** $c^2\, \partial^2 u/\partial x^2 = \partial^2 u/\partial t^2$ and the **heat equation** $c\, \partial^2 u/\partial x^2 = \partial u/\partial t$ are two of the most important equations in physics (c is a constant). These are called **partial differential equations**. Show each of the following:

(a) $u = \cos x \cos ct$ and $u = e^x \cosh ct$ satisfy the wave equation.

(b) $u = e^{-ct} \sin x$ and $u = t^{-1/2}e^{-x^2/(4ct)}$ satisfy the heat equation.

47. For the contour map for $z = f(x, y)$ shown in Figure 4, estimate each value.

(a) $f_y(1, 1)$ (b) $f_x(-4, 2)$
(c) $f_x(-5, -2)$ (d) $f_y(0, -2)$

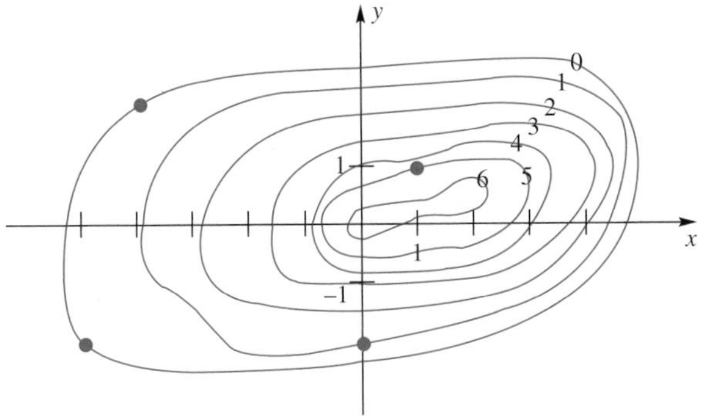

Figure 4

CAS **48.** A CAS can be used to calculate and graph partial derivatives. Draw the graphs of each of the following:

(a) $\sin(x + y^2)$ (b) $D_x \sin(x + y^2)$
(c) $D_y \sin(x + y^2)$ (d) $D_x(D_y \sin(x + y^2))$

49. Give definitions in terms of limits for the following partial derivatives:

(a) $f_y(x, y, z)$ (b) $f_z(x, y, z)$
(c) $G_x(w, x, y, z)$ (d) $\frac{\partial}{\partial z}\lambda(x, y, z, t)$
(e) $\frac{\partial}{\partial b_2} S(b_0, b_1, b_2, \ldots, b_n)$

50. Find each partial derivative.

(a) $\frac{\partial}{\partial w}(\sin w \sin x \cos y \cos z)$ (b) $\frac{\partial}{\partial x}[x \ln(wxyz)]$
(c) $\lambda_t(x, y, z, t)$, where $\lambda(x, y, z, t) = \frac{t \cos x}{1 + xyzt}$

Answers to Concepts Review:
1. $\lim_{\Delta x \to 0} [f(x_0 + \Delta x, y_0) - f(x_0, y_0)]/\Delta x$; partial derivative of f with respect to x **2.** 5; 1 **3.** $\partial^2 f/\partial y\, \partial x$ **4.** 0

12.3 Limits and Continuity

Our aim in this section is to give meaning to the statement

$$\lim_{(x,y)\to(a,b)} f(x, y) = L$$

It may seem odd that we covered partial derivatives before limits for functions of two or more variables. After all, we covered limits in Chapter 1 and derivatives in Chapter 2. However, partial differentiation is actually a simpler idea because all variables but one are held fixed. The only concept necessary for defining the partial derivative is that of the limit of a function of one variable, which goes back to Chapter 1. On the other hand, the limit of a function of two (or more) variables is a deeper concept because we must account for *all* ways that (x, y) approaches (a, b). This cannot be reduced to treating "one variable at a time" like partial differentiation.

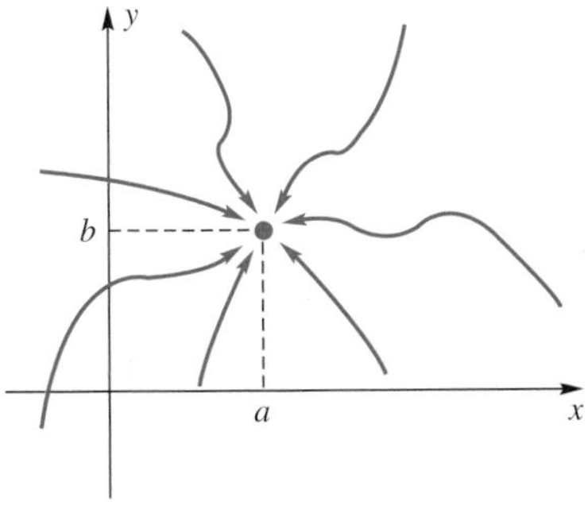

Figure 1

The limit of a function of two variables has the usual intuitive meaning: The values of $f(x, y)$ get closer and closer to the number L as (x, y) approaches (a, b). The problem is that (x, y) can approach (a, b) in infinitely many ways (Figure 1). We want a definition that gives the same L no matter what path (x, y) takes in approaching (a, b). Fortunately, the formal definition given first for real-valued functions of one variable (Section 1.1) and then for vector-valued functions (Section 11.5) are similar to what we need here.

Definition Limit of a Function of Two Variables

To say that $\lim_{(x,y)\to(a,b)} f(x, y) = L$ means that for each $\varepsilon > 0$ (no matter how small) there is a corresponding $\delta > 0$ such that $|f(x, y) - L| < \varepsilon$, provided that $0 < \|(x, y) - (a, b)\| < \delta$.

To interpret $\|(x, y) - (a, b)\|$, think of (x, y) and (a, b) as vectors. Then

$$\|(x, y) - (a, b)\| = \sqrt{(x - a)^2 + (y - b)^2}$$

and the points satisfying $0 < \|(x, y) - (a, b)\| < \delta$ are those points inside a circle of radius δ, excluding the center (a, b) (see Figure 2). The essence of the definition is this: We can make $f(x, y)$ as close as we like to L (within ε, where distance is measured by $|f(x, y) - L|$) as long as we take (x, y) sufficiently close to (a, b)

(within δ, with distance being measured by $\|(x, y) - (a, b)\|$). Compare this definition to the definition of limit given in Chapter 1 and the definition of a vector-valued function given in Chapter 11; the similarities will be obvious.

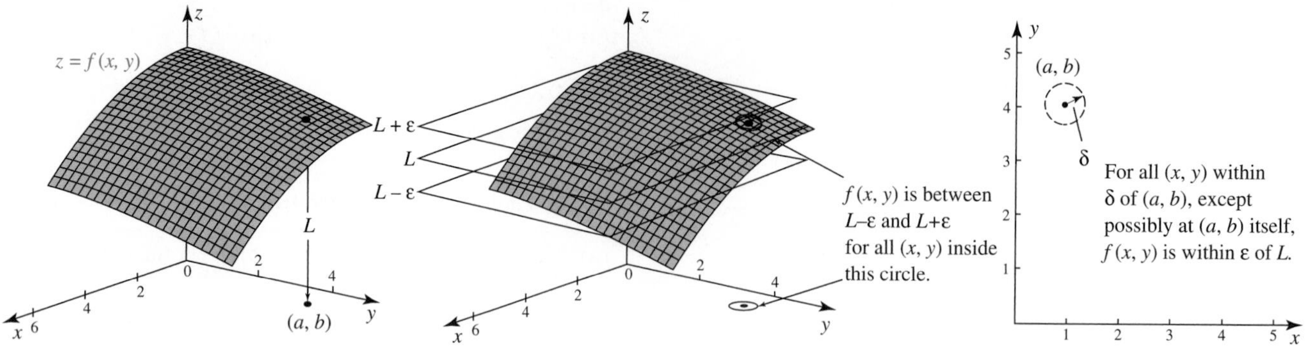

Figure 2

Note several aspects of this definition.

1. The path of approach to (a, b) is irrelevant. This means that if different paths of approach lead to different L-values then the limit does not exist.
2. The behavior of $f(x, y)$ at (a, b) is irrelevant; the function does not even have to be defined at (a, b). This follows from the restriction $0 < \|(x, y) - (a, b)\|$.
3. The definition is phrased so that it immediately extends to functions of three (or more) variables. Simply replace (x, y) and (a, b) by (x, y, z) and (a, b, c) wherever they occur.

We might expect that limits for many functions can be obtained by substitution. This was true for many (but certainly not all) functions of one variable. Before we state a theorem that justifies evaluating limits by substitution, we give a few definitions. A **polynomial** in the variables x and y is a function of the form

$$f(x, y) = \sum_{i=1}^{n}\sum_{j=1}^{m} c_{ij}x^i y^j$$

and a **rational function** in the variables x and y is a function of the form

$$f(x, y) = \frac{p(x, y)}{q(x, y)}$$

where p and q are polynomials in x and y, assuming q is not identically zero. The following theorem is analogous to Theorem 1.3B.

Theorem A

If $f(x, y)$ is a polynomial, then

$$\lim_{(x,y)\to(a,b)} f(x, y) = f(a, b)$$

and if $f(x, y) = p(x, y)/q(x, y)$, where p and q are polynomials, then

$$\lim_{(x,y)\to(a,b)} f(x, y) = \frac{p(a, b)}{q(a, b)}$$

provided $q(a, b) \neq 0$. Furthermore, if

$$\lim_{(x,y)\to(a,b)} p(x, y) = L \neq 0 \quad \text{and} \quad \lim_{(x,y)\to(a,b)} q(x, y) = 0$$

then

$$\lim_{(x,y)\to(a,b)} \frac{p(x, y)}{q(x, y)}$$

does not exist.

EXAMPLE 1 Evaluate the following limits if they exist:

$$\text{(a)} \lim_{(x,y)\to(1,2)} (x^2y + 3y) \quad \text{and} \quad \text{(b)} \lim_{(x,y)\to(0,0)} \frac{x^2 + y^2 + 1}{x^2 - y^2}$$

SOLUTION

(a) The function whose limit we seek is a polynomial, so by Theorem A

$$\lim_{(x,y)\to(1,2)} (x^2y + 3y) = 1^2 \cdot 2 + 3 \cdot 2 = 8$$

(b) The second function is a rational function, but the limit of the denominator is equal to 0, while the limit of the numerator is 1. Thus, by Theorem A, this limit does not exist. ■

EXAMPLE 2 Show that the function f defined by

$$f(x, y) = \frac{x^2 - y^2}{x^2 + y^2}$$

has no limit at the origin (Figure 3).

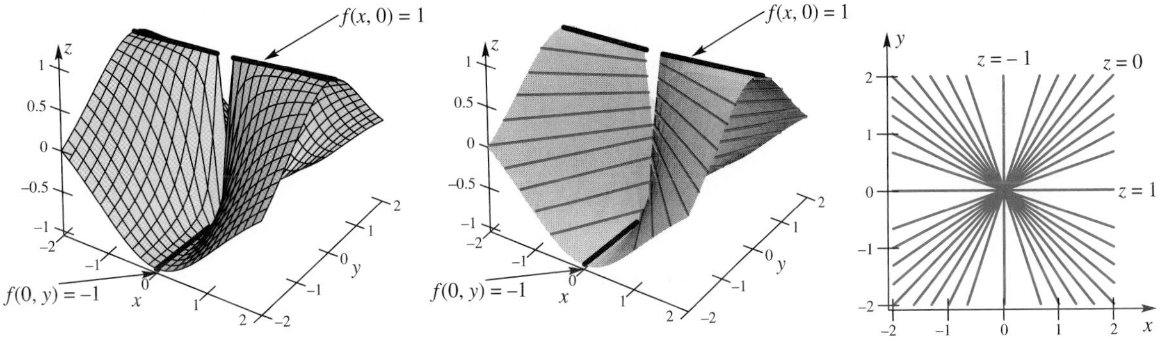

Figure 3

SOLUTION The function f is defined everywhere in the xy-plane except at the origin. At all points on the x-axis different from the origin, the value of f is

$$f(x, 0) = \frac{x^2 - 0}{x^2 + 0} = 1$$

Thus, the limit of $f(x, y)$ as (x, y) approaches $(0, 0)$ along the x-axis is

$$\lim_{(x,0)\to(0,0)} f(x, 0) = \lim_{(x,0)\to(0,0)} \frac{x^2 - 0}{x^2 + 0} = +1$$

Similarly, the limit of $f(x, y)$ as (x, y) approaches $(0, 0)$ along the y-axis is

$$\lim_{(0,y)\to(0,0)} f(0, y) = \lim_{(0,y)\to(0,0)} \frac{0 - y^2}{0 + y^2} = -1$$

Thus, we get different values depending on how $(x, y) \to (0, 0)$. In fact, there are points arbitrarily close to $(0, 0)$ at which the value of f is 1 and other points equally close at which the value of f is -1. Therefore, the limit cannot exist at $(0, 0)$. ■

Polar Coordinates for Example 2

We can use polar coordinates to show that the limit in Example 2 doesn't exist.

$$\lim_{(x,y)\to(0,0)} \frac{x^2 - y^2}{x^2 + y^2}$$

$$= \lim_{r\to 0} \frac{r^2\cos^2\theta - r^2\sin^2\theta}{r^2}$$

$$= \lim_{r\to 0} \cos 2\theta$$

$$= \cos 2\theta$$

which takes on all values between -1 and 1 in every neighborhood of $(0, 0)$. We conclude that the limit does not exist.

It is often easier to analyze limits of functions of two variables, especially limits at the origin, by changing to polar coordinates. The important point is that $(x, y) \to (0, 0)$ if and only if $r = \sqrt{x^2 + y^2} \to 0$. Thus, limits for functions of two variables can sometimes be expressed as limits involving just one variable, r.

EXAMPLE 3 Evaluate the following limits if they exist:

$$\text{(a)} \lim_{(x,y)\to(0,0)} \frac{\sin(x^2 + y^2)}{3x^2 + 3y^2} \quad \text{and} \quad \text{(b)} \lim_{(x,y)\to(0,0)} \frac{xy}{x^2 + y^2}$$

SOLUTION

(a) Changing to polar coordinates and using L'Hôpital's Rule, we have

$$\lim_{(x,y)\to(0,0)} \frac{\sin(x^2 + y^2)}{3x^2 + 3y^2} = \lim_{r\to 0} \frac{\sin r^2}{3r^2} = \frac{1}{3}\lim_{r\to 0} \frac{2r\cos r^2}{2r} = \frac{1}{3}$$

(b) Again, changing to polar coordinates gives

$$\lim_{(x,y)\to(0,0)} \frac{xy}{x^2 + y^2} = \lim_{r\to 0} \frac{r\cos\theta\, r\sin\theta}{r^2} = \cos\theta\sin\theta$$

Since this limit depends on θ, straight line paths to the origin will lead to different limits. Thus, this limit does not exist. ■

Continuity at a Point To say that $f(x, y)$ is **continuous** at the point (a, b), we require the following: (1) f has a value at (a, b), (2) f has a limit at (a, b), and (3) the value of f at (a, b) is equal to the limit there. In summary, we require that

$$\lim_{(x,y)\to(a,b)} f(x, y) = f(a, b)$$

This is essentially the same requirement for continuity of a function of one variable. Intuitively, this again means that f has no jumps, wild fluctuations, or unbounded behavior at (a, b).

Theorem A can be used to say that polynomial functions are continuous for all (x, y) and that rational functions are continuous everywhere except where the denominator is equal to 0. Furthermore, sums, differences, products, and quotients of continuous functions are continuous (provided, in the latter case that we avoid division by 0). These results, along with the next theorem, can be used to establish the continuity of many functions of two variables.

Theorem B Composition of Functions

If a function g of two variables is continuous at (a, b) and a function f of one variable is continuous at $g(a, b)$, then the composite function $f \circ g$, defined by $(f \circ g)(x, y) = f(g(x, y))$, is continuous at (a, b).

The proof of this theorem is similar to the proof of Theorem 1.6E.

EXAMPLE 4 Describe the points (x, y) for which the following functions are continuous.

(a) $H(x, y) = \dfrac{2x + 3y}{y - 4x^2}$, (b) $F(x, y) = \cos(x^3 - 4xy + y^2)$

SOLUTION

(a) $H(x, y)$ is a rational function, so it is continuous at every point where the denominator is not 0. The denominator, $y - 4x^2$ is equal to zero along the parabola $y = 4x^2$. Thus, $H(x, y)$ is continuous for all (x, y) except those along the parabola $y = 4x^2$.

(b) The function $g(x, y) = x^3 - 4xy + y^2$, being a polynomial, is continuous for all (x, y). Also, $f(t) = \cos t$ is continuous for every real number t. We conclude from Theorem B that $F(x, y)$ is continuous for all (x, y). ■

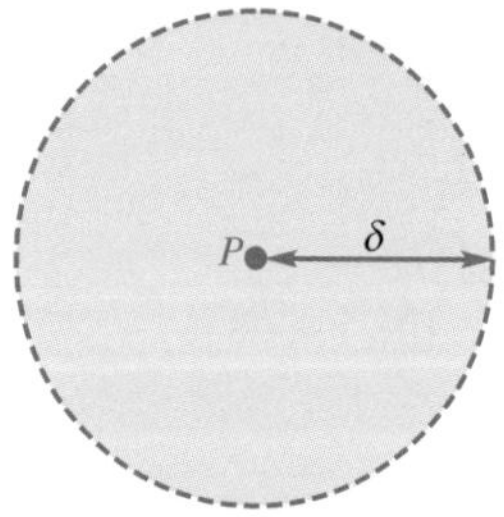

A neighborhood in two-space

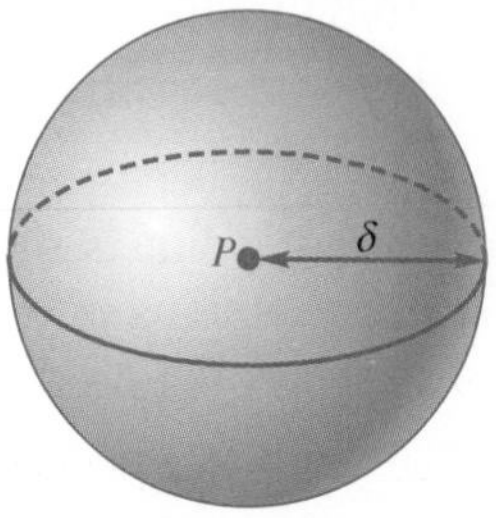

A neighborhood in three-space

Figure 4

Continuity on a Set To say that $f(x, y)$ is continuous on a set S ought to mean that $f(x, y)$ is continuous at every point of the set. It does mean that, but there are some subtleties connected with this statement that need to be cleared up.

First we need to introduce some language relative to sets in the plane (and higher-dimensional spaces). By a **neighborhood** of radius δ of a point P, we mean the set of all points Q satisfying $\|Q - P\| < \delta$. In two-space, a neighborhood is the "inside" of a circle; in three-space, it is the inside of a sphere (Figure 4). A point P

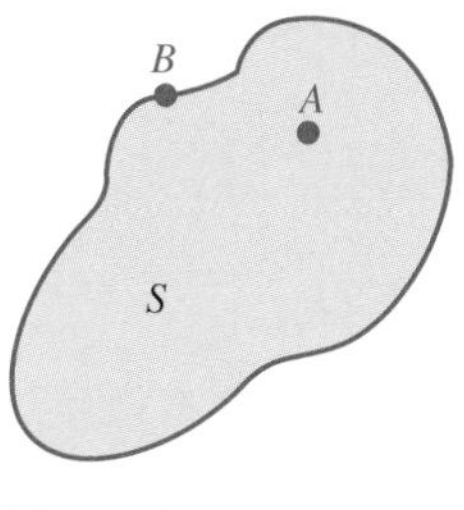

Figure 5

is an **interior point** of a set S if there is a neighborhood of P contained in S. The set of all interior points of S is the **interior** of S. On the other hand, P is a **boundary point** of S if every neighborhood of P contains points that are in S and points that are not in S. The set of all boundary points of S is called the **boundary** of S. In Figure 5, A is an interior point and B is a boundary point of S. A set is **open** if all its points are interior points, and it is **closed** if it contains all its boundary points. It is possible for a set to be neither open nor closed. This, incidentally, explains the use of "open intervals" and "closed intervals" in one-dimensional space. Finally, a set S is **bounded** if there exists an $R > 0$ such that all ordered pairs in S are inside a circle of radius R centered at the origin.

If S is an open set, to say that f is continuous on S means precisely that f is continuous at every point of S. On the other hand, if S contains some or all of its boundary points, we must be careful to give the right interpretation of continuity at such points (recall that in one-space we had to talk about left and right continuity at the end points of an interval). To say that f is continuous at a boundary point P of S means that $f(Q)$ must approach $f(P)$ as Q approaches P through points of S.

Here is an example that will help to clarify what we have said (see Figure 6). Let

$$f(x, y) = \begin{cases} 0 & \text{if } x^2 + y^2 \le 1 \\ 4 & \text{otherwise} \end{cases}$$

If S is the set $\{(x, y): x^2 + y^2 \le 1\}$, it is correct to say that $f(x, y)$ is continuous on S. On the other hand, it would be incorrect to say that $f(x, y)$ is continuous on the whole plane.

We said in Section 12.2 that for most functions of two variables studied in a first course $f_{xy} = f_{yx}$; that is, the order of differentiation in mixed partial derivatives is immaterial. Now that continuity is defined, conditions for this to be true can be simply stated.

> **The Boundary of a Set**
>
> If you are standing on the boundary (i.e., the border) between the United States and Canada, then you can reach into both countries, *no matter how short your reach is*. This is the essence of our definition of a boundary point. Any neighborhood (your reach) of a boundary point will include points in S and points outside of S, no matter how small the neighborhood is.
>
> The boundary of a set will play an important role later in this chapter when we consider optimization of functions, and in Chapter 13 and 14 when we study multiple integrals.

Theorem C Equality of Mixed Partials

If f_{xy} and f_{yx} are continuous on an open set S, then $f_{xy} = f_{yx}$ at each point of S.

A proof of this theorem is given in books on advanced calculus. A counterexample for which continuity of f_{xy} is lacking is given in Problem 42.

Our discussion of continuity has dealt mainly with functions of two variables. We believe you can make the simple changes that are required to describe continuity for functions of three or more variables.

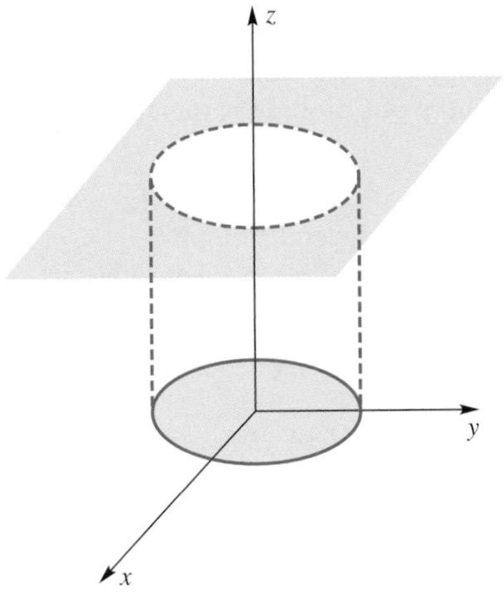

Figure 6

Concepts Review

1. In intuitive language, to say that $\lim_{(x,y)\to(1,2)} f(x, y) = 3$ means that $f(x, y)$ gets close to _____ when _____.

2. For $f(x, y)$ to be continuous at (1, 2) means that _____.

3. The point P is an interior point of set S if there is a neighborhood of P that is _____.

4. The set S is open if every point of S is _____; S is closed if S contains all its _____.

Problem Set 12.3

In Problems 1–16, find the indicated limit or state that it does not exist.

1. $\lim_{(x,y)\to(1,3)} (3x^2y - xy^3)$

2. $\lim_{(x,y)\to(-2,1)} (xy^3 - xy + 3y^2)$

3. $\lim_{(x,y)\to(2,\pi)} [x\cos^2(xy) - \sin(xy/3)]$

4. $\lim_{(x,y)\to(1,2)} \dfrac{x^3 - 3x^2y + 3xy^2 - y^3}{y - 2x^2}$

5. $\lim_{(x,y)\to(-1,2)} \dfrac{xy - y^3}{(x + y + 1)^2}$

6. $\lim_{(x,y)\to(0,0)} \dfrac{xy + \cos x}{xy - \cos x}$

7. $\lim_{(x,y)\to(0,0)} \dfrac{\sin(x^2 + y^2)}{x^2 + y^2}$

8. $\lim_{(x,y)\to(0,0)} \dfrac{\tan(x^2 + y^2)}{x^2 + y^2}$

9. $\lim_{(x,y)\to(0,0)} \dfrac{x^2 + y^2}{x^4 - y^4}$

10. $\lim_{(x,y)\to(0,0)} \dfrac{x^4 - y^4}{x^2 + y^2}$

11. $\lim_{(x,y)\to(0,0)} \dfrac{xy}{\sqrt{x^2 + y^2}}$

12. $\lim_{(x,y)\to(0,0)} \dfrac{xy}{(x^2 + y^2)^2}$

13. $\lim_{(x,y)\to(0,0)} \dfrac{x^{7/3}}{x^2 + y^2}$

14. $\lim_{(x,y)\to(0,0)} xy\dfrac{x^2 - y^2}{x^2 + y^2}$

15. $\lim_{(x,y)\to(0,0)} \dfrac{x^2y^2}{x^2 + y^4}$

16. $\lim_{(x,y)\to(0,0)} \dfrac{xy^2}{x^2 + y^4}$

In Problems 17–26, describe the largest set S on which it is correct to say that f is continuous.

17. $f(x, y) = \dfrac{x^2 + xy - 5}{x^2 + y^2 + 1}$

18. $f(x, y) = \ln(1 + x^2 + y^2)$

19. $f(x, y) = \ln(1 - x^2 - y^2)$

20. $f(x, y) = \dfrac{1}{\sqrt{1 + x + y}}$

21. $f(x, y) = \dfrac{x^2 + 3xy + y^2}{y - x^2}$

22. $f(x, y) = \begin{cases} \dfrac{\sin(xy)}{xy}, & \text{if } xy \neq 0 \\ 1, & \text{if } xy = 0 \end{cases}$

23. $f(x, y) = \sqrt{x - y + 1}$

24. $f(x, y) = (4 - x^2 - y^2)^{-1/2}$

25. $f(x, y, z) = \dfrac{1 + x^2}{x^2 + y^2 + z^2}$

26. $f(x, y, z) = \ln(4 - x^2 - y^2 - z^2)$

In Problems 27–32, sketch the indicated set. Describe the boundary of the set. Finally, state whether the set is open, closed, or neither.

27. $\{(x, y): 2 \le x \le 4, 1 \le y \le 5\}$

28. $\{(x, y): x^2 + y^2 < 4\}$

29. $\{(x, y): 0 < x^2 + y^2 \le 1\}$

30. $\{(x, y): 1 < x \le 4\}$

31. $\{(x, y): x > 0, y < \sin(1/x)\}$

32. $\{(x, y): x = 0, y = 1/n, n \text{ a positive integer}\}$

33. Let

$$f(x, y) = \begin{cases} \dfrac{x^2 - 4y^2}{x - 2y}, & \text{if } x \neq 2y \\ g(x), & \text{if } x = 2y \end{cases}$$

If f is continuous in the whole plane, find a formula for $g(x)$.

34. Prove that

$$\lim_{(x,y)\to(a,b)} [f(x, y) + g(x, y)] = \lim_{(x,y)\to(a,b)} f(x, y) + \lim_{(x,y)\to(a,b)} g(x, y)$$

provided that the latter two limits exist.

35. Show that

$$\lim_{(x,y)\to(0,0)} \frac{xy}{x^2 + y^2}$$

does not exist by considering one path to the origin along the x-axis and another path along the line $y = x$.

36. Show that

$$\lim_{(x,y)\to(0,0)} \frac{xy + y^3}{x^2 + y^2}$$

does not exist.

37. Let $f(x, y) = x^2y/(x^4 + y^2)$.

(a) Show that $f(x, y) \to 0$ as $(x, y) \to (0, 0)$ along any straight line $y = mx$.

(b) Show that $f(x, y) \to \frac{1}{2}$ as $(x, y) \to (0, 0)$ along the parabola $y = x^2$.

(c) What conclusion do you draw?

38. Let $f(x, y)$ be the shortest distance that a raindrop landing at latitude x and longitude y in the state of Colorado must travel to reach an ocean. Where in Colorado is this function discontinuous?

39. Let H be the hemispherical shell $x^2 + y^2 + (z - 1)^2 = 1, 0 \le z < 1$, shown in Figure 7, and let $D = \{(x, y, z): 1 \le z \le 2\}$. For each function defined below, determine its set of discontinuities within D.

(a) $f(x, y, z)$ is the time required for a particle dropped from (x, y, z) to reach the level $z = 0$.

(b) $f(x, y, z)$ is the area of the inside of H (assumed opaque) that can be seen from (x, y, z).

(c) $f(x, y, z)$ is the area of the shadow of H on the xy-plane due to a point light source at (x, y, z).

(d) $f(x, y, z)$ is the distance along the shortest path from (x, y, z) to $(0, 0, 0)$ that does not penetrate H.

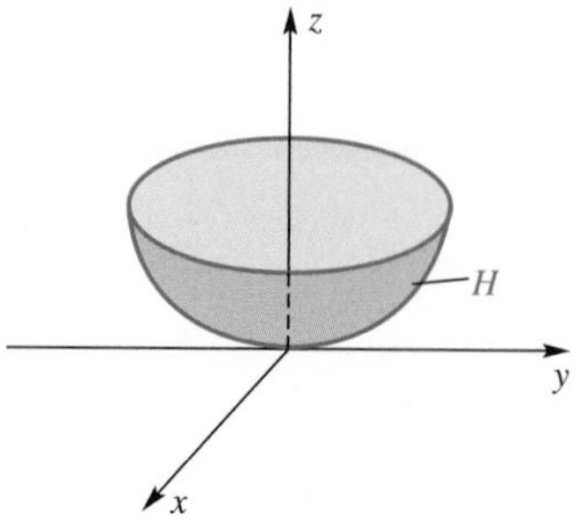

Figure 7

40. Let f, a function of n variables, be continuous on an open set D, and suppose that P_0 is in D with $f(P_0) > 0$. Prove that there is a $\delta > 0$ such that $f(P) > 0$ in a neighborhood of P_0 with radius δ.

41. The French Railroad Suppose that Paris is located at the origin of the xy-plane. Rail lines emanate from Paris along all rays, and these are the only rail lines. Determine the set of discontinuities of the following functions.

(a) $f(x, y)$ is the distance from (x, y) to $(1, 0)$ on the French railroad.

(b) $g(u, v, x, y)$ is the distance from (u, v) to (x, y) on the French railroad.

42. Let $f(x, y) = xy\dfrac{x^2 - y^2}{x^2 + y^2}$ if $(x, y) \neq (0, 0)$ and $f(0, 0) = 0$.

Show that $f_{xy}(0, 0) \neq f_{yx}(0, 0)$ by completing the following steps:

(a) Show that $f_x(0, y) = \lim_{h \to 0} \dfrac{f(0 + h, y) - f(0, y)}{h} = -y$ for all y.

(b) Similarly, show that $f_y(x, 0) = x$ for all x.

(c) Show that $f_{yx}(0, 0) = \lim_{h \to 0} \dfrac{f_y(0 + h, 0) - f_y(0, 0)}{h} = 1$.

(d) Similarly, show that $f_{xy}(0, 0) = -1$.

CAS **43.** Plot the graph of the function mentioned in Problem 42. Do you see why this surface is sometimes called the *dog saddle*?

CAS **44.** Plot the graphs of each of the following functions on $-2 \le x \le 2$, $-2 \le y \le 2$, and determine where on this set they are discontinuous.

(a) $f(x, y) = x^2/(x^2 + y^2)$, $f(0, 0) = 0$

(b) $f(x, y) = \tan(x^2 + y^2)/(x^2 + y^2)$, $f(0, 0) = 0$

CAS **45.** Plot the graph of $f(x, y) = x^2y/(x^4 + y^2)$ in an orientation that illustrates its unusual characteristics (see Problem 37).

46. Give definitions of continuity at a point and continuity on a set for a function of three variables.

47. Show that the function defined by

$$f(x, y, z) = \frac{xyz}{x^3 + y^3 + z^3} \quad \text{for } (x, y, z) \neq (0, 0, 0)$$

and $f(0, 0, 0) = 0$ is not continuous at $(0, 0, 0)$.

48. Show that the function defined by

$$f(x, y, z) = (y + 1)\frac{x^2 - z^2}{x^2 + z^2} \quad \text{for } (x, y, z) \neq (0, 0, 0)$$

and $f(0, 0, 0) = 0$ is not continuous at $(0, 0, 0)$.

Answers to Concepts Review: **1.** 3; (x, y) approaches $(1, 2)$ **2.** $\lim_{(x,y) \to (1,2)} f(x, y) = f(1, 2)$ **3.** contained in S **4.** an interior point of S; boundary points

---

12.4 Differentiability

For a function of a single variable, differentiability of f at x meant the existence of the derivative $f'(x)$. This, in turn, was equivalent to the graph of f having a nonvertical tangent line at x.

Now we ask: What is the right concept of differentiability for a function of two variables? Surely it must correspond in a natural way to the existence of a tangent plane, and clearly this requires more than the mere existence of the partial derivatives of f, for they reflect the behavior of f in only two directions. To emphasize this point, consider

$$f(x, y) = -10\sqrt{|xy|}$$

which is shown in Figure 1. Note that $f_x(0, 0)$ and $f_y(0, 0)$ both exist and equal 0; yet no one would claim that the graph has a tangent plane at the origin. The reason is, of course, that the graph of f is not well approximated there by any plane (in particular, the xy-plane) except in two directions. A tangent plane ought to approximate the graph very well in all directions.

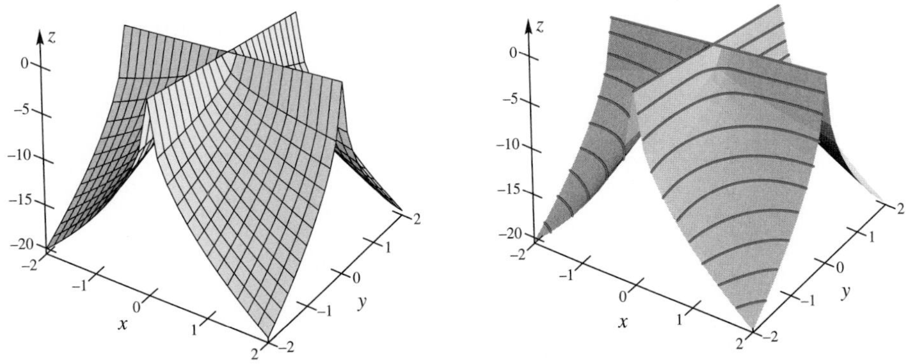

Figure 1

$z = -10\sqrt{|xy|}$

Consider a second question. What plays the role of the derivative for a function of two variables? Again the partial derivatives fall short, if for no other reason than because there are two of them.

To answer these two questions, we start by downplaying the distinction between the point (x, y) and the vector $\langle x, y \rangle$. Thus, we write $\mathbf{p} = (x, y) = \langle x, y \rangle$ and $f(\mathbf{p}) = f(x, y)$. Recall that

$$(1) \qquad f'(a) = \lim_{x \to a} \frac{f(x) - f(a)}{x - a} = \lim_{h \to 0} \frac{f(a + h) - f(a)}{h}$$

The analog would seem to be

$$(2) \qquad f'(\mathbf{p}_0) = \lim_{\mathbf{p} \to \mathbf{p}_0} \frac{f(\mathbf{p}) - f(\mathbf{p}_0)}{\mathbf{p} - \mathbf{p}_0} = \lim_{\mathbf{h} \to \mathbf{0}} \frac{f(\mathbf{p}_0 + \mathbf{h}) - f(\mathbf{p}_0)}{\mathbf{h}}$$

but, unfortunately, the division by a vector makes no sense.

But let us not give up too quickly. Another way to look at differentiability of a function of a single variable is as follows. If f is differentiable at a, then there exists a tangent line through $(a, f(a))$ that approximates the function for values of x near a. In other words, f is almost *linear* near a. Figure 2 illustrates this for a function of a single variable; as we zoom in on the graph of $y = f(x)$, we see that the tangent line and the function become almost indistinguishable.

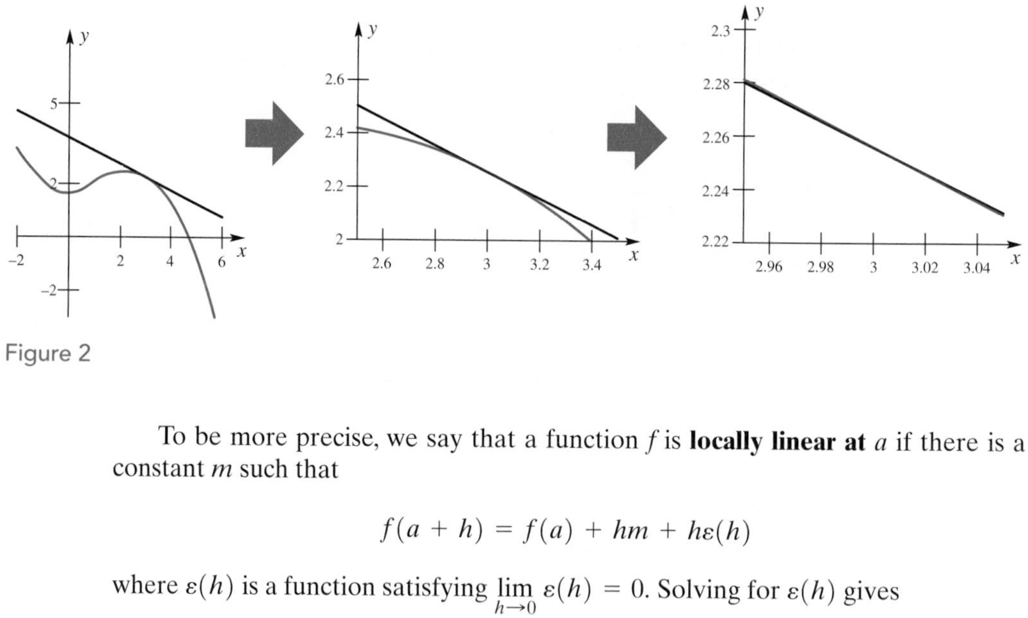

Figure 2

To be more precise, we say that a function f is **locally linear at** a if there is a constant m such that

$$f(a + h) = f(a) + hm + h\varepsilon(h)$$

where $\varepsilon(h)$ is a function satisfying $\lim_{h \to 0} \varepsilon(h) = 0$. Solving for $\varepsilon(h)$ gives

$$\varepsilon(h) = \frac{f(a + h) - f(a)}{h} - m$$

The function $\varepsilon(h)$ is the difference between the slope of the secant line through the points $(a, f(a))$ and $(a + h, f(a + h))$ and the slope of the tangent line through $(a, f(a))$. If f is locally linear at a, then

$$\lim_{h \to 0} \varepsilon(h) = \lim_{h \to 0} \left[\frac{f(a + h) - f(a)}{h} - m \right] = 0$$

which means that

$$\lim_{h \to 0} \frac{f(a + h) - f(a)}{h} = m$$

We conclude that f must be differentiable at a and that m must equal $f'(a)$.

Conversely, if f is differentiable at a, then $\lim_{h\to 0}\dfrac{f(a+h)-f(a)}{h} = f'(a) = m$; hence, f is locally linear. Therefore, in the one-variable case, f is locally linear at a if and only if f is differentiable at a.

This concept of local linearity *does* carry over to the situation in which f is a function of two variables, and we will use this characteristic to define differentiability of a function of two variables. First, we define local linearity.

Definition Local Linearity for a Function of Two Variables

We say that f is **locally linear** at (a, b) if

$$f(a + h_1, b + h_2) \\ = f(a, b) + h_1 f_x(a, b) + h_2 f_y(a, b) + h_1\varepsilon_1(h_1, h_2) + h_2\varepsilon_2(h_1, h_2)$$

where $\varepsilon_1(h_1, h_2) \to 0$ as $(h_1, h_2) \to 0$ and $\varepsilon_2(h_1, h_2) \to 0$ as $(h_1, h_2) \to 0$.

Just as h was a small increment in x for the one-variable case, we can think of h_1 and h_2 as small increments in x and y, respectively, for the two-variable case.

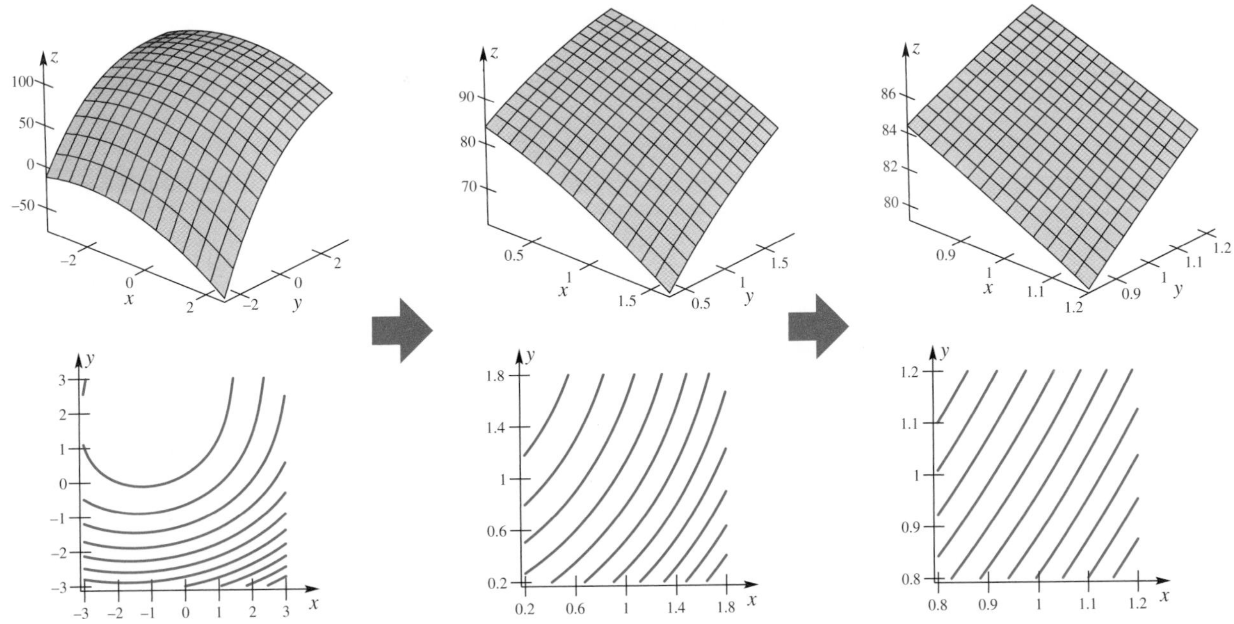

Figure 3

Figure 3 shows what can happen when we zoom in on the graph of a function of two variables. (In Figure 3 we zoom in on the graph at the point $(x, y) = (1, 1)$.) If we zoom in far enough, the surface resembles a plane, and the contour plot appears to consist of parallel lines. We can simplify the above definition by defining $\mathbf{p}_0 = (a, b)$, $\mathbf{h} = (h_1, h_2)$, and $\varepsilon(\mathbf{h}) = (\varepsilon_1(h_1, h_2), \varepsilon_2(h_1, h_2))$. (The function $\varepsilon(\mathbf{h})$ is a vector-valued function of a vector variable.) Thus,

$$f(\mathbf{p}_0 + \mathbf{h}) = f(\mathbf{p}_0) + (f_x(\mathbf{p}_0), f_y(\mathbf{p}_0))\cdot\mathbf{h} + \varepsilon(\mathbf{h})\cdot\mathbf{h}$$

This formulation easily carries over to the case where f is a function of three (or more) variables. We now define differentiability to be synonymous with local linearity.

Definition Differentiability for a Function of Two or More Variables

The function f is **differentiable** at $\mathbf{p}$ if it is locally linear at $\mathbf{p}$. The function f is differentiable on an open set R if it is differentiable at every point in R.

The vector $(f_x(\mathbf{p}), f_y(\mathbf{p})) = f_x(\mathbf{p})\mathbf{i} + f_y(\mathbf{p})\mathbf{j}$ is denoted $\nabla f(\mathbf{p})$ and is called the **gradient** of f. Thus, f is differentiable at $\mathbf{p}$ if and only if

$$f(\mathbf{p} + \mathbf{h}) = f(\mathbf{p}) + \nabla f(\mathbf{p}) \cdot \mathbf{h} + \varepsilon(\mathbf{h}) \cdot \mathbf{h}$$

where $\varepsilon(\mathbf{h}) \to \mathbf{0}$ as $\mathbf{h} \to \mathbf{0}$. The operator ∇ is read "del" and is often called the **del operator.**

In the sense described above, *the gradient becomes the analog of the derivative.* We point out several aspects of our definitions.

1. The derivative $f'(x)$ is a number, whereas the gradient $\nabla f(\mathbf{p})$ is a vector.
2. The products $\nabla f(\mathbf{p}) \cdot \mathbf{h}$ and $\varepsilon(\mathbf{h}) \cdot \mathbf{h}$ are dot products.
3. The definitions of differentiability and gradient are easily extended to any number of dimensions.

The following theorem gives a condition that guarantees the differentiability of a function at a point.

Theorem A

If $f(x, y)$ has continuous partial derivatives $f_x(x, y)$ and $f_y(x, y)$ on a disk D whose interior contains (a, b), then $f(x, y)$ is differentiable at (a, b).

Interval Notation in the Proof

The interval notation used in the proof, such as $[a, a + h_1]$, would suggest that $h_1 > 0$. This need not be the case, as h_1 and h_2 can be negative. In this proof, we must interpret intervals to mean all those points between the two end points (regardless of which is the larger). The interval *includes* the endpoints in the case of a closed interval and *excludes* the endpoints in the case of an open interval.

Proof Let h_1 and h_2 be increments in x and y, respectively, that are so small that $(a + h_1, b + h_2)$ is in the interior of the disk D. (That such values h_1 and h_2 exist is a consequence of the fact that the interior of the disk D is an *open* set.) The difference between $f(a + h_1, b + h_2)$ and $f(a, b)$ is

(3) $$f(a + h_1, b + h_2) - f(a, b) = [f(a + h_1, b) - f(a, b)] + [f(a + h_1, b + h_2) - f(a + h_1, b)]$$

We now apply the Mean Value Theorem for Derivatives (Theorem 3.6A) twice: once to the difference $f(a + h_1, b) - f(a, b)$, and once to the difference $f(a + h_1, b + h_2) - f(a + h_1, b)$. In the first case, we define $g_1(x) = f(x, b)$ for x in the interval $[a, a + h_1]$, and from the Mean Value Theorem for Derivatives we conclude that there exists a value c_1 in $(a, a + h_1)$ such that

$$g_1(a + h_1) - g_1(a) = f(a + h_1, b) - f(a, b) = h_1 g_1'(c_1) = h_1 f_x(c_1, b)$$

For the second case, we define $g_2(y) = f(a + h_1, y)$ for y in the interval $[b, b + h_2]$. There exists a c_2 in the interval $(b, b + h_2)$ such that

$$g_2(b + h_2) - g_2(b) = h_2 g_2'(c_2)$$

This gives

$$g_2(b + h_2) - g_2(b) = f(a + h_1, b + h_2) - f(a + h_1, b)$$

$$= h_2 g_2'(c_2) = h_2 f_y(a + h_1, c_2)$$

Equation (3) becomes

$$\begin{aligned}f(a + h_1, b + h_2) - f(a, b) &= h_1 f_x(c_1, b) + h_2 f_y(a + h_1, c_2)\\ &= h_1\left[f_x(c_1, b) + f_x(a, b) - f_x(a, b)\right]\\ &\quad + h_2\left[f_y(a + h_1, c_2) + f_y(a, b) - f_y(a, b)\right]\\ &= h_1 f_x(a, b) + h_2 f_y(a, b)\\ &\quad + h_1\left[f_x(c_1, b) - f_x(a, b)\right]\\ &\quad + h_2\left[f_y(a + h_1, c_2) - f_y(a, b)\right]\end{aligned}$$

Now, let $\varepsilon_1(h_1, h_2) = f_x(c_1, b) - f_x(a, b)$ and $\varepsilon_2(h_1, h_2) = f_y(a + h_1, c_2) - f_y(a, b)$. Since $c_1 \in (a, a + h_1)$ and $c_2 \in (b, b + h_2)$, we conclude that $c_1 \to a$ and $c_2 \to b$ as $h_1, h_2 \to 0$. Thus,

$$\begin{aligned}f(a + h_1, b + h_2) - f(a, b) &= h_1 f_x(a, b) + h_2 f_y(a, b)\\ &\quad + h_1\varepsilon_1(h_1, h_2) + h_2\varepsilon_2(h_1, h_2)\end{aligned}$$

where $\varepsilon_1(h_1, h_2) \to 0$ and $\varepsilon_2(h_1, h_2) \to 0$ as $(h_1, h_2) \to (0, 0)$. Therefore, f is locally linear and hence differentiable at (a, b). ■

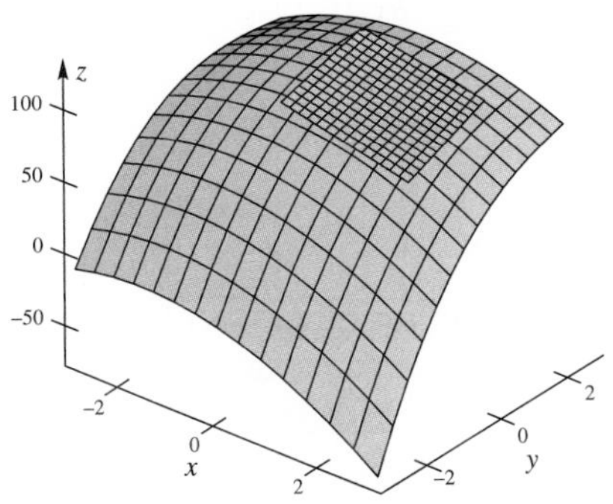

Figure 4

If the function f is differentiable at $\mathbf{p}_0$, then, when $\mathbf{h}$ has small magnitude

$$f(\mathbf{p}_0 + \mathbf{h}) \approx f(\mathbf{p}_0) + \nabla f(\mathbf{p}_0) \cdot \mathbf{h}$$

Letting $\mathbf{p} = \mathbf{p}_0 + \mathbf{h}$, we find that the function T defined by

$$T(\mathbf{p}) = f(\mathbf{p}_0) + \nabla f(\mathbf{p}_0) \cdot (\mathbf{p} - \mathbf{p}_0)$$

should be a good approximation to $f(\mathbf{p})$ if $\mathbf{p}$ is close to $\mathbf{p}_0$. The equation $z = T(\mathbf{p})$ defines a plane that approximates f near $\mathbf{p}_0$. Naturally, this plane is called the **tangent plane.** See Figure 4.

EXAMPLE 1 Show that $f(x, y) = xe^y + x^2 y$ is differentiable everywhere and calculate its gradient. Then find the equation of the tangent plane at (2, 0).

SOLUTION We note first that

$$\frac{\partial f}{\partial x} = e^y + 2xy \qquad \text{and} \qquad \frac{\partial f}{\partial y} = xe^y + x^2$$

Both of these functions are continuous everywhere and so, by Theorem A, f is differentiable everywhere. The gradient is

$$\nabla f(x, y) = (e^y + 2xy)\mathbf{i} + (xe^y + x^2)\mathbf{j} = \langle e^y + 2xy, xe^y + x^2\rangle$$

Thus,

$$\nabla f(2, 0) = \mathbf{i} + 6\mathbf{j} = \langle 1, 6\rangle$$

and the equation of the tangent plane is

$$\begin{aligned}z &= f(2, 0) + \nabla f(2, 0) \cdot \langle x - 2, y\rangle\\ &= 2 + \langle 1, 6\rangle \cdot \langle x - 2, y\rangle\\ &= 2 + x - 2 + 6y = x + 6y\end{aligned}$$

■

EXAMPLE 2 For $f(x, y, z) = x \sin z + x^2 y$, find $\nabla f(1, 2, 0)$.

SOLUTION The partial derivatives are

$$\frac{\partial f}{\partial x} = \sin z + 2xy, \qquad \frac{\partial f}{\partial y} = x^2, \qquad \frac{\partial f}{\partial z} = x \cos z$$

At (1, 2, 0), these partials have the values 4, 1, and 1, respectively. Thus,

$$\nabla f(1, 2, 0) = 4\mathbf{i} + \mathbf{j} + \mathbf{k}$$

■

Rules for Gradients In many respects, gradients behave like derivatives. Recall that D, considered as an operator, is linear. The operator ∇ is also linear.

Theorem B Properties of ∇

The gradient operator ∇ satisfies

1. $\nabla[f(\mathbf{p}) + g(\mathbf{p})] = \nabla f(\mathbf{p}) + \nabla g(\mathbf{p})$
2. $\nabla[\alpha f(\mathbf{p})] = \alpha \nabla f(\mathbf{p})$
3. $\nabla[f(\mathbf{p})g(\mathbf{p})] = f(\mathbf{p})\nabla g(\mathbf{p}) + g(\mathbf{p})\nabla f(\mathbf{p})$

Proof All three results follow from the corresponding facts for partial derivatives. We prove (3) in the two-variable case, suppressing the point $\mathbf{p}$ for brevity.

$$\begin{aligned}\nabla fg &= \frac{\partial(fg)}{\partial x}\mathbf{i} + \frac{\partial(fg)}{\partial y}\mathbf{j}\\ &= \left(f\frac{\partial g}{\partial x} + g\frac{\partial f}{\partial x}\right)\mathbf{i} + \left(f\frac{\partial g}{\partial y} + g\frac{\partial f}{\partial y}\right)\mathbf{j}\\ &= f\left(\frac{\partial g}{\partial x}\mathbf{i} + \frac{\partial g}{\partial y}\mathbf{j}\right) + g\left(\frac{\partial f}{\partial x}\mathbf{i} + \frac{\partial f}{\partial y}\mathbf{j}\right)\\ &= f\nabla g + g\nabla f\end{aligned}$$

■

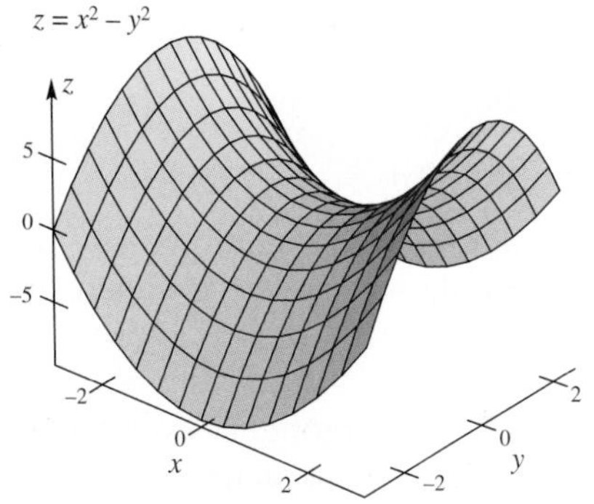

Figure 5

Continuity versus Differentiability Recall that for functions of one variable, differentiability implies continuity, but not vice versa. The same is true here.

Theorem C

If f is differentiable at $\mathbf{p}$, then f is continuous at $\mathbf{p}$.

Proof Since f is differentiable at $\mathbf{p}$,

$$f(\mathbf{p} + \mathbf{h}) - f(\mathbf{p}) = \nabla f(\mathbf{p})\cdot\mathbf{h} + \boldsymbol{\varepsilon}(\mathbf{h})\cdot\mathbf{h}$$

Thus,

$$\begin{aligned}|f(\mathbf{p} + \mathbf{h}) - f(\mathbf{p})| &\le |\nabla f(\mathbf{p})\cdot\mathbf{h}| + |\boldsymbol{\varepsilon}(\mathbf{h})\cdot\mathbf{h}|\\ &= \|\nabla f(\mathbf{p})\|\|\mathbf{h}\||\cos\theta| + |\boldsymbol{\varepsilon}(\mathbf{h})\cdot\mathbf{h}|\end{aligned}$$

Both of the latter terms approach 0 as $\mathbf{h}\to\mathbf{0}$, and so

$$\lim_{\mathbf{h}\to\mathbf{0}} f(\mathbf{p} + \mathbf{h}) = f(\mathbf{p})$$

This last equality is one way of formulating the continuity of f at $\mathbf{p}$. ■

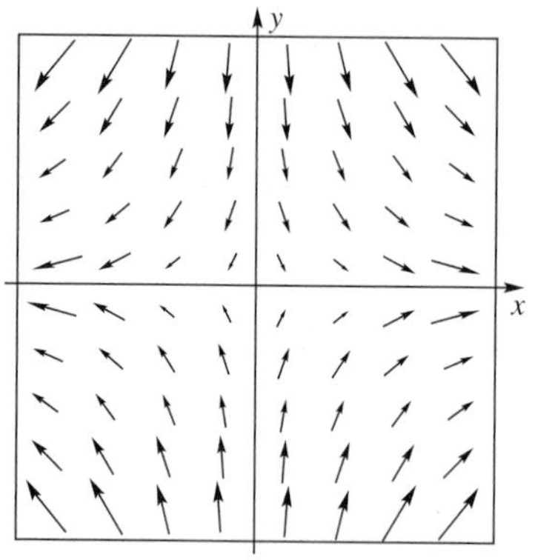

Figure 6

The Gradient Field The gradient ∇f associates with each point $\mathbf{p}$ in the domain of f a vector $\nabla f(\mathbf{p})$. The set of all these vectors is called the **gradient field** for f. In Figures 5 and 6, we show graphs of the surface $z = x^2 - y^2$ and the corresponding gradient field. Do these figures suggest something about the direction in which the gradient vectors point? We explore this subject in the next section.

Concepts Review

1. The analog of the derivative $f'(x)$ for a function of more than one variable is the _____, denoted by $\nabla f(\mathbf{p})$.

2. The function $f(x, y)$ is differentiable at (a, b) if and only if f is _____ at (a, b).

3. For a function f of two variables, the gradient is $\nabla f(\mathbf{p}) =$ _____. Thus, if $f(x, y) = xy^2$, $\nabla f(x, y) =$ _____.

4. $f(x, y)$ being differentiable at (x_0, y_0) is equivalent to the existence of a _____ to the graph at this point.

Problem Set 12.4

In Problems 1–10, find the gradient ∇f.

1. $f(x, y) = x^2y + 3xy$
2. $f(x, y) = x^3y - y^3$
3. $f(x, y) = xe^{xy}$
4. $f(x, y) = x^2y \cos y$
5. $f(x, y) = x^2y/(x + y)$
6. $f(x, y) = \sin^3(x^2y)$
7. $f(x, y, z) = \sqrt{x^2 + y^2 + z^2}$
8. $f(x, y, z) = x^2y + y^2z + z^2x$
9. $f(x, y, z) = x^2ye^{x-z}$
10. $f(x, y, z) = xz \ln(x + y + z)$

In Problems 11–14, find the gradient vector of the given function at the given point **p**. *Then find the equation of the tangent plane at* **p** *(see Example 1).*

11. $f(x, y) = x^2y - xy^2$, $\mathbf{p} = (-2, 3)$
12. $f(x, y) = x^3y + 3xy^2$, $\mathbf{p} = (2, -2)$
13. $f(x, y) = \cos \pi x \sin \pi y + \sin 2\pi y$, $\mathbf{p} = (-1, \frac{1}{2})$
14. $f(x, y) = \dfrac{x^2}{y}$, $\mathbf{p} = (2, -1)$

In Problems 15 and 16, find the equation $w = T(x, y, z)$ of the tangent "hyperplane" at **p**.

15. $f(x, y, z) = 3x^2 - 2y^2 + xz^2$, $\mathbf{p} = (1, 2, -1)$
16. $f(x, y, z) = xyz + x^2$, $\mathbf{p} = (2, 0, -3)$

17. Show that

$$\nabla\left(\frac{f}{g}\right) = \frac{g\nabla f - f\nabla g}{g^2}$$

18. Show that

$$\nabla(f^r) = rf^{r-1}\nabla f$$

19. Find all points (x, y) at which the tangent plane to the graph of $z = x^2 - 6x + 2y^2 - 10y + 2xy$ is horizontal.

20. Find all points (x, y) at which the tangent plane to the graph of $z = x^3$ is horizontal.

21. Find parametric equations of the line tangent to the surface $z = y^2 + x^3y$ at the point $(2, 1, 9)$ whose projection on the xy-plane is
(a) parallel to the x-axis; (b) parallel to the y-axis;
(c) parallel to the line $x = y$.

22. Find parametric equations of the line tangent to the surface $z = x^2y^3$ at the point $(3, 2, 72)$ whose projection on the xy-plane is
(a) parallel to the x-axis; (b) parallel to the y-axis;
(c) parallel to the line $x = -y$.

23. Refer to Figure 1. Find the equation of the tangent plane to $z = -10\sqrt{|xy|}$ at $(1, -1)$. *Recall:* $d|x|/dx = |x|/x$ for $x \neq 0$.

24. **Mean Value Theorem for Several Variables** If f is differentiable at each point of the line segment from **a** to **b**, then there exists on that line segment a point **c** between **a** and **b** such that

$$f(\mathbf{b}) - f(\mathbf{a}) = \nabla f(\mathbf{c}) \cdot (\mathbf{b} - \mathbf{a})$$

Assuming that this result is true, show that, if f is differentiable on a convex set S and if $\nabla f(\mathbf{p}) = \mathbf{0}$ on S, then f is constant on S. *Note:* A set S is convex if each pair of points in S can be connected by a line segment in S.

25. Find all values of **c** that satisfy the Mean Value Theorem for Several Variables (see Problem 24) for the function $f(x, y) = 9 - x^2 - y^2$ where $\mathbf{a} = \langle 0, 0\rangle$ and $\mathbf{b} = \langle 2, 1\rangle$.

26. Find all values of **c** that satisfy the Mean Value Theorem for Several Variables (see Problem 24) for the function $f(x, y) = \sqrt{4 - x^2}$ where $\mathbf{a} = \langle 0, 0\rangle$ and $\mathbf{b} = \langle 2, 6\rangle$.

27. Use the result of Problem 24 to show that if $\nabla f(\mathbf{p}) = \nabla g(\mathbf{p})$ for all **p** in a convex set S then f and g differ by a constant on S.

28. Find the most general function $f(\mathbf{p})$ satisfying $\nabla f(\mathbf{p}) = \mathbf{p}$.

CAS 29. Plot the graph of $f(x, y) = -|xy|$ together with its gradient field.
(a) Based on this and Figures 5 and 6, make a conjecture about the direction in which a gradient vector points.
(b) Is f differentiable at the origin? Justify your answer.

CAS 30. Plot the graph of $f(x, y) = \sin x + \sin y - \sin(x + y)$ on $0 \le x \le 2\pi, 0 \le y \le 2\pi$. Also draw the gradient field to see if your conjecture in Problem 29 (a) holds up.

31. Prove Theorem B for
(a) the three-variable case and
(b) the n-variable case. *Hint:* Denote the standard unit vectors by $\mathbf{i}_1, \mathbf{i}_2, \ldots, \mathbf{i}_n$.

Answers to Concepts Review: 1. gradient 2. locally linear 3. $\dfrac{\partial f(\mathbf{p})}{\partial x}\mathbf{i} + \dfrac{\partial f(\mathbf{p})}{\partial y}\mathbf{j}$; $y^2\mathbf{i} + 2xy\,\mathbf{j}$ 4. tangent plane

12.5 Directional Derivatives and Gradients

Consider again a function $f(x, y)$ of two variables. The partial derivatives $f_x(x, y)$ and $f_y(x, y)$ measure the rate of change (and the slope of the tangent line) in directions parallel to the x- and y-axes. Our goal now is to study the rate of change of f in an arbitrary direction. This leads to the concept of the directional derivative, which in turn is related to the gradient.

It will be convenient to use vector notation. Let $\mathbf{p} = (x, y)$, and let $\mathbf{i}$ and $\mathbf{j}$ be the unit vectors in the positive x- and y-directions. Then the two partial derivatives at $\mathbf{p}$ may be written as follows:

$$f_x(\mathbf{p}) = \lim_{h \to 0} \frac{f(\mathbf{p} + h\mathbf{i}) - f(\mathbf{p})}{h}$$

$$f_y(\mathbf{p}) = \lim_{h \to 0} \frac{f(\mathbf{p} + h\mathbf{j}) - f(\mathbf{p})}{h}$$

To get the concept we are after, all we have to do is replace $\mathbf{i}$ or $\mathbf{j}$ by an arbitrary unit vector $\mathbf{u}$.

Definition

For any unit vector $\mathbf{u}$, let

$$D_{\mathbf{u}} f(\mathbf{p}) = \lim_{h \to 0} \frac{f(\mathbf{p} + h\mathbf{u}) - f(\mathbf{p})}{h}$$

This limit, if it exists, is called the **directional derivative** of f at $\mathbf{p}$ in the direction $\mathbf{u}$.

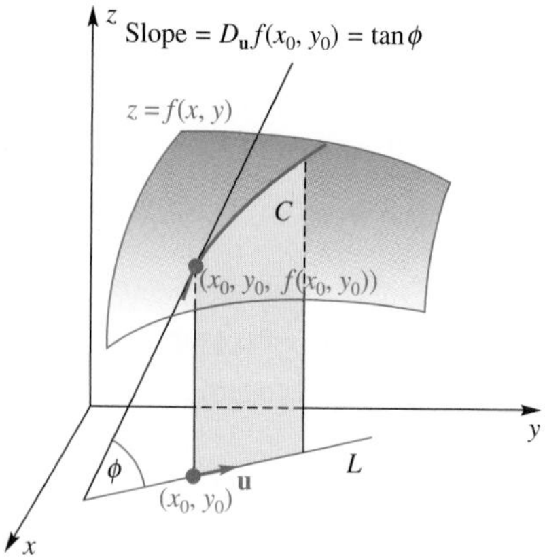

Figure 1

Thus, $D_{\mathbf{i}} f(\mathbf{p}) = f_x(\mathbf{p})$ and $D_{\mathbf{j}} f(\mathbf{p}) = f_y(\mathbf{p})$. Since $\mathbf{p} = (x, y)$, we also use the notation $D_{\mathbf{u}} f(x, y)$. Figure 1 gives the geometric interpretation of $D_{\mathbf{u}} f(x_0, y_0)$. The vector $\mathbf{u}$ determines a line L in the xy-plane through (x_0, y_0). The plane through L perpendicular to the xy-plane intersects the surface $z = f(x, y)$ in a curve C. Its tangent at the point $(x_0, y_0, f(x_0, y_0))$ has slope $D_{\mathbf{u}} f(x_0, y_0)$. Another useful interpretation is that $D_{\mathbf{u}} f(x_0, y_0)$ measures the rate of change of f with respect to distance in the direction $\mathbf{u}$.

Connection with the Gradient

Recall from Section 12.4 that $\nabla f(\mathbf{p})$ is

$$\nabla f(\mathbf{p}) = f_x(\mathbf{p})\mathbf{i} + f_y(\mathbf{p})\mathbf{j}$$

Theorem A

Let f be differentiable at $\mathbf{p}$. Then f has a directional derivative at $\mathbf{p}$ in the direction of the unit vector $\mathbf{u} = u_1\mathbf{i} + u_2\mathbf{j}$ and

$$D_{\mathbf{u}} f(\mathbf{p}) = \mathbf{u} \cdot \nabla f(\mathbf{p})$$

That is,

$$D_{\mathbf{u}} f(x, y) = u_1 f_x(x, y) + u_2 f_y(x, y)$$

Proof Since f is differentiable at $\mathbf{p}$,

$$f(\mathbf{p} + h\mathbf{u}) - f(\mathbf{p}) = \nabla f(\mathbf{p}) \cdot (h\mathbf{u}) + \boldsymbol{\varepsilon}(h\mathbf{u}) \cdot (h\mathbf{u})$$

where $\boldsymbol{\varepsilon}(h\mathbf{u}) \to \mathbf{0}$ as $h \to 0$. Thus,

$$\frac{f(\mathbf{p} + h\mathbf{u}) - f(\mathbf{p})}{h} = \nabla f(\mathbf{p}) \cdot \mathbf{u} + \boldsymbol{\varepsilon}(h\mathbf{u}) \cdot \mathbf{u}$$

The conclusion follows by taking limits as $h \to 0$. ■

EXAMPLE 1 If $f(x, y) = 4x^2 - xy + 3y^2$, find the directional derivative of f at $(2, -1)$ in the direction of the vector $\mathbf{a} = 4\mathbf{i} + 3\mathbf{j}$.

SOLUTION The unit vector $\mathbf{u}$ in the direction of $\mathbf{a}$ is $\left(\frac{4}{5}\right)\mathbf{i} + \left(\frac{3}{5}\right)\mathbf{j}$. Also, $f_x(x, y) = 8x - y$ and $f_y(x, y) = -x + 6y$; thus, $f_x(2, -1) = 17$ and $f_y(2, -1) = -8$. Consequently, by Theorem A,

$$D_{\mathbf{u}} f(2, -1) = \left\langle \tfrac{4}{5}, \tfrac{3}{5} \right\rangle \cdot \langle 17, -8 \rangle = \tfrac{4}{5}(17) + \tfrac{3}{5}(-8) = \tfrac{44}{5}$$ ■

Although we will not go through the details, we assert that what we have done is valid for functions of three or more variables, with obvious modifications.

EXAMPLE 2 Find the directional derivative of the function $f(x, y, z) = xy \sin z$ at the point $(1, 2, \pi/2)$ in the direction of the vector $\mathbf{a} = \mathbf{i} + 2\mathbf{j} + 2\mathbf{k}$.

SOLUTION The unit vector $\mathbf{u}$ in the direction of $\mathbf{a}$ is $\frac{1}{3}\mathbf{i} + \frac{2}{3}\mathbf{j} + \frac{2}{3}\mathbf{k}$. Also, $f_x(x, y, z) = y \sin z$, $f_y(x, y, z) = x \sin z$, and $f_z(x, y, z) = xy \cos z$, and so $f_x(1, 2, \pi/2) = 2$, $f_y(1, 2, \pi/2) = 1$, and $f_z(1, 2, \pi/2) = 0$. We conclude that

$$D_{\mathbf{u}} f\left(1, 2, \frac{\pi}{2}\right) = \frac{1}{3}(2) + \frac{2}{3}(1) + \frac{2}{3}(0) = \frac{4}{3}$$ ■

Maximum Rate of Change For a given function f at a given point $\mathbf{p}$, it is natural to ask in what direction the function is changing most rapidly, that is, in what direction is $D_{\mathbf{u}} f(\mathbf{p})$ the largest? From the geometric formula for the dot product (Section 11.3), we may write

$$D_{\mathbf{u}} f(\mathbf{p}) = \mathbf{u} \cdot \nabla f(\mathbf{p}) = \|\mathbf{u}\| \|\nabla f(\mathbf{p})\| \cos \theta = \|\nabla f(\mathbf{p})\| \cos \theta$$

where θ is the angle between $\mathbf{u}$ and $\nabla f(\mathbf{p})$. Thus, $D_{\mathbf{u}} f(\mathbf{p})$ is maximized when $\theta = 0$ and minimized when $\theta = \pi$. We summarize as follows.

Theorem B

A function increases most rapidly at $\mathbf{p}$ in the direction of the gradient (with rate $\|\nabla f(\mathbf{p})\|$) and decreases most rapidly in the opposite direction (with rate $-\|\nabla f(\mathbf{p})\|$).

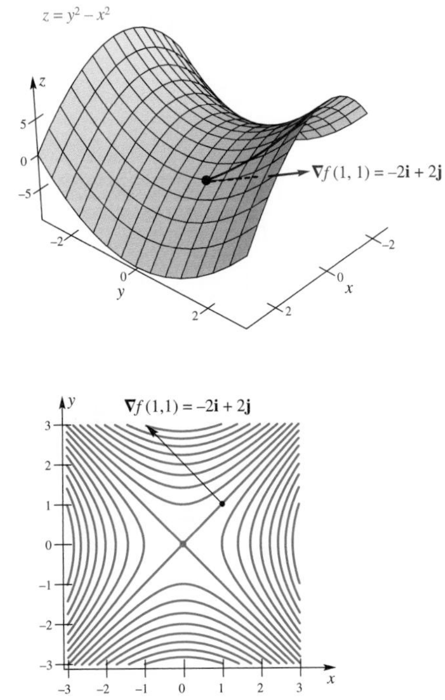

Figure 2

EXAMPLE 3 Suppose that a bug is located on the hyperbolic paraboloid $z = y^2 - x^2$ at the point $(1, 1, 0)$, as in Figure 2. In what direction should it move for the steepest climb and what is the slope as it starts out?

SOLUTION Let $f(x, y) = y^2 - x^2$. Since $f_x(x, y) = -2x$ and $f_y(x, y) = 2y$,

$$\nabla f(1, 1) = f_x(1, 1)\mathbf{i} + f_y(1, 1)\mathbf{j} = -2\mathbf{i} + 2\mathbf{j}$$

Thus, the bug should move from $(1, 1, 0)$ in the direction $-2\mathbf{i} + 2\mathbf{j}$, where the slope will be $\|-2\mathbf{i} + 2\mathbf{j}\| = \sqrt{8} = 2\sqrt{2}$. ■

Level Curves and Gradients Recall from Section 12.1 that the *level curves* of a surface $z = f(x, y)$ are the projections onto the xy-plane of the curves of intersection of the surface with planes $z = k$ that are parallel to the xy-plane. The value of the function at all points on the same level curve is constant (Figure 3).

Denote by L the level curve of $f(x, y)$ that passes through an arbitrarily chosen point $P(x_0, y_0)$ in the domain of f, and let the unit vector $\mathbf{u}$ be tangent to L at P. Since the value of f is the same at all points on the level curve L, its directional derivative $D_{\mathbf{u}} f(x_0, y_0)$, which is the rate of change of $f(x, y)$ in the direction $\mathbf{u}$, is zero when $\mathbf{u}$ is tangent to L. (This statement, which seems very clear intuitively, requires justification, which we omit since the result we want also follows from an argument to be given in Section 12.7.) Since

$$0 = D_{\mathbf{u}} f(x_0, y_0) = \nabla f(x_0, y_0) \cdot \mathbf{u}$$

we conclude that ∇f and $\mathbf{u}$ are perpendicular, a result worthy of theorem status.

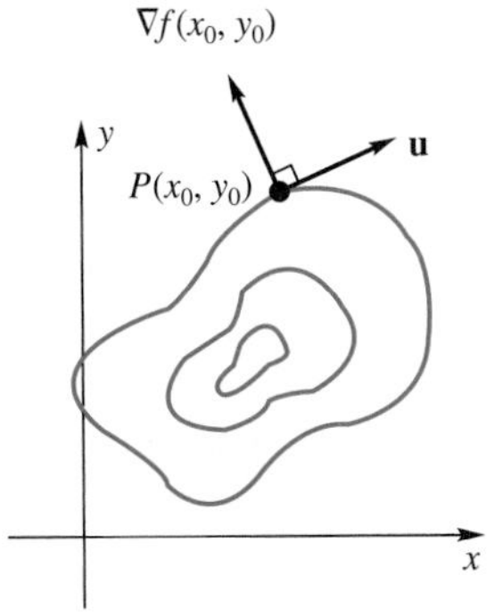

Figure 3

Theorem C

The gradient of f at a point P is perpendicular to the level curve of f that goes through P.

EXAMPLE 4 For the paraboloid $z = x^2/4 + y^2$, find the equation of its level curve that passes through the point $P(2, 1)$ and sketch it. Find the gradient vector of the paraboloid at P, and draw the gradient with its initial point at P.

SOLUTION The level curve of the paraboloid that corresponds to the plane $z = k$ has the equation $x^2/4 + y^2 = k$. To find the value of k belonging to the level curve through P, we substitute $(2, 1)$ for (x, y) and obtain $k = 2$. Thus, the equation of the level curve that goes through P is that of the ellipse

$$\frac{x^2}{8} + \frac{y^2}{2} = 1$$

Next let $f(x, y) = x^2/4 + y^2$. Since $f_x(x, y) = x/2$ and $f_y(x, y) = 2y$, the gradient of the paraboloid at $P(2, 1)$ is

$$\nabla f(2, 1) = f_x(2, 1)\mathbf{i} + f_y(2, 1)\mathbf{j} = \mathbf{i} + 2\mathbf{j}$$

The level curve and the gradient at P are shown in Figure 4. ■

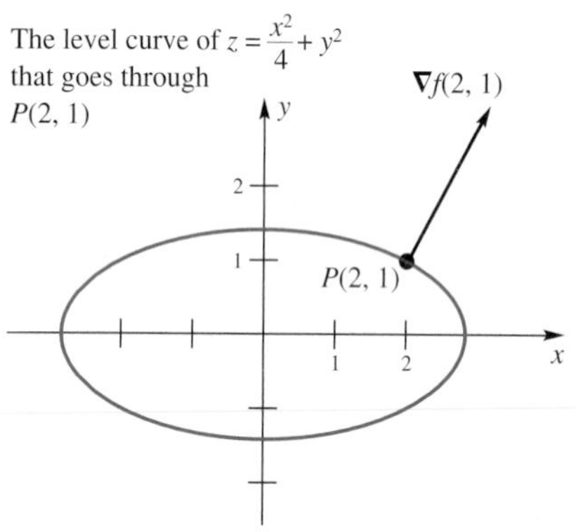

Figure 4

To provide additional illustration of Theorems B and C, we asked our computer to draw the surface $z = |xy|$, together with its contour map and gradient field. The results are shown in Figure 5. Note that the gradient vectors are perpendicular to the level curves and that they do point in the direction of greatest increase of z.

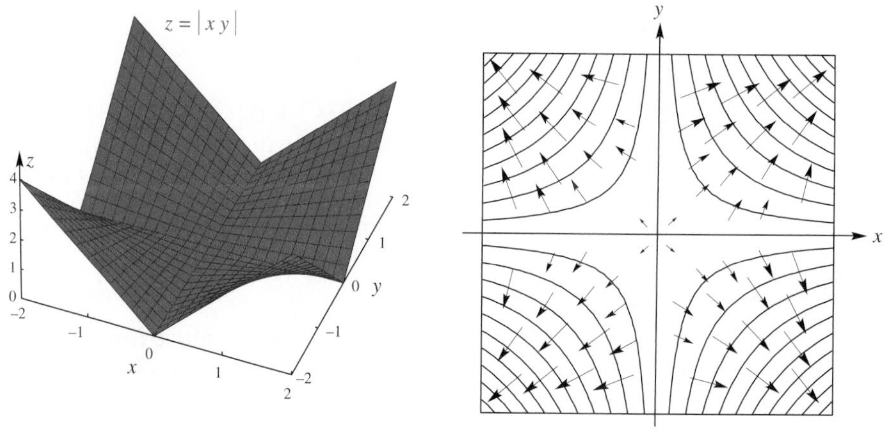

Figure 5

Higher Dimensions The concept of level curves for functions of two variables generalizes to level surfaces for functions of three variables. If f is a function of three variables, the surface $f(x, y, z) = k$, where k is a constant, is a level surface for f. At all points on a level surface, the value of the function is the same, and the gradient vector of $f(x, y, z)$ at a point $P(x, y, z)$ in its domain is normal to the level surface of f that goes through P.

In problems of heat conduction in a homogeneous body, where $w = f(x, y, z)$ gives the temperature at the point (x, y, z), the level surface $f(x, y, z) = k$ is called an *isothermal surface* because all points on it have the same temperature k. At any given point of the body, heat flows in the direction opposite to the gradient (i.e., in the direction of the greatest *decrease* in temperature) and therefore perpendicular to the isothermal surface through the point. If $w = f(x, y, z)$ gives the electrostatic potential (voltage) at any point in an electric potential field, the level surfaces of the function are called *equipotential surfaces*. All points on an equipotential surface have the same electrostatic potential, and the direction of flow of electricity is along the negative gradient, that is, in the direction of greatest drop in potential.

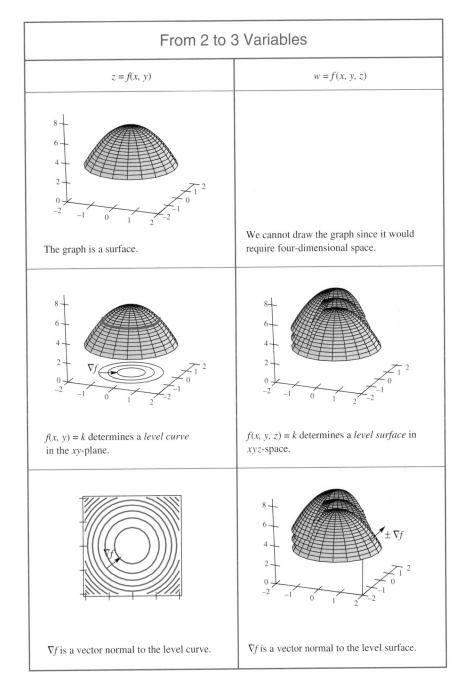

From 2 to 3 Variables

$z = f(x, y)$	$w = f(x, y, z)$
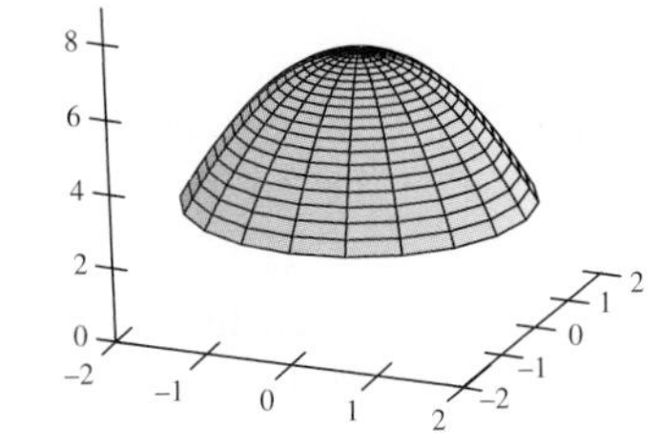 The graph is a surface.	We cannot draw the graph since it would require four-dimensional space.
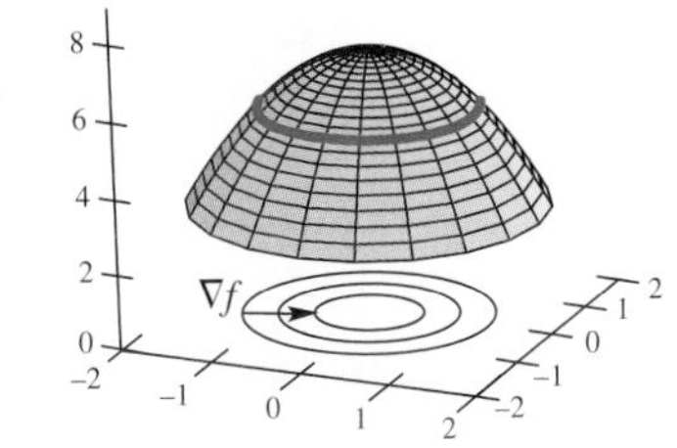 $f(x, y) = k$ determines a *level curve* in the xy-plane.	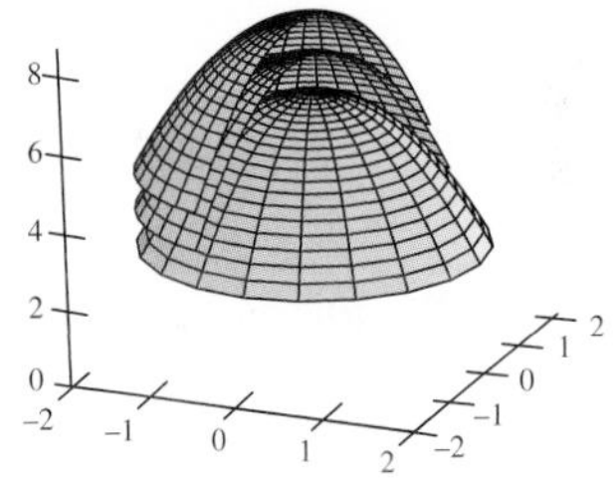 $f(x, y, z) = k$ determines a *level surface* in xyz-space.
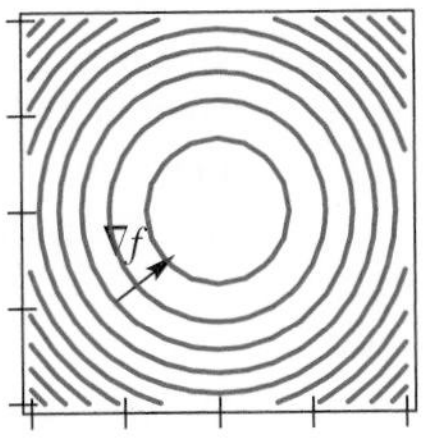 ∇f is a vector normal to the level curve.	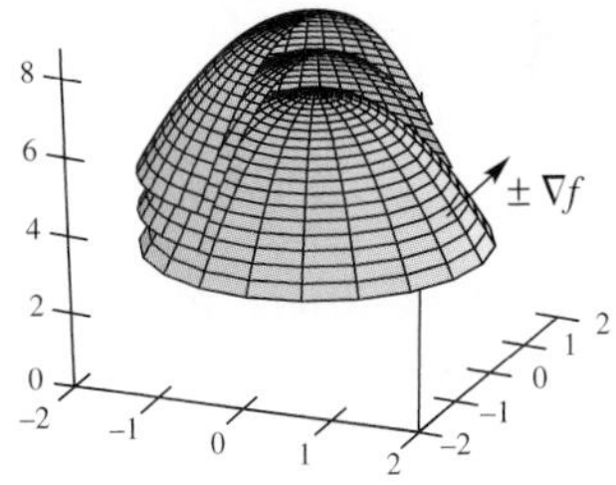 ∇f is a vector normal to the level surface.

EXAMPLE 5 If the temperature at any point in a homogeneous body is given by $T = e^{xy} - xy^2 - x^2yz$, what is the direction of the greatest drop in temperature at the point $(1, -1, 2)$?

SOLUTION The greatest decrease in temperature at $(1, -1, 2)$ is in the direction of the negative gradient at that point.

Since $\nabla T = (ye^{xy} - y^2 - 2xyz)\mathbf{i} + (xe^{xy} - 2xy - x^2z)\mathbf{j} + (-x^2y)\mathbf{k}$, we find that $-\nabla T$ at $(1, -1, 2)$ is

$$(e^{-1} - 3)\mathbf{i} - e^{-1}\mathbf{j} - \mathbf{k}$$

■

Concepts Review

1. The directional derivative of f at $\mathbf{p}$ in the direction of the unit vector $\mathbf{u}$ is denoted by $D_{\mathbf{u}}f(\mathbf{p})$ and is defined as $\lim_{h\to 0}$ ______.

2. If $\mathbf{u} = u_1\mathbf{i} + u_2\mathbf{j}$ is a unit vector, then we may calculate $D_{\mathbf{u}}f(x, y)$ from the formula $D_{\mathbf{u}}f(x, y) =$ ______.

3. The gradient vector ∇f always points in the direction of ______ of f.

4. The gradient vector of f at P is always perpendicular to the ______ of f through P.

Problem Set 12.5

In Problems 1–8, find the directional derivative of f at the point **p** *in the direction of* **a**.

1. $f(x, y) = x^2y$; $\mathbf{p} = (1, 2)$; $\mathbf{a} = 3\mathbf{i} - 4\mathbf{j}$

2. $f(x, y) = y^2 \ln x$; $\mathbf{p} = (1, 4)$; $\mathbf{a} = \mathbf{i} - \mathbf{j}$

3. $f(x, y) = 2x^2 + xy - y^2$; $\mathbf{p} = (3, -2)$; $\mathbf{a} = \mathbf{i} - \mathbf{j}$

4. $f(x, y) = x^2 - 3xy + 2y^2$; $\mathbf{p} = (-1, 2)$; $\mathbf{a} = 2\mathbf{i} - \mathbf{j}$

5. $f(x, y) = e^x \sin y$; $\mathbf{p} = (0, \pi/4)$; $\mathbf{a} = \mathbf{i} + \sqrt{3}\mathbf{j}$

6. $f(x, y) = e^{-xy}$; $\mathbf{p} = (1, -1)$; $\mathbf{a} = -\mathbf{i} + \sqrt{3}\mathbf{j}$

7. $f(x, y, z) = x^3y - y^2z^2$; $\mathbf{p} = (-2, 1, 3)$; $\mathbf{a} = \mathbf{i} - 2\mathbf{j} + 2\mathbf{k}$

8. $f(x, y, z) = x^2 + y^2 + z^2$; $\mathbf{p} = (1, -1, 2)$; $\mathbf{a} = \sqrt{2}\mathbf{i} - \mathbf{j} - \mathbf{k}$

In Problems 9–12, find a unit vector in the direction in which f increases most rapidly at **p**. *What is the rate of change in this direction?*

9. $f(x, y) = x^3 - y^5$; $\mathbf{p} = (2, -1)$

10. $f(x, y) = e^y \sin x$; $\mathbf{p} = (5\pi/6, 0)$

11. $f(x, y, z) = x^2yz$; $\mathbf{p} = (1, -1, 2)$

12. $f(x, y, z) = xe^{yz}$; $\mathbf{p} = (2, 0, -4)$

13. In what direction $\mathbf{u}$ does $f(x, y) = 1 - x^2 - y^2$ decrease most rapidly at $\mathbf{p} = (-1, 2)$?

14. In what direction $\mathbf{u}$ does $f(x, y) = \sin(3x - y)$ decrease most rapidly at $\mathbf{p} = (\pi/6, \pi/4)$?

15. Sketch the level curve of $f(x, y) = y/x^2$ that goes through $\mathbf{p} = (1, 2)$. Calculate the gradient vector $\nabla f(\mathbf{p})$ and draw this vector, placing its initial point at $\mathbf{p}$. What should be true about $\nabla f(\mathbf{p})$?

16. Follow the instructions of Problem 15 for $f(x, y) = x^2 + 4y^2$ and $\mathbf{p} = (2, 1)$.

17. Find the directional derivative of $f(x, y, z) = xy + z^2$ at $(1, 1, 1)$ in the direction toward $(5, -3, 3)$.

18. Find the directional derivative of $f(x, y) = e^{-x} \cos y$ at $(0, \pi/3)$ in the direction toward the origin.

19. The temperature at (x, y, z) of a solid sphere centered at the origin is given by

$$T(x, y, z) = \frac{200}{5 + x^2 + y^2 + z^2}$$

(a) By inspection, decide where the solid sphere is hottest.

(b) Find a vector pointing in the direction of greatest increase of temperature at $(1, -1, 1)$.

(c) Does the vector of part (b) point toward the origin?

20. The temperature at (x, y, z) of a solid sphere centered at the origin is $T(x, y, z) = 100e^{-(x^2+y^2+z^2)}$. Note that it is hottest at the origin. Show that the direction of greatest decrease in temperature is always a vector pointing away from the origin.

21. Find the gradient of $f(x, y, z) = \sin\sqrt{x^2 + y^2 + z^2}$. Show that the gradient always points directly toward the origin or directly away from the origin.

22. Suppose that the temperature T at the point (x, y, z) depends only on the distance from the origin. Show that the direction of greatest increase in T is either directly toward the origin or directly away from the origin.

23. The elevation of a mountain above sea level at the point (x, y) is $f(x, y)$. A mountain climber at $\mathbf{p}$ notes that the slope in the easterly direction is $-\frac{1}{2}$ and the slope in the northerly direction is $-\frac{1}{4}$. In what direction should he move for fastest descent?

24. Given that $f_x(2, 4) = -3$ and $f_y(2, 4) = 8$, find the directional derivative of f at $(2, 4)$ in the direction toward $(5, 0)$.

25. The elevation of a mountain above sea level at (x, y) is $3000e^{-(x^2+2y^2)/100}$ meters. The positive x-axis points east and the positive y-axis points north. A climber is directly above $(10, 10)$. If the climber moves northwest, will she ascend or descend and at what slope?

26. If the temperature of a plate at the point (x, y) is $T(x, y) = 10 + x^2 - y^2$, find the path a heat-seeking particle (which always moves in the direction of greatest increase in temperature) would follow if it starts at $(-2, 1)$. *Hint:* The particle moves in the direction of the gradient

$$\nabla T = 2x\mathbf{i} - 2y\mathbf{j}$$

We may write the path in parametric form as

$$\mathbf{r}(t) = x(t)\mathbf{i} + y(t)\mathbf{j}$$

and we want $x(0) = -2$ and $y(0) = 1$. To move in the required direction means that $\mathbf{r}'(t)$ should be parallel to ∇T. This will be satisfied if

$$\frac{x'(t)}{2x(t)} = -\frac{y'(t)}{2y(t)}$$

together with the conditions $x(0) = -2$ and $y(0) = 1$. Now solve this differential equation and evaluate the arbitrary constant of integration.

27. Do Problem 26 assuming that $T(x, y) = 20 - 2x^2 - y^2$.

28. The point $P(1, -1, -10)$ is on the surface $z = -10\sqrt{|xy|}$ (see Figure 1 of Section 12.4). Starting at P, in what direction $\mathbf{u} = u_1\mathbf{i} + u_2\mathbf{j}$ should one move in each case?

(a) To climb most rapidly.

(b) To stay at the same level.

(c) To climb at slope 1.

29. The temperature T in degrees Celsius at (x, y, z) is given by $T = 10/(x^2 + y^2 + z^2)$, where distances are in meters. A bee is flying away from the hot spot at the origin on a spiral path so that its position vector at time t seconds is $\mathbf{r}(t) = t\cos \pi t\,\mathbf{i} + t\sin \pi t\,\mathbf{j} + t\,\mathbf{k}$. Determine the rate of change of T in each case.

(a) With respect to distance traveled at $t = 1$.

(b) With respect to time at $t = 1$. (Think of two ways to do this.)

30. Let $\mathbf{u} = (3\mathbf{i} - 4\mathbf{j})/5$ and $\mathbf{v} = (4\mathbf{i} + 3\mathbf{j})/5$ and suppose that at some point P, $D_{\mathbf{u}}f = -6$ and $D_{\mathbf{v}}f = 17$.

(a) Find ∇f at P.

(b) Note that $\|\nabla f\|^2 = (D_{\mathbf{u}}f)^2 + (D_{\mathbf{v}}f)^2$ in part (a). Show that this relation always holds if $\mathbf{u}$ and $\mathbf{v}$ are perpendicular.

31. Figure 6 shows the contour map for a hill 60 feet high, which we assume has equation $z = f(x, y)$.

(a) A raindrop landing on the hill above point A will reach the xy-plane at A' by following the path of steepest descent from A. Draw this path and use it to estimate A'.

(b) Do the same for point B.

(c) Estimate f_x at C, f_y at D, and $D_{\mathbf{u}}f$ at E, where $\mathbf{u} = (\mathbf{i} + \mathbf{j})/\sqrt{2}$.

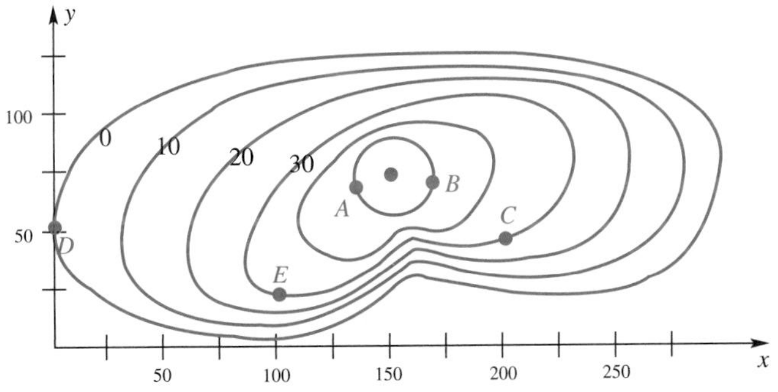

Figure 6

32. According to Theorem A, the differentiability of f at $\mathbf{p}$ implies the existence of $D_{\mathbf{u}}f(\mathbf{p})$ in all directions. Show that the converse is false by considering

$$f(x, y) = \begin{cases} 1 & \text{if } 0 < y < x^2 \\ 0 & \text{otherwise} \end{cases}$$

at the origin.

CAS **33.** Plot the graph of

$$z = x^2 - y^2$$

on $-5 \le x \le 5, -5 \le y \le 5$; also plot its contour map and gradient field, thus illustrating Theorems B and C. Then estimate the xy-coordinates of the point where a raindrop landing above the point $(-5, -0.1)$ will leave this surface.

CAS **34.** Follow the directions of Problem 33 for

$$z = x - x^3/9 - y^2$$

CAS **35.** For the monkey saddle

$$z = x^3 - 3xy^2$$

on $-5 \le x \le 5, -5 \le y \le 5$, estimate the xy-coordinates of the point where a raindrop landing above the point $(5, -0.2)$ will leave the surface.

CAS **36.** Where will a raindrop landing above the point $(4, 1)$ on the surface

$$z = \sin x + \sin y - \sin(x + y)$$

$0 \le x \le 2\pi, 0 \le y \le 2\pi$, come to rest?

Answers to Concepts Review: **1.** $[f(\mathbf{p} + h\mathbf{u}) - f(\mathbf{p})]/h$ **2.** $u_1 f_x(x, y) + u_2 f_y(x, y)$ **3.** greatest increase **4.** level curve

---

12.6 The Chain Rule

The Chain Rule for composite functions of one variable is by now familiar to all our readers. If $y = f(x(t))$, where both f and x are differentiable functions, then

$$\frac{dy}{dt} = \frac{dy}{dx}\frac{dx}{dt}$$

Our goal is to obtain generalizations for functions of several variables.

Beauty and Generality

Does the general analog of the one variable Chain Rule (Theorem A, Section 2.5) hold? Yes, and here is a particularly elegant statement of it. Let $\mathbb{R}$ denote the real numbers and $\mathbb{R}^n$ denote Euclidean n-space, let g be a function from $\mathbb{R}$ to $\mathbb{R}^n$, and let f be a function from $\mathbb{R}^n$ to $\mathbb{R}$. If g is differentiable at t and if f is differentiable at $g(t)$, then the composite function $f \circ g$ is differentiable at t and

$$(f \circ g)'(t) = \nabla f(g(t)) \cdot g'(t)$$

First Version If $z = f(x, y)$, where x and y are functions of t, then it makes sense to ask for dz/dt, and there ought to be a formula for it.

Theorem A Chain Rule

Let $x = x(t)$ and $y = y(t)$ be differentiable at t, and let $z = f(x, y)$ be differentiable at $(x(t), y(t))$. Then $z = f(x(t), y(t))$ is differentiable at t and

$$\frac{dz}{dt} = \frac{\partial z}{\partial x}\frac{dx}{dt} + \frac{\partial z}{\partial y}\frac{dy}{dt}$$

Proof We mimic the one-variable proof of Appendix A.2, Theorem B. To simplify notation, let $\mathbf{p} = (x, y)$, $\Delta\mathbf{p} = (\Delta x, \Delta y)$, and $\Delta z = f(\mathbf{p} + \Delta\mathbf{p}) - f(\mathbf{p})$. Then, since f is differentiable,

$$\begin{aligned}\Delta z &= f(\mathbf{p} + \Delta\mathbf{p}) - f(\mathbf{p}) = \nabla f(\mathbf{p}) \cdot \Delta\mathbf{p} + \boldsymbol{\varepsilon}(\Delta\mathbf{p}) \cdot \Delta\mathbf{p} \\ &= f_x(\mathbf{p})\,\Delta x + f_y(\mathbf{p})\,\Delta y + \boldsymbol{\varepsilon}(\Delta\mathbf{p}) \cdot \Delta\mathbf{p}\end{aligned}$$

with $\boldsymbol{\varepsilon}(\Delta\mathbf{p}) \to \mathbf{0}$ as $\Delta\mathbf{p} \to \mathbf{0}$.

When we divide both sides by Δt, we obtain

$$\text{(1)} \qquad \frac{\Delta z}{\Delta t} = f_x(\mathbf{p})\frac{\Delta x}{\Delta t} + f_y(\mathbf{p})\frac{\Delta y}{\Delta t} + \boldsymbol{\varepsilon}(\Delta\mathbf{p}) \cdot \left\langle \frac{\Delta x}{\Delta t}, \frac{\Delta y}{\Delta t} \right\rangle$$

Now, $\left\langle \frac{\Delta x}{\Delta t}, \frac{\Delta y}{\Delta t} \right\rangle$ approaches $\left\langle \frac{dx}{dt}, \frac{dy}{dt} \right\rangle$ as $\Delta t \to 0$. Also, when $\Delta t \to 0$, both Δx and Δy approach 0 (remember that $x(t)$ and $y(t)$ are continuous, being differentiable). It follows that $\Delta\mathbf{p} \to \mathbf{0}$, and hence $\boldsymbol{\varepsilon}(\Delta\mathbf{p}) \to \mathbf{0}$ as $\Delta t \to 0$. Consequently, when we let $\Delta t \to 0$ in (1), we get

$$\frac{dz}{dt} = f_x(\mathbf{p})\frac{dx}{dt} + f_y(\mathbf{p})\frac{dy}{dt}$$

a result equivalent to the claimed assertion. ■

The Chain Rule: Two-Variable Case

Here is a device that may help you to remember the Chain Rule.

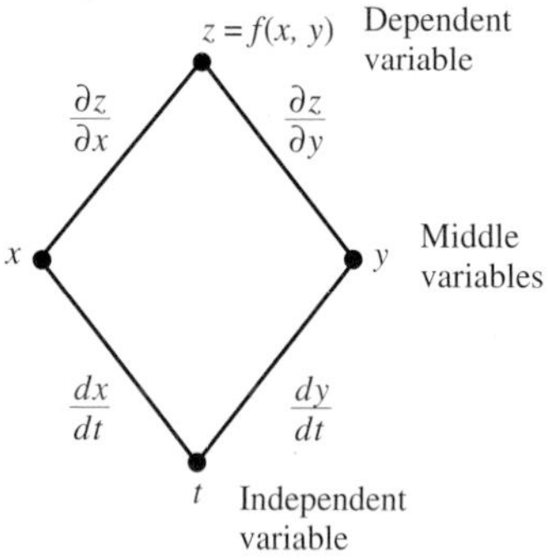

$$\frac{dz}{dt} = \frac{\partial z}{\partial x}\frac{dx}{dt} + \frac{\partial z}{\partial y}\frac{dy}{dt}$$

EXAMPLE 1 Suppose that $z = x^3 y$, where $x = 2t$ and $y = t^2$. Find dz/dt.

SOLUTION

$$\begin{aligned}\frac{dz}{dt} &= \frac{\partial z}{\partial x}\frac{dx}{dt} + \frac{\partial z}{\partial y}\frac{dy}{dt} \\ &= (3x^2 y)(2) + (x^3)(2t) \\ &= 6(2t)^2(t^2) + 2(2t)^3(t) \\ &= 40t^4\end{aligned}$$

■

We could have done Example 1 without use of the Chain Rule. By direct substitution,

$$z = x^3 y = (2t)^3 t^2 = 8t^5$$

and so $dz/dt = 40t^4$. However, the direct substitution method is often not available or not convenient—witness the next example.

EXAMPLE 2 As a solid right circular cylinder is heated, its radius r and height h increase; hence, so does its surface area S. Suppose that at the instant when $r = 10$ centimeters and $h = 100$ centimeters, r is increasing at 0.2 centimeter per hour and h is increasing at 0.5 centimeter per hour. How fast is S increasing at this instant?

SOLUTION The formula for the total surface area of a cylinder (Figure 1) is

$$S = 2\pi rh + 2\pi r^2$$

Thus,

$$\begin{aligned}\frac{dS}{dt} &= \frac{\partial S}{\partial r}\frac{dr}{dt} + \frac{\partial S}{\partial h}\frac{dh}{dt} \\ &= (2\pi h + 4\pi r)(0.2) + (2\pi r)(0.5)\end{aligned}$$

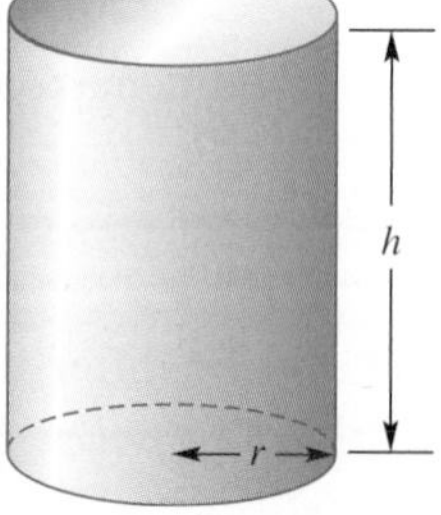

Figure 1

At $r = 10$ and $h = 100$,

$$\begin{aligned}\frac{dS}{dt} &= (2\pi \cdot 100 + 4\pi \cdot 10)(0.2) + (2\pi \cdot 10)(0.5) \\ &= 58\pi \text{ square centimeters per hour}\end{aligned}$$

■

The result in Theorem A extends readily to a function of three variables, as we now illustrate.

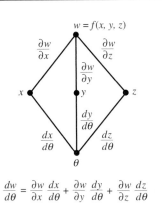

The Chain Rule: Three-Variable Case

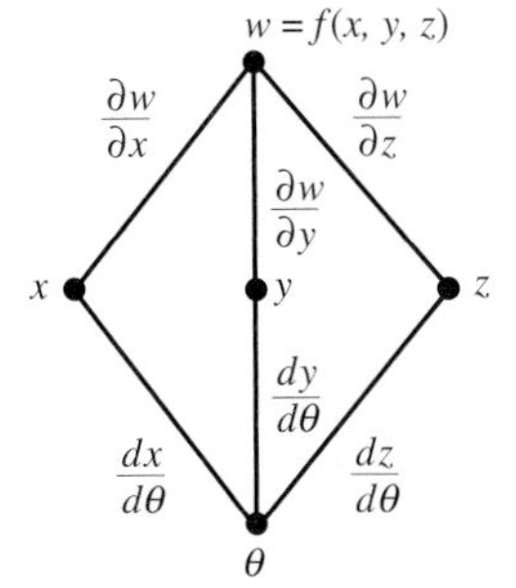

$$\frac{dw}{d\theta} = \frac{\partial w}{\partial x}\frac{dx}{d\theta} + \frac{\partial w}{\partial y}\frac{dy}{d\theta} + \frac{\partial w}{\partial z}\frac{dz}{d\theta}$$

EXAMPLE 3 Suppose that $w = x^2y + y + xz$, where $x = \cos\theta$, $y = \sin\theta$, and $z = \theta^2$. Find $dw/d\theta$ and evaluate it at $\theta = \pi/3$.

SOLUTION

$$\begin{aligned}\frac{dw}{d\theta} &= \frac{\partial w}{\partial x}\frac{dx}{d\theta} + \frac{\partial w}{\partial y}\frac{dy}{d\theta} + \frac{\partial w}{\partial z}\frac{dz}{d\theta} \\ &= (2xy + z)(-\sin\theta) + (x^2 + 1)(\cos\theta) + (x)(2\theta) \\ &= -2\cos\theta\sin^2\theta - \theta^2\sin\theta + \cos^3\theta + \cos\theta + 2\theta\cos\theta\end{aligned}$$

At $\theta = \pi/3$,

$$\begin{aligned}\frac{dw}{d\theta} &= -2\cdot\frac{1}{2}\cdot\frac{3}{4} - \frac{\pi^2}{9}\cdot\frac{\sqrt{3}}{2} + \left(\frac{1}{4} + 1\right)\frac{1}{2} + \frac{2\pi}{3}\cdot\frac{1}{2} \\ &= -\frac{1}{8} - \frac{\pi^2\sqrt{3}}{18} + \frac{\pi}{3}\end{aligned}$$

■

Second Version Suppose next that $z = f(x, y)$, where $x = x(s, t)$ and $y = y(s, t)$. Then it makes sense to ask for $\partial z/\partial s$ and $\partial z/\partial t$.

Theorem B Chain Rule

Let $x = x(s, t)$ and $y = y(s, t)$ have first partial derivatives at (s, t), and let $z = f(x, y)$ be differentiable at $(x(s, t), y(s, t))$. Then $z = f(x(s, t), y(s, t))$ has first partial derivatives given by

1. $\dfrac{\partial z}{\partial s} = \dfrac{\partial z}{\partial x}\dfrac{\partial x}{\partial s} + \dfrac{\partial z}{\partial y}\dfrac{\partial y}{\partial s}$; 2. $\dfrac{\partial z}{\partial t} = \dfrac{\partial z}{\partial x}\dfrac{\partial x}{\partial t} + \dfrac{\partial z}{\partial y}\dfrac{\partial y}{\partial t}$.

Proof If s is held fixed, then $x(s, t)$ and $y(s, t)$ become functions of t alone, which means that Theorem A applies. When we use this theorem with ∂ replacing d to indicate that s is fixed, we obtain the formula in (2) for $\partial z/\partial t$. The formula for $\partial z/\partial s$ is obtained in a similar way by holding t fixed. ■

EXAMPLE 4 If $z = 3x^2 - y^2$, where $x = 2s + 7t$ and $y = 5st$, find $\partial z/\partial t$ and express it in terms of s and t.

SOLUTION

$$\begin{aligned}\frac{\partial z}{\partial t} &= \frac{\partial z}{\partial x}\frac{\partial x}{\partial t} + \frac{\partial z}{\partial y}\frac{\partial y}{\partial t} \\ &= (6x)(7) + (-2y)(5s) \\ &= 42(2s + 7t) - 10st(5s) \\ &= 84s + 294t - 50s^2t\end{aligned}$$

Of course, if we substitute the expressions for x and y into the formula for z and then take the partial derivative with respect to t, we get the same answer:

$$\begin{aligned}\frac{\partial z}{\partial t} &= \frac{\partial}{\partial t}[3(2s + 7t)^2 - (5st)^2] \\ &= \frac{\partial}{\partial t}[12s^2 + 84st + 147t^2 - 25s^2t^2] \\ &= 84s + 294t - 50s^2t\end{aligned}$$

■

Here is the corresponding result for three intermediate variables illustrated in an example.

EXAMPLE 5 If $w = x^2 + y^2 + z^2 + xy$, where $x = st$, $y = s - t$, and $z = s + 2t$, find $\partial w/\partial t$.

SOLUTION

$$\begin{aligned}\frac{\partial w}{\partial t} &= \frac{\partial w}{\partial x}\frac{\partial x}{\partial t} + \frac{\partial w}{\partial y}\frac{\partial y}{\partial t} + \frac{\partial w}{\partial z}\frac{\partial z}{\partial t}\\ &= (2x + y)(s) + (2y + x)(-1) + (2z)(2)\\ &= (2st + s - t)(s) + (2s - 2t + st)(-1) + (2s + 4t)2\\ &= 2s^2t + s^2 - 2st + 2s + 10t\end{aligned}$$

■

Implicit Functions Suppose that $F(x, y) = 0$ defines y implicitly as a function of x, for example, $y = g(x)$, but that the function g is difficult or impossible to determine. We can still find dy/dx. One method for doing this, implicit differentiation, was discussed in Section 2.7. Here is another method.

Let's differentiate both sides of $F(x, y) = 0$ with respect to x using the Chain Rule. We obtain

$$\frac{\partial F}{\partial x}\frac{dx}{dx} + \frac{\partial F}{\partial y}\frac{dy}{dx} = 0$$

Solving for dy/dx yields the formula

$$\boxed{\frac{dy}{dx} = -\frac{\partial F/\partial x}{\partial F/\partial y}}$$

EXAMPLE 6 Find dy/dx if $x^3 + x^2y - 10y^4 = 0$ using

(a) the Chain Rule, and (b) implicit differentiation.

SOLUTION

(a) Let $F(x, y) = x^3 + x^2y - 10y^4$. Then

$$\frac{dy}{dx} = -\frac{\partial F/\partial x}{\partial F/\partial y} = -\frac{3x^2 + 2xy}{x^2 - 40y^3}$$

(b) Differentiate both sides with respect to x to obtain

$$3x^2 + x^2\frac{dy}{dx} + 2xy - 40y^3\frac{dy}{dx} = 0$$

Solving for dy/dx gives the same result as we obtained with the Chain Rule. ■

If z is an implicit function of x and y defined by the equation $F(x, y, z) = 0$, then differentiation of both sides with respect to x, holding y fixed, yields

$$\frac{\partial F}{\partial x}\frac{\partial x}{\partial x} + \frac{\partial F}{\partial y}\frac{\partial y}{\partial x} + \frac{\partial F}{\partial z}\frac{\partial z}{\partial x} = 0$$

If we solve for $\partial z/\partial x$ and note that $\partial y/\partial x = 0$, we get the first of the formulas below. A similar calculation holding x fixed and differentiating with respect to y produces the second formula.

$$\boxed{\frac{\partial z}{\partial x} = -\frac{\partial F/\partial x}{\partial F/\partial z}, \quad \frac{\partial z}{\partial y} = -\frac{\partial F/\partial y}{\partial F/\partial z}}$$

EXAMPLE 7 If $F(x, y, z) = x^3e^{y+z} - y\sin(x - z) = 0$ defines z implicitly as a function of x and y, find $\partial z/\partial x$.

SOLUTION

$$\frac{\partial z}{\partial x} = -\frac{\partial F/\partial x}{\partial F/\partial z} = -\frac{3x^2e^{y+z} - y\cos(x - z)}{x^3e^{y+z} + y\cos(x - z)}$$ ■

Concepts Review

1. If $z = f(x, y)$, where $x = g(t)$ and $y = h(t)$, then the Chain Rule says that $dz/dt =$ ______.

2. Thus, if $z = xy^2$, where $x = \sin t$ and $y = \cos t$, then $dz/dt =$ ______.

3. If $z = f(x, y)$, where $x = g(s, t)$ and $y = h(s, t)$, then the Chain Rule says that $\partial z/\partial t =$ ______.

4. Thus, if $z = xy^2$, where $x = st$ and $y = s^2 + t^2$, then $\partial z/\partial t$ at $s = 1$ and $t = 1$ has the value ______.

Problem Set 12.6

In Problems 1–6, find dw/dt by using the Chain Rule. Express your final answer in terms of t.

1. $w = x^2y^3$; $x = t^3$, $y = t^2$

2. $w = x^2y - y^2x$; $x = \cos t$, $y = \sin t$

3. $w = e^x \sin y + e^y \sin x$; $x = 3t$, $y = 2t$

4. $w = \ln(x/y)$; $x = \tan t$, $y = \sec^2 t$

5. $w = \sin(xyz^2)$; $x = t^3$, $y = t^2$, $z = t$

6. $w = xy + yz + xz$; $x = t^2$, $y = 1 - t^2$, $z = 1 - t$

In Problems 7–12, find $\partial w/\partial t$ by using the Chain Rule. Express your final answer in terms of s and t.

7. $w = x^2y$; $x = st$, $y = s - t$

8. $w = x^2 - y\ln x$; $x = s/t$, $y = s^2t$

9. $w = e^{x^2+y^2}$; $x = s\sin t$, $y = t\sin s$

10. $w = \ln(x + y) - \ln(x - y)$; $x = te^s$, $y = e^{st}$

11. $w = \sqrt{x^2 + y^2 + z^2}$; $x = \cos st$, $y = \sin st$, $z = s^2t$

12. $w = e^{xy+z}$; $x = s + t$, $y = s - t$, $z = t^2$

13. If $z = x^2y$, $x = 2t + s$, and $y = 1 - st^2$, find

$$\left.\frac{\partial z}{\partial t}\right|_{s=1,\,t=-2}$$

14. If $z = xy + x + y$, $x = r + s + t$, and $y = rst$, find

$$\left.\frac{\partial z}{\partial s}\right|_{r=1,\,s=-1,\,t=2}$$

15. If $w = u^2 - u\tan v$, $u = x$, and $v = \pi x$, find

$$\left.\frac{dw}{dx}\right|_{x=1/4}$$

16. If $w = x^2y + z^2$, $x = \rho\cos\theta\sin\phi$, $y = \rho\sin\theta\sin\phi$, and $z = \rho\cos\phi$, find

$$\left.\frac{\partial w}{\partial \theta}\right|_{\rho=2,\,\theta=\pi,\,\phi=\pi/2}$$

17. The part of a tree normally sawed into lumber is the trunk, a solid shaped approximately like a right circular cylinder. If the radius of the trunk of a certain tree is growing $\frac{1}{2}$ inch per year and the height is increasing 8 inches per year, how fast is the volume increasing when the radius is 20 inches and the height is 400 inches? Express your answer in board feet per year (1 board foot = 1 inch by 12 inches by 12 inches).

18. The temperature of a metal plate at (x, y) is e^{-x-3y} degrees. A bug is walking northeast at a rate of $\sqrt{8}$ feet per minute (i.e., $dx/dt = dy/dt = 2$). From the bug's point of view, how is the temperature changing with time as it crosses the origin?

19. A boy's toy boat slips from his grasp at the edge of a straight river. The stream carries it along at 5 feet per second. A crosswind blows it toward the opposite bank at 4 feet per second. If the boy runs along the shore at 3 feet per second following his boat, how fast is the boat moving away from him when $t = 3$ seconds?

20. Sand is pouring onto a conical pile in such a way that at a certain instant the height is 100 inches and increasing at 3 inches per minute and the base radius is 40 inches and increasing at 2 inches per minute. How fast is the volume increasing at that instant?

In Problems 21–24, use the method of Example 6a to find dy/dx.

21. $x^3 + 2x^2y - y^3 = 0$

22. $ye^{-x} + 5x - 17 = 0$

23. $x\sin y + y\cos x = 0$

24. $x^2\cos y - y^2\sin x = 0$

25. If $3x^2z + y^3 - xyz^3 = 0$, find $\partial z/\partial x$ (Example 7).

26. If $ye^{-x} + z\sin x = 0$, find $\partial x/\partial z$ (Example 7).

27. If $T = f(x, y, z, w)$ and x, y, z, and w are each functions of s and t, write a chain rule for $\partial T/\partial s$.

28. Let $z = f(x, y)$, where $x = r\cos\theta$ and $y = r\sin\theta$. Show that

$$\left(\frac{\partial z}{\partial x}\right)^2 + \left(\frac{\partial z}{\partial y}\right)^2 = \left(\frac{\partial z}{\partial r}\right)^2 + \frac{1}{r^2}\left(\frac{\partial z}{\partial \theta}\right)^2$$

29. The wave equation of physics is the partial differential equation

$$\frac{\partial^2 y}{\partial t^2} = c^2 \frac{\partial^2 y}{\partial x^2}$$

where c is a constant. Show that if f is any twice differentiable function then

$$y(x, t) = \tfrac{1}{2}[f(x - ct) + f(x + ct)]$$

satisfies this equation.

30. Show that if $w = f(r - s, s - t, t - r)$ then

$$\frac{\partial w}{\partial r} + \frac{\partial w}{\partial s} + \frac{\partial w}{\partial t} = 0$$

31. Let $F(t) = \int_{g(t)}^{h(t)} f(u)\, du$, where f is continuous and g and h are differentiable. Show that

$$F'(t) = f(h(t))h'(t) - f(g(t))g'(t)$$

and use this result to find $F'(\sqrt{2})$, where

$$F(t) = \int_{\sin\sqrt{2}\,\pi t}^{t^2} \sqrt{9 + u^4}\, du$$

32. Call a function $f(x, y)$ *homogeneous of degree* 1 if $f(tx, ty) = tf(x, y)$ for all $t > 0$. For example, $f(x, y) = x + ye^{y/x}$ satisfies this criterion. Prove **Euler's Theorem** that such a function satisfies

$$f(x, y) = x\frac{\partial f}{\partial x} + y\frac{\partial f}{\partial y}$$

Note: Let $f(x, y)$ denote the value of production from x units of capital and y units of labor. Then f is a homogeneous function (e.g., doubling capital and labor doubles production). Euler's Theorem then asserts an important law of economics that may be phrased as follows: The value of production $f(x, y)$ equals the cost of capital plus the cost of labor provided that they are paid for at their respective marginal rates $\partial f/\partial x$ and $\partial f/\partial y$.

C **33.** Leaving from the same point P, airplane A flies due east while airplane B flies N 50° E. At a certain instant, A is 200 miles from P flying at 450 miles per hour, and B is 150 miles from P flying at 400 miles per hour. How fast are they separating at that instant?

34. Recall Newton's Law of Gravitation, which asserts that the magnitude F of the force of attraction between objects of masses M and m is $F = GMm/r^2$, where r is the distance between them and G is a universal constant. Let an object of mass M be located at the origin, and suppose that a second object of changing mass m (say from fuel consumption) is moving away from the origin so that its position vector is $\mathbf{r} = x\mathbf{i} + y\mathbf{j} + z\mathbf{k}$. Obtain a formula for dF/dt in terms of the time derivatives of m, x, y, and z.

Answers to Concepts Review: **1.** $\frac{\partial z}{\partial x}\frac{dx}{dt} + \frac{\partial z}{\partial y}\frac{dy}{dt}$ **2.** $y^2 \cos t + 2xy(-\sin t) = \cos^3 t - 2\sin^2 t \cos t$ **3.** $\frac{\partial z}{\partial x}\frac{\partial x}{\partial t} + \frac{\partial z}{\partial y}\frac{\partial y}{\partial t}$ **4.** 12

12.7 Tangent Planes and Approximations

We introduced the notion of a tangent plane to a surface in Section 12.4, but dealt only with surfaces determined by equations of the form $z = f(x, y)$ (Figure 1). Now we want to consider the more general situation of a surface determined by $F(x, y, z) = k$. (Note that $z = f(x, y)$ can be written as $F(x, y, z) = f(x, y) - z = 0$.) Consider a curve on this surface passing through the point (x_0, y_0, z_0). If $x = x(t)$, $y = y(t)$, and $z = z(t)$ are parametric equations for this curve, then for all t,

$$F(x(t), y(t), z(t)) = k$$

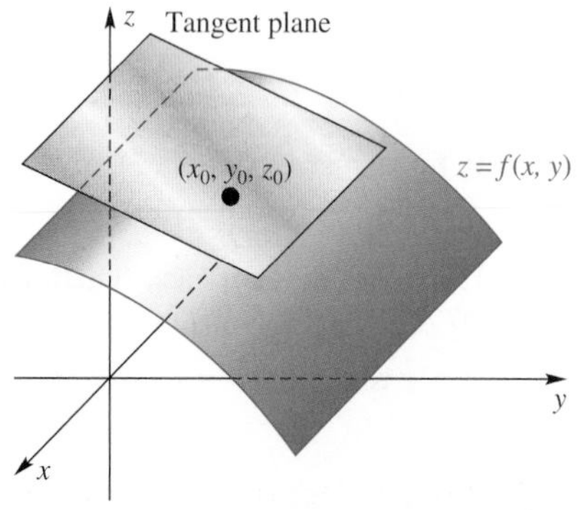

Figure 1

By the Chain Rule,

$$\frac{dF}{dt} = \frac{\partial F}{\partial x}\frac{dx}{dt} + \frac{\partial F}{\partial y}\frac{dy}{dt} + \frac{\partial F}{\partial z}\frac{dz}{dt} = \frac{d}{dt}(k) = 0$$

We can express this in terms of the gradient of F and the derivative of the vector expression for the curve $\mathbf{r}(t) = x(t)\mathbf{i} + y(t)\mathbf{j} + z(t)\mathbf{k}$ as

$$\nabla F \cdot \frac{d\mathbf{r}}{dt} = 0$$

As we learned earlier (Section 11.5), $d\mathbf{r}/dt$ is tangent to the curve. In summary, the gradient at (x_0, y_0, z_0) is perpendicular to the tangent line at this point.

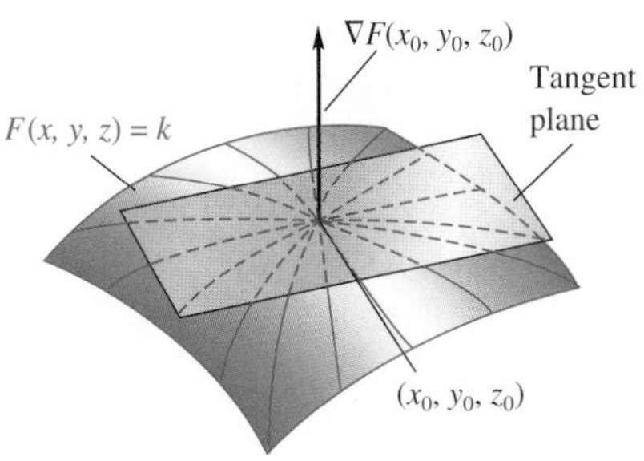

Figure 2

The argument just given is valid for any curve through (x_0, y_0, z_0) that lies in the surface $F(x, y, z) = k$ (Figure 2). This suggests the following general definition.

Definition

Let $F(x, y, z) = k$ determine a surface, and suppose that F is differentiable at a point $P(x_0, y_0, z_0)$ of this surface, with $\nabla F(x_0, y_0, z_0) \neq \mathbf{0}$. Then the plane through P perpendicular to $\nabla F(x_0, y_0, z_0)$ is called the **tangent plane** to the surface at P.

As a consequence of this definition and Section 11.3, we can write the equation of the tangent plane.

Theorem A Tangent Planes

For the surface $F(x, y, z) = k$, the equation of the tangent plane at (x_0, y_0, z_0) is $\nabla F(x_0, y_0, z_0) \cdot \langle x - x_0, y - y_0, z - z_0 \rangle = 0$; that is,

$$F_x(x_0, y_0, z_0)(x - x_0) + F_y(x_0, y_0, z_0)(y - y_0) + F_z(x_0, y_0, z_0)(z - z_0) = 0$$

In particular, for the surface $z = f(x, y)$, the equation of the tangent plane at $(x_0, y_0, f(x_0, y_0))$ is

$$z - z_0 = f_x(x_0, y_0)(x - x_0) + f_y(x_0, y_0)(y - y_0)$$

Proof The first statement is immediate, and the second follows from it by considering $F(x, y, z) = f(x, y) - z$. ■

If z is a function of x and y, say $z = f(x, y)$, then from the second part of Theorem A, we can write the equation of the tangent plane as

$$z - f(x_0, y_0) = f_x(x_0, y_0)(x - x_0) + f_y(x_0, y_0)(y - y_0)$$

Letting $\mathbf{p} = (x, y)$ and $\mathbf{p}_0 = (x_0, y_0)$, we see that the equation of the tangent plane is

$$\begin{aligned} z &= f(x_0, y_0) + \langle f_x(x_0, y_0), f_y(x_0, y_0) \rangle \cdot \langle x - x_0, y - y_0 \rangle \\ &= f(\mathbf{p}_0) + \nabla f(\mathbf{p}_0) \cdot (\mathbf{p} - \mathbf{p}_0) \end{aligned}$$

Thus, our definition in this section agrees with the definition of a tangent plane given in Section 12.4.

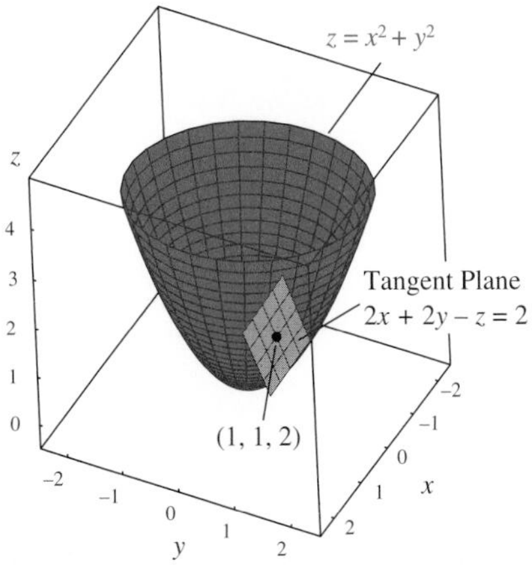

Figure 3

EXAMPLE 1 Find the equation of the tangent plane (Figure 3) to $z = x^2 + y^2$ at the point $(1, 1, 2)$.

SOLUTION Let $f(x, y) = x^2 + y^2$, and note that $\nabla f(x, y) = 2x\mathbf{i} + 2y\mathbf{j}$. Thus, $\nabla f(1, 1) = 2\mathbf{i} + 2\mathbf{j}$, and from Theorem A, the required equation is

$$z - 2 = 2(x - 1) + 2(y - 1)$$

or

$$2x + 2y - z = 2$$ ■

EXAMPLE 2 Find the equation of the tangent plane and the normal line to the surface $x^2 + y^2 + 2z^2 = 23$ at $(1, 2, 3)$.

SOLUTION Let $F(x, y, z) = x^2 + y^2 + 2z^2 - 23$ so that $\nabla F(x, y, z) = 2x\,\mathbf{i} + 2y\,\mathbf{j} + 4z\,\mathbf{k}$ and $\nabla F(1, 2, 3) = 2\mathbf{i} + 4\mathbf{j} + 12\mathbf{k}$. According to Theorem A, the equation of the tangent plane at $(1, 2, 3)$ is

$$2(x - 1) + 4(y - 2) + 12(z - 3) = 0$$

Similarly, the symmetric equations of the normal line through $(1, 2, 3)$ are

$$\frac{x - 1}{2} = \frac{y - 2}{4} = \frac{z - 3}{12}$$

Differentials and Approximations We suggest that you review Section 2.9, where the topics of differentials and approximations are treated for functions of one variable.

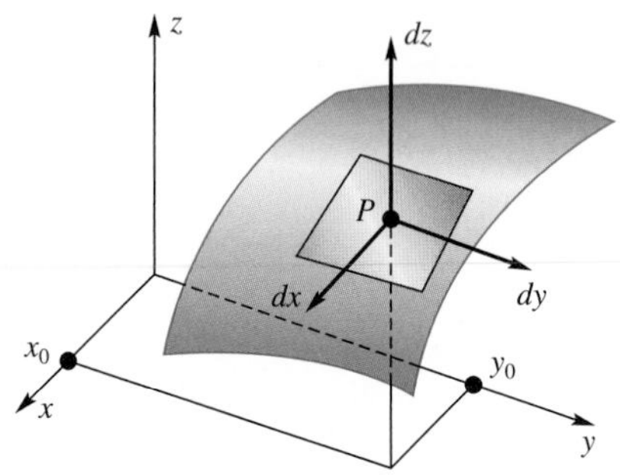

Figure 4

Let $z = f(x, y)$, and let $P(x_0, y_0, z_0)$ be a fixed point on the corresponding surface. Introduce new coordinate axes (the dx-, dy-, and dz-axes) parallel to the old axes, with P as origin (Figure 4). In the old system, the tangent plane at P has equation

$$z - z_0 = f_x(x_0, y_0)(x - x_0) + f_y(x_0, y_0)(y - y_0)$$

but in the new system this takes the simple form

$$dz = f_x(x_0, y_0)\,dx + f_y(x_0, y_0)\,dy$$

This suggests a definition.

Definition

Let $z = f(x, y)$, where f is a differentiable function, and let dx and dy (called the differentials of x and y) be variables. The **differential of the dependent variable,** dz, also called the total **differential of f** and written $df(x, y)$, is defined by

$$dz = df(x, y) = f_x(x, y)\,dx + f_y(x, y)\,dy = \nabla f \cdot \langle dx, dy \rangle$$

The significance of dz arises from the fact that if $dx = \Delta x$ and $dy = \Delta y$ represent small changes in x and y, respectively, then dz will be a good approximation to Δz, the corresponding change in z. This is illustrated in Figure 5 and, while dz does not appear to be a very good approximation to Δz, you can see that it will get better and better as Δx and Δy get smaller.

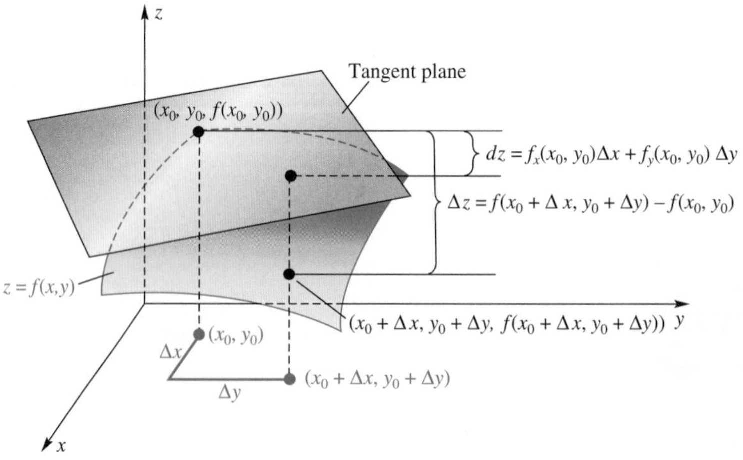

Figure 5

EXAMPLE 3 Let $z = f(x, y) = 2x^3 + xy - y^3$. Compute Δz and dz as (x, y) changes from $(2, 1)$ to $(2.03, 0.98)$.

SOLUTION

$$\begin{aligned}
\Delta z &= f(2.03, 0.98) - f(2, 1) \\
&= 2(2.03)^3 + (2.03)(0.98) - (0.98)^3 - [2(2)^3 + 2(1) - 1^3] \\
&= 0.779062 \\
dz &= f_x(x, y)\,\Delta x + f_y(x, y)\,\Delta y \\
&= (6x^2 + y)\,\Delta x + (x - 3y^2)\,\Delta y
\end{aligned}$$

At $(2, 1)$ with $\Delta x = 0.03$ and $\Delta y = -0.02$,

$$dz = (25)(0.03) + (-1)(-0.02) = 0.77$$

■

EXAMPLE 4 The formula $P = k(T/V)$, where k is a constant, gives the pressure P of a confined gas of volume V and temperature T. Find, approximately, the maximum percentage error in P introduced by an error of $\pm 0.4\%$ in measuring the temperature and an error of $\pm 0.9\%$ in measuring the volume.

SOLUTION The error in P is ΔP, which we will approximate by dP. Thus,

$$\begin{aligned}
|\Delta P| \approx |dP| &= \left|\frac{\partial P}{\partial T}\Delta T + \frac{\partial P}{\partial V}\Delta V\right| \\
&\le \left|\frac{k}{V}(\pm 0.004T)\right| + \left|-\frac{kT}{V^2}(\pm 0.009V)\right| \\
&= \frac{kT}{V}(0.004 + 0.009) = 0.013\frac{kT}{V} = 0.013P
\end{aligned}$$

The maximum relative error, $|\Delta P|/P$, is approximately 0.013, and the maximum percentage error is approximately 1.3%. ■

Taylor Polynomials for Functions of Two or More Variables Recall that for functions of one variable we could approximate the function $f(x)$ using a Taylor polynomial $P_n(x)$. The Taylor polynomials of order one and two are

$$P_1(x) = f(x_0) + f'(x_0)(x - x_0)$$

$$P_2(x) = f(x_0) + f'(x_0)(x - x_0) + \frac{1}{2}f''(x_0)(x - x_0)^2$$

The first is the tangent line at the point $(x_0, f(x_0))$. The analogous quantities for a function $f(x, y)$ of two variables are

$$P_1(x, y) = f(x_0, y_0) + [f_x(x_0, y_0)(x - x_0) + f_y(x_0, y_0)(y - y_0)]$$

which is, of course, the tangent plane at $(x_0, y_0, f(x_0, y_0))$, and

$$\begin{aligned}
P_2(x, y) &= f(x_0, y_0) + [f_x(x_0, y_0)(x - x_0) + f_y(x_0, y_0)(y - y_0)] \\
&\quad + \frac{1}{2}[f_{xx}(x_0, y_0)(x - x_0)^2 + 2f_{xy}(x_0, y_0)(x - x_0)(y - y_0) + f_{yy}(x_0, y_0)(y - y_0)^2]
\end{aligned}$$

These results generalize to nth-order Taylor polynomials and to functions of more than two variables.

EXAMPLE 5 Find the first- and second-order Taylor polynomials to the function $f(x, y) = 1 - e^{-x^2-2y^2}$ at $(0, 0)$, and use them to approximate $f(0.05, -0.06)$.

SOLUTION

$$f_x(x, y) = 2xe^{-x^2-2y^2}$$
$$f_y(x, y) = 4ye^{-x^2-2y^2}$$
$$f_{xx}(x, y) = (2 - 4x^2)e^{-x^2-2y^2}$$
$$f_{yy}(x, y) = (4 - 16y^2)e^{-x^2-2y^2}$$
$$f_{xy}(x, y) = -8xye^{-x^2-2y^2}$$

Thus,

$$P_1(x, y) = f(0,0) + [f_x(0,0)(x - 0) + f_y(0,0)(y - 0)]$$
$$= (1 - e^0) + (0x + 0y) = 0$$

and

$$P_2(x, y) = f(0,0) + [f_x(0,0)(x - 0) + f_y(0,0)(y - 0)]$$
$$+ \frac{1}{2}[f_{xx}(0,0)(x - 0)^2 + 2f_{xy}(0,0)(x - 0)(y - 0) + f_{yy}(0,0)(y - 0)^2]$$
$$= (1 - e^0) + (0x + 0y) + \frac{1}{2}[2x^2 + 2\cdot 0xy + 4y^2]$$
$$= x^2 + 2y^2$$

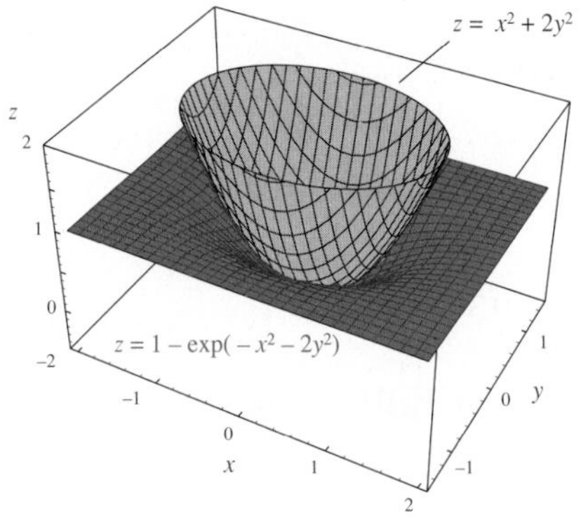

Figure 6

The first-order approximation to $f(0.05, -0.06)$, is

$$f(0.05, -0.06) \approx P_1(0.05, -0.06) = 0$$

and the second-order approximation is

$$f(0.05, -0.06) \approx P_2(0.05, -0.06) = 0.05^2 + 2(-0.06)^2 = 0.00970$$

Figure 6 shows the second-order polynomial (the first-order polynomial is simply $P_1(x, y) = 0$) along with the function $f(x, y)$. The true value for $f(0.05, -0.06)$ is

$$f(0.05, -0.06) = 1 - e^{-0.05^2-2(0.06)^2} = 1 - e^{-0.0097} \approx 0.00965$$ ■

Concepts Review

1. Let $F(x, y, z) = k$ determine a surface. The direction of the gradient vector ∇F is _____ to the surface.

2. Let $z = x^2 + xy$ determine a surface. A vector at $(1, 1, 2)$ perpendicular to this surface is _____.

3. Let $xy^2z^3 = 2$ determine a surface. An equation for the tangent plane at $(2, 1, 1)$ is _____.

4. We define the total differential of $f(x, y)$ by $df(x, y) =$ _____.

Problem Set 12.7

In Problems 1–8, find the equation of the tangent plane to the given surface at the indicated point.

1. $x^2 + y^2 + z^2 = 16; \left(2, 3, \sqrt{3}\right)$
2. $8x^2 + y^2 + 8z^2 = 16; \left(1, 2, \sqrt{2}/2\right)$
3. $x^2 - y^2 + z^2 + 1 = 0; \left(1, 3, \sqrt{7}\right)$
4. $x^2 + y^2 - z^2 = 4; (2, 1, 1)$
5. $z = \frac{x^2}{4} + \frac{y^2}{4}; (2, 2, 2)$
6. $z = xe^{-2y}; (1, 0, 1)$
7. $z = 2e^{3y}\cos 2x; (\pi/3, 0, -1)$
8. $z = x^{1/2} + y^{1/2}; (1, 4, 3)$

C *In Problems 9–12, use the total differential dz to approximate the change in z as (x, y) moves from P to Q. Then use a calculator to find the corresponding exact change Δz (to the accuracy of your calculator). See Example 3.*

9. $z = 2x^2y^3; P(1, 1), Q(0.99, 1.02)$
10. $z = x^2 - 5xy + y; P(2, 3), Q(2.03, 2.98)$
11. $z = \ln(x^2y); P(-2, 4), Q(-1.98, 3.96)$

12. $z = \tan^{-1} xy$; $P(-2, -0.5)$, $Q(-2.03, -0.51)$

13. Find all points on the surface

$$z = x^2 - 2xy - y^2 - 8x + 4y$$

where the tangent plane is horizontal.

14. Find a point on the surface $z = 2x^2 + 3y^2$ where the tangent plane is parallel to the plane $8x - 3y - z = 0$.

15. Show that the surfaces $x^2 + 4y + z^2 = 0$ and $x^2 + y^2 + z^2 - 6z + 7 = 0$ are tangent to each other at $(0, -1, 2)$; that is, show that they have the same tangent plane at $(0, -1, 2)$.

16. Show that the surfaces $z = x^2 y$ and $y = \frac{1}{4}x^2 + \frac{3}{4}$ intersect at $(1, 1, 1)$ and have perpendicular tangent planes there.

17. Find a point on the surface $x^2 + 2y^2 + 3z^2 = 12$ where the tangent plane is perpendicular to the line with parametric equations: $x = 1 + 2t$, $y = 3 + 8t$, $z = 2 - 6t$.

18. Show that the equation of the tangent plane to the ellipsoid

$$\frac{x^2}{a^2} + \frac{y^2}{b^2} + \frac{z^2}{c^2} = 1$$

at (x_0, y_0, z_0) can be written in the form

$$\frac{x_0 x}{a^2} + \frac{y_0 y}{b^2} + \frac{z_0 z}{c^2} = 1$$

19. Find the parametric equations of the line that is tangent to the curve of intersection of the surfaces

$$f(x, y, z) = 9x^2 + 4y^2 + 4z^2 - 41 = 0$$

and

$$g(x, y, z) = 2x^2 - y^2 + 3z^2 - 10 = 0$$

at the point $(1, 2, 2)$. *Hint:* This line is perpendicular to $\nabla f(1, 2, 2)$ and $\nabla g(1, 2, 2)$.

20. Find the parametric equations of the line that is tangent to the curve of intersection of the surfaces $x = z^2$ and $y = z^3$ at $(1, 1, 1)$ (see Problem 19).

21. In determining the specific gravity of an object, its weight in air is found to be $A = 36$ pounds and its weight in water is $W = 20$ pounds, with a possible error in each measurement of 0.02 pound. Find, approximately, the maximum possible error in calculating its specific gravity S, where $S = A/(A - W)$.

22. Use differentials to find the approximate amount of copper in the four sides and bottom of a rectangular copper tank that is 6 feet long, 4 feet wide, and 3 feet deep *inside*, if the sheet copper is $\frac{1}{4}$ inch thick. *Hint:* Make a sketch.

23. The radius and height of a right circular cone are measured with errors of at most 2% and 3%, respectively. Use differentials to estimate the maximum percentage error in the calculated volume (see Example 4).

24. The period T of a pendulum of length L is given by $T = 2\pi\sqrt{L/g}$, where g is the acceleration of gravity. Show that $dT/T = \frac{1}{2}[dL/L - dg/g]$, and use this result to estimate the maximum percentage error in T due to an error of 0.5% in measuring L and 0.3% in measuring g.

25. The formula $1/R = 1/R_1 + 1/R_2$ determines the combined resistance R when resistors of resistance R_1 and R_2 are connected in parallel. Suppose that R_1 and R_2 were measured at 25 and 100 ohms, respectively, with possible errors in each measurement of 0.5 ohm. Calculate R and give an estimate for the maximum error in this value.

26. A bee sat at the point $(1, 2, 1)$ on the ellipsoid $x^2 + y^2 + 2z^2 = 6$ (distances in feet). At $t = 0$, it took off along the normal line at a speed of 4 feet per second. Where and when did it hit the plane $2x + 3y + z = 49$?

27. Show that a plane tangent at any point of the surface $xyz = k$ forms with the coordinate planes a tetrahedron of fixed volume and find this volume.

28. Find and simplify the equation of the tangent plane at (x_0, y_0, z_0) to the surface $\sqrt{x} + \sqrt{y} + \sqrt{z} = a$. Then show that the sum of the intercepts of this plane with the coordinate axes is a^2.

C **29.** For the function $f(x, y) = \sqrt{x^2 + y^2}$, find the second-order Taylor approximation based at $(x_0, y_0) = (3, 4)$. Then estimate $f(3.1, 3.9)$ using

(a) the first-order approximation,

(b) the second-order approximation, and

(c) your calculator directly.

C **30.** For the function $f(x, y) = \tan((x^2 + y^2)/64)$, find the second-order Taylor approximation based at $(x_0, y_0) = (0, 0)$. Then estimate $f(0.2, -0.3)$ using

(a) the first-order approximation,

(b) the second-order approximation, and

(c) your calculator directly.

Answers to Concepts Review: **1.** perpendicular **2.** $\langle 3, 1, -1 \rangle$ **3.** $(x - 2) + 4(y - 1) + 6(z - 1) = 0$ **4.** $\frac{\partial f}{\partial x} dx + \frac{\partial f}{\partial y} dy$

12.8 Maxima and Minima

Our goal is to extend the notions of Chapter 3 to functions of several variables; a quick review of that chapter, especially Sections 3.1 and 3.3, will be helpful. The definitions given there extend almost without change, but for clarity we repeat them. In what follows, let $\mathbf{p} = (x, y)$ and $\mathbf{p}_0 = (x_0, y_0)$ be a variable point and a fixed point, respectively, in two-space (they could just as well be points in n-space).

Definition

Let f be a function with domain S, and let $\mathbf{p}_0$ be a point in S.

(i) $f(\mathbf{p}_0)$ is a **global maximum value** of f on S if $f(\mathbf{p}_0) \geq f(\mathbf{p})$ for all $\mathbf{p}$ in S.

(ii) $f(\mathbf{p}_0)$ is a **global minimum value** of f on S if $f(\mathbf{p}_0) \leq f(\mathbf{p})$ for all $\mathbf{p}$ in S.

(iii) $f(\mathbf{p}_0)$ is a **global extreme value** of f on S if $f(\mathbf{p}_0)$ is either a global maximum value or a global minimum value.

We obtain definitions for **local maximum value** and **local minimum value** if in (i) and (ii) we require only that the inequalities hold on $N \cap S$, where N is some neighborhood of $\mathbf{p}_0$. $f(\mathbf{p}_0)$ is a **local extreme value** of f on S if $f(\mathbf{p}_0)$ is either a local maximum value or a local minimum value.

Figure 1 gives a geometric interpretation of the concepts we have defined. Note that a global maximum (or minimum) is automatically a local maximum (or minimum).

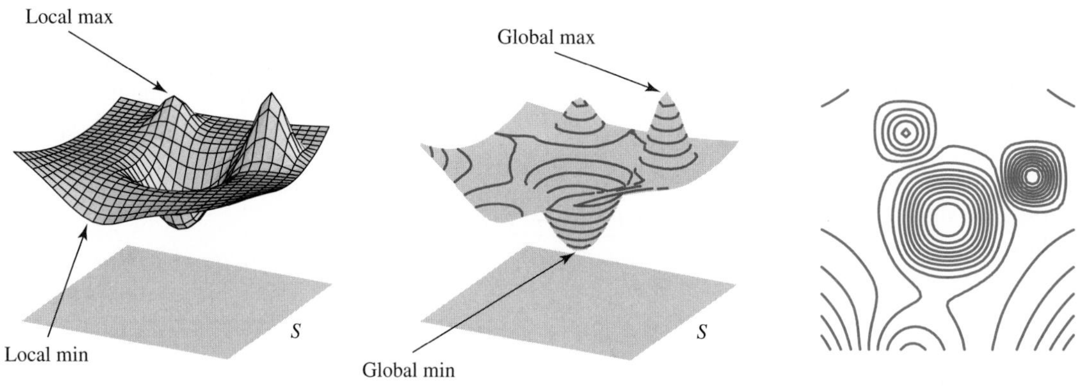

Figure 1

Our first theorem is a big one—difficult to prove, but intuitively clear.

Theorem A Max–Min Existence Theorem

If f is continuous on a closed bounded set S, then f attains both a (global) maximum value and a (global) minimum value there.

The proof may be found in most books on advanced calculus.

Where Do Extreme Values Occur? The situation is analogous to the one-variable case. The **critical points** of f on S are of three types.

1. **Boundary points.** See Section 12.3.
2. **Stationary points.** We call $\mathbf{p}_0$ a stationary point if $\mathbf{p}_0$ is an interior point of S where f is differentiable and $\nabla f(\mathbf{p}_0) = \mathbf{0}$. At such a point, the tangent plane is horizontal.
3. **Singular points.** We call $\mathbf{p}_0$ a singular point if $\mathbf{p}_0$ is an interior point of S where f is not differentiable, for example, a point where the graph of f has a sharp corner.

Now we can state another big theorem; we can actually prove this one.

Theorem B Critical Point Theorem

Let f be defined on a set S containing $\mathbf{p}_0$. If $f(\mathbf{p}_0)$ is an extreme value, then $\mathbf{p}_0$ must be a critical point; that is, either $\mathbf{p}_0$ is

1. a boundary point of S; or
2. a stationary point of f; or
3. a singular point of f.

Proof Suppose that $\mathbf{p}_0$ is neither a boundary point nor a singular point (so that $\mathbf{p}_0$ is an interior point where ∇f exists). We will be done if we can show that $\nabla f(\mathbf{p}_0) = \mathbf{0}$. For simplicity, set $\mathbf{p}_0 = (x_0, y_0)$; the higher-dimensional cases will follow in a similar fashion.

Since f has an extreme value at (x_0, y_0), the function $g(x) = f(x, y_0)$ has an extreme value at x_0. Moreover, g is differentiable at x_0 since f is differentiable at (x_0, y_0) and therefore, by the Critical Point Theorem for functions of one variable (Theorem 3.1 B),

$$g'(x_0) = f_x(x_0, y_0) = 0$$

Similarly, the function $h(y) = f(x_0, y)$ has an extreme value at y_0 and satisfies

$$h'(y_0) = f_y(x_0, y_0) = 0$$

The gradient is $\mathbf{0}$ since both partials are 0. ■

The theorem and its proof are valid whether the extreme values are global or local extreme values.

EXAMPLE 1 Find the local maximum or minimum values of $f(x, y) = x^2 - 2x + y^2/4$.

SOLUTION The given function is differentiable throughout its domain, the xy-plane. Thus, the only possible critical points are the stationary points obtained by setting $f_x(x, y)$ and $f_y(x, y)$ equal to zero. But $f_x(x, y) = 2x - 2$ and $f_y(x, y) = y/2$ are zero only when $x = 1$ and $y = 0$. It remains to decide whether $(1, 0)$ gives a maximum or a minimum or neither. We will develop a simple tool for this soon, but for now we must use a little ingenuity. Note that $f(1, 0) = -1$ and

$$\begin{aligned} f(x, y) &= x^2 - 2x + \frac{y^2}{4} = x^2 - 2x + 1 + \frac{y^2}{4} - 1 \\ &= (x - 1)^2 + \frac{y^2}{4} - 1 \geq -1 \end{aligned}$$

Thus, $f(1, 0)$ is actually a global minimum for f. There are no local maximum values. ■

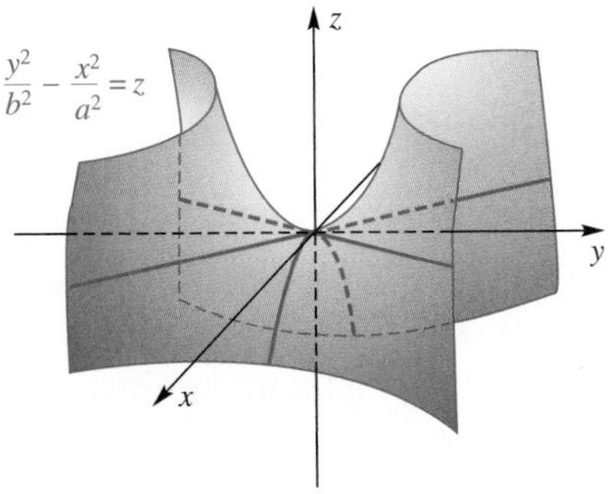

Figure 2

EXAMPLE 2 Find the local minimum or maximum values of $f(x, y) = -x^2/a^2 + y^2/b^2$.

SOLUTION The only critical points are obtained by setting $f_x(x, y) = -2x/a^2$ and $f_y(x, y) = 2y/b^2$ equal to zero. This yields the point $(0, 0)$, which gives neither a maximum nor minimum (see Figure 2). It is called a **saddle point.** The given function has no local extrema. ■

Example 2 illustrates the troublesome fact that $\nabla f(x_0, y_0) = \mathbf{0}$ does not guarantee that there is a local extremum at (x_0, y_0). Fortunately, there is a nice criterion for deciding what is happening at a stationary point—our next topic.

Sufficient Conditions for Extrema The next theorem is analogous to the Second Derivative Test for functions of one variable (Theorem 3.3B). A rigorous proof is beyond the scope of this book, but we provide a sketch of the proof that uses the Taylor polynomial for functions of two variables introduced in the previous section.

Theorem C Second Partials Test

Suppose that $f(x, y)$ has continuous second partial derivatives in a neighborhood of (x_0, y_0) and that $\nabla f(x_0, y_0) = \mathbf{0}$. Let

$$D = D(x_0, y_0) = f_{xx}(x_0, y_0)f_{yy}(x_0, y_0) - f_{xy}^2(x_0, y_0)$$

Then

1. if $D > 0$ and $f_{xx}(x_0, y_0) < 0$, then $f(x_0, y_0)$ is a local maximum value;
2. if $D > 0$ and $f_{xx}(x_0, y_0) > 0$, then $f(x_0, y_0)$ is a local minimum value;
3. if $D < 0$, then $f(x_0, y_0)$ is not an extreme value ((x_0, y_0) is a saddle point);
4. if $D = 0$, then the test is inconclusive.

Sketch of Proof We will assume that $f(0, 0) = 0$ and that $x_0 = y_0 = 0$. (If these conditions do not hold, we can translate the graph, without altering its shape, to make these conditions true and then translate back.) For (x, y) near $(0, 0)$, the function f behaves much like the second-order Taylor polynomial about $(0, 0)$

$$P_2(x, y) =$$
$$f(0,0) + f_x(0,0)x + f_y(0,0)y + \frac{1}{2}[f_{xx}(0,0)x^2 + 2f_{xy}(0,0)xy + f_{yy}(0,0)y^2]$$

(A rigorous proof would take into account the remainder in using $P_2(x, y)$ to approximate $f(x, y)$.) Under the condition that $\nabla f(0, 0) = \langle f_x(0, 0), f_y(0, 0)\rangle = \mathbf{0}$, and the condition $f(0, 0) = 0$, the second-order Taylor polynomial reduces to

$$P_2(x, y) = \frac{1}{2}[f_{xx}(0,0)x^2 + 2f_{xy}(0,0)xy + f_{yy}(0,0)y^2]$$

Let $A = f_{xx}(0, 0)$, $B = f_{xy}(0, 0)$, and $C = f_{yy}(0, 0)$. This gives

$$P_2(x, y) = \frac{1}{2}[Ax^2 + 2Bxy + Cy^2]$$

Completing the square on x gives

$$P_2(x, y) = \frac{A}{2}\left[x^2 + 2\frac{By}{A}x + \left(\frac{By}{A}\right)^2 + \frac{C}{A}y^2 - \left(\frac{By}{A}\right)^2\right]$$
$$= \frac{A}{2}\left[\left(x + \frac{B}{A}y\right)^2 + \left(\frac{C}{A} - \frac{B^2}{A^2}\right)y^2\right]$$

The expression $\left(x + \frac{B}{A}y\right)^2$ is positive for all (x, y) except $(0, 0)$. If $\frac{C}{A} - \frac{B^2}{A^2} > 0$, that is, if $AC - B^2 = f_{xx}(0,0)f_{yy}(0,0) - f_{xy}^2(0,0) = D > 0$, then the expression in brackets will be positive for all $(x, y) \neq (0, 0)$. If, in addition, $A > 0$, then $P_2(x, y) > 0$ for $(x, y) \neq (0, 0)$, in which case $f(0, 0) = 0$ is a local minimum. Similarly, if $D > 0$ and $A < 0$ then $P_2(x, y) < 0$ for $(x, y) \neq (0, 0)$, in which case $f(0, 0)$ is a local maximum. When $D > 0$, then the graph of $P_2(x, y)$ is a (rotated) paraboloid with vertex at $(0, 0)$ opening upward if $A > 0$ and downward if $A < 0$.

When $D < 0$, the graph of $P_2(x, y)$ is a rotated hyperbolic paraboloid with a saddle point at $(0, 0)$. (See Figure 11 in Section 11.8.)

Finally, when $D = 0$, then all terms in $P_2(x, y)$ are zero, and so $P_2(x, y) = 0$. In this case, higher order terms would be required to determine the behavior of $f(x, y)$ near $(0, 0)$. Since the theorem makes no assumptions about these higher-order terms, we can draw no conclusion about whether $f(0, 0)$ is a local minimum or maximum. ■

EXAMPLE 3 Find the extrema, if any, of the function F defined by $F(x, y) = 3x^3 + y^2 - 9x + 4y$.

SOLUTION Since $F_x(x, y) = 9x^2 - 9$ and $F_y(x, y) = 2y + 4$, the critical points, obtained by solving the simultaneous equations $F_x(x, y) = F_y(x, y) = 0$, are $(1, -2)$ and $(-1, -2)$.

Now $F_{xx}(x, y) = 18x$, $F_{yy}(x, y) = 2$, and $F_{xy} = 0$. Thus, at the critical point $(1, -2)$,

$$D = F_{xx}(1, -2) \cdot F_{yy}(1, -2) - F_{xy}^2(1, -2) = 18(2) - 0 = 36 > 0$$

Furthermore, $F_{xx}(1, -2) = 18 > 0$ and so, by Theorem C(2), $F(1, -2) = -10$ is a local minimum value of F.

In testing the given function at the other critical point, $(-1, -2)$, we find that $F_{xx}(-1, -2) = -18$, $F_{yy}(-1, -2) = 2$, and $F_{xy}(-1, -2) = 0$, which makes $D = -36 < 0$. Thus, by Theorem C(3), $(-1, -2)$ is a saddle point and $F(-1, -2)$ is not an extremum. ■

EXAMPLE 4 Find the minimum distance between the origin and the surface $z^2 = x^2y + 4$.

SOLUTION Let $P(x, y, z)$ be any point on the surface. The square of the distance between the origin and P is $d^2 = x^2 + y^2 + z^2$. We seek the coordinates of P that make d^2 (and hence d) a minimum.

Since P is on the surface, its coordinates satisfy the equation of the surface. Substituting $z^2 = x^2y + 4$ in $d^2 = x^2 + y^2 + z^2$, we obtain d^2 as a function of two variables x and y.

$$d^2 = f(x, y) = x^2 + y^2 + x^2y + 4$$

To find the critical points, we set $f_x(x, y) = 0$ and $f_y(x, y) = 0$, obtaining

$$2x + 2xy = 0 \quad \text{and} \quad 2y + x^2 = 0$$

By eliminating y between these equations, we get

$$2x - x^3 = 0$$

Thus, $x = 0$ or $x = \pm\sqrt{2}$. Substituting these values in the second of the equations, we obtain $y = 0$ and $y = -1$. Therefore, the critical points are $(0, 0)$, $(\sqrt{2}, -1)$, and $(-\sqrt{2}, -1)$. (There are no boundary points.)

To test each of these, we need $f_{xx}(x, y) = 2 + 2y$, $f_{yy}(x, y) = 2$, $f_{xy}(x, y) = 2x$, and

$$D(x, y) = f_{xx}f_{yy} - f_{xy}^2 = 4 + 4y - 4x^2$$

Since $D(\pm\sqrt{2}, -1) = -8 < 0$, neither $(\sqrt{2}, -1)$ nor $(-\sqrt{2}, -1)$ yields an extremum. However, $D(0, 0) = 4 > 0$ and $f_{xx}(0, 0) = 2 > 0$; so $(0, 0)$ yields the minimum distance. Substituting $x = 0$ and $y = 0$ in the expression for d^2, we find $d^2 = 4$.

The minimum distance between the origin and the given surface is 2. ■

It is easy to check the boundary points when we are maximizing (or minimizing) a function of one variable, because the boundary usually consists of just the two endpoints. For functions of two or more variables, it is a more difficult problem. In some cases, such as the next example, the entire boundary can be parameterized and then the methods of Chapter 3 can be used to find the maximum and minimum. In other cases, such as Example 6, pieces of the boundary can be parameterized and then the function can be maximized or minimized on each piece. We will see another method, Lagrange multipliers, in the next section.

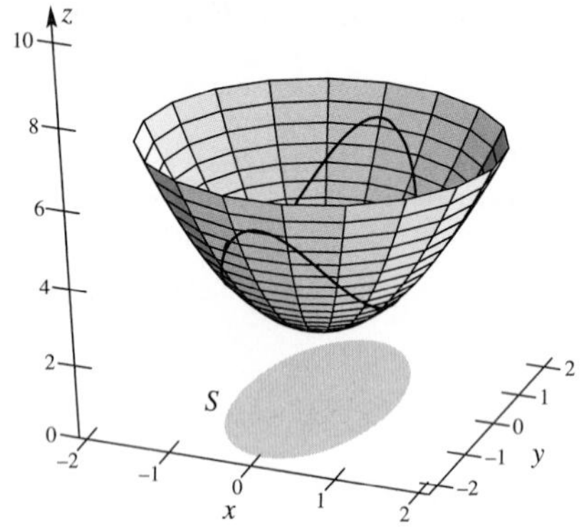

Figure 3

EXAMPLE 5 Find the maximum and minimum values of $f(x, y) = 2 + x^2 + y^2$ on the closed and bounded set $S = \left\{(x, y): x^2 + \frac{1}{4}y^2 \le 1\right\}$.

SOLUTION Figure 3 shows the surface $z = f(x, y)$ along with the set S, shown in the xy-plane. The first partial derivatives are $f_x(x, y) = 2x$ and $f_y(x, y) = 2y$. Thus, the only interior critical point is $(0, 0)$. Since

$$D(0,0) = f_{xx}(0,0)f_{yy}(0,0) - f_{xy}^2(0,0) = 2\cdot 2 - 0 = 4 > 0$$

and $f_{xx}(0, 0) = 2 > 0$, we know that $f(0, 0) = 2$ is a minimum.

The global maximum must then occur on the boundary of S. Figure 3 also shows the boundary of S projected upward to the surface $z = f(x, y)$; somewhere along this curve, f should achieve a maximum. We can describe parametrically the boundary of S by

$$x = \cos t, \qquad y = 2 \sin t, \qquad 0 \le t \le 2\pi$$

The optimization problem then reduces to one of optimizing the function of *one* variable

$$g(t) = f(\cos t, 2 \sin t), \qquad 0 \le t \le 2\pi$$

By the Chain Rule (Theorem 12.6A),

$$\begin{aligned} g'(t) &= \frac{\partial f}{\partial x}\frac{dx}{dt} + \frac{\partial f}{\partial y}\frac{dy}{dt} \\ &= 2x(-\sin t) + 2y(2\cos t) \\ &= -2\sin t\cos t + 8\sin t\cos t \\ &= 6\sin t\cos t = 3\sin 2t \end{aligned}$$

Setting $g'(t) = 0$ yields $t = 0, \frac{\pi}{2}, \pi, \frac{3\pi}{2}$, and 2π. Thus, g has five critical points on $[0, 2\pi]$. These five values for t determine the five critical points $(1, 0)$, $(0, 2)$, $(-1, 0)$, $(0, -2)$, and $(1, 0)$ for f; the last point is the same as the first because an angle of 2π yields the same point as an angle of 0. The corresponding values of f are

$$f(1,0) = 3 \qquad f(0,2) = 6$$
$$f(-1,0) = 3 \qquad f(0,-2) = 6$$

At the critical point interior to S, we have $f(0, 0) = 2$. Therefore, we conclude that the minimum value of f on S is 2, and the maximum value is 6. ■

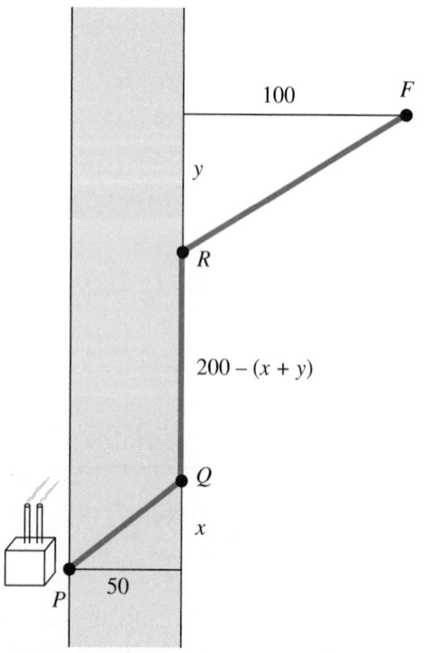

Figure 4

EXAMPLE 6 A power cable must be laid from a power plant to a new factory located across a shallow river. The river is 50 feet wide and the factory is 200 feet down stream and 100 feet from the bank as shown in Figure 4. The cable costs \$600 per foot to lay under water, \$100 per foot to lay along the bank, and \$200 per foot to lay from the bank to the factory. What path should be taken to minimize the cost and what is the minimum cost?

SOLUTION Let P, Q, R, and F denote the points as shown in Figure 4. Let x denote the distance from the point directly across from the power plant to Q, and let y denote the distance from R to the point on the bank nearest the factory. The lengths and costs of the cable are shown in the table below.

Type of Cable	Length	Cost
Underwater	$\sqrt{x^2 + 50^2}$	\$600/foot
Along Bank	$200 - (x + y)$	\$100/foot
Across Land	$\sqrt{y^2 + 100^2}$	\$200/foot

The total cost is therefore

$$C(x, y) = 600\sqrt{x^2 + 50^2} + 100(200 - x - y) + 200\sqrt{y^2 + 100^2}$$

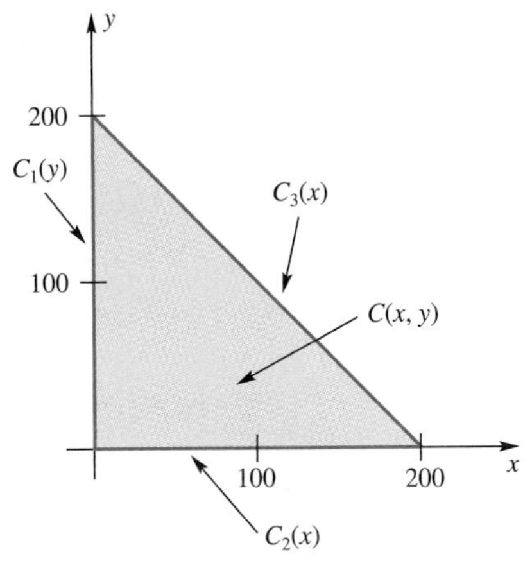

Figure 5

The values of (x, y) must satisfy $x \geq 0$, $y \geq 0$, $x + y \leq 200$ (see Figure 5). Taking partial derivatives and setting them equal to 0 gives

$$C_x(x, y) = 300(x^2 + 50^2)^{-1/2}(2x) - 100 = \frac{600x}{\sqrt{x^2 + 50^2}} - 100 = 0$$

$$C_y(x, y) = 100(y^2 + 100^2)^{-1/2}(2y) - 100 = \frac{200y}{\sqrt{y^2 + 100^2}} - 100 = 0$$

The solution to this system of equations is

$$x = \frac{10}{7}\sqrt{35} \approx 8.4515$$

$$y = \frac{100}{3}\sqrt{3} \approx 57.735$$

We now apply the second partials test:

$$C_{xx}(x, y) = \frac{600\sqrt{x^2 + 50^2} - 600x^2/\sqrt{x^2 + 50^2}}{x^2 + 50^2}$$

$$C_{yy}(x, y) = \frac{200\sqrt{y^2 + 100^2} - 200y^2/\sqrt{y^2 + 100^2}}{y^2 + 100^2}$$

$$C_{xy}(x, y) = 0$$

Evaluating D at $x = \frac{10}{7}\sqrt{35}$ and $y = \frac{100}{3}\sqrt{3}$, gives

$$D = C_{xx}\left(\tfrac{10}{7}\sqrt{35}, \tfrac{100}{3}\sqrt{3}\right)C_{yy}\left(\tfrac{10}{7}\sqrt{35}, \tfrac{100}{3}\sqrt{3}\right) - \left[C_{xy}\left(\tfrac{10}{7}\sqrt{35}, \tfrac{100}{3}\sqrt{3}\right)\right]^2$$

$$= \frac{35\sqrt{35}}{18}\frac{3\sqrt{3}}{4} - 0^2$$

$$= \frac{35}{24}\sqrt{105} > 0$$

Thus, $x = \frac{10}{7}\sqrt{35}$ and $y = \frac{100}{3}\sqrt{3}$ yields a local minimum, which is

$$C\left(\tfrac{10}{7}\sqrt{35}, \tfrac{100}{3}\sqrt{3}\right) =$$

$$36{,}000\sqrt{\frac{5}{7}} + 100\left(200 - \tfrac{10}{7}\sqrt{35} - \tfrac{100}{3}\sqrt{3}\right) + \frac{40{,}000}{\sqrt{3}} \approx \$66{,}901$$

We must also check the boundary. When $x = 0$, the cost function is

$$C_1(y) = C(0, y) = 30{,}000 + 100(200 - y) + 200\sqrt{y^2 + 100^2}$$

(The function $C_1(y)$ agrees with $C(x, y)$ on the left boundary of the triangular domain for $C(x, y)$. Similarly, $C_2(x)$ and $C_3(x)$, defined below, agree with $C(x, y)$ on the lower and upper boundaries, respectively. See Figure 5.) Using the methods from Chapter 3 (details are left to the reader), we find that C_1 achieves a minimum of approximately \$67,321 when $y = 100/\sqrt{3}$. On the boundary where $y = 0$, the cost function is

$$C_2(x) = C(x, 0) = 20{,}000 + 600\sqrt{x^2 + 50^2} + 100(200 - x)$$

Again, using the methods of Chapter 3, we find that C_2 reaches a minimum of approximately \$69,580 when $x = 10\sqrt{5/7}$. Finally, we must address the boundary where $x + y = 200$. We can substitute $200 - x$ for y to get the cost in terms of x alone:

$$\begin{aligned} C_3(x) &= C(x, 200 - x) \\ &= 600\sqrt{x^2 + 50^2} + 200\sqrt{(200 - x)^2 + 100^2} \end{aligned}$$

This function reaches a minimum of approximately \$73,380 when $x \approx 15.3292$.

Therefore, the minimum cost path is the path where $x = \frac{10}{7}\sqrt{35} \approx 8.4515$ and $y = \frac{100}{3}\sqrt{3} \approx 57.735$, which yields a cost of \$66,901. ■

Concepts Review

1. If $f(x, y)$ is continuous on a(n) _____ set S, then f attains both a maximum value and a minimum value on S.

2. If $f(x, y)$ attains a maximum value at a point (x_0, y_0), then (x_0, y_0) is either a(n) _____ point or a(n) _____ point or a(n) _____ point.

3. If (x_0, y_0) is a stationary point for f, then f is differentiable there and _____.

4. In the Second Partials Test for a function f of two variables, the number $D =$ _____ plays a crucial role.

Problem Set 12.8

In Problems 1–10, find all critical points. Indicate whether each such point gives a local maximum or a local minimum, or whether it is a saddle point. Hint: Use Theorem C.

1. $f(x, y) = x^2 + 4y^2 - 4x$
2. $f(x, y) = x^2 + 4y^2 - 2x + 8y - 1$
3. $f(x, y) = 2x^4 - x^2 + 3y^2$
4. $f(x, y) = xy^2 - 6x^2 - 3y^2$
5. $f(x, y) = xy$
6. $f(x, y) = x^3 + y^3 - 6xy$
7. $f(x, y) = xy + \dfrac{2}{x} + \dfrac{4}{y}$
8. $f(x, y) = e^{-(x^2+y^2-4y)}$
9. $f(x, y) = \cos x + \cos y + \cos(x + y)$; $0 < x < \pi/2, 0 < y < \pi/2$
10. $f(x, y) = x^2 + a^2 - 2ax \cos y$; $-\pi < y < \pi$

In Problems 11–14, find the global maximum value and global minimum value of f on S and indicate where each occurs.

11. $f(x, y) = 3x + 4y$; $S = \{(x, y): 0 \le x \le 1, -1 \le y \le 1\}$
12. $f(x, y) = x^2 + y^2$; $S = \{(x, y): -1 \le x \le 3, -1 \le y \le 4\}$
13. $f(x, y) = x^2 - y^2 + 1$; $S = \{(x, y): x^2 + y^2 \le 1\}$ (See Example 5.)
14. $f(x, y) = x^2 - 6x + y^2 - 8y + 7$; $S = \{(x, y): x^2 + y^2 \le 1\}$

15. Express a positive number N as a sum of three positive numbers such that the product of these three numbers is a maximum.

16. Use the methods of this section to find the shortest distance from the origin to the plane $x + 2y + 3z = 12$.

17. Find the dimensions of the closed rectangular box of volume V_0 with minimum surface area.

18. Find the dimensions of the rectangular box of volume V_0 for which the sum of the edge lengths is least.

19. A rectangular metal tank with open top is to hold 256 cubic feet of liquid. What are the dimensions of the tank that require the least material to build?

20. A rectangular box, whose edges are parallel to the coordinate axes, is inscribed in the ellipsoid $96x^2 + 4y^2 + 4z^2 = 36$. What is the greatest possible volume for such a box?

21. Find the three-dimensional vector with length 9, the sum of whose components is a maximum.

22. Find the point on the plane $2x + 4y + 3z = 12$ that is closest to the origin. What is the minimum distance?

CAS 23. Find the point on the paraboloid $z = x^2 + y^2$ that is closest to $(1, 2, 0)$. What is the minimum distance?

24. Find the minimum distance between the point $(1, 2, 0)$ and the quadric cone $z^2 = x^2 + y^2$.

25. An open gutter with cross section in the form of a trapezoid with equal base angles is to be made by bending up equal strips along both sides of a long piece of metal 12 inches wide. Find the base angles and the width of the sides for maximum carrying capacity.

26. Find the minimum distance between the lines having parametric equations $x = t - 1$, $y = 2t$, $z = t + 3$ and $x = 3s$, $y = s + 2$, $z = 2s - 1$.

27. Convince yourself that the maximum and minimum values of a linear function $f(x, y) = ax + by + c$ over a closed polygonal set (i.e., a polygon and its interior) will always occur at a vertex of the polygon. Then use this fact to find each of the following:

(a) maximum value of $2x + 3y + 4$ on the closed polygon with vertices $(-1, 2)$, $(0, 1)$, $(1, 0)$, $(-3, 0)$, and $(0, -4)$

(b) minimum value of $-3x + 2y + 1$ on the closed polygon with vertices $(-3, 0)$, $(0, 5)$, $(2, 3)$, $(4, 0)$, and $(1, -4)$

28. Use the result of Problem 27 to maximize $2x + y$ subject to the constraints $4x + y \le 8$, $2x + 3y \le 14$, $x \ge 0$, and $y \ge 0$. *Hint:* Begin by graphing the set determined by the constraints.

29. Find the maximum and minimum values of $z = y^2 - x^2$ (Figure 2) on the closed triangle with vertices $(0, 0)$, $(1, 2)$, and $(2, -2)$.

30. Least Squares Given n points $(x_1, y_1), (x_2, y_2), \ldots, (x_n, y_n)$ in the xy-plane, we wish to find the line $y = mx + b$ such that the sum of the squares of the vertical distances from the points to the line is a minimum; that is, we wish to minimize

$$f(m, b) = \sum_{i=1}^{n} (y_i - mx_i - b)^2$$

(See Figure 6. Also, remember that the x_i's and the y_i's are *fixed*.)

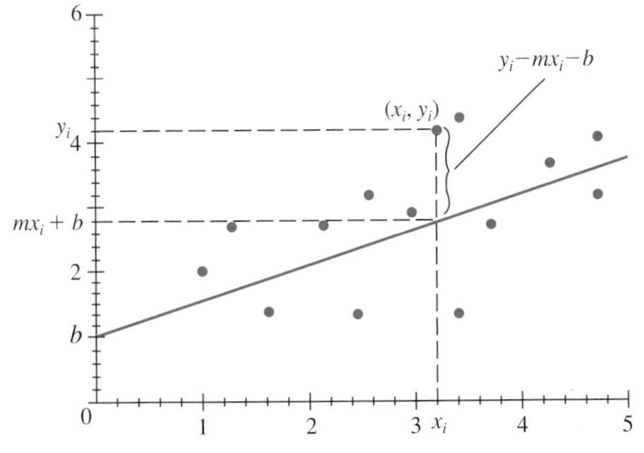

Figure 6

(a) Find $\partial f/\partial m$ and $\partial f/\partial b$, and set these results equal to zero. Show that this leads to the system of equations

$$m\sum_{i=1}^{n} x_i^2 + b\sum_{i=1}^{n} x_i = \sum_{i=1}^{n} x_i y_i$$

$$m\sum_{i=1}^{n} x_i + nb = \sum_{i=1}^{n} y_i$$

(b) Solve this system for m and b.

(c) Use the Second Partials Test (Theorem C) to show that f is *minimized* for this choice of m and b.

31. Find the least-squares line (Problem 30) for the data $(3, 2)$, $(4, 3)$, $(5, 4)$, $(6, 4)$, and $(7, 5)$.

32. Find the maximum and minimum values of $z = 2x^2 + y^2 - 4x - 2y + 5$ (Figure 3) on the set bounded by the closed triangle with vertices $(0, 0)$, $(4, 0)$, and $(0, 1)$.

33. Suppose that in Example 6, the costs were as follows: underwater \$400/foot; along the bank \$200/foot; and across land \$300/foot. What path should be taken to minimize the cost and what is the minimum cost?

34. Suppose that in Example 6, the costs were as follows: underwater \$500/foot; along the bank \$200/foot; and across land \$100/foot. What path should be taken to minimize the cost and what is the minimum cost?

35. Find the maximum and minimum values of $f(x, y) = 10 + x + y$ on the disk $x^2 + y^2 \le 9$. *Hint:* Parametrize the boundary by $x = 3\cos t$, $y = 3\sin t$, $0 \le t \le 2\pi$.

36. Find the maximum and minimum values of $f(x, y) = x^2 + y^2$ on the ellipse with interior $x^2/a^2 + y^2/b^2 \le 1$, where $a > b$. *Hint:* Parametrize the boundary by $x = a\cos t$, $y = b\sin t$, $0 \le t \le 2\pi$.

37. A box is to be made where the material for the sides and the lid cost \$0.25 per square foot and the cost for the bottom is \$0.40 per square foot. Find the dimensions of a box with volume 2 cubic feet that has minimum cost.

CAS **38.** A steel box without a lid having volume 60 cubic feet is to be made from material that costs \$4 per square foot for the bottom and \$1 per square foot for the sides. Welding the sides to the bottom costs \$3 per linear foot and welding the sides together costs \$1 per linear foot. Find the dimensions of the box that has minimum cost and find the minimum cost. *Hint:* Use symmetry to obtain one equation in one unknown and use a CAS or Newton's Method to approximate the solution.

39. Suppose that the temperature T on the circular plate $\{(x, y): x^2 + y^2 \le 1\}$ is given by $T = 2x^2 + y^2 - y$. Find the hottest and coldest spots on the plate.

40. A wire of length k is to be cut into (at most) three pieces to form a circle and two squares, any of which may be degenerate. How should this be done to maximize and minimize the area thus enclosed? *Hint:* Reduce the problem to that of optimizing $x^2 + y^2 + z^2$ on the part of the plane $2\sqrt{\pi}x + 4y + 4z = k$ in the first octant. Then reason geometrically.

41. Find the shape of the triangle of largest area that can be inscribed in a circle of radius r. *Hint:* Let α, β, and γ be the central angles that subtend the three sides of the triangle. Show that the area of the triangle is $\frac{1}{2}r^2[\sin\alpha + \sin\beta - \sin(\alpha + \beta)]$. Maximize.

42. Let (a, b, c) be a fixed point in the first octant. Find the plane through this point that cuts off from the first octant the tetrahedron of minimum volume, and determine the resulting volume.

CAS *Sometimes finding the extrema for a function of two variables can best be handled by commonsense methods using a computer. To illustrate, look at the pictures of the surfaces and the corresponding contour maps for the five functions graphed in Figures 15–19 of Section 12.1. Note that these graphs suggest that we can locate the extrema visually. With the additional ability to evaluate the function at points, we can experimentally approximate maxima and minima with good accuracy. In Problems 43–53, use your technology to find the point where the indicated maximum or minimum occurs and give the functional value at this point. Note that Problems 43–47 refer to the five functions from Section 12.1.*

43. $f(x, y) = x - x^3/9 - y^2/2$; $-3.8 \le x \le 3.8$, $-3.8 \le y \le 3.8$; local maximum point near $(2, 0)$; also global maximum. Check using calculus.

44. $f(x, y) = y/(1 + x^2 + y^2)$; $-5 \le x \le 5, -5 \le y \le 5$; global maximum point and global minimum. Check using calculus.

45. $f(x, y) = -1 + \cos(y/(1 + x^2 + y^2))$; $-3.8 \le x \le 3.8$, $-3.8 \le y \le 3.8$; global minimum.

46. $f(x, y) = \exp(-x^2 - y^2 + xy/4)$; $-2 \le x \le 2$, $-2 \le y \le 2$; global maximum and global minimum. Check using calculus.

47. $f(x, y) = \exp(-(x^2 + y^2)/4)\sin\left(x\sqrt{|y|}\right)$; $-5 \le x \le 5$, $-5 \le y \le 5$; global maximum and global minimum.

48. $f(x, y) = -x/(x^2 + y^2)$, $f(0, 0) = 0$; $-1 \le x \le 1$, $-1 \le y \le 1$; global maximum and global minimum. Be careful.

49. $f(x, y) = 8\cos(xy + 2x) + x^2y^2$; $-3 \le x \le 3$, $-3 \le y \le 3$; global maximum and global minimum.

50. $f(x, y) = (\sin x)/(6 + x + |y|)$; $-3 \le x \le 3$, $-3 \le y \le 3$; global maximum and global minimum.

51. $f(x, y) = \cos(|x| + y^2) + 10x\exp(-x^2 - y^2)$; $-2 \le x \le 2, -2 \le y \le 2$; global maximum and global minimum.

52. $f(x, y) = (x^2 - x - 5)(1 - 9y)\sin x \sin y$; $-6 \le x \le 6, -6 \le y \le 6$; global maximum and global minimum.

53. $f(x, y) = 2\sin x + \sin y - \sin(x + y)$; $0 \le x \le 2\pi, 0 \le y \le 2\pi$; global maximum and global minimum.

54. Let three arms of lengths 6, 8, and 10 emanate from N, as shown in Figure 7. Let $K(\alpha, \beta)$ and $L(\alpha, \beta)$ denote the area and perimeter, respectively, of the triangle ABC determined by these arms.

(a) Find formulas for $K(\alpha, \beta)$ and $L(\alpha, \beta)$.

(b) Determine (α, β) in $D = \{(\alpha, \beta): 0 \le \alpha \le \pi, 0 \le \beta \le \pi\}$ that maximizes $K(\alpha, \beta)$.

(c) Determine (α, β) in D that maximizes $L(\alpha, \beta)$

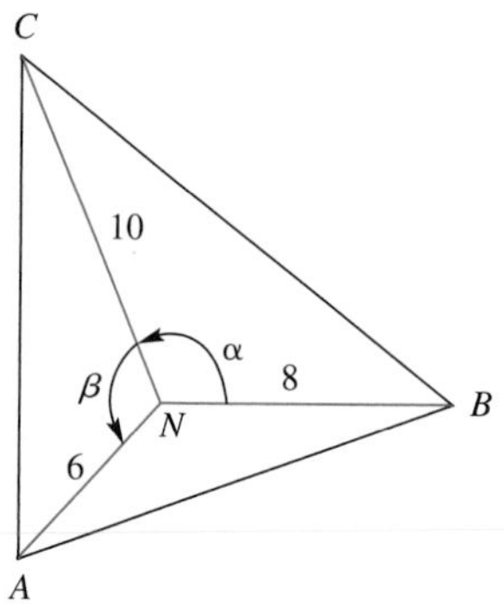

Figure 7

Answers to Concepts Review: **1.** closed and bounded **2.** boundary; stationary; singular **3.** $\nabla f(x_0, y_0) = \mathbf{0}$ **4.** $f_{xx}(x_0, y_0)f_{yy}(x_0, y_0) - f_{xy}^2(x_0, y_0)$

12.9 The Method of Lagrange Multipliers

We begin by distinguishing between two kinds of problems. To find the minimum value of $x^2 + 2y^2 + z^4 + 4$ is a *free extremum* problem. To find the minimum of $x^2 + 2y^2 + z^4 + 4$ subject to the condition that $x + 3y - z = 7$ is a *constrained extremum* problem. Many of the problems of the real world, especially those in economics, are of the latter type. For example, a manufacturer may wish to maximize profits, but is likely to be constrained by the amount of raw materials available, the size of its labor force, and so on.

Example 4 of the previous section was a constrained extremum problem. We were asked to find the minimum distance from the surface $z^2 = x^2y + 4$ to the origin. We formulated the problem as that of minimizing $d^2 = x^2 + y^2 + z^2$, subject to the constraint $z^2 = x^2y + 4$. We handled this problem by substituting the value for z^2 from the constraint in the expression for d^2 and then solving the resulting *free* (i.e., the unconstrained) extremum problem. Example 5 of the previous section was also a constrained optimization problem. We knew that the maximum had to occur on the boundary of the region S, so we were led to the problem of maximizing $z = 2 + x^2 + y^2$ subject to the constraint that $x^2 + \frac{1}{4}y^2 = 1$. This problem was solved by finding a parametrization for the constraint and then maximizing a function of one variable (the variable being the parameter in the constraint). It often happens, however, that the constraint equation is not easily solved for one of the variables or that the constraint cannot be parametrized in terms of one variable. Even when one of these techniques can be applied, another method may be simpler; this is the method of **Lagrange multipliers.**

Geometric Interpretation of the Method

Part of the problem in Example 5 of the previous section was to maximize the objective function $f(x, y) = 2 + x^2 + y^2$ subject to the constraint $g(x, y) = 0$, where $g(x, y) = x^2 + \frac{1}{4}y^2 - 1$. Figure 1 shows the surface $z = f(x, y)$ along with the constraint.

Here, the elliptical cylinder represents the constraint. The second part of Figure 1 shows the intersection of the constraint and the surface $z = f(x, y)$. The optimization problem is to find where, along this curve of intersection, the function is a maximum and where it is a minimum. Both the second and third parts of Figure 1 suggest that the maximum and minimum will occur when a level curve of the objective function f is *tangent* to the constraint curve. This is the key idea behind the method of Lagrange multipliers.

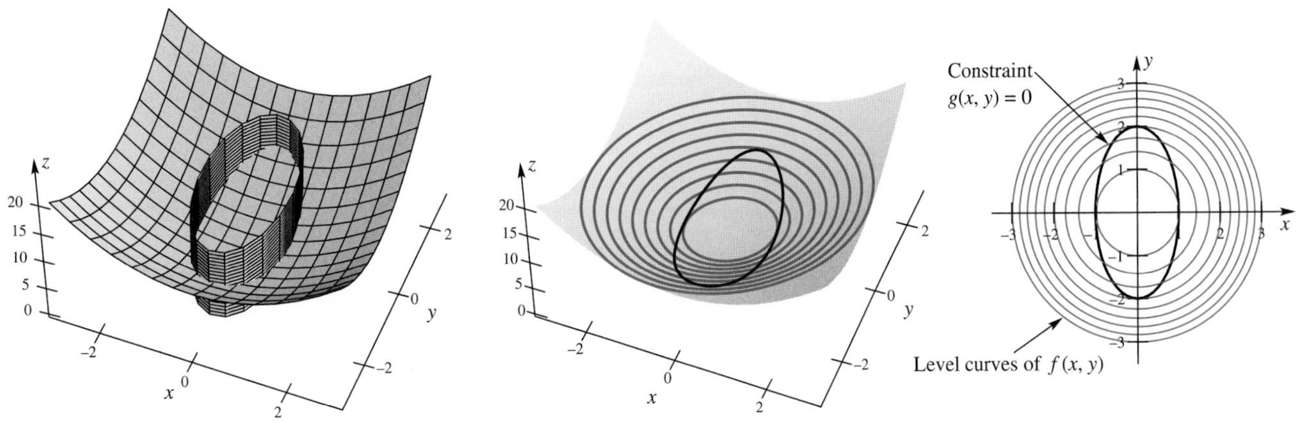

Figure 1

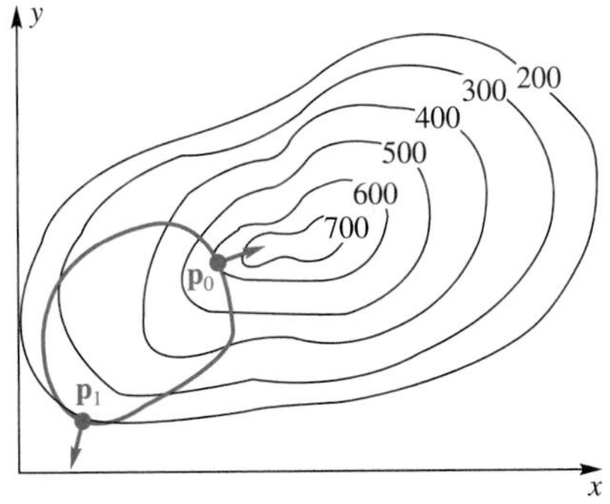

Figure 2

Next we consider the general problem of optimizing $f(x, y)$ subject to the constraint $g(x, y) = 0$. The level curves of f are the curves $f(x, y) = k$, where k is a constant. They are shown as black curves in Figure 2 for $k = 200, 300, \ldots, 700$. The graph of the constraint $g(x, y) = 0$ is also a curve; it is shown in color in Figure 2. To maximize f subject to the constraint $g(x, y) = 0$ means to find the level curve with the greatest possible k that intersects the constraint curve. It is evident from Figure 2 that such a level curve is *tangent* to the constraint curve at a point $\mathbf{p}_0 = (x_0, y_0)$ and therefore that the maximum value of f subject to the constraint $g(x, y) = 0$ is $f(x_0, y_0)$. The other point of tangency, $\mathbf{p}_1 = (x_1, y_1)$, gives the minimum value $f(x_1, y_1)$ of f subject to the constraint $g(x, y) = 0$.

Lagrange's method provides an algebraic procedure for finding the points $\mathbf{p}_0$ and $\mathbf{p}_1$. Since, at such a point, the level curve and the constraint curve are tangent (i.e., have a common tangent line), the two curves have a common perpendicular line. But at any point of a level curve the gradient vector ∇f is perpendicular to the level curve (Section 12.5), and similarly, ∇g is perpendicular to the constraint curve. Thus, ∇f and ∇g are parallel at $\mathbf{p}_0$ and also at $\mathbf{p}_1$; that is,

$$\nabla f(\mathbf{p}_0) = \lambda_0 \nabla g(\mathbf{p}_0) \quad \text{and} \quad \nabla f(\mathbf{p}_1) = \lambda_1 \nabla g(\mathbf{p}_1)$$

for some nonzero numbers λ_0 and λ_1.

The argument just given is admittedly an intuitive one, but it can be made completely rigorous under appropriate hypotheses. Moreover, this argument works just as well for the problem of maximizing or minimizing $f(x, y, z)$ subject to the constraint $g(x, y, z) = 0$. We simply consider level *surfaces* rather than level curves. In fact, the result is valid in any number of variables.

All this suggests the following formulation of the method of Lagrange multipliers.

Theorem A Lagrange's Method

To maximize or minimize $f(\mathbf{p})$ subject to the constraint $g(\mathbf{p}) = 0$, solve the system of equations

$$\nabla f(\mathbf{p}) = \lambda \nabla g(\mathbf{p}) \quad \text{and} \quad g(\mathbf{p}) = 0$$

for $\mathbf{p}$ and λ. Each such point $\mathbf{p}$ is a critical point for the constrained extremum problem, and the corresponding λ is called a Lagrange multiplier.

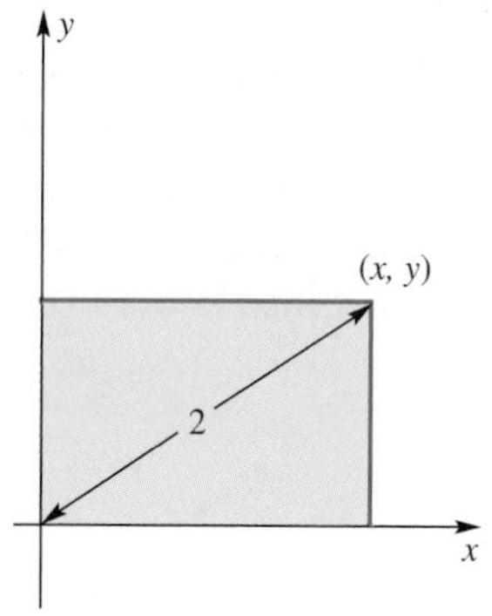

Figure 3

Applications We illustrate the method with several examples.

EXAMPLE 1 What is the greatest area that a rectangle can have if the length of its diagonal is 2?

SOLUTION Place the rectangle in the first quadrant with two of its sides along the coordinate axes; then the vertex opposite the origin has coordinates (x, y), with x and y positive (Figure 3). The length of its diagonal is $\sqrt{x^2 + y^2} = 2$, and its area is xy.

Thus, we may formulate the problem to be that of maximizing $f(x, y) = xy$ subject to the constraint $g(x, y) = x^2 + y^2 - 4 = 0$. The corresponding gradients are

$$\nabla f(x, y) = f_x(x, y)\mathbf{i} + f_y(x, y)\mathbf{j} = y\,\mathbf{i} + x\,\mathbf{j}$$

$$\nabla g(x, y) = g_x(x, y)\mathbf{i} + g_y(x, y)\mathbf{j} = 2x\,\mathbf{i} + 2y\,\mathbf{j}$$

Lagrange's equations thus become

(1) $$y = \lambda(2x)$$

(2) $$x = \lambda(2y)$$

(3) $$x^2 + y^2 = 4$$

which we must solve simultaneously. If we multiply the first equation by y and the second by x, we get $y^2 = 2\lambda xy$ and $x^2 = 2\lambda xy$, from which

(4) $$y^2 = x^2$$

From (3) and (4), we find that $x = \sqrt{2}$ and $y = \sqrt{2}$; and by substituting these values in (1), we obtain $\lambda = \frac{1}{2}$. Thus, the solution to equations (1) through (3), keeping x and y positive, is $x = \sqrt{2}$, $y = \sqrt{2}$, and $\lambda = \frac{1}{2}$.

We conclude that the rectangle of greatest area with diagonal 2 is the square having sides of length $\sqrt{2}$. Its area is 2. A geometric interpretation of this problem is shown in Figure 4. ■

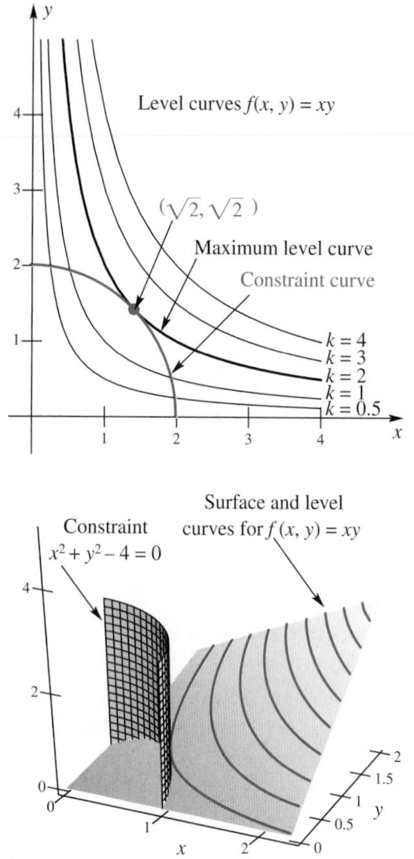

Figure 4

EXAMPLE 2 Use Lagrange's method to find the maximum and minimum values of

$$f(x, y) = y^2 - x^2$$

on the ellipse $x^2/4 + y^2 = 1$.

SOLUTION Refer to Figure 2 of Section 12.8 for a graph of the hyperbolic paraboloid $z = f(x, y) = y^2 - x^2$. From this figure, we would certainly guess that the minimum value occurs at $(\pm 2, 0)$ and the maximum value at $(0, \pm 1)$. But let us justify this conjecture.

We may write the constraint as $g(x, y) = x^2 + 4y^2 - 4 = 0$. Now

$$\nabla f(x, y) = -2x\,\mathbf{i} + 2y\,\mathbf{j}$$

and

$$\nabla g(x, y) = 2x\,\mathbf{i} + 8y\,\mathbf{j}$$

The Lagrange equations are

(1) $$-2x = \lambda 2x$$

(2) $$2y = \lambda 8y$$

(3) $$x^2 + 4y^2 = 4$$

Note from the third equation that x and y cannot both be 0. If $x \neq 0$, the first equation implies that $\lambda = -1$, and the second equation then requires that $y = 0$. We conclude from the third equation that $x = \pm 2$. We have thus obtained the critical points $(\pm 2, 0)$.

Exactly the same argument with $y \neq 0$ yields $\lambda = \frac{1}{4}$ from the second equation, then $x = 0$ from the first equation, and finally $y = \pm 1$ from the third equation. We conclude that $(0, \pm 1)$ are also critical points.

Now, for $f(x, y) = y^2 - x^2$,

$$f(2, 0) = -4$$

$$f(-2, 0) = -4$$

$$f(0, 1) = 1$$

$$f(0, -1) = 1$$

The minimum value of $f(x, y)$ on the given ellipse is -4; the maximum value is 1. ■

EXAMPLE 3 Find the minimum of $f(x, y, z) = 3x + 2y + z + 5$ subject to the constraint $g(x, y, z) = 9x^2 + 4y^2 - z = 0$.

SOLUTION The gradients of f and g are $\nabla f(x, y, z) = 3\mathbf{i} + 2\mathbf{j} + \mathbf{k}$ and $\nabla g(x, y, z) = 18x\mathbf{i} + 8y\mathbf{j} - \mathbf{k}$. To find the critical points, we solve the equations

$$\nabla f(x, y, z) = \lambda \nabla g(x, y, z) \quad \text{and} \quad g(x, y, z) = 0$$

for (x, y, z, λ), in which λ is a Lagrange multiplier. This is equivalent, in the present problem, to solving the following system of four simultaneous equations in the four variables x, y, z, and λ.

(1) $$3 = 18x\lambda$$

(2) $$2 = 8y\lambda$$

(3) $$1 = -\lambda$$

(4) $$9x^2 + 4y^2 - z = 0$$

From (3), $\lambda = -1$. Substituting this result in equations (1) and (2), we get $x = -\frac{1}{6}$ and $y = -\frac{1}{4}$. By putting these values for x and y in equation (4), we obtain $z = \frac{1}{2}$. Thus, the solution of the foregoing system of four simultaneous equations is $\left(-\frac{1}{6}, -\frac{1}{4}, \frac{1}{2}, -1\right)$, and the only critical point is $\left(-\frac{1}{6}, -\frac{1}{4}, \frac{1}{2}\right)$. Therefore, the minimum of $f(x, y, z)$, subject to the constraint $g(x, y, z) = 0$, is $f\left(-\frac{1}{6}, -\frac{1}{4}, \frac{1}{2}\right) = \frac{9}{2}$. (How do we know that this value is a minimum rather than a maximum?) ■

Two or More Constraints When more than one constraint is imposed on the variables of a function that is to be maximized or minimized, additional Lagrange multipliers are used (one for each constraint). For example, if we seek the extrema of a function f of three variables subject to the two constraints $g(x, y, z) = 0$ and $h(x, y, z) = 0$, we solve the equations

$$\nabla f(x, y, z) = \lambda \nabla g(x, y, z) + \mu \nabla h(x, y, z), \quad g(x, y, z) = 0, \quad h(x, y, z) = 0$$

for x, y, z, λ, and μ, where λ and μ are Lagrange multipliers. This is equivalent to finding the solutions of the system of five simultaneous equations in the variables x, y, z, λ, and μ.

(1) $$f_x(x, y, z) = \lambda g_x(x, y, z) + \mu h_x(x, y, z)$$

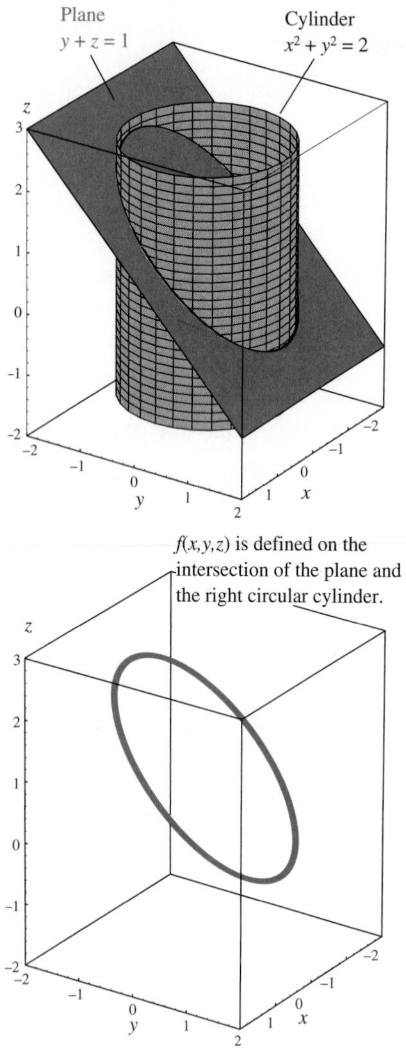
Plane
$y + z = 1$
Cylinder
$x^2 + y^2 = 2$
f(x,y,z) is defined on the intersection of the plane and the right circular cylinder.

Figure 5

(2) $$f_y(x, y, z) = \lambda g_y(x, y, z) + \mu h_y(x, y, z)$$

(3) $$f_z(x, y, z) = \lambda g_z(x, y, z) + \mu h_z(x, y, z)$$

(4) $$g(x, y, z) = 0$$

(5) $$h(x, y, z) = 0$$

From the solutions of this system, we obtain the critical points.

EXAMPLE 4 Find the maximum and minimum values of $f(x, y, z) = x + 2y + 3z$ on the ellipse that is the intersection of the cylinder $x^2 + y^2 = 2$ and the plane $y + z = 1$ (see Figure 5).

SOLUTION We want to maximize and minimize $f(x, y, z)$ subject to $g(x, y, z) = x^2 + y^2 - 2 = 0$ and $h(x, y, z) = y + z - 1 = 0$. The corresponding Lagrange equations are

(1) $$1 = 2\lambda x$$

(2) $$2 = 2\lambda y + \mu$$

(3) $$3 = \mu$$

(4) $$x^2 + y^2 - 2 = 0$$

(5) $$y + z - 1 = 0$$

From (1), $x = 1/(2\lambda)$; from (2) and (3), $y = -1/(2\lambda)$. Thus, from (4), $(1/(2\lambda))^2 + (-1/(2\lambda))^2 = 2$, which implies that $\lambda = \pm\frac{1}{2}$. The solution $\lambda = \frac{1}{2}$ yields the critical point $(x, y, z) = (1, -1, 2)$, and $\lambda = -\frac{1}{2}$ yields the critical point $(x, y, z) = (-1, 1, 0)$. We conclude that $f(1, -1, 2) = 5$ is the maximum value and $f(-1, 1, 0) = 1$ is the minimum value. ■

Optimizing a Function over a Closed and Bounded Set We can find the maximum or minimum of a function $f(x, y)$ over a closed and bounded set S using the following steps. First, use the methods of Section 12.8 to find the maximum or minimum on the interior of S. Second, use Lagrange multipliers to find the points along the boundary that give a local maximum or minimum. Finally, evaluate the function at these points to find the maximum and minimum over S.

EXAMPLE 5 Find the maximum and minimum for the function $f(x, y) = 4 + xy - x^2 - y^2$ over the set $S = \{(x, y): x^2 + y^2 \le 1\}$

SOLUTION Figure 6 shows the graph of $z = 4 + xy - x^2 - y^2$. The set S is the circle with interior centered at the origin having radius 1. Thus, we are to find the maximum and minimum of f over that region that is on or inside the curve drawn on the top of Figure 6. We begin by finding all critical points on the interior of S:

$$\frac{\partial f}{\partial x} = y - 2x = 0$$

$$\frac{\partial f}{\partial y} = x - 2y = 0$$

The only solution, and thus the only interior critical point, is $(0, 0)$. Next we apply the method of Lagrange multipliers to find points along the boundary where the

Thinking of Symmetry

Figure 6 suggests that four of the points to check will be symmetric about the origin. This turned out to be the case.

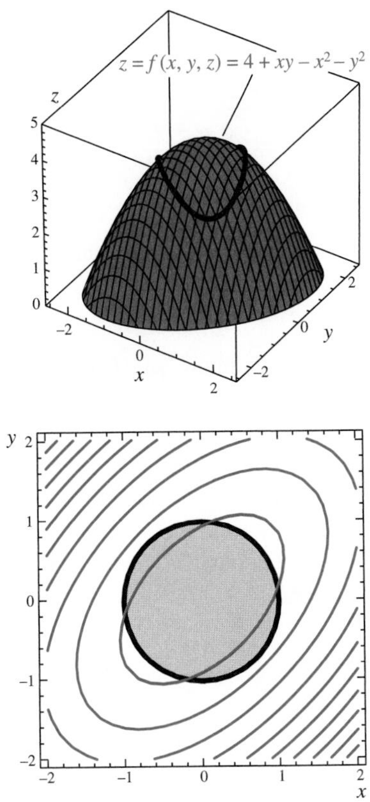

Figure 6

function is a maximum or minimum. A point on the boundary will satisfy the constraint $x^2 + y^2 - 1 = 0$, so we let $g(x, y) = x^2 + y^2 - 1$. Then

$$\nabla f(x, y) = (y - 2x)\mathbf{i} + (x - 2y)\mathbf{j}$$
$$\nabla g(x, y) = 2x\,\mathbf{i} + 2y\,\mathbf{j}$$

Setting $\nabla f(x, y) = \lambda \nabla g(x, y)$ leads to

$$y - 2x = \lambda 2x$$
$$x - 2y = \lambda 2y$$

Solving these two equations for λ gives

$$\frac{y}{2x} - 1 = \lambda = \frac{x}{2y} - 1$$

which leads to $x = \pm y$. This, together with the constraint $x^2 + y^2 - 1 = 0$, leads to $x = \pm\sqrt{2}/2$, $y = \pm\sqrt{2}/2$. We must therefore evaluate f at the five points $(0, 0)$, $\left(\frac{\sqrt{2}}{2}, \frac{\sqrt{2}}{2}\right)$, $\left(-\frac{\sqrt{2}}{2}, \frac{\sqrt{2}}{2}\right)$, $\left(\frac{\sqrt{2}}{2}, -\frac{\sqrt{2}}{2}\right)$, and $\left(-\frac{\sqrt{2}}{2}, -\frac{\sqrt{2}}{2}\right)$:

$$f(0, 0) = 4 \qquad f\left(\frac{\sqrt{2}}{2}, \frac{\sqrt{2}}{2}\right) = \frac{7}{2} \qquad f\left(-\frac{\sqrt{2}}{2}, \frac{\sqrt{2}}{2}\right) = \frac{5}{2}$$
$$f\left(\frac{\sqrt{2}}{2}, -\frac{\sqrt{2}}{2}\right) = \frac{5}{2} \qquad f\left(-\frac{\sqrt{2}}{2}, -\frac{\sqrt{2}}{2}\right) = \frac{7}{2}$$

The maximum that f attains over S is 4, and this occurs at $(x, y) = (0, 0)$. The minimum that f attains over S is $\frac{5}{2}$, and this occurs at the two points $\left(-\frac{\sqrt{2}}{2}, \frac{\sqrt{2}}{2}\right)$ and $\left(\frac{\sqrt{2}}{2}, -\frac{\sqrt{2}}{2}\right)$. ■

Concepts Review

1. To maximize $f(x, y)$ is a(n) _____ extremum problem; to maximize $f(x, y)$ subject to $g(x, y) = 0$ is a(n) _____ extremum problem.

2. The method of Lagrange multipliers depends on the fact that at an extreme value the vectors ∇f and ∇g are _____.

3. Thus, to use the method of Lagrange multipliers, we attempt to solve the equations $\nabla f(x, y) = \lambda \nabla g(x, y)$ and _____ simultaneously.

4. Sometimes simple geometric reasoning yields a solution. The maximum value of $f(x, y) = x^4 + y^4$ on the circle $(x - 1)^2 + (y - 1)^2 = 2$ clearly occurs at _____.

Problem Set 12.9

1. Find the minimum of $f(x, y) = x^2 + y^2$ subject to the constraint $g(x, y) = xy - 3 = 0$.

2. Find the maximum of $f(x, y) = xy$ subject to the constraint $g(x, y) = 4x^2 + 9y^2 - 36 = 0$.

3. Find the maximum of $f(x, y) = 4x^2 - 4xy + y^2$ subject to the constraint $x^2 + y^2 = 1$.

4. Find the minimum of $f(x, y) = x^2 + 4xy + y^2$ subject to the constraint $x - y - 6 = 0$.

5. Find the minimum of $f(x, y, z) = x^2 + y^2 + z^2$ subject to the constraint $x + 3y - 2z = 12$.

6. Find the minimum of $f(x, y, z) = 4x - 2y + 3z$ subject to the constraint $2x^2 + y^2 - 3z = 0$.

7. What are the dimensions of the rectangular box, open at the top, that has maximum volume when the surface area is 48?

8. Find the minimum distance between the origin and the plane $x + 3y - 2z = 4$.

9. The material for the bottom of a rectangular box costs three times as much per square foot as the material for the sides and top. Find the greatest volume that such a box can have if the total amount of money available for material is \$12 and the material for the bottom costs \$0.60 per square foot.

10. Find the minimum distance between the origin and the surface $x^2y - z^2 + 9 = 0$.

11. Find the maximum volume of a closed rectangular box with faces parallel to the coordinate planes inscribed in the ellipsoid

$$\frac{x^2}{a^2} + \frac{y^2}{b^2} + \frac{z^2}{c^2} = 1$$

12. Find the maximum volume of the first-octant rectangular box with faces parallel to the coordinate planes, one vertex at $(0, 0, 0)$, and diagonally opposite vertex on the plane

$$\frac{x}{a} + \frac{y}{b} + \frac{z}{c} = 1$$

In Problems 13–20, use the method of Lagrange multipliers to solve these problems from Section 12.8.

13. Problem 21

14. Problem 22

15. Problem 23

16. Problem 24

17. Problem 37

18. Problem 38

19. Problem 40 (minimum only)

20. Problem 42; *Hint:* Let the plane be $\frac{x}{A} + \frac{y}{B} + \frac{z}{C} = 1$.

In Problems 21–25, find the maximum and minimum of the function f over the closed and bounded set S. Use the methods of Section 12.8 to find the maximum and minimum on the the interior of S; then use Lagrange multipliers to find the maximum and minimum over the boundary of S.

21. $f(x, y) = 10 + x + y$; $S = \{(x, y): x^2 + y^2 \le 1\}$

22. $f(x, y) = x + y - xy$; $S = \{(x, y): x^2 + y^2 \le 9\}$

23. $f(x, y) = x^2 + y^2 + 3x - xy$; $S = \{(x, y): x^2 + y^2 \le 9\}$

24. $f(x, y) = \frac{x}{1 + y^2}$; $S = \left\{(x, y): \frac{x^2}{4} + \frac{y^2}{9} \le 1\right\}$

25. $f(x, y) = (1 + x + y)^2$; $S = \left\{(x, y): \frac{x^2}{4} + \frac{y^2}{16} \le 1\right\}$

26. Find the shape of the triangle of maximum perimeter that can be inscribed in a circle of radius r. *Hint:* Let α, β, and γ be as in Figure 7 and reduce the problem to maximizing $P = 2r(\sin \alpha/2 + \sin \beta/2 + \sin \gamma/2)$ subject to $\alpha + \beta + \gamma = 2\pi$.

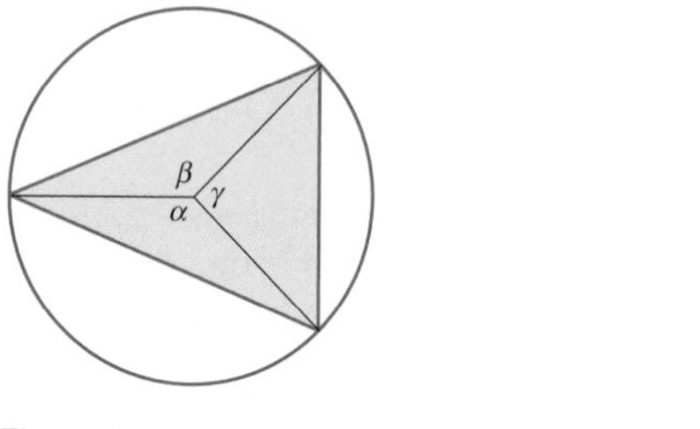

Figure 7

27. Consider the Cobb–Douglas production model for a manufacturing process depending on three inputs x, y, and z with unit costs a, b, and c, respectively, given by

$$P = kx^\alpha y^\beta z^\gamma, \quad \alpha > 0, \beta > 0, \gamma > 0, \alpha + \beta + \gamma = 1$$

subject to the cost constraint $ax + by + cz = d$. Determine x, y, and z to maximize the production P.

28. Find the minimum distance from the origin to the line of intersection of the two planes

$$x + y + z = 8 \quad \text{and} \quad 2x - y + 3z = 28$$

29. Find the maximum and minimum of $f(x, y, z) = -x + 2y + 2z$ on the ellipse $x^2 + y^2 = 2$, $y + 2z = 1$ (see Example 4).

30. Let $w = x_1 x_2 \cdots x_n$.

(a) Maximize w subject to $x_1 + x_2 + \cdots + x_n = 1$ and all $x_i > 0$.

(b) Use part (a) to deduce the famous **Geometric Mean–Arithmetic Mean Inequality** for positive numbers $a_1, a_2, \ldots, a_n$; that is,

$$\sqrt[n]{a_1 a_2 \cdots a_n} \le \frac{a_1 + a_2 + \cdots + a_n}{n}$$

31. Maximize $w = a_1 x_1 + a_2 x_2 + \cdots + a_n x_n$, all $a_i > 0$, subject to $x_1^2 + x_2^2 + \cdots + x_n^2 = 1$.

CAS *Drawing surfaces and level curves, plus using a little common sense, can allow us to solve some constrained extremum problems. Solve the following, which are based on the functions of Figures 6 through 9 of Section 12.1.*

32. Maximize $z = -4x^3y^2$ subject to $x^2 + y^2 = 1$.

33. Minimize $z = x - x^3/8 - y^2/3$ subject to $x^2/16 + y^2 = 1$.

34. Maximize $z = xy \exp(-x^2 - y^2)$ subject to $xy = 2$.

35. Minimize $z = \exp(-|x|) \cos\sqrt{x^2 + y^2}$ subject to $x^2 + y^2/9 = 1$.

Answers to Concepts Review: **1.** free; constrained **2.** parallel **3.** $g(x, y) = 0$ **4.** $(2, 2)$

12.10 Chapter Review

Concepts Test

Respond with true or false to each of the following assertions. Be prepared to justify your answer.

1. The level curves of $z = 2x^2 + 3y^2$ are ellipses.

2. If $f_x(0, 0) = f_y(0, 0)$, then $f(x, y)$ is continuous at the origin.

3. If $f_x(0, 0)$ exists, then $g(x) = f(x, 0)$ is continuous at $x = 0$.

4. If $\lim_{(x,y)\to(0,0)} f(x, y) = L$, then $\lim_{y\to 0} f(y, y) = L$.

5. If $f(x, y) = g(x)h(y)$, where g and h are continuous for all x and y, respectively, then f is continuous on the whole xy-plane.

6. If $f(x, y) = g(x)h(y)$, where both g and h are twice differentiable, then

$$\frac{\partial^2 f}{\partial x^2} + \frac{\partial^2 f}{\partial y^2} = g''(x)h(y) + g(x)h''(y)$$

7. If $f(x, y)$ and $g(x, y)$ have the same gradient, then they are identical functions.

8. The gradient of f is perpendicular to the graph of $z = f(x, y)$.

9. If f is differentiable and $\nabla f(a, b) = \mathbf{0}$, then the graph of $z = f(x, y)$ has a horizontal tangent plane at (a, b).

10. If $\nabla f(\mathbf{p}_0) = \mathbf{0}$, then f has an extreme value at $\mathbf{p}_0$.

11. If $T = e^y \sin x$ gives the temperature at a point (x, y) in the plane, then a heat-seeking object would move away from the origin in the direction $\mathbf{i}$.

12. The function $f(x, y) = \sqrt[3]{x^2 + y^4}$ has a global minimum value at the origin.

13. The function $f(x, y) = \sqrt[3]{x + y^4}$ has neither a global minimum nor a global maximum value.

14. If $f(x, y) = 4x + 4y$, then $|D_{\mathbf{u}}f(x, y)| \le 4$.

15. If $D_{\mathbf{u}}f(x, y)$ exists, then $D_{-\mathbf{u}}f(x, y) = -D_{\mathbf{u}}f(x, y)$.

16. The set $\{(x, y): y = x, 0 \le x \le 1\}$ is a closed set in the plane.

17. If $f(x, y)$ is continuous on a closed bounded set S, then f attains a maximum value on S.

18. If $f(x, y)$ attains its maximum value at an interior point (x_0, y_0) of S, then $\nabla f(x_0, y_0) = \mathbf{0}$.

19. The function $f(x, y) = \sin(xy)$ does not attain a maximum value on the set $\{(x, y): x^2 + y^2 < 4\}$.

20. If $f_x(x_0, y_0)$ and $f_y(x_0, y_0)$ both exist, then f is differentiable at (x_0, y_0).

Sample Test Problems

1. Find and sketch the domain of each indicated function of two variables, showing clearly any points on the boundary of the domain that belong to the domain.

(a) $z = \sqrt{x^2 + 4y^2 - 100}$ (b) $z = -\sqrt{2x - y - 1}$

2. Sketch the level curves of $f(x, y) = (x + y^2)$ for $k = 0, 1, 2, 4$.

In Problems 3–6, find $\partial f/\partial x$, $\partial^2 f/\partial x^2$, and $\partial^2 f/\partial y\, \partial x$.

3. $f(x, y) = 3x^4y^2 + 7x^2y^7$ **4.** $f(x, y) = \cos^2 x - \sin^2 y$

5. $f(x, y) = e^{-y} \tan x$ **6.** $f(x, y) = e^{-x} \sin y$

7. If $F(x, y) = 5x^3y^6 - xy^7$, find $\partial^3 F(x, y)/\partial x\, \partial y^2$.

8. If f is the function of three variables defined by $f(x, y, z) = xy^3 - 5x^2yz^4$, find $f_x(2, -1, 1)$, $f_y(2, -1, 1)$, and $f_z(2, -1, 1)$.

9. Find the slope of the tangent to the curve of intersection of the surface $z = x^2 + y^2/4$ and the plane $x = 2$ at the point $(2, 2, 5)$.

10. For what points is the function defined by $f(x, y) = xy/(x^2 - y)$ continuous?

11. Does $\displaystyle\lim_{(x,y)\to(0,0)} \frac{x - y}{x + y}$ exist? Explain.

12. In each case, find the indicated limit or state that it does not exist.

(a) $\displaystyle\lim_{(x,y)\to(2,2)} \frac{x^2 - 2y}{x^2 + 2y}$ (b) $\displaystyle\lim_{(x,y)\to(2,2)} \frac{x^2 + 2y}{x^2 - 2y}$

(c) $\displaystyle\lim_{(x,y)\to(0,0)} \frac{x^4 - 4y^4}{x^2 + 2y^2}$

13. Find $\nabla f(1, 2, -1)$.

(a) $f(x, y, z) = x^2yz^3$ (b) $f(x, y, z) = y^2 \sin xz$

14. Find the directional derivative of $f(x, y) = \tan^{-1}(3xy)$. What is its value at the point $(4, 2)$ in the direction $\mathbf{u} = \left(\sqrt{3}/2\right)\mathbf{i} - (1/2)\mathbf{j}$?

15. Find the slope of the tangent line to the curve of intersection of the vertical plane $x - \sqrt{3}y + 2\sqrt{3} - 1 = 0$ and the surface $z = x^2 + y^2$ at the point $(1, 2, 5)$.

16. In what direction is $f(x, y) = 9x^4 + 4y^2$ increasing most rapidly at $(1, 2)$?

17. For $f(x, y) = x^2/2 + y^2$,

(a) find the equation of its level curve that goes through the point $(4, 1)$ in its domain;

(b) find the gradient vector ∇f at $(4, 1)$;

(c) draw the level curve and draw the gradient vector with its initial point at $(4, 1)$.

18. If $F(u, v) = \tan^{-1}(uv)$, $u = \sqrt{xy}$, and $v = \sqrt{x} - \sqrt{y}$, find $\partial F/\partial x$ and $\partial F/\partial y$ in terms of u, v, x, and y.

19. If $f(u, v) = u/v$, $u = x^2 - 3y + 4z$, and $v = xyz$, find f_x, f_y, and f_z in terms of x, y, and z.

20. If $F(x, y) = x^3 - xy^2 - y^4$, $x = 2\cos 3t$, and $y = 3\sin t$, find dF/dt at $t = 0$.

21. If $F(x, y, z) = (5x^2y/z^3)$, $x = t^{3/2} + 2$, $y = \ln 4t$, and $z = e^{3t}$, find dF/dt in terms of x, y, z, and t.

22. A triangle has vertices A, B, and C. The length of the side $c = AB$ is increasing at the rate of 3 inches per second, the side $b = AC$ is decreasing at 1 inch per second, and the included angle α is increasing at 0.1 radian per second. If $c = 10$ inches and $b = 8$ inches when $\alpha = \pi/6$, how fast is the area changing?

23. Find the gradient vector of $F(x, y, z) = 9x^2 + 4y^2 + 9z^2 - 34$ at the point $P(1, 2, -1)$. Write the equation of the tangent plane to the surface $F(x, y, z) = 0$ at P.

24. A right circular cylinder is measured to have a radius of 10 ± 0.02 inches and a height of 6 ± 0.01 inches. Calculate its volume and use differentials to give an estimate of the possible error.

25. If $f(x, y, z) = xy^2/(1 + z^2)$, use differentials to estimate $f(1.01, 1.98, 2.03)$.

26. Find the extrema of $f(x, y) = x^2y - 6y^2 - 3x^2$.

27. A rectangular box whose edges are parallel to the coordinate axes is inscribed in the ellipsoid $36x^2 + 4y^2 + 9z^2 = 36$. What is the greatest possible volume for such a box?

28. Use Lagrange multipliers to find the maximum and the minimum of $f(x, y) = xy$ subject to the constraint $x^2 + y^2 = 1$.

29. Use Lagrange multipliers to find the dimensions of the right circular cylinder with maximum volume if its surface area is 24π.

REVIEW & PREVIEW PROBLEMS

In Problems 1–6, sketch a graph of the given function.

1. $f(x, y) = \sqrt{64 - x^2 - y^2}$

2. $f(x, y) = 9 - x^2 - y^2$

3. $f(x, y) = x^2 + 4y^2$

4. $f(x, y) = x^2 - y^2$

5. $f(x, y) = x^2$

6. $f(x, y) = \sqrt{9 - y^2}$

In Problems 7–14, sketch the graph of the given cylindrical or spherical equation.

7. $r = 2$

8. $\rho = 2$

9. $\phi = \pi/4$

10. $\theta = \pi/4$

11. $r^2 + z^2 = 9$

12. $r = \cos\theta$

13. $r = 2\sin\theta$

14. $z = 9 - r^2$

Evaluate the integrals in Problems 15–26.

15. $\int e^{-2x}\,dx$

16. $\int xe^{-2x}\,dx$

17. $\int_{-a/2}^{a/2} \cos\left(\frac{\pi x}{a}\right) dx$

18. $\int_0^2 (a + bx + c^2x^2)\,dx$

19. $\int_0^{\pi} \sin^2 x\,dx$

20. $\int_{1/4}^{3/4} \frac{1}{1 - x^2}\,dx$

21. $\int_0^1 \frac{x}{1 + x^2}\,dx$

22. $\int_0^4 \frac{e^x}{1 + e^{2x}}\,dx$

23. $\int_0^3 r\sqrt{4r^2 + 1}\,dr$

24. $\int_0^{a/2} \frac{a\,r}{\sqrt{a^2 - r^2}}\,dr$

25. $\int_0^{\pi/2} \cos^2\theta\,d\theta$

26. $\int_0^{\pi/2} \cos^4\theta\,d\theta$

27. Without using the Second Fundamental Theorem of Calculus, evaluate

$$\int_0^{2\pi} \left[\sqrt{a^2 - b^2} - \sqrt{a^2 - c^2}\right] d\theta$$

28. Find the area of that part of the plane $x + y + z = 1$ that is in the first octant.

In Problems 29–34, find the volume of the indicated solid in three-space using basic properties of geometry or the methods of Chapter 5.

29. The solid bounded above by $z = \sqrt{9 - y^2}$, below by the xy-plane, and laterally by the planes $x = 0$ and $x = 8$.

30. The solid in three-space consisting of those points whose spherical coordinates satisfy $\rho \le 7$.

31. The solid obtained when the graph of $y = \sin x$, $0 \le x \le \pi$ is revolved about the x-axis.

32. The solid in three-space consisting of those points whose cylindrical coordinates satisfy $r \le 7$ and $0 \le z \le 100$.

33. The solid in three-space bounded above by $z = 9 - x^2 - y^2$ and below by the xy-plane. *Hint:* Interpret this as a solid of revolution.

34. The solid in three-space consisting of those points whose spherical coordinates satisfy $1 \le \rho \le 4$ and $0 \le \phi \le \pi/2$.

CHAPTER 13 Multiple Integrals

13.1 Double Integrals over Rectangles
13.2 Iterated Integrals
13.3 Double Integrals over Nonrectangular Regions
13.4 Double Integrals in Polar Coordinates
13.5 Applications of Double Integrals
13.6 Surface Area
13.7 Triple Integrals in Cartesian Coordinates
13.8 Triple Integrals in Cylindrical and Spherical Coordinates
13.9 Change of Variables in Multiple Integrals

13.1 Double Integrals over Rectangles

Differentiation and integration are the major processes of calculus. We have studied differentiation in two- and three-dimensional space (Chapter 12); it is time to consider integration in two- and three-dimensional space. The theory and the applications of single (Riemann) integrals are to be generalized to multiple integrals. In Chapter 5 we used single integrals to calculate the area of curved planar regions, to find the length of planar curves, and to determine the center of mass of straight wires of variable density. In this chapter we use multiple integrals to find the volume of general solids, the area of general surfaces, and the center of mass of laminas and solids of variable density.

The intimate connection between integration and differentiation was enunciated in the Fundamental Theorems of Calculus; these theorems provided the principal theoretical tools for evaluating single integrals. Here we reduce multiple integration to a succession of single integrations where again the Second Fundamental Theorem will play a central role. The integration skills that you learned in Chapters 4 through 7 will be tested.

The Riemann integral for a function of one variable was defined in Section 4.2, a section worth reviewing. Recall that we formed a partition P of the interval $[a, b]$ into subintervals of length Δx_k, $k = 1, 2, \ldots, n$, picked a sample point $\overline{x}_k$ from the kth subinterval, and then wrote

$$\int_a^b f(x)\,dx = \lim_{\|P\|\to 0} \sum_{k=1}^{n} f(\overline{x}_k)\,\Delta x_k$$

We proceed in a very similar fashion to define the integral for a function of two variables.

Let R be a rectangle with sides parallel to the coordinate axes; that is, let

$$R = \{(x, y): a \le x \le b, c \le y \le d\}$$

Form a partition P of R by means of lines parallel to the x- and y-axes, as in Figure 1. This divides R into subrectangles, say n of them, which we denote by R_k, $k = 1, 2, \ldots, n$. Let Δx_k and Δy_k be the lengths of the sides of R_k, and let $\Delta A_k = \Delta x_k \Delta y_k$ be its area. In R_k, pick a sample point $(\overline{x}_k, \overline{y}_k)$ and form the Riemann sum

$$\sum_{k=1}^{n} f(\overline{x}_k, \overline{y}_k)\,\Delta A_k$$

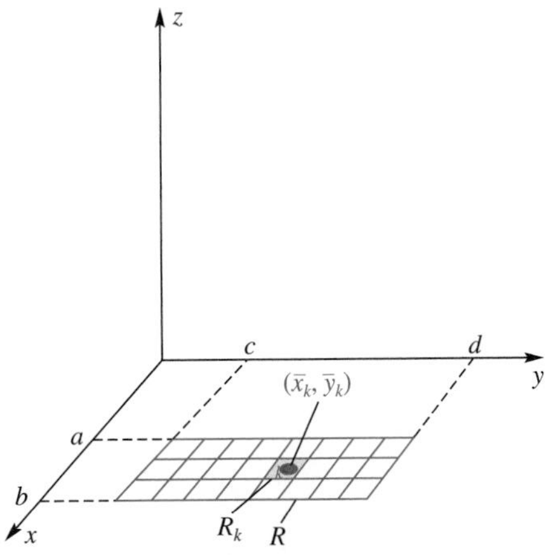

Figure 1

which corresponds (if $f(x, y) \geq 0$) to the sum of the volumes of n boxes (Figures 2 and 3). Letting the partition get finer and finer in such a way that all the R_k's get smaller will lead to the concept that we want.

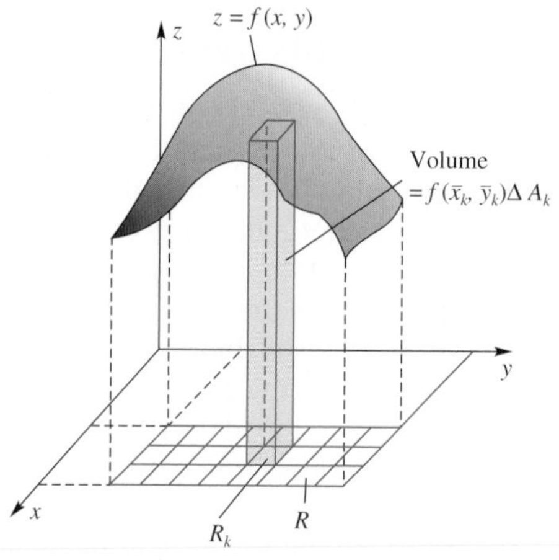

Figure 2

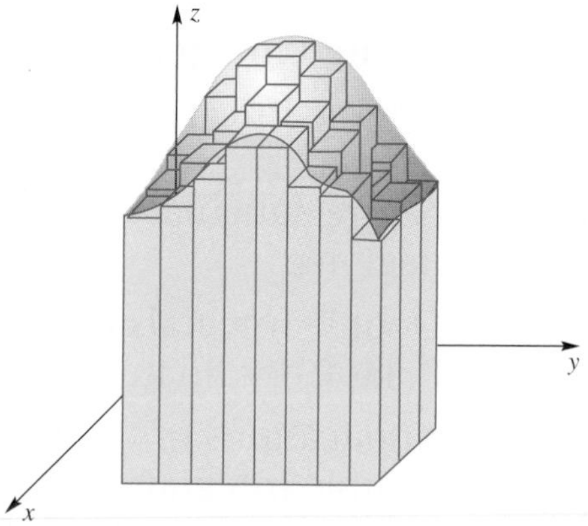

Figure 3

We are ready for a formal definition. We use the notation introduced above, with the additional proviso that the norm of the partition P, denoted by $\|P\|$, is the length of the longest diagonal of any subrectangle in the partition.

Definition **The Double Integral**

Let f be a function of two variables that is defined on a closed rectangle R. If

$$\lim_{\|P\|\to 0} \sum_{k=1}^{n} f(\overline{x}_k, \overline{y}_k)\, \Delta A_k$$

exists, we say that f is integrable on R. Moreover, $\iint_R f(x, y)\, dA$, called the **double integral** of f over R, is then given by

$$\iint_R f(x, y)\, dA = \lim_{\|P\|\to 0} \sum_{k=1}^{n} f(\overline{x}_k, \overline{y}_k)\, \Delta A_k$$

This definition of the double integral contains the limit as $\|P\| \to 0$. This is not a limit in the sense of Chapter 1, so we should clarify what this really means. We say that $\lim_{\|P\|\to 0} \sum_{k=1}^{n} f(\overline{x}_k, \overline{y}_k)\, \Delta A_k = L$ if for every $\varepsilon > 0$ there exists a $\delta > 0$ such that, for every partition P of the rectangle R by lines parallel to the x- and y-axes that satisfies $\|P\| < \delta$ and for any choice of the sample points $(\overline{x}_k, \overline{y}_k)$ in the kth rectangle, we have $\left| \sum_{k=1}^{n} f(\overline{x}_k, \overline{y}_k)\, \Delta A_k - L \right| < \varepsilon$.

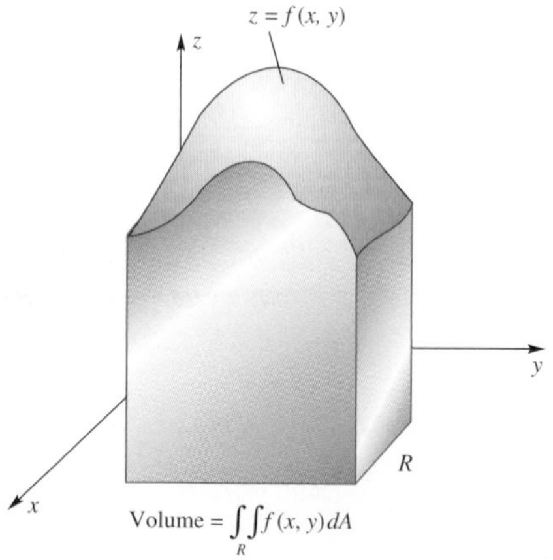

Figure 4

Recall that if $f(x) \geq 0$, $\int_a^b f(x)\, dx$ represents the area of the region under the curve $y = f(x)$ between a and b. In a similar manner, if $f(x, y) \geq 0$, $\iint_R f(x, y)\, dA$ represents the **volume** of the solid under the surface $z = f(x, y)$ and above the rectangle R (Figure 4). In fact, we take this integral as the definition of the volume of such a solid.

The Existence Question Not every function of two variables is integrable on a given rectangle R. The reasons are the same as in the one-variable case (Section 4.2). In particular, a function that is unbounded on R will always fail to be integrable. Fortunately, there is a natural generalization of Theorem 4.2A, although its proof is beyond the level of a first course.

Theorem A Integrability Theorem

If f is bounded on the closed rectangle R and if it is continuous there except on a finite number of smooth curves, then f is integrable on R. In particular, if f is continuous on all of R, then f is integrable there.

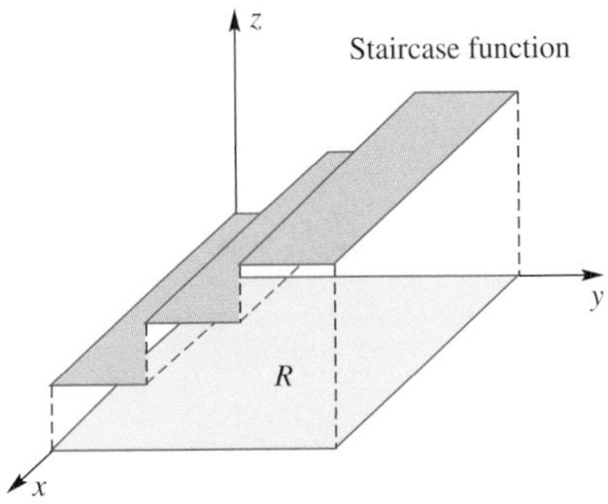

Figure 5

As a consequence, most of the common functions (provided they are bounded) are integrable on every rectangle. For example,

$$f(x, y) = e^{\sin(xy)} - y^3 \cos(x^2 y)$$

is integrable on every rectangle. On the other hand,

$$g(x, y) = \frac{x^2 y - 2x}{y - x^2}$$

would fail to be integrable on any rectangle that intersected the parabola $y = x^2$. The *staircase* function of Figure 5 is integrable on R because its discontinuities occur along two line segments.

Properties of the Double Integral The double integral inherits most of the properties of the single integral.

1. The double integral is linear; that is,

   a. $\displaystyle\iint_R kf(x, y)\,dA = k\iint_R f(x, y)\,dA;$

   b. $\displaystyle\iint_R [f(x, y) + g(x, y)]\,dA = \iint_R f(x, y)\,dA + \iint_R g(x, y)\,dA.$

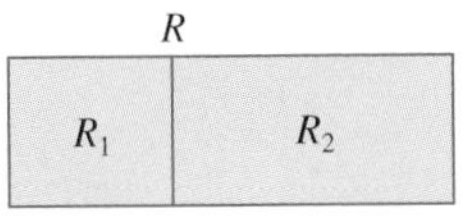

Figure 6

2. The double integral is additive on rectangles (Figure 6) that overlap only on a line segment.

$$\iint_R f(x, y)\,dA = \iint_{R_1} f(x, y)\,dA + \iint_{R_2} f(x, y)\,dA$$

3. The comparison property holds. If $f(x, y) \le g(x, y)$ for all (x, y) in R, then

$$\iint_R f(x, y)\,dA \le \iint_R g(x, y)\,dA$$

All of these properties hold on more general sets than rectangles, but that is a matter we take up in Section 13.3.

Evaluation of Double Integrals This topic will receive major attention in the next section, where we will develop a powerful tool for evaluating double integrals. However, we can already evaluate a few integrals, and we can approximate others.

Note first that if $f(x, y) = 1$ on R then the double integral is the area of R, and from this it follows that

$$\iint_R k\,dA = k\iint_R 1\,dA = kA(R)$$

EXAMPLE 1 Let f be the staircase function of Figure 5; that is, let

$$f(x, y) = \begin{cases} 1, & \text{if } 0 \le x \le 3, 0 \le y \le 1 \\ 2, & \text{if } 0 \le x \le 3, 1 < y \le 2 \\ 3, & \text{if } 0 \le x \le 3, 2 < y \le 3 \end{cases}$$

Calculate $\iint_R f(x, y)\, dA$, where $R = \{(x, y): 0 \le x \le 3, 0 \le y \le 3\}$.

SOLUTION Introduce rectangles R_1, R_2, and R_3 as follows:

$$R_1 = \{(x, y): 0 \le x \le 3, 0 \le y \le 1\}$$
$$R_2 = \{(x, y): 0 \le x \le 3, 1 \le y \le 2\}$$
$$R_3 = \{(x, y): 0 \le x \le 3, 2 \le y \le 3\}$$

Then, using the additivity property of the double integral, we obtain

$$\iint_R f(x, y)\, dA = \iint_{R_1} f(x, y)\, dA + \iint_{R_2} f(x, y)\, dA + \iint_{R_3} f(x, y)\, dA$$
$$= 1A(R_1) + 2A(R_2) + 3A(R_3)$$
$$= 1 \cdot 3 + 2 \cdot 3 + 3 \cdot 3 = 18$$

In this derivation, we also used the fact that the value of f on the boundary of a rectangle does not affect the value of the integral. ■

Example 1 was a minor accomplishment, and to be honest we cannot do much more without more tools. However, we can always approximate a double integral by calculating a Riemann sum. In general, we can expect the approximation to be better the finer the partition we use.

EXAMPLE 2 Approximate $\iint_R f(x, y)\, dA$, where

$$f(x, y) = \frac{64 - 8x + y^2}{16}$$

and

$$R = \{(x, y): 0 \le x \le 4, 0 \le y \le 8\}$$

Do this by calculating the Riemann sum obtained by dividing R into eight equal squares and using the center of each square as the sample point (Figure 7).

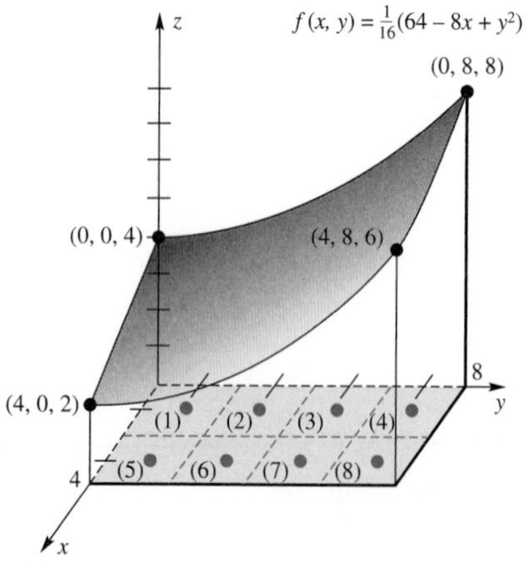

Figure 7

SOLUTION The values of the function at the required sample points are as follows:

(1) $f(x_1, y_1) = f(1, 1) = \frac{57}{16}$; (5) $f(x_5, y_5) = f(3, 1) = \frac{41}{16}$

(2) $f(x_2, y_2) = f(1, 3) = \frac{65}{16}$; (6) $f(x_6, y_6) = f(3, 3) = \frac{49}{16}$

(3) $f(x_3, y_3) = f(1, 5) = \frac{81}{16}$; (7) $f(x_7, y_7) = f(3, 5) = \frac{65}{16}$

(4) $f(x_4, y_4) = f(1, 7) = \frac{105}{16}$; (8) $f(x_8, y_8) = f(3, 7) = \frac{89}{16}$

Thus, since $\Delta A_k = 4$,

$$\iint_R f(x, y)\, dA \approx \sum_{k=1}^{8} f(\overline{x}_k, \overline{y}_k)\, \Delta A_k$$
$$= 4\sum_{k=1}^{8} f(\overline{x}_k, \overline{y}_k)$$
$$= \frac{4(57 + 65 + 81 + 105 + 41 + 49 + 65 + 89)}{16} = 138$$

In Section 13.2, we shall learn how to find the exact value of this integral. It is $138\frac{2}{3}$. ■

Concepts Review

1. Assume that the rectangle R has been partitioned into n subrectangles of area ΔA_k with sample points $(\overline{x}_k, \overline{y}_k)$, $k = 1, 2, \ldots, n$. Then $\iint_R f(x, y)\, dA = \lim_{\|P\|\to 0}$ _____.

2. If $f(x, y) \geq 0$ on R, then $\iint_R f(x, y)\, dA$ can be interpreted geometrically as _____.

3. If f is _____ on R, then f is integrable there.

4. If $f(x, y) = 6$ on the rectangle $R = \{(x, y): 1 \leq x \leq 2, 0 \leq y \leq 2\}$, then $\iint_R f(x, y)\, dA$ has the value _____.

Problem Set 13.1

In Problems 1–4, let $R = \{(x, y): 1 \leq x \leq 4, 0 \leq y \leq 2\}$. Evaluate $\iint_R f(x, y)\, dA$, where f is the given function (see Example 1).

1. $f(x, y) = \begin{cases} 2 & 1 \leq x < 3, 0 \leq y \leq 2 \\ 3 & 3 \leq x \leq 4, 0 \leq y \leq 2 \end{cases}$

2. $f(x, y) = \begin{cases} -1 & 1 \leq x \leq 4, 0 \leq y < 1 \\ 2 & 1 \leq x \leq 4, 1 \leq y \leq 2 \end{cases}$

3. $f(x, y) = \begin{cases} 2 & 1 \leq x < 3, 0 \leq y < 1 \\ 1 & 1 \leq x < 3, 1 \leq y \leq 2 \\ 3 & 3 \leq x \leq 4, 0 \leq y \leq 2 \end{cases}$

4. $f(x, y) = \begin{cases} 2 & 1 \leq x \leq 4, 0 \leq y < 1 \\ 3 & 1 \leq x < 3, 1 \leq y \leq 2 \\ 1 & 3 \leq x \leq 4, 1 \leq y \leq 2 \end{cases}$

Suppose that $R = \{(x, y): 0 \leq x \leq 2, 0 \leq y \leq 2\}$, $R_1 = \{(x, y): 0 \leq x \leq 2, 0 \leq y \leq 1\}$, and $R_2 = \{(x, y): 0 \leq x \leq 2, 1 \leq y \leq 2\}$. Suppose, in addition, that $\iint_R f(x, y)\, dA = 3$, $\iint_R g(x, y)\, dA = 5$, and $\iint_{R_1} g(x, y)\, dA = 2$.

Use the properties of integrals to evaluate the integrals in Problems 5–8

5. $\iint_R [3f(x, y) - g(x, y)]\, dA$

6. $\iint_R [2f(x, y) + 5g(x, y)]\, dA$

7. $\iint_{R_2} g(x, y)\, dA$

8. $\iint_{R_1} [2g(x, y) + 3]\, dA$

In Problems 9–14, $R = \{(x, y): 0 \leq x \leq 6, 0 \leq y \leq 4\}$ and P is the partition of R into six equal squares by the lines $x = 2$, $x = 4$, and $y = 2$. Approximate $\iint_R f(x, y)\, dA$ by calculating the corresponding Riemann sum $\sum_{k=1}^{6} f(\overline{x}_k, \overline{y}_k)\, \Delta A_k$, assuming that $(\overline{x}_k, \overline{y}_k)$ are the centers of the six squares (see Example 2).

9. $f(x, y) = 12 - x - y$

10. $f(x, y) = 10 - y^2$

11. $f(x, y) = x^2 + 2y^2$

12. $f(x, y) = \frac{1}{6}(48 - 4x - 3y)$

C **13.** $f(x, y) = \sqrt{x + y}$

C **14.** $f(x, y) = e^{xy}$

In Problems 15–20, sketch the solid whose volume is given by the following double integrals over the rectangle $R = \{(x, y): 0 \leq x \leq 2, 0 \leq y \leq 3\}$.

15. $\iint_R 3\, dA$

16. $\iint_R (x + 1)\, dA$

17. $\iint_R (y + 1)\, dA$

18. $\iint_R (x - y + 4)\, dA$

19. $\iint_R (x^2 + y^2)\, dA$

20. $\iint_R (25 - x^2 - y^2)\, dA$

21. Calculate $\iint_R (6 - y)\, dA$, where $R = \{(x, y): 0 \leq x \leq 1, 0 \leq y \leq 1\}$. *Hint:* This integral represents the volume of a certain solid. Sketch this solid and calculate its volume from elementary principles.

22. Calculate $\iint_R (1 + x)\,dA$, where $R = \{(x, y): 0 \le x \le 2, 0 \le y \le 1\}$. See the hint in Problem 21.

23. Use the comparison property of double integrals to show that if $f(x, y) \ge 0$ on R then $\iint_R f(x, y)\,dA \ge 0$.

24. Suppose that $m \le f(x, y) \le M$ on R. Show that

$$mA(R) \le \iint_R f(x, y)\,dA \le MA(R)$$

C **25.** Let R be the rectangle shown in Figure 8. For the indicated partition into 12 equal squares, calculate the smallest and largest Riemann sums for $\iint_R \sqrt{x^2 + y^2}\,dA$ and thereby obtain numbers c and C such that

$$c \le \iint_R \sqrt{x^2 + y^2}\,dA \le C$$

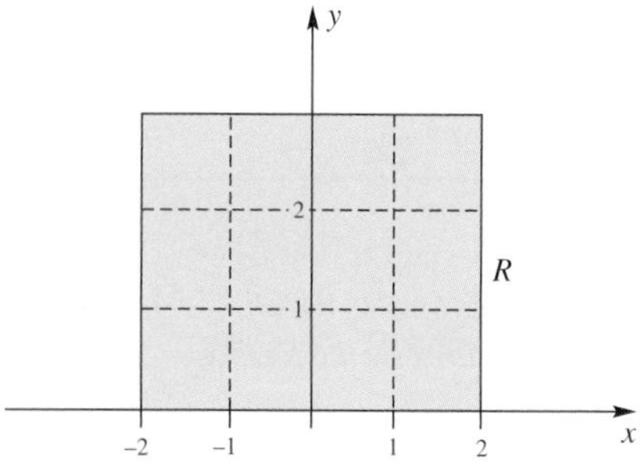

Figure 8

26. Evaluate $\iint_R x\cos^2(xy)\,dA$, where R is the rectangle of Figure 8. *Hint:* Does the graph of the integrand have any kind of symmetry?

27. Recall that $[\![x]\!]$ is the greatest integer function. For R of Figure 8, evaluate:

(a) $\iint_R [\![x]\!][\![y]\!]\,dA$ (b) $\iint_R ([\![x]\!] + [\![y]\!])\,dA$

28. Suppose that the rectangle of Figure 8 represents a thin plate (lamina) whose mass density at (x, y) is $\delta(x, y)$, say in grams per square centimeter. What does $\iint_R \delta(x, y)\,dA$ represent?

29. Colorado is a rectangular state (if we ignore the curvature of the earth). Let $f(x, y)$ be the number of inches of rainfall during 2005 at the point (x, y) in that state. What does $\iint_{\text{Colorado}} f(x, y)\,dA$ represent? What does this number divided by the area of Colorado represent?

30. Let $f(x, y) = 1$ if both x and y are rational numbers, and let $f(x, y) = 0$ otherwise. Show that $f(x, y)$ is not integrable over the rectangle R in Figure 8.

31. Use the two graphs in Figure 9 to approximate

$$\iint_R f(x, y)\,dA;\quad R = \{(x, y): 0 \le x \le 4, 0 \le y \le 4\}$$

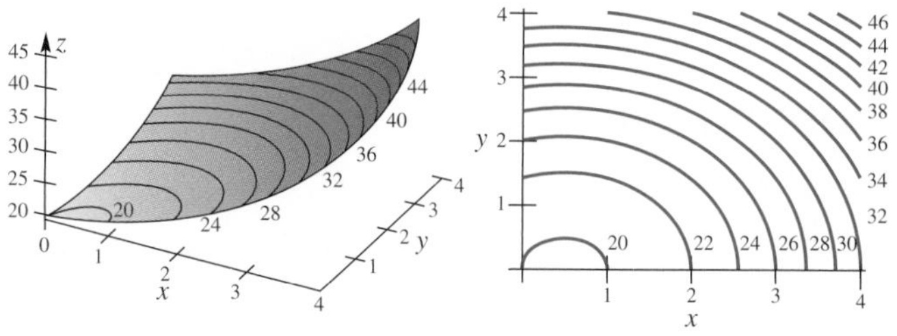

Figure 9

Answers to Concepts Review: **1.** $\sum_{k=1}^{n} f(\bar{x}_k, \bar{y}_k)\,\Delta A_k$ **2.** the volume of the solid under $z = f(x, y)$ and above R **3.** continuous **4.** 12

13.2 Iterated Integrals

Now we face in earnest the problem of evaluating $\iint_R f(x, y)\,dA$, where R is the rectangle

$$R = \{(x, y): a \le x \le b, c \le y \le d\}$$

Suppose for the time being that $f(x, y) \ge 0$ on R so that we may interpret the double integral as the volume V of the solid under the surface of Figure 1.

$$V = \iint_R f(x, y)\,dA \tag{1}$$

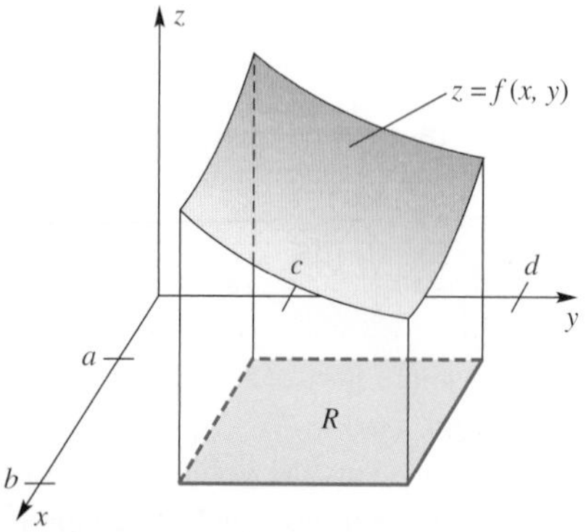

Figure 1

There is another way to calculate the volume of this solid, which at least intuitively seems just as valid. Slice the solid into thin slabs by means of planes parallel to the xz-plane. A typical such slab is shown in Figure 2a. The area of the face of this slab depends on how far it is from the xz-plane; that is, it depends on y. Therefore, we denote this area by $A(y)$ (see Figure 2b).

The volume ΔV of the slab is given approximately by

$$\Delta V \approx A(y)\,\Delta y$$

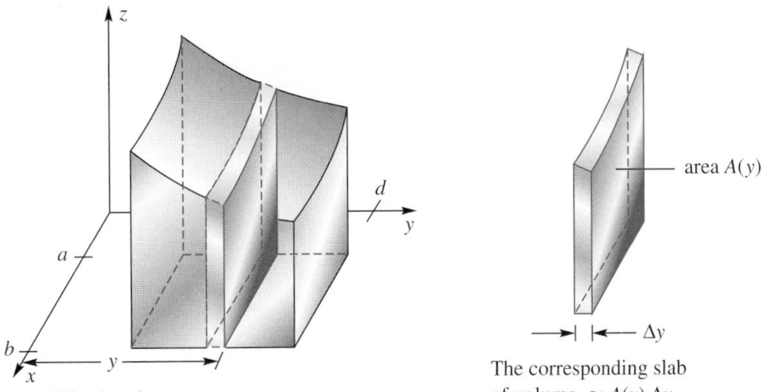

Slice by planes y = constant

(a)

The corresponding slab of volume $\approx A(y)\,\Delta y$

(b)

Figure 2

and, recalling our old motto (*slice, approximate, integrate*), we may write

$$V = \int_c^d A(y)\,dy$$

On the other hand, for fixed y we may calculate $A(y)$ by means of an ordinary single integral; in fact,

$$A(y) = \int_a^b f(x, y)\,dx$$

Thus, we have a solid whose cross sectional areas are known to be $A(y)$. The problem of finding the volume of a region whose cross sections are known was treated in Section 5.2. We conclude that

(2) $$V = \int_c^d A(y)\,dy = \int_c^d \left[\int_a^b f(x, y)\,dx\right] dy$$

The last expression is called an **iterated integral.**

When we equate the expressions for V from (1) and (2), we obtain the result that we want.

$$\iint_R f(x, y)\,dA = \int_c^d \left[\int_a^b f(x, y)\,dx\right] dy$$

If we had begun the process above by slicing the solid with planes parallel to the yz-plane, we would have obtained another iterated integral, with the integrations occurring in the opposite order.

$$\iint_R f(x, y)\,dA = \int_a^b \left[\int_c^d f(x, y)\,dy\right] dx$$

Two remarks are in order. First, while the two boxed results were derived under the assumption that f was nonnegative, they are valid in general. Second, the whole exercise would be rather pointless unless iterated integrals can be evaluated. Fortunately, iterated integrals are often easy to evaluate, as we demonstrate next.

> **What if f is Negative?**
>
> If $f(x, y)$ is negative on part of R, then $\iint_R f(x, y)\,dA$ gives the *signed* volume of the solid between the surface $z = f(x, y)$ and the rectangle R of the xy-plane.
>
> 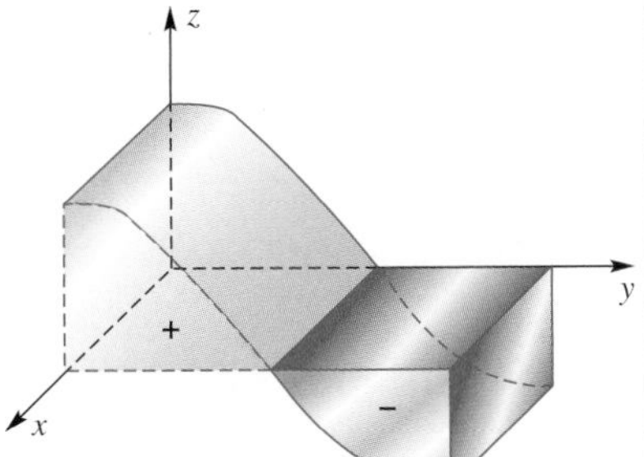
>
>
> The actual volume of this solid is
>
> $$\iint_R |f(x, y)|\,dA$$

Evaluating Iterated Integrals We begin with a simple example.

EXAMPLE 1 Evaluate $\int_0^3 \left[\int_1^2 (2x + 3y)\,dx\right] dy.$

SOLUTION In the inner integration, y is a constant, so

$$\int_1^2 (2x + 3y)\,dx = \left[x^2 + 3yx\right]_1^2 = 4 + 6y - (1 + 3y) = 3 + 3y$$

Consequently,

$$\int_0^3 \left[\int_1^2 (2x + 3y)\,dx\right] dy = \int_0^3 [3 + 3y]\,dy = \left[3y + \frac{3}{2}y^2\right]_0^3$$
$$= 9 + \frac{27}{2} = \frac{45}{2}$$

■

EXAMPLE 2 Evaluate $\int_1^2 \left[\int_0^3 (2x + 3y)\,dy\right] dx$.

SOLUTION Note that we have simply reversed the order of integration from Example 1; we expect the same answer as in that example.

$$\int_0^3 (2x + 3y)\,dy = \left[2xy + \frac{3}{2}y^2\right]_0^3$$
$$= 6x + \frac{27}{2}$$

Thus,

$$\int_1^2 \left[\int_0^3 (2x + 3y)\,dy\right] dx = \int_1^2 \left[6x + \frac{27}{2}\right] dx = \left[3x^2 + \frac{27}{2}x\right]_1^2$$
$$= 12 + 27 - \left(3 + \frac{27}{2}\right) = \frac{45}{2}$$

From now on, we shall usually omit the brackets in the iterated integral. ■

A Note on Notation

The order of dx and dy is important because it specifies which integration is to be done first. The first integration involves the integrand, the integral symbol closest to it on the left, and the first dx or dy symbol on its right. We will sometimes refer to this integral as the *inner integral* and its value as the *inner integration*.

EXAMPLE 3 Evaluate $\int_0^8 \int_0^4 \frac{1}{16}[64 - 8x + y^2]\,dx\,dy$.

SOLUTION Note that this iterated integral corresponds to the double integral of Example 2 of Section 13.1 for which we claimed the answer $138\frac{2}{3}$. We will often omit a separate consideration of the inner integral; instead, we will work our way from the inside out.

$$\int_0^8 \int_0^4 \frac{1}{16}\left[64 - 8x + y^2\right] dx\,dy = \frac{1}{16}\int_0^8 \left[64x - 4x^2 + xy^2\right]_0^4 dy$$
$$= \frac{1}{16}\int_0^8 \left[256 - 64 + 4y^2\right] dy$$
$$= \int_0^8 \left(12 + \frac{1}{4}y^2\right) dy$$
$$= \left[12y + \frac{y^3}{12}\right]_0^8$$
$$= 96 + \frac{512}{12} = 138\frac{2}{3}$$

■

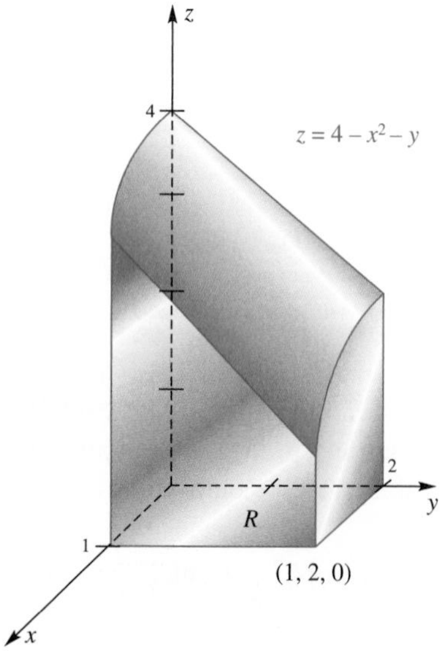

Figure 3

Calculating Volumes Now we can calculate volumes for a wide variety of solids.

EXAMPLE 4 Find the volume V of the solid under the surface $z = 4 - x^2 - y$ and over the rectangle $R = \{(x, y): 0 \le x \le 1, 0 \le y \le 2\}$ (see Figure 3).

SOLUTION ≈ Let's estimate this volume by assuming that the solid has constant height 2.5, giving it a volume of $(2.5)(2) = 5$. If the following calculation gives an answer that is not close to 5, we will know we have made a mistake.

$$V = \iint_R (4 - x^2 - y)\,dA = \int_0^2 \int_0^1 (4 - x^2 - y)\,dx\,dy$$
$$= \int_0^2 \left[4x - \frac{x^3}{3} - yx\right]_0^1 dy = \int_0^2 \left[4 - \frac{1}{3} - y\right] dy$$
$$= \left[\frac{11}{3}y - \frac{1}{2}y^2\right]_0^2 = \frac{22}{3} - 2 = \frac{16}{3}$$ ■

Concepts Review

1. The expression $\int_a^b \left[\int_c^d f(x, y)\,dy\right] dx$ is called a(n) _____ integral.

2. Let $R = \{(x, y): -1 \le x \le 2, 0 \le y \le 2\}$. Then $\iint_R f(x, y)\,dA$ can be expressed as an iterated integral either as _____ or as _____.

3. For a general function f defined on R, $\iint_R f(x, y)\,dA$ can be interpreted as the _____ volume of the solid between the surface $z = f(x, y)$ and the xy-plane; the part above this plane gets a _____ sign; the part below, a _____ sign.

4. Thus, if a double integral turns out to have a negative value, we know that more than half of the solid _____.

Problem Set 13.2

In Problems 1–16, evaluate each of the iterated integrals.

1. $\int_0^2 \int_0^3 (9 - x)\,dy\,dx$

2. $\int_{-2}^2 \int_0^1 (9 - x^2)\,dy\,dx$

3. $\int_0^2 \int_1^3 x^2 y\,dy\,dx$

4. $\int_{-1}^4 \int_1^2 (x + y^2)\,dy\,dx$

5. $\int_1^2 \int_0^3 (xy + y^2)\,dx\,dy$

6. $\int_{-1}^1 \int_1^2 (x^2 + y^2)\,dx\,dy$

7. $\int_0^{\pi} \int_0^1 x \sin y\,dx\,dy$

8. $\int_0^{\ln 3} \int_0^{\ln 2} e^{x+y}\,dy\,dx$

9. $\int_0^{\pi/2} \int_0^1 x \sin xy\,dy\,dx$

10. $\int_0^1 \int_0^1 xe^{xy}\,dy\,dx$

11. $\int_0^3 \int_0^1 2x\sqrt{x^2 + y}\,dx\,dy$

12. $\int_0^1 \int_0^1 \frac{y}{(xy + 1)^2}\,dx\,dy$

13. $\int_0^{\ln 3} \int_0^1 xy\,e^{xy^2}\,dy\,dx$

14. $\int_0^1 \int_0^2 \frac{y}{1 + x^2}\,dy\,dx$

15. $\int_0^{\pi} \int_0^3 y \cos^2 x\,dy\,dx$

16. $\int_{-1}^1 \int_0^1 xe^{x^2}\,dx\,dy$

In Problems 17–20, evaluate the indicated double integral over R.

17. $\iint_R xy^3\,dA$; $R = \{(x, y): 0 \le x \le 1, -1 \le y \le 1\}$

18. $\iint_R (x^2 + y^2)\,dA$; $R = \{(x, y): -1 \le x \le 1, 0 \le y \le 2\}$

19. $\iint_R \sin(x + y)\,dA$;
$R = \{(x, y): 0 \le x \le \pi/2, 0 \le y \le \pi/2\}$

20. $\iint_R xy\sqrt{1 + x^2}\,dA$;
$R = \{(x, y): 0 \le x \le \sqrt{3}, 1 \le y \le 2\}$

In Problems 21–24, find the volume under the surface in each figure.

21.
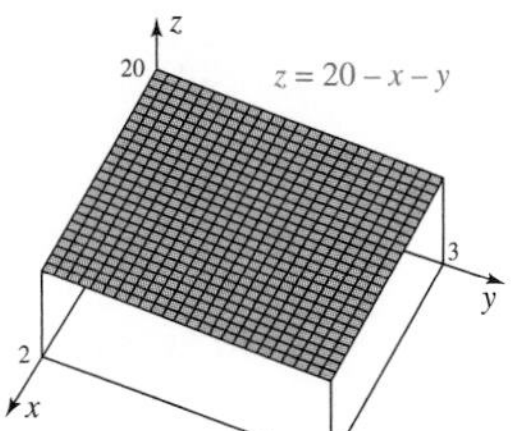

22.
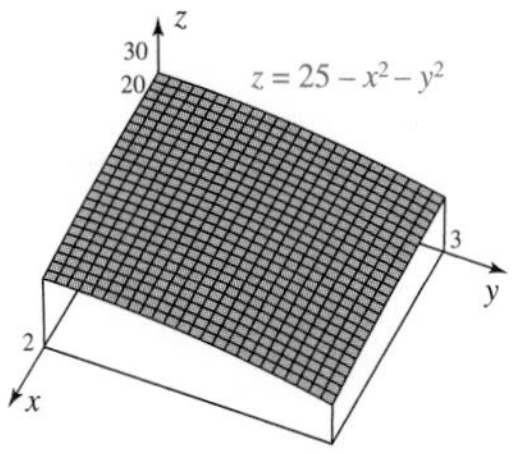

23.
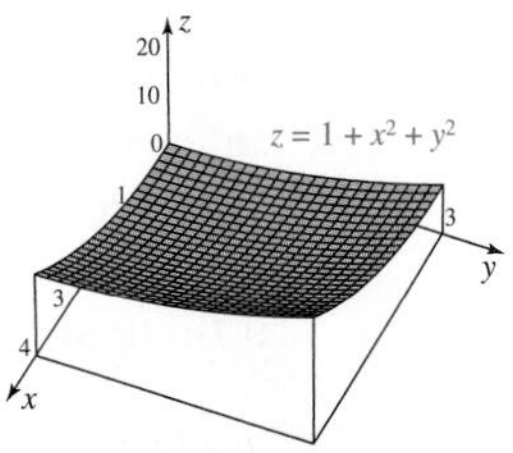

24.
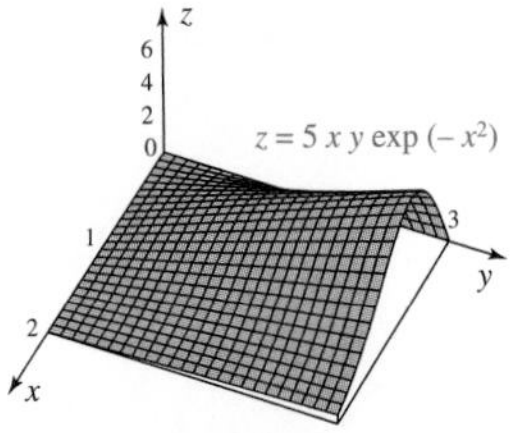

In Problems 25–28, sketch the solid whose volume is the indicated iterated integral.

25. $\int_0^1 \int_0^2 \frac{x}{2}\, dx\, dy$ **26.** $\int_0^1 \int_0^1 (2 - x - y)\, dy\, dx$

27. $\int_0^2 \int_0^2 (x^2 + y^2)\, dy\, dx$ **28.** $\int_0^2 \int_0^2 (4 - y^2)\, dy\, dx$

In Problems 29–32, find the volume of the given solid. First, sketch the solid; then estimate its volume; finally, determine its exact volume.

29. Solid under the plane $z = x + y + 1$ over $R = \{(x, y): 0 \le x \le 1, 1 \le y \le 3\}$

30. Solid under the plane $z = 2x + 3y$ and over $R = \{(x, y): 1 \le x \le 2, 0 \le y \le 4\}$

31. Solid between $z = x^2 + y^2 + 2$ and $z = 1$ and lying above $R = \{(x, y): -1 \le x \le 1, 0 \le y \le 1\}$

32. Solid in the first octant enclosed by $z = 4 - x^2$ and $y = 2$

33. Show that if $f(x, y) = g(x)h(y)$ then

$$\int_a^b \int_c^d f(x, y)\, dy\, dx = \left[\int_a^b g(x)\, dx\right]\left[\int_c^d h(y)\, dy\right]$$

34. Use Problem 33 to evaluate

$$\int_0^{\sqrt{\ln 2}} \int_0^1 \frac{xye^{x^2}}{1 + y^2}\, dy\, dx$$

35. Evaluate

$$\int_0^1 \int_0^1 xye^{x^2+y^2}\, dy\, dx$$

36. Find the volume of the solid trapped between the surface $z = \cos x \cos y$ and the xy-plane, where $-\pi \le x \le \pi$, $-\pi \le y \le \pi$.

In Problems 37–39, evaluate each iterated integral.

37. $\int_{-2}^2 \int_{-1}^1 |x^2y^3|\, dy\, dx$ **38.** $\int_{-2}^2 \int_{-1}^1 [\![x^2]\!]y^3\, dy\, dx$

39. $\int_{-2}^2 \int_{-1}^1 [\![x^2]\!]|y^3|\, dy\, dx$

40. Evaluate $\int_0^{\sqrt{3}} \int_0^1 \frac{8x}{(x^2 + y^2 + 1)^2}\, dy\, dx$. *Hint:* Reverse the order of integration.

41. Prove the **Cauchy–Schwarz Inequality for Integrals:**

$$\left[\int_a^b f(x)g(x)\, dx\right]^2 \le \int_a^b f^2(x)\, dx \int_a^b g^2(x)\, dx$$

Hint: Consider the double integral of

$$F(x, y) = [f(x)g(y) - f(y)g(x)]^2$$

over the rectangle $R = \{(x, y): a \le x \le b, a \le y \le b\}$.

42. Suppose that f is increasing on $[a, b]$ and $\int_a^b f(x)\, dx > 0$. Prove that

$$\frac{\int_a^b xf(x)\, dx}{\int_a^b f(x)\, dx} > \frac{a + b}{2}$$

and give a physical interpretation of this result. *Hint:* Let $F(x, y) = [y - x][f(y) - f(x)]$ and use the hint of Problem 41.

Answers to Concepts Review: **1.** iterated **2.** $\int_{-1}^2\left[\int_0^2 f(x, y)\, dy\right] dx$; $\int_0^2\left[\int_{-1}^2 f(x, y)\, dx\right] dy$ **3.** signed; plus; minus **4.** is below the xy-plane

13.3 Double Integrals over Nonrectangular Regions

Consider an arbitrary closed bounded set S in the plane. Surround S by a rectangle R with sides parallel to the coordinate axes (Figure 1). Suppose that $f(x, y)$ is defined on S, and define (or redefine, if necessary) $f(x, y) = 0$ on the part of R outside of S (Figure 2). We say that f is integrable on S if it is integrable on R and write

$$\iint_S f(x, y)\, dA = \iint_R f(x, y)\, dA$$

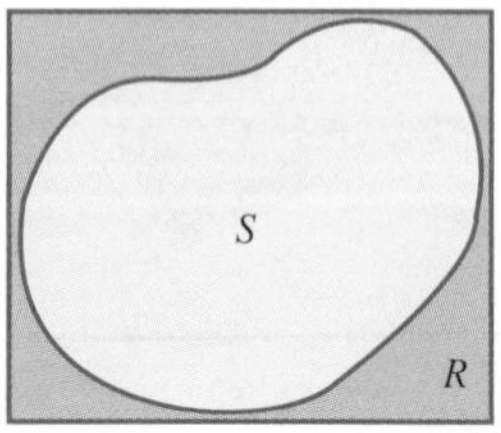

Figure 1

We assert that the double integral on a general set S is (1) linear, (2) additive on sets that overlap only on smooth curves, and (3) satisfies the comparison property (see Section 13.1).

Evaluation of Double Integrals over General Sets

Sets with curved boundaries can be very complicated. For our purposes, it will be sufficient to consider x-simple sets and y-simple sets (and finite unions of such sets). A set S is y-simple if it is simple in the y-direction, meaning that a line in this direction intersects S in a single interval (or point or not at all). Thus, a set S is ***y*-simple** (Figure 3) if there are functions ϕ_1 and ϕ_2 on $[a, b]$ such that

$$S = \{(x, y): \phi_1(x) \le y \le \phi_2(x), a \le x \le b\}$$

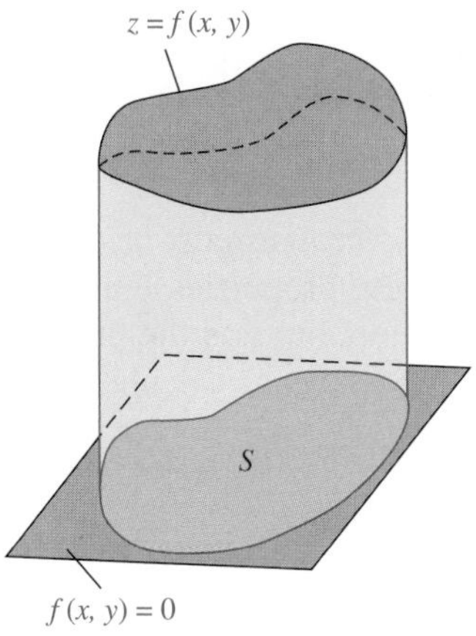

Figure 2

A set S is ***x*-simple** (Figure 4) if there are functions ψ_1 (ψ is the Greek letter psi) and ψ_2 on $[c, d]$ such that

$$S = \{(x, y): \psi_1(y) \leq x \leq \psi_2(y), c \leq y \leq d\}$$

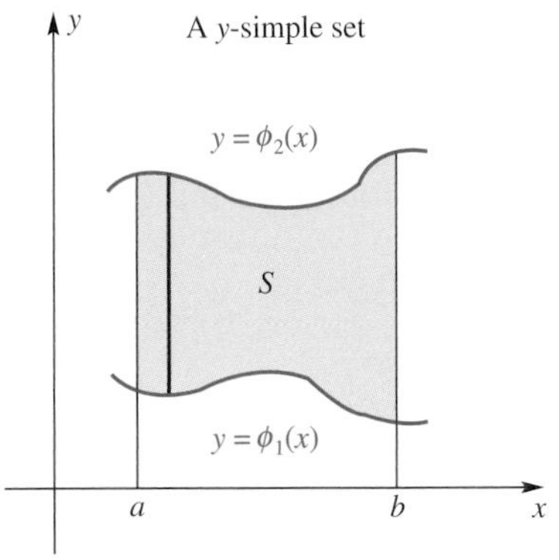

Figure 3

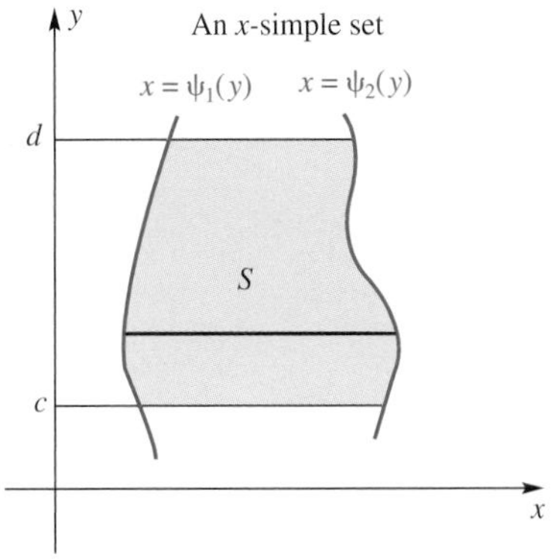

Figure 4

Figure 5 exhibits a set that is neither x-simple nor y-simple.

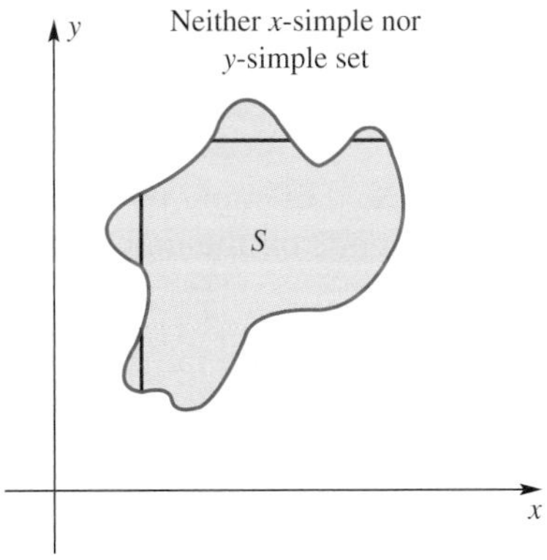

Figure 5

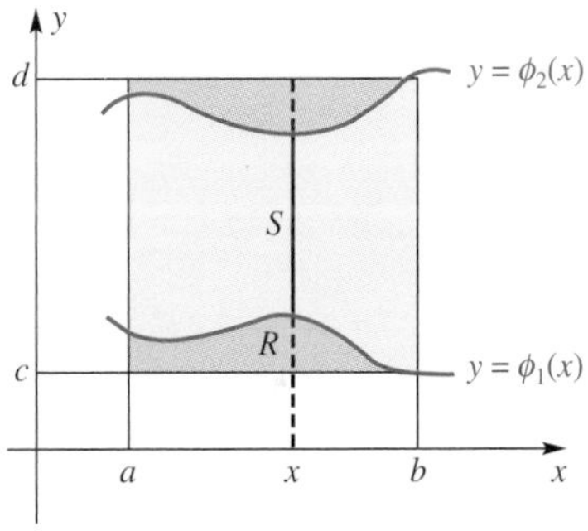

Figure 6

Now suppose that we wish to evaluate the double integral of a function $f(x, y)$ over a y-simple set S. We enclose S in a rectangle R (Figure 6) and make $f(x, y) = 0$ outside S. Then

$$\iint_S f(x, y)\, dA = \iint_R f(x, y)\, dA = \int_a^b \left[\int_c^d f(x, y)\, dy \right] dx$$

$$= \int_a^b \left[\int_{\phi_1(x)}^{\phi_2(x)} f(x, y)\, dy \right] dx$$

In summary,

$$\boxed{\iint_S f(x, y)\, dA = \int_a^b \int_{\phi_1(x)}^{\phi_2(x)} f(x, y)\, dy\, dx}$$

In the inner integration, x is held fixed; thus, this integration is along the heavy vertical line of Figure 6. This integration yields the area $A(x)$ of the cross section shown in Figure 7. Finally, $A(x)$ is integrated from a to b.

If the set S is x-simple (Figure 4), similar reasoning leads to the formula

$$\iint_S f(x, y)\, dA = \int_c^d \int_{\psi_1(y)}^{\psi_2(y)} f(x, y)\, dx\, dy$$

If the set S is neither x-simple nor y-simple (Figure 5), it can usually be considered as a union of pieces that have one or the other of these properties. For example, the annulus of Figure 8 is not simple in either direction, but it is the union of the two y-simple sets S_1 and S_2. The integrals on these pieces can be calculated and added together to obtain the integral over S.

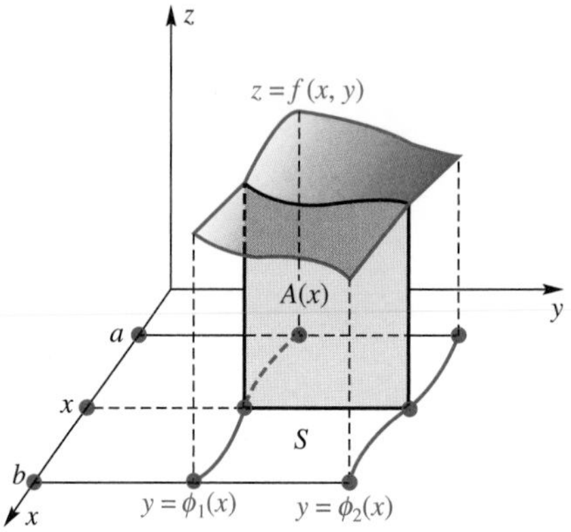

Figure 7

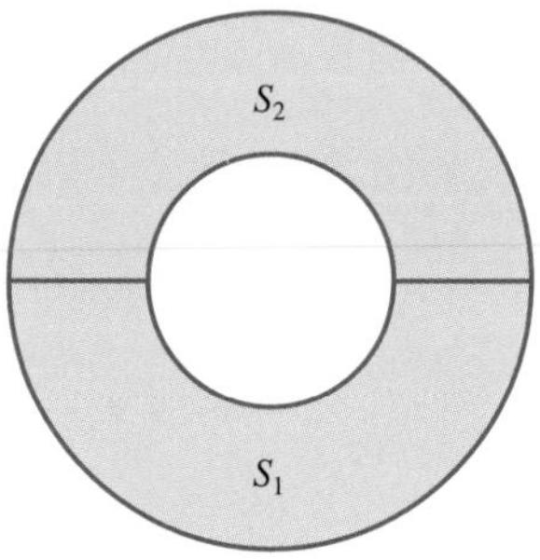

Figure 8

Some Examples For some preliminary practice, we evaluate two iterated integrals, where the limits on the inner integral sign are variables.

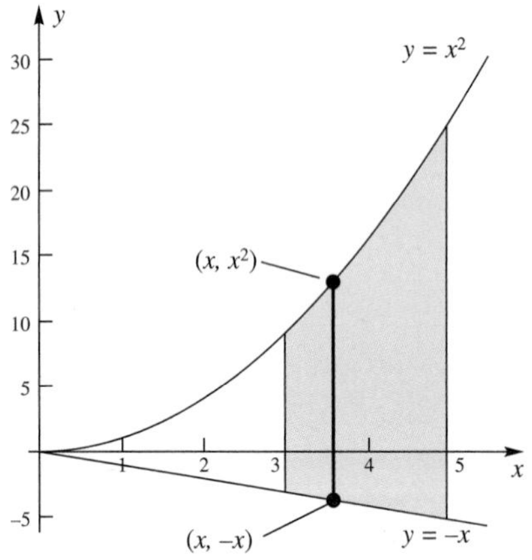

Figure 9

EXAMPLE 1 Evaluate the iterated integral

$$\int_3^5 \int_{-x}^{x^2} (4x + 10y)\, dy\, dx$$

SOLUTION We first perform the inner integration with respect to y, temporarily thinking of x as constant (see Figure 9), and obtain

$$\begin{aligned}\int_3^5 \int_{-x}^{x^2} (4x + 10y)\, dy\, dx &= \int_3^5 \left[4xy + 5y^2\right]_{-x}^{x^2} dx \\ &= \int_3^5 \left[(4x^3 + 5x^4) - (-4x^2 + 5x^2)\right] dx \\ &= \int_3^5 (5x^4 + 4x^3 - x^2)\, dx = \left[x^5 + x^4 - \frac{x^3}{3}\right]_3^5 \\ &= \frac{10{,}180}{3} = 3393\frac{1}{3}\end{aligned}$$

■

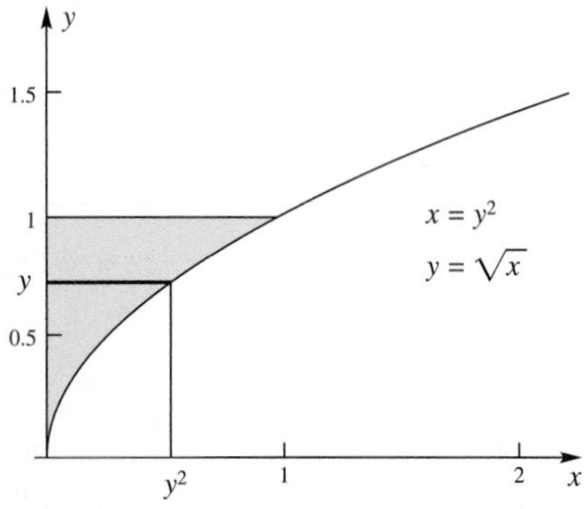

Figure 10

Notice that for iterated integrals, the outer integral cannot have limits that depend on either variable of integration.

EXAMPLE 2 Evaluate the iterated integral

$$\int_0^1 \int_0^{y^2} 2ye^x\, dx\, dy$$

SOLUTION The region of integration is shown in Figure 10.

$$\int_0^1 \int_0^{y^2} 2ye^x \, dx \, dy = \int_0^1 \left[\int_0^{y^2} 2ye^x \, dx \right] dy$$
$$= \int_0^1 \left[2ye^x \right]_0^{y^2} dy = \int_0^1 (2ye^{y^2} - 2ye^0) \, dy$$
$$= \int_0^1 e^{y^2}(2y \, dy) - 2\int_0^1 y \, dy$$
$$= \left[e^{y^2} \right]_0^1 - 2\left[\frac{y^2}{2} \right]_0^1 = e - 1 - 2\left(\frac{1}{2} \right) = e - 2 \qquad \blacksquare$$

We turn to the problem of calculating volumes by means of iterated integrals.

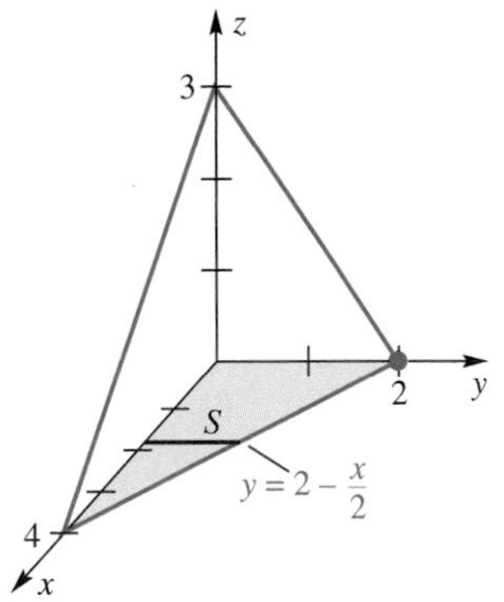

Figure 11

EXAMPLE 3 Use double integration to find the volume of the tetrahedron bounded by the coordinate planes and the plane $3x + 6y + 4z - 12 = 0$.

SOLUTION Denote by S the triangular region in the xy-plane that forms the base of the tetrahedron (Figures 11 and 12). We seek the volume of the solid under the surface $z = \frac{3}{4}(4 - x - 2y)$ and above the region S.

The given plane intersects the xy-plane in the line $x + 2y - 4 = 0$, a segment of which belongs to the boundary of S. Since this equation can be written $y = 2 - x/2$ and $x = 4 - 2y$, S can be thought of as the y-simple set

$$S = \left\{ (x, y): 0 \le x \le 4, 0 \le y \le 2 - \frac{x}{2} \right\}$$

or as the x-simple set

$$S = \{(x, y): 0 \le x \le 4 - 2y, 0 \le y \le 2\}$$

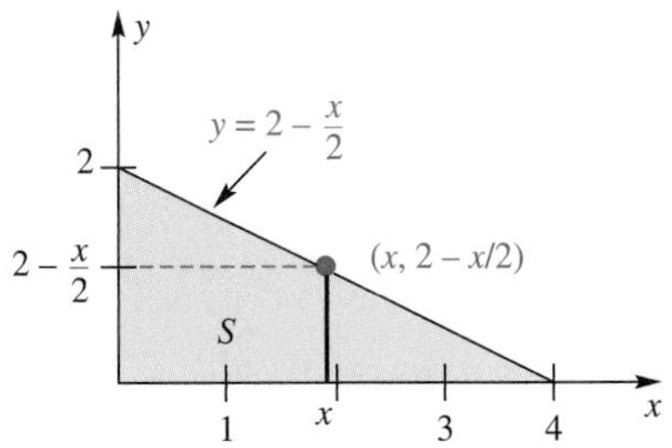

Figure 12

We will treat S as a y-simple set; the final result would be the same either way, as you should verify.

The volume V of the solid is

$$V = \iint_S \frac{3}{4}(4 - x - 2y) \, dA$$

In writing this as an iterated integral, we fix x and integrate along a line (Figure 11 and 12) from $y = 0$ to $y = 2 - x/2$, and then integrate the result from $x = 0$ to $x = 4$. Thus,

$$V = \int_0^4 \int_0^{2-x/2} \frac{3}{4}(4 - x - 2y) \, dy \, dx$$
$$= \int_0^4 \left[\frac{3}{4} \int_0^{2-x/2} (4 - x - 2y) \, dy \right] dx$$
$$= \int_0^4 \frac{3}{4} \left[4y - xy - y^2 \right]_0^{2-x/2} dx$$
$$= \frac{3}{16} \int_0^4 (16 - 8x + x^2) \, dx$$
$$= \frac{3}{16} \left[16x - 4x^2 + \frac{x^3}{3} \right]_0^4 = 4$$

You may recall that the volume of a tetrahedron is one-third the area of the base times the height. In the case at hand, $V = \frac{1}{3}(4)(3) = 4$. This confirms our answer. $\blacksquare$

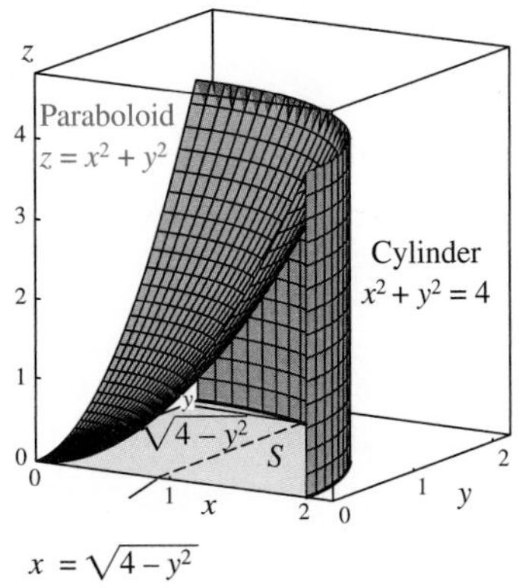

Figure 13

EXAMPLE 4 Find the volume of the solid in the first octant ($x \ge 0, y \ge 0, z \ge 0$) bounded by the circular paraboloid $z = x^2 + y^2$, the cylinder $x^2 + y^2 = 4$, and the coordinate planes (Figure 13).

SOLUTION The region S in the first quadrant of the xy-plane is bounded by a quarter of the circle $x^2 + y^2 = 4$ and the lines $x = 0$ and $y = 0$. Although S can be thought of as either a y-simple or an x-simple region, we shall treat S as the latter and write its boundary curves as $x = \sqrt{4 - y^2}$, $x = 0$, and $y = 0$. Thus,

$$S = \left\{(x, y): 0 \le x \le \sqrt{4 - y^2}, 0 \le y \le 2\right\}$$

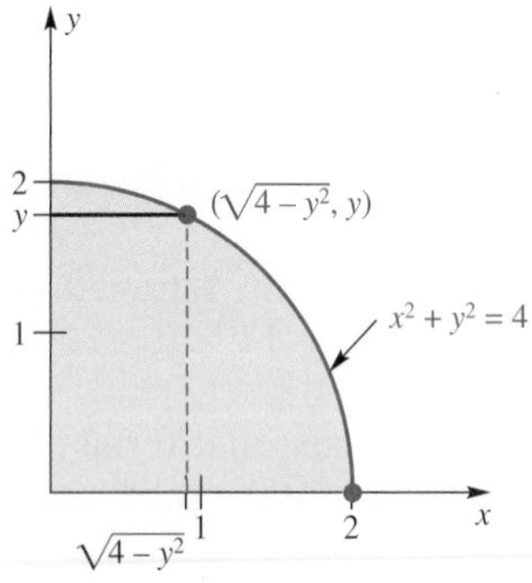

Figure 14

Figure 14 shows the region S in the xy-plane. Now our goal is to calculate

$$V = \iint_S (x^2 + y^2)\, dA$$

by means of an iterated integral. This time we first fix y and integrate along a line (Figure 14) from $x = 0$ to $x = \sqrt{4 - y^2}$ and then integrate the result from $y = 0$ to $y = 2$.

$$V = \iint_S (x^2 + y^2)\, dA = \int_0^2 \int_0^{\sqrt{4-y^2}} (x^2 + y^2)\, dx\, dy$$

$$= \int_0^2 \left[\frac{1}{3}(4 - y^2)^{3/2} + y^2\sqrt{4 - y^2}\right] dy$$

By the trigonometric substitution $y = 2 \sin \theta$, the latter integral can be rewritten as

$$\int_0^{\pi/2} \left[\frac{8}{3}\cos^3\theta + 8\sin^2\theta\cos\theta\right] 2\cos\theta\, d\theta$$

$$= \int_0^{\pi/2} \left[\frac{16}{3}\cos^4\theta + 16\sin^2\theta\cos^2\theta\right] d\theta$$

$$= \frac{16}{3}\int_0^{\pi/2} \cos^2\theta\,(1 - \sin^2\theta + 3\sin^2\theta)\, d\theta$$

$$= \frac{16}{3}\int_0^{\pi/2} (\cos^2\theta + 2\sin^2\theta\cos^2\theta)\, d\theta$$

$$= \frac{16}{3}\int_0^{\pi/2} \left(\cos^2\theta + \frac{1}{2}\sin^2 2\theta\right) d\theta$$

$$= \frac{16}{3}\int_0^{\pi/2} \left(\frac{1 + \cos 2\theta}{2} + \frac{1 - \cos 4\theta}{4}\right) d\theta = 2\pi$$

≈ Is this answer reasonable? Note that the volume of the complete quarter-cylinder in Figure 13 is $\frac{1}{4}\pi r^2 h = \frac{1}{4}\pi(2^2)(4) = 4\pi$. One-half this number is certainly a reasonable value for the required volume. ■

EXAMPLE 5 By changing the order of integration, evaluate

$$\int_0^4 \int_{x/2}^2 e^{y^2}\, dy\, dx$$

SOLUTION The inner integral cannot be evaluated as it stands because e^{y^2} does not have an antiderivative in terms of elementary functions. However, we recognize that the given iterated integral is equal to

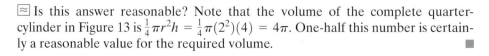

$$\iint_S e^{y^2}\, dA$$

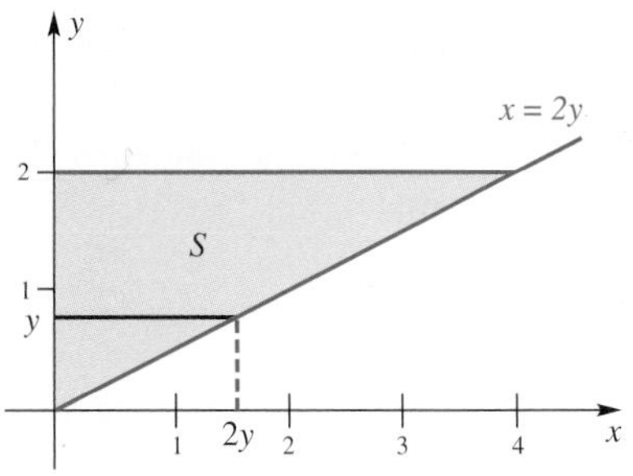

Figure 15

where $S = \{(x, y): x/2 \le y \le 2, 0 \le x \le 4\} = \{(x, y): 0 \le x \le 2y, 0 \le y \le 2\}$ (see Figure 15). If we write this double integral as an iterated integral with the x-integration performed first, we get

$$\int_0^2 \int_0^{2y} e^{y^2}\, dx\, dy = \int_0^2 \left[x e^{y^2} \right]_0^{2y} dy$$

$$= \int_0^2 2y e^{y^2}\, dy = \left[e^{y^2} \right]_0^2 = e^4 - 1 \qquad \blacksquare$$

Concepts Review

1. For an arbitrary set S, we define $\iint_S f(x, y)\, dA$ as $\iint_R f(x, y)\, dA$, where R is ____ and $f(x, y) =$ ____ outside of the set S.

2. A set S is called y-simple if there are functions ϕ_1 and ϕ_2 on $[a, b]$ such that $S = \{(x, y):$ ____ $, a \le x \le b\}$.

3. If S is a y-simple set, as in Question 2, then the double integral over S can be written as the iterated integral $\iint_S f(x, y)\, dA =$ ____.

4. If S is the triangle in the first quadrant bounded by $x + y = 1$, then $\iint_S 2x\, dA$ can be written as the iterated integral ____, which has value ____.

Problem Set 13.3

Evaluate the iterated integrals in Problems 1–14.

1. $\int_0^1 \int_0^{3x} x^2\, dy\, dx$

2. $\int_1^2 \int_0^{x-1} y\, dy\, dx$

3. $\int_{-1}^3 \int_0^{3y} (x^2 + y^2)\, dx\, dy$

4. $\int_{-3}^1 \int_0^x (x^2 - y^3)\, dy\, dx$

5. $\int_1^3 \int_{-y}^{2y} x e^{y^3}\, dx\, dy$

6. $\int_1^5 \int_0^x \dfrac{3}{x^2 + y^2}\, dy\, dx$

7. $\int_{1/2}^1 \int_0^{2x} \cos(\pi x^2)\, dy\, dx$

8. $\int_0^{\pi/4} \int_{\sqrt{2}}^{\sqrt{2}\cos\theta} r\, dr\, d\theta$

9. $\int_0^{\pi/9} \int_{\pi/4}^{3r} \sec^2\theta\, d\theta\, dr$

10. $\int_0^2 \int_{-x}^x e^{-x^2}\, dy\, dx$

11. $\int_0^{\pi/2} \int_0^{\sin y} e^x \cos y\, dx\, dy$

12. $\int_1^2 \int_0^{x^2} \dfrac{y^2}{x}\, dy\, dx$

13. $\int_0^2 \int_0^{\sqrt{4-x^2}} (x + y)\, dy\, dx$

14. $\int_{\pi/6}^{\pi/2} \int_0^{\sin\theta} 6r\cos\theta\, dr\, d\theta$

In Problems 15–20, evaluate the given double integral by changing it to an iterated integral.

15. $\iint_S xy\, dA$; S is the region bounded by $y = x^2$ and $y = 1$.

16. $\iint_S (x + y)\, dA$; S is the triangular region with vertices $(0, 0)$, $(0, 4)$, and $(1, 4)$.

17. $\iint_S (x^2 + 2y)\, dA$; S is the region between $y = x^2$ and $y = \sqrt{x}$.

18. $\iint_S (x^2 - xy)\, dA$; S is the region between $y = x$ and $y = 3x - x^2$.

19. $\iint_S \dfrac{2}{1 + x^2}\, dA$; S is the triangular region with vertices at $(0, 0)$, $(2, 2)$, and $(0, 2)$.

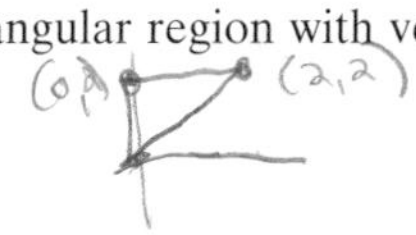

20. $\iint_S x\, dA$; S is the region between $y = x$ and $y = x^3$. (Note that S has two parts.)

In Problems 21–32, sketch the indicated solid. Then find its volume by an iterated integration.

21. Tetrahedron bounded by the coordinate planes and the plane $z = 6 - 2x - 3y$

22. Tetrahedron bounded by the coordinate planes and the plane $3x + 4y + z - 12 = 0$

23. Wedge bounded by the coordinate planes and the planes $x = 5$ and $y + 2z - 4 = 0$

24. Solid in the first octant bounded by the coordinate planes and the planes $2x + y - 4 = 0$ and $8x + y - 4z = 0$

25. Solid in the first octant bounded by the surface $9x^2 + 4y^2 = 36$ and the plane $9x + 4y - 6z = 0$

26. Solid in the first octant bounded by the surface $z = 9 - x^2 - y^2$ and the coordinate planes

27. Solid in the first octant bounded by the cylinder $y = x^2$ and the planes $x = 0$, $z = 0$, and $y + z = 1$

28. Solid bounded by the parabolic cylinder $x^2 = 4y$ and the planes $z = 0$ and $5y + 9z - 45 = 0$

29. Solid in the first octant bounded by the cylinder $z = \tan x^2$ and the planes $x = y$, $x = 1$, and $y = 0$

30. Solid in the first octant bounded by the surface $z = e^{x-y}$, the plane $x + y = 1$, and the coordinate planes

31. Solid in the first octant bounded by the surface $9z = 36 - 9x^2 - 4y^2$ and the coordinate planes

32. Solid in the first octant bounded by the circular cylinders $x^2 + z^2 = 16$ and $y^2 + z^2 = 16$ and the coordinate planes

In Problems 33–38, write the given iterated integral as an iterated integral with the order of integration interchanged. Hint: Begin by sketching a region S and representing it in two ways, as in Example 5.

33. $\int_0^1 \int_0^x f(x, y)\, dy\, dx$ **34.** $\int_0^2 \int_{y^2}^{2y} f(x, y)\, dx\, dy$

35. $\int_0^1 \int_{x^2}^{x^{1/4}} f(x, y)\, dy\, dx$ **36.** $\int_{1/2}^1 \int_{x^3}^{x} f(x, y)\, dy\, dx$

37. $\int_0^1 \int_{-y}^{y} f(x, y)\, dx\, dy$ **38.** $\int_{-1}^0 \int_{-\sqrt{y+1}}^{\sqrt{y+1}} f(x, y)\, dx\, dy$

39. Evaluate $\iint_S xy^2\, dA$, where S is the region shown in Figure 16.

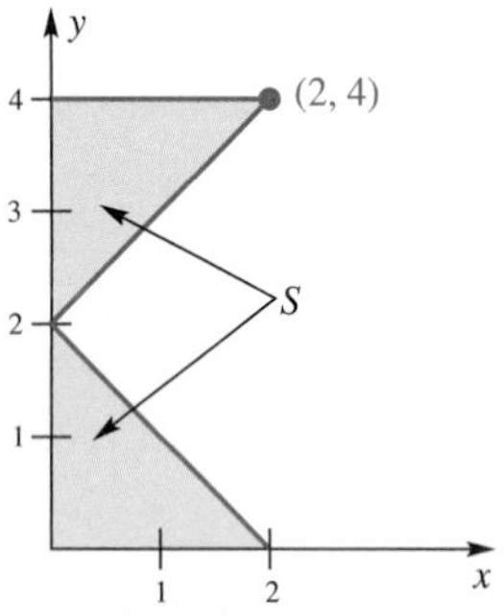

Figure 16

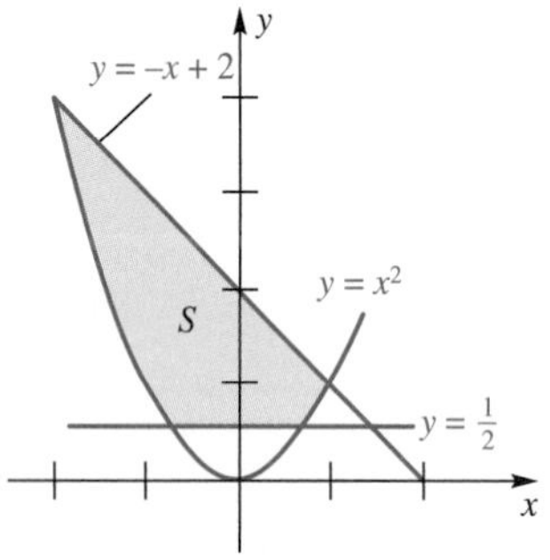

Figure 17

40. Evaluate $\iint_S xy\, dA$, where S is the region in Figure 17.

41. Evaluate $\iint_S (x^2 + x^4 y)\, dA$, where $S = \{(x, y): 1 \le x^2 + y^2 \le 4\}$. *Hint:* Use symmetry to reduce the problem to evaluating $4\left[\iint_{S_1} x^2\, dA + \iint_{S_2} x^2\, dA\right]$, where S_1 and S_2 are as in Figure 18.

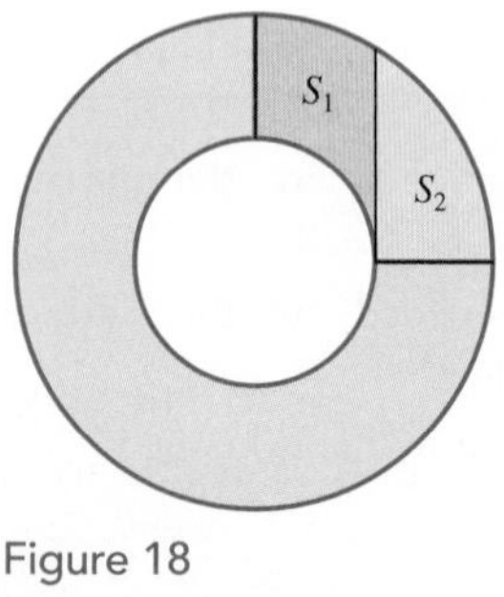

Figure 18

42. Evaluate $\iint_S \sin(xy^2)\, dA$, where S is the annulus $\{(x, y): 1 \le x^2 + y^2 \le 4\}$. *Hint:* Done without thinking, this problem is hard; using symmetry, it is trivial.

43. Evaluate $\iint_S \sin(y^3)\, dA$, where S is the region bounded by $y = \sqrt{x}$, $y = 2$, and $x = 0$. *Hint:* If one order of integration does not work, try the other.

44. Evaluate $\iint_S x^2\, dA$, where S is the region between the ellipse $x^2 + 2y^2 = 4$ and the circle $x^2 + y^2 = 4$.

45. Figure 19 shows a contour map for the depth of a river between a dam and a bridge. Approximate the volume of water between the dam and the bridge. *Hint:* Slice the river into eleven 100-feet sections parallel to the bridge and assume that cross-sections are isosceles triangles. The river is approximately 300 feet wide by the dam and 175 feet wide by the bridge.

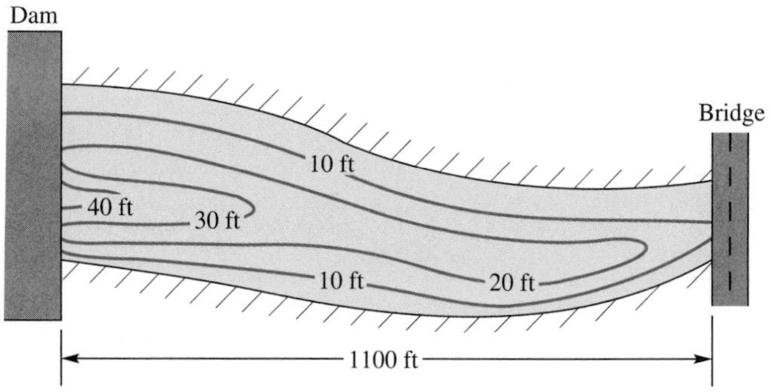

Figure 19

46. Suppose that $f(x, y)$ is a continuous function defined on a region R that is closed and bounded. Show that there is an ordered pair (a, b) in R such that

$$\iint_R f(x, y)\, dA = f(a, b)\, A(R)$$

This result is called the **Mean Value Theorem for Double Integrals.** *Hint:* You will need the Intermediate Value Theorem (Theorem 1.6F).

Answers to Concepts Review: **1.** a rectangle containing S; 0 **2.** $\phi_1(x) \le y \le \phi_2(x)$ **3.** $\int_a^b \int_{\phi_1(x)}^{\phi_2(x)} f(x, y)\, dy\, dx$ **4.** $\int_0^1 \int_0^{1-x} 2x\, dy\, dx$; $\frac{1}{3}$

13.4 Double Integrals in Polar Coordinates

Certain curves in the plane, such as circles, cardioids, and roses, are easier to describe in terms of polar coordinates than in Cartesian (rectangular) coordinates. Thus, we can expect that double integrals over regions enclosed by such curves are more easily evaluated using polar coordinates. In Section 13.9, we will see how to make more general transformations. For now, we study in depth just one particular transformation, from rectangular to polar coordinates, because this technique is so useful.

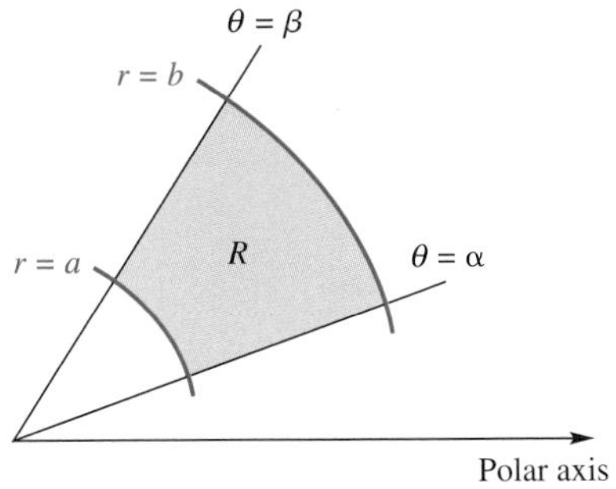

Figure 1

Let R have the shape shown in Figure 1, which we call a *polar rectangle* and will describe analytically in a moment. Let $z = f(x, y)$ determine a surface over R and suppose that f is continuous and nonnegative. Then the volume V of the solid under this surface and above R (Figure 2) is given by

$$(1) \qquad V = \iint_R f(x, y)\, dA$$

In polar coordinates, a polar rectangle R has the form

$$R = \{(r, \theta): a \le r \le b, \alpha \le \theta \le \beta\}$$

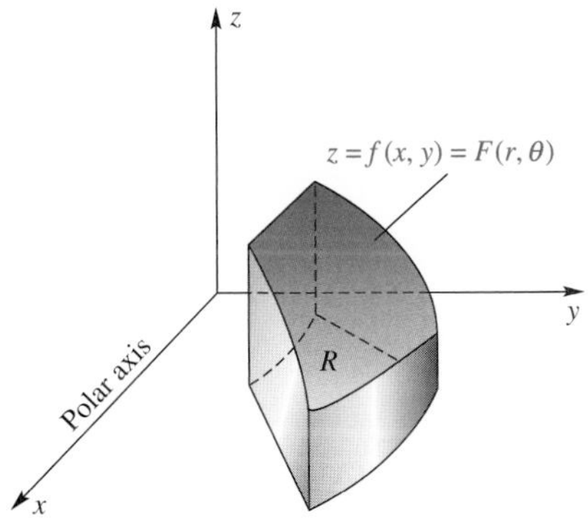

Figure 2

where $a \ge 0$ and $\beta - \alpha \le 2\pi$. Also, the equation of the surface can be written as

$$z = f(x, y) = f(r\cos\theta, r\sin\theta) = F(r, \theta)$$

We are going to calculate the volume V in a new way using polar coordinates.

Partition R into smaller polar rectangles $R_1, R_2, \ldots, R_n$ by means of a polar grid and let Δr_k and $\Delta\theta_k$ denote the dimensions of the typical piece R_k, as shown in Figure 3. The area $A(R_k)$ is given by (see Problem 38)

$$A(R_k) = \bar{r}_k\, \Delta r_k\, \Delta\theta_k$$

where $\bar{r}_k$ is the average radius of R_k. Thus,

$$V \approx \sum_{k=1}^{n} F(\bar{r}_k, \bar{\theta}_k)\bar{r}_k\, \Delta r_k\, \Delta\theta_k$$

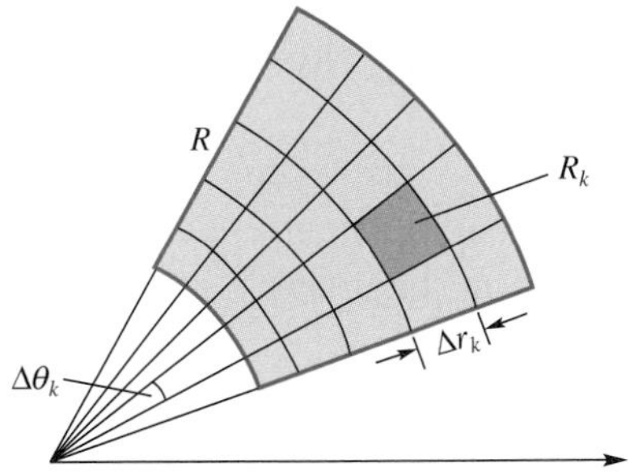

Figure 3

When we take the limit as the norm of the partition approaches zero, we ought to get the actual volume. This limit is a double integral.

$$(2) \qquad V = \iint_R F(r, \theta)\, r\, dr\, d\theta = \iint_R f(r\cos\theta, r\sin\theta)\, r\, dr\, d\theta$$

Now we have two expressions for V, that is, (1) and (2). Equating them yields

$$\boxed{\iint_R f(x, y)\, dA = \iint_R f(r\cos\theta, r\sin\theta)\, r\, dr\, d\theta}$$

The boxed result was derived under the assumption that f was nonnegative, but it is valid for very general functions, in particular for continuous functions of arbitrary sign.

Iterated Integrals The result announced above becomes useful when we write the polar double integral as an iterated integral, a statement we now illustrate.

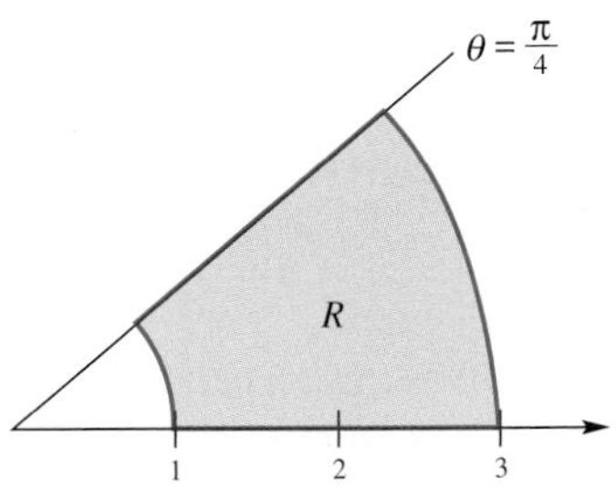

Figure 4

EXAMPLE 1 Find the volume V of the solid above the polar rectangle $R = \{(r, \theta): 1 \le r \le 3, 0 \le \theta \le \pi/4\}$ (Figure 4) and under the surface $z = e^{x^2+y^2}$.

SOLUTION Since $x^2 + y^2 = r^2$,

$$V = \iint_R e^{x^2+y^2}\, dA = \int_0^{\pi/4}\left[\int_1^3 e^{r^2} r\, dr\right] d\theta = \int_0^{\pi/4}\left[\frac{1}{2}e^{r^2}\right]_1^3 d\theta = \int_0^{\pi/4}\frac{1}{2}(e^9 - e)\, d\theta = \frac{\pi}{8}(e^9 - e) \approx 3181$$

Without the help of polar coordinates, we could not have done this problem. Note how the extra factor of r was just what we needed in order to antidifferentiate e^{r^2}. ■

General Regions Recall how we extended the double integral over an ordinary rectangle R to the integral over a general set S. We simply enclosed S in a rectangle and gave the function to be integrated the value zero outside S. We can do the same thing for double integrals in polar coordinates, except that we use polar rectangles rather than ordinary rectangles. Omitting the details, we simply assert that the boxed result stated earlier holds for general sets S.

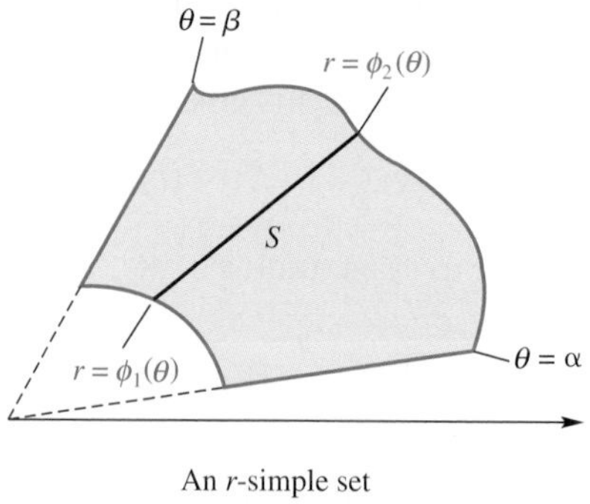

Figure 5

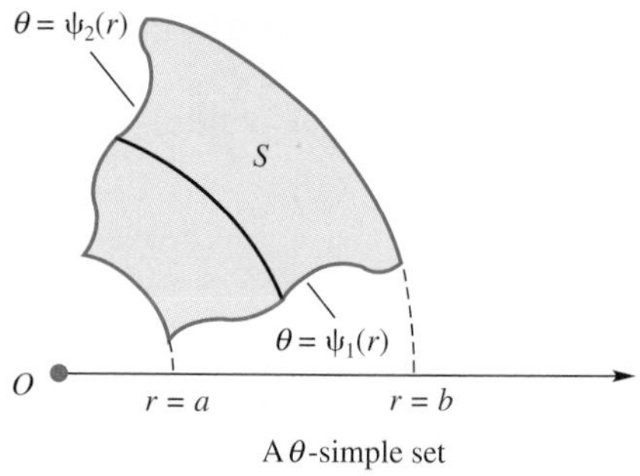

Figure 6

Of special interest for polar integration are what we shall call r-simple and θ-simple sets. Call a set S an ***r*-simple** set if it has the form (Figure 5)

$$S = \{(r, \theta): \phi_1(\theta) \le r \le \phi_2(\theta), \alpha \le \theta \le \beta\}$$

and call it ***θ*-simple** if it has the form (Figure 6)

$$S = \{(r, \theta): a \le r \le b, \psi_1(r) \le \theta \le \psi_2(r)\}$$

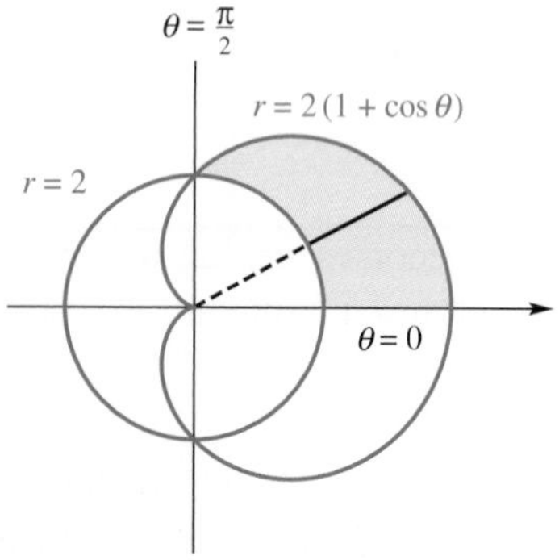

Figure 7

EXAMPLE 2 Evaluate

$$\iint_S y\, dA$$

where S is the region in the first quadrant that is outside the circle $r = 2$ and inside the cardioid $r = 2(1 + \cos\theta)$ (see Figure 7).

SOLUTION Since S is an r-simple set, we write the given integral as an iterated polar integral, with r as the inner variable of integration. In this inner integration, θ is held fixed; the integration is along the heavy line of Figure 7 from $r = 2$ to $r = 2(1 + \cos\theta)$.

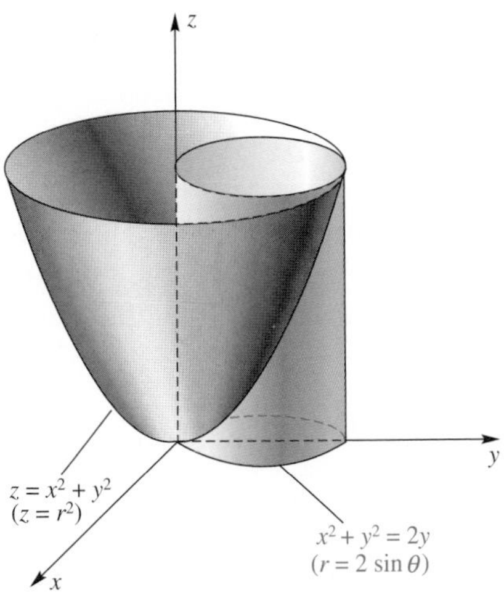

Figure 8

$$\iint_S y\,dA = \int_0^{\pi/2}\int_2^{2(1+\cos\theta)} (r\sin\theta)r\,dr\,d\theta$$

$$= \int_0^{\pi/2}\left[\frac{r^3}{3}\sin\theta\right]_2^{2(1+\cos\theta)} d\theta$$

$$= \frac{8}{3}\int_0^{\pi/2}\left[(1+\cos\theta)^3\sin\theta - \sin\theta\right]d\theta$$

$$= \frac{8}{3}\left[-\frac{1}{4}(1+\cos\theta)^4 + \cos\theta\right]_0^{\pi/2}$$

$$= \frac{8}{3}\left[-\frac{1}{4} + 0 - (-4+1)\right] = \frac{22}{3}$$ ■

EXAMPLE 3 Find the volume of the solid under the surface $z = x^2 + y^2$, above the xy-plane, and inside the cylinder $x^2 + y^2 = 2y$ (Figure 8).

SOLUTION From symmetry, we can double the volume in the first octant. When we use $x = r\cos\theta$ and $y = r\sin\theta$, the equation of the surface becomes $z = r^2$ and that of the cylinder, $r = 2\sin\theta$. Let S denote the region shown in Figure 9. The required volume V is given by

$$V = 2\iint_S (x^2+y^2)\,dA = 2\int_0^{\pi/2}\int_0^{2\sin\theta} r^2 r\,dr\,d\theta$$

$$= 2\int_0^{\pi/2}\left[\frac{r^4}{4}\right]_0^{2\sin\theta} d\theta = 8\int_0^{\pi/2}\sin^4\theta\,d\theta$$

$$= 8\left(\frac{3}{8}\cdot\frac{\pi}{2}\right) = \frac{3\pi}{2}$$

The last integral was evaluated by means of Formula 113 in the table of integrals at the end of the book. ■

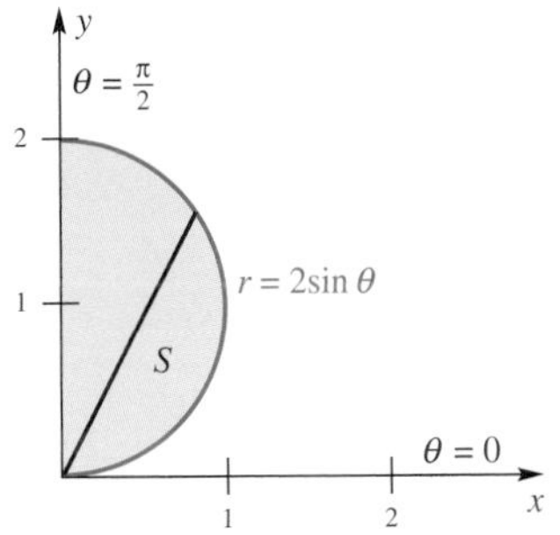

Figure 9

Common Sense

To estimate the volume in Example 3, note that the height of the cylinder displayed in Figure 8 is 4 (let $x = 0$ and $y = 2$ in $z = x^2 + y^2$). Thus, the desired volume is somewhat less than half the volume of a cylinder of radius 1 and height 4, that is, less than $\left(\frac{1}{2}\right)\pi(1^2)4 = 2\pi$. The answer we got, $3\pi/2$, is reasonable.

A Probability Integral

In Chapter 8 we discussed the standard normal probability density function

$$f(x) = \frac{1}{\sqrt{2\pi}}e^{-x^2/2}$$

At that time, we claimed, but were unable to prove, that $\int_{-\infty}^{\infty} f(x)\,dx = 1$. In the next two examples, we will prove this result.

EXAMPLE 4 Show that $I = \int_0^\infty e^{-x^2}\,dx = \frac{\sqrt{\pi}}{2}$.

SOLUTION We are going to sneak up on this problem in a roundabout, but decidedly ingenious, way. First recall that

$$I = \int_0^\infty e^{-x^2}\,dx = \lim_{b\to\infty}\int_0^b e^{-x^2}\,dx$$

Now let V_b be the volume of the solid (Figure 10) that lies under the surface $z = e^{-x^2-y^2}$ and above the square with vertices $(\pm b, \pm b)$. Then

$$V_b = \int_{-b}^{b}\int_{-b}^{b} e^{-x^2-y^2}\,dy\,dx = \int_{-b}^{b} e^{-x^2}\left[\int_{-b}^{b} e^{-y^2}\,dy\right]dx$$

$$= \int_{-b}^{b} e^{-x^2}\,dx\int_{-b}^{b} e^{-y^2}\,dy = \left[\int_{-b}^{b} e^{-x^2}\,dx\right]^2 = 4\left[\int_0^b e^{-x^2}\,dx\right]^2$$

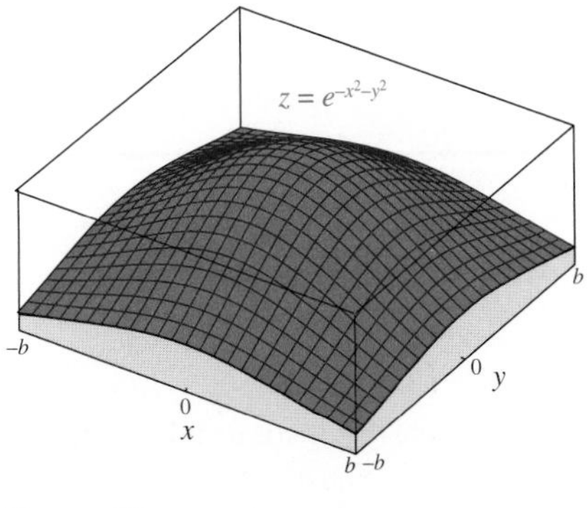

Figure 10

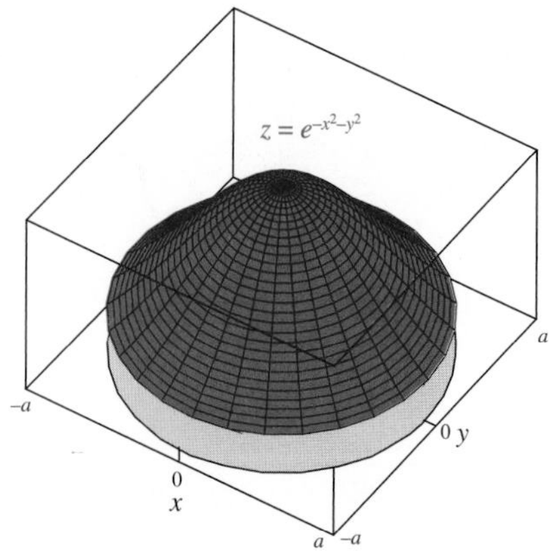

Figure 11

It follows that the volume of the region under $z = e^{-x^2-y^2}$ above the whole xy-plane is

$$(1) \qquad V = \lim_{b\to\infty} V_b = \lim_{b\to\infty} 4\left[\int_0^b e^{-x^2}\,dx\right]^2 = 4\left[\int_0^\infty e^{-x^2}\,dx\right]^2 = 4I^2$$

On the other hand, we can also calculate V using polar coordinates. Here V is the limit as $a \to \infty$ of V_a, the volume of the solid under the surface $z = e^{-x^2-y^2} = e^{-r^2}$ above the circular region of radius a centered at the origin (Figure 11).

$$(2) \qquad V = \lim_{a\to\infty} V_a = \lim_{a\to\infty} \int_0^{2\pi}\int_0^a e^{-r^2} r\,dr\,d\theta = \lim_{a\to\infty}\int_0^{2\pi}\left[-\frac{1}{2}e^{-r^2}\right]_0^a d\theta$$

$$= \lim_{a\to\infty} \frac{1}{2}\int_0^{2\pi}\left[1 - e^{-a^2}\right]d\theta = \lim_{a\to\infty} \pi\left[1 - e^{-a^2}\right] = \pi$$

Equating the two values obtained for V in (1) and (2) yields $4I^2 = \pi$, or $I = \frac{1}{2}\sqrt{\pi}$, as desired. ■

EXAMPLE 5 Show that $\displaystyle\int_{-\infty}^{\infty} \frac{1}{\sqrt{2\pi}} e^{-x^2/2}\,dx = 1.$

SOLUTION By symmetry,

$$\int_{-\infty}^{\infty} \frac{1}{\sqrt{2\pi}} e^{-x^2/2}\,dx = 2\int_0^{\infty} \frac{1}{\sqrt{2\pi}} e^{-x^2/2}\,dx$$

Now we make the substitution $u = x/\sqrt{2}$, so $dx = \sqrt{2}\,du$. The limits on the integral remain the same, so we have

$$\int_{-\infty}^{\infty} \frac{1}{\sqrt{2\pi}} e^{-x^2/2}\,dx = 2\int_0^{\infty} \frac{1}{\sqrt{2\pi}} e^{-u^2}\sqrt{2}\,du$$

$$= \frac{2\sqrt{2}}{\sqrt{2\pi}}\int_0^{\infty} e^{-u^2}\,du$$

$$= \frac{2\sqrt{2}}{\sqrt{2\pi}}\frac{\sqrt{\pi}}{2} = 1$$

To get the last line, we used the result of Example 4. ■

Concepts Review

1. A polar rectangle R has the form $R = \{(r, \theta)\text{: \_\_\_\_\_}\}$.

2. The $dy\,dx$ of integrals in Cartesian (rectangular) coordinates transforms to _____ for integrals in polar coordinates.

3. The integral $\iint_S (x^2 + y^2)\,dA$, where S is the semicircle bounded by $y = \sqrt{4 - x^2}$ and $y = 0$, becomes the iterated integral _____ in polar coordinates.

4. The value of the integral in Question 3 is _____.

Problem Set 13.4

In Problems 1–6, evaluate the iterated integrals.

1. $\displaystyle\int_0^{\pi/2}\int_0^{\cos\theta} r^2 \sin\theta\,dr\,d\theta$

2. $\displaystyle\int_0^{\pi/2}\int_0^{\sin\theta} r\,dr\,d\theta$

3. $\displaystyle\int_0^{\pi}\int_0^{\sin\theta} r^2\,dr\,d\theta$

4. $\displaystyle\int_0^{\pi}\int_0^{1-\cos\theta} r\sin\theta\,dr\,d\theta$

5. $\displaystyle\int_0^{\pi}\int_0^{2} r\cos\frac{\theta}{4}\,dr\,d\theta$

6. $\displaystyle\int_0^{2\pi}\int_0^{\theta} r\,dr\,d\theta$

In Problems 7–12, find the area of the given region S by calculating $\iint_S r\,dr\,d\theta$. Be sure to make a sketch of the region first.

7. S is the region inside the circle $r = 4\cos\theta$ and outside the circle $r = 2$.

8. S is the smaller region bounded by $\theta = \pi/6$ and $r = 4\sin\theta$.

9. S is one leaf of the four-leaved rose $r = a \sin 2\theta$.

10. S is the region inside the cardioid $r = 6 - 6 \sin \theta$.

11. S is the region inside the larger loop of the limaçon $r = 2 - 4 \sin \theta$.

12. S is the region outside the circle $r = 2$ and inside the lemniscate $r^2 = 9 \cos 2\theta$.

In Problems 13–18, an iterated integral in polar coordinates is given. Sketch the region whose area is given by the iterated integral and evaluate the integral, thereby finding the area of the region.

13. $\int_0^{\pi/4} \int_0^2 r\,dr\,d\theta$ **14.** $\int_0^{2\pi} \int_1^3 r\,dr\,d\theta$

15. $\int_0^{\pi/2} \int_0^{\theta} r\,dr\,d\theta$ **16.** $\int_0^{\pi/2} \int_0^{\cos\theta} r\,dr\,d\theta$

17. $\int_0^{\pi} \int_0^{\sin\theta} r\,dr\,d\theta$ **18.** $\int_0^{3\pi/2} \int_0^{\theta^2} r\,dr\,d\theta$

In Problems 19–26, evaluate by using polar coordinates. Sketch the region of integration first.

19. $\iint_S e^{x^2+y^2}\,dA$, where S is the region enclosed by $x^2 + y^2 = 4$

20. $\iint_S \sqrt{4 - x^2 - y^2}\,dA$, where S is the first quadrant sector of the circle $x^2 + y^2 = 4$ between $y = 0$ and $y = x$

21. $\iint_S \frac{1}{4 + x^2 + y^2}\,dA$, where S is as in Problem 20

22. $\iint_S y\,dA$, where S is the first quadrant polar rectangle inside $x^2 + y^2 = 4$ and outside $x^2 + y^2 = 1$

23. $\int_0^1 \int_0^{\sqrt{1-x^2}} (4 - x^2 - y^2)^{-1/2}\,dy\,dx$

24. $\int_0^1 \int_0^{\sqrt{1-y^2}} \sin(x^2 + y^2)\,dx\,dy$

25. $\int_0^1 \int_x^1 x^2\,dy\,dx$

26. $\int_1^2 \int_0^{\sqrt{2x-x^2}} (x^2 + y^2)^{-1/2}\,dy\,dx$

27. Find the volume of the solid in the first octant under the paraboloid $z = x^2 + y^2$ and inside the cylinder $x^2 + y^2 = 9$ by using polar coordinates.

28. Using polar coordinates, find the volume of the solid bounded above by $2x^2 + 2y^2 + z^2 = 18$, below by $z = 0$, and laterally by $x^2 + y^2 = 4$.

29. Switch to rectangular coordinates and then evaluate

$$\int_{3\pi/4}^{4\pi/3} \int_0^{-5 \sec\theta} r^3 \sin^2\theta\,dr\,d\theta$$

30. Let $V = \iint_S \sin\sqrt{x^2 + y^2}\,dA$ and $W = \iint_S |\sin\sqrt{x^2 + y^2}|\,dA$, where S is the region inside the circle $x^2 + y^2 = 4\pi^2$.

(a) Without calculation, determine the sign of V.

(b) Evaluate V. (c) Evaluate W.

31. The centers of two spheres of radius a are $2b$ units apart with $b \le a$. Find the volume of their intersection in terms of $d = a - b$.

32. The depth (in feet) of water distributed by a rotating lawn sprinkler in an hour is $ke^{-r/10}$, $0 \le r \le 10$, where r is the distance from the sprinkler and k is a constant. Determine k if 100 cubic feet of water is distributed in 1 hour.

33. Find the volume of the solid cut from the sphere $r^2 + z^2 \le a^2$ by the cylinder $r = a \sin\theta$.

34. Find the volume of the wedge cut from a tall right circular cylinder of radius a by a plane through a diameter of its base and making an angle α $(0 < \alpha < \pi/2)$ with the base (compare Problem 37, Section 5.2).

35. Consider the ring A of height $2b$ obtained from a sphere of radius a when a hole of radius $c(c < a)$ is bored through the center of the sphere (Figure 12). Show that the volume of A is $4\pi b^3/3$, which is remarkable for two reasons. It is independent of the radius a, and it is the same as the volume of a sphere of radius b.

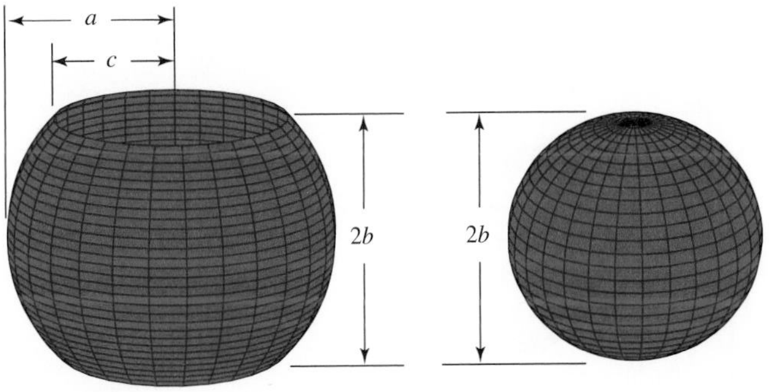

Figure 12

36. There is a simple explanation for the remarkable result in Problem 35. Show that a horizontal plane that intersects the region in Figure 12 and a sphere of radius b next to it will intersect in equal areas. Then apply Cavalieri's Principle for volume. (See Problem 40 in Section 5.2.)

37. Show that

$$\int_0^{\infty} \int_0^{\infty} \frac{1}{(1 + x^2 + y^2)^2}\,dy\,dx = \frac{\pi}{4}$$

38. Recall the formula $A = \frac{1}{2}r^2\theta$ for the area of the sector of a circle of radius r and central angle θ radians (Section 10.7). Use this to obtain the formula

$$A = \frac{r_1 + r_2}{2}(r_2 - r_1)(\theta_2 - \theta_1)$$

for the area of the polar rectangle $\{(r, \theta): r_1 \le r \le r_2, \theta_1 \le \theta \le \theta_2\}$.

39. Show that

$$\int_{-\infty}^{\infty} \frac{1}{\sigma\sqrt{2\pi}} e^{-(x-\mu)^2/2\sigma^2}\,dx = 1$$

for all μ and for all $\sigma > 0$. *Hint:* Use the result of Example 5.

Answers to Concepts Review: **1.** $a \le r \le b; \alpha \le \theta \le \beta$ **2.** $r\,dr\,d\theta$ **3.** $\int_0^{\pi} \int_0^2 r^3\,dr\,d\theta$ **4.** 4π

13.5 Applications of Double Integrals

The most obvious application of double integrals is in calculating volumes of solids. This use of double integrals has been amply illustrated, so now we turn to other applications (mass, center of mass, moment of inertia, and radius of gyration).

Consider a flat sheet that is so thin that we may consider it to be two-dimensional. In Section 5.6, we called such a sheet a lamina, but there we considered only laminas of constant density. Here we wish to study laminas of variable density, that is, laminas made of nonhomogeneous material (Figure 1).

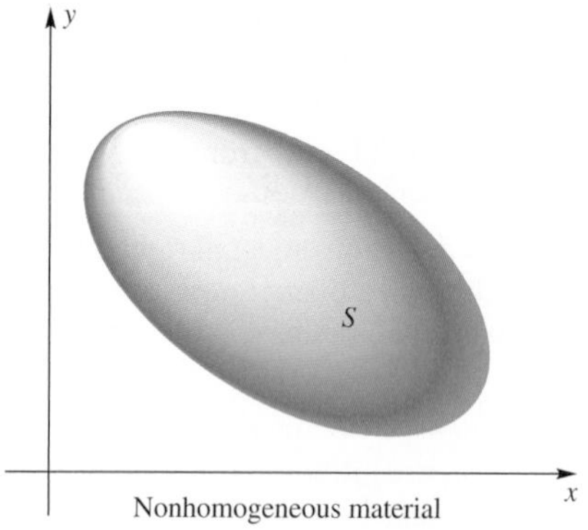

Figure 1

Suppose that a lamina covers a region S in the xy-plane, and let the density (mass per unit area) at (x, y) be denoted by $\delta(x, y)$. Partition S into small rectangles $R_1, R_2, \ldots, R_k$, as shown in Figure 2. Pick a point $(\overline{x}_k, \overline{y}_k)$ in R_k. Then the mass of R_k is approximately $\delta(\overline{x}_k, \overline{y}_k)A(R_k)$, and the total mass of the lamina is approximately

$$m \approx \sum_{k=1}^{n} \delta(\overline{x}_k, \overline{y}_k)A(R_k)$$

The actual mass m is obtained by taking the limit of the above expression as the norm of the partition approaches zero, which is, of course, a double integral.

$$\boxed{m = \iint_S \delta(x, y)\, dA}$$

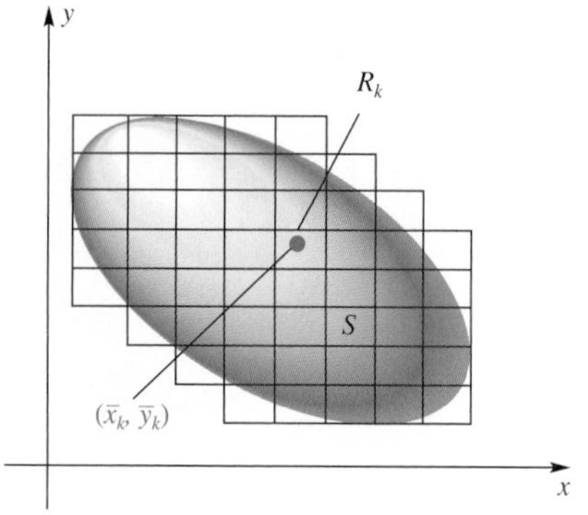

Figure 2

EXAMPLE 1 A lamina with density $\delta(x, y) = xy$ is bounded by the x-axis, the line $x = 8$, and the curve $y = x^{2/3}$ (Figure 3). Find its total mass.

SOLUTION

$$\begin{aligned} m &= \iint_S xy\, dA = \int_0^8 \int_0^{x^{2/3}} xy\, dy\, dx \\ &= \int_0^8 \left[\frac{xy^2}{2}\right]_0^{x^{2/3}} dx = \frac{1}{2}\int_0^8 x^{7/3}\, dx \\ &= \frac{1}{2}\left[\frac{3}{10}x^{10/3}\right]_0^8 = \frac{768}{5} = 153.6 \end{aligned}$$

■

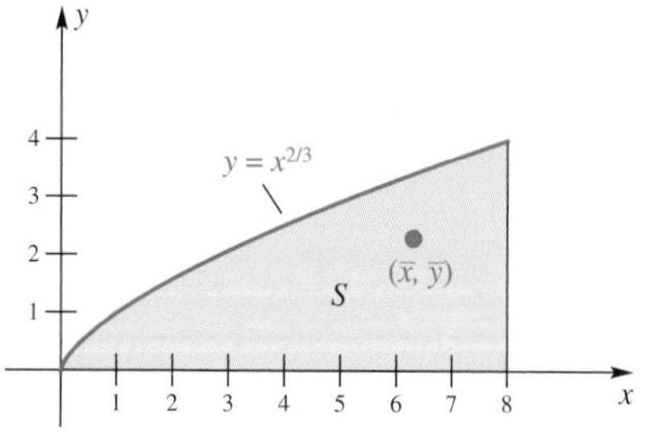

Figure 3

Center of Mass We suggest that you review the concept of center of mass from Section 5.6. There we learned that if $m_1, m_2, \ldots, m_n$ is a collection of point masses situated at $(x_1, y_1), (x_2, y_2), \ldots, (x_n, y_n)$, respectively, then the total moments with respect to the y-axis and the x-axis are given by

$$M_y = \sum_{k=1}^{n} x_k m_k \qquad M_x = \sum_{k=1}^{n} y_k m_k$$

Moreover, the coordinates $(\overline{x}, \overline{y})$ of the center of mass (balance point) are

$$\overline{x} = \frac{M_y}{m} = \frac{\sum_{k=1}^{n} x_k m_k}{\sum_{k=1}^{n} m_k} \qquad \overline{y} = \frac{M_x}{m} = \frac{\sum_{k=1}^{n} y_k m_k}{\sum_{k=1}^{n} m_k}$$

Consider now a lamina of variable density $\delta(x, y)$ covering a region S in the xy-plane, as in Figure 1. Partition this lamina as in Figure 2 and assume as an approximation that the mass of each R_k is concentrated at $(\overline{x}_k, \overline{y}_k)$, $k = 1, 2, \ldots, n$. Finally, take the limit as the norm of the partition tends to zero. This leads to the formulas

$$\bar{x} = \frac{M_y}{m} = \frac{\displaystyle\iint_S x\delta(x, y)\, dA}{\displaystyle\iint_S \delta(x, y)\, dA} \qquad \bar{y} = \frac{M_x}{m} = \frac{\displaystyle\iint_S y\delta(x, y)\, dA}{\displaystyle\iint_S \delta(x, y)\, dA}$$

EXAMPLE 2 Find the center of mass of the lamina of Example 1.

SOLUTION In Example 1, we showed that the mass m of this lamina is $\frac{768}{5}$. The moments M_y and M_x with respect to the y-axis and x-axis are

$$M_y = \iint_S x\delta(x, y)\, dA = \int_0^8 \int_0^{x^{2/3}} x^2 y\, dy\, dx$$

$$= \frac{1}{2}\int_0^8 x^{10/3}\, dx = \frac{12{,}288}{13} \approx 945.23$$

$$M_x = \iint_S y\delta(x, y)\, dA = \int_0^8 \int_0^{x^{2/3}} x y^2\, dy\, dx$$

$$= \frac{1}{3}\int_0^8 x^3\, dx = \frac{1024}{3} \approx 341.33$$

We conclude that

$$\bar{x} = \frac{M_y}{m} = \frac{80}{13} \approx 6.15, \qquad \bar{y} = \frac{M_x}{m} = \frac{20}{9} \approx 2.22$$

≅ Notice in Figure 3 that the center of mass $(\bar{x}, \bar{y})$ is in the upper-right portion of S; but this is to be expected since a lamina with density $\delta(x, y) = xy$ gets heavier as the distance from the x- and y-axes increases. ■

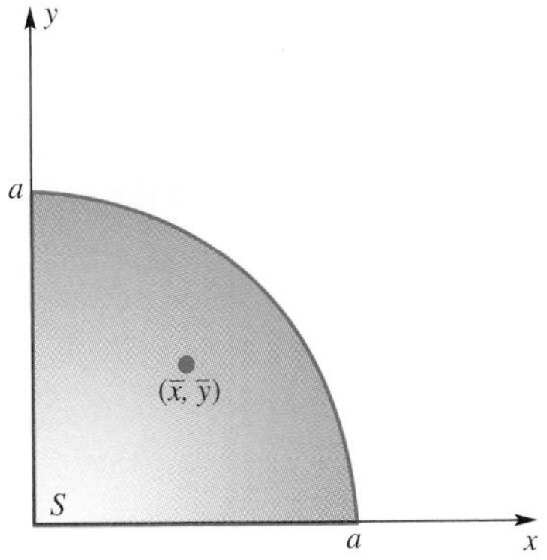

Figure 4

EXAMPLE 3 Find the center of mass of a lamina in the shape of a quarter-circle of radius a whose density is proportional to the distance from the center of the circle (Figure 4).

SOLUTION By hypothesis, $\delta(x, y) = k\sqrt{x^2 + y^2}$, where k is a constant. The shape of S suggests the use of polar coordinates.

$$m = \iint_S k\sqrt{x^2 + y^2}\, dA = k\int_0^{\pi/2} \int_0^a r r\, dr\, d\theta$$

$$= k\int_0^{\pi/2} \frac{a^3}{3}\, d\theta = \frac{k\pi a^3}{6}$$

Also,

$$M_y = \iint_S xk\sqrt{x^2 + y^2}\, dA = k\int_0^{\pi/2} \int_0^a (r\cos\theta) r^2\, dr\, d\theta$$

$$= k\int_0^{\pi/2} \frac{a^4}{4}\cos\theta\, d\theta = \left[\frac{ka^4}{4}\sin\theta\right]_0^{\pi/2} = \frac{ka^4}{4}$$

We conclude that

$$\bar{x} = \frac{M_y}{m} = \frac{ka^4/4}{k\pi a^3/6} = \frac{3a}{2\pi}$$

Because of the symmetry of the lamina, we recognize that $\bar{y} = \bar{x}$, so no further calculation is needed. ■

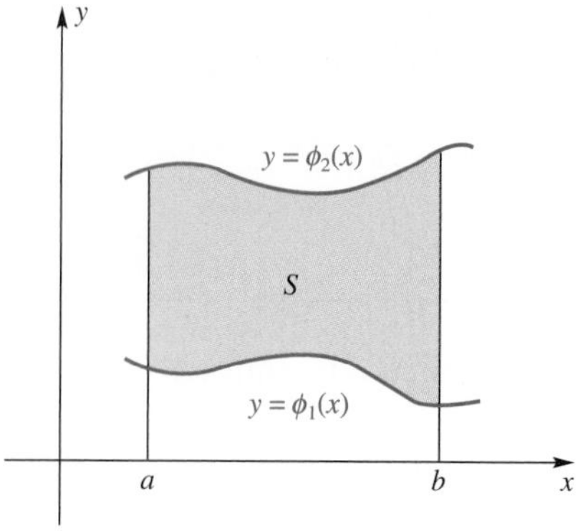

Figure 5

A perceptive reader might well ask a question at this point. What if a lamina is homogeneous; that is, what if $\delta(x, y) = k$, a constant? Will the formulas derived in this section, which involve double integrals, agree with those of Section 5.6, which involved only single integrals? The answer is yes. To give a partial justification, consider calculating M_y for a y-simple region S (Figure 5).

$$M_y = \iint_S xk\,dA = k\int_a^b \int_{\phi_1(x)}^{\phi_2(x)} x\,dy\,dx = k\int_a^b x[\phi_2(x) - \phi_1(x)]\,dx$$

The single integral on the right is the one given in Section 5.6.

Moment of Inertia From physics, we learn that the kinetic energy, KE, of a particle of mass m and velocity v, moving in a straight line, is

$$KE = \tfrac{1}{2}mv^2 \tag{1}$$

If, instead of moving in a straight line, the particle rotates about an axis with an angular velocity of ω radians per unit of time, its linear velocity is $v = r\omega$, where r is the radius of its circular path. When we substitute this in (1), we obtain

$$KE = \tfrac{1}{2}(r^2m)\omega^2$$

The expression r^2m is called the **moment of inertia** of the particle and is denoted by I. Thus, for a rotating particle,

$$KE = \tfrac{1}{2}I\omega^2 \tag{2}$$

We conclude from (1) and (2) that the moment of inertia of a body in circular motion plays a role similar to the mass of a body in linear motion.

For a system of n particles in a plane with masses $m_1, m_2, \ldots, m_n$ and at distances $r_1, r_2, \ldots, r_n$ from a line L, the moment of inertia of the system about L is defined to be

$$I = m_1r_1^2 + m_2r_2^2 + \cdots + m_nr_n^2 = \sum_{k=1}^{n} m_kr_k^2$$

In other words, we add the moments of inertia of the individual particles.

Now consider a lamina with density $\delta(x, y)$ covering a region S of the xy-plane (Figure 1). If we partition S as in Figure 2, approximate the moments of inertia of each piece R_k, add, and take the limit, we are led to the following formulas. The **moments of inertia** (also called the second moments) of the lamina about the x-, y-, and z-axes are given by

$$I_x = \iint_S y^2\delta(x, y)\,dA \qquad I_y = \iint_S x^2\delta(x, y)\,dA$$

$$I_z = \iint_S (x^2 + y^2)\,\delta(x, y)\,dA = I_x + I_y$$

EXAMPLE 4 Find the moments of inertia about the x-, y-, and z-axes of the lamina of Example 1.

SOLUTION

$$I_x = \iint_S xy^3\,dA = \int_0^8 \int_0^{x^{2/3}} xy^3\,dy\,dx = \frac{1}{4}\int_0^8 x^{11/3}\,dx = \frac{6144}{7} \approx 877.71$$

$$I_y = \iint_S x^3y\,dA = \int_0^8 \int_0^{x^{2/3}} x^3y\,dy\,dx = \frac{1}{2}\int_0^8 x^{13/3}\,dx = 6144$$

$$I_z = I_x + I_y = \frac{49{,}152}{7} \approx 7021.71$$

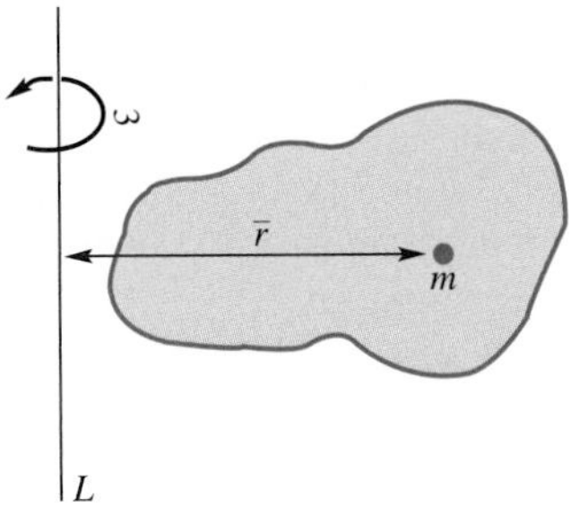

Figure 6

Consider the problem of replacing a general mass system of total mass m by a single point mass m with the same moment of inertia I with respect to a line L (Figure 6). How far should this point be from L? The answer is $\bar{r}$, where $m\bar{r}^2 = I$. The number

$$\bar{r} = \sqrt{\frac{I}{m}}$$

is called the **radius of gyration** of the system. Thus, the kinetic energy of the system rotating about L with angular velocity ω is

$$KE = \tfrac{1}{2}m\bar{r}^2\omega^2$$

Concepts Review

1. If the density at (x, y) is x^2y^4, then the mass m of the lamina S is given by $m =$ _____.

2. The y-coordinate of the center of mass of the lamina of Question 1 is given by $\bar{y} =$ _____.

3. The moment of inertia with respect to the y-axis of the lamina S of Question 1 is given by $I_y =$ _____.

4. If $S = \{(x, y): 0 \le x \le 1, 0 \le y \le 1\}$, then geometric reasoning says that if $\delta(x, y) = x^2y^4$, then both $\bar{x}$ and $\bar{y}$ are _____ than $\frac{1}{2}$.

Problem Set 13.5

In Problems 1–10, find the mass m and center of mass $(\bar{x}, \bar{y})$ of the lamina bounded by the given curves and with the indicated density.

1. $x = 0, x = 4, y = 0, y = 3; \delta(x, y) = y + 1$
2. $y = 0, y = \sqrt{4 - x^2}; \delta(x, y) = y$
3. $y = 0, y = \sin x, 0 \le x \le \pi; \delta(x, y) = y$
4. $y = 1/x, y = x, y = 0, x = 2; \delta(x, y) = x$
5. $y = e^{-x}, y = 0, x = 0, x = 1; \delta(x, y) = y^2$
6. $y = e^x, y = 0, x = 0, x = 1; \delta(x, y) = 2 - x + y$
7. $r = 2\sin\theta; \delta(r, \theta) = r$
8. $r = 1 + \cos\theta; \delta(r, \theta) = r$
9. $r = 1, r = 2, \theta = 0, \theta = \pi, (0 \le \theta \le \pi); \delta(r, \theta) = 1/r$
10. $r = 2 + 2\cos\theta; \delta(r, \theta) = r$

In Problems 11–14, find the moments of inertia I_x, I_y, and I_z for the lamina bounded by the given curves and with the indicated density δ.

11. $y = \sqrt{x}, x = 9, y = 0; \delta(x, y) = x + y$
12. $y = x^2, y = 4; \delta(x, y) = y$
13. Square with vertices $(0, 0), (0, a), (a, a), (a, 0); \delta(x, y) = x + y$
14. Triangle with vertices $(0, 0), (0, a), (a, 0); \delta(x, y) = x^2 + y^2$

In Problems 15–20, an iterated integral is given either in rectangular or polar coordinates. The double integral gives the mass of some lamina R. Sketch the lamina R and determine the density δ. Then find the mass and center of mass.

15. $\int_0^2 \int_0^x k\,dy\,dx$
16. $\int_0^1 \int_x^1 ky\,dy\,dx$
17. $\int_{-3}^3 \int_0^{9-x^2} k(x^2 + y^2)\,dy\,dx$
18. $\int_{-\pi/2}^{\pi/2} \int_0^{\cos x} k\,dy\,dx$
19. $\int_0^\pi \int_1^3 kr^2\,dr\,d\theta$
20. $\int_0^{\pi/2} \int_0^\theta kr\,dr\,d\theta$

21. Find the radius of gyration of the lamina of Problem 13 with respect to the x-axis.

22. Find the radius of gyration of the lamina of Problem 14 with respect to the y-axis.

23. Find the moment of inertia and radius of gyration of a homogeneous (δ a constant) circular lamina of radius a with respect to a diameter.

24. Show that the moment of inertia of a homogeneous rectangular lamina with sides of length a and b about a perpendicular axis through its center of mass is

$$I = \tfrac{1}{12}(a^3b + ab^3)$$

Here k is the constant density.

25. Find the moment of inertia of the lamina of Problem 23 about a line tangent to its boundary. *Hint:* Let the circle be $r = 2a\sin\theta$; then the tangent line is the x-axis. Formula 113 at the back of the book may help with the integration.

26. Consider the lamina S of constant density k bounded by the cardioid $r = a(1 + \sin\theta)$, as shown in Figure 7. Find its center of mass and moment of inertia with respect to the x-axis. *Hint:* Problem 7 of Section 10.7 suggests the useful fact that S has area $3\pi a^2/2$; also, Formula 113 at the back of the book may prove helpful.

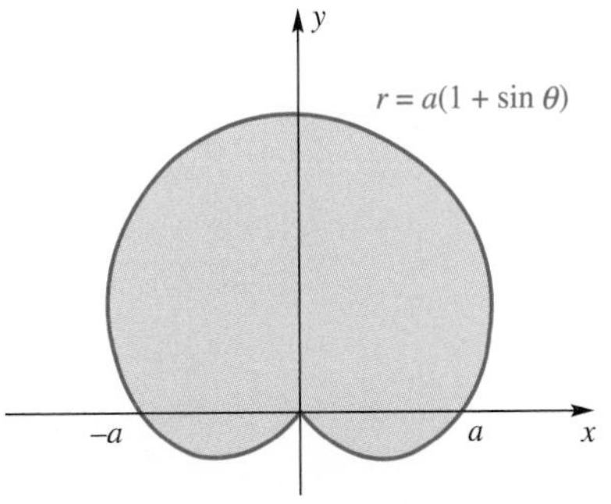

Figure 7

27. Find the center of mass of that part of the cardioid of Problem 26 that is outside the circle $r = a$.

28. Parallel Axis Theorem Consider a lamina S of mass m together with parallel lines L and L' in the plane of S, the line L passing through the center of mass of S. Show that if I and I' are the moments of inertia of S about L and L', respectively, then $I' = I + d^2m$, where d is the distance between L and L'. *Hint:* Assume that S lies in the xy-plane, L is the y-axis, and L' is the line $x = -d$.

29. Refer to the lamina of Problem 13, for which we found $I_y = 5a^5/12$. Find

(a) m (b) $\bar{x}$ (c) I_L

where L is a line through $(\bar{x}, \bar{y})$ parallel to the y-axis (see Problem 28).

30. Use the Parallel Axis Theorem together with Problem 23 to solve Problem 25 another way.

31. Find I_x, I_y, and I_z for the two-piece lamina of constant density k shown in Figure 8 (see Problems 23 and 28).

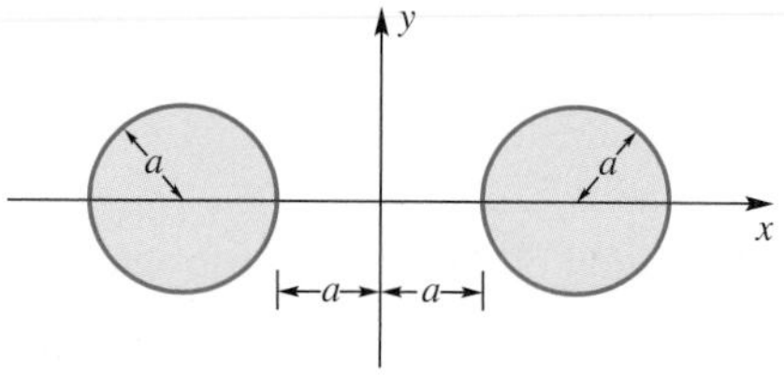

Figure 8

32. The Parallel Axis Theorem also holds for lines that are perpendicular to a lamina. Use this fact to find the moment of inertia of the rectangular lamina of Problem 24 about an axis perpendicular to the lamina and through a corner.

33. Let S_1 and S_2 be disjoint laminas in the xy-plane of mass m_1 and m_2 with centers of mass $(\bar{x}_1, \bar{y}_1)$ and $(\bar{x}_2, \bar{y}_2)$. Show that the center of mass $(\bar{x}, \bar{y})$ of the combined lamina $S_1 \cup S_2$ satisfies

$$\bar{x} = \bar{x}_1 \frac{m_1}{m_1 + m_2} + \bar{x}_2 \frac{m_2}{m_1 + m_2}$$

with a similar formula for $\bar{y}$. Conclude that in finding $(\bar{x}, \bar{y})$ the two laminas can be treated as if they were point masses at $(\bar{x}_1, \bar{y}_1)$ and $(\bar{x}_2, \bar{y}_2)$.

34. Let S_1 and S_2 be the homogeneous circular laminas of radius a and ta $(t > 0)$ centered at $(-a, a)$ and $(ta, 0)$, respectively. Use Problem 17 to find the center of mass of $S_1 \cup S_2$.

35. Let S be a lamina in the xy-plane with center of mass at the origin, and let L be the line $ax + by = 0$, which goes through the origin. Show that the (signed) distance d of a point (x, y) from L is $d = (ax + by)/\sqrt{a^2 + b^2}$, and use this to conclude that the moment of S with respect to L is 0. *Note:* This shows that a lamina will balance on any line through its center of mass.

36. For the lamina of Example 3, find the equation of the balance line that makes an angle of 135° with the positive x-axis (see Problem 35). Write your answer in the form $Ax + By = C$.

Answers to Concepts Review: **1.** $\iint_S x^2y^4\,dA$ **2.** $\iint_S x^2y^5\,dA/m$ **3.** $\iint_S x^4y^4\,dA$ **4.** greater

13.6 Surface Area

We have seen some special cases of surface area. For example, in Example 3 of Section 11.4 we found the area of a parallelogram in space. We have also seen (Problems 29 and 30 of Section 5.4) that the surface area of a sphere is $4\pi r^2$. In this section, we develop a formula for the area of a surface defined by $z = f(x, y)$ over a specified region.

Suppose that G is such a surface over the closed and bounded region S in the xy-plane. Assume that f has continuous first partial derivatives f_x and f_y. We begin by creating a partition P of the region S with lines parallel to the x- and y-axes (see Figure 1). Let R_m, $m = 1, 2, \ldots, n$, denote the resulting rectangles that lie completely within S. For each m, let G_m be that part of the surface that projects onto R_m, and

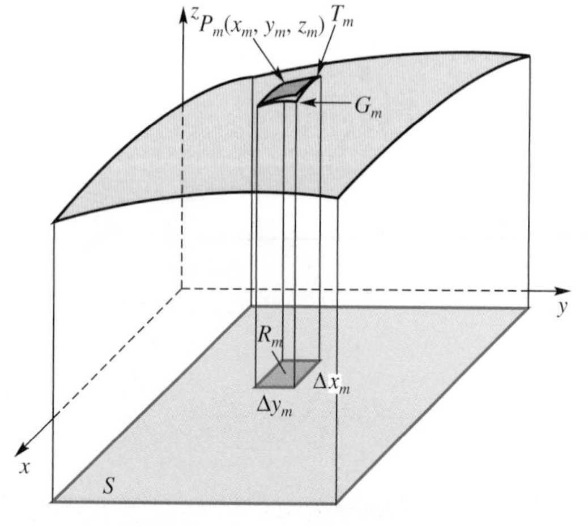

Figure 1

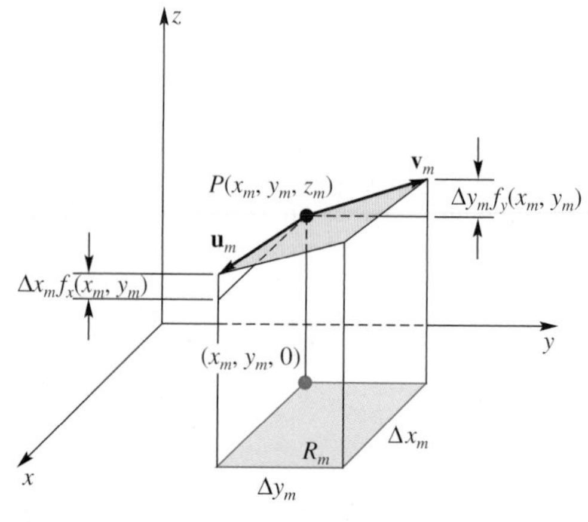

Figure 2

let P_m be the point of G_m that projects onto the corner of R_m with the smallest x- and y-coordinates. Finally, let T_m denote the parallelogram from the tangent plane at P_m that projects onto R_m, as shown in Figure 1, and then in more detail in Figure 2.

We next find the area of the parallelogram T_m whose projection is R_m. Let $\mathbf{u}_m$ and $\mathbf{v}_m$ denote the vectors that form the sides of T_m. Then

$$\mathbf{u}_m = \Delta x_m \mathbf{i} + f_x(x_m, y_m)\, \Delta x_m \mathbf{k}$$
$$\mathbf{v}_m = \Delta y_m \mathbf{j} + f_y(x_m, y_m)\, \Delta y_m \mathbf{k}$$

From Section 11.4, we know that the area of the parallelogram T_m is $\|\mathbf{u}_m \times \mathbf{v}_m\|$, where

$$\begin{aligned}\mathbf{u}_m \times \mathbf{v}_m &= \begin{vmatrix} \mathbf{i} & \mathbf{j} & \mathbf{k} \\ \Delta x_m & 0 & f_x(x_m, y_m)\, \Delta x_m \\ 0 & \Delta y_m & f_y(x_m, y_m)\, \Delta y_m \end{vmatrix} \\ &= (0 - f_x(x_m, y_m)\, \Delta x_m\, \Delta y_m)\mathbf{i} - (f_y(x_m, y_m)\, \Delta x_m\, \Delta y_m - 0)\mathbf{j} \\ &\quad + (\Delta x_m\, \Delta y_m - 0)\mathbf{k} \\ &= \Delta x_m\, \Delta y_m[-f_x(x_m, y_m)\mathbf{i} - f_y(x_m, y_m)\mathbf{j} + \mathbf{k}] \\ &= A(R_m)[-f_x(x_m, y_m)\mathbf{i} - f_y(x_m, y_m)\mathbf{j} + \mathbf{k}]\end{aligned}$$

The area of T_m is therefore

$$A(T_m) = \|\mathbf{u}_m \times \mathbf{v}_m\| = A(R_m)\sqrt{[f_x(x_m, y_m)]^2 + [f_y(x_m, y_m)]^2 + 1}$$

We then add the areas of these tangent parallelograms T_m, $m = 1, 2, \ldots, n$, and take the limit to arrive at the surface area of G.

$$\begin{aligned}A(G) &= \lim_{\|P\|\to 0} \sum_{m=1}^{n} A(T_m) \\ &= \lim_{\|P\|\to 0} \sum_{m=1}^{n} \sqrt{[f_x(x_m, y_m)]^2 + [f_y(x_m, y_m)]^2 + 1}\, A(R_m) \\ &= \iint_S \sqrt{[f_x(x, y)]^2 + [f_y(x, y)]^2 + 1}\, dA\end{aligned}$$

or, more concisely,

$$\boxed{A(G) = \iint_S \sqrt{f_x^2 + f_y^2 + 1}\, dA}$$

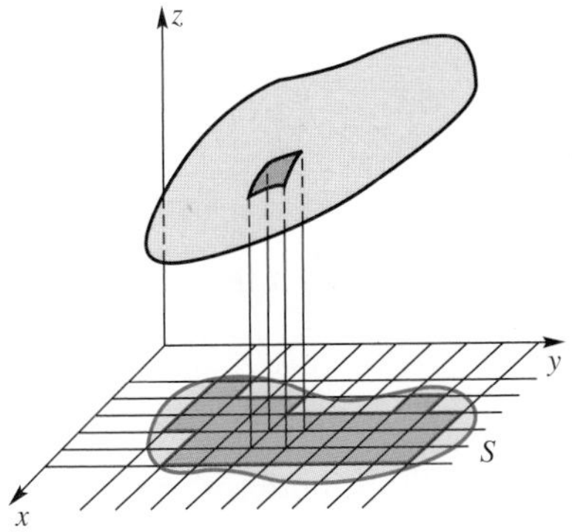

Figure 3

Figure 1 is drawn as if the region S in the xy-plane were a rectangle; this need not be the case. Figure 3 shows what happens when S is not a rectangle.

Some Examples We illustrate the boxed formula for surface area with four examples.

EXAMPLE 1 If S is the rectangular region in the xy-plane that is bounded by the lines $x = 0$, $x = 1$, $y = 0$, and $y = 2$, find the area of the part of the cylindrical surface $z = \sqrt{4 - x^2}$ that projects onto S (Figure 4).

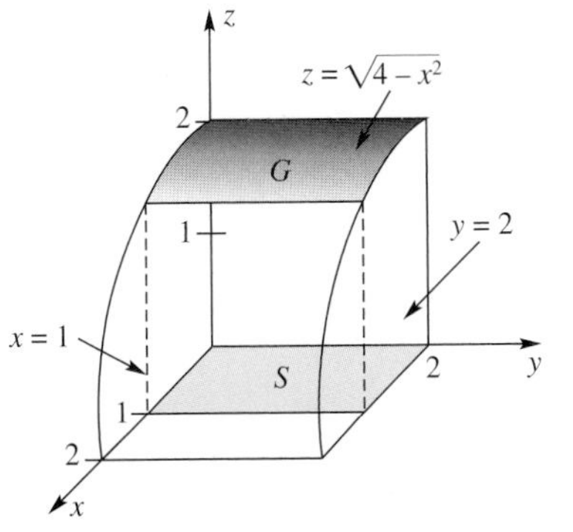

Figure 4

SOLUTION Let $f(x, y) = \sqrt{4 - x^2}$. Then $f_x = -x/\sqrt{4 - x^2}$, $f_y = 0$, and

$$\begin{aligned}A(G) &= \iint_S \sqrt{f_x^2 + f_y^2 + 1}\, dA = \iint_S \sqrt{\frac{x^2}{4 - x^2} + 1}\, dA = \iint_S \frac{2}{\sqrt{4 - x^2}}\, dA \\ &= \int_0^1 \int_0^2 \frac{2}{\sqrt{4 - x^2}}\, dy\, dx = 4\int_0^1 \frac{1}{\sqrt{4 - x^2}}\, dx = 4\left[\sin^{-1}\frac{x}{2}\right]_0^1 = \frac{2\pi}{3}\end{aligned}$$ ■

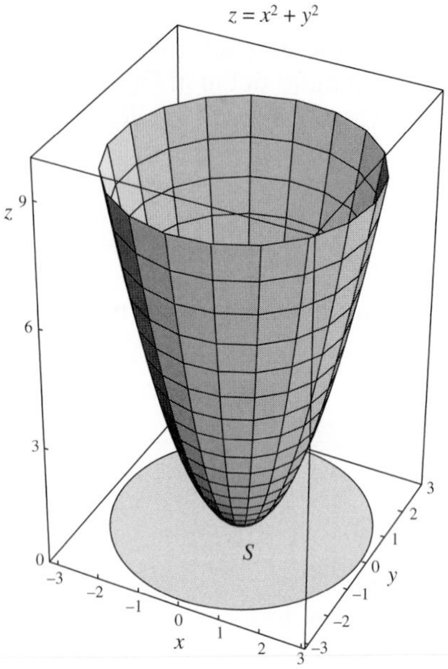

Figure 5

EXAMPLE 2 Find the area of the surface $z = x^2 + y^2$ below the plane $z = 9$.

SOLUTION The designated part G of the surface projects onto the circular region S inside the circle $x^2 + y^2 = 9$ (Figure 5). Let $f(x, y) = x^2 + y^2$. Then $f_x = 2x$, $f_y = 2y$, and

$$A(G) = \iint_S \sqrt{4x^2 + 4y^2 + 1}\, dA$$

The shape of S suggests use of polar coordinates.

$$\begin{aligned} A(G) &= \int_0^{2\pi} \int_0^3 \sqrt{4r^2 + 1}\, r\, dr\, d\theta \\ &= \int_0^{2\pi} \frac{1}{8}\left[\frac{2}{3}(4r^2 + 1)^{3/2}\right]_0^3 d\theta \\ &= \int_0^{2\pi} \frac{1}{12}(37^{3/2} - 1)\, d\theta = \frac{\pi}{6}(37^{3/2} - 1) \approx 117.32 \end{aligned}$$

A right circular cylinder (with height equal to diameter) and an inscribed sphere have the remarkable property that the surfaces between two parallel planes (perpendicular to the axis of the cylinder) have equal area. Our next example demonstrates this property for a hemisphere, showing that the two surfaces in Figure 6 have equal area. The steps easily extend to show that the property holds for a sphere.

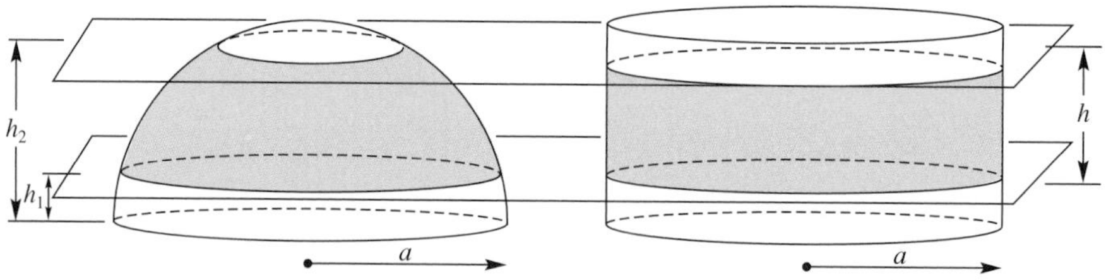

Figure 6

EXAMPLE 3 Show that the area of the surface G cut from the hemisphere $x^2 + y^2 + z^2 = a^2$, $z \geq 0$, by the planes $z = h_1$ and $z = h_2$ $(0 \leq h_1 \leq h_2 \leq a)$ is

$$A(G) = 2\pi a(h_2 - h_1)$$

Show that this is also the surface area on the right circular cylinder $x^2 + y^2 = a^2$ between the planes $z = h_1$ and $z = h_2$.

SOLUTION Let $h = h_2 - h_1$. The surface of the hemisphere is defined by

$$z = \sqrt{a^2 - x^2 - y^2}$$

and its projection S in the xy-plane is the annulus $b \leq x^2 + y^2 \leq c$, where $b = \sqrt{a^2 - h_2^2}$ and $c = \sqrt{a^2 - h_1^2}$ (see Figure 7). The surface area of the hemisphere between the two horizontal planes is

$$\begin{aligned} A(G) &= \iint_S \sqrt{\left[\frac{\partial}{\partial x}\sqrt{a^2 - x^2 - y^2}\right]^2 + \left[\frac{\partial}{\partial y}\sqrt{a^2 - x^2 - y^2}\right]^2 + 1}\, dA \\ &= \iint_S \sqrt{\frac{x^2}{a^2 - x^2 - y^2} + \frac{y^2}{a^2 - x^2 - y^2} + 1}\, dA \\ &= \iint_S \frac{a}{\sqrt{a^2 - x^2 - y^2}}\, dA \end{aligned}$$

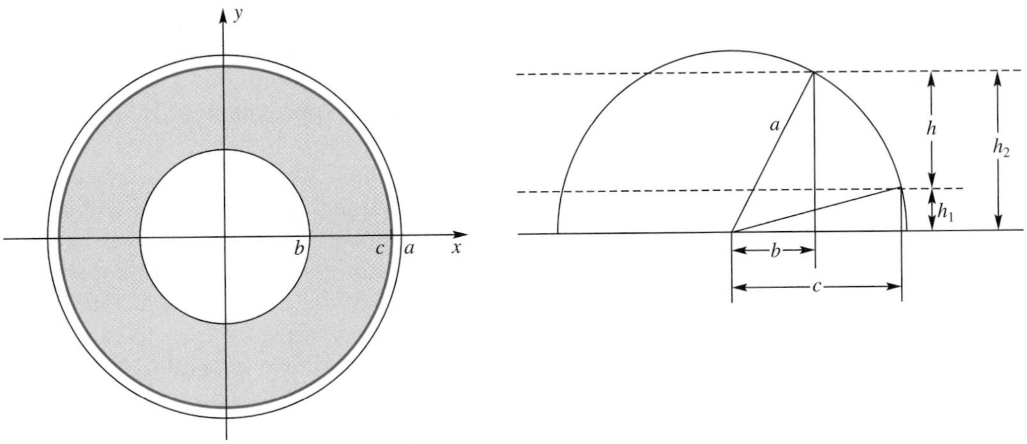

Figure 7

This integral is most easily evaluated using polar coordinates.

$$A(G) = \int_0^{2\pi}\int_b^c \frac{a}{\sqrt{a^2 - r^2}}\, r\, dr\, d\theta = \int_0^{2\pi} a\left[-\sqrt{a^2 - c^2} + \sqrt{a^2 - b^2}\right] d\theta$$
$$= 2\pi a\left[\sqrt{a^2 - b^2} - \sqrt{a^2 - c^2}\right] = 2\pi a(h_2 - h_1) = 2\pi ah$$

Because the surface area for the cylinder is the circumference $(2\pi a)$ of the circle times the height h, the surface area for that part of the cylinder between the two planes is $2\pi ah$, which, of course, agrees with the surface area for the hemisphere. ■

EXAMPLE 4 Find the surface area of the hyperbolic paraboloid $z = x^2 - y^2$ over the triangle with vertices $(0, 0)$, $(2, 0)$, and $(0, 2)$.

SOLUTION Let $f(x, y) = x^2 - y^2$. Then, $f_x(x, y) = 2x$ and $f_y(x, y) = -2y$. The area is given by the iterated integral

$$A = \int_0^2\int_0^{2-x} \sqrt{f_x^2 + f_y^2 + 1}\, dy\, dx$$
$$= \int_0^2\int_0^{2-x} \sqrt{(2x)^2 + (-2y)^2 + 1}\, dy\, dx$$
$$= 2\int_0^2\left(\int_0^{2-x} \sqrt{y^2 + \left(x^2 + \frac{1}{4}\right)}\, dy\right) dx$$

[Use Formula 44 from the table of integrals]

$$= 2\int_0^2\left[\frac{y}{2}\sqrt{y^2 + \left(x^2 + \frac{1}{4}\right)} + \frac{\left(x^2 + \frac{1}{4}\right)}{2}\ln\left|y + \sqrt{y^2 + \left(x^2 + \frac{1}{4}\right)}\right|\right]_0^{2-x} dx$$
$$= 2\int_0^2\left[\frac{2-x}{2}\sqrt{(2-x)^2 + \left(x^2 + \frac{1}{4}\right)}\right.$$
$$+ \frac{1}{2}\left(x^2 + \frac{1}{4}\right)\ln\left|(2-x) + \sqrt{(2-x)^2 + \left(x^2 + \frac{1}{4}\right)}\right|$$
$$\left. - \frac{1}{2}\left(x^2 + \frac{1}{4}\right)\ln\left|\sqrt{x^2 + \frac{1}{4}}\right|\right] dx$$
$$= 2\int_0^2\left[\frac{2-x}{2}\sqrt{2x^2 - 4x + \frac{17}{4}} + \frac{4x^2 + 1}{8}\ln\left((2-x) + \sqrt{2x^2 - 4x + \frac{17}{4}}\right)\right.$$
$$\left. - \frac{4x^2 + 1}{16}\ln\left(x^2 + \frac{1}{4}\right)\right] dx$$

Surface Area Problems

For most surface area problems it is easy to set up the double integral. This is just a matter of substituting the required derivatives in the formula. However, it is often difficult or impossible to evaluate these integrals using the Second Fundamental Theorem of Calculus because of the difficulty of finding antiderivatives.

This last integral is too complicated to evaluate using the Second Fundamental Theorem of Calculus, so we rely on a numerical method. The Parabolic Rule with $n = 10$ gives an approximation of 4.8363 to this last integral. (Larger values of n give virtually the same approximation.) ■

In this last example, we were able to evaluate the inner integral by finding an antiderivative and applying the Second Fundamental Theorem of Calculus. Then at the end, a numerical approximation was needed to evaluate the integral. Although there exist numerical methods for approximating double integrals, they are rather cumbersome to use and require evaluation of the function at a large number of points. It is always advantageous to evaluate the inner integral, if this is possible, in order to leave a single integral to be approximated.

Concepts Review

1. The area of a parallelogram with sides equal to the vectors **u** and **v** is ____.

2. More generally, if $z = f(x, y)$ determines a surface G that projects onto the region S in the xy-plane, then the area of G is given by the formula $A(G) =$ ____.

3. Applying the result of Question 2 with $z = (a^2 - x^2 - y^2)^{1/2}$ leads to the integral formula $A =$ ____ for the area of a hemisphere of radius a. When this integral is evaluated, we obtain the familiar formula $A =$ ____.

4. Consider a sphere inscribed in a cylindrical can of radius a. Two planes, both perpendicular to the axis of the cylinder and separated by distance h, will cut off regions on both the cylinder and the sphere that have area ____.

Problem Set 13.6

In Problems 1–17, find the area of the indicated surface. Make a sketch in each case.

1. The part of the plane $3x + 4y + 6z = 12$ that is above the rectangle in the xy-plane with vertices $(0, 0)$, $(2, 0)$, $(2, 1)$, and $(0, 1)$

2. The part of the plane $3x - 2y + 6z = 12$ that is bounded by the planes $x = 0$, $y = 0$, and $3x + 2y = 12$

3. The part of the surface $z = \sqrt{4 - y^2}$ that is directly above the square in the xy-plane with vertices $(1, 0)$, $(2, 0)$, $(2, 1)$, and $(1, 1)$

4. The part of the surface $z = \sqrt{4 - y^2}$ in the first octant that is directly above the circle $x^2 + y^2 = 4$ in the xy-plane

5. The part of the cylinder $x^2 + z^2 = 9$ that is directly over the rectangle in the xy-plane with vertices $(0, 0)$, $(2, 0)$, $(2, 3)$, and $(0, 3)$

6. The part of the paraboloid $z = x^2 + y^2$ that is cut off by the plane $z = 4$

7. The part of the conical surface $x^2 + y^2 = z^2$ that is directly over the triangle in the xy-plane with vertices $(0, 0)$, $(4, 0)$, and $(0, 4)$

8. The part of the surface $z = x^2/4 + 4$ that is cut off by the planes $x = 0$, $x = 1$, $y = 0$, and $y = 2$

9. The part of the sphere $x^2 + y^2 + z^2 = a^2$ inside the circular cylinder $x^2 + y^2 = b^2$, where $0 < b \le a$

10. The part of the sphere $x^2 + y^2 + z^2 = a^2$ inside the elliptic cylinder $b^2x^2 + a^2y^2 = a^2b^2$, where $0 < b \le a$

11. The part of the sphere $x^2 + y^2 + z^2 = a^2$ inside the circular cylinder $x^2 + y^2 = ay$ ($r = a \sin\theta$ in polar coordinates), $a > 0$

12. The part of the cylinder $x^2 + y^2 = ay$ inside the sphere $x^2 + y^2 + z^2 = a^2$, $a > 0$. *Hint:* Project to the yz-plane to get the region of integration.

13. The part of the saddle $az = x^2 - y^2$ inside the cylinder $x^2 + y^2 = a^2$, $a > 0$

14. The surface of the solid that is the intersection of the two solid cylinders $x^2 + z^2 \le a^2$ and $x^2 + y^2 \le a^2$. *Hint:* You may need the integration formula $\int (1 + \sin\theta)^{-1}\, d\theta = -\tan[(\pi - 2\theta)/4] + C$.

15. The part of $z = 9 - x^2 - y^2$ above the plane $z = 5$.

16. The part of $z = 9 - x^2$ above the xy-plane with $0 \le x \le 20$

17. The part of the plane $Ax + By + Cz = D$ (where A, B, C, and D are all positive) that lies in the first octant.

18. Figure 8 shows the Engineering Building at Southern Illinois University Edwardsville. The spiral staircase, visible in the

Figure 8

middle of the photo, is in the shape of a right circular cylinder with diameter 36 feet. The roof is slanted at a 45-degree angle. What is the surface area of the roof?

Problems 19–21 are related to Example 3.

19. Consider that part of the sphere $x^2 + y^2 + z^2 = a^2$ between the planes $z = h_1$ and $z = h_2$, where $0 \le h_1 < h_2 \le a$. Find that value of h such that the plane $z = h$ cuts the surface area in half.

20. Show that the polar cap (Figure 9) on a sphere of radius a determined by the spherical angle ϕ has area $2\pi a^2(1 - \cos\phi)$.

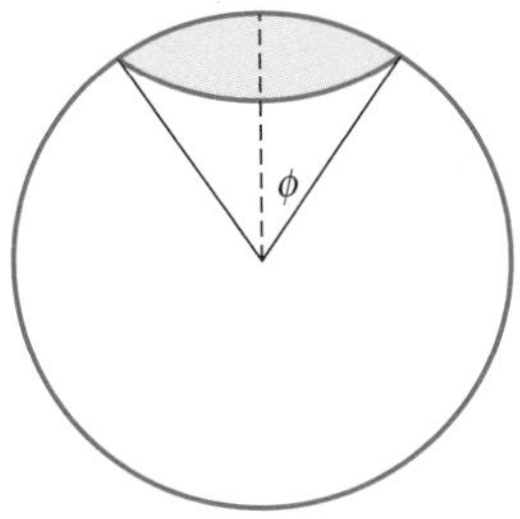

Figure 9

21. Another Old Goat Problem (see Problem Set 10.7) Four goats have grazing areas A, B, C, and D, respectively. The first three goats are each tethered by ropes of length b, the first on a flat plane, the second on the outside of a sphere of radius a, and the third on the inside of a sphere of radius a. The fourth goat must stay inside a ring of radius b that has been dropped over a sphere of radius a. Determine formulas for A, B, C, and D and arrange them in order of size. Assume that $b < a$.

22. Let S be a planar region in three-space, and let S_{xy}, S_{xz}, and S_{yz} be the projections on the three coordinate planes (Figure 10). Show that

$$[A(S)]^2 = [A(S_{xy})]^2 + [A(S_{xz})]^2 + [A(S_{yz})]^2$$

23. Assume that the region S of Figure 10 lies in the plane $z = f(x, y) = ax + by + c$ and that S is above the xy-plane. Show that the volume of the solid cylinder under S is $A(S_{xy})f(\bar{x}, \bar{y})$, where $(\bar{x}, \bar{y})$ is the centroid of S_{xy}.

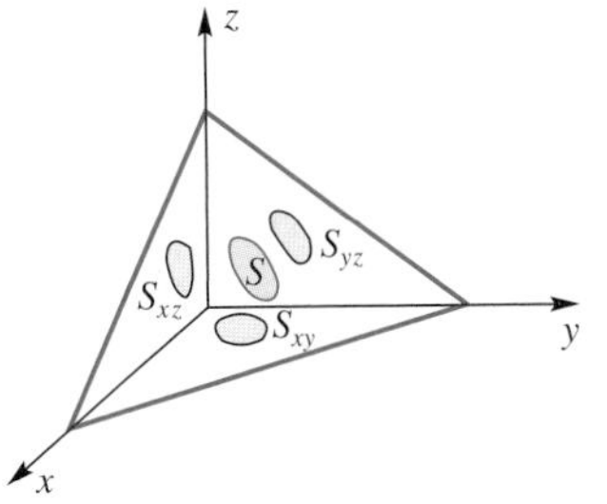

Figure 10

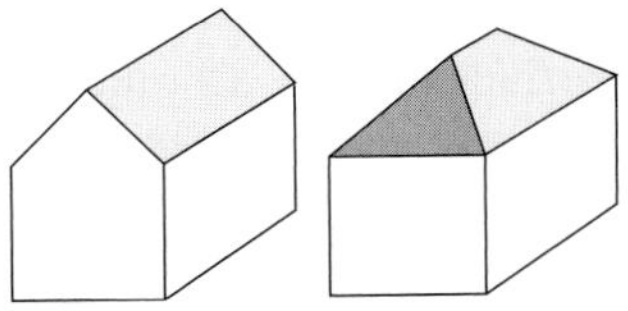

Figure 11

24. Joe's house has a rectangular base with a gable roof, and Alex's house has the same base with a pyramid-type roof (see Figure 11). The slopes of all parts of both roofs are the same. Whose roof has the smaller area?

25. Show that the surface area of a nonvertical plane over a region S in the xy-plane is $A(S)\sec\gamma$ where γ is the acute angle between a normal vector to the plane and the positive z-axis.

26. Let $\gamma = \gamma(x, y, f(x, y))$ be the acute angle between the z-axis and a normal vector to the surface $z = f(x, y)$ at the point $(x, y, f(x, y))$ on the surface. Show that $\sec\gamma = \sqrt{f_x^2 + f_y^2 + 1}$. (Note that this gives another formula for surface area: $A(G) = \iint_S \sec\gamma\, dA$.)

In Problems 27–28, find the surface area of the given surface. If an integral cannot be evaluated using the Second Fundamental Theorem of Calculus, then use the Parabolic Rule with $n = 10$.

27. The paraboloid $z = x^2 + y^2$ over the region

(a) in the first quadrant and inside the circle $x^2 + y^2 = 9$

(b) inside the triangle with vertices $(0, 0), (3, 0), (0, 3)$

28. The hyperbolic paraboloid $z = y^2 - x^2$ over the region

(a) in the first quadrant and inside the circle $x^2 + y^2 = 9$

(b) inside the triangle with vertices $(0, 0), (3, 0), (0, 3)$

29. Six surfaces are given below. Without performing any integration, rank the surfaces in order of their surface area from smallest to largest. *Hint:* There may be some "ties."

(a) The paraboloid $z = x^2 + y^2$ over the region in the first quadrant and inside the circle $x^2 + y^2 = 1$

(b) The hyperbolic paraboloid $z = x^2 - y^2$ over the region in the first quadrant and inside the circle $x^2 + y^2 = 1$

(c) The paraboloid $z = x^2 + y^2$ over the region inside the rectangle with vertices $(0, 0), (1, 0), (1, 1)$, and $(0, 1)$

(d) The hyperbolic paraboloid $z = x^2 - y^2$ over the region inside the rectangle with vertices $(0, 0), (1, 0), (1, 1)$, and $(0, 1)$

(e) The paraboloid $z = x^2 + y^2$ over the region inside the triangle with vertices $(0, 0), (1, 0)$, and $(0, 1)$

(f) The hyperbolic paraboloid $z = x^2 - y^2$ over the region inside the triangle with vertices $(0, 0), (1, 0)$, and $(0, 1)$

Answers to Concepts Review: **1.** $\|\mathbf{u} \times \mathbf{v}\|$ **2.** $\iint_S \sqrt{f_x^2 + f_y^2 + 1}\, dA$ **3.** $\int_{-a}^{a}\int_{-\sqrt{a^2-x^2}}^{\sqrt{a^2-x^2}} \left(a/\sqrt{a^2 - x^2 - y^2}\right) dy\, dx = \int_0^{2\pi}\int_0^{a} \left(ar/\sqrt{a^2 - r^2}\right) dr\, d\theta; 2\pi a^2$ **4.** $2\pi ah$

13.7 Triple Integrals in Cartesian Coordinates

The concept embodied in single and double integrals extends in a natural way to triple and even n-dimensional integrals.

Consider a function f of three variables defined over a box-shaped region B with faces parallel to the coordinate planes. We can no longer graph f (four dimensions would be required), but we can picture B (Figure 1). Form a partition P of B by passing planes through B parallel to the coordinate planes, thus cutting B into small subboxes $B_1, B_2, \ldots, B_n$; a typical subbox, B_k, is shown in Figure 1. On B_k, pick a sample point $(\overline{x}_k, \overline{y}_k, \overline{z}_k)$ and consider the Riemann sum

$$\sum_{k=1}^{n} f(\overline{x}_k, \overline{y}_k, \overline{z}_k)\,\Delta V_k$$

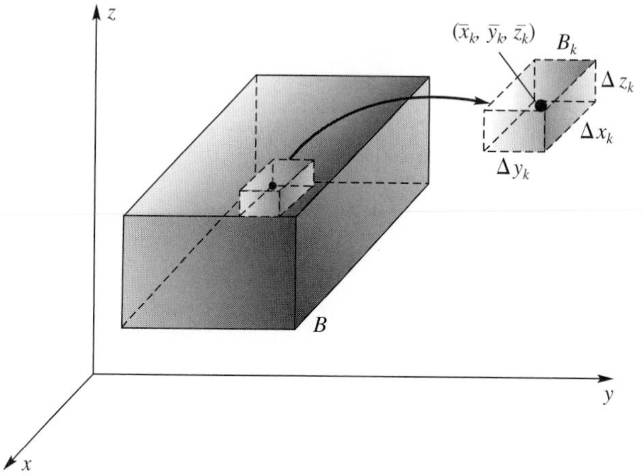

Figure 1

where $\Delta V_k = \Delta x_k\,\Delta y_y\,\Delta z_k$ is the volume of B_k. Let the norm of the partition $\|P\|$ be the length of the longest diagonal of all the subboxes. Then we define the **triple integral** by

$$\iiint_B f(x, y, z)\,dV = \lim_{\|P\|\to 0} \sum_{k=1}^{n} f(\overline{x}_k, \overline{y}_k, \overline{z}_k)\,\Delta V_k$$

provided that this limit exists.

The question of what kind of functions are integrable arises here, as it did for single and double integrals. It is certainly sufficient that f be continuous on B. Actually, we can allow some discontinuities, for example, on a finite number of smooth surfaces. We do not prove this (a very difficult task), but we assert that it is true.

As you would expect, the triple integral has the standard properties: linearity, additivity on sets that overlap only on a boundary surface, and the comparison property. Finally, *triple integrals can be written as triple iterated integrals*, as we now illustrate.

EXAMPLE 1 Evaluate $\iiint_B x^2yz\,dV$, where B is the box

$$B = \{(x, y, z): 1 \le x \le 2, 0 \le y \le 1, 0 \le z \le 2\}$$

SOLUTION

$$\iiint_B x^2yz\,dV = \int_0^2\int_0^1\int_1^2 x^2yz\,dx\,dy\,dz$$

$$= \int_0^2\int_0^1\left[\frac{1}{3}x^3yz\right]_1^2 dy\,dz = \int_0^2\int_0^1 \frac{7}{3}yz\,dy\,dz$$

$$= \frac{7}{3}\int_0^2 \left[\frac{1}{2}y^2 z\right]_0^1 dz = \frac{7}{3}\int_0^2 \frac{1}{2}z\,dz$$

$$= \frac{7}{6}\left[\frac{z^2}{2}\right]_0^2 = \frac{7}{3}$$

There are six possible orders of integration. Every one of them will yield the answer $\frac{7}{3}$. ■

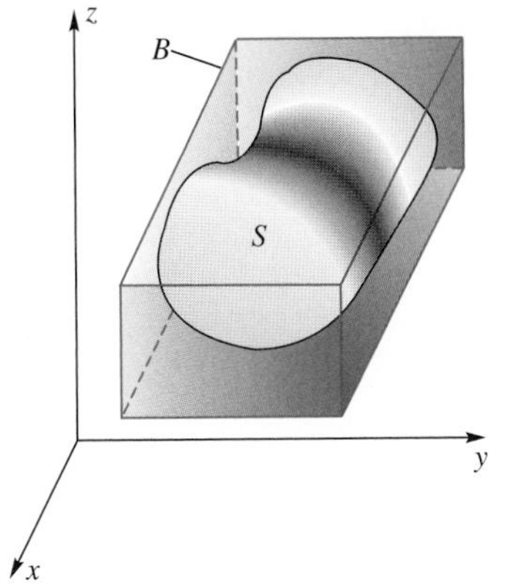

Figure 2

General Regions Consider a closed bounded set S in three-space and enclose it in a box B, as shown in Figure 2. Let $f(x, y, z)$ be defined on S, and give f the value zero outside S. Then we define

$$\iiint_S f(x, y, z)\,dV = \iiint_B f(x, y, z)\,dV$$

The integral on the right was defined in our opening remarks, but that does not mean that it is easy to evaluate. In fact, if the set S is sufficiently complicated, we may not be able to make the evaluation.

Let S be a z-simple set (vertical lines intersect S in a single line segment), and let S_{xy} be its projection in the xy-plane (Figure 3). Then

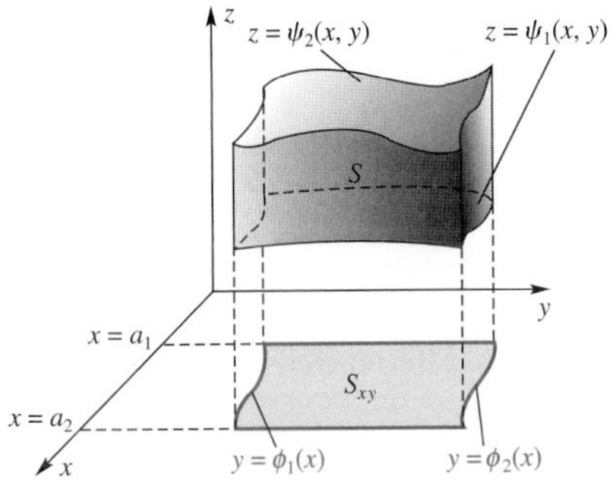

Figure 3

$$\iiint_S f(x, y, z)\,dV = \iint_{S_{xy}} \left[\int_{\psi_1(x, y)}^{\psi_2(x, y)} f(x, y, z)\,dz\right] dA$$

If, in addition, S_{xy} is a y-simple set (as shown in Figure 3), we can rewrite the outer double integral as an iterated integral.

$$\iiint_S f(x, y, z)\,dV = \int_{a_1}^{a_2}\int_{\phi_1(x)}^{\phi_2(x)}\int_{\psi_1(x, y)}^{\psi_2(x, y)} f(x, y, z)\,dz\,dy\,dx$$

Other orders of integration may be possible, depending on the shape of S, but in each case we should expect the limits on the inner integral to be functions of two variables, those on the middle integral to be functions of one variable, and those on the outer integral to be constants.

We give several examples. The first simply illustrates evaluation of a triple iterated integral.

> **Limits of Integration**
>
> The limits of integration on the innermost integral may depend on *both* of the other variables of integration. The limits of integration on the middle integral may depend *only* on the outermost variable of integration. Finally, the limits of integration for the outermost integral may not depend on any of the variables of integration.

EXAMPLE 2 Evaluate the iterated integral

$$\int_{-2}^{5}\int_0^{3x}\int_y^{x+2} 4\,dz\,dy\,dx$$

SOLUTION

$$\int_{-2}^{5}\int_0^{3x}\int_y^{x+2} 4\,dz\,dy\,dx = \int_{-2}^{5}\int_0^{3x}\left(\int_y^{x+2} 4\,dz\right)dy\,dx$$

$$= \int_{-2}^{5}\int_0^{3x} [4z]_y^{x+2}\,dy\,dx$$

$$= \int_{-2}^{5}\int_0^{3x} (4x - 4y + 8)\,dy\,dx$$

$$= \int_{-2}^{5} [4xy - 2y^2 + 8y]_0^{3x}\,dx$$

$$= \int_{-2}^{5} (-6x^2 + 24x)\,dx = -14$$ ■

EXAMPLE 3 Evaluate the triple integral of $f(x, y, z) = 2xyz$ over the solid region S in the first octant that is bounded by the parabolic cylinder $z = 2 - \frac{1}{2}x^2$ and the planes $z = 0$, $y = x$, and $y = 0$.

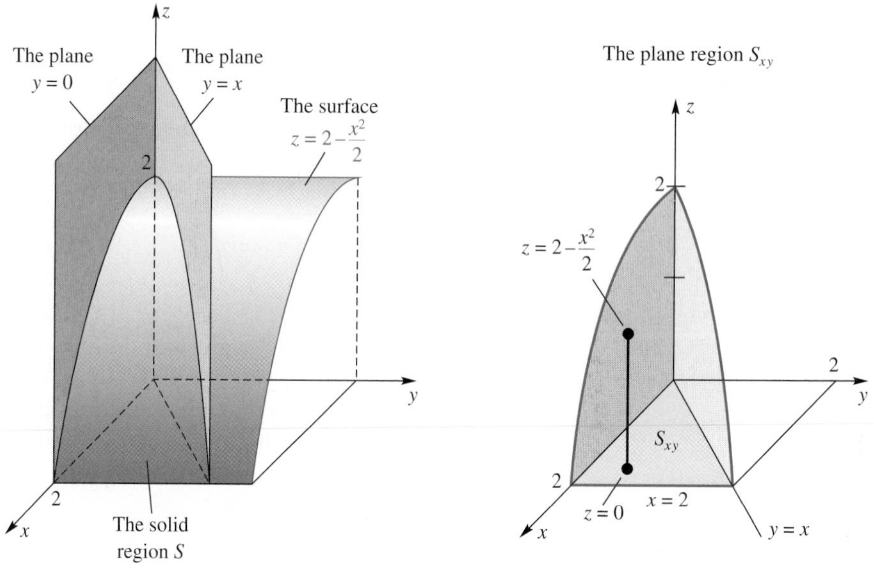

Figure 4

SOLUTION The solid region S is shown in Figure 4. The triple integral

$$\iiint_S 2xyz\, dV$$

can be evaluated by an iterated integral.

Note first that S is a z-simple set and that its projection S_{xy} in the xy-plane is y-simple (also x-simple). In the first integration, x and y are fixed; we integrate along a vertical line from $z = 0$ to $z = 2 - x^2/2$. The result is then integrated over the set S_{xy}.

$$\begin{aligned}\iiint_S 2xyz\, dV &= \int_0^2 \int_0^x \int_0^{2-x^2/2} 2xyz\, dz\, dy\, dx \\ &= \int_0^2 \int_0^x [xyz^2]_0^{2-x^2/2}\, dy\, dx \\ &= \int_0^2 \int_0^x \left(4xy - 2x^3y + \frac{1}{4}x^5y\right) dy\, dx \\ &= \int_0^2 \left(2x^3 - x^5 + \frac{1}{8}x^7\right) dx = \frac{4}{3}\end{aligned}$$ ■

Many different orders of integration are possible in Example 3. We illustrate another way to do this problem.

EXAMPLE 4 Evaluate the integral of Example 3 by doing the integration in the order $dy\, dx\, dz$.

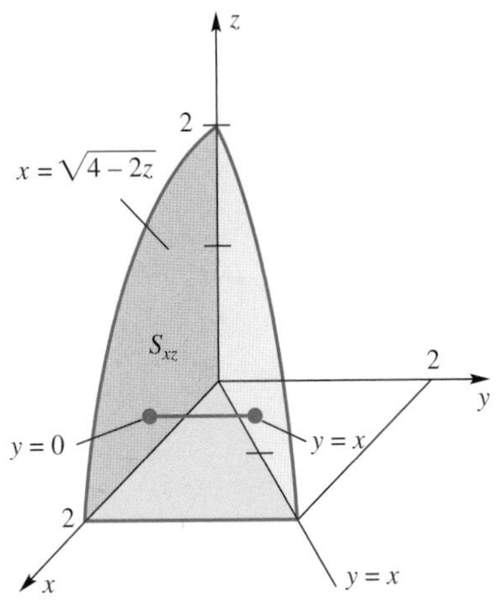

Figure 5

SOLUTION Note that the solid S is y-simple and that it projects onto the plane set S_{xz} shown in Figure 5. We first integrate along a horizontal line from $y = 0$ to $y = x$; then we integrate the result over S_{xz}.

$$\iiint_S 2xyz\,dV = \int_0^2 \int_0^{\sqrt{4-2z}} \int_0^x 2xyz\,dy\,dx\,dz$$
$$= \int_0^2 \int_0^{\sqrt{4-2z}} x^3 z\,dx\,dz = \frac{1}{4}\int_0^2 \left(\sqrt{4-2z}\right)^4 z\,dz$$
$$= \frac{1}{4}\int_0^2 (16z - 16z^2 + 4z^3)\,dz = \frac{4}{3}$$ ■

Mass and Center of Mass The concepts of mass and center of mass generalize easily to solid regions. By now, the process that leads to the correct formula is very familiar and can be summarized in our motto: *slice, approximate, integrate.* Figure 6 gives away the whole idea. The symbol $\delta(x, y, z)$ denotes the density (mass per unit volume) at (x, y, z).

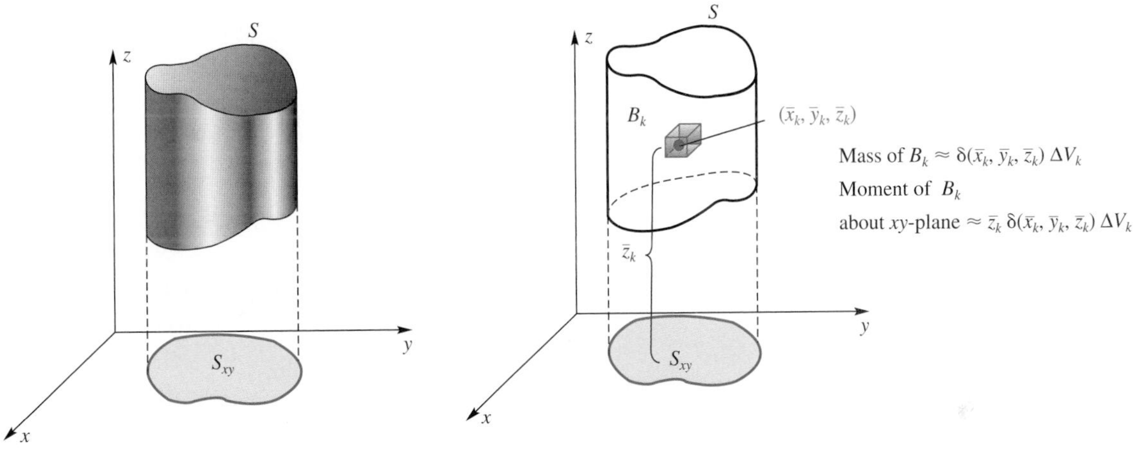

Figure 6

The corresponding integral formulas for the mass m of the solid S, moment M_{xy} of S with respect to the xy-plane, and z-coordinate $\bar{z}$ of the center of mass are

$$m = \iiint_S \delta(x, y, z)\,dV$$
$$M_{xy} = \iiint_S z\delta(x, y, z)\,dV$$
$$\bar{z} = \frac{M_{xy}}{m}$$

There are similar formulas for M_{yz}, M_{xz}, $\bar{x}$, and $\bar{y}$.

EXAMPLE 5 Find the mass and center of mass of the solid S of Example 3, assuming that its density is proportional to the distance from its base in the xy-plane.

SOLUTION By hypothesis, $\delta(x, y, z) = kz$, where k is a constant. Thus,

$$\begin{aligned}
m &= \iiint_S kz\, dV = \int_0^2 \int_0^x \int_0^{2-x^2/2} kz\, dz\, dy\, dx \\
&= k \int_0^2 \int_0^x \frac{1}{2}\left(2 - \frac{x^2}{2}\right)^2 dy\, dx = k \int_0^2 \int_0^x \left(2 - x^2 + \frac{1}{8}x^4\right) dy\, dx \\
&= k \int_0^2 \left(2x - x^3 + \frac{1}{8}x^5\right) dx = k\left[x^2 - \frac{x^4}{4} + \frac{x^6}{48}\right]_0^2 = \frac{4}{3}k \\
M_{xy} &= \iiint_S kz^2\, dV = \int_0^2 \int_0^x \int_0^{2-x^2/2} kz^2\, dz\, dy\, dx \\
&= \frac{k}{3} \int_0^2 \int_0^x \left(2 - \frac{x^2}{2}\right)^3 dy\, dx \\
&= \frac{k}{3} \int_0^2 \int_0^x \left(8 - 6x^2 + \frac{3}{2}x^4 - \frac{1}{8}x^6\right) dy\, dx \\
&= \frac{k}{3} \int_0^2 \left(8x - 6x^3 + \frac{3}{2}x^5 - \frac{1}{8}x^7\right) dx \\
&= \frac{k}{3}\left[4x^2 - \frac{3}{2}x^4 + \frac{1}{4}x^6 - \frac{1}{64}x^8\right]_0^2 = \frac{4}{3}k \\
M_{xz} &= \iiint_S kyz\, dV = \int_0^2 \int_0^x \int_0^{2-x^2/2} kyz\, dz\, dy\, dx \\
&= k \int_0^2 \int_0^x \frac{1}{2}y\left(2 - \frac{x^2}{2}\right)^2 dy\, dx = k \int_0^2 \frac{1}{4}x^2\left(2 - \frac{x^2}{2}\right)^2 dx \\
&= k \int_0^2 \left(x^2 - \frac{1}{2}x^4 + \frac{1}{16}x^6\right) dx = \frac{64}{105}k \\
M_{yz} &= \iiint_S kxz\, dV = \int_0^2 \int_0^x \int_0^{2-x^2/2} kxz\, dz\, dy\, dx = \frac{128}{105}k \\
\bar{z} &= \frac{M_{xy}}{m} = \frac{4k/3}{4k/3} = 1 \\
\bar{y} &= \frac{M_{xz}}{m} = \frac{64k/105}{4k/3} = \frac{16}{35} \\
\bar{x} &= \frac{M_{yz}}{m} = \frac{128k/105}{4k/3} = \frac{32}{35}
\end{aligned}$$

■

Multiple Random Variables We saw in Section 5.7 how probabilities for random variables can be computed as areas under the probability density function and how expectations can be computed like moments. These concepts are easily generalized to the case of a pair (or triple, etc.) of random variables. A function $f(x, y, z)$ is a **joint probability density function** (PDF) for the random variables (X, Y, Z) if $f(x, y, z) \geq 0$ for all (x, y, z) in S and

$$\iiint_S f(x, y, z)\, dz\, dy\, dx = 1$$

where S is the region of all possible values for (X, Y, Z). A probability involving (X, Y, Z) can then be computed as the triple integral over the appropriate region. The expected value of some function $g(X, Y, Z)$ is defined to be

$$E(g(X, Y, Z)) = \iiint_S g(x, y, z)\, f(x, y, z)\, dz\, dy\, dx$$

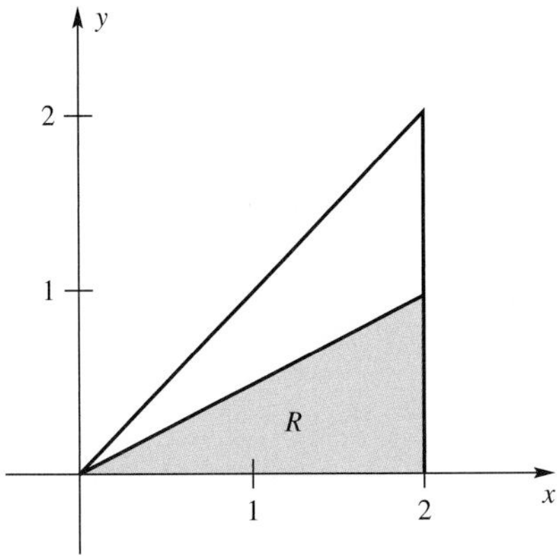

Figure 7

With obvious modifications, this discussion applies to pairs (or n-tuples) of random variables.

EXAMPLE 6 The joint PDF for the random variables (X, Y, Z) is of the form

$$f(x, y, z) = \begin{cases} \frac{1}{2}, & \text{if } 0 \le x \le 2; 0 \le y \le x; 0 \le z \le 1 \\ 0, & \text{otherwise} \end{cases}$$

Find (a) $P(Y \le X/2)$, and (b) $E(Y)$.

SOLUTION

(a) We notice that $Y \le X/2$ if and only if (X, Y) is in the shaded region R of Figure 7 and $0 \le Z \le 1$. Thus,

$$P\left(Y \le \frac{X}{2}\right) = \int_0^2 \int_0^{x/2} \int_0^1 \frac{1}{2}\, dz\, dy\, dx = \int_0^2 \int_0^{x/2} \frac{1}{2}\, dy\, dx = \int_0^2 \frac{x}{4}\, dx = \left[\frac{x^2}{8}\right]_0^2 = \frac{1}{2}$$

(b) The expectation of Y is

$$E(Y) = \iiint_S y\, f(x, y, z)\, dz\, dy\, dx = \int_0^2 \int_0^x \int_0^1 \frac{y}{2}\, dz\, dy\, dx$$

$$= \int_0^2 \int_0^x \frac{y}{2}\, dy\, dx = \int_0^2 \left[\frac{y^2}{4}\right]_0^x dx$$

$$= \int_0^2 \frac{x^2}{4}\, dx = \left[\frac{x^3}{12}\right]_0^2 = \frac{2}{3}$$ ■

Concepts Review

1. $\iiint_S 1\, dV$ gives the _____ of the solid S.

2. If the density at (x, y, z) is $|xyz|$, then the mass of S is _____.

3. $\int_0^1 \int_0^1 \int_{x^2}^x f(x, y, z)\, dy\, dx\, dz = \int_0^1 \int_0^1 \int_{g(y)}^{h(y)} f(x, y, z)\, dx\, dy\, dz$, where $g(y) =$ _____ and $h(y) =$ _____.

4. Let S be the solid unit sphere centered at the origin. Then, from symmetry, we conclude that $\iiint_S (x + y + z)\, dV =$ _____.

Problem Set 13.7

In Problems 1–10, evaluate the iterated integrals.

1. $\int_{-3}^7 \int_0^{2x} \int_y^{x-1} dz\, dy\, dx$

2. $\int_0^2 \int_{-1}^4 \int_0^{3y+x} dz\, dy\, dx$

3. $\int_1^4 \int_{z-1}^{2z} \int_0^{y+2z} dx\, dy\, dz$

4. $\int_0^5 \int_{-2}^4 \int_1^2 6xy^2z^3\, dx\, dy\, dz$

5. $\int_4^{24} \int_0^{24-x} \int_0^{24-x-y} \frac{y+z}{x}\, dz\, dy\, dx$

6. $\int_0^5 \int_0^3 \int_{z^2}^9 xyz\, dx\, dz\, dy$

7. $\int_0^2 \int_1^z \int_0^{\sqrt{x/z}} 2xyz\, dy\, dx\, dz$

8. $\int_0^{\pi/2} \int_0^z \int_0^y \sin(x + y + z)\, dx\, dy\, dz$

9. $\int_{-2}^4 \int_{x-1}^{x+1} \int_0^{\sqrt{2y/x}} 3xyz\, dz\, dy\, dx$

10. $\int_0^{\pi/2} \int_{\sin 2z}^0 \int_0^{2yz} \sin\left(\frac{x}{y}\right) dx\, dy\, dz$

In Problems 11–20, sketch the solid S. Then write an iterated integral for

$$\iiint_S f(x, y, z)\, dV$$

11. $S = \{(x, y, z): 0 \le x \le 1, 0 \le y \le 3, 0 \le z \le \frac{1}{6}(12 - 3x - 2y)\}$

12. $S = \{(x, y, z): 0 \le x \le \sqrt{4 - y^2}, 0 \le y \le 2, 0 \le z \le 3\}$

13. $S = \{(x, y, z): 0 \le x \le \frac{1}{2}y, 0 \le y \le 4, 0 \le z \le 2\}$

14. $S = \{(x, y, z): 0 \le x \le \sqrt{y}, 0 \le y \le 4, 0 \le z \le \frac{3}{2}x\}$

15. $S = \{(x, y, z): 0 \le x \le 3z, 0 \le y \le 4 - x - 2z, 0 \le z \le 2\}$

16. $S = \{(x, y, z): 0 \le x \le y^2, 0 \le y \le \sqrt{z}, 0 \le z \le 1\}$

17. S is the tetrahedron with vertices $(0, 0, 0)$, $(3, 2, 0)$, $(0, 3, 0)$, and $(0, 0, 2)$.

18. S is the region in the first octant bounded by the surface $z = 9 - x^2 - y^2$ and the coordinate planes.

19. S is the region in the first octant bounded by the cylinder $y^2 + z^2 = 1$ and the planes $x = 1$ and $x = 4$.

20. S is the smaller region bounded by the cylinder $x^2 + y^2 - 2y = 0$ and the planes $x - y = 0$, $z = 0$, and $z = 3$.

In Problems 21–28, use triple iterated integrals to find the indicated quantities.

21. Volume of the solid in the first octant bounded by $y = 2x^2$ and $y + 4z = 8$

22. Volume of the solid in the first octant bounded by the elliptic cylinder $y^2 + 64z^2 = 4$ and the plane $y = x$

23. Volume of the solid bounded by the cylinders $x^2 = y$ and $z^2 = y$ and the plane $y = 1$

24. Volume of the solid bounded by the cylinder $y = x^2 + 2$ and the planes $y = 4$, $z = 0$, and $3y - 4z = 0$

25. Center of mass of the tetrahedron bounded by the planes $x + y + z = 1$, $x = 0$, $y = 0$, and $z = 0$ if the density is proportional to the sum of the coordinates of the point

26. Center of mass of the solid bounded by the cylinder $x^2 + y^2 = 9$ and the planes $z = 0$ and $z = 4$ if the density is proportional to the square of the distance from the origin

27. Center of mass of that part of the solid sphere $\{(x, y, z): x^2 + y^2 + z^2 \le a^2\}$ that lies in the first octant, assuming that it has constant density

28. Moment of inertia I_x about the x-axis of the solid bounded by the cylinder $y^2 + z^2 = 4$ and the planes $x - y = 0$, $x = 0$, and $z = 0$ if the density $\delta(x, y, z) = z$. *Hint:* You will need to develop your own formula; *slice, approximate, integrate*.

In Problems 29–32, write the given iterated integral as an iterated integral with the indicated order of integration.

29. $\int_0^1 \int_0^{\sqrt{1-y^2}} \int_0^{\sqrt{1-y^2-z^2}} f(x, y, z)\, dx\, dz\, dy$; $dz\, dy\, dx$

30. $\int_0^2 \int_0^{4-2y} \int_0^{4-2y-z} f(x, y, z)\, dx\, dz\, dy$; $dz\, dy\, dx$

31. $\int_0^2 \int_0^{9-x^2} \int_0^{2-x} f(x, y, z)\, dz\, dy\, dx$; $dy\, dx\, dz$

32. $\int_0^2 \int_0^{9-x^2} \int_0^{2-x} f(x, y, z)\, dz\, dy\, dx$; $dz\, dx\, dy$

33. Consider the solid (Figure 8) in the first octant cut off from the square cylinder with sides $x = 0$, $x = 1$, $z = 0$, and $z = 1$ by the plane $2x + y + 2z = 6$. Find its volume in three ways.

(a) Hard way: by a $dz\, dy\, dx$ integration
(b) Easier way: by a $dy\, dx\, dz$ integration
(c) Easiest way: by Problem 23 of Section 13.6

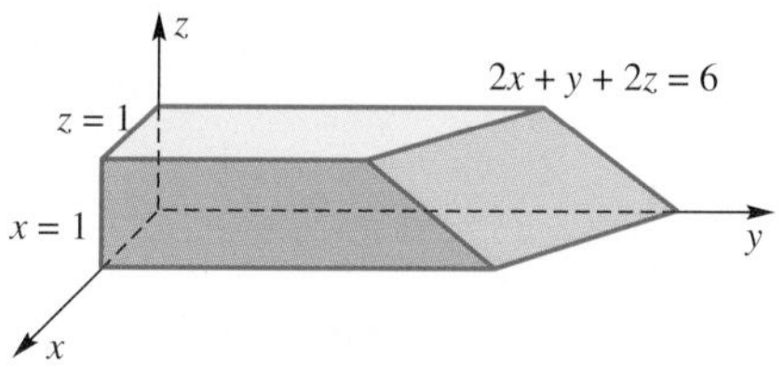

Figure 8

34. Assuming that the density of the solid of Figure 8 is a constant k, find the moment of inertia of the solid with respect to the y-axis.

35. If the temperature at (x, y, z) is $T(x, y, z) = 30 - z$ degrees, find the average temperature of the solid of Figure 8.

36. Assuming that the temperature of the solid in Figure 8 is $T(x, y, z) = 30 - z$, find all points in the solid where the actual temperature equals the average temperature.

37. Find the center of mass of the homogeneous solid in Figure 8.

38. Consider the solid (Figure 9) in the first octant cut off from the square cylinder with sides $x = 0$, $x = 1$, $y = 0$, and $y = 1$ by the plane $x + y + z = 4$. Find its volume in three ways.

(a) Hard way: by a $dx\, dz\, dy$ integration
(b) Easier way: by a $dz\, dy\, dx$ integration
(c) Easiest way: by Problem 23 of Section 13.6

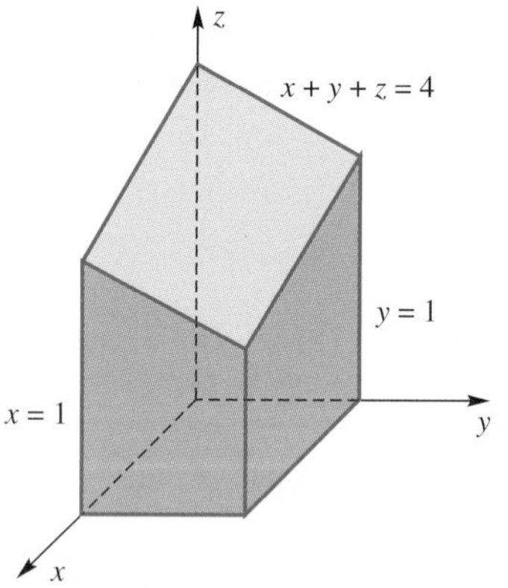

Figure 9

39. Find the center of mass of the homogeneous solid in Figure 9.

40. Suppose the temperature of the solid in Figure 9 begins at 40° at the bottom (the xy-plane) and increases (continuously) 5° for each unit above the xy-plane. Find the average temperature in the solid.

41. Soda Can Problem A full soda can of height h stands on the xy-plane. Punch a hole in the base and watch $\bar{z}$ (the z-coordinate of the center of mass) as the soda leaks away. Starting at $h/2$, $\bar{z}$ gradually drops to a minimum and then rises back to $h/2$ when the can is empty. Show that $\bar{z}$ is least when it coincides with the height of the soda. (Do not neglect the mass of the can itself.)

Would the same conclusion hold for a soda bottle? *Hint:* Don't calculate; think geometrically.

CAS **42.** Let $S = \{(x, y, z): x^2/a^2 + y^2/b^2 + z^2/c^2 \le 1\}$. Evaluate $\iiint_S (xy + xz + yz)\, dV$.

43. Suppose that the random variables (X, Y) have joint PDF

$$f(x, y) = \begin{cases} ky, & \text{if } 0 \le x \le 12; 0 \le y \le x \\ 0, & \text{otherwise} \end{cases}$$

Find each of the following:

(a) k (b) $P(Y > 4)$ (c) $E(X)$

44. Suppose that the random variables (X, Y, Z) have joint PDF

$$f(x, y) = \begin{cases} kxy, & \text{if } 0 \le x \le y; 0 \le y \le 4; 0 \le z \le 2 \\ 0, & \text{otherwise} \end{cases}$$

Find each of the following:

(a) k (b) $P(X > 2)$ (c) $E(X)$

45. Suppose that the random variables (X, Y) have joint PDF

$$f(x, y) = \begin{cases} \frac{3}{256}(x^2 + y^2), & \text{if } 0 \le x \le y; 0 \le y \le 4 \\ 0, & \text{otherwise} \end{cases}$$

Find each of the following:

(a) $P(X > 2)$ (b) $P(X + Y \le 4)$ (c) $E(X + Y)$

46. Suppose that the random variables (X, Y) have joint probability density function $f(x, y)$. The **marginal probability density function** of X is defined to be

$$f_X(x) = \int_{a(x)}^{b(x)} f(x, y)\, dy$$

where $a(x)$ and $b(x)$ are the smallest and largest possible values, respectively, that y can be for the given x. Show that

(a) $P(a < X < b) = \int_a^b f_X(x)\, dx$

(b) $E(X) = \int_a^b x\, f_X(x)\, dx$

47. Find the marginal PDF for the random variable X in Problem 43 and use it to calculate $E(X)$.

48. Give a reasonable definition for the marginal PDF for Y and use it to calculate the marginal PDF for Y in Problem 44.

Answers to Concepts Review: **1.** volume **2.** $\iiint_S |xyz|\, dV$ **3.** y; $\sqrt{y}$ **4.** 0

13.8 Triple Integrals in Cylindrical and Spherical Coordinates

When a solid region S in three-space has an axis of symmetry, the evaluation of triple integrals over S is often facilitated by using cylindrical coordinates. Similarly if S is symmetric with respect to a point, spherical coordinates may be helpful. Cylindrical and spherical coordinates were introduced in Section 11.9, a section you may wish to review before going on. Both of these are special cases of transformations of variables for multiple integrals, the topic of Section 13.9.

Cylindrical Coordinates Figure 1 serves to remind us of the meaning of cylindrical coordinates and displays the symbols that we use. Cylindrical and Cartesian (rectangular) coordinates are related by the equations

$$x = r\cos\theta, \qquad y = r\sin\theta, \qquad x^2 + y^2 = r^2$$

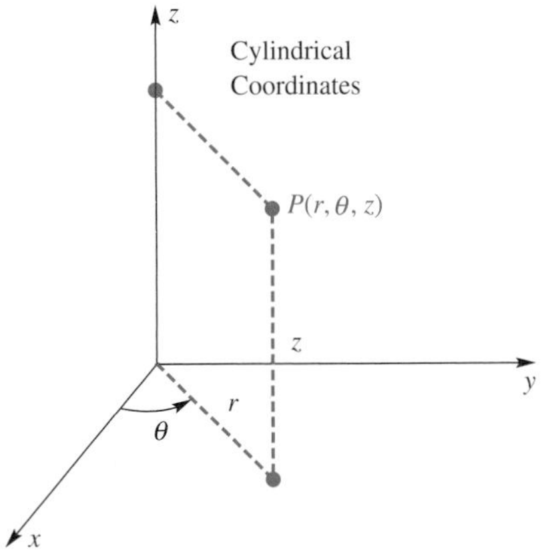

Figure 1

As a result, the function $f(x, y, z)$ transforms to

$$f(x, y, z) = f(r\cos\theta, r\sin\theta, z) = F(r, \theta, z)$$

when written in cylindrical coordinates.

Suppose now that we wish to evaluate $\iiint_S f(x, y, z)\, dV$, where S is a solid region. Consider partitioning S by means of a cylindrical grid, where the typical volume element has the shape shown in Figure 2. Since this piece (called a *cylindrical wedge*) has volume $\Delta V_k = \bar{r}_k\, \Delta r_k\, \Delta\theta_k\, \Delta z_k$, the sum that approximates the integral has the form

$$\sum_{k=1}^{n} F(\bar{r}_k, \bar{\theta}_k, \bar{z}_k)\bar{r}_k\, \Delta z_k\, \Delta r_k\, \Delta\theta_k$$

Taking the limit as the norm of the partition tends to zero leads to a new integral and suggests an important formula for changing from Cartesian to cylindrical coordinates in a triple integral.

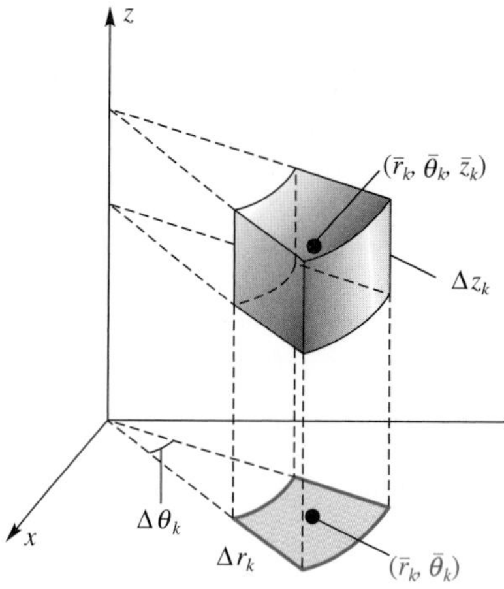

Figure 2

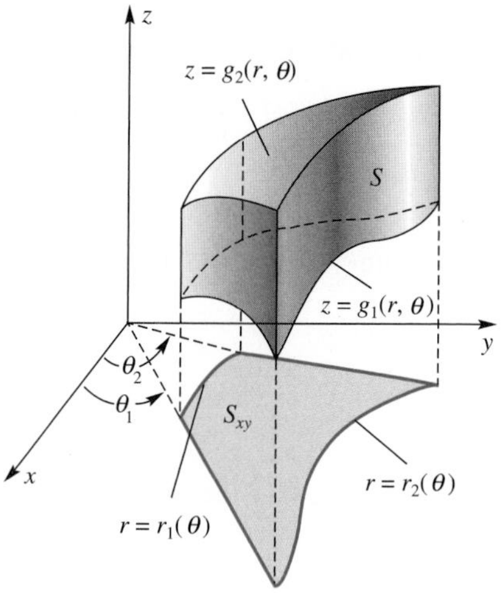

Figure 3

Cylindrical Coordinates and Polar Coordinates

The transformation from Cartesian coordinates (in 2-space) to polar coordinates is

$$x = r\cos\theta \qquad y = r\sin\theta$$

whereas the transformation from Cartesian coordinates (in 3-space) to cylindrical coordinates is

$$x = r\cos\theta \qquad y = r\sin\theta \qquad z = z$$

In other words, to identify a point in three-space using cylindrical coordinates, we specify the polar coordinates for the ordered pair (x, y) and then "tack on" the z component. Since

$$dx\,dy = r\,dr\,d\theta$$

we shouldn't be surprised that

$$dx\,dy\,dz = r\,dz\,dr\,d\theta$$

Let S be a z-simple solid and suppose that its projection S_{xy} in the xy-plane is r-simple, as shown in Figure 3. If f is continuous on S, then

$$\iiint_S f(x, y, z)\,dV = \int_{\theta_1}^{\theta_2}\int_{r_1(\theta)}^{r_2(\theta)}\int_{g_1(r,\theta)}^{g_2(r,\theta)} f(r\cos\theta, r\sin\theta, z)r\,dz\,dr\,d\theta$$

The key fact to note is that the $dz\,dy\,dx$ of Cartesian coordinates becomes $r\,dz\,dr\,d\theta$ in cylindrical coordinates.

EXAMPLE 1 Find the mass and center of mass of a solid right circular cylinder S, assuming that the density is proportional to the distance from the base.

SOLUTION With S oriented as shown in Figure 4, we can write the density function as $\delta(x, y, z) = kz$, where k is a constant. Then

$$m = \iiint_S \delta(x, y, z)\,dV = k\int_0^{2\pi}\int_0^a\int_0^h zr\,dz\,dr\,d\theta$$

$$= k\int_0^{2\pi}\int_0^a \frac{1}{2}h^2 r\,dr\,d\theta = \frac{1}{2}kh^2\int_0^{2\pi}\int_0^a r\,dr\,d\theta$$

$$= \frac{1}{2}kh^2\int_0^{2\pi}\frac{1}{2}a^2\,d\theta = \frac{1}{2}kh^2\pi a^2$$

$$M_{xy} = \iiint_S z\delta(x, y, z)\,dV = k\int_0^{2\pi}\int_0^a\int_0^h z^2 r\,dz\,dr\,d\theta$$

$$= k\int_0^{2\pi}\int_0^a \frac{1}{3}h^3 r\,dr\,d\theta = \frac{1}{3}kh^3\int_0^{2\pi}\int_0^a r\,dr\,d\theta$$

$$= \frac{1}{3}kh^3\pi a^2$$

$$\bar{z} = \frac{M_{xy}}{m} = \frac{\frac{1}{3}kh^3\pi a^2}{\frac{1}{2}kh^2\pi a^2} = \frac{2}{3}h$$

By symmetry, $\bar{x} = \bar{y} = 0$. ■

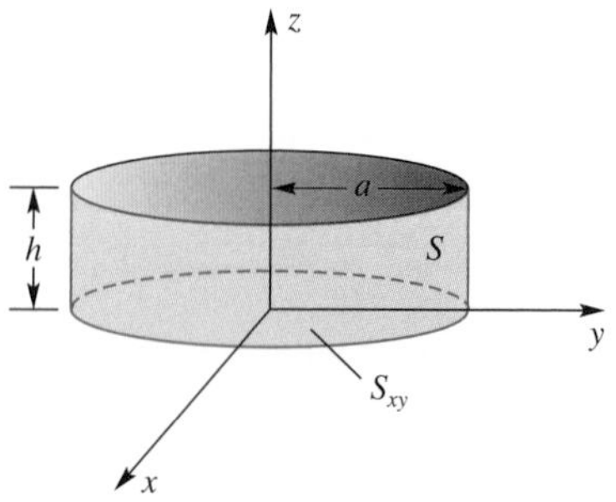

Figure 4

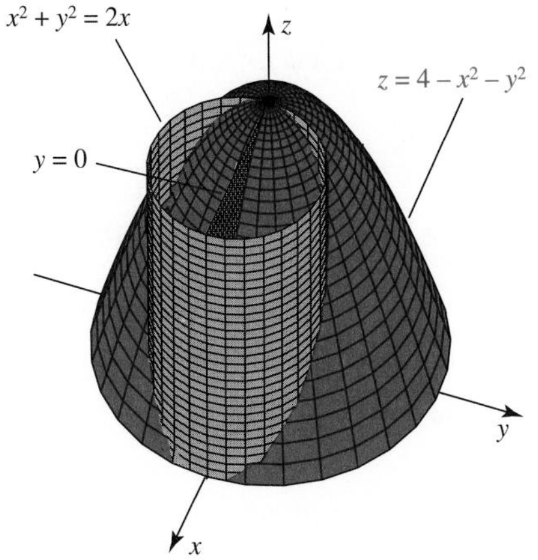

Figure 5

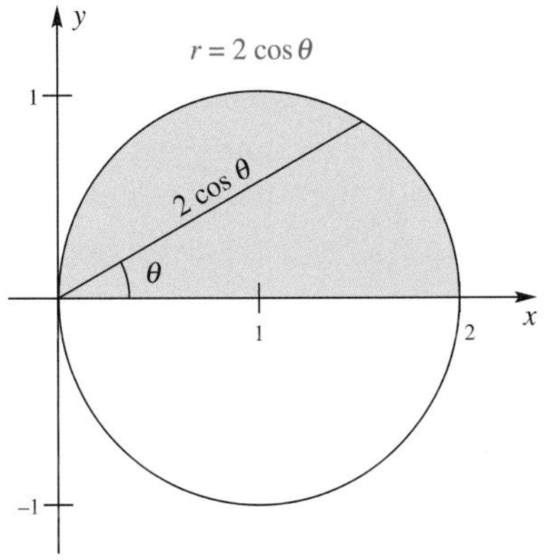

Figure 6

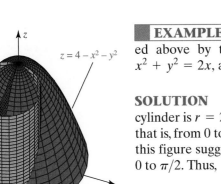

EXAMPLE 2 Find the volume of the solid region S in the first octant bounded above by the paraboloid $z = 4 - x^2 - y^2$, and laterally by the cylinder $x^2 + y^2 = 2x$, as shown in Figure 5.

SOLUTION In cylindrical coordinates, the paraboloid is $z = 4 - r^2$ and the cylinder is $r = 2 \cos \theta$. The z-variable runs from the xy-plane up to the paraboloid, that is, from 0 to $4 - r^2$. Figure 6 shows the "footprint" of the solid in the xy-plane; this figure suggests that for a fixed θ, r goes from 0 to $2 \cos \theta$. Finally, θ goes from 0 to $\pi/2$. Thus,

$$
\begin{aligned}
V &= \iiint_S 1\, dV = \int_0^{\pi/2} \int_0^{2\cos\theta} \int_0^{4-r^2} r\, dz\, dr\, d\theta \\
&= \int_0^{\pi/2} \int_0^{2\cos\theta} r(4 - r^2)\, dr\, d\theta = \int_0^{\pi/2} \left[2r^2 - \frac{1}{4} r^4 \right]_0^{2\cos\theta} d\theta \\
&= \int_0^{\pi/2} (8 \cos^2 \theta - 4 \cos^4 \theta)\, d\theta \\
&= 8 \cdot \frac{1}{2} \cdot \frac{\pi}{2} - 4 \cdot \frac{3}{8} \cdot \frac{\pi}{2} = \frac{5\pi}{4}
\end{aligned}
$$

We used Formula 113 from the table of integrals at the end of the book to make the last calculation. ■

Spherical Coordinates Figure 7 serves to remind us of the meaning of spherical coordinates, which were introduced in Section 11.9. There we learned that the equations

$$x = \rho \sin \phi \cos \theta, \qquad y = \rho \sin \phi \sin \theta, \qquad z = \rho \cos \phi$$

relate spherical coordinates and Cartesian coordinates. Figure 8 exhibits the volume element in spherical coordinates (called a *spherical wedge*). Though we omit the details, it can be shown that the volume of the indicated spherical wedge is

$$\Delta V = \bar{\rho}^2 \sin \bar{\phi}\, \Delta\rho\, \Delta\theta\, \Delta\phi$$

where $(\bar{\rho}, \bar{\theta}, \bar{\phi})$ is an appropriately chosen point in the wedge.

Partitioning a solid S by means of a spherical grid, forming the appropriate sum, and taking the limit leads to an iterated integral in which $dz\, dy\, dx$ is replaced by $\rho^2 \sin \phi\, d\rho\, d\theta\, d\phi$.

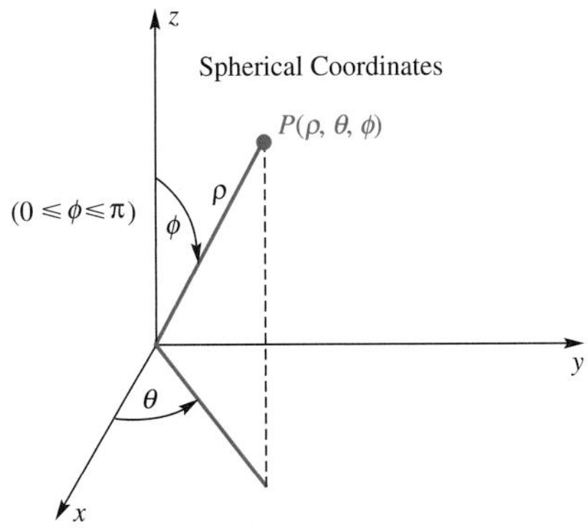

Figure 7

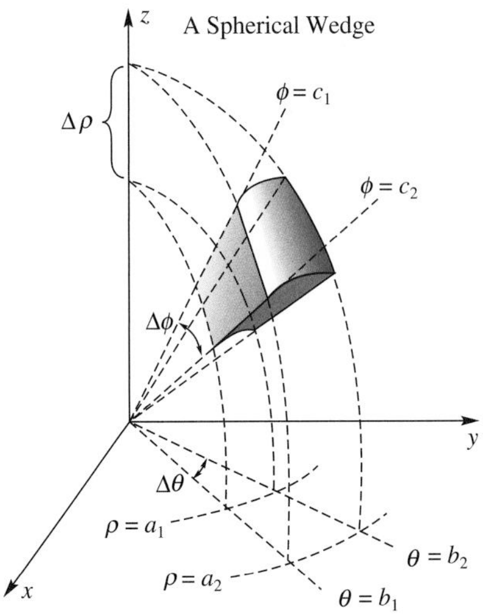

Figure 8

$$\iiint_S f(x, y, z)\, dV = \iiint_{\substack{\text{appropriate}\\ \text{limits}}} f(\rho \sin\phi \cos\theta, \rho \sin\phi \sin\theta, \rho\cos\phi)\rho^2 \sin\phi\, d\rho\, d\theta\, d\phi$$

EXAMPLE 3 Find the mass of a solid sphere S if its density δ is proportional to the distance from the center.

SOLUTION Center the sphere at the origin and let its radius be a. The density δ is given by $\delta = k\sqrt{x^2 + y^2 + z^2} = k\rho$. Thus, the mass m is given by

$$\begin{aligned} m &= \iiint_S \delta\, dV = k\int_0^{\pi}\int_0^{2\pi}\int_0^a \rho\rho^2 \sin\phi\, d\rho\, d\theta\, d\phi \\ &= k\frac{a^4}{4}\int_0^{\pi}\int_0^{2\pi} \sin\phi\, d\theta\, d\phi = \frac{1}{2}k\pi a^4 \int_0^{\pi} \sin\phi\, d\phi \\ &= k\pi a^4 \end{aligned}$$

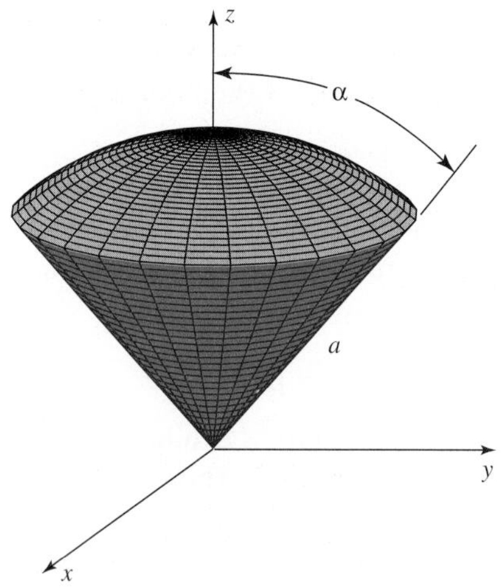

Figure 9

EXAMPLE 4 Find the volume and center of mass of the homogeneous solid S that is bounded above by the sphere $\rho = a$ and below by the cone $\phi = \alpha$, where a and α are constants (Figure 9).

SOLUTION The volume V is given by

$$\begin{aligned} V &= \int_0^{\alpha}\int_0^{2\pi}\int_0^a \rho^2 \sin\phi\, d\rho\, d\theta\, d\phi \\ &= \int_0^{\alpha}\int_0^{2\pi}\left(\frac{a^3}{3}\right)\sin\phi\, d\theta\, d\phi \\ &= \frac{2\pi a^3}{3}\int_0^{\alpha} \sin\phi\, d\phi = \frac{2\pi a^3}{3}(1 - \cos\alpha) \end{aligned}$$

It follows that the mass m of the solid is

$$m = kV = \frac{2\pi a^3 k}{3}(1 - \cos\alpha)$$

where k is the constant density.

From symmetry, the center of mass is on the z-axis; that is, $\bar{x} = \bar{y} = 0$. To find $\bar{z}$, we first calculate M_{xy}.

$$\begin{aligned} M_{xy} &= \iiint_S kz\, dV = \int_0^{\alpha}\int_0^{2\pi}\int_0^a k(\rho\cos\phi)\rho^2 \sin\phi\, d\rho\, d\theta\, d\phi \\ &= \int_0^{\alpha}\int_0^{2\pi}\int_0^a k\rho^3 \sin\phi\cos\phi\, d\rho\, d\theta\, d\phi \\ &= \int_0^{\alpha}\int_0^{2\pi} \frac{1}{4}ka^4 \sin\phi\cos\phi\, d\theta\, d\phi \\ &= \int_0^{\alpha} \frac{1}{2}\pi k a^4 \sin\phi\cos\phi\, d\phi = \frac{1}{4}\pi a^4 k \sin^2\alpha \end{aligned}$$

Thus,

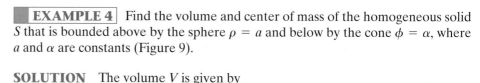

$$\begin{aligned} \bar{z} &= \frac{\frac{1}{4}\pi a^4 k \sin^2\alpha}{\frac{2}{3}\pi a^3 k(1 - \cos\alpha)} = \frac{3a\sin^2\alpha}{8(1 - \cos\alpha)} \\ &= \frac{3}{8}a(1 + \cos\alpha) \end{aligned}$$

Concepts Review

1. $dz\,dy\,dx$ takes the form _____ in cylindrical coordinates and the form _____ in spherical coordinates.

2. $\int_0^1 \int_0^{\sqrt{1-x^2}} \int_0^3 xy\,dz\,dy\,dx$ becomes _____ in cylindrical coordinates.

3. If S is the unit sphere centered at the origin, then $\iiint_S z^2\,dV$, when written as an iterated integral in spherical coordinates, becomes _____.

4. The value of the integral in Question 3 is _____.

Problem Set 13.8

In Problems 1–6, evaluate the integral which is given in cylindrical or spherical coordinates, and describe the region R of integration.

1. $\int_0^{2\pi} \int_0^3 \int_0^{12} r\,dz\,dr\,d\theta$

2. $\int_0^{2\pi} \int_1^3 \int_0^{12} r\,dz\,dr\,d\theta$

3. $\int_0^{\pi/4} \int_0^3 \int_0^{9-r^2} zr\,dz\,dr\,d\theta$

4. $\int_0^{\pi} \int_0^{\sin\theta} \int_0^{2} r\,dz\,dr\,d\theta$

5. $\int_0^{\pi} \int_0^{2\pi} \int_0^{a} \rho^2 \sin\phi\,d\rho\,d\theta\,d\phi$

6. $\int_0^{\pi/2} \int_0^{\pi/2} \int_0^{a} \rho^2 \cos^2\phi \sin\phi\,d\rho\,d\theta\,d\phi$

In Problems 7–14, use cylindrical coordinates to find the indicated quantity.

7. Volume of the solid bounded by the paraboloid $z = x^2 + y^2$ and the plane $z = 4$

8. Volume of the solid bounded above by the sphere $x^2 + y^2 + z^2 = 9$, below by the plane $z = 0$, and laterally by the cylinder $x^2 + y^2 = 4$

9. Volume of the solid bounded above by the sphere centered at the origin having radius 5 and below by the plane $z = 4$.

10. Volume of the solid bounded above by the plane $z = y + 4$, below by the xy-plane, and laterally by the right circular cylinder having radius 4 and whose axis is the z-axis.

11. Volume of the solid bounded above by the sphere $r^2 + z^2 = 5$ and below by the paraboloid $r^2 = 4z$

12. Volume of the solid under the surface $z = xy$, above the xy-plane, and within the cylinder $x^2 + y^2 = 2x$

13. Center of mass of the homogeneous solid bounded above by $z = 12 - 2x^2 - 2y^2$ and below by $z = x^2 + y^2$

14. Center of mass of the homogeneous solid inside $x^2 + y^2 = 4$, outside $x^2 + y^2 = 1$, below $z = 12 - x^2 - y^2$, and above $z = 0$

In Problems 15–22, use spherical coordinates to find the indicated quantity.

15. Mass of the solid inside the sphere $\rho = b$ and outside the sphere $\rho = a\ (a < b)$ if the density is proportional to the distance from the origin

16. Mass of a solid inside a sphere of radius $2a$ and outside a circular cylinder of radius a whose axis is a diameter of the sphere, if the density is proportional to the square of the distance from the center of the sphere

17. Center of mass of a solid hemisphere of radius a, if the density is proportional to the distance from the center of the sphere

18. Center of mass of a solid hemisphere of radius a, if the density is proportional to the distance from the axis of symmetry

19. Moment of inertia of the solid of Problem 18 with respect to its axis of symmetry

20. Volume of the solid within the sphere $x^2 + y^2 + z^2 = 16$, outside the cone $z = \sqrt{x^2 + y^2}$, and above the xy-plane

21. Volume of the smaller wedge cut from the unit sphere by two planes that meet at a diameter at an angle of 30°

22. $\int_{-3}^{3} \int_{-\sqrt{9-x^2}}^{\sqrt{9-x^2}} \int_{-\sqrt{9-x^2-z^2}}^{\sqrt{9-x^2-z^2}} (x^2 + y^2 + z^2)^{3/2}\,dy\,dz\,dx$

23. Find the volume of the solid bounded above by the plane $z = y$ and below by the paraboloid $z = x^2 + y^2$. *Hint:* In cylindrical coordinates the plane has equation $z = r\sin\theta$ and the paraboloid has equation $z = r^2$. Solve simultaneously to get the projection in the xy-plane.

24. Find the volume of the solid inside both of the spheres $\rho = 2\sqrt{2}\cos\phi$ and $\rho = 2$.

25. For a solid sphere of radius a, find each average distance.

(a) From its center

(b) From a diameter

(c) From a point on its boundary (consider $\rho = 2a\cos\phi$)

26. For any homogeneous solid S, show that the average value of the linear function $f(x, y, z) = ax + by + cz + d$ on S is $f(\bar{x}, \bar{y}, \bar{z})$, where $(\bar{x}, \bar{y}, \bar{z})$ is the center of mass.

27. A homogeneous solid sphere of radius a is centered at the origin. For the section S bounded by the half-planes $\theta = -\alpha$ and $\theta = \alpha$ (like a section of an orange), find each value.

(a) x-coordinate of the center of mass

(b) Average distance from the z-axis

28. All spheres in this problem have radius a, constant density k, and mass m. Find in terms of a and m the moment of inertia of each of the following:

(a) A solid sphere about a diameter

(b) A solid sphere about a tangent line to its boundary (the Parallel Axis Theorem holds also for solids; see Problem 28 of Section 13.5)

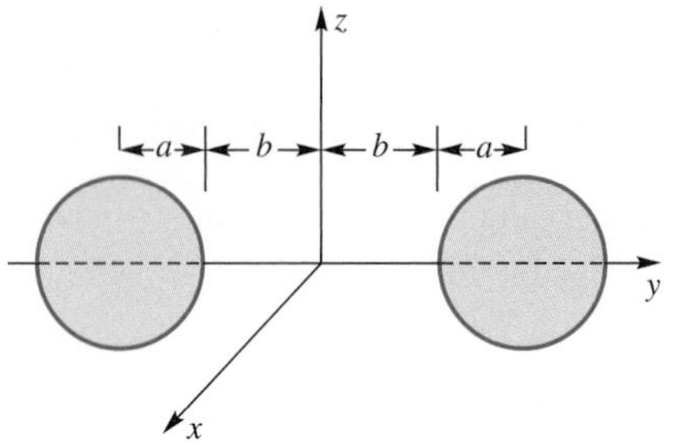

Figure 10

(c) The two-sphere solid of Figure 10 about the z-axis

29. Suppose that the left sphere in Figure 10 has density k and the right sphere density ck. Find the y-coordinate of the center of mass of this two-sphere solid (convince yourself that the analogue of Problem 33 of Section 13.5 is valid).

Answers to Concepts Review: **1.** $r\,dz\,dr\,d\theta$; $\rho^2 \sin\phi\,d\rho\,d\theta\,d\phi$ **2.** $\int_0^{\pi/2}\int_0^1\int_0^3 r^3 \cos\theta \sin\theta\,dz\,dr\,d\theta$ **3.** $\int_0^{\pi}\int_0^{2\pi}\int_0^1 \rho^4 \cos^2\phi \sin\phi\,d\rho\,d\theta\,d\phi$ **4.** $4\pi/15$

13.9 Change of Variables in Multiple Integrals

The formulas

$$dx\,dy = r\,dr\,d\theta$$
$$dx\,dy\,dz = r\,dz\,dr\,d\theta$$
$$dx\,dy\,dz = \rho^2 \sin\phi\,d\rho\,d\theta\,d\phi$$

are specific cases of a change of variable formula. They illustrate a general result that we discuss in this section. Before presenting the result for multiple integrals, we review the concept of change of variables, or substitutions, for single integrals.

Terminology and Notation

A function f from a set A to a set B is said to be **one-to-one** if distinct elements x and y in A get mapped to distinct elements $f(x)$ and $f(y)$ in B. The function f is **onto** if its range consists of the set B. A function f that is one-to-one and onto is guaranteed to have an inverse f^{-1}. $\mathbb{R}^2$ denotes the set of all ordered pairs of real numbers.

If g is a one-to-one function of a single variable, then g has an inverse g^{-1} and we know from Chapter 4 that

$$\int_a^b f(g(x))\,g'(x)\,dx = \int_{g(a)}^{g(b)} f(u)\,du$$

Interchanging the roles of x and u allows us to write this as

$$\int_a^b f(x)\,dx = \int_{g^{-1}(a)}^{g^{-1}(b)} f(g(u))\,g'(u)\,du$$

This last formula can be seen as the result of making the substitution $x = g(u)$. This function, or mapping, is illustrated in Figure 1. In this section we will develop an analogous formula for change of variables in multiple integrals. We begin by studying transformations from $\mathbb{R}^2$ to $\mathbb{R}^2$.

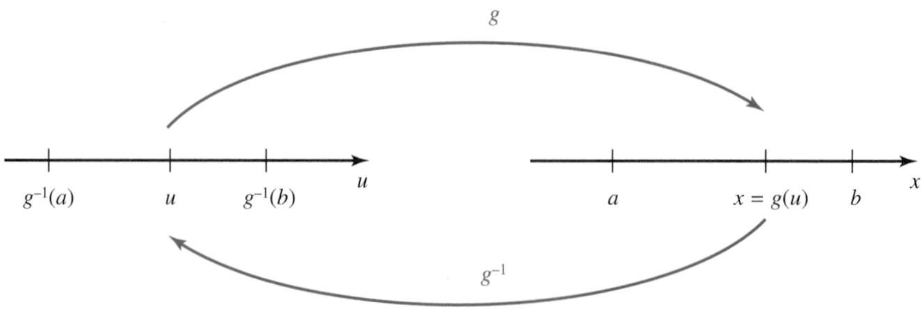

Figure 1

Transformations from the *uv*-Plane to the *xy*-Plane Let

$$x = x(u, v) \quad \text{and} \quad y = y(u, v)$$

and let

$$G(u, v) = (x(u, v), y(u, v))$$

The function G is a vector-valued function with a vector input. Such a function is called a **transformation** from $\mathbb{R}^2$ to $\mathbb{R}^2$. The ordered pair $(x, y) = G(u, v)$ is called the **image** of (u, v) under the transformation G, and (u, v) is called the **preimage** of (x, y). The **image of a set** S in the uv-plane is equal to that set of points (x, y) in the xy-plane satisfying $(x, y) = G(u, v)$, where (u, v) is in S. The function G cannot be graphed in the ordinary way because it would require four dimensions. Instead, we illustrate the function as a mapping from points in the uv-plane to points in the xy-plane. The situation is illustrated in Figure 2. This figure shows a grid in the uv-plane with lines parallel to the u- and v-axes, and its image in the xy-plane. The images of vertical lines in the uv-plane are called ***u*-curves** of G (vertical lines in the uv-plane are of the form $u =$ constant). Analogously, the images of horizontal lines are called ***v*-curves** of G.

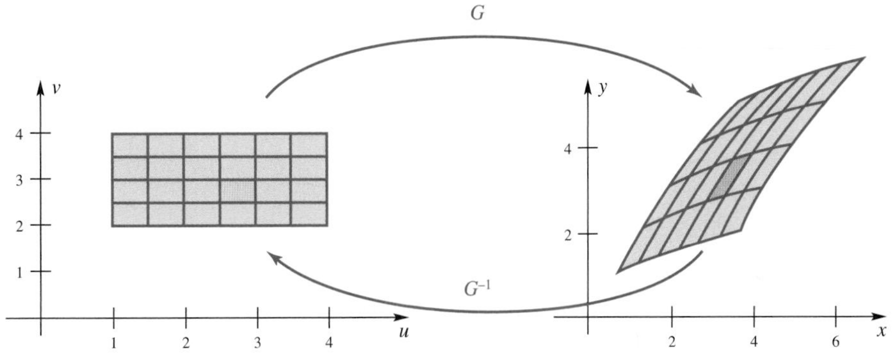

Figure 2

EXAMPLE 1 Let

$$x = x(u, v) = u + v$$
$$y = y(u, v) = u - v$$

and $G(u, v) = (x(u, v), y(u, v))$. Find and graph the u-curves and v-curves for G for the grid $\{(u, v): (u = 3, 4, 5 \text{ and } 1 \le v \le 4) \text{ or } (v = 1, 2, 3, 4 \text{ and } 3 \le u \le 5)\}$.

SOLUTION If we solve the system

$$x = u + v$$
$$y = u - v$$

for u and v, we obtain

$$u = \frac{1}{2}x + \frac{1}{2}y$$
$$v = \frac{1}{2}x - \frac{1}{2}y$$

The u-curves are determined by

$$C = \frac{1}{2}x + \frac{1}{2}y, \qquad C = 3, 4, 5$$

This leads to the curves

$$y = 2C - x, \quad C = 3, 4, 5$$

These are parallel lines, each having a slope of -1. Similarly, the v-curves are obtained by solving the equations

$$C = \frac{1}{2}x - \frac{1}{2}y, \qquad C = 1, 2, 3, 4$$

for y when $C = 1, 2, 3,$ and 4. The solution is

$$y = -2C + x, \qquad C = 1, 2, 3, 4$$

These are also parallel lines, each having slope +1. Figure 3 shows these curves. The u-curve for $u = 3$ and the v-curve for $v = 2$ are dashed. ■

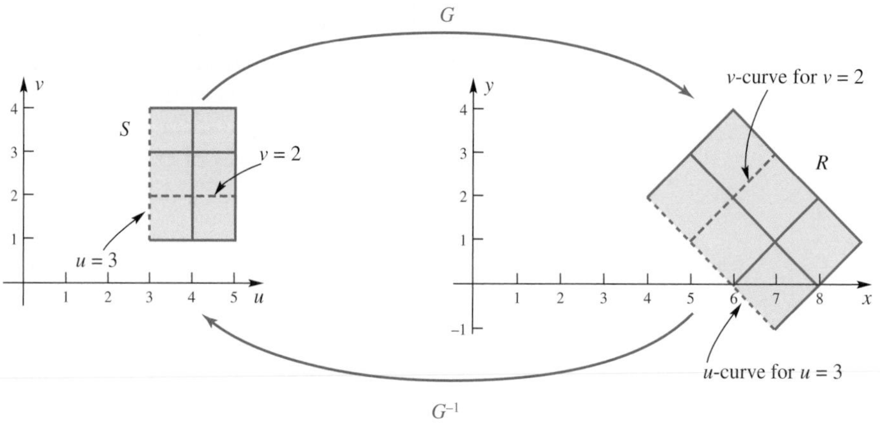

Figure 3

EXAMPLE 2 For $u > 0$ and $v > 0$, let

$$x = x(u, v) = u^2 - v^2$$
$$y = y(u, v) = uv$$

and

$$G(u, v) = (x(u, v), y(u, v))$$

Find and graph the u-curves and v-curves for G for the grid $\{(u, v)$: $(u = 0, 1, 2, 3, 4, 5$ and $0 \le v \le 5)$ or $(v = 0, 1, 2, 3, 4, 5$ and $0 \le u \le 5)\}$, and identify the u-curve for $u = 4$.

SOLUTION In order to solve the system

$$x = u^2 - v^2$$
$$y = uv$$

for u and v, we solve the second equation for v (in terms of u), getting $v = y/u$. Substituting this result into the first equation gives

$$x = u^2 - y^2/u^2$$

which is equivalent to

$$u^4 - xu^2 - y^2 = 0$$

This is a quadratic equation in u^2, so we can apply the quadratic formula to obtain

$$u^2 = \frac{x + \sqrt{x^2 + 4y^2}}{2}$$

(We must take the positive sign in the quadratic formula, otherwise the expression on the right side will be negative.) Thus,

$$u = \sqrt{\frac{1}{2}\left(x + \sqrt{x^2 + 4y^2}\right)}$$

$$v = \frac{y}{u} = \frac{y}{\sqrt{\frac{1}{2}\left(x + \sqrt{x^2 + 4y^2}\right)}}$$

These formulas apply so long as $(x, y) \neq (0, 0)$; we leave it as an exercise to show that $(x, y) = (0, 0)$ if and only if $(u, v) = (0, 0)$. The u-curves are determined by

$$C = \sqrt{\frac{x + \sqrt{x^2 + 4y^2}}{2}}, \qquad C = 0, 1, 2, 3, 4, 5$$

Which simplifies as follows:

$$\begin{aligned} 2C^2 &= x + \sqrt{x^2 + 4y^2} \\ 4C^4 - 4C^2 x + x^2 &= x^2 + 4y^2 \\ x &= -\frac{y^2 - C^4}{C^2} \end{aligned}$$

for $C = 0, 1, 2, 3, 4, 5$. These are horizontal parabolas opening to the left. Similarly, the v-curves are determined by

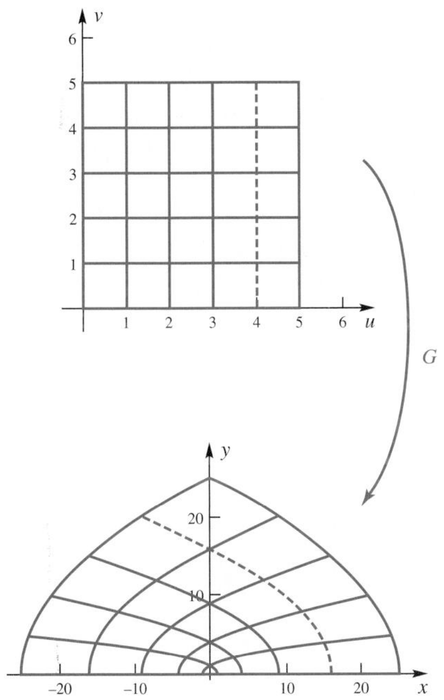

Figure 4

$$\begin{aligned} C &= \frac{y}{\sqrt{\left(x + \sqrt{x^2 + 4y^2}\right)/2}} \\ C^2\left(x + \sqrt{x^2 + 4y^2}\right) &= 2y^2 \\ x^2 + 4y^2 &= \frac{4y^4}{C^4} - \frac{4xy^2}{C^2} + x^2 \\ x &= \frac{y^2 - C^4}{C^2} \end{aligned}$$

These are horizontal parabolas opening to the right. The u- and v-curves are shown in Figure 4.

The u-curve corresponding to $u = 4$ $(0 \leq v \leq 5)$ is

$$x = -\frac{y^2 - 4^4}{4^2} = -\frac{1}{16}y^2 + 16, \qquad 0 \leq y \leq 20$$

This u-curve is the dashed curve in Figure 4. ■

The Change of Variable Formula for Double Integrals

When making a change of variable in a single integral, such as $\int_a^b f(x)\, dx$, we must take into account

1. the integrand $f(x)$,
2. the differential dx, and
3. the limits of integration.

For double integrals, such as $\iint_R f(x, y)\, dx\, dy$, the procedure is similar: We must take into account

1. the integrand $f(x, y)$,
2. the differential $dx\, dy$, and
3. the region of integration.

The main result is given in the next theorem.

Theorem A Change of Variables for Double Integrals

Suppose G is a one-to-one transformation from $\mathbb{R}^2$ to $\mathbb{R}^2$ which maps the bounded region S in the uv-plane onto the bounded region R in the xy-plane. If G is of the form $G(u, v) = (x(u, v), y(u, v))$, then

$$\iint_R f(x, y)\, dx\, dy = \iint_S f(x(u, v), y(u, v))|J(u, v)|\, du\, dv$$

where $J(u, v)$, called the **Jacobian,** is equal to the determinant

$$J(u, v) = \begin{vmatrix} \dfrac{\partial x}{\partial u} & \dfrac{\partial x}{\partial v} \\ \dfrac{\partial y}{\partial u} & \dfrac{\partial y}{\partial v} \end{vmatrix} = \frac{\partial x}{\partial u}\frac{\partial y}{\partial v} - \frac{\partial x}{\partial v}\frac{\partial y}{\partial u}$$

Sketch of Proof We begin in the uv-plane by taking a regular partition (i.e., a partition with a constant Δu and Δv) of a rectangle containing S. The image of this partition will be a partition of the region R in the xy-plane, although in general the u-curves and v-curves are not parallel to the x- and y-axes. (In fact, the u-curves and the v-curves are usually not lines.) Let $(u_i, v_i), i = 1, 2, \ldots, n$, be the lower left corner of the ith rectangle, and let (x_i, y_i) be the image of (u_i, v_i) under the transformation G. Let S_k denote the kth rectangle in the partition of the region S, and let R_k be its image in the xy-plane. See Figure 5. The double integral of f over the region R is then

$$\iint_R f(x, y)\, dx\, dy \approx \sum_{k=1}^{n} f(x_k, y_k)\, \Delta A_k$$

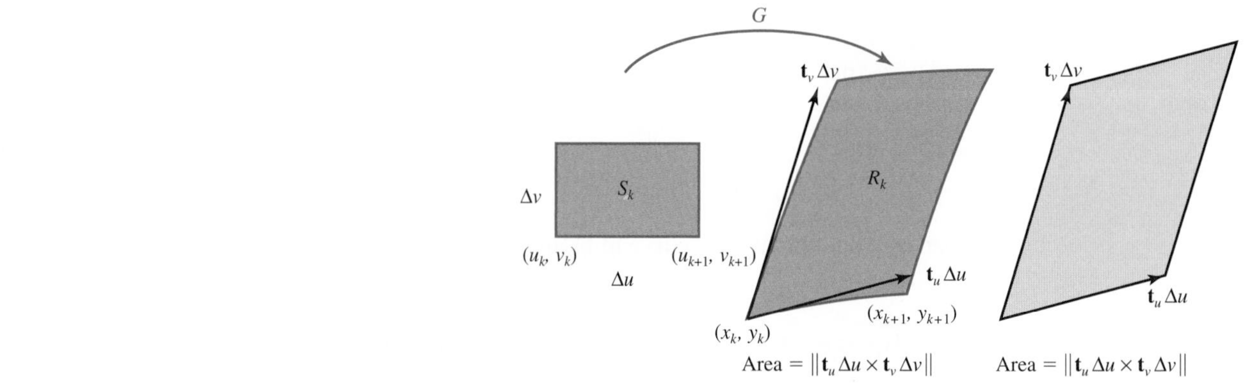

Figure 5

where ΔA_k is the area of R_k. Although the region R_k is not a rectangle, it closely resembles a parallelogram, so its area is roughly that of a parallelogram. In Section 11.4 we showed how to get the area of a parallelogram using the cross product of the two vectors that make up two sides. We must therefore find the two vectors that are tangent to the u-curve and the v-curve at (x_k, y_k). We will show how the tangent vector to the u-curve is obtained; the tangent to the v-curve is obtained similarly. Suppose that (x_{k+1}, y_{k+1}) is the image of (u_{k+1}, v_{k+1}) as shown in Figure 5. The vector from (x_k, y_k) to (x_{k+1}, y_{k+1}) is then

$$\begin{aligned}(x_{k+1} - x_k)\mathbf{i} + (y_{k+1} - y_k)\mathbf{j} &= [x(u_{k+1}, v_k) - x(u_k, v_k)]\,\mathbf{i} \\ &\quad + [y(u_{k+1}, v_k) - y(u_k, v_k)]\,\mathbf{j} \\ &\approx \Delta u \frac{\partial x}{\partial u}(u_k, v_k)\,\mathbf{i} + \Delta u \frac{\partial y}{\partial u}(u_k, v_k)\,\mathbf{j} \\ &= \Delta u\left(\frac{\partial x}{\partial u}(u_k, v_k)\,\mathbf{i} + \frac{\partial y}{\partial u}(u_k, v_k)\,\mathbf{j}\right)\end{aligned}$$

The vector in parentheses, which we will call $\mathbf{t}_u$, is tangent to the u-curve through (x_k, y_k). Similarly, the vector

$$\mathbf{t}_v = \frac{\partial x}{\partial v}(u_k, v_k)\,\mathbf{i} + \frac{\partial y}{\partial v}(u_k, v_k)\,\mathbf{j}$$

is tangent to the v-curve through (x_k, y_k). The area ΔA_k of the region R_k is therefore

$$\begin{aligned}
\Delta A_k &\approx \|\Delta u\,\mathbf{t}_u \times \Delta v\,\mathbf{t}_v\| \\
&= \left\| \begin{vmatrix} \mathbf{i} & \mathbf{j} & \mathbf{k} \\ \Delta u \dfrac{\partial x}{\partial u}(u_k, v_k) & \Delta u \dfrac{\partial y}{\partial u}(u_k, v_k) & 0 \\ \Delta v \dfrac{\partial x}{\partial v}(u_k, v_k) & \Delta v \dfrac{\partial y}{\partial v}(u_k, v_k) & 0 \end{vmatrix} \right\| \\
&= \Delta u\,\Delta v \left\| \begin{vmatrix} \dfrac{\partial x}{\partial u} & \dfrac{\partial y}{\partial u} \\ \dfrac{\partial x}{\partial v} & \dfrac{\partial y}{\partial v} \end{vmatrix}_{(u_k, v_k)} \mathbf{k} \right\| \\
&= \Delta u\,\Delta v \left| \left[\frac{\partial x}{\partial u}\frac{\partial y}{\partial v} - \frac{\partial y}{\partial u}\frac{\partial x}{\partial v} \right]_{(u_k, v_k)} \right| \|\mathbf{k}\| \\
&= |J(u_k, v_k)|\,\Delta u\,\Delta v
\end{aligned}$$

Thus, we have

$$\begin{aligned}
\iint_R f(x, y)\,dx\,dy &\approx \sum_{k=1}^{n} f(x_k, y_k)\,\Delta A_k \\
&\approx \sum_{k=1}^{n} f(x(u_k, v_k), y(u_k, v_k))|J(u_k, v_k)|\,\Delta u\,\Delta v \\
&\approx \iint_S f(x(u, v), y(u, v))|J(u, v)|\,du\,dv
\end{aligned}$$

This completes the sketch of the proof. ■

EXAMPLE 3 Evaluate $\displaystyle\iint_R \cos(x - y)\sin(x + y)\,dA$, where R is the triangle with vertices $(0, 0)$, $(\pi, -\pi)$, and (π, π).

SOLUTION Let $u = x - y$ and $v = x + y$. Solving for x and y gives $x = \frac{1}{2}(u + v)$ and $y = \frac{1}{2}(v - u)$. The region R can be specified as

$$-x \le y \le x$$
$$0 \le x \le \pi$$

Substituting u and v gives

$$-\frac{1}{2}(u + v) \le \frac{1}{2}(v - u) \le \frac{1}{2}(u + v)$$
$$0 \le \frac{1}{2}(u + v) \le \pi$$

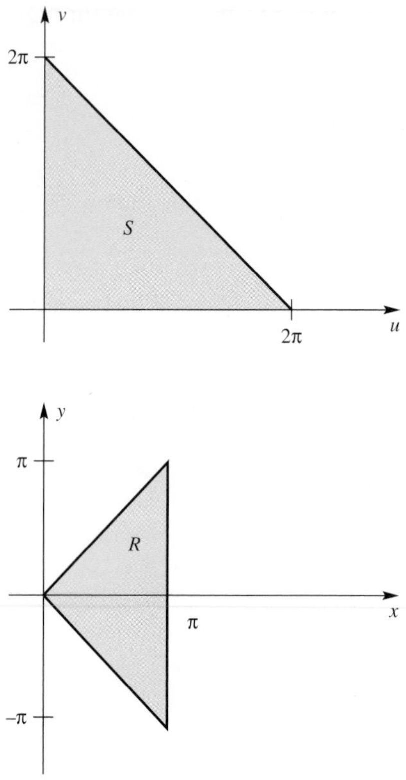

Figure 6

which reduces to

$$u \geq 0, \quad v \geq 0$$
$$0 \leq u + v \leq 2\pi$$

This is the region S in the uv-plane (see Figure 6). The Jacobian for this transformation is

$$J = \begin{vmatrix} \dfrac{\partial x}{\partial u} & \dfrac{\partial x}{\partial v} \\ \dfrac{\partial y}{\partial u} & \dfrac{\partial y}{\partial v} \end{vmatrix} = \begin{vmatrix} \dfrac{1}{2} & \dfrac{1}{2} \\ -\dfrac{1}{2} & \dfrac{1}{2} \end{vmatrix} = \frac{1}{2}$$

Thus

$$\begin{aligned}
\iint_R \cos(x - y)\sin(x + y)\,dA &= \iint_S \cos u \sin v \left|\frac{1}{2}\right| dv\,du \\
&= \frac{1}{2}\int_0^{2\pi}\int_0^{2\pi - u} \cos u \sin v\,dv\,du \\
&= \frac{1}{2}\int_0^{2\pi} \cos u\,(1 - \cos(2\pi - u))\,du \\
&= \frac{1}{2}\int_0^{2\pi} \cos u\,(1 - \cos u)\,du \\
&= \frac{1}{2}\int_0^{2\pi} (\cos u - \cos^2 u)\,du \\
&= \frac{1}{2}\int_0^{2\pi} \left(\cos u - \frac{1 + \cos 2u}{2}\right) du \\
&= \frac{1}{2}\int_0^{2\pi} \left(\cos u - \frac{1}{2} - \frac{1}{2}\cos 2u\right) du \\
&= \frac{1}{2}\left[\sin u - \frac{1}{2}u - \frac{1}{4}\sin 2u\right]_0^{2\pi} = -\frac{1}{2}\pi
\end{aligned}$$

■

The region of integration often suggests a transformation, as the next example illustrates.

EXAMPLE 4 Find the center of mass of the region R in the first quadrant bounded by

$$x^2 + y^2 = 9 \qquad y^2 - x^2 = 1$$
$$x^2 + y^2 = 16 \qquad y^2 - x^2 = 9$$

if the density is proportional to the square of the distance from the origin.

SOLUTION The mass is $\iint_R k(x^2 + y^2)\,dx\,dy$. Although the integrand is simple, this is a difficult integral to evaluate because the limits are complicated. However, the substitutions $u = x^2 + y^2$ and $v = y^2 - x^2$ transform the region R to the region S, which, in the uv-plane, is the rectangle

$$9 \leq u \leq 16 \quad \text{and} \quad 1 \leq v \leq 9$$

(See Figure 7.) Solving for x and y in the system $u = x^2 + y^2$ and $v = y^2 - x^2$ gives

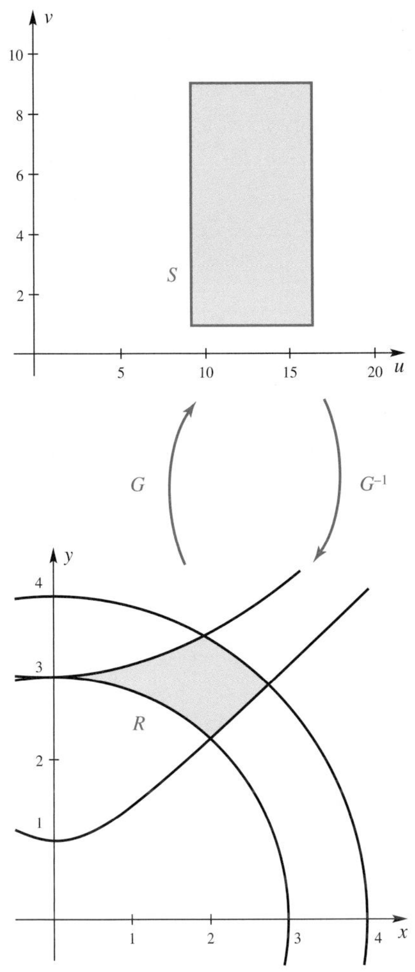

Figure 7

$$x = \sqrt{\frac{u-v}{2}} = \frac{1}{\sqrt{2}}(u-v)^{1/2}$$
$$y = \sqrt{\frac{u+v}{2}} = \frac{1}{\sqrt{2}}(u+v)^{1/2}$$

The Jacobian for this transformation is therefore

$$J(u,v) = \begin{vmatrix} \frac{\partial x}{\partial u} & \frac{\partial x}{\partial v} \\ \frac{\partial y}{\partial u} & \frac{\partial y}{\partial v} \end{vmatrix} = \begin{vmatrix} \frac{1}{2\sqrt{2}}(u-v)^{-1/2} & -\frac{1}{2\sqrt{2}}(u-v)^{-1/2} \\ \frac{1}{2\sqrt{2}}(u+v)^{-1/2} & \frac{1}{2\sqrt{2}}(u+v)^{-1/2} \end{vmatrix} = \frac{1}{4\sqrt{u^2-v^2}}$$

The mass is then

$$\begin{aligned} m &= \iint_R k(x^2+y^2)\,dx\,dy = k\iint_S u|J(u,v)|\,du\,dv \\ &= k\int_1^9\int_9^{16} \frac{u}{4\sqrt{u^2-v^2}}\,du\,dv \\ &= \frac{k}{4}\int_1^9 \left[\sqrt{u^2-v^2}\,\right]_{u=9}^{u=16} dv \\ &= \frac{k}{4}\int_1^9 \left(\sqrt{256-v^2}-\sqrt{81-v^2}\right) dv \\ &= \frac{k}{4}\left(\frac{45}{2}\sqrt{7} + 128\arcsin\frac{9}{16} - \frac{81}{4}\pi - \frac{1}{2}\sqrt{255}\right. \\ &\qquad \left. - 128\arcsin\frac{1}{16} + 2\sqrt{5} + \frac{81}{2}\arcsin\frac{1}{9}\right) \approx 16.343k \end{aligned}$$

The integral on the third last line could be evaluated using Formula 54 from the table of integrals at the back of the book, or with a CAS. The moments are

$$\begin{aligned} M_y &= \iint_R xk(x^2+y^2)\,dx\,dy = k\iint_S \sqrt{\frac{u-v}{2}}\,u|J(u,v)|\,du\,dv \\ &= \frac{k}{4\sqrt{2}}\int_9^{16}\int_1^9 \frac{u\sqrt{u-v}}{\sqrt{u^2-v^2}}\,dv\,du \\ &= \frac{k}{4\sqrt{2}}\int_9^{16}\int_1^9 \frac{u}{\sqrt{u+v}}\,dv\,du \\ &= \frac{k}{2\sqrt{2}}\int_9^{16}\left(u\sqrt{u+9}-u\sqrt{u+1}\right) du \\ &= \frac{k\sqrt{2}}{4}\left(500 - \frac{1564}{15}\sqrt{17} - \frac{324}{5}\sqrt{2} + \frac{100}{3}\sqrt{10}\right) \approx 29.651k \end{aligned}$$

and

$$\begin{aligned} M_x &= \iint_R yk(x^2+y^2)\,dx\,dy = k\iint_S \sqrt{\frac{u+v}{2}}\,u|J(u,v)|\,du\,dv \\ &= \frac{k}{4\sqrt{2}}\int_9^{16}\int_1^9 \frac{u\sqrt{u+v}}{\sqrt{u^2-v^2}}\,dv\,du \\ &= \frac{k}{4\sqrt{2}}\int_9^{16}\int_1^9 \frac{u}{\sqrt{u-v}}\,dv\,du \\ &= \frac{k}{2\sqrt{2}}\int_9^{16}\left(u\sqrt{u-1}-u\sqrt{u-9}\right) du \\ &= \frac{k\sqrt{2}}{4}\left(100\sqrt{15} - \frac{308}{5}\sqrt{7} - \frac{928}{15}\sqrt{2}\right) \approx 48.376k \end{aligned}$$

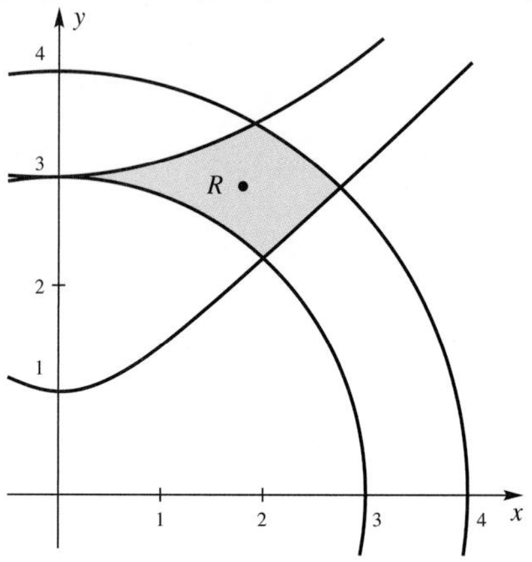

Figure 8

The integrals on the third last lines can be evaluated using Formula 96 from the table of integrals, or by using a CAS. The coordinates of the center of mass are thus

$$\bar{x} = \frac{M_y}{m} \approx \frac{29.651k}{16.343k} \approx 1.814$$

$$\bar{y} = \frac{M_x}{m} \approx \frac{48.376k}{16.343k} \approx 2.960$$

The point $(\bar{x}, \bar{y}) = (1.814, 2.960)$ is shown in Figure 8. ■

The Change of Variable Formula for Triple Integrals Theorem A generalizes to triple (and even higher-dimensional) integrals. If G is a one-to-one transformation from $\mathbb{R}^3$ to $\mathbb{R}^3$ which maps the bounded region S in uvw-space onto the bounded region R in the xyz-plane, and if G is of the form $G(u, v) = (x(u, v, w), y(u, v, w), z(u, v, w))$, then

$$\iiint_R f(x, y, z)\, dx\, dy\, dz = \iiint_S f(x(u, v, w), y(u, v, w), z(u, v, w)) \times |J(u, v, w)|\, du\, dv\, dw$$

where $J(u, v, w)$ is the determinant

$$J(u, v, w) = \begin{vmatrix} \frac{\partial x}{\partial u} & \frac{\partial x}{\partial v} & \frac{\partial x}{\partial w} \\ \frac{\partial y}{\partial u} & \frac{\partial y}{\partial v} & \frac{\partial y}{\partial w} \\ \frac{\partial z}{\partial u} & \frac{\partial z}{\partial v} & \frac{\partial z}{\partial w} \end{vmatrix}.$$

EXAMPLE 5 Derive the change of variable formula $dx\, dy\, dz = r\, dr\, d\theta\, dz$ for the transformation to cylindrical coordinates.

SOLUTION Since the change of variables is $x = r\cos\theta$, $y = r\sin\theta$ and $z = z$, the Jacobian is

$$J(r, \theta, z) = \begin{vmatrix} \frac{\partial x}{\partial r} & \frac{\partial x}{\partial \theta} & \frac{\partial x}{\partial z} \\ \frac{\partial y}{\partial r} & \frac{\partial y}{\partial \theta} & \frac{\partial y}{\partial z} \\ \frac{\partial z}{\partial r} & \frac{\partial z}{\partial \theta} & \frac{\partial z}{\partial z} \end{vmatrix} = \begin{vmatrix} \cos\theta & -r\sin\theta & 0 \\ \sin\theta & r\cos\theta & 0 \\ 0 & 0 & 1 \end{vmatrix}$$

$$= 0\begin{vmatrix} -r\sin\theta & 0 \\ r\cos\theta & 0 \end{vmatrix} - 0\begin{vmatrix} \cos\theta & 0 \\ \sin\theta & 0 \end{vmatrix} + 1\begin{vmatrix} \cos\theta & -r\sin\theta \\ \sin\theta & r\cos\theta \end{vmatrix}$$

$$= r\cos^2\theta + r\sin^2\theta = r$$

Thus,

$$dx\, dy\, dz = |J(r, \theta, z)|\, dr\, d\theta\, dz = r\, dr\, d\theta\, dz$$ ■

We leave it as an exercise (Problem 21) to derive the relationship $dx\, dy\, dz = \rho^2 \sin\phi\, d\rho\, d\theta\, d\phi$ for spherical coordinates.

Concepts Review

1. Under a transformation from the uv-plane to the xy-plane, the image of a vertical line is called a _____ and the image of a horizontal line is called a _____.

2. A change of variable for a double integral must take into account _____, _____, and _____.

3. The determinant $\begin{vmatrix} \frac{\partial x}{\partial u} & \frac{\partial x}{\partial v} \\ \frac{\partial y}{\partial u} & \frac{\partial y}{\partial v} \end{vmatrix}$ is called the _____.

4. The formula for a change of variable in a double integral is $\iint_R f(x, y)\, dx\, dy = \iint_S f(x(u, v), y(u, v))$ _____ $du\, dv$.

Problem Set 13.9

1. For the transformation $x = u + v$, $y = v - u$, sketch the u-curves and v-curves for the grid $\{(u, v): (u = 2, 3, 4, 5 \text{ and } 1 \le v \le 3) \text{ or } (v = 1, 2, 3 \text{ and } 2 \le u \le 5)\}$.

2. For the transformation $x = 2u + v$, $y = v - u$, sketch the u-curves and v-curves for the grid $\{(u, v): (u = 2, 3, 4, 5 \text{ and } 1 \le v \le 3) \text{ or } (v = 1, 2, 3 \text{ and } 2 \le u \le 5)\}$.

3. For the transformation $x = u \sin v$, $y = u \cos v$, sketch the u-curves and v-curves for the grid $\{(u, v): (u = 0, 1, 2, 3 \text{ and } 0 \le v \le \pi) \text{ or } (v = 0, \pi/2, \pi \text{ and } 0 \le u \le 3)\}$.

4. For the transformation $x = u \cos v$, $y = u \sin v$, sketch the u-curves and v-curves for the grid $\{(u, v): (u = 0, 1, 2, 3 \text{ and } 0 \le v \le 2\pi) \text{ or } (v = 0, \pi, 2\pi \text{ and } 0 \le u \le 3)\}$.

5. For the transformation $x = u/(u^2 + v^2)$, $y = -v/(u^2 + v^2)$, sketch the u-curves and v-curves for the grid $\{(u, v): (u = 0, 1, 2, 3 \text{ and } 1 \le v \le 3) \text{ or } (v = 1, 2, 3 \text{ and } 0 \le u \le 3)\}$.

6. For the transformation $x = u + u/(u^2 + v^2)$, $y = v - v/(u^2 + v^2)$, sketch the u-curves and v-curves for the grid $\{(u, v): (u = -2, -1, 0, 1, 2 \text{ and } 1 \le v \le 3) \text{ or } (v = 1, 2, 3 \text{ and } -2 \le u \le 2)\}$.

In Problems 7–10, find the image of the rectangle with the given corners and find the Jacobian of the transformation.

7. $x = u + 2v$, $y = u - 2v$; $(0, 0), (2, 0), (2, 1), (0, 1)$

8. $x = 2u + 3v$, $y = u - v$; $(0, 0), (3, 0), (3, 1), (0, 1)$

9. $x = u^2 + v^2$, $y = v$; $(0, 0), (1, 0), (1, 1), (0, 1)$

10. $x = u$, $y = u^2 - v^2$; $(0, 0), (3, 0), (3, 1), (0, 1)$

In Problems 11–16, find the transformation from the uv-plane to the xy-plane and find the Jacobian. Assume that $x \ge 0$ and $y \ge 0$.

11. $u = x + 2y$, $v = x - 2y$

12. $u = 2x - 3y$, $v = 3x - 2y$

13. $u = x^2 + y^2$, $v = x$

14. $u = x^2 - y^2$, $v = x + y$

15. $u = xy$, $v = x$

16. $u = x^2$, $v = xy$

In Problems 17–20, use a transformation to evaluate the given double integral over the region R which is the triangle with vertices (1, 0), (4, 0), and (4, 3).

17. $\displaystyle\iint_R \ln \frac{x + y}{x - y}\, dA$

18. $\displaystyle\iint_R \sqrt{\frac{x + y}{x - y}}\, dA$

19. $\displaystyle\iint_R \sin(\pi(2x - y)) \cos(\pi(y - 2x))\, dA$

20. $\displaystyle\iint_R (2x - y) \cos(y - 2x)\, dA$

21. Find the Jacobian for the transformation from rectangular coordinates to spherical coordinates.

22. Find the volume of the ellipsoid $x^2/a^2 + y^2/b^2 + z^2/c^2 = 1$ by making the change of variables $x = ua$, $y = vb$, and $z = cw$. Also, find the moment of inertia of this solid about the z-axis assuming that it has constant density k.

23. Suppose X and Y are continuous random variables with joint PDF $f(x, y)$ and suppose U and V are random variables that are functions of X and Y such that the transformation

$$X = x(U, V) \quad \text{and} \quad Y = y(U, V)$$

is one-to-one. Show that the joint PDF of U and V is

$$g(u, v) = f(x(u, v), y(u, v))|J(u, v)|$$

Hint: Let R be a region in the xy-plane and let S be its preimage. Show that $P((X, Y) \in R) = P((U, V) \in S)$ and get a double integral for each of these.

24. Suppose that the random variables X and Y have joint PDF

$$f(x, y) = \begin{cases} \frac{1}{4}, & \text{if } 0 \le x \le 2, 0 \le y \le 2 \\ 0, & \text{otherwise} \end{cases}$$

that is, X and Y are uniformly distributed over the square $0 \le x \le 2$, $0 \le y \le 2$. Find

(a) the joint PDF of $U = X + Y$ and $V = X - Y$, and

(b) the marginal PDF of U.

25. Suppose X and Y have joint PDF

$$f(x, y) = \begin{cases} e^{-x-y}, & \text{if } x \ge 0, y \ge 0 \\ 0, & \text{otherwise} \end{cases}$$

Find

(a) the joint PDF of $U = X + Y$ and $V = X$

(b) the marginal PDF of U.

Answers to Concepts Review: **1.** u-curve; v-curve **2.** the integrand; the differential $dx\, dy$; the region of integration **3.** Jacobian **4.** $|J(u, v)|$

13.10 Chapter Review

Concepts Test

Respond with true or false to each of the following assertions. Be prepared to defend your answer.

1. $\int_a^b \int_a^b f(x)f(y)\,dy\,dx = \left[\int_a^b f(x)\,dx\right]^2$

2. $\int_0^1 \int_0^x f(x, y)\,dy\,dx = \int_0^1 \int_0^y f(x, y)\,dx\,dy$

3. $\int_0^2 \int_{-1}^1 \sin(x^3y^3)\,dx\,dy = 0$

4. $\int_{-1}^1 \int_{-1}^1 e^{x^2+2y^2}\,dy\,dx = 4\int_0^1 \int_0^1 e^{x^2+2y^2}\,dy\,dx$

5. $\int_1^2 \int_0^2 \sin^2(x/y)\,dx\,dy \le 2$

6. If f is continuous and nonnegative on R and $f(x_0, y_0) > 0$, where (x_0, y_0) is an interior point of R, then $\iint_R f(x, y)\,dA > 0$.

7. If $\iint_R f(x, y)\,dA \le \iint_R g(x, y)\,dA$, then $f(x, y) \le g(x, y)$ on R.

8. If $f(x, y) \ge 0$ on R and $\iint_R f(x, y)\,dA = 0$, then $f(x, y) = 0$ for all (x, y) in R.

9. If $\delta(x, y) = k$ gives the density of a lamina at (x, y), the coordinates of the center of mass of the lamina do not involve k.

10. If $\delta(x, y) = y^2/(1 + x^2)$ gives the density of the lamina $\{(x, y): 0 \le x \le 1, 0 \le y \le 1\}$, we know without calculating that $\bar{x} < \frac{1}{2}$ and $\bar{y} > \frac{1}{2}$.

11. If $S = \{(x, y, z): 1 \le x^2 + y^2 + z^2 \le 16\}$, then

$$\iiint_S dV = 84\pi$$

12. If the top of a right circular cylinder of radius 1 is sliced off by a plane that makes an angle of 30° with the base of the cylinder, the area of the resulting slanted top is $2\sqrt{3}\pi/3$.

13. There are eight possible orders of integration for a triple iterated integral.

14. $\int_0^2 \int_0^{2\pi} \int_0^1 dr\,d\theta\,dz$ represents the volume of a right circular cylinder of radius 1 and height 2.

15. If $|f_x| \le 2$ and $|f_y| \le 2$, then the surface G determined by $z = f(x, y)$, $0 \le x \le 1$, $0 \le y \le 1$, has area at most 3.

16. For the transformation from Cartesian to polar coordinates, the Jacobian is $J(r, \theta) = r$.

17. For the transformation, $x = 2u$, $y = 2v$, the Jacobian is $J(u, v) = 2$.

Sample Test Problems

In Problems 1–4, evaluate each integral.

1. $\int_0^1 \int_x^{\sqrt{x}} xy\,dy\,dx$

2. $\int_{-2}^2 \int_{-\sqrt{4-y^2}}^{\sqrt{4-y^2}} 2xy^2\,dx\,dy$

3. $\int_0^{\pi/2} \int_0^{2\sin\theta} r\cos\theta\,dr\,d\theta$

4. $\int_1^2 \int_3^x \int_0^{\sqrt{3}y} \frac{y}{y^2 + z^2}\,dz\,dy\,dx$

In Problems 5–8, rewrite the iterated integral with the indicated order of integration. Make a sketch first.

5. $\int_0^1 \int_x^1 f(x, y)\,dy\,dx$; $dx\,dy$

6. $\int_0^1 \int_0^{\cos^{-1} y} f(x, y)\,dx\,dy$; $dy\,dx$

7. $\int_0^1 \int_0^{(1-x)/2} \int_0^{1-x-2y} f(x, y, z)\,dz\,dy\,dx$; $dx\,dz\,dy$

8. $\int_0^2 \int_{x^2}^4 \int_0^{4-y} f(x, y, z)\,dz\,dy\,dx$; $dx\,dy\,dz$

9. Write the triple iterated integrals for the volume of a sphere of radius a in each case.

(a) Cartesian coordinates

(b) Cylindrical coordinates

(c) Spherical coordinates

10. Evaluate $\iint_S (x + y)\,dA$, where S is the region bounded by $y = \sin x$ and $y = 0$ between $x = 0$ and $x = \pi$.

11. Evaluate $\iiint_S z^2\,dV$, where S is the region bounded by $x^2 + z = 1$ and $y^2 + z = 1$ and the xy-plane.

12. Evaluate $\iint_S \frac{1}{x^2 + y^2}\,dA$, where S is the region between the circles $x^2 + y^2 = 4$ and $x^2 + y^2 = 9$.

13. Find the center of mass of the rectangular lamina bounded by $x = 1$, $x = 3$, $y = 0$, and $y = 2$ if the density is $\delta(x, y) = xy^2$.

14. Find the moment of inertia of the lamina of Problem 13 with respect to the x-axis.

15. Find the area of the surface of the cylinder $z^2 + y^2 = 9$ lying in the first octant between the planes $y = x$ and $y = 3x$.

16. Evaluate by changing to cylindrical or spherical coordinates.

(a) $\int_0^3 \int_0^{\sqrt{9-x^2}} \int_0^2 \sqrt{x^2 + y^2}\,dz\,dy\,dx$

(b) $\int_0^2 \int_0^{\sqrt{4-x^2}} \int_0^{\sqrt{4-x^2-y^2}} z\sqrt{4 - x^2 - y^2}\, dz\, dy\, dx$

17. Find the mass of the solid between the spheres $x^2 + y^2 + z^2 = 1$ and $x^2 + y^2 + z^2 = 9$ if the density is proportional to the distance from the origin.

18. Find the center of mass of the homogeneous lamina bounded by the cardioid $r = 4(1 + \sin\theta)$.

19. Find the mass of the solid in the first octant under the plane $x/a + y/b + z/c = 1$ (a, b, c positive) if the density is $\delta(x, y, z) = kx$.

20. Compute the volume of the solid bounded by $z = x^2 + y^2$, $z = 0$, and $x^2 + (y - 1)^2 = 1$.

21. Use a transformation to evaluate the integral

$$\iint_R \sin(x - y)\cos(x + y)\, dA$$

where R is the rectangle with vertices $(0, 0)$, $(\pi/2, -\pi/2)$, $(\pi, 0)$, and $(\pi/2, \pi/2)$.

REVIEW & PREVIEW PROBLEMS

In Problems 1–9, find parametric equations for the given curve. (Be sure to give the domain for the parameter t.)

1. The circle centered at the origin having radius 3

2. The circle centered at $(2, 1)$ having radius 1

3. The semicircle $x^2 + y^2 = 4$ with $y > 0$

4. The semicircle $x^2 + y^2 = a^2$ with $y \leq 0$ having a clockwise orientation

5. That part of the line $y = 2$ between the points $(-2, 2)$ and $(3, 2)$

6. That part of the line $y = 9 - x$ that is in the first quadrant with an orientation that is down and to the right

7. That part of the line $y = 9 - x$ that is in the first quadrant with an orientation that is up and to the left

8. That part of the parabola $y = 9 - x^2$ that is above the x-axis having an orientation that is left to right

9. That part of the parabola $y = 9 - x^2$ that is above the x-axis having an orientation that is right to left

10. Use the arc length formula to find the length of the curve in Problem 6.

In Problems 11–16, find the gradient of the given function.

11. $f(x, y) = x \sin x + y \cos y$

12. $f(x, y) = xe^{-xy} + ye^{xy}$

13. $f(x, y, z) = x^2 + y^2 + z^2$

14. $f(x, y, z) = \dfrac{1}{x^2 + y^2 + z^2}$

15. $f(x, y, z) = xy + xz + yz$

16. $f(x, y, z) = \dfrac{1}{\sqrt{x^2 + y^2 + z^2}}$

Evaluate the integrals in Problems 17–22.

17. $\displaystyle\int_0^{\pi} \sin^2 t \, dt$

18. $\displaystyle\int_0^{\pi} \sin t \cos t \, dt$

19. $\displaystyle\int_0^1 \int_1^2 xy \, dy \, dx$

20. $\displaystyle\int_{-1}^1 \int_1^4 (x^2 + 2y) \, dy \, dx$

21. $\displaystyle\int_0^{2\pi} \int_1^2 r^2 \, dr \, d\theta$

22. $\displaystyle\int_0^{2\pi} \int_0^{\pi} \int_1^2 \rho^2 \sin \phi \, d\rho \, d\phi \, d\theta$

23. The integral in Problem 22 represents the volume of some region in three-space. What is this region?

24. Find the surface area of that part of the paraboloid $z = 144 - x^2 - y^2$ that lies above the plane $z = 36$.

25. Find a unit normal vector to the graph of $x^2 + y^2 + z^2 = 169$ at the point $(3, 4, 12)$.

CHAPTER 14 Vector Calculus

14.1 Vector Fields
14.2 Line Integrals
14.3 Independence of Path
14.4 Green's Theorem in the Plane
14.5 Surface Integrals
14.6 Gauss's Divergence Theorem
14.7 Stokes's Theorem

14.1 Vector Fields

The concept of a function has played a central role in calculus. This concept, and the associated calculus, has been steadily generalized. Most of the first two-thirds of this book deals with functions where the input is a real number and the output is a real number. In Chapter 11 we introduced vector-valued functions, that is, functions whose input is a real number and whose output is a vector. Then in Chapter 12 we introduced real-valued functions of several variables, that is, functions whose input is a pair or triple (or n-tuple) of real numbers and whose output is a real number. The natural next step is to study functions whose input is a vector and whose output is a vector. This is the final step in the usual calculus sequence.

Consider then a function $\mathbf{F}$ that associates with each point $\mathbf{p}$ in n-space a vector $\mathbf{F}(\mathbf{p})$. A typical example in two-space is

$$\mathbf{F}(\mathbf{p}) = \mathbf{F}(x, y) = -\tfrac{1}{2}y\mathbf{i} + \tfrac{1}{2}x\mathbf{j}$$

For historical reasons, we refer to such a function as a **vector field,** a name arising from a visual image that we now describe. Imagine that to each point $\mathbf{p}$ in a region of space is attached a vector $\mathbf{F}(\mathbf{p})$ emanating from $\mathbf{p}$. We cannot draw all these vectors, but a representative sample can give us a good intuitive picture of a field. Figure 1 is just such a picture for the vector field $\mathbf{F}(x, y) = -\tfrac{1}{2}y\mathbf{i} + \tfrac{1}{2}x\mathbf{j}$ mentioned earlier. It is the velocity field of a wheel spinning at a constant rate of $\tfrac{1}{2}$ radian per unit of time (see Example 2). Figure 2 might represent the velocity field for water flowing in a curved pipe.

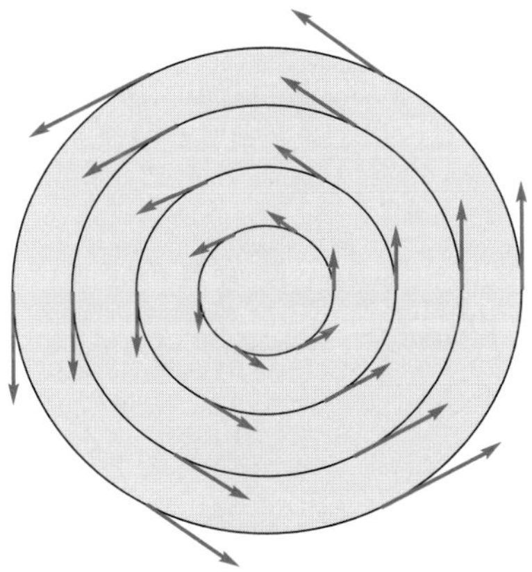

Figure 1

Other vector fields that arise naturally in science are electric fields, magnetic fields, force fields, and gravitational fields. We consider only the case in which these fields are independent of time, which we call **steady vector fields.** In contrast to a vector field, a function F that attaches a number to each point in space is called a **scalar field.** The function that gives the temperature at each point would be a good physical example of a scalar field.

EXAMPLE 1 Sketch a representative sample of vectors from the vector field

$$\mathbf{F}(x, y) = \frac{x\mathbf{i} + y\mathbf{j}}{\sqrt{x^2 + y^2}}$$

SOLUTION $\mathbf{F}(x, y)$ is a unit vector pointing in the same direction as $x\mathbf{i} + y\mathbf{j}$, that is, away from the origin. Several such vectors are shown in Figure 3. ■

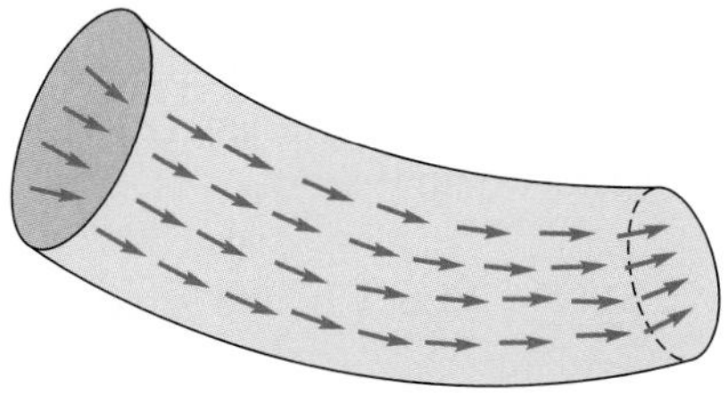

Figure 2

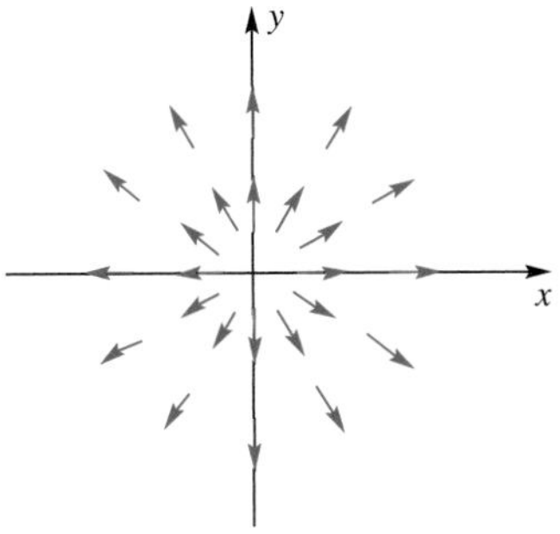

Figure 3

EXAMPLE 2 Sketch a representative sample of vectors from the vector field

$$\mathbf{F}(x, y) = -\tfrac{1}{2}y\mathbf{i} + \tfrac{1}{2}x\mathbf{j}$$

and show that each vector is tangent to a circle centered at the origin and has length equal to one-half the radius of that circle (see Figure 1).

SOLUTION Figure 4 shows a plot of the vector field. If $\mathbf{r} = x\mathbf{i} + y\mathbf{j}$ is the position vector of the point (x, y), then

$$\mathbf{r} \cdot \mathbf{F}(x, y) = -\tfrac{1}{2}xy + \tfrac{1}{2}xy = 0$$

Thus, $\mathbf{F}(x, y)$ is perpendicular to $\mathbf{r}$ and is therefore tangent to the circle of radius $\|\mathbf{r}\|$. Finally,

$$\|\mathbf{F}(x, y)\| = \sqrt{\left(-\tfrac{1}{2}y\right)^2 + \left(\tfrac{1}{2}x\right)^2} = \tfrac{1}{2}\|\mathbf{r}\|$$ ■

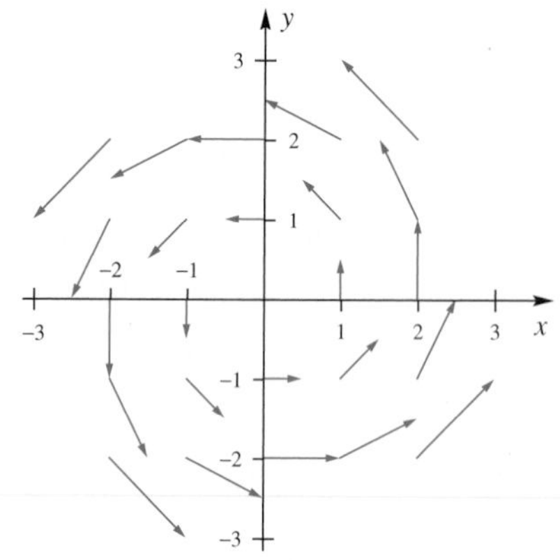

Figure 4

According to Isaac Newton, the magnitude of the force of attraction between objects of mass M and m, respectively, is given by GMm/d^2, where d is the distance between the objects and G is a universal constant. This is the famous Inverse Square Law of Gravitational Attraction. It supplies us with an important example of a vector field. Since the vectors represent forces, we will call such a field a **force field.**

EXAMPLE 3 Suppose that a spherical object of mass M (e.g., the earth) is centered at the origin. Derive the formula for the gravitational field of force $\mathbf{F}(x, y, z)$ exerted by this mass on an object of mass m located at a point (x, y, z) in space. Then sketch this field.

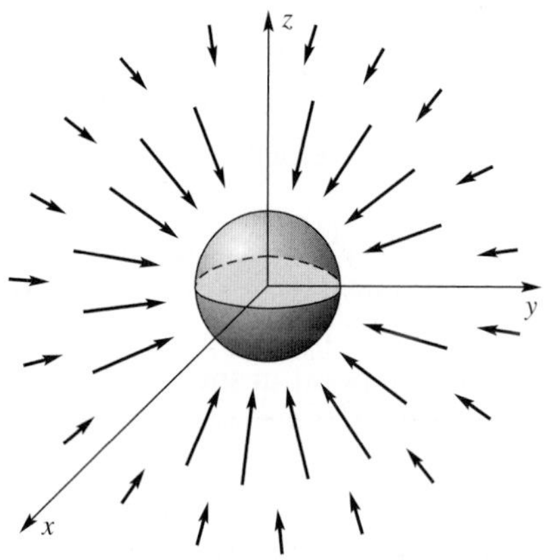

Figure 5

SOLUTION We assume that we may treat the object of mass M as a point mass located at the origin. Let $\mathbf{r} = x\mathbf{i} + y\mathbf{j} + z\mathbf{k}$. Then $\mathbf{F}$ has magnitude

$$\|\mathbf{F}\| = \frac{GMm}{\|\mathbf{r}\|^2}$$

The direction of $\mathbf{F}$ is toward the origin; that is, $\mathbf{F}$ has the direction of the unit vector $-\mathbf{r}/\|\mathbf{r}\|$. We conclude that

$$\mathbf{F}(x, y, z) = \frac{GMm}{\|\mathbf{r}\|^2}\left(\frac{-\mathbf{r}}{\|\mathbf{r}\|}\right) = -GMm\frac{\mathbf{r}}{\|\mathbf{r}\|^3}$$

This field is sketched in Figure 5. ■

The Gradient of a Scalar Field Let $f(x, y, z)$ determine a scalar field and suppose that f is differentiable. Then the gradient of f, denoted by ∇f, is the vector field given by

$$\mathbf{F}(x, y, z) = \nabla f(x, y, z) = \frac{\partial f}{\partial x}\mathbf{i} + \frac{\partial f}{\partial y}\mathbf{j} + \frac{\partial f}{\partial z}\mathbf{k}$$

We first met **gradient fields** in Sections 12.4 and 12.5. There we learned that $\nabla f(x, y, z)$ points in the direction of greatest increase of $f(x, y, z)$. A vector field $\mathbf{F}$ that is the gradient of a scalar field f is called a **conservative vector field,** and f is its **potential function** (the origin of these names will be clarified in Section 14.3). Such fields and their potential functions are important in physics. In particular, fields that obey the inverse square law (e.g., electric fields and gravitational fields) are conservative, as we now show.

EXAMPLE 4 Let $\mathbf{F}$ be the force resulting from an inverse square law; that is, let

$$\mathbf{F}(x, y, z) = -c\frac{\mathbf{r}}{\|\mathbf{r}\|^3} = -c\frac{x\mathbf{i} + y\mathbf{j} + z\mathbf{k}}{(x^2 + y^2 + z^2)^{3/2}}$$

where c is a constant (see Example 3). Show that

$$f(x, y, z) = \frac{c}{\sqrt{x^2 + y^2 + z^2}} = c(x^2 + y^2 + z^2)^{-1/2}$$

is a potential function for $\mathbf{F}$ and therefore that $\mathbf{F}$ is conservative (for $\mathbf{r} \neq \mathbf{0}$).

SOLUTION

$$\begin{aligned}\nabla f(x, y, z) &= \frac{\partial f}{\partial x}\mathbf{i} + \frac{\partial f}{\partial y}\mathbf{j} + \frac{\partial f}{\partial z}\mathbf{k} \\ &= -\frac{c}{2}(x^2 + y^2 + z^2)^{-3/2}(2x\mathbf{i} + 2y\mathbf{j} + 2z\mathbf{k}) \\ &= \mathbf{F}(x, y, z)\end{aligned}$$

■

Example 4 was really too easy since we gave the function f. A much harder and more significant problem is this. Given a vector field $\mathbf{F}$, decide whether it is conservative, and, if so, find its potential function. We discuss this problem in Section 14.3.

The Divergence and Curl of a Vector Field

With a given vector field

$$\mathbf{F}(x, y, z) = M(x, y, z)\,\mathbf{i} + N(x, y, z)\,\mathbf{j} + P(x, y, z)\,\mathbf{k}$$

are associated two other important fields. The first, called the **divergence** of $\mathbf{F}$, is a scalar field; the second, called the **curl** of $\mathbf{F}$, is a vector field.

Definition div and curl

Let $\mathbf{F} = M\mathbf{i} + N\mathbf{j} + P\mathbf{k}$ be a vector field for which the first partial derivatives of M, N, and P exist. Then

$$\operatorname{div} \mathbf{F} = \frac{\partial M}{\partial x} + \frac{\partial N}{\partial y} + \frac{\partial P}{\partial z}$$

$$\operatorname{curl} \mathbf{F} = \left(\frac{\partial P}{\partial y} - \frac{\partial N}{\partial z}\right)\mathbf{i} + \left(\frac{\partial M}{\partial z} - \frac{\partial P}{\partial x}\right)\mathbf{j} + \left(\frac{\partial N}{\partial x} - \frac{\partial M}{\partial y}\right)\mathbf{k}$$

At this point it is hard to see the significance of these fields; this will be apparent later. Our interest now is in learning to calculate divergence and curl easily and in relating them to the gradient operator ∇. Recall that ∇ is the operator

$$\nabla = \frac{\partial}{\partial x}\mathbf{i} + \frac{\partial}{\partial y}\mathbf{j} + \frac{\partial}{\partial z}\mathbf{k}$$

When ∇ operates on a function f, it produces the gradient ∇f, which we will also write as grad f. By a slight (but very helpful) abuse of notation, we can write

$$\begin{aligned}\nabla \cdot \mathbf{F} &= \left(\frac{\partial}{\partial x}\mathbf{i} + \frac{\partial}{\partial y}\mathbf{j} + \frac{\partial}{\partial z}\mathbf{k}\right) \cdot (M\mathbf{i} + N\mathbf{j} + P\mathbf{k}) \\ &= \frac{\partial M}{\partial x} + \frac{\partial N}{\partial y} + \frac{\partial P}{\partial z} = \operatorname{div} \mathbf{F}\end{aligned}$$

$$\begin{aligned}\nabla \times \mathbf{F} &= \begin{vmatrix} \mathbf{i} & \mathbf{j} & \mathbf{k} \\ \dfrac{\partial}{\partial x} & \dfrac{\partial}{\partial y} & \dfrac{\partial}{\partial z} \\ M & N & P \end{vmatrix} \\ &= \left(\frac{\partial P}{\partial y} - \frac{\partial N}{\partial z}\right)\mathbf{i} - \left(\frac{\partial P}{\partial x} - \frac{\partial M}{\partial z}\right)\mathbf{j} + \left(\frac{\partial N}{\partial x} - \frac{\partial M}{\partial y}\right)\mathbf{k} = \operatorname{curl} \mathbf{F}\end{aligned}$$

Thus, grad f, div $\mathbf{F}$, and curl $\mathbf{F}$ can all be written in terms of the operator ∇; this is the way to remember how these fields are defined.

What Do They Mean?

To help you visualize the divergence and curl, we offer this physical interpretation. If $\mathbf{F}$ denotes the velocity field for a fluid, then div $\mathbf{F}$ at a point $\mathbf{p}$ measures the tendency of that fluid to diverge away from $\mathbf{p}$ (div $\mathbf{F} > 0$) or accumulate toward $\mathbf{p}$ (div $\mathbf{F} < 0$). On the other hand, curl $\mathbf{F}$ picks out the direction of the axis about which the fluid rotates (curls) most rapidly, and $\|\text{curl } \mathbf{F}\|$ is a measure of the speed of this rotation. The direction of rotation is according to the right-hand rule. We will expand on this discussion later in the chapter.

EXAMPLE 5 Let

$$\mathbf{F}(x, y, z) = x^2yz\,\mathbf{i} + 3xyz^3\,\mathbf{j} + (x^2 - z^2)\,\mathbf{k}$$

Find div **F** and curl **F**.

SOLUTION

$$\text{div}\,\mathbf{F} = \nabla\cdot\mathbf{F} = 2xyz + 3xz^3 - 2z$$

$$\text{curl}\,\mathbf{F} = \nabla\times\mathbf{F} = \begin{vmatrix} \mathbf{i} & \mathbf{j} & \mathbf{k} \\ \dfrac{\partial}{\partial x} & \dfrac{\partial}{\partial y} & \dfrac{\partial}{\partial z} \\ x^2yz & 3xyz^3 & x^2 - z^2 \end{vmatrix}$$

$$= -(9xyz^2)\mathbf{i} - (2x - x^2y)\mathbf{j} + (3yz^3 - x^2z)\mathbf{k}$$

■

Concepts Review

1. A function that associates with each point (x, y, z) in space a vector $\mathbf{F}(x, y, z)$ is called a _____.

2. In particular, the function that associates with the scalar function $f(x, y, z)$ the vector $\nabla f(x, y, z)$ is called a _____.

3. Two important examples of vector fields in physics that arise as gradients of scalar fields are _____ and _____.

4. Given a vector field $\mathbf{F} = M\mathbf{i} + N\mathbf{j} + P\mathbf{k}$, we introduce a corresponding scalar field div **F** and a vector field curl **F**. They can be defined symbolically by div **F** = _____ and curl **F** = _____.

Problem Set 14.1

In Problems 1–6, sketch a sample of vectors for the given vector field **F**.

1. $\mathbf{F}(x, y) = x\mathbf{i} + y\mathbf{j}$
2. $\mathbf{F}(x, y) = x\mathbf{i} - y\mathbf{j}$
3. $\mathbf{F}(x, y) = -x\mathbf{i} + 2y\mathbf{j}$
4. $\mathbf{F}(x, y) = 3x\mathbf{i} + y\mathbf{j}$
5. $\mathbf{F}(x, y, z) = x\mathbf{i} + 0\mathbf{j} + \mathbf{k}$
6. $\mathbf{F}(x, y, z) = -z\mathbf{k}$

In Problems 7–12, find ∇f.

7. $f(x, y, z) = x^2 - 3xy + 2z$
8. $f(x, y, z) = \sin(xyz)$
9. $f(x, y, z) = \ln|xyz|$
10. $f(x, y, z) = \frac{1}{2}(x^2 + y^2 + z^2)$
11. $f(x, y, z) = xe^y \cos z$
12. $f(x, y, z) = y^2e^{-2z}$

In Problems 13–18, find div **F** *and curl* **F**.

13. $\mathbf{F}(x, y, z) = x^2\mathbf{i} - 2xy\mathbf{j} + yz^2\mathbf{k}$
14. $\mathbf{F}(x, y, z) = x^2\mathbf{i} + y^2\mathbf{j} + z^2\mathbf{k}$
15. $\mathbf{F}(x, y, z) = yz\mathbf{i} + xz\mathbf{j} + xy\mathbf{k}$
16. $\mathbf{F}(x, y, z) = \cos x\,\mathbf{i} + \sin y\,\mathbf{j} + 3\mathbf{k}$
17. $\mathbf{F}(x, y, z) = e^x \cos y\,\mathbf{i} + e^x \sin y\,\mathbf{j} + z\mathbf{k}$
18. $\mathbf{F}(x, y, z) = (y + z)\mathbf{i} + (x + z)\mathbf{j} + (x + y)\mathbf{k}$

19. Let f be a scalar field and **F** a vector field. Indicate which of the following are scalar fields, vector fields, or meaningless.

(a) div f (b) grad f
(c) curl **F** (d) div (grad f)
(e) curl (grad f) (f) grad (div **F**)
(g) curl (curl **F**) (h) div (div **F**)
(i) grad (grad f) (j) div (curl (grad f))
(k) curl (div(grad f))

20. Assuming that the required partial derivatives exist and are continuous, show that

(a) div(curl **F**) = 0; (b) curl(grad f) = **0**;
(c) div(f**F**) = (f)(div **F**) + (grad f)·**F**;
(d) curl (f**F**) = (f)(curl **F**) + (grad f) × **F**.

21. Let $\mathbf{F}(x, y, z) = c\mathbf{r}/\|\mathbf{r}\|^3$ be an inverse square law field (see Examples 3 and 4). Show that curl **F** = **0** and div **F** = 0. *Hint:* Use Problem 20 with $f = -c/\|\mathbf{r}\|^3$.

22. Let $\mathbf{F}(x, y, z) = c\mathbf{r}/\|\mathbf{r}\|^m$, $c \neq 0$, $m \neq 3$. Show in contrast to Problem 21 that div **F** ≠ 0, though curl **F** = **0**.

23. Let $\mathbf{F}(x, y, z) = f(r)\mathbf{r}$, where $r = \|\mathbf{r}\| = \sqrt{x^2 + y^2 + z^2}$ and f is a differentiable scalar function (except possibly at $r = 0$). Show that curl **F** = **0** (except at $r = 0$). *Hint:* First show that grad $f = f'(r)\mathbf{r}/r$ and then apply Problem 20d.

24. Let $\mathbf{F}(x, y, z)$ be as in Problem 23. Show that if div **F** = 0 then $f(r) = cr^{-3}$, where c is a constant.

25. This problem relates to the interpretation of div and curl given in the margin box just after their definition. Consider the four velocity fields **F**, **G**, **H**, and **L**, which have for every z the configuration illustrated in Figure 6. Determine each of the following by geometric reasoning.

(a) Is the divergence at **p** positive, negative, or zero?
(b) Will a paddle wheel with vertical axis at **p** (Figure 4, Section 14.7) rotate clockwise, counterclockwise, or not at all?
(c) Now suppose $\mathbf{F} = c\mathbf{j}$, $\mathbf{G} = e^{-y^2}\mathbf{j}$, $\mathbf{H} = e^{-x^2}\mathbf{j}$, and $\mathbf{L} = (x\mathbf{i} + y\mathbf{j})/\sqrt{x^2 + y^2}$, which might be modeled as in

Figure 6. Calculate the divergence and curl for each of these fields and thereby confirm your answers in parts (a) and (b).

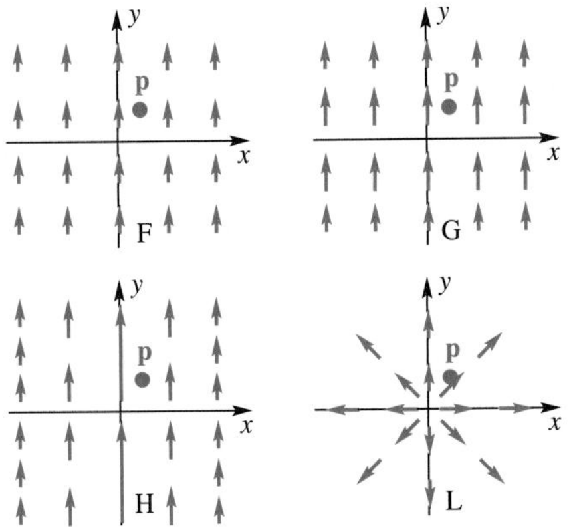

Figure 6

CAS **26.** Sketch a plot of the vector field $\mathbf{F} = y\mathbf{i}$ for (x, y) in the rectangle $1 \le x \le 2, 0 \le y \le 2$. From the plot, use the marginal box that describes the interpretation of div and curl to determine whether div is positive, negative, or zero at the point $(1, 1)$, and whether a paddle wheel placed at $(1, 1)$ would rotate clockwise, counterclockwise, or not at all.

CAS **27.** Sketch a plot of the vector field

$$\mathbf{F} = -\frac{x}{(1 + x^2 + y^2)^{3/2}}\mathbf{i} - \frac{y}{(1 + x^2 + y^2)^{3/2}}\mathbf{j}$$

for (x, y) in the rectangle $-1 \le x \le 1, -1 \le y \le 1$. From the plot, use the marginal box that describes the interpretation of div and curl to determine whether div is positive, negative, or zero at the origin, and whether a paddle wheel placed at the origin would rotate clockwise, counterclockwise, or not at all. (For the curl, think of $\mathbf{F}$ as being a vector field in 3-space with z-component equal to 0.)

28. Consider the velocity field $\mathbf{v}(x, y, z) = -\omega y\mathbf{i} + \omega x\mathbf{j}$, $\omega > 0$ (see Example 2 and Figure 1). Note that $\mathbf{v}$ is perpendicular to $x\mathbf{i} + y\mathbf{j}$ and that $\|\mathbf{v}\| = \omega\sqrt{x^2 + y^2}$. Thus, $\mathbf{v}$ describes a fluid that is rotating (like a solid) about the z-axis with constant angular velocity ω. Show that div $\mathbf{v} = 0$ and curl $\mathbf{v} = 2\omega\mathbf{k}$.

29. An object of mass m, which is revolving in a circular orbit with constant angular velocity ω, is subject to the centrifugal force given by

$$\mathbf{F}(x, y, z) = m\omega^2\mathbf{r} = m\omega^2(x\mathbf{i} + y\mathbf{j} + z\mathbf{k})$$

Show that $f(x, y, z) = \frac{1}{2}m\omega^2(x^2 + y^2 + z^2)$ is a potential function for $\mathbf{F}$.

30. The scalar function div(grad f) $= \nabla \cdot \nabla f$ (also written $\nabla^2 f$) is called the *Laplacian*, and a function f satisfying $\nabla^2 f = 0$ is said to be *harmonic*, concepts important in physics. Show that $\nabla^2 f = f_{xx} + f_{yy} + f_{zz}$. Then find $\nabla^2 f$ for each of the following functions and decide which are harmonic.

(a) $f(x, y, z) = 2x^2 - y^2 - z^2$

(b) $f(x, y, z) = xyz$

(c) $f(x, y, z) = x^3 - 3xy^2 + 3z$

(d) $f(x, y, z) = (x^2 + y^2 + z^2)^{-1/2}$

31. Show that

(a) $\text{div}(\mathbf{F} \times \mathbf{G}) = \mathbf{G} \cdot \text{curl}\,\mathbf{F} - \mathbf{F} \cdot \text{curl}\,\mathbf{G}$

(b) $\text{div}(\nabla f \times \nabla g) = 0$

32. By analogy with earlier definitions, define each of the following:

(a) $\lim_{(x, y, z) \to (a, b, c)} \mathbf{F}(x, y, z) = \mathbf{L}$

(b) $\mathbf{F}(x, y, z)$ is continuous at (a, b, c)

Answers to Concepts Review: **1.** vector-valued function of three real variables; or a vector field **2.** gradient field **3.** gravitational fields; electric fields **4.** $\nabla \cdot \mathbf{F}$; $\nabla \times \mathbf{F}$

14.2 Line Integrals

One kind of generalization of the definite integral $\int_a^b f(x)\,dx$ is obtained by replacing the set $[a, b]$ over which we integrate by two- and three-dimensional sets. This led us to the double and triple integrals of Chapter 13. A very different generalization is obtained by replacing $[a, b]$ with a curve C in the xy-plane. The resulting integral $\int_C f(x, y)\,ds$ is called a **line integral,** but would more properly be called a *curve integral.*

Let C be a smooth plane curve; that is, let C be given parametrically by

$$x = x(t), \qquad y = y(t), \qquad a \le t \le b$$

where x' and y' are continuous and not simultaneously zero on (a, b). We say that C is **positively oriented** if its direction corresponds to increasing values of t. We suppose that C is positively oriented and that C is traced only once as t varies from a to b. Thus, C has initial point $A = (x(a), y(a))$ and terminal point $B = (x(b), y(b))$. Consider the partition P of the parameter interval $[a, b]$ obtained by inserting the points

$$a = t_0 < t_1 < t_2 < \cdots < t_n = b$$

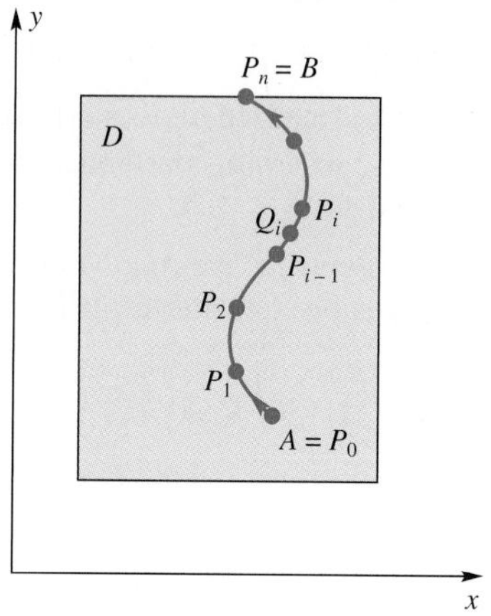

Figure 1

This partition of $[a, b]$ results in a division of the curve C into n subarcs $P_{i-1}P_i$ in which the point P_i corresponds to t_i. Let Δs_i denote the length of the arc $P_{i-1}P_i$, and let $\|P\|$ be the norm of the partition P; that is, let $\|P\|$ be the largest $\Delta t_i = t_i - t_{i-1}$. Finally, choose a sample point $Q_i(\overline{x}_i, \overline{y}_i)$ on the subarc $P_{i-1}P_i$ (see Figure 1).

Now consider the Riemann sum

$$\sum_{i=1}^{n} f(\overline{x}_i, \overline{y}_i)\, \Delta s_i$$

If f is nonnegative, this sum approximates the area of the curved vertical curtain shown in Figure 2. If f is continuous on a region D containing the curve C, then this Riemann sum has a limit as $\|P\| \to 0$. This limit is called the **line integral of f along C from A to B with respect to arc length**; that is,

$$\int_C f(x, y)\, ds = \lim_{\|P\| \to 0} \sum_{i=1}^{n} f(\overline{x}_i, \overline{y}_i)\, \Delta s_i$$

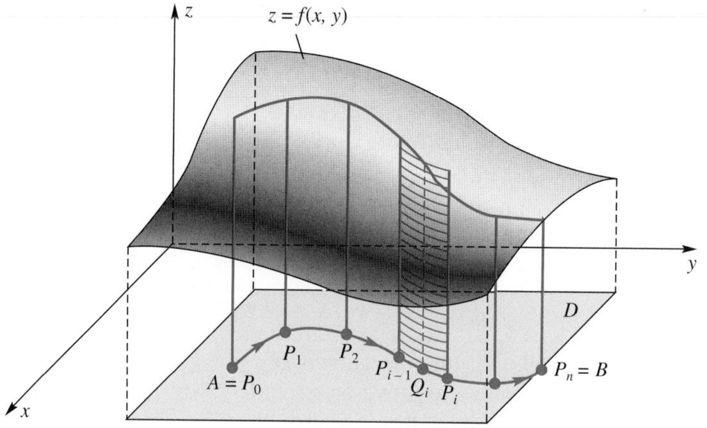

Figure 2

It represents, for $f(x, y) \ge 0$, the exact area of the curved curtain of Figure 2.

The definition does not provide a very good way of evaluating $\int_C f(x, y)\, ds$. That is best accomplished by expressing everything in terms of the parameter t and leads to an ordinary definite integral. Using $ds = \sqrt{[x'(t)]^2 + [y'(t)]^2}\, dt$ (see Section 5.4) gives

$$\int_C f(x, y)\, ds = \int_a^b f(x(t), y(t)) \sqrt{[x'(t)]^2 + [y'(t)]^2}\, dt$$

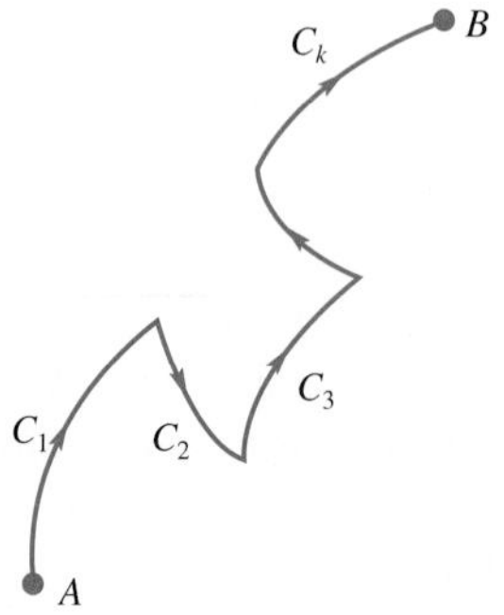

Figure 3

Of course, a curve can be parametrized in many different ways; fortunately, it can be proved that any parametrization results in the same value for $\int_C f(x, y)\, ds$.

The definition of a line integral can be extended to the case where C, though not smooth itself, is piecewise smooth, that is, consists of several smooth curves $C_1, C_2, \ldots, C_k$ joined together, as shown in Figure 3. We simply define the integral over C to be the sum of the integrals over the individual curves.

Examples and Applications We begin with two examples where C is part of a circle.

EXAMPLE 1 Evaluate $\int_C x^2 y\, ds$, where C is determined by the parametric equations $x = 3\cos t$, $y = 3\sin t$, $0 \le t \le \pi/2$. Also show that the parametrization $x = \sqrt{9 - y^2}$, $y = y$, $0 \le y \le 3$, gives the same value.

SOLUTION Using the first parametrization, we obtain

$$\begin{aligned}\int_C x^2 y\, ds &= \int_0^{\pi/2} (3\cos t)^2(3\sin t)\sqrt{(-3\sin t)^2 + (3\cos t)^2}\, dt \\ &= 81\int_0^{\pi/2} \cos^2 t \sin t\, dt \\ &= \left[-\frac{81}{3}\cos^3 t\right]_0^{\pi/2} = 27\end{aligned}$$

For the second parametrization, we use another formula for ds as given in Section 5.4. This gives

$$ds = \sqrt{1 + \left(\frac{dx}{dy}\right)^2}\, dy = \sqrt{1 + \frac{y^2}{9 - y^2}}\, dy = \frac{3}{\sqrt{9 - y^2}}\, dy$$

and

$$\begin{aligned}\int_C x^2 y\, ds &= \int_0^3 (9 - y^2)y\frac{3}{\sqrt{9 - y^2}}\, dy = 3\int_0^3 \sqrt{9 - y^2}\, y\, dy \\ &= -\left[(9 - y^2)^{3/2}\right]_0^3 = 27\end{aligned}$$

■

EXAMPLE 2 A thin wire is bent in the shape of the semicircle

$$x = a\cos t, \qquad y = a\sin t, \qquad 0 \le t \le \pi, \qquad a > 0$$

If the density of the wire at a point is proportional to its distance from the x-axis, find the mass and center of mass of the wire.

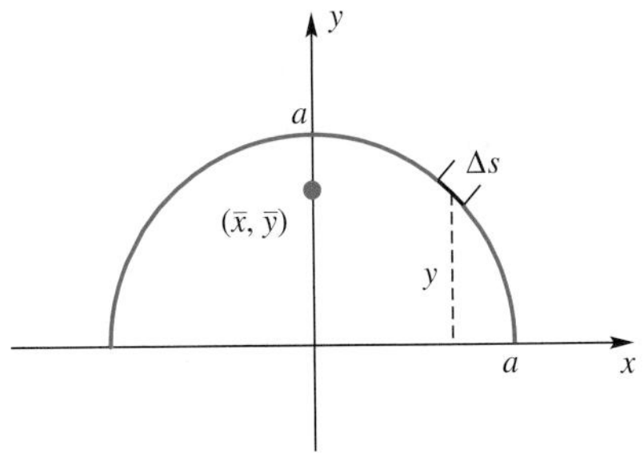

Figure 4

SOLUTION Our old motto, *slice, approximate, integrate*, is still appropriate. The mass of a small piece of wire of length Δs (Figure 4) is approximately $\delta(x, y)\,\Delta s$, where $\delta(x, y) = ky$ is the density at (x, y) (k is a constant). Thus, the mass m of the whole wire is

$$\begin{aligned}m &= \int_C ky\, ds = \int_0^{\pi} ka\sin t\sqrt{a^2\sin^2 t + a^2\cos^2 t}\, dt \\ &= ka^2\int_0^{\pi} \sin t\, dt \\ &= [-ka^2\cos t]_0^{\pi} = 2ka^2\end{aligned}$$

The moment of the wire with respect to the x-axis is given by

$$\begin{aligned}M_x &= \int_C y\, ky\, ds = \int_0^{\pi} ka^3\sin^2 t\, dt \\ &= \frac{ka^3}{2}\int_0^{\pi} (1 - \cos 2t)\, dt \\ &= \frac{ka^3}{2}\left[t - \frac{1}{2}\sin 2t\right]_0^{\pi} = \frac{ka^3\pi}{2}\end{aligned}$$

Thus,

$$\overline{y} = \frac{M_x}{m} = \frac{\frac{1}{2}ka^3\pi}{2ka^2} = \frac{1}{4}\pi a$$

From symmetry, $\overline{x} = 0$, so the center of mass is at $(0, \pi a/4)$. ■

All that we have done extends easily to a smooth curve C in three-space. In particular, if C is given parametrically by

$$x = x(t), \quad y = y(t), \quad z = z(t), \quad a \le t \le b$$

then

$$\int_C f(x, y, z)\, ds = \int_a^b f(x(t), y(t), z(t))\sqrt{[x'(t)]^2 + [y'(t)]^2 + [z'(t)]^2}\, dt$$

EXAMPLE 3 Find the mass of a wire of density $\delta(x, y, z) = kz$ if it has the shape of the helix C with parametrization

$$x = 3\cos t, \quad y = 3\sin t, \quad z = 4t, \quad 0 \le t \le \pi$$

SOLUTION

$$m = \int_C kz\, ds = k\int_0^\pi (4t)\sqrt{9\sin^2 t + 9\cos^2 t + 16}\, dt$$

$$= 20k\int_0^\pi t\, dt = \left[20k\frac{t^2}{2}\right]_0^\pi = 10k\pi^2$$

The units for m depend on those for length and density. ■

Work Suppose that the force acting at a point (x, y, z) in space is given by the vector field

$$\mathbf{F}(x, y, z) = M(x, y, z)\,\mathbf{i} + N(x, y, z)\,\mathbf{j} + P(x, y, z)\,\mathbf{k}$$

where M, N, and P are continuous. We want to find the work W done by $\mathbf{F}$ in moving a particle along a smooth oriented curve C. Let $\mathbf{r} = x\mathbf{i} + y\mathbf{j} + z\mathbf{k}$ be the position vector for a point $Q(x, y, z)$ on the curve (Figure 5). If $\mathbf{T}$ is the unit tangent vector at Q, then $\mathbf{F}\cdot\mathbf{T}$ is the tangential component of $\mathbf{F}$ at Q. The work done by $\mathbf{F}$ in moving the particle from Q a short distance Δs along the curve is approximately $\mathbf{F}\cdot\mathbf{T}\,\Delta s$, and consequently the work done in moving the particle from A to B along C is defined to be $\int_C \mathbf{F}\cdot\mathbf{T}\, ds$. Work is a *scalar* quantity, but it can be positive or negative. It is positive when the component of force along the curve is in the direction of the object's motion, and it is negative when the component of force along the curve is in the direction opposite the object's motion. From Section 11.7, we know that $\mathbf{T} = (d\mathbf{r}/dt)(dt/ds) = d\mathbf{r}/ds$, and so we have the following alternative formulas for work.

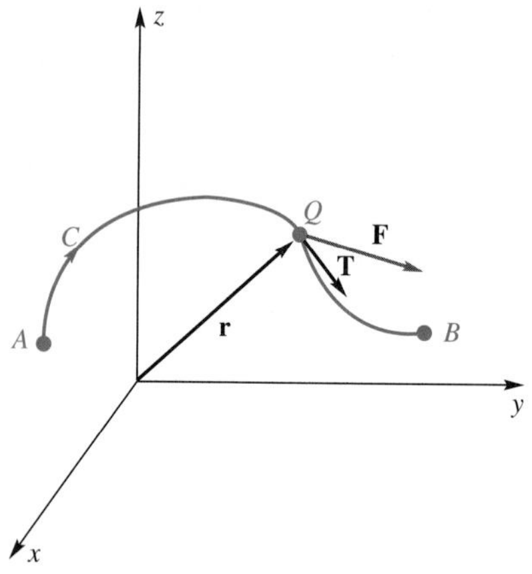

Figure 5

$$W = \int_C \mathbf{F}\cdot\mathbf{T}\, ds = \int_C \mathbf{F}\cdot\frac{d\mathbf{r}}{dt}\, dt = \int_C \mathbf{F}\cdot d\mathbf{r}$$

To interpret the last expression, think of $\mathbf{F}\cdot d\mathbf{r}$ as representing the work done by $\mathbf{F}$ in moving a particle along the "infinitesimal" tangent vector $d\mathbf{r}$, a formulation preferred by many physicists and applied mathematicians.

There is still another expression for work that is often useful in calculations. If we agree to write $d\mathbf{r} = dx\,\mathbf{i} + dy\,\mathbf{j} + dz\,\mathbf{k}$, then

$$\mathbf{F}\cdot d\mathbf{r} = (M\mathbf{i} + N\mathbf{j} + P\mathbf{k})\cdot(dx\,\mathbf{i} + dy\,\mathbf{j} + dz\,\mathbf{k}) = M\,dx + N\,dy + P\,dz$$

and

$$W = \int_C \mathbf{F} \cdot d\mathbf{r} = \int_C M\,dx + N\,dy + P\,dz$$

The integrals $\int_C M\,dx$, $\int_C N\,dy$, and $\int_C P\,dz$ are a special kind of line integral. They are defined just as $\int_C f\,ds$ was defined at the beginning of the section, except that Δs_i is replaced by Δx_i, Δy_i, and Δz_i, respectively. However, we point out that, while Δs_i is always taken to be positive, Δx_i, Δy_i, and Δz_i may well be negative on a path C. The result of this is that a change in the orientation of C switches the sign of $\int_C M\,dx$, $\int_C N\,dy$, and $\int_C P\,dz$ while leaving that of $\int_C f\,ds$ unchanged (see Problem 33).

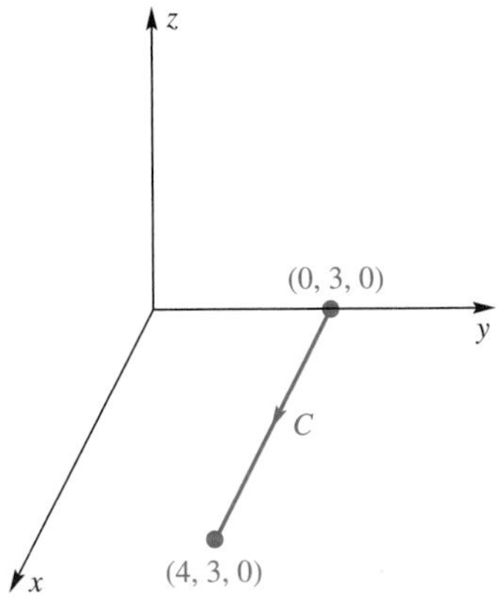

Figure 6

EXAMPLE 4 Find the work done by the inverse square law force field

$$\mathbf{F}(x, y, z) = -c\frac{\mathbf{r}}{\|\mathbf{r}\|^3} = \frac{-c(x\,\mathbf{i} + y\,\mathbf{j} + z\,\mathbf{k})}{(x^2 + y^2 + z^2)^{3/2}} = M\mathbf{i} + N\mathbf{j} + P\mathbf{k}$$

in moving a particle along the straight-line curve C from $(0, 3, 0)$ to $(4, 3, 0)$ shown in Figure 6.

SOLUTION Along C, $y = 3$ and $z = 0$, so $dy = dz = 0$. Using x as the parameter, we obtain

$$W = \int_C M\,dx + N\,dy + P\,dz = -c\int_C \frac{x\,dx + y\,dy + z\,dz}{(x^2 + y^2 + z^2)^{3/2}}$$

$$= -c\int_0^4 \frac{x}{(x^2 + 9)^{3/2}}\,dx = \left[\frac{c}{(x^2 + 9)^{1/2}}\right]_0^4 = -\frac{2c}{15}$$

Of course, appropriate units must be assigned, depending on those for length and force. If $c > 0$, then the work done by the force field **F** is *negative*. Does this make sense? In this problem, the force always points toward the origin, so the component of force along the curve is always in the direction *opposite* the path of the particle's motion (see Figure 7). When this happens, the work is negative. ■

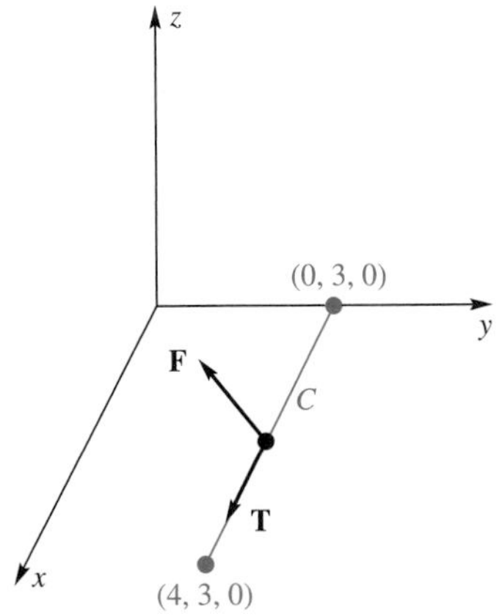

Figure 7

Here is a planar version of this type of line integral.

EXAMPLE 5 Evaluate the line integral

$$\int_C (x^2 - y^2)\,dx + 2xy\,dy$$

along the curve C whose parametric equations are $x = t^2$, $y = t^3$, $0 \le t \le \frac{3}{2}$.

SOLUTION Since $dx = 2t\,dt$ and $dy = 3t^2\,dt$,

$$\int_C (x^2 - y^2)\,dx + 2xy\,dy = \int_0^{3/2} \left[(t^4 - t^6)2t + 2t^5(3t^2)\right] dt$$

$$= \int_0^{3/2} (2t^5 + 4t^7)\,dt = \frac{8505}{512} \approx 16.61$$ ■

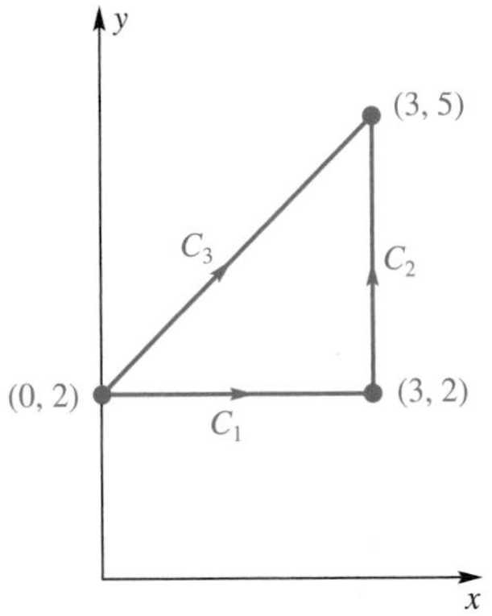

Figure 8

EXAMPLE 6 Evaluate $\int_C xy^2\,dx + xy^2\,dy$ along the path $C = C_1 \cup C_2$ shown in Figure 8. Also evaluate this integral along the straight path C_3 from $(0, 2)$ to $(3, 5)$.

SOLUTION On C_1, $y = 2$, $dy = 0$, and

$$\int_{C_1} xy^2\,dx + xy^2\,dy = \int_0^3 4x\,dx = [2x^2]_0^3 = 18$$

On C_2, $x = 3$, $dx = 0$, and

$$\int_{C_2} xy^2\,dx + xy^2\,dy = \int_2^5 3y^2\,dy = [y^3]_2^5 = 117$$

We conclude that

$$\int_C xy^2\,dx + xy^2\,dy = 18 + 117 = 135$$

On C_3, $y = x + 2$, $dy = dx$, and so

$$\begin{aligned}\int_{C_3} xy^2\,dx + xy^2\,dy &= 2\int_0^3 x(x+2)^2\,dx \\ &= 2\int_0^3 (x^3 + 4x^2 + 4x)\,dx \\ &= 2\left[\frac{x^4}{4} + \frac{4x^3}{3} + 2x^2\right]_0^3 = \frac{297}{2}\end{aligned}$$

Note that the two paths from (0, 2) to (3, 5) give different values for the integral. ■

Concepts Review

1. A curve C given parametrically by $x = x(t)$, $y = y(t)$, $a \le t \le b$, is said to be positively oriented if its positive direction corresponds to _____.

2. The line integral $\int_C f(x, y)\,ds$, where C is the positively oriented curve of Question 1, is defined as $\lim_{\|P\|\to 0}$ _____.

3. The line integral in Question 2 transforms to the ordinary integral $\int_a^b$ _____ dt.

4. If $\mathbf{r} = x(t)\mathbf{i} + y(t)\mathbf{j}$ is the position vector of a point on the curve C of Question 1 and if $\mathbf{F} = M(x, y)\mathbf{i} + N(x, y)\mathbf{j}$ is a force field in the plane, then the work W done by $\mathbf{F}$ in moving an object along C is given by $\int_C$ _____ dt.

Problem Set 14.2

In Problems 1–16, evaluate each line integral.

1. $\int_C (x^3 + y)\,ds$; C is the curve $x = 3t$, $y = t^3$, $0 \le t \le 1$.

2. $\int_C xy^{2/5}\,ds$; C is the curve $x = \frac{1}{2}t$, $y = t^{5/2}$, $0 \le t \le 1$.

3. $\int_C (\sin x + \cos y)\,ds$; C is the line segment from $(0, 0)$ to $(\pi, 2\pi)$.

4. $\int_C xe^y\,ds$; C is the line segment from $(-1, 2)$ to $(1, 1)$.

5. $\int_C (2x + 9z)\,ds$; C is the curve $x = t$, $y = t^2$, $z = t^3$, $0 \le t \le 1$.

6. $\int_C (x^2 + y^2 + z^2)\,ds$; C is the curve $x = 4\cos t$, $y = 4\sin t$, $z = 3t$, $0 \le t \le 2\pi$.

7. $\int_C y\,dx + x^2\,dy$; C is the curve $x = 2t$, $y = t^2 - 1$, $0 \le t \le 2$.

8. $\int_C y\,dx + x^2\,dy$; C is the right-angle curve from $(0, -1)$ to $(4, -1)$ to $(4, 3)$.

9. $\int_C y^3\,dx + x^3\,dy$; C is the right-angle curve from $(-4, 1)$ to $(-4, -2)$ to $(2, -2)$.

10. $\int_C y^3\,dx + x^3\,dy$; C is the curve $x = 2t$, $y = t^2 - 3$, $-2 \le t \le 1$.

11. $\int_C (x + 2y)\,dx + (x - 2y)\,dy$; C is the line segment from $(1, 1)$ to $(3, -1)$.

12. $\int_C y\,dx + x\,dy$; C is the curve $y = x^2, 0 \le x \le 1$.

13. $\int_C (x + y + z)\,dx + x\,dy - yz\,dz$; C is the line segment from $(1, 2, 1)$ to $(2, 1, 0)$.

14. $\int_C xz\,dx + (y + z)\,dy + x\,dz$; C is the curve $x = e^t$, $y = e^{-t}, z = e^{2t}, 0 \le t \le 1$.

15. $\int_C (x + y + z)\,dx + (x - 2y + 3z)\,dy +$ $(2x + y - z)\,dz$; C is the line-segment path from $(0, 0, 0)$ to $(2, 0, 0)$ to $(2, 3, 0)$ to $(2, 3, 4)$.

16. Same integral as in Problem 15; C is the line segment from $(0, 0, 0)$ to $(2, 3, 4)$.

17. Find the mass of a wire with the shape of the curve $y = x^2$ between $(-2, 4)$ and $(2, 4)$ if the density is given by $\delta(x, y) = k|x|$.

18. A wire of constant density has the shape of the helix $x = a\cos t$, $y = a\sin t$, $z = bt$, $0 \le t \le 3\pi$. Find its mass and center of mass.

In Problems 19–24, find the work done by the force field **F** *in moving a particle along the curve C.*

19. $\mathbf{F}(x, y) = (x^3 - y^3)\mathbf{i} + xy^2\mathbf{j}$; C is the curve $x = t^2$, $y = t^3, -1 \le t \le 0$.

20. $\mathbf{F}(x, y) = e^x\mathbf{i} - e^{-y}\mathbf{j}$; C is the curve $x = 3\ln t$, $y = \ln 2t$, $1 \le t \le 5$.

21. $\mathbf{F}(x, y) = (x + y)\mathbf{i} + (x - y)\mathbf{j}$; C is the quarter-ellipse, $x = a\cos t, y = b\sin t, 0 \le t \le \pi/2$.

22. $\mathbf{F}(x, y, z) = (2x - y)\mathbf{i} + 2z\mathbf{j} + (y - z)\mathbf{k}$; C is the line segment from $(0,0,0)$ to $(1,1,1)$.

23. Same **F** as in Problem 22; C is the curve $x = \sin(\pi t/2)$, $y = \sin(\pi t/2), z = t, 0 \le t \le 1$.

24. $\mathbf{F}(x, y, z) = y\mathbf{i} + z\mathbf{j} + x\mathbf{k}$; C is the curve $x = t, y = t^2$, $z = t^3, 0 \le t \le 2$.

25. Figure 9 shows a plot of a vector field **F** along with three curves, C_1, C_2, and C_3. Determine whether each line integral $\int_{C_i} \mathbf{F} \cdot d\mathbf{r}$, $i = 1, 2, 3$, is positive, negative, or zero, and justify your answers.

26. Figure 10 shows a plot of a vector field **F** along with three curves, C_1, C_2, and C_3. Determine whether each line integral $\int_{C_i} \mathbf{F} \cdot d\mathbf{r}$, $i = 1, 2, 3$, is positive, negative, or zero, and justify your answers.

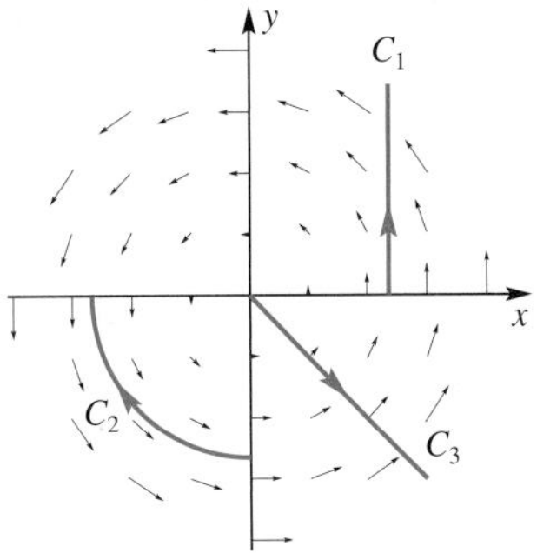

Figure 9

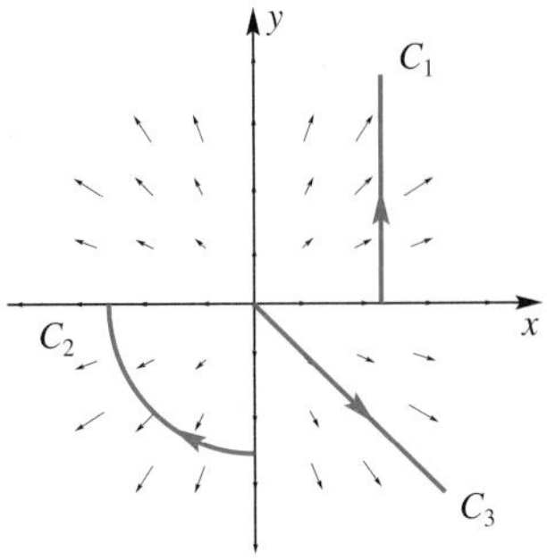

Figure 10

27. Christy plans to paint both sides of a fence whose base is in the xy-plane with shape $x = 30\cos^3 t$, $y = 30\sin^3 t$, $0 \le t \le \pi/2$, and whose height at (x, y) is $1 + \frac{1}{3}y$, all measured in feet. Sketch a picture of the fence and decide how much paint she will need if a gallon covers 200 square feet.

28. A squirrel weighing 1.2 pounds climbed a cylindrical tree by following the helical path $x = \cos t$, $y = \sin t$, $z = 4t$, $0 \le t \le 8\pi$ (distance measured in feet). How much work did it do? Use a line integral, but then think of a trivial way to answer this question.

29. Use a line integral to find the area of the part cut out of the vertical square cylinder $|x| + |y| = a$ by the sphere $x^2 + y^2 + z^2 = a^2$. Check your answer by finding a trivial way to do this problem.

30. A wire of constant density k has the shape $|x| + |y| = a$. Find its moment of inertia with respect to the y-axis and with respect to the z-axis.

31. Use a line integral to find the area of that part of the cylinder $x^2 + y^2 = ay$ inside the sphere $x^2 + y^2 + z^2 = a^2$ (compare with Problem 12, Section 13.6). *Hint:* Use polar coordinates where $ds = [r^2 + (dr/d\theta)^2]^{1/2}\,d\theta$.

32. Two circular cylinders of radius a intersect so that their axes meet at right angles. Use a line integral to find the area of the part from one cut off by the other (compare with Problem 14, Section 13.6). See Figure11.

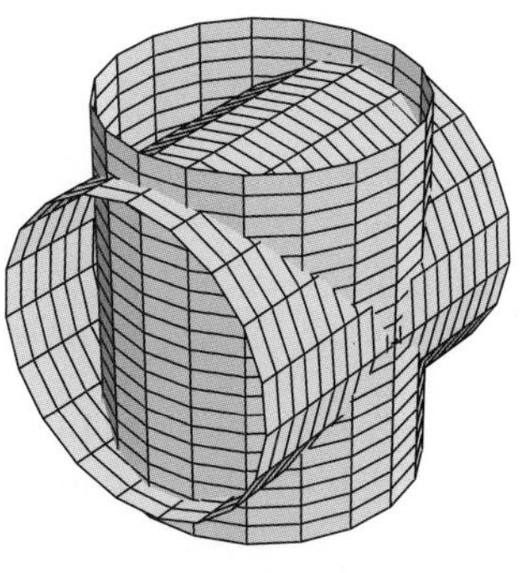

Figure 11

33. Evaluate

(a) $\int_C x^2 y\,ds$ using the parametrization $x = 3\sin t$, $y = 3\cos t$, $0 \le t \le \pi/2$, which reverses the orientation of C in Example 1, and

(b) $\int_{C_4} xy^2\,dx + xy^2\,dy$ using the parametrization $x = 3 - t$, $y = 5 - t, 0 \le t \le 3$, and note that C_4 has the reverse orientation of C_3 in Example 6.

Orientation-reversing parametrizations do not change the sign of $\int_C f\,ds$, but do change the sign of the other types of line integrals considered in this section.

Answers to Concepts Review: **1.** increasing values of t **2.** $\sum_{i=1}^{n} f(\bar{x}_i, \bar{y}_i)\,\Delta s_i$ **3.** $f(x(t), y(t))\sqrt{[x'(t)]^2 + [y'(t)]^2}$ **4.** $\mathbf{F}\cdot d\mathbf{r}/dt$

14.3 Independence of Path

The basic tool in evaluating ordinary definite integrals is the Second Fundamental Theorem of Calculus. In symbols, it says that

$$\int_a^b f'(x)\,dx = f(b) - f(a)$$

Now we ask the question: Is there an analogous theorem for line integrals? The answer is yes.

In what follows, interpret $\mathbf{r}(t)$ as $x(t)\mathbf{i} + y(t)\mathbf{j}$ if the context is two-space and as $x(t)\mathbf{i} + y(t)\mathbf{j} + z(t)\mathbf{k}$ if it is three-space. Correspondingly, $f(\mathbf{r})$ will mean $f(x, y)$ in the first case and $f(x, y, z)$ in the second.

Theorem A Fundamental Theorem for Line Integrals

Let C be a piecewise smooth curve given parametrically by $\mathbf{r} = \mathbf{r}(t), a \le t \le b$, which begins at $\mathbf{a} = \mathbf{r}(a)$ and ends at $\mathbf{b} = \mathbf{r}(b)$. If f is continuously differentiable on an open set containing C, then

$$\int_C \nabla f(\mathbf{r})\cdot d\mathbf{r} = f(\mathbf{b}) - f(\mathbf{a})$$

Proof We suppose first that C is smooth. Then

$$\begin{aligned}\int_C \nabla f(\mathbf{r})\cdot d\mathbf{r} &= \int_a^b [\nabla f(\mathbf{r}(t))\cdot \mathbf{r}'(t)]\,dt\\ &= \int_a^b \frac{d}{dt} f(\mathbf{r}(t))\,dt = f(\mathbf{r}(b)) - f(\mathbf{r}(a))\\ &= f(\mathbf{b}) - f(\mathbf{a})\end{aligned}$$

Note how we first wrote the line integral as an ordinary definite integral, then applied the Chain Rule, and finally used the Second Fundamental Theorem of Calculus.

If C is not smooth but only piecewise smooth, we simply apply the above result to the individual pieces. We leave the details to you. ■

EXAMPLE 1 Recall from Example 4 of Section 14.1 that

$$f(x, y, z) = f(\mathbf{r}) = \frac{c}{\|\mathbf{r}\|} = \frac{c}{\sqrt{x^2 + y^2 + z^2}}$$

is a potential function for the inverse square law field $\mathbf{F}(\mathbf{r}) = -c\mathbf{r}/\|\mathbf{r}\|^3$. Calculate $\int_C \mathbf{F}(\mathbf{r}) \cdot d\mathbf{r}$, where C is any simple piecewise smooth curve from $(0, 3, 0)$ to $(4, 3, 0)$ that misses the origin.

SOLUTION Since $\mathbf{F}(\mathbf{r}) = \nabla f(\mathbf{r})$,

$$\int_C \mathbf{F}(\mathbf{r}) \cdot d\mathbf{r} = \int_C \nabla f(\mathbf{r}) \cdot d\mathbf{r} = f(4, 3, 0) - f(0, 3, 0)$$

$$= \frac{c}{\sqrt{16 + 9}} - \frac{c}{\sqrt{9}} = -\frac{2c}{15}$$ ■

Now compare Example 1 with Example 4 of the previous section. There we calculated the same integral, but for a specific curve C, the line segment from $(0, 3, 0)$ to $(4, 3, 0)$. Surprisingly, we will get the same answer no matter what curve we take from $(0, 3, 0)$ to $(4, 3, 0)$. We say that the given line integral is independent of path.

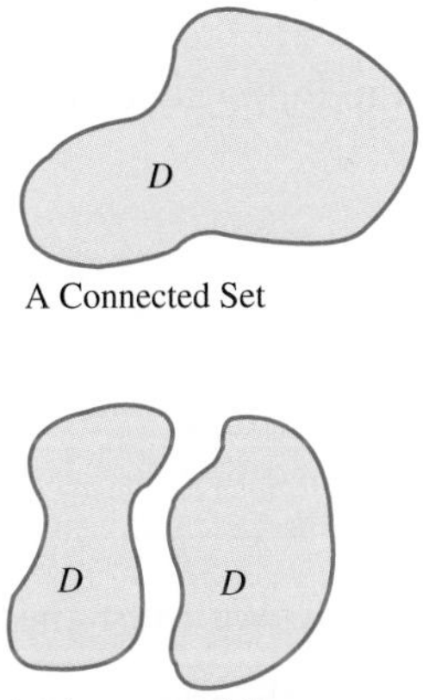

Figure 1

Criteria for Independence of Path Call a set D **connected** if any two points in D can be joined by a piecewise smooth curve lying entirely in D (Figure 1). Then call $\int_C \mathbf{F}(\mathbf{r}) \cdot d\mathbf{r}$ **independent of path in D** if for any two points A and B in D the line integral has the same value for every path C in D that is positively oriented from A to B.

One consequence of Theorem A is that if $\mathbf{F}$ is the gradient of another function f then $\int_C \mathbf{F}(\mathbf{r}) \cdot d\mathbf{r}$ is independent of path. The converse is also true.

Theorem B Independence of Path Theorem

Let $\mathbf{F}(\mathbf{r})$ be continuous on an open connected set D. Then the line integral $\int_C \mathbf{F}(\mathbf{r}) \cdot d\mathbf{r}$ is independent of path in D if and only if $\mathbf{F}(\mathbf{r}) = \nabla f(\mathbf{r})$ for some scalar function f; that is, if and only if $\mathbf{F}$ is a conservative vector field on D.

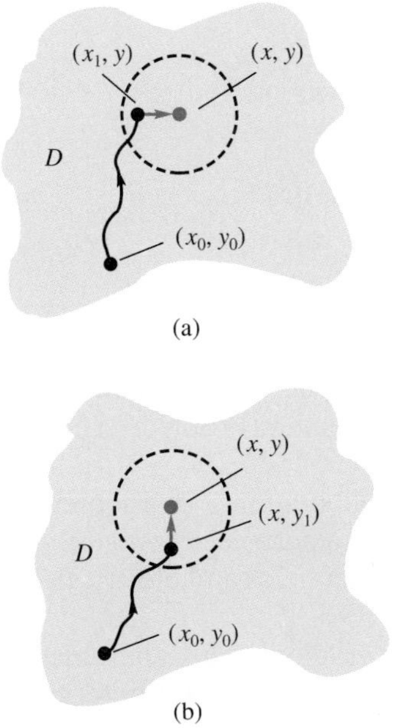

Figure 2

Proof Theorem A takes care of the "if" statement. Suppose then that $\int_C \mathbf{F}(\mathbf{r}) \cdot d\mathbf{r}$ is independent of path in D. Our task is to construct a function f satisfying $\nabla f = \mathbf{F}$; that is, we must find a potential function for the vector field $\mathbf{F}$. For simplicity, we restrict ourselves to the two-dimensional case, where D is a plane set and $\mathbf{F}(\mathbf{r}) = M(x, y)\mathbf{i} + N(x, y)\mathbf{j}$.

Let (x_0, y_0) be a fixed point of D, and let (x, y) be any other point of D. Choose a third point (x_1, y) in D and slightly to the left of (x, y), and join it to (x, y) by a horizontal segment in D. Then join (x_0, y_0) to (x_1, y) by a curve in D. (All this is possible because D is both open and connected; see Figure 2a). Finally, let C denote the path from (x_0, y_0) to (x, y) composed of these two pieces, and define f by

$$f(x, y) = \int_C \mathbf{F}(\mathbf{r}) \cdot d\mathbf{r} = \int_{(x_0,y_0)}^{(x_1,y)} \mathbf{F}(\mathbf{r}) \cdot d\mathbf{r} + \int_{(x_1,y)}^{(x,y)} \mathbf{F}(\mathbf{r}) \cdot d\mathbf{r}$$

That we get a unique value is clear from the assumed independence of path.

The first integral on the right above does not depend on x; the second, which has y fixed, can be written as an ordinary definite integral using, for example, t as a parameter. It follows that

$$\frac{\partial f}{\partial x} = 0 + \frac{\partial}{\partial x}\int_{x_1}^{x} M(t, y)\,dt = M(x, y)$$

The last equality is a consequence of the First Fundamental Theorem of Calculus (Theorem 4.4A).

A similar argument using Figure 2b shows that $\partial f/\partial y = N(x, y)$. We conclude that $\nabla f = M(x, y)\mathbf{i} + N(x, y)\mathbf{j} = \mathbf{F}$, as desired. ■

The results of this section culminate in the next theorem, which draws a connection between the ideas of a conservative vector field, path independence of a line integral, and the line integral over all closed paths being zero.

Theorem C Equivalent Conditions for Line Integrals

Let $\mathbf{F}(\mathbf{r})$ be continuous on an open connected set D. Then the following conditions are equivalent:

(1) $\mathbf{F} = \nabla f$ for some function f. ($\mathbf{F}$ is conservative on D.)

(2) $\int_C \mathbf{F}(\mathbf{r}) \cdot d\mathbf{r}$ is independent of path in D.

(3) $\int_C \mathbf{F}(\mathbf{r}) \cdot d\mathbf{r} = 0$ for every closed path in D.

Proof Theorem B establishes that (1) and (2) are equivalent. We must then show that (2) and (3) are equivalent. Suppose that $\int_C \mathbf{F}(\mathbf{r}) \cdot d\mathbf{r}$ is path independent. We must show that $\int_C \mathbf{F}(\mathbf{r}) \cdot d\mathbf{r} = 0$ for every closed path in D. Let C be a closed path in D and let A and B be distinct points on C as shown in the first part of Figure 3. Suppose C is composed of two curves, C_1, going from A to B, and C_2 going from B to A. Let $-C_2$ denote the curve C_2 with opposite orientation, as shown in the second part of Figure 3. Since C_1 and $-C_2$ have the same initial and terminal points, namely A and B, the independence of path implies that

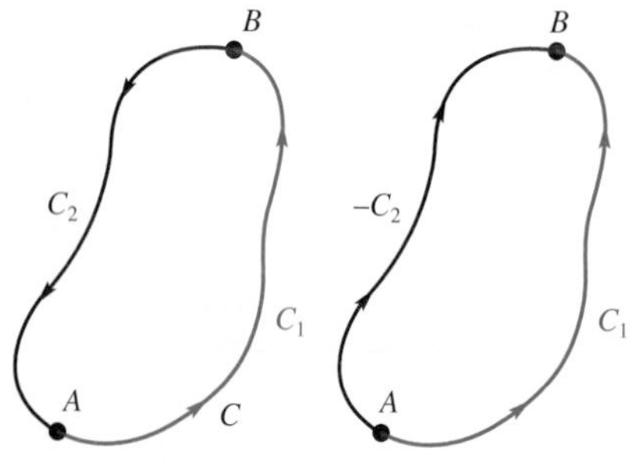

Figure 3

$$\begin{aligned}\int_C \mathbf{F}(\mathbf{r}) \cdot d\mathbf{r} &= \int_{C_1} \mathbf{F}(\mathbf{r}) \cdot d\mathbf{r} + \int_{C_2} \mathbf{F}(\mathbf{r}) \cdot d\mathbf{r} \\ &= \int_{C_1} \mathbf{F}(\mathbf{r}) \cdot d\mathbf{r} - \int_{-C_2} \mathbf{F}(\mathbf{r}) \cdot d\mathbf{r} \\ &= \int_{C_1} \mathbf{F}(\mathbf{r}) \cdot d\mathbf{r} - \int_{C_1} \mathbf{F}(\mathbf{r}) \cdot d\mathbf{r} = 0\end{aligned}$$

This shows that (2) implies (3). The argument that (3) implies (2) is essentially the reverse of that given above. We leave the details to the reader (Problem 31). ■

There is an interesting physical interpretation of Condition 3 of Theorem C. The work done by a conservative force field as it moves a particle around a closed path is zero. In particular, this is true of both gravitational fields and electric fields, since they are conservative.

While Conditions 2 and 3 each imply that $\mathbf{F}$ is the gradient of a scalar function f, they are not particularly useful in this connection. A more useful criterion is given in the following theorem. We need, however, to impose the additional condition on D, that it is **simply connected.** In two-space, this means that D has no "holes" and in three-space that it has no "tunnels" all the way through D. (For the technical definition, see any advanced calculus book.)

Theorem D

Let $\mathbf{F} = M\mathbf{i} + N\mathbf{j} + P\mathbf{k}$, where M, N, and P are continuous together with their first-order partial derivatives in an open connected set D, which is also simply connected. Then $\mathbf{F}$ is conservative ($\mathbf{F} = \nabla f$) if and only if curl $\mathbf{F} = \mathbf{0}$, that is, if and only if

$$\frac{\partial M}{\partial y} = \frac{\partial N}{\partial x}, \quad \frac{\partial M}{\partial z} = \frac{\partial P}{\partial x}, \quad \frac{\partial N}{\partial z} = \frac{\partial P}{\partial y}$$

In the two-variable case, where $\mathbf{F} = M\mathbf{i} + N\mathbf{j}$, $\mathbf{F}$ is conservative if and only if

$$\frac{\partial M}{\partial y} = \frac{\partial N}{\partial x}$$

The "only if" statement is easy to prove (Problem 21). The "if" statement follows from Green's Theorem (Theorem 14.4A) in the two-variable case and from Stokes's Theorem in the three-variable case (see Example 4 of Section 14.7). Problem 29 shows the need for simple connectedness.

Recovering a Function from Its Gradient Suppose that we are given a vector field $\mathbf{F}$ satisfying the conditions of Theorem D. Then we know there is a function f satisfying $\nabla f = \mathbf{F}$. But how can we find f? We illustrate the answer first for a two-dimensional vector field.

EXAMPLE 2 Determine whether $\mathbf{F} = (4x^3 + 9x^2y^2)\mathbf{i} + (6x^3y + 6y^5)\mathbf{j}$ is conservative, and if so, find the function f of which it is the gradient.

SOLUTION $M(x, y) = 4x^3 + 9x^2y^2$ and $N(x, y) = 6x^3y + 6y^5$. In this, the two-variable case, the conditions of Theorem D reduce to showing that

$$\frac{\partial M}{\partial y} = \frac{\partial N}{\partial x}$$

Now

$$\frac{\partial M}{\partial y} = 18x^2y, \quad \frac{\partial N}{\partial x} = 18x^2y$$

so the condition is satisfied and f must exist.

To find f, we first note that

$$\nabla f = \frac{\partial f}{\partial x}\mathbf{i} + \frac{\partial f}{\partial y}\mathbf{j} = M\mathbf{i} + N\mathbf{j}$$

Thus,

$$(1) \qquad \frac{\partial f}{\partial x} = 4x^3 + 9x^2y^2, \quad \frac{\partial f}{\partial y} = 6x^3y + 6y^5$$

If we antidifferentiate the left equation with respect to x, we obtain

$$(2) \qquad f(x, y) = x^4 + 3x^3y^2 + C_1(y)$$

in which the "constant" of integration C_1 may depend on y. But the partial with respect to y of the expression in (2) must equal $6x^3y + 6y^5$; thus

$$\frac{\partial f}{\partial y} = 6x^3y + C_1'(y) = 6x^3y + 6y^5$$

We conclude that $C_1'(y) = 6y^5$. Another antidifferentiation gives

$$C_1(y) = y^6 + C$$

where C is a constant (independent of both x and y). Substitution of this result in (1) yields

$$f(x, y) = x^4 + 3x^3y^2 + y^6 + C$$ ■

Next we use the result of Example 2 to calculate a line integral.

Notation for Line Integrals That Are Independent of Path

If a line integral

$$\int_C P(x, y)\,dx + Q(x, y)\,dy$$

is independent of path, then we will often write it as

$$\int_{(a,b)}^{(c,d)} P(x, y)\,dx + Q(x, y)\,dy$$

indicating only the initial point (a, b) and the terminal point (c, d) for the path C.

Similarly, we will write

$$\left[f(x, y)\right]_{(a,b)}^{(c,d)}$$

to mean

$$f(c, d) - f(a, b)$$

EXAMPLE 3 Let $\mathbf{F}(\mathbf{r}) = \mathbf{F}(x, y) = (4x^3 + 9x^2y^2)\mathbf{i} + (6x^3y + 6y^5)\mathbf{j}$. Calculate $\int_C \mathbf{F}(\mathbf{r}) \cdot d\mathbf{r} = \int_C (4x^3 + 9x^2y^2)\,dx + (6x^3y + 6y^5)\,dy$, where C is any path from $(0, 0)$ to $(1, 2)$.

SOLUTION Example 1 shows that $\mathbf{F} = \nabla f$, where

$$f(x, y) = x^4 + 3x^3y^2 + y^6 + C$$

and thus the given line integral is independent of path. In fact, by Theorem A,

$$\int_C \mathbf{F}(\mathbf{r}) \cdot d\mathbf{r} = \int_{(0,0)}^{(1,2)} (4x^3 + 9x^2y^2)\,dx + (6x^3y + 6y^5)\,dy$$

$$= \left[x^4 + 3x^3y^2 + y^6 + C\right]_{(0,0)}^{(1,2)} = 1 + 12 + 64 = 77$$ ■

EXAMPLE 4 Show that $\mathbf{F} = (e^x \cos y + yz)\mathbf{i} + (xz - e^x \sin y)\mathbf{j} + xy\,\mathbf{k}$ is conservative, and find f such that $\mathbf{F} = \nabla f$.

SOLUTION

$$M = e^x \cos y + yz, \qquad N = xz - e^x \sin y, \qquad P = xy$$

and so

$$\frac{\partial M}{\partial y} = -e^x \sin y + z = \frac{\partial N}{\partial x}, \qquad \frac{\partial M}{\partial z} = y = \frac{\partial P}{\partial x}, \qquad \frac{\partial N}{\partial z} = x = \frac{\partial P}{\partial y}$$

which are the conditions of Theorem D. Now

(3)
$$\frac{\partial f}{\partial x} = e^x \cos y + yz$$
$$\frac{\partial f}{\partial y} = xz - e^x \sin y$$
$$\frac{\partial f}{\partial z} = xy$$

When we antidifferentiate the first of these with respect to x, we get

(4) $$f(x, y, z) = e^x \cos y + xyz + C_1(y, z)$$

Now differentiate (4) with respect to y and set the result equal to the second expression in (3).

$$-e^x \sin y + xz + \frac{\partial C_1}{\partial y} = xz - e^x \sin y$$

or

$$\frac{\partial C_1(y, z)}{\partial y} = 0$$

Antidifferentiating the latter with respect to y gives

$$C_1(y, z) = C_2(z)$$

which we in turn substitute into (4).

$$f(x, y, z) = e^x \cos y + xyz + C_2(z) \tag{5}$$

When we differentiate (5) with respect to z and equate the result to the third expression in (3), we get

$$\frac{\partial f}{\partial z} = xy + C_2'(z) = xy$$

or $C_2'(z) = 0$ and $C_2(z) = C$. We conclude that

$$f(x, y, z) = e^x \cos y + xyz + C$$

■

Conservation of Energy Let us make an application to physics and at the same time offer a reason for the name *conservative* force field. We will establish the Law of Conservation of Energy, which says that the sum of the kinetic energy and the potential energy of an object due to a conservative force is constant.

Suppose that an object of mass m is moving along a smooth curve C given by

$$\mathbf{r} = \mathbf{r}(t) = x(t)\mathbf{i} + y(t)\mathbf{j} + z(t)\mathbf{k}, \qquad a \le t \le b$$

under the influence of a conservative force $\mathbf{F}(\mathbf{r}) = \nabla f(\mathbf{r})$. From physics, we learn three facts about the object at time t.

1. $\mathbf{F}(\mathbf{r}(t)) = m\mathbf{a}(t) = m\mathbf{r}''(t)$ (Newton's Second Law)
2. $\text{KE} = \frac{1}{2}m\|\mathbf{r}'(t)\|^2$ (KE = kinetic energy)
3. $\text{PE} = -f(\mathbf{r})$ (PE = potential energy)

Thus,

$$\begin{aligned}
\frac{d}{dt}(\text{KE} + \text{PE}) &= \frac{d}{dt}\left[\frac{1}{2}m\|\mathbf{r}'(t)\|^2 - f(\mathbf{r})\right] \\
&= \frac{m}{2}\frac{d}{dt}[\mathbf{r}'(t)\cdot\mathbf{r}'(t)] - \left[\frac{\partial f}{\partial x}\frac{dx}{dt} + \frac{\partial f}{\partial y}\frac{dy}{dt} + \frac{\partial f}{\partial z}\frac{dz}{dt}\right] \\
&= m\mathbf{r}''(t)\cdot\mathbf{r}'(t) - \nabla f(\mathbf{r})\cdot\mathbf{r}'(t) \\
&= [m\mathbf{r}''(t) - \nabla f(\mathbf{r})]\cdot\mathbf{r}'(t) \\
&= [\mathbf{F}(\mathbf{r}) - \mathbf{F}(\mathbf{r})]\cdot\mathbf{r}'(t) = 0
\end{aligned}$$

We conclude that KE + PE is constant.

Concepts Review

1. Let C be determined by $\mathbf{r} = \mathbf{r}(t)$, $a \le t \le b$, and let $\mathbf{a} = \mathbf{r}(a)$ and $\mathbf{b} = \mathbf{r}(b)$. Then, by the Fundamental Theorem for Line Integrals, $\int_C \nabla f(\mathbf{r})\cdot d\mathbf{r} =$ _____.

2. $\int_C \mathbf{F}(\mathbf{r})\cdot d\mathbf{r}$ is independent of path if and only if $\mathbf{F}$ is a _____ vector field, that is, if and only if $\mathbf{F}(\mathbf{r}) =$ _____ for some scalar function f.

3. If curl $\mathbf{F} =$ _____ in an open connected and simply connected set D, then $\mathbf{F} = \nabla f$ for some f defined on D. Conversely, curl $(\nabla f) =$ _____.

4. Let $\mathbf{F} = f(x)\mathbf{i} + g(y)\mathbf{j}$ be a two-dimensional vector field. If $\partial f/\partial y = \partial g/\partial x$, we conclude that _____.

Problem Set 14.3

In Problems 1–12, determine whether the given field $\mathbf{F}$ *is conservative. If so, find* f *so that* $\mathbf{F} = \nabla f$*; if not, state that* $\mathbf{F}$ *is not conservative. See Examples 2 and 4.*

1. $\mathbf{F}(x, y) = (10x - 7y)\mathbf{i} - (7x - 2y)\mathbf{j}$

2. $\mathbf{F}(x, y) = (12x^2 + 3y^2 + 5y)\mathbf{i} + (6xy - 3y^2 + 5x)\mathbf{j}$

3. $\mathbf{F}(x, y) = (45x^4y^2 - 6y^6 + 3)\mathbf{i} + (18x^5y - 12xy^5 + 7)\mathbf{j}$

4. $\mathbf{F}(x, y) = (35x^4 - 3x^2y^4 + y^9)\mathbf{i} - (4x^3y^3 - 9xy^8)\mathbf{j}$

5. $\mathbf{F}(x, y) = \left(\frac{6x^2}{5y^2}\right)\mathbf{i} - \left(\frac{4x^3}{5y^3}\right)\mathbf{j}$

6. $\mathbf{F}(x, y) = 4y^2 \cos(xy^2)\mathbf{i} + 8x \cos(xy^2)\mathbf{j}$

7. $\mathbf{F}(x, y) = (2e^y - ye^x)\mathbf{i} + (2xe^y - e^x)\mathbf{j}$

8. $\mathbf{F}(x, y) = -e^{-x} \ln y\, \mathbf{i} + e^{-x}y^{-1}\mathbf{j}$

9. $\mathbf{F}(x, y, z) = 3x^2\mathbf{i} + 6y^2\mathbf{j} + 9z^2\mathbf{k}$

10. $\mathbf{F}(x, y, z) = (2xy + z^2)\mathbf{i} + x^2\mathbf{j} + (2xz + \pi \cos \pi z)\mathbf{k}$

11. $\mathbf{F}(x, y, z) = \frac{-2x}{x^2 + z^2}\mathbf{i} + \frac{-2z}{x^2 + z^2}\mathbf{k}$

12. $\mathbf{F}(x, y, z) = (1 + 2yz^2)\mathbf{j} + (1 + 2y^2z)\mathbf{k}$

In Problems 13–20, show that the given line integral is independent of path (use Theorem C) and then evaluate the integral (either by choosing a convenient path or, if you prefer, by finding a potential function f and applying Theorem A).

13. $\int_{(-1,2)}^{(3,1)} (y^2 + 2xy)\, dx + (x^2 + 2xy)\, dy$

14. $\int_{(0,0)}^{(1,\pi/2)} e^x \sin y\, dx + e^x \cos y\, dy$

15. $\int_{(2,1)}^{(6,3)} \frac{x^3}{(x^4 + y^4)^2}\, dx + \frac{y^3}{(x^4 + y^4)^2}\, dy$

16. $\int_{(-1,1)}^{(4,2)} \left(y - \frac{1}{x^2}\right) dx + \left(x - \frac{1}{y^2}\right) dy$

17. $\int_{(0,0,0)}^{(1,1,1)} (6xy^3 + 2z^2)\, dx + 9x^2y^2\, dy + (4xz + 1)\, dz$

Hint: Try the path consisting of line segments from $(0, 0, 0)$ to $(1, 0, 0)$ to $(1, 1, 0)$ to $(1, 1, 1)$.

18. $\int_{(0,1,0)}^{(1,1,1)} (yz + 1)\, dx + (xz + 1)\, dy + (xy + 1)\, dz$

19. $\int_{(0,0,0)}^{(-1,0,\pi)} (y + z)\, dx + (x + z)\, dy + (x + y)\, dz$

20. $\int_{(0,0,0)}^{(\pi,\pi,0)} (\cos x + 2yz)\, dx + (\sin y + 2xz)\, dy + (z + 2xy)\, dz$

21. Suppose that $\nabla f(x, y, z) = M(x, y, z)\mathbf{i} + N(x, y, z)\mathbf{j} + P(x, y, z)\mathbf{k}$, where M, N, and P have continuous first-order partial derivatives in an open set D. Prove that

$$\frac{\partial M}{\partial y} = \frac{\partial N}{\partial x}, \quad \frac{\partial M}{\partial z} = \frac{\partial P}{\partial x}, \quad \frac{\partial N}{\partial z} = \frac{\partial P}{\partial y}$$

in D. *Hint:* Use Theorem 12.3C on f.

22. For each (x, y, z), let $\mathbf{F}(x, y, z)$ be a vector pointed toward the origin with magnitude inversely proportional to the distance from the origin; that is, let

$$\mathbf{F}(x, y, z) = \frac{-k(x\mathbf{i} + y\mathbf{j} + z\mathbf{k})}{x^2 + y^2 + z^2}$$

Show that $\mathbf{F}$ is conservative by finding a potential function for $\mathbf{F}$. *Hint:* If this looks like hard work, see Problem 24.

23. Follow the directions of Problem 22 for $\mathbf{F}(x, y, z)$ directed away from the origin with magnitude that is proportional to the distance from the origin.

24. Generalize Problems 22 and 23 by showing that if

$$\mathbf{F}(x, y, z) = [g(x^2 + y^2 + z^2)](x\mathbf{i} + y\mathbf{j} + z\mathbf{k})$$

where g is a continuous function of one variable, then $\mathbf{F}$ is conservative. *Hint:* Show that $\mathbf{F} = \nabla f$, where $f(x, y, z) = \frac{1}{2}h(x^2 + y^2 + z^2)$ and $h(u) = \int g(u)\, du$.

25. Suppose that an object of mass m is moved along a smooth curve C described by

$$\mathbf{r} = \mathbf{r}(t) = x(t)\mathbf{i} + y(t)\mathbf{j} + z(t)\mathbf{k}, \quad a \le t \le b$$

while subject only to the continuous force $\mathbf{F}$. Show that the work done is equal to the change in the kinetic energy of the object; that is, show that

$$\int_C \mathbf{F} \cdot d\mathbf{r} = \frac{m}{2}[\|\mathbf{r}'(b)\|^2 - \|\mathbf{r}'(a)\|^2]$$

Hint: $\mathbf{F}(\mathbf{r}(t)) = m\mathbf{r}''(t)$.

26. Matt moved a heavy object along the ground from A to B. The object was at rest at the beginning and at the end. Does Problem 25 imply that Matt did no work? Explain.

27. We normally consider the gravitational force of the earth on an object of mass m to be given by the constant $\mathbf{F} = -gm\mathbf{k}$, but, of course, this is valid only in regions near the earth's surface. Find the potential function f for $\mathbf{F}$ and use it to show that the work done by $\mathbf{F}$ when an object is moved from (x_1, y_1, z_1) to a nearby point (x_2, y_2, z_2) is $mg(z_1 - z_2)$.

C **28.** The distance from the earth (mass m) to the sun (mass M) varies from a maximum (aphelion) of 152.1 million kilometers to a minimum (perihelion) of 147.1 million kilometers. Assume that Newton's Inverse Square Law $\mathbf{F} = -GMm\mathbf{r}/\|\mathbf{r}\|^3$ holds, with $G = 6.67 \times 10^{-11}$ newton-meter^2/kilogram^2, $M = 1.99 \times 10^{30}$ kilograms, and $m = 5.97 \times 10^{24}$ kilograms. How much work does $\mathbf{F}$ do in moving the earth in each case?

(a) From aphelion to perihelion

(b) Around a complete orbit

29. This problem shows the need for simple connectedness in the "if" statement of Theorem C. Let $\mathbf{F} = (y\mathbf{i} - x\mathbf{j})/(x^2 + y^2)$ on the set $D = \{(x, y): x^2 + y^2 \ne 0\}$. Show each of the following.

(a) The condition $\partial M/\partial y = \partial N/\partial x$ holds on D.

(b) $\mathbf{F}$ is not conservative on D.

Hint: To establish part (b), show that $\int_C \mathbf{F} \cdot d\mathbf{r} = -2\pi$ where C is the circle with parametric equations $x = \cos t$, $y = \sin t$, $0 \le t \le 2\pi$.

30. Let $f(x, y) = \tan^{-1}(y/x)$. Show that $\nabla f = (y\mathbf{i} - x\mathbf{j})/(x^2 + y^2)$, which is the vector function of Problem 29. Why doesn't the hint to that problem violate Theorem A?

31. Prove that in Theorem C, Condition (3) implies Condition (2).

Answers to Concepts Review: **1.** $f(\mathbf{b}) - f(\mathbf{a})$ **2.** gradient or conservative; $\nabla f(\mathbf{r})$ **3.** 0; 0 **4.** $\mathbf{F}$ is conservative

14.4 Green's Theorem in the Plane

We begin with another look at the Second Fundamental Theorem of Calculus,

$$\int_a^b f'(x)\,dx = f(b) - f(a)$$

It says that the integral of a function over a set $S = [a, b]$ is equal to a related function (the antiderivative) evaluated in a certain way on the boundary of S, which in this case consists of just the two points a and b. In the remainder of this chapter, we are going to give three generalizations of this result: the theorems of Green, Gauss, and Stokes. These are generalizations of the Second Fundamental Theorem of Calculus in the sense that some integral (double or triple integral, or surface integral, defined in the next section) is expressed as some quantity evaluated on the boundary of the region of integration. These theorems are applied in physics, particularly in the study of heat, electricity, magnetism, and fluid flow. The first of these theorems is due to George Green (1793–1841), a self-taught English mathematical physicist.

We suppose that C is a simple closed curve (Section 10.4) that forms the boundary of a region S in the xy-plane. Let C be oriented so that traversing C in its positive direction keeps S to the left (the counterclockwise orientation). The corresponding line integral of $\mathbf{F}(x, y) = M(x, y)\mathbf{i} + N(x, y)\mathbf{j}$ around C is denoted by

$$\oint_C M\,dx + N\,dy$$

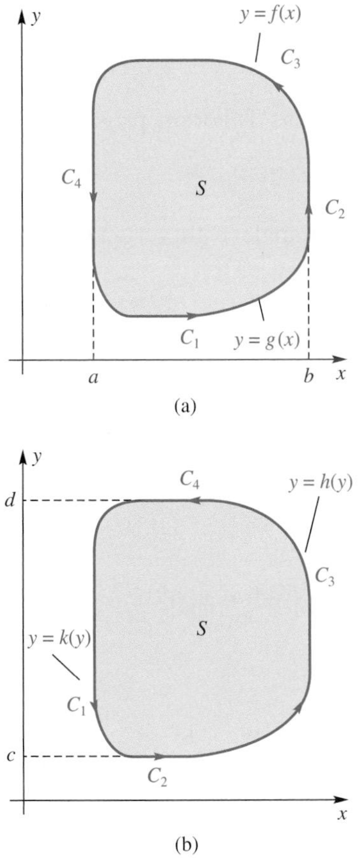

Figure 1

Theorem A **Green's Theorem**

Let C be a piecewise smooth, simple closed curve that forms the boundary of a region S in the xy-plane. If $M(x, y)$ and $N(x, y)$ are continuous and have continuous partial derivatives on S and its boundary C, then

$$\iint_S \left(\frac{\partial N}{\partial x} - \frac{\partial M}{\partial y}\right) dA = \oint_C M\,dx + N\,dy$$

Proof We prove the theorem for the case where S is both an x-simple and a y-simple set and then discuss extensions to the general case.

Since S is y-simple, it has the shape of Figure 1a; that is,

$$S = \{(x, y): g(x) \le y \le f(x), a \le x \le b\}$$

Its boundary C consists of four arcs C_1, C_2, C_3, and C_4 (although C_2 or C_4 could be degenerate) and

$$\oint_C M\,dx = \int_{C_1} M\,dx + \int_{C_2} M\,dx + \int_{C_3} M\,dx + \int_{C_4} M\,dx$$

The integrals over C_2 and C_4 are zero, since on these curves x is constant, so $dx = 0$. Thus,

$$\begin{aligned}\oint_C M\,dx &= \int_a^b M(x, g(x))\,dx + \int_b^a M(x, f(x))\,dx \\ &= -\int_a^b [M(x, f(x)) - M(x, g(x))]\,dx \\ &= -\int_a^b \int_{g(x)}^{f(x)} \frac{\partial M(x, y)}{\partial y}\,dy\,dx \\ &= -\iint_S \frac{\partial M}{\partial y}\,dA\end{aligned}$$

A Promised Result

Suppose that $\partial N/\partial x = \partial M/\partial y$. Then Green's Theorem tells us that

$$\oint_C M\,dx + N\,dy = 0$$

This in turn implies that the field $\mathbf{F} = M\mathbf{i} + N\mathbf{j}$ is conservative. This is part of what we claimed in Theorem 14.3D for the two-variable case

Similarly, by treating S as an x-simple set, we obtain

$$\oint_C N\,dy = \iint_S \frac{\partial N}{\partial x}\,dA$$

although the curves C_1, C_2, C_3, and C_4 must be redefined as in Figure 1b. We conclude that Green's Theorem holds on a set that is both x-simple and y-simple.

The result extends easily to a region S that decomposes into a union of regions $S_1, S_2, \ldots, S_k$, which are both x- and y-simple (Figure 2). We simply apply the theorem in the form that we have proved to each of these sets and then add the results. Note that the contributions of the line integrals cancel on boundaries shared by adjoining regions, since these boundaries are traversed twice, but in opposite directions. ■

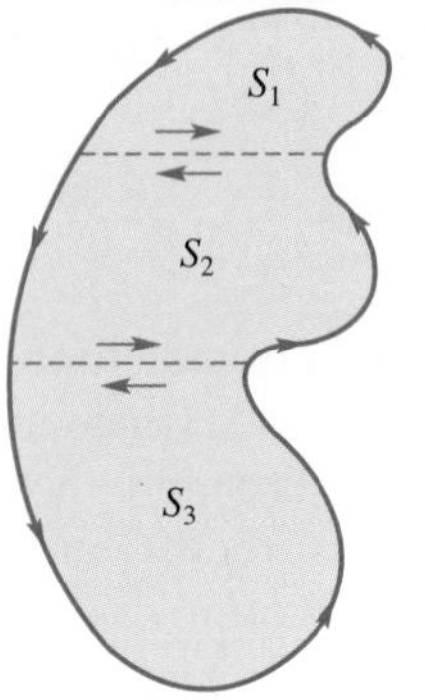

Figure 2

Green's Theorem even holds for a region S with one or more holes (Figure 3), provided that each part of the boundary is oriented so that S is always on the left as one traverses the curve in its positive direction. We simply decompose it into ordinary regions in the manner shown in Figure 4.

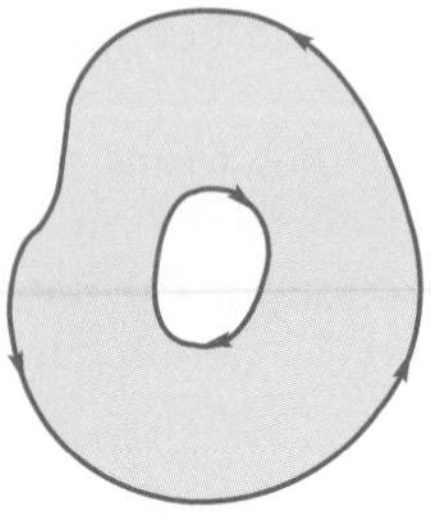

Figure 3

Examples and Applications Sometimes, Green's Theorem provides the simplest way of evaluating a line integral.

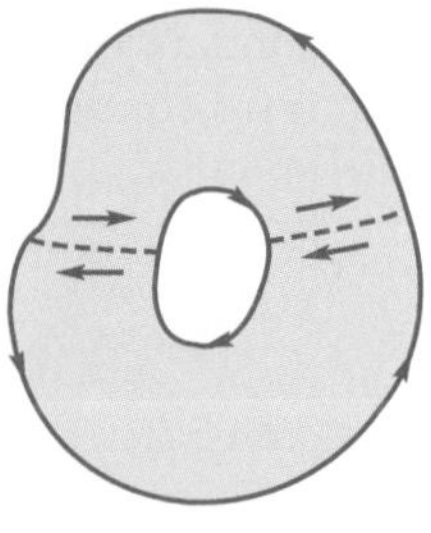

Figure 4

EXAMPLE 1 Let C be the boundary of the triangle with vertices $(0, 0)$, $(1, 2)$, and $(0, 2)$ (Figure 5). Calculate

$$\oint_C 4x^2y\,dx + 2y\,dy$$

(a) by the direct method, and (b) by Green's Theorem.

SOLUTION

(a) On C_1, $y = 2x$ and $dy = 2\,dx$, so

$$\int_{C_1} 4x^2y\,dx + 2y\,dy = \int_0^1 8x^3\,dx + 8x\,dx = [2x^4 + 4x^2]_0^1 = 6$$

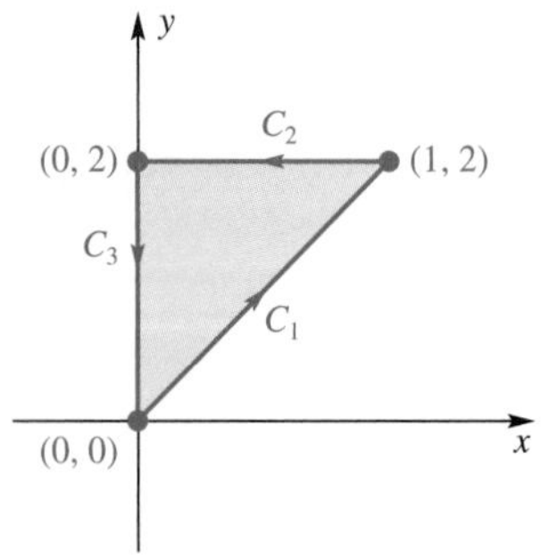

Figure 5

Also,

$$\int_{C_2} 4x^2y\,dx + 2y\,dy = \int_1^0 8x^2\,dx = \left[\frac{8x^3}{3}\right]_1^0 = -\frac{8}{3}$$

$$\int_{C_3} 4x^2y\,dx + 2y\,dy = \int_2^0 2y\,dy = [y^2]_2^0 = -4$$

Thus,

$$\oint_C 4x^2y\,dx + 2y\,dy = 6 - \frac{8}{3} - 4 = -\frac{2}{3}$$

(b) By Green's Theorem,

$$\begin{aligned}\oint_C 4x^2y\,dx + 2y\,dy &= \int_0^1 \int_{2x}^2 (0 - 4x^2)\,dy\,dx \\ &= \int_0^1 [-4x^2y]_{2x}^2\,dx = \int_0^1 (-8x^2 + 8x^3)\,dx \\ &= \left[\frac{-8x^3}{3} + 2x^4\right]_0^1 = -\frac{2}{3}\end{aligned}$$

■

EXAMPLE 2 Show that if a region S in the plane has boundary C, where C is a piecewise smooth, simple closed curve, then the area of S is given by

$$A(S) = \frac{1}{2}\oint_C (-y\,dx + x\,dy)$$

SOLUTION Let $M(x, y) = -y/2$ and $N(x, y) = x/2$, and apply Green's Theorem.

$$\oint_C \left(-\frac{y}{2}\,dx + \frac{x}{2}\,dy\right) = \iint_S \left(\frac{1}{2} + \frac{1}{2}\right) dA = A(S) \qquad \blacksquare$$

EXAMPLE 3 Use the result of Example 2 to find the area enclosed by the ellipse $\dfrac{x^2}{a^2} + \dfrac{y^2}{b^2} = 1$.

SOLUTION The given ellipse has parametric equations

$$x = a\cos t, \qquad y = b\sin t, \qquad 0 \le t \le 2\pi$$

Thus,

$$\begin{aligned} A(S) &= \frac{1}{2}\oint_C (-y\,dx + x\,dy) \\ &= \frac{1}{2}\int_0^{2\pi} (-(b\sin t)(-a\sin t\,dt) + (a\cos t)(b\cos t\,dt)) \\ &= \frac{1}{2}\int_0^{2\pi} ab(\sin^2 t + \cos^2 t)\,dt \\ &= \frac{1}{2}ab\int_0^{2\pi} dt = \pi ab \end{aligned} \qquad \blacksquare$$

EXAMPLE 4 Use Green's Theorem to evaluate the line integral

$$\oint_C (x^3 + 2y)\,dx + (4x - 3y^2)\,dy$$

where C is the ellipse $\dfrac{x^2}{a^2} + \dfrac{y^2}{b^2} = 1$.

SOLUTION Let $M(x, y) = x^3 + 2y$ and $N(x, y) = 4x - 3y^2$ so that $\partial M/\partial y = 2$ and $\partial N/\partial x = 4$. By Green's Theorem and Example 3,

$$\oint_C (x^3 + 2y)\,dx + (4x - 3y^2)\,dy = \iint_S (4 - 2)\,dA = 2A(S) = 2\pi ab \qquad \blacksquare$$

Vector Forms of Green's Theorem Our next goal is to restate Green's Theorem for the plane in its vector form in two different ways. It is these forms that we will generalize later to two important theorems in three-space.

We suppose that C is a smooth, simple closed curve in the xy-plane and that it has been given a counterclockwise orientation by means of its arc length parametrization $x = x(s)$ and $y = y(s)$. Then

$$\mathbf{T} = \frac{dx}{ds}\,\mathbf{i} + \frac{dy}{ds}\,\mathbf{j}$$

is a unit tangent vector and

$$\mathbf{n} = \frac{dy}{ds}\,\mathbf{i} - \frac{dx}{ds}\,\mathbf{j}$$

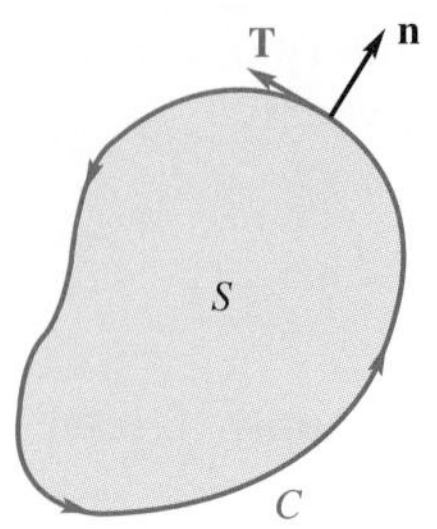

Figure 6

is a unit normal vector pointing out of the region S bounded by C (Figure 6). (Note that $\mathbf{T}\cdot\mathbf{n} = 0$.) If $\mathbf{F}(x, y) = M(x, y)\mathbf{i} + N(x, y)\mathbf{j}$ is a vector field, then

$$\oint_C \mathbf{F}\cdot\mathbf{n}\,ds = \oint_C (M\mathbf{i} + N\mathbf{j})\cdot\left(\frac{dy}{ds}\mathbf{i} - \frac{dx}{ds}\mathbf{j}\right)ds = \oint_C (-N\,dx + M\,dy)$$
$$= \iint_S \left(\frac{\partial M}{\partial x} + \frac{\partial N}{\partial y}\right)dA$$

The last equality comes from Green's Theorem. On the other hand,

$$\operatorname{div}\mathbf{F} = \nabla\cdot\mathbf{F} = \frac{\partial M}{\partial x} + \frac{\partial N}{\partial y}$$

We conclude that

$$\oint_C \mathbf{F}\cdot\mathbf{n}\,ds = \iint_S \operatorname{div}\mathbf{F}\,dA = \iint_S \nabla\cdot\mathbf{F}\,dA$$

a result sometimes called Gauss's Divergence Theorem in the plane.

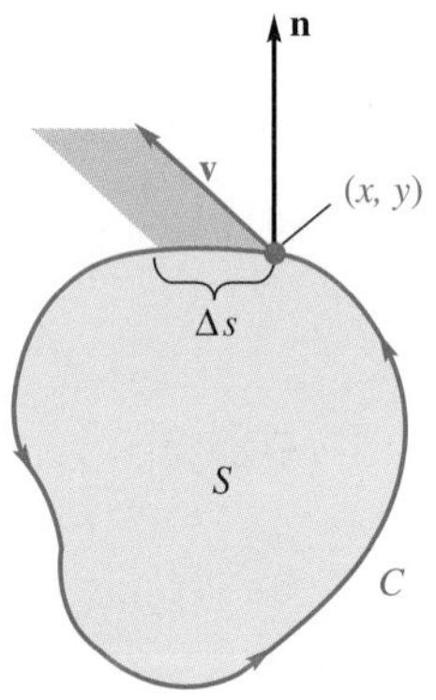

Figure 7

We give a physical interpretation to this last formula and thereby also understand the origin of the term *divergence*. Imagine a uniform layer of a fluid of constant density moving across the xy-plane, a layer so thin that we may consider it to be two-dimensional. We wish to compute the rate at which the fluid in a region S crosses its boundary curve C (Figure 7).

Let $\mathbf{F}(x, y) = \mathbf{v}(x, y)$ denote the velocity vector of the fluid at (x, y), and let Δs be the length of a short segment of the curve with initial point (x, y). The amount of fluid crossing this segment per unit of time is approximately the area of the parallelogram of Figure 7, that is $\mathbf{v}\cdot\mathbf{n}\,\Delta s$. The (net) amount of fluid leaving S per unit of time, called the **flux** of the vector field $\mathbf{F}$ across the curve C in the outward direction, is therefore

$$\text{flux of }\mathbf{F}\text{ across }C = \oint_C \mathbf{F}\cdot\mathbf{n}\,ds$$

Now consider a fixed point (x_0, y_0) in S and a small circle C_r of radius r around it. On S_r, the circular region with boundary C_r, div $\mathbf{F}$ will be approximately equal to its value div $\mathbf{F}(x_0, y_0)$ at the center (we are assuming that div $\mathbf{F}$ is continuous); so, by Green's Theorem,

$$\text{flux of }\mathbf{F}\text{ across }C_r = \oint_C \mathbf{F}\cdot\mathbf{n}\,ds = \iint_{S_r} \operatorname{div}\mathbf{F}\,dA \approx \operatorname{div}\mathbf{F}(x_0, y_0)(\pi r^2)$$

We conclude that div $\mathbf{F}(x_0, y_0)$ measures the rate at which the fluid is "diverging away" from (x_0, y_0). If div $\mathbf{F}(x_0, y_0) > 0$, there is a *source* of fluid at (x_0, y_0); if div $\mathbf{F}(x_0, y_0) < 0$, there is a *sink* for the fluid at (x_0, y_0). If the flux across the boundary of a region is zero, then the sources and sinks in the region must balance each other. On the other hand, if there are no sources or sinks in a region S, then div $\mathbf{F} = 0$ and, by Green's Theorem, there is a net flow of zero across the boundary of S.

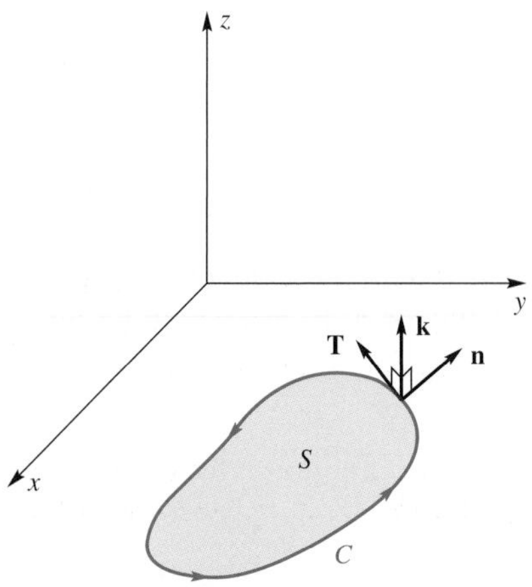

Figure 8

There is another vector form for Green's Theorem. We redraw Figure 6, but now as a subset of three-space (Figure 8). If $\mathbf{F} = M\mathbf{i} + N\mathbf{j} + 0\mathbf{k}$, then Green's Theorem says that

$$\oint_C \mathbf{F}\cdot\mathbf{T}\,ds = \oint_C M\,dx + N\,dy = \iint_S \left(\frac{\partial N}{\partial x} - \frac{\partial M}{\partial y}\right)dA$$

On the other hand,

$$\text{curl } \mathbf{F} = \nabla \times \mathbf{F} = \begin{vmatrix} \mathbf{i} & \mathbf{j} & \mathbf{k} \\ \dfrac{\partial}{\partial x} & \dfrac{\partial}{\partial y} & \dfrac{\partial}{\partial z} \\ M & N & 0 \end{vmatrix} = \left(\frac{\partial N}{\partial x} - \frac{\partial M}{\partial y}\right)\mathbf{k}$$

so that

$$(\text{curl } \mathbf{F}) \cdot \mathbf{k} = \left(\frac{\partial N}{\partial x} - \frac{\partial M}{\partial y}\right)$$

Green's Theorem thus takes the form

$$\oint_C \mathbf{F} \cdot \mathbf{T}\, ds = \iint_S (\text{curl } \mathbf{F}) \cdot \mathbf{k}\, dA$$

which is sometimes called Stokes's Theorem in the plane.

If we apply this result to a small circle C_r centered at (x_0, y_0), we obtain

$$\oint_{C_r} \mathbf{F} \cdot \mathbf{T}\, ds \approx (\text{curl } \mathbf{F}(x_0, y_0)) \cdot \mathbf{k}(\pi r^2)$$

This says that the flow in the direction of the tangent to C_r (the *circulation* of **F** around C_r) is measured by the curl of **F**. In other words, curl **F** measures the tendency of the fluid to rotate about (x_0, y_0). If curl $\mathbf{F} = \mathbf{0}$ in a region S, the corresponding fluid flow is said to be *irrotational.*

EXAMPLE 5 The vector field $\mathbf{F}(x, y) = -\frac{1}{2}y\mathbf{i} + \frac{1}{2}x\mathbf{j} = M\mathbf{i} + N\mathbf{j}$ is the velocity field of a steady counterclockwise rotation of a wheel about the z-axis (see Example 2 of Section 14.1). Calculate $\oint_C \mathbf{F} \cdot \mathbf{n}\, ds$ and $\oint_C \mathbf{F} \cdot \mathbf{T}\, ds$ for any closed curve C in the xy-plane.

SOLUTION If S is the region enclosed by C,

$$\oint_C \mathbf{F} \cdot \mathbf{n}\, ds = \iint_S \text{div } \mathbf{F}\, dA = \iint_S \left(\frac{\partial M}{\partial x} + \frac{\partial N}{\partial y}\right) dA = 0$$

$$\oint_C \mathbf{F} \cdot \mathbf{T}\, ds = \iint_S (\text{curl } \mathbf{F}) \cdot \mathbf{k}\, dA = \iint_S \left(\frac{\partial N}{\partial x} - \frac{\partial M}{\partial y}\right) dA$$

$$= \iint_S \left(\frac{1}{2} + \frac{1}{2}\right) dA = A(S)$$

■

Concepts Review

1. Let C be a simple closed curve bounding a region S in the xy-plane. Then, by Green's Theorem, $\oint_C M\, dx + N\, dy = \iint_S$ ____ dA.

2. Thus, if C is the boundary of the square $S = \{(x, y): 0 \le x \le 1,\ 0 \le y \le 1\}$, $\oint_C y\, dx - x\, dy = \iint_S$ ____ $dA =$ ____.

3. The div $\mathbf{F}(x, y)$ measures the rate at which a homogeneous fluid flow with velocity field **F** diverges away from (x, y). If div $\mathbf{F}(x, y) > 0$, there is a(n) ____ of fluid at (x, y): if div $\mathbf{F}(x, y) < 0$, there is a(n) ____ at (x, y).

4. On the other hand, curl $\mathbf{F}(x, y)$ measures the tendency of the fluid to ____ about (x, y). If curl $\mathbf{F}(x, y) = \mathbf{0}$ in a region, the flow is ____.

Problem Set 14.4

In Problems 1–6, use Green's Theorem to evaluate the given line integral. Begin by sketching the region S.

1. $\oint_C 2xy\,dx + y^2\,dy$, where C is the closed curve formed by $y = x/2$ and $y = \sqrt{x}$ between $(0,0)$ and $(4,2)$

2. $\oint_C \sqrt{y}\,dx + \sqrt{x}\,dy$, where C is the closed curve formed by $y = 0$, $x = 2$, and $y = x^2/2$

3. $\oint_C (2x + y^2)\,dx + (x^2 + 2y)\,dy$, where C is the closed curve formed by $y = 0$, $x = 2$, and $y = x^3/4$

4. $\oint_C xy\,dx + (x + y)\,dy$, where C is the triangle with vertices $(0,0)$, $(2,0)$, and $(0,1)$

5. $\oint_C (x^2 + 4xy)\,dx + (2x^2 + 3y)\,dy$, where C is the ellipse $9x^2 + 16y^2 = 144$

6. $\oint_C (e^{3x} + 2y)\,dx + (x^2 + \sin y)\,dy$, where C is the rectangle with vertices $(2,1)$, $(6,1)$, $(6,4)$, and $(2,4)$.

In Problems 7 and 8, use the result of Example 2 to find the area of the indicated region S. Make a sketch.

7. S is bounded by the curves $y = 4x$ and $y = 2x^2$.

8. S is bounded by the curves $y = \frac{1}{2}x^3$ and $y = x^2$.

In Problems 9–12, use the vector forms of Green's Theorem to calculate (a) $\oint_C \mathbf{F}\cdot\mathbf{n}\,ds$ *and (b)* $\oint_C \mathbf{F}\cdot\mathbf{T}\,ds$.

9. $\mathbf{F} = y^2\mathbf{i} + x^2\mathbf{j}$; C is the boundary of unit square with vertices $(0,0)$, $(1,0)$, $(1,1)$, and $(0,1)$.

10. $\mathbf{F} = ay\mathbf{i} + bx\mathbf{j}$; C as in Problem 9.

11. $\mathbf{F} = y^3\mathbf{i} + x^3\mathbf{j}$; C is the unit circle.

12. $\mathbf{F} = x\mathbf{i} + y\mathbf{j}$; C is the unit circle.

13. Suppose that the integrals $\oint \mathbf{F}\cdot\mathbf{T}\,ds$ taken counterclockwise around the circles $x^2 + y^2 = 36$ and $x^2 + y^2 = 1$ are 30 and -20, respectively. Calculate $\iint_S (\text{curl }\mathbf{F})\cdot\mathbf{k}\,dA$, where S is the region between the circles.

14. If $\mathbf{F} = (x^2 + y^2)\mathbf{i} + 2xy\,\mathbf{j}$, find the flux of $\mathbf{F}$ across the boundary C of the unit square with vertices $(0,0)$, $(1,0)$, $(1,1)$, and $(0,1)$; that is, calculate $\oint_C \mathbf{F}\cdot\mathbf{n}\,ds$.

15. Find the work done by $\mathbf{F} = (x^2 + y^2)\mathbf{i} - 2xy\,\mathbf{j}$ in moving a body counterclockwise around the curve C of Problem 14.

16. If $\mathbf{F} = (x^2 + y^2)\mathbf{i} + 2xy\,\mathbf{j}$, calculate the circulation of $\mathbf{F}$ around C of Problem 14; that is, calculate $\oint_C \mathbf{F}\cdot\mathbf{T}\,ds$.

17. Show that the work done by a constant force $\mathbf{F}$ in moving a body around a simple closed curve is 0.

18. Use Green's Theorem to prove the plane case of Theorem 14.3D; that is, show that $\partial N/\partial x = \partial M/\partial y$ implies that $\oint_C M\,dx + N\,dy = 0$, which implies that $\mathbf{F} = M\mathbf{i} + N\mathbf{j}$ is conservative.

19. Let

$$\mathbf{F} = \frac{y}{x^2 + y^2}\mathbf{i} - \frac{x}{x^2 + y^2}\mathbf{j} = M\mathbf{i} + N\mathbf{j}$$

(a) Show that $\partial N/\partial x = \partial M/\partial y$.

(b) Show, by using the parametrization $x = \cos t$, $y = \sin t$, that $\oint_C M\,dx + N\,dy = -2\pi$, where C is the unit circle.

(c) Why doesn't this contradict Green's Theorem?

20. Let $\mathbf{F}$ be as in Problem 19. Calculate $\oint_C M\,dx + N\,dy$, where

(a) C is the ellipse $x^2/9 + y^2/4 = 1$

(b) C is the square with vertices $(1,-1)$, $(1,1)$, $(-1,1)$, and $(-1,-1)$

(c) C is the triangle with vertices $(1,0)$, $(2,0)$, and $(1,1)$.

21. Let the piecewise smooth, simple closed curve C be the boundary of a region S in the xy-plane. Modify the argument in Example 2 to show that

$$A(S) = \oint_C (-y)\,dx = \oint_C x\,dy$$

22. Let S and C be as in Problem 21. Show that the moments M_x and M_y about the x- and y-axes are given by

$$M_x = -\frac{1}{2}\oint_C y^2\,dx, \qquad M_y = \frac{1}{2}\oint_C x^2\,dy$$

23. Calculate the area of the asteroid $x^{2/3} + y^{2/3} = a^{2/3}$. *Hint:* Parametrize by $x = a\cos^3 t$, $y = a\sin^3 t$, $0 \le t \le 2\pi$.

24. Calculate the work done by $\mathbf{F} = 2y\mathbf{i} - 3x\mathbf{j}$ in moving an object around the asteroid of Problem 23.

25. Let $\mathbf{F}(\mathbf{r}) = \mathbf{r}/\|\mathbf{r}\|^2 = (x\,\mathbf{i} + y\,\mathbf{j})/(x^2 + y^2)$.

(a) Show that $\int_C \mathbf{F}\cdot\mathbf{n}\,ds = 2\pi$, where C is the circle centered at the origin of radius a and $\mathbf{n} = (x\mathbf{i} + y\,\mathbf{j})/\sqrt{x^2 + y^2}$ is the exterior unit normal to C.

(b) Show that div $\mathbf{F} = 0$.

(c) Explain why the results of parts (a) and (b) do not contradict the vector form of Green's Theorem.

(d) Show that if C is a smooth simple closed curve then $\int_C \mathbf{F}\cdot\mathbf{n}\,ds$ equals 2π or 0 accordingly as the origin is inside or outside C.

26. Area of a Polygon Let $V_0(x_0, y_0), V_1(x_1, y_1), \ldots, V_n(x_n, y_n)$, be the vertices of a simple polygon P, labeled counterclockwise and with $V_0 = V_n$. Show each of the following:

(a) $\int_C x\,dy = \frac{1}{2}(x_1 + x_0)(y_1 + y_0)$, where C is the edge V_0V_1

(b) Area $(P) = \sum_{i=1}^{n} \frac{x_i + x_{i-1}}{2}(y_i - y_{i-1})$

(c) The area of a polygon with vertices having integral coordinates is always a multiple of $\frac{1}{2}$.

(d) The formula in part (b) gives the correct answer for the polygon with vertices $(2, 0)$, $(2, -2)$, $(6, -2)$, $(6, 0)$, $(10, 4)$, and $(-2, 4)$.

CAS *In each of the following problems, plot the graph of $f(x, y)$ and the corresponding gradient field $\mathbf{F} = \nabla f$ on $S = \{(x, y): -3 \le x \le 3, -3 \le y \le 3\}$. Note that, in each case,* curl $\mathbf{F} = \mathbf{0}$ *(Theorem 14.3D), and so there is no tendency for rotation around any point.*

27. Let $f(x, y) = x^2 + y^2$.

(a) By visually examining the field $\mathbf{F}$, convince yourself that div $\mathbf{F} > 0$ on S. Then calculate div $\mathbf{F}$.

(b) Calculate the flux of $\mathbf{F}$ across the boundary of S.

28. Let $f(x, y) = \ln(\cos(x/3)) - \ln(\cos(y/3))$.

(a) Guess whether div $\mathbf{F}$ is positive or negative at a few points and then calculate div $\mathbf{F}$ to check on your guesses.

(b) Calculate the flux of $\mathbf{F}$ across the boundary of S.

29. Let $f(x, y) = \sin x \sin y$.

(a) By visually examining the field $\mathbf{F}$, guess where div $\mathbf{F}$ is positive and where it is negative. Then calculate div $\mathbf{F}$ to check on your guesses.

(b) Calculate the flux of $\mathbf{F}$ across the boundary of S; then calculate it across the boundary of $T = \{(x, y): 0 \le x \le 3, 0 \le y \le 3\}$.

30. Let $f(x, y) = \exp(-(x^2 + y^2)/4)$. Guess where div $\mathbf{F}$ is positive and where it is negative. Then determine this analytically.

Answers to Concepts Review: **1.** $\frac{\partial N}{\partial x} - \frac{\partial M}{\partial y}$ **2.** -2; -2 **3.** source; sink **4.** rotate; irrotational

---

14.5 Surface Integrals

A line integral generalizes the ordinary definite integral; in a similar way, a surface integral generalizes a double integral.

Let the surface G be the graph of $z = f(x, y)$, where (x, y) ranges over a rectangle R in the xy-plane. Let P be a partition of R into n subrectangles R_i; this results in a corresponding partition of the surface G into n pieces G_i (Figure 1). Choose a sample point $(\bar{x}_i, \bar{y}_i)$ in R_i, and let $(\bar{x}_i, \bar{y}_i, \bar{z}_i) = (\bar{x}_i, \bar{y}_i, f(\bar{x}_i, \bar{y}_i))$ be the corresponding point on G_i. Then define the **surface integral** of g over G by

$$\iint_G g(x, y, z)\,dS = \lim_{\|P\| \to 0} \sum_{i=1}^{n} g(\bar{x}_i, \bar{y}_i, \bar{z}_i)\,\Delta S_i$$

where ΔS_i is the area of G_i. Finally, extend the definition to the case where R is a general closed, bounded set in the xy-plane in the usual way (by giving g the value 0 outside R).

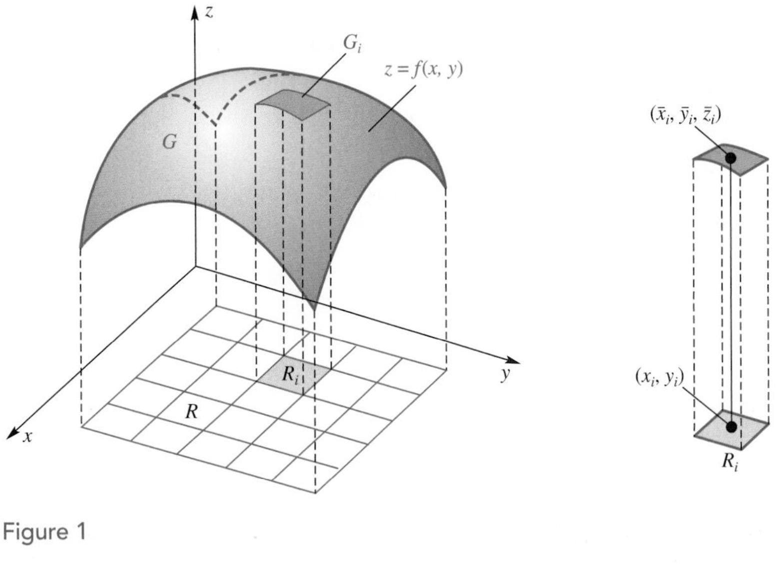

Figure 1

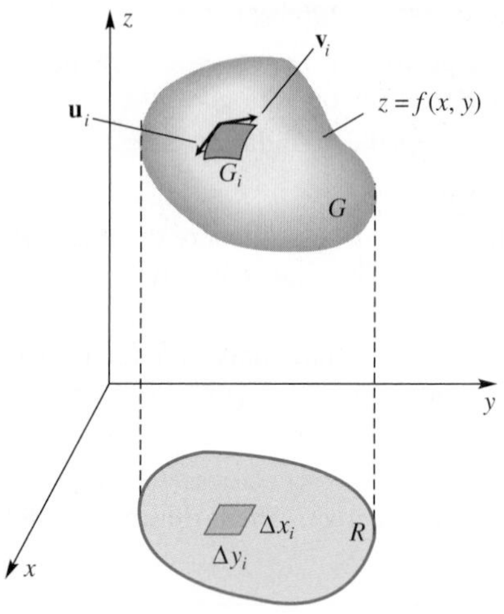

Figure 2

Evaluating Surface Integrals A definition is not enough; we need a practical way to evaluate a surface integral. The development in Section 13.6 suggests the correct result. There we showed that under appropriate hypotheses, the area of a small patch G_i (Figure 2) of the surface is approximately $\|\mathbf{u}_i \times \mathbf{v}_i\|$, where $\mathbf{u}_i$ and $\mathbf{v}_i$ are sides of a parallelogram that is tangent to the surface. Thus,

$$A(G_i) \approx \|\mathbf{u}_i \times \mathbf{v}_j\| \approx \sqrt{(f_x(x_i, y_i))^2 + (f_y(x_i, y_i))^2 + 1}\,\Delta y_i\,\Delta x_i$$

This leads to the following theorem.

Theorem A

Let G be a surface given by $z = f(x, y)$, where (x, y) is in R. If f has continuous first-order partial derivatives and $g(x, y, z) = g(x, y, f(x, y))$ is continuous on R, then

$$\iint_G g(x, y, z)\,dS = \iint_R g(x, y, f(x, y))\sqrt{f_x^2 + f_y^2 + 1}\,dy\,dx$$

Note that when $g(x, y, z) = 1$ Theorem A gives the formula for surface area given in Section 13.6.

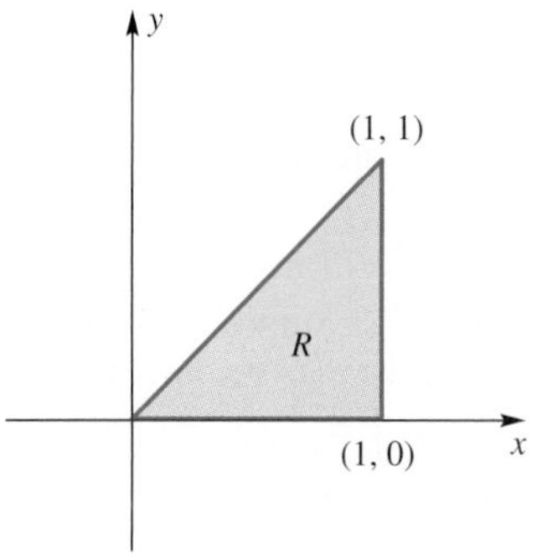

Figure 3

EXAMPLE 1 Evaluate $\iint_G (xy + z)\,dS$, where G is the part of the plane $2x - y + z = 3$ above the triangle R sketched in Figure 3.

SOLUTION In this case, $z = 3 + y - 2x = f(x, y)$, $f_x = -2$, $f_y = 1$, and $g(x, y, z) = xy + 3 + y - 2x$. Thus,

$$\iint_G (xy + z)\,dS = \int_0^1 \int_0^x (xy + 3 + y - 2x)\sqrt{(-2)^2 + 1^2 + 1}\,dy\,dx$$

$$= \sqrt{6}\int_0^1 \left[\frac{xy^2}{2} + 3y + \frac{y^2}{2} - 2xy\right]_0^x dx$$

$$= \sqrt{6}\int_0^1 \left[\frac{x^3}{2} + 3x - \frac{3x^2}{2}\right] dx = \frac{9\sqrt{6}}{8}$$

■

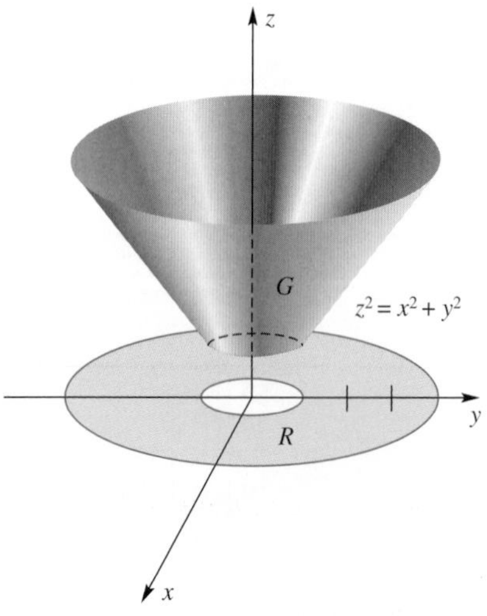

Figure 4

EXAMPLE 2 Evaluate $\iint_G xyz\,dS$, where G is the portion of the cone $z^2 = x^2 + y^2$ between the planes $z = 1$ and $z = 4$ (Figure 4).

SOLUTION We may write

$$z = (x^2 + y^2)^{1/2} = f(x, y)$$

from which

$$f_x^2 + f_y^2 + 1 = \frac{x^2}{x^2 + y^2} + \frac{y^2}{x^2 + y^2} + 1 = 2$$

Thus,

$$\iint_G xyz\,dS = \iint_R xy\sqrt{x^2 + y^2}\sqrt{2}\,dy\,dx$$

After a change to polar coordinates, this becomes

$$\sqrt{2}\int_0^{2\pi}\int_1^4 (r\cos\theta)(r\sin\theta)r^2\,dr\,d\theta = \sqrt{2}\int_0^{2\pi}\left[\sin\theta\cos\theta\frac{r^5}{5}\right]_1^4 d\theta$$
$$= \frac{1023\sqrt{2}}{5}\left[\frac{\sin^2\theta}{2}\right]_0^{2\pi} = 0 \quad ■$$

EXAMPLE 3 The portion of the spherical surface G with equation

$$z = f(x, y) = \sqrt{9 - x^2 - y^2}$$

where x and y satisfy $x^2 + y^2 \leq 4$ has a thin metal covering whose density at (x, y, z) is $\delta(x, y, z) = z$. Find the mass of this covering.

SOLUTION Let R be the projection of G in the xy-plane; that is, let $R = \{(x, y): x^2 + y^2 \leq 4\}$. Then

$$m = \iint_G \delta(x, y, z)\,dS = \iint_R z\sqrt{f_x^2 + f_y^2 + 1}\,dA$$
$$= \iint_R z\sqrt{\frac{x^2}{9 - x^2 - y^2} + \frac{y^2}{9 - x^2 - y^2} + 1}\,dA$$
$$= \iint_R z\frac{3}{z}\,dA = 3(\pi 2^2) = 12\pi \quad ■$$

Let the surface G be given by an equation of the form $y = h(x, z)$, and let R be its projection in the xz-plane. Then the appropriate formula for the surface integral is

$$\iint_G g(x, y, z)\,dS = \iint_R g(x, h(x, z), z)\sqrt{h_x^2 + h_z^2 + 1}\,dx\,dz$$

There is a corresponding formula when the surface G is given by $x = k(y, z)$.

EXAMPLE 4 Evaluate $\iint_G (x^2 + z^2)\,dS$, where G is the part of the paraboloid $y = 1 - x^2 - z^2$ that projects onto $R = \{(x, z): x^2 + z^2 \leq 1\}$.

SOLUTION

$$\iint_G (x^2 + z^2)\,dS = \iint_R (x^2 + z^2)\sqrt{4x^2 + 4z^2 + 1}\,dA$$

If we use polar coordinates, this becomes

$$\int_0^{2\pi}\int_0^1 r^2\sqrt{4r^2 + 1}\,r\,dr\,d\theta$$

In the inner integral, let $u = \sqrt{4r^2 + 1}$, so $u^2 = 4r^2 + 1$ and $u\,du = 4r\,dr$. We obtain

$$\frac{1}{16}\int_0^{2\pi}\int_1^{\sqrt{5}} (u^2 - 1)u^2\,du\,d\theta = \frac{\left(25\sqrt{5} + 1\right)\pi}{60} \approx 2.979 \quad ■$$

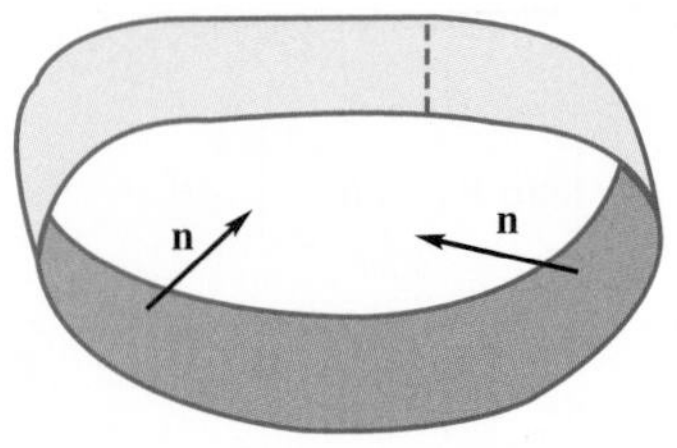

Figure 5

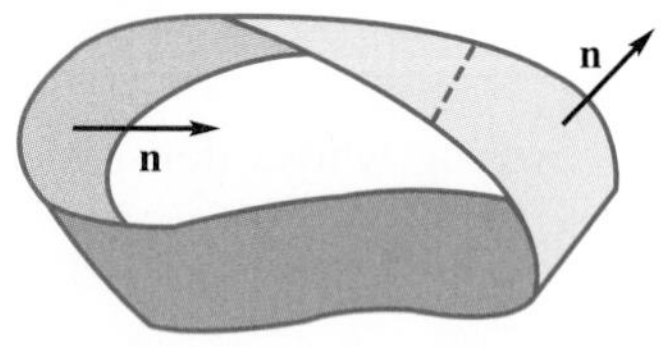

Figure 6

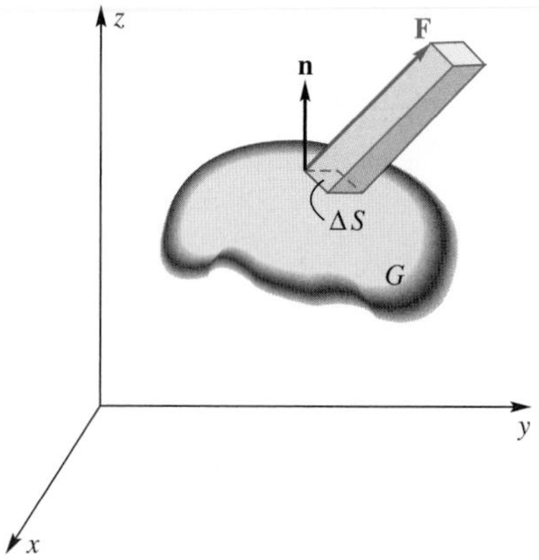

Figure 7

The Flux of a Vector Field Through a Surface For what we discuss now and for later applications, we need to limit further the kinds of surfaces that we consider. Most surfaces that arise in practice have two sides. However, it is surprisingly easy to construct a surface with just one side. Take a paper band (Figure 5), slit it at the dotted line, give one end a half twist, and paste it back together (Figure 6). You obtain a one-sided surface, called a Möbius band.

From now on, we consider only two-sided surfaces, so it will make sense to talk about a fluid flowing through the surface from one side to the other as if the surface were a screen. We also suppose that the surface is *smooth*, meaning that it has a continuously varying unit normal $\mathbf{n}$. Let G be such a smooth, two-sided surface, and assume that it is submerged in a fluid with a continuous velocity field $\mathbf{F}(x, y, z)$. If ΔS is the area of a small piece of G, then $\mathbf{F}$ is almost constant there, and the volume ΔV of fluid crossing this piece in the direction of the unit normal $\mathbf{n}$ (Figure 7) is

$$\Delta V \approx \mathbf{F} \cdot \mathbf{n}\, \Delta S$$

We conclude that

$$\boxed{\text{flux of } \mathbf{F} \text{ across } G = \iint_G \mathbf{F} \cdot \mathbf{n}\, dS}$$

EXAMPLE 5 Find the upward flux of $\mathbf{F} = -y\,\mathbf{i} + x\,\mathbf{j} + 9\,\mathbf{k}$ across the part of the spherical surface G determined by

$$z = f(x, y) = \sqrt{9 - x^2 - y^2}, \qquad 0 \le x^2 + y^2 \le 4$$

SOLUTION Note that the field $\mathbf{F}$ is a rotating stream flowing in the direction of the positive z-axis.

The equation of the surface may be written as

$$H(x, y, z) = z - \sqrt{9 - x^2 - y^2} = z - f(x, y) = 0$$

and thus

$$\mathbf{n} = \frac{\nabla H}{\|\nabla H\|} = \frac{-f_x\mathbf{i} - f_y\,\mathbf{j} + \mathbf{k}}{\sqrt{f_x^2 + f_y^2 + 1}} = \frac{(x/z)\mathbf{i} + (y/z)\mathbf{j} + \mathbf{k}}{\sqrt{(x/z)^2 + (y/z)^2 + 1}}$$

is a unit vector normal to the surface. The vector $-\mathbf{n}$ is also normal to the surface, but, since we desire the normal unit vector that points upward, $\mathbf{n}$ is the right choice. A straightforward computation using the fact that $x^2 + y^2 + z^2 = 9$ gives

$$\mathbf{n} = \frac{(x/z)\mathbf{i} + (y/z)\mathbf{j} + \mathbf{k}}{3/z} = \frac{x}{3}\,\mathbf{i} + \frac{y}{3}\,\mathbf{j} + \frac{z}{3}\,\mathbf{k}$$

(A simple geometric argument will also give this result; the normal must point directly away from the origin.)

The flux of $\mathbf{F}$ across G is given by

$$\begin{aligned}\text{flux} &= \iint_G \mathbf{F} \cdot \mathbf{n}\, dS \\ &= \iint_G (-y\,\mathbf{i} + x\,\mathbf{j} + 9\,\mathbf{k}) \cdot \left(\frac{x}{3}\,\mathbf{i} + \frac{y}{3}\,\mathbf{j} + \frac{z}{3}\,\mathbf{k}\right) dS \\ &= \iint_G 3z\, dS\end{aligned}$$

Finally, we write this surface integral as a double integral, using the fact that R is a circle of radius 2 and that $\sqrt{f_x^2 + f_y^2 + 1} = 3/\sqrt{9 - x^2 - y^2} = 3/z$.

$$\text{flux} = \iint_G 3z\, dS = \iint_R 3z \frac{3}{z}\, dA = 9(\pi \cdot 2^2) = 36\pi$$

The total flux across G in one unit of time is 36π cubic units. ■

An observant reader, after noting the cancellation that occurred in Example 5, will suspect that a theorem is lurking near.

Theorem B

Let G be a smooth, two-sided surface given by $z = f(x, y)$, where (x, y) is in R, and let $\mathbf{n}$ denote the upward unit normal on G. If f has continuous first-order partial derivatives and $\mathbf{F} = M\mathbf{i} + N\mathbf{j} + P\mathbf{k}$ is a continuous vector field, then the flux of $\mathbf{F}$ across G is given by

$$\text{flux } \mathbf{F} = \iint_G \mathbf{F} \cdot \mathbf{n}\, dS = \iint_R [-Mf_x - Nf_y + P]\, dx\, dy$$

Proof If we write $H(x, y, z) = z - f(x, y)$, we obtain

$$\mathbf{n} = \frac{\nabla H}{\|\nabla H\|} = \frac{-f_x\mathbf{i} - f_y\mathbf{j} + \mathbf{k}}{\sqrt{f_x^2 + f_y^2 + 1}}$$

It follows from Theorem A that

$$\iint_G \mathbf{F} \cdot \mathbf{n}\, dS = \iint_R (M\,\mathbf{i} + N\,\mathbf{j} + P\,\mathbf{k}) \cdot \frac{-f_x\mathbf{i} - f_y\mathbf{j} + \mathbf{k}}{\sqrt{f_x^2 + f_y^2 + 1}} \sqrt{f_x^2 + f_y^2 + 1}\, dx\, dy$$

$$= \iint_R (-Mf_x - Nf_y + P)\, dx\, dy$$ ■

You might try reworking Example 5 using Theorem B. We offer a different example.

EXAMPLE 6 Evaluate the flux for the vector field $\mathbf{F} = x\,\mathbf{i} + y\,\mathbf{j} + z\,\mathbf{k}$ across the part G of the paraboloid $z = 1 - x^2 - y^2$ that lies above the xy-plane, taking $\mathbf{n}$ to be the upward normal.

SOLUTION

$$f(x, y) = 1 - x^2 - y^2, \qquad f_x = -2x, \qquad f_y = -2y$$

$$\begin{aligned} -Mf_x - Nf_y + P &= 2x^2 + 2y^2 + z \\ &= 2x^2 + 2y^2 + 1 - x^2 - y^2 = 1 + x^2 + y^2 \end{aligned}$$

$$\iint_G \mathbf{F} \cdot \mathbf{n}\, dS = \iint_R (1 + x^2 + y^2)\, dx\, dy = \int_0^{2\pi} \int_0^1 (1 + r^2) r\, dr\, d\theta = \frac{3}{2}\pi$$ ■

Parametrized Surfaces We have seen that curves in space can be expressed as $\mathbf{r}(t) = x(t)\mathbf{i} + y(t)\mathbf{j} + z(t)\mathbf{k}$ where $a \le t \le b$. What if $\mathbf{r}$ is a function of two parameters, say u and v? It shouldn't be too surprising that a relationship like

$$\mathbf{r}(u, v) = x(u, v)\mathbf{i} + y(u, v)\mathbf{j} + z(u, v)\mathbf{k}, \quad (u, v) \in R$$

yields a surface. For each (u, v) in the set R, we obtain a vector $\mathbf{r}$ in three-space. The set of points that are the terminal points of the vectors $\mathbf{r}(u, v)$ (emanating from the origin) is called a **parametrized surface.**

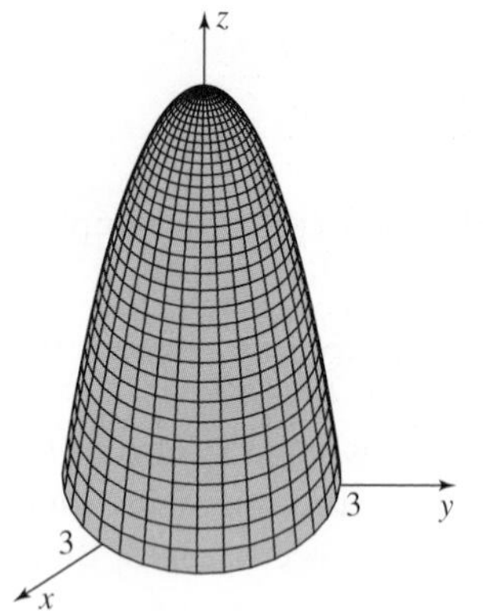

Figure 8

EXAMPLE 7 Describe and sketch the surfaces defined parametrically by

(a) $\mathbf{r}(u, v) = u\,\mathbf{i} + v\,\mathbf{j} + (9 - u^2 - v^2)\mathbf{k}, u^2 + v^2 \le 9$

(b) $\mathbf{r}(u, v) = u\cos v\,\mathbf{i} + u\sin v\,\mathbf{j} + (9 - u^2)\mathbf{k}, 0 \le u \le 3, 0 \le v \le 2\pi$

(c) $\mathbf{r}(u, v) = 3\cos u \sin v\,\mathbf{i} + 3\sin u \sin v\,\mathbf{j} + 3\cos v\,\mathbf{k}, 0 \le u \le 2\pi, 0 \le v \le \pi$.

SOLUTION

(a) For this $\mathbf{r}$ we see that the x and y components are simply u and v, and that z is $9 - u^2 - v^2$. This is just the graph of the function $f(x, y) = 9 - x^2 - y^2$ over the disk $x^2 + y^2 \le 9$. The graph, a paraboloid with vertex at $(0, 0, 9)$ opening downward, is shown in Figure 8.

(b) The x and y components look like the formulas for polar coordinates, except that u and v are substituted for r and θ, respectively. Since $z = 9 - u^2 = 9 - x^2 - y^2$, this surface is the same as part (a).

(c) We recognize the components of $\mathbf{r}(u, v)$ as the spherical coordinates of the points on a sphere of radius 3 centered at the origin. As u ranges from 0 to 2π and v ranges from 0 to π, we obtain the full sphere, shown in Figure 9. ■

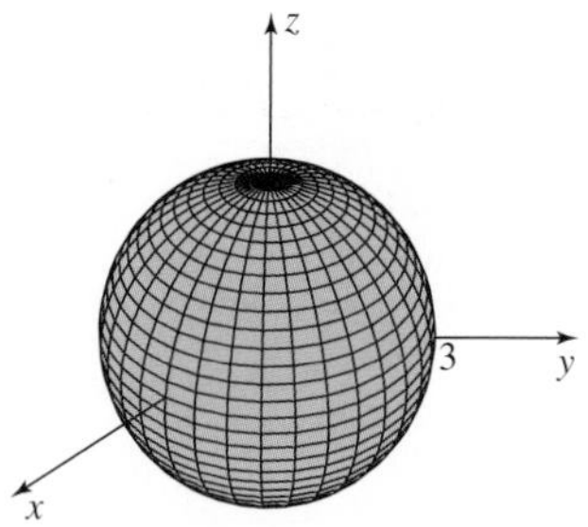

Figure 9

There are many situations where we must describe a known surface as a parameterized surface. There is often more than one way to do this, as the next example suggests.

EXAMPLE 8 Find parametric equations for the surfaces (a) the right circular cylinder of radius 2 with axis along the y-axis for $-4 \le y \le 4$, and (b) the hemisphere of radius 2 centered at the origin lying above the xy-plane.

SOLUTION

(a) If we think of polar coordinates in the xz-plane, we have $x = 2\cos v$ and $z = 2\sin v$. The other parameter, u, is the distance from the yz-plane. The parametric equation is thus $\mathbf{r}(u, v) = 2\cos v\,\mathbf{i} + u\,\mathbf{j} + 2\sin v\,\mathbf{j}$ and the domain for (u, v) is $-4 \le u \le 4, 0 \le v \le 2\pi$.

(b) For a hemisphere, we can use cylindrical coordinates $x = u\cos v$ and $y = u\sin v$, in which case the hemisphere $x^2 + y^2 + z^2 = 4, z \ge 0$, becomes $z = \sqrt{4 - x^2 - y^2} = \sqrt{4 - u^2}$. The parametric equation is thus $\mathbf{r}(u, v) = u\cos v\,\mathbf{i} + u\sin v\,\mathbf{j} + \sqrt{4 - u^2}\,\mathbf{k}$ with domain $0 \le u \le 2, 0 \le v \le 2\pi$. Alternatively, we could think in terms of spherical coordinates and allow the angle measured from the positive z-axis (usually called ϕ) to range from 0 to $\pi/2$ (rather than from 0 to π as we would to get the full sphere). The parametric equation would then be $\mathbf{r}(u, v) = 2\cos u \sin v\,\mathbf{i} + 2\sin u \sin v\,\mathbf{j} + 2\cos v\,\mathbf{k}$, $0 \le u \le 2\pi, 0 \le v \le \pi/2$. ■

Surface Area for a Parametrized Surface Figure 10 shows the mapping from the rectangle R to the surface G. (In general, the domain needn't be a rectangle, but to simplify the derivation we assume that it is a rectangle.) If we partition the rectangle R, we see that R_i, the ith rectangle, gets mapped to a curved patch G_i, and that the sides of the rectangles in the partition of R get mapped to a curved part of the surface. However, if Δu_i and Δv_i are small, the patch will closely resemble a parallelogram having as sides the vectors $\Delta u_i \mathbf{r}_u(u_i, v_i)$ and $\Delta v_i \mathbf{r}_v(u_i, v_i)$, where (u_i, v_i) is the lower left corner of the ith rectangle, and $\mathbf{r}_u$ and $\mathbf{r}_v$ denote the partial derivatives $\dfrac{\partial \mathbf{r}}{\partial u}$ and $\dfrac{\partial \mathbf{r}}{\partial v}$, respectively.

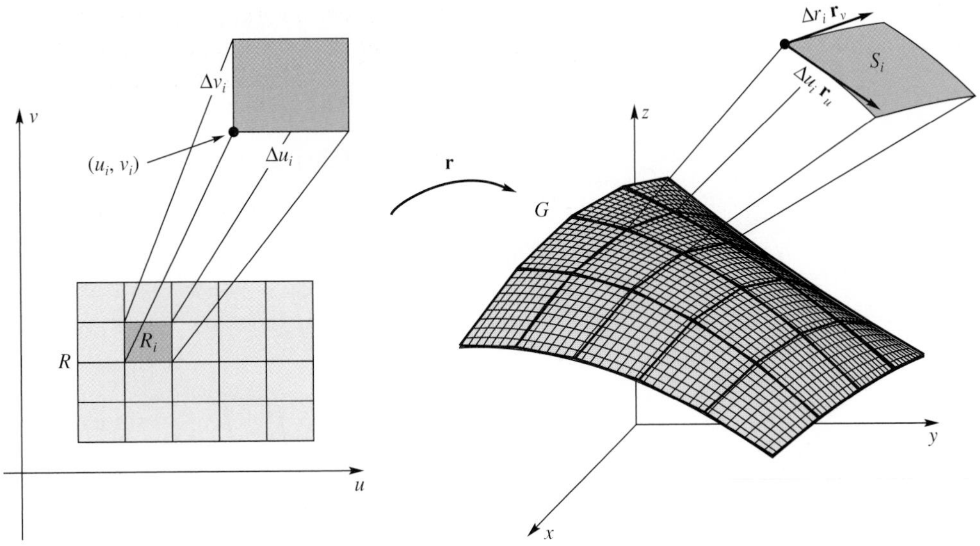

Figure 10

The surface area of the patch G_i is approximately

$$\Delta S_i \approx \|(\Delta u_i \mathbf{r}_u(u_i, v_i)) \times (\Delta v_i \mathbf{r}_v(u_i, v_i))\| = \|\mathbf{r}_u(u_i, v_i) \times \mathbf{r}_v(u_i, v_i)\|\, \Delta u_i\, \Delta v_i$$

The surface area of the parametrized surface is therefore

$$SA = \iint_R \|\mathbf{r}_u(u, v) \times \mathbf{r}_v(u, v)\|\, dA$$

The differential of surface area is

$$dS = \|\mathbf{r}_u(u, v) \times \mathbf{r}_v(u, v)\|\, dA$$

A surface integral for a parametrized surface is then

$$\iint_G f(x, y, z)\, dS = \iint_R f(\mathbf{r}(u, v))\|\mathbf{r}_u(u, v) \times \mathbf{r}_v(u, v)\|\, dA$$

EXAMPLE 9 A thin spherical shell of radius 5 centered at the origin has a hole of radius 3 removed from the top (Figure 11). Find the surface area, mass, and center of mass of this shell assuming that the density is proportional to the square of the distance from the z-axis.

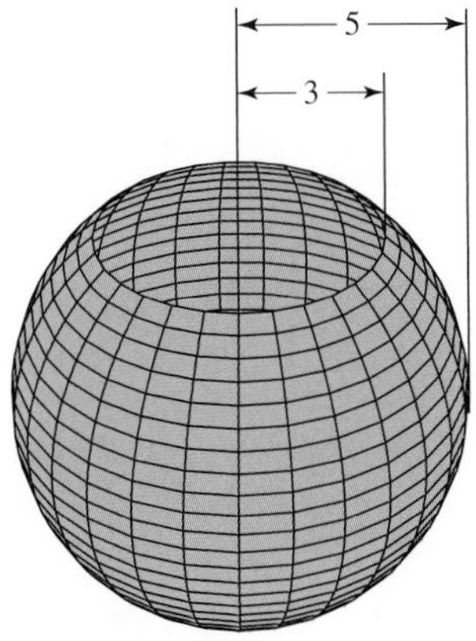

Figure 11

SOLUTION We can parametrize the surface as

$$\mathbf{r}(u, v) = 5 \cos u \sin v\, \mathbf{i} + 5 \sin u \sin v\, \mathbf{j} + 5 \cos v\, \mathbf{k}$$

for $0 \le u \le 2\pi$, $\sin^{-1}\frac{3}{5} \le v \le \pi$. The required derivatives are

$$\mathbf{r}_u(u, v) = -5 \sin u \sin v\, \mathbf{i} + 5 \cos u \sin v\, \mathbf{j} + 0\mathbf{k}$$
$$\mathbf{r}_v(u, v) = 5 \cos u \cos v\, \mathbf{i} + 5 \sin u \cos v\, \mathbf{j} - 5 \sin v\, \mathbf{k}$$

$$\mathbf{r}_u(u, v) \times \mathbf{r}_v(u, v) = \begin{vmatrix} \mathbf{i} & \mathbf{j} & \mathbf{k} \\ \dfrac{\partial x}{\partial u} & \dfrac{\partial y}{\partial u} & \dfrac{\partial z}{\partial u} \\ \dfrac{\partial x}{\partial v} & \dfrac{\partial y}{\partial v} & \dfrac{\partial z}{\partial v} \end{vmatrix}$$

$$= \begin{vmatrix} \mathbf{i} & \mathbf{j} & \mathbf{k} \\ -5 \sin u \sin v & 5 \cos u \sin v & 0 \\ 5 \cos u \cos v & 5 \sin u \cos v & -5 \sin v \end{vmatrix}$$

$$= -25\cos u\sin^2 v\,\mathbf{i} - 25\sin u\sin^2 v\,\mathbf{j}$$
$$-25\sin v\cos v(\sin^2 u + \cos^2 u)\,\mathbf{k}$$
$$= -25\cos u\sin^2 v\,\mathbf{i} - 25\sin u\sin^2 v\,\mathbf{j} - 25\sin v\cos v\,\mathbf{k}$$

The magnitude of this cross product is found to be (see Problem 27)

$$\|\mathbf{r}_u(u,v)\times\mathbf{r}_v(u,v)\| = 25\,|\sin v|$$

The surface area is

$$SA = \iint_R \|\mathbf{r}_u(u,v)\times\mathbf{r}_v(u,v)\|\,dA = \iint_R 25\,|\sin v|\,dv\,du$$
$$= 25\int_0^{2\pi}\int_{\sin^{-1}(3/5)}^{\pi} \sin v\,dv\,du$$
$$= 25\int_0^{2\pi} [-\cos v]_{\sin^{-1}(3/5)}^{\pi}\,du$$
$$= 25(2\pi)\frac{9}{5} = 90\pi \approx 282.74$$

The mass is equal to the value of the surface integral

$$m = \iint_G \delta(x,y,z)\,dS = \iint_G k(x^2+y^2)\,dS$$
$$= \iint_R (25k\sin^2 v)\|\mathbf{r}_u(u,v)\times\mathbf{r}_v(u,v)\|\,dA$$
$$= 25^2k\int_0^{2\pi}\int_{\sin^{-1}(3/5)}^{\pi} \sin^2 v\,|\sin v|\,dv\,du$$
$$= 625k\int_0^{2\pi}\int_{\sin^{-1}(3/5)}^{\pi} \sin^3 v\,dv\,du$$
$$= 625k\int_0^{2\pi}\left[-\frac{1}{3}\sin^2 v\cos v - \frac{2}{3}\cos v\right]_{\sin^{-1}(3/5)}^{\pi} du$$
$$= 625k\int_0^{2\pi}\frac{162}{125}\,du$$
$$= 1620\pi k \approx 5089.4k$$

By symmetry $\overline{x} = \overline{y} = 0$. The moment about the xy-plane is

$$M_{xy} = \iint_G z\delta(x,y,z)\,dS$$
$$= \iint_R (5\cos v)(25k\sin^2 v)\|\mathbf{r}_u(u,v)\times\mathbf{r}_v(u,v)\|\,dA$$
$$= 5\cdot 25^2k\int_0^{2\pi}\int_{\sin^{-1}(3/5)}^{\pi} \cos v\sin^3 v\,dv\,du$$
$$= 3125k\int_0^{2\pi}\left[\frac{1}{4}\sin^4 v\right]_{\sin^{-1}(3/5)}^{\pi} du$$
$$= 3125k\int_0^{2\pi}\left[-\frac{81}{2500}\right] du$$
$$= -\frac{81\cdot 3125k}{2500}2\pi = -\frac{405}{2}k\pi$$

The z-component of the center of mass is

$$\bar{z} = \frac{M_{xy}}{m} = \frac{-\frac{405}{2}k\pi}{1620\pi k} = -\frac{1}{8}$$

Thus, the center of mass is $\left(0, 0, -\frac{1}{8}\right)$. We would expect the z-component of the center of mass to be negative because some material is removed from the top of the sphere but not the bottom. ■

Concepts Review

1. A _____ generalizes the ordinary double integral similar to the way a line integral generalizes the definite integral.

2. If G is a surface, $\iint_G g(x, y, z)\, dS = \lim_{\|P\| \to 0}$ _____.

3. Let G be a surface given by $z = f(x, y)$, where (x, y) is in R. Then $\iint_G g(x, y, z)\, dS = \iint_R g(x, y, f(x, y))$ _____ $dy\, dx$.

4. Consider the cone with axis along the z-axis, with vertex at the origin, and making an angle of 30° with the z-axis. If G is the portion of this cone above the set $R = \{(x, y): x^2 + y^2 \le 9\}$, then $\iint_G dS = \iint_R$ _____ $dy\, dx =$ _____.

Problem Set 14.5

In Problems 1–8, evaluate $\iint_G g(x, y, z)\, dS$.

1. $g(x, y, z) = x^2 + y^2 + z$; $G: z = x + y + 1$, $0 \le x \le 1$, $0 \le y \le 1$

2. $g(x, y, z) = x$; $G: x + y + 2z = 4$, $0 \le x \le 1$, $0 \le y \le 1$

3. $g(x, y, z) = x + y$; $G: z = \sqrt{4 - x^2}$, $0 \le x \le \sqrt{3}$, $0 \le y \le 1$

4. $g(x, y, z) = 2y^2 + z$; $G: z = x^2 - y^2$, $0 \le x^2 + y^2 \le 1$

5. $g(x, y, z) = \sqrt{4x^2 + 4y^2 + 1}$; G is the part of $z = x^2 + y^2$ below $y = z$

6. $g(x, y, z) = y$; $G: z = 4 - y^2$, $0 \le x \le 3$, $0 \le y \le 2$

7. $g(x, y, z) = x + y$; G is the surface of the cube $0 \le x \le 1$, $0 \le y \le 1$, $0 \le z \le 1$.

8. $g(x, y, z) = z$; G is the tetrahedron bounded by the coordinate planes and the plane $4x + 8y + 2z = 16$.

In Problems 9–12, use Theorem B to calculate the flux of $\mathbf{F}$ *across* G.

9. $\mathbf{F}(x, y, z) = -y\,\mathbf{i} + x\,\mathbf{j}$; G is the part of the plane $z = 8x - 4y - 5$ above the triangle with vertices $(0, 0, 0)$, $(0, 1, 0)$, and $(1, 0, 0)$.

10. $\mathbf{F}(x, y, z) = (9 - x^2)\mathbf{j}$; G is the part of the plane $2x + 3y + 6z = 6$ in the first octant.

11. $\mathbf{F}(x, y, z) = y\,\mathbf{i} - x\,\mathbf{j} + 2\,\mathbf{k}$; G is the surface determined by $z = \sqrt{1 - y^2}$, $0 \le x \le 5$.

12. $\mathbf{F}(x, y, z) = 2\mathbf{i} + 5\mathbf{j} + 3\mathbf{k}$; G is the part of the cone $z = (x^2 + y^2)^{1/2}$ that is inside the cylinder $x^2 + y^2 = 1$.

13. Find the mass of the triangle with vertices $(a, 0, 0)$, $(0, a, 0)$, and $(0, 0, a)$ if its density δ satisfies $\delta(x, y, z) = kx^2$.

14. Find the mass of the surface $z = 1 - (x^2 + y^2)/2$ over $0 \le x \le 1$, $0 \le y \le 1$, if $\delta(x, y, z) = kxy$.

15. Find the center of mass of the homogeneous triangle with vertices $(a, 0, 0)$, $(0, a, 0)$, and $(0, 0, a)$.

16. Find the center of mass of the homogeneous triangle with vertices $(a, 0, 0)$, $(0, b, 0)$, and $(0, 0, c)$, where a, b, and c are all positive.

In Problems 17–20, plot the parametric surface over the indicated domain.

17. $\mathbf{r}(u, v) = u\,\mathbf{i} + 3v\,\mathbf{j} + (4 - u^2 - v^2)\mathbf{k}$; $0 \le u \le 2$, $0 \le v \le 1$

18. $\mathbf{r}(u, v) = 2u\,\mathbf{i} + 3v\,\mathbf{j} + (u^2 + v^2)\mathbf{k}$; $-1 \le u \le 1$, $-2 \le v \le 2$

19. $\mathbf{r}(u, v) = 2\cos v\,\mathbf{i} + 3\sin v\,\mathbf{j} + u\mathbf{k}$; $-6 \le u \le 6$, $0 \le v \le 2\pi$

20. $\mathbf{r}(u, v) = u\,\mathbf{i} + 3\sin v\,\mathbf{j} + 5\cos v\,\mathbf{k}$; $-6 \le u \le 6$, $0 \le v \le 2\pi$

CAS *In Problems 21–24, use a CAS to plot the parametric surface over the indicated domain and find the surface area of the resulting surface.*

21. $\mathbf{r}(u, v) = u\sin v\,\mathbf{i} + u\cos v\,\mathbf{j} + v\,\mathbf{k}$; $-6 \le u \le 6$, $0 \le v \le \pi$

22. $\mathbf{r}(u, v) = \sin u \sin v\, \mathbf{i} + \cos u \sin v\, \mathbf{j} + \sin v\, \mathbf{k}$; $0 \le u \le 2\pi, 0 \le v \le 2\pi$

23. $\mathbf{r}(u, v) = u^2 \cos v\, \mathbf{i} + u^2 \sin v\, \mathbf{j} + 5u\, \mathbf{k}; 0 \le u \le 2\pi$, $0 \le v \le 2\pi$

24. $\mathbf{r}(u, v) = \cos u \cos v\, \mathbf{i} + \cos u \sin v\, \mathbf{j} + \cos u\, \mathbf{k}$; $0 \le u \le \pi/2, 0 \le v \le 2\pi$

25. Find the mass of the surface in Problem 23 if the density is proportional to the distance from the xy-plane.

26. Find the mass of the surface in Problem 24 if the density is proportional to (a) the distance from the z-axis, and (b) the distance from the xy-plane.

27. Show that the magnitude of the cross product $\|\mathbf{r}_u(u, v) \times \mathbf{r}_v(u, v)\|$ in Example 9 is equal to $25\,|\sin v|$.

28. Refer to Example 3. The hemispherical surface $z = f(x, y) = \sqrt{9 - x^2 - y^2}$ has a thin metal covering with density $\delta(x, y, z) = z$. Find the mass of this covering. Note that Theorem A does not apply directly, since f_x and f_y are undefined on the boundary $x^2 + y^2 = 9$ of R. Therefore, proceed by letting R_ε be the region $0 \le x^2 + y^2 \le (3 - \varepsilon)^2$, make the calculation, and then let $\varepsilon \to 0$. Discover that you get the same answer as you would if you ignored this subtle point.

29. Let G be the sphere $x^2 + y^2 + z^2 = a^2$. Evaluate each of the following:

(a) $\iint_G z\, dS$

(b) $\iint_G \frac{x + y^3 + \sin z}{1 + z^4}\, dS$

(c) $\iint_G (x^2 + y^2 + z^2)\, dS$

(d) $\iint_G x^2\, dS$

(e) $\iint_G (x^2 + y^2)\, dS$

Hint: Use symmetry properties to make this a trivial problem.

30. The sphere $x^2 + y^2 + z^2 = a^2$ has constant area density k. Find each moment of inertia.

(a) About a diameter

(b) About a tangent line (assume the Parallel Axis Theorem from Problem 28 of Section 13.5).

31. Find the total force against the surface of a tank full of a liquid of weight density k for each tank shape.

(a) Sphere of radius a

(b) Hemisphere of radius a with a flat base

(c) Vertical cylinder of radius a and height h

Hint: The force against a small patch of area ΔG is approximately $kd\, \Delta G$, where d is the depth of the water at the patch.

32. Find the center of mass of that part of the sphere $x^2 + y^2 + z^2 = a^2$ between the planes $z = h_1$ and $z = h_2$, where $0 \le h_1 \le h_2 \le a$. Do this by the methods of this section and then compare with Problem 19 of Section 13.6.

Answers to Concepts Review: **1.** surface integral **2.** $\sum_{i=1}^{n} g(\overline{x}_i, \overline{y}_i, \overline{z}_i)\, \Delta S_i$ **3.** $\sqrt{f_x^2 + f_y^2 + 1}$ **4.** 2; 18π

14.6 Gauss's Divergence Theorem

The Boundary of a Set

Recall from Section 12.3 that a point P is a boundary point for a set S if every neighborhood of P contains points that are in S and points that are not in S. The boundary of a set is the set of all its boundary points.

The theorems of Green, Gauss, and Stokes all relate an integral over a set S to another integral over the boundary of S. To emphasize the similarity in these theorems, we introduce the notation ∂S to stand for the boundary of S. Thus, one form of Green's Theorem (Section 14.4) can be written as

$$\oint_{\partial S} \mathbf{F} \cdot \mathbf{n}\, ds = \iint_S \operatorname{div} \mathbf{F}\, dA$$

It says that the flux of $\mathbf{F}$ across the boundary ∂S of a closed bounded plane region S is equal to the double integral of div $\mathbf{F}$ over that region. Gauss's Theorem (also called the *Divergence Theorem*) lifts this result up one dimension.

Gauss's Theorem Let S be a closed bounded solid in three-space that is completely enclosed by a piecewise smooth surface ∂S (Figure 1).

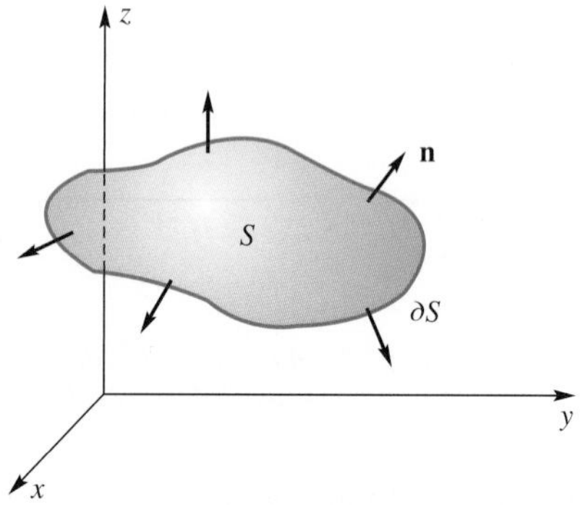

Figure 1

Theorem A Gauss's Theorem

Let $\mathbf{F} = M\mathbf{i} + N\mathbf{j} + P\mathbf{k}$ be a vector field such that M, N, and P have continuous first-order partial derivatives on a solid S with boundary ∂S. If $\mathbf{n}$ denotes the outer unit normal to ∂S, then

$$\iint_{\partial S} \mathbf{F} \cdot \mathbf{n}\, dS = \iiint_S \operatorname{div} \mathbf{F}\, dV$$

In other words, the flux of $\mathbf{F}$ across the boundary of a closed region in three-space is the triple integral of its divergence over that region.

It is useful both for some applications and for the proof to state the conclusion to Gauss's Theorem in its Cartesian (nonvector) form. We may write

$$\mathbf{n} = \cos\alpha\,\mathbf{i} + \cos\beta\,\mathbf{j} + \cos\gamma\,\mathbf{k}$$

where α, β, and γ are the direction angles for $\mathbf{n}$. Thus

$$\mathbf{F}\cdot\mathbf{n} = M\cos\alpha + N\cos\beta + P\cos\gamma$$

and so Gauss's formula becomes

$$\iint_{\partial S}(M\cos\alpha + N\cos\beta + P\cos\gamma)\,dS = \iiint_{S}\left(\frac{\partial M}{\partial x} + \frac{\partial N}{\partial y} + \frac{\partial P}{\partial z}\right)dV$$

Proof of Gauss's Theorem We first consider the case where the region S is x-simple, y-simple, and z-simple. It will be sufficient to show that

$$\iint_{\partial S} M\cos\alpha\,dS = \iiint_{S}\frac{\partial M}{\partial x}\,dV$$

$$\iint_{\partial S} N\cos\beta\,dS = \iiint_{S}\frac{\partial N}{\partial y}\,dV$$

$$\iint_{\partial S} P\cos\gamma\,dS = \iiint_{S}\frac{\partial P}{\partial z}\,dV$$

Since these demonstrations are similar, we show only the third.

Since S is z-simple, it can be described by the inequalities $f_1(x, y) \le z \le f_2(x, y)$. As in Figure 2, ∂S consists of three parts: S_1, corresponding to $z = f_1(x, y)$; S_2, corresponding to $z = f_2(x, y)$; and the lateral surface S_3, which may be empty. On S_3, $\cos\gamma = \cos 90° = 0$, so we can ignore its contribution. Also, from Problem 26 of Section 13.6 and Theorem 14.5A,

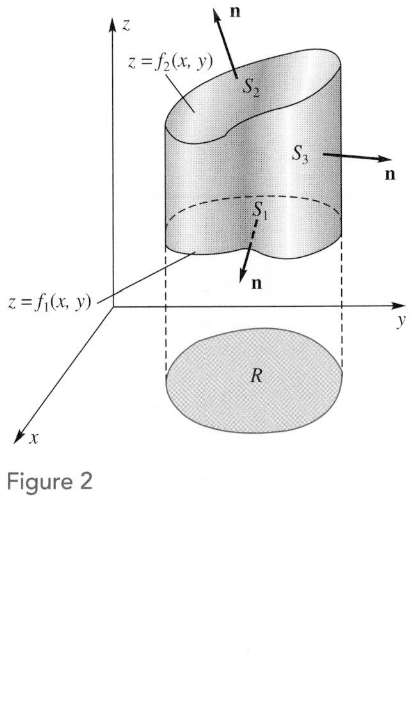

Figure 2

$$\iint_{S_2} P\cos\gamma\,dS = \iint_{R} P(x, y, f_2(x, y))\,dx\,dy$$

The result to which we just referred assumes that the normal $\mathbf{n}$ points upward. Hence, when we apply it to S_1, where $\mathbf{n}$ is a lower normal (Figure 2), we must reverse the sign.

$$\iint_{S_1} P\cos\gamma\,dS = -\iint_{R} P(x, y, f_1(x, y))\,dx\,dy$$

It follows that

$$\begin{aligned}\iint_{\partial S} P\cos\gamma\,dS &= \iint_{R}\left[P(x, y, f_2(x, y)) - P(x, y, f_1(x, y))\right]dx\,dy \\ &= \iint_{R}\left[\int_{f_1(x,y)}^{f_2(x,y)}\frac{\partial P}{\partial z}\,dz\right]dx\,dy \\ &= \iiint_{S}\frac{\partial P}{\partial z}\,dV\end{aligned}$$

The result just proved extends easily to regions that are finite unions of the type considered. We omit the details. ■

EXAMPLE 1 Verify Gauss's Theorem for $\mathbf{F} = x\,\mathbf{i} + y\,\mathbf{j} + z\,\mathbf{k}$ and $S = \{(x, y, z): x^2 + y^2 + z^2 \le a^2\}$ by independently calculating (a) $\iint_{\partial S} \mathbf{F}\cdot\mathbf{n}\,dS$ and (b) $\iiint_S \operatorname{div}\mathbf{F}\,dV$.

SOLUTION

(a) On ∂S, $\mathbf{n} = (x\,\mathbf{i} + y\,\mathbf{j} + z\,\mathbf{k})/a$, and so $\mathbf{F}\cdot\mathbf{n} = (x^2 + y^2 + z^2)/a = a$. Thus,

$$\iint_{\partial S} \mathbf{F}\cdot\mathbf{n}\,dS = a\iint_{\partial S} dS = a(4\pi a^2) = 4\pi a^3$$

(b) Since $\operatorname{div}\mathbf{F} = 3$,

$$\iiint_S \operatorname{div}\mathbf{F}\,dV = 3\iiint_S dV = 3\,\frac{4\pi a^3}{3} = 4\pi a^3$$

■

EXAMPLE 2 Compute the flux of the vector field $\mathbf{F} = x^2y\,\mathbf{i} + 2xz\,\mathbf{j} + yz^3\,\mathbf{k}$ across the surface of the rectangular solid S determined by (Figure 3)

$$0 \le x \le 1, \qquad 0 \le y \le 2, \qquad 0 \le z \le 3$$

(a) by a direct method and (b) by Gauss's Theorem.

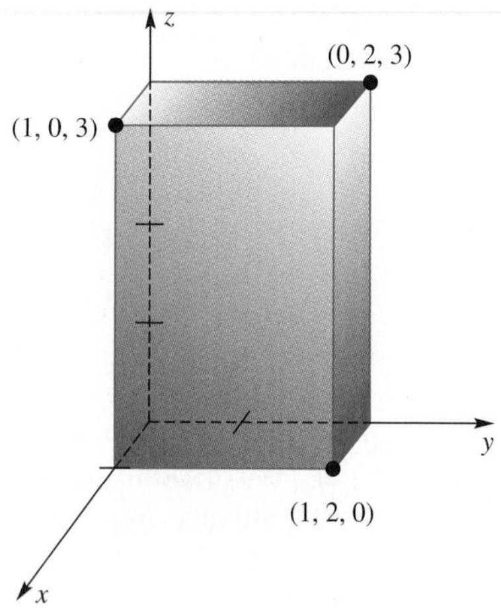

Figure 3

SOLUTION

(a) To calculate $\iint_{\partial S} \mathbf{F}\cdot\mathbf{n}\,dS$ directly, we calculate this integral over the six faces and add the results. On the face $x = 1$, $\mathbf{n} = \mathbf{i}$, and $\mathbf{F}\cdot\mathbf{n} = x^2y = 1^2y = y$, so $\iint_{x=1} \mathbf{F}\cdot\mathbf{n}\,dS = \int_0^3\int_0^2 y\,dy\,dz = 6$. By similar calculations, we can construct the following table:

Face	$\mathbf{n}$	$\mathbf{F}\cdot\mathbf{n}$	$\iint_{\text{face}} \mathbf{F}\cdot\mathbf{n}\,dS$
$x = 1$	$\mathbf{i}$	y	6
$x = 0$	$-\mathbf{i}$	0	0
$y = 2$	$\mathbf{j}$	$2xz$	9/2
$y = 0$	$-\mathbf{j}$	$-2xz$	−9/2
$z = 3$	$\mathbf{k}$	$27y$	54
$z = 0$	$-\mathbf{k}$	0	0

Thus,

$$\iint_{\partial S} \mathbf{F}\cdot\mathbf{n}\,dS = 6 + 0 + \frac{9}{2} - \frac{9}{2} + 54 + 0 = 60$$

(b) By Gauss's Theorem,

$$\begin{aligned}
\iint_{\partial S} \mathbf{F}\cdot\mathbf{n}\,dS &= \iiint_S (2xy + 0 + 3yz^2)\,dV \\
&= \int_0^1\int_0^2\int_0^3 (2xy + 3yz^2)\,dz\,dy\,dx = \int_0^1\int_0^2 (6xy + 27y)\,dy\,dx \\
&= \int_0^1 (12x + 54)\,dx = \left[6x^2 + 54x\right]_0^1 = 60
\end{aligned}$$

■

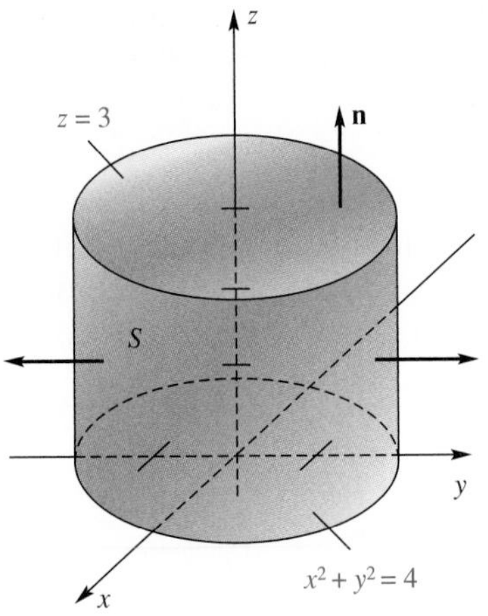

Figure 4

EXAMPLE 3 Let S be the solid cylinder bounded by $x^2 + y^2 = 4$, $z = 0$, and $z = 3$, and let $\mathbf{n}$ be the outer unit normal to the boundary ∂S (Figure 4). If $\mathbf{F} = (x^3 + \tan yz)\mathbf{i} + (y^3 - e^{xz})\mathbf{j} + (3z + x^3)\mathbf{k}$, find the flux of $\mathbf{F}$ across ∂S.

SOLUTION Imagine the difficulty in trying to evaluate $\iint_{\partial S} \mathbf{F} \cdot \mathbf{n}\, dS$ directly.

However,

$$\operatorname{div} \mathbf{F} = 3x^2 + 3y^2 + 3 = 3(x^2 + y^2 + 1)$$

and so, by Gauss's Theorem and a shift to cylindrical coordinates,

$$\begin{aligned}
\iint_{\partial S} \mathbf{F} \cdot \mathbf{n}\, dS &= 3 \iiint_S (x^2 + y^2 + 1)\, dV \\
&= 3 \int_0^{2\pi} \int_0^2 \int_0^3 (r^2 + 1) r\, dz\, dr\, d\theta \\
&= 9 \int_0^{2\pi} \int_0^2 (r^3 + r)\, dr\, d\theta \\
&= 9 \int_0^{2\pi} 6\, d\theta = 108\pi
\end{aligned}$$

■

Extensions and Applications So far we have implicitly assumed that the solid S has no holes in its interior and that its boundary ∂S consists of one connected surface. In fact, Gauss's Theorem holds for a solid with holes, like a chunk of Swiss cheese, provided that we always require $\mathbf{n}$ to point away from the interior of the solid. For example, let S be the solid shell between two concentric spheres centered at the origin. Gauss's Theorem applies, provided that we recognize that ∂S now consists of two surfaces (an outer surface where $\mathbf{n}$ points away from the origin and an inner surface where $\mathbf{n}$ points toward the origin).

EXAMPLE 4 Let S be the solid determined by

$$1 \le x^2 + y^2 + z^2 \le 4$$

and let $\mathbf{F} = x\,\mathbf{i} + (2y + z)\mathbf{j} + (z + x^2)\mathbf{k}$. Evaluate

$$\iint_{\partial S} \mathbf{F} \cdot \mathbf{n}\, dS$$

SOLUTION

$$\begin{aligned}
\iint_{\partial S} \mathbf{F} \cdot \mathbf{n}\, dS &= \iiint_S \operatorname{div} \mathbf{F}\, dV \\
&= \iiint_S (1 + 2 + 1)\, dV \\
&= 4\left[\frac{4}{3}\pi(2^3) - \frac{4}{3}\pi(1^3)\right] = \frac{112\pi}{3}
\end{aligned}$$

■

Recall from Section 14.1 that the gravitational field $\mathbf{F}$ due to a point mass M at the origin has the form

$$\mathbf{F}(x, y, z) = -cM \frac{\mathbf{r}}{\|\mathbf{r}\|^3}$$

where $\mathbf{r} = x\,\mathbf{i} + y\,\mathbf{j} + z\,\mathbf{k}$ and c is a constant.

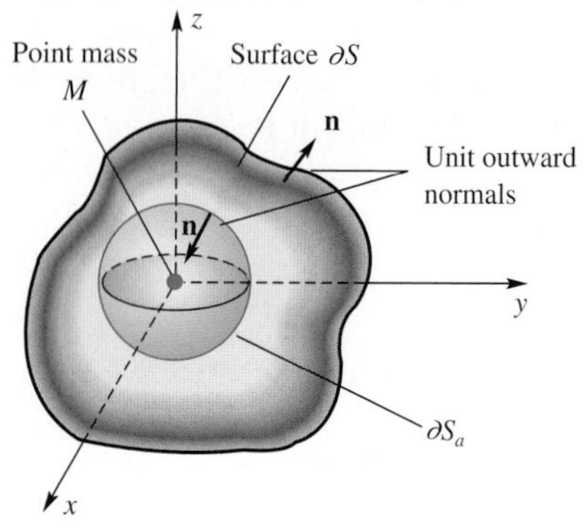

Figure 5

EXAMPLE 5 Let S be a solid region containing a point mass M at the origin in its interior and with the corresponding field $\mathbf{F} = -cM\mathbf{r}/\|\mathbf{r}\|^3$. Show that the flux of $\mathbf{F}$ across ∂S is $-4\pi cM$, regardless of the shape of S.

SOLUTION Since $\mathbf{F}$ is discontinuous at the origin, Gauss's Theorem does not apply directly. However, let us imagine that a small solid sphere S_a centered at the origin and of radius a has been removed from S, leaving a solid W with outer boundary ∂S and inner boundary ∂S_a (Figure 5). When we apply Gauss's Theorem to W, we get

$$\iint_{\partial S} \mathbf{F}\cdot\mathbf{n}\,dS + \iint_{\partial S_a} \mathbf{F}\cdot\mathbf{n}\,dS = \iint_{\partial W} \mathbf{F}\cdot\mathbf{n}\,dS = \iiint_W \operatorname{div}\mathbf{F}\,dV$$

But div $\mathbf{F} = 0$, which is easy to check (Problem 21 of Section 14.1), and so

$$\iint_{\partial S} \mathbf{F}\cdot\mathbf{n}\,dS = -\iint_{\partial S_a} \mathbf{F}\cdot\mathbf{n}\,dS$$

On the surface ∂S_a, $\mathbf{n} = -\mathbf{r}/\|\mathbf{r}\|$ and $\|\mathbf{r}\| = a$. Consequently,

$$\begin{aligned}
-\iint_{\partial S_a} \mathbf{F}\cdot\mathbf{n}\,dS &= -\iint_{\partial S_a}\left(-cM\frac{\mathbf{r}}{\|\mathbf{r}\|^3}\right)\cdot\left(-\frac{\mathbf{r}}{\|\mathbf{r}\|}\right)dS \\
&= -cM\iint_{\partial S_a}\frac{\mathbf{r}\cdot\mathbf{r}}{a^4}\,dS \\
&= -cM\iint_{\partial S_a}\frac{1}{a^2}\,dS \\
&= \frac{-cM}{a^2}(4\pi a^2) = -4\pi cM
\end{aligned}$$

■

We can extend the result of Example 5 to the case where a solid S contains k point masses $M_1, M_2, \ldots, M_k$ in its interior. The result, known as *Gauss's Law*, gives the flux of $\mathbf{F}$ across ∂S as

$$\iint_{\partial S} \mathbf{F}\cdot\mathbf{n}\,dS = -4\pi c(M_1 + M_2 + \cdots + M_k)$$

Finally, Gauss's Law can be extended to a body B with a continuously distributed mass of size M by subdividing it into small pieces and approximating these pieces by point masses. The result for any region S containing B is

$$\iint_{\partial S} \mathbf{F}\cdot\mathbf{n}\,dS = -4\pi cM$$

Concepts Review

1. The theorems of Green, Gauss, and Stokes all relate an integral over S to another integral over the _____ of S, which is denoted by _____.

2. In particular, the theorem of Gauss says that $\iiint_S \operatorname{div}\mathbf{F}\,dV = \iint_{\partial S}$ _____ dS.

3. Another way to state Gauss's Theorem is to say that the flux of $\mathbf{F}$ across the boundary of S equals $\iiint_S$ _____ dV.

4. A consequence of Gauss's Theorem is that the _____ of the gravitational field due to a mass M across the boundary of any solid S containing M is $-4\pi cM$; that is, it is independent of _____ of S.

Problem Set 14.6

In Problems 1–14, use Gauss's Divergence Theorem to calculate $\iint_{\partial S} \mathbf{F}\cdot\mathbf{n}\,dS$.

1. $\mathbf{F}(x, y, z) = z\,\mathbf{i} + x\,\mathbf{j} + y\,\mathbf{k}$; S is the hemisphere $0 \le z \le \sqrt{9 - x^2 - y^2}$.

2. $\mathbf{F}(x, y, z) = x\,\mathbf{i} + 2y\,\mathbf{j} + 3z\,\mathbf{k}$; S is the cube $0 \le x \le 1, 0 \le y \le 1, 0 \le z \le 1$.

3. $\mathbf{F}(x, y, z) = \cos z^2\mathbf{i} + y\,\mathbf{j} + \cos x^2\mathbf{k}$; S is the cube $-1 \le x \le 1, -1 \le y \le 1, -1 \le z \le 1$.

4. $\mathbf{F}(x, y, z) = x^3\,\mathbf{i} + y^3\,\mathbf{j} + z^3\,\mathbf{k}$; S is the hemisphere $0 \le z \le \sqrt{a^2 - x^2 - y^2}$.

5. $\mathbf{F}(x, y, z) = x^2yz\,\mathbf{i} + xy^2z\,\mathbf{j} + xyz^2\,\mathbf{k}$; S is the box $0 \le x \le a, 0 \le y \le b, 0 \le z \le c$.

6. $\mathbf{F}(x, y, z) = 3x\,\mathbf{i} - 2y\,\mathbf{j} + 4z\,\mathbf{k}$; S is the solid sphere $x^2 + y^2 + z^2 \le 9$.

7. $\mathbf{F}(x, y, z) = x^2\,\mathbf{i} + y^2\mathbf{j} + z^2\,\mathbf{k}$; S is the parabolic solid $0 \le z \le 4 - x^2 - y^2$.

8. $\mathbf{F}(x, y, z) = (x^2 + \cos yz)\mathbf{i} + (y - e^z)\mathbf{j} + (z^2 + x^2)\mathbf{k}$; S is the solid bounded by $x^2 + y^2 = 4$, $x + z = 2$, $z = 0$.

9. $\mathbf{F}(x, y, z) = (x + z^2)\mathbf{i} + (y - z^2)\mathbf{j} + x\,\mathbf{k}$; S is the solid $0 \le y^2 + z^2 \le 1, 0 \le x \le 2$.

10. $\mathbf{F}(x, y, z) = x^2\,\mathbf{i} + y^2\,\mathbf{j} + z^2\,\mathbf{k}$; S is the solid enclosed by $x + y + z = 4$, $x = 0$, $y = 0$, $z = 0$.

11. $\mathbf{F}(x, y, z) = 2x\,\mathbf{i} + 3y\,\mathbf{j} + 4z\,\mathbf{k}$; S is the solid spherical shell $9 \le x^2 + y^2 + z^2 \le 25$.

12. $\mathbf{F}(x, y, z) = 2z\,\mathbf{i} + x\,\mathbf{j} + z^2\,\mathbf{k}$; S is the solid cylindrical shell $1 \le x^2 + y^2 \le 4, 0 \le z \le 2$.

13. $\mathbf{F}(x, y, z) = z^2\,\mathbf{i} + y^2\,\mathbf{j} + x^2\,\mathbf{k}$; S is the cylindrical region $x^2 + z^2 \le 1, 0 \le y \le 10$. *Hint:* Make the transformation $x = r\cos\theta, z = r\sin\theta, y = y$ (similar to cylindrical coordinates) and use the methods of Section 13.9 to get the Jacobian.

14. $\mathbf{F}(x, y, z) = (x^3 + y)\mathbf{i} + (y^3 + z)\mathbf{j} + (x + z^3)\mathbf{k}$; S is the region $x^2 + z^2 \le y^2, x^2 + y^2 + z^2 \le 1, y \ge 0$. *Hint:* Make a transformation similar to spherical coordinates and use the methods of Section 13.9 to get the Jacobian.

15. Let $\mathbf{F}(x, y, z) = x\,\mathbf{i} + y\,\mathbf{j} + z\,\mathbf{k}$, and let S be a solid for which Gauss's Divergence Theorem applies. Show that the volume of S is given by

$$V(S) = \frac{1}{3}\iint_{\partial S} \mathbf{F}\cdot\mathbf{n}\,dS$$

16. Use the result of Problem 15 to verify the formula for the volume of a right circular cylinder of height h and radius a.

17. Consider the plane $ax + by + cz = d$, where a, b, c, and d are all positive. Use Problem 15 to show that the volume of the tetrahedron cut from the first octant by this plane is $dD/\left(3\sqrt{a^2 + b^2 + c^2}\right)$, where D is the area of that part of the plane in the first octant.

18. Let $\mathbf{F}$ be a constant vector field. Show that

$$\iint_{\partial S} \mathbf{F}\cdot\mathbf{n}\,dS = 0$$

for any "nice" solid S. What should we mean by "nice"?

19. Calculate $\iint_{\partial S} \mathbf{F}\cdot\mathbf{n}\,dS$ for each of the following. Looked at the right way, all are quite easy and some are even trivial.

(a) $\mathbf{F} = (2x + yz)\mathbf{i} + 3y\,\mathbf{j} + z^2\,\mathbf{k}$;
S is the solid sphere $x^2 + y^2 + z^2 \le 1$.

(b) $\mathbf{F} = (x^2 + y^2 + z^2)^{5/3}(x\,\mathbf{i} + y\,\mathbf{j} + z\,\mathbf{k})$; S as in part (a).

(c) $\mathbf{F} = x^2\mathbf{i} + y^2\mathbf{j} + z^2\mathbf{k}$;
S is the solid sphere $(x - 2)^2 + y^2 + z^2 \le 1$.

(d) $\mathbf{F} = x^2\mathbf{i}$; S is the cube $0 \le x \le 1, 0 \le y \le 1, 0 \le z \le 1$.

(e) $\mathbf{F} = (x + z)\mathbf{i} + (y + x)\mathbf{j} + (z + y)\mathbf{k}$; S is the tetrahedron cut from the first octant by the plane $3x + 4y + 2z = 12$.

(f) $\mathbf{F} = x^3\mathbf{i} + y^3\mathbf{j} + z^3\mathbf{k}$; S as in part (a).

(g) $\mathbf{F} = (x\,\mathbf{i} + y\,\mathbf{j})\ln(x^2 + y^2)$;
S is the solid cylinder $x^2 + y^2 \le 4, 0 \le z \le 2$.

20. Calculate $\iint_{\partial S} \mathbf{F}\cdot\mathbf{n}\,dS$. In each case, $\mathbf{r} = x\,\mathbf{i} + y\,\mathbf{j} + z\,\mathbf{k}$.

(a) $\mathbf{F} = \mathbf{r}/\|\mathbf{r}\|^3$; S is the solid sphere $(x - 2)^2 + y^2 + z^2 \le 1$.

(b) $\mathbf{F} = \mathbf{r}/\|\mathbf{r}\|^3$; S is the solid sphere $x^2 + y^2 + z^2 \le a^2$.

(c) $\mathbf{F} = \mathbf{r}/\|\mathbf{r}\|^2$; S as in part (b).

(d) $\mathbf{F} = f(\|\mathbf{r}\|)\mathbf{r}$, f any scalar function; S as in part (b).

(e) $\mathbf{F} = \|\mathbf{r}\|^n\mathbf{r}$, $n \ge 0$; S is the solid sphere $x^2 + y^2 + z^2 \le az$ ($\rho \le a\cos\phi$ in spherical coordinates).

21. We have defined the Laplacian of a scalar field by

$$\nabla^2 f = \frac{\partial^2 f}{\partial x^2} + \frac{\partial^2 f}{\partial y^2} + \frac{\partial^2 f}{\partial z^2}$$

Show that if $D_{\mathbf{n}}f$ is the directional derivative in the direction of the unit normal vector $\mathbf{n}$, then

$$\iint_{\partial S} D_{\mathbf{n}} f\,dS = \iiint_S \nabla^2 f\,dV$$

22. Suppose that $\nabla^2 f$ is identically zero in a region S. Show that

$$\iint_{\partial S} f D_{\mathbf{n}} f\,dS = \iiint_S \|\nabla f\|^2\,dV$$

23. Establish Green's First Identity

$$\iint_{\partial S} f D_{\mathbf{n}} g\,dS = \iiint_S (f\,\nabla^2 g + \nabla f\cdot\nabla g)\,dV$$

by applying Gauss's Divergence Theorem to $\mathbf{F} = f\,\nabla g$.

24. Establish Green's Second Identity:

$$\iint_{\partial S} (f D_{\mathbf{n}} g - g D_{\mathbf{n}} f)\,dS = \iiint_S (f\,\nabla^2 g - g\,\nabla^2 f)\,dV$$

Answers to Concepts Review: **1.** boundary; ∂S **2.** $\mathbf{F}\cdot\mathbf{n}$ **3.** div $\mathbf{F}$ **4.** flux; the shape

14.7 Stokes's Theorem

We showed in Section 14.4 that the conclusion to Green's Theorem could be written as

$$\oint_{\partial S} \mathbf{F} \cdot \mathbf{T}\, ds = \iint_S (\text{curl } \mathbf{F}) \cdot \mathbf{k}\, dA$$

As stated, it was a theorem for a plane set S bounded by a simple closed curve ∂S. We are going to generalize this result to the case where S is a curved surface in three-space. In this form, the theorem is due to the Irish scientist George Gabriel Stokes (1819–1903).

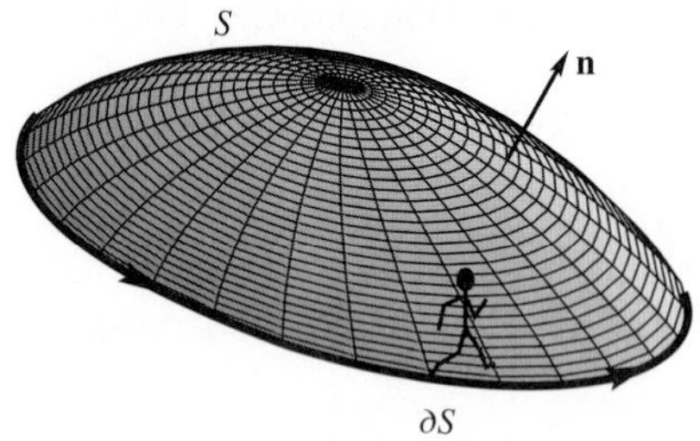

Figure 1

We will need to put some restrictions on the surface S. First, we suppose that S is two-sided with a continuously varying unit normal $\mathbf{n}$ (the one-sided Möbius band of Section 14.5 is thereby eliminated from our discussion). Second, we require that the boundary ∂S be a piecewise smooth, simple closed curve, oriented consistently with $\mathbf{n}$. This means that, if you stand near the edge of the surface with your head in the direction $\mathbf{n}$ and your eyes looking in the direction of the curve, the surface is to your left (Figure 1).

Theorem A Stokes's Theorem

Let $S, \partial S$, and $\mathbf{n}$ be as indicated above, and suppose that $\mathbf{F} = M\mathbf{i} + N\mathbf{j} + P\mathbf{k}$ is a vector field, with M, N, and P having continuous first-order partial derivatives on S and its boundary ∂S. If $\mathbf{T}$ denotes the unit tangent vector to ∂S, then

$$\oint_{\partial S} \mathbf{F} \cdot \mathbf{T}\, ds = \iint_S (\text{curl } \mathbf{F}) \cdot \mathbf{n}\, dS$$

Examples and Applications

The proof of Stokes's Theorem is more appropriate for a course in advanced calculus. However, we can at least verify the theorem in an example.

EXAMPLE 1 Verify Stokes's Theorem for $\mathbf{F} = y\,\mathbf{i} - x\,\mathbf{j} + yz\,\mathbf{k}$ if S is the paraboloid $z = x^2 + y^2$ with the circle $x^2 + y^2 = 1, z = 1$ as its boundary (Figure 2).

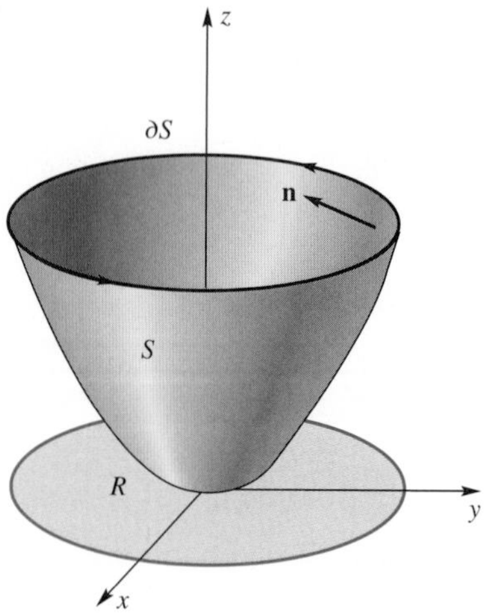

Figure 2

SOLUTION We may describe ∂S by the parametric equations

$$x = \cos t, \qquad y = \sin t, \qquad z = 1$$

Then $dz = 0$ and (see Section 14.2)

$$\oint_{\partial S} \mathbf{F} \cdot \mathbf{T}\, ds = \oint_{\partial S} y\, dx - x\, dy = \int_0^{2\pi} [\sin t(-\sin t)\, dt - \cos t \cos t\, dt]$$

$$= -\int_0^{2\pi} [\sin^2 t + \cos^2 t]\, dt = -2\pi$$

On the other hand, to calculate $\iint_S (\text{curl } \mathbf{F}) \cdot \mathbf{n}\, dS$ we first obtain

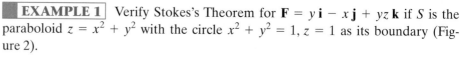

$$\text{curl } \mathbf{F} = \nabla \times \mathbf{F} = \begin{vmatrix} \mathbf{i} & \mathbf{j} & \mathbf{k} \\ \dfrac{\partial}{\partial x} & \dfrac{\partial}{\partial y} & \dfrac{\partial}{\partial z} \\ y & -x & yz \end{vmatrix} = z\,\mathbf{i} + 0\,\mathbf{j} - 2\,\mathbf{k}$$

Then, by Theorem 14.5B,

$$\begin{aligned}\iint_S (\text{curl }\mathbf{F})\cdot\mathbf{n}\,dS &= \iint_R [-z(2x) - 0(2y) - 2]\,dx\,dy\\ &= -2\iint_R [xz + 1]\,dx\,dy\\ &= -2\iint_R [x(x^2 + y^2) + 1]\,dx\,dy\\ &= -2\int_0^{2\pi}\int_0^1 [r^3\cos\theta + 1]r\,dr\,d\theta\\ &= -2\int_0^{2\pi}\left[\frac{1}{5}\cos\theta + \frac{1}{2}\right]d\theta = -2\pi\end{aligned}$$

■

EXAMPLE 2 Let S be that part of the spherical surface $x^2 + y^2 + (z - 4)^2 = 10$ below the plane $z = 1$, and let $\mathbf{F} = y\,\mathbf{i} - x\,\mathbf{j} + yz\,\mathbf{k}$. Use Stokes's Theorem to calculate

$$\iint_S (\text{curl }\mathbf{F})\cdot\mathbf{n}\,dS$$

where $\mathbf{n}$ is the upward unit normal.

SOLUTION Note that the field $\mathbf{F}$ is the same as that of Example 1 and also that S has the same circle as its boundary curve. We conclude that

$$\iint_S (\text{curl }\mathbf{F})\cdot\mathbf{n}\,dS = \oint_{\partial S}\mathbf{F}\cdot\mathbf{n}\,ds = -2\pi$$

In fact, we conclude that the flux of curl $\mathbf{F}$ is -2π for all surfaces S that have the circle ∂S of Figure 2 as their oriented boundary. ■

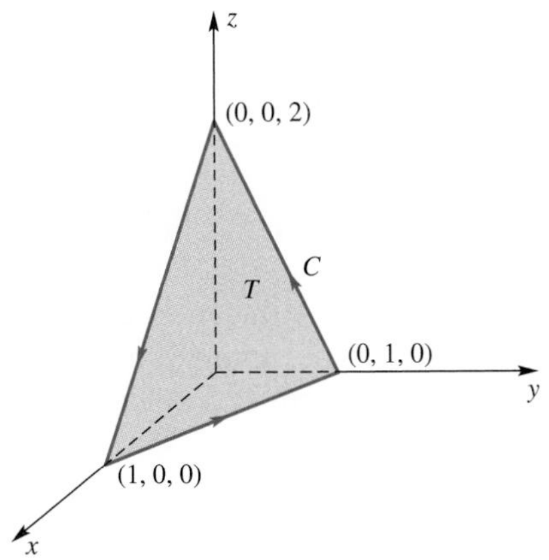

Figure 3

EXAMPLE 3 Use Stokes's Theorem to evaluate $\oint_C \mathbf{F}\cdot\mathbf{T}\,ds$, where $\mathbf{F} = 2z\,\mathbf{i} + (8x - 3y)\mathbf{j} + (3x + y)\mathbf{k}$ and C is the triangular curve of Figure 3.

SOLUTION We could let S be any surface with C as its oriented boundary, but it is to our advantage to choose the simplest such surface, the flat planar triangle T. To determine $\mathbf{n}$ for this surface, we note that the vectors

$$\mathbf{A} = (0 - 1)\mathbf{i} + (0 - 0)\mathbf{j} + (2 - 0)\mathbf{k} = -\mathbf{i} + 2\mathbf{k}$$
$$\mathbf{B} = (0 - 1)\mathbf{i} + (1 - 0)\mathbf{j} + (0 - 0)\mathbf{k} = -\mathbf{i} + \mathbf{j}$$

lie on this surface and hence

$$\mathbf{N} = \mathbf{A}\times\mathbf{B} = \begin{vmatrix}\mathbf{i} & \mathbf{j} & \mathbf{k}\\ -1 & 0 & 2\\ -1 & 1 & 0\end{vmatrix} = -2\mathbf{i} - 2\mathbf{j} - \mathbf{k}$$

is perpendicular to it. The upward unit normal $\mathbf{n}$ is therefore

$$\mathbf{n} = \frac{2\mathbf{i} + 2\mathbf{j} + \mathbf{k}}{\sqrt{4 + 4 + 1}} = \frac{2}{3}\mathbf{i} + \frac{2}{3}\mathbf{j} + \frac{1}{3}\mathbf{k}$$

Also,

$$\text{curl } \mathbf{F} = \begin{vmatrix} \mathbf{i} & \mathbf{j} & \mathbf{k} \\ \dfrac{\partial}{\partial x} & \dfrac{\partial}{\partial y} & \dfrac{\partial}{\partial z} \\ 2z & 8x - 3y & 3x + y \end{vmatrix} = \mathbf{i} - \mathbf{j} + 8\mathbf{k}$$

and curl $\mathbf{F} \cdot \mathbf{n} = \frac{8}{3}$. We conclude that

$$\oint_C \mathbf{F} \cdot \mathbf{T}\, ds = \iint_T (\text{curl } \mathbf{F}) \cdot \mathbf{n}\, dS = \frac{8}{3}(\text{area of } T) = \frac{8}{3}\left(\frac{3}{2}\right) = 4 \qquad \blacksquare$$

EXAMPLE 4 Let the vector field $\mathbf{F}$ and the region D satisfy the hypotheses of Theorem 14.3D. Show that if curl $\mathbf{F} = \mathbf{0}$ in D then $\mathbf{F}$ is conservative there.

SOLUTION From the discussion in Section 14.3, we conclude that it is enough to show that $\oint_C \mathbf{F} \cdot d\mathbf{r} = 0$ for any simple closed path C in D. Let S be a surface having C as its boundary and oriented consistently with C (the simple connectedness of D can be shown to guarantee the existence of such a surface). Then, from Stokes's Theorem,

$$\oint_C \mathbf{F} \cdot d\mathbf{r} = \oint_C \mathbf{F} \cdot \mathbf{T}\, ds = \iint_S (\text{curl } \mathbf{F}) \cdot \mathbf{n}\, dS = 0 \qquad \blacksquare$$

Physical Interpretation of the Curl We offered an interpretation of the curl in Section 14.4. Now we can amplify that discussion. Let C be a circle of radius a centered at the point P. Then

$$\oint_C \mathbf{F} \cdot \mathbf{T}\, ds$$

is called the *circulation* of $\mathbf{F}$ around C and measures the tendency of a fluid with velocity field $\mathbf{F}$ to circulate around C. Now, if $\mathbf{F}$ is continuous and C is very small, Stokes's Theorem gives

$$\oint_C \mathbf{F} \cdot \mathbf{T}\, ds = \iint_S (\text{curl } \mathbf{F}) \cdot \mathbf{n}\, dS \approx [\text{curl } \mathbf{F}(P)] \cdot \mathbf{n}\, (\pi a^2)$$

The expression on the right will have the largest magnitude if $\mathbf{n}$ has the same direction as curl $\mathbf{F}(P)$.

Suppose that a small paddle wheel is placed in the fluid with center at P and axis having direction $\mathbf{n}$ (Figure 4). This wheel will rotate most rapidly if $\mathbf{n}$ has the direction of curl $\mathbf{F}$. The direction of rotation will be that determined by the right-hand rule.

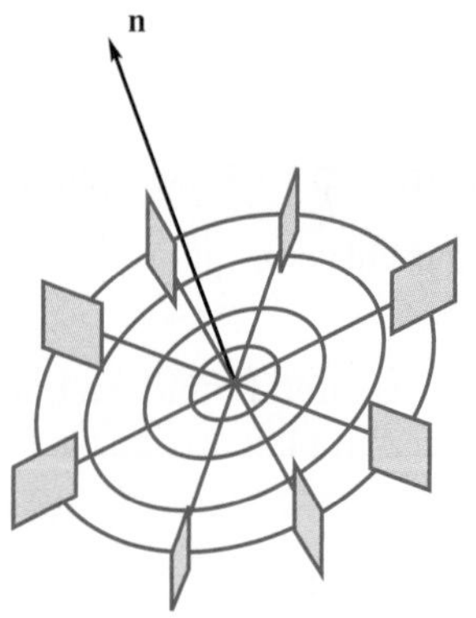

Figure 4

Concepts Review

1. Stokes's Theorem in three-space says that under appropriate hypotheses $\int_{\partial S} \mathbf{F} \cdot \mathbf{T}\, dS = \iint_S$ _____ dS. Here S is a surface and ∂S is its boundary.

2. One of these hypotheses is that S be two-sided. An important example of a one-sided surface is the _____, obtained by cutting an ordinary cylindrical band, giving it a half twist, and pasting it back together.

3. It follows from Stokes's Theorem that all two-sided surfaces with the same boundary ∂S give the same value for _____.

4. A paddle wheel centered at P and immersed in a fluid with velocity field $\mathbf{F}$ will rotate most rapidly about P if $\mathbf{n}$ has the direction of _____.

Problem Set 14.7

In Problems 1–6, use Stokes's Theorem to calculate

$$\iint_S (\text{curl } \mathbf{F}) \cdot \mathbf{n}\, dS$$

1. $\mathbf{F} = x^2\,\mathbf{i} + y^2\,\mathbf{j} + z^2\,\mathbf{k}$; S is the hemisphere $z = \sqrt{1 - x^2 - y^2}$ and $\mathbf{n}$ is the upper normal.

2. $\mathbf{F} = xy\,\mathbf{i} + yz\,\mathbf{j} + xz\,\mathbf{k}$; S is the triangular surface with vertices $(0, 0, 0)$, $(1, 0, 0)$, and $(0, 2, 1)$ and $\mathbf{n}$ is the upper normal.

3. $\mathbf{F} = (y + z)\mathbf{i} + (x^2 + z^2)\mathbf{j} + y\,\mathbf{k}$; S is the half-cylinder $z = \sqrt{1 - x^2}$ between $y = 0$ and $y = 1$ and $\mathbf{n}$ is the upper normal.

4. $\mathbf{F} = xz^2\,\mathbf{i} + x^3\mathbf{j} + \cos xz\,\mathbf{k}$; S is the part of the ellipsoid $x^2 + y^2 + 3z^2 = 1$ below the xy-plane and $\mathbf{n}$ is the lower normal.

5. $\mathbf{F} = yz\,\mathbf{i} + 3xz\,\mathbf{j} + z^2\mathbf{k}$; S is the part of the sphere $x^2 + y^2 + z^2 = 16$ below the plane $z = 2$ and $\mathbf{n}$ is the outward normal.

6. $\mathbf{F} = (z - y)\mathbf{i} + (z + x)\mathbf{j} - (x + y)\mathbf{k}$; S is the part of the paraboloid $z = 1 - x^2 - y^2$ above the xy-plane and $\mathbf{n}$ is the upward normal.

In Problems 7–12, use Stokes's Theorem to calculate $\oint_C \mathbf{F} \cdot \mathbf{T}\, ds$.

7. $\mathbf{F} = 2z\,\mathbf{i} + x\,\mathbf{j} + 3y\,\mathbf{k}$; C is the ellipse that is the intersection of the plane $z = x$ and the cylinder $x^2 + y^2 = 4$, oriented clockwise as viewed from above.

8. $\mathbf{F} = y\,\mathbf{i} + z\,\mathbf{j} + x\,\mathbf{k}$; C is the triangular curve with vertices $(0, 0, 0)$, $(2, 0, 0)$, and $(0, 2, 2)$, oriented counterclockwise as viewed from above.

9. $\mathbf{F} = (y - x)\mathbf{i} + (x - z)\mathbf{j} + (x - y)\mathbf{k}$; C is the boundary of the plane $x + 2y + z = 2$ in the first octant, oriented clockwise as viewed from above.

10. $\mathbf{F} = y(x^2 + y^2)\mathbf{i} - x(x^2 + y^2)\mathbf{j}$; C is the rectangular path from $(0, 0, 0)$ to $(1, 0, 0)$ to $(1, 1, 1)$ to $(0, 1, 1)$ to $(0, 0, 0)$.

11. $\mathbf{F} = (z - y)\mathbf{i} + y\,\mathbf{j} + x\,\mathbf{k}$; C is the intersection of the cylinder $x^2 + y^2 = x$ with the sphere $x^2 + y^2 + z^2 = 1$, oriented counterclockwise as viewed from above.

12. $\mathbf{F} = (y - z)\mathbf{i} + (z - x)\mathbf{j} + (x - y)\mathbf{k}$; C is the ellipse which is the intersection of the plane $x + z = 1$ and the cylinder $x^2 + y^2 = 1$, oriented clockwise as viewed from above.

13. Suppose that the surface S is determined by the formula $z = g(x, y)$. Show that the surface integral in Stokes's Theorem can be written as a double integral in the following way:

$$\iint_S (\text{curl } \mathbf{F}) \cdot \mathbf{n}\, dS = \iint_{S_{xy}} (\text{curl } \mathbf{F}) \cdot (-g_x\mathbf{i} - g_y\mathbf{j} + \mathbf{k})\, dA$$

where $\mathbf{n}$ is the upward normal to S and S_{xy} is the projection of S in the xy-plane.

14. Let $\mathbf{F} = x^2\,\mathbf{i} - 2xy\,\mathbf{j} + yz^2\,\mathbf{k}$ and ∂S be the boundary of the surface $z = xy$, $0 \le x \le 1$, $0 \le y \le 1$, oriented counterclockwise as viewed from above. Use Stokes's Theorem and Problem 13 to evaluate $\oint_{\partial S} \mathbf{F} \cdot \mathbf{T}\, ds$.

15. Let $\mathbf{F} = 2\,\mathbf{i} + xz\,\mathbf{j} + z^3\mathbf{k}$ and ∂S be the boundary of the surface $z = xy^2$, $0 \le x \le 1$, $0 \le y \le 1$, oriented counterclockwise as viewed from above. Evaluate $\oint_{\partial S} \mathbf{F} \cdot \mathbf{T}\, ds$.

16. Let $\mathbf{F} = 2\,\mathbf{i} + xz\,\mathbf{j} + z^3\mathbf{k}$ and ∂S be the boundary of the surface $z = x^2y^2$, $x^2 + y^2 \le a^2$, oriented counterclockwise as viewed from above. Evaluate $\oint_{\partial S} \mathbf{F} \cdot \mathbf{T}\, ds$.

17. Let $\mathbf{F} = 2z\,\mathbf{i} + 2y\,\mathbf{k}$, and let ∂S be the intersection of the cylinder $x^2 + y^2 = ay$ with the hemisphere $z = \sqrt{a^2 - x^2 - y^2}$, $a > 0$. Assuming distances in meters and force in newtons, find the work done by the force $\mathbf{F}$ in moving an object around ∂S in the counterclockwise direction as viewed from above.

18. A central force is one of the form $\mathbf{F} = f(\|\mathbf{r}\|)\mathbf{r}$, where f has a continuous derivative (except possibly at $\|\mathbf{r}\| = 0$). Show that the work done by such a force in moving an object around a closed path that misses the origin is 0.

19. Let S be a solid sphere (or any solid enclosed by a "nice" surface ∂S). Show that

$$\iint_{\partial S} (\text{curl } \mathbf{F}) \cdot \mathbf{n}\, dS = 0$$

(a) By using Stokes's Theorem.
(b) By using Gauss's Theorem. *Hint:* Show $\text{div}(\text{curl } \mathbf{F}) = 0$.

20. Show that

$$\oint_{\partial S} (f\,\nabla g) \cdot \mathbf{T}\, ds = \iint_S (\nabla f \times \nabla g) \cdot \mathbf{n}\, dS$$

Answers to Concept Review: **1.** curl $\mathbf{F} \cdot \mathbf{n}$ **2.** Möbius band **3.** $\iint_S (\text{curl } \mathbf{F}) \cdot \mathbf{n}\, dS$ **4.** curl $\mathbf{F}$

14.8 Chapter Review

Concepts Test

Respond with true or false to each of the following assertions. Be prepared to justify your answer.

1. Inverse square law fields are conservative.

2. The divergence of a vector field is another vector field.

3. A physicist might be interested in both curl (grad f) and grad (curl $\mathbf{F}$).

4. If f has continuous second-order partial derivatives, then curl (grad f) $= \mathbf{0}$.

5. The work done by a conservative force field as it moves an object around a closed path is zero.

6. If $\int_C \mathbf{F}(\mathbf{r}) \cdot d\mathbf{r} = 0$ for every closed path in an open connected set D, then there is a function f such that $\nabla f = \mathbf{F}$ in D.

7. The field $\mathbf{F}(x, y, z) = (2x + 2y)\mathbf{i} + 2x\,\mathbf{j} + yz^2\mathbf{k}$ is conservative.

8. Green's Theorem holds for a region S with a hole provided that the complete boundary of S is oriented correctly.

9. The double integral is a special case of a surface integral.

10. A surface always has two sides.

11. If there are no sources or sinks in a region, then the net flow across the boundary of the region is zero.

12. If S is a sphere with outward normal $\mathbf{n}$ and $\mathbf{F}$ is a constant vector field, then

$$\iint_S (\mathbf{F} \cdot \mathbf{n})\, dS = 0$$

Sample Test Problems

1. Sketch a sample of vectors from the vector field $\mathbf{F}(x, y) = x\,\mathbf{i} + 2y\,\mathbf{j}$.

2. Find div $\mathbf{F}$, curl $\mathbf{F}$, grad(div $\mathbf{F}$), and div(curl $\mathbf{F}$) if $\mathbf{F}(x, y, z) = 2xyz\,\mathbf{i} - 3y^2\mathbf{j} + 2y^2z\,\mathbf{k}$.

3. We showed in Problem 20, Section 14.1, that

$$\text{curl}(f\mathbf{F}) = f(\text{curl } \mathbf{F}) + \nabla f \times \mathbf{F}$$

and in Problem 21, Section 14.3, that

$$\text{curl}(\nabla f) = \mathbf{0}$$

Use these facts to show that

$$\text{curl}(f\,\nabla f) = \mathbf{0}$$

4. Find a function f satisfying

(a) $\nabla f = (2xy + y)\mathbf{i} + (x^2 + x + \cos y)\mathbf{j}$;

(b) $\nabla f = (yz - e^{-x})\mathbf{i} + (xz + e^y)\mathbf{j} + xy\,\mathbf{k}$.

5. Evaluate:

(a) $\int_C (1 - y^2)\, ds$; C is the quarter circle from $(0, -1)$ to $(1, 0)$, centered at the origin.

(b) $\int_C xy\, dx + z \cos x\, dy + z\, dz$; C is the curve $x = t$, $y = \cos t$, $z = \sin t$, $0 \le t \le \pi/2$.

6. Show that $\int_C y^2\, dx + 2xy\, dy$ is independent of path, and use this to calculate the integral on any path from $(0, 0)$ to $(1, 2)$.

7. Find the work done by $\mathbf{F} = y^2\mathbf{i} + 2xy\,\mathbf{j}$ in moving an object from $(1, 1)$ to $(3, 4)$ (see Problem 6).

8. Evaluate (see Problem 4b).

$$\int_{(0,0,0)}^{(1,1,4)} (yz - e^{-x})\, dx + (xy + e^y)\, dy + xy\, dz$$

9. Evaluate $\oint_C xy\, dx + (x^2 + y^2)\, dy$ if

(a) C is the square path $(0, 0)$ to $(1, 0)$ to $(1, 1)$ to $(0, 1)$ to $(0, 0)$;

(b) C is the triangular path $(0, 0)$ to $(2, 0)$ to $(2, 1)$ to $(0, 0)$;

(c) C is the circle $x^2 + y^2 = 1$ traversed in the clockwise direction.

10. Calculate the flux of $\mathbf{F} = x\,\mathbf{i} + y\,\mathbf{j}$ across the square curve C with vertices $(1, 1)$, $(-1, 1)$, $(-1, -1)$, and $(1, -1)$; that is, calculate $\oint_C \mathbf{F} \cdot \mathbf{n}\, ds$.

11. Calculate the flux of $\mathbf{F} = x\,\mathbf{i} + y\,\mathbf{j} + 3\,\mathbf{k}$ across the sphere $x^2 + y^2 + z^2 = 1$.

12. Evaluate $\iint_G xyz\, dS$, where G is the part of the plane $z = x + y$ above the triangular region with vertices $(0, 0, 0)$, $(1, 0, 0)$, and $(0, 2, 0)$.

13. Evaluate $\iint_G (\text{curl } \mathbf{F}) \cdot \mathbf{n}\, dS$, where

$$\mathbf{F} = x^3y\,\mathbf{i} + e^y\mathbf{j} + z \tan\left(\frac{xyz}{4}\right)\mathbf{k}$$

and G is the part of the sphere $x^2 + y^2 + z^2 = 2$ above the plane $z = 1$ and $\mathbf{n}$ is the upward unit normal.

14. Evaluate $\iint_G \mathbf{F} \cdot \mathbf{n}\, dS$, where

$$\mathbf{F} = \sin x\,\mathbf{i} + (1 - \cos x)y\,\mathbf{j} + 4z\,\mathbf{k}$$

and G is the closed surface bounded by $z = \sqrt{9 - x^2 - y^2}$, and $z = 0$ with outward unit normal $\mathbf{n}$.

15. Let C be the circle that is the intersection of the plane $ax + by + z = 0$ $(a \ge 0, b \ge 0)$ and the sphere $x^2 + y^2 + z^2 = 9$. For $\mathbf{F} = y\,\mathbf{i} - x\,\mathbf{j} + 3y\,\mathbf{k}$, evaluate

$$\oint_C \mathbf{F} \cdot \mathbf{T}\, ds$$

Hint: Use Stokes's Theorem.

Appendix

A.1 Mathematical Induction

A.2 Proofs of Several Theorems

Theorem A
Main Limit Theorem

Theorem B
Chain Rule

Theorem C
Power Rule

Theorem D
Vector Limits

A.1 Mathematical Induction

Often in mathematics we are faced with the task of wanting to establish that a certain proposition P_n is true for every integer $n \geq 1$ (or perhaps every integer $n \geq N$). Here are three examples:

1. P_n: $1^2 + 2^2 + 3^2 + \cdots + n^2 = \dfrac{n(n+1)(2n+1)}{6}$
2. Q_n: $2^n > n + 20$
3. R_n: $n^2 - n + 41$ is prime

Proposition P_n is true for every positive integer, and Q_n is true for every integer greater than or equal to 5 (as we will show soon). The third proposition, R_n, is interesting. Note that for $n = 1, 2, 3, \ldots$, the values of $n^2 - n + 41$ are $41, 43, 47, 53, 61, \ldots$ (prime numbers so far). In fact, we will get a prime number for all n's through 40; but at $n = 41$, the formula yields the composite number $1681 = (41)(41)$. Showing the truth of a proposition for 40 (or 40 million) individual cases may make a proposition plausible, but it most certainly does not prove it is true for all n. The chasm between any finite number of cases and *all* cases is infinitely wide.

What is to be done? Is there a procedure for establishing that a proposition P_n is true for *all* n? An affirmative answer is provided by the **Principle of Mathematical Induction.**

Principle of Mathematical Induction

Let $\{P_n\}$ be a sequence of propositions (statements) satisfying these two conditions:

(i) P_N is true (usually N will be 1).
(ii) The truth of P_i implies the truth of P_{i+1}, $i \geq N$.

Then, P_n is true for every integer $n \geq N$.

We do not prove this principle; it is often taken as an axiom, and we hope it seems obvious. After all, if the first domino falls and if each domino knocks over the next one, then the whole row of dominoes will fall. Our efforts will be directed toward illustrating how we use mathematical induction.

EXAMPLE 1 Prove that

$$P_n:\ 1^2 + 2^2 + 3^2 + \cdots + n^2 = \frac{n(n+1)(2n+1)}{6}$$

is true for all $n \geq 1$.

SOLUTION First, we note that

$$P_1:\ 1^2 = \frac{1(1+1)(2+1)}{6}$$

is a true statement.

Second, we demonstrate implication (ii). We begin by writing the statements P_i and P_{i+1}.

$$P_i:\ 1^2 + 2^2 + \cdots + i^2 = \frac{i(i+1)(2i+1)}{6}$$

$$P_{i+1}:\ 1^2 + 2^2 + \cdots + i^2 + (i+1)^2 = \frac{(i+1)(i+2)(2i+3)}{6}$$

We must show that P_i implies P_{i+1}, so we assume that P_i is true. Then the left side of P_{i+1} can be written as follows (* indicates where P_i is used):

$$[1^2 + 2^2 + \cdots + i^2] + (i+1)^2 \overset{*}{=} \frac{i(i+1)(2i+1)}{6} + (i+1)^2$$
$$= (i+1)\frac{2i^2 + i + 6i + 6}{6}$$
$$= \frac{(i+1)(i+2)(2i+3)}{6}$$

This chain of equalities leads to the statement P_{i+1}. Thus, the truth of P_i does imply the truth of P_{i+1}. By the Principle of Mathematical Induction, P_n is true for each positive integer n. ■

EXAMPLE 2 Prove that P_n: $2^n > n + 20$ is true for each integer $n \geq 5$.

SOLUTION First, we note that P_5: $2^5 > 5 + 20$ is true. Second, we suppose that P_i: $2^i > i + 20$ is true and attempt to deduce from this that P_{i+1}: $2^{i+1} > i + 1 + 20$ is true. But

$$2^{i+1} = 2 \cdot 2^i \overset{*}{>} 2(i + 20) = 2i + 40 > i + 21$$

Read from left to right, this is proposition P_{i+1}. Thus, P_n is true for $n \geq 5$. ■

EXAMPLE 3 Prove that

$$P_n: \quad x - y \text{ is a factor of } x^n - y^n$$

is true for each integer $n \geq 1$.

SOLUTION Trivially, $x - y$ is a factor of $x - y$, so P_1 is true. Suppose that $x - y$ is a factor of $x^i - y^i$; that is,

$$x^i - y^i = Q(x, y)(x - y)$$

for some polynomial $Q(x, y)$. Then

$$x^{i+1} - y^{i+1} = x^{i+1} - x^i y + x^i y - y^{i+1}$$
$$= x^i(x - y) + y(x^i - y^i)$$
$$\overset{*}{=} x^i(x - y) + y\,Q(x, y)(x - y)$$
$$= [x^i + yQ(x, y)](x - y)$$

Thus, the truth of P_i does imply the truth of P_{i+1}. We conclude by the Principle of Mathematical Induction that P_n is true for all $n \geq 1$. ■

Problem Set A.1

In Problems 1–8, use the Principle of Mathematical Induction to prove that the given proposition is true for each integer $n \geq 1$.

1. $1 + 2 + 3 + \cdots + n = \dfrac{n(n+1)}{2}$

2. $1 + 3 + 5 + \cdots + (2n - 1) = n^2$

3. $1 \cdot 2 + 2 \cdot 3 + 3 \cdot 4 + \cdots + n(n+1) = \dfrac{n(n+1)(n+2)}{3}$

4. $1^2 + 3^2 + 5^2 + \cdots + (2n - 1)^2 = \dfrac{n(2n-1)(2n+1)}{3}$

5. $1^3 + 2^3 + 3^3 + \cdots + n^3 = \left[\dfrac{n(n+1)}{2}\right]^2$

6. $1^4 + 2^4 + 3^4 + \cdots + n^4 = \dfrac{n(n+1)(6n^3 + 9n^2 + n - 1)}{30}$

7. $n^3 - n$ is divisible by 6.

8. $n^3 + (n+1)^3 + (n+2)^3$ is divisible by 9.

In Problems 9–12, make a conjecture about the first integer N for which the proposition is true for all $n \geq N$, and then prove the proposition for all $n \geq N$.

9. $3n + 25 < 3^n$

10. $n - 100 > \log_{10} n$

11. $n^2 \leq 2^n$

12. $|\sin nx| \leq n|\sin x|$ for all x

In Problems 13–20, indicate what conclusion about P_n can be drawn from the given information.

13. P_5 is true, and P_i true implies P_{i+2} true.

14. P_1 and P_2 are true, and P_i true implies P_{i+2} true.

15. P_{30} is true, and P_i true implies P_{i-1} true.

16. P_{30} is true, and P_i true implies both P_{i+1} and P_{i-1} true.

17. P_1 is true, and P_i true implies both P_{4i} and P_{i-1} true.

18. P_1 is true, and P_{2i} true implies P_{2i+1} true.

19. P_1 and P_2 are true, and P_i and P_{i+1} true imply P_{i+2} true.

20. P_1 is true, and P_j true for $j \leq i$ implies P_{i+1} true.

In Problems 21–27, decide for what n's the given proposition is true and then use mathematical induction (perhaps in one of the alternative forms that you may have discovered in Problems 13–20) to prove each of the following.

21. $x + y$ is a factor of $x^n + y^n$.

22. The sum of the measures of the interior angles of an n-sided convex (no holes or dents) polygon is $(n - 2)\pi$.

23. The number of diagonals of an n-sided convex polygon is $\dfrac{n(n-3)}{2}$.

24. $\dfrac{1}{n+1} + \dfrac{1}{n+2} + \dfrac{1}{n+3} + \cdots + \dfrac{1}{2n} > \dfrac{3}{5}$

25. $\left(1 - \frac{1}{4}\right)\left(1 - \frac{1}{9}\right)\left(1 - \frac{1}{16}\right)\cdots\left(1 - \frac{1}{n^2}\right) = \dfrac{n+1}{2n}$

26. Let $f_0 = 0, f_1 = 1$, and $f_{n+2} = f_{n+1} + f_n$ for $n \geq 0$ (this is the Fibonacci sequence). Then

$$f_n = \frac{1}{\sqrt{5}}\left[\left(\frac{1+\sqrt{5}}{2}\right)^n - \left(\frac{1-\sqrt{5}}{2}\right)^n\right]$$

27. Let $a_0 = 0, a_1 = 1$, and $a_{n+2} = (a_{n+1} + a_n)/2$ for $n \geq 0$. Then

$$a_n = \frac{2}{3}\left[1 - \left(-\frac{1}{2}\right)^n\right]$$

28. What is wrong with the following argument, which purports to show that all people in any set of n people are the same age? The statement is certainly true for a set consisting of one person. Suppose that it is true for any set of i people, and consider a set W of $i + 1$ people. We may think of W as the union of sets X and Y, each consisting of i people (draw a picture, for example, when W has 6 people). By supposition, each of these sets consists of identically aged people. But X and Y overlap (in $X \cap Y$), and so all members of $W = X \cup Y$ also are the same age.

A.2 Proofs of Several Theorems

Theorem A Main Limit Theorem

Let n be a positive integer, k be a constant, and f and g be functions that have limits at c. Then

1. $\lim_{x\to c} k = k$
2. $\lim_{x\to c} x = c$
3. $\lim_{x\to c} kf(x) = k \lim_{x\to c} f(x)$
4. $\lim_{x\to c}[f(x) + g(x)] = \lim_{x\to c} f(x) + \lim_{x\to c} g(x)$
5. $\lim_{x\to c}[f(x) - g(x)] = \lim_{x\to c} f(x) - \lim_{x\to c} g(x)$
6. $\lim_{x\to c}[f(x)\cdot g(x)] = \lim_{x\to c} f(x)\cdot \lim_{x\to c} g(x)$
7. $\lim_{x\to c}\dfrac{f(x)}{g(x)} = \dfrac{\lim_{x\to c} f(x)}{\lim_{x\to c} g(x)}$, provided $\lim_{x\to c} g(x) \neq 0$
8. $\lim_{x\to c}[f(x)]^n = \left[\lim_{x\to c} f(x)\right]^n$
9. $\lim_{x\to c}\sqrt[n]{f(x)} = \sqrt[n]{\lim_{x\to c} f(x)}$, provided $\lim_{x\to c} f(x) > 0$ when n is even

Proof We proved parts 1 through 5 near the end of Section 1.3, so we should start with part 6. However, we choose first to prove a special case of part 8:

$$\lim_{x\to c}[g(x)]^2 = \left[\lim_{x\to c} g(x)\right]^2$$

To see this, recall that we have proved that $\lim_{x\to c} x^2 = c^2$ (Example 7 of Section 1.2), and so $f(x) = x^2$ is continuous everywhere. Thus, by the Composite Limit Theorem (Theorem 1.6E),

$$\lim_{x \to c}[g(x)]^2 = \lim_{x \to c} f(g(x)) = f\left[\lim_{x \to c} g(x)\right] = \left[\lim_{x \to c} g(x)\right]^2$$

Next, write

$$f(x)g(x) = \frac{1}{4}\{[f(x) + g(x)]^2 - [f(x) - g(x)]^2\}$$

and apply parts 3, 4, and 5, plus what we have just proved. Part 6 is proved.

To prove part 7, apply the Composite Limit Theorem with $f(x) = 1/x$ and use Example 8 of Section 1.2. Then

$$\lim_{x \to c} \frac{1}{g(x)} = \lim_{x \to c} f(g(x)) = f\left(\lim_{x \to c} g(x)\right) = \frac{1}{\lim_{x \to c} g(x)}$$

Finally, by part 6,

$$\lim_{x \to c} \frac{f(x)}{g(x)} = \lim_{x \to c}\left[f(x) \cdot \frac{1}{g(x)}\right] = \lim_{x \to c} f(x) \cdot \lim_{x \to c} \frac{1}{g(x)}$$

from which the result follows.

Part 8 follows from repeated use of part 6 (technically, by mathematical induction).

We prove part 9 only for square roots. Let $f(x) = \sqrt{x}$, which is continuous for positive numbers by Example 5 of Section 1.2. By the Composite Limit Theorem,

$$\lim_{x \to c} \sqrt{g(x)} = \lim_{x \to c} f(g(x)) = f\left(\lim_{x \to c} g(x)\right) = \sqrt{\lim_{x \to c} g(x)}$$

which is equivalent to the desired result. ■

Theorem B Chain Rule

If g is differentiable at a and f is differentiable at $g(a)$, then $f \circ g$ is differentiable at a and

$$(f \circ g)'(a) = f'(g(a))g'(a)$$

Proof We offer a proof that generalizes easily to higher dimensions (see Section 12.6). By hypothesis, f is differentiable at $b = g(a)$; that is, there is a number $f'(b)$ such that

(1) $$\lim_{\Delta u \to 0} \frac{f(b + \Delta u) - f(b)}{\Delta u} = f'(b)$$

Define a function ε depending on Δu by

$$\varepsilon(\Delta u) = \frac{f(b + \Delta u) - f(b)}{\Delta u} - f'(b)$$

and multiply both sides by Δu to obtain

(2) $$f(b + \Delta u) - f(b) = f'(b)\,\Delta u + \Delta u\,\varepsilon(\Delta u)$$

The existence of the limit in (1) is equivalent to $\varepsilon(\Delta u) \to 0$ as $\Delta u \to 0$ in (2). If, in (2), we replace Δu by $g(a + \Delta x) - g(a)$ and b by $g(a)$, we get

$$f(g(a + \Delta x)) - f(g(a)) = f'(g(a))[g(a + \Delta x) - g(a)] + [g(a + \Delta x) - g(a)]\varepsilon(\Delta u)$$

or, upon dividing both sides by Δx,

(3) $$\frac{f(g(a + \Delta x)) - f(g(a))}{\Delta x} = f'(g(a))\frac{g(a + \Delta x) - g(a)}{\Delta x} + \frac{g(a + \Delta x) - g(a)}{\Delta x}\varepsilon(\Delta u)$$

In (3), let $\Delta x \to 0$. Since g is differentiable at a, it is continuous there, so $\Delta x \to 0$ forces $\Delta u \to 0$; this, in turn, makes $\varepsilon(\Delta u) \to 0$. We conclude that

$$\lim_{\Delta x \to 0} \frac{f(g(a+\Delta x)) - f(g(a))}{\Delta x} = f'(g(a)) \lim_{\Delta x \to 0} \frac{g(a+\Delta x) - g(a)}{\Delta x} + 0$$

That is, $f \circ g$ is differentiable at a and

$$(f \circ g)'(a) = f'(g(a))g'(a)$$ ■

Theorem C Power Rule

If r is rational, then x^r is differentiable at any x that is in an open interval on which x^{r-1} is real and

$$D_x(x^r) = rx^{r-1}$$

Proof Consider first the case where $r = 1/q$, q a positive integer. Recall that $a^q - b^q$ factors as

$$a^q - b^q = (a-b)(a^{q-1} + a^{q-2}b + \cdots + ab^{q-2} + b^{q-1})$$

so

$$\frac{a-b}{a^q - b^q} = \frac{1}{a^{q-1} + a^{q-2}b + \cdots + ab^{q-2} + b^{q-1}}$$

Thus, if $f(t) = t^{1/q}$,

$$\begin{aligned} f'(x) &= \lim_{t \to x} \frac{t^{1/q} - x^{1/q}}{t - x} = \lim_{t \to x} \frac{t^{1/q} - x^{1/q}}{(t^{1/q})^q - (x^{1/q})^q} \\ &= \lim_{t \to x} \frac{1}{t^{(q-1)/q} + t^{(q-2)/q}x^{1/q} + \cdots + x^{(q-1)/q}} \\ &= \frac{1}{qx^{(q-1)/q}} = \frac{1}{q}x^{1/q-1} \end{aligned}$$

Now, by the Chain Rule, and with p an integer,

$$D_x(x^{p/q}) = D_x[(x^{1/q})^p] = p(x^{1/q})^{p-1} D_x(x^{1/q}) = px^{p/q-1/q}\frac{1}{q}x^{1/q-1} = \frac{p}{q}x^{p/q-1}$$ ■

Theorem D Vector Limits

Let $\mathbf{F}(t) = f(t)\mathbf{i} + g(t)\mathbf{j}$. Then $\mathbf{F}$ has a limit at c if and only if f and g have limits at c. In that case,

$$\lim_{t \to c} \mathbf{F}(t) = \left[\lim_{t \to c} f(t)\right]\mathbf{i} + \left[\lim_{t \to c} g(t)\right]\mathbf{j}$$

Proof First, note that for any vector $\mathbf{u} = u_1\mathbf{i} + u_2\mathbf{j}$,

$$|u_1| \le \|\mathbf{u}\| \le |u_1| + |u_2|$$

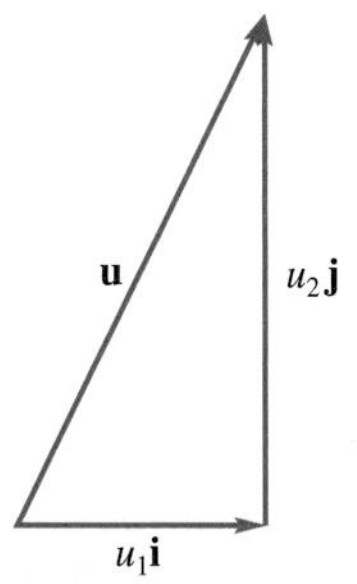

Figure 1

This fact is readily seen from Figure 1.

Now suppose that $\lim_{t \to c} \mathbf{F}(t) = \mathbf{L} = a\,\mathbf{i} + b\,\mathbf{j}$. This means that for any $\varepsilon > 0$ there is a corresponding $\delta > 0$ such that

$$0 < |t - c| < \delta \Rightarrow \|\mathbf{F}(t) - \mathbf{L}\| < \varepsilon$$

But, by the left part of the boxed inequality,

$$|f(t) - a| \leq \|\mathbf{F}(t) - \mathbf{L}\|$$

and so

$$0 < |t - c| < \delta \Rightarrow |f(t) - a| < \varepsilon$$

This shows that $\lim_{t \to c} f(t) = a$. A similar argument establishes that $\lim_{t \to c} g(t) = b$. The first half of our theorem is complete.

Conversely, suppose that

$$\lim_{t \to c} f(t) = a \quad \text{and} \quad \lim_{t \to c} g(t) = b$$

and let $\mathbf{L} = a\,\mathbf{i} + b\,\mathbf{j}$. For any given $\varepsilon > 0$, there is a corresponding $\delta > 0$ such that $0 < |t - c| < \delta$ implies that both

$$|f(t) - a| < \frac{\varepsilon}{2} \quad \text{and} \quad |g(t) - b| < \frac{\varepsilon}{2}$$

Hence, by the right part of the boxed inequality,

$$0 < |t - c| < \delta \Rightarrow \|\mathbf{F}(t) - \mathbf{L}\| \leq \frac{\varepsilon}{2} + \frac{\varepsilon}{2} = \varepsilon$$

Thus,

$$\lim_{t \to c} \mathbf{F}(t) = \mathbf{L} = a\,\mathbf{i} + b\,\mathbf{j} = \lim_{t \to c} f(t)\,\mathbf{i} + \lim_{t \to c} g(t)\,\mathbf{j}$$

■

Answers to Odd-Numbered Problems

Problem Set 0.1

1. 16 **3.** −148 **5.** $\frac{58}{91}$ **7.** $\frac{1}{24}$ **9.** $\frac{6}{49}$ **11.** $\frac{7}{15}$
13. $\frac{1}{3}$ **15.** 2 **17.** $3x^2 - x - 4$ **19.** $6x^2 - 15x - 9$
21. $9t^4 - 6t^3 + 7t^2 - 2t + 1$ **23.** $x + 2, x \neq 2$
25. $t - 7, t \neq -3$ **27.** $\dfrac{2(3x + 10)}{x(x + 2)}$
29. **(a)** 0; **(b)** Undefined; **(c)** 0; **(d)** Undefined; **(e)** 0; **(f)** 1
31. 0.08333... **33.** 0.142857... **35.** 3.6666...
37. $\frac{41}{333}$ **39.** $\frac{254}{99}$ **41.** $\frac{1}{5}$
43. Those rational numbers that can be expressed by a terminating decimal followed by zeros
49. Irrational **51.** 20.39230485 **53.** 0.00028307388
55. 0.000691744752 **59.** 132,700,874 ft
61. 651,441 board ft
63. **(a)** If I stay home from work today then it rains. If I do not stay home from work, then it does not rain.
(b) If the candidate will be hired then she meets all the qualifications. If the candidate will not be hired then she does not meet all the qualifications.
65. **(a)** If a triangle is a right triangle, then $a^2 + b^2 = c^2$. If a triangle is not a right triangle, then $a^2 + b^2 \neq c^2$.
(b) If the measure of angle ABC is greater than 0° and less than 90°, it is acute. If the measure of angle ABC is less than 0° or greater than 90°, then it is not acute.
67. **(a)** The statement, converse, and contrapositive are all true.
(b) The statement, converse, and contrapositive are all true.
69. **(a)** Some isosceles triangles are not equilateral. The negation is true.
(b) All real numbers are integers. The original statement is true.
(c) Some natural number is larger than its square. The original statement is true.
71. **(a)** True; **(b)** False; **(c)** False; **(d)** True; **(e)** True
75. **(a)** $3 \cdot 3 \cdot 3 \cdot 3 \cdot 3$ or 3^5; **(b)** $2 \cdot 2 \cdot 31$ or $2^2 \cdot 31$
(c) $2 \cdot 2 \cdot 3 \cdot 5 \cdot 5 \cdot 17$ or $2^2 \cdot 3 \cdot 5^2 \cdot 17$
81. **(a)** Rational; **(b)** Rational; **(c)** Rational; **(d)** Irrational

Problem Set 0.2

1. **(a)** 

(b) 

(c) 

(d) 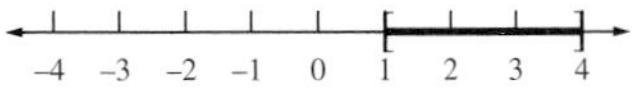

(e) 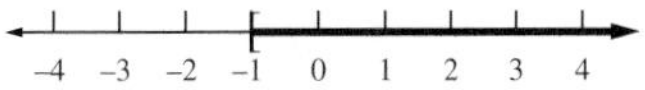

(f) 

3. $(-2, \infty)$; 

5. $\left[-\frac{5}{2}, \infty\right)$; 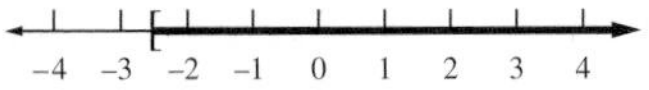

7. $(-2, 1)$; 

9. $\left[-\dfrac{1}{2}, \dfrac{2}{3}\right)$; 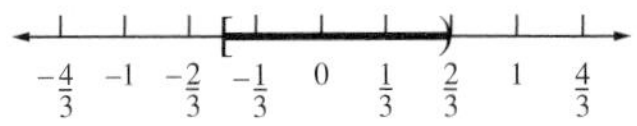

11. $\left(-1 - \sqrt{13}, -1 + \sqrt{13}\right)$; 

13. $(-\infty, -3) \cup \left(\frac{1}{2}, \infty\right)$; 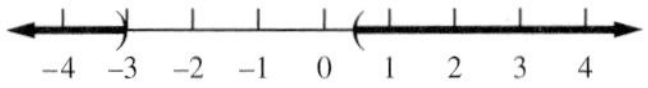

15. $[-4, 3)$; 

17. $(-\infty, 0) \cup \left(\frac{2}{5}, \infty\right)$; 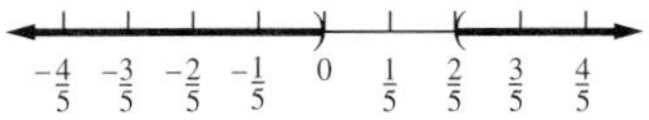

19. $\left(-\infty, \frac{2}{3}\right) \cup \left[\frac{3}{4}, \infty\right)$; 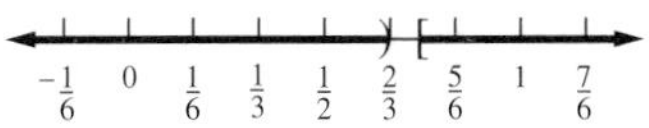

21. $(-2, 1) \cup (3, \infty)$; 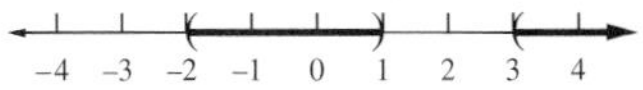

23. $\left(-\infty, \frac{3}{2}\right] \cup [3, \infty)$; 

25. $(-\infty, -1) \cup (0, 6)$; 

27. **(a)** False; **(b)** True; **(c)** False
31. **(a)** $(-2, 1)$; **(b)** $(-2, \infty)$; **(c)** No values
33. **(a)** $[-3, -1] \cup [2, \infty)$; **(b)** $(-\infty, -2] \cup [2, \infty)$;
(c) $(-2, -1) \cup (1, 2)$
35. $(-\infty, -3] \cup [7, \infty)$ **37.** $\left[-\dfrac{15}{4}, \dfrac{5}{4}\right]$
39. $(-\infty, -7] \cup [42, \infty)$ **41.** $(-\infty, 1) \cup \left(\dfrac{7}{5}, \infty\right)$
43. $\left(-\dfrac{1}{3}, 0\right) \cup \left(0, \dfrac{1}{9}\right)$ **45.** $(-\infty, -1] \cup [4, \infty)$
47. $(-\infty, -6) \cup \left(\frac{1}{3}, \infty\right)$ **53.** $\frac{\varepsilon}{3}$ **55.** $\frac{\varepsilon}{6}$ **57.** 0.0064 in.
59. $\left(-\infty, \frac{7}{3}\right) \cup (5, \infty)$ **61.** $\left(-\frac{4}{5}, \frac{16}{3}\right)$ **77.** $\dfrac{60}{11} \leq R \leq \dfrac{120}{13}$

Problem Set 0.3

1. 2 **3.** $\sqrt{170}$

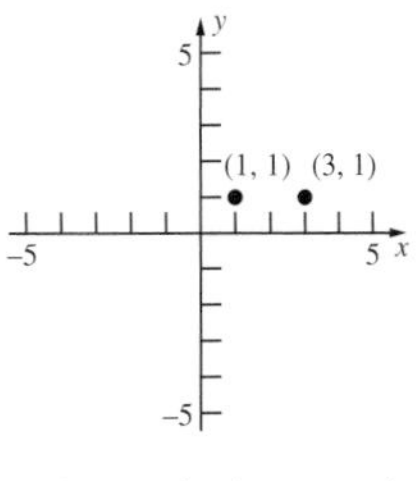

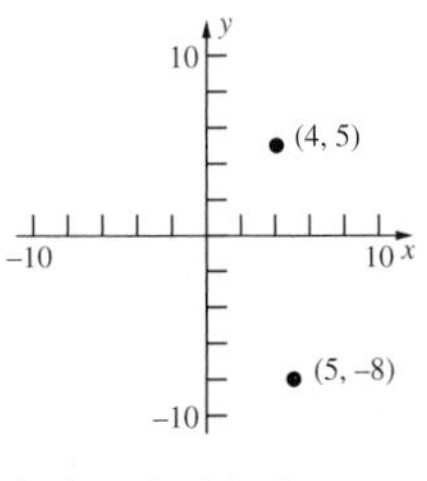

7. $(-1, 3), (-1, -1); (7, 3), (7, -1); (1, 1), (5, 1)$
9. $\dfrac{\sqrt{61}}{2}$ **11.** $(x - 1)^2 + (y - 1)^2 = 1$

13. $(x-2)^2+(y+1)^2=25$ **15.** $(x-2)^2+(y-5)^2=5$
17. Center $=(-1,3)$; radius $=\sqrt{10}$
19. Center $=(6,0)$; radius $=1$

21. Center $=\left(-2,-\frac{3}{4}\right)$; radius $=\dfrac{\sqrt{13}}{4}$

23. 1 **25.** $\frac{9}{7}$ **27.** $-\frac{5}{3}$ **29.** $y=-x+4$; $x+y-4=0$
31. $y=2x+3$; $2x-y+3=0$
33. $y=\frac{5}{2}x-2$; $5x-2y-4=0$
35. Slope $=-\frac{2}{3}$; y-intercept $=\frac{1}{3}$
37. Slope $=-5$; y-intercept $=4$
39. **(a)** $y=2x-9$; **(b)** $y=-\frac{1}{2}x-\frac{3}{2}$; **(c)** $y=-\frac{2}{3}x-1$; **(d)** $y=\frac{3}{2}x-\frac{15}{2}$; **(e)** $y=-\frac{3}{4}x-\frac{3}{4}$; **(f)** $x=3$; **(g)** $y=-3$
41. $y=\frac{3}{2}x+2$ **43.** It lies above the line.
45. $(-1,2)$; $y=\frac{3}{2}x+\frac{7}{2}$ **47.** $(3,1)$; $y=-\frac{4}{3}x+5$
49. Inscribed: $(x-4)^2+(y-1)^2=4$; circumscribed: $(x-4)^2+(y-1)^2=8$
55. $d=2\sqrt{3}+4$ **61.** $18+2\sqrt{17}+4\pi\approx 38.8$
63. $\dfrac{7}{5}$ **65.** $\dfrac{18}{13}$ **67.** $\dfrac{\sqrt{5}}{5}$ **69.** $y=\frac{3}{5}x+\frac{4}{5}$ **71.** $r=1$
73. $x+\sqrt{3}y=12$ and $x-\sqrt{3}y=12$ **77.** 8

Problem Set 0.4

1.
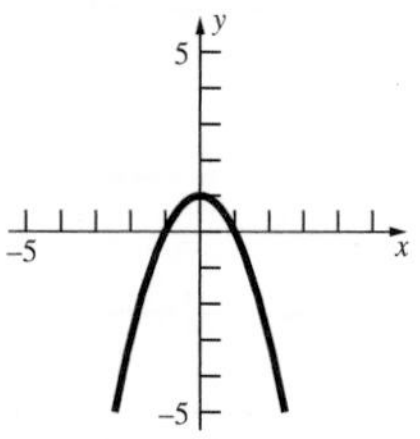

3.
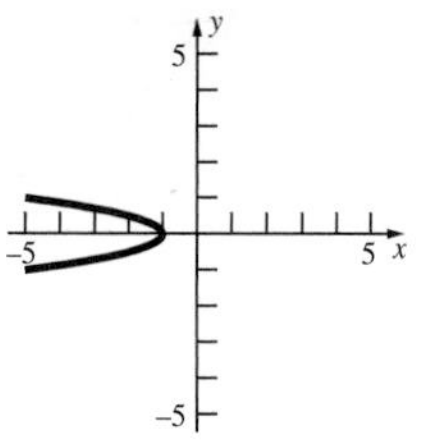

5.
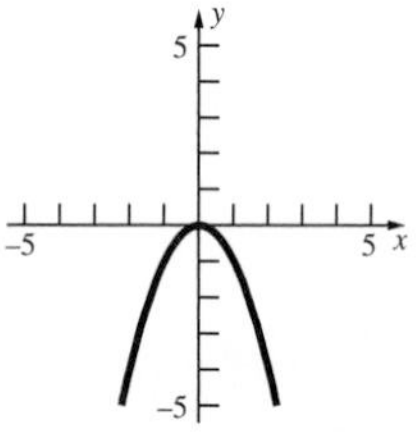

7.
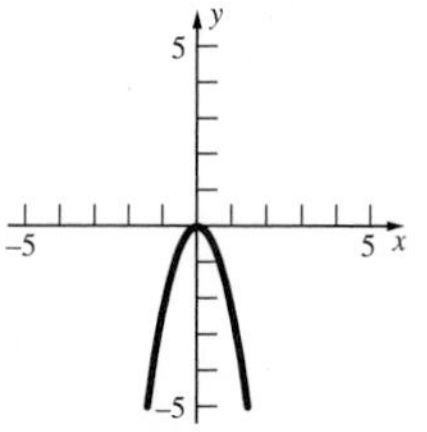

9.
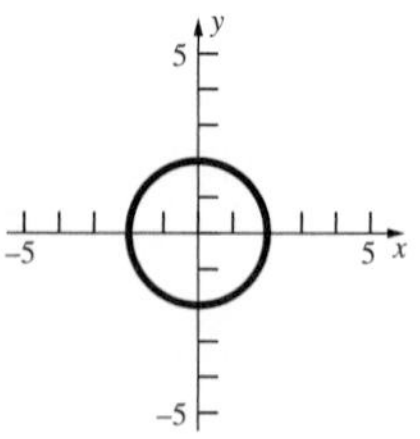

11.
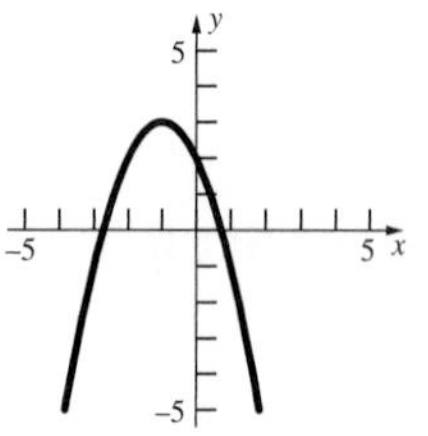

13.
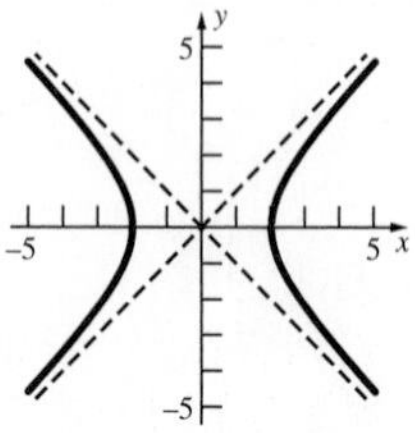

15.
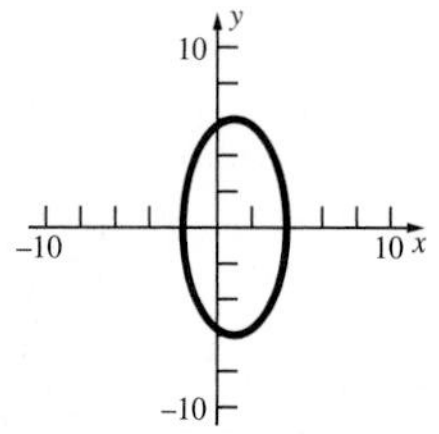

17.
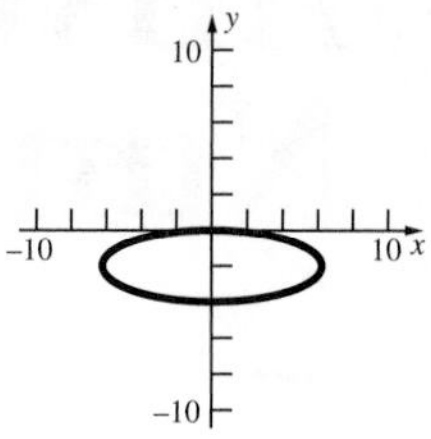

19.
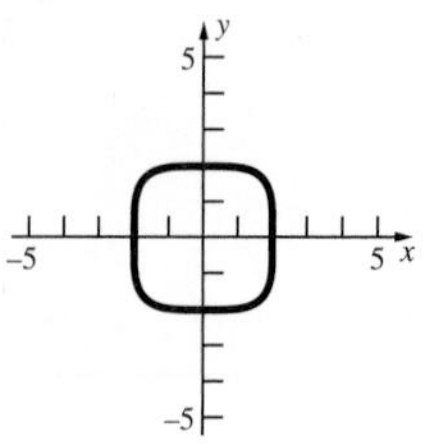

21.
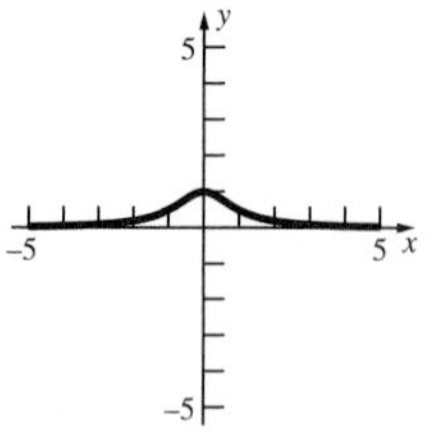

23.
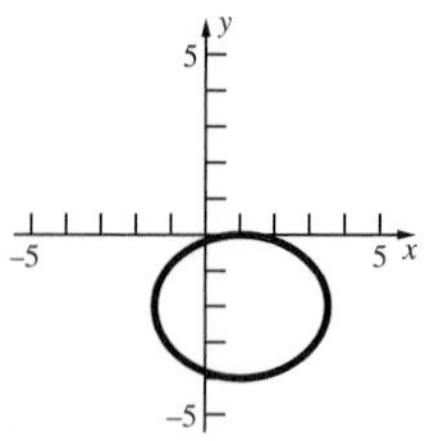

25.
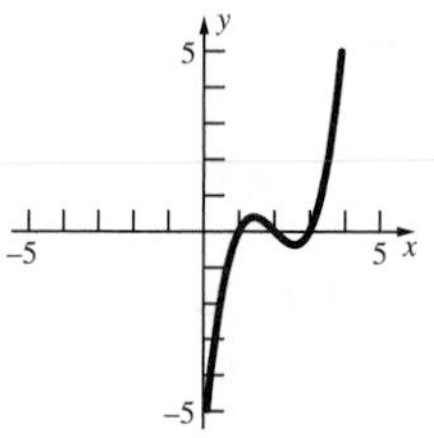

27.
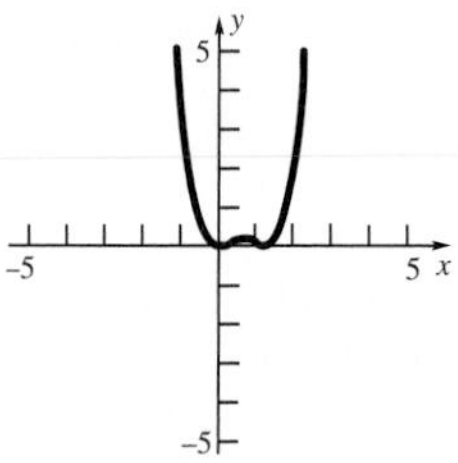

29.
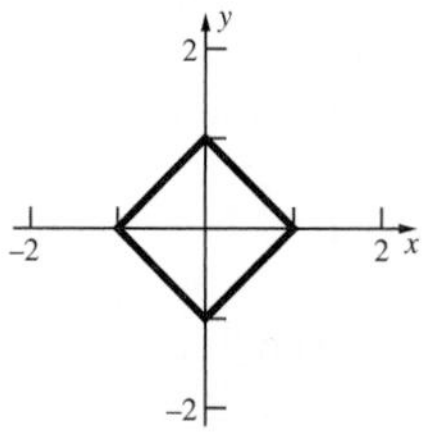

31.
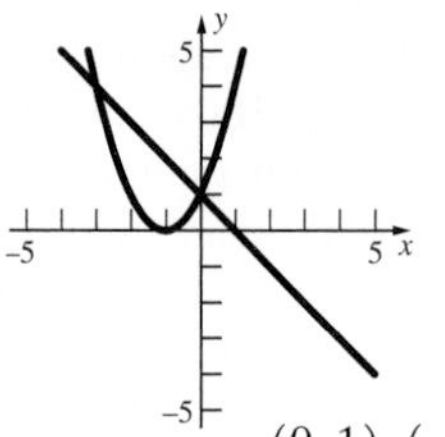

$(0,1)$, $(-3,4)$

33.
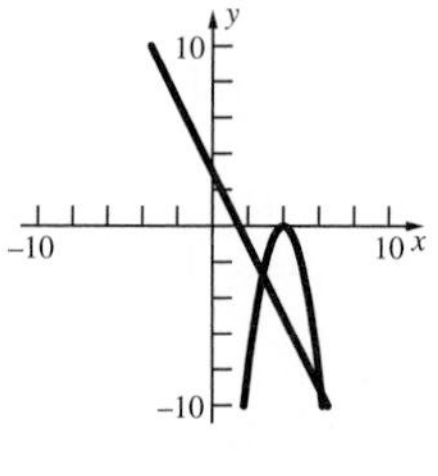

$\left(\frac{9}{2}-\frac{1}{2}\sqrt{11}, -6+\sqrt{11}\right)$,
$\left(\frac{9}{2}+\frac{1}{2}\sqrt{11}, -6-\sqrt{11}\right)$

35.
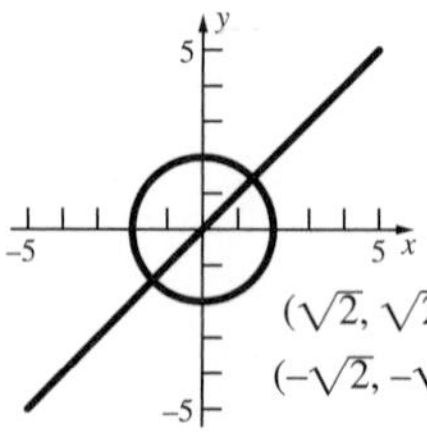

$(\sqrt{2},\sqrt{2})$,
$(-\sqrt{2},-\sqrt{2})$

37.
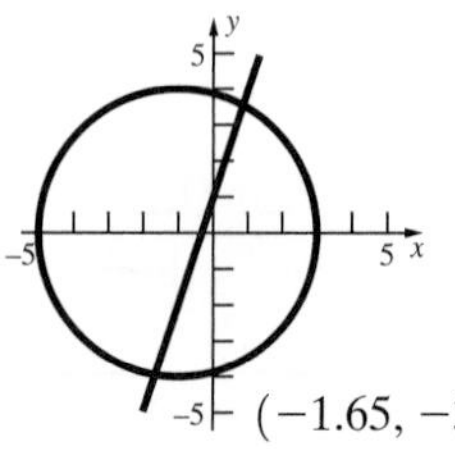

$(-1.65,-3.95)$,
$(0.85, 3.55)$

39. **(a)** (2) **(b)** (1) **(c)** (3) **(d)** (4)
41. Four distinct distances

Problem Set 0.5

1. **(a)** 0; **(b)** -3; **(c)** 1; **(d)** $1-k^2$; **(e)** -24; **(f)** $\frac{15}{16}$; **(g)** $-2h-h^2$; **(h)** $-2h-h^2$; **(i)** $-4h-h^2$
3. **(a)** -1; **(b)** -1000; **(c)** 100; **(d)** $\dfrac{1}{y^2-1}$; **(e)** $-\dfrac{1}{x+1}$; **(f)** $\dfrac{x^2}{1-x^2}$

5. **(a)** Undefined; **(b)** 2.658; **(c)** 0.841

7. **(a)** Not a function; **(b)** $f(x) = \dfrac{1-x}{x+1}$;
(c) $f(x) = \frac{1}{2}(x^2 - 1)$; **(d)** $f(x) = \dfrac{x}{1-x}$

9. $4a + 2h$ **11.** $-\dfrac{3}{x^2 - 4x + hx - 2h + 4}$

13. **(a)** $\{z \in \text{reals: } z \geq -\frac{3}{2}\}$; **(b)** $\{v \in \text{reals: } v \neq \frac{1}{4}\}$;
(c) $\{x \in \text{reals: } |x| \geq 3\}$; **(d)** $\{y \in \text{reals: } |y| \leq 5\}$

15. Even

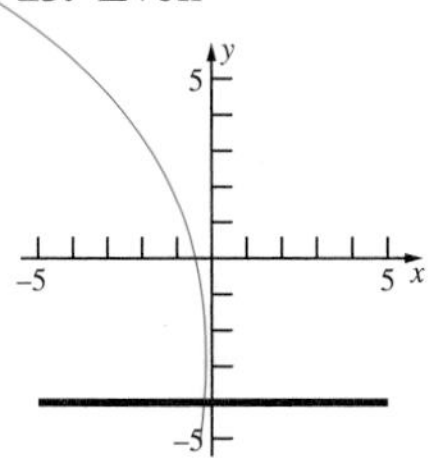

17. Neither

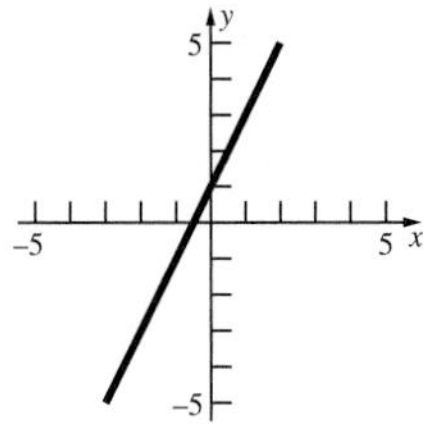

19. Neither

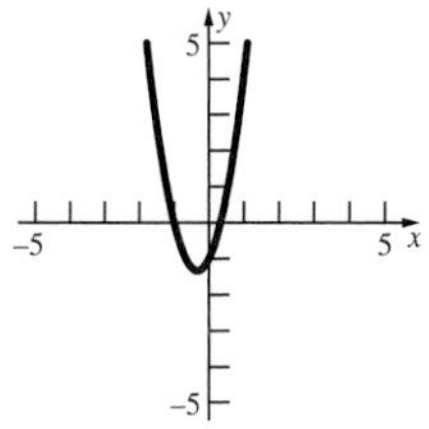

21. Odd

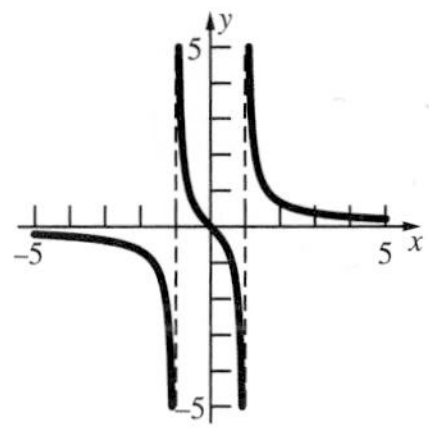

23. Neither

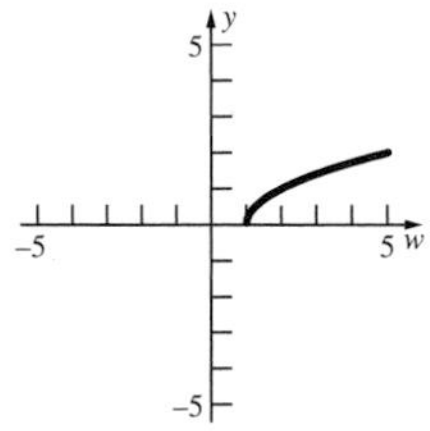

25. Even

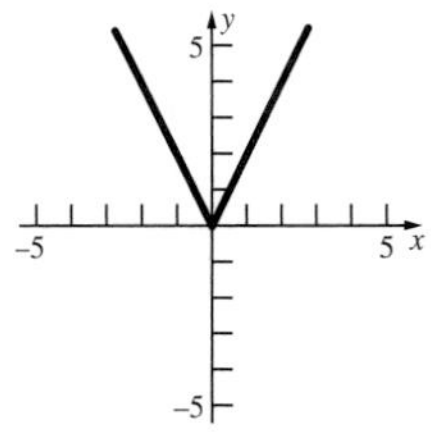

27. Neither

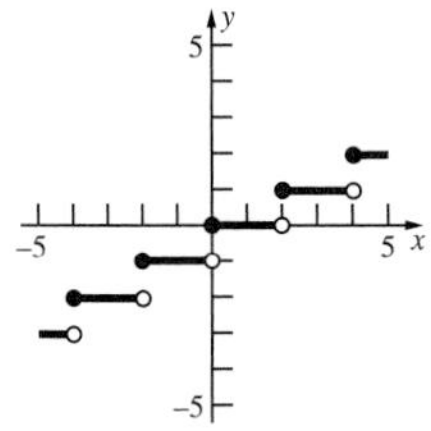

29. Neither

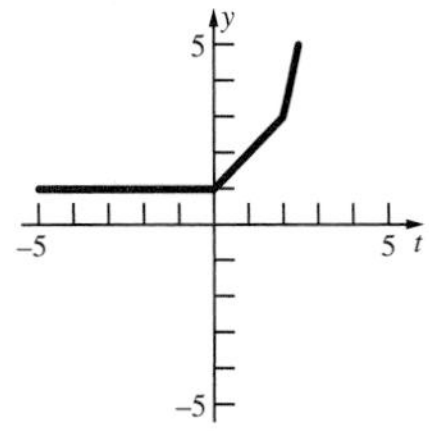

31. $T(x) = 5000 + 805x$, $\{x \in \text{integers: } 0 \leq x \leq 100\}$;
$u(x) = \frac{5000}{x} + 805$, $\{x \in \text{integers: } 0 < x \leq 100\}$

33. $E(x) = x - x^2$

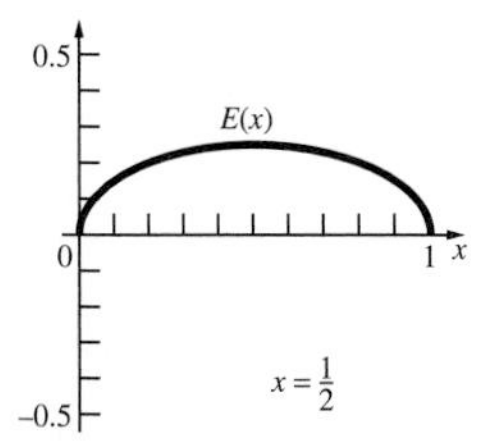

35. $L(x) = \sqrt{h^2 - x^2}$

37. **(a)** $E(x) = 24 + 0.40x$; **(b)** 240 miles

39. $A(d) = \dfrac{2d - \pi d^2}{4}$, $\left\{d \in \text{reals: } 0 < d < \dfrac{1}{\pi}\right\}$

41. **(a)** $B(0) = 0$ **(b)** $B(\frac{1}{2}) = \frac{1}{2}B(1) = \frac{1}{2} \cdot \frac{1}{6} = \frac{1}{12}$

(c)

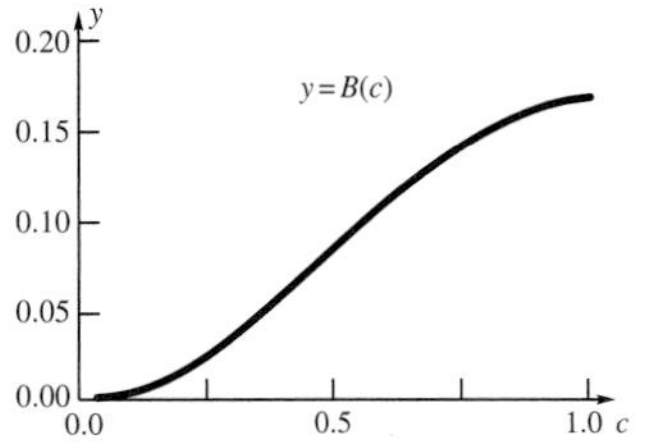

45. **(a)** $f(1.38) \approx 0.2994$, $f(4.12) \approx 3.6852$

(b)

x	$f(x)$
−4	−4.05
−3	−3.1538
−2	−2.375
−1	−1.8
0	−1.25
1	−0.2
2	1.125
3	2.3846
4	3.55

47.

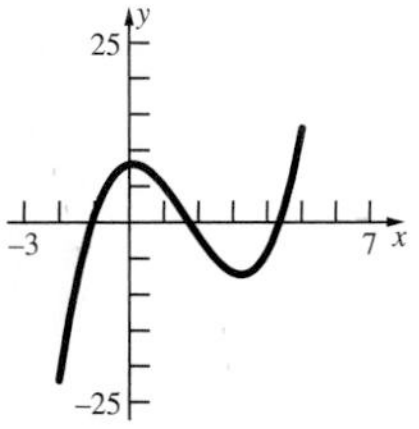

(a) $\{y \in \text{reals: } -22 \leq y \leq 13\}$;
(b) $[-1.1, 1.7] \cup [4.3, 5]$

49.

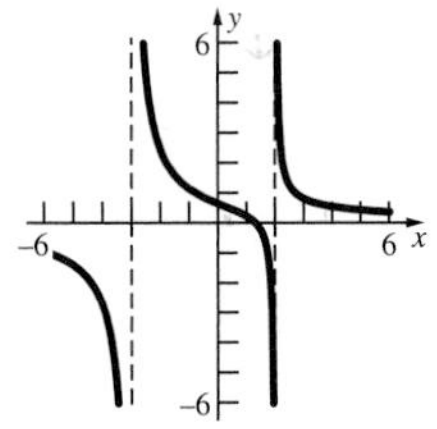

(a) x-intercept $\frac{4}{3}$, y-intercept $\frac{2}{3}$;
(b) all reals:
(c) $x = -3, x = 2$; **(d)** $y = 0$

Problem Set 0.6

1. **(a)** 9; **(b)** 0; **(c)** $\frac{3}{2}$; **(d)** 4; **(e)** 16; **(f)** 25

3. **(a)** $t^3 + 1 + \dfrac{1}{t}$; **(b)** $\dfrac{1}{r^3} + 1$; **(c)** $\dfrac{1}{r^3 + 1}$; **(d)** $(z^3 + 1)^3$;
(e) $125t^3 + 1 - \dfrac{1}{5t}$; **(f)** $\dfrac{1}{t^3} + 1 - t$

5. $(f \circ g)(x) = \sqrt{x^2 + 2x - 3}$; $(g \circ f)(x) = 1 + \sqrt{x^2 - 4}$

7. 1.188 **9.** 4.789

11. **(a)** $g(x) = \sqrt{x}, f(x) = x + 7$;
(b) $g(x) = x^{15}, f(x) = x^2 + x$

13. $p = f \circ g \circ h$ if $f(x) = 1/x$, $g(x) = \sqrt{x}$, $h(x) = x^2 + 1$;
$p = f \circ g \circ h$ if $f(x) = 1/\sqrt{x}$, $g(x) = x + 1$, $h(x) = x^2$

15.

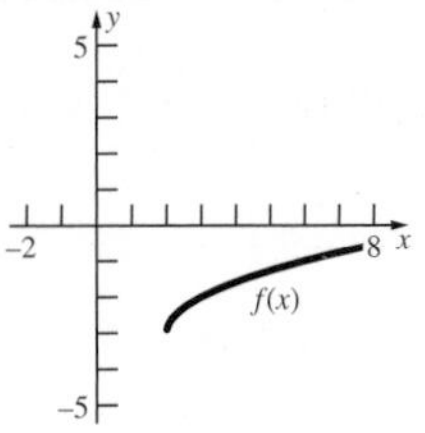

17.

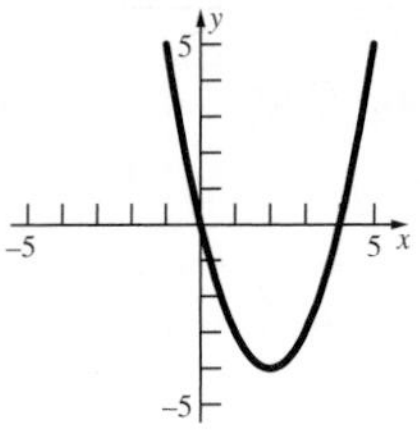

19.

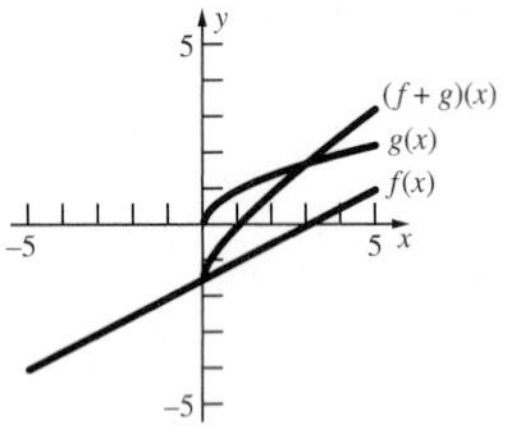

21.

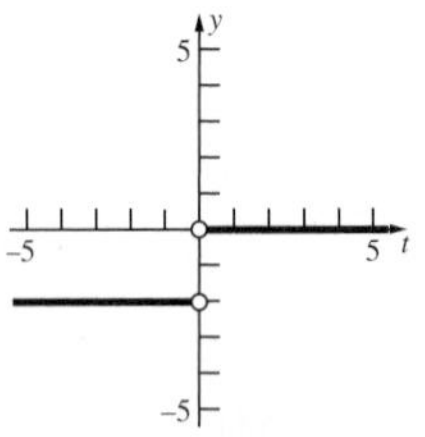

23. **(a)** Even; **(b)** Odd; **(c)** Even; **(d)** Even; **(e)** Odd

25. No, in both cases. (Consider $f(x) = x^2 + x$ and $f(x) = x^3 + 1$.)

27. **(a)** $P = \sqrt{t + \sqrt{t} + 27}$; **(b)** $P \approx 7$

29. $D(t) = \begin{cases} 400t & \text{if } 0 \le t \le 1 \\ \sqrt{250{,}000t^2 - 180{,}000t + 90{,}000} & \text{if } t > 1 \end{cases}$

33. **(a)** $\dfrac{1}{1 - x}$; **(b)** x; **(c)** $1 - x$

37.

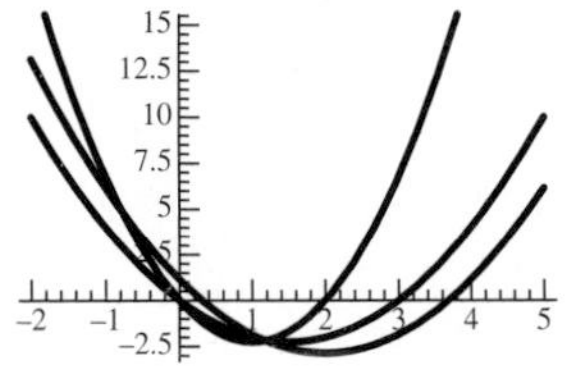

39.

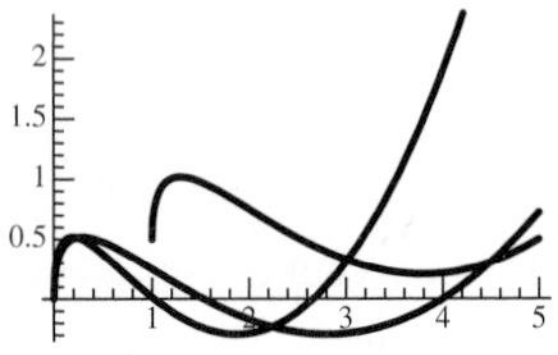

41. **(a)**

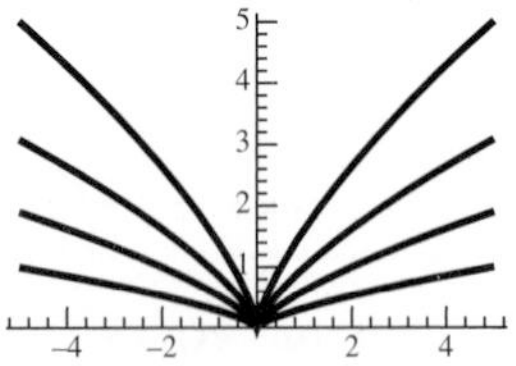

(b)

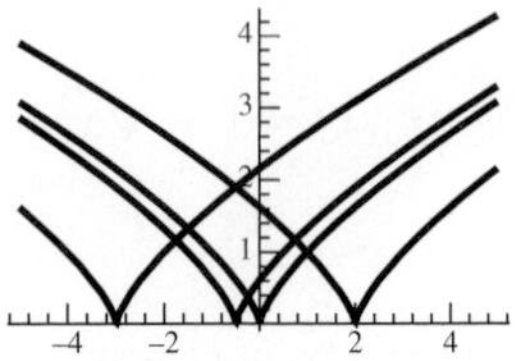

(c)

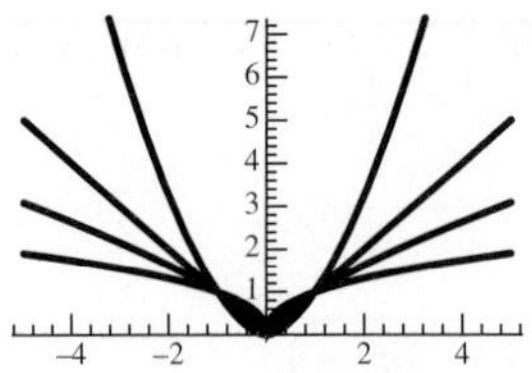

Problem Set 0.7

1. **(a)** $\frac{\pi}{6}$; **(b)** $\frac{\pi}{4}$; **(c)** $-\frac{\pi}{3}$; **(d)** $\frac{4\pi}{3}$; **(e)** $-\frac{37\pi}{18}$; **(f)** $\frac{\pi}{18}$;

3. **(a)** 0.5812; **(b)** 0.8029; **(c)** −1.1624; **(d)** 4.1907; **(e)** −6.4403; **(f)** 0.1920;

5. **(a)** 68.37; **(b)** 0.8845; **(c)** 0.4855; **(d)** −0.3532;

7. **(a)** 46.097; **(b)** 0.0789

9. **(a)** $\dfrac{\sqrt{3}}{3}$; **(b)** −1; **(c)** $-\sqrt{2}$; **(d)** 1; **(e)** 1; **(f)** −1

15. **(a)**

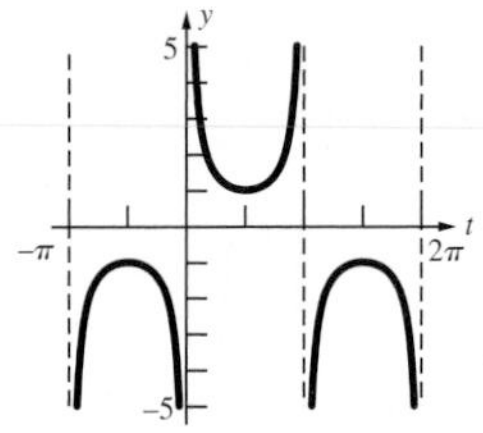

(b)

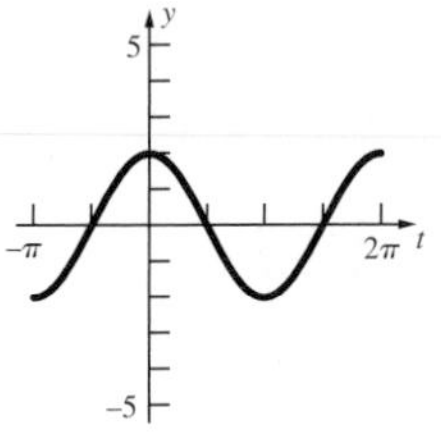

(c)

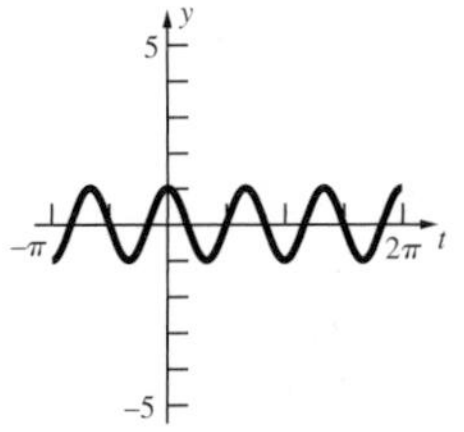

(d)

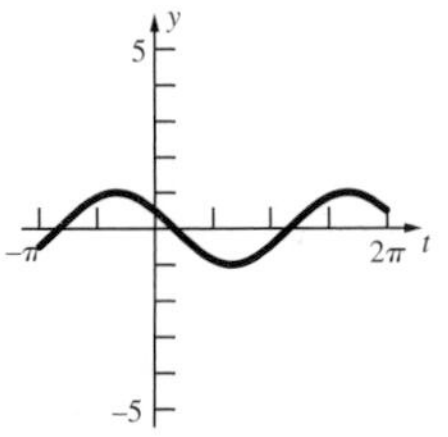

17. Period = π; Amplitude = 2

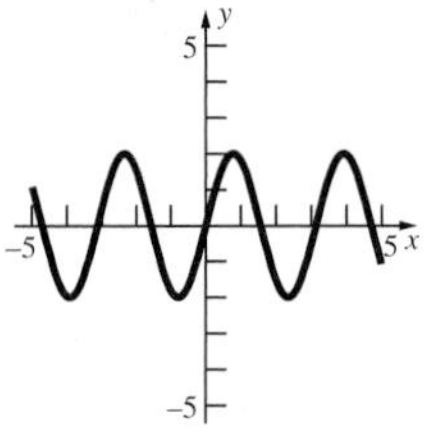

19. Period = $\frac{\pi}{2}$; shift: 2 units up

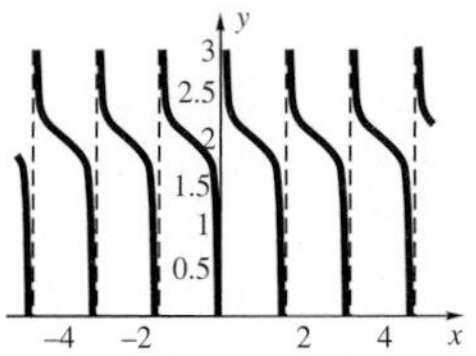

21. Period $= \pi$; amplitude $= 7$; shift: 21 units up, $\frac{3}{2}$ units left

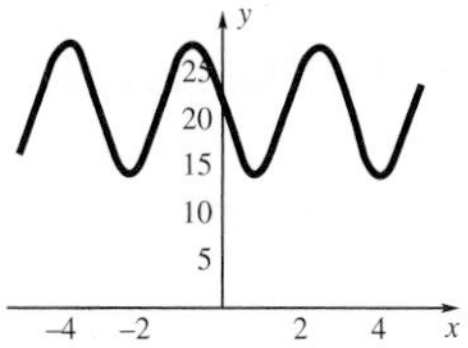

23. Period $= \frac{\pi}{2}$; shift: $\frac{\pi}{6}$ units right

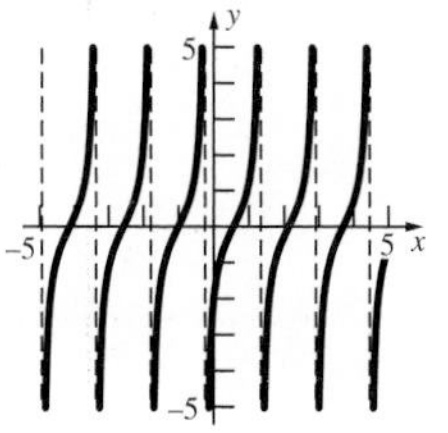

25. **(a)** Even; **(b)** Even; **(c)** Odd; **(d)** Even; **(e)** Even; **(f)** Odd

27. $\frac{1}{4}$ **29.** $\frac{1}{8}$ **31.** $\dfrac{2-\sqrt{2}}{4}$

35. 336 rev/min **37.** 28 rev/sec

39. **(a)** $\frac{\pi}{3}$; **(b)** $\frac{5\pi}{6}$

41. **(a)** 0.1419; **(b)** 1.8925; **(c)** 1.7127

43. 25 cm^2 **45.** $r^2 \sin\frac{t}{2}\cos\frac{t}{2} + \frac{\pi r^2}{2}\sin^2\frac{t}{2}$

47. 67.5°F

49. As t increases, the point on the rim of the wheel will move around the circle of radius 2.
(a) $x(2) \approx 1.902$; $y(2) \approx 0.618$; $x(6) \approx -1.176$; $y(6) \approx -1.618$; $x(10) = 0$; $y(10) = 2$; $x(0) = 0$; $y(0) = 2$
(b) $x(t) = -2\sin\left(\frac{\pi}{5}t\right)$, $y(t) = 2\cos\left(\frac{\pi}{5}t\right)$
(c) The point is at $(2, 0)$ when $\frac{\pi}{5}t = \frac{\pi}{2}$; that is, when $t = \frac{5}{2}$.

51. **(c)** $A_1\sin(\omega t + \phi_1) + A_2\sin(\omega t + \phi_2) + A_3\sin(\omega t + \phi_3)$
$= (A_1\cos\phi_1 + A_2\cos\phi_2 + A_3\cos\phi_3)\sin\omega t$
$+ (A_1\sin\phi_1 + A_2\sin\phi_2 + A_3\sin\phi_3)\cos\omega t$

53. **(a)**

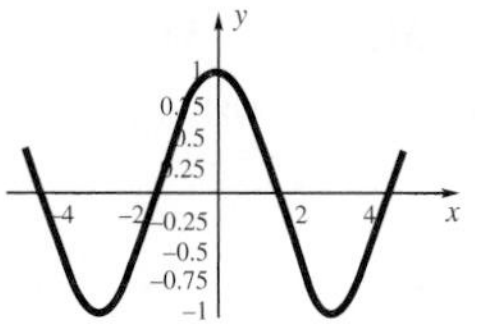

(b)

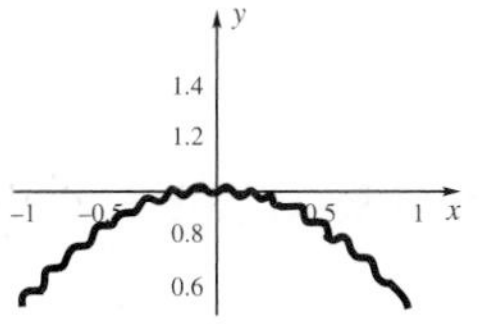

(c)

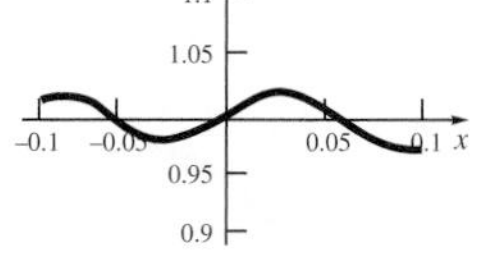

Chapter Review 0.8

Concepts Test

1. False **3.** False **5.** False **7.** False **9.** True
11. True **13.** True **15.** False **17.** True **19.** True
21. True **23.** True **25.** True **27.** True **29.** True
31. True **33.** True **35.** True **37.** False **39.** False
41. True **43.** True **45.** False **47.** True **49.** True
51. False **53.** True **55.** False **57.** True **59.** False
61. True **63.** False

Sample Test Problems

1. **(a)** $2, \frac{25}{4}, \frac{4}{25}$; **(b)** $1, 9, 49$; **(c)** $64, 8, \frac{1}{8}$; **(d)** $1, \dfrac{\sqrt{2}}{2}, \sqrt{2}$

7. 2.66

9. $\{x: x < \frac{1}{3}\}$; $\left(-\infty; \frac{1}{3}\right)$;

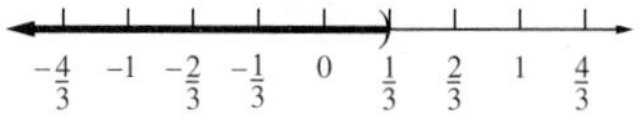

11. $\{x: \frac{1}{3} \le x \le 3\}$; $\left[\frac{1}{3}, 3\right]$;

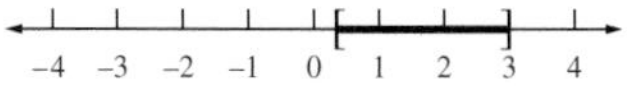

13. $\{t: \frac{3}{7} \le t \le \frac{5}{3}\}$; $\left[\frac{3}{7}, \frac{5}{3}\right]$;

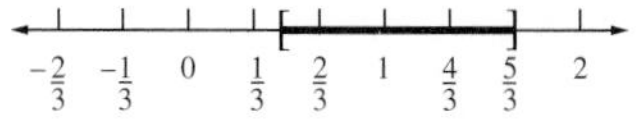

15. $\{x: -4 \le x \le 3\}$; $[-4, 3]$;

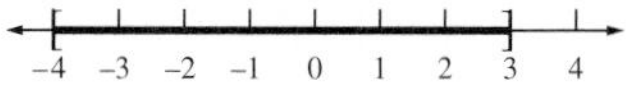

17. $\{x: x \le -\frac{1}{2} \text{ or } x > 1\}$; $\left(-\infty, -\frac{1}{2}\right] \cup (1, \infty)$;

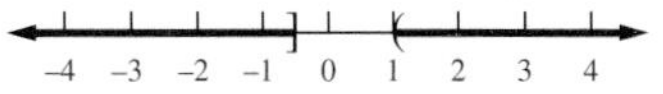

19. Any negative number **21.** $t \le 5$

25.

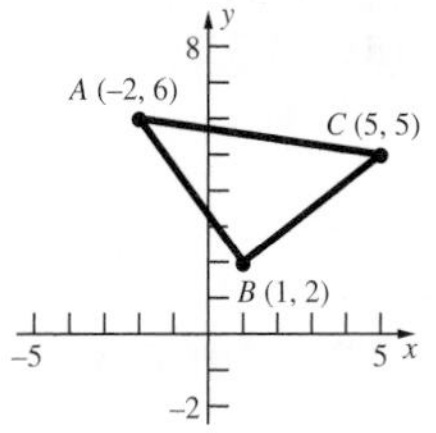

27. $(x - 6)^2 + (y - 2)^2 = 20$ **29.** 5

31. **(a)** $y = \frac{2}{9}x + \frac{13}{9}$; **(b)** $y = \frac{3}{2}x + 4$; **(c)** $y = \frac{4}{3}x + \frac{11}{3}$; **(d)** $x = -2$; **(e)** $y = x + 3$

33. (b)

35.

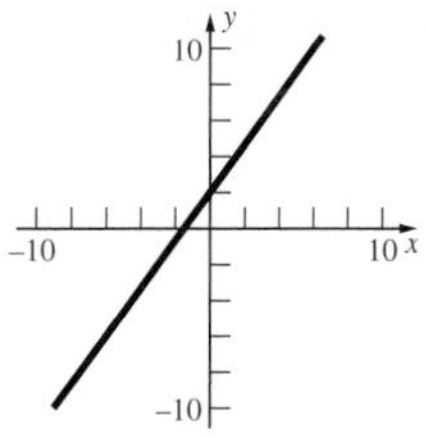

37.

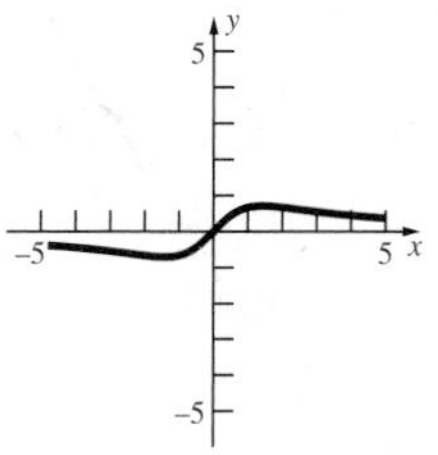

39. (0, 4) and (3, 7)

41. **(a)** $-\frac{1}{2}$; **(b)** 4; **(c)** Does not exist; **(d)** $\dfrac{1}{t} - \dfrac{1}{t-1}$; **(e)** $\dfrac{t}{1+t} - t$

43. **(a)** $\{x \in \text{reals}: x \ne -1, 1\}$; **(b)** $\{x \in \text{reals}: |x| \le 2\}$;

45. **(a)**

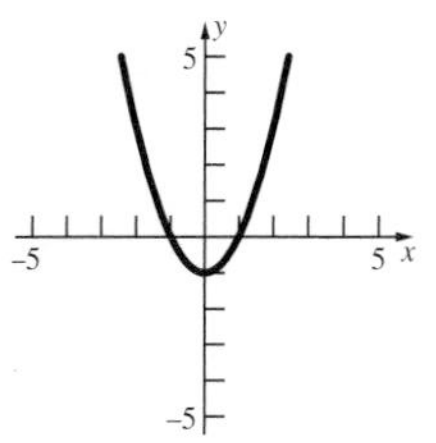

(b)

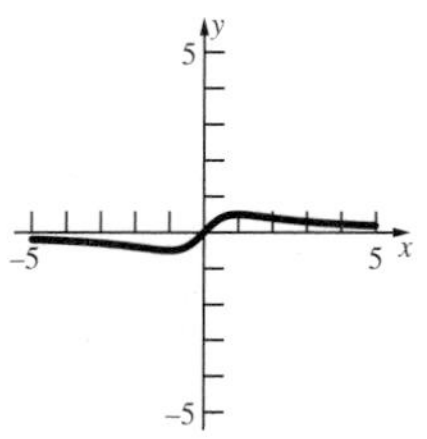

(c)

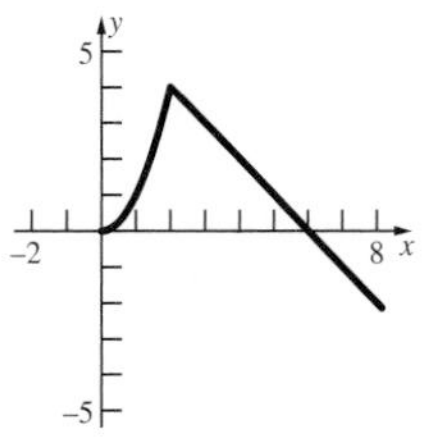

47. $V(x) = x(32 - 2x)(24 - 2x)$, $\{x \in \text{reals}: 0 \le x \le 12\}$

49. (a)

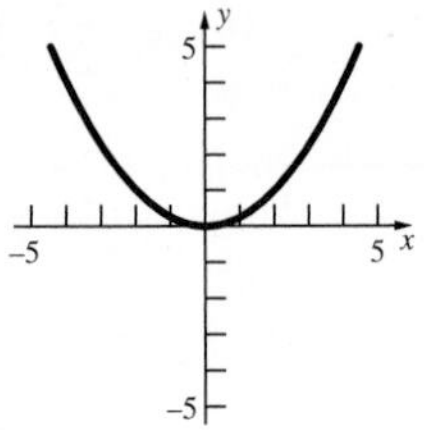

(b)

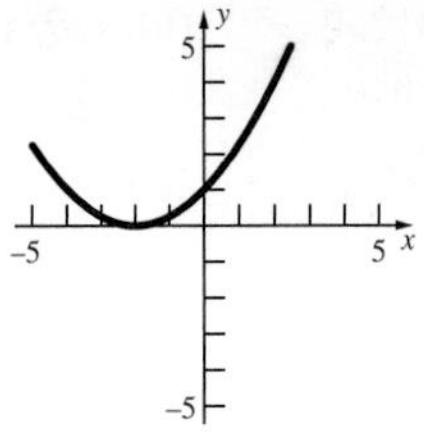

(c)

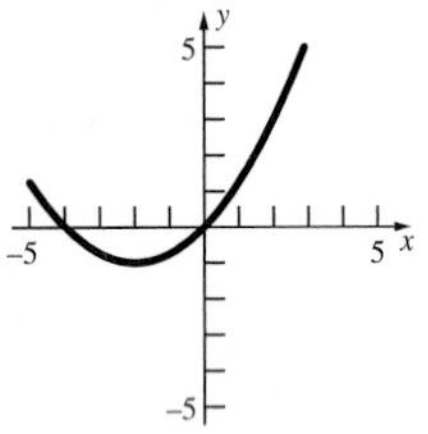

51. $f(x) = \sqrt{x}, g(x) = 1 + x, h(x) = x^2, k(x) = \sin x$

53. (a) -0.8; **(b)** -0.6; **(c)** -0.96; **(d)** -1.333;
(e) 0.8; **(f)** -0.8

55. 18.85 in.

Chapter 1 Review and Preview Problems

1. (a) $0 < x < 2$; **(b)** $-6 < x < 8$

3. 4,10 **5.** 4,10

7. (a) $4 < x < 10$; **(b)** $4 \le x \le 10$; **(c)** $6 \le x \le 8$;
(d) $6.9 < x < 7.1$

9. (a) $x \ne 1$; **(b)** $x \ne 1, -0.5$

11. 1, 1.9, 1.99, 1.999, 2.001, 2.01, 2.1, 3; -1, -0.0357143, -0.0033557, -0.000333556, 0.000333111, 0.00331126, 0.03125, 0.2

13. $4.9 < x < 5.1$

15. (a) True; **(b)** False; **(c)** True; **(d)** True

Problem Set 1.1

1. -2 **3.** -1 **5.** 0 **7.** 4 **9.** 12 **11.** $-2t$

13. $\dfrac{\sqrt{6}}{9}$ **15.** 36 **17.** 4 **19.** 0.5 **21.** 0 **23.** 2

25. 0 **27.** 0.25

29. (a) 2; **(b)** 1; **(c)** Does not exist; **(d)** $\frac{5}{2}$; **(e)** 2;
(f) Does not exist; **(g)** 2; **(h)** 1; **(i)** 2.5

31. (a) 2; **(b)** undefined; **(c)** 2; **(d)** 4; **(e)** does not exist;
(f) does not exist

33.

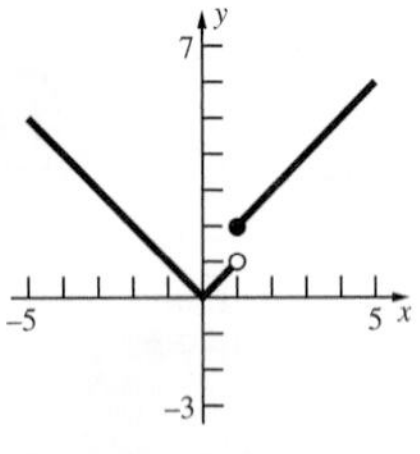

(a) 0; **(b)** Does not exist;
(c) 2; **(d)** 2

35.

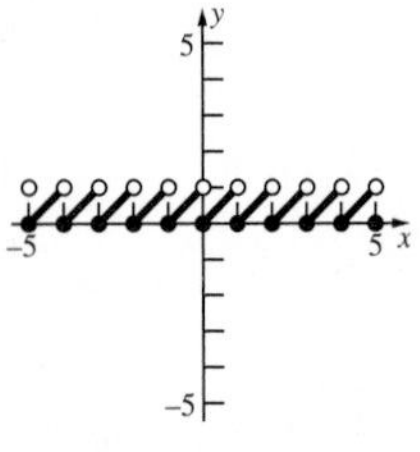

(a) 0; **(b)** Does not exist; **(c)** 1;
(d) $\frac{1}{2}$

37. Does not exist

39. (a) Does not exist; **(b)** 0

41. $a = -1, 0, 1$

43. (a) Does not exist; **(b)** -1; **(c)** -3; **(d)** Does not exist

45. (a) 1; **(b)** 0; **(c)** -1; **(d)** -1

47. Does not exist **49.** 0 **51.** $\frac{1}{2}$ **53.** Does not exist

55. 6 **57.** -3

Problem Set 1.2

1. $0 < |t - a| < \delta \Rightarrow |f(t) - M| < \varepsilon$

3. $0 < |z - d| < \delta \Rightarrow |h(z) - P| < \varepsilon$

5. $0 < c - x < \delta \Rightarrow |f(x) - L| < \varepsilon$

7. 0.001

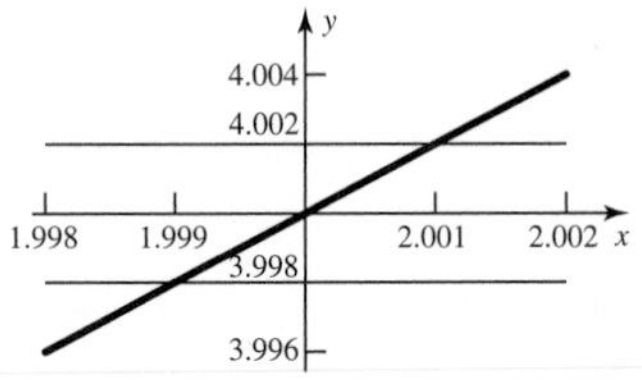

9. 0.0019

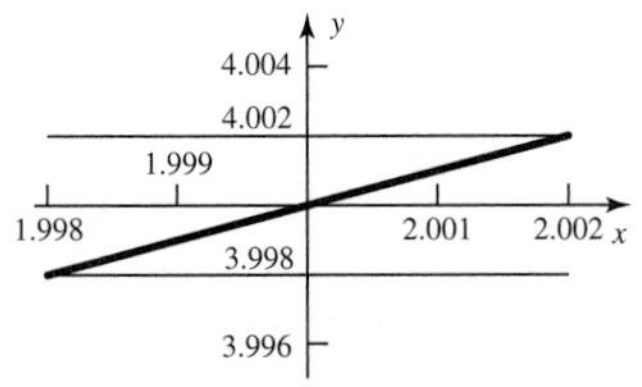

31. (b), (c)

33. (a) $\dfrac{x^3 - x^2 - 2x - 4}{x^4 - 4x^3 + x^2 + x + 6}$; **(b)** No; **(c)** 3

Problem Set 1.3

1. 3 **3.** -3 **5.** -5 **7.** 2 **9.** -1 **11.** 2 **13.** 0

15. -4 **17.** $-\frac{2}{3}$ **19.** $\frac{3}{2}$ **21.** $\dfrac{x+2}{5}$ **23.** -1

25. $\sqrt{10}$ **27.** -6 **29.** 6 **31.** 12 **33.** $-\frac{1}{4}$ **41.** 0

43. 0 **45.** $\frac{2}{5}$ **47.** -1 **51. (a)** 1; **(b)** 0

Problem Set 1.4

1. 1 **3.** 1 **5.** $\frac{1}{2}$ **7.** 3 **9.** $\frac{1}{2\pi}$ **11.** 0 **13.** 7

15. 0

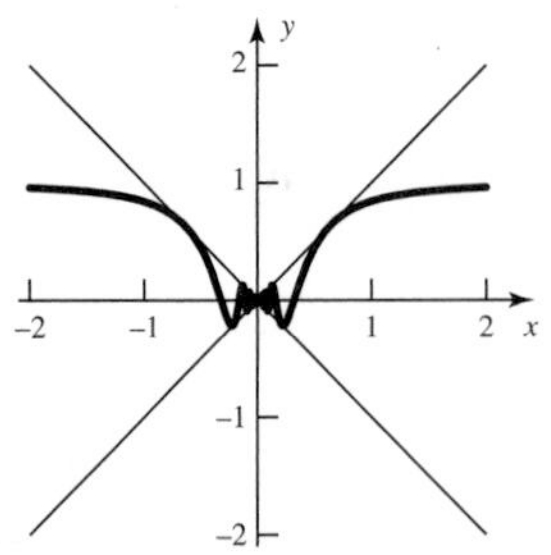

17. 0

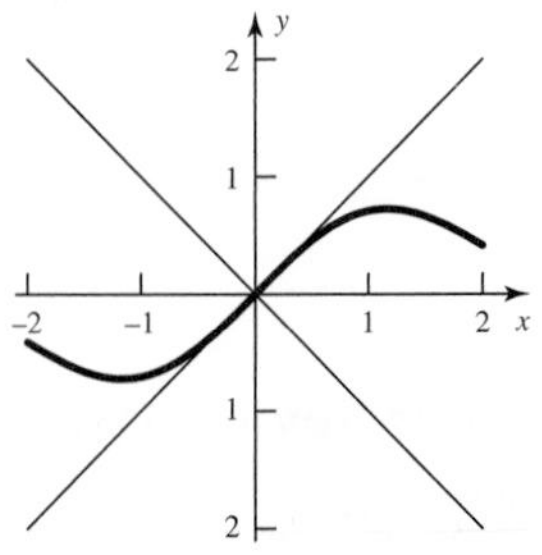

19. 2

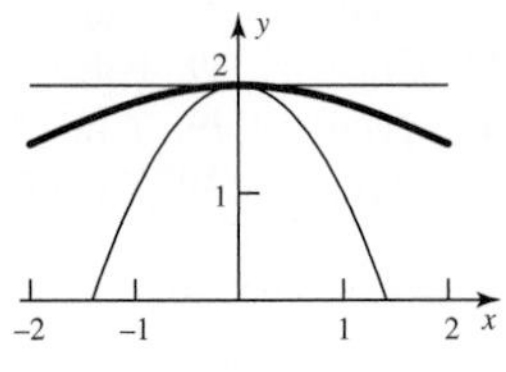

Problem Set 1.5

1. 1 **3.** −1 **5.** −1 **7.** $\frac{1}{2}$ **9.** $\frac{3}{\pi}$ **11.** $\frac{3}{\sqrt{2}}$

13. 2 **15.** $\frac{1}{2}$ **17.** ∞ **19.** 2 **21.** 0 **23.** $-\infty$

25. 1 **27.** ∞ **29.** ∞ **31.** ∞ **33.** $-\infty$ **35.** 5

37. 0 **39.** −1 **41.** $-\infty$

43. Horizontal asymptote $y = 0$
Vertical asymptote $x = -1$

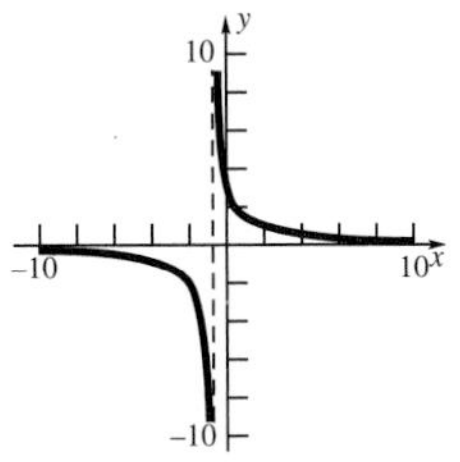

45. Horizontal asymptote $y = 2$
Vertical asymptote $x = 3$

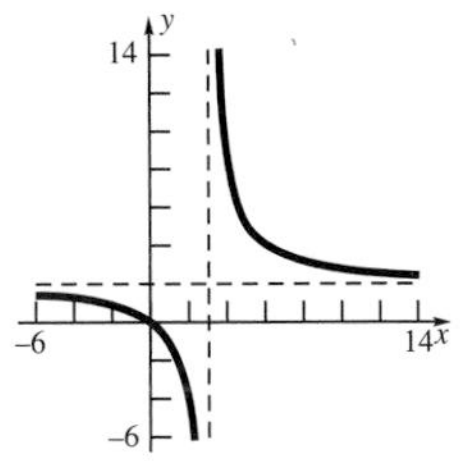

47. Horizontal asymptote $y = 0$
No vertical asymptotes

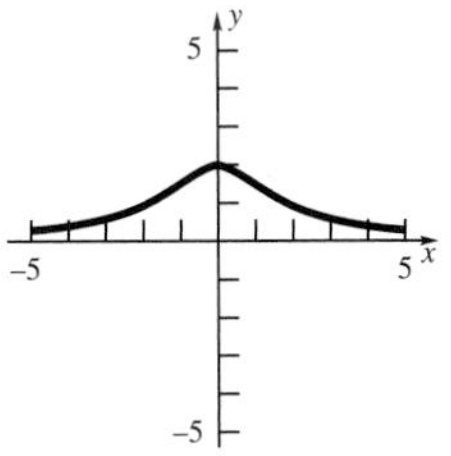

49. The oblique asymptote is $y = 2x + 3$.

51. (a) We say that $\lim_{x \to c^+} f(x) = -\infty$ if for each negative number M there corresponds a $\delta > 0$ such that $0 < x - c < \delta \Rightarrow f(x) < M$.
(b) We say that $\lim_{x \to c^-} f(x) = \infty$ if for each positive number M there corresponds a $\delta > 0$ such that $0 < c - x < \delta \Rightarrow f(x) > M$.

55. (a) Does not exist. **(b)** 0 **(c)** 1 **(d)** ∞ **(e)** 0
(f) $\frac{1}{2}$ **(g)** Does not exist. **(h)** 0

57. $\frac{3}{2}$ **59.** $-\frac{3}{2\sqrt{2}}$ **61.** 1 **63.** ∞ **65.** −1

67. $-\infty$ **69.** e **71.** 1

Problem Set 1.6

1. Continuous

3. Not continuous; $\lim_{x \to 3} \frac{3}{x - 3}$ and $h(3)$ do not exist.

5. Not continuous; $\lim_{t \to 3} \frac{|t - 3|}{t - 3}$ and $h(3)$ do not exist.

7. Continuous **9.** Not continuous; $h(3)$ does not exist.

11. Continuous **13.** Continuous **15.** Continuous

17. $(-\infty, -5), [-5, 4], (4, 6), [6, 8], (8, \infty)$

19. Define $f(3) = -12$. **21.** Define $H(1) = \frac{1}{2}$.

23. Define $F(-1) = -\sin 2$. **25.** 3, π

27. Every $\theta = n\pi + \frac{\pi}{2}$ where n is any integer. **29.** −1

31. $(-\infty, -2] \cup [2, \infty)$ **33.** 1

35. Every $t = n + \frac{1}{2}$ where n is any integer.

37.

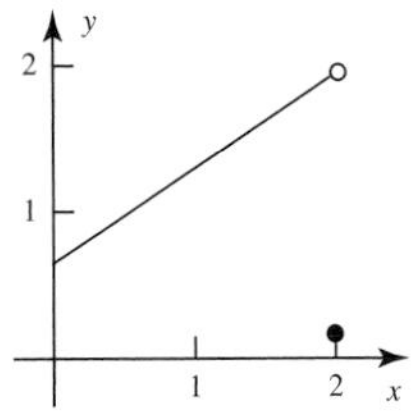

39.

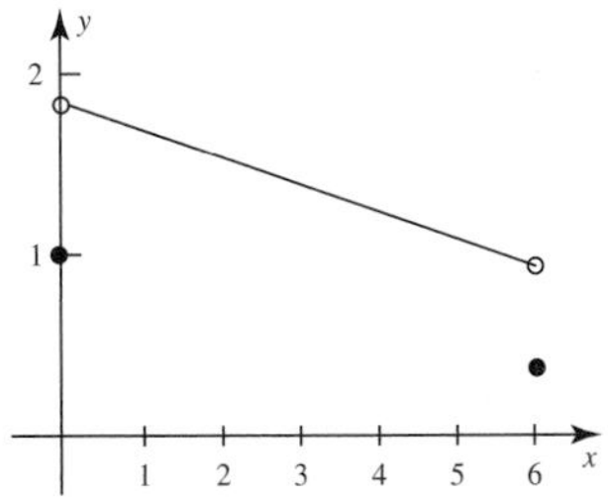

41. Continuous.

43. Discontinuous: removable, define $f(0) = 1$

45. Discontinuous, removable, redefine $g(0) = 1$

47. Discontinuous: nonremovable.

49. The function is continuous on the intervals $(0, 1], (1, 2], (2, 3], \ldots$

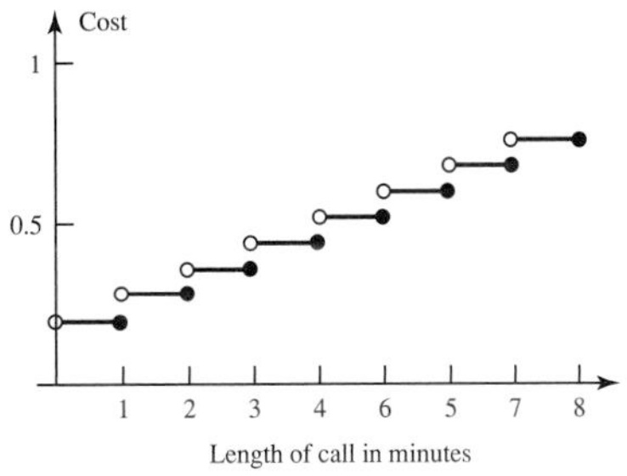

51. The function is continuous on the intervals $(0, 0.25], (0.25, 0.375], (0.375, 0.5], \ldots$

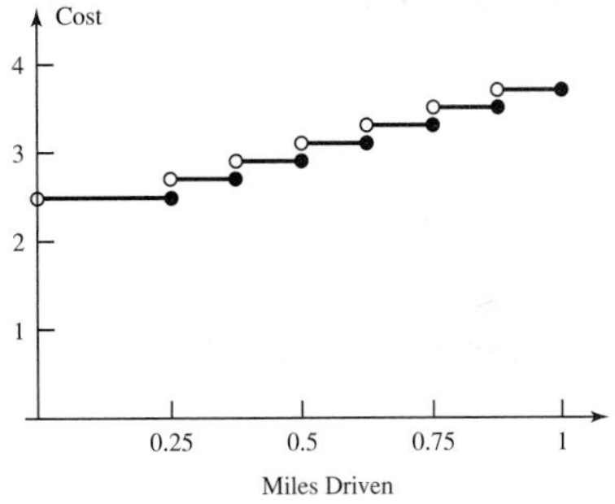

55. The interval [0.6,0.7] contains the solution.

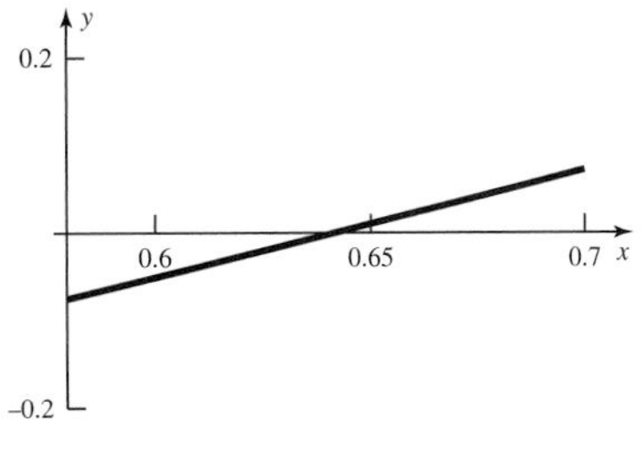

65. Yes, g is continuous.

71. (a) Domain $\left[-\frac{3}{4}, \frac{3}{4}\right]$, Range $\{-3/4, 0, 3/4\}$

(b) Discontinuous at $x = 0$ **(c)** $-\frac{3}{4}, 0, \frac{3}{4}$

Chapter Review 1.7

Concepts Test

1. False **3.** False **5.** False **7.** True **9.** False
11. True **13.** True **15.** False **17.** False **19.** False
21. True **23.** True **25.** True **27.** True **29.** False
31. True

Sample Test Problems

1. 0 **3.** 2 **5.** $\frac{1}{8}$ **7.** $\frac{1}{2}$ **9.** 4 **11.** -1 **13.** -1
15. $\frac{5}{3}$ **17.** 1 **19.** ∞ **21.** ∞
25. **(a)** $x = -1, 1$ **(b)** $f(-1) = -1$
27. **(a)** 14 **(b)** -12 **(c)** -2 **(d)** -2 **(e)** 5 **(f)** 0
29. $a = 2, b = -1$ **31.** Vertical: none, Horizontal $y = 0$
33. Vertical: $x = -1, 1$, Horizontal: $y = 1$
35. Vertical: $x = \pm\pi/4, \pm 3\pi/4, \pm 5\pi/4, \ldots$, Horizontal: none

Chapter 2 Review and Preview Problems

1. **(a)** 4 **(b)** 4.41 **(c)** 0.41 **(d)** 4.1 **(e)** $a^2 + 2ah + h^2$ **(f)** $2ah + h^2$ **(g)** $2a + h$ **(h)** $2a$
3. **(a)** $\sqrt{2} \approx 1.41$ **(b)** $\sqrt{2.1} \approx 1.45$ **(c)** 0.035 **(d)** 0.35 **(e)** $\sqrt{a+h}$ **(f)** $\sqrt{a+h} - \sqrt{a}$ **(g)** $\left(\sqrt{a+h} - \sqrt{a}\right)/h$ **(h)** $\frac{1}{2\sqrt{a}}$
5. **(a)** $a^3 + 3a^2b$ **(b)** $a^4 + 4a^3b$ **(c)** $a^5 + 5a^4b$
7. $\sin(x + h) = \sin x \cos h + \cos x \sin h$
9. **(a)** $(10, 0), (10, 0), (10, 0)$ **(b)** $t = 1/4$
11. **(a)** North plane has traveled 600 miles. East plane has traveled 400 miles. **(b)** 721 miles **(c)** 840 miles

Problem Set 2.1

1. 4
3. -2 **5.** $\frac{5}{2}$

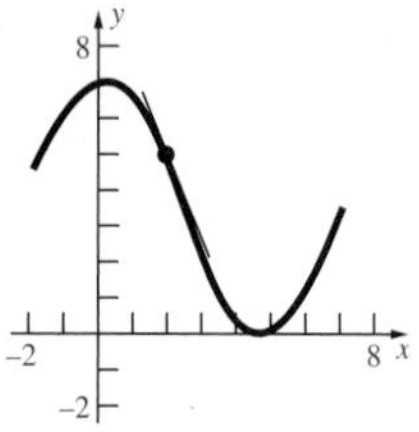

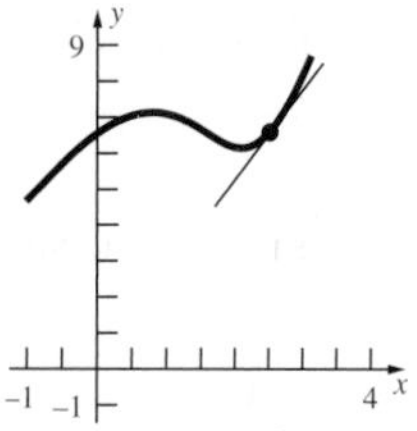

7. **(a), (b)**

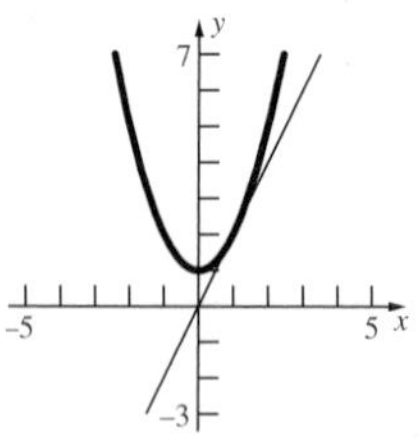

(c) 2; **(d)** 2.01; **(e)** 2

9. $-4, -2, 0, 2, 4$

11.

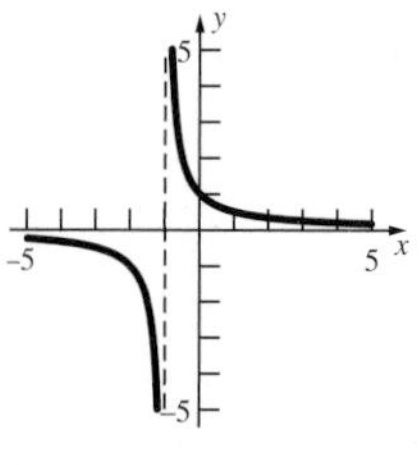

$y - \frac{1}{2} = -\frac{1}{4}(x - 1)$

13. **(a)** 16 ft; **(b)** 48 ft; **(c)** 80 ft/s; **(d)** 96.16 ft/s; **(e)** 96 ft/s
15. **(a)** $\frac{1}{\sqrt{2\alpha + 1}}$ ft/s; **(b)** 1.5 sec
17. **(a)** 0.02005 g; **(b)** 2.005 g/h; **(c)** 2 g/h
19. **(a)** 49 g/cm; **(b)** 27 g/cm
21. 4 **23.** 29,167 gal/h; 75,000 gal/h
25. **(a)** 0.5 °F/day **(b)** 0.067 °F/day **(c)** January and July **(d)** March and November
27. **(a)** Increasing **(b)** Decreasing
29. 24π km^2/day
31.

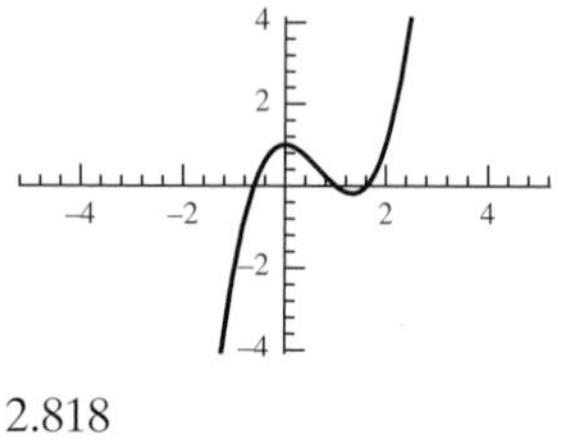

(a) 7; **(b)** 0; **(c)** -1; **(d)** 17.92

33. 2.818

Problem Set 2.2

1. 2 **3.** 5 **5.** 2 **7.** $6x$ **9.** $2ax + b$
11. $3x^2 + 4x$ **13.** $-\frac{2}{x^2}$ **15.** $-\frac{12x}{(x^2 + 1)^2}$
17. $-\frac{7}{(x - 4)^2}$ **19.** $\frac{3}{2\sqrt{3x}}$ **21.** $-\frac{3}{2(x - 2)^{3/2}}$
23. $2x - 3$ **25.** $-\frac{5}{(x - 5)^2}$ **27.** $f(x) = 2x^3$ at $x = 5$
29. $f(x) = x^2$ at $x = 2$ **31.** $f(x) = x^2$ at x
33. $f(t) = \frac{2}{t}$ at t **35.** $f(x) = \cos x$ at x

37.

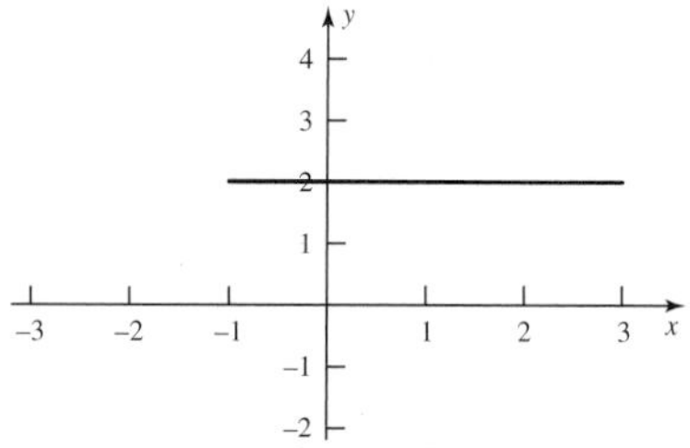

39.

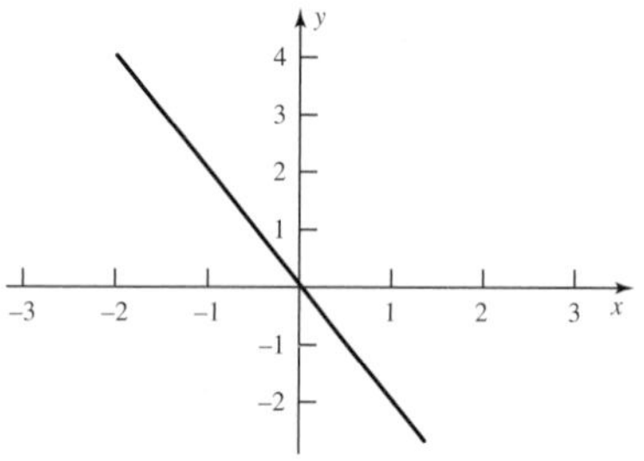

41.

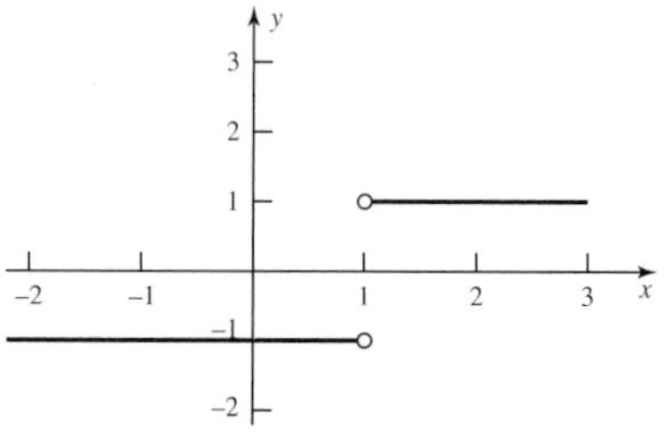

43.

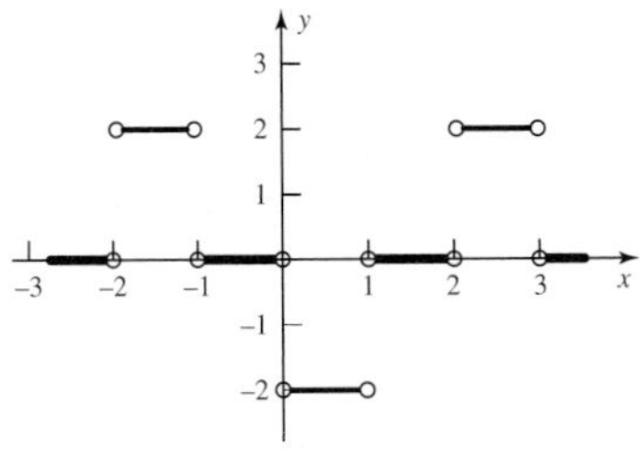

45. 1.5 **47.** −0.1667 **49.** 0.0081 **51.** $2x$

53. $-1/(x+1)^2$ **55.** $2/(x+1)^2$ **57.** $-\frac{1}{2}, 1, \frac{2}{3}, -3$

59.

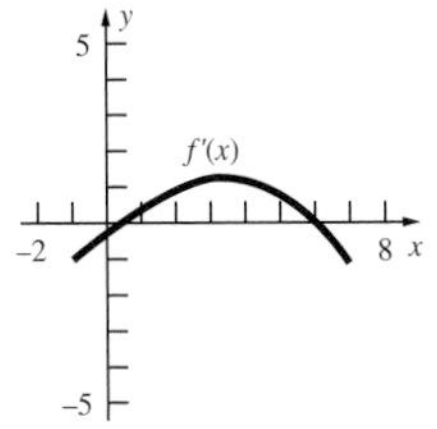

61. (a) $\frac{5}{2}, \frac{3}{2}$, 1.8, −0.6; **(b)** 0.5; **(c)** 5; **(d)** 3, 5; **(e)** 1, 3, 5; **(f)** 0; **(g)** −0.7, 1.5, (5, 7)

63.

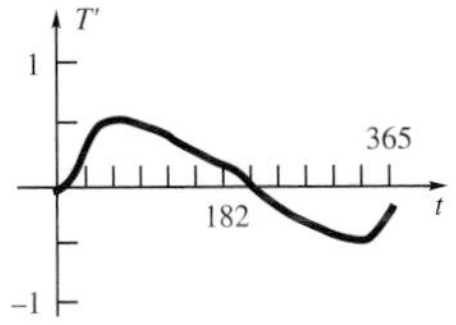

65. f is short-dashed; $g = f'$ is solid; g' is long-dashed

67. $m = 4, b = -4$

69. (a) m; **(b)** $-m$

71.

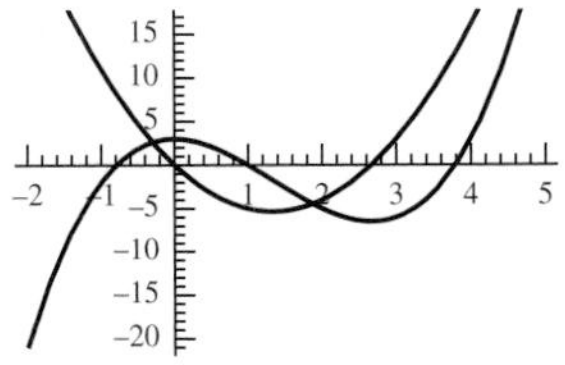

(a) $\left(0, \frac{8}{3}\right)$; **(b)** $\left[0, \frac{8}{3}\right]$; **(c)** $f(x)$ decreases as x increases when $f'(x) < 0$.

Problem Set 2.3

1. $4x$ **3.** π **5.** $-4x^{-3}$ **7.** $-\dfrac{\pi}{x^2}$ **9.** $-\dfrac{500}{x^6}$

11. $2x + 2$ **13.** $4x^3 + 3x^2 + 2x + 1$

15. $7\pi x^6 - 10x^4 + 10x^{-3}$ **17.** $-\dfrac{9}{x^4} - 4x^{-5}$

19. $-\dfrac{2}{x^2} + \dfrac{2}{x^3}$ **21.** $-\dfrac{1}{2x^2} + 2$ **23.** $3x^2 + 1$

25. $8x + 4$ **27.** $5x^4 + 6x^2 + 2x$

29. $5x^4 + 42x^2 + 2x - 51$ **31.** $60x^3 - 30x^2 - 32x + 14$

33. $-\dfrac{6x}{(3x^2+1)^2}$ **35.** $\dfrac{-8x+3}{(4x^2-3x+9)^2}$ **37.** $\dfrac{2}{(x+1)^2}$

39. $\dfrac{6x^2+20x+3}{(3x+5)^2}$ **41.** $\dfrac{4x^2+4x-5}{(2x+1)^2}$ **43.** $\dfrac{x^2-1}{(x^2+1)^2}$

45. (a) 23; **(b)** 4; **(c)** $-\frac{17}{9}$

49. $y = 1$ **51.** $(0, 0)$ and $\left(\frac{2}{3}, -\frac{4}{27}\right)$

53. (2.817, 0.563) and (−2.817, −0.563)

55. (a) −24 ft/s; **(b)** 1.25 s

57. $y = 2x + 1, y = -2x + 9$ **59.** $3\sqrt{5}$

61. 681 cm^3 per week

Problem Set 2.4

1. $2\cos x - 3\sin x$ **3.** 0 **5.** $\sec x \tan x$ **7.** $\sec^2 x$

9. $\sec^2 x$ **11.** $\cos^2 x - \sin^2 x$ **13.** $\dfrac{x\cos x - \sin x}{x^2}$

15. $-x^2 \sin x + 2x\cos x$ **17.** $2\tan x \sec^2 x$

19. $y - 0.5403 = -0.8415(x - 1)$ **21.** $-2\sin^2 x + 2\cos^2 x$

23. $30\sqrt{3}$ ft/sec **25.** $y = x$

27. $x = \frac{\pi}{4} + k\frac{\pi}{2}$ where k is an integer.

33. (a)

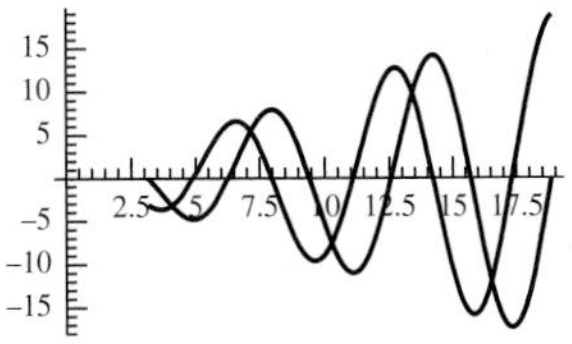

(b) 6; 5; **(c)** $f(x) = x\sin x$ with $a = 0$ and $b = \pi$ is a counterexample; **(d)** 24.93

Problem Set 2.5

1. $15(1+x)^{14}$ **3.** $-10(3-2x)^4$

5. $11(3x^2 - 4x + 3)(x^3 - 2x^2 + 3x + 1)^{10}$

7. $-\dfrac{5}{(x+3)^6}$ **9.** $(2x+1)\cos(x^2+x)$

11. $-3\sin x\cos^2 x$ **13.** $-\dfrac{6(x+1)^2}{(x-1)^4}$

15. $-\dfrac{3x^2+12x}{(x+2)^2}\sin\left(\dfrac{3x^2}{x+2}\right)$

17. $2(3x-2)(3-x^2)(9+4x-9x^2)$

19. $\dfrac{(x+1)(3x-11)}{(3x-4)^2}$ **21.** $4x(x^2+4)$

23. $\dfrac{51(3t-2)^2}{(t+5)^4}$ **25.** $\dfrac{(6t+47)(3t-2)^2}{(t+5)^2}$

27. $\dfrac{3\sin^2 x(\cos x\cos 2x + 2\sin x\sin 2x)}{\cos^4 2x}$ **29.** 9.6

31. 1.4183 **33.** $4(2x+3)\sin^3(x^2+3x)\cos(x^2+3x)$

35. $-3\sin t\sin^2(\cos t)\cos(\cos t)$

37. $-8\theta\cos^3(\sin\theta^2)\sin(\sin\theta^2)(\cos\theta^2)$

39. $-2\cos[\cos(\sin 2x)]\sin(\sin 2x)(\cos 2x)$

41. 2 **43.** 1 **45.** −1 **47.** $2F'(2x)$

49. $-2(F(t))^{-3}F'(t)$ **51.** $4(1 + F(2z))F'(2z)$

53. $-\sin x F'(\cos x)$ **55.** $2F'(2x)\sec^2(F(2x))$

57. $2F(x)F'(x)\sin F(x)\cos F(x) + F'(x)\sin^2 F(x)$

59. $-2\sin 1$ **61.** −1 **63.** $x = \pi/4 + k\pi, k = 0, \pm 1, \pm 2, \ldots$

65. $y = -\dfrac{1}{2}x + \dfrac{3}{4}$ **67.** $x = 3/2$

69. (a) $(10\cos 8\pi t, 10\sin 8\pi t)$; **(b)** 80π cm/s

71. (a) $(\cos 2\pi t, \sin 2\pi t)$; **(b)** $\sin 2\pi t + \sqrt{25 - \cos^2 2\pi t}$;

(c) $2\pi\cos 2\pi t\left(1 + \dfrac{\sin 2\pi t}{\sqrt{25 - \cos^2 2\pi t}}\right)$

73. 0.38 in/min **75.** $x_0 = \pi/3; \theta = 1.25$ rad.

79. $\cot x|\sin x|$ **81.** 16

Problem Set 2.6

1. 6 **3.** 162 **5.** $-343\cos(7x)$ **7.** $-\dfrac{6}{(x-1)^4}$

9. 2 **11.** $\frac{1}{2}$ **13.** $2\pi^2$ **15.** −900

19. **(a)** 0; **(b)** 0; **(c)** 0

21. $f''(-5) = -24$; $f''(3) = 24$

23. **(a)** $v(t) = 12 - 4t$; $a(t) = -4$ **(b)** $(-\infty, 3)$; **(c)** $(3, \infty)$; **(d)** All t; **(e)**

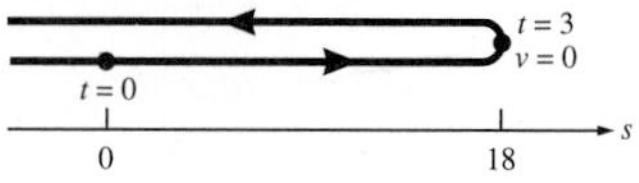

25. **(a)** $v(t) = 3t^2 - 18t + 24$; $a(t) = 6t - 18$; **(b)** $(-\infty, 2) \cup (4, \infty)$; **(c)** $(2, 4)$; **(d)** $(-\infty, 3)$;

(e)

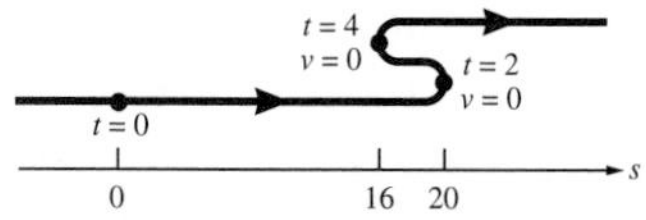

27. **(a)** $v(t) = 2t - \dfrac{16}{t^2}$; $a(t) = 2 + \dfrac{32}{t^3}$; **(b)** $(2, \infty)$; **(c)** $(0, 2)$; **(d)** No t;

(e)

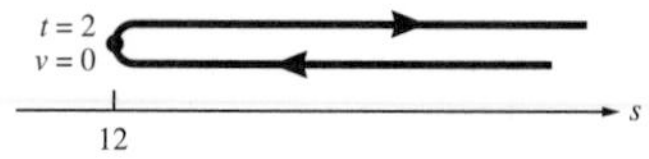

29. $v(1) = 11$; $v(4) = -16$

31. **(a)** $\frac{3}{4}$ s; **(b)** $\frac{1}{2}$ s, $\frac{3}{4}$ s; **(c)** 0 s, $\frac{3}{2}$ s

33. **(a)** 48 ft/s; **(b)** $\frac{3}{2}$ s; **(c)** 292 ft; **(d)** 5.77 s; **(e)** 137 ft/s

35. 581 ft/s **37.** $(-\infty, -2) \cup (1, 4)$

39. $D_x^n(uv) = \sum_{k=0}^{n} \binom{n}{k} D_x^{n-k}(u) D_x^k(v)$ where $\binom{n}{k}$ is the binomial coefficient $\dfrac{n!}{(n-k)!k!}$.

41. **(a)**

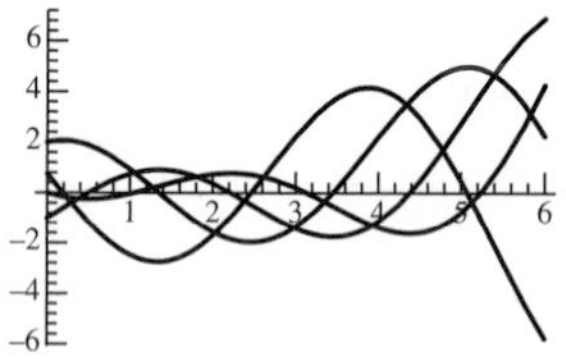

(b) -1.2826

Problem Set 2.7

1. $\dfrac{x}{y}$ **3.** $-\dfrac{y}{x}$ **5.** $\dfrac{1 - y^2}{2xy}$ **7.** $\dfrac{12x^2 + 7y^2}{6y^2 - 14xy}$

9. $\dfrac{y^3 - \dfrac{5y}{2\sqrt{5xy}}}{\dfrac{5x}{2\sqrt{5xy}} + 2 - 2y - 3xy^2}$ **11.** $-\dfrac{y}{x}$

13. $y - 3 = -\frac{9}{7}(x - 1)$ **15.** $y = 1$

17. $y + 1 = \frac{1}{2}(x - 1)$ **19.** $5x^{2/3} + \dfrac{1}{2\sqrt{x}}$

21. $\dfrac{1}{3\sqrt[3]{x^2}} - \dfrac{1}{3\sqrt[3]{x^4}}$ **23.** $\dfrac{3x - 2}{2\sqrt[4]{(3x^2 - 4x)^3}}$

25. $-\dfrac{6x^2 + 4}{3\sqrt[3]{(x^3 + 2x)^5}}$ **27.** $\dfrac{2x + \cos x}{2\sqrt{x^2 + \sin x}}$

29. $-\dfrac{x^2 \cos x + 2x \sin x}{3\sqrt[3]{(x^2 \sin x)^4}}$ **31.** $-\dfrac{(x + 1)\sin(x^2 + 2x)}{2\sqrt[4]{[1 + \cos(x^2 + 2x)]^3}}$

33. $\dfrac{ds}{dt} = -\dfrac{s^2 + 3t^2}{2st}$; $\dfrac{dt}{ds} = -\dfrac{2st}{s^2 + 3t^2}$

35. $\sqrt{3}y + x = 0$, $\sqrt{3}y - x = 0$

37. **(a)** $y' = -\dfrac{y}{x + 3y^2}$; **(b)** $y'' = \dfrac{2xy}{(x + 3y^2)^3}$

39. -15; **45.** $\theta \approx 2.0344$

47. $y = 2(x + 4)$; $y = 2(x - 4)$ **49.** $\frac{13}{3}$

Problem Set 2.8

1. 1296 in.^{3}/s **3.** 392 mi/h **5.** 471 mi/h **7.** 0.258 ft/s

9. 0.0796 ft/s **11.** $\frac{1}{12}$ ft/min **13.** 1.018 in.^{2}/s

15. 15.71 km/min

17. **(a)** $\frac{1}{2}$ ft/s; **(b)** $\frac{5}{2}$ ft/s **(c)** $\frac{1}{24}$ rad/s

19. 110 ft/s **21.** -0.016 ft/h **23.** 13.33 ft/s

25. 4049 ft^3/hr

27. **(a)** -1.125 ft/s; **(b)** -0.08 ft/s^2

29. **(b)** 3 hours

31. $\frac{16}{3}$ ft/s when the girl is at least 30 ft from the light pole and $\frac{80}{17}$ ft/s when she is less than 30 ft from the pole.

Problem Set 2.9

1. $dy = (2x + 1)\,dx$ **3.** $dy = -8(2x + 3)^{-5}\,dx$

5. $dy = 3(\sin x + \cos x)^2(\cos x - \sin x)\,dx$

7. $dy = -\frac{3}{2}(14x + 3)(7x^2 + 3x - 1)^{-5/2}\,dx$

9. $ds = \frac{3}{2}(2t + \csc^2 t)\sqrt{t^2 - \cot t + 2}\,dt$

11.

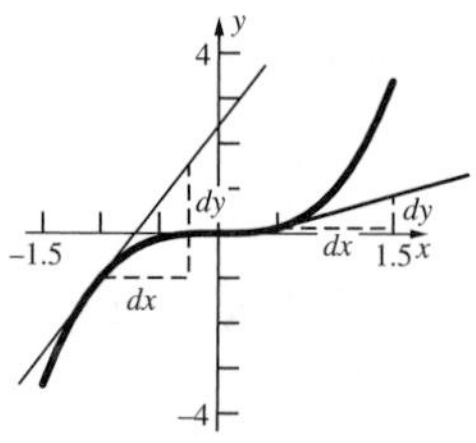

13.

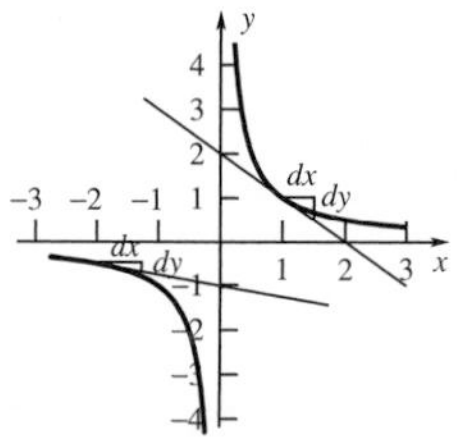

15. **(a)** $\Delta y = -\frac{1}{3}$ **(b)** $\Delta y = -0.3$

17. **(a)** $\Delta y = 67$ $dy = 34$ **(b)** $\Delta y \approx 0.1706$ $dy = 0.17$

19. 5.9917 **21.** 39.27 cm^3 **23.** 893 ft^3 **25.** 12.6 ft

27. 4189 ± 62.8 cm^3; relative error ≈ 0.015

29. 79.097 ± 0.729 cm; relative error ≈ 0.0092

31. $dy = 0.01$; $|\Delta y - dy| \le 0.000003$ **33.** 8.0125

35. 754 cm^3

37. $L(x) = 4x - 4$ **39.** $L(x) = x$

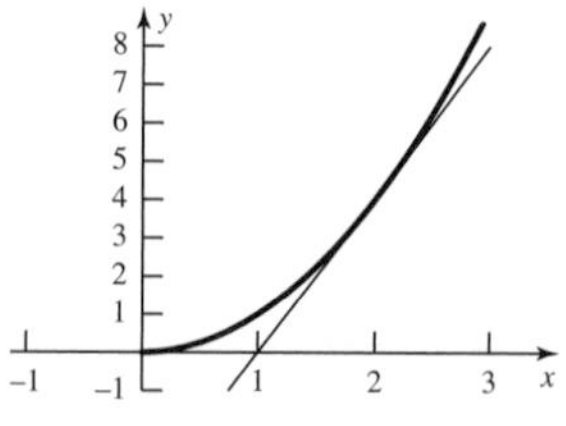

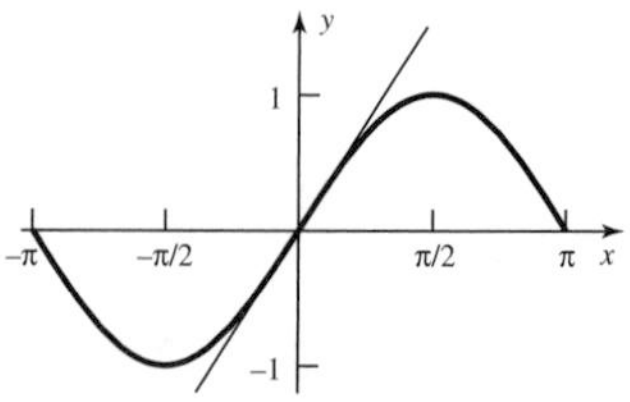

41. $L(x) = 1$ **43.** $L(x) = x$

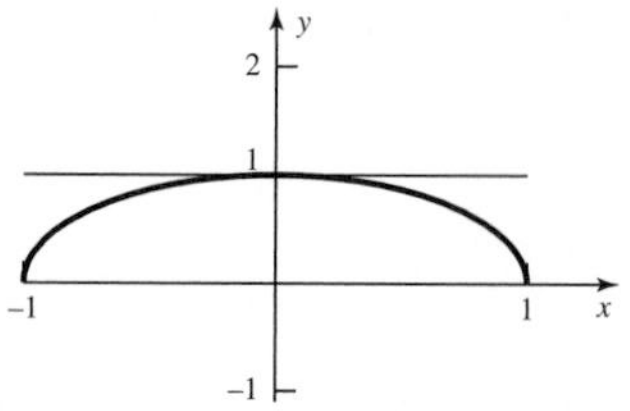

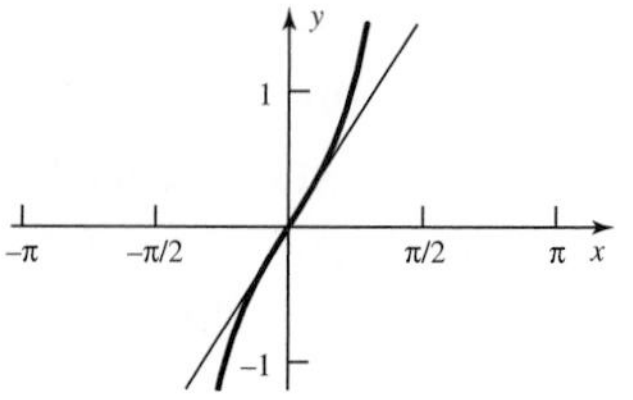

45. $L(x) = f(x)$

Chapter Review 2.10

Concepts Test

1. False **3.** True **5.** True **7.** True **9.** True **11.** True **13.** False **15.** True **17.** False **19.** True **21.** True **23.** True **25.** True **27.** False **29.** True **31.** True **33.** True **35.** True **37.** False

Sample Test Problems

1. (a) $9x^2$; **(b)** $10x^4 + 3$; **(c)** $-\frac{1}{3x^2}$; **(d)** $-\frac{6x}{(3x^2+2)^2}$; **(e)** $\frac{3}{2\sqrt{3x}}$; **(f)** $3\cos 3x$; **(g)** $\frac{x}{\sqrt{x^2+5}}$; **(h)** $-\pi \sin \pi x$

3. (a) $f(x) = 3x$ at $x = 1$; **(b)** $f(x) = 4x^3$ at $x = 2$; **(c)** $f(x) = \sqrt{x^3}$ at $x = 1$; **(d)** $f(x) = \sin x$ at $x = \pi$; **(e)** $f(x) = \frac{4}{x}$ at x; **(f)** $f(x) = -\sin 3x$ at x; **(g)** $f(x) = \tan x$ at $x = \frac{\pi}{4}$; **(h)** $f(x) = \frac{1}{\sqrt{x}}$ at $x = 5$

5. $15x^4$ **7.** $3z^2 + 8z + 2$ **9.** $\frac{-24t^2 + 60t + 10}{(6t^2 + 2t)^2}$

11. $\frac{-4x^4 + 10x^2 + 2}{(x^3 + x)^2}$ **13.** $-\frac{x}{\sqrt{(x^2+4)^3}}$

15. $-\sin\theta + 6\sin^2\theta\cos\theta - 3\cos^3\theta$ **17.** $2\theta\cos(\theta^2)$

19. $2\pi\sin(\sin(\pi\theta))\cos(\sin(\pi\theta))\cos(\pi\theta)$ **21.** $3\sec^2 3\theta$

23. 672 **25.** $\frac{-\csc^2 x - 2x\cot x\tan x^2}{\sec x^2}$ **27.** $16 - 4\pi$

29. 458.8

31. $F'(r(x) + s(x))(r''(x) + s''(x)) + (r'(x) + s'(x))^2 F''(r(x) + s(x)) + s''(x)$

33. $27z^2\cos(9z^3)$

35. 314 m^3 per meter increase in the radius. **37.** 0.167 ft/min

39. (a) $(1, 3)$ **(b)** $a(1) = -6, a(3) = 6$; **(c)** $(2, \infty)$

41. (a) $\frac{1-x}{y}$; **(b)** $-\frac{y^2 + 2xy}{x^2 + 2xy}$; **(c)** $\frac{x^2y^3 - x^2}{y^2 - x^3y^2}$; **(d)** $\frac{2x - \sin(xy) - xy\cos(xy)}{x^2\cos(xy)}$; **(e)** $-\frac{\tan(xy) + xy\sec^2(xy)}{x^2\sec^2(xy)}$

43. 0.0714

45. (a) 84; **(b)** 23; **(c)** 20; **(d)** 26

47. 104 mi/h

49. (a) $\cot\theta\,|\sin\theta|$; **(b)** $-\tan\theta\,|\cos\theta|$

Chapter 3 Review and Preview Problems

1. $(2, 3)$ **3.** $(-\infty, 0] \cup [1, 2]$

5. $(-\infty, -2) \cup [0, 2) \cup (2, \infty)$

7. $8(2x + 1)^3$ **9.** $-2(x^2 - 1)\sin 2x + 2x\cos 2x$

11. $6(\sec^2 3x)(\tan 3x)$ **13.** $\frac{\cos\sqrt{x}}{2\sqrt{x}}$

15. $x = k\pi$, where k is an integer

17. $x = (2k + 1)\pi/2$, where k is an integer

19. $\frac{\sqrt{x^2+1}}{4} + \frac{4-x}{10}$

21. (a) $x^2 + 3$ is one such function
(b) $-\cos x + 8$ is one such function
(c) $\frac{1}{3}x^3 + \frac{1}{2}x^2 + x + 2$ is one such function

Problem Set 3.1

1. Critical points: $-2, 0, 2, 4$; maximum value 10; minimum value 1

3. Critical points: $-2, -1, 0, 1, 2, 3, 4$; maximum value 3; minimum value 1

5. Critical points: $-4, -2, 0$; maximum value 4, minimum value 0

7. Critical points: $-2, -\frac{3}{2}, 1$; maximum value 4, minimum value $-\frac{9}{4}$

9. Critical points: $-1, 1$; No maximum value, minimum value -1

11. Critical points: $-1, 3$; No maximum value, no minimum value

13. Critical points: $-2, -1, 0, 1, 2$; maximum value 10; minimum value 1

15. Critical point: 0; maximum value 1, no minimum value

17. Critical points: $-\frac{\pi}{4}, \frac{\pi}{6}$; Maximum value $\frac{1}{2}$, minimum value $-\frac{1}{\sqrt{2}}$

19. Critical points: 0, 1, 3; maximum value 2, minimum value 0

21. Critical points: $-1, 0, 27$; Maximum value 3, minimum value -1

23. Critical points: $0, \pi, 2\pi, 3\pi, 4\pi, 5\pi, 6\pi, 7\pi, 8\pi$; maximum value 1; minimum value -1

25. Critical points: $-\frac{\pi}{4}, 0, \frac{\pi}{4}$; maximum value $\frac{\pi^2\sqrt{2}}{16}$; minimum value: 0

27. (a) Critical points: $-1, 2 - \frac{\sqrt{33}}{3}, 2 + \frac{\sqrt{33}}{3}, 5$; maximum value ≈ 2.04; minimum value ≈ -26.04
(b) Critical points: $-1, -0.4836, 2 - \frac{\sqrt{33}}{3}, 0.7172, 2 + \frac{\sqrt{33}}{3}, 5$; maximum value ≈ 26.04; minimum value $= 0$

29. Answers will vary. One possibility:

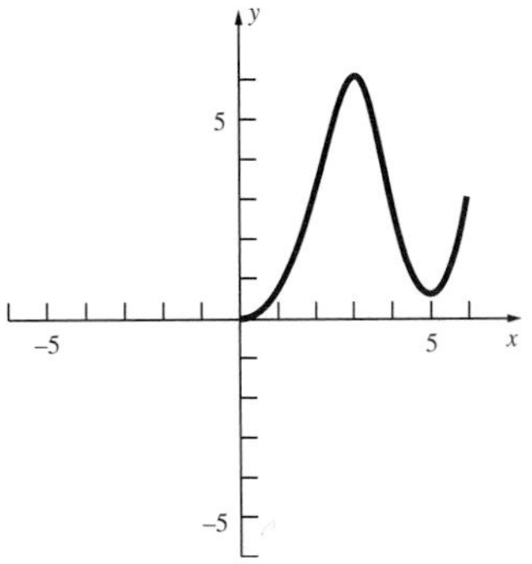

31. Answers will vary. One possibility:

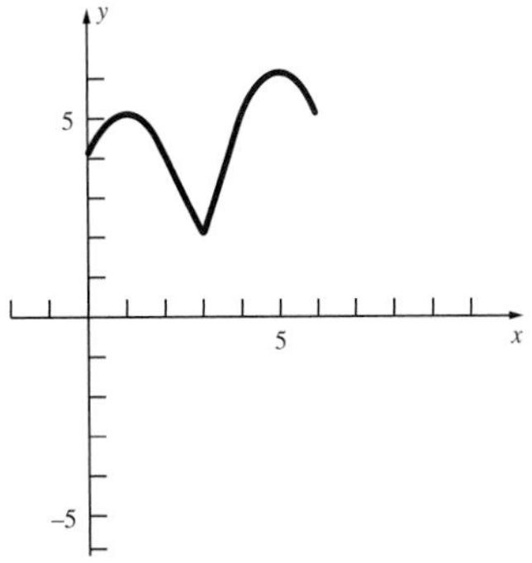

33. Answers will vary. One possibility:

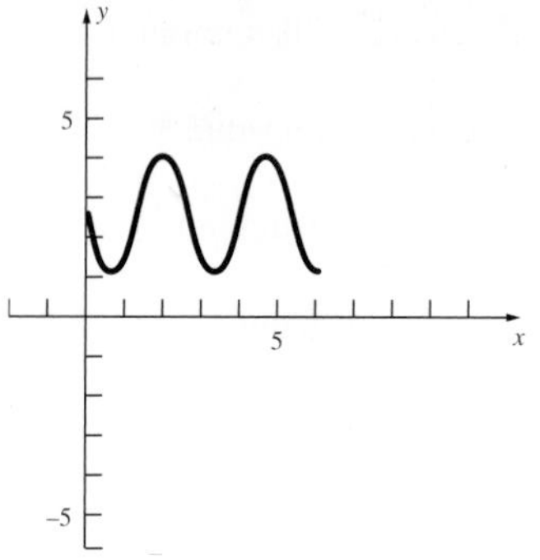

35. Answers will vary. One possibility:

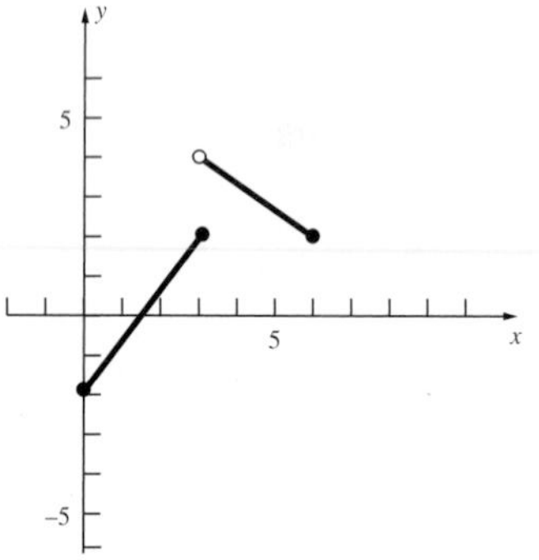

Problem Set 3.2

1. Increasing on $(-\infty, \infty)$

3. Increasing on $[-1, \infty)$, decreasing on $(-\infty, -1]$

5. Increasing on $(-\infty, 1] \cup [2, \infty)$, decreasing on $[1, 2]$

7. Increasing on $[2, \infty)$, decreasing on $(-\infty, 2]$

9. Increasing on $\left[0, \frac{\pi}{2}\right] \cup \left[\frac{3\pi}{2}, 2\pi\right]$, decreasing on $\left[\frac{\pi}{2}, \frac{3\pi}{2}\right]$

11. Concave up for all x; no inflection points

13. Concave up on $(0, \infty)$, concave down on $(-\infty, 0)$; inflection point $(0, 0)$

15. Concave up on $(-\infty, -1) \cup (4, \infty)$, concave down on $(-1, 4)$; inflection points $(-1, -19)$ and $(4, -499)$

17. Concave up for all x; no inflection points

19. Increasing on $(-\infty, -2] \cup [2, \infty)$, decreasing on $[-2, 2]$; concave up on $(0, \infty)$, concave down on $(-\infty, 0)$

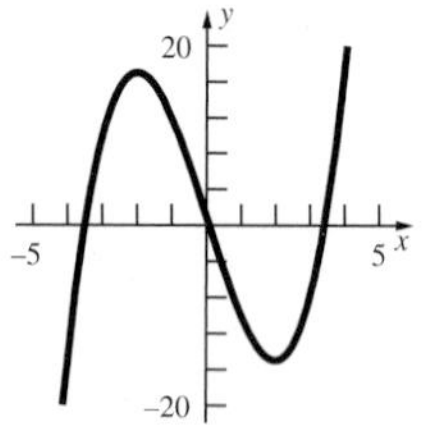

21. Increasing on $[1, \infty)$, decreasing on $(-\infty, 1]$; concave up on $(-\infty, 0) \cup \left(\frac{2}{3}, \infty\right)$, concave down on $\left(0, \frac{2}{3}\right)$

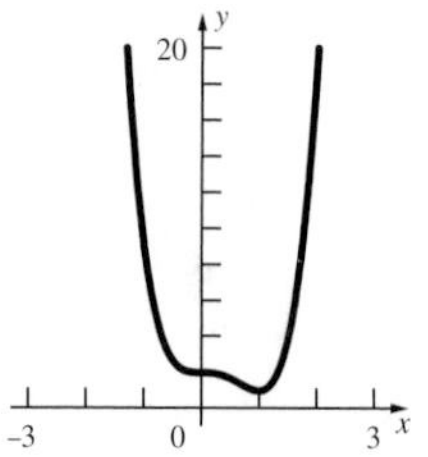

23. Increasing on $(-\infty, -1] \cup [1, \infty)$, decreasing on $[-1, 1]$; concave up on $\left(-\frac{1}{\sqrt{2}}, 0\right) \cup \left(\frac{1}{\sqrt{2}}, \infty\right)$, concave down on $\left(-\infty, -\frac{1}{\sqrt{2}}\right) \cup \left(0, \frac{1}{\sqrt{2}}\right)$.

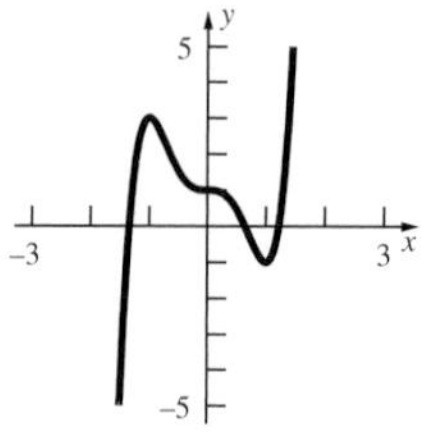

25. Increasing on $\left[0, \frac{\pi}{2}\right]$, decreasing on $\left[\frac{\pi}{2}, \pi\right]$; concave down on $(0, \pi)$.

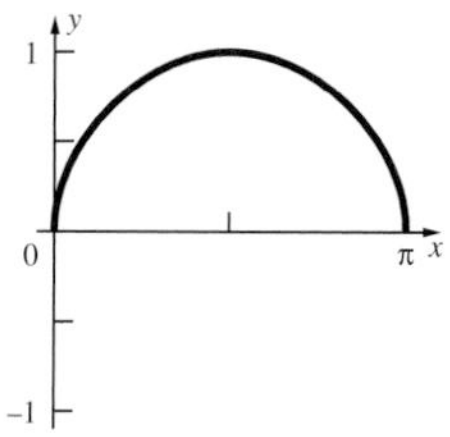

27. Increasing on $\left[0, \frac{2}{5}\right]$, decreasing on $(-\infty, 0] \cup \left[\frac{2}{5}, \infty\right)$; concave up on $\left(-\infty, -\frac{1}{5}\right)$, concave down on $\left(-\frac{1}{5}, 0\right) \cup (0, \infty)$.

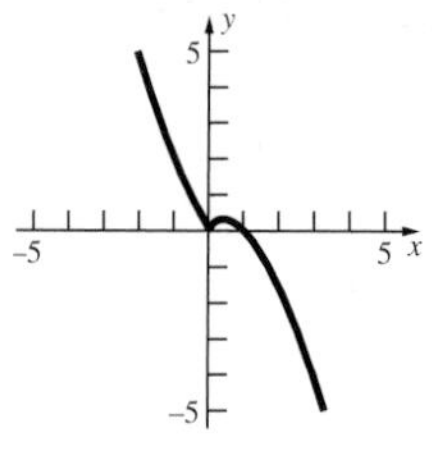

29.

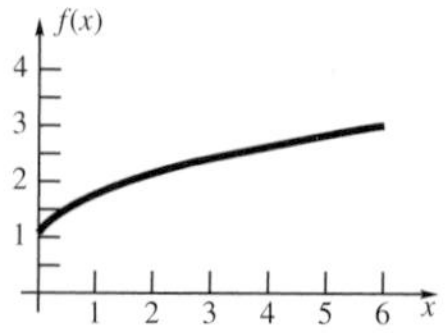

31.

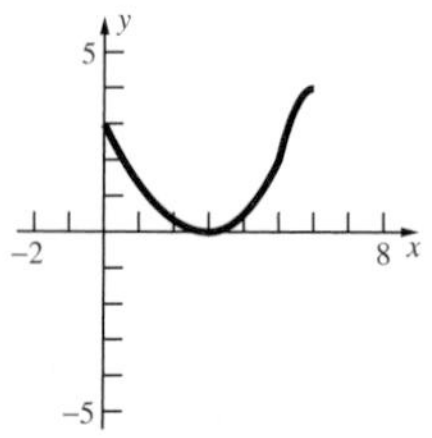

33.

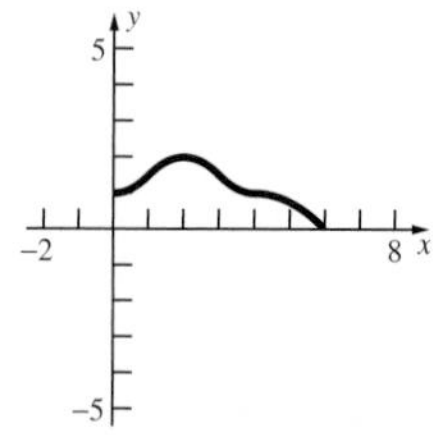

41. $a = \frac{39}{8}, b = \frac{13}{2}$

43. (a) No conditions needed;

(b) $f(x) > -\dfrac{f'(x)}{g'(x)}g(x)$ for all x;

(c) No conditions needed

45. (a)

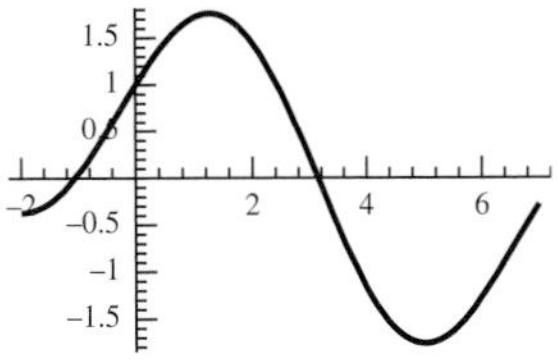

(b) $(1.3, 5)$; **(c)** $(-0.25, 3.1) \cup (6.5, 7]$

(d)

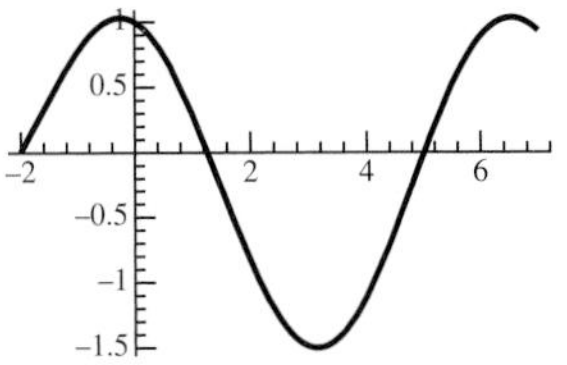

(e)

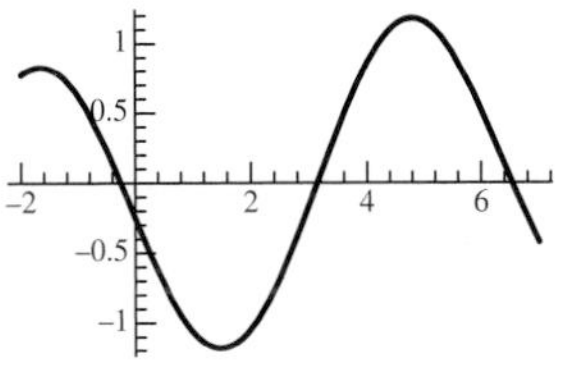

47. $[-0.598, 0.680]$

49. (a) $\dfrac{ds}{dt} = ks$, k a constant; **(b)** $\dfrac{d^2s}{dt^2} > 0$

(c) $\dfrac{d^3s}{dt^3} < 0, \dfrac{d^2s}{dt^2} > 0$ **(d)** $\dfrac{d^2s}{dt^2} = 10$ mph/min

(e) $\dfrac{ds}{dt}$ and $\dfrac{d^2s}{dt^2}$ are approaching zero. **(f)** $\dfrac{ds}{dt}$ is constant.

51. (a) $\dfrac{dC}{dt} > 0, \dfrac{d^2C}{dt^2} > 0$, where C is the car's cost. Concave up.

(b) $f(t)$ is oil consumption at time t. $\dfrac{df}{dt} < 0, \dfrac{d^2f}{dt^2} > 0$. Concave up.

(c) $\dfrac{dP}{dt} > 0, \dfrac{d^2P}{dt^2} < 0$, where P is world population. Concave down.

(d) $\dfrac{d\theta}{dt} > 0, \dfrac{d^2\theta}{dt^2} > 0$, where θ is the angle that the tower makes with the vertical. Concave up.

(e) $P = f(t)$ is profit at time t. $\dfrac{dP}{dt} > 0, \dfrac{d^2P}{dt^2} < 0$. Concave down.

(f) R is revenue at time t. $R < 0, \dfrac{dR}{dt} > 0$. Could be either concave up or down.

53. $h(t) = \sqrt[3]{\dfrac{2400}{\pi}t + 27000} - 30$

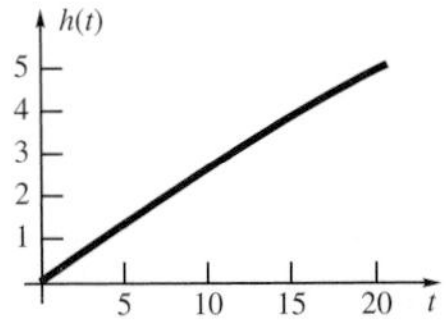

55.

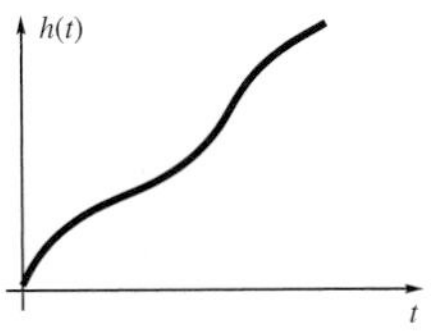

57. (a)

Depth	V	$A \approx \Delta V$	$r \approx \sqrt{\Delta V/\pi}$
1	4	4	1.13
2	8	4	1.13
3	11	3	0.98
4	14	3	0.98
5	20	6	1.38
6	28	8	1.60

(b)

Depth	V	$A \approx \Delta V$	$r \approx \sqrt{\Delta V/\pi}$
1	4	4	1.13
2	9	5	1.26
3	12	3	0.98
4	14	2	0.80
5	20	6	1.38
6	28	8	1.60

Problem Set 3.3

1. Critical points: 0, 4; local minimum at $x = 4$; local maximum at $x = 0$

3. No critical points; no local minima or maxima on $\left(0, \frac{\pi}{4}\right)$

5. Critical point: 0; local minimum at $\theta = 0$

7. Critical points $-2, 2$; local minimum at $x = -2$, local maximum at $x = 2$

9. Critical point $-\dfrac{\sqrt[3]{4}}{2}$; local minimum at $-\dfrac{\sqrt[3]{4}}{2}$

11. Critical points: $-1, 1$; local minimum value $f(1) = -2$; local maximum value $f(-1) = 2$

13. Critical points $0, \frac{3}{2}$; local minimum value $H\left(\frac{3}{2}\right) = -\frac{27}{16}$; no local maximum

15. Critical point: 2; no local minimum values; local maximum value $g(2) = \pi$

17. No critical points
No local minimum or maximum values

19. No critical points
No local minimum or maximum values

21. Maximum value $f(\pi/4) = 1$; minimum value $f(0) = f(\pi/2) = 0$

23. Maximum value $g(4) = \dfrac{1}{6}$; minimum value $g(0) = 0$

25. Maximum value $F(9/16) = 9/4$; minimum value $F(4) = -4$

27. Minimum value $f(\tan^{-1}(4/3)) = 125$; no maximum value

29. Maximum value $H(-2) = H(2) = 3$; minimum value $H(-1) = H(1) = 0$

31. Local minimum at $x = 0$

33. Local minimum at $x = 4$; local maximum at $x = 3$

35. No local extrema

37. Answers will vary. One possibility:

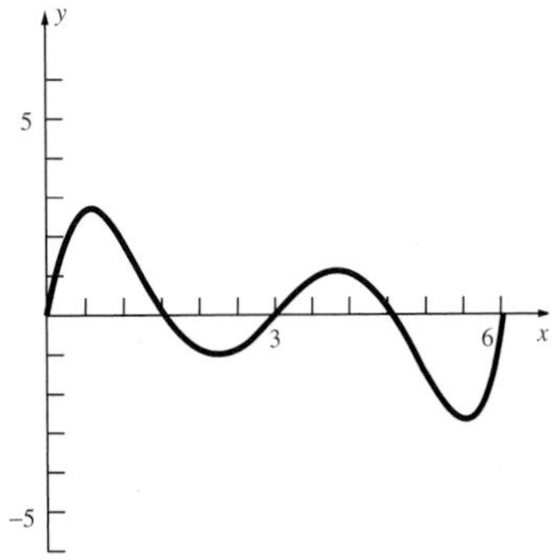

39. Answers will vary. One possibility:

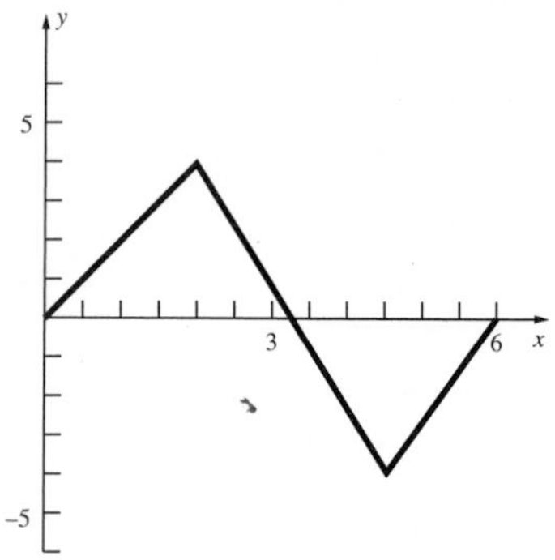

41. Answers will vary. One possibility:

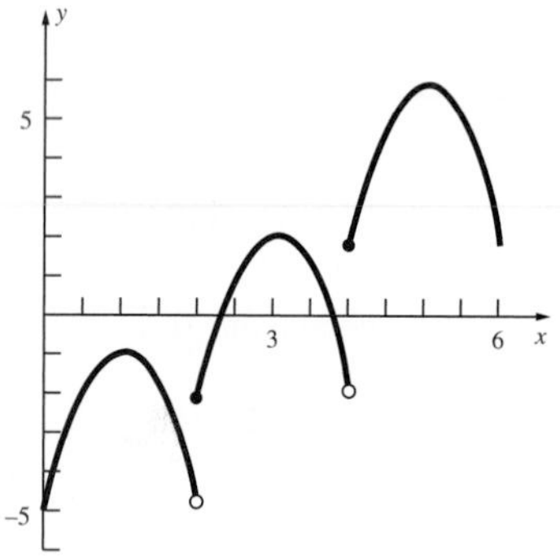

45. f has an inflection point at c.

Problem Set 3.4

1. −4 and 4 **3.** $\frac{1}{16}$ **5.** $\left(-\frac{3}{\sqrt{2}}, \frac{9}{2}\right), \left(\frac{3}{\sqrt{2}}, \frac{9}{2}\right)$ **7.** $\frac{1}{2}$

9. 1024 in^3 **11.** $x = 10$ ft, $y = 40$ ft

13. $x = 15\sqrt{3}$ ft, $y = 20\sqrt{3}$ ft

15. $x = \dfrac{10\sqrt{5}}{\sqrt{3}}$ ft, $y = 6\sqrt{15}$ ft **17.** $P(2\sqrt{2}, 2), Q(0, 0)$

19. $\dfrac{6}{\sqrt{7}}$ miles down the shore from P **21.** At the town

23. about 8:09 A.M. **25.** $\dfrac{4\pi\sqrt{3}}{9}r^3$

27. $h = \sqrt{2}r$, $x = \dfrac{r}{\sqrt{2}}$ where h = height of the cylinder, x = radius of the cylinder, r = radius of the sphere

29. (a) 43.50 cm from one end; shorter length bent to form square
(b) No cut, wire bent to form square

31. height $= \left(\dfrac{3V}{\pi}\right)^{1/3}$, radius $= \frac{1}{2}\left(\dfrac{3V}{\pi}\right)^{1/3}$

33. $r = \sqrt{A}, \theta = 2$ **35.** 4 by 8

37. $r = \sqrt{A/(6\pi)}, \quad h = 2r$

39. Maximum area is for a square. **41.** $\pi/3$

43. $x = 1, y = 3, z = 3$

45. (a) $x = 2a/3$ maximizes area of A.
(b) $x = 2a/3$ minimizes area of B.
(c) $x = 3a/4$ minimizes length z.

47. (a) $L' = 3, L = 4, \phi = 90°$; **(b)** $L' = 5, L = 12, \phi = 90°$;
(c) $\phi = 90°, L = \sqrt{m^2 - h^2}, L' = h$

49. $t \approx 13.8279$, distance ≈ 0.047851 million miles

51. $5\sqrt{5}$ ft

53. (a) $b = \left(\sum_{i=1}^{n} x_i y_i - 5\sum_{i=1}^{n} x_i\right) \Big/ \sum_{i=1}^{n} x_i^2$ **(b)** $b \approx 3.0119$
(c) 50.179 hours

55. $p(n) = 300 - \dfrac{n}{2}$; $R(n) = 300n - \dfrac{n^2}{2}$

57. $n = 200$

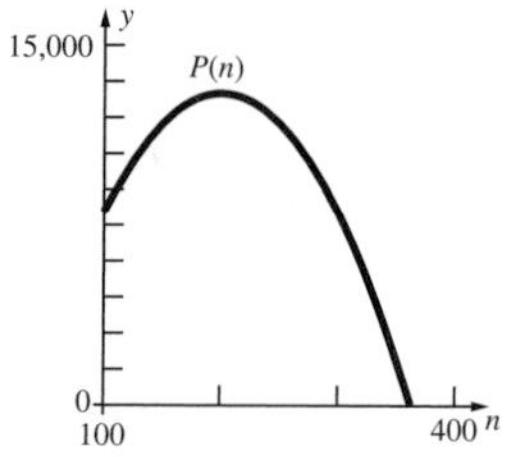

59. \$1.92 per unit; \$1.33

61. (a) $R(x) = 20x + 4x^2 - \dfrac{x^3}{3}$; $\dfrac{dR}{dx} = 20 + 8x - x^2$
(b) $0 \le x \le 10$ **(c)** 4

63. $x_1 = 25$, $\dfrac{dR}{dx} = 0$ at x_1

65. (a) No.
(b) $x = 500$.

67. $P(300) = \$2410$

Problem Set 3.5

1.

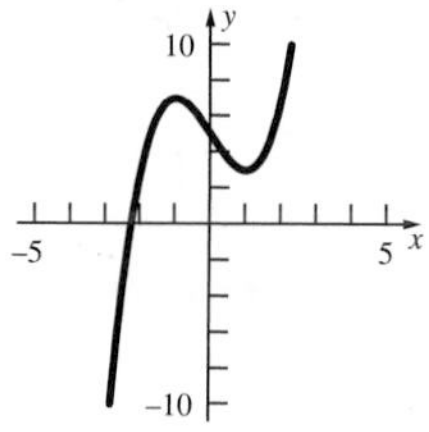

3.

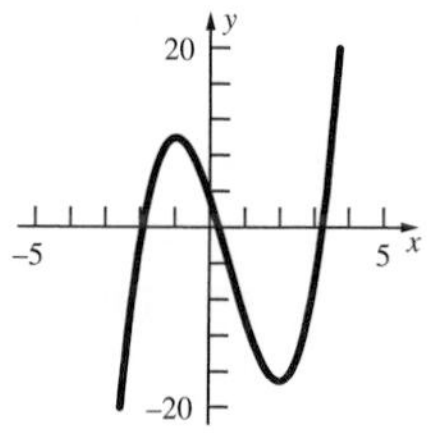

5.

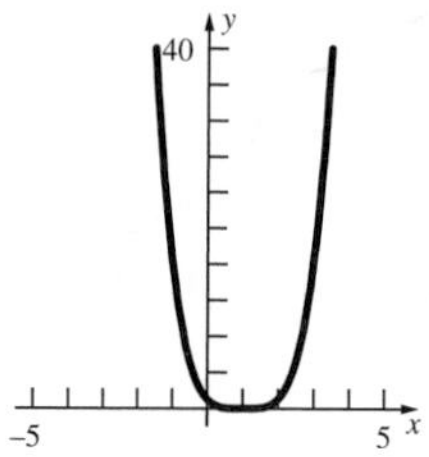

7.

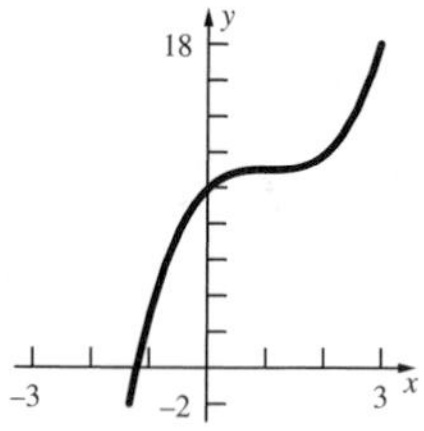

9.

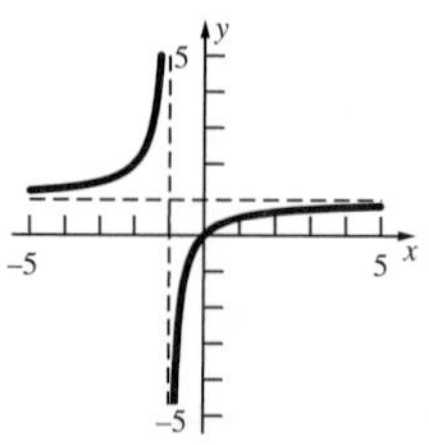

11.

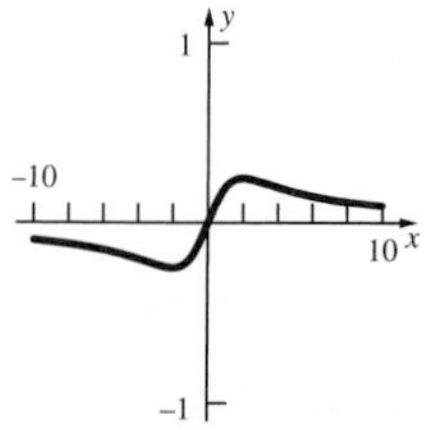

13.

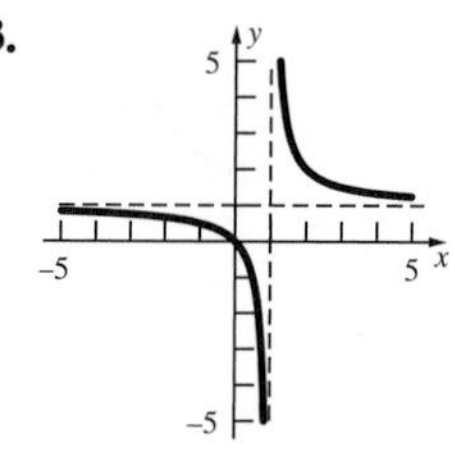

15.

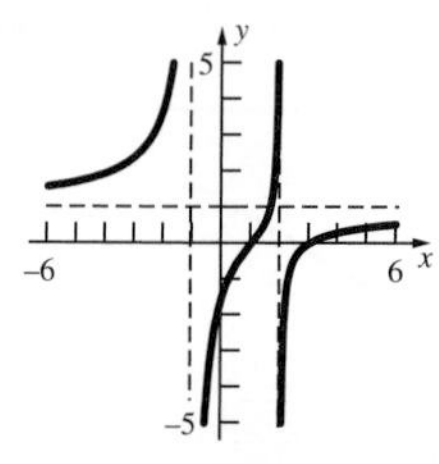

17.

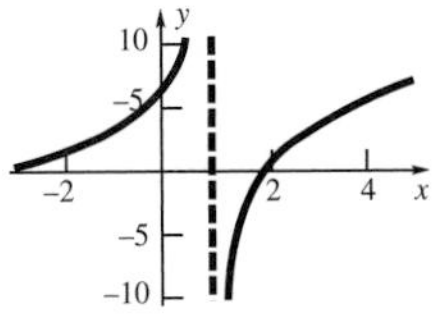

19.

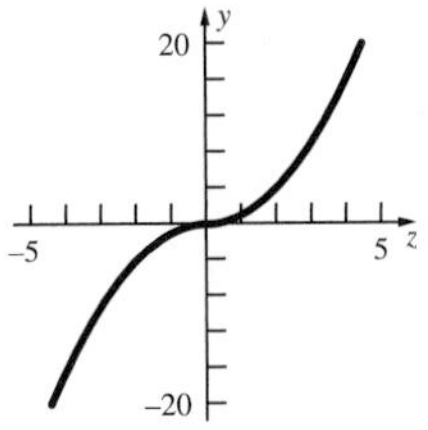

45.

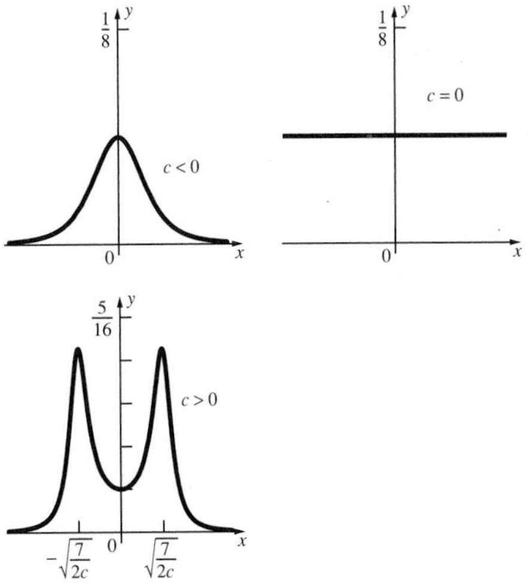

21.

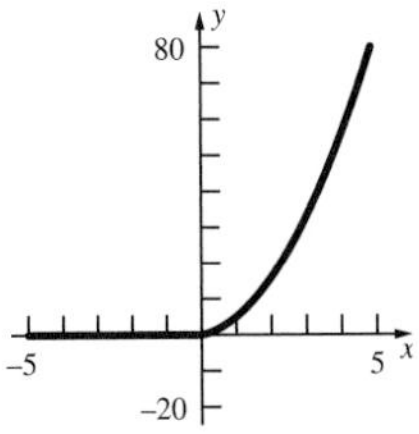

23.

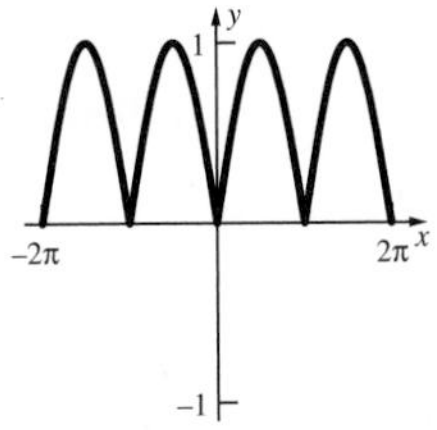

25.

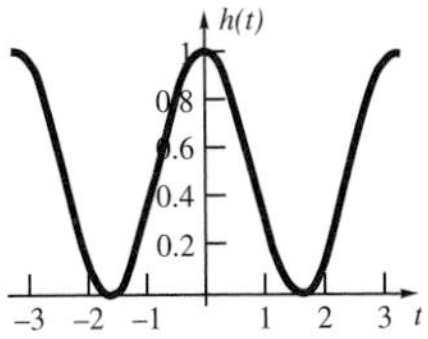

27.

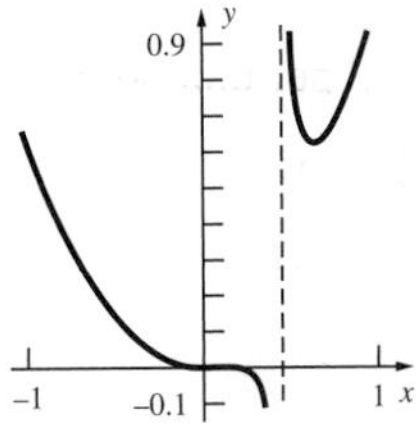

47.

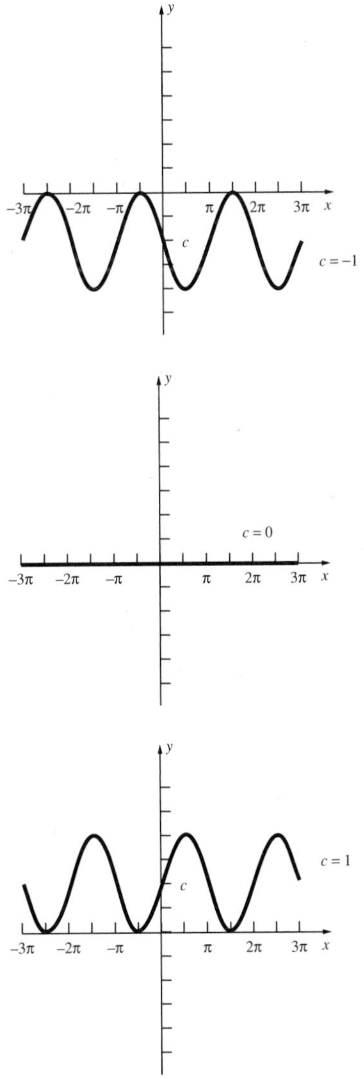

29.

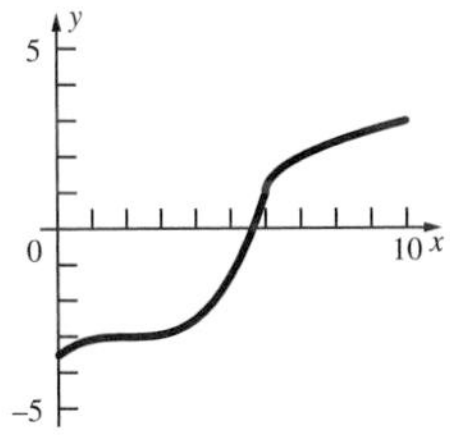

31.

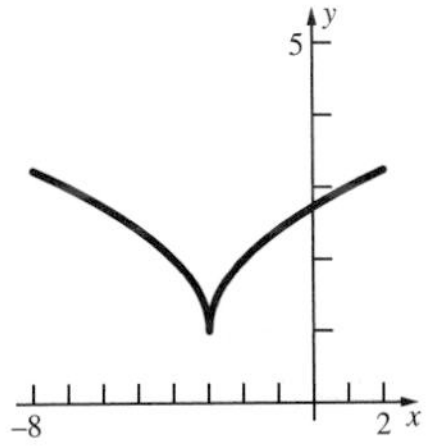

33.

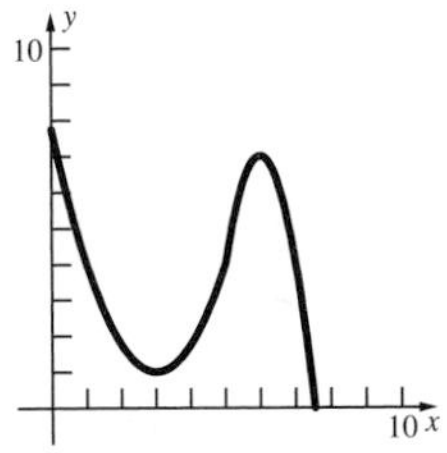

35.

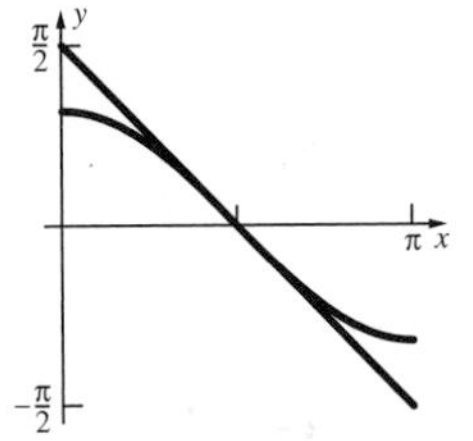

37.

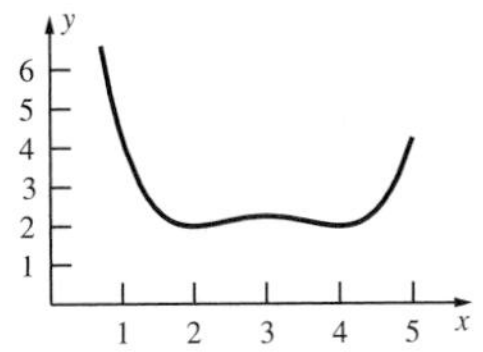

39.

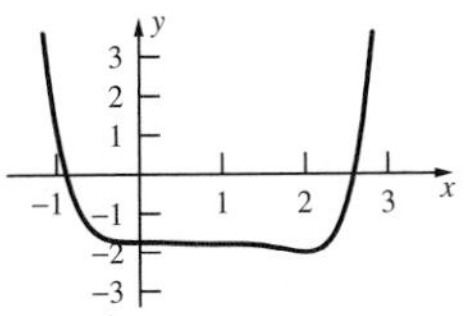

43.

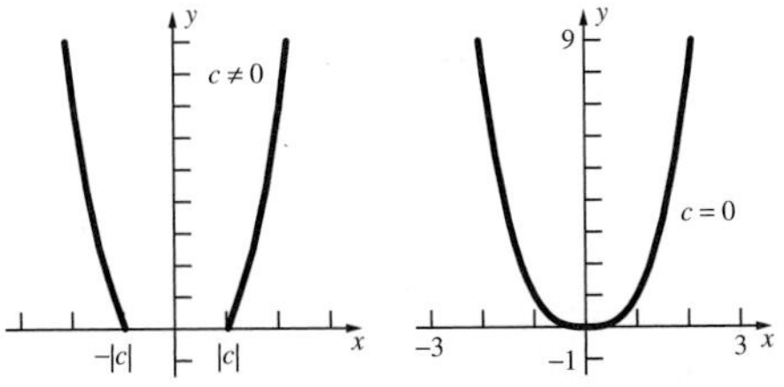

49.

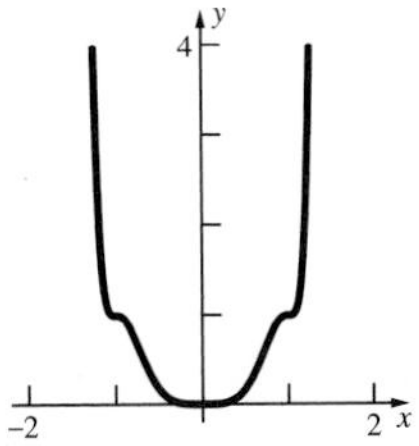

51. (a) Not possible; **(b)** Not possible;
(c)

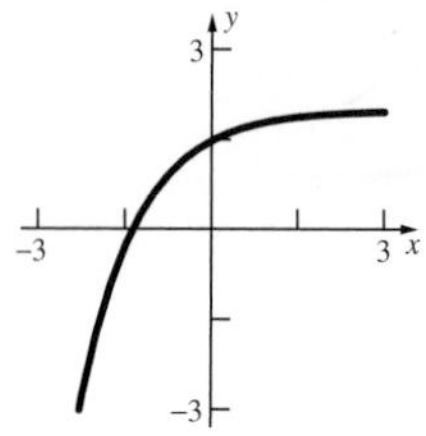

53. (a)

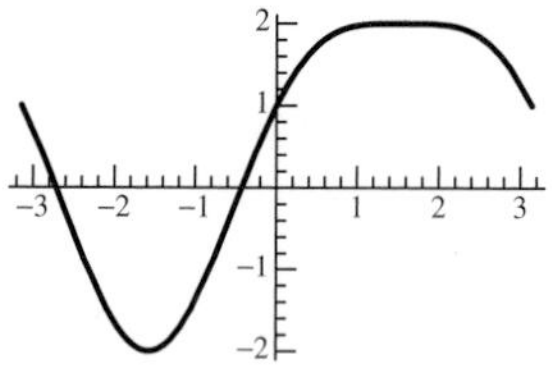

Global minimum: $f\left(-\frac{\pi}{2}\right) = -2$

Global maximum: $f\left(\frac{\pi}{2}\right) = 2$

Inflection points: $\left(-\frac{\pi}{6}, -\frac{1}{4}\right), \left(-\frac{5\pi}{6}, -\frac{1}{4}\right)$

(b)

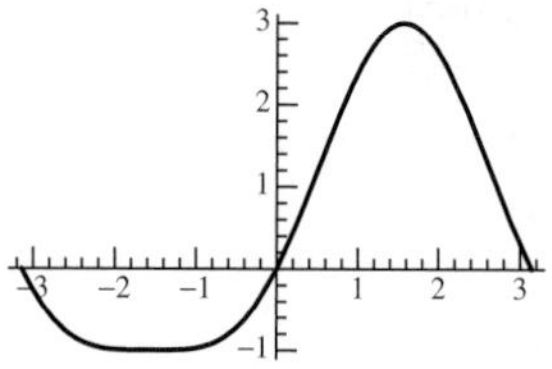

Global minimum: $f\left(-\frac{\pi}{2}\right) = -1$

Global maximum: $f\left(\frac{\pi}{2}\right) = 3$

Inflection points: $\left(\frac{\pi}{6}, \frac{5}{4}\right), \left(\frac{5\pi}{6}, \frac{5}{4}\right)$

(c)

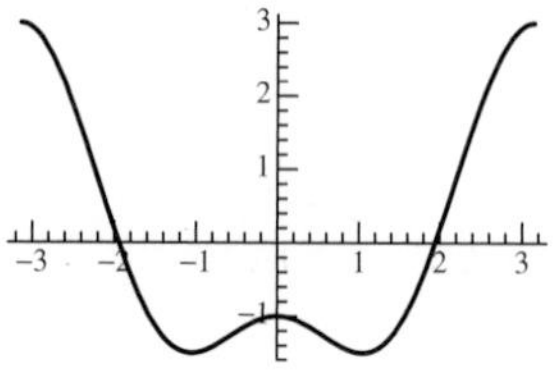

Global minimum: $f\left(-\frac{\pi}{3}\right) = f\left(\frac{\pi}{3}\right) = -1.5$

Global maximum: $f(-\pi) = f(\pi) = 3$

Inflection points: $\approx(-2.206, 0.890), (-0.568, -1.265), (0.568, -1.265), (2.206, 0.890)$

(d)

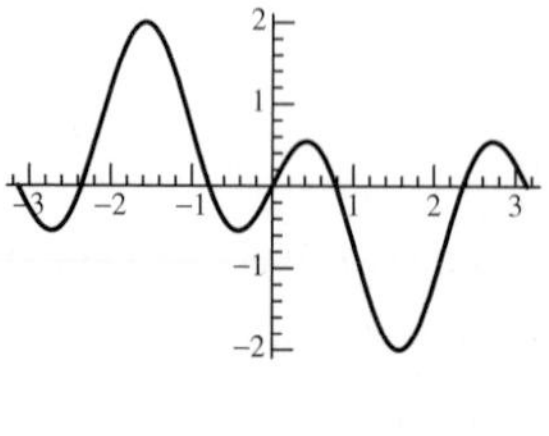

Global minimum: $f\left(\frac{\pi}{2}\right) = -2$

Global maximum: $f\left(-\frac{\pi}{2}\right) = 2$

Inflection points: $(0, 0), \approx(-2.126, 0.755), (-1.016, 0.755), (1.016, -0.755), (2.126, -0.755)$

(e)

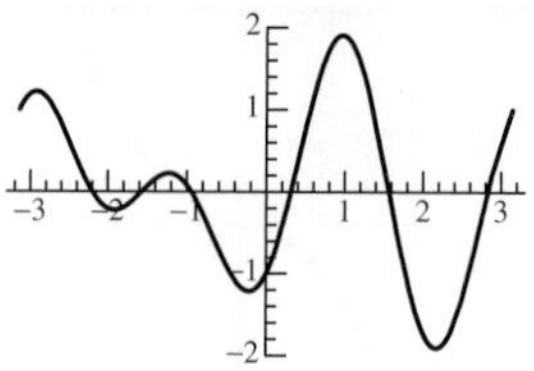

Global minimum: $f(2.17) \approx -1.9$

Global maximum: $f(0.97) \approx 1.9$

Inflection points: $\left(-\frac{\pi}{2}, 0\right), \left(\frac{\pi}{2}, 0\right), \approx(-2.469, 0.542), (-0.673, -0.542), (0.413, 0.408), (2.729, -0.408)$

55. (a) Increasing on $(-\infty, -3] \cup [-1, 0]$: decreasing on $[-3, -1] \cup [0, \infty)$;

(b) Concave up on $(-2, 0) \cup (0, 2)$; concave down on $(-\infty, -2) \cup (2, \infty)$;

(c) Local maximum at $x = -3$; local minimum at $x = -1$;
(d) $x = -2, 2$

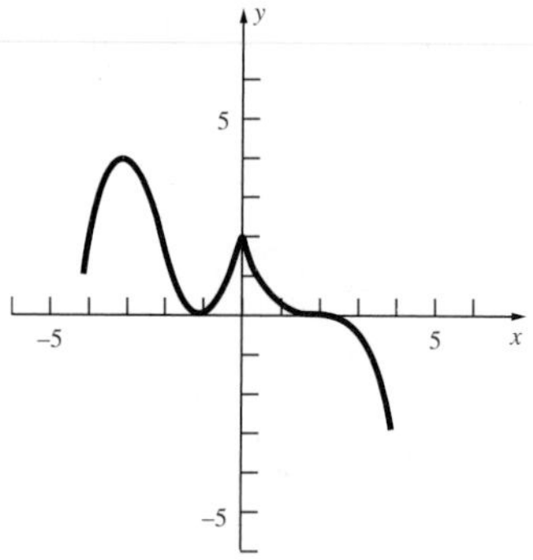

57.

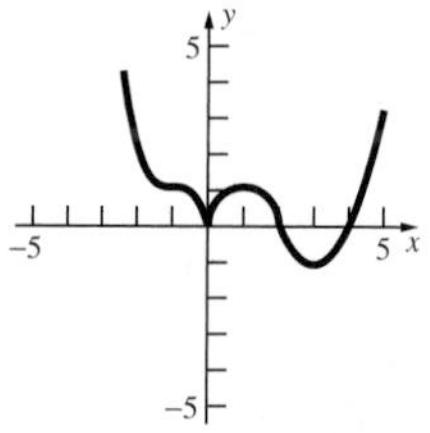

59. (a)

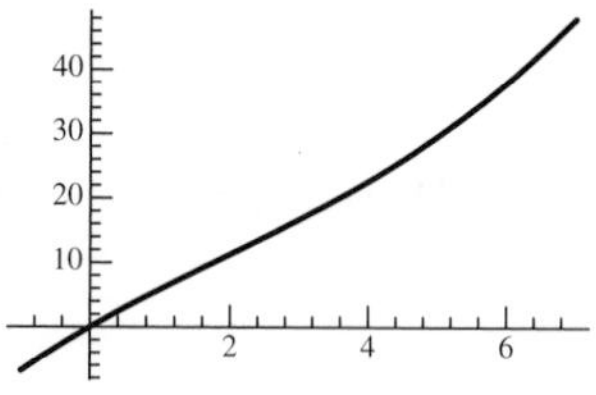

Global minimum: $f(-1) \approx -6.9$

Global maximum: $f(7) \approx 48.0$

Inflection point: $\approx(2.02, 11.4)$

(b)

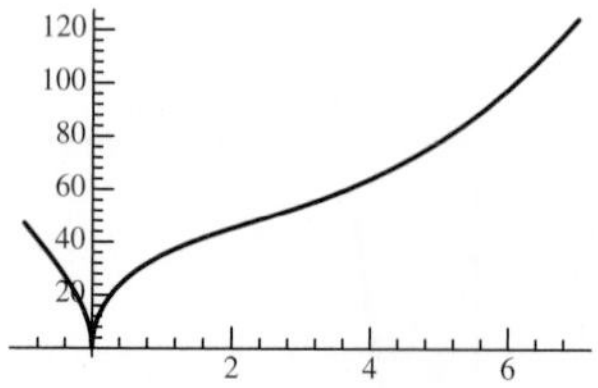

Global minimum: $f(0) = 0$

Global maximum: $f(7) \approx 124.4$

Inflection point: $\approx(2.34, 48.09)$

(c)

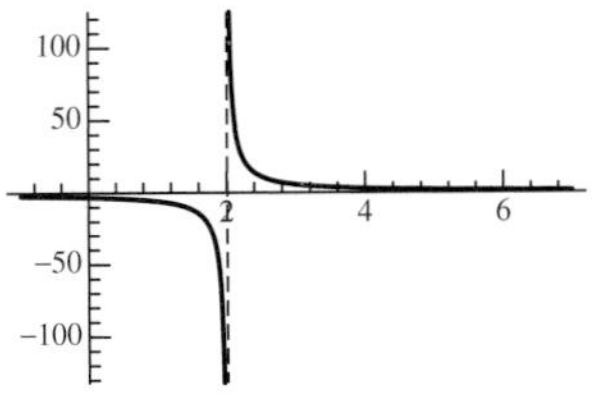

No global minimum or maximum.
No inflections points.

(d)

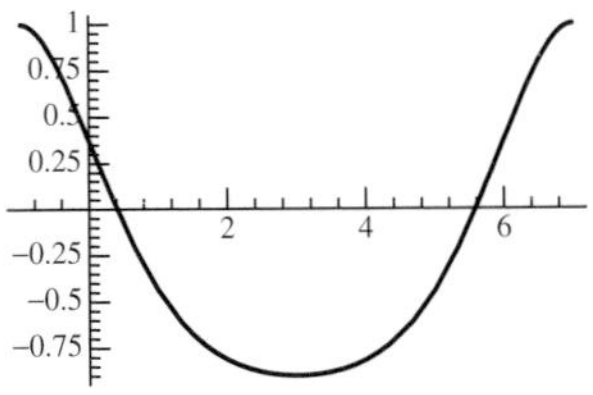

Global minimum: $f(3) \approx -0.9$
Global maximum: $f(-1) = f(7) \approx 1.0$
Inflection points: $\approx (0.05, 0.3), (5.9, 0.3)$

Problem Set 3.6

1. $1 < c < 2$

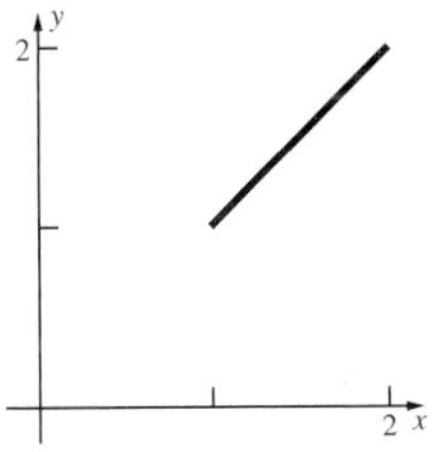

3. $c = 0$

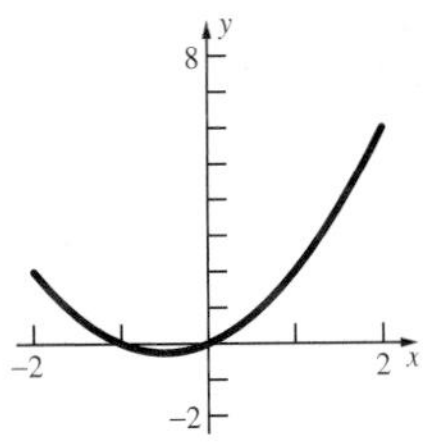

5. $c = -1$

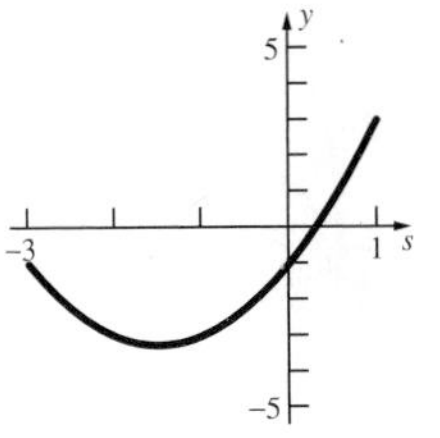

7. $c = 1$

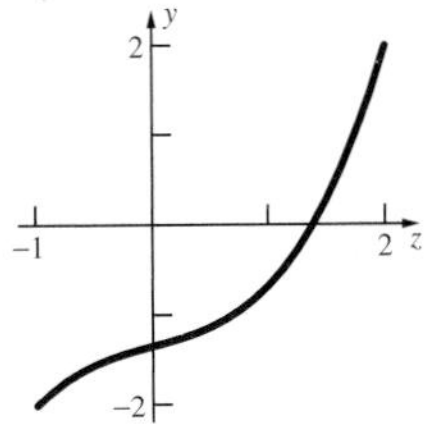

9. $c = 3 - \sqrt{3} \approx 1.27$

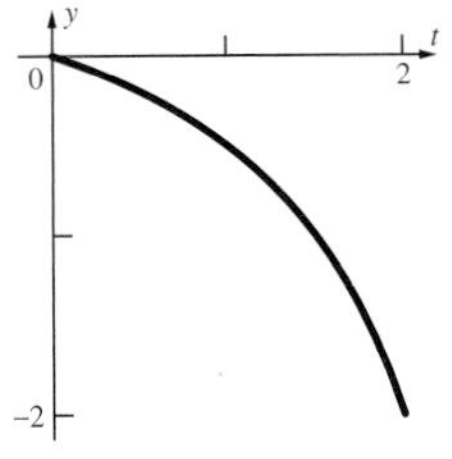

11. $c = \frac{16}{27} \approx 0.59$

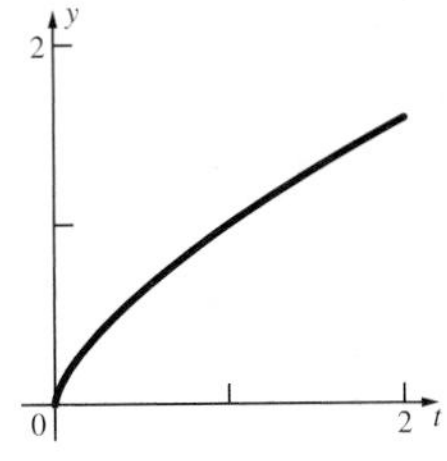

13. $c = \left(\frac{3}{5}\right)^{3/2} \approx 0.46$

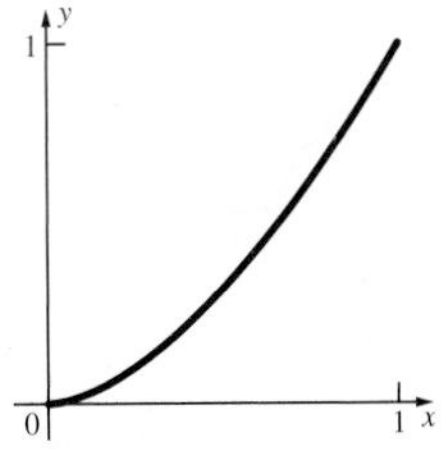

15. $c = \pm\frac{\pi}{2}$

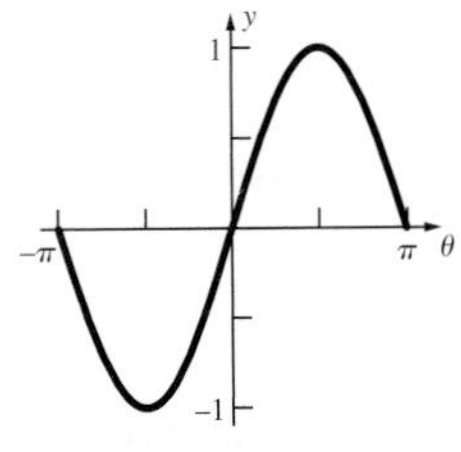

17. Does not apply, $T(\theta)$ not continuous at $\theta = \frac{\pi}{2}$

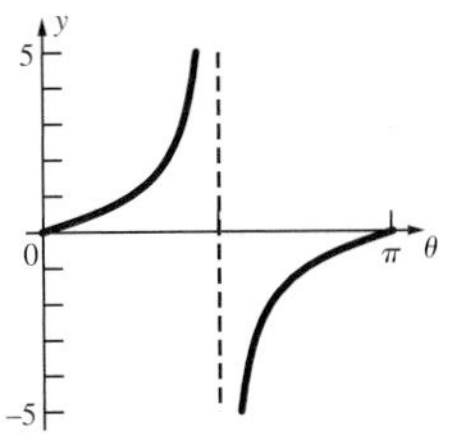

19. $c = \sqrt{2} \approx 1.41$

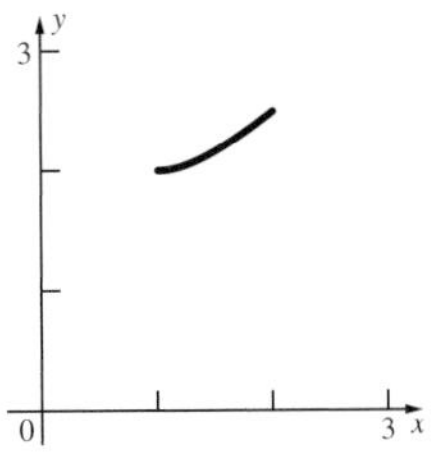

21. Does not apply, $f(x)$ is not differentiable at $x = 0$

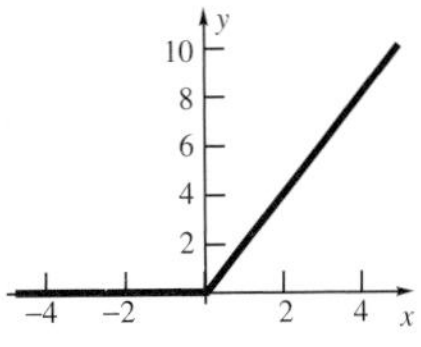

23. $\approx 1.5, 3.75, 7$

Problem Set 3.7

1. 1.46 **3.** 1.11 **5.** −0.12061 **7.** 3.69815 **9.** 0.45018 **11.** 2, 0.58579, 3.41421 **13.** 0.48095 **15.** 1.81712
17. Minimum $f(-0.60583) \approx -0.32645$; Maximum $f(1) = 4$
19. Minimum $f(4.493409) \approx -0.21723$
Maximum $f(7.725252) \approx 0.128375$
21. 0.9643 **23.** **(c)** $i = 0.0151308$; $r = 18.157\%$
25. 0.91486 **27.** 2.21756

29. **(a)**

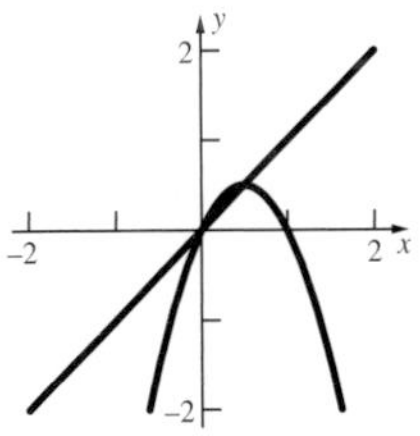

(b) 0.5; **(c)** $\frac{1}{2}$;

31. **(a)** $x_1 = 0, x_2 = 1, x_3 = 1.4142136, x_4 = 1.553774,$ $x_5 = 1.5980532$; **(b)** $x = \frac{1}{2}\left(1 + \sqrt{5}\right) \approx 1.618034$
(c) $x = 1.618034$
33. **(a)** $x_1 = 1, x_2 = 2, x_3 = 1.5, x_4 \approx 1.6666667, x_5 = 1.6$
(b) $x = \dfrac{1 + \sqrt{5}}{2} \approx 1.618034$. **(c)** $\left(1 + \sqrt{5}\right)/2 \approx 1.618034$.
35. **(a)** The algorithm computes the root of $\frac{1}{x} - a = 0$ for x_1 close to $\frac{1}{a}$. **37.** 20.84 ft.
39. **(a)** (28.0279, 7.1828) **(b)** (6.7728, 45.1031)

Problem Set 3.8

1. $5x + C$ **3.** $\frac{1}{3}x^3 + \pi x + C$ **5.** $\frac{4}{9}x^{9/4} + C$
7. $3\sqrt[3]{x} + C$ **9.** $\frac{1}{3}x^3 - \frac{1}{2}x^2 + C$ **11.** $\frac{2}{3}x^6 - \frac{1}{4}x^4 + C$
13. $\dfrac{27}{8}x^8 + \dfrac{1}{2}x^6 - \dfrac{45}{4}x^4 + \dfrac{\sqrt{2}}{2}x^2 + C$ **15.** $-\dfrac{3}{x} + \dfrac{1}{x^2} + C$
17. $x^4 + \frac{3}{2}x^2 + C$ **19.** $\frac{1}{3}x^3 + \frac{1}{2}x^2 + C$ **21.** $\frac{1}{3}(x + 1)^3 + C$
23. $\frac{2}{9}z^{9/2} + \frac{4}{5}z^{5/2} + 2z^{1/2} + C$ **25.** $-\cos\theta - \sin\theta + C$
27. $\frac{1}{4}\left(\sqrt{2}x + 1\right)^4 + C$ **29.** $\frac{1}{21}(5x^3 + 3x - 8)^7 + C$
31. $\frac{9}{16}\sqrt[3]{(2t^2 - 11)^4} + C$ **33.** $\frac{2}{9}(x^3 + 4)^{3/2} + C$
35. $-\frac{1}{5}(1 + \cos x)^5 + C$ **37.** $\frac{1}{2}x^3 + \frac{1}{2}x^2 + C_1x + C_2$

39. $\frac{4}{15}x^{5/2} + C_1x + C_2$ **41.** $\frac{1}{6}x^3 + \frac{1}{2x} + C_1x + C_2$

45. $x^2\sqrt{x-1} + C$ **47.** $\dfrac{5x^3+2}{2\sqrt{x^3+1}} + C$

51. $\frac{1}{2}x^2 + C$ if $x \ge 0$, $-\frac{1}{2}x^2 + C$ if $x < 0$

53. **(a)** $-2\cos(3(x-2)) + C$ **(b)** $\dfrac{1}{2}\cos\dfrac{x}{2} - \dfrac{9}{2}\cos\dfrac{x}{6} + C$
(c) $\dfrac{1}{2}x^2\sin 2x + C$

Problem Set 3.9

5. $y = \frac{1}{3}x^3 + x + C$; $y = \frac{1}{3}x^3 + x - \frac{1}{3}$

7. $y = \pm\sqrt{x^2 + C}$; $y = \sqrt{x^2}$ **9.** $z = \dfrac{3}{C - t^3}$; $z = \dfrac{3}{10 - t^3}$

11. $s = \frac{16}{3}t^3 + 2t^2 - t + C$; $s = \frac{16}{3}t^3 + 2t^2 - t + 100$

13. $y = \frac{1}{10}(2x+1)^5 + C$; $y = \frac{1}{10}(2x+1)^5 + \frac{59}{10}$

15. $y = \frac{3}{2}x^2 + \frac{1}{2}$ **17.** $v = 5$ cm/s; $s = \frac{22}{3}$ cm

19. $v \approx 2.83$ cm/s; $s \approx 12.6$ cm **21.** 144 ft

23. $v = 32.24$ ft/s; $s = 1198.54$ ft

27. Moon: ≈ 1.470 mi/s; Venus: ≈ 6.257 mi/s
Jupiter: ≈ 36.812 mi/s; Sun: ≈ 382.908 mi/s

29. 2.2 ft/s^2 **31.** 5500 m

33. **(a)**

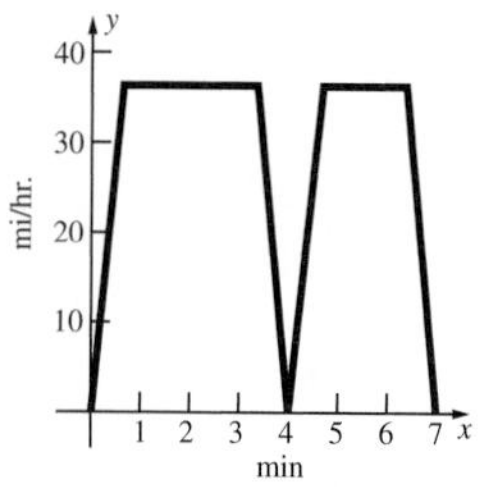

(b) 36 mi/h; **(c)** 0.9 mi/min^2

35. **(a)** $\dfrac{dV}{dt} = C_1\dfrac{\sqrt{V}}{10}$, $V(0) = 1600$, $V(40) = 0$;
(b) $V = \frac{1}{400}(-20t + 800)^2$; **(c)** 900 cm^3

37. **(a)** $v(t) = \begin{cases} -32t & \text{for } 0 \le t < 1 \\ -32(t-1) + 24 & \text{for } 1 < t \le 2.5 \end{cases}$
(b) $t \approx 0.66, 1.75$ s

Chapter Review 3.10

Concepts Test

1. True **3.** True **5.** True **7.** True **9.** True
11. False **13.** True **15.** True **17.** True **19.** False
21. False **23.** False **25.** True **27.** True **29.** True
31. False **33.** True **35.** True **37.** False **39.** True
41. True **43.** True **45.** False **47.** True

Sample Test Problems

1. Critical points: 0, 1, 4; minimum value $f(1) = -1$; maximum value $f(4) = 8$

3. Critical points: $-2, -\frac{1}{2}$; minimum value $f(-2) = \frac{1}{4}$; maximum value $f\left(-\frac{1}{2}\right) = 4$

5. Critical points: $-\frac{1}{2}, 0, 1$; minimum value $f(0) = 0$; maximum value $f(1) = 1$

7. Critical points: $-2, 0, 1, 3$; minimum value $f(1) = -1$; maximum value $f(3) = 135$

9. Critical points: $-1, 0, 2, 3$; minimum value $f(2) = -9$; maximum value $f(3) = 88$

11. Critical points: $\frac{\pi}{4}, \frac{\pi}{2}, \frac{4\pi}{3}$; minimum value $f\left(\frac{4\pi}{3}\right) \approx -0.87$; maximum value $f\left(\frac{\pi}{2}\right) = 1$

13. Increasing: $\left(-\infty, \frac{3}{2}\right]$; concave down: $(-\infty, \infty)$

15. Increasing: $(-\infty, -1] \cup [1, \infty)$; concave down: $(-\infty, 0)$

17. Increasing: $\left[0, \frac{1}{5}\right]$; concave down: $\left(\frac{3}{20}, \infty\right)$

19. Increasing: $\left(-\infty, \frac{3}{4}\right]$; concave down: $(-\infty, 0) \cup \left(\frac{1}{2}, \infty\right)$

21. Increasing: $\left(-\infty, 0\right] \cup \left[\frac{8}{3}, \infty\right)$; decreasing: $\left[0, \frac{8}{3}\right]$;
Local minimum value $f\left(\frac{8}{3}\right) = -\frac{256}{27}$
Local maximum value $f(0) = 0$
Inflection point: $\left(\frac{4}{3}, -\frac{128}{27}\right)$

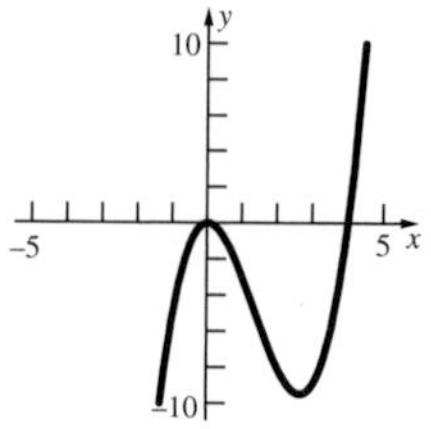

23.

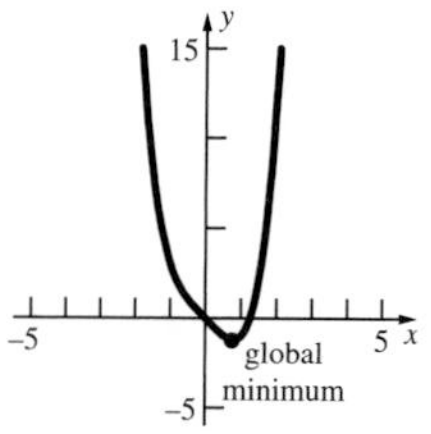

25.

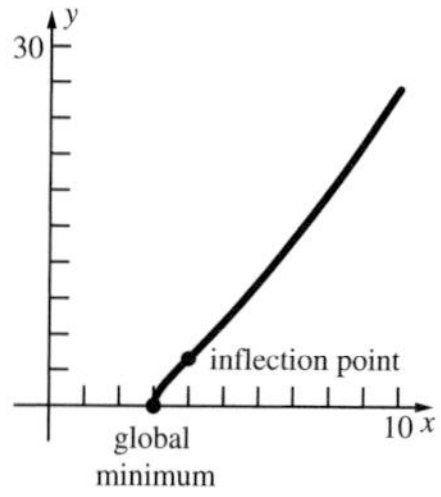

27.

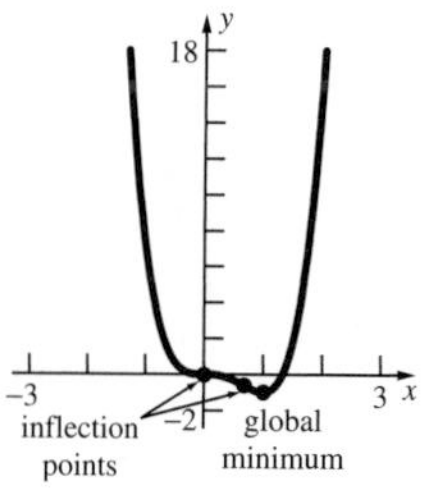

29.

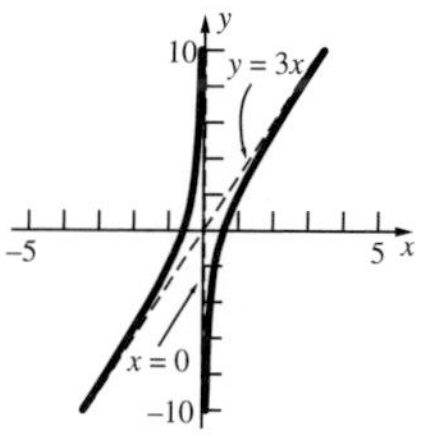

31.

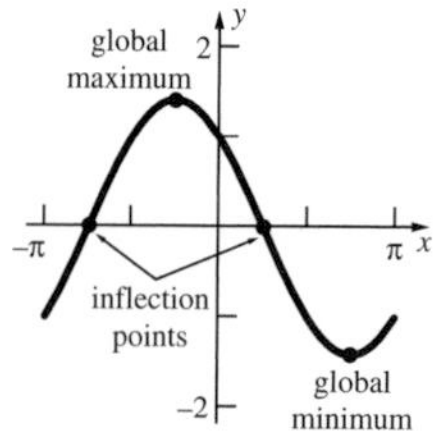

33.

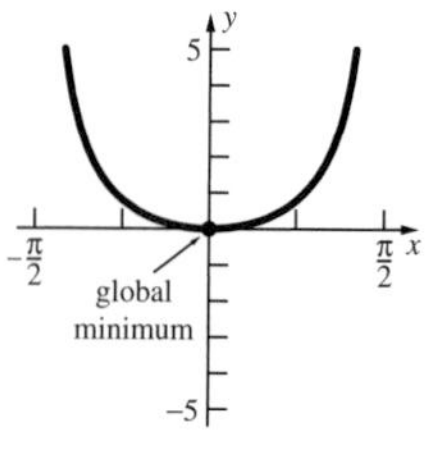

35.

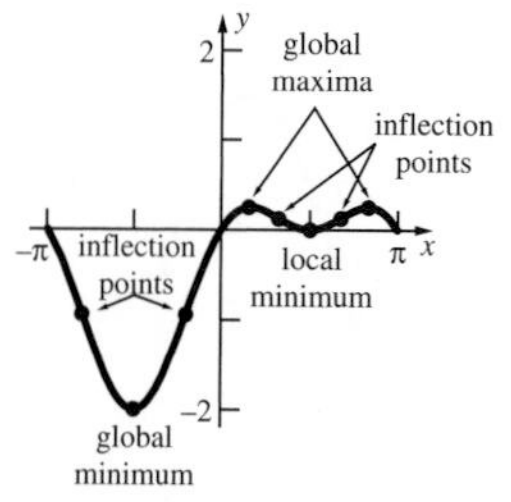

37.

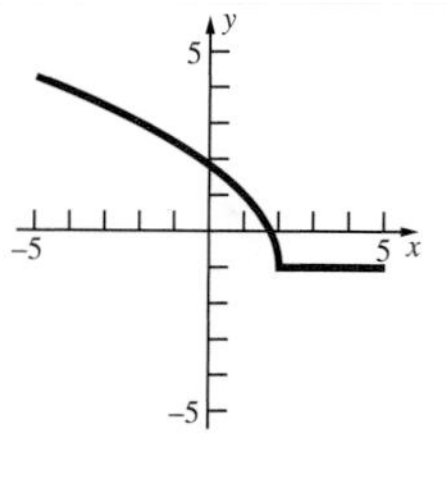

39.

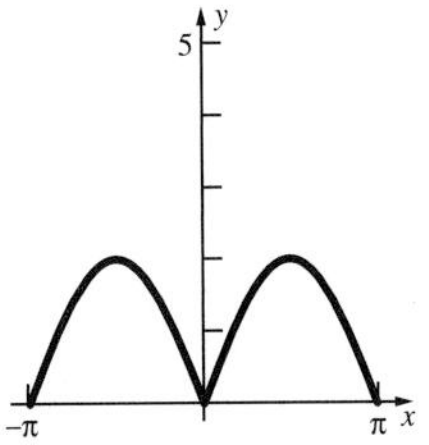

41. 11.18 ft **43.** $r = 4\sqrt[3]{2}, h = 8\sqrt[3]{2}$

45. **(a)** $c = \pm\sqrt{3}$

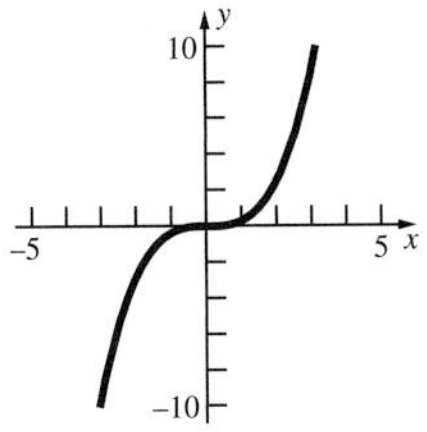

(b) Does not apply, $F'(0)$ does not exist.

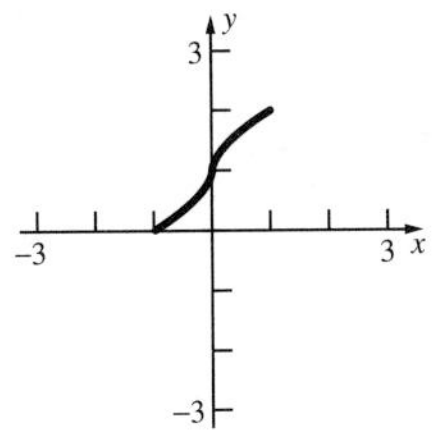

(c) $c = 1 + \sqrt{2}$

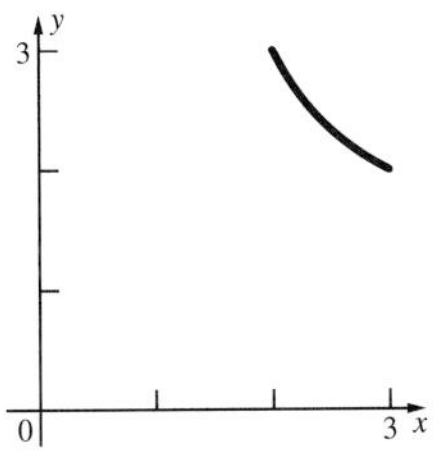

47.

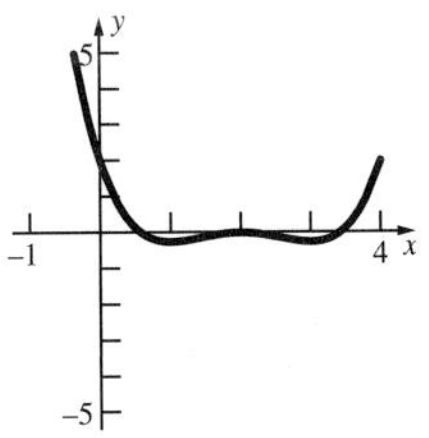

49. 0.281785 **51.** 0.281785 **53.** $\frac{1}{4}x^4 - x^3 + 2x^{3/2} + C$

55. $\frac{1}{3}y^3 + 9\cos y - \frac{26}{y} + C$ **57.** $\frac{3}{16}(2z^2 - 3)^{4/3} + C$

59. $\frac{1}{18}\tan^3(3x^2 + 6x) + C$ **61.** $\frac{3}{25}(t^5 + 5)^{5/3} + C$

63. $\frac{2}{3}\sqrt{x^3 + 9} + C$ **65.** $-\frac{1}{2(2y-1)^2} + C$

67. $\frac{5}{24}(2y^3 + 3y^2 + 6y)^{4/5} + C$ **69.** $y = 2\sqrt{x+1} + 14$

71. $y = \frac{1}{3}(2t-1)^{3/2} - 1$ **73.** $y = \sqrt{3x^2 - \frac{1}{4}x^4 + 9}$

75. 7 s; −176 ft/s

Chapter 4 Review and Preview Problems

1. $\frac{\sqrt{3}}{4}a^2$ **3.** $\frac{5}{4}a^2 \cot 36°$ **5.** $3.6 \cdot 5.8 + \frac{1}{2}\pi(1.8)^2 \approx 25.97$

7. 3.5 **9.** $\frac{1}{2}x^2 + x$ **11.** 6

Problem Set 4.1

1. 15 **3.** $\frac{481}{280}$ **5.** $\frac{85}{2}$ **7.** 3 **9.** $\sum_{i=1}^{41} i$ **11.** $\sum_{i=1}^{100} \frac{1}{i}$

13. $\sum_{i=1}^{50} a_{2i-1}$ **15.** 90 **17.** −10 **19.** 14,950

21. 2640 **23.** $\frac{4n^3 - 3n^2 - n}{6}$

27. **(a)** $1 - \left(\frac{1}{2}\right)^{10}$; **(b)** $2^{11} - 2$

33. $\bar{x} = 55/7 \approx 7.86; s^2 \approx 12.41$ **37.** $c = \bar{x}$ **39.** 715

41. $S = \frac{m(m+1)(3n - m + 1)}{6}$ **43.** $\frac{7}{2}$ **45.** $\frac{9}{2}$ **47.** $\frac{23}{8}$

49. $A = 6$

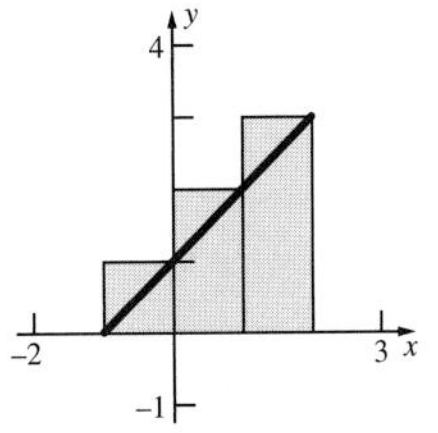

51. $A = \frac{1243}{216}$

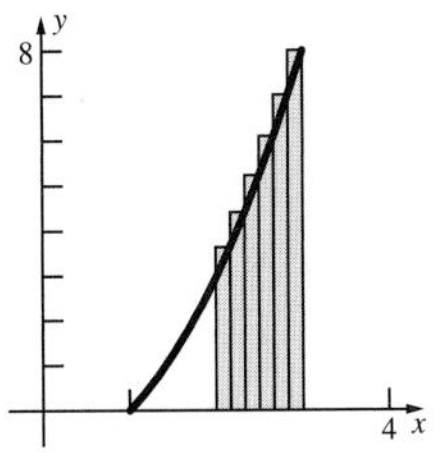

53. $\frac{5}{2}$ **55.** 4 **57.** $\frac{1}{4}$ **59.** $2\frac{1}{2}$ ft

63. **(a)** $\frac{125}{3}$; **(b)** 21; **(c)** 39

65. **(a)** 4; **(b)** $\frac{15}{4}$; **(c)** 10.5; **(d)** 102.4

Problem Set 4.2

1. 5.625 **3.** 15.6875 **5.** 2.625 **7.** $\int_1^3 x^3\,dx$

9. $\int_{-1}^{1} \frac{x^2}{1+x}\,dx$ **11.** 4 **13.** $3\pi - 3$ **15.** $\frac{35}{2}$

17. $\frac{27}{2}$ **19.** $\frac{1}{2} + \frac{\pi}{4}$ **21.** $\frac{1}{2}\pi A^2$ **23.** $\frac{2}{15}$ **25.** 3

27. 40, 80, 120, 160, 200, 240 **29.** 20, 80, 160, 240, 320, 400

31. **(a)** −3; **(b)** 19; **(c)** 3; **(d)** 2; **(e)** 9; **(f)** 0; **(g)** 1; **(h)** 2

35. Left: 5.24; Right: 6.84; Midpoint: 5.98

37. Left: 0.8638; Right: 0.8178; Midpoint: 0.8418

Problem Set 4.3

1. $A(x) = 2x$

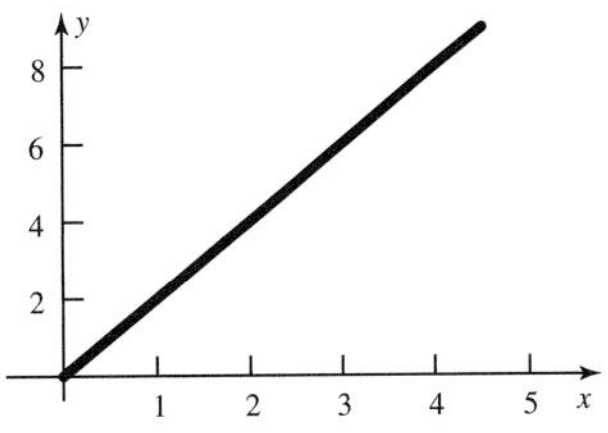

3. $A(x) = \frac{1}{2}(x - 1)(-1 + x), x > 1$

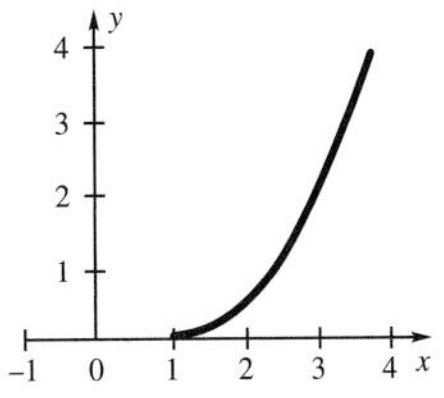

5. $A(x) = ax^2/2$

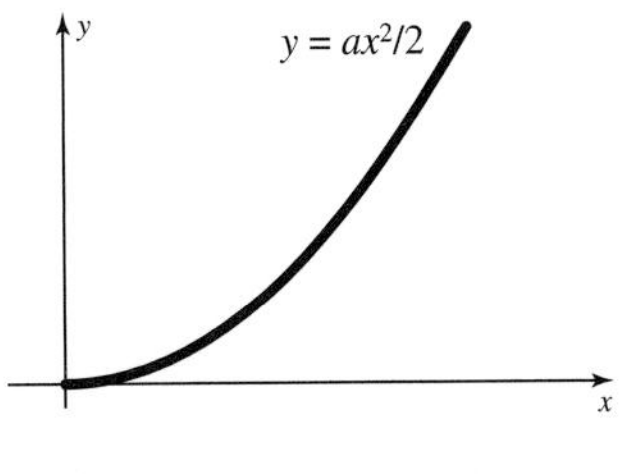

7. $A(x) = \begin{cases} 2x & \text{if } 0 \le x \le 1 \\ 2 + (x - 1) & \text{if } 1 < x \le 2 \\ 3 + 2(x - 2) & \text{if } 2 < x \le 3 \\ 5 + (x - 3) & \text{if } 3 < x \le 4 \\ \text{etc.} & \end{cases}$

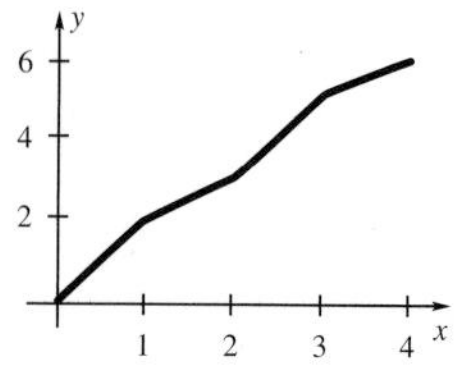

9. 6 **11.** 14 **13.** −31 **15.** 23 **17.** $2x$
19. $2x^2 + \sqrt{x}$ **21.** $-(x - 2)\cot 2x$ **23.** $2x\sin(x^2)$

25. $\dfrac{2x^5}{1 + x^4} + \dfrac{x^2}{1 + x^2}$

27. $f(x)$ is increasing on $[0, \infty)$ and concave up on $(0, \infty)$.

29. $f(x)$ is increasing on $\left[0, \frac{\pi}{2}\right], \left[\frac{3\pi}{2}, \frac{5\pi}{2}\right], \ldots$ and concave up on $(\pi, 2\pi), (3\pi, 4\pi), \ldots$

31. $f(x)$ is increasing on $(0, \infty)$ and never concave up.

33. 10;

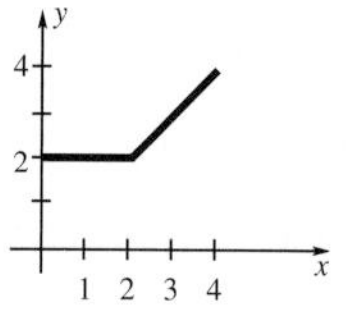

35. 4;

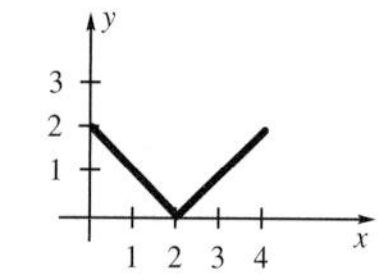

37. (a) Local minima at 0, ≈3.8, ≈5.8, ≈7.9, ≈9.9, 10; local maxima at ≈3.1, ≈5.0, ≈7.1, ≈9.0
(b) $G(0) = 0$ is global minimum, $G(9)$ is global maximum
(c) G is concave down on ≈(0.7, 1.5), (2.5, 3.5), (4.5, 5.5), (6.5, 7.5), (8.5, 9.5)
(d)

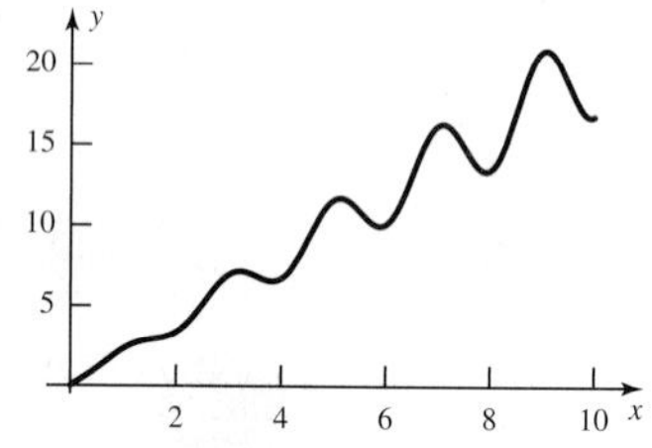

39. (a) 0 **(b)** $\frac{1}{5}x^5 + x + C$ **(c)** $\frac{1}{5}x^5 + x$ **(d)** $\frac{6}{5}$

43. Lower bound 20; upper bound 276

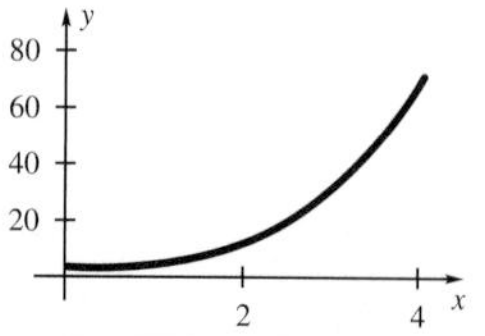

45. Lower bound $\frac{68}{5}$; upper bound 20

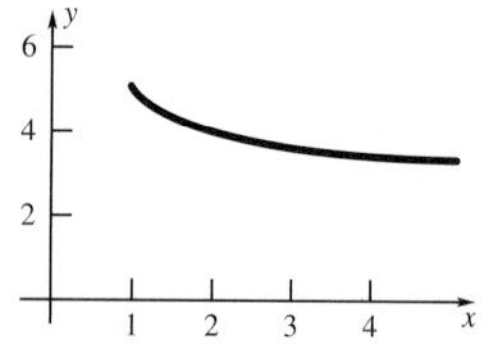

47. Lower bound 20π; upper bound $\frac{101}{5}\pi$

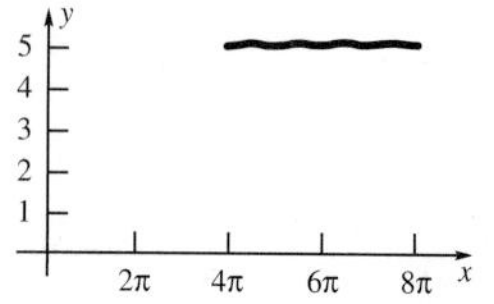

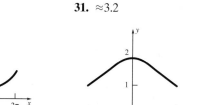

49. $\frac{1}{2}$ **51.** 2 **53.** $\sqrt{x}/2$ **55.** True **57.** False
59. True

61. $s(t) = \begin{cases} t^2/2, & 0 \le t \le 2 \\ -4 + 4t - t^2/2, & t > 2 \end{cases}$
$t = 4 + 2\sqrt{2} \approx 6.83$

Problem Set 4.4

1. 4 **3.** 15 **5.** $\frac{3}{4}$ **7.** $\frac{16}{3}$ **9.** $\frac{1783}{96}$ **11.** 1 **13.** $\frac{22}{5}$
15. $\frac{2}{9}(3x + 2)^{3/2} + C$ **17.** $\frac{1}{3}\sin(3x + 2) + C$
19. $-\frac{1}{6}\cos(6x - 7) + C$ **21.** $\frac{1}{3}(x^2 + 4)^{3/2} + C$
23. $-\frac{7}{10}(x^2 + 3)^{-5/7} + C$ **25.** $-\frac{1}{2}\cos(x^2 + 4) + C$
27. $-\cos\sqrt{x^2 + 4} + C$ **29.** $\frac{1}{27}\sin[(x^3 + 5)^9] + C$
31. $\frac{1}{3}[\sin(x^2 + 4)]^{3/2} + C$ **33.** $-\frac{1}{30}\cos^{10}(x^3 + 5) + C$
35. $\frac{2047}{11}$ **37.** $\frac{4}{5}$ **39.** $\frac{122}{9}$ **41.** 0 **43.** $\frac{1}{3}$ **45.** $\frac{9}{2}$
47. $\frac{1}{64}$ **49.** $\frac{\sin 3}{3}$ **51.** $\frac{1}{\pi}$ **53.** 1 **55.** $1 - \cos 1$
57. $\dfrac{1 - \cos^4 1}{8}$
59. (a) positive, **(b)** negative, **(c)** negative, **(d)** positive
61. 50 gallons; 20 hours **63.** 86 gallons **65.** 134
69. 9 **71.** 2

Problem Set 4.5

1. 40 **3.** $\frac{1}{3}$ **5.** $\frac{17}{6}$ **7.** 0 **9.** 0 **11.** $\frac{609}{8}$
13. $\frac{8}{\pi}\left(-\cos\sqrt{\frac{\pi}{2}} + \cos\sqrt{\frac{\pi}{4}}\right)$ **15.** $\frac{115}{81}$ **17.** $\frac{\sqrt{39}}{3}$
19. $c = 1$ **21.** $c = 0$ **23.** $c = \frac{\sqrt{21} + 3}{6}$ **25.** $c = \frac{5}{2}$
27. $(A + B)/2$

29. $\approx 1250\pi$

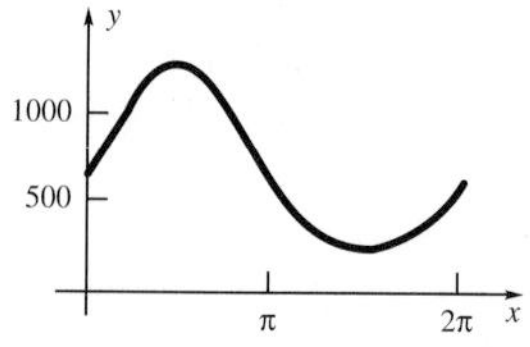

31. ≈3.2

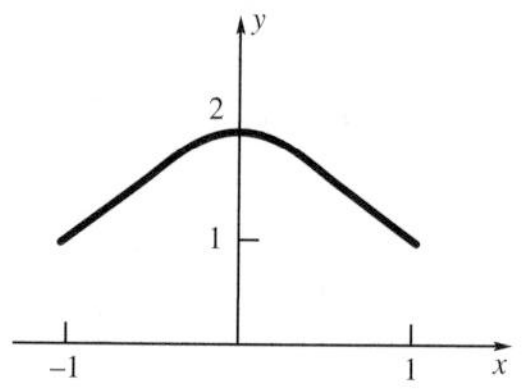

33. ≈25 **35.** 0 **37.** 0 **39.** 2π **41.** $\frac{8}{3}$ **43.** $\frac{1}{2}$

45. Even: $\int_{-b}^{-a} f(x)\,dx = \int_{a}^{b} f(x)\,dx$;
Odd: $\int_{-b}^{-a} f(x)\,dx = -\int_{a}^{b} f(x)\,dx$

47. 8 **49.** 2 **51.** 2
57. (a) Even; **(b)** 2π
(c)

Interval	Value of Integral
$\left[0, \frac{\pi}{2}\right]$	0.46
$\left[-\frac{\pi}{2}, \frac{\pi}{2}\right]$	0.92
$\left[0, \frac{3\pi}{2}\right]$	−0.46
$\left[-\frac{3\pi}{2}, \frac{3\pi}{2}\right]$	−0.92
$[0, 2\pi]$	0
$\left[\frac{\pi}{6}, \frac{13\pi}{6}\right]$	0
$\left[\frac{\pi}{6}, \frac{4\pi}{3}\right]$	−0.44
$\left[\frac{13\pi}{6}, \frac{10\pi}{3}\right]$	−0.44

Problem Set 4.6

1. 0.7877, 0.5654, 0.6766, 0.6671, $\frac{2}{3}$

3. 1.6847, 2.0382, 1.8615, 1.8755, $\frac{4\sqrt{2}}{3}$

5. 3.4966, 7.4966, 5.4966, 5.2580, 5.25

7.

	LRS	RRS	MRS	Trap	Parabolic
$n = 4$	0.5728	0.3728	0.4590	0.4728	0.4637
$n = 8$	0.5159	0.4159	0.4625	0.4659	0.4636
$n = 16$	0.4892	0.4392	0.4634	0.4642	0.4636

9.

	LRS	RRS	MRS	Trap	Parabolic
$n = 4$	2.6675	3.2856	2.9486	2.9765	2.9580
$n = 8$	2.8080	3.1171	2.9556	2.9625	2.9579
$n = 16$	2.8818	3.0363	2.9573	2.9591	2.9579

11. 12, 1.1007 **13.** 8, 4.6637 **15.** 6, 1.0989 **19.** smaller
21. larger **25.** LRS < MRS < Parabolic < Trap < RRS
27. 4570 ft^2 **29.** 1,074,585,600 ft^3
31. Using a right Riemann sum ≈13,740 gallons

Chapter Review 4.7

Concepts Test

1. True **3.** True **5.** False **7.** True **9.** True
11. True **13.** True **15.** True **17.** True **19.** False
21. True **23.** True **25.** False **27.** False **29.** False
31. True **33.** False **35.** True **37.** False **39.** True
41. True **43.** True **45.** False

Sample Test Problems

1. $\frac{5}{4}$ **3.** $\frac{50}{3} - \frac{26}{\pi} + \frac{\pi^3}{3} - 9\cos 1$

5. $\frac{1}{16}\left[-15\left(-125 + \sqrt[3]{5}\right)\right]$ **7.** $\frac{1}{18}\tan^3(3\pi^2 + 6\pi)$

9. 46.9 **11.** $-\frac{1}{2}\cos(x^2 + 2x + 3) + C$

13. $\frac{7}{4}$

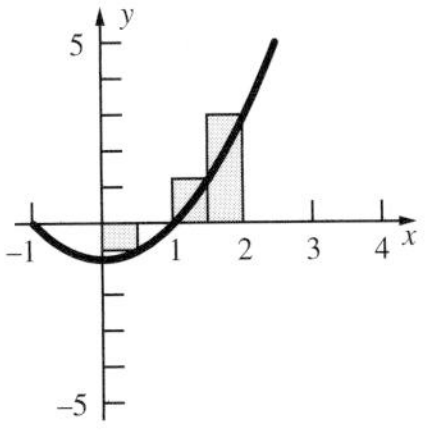

15. $\frac{5}{6}$ **17.** $\frac{39}{4}$ **19.** 1870

21. (a) $\sum_{n=2}^{78} \frac{1}{n}$; **(b)** $\sum_{n=1}^{50} nx^{2n}$

23. (a) −2; **(b)** −4; **(c)** 6; **(d)** −12; **(e)** −2

25. (a) −8; **(b)** 8; **(c)** 0; **(d)** −16; **(e)** −2 **(f)** −5

27. $c = -\sqrt{7}$

29. (a) $\sin^2 x$; **(b)** $f(x+1) - f(x)$

(c) $-\frac{1}{x^2}\int_0^x f(z)\,dz + \frac{1}{x}f(x)$; **(d)** $\int_0^x f(t)\,dt$;

(e) $g'(g(x))g'(x)$; **(f)** $-f(x)$

33. 0.2043 **35.** 372 **37.** MRS < Trap < LRS

Chapter 5 Review and Preview Problems

1. $\frac{1}{4}$ **3.** $\sqrt[3]{4} - 1$ **5.** $\sqrt{10}$ **7.** 1.6π

9. $[\pi(r_2^2 - r_1^2)]\,\Delta x$ **11.** $\frac{51}{10}$ **13.** $\frac{16}{15}$

Problem Set 5.1

1. 6 **3.** $\frac{40}{3}$ **5.** $\frac{9}{2}$ **7.** $\frac{253}{12}$ **9.** $\frac{9}{2}$

11. 6

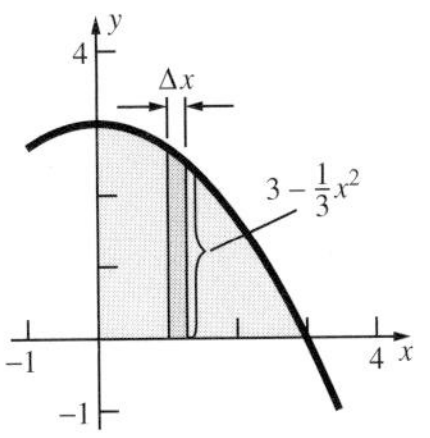

13. 24

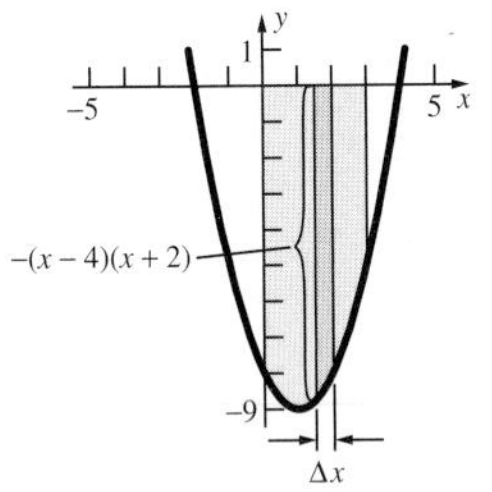

15. $\frac{17}{6}$

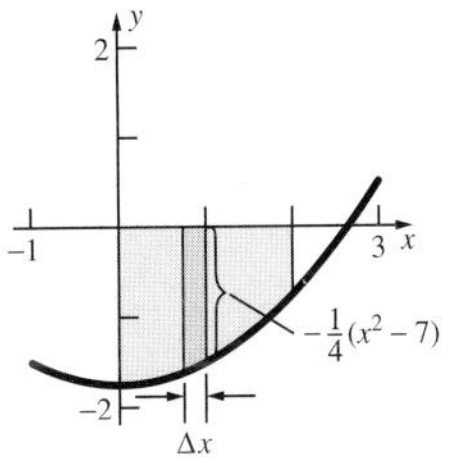

17. $3\sqrt[3]{2}$

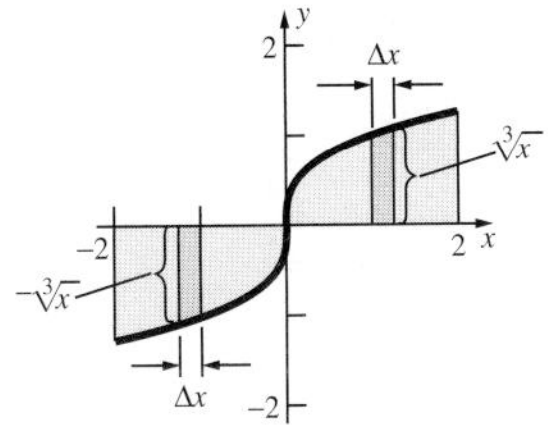

19. $\frac{13\sqrt{13}}{6}$

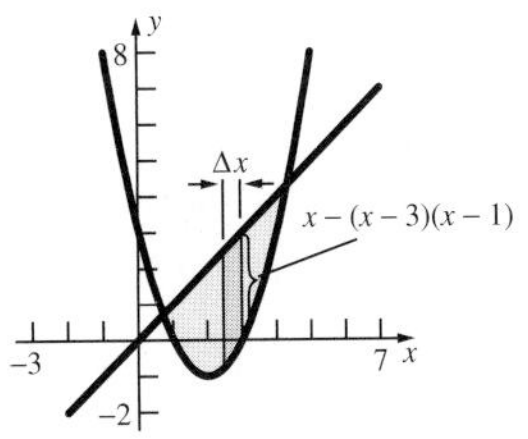

21. $\frac{1}{3}$

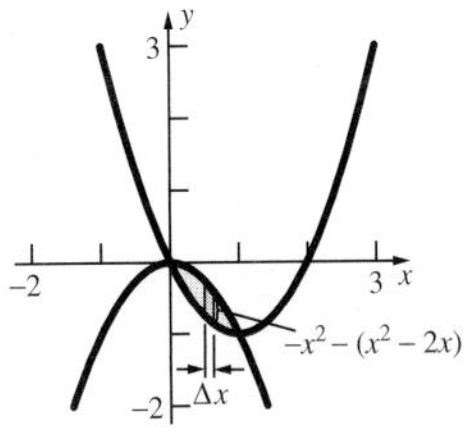

23. $\frac{256}{3}$

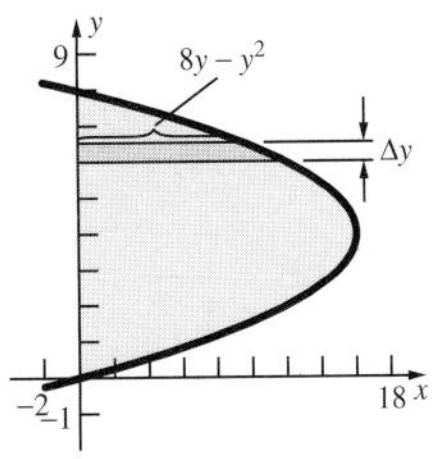

25. $\frac{1}{216}$

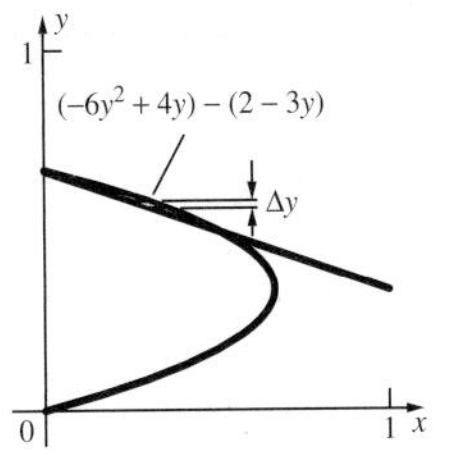

27. 4

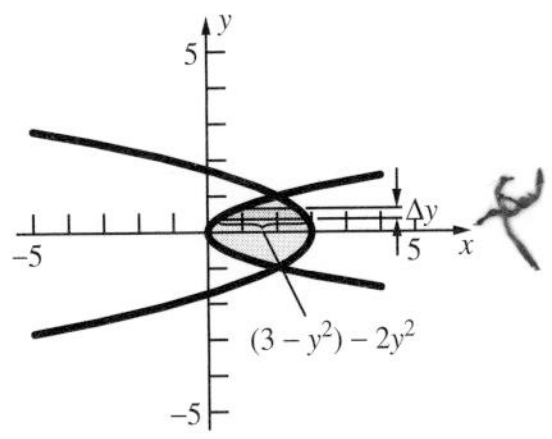

29. 22

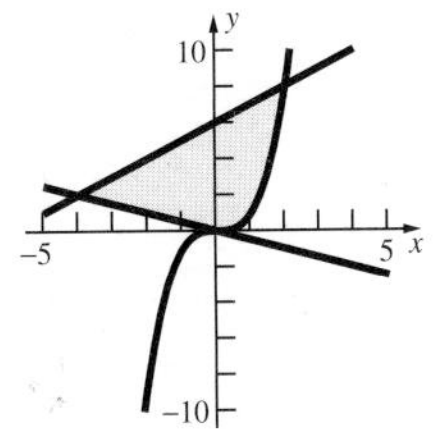

31. 130 ft; 194 ft **33.** 6 s; $2 + 2\sqrt{2}$ s

35. Area $(A) = 9$; $A(B) = \frac{37}{6}$; $A(C) = \frac{37}{6}$; $A(D) = \frac{44}{3}$; $A(A + B + C + D) = 36$

Problem Set 5.2

1. $\frac{206\pi}{15}$

3. (a) $\frac{256\pi}{15}$; **(b)** 8π

5. $\frac{1024}{5\pi}$

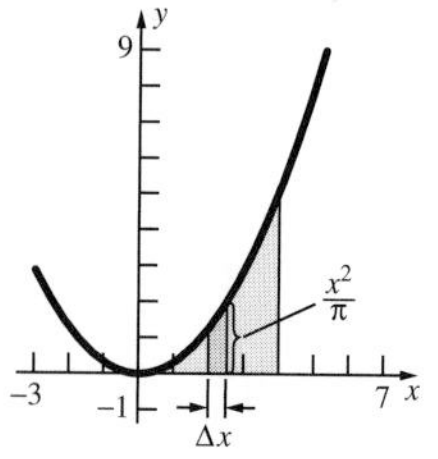

7. $\frac{\pi}{4}$

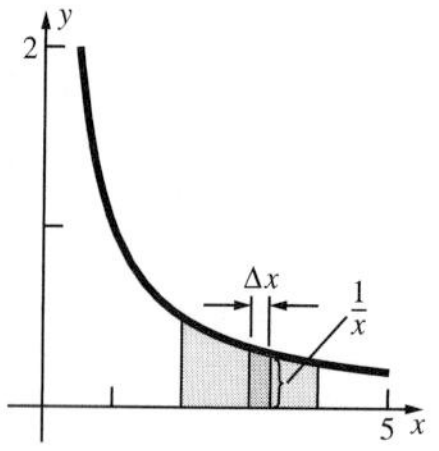

9. $\frac{100\pi}{3}$

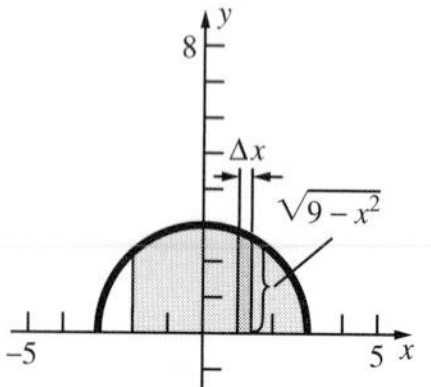

11. $\frac{243\pi}{5}$

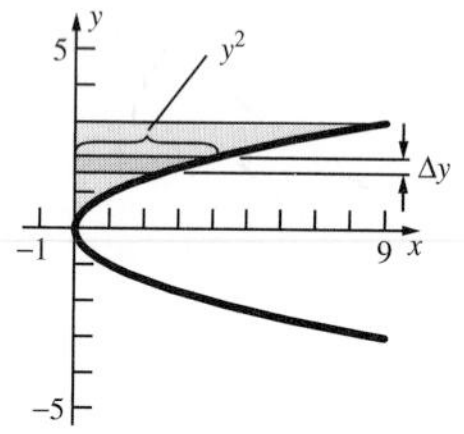

13. 32π

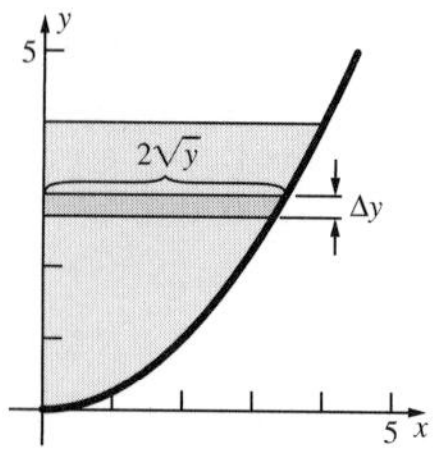

15. $\frac{6561\pi}{4}$

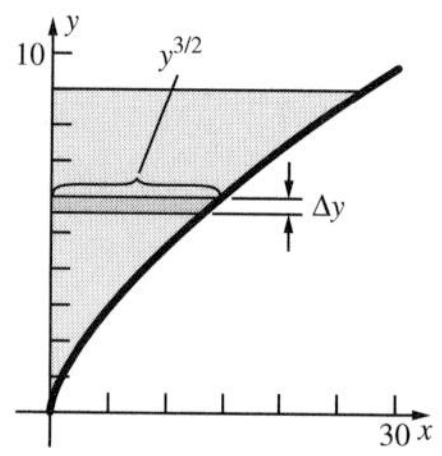

17. $\frac{4}{3}ab^2\pi$ **19.** $\frac{512\pi}{3}$ **21.** $\frac{2\pi}{3}$ **23.** $\frac{128}{3}$ **25.** 2 **27.** $\frac{2}{3}$

29. $2\pi r^2L - \frac{16}{3}r^3$ **31.** $\pi r^2(L_1 + L_2) - \frac{8}{3}r^3$

33. (a) $\frac{1024\pi}{35}$; **(b)** $\frac{704\pi}{5}$

35. $2\pi + \frac{16}{3}$ **37.** $\frac{2}{3}r^3 \tan\theta$

39. (a) $\frac{1}{3}\pi r^2 h$; **(b)** $\frac{\sqrt{2}}{12}r^3$

41. $\frac{2}{3}\pi r^3$

Problem Set 5.3

1. (a), (b)

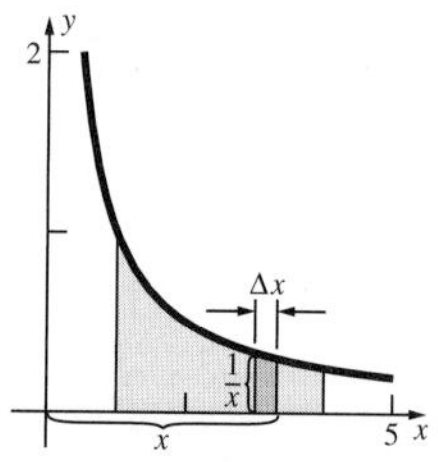

(c) $\Delta V \approx 2\pi\Delta x$;

(d) $2\pi\int_1^4 dx$; **(e)** 6π

3. (a), (b)

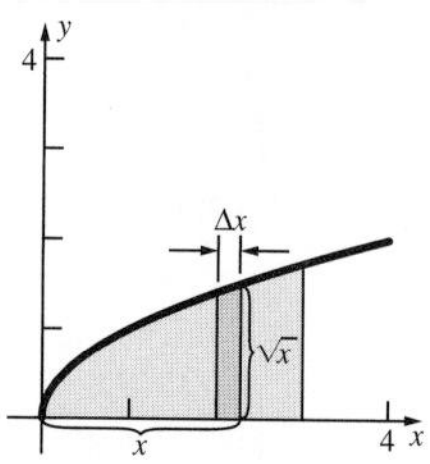

(c) $\Delta V \approx 2\pi x^{3/2}\Delta x$;

(d) $2\pi\int_0^3 x^{3/2}\,dx$;

(e) $\frac{36\sqrt{3}}{5}\pi$

5. (a), (b)

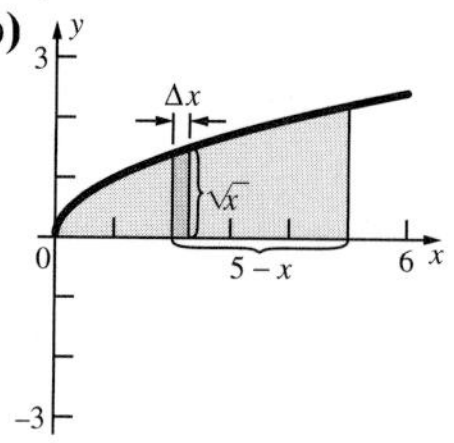

(c) $\Delta V \approx 2\pi(5x^{1/2} - x^{3/2})\Delta x$

(d) $2\pi\int_0^5 (5x^{1/2} - x^{3/2})\,dx$

(e) $\frac{40\sqrt{5}}{3}\pi$

7. (a), (b)

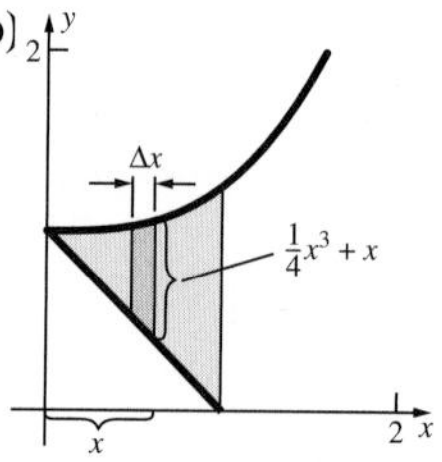

(c) $\Delta V \approx 2\pi\left(\frac{1}{4}x^4 + x^2\right)\Delta x$;

(d) $2\pi\int_0^1 \left(\frac{1}{4}x^4 + x^2\right)dx$

(e) $\frac{23\pi}{30}$

9. (a), (b)

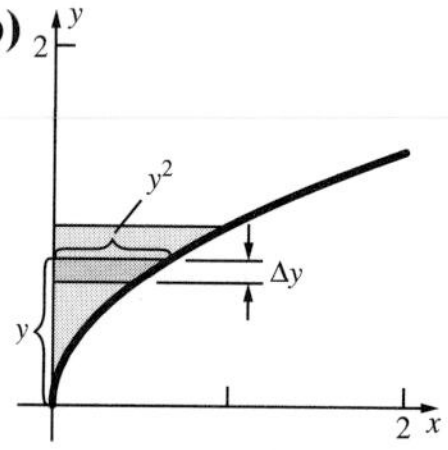

(c) $\Delta V \approx 2\pi y^3\Delta y$;

(d) $2\pi\int_0^1 y^3\,dy$

(e) $\frac{\pi}{2}$

11. (a), (b)

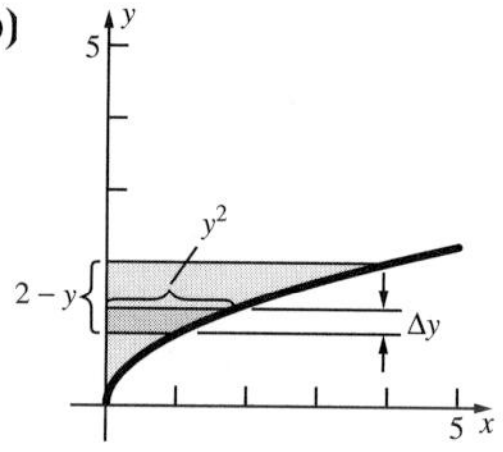

(c) $\Delta V \approx 2\pi(2y^2 - y^3)\Delta y$

(d) $2\pi\int_0^2 (2y^2 - y^3)\,dy$;

(e) $\frac{8\pi}{3}$

13. (a) $\pi\int_a^b [f(x)^2 - g(x)^2]\,dx$;

(b) $2\pi\int_a^b x[f(x) - g(x)]\,dx$;

(c) $2\pi\int_a^b (x - a)[f(x) - g(x)]\,dx$;

(d) $2\pi\int_a^b (b - x)[f(x) - g(x)]\,dx$

15.

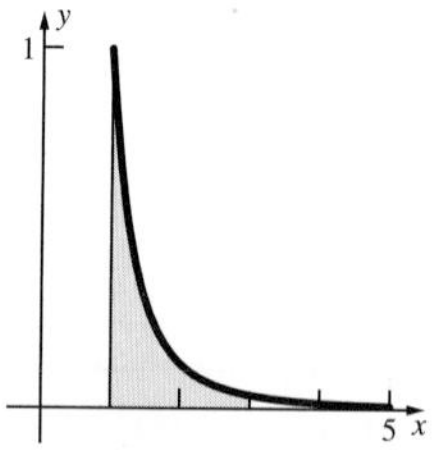

(a) $\int_1^3 \frac{1}{x^3}\,dx$;

(b) $2\pi\int_1^3 \frac{1}{x^2}\,dx$;

(c) $\pi\int_1^3 \left(\frac{1}{x^6} + \frac{2}{x^3}\right)dx$;

(d) $2\pi\int_1^3 \left(\frac{4}{x^3} - \frac{1}{x^2}\right)dx$

17. $\frac{64\pi}{5}$ **19.** $\frac{4\pi}{3}(b^2 - a^2)^{3/2}$ **21.** $\pi\left(\sqrt{2} - 1\right)$

23. (a) $\frac{2\pi}{15}$; **(b)** $\frac{\pi}{6}$; **(c)** $\frac{\pi}{60}$

25. $\frac{1}{3}rS$

Problem Set 5.4

1. $\frac{1}{54}\left(181\sqrt{181} - 13\sqrt{13}\right)$ **3.** 9 **5.** $\frac{595}{144}$

7. $\frac{1}{3}\left(2\sqrt{2} - 1\right)$ **9.** 4π

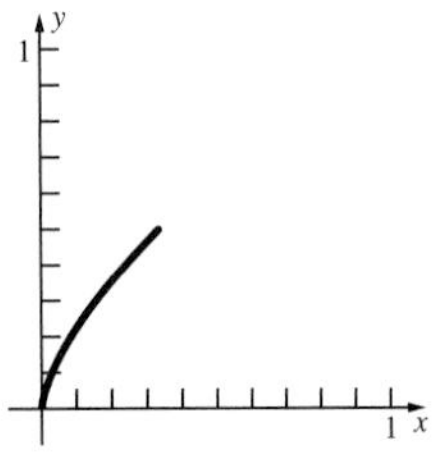

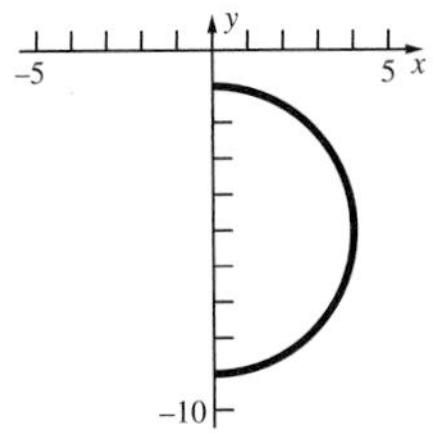

11. $2\sqrt{5}$ **13.** $\int_0^2 \sqrt{1 + 4t^2}\, dt \approx 4.6468$

15. $\int_0^{\pi/2} \sqrt{\cos^2 t + 4\sin^2 2t}\, dt \approx 2.3241$

17. $6a$

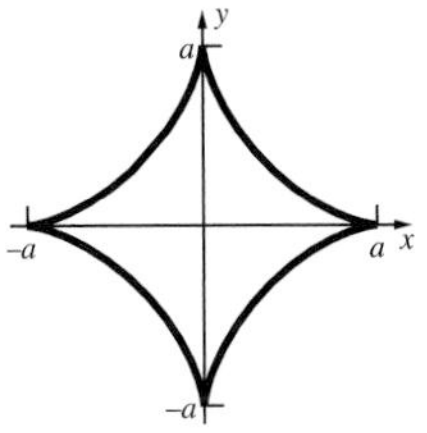

19. $8a$

21. **(a)** $\frac{2}{5}\left(4\sqrt{2} - 1\right)$; **(b)** 16

23. $6\sqrt{37}\pi$ **25.** $248\sqrt{2}\pi/9$ **27.** $\frac{\pi}{27}\left(10\sqrt{10} - 1\right)$

29. $4\pi r^2$

33. **(b)** $\frac{64}{3}\pi a^2$

35. **(a)**

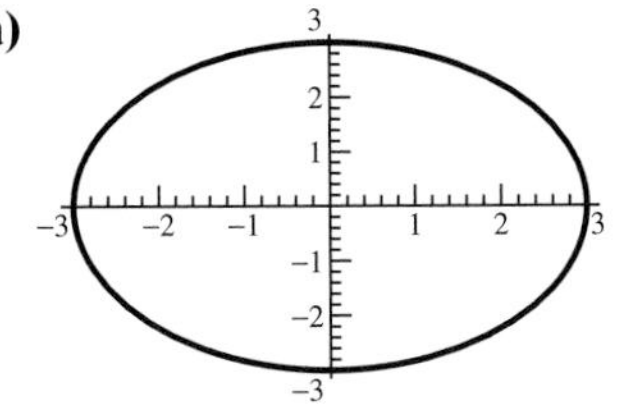

(b)

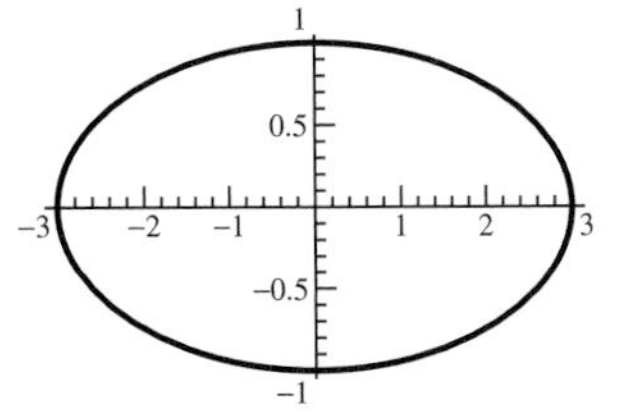

(c)

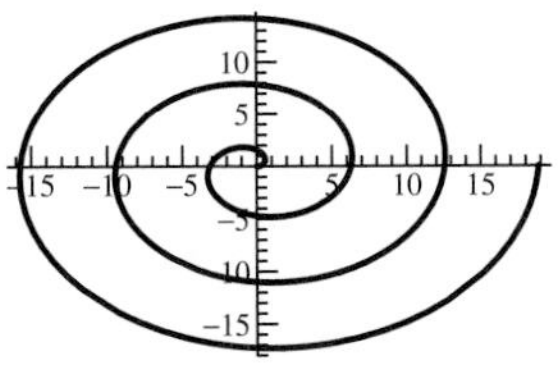

(d)

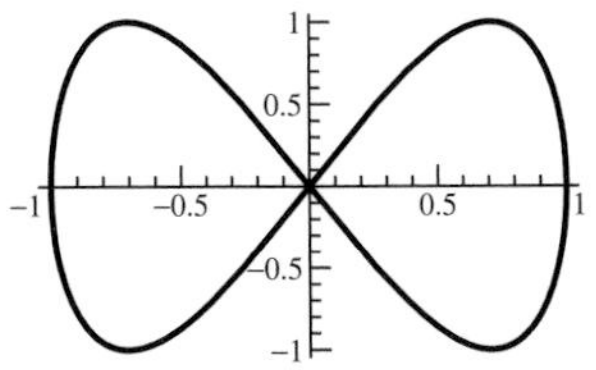

(e)

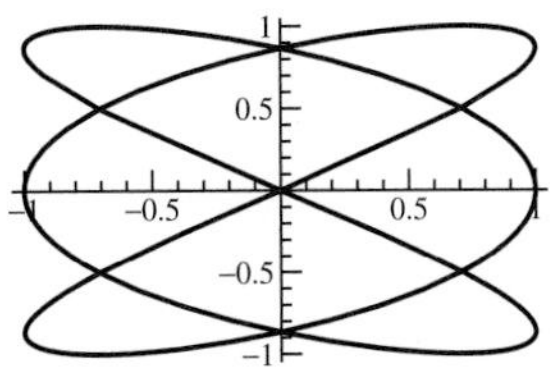

(f)

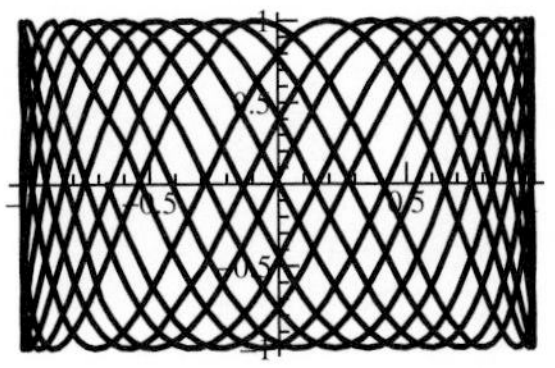

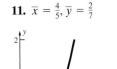

37. $n = 1$: $L \approx 1.41$; $n = 2$: $L \approx 1.48$; $n = 4$: $L \approx 1.60$
$n = 10$: $L \approx 1.75$; $n = 100$: $L \approx 1.95$; $n = 10{,}000$: $L \approx 2$

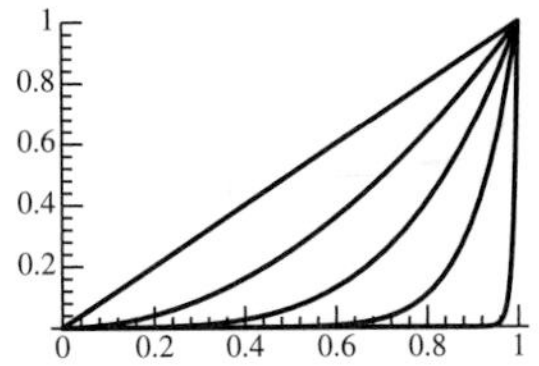

Problem Set 5.5

1. 1.5 ft-lb **3.** 0.012 Joules **7.** 18 ft-lb **9.** 52,000 ft-lb
11. 76.128 ft-lb **13.** 125.664 ft-lb **17.** 2075.83 in.-lb
19. 350,000 ft-lb **21.** 952.381 mi-lb **23.** 43,200 ft-lb
25. 1684.8 pounds **27.** 1684.8 pounds **29.** 16.64 pounds
33. 74,880 pounds **35.** $\frac{3mh}{4} + 15m$ **37.** $\frac{8475}{32}$ ft-lb

Problem Set 5.6

1. $\frac{5}{21}$ **3.** $\frac{21}{5}$ **5.** $M_y = 17, M_x = -3; \bar{x} = 1, \bar{y} = -\frac{3}{17}$

9. $\bar{x} = 0, \bar{y} = \frac{4}{5}$ **11.** $\bar{x} = \frac{4}{5}, \bar{y} = \frac{2}{7}$

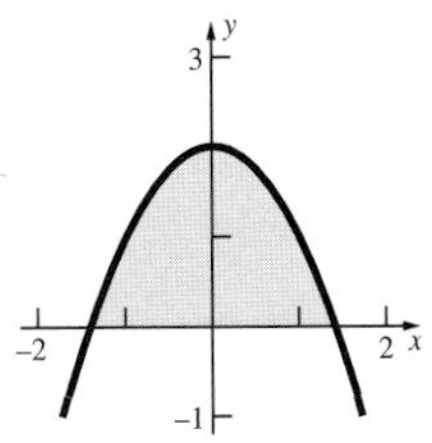

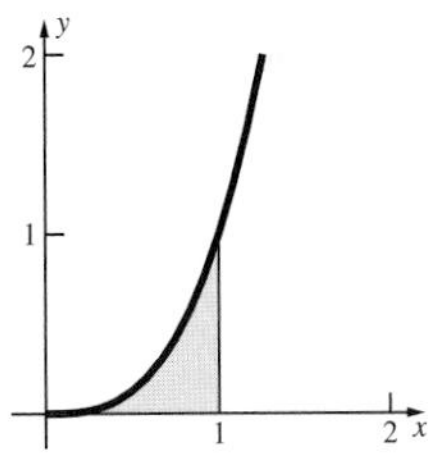

13. $\bar{x} = \frac{192}{95}, \bar{y} = \frac{27}{19}$ **15.** $\bar{x} = \frac{6}{5}, \bar{y} = 0$

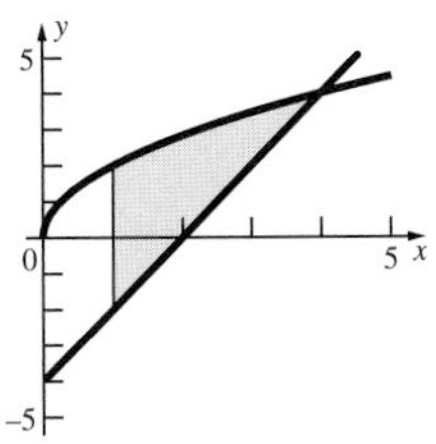

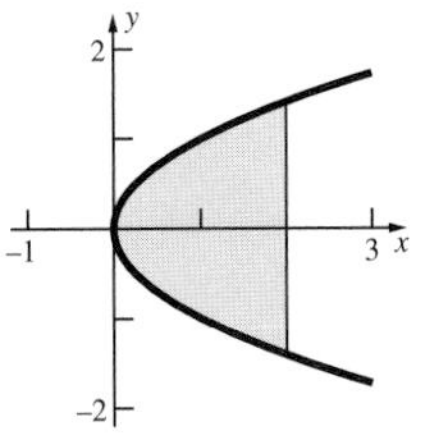

17. $m(R_1) = \frac{1}{2}\delta, \bar{x}_1 = \frac{2}{3}, \bar{y} = \frac{1}{3}, M_y(R_1) = \frac{1}{3}\delta, M_x(R_1) = \frac{1}{6}\delta$;
$m(R_2) = 2\delta, \bar{x}_2 = 2, \bar{y}_2 = \frac{1}{2}, M_y(R_2) = 4\delta, M_x(R_2) = \delta$.

21. $\bar{x} = -\frac{3}{14}, \bar{y} = \frac{1}{14}$ **23.** $\bar{x} = \frac{9}{16}, \bar{y} = \frac{31}{16}$ **25.** $\frac{2\pi}{5}$

27. The centroid is $\frac{4a}{3\pi}$ units perpendicular from the center of the diameter. $\left(\bar{y} = \frac{4a}{3\pi}, \bar{x} = 0\right)$

29. **(a)** $V = 2\pi \int_c^d (K - y)w(y)\, dy$

31. **(a)** $4\pi r^3 n \sin\frac{\pi}{2n}\cos^2\frac{\pi}{2n}$

35. $\overline{x} \approx 7.00$ cm above the center of the hole; $\overline{y} \approx 0.669$ cm to the right of the center of the hole

Problem Set 5.7

1. **(a)** 0.1 **(b)** 0.35
3. **(a)** 0.2 **(b)** 0
5. **(a)** 0.6 **(b)** 2.2
7. **(a)** 0.6 **(b)** 2

9. **(a)** 0.9 **(b)** 10 **(c)** $F(x) = \begin{cases} 0, & x < 0 \\ x/20, & 0 \le x \le 20 \\ 1, & x > 20 \end{cases}$

11. **(a)** $\frac{27}{32}$ **(b)** 4

(c) $F(x) = \begin{cases} 0, & x < 0 \\ \frac{3}{64}x^2 - \frac{1}{256}x^3, & 0 \le x \le 8 \\ 1, & x > 8 \end{cases}$

13. **(a)** 0.6875 **(b)** 2.4

(c) $F(x) = \begin{cases} 0, & x < 0 \\ \frac{1}{16}x^3 - \frac{3}{256}x^4, & 0 \le x \le 4 \\ 1, & x > 4 \end{cases}$

15. **(a)** $\frac{1}{2}$ **(b)** 2

(c) $F(x) = \begin{cases} 0, & x < 0 \\ \frac{1}{2} - \frac{1}{2}\cos\frac{\pi x}{4}, & 0 \le x \le 4 \\ 1, & x > 4 \end{cases}$

17. **(a)** $\frac{1}{3}$ **(b)** $\frac{4}{3}\ln 4$ **(c)** $F(x) = \begin{cases} 0, & x < 1 \\ \frac{4x - 4}{3x}, & 1 \le x \le 4 \\ 1, & x > 4 \end{cases}$

21. $\frac{a + b}{2}$ **23.** $k = \frac{6}{125}$

25. **(a)** $\frac{1}{4}$ **(b)** $\frac{1}{8}$ **(c)** 2

(d) $F(x) = \begin{cases} 0, & \text{if } x < 0 \\ x^2/8, & \text{if } 0 \le x \le 2 \\ -x^2/8 + x - 1, & \text{if } 2 < x \le 4 \\ 1, & \text{if } x > 4 \end{cases}$

(e) $F(y) = \begin{cases} 0, & \text{if } y < 0 \\ y^2/28800, & \text{if } 0 \le y \le 120 \\ -y^2/28800 + y/60 - 1, & \text{if } 120 < y \le 240 \\ 1, & \text{if } y > 240 \end{cases}$

27. **(a)** $\approx$95,802,719 **(b)** $\approx$0.884 **(c)** 0.2625
(d) For $0 \le x \le 0.6$, $F(x) \approx 6.3868 \times 10^6 x^{15} - 3.2847 \times 10^7 x^{14} + 7.4284 \times 10^7 x^{13} - 9.6569 \times 10^7 x^{12} + 7.9011 \times 10^7 x^{11} - 4.1718 \times 10^7 x^{10} + 1.3906 \times 10^7 x^9 - 2.6819 \times 10^6 x^8 + 2.2987 \times 10^5 x^7$
(e) $F(25.4y)$ where F is as in (d)

29. $G(y) = \begin{cases} 0, & y < 0 \\ \sqrt{y^2 - 1}, & 0 \le y \le \sqrt{2} \\ 1, & y > \sqrt{2} \end{cases}$

$g(y) = y/\sqrt{y^2 - 1}, 0 \le y \le \sqrt{2}$.

33. $F(x) = \begin{cases} 0, & x < 0 \\ 0.8, & 0 \le x < 1 \\ 0.9, & 1 \le x < 2 \\ 0.95, & 2 \le x < 3 \\ 1, & x \ge 3 \end{cases}$

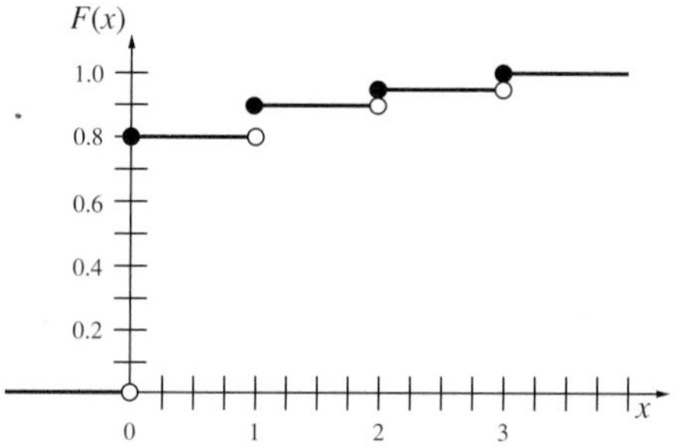

35. **(a)** 1 **(b)** $\frac{1}{12}$ **(c)** $\frac{2}{(y + 1)^2}$ for $0 \le y \le 1$ **(d)** 0.38625

37. $2, \frac{32}{7}$ **39.** $\frac{4}{7}$

Chapter Review 5.8

Concepts Test

1. False **3.** False **5.** True **7.** False **9.** False **11.** False **13.** True **15.** True **17.** True **19.** True **21.** True **23.** True

Sample Test

1. $\frac{1}{6}$ **3.** $\frac{\pi}{6}$ **5.** $\frac{5\pi}{6}$

7. $V(S_1) = \frac{\pi}{30}$; $V(S_2) = \frac{\pi}{6}$; $V(S_3) = \frac{7\pi}{10}$; $V(S_4) = \frac{5\pi}{6}$

9. 205,837 ft-lb **11.** **(a), (b)** $\frac{32}{3}$ **13.** $\frac{2048\pi}{15}$

15. $\frac{53}{6}$ **17.** 36 **19.** $\pi\int_a^b [f^2(x) - g^2(x)]\,dx$

21. $M_y = \delta\int_a^b x[f(x) - g(x)]\,dx$

$M_x = \frac{\delta}{2}\int_a^b [f^2(x) - g^2(x)]\,dx$

23. $2\pi\int_a^b f(x)\sqrt{1 + [f'(x)]^2}\,dx$

$+ 2\pi\int_a^b g(x)\sqrt{1 + [g'(x)]^2}\,dx$

$+ \pi[f^2(a) - g^2(a)] + \pi[f^2(b) - g^2(b)]$

25. **(a)** $\frac{3}{4}$ **(b)** $\frac{6 - x}{18}$ for $0 \le x \le 6$ **(c)** 2

Chapter 6 Review and Preview Problems

1. $-\frac{1}{x} + C$ **3.** $-\frac{100}{x^{0.01}} + C$ **5.** 0 **7.** $\frac{2}{x}$

9. **(a)** 2; **(b)** 2.48832; **(c)** 2.593742; **(d)** 2.691588; **(e)** 2.704814

11. **(a)** 2.25; **(b)** 2.593742; **(c)** 2.6533; **(d)** 2.70481; **(e)** 2.71152

13. $x = \frac{12k + 1}{6}\pi$ or $x = \frac{12k + 5}{6}\pi$ **15.** $x = \frac{4k + 1}{4}\pi$

17. $\sin\theta = \frac{\sqrt{x^2-1}}{x}$; $\cos\theta = \frac{1}{x}$; $\tan\theta = \sqrt{x^2-1}$

$\cot\theta = \frac{1}{\sqrt{x^2-1}}$; $\sec\theta = x$; $\csc\theta = \frac{x}{\sqrt{x^2-1}}$

19. $\sin\theta = \frac{1}{\sqrt{1+x^2}}$; $\cos\theta = \frac{x}{\sqrt{1+x^2}}$; $\tan\theta = \frac{1}{x}$

$\cot\theta = x$; $\sec\theta = \frac{\sqrt{1+x^2}}{x}$; $\csc\theta = \sqrt{1+x^2}$

21. $y = \frac{-2}{x^2-2}$

Problem Set 6.1

1. **(a)** 1.792; **(b)** 0.406; **(c)** 4.396; **(d)** 0.3465; **(e)** −3.584; **(f)** 3.871

3. $\frac{2x+3}{x^2+3x+\pi}$ **5.** $\frac{3}{x-4}$ **7.** $\frac{3}{x}$

9. $2x + 4x\ln x + \frac{3}{x}(\ln x)^2$ **11.** $\frac{1}{\sqrt{x^2+1}}$ **13.** $\frac{1}{243}$

15. $\frac{1}{2}\ln|2x+1| + C$ **17.** $\ln|3v^2+9v| + C$

19. $(\ln x)^2 + C$ **21.** $\frac{1}{10}[\ln(486+\pi) - \ln\pi]$

23. $\frac{x^2}{2} + x + \ln|x-1| + C$

25. $\frac{x^4}{4} - \frac{4x^3}{3} + 8x^2 - 64x + 256\ln|x+4| + C$

27. $\ln\frac{(x+1)^2}{x}$ **29.** $\ln\frac{x^2(x-2)}{x+2}$ **31.** $-\frac{x^3+33x^2+8}{2(x^3-4)^{3/2}}$

33. $-\frac{10x^2+219x-118}{6(x-4)^2(x+13)^{1/2}(2x+1)^{4/3}}$

35.

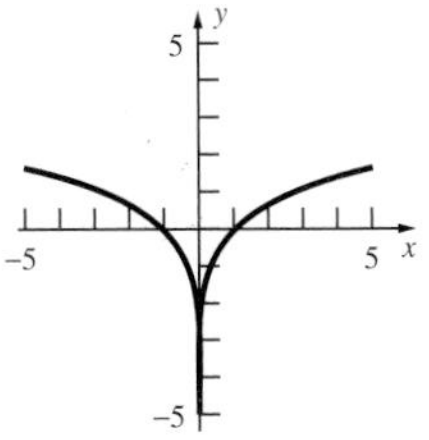

37.

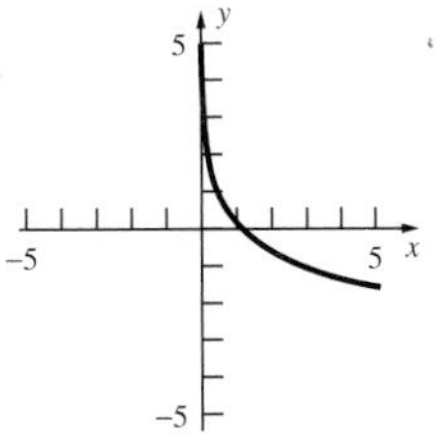

39.

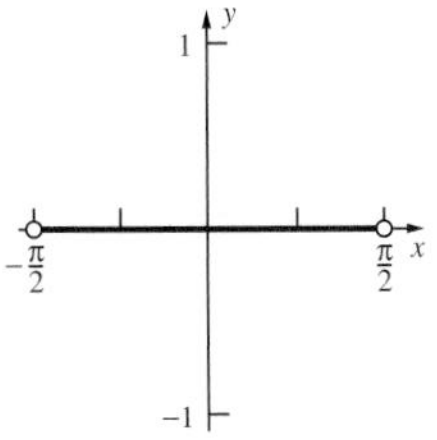

41. Minimum $f(1) = -1$ **43.** $\lim_{x\to\infty}\ln x = \infty$

45. $x = 3$ **47.** ln 2

49. **(a)** 1 **(b)** 3

51. $\ln\sqrt{3} \approx 0.5493$ **53.** $\pi\ln 4 \approx 4.355$

57. **(a)** Maxima: $\left(\frac{\pi}{2}, 0.916\right)$, $\left(\frac{5\pi}{2}, 0.916\right)$; minimum: $\left(\frac{3\pi}{2}, -0.693\right)$; **(b)** (3.871, −0.182), (5.553, −0.183); **(c)** 4.042

59.

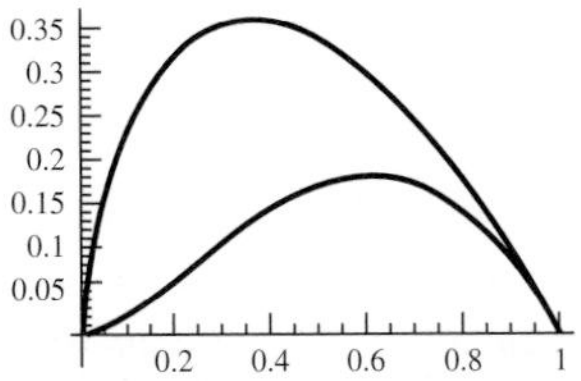

(a) 0.139; **(b)** 0.260

Problem Set 6.2

1. $f^{-1}(2) = 4$ **3.** No inverse **5.** $f^{-1}(2) \approx -1.3$

15. $f^{-1}(x) = x - 1$ **17.** $f^{-1}(x) = x^2 - 1, x \geq 0$

19. $f^{-1}(x) = 3 - \frac{1}{x}$ **21.** $f^{-1}(x) = -\frac{\sqrt{x}}{2}$

23. $f^{-1}(x) = 1 + \sqrt[3]{x}$ **25.** $f^{-1}(x) = \frac{1+x}{1-x}$

27. $f^{-1}(x) = \left(\frac{2-x}{x-1}\right)^{1/3}$ **29.** $V = \frac{4\pi h^3}{27}$; $h = 3\sqrt[3]{\frac{V}{4\pi}}$

31. $(-\infty, -0.25]$ or $[-0.25, \infty)$; then $f^{-1}(x) = \frac{1}{4}\left(-1 - \sqrt{8x+33}\right)$ or $f^{-1}(x) = \frac{1}{4}\left(-1 + \sqrt{8x+33}\right)$

33. $(f^{-1})'(3) \approx \frac{1}{3}$ **35.** $(f^{-1})'(3) \approx -\frac{1}{3}$

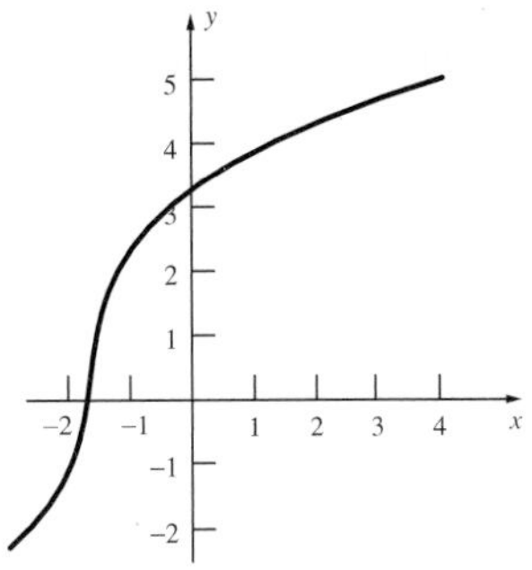

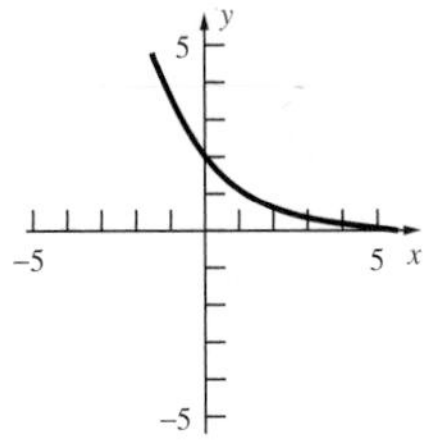

37. $\frac{1}{16}$ **39.** $\frac{1}{4}$

43. **(a)** 1 **(b)** $\frac{2}{\sqrt{7}}$ **(c)** $\frac{1}{\sqrt{2}}$

45. $\frac{3}{5}$

Problem Set 6.3

1. **(a)** 20.086; **(b)** 8.1662; **(c)** 4.1; **(d)** 1.20

3. x^3 **5.** cos x **7.** $3\ln x - 3x$ **9.** $3x^2$ **11.** e^{x+2}

13. $\frac{e^{\sqrt{x+2}}}{2\sqrt{x+2}}$ **15.** $2x$ **17.** $x^2e^x(x+3)$

19. $x\sqrt{e^{x^2}} + \frac{x}{|x|}e^{\sqrt{x^2}}$ **21.** $-\frac{y}{x}$

23. **(a)**

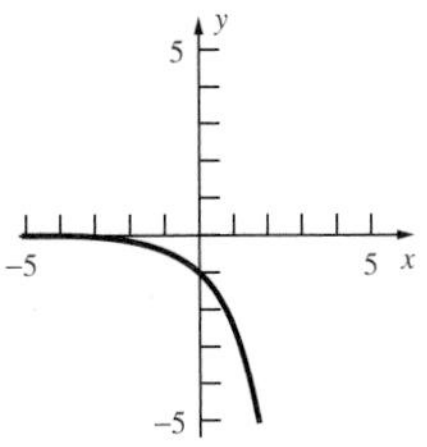

(b)

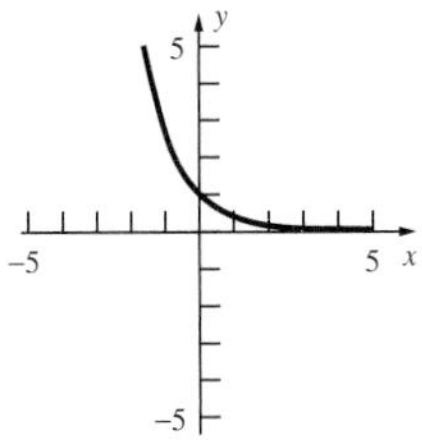

25. Domain $= (-\infty, \infty)$; increasing on $(-\infty, \infty)$; concave up on $(-\infty, \infty)$; no extreme values or points of inflection.

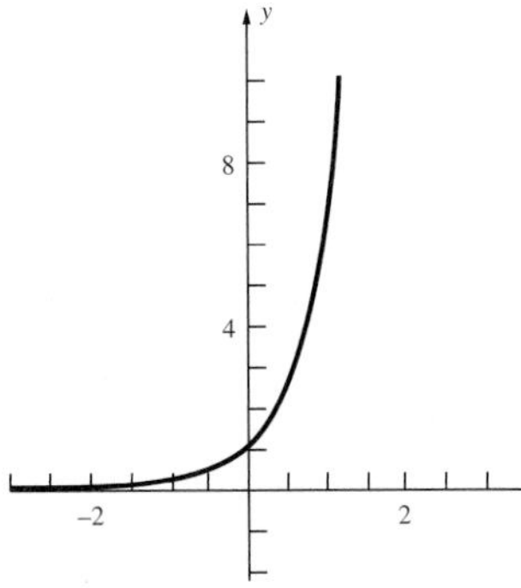

27. Domain $= (-\infty, \infty)$; increasing on $(-\infty, 1)$ and decreasing on $(1, \infty)$; maximum at $(1, 1/e)$; concave up on $(2, \infty)$ and concave down on $(-\infty, 2)$; point of inflection at $(2, 2/e^2)$

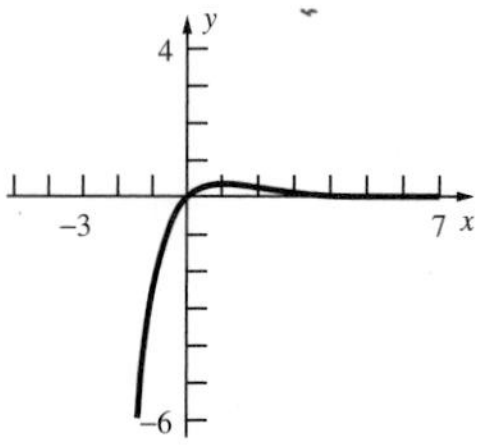

29. Domain $= (-\infty, \infty)$; increasing on $(0, \infty)$ and decreasing on $(-\infty, 0)$; minimum at $(0, 0)$; concave up on $(-1, 1)$ and concave down on $(-\infty, -1) \cup (1, \infty)$; points of inflection at $(-1, \ln 2)$ and $(1, \ln 2)$

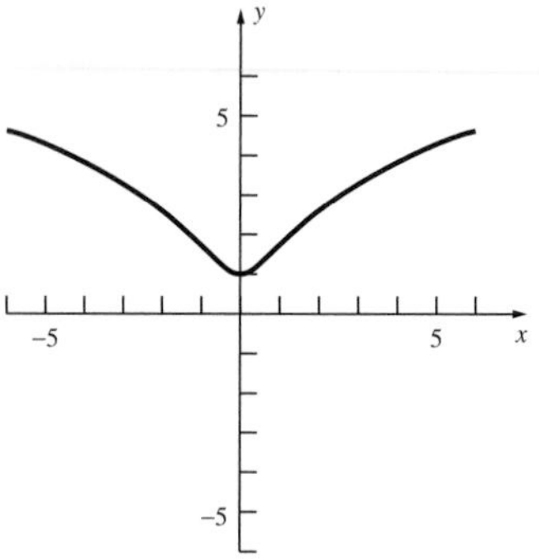

31. Domain $= (-\infty, \infty)$; increasing on $(-\infty, \infty)$; concave up on $(-\infty, \infty)$; no extreme values or points of inflection.

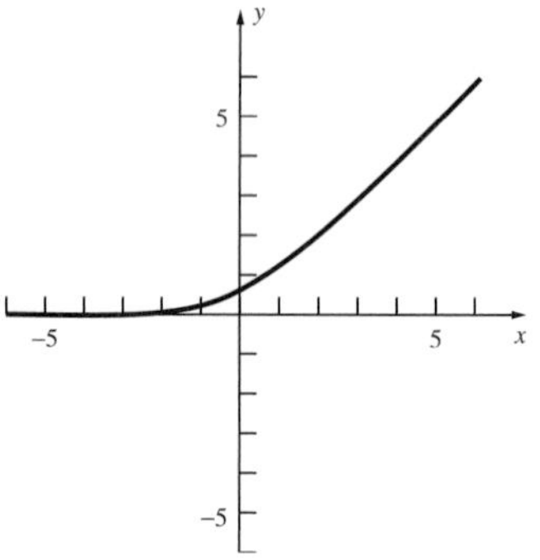

33. Domain $= (-\infty, \infty)$; increasing on$(-\infty, 2)$ and decreasing on $(2, \infty)$; maximum at $(2, 1)$; concave up on $\left(-\infty, \frac{4 - \sqrt{2}}{2}\right) \cup \left(\frac{4 + \sqrt{2}}{2}, \infty\right)$ and concave down on $\left(\frac{4 - \sqrt{2}}{2}, \frac{4 + \sqrt{2}}{2}\right)$; points of inflection at $\left(\frac{4 - \sqrt{2}}{2}, \frac{1}{\sqrt{e}}\right)$ and $\left(\frac{4 + \sqrt{2}}{2}, \frac{1}{\sqrt{e}}\right)$

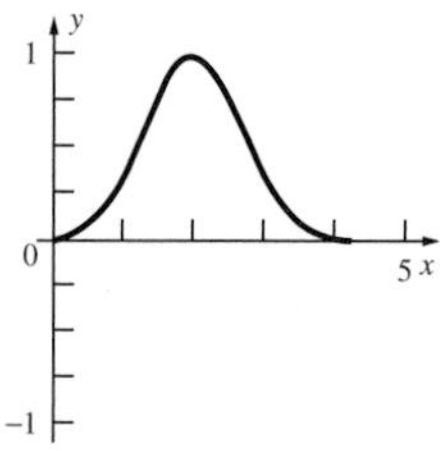

35. Domain $= (-\infty, \infty)$; increasing on $(-\infty, \infty)$; concave up on $(-\infty, 0)$ and concave down on $(0, \infty)$; point of inflection at $(0, 0)$; no extreme values.

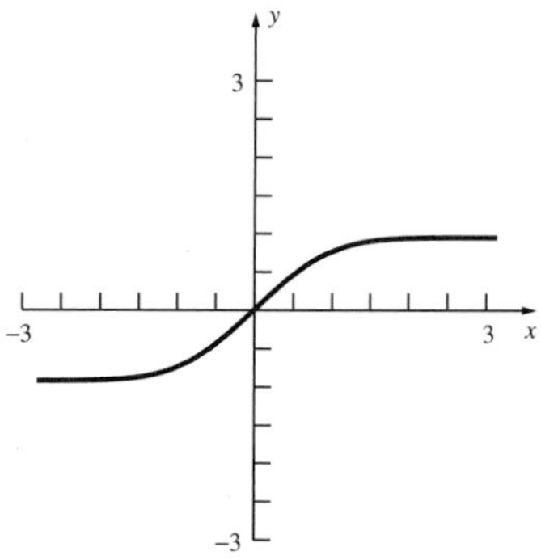

37. $\frac{1}{3}e^{3x+1} + C$ **39.** $\frac{1}{2}e^{x^2+6x} + C$ **41.** $e^{-1/x} + C$

43. $\frac{1}{2}e^3(e^2 - 1)$ **45.** 4π **47.** $\dfrac{3 - e}{2e}$

49. **(a)** 3,628,800; 3,598,696; **(b)** 8.31×10^{81}

51. $\sqrt{2}(e^\pi - 1)$

53. **(a)** 0; 0 **(b)** Maximum: $\left(e, \frac{1}{2}\right)$; minimum: $\left(\frac{1}{e}, -\frac{1}{2}\right)$ **(c)** $\sqrt{e}$

55. **(a)** 3.11; **(b)** 0.910

57. 4.2614 **59.** Behaves like $-x$; behaves like $2 \ln x$

Problem Set 6.4

1. 3 **3.** 8 **5.** 9 **7.** 1 **9.** 1.544 **11.** 0.1747

13. 4.08746 **15.** 1.9307 **17.** $2 \cdot 6^{2x} \ln 6$ **19.** $\dfrac{1}{\ln 3}$

21. $3^z\left[\dfrac{1}{z + 5} + \ln(z + 5) \ln 3\right]$ **23.** $\dfrac{2^{x^2-1}}{\ln 2} + C$

25. $\dfrac{40}{\ln 5}$ **27.** $10^{x^2} 2x \ln 10 + 20x^{19}$

29. $(\pi + 1)x^\pi + (\pi + 1)^x \ln(\pi + 1)$

31. $(x^2 + 1)^{\ln x}\left(\dfrac{\ln(x^2 + 1)}{x} + \dfrac{2x \ln x}{x^2 + 1}\right)$ **33.** sin 1

35. Domain $= (-\infty, \infty)$; decreasing on $(-\infty, \infty)$; concave upward on $(-\infty, \infty)$; no extreme values or points of inflection.

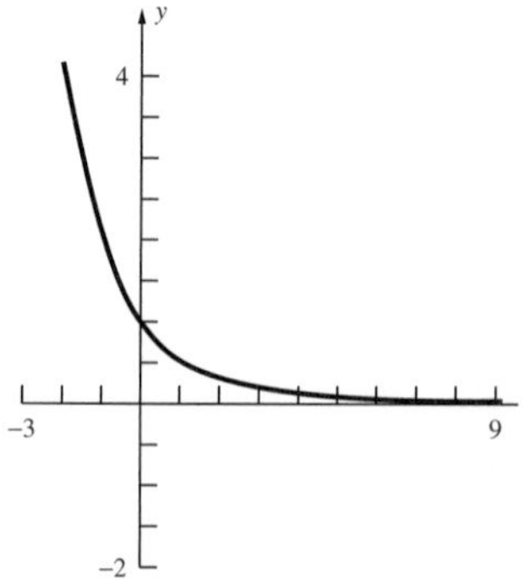

37. Domain $= (-\infty, \infty)$; increasing on $(0, \infty)$ and decreasing on $(-\infty, 0)$; concave up on $(-1, 1)$ and concave down on

$(-\infty, -1) \cup (1, \infty)$; minimum at $(0, 0)$; points of inflection at $(-1, 1)$ and $(1, 1)$

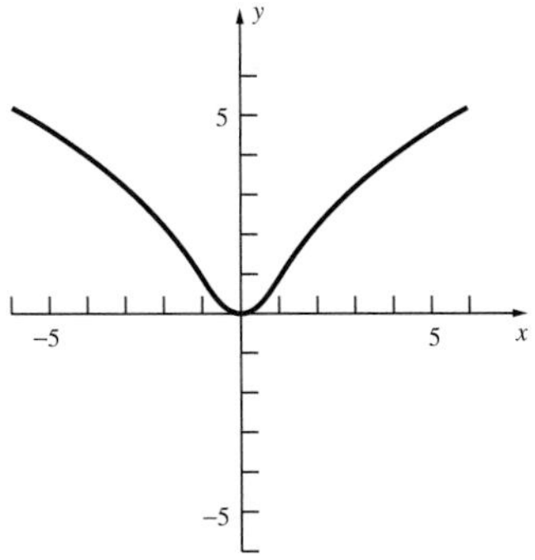

39. Domain $= (-\infty, \infty)$; increasing on $(-\infty, \infty)$; concave up on $(-\infty, 0)$ and concave down on $(0, \infty)$; no extreme values; point of inflection at $\left(0, \int_1^0 2^{-t^2}\,dt\right) \approx (0, -0.81)$

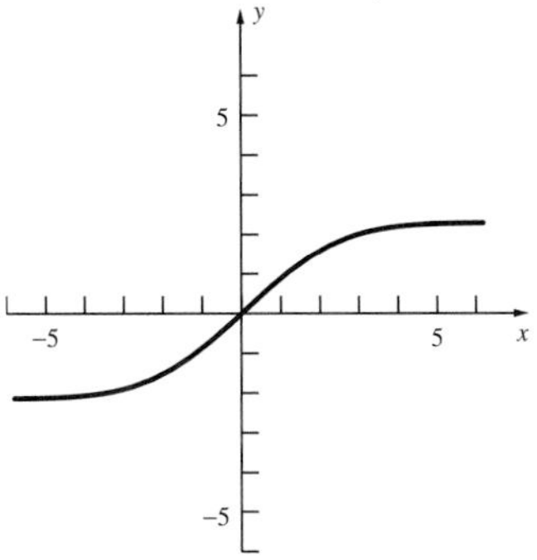

41. $\log_{1/2} x = -\log_2 x$

43. $E \approx 5.017 \times 10^8$ kW-h for magnitude 7;
$E \approx 1.560 \times 10^{10}$ kW-h for magnitude 8

45. $r = 2^{1/12} \approx 1.0595$; frequency of $\overline{C} = 440\sqrt[4]{2} \approx 523.25$

47. If $y = A \cdot b^x$, then $\ln y = \ln A + x \ln b$, so the $\ln y$ vs. x plot will be linear. If $y = C \cdot x^d$, then $\ln y = \ln C + d \ln x$, so the $\ln y$ vs $\ln x$ plot will be linear.

49. $f'(x) = x^{(x^2)}(2x \ln x + x)$

$g'(x) = x^{x^x + x}\left[\ln x + (\ln x)^2 + \dfrac{1}{x}\right]$

53. $\lim_{x \to 0^+} x^x = 1$; minimum: $(e^{-1}, e^{-1/e})$ **55.** 20.2259

57. $b \approx 2^{5/2}, C \approx 2^{3/2}$

Problem Set 6.5

1. $y = 4e^{-6t}$ **3.** $y = 2e^{0.005(t-10)}$ **5.** 56,569

7. 15.8 days **9.** 4.64 million; 4.79 million; 6.17 million; 105 million

11. 126,822 **13.** 7.43 g

15. $t_c \approx 201$ yrs(2187) $t_s \approx 191$ yrs(2177)

17. 2950 years ago **19.** 81.6ºF **21.** 83.7ºC **23.** 8:45 pm

25. (a) \$401.71 **(b)** \$402.15 **(c)** \$402.19 **(d)** \$402.19

27. (a) 11.58 yrs. **(b)** 11.55 yrs.

29. \$133.6 billion **31.** \$1051.27 **33.** $t = \dfrac{100 \ln 2}{p}$

35.

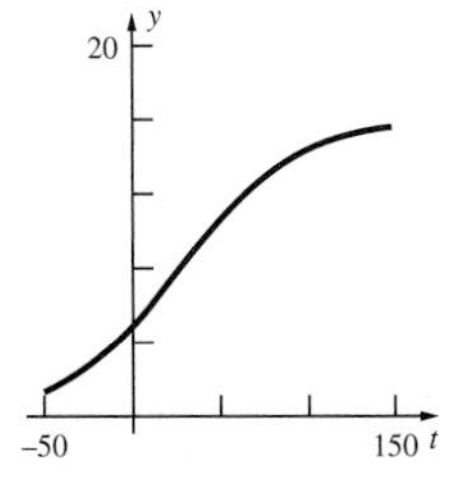

37. (a) $\frac{1}{e}$; **(b)** e^3; **(c)** e^2; **(d)** $\frac{1}{e^2}$

39. 15.25 million **45.** 75.25 years from 2004

47. (a) $k = 0.0132 - 0.0001t$ **(b)** $y' = (0.0132 - 0.0001t)y$
(c) $y = 6.4^{0.0132t - 0.00005t^2}$

(d)

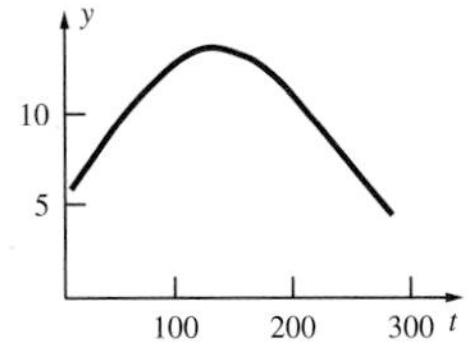

(e) The maximum population will occur when $t = 132$, which is year 2136. The model predicts that the population will return to the 2004 level in year 2268.

49.

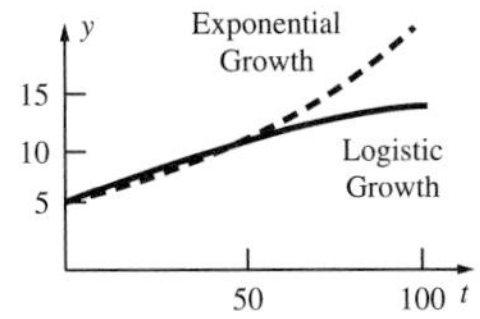

Exponential growth: 6.93 billion in 2010; 10.29 billion in 2040; 19.92 billion in 2090:
Logistic growth: 7.13 billion in 2010; 10.90 billion in 2040; 15.15 billion in 2090

Problem Set 6.6

1. $y = e^{-x}(x + C)$ **3.** $y = a + C(1 - x^2)^{1/2}$

5. $y = xe^x + Cx$ **7.** $y = 1 + Cx^{-1}$

9. $y = 1 + Ce^{-\int f(x)\,dx}$ **11.** $y = x^4 + 2x$ goes through $(1, 3)$.

13. $y = e^{-x}(1 - x^{-1})$ goes through $(1, 0)$. **15.** 38.506 lb.

17. $y(t) = 2(60 - t) - \left(\dfrac{1}{1800}\right)(60 - t)^3$

19. $I(t) = 10^{-6}(1 - \exp(-10^6 t))$

21. $I(t) = 0.12 \sin 377t$

23. (a) 21.97 min **(b)** 26.67 min **(c)** $c > 7.7170$
(d) $400e^{-0.04T} + T = 150$.

25. (a) 200.32 ft **(b)** $95 - 4T - 95e^{-0.05T} = 0$

Problem Set 6.7

1. $\lim_{t \to \infty} y(t) = 12$ and $y(2) \approx 10.5$

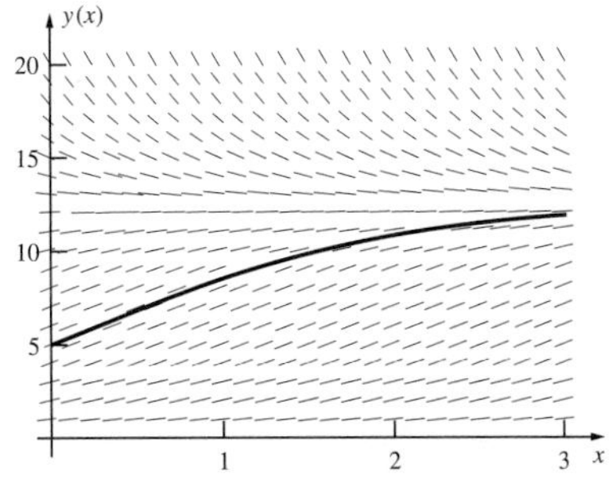

3. $\lim_{t\to\infty} y(t) = 0$ and $y(2) \approx 6$

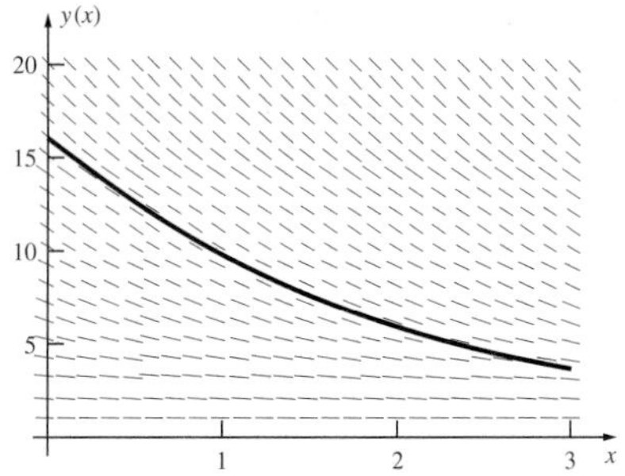

5. The oblique asymptote is $y = x$.

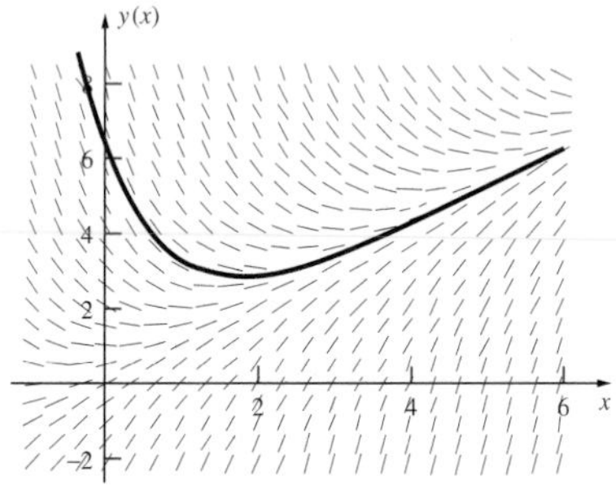

7. $y = \frac{1}{2}e^{x/2}$

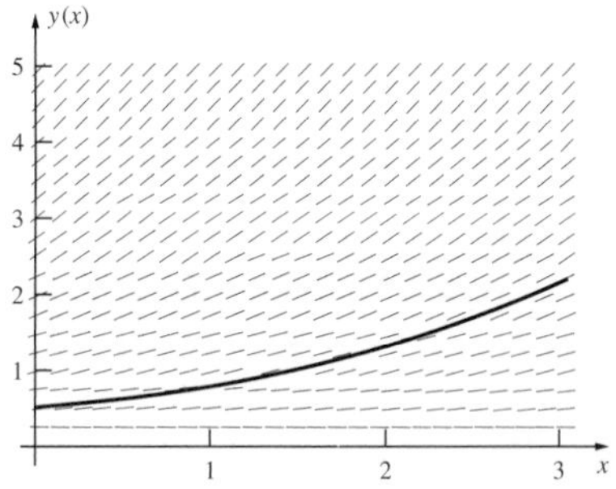

9. $y = x + 1 + 3e^{-x}$.

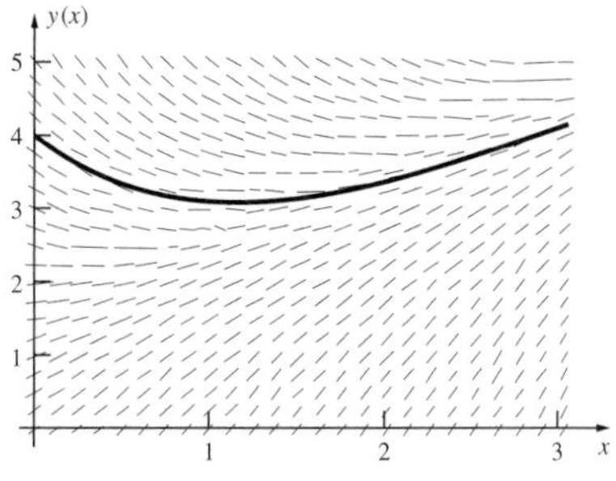

11.

x_n	Euler's Method y_n
0.0	3.0
0.2	4.2
0.4	5.88
0.6	8.232
0.8	11.5248
1.0	16.1347

13.

x_n	Euler's Method y_n
0.0	0.0
0.2	0.0
0.4	0.04
0.6	0.12
0.8	0.24
1.0	0.40

15.

x_n	Euler's Method y_n
1.0	1.0
1.2	1.2
1.4	1.488
1.6	1.90464
1.8	2.51412
2.0	3.41921

19. (a) $y(x_1) \approx 0$ **(b)** $y(x_2) \approx 0.00099998$
(c) $y(x_{10}) \approx 0.269097$

21. (a) $\Delta y = \frac{1}{2}[f(x_0, y_0) + f(x_1, \hat{y}_1)]$
(c) $x_n = x_{n-1} + h$
$\hat{y}_n = y_{n-1} + h \cdot f(x_{n-1}, y_{n-1})$
$y_n = y_{n-1} + \frac{h}{2}[f(x_{n-1}, y_{n-1}) + f(x_n, \hat{y}_n)]$

23.

x_n	y_n
0.0	2.0
0.2	1.64
0.4	1.3448
0.6	1.10274
0.8	0.90424
1.0	0.74148

25.

x_n	y_n
0.0	0.0
0.2	0.004
0.4	0.024
0.6	0.076
0.8	0.176
1.0	0.340

27.

x_n	y_n
1.0	2.0
1.2	1.312
1.4	0.80609
1.6	0.46689
1.8	0.25698
2.0	0.13568

Problem Set 6.8

1. $\frac{\pi}{4}$ **3.** $-\frac{\pi}{3}$ **5.** $\frac{\pi}{3}$ **7.** $-\frac{\pi}{6}$ **9.** 0.4567 **11.** 0.1115
13. 0.9548 **15.** 2.038 **17.** 0.6259 **19.** $\theta = \sin^{-1}\frac{x}{8}$
21. $\theta = \sin^{-1}\frac{5}{x}$ **23.** $\theta = \tan^{-1}\frac{3}{x} - \tan^{-1}\frac{1}{x}$ **25.** $\frac{1}{9}$ **27.** $\frac{56}{65}$
33. (a) $\frac{\pi}{2}$; **(b)** $-\frac{\pi}{2}$
35. (a) $\frac{\pi}{2}$; **(b)** $-\frac{\pi}{2}$
37. The tangent lines approach the vertical
39. $\frac{\cos x}{2 + \sin x}$ **41.** $\sec x$ **43.** $\frac{4x}{\sqrt{1 - 4x^4}}$

45. $x^2\left[\dfrac{xe^x}{1+e^{2x}} + 3\tan^{-1}(e^x)\right]$ **47.** $\dfrac{3(\tan^{-1}x)^2}{1+x^2}$

49. $\dfrac{3}{|x|\sqrt{x^6-1}}$ **51.** $\dfrac{3(1+\sin^{-1}x)^2}{\sqrt{1-x^2}}$ **53.** $\dfrac{2}{x[1+(\ln x^2)^2]}$

55. $\frac{1}{3}\sin 3x + C$ **57.** $\frac{1}{4}\sin^2 2x + C$ **59.** $\dfrac{\sin e^2 - \sin 1}{2}$

61. $\dfrac{\pi}{4}$ **63.** $\dfrac{\pi}{2}$ **65.** $\frac{1}{2}\arctan 2x + C$

67. $\dfrac{1}{3}\sin^{-1}\left(\dfrac{\sqrt{3}}{2}x\right) + C$ **69.** $\dfrac{1}{2}\tan^{-1}\left(\dfrac{x-3}{2}\right) + C$

71. $\dfrac{1}{3}\sec^{-1}\left(\dfrac{2|x|}{3}\right) + C$

73. $\theta = \tan^{-1}\dfrac{7.6}{b} - \tan^{-1}\dfrac{2.6}{b}$; if $b = 12.9$, $\theta \approx 0.3335$

77. $\pi b^2 - b^2\cos^{-1}\dfrac{b}{2a} - 2a^2\sin^{-1}\dfrac{b}{2a} + \dfrac{1}{2}b\sqrt{4a^2-b^2}$

79.

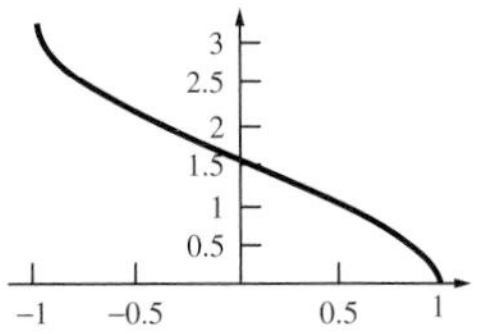

$\dfrac{\pi}{2} - \arcsin x = \arccos x$

87. 4.9 ft **89.** $\frac{1}{13}$ rad/s **91.** 1 rev/min

93. 3.96×10^{-4} rad/s

Problem Set 6.9

13. $2\sinh x\cosh x = \sinh 2x$

15. $10\sinh x\cosh x = 5\sinh 2x$ **17.** $3\sinh(3x+1)$

19. $\coth x$ **21.** $x^2\sinh x + 2x\cosh x$

23. $\cosh 3x\cosh x + 3\sinh 3x\sinh x$

25. $2\tanh x\cosh 2x + \sinh 2x\operatorname{sech}^2 x$

27. $\dfrac{2x}{\sqrt{x^4+1}}$ **29.** $-\dfrac{1}{2(x^2-3x+2)}$

31. $\dfrac{3x}{\sqrt{9x^2-1}} + \cosh^{-1}3x$ **33.** $\dfrac{1}{\sqrt{x^2-1}\cosh^{-1}x}$

35. $-\csc^2 x\operatorname{sech}^2(\cot x)$ **37.** $\frac{20}{9}$

39. $\frac{1}{2\pi}\sinh(\pi x^2+5) + C$ **41.** $2\cosh(2z^{1/4}) + C$

43. $\cosh(\sin x) + C$ **45.** $\frac{1}{4}[\ln(\sinh x^2)]^2 + C$ **47.** $\frac{1}{4}$

49. $\frac{\pi}{2} + \frac{\pi}{4}\sinh 2$ **51.** $\pi + \frac{\pi}{2}\sinh 2$

55. (a)

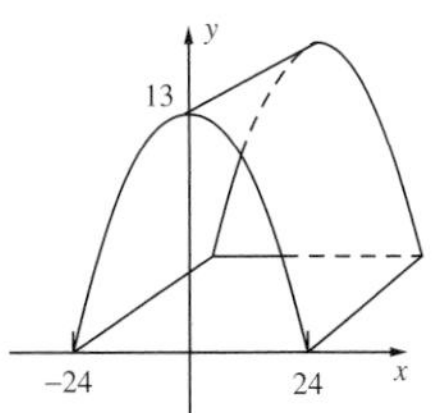

(b) 42,200 ft^3; **(c)** 5640 ft^2

61.

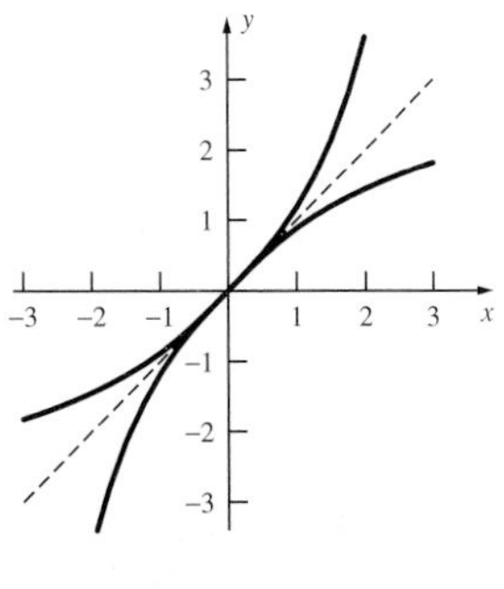

$y = \sinh x$ and $y = \ln\left(x + \sqrt{x^2+1}\right)$ are inverse functions.

Chapter Review 6.10

Concepts Test

1. False **3.** True **5.** True **7.** False **9.** True
11. True **13.** True **15.** True **17.** False **19.** True
21. False **23.** True **25.** False **27.** False **29.** False
31. True **33.** False **35.** True **37.** True **39.** True
41. True **43.** True **45.** False **47.** True

Sample Test Problems

1. $\dfrac{4}{x}$ **3.** $(2x-4)e^{x^2-4x}$ **5.** $\sec^2 x$ **7.** $\dfrac{\operatorname{sech}^2\sqrt{x}}{\sqrt{x}}$

9. $|\sec x|$ **11.** $\dfrac{1}{\sqrt{e^{2x}-1}}$ **13.** $\dfrac{15e^{5x}}{e^{5x}+1}$

15. $-\dfrac{e^{\sqrt{x}}\sin e^{\sqrt{x}}}{2\sqrt{x}}$ **17.** $-\dfrac{1}{\sqrt{x-x^2}}$ **19.** $-\dfrac{\csc\sqrt{x}\cot\sqrt{x}}{\sqrt{x}}$

21. $20\sec 5x(2\sec^2 5x - 1)$ **23.** $x^{1+x}\left(\ln x + 1 + \frac{1}{x}\right)$

25. $\frac{1}{3}e^{3x-1} + C$ **27.** $-\cos e^x + C$ **29.** $\dfrac{\ln(e^{x+3}+1)}{e} + C$

31. $2\sin^{-1}2x + C$ **33.** $-\tan^{-1}(\ln x) + C$

35. Increasing: $[-\frac{\pi}{2}, \frac{\pi}{4}]$; decreasing: $[\frac{\pi}{4}, \frac{\pi}{2}]$; concave up: $(-\frac{\pi}{2}, -\frac{\pi}{4})$; concave down: $(-\frac{\pi}{4}, \frac{\pi}{2})$; inflection point: $(-\frac{\pi}{4}, 0)$; global minimum: $(-\frac{\pi}{2}, -1)$ global maximum: $\left(\frac{\pi}{4}, \sqrt{2}\right)$

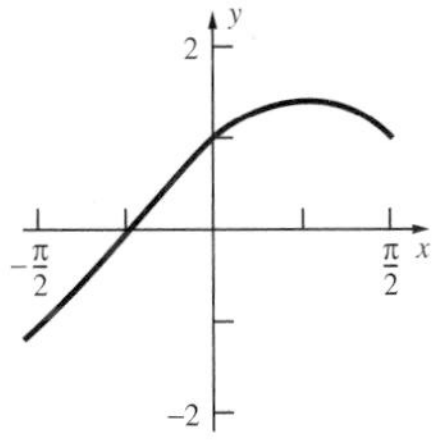

37. (b) 1 **(c)** $\frac{1}{15}$

39.

x_n	y_n
1.0	2.0
1.2	2.4
1.4	2.976
1.6	3.80928
1.8	5.02825
2.0	6.83842

41. $y = 1$ **43.** $y = Cx^{-1}$ **45.** $y = 1 + 2e^{-x^2}$

47. $y = -e^x + Ce^{2x}$

Chapter 7 Review and Preview Problems

1. $-\frac{1}{2}\cos 2x + C$ **3.** $-\frac{1}{2}\cos x^2 + C$ **5.** $\ln|\sec t| + C$

7. $\frac{1}{3}(x^2+2)^{3/2} + C$ **9.** $\ln x$ **11.** $x^2\sin x$

13. $\sin^2 x = \dfrac{1-\cos 2x}{2}$ **15.** $\cos^4 x = \left(\dfrac{1+\cos 2x}{2}\right)^2$

17. $\cos 3x\cos 5x = \dfrac{\cos 8x + \cos 2x}{2}$ **19.** $|a|\cos t$

21. $|a|\cdot|\tan t|$ **23.** $\dfrac{2x-1}{x(1-x)}$ **25.** $\dfrac{5x+3}{x(x+1)(x-3)}$

Problem Set 7.1

1. $\frac{1}{6}(x-2)^6 + C$ **3.** 1302 **5.** $\frac{1}{2}\tan^{-1}\left(\frac{x}{2}\right) + C$

7. $\frac{1}{2}\ln(x^2+4) + C$ **9.** $2(4+z^2)^{3/2} + C$

11. $\frac{1}{2}\tan^2 z + C$ **13.** $-2\cos\sqrt{t} + C$ **15.** $\tan^{-1}\frac{\sqrt{2}}{2}$

17. $\frac{3}{2}x^2 - x + \ln|x + 1| + C$ **19.** $-\frac{1}{2}\cos(\ln 4x^2) + C$

21. $6\sin^{-1}(e^x) + C$ **23.** $-3\sqrt{1 - e^{2x}} + C$ **25.** $1/\ln 3$

27. $x - \ln|\sin x| + C$ **29.** $\ln|\sec e^x + \tan e^x| + C$

31. $\tan x + e^{\sin x} + C$ **33.** $-\frac{1}{3\sin(t^3 - 2)} + C$

35. $-\frac{1}{3}[\cot(t^3 - 2) + t^3] + C$ **37.** $\frac{1}{2}e^{\tan^{-1}2t} + C$

39. $\frac{1}{6}\sin^{-1}\left(\frac{3y^2}{4}\right) + C$ **41.** $\frac{1}{3}\cosh x^3 + C$

43. $\frac{1}{3}\sin^{-1}\left(\frac{e^{3t}}{2}\right) + C$ **45.** $\frac{1}{4}\tan^{-1}\left(\frac{1}{4}\right)$

47. $\frac{1}{2}\tan^{-1}\left(\frac{x + 1}{2}\right) + C$ **49.** $\frac{1}{3}\tan^{-1}(3x + 3) + C$

51. $\frac{1}{18}\ln|9x^2 + 18x + 10| + C$ **53.** $\frac{1}{3}\sec^{-1}\left(\frac{|\sqrt{2}t|}{3}\right) + C$

55. $\ln(\sqrt{2} + 1)$ **57.** π^2

Problem Set 7.2

1. $xe^x - e^x + C$ **3.** $\frac{1}{5}te^{5t+\pi} - \frac{1}{25}e^{5t+\pi} + C$

5. $x\sin x + \cos x + C$

7. $(t - 3)\sin(t - 3) + \cos(t - 3) + C$

9. $\frac{2}{3}t(t + 1)^{3/2} - \frac{4}{15}(t + 1)^{5/2} + C$ **11.** $x\ln 3x - x + C$

13. $x\arctan x - \frac{1}{2}\ln(1 + x^2) + C$ **15.** $-\frac{\ln x}{x} - \frac{1}{x} + C$

17. $\frac{2}{9}(e^{3/2} + 2)$ **19.** $\frac{1}{4}z^4\ln z - \frac{1}{16}z^4 + C$

21. $t\arctan\left(\frac{1}{t}\right) + \frac{1}{2}\ln(1 + t^2) + C$ **23.** $\frac{\pi}{2\sqrt{3}} + \ln 2$

25. $\frac{2}{9}x^3(x^3 + 4)^{3/2} - \frac{4}{45}(x^3 + 4)^{5/2} + C$

27. $\frac{t^4}{6(7 - 3t^4)^{1/2}} + \frac{1}{9}(7 - 3t^4)^{1/2} + C$

29. $\frac{z^4}{4(4 - z^4)} + \frac{1}{4}\ln|4 - z^4| + C$

31. $x\cosh x - \sinh x + C$

33. $\frac{x}{150}(3x + 10)^{50} - \frac{1}{22950}(3x + 10)^{51} + C$

35. $\frac{x}{\ln 2}2^x - \frac{1}{(\ln 2)^2}2^x + C$ **37.** $x^2e^x - 2xe^x + 2e^x + C$

39. $z\ln^2 z - 2z\ln z + 2z + C$ **41.** $\frac{1}{2}e^t(\sin t + \cos t) + C$

43. $x^2\sin x + 2x\cos x - 2\sin x + C$

45. $\frac{x}{2}[\sin(\ln x) - \cos(\ln x)] + C$

47. $x\ln^3 x - 3x\ln^2 x + 6x\ln x - 6x + C$ **65.** 1

67. $9 - \frac{9}{e^3} \approx 8.552$

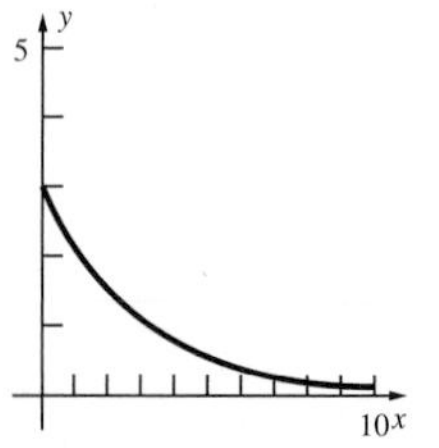

69. $\frac{\sqrt{2}\pi}{4} - 1$ **71.** $\bar{x} = \frac{e^2 + 1}{4}, \bar{y} = \frac{e - 2}{4}$

73. **(a)** $(x^3 - 2x)e^x - (3x^2 - 2)e^x + 6xe^x - 6e^x + C$

(b) $(x^2 - 3x + 1)(-\cos x) - (2x - 3)(-\sin x) + 2\cos x + C$

87. $e^x(3x^4 - 12x^3 + 38x^2 - 76x + 76)$

Problem Set 7.3

1. $\frac{1}{2}x - \frac{1}{4}\sin 2x + C$ **3.** $-\cos x + \frac{1}{3}\cos^3 x + C$ **5.** $\frac{8}{15}$

7. $-\frac{1}{12}\cos^3 4x + \frac{1}{10}\cos^5 4x - \frac{1}{28}\cos^7 4x + C$

9. $-\frac{1}{3}\csc 3\theta - \frac{1}{3}\sin 3\theta + C$

11. $\frac{3}{128}t - \frac{1}{384}\sin 12t + \frac{1}{3072}\sin 24t + C$

13. $\frac{1}{2}\cos y - \frac{1}{18}\cos 9y + C$

15. $\frac{1}{16}w - \frac{1}{32}\sin 2w - \frac{1}{24}\sin^3 w + C$

17. $\frac{1}{3}\left[-x\cos^3 x + \sin x - \frac{1}{3}\sin^3 x\right] + C$

19. $\frac{1}{3}\tan^3 x - \tan x + x + C$ **21.** $\frac{1}{2}\tan^2 x + \ln|\cos x| + C$

23. $\frac{1}{2}\tan^4\left(\frac{\theta}{2}\right) - \tan^2\left(\frac{\theta}{2}\right) - 2\ln\left|\cos\frac{\theta}{2}\right| + C$

25. $-\frac{1}{2}\tan^{-2} x + \ln|\tan x| + C$ **27.** $\frac{1}{4}\sec^4 x - \frac{1}{2}\sec^2 x + C$

29. 0 for $m \neq n$, since $\sin k\pi = 0$ for all integers k.

31. $\frac{\pi^4}{3} + \frac{5\pi^2}{2}$

Problem Set 7.4

1. $\frac{2}{5}(x + 1)^{5/2} - \frac{2}{3}(x + 1)^{3/2} + C$

3. $\frac{2}{27}(3t + 4)^{3/2} - \frac{8}{9}(3t + 4)^{1/2} + C$

5. $2\sqrt{2} - 2 - 2e\ln\left(\frac{\sqrt{2} + e}{1 + e}\right)$

7. $\frac{2}{63}(3t + 2)^{7/2} - \frac{4}{45}(3t + 2)^{5/2} + C$

9. $2\ln\left|\frac{2 - \sqrt{4 - x^2}}{x}\right| + \sqrt{4 - x^2} + C$

11. $\frac{x}{4\sqrt{x^2 + 4}} + C$ **13.** $-\frac{\sqrt{2}}{9} - \frac{1}{2}\sec^{-1}(-3) + \frac{\sqrt{3}}{8} + \frac{\pi}{3}$

15. $-2\sqrt{1 - z^2} - 3\sin^{-1} z + C$

17. $\ln|\sqrt{x^2 + 2x + 5} + x + 1| + C$

19. $3\sqrt{x^2 + 2x + 5} - 3\ln|\sqrt{x^2 + 2x + 5} + x + 1| + C$

21. $\frac{9}{2}\sin^{-1}\left(\frac{x + 2}{3}\right) + \frac{x + 2}{2}\sqrt{5 - 4x - x^2} + C$

23. $\sin^{-1}\left(\frac{x - 2}{2}\right) + C$

25. $\ln|x^2 + 2x + 2| - \tan^{-1}(x + 1) + C$

27. $\frac{\pi}{16}\left(\frac{1}{10} + \frac{\pi}{4} - \tan^{-1}\frac{1}{2}\right)$ **29.** $\frac{1}{2}\ln|x^2 + 9| + C$

31. $2\ln\left|\frac{2 - \sqrt{4 - x^2}}{x}\right| + \sqrt{4 - x^2} + C$

35. $\frac{dy}{dx} = -\frac{\sqrt{a^2 - x^2}}{x}, y = -\sqrt{a^2 - x^2} - a\ln\left|\frac{a - \sqrt{a^2 - x^2}}{x}\right|$

Problem Set 7.5

1. $\ln|x| - \ln|x + 1| + C$

3. $-\frac{3}{2}\ln|x + 1| + \frac{3}{2}\ln|x - 1| + C$

5. $3\ln|x + 4| - 2\ln|x - 1| + C$

7. $4\ln|x + 5| - \ln|x - 2| + C$

9. $2\ln|2x - 1| - \ln|x + 5| + C$

11. $\frac{5}{3}\ln|3x - 2| + 4\ln|x + 1| + C$

13. $2\ln|x| - \ln|x + 1| + \ln|x - 2| + C$

15. $\ln|2x - 1| - \ln|x + 3| + 3\ln|x - 2| + C$

17. $\frac{1}{2}x^2 - x + \frac{8}{3}\ln|x+2| + \frac{1}{3}\ln|x-1| + C$

19. $\frac{1}{2}x^2 - 2\ln|x| + 7\ln|x+2| + 7\ln|x-2| + C$

21. $\ln|x-3| - \frac{4}{x-3} + C$

23. $-\frac{3}{x+1} + \frac{1}{2(x+1)^2} + C$

25. $2\ln|x| + \ln|x-4| + \frac{1}{x-4} + C$

27. $-2\ln|x| + \frac{1}{2}\tan^{-1}\left(\frac{x}{2}\right) + 2\ln|x^2+4| + C$

29. $-2\ln|2x-1| + \frac{3}{2}\ln|x^2+9| + C$

31. $-\frac{2}{125}\ln|x-1| - \frac{1}{25(x-1)} + \frac{2}{125}\ln|x+4| - \frac{1}{25(x+4)} + C$

33. $\sin t - \frac{50}{13}\ln|\sin t + 3| - \frac{68}{13}\tan^{-1}(\sin t - 2) - \frac{41}{26}\ln|\sin^2 t - 4\sin t + 5| + C$

35. $\frac{1}{2}\ln|x^2+1| + \frac{5}{2(x^2+1)} + C$

37. $\frac{3}{2}\tan^{-1}\frac{x}{2} + \frac{2x-5}{2(x^2+4)} + C$

39. $\frac{1}{8}\ln\left(\frac{\sqrt{2}+1}{\sqrt{2}-1}\right) + \frac{1}{2}\tan^{-1}\frac{1}{\sqrt{2}} + \frac{1}{6\sqrt{2}}$

41. $y(t) = \frac{e^t}{1+e^t}$; $y(3) \approx 0.953$

43. $y(t) = \frac{8000e^{2.4t}}{7+e^{2.4t}}$; $y(3) \approx 7958.4$

45. $y(t) = \frac{Le^{KLt}}{\left(\frac{L-y_0}{y_0}\right) + e^{KLt}}$

47. If $y_0 < L$, then $y'(0) = Ky_0(L-y_0) > 0$ and the population is increasing initially.

49. **(a)** $y = \frac{16}{1+7e^{-\left(\frac{1}{50}\ln\frac{7}{3}\right)t}}$ **(b)** $y(90) \approx 6.34$ billion
(c) The population will be 9 billion in 2055.

51. **(a)** $x(t) = \frac{ab(1-e^{(a-b)kt})}{b-ae^{(a-b)kt}}$ **(c)** 1.65 grams
(d) $x(t) = a\left(\frac{akt}{akt+1}\right)$

53. $y(t) = \frac{ACe^{(A+B)kt} - B}{1+Ce^{(A+B)kt}}$

Problem Set 7.6

1. $-\frac{1}{5}e^{-5x}\left(\frac{1}{5}+x\right) + C$ **3.** $\frac{1}{2}[\ln 2]^2$

5. $\frac{1}{64}[24x + 8\sin 4x + \sin 8x] + C$

7. $\frac{1}{2}\left(\ln\frac{2}{3} - \ln\frac{3}{5}\right) \approx 0.0527$ **9.** $\frac{2}{15}\left[77\sqrt{7} + 8\sqrt{2}\right] \approx 28.67$

11. 0

13. **(a)** $\frac{2}{135}(9x-2)(3x+1)^{3/2} + C$
(b) $\frac{2}{135}(9e^x-2)(3e^x+1)^{3/2} + C$

15. **(a)** $\frac{1}{24}\ln\left|\frac{4x+3}{4x-3}\right| + C$ **(b)** $\frac{1}{24}\ln\left|\frac{4e^x+3}{4e^x-3}\right| + C$

17. **(a)** $\frac{1}{16}\left[x(4x^2-9)\sqrt{9-2x^2} + \frac{81}{\sqrt{2}}\sin^{-1}\left(\frac{\sqrt{2}x}{3}\right)\right] + C$
(b) $\frac{1}{16}\left[\sin x(4\sin^2 x - 9)\sqrt{9-2\sin^2 x} + \frac{81}{\sqrt{2}}\sin^{-1}\left(\frac{\sqrt{2}\sin x}{3}\right)\right] + C$

19. **(a)** $\frac{\sqrt{3}}{3}\ln|\sqrt{3}x + \sqrt{5+3x^2}| + C$
(b) $\frac{\sqrt{3}}{6}\ln|\sqrt{3}x^2 + \sqrt{5+3x^4}| + C$

21. **(a)** $\ln|(t+1) + \sqrt{t^2+2t-3}| + C$
(b) $\ln\left|\left(t+\frac{3}{2}\right) + \sqrt{t^2+3t-5}\right| + C$

23. **(a)** $\frac{2}{27}(3y-10)\sqrt{3y+5} + C$
(b) $\frac{2}{27}(3\sin t - 10)\sqrt{3\sin t + 5} + C$

25. $\frac{1}{12}(\sinh 6t - 6t) + C$

27. $\frac{1}{3}(1-\cos t)\sqrt{2\cos t + 1} + C$

29. $-\frac{2}{5}\sqrt{\cos t + 1}\left[\cos^2 t - \frac{4}{3}(\cos t - 2)\right] + C$

31. $\pi - 2 \approx 1.14159$ **33.** $\frac{231\pi}{2048} \approx 0.35435$

35. 0.11083 **37.** 1.10577 **39.** $4\ln 2 + 2 \approx 4.77259$

41. $e - 1 \approx 1.71828$ **43.** $e - 1 \approx 1.71828$

45. $c \approx 0.59601$ **47.** $c \approx 0.16668$ **49.** $c \approx 9.2365$

51. $\bar{x} = \frac{8}{3(c+1)}$; $c = \frac{1}{3}$

53. $\bar{x} = \frac{cu}{u+18} + 3$ where $u = -18e^{-c/3}$; $c \approx 5.7114$

55. **(a)** $\frac{2}{\sqrt{\pi}}e^{-x^2}$ **(b)** $\frac{\sin x}{x}$

57. **(a)** erf(x) is increasing on $(0, \infty)$.
(b) erf(x) is not concave up on $(0, \infty)$.

59. **(a)** $C(x)$ is increasing on $(0,1) \cup (\sqrt{3}, 2)$.
(b) $C(x)$ is concave up on $(\sqrt{2}, 2)$.

Chapter Review 7.7

Concepts Test

1. True **3.** False **5.** True **7.** True **9.** True
11. False **13.** True **15.** True **17.** False **19.** True
21. False **23.** True **25.** False **27.** True

Sample Test Problems

1. 2 **3.** $e - 1$ **5.** $\frac{1}{3}y^3 - \frac{1}{2}y^2 + 2y - 2\ln|1+y| + C$

7. $\frac{1}{2}\ln|y^2 - 4y + 2| + C$ **9.** $e^t + 2\ln|e^t - 2| + C$

11. $\frac{1}{\sqrt{2}}\sin^{-1}\left(\frac{x-1}{3}\right) + C$ **13.** $\frac{1}{\sqrt{3}}\ln\left|\sqrt{y^2+\frac{2}{3}} + y\right| + C$

15. $-\ln|\ln|\cos x|| + C$ **17.** $\cosh x + C$

19. $-x\cot x - \frac{1}{2}x^2 + \ln|\sin x| + C$ **21.** $\frac{1}{4}[\ln(t^2)]^2 + C$

23. $-\frac{3}{82}e^{t/3}(9\cos 3t - \sin 3t) + C$

25. $-\frac{1}{2}\cos x - \frac{1}{4}\cos 2x + C$ **27.** $\frac{1}{6}\sec^3(2x) - \frac{1}{2}\sec(2x) + C$

29. $\frac{2}{5}\tan^{5/2} x + \frac{2}{9}\tan^{9/2} x + C$ **31.** $-\sqrt{9 - e^{2y}} + C$

33. $3\sin x + C$ **35.** $\frac{1}{4}\tan^{-1}(e^{4x}) + C$

37. $\frac{2}{3}(w+5)^{3/2} - 10(w+5)^{1/2} + C$

39. $-\frac{1}{6}\tan^{-1}\left(\frac{\cos^2 y}{3}\right) + C$

41. $\ln|x| - \dfrac{2}{x} - \dfrac{1}{2}\ln|x^2+3| + \dfrac{2}{\sqrt{3}}\tan^{-1}\left(\dfrac{x}{\sqrt{3}}\right) + C$

43. (a) $\dfrac{A}{2x+1} + \dfrac{B}{(2x+1)^2} + \dfrac{C}{(2x+1)^3}$

(b) $\dfrac{A}{x-1} + \dfrac{B}{(x-1)^2} + \dfrac{C}{2-x} + \dfrac{D}{(2-x)^2} + \dfrac{E}{(2-x)^3}$

(c) $\dfrac{Ax+B}{x^2+x+10} + \dfrac{Cx+D}{(x^2+x+10)^2}$

(d) $\dfrac{A}{1-x} + \dfrac{B}{(1-x)^2} + \dfrac{C}{1+x} + \dfrac{D}{(1+x)^2}$
$+ \dfrac{Ex+F}{x^2-x+10} + \dfrac{Gx+H}{(x^2-x+10)^2}$

(e) $\dfrac{A}{x+3} + \dfrac{B}{(x+3)^2} + \dfrac{C}{(x+3)^3} + \dfrac{D}{(x+3)^4}$
$+ \dfrac{Ex+F}{x^2+2x+10} + \dfrac{Gx+H}{(x^2+2x+10)^2}$

(f) $\dfrac{Ax+B}{2x^2+x+10} + \dfrac{Cx+D}{(2x^2+x+10)^2} + \dfrac{Ex+F}{(2x^2+x+10)^3}$

45. $\sqrt{5} + 4\ln\left(\dfrac{1+\sqrt{5}}{2}\right)$

47. $2\pi\ln\frac{32}{25}$ **49.** $4\pi[2 - \ln 3 - \frac{1}{2}(\ln 3)^2]$ **51.** $\ln 7 - \frac{6}{7}$

53. $\ln\left(\dfrac{2\sqrt{3}+3}{3}\right)$

55. (a) $\dfrac{\sin x}{2}\sqrt{\sin^2 x + 4} + 2\ln\left|\sin x + \sqrt{\sin^2 x + 4}\right| + C$

(b) $\dfrac{1}{4}\ln\left|\dfrac{1+2x}{1-2x}\right| + C$

57. $c \approx 0.5165$

Chapter 8 Review and Preview Problems

1. $\dfrac{5}{3}$ **3.** 6 **5.** 2 **7.** 1 **9.** 0 **11.** ∞ **13.** $\dfrac{\pi}{2}$

15.

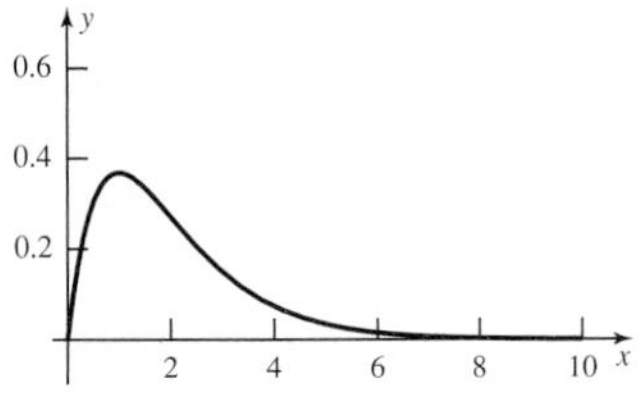

$\lim_{x\to\infty} xe^{-x} = 0.$

17.

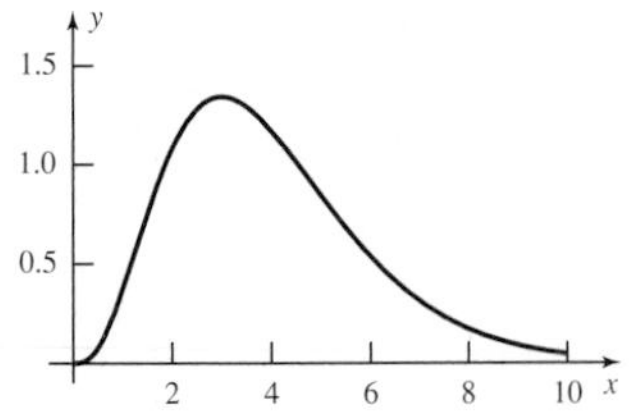

$\lim_{x\to\infty} x^3e^{-x} = 0.$

19.

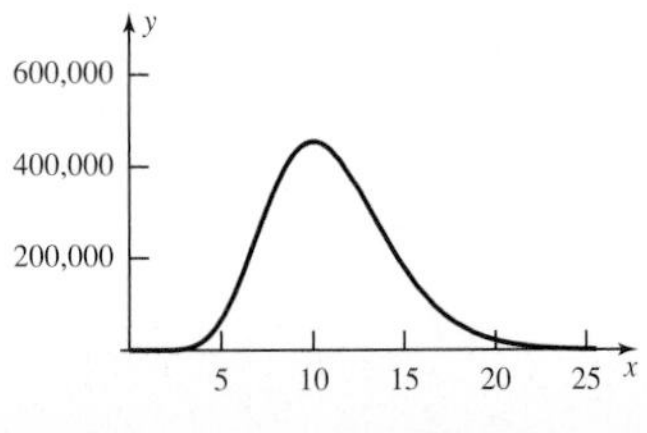

$\lim_{x\to\infty} x^{10}e^{-x} = 0.$

21.

a	1	2	4	8	16
$1 - e^{-a}$	0.632	0.865	0.982	0.99966	0.999999887

23.

a	1	2	4	8	16
$\ln\left(\sqrt{1+a^2}\right)$	0.3466	0.8047	1.4166	2.0872	2.7745

25.

a	2	4	8	16
$1 - \dfrac{1}{a}$	0.5	0.75	0.875	0.9375

27.

a	1	1/2	1/4	1/8	1/16
$4 - 2\sqrt{a}$	2	2.58579	3	3.29289	3.5

Problem Set 8.1

1. 1 **3.** -1 **5.** $-\frac{2}{7}$ **7.** $-\infty$ **9.** 0
11. $-\frac{3}{2}$ **13.** $-\frac{2}{7}$ **15.** $-\frac{1}{4}$ **17.** $-\infty$ **19.** $-\frac{1}{24}$
21. $-\infty$ **23.** 1
27. (a) $\frac{3}{4}$; **(b)** $\frac{1}{2}$
29. $c = 1$ **31.** $4\pi b^2$ **35.** $\frac{1}{24}$ **37.** 2

39. The ratio of the slopes is 1/2, indicating that the limit of the ratio should be about 1/2.

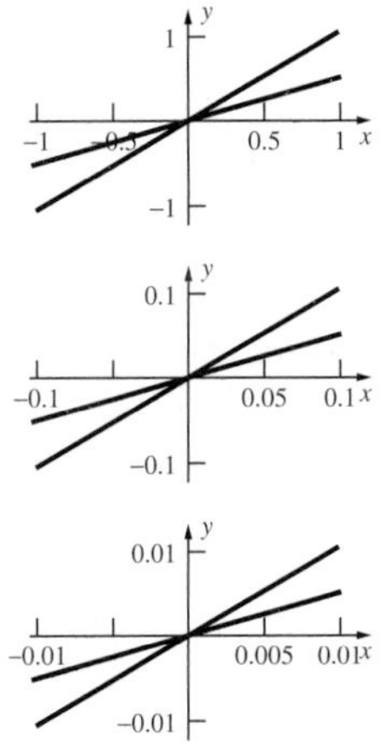

41. The ratio of the slopes is $-1/1 = -1$, indicating that the limit of the ratio should be about -1.

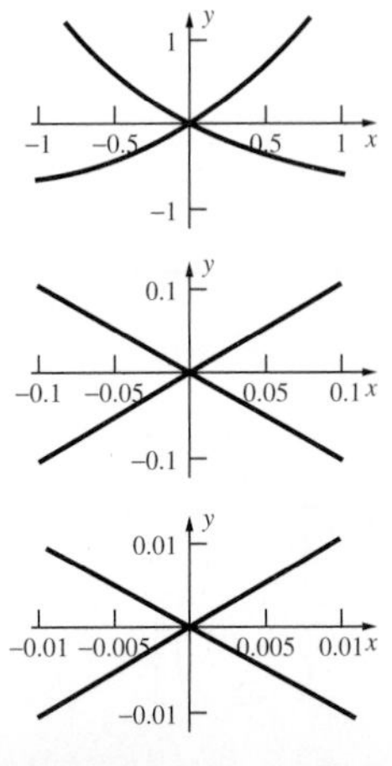

Problem Set 8.2

1. 0 **3.** 0 **5.** 3 **7.** 0 **9.** ∞ **11.** 0 **13.** 1 **15.** 1 **17.** 0 **19.** e^4 **21.** 1 **23.** 1 **25.** 0 **27.** 1 **29.** 0 **31.** ∞ **33.** 1 **35.** Limit does not exist. **37.** 0 **39.** 1

41. (a) 1; **(b)** 1; **(c)** $\ln a$; **(d)** ∞

43. As $x \to 0^+$, $y \to 0$. As $x \to \infty$, $y \to 1$. Maximum value $e^{1/e}$ at $x = e$.

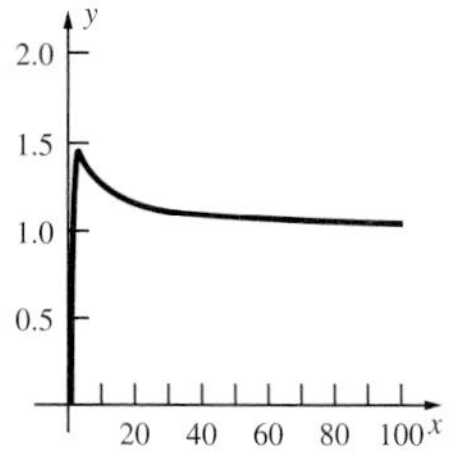

45. $1/(k + 1)$

47. (a) 3.162; **(b)** 4.163; **(c)** 4.562

49. No absolute minimum; absolute maximum at $x \approx 25$

Problem Set 8.3

1. Diverges **3.** $\frac{1}{e}$ **5.** Diverges **7.** 100,000 **9.** Diverges **11.** Diverges **13.** $\frac{1}{2}(\ln 2 + 1)$ **15.** $-\frac{1}{4}$ **17.** Diverges **19.** $\frac{\pi}{3}$ **21.** π **23.** $\frac{1}{2}$ **25.** $\frac{1}{2}\ln 3$ **29.** \$1,250,000

31. (b) $\mu = \dfrac{a + b}{2}$; $\sigma^2 = \dfrac{(b - a)^2}{12}$; **(c)** $\dfrac{1}{5}$

35. (a) $C = 3$ and $M = \dfrac{4 \times 10^4}{3}$; **(b)** $\sigma^2 = \dfrac{4 \times 10^8}{3}$

(c) $^6/_{25}$ of one percent earn over \$100,000.

41. $\displaystyle\int_1^{100} \frac{1}{x^2}\,dx = 0.99$; $\displaystyle\int_1^{100} \frac{1}{x^{1.1}}\,dx \approx 3.69$

$\displaystyle\int_1^{100} \frac{1}{x^{1.01}}\,dx \approx 4.50$; $\displaystyle\int_1^{100} \frac{1}{x}\,dx = \ln 100 \approx 4.61$;

$\displaystyle\int_1^{100} \frac{1}{x^{0.99}}\,dx \approx 4.71$

43. $\displaystyle\int_{-1}^{1} \frac{1}{\sqrt{2\pi}} \exp(-0.5x^2)\,dx \approx 0.6827$;

$\displaystyle\int_{-2}^{2} \frac{1}{\sqrt{2\pi}} \exp(-0.5x^2)\,dx \approx 0.9545$;

$\displaystyle\int_{-3}^{3} \frac{1}{\sqrt{2\pi}} \exp(-0.5x^2)\,dx \approx 0.9973$;

$\displaystyle\int_{-4}^{4} \frac{1}{\sqrt{2\pi}} \exp(-0.5x^2)\,dx \approx 0.9999$

Problem Set 8.4

1. $\dfrac{3}{\sqrt[3]{2}}$ **3.** $2\sqrt{7}$ **5.** $\frac{\pi}{2}$ **7.** Diverges **9.** $\frac{21}{2}$

11. $\frac{1}{2}(2^{2/3} - 10^{2/3})$ **13.** Diverges **15.** Diverges **17.** Diverges **19.** Diverges **21.** Diverges **23.** Diverges **25.** Diverges **27.** $2\sqrt{2}$ **29.** Diverges **31.** $\ln(2 + \sqrt{3})$ **35.** 0 **37.** Diverges **41.** 6 **43. (a)** 3 **45.** No **49.** Converges

55. (a) $C = \beta^\alpha/\Gamma(\alpha)$; **(b)** $\mu = \alpha/\beta$; **(c)** $\sigma^2 = \alpha/\beta^2$

57. (a) $\frac{\pi}{2}$; **(b)** π

Chapter Review 8.5

Concepts Test

1. True **3.** False **5.** False **7.** True **9.** True **11.** False **13.** True **15.** True **17.** False **19.** True **21.** True **23.** True **25.** False

Sample Test Problems

1. 4 **3.** 0 **5.** 2 **7.** 0 **9.** 0 **11.** 1 **13.** 0 **15.** 0 **17.** 1 **19.** 1 **21.** $\frac{1}{2}e^2$ **23.** Diverges **25.** $1 - \frac{\pi}{4}$ **27.** Diverges **29.** $\frac{1}{\ln 2}$ **31.** 6 **33.** Diverges **35.** $\frac{\pi}{4}$ **37.** 0

39. Converges: $p > 1$; diverges: $p \le 1$

41. Converges **43.** Diverges

Chapter 9 Review and Preview Problems

1. Original: If $x > 0$, then $x^2 > 0$. (AT)
Converse: If $x^2 > 0$, then $x > 0$.
Contrapositive: If $x^2 \le 0$, then $x \le 0$. (AT)

3. Original: f differentiable at $c \Rightarrow f$ continuous at c (AT)
Converse: f continuous at $c \Rightarrow f$ differentiable at c
Contrapositive: f discontinuous at $c \Rightarrow f$ non-differentiable at c (AT)

5. Original: f right continuous at $c \Rightarrow f$ continuous at c
Converse: f continuous at $c \Rightarrow f$ right continuous at c (AT)
Contrapositive: f discontinuous at $c \Rightarrow f$ not right continuous at c

7. Original: $f(x) = x^2 \Rightarrow f'(x) = 2x$ (AT)
Converse: $f'(x) = 2x \Rightarrow f(x) = x^2$
Contrapositive: $f'(x) \ne 2x \Rightarrow f(x) \ne x^2$ (AT)

9. $\dfrac{7}{4}$ **11.** $\dfrac{25}{12}$ **13.** $\dfrac{1}{2}$ **15.** 0 **17.** diverges **19.** converges **21.** diverges

Problem Set 9.1

1. $\frac{1}{3}$ **3.** 4 **5.** 1 **7.** Diverges **9.** 0 **11.** Diverges **13.** 0 **15.** 2 **17.** 0 **19.** e **21.** $a_n = \dfrac{n}{n + 1}$; 1

23. $a_n = (-1)^n \dfrac{n}{2n - 1}$; diverges **25.** $a_n = \dfrac{n}{2n - 1}$; $\dfrac{1}{2}$

27. $a_n = n \sin\dfrac{1}{n}$; 1 **29.** $a_n = \dfrac{2^n}{n^2}$; diverges **31.** $\frac{1}{2}, \frac{5}{4}, \frac{9}{8}, \frac{13}{16}$

33. $\frac{3}{4}, \frac{2}{3}, \frac{5}{8}, \frac{3}{5}$ **35.** $1, \frac{3}{2}, \frac{7}{4}, \frac{15}{8}$ **37.** 2.3028 **39.** $\frac{1}{2}(1 + \sqrt{13})$

41. 1.1118 **43.** $1 - \cos 1$ **51.** No **53.** $\dfrac{\pi}{2\sqrt{3}}$ **55.** $e^{1/2}$ **57.** e^{-2} **59.** e^{-1}

Problem Set 9.2

1. $\frac{1}{6}$ **3.** $\frac{31}{6}$ **5.** Diverges **7.** -1 **9.** Diverges

11. $\dfrac{e^2}{\pi(\pi - e)}$ **13.** 3 **15.** $\frac{2}{9}$ **17.** $\frac{13}{999}$ **19.** $\frac{1}{2}$

21. 1 **25.** 500 ft **27.** \$4 billion **29.** $\frac{1}{4}$ **31.** $\frac{4}{5}$; No

33. (a) Perimeter is infinite. **(b)** $A = \dfrac{8}{5}\left(\dfrac{81\sqrt{3}}{4}\right)$

35. $111\frac{1}{9}$ yd **37.** $\dfrac{3}{5}$ **39.** $\Pr(X = n) = \left(\dfrac{5}{6}\right)^{n-1}\left(\dfrac{1}{6}\right)$

43. (b) Indefinitely
47. (a) 2; **(b)** 1
49. (a) $\frac{Ce^{kt}}{e^{kt} - 1}$; **(b)** $\frac{8}{3}$mg
51. 1

Problem Set 9.3

1. Diverges **3.** Diverges **5.** Diverges **7.** Diverges
9. Converges **11.** Converges **13.** Diverges
15. Diverges **17.** Diverges **19.** Converges
21. Converges **23.** 0.0404 **25.** 0.1974 **27.** $n > 5000$
29. $n > 5000$ **31.** $n > 50$ **33.** $p > 1$
39. 272,404,866

Problem Set 9.4

1. Diverges **3.** Converges **5.** Converges
7. Diverges **9.** Converges **11.** Diverges; nth-Term Test
13. Converges; Limit Comparison Test
15. Converges; Ratio Test
17. Converges; Limit Comparison Test
19. Converges; Limit Comparison Test
21. Converges; Limit Comparison Test
23. Converges; Ratio Test
25. Converges; Integral Test
27. Diverges; nth Term Test
29. Converges; Comparison Test
31. Converges; Ratio Test
33. Converges; Ratio Test
43. (a) Diverges; **(b)** Converges; **(c)** Converges; **(d)** Converges; **(e)** Diverges **(f)** Converges
45. Converges for $p > 1$, diverges for $p \le 1$.

Problem Set 9.5

1. $|S - S_9| \le 0.065$ **3.** $|S - S_9| \le 0.417$
5. $|S - S_9| \le 0.230$ **13.** Conditionally convergent
15. Divergent **17.** Conditionally convergent
19. Absolutely convergent **21.** Conditionally convergent
23. Conditionally convergent **25.** Absolutely convergent
27. Conditionally convergent **29.** Divergent
35. (a) $1 + \frac{1}{3} \approx 1.33$; **(b)** $1 + \frac{1}{3} - \frac{1}{2} \approx 0.833$
45. ln 2

Problem Set 9.6

1. All x **3.** $-1 \le x \le 1$ **5.** $-1 \le x \le 1$
7. $1 < x \le 3$ **9.** $-1 \le x \le 1$ **11.** All x
13. $-1 < x < 1$ **15.** $-1 < x \le 1$ **17.** $-1 \le x \le 1$
19. $-2 < x < 2$ **21.** All x **23.** $0 \le x < 2$
25. $-3 < x < 1$ **27.** $-6 \le x \le -4$
29. If $\lim_{n\to\infty} \frac{x_0^n}{n!} \ne 0$, then $\sum \frac{x_0^n}{n!}$ will not converge.
31. $\sqrt{2}$ **33.** $\frac{1}{4 - x}$; $2 < x < 4$
35. (a) $-1 \le x < \frac{1}{3}$; **(b)** $-\frac{1}{2} < x \le \frac{7}{2}$
37. $S(x) = \frac{a_0 + a_1x + a_2x^2}{1 - x^3}$, $|x| < 1$

Problem Set 9.7

1. $1 - x + x^2 - x^3 + x^4 - x^5 + \cdots; 1$
3. $1 + 3x + 6x^2 + 10x^3 + \cdots; 1$
5. $\frac{1}{2} + \frac{3x}{4} + \frac{9x^2}{8} + \frac{27x^3}{16} + \cdots; \frac{2}{3}$
7. $x^2 + x^6 + x^{10} + x^{14} + \cdots; 1$
9. $\frac{x^2}{2} - \frac{x^3}{6} + \frac{x^4}{12} - \frac{x^5}{20} + \cdots; 1$
11. $2x + \frac{2x^3}{3} + \frac{2x^5}{5} + \cdots; 1$
13. $1 - x + \frac{x^2}{2!} - \frac{x^3}{3!} + \frac{x^4}{4!} - \frac{x^5}{5!} + \cdots$
15. $2 + \frac{2x^2}{2!} + \frac{2x^4}{4!} + \frac{2x^6}{6!} + \cdots$
17. $1 + \frac{x^2}{2} + \frac{x^3}{3} + \frac{3x^4}{8} + \frac{11x^5}{30} + \cdots$
19. $x - x^2 + \frac{x^3}{6} + \frac{x^4}{6} + \frac{3x^5}{40} + \cdots$
21. $x + \frac{2x^3}{3} + \frac{13x^5}{15} - \frac{29x^7}{105} + \cdots$
23. $x + \frac{x^3}{6} - \frac{x^4}{12} + \frac{3x^5}{40} - \cdots$
25. (a) $\frac{x}{1 + x}$; **(b)** $\frac{e^x - (1 + x)}{x^2}$; **(c)** $-\ln(1 - 2x)$
27. $\frac{x}{(1 - x)^2}$, $-1 < x < 1$
29. (a) $x + \frac{x^2}{2} - \frac{x^3}{6} - \cdots$; **(b)** $1 + x + x^2 + \frac{5x^3}{6} + \cdots$
31. $\frac{x}{2} + \frac{3x^2}{4} + \frac{7x^3}{8} + \cdots$ **33.** $\frac{x}{1 - x - x^2}$ **35.** 3.14159

Problem Set 9.8

1. $x + \frac{x^3}{3} + \frac{2x^5}{15}$ **3.** $x + x^2 + \frac{x^3}{3} - \frac{x^5}{30}$
5. $x - \frac{x^2}{2} - \frac{x^3}{6} + \frac{3x^5}{40}$ **7.** $1 + 3x + \frac{x^2}{2} + \frac{x^4}{24} + \frac{x^5}{60}$
9. $1 + x + \frac{3x^2}{2} + \frac{3x^3}{2} + \frac{37x^4}{24} + \frac{37x^5}{24}$ **11.** $1 - x + x^3 - x^4$
13. $x^3 - \frac{x^5}{2}$ **15.** $2x - \frac{x^3}{6} + \frac{61x^5}{120}$
17. $1 + \frac{3x}{2} + \frac{3x^2}{8} - \frac{x^3}{16} + \frac{3x^4}{128} - \frac{3x^5}{256}$
19. $e + e(x - 1) + \frac{e}{2}(x - 1)^2 + \frac{e}{6}(x - 1)^3$
21. $\frac{1}{2} - \frac{\sqrt{3}}{2}\left(x - \frac{\pi}{3}\right) - \frac{1}{4}\left(x - \frac{\pi}{3}\right)^2 + \frac{\sqrt{3}}{12}\left(x - \frac{\pi}{3}\right)^3$
23. $3 + 5(x - 1) + 4(x - 1)^2 + (x - 1)^3$
27. $x + \frac{x^3}{6} + \frac{3x^5}{40} + \frac{5x^7}{112}$ **29.** 0.9045
31. $1 - (x - 1) + (x - 1)^2 - (x - 1)^3 + \cdots$
33. (a) 25; **(b)** -3; **(c)** 0; **(d)** $4e$; **(e)** -4
35. $x - \frac{x^3}{3} + \frac{2x^5}{15}$ **41.** $x - \frac{x^3}{6} + \frac{x^5}{120} - \frac{x^7}{5040}$
43. $-2 + x - x^2 - \frac{5x^3}{6}$ **45.** $x + \frac{x^2}{2} - \frac{5x^4}{24} - \frac{23x^5}{120}$
47. $x + x^2 + \frac{x^3}{3} - \frac{x^5}{30}$

Problem Set 9.9

1. $1 + 2x + 2x^2 + \frac{4}{3}x^3 + \frac{2}{3}x^4$; 1.2712 **3.** $2x - \frac{4}{3}x^3$; 0.2377
5. $x - \frac{1}{2}x^2 + \frac{1}{3}x^3 - \frac{1}{4}x^4$; 0.1133 **7.** $x - \frac{1}{3}x^3$; 0.1194

9. $e + e(x-1) + \frac{e}{2}(x-1)^2 + \frac{e}{6}(x-1)^3$

11. $\frac{\sqrt{3}}{3} + \frac{4}{3}\left(x - \frac{\pi}{6}\right) + \frac{4\sqrt{3}}{9}\left(x - \frac{\pi}{6}\right)^2 + \frac{8}{9}\left(x - \frac{\pi}{6}\right)^3$

13. $\frac{\pi}{4} - \frac{1}{2}(x-1) + \frac{1}{4}(x-1)^2 - \frac{1}{12}(x-1)^3$

15. $7 + 2(x-1) + (x-1)^2 + (x-1)^3$

17. $f(x) \approx 1 + x + x^2 + x^3 + x^4$
(a) 1.1111; **(b)** 1.9375; **(c)** 4.0951; **(d)** 31

19.
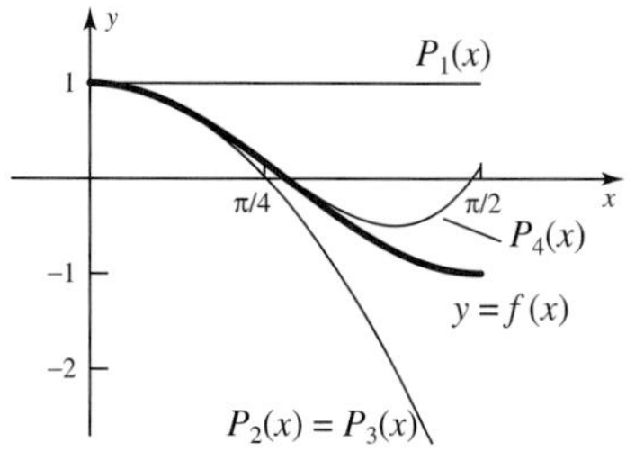

21.
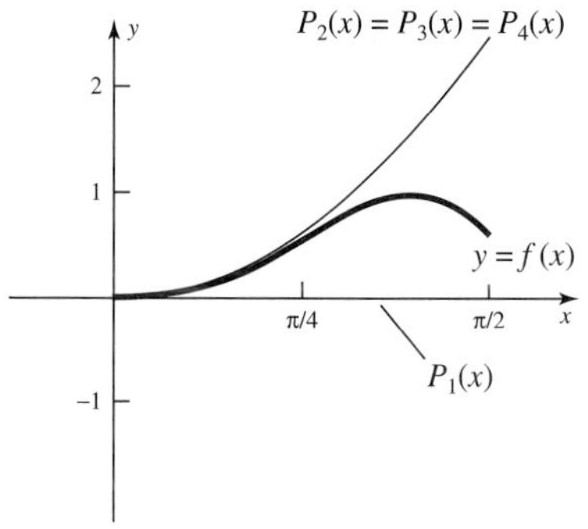

23.
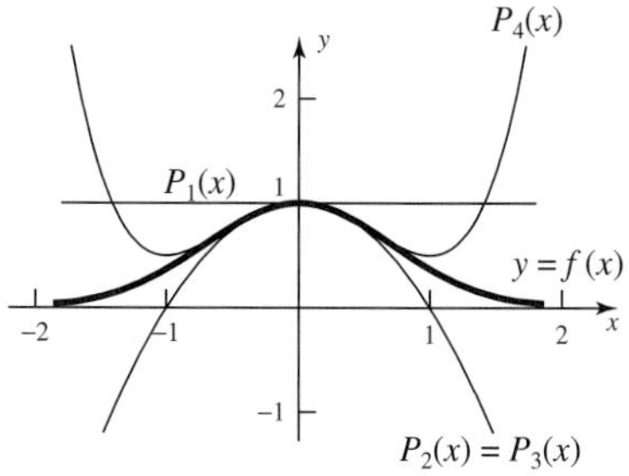

25.
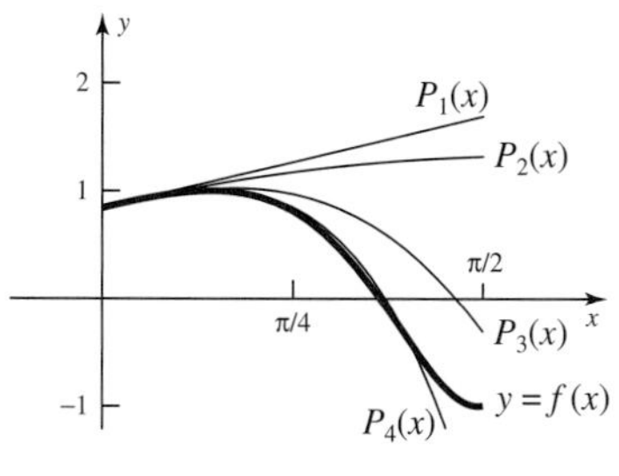

27.
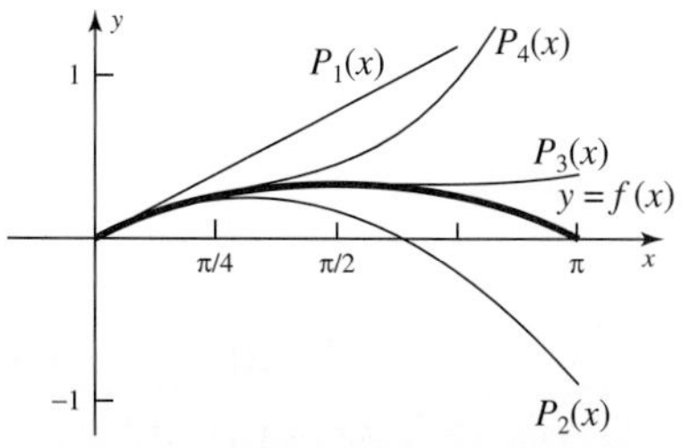

29. $e^6 + 1$ **31.** $2\sqrt{2}\pi$ **33.** $\frac{e^4}{3}$ **35.** $\frac{17}{10 \ln 2}$

37. $R_6(x) = \frac{x^7}{7(2+c)^7}$; 8.719×10^{-6}

39. $R_6(x) = -\frac{\cos c}{5040}\left(x - \frac{\pi}{4}\right)^7$; 2.685×10^{-8}

41. $R_6(x) = \frac{-(x-1)^7}{c^8}$; 2 **43.** $n \ge 9$

45. $1 + \frac{x}{2} - \frac{x^2}{8} + \frac{x^3}{16}$; $|R_3(x)| \le 0.0276$

47. $1 - \frac{1}{2}x + \frac{3}{8}x^2 - \frac{5}{16}x^3$; $|R_3(x)| \le 2.15 \times 10^{-6}$

49. 0.1224; $|\text{Error}| \le 0.00013025$ **51.** $n > 42$

53. $A = \frac{1}{2}tr^2 - \frac{1}{2}r^2 \sin t$; $A \approx \frac{1}{12}r^2t^3$

55. (c)

r	n (exact)	n (approx.)	n (rule 72)
0.05	13.892	13.889	14.4
0.10	6.960	6.959	7.2
0.15	4.650	4.649	4.8
0.20	3.495	3.494	3.6

57. $-1 - (x-1)^2 + (x-1)^3 + (x-1)^4$

59. 0.681998; $|R_3| \le 6.19 \times 10^{-8}$

Chapter Review 9.10

Concepts Test

1. False **3.** True **5.** False **7.** False **9.** True
11. True **13.** False **15.** True **17.** False **19.** True
21. True **23.** True **25.** True **27.** True **29.** True
31. True **33.** False **35.** True **37.** True **39.** True
41. True

Sample Test Problems

1. 3 **3.** e^4 **5.** 1 **7.** 0 **9.** 1 **11.** Diverges

13. $\frac{e^2}{e^2 - 1}$ **15.** $\frac{91}{99}$ **17.** cos 2 **19.** Diverges

21. Converges **23.** Converges **25.** Diverges
27. Converges **29.** Diverges **31.** Converges
33. Conditionally convergent **35.** Diverges

37. $-1 \le x \le 1$ **39.** $3 < x \le 5$ **41.** $1 < x < 5$

43. $1 - 2x + 3x^2 - 4x^3 + \cdots; -1 < x < 1$

45. $x^2 - \frac{x^4}{3} + \frac{2x^6}{45} - \frac{x^8}{315} + \cdots$; all x

47. $1 + x - \frac{x^2}{2!} - \frac{x^3}{3!} + \frac{x^4}{4!} + \frac{x^5}{5!} - \cdots$; all x **49.** $n > 3$

51. (a) $1 + x^3 + x^6$ **(b)** $1 + \frac{1}{2}x^2 - \frac{1}{8}x^4$ **(c)** $\frac{x^2}{2!} - \frac{x^3}{3!} + \frac{x^4}{4!}$
(d) $x + \frac{x^3}{2} + \frac{5x^5}{4!}$ **(e)** $x - x^2 + \frac{x^3}{3}$ **(f)** $1 - x + x^2$

53. $P(x) = x$; 0.2

55. $P_4(x) = 3 + 9(x-2) + 4(x-2)^2 + (x-2)^3$

57. $f(x) \approx \frac{1}{2} - \frac{1}{4}(x-1) + \frac{1}{8}(x-1)^2 - \frac{1}{16}(x-1)^3 + \frac{1}{32}(x-1)^4$

59. $\sin^2 x \approx x^2 - \frac{1}{3}x^4$; $|R_4(x)| < 2.85 \times 10^{-6}$

61. -0.00269867; Error $< 1.63 \times 10^{-5}$

Chapter 10 Review and Preview Problems

1. **(a)** $y = x - 1$ **(b)** $y = -x + 3$

3. $(2.4, 2.4), (-2.4, 2.4), (2.4, -2.4), (-2.4, -2.4)$

5. $y = 2\sqrt{3}x + 4$

7. $\left(5, 4\sqrt{3}\right), \left(-5, 4\sqrt{3}\right), \left(5, -4\sqrt{3}\right), \left(-5, -4\sqrt{3}\right)$;

T_1: $4\sqrt{3}x + 15y = 80\sqrt{3}$; T_2: $5\sqrt{3}x - 4y = 9\sqrt{3}$; $\alpha = 90°$

9. $r = 5$; $\theta = \sin^{-1}(0.6)$ **11.** $x = y = 4\sqrt{2}$

Problem Set 10.1

1. Focus at $(1, 0)$; directrix $x = -1$

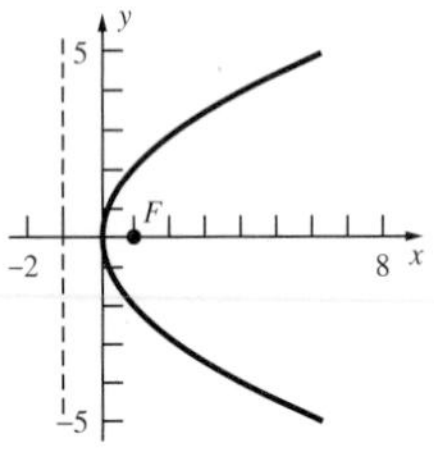

3. Focus at $(0, -3)$; directrix $y = 3$

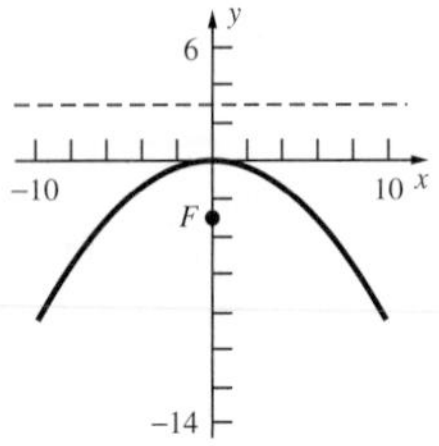

5. Focus at $\left(\frac{1}{4}, 0\right)$; directrix $x = -\frac{1}{4}$

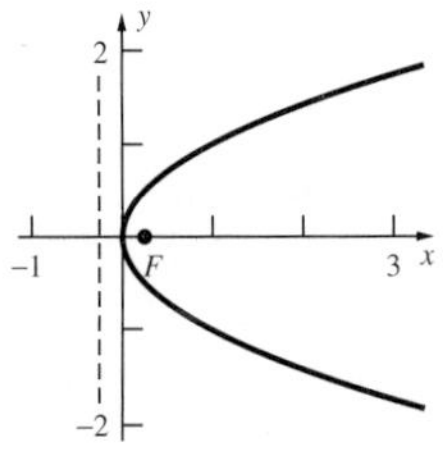

7. Focus at $\left(0, \frac{3}{4}\right)$; directrix $y = -\frac{3}{4}$

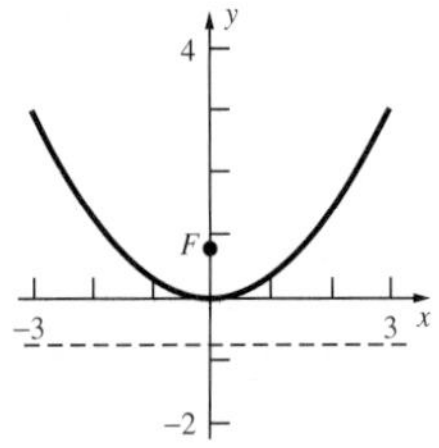

9. $y^2 = 8x$ **11.** $x^2 = -8y$ **13.** $y^2 = -16x$

15. $y^2 = \frac{1}{3}x$

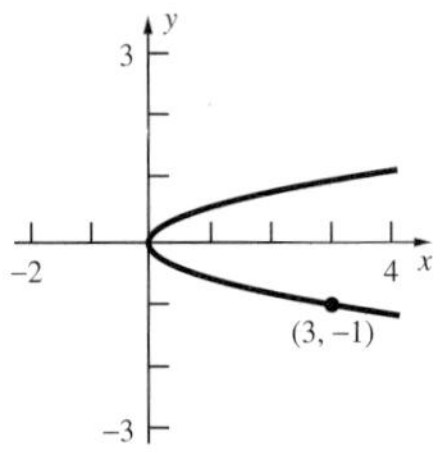

17. $x^2 = -\frac{36}{5}y$

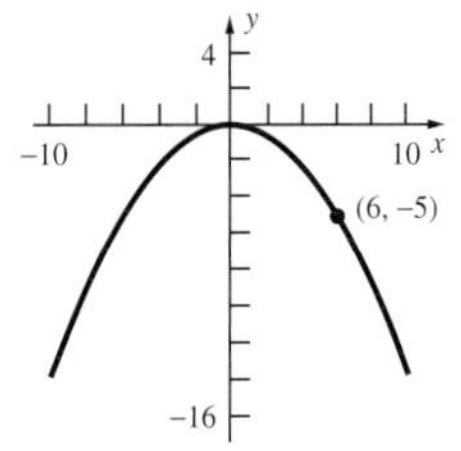

19. $y = -2x - 2$; $y = \frac{1}{2}x - \frac{9}{2}$

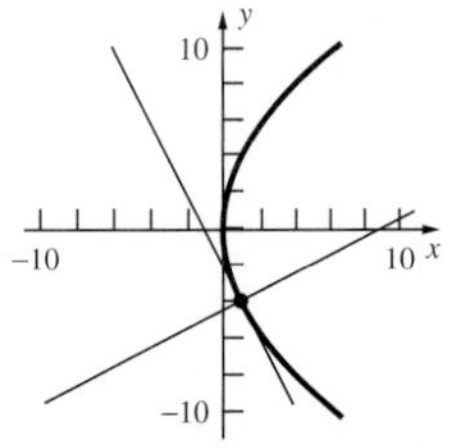

21. $y = 4x - 8$; $y = -\frac{1}{4}x + 9$

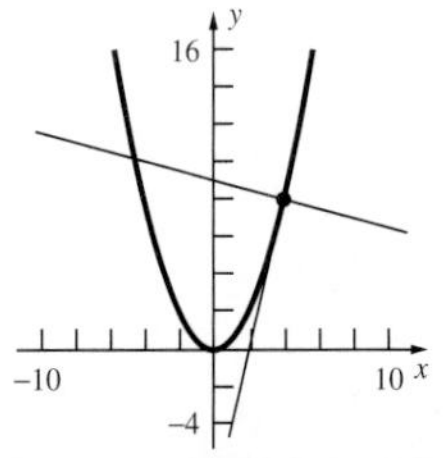

23. $y = \dfrac{\sqrt{5}}{2}x - \dfrac{3\sqrt{5}}{2}$; $y = -\dfrac{2\sqrt{5}}{5}x - \dfrac{21\sqrt{5}}{5}$

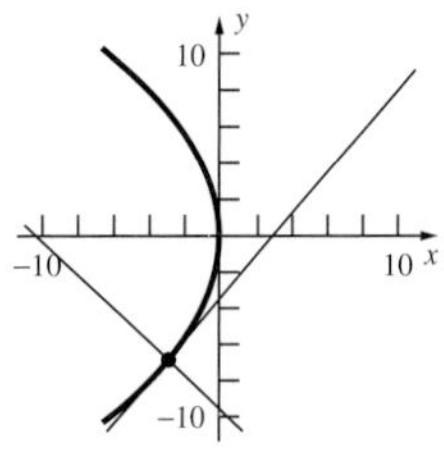

25. $y = -\sqrt{2}x + 3$; $y = \dfrac{\sqrt{2}}{2}x - 6$

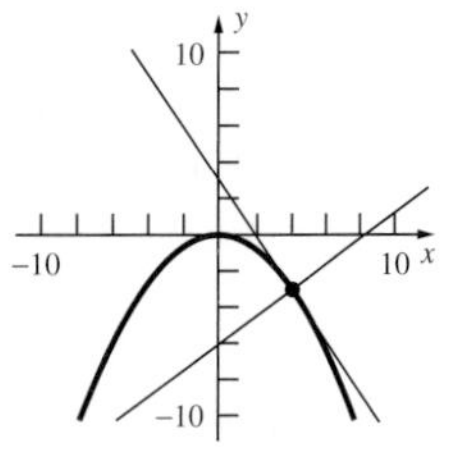

27. $\left(4, 2\sqrt{5}\right)$

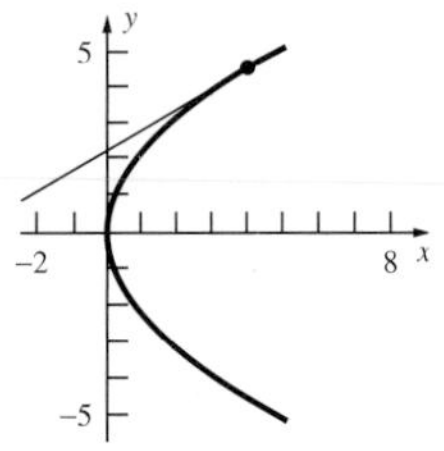

29. $y = \frac{3}{2}x - 3$ **35.** 14.8 million mi **37.** $2p$

39. $L = 4p$ **41.** $y = \dfrac{\delta x^2}{2H}$

43.

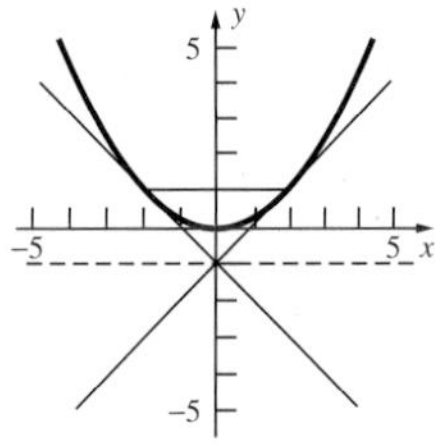

Problem Set 10.2

1. Horizontal ellipse **3.** Vertical hyperbola

5. Vertical parabola (opens up) **7.** Vertical ellipse

9.

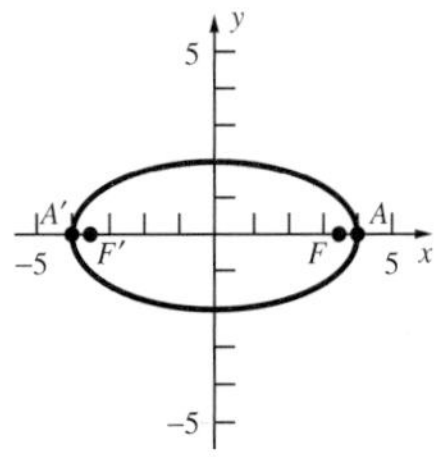

11.

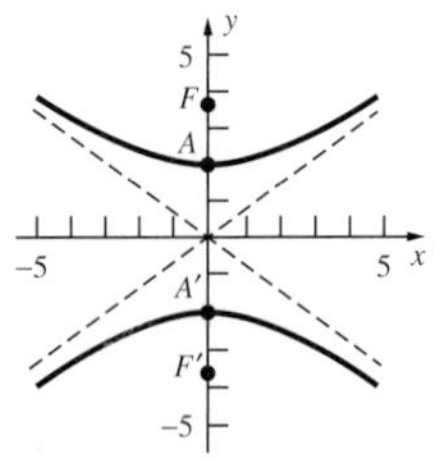

13.

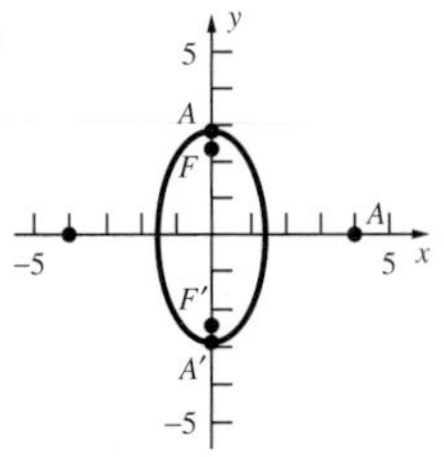

15.

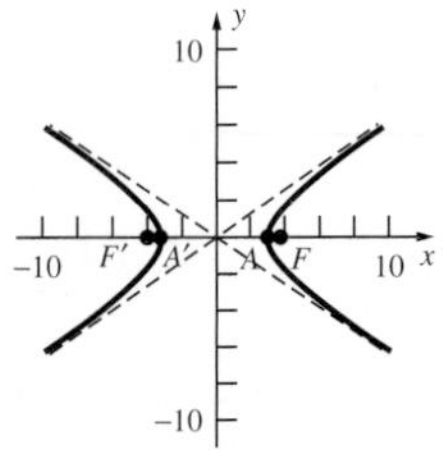

17. $\dfrac{x^2}{36} + \dfrac{y^2}{27} = 1$ **19.** $\dfrac{x^2}{200} + \dfrac{y^2}{225} = 1$ **21.** $\dfrac{x^2}{25} + \dfrac{y^2}{\frac{225}{21}} = 1$

23. $\frac{y^2}{16} - \frac{x^2}{9} = 1$ **25.** $\frac{x^2}{64} - \frac{y^2}{16} = 1$ **27.** $\frac{x^2}{16} + \frac{y^2}{12} = 1$

29. $\frac{y^2}{5} - \frac{x^2}{20} = 1$ **31.** $\frac{x^2}{88} + \frac{y^2}{169} = 1$ **33.** $\frac{x^2}{36} - \frac{y^2}{13} = 1$

35. $x + \sqrt{6}y = 9$ **37.** $x - \sqrt{6}y = 9$

39. $5x + 12y = 169$ **41.** $y = 13$ **43.** 8.66 ft **45.** $\frac{2b^2}{a}$

47. 0.58 AU **49.** 0.05175 **51.** $\left(-\sqrt{3}, \frac{3}{2}\right), \left(\sqrt{3}, \frac{3}{2}\right)$

53. $(-7, 3), (7, -3)$ **55.** πab

57. $\frac{\pi b^2}{3a^2}\left[(a^2 + b^2)^{3/2} - 3a^2\sqrt{a^2 + b^2} + 2a^3\right]$

59. $a\sqrt{2}$ by $b\sqrt{2}$ **61.** $\left(6, 5\sqrt{3}\right)$

63.

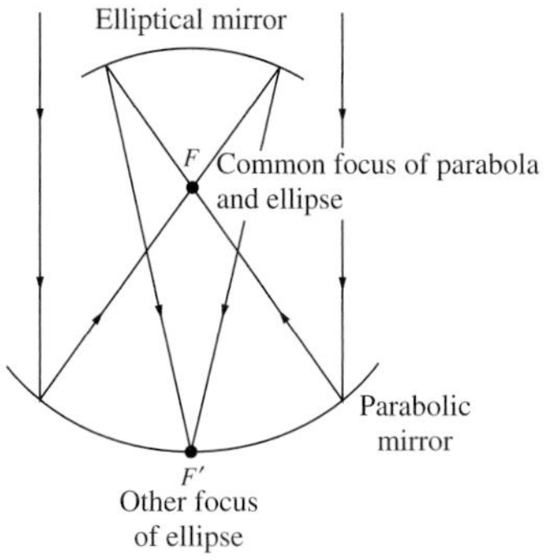

69. $\left(\sqrt{\frac{17}{3}}, 5\right)$ **73.** $\frac{x^2}{a^2} + \frac{y^2}{b^2} = 1$

77.

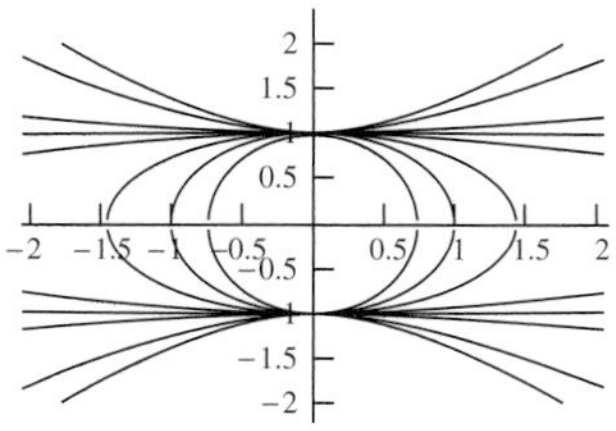

Problem Set 10.3

1. Circle **3.** Ellipse **5.** Point **7.** Parabola
9. Empty set **11.** Intersecting lines **13.** Line

15.

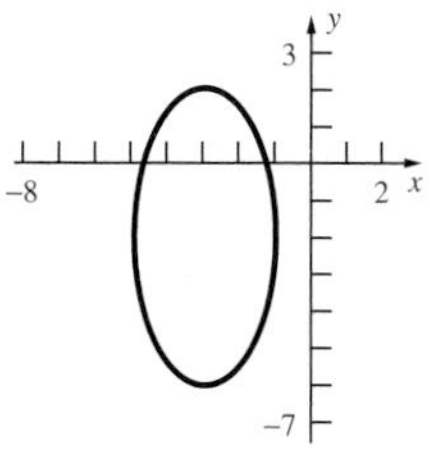

17.

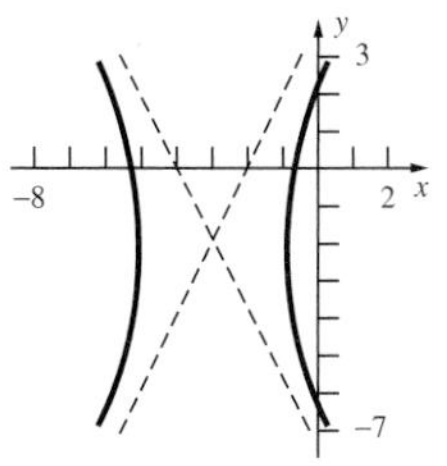

19.

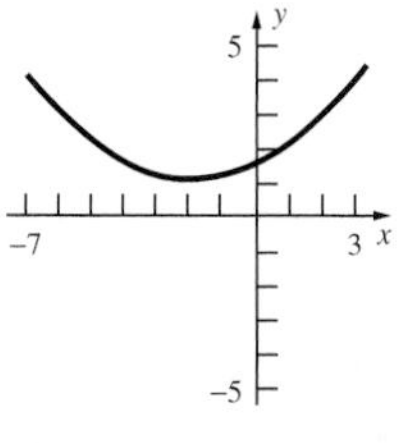

21.

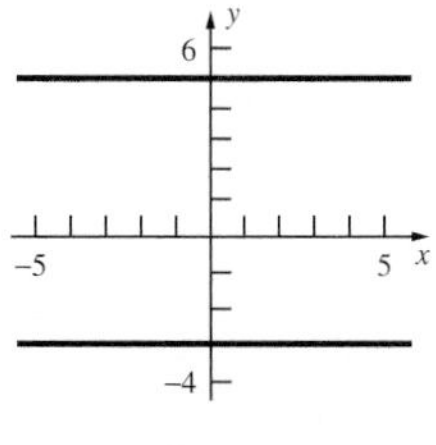

23.

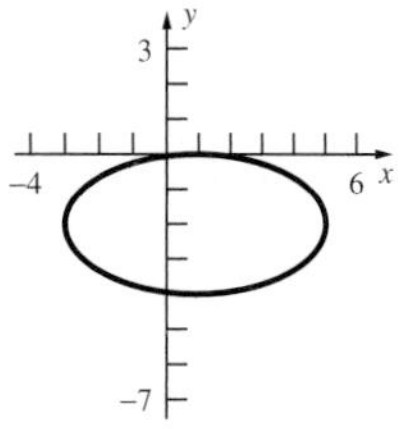

25.

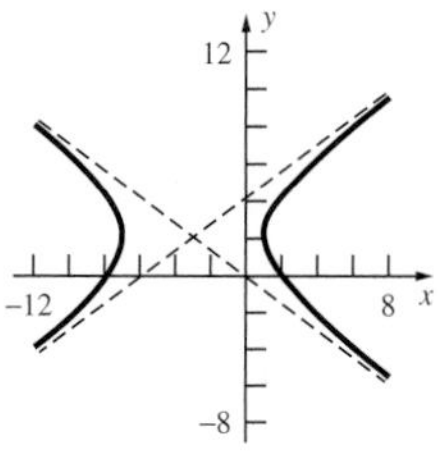

27.

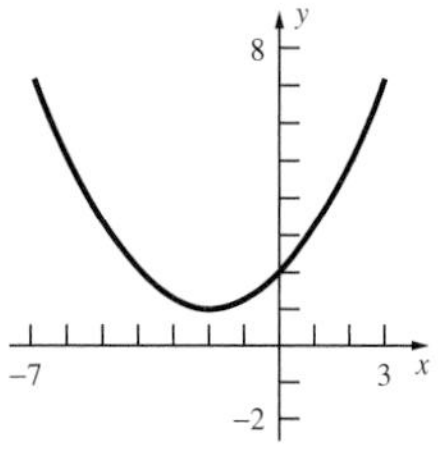

29. Focus at $\left(\frac{21}{20}, 1\right)$; directrix $x = -\frac{29}{20}$ **31.** $(-2, 2), (4, -2)$

33. $\frac{(x-5)^2}{25} + \frac{(y-1)^2}{16} = 1$ **35.** $(x-2)^2 = 8(y-3)$

37. $\frac{(y-3)^2}{9} - \frac{x^2}{16} = 1$ **39.** $(y-5)^2 = -16(x-6)$

41. $\frac{x^2}{8} + \frac{(y-2)^2}{4} = 1$

43. $\frac{u^2}{4} + \frac{v^2}{12} = 1$

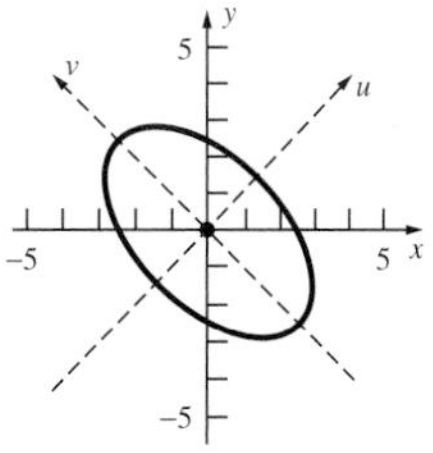

45. $\frac{u^2}{\frac{112}{9}} + \frac{v^2}{16} = 1$

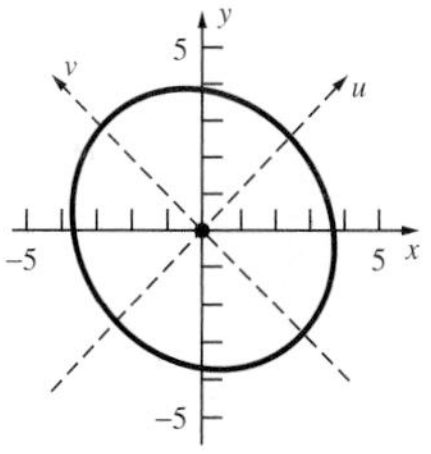

47. $\frac{(u-2)^2}{4} - \frac{v^2}{3} = 1$

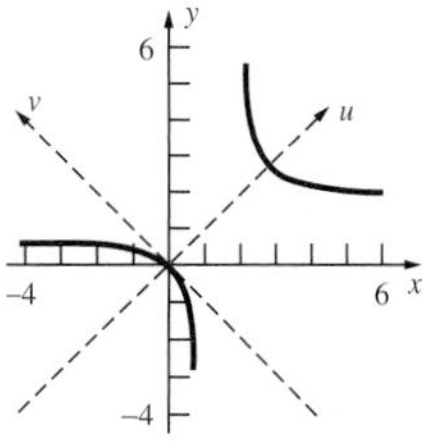

49. $\frac{v^2}{4} - \frac{u^2}{36} = 1$

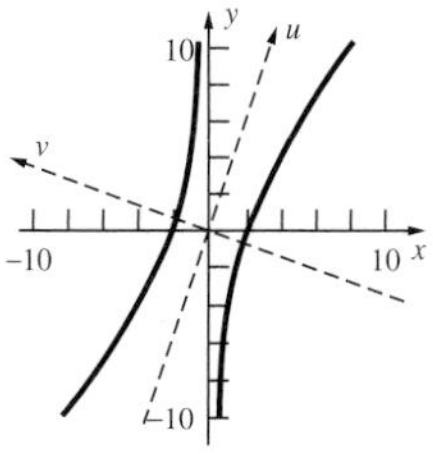

51. $\frac{(u+2)^2}{2} + \frac{(v+3)^2}{4} = 1$

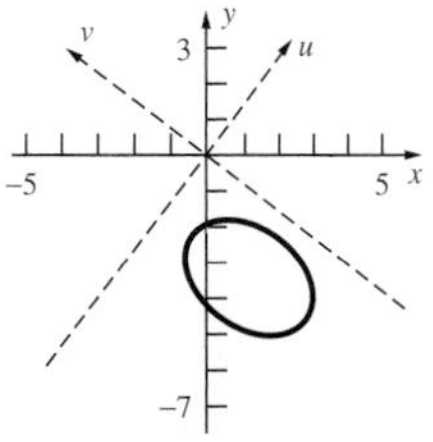

53. **(a)** $y = x^2 - x$; **(b)** $x = \frac{1}{4}y^2 - y$;
(c) $\left(x - \frac{5}{2}\right)^2 + \left(y - \frac{5}{2}\right)^2 = \frac{25}{2}$

55. If $K < -1$, the conic is a vertical ellipse. If $K = -1$, the conic is a circle. If $-1 < K < 0$, the conic is a horizontal ellipse. If $K = 0$, the conic is a horizontal parabola. If $K > 0$, the conic is a horizontal hyperbola.

59. $u = x\cos\theta + y\sin\theta$, $v = -x\sin\theta + y\cos\theta$

61. $\left(-\frac{1}{5}, -\frac{7}{5}\right), \left(\frac{1}{5}, \frac{7}{5}\right)$

67. **(a)** $-2 < B < 2$; **(b)** $B = 0$; **(c)** $B < -2$ or $B > 2$; **(d)** $B = \pm 2$

Problem Set 10.4

1. **(a)**

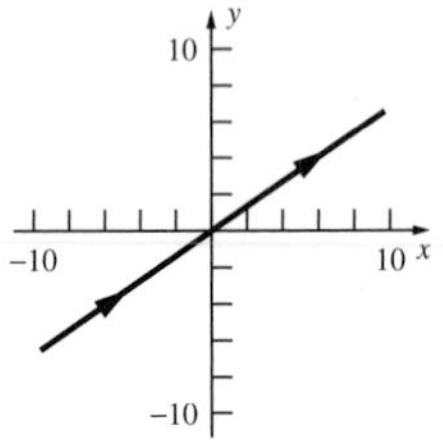

(b) Simple; not closed
(c) $y = \frac{2}{3}x$

3. **(a)**

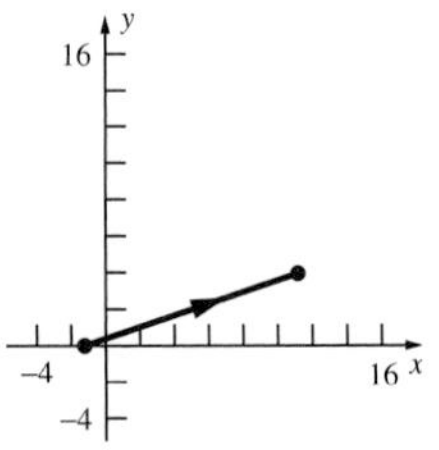

(b) Simple, not closed
(c) $y = \frac{1}{3}(x + 1)$

5. **(a)**

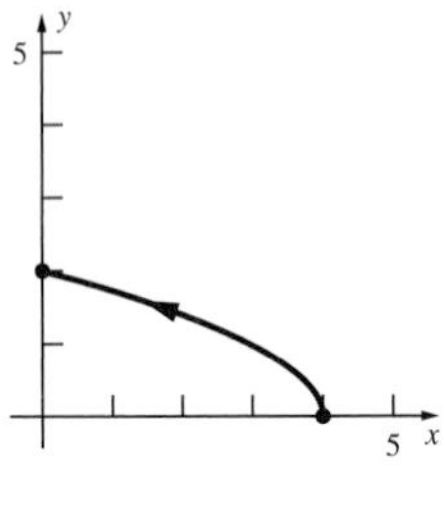

(b) Simple; not closed
(c) $y = \sqrt{4 - x}$

7. **(a)**

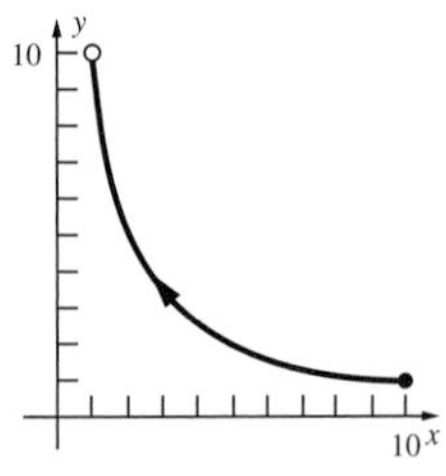

(b) Simple; not closed
(c) $y = \frac{1}{x}$

9. **(a)**

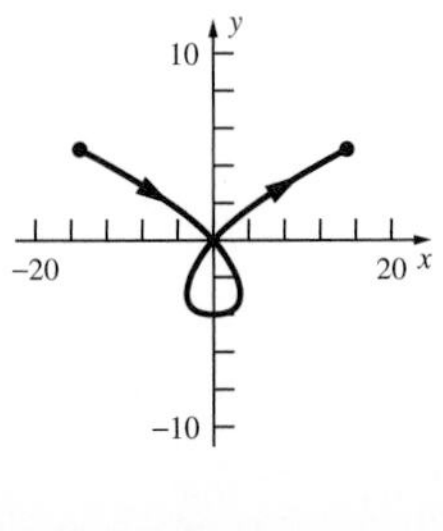

(b) Not simple; not closed
(c) $x^2 = y^3 + 4y^2$

11. **(a)**

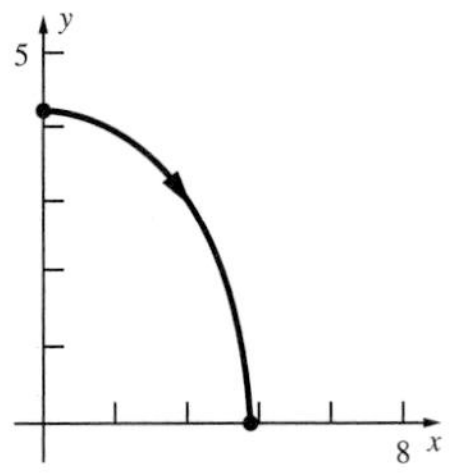

(b) Simple; not closed
(c) $\frac{x^2}{8} + \frac{y^2}{18} = 1$

13. **(a)**

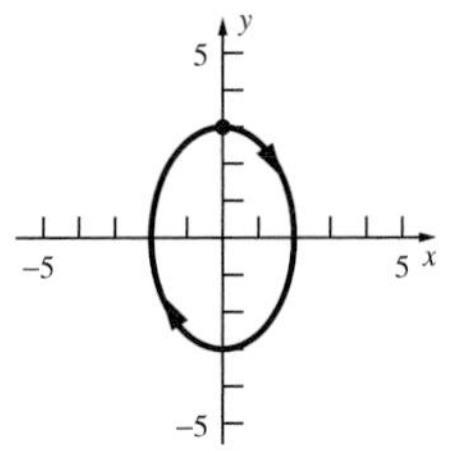

(b) Simple; closed
(c) $\frac{x^2}{4} + \frac{y^2}{9} = 1$

15. **(a)**

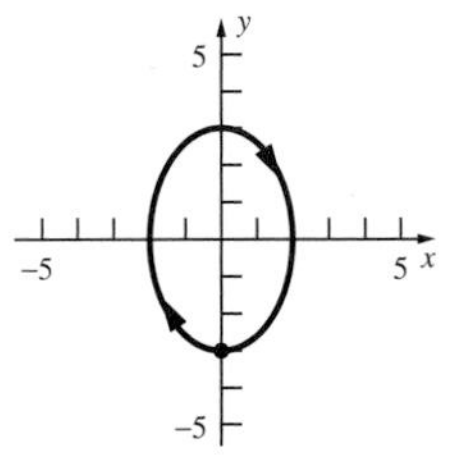

(b) Not simple; closed
(c) $\frac{x^2}{4} + \frac{y^2}{9} = 1$

17. **(a)**

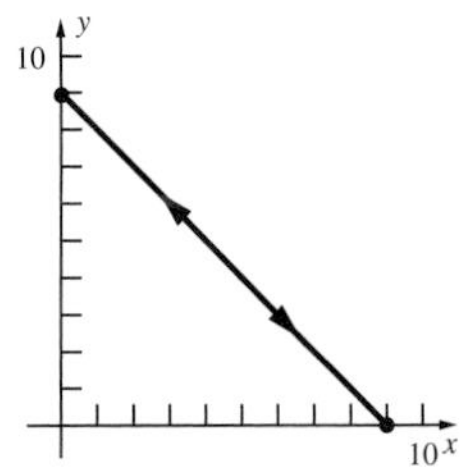

(b) Not simple; closed
(c) $x + y = 9$

19. **(a)**

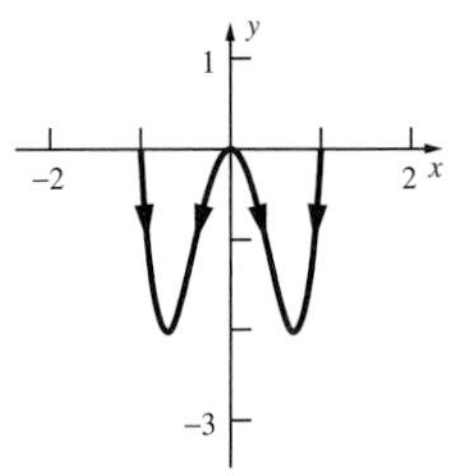

(b) Not simple; not closed
(c) $y = -8x^2(1 - x^2)$

21. $\frac{dy}{dx} = 2\tau$; $\frac{d^2y}{dx^2} = \frac{1}{3\tau}$ **23.** $\frac{dy}{dx} = \frac{3\sqrt{5}}{4}\theta$; $\frac{d^2y}{dx^2} = \frac{3\sqrt{5}}{16\theta}$

25. $\frac{dy}{dx} = \cot t$; $\frac{d^2y}{dx^2} = -\csc^3 t$

27. $\frac{dy}{dx} = \frac{5}{3}\sin t$; $\frac{d^2y}{dx^2} = \frac{5}{9}\cos^3 t$

29. $\frac{dy}{dx} = \frac{(1 - 2t)(1 + t^2)^2}{2t^3(1 - t)^2}$;

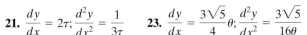

$$\frac{d^2y}{dx^2} = \frac{(3t^5 + 7t^4 - 6t^3 + 10t^2 - 9t + 3)(1 + t^2)^2}{4t^5(1 - t)^3}$$

31. $y - 8 = 3(x - 4)$

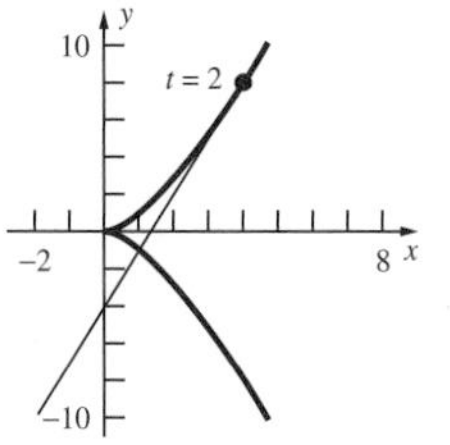

33. $y + \dfrac{2}{\sqrt{3}} = -2\left(x - \dfrac{4}{\sqrt{3}}\right)$

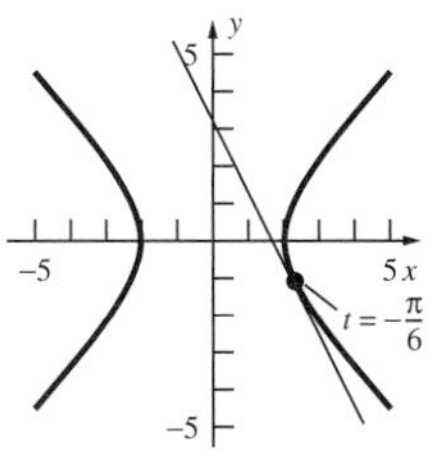

35. $3\sqrt{13}$ **37.** $\frac{1}{27}\left(31\sqrt{31} - 8\right)$ **39.** $16\sqrt{2} - 8$

41. $\dfrac{713\sqrt{713} - 227\sqrt{227}}{243}$ **43.** $\frac{39}{16}$ **45.** $\frac{1}{2}\ln 2$

47. **(a)** 2π; **(b)** 6π;
(c) The curve in part **(a)** goes around the unit circle once. The curve in part **(b)** goes around the unit circle three times.

49. $4\pi^2$ **51.** $4\pi^2$ **53.** $\frac{2\pi}{3}\left(29\sqrt{29} - 1\right)$ **55.** $-\frac{44}{3}$

57. 8

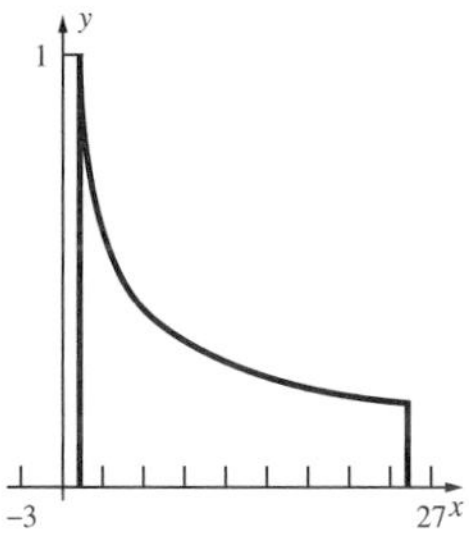

59.

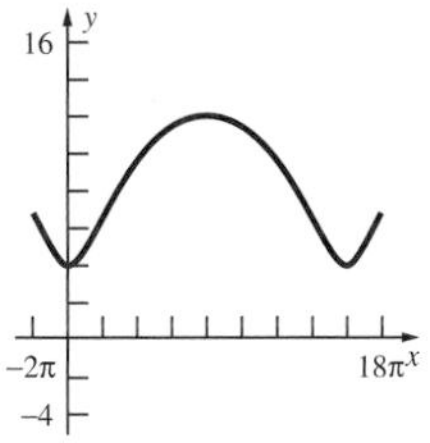

61. $x = (a - b)\cos t + b\cos\left(\dfrac{a - b}{b}t\right)$,

$y = (a - b)\sin t - b\sin\left(\dfrac{a - b}{a}t\right)$

65. $L = \dfrac{16a}{3}$

67. **(a)**

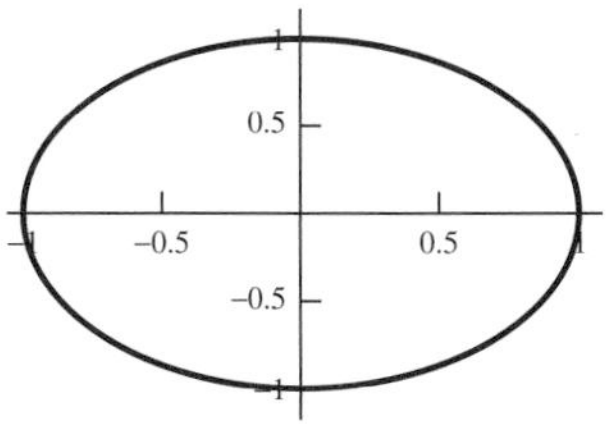

(b)

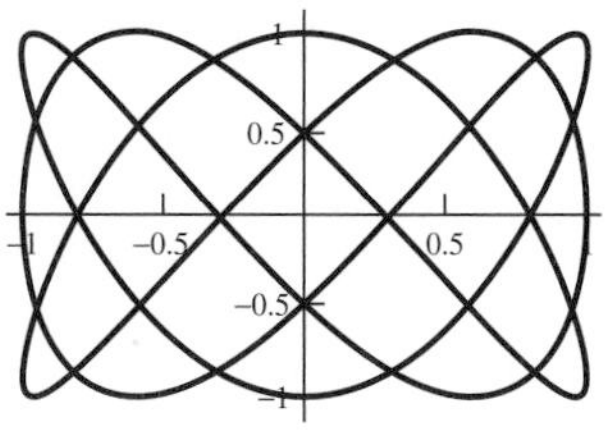

(c)

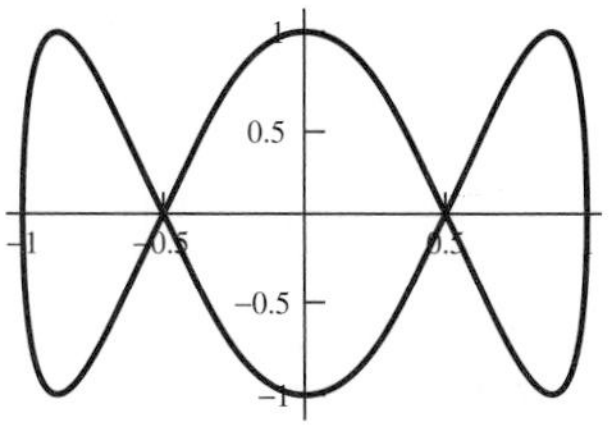

(d)

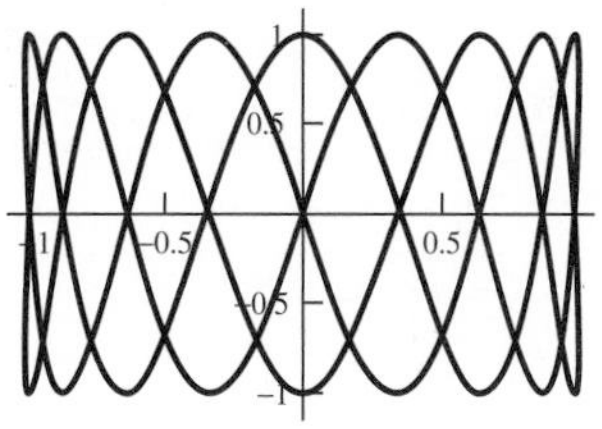

69. a, b, c

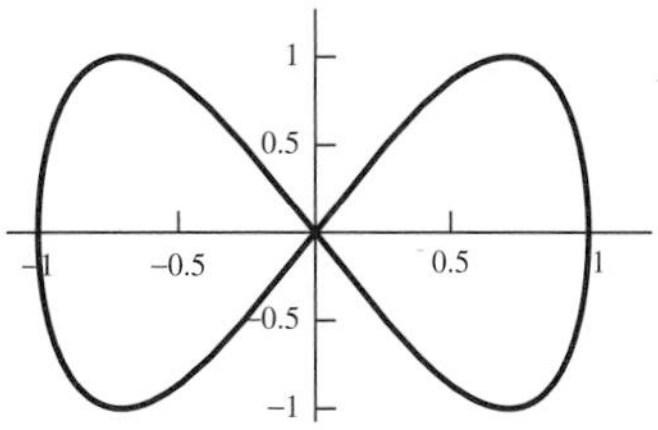

d, e, f

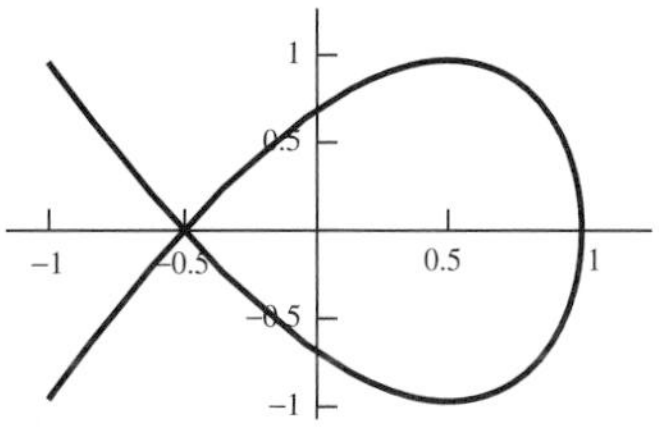

71. **(a)** $0 \le t \le 2$

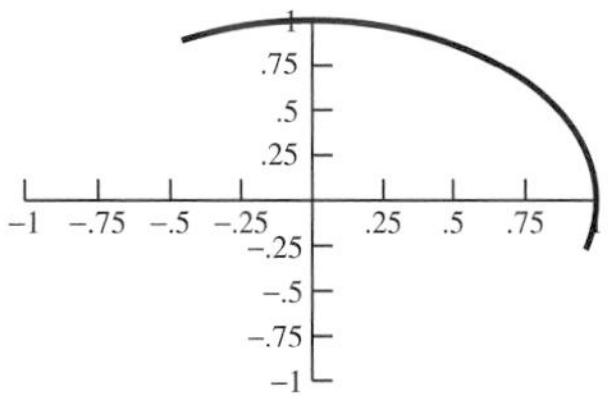

(b) $0 \le t \le 1$

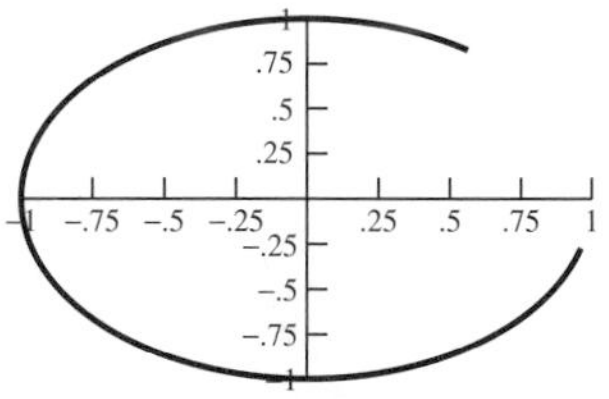

(c) $0.25 \le t \le 2$

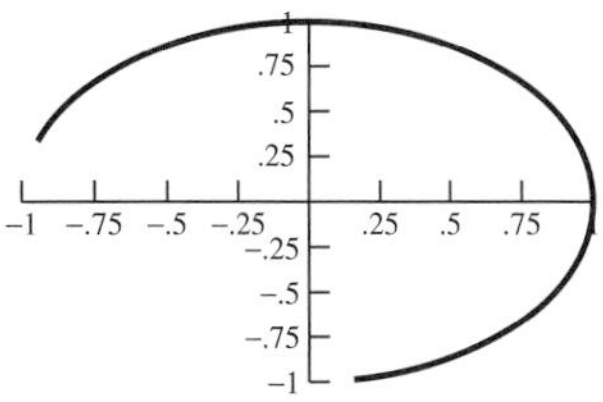

(d) $0 \le t \le 2\pi$

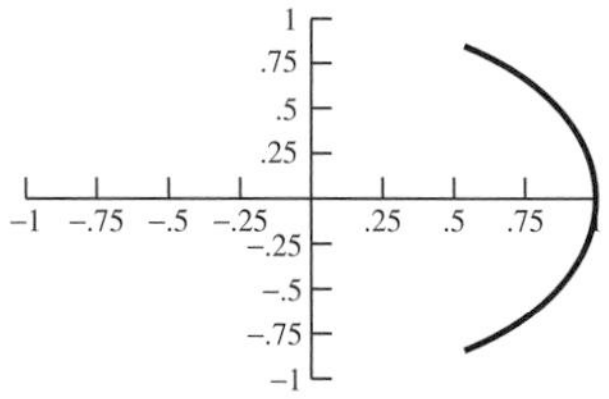

73. **(a)**

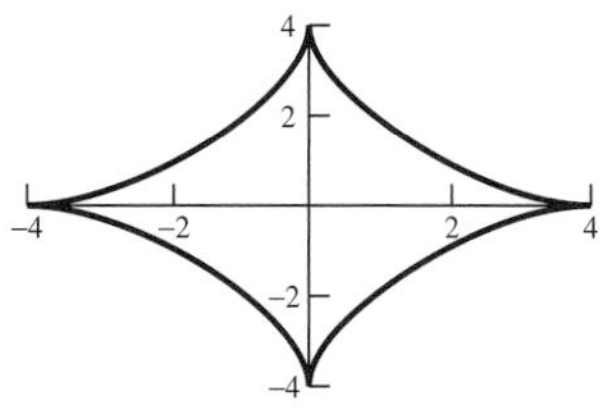

(b)

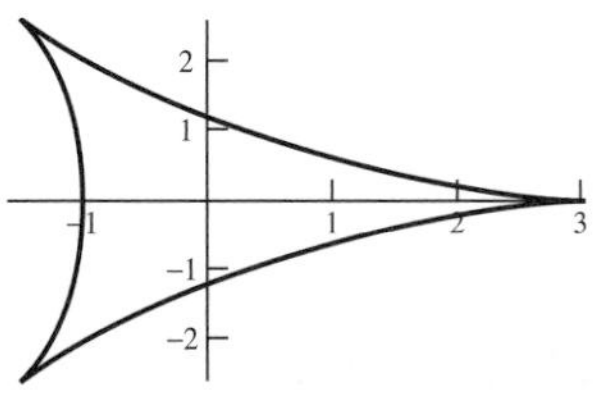

(c)

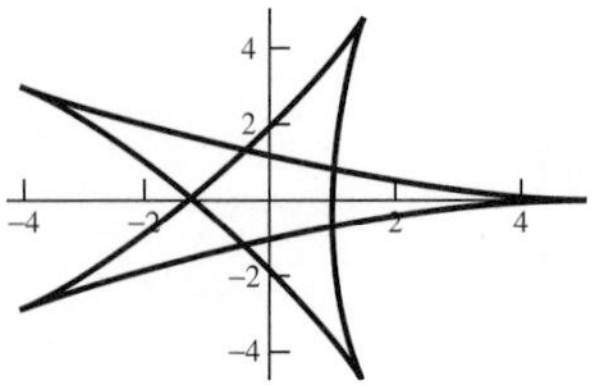

(d)

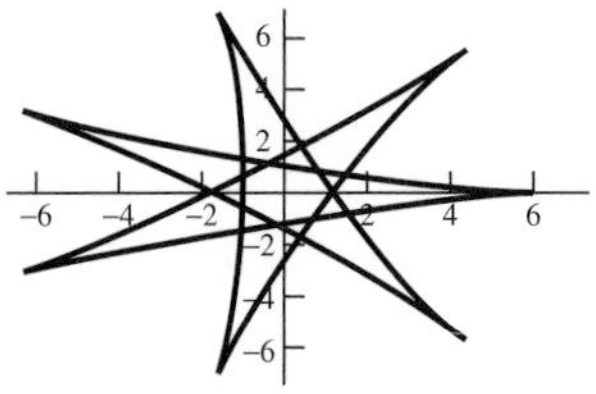

75. Quadrant I for $t > 0$, quadrant II for $-1 < t < 0$, quadrant III for no t, quadrant IV for $t < -1$.

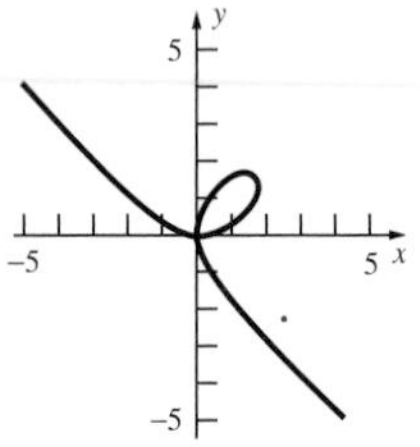

Problem Set 10.5

1.

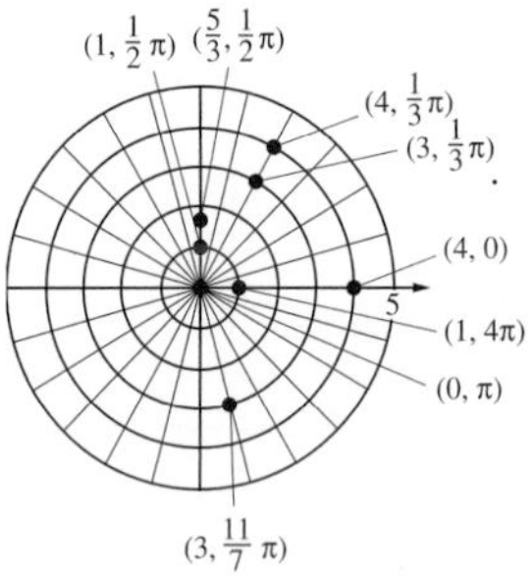

3.

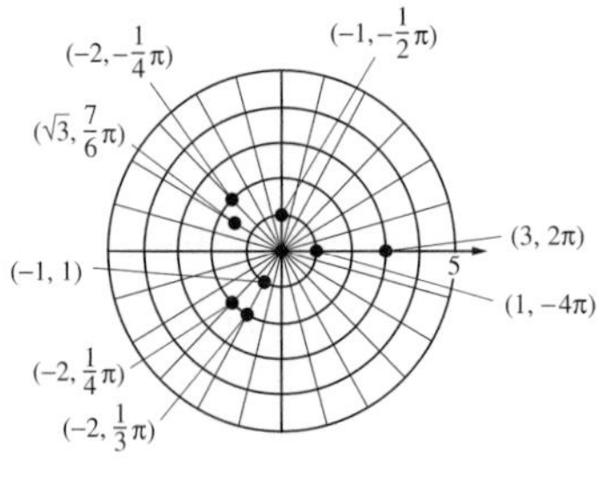

5.

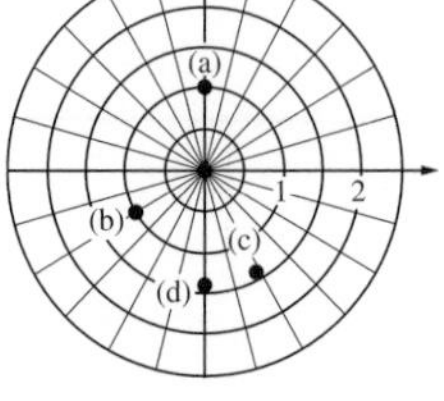

(a) $\left(1, -\frac{3}{2}\pi\right), \left(1, \frac{5}{2}\pi\right), \left(-1, -\frac{1}{2}\pi\right), \left(-1, \frac{3}{2}\pi\right)$

(b) $\left(1, -\frac{3}{4}\pi\right), \left(1, \frac{5}{4}\pi\right), \left(-1, -\frac{7}{4}\pi\right), \left(-1, \frac{9}{4}\pi\right)$

(c) $\left(\sqrt{2}, -\frac{7}{3}\pi\right), \left(\sqrt{2}, \frac{5}{3}\pi\right), \left(-\sqrt{2}, -\frac{4}{3}\pi\right), \left(-\sqrt{2}, \frac{2}{3}\pi\right)$

(d) $\left(\sqrt{2}, -\frac{1}{2}\pi\right), \left(\sqrt{2}, \frac{3}{2}\pi\right), \left(-\sqrt{2}, -\frac{3}{2}\pi\right), \left(-\sqrt{2}, \frac{1}{2}\pi\right)$

7. (a) $(0, 1)$; **(b)** $\left(-\frac{\sqrt{2}}{2}, -\frac{\sqrt{2}}{2}\right)$; **(c)** $\left(\frac{\sqrt{2}}{2}, -\frac{\sqrt{6}}{2}\right)$; **(d)** $\left(0, -\sqrt{2}\right)$

9. (a) $\left(6, \frac{1}{6}\pi\right)$; **(b)** $\left(4, \frac{5}{6}\pi\right)$; **(c)** $\left(2, \frac{5}{4}\pi\right)$; **(d)** $(0, 0)$

11. $r = \dfrac{2}{3 \sin \theta - \cos \theta}$

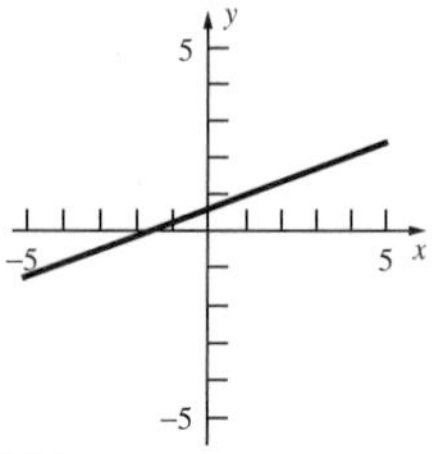

13. $r = -2 \csc \theta$

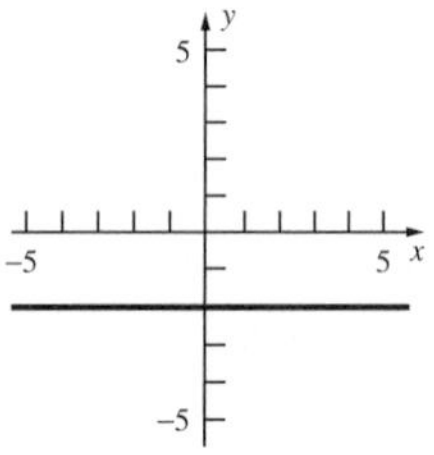

15. $r = 2$

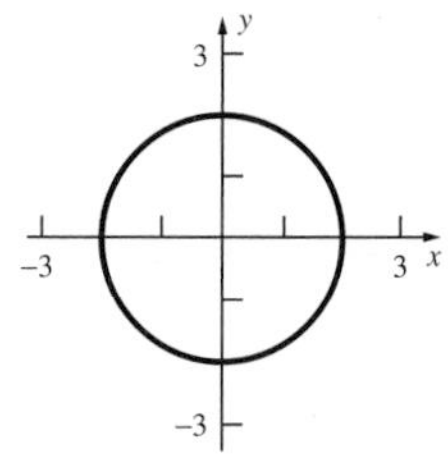

17. $x = 0$ **19.** $x = -3$ **21.** $y = 1$

23. Circle

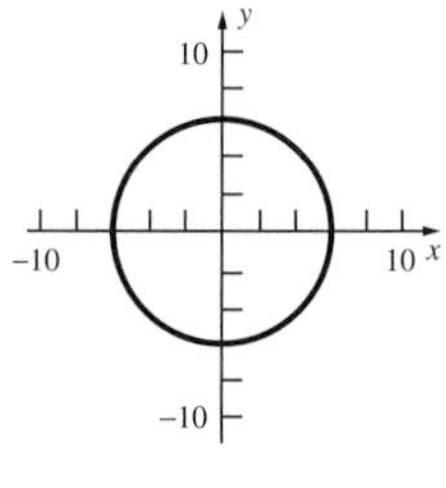

25. Line

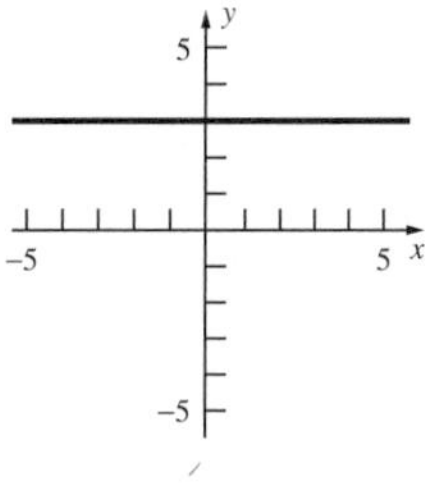

27. Circle

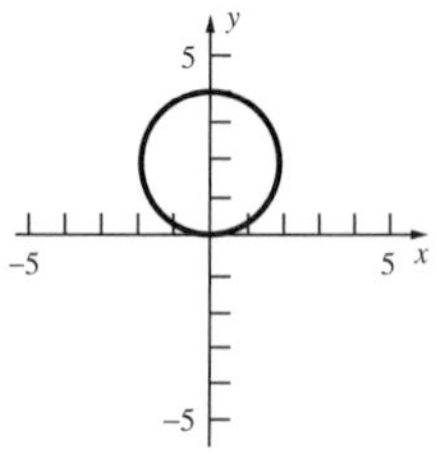

29. Parabola; $e = 1$

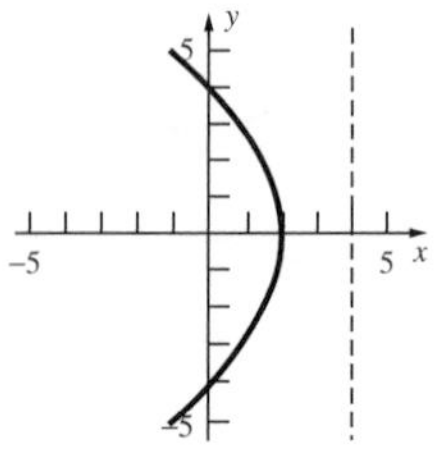

31. Ellipse; $e = \frac{1}{2}$

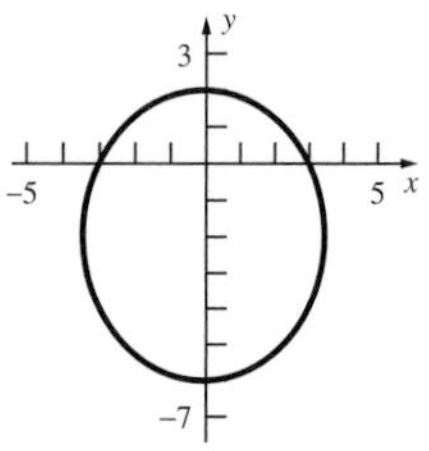

33. Parabola; $e = 1$

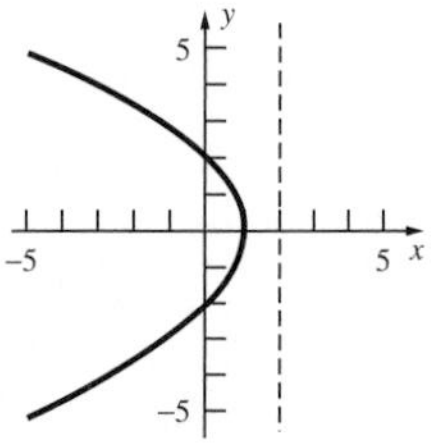

35. Hyperbola; $e = 2$

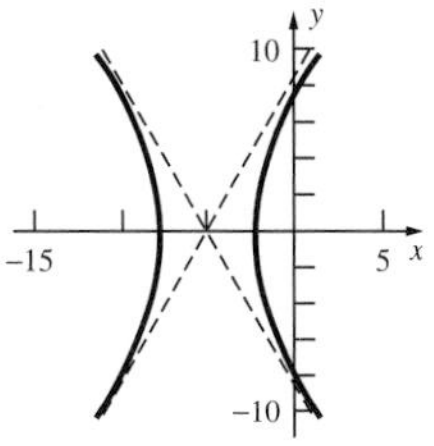

39. $2ed$ **41.** 0.83 **43.** 25 million mi

45. $e = 0.1$

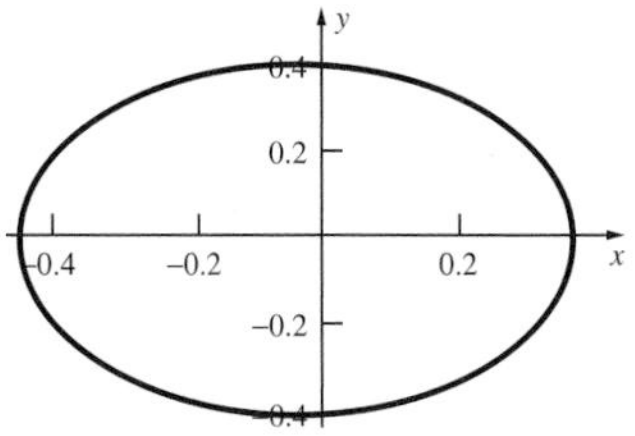

$e = 0.5$

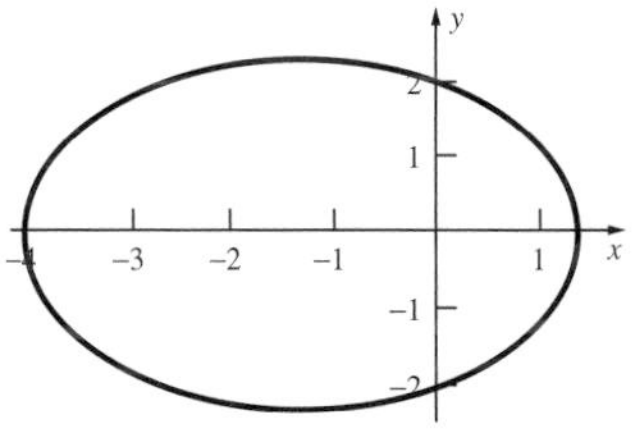

$e = 0.9$

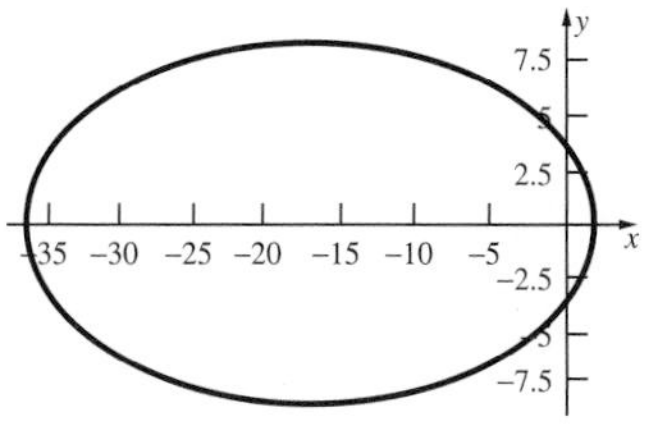

$e = 1$

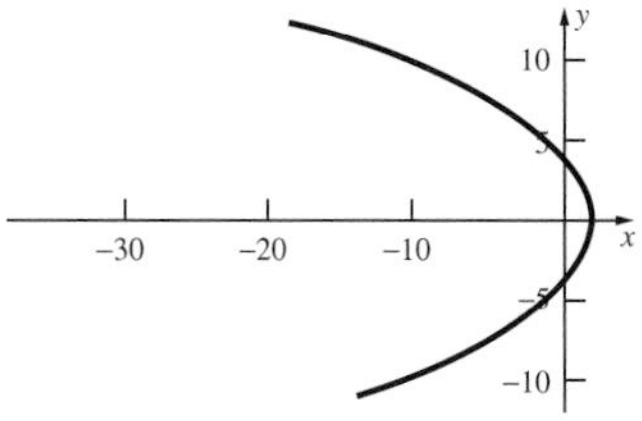

$e = 1.1$

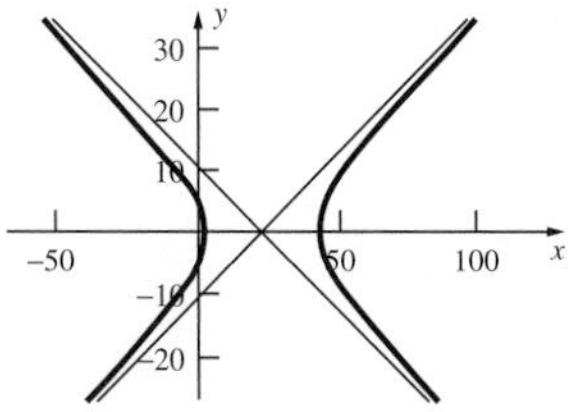

$e = 1.3$

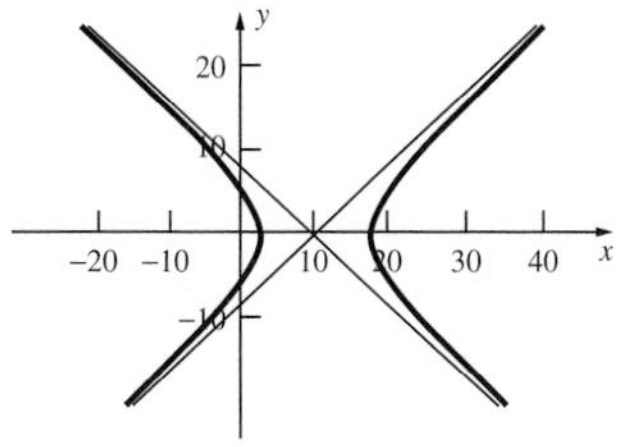

Problem Set 10.6

1.

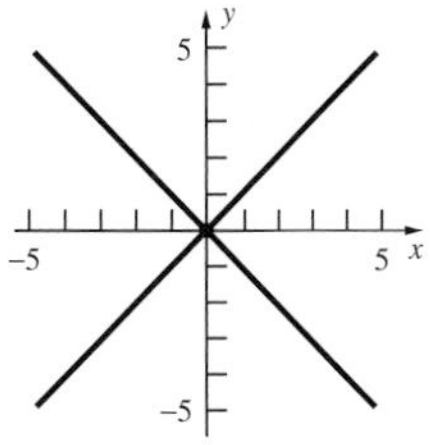

3.

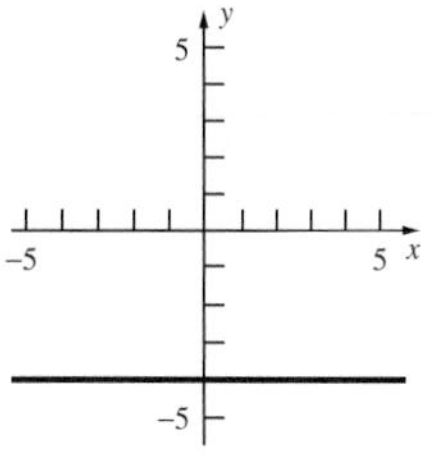

5.

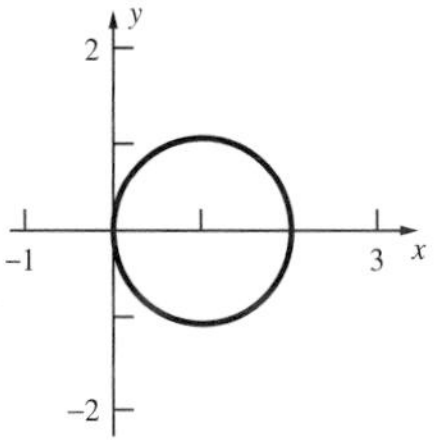

7.

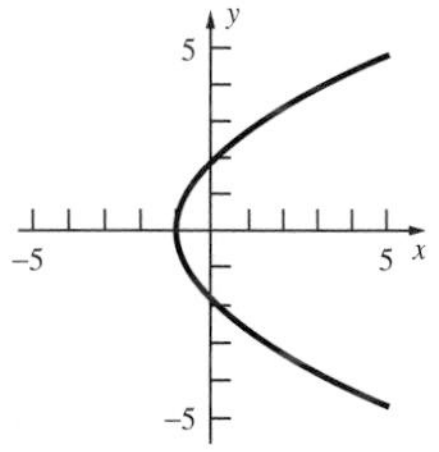

9.

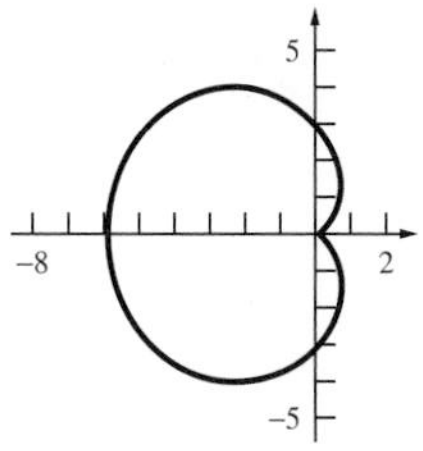

11.

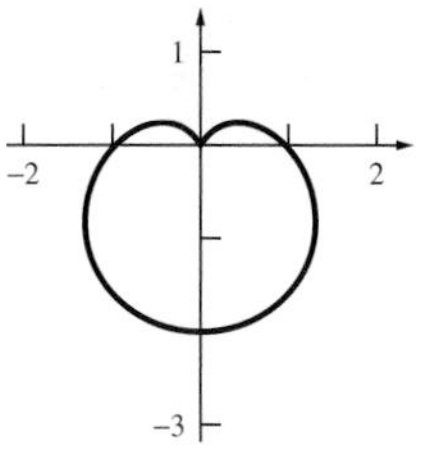

13.

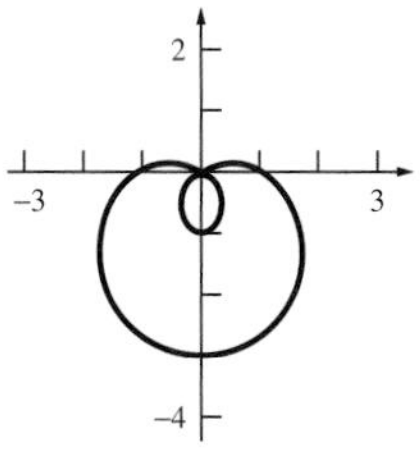

15.

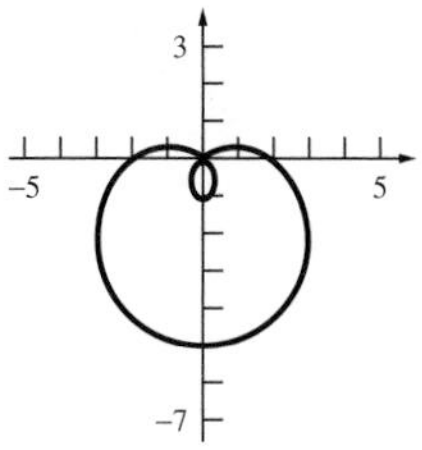

17.

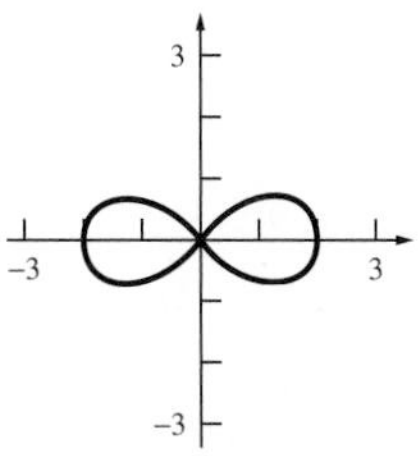

19.

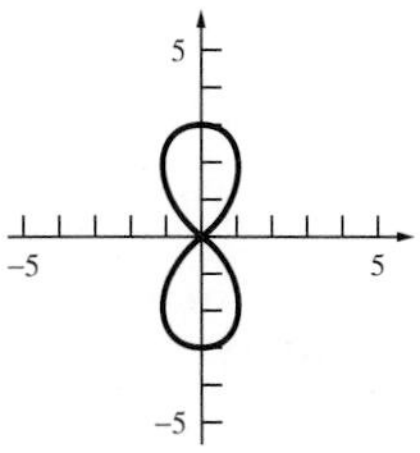

21.

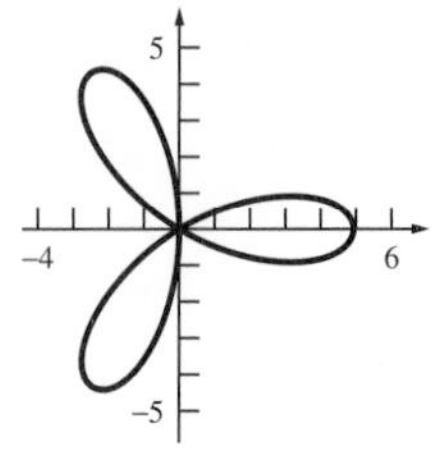

23.

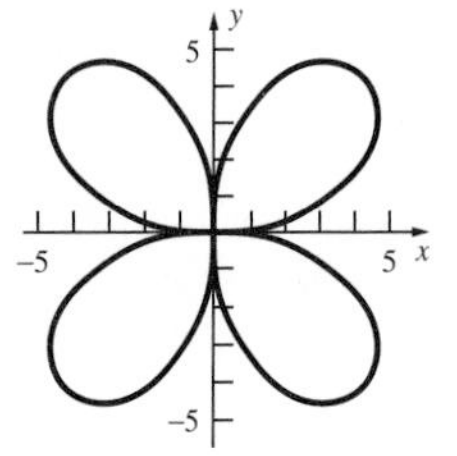

25.

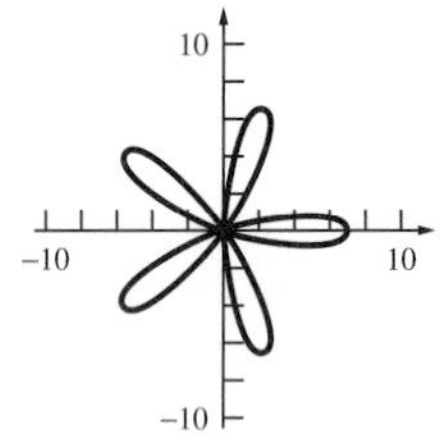

27.

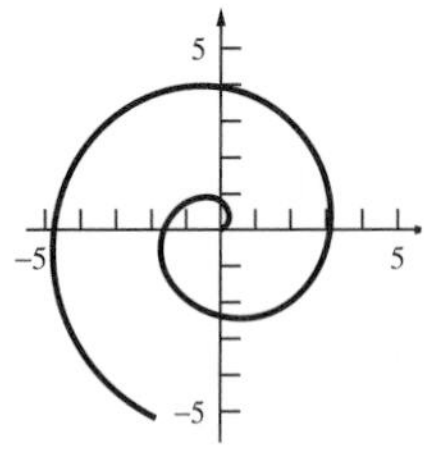

29.

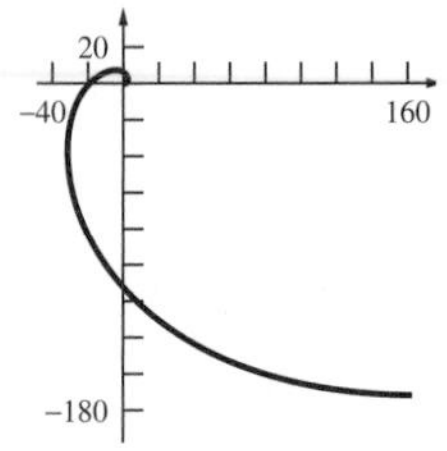

31.

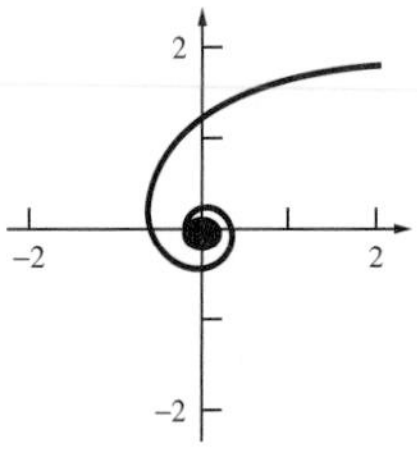

33. $\left(6, \frac{\pi}{3}\right), \left(6, \frac{5\pi}{3}\right)$

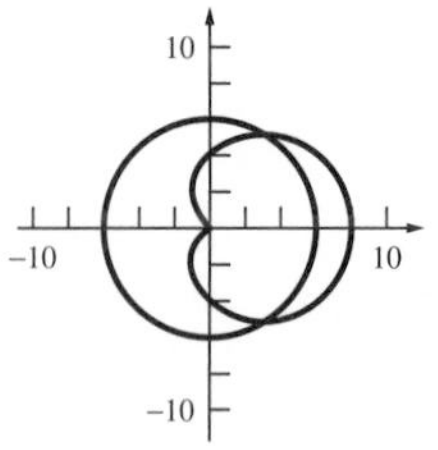

35. $(0, 0), \left(\frac{3\sqrt{3}}{2}, \frac{\pi}{3}\right)$

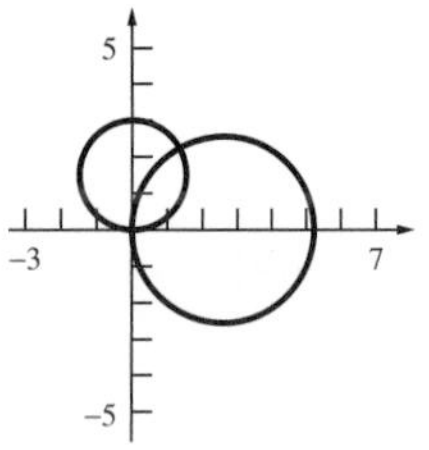

37. $\left(3, \frac{\pi}{6}\right), \left(3, \frac{5\pi}{6}\right), \left(6, \frac{\pi}{2}\right)$

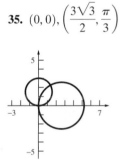

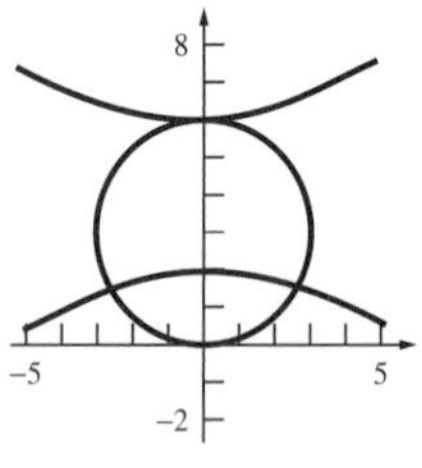

43. (a) $r = \frac{45}{\sin\theta}$; **(b)** $r = 6$; **(c)** $r = \pm\frac{1}{\sqrt{\cos 2\theta}}$
(d) $r = \pm\frac{1}{\sqrt{2\sin 2\theta}}$; **(e)** $r = \frac{2}{\sin\theta - 3\cos\theta}$;
(f) $r = \frac{-2\sin\theta \pm \sqrt{4\sin^2\theta + 6\cos^2\theta}}{3\cos^2\theta}$;
(g) $r = -\cos\theta + 2\sin\theta \pm \sqrt{(\cos\theta - 2\sin\theta)^2 + 25}$;

45. (a) VII **(b)** I **(c)** VIII **(d)** III **(e)** V **(f)** II **(g)** VI **(h)** IV

47.

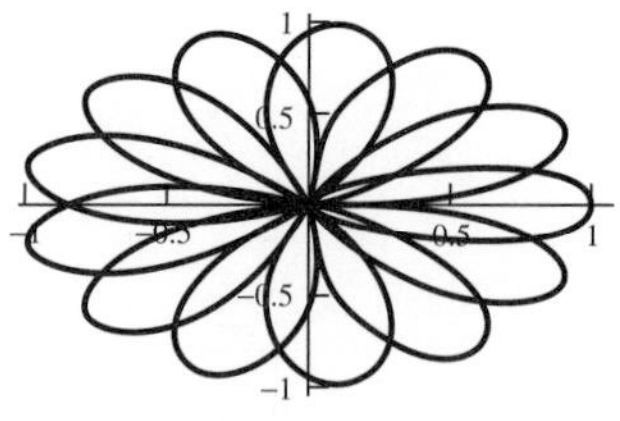

49.

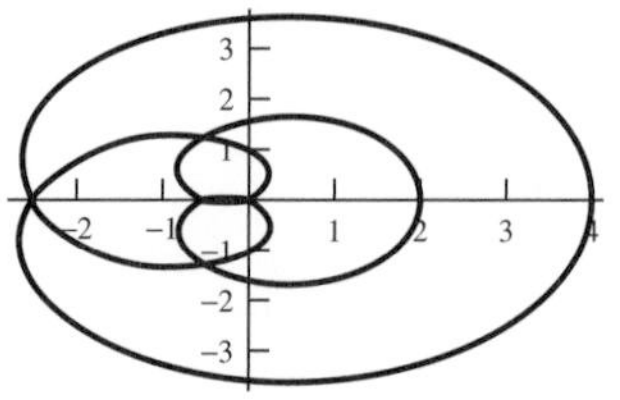

51. (a) The graph for $\phi = 0$ is the graph for $\phi \neq 0$ rotated by ϕ counterclockwise about the pole.
(b) As n increases, the number of "leaves" increases.
53. The spiral will unwind clockwise for $c < 0$. The spiral will unwind counterclockwise for $c > 0$.
55. (a) III; **(b)** IV; **(c)** I; **(d)** II; **(e)** VI; **(f)** V

Problem Set 10.7

1. πa^2

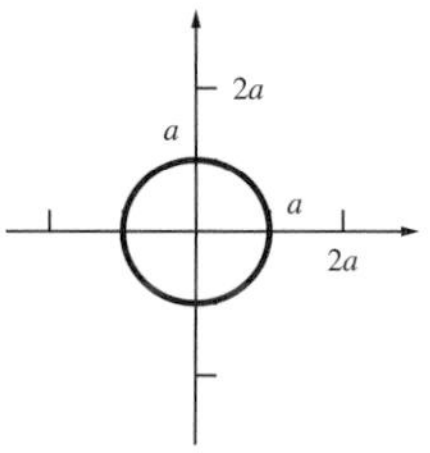

3. $\frac{9}{2}\pi$

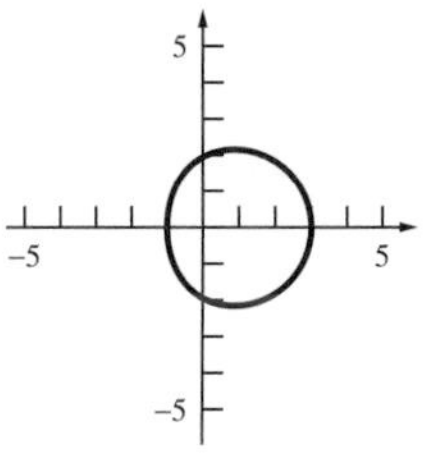

5. $\frac{27}{2}\pi$

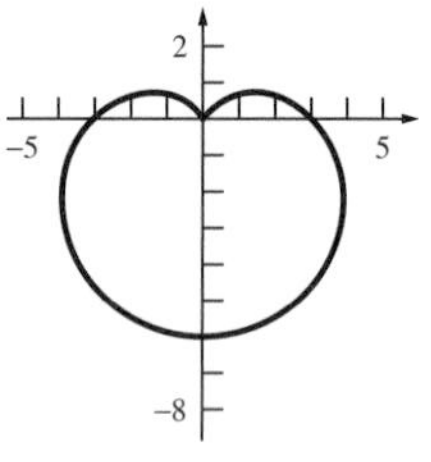

7. $\frac{3}{2}\pi a^2$

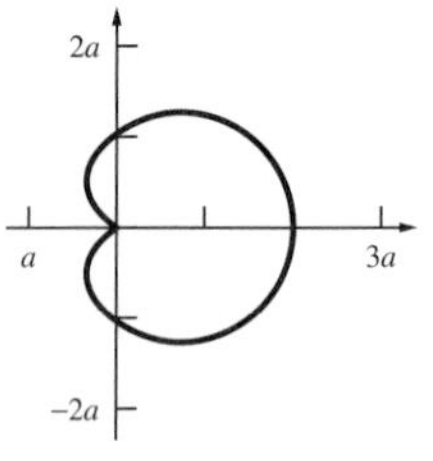

9. 9

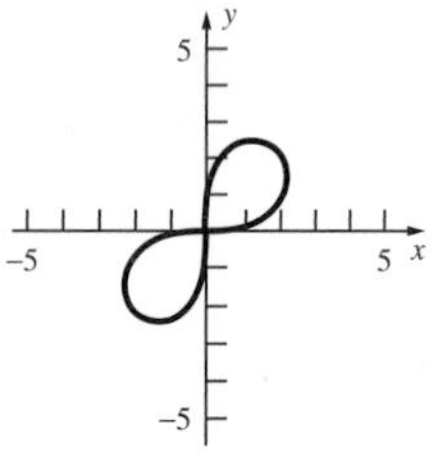

11. $\frac{17}{2}\pi - 17\sin^{-1}\frac{3}{4} - \frac{9\sqrt{7}}{2}$

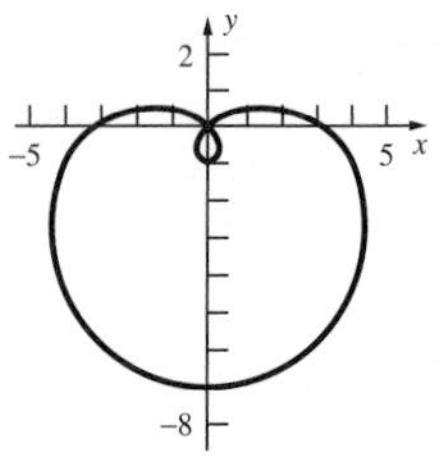

13. $\frac{17}{2}\pi - \frac{17}{2}\cos^{-1}\frac{2}{3} + 3\sqrt{5}$

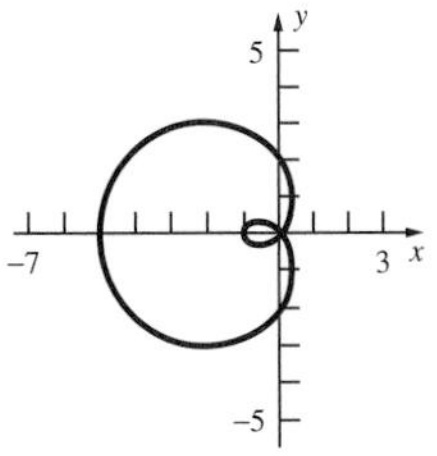

15. 4π

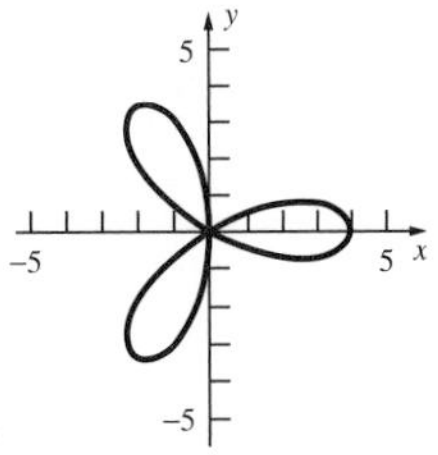

17. 51π

19. $4\sqrt{3} - \frac{4}{3}\pi$

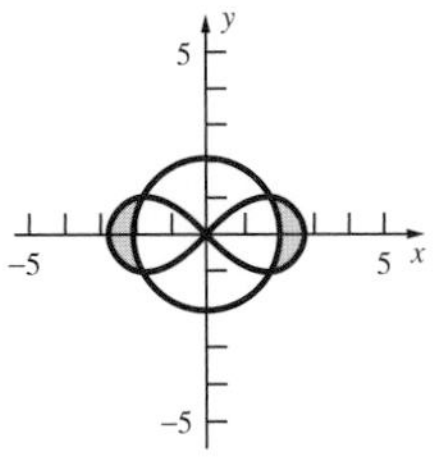

21. $9\sqrt{2} - \frac{27}{4}$

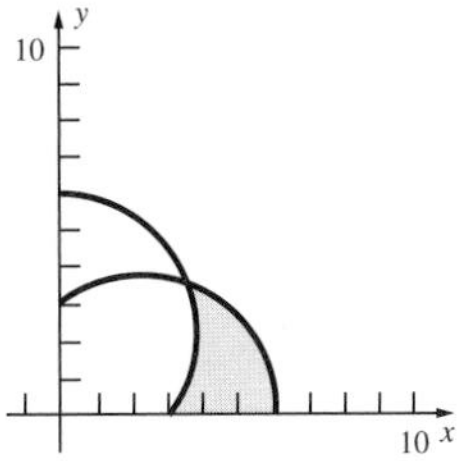

23. (a) $\frac{1}{\sqrt{3}}$; **(b)** -1 **(c)** $\frac{\sqrt{3}}{5}$ **(d)** $-\frac{7}{\sqrt{3}}$

25. $\left(-1, \frac{\pi}{2}\right), \left(3, \frac{3\pi}{2}\right), \left(\frac{1}{2}, \sin^{-1}\frac{1}{4}\right), \left(\frac{1}{2}, \pi - \sin^{-1}\frac{1}{4}\right)$ **27.** $8a$

29. $\frac{1}{2}\pi a^2$ if n is even. $\frac{1}{4}\pi a^2$ if n is odd.

31. (a) $a^2 \tan^{-1}\frac{b}{a} + b^2\left(\frac{\pi}{2} - \tan^{-1}\frac{b}{a}\right) - ab$

33. $a^2\left[(k^2 - 1)\pi + (2 - k^2)\cos^{-1}\left(\frac{k}{2}\right) + \frac{k\sqrt{4 - k^2}}{2}\right]$

35. $1.26a$ **37.** 4π; 26.73

39. 63.46

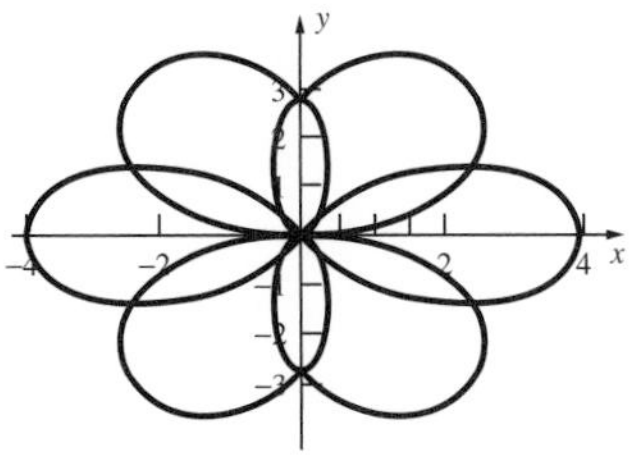

Chapter Review 10.8

Concepts Test

1. False **3.** False **5.** True **7.** True **9.** False
11. True **13.** False **15.** True **17.** True **19.** False
21. False **23.** False **25.** False **27.** False **29.** True
31. False **33.** True

Sample Test Problems

1. (a) (5); **(b)** (9); **(c)** (4); **(d)** (3); **(e)** (2); **(f)** (8); **(g)** (8); **(h)** (1); **(i)** (7); **(j)** (6)

3. Ellipse
Foci at $(0, \pm\sqrt{5})$
Vertices at $(0, \pm 3)$

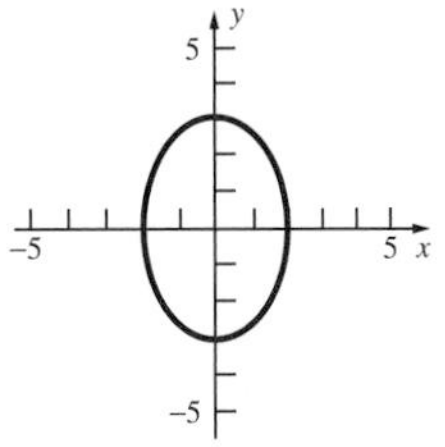

5. Parabola
Focus at $\left(0, -\frac{9}{4}\right)$
Vertex at $(0, 0)$

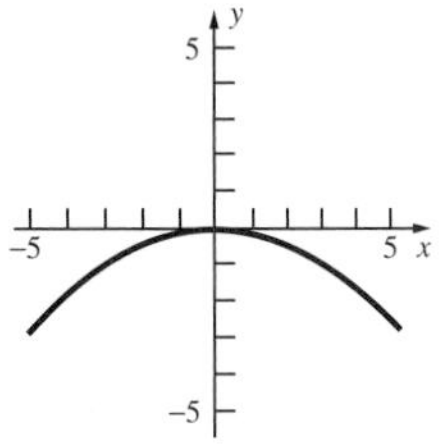

7. Ellipse
Foci at $(\pm 4, 0)$
Vertices at $(\pm 5, 0)$

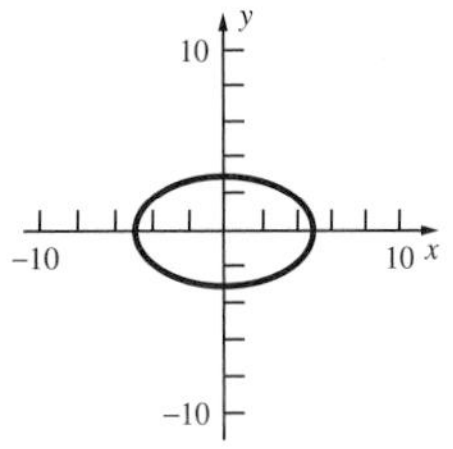

9. Parabola
Focus at $(0, 0)$
Vertex at $\left(0, \frac{5}{4}\right)$

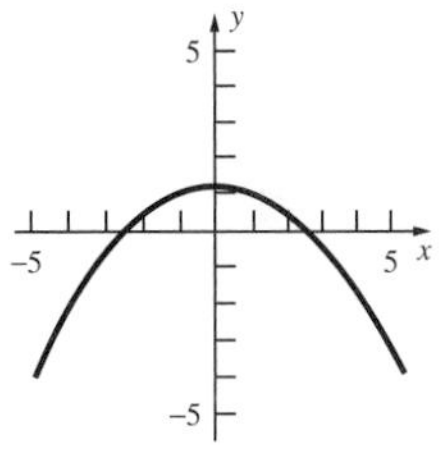

11. $\frac{x^2}{16} + \frac{y^2}{12} = 1$ **13.** $y^2 = -9x$ **15.** $\frac{x^2}{4} - y^2 = 1$

17. $\frac{(x - 1)^2}{25} + \frac{(y - 2)^2}{16} = 1$

19. Circle

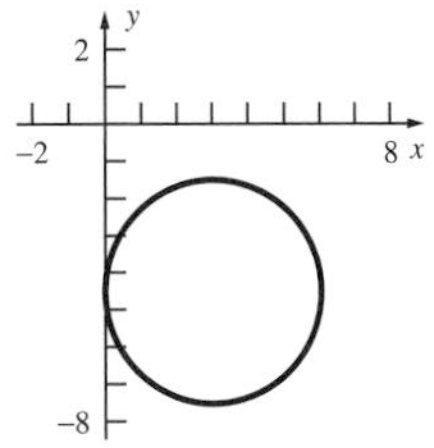

21. Parabola

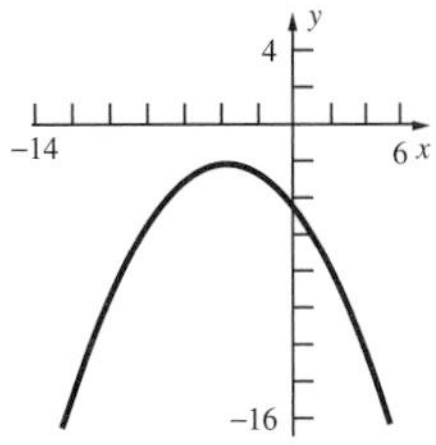

23. $r = \frac{5}{2}$; $s = -\frac{1}{2}$; hyperbola; $4\sqrt{6}$

25. $y = \frac{1}{3}(x - 2)$

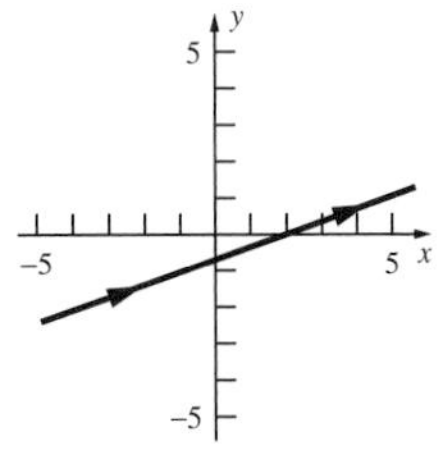

27. $\frac{(x + 2)^2}{16} + \frac{(y - 1)^2}{9} = 1$

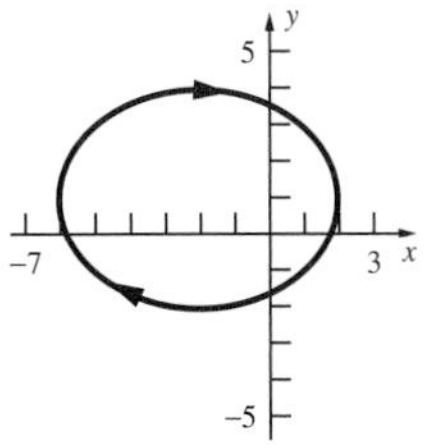

29. $y = -\frac{1}{2}(x - 7)$ **31.** $27\sqrt{2}$

33.

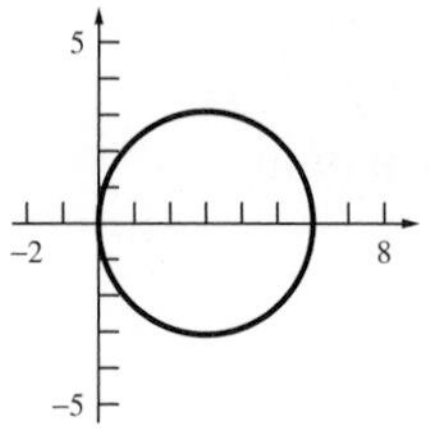

35.

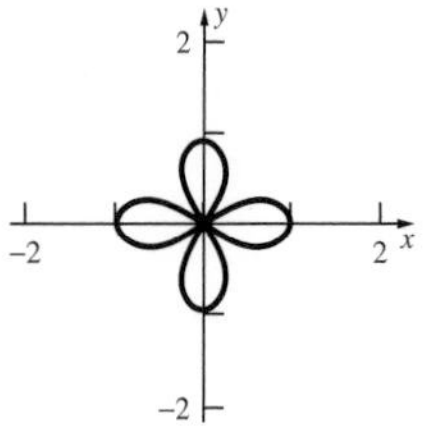

37.

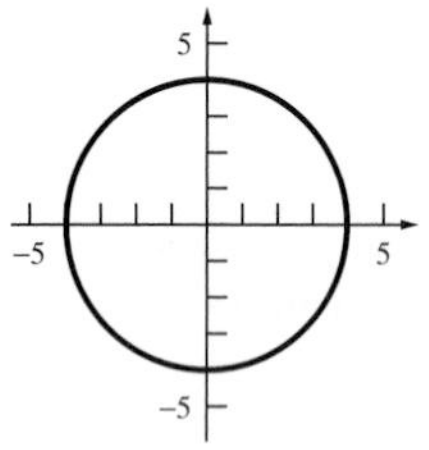

39.

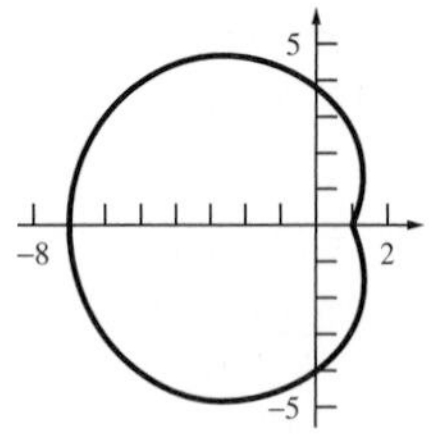

41.

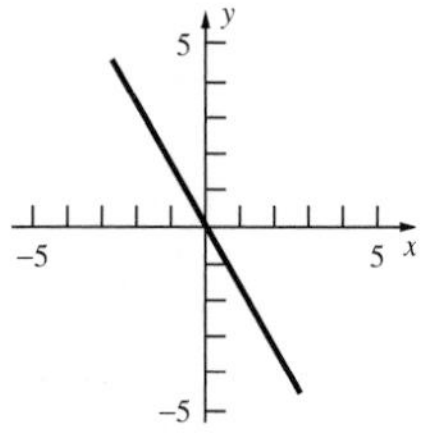

43.

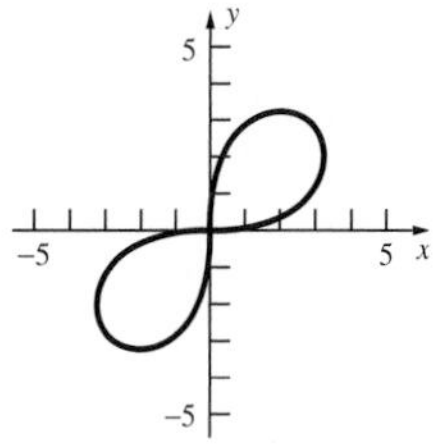

45. $(x-3)^2+(y-3)^2=9$

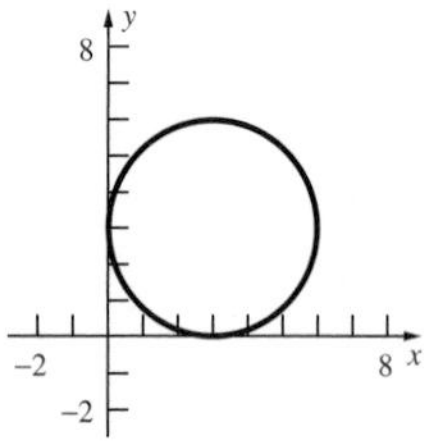

47. -1 **49.** $\frac{75}{2}\pi$ **51.** $\frac{22}{3}$

53. **(a)** I; **(b)** IV; **(c)** III; **(d)** II; **(e)** V

Chapter 11 Review and Preview Problems

1.

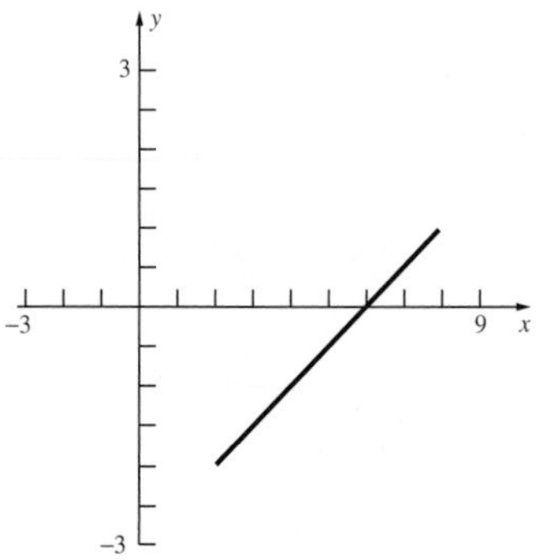

3.

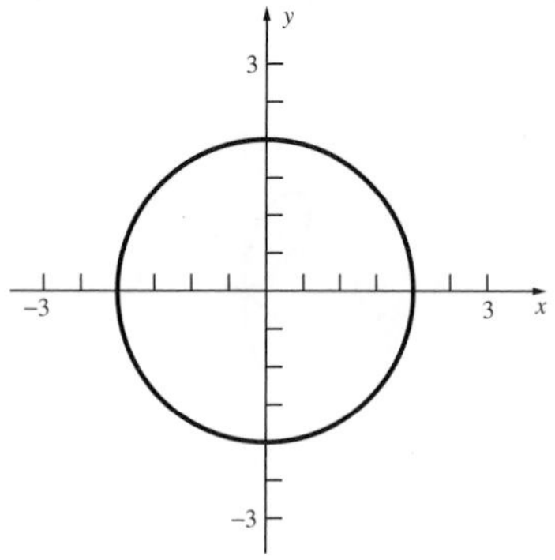

5.

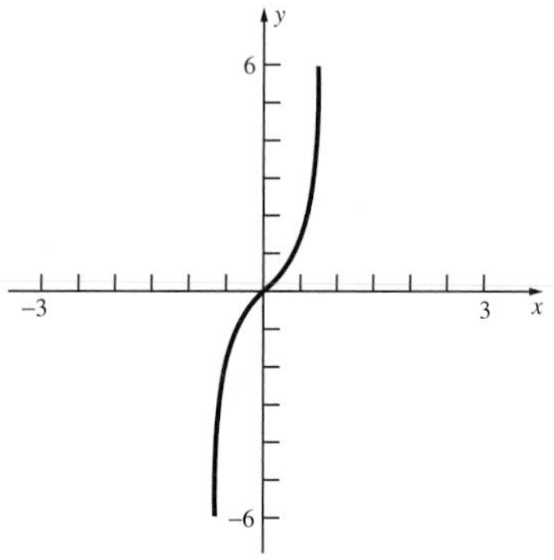

7. $x=h\cdot\cos\theta$
$y=h\cdot\sin\theta$

9. $\frac{1}{243}\left[(328)^{3/2}-8\right]\approx 24.4129$

11. $\pi|a|$ **13.** $(0.8, 2.6)$, distance is $\sqrt{0.8}$

15. **(a)** $v(t)=2t-6$; $a(t)=2$ **(b)** $t>3$

17.

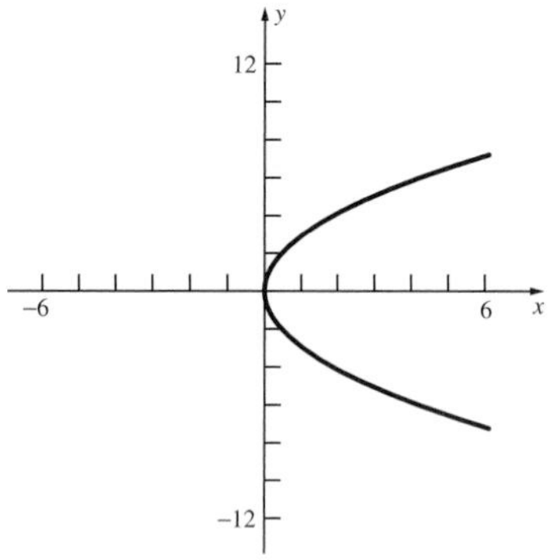

19.

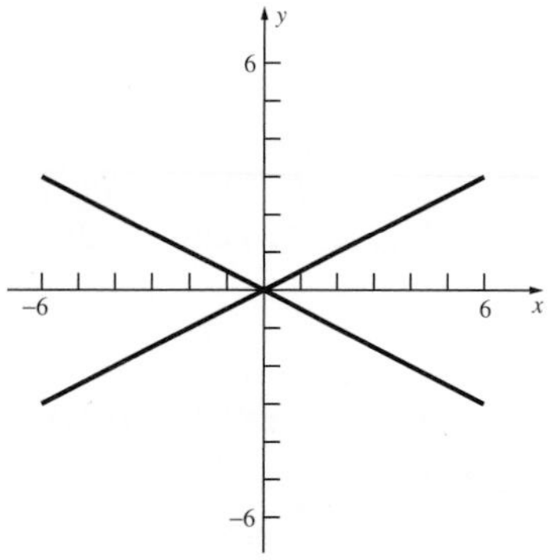

21.

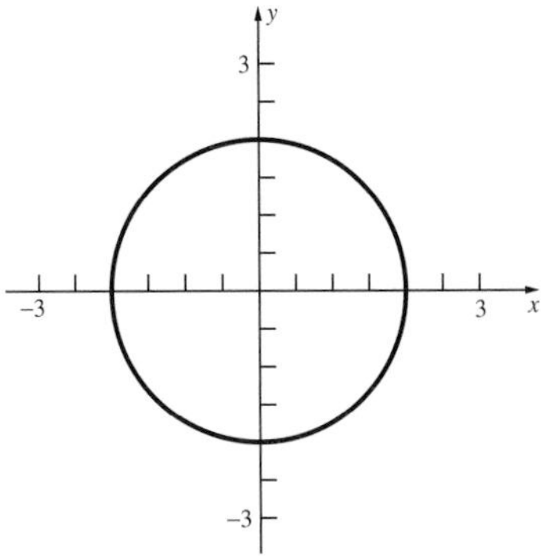

23.

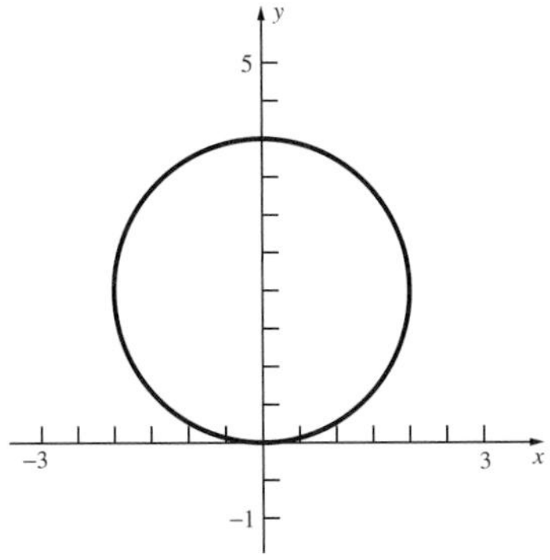

Problem Set 11.1

1. $A(1, 2, 3)$, $B(2, 0, 1)$, $C(-2, 4, 5)$, $D(0, 3, 0)$, $E(-1, -2, -3)$

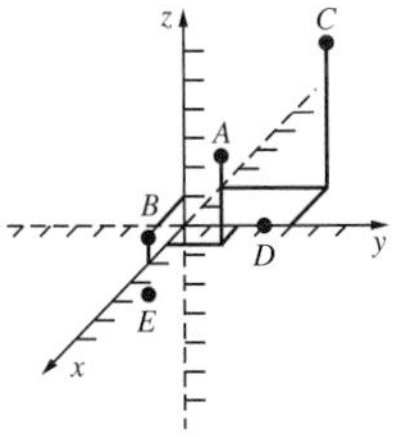

3. $x = 0; x = 0, y = 0$

5. (a) $\sqrt{43}$; **(b)** 5; **(c)** $\sqrt{(e + \pi)^2 + (\pi + 4)^2 + 3}$

9.

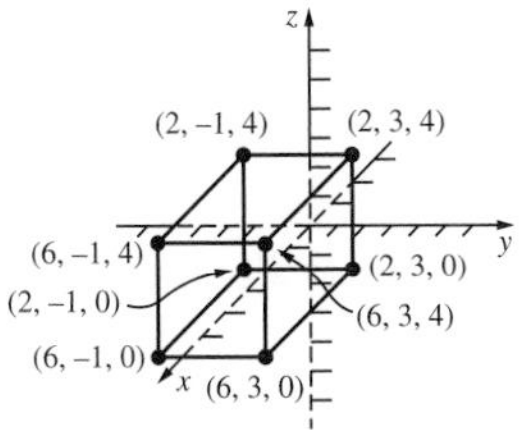

11. (a) $(x - 1)^2 + (y - 2)^2 + (z - 3)^2 = 25$;
(b) $(x + 2)^2 + (y + 3)^2 + (z + 6)^2 = 5$;
(c) $(x - \pi)^2 + (y - e)^2 + \left(z - \sqrt{2}\right)^2 = \pi$

13. $(6, -7, 4); 10$ **15.** $\left(\frac{1}{2}, -1, -2\right); \sqrt{\frac{17}{2}}$

17.

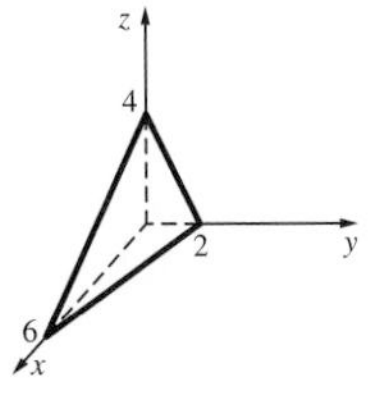

19.

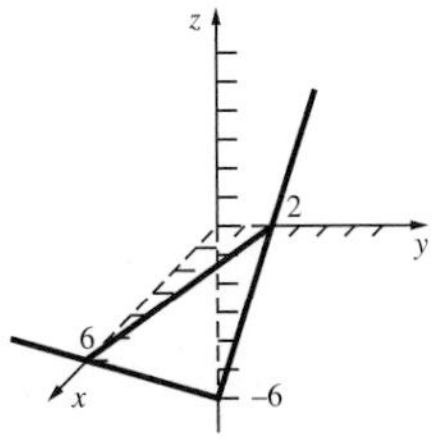

21.

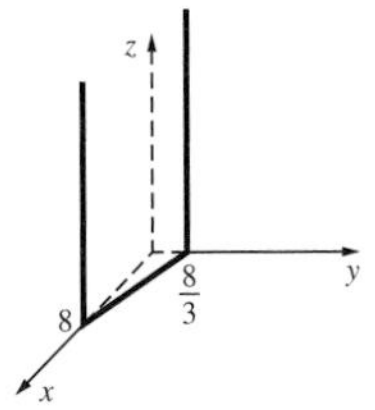

23.

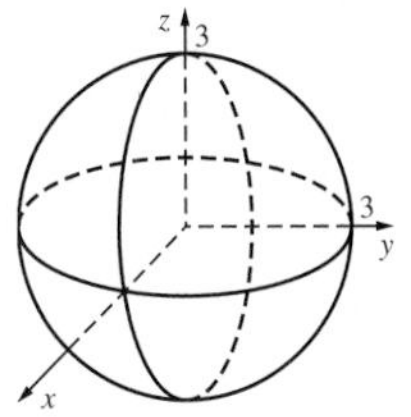

25. $2\sqrt{6}$ **27.** 16.59 **29.** 72 **31.** $2\pi\sqrt{13}$

33. 7.2273 **35.** 34.8394

37. $(x - 1)^2 + (y - 1)^2 + \left(z - \frac{11}{2}\right)^2 = \frac{53}{4}$

39. $(x - 6)^2 + (y - 6)^2 + (z - 6)^2 = 36$

41. (a) Plane parallel to and 2 units above the xy-plane;
(b) Plane perpendicular to the xy-plane, whose trace in the xy-plane is the line $x = y$;
(c) Union of the yz-plane ($x = 0$) and the xz-plane ($y = 0$);
(d) Union of the three coordinate planes;
(e) Cylinder of radius 2, parallel to the z-axis;
(f) Top half of the sphere with center $(0, 0, 0)$ and radius 3

43. Center $(1, 2, 5)$, radius 4 **45.** $\dfrac{11\pi}{12}$

Problem Set 11.2

1.

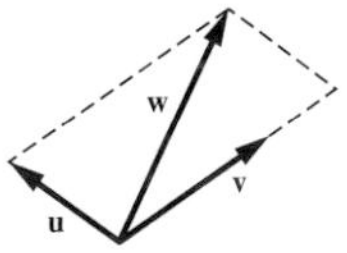

3.

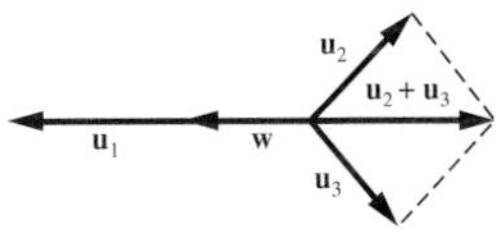

5. $\frac{1}{2}\mathbf{u} + \frac{1}{2}\mathbf{v}$ **7.** 1

9. $\mathbf{u} + \mathbf{v} = \langle 2, 4\rangle$; $\mathbf{u} - \mathbf{v} = \langle -4, -4\rangle$; $\|\mathbf{u}\| = 1$; $\|\mathbf{v}\| = 5$

11. $\mathbf{u} + \mathbf{v} = \langle 10, 14\rangle$; $\mathbf{u} - \mathbf{v} = \langle 14, 10\rangle$; $\|\mathbf{u}\| = 12\sqrt{2}$; $\|\mathbf{v}\| = 2\sqrt{2}$

13. $\mathbf{u} + \mathbf{v} = \langle 2, 4, 0\rangle$; $\mathbf{u} - \mathbf{v} = \langle -4, -4, 0\rangle$; $\|\mathbf{u}\| = 1$; $\|\mathbf{v}\| = 5$

15. $\mathbf{u} + \mathbf{v} = \langle -4, 0, 1\rangle$; $\mathbf{u} - \mathbf{v} = \langle 6, 0, 1\rangle$; $\|\mathbf{u}\| = \sqrt{2}$; $\|\mathbf{v}\| = 5$

17. $\|\mathbf{w}\| \approx 79.34$; S 7.5°W **19.** 150 N

21. N 2.08° E; 467 mi/h **23.** 80 mi/h

33. $\alpha + \beta = 143.13°$, $\beta + \gamma = 126.87°$, $\alpha + \gamma = 90°$

35. $50/\sqrt{2}$ lbs.

Problem Set 11.3

1. (a) $-12\mathbf{i} + 18\mathbf{j}$; **(b)** -13; **(c)** -28; **(d)** 375; **(e)** $-15\sqrt{13}$ **(f)** $13 - \sqrt{13}$

3. (a) $-\dfrac{7}{5\sqrt{2}}$ **(b)** $-\dfrac{1}{\sqrt{5}}$

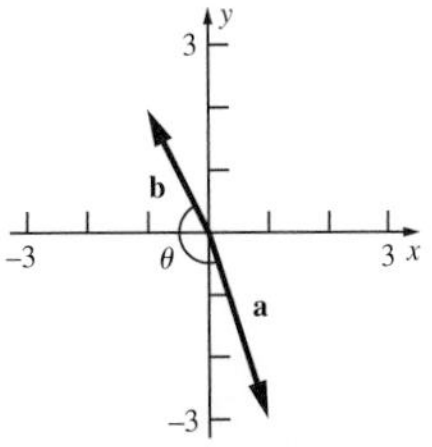

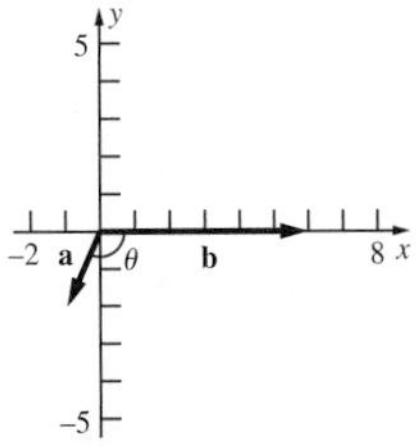

(c) 0 **(d)** $-\dfrac{51}{\sqrt{2665}}$

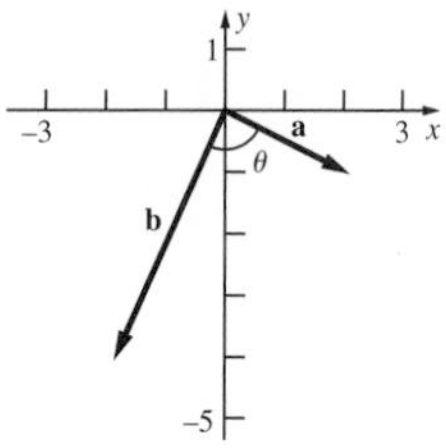

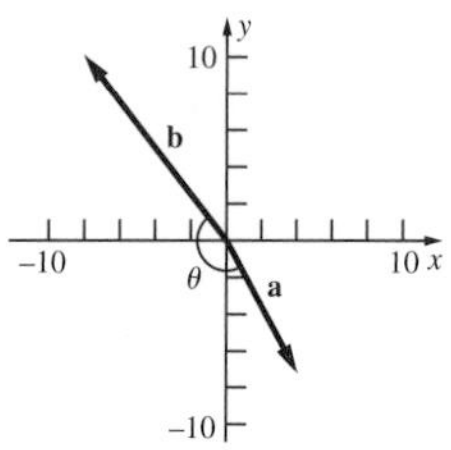

5. (a) 1; **(b)** 4; **(c)** $\dfrac{\sqrt{6}}{6}\mathbf{i} + \dfrac{\sqrt{6}}{3}\mathbf{j} - \dfrac{\sqrt{6}}{6}\mathbf{k}$; **(d)** 2; **(e)** $\sqrt{3}/6$; **(f)** 0

7. $\theta_{a,b} = 90°$; $\theta_{a,c} = 90°$; $\theta_{b,c} = 125.26°$

9. (a) $\alpha_a = 45°$, $\beta_a = 45°$, $\gamma_a = 90°$;
(b) $\alpha_b \approx 54.74°$, $\beta_b \approx 125.26°$, $\gamma_b \approx 54.74°$
(c) $\alpha_c \approx 131.81°$, $\beta_c \approx 48.19°$, $\gamma_c \approx 70.53°$

15. $\dfrac{10}{\sqrt{593}}\mathbf{i} - \dfrac{40}{\sqrt{593}}\mathbf{j} + \dfrac{240}{\sqrt{593}}\mathbf{k}$; $-\dfrac{10}{\sqrt{593}}\mathbf{i} + \dfrac{40}{\sqrt{593}}\mathbf{j} - \dfrac{240}{\sqrt{593}}\mathbf{k}$

17. $\cos^{-1}\dfrac{11}{\sqrt{129}}$ **19.** $\pm 2\sqrt{6}$ **21.** c is any number; $d = -2$

23. 0 **25.** $\left\langle \dfrac{11}{5}, \dfrac{22}{5} \right\rangle$ **27.** $\langle 0, 2\rangle$ **29.** $\langle 2, 0, -1\rangle$

31. $\left\langle \dfrac{15}{7}, \dfrac{10}{7}, \dfrac{5}{7} \right\rangle$ **33.** $\langle 0, 0, 1\rangle$

35. (a) $\mathbf{u}$; **(b)** $\mathbf{u}$

37. $\sqrt{3}$ **39.** $2\sqrt{3}$

41. infinitely many answers; one is $\mathbf{u} = \langle 1, 2, 0\rangle$, $\mathbf{v} = \langle -2, 1, -2\rangle$

43. a, b **51.** $k = \dfrac{2}{3}$, $m = -\dfrac{5}{3}$ **57.** 37.86°

59. 100 joules **61.** 94 ft-lb **63.** 32 joules

65. $2x - 4y + 3z = -15$ **67.** $x + 4y + 4z = 13$

69. 56.91°

71. (a) $z = 2$; **(b)** $2x - 3y - 4z = -13$ **73.** $7/\sqrt{11}$

75. $37/\sqrt{56}$ **77.** 0 **81.** $a = 5$, $b = -2$, $c = 5$

85. (4, 2, 3)

87. $a\mathbf{i} + b\mathbf{j} - c\mathbf{k}$; $a\mathbf{i} - b\mathbf{j} - c\mathbf{k}$; $-a\mathbf{i} - b\mathbf{j} - c\mathbf{k}$, the opposite of the original direction.

Problem Set 11.4

1. (a) $-4\mathbf{i} - 10\mathbf{j} - 4\mathbf{k}$; **(b)** $-6\mathbf{i} - 36\mathbf{j} - 27\mathbf{k}$; **(c)** 8; **(d)** $-98\mathbf{i} - 59\mathbf{j} + 88\mathbf{k}$

3. $c(-14\mathbf{i} - 2\mathbf{j} + 6\mathbf{k})$, c in $\mathbb{R}$ **5.** $\pm\left\langle \dfrac{7}{\sqrt{86}}, -\dfrac{1}{\sqrt{86}}, \dfrac{6}{\sqrt{86}} \right\rangle$

7. $2\sqrt{74}$ **9.** $4\sqrt{6}$ **11.** $2x - y - z = -3$

13. $15x + 35y + 21z = 105$ **15.** $x - y + 2z = -1$

17. $7x + 5y + 4z = -5$ **19.** $x - 2y - 2z = 4$

21. $-x + 10y + 17z = -3$ **23.** 69

25. (a) 9: **(b)** $\sqrt{35}$: **(c)** 40.01°

27. (c),(d) **35.** $\frac{1}{2}\sqrt{a^2b^2 + a^2c^2 + b^2c^2}$

Problem Set 11.5

1. $2\mathbf{i} - \mathbf{j}$ **3.** $\frac{1}{2}\mathbf{i} - 4\mathbf{j}$ **5.** $\mathbf{i}$ **7.** Does not exist

9. (a) $\{t \in \mathbb{R}: t \le 3\}$; **(b)** $\{t \in \mathbb{R}: t \le 20\}$
(c) $\{t \in \mathbb{R}: -3 \le t \le 3\}$

11. (a) $\{t \in \mathbb{R}: t \le 3\}$; **(b)** $\{t \in \mathbb{R}: t < 20, t^2 \text{ not an integer}\}$
(c) $\{t \in \mathbb{R}: -3 \le t \le 3\}$

13. (a) $9(3t + 4)^2\mathbf{i} + 2te^{t^2}\mathbf{j}$; $54(3t + 4)\mathbf{i} + 2(2t^2 + 1)e^{t^2}\mathbf{j}$;
(b) $\sin 2t\mathbf{i} - 3\sin 3t\mathbf{j} + 2t\mathbf{k}$; $2\cos 2t\mathbf{i} - 9\cos 3t\mathbf{j} + 2\mathbf{k}$

15. $-2e^{-2t} - \dfrac{4}{t^3} + \dfrac{4}{t^3}\ln t^2$

17. $-\dfrac{e^{-3t}}{2}\left(\dfrac{6t - 7}{\sqrt{t - 1}}\right)\mathbf{i} + e^{-3t}\left(\dfrac{2}{t} - 3\ln(2t^2)\right)\mathbf{j}$

19. $\mathbf{v}(1) = 4\mathbf{i} + 10\mathbf{j} + 2\mathbf{k}$; $\mathbf{a}(1) = 10\mathbf{j}$; $s(1) = 2\sqrt{30}$

21. $\mathbf{v}(2) = -\frac{1}{4}\mathbf{i} - \frac{4}{9}\mathbf{j} + 80\mathbf{k}$; $\mathbf{a}(2) = \frac{1}{4}\mathbf{i} + \frac{26}{27}\mathbf{j} + 160\mathbf{k}$;
$s(2) = \dfrac{\sqrt{8{,}294{,}737}}{36}$

23. $\mathbf{v}(2) = 4\mathbf{j} + \dfrac{2^{2/3}}{3}\mathbf{k}$; $\mathbf{a}(2) = 4\mathbf{j} - \dfrac{1}{9\sqrt[3]{2}}\mathbf{k}$;
$s(2) = \sqrt{16 + \dfrac{2^{4/3}}{9}}$

25. $\mathbf{v}(\pi) = -\mathbf{j} + \mathbf{k}$; $\mathbf{a}(\pi) = \mathbf{i}$, $s(\pi) = \sqrt{2}$

27. $\mathbf{v}\left(\frac{\pi}{4}\right) = 2\mathbf{i} + 3e^{\pi/4}\mathbf{j}$; $\mathbf{a}\left(\frac{\pi}{4}\right) = 4\mathbf{i} + 3e^{\pi/4}\mathbf{j} + 16\mathbf{k}$;
$s\left(\frac{\pi}{4}\right) = \sqrt{4 + 9e^{\pi/2}}$

29. $\mathbf{v}(2) = 2\pi\mathbf{i} + \mathbf{j} - e^{-2}\mathbf{k}$; $\mathbf{a}(2) = 2\pi\mathbf{i} - 2\pi^2\mathbf{j} + e^{-2}\mathbf{k}$;
$s(2) = \sqrt{4\pi^2 + 1 + e^{-4}}$

33. $2\sqrt{2}$ **35.** 144 **37.** $\sqrt{41}$

39. $-6t\sin(3t^2 - 4)\mathbf{i} + 18te^{9t^2 - 12}\mathbf{j}$

41. $(e - 1)\mathbf{i} + (1 - e^{-1})\mathbf{j}$

43. $\mathbf{r}(t) = 5\cos(6t)\mathbf{i} + 5\sin(6t)\mathbf{j}$;
$\mathbf{v}(t) = -30\sin(6t)\mathbf{i} + 30\cos(6t)\mathbf{j}$;
$\|\mathbf{v}(t)\| = 30$; $\mathbf{a}(t) = -180\cos(6t)\mathbf{i} - 180\sin(6t)\mathbf{j}$

45. (b) $R_p = 10R_m$; $t = \frac{\pi}{9}$

47. (a) Winding upward around the right circular cylinder $x = \sin t$, $y = \cos t$, as t increases.
(b) Same as part (a), but winding much faster by a factor of $3t^2$.
(c) With standard orientation of the axes, the motion is winding to the right around the right circular cylinder $x = \sin t$, $z = \cos t$.
(d) Spiraling upward, with increasing radius, along the spiral $x = t\sin t$, $y = t\cos t$.
(e) Spiraling upward, with decreasing radius, along the spiral $x = \dfrac{1}{t^2}\sin t$, $y = \dfrac{1}{t^2}\cos t$.
(f) Spiraling to the right, with increasing radius along the spiral $x = t^2\sin(\ln t)$, $z = t^2\cos(\ln t)$.

Problem Set 11.6

1. $x = 1 + 3t$, $y = -2 + 7t$, $z = 3 + 3t$

3. $x = 4 + t$, $y = 2$, $z = 3 - 2t$

5. $x = 4 + 3t$, $y = 5 + 2t$, $z = 6 + t$; $\dfrac{x - 4}{3} = \dfrac{y - 5}{2} = \dfrac{z - 6}{1}$

7. $x = 1 + t$, $y = 1 + 10t$, $z = 1 + 100t$;
$\dfrac{x - 1}{1} = \dfrac{y - 1}{10} = \dfrac{z - 1}{100}$

9. $\dfrac{x - 4}{27} = \dfrac{y + 5}{-50} = \dfrac{z}{-6}$ **11.** $\dfrac{x + 8}{10} = \dfrac{y}{2} = \dfrac{z + \frac{21}{2}}{9}$

13. $\dfrac{x - 4}{1} = \dfrac{y}{-5} = \dfrac{z - 6}{2}$ **15.** $x = 5t$, $y = -3t$, $z = 4$

17. $x + y + 6z = 11$ **19.** $3x - 2y = 5$

21. **(b)** $2x + y - z = 7$; **(c)** $(-1, 2, -1)$; **(d)** $\sqrt{6}$

23. $\dfrac{x-1}{-\sqrt{3}} = \dfrac{y - 3\sqrt{3}}{3} = \dfrac{z - \frac{\pi}{3}}{1}$ **25.** $3x - 4y + 5z = -22$

27. $\left(-\dfrac{1}{2}, 0, \dfrac{37}{4\sqrt{7}}\right)$ **29.** $\left(\dfrac{5}{2}, 1, 0\right)$

31. **(a)** $\dfrac{8\sqrt{2}}{3}$; **(b)** $\dfrac{3\sqrt{26}}{7}$

Problem Set 11.7

1. $\mathbf{v}(1) = \langle 1, 2\rangle$; $\mathbf{a}(1) = \langle 0, 2\rangle$

$\mathbf{T}(1) = \left\langle \dfrac{1}{\sqrt{5}}, \dfrac{2}{\sqrt{5}}\right\rangle$; $\kappa = \dfrac{2}{5^{3/2}}$

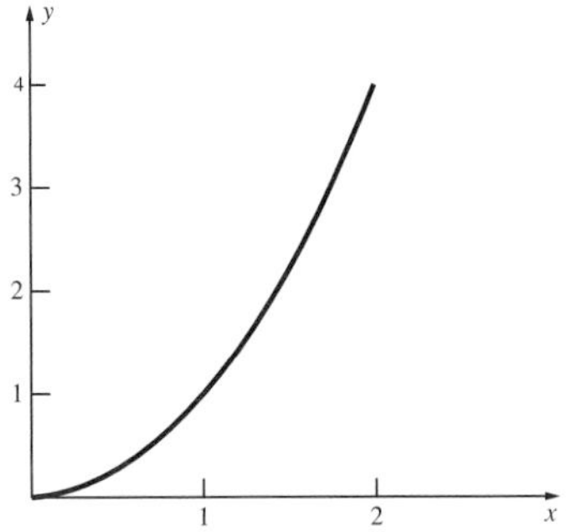

3. $\mathbf{v}(\pi) = \langle 1, 0, -2\rangle$; $\mathbf{a}(\pi) = \langle 0, 2, 0\rangle$

$\mathbf{T}(\pi) = \left\langle \dfrac{1}{\sqrt{5}}, 0, -\dfrac{2}{\sqrt{5}}\right\rangle$; $\kappa = \dfrac{2}{5}$

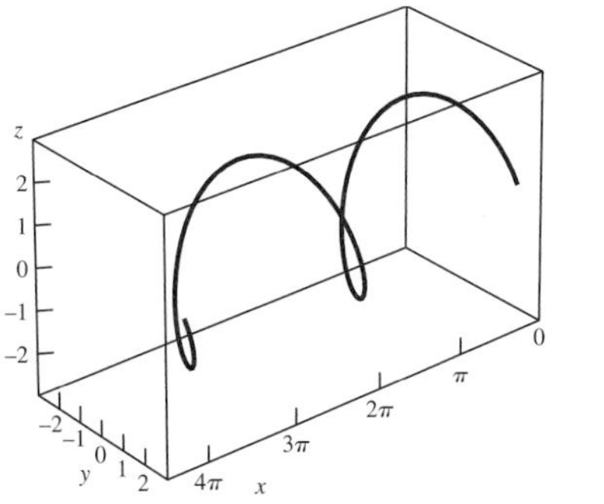

5. $\mathbf{v}(\pi) = \left\langle \dfrac{\pi}{4}, 0, -5\right\rangle$; $\mathbf{a}(\pi) = \left\langle \dfrac{1}{4}, 5, 0\right\rangle$

$\mathbf{T}(\pi) = \left\langle \dfrac{\pi}{\sqrt{400 + \pi^2}}, 0, -\dfrac{20}{\sqrt{400 + \pi^2}}\right\rangle$; $\kappa \approx 0.195422$

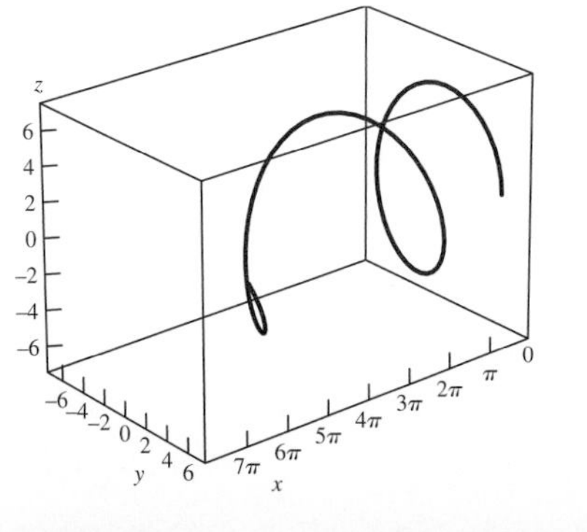

7. $\dfrac{1}{\sqrt{2}}\mathbf{i} + \dfrac{1}{\sqrt{2}}\mathbf{j}$; $\dfrac{1}{4\sqrt{2}}$ **9.** $-\dfrac{3}{5}\mathbf{i} + \dfrac{4}{5}\mathbf{j}$; $\dfrac{24\sqrt{2}}{125}$

11. $-\dfrac{2}{\sqrt{13}}\mathbf{i} - \dfrac{3}{\sqrt{13}}\mathbf{j}$; $\dfrac{6}{13\sqrt{13}}$ **13.** $-\dfrac{1}{\sqrt{2}}\mathbf{i} + \dfrac{1}{\sqrt{2}}\mathbf{j}$; $\dfrac{1}{\sqrt{2}}$

15. $\kappa = \dfrac{4}{17\sqrt{17}}$; $R = \dfrac{17\sqrt{17}}{4}$ **17.** $\kappa = \dfrac{2}{3\sqrt{3}}$; $R = \dfrac{3\sqrt{3}}{2}$

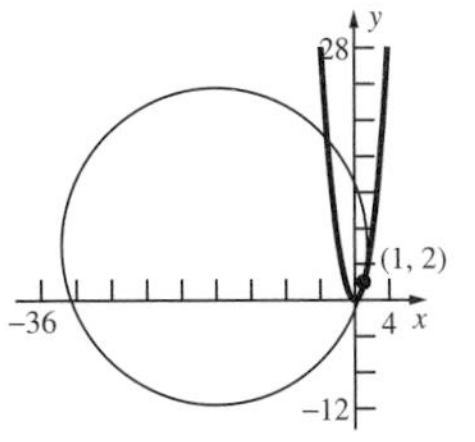

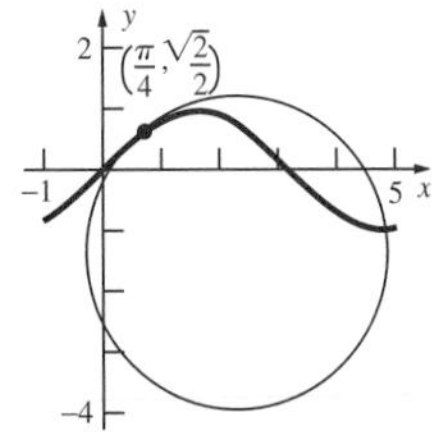

19. $\kappa = \frac{2}{25}$; $R = \frac{25}{2}$ **21.** $\kappa = \frac{1}{4}$; $R = 4$

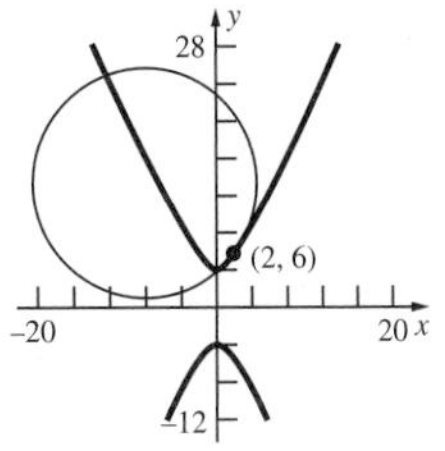

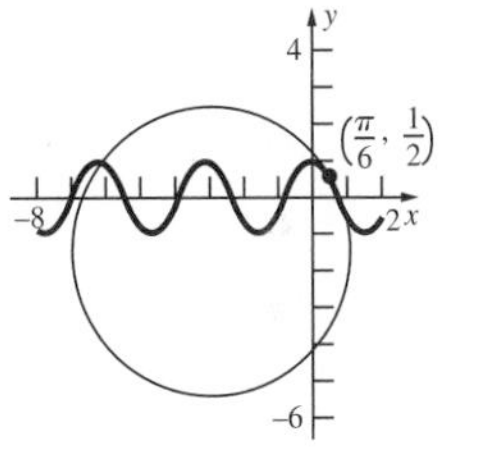

23. $\kappa = \dfrac{4}{5\sqrt{5}}$; $R = \dfrac{5\sqrt{5}}{4}$ **25.** $\kappa = \dfrac{3}{5\sqrt{10}}$; $R = \dfrac{5\sqrt{10}}{3}$

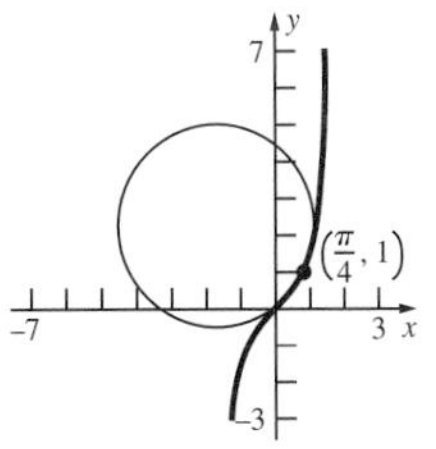

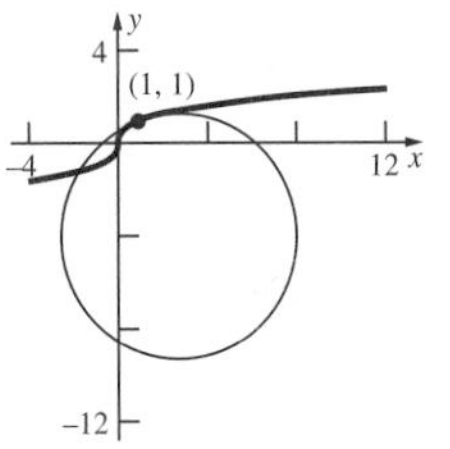

27. $\dfrac{\sqrt{11}}{21\sqrt{7}}$; $\mathbf{T} = \dfrac{2}{\sqrt{21}}\mathbf{i} + \dfrac{1}{\sqrt{21}}\mathbf{j} + \dfrac{4}{\sqrt{21}}\mathbf{k}$;

$\mathbf{N} = -\dfrac{5}{\sqrt{77}}\mathbf{i} - \dfrac{6}{\sqrt{77}}\mathbf{j} + \dfrac{4}{\sqrt{77}}\mathbf{k}$;

$\mathbf{B} = \dfrac{4}{\sqrt{33}}\mathbf{i} - \dfrac{4}{\sqrt{33}}\mathbf{j} - \dfrac{1}{\sqrt{33}}\mathbf{k}$

29. $\kappa = \frac{9}{91}$; $\mathbf{T} = \left\langle -\dfrac{3}{\sqrt{13}}, 0, \dfrac{2}{\sqrt{13}}\right\rangle$; $\mathbf{N} = \langle 0, 1, 0\rangle$;

$\mathbf{B} = \left\langle -\dfrac{2}{\sqrt{13}}, 0, -\dfrac{3}{\sqrt{13}}\right\rangle$

31. $\kappa = \frac{1}{3}\operatorname{sech}^2\frac{1}{3}$; $\mathbf{T} = \tanh\frac{1}{3}\mathbf{i} + \operatorname{sech}\frac{1}{3}\mathbf{j}$;

$\mathbf{N} = \operatorname{sech}\frac{1}{3}\mathbf{i} - \tanh\frac{1}{3}\mathbf{j}$; $\mathbf{B} = -\mathbf{k}$

33. $\kappa = \dfrac{1}{2\sqrt{2}}$; $\mathbf{T} = -\dfrac{1}{2}\mathbf{i} + \dfrac{1}{2}\mathbf{j} + \dfrac{1}{\sqrt{2}}\mathbf{k}$;

$\mathbf{N} = \dfrac{1}{\sqrt{2}}\mathbf{i} + \dfrac{1}{\sqrt{2}}\mathbf{j}$; $\mathbf{B} = -\dfrac{1}{2}\mathbf{i} + \dfrac{1}{2}\mathbf{j} - \dfrac{1}{\sqrt{2}}\mathbf{k}$

35. $\left(\dfrac{1}{\sqrt{2}}, -\dfrac{\ln 2}{2}\right)$ **37.** $(0, 1)$ **39.** $\left(-\dfrac{1}{2}\ln 2, \dfrac{1}{\sqrt{2}}\right)$

41. $a_T = \dfrac{12}{\sqrt{13}}, a_N = \dfrac{18}{\sqrt{13}}$ **43.** $a_T = -\sqrt{2}, a_N = \sqrt{2}$

45. $a_T = \dfrac{40a}{3\sqrt{41}}, a_N = \dfrac{3a}{\sqrt{41}}$

47. $a_T(1) = \dfrac{4}{\sqrt{14}};\quad a_N(1) = 2\sqrt{\dfrac{5}{7}}$

49. $a_T(0) = 0;\quad a_N(0) = \sqrt{2}$

51. $a_T(3) = 36\sqrt{\dfrac{3}{55}};\quad a_N(3) = 6\sqrt{\dfrac{2}{55}}$

53. $(0, 0); (1, 0), (-1, 0)$

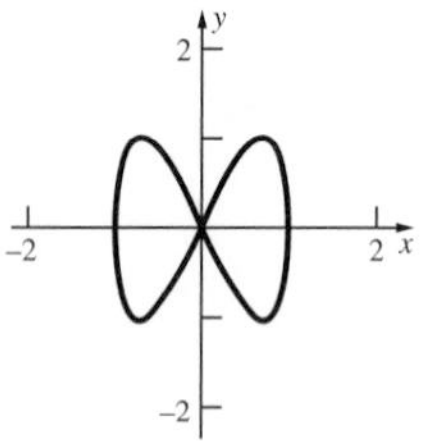

55. The speed is constant; the curvature is zero.

57. $(\cos 5)\mathbf{i} - (\sin 5)\mathbf{j} + 7\mathbf{k}$ **59.** $5\mathbf{T} + 5\mathbf{N}; -\mathbf{i} - 7\mathbf{j}$

61. 72 ft/s **67.** $P_5(x) = 10x^3 - 15x^4 + 6x^5$ **71.** $\frac{3}{4}$

73. $\dfrac{3}{8\sqrt{2}}$ **75.** $\frac{3}{16}$

79. max ≈ 0.7606; min ≈ 0.1248

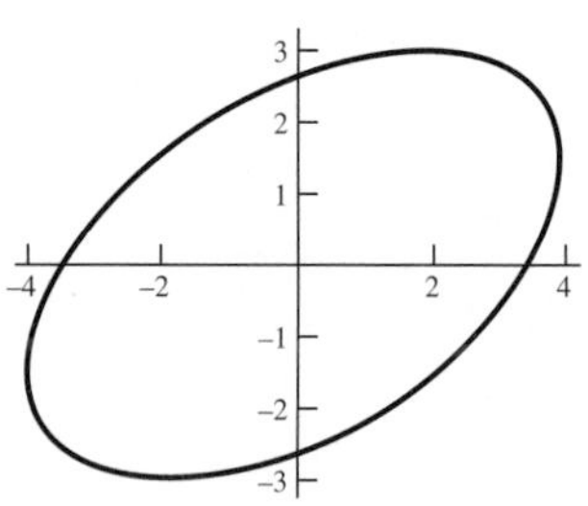

85. $(6, 0, 8); 8\sqrt{9\pi^2 + 1}$

Problem Set 11.8

1. Elliptic cylinder

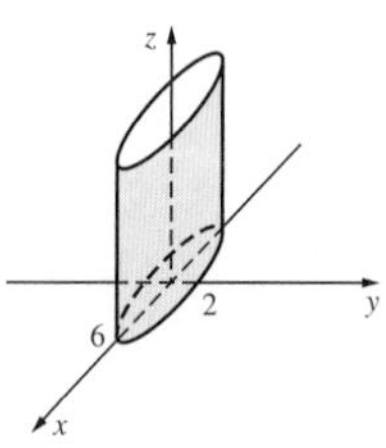

3. Plane

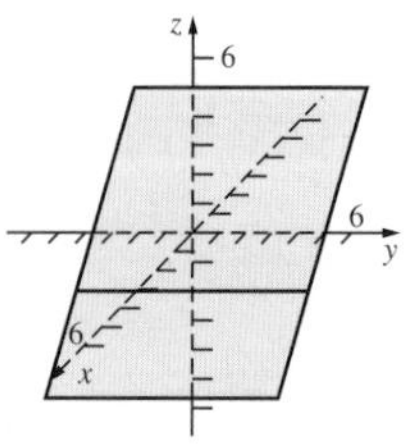

5. Circular cylinder

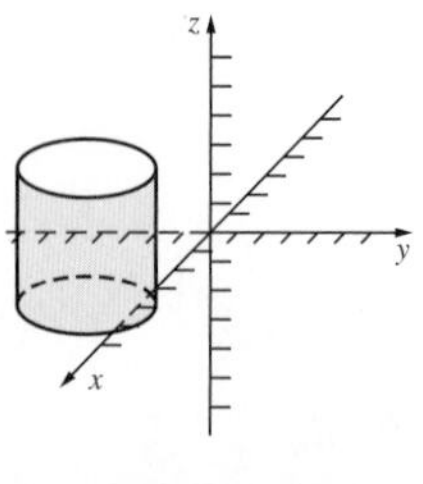

7. Ellipsoid

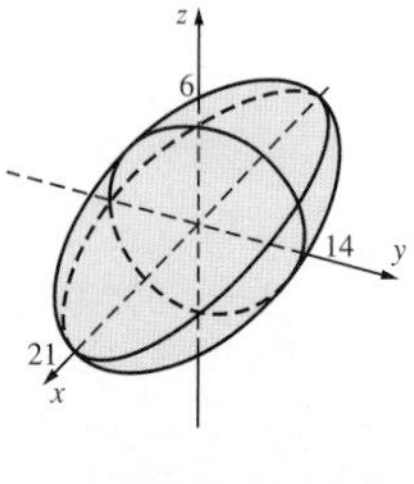

9. Elliptic paraboloid

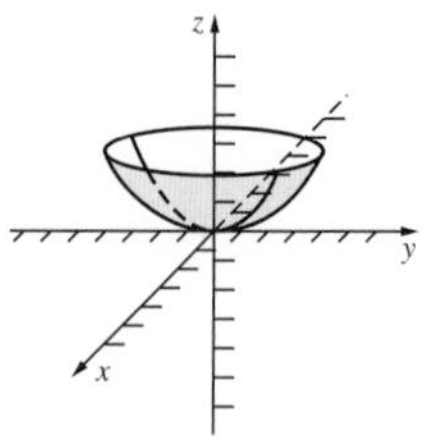

11. Cylinder

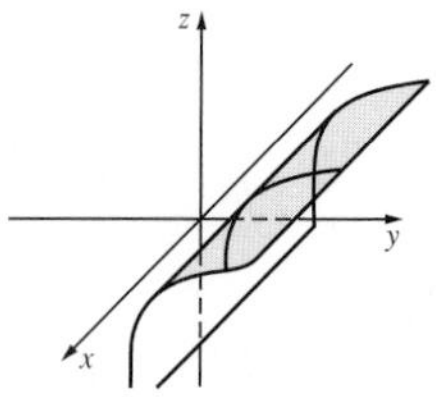

13. Hyperbolic paraboloid

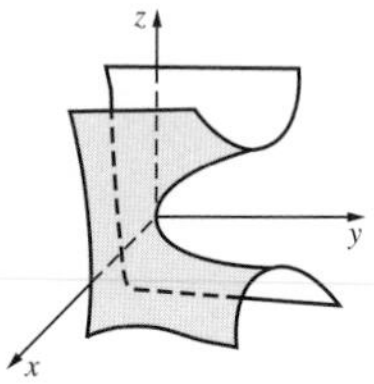

15. Elliptic paraboloid

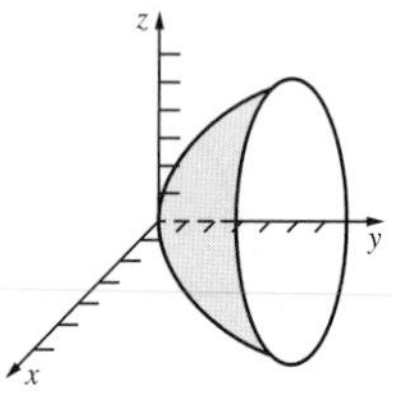

17. Plane

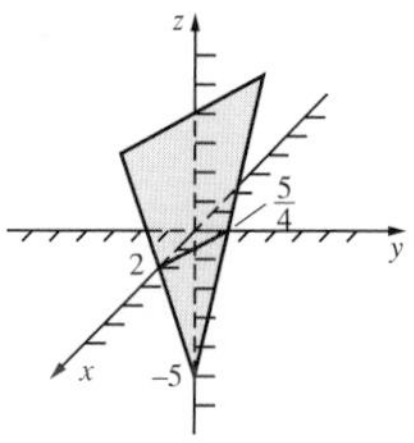

19. Hemisphere

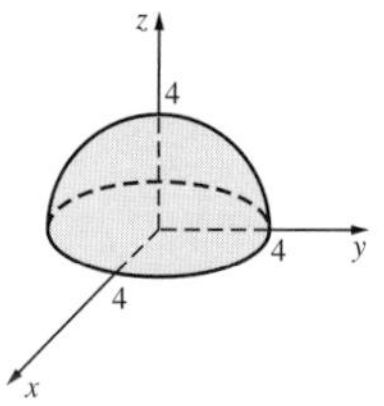

21. (a) Replacing x by $-x$ results in an equivalent equation.
(b) Replacing x by $-x$ and y by $-y$ results in an equivalent equation.
(c) Replacing x by $-x$, y by $-y$, and z by $-z$ results in an equivalent equation.

23. All central ellipsoids are symmetric with respect to (a) the origin, (b) the x-axis, and the (c) xy-plane.

25. All central hyperboloids of two sheets are symmetric with respect to (a) the origin, (b) the z-axis, and (c) the yz-plane.

27. $y = 2x^2 + 2z^2$ **29.** $4x^2 + 3y^2 + 4z^2 = 12$

31. $\left(0, \pm 2\sqrt{5}, 4\right)$ **33.** $\dfrac{\pi ab(c^2 - h^2)}{c^2}$

35. Major diameter 4; minor diameter $2\sqrt{2}$

37. $x^2 + 9y^2 - 9z^2 = 0$

Problem Set 11.9

1. Cylindrical to Spherical: $\rho = \sqrt{r^2 + z^2}, \cos\phi = \dfrac{z}{\sqrt{r^2 + z^2}},$ $\theta = \theta$

Spherical to Cylindrical: $r = \rho\sin\phi, z = \rho\cos\phi, \theta = \theta$

3. (a) $\left(3\sqrt{3}, 3, -2\right)$; **(b)** $\left(-2, -2\sqrt{3}, -8\right)$

5. (a) $\left(4\sqrt{2}, \frac{5\pi}{3}, \frac{\pi}{4}\right)$; **(b)** $\left(4, \frac{3\pi}{4}, \frac{\pi}{6}\right)$

7.

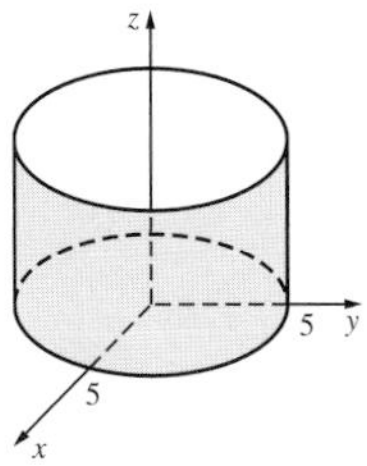

9.

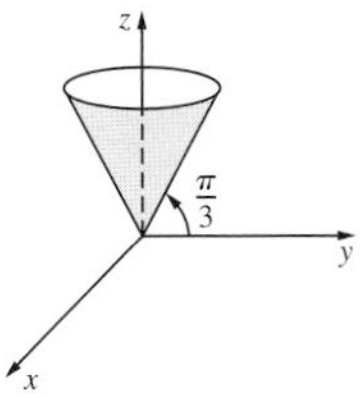

11.

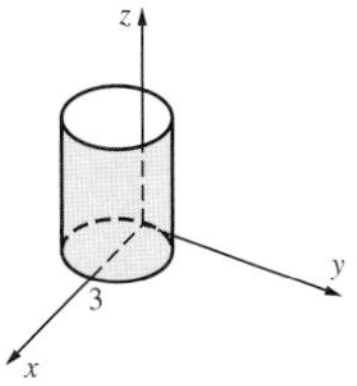

13.

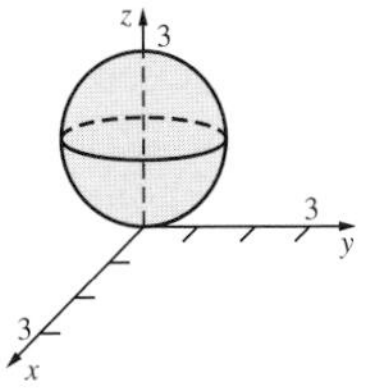

15.

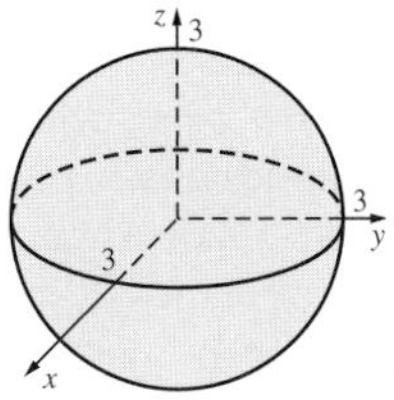

17. $r = 3$ **19.** $r^2 + 4z^2 = 10$ **21.** $\cos^2 \phi = \frac{1}{3}$

23. $\rho^2 = \dfrac{4}{1 + \cos^2 \phi}$ **25.** $r = \dfrac{4}{\sin \theta + \cos \theta}$

27. $\rho \sin \phi = 3$ **29.** $x^2 - y^2 = z$ **31.** $z = 2r^2$

33. 4029 mi **35.** 4552 mi **37.** 2252 mi

41. (a) 3485 mi; **(b)** 4552 mi; **(c)** 9331 mi; **(d)** 7798 mi; **(e)** 12,441 mi

Chapter Review 11.10

Concepts Test

1. True **3.** True **5.** False **7.** True **9.** True
11. False **13.** True **15.** True **17.** True **19.** True
21. True **23.** False **25.** True **27.** True **29.** True
31. True **33.** False **35.** False **37.** False **39.** False
41. False **43.** True **45.** True **47.** False **49.** True
51. True **53.** False **55.** False **57.** False

Sample Test Problems

1. $(x - 1)^2 + (y - 2)^2 + (z - 4)^2 = 11$

3. (a) $\langle 4, -17 \rangle$ **(b)** -3 **(c)** -15 **(d)** -234 **(e)** -36 **(f)** 30

5. (a) $2\mathbf{i} + \mathbf{j} + 4\mathbf{k}$; **(b)** -9; **(c)** -14; **(d)** does not exist; **(e)** $\sqrt{17}$; **(f)** 7

7.

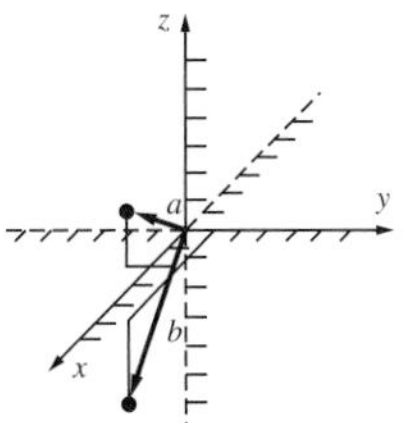

(a) $3; \sqrt{35}$; **(b)** $\dfrac{2}{3}, -\dfrac{1}{3}, \dfrac{2}{3}; \dfrac{5}{\sqrt{35}}, \dfrac{1}{\sqrt{35}}, -\dfrac{3}{\sqrt{35}}$;

(c) $\frac{2}{3}\mathbf{i} - \frac{1}{3}\mathbf{j} + \frac{2}{3}\mathbf{k}$; **(d)** $\cos^{-1}\dfrac{1}{\sqrt{35}}$

9. $c\langle 10, -11, -3 \rangle$, c in $\mathbb{R}$

11. (a) $y = 7$; **(b)** $x = -5$; **(c)** $z = -2$;

(d) $3x - 4y + z = -45$

13. 1 **15.** $x = -2 + 8t, y = 1 + t, z = 5 - 8t$

17. $x = 2t, y = 25 + t, z = 16$

19. $\mathbf{r}(t) = \langle 2, -2, 1 \rangle + t\langle 5, -4, -3 \rangle$

21. Tangent line: $\dfrac{x - 2}{1} = \dfrac{y - 2}{2} = \dfrac{z - \frac{8}{3}}{4}$

Normal Plane: $3x + 6y + 12z = 50$

23. $\sqrt{3}(e^5 - e)$ **25.** N 12.22°W; 409.27 mi/h

27. (a) $\left\langle \dfrac{1}{t}, -6t \right\rangle; \left\langle -\dfrac{1}{t^2}, -6 \right\rangle$;

(b) $\langle \cos t, -2 \sin 2t \rangle; \langle -\sin t, -4 \cos 2t \rangle$;

(c) $\langle \sec^2 t, -4t^3 \rangle; \langle 2 \sec^2 t \tan t, -12t^2 \rangle$;

29. $a_T = \dfrac{22}{\sqrt{14}}; a_N = \dfrac{2\sqrt{19}}{\sqrt{14}}$

31. Sphere

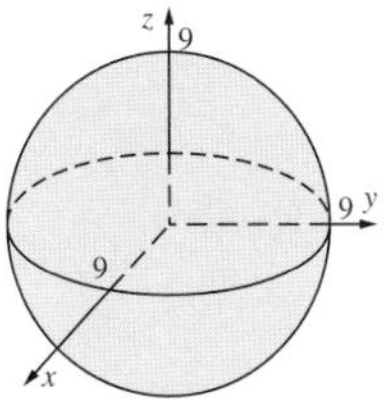

33. Circular paraboloid

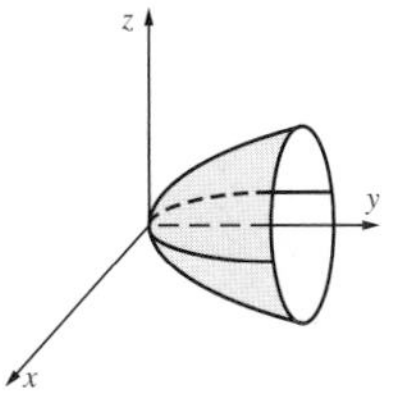

35. Plane

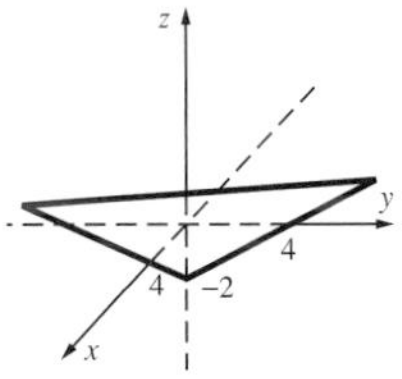

37. Ellipsoid

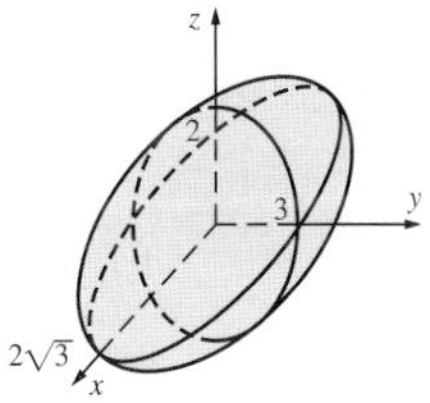

39. (a) $r = 3$; **(b)** $r^2 = \dfrac{16}{1 + 3 \sin^2 \theta}$; **(c)** $r^2 = 9z$;

(d) $r^2 + 4z^2 = 10$

41. (a) $\rho = 2$; **(b)** $\cos^2 \phi = \frac{1}{2}$ (Other forms are possible.);

(c) $\rho^2 = \dfrac{1}{2 \sin^2 \phi \cos^2 \theta - 1}$; **(d)** $r = \cot \phi \csc \phi$

43. 1.25

Chapter 12 Review and Preview Problems

1. $x^2 + y^2 + z^2 = 64$

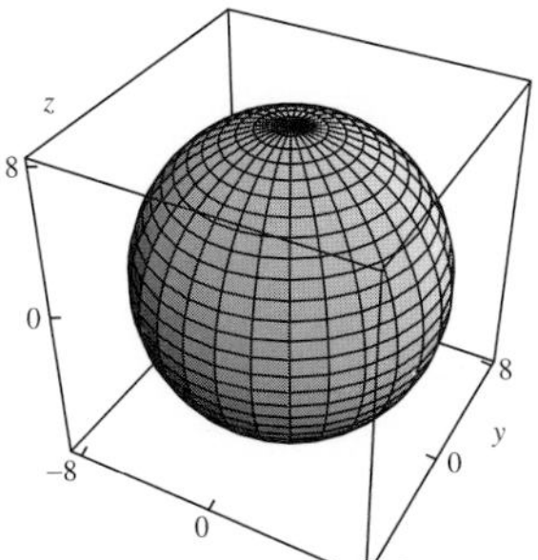

3. $z = x^2 + 4y^2$

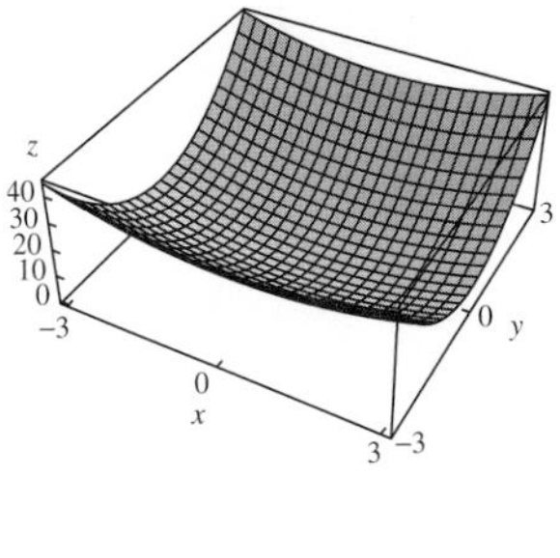

5. (a) $6x^2$; **(b)** $15x^2$; **(c)** $3kx^2$; **(d)** $3ax^2$

7. (a) $2 \cos 2a$; **(b)** $17 \cos 17a$; **(c)** $t \cos ta$; **(d)** $s \cos sa$

9. Continuous and differentiable at $x = 2$

11. Continuous at $x = 4$; not differentiable at $x = 4$

13. maximum value of f on $[0, 4]$ is 5; minimum value is -15

15. $S(r) = 2\pi r^2 + \dfrac{16}{r}$

Problem Set 12.1

1. (a) 5; **(b)** 0; **(c)** 6; **(d)** $a^6 + a^2$; **(e)** $2x^2$;
(f) $(2, -4)$ is not in the domain of f. Domain is set of all (x, y) such that $y > 0$.

3. (a) 0; **(b)** 2; **(c)** 16; **(d)** -4.2469;

5. t^2

7.

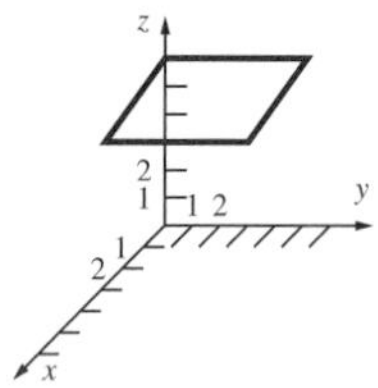

9.

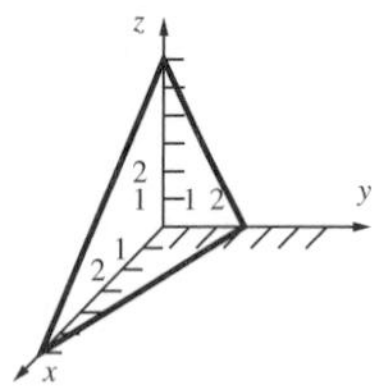

11.

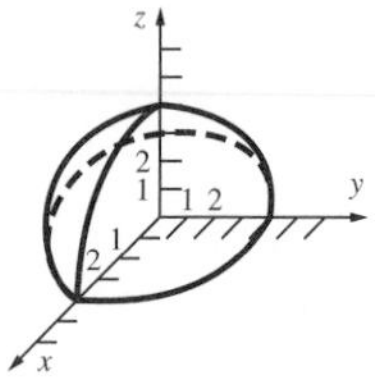

13.

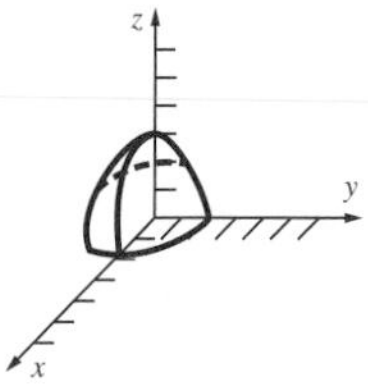

15.

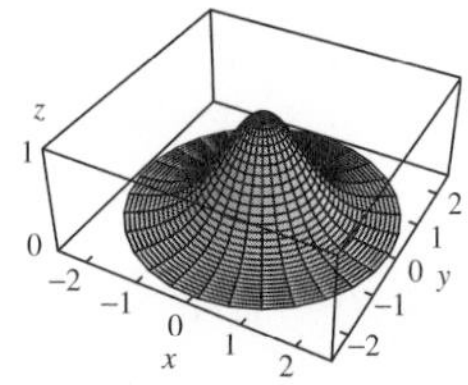

17.

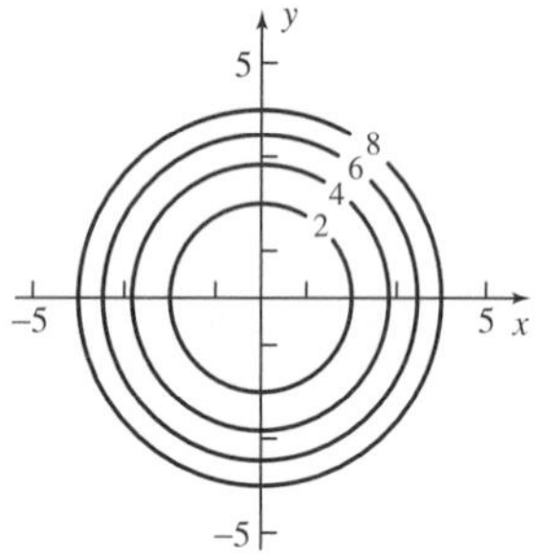

19.

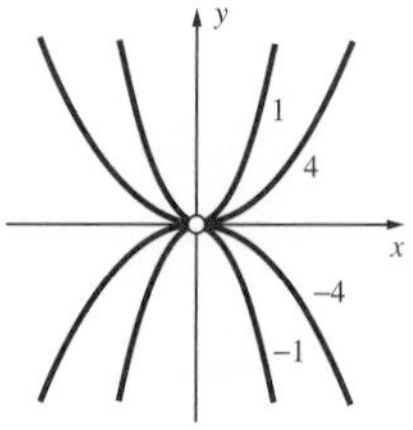

21.

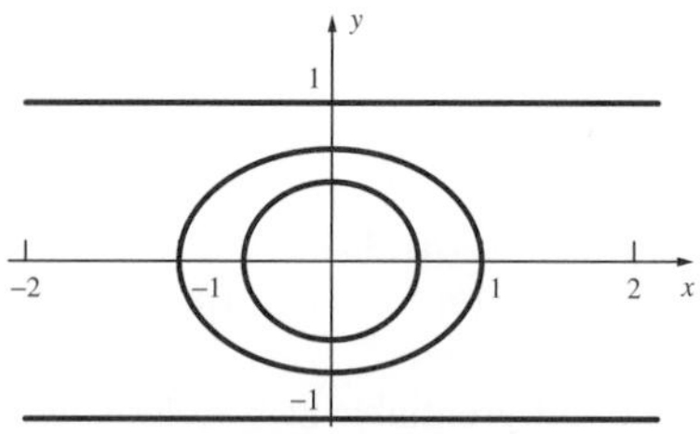

23.

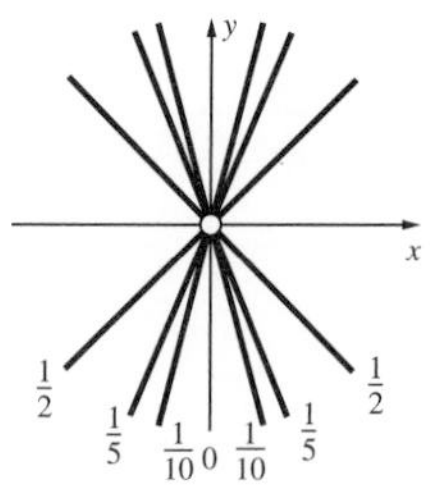

25. (a) San Francisco **(b)** northwest: southeast
(c) southwest or northeast

27. The set of all points on and outside the sphere $x^2 + y^2 + z^2 = 16$.

29. The set of all points on and inside the ellipsoid $x^2/9 + y^2/16 + z^2/1 = 1$.

31. All points in $\mathbb{R}^3$ except the origin $(0, 0, 0)$.

33. The set of all spheres with centers at the origin.

35. A set of hyperboloids of revolution about the z-axis when $k = 0$. When $k \neq 0$, the level surface is an elliptic cone.

37. A set of hyperbolic cylinders parallel to the z-axis when $k \neq 0$. When $k = 0$, the level surface is a pair of planes.

39. (a) All points in $\mathbb{R}^4$ except the origin $(0, 0, 0, 0)$.
(b) All points in $\mathbb{R}^n$.
(c) All points in $\mathbb{R}^n$ that satisfy $x_1^2 + x_2^2 + \cdots + x_n^2 \leq 1$.

41. (a) gentle climb, steep climb; **(b)** 6490 ft, 3060 ft

43.

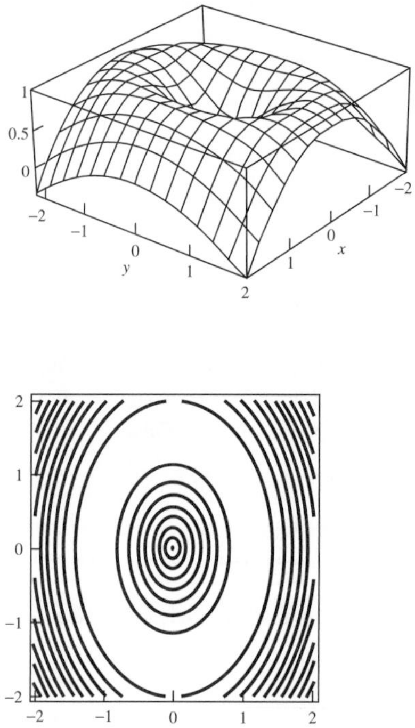

45.

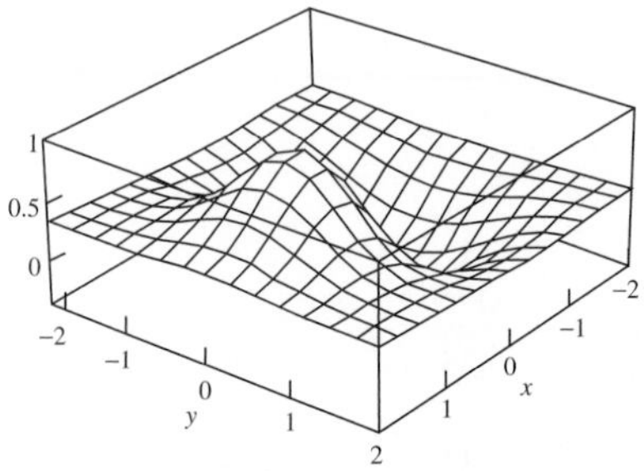

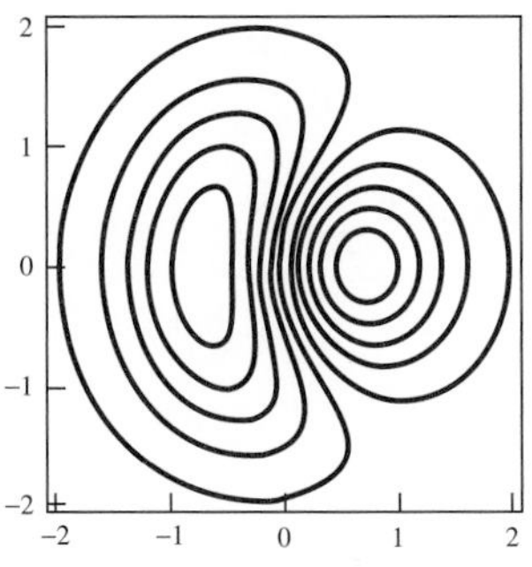

Problem Set 12.2

1. $f_x(x, y) = 8(2x - y)^3; f_y(x, y) = -4(2x - y)^3$

3. $f_x(x, y) = (x^2 + y^2)/(x^2y); f_y(x, y) = -(x^2 + y^2)/(xy^2)$

5. $f_x(x, y) = e^y \cos x; f_y(x, y) = e^y \sin x$

7. $f_x(x, y) = x(x^2 - y^2)^{-1/2}; f_y(x, y) = -y(x^2 - y^2)^{-1/2}$

9. $g_x(x, y) = -ye^{-xy}; g_y(x, y) = -xe^{-xy}$

11. $f_x(x, y) = 4/[1 + (4x - 7y)^2];$
$f_y(x, y) = -7/[1 + (4x - 7y)^2]$

13. $f_x(x, y) = -2xy \sin(x^2 + y^2);$
$f_y(x, y) = -2y^2 \sin(x^2 + y^2) + \cos(x^2 + y^2)$

15. $F_x(x, y) = 2 \cos x \cos y; F_y(x, y) = -2 \sin x \sin y$

17. $f_{xy}(x, y) = 12xy^2 - 15x^2y^4 = f_{yx}(x, y)$

19. $f_{xy}(x, y) = -6e^{2x} \sin y = f_{yx}(x, y)$

21. $F_x(3, -2) = \frac{1}{9}; F_y(3, -2) = -\frac{1}{2}$

23. $f_x(\sqrt{5}, -2) = -\frac{4}{21}; f_y(\sqrt{5}, -2) = -4\sqrt{5}/21$

25. 1 **27.** 3 **29.** 120π **31.** $k/100$

33. $\partial^2 f/\partial x^2 = 6xy; \partial^2 f/\partial y^2 = -6xy$

35. $180x^4y^2 - 12x^2$

37. (a) $\partial^3 f/\partial y^3$; **(b)** $\partial^3 f/\partial y\, \partial x^2$; **(c)** $\partial^4 f/\partial y^3\, \partial x$

39. (a) $6xy - yz$; **(b)** 8; **(c)** $6x - z$

41. $-yze^{-xyz} - y(xy - z^2)^{-1}$ **43.** $(1, 0, 29)$

45. $\left\{(x, y): x < \frac{1}{2}, y > \frac{1}{2}, y < x + \frac{1}{2}\right\}$
$\cup \left\{(x, y): x > \frac{1}{2}, y < \frac{1}{2}, x < y + \frac{1}{2}\right\}, \left\{z: 0 < z \le \sqrt{3}/36\right\}$

47. (a) -4; **(b)** $\frac{2}{3}$; **(c)** $\frac{2}{5}$; **(d)** $\frac{8}{3}$

49. (a) $f_y(x, y, z) = \lim_{\Delta y \to 0} \dfrac{f(x, y + \Delta y, z) - f(x, y, z)}{\Delta y}$

(b) $f_z(x, y, z) = \lim_{\Delta z \to 0} \dfrac{f(x, y, z + \Delta z) - f(x, y, z)}{\Delta z}$

(c) $G_x(w, x, y, z) = \lim_{\Delta x \to 0} \dfrac{G(w, x + \Delta x, y, z) - G(w, x, y, z)}{\Delta x}$

(d) $\dfrac{\partial}{\partial z}\lambda(x, y, z, t) = \lim_{\Delta z \to 0} \dfrac{\lambda(x, y, z + \Delta z, t) - \lambda(x, y, z, t)}{\Delta z}$

(e) $\dfrac{\partial}{\partial b_2} S(b_0, b_1, b_2, \ldots, b_n)$
$= \lim_{\Delta b_2 \to 0} \left(\dfrac{S(b_0, b_1, b_2 + \Delta b_2, \ldots, b_n) - S(b_0, b_1, b_2, \ldots, b_n)}{\Delta b_2} \right)$

Problem Set 12.3

1. -18 **3.** $2 - \frac{1}{2}\sqrt{3}$ **5.** $-\dfrac{5}{2}$; **7.** 1;

9. Does not exist; **11.** 0 **13.** 0 **15.** 0

17. Entire plane **19.** $\{(x, y): x^2 + y^2 < 1\}$

21. $\{(x, y): y \ne x^2\}$ **23.** $\{(x, y): y \le x + 1\}$

25. All (x, y, z), except $(0, 0, 0)$.

27. The boundary consists of the line segments that form the outer edges of the given rectangle; the set is closed.

29. Boundary: $\{(x, y): x^2 + y^2 = 1\} \cup \{(0, 0)\}$; the set is neither open nor closed.

31. Boundary: $\{(x, y): y = \sin(1/x), x > 0\} \cup \{(x, y): x = 0, y \le 1\}$; the set is open.

33. $g(x) = 2x$

35. $\lim_{x \to 0} f(x, 0) = \lim_{x \to 0} [0/(x^2 + 0)] = 0;$
$\lim_{x \to 0} f(x, x) = \lim_{x \to 0} [x^2/(x^2 + x^2)] = \frac{1}{2}$

37. (a) $\lim_{x \to 0} f(x, mx) = \lim_{x \to 0} mx^3/(x^4 + m^2x^2)$
$= \lim_{x \to 0} mx/(x^2 + m^2) = 0;$

(b) $\lim_{x \to 0} f(x, x^2) = \lim_{x \to 0} x^4/(x^4 + x^4) = \frac{1}{2};$

(c) $\lim_{(x, y) \to (0, 0)} f(x, y)$ does not exist.

39. (a) $\{(x, y, z): x^2 + y^2 = 1, 1 \le z \le 2\}$;
(b) $\{(x, y, z): x^2 + y^2 = 1, z = 1\}$; **(c)** $\{(x, y, z): z = 1\}$;
(d) empty set.

41. (a) $\{(x, y): x > 0, y = 0\}$;
(b) $\{(u, v, x, y): \langle x, y\rangle = k\langle u, v\rangle, k > 0, \langle u, v\rangle \ne \langle 0, 0\rangle\}$

43.

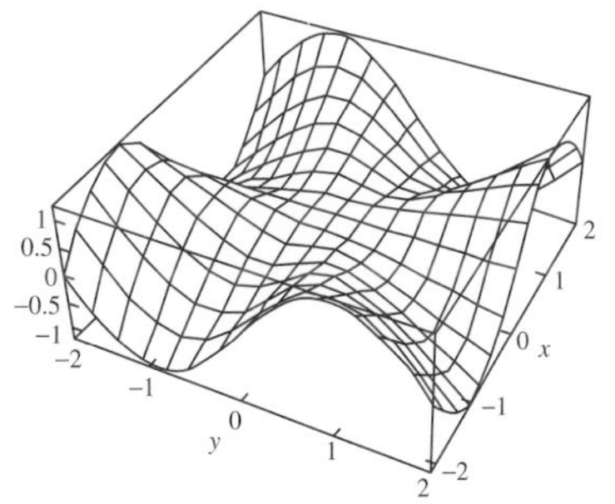

45.

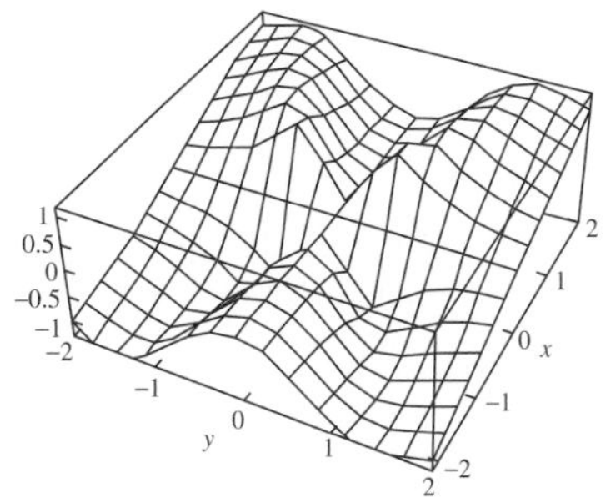

Problem Set 12.4

1. $(2xy + 3y)\mathbf{i} + (x^2 + 3x)\mathbf{j}$ **3.** $e^{xy}(1 + xy)\mathbf{i} + x^2e^{xy}\mathbf{j}$

5. $(x + y)^{-2}[(x^2y + 2xy^2)\mathbf{i} + x^3\mathbf{j}]$

7. $(x^2 + y^2 + z^2)^{-1/2}(x\mathbf{i} + y\mathbf{j} + z\mathbf{k})$

9. $xe^{x-z}[(yx + 2y)\mathbf{i} + x\mathbf{j} - xy\mathbf{k}]$

11. $\langle -21, 16\rangle, z = -21x + 16y - 60$

13. $\langle 0, -2\pi\rangle, z = -2\pi y + \pi - 1$

15. $w = 7x - 8y - 2z + 3$ **19.** $(1, 2)$

21. (a) $x = 2 + t, y = 1, z = 9 + 12t$
(b) $x = 2, y = 1 + 10t, z = 9 + 10t$
(c) $x = 2 - t, y = 1 - t, z = 9 - 22t$

23. $z = -5x + 5y$ **25.** $\mathbf{c} = \langle 1, \frac{1}{2}\rangle$

29.

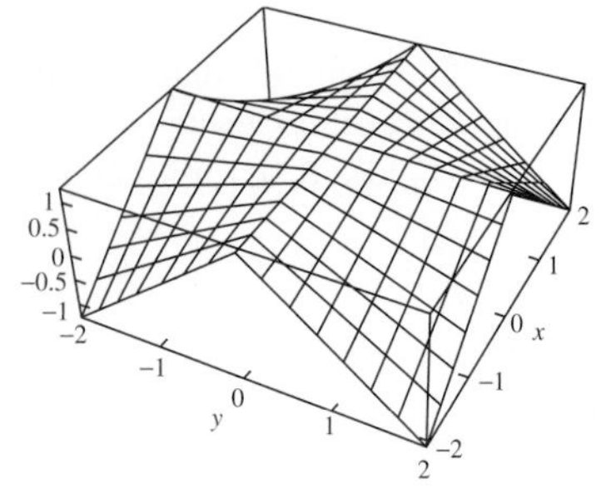

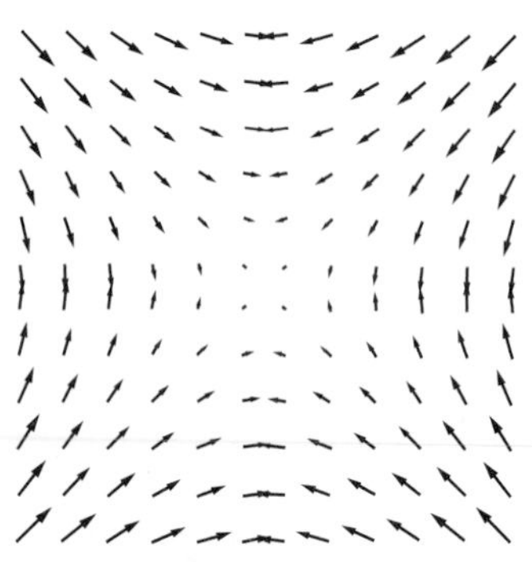

(a) The gradient points in the direction of greatest increase of the function.
(b) No.

Problem Set 12.5

1. $\frac{8}{5}$ **3.** $3\sqrt{2}/2$ **5.** $(\sqrt{2}+\sqrt{6})/4$ **7.** $\frac{52}{3}$
9. $\frac{12}{13}\mathbf{i}-\frac{5}{13}\mathbf{j}$; 13 **11.** $-\frac{4}{\sqrt{21}}\mathbf{i}+\frac{2}{\sqrt{21}}\mathbf{j}-\frac{1}{\sqrt{21}}\mathbf{k}$; $\sqrt{21}$
13. $(1/\sqrt{5})(-\mathbf{i}+2\mathbf{j})$
15. $\nabla f(\mathbf{p})=-4\mathbf{i}+\mathbf{j}$ is perpendicular to the tangent line at **p**.
17. $\frac{2}{3}$
19. (a) $(0,0,0)$; **(b)** $-\mathbf{i}+\mathbf{j}-\mathbf{k}$; **(c)** yes.
21. $(x^2+y^2+z^2)^{-1/2}\cos\sqrt{x^2+y^2+z^2}\,\langle x,y,z\rangle$
23. N 63.43°E **25.** Descend: $-300\sqrt{2}e^{-3}$ **27.** $x=-2y^2$
29. (a) $-10/\sqrt{2+\pi^2}$ deg/m; **(b)** -10 deg/s
31. (a) $(100,120)$; **(b)** $(190,25)$; **(c)** $-\frac{1}{3},0,\frac{2}{5}$
33. Leave at about $(-0.1,-5)$. **35.** Leave at about $(3,5)$.

Problem Set 12.6

1. $12t^{11}$ **3.** $e^{3t}(3\sin 2t+2\cos 2t)+e^{2t}(3\cos 3t+2\sin 3t)$
5. $7t^6\cos(t^7)$ **7.** $2s^3t-3s^2t^2$
9. $2(s^2\sin t\cos t+t\sin^2 s)\exp(s^2\sin^2 t+t^2\sin^2 s)$
11. $s^4t(1+s^4t^2)^{-1/2}$ **13.** 72 **15.** $-\frac{1}{2}(\pi+1)$
17. 244.35 board ft per year **19.** $\sqrt{20}$ ft/s
21. $(3x^2+4xy)/(3y^2-2x^2)$
23. $(y\sin x-\sin y)/(x\cos y+\cos x)$
25. $(yz^3-6xz)/(3x^2-3xyz^2)$
27. $\partial T/\partial s=(\partial T/\partial x)(\partial x/\partial s)+(\partial T/\partial y)(\partial y/\partial s)+(\partial T/\partial z)(\partial z/\partial s)+(\partial T/\partial w)(\partial w/\partial s)$
31. $10\sqrt{2}-3\pi\sqrt{2}$ **33.** 288 mi/h

Problem Set 12.7

1. $2(x-2)+3(y-3)+\sqrt{3}(z-\sqrt{3})=0$
3. $(x-1)-3(y-3)+\sqrt{7}(z-\sqrt{7})=0$
5. $x+y-z=2$ **7.** $z+1=-2\sqrt{3}(x-\frac{1}{3}\pi)-3y$
9. 0.08; 0.08017992 **11.** -0.03; -0.03015101
13. $(3,-1,-14)$
15. $\langle 0,1,1\rangle$ is normal to both surfaces at $(0,-1,2)$
17. $(1,2,-1)$ and $(-1,-2,1)$
19. $x=1+32t$; $y=2-19t$; $z=2-17t$ **21.** 0.004375 lb
23. 7% **25.** 20 ± 0.34 **27.** $V=9|k|/2$
29. (a) 4.98; **(b)** 4.98196; **(c)** 4.9819675

Problem Set 12.8

1. $(2,0)$; local minimum point.
3. $(0,0)$; saddle point; $(\pm\frac{1}{2},0)$; local minimum points.
5. $(0,0)$; saddle point.
7. $(1,2)$; local minimum point. **9.** No critical points.
11. Global maximum of 7 at $(1,1)$; global minimum of -4 at $(0,-1)$.
13. Global maximum of 2 at $(\pm1,0)$; global minimum of 0 at $(0,\pm1)$.
15. Each of the three numbers is $N/3$. **17.** A cube.
19. Base 8 ft by 8 ft; depth 4 ft. **21.** $3\sqrt{3}(\mathbf{i}+\mathbf{j}+\mathbf{k})$
23. $(0.393,0.786,0.772)$; 1.56
25. Width of turned-up sides is 4″; base angle $^{2\pi}/_{3}$
27. (a) maximum value of 8 occurs at $(-1,2)$
(b) minimum value of -11 occurs at $(4,0)$
29. Maximum of 3 at $(1,2)$; minimum of $-\frac{12}{5}$ at $(\frac{8}{5},-\frac{2}{5})$.
31. $y=\frac{7}{10}x+\frac{1}{10}$ **33.** $x=50/\sqrt{3}$, $y=100/\sqrt{1.25}$; \$79,681
35. Maximum of $10+3\sqrt{2}$ at $(3/\sqrt{2},3/\sqrt{2})$; Minimum of $10-3\sqrt{2}$ at $(-3/\sqrt{2},-3/\sqrt{2})$
37. Length 1.1544 ft, Width 1.1544 ft, Height 1.501 ft
39. $(\pm\sqrt{3}/2,-\frac{1}{2})$ where $T=9/4$; $(0,1/2)$, where $T=-1/4$
41. Equilateral triangle.
43. Local maximum: $f(1.75,0)=1.15$; global maximum: $f(-3.8,0)=2.30$
45. Global minimum: $f(0,1)=f(0,-1)=-0.12$.
47. Global maximum $f(1.13,0.79)=f(1.13,-0.79)=0.53$ global minimum $f(-1.13,0.79)=f(-1.13,-0.79)=-0.53$.
49. Global maximum $f(3,3)=f(-3,3)\approx 74.9225$ global minimum $f(1.5708,0)=f(-1.5708,0)=-8$.
51. Global maximum: $f(0.67,0)=5.06$; global minimum: $f(-0.75,0)=-3.54$.
53. Global maximum: $f(2.1,2.1)=3.5$; global minimum: $f(4.2,4.2)=-3.5$.

Problem Set 12.9

1. $f(\sqrt{3},\sqrt{3})=f(-\sqrt{3},-\sqrt{3})=6$
3. $f(2/\sqrt{5},-1/\sqrt{5})=f(-2/\sqrt{5},1/\sqrt{5})=5$
5. $f(\frac{6}{7},\frac{18}{7},-\frac{12}{7})=\frac{72}{7}$ **7.** Base is 4 by 4; depth is 2.
9. $10\sqrt{5}$ ft³ **11.** $8abc/(3\sqrt{3})$
13. Maximum is $9\sqrt{3}$ when $\langle x,y,z\rangle=\langle 3\sqrt{3},3\sqrt{3},3\sqrt{3}\rangle$.
15. Minimum distance is 1.5616 at pt $(0.393,0.786,0.772)$
17. Length = Width = 1.1544 ft, Height = 1.501 ft
19. $c_0=\dfrac{\pi k}{8+\pi}$; $p_0=\dfrac{4k}{8+\pi}$; $q_0=\dfrac{4k}{8+\pi}$;
$A(c_0,p_0,q_0)=\dfrac{k}{4(8+\pi)}\approx 0.224k^2$ is a minimum value.

21. $f\left(\frac{1}{\sqrt{2}}, \frac{1}{\sqrt{2}}\right) = 10 + \sqrt{2}$ is the maximum value;
$f\left(-\frac{1}{\sqrt{2}}, -\frac{1}{\sqrt{2}}\right) = 10 - \sqrt{2}$ is the minimum value

23. $f\left(\frac{3\sqrt{3}}{2}, -\frac{3}{2}\right) \approx 20.6913$ is the maximum value;
$f(-2, -1) = -3$ is the minimum value

25. $f\left(\frac{2}{\sqrt{5}}, \frac{8}{\sqrt{5}}\right) \approx 29.9443$ is the maximum value;
$f(x, -1-x) = 0$ for $-\frac{1}{5} - \frac{2}{5}\sqrt{19} \le x \le -\frac{1}{5} + \frac{2}{5}\sqrt{19}$ is the minimum value

27. $x = \alpha d/a,\ y = \beta d/b,\ z = \gamma d/c$

29. $f(-1, 1, 0) = 3,\ f(1, -1, 1) = -1$

31. $\sqrt{A}$ is the maximum value of w, where $A = a_1^2 + a_2^2 + \cdots + a_n^2$

33. $f(4, 0) = -4$ **35.** $f(0, 3) = f(0, -3) = -0.99$

Chapter Review 12.10

Concepts Test

1. True **3.** True **5.** True **7.** False **9.** True
11. True **13.** True **15.** True **17.** True **19.** False

Sample Test Problems

1. **(a)** $\{(x, y): x^2 + 4y^2 \ge 100\}$ **(b)** $\{(x, y): 2x - y \ge 1\}$

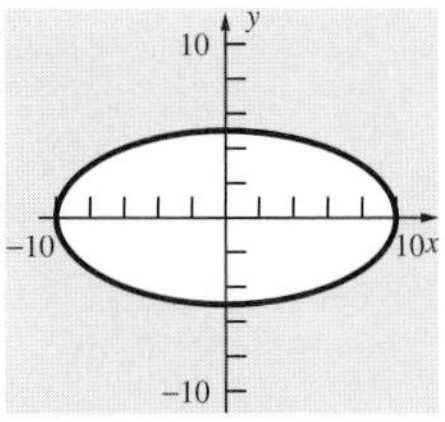

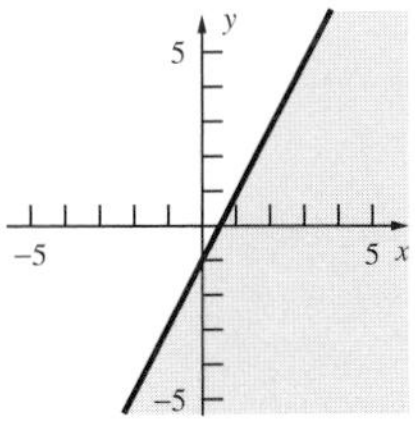

3. $12x^3y^2 + 14xy^7;\ 36x^2y^2 + 14y^7;\ 24x^3y + 98xy^6$

5. $e^{-y}\sec^2 x;\ 2e^{-y}\sec^2 x \tan x;\ -e^{-y}\sec^2 x$

7. $450x^2y^4 - 42y^5$ **9.** 1 **11.** Does not exist.

13. **(a)** $-4\mathbf{i} - \mathbf{j} + 6\mathbf{k}$; **(b)** $-4(\cos 1\mathbf{i} + \sin 1\mathbf{j} - \cos 1\mathbf{k})$

15. $\sqrt{3} + 2$

17. **(a)** $x^2 + 2y^2 = 18$; **(b)** $4\mathbf{i} + 2\mathbf{j}$;

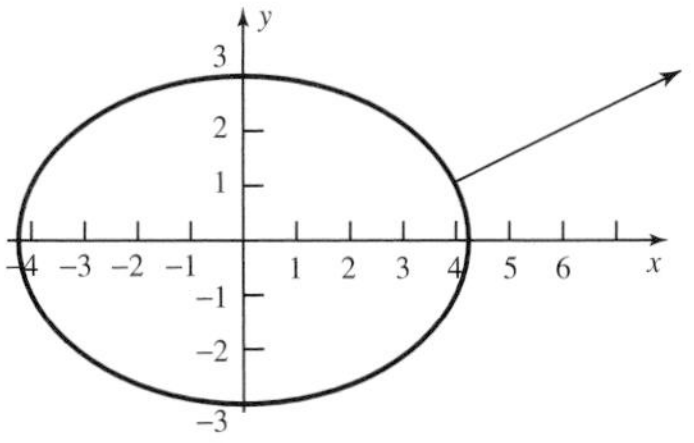

19. $(x^2 + 3y - 4z)/x^2yz;\ (-x^2 - 4x)/xy^2z;\ (3y - x^2)/xyz^2$

21. $15xy\sqrt{t}/z^3 + 5x^2/tz^3 - 45x^2ye^{3t}/z^4$

23. $18\mathbf{i} + 16\mathbf{j} - 18\mathbf{k};\ 9x + 8y - 9z = 34$ **25.** 0.7728

27. $16\sqrt{3}/3$ **29.** Radius 2; height 4.

Chapter 13 Review and Preview Problems

1.

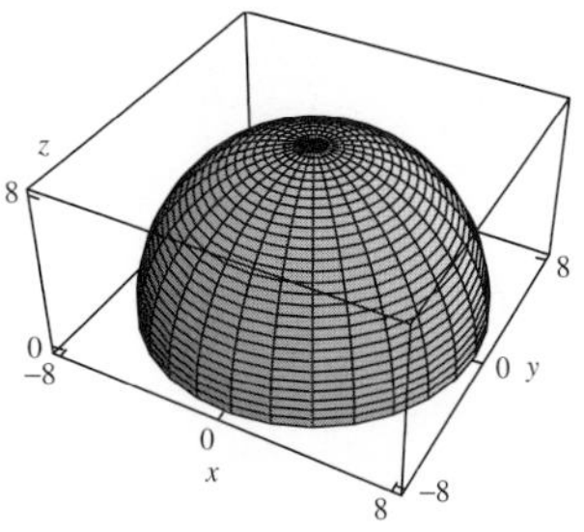

3.

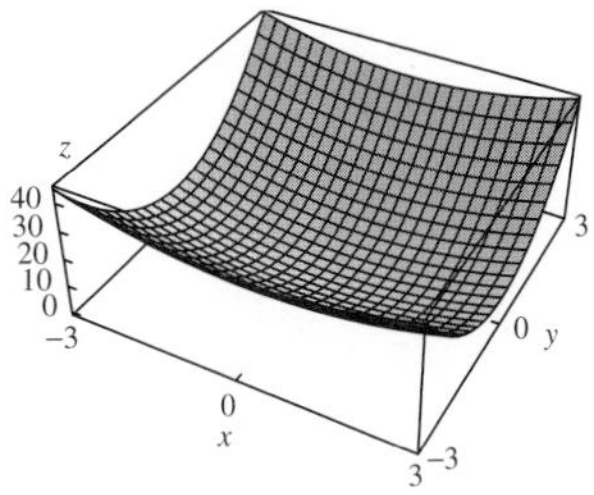

5.

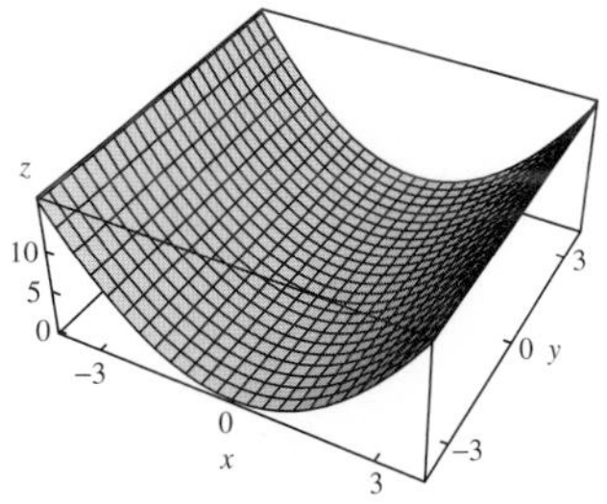

7.

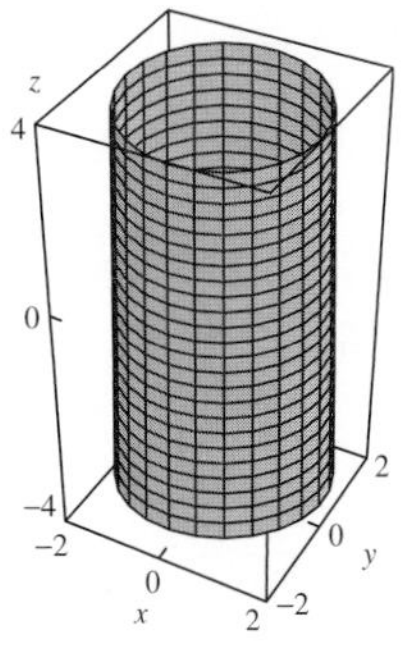

9.

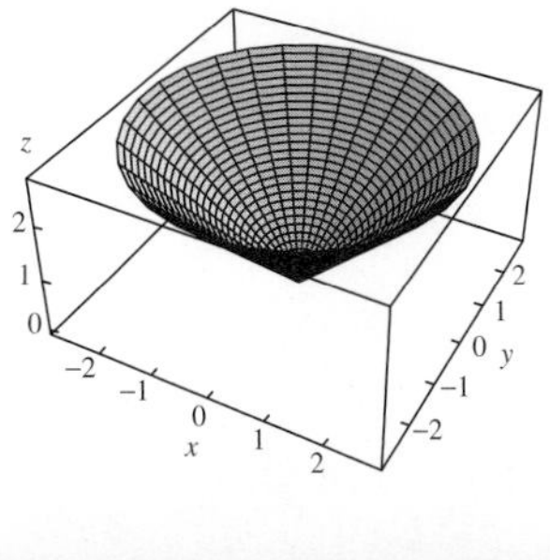

11.

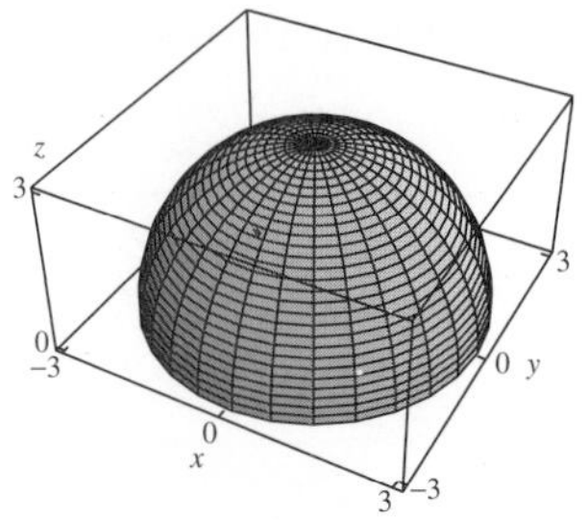

13.

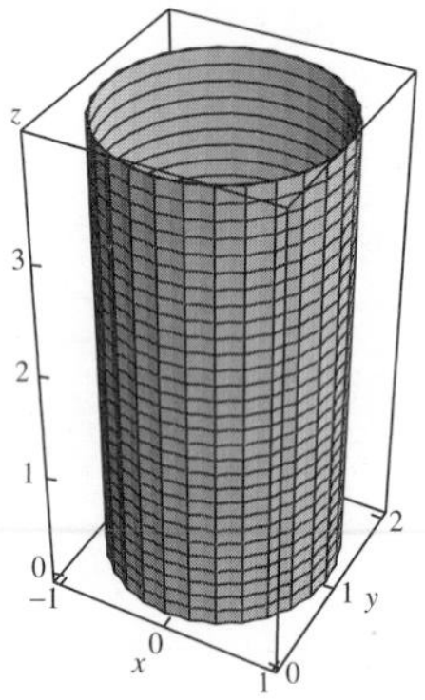

15. $-\frac{1}{2}e^{-2x}+C$ **17.** $\frac{2a}{\pi}$ **19.** $\frac{\pi}{2}$ **21.** $\frac{1}{2}\ln 2$

23. $\frac{-1+37^{3/2}}{12}$ **25.** $\frac{\pi}{4}$ **27.** $2\pi\left(\sqrt{a^2-b^2}-\sqrt{a^2-c^2}\right)$

29. 36π **31.** $\frac{\pi^2}{2}$ **33.** $\frac{81\pi}{2}$

Problem Set 13.1

1. 14 **3.** 12 **5.** 4 **7.** 3 **9.** 168 **11.** 520
13. 52.57

15.

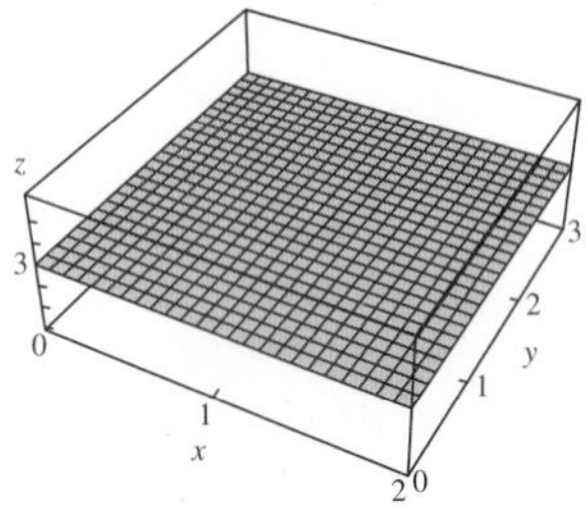

17.

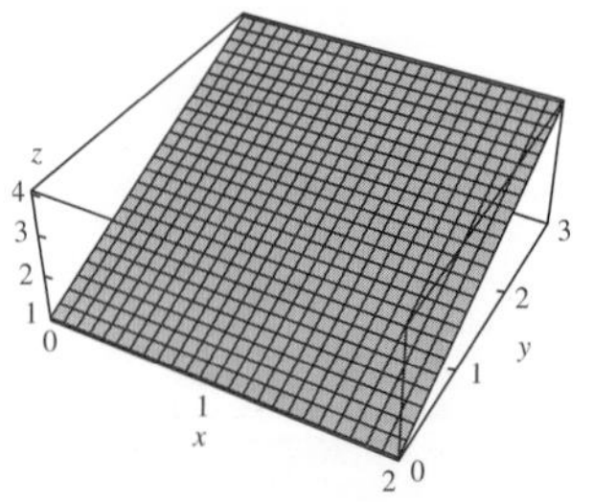

19.

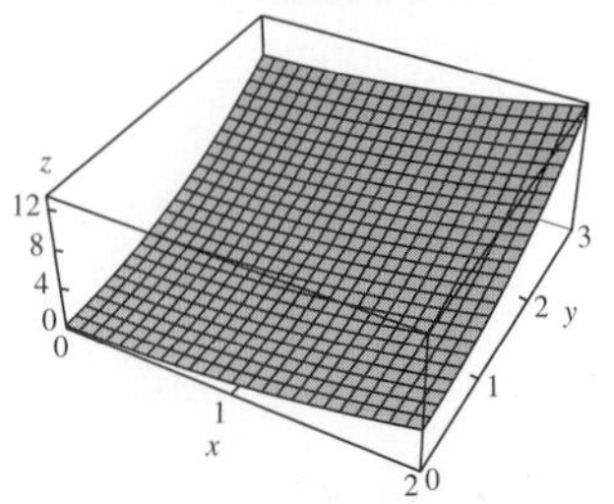

21. 5.5 **25.** $c = 15.30$, $C = 30.97$
27. **(a)** -6; **(b)** 6
29. Number of cubic inches of rain that fell on all Colorado in 1999; average rainfall in Colorado during 1999.
31. Approximately 458.

Problem Set 13.2

1. 48 **3.** $\frac{32}{3}$ **5.** $\frac{55}{4}$ **7.** 1 **9.** $\pi/2 - 1$

11. $\frac{4}{15}\left[31 - 9\sqrt{3}\right] \approx 4.110$ **13.** $1 - \frac{1}{2}\ln 3 \approx 0.4507$

15. $\frac{9\pi}{4}$ **17.** 0 **19.** 2 **21.** 105 **23.** 112

25.

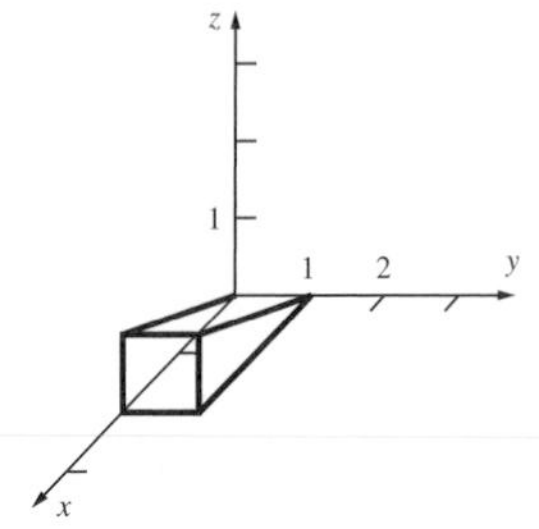

27.

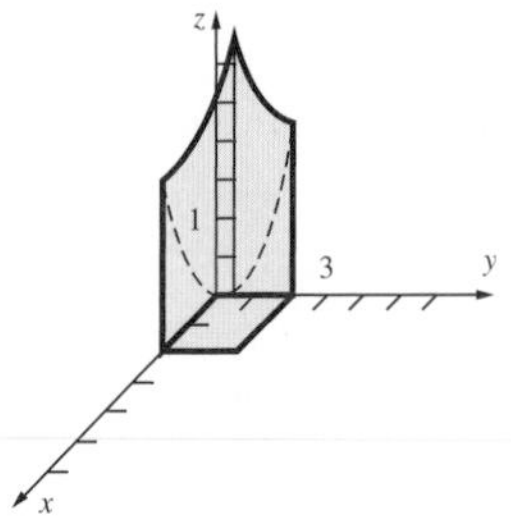

29. 7 **31.** $\frac{10}{3}$ **35.** $\frac{1}{4}(e-1)^2$

37. **(a)** $\frac{8}{3}$ **39.** $5 - \sqrt{3} - \sqrt{2}$

Problem Set 13.3

1. $\frac{3}{4}$ **3.** 240 **5.** $\frac{1}{2}(e^{27} - e)$ **7.** $-\sqrt{2}/(2\pi)$

9. $(3\ln 2 - \pi)/9$ **11.** $e - 2$ **13.** $\frac{16}{3}$ **15.** 0 **17.** $\frac{27}{70}$

19. $4\tan^{-1} 2 - \ln 5$ **21.** 6 **23.** 20 **25.** 10 **27.** $\frac{4}{15}$

29. $-\frac{1}{2}\ln(\cos 1)$ **31.** 3π **33.** $\int_0^1\int_y^1 f(x, y)\,dx\,dy$

35. $\int_0^1\int_{y^4}^{\sqrt{y}} f(x, y)\,dx\,dy$

37. $\int_{-1}^0\int_{-x}^1 f(x, y)\,dy\,dx + \int_0^1\int_x^1 f(x, y)\,dy\,dx$ **39.** $\frac{256}{15}$

41. $15\pi/4$ **43.** $\frac{1}{3}(1 - \cos 8)$

45. approximately 4,133,000 ft^3

Problem Set 13.4

1. $\frac{1}{12}$ **3.** $\frac{4}{9}$ **5.** $4\sqrt{2}$ **7.** $2\sqrt{3} + \frac{4}{3}\pi \approx 7.653$

9. $\pi a^2/8$ **11.** $8\pi + 6\sqrt{3} \approx 35.525$ **13.** $\frac{\pi}{2}$ **15.** $\frac{\pi^3}{48}$

17. $\frac{\pi}{4}$ **19.** $\pi(e^4 - 1) \approx 168.384$ **21.** $(\pi \ln 2)/8 \approx 0.272$

23. $\pi\left(2 - \sqrt{3}\right)/2 \approx 0.421$ **25.** $\frac{1}{12}$ **27.** $81\pi/8 \approx 31.809$

29. $625\left(3\sqrt{3} + 1\right)/12 \approx 322.716$ **31.** $\frac{2}{3}\pi d^2(3a - d)$

33. $\frac{2}{9}a^3(3\pi - 4)$

Problem Set 13.5

1. $m = 30$; $\bar{x} = 2$; $\bar{y} = 1.8$
3. $m = \pi/4$; $\bar{x} = \pi/2$; $\bar{y} = 16/(9\pi)$
5. $m \approx 0.1056$; $\bar{x} \approx 0.281$; $\bar{y} \approx 0.581$
7. $m = 32/9$; $\bar{x} = 0$; $\bar{y} = 6/5$ **9.** $m = \pi$; $\bar{x} = 0$; $\bar{y} = \frac{3}{\pi}$
11. $I_x \approx 269$; $I_y \approx 5194$; $I_z \approx 5463$
13. $I_x = I_y = 5a^5/12$; $I_z \approx 5a^5/6$

15. $k; 2k; \left(\frac{4}{3}, \frac{2}{3}\right)$

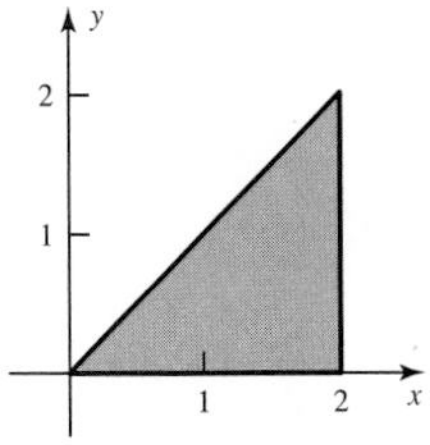

17. The density is proportional to the squared distance from the origin; $\frac{25596k}{35}$; $\left(0, \frac{450}{79}\right)$

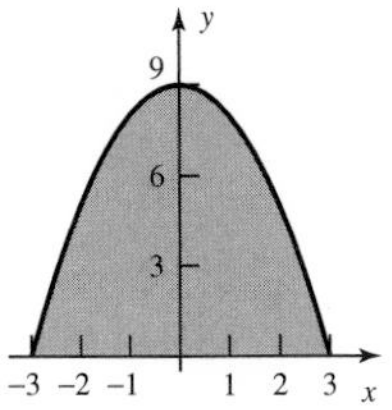

19. The density is proportional to the distance from the origin; $\frac{26k\pi}{3}$; $\left(0, \frac{60}{13\pi}\right)$

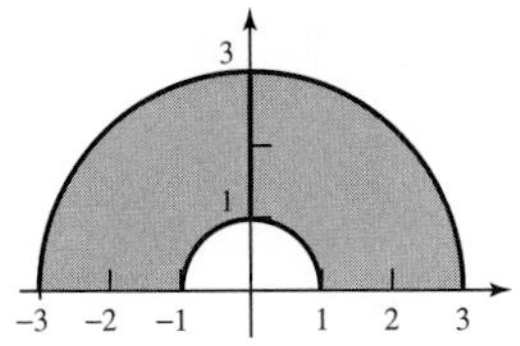

21. $\bar{r} = \sqrt{5/12}a \approx 0.6455a$ **23.** $I_x = \pi\delta a^4/4; \bar{r} = a/2$
25. $5\pi\delta a^4/4$ **27.** $\bar{x} = 0, \bar{y} = (15\pi + 32)a/(6\pi + 48)$
29. **(a)** a^3; **(b)** $7a/12$; **(c)** $11a^5/144$
31. $I_x = \pi k a^4/2, I_y = 17k\pi a^4/2, I_z = 9\pi k a^4$

Problem Set 13.6

1. $\sqrt{61}/3$ **3.** $\pi/3$ **5.** $9\sin^{-1}\left(\frac{2}{3}\right)$ **7.** $8\sqrt{2}$
9. $4\pi a\left(a - \sqrt{a^2 - b^2}\right)$ **11.** $2a^2(\pi - 2)$
13. $\frac{1}{6}\pi a^2\left(5\sqrt{5} - 1\right)$ **15.** $\frac{(17^{3/2} - 1)\pi}{6}$
17. $\frac{D^2\sqrt{A^2 + B^2 + C^2}}{2ABC}$
19. $(h_1 + h_2)/2$
21. $A = \pi b^2, B = 2\pi a^2[1 - \cos(b/a)], C = \pi b^2,$
$D = \pi b^2\left[2a/\left(a + \sqrt{a^2 - b^2}\right)\right], B < A = C < D$
27. **(a)** 29.3297 **(b)** 15.4233
29. E/F (tie), A/B (tie), C/D (tie)

Problem Set 13.7

1. -40 **3.** $\frac{189}{2}$ **5.** 1927.54 **7.** $\frac{2}{3}$ **9.** 156

11. $\int_0^1 \int_0^3 \int_0^{(1/6)(12-3x-2y)} f(x, y, z)\, dz\, dy\, dx$

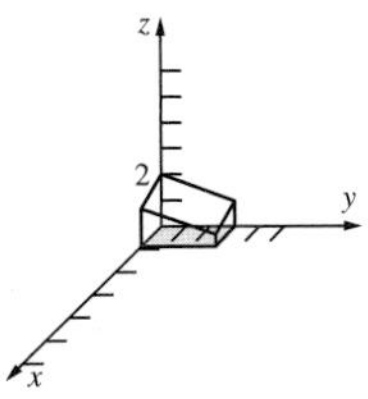

13. $\int_0^2 \int_0^4 \int_0^{y/2} f(x, y, z)\, dx\, dy\, dz$

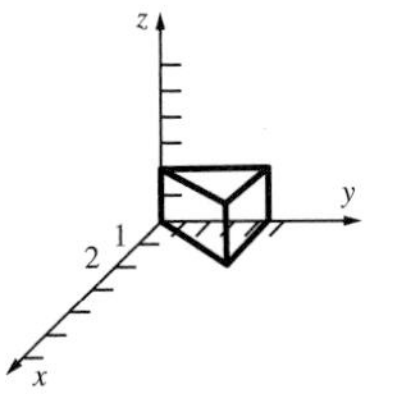

15. $\int_0^{12/5} \int_{x/3}^{(4-x)/2} \int_0^{4-x-2z} f(x, y, z)\, dy\, dz\, dx$

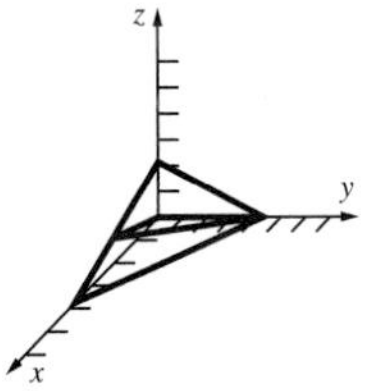

17. $\int_0^3 \int_{2x/3}^{(9-x)/3} \int_0^{(18-2x-6y)/9} f(x, y, z)\, dz\, dy\, dx$

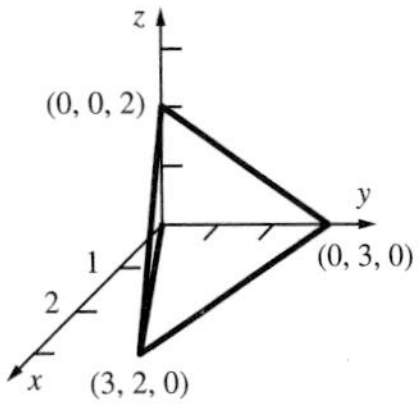

19. $\int_1^4 \int_0^1 \int_0^{\sqrt{1-z^2}} f(x, y, z)\, dy\, dz\, dx$

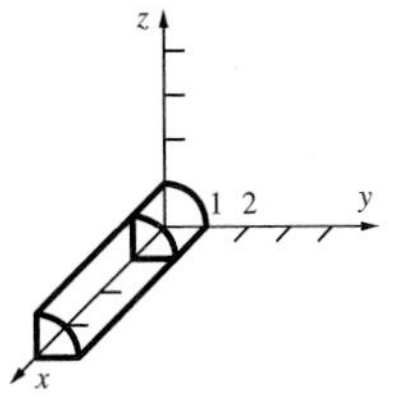

21. $\frac{128}{15}$ **23.** $4\int_0^1 \int_{x^2}^1 \int_0^{\sqrt{y}} dz\, dy\, dx = 2$

25. $\bar{x} = \bar{y} = \bar{z} = \frac{4}{15}$ **27.** $\bar{x} = \bar{y} = \bar{z} = 3a/8$

29. $\int_0^1 \int_0^{\sqrt{1-x^2}} \int_0^{\sqrt{1-x^2-y^2}} f(x, y, z)\, dz\, dy\, dx$

31. $\int_0^2 \int_0^{2-z} \int_0^{9-x^2} f(x, y, z)\, dy\, dx\, dz$ **33.** 4

35. Ave $T = 29.54$ **37.** $(\bar{x}, \bar{y}, \bar{z}) = \left(\frac{11}{24}, \frac{25}{12}, \frac{11}{24}\right)$

39. $(\bar{x}, \bar{y}, \bar{z}) = \left(\frac{17}{36}, \frac{17}{36}, \frac{55}{36}\right)$

43. (a) $k = \frac{1}{288}$ **(b)** $\frac{26}{27}$ **(c)** 9

45. (a) $\frac{7}{16}$ **(b)** $\frac{1}{4}$ **(c)** 5

47. $x^2/576, 0 \le x \le 12, 9$

Problem Set 13.8

1. Right circular cylinder about the z-axis with radius 3 and height 12; $V = 108\pi$

3. Region under the paraboloid $z = 9 - r^2$ above the xy-plane in that part of the first quadrant satisfying $0 \le \theta \le \frac{\pi}{4}$; $V = \frac{243\pi}{16}$

5. Sphere centered at the origin with radius a; $V = \frac{4}{3}\pi a^3$

7. 8π **9.** $14\pi/3$ **11.** $2\pi(5\sqrt{5} - 4)/3 \approx 15.038$

13. $\bar{x} = \bar{y} = 0; \bar{z} = \frac{16}{3}$ **15.** $k\pi(b^4 - a^4)$

17. $\bar{x} = \bar{y} = 0; \bar{z} = 2a/5$ **19.** $k\pi^2 a^6/16$ **21.** $\pi/9$

23. $\pi/32$

25. (a) $3a/4$; **(b)** $3\pi a/16$; **(c)** $6a/5$

27. (a) $3\pi a \sin\alpha/16\alpha$; **(b)** $3\pi a/16$

29. $(a + b)(c - 1)/(c + 1)$

Problem Set 13.9

1.

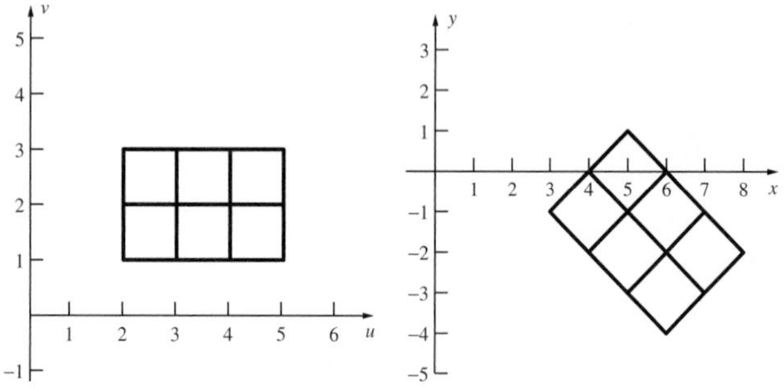

3.

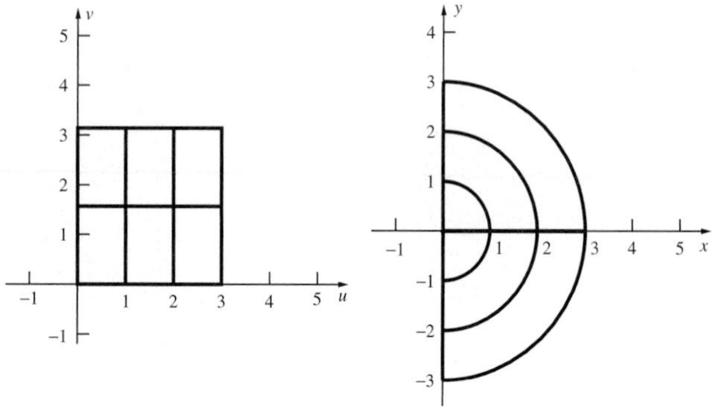

5.

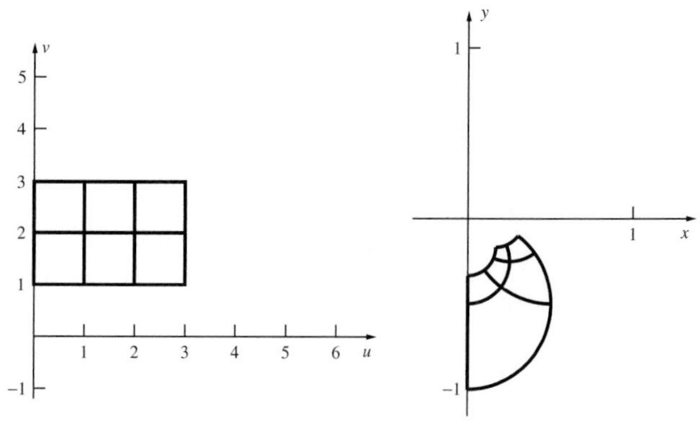

7. Image is the square with corners $(0, 0)$, $(2, 2)$, $(4, 0)$, and $(2, -2)$; $J = -4$.

9. Image is the set of (x, y) that satisfy $y^2 \le x \le y^2 + 1, 0 \le y \le 1$; $J = 2u$.

11. $x = u/2 + v/2; y = u/4 - v/4; J = -\frac{1}{4}$

13. $x = v; y = \sqrt{u - v^2}; J = -\frac{1}{2\sqrt{u - v^2}}$

15. $x = v; y = u/v; J = -\frac{1}{v}$ **17.** 3.15669 **19.** 0

21. $-\rho^2 \sin\phi$

25. (a) $g(u, v) = \begin{cases} e^{-u}, & \text{if } 0 \le v \le u \\ 0, & \text{otherwise} \end{cases}$

(b) $g_U(u) = \begin{cases} ue^{-u}, & \text{if } 0 \le u \\ 0, & \text{otherwise} \end{cases}$

Chapter Review 13.10

Concepts Test

1. True **3.** True **5.** True **7.** False **9.** True
11. True **13.** False **15.** True **17.** False

Sample Test Problems

1. $\frac{1}{24}$ **3.** $\frac{2}{3}$ **5.** $\int_0^1 \int_0^y f(x, y)\, dx\, dy$

7. $\int_0^{1/2} \int_0^{1-2y} \int_0^{1-2y-z} f(x, y, z)\, dx\, dz\, dy$

9. (a) $8\int_0^a \int_0^{\sqrt{a^2-x^2}} \int_0^{\sqrt{a^2-x^2-y^2}} dz\, dy\, dx$;

(b) $8\int_0^{\pi/2} \int_0^a \int_0^{\sqrt{a^2-r^2}} r\, dz\, dr\, d\theta$;

(c) $8\int_0^{\pi/2} \int_0^{\pi/2} \int_0^a \rho^2 \sin\phi\, d\rho\, d\phi\, d\theta$

11. 0.8857 **13.** $\bar{x} = \frac{13}{6}; \bar{y} = \frac{3}{2}$ **15.** 6 **17.** $80\pi k$

19. $ka^2bc/24$ **21.** 0

Chapter 14 Review and Preview Problems

1. $x = 3\cos t, y = 3\sin t, 0 \le t < 2\pi$ is one possibility.
3. $x = 2\cos t, y = 2\sin t, 0 < t < \pi$ is one possibility.
5. $x = -2 + 5t, y = 2, 0 \le t \le 1$ is one possibility.
7. $x = 9 - t, y = t, 0 < t < 9$ is one possibility.
9. $x = -t, y = 9 - t^2, -3 \le t \le 3$ is one possibility.

11. $\nabla f(x, y) = (x \cos x + \sin x)\mathbf{i} + (\cos y - y \sin y)\mathbf{j}$
13. $\nabla f(x, y, z) = 2x\mathbf{i} + 2y\mathbf{j} + 2z\mathbf{k}$
15. $\nabla f(x, y, z) = (y + z)\mathbf{i} + (x + z)\mathbf{j} + (x + y)\mathbf{k}$

17. $\frac{\pi}{2}$ **19.** $\frac{3}{4}$ **21.** $\frac{14\pi}{3}$

23. The volume in problem 22 is that of a spherical shell centered at $(0, 0, 0)$ with outer radius $= 2$ and inner radius $= 1$.

25. $\left\langle \frac{3}{13}, \frac{4}{13}, \frac{12}{13} \right\rangle$

Problem Set 14.1

1.

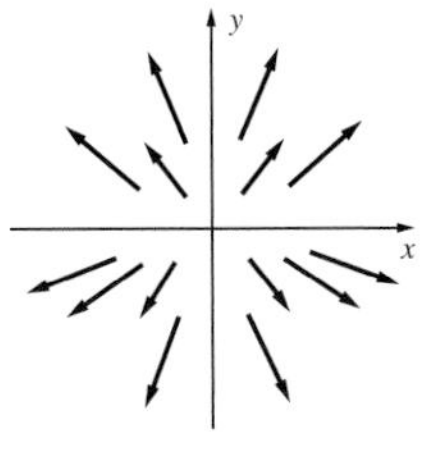

3.

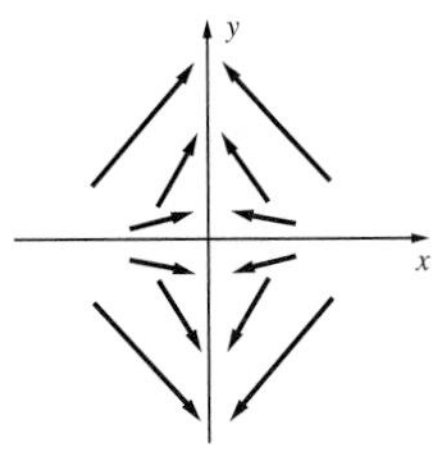

5.

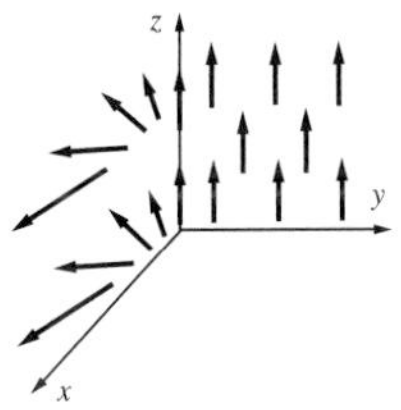

7. $(2x - 3y)\mathbf{i} - 3x\mathbf{j} + 2\mathbf{k}$ **9.** $x^{-1}\mathbf{i} + y^{-1}\mathbf{j} + z^{-1}\mathbf{k}$
11. $e^y \cos z\mathbf{i} + xe^y \cos z\mathbf{j} - xe^y \sin z\mathbf{k}$ **13.** $2yz$; $z^2\mathbf{i} - 2y\mathbf{k}$
15. 0;$\mathbf{0}$ **17.** $2e^x \cos y + 1$; $2e^x \sin y\mathbf{k}$
19. (a) Meaningless; **(b)** vector field; **(c)** vector field; **(d)** scalar field; **(e)** vector field; **(f)** vector field; **(g)** vector field; **(h)** meaningless; **(i)** meaningless; **(j)** scalar field; **(k)** meaningless.
25. (a) div $\mathbf{F} = 0$, div $\mathbf{G} < 0$, div $\mathbf{H} = 0$, div $\mathbf{L} > 0$;
(b) clockwise for $\mathbf{H}$, not at all for others.
(c) div $\mathbf{F} = 0$, curl $\mathbf{F} = \mathbf{0}$, div $\mathbf{G} = -2ye^{-y^2}$, curl $\mathbf{G} = \mathbf{0}$, div $\mathbf{H} = 0$, curl $\mathbf{H} = -2xe^{-x^2}\mathbf{k}$. div $\mathbf{L} = 1/\sqrt{x^2 + y^2}$, curl $\mathbf{L} = \mathbf{0}$

27.

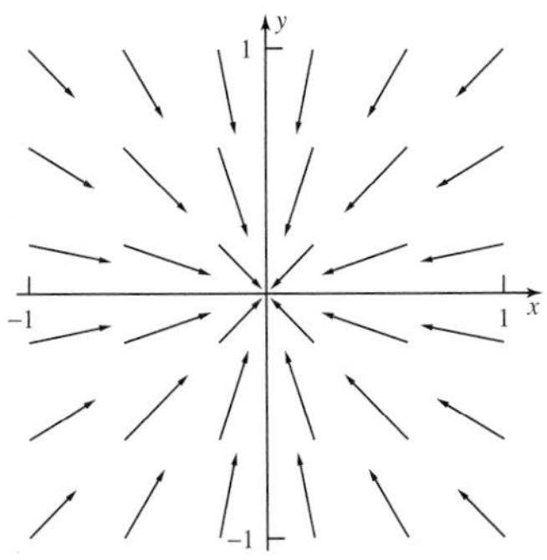

div $\mathbf{F} > 0$. A paddle wheel at the origin will not rotate.

Problem Set 14.2

1. $14(2\sqrt{2} - 1)$ **3.** $2\sqrt{5}$ **5.** $\frac{1}{6}(14\sqrt{14} - 1)$ **7.** $\frac{100}{3}$
9. 144 **11.** 0 **13.** $\frac{17}{6}$ **15.** 19 **17.** $k(17\sqrt{17} - 1)/6$
19. $-\frac{7}{44}$ **21.** $-\frac{1}{2}(a^2 + b^2)$ **23.** $2 - 2/\pi$
25. Work along C_1 is positive; work along C_2 is negative; work along C_3 is zero.
27. 2.25 gal **29.** $2\pi a^2$ **31.** $4a^2$
33. (a) 27; **(b)** $-297/2$

Problem Set 14.3

1. $f(x, y) = 5x^2 - 7xy + y^2 + C$ **3.** Not conservative.
5. $f(x, y) = \frac{2}{5}x^3y^{-2} + C$ **7.** $f(x, y) = 2xe^y - ye^x + C$
9. $f(x, y, z) = x^3 + 2y^3 + 3z^3 + C$ **11.** $\ln\left(\frac{1}{x^2 + z^2}\right) + C$
13. 14 **15.** $\frac{20}{1377}$ **17.** 6 **19.** $-\pi$
23. $f(x, y, z) = \frac{1}{2}k(x^2 + y^2 + z^2)$

25. $$\int_C \mathbf{F} \cdot d\mathbf{r} = \int_a^b m\mathbf{r}''(t) \cdot \mathbf{r}'(t)\, dt$$
$$= \frac{1}{2}m \int_a^b (d/dt)[\mathbf{r}'(t) \cdot \mathbf{r}'(t)]\, dt = \frac{1}{2}m \int_a^b (d/dt)|\mathbf{r}'(t)|^2\, dt$$
$$= \left[\frac{1}{2}m|\mathbf{r}'(t)|^2\right]_a^b = \frac{1}{2}m[|\mathbf{r}'(b)|^2 - |\mathbf{r}'(a)|^2]$$

27. $f(x, y, z) = -gmz$

Problem Set 14.4

1. $-\frac{64}{15}$ **3.** $\frac{72}{35}$ **5.** 0 **7.** $\frac{8}{3}$
9. (a) 0; **(b)** 0
11. (a) 0; **(b)** 0
13. 50 **15.** -2
19. (c) M and N have a discontinuity at $(0, 0)$.
23. $3\pi a^2/8$
27. (a) div $\mathbf{F} = 4$; **(b)** 144
29. (a) div $\mathbf{F} < 0$ in quadrants I and III; div $\mathbf{F} > 0$ in quadrants II and IV;
(b) 0; $-2(1 - \cos 3)^2$

Problem Set 14.5

1. $8\sqrt{3}/3$ **3.** $2 + \pi/3$ **5.** $5\pi/8$ **7.** 6 **9.** 2
11. 20 **13.** $\sqrt{3}ka^4/12$ **15.** $\bar{x} = \bar{y} = \bar{z} = a/3$

17.

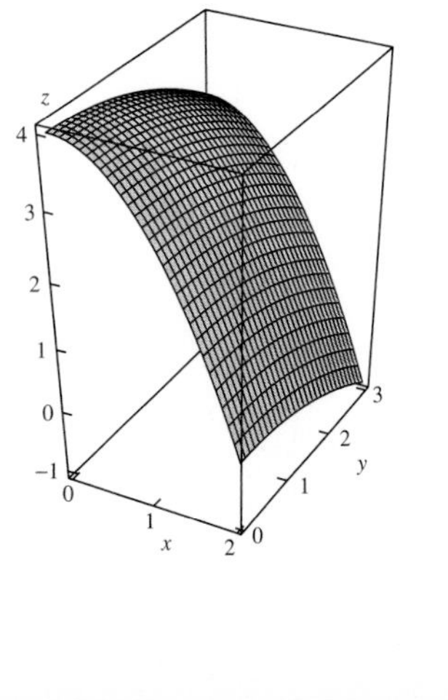

19.

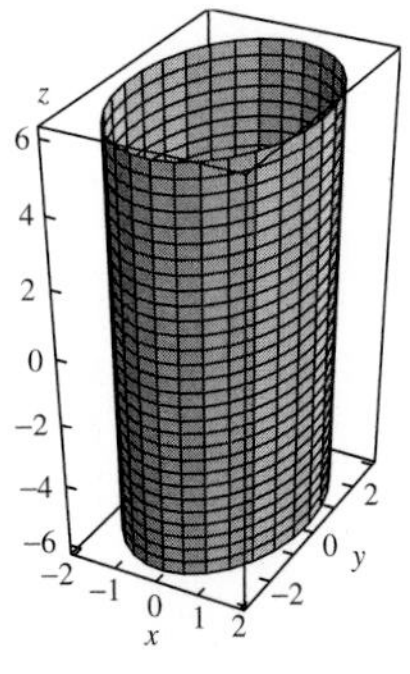

21. $\pi\left[6\sqrt{37} + \ln\sqrt{\dfrac{\sqrt{37}+6}{\sqrt{37}-6}}\right]$

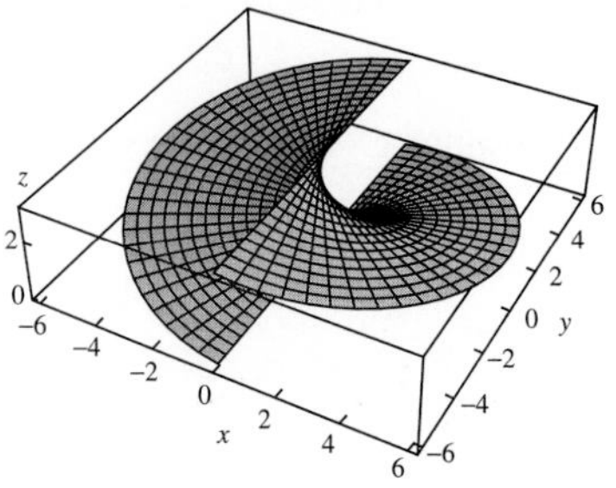

23. $\dfrac{\pi}{4}\left[\dfrac{\pi}{2}(32\pi^2 + 25)\sqrt{16\pi^2 + 25} - \dfrac{625}{8}\ln\left|4\pi + \sqrt{16\pi^2 + 25}\right| + \dfrac{625}{8}\ln 5\right]$

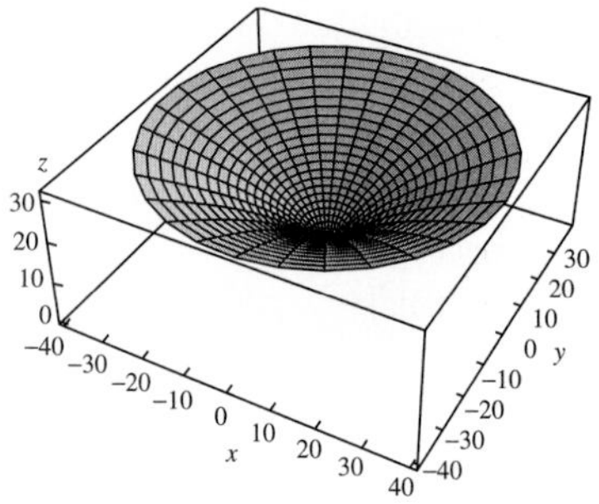

25. $\dfrac{702965k}{32}$

29. **(a)** 0; **(b)** 0; **(c)** $4\pi a^4$; **(d)** $4\pi a^4/3$; **(e)** $8\pi a^4/3$

31. **(a)** $4k\pi a^3$; **(b)** $2k\pi a^3$; **(c)** $hk\pi a(a + h)$

Problem Set 14.6

1. 0 **3.** 8 **5.** $3a^2b^2c^2/4$ **7.** $64\pi/3$ **9.** 4π
11. 1176π **13.** 100π

15. $\nabla \cdot \mathbf{F} = 3$ and so $\displaystyle\iint_{\partial S} \mathbf{F} \cdot \mathbf{n}\, dS = \iiint_S 3\, dV = 3V(S)$.

19. **(a)** $20\pi/3$; **(b)** 4π; **(c)** $16\pi/3$; **(d)** 1; **(e)** 36; **(f)** $12\pi/5$; **(g)** $32\pi \ln 2$

Problem Set 14.7

1. 0 **3.** -2 **5.** -48π **7.** 8π **9.** 2 **11.** $\pi/4$
15. 1/3 **17.** $\frac{4}{3}a^2$ joules

Chapter Review 14.8

Concepts Test

1. True **3.** False **5.** True **7.** False **9.** True
11. True.

Sample Test Problems

3. $\text{curl}(f\,\nabla f) = f\,\text{curl}(\nabla f) + \nabla f \times \nabla f = \mathbf{0} + \mathbf{0} = \mathbf{0}$

5. **(a)** $\pi/4$; **(b)** $(3\pi - 5)/6$

7. 47

9. **(a)** $\frac{1}{2}$; **(b)** $\frac{4}{3}$; **(c)** 0

11. 6π **13.** 0 **15.** $9\pi(3a - 2)/\sqrt{a^2 + b^2 + 1}$

Index

Symbols
θ-simple set, 692

A
Abel, Niels Henrik, 485
Absolute convergence, 477
Absolute convergence test, 476
Absolute error, 145
Absolute ratio test, 477
Absolute value function, 32
Absolute values, 11
  continuity of, 84
  as distance, 62
  inequalities involving, 11–12
  properties of, 11
Acceleration, 98, 127
  components of, 598–601
Accumulation function, 233, 319
Addition identities, 47
Algorithms, 193
Alternating harmonic series, 474–475
Alternating series, 474–478
  absolute convergence test, 476
  absolute ratio test, 477
  alternating series test, 475–476
  conditionally convergent series, 477–478
  convergence test, 475–476
  Rearrangement Theorem, 478
Amplitude, of trigonometric functions, 43–45
Angle of inclination, 49
Angles, 46–47
  direction, 569
  of inclination, 49
  rotation of axes of conics, 526
Angular velocity, 117
Anticommutative Law, 575
Antiderivatives, 197–202
  general, 198
  Generalized Power Rule, 200–201
  notation for, 198–199
Antidifferentiation, 411
Aphelion, 542
Approximations, 144–146, 413
  derivatives, 144–146
  linear, 145–146
  tangent planes, 654–655
Arc length, 295–298, 534, 582
  differential of, 298–299
Archimedes, 93, 232, 513
Area, 215–221
  by circumscribed polygons, 216
  by inscribed polygons, 216
  of a plane region, 275–279
    distance and displacement, 278
    region above the x-axis, 275
    region below the x-axis, 275–276
    region between two curves, 277–278
  surface, 700–705
  of a surface of revolution, 299
Asymptote, 31, 80–81
  horizontal, 80
  oblique, 82
  vertical, 80
Average value, of a function, 253
Average velocity, 95
  derivatives, 95
Axiom of Completeness, 8
Axis, 282, 510–511
  general equation of a conic section, 525–526
  rotations, 526–528
    determining the angle θ, 527–528
  translations, 523–524

B
Ball, 556
Barrow, Isaac, 232
Basis vectors, 563
Best-fit line, 171
Beta function, 393
Binomial Formula, 452
Binomial Series, 493–495
Binormal vector, 599–600
Bisection Method, 87, 190–192
Boundary of a set, 764
Boundary points, 633, 658
Bounded partial sums, 464
Bounded sum test, 464, 473
Boundedness property, 236
Bouyer, Martine, 488
Boyle's Law, 142
Brahe, Tycho, 585

C
Calculators, 3, 413–416
Calculus:
  defined, 55, 66
  differential, 100
  First Fundamental Theorem of, 232–240
  graphing functions using, 178–182
  in polar coordinate system, 547–550
  Second Fundamental Theorem of, 243
Carbon dating, 353
Cardiod, 543
Cartesian coordinates, 16
  in three-space, 555–559
    curves in three-space, 558–559
    Distance Formula, 556
    graphs in three-space, 557–558
    Midpoint Formula, 557
    spheres, 556–557
  triple integrals in, 706–711
Cartesian (rectangular) coordinate system, 537
Catenary, 378–379
Cauchy, Augustin Louis, 66
Cauchy's Mean Value Theorem, 425–426
Cauchy–Schwarz Inequality for Integrals, 684
Cavalieri's Principle, 288, 695
Center of curvature, for a plane curve, 595–596
Center of mass, 309
  and double integrals, 696–698
  and triple integrals in Cartesian coordinates, 709–710
Central conics, 513
Central quadrics, 605
Centroid, 311
Chain Rule, 118–123, 130, 132, 238, 245, 338–339, 343, 345, 411, 532, 647–651, 742
  applications of, 119–121
  applying more than once, 121–122
  derivatives, 118–123
  first version, for functions of two variables, 647–649
  implicit functions, 650–651
  partial proof of, 122–123
  proof of, 778
  second version, for functions of two variables, 649–650
  three-variable case, 649
  two-variable case, 648
Change of variable formula:
  for double integrals, 721–726
  for triple integrals, 726
Circle of curvature, 595
Circles:
  defined, 17
  equation of, 17–18
  polar equations for, 540–541
Circulation, 753, 772
Closed curve, 530
Closed interval, 8–9
Closed set, 633
Coefficient of friction, 602
Cofunction identities, 47
Coin, volume of, 281
Collapsing series, 458
Common logarithms, 344
Comparison properties, 235–236
Comparison test, 469
Completeness property of the real numbers, 454
Completing the square, 524
Complex numbers, 2
Components of acceleration, 598–601
  vector forms for, 599
Composite limit theorem, 85–86
Composition, of functions, 36–37
Compounded continuously, use of term, 351–352
Computer algebra systems (CAS), 24
  and calculators, 413–416
Computer graphs, 618–619
  and level curves, 619–621
Computers, 3
Concave down, 156–157
Concave side of a curve, 582
Concave up, 156–157
Concavity, 156–160
  inflection points, 159–160
Concavity Theorem, 157
Conchoid, 545

Conditionally convergent series, 477–478
Conic sections (conics), 509–511
central, 513
polar equations for, 540–541
Connected set, 743
Conservation of energy, 747
Conservative vector fields, 732
Constant function, 38
Constant Function Rule, for derivatives, 107–108
Constant Multiple Rule, for derivatives, 108
Constrained extremum problem, 666
Continued fraction, 196
Continuity:
of absolute value and *n*th root functions, 84
of familiar functions, 83–84
of a function of two or more variables, 632
under function operations, 84
of functions, 82–88
on an interval, 86–87
at a point, 83, 632
of polynomial and rational functions, 83
on a set, 632–633
of trigonometric functions, 84
Continuous functions, 83
Continuous mass distribution along a line, 309–310
Continuous on an open interval, 86
Continuous on a closed interval, 86
Continuous random variables, 318
Contour map, 619
Contour plots, 619
applications of, 621
Contrapositive, 4, 457
Converge, use of term, 434, 450, 456, 463
Convergence, 449–450
of Taylor series, 489–493
Convergence set, 480–481
Convergence tests:
Absolute Convergence Test, 476
Absolute Ratio Test, 477
Alternating Series Test, 475
Bounded Sum Test, 464
Limit Comparison Test, 470
Ordinary Comparison Test, 469
p-Series Test, 465
Ratio Test, 471
Convergent sequence, 79, 449–454
Convergent series:
linearity of, 459
properties of, 459–460
Converse, 4, 457
Coordinate, 2
Coordinate axes, 16
Cosecant, 45–46
Cosine, 41–42
basic properties of, 41–42
graphs of, 42–43
Cotangent, 45–46
Coulomb's Law, 307
CRC Standard Mathematical Tables and Formulae, 412
Critical Point Theorem, 152, 659
Critical points, 152, 658
Cross product, 574–577
algebraic properties, 577
anticommutative law, 575
applications, 576–577
geometric interpretation of **u** × **v**, 575
torque, 575
Cross sections, 603
Cumulative distribution function (CDF), 319
Curl, of a vector field, 733–734
Current, 99
Curtate cycloid, 536
Curvature:
binormal, 599–600
center of, for a plane curve, 595–596
circle of curvature, 595
components of acceleration, 598–601
vector forms for, 599
defined, 593
examples of, 594–595
osculating circle, 595
radius of, for a plane curve, 595–596
Curve integral, 735
Curves:
calculus for curves defined parametrically, 532–534
parametric representation of, 530–534
cycloid, 531–532
parameter, eliminating, 530–531
in three-space, 558–559
Curvilinear motion, 581–582, 581–585
Cycloid, 295, 531–532, 536
Cylinders, in three-space, 604
Cylindrical coordinate system, 609
Cylindrical coordinates, 609–610
and polar coordinates, 714
triple integrals in, 713–715
Cylindrical shells, 288–289

D

D notation, 125
Definite integrals, 215–273, 413
area, 215–221
boundedness property, 236
comparison property, 235–236
defined, 225–227
evaluating, 240
Integrability Theorem, 227
interval additive property, 229–230
linearity of, 236–237
numerical integration, 260–268
Riemann sum and, 224–225
Substitution Rule for, 248
use of symmetry in evaluating, 255–256
velocity and position, 230
Degree of polynomial function, 39
Del operator, 638
Delta, finding, 12
Deltoid, 536
Denseness, 3
Dependent variables, 30, 617
Derivatives, 93–149, *See also* Antiderivatives; Chain Rule; Differentials; Higher-order derivatives
applications of, 151–213
approximations, 144–146
average velocity, 95
Chain Rule, 118–123
concavity, 156–159
Constant Function Rule, 107–108
Constant Multiple Rule, 108
defined, 100
Difference Rule, 109
differentiability and continuity, 102–103
differential equations, 203–208
differentials, 142–146
differentiation, 100
economic applications, 172
equivalent forms for, 101–102
existence question, 151–152
finding, 100–101
rules for, 107–110
First Derivative Test, 163–164
formulas, 114–117
of functions of two or more variables, 617–672
graph of, 104
summary of the method, 181–182
using to graph a function, 182–183
higher-order, 125–129
Identity Function Rule, 108
implicit differentiation, 130–134
increments, 103–104
instantaneous velocity, 93, 95–96
Leibniz notation for, 104
linear operator, 109
Mean Value Theorem, 185–188
monotonicity, 155–156
Power Rule, 108, 110
practical problems, 167–174
Product Rule, 111–112, 116
Quotient Rule, 112, 116
rate of change, 97
related rates, 135–140
Second Derivative Test, 164, 171
Sum Rule, 109
tangent line, 93–95
of trigonometric functions, 114–117
Derive, 413
Descartes, René, 537
Determinants, 574
Difference Rule, for derivatives, 109
Differences, 35–36
Differentiability, 635–640
continuity versus, 640
for a function of two or more variables, 638
Differential calculus, 100
Differential equations, 203–208
approximations for, 359–363
Euler's Method, 360–363
slope fields, 359–360
defined, 203–204
first-order linear, 355–359
first-order separable, 204
motion problems, 206–208
separation of variables, 205–206
Differential of arc length, 298–299
Differential of the dependent variable, 143
Differential of the independent variable, 143
Differentials, 142–146
absolute error, 145
approximations, 144–146
defined, 143
estimating errors, 144–145
relative error, 145
tangent planes, 654–655
Differentiation, 100
formulas, 581
Directed distance, 1–2
Direction:
angles, 569
cosines, 569

numbers, 589
vectors, 560, 589
Directional derivatives and gradients, 642–645
connection with the gradient, 642–643
higher dimensions, 644–645
level curves and gradients, 643–644
level surface, 644
maximum rate of change, 643
Directrix, 509
Disconnected set, 743
Discontinuous functions, 83
Discrete random variables, 318
Discriminant, quadratic equation, 13
Disjoint events, 316
Disks, method of, 281–288
Displacement, 279
Distance, absolute values as, 62
Distance Formula, 16–17, 556
Diverge, use of term, 434, 450, 456
Divergence, 752
*n*th-term test for, 457–458
of a vector field, 733–734
Divergence Theorem, 764
Dog saddle, 635
Domain:
functions, 617
natural, 30
and range, 30
restricting the, 365
Dot product, 566–572
defined, 566
orthogonal vectors, 568
perpendicularity criterion, 567
planes, 570–572
projections, 569–570
properties of, 567
Double-angle identities, 47
Double integrals, 676–679
applications of, 696–699
center of mass, 696–698
moment of inertia, 698–699
change of variable formula for, 721–726
sketch of proof, 721–723
evaluation of, 677–679
examples of, 686–689
over general sets, 684–686
over general sets, evaluation of, 684–686
over nonrectangular regions, 684–689
over rectangular regions, 684–689
in polar coordinates, 691–694
iterated integrals, 691–692
properties of, 677
Doubling time, 349

E

e, 337, 352
Eccentricity, 509
Einstein's Special Theory of Relativity, 82, 147
Ellipse, 509, 513–520
horizontal, 516
optical properties, 519–520
polar equation for, 541
reflecting property of, applications, 520
standard equation of, 515
string properties of, 517–519
whispering gallery, 520
Ellipsoid, in three-space, 605
Elliptic cone, in three-space, 606
Elliptic paraboloid, in three-space, 606
Epicycloid, 536
Equations:
of a vertical line, 20
parametric, 294
solving numerically, 190–195
Bisection Method, 190–192
Fixed-Point Algorithm, 194–195
Newton's Method, 192–194
Equipotential curves, 623
Equipotential surfaces, 644
Error analysis, 266–268
Error of calculation, 499, 501–502
Error of the method, 499–500
Escape velocity, 209
Estimation, 3–4
Euclidean plane, 555
Euler, Leonhard, 337, 361
Euler's constant, 468
Euler's Method, 360–363
Euler's Theorem, 652
Even function, 31–32
Expected value, 318
Explicit algebraic functions, 39
Explicit formula, for infinite series, 449
Exponential distribution, 436–437
Exponential function rules, 343
Exponential function to the base *a*, 342
Exponential functions, 344
Exponential growth and decay, 347–352
compound interest, 351–352
differential equation, solving, 348–349
doubling time, 349
exponential decay, defined, 348
exponential growth, defined, 348
logistic model, 349
Newton's Law of Cooling, 350–351
radioactive decay, 350
Exponents, properties of, 342
Extrema:
on open intervals, 165
sufficient conditions for, 660–664
Extreme value, 151, 152–154
defined, 153–154
intervals of occurrence, 152

F

Falling-body problems, 128–129
Familiar functions, continuity of, 83–84
Family of functions, 233
Fermat, Pierre de, 537
Fibonacci, Leonardo, 454
Fibonacci Sequence, 454
Final end points, 530
First derivative, 125
and monotonicity, 155–156
First Derivative Test, 163–164
First Fundamental Theorem of Calculus, 232–240, 582, 744
position as accumulated velocity, 239
proof of, 237–239
sketch of proof, 235
First-order linear differential equations, 355–357
applications, 356–357
defined, 355
general solution, 355
initial condition, 355
integrating factor, 355
particular solution, 355
solving, 355–356
First-order separable differential equations, 204
Fixed cost, 172
Fixed-Point Algorithm, 194–195
Fluid force, 304–305
Flux, 752
of a vector field through a surface, 758–759
Focus, 509
Folium of Descartes, 537
Foot-pounds, 301
Force field, 731
Formulas, derivatives, 114–117
Fourth derivative, 125
Fourth-Order Runge-Kutta Method, 363
Fractional rate of change, 354
Free extremum problem, 666
Frustum of a cone, 299
Function notation, 29
Function operations, continuity under, 84
Functions, 29, 233
absolute value, 32
average value, 253
composition of, 36–37, 632
computer graphs, 618–619
constant, 38
continuity, 82–88
at a point, 632
on a set, 632–633
cosine, 41
discontinuous, 83
domain, 617
even and odd, 31–32
explicit algebraic, 39
graphs of, 31, 617–618
greatest integer, 32
harmonic, 628
identity, 38
involving roots, 180–181
level curves, 619–620
limits, 629–632
linear, 39
locally linear, 636
objective, 151
operations on, 35–39
partial catalog of, 38–39
periodic, 43–44
polynomial, 38
quadratic, 39
range, 29, 617
rational, 39, 630
sine, 41
special, 32
of three or more variables, 622
translations, 37–38
trigonometric, 41–48
of two or more variables, 617–622
of two variables:
differentiability for, 638
limit of, 629
local linearity for, 637
Taylor polynomials, 655–656
Fundamental Theorem for Line Integrals, 742
Fundamental Theorem of Arithmetic, 8
Fundamental Theorem of Calculus, *See* First Fundamental Theorem of Calculus or Second Fundamental Theorem of Calculus

Fundamental theorems, 232
Fundamental triangle, 516
Future value, 354

G

Gabriel's horn, 439–440
Galileo, Galilei, 93
Gamma probability density function, 446
Gauss's Divergence Theorem, 752, 764–768
  extensions and applications, 767–768
  proof of, 765
General antiderivative, 198
General exponential and logarithmic functions, 342–345
  function log a, 343–344
General linear equation, 21
General solution, of a differential equation, 355
Generalized Power Rule, 200–202
Geographic center, 315
Geometric mean–arithmetic mean inequality, 15
Geometric series, 456–457
Geometry in space and vectors, 555–615, 556
Global extreme value, 658
Global maximum value, 162, 658
Global minimum value, 658
Golden ratio, 454
Grade, 19
Gradient fields, 640, 732–733
Gradient, 638, 641–645
  and level curves, 643–644
  recovering function from its, 745–747
  rules for, 640
Graph of a function, 31
Graphing calculator, 24, 26
Graphs:
  computer, 618–619
  of derivatives, 104
    summary of the method, 181–182
    using to graph a function, 182–183
  of functions, 617–618
  of polar equations, 542–544
    cardioids, 543
    lemniscates, 543
    limaçons, 543
    roses, 543–544
    spirals, 544
  use of symmetry, 542–543
Graphs in three-space, 557–558
Graphs of equations, 24–27
  graphing procedure, 24–25
  intersections of, 26–27
  symmetry of, 25–26
Greatest integer function, 32
Green, George, 749
Green's Theorem, 745, 749–753, 764
  examples and applications, 750–755
  proof of, 749–750
  vector forms of, 751–753
Guilloud, Jean, 488

H

Half-angle identities, 47
Half-life, 350
Handbook of Mathematical Functions, 412, 417–418
Harmonic functions, 628, 735
Harmonic series, 458
Helix, 559, 583
Higher-order derivatives, 125–129
  acceleration, 126–128
  D notation, 125
  falling-body problems, 128–129
  first derivative, 125
  fourth derivative, 125
  Leibniz notation, 125–126
  prime notation, 125
  second derivative, 125–126
  third derivative, 125
  velocity, 126–128
Homogeneous of degree 1, 652
Hooke's Law, 170, 302, 306, 322
Horizontal asymptote, 80
Horizontal ellipses, 516
Horizontal slicing, 278
Huygens, Christian, 532
Hyperbola, 509, 513–520
  optical properties, 519–520
  polar equation for, 541
  standard equation of, 516–517
  string properties of, 517–519
    applications, 520
Hyperbolic functions, 374–378
  applications, 378
  catenary, 378
  derivatives of, 374–376
  inverse, 376–377
Hyperbolic paraboloid, in three-space, 606
Hyperboloid of one sheet, in three-space, 605
Hyperboloid of two sheets, in three-space, 606
Hypocycloid, 536
  of four cusps, 536

I

Identity function, 38
Identity Function Rule, for derivatives, 108
Image of a set, 719
Implicit differentiation, 130–134
  Power Rule, 133
Improper integrals, infinite limits, 433–440
  infinite integrands, 442–444
Increments, 103–104
  derivatives, 103–104
Indefinite integrals, 244
  as linear operator, 199–200
  Substitution Rule for, 246
  use of term, 199
Independence of path, 742–747
  conservation of energy, 747
  criteria for, 743–745
  notation for line integrals, 746
  recovering a function from its gradient, 745–747
  theorem, 743
Independent variables, 30, 617
Indeterminate forms, 423–447
  of type ∞/∞, 429
  of type 0^0, ∞^0, 1^∞, 431–432
  of type $0/0$, 423–427
    l'Hôpital's rule for, 429
  of types $0 \cdot \infty$ and $\infty - \infty$, 430–431
Inequalities, 8–14
  involving absolute values, 11–12
  solving, 9–11
Infinite integrands, 442–444
Infinite limits, 79–80
Infinite sequences, 79, 449
  convergence, 449–450
  limits of sequences, properties of, 450
  monotonic sequences, 452–453
Infinite series, 448, 455–460
  grouping terms in, 460
Infinity, limits at, 77–78
Inflection points, 159–160
Initial end points, 530
Inner integral, 682
Inner integration, 682
Instantaneous velocity, 93, 95–96, 100, 126
Integers, 1
Integrability Theorem, 227, 677
Integral sign, 199
Integral Test, 464, 473
Integrals:
  applications of, 275–323
    area of a plane region, 275–279
    moments and center of mass, 308–313
    plane curve, length of, 294–299
    probability and random variables, 316–320
    solids of revolution, volumes of, 288–292
    solids, volumes of, 281–286
    work and fluid force, 301–306
  definite, 224–232
  double integrals over nonrectangular regions, 684–689
  estimating, 255
  improper, 433–444
  iterated, 680–683
  Mean Value Theorem for, 253–258
  multiple, 675–729
  table of, 412–413
Integrands, 199
  infinite at an end point, 442–444
  infinite at an interior point, 444
Integration, *See* Antiderivatives
  basic rules, 383–385
  both limits infinite, 435–437
  computer algebra systems and calculators, 413–416
  converge, use of term, 434, 450, 456
  diverge, use of term, 434, 450, 456
  functions defined by tables, 416–417
  infinite limits of, 433–440
  inner, 682
  limits of, 707
  numerical, 260–268
  one infinite limit, 433–435
  by parts, 383, 387–391, 411–412
    definite integrals, 387–389
    indefinite integrals, 387
    reduction formula, 390
    repeated, 389–390
  probability density functions (PDFs), 436–439
  of rational functions, 404–410
    logistic differential equation, 408–410
    partial fraction decomposition (linear factors), 405–407
    partial fraction decomposition (quadratic factors), 407–408
  rationalizing substitutions, 399–402
  repeated, by parts, 389–390
  special functions, 417–418
  standard forms, 383–384
  strategies for, 411–418
  substitution, 383

in definite integrals, 385
in indefinite integrals, 384–385
techniques, 383–421
trigonometric integrals, 393–398
Interior point, 633
Intermediate Value Theorem, 87–88
Interval, continuity on, 86–87
Interval additive property, 229–230
Interval notation, 638
Intervals, 8–11
Inverse, defined, 332
Inverse Function Theorem, 335, 369, 532
Inverse functions, 331–335
derivatives of, 334–335
existence of, 332–333
Inverse hyperbolic functions, 376–377
Inverse Square Law of Gravitational Attraction, 731, 748
Inverse trigonometric functions:
derivatives of, 368–369
inverse sine and inverse cosine, 365–366
inverse tangent and inverse secant, 366–367
manipulating the integrand, 371
restricting the domain, 365
useful identities, 368
Irrational numbers, 1
decimal representations of, 3
Irrotational flow, 753
Isobars, 623
Isoseismic curves, 621
Isothermal curves, 621
Isothermal surface, 644
Isotherms, 621
Iterated integrals, 680–683
defined, 681
evaluating, 681–683
inner integral, 682
inner integration, 682
Iteration scheme, 192

J

Jacobian of a transformation, 722–725
Joint probability density function, 710–711
Joules, 301

K

Kepler, Johannes, 93, 517, 585
Kepler's Laws of Planetary Motion, 585–587
First Law, 585
Second Law, 585–586
Third Law, 586–587
Koch snowflake, 462

L

Lagrange, Joseph Louis, 500
Lagrange multipliers:
defined, 667
method of, 666–671
applications, 668–669
defined, 666
geometric interpretation of, 666–667
optimizing a function over a closed and bounded set, 670–671
two or more constraints, 669–670
Lagrange's Identity, 578
Lamina, 311, 696–698
center of mass, 311, 696–698
mass, 311, 696
Laplace, Pierre-Simon de, 446
Laplace transform, 446
Laplace's Equation, 628
Laplacian, 735
Last Judgement (Memling), 440
Latus rectum, 512
Law of Conservation of Energy, 747
Law of Cosines, 540
Law of the Excluded Middle, 5
Least squares, 170–171
Least squares line through the origin, 171
Least upper bound, 8
Left continuous function, 86
Left Riemann Sum, 266
Left-hand limit, 58
Leibniz, Gottfried Wilhelm von, 66, 104, 110, 232
Leibniz notation, 104, 125–126
Lemniscate, 543
Lenses, 519
Level curves, 619–620
and computer graphs, 619–621
and gradients, 643–644
Level surfaces, 622, 644
l'Hôpital, Guillaume François Antoine de, 423
l'Hôpital's Rule, 423–424, 451, 476
geometric interpretation of, 423
Limaçon, 543
Limit Comparison Test, 470, 473
Limits, 55, 61–67, 579
asymptotes, 80–81
composite limit theorem, 85–86
continuity of functions, 82–88
examples, 57
infinite, 79–80
at infinity, 77–78
intuitive meaning of, 57
involving trigonometric functions, 73–76
left-hand, 58
limit proofs, 63–66
limit theorems, 68–72
Main Limit Theorem:
applications of, 68–69
proof of, 71–72
one-sided, 58–59, 66–68
precise meaning of, 61–63
preliminary analysis, 63–66
problems leading to the concept of, 55
right-hand, 58, 66–67
rigorous definitions of limits as $x \to \pm\infty$, 78–79
rigorous study of, 61–67
of sequences, 79
Squeeze Theorem, 72, 75
Substitution Theorem, 69–71
warning flags, 57–58
Limits of integration, 707
Line integrals, 735–740
equivalent conditions for, 744
examples and applications, 737–738
fundamental theorem for, 742
work, 738–740
Linear approximations, 145–146
Linear equation, 557
Linear functions, 39, 109
Linear operator, 109
Lines, 18–19
point-slope form, 20
polar equations for, 540–541
slope-intercept form, 20
slope of, 18–19
in three-space, 589–591
Lissajous figure, 536
Local extreme values, 162–163, 658
where they occur, 163–165
Local maximum value, 162, 658
Local minimum value, 162, 658
Locally linear functions, 636
Logarithmic differentiation, 329, 345
Logarithmic function, 325–330
to the base *a*, 343
Logarithmic spiral, 544
Logistic differential equation, 349, 354, 408–410
LORAN, 520

M

Machin, John, 488
Maclaurin, Colin, 489
Maclaurin polynomials, 497, 498–499
Maclaurin series, 491–492, 494–495
defined, 489
Maclaurin's Formula, 496, 500
Magnitude, vectors, 560, 563
Main Limit Theorem, 423
applications of, 68–69
proof of, 71–72
Major axis, 509–510, 515
Major diameter, 515
Maple, 3, 31, 413
three-dimensional graphs, 618
Marginal cost, 173
Marginal price, 173
Marginal profit, 173
Marginal revenue, 173–174
Mass, and triple integrals in Cartesian coordinates, 709–710
Mass distributions in the plane, 310–311
Mathematica, 3, 413
three-dimensional graphs, 618
Mathematical induction, 5, A-1–A-3
Mathematical modeling, 172
Maximum value, 151
Max–Min Existence Theorem, 151, 658
Mean, 317–318
Mean Value Theorem, 185–188
for derivatives, 185–188, 254, 425–426, 490–491, 638
for double integrals, 690
for integrals, 253–258
Median, 321
of a continuous random variable, 322
Memling, Hans, 440
Memorization, 111
Method of equating coefficients, 496
Method of iterations, 190
Method of successive approximations, 190
Midpoint Formula, 18, 557
Midpoint Riemann Sum, 266
Minimum value, 151
Minkowski's Inequality, 337
Minor axis, 515
Minor diameter, 515
Mixed partial derivatives, 627
equality of, 633
Möbius band, 758, 770
Models, 170
Moment, 308

Moment of inertia, and double integrals, 698–699
Monotonic Sequence Theorem, 453, 464
Monotonic sequences, 452–453
Monotonicity, 155–156
  and first derivative, 155–156
Monotonicity Theorem, 155
Motion problems, 206–208
Multiple integrals, 675–729
  change of variables in, 718–726
    terminology and notation, 718
  double integrals, 676
    evaluation of, 677–679
    over nonrectangular regions, 684–689
    in polar coordinates, 691–694
    properties of, 677
  existence question, 677
  iterated integrals, 680–683
  surface area, 700–704
  triple integrals:
    in Cartesian coordinates, 706–711
    in cylindrical and spherical coordinates, 713–717
Multiplier effect, 461

N

Napier, John, 327
Napier's Inequality, 331
Natural domain, 30
Natural exponential function, e^x, 337–340
  derivative of e^x, 338–339
  properties of, 337–338
Natural logarithm function, 325–330
  defined, 325
  derivative of, 326–327
  graph of, 329
  logarithmic differentiation, 329
  properties of, 327–329
Natural numbers, 1
Negation, 4, 6
Neighborhood, 632–633
Newton, Isaac, 66, 93, 232
Newton-meters, 301
Newton's Inverse Square Law, 435
Newton's Law of Cooling, 350–351
Newton's Law of Gravitation, 652
Newton's Method, 192–194
  of solving equations numerically, 192–194
Newton's Second Law, 170, 586
Nondecreasing sequence, 452
Nonremovable point of discontinuity, 83
Nonrepeating decimals, 2–3
Normal components of acceleration, 598
Normal distribution, 438
Notation:
  for antiderivatives, 638
  D, 125
  interval, 638
  Leibniz, 125–126
  for line integrals, 746
  prime, 125
  for square roots, 13
nth partial sum, 455–456, 463
nth root functions, continuity of, 84
nth-Term Test for Divergence, 473
Number, 159, 233
Numerical integration, 260–268
  definite integrals, 260–268
  error analysis, 266–268
  functions defined by a table, 268
  Parabolic Rule (Simpson's Rule), 265
  Riemann sums, 260
  Trapezoidal Rule, 264

O

Objective function, 151
Oblique asymptote, 82, 180
Odd-even trigonometric identities, 47
Odd function, 32
Old Goat Problem, 705
One-sided limits, 58–59, 66–68
One-to-one function, 332
Open interval, 8
Open set, 633
Operator, 107
Optical property:
  of ellipse, 519–520
  of hyperbola, 519–520
  of parabola, 511
Ordered pair, 16, 159
Ordinary Comparison Test, 469–471, 473
Orientation, curves, 295
Origin, 2, 16, 555
Orthogonal vectors, 568
Osculating plane, 600

P

Pappus, 312
Pappus's Theorem, 312–313, 322
Parabola, 26, 509–511
  applications, 511
  defined, 510
  optical properties, 511
    applications, 520
  polar equation for, 541
  standard equation, 510
Parabolic Rule (Simpson's Rule), 265–268, 363, 413
Paradox of Gabriel's horn, 439–440
Parallel Axis Theorem, 700
Parallel lines, 21
Parallelepided, 556
Parameter, 530
  eliminating, 530–531
Parameters, 294
Parametric equations, 294, 589
Parametric representation:
  defined, 49
  of curves in the plane, 530–534
    cycloid, 531–532
    parameter, eliminating, 530–531
Parametrization of a curve, 530
Parametrized surface:
  defined, 760
  surface area for, 760–763
Pareto probability density function, 441
Partial derivative symbol, 633
Partial derivatives, 624–627
  geometric and physical interpretations, 625–626
  mixed, 627
  partial derivative of f with respect to x, 624, 627
  second, 626
  third, 626
Partial differential equations, 628
Partial fraction decomposition, 404–412
Partial sum, 455
Partition, 224–229
  regular, 227–229
Pascal, Blaise, 232, 304–305
Perihelion, 542
Period, 43
Periodic, of trigonometric functions, 43–45
Periodicity, 257
Perpendicular lines, 21–22
Perpendicularity criterion, 567
Pitch, 19
Plane curve, 295, 530
  arc length, 294–299
    differential of, 298–299
  orientation, 295
Planes, 570–572
  dot product, 570–572
  linear equation, 571
  osculating, 600
  standard form for the equation of a plane, 570
Point–slope form, 19–20
Polar coordinate system, 537–541
  area in, 547–548
  calculus in, 547–550
  intersection of curves in, 544
  polar axis, 537
  polar coordinates, defined, 537
  polar equations, 538
    for lines, circles, and conics, 540–541
  pole (origin), 537
  relation to Cartesian coordinates, 539–540
  tangents in, 549–550
Polar coordinates:
  and cylindrical coordinates, 714
  double integrals in, 691–694
    iterated integrals, 691–692
Polar equations:
  graphs of, 542–544
    cardioids, 543
    lemniscates, 543
    limaçons, 543
    roses, 543–544
    spirals, 544
  intersection of curves in polar coordinates, 544
Polar rectangle, 691
Polynomial functions, 38–39, 178–179, 630
  continuity of, 83
Polynomials:
  Maclaurin, 497–499
  Taylor, 497–499, 655–656
Position, 227
  as accumulated velocity, 239
  definite integrals, 230
Position vector, 582
Positive series, 468–473
  comparing a series with itself, 471–473
  comparing one series with another, 469–471
  Limit Comparison Test, 470
  Ordinary Comparison Test, 469–470
  Ratio Test, 471–473
Positively oriented, use of term, 735
Potential function, 732
Power function, 344
Power Rule, for derivatives, 108, 110, 133
Power series, 455, 479–482
  convergence set, 480–481
  operations on, 484–487
    algebraic operations, 486–487

power series in $x - a$, 487
term-by-term differentiation and integration, 484–486
in x, 480
in $x - a$, 481–482
Powers, 35–36
Precalculus, 55
Preimage, 719
Present value, 354
Price, 172
Prime notation, 125
Prime number, 8
Principal square root, 13
Principal unit normal vector, 598
Probability:
continuous random variables, 318
cumulative distribution function (CDF), 319
discrete random variables, 318
disjoint events, 316
expected value, 318
mean, 317–318
probability distribution, 316
random outcome, 316
random variables, 316
expectation of, 317
sample space, 316
Probability density function (PDF), 318, 436–439
Probability distribution, 316
Probability integral, 693
Product identities, trigonometric, 48
Product Rule, for derivatives, 111–112, 116, 345, 411, 598
Products, 35–36
Projections, 569–570
of **u** on **v**, 570
Prolate cycloid, 536
Prolate spheroid, 287
Proof:
Chain Rule, partial, 122–123
by contradiction, 5
First Fundamental Theorem of Calculus, 237–239
Gauss's Divergence Theorem, 765
Green's Theorem, 749–750
key to, 186
Main Limit Theorem, 71–72
p-series, 465–466
Pythagorean identities, trigonometric, 47
Pythagorean Theorem, 4, 16, 136, 556
in three-space, 578

Q

Quadrants, 16
Quadratic Formula, 13
Quadratic functions, 39
Quadric surfaces, in three-space, 605–607
Quantifiers, 5–6
Quotient Rule, for derivatives, 112, 116, 411
Quotients, 35–36

R

Radioactive decay, 350
Radius and center of curvature, for a plane curve, 595–596
Radius of gyration, 699
Ramanujan, Srinivasa, 487
Random outcome, 316
Random variables, 316
expectation of, 317
pairs of, 710–711
Range:
and domain, 30
functions, 29, 617
Rates of change, 97
derivatives, 97
Ratio Test, 471–473
Rational functions, 39, 179–180, 630
continuity of, 83
integration of, 404–410
proper, 404
Rational numbers, 1
Rationalizing substitutions, 399–402
Real line, 2
Real numbers, 1–2
Real-valued function of two real variables, 617
Rearrangement Theorem, 478
Reciprocal spiral, 546
Rectangular coordinate system, 16–22
Cartesian coordinates, 16
coordinate axes, 16
distance formula, 16–17
equation of a circle, 17–18
equation of a vertical line, 20
form $Ax + By + C = 0$, 20–21
lines, 18–19
midpoint formula, 18
ordered pair, 16
origin, 16
parallel lines, 21
perpendicular lines, 21–22
point–slope form, 19–20
quadrants, 16
slope–intercept form, 20
x-axis, 16
x-coordinate, 16
y-axis, 16
y-coordinate, 16
Recursion formula, 192, 449
Reduction formula, 390
Related rates, 135–140
derivatives, 135–140
graphical problem, 140
simple examples, 135–137
systematic procedure, 137–139
Relative error, 145
Relative maximum value, 162
Removable point of discontinuity, 83
Repeated integration by parts, 389–390
Repeating decimals, 2–3
Restricting the domain, 365
Riemann sum, 224–225, 234, 260–264, 547, 706, 736
Riemann Sum Rule, 363
Right- and left-hand limits, 58
Right continuous function, 86
Right cylinders, 281
Right Riemann Sum, 266
Right-hand limit, 58, 66–67
Right-handed system, 555
Rise, 18–19
Roots, functions involving, 180–181
Rose, 543–544
Rotation of axes, for conic, 526–528
determining the angle θ, 527–528
r-simple set, 692
Run, 18–19

S

Saddle point, 659
Sample point, 282
Sample space, 316
Scalar field, 731
gradient of, 732–733
Scalar product, *See* Dot product
Scalar projection of **u** on **v**, 570
Scalars, 560
dividing a vector by, 564
Scatter plot, 170
Secant, 45–46
inverse, 366–367
Secant line, 93
Second derivative, 125–126
Second Derivative Test, 164, 171
Second Fundamental Theorem of Calculus, 243–250, 260, 340, 413–415, 417, 704, 742, 749
accumulated rate of change, 249–250
method of substitution, 245–249
Second moments, *See* Moment of inertia
Second Partials Test, 660
Separation of variables, 205–206
Sequence:
defined, 463
of partial sums, 463
Series:
approximating the sum of, 466–467
of constants, 479
defined, 463
of functions, 479
and improper integrals, 464–465
tail of, 466
Shells, 288–292
method of, 289–291
Sigma (Σ) notation, 170–171, 216–217, 455
Simple curve, 530
Simply connected set, 745–746
Simpson's Rule (Parabolic Rule), 265–268
Sine, 41–42
basic properties of, 41–42
graphs of, 42
Singular points, 152, 658
Slabs, 281
Slice, approximate, and integrate strategy, 276–277, 280–282, 286, 289–290, 302, 309–311, 547, 681, 709, 737
Slope, 18–19
Slope fields, 359–360
Slope of the tangent line, 93, 100
Slope–intercept form, 20
Smooth curve, 295, 558
Soda Can Problem, 712–713
Solid of revolution, defined, 282
Solid sphere, 556
Solids with known cross sections, 285–286
Solving equations, 8
Speed, 127, 582
Spheres, 556–557
defined, 556
standard equation of, 556
Spherical coordinates, 609–611
in geography, 611–612
triple integrals in, 715–717
Spherical wedge, 715
Spiral of Archimedes, 544, 546
Spirals, 544
reciprocal, 546

Split points, 10
Spring constant, 302
Square roots, notation for, 13
Squares, 13–14
Squeeze Theorem, 72, 75, 451–452, 547
Standard equation:
  circle, 17
  ellipse, 515
  hyperbola, 516–517
  parabola, 510
  sphere, 556
Standard form for the equation of a plane, 570
Standard normal distribution, 438
Standard unit vectors, 563
Stationary points, 152, 658
Steady vector fields, 731
Stirling's Formula, 341, 468
Stokes, George Gabriel, 770
Stokes's Theorem, 745, 753, 770–772
  examples and applications, 770–772
  physical interpretation of the curl, 772
Strictly monotonic function, 332
String properties, of conics, 517–519
Substitution Rule:
  for definite integrals, 248
  for indefinite integrals, 246
Substitution Theorem, for limits, 69–71
Substitution(s), 411
  rationalizing, 399–402
Sum, 35–36
  of a series, 463
Sum identities, 47
Sum Rule, for derivatives, 109, 411
Surface area, 700–704
  problems, 701
Surface integrals, 755–763
  evaluating, 756–757
  flux of a vector field through a surface, 758–759
  parameterized surface:
    defined, 760
    surface area for, 760–763
Surfaces in three-space, 603–607
  cylinders, 604
  quadric surfaces, 605–607
Symmetric derivative, 104
Symmetric equations, 590
Symmetric with respect to the origin, 25
Symmetric with respect to the *x*-axis, 25
Symmetric with respect to the *y*-axis, 25
Symmetry, 253–258
  and graphs, 542–543
Symmetry Theorem, 256

T

Tables of integrals, 412–413
Tail, vectors, 560–561
Tangent lines, 93–95
  to a curve, 591
  defined, 93
  in three-space, 589–591
Tangent planes, 639, 652–656
  approximations, 654–655
  defined, 653
  differentials, 654–655
  Taylor polynomials for functions of two or more variables, 655–656
Tangential components of acceleration, 598
Tangents, 45–46
  in polar coordinate system, 549–550
Taylor approximation to a function, 497–502
Taylor, Brook, 489
Taylor polynomials, 497–499, 655–656
  for functions of two or more variables, 655–656
  of order 1, 497–499
  of order *n*, 497–498
Taylor series:
  convergence of, 489–493
  defined, 489
  Taylor's Formula with Remainder, 490–491
  Taylor's Theorem, 491
Test points, 10
Theorems, 4–5
Third derivative, 125
Three-space:
  Cartesian coordinates in, 555–559
    curves in three-space, 558–559
    Distance Formula, 556
    graphs in three-space, 557–558
    Midpoint Formula, 557
    spheres, 556–557
  curves in, 558–559
  cylinders in, 604
  ellipsoid in, 605
  elliptic cone in, 606
  elliptic paraboloid in, 606
  lines in, 589–591
  Pythagorean Theorem in, 578
  tangent lines in, 589–591
  vectors in, 560–564
Time, measuring, 127
Time rate of change, 135
Torque, 575
Torricelli's Law, 209
Total distance, 279
Total profit, 172
Total revenue, 172
Traces, 557, 603
Transcendental functions, 325–381
  exponential growth and decay, 347–352
  first-order linear differential equations, 355–357
  general exponential and logarithmic functions, 342–345
  hyperbolic functions, 374–378
  inverse functions, 331–335
  inverse trigonometric functions, 365–371
  natural exponential function, 337–340
  natural logarithm function, 325–330
Transformation, 719
Translations:
  axes, for conic, 523–524
  of functions, 37–38
Trapezoidal Rule, 264, 266, 363, 417
Triangle Inequality, 11, 574
Trigonometric functions, 41–48
  amplitude of, 43–45
  angles, 46–47
  continuity of, 84
  cosecant, 45–46
  cosine, 41–42
  cotangent, 45–46
  derivatives of, 114–117, 368
  important identities, list of, 47–48
  inverse, 365–368
  limits involving, 73–76
  periodic of, 43–45
  secant, 45–46
  sine, 41–42
  tangent, 45–46
  trigonometric identities, 47
Trigonometric integrals, 393–398
  natural logarithm function, 329–330
Trigonometric substitutions, 412
Trihedral, 600
Triple integrals:
  in Cartesian coordinates, 706–711
    general regions, 707–709
    mass and center of mass, 709–710
    pairs of random vectors, 710–711
  change of variable formula for, 726
  in cylindrical and spherical coordinates, 713–717
  defined, 706
Two-space, 555
  vectors in, 563

U

u-curves, 719–720
Unbounded integrand, 442
Uniform circular motion, 582–583
Uniform distribution, 321
Union, 10
Uniqueness Theorem, 489
Unit vectors, 564
Upper bound, 8

V

Variable cost, 172
Variance, 322
v-curves, 719–720
Vector calculus, 731–774
  line integrals, 735–740
  surface integrals, 755–763
  vector fields, 731–733
Vector fields, 731–733
  conservative, 732
  defined, 731
  divergence and curl of, 733–734
  steady, 731
Vector forms for components of acceleration, 599
Vector product, *See* Cross product
Vector projection of **u** on **v**, 570
Vectors, 560–564
  algebraic approach to, 562–563
  application, 562
  basis, 563
  components, 562
  direction, 560
  equivalent, 561
  head (tip), 561
  magnitude, 560, 563
  operations on, 561–562
  orthogonal, 568
  scalars, 560
  standard unit, 563
  tail, 560–561
  in two dimensions, 580
  in two-space, 563
  unit, 564
  zero, 563

Vector-valued functions, 579–587
curvilinear motion, 581–585
differentiation formulas, 581
Kepler's Laws of planetary motion, 585–587
limit of, 579
Velocity, 126–128, 227, 582
definite integrals, 230
instantaneous, 93, 95–96
Vertex, 510
Vertical asymptote, 80
Vertical line, equation of, 20
Volume, 676
calculating, 682–683
of a coin, 281
solid of revolution, 288–292
solids, 282

W

Washers, method of, 284–285
Wave equation, 628
Weibull distribution, 441
Weierstrass, Karl, 67
Whispering gallery, 520
Work, 301–304, 738–740
application to pumping a liquid, 303–304
application to springs, 302–303

X

x-axis, 16
x-coordinate, 16
x-intercepts, 26
x-simple set, 685

Y

y-axis, 16
y-coordinate, 16
y-intercepts, 26
Young's Inequality, 337
y-simple set, 684–685

Z

Zeno's paradox, 462
Zero vectors, 563
z-simple set, 707

Photo Credits

Front Endpapers

Descartes	Frans Hals/Louvre
Newton	Library of Congress
Leibniz	The Granger Collection, New York
Euler	Library of Congress
Kepler	Library of Congress
Pascal	Courtesy of International Business Machines Corporation. Unauthorized use not permitted.
L'Höpital	The Granger Collection, New York
Bernoulli	The Granger Collection, New York
Lagrange	New York Public Library
Gauss	The Granger Collection, New York
Cauchy	Corbis
Riemann	The Granger Collection, New York
Lebesgue	The Granger Collection, New York
Agnesi	Library of Congress
Weierstrass	Corbis
Kovalevsky	Library of Congress
Gibbs	Corbis

Text

p. 219	Kennedy Space Center/NASA
p. 224	Susan Van Etten/PhotoEdit
p. 362	David Frazier/Photo Researchers, Inc.
p. 418	Scala/Art Resource, NY
p. 704	S. Rigdon

Table of Integrals

ELEMENTARY FORMS

1 $\int u\,dv = uv - \int v\,du$

2 $\int u^n\,du = \frac{1}{n+1}u^{n+1} + C \quad \text{if } n \neq -1$

3 $\int \frac{du}{u} = \ln|u| + C$

4 $\int e^u\,du = e^u + C$

5 $\int a^u\,du = \frac{a^u}{\ln a} + C$

6 $\int \sin u\,du = -\cos u + C$

7 $\int \cos u\,du = \sin u + C$

8 $\int \sec^2 u\,du = \tan u + C$

9 $\int \csc^2 u\,du = -\cot u + C$

10 $\int \sec u \tan u\,du = \sec u + C$

11 $\int \csc u \cot u\,du = -\csc u + C$

12 $\int \tan u\,du = -\ln|\cos u| + C$

13 $\int \cot u\,du = \ln|\sin u| + C$

14 $\int \sec u\,du = \ln|\sec u + \tan u| + C$

15 $\int \csc u\,du = \ln|\csc u - \cot u| + C$

16 $\int \frac{du}{\sqrt{a^2 - u^2}} = \sin^{-1}\frac{u}{a} + C$

17 $\int \frac{du}{a^2 + u^2} = \frac{1}{a}\tan^{-1}\frac{u}{a} + C$

18 $\int \frac{du}{a^2 - u^2} = \frac{1}{2a}\ln\left|\frac{u+a}{u-a}\right| + C$

19 $\int \frac{du}{u\sqrt{u^2 - a^2}} = \frac{1}{a}\sec^{-1}\left|\frac{u}{a}\right| + C$

TRIGONOMETRIC FORMS

20 $\int \sin^2 u\,du = \frac{1}{2}u - \frac{1}{4}\sin 2u + C$

21 $\int \cos^2 u\,du = \frac{1}{2}u + \frac{1}{4}\sin 2u + C$

22 $\int \tan^2 u\,du = \tan u - u + C$

23 $\int \cot^2 u\,du = -\cot u - u + C$

24 $\int \sin^3 u\,du = -\frac{1}{3}(2 + \sin^2 u)\cos u + C$

25 $\int \cos^3 u\,du = \frac{1}{3}(2 + \cos^2 u)\sin u + C$

26 $\int \tan^3 u\,du = \frac{1}{2}\tan^2 u + \ln|\cos u| + C$

27 $\int \cot^3 u\,du = -\frac{1}{2}\cot^2 u - \ln|\sin u| + C$

28 $\int \sec^3 u\,du = \frac{1}{2}\sec u \tan u + \frac{1}{2}\ln|\sec u + \tan u| + C$

29 $\int \csc^3 u\,du = -\frac{1}{2}\csc u \cot u + \frac{1}{2}\ln|\csc u - \cot u| + C$

30 $\int \sin au \sin bu\,du = \frac{\sin(a-b)u}{2(a-b)} - \frac{\sin(a+b)u}{2(a+b)} + C \quad \text{if } a^2 \neq b^2$

31 $\int \cos au \cos bu\,du = \frac{\sin(a-b)u}{2(a-b)} + \frac{\sin(a+b)u}{2(a+b)} + C \quad \text{if } a^2 \neq b^2$

32 $\int \sin au \cos bu\,du = -\frac{\cos(a-b)u}{2(a-b)} - \frac{\cos(a+b)u}{2(a+b)} + C \quad \text{if } a^2 \neq b^2$

33 $\int \sin^n u\,du = -\frac{1}{n}\sin^{n-1} u \cos u + \frac{n-1}{n}\int \sin^{n-2} u\,du$

34 $\int \cos^n u\,du = \frac{1}{n}\cos^{n-1} u \sin u + \frac{n-1}{n}\int \cos^{n-2} u\,du$

35 $\int \tan^n u\,du = \frac{1}{n-1}\tan^{n-1} u - \int \tan^{n-2} u\,du \quad \text{if } n \neq 1$

36 $\int \cot^n u\,du = \frac{-1}{n-1}\cot^{n-1} u - \int \cot^{n-2} u\,du \quad \text{if } n \neq 1$

37 $\int \sec^n u\,du = \frac{1}{n-1}\sec^{n-2} u \tan u + \frac{n-2}{n-1}\int \sec^{n-2} u\,du \quad \text{if } n \neq 1$

38 $\int \csc^n u\,du = \frac{-1}{n-1}\csc^{n-2} u \cot u + \frac{n-2}{n-1}\int \csc^{n-2} u\,du \quad \text{if } n \neq 1$

39a $\int \sin^n u \cos^m u\,du = -\frac{\sin^{n-1} u \cos^{m+1} u}{n+m} + \frac{n-1}{n+m}\int \sin^{n-2} u \cos^m u\,du \quad \text{if } n \neq -m$

39b $\int \sin^n u \cos^m u\,du = \frac{\sin^{n+1} u \cos^{m-1} u}{n+m} + \frac{m-1}{n+m}\int \sin^n u \cos^{m-2} u\,du \quad \text{if } m \neq -n$

40 $\int u \sin u\,du = \sin u - u\cos u + C$

41 $\int u \cos u\,du = \cos u + u \sin u + C$

42 $\int u^n \sin u\,du = -u^n \cos u + n\int u^{n-1}\cos u\,du$

43 $\int u^n \cos u\,du = u^n \sin u - n\int u^{n-1}\sin u\,du$

FORMS INVOLVING $\sqrt{u^2 \pm a^2}$

44 $\displaystyle\int \sqrt{u^2 \pm a^2}\, du = \frac{u}{2}\sqrt{u^2 \pm a^2} \pm \frac{a^2}{2}\ln\left|u + \sqrt{u^2 \pm a^2}\right| + C$

45 $\displaystyle\int \frac{du}{\sqrt{u^2 \pm a^2}} = \ln\left|u + \sqrt{u^2 \pm a^2}\right| + C$

46 $\displaystyle\int \frac{\sqrt{u^2 + a^2}}{u}\, du = \sqrt{u^2 + a^2} - a\ln\left(\frac{a + \sqrt{u^2 + a^2}}{u}\right) + C$

47 $\displaystyle\int \frac{\sqrt{u^2 - a^2}}{u}\, du = \sqrt{u^2 - a^2} - a\sec^{-1}\frac{u}{a} + C$

48 $\displaystyle\int u^2\sqrt{u^2 \pm a^2}\, du = \frac{u}{8}(2u^2 \pm a^2)\sqrt{u^2 \pm a^2} - \frac{a^4}{8}\ln\left|u + \sqrt{u^2 \pm a^2}\right| + C$

49 $\displaystyle\int \frac{u^2\, du}{\sqrt{u^2 \pm a^2}} = \frac{u}{2}\sqrt{u^2 \pm a^2} \mp \frac{a^2}{2}\ln\left|u + \sqrt{u^2 \pm a^2}\right| + C$

50 $\displaystyle\int \frac{du}{u^2\sqrt{u^2 \pm a^2}} = \mp\frac{\sqrt{u^2 \pm a^2}}{a^2u} + C$

51 $\displaystyle\int \frac{\sqrt{u^2 \pm a^2}}{u^2}\, du = -\frac{\sqrt{u^2 \pm a^2}}{u} + \ln\left|u + \sqrt{u^2 \pm a^2}\right| + C$

52 $\displaystyle\int \frac{du}{(u^2 \pm a^2)^{3/2}} = \frac{\pm u}{a^2\sqrt{u^2 \pm a^2}} + C$

53 $\displaystyle\int (u^2 \pm a^2)^{3/2}\, du = \frac{u}{8}(2u^2 \pm 5a^2)\sqrt{u^2 \pm a^2} + \frac{3a^4}{8}\ln\left|u + \sqrt{u^2 \pm a^2}\right| + C$

FORMS INVOLVING $\sqrt{a^2 - u^2}$

54 $\displaystyle\int \sqrt{a^2 - u^2}\, du = \frac{u}{2}\sqrt{a^2 - u^2} + \frac{a^2}{2}\sin^{-1}\frac{u}{a} + C$

55 $\displaystyle\int \frac{\sqrt{a^2 - u^2}}{u}\, du = \sqrt{a^2 - u^2} - a\ln\left|\frac{a + \sqrt{a^2 - u^2}}{u}\right| + C$

56 $\displaystyle\int \frac{u^2\, du}{\sqrt{a^2 - u^2}} = -\frac{u}{2}\sqrt{a^2 - u^2} + \frac{a^2}{2}\sin^{-1}\frac{u}{a} + C$

57 $\displaystyle\int u^2\sqrt{a^2 - u^2}\, du = \frac{u}{8}(2u^2 - a^2)\sqrt{a^2 - u^2} + \frac{a^4}{8}\sin^{-1}\frac{u}{a} + C$

58 $\displaystyle\int \frac{du}{u^2\sqrt{a^2 - u^2}} = -\frac{\sqrt{a^2 - u^2}}{a^2u} + C$

59 $\displaystyle\int \frac{\sqrt{a^2 - u^2}}{u^2}\, du = -\frac{\sqrt{a^2 - u^2}}{u} - \sin^{-1}\frac{u}{a} + C$

60 $\displaystyle\int \frac{du}{u\sqrt{a^2 - u^2}} = -\frac{1}{a}\ln\left|\frac{a + \sqrt{a^2 - u^2}}{u}\right| + C$

61 $\displaystyle\int \frac{du}{(a^2 - u^2)^{3/2}} = \frac{u}{a^2\sqrt{a^2 - u^2}} + C$

62 $\displaystyle\int (a^2 - u^2)^{3/2}\, du = \frac{u}{8}(5a^2 - 2u^2)\sqrt{a^2 - u^2} + \frac{3a^4}{8}\sin^{-1}\frac{u}{a} + C$

EXPONENTIAL AND LOGARITHMIC FORMS

63 $\displaystyle\int ue^u\, du = (u - 1)e^u + C$

64 $\displaystyle\int u^n e^u\, du = u^n e^u - n\int u^{n-1}e^u\, du$

65 $\displaystyle\int \ln u\, du = u\ln u - u + C$

66 $\displaystyle\int u^n \ln u\, du = \frac{u^{n+1}}{n+1}\ln u - \frac{u^{n+1}}{(n+1)^2} + C$

67 $\displaystyle\int e^{au}\sin bu\, du = \frac{e^{au}}{a^2 + b^2}(a\sin bu - b\cos bu) + C$

68 $\displaystyle\int e^{au}\cos bu\, du = \frac{e^{au}}{a^2 + b^2}(a\cos bu + b\sin bu) + C$

INVERSE TRIGONOMETRIC FORMS

69 $\displaystyle\int \sin^{-1}u\, du = u\sin^{-1}u + \sqrt{1 - u^2} + C$

70 $\displaystyle\int \tan^{-1}u\, du = u\tan^{-1}u - \frac{1}{2}\ln(1 + u^2) + C$

71 $\displaystyle\int \sec^{-1}u\, du = u\sec^{-1}u - \ln\left|u + \sqrt{u^2 - 1}\right| + C$

72 $\displaystyle\int u\sin^{-1}u\, du = \frac{1}{4}(2u^2 - 1)\sin^{-1}u + \frac{u}{4}\sqrt{1 - u^2} + C$

73 $\displaystyle\int u\tan^{-1}u\, du = \frac{1}{2}(u^2 + 1)\tan^{-1}u - \frac{u}{2} + C$

74 $\displaystyle\int u\sec^{-1}u\, du = \frac{u^2}{2}\sec^{-1}u - \frac{1}{2}\sqrt{u^2 - 1} + C$

75 $\displaystyle\int u^n\sin^{-1}u\, du = \frac{u^{n+1}}{n+1}\sin^{-1}u - \frac{1}{n+1}\int \frac{u^{n+1}}{\sqrt{1 - u^2}}\, du \quad \text{if } n \neq -1$

76 $\displaystyle\int u^n\tan^{-1}u\, du = \frac{u^{n+1}}{n+1}\tan^{-1}u - \frac{1}{n+1}\int \frac{u^{n+1}}{1 + u^2}\, du \quad \text{if } n \neq -1$

77 $\displaystyle\int u^n\sec^{-1}u\, du = \frac{u^{n+1}}{n+1}\sec^{-1}u - \frac{1}{n+1}\int \frac{u^n}{\sqrt{u^2 - 1}}\, du \quad \text{if } n \neq -1$